TABLE 3.1 RELATIVE STRENGTH OF SELECTED ACIDS AND THEIR CONJUGATE BASES

	Acid	Approximate pK_a	Conjugate Base	
Strongest acid	$HSbF_6$	< -12	SbF_6^-	Weakest base
	HI	-10	I^-	
	H_2SO_4	-9	HSO_4^-	
	HBr	-9	Br^-	
	HCl	-7	Cl^-	
	$C_6H_5SO_3H$	-6.5	$C_6H_5SO_3^-$	
	$(CH_3)_2\overset{+}{O}H$	-3.8	$(CH_3)_2O$	
	$(CH_3)_2C=\overset{+}{O}H$	-2.9	$(CH_3)_2C=O$	
	$CH_3\overset{+}{O}H_2$	-2.5	CH_3OH	
	H_3O^+	-1.74	H_2O	
	HNO_3	-1.4	NO_3^-	
	CF_3CO_2H	0.18	$CF_3CO_2^-$	
	HF	3.2	F^-	
	$C_6H_5CO_2H$	4.21	$C_6H_5CO_2^-$	
	$C_6H_5NH_3^+$	4.63	$C_6H_5NH_2$	
	CH_3CO_2H	4.75	$CH_3CO_2^-$	
	H_2CO_3	6.35	HCO_3^-	
	$CH_3COCH_2COCH_3$	9.0	$CH_3CO\overline{C}HCOCH_3$	
	NH_4^+	9.2	NH_3	
	C_6H_5OH	9.9	$C_6H_5O^-$	
	HCO_3^-	10.2	CO_3^{2-}	
	$CH_3NH_3^+$	10.6	CH_3NH_2	
	H_2O	15.7	HO^-	
	CH_3CH_2OH	16	$CH_3CH_2O^-$	
	$(CH_3)_3COH$	18	$(CH_3)_3CO^-$	
	CH_3COCH_3	19.2	$^-CH_2COCH_3$	
	$HC\equiv CH$	25	$HC\equiv C^-$	
	$C_6H_5NH_2$	31	$C_6H_5NH^-$	
	H_2	35	H^-	
	$(i\text{-}Pr)_2NH$	36	$(i\text{-}Pr)_2N^-$	
	NH_3	38	$^-NH_2$	
	$CH_2=CH_2$	44	$CH_2=CH^-$	
Weakest acid	CH_3CH_3	50	$CH_3CH_2^-$	Strongest base

Increasing acid strength

Increasing base strength

ORGANIC CHEMISTRY 11E

ORGANIC CHEMISTRY

T.W. Graham Solomons
University of South Florida

Craig B. Fryhle
Pacific Lutheran University

Scott A. Snyder
Columbia University

11E

WILEY

In memory of my beloved son, John Allen Solomons. TWGS

For my family. CBF

For Cathy, who has always inspired me. SAS

VICE PRESIDENT, PUBLISHER Petra Recter
SPONSORING EDITOR Joan Kalkut
PROJECT EDITOR Jennifer Yee
MARKETING MANAGER Kristine Ruff
MARKETING ASSISTANT Andrew Ginsberg
SENIOR PRODUCTION EDITOR Elizabeth Swain
SENIOR DESIGNER Maureen Eide
SENIOR PRODUCT DESIGNERS Bonnie Roth, Geraldine Osnato
CONTENT EDITOR Veronica Armour
MEDIA SPECIALIST Svetlana Barskaya
SENIOR PHOTO EDITOR Lisa Gee
DESIGN DIRECTOR Harry Nolan
TEXT AND COVER DESIGNER Maureen Eide
COVER IMAGE © Gerhard Schulz/Age Fotostock America, Inc.
This book is printed on acid-free paper.

Founded in 1807, John Wiley & Sons, Inc. has been a valued source of knowledge and understanding for more than 200 years, helping people around the world meet their needs and fulfill their aspirations. Our company is built on a foundation of principles that include responsibility to the communities we serve and where we live and work. In 2008, we launched a Corporate Citizenship Initiative, a global effort to address the environmental, social, economic, and ethical challenges we face in our business. Among the issues we are addressing are carbon impact, paper specifications and procurement, ethical conduct within our business and among our vendors, and community and charitable support. For more information, please visit our website: www.wiley.com/go/citizenship.

Evaluation copies are provided to qualified academics and professionals for review purposes only, for use in their courses during the next academic year. These copies are licensed and may not be sold or transferred to a third party. Upon completion of the review period, please return the evaluation copy to Wiley. Return instructions and a free of charge return shipping label are available at **www.wiley.com/go/returnlabel**. Outside of the United States, please contact your local representative.

ISBN 978-1-118-13357-6 (cloth)
Binder-ready version ISBN 978-1-118-14739-9

Printed in the United States of America
10 9 8 7 6 5 4 3 2 1

[BRIEF CONTENTS]

v

[CONTENTS]

3
Acids and Bases
AN INTRODUCTION TO ORGANIC REACTIONS AND THEIR MECHANISMS   **104**

3.1 Acid–Base Reactions   105

3.2 **HOW TO** Use Curved Arrows in Illustrating Reactions   107

[**A MECHANISM FOR THE REACTION**] Reaction of Water with Hydrogen Chloride: The Use of Curved Arrows   107

3.3 Lewis Acids and Bases   109

3.4 Heterolysis of Bonds to Carbon: Carbocations and Carbanions   111

3.5 The Strength of Brønsted–Lowry Acids and Bases: K_a and pK_a 113

3.6 **HOW TO** Predict the Outcome of Acid–Base Reactions   118

3.7 Relationships Between Structure and Acidity   120

3.8 Energy Changes   123

3.9 The Relationship Between the Equilibrium Constant and the Standard Free-Energy Change, $\Delta G°$ 125

3.10 Acidity: Carboxylic Acids versus Alcohols   126

3.11 The Effect of the Solvent on Acidity   130

3.12 Organic Compounds as Bases   130

3.13 A Mechanism for an Organic Reaction   132

[**A MECHANISM FOR THE REACTION**] Reaction of *tert*-Butyl Alcohol with Concentrated Aqueous HCl   132

3.14 Acids and Bases in Nonaqueous Solutions   133

3.15 Acid–Base Reactions and the Synthesis of Deuterium- and Tritium-Labeled Compounds   134

3.16 Applications of Basic Principles   135

[**WHY DO THESE TOPICS MATTER?**]   136

4
Nomenclature and Conformations of Alkanes and Cycloalkanes   **142**

4.1 Introduction to Alkanes and Cycloalkanes   143

THE CHEMISTRY OF... Petroleum Refining   143

4.2 Shapes of Alkanes   144

4.3 **HOW TO** Name Alkanes, Alkyl Halides, and Alcohols: The IUPAC System   146

4.4 **HOW TO** Name Cycloalkanes   153

4.5 **HOW TO** Name Alkenes and Cycloalkenes   156

4.6 **HOW TO** Name Alkynes   158

4.7 Physical Properties of Alkanes and Cycloalkanes   159

THE CHEMISTRY OF... Pheromones: Communication by Means of Chemicals   161

4.8 Sigma Bonds and Bond Rotation   162

4.9 Conformational Analysis of Butane   164

THE CHEMISTRY OF... Muscle Action   166

4.10 The Relative Stabilities of Cycloalkanes: Ring Strain   167

4.11 Conformations of Cyclohexane: The Chair and the Boat   168

THE CHEMISTRY OF... Nanoscale Motors and Molecular Switches   170

4.12 Substituted Cyclohexanes: Axial and Equatorial Hydrogen Groups   171

4.13 Disubstituted Cycloalkanes: Cis–Trans Isomerism   175

4.14 Bicyclic and Polycyclic Alkanes   179

4.15 Chemical Reactions of Alkanes   180

4.16 Synthesis of Alkanes and Cycloalkanes   180

4.17 **HOW TO** Gain Structural Information from Molecular Formulas and the Index of Hydrogen Deficiency   182

4.18 Applications of Basic Principles   184

[**WHY DO THESE TOPICS MATTER?**]   185

See **SPECIAL TOPIC A:** ^{13}C NMR Spectroscopy—A Practical Introduction in *WileyPLUS*

5
Stereochemistry
CHIRAL MOLECULES   **191**

5.1 Chirality and Stereochemistry   192

5.2 Isomerism: Constitutional Isomers and Stereoisomers   193

5.3 Enantiomers and Chiral Molecules   195

5.4 Molecules Having One Chirality Center are Chiral   196

5.5 More about the Biological Importance of Chirality   199

5.6 **HOW TO** Test for Chirality: Planes of Symmetry   201

5.7 Naming Enantiomers: The *R,S*-System   202

5.8 Properties of Enantiomers: Optical Activity   206

5.9 The Origin of Optical Activity   211

6
Ionic Reactions
NUCLEOPHILIC SUBSTITUTION AND ELIMINATION REACTIONS OF ALKYL HALIDES **239**

7
Alkenes and Alkynes I
PROPERTIES AND SYNTHESIS. ELIMINATION REACTIONS OF ALKYL HALIDES **291**

8

Alkenes and Alkynes II
ADDITION REACTIONS   337

9

Nuclear Magnetic Resonance and Mass Spectrometry
TOOLS FOR STRUCTURE DETERMINATION   391

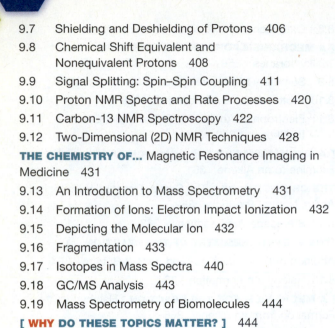

10

Radical Reactions   457

See SPECIAL TOPIC B: Chain-Growth Polymers in WileyPLUS

11

Alcohols and Ethers
SYNTHESIS AND REACTIONS   498

12

Alcohols from Carbonyl Compounds
OXIDATION–REDUCTION AND ORGANOMETALLIC COMPOUNDS   **542**

13

Conjugated Unsaturated Systems   581

14

Aromatic Compounds   626

15

Reactions of Aromatic Compounds 669

16

Aldehydes and Ketones

NUCLEOPHILIC ADDITION TO THE CARBONYL GROUP 720

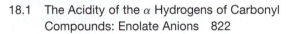

17

Carboxylic Acids and Their Derivatives

NUCLEOPHILIC ADDITION–ELIMINATION AT THE ACYL CARBON   771

18

Reactions at the α Carbon of Carbonyl Compounds

ENOLS AND ENOLATES   821

19
Condensation and Conjugate Addition Reactions of Carbonyl Compounds
MORE CHEMISTRY OF ENOLATES 858

20
Amines 897

21
Phenols and Aryl Halides
NUCLEOPHILIC AROMATIC SUBSTITUTION 944

22
Carbohydrates   979

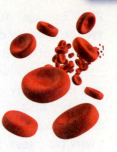

23
Lipids   1027

24

Amino Acids and Proteins 1060

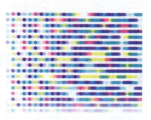

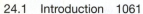

25

Nucleic Acids and Protein Synthesis 1105

[A MECHANISM FOR THE REACTION]

THE CHEMISTRY OF...

HOW TO...

[PREFACE]

"IT'S ORGANIC CHEMISTRY!"

That's what we want students to exclaim after they become acquainted with our subject. Our lives revolve around organic chemistry, whether we all realize it or not. When we understand organic chemistry, we see how life itself would be impossible without it, how the quality of our lives depends upon it, and how examples of organic chemistry leap out at us from every direction. That's why we can envision students enthusiastically exclaiming "It's organic chemistry!" when, perhaps, they explain to a friend or family member how one central theme—organic chemistry— pervades our existence. We want to help students experience the excitement of seeing the world through an organic lens, and how the unifying and simplifying nature of organic chemistry helps make many things in nature comprehensible.

Our book makes it possible for students to learn organic chemistry well and to see the marvelous ways that organic chemistry touches our lives on a daily basis. Our book helps students develop their skills in critical thinking, problem solving, and analysis—skills that are so important in today's world, no matter what career paths they choose. The richness of organic chemistry lends itself to solutions for our time, from the fields of health care, to energy, sustainability, and the environment. After all, it's organic chemistry!

Guided by these goals, and by wanting to make our book even more **accessible to students** than it has ever been before, we have brought many changes to this edition.

NEW TO THIS EDITION

With this edition we bring Scott Snyder on board as a co-author. We're very excited to have Scott join our team. Scott brings a rich resource of new perspectives to the book, particularly in the arena of complex molecule synthesis. Scott has infused new examples and applications of exciting chemistry that help achieve our goals. In addition to adding his perspectives to the presentation of core chemistry throughout the book, Scott's work is manifest in most of this edition's chapter openers and in all of the chapter closers, couched in a new feature called "Why do these topics matter?".

"Why do these topics matter?" is a new feature that bookends each chapter with a teaser in the opener and a captivating example of organic chemistry in the closer. The chapter opener seeks to whet the student's appetite both for the core chemistry in that chapter as well as a prize that comes at the end of the chapter in the form of a "Why do these topics matter?" vignette. These new closers consist of fascinating nuggets of organic chemistry that stem from research relating to medical, environmental, and other aspects of organic chemistry in the world around us, as well as the history of the science. They show the rich relevance of what students have learned to applications that have direct bearing on our lives and wellbeing. For example, in Chapter 6, the opener talks about the some of the benefits and drawbacks of making substitutions in a recipe, and then compares such changes to the nucleophilic displacement reactions that similarly allow chemists to change molecules and their properties. The closer then shows how exactly such reactivity has enabled scientists to convert simple table sugar into the artificial sweetener Splenda which is 600 times as sweet, but has no calories!

Laying the foundation earlier Certain tools are absolutely key to success in organic chemistry. Among them is the ability to draw structural formulas quickly and correctly. In this edition, we help students learn these skills even sooner than ever before by moving coverage of structural formulas and the use curved arrows earlier in the text (Section 3.2). We have woven together instruction about Lewis structures, covalent bonds, and dash structural formulas, so that students build their skills in these areas as a coherent unit, using organic examples that include alkanes, alkenes, alkynes, and alkyl halides. One could say that it's a "use organic to teach organic" approach.

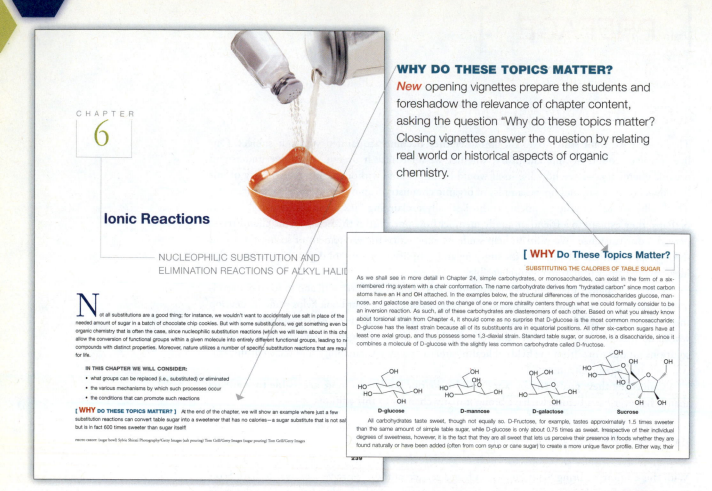

WHY DO THESE TOPICS MATTER?

New opening vignettes prepare the students and foreshadow the relevance of chapter content, asking the question "Why do these topics matter? Closing vignettes answer the question by relating real world or historical aspects of organic chemistry.

Getting to the heart of the matter quicker Acid-base chemistry, and electrophiles and nucleophiles are at the heart of organic chemistry. Students cannot master the subject if they do not have a firm and early grasp of these topics. In this edition, we cut to the chase with these topics earlier in Chapter 3 than ever before, providing a streamlined and highly efficient route to student mastery of these critical concepts.

Improving a core area: substitution reactions All organic instructors know how important it is for their students to have a solid understanding of substitution reactions. This is one reason our text has proven its lasting value. In this edition we have even further enhanced the presentation of substitution reactions in several ways, including a revised introduction of S_N1 reactions (Section 6.10) through the classic hydrolysis experiments of Hughes, and a newly organized presentation of solvent effects on the rate of substitution reactions.

Striking a strong balance of synthetic methods Students need to learn methods of organic synthesis that are useful, as environmentally friendly as possible, and that are placed in the best overall contextual framework. In this edition we incorporate the Swern oxidation (Section 12.4), long held as a useful oxidation method and one that provides a less toxic alternative to chromate oxidations in some cases. We also restore coverage of the Wolff-Kishner reduction (Section 16.8C) and the Baeyer-Villiger oxidation (Section 16.12), two methods whose importance has been proven by the test of time. The chemistry of radical reactions has also been refocused and streamlined by reducing thermochemistry content and by centralizing the coverage of allylic and benzylic radical substitutions (including NBS reactions) in one chapter (Sections 10.8 and 10.9), instead of distributing it between two, as before. The addition of sulfuric acid to alkenes and the Kolbe reaction have been deleted from the text, since these have little practical use in the laboratory. Toward the inclusion of modern, though mechanistically complex, methods of organic synthesis, we introduce catalytic oxidation methods (e.g., Sharpless and others) in special boxes, and provide coverage of transition metal organometallic reactions (Heck, Suzuki, and others) in Special Topic G.

Maintaining an eye for clarity With every edition we improve the presentation of topics, reactions, and diagrams where the opportunity arises. In this edition some examples include improved discussion and diagrams regarding endo and exo Diels-Alder transition states, the effect of diene stereochemistry in Diels-Alder reactions (Section 13.10B), and improved mechanism depictions for aromatic sulfonation and thionyl chloride substitution.

Resonating with topics in spectroscopy The authors have incorporated new figures to depict shielding and deshielding of alkenyl and alkynyl hydrogens by magnetic anisotropy, and clarified the discussion of shielding and deshielding in NMR chemical shifts (no longer invoking the terms upfield and downfield). The discussion of chlorine and bromine isotopic signatures in mass spectra has been enhanced, and presentation of mass spectrometer designs has been refocused.

Showing how things work A mechanistic understanding of organic chemistry is key to student success in organic chemistry. Mechanisms have always been central to the book, and in this edition the authors have added a mechanistic framework for the Swern and chromate alcohol oxidations (Section 12.4) by presenting elimination of the carbinol hydrogen and a leaving group from oxygen as the common theme.

TRADITIONAL PEDAGOGICAL STRENGTHS

Solved Problems Knowing "where to begin" to solve organic chemistry problems is one of the greatest challenges faced by today's students. By modeling problem solving strategies, students begin to understand the patterns inherent in organic chemistry and learn to apply that knowledge to new situations. In this edition we have added even more Solved Problems. Now over 165 Solved Problems guide students in their strategies for problem solving. Solved Problems are **usually paired with a related Practice Problem**.

Practice Problems Students need ample opportunities to practice and apply their new found strategies for solving organic chemistry problems. We've added to our rich array of in-text Practice Problems to provide students with even more opportunities to check their progress as they study. If they can work the practice problem, they should move on. If not, they should review the preceding presentation.

SOLVED PROBLEMS model problem solving strategies.

PRACTICE PROBLEMS provides opportunities to check progress.

●●● SOLVED PROBLEM 3.3

Identify the electrophile and the nucleophile in the following reaction, and add curved arrows to indicate the flow of electrons for the bond-forming and bond-breaking steps.

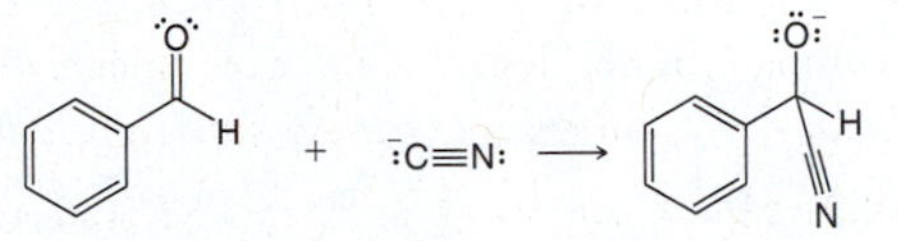

STRATEGY AND ANSWER: The aldehyde carbon is electrophilic due to the electronegativity of the carbonyl oxygen. The cyanide anion acts as a Lewis base and is the nucleophile, donating an electron pair to the carbonyl carbon, and causing an electron pair to shift to the oxygen so that no atom has more than an octet of electrons.

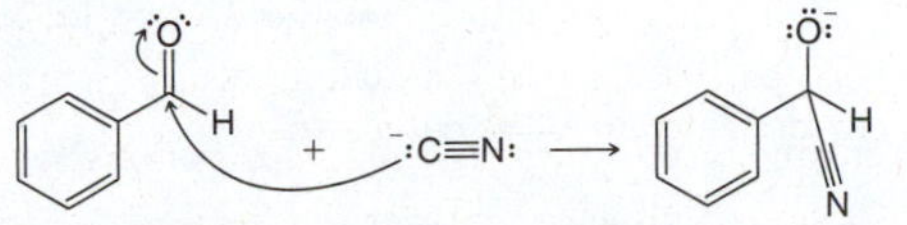

●●● PRACTICE PROBLEM 3.4 Use the curved-arrow notation to write the reaction that would take place between dimethylamine $(CH_3)_2NH$ and boron trifluoride. Identify the Lewis acid, Lewis base, nucleophile, and electrophile and assign appropriate formal charges.

End-of-Chapter Problems As athletes and musicians know, practice makes perfect. The same is true with organic chemistry. The End of Chapter problems, categorized by topic, provide essential practice for students and help them build mastery of both concepts and skills presented throughout the chapter. Many of the End of Chapter problems are cast in a visual format using structures, equations, and schemes. In addition, we still provide **Challenge Problems** and **Learning Group Problems** to address myriad teaching goals and styles. Learning Group Problems engage students in synthesizing information and concepts from throughout a chapter. They can be used to facilitate collaborative learning in small groups, and can serve as a culminating activity that demonstrates student mastery over an integrated set of principles. Supplementary material provided to instructors includes suggestions about how to orchestrate the use of learning groups.

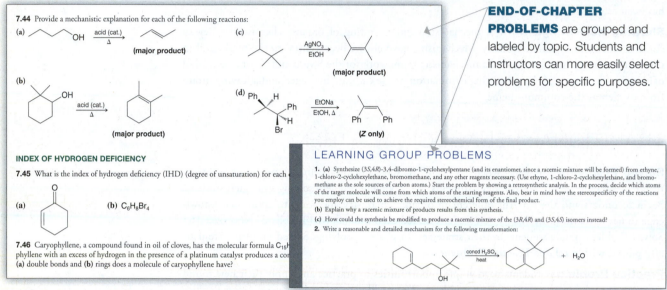

7.44 Provide a mechanistic explanation for each of the following reactions:

(a) (acid (cat.), Δ) → (major product)

(b) (acid (cat.), Δ) → (major product)

(c) (AgNO₃, EtOH) → (major product)

(d) (EtONa, EtOH, Δ) → (Z only)

INDEX OF HYDROGEN DEFICIENCY

7.45 What is the index of hydrogen deficiency (IHD) (degree of unsaturation) for each

(a)

(b) $C_6H_8Br_4$

7.46 Caryophyllene, a compound found in oil of cloves, has the molecular formula C_{15} phyllene with an excess of hydrogen in the presence of a platinum catalyst produces a co (a) double bonds and (b) rings does a molecule of caryophyllene have?

END-OF-CHAPTER PROBLEMS are grouped and labeled by topic. Students and instructors can more easily select problems for specific purposes.

LEARNING GROUP PROBLEMS

1. (a) Synthesize (3S,4R)-3,4-dibromo-1-cyclohexylpentane (and its enantiomer, since a racemic mixture will be formed) from ethyne, 1-chloro-2-cyclohexylethane, bromomethane, and any other reagents necessary. (Use ethyne, 1-chloro-2-cyclohexylethane, and bromomethane as the sole sources of carbon atoms.) Start the problem by showing a retrosynthetic analysis. In the process, decide which atoms of the target molecule will come from which atoms of the starting reagents. Also, bear in mind how the stereospecificity of the reactions you employ can be used to achieve the required stereochemical form of the final product.

(b) Explain why a racemic mixture of products results from this synthesis.

(c) How could the synthesis be modified to produce a racemic mixture of the (3R,4R) and (3S,4S) isomers instead?

2. Write a reasonable and detailed mechanism for the following transformation:

(concd H_2SO_4, heat) → + H_2O

A Mechanism for the Reaction Understanding mechanisms and the ability to recognize patterns among them is a key component in determining student success in organic chemistry. We provide A Mechanism for the Reaction boxes that show step-by-step details about how reactions take place so that students have the tools to understand rather than memorize organic reactions.

A MECHANISM FOR THE REACTION Stepped out reactions with just the right amount of detail provides the tools for students to understand rather than memorize reaction mechanisms

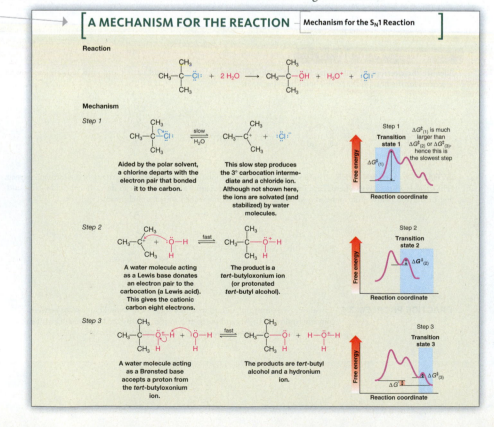

[A MECHANISM FOR THE REACTION — Mechanism for the S_N1 Reaction **]**

Reaction

$$CH_3-C(CH_3)_2-\ddot{C}l: + 2 H_2O \longrightarrow CH_3-C(CH_3)_2-\ddot{O}H + H_3O^+ + :\ddot{C}l:^-$$

Mechanism

Step 1

Aided by the polar solvent, a chlorine departs with the electron pair that bonded it to the carbon.

This slow step produces the 3° carbocation intermediate and a chloride ion. Although not shown here, the ions are solvated (and stabilized) by water molecules.

Step 1 — Transition state 1. $\Delta G^{\ddagger}_{(1)}$ is much larger than $\Delta G^{\ddagger}_{(2)}$ or $\Delta G^{\ddagger}_{(3)}$, hence this is the slowest step

Step 2

A water molecule acting as a Lewis base donates an electron pair to the carbocation (a Lewis acid). This gives the cationic carbon eight electrons.

The product is a *tert*-butyloxonium ion (or protonated *tert*-butyl alcohol).

Step 2 — Transition state 2. $\Delta G^{\ddagger}_{(2)}$

Step 3

A water molecule acting as a Brønsted base accepts a proton from the *tert*-butyloxonium ion.

The products are *tert*-butyl alcohol and a hydronium ion.

Step 3 — Transition state 3. $\Delta G^{\ddagger}_{(3)}$, $\Delta G^{\ddagger}$

Key Ideas as Bullet Points The amount of content covered in organic chemistry can be overwhelming to students. To help students focus on the most essential topics, key ideas are emphasized as bullet points in every section. In preparing bullet points, we have distilled appropriate concepts into simple declarative statements that convey core ideas accurately and clearly. No topic is ever presented as a bullet point if its integrity would be diminished by oversimplification, however.

"How to" Sections Students need to master important skills to support their conceptual learning. "How to" Sections throughout the text give step-by-step instructions to guide students in performing important tasks, such as using curved arrows, drawing chair conformations, planning a Grignard synthesis, determining formal charges, writing Lewis structures, and using ^{13}C and ^{1}H NMR spectra to determine structure.

The Chemistry of . . . . Virtually every instructor has the goal of showing students how organic chemistry relates to their field of study and to their everyday life experience. The authors assist their colleagues in this goal by providing boxes titled "The Chemistry of. . ." that provide interesting and targeted examples that engage the student with chapter content.

Summary and Review Tools At the end of each chapter, Summary and Review Tools provide visually oriented roadmaps and frameworks that students can use to help organize and assimilate concepts as they study and review chapter content. Intended to accommodate diverse learning styles, these include Synthetic Connections, Concept Maps, Thematic Mechanism Review Summaries, and the detailed Mechanism for the Reaction boxes already mentioned. We also provide Helpful Hints and richly annotated illustrations throughout the text.

SUMMARY AND REVIEW TOOLS Visually oriented study tools accommodate diverse learning styles.

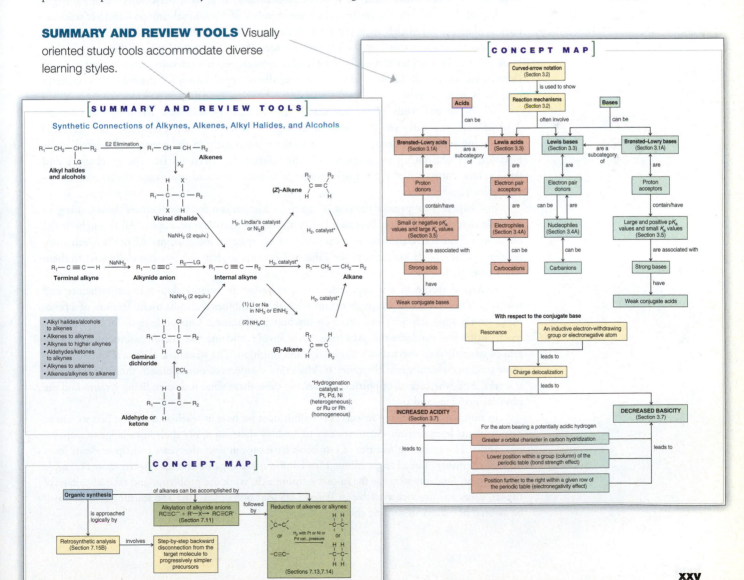

COVERAGE

Throughout the book, we have streamlined or reduced content to match the modern practice of organic chemistry, and we have provided new coverage of current reactions, while maintaining our commitment to an appropriate level and breadth of coverage.

- Chapters on carbonyl chemistry that are organized to emphasize mechanistic themes of nucleophilic addition, acyl substitution, and reactivity at the α-carbon.
- Presentation of the important modern synthetic methods of the Grubbs, Heck, Sonogashira, Stille, and Suzuki transition metal catalyzed carbon-carbon bond-forming reactions in a practical and student-oriented way that includes review problems and mechanistic context (Special Topic G).

ORGANIZATION — An Emphasis on the Fundamentals

So much of organic chemistry makes sense and can be generalized if students master and apply a few fundamental concepts. Therein lays the beauty of organic chemistry. If students learn the essential principles, they will see that memorization is not needed to succeed.

Most important is for students to have a solid understanding of structure—of hybridization and geometry, steric hindrance, electronegativity, polarity, formal charges, and resonance —so that they can make intuitive sense of mechanisms. It is with these topics that we begin in Chapter 1. In Chapter 2 we introduce the families of functional groups—so that students have a platform on which to apply these concepts. We also introduce intermolecular forces, and infrared (IR) spectroscopy—a key tool for identifying functional groups. Throughout the book we include calculated models of molecular orbitals, electron density surfaces, and maps of electrostatic potential. These models enhance students' appreciation for the role of structure in properties and reactivity.

We begin our study of mechanisms in the context of acid-base chemistry in Chapter 3. Acid-base reactions are fundamental to organic reactions, and they lend themselves to introducing several important topics that students need early in the course: (1) curved arrow notation for illustrating mechanisms, (2) the relationship between free-energy changes and equilibrium constants, and (3) the importance of inductive and resonance effects and of solvent effects.

In Chapter 3 we present the first of many "A Mechanism for the Reaction" boxes, using an example that embodies both Brønsted-Lowry and Lewis acid-base principles. All throughout the book, we use boxes like these to show the details of key reaction mechanisms. All of the Mechanism for the Reaction boxes are listed in the Table of Contents so that students can easily refer to them when desired.

A central theme of our approach is to emphasize the relationship between structure and reactivity. This is why we choose an organization that combines the most useful features of a functional group approach with one based on reaction mechanisms. Our philosophy is to emphasize mechanisms and fundamental principles, while giving students the anchor points of functional groups to apply their mechanistic knowledge and intuition. The structural aspects of our approach show students what organic chemistry is. Mechanistic aspects of our approach show students how it works. And wherever an opportunity arises, we show them what it does in living systems and the physical world around us.

In summary, our writing reflects the commitment we have as teachers to do the best we can to help students learn organic chemistry and to see how they can apply their knowledge to improve our world. The enduring features of our book have proven over the years to help students learn organic chemistry. The changes in our 11th edition make organic chemistry even more accessible and relevant. Students who use the in-text learning aids, work the problems, and take advantage of the resources and practice available in *WileyPLUS* (our online teaching and learning solution) will be assured of success in organic chemistry.

WILEYPLUS FOR ORGANIC CHEMISTRY—
A Powerful Teaching and Learning Solution

WileyPLUS is an innovative, research-based online environment for effective teaching and learning. *WileyPLUS* builds student confidence because it takes the guesswork out of studying by providing students with a clear roadmap: what to do, how to do it, if they did it right. Students will take more initiative so instructors will have greater impact on their achievement in the classroom and beyond.

Breadth of Depth of Assessment: Four unique silos of assessment are available to instructors for creating online homework and quizzes and are designed to enable and support problem-solving skill development and conceptual understanding

WILEYPLUS ASSESSMENT ········ | FOR ORGANIC CHEMISTRY

REACTION EXPLORER	MEANINGFUL PRACTICE OF MECHANISMS AND SYNTHESIS PROBLEMS (A DATABASE OF OVER 100,000 QUESTIONS)
IN CHAPTER/EOC ASSESSMENT	90-100% OF REVIEW PROBLEMS AND END OF CHAPTER (EOC) QUESTIONS ARE CODED FOR ONLINE ASSESSMENT
CONCEPT MASTERY	PRE-BUIT CONCEPT MASTERY ASSIGNMENTS (FROM DATABASE OF OVER 25,000 QUESTIONS)
TEST BANK	RICH TESTBANK CONSISTING OF OVER 3,000 QUESTIONS

Reaction Explorer Students ability to understand mechanisms and predict syntheis reactions greatly impacts their level of success in the course. Reaction Explorer is an interactive system for **learning and practicing reactions**, **syntheses** and **mechanisms** in organic chemistry with advanced support for the automatic generation of random problems and curved arrow mechanism diagrams.

MECHANISM EXPLORER:
valuable practice with
reactions and mechanisms

SYNTHESIS EXPLORER:
meaningful practice with single
and multi-step synthesis

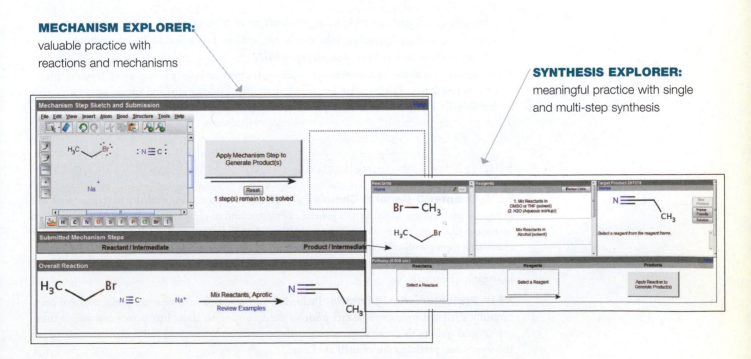

End of Chapter Problems. Approximately 90% of the end of chapter problems are included in *WileyPLUS*. Many of the problems are algorithmic and feature structure drawing/assessment functionality using MarvinSketch, with immediate answer feedback and video question assistance. A subset of these end of chapter problems is linked to **Guided Online tutorials** which are stepped-out problem-solving tutorials that walk the student through the problem, offering individualized feedback at each step.

Prebuilt concept mastery assignments Students must continuously practice and work organic chemistry in order to master the concepts and skills presented in the course. Prebuilt concept mastery assignments offer students ample opportunities for practice, covering all the major topics and concepts within an organic chemistry course. Each assignment is organized by topic and features **feedback for incorrect answers**. These assignments are drawn from a unique database of over 25,000 questions, over half of which require students to draw a structure using MarvinSketch.

PREBUILT CONCEPT MASTERY ASSIGNMENTS

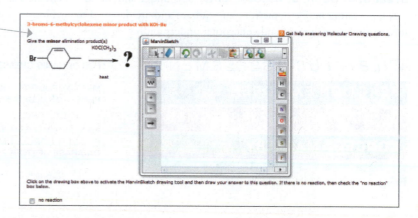

WHAT DO STUDENTS RECEIVE WITH *WILEYPLUS*?

- The complete digital textbook, saving students up to 60% off the cost of a printed text.
- Question assistance, including links to relevant sections in the online digital textbook.
- Immediate feedback and proof of progress, 24/7.
- Integrated, multi-media resources that address students' unique learning styles, levels of proficiency, and levels of preparation by providing multiple study paths and encourage more active learning.

WILEYPLUS STUDENT RESOURCES

NEW Chapter 0 General Chemistry Refresher. To ensure students have mastered the necessary prerequisite content from general chemistry, and to eliminate the burden on instructors to review this material in lecture, *WileyPLUS* now includes a complete chapter of core general chemistry topics with corresponding assignments. Chapter 0 is available to students and can be assigned in *WileyPLUS* to ensure and gauge understanding of the core topics required to succeed in organic chemistry.

NEW Prelecture Assignments. Preloaded and ready to use, these assignments have been carefully designed to assess students prior to their coming to class. Instructors can assign these pre-created quizzes to gauge student preparedness prior to lecture and tailor class time based on the scores and participation of their students.

Video Mini-Lectures, Office Hour Videos, and Solved Problem Videos In each chapter, several types of video assistance are included to help students with conceptual understanding and problem solving strategies. The video mini-lectures focus on challenging concepts; the office hours videos take these concepts and apply them to example problems, emulating the experience that a student would get if she or he were to attend office hours and ask for assistance in working a problem. The Solved Problem videos demonstrate good problems solving strategies for the student by walking through in text solved problems using audio and a whiteboard. The goal is to illustrate good problem solving strategies.

Skill Building Exercises are animated exercises with instant feedback to reinforce the key skills required to succeed in organic chemistry.

3D Molecular Visualizations use the latest visualization technologies to help students visualize concepts with audio. Instructors can assign quizzes based on these visualizations in *WileyPLUS*.

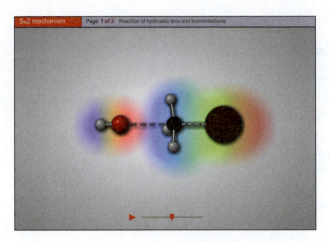

WHAT DO INSTRUCTORS RECEIVE WITH *WILEYPLUS*?

- Reliable resources that reinforce course goals inside and outside of the classroom.
- The ability to easily identify students who are falling behind by tracking their progress and offering assistance easily, even before they come to office hours. *WileyPLUS* simplifies and automates such tasks as student performance assessment, creating assignments, scoring student work, keeping grades, and more.
- Media-rich course materials and assessment content that allow you to customize your classroom presentation with a wealth of resources and functionality from PowerPoint slides to a database of rich visuals. You can even add your own materials to your *WileyPLUS* course.

ADDITIONAL INSTRUCTOR RESOURCES

All Instructor Resources are available within *WileyPLUS* or they can be accessed by contacting your local Wiley Sales Representative. Many of the assets are located on the book companion site, www.wiley.com/college/solomons

Test Bank Authored by Robert Rossi, of Gloucester County College, Jeffrey Allison, of Austin Community College, and Gloria Silva, of Carnegie Mellon University, the Test Bank for this edition has been completely revised and updated to include over 3,000 short answer, multiple choice, and essay/drawing questions. The Test Bank files, along with a software tool for managing and creating exams, are available online.

PowerPoint Lecture Slides PowerPoint Lecture Slides have been prepared by Professor William Tam, of the University of Guelph, Dr. Phillis Chang, and Gary Porter, of Bergen Community College. The PowerPoint slides include additional examples, illustrations, and presentations that help reinforce and test students' grasp of organic chemistry concepts. An additional set of PowerPoint slides features the illustrations, figures, and tables from the text. All PowerPoint slide presentations are customizable to fit your course.

Personal Response System ("Clicker") Questions A bank of questions is available for anyone using personal response system technology in their classroom. The clicker questions are also available in a separate set of PowerPoint slides.

Digital Image Library Images from the text are available online in JPEG format. Instructors may use these images to customize their presentations and to provide additional visual support for quizzes and exams.

ADDITIONAL STUDENT RESOURCES

Study Guide and Solutions Manual (978-1-118-14790-0)

The Study Guide and Solutions Manual for *Organic Chemistry, Eleventh Edition*, authored by Jon Antilla, of the University of South Florida, Robert Johnson, of Xavier University, Craig Fryhle, Graham Solomons, and Scott Snyder **contains explained solutions to all of the problems in the text**. The Study Guide also contains:

- An introductory essay "Solving the Puzzle—or—Structure is Everything" that serves as a bridge from general to organic chemistry
- Summary tables of reactions by mechanistic type and functional group
- A review quiz for each chapter
- A set of hands-on molecular model exercises
- Solutions to the problems in the Special Topics sections (many of the Special Topics are only available within *WileyPLUS*)

MOLECULAR VISIONS™ MODEL KITS

We believe that the tactile and visual experience of manipulating physical models is key to students' understanding that organic molecules have shape and occupy space. To support our pedagogy, we have arranged with the Darling Company to bundle a special ensemble of Molecular Visions™ model kits with our book (for those who choose that option). We use Helpful Hint icons and margin notes to frequently encourage students to use hand-held models to investigate the three-dimensional shape of molecules we are discussing in the book.

CUSTOMIZATION AND FLEXIBLE OPTIONS TO MEET YOUR NEEDS

Wiley Custom Select allows you to create a textbook with precisely the content you want, in a simple, three-step online process that brings your students a cost-efficient alternative to a traditional textbook. Select from an extensive collection of content at **http://customselect.wiley.com,** upload your own materials as well, and select from multiple delivery formats—full color or black and white print with a variety of binding options, or eBook. Preview the full text online, get an instant price quote, and submit your order; we'll take it from there.

WileyFlex offers content in flexible and cost-saving options to students. Our goal is to deliver our learning materials to our customers in the formats that work best for them, whether it's a traditional text, eTextbook, *WileyPLUS*, loose-leaf binder editions, or customized content through Wiley Custom Select.

[ACKNOWLEDGMENTS]

We are especially grateful to the following people who provided detailed reviews and participated in focus groups that helped us prepare this new edition of Organic Chemistry.

ARIZONA

Cindy Browder, *Northern Arizona University*
Tony Hascall, *Northern Arizona University*

ARKANSAS

David Bateman, *Henderson State University*
Kenneth Nolan Carter, *University of Central Arkansas*

CALIFORNIA

Thomas Bertolini, *University of Southern California*
David Brook, *San Jose State University*
Rebecca Broyer, *University of Southern California*
Paul Buonora, *California State University-Long Beach*
Steven Farmer, *Sonoma State University*
Amelia Fuller, *Santa Clara University*
Andreas F. Franz, *University of the Pacific*
Karl Haushalter, *Harvey Mudd College*
Jianhua Ren, *The University of the Pacific*
Harold Rogers, *California State University-Fullerton*
Douglas Smith, *California State University-San Bernardino*
Daniel Wellmanm, *Chapman University*
Liang Xue, *University of the Pacific*

CONNECTICUT

Andrew Karatjas, *Southern Connecticut State University*
Erika Taylor, *Wesleyan University*

GEORGIA

Koushik Banerjee, *Georgia College & State University*
Stefan France, *Georgia Institute of Technology*
Arthur Ragauskas, *Georgia Institute of Technology*
Laren Tolbert, *Georgia Institute of Technology*
Christine Whitlock, *Georgia Southern University*

ILLINOIS

Eugene Losey, *Elmhurst College*
Valerie Keller, *University of Chicago*
Richard Nagorski, *Illinois State University*

IDAHO

Owen McDougal, *Boise State University*
Joshua Pak, *Idaho State University*

INDIANA

Adam Azman, *Butler University*
Paul Morgan, *Butler University*

IOWA

Ned Bowden, *University of Iowa*
William Jenks, *Iowa State University*
Olga Rinco, *Luther College*

KANSAS

Loretta Dorn, *Fort Hays State University*

KENTUCKY

Mark Blankenbuehler, *Morehead State University*

LOUISIANA

Marilyn Cox, *Louisiana Tech University*
August Gallo, *University of Louisiana-Lafayette*
Sean Hickey, *University of New Orleans*

MASSACHUSETTS

Samuel Thomas, *Tufts University*

MICHIGAN

Bal Barot, *Lake Michigan College*
Kimberly Greve, *Kalamazoo Valley Community College*
Scott Ratz, *Alpena Community College*
Ronald Stamper, *University of Michigan*
Gregg Wilmes, *Eastern Michigan University*

MINNESOTA

Eric Fort, *University of St. Thomas*

MISSISSIPPI

Tim Ross, *Itawamba Community College*
Douglas Masterson, *University of Southern Mississippi*

NEW JERSEY

Heba Abourahma, *The College of New Jersey*
Bruce Hietbrink, *Richard Stockton College*
David Hunt, *The College of New Jersey*
Subash Jonnalagadda, *Rowan University*

NEW MEXICO

Donald Bellew, *University of New Mexico*

NEW YORK

Brahmadeo Dewprashad, *Borough of Manhattan Community College*

Galina Melman, *Clarkson University*
Gloria Proni, *John Jay College*

NORTH CAROLINA

Nicole Bennett, *Appalachian State University*
Brian Love, *East Carolina University*
Andrew Morehead, *East Carolina University*
Jim Parise, *Duke University*
Cornelia Tirla, *University of North Carolina-Pembroke*
Wei You, *University of North Carolina-Chapel Hill*

NORTH DAKOTA

Karla Wohlers, *North Dakota State University*

OHIO

Neil Ayres, *University of Cincinnati*
Benjamin Gung, *Miami University-Oxford*
Debbie Lieberman, *University of Cincinnati*
Michael Nichols, *John Carroll University*
Allan Pinhas, *University of Cincinnati*
Joel Shulman, *University of Cincinnati*

OKLAHOMA

Donna Nelson, *University of Oklahoma*

PENNSYLVANIA

Melissa Cichowicz, *West Chester University*
Dian He, *Holy Family University*
Joel Ressner, *West Chester University*

SOUTH CAROLINA

Carl Heltzel, *Clemson University*

SOUTH DAKOTA

Grigoriy Sereda, *University of South Dakota*

TENNESSEE

Ramez Elgammal, *University of Tennessee*
Scott Handy, *Middle Tennessee State University*
Aleksey Vasiliev, *East Tennessee State University*

TEXAS

Jeff Allison, *Austin Community College-South Austin Campus*
Shawn Amorde, *Austin Community College*
Jennifer Irvin, *Texas State University*

VIRGINIA

Suzanne Slayden, *George Mason University*

WISCONSIN

Elizabeth Glogowski, *University of Wisconsin-Eau Claire*
Tehshik Yoon, *University of Wisconsin-Madison*

CANADA

Jeremy Wulff, *University of Victoria*
France-Isabelle Auzanneau, *University of Guelph*

We are also grateful to the many people who provided reviews that guided preparation of the earlier editions of our book:

ALABAMA

Wayne Brouillette, *University of Alabama*
Stephen A. Woski, *University of Alabama*

ALASKA

John W. Keller, *University of Alaska-Fairbanks*

ARIZONA

Kaiguo Chang, *University of Arkansas at Fort Smith*
Michael J. Panigot, *Arkansas State University-Jonesboro*
Lee Harris, *University of Arizona*
Colleen Kelley, *Pima Community College*
Christine A. Pruis, *Arizona State University*

CALIFORNIA

David Ball, *California State University*
Stuart R. Berryhill, *California State University-Long Beach*
Jeff Charonnat, *California State University-Northridge*
Phillip Crews, *University of California-Santa Cruz*
Peter deLijser, *California State University-Fullerton*
James Ellern, *University of Southern California*
Ihsan Erden, *San Francisco State University*
Andreas Franz, *University of the Pacific*
Steven A. Hardinger, *University of California-Los Angeles*
Carl A. Hoeger, *University of California-San Diego*
Stanley N. Johnson, *Orange Coast College*
Michael Kzell, *Orange Coast College*
Shelli R. McAlpine, *San Diego State University*
Stephen Rodemeyer, *California State University-Fresno*
Harold R. Rogers, *California State University-Fullerton*
Vyacheslav V. Samoshin, *University of the Pacific*
Donald Wedegaertner, *University of the Pacific*

CONNECTICUT

Ed O'Connell, *Fairfield University*

DISTRICT OF COLUMBIA

Jesse M. Nicholson, *Howard University*
Richard Tarkka, *George Washington University*

DELAWARE

James Damewood, *University of Delaware*
Kraig Wheeler, *Delaware State University*

FLORIDA

George Fisher, *Barry University*
Wayne Guida, *Eckerd College*
John Helling, *University of Florida*
Paul Higgs, *Barry University*
Cyril Parkanyi, *Florida Atlantic University*
Bill J. Baker, *University of South Florida, Chico*

GEORGIA

Winfield M. Baldwin, *University of Georgia*
Edward M. Burgess, *Georgia Institute of Technology*
David Collard, *Georgia Institute of Technology*
D. Scott Davis, *Mercer University*
Leyte L. Winfield, *Spelman College*

IOWA

Janelle Torres y Torres, *Muscatine Community College*

IDAHO

Lyle W. Castle, *Idaho State University*

ILLINOIS

Ronald Baumgarten, *University of Illinois-Chicago*
Robert C. Duty, *Illinois State University*
Eugene Losey, *Elmhurst College*
Christine Russell, *College of DuPage*
Warren Sherman, *Chicago State University*

INDIANA

Jeremiah P. Freeman, *University of Notre Dame*
Catherine Reck, *Indiana University-Bloomington*
Joseph Wolinsky, *Purdue University*
Anne M. Wilson, *Butler University*

KANSAS

John A. Landgrebe, *University of Kansas*
Dilip K. Paul, *Pittsburg State University*
Robert Pavlis, *Pittsburg State University*

KENTUCKY

Arthur Cammers, *University of Kentucky*
Frederick A. Luzzio, *University of Louisville*
John L. Meisenheimer, *Eastern Kentucky University*

LOUISIANA

Sean Hickey, *University of New Orleans*
Cynthia M. Lamberty, *Nicholls State University*
William A. Pryor, *Louisiana State University*
John Sevenair, *Xavier University of Louisiana*
James G. Traynham, *Louisiana State University*

MASSACHUSETTS

Ed Brusch, *Tufts University*
Michael Hearn, *Wellesley College*
Philip W. LeQuesne, *Northeastern University*
James W. Pavlik, *Worcester Polytechnic Institute*
Ralph Salvatore, *University of Massachusetts-Boston*
Jean Stanley, *Wellesley College*
Robert Stolow, *Tufts University*
Arthur Watterson, *University of Massachusetts-Lowell*

MAINE

Jennifer Koviach-Cote, *Bates College*

MARYLAND

Phillip DeShong, *University of Maryland*
Tiffany Gierasch, *University of Maryland-Baltimore County*
Ray A. Goss Jr. *Prince George's Community College*
Thomas Lectka, *Johns Hopkins University*
Jesse More, *Loyola College*
Andrew Morehead, *University of Maryland*

MICHIGAN

Angela J. Allen, *University of Michigan-Dearborn*
James Ames, *University of Michigan-Flint*
Todd A. Carlson, *Grand Valley State University*
Brian Coppola, *University of Michigan*
Roman Dembinski, *Oakland University*
David H. Kenny, *Michigan Technological University*
Renee Muro, *Oakland Community College*
Everett Nienhouse, *Ferris State College*
Thomas R. Riggs, *University of Michigan*
Darrell Watson, *GMI Engineering and Management Institure*
Regina Zibuck, *Wayne State University*

MINNESOTA

Paul A. Barks, *North Hennepin State Junior College*
Robert Carlson, *University of Minnesota*
Stuart Fenton, *University of Minnesota*
Brad Glorvigen, *University of St. Thomas*
Frank Guziec, *New Mexico State University*

John Holum, *Augsburg College*
Julie E. Larson, *Bemidji State University*
Rita Majerle, *Hamline University*
Viktor V. Zhdankin, *University of Minnesota—Duluth*

MISSOURI

Eric Bosch, *Southwest Missouri State University*
Peter Gaspar, *Washington University-St. Louis*
Bruce A. Hathaway, *Southeast Missouri State University*
Ekaterina N. Kadnikova, *University of Missouri*
Shon Pulley, *University of Missouri-Columbia*

NORTH CAROLINA

Nina E. Heard, *University of North Carolina-Greensboro*
Shouquan Huo, *East Carolina University*
Paul J. Kropp, *University of North Carolina-Chapel Hill*
Samuel G. Levine, *North Carolina State University*
Andrew T. Morehead Jr., *East Carolina University*
Yousry Sayed, *University of North Carolina-Wilmington*
George Wahl, *North Carolina State University*
Mark Welker, *Wake Forest University*
Michael Wells, *Campbell University*
Justin Wyatt, *College of Charleston*

NORTH DAKOTA

A. William Johnson, *University of North Dakota*

NEBRASKA

Desmond M. S. Wheeler, *University of Nebraska*

NEW HAMPSHIRE

Carolyn Kraebel Weinreb, *Saint Anselm College*

NEW JERSEY

Bruce S.Burnham, *Rider University*
Jerry A. Hirsch, *Seton Hall University*
John L. Isidor, *Montclair State University*
Allan K. Lazarus, *Trenton State College*
Kenneth R. Overly, *Richard Stockton College*
Alan Rosan, *Drew University*
John Sowa, *Seton Hall University*
Daniel Trifan, *Fairleigh Dickinson University*

NEW MEXICO

Paul Papadopoulos, *University of New Mexico*

NEW YORK

W. Lawrence Armstrong. *SUNY College-Oneonta*
Karen Aubrecht, *State University of New York-Stonybrook*
William D. Closson, *State University of New York-Albany*
Jeremy Cody, *Rochester Institute of Technology*
Sidney Cohen, *Buffalo State College*
Bill Fowler, *State University of New York-Stonybrook*
Cristina H.Geiger, *SUNY Geneseo*
William H. Hersh, *Queens College*
Robert C. Kerber, *State University of New York-Stony Brook*
James Leighton, *Columbia University*
Patricia Lutz, *Wagner College*
Jerry March, *Adelphi University*
Joseph J. Tufariello, *State University of New York-Buffalo*
Kay Turner, *Rochester Institute of Technology*
James Van Verth, *Canisius College*
Herman E. Zieger, *Brooklyn College.*

OHIO

Jovica Badjic, *The Ohio State University*
Kenneth Berlin, *Oklahoma State University*
Christopher Callam, *The Ohio State University*
George Clemans, *Bowling Green State University*
Clarke W. Earley, *Kent State University*
Gideon Fraenkel, *The Ohio State University*
Christopher M. Hadad, *The Ohio State University*
James W. Hershberger, *Miami University-Oxford*
Robert G. Johnson, *Xavier University*
Adam I. Keller, *Columbus State Community College*
Chase Smith, *Ohio Northern University*
Doug Smith, *University of Toledo*
Frank Switzer, *Xavier University*
Mark C. McMills, *Ohio University*

OKLAHOMA

O. C. Dermer, *Oklahoma State University*
John DiCesare, *University of Tulsa*
Kirk William Voska, *Rogers State University*

OREGON

Bruce Branchaud, *University of Oregon*
Arlene R. Courtney, *Western Oregon University*
M. K. Gleicher, *Oregon State University*
John F. Keana, *University of Oregon*
James W. Long, *University of Oregon*

PENNSYLVANIA

George Bandik, *University of Pittsburgh*
Trudy Dickneider, *University of Scranton*

Paul Dowd, *University of Pittsburgh*
Mark Forman, *Saint Joseph's University*
Felix Goodson, *West Chester University*
Kenneth Hartman, *Geneva College*
Michael S. Leonard, *Washington and Jefferson College*
Robert Levine, *University of Pittsburgh*
John Mangravite, *West Chester University*
Przemyslaw Maslak, *Pennsylvania State University*
James McKee, *University of the Sciences Philadelphia*
Joel M. Ressner, *West Chester University*
Don Slavin, *Community College of Philadelphia*
Jennifer A. Tripp, *University of Scranton*
John Williams, *Temple University*

PUERTO RICO

Gerado Molina, *Universidad de Puerto Rico*

SOUTH CAROLINA

Marion T. Doig III, *College of Charleston*
Rick Heldrich, *College of Charleston*
Neal Tonks, *College of Charleston*

SOUTH DAKOTA

David C. Hawkinson, *University of South Dakota*

TENNESSEE

Newell S. Bowman, *The University of Tennessee*
Mohammad R. Karim, *Tennessee State University*
Paul Langford, *David Lipscomb University*
Ronald M. Magid, *University of Tennessee*
Aleksey Vasiliev, *East Tennessee State University*
Steven Bachrach. *Trinity University*

TEXAS

Ed Biehl, *Southern Methodist University*
Brian M. Bocknack, *University of Texas-Austin*
John Hogg, *Texas A & M University*
Javier Macossay, *The University of Texas-Pan American*
Janet Maxwell, *Angelo State University*
Gary Miracle, *Texas Tech University*
Michael Richmond, *University of North Texas*
Jonathan Sessler, *University of Texas-Austin*
Rueben Walter, *Tarleton State University*
James K. Whitesell, *The University of Texas-Austin*
David Wiedenfeld, *University of North Texas*
Carlton Willson, *University of Texas-Austin*

UTAH

Merritt B. Andrus, *Brigham Young University*
Eric Edstrom, *Utah State University*
Richard Steiner, *University of Utah*
Heidi Vollmer-Snarr, *Brigham Young University*

VIRGINIA

Chris Abelt, *College of William & Mary*
Harold Bell, *Virginia Polytechnic Institute and State University*
Randolph Coleman, *College of William & Mary*
Roy Gratz, *Mary Washington College*
Philip L. Hall, *Virginia Polytechnic Institute and State University*

Eric Remy, *Virginia Polytechnic Institute and State University*
John Jewett, *University of Vermont*

WASHINGTON

Kevin Bartlett, *Seattle Pacific University*
Jeffrey P. Jones, *Washington State University*
Tomikazu Sasaki, *University of Washington*
Darrell J. Woodman, *University of Washington*

WEST VIRGINIA

John H. Penn, *West Virginia University*

WISCONSIN

Ronald Starkey, *University of Wisconsin-Green Bay*
Linfeng Xie, *University of Wisconsin-Oshkosh*

CANADA

Edward V. Blackburn, *University of Alberta*
Christine Brzezowski, *University of Alberta*
Shadi Dalili, *University of Toronto*
Sandro Gambarotta, *University of Ottawa*
Dennis Hall, *University of Alberta*
David Harpp, *McGill University*
Ian Hunt, *University of Calgary*
Karl R. Kopecky, *University of Alberta*
John Otto Olson, *University of Alberta*
Frank Robinson, *University of Victoria—British Columbia*
Adrian L. Schwan, *University of Guelph*
Rik R. Tykwinski, *University of Alberta*

Many people have helped with this edition, and we owe a great deal of thanks to each one of them. We thank Sean Hickey (*University of New Orleans*) and Justin Wyatt (*College of Charleston*) for their reviews of the manuscript and problems. We are grateful to Alan Shusterman (*Reed College*) and Warren Hehre (*Wavefunction, Inc.*) for assistance in prior editions regarding explanations of electrostatic potential maps and other calculated molecular models. We would also like to thank those scientists who allowed us to use or adapt figures from their research as illustrations for a number of the topics in our book.

A book of this scope could not be produced without the excellent support we have had from many people at John Wiley and Sons, Inc. Photo Editor Lisa Gee obtained photographs that so aptly illustrate examples in our book. Maureen Eide led development of the striking new design of the 11th edition. Jennifer Yee ensured coordination and cohesion among many aspects of this project, especially regarding reviews, supplements, and the Study Guide and Solutions Manual. Joan Kalkut, Sponsoring Editor, provided advice, ideas, and greatly assisted with development and production of the manuscript throughout the process. Elizabeth Swain brought the book to print through her incredible skill in orchestrating the production process and converting manuscript to final pages. Publisher Petra Recter led the project from the outset and provided careful oversight and encouragement through all stages of work on the 11th edition, even as she prepared to welcome twins into the world. (Congratulations, Petra!) Kristine Ruff enthusiastically and effectively helped tell the 'story' of our book to the many people we hope will consider using it. We are thankful to all of these people and others behind the scenes at Wiley for the skills and dedication that they provided to bring this book to fruition.

TWGS would like to thank his wife Judith for her support over ten editions of this book. She joins me in dedicating this edition to the memory of our beloved son, Allen.

CBF would like to thank his colleagues, students, and mentors for what they have taught him over the years. Most of all, he would like to thank his wife Deanna for the support and patience she gives to make this work possible.

SAS would like to thank his parents, his mentors, his colleagues, and his students for all that they have done to inspire him. Most of all, he would like to thank his wife Cathy for all that she does and her unwavering support.

T. W. Graham Solomons
Craig B. Fryhle
Scott A. Snyder

[ABOUT THE AUTHORS]

T. W. GRAHAM SOLOMONS did his undergraduate work at The Citadel and received his doctorate in organic chemistry in 1959 from Duke University where he worked with C. K. Bradsher. Following this he was a Sloan Foundation Postdoctoral Fellow at the University of Rochester where he worked with V. Boekelheide. In 1960 he became a charter member of the faculty of the University of South Florida and became Professor of Chemistry in 1973. In 1992 he was made Professor Emeritus. In 1994 he was a visiting professor with the Faculté des Sciences Pharmaceutiques et Biologiques, Université René Descartes (Paris V). He is a member of Sigma Xi, Phi Lambda Upsilon, and Sigma Pi Sigma. He has received research grants from the Research Corporation and the American Chemical Society Petroleum Research Fund. For several years he was director of an NSF-sponsored Undergraduate Research Participation Program at USF. His research interests have been in the areas of heterocyclic chemistry and unusual aromatic compounds. He has published papers in the *Journal of the American Chemical Society*, the *Journal of Organic Chemistry*, and the *Journal of Heterocyclic Chemistry*. He has received several awards for distinguished teaching. His organic chemistry textbooks have been widely used for 30 years and have been translated into French, Japanese, Chinese, Korean, Malaysian, Arabic, Portuguese, Spanish, Turkish, and Italian. He and his wife Judith have a daughter who is a building conservator and a son who is a research biochemist.

CRAIG BARTON FRYHLE is Chair and Professor of Chemistry at Pacific Lutheran University. He earned his B.A. degree from Gettysburg College and Ph.D. from Brown University. His experiences at these institutions shaped his dedication to mentoring undergraduate students in chemistry and the liberal arts, which is a passion that burns strongly for him. His research interests have been in areas relating to the shikimic acid pathway, including molecular modeling and NMR spectrometry of substrates and analogues, as well as structure and reactivity studies of shikimate pathway enzymes using isotopic labeling and mass spectrometry. He has mentored many students in undergraduate research, a number of who have later earned their Ph.D. degrees and gone on to academic or industrial positions. He has participated in workshops on fostering undergraduate participation in research, and has been an invited participant in efforts by the National Science Foundation to enhance undergraduate research in chemistry. He has received research and instrumentation grants from the National Science Foundation, the M J. Murdock Charitable Trust, and other private foundations. His work in chemical education, in addition to textbook coauthorship, involves incorporation of student-led teaching in the classroom and technology-based strategies in organic chemistry. He has also developed experiments for undergraduate students in organic laboratory and instrumental analysis courses. He has been a volunteer with the hands-on science program in Seattle public schools, and Chair of the Puget Sound Section of the American Chemical Society. His passion for climbing has led to ascents of high peaks in several parts of the world. He resides in Seattle with his wife, where both enjoy following the lives of their two daughters as they unfold in new places.

SCOTT A. SNYDER is Associate Professor of Chemistry at Columbia University. He grew up in the suburbs of Buffalo NY and was an undergraduate at Williams College, where he graduated summa cum laude in 1999, before pursuing his doctoral studies at The Scripps Research Institute under the tutelege of K. C. Nicolaou as an NSF, Pfizer, and Bristol-Myers-Squibb predoctoral fellow. While there, he co-authored the graduate textbook *Classics in Total Synthesis II* with his doctoral mentor. Scott was then an NIH postdoctoral fellow in the laboratories of E. J. Corey at Harvard University before assuming his current position in 2006. His research interests lie in the arena of natural products total synthesis, especially in the realm of unique polyphenols and halogenated materials, and to date he has trained more than 60 students at the high school, undergraduate, graduate, and postdoctoral levels and co-authored more than 40 research and review articles. Scott has received a number of awards and honors, including a Camille and Henry Dreyfus New Faculty Award, Amgen New Faculty and Young Investigator Awards, Eli Lilly New Faculty and Grantee Awards, a Bristol-Myers Squibb Unrestricted Grant Award, an NSF CAREER Award, an Alfred P. Sloan Foundation Fellowship, a DuPont Young Professor Award, and an Arthur C. Cope Scholar Award from the American Chemical Society. He has also received recognition for his teaching through a Cottrell Scholar Award from the Research Corporation for Science Advancement and a Columbia Presidential Teaching Award. He is a member of the international advisory board for *The Chemical Record* and the editorial board of *Chirality*. He lives north of New York City with his wife Cathy where he enjoys gardening, cooking, and watching movies.

[TO THE STUDENT]

Contrary to what you may have heard, organic chemistry does not have to be a difficult course. It will be a rigorous course, and it will offer a challenge. But you will learn more in it than in almost any course you will take—and what you learn will have a special relevance to life and the world around you. However, because organic chemistry can be approached in a logical and systematic way, you will find that with the right study habits, mastering organic chemistry can be a deeply satisfying experience. Here, then, are some suggestions about how to study:

1. **Keep up with your work from day to day—never let yourself get behind.** Organic chemistry is a course in which one idea almost always builds on another that has gone before. It is essential, therefore, that you keep up with, or better yet, be a little ahead of your instructor. Ideally, you should try to stay one day ahead of your instructor's lectures in your own class preparations. The lecture, then, will be much more helpful because you will already have some understanding of the assigned material. Your time in class will clarify and expand ideas that are already familiar ones.

2. **Study material in small units, and be sure that you understand each new section before you go on to the next.** Again, because of the cumulative nature of organic chemistry, your studying will be much more effective if you take each new idea as it comes and try to understand it completely before you move on to the next concept.

3. **Work all of the in-chapter and assigned problems.** One way to check your progress is to work each of the in-chapter problems when you come to it. These problems have been written just for this purpose and are designed to help you decide whether or not you understand the material that has just been explained. You should also carefully study the Solved Problems. If you understand a Solved Problem and can work the related in-chapter problem, then you should go on; if you cannot, then you should go back and study the preceding material again. Work all of the problems assigned by your instructor from the end of the chapter, as well. Do all of your problems in a notebook and bring this book with you when you go to see your instructor for extra help.

4. **Write when you study.** Write the reactions, mechanisms, structures, and so on, over and over again. Organic chemistry is best assimilated through the fingertips by writing, and not through the eyes by simply looking, or by highlighting material in the text, or by referring to flash cards. There is a good reason for this. Organic structures, mechanisms, and reactions are complex. If you simply examine them, you may think you understand them thoroughly, but that will be a misperception. The reaction mechanism may make sense to you in a certain way, but you need a deeper understanding than this. You need to know the material so thoroughly that you can explain it to someone else. This level of understanding comes to most of us (those of us without photographic memories) through writing. Only by writing the reaction mechanisms do we pay sufficient attention to their details, such as which atoms are connected to which atoms, which bonds break in a reaction and which bonds form, and the three-dimensional aspects of the structures. When we write reactions and mechanisms, connections are made in our brains that provide the long-term memory needed for success in organic chemistry. We virtually guarantee that your grade in the course will be directly proportional to the number of pages of paper that your fill with your own writing in studying during the term.

5. **Learn by teaching and explaining.** Study with your student peers and practice explaining concepts and mechanisms to each other. Use the Learning Group Problems and other exercises your instructor may assign as vehicles for teaching and learning interactively with your peers.

6. **Use the answers to the problems in the Study Guide in the proper way.** Refer to the answers only in two circumstances: (1) When you have finished a problem, use the Study Guide to check your answer. (2) When, after making a real effort to solve the problem, you find that you are completely stuck, then look at the answer for a clue and go back to work out the problem on your own. The value of a problem is in solving it. If you simply read the problem and look up the answer, you will deprive yourself of an important way to learn.

7. **Use molecular models when you study.** Because of the three-dimensional nature of most organic molecules, molecular models can be an invaluable aid to your understanding of them. When you need to see the three-dimensional aspect of a particular topic, use the Molecular Visions™ model set that may have been packaged with your textbook, or buy a set of models separately. An appendix to the *Study Guide* that accompanies this text provides a set of highly useful molecular model exercises.

8. **Make use of the rich online teaching resources in *WileyPLUS*** and do any online exercises that may be assigned by your instructor.

CHAPTER

1

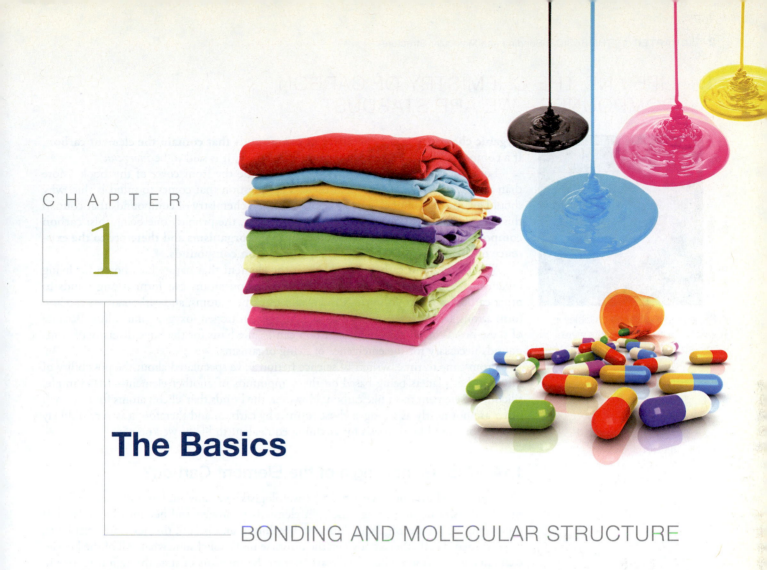

The Basics

BONDING AND MOLECULAR STRUCTURE

Organic chemistry plays a role in all aspects of our lives, from the clothing we wear, to the pixels of our television and computer screens, to preservatives in food, to the inks that color the pages of this book. If you take the time to understand organic chemistry, to learn its overall logic, then you will truly have the power to change society. Indeed, organic chemistry provides the power to synthesize new drugs, to engineer molecules that can make computer processors run more quickly, to understand why grilled meat can cause cancer and how its effects can be combated, and to design ways to knock the calories out of sugar while still making food taste deliciously sweet. It can explain biochemical processes like aging, neural functioning, and cardiac arrest, and show how we can prolong and improve life. It can do almost anything.

IN THIS CHAPTER WE WILL CONSIDER:

- what kinds of atoms make up organic molecules
- the principles that determine how the atoms in organic molecules are bound together
- how best to depict organic molecules

[**WHY** DO THESE TOPICS MATTER?]   At the end of the chapter, we will see how some of the unique organic structures that nature has woven together possess amazing properties that we can harness to aid human health.

1.1 LIFE AND THE CHEMISTRY OF CARBON COMPOUNDS—WE ARE STARDUST

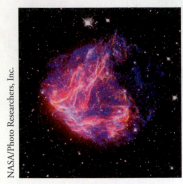

NASA/Photo Researchers, Inc.

Supernovae were the crucibles in which the heavy elements were formed.

Organic chemistry is the chemistry of compounds that contain the element carbon. If a compound does not contain the element carbon, it is said to be *inorganic*.

Look for a moment at the periodic table inside the front cover of this book. More than a hundred elements are listed there. The question that comes to mind is this: why should an entire field of chemistry be based on the chemistry of compounds that contain this one element, carbon? There are several reasons, the primary one being this: **carbon compounds are central to the structure of living organisms and therefore to the existence of life on Earth. We exist because of carbon compounds.**

What is it about carbon that makes it the element that nature has chosen for living organisms? There are two important reasons: carbon atoms can form strong bonds to other carbon atoms to form rings and chains of carbon atoms, and carbon atoms can also form strong bonds to elements such as hydrogen, nitrogen, oxygen, and sulfur. Because of these bond-forming properties, carbon can be the basis for the huge diversity of compounds necessary for the emergence of living organisms.

From time to time, writers of science fiction have speculated about the possibility of life on other planets being based on the compounds of another element—for example, silicon, the element most like carbon. However, the bonds that silicon atoms form to each other are not nearly as strong as those formed by carbon, and therefore it is very unlikely that silicon could be the basis for anything equivalent to life as we know it.

1.1A What Is the Origin of the Element Carbon?

Through the efforts of physicists and cosmologists, we now understand much of how the elements came into being. The light elements hydrogen and helium were formed at the beginning, in the Big Bang. Lithium, beryllium, and boron, the next three elements, were formed shortly thereafter when the universe had cooled somewhat. All of the heavier elements were formed millions of years later in the interiors of stars through reactions in which the nuclei of lighter elements fuse to form heavier elements.

The energy of stars comes primarily from the fusion of hydrogen nuclei to produce helium nuclei. This nuclear reaction explains why stars shine. Eventually some stars begin to run out of hydrogen, collapse, and explode—they become supernovae. Supernovae explosions scatter heavy elements throughout space. Eventually, some of these heavy elements drawn by the force of gravity became part of the mass of planets like the Earth.

1.1B How Did Living Organisms Arise?

This question is one for which an adequate answer cannot be given now because there are many things about the emergence of life that we do not understand. However, we do know this. Organic compounds, some of considerable complexity, are detected in outer space, and meteorites containing organic compounds have rained down on Earth since it was formed. A meteorite that fell near Murchison, Victoria, Australia, in 1969 was found to contain over 90 different amino acids, 19 of which are found in living organisms on Earth. While this does not mean that life arose in outer space, it does suggest that events in outer space may have contributed to the emergence of life on Earth.

In 1924 Alexander Oparin, a biochemist at the Moscow State University, postulated that life on Earth may have developed through the gradual evolution of carbon-based molecules in a "primordial soup" of the compounds that were thought to exist on a prebiotic Earth: methane, hydrogen, water, and ammonia. This idea was tested by experiments carried out at the University of Chicago in 1952 by Stanley Miller and Harold Urey. They showed that amino acids and other complex organic compounds are synthesized when an electric spark (think of lightning) passes through a flask containing a mixture of these four compounds (think of the early atmosphere). Miller and Urey in their 1953 publication reported that five amino acids (essential constituents of proteins) were formed. In 2008, examination of archived solutions from Miller and Urey's original

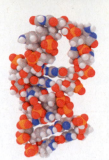

experiments have shown that 22 amino acids, rather than the 5 amino acids originally reported, were actually formed.

Similar experiments have shown that other precursors of biomolecules can also arise in this way—compounds such as ribose and adenine, two components of RNA. Some RNA molecules can not only store genetic information as DNA does, they can also act as catalysts, as enzymes do.

There is much to be discovered to explain exactly how the compounds in this soup became living organisms, but one thing seems certain. The carbon atoms that make up our bodies were formed in stars, so, in a sense, we are stardust.

An RNA molecule

1.1C Development of the Science of Organic Chemistry

The science of organic chemistry began to flower with the demise of a nineteenth century theory called vitalism. According to vitalism, organic compounds were only those that came from living organisms, and only living things could synthesize organic compounds through intervention of a vital force. Inorganic compounds were considered those compounds that came from nonliving sources. Friedrich Wöhler, however, discovered in 1828 that an organic compound called urea (a constituent of urine) could be made by evaporating an aqueous solution of the inorganic compound ammonium cyanate. With this discovery, the synthesis of an organic compound, began the evolution of organic chemistry as a scientific discipline.

$$NH_4^+NCO^- \xrightarrow{\text{heat}} \underset{\substack{\\ \text{Urea}}}{H_2N\overset{\displaystyle\overset{O}{\|}}{C}NH_2}$$

Ammonium cyanate   **Urea**

THE CHEMISTRY OF... Natural Products

Despite the demise of vitalism in science, the word "organic" is still used today by some people to mean "coming from living organisms" as in the terms "organic vitamins" and "organic fertilizers." The commonly used term "organic food" means that the food was grown without the use of synthetic fertilizers and pesticides. An "organic vitamin" means to these people that the vitamin was isolated from a natural source and not synthesized by a chemist. While there are sound arguments to be made against using food contaminated with certain pesticides, while there may be environmental benefits to be obtained from organic farming, and while "natural" vitamins may contain beneficial substances not present in synthetic vitamins, it is impossible to argue that pure "natural" vitamin C, for example, is healthier than pure "synthetic" vitamin C, since the two substances are identical in all respects. In science today, the study of compounds from living organisms is called natural products chemistry. In the closer to this chapter we will consider more about why natural products chemistry is important.

Vitamin C

Vitamin C is found in various citrus fruits.

1.2 ATOMIC STRUCTURE

Before we begin our study of the compounds of carbon we need to review some basic but familiar ideas about the chemical elements and their structure.

- The **compounds** we encounter in chemistry are made up of **elements** combined in different proportions.
- **Elements** are made up of **atoms**. An atom (Fig. 1.1) consists of a dense, positively charged *nucleus* containing **protons** and **neutrons** and a surrounding cloud of **electrons**.

Each proton of the nucleus bears one positive charge; electrons bear one negative charge. Neutrons are electrically neutral; they bear no charge. Protons and neutrons have

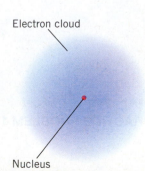

Electron cloud

Nucleus

FIGURE 1.1 An atom is composed of a tiny nucleus containing protons and neutrons and a large surrounding volume containing electrons. The diameter of a typical atom is about 10,000 times the diameter of its nucleus.

nearly equal masses (approximately 1 atomic mass unit each) and are about 1800 times as heavy as electrons. Most of the **mass** of an atom, therefore, comes from the mass of the nucleus; the atomic mass contributed by the electrons is negligible. Most of the **volume** of an atom, however, comes from the electrons; the volume of an atom occupied by the electrons is about 10,000 times larger than that of the nucleus.

The elements commonly found in organic molecules are carbon, hydrogen, nitrogen, oxygen, phosphorus, and sulfur, as well as the halogens: fluorine, chlorine, bromine, and iodine.

Each **element** is distinguished by its **atomic number** (**Z**), a **number equal to the number of protons in its nucleus**. Because an atom is electrically neutral, **the atomic number also equals the number of electrons surrounding the nucleus.**

1.2A Isotopes

Before we leave the subject of atomic structure and the periodic table, we need to examine one other observation: **the existence of atoms of the same element that have different masses**.

For example, the element carbon has six protons in its nucleus giving it an atomic number of 6. Most carbon atoms also have six neutrons in their nuclei, and because each proton and each neutron contributes one atomic mass unit (1 amu) to the mass of the atom, carbon atoms of this kind have a mass number of 12 and are written as ^{12}C.

- **Although all the nuclei of all atoms of the same element will have the same number of protons**, some atoms of the same element **may have different masses** because they have **different numbers of neutrons**. Such atoms are called **isotopes**.

For example, about 1% of the atoms of elemental carbon have nuclei containing 7 neutrons, and thus have a mass number of 13. Such atoms are written ^{13}C. A tiny fraction of carbon atoms have 8 neutrons in their nucleus and a mass number of 14. Unlike atoms of carbon-12 and carbon-13, atoms of carbon-14 are radioactive. The ^{14}C isotope is used in *carbon dating*. The three forms of carbon, ^{12}C, ^{13}C, and ^{14}C, are isotopes of one another.

Most atoms of the element hydrogen have one proton in their nucleus and have no neutron. They have a mass number of 1 and are written ^{1}H. A very small percentage (0.015%) of the hydrogen atoms that occur naturally, however, have one neutron in their nucleus. These atoms, called *deuterium* atoms, have a mass number of 2 and are written ^{2}H. An unstable (and radioactive) isotope of hydrogen, called *tritium* (^{3}H), has two neutrons in its nucleus.

PRACTICE PROBLEM 1.1   There are two stable isotopes of nitrogen, ^{14}N and ^{15}N. How many protons and neutrons does each isotope have?

1.2B Valence Electrons

We discuss the electron configurations of atoms in more detail in Section 1.10. For the moment we need only to point out that the electrons that surround the nucleus exist in **shells** of increasing energy and at increasing distances from the nucleus. The most important shell, called the **valence shell**, is the outermost shell because the electrons of this shell are the ones that an atom uses in making chemical bonds with other atoms to form compounds.

- How do we know how many electrons an atom has in its valence shell? We look at the periodic table. The number of electrons in the valence shell (called **valence electrons**) is equal to the group number of the atom. For example, carbon is in group **IVA** and carbon has *four* valence electrons; oxygen is in group **VIA** and oxygen has *six* valence electrons. The halogens of group **VIIA** all have *seven* electrons.

PRACTICE PROBLEM 1.2   How many valence electrons does each of the following atoms have?

(a) Na   **(b)** Cl   **(c)** Si   **(d)** B   **(e)** Ne   **(f)** N

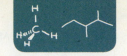

1.3 CHEMICAL BONDS: THE OCTET RULE

The first explanations of the nature of chemical bonds were advanced by G. N. Lewis (of the University of California, Berkeley) and W. Kössel (of the University of Munich) in 1916. Two major types of chemical bonds were proposed:

1. **Ionic** (or electrovalent) bonds are formed by the transfer of one or more electrons from one atom to another to create ions.
2. **Covalent** bonds result when atoms share electrons.

The central idea in their work on bonding is that atoms without the electronic configuration of a noble gas generally react to produce such a configuration because these configurations are known to be highly stable. For all of the noble gases except helium, this means achieving an octet of electrons in the valence shell.

- The **valence shell** is the outermost shell of electrons in an atom.
- The tendency for an atom to achieve a configuration where its valence shell contains eight electrons is called the **octet rule**.

The concepts and explanations that arise from the original propositions of Lewis and Kössel are satisfactory for explanations of many of the problems we deal with in organic chemistry today. For this reason we shall review these two types of bonds in more modern terms.

1.3A Ionic Bonds

Atoms may gain or lose electrons and form charged particles called **ions**.

- An **ionic bond** is an attractive force between oppositely charged ions.

One source of such ions is a reaction between atoms of widely differing electronegativities (Table 1.1).

- **Electronegativity** *is a measure of the ability of an atom to attract electrons.*
- Electronegativity increases as we go across a horizontal row of the periodic table from left to right and it increases as we go up a vertical column (Table 1.1).

An example of the formation of an ionic bond is the reaction of lithium and fluorine atoms:

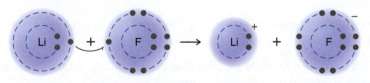

Lithium, a typical metal, has a very low electronegativity; fluorine, a nonmetal, is the most electronegative element of all. The loss of an electron (a negatively charged species)

Helpful Hint

Terms and concepts that are fundamentally important to your learning organic chemistry are set in bold blue type. You should learn them as they are introduced. These terms are also defined in the glossary.

Helpful Hint

We will use electronegativity frequently as a tool for understanding the properties and reactivity of organic molecules.

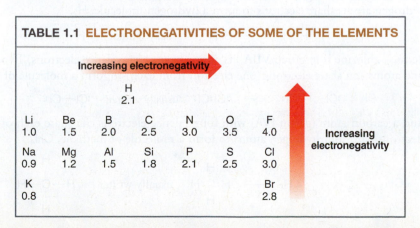

TABLE 1.1 **ELECTRONEGATIVITIES OF SOME OF THE ELEMENTS**

Increasing electronegativity →

			H 2.1			
Li 1.0	Be 1.5	B 2.0	C 2.5	N 3.0	O 3.5	F 4.0
Na 0.9	Mg 1.2	Al 1.5	Si 1.8	P 2.1	S 2.5	Cl 3.0
K 0.8						Br 2.8

Increasing electronegativity ↑

by the lithium atom leaves a lithium cation (Li⁺); the gain of an electron by the fluorine atom gives a fluoride anion (F⁻).

- Ions form because atoms can achieve the electronic configuration of a noble gas by gaining or losing electrons.

The lithium cation with two electrons in its valence shell is like an atom of the noble gas helium, and the fluoride anion with eight electrons in its valence shell is like an atom of the noble gas neon. Moreover, crystalline lithium fluoride forms from the individual lithium and fluoride ions. In this process negative fluoride ions become surrounded by positive lithium ions, and positive lithium ions by negative fluoride ions. In this crystalline state, the ions have substantially lower energies than the atoms from which they have been formed. Lithium and fluorine are thus "stabilized" when they react to form crystalline lithium fluoride.

We represent the formula for lithium fluoride as LiF, because that is the simplest formula for this ionic compound.

Ionic substances, because of their strong internal electrostatic forces, are usually very high melting solids, often having melting points above 1000 °C. In polar solvents, such as water, the ions are solvated (see Section 2.13D), and such solutions usually conduct an electric current.

- Ionic compounds, often called **salts**, form only when atoms of very different electronegativities transfer electrons to become ions.

PRACTICE PROBLEM 1.3 Using the periodic table, which element in each pair is more electronegative?
(a) Si, O **(b)** N, C **(c)** Cl, Br **(d)** S, P

1.3B Covalent Bonds and Lewis Structures

When two or more atoms of the same or similar electronegativities react, a complete transfer of electrons does not occur. In these instances the atoms achieve noble gas configurations by *sharing electrons*.

- **Covalent bonds** form by sharing of electrons between atoms of similar electronegativities to achieve the configuration of a noble gas.
- **Molecules** are composed of atoms joined exclusively or predominantly by covalent bonds.

Molecules may be represented by electron-dot formulas or, more conveniently, by formulas where each pair of electrons shared by two atoms is represented by a line.

- A **dash structural formula** has lines that show bonding electron pairs and includes elemental symbols for the atoms in a molecule.

Some examples are shown here:

1. Hydrogen, being in group IA of the periodic table, has one valence electron. Two hydrogen atoms share electrons to form a hydrogen molecule, H₂.

   H₂   H· + ·H ⟶ H:H   usually written   H—H

2. Because chlorine is in group VIIA, its atoms have seven valence electrons. Two chlorine atoms can share electrons (one electron from each) to form a molecule of Cl₂.

   Cl₂   :Cl· + ·Cl: ⟶ :Cl:Cl: usually written :Cl—Cl:

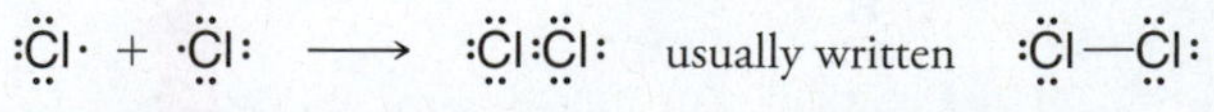

3. And a carbon atom (group IVA) with four valence electrons can share each of these electrons with four hydrogen atoms to form a molecule of methane, CH₄.

   CH₄   ·Ċ· + 4 H· ⟶ H:C:H usually written H—C—H (with H above and below)

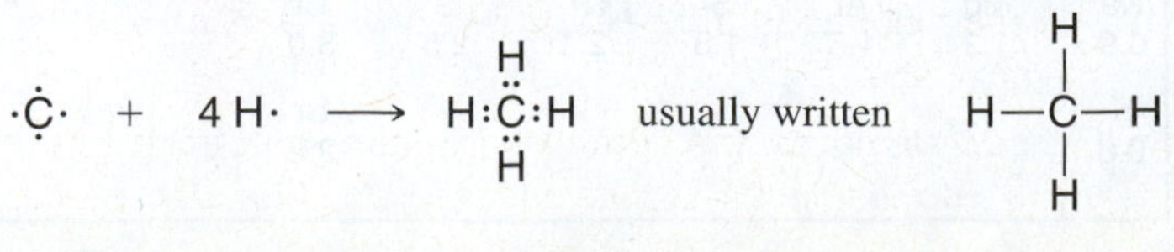

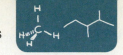

Two carbon atoms can use one electron pair between them to form a **carbon–carbon single bond** while also bonding hydrogen atoms or other groups to achieve an octet of valence electrons. Consider the example of ethane below.

C_2H_6 H:C:C:H   and as a dash formula

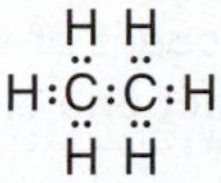

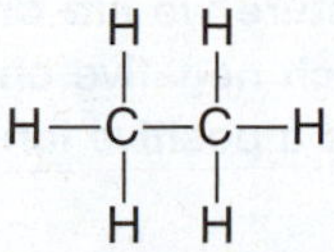

Ethane

These formulas are often called **Lewis structures**; in writing them we show all of the valence electrons. Unshared electron pairs are shown as dots, and in dash structural formulas, bonding electron pairs are shown as lines.

4. Atoms can share *two or more pairs of electrons* to form **multiple covalent bonds**. For example, two nitrogen atoms possessing five valence electrons each (because nitrogen is in group VA) can share electrons to form a **triple bond** between them.

N_2 :N::N: and as a dash formula   :N≡N:

Carbon atoms can also share more than one electron pair with another atom to form a multiple covalent bond. Consider the examples of a **carbon–carbon double bond** in ethene (ethylene) and a **carbon–carbon triple bond** in ethyne (acetylene).

C_2H_4 :C::C: and as a dash formula

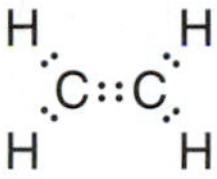

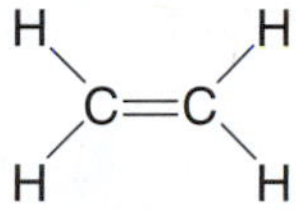

Ethene

C_2H_2 H:C::C:H   and as a dash formula   H—C≡C—H

Ethyne

5. Ions, themselves, may contain covalent bonds. Consider, as an example, the ammonium ion.

NH_4^+ H:N:H   and as a dash formula

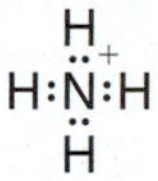

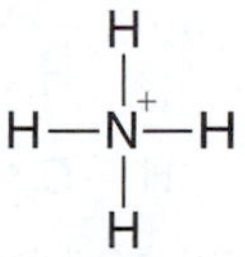

PRACTICE PROBLEM 1.4

Consider the following compounds and decide whether the bond in them would be ionic or covalent.

(a) KCl   (b) F_2 (c) PH_3 (d) CBr_4

1.4 HOW TO WRITE LEWIS STRUCTURES

Several simple rules allow us to draw proper Lewis structures:

1. **Lewis structures show the connections between atoms in a molecule or ion using only the valence electrons of the atoms involved.** Valence electrons are those of an atom's outermost shell.

2. **For main group elements, the number of valence electrons a neutral atom brings to a Lewis structure is the same as its group number in the periodic table.**

Helpful Hint

The ability to write proper **Lewis structures** is one of the most important tools for learning organic chemistry.

Carbon, for example, is in group IVA and has four valence electrons; the halogens (e.g., fluorine) are in group VIIA and each has seven valence electrons; hydrogen is in group IA and has one valence electron.

3. **If the structure we are drawing is a negative ion (an anion), we add one electron for each negative charge to the original count of valence electrons. If the structure is a positive ion (a cation), we subtract one electron for each positive charge.**

4. **In drawing Lewis structures we try to give each atom the electron configuration of a noble gas.** To do so, we draw structures where atoms share electrons to form covalent bonds or transfer electrons to form ions.

    a. Hydrogen forms one covalent bond by sharing its electron with an electron of another atom so that it can have two valence electrons, the same number as in the noble gas helium.

    b. Carbon forms four covalent bonds by sharing its four valence electrons with four valence electrons from other atoms, so that it can have eight electrons (the same as the electron configuration of neon, satisfying the octet rule).

    c. To achieve an octet of valence electrons, elements such as nitrogen, oxygen, and the halogens typically share only some of their valence electrons through covalent bonding, leaving others as unshared electron pairs.

The following problems illustrate the rules above.

●●● SOLVED PROBLEM 1.1

Write the Lewis structure of CH_3F.

STRATEGY AND ANSWER:

1. We find the total number of valence electrons of all the atoms:

$$4 + 3(1) + 7 = 14$$
$$\quad\uparrow\qquad\uparrow\qquad\uparrow$$
$$\quad C\quad\ 3H\quad F$$

2. We use pairs of electrons to form bonds between all atoms that are bonded to each other. We represent these bonding pairs with lines. In our example this requires four pairs of electrons (8 of the 14 valence electrons).

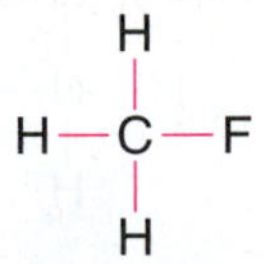

$$\begin{array}{c} \text{H} \\ | \\ \text{H}-\text{C}-\text{F} \\ | \\ \text{H} \end{array}$$

3. We then add the remaining electrons in pairs so as to give each hydrogen 2 electrons (a duet) and every other atom 8 electrons (an octet). In our example, we assign the remaining 6 valence electrons to the fluorine atom in three nonbonding pairs.

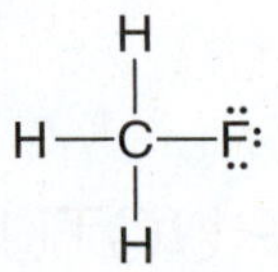

$$\begin{array}{c} \text{H} \\ | \\ \text{H}-\text{C}-\ddot{\text{F}}\text{:} \\ | \\ \text{H} \end{array}$$

●●●

PRACTICE PROBLEM 1.5   Write the Lewis structure of (a) CH_2Fl_2 (difluoromethane) and (b) $CHCl_3$ (chloroform).

Write a Lewis structure for methylamine (CH_3NH_2).

STRATEGY AND ANSWER:

1. We find the total number of valence electrons for all the atoms.

$$\begin{array}{ccc} 4 & 5 & 5(1) = 14 = 7 \text{ pairs} \\ \uparrow & \uparrow & \uparrow \\ C & N & 5H \end{array}$$

2. We use one electron pair to join the carbon and nitrogen.

$$C-N$$

3. We use three pairs to form single bonds between the carbon and three hydrogen atoms.
4. We use two pairs to form single bonds between the nitrogen atom and two hydrogen atoms.
5. This leaves one electron pair, which we use as a lone pair on the nitrogen atom.

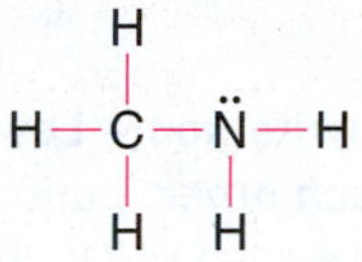

$$\begin{array}{c} H \\ | \\ H-C-\ddot{N}-H \\ | \quad | \\ H \quad H \end{array}$$

●●●

Write the Lewis structure of CH_3OH.

PRACTICE PROBLEM 1.6

5. **If necessary, we use multiple bonds to satisfy the octet rule (i.e., give atoms the noble gas configuration).** The carbonate ion (CO_3^{2-}) illustrates this:

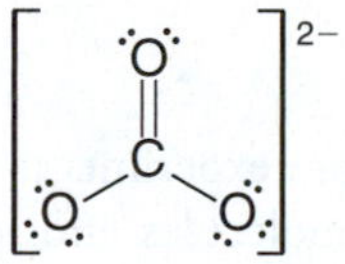

$$\left[\begin{array}{c} \ddot{O} \\ \| \\ C \\ \diagup \quad \diagdown \\ \ddot{O}: \quad :\ddot{O}: \end{array} \right]^{2-}$$

The organic molecules ethene (C_2H_4) and ethyne (C_2H_2), as mentioned earlier, have a double and triple bond, respectively:

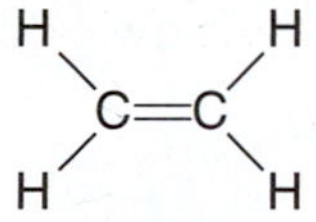

$$\begin{array}{cc} H \qquad H \\ \diagdown \quad \diagup \\ C=C \\ \diagup \quad \diagdown \\ H \qquad H \end{array} \qquad \text{and} \qquad H-C\equiv C-H$$

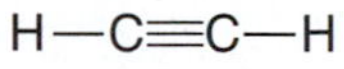

●●● SOLVED PROBLEM 1.3

Write the Lewis structure of CH_2O (formaldehyde).

STRATEGY AND ANSWER:

1. Find the total number of valence electrons of all the atoms:

$$\begin{array}{ccc} 2(1) + 1(4) + 1(6) = 12 \\ \uparrow \qquad \uparrow \qquad \uparrow \\ 2H \qquad 1C \qquad 1O \end{array}$$

2. **(a)** Use pairs of electrons to form single bonds.

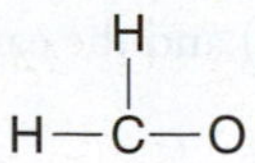

$$\begin{array}{c} H \\ | \\ H-C-O \end{array}$$

(continues on next page)

(b) Determine which atoms already have a full valence shell and which ones do not, and how many valence electrons we have used so far. In this case, we have used 6 valence electrons, and the valence shell is full for the hydrogen atoms but not for the carbon and oxygen.

(c) We use the remaining electrons as bonds or unshared electron pairs, to fill the valence shell of any atoms whose valence shell is not yet full, taking care not to exceed the octet rule. In this case 6 of the initial 12 valence electrons are left to use. We use 2 electrons to fill the valence shell of the carbon by another bond to the oxygen, and the remaining 4 electrons as two unshared electron pairs with the oxygen, filling its valence shell.

$$\begin{array}{c} H \\ | \\ H-C=\ddot{O}: \end{array}$$

• • •

PRACTICE PROBLEM 1.7 Write a dash structural formula showing all valence electrons for CH_3CHO (acetaldehyde).

6. **Before we can write some Lewis structures, *we must know how the atoms are connected to each other***. Consider nitric acid, for example. Even though the formula for nitric acid is often written HNO_3, the hydrogen is actually connected to an oxygen, not to the nitrogen. The structure is $HONO_2$ and not HNO_3. Thus the correct Lewis structure is:

$$\begin{array}{ccccc} & & \ddot{O}: & & \\ & & \| & & \\ H-\ddot{O}-N & & \text{and not} & H-N-\ddot{O}-\ddot{O}: \\ & & \overset{..}{\underset{..}{O}}: & & \overset{\|}{\underset{..}{:O}}. \end{array}$$

Helpful Hint

Check your progress by doing each Practice Problem as you come to it in the text.

This knowledge comes ultimately from experiments. If you have forgotten the structures of some of the common inorganic molecules and ions (such as those listed in Practice Problem 1.8), this may be a good time for a review of the relevant portions of your general chemistry text.

• • • **SOLVED PROBLEM 1.4**

Assume that the atoms are connected in the same way they are written in the formula, and write a Lewis structure for the toxic gas hydrogen cyanide (HCN).

STRATEGY AND ANSWER:

1. We find the total number of valence electrons on all of the atoms:

$$1 + 4 + 5 = 10$$
$$\uparrow \quad \uparrow \quad \uparrow$$
$$H \quad C \quad N$$

2. We use one pair of electrons to form a single bond between the hydrogen atom and the carbon atom (see below), and we use three pairs to form a triple bond between the carbon atom and the nitrogen atom. This leaves two electrons. We use these as an unshared pair on the nitrogen atom. Now each atom has the electronic structure of a noble gas. The hydrogen atom has two electrons (like helium) and the carbon and nitrogen atoms each have eight electrons (like neon).

$$H-C\equiv N:$$

PRACTICE PROBLEM 1.8

Write a Lewis structure for each of the following:

(a) HF **(c)** CH_3F **(e)** H_2SO_3 **(g)** H_3PO_4
(b) F_2 **(d)** HNO_2 **(f)** BH_4^- **(h)** H_2CO_3

1.4A Exceptions to the Octet Rule

Atoms share electrons, not just to obtain the configuration of an inert gas, but because sharing electrons produces increased electron density between the positive nuclei. The resulting attractive forces of nuclei for electrons is the "glue" that holds the atoms together (cf. Section 1.11).

- Elements of the second period of the periodic table can have a maximum of four bonds (i.e., have eight electrons around them) because these elements have only one 2*s* and three 2*p* orbitals available for bonding.

Each orbital can contain two electrons, and a total of eight electrons fills these orbitals (Section 1.10A). The **octet rule**, therefore, only applies to these elements, and even here, as we shall see in compounds of beryllium and boron, fewer than eight electrons are possible.

- Elements of the third period and beyond have *d* orbitals that can be used for bonding.

These elements can accommodate more than eight electrons in their valence shells and therefore can form more than four covalent bonds. Examples are compounds such as PCl_5 and SF_6. Bonds written as (dashed wedges) project behind the plane of the paper. Bonds written as (solid wedges) project in front of the paper.

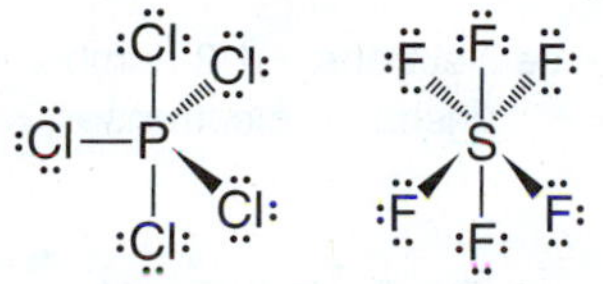

SOLVED PROBLEM 1.5

Write a Lewis structure for the sulfate ion (SO_4^{2-}). (*Note*: The sulfur atom is bonded to all four oxygen atoms.)

STRATEGY AND ANSWER:

1. We find the total number of valence electrons including the extra 2 electrons needed to give the ion the double negative charge:

$$6 + 4(6) + 2 = 32$$

↑ S ↑ 4O ↑ $2e^-$

2. We use four pairs of electrons to form bonds between the sulfur atom and the four oxygen atoms:

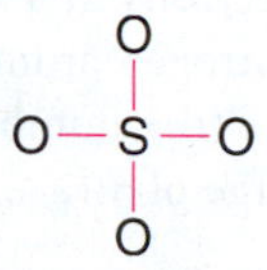

3. We add the remaining 24 electrons as unshared pairs on oxygen atoms and as **double bonds** between the sulfur atom and two oxygen atoms. This gives each oxygen 8 electrons and the sulfur atom 12:

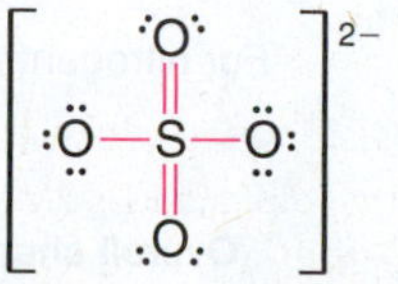

PRACTICE PROBLEM 1.9

Write a Lewis structure for the phosphate ion (PO_4^{3-}).

Some highly reactive molecules or ions have atoms with fewer than eight electrons in their outer shell. An example is boron trifluoride (BF_3). In a BF_3 molecule the central boron atom has only six electrons around it:

$$:\ddot{F}:$$
$$\underset{:\ddot{F}\quad\quad\ddot{F}:}{B}$$

1.5 FORMAL CHARGES AND **HOW TO** CALCULATE THEM

Helpful Hint

Proper assignment of **formal charges** is another essential tool for learning organic chemistry.

Many **Lewis structures** are incomplete until we decide whether any of their atoms have a **formal charge**. Calculating the formal charge on an atom in a Lewis structure is simply a bookkeeping method for its valence electrons.

- First, we examine each atom and, using the periodic table, we determine how many **valence electrons** it would have if it were an atom not bonded to any other atoms. **This is equal to the group number of the atom in the periodic table.** For hydrogen this number equals 1, for carbon it equals 4, for nitrogen it equals 5, and for oxygen it equals 6.

Next, we examine the atom in the Lewis structure and we assign the valence electrons in the following way:

- **We assign to each atom half of the electrons it is sharing with another atom and all of its unshared (lone) electron pairs.**

Then we do the following calculation for the atom:

Formal charge = number of valence electrons − 1/2 number of shared electrons − number of unshared electrons

or

$$F = Z - (1/2)S - U$$

where F is the formal charge, Z is the group number of the element, S equals the number of shared electrons, and U is the number of unshared electrons.

- It is important to note, too, that **the arithmetic sum of all the formal charges in a molecule or ion will equal the overall charge on the molecule or ion**.

Let us consider several examples showing how this is done.

The Ammonium Ion (NH_4^+) As we see below, the ammonium ion has no unshared electron pairs. We divide all of the electrons in bonds equally between the atoms that share them. Thus, each hydrogen is assigned one electron. We subtract this from one (the number of valence electrons in a hydrogen atom) to give each hydrogen atom a formal charge of zero. The nitrogen atom is assigned four electrons (one from each bond). We subtract four from five (the number of valence electrons in a nitrogen atom) to give the nitrogen a formal charge of +1.

$$\begin{matrix} & H & + \\ & \ddot{} & \\ H & :N: & H \\ & \ddot{} & \\ & H & \end{matrix}$$

For hydrogen:	valence electrons of free atom	=	1
	subtract assigned electrons	=	− 1
	Formal charge on each hydrogen	=	0
For nitrogen:	valence electrons of free atom	=	5
	subtract assigned electrons	=	− (1/2)8
	Formal charge on nitrogen	=	+1

Overall charge on ion = 4(0) + 1 = +1

The Nitrate Ion (NO_3^-) Let us next consider the nitrate ion (NO_3^-), an ion that has oxygen atoms with unshared electron pairs. Here we find that the nitrogen atom has a formal charge of +1, that two oxygen atoms have formal charges of −1, and that one oxygen has a formal charge equal to 0.

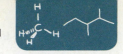

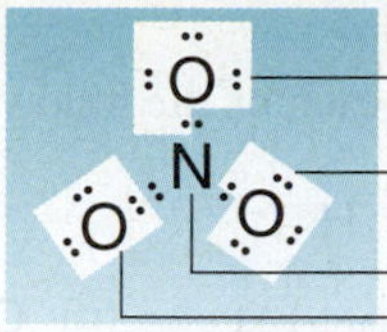

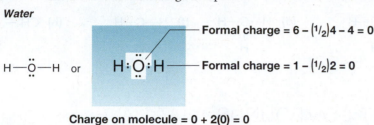

Formal charge = $6 - (^1/_2)2 - 6 = -1$
Formal charge = $6 - (^1/_2)2 - 6 = -1$
Formal charge = $5 - (^1/_2)8 = +1$
Formal charge = $6 - (^1/_2)4 - 4 = 0$
Charge on ion = $2(-1) + 1 + 0 = -1$

Water and Ammonia The sum of the formal charges on each atom making up a molecule must be zero. Consider the following examples:

Water

H—Ö—H or H:Ö:H

Formal charge = $6 - (^1/_2)4 - 4 = 0$
Formal charge = $1 - (^1/_2)2 = 0$

Charge on molecule = $0 + 2(0) = 0$

Ammonia

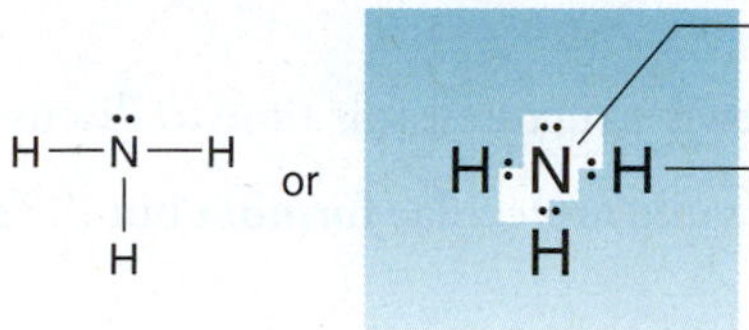

H—N—H or H:N:H
    | |
    H   H

Formal charge = $5 - (^1/_2)6 - 2 = 0$
Formal charge = $1 - (^1/_2)2 = 0$

Charge on molecule = $0 + 3(0) = 0$

Write a Lewis structure for each of the following negative ions, and assign the formal negative charge to the correct atom:

(a) CH_3O^- (b) NH_2^- (c) CN^- (d) HCO_2^- (e) HCO_3^- (f) HC_2^-

PRACTICE PROBLEM 1.10

1.5A A Summary of Formal Charges

With this background, it should now be clear that each time an oxygen atom of the type —Ö: appears in a molecule or ion, it will have a formal charge of -1, and that each time an oxygen atom of the type =Ö or —Ö— appears, it will have a formal charge of 0. Similarly, —N— will be $+1$, and —N̈— will be zero. These and other common structures are summarized in Table 1.2.

Helpful Hint

In later chapters, when you are evaluating how reactions proceed and what products form, you will find it essential to keep track of formal charges.

TABLE 1.2 A SUMMARY OF FORMAL CHARGES			
Group	Formal Charge of +1	Formal Charge of 0	Formal Charge of −1
IIIA		∖B∕	—B̈—
IVA	∖Ċ⁺∕ =Ċ⁺— ≡Ċ⁺	—C̈— =C̈∕ ≡C—	—C̈⁻— =C̈·⁻ ≡C:⁻
VA	—N⁺— =N⁺∕ =N⁺—	—N̈— ∕N̈∖ ≡N:	—N̈⁻— =N̈·⁻
VIA	—Ö⁺— ∕Ö⁺∖	—Ö— =Ö:	—Ö:⁻
VIIA	—Ẍ⁺—	—Ẍ: (X = F, Cl, Br, or I)	:Ẍ:⁻

PRACTICE PROBLEM 1.11 Assign the proper formal charge to the colored atom in each of the following structures:

(a)
```
      H    H
      |    |
  H — C —— C
      |    |
      H    H
```

(c)
```
       ··
      :O:
      ‖
      C
     / \
    H   O:
        ··
```

(e)
```
      H    H
      |    |
  H — C —— N — H
      |    |
      H    H
```

(g) CH_3—C≡N:

(b)
```
  H — Ö — H
       |
       H
```

(d)
```
      :Ö:
      |
  H — C — H
      |
      H
```

(f)
```
   H — Ö — H
       |
   H — C — H
       |
       H
```

(h) CH_3—N≡N:

1.6 ISOMERS: DIFFERENT COMPOUNDS THAT HAVE THE SAME MOLECULAR FORMULA

Now that we have had an introduction to Lewis structures, it is time to discuss isomers.

- **Isomers** are compounds that have the same **molecular formula** but different structures.

We will learn about several kinds of isomers during the course of our study. For now, let us consider a type called constitutional isomers.

- **Constitutional isomers** are different compounds that have the same molecular formula but differ in the sequence in which their atoms are bonded—that is, their **connectivity**.

CONTENTS: ACETONE, WATER, GLYCERIN, DIGLYCEROL, GELATIN, FRAGRANCE, YELLOW 11.

Acetone is used in some nailpolish removers.

Acetone, used in nail polish remover and as a paint solvent, and propylene oxide, used with seaweed extracts to make food-grade thickeners and foam stabilizers for beer (among other applications) are isomers. Both of these compounds have the molecular formula C_3H_6O and therefore the same molecular weight. Yet acetone and propylene oxide have distinctly different boiling points and chemical reactivity that, as a result, lend themselves to distinctly different practical applications. Their shared molecular formula simply gives us no basis for understanding the differences between them. We must, therefore, move to a consideration of their structural formulas.

On examining the structures of acetone and propylene oxide several key aspects are clearly different (Fig. 1.2). Acetone contains a double bond between the oxygen atom and the central carbon atom. Propylene oxide does not contain a double bond, but has three atoms joined in a ring. The connectivity of the atoms is clearly different in acetone

Propylene oxide alginates, made from propylene oxide and seaweed extracts, are used as food thickeners.

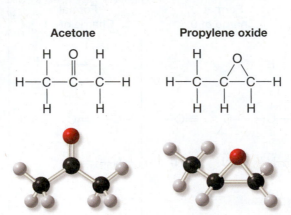

Acetone
```
   H   O   H
   |   ‖   |
H—C — C — C—H
   |       |
   H       H
```

Propylene oxide
```
   H        H
   |       O
H—C — C  C—H
   |   \ /  |
   H    C   H
   H    H
```

FIGURE 1.2 Ball-and-stick models and chemical formulas show the different structures of acetone and propylene oxide.

Helpful Hint

Build handheld models of these compounds and compare their structures.

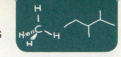

and propylene oxide. Their structures have the same molecular formula but a different constitution. They are constitutional isomers.*

- Constitutional isomers usually have different physical properties (e.g., melting point, boiling point, and density) and different chemical properties (reactivity).

●●● **SOLVED PROBLEM 1.6**

There are two constitutional isomers with the formula C_2H_6O. Write structural formulas for these isomers.

STRATEGY AND ANSWER: If we recall that carbon can form four covalent bonds, oxygen can form two, and hydrogen only one, we can arrive at the following constitutional isomers.

$$\underset{\textbf{Dimethyl ether}}{H-\overset{\overset{\textstyle H}{|}}{\underset{\underset{\textstyle H}{|}}{C}}-O-\overset{\overset{\textstyle H}{|}}{\underset{\underset{\textstyle H}{|}}{C}}-H} \qquad \underset{\textbf{Ethanol}}{H-\overset{\overset{\textstyle H}{|}}{\underset{\underset{\textstyle H}{|}}{C}}-\overset{\overset{\textstyle H}{|}}{\underset{\underset{\textstyle H}{|}}{C}}-O-H}$$

It should be noted that these two isomers are clearly different in their physical properties. At room temperature and 1 atm pressure, dimethyl ether is a gas. Ethanol is a liquid.

●●● **SOLVED PROBLEM 1.7**

Which of the following compounds are constitutional isomers of one another?

ANSWER: First determine the molecular formula for each compound. You will then see that **B** and **D** have the same molecular formula (C_4H_8O) but have different connectivities. They are, therefore, constitutional isomers of each other. **A**, **C**, and **E** also have the same molecular formula (C_3H_6O) and are constitutional isomers of one another.

1.7 HOW TO WRITE AND INTERPRET STRUCTURAL FORMULAS

Organic chemists use a variety of formats to write **structural formulas**. We have already used electron-dot formulas and dash formulas in previous sections. Two other important types of formulas are condensed formulas and bond-line (skeletal) formulas. Examples of these four types of structural formulas are shown in Fig. 1.3 using propyl alcohol as an example.

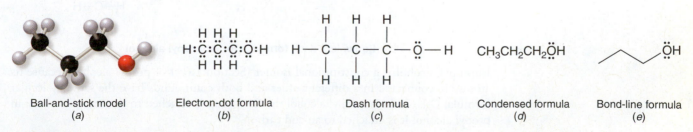

| Ball-and-stick model | Electron-dot formula | Dash formula | Condensed formula | Bond-line formula |
| (a) | (b) | (c) | (d) | (e) |

FIGURE 1.3 Structural formulas for propyl alcohol.

*An older term for isomers of this type was **structural isomers**. The International Union of Pure and Applied Chemistry (IUPAC) now recommends that use of the term "structural" when applied to constitutional isomers be abandoned.

Although electron-dot formulas account explicitly for all of the valence electrons in a molecule, they are tedious and time-consuming to write. Dash, condensed, and bond-line formulas are therefore used more often.

Generally it is best to draw unshared electron pairs in chemical formulas, though sometimes they are omitted if we are not considering the chemical properties or reactivity of a compound. When we write chemical reactions, however, we shall see that it is necessary to include the unshared electron pairs when they participate in a reaction. It is a good idea, therefore, to be in the habit of writing unshared electrons pairs.

1.7A More About Dash Structural Formulas

- **Dash structural formulas** have lines that show bonding electron pairs, and include elemental symbols for all of the atoms in a molecule.

If we look at the ball-and-stick model for propyl alcohol given in Fig. 1.3*a* and compare it with the electron-dot, dash, and condensed formulas in Figs. 1.3*b–d* we find that the chain of atoms is straight in those formulas. In the model, which corresponds more accurately to the actual shape of the molecule, the chain of atoms is not at all straight. Also of importance is this: ***Atoms joined by single bonds can rotate relatively freely with respect to one another***. (We shall discuss the reason for this in Section 1.12B.) This relatively free rotation means that the chain of atoms in propyl alcohol can assume a variety of arrangements like these:

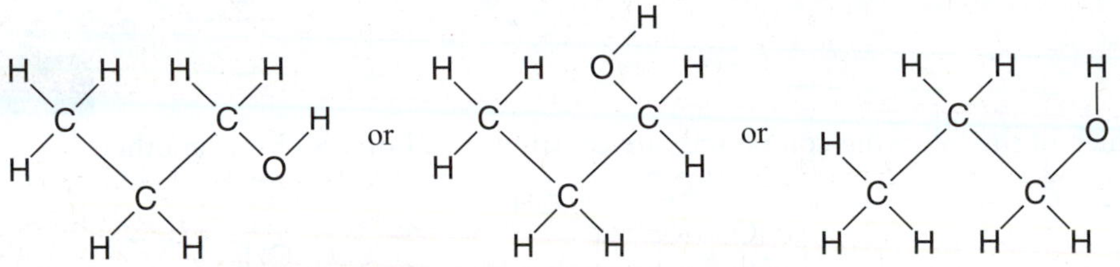

Equivalent dash formulas for propyl alcohol

Helpful Hint

It is important that you be able to recognize when a set of structural formulas has the same connectivity versus when they are constitutional isomers.

It also means that all of the structural formulas above are *equivalent* and all represent propyl alcohol. Dash structural formulas such as these indicate the way in which the atoms are attached to each other and *are not* representations of the actual shapes of the molecule. (Propyl alcohol does not have 90° **bond angles**. It has tetrahedral bond angles.) Dash structural formulas show what is called the **connectivity** of the atoms. *Constitutional isomers (Section 1.6A) have different connectivities and, therefore, must have different structural formulas.*

Consider the compound called isopropyl alcohol, whose formula we might write in a variety of ways:

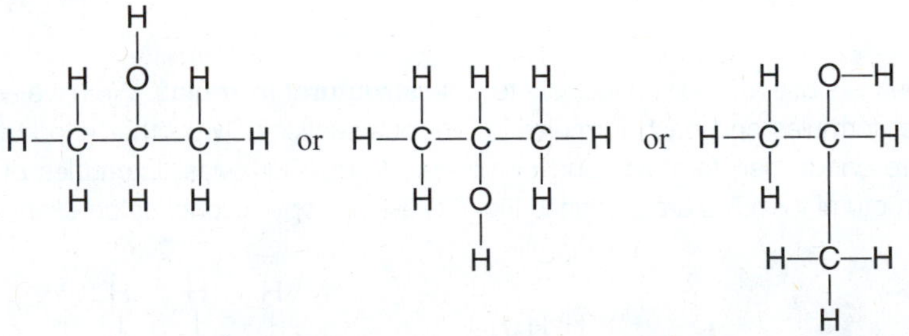

Equivalent dash formulas for isopropyl alcohol

Isopropyl alcohol is a constitutional isomer (Section 1.6A) of propyl alcohol because its atoms are connected in a different order and both compounds have the same molecular formula, C_3H_8O. In isopropyl alcohol the OH group is attached to the central carbon; in propyl alcohol it is attached to an end carbon.

- In problems you will often be asked to write structural formulas for all the isomers that have a given molecular formula. Do not make the error of writing several equivalent formulas, like those that we have just shown, mistaking them for different constitutional isomers.

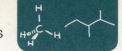

There are actually three constitutional isomers with the molecular formula C_3H_8O. We have seen two of them in propyl alcohol and isopropyl alcohol. Write a dash formula for the third isomer.

PRACTICE PROBLEM 1.12

1.7B Condensed Structural Formulas

Condensed structural formulas are somewhat faster to write than dash formulas and, when we become familiar with them, they will impart all the information that is contained in the dash structure. In condensed formulas all of the hydrogen atoms that are attached to a particular carbon are usually written immediately after the carbon. In fully condensed formulas, all of the atoms that are attached to the carbon are usually written immediately after that carbon, listing hydrogens first. For example,

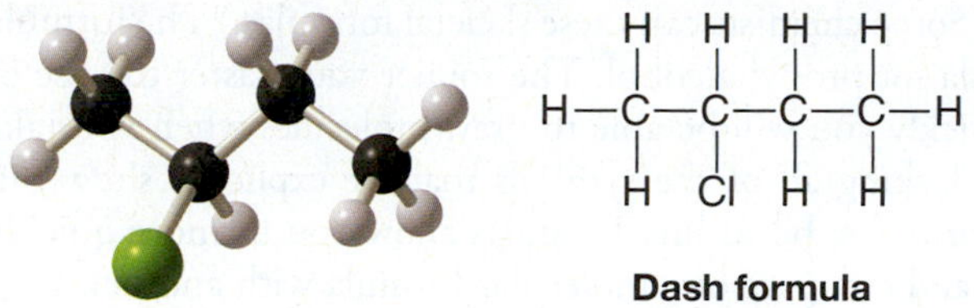

$$H-\underset{\underset{H}{|}}{\overset{\overset{H}{|}}{C}}-\underset{\underset{Cl}{|}}{\overset{\overset{H}{|}}{C}}-\underset{\underset{H}{|}}{\overset{\overset{H}{|}}{C}}-\underset{\underset{H}{|}}{\overset{\overset{H}{|}}{C}}-H$$

Dash formula

$CH_3CHCH_2CH_3$ or $CH_3CHClCH_2CH_3$

with Cl below the second carbon

Condensed formulas

The condensed formula for isopropyl alcohol can be written in four different ways:

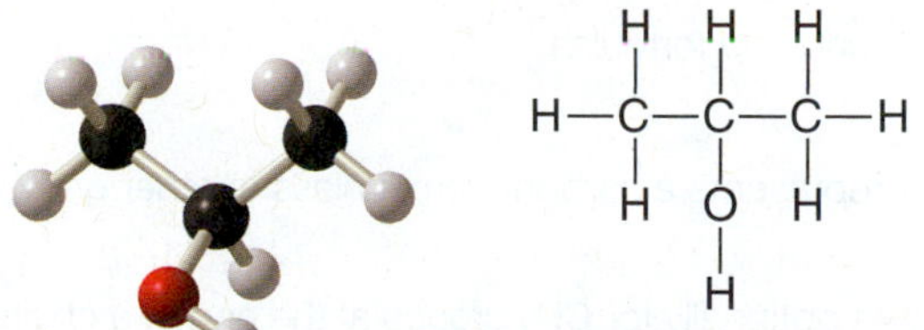

$$H-\underset{\underset{H}{|}}{\overset{\overset{H}{|}}{C}}-\underset{\underset{O}{|}}{\overset{\overset{H}{|}}{C}}-\underset{\underset{H}{|}}{\overset{\overset{H}{|}}{C}}-H$$

Dash formula

CH_3CHCH_3 or $CH_3CH(OH)CH_3$
with OH below the middle carbon

$CH_3CHOHCH_3$ or $(CH_3)_2CHOH$

Condensed formulas

●●● **SOLVED PROBLEM 1.8**

Write a condensed structural formula for the compound that follows:

$$H-\underset{\underset{H}{|}}{\overset{\overset{H}{|}}{C}}-\underset{\underset{\underset{\underset{H}{|}}{C}-H}{|}}{\overset{\overset{H}{|}}{C}}-\underset{\underset{H}{|}}{\overset{\overset{H}{|}}{C}}-\underset{\underset{H}{|}}{\overset{\overset{H}{|}}{C}}-H$$

ANSWER:

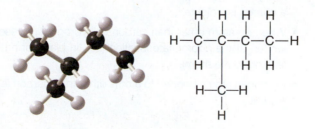

$CH_3CHCH_2CH_3$ or $CH_3CH(CH_3)CH_2CH_3$ or $(CH_3)_2CHCH_2CH_3$
with CH_3 below

or $CH_3CH_2CH(CH_3)_2$ or $CH_3CH_2CHCH_3$
with CH_3 below

● ● ●

PRACTICE PROBLEM 1.13 Write a condensed structural formula for the following compound.

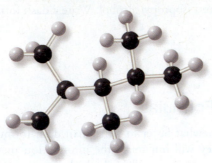

1.7C Bond-Line Formulas

The most common type of structural formula used by organic chemists, and the fastest to draw, is the **bond-line formula**. (Some chemists call these skeletal formulas.) The formula in Fig. 1.3e is a bond-line formula for propyl alcohol. The sooner you master the use of bond-line formulas, the more quickly you will be able to draw molecules when you take notes and work problems. And, lacking all of the symbols that are explicitly shown in dash and condensed structural formulas, bond-line formulas allow you to more quickly interpret molecular connectivity and compare one molecular formula with another.

HOW TO DRAW BOND-LINE FORMULAS

We apply the following rules when we draw bond-line formulas:

- Each line represents a bond.
- Each **bend** in a line or **terminus** of a line represents a carbon atom, unless another group is shown explicitly.
- No Cs are written for carbon atoms, except optionally for CH_3 groups at the end of a chain or branch.
- No Hs are shown for hydrogen atoms, unless they are needed to give a three-dimensional perspective, in which case we use dashed or solid wedges (as explained in the next section).
- The number of hydrogen atoms bonded to each carbon is inferred by assuming that as many hydrogen atoms are present as needed to fill the valence shell of the carbon, unless a charge is indicated.
- When an atom other than carbon or hydrogen is present, the symbol for that element is written at the appropriate location (i.e., in place of a bend or at the terminus of the line leading to the atom).
- Hydrogen atoms bonded to atoms other than carbon (e.g., oxygen or nitrogen) are written explicitly.

Consider the following examples of molecules depicted by bond-line formulas.

$$CH_3CHClCH_2CH_3 = \begin{array}{ccc} CH_3 & CH_2 \\ | & | \\ CH & CH_3 \\ | \\ Cl \end{array} = $$

Bond-line formulas

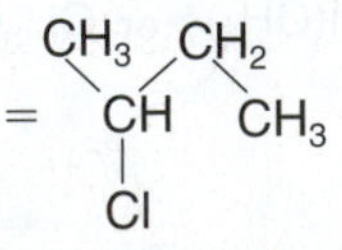

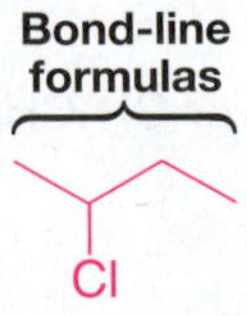

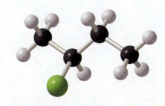

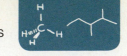

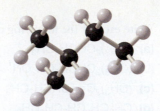

CH₃CH(CH₃)CH₂CH₃ = (structural formula with CH₃, CH₂, CH, CH₃, Cl) =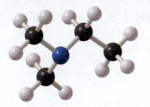

(CH₃)₂NCH₂CH₃ = (structural formula with CH₃, CH₂, N, CH₃, CH₃) = (bond-line with N)

Helpful Hint

As you become more familiar with organic molecules, you will find bond-line formulas to be very useful tools for representing structures.

Bond-line formulas are easy to draw for molecules with multiple bonds and for cyclic molecules, as well. The following are some examples.

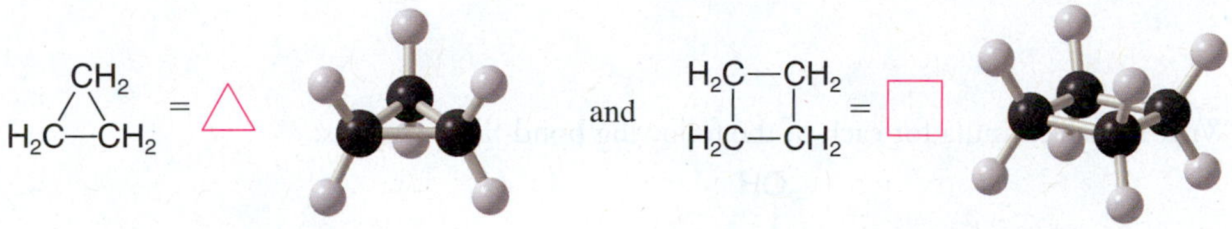

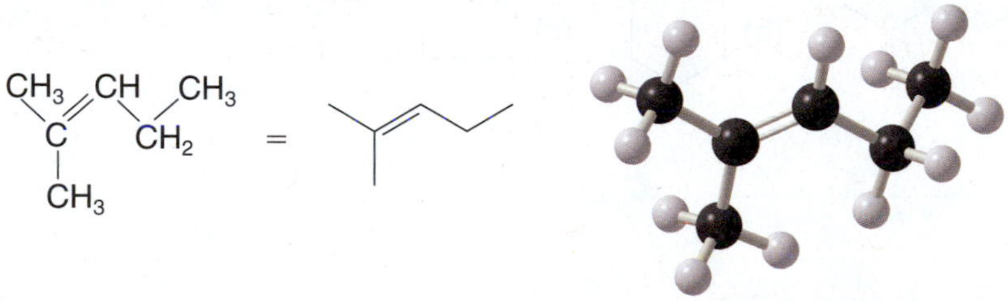

CH₂ / H₂C—CH₂ = △   and   H₂C—CH₂ / H₂C—CH₂ = ☐

CH₃ CH CH₃ / CH₂ / CH₃ = (bond-line structure)

CH₂=CHCH₂OH = (bond-line structure with OH)

(bond-line structure with triple bond and CH₃ ball-and-stick model)

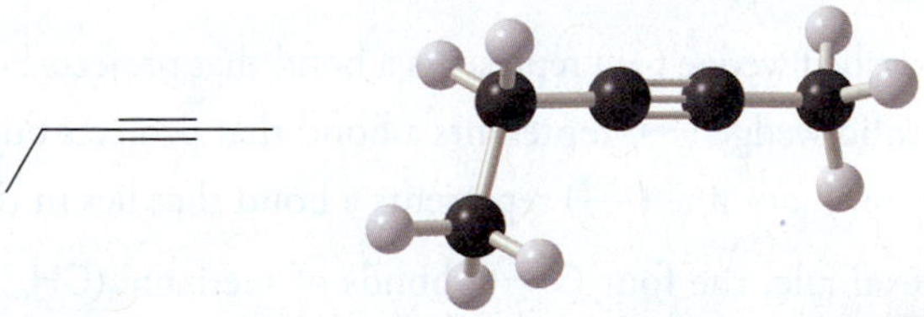

●●● **SOLVED PROBLEM 1.9**

Write the bond-line formula for

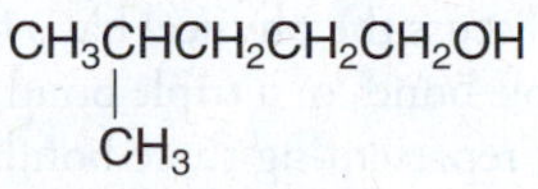

CH₃CHCH₂CH₂CH₂OH
|
CH₃

STRATEGY AND ANSWER: First, for the sake of practice, we outline the carbon skeleton, including the OH group, as follows:

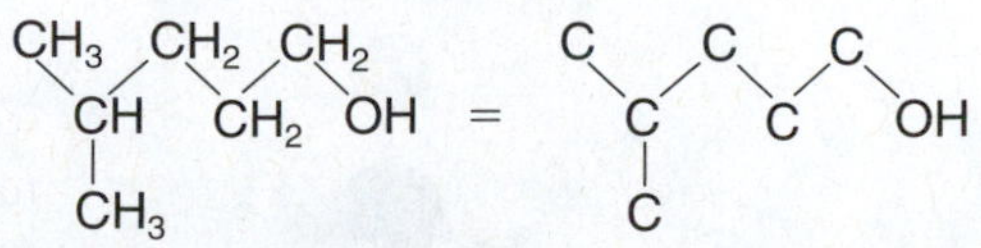

CH₃ CH₂ CH₂ / CH CH₂ OH / CH₃   =   C C C / C C OH / C

Then we write the bond-line formula as (bond-line structure) OH. As you gain experience you will likely skip the intermediate steps shown above and proceed directly to writing bond-line formulas.

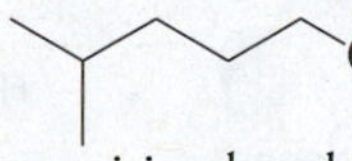

PRACTICE PROBLEM 1.14 Write each of the following condensed structural formulas as a bond-line formula:

(a) $(CH_3)_2CHCH_2CH_3$

(b) $(CH_3)_2CHCH_2CH_2OH$

(c) $(CH_3)_2C{=}CHCH_2CH_3$

(d) $CH_3CH_2CH_2CH_2CH_3$

(e) $CH_3CH_2CH(OH)CH_2CH_3$

(f) $CH_2{=}C(CH_2CH_3)_2$

(g) $CH_3\overset{\overset{\displaystyle O}{\|}}{C}CH_2CH_2CH_2CH_3$

(h) $CH_3CHClCH_2CH(CH_3)_2$

PRACTICE PROBLEM 1.15 Which molecules in Practice Problem 1.14 form sets of constitutional isomers?

PRACTICE PROBLEM 1.16 Write a dash formula for each of the following bond-line formulas:

(a) (b) (c)

1.7D Three-Dimensional Formulas

None of the formulas that we have described so far convey any information about how the atoms of a molecule are arranged in space. Molecules exist in three dimensions. We can depict three-dimensional geometry in molecules using bonds represented by dashed wedges, solid wedges, and lines.

- A dashed wedge (⫶⫶⫶) represents a bond that projects behind the plane of the paper.
- A solid wedge (◄) represents a bond that projects out of the plane of the paper.
- An ordinary line (—) represents a bond that lies in the plane of the paper.

For example, the four C—H bonds of methane (CH_4) are oriented toward the corners of a regular tetrahedron, with the carbon in the center and an approximately 109° angle between each C—H bond, as was originally postulated by J. H. van't Hoff and L. A. Le Bel in 1874. Figure 1.4 shows the tetrahedral structure of methane.

We will discuss the physical basis for the geometries of carbon when it has only single bonds, a double bond, or a triple bond in Sections 1.12–14. For now, let us consider some guidelines for representing these bonding patterns in three dimensions using dashed and solid wedge bonds.

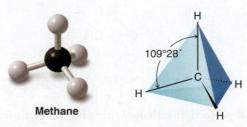

Methane

FIGURE 1.4 The tetrahedral structure of methane.

In general for carbon atoms that have only single bonds:

- A carbon atom with **four single bonds** has tetrahedral geometry (Section 1.12) and can be drawn with two bonds in the plane of the paper separated by approximately 109°, one bond behind the plane using a dashed wedge, and one bond in front of the plane using a solid wedge.
- The dashed wedge and solid wedge bonds in tetrahedral geometry nearly eclipse each other when drawn in proper three-dimensional perspective.

For carbon atoms with a double or a triple bond:

- A carbon atom with a **double bond** has trigonal planar geometry (Section 1.13) and can be depicted with bonds that are all in the plane of the paper and separated by 120°.
- A carbon atom with a **triple bond** has linear geometry (Section 1.14) and can be depicted with its bonds in the plane of the paper and separated by a 180° angle.

Last, when drawing three-dimensional formulas for molecules:

- Draw as many carbon atoms in the plane of the paper as possible using ordinary lines, then use dashed or solid wedge bonds for substituent groups or hydrogen atoms that are needed to show three dimensions.

Some examples of three-dimensional formulas are shown below.

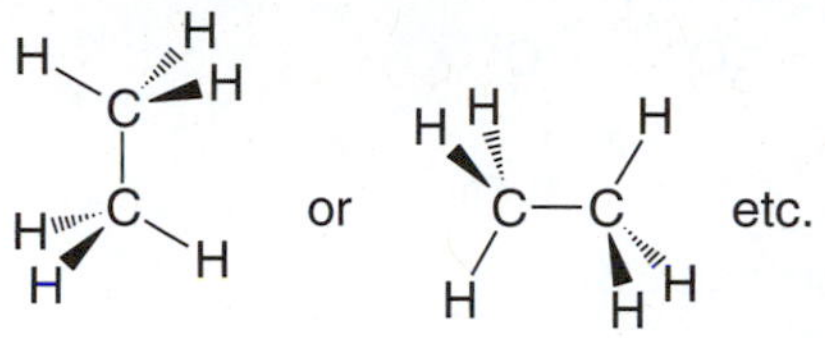

Ethane 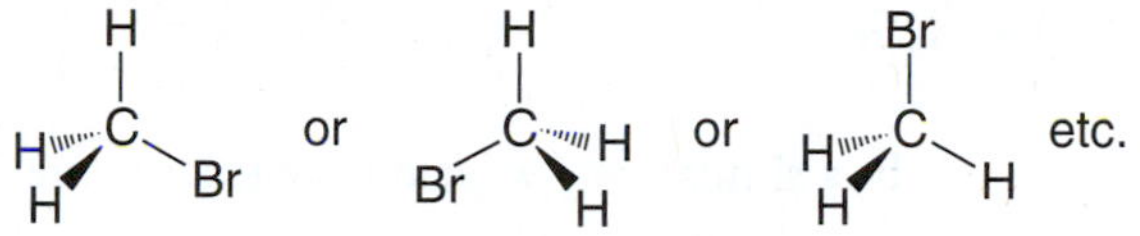   **Bromomethane**

Examples of bond-line formulas that include three-dimensional representations	An example involving trigonal planar geometry	An example involving linear geometry
The carbon chains are shown in the plane of the paper. The dashed and solid wedge bonds nearly eclipse each other.	Bonds to the carbon with the double bond are in the plane of the paper and separated by 120°.	Bonds to the carbon with the triple bond are in the plane of the paper and separated by 180°.

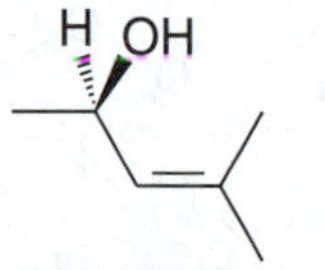

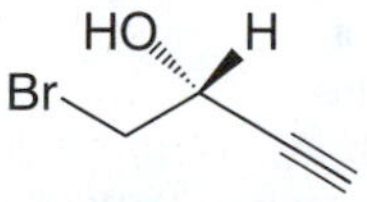

●●● SOLVED PROBLEM 1.10

Write a bond-line formula for the following compound showing three dimensions at the carbon bearing the chlorine atom.

$$CH_3CH_2CHCH_2CH_3$$
$$|$$
$$Cl$$

STRATEGY AND ANSWER: First draw the carbon skeleton, placing as many carbon atoms in the plane of the paper as possible (which is all of them, in this case).

Then add the chlorine atom at the appropriate carbon using a three-dimensional representation.

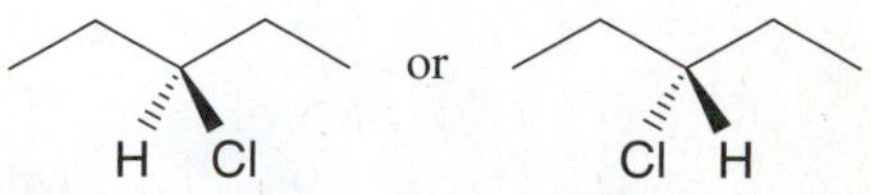

●●● PRACTICE PROBLEM 1.17

Write three-dimensional (wedge–dashed wedge–line) representations for each of the following:

(a) CH_3Cl **(b)** CH_2Cl_2 **(c)** CH_2BrCl **(d)** CH_3CH_2Cl

1.8 RESONANCE THEORY

Often more than one *equivalent* Lewis structure can be written for a molecule or ion. Consider, for example, the carbonate ion (CO_3^{2-}). We can write three *different* but *equivalent* structures, **1–3**:

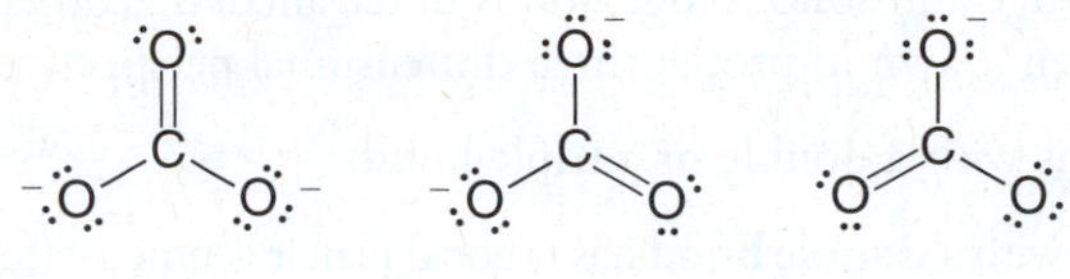

Notice two important features of these structures. First, each atom has the noble gas configuration. Second, *and this is especially important*, we can convert one structure into any other by *changing only the positions of the electrons*. We do not need to change the relative positions of the atomic nuclei. For example, if we move the electron pairs in the manner indicated by the **curved arrows** in structure **1**, we change structure **1** into structure **2**:

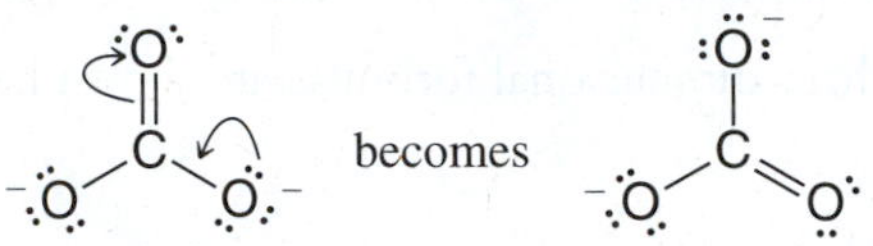

In a similar way we can change structure **2** into structure **3**:

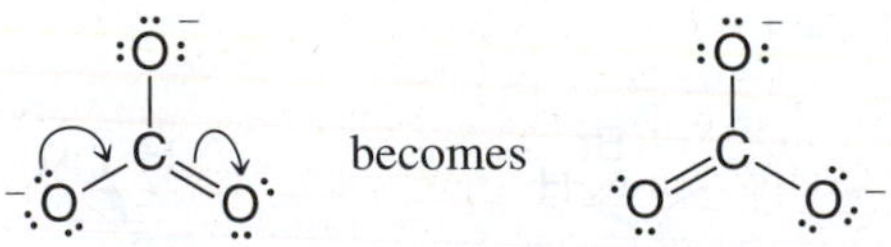

Helpful Hint

Curved arrows (Section 3.2) show movement of electron pairs, not atoms. The tail of the arrow begins at the current position of the electron pair. The *head* of the arrow points to the location where the electron pair will be in the next structure. Curved-arrow notation is one of the most important tools that you will use to understand organic reactions.

Structures **1–3**, although not identical on paper, *are equivalent*. None of them alone, however, fits important data about the carbonate ion.

X-ray studies have shown that carbon–oxygen double bonds are shorter than single bonds. The same kind of study of the carbonate ion shows, however, that all of its carbon–oxygen bonds *are of equal length*. One is not shorter than the others as would be expected from representations **1**, **2**, and **3**. Clearly none of the three structures agrees with this evidence. In each structure, **1–3**, one carbon–oxygen bond is a double bond and the other two are single bonds. None of the structures, therefore, is correct. How, then, should we represent the carbonate ion?

One way is through a theory called **resonance theory**. This theory states that whenever a molecule or ion can be represented by two or more Lewis structures *that differ only in the positions of the electrons*, two things will be true:

1. None of these structures, which we call **resonance structures** or **resonance contributors**, will be a realistic representation for the molecule or ion. None will be in complete accord with the physical or chemical properties of the substance.

2. The actual molecule or ion will be better represented by a *hybrid (average) of these structures*.

   - *Resonance structures, then, are not real structures for the actual molecule or ion; they exist only on paper*. As such, they can never be isolated. No single contributor adequately represents the molecule or ion. In resonance theory we view the carbonate ion, which is, of course, a real entity, as having a structure that is a **hybrid** of the three **hypothetical** resonance structures.

What would a hybrid of structures **1–3** be like? Look at the structures and look especially at a particular carbon–oxygen bond, say, the one at the top. This carbon–oxygen bond is a double bond in one structure (**1**) and a single bond in the other two (**2** and **3**). The actual carbon–oxygen bond, since it is a hybrid, must be something in between a double bond and a single bond. Because the carbon–oxygen bond is a single bond in two of the structures and a double bond in only one, it must be more like a single bond than a double bond. It must be like a one and one-third bond. We could call it a partial double bond. And, of course, what we have just said about any one carbon–oxygen bond will be equally true of the other two. Thus all of the carbon–oxygen bonds of the carbonate ion are partial double bonds, and *all are equivalent*. All of them *should be* the same length, and this is exactly what experiments tell us. The bonds are all 1.28 Å long, a distance which is intermediate between that of a carbon–oxygen single bond (1.43 Å) and that of a carbon–oxygen double bond (1.20 Å). One angstrom equals 1×10^{-10} meter.

- One other important point: by convention, when we draw resonance structures, we connect them by double-headed arrows ($\leftrightarrow$) to indicate clearly that they are hypothetical, not real. For the carbonate ion we write them this way:

We should not let these arrows, or the word "resonance," mislead us into thinking that the carbonate ion fluctuates between one structure and another. These structures individually do not represent reality and exist only on paper; therefore, the carbonate ion cannot fluctuate among them because it is a hybrid of them.

- Resonance structures do not represent an **equilibrium**.

In an equilibrium between two or more species, it is quite correct to think of different structures and moving (or fluctuating) atoms, *but not in the case of resonance* (as in the carbonate ion). Here the atoms do not move, and the "structures" exist only on paper. **An equilibrium is indicated by $\rightleftharpoons$ and resonance by $\leftrightarrow$.**

How can we write the structure of the carbonate ion in a way that will indicate its actual structure? We may do two things: we may write all of the resonance structures as we have just done and let the reader mentally fashion the hybrid, or we may write a non-Lewis structure that attempts to represent the hybrid. For the carbonate ion we might do the following:

Hybrid   **Contributing resonance structures**

The bonds in the structure on the left are indicated by a combination of a solid line and a dashed line. This is to indicate that the bonds are something in between a single bond and a double bond. As a rule, we use a solid line whenever a bond appears in all structures, and a dashed line when a bond exists in one or more but not all. We also place a $\delta-$ (read partial minus) beside each oxygen to indicate that something less than a full negative charge resides on each oxygen atom. In this instance, each oxygen atom has two-thirds of a full negative charge.

Calculations from theory show the equal charge density at each oxygen in the carbonate anion. Figure 1.5 shows a calculated **electrostatic potential map** of the electron density in the carbonate ion. In an electrostatic potential map, regions of relatively more negative charge are red, while more positive regions (i.e., less negative regions) are indicated by colors trending toward blue. Equality of the bond lengths in the carbonate anion (partial double bonds as shown in the resonance hybrid above) is also evident in this model.

Helpful Hint

Each type of arrow in organic chemistry (e.g., $\frown$, $\rightleftharpoons$, and $\leftrightarrow$) has a specific meaning. It is important that you use each type of arrow only for the purpose for which it is defined.

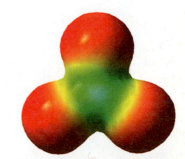

FIGURE 1.5 A calculated electrostatic potential map for the carbonate anion, showing the equal charge distribution at the three oxygen atoms. In electrostatic potential maps like this one, colors trending toward red mean increasing concentration of negative charge, while those trending toward blue mean less negative (or more positive) charge.

1.8A The Use of Curved Arrows: HOW TO Write Resonance Structures

As we have mentioned earlier, curved arrows are often used in writing **resonance structures**, and as we shall see in Section 3.2 they are essential in writing reaction mechanisms. Let us now point out several important things to remember about their use.

- Curved arrows are used to show the movement of both **bonding** and **unshared** electrons.
- A double-barbed curved arrow (⌒) shows the movement of **two** electrons (an electron pair). [Later, we will see that a single-barbed arrow (⌒) can be used to show the movement of a single electron.]
- A curved arrow should originate precisely at the location of the relevant electrons in the initial formula and point precisely to where those electrons will be drawn in the new formula.
- A new formula should be drawn to show the result of the electron shift(s). All formulas should be proper Lewis structures and should include formal charges as appropriate. The maximum number of valence electrons should not be exceeded for any atom in a formula.

1.8B Rules for Writing Resonance Structures

1. **Resonance structures exist only on paper**. Although they have no real existence of their own, **resonance structures** are useful because they allow us to describe molecules and ions for which a single Lewis structure is inadequate. We write two or more Lewis structures, calling them resonance structures or resonance contributors. We connect these structures by double-headed arrows (↔), and we say that the real molecule or ion is a hybrid of all of them.

2. **We are only allowed to move electrons in writing resonance structures**. The positions of the nuclei of the atoms must remain the same in all of the structures. Structure **3** is not a resonance structure of **1** or **2**, for example, because in order to form it we would have to move a hydrogen atom and this is not permitted:

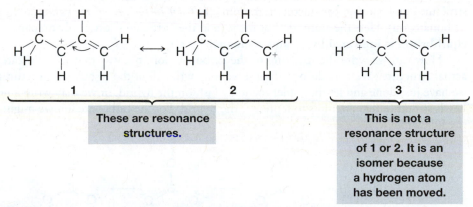

These are resonance structures.

This is not a resonance structure of 1 or 2. It is an isomer because a hydrogen atom has been moved.

Generally speaking, when we move electrons, we move only those of **multiple bonds** (as in the example above) and those of **nonbonding electron pairs**.

3. **All of the structures must be proper Lewis structures**. We should not write structures in which carbon has five bonds, for example:

$$H-\overset{\underset{|}{H}}{\underset{H}{C}}-\overset{+}{\underset{..}{O}}-H$$

This is not a proper resonance structure for methanol because carbon has five bonds. Elements of the first major row of the periodic table cannot have more than eight electrons in their valence shell.

4. **The energy of the resonance hybrid is lower than the energy of any contributing structure**. Resonance stabilizes a molecule or ion. This is especially true when the resonance structures are equivalent. Chemists call this stabilization *resonance stabilization*. **If the resonance structures are equivalent, then the resonance stabilization is large.**

In Chapter 14 we shall find that benzene is highly resonance stabilized because it is a hybrid of the two equivalent forms that follow:

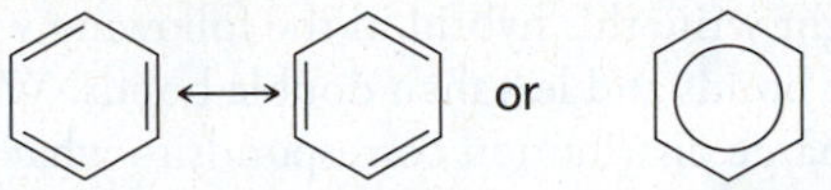

Resonance structures for benzene — Representation of hybrid

5. **The more stable a structure is (when taken by itself), the greater is its contribution to the hybrid.**

1.8C How Do We Decide When One Resonance Structure Contributes More to the Hybrid Than Another?

The following rules will help us:

1. **The more covalent bonds a structure has, the more stable it is**. Consider the resonance structures for formaldehyde below. (Formaldehyde is a chemical used to preserve biological specimens.) Structure **A** has more covalent bonds, and therefore makes a larger contribution to the hybrid. In other words, the hybrid is more like structure **A** than structure **B**.

Four covalent bonds

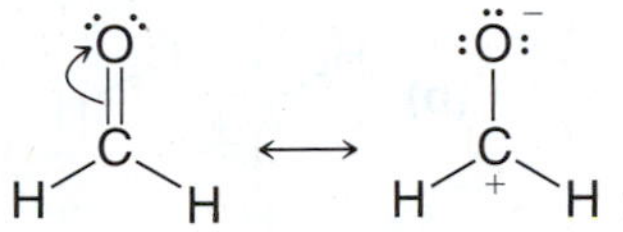

Three covalent bonds

A — **B**

More stable — **Less stable**

Resonance structures for formaldehyde

These structures also illustrate two other considerations:

2. **Charge separation decreases stability**. It takes energy to separate opposite charges, and therefore a structure with separated charges is less stable. Structure **B** for formaldehyde has separated plus and minus charges; therefore, on this basis, too, it is the less stable contributor and makes a smaller contribution to the hybrid.
3. **Structures in which all the atoms have a complete valence shell of electrons (i.e., the noble gas structure) are more stable**. Look again at structure **B**. The carbon atom has only six electrons around it, whereas in **A** it has eight. On this basis we can conclude that **A** is more stable and makes a larger contribution.

SOLVED PROBLEM 1.11

The following is one way of writing the structure of the nitrate ion:

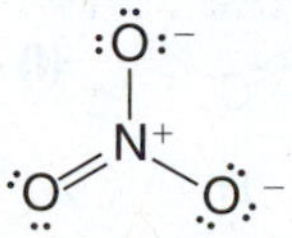

However, considerable physical evidence indicates that all three nitrogen–oxygen bonds are equivalent and that they have the same length, a bond distance between that expected for a nitrogen–oxygen single bond and a nitrogen–oxygen double bond. Explain this in terms of resonance theory.

STRATEGY AND ANSWER: We recognize that if we move the electron pairs in the following way, we can write three *different* but *equivalent* structures for the nitrate ion:

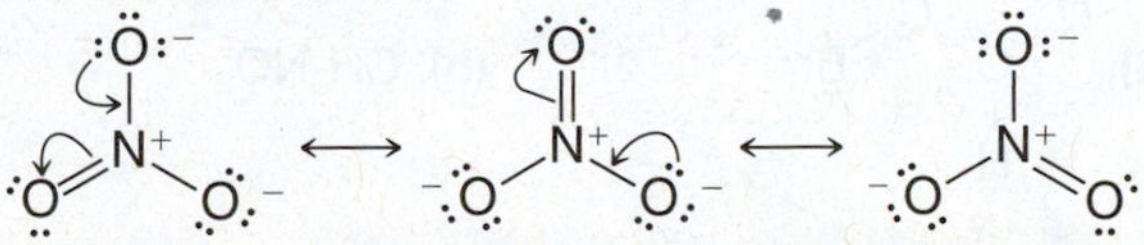

(continues on the next page)

Since these structures differ from one another *only in the positions of their electrons*, they are *resonance structures* or *resonance contributors*. As such, no single structure taken alone will adequately represent the nitrate ion. The actual molecule will be best represented by a *hybrid of these three structures*. We might write this hybrid in the following way to indicate that all of the bonds are equivalent and that they are more than single bonds and less than double bonds. We also indicate that each oxygen atom bears an equal partial negative charge. This charge distribution corresponds to what we find experimentally.

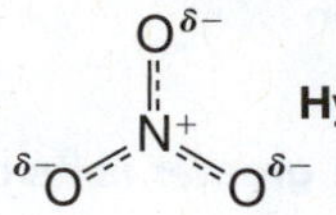

Hybrid structure for the nitrate ion

PRACTICE PROBLEM 1.18 **(a)** Write two resonance structures for the formate ion HCO_2^-. (*Note*: The hydrogen and oxygen atoms are bonded to the carbon.) **(b)** Explain what these structures predict for the carbon–oxygen bond lengths of the formate ion, and **(c)**, for the electrical charge on the oxygen atoms.

PRACTICE PROBLEM 1.19 Write the resonance structure that would result from moving the electrons as the curved arrows indicate. Be sure to include formal charges if needed.

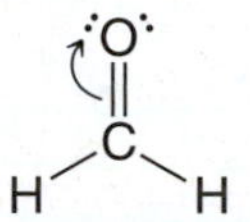

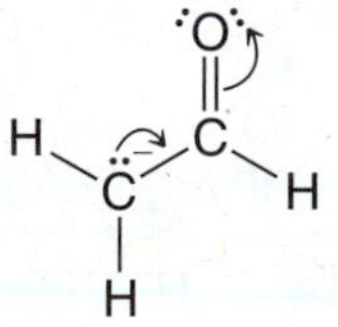

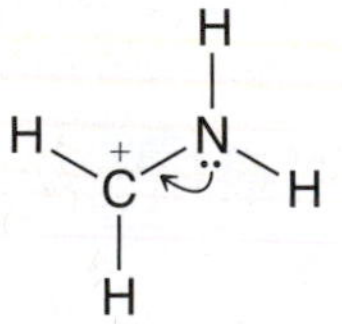

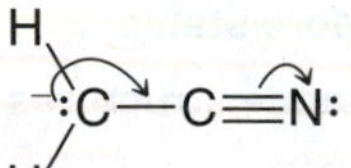

PRACTICE PROBLEM 1.20 Add any missing unshared electron pairs (if any), then, using curved arrows to show the shifts in electrons, write the contributing resonance structures and resonance hybrid for each of the following:

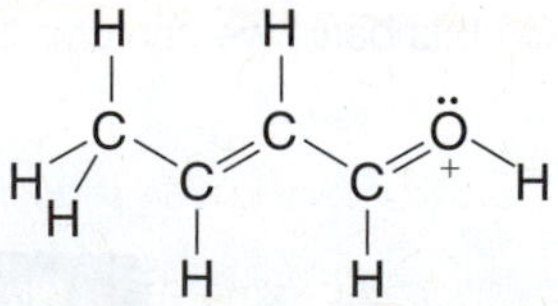

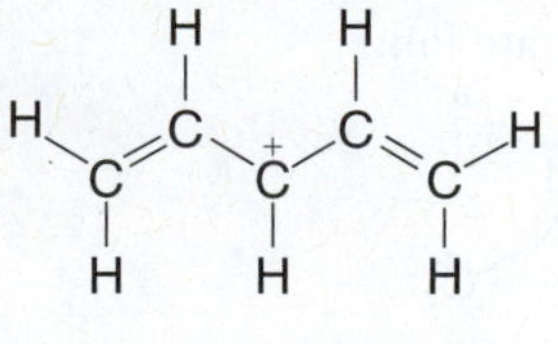

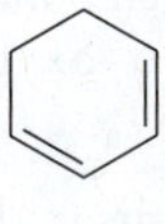

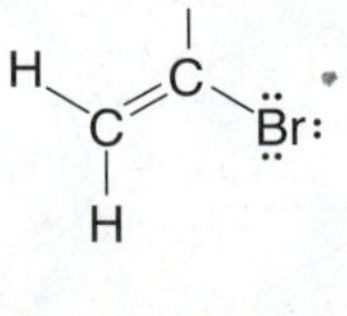

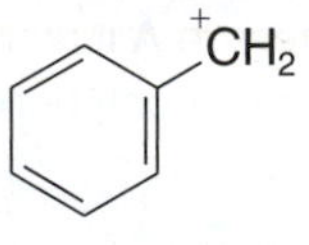

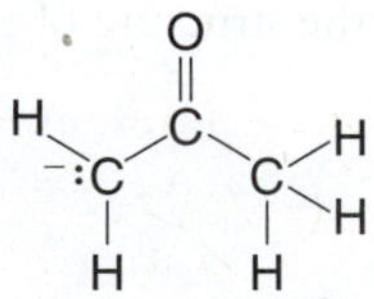

(g) $CH_3\overset{+}{S}CH_2$

(h) CH_3NO_2

For each set of resonance structures that follow, add a curved arrow that shows how electrons in the left formula shift to become the right formula, and designate the formula that would contribute most to the hybrid. Explain your choice:

PRACTICE PROBLEM 1.21

(a)

(b) CH_3-C ⟷ CH_3-C

(c) $:NH_2-C≡N:$ ⟷ $\overset{+}{N}H_2=C=\overset{..}{N}:^-$

1.9 QUANTUM MECHANICS AND ATOMIC STRUCTURE

A theory of atomic and molecular structure was advanced independently and almost simultaneously by three people in 1926: Erwin Schrödinger, Werner Heisenberg, and Paul Dirac. This theory, called **wave mechanics** by Schrödinger and **quantum mechanics** by Heisenberg, has become the basis from which we derive our modern understanding of bonding in molecules. At the heart of quantum mechanics are equations called wave functions (denoted by the Greek letter **psi**, ψ).

- Each **wave function** (ψ) corresponds to a different *energy state* for an electron.
- Each *energy state* is a sublevel where one or two electrons can reside.
- The **energy** associated with the state of an electron can be calculated from the wave function.
- The **relative probability** of finding an electron in a given region of space can be calculated from the wave function (Section 1.10).
- The solution to a wave function can be positive, negative, or zero (Fig. 1.6).
- The **phase sign** of a wave equation indicates whether the solution is positive or negative when calculated for a given point in space relative to the nucleus.

Wave functions, whether they are for sound waves, lake waves, or the energy of an electron, have the possibility of constructive interference and destructive interference.

- **Constructive interference** occurs when wave functions with the same phase sign interact. There is a *reinforcing effect* and the amplitude of the wave function increases.
- **Destructive interference** occurs when wave functions with opposite phase signs interact. There is a *subtractive effect* and the amplitude of the wave function goes to zero or changes sign.

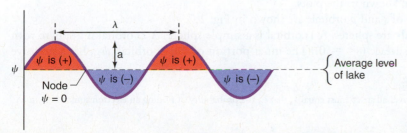

FIGURE 1.6 A wave moving across a lake is viewed along a slice through the lake. For this wave the wave function, ψ, is plus (+) in crests and minus (−) in troughs. At the average level of the lake it is zero; these places are called nodes. The magnitude of the crests and troughs is the amplitude (a) of the wave. The distance from the crest of one wave to the crest of the next is the wavelength (λ, or lambda).

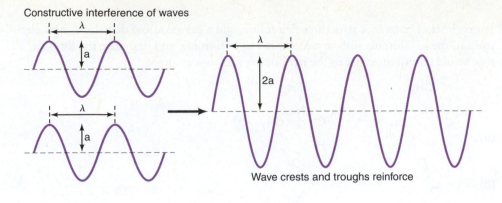

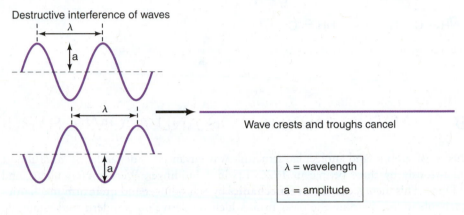

Experiments have shown that electrons have properties of waves and particles, which was an idea first put forth by Louis de Broglie in 1923. Our discussion focuses on the wavelike properties of electrons, however.

1.10 ATOMIC ORBITALS AND ELECTRON CONFIGURATION

A physical interpretation related to the electron wave function was put forth by Max Born in 1926:

- The square of a wave function (ψ^2) for a particular x, y, z location expresses the probability of finding an electron at that location in space.

If the value of ψ^2 is large in a unit volume of space, the probability of finding an electron in that volume is high—we say that the **electron probability density** is large. Conversely, if ψ^2 for some other volume of space is small, the probability of finding an electron there is low.* This leads to the general definition of an orbital and, by extension, to the familiar shapes of atomic orbitals.

- An **orbital** is a region of space where the probability of finding an electron is high.
- **Atomic orbitals** are plots of ψ^2 in three dimensions. These plots generate the familiar s, p, and d orbital shapes.

The volumes that we show are those that would contain the electron 90–95% of the time. There is a finite, but very small, probability of finding an electron at greater distance from the nucleus than shown in the plots.

The shapes of s and p orbitals are shown in Fig. 1.7.

All *s orbitals* are spheres. A 1s orbital is a simple sphere. A 2s orbital is a sphere with an inner nodal surface ($\psi^2 = 0$). The inner portion of the 2s orbital, ψ_{2s}, has a negative phase sign.

*Integration of ψ^2 over all space must equal 1; that is, the probability of finding an electron somewhere in all of space is 100%.

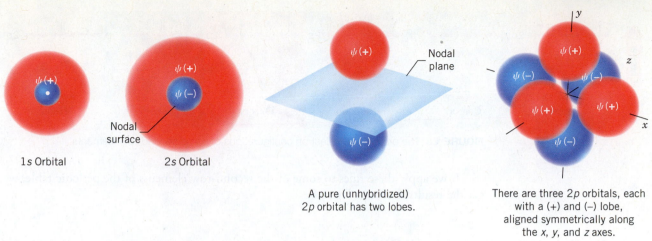

1s Orbital   2s Orbital   Nodal surface

A pure (unhybridized) 2p orbital has two lobes.   Nodal plane

There are three 2p orbitals, each with a (+) and (−) lobe, aligned symmetrically along the x, y, and z axes.

FIGURE 1.7 The shapes of some s and p orbitals. Pure, unhybridized p orbitals are almost-touching spheres. The p orbitals in hybridized atoms are lobe-shaped (Section 1.13).

The shape of a *p* **orbital** is like that of almost-touching spheres or lobes. The phase sign of a 2p wave function, ψ_{2p}, is positive in one lobe and negative in the other. A nodal plane separates the two lobes of a p orbital, and the three p orbitals of a given energy level are arranged in space along the x, y, and z axes in a Cartesian coordinate system.

- The + and − signs of wave functions do not imply positive or negative charge or greater or lesser probability of finding an electron.
- ψ^2 (the probability of finding an electron) is always positive, because squaring either a positive or negative solution to ψ leads to a positive value.

Thus, the probability of finding an electron in either lobe of a p orbital is the same. We shall see the significance of the + and − signs later when we see how atomic orbitals combine to form molecular orbitals.

1.10A Electron Configurations

The relative energies of atomic orbitals in the first and second principal shells are as follows:

- Electrons in 1s orbitals have the lowest energy because they are closest to the positive nucleus.
- Electrons in 2s orbitals are next lowest in energy.
- Electrons of the three 2p orbitals have equal but higher energy than the 2s orbital.
- Orbitals of equal energy (such as the three 2p orbitals) are called **degenerate orbitals**.

We can use these relative energies to arrive at the electron configuration of any atom in the first two rows of the periodic table. We need follow only a few simple rules.

1. **Aufbau principle**: Orbitals are filled so that those of lowest energy are filled first. (*Aufbau* is German for "building up.")

2. **Pauli exclusion principle**: A maximum of two electrons may be placed in each orbital *but only when the spins of the electrons are paired*. An electron spins about its own axis. For reasons that we cannot develop here, an electron is permitted only one or the other of just two possible spin orientations. We usually show these orientations by arrows, either ↑ or ↓. Thus two spin-paired electrons would be designated ↓↑. Unpaired electrons, which are not permitted in the same orbital, are designated ↑↑ (or ↓↓).

3. **Hund's rule**: When we come to orbitals of equal energy (degenerate orbitals) such as the three p orbitals, we add one electron to each *with their spins unpaired* until each of the degenerate orbitals contains one electron. (This allows the electrons, which repel each other, to be farther apart.) Then we begin adding a second electron to each degenerate orbital so that the spins are paired.

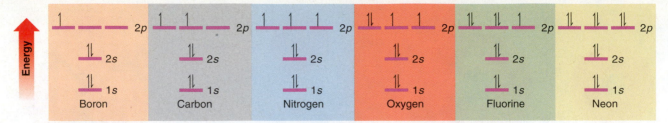

FIGURE 1.8 The ground state electron configurations of some second-row elements.

If we apply these rules to some of the second-row elements of the periodic table, we get the results shown in Fig. 1.8.

1.11 MOLECULAR ORBITALS

Atomic orbitals provide a means for understanding how atoms form covalent bonds. Let us consider a very simple case—formation of a bond between two hydrogen atoms to form a hydrogen molecule (Fig. 1.9).

When two hydrogen atoms are relatively far apart their total energy is simply that of two isolated hydrogen atoms (**I**). Formation of a covalent bond reduces the overall energy of the system, however. As the two hydrogen atoms move closer together (**II**), each nucleus increasingly attracts the other's electron. This attraction more than compensates for the repulsive force between the two nuclei (or the two electrons). The result is a covalent bond (**III**), such that the internuclear distance is an ideal balance that allows the two electrons to be shared between both atoms while at the same time avoiding repulsive interactions between their nuclei. This ideal internuclear distance between hydrogen atoms is 0.74 Å, and we call this the **bond length** in a hydrogen molecule. If the nuclei are moved closer together (**IV**) the repulsion of the two positively charged nuclei predominates, and the energy of the system rises.

Notice that each H· has a shaded area around it, indicating that its precise position is uncertain. Electrons are constantly moving.

- According to the **Heisenberg uncertainty principle**, we cannot simultaneously know the position and momentum of an electron.

These shaded areas in our diagram represent orbitals, and they result from applying the principles of quantum mechanics. Plotting the square of the wave function (ψ^2) gives us a three-dimensional region called an orbital where finding an electron is highly probable.

- An **atomic orbital** represents the region of space where one or two electrons of an isolated atom are likely to be found.

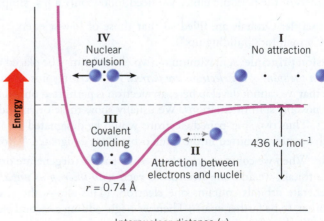

FIGURE 1.9 The potential energy of the hydrogen molecule as a function of internuclear distance.

FIGURE 1.10 (*a*) The overlapping of two hydrogen 1*s* atomic orbitals with the same phase sign (indicated by their identical color) to form a bonding molecular orbital. (*b*) The analogous overlapping of two waves with the same phase, resulting in constructive interference and enhanced amplitude.

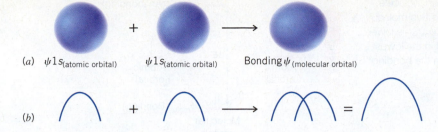

In the case of our hydrogen model above, the shaded spheres represent the 1*s* orbital of each hydrogen atom. As the two hydrogen atoms approach each other their 1*s* orbitals begin to overlap until their atomic orbitals combine to form molecular orbitals.

- A **molecular orbital (MO)** represents the region of space where one or two electrons of a molecule are likely to be found.
- An orbital (atomic or molecular) can contain a maximum of two spin-paired electrons (Pauli exclusion principle).
- When atomic orbitals combine to form molecular orbitals, **the number of molecular orbitals that result always equals the number of atomic orbitals that combine.**

Thus, in the formation of a hydrogen molecule the two ψ_{1s} atomic orbitals combine to produce two molecular orbitals. Two orbitals result because the mathematical properties of wave functions permit them to be combined by either addition or subtraction. That is, they can combine either in or out of phase.

- A **bonding molecular orbital** (ψ_{molec}) results when two orbitals of the same phase overlap (Fig. 1.10).
- An **antibonding molecular orbital** (ψ^*_{molec}) results when two orbitals of opposite phase overlap (Fig. 1.11).

The bonding molecular orbital of a hydrogen molecule in its lowest energy (ground) state contains both electrons from the individual hydrogen atoms. The value of ψ (and therefore also ψ^2) is large between the nuclei, precisely as expected since the electrons are shared by both nuclei to form the covalent bond.

The antibonding molecular orbital contains no electrons in the ground state of a hydrogen molecule. Furthermore, the value of ψ (and therefore also ψ^2) goes to zero between the nuclei, creating a node ($\psi = 0$). The antibonding orbital does not provide for electron density between the atoms, and thus it is not involved in bonding.

What we have just described has its counterpart in a mathematical treatment called the **LCAO (linear combination of atomic orbitals)** method. In the LCAO treatment, wave functions for the atomic orbitals are combined in a linear fashion (by addition or subtraction) in order to obtain new wave functions for the molecular orbitals.

FIGURE 1.11 (*a*) The overlapping of two hydrogen 1*s* atomic orbitals with opposite phase signs (indicated by their different colors) to form an antibonding molecular orbital. (*b*) The analogous overlapping of two waves with the opposite sign, resulting in destructive interference and decreased amplitude. A node exists where complete cancellation by opposite phases makes the value of the combined wave function zero.

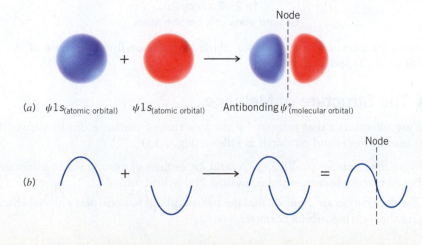

FIGURE 1.12 Energy diagram for the hydrogen molecule. Combination of two atomic orbitals, ψ_{1s}, gives two molecular orbitals, ψ_{molec} and ψ^*_{molec}. The energy of ψ_{molec} is lower than that of the separate atomic orbitals, and in the lowest electronic energy state of molecular hydrogen the bonding MO contains both electrons.

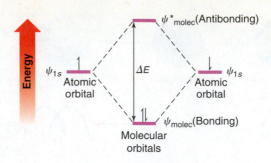

Molecular orbitals, like atomic orbitals, correspond to particular energy states for an electron. Calculations show that the relative energy of an electron in the bonding molecular orbital of the hydrogen molecule is substantially less than its energy in a ψ_{1s} atomic orbital. These calculations also show that the energy of an electron in the antibonding molecular orbital is substantially greater than its energy in a ψ_{1s} atomic orbital.

An energy diagram for the molecular orbitals of the hydrogen molecule is shown in Fig. 1.12. Notice that electrons are placed in molecular orbitals in the same way that they are in atomic orbitals. Two electrons (with their spins opposed) occupy the bonding molecular orbital, where their total energy is less than in the separate atomic orbitals. This is, as we have said, the *lowest electronic state* or *ground state* of the hydrogen molecule. An electron may occupy the antibonding molecular orbital in what is called an *excited state* for the molecule. This state forms when the molecule in the ground state (Fig. 1.12) absorbs a photon of light having the proper energy (ΔE).

1.12 THE STRUCTURE OF METHANE AND ETHANE: sp^3 HYBRIDIZATION

The s and p orbitals used in the quantum mechanical description of the carbon atom, given in Section 1.10, were based on calculations for hydrogen atoms. These simple s and p orbitals do not, when taken alone, provide a satisfactory model for the *tetravalent–tetrahedral* carbon of methane (CH_4, see Practice Problem 1.22). However, a satisfactory model of methane's structure that is based on quantum mechanics *can* be obtained through an approach called **orbital hybridization**. Orbital hybridization, in its simplest terms, is nothing more than a mathematical approach that involves the combining of individual wave functions for s and p orbitals to obtain wave functions for new orbitals. The new orbitals have, *in varying proportions*, the properties of the original orbitals taken separately. These new orbitals are called **hybrid atomic orbitals**.

According to quantum mechanics, the electronic configuration of a carbon atom in its lowest energy state—called the **ground state**—is that given here:

$$C \quad \underset{1s}{\underline{\uparrow\downarrow}} \quad \underset{2s}{\underline{\uparrow\downarrow}} \quad \underset{2p_x}{\underline{\uparrow}} \quad \underset{2p_y}{\underline{\uparrow}} \quad \underset{2p_z}{\underline{}}$$

Ground state of a carbon atom

The valence electrons of a carbon atom (those used in bonding) are those of the *outer level*, that is, the $2s$ and $2p$ electrons.

1.12A The Structure of Methane

Hybrid atomic orbitals that account for the structure of methane can be derived from carbon's second-shell s and p orbitals as follows (Fig. 1.13):

- Wave functions for the $2s$, $2p_x$, $2p_y$, and $2p_z$ orbitals of ground state carbon are mixed to form four new and equivalent $2sp^3$ hybrid orbitals.
- The designation sp^3 signifies that the hybrid orbital has one part s orbital character and three parts p orbital character.

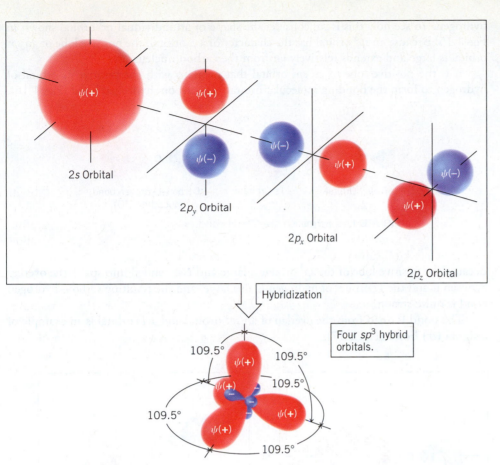

FIGURE 1.13 Hybridization of pure atomic orbitals of a carbon atom to produce sp^3 hybrid orbitals.

- The mathematical result is that the four $2sp^3$ orbitals are oriented at angles of 109.5° with respect to each other. This is precisely the orientation of the four hydrogen atoms of methane. Each H—C—H bond angle is 109.5°.

If, in our imagination, we visualize the hypothetical formation of methane from an sp^3-hybridized carbon atom and four hydrogen atoms, the process might be like that shown in Fig. 1.14. For simplicity we show only the formation of the **bonding molecular orbital** for each carbon–hydrogen bond. We see that an sp^3-hybridized carbon gives a *tetrahedral structure for methane, and one with four equivalent C—H bonds.*

In addition to accounting properly for the shape of methane, the orbital hybridization model also explains the very strong bonds that are formed between carbon and

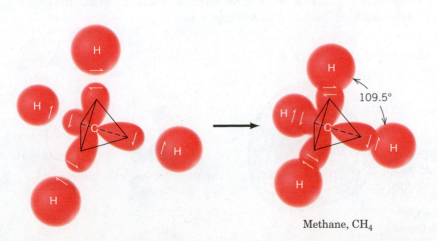

Methane, CH_4

FIGURE 1.14 The hypothetical formation of methane from an sp^3-hybridized carbon atom and four hydrogen atoms. In orbital hybridization we combine orbitals, *not* electrons. The electrons can then be placed in the hybrid orbitals as necessary for bond formation, but always in accordance with the Pauli principle of no more than two electrons (with opposite spin) in each orbital. In this illustration we have placed one electron in each of the hybrid carbon orbitals. In addition, we have shown only the bonding molecular orbital of each C—H bond because these are the orbitals that contain the electrons in the lowest energy state of the molecule.

FIGURE 1.15 The shape of an sp^3 orbital.

hydrogen. To see how this is so, consider the shape of an individual sp^3 orbital shown in Fig. 1.15. Because an sp^3 orbital has the character of a p orbital, the positive lobe of an sp^3 orbital is large and extends relatively far from the carbon nucleus.

It is the positive lobe of an sp^3 orbital that overlaps with the positive $1s$ orbital of hydrogen to form the bonding molecular orbital of a carbon–hydrogen bond (Fig. 1.16).

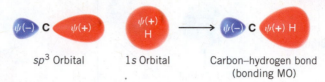

sp^3 Orbital $1s$ Orbital   Carbon–hydrogen bond
(bonding MO)

FIGURE 1.16 Formation of a C—H bond.

Because the positive lobe of the sp^3 orbital is large and is extended into space, the overlap between it and the $1s$ orbital of hydrogen is also large, and the resulting carbon–hydrogen bond is quite strong.

The bond formed from the overlap of an sp^3 orbital and a $1s$ orbital is an example of a **sigma (σ) bond** (Fig. 1.17).

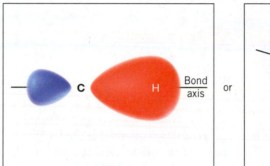

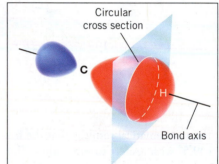

FIGURE 1.17 A σ (sigma) bond.

- A **sigma (σ) bond** has a circularly symmetrical orbital cross section when viewed along the bond between two atoms.
- All purely **single bonds** are sigma bonds.

From this point on we shall often show only the bonding molecular orbitals because they are the ones that contain the electrons when the molecule is in its lowest energy state. Consideration of antibonding orbitals is important when a molecule absorbs light and in explaining certain reactions. We shall point out these instances later.

In Fig. 1.18 we show a calculated structure for methane where the tetrahedral geometry derived from orbital hybridization is clearly apparent.

FIGURE 1.18 (a) In this structure of methane, based on quantum mechanical calculations, the inner solid surface represents a region of high electron density. High electron density is found in each bonding region. The outer mesh surface represents approximately the furthest extent of overall electron density for the molecule. (b) This ball-and-stick model of methane is like the kind you might build with a molecular model kit. (c) This structure is how you would draw methane. Ordinary lines are used to show the two bonds that are in the plane of the paper, a solid wedge is used to show the bond that is in front of the paper, and a dashed wedge is used to show the bond that is behind the plane of the paper.

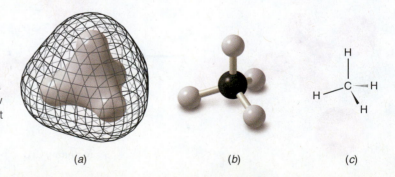

(a)   (b)   (c)

1.12B The Structure of Ethane

The bond angles at the carbon atoms of ethane, and of all alkanes, are also tetrahedral like those in methane. A satisfactory model for ethane can be provided by sp^3-hybridized carbon atoms. Figure 1.19 shows how we might imagine the bonding molecular orbitals of an ethane molecule being constructed from two sp^3-hybridized carbon atoms and six hydrogen atoms.

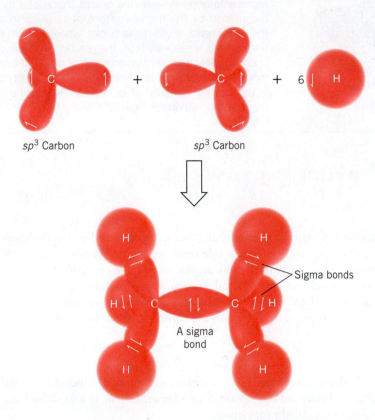

sp³ Carbon   sp³ Carbon

Sigma bonds

A sigma bond

FIGURE 1.19 The hypothetical formation of the bonding molecular orbitals of ethane from two sp^3-hybridized carbon atoms and six hydrogen atoms. All of the bonds are sigma bonds. (Antibonding sigma molecular orbitals—called σ^* orbitals—are formed in each instance as well, but for simplicity these are not shown.)

The carbon–carbon bond of ethane is a *sigma bond* with cylindrical symmetry, formed by two overlapping sp^3 orbitals. (The carbon–hydrogen bonds are also sigma bonds. They are formed from overlapping carbon sp^3 orbitals and hydrogen s orbitals.)

- Rotation of groups joined by a single bond does not usually require a large amount of energy.

Consequently, groups joined by single bonds rotate relatively freely with respect to one another. (We discuss this point further in Section 4.8.) In Fig. 1.20 we show a calculated structure for ethane in which the tetrahedral geometry derived from orbital hybridization is clearly apparent.

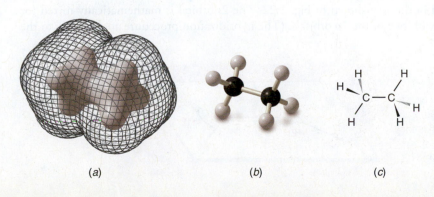

(a)   (b)   (c)

FIGURE 1.20 (a) In this structure of ethane, based on quantum mechanical calculations, the inner solid surface represents a region of high electron density. High electron density is found in each bonding region. The outer mesh surface represents approximately the furthest extent of overall electron density for the molecule. (b) A ball-and-stick model of ethane, like the kind you might build with a molecular model kit. (c) A structural formula for ethane as you would draw it using lines, wedges, and dashed wedges to show in three dimensions its tetrahedral geometry at each carbon.

THE CHEMISTRY OF... Calculated Molecular Models: Electron Density Surfaces

In this book we make frequent use of molecular models derived from quantum mechanical calculations. These models will help us visualize the shapes of molecules as well as understand their properties and reactivity. A useful type of model is one that shows a calculated three-dimensional surface at which a chosen value of electron density is the same all around a molecule, called an **electron density surface**. If we make a plot where the value chosen is for low electron density, the result is a van der Waals surface, the surface that represents approximately the overall shape of a molecule as determined by the furthest extent of its electron cloud. On the other hand, if we make a plot where the value of electron density is relatively high, the resulting surface is one that approximately represents the region of covalent bonding in a molecule. Surfaces of low and high electron density are shown in this box for dimethyl ether. Similar models are shown for methane and ethane in Figs. 1.18 and 1.20.

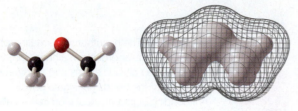

Dimethyl ether

1.13 THE STRUCTURE OF ETHENE (ETHYLENE): sp^2 HYBRIDIZATION

The carbon atoms of many of the molecules that we have considered so far have used their four valence electrons to form four single covalent (sigma) bonds to four other atoms. We find, however, that many important organic compounds exist in which carbon atoms share more than two electrons with another atom. In molecules of these compounds some bonds that are formed are multiple covalent bonds. When two carbon atoms share two pairs of electrons, for example, the result is a carbon–carbon double bond:

$$:C::C: \quad \text{or} \quad \text{\Large{}C{=}C}$$

Hydrocarbons whose molecules contain a carbon–carbon double bond are called **alkenes**. Ethene (C_2H_4) and propene (C_3H_6) are both alkenes. Ethene is also called ethylene, and propene is sometimes called propylene.

Ethene	**Propene**

In ethene the only carbon–carbon bond is a double bond. Propene has one carbon–carbon single bond and one carbon–carbon double bond.

The spatial arrangement of the atoms of alkenes is different from that of alkanes. The six atoms of ethene are coplanar, and the arrangement of atoms around each carbon atom is triangular (Fig. 1.21).

● Carbon–carbon double bonds are comprised of sp^2-hybridized carbon atoms.

The mathematical mixing of orbitals that furnish the sp^2 **orbitals** for our model can be visualized in the way shown in Fig. 1.22. The 2s orbital is mathematically mixed (or hybridized) with two of the 2p orbitals. (The hybridization procedure applies only to the

FIGURE 1.21 The structure and bond angles of ethene. The plane of the atoms is perpendicular to the paper. The dashed wedge bonds project behind the plane of the paper, and the solid wedge bonds project in front of the paper.

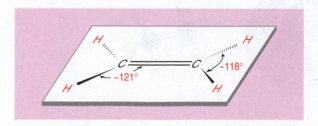

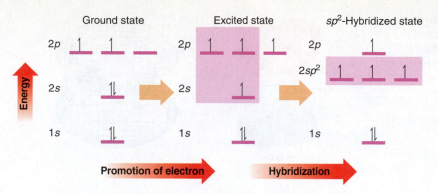

Ground state   Excited state sp^2-Hybridized state

FIGURE 1.22 A process for deriving sp^2-hybridized carbon atoms.

Promotion of electron   Hybridization

orbitals, not to the electrons.) One $2p$ orbital is left unhybridized. One electron is then placed in each of the sp^2 hybrid orbitals and one electron remains in the $2p$ orbital.

The three sp^2 orbitals that result from hybridization are directed toward the corners of a regular triangle (with angles of 120° between them). The carbon p orbital that is not hybridized is perpendicular to the plane of the triangle formed by the hybrid sp^2 orbitals (Fig. 1.23).

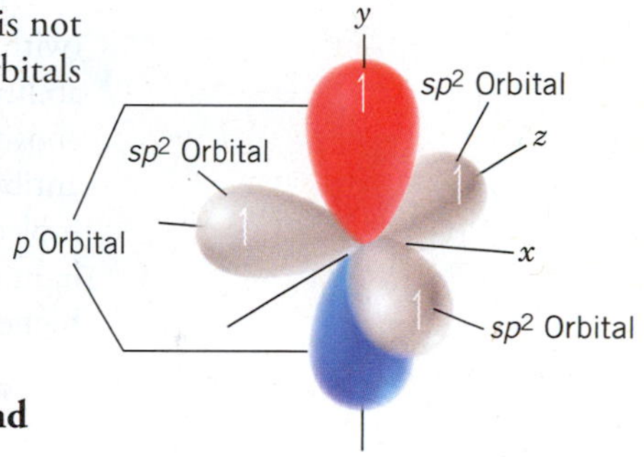

In our model for ethene (Fig. 1.24) we see the following:

- Two sp^2-hybridized carbon atoms form a sigma (σ) bond between them by overlap of one sp^2 orbital from each carbon. The remaining carbon sp^2 orbitals form σ bonds to four hydrogens through overlap with the hydrogen $1s$ orbitals. These five σ bonds account for 10 of the 12 valence electrons contributed by the two carbons and four hydrogens, and comprise the **σ-bond framework** of the molecule.

- The remaining two bonding electrons are each located in an unhybridized p orbital of each carbon. Sideways overlap of these p orbitals and sharing of the two electrons between the carbons leads to a **pi (π) bond**. The overlap of these orbitals is shown schematically in Fig. 1.25.

The bond angles that we would predict on the basis of sp^2-hybridized carbon atoms (120° all around) are quite close to the bond angles that are actually found (Fig. 1.21).

We can better visualize how these p orbitals interact with each other if we view a structure showing calculated molecular orbitals for ethene (Fig. 1.25). We see that the parallel p orbitals *overlap above and below the plane of the σ framework*.

Note the difference in shape of the bonding molecular orbital of a π bond as contrasted to that of a σ bond. A σ bond has cylindrical symmetry about a line connecting the two bonded nuclei. A π bond has a nodal plane passing through the two bonded nuclei and between the π molecular orbital lobes.

- When two p atomic orbitals combine to form a π bond, two **π molecular orbital** molecular orbitals form: one is a bonding molecular orbital and the other is an **antibonding molecular orbital**.

FIGURE 1.23 An sp^2-hybridized carbon atom.

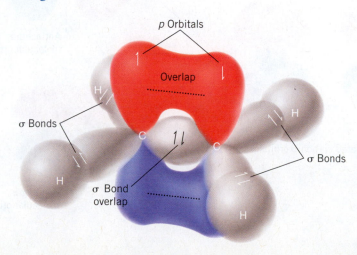

FIGURE 1.24 A model for the bonding molecular orbitals of ethene formed from two sp^2-hybridized carbon atoms and four hydrogen atoms.

FIGURE 1.25 (a) A wedge–dashed wedge formula for the sigma bonds in ethene and a schematic depiction of the overlapping of adjacent *p* orbitals that form the π bond. (b) A calculated structure for ethene. The blue and red colors indicate opposite phase signs in each lobe of the π molecular orbital. A ball-and-stick model for the σ bonds in ethene can be seen through the mesh that indicates the π bond.

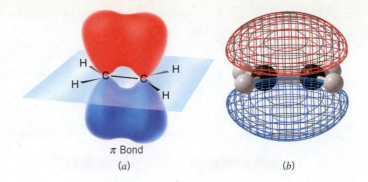

π Bond

(a)   (b)

The bonding π molecular orbital results when *p*-orbital lobes of like signs overlap; the antibonding π molecular orbital results when opposite signs overlap (Fig. 1.26).

The bonding π orbital is the lower energy orbital and contains both π electrons (with opposite spins) in the ground state of the molecule. The region of greatest probability of finding the electrons in the bonding π orbital is a region generally situated above and below the plane of the σ-bond framework between the two carbon atoms. The antibonding π* orbital is of higher energy, and it is not occupied by electrons when the molecule is in the ground state. It can become occupied, however, if the molecule absorbs light of the right frequency and an electron is promoted from the lower energy level to the higher one. The antibonding π* orbital has a nodal plane between the two carbon atoms.

- To summarize, a carbon–carbon double bond consists of one σ bond and one π bond.

The σ bond results from two *sp²* orbitals overlapping end to end and is symmetrical about an axis linking the two carbon atoms. The π bond results from a sideways overlap of two *p* orbitals; it has a nodal plane like a *p* orbital. In the ground state the electrons of the π bond are located between the two carbon atoms but generally above and below the plane of the σ-bond framework.

Electrons of the π bond have greater energy than electrons of the σ bond. The relative energies of the σ and π molecular orbitals (with the electrons in the ground state) are shown in the margin diagram. The σ* orbital is the antibonding **sigma orbital**.

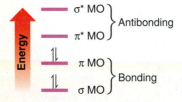

The relative energies of electrons involved in σ and π bonds

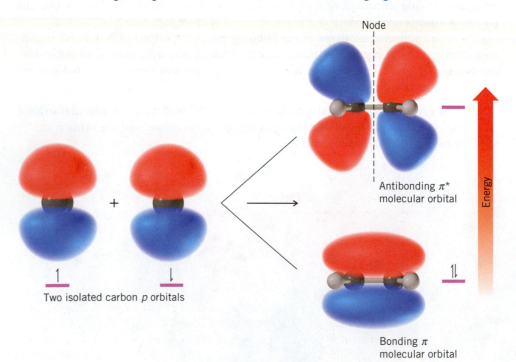

FIGURE 1.26 How two isolated carbon *p* orbitals combine to form two π (pi) molecular orbitals. The bonding MO is of lower energy. The higher energy antibonding MO contains an additional node. Both orbitals have a node in the plane containing the **C** and **H** atoms.

1.13A Restricted Rotation and the Double Bond

The σ–π model for the carbon–carbon double bond also accounts for an important property of the **double bond**:

- There is a large energy barrier to rotation associated with groups joined by a double bond.

Maximum overlap between the p orbitals of a π bond occurs when the axes of the p orbitals are exactly parallel. Rotating one carbon of the double bond 90° (Fig. 1.27) breaks the π bond, for then the axes of the p orbitals are perpendicular and there is no net overlap between them. Estimates based on thermochemical calculations indicate that the strength of the π bond is 264 kJ mol⁻¹. This, then, is the barrier to rotation of the double bond. It is markedly higher than the rotational barrier of groups joined by carbon–carbon single bonds (13–26 kJ mol⁻¹). While groups joined by single bonds rotate relatively freely at room temperature, those joined by double bonds do not.

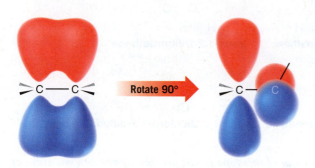

Rotate 90°

FIGURE 1.27 A stylized depiction of how rotation of a carbon atom of a double bond through an angle of 90° results in breaking of the π bond.

1.13B Cis–Trans Isomerism

Restricted rotation of groups joined by a double bond causes a new type of isomerism that we illustrate with the two dichloroethenes written as the following structures:

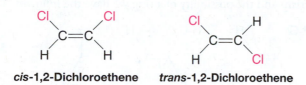

cis-1,2-Dichloroethene   *trans*-1,2-Dichloroethene

- These two compounds are isomers; they are different compounds that have the same molecular formula.

We can tell that they are different compounds by trying to place a model of one compound on a model of the other so that all parts coincide, that is, to try to **superpose** one on the other. We find that it cannot be done. Had one been **superposable** on the other, all parts of one model would correspond in three dimensions exactly with the other model. (*The notion of superposition is different from simply superimposing one thing on another*. The latter means only to lay one on the other without the necessary condition that all parts coincide.)

- We indicate that they are different isomers by attaching the prefix cis or trans to their names (*cis*, Latin: on this side; *trans*, Latin: across).

cis-1,2-Dichloroethene and *trans*-1,2-dichloroethene are not constitutional isomers because the connectivity of the atoms is the same in each. The two compounds **differ only in the arrangement of their atoms in space**. Isomers of this kind are classified formally as **stereoisomers**, but often they are called simply cis–trans isomers. (We shall study stereoisomerism in detail in Chapters 4 and 5.)

The structural requirements for **cis–trans isomerism** will become clear if we consider a few additional examples. 1,1-Dichloroethene and 1,1,2-trichloroethene do not show this type of isomerism.

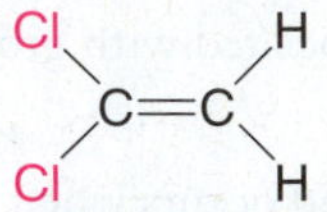

1,1-Dichloroethene
(no cis-trans isomerism)

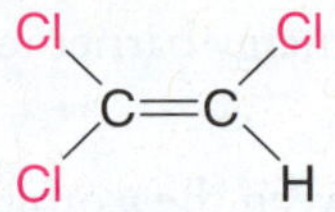

1,1,2-Trichloroethene
(no cis-trans isomerism)

1,2-Difluoroethene and 1,2-dichloro-1,2-difluoroethene do exist as cis–trans isomers. Notice that we designate the isomer with two identical groups on the same side as being cis:

cis-1,2-Difluoroethene   *trans*-1,2-Difluoroethene

cis-1,2-Dichloro-1,2-difluoroethene   *trans*-1,2-Dichloro-1,2-difluoroethene

Clearly, then, *cis–trans isomerism of this type is not possible if one carbon atom of the double bond bears two identical groups*.

● ● ● **SOLVED PROBLEM 1.12**

Write structures of all the isomers of C_2H_5F.

ANSWER: Taking into account cis–trans isomerism and the possibility of a ring we have the following four possibilities.

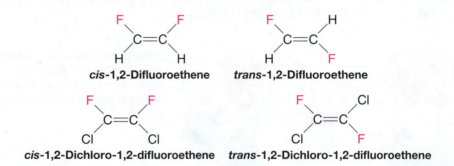

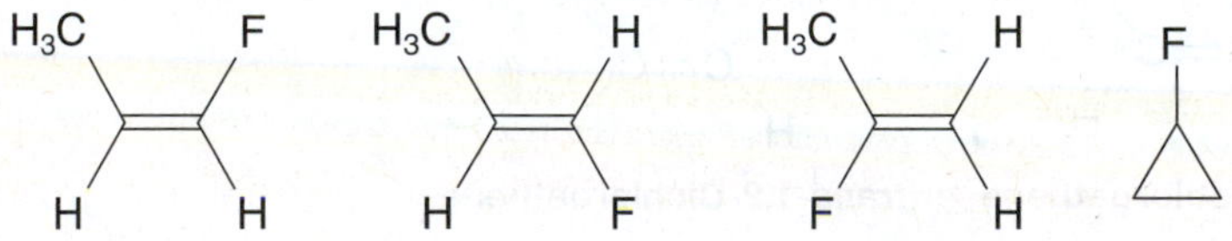

● ● ●

PRACTICE PROBLEM 1.22 Which of the following alkenes can exist as cis–trans isomers? Write their structures. Build hand-held models to prove that one isomer is not superposable on the other.

(a) $CH_2\!=\!CHCH_2CH_3$ **(c)** $CH_2\!=\!C(CH_3)_2$

(b) $CH_3CH\!=\!CHCH_3$ **(d)** $CH_3CH_2CH\!=\!CHCl$

1.14 THE STRUCTURE OF ETHYNE (ACETYLENE): *sp* HYBRIDIZATION

Hydrocarbons in which two carbon atoms share three pairs of electrons between them, and are thus bonded by a triple bond, are called **alkynes**. The two simplest alkynes are ethyne and propyne.

$$H\!-\!C\!\equiv\!C\!-\!H \qquad CH_3\!-\!C\!\equiv\!C\!-\!H$$

Ethyne
(acetylene)
(C_2H_2)

Propyne
(C_3H_4)

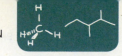

Ethyne, a compound that is also called **acetylene**, consists of a linear arrangement of atoms. The H—C≡C bond angles of ethyne molecules are 180°:

$$H-C \equiv C-H$$
$$180° \quad 180°$$

We can account for the structure of ethyne on the basis of orbital hybridization as we did for ethane and ethene. In our model for ethane (Section 1.12B) we saw that the carbon orbitals are sp^3 hybridized, and in our model for ethene (Section 1.13) we saw that they are sp^2 hybridized. In our model for ethyne we shall see that the carbon atoms are *sp hybridized*.

The mathematical process for obtaining the *sp* hybrid orbitals of ethyne can be visualized in the following way (Fig. 1.28).

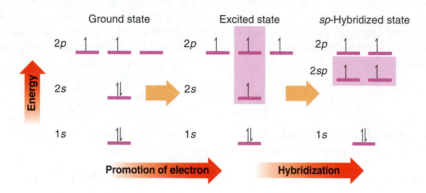

FIGURE 1.28 A process for deriving *sp*-hybridized carbon atoms.

- The 2s orbital and one 2p orbital of carbon are hybridized to form two **sp orbitals**.
- The remaining two 2p orbitals are not hybridized.

Calculations show that the *sp* hybrid orbitals have their large positive lobes oriented at an angle of 180° with respect to each other. The two 2p orbitals that were not hybridized are each perpendicular to the axis that passes through the center of the two *sp* orbitals (Fig. 1.29). We place one electron in each orbital.

We envision the bonding molecular orbitals of ethyne being formed in the following way (Fig. 1.30).

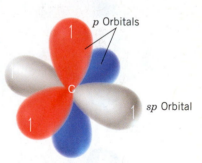

FIGURE 1.29 An *sp*-hybridized carbon atom.

- Two carbon atoms overlap *sp* orbitals to form a sigma bond between them (this is one bond of the triple bond). The remaining two *sp* orbitals at each carbon atom overlap with *s* orbitals from hydrogen atoms to produce two sigma C—H bonds.

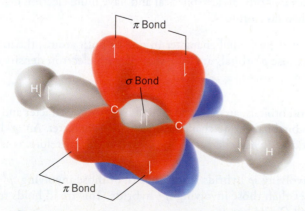

FIGURE 1.30 Formation of the bonding molecular orbitals of ethyne from two *sp*-hybridized carbon atoms and two hydrogen atoms. (Antibonding orbitals are formed as well, but these have been omitted for simplicity.)

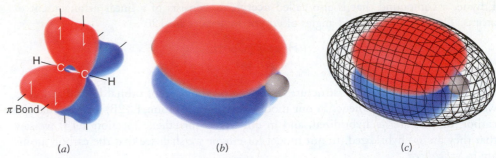

(a)   (b)   (c)

FIGURE 1.31 (a) The structure of ethyne (acetylene) showing the sigma-bond framework and a schematic depiction of the two pairs of *p* orbitals that overlap to form the two π bonds in ethyne. (b) A structure of ethyne showing calculated π molecular orbitals. Two pairs of π molecular orbital lobes are present, one pair for each π bond. The red and blue lobes in each π bond represent opposite phase signs. The hydrogen atoms of ethyne (white spheres) can be seen at each end of the structure (the carbon atoms are hidden by the molecular orbitals). (c) The mesh surface in this structure represents approximately the furthest extent of overall electron density in ethyne. Note that the overall electron density (but not the π-bonding electrons) extends over both hydrogen atoms.

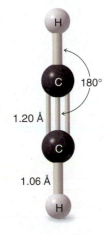

- The two *p* orbitals on each carbon atom also overlap side to side to form two π bonds. These are the other two bonds of the triple bond.

- The carbon–carbon triple bond consists of two π bonds and one σ bond.

Structures for ethyne based on calculated molecular orbitals and electron density are shown in Fig. 1.31. Circular symmetry exists along the length of a triple bond (Fig. 1.31*b*). As a result, there is no restriction of rotation for groups joined by a triple bond (as compared with alkenes), and if rotation would occur, no new compound would form.

1.14A Bond Lengths of Ethyne, Ethene, and Ethane

The carbon–carbon triple bond of ethyne is shorter than the carbon–carbon double bond of ethene, which in turn is shorter than the carbon–carbon single bond of ethane. The reason is that **bond lengths** are affected by the hybridization states of the carbon atoms involved.

- The greater the *s* orbital character in one or both atoms, the shorter is the bond. This is because *s* orbitals are spherical and have more electron density closer to the nucleus than do *p* orbitals.

- The greater the *p* orbital character in one or both atoms, the longer is the bond. This is because *p* orbitals are lobe-shaped with electron density extending away from the nucleus.

In terms of hybrid orbitals, an *sp* hybrid orbital has 50% *s* character and 50% *p* character. An *sp²* hybrid orbital has 33% *s* character and 67% *p* character. An *sp³* hybrid orbital has 25% *s* character and 75% *p* character. The overall trend, therefore, is as follows:

- Bonds involving *sp* hybrids are shorter than those involving *sp²* hybrids, which are shorter than those involving *sp³* hybrids. This trend holds true for both C—C and C—H bonds.

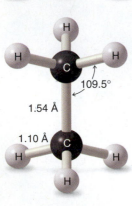

FIGURE 1.32 Bond angles and bond lengths of ethyne, ethene, and ethane.

The bond lengths and bond angles of ethyne, ethene, and ethane are summarized in Fig. 1.32.

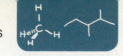

1.15 A SUMMARY OF IMPORTANT CONCEPTS THAT COME FROM QUANTUM MECHANICS

1. An **atomic orbital** (**AO**) corresponds to a region of space about the nucleus of a single atom where there is a high probability of finding an electron. Atomic orbitals called *s* orbitals are spherical; those called *p* orbitals are like two almost-tangent spheres. Orbitals can hold a maximum of two electrons when their spins are paired. Orbitals are described by the square of a wave function, ψ^2, and each orbital has a characteristic energy. The phase signs associated with an orbital may be $+$ or $-$.

2. When atomic orbitals overlap, they combine to form **molecular orbitals** (**MOs**). Molecular orbitals correspond to regions of space encompassing two (or more) nuclei where electrons are to be found. Like atomic orbitals, molecular orbitals can hold up to two electrons if their spins are paired.

3. When atomic orbitals with the same phase sign interact, they combine to form a **bonding molecular orbital**:

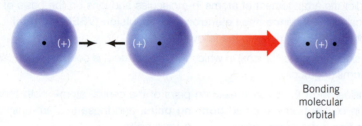

Bonding
molecular
orbital

The electron probability density of a bonding molecular orbital is large in the region of space between the two nuclei where the negative electrons hold the positive nuclei together.

4. An **antibonding molecular orbital** forms when orbitals of opposite phase sign overlap:

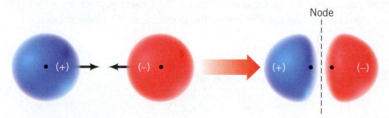

Node

An antibonding orbital has higher energy than a bonding orbital. The electron probability density of the region between the nuclei is small and it contains a **node**—a region where $\psi = 0$. Thus, having electrons in an antibonding orbital does not help hold the nuclei together. The internuclear repulsions tend to make them fly apart.

5. The **energy of electrons** in a bonding *molecular* orbital is less than the energy of the electrons in their separate *atomic* orbitals. The energy of electrons in an antibonding orbital is greater than that of electrons in their separate atomic orbitals.

6. The **number of molecular orbitals** always equals the number of atomic orbitals from which they are formed. Combining two atomic orbitals will always yield two molecular orbitals—one bonding and one antibonding.

7. **Hybrid atomic orbitals** are obtained by mixing (hybridizing) the wave functions for orbitals of different types (i.e., *s* and *p* orbitals) but from the same atom.

8. Hybridizing three *p* orbitals with one *s* orbital yields four sp^3 **orbitals**. Atoms that are sp^3 hybridized direct the axes of their four sp^3 orbitals toward the corners of a tetrahedron. The carbon of methane is sp^3 hybridized and **tetrahedral**.

9. Hybridizing two *p* orbitals with one *s* orbital yields three sp^2 **orbitals**. Atoms that are sp^2 hybridized point the axes of their three sp^2 orbitals toward the corners of an equilateral triangle. The carbon atoms of ethene are sp^2 hybridized and **trigonal planar**.

Helpful Hint

A summary of sp^3, sp^2, and sp hybrid orbital geometries.

10. Hybridizing one *p* orbital with one *s* orbital yields two ***sp* orbitals**. Atoms that are *sp* hybridized orient the axes of their two *sp* orbitals in opposite directions (at an angle of 180°). The carbon atoms of ethyne are *sp* hybridized and ethyne is a **linear** molecule.

11. A **sigma (σ) bond** (a type of single bond) is one in which the electron density has circular symmetry when viewed along the bond axis. In general, the skeletons of organic molecules are constructed of atoms linked by sigma bonds.

12. A **pi (π) bond**, part of double and triple carbon–carbon bonds, is one in which the electron densities of two adjacent parallel *p* orbitals overlap sideways to form a bonding pi molecular orbital.

1.16 HOW TO PREDICT MOLECULAR GEOMETRY: THE VALENCE SHELL ELECTRON PAIR REPULSION MODEL

We can predict the arrangement of atoms in molecules and ions on the basis of a relatively simple idea called the **valence shell electron pair repulsion (VSEPR) model**.

We apply the **VSEPR** model in the following way:

1. We consider molecules (or ions) in which the central atom is covalently bonded to two or more atoms or groups.

2. We consider all of the valence electron pairs of the central atom—both those that are shared in covalent bonds, called **bonding pairs**, and those that are unshared, called **nonbonding pairs** or **unshared pairs** or **lone pairs**.

3. Because electron pairs repel each other, the electron pairs of the valence shell tend to stay as far apart as possible. The repulsion between nonbonding pairs is generally greater than that between bonding pairs.

4. We arrive at the *geometry* of the molecule by considering all of the electron pairs, bonding and nonbonding, but we describe the *shape* of the molecule or ion by referring to the positions of the nuclei (or atoms) and not by the positions of the electron pairs.

In the following sections we consider several examples.

FIGURE 1.33 A tetrahedral shape for methane allows the maximum separation of the four bonding electron pairs.

1.16A Methane

The valence shell of methane contains four pairs of bonding electrons. Only a tetrahedral orientation will allow four pairs of electrons to have equal and maximum possible separation from each other (Fig. 1.33). Any other orientation, for example, a square planar arrangement, places some electron pairs closer together than others. Thus, methane has a tetrahedral shape.

The bond angles for any atom that has a regular tetrahedral structure are 109.5°. A representation of these angles in methane is shown in Fig. 1.34.

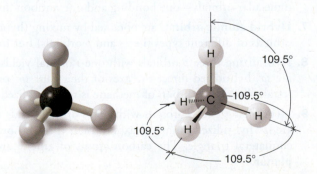

FIGURE 1.34 The bond angles of methane are 109.5°.

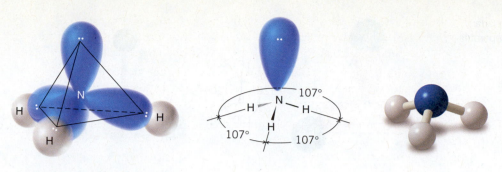

FIGURE 1.35 The tetrahedral arrangement of the electron pairs of an ammonia molecule that results when the nonbonding electron pair is considered to occupy one corner. This arrangement of electron pairs explains the trigonal pyramidal shape of the NH_3 molecule. Ball-and-stick models do not show unshared electrons.

1.16B Ammonia

The shape of a molecule of ammonia (NH_3) is a **trigonal pyramid**. There are three bonding pairs of electrons and one nonbonding pair. The bond angles in a molecule of ammonia are 107°, a value very close to the tetrahedral angle (109.5°). We can write a general tetrahedral structure for the electron pairs of ammonia by placing the nonbonding pair at one corner (Fig. 1.35). A *tetrahedral arrangement* of the electron pairs explains the *trigonal pyramidal* arrangement of the four atoms. The bond angles are 107° (not 109.5°) because the nonbonding pair occupies more space than the bonding pairs.

What do the bond angles of ammonia suggest about the hybridization state of the nitrogen atom of ammonia?

PRACTICE PROBLEM 1.23

1.16C Water

A molecule of water has an **angular** or **bent** shape. The H—O—H bond angle in a molecule of water is 104.5°, an angle that is also quite close to the 109.5° bond angles of methane.

We can write a general tetrahedral structure for the electron pairs of a molecule of water *if we place the two bonding pairs of electrons and the two nonbonding electron pairs at the corners of the tetrahedron.* Such a structure is shown in Fig. 1.36. A *tetrahedral arrangement* of the electron pairs accounts for the *angular arrangement* of the three atoms. The bond angle is less than 109.5° because the nonbonding pairs are effectively "larger" than the bonding pairs and, therefore, the structure is not perfectly tetrahedral.

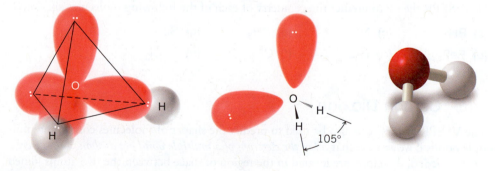

FIGURE 1.36 An approximately tetrahedral arrangement of the electron pairs of a molecule of water that results when the pairs of nonbonding electrons are considered to occupy corners. This arrangement accounts for the angular shape of the H_2O molecule.

What do the bond angles of water suggest about the hybridization state of the oxygen atom of water?

PRACTICE PROBLEM 1.24

1.16D Boron Trifluoride

Boron, a group IIIA element, has only three valence electrons. In the compound boron trifluoride (BF_3) these three electrons are shared with three fluorine atoms. As a result, the boron atom in BF_3 has only six electrons (three bonding pairs) around it. Maximum separation of three bonding pairs occurs when they occupy the corners of an equilateral

FIGURE 1.37 The triangular (trigonal planar) shape of boron trifluoride maximally separates the three bonding pairs.

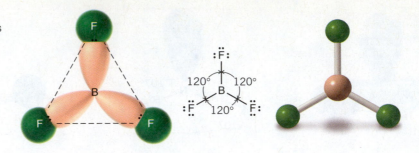

triangle. Consequently, in the boron trifluoride molecule the three fluorine atoms lie in a plane at the corners of an equilateral triangle (Fig. 1.37). Boron trifluoride is said to have a *trigonal planar structure*. The bond angles are 120°.

● ● ●

PRACTICE PROBLEM 1.25   What do the bond angles of boron trifluoride suggest about the hybridization state of the boron atom?

1.16E Beryllium Hydride

The central beryllium atom of BeH_2 has only two electron pairs around it; both electron pairs are bonding pairs. These two pairs are maximally separated when they are on opposite sides of the central atom, as shown in the following structures. This arrangement of the electron pairs accounts for the *linear geometry* of the BeH_2 molecule and its bond angle of 180°.

$$H : Be : H \quad \text{or} \quad H \overset{180°}{\frown} Be \text{---} H$$

Linear geometry of BeH_2

● ● ●

PRACTICE PROBLEM 1.26   What do the bond angles of beryllium hydride suggest about the hybridization state of the beryllium atom?

● ● ●

PRACTICE PROBLEM 1.27   Use VSEPR theory to predict the geometry of each of the following molecules and ions:

(a) $\bar{B}H_4$ (c) $\overset{+}{N}H_4$ (e) BH_3 (g) SiF_4

(b) BeF_2 (d) H_2S (f) CF_4 (h) $\bar{:}CCl_3$

1.16F Carbon Dioxide

The VSEPR method can also be used to predict the shapes of molecules containing multiple bonds if we assume that *all of the electrons of a multiple bond act as though they were a single unit* and, therefore, are located in the region of space between the two atoms joined by a multiple bond.

This principle can be illustrated with the structure of a molecule of carbon dioxide (CO_2). The central carbon atom of carbon dioxide is bonded to each oxygen atom by a double bond. Carbon dioxide is known to have a linear shape; the bond angle is 180°.

$$:\overset{180°}{O} = C = O: \quad \text{or} \quad :O :: C :: O:$$

The four electrons of each double bond act as a single unit and are maximally separated from each other.

Such a structure is consistent with a maximum separation of the two groups of four bonding electrons. The nonbonding pairs associated with the oxygen atoms have no effect on the shape.

TABLE 1.3 SHAPES OF MOLECULES AND IONS FROM VSEPR THEORY

Number of Electron Pairs at Central Atom			Hybridization State of Central Atom	Shape of Molecule or Ion[a]	Examples
Bonding	Nonbonding	Total			
2	0	2	sp	Linear	BeH_2
3	0	3	sp^2	Trigonal planar	BF_3, $\overset{+}{C}H_3$
4	0	4	sp^3	Tetrahedral	CH_4, $\overset{+}{N}H_4$
3	1	4	$\sim sp^3$	Trigonal pyramidal	NH_3, $\bar{C}H_3$
2	2	4	$\sim sp^3$	Angular	H_2O

[a]Referring to positions of atoms and excluding nonbonding pairs.

Predict the bond angles of

(a) $F_2C{=}CF_2$ (b) $CH_3C{\equiv}CCH_3$ (c) $HC{\equiv}N$

PRACTICE PROBLEM 1.28

The shapes of several simple molecules and ions as predicted by VSEPR theory are shown in Table 1.3. In this table we have also included the hybridization state of the central atom.

1.17 APPLICATIONS OF BASIC PRINCIPLES

Throughout the early chapters of this book we review certain basic principles that underlie and explain much of the chemistry we shall be studying. Consider the following principles and how they apply in this chapter.

Opposite Charges Attract We see this principle operating in our explanations for covalent and ionic bonds (Section 1.3A). It is the attraction of the *positively* charged nuclei for the *negatively* charged electrons that underlies our explanation for the covalent bond. It is the attraction of the oppositely charged ions in crystals that explains the ionic bond.

Like Charges Repel It is the repulsion of the electrons in covalent bonds of the valence shell of a molecule that is central to the valence shell electron pair repulsion model for explaining molecular geometry. And, although it is not so obvious, this same factor underlies the explanations of molecular geometry that come from orbital hybridization because these repulsions are taken into account in calculating the orientations of the hybrid orbitals.

Nature Tends toward States of Lower Potential Energy This principle explains so much of the world around us. It explains why water flows downhill: the potential energy of the water at the bottom of the hill is lower than that at the top. (We say that water is in a more stable state at the bottom.) This principle underlies the aufbau principle (Section 1.10A): in its lowest energy state, the electrons of an atom occupy the lowest energy orbitals available [but Hund's rule still applies, as well as the Pauli exclusion principle (Section 1.10A), allowing only two electrons per orbital]. Similarly in molecular orbital theory (Section 1.11), electrons fill lower energy bonding molecular orbitals first because this gives the molecule lower potential energy (or greater stability). Energy has to be provided to move an electron to a higher orbital and provide an excited (less stable) state.

Orbital Overlap Stabilizes Molecules This principle is part of our explanation for covalent bonds. When orbitals of the same phase from different nuclei overlap, the electrons in these orbitals can be shared by both nuclei, resulting in stabilization. The result is a covalent bond.

WHY Do These Topics Matter?]

NATURAL PRODUCTS THAT CAN TREAT DISEASE

Everywhere on Earth, organisms make organic molecules comprised almost exclusively of carbon, hydrogen, nitrogen, and oxygen. Sometimes a few slightly more exotic atoms, such as halogens and sulfur, are present. Globally, these compounds aid in day-to-day functioning of these organisms and/or their survival against predators. Organic molecules include the chlorophyll in green plants, which harnesses the energy of sunlight; vitamin C is synthesized by citrus trees, protecting them against oxidative stress; capsaicin, a molecule synthesized by pepper plants (and makes peppers taste hot), serves to ward off insects and birds that might try to eat them; salicylic acid, made by willow trees, is a signaling hormone; lovastatin, found in oyster mushrooms, protects against bacterial attacks.

Capsaicin

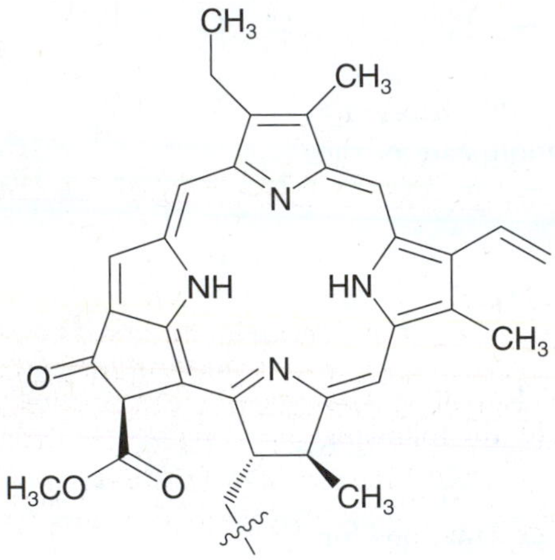

Chlorophyll core

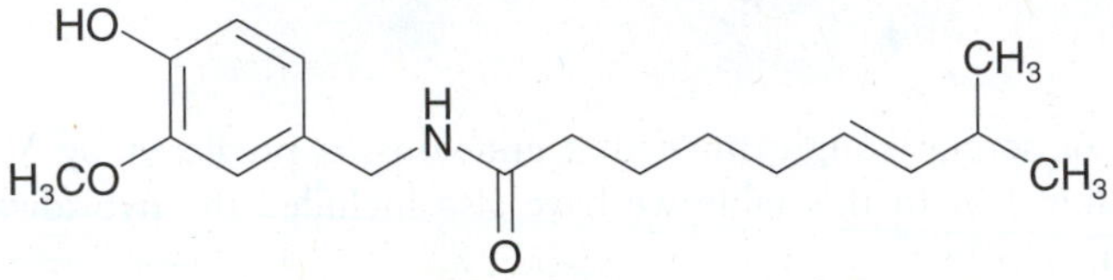

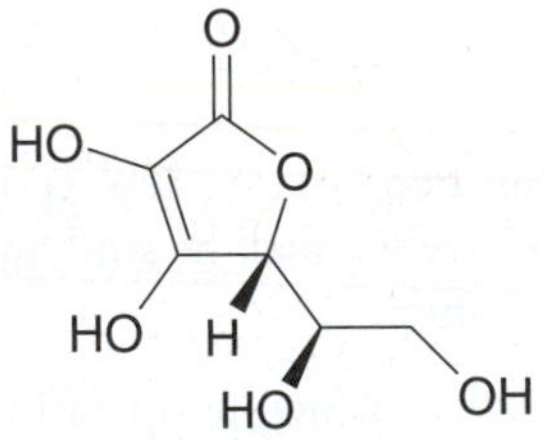

Vitamin C

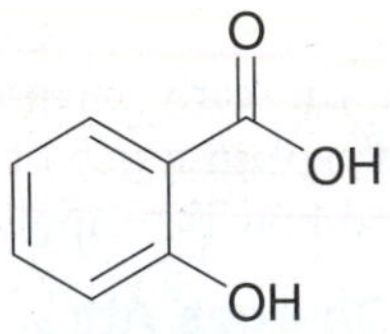

Salicylic acid

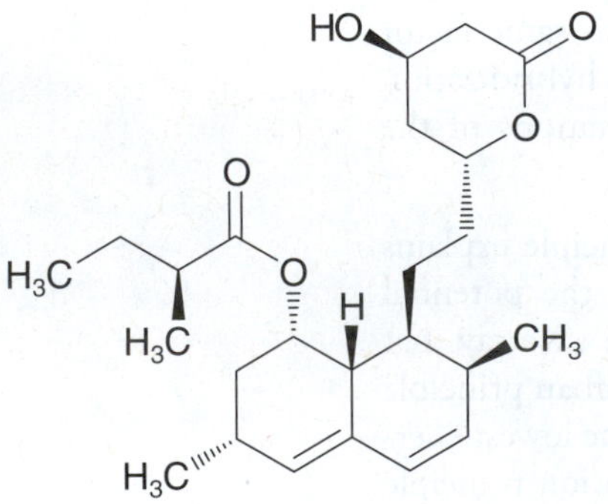

Lovastatin

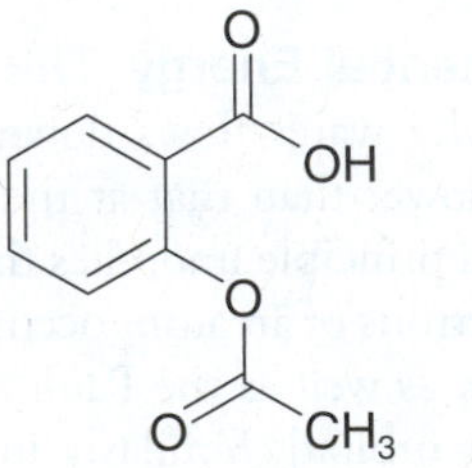

Aspirin

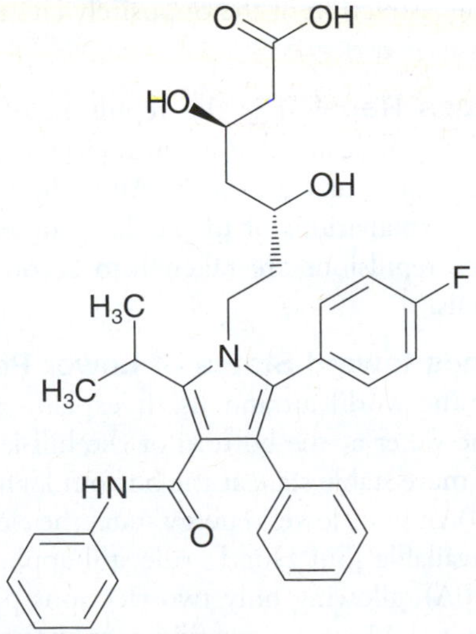

Lipitor

These compounds are all natural products, and many advances in modern society are the result of their study and use. Capsaicin, it turns out, is an effective analgesic. It can modulate pain when applied to the skin and is currently sold under the tradename Capzacin. Salicylic acid is a painkiller as well as an anti-acne medication, while lovastatin is used as a drug to decrease levels of cholesterol in human blood. The power of modern organic chemistry lies in the ability to take such molecules, sometimes found in trace quantities in nature, and make them from readily available and inexpensive starting materials on a large scale so that all members of society can benefit from them. For instance, although we can obtain vitamin C from eating certain fruits, chemists can make large quantities in the laboratory for use in daily supplements; while some may think that "natural" vitamin C is healthier, the "syn-thetic" compound is equally effective since they are exactly the same chemically.

Perhaps more important, organic chemistry also provides the opportunity to change the structures of these and other natural products to make molecules with different, and potentially even more impressive, properties. For example, the addition of a few atoms to salicylic acid through a chemical reaction is what led to the discovery of aspirin (see Chapter 17), a molecule with far greater potency as a painkiller and fewer side effects than nature's compound. Similarly, scientists at Parke–Davis Warner–Lambert (now Pfizer) used the structure and activity of lovastatin as inspiration to develop Lipitor, a molecule that has saved countless lives by lowering levels of cholesterol in human serum. In fact, of the top 20 drugs based on gross sales, slightly over half are either natural products or their derivatives.

To learn more about these topics, see:

1. Nicolaou, K. C.; Montagnon, T. *Molecules that Changed the World*. Wiley-VCH: Weinheim, **2008**, p. 366.

2. Nicolaou, K. C.; Sorensen, E. J.; Winssinger, N, "The Art and Science of Organic and Natural Products Synthesis" in *J. Chem. Educ.* **1998**, *75*, 1225–1258.

SUMMARY AND REVIEW TOOLS

In Chapter 1 you have studied concepts and skills that are absolutely essential to your success in organic chemistry. You should now be able to use the periodic table to determine the number of valence electrons an atom has in its neutral state or as an ion. You should be able to use the periodic table to compare the relative electronegativity of one element with another, and determine the formal charge of an atom or ion. Electronegativity and formal charge are key concepts in organic chemistry.

You should be able to draw chemical formulas that show all of the valence electrons in a molecule (Lewis structures), using lines for bonds and dots to show unshared electrons. You should be proficient in representing structures as dash structural formulas, condensed structural formulas, and bond-line structural formulas. In particular, the more quickly you become skilled at using and interpreting bond-line formulas, the faster you will be able to process structural information in organic chemistry. You have also learned about resonance structures, the use of which will help us in understanding a variety of concepts in later chapters.

Last, you have learned to predict the three-dimensional structure of molecules using the valence shell electron pair repulsion (VSEPR) model and molecular orbital (MO) theory. An ability to predict three-dimensional structure is critical to understanding the properties and reactivity of molecules.

We encourage you to do all of the problems that your instructor has assigned. We also recommend that you use the summary and review tools in each chapter, such as the concept map that follows. Concept maps can help you see the flow of concepts in a chapter and also help remind you of key points. In fact, we encourage you to build your own concept maps for review when the opportunity arises.

Work especially hard to solidify your knowledge from this and other early chapters in the book. These chapters have everything to do with helping you learn basic tools you need for success throughout organic chemistry.

The study aids for this chapter include key terms and concepts (which are hyperlinked to the glossary from the bold, blue terms in the *WileyPLUS* version of the book at wileyplus.com) and a Concept Map after the end-of-chapter problems.

KEY TERMS AND CONCEPTS

The key terms and concepts that are highlighted in bold, blue text within the chapter are defined in the glossary (at the back of the book) and have hyperlinked definitions in the accompanying *WileyPLUS* course (www.wileyplus.com).

PROBLEMS PLUS

Note to Instructors: Many of the homework problems are available for assignment via *WileyPlus*, an online teaching and learning solution.

ELECTRON CONFIGURATION

1.29 Which of the following ions possess the electron configuration of a noble gas?

(a) Na^+ **(b)** Cl^- **(c)** F^+ **(d)** H^- **(e)** Ca^{2+} **(f)** S^{2-} **(g)** O^{2-} **(h)** Br^+

LEWIS STRUCTURES

1.30 Write a Lewis structure for each of the following:

(a) $SOCl_2$ **(b)** $POCl_3$ **(c)** PCl_5 **(d)** $HONO_2$ (HNO_3)

1.31 Give the formal charge (if one exists) on each atom of the following:

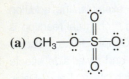

1.32 Add any unshared electrons to give each element an octet in its valence shell in the formulas below and indicate any formal charges. Note that all of the hydrogen atoms that are attached to heteroatoms have been drawn if they are present.

(a) 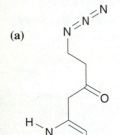   **(b)** 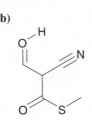   **(c)** 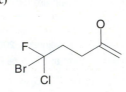   **(d)** 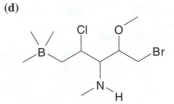   **(e)**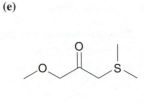

STRUCTURAL FORMULAS AND ISOMERISM

1.33 Write a condensed structural formula for each compound given here.

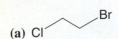

(a)   **(b)**   **(c)**   **(d)**

1.34 What is the molecular formula for each of the compounds given in Exercise 1.33?

1.35 Consider each pair of structural formulas that follow and state whether the two formulas represent the same compound, whether they represent different compounds that are constitutional isomers of each other, or whether they represent different compounds that are not isomeric.

(a)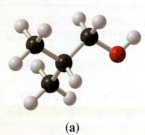

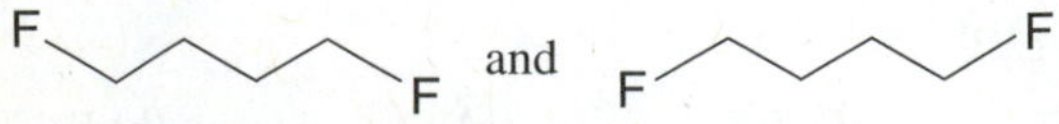

(e)

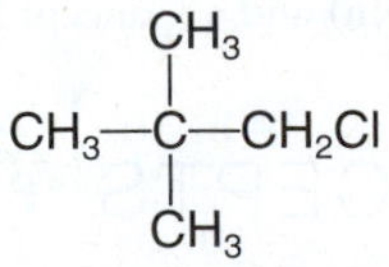

(b)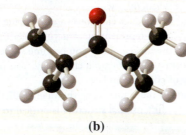

(f) CH_2=$CHCH_2CH_3$ and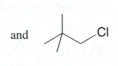

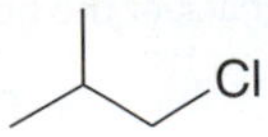

(c) H—C—Cl and Cl—C—Cl

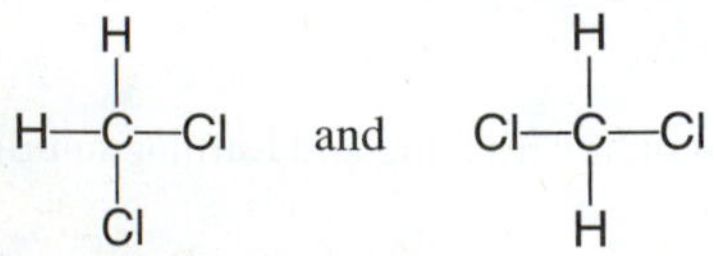

(g)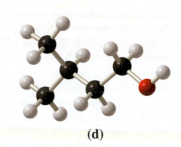

(d) F ... F and F ... F

(h) CH_3CH_2 and
   CH_2CH_3

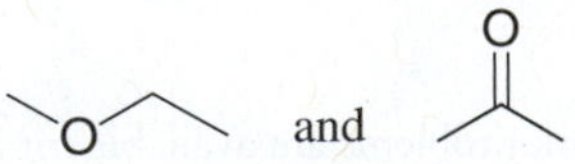

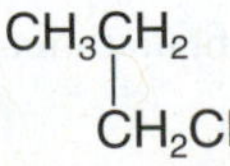

(i) $CH_3OCH_2CH_3$ and

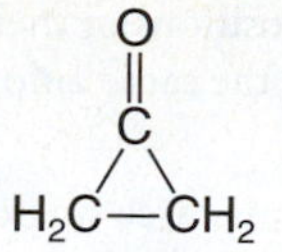

(m) H—C—Br and CH_3—C—Br

(j) $CH_2ClCHClCH_3$ and $CH_3CHClCH_2Cl$

(n) CH_3—C—H and CH_3—C—CH_3

(k) $CH_3CH_2CHClCH_2Cl$ and CH_3CHCH_2Cl
$\phantom{CH_3CH_2CHClCH_2Cl \text{ and } }CH_2Cl$

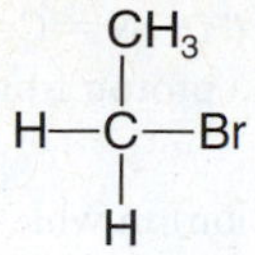

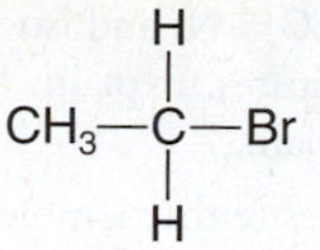

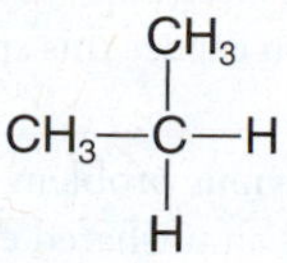

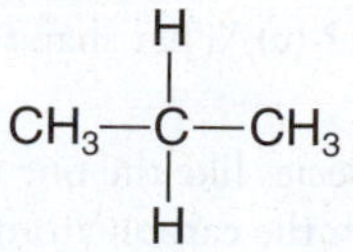

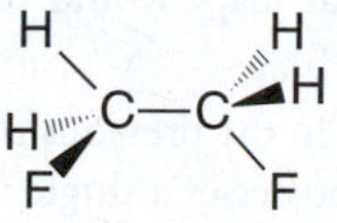

(o)

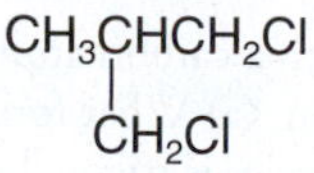

(l)

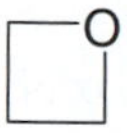

and

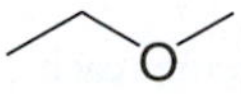

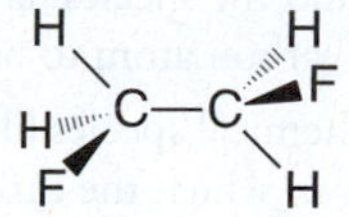

(p)

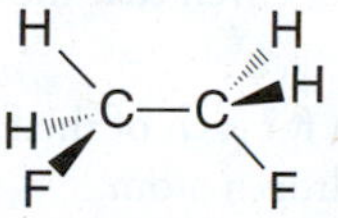

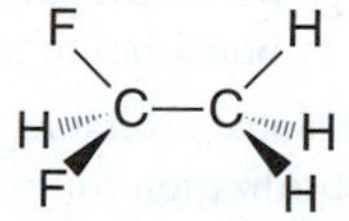

1.36 Rewrite each of the following using bond-line formulas:

(a) 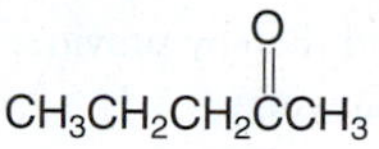 $CH_3CH_2CH_2\overset{\displaystyle O}{\overset{\|}{C}}CH_3$

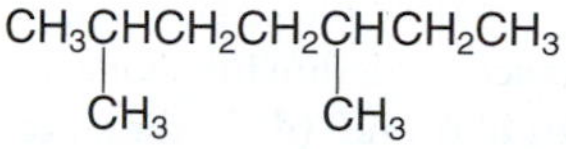

(c) $(CH_3)_3CCH_2CH_2CH_2OH$

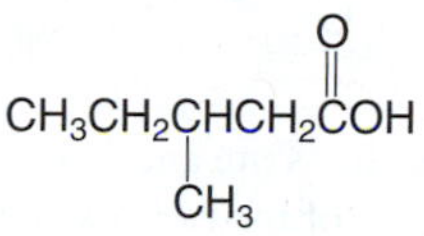

(e) $CH_2{=}CHCH_2CH_2CH{=}CHCH_3$

(b) $CH_3CHCH_2CH_2CHCH_2CH_3$
CH_3CH_3

(d) $CH_3CH_2CHCH_2\overset{\displaystyle O}{\overset{\|}{C}}OH$
CH_3

(f)

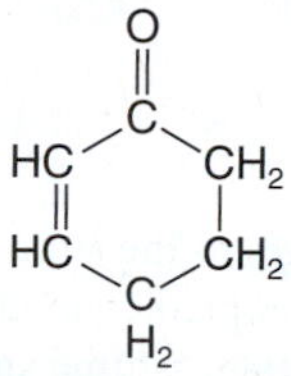

1.37 Write bond-line formulas for all of the constitutional isomers with the molecular formula C_4H_8.

1.38 Write structural formulas for at least three constitutional isomers with the molecular formula CH_3NO_2. (In answering this question you should assign a formal charge to any atom that bears one.)

RESONANCE STRUCTURES

1.39 Write the resonance structure that would result from moving the electrons in the way indicated by the curved arrows.

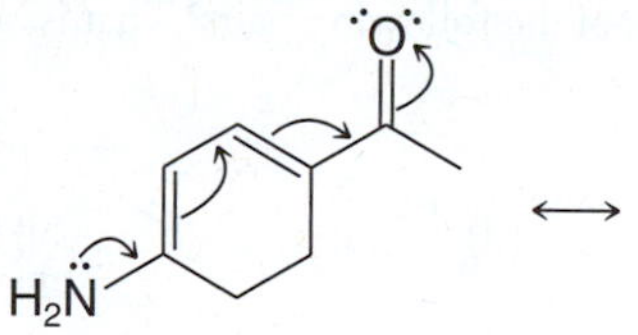

1.40 Show the curved arrows that would convert **A** into **B**.

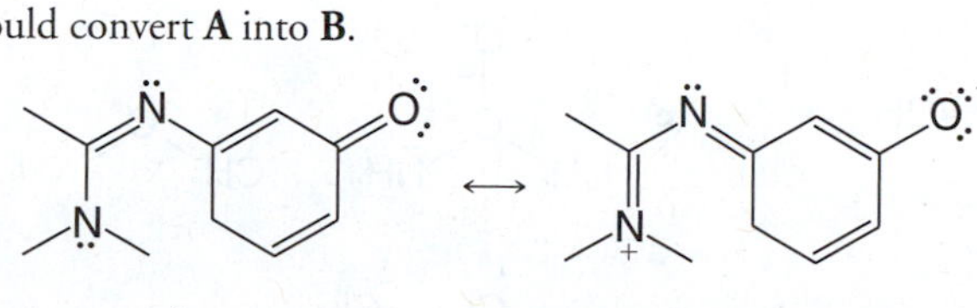

 **A** **B**

1.41 For the following write all possible resonance structures. Be sure to include formal charges where appropriate.

(a)

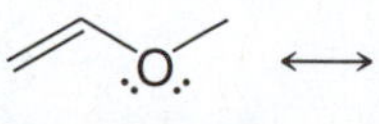

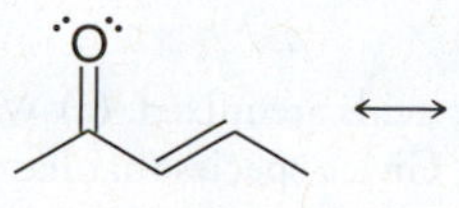

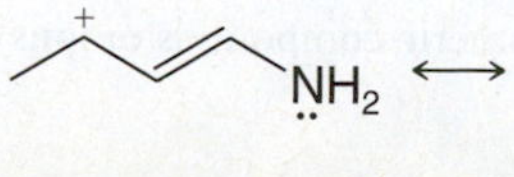

(d)

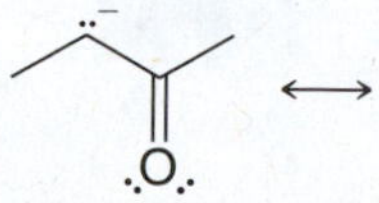

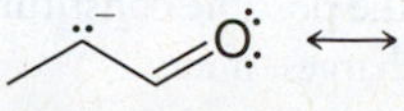

(g)

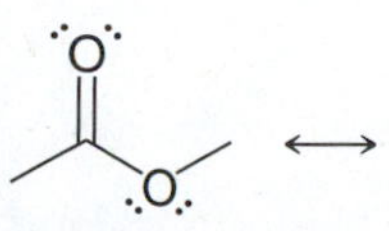

(b)

(e)

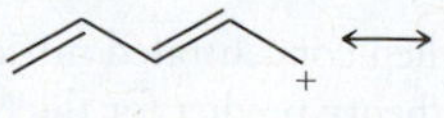

(h)

(c)

(f)

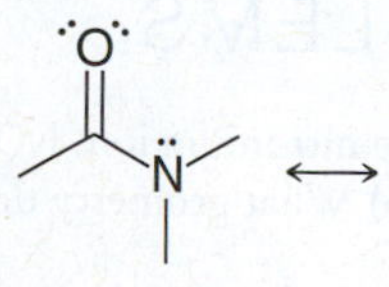

(i)

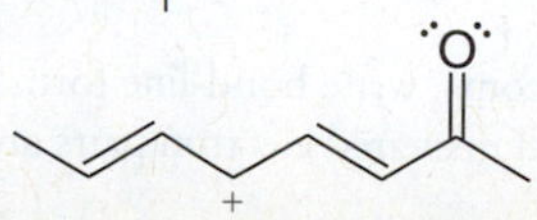

1.42 **(a)** Cyanic acid (H—O—C≡N) and isocyanic acid (H—N=C=O) differ in the positions of their electrons but their structures do not represent resonance structures. Explain. **(b)** Loss of a proton from cyanic acid yields the same anion as that obtained by loss of a proton from isocyanic acid. Explain.

1.43 Consider a chemical species (either a molecule or an ion) in which a carbon atom forms three single bonds to three hydrogen atoms and in which the carbon atom possesses no other valence electrons. **(a)** What formal charge would the carbon atom have? **(b)** What total charge would the species have? **(c)** What shape would you expect this species to have? **(d)** What would you expect the hybridization state of the carbon atom to be?

1.44 Consider a chemical species like the one in the previous problem in which a carbon atom forms three single bonds to three hydrogen atoms, but in which the carbon atom possesses an unshared electron pair. **(a)** What formal charge would the carbon atom have? **(b)** What total charge would the species have? **(c)** What shape would you expect this species to have? **(d)** What would you expect the hybridization state of the carbon atom to be?

1.45 Consider another chemical species like the ones in the previous problems in which a carbon atom forms three single bonds to three hydrogen atoms but in which the carbon atom possesses a single unpaired electron. **(a)** What formal charge would the carbon atom have? **(b)** What total charge would the species have? **(c)** Given that the shape of this species is trigonal planar, what would you expect the hybridization state of the carbon atom to be?

1.46 Draw a three-dimensional orbital representation for each of the following molecules, indicate whether each bond in it is a σ or π bond, and provide the hybridization for each non-hydrogen atom.

(a) CH_2O

(b) $H_2C=CHCH=CH_2$

(c) $H_2C=C=C=CH_2$

1.47 Ozone (O_3) is found in the upper atmosphere where it absorbs highly energetic ultraviolet (UV) radiation and thereby provides the surface of Earth with a protective screen (cf. Section 10.11E). One possible resonance structure for ozone is the following:

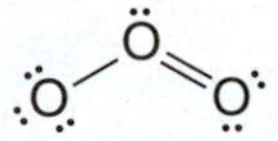

(a) Assign any necessary formal charges to the atoms in this structure. **(b)** Write another equivalent resonance structure for ozone. **(c)** What do these resonance structures predict about the relative lengths of the two oxygen–oxygen bonds of ozone? **(d)** In the structure above, and the one you have written, assume an angular shape for the ozone molecule. Is this shape consistent with VSEPR theory? Explain your answer.

1.48 Write resonance structures for the azide ion, N_3^-. Explain how these resonance structures account for the fact that both bonds of the azide ion have the same length.

1.49 Write structural formulas of the type indicated: **(a)** bond-line formulas for seven constitutional isomers with the formula $C_4H_{10}O$; **(b)** condensed structural formulas for two constitutional isomers with the formula C_2H_7N; **(c)** condensed structural formulas for four constitutional isomers with the formula C_3H_9N; **(d)** bond-line formulas for three constitutional isomers with the formula C_5H_{12}.

1.50 What is the relationship between the members of the following pairs? That is, are they constitutional isomers, the same, or something else (specify)?

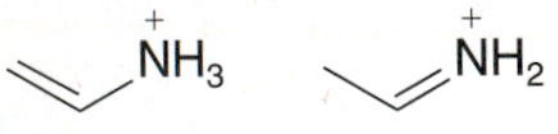

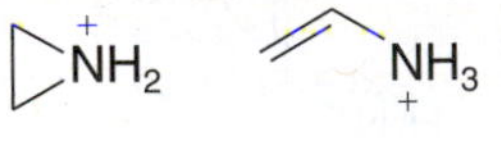

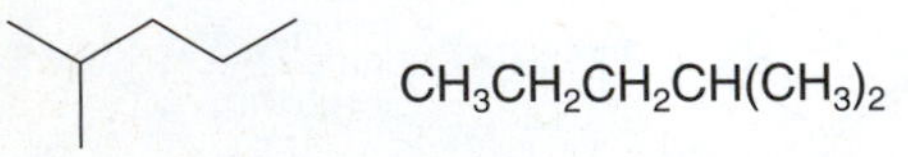

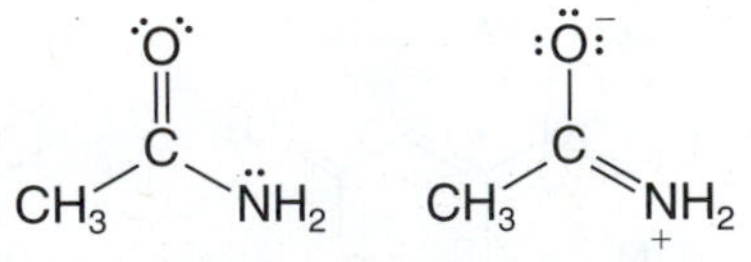

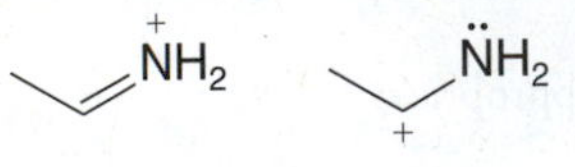

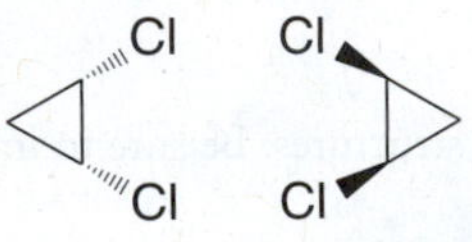

CHALLENGE PROBLEMS

1.51 In Chapter 15 we shall learn how the nitronium ion, NO_2^+, forms when concentrated nitric and sulfuric acids are mixed. **(a)** Write a Lewis structure for the nitronium ion. **(b)** What geometry does VSEPR theory predict for the NO_2^+ ion? **(c)** Give a species that has the same number of electrons as NO_2^+.

1.52 Given the following sets of atoms, write bond-line formulas for all of the possible constitutionally isomeric compounds or ions that could be made from them. Show all unshared electron pairs and all formal charges, if any.

Set	C atoms	H atoms	Other
A	3	6	2 Br atoms
B	3	9	1 N atom and 1 O atom (not on same C)
C	3	4	1 O atom
D	2	7	1 N atom and 1 proton
E	3	7	1 extra electron

1.53 (a) Consider a carbon atom in its ground state. Would such an atom offer a satisfactory model for the carbon of methane? If not, why not? (*Hint:* Consider whether a ground state carbon atom could be tetravalent, and consider the bond angles that would result if it were to combine with hydrogen atoms.)

(b) Consider a carbon atom in the excited state:

$$\text{C} \quad \frac{\uparrow\downarrow}{1s} \ \frac{\uparrow}{2s} \ \frac{\uparrow}{2p_x} \ \frac{\uparrow}{2p_y} \ \frac{\uparrow}{2p_z}$$

Excited state of a carbon atom

Would such an atom offer a satisfactory model for the carbon of methane? If not, why not?

1.54 Open computer molecular models for dimethyl ether, dimethylacetylene, and *cis*-1,2-dichloro-1,2-difluoroethene from the 3D Molecular Models section of the book's website. By interpreting the computer molecular model for each one, draw **(a)** a dash formula, **(b)** a bond-line formula, and **(c)** a three-dimensional dashed-wedge formula. Draw the models in whatever perspective is most convenient—generally the perspective in which the most atoms in the chain of a molecule can be in the plane of the paper.

1.55 Boron is a group IIIA element. Open the molecular model for boron trifluoride from the 3D Molecular Models section of the book's website. Near the boron atom, above and below the plane of the atoms in BF_3, are two relatively large lobes. Considering the position of boron in the periodic table and the three-dimensional and electronic structure of BF_3, what type of orbital does this lobe represent? Is it a hybridized orbital or not?

1.56 There are two contributing resonance structures for an anion called acetaldehyde enolate, whose condensed molecular formula is CH_2CHO^-. Draw the two resonance contributors and the resonance hybrid, then consider the map of electrostatic potential (MEP) shown below for this anion. Comment on whether the MEP is consistent or not with predominance of the resonance contributor you would have predicted to be represented most strongly in the hybrid.

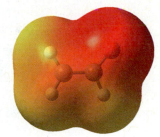

LEARNING GROUP PROBLEMS

Consider the compound with the following condensed molecular formula:

$$CH_3CHOHCH=CH_2$$

1. Write a full dash structural formula for the compound.

2. Show all nonbonding electron pairs on your dash structural formula.

3. Indicate any formal charges that may be present in the molecule.

4. Label the hybridization state at every carbon atom and the oxygen.

5. Draw a three-dimensional perspective representation for the compound showing approximate bond angles as clearly as possible. Use ordinary lines to indicate bonds in the plane of the paper, solid wedges for bonds in front of the paper, and dashed wedges for bonds behind the paper.

6. Label all the bond angles in your three-dimensional structure.

7. Draw a bond-line formula for the compound.

8. Devise two structures, each having two *sp*-hybridized carbons and the molecular formula C_4H_6O. Create one of these structures such that it is linear with respect to all carbon atoms. Repeat parts 1–7 above for both structures.

Helpful Hint

Your instructor will tell you how to work these problems as a Learning Group.

[**C O N C E P T   M A P**]

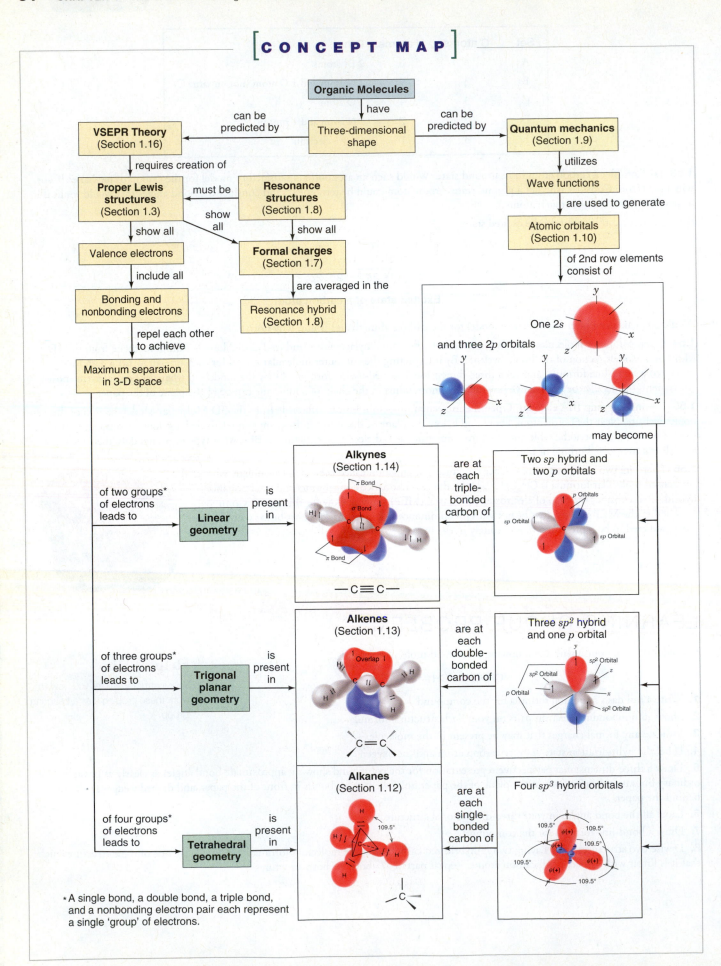

CHAPTER

2

Families of Carbon Compounds

FUNCTIONAL GROUPS, INTERMOLECULAR FORCES, AND INFRARED (IR) SPECTROSCOPY

In this chapter we introduce one of the great simplifying concepts of organic chemistry—the functional group. Functional groups are common and specific arrangements of atoms that impart predictable reactivity and properties to a molecule. Even though there are millions of organic compounds, you may be relieved to know that we can readily understand much about whole families of compounds simply by learning about the properties of the common functional groups.

For example, all alcohols contain an —OH (hydroxyl) functional group attached to a saturated carbon bearing nothing else but carbon or hydrogen. Alcohols as simple as ethanol in alcoholic beverages and as complex as ethinyl estradiol (Section 2.1C) in birth control pills have this structural unit in common. All aldehydes have a —C(=O)— (carbonyl) group with one bond to a hydrogen and the other to one or more carbons, such as in benzaldehyde (which comes from almonds). All ketones include a carbonyl group bonded by its carbon to one or more other carbons on each side, as in the natural oil menthone, found in geraniums and spearmint.

Ethanol   **Benzaldehyde**   **Menthone**

PHOTO CREDIT: © ValentynVolkov/iStockphoto

Members of each functional group family share common chemical properties and reactivity, and this fact helps greatly in organizing our knowledge of organic chemistry. As you progress in this chapter it will serve you well to learn the arrangements of atoms that define the common functional groups. This knowledge will be invaluable to your study of organic chemistry.

IN THIS CHAPTER WE WILL CONSIDER:

• the major functional groups

• the correlation between properties of functional groups and molecules and intermolecular forces

• infrared (IR) spectroscopy, which can be used to determine what functional groups are present in a molecule

[WHY DO THESE TOPICS MATTER?]   At the end of the chapter, we will see how these important concepts merge together to explain how the world's most powerful antibiotic behaves and how bacteria have evolved to escape its effects.

2.1 HYDROCARBONS: REPRESENTATIVE ALKANES, ALKENES, ALKYNES, AND AROMATIC COMPOUNDS

We begin this chapter by introducing the class of compounds that contains only carbon and hydrogen, and we shall see how the -ane, -ene, or -yne ending in a name tells us what kinds of carbon–carbon bonds are present.

• **Hydrocarbons** are compounds that contain only carbon and hydrogen atoms.

Methane (CH_4) and ethane (C_2H_6) are hydrocarbons, for example. They also belong to a subgroup of compounds called alkanes.

 Propane (an alkane)

• **Alkanes** are hydrocarbons that do not have multiple bonds between carbon atoms, and we can indicate this in the family name and in names for specific compounds by the **-ane** ending.

Other hydrocarbons may contain double or triple bonds between their carbon atoms.

 Propene (an alkene)

• **Alkenes** contain at least one carbon–carbon double bond, and this is indicated in the family name and in names for specific compounds by an **-ene** ending.

 Propyne (an alkyne)

• **Alkynes** contain at least one carbon–carbon triple bond, and this is indicated in the family name and in names for specific compounds by an **-yne** ending.

 Benzene (an aromatic compound)

• **Aromatic compounds** contain a special type of ring, the most common example of which is a benzene ring. There is no special ending for the general family of aromatic compounds.

We shall introduce representative examples of each of these classes of hydrocarbons in the following sections.

Generally speaking, compounds such as alkanes, whose molecules contain only single bonds, are referred to as **saturated compounds** because these compounds contain the maximum number of hydrogen atoms that the carbon compound can possess. Compounds with multiple bonds, such as alkenes, alkynes, and aromatic hydrocarbons, are called **unsaturated compounds** because they possess fewer than the maximum number of hydrogen atoms, and they are capable of reacting with hydrogen under the proper conditions. We shall have more to say about this in Chapter 7.

Methane

2.1A Alkanes

The primary sources of alkanes are natural gas and petroleum. The smaller alkanes (methane through butane) are gases under ambient conditions. Methane is the principal component of natural gas. Higher molecular weight alkanes are obtained largely by refining petroleum. Methane, the simplest alkane, was one major component of the early atmosphere of this planet. Methane is still found in Earth's atmosphere, but no longer in appreciable amounts. It is, however, a major component of the atmospheres of Jupiter, Saturn, Uranus, and Neptune.

Some living organisms produce methane from carbon dioxide and hydrogen. These very primitive creatures, called *methanogens*, may be Earth's oldest organisms, and they

may represent a separate form of evolutionary development. Methanogens can survive only in an anaerobic (i.e., oxygen-free) environment. They have been found in ocean trenches, in mud, in sewage, and in cows' stomachs.

2.1B Alkenes

Ethene

Ethene and propene, the two simplest alkenes, are among the most important industrial chemicals produced in the United States. Each year, the chemical industry produces more than 30 billion pounds of ethene and about 15 billion pounds of propene. Ethene is used as a starting material for the synthesis of many industrial compounds, including ethanol, ethylene oxide, ethanal, and the polymer polyethylene (Section 10.10). Propene is used in making the polymer polypropylene (Section 10.10 and Special Topic B*), and, in addition to other uses, propene is the starting material for a synthesis of acetone and cumene (Section 21.4B).

Ethene also occurs in nature as a plant hormone. It is produced naturally by fruits such as tomatoes and bananas and is involved in the ripening process of these fruits. Much use is now made of ethene in the commercial fruit industry to bring about the ripening of tomatoes and bananas picked green because the green fruits are less susceptible to damage during shipping.

There are many naturally occurring alkenes. Two examples are the following:

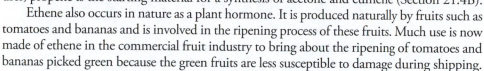

β-Pinene
(a component of turpentine)   **An aphid alarm pheromone**

●●● SOLVED PROBLEM 2.1

Propene, $CH_3CH=CH_2$, is an alkene. Write the structure of a constitutional isomer of propene that is not an alkene. (*Hint*: It does not have a double bond.)

STRATEGY AND ANSWER: A compound with a ring of *n* carbon atoms will have the same molecular formula as an alkene with the same number of carbons.

 is a constitutional isomer of

Cyclopropane   **Propene**
C_3H_6 C_3H_6

Cyclopropane has anesthetic properties.

2.1C Alkynes

Ethyne

The simplest alkyne is ethyne (also called acetylene). Alkynes occur in nature and can be synthesized in the laboratory.

Two examples of alkynes among thousands that have a biosynthetic origin are capillin, an antifungal agent, and dactylyne, a marine natural product that is an inhibitor of pentobarbital metabolism. Ethinyl estradiol is a synthetic alkyne whose estrogen-like properties have found use in oral contraceptives.

Capillin   **Dactylyne**   **Ethinyl estradiol**
[17α-ethynyl-1,3,5(10)-estratriene-3,17β-diol]

*Special Topics A–F and H are in *WileyPLUS*; Special Topic G can be found later in this volume.

2.1D Benzene: A Representative Aromatic Hydrocarbon

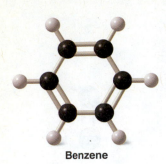

Benzene

In Chapter 14 we shall study in detail a group of unsaturated cyclic hydrocarbons known as **aromatic compounds**. The compound known as **benzene** is the prototypical aromatic compound. Benzene can be written as a six-membered ring with alternating single and double bonds, called a **Kekulé structure** after August Kekulé, who first conceived of this representation:

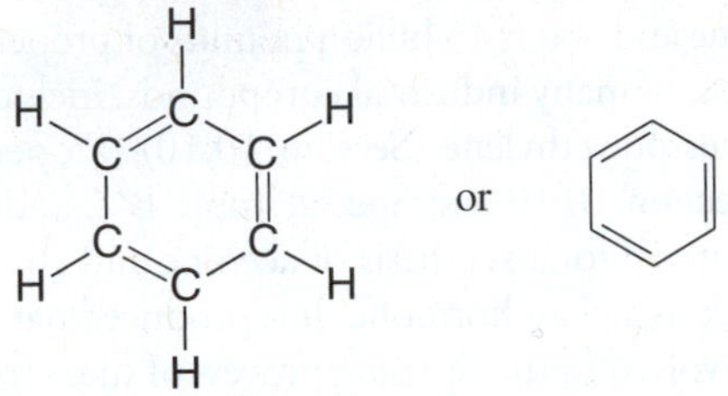

**Kekulé structure
for benzene**

**Bond-line representation
of Kekulé structure**

Even though the Kekulé structure is frequently used for benzene compounds, there is much evidence that this representation is inadequate and incorrect. For example, if benzene had alternating single and double bonds as the Kekulé structure indicates, we would expect the lengths of the carbon–carbon bonds around the ring to be alternately longer and shorter, as we typically find with carbon–carbon single and double bonds (Fig. 1.31). In fact, the carbon–carbon bonds of benzene are all the same length (1.39 Å), a value in between that of a carbon–carbon single bond and a carbon–carbon double bond. There are two ways of dealing with this problem: with resonance theory or with molecular orbital theory.

If we use resonance theory, we visualize benzene as being represented by either of two equivalent Kekulé structures:

**Two contributing Kekulé structures
for benzene**

**A representation of the
resonance hybrid**

Based on the principles of resonance theory (Section 1.8) we recognize that benzene cannot be represented adequately by either structure, but that, instead, *it should be visualized as a hybrid of the two structures*. We represent this hybrid by a hexagon with a circle in the middle. Resonance theory, therefore, solves the problem we encountered in understanding how all of the carbon–carbon bonds are the same length. According to resonance theory, the bonds are not alternating single and double bonds, they are a resonance hybrid of the two. Any bond that is a single bond in the first contributor is a double bond in the second, and vice versa.

- All of the carbon–carbon bonds in benzene are one and one-half bonds, have a bond length in between that of a single bond and a double bond, and have bond angles of 120°.

In the molecular orbital explanation, which we shall describe in much more depth in Chapter 14, we begin by recognizing that the carbon atoms of the benzene ring are sp^2 hybridized. Therefore, each carbon has a p orbital that has one lobe above the plane of the ring and one lobe below, as shown on the next page in the schematic and calculated p orbital representations.

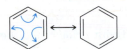

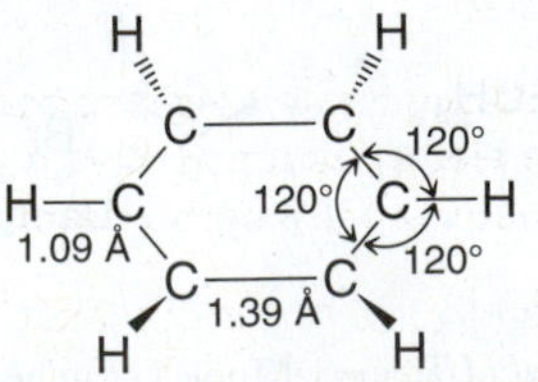

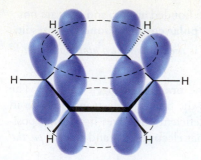

Schematic representation
of benzene *p* orbitals

Calculated *p* orbital
shapes in benzene

Calculated benzene molecular
orbital resulting from favorable
overlap of *p* orbitals above and
below plane of benzene ring

The lobes of each *p* orbital above and below the ring overlap with the lobes of *p* orbitals on the atoms to either side of it. This kind of overlap of *p* orbitals leads to a set of bonding molecular orbitals that encompass all of the carbon atoms of the ring, as shown in the calculated molecular orbital. Therefore, the six electrons associated with these *p* orbitals (one electron from each orbital) are **delocalized** about all six carbon atoms of the ring. This delocalization of electrons explains how all the carbon–carbon bonds are equivalent and have the same length. In Section 14.7B, when we study nuclear magnetic resonance spectroscopy, we shall present convincing physical evidence for this delocalization of the electrons.

Cyclobutadiene (below) is like benzene in that it has alternating single and double bonds in a ring. However, its bonds are not the same length, the double bonds being shorter than the single bonds; the molecule is rectangular, not square. Explain why it would be incorrect to write resonance structures as shown.

● ● ● · · · · · · · · ·

PRACTICE PROBLEM 2.1

2.2 POLAR COVALENT BONDS

In our discussion of chemical bonds in Section 1.3, we examined compounds such as LiF in which the bond is between two atoms with very large electronegativity differences. In instances like these, a complete transfer of electrons occurs, giving the compound an **ionic bond**:

$$\text{Li}^+ \quad :\ddot{\text{F}}:^-$$

Lithium fluoride has an ionic bond.

We also described molecules in which electronegativity differences are not large, or in which they are the same, such as the carbon–carbon bond of ethane. Here the electrons are shared equally between the atoms.

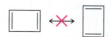

Ethane has a covalent bond.
The electrons are shared equally
between the carbon atoms.

Lithium fluoride crystal model

Until now, we have not considered the possibility that the electrons of a covalent bond might be shared unequally.

- If electronegativity differences exist between two bonded atoms, and they are not large, the electrons are not shared equally and a **polar covalent bond** is the result.
- Remember: one definition of **electronegativity** is *the ability of an atom to attract electrons that it is sharing in a covalent bond.*

An example of such a polar covalent bond is the one in hydrogen chloride. The chlorine atom, with its greater electronegativity, pulls the bonding electrons closer to it. This makes the hydrogen atom somewhat electron deficient and gives it a *partial* positive charge ($\delta+$). The chlorine atom becomes somewhat electron rich and bears a *partial* negative charge ($\delta-$):

$$\overset{\delta+}{H} : \overset{\delta-}{\underset{\cdot\cdot}{\overset{\cdot\cdot}{Cl}}} :$$

Because the hydrogen chloride molecule has a partially positive end and a partially negative end, it is a **dipole**, and it has a **dipole moment**.

The direction of polarity of a polar bond can be symbolized by a vector quantity $\longmapsto$. The crossed end of the arrow is the positive end and the arrowhead is the negative end:

$$(\text{positive end}) \longmapsto (\text{negative end})$$

In HCl, for example, we would indicate the direction of the dipole moment in the following way:

$$H\!-\!Cl$$
$$\longmapsto$$

The dipole moment is a physical property that can be measured experimentally. It is defined as the product of the magnitude of the charge in electrostatic units (esu) and the distance that separates them in centimeters (cm):

$$\text{Dipole moment} = \text{charge (in esu)} = \text{distance (in cm)}$$
$$\mu = e \times d$$

The charges are typically on the order of 10^{-10} esu and the distances are on the order of 10^{-8} cm. Dipole moments, therefore, are typically on the order of 10^{-18} esu cm. For convenience, this unit, 1×10^{-18} esu cm, is defined as one **debye** and is abbreviated D. (The unit is named after Peter J. W. Debye, a chemist born in the Netherlands and who taught at Cornell University from 1936 to 1966. Debye won the Nobel Prize in Chemistry in 1936.) In SI units $1\ D = 3.336 \times 10^{-30}$ coulomb meter (C · m).

If necessary, the length of the arrow can be used to indicate the magnitude of the dipole moment. Dipole moments, as we shall see in Section 2.3, are very useful quantities in accounting for physical properties of compounds.

PRACTICE PROBLEM 2.2   Write $\delta+$ and $\delta-$ by the appropriate atoms and draw a dipole moment vector for any of the following molecules that are polar:

(a) HF   **(b)** IBr   **(c)** Br_2 **(d)** F_2

Polar covalent bonds strongly influence the physical properties and reactivity of molecules. In many cases, these polar covalent bonds are part of **functional groups**, which we shall study shortly (Sections 2.5–2.13). Functional groups are defined groups of atoms in a molecule that give rise to the function (reactivity or physical properties) of the molecule. Functional groups often contain atoms having different electronegativity values and unshared electron pairs. (Atoms such as oxygen, nitrogen, and sulfur that form covalent bonds and have unshared electron pairs are called **heteroatoms**.)

2.2A Maps of Electrostatic Potential

One way to visualize the distribution of charge in a molecule is with a **map of electrostatic potential (MEP)**. Regions of an electron density surface that are more negative than others in an MEP are colored red. These regions would attract a positively charged species (or repel a negative charge). Regions in the MEP that are less negative (or are positive) are blue. Blue regions are likely to attract electrons from another molecule. The spectrum of colors from red to blue indicates the trend in charge from most negative to least negative (or most positive).

Figure 2.1 shows a map of electrostatic potential for the low-electron-density surface of hydrogen chloride. We can see clearly that negative charge is concentrated near the chlorine atom and that positive charge is localized near the hydrogen atom, as we predict based on the difference in their electronegativity values. Furthermore, because this MEP is plotted at the low-electron-density surface of the molecule (the van der Waals surface, Section 2.13B), it also gives an indication of the molecule's overall shape.

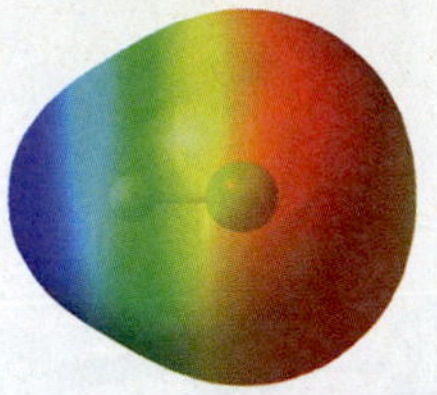

2.3 POLAR AND NONPOLAR MOLECULES

In the discussion of dipole moments in the previous section, our attention was restricted to simple diatomic molecules. Any *diatomic* molecule in which the two atoms are *different* (and thus have different electronegativities) will, of necessity, have a dipole moment. In general, a molecule with a dipole moment is a **polar molecule**. If we examine Table 2.1, however, we find that a number of molecules (e.g., CCl_4, CO_2) consist of more than two atoms, have *polar* bonds, *but have no dipole moment*. With our knowledge of the shapes of molecules (Sections 1.12–1.16) we can understand how this can occur.

FIGURE 2.1 A calculated map of electrostatic potential for hydrogen chloride showing regions of relatively more negative charge in red and more positive charge in blue. Negative charge is clearly localized near the chlorine, resulting in a strong dipole moment for the molecule.

TABLE 2.1 DIPOLE MOMENTS OF SOME SIMPLE MOLECULES

Formula	μ (D)	Formula	μ (D)
H_2	0	CH_4	0
Cl_2	0	CH_3Cl	1.87
HF	1.83	CH_2Cl_2	1.55
HCl	1.08	$CHCl_3$	1.02
HBr	0.80	CCl_4	0
HI	0.42	NH_3	1.47
BF_3	0	NF_3	0.24
CO_2	0	H_2O	1.85

Consider a molecule of carbon tetrachloride (CCl_4). Because the electronegativity of chlorine is greater than that of carbon, each of the carbon–chlorine bonds in CCl_4 is polar. Each chlorine atom has a partial negative charge, and the carbon atom is considerably positive. Because a molecule of carbon tetrachloride is tetrahedral (Fig. 2.2), however, *the center of positive charge and the center of negative charge coincide, and the molecule has no net dipole moment.*

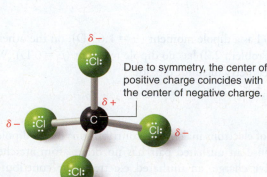

Due to symmetry, the center of positive charge coincides with the center of negative charge.

FIGURE 2.2 Charge distribution in carbon tetrachloride. The molecule has no net dipole moment.

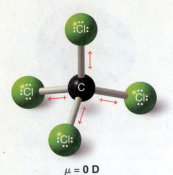

$\mu = 0\ D$

FIGURE 2.3 A tetrahedral orientation of equal bond moments causes their effects to cancel.

This result can be illustrated in a slightly different way: if we use arrows ($\longleftrightarrow$) to represent the direction of polarity of each bond, we get the arrangement of bond moments shown in Fig. 2.3. Since the bond moments are vectors of equal magnitude arranged tetrahedrally, their effects cancel. Their vector sum is zero. The molecule has *no net dipole moment*.

The chloromethane molecule (CH_3Cl) has a net dipole moment of 1.87 D. Since carbon and hydrogen have electronegativities (Table 1.1) that are nearly the same, the contribution of three C—H bonds to the net dipole is negligible. The electronegativity difference between carbon and chlorine is large, however, and the highly polar C—Cl bond accounts for most of the dipole moment of CH_3Cl (Fig. 2.4).

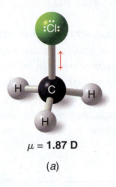

$\mu = 1.87\ D$

(a)

(b)

FIGURE 2.4 (a) The dipole moment of chloromethane arises mainly from the highly polar carbon–chlorine bond. (b) A map of electrostatic potential illustrates the polarity of chloromethane.

●●● **SOLVED PROBLEM 2.2**

Although molecules of CO_2 have polar bonds (oxygen is more electronegative than carbon), carbon dioxide (Table 2.1) has no dipole moment. What can you conclude about the geometry of a carbon dioxide molecule?

STRATEGY AND ANSWER: For a CO_2 molecule to have a zero dipole moment, the bond moments of the two carbon–oxygen bonds must cancel each other. This can happen only if molecules of carbon dioxide are linear.

$$\overset{..}{O}=C=\overset{..}{O}$$
$$\longleftarrow \quad \longrightarrow$$
$$\mu = 0\ D$$

●●●

PRACTICE PROBLEM 2.3   Boron trifluoride (BF_3) has no dipole moment ($\mu = 0$ D). Explain how this observation confirms the geometry of BF_3 predicted by VSEPR theory.

●●●

PRACTICE PROBLEM 2.4   Tetrachloroethene (CCl_2═CCl_2) does not have a dipole moment. Explain this fact on the basis of the shape of CCl_2═CCl_2.

●●●

PRACTICE PROBLEM 2.5   Sulfur dioxide (SO_2) has a dipole moment ($\mu = 1.63$ D); on the other hand, carbon dioxide (see Solved Problem 2.2) has no dipole moment ($\mu = 0$ D). What do these facts indicate about the geometry of sulfur dioxide?

Unshared pairs of electrons make large contributions to the dipole moments of water and ammonia. Because an unshared pair has no other atom attached to it to partially neutralize its negative charge, an unshared electron pair contributes a large moment directed away from the central atom (Fig. 2.5). (The O—H and N—H moments are also appreciable.)

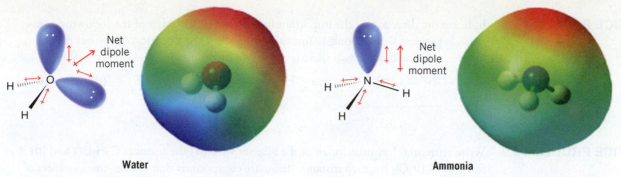

FIGURE 2.5 Bond moments and the resulting dipole moments of water and ammonia.

Using a three-dimensional formula, show the direction of the dipole moment of CH_3OH.
Write $\delta+$ and $\delta-$ signs next to the appropriate atoms.

PRACTICE PROBLEM 2.6

Trichloromethane ($CHCl_3$, also called *chloroform*) has a larger dipole moment than $CFCl_3$.
Use three-dimensional structures and bond moments to explain this fact.

PRACTICE PROBLEM 2.7

2.3A Dipole Moments in Alkenes

Cis–trans isomers of alkenes (Section 1.13B) have different physical properties. They have
different melting points and boiling points, and often cis–trans isomers differ markedly
in the magnitude of their dipole moments. Table 2.2 summarizes some of the physical
properties of two pairs of cis–trans isomers.

TABLE 2.2 PHYSICAL PROPERTIES OF SOME CIS–TRANS ISOMERS

Compound	Melting Point (°C)	Boiling Point (°C)	Dipole Moment (D)
cis-1,2-Dichloroethene	−80	60	1.90
trans-1,2-Dichloroethene	−50	48	0
cis-1,2-Dibromoethene	−53	112	1.35
trans-1,2-Dibromoethene	−6	108	0

SOLVED PROBLEM 2.3

Explain why *cis*-1,2-dichloroethene (Table 2.2) has a large dipole moment whereas *trans*-1,2-dichloroethene has a dipole
moment equal to zero.

STRATEGY AND ANSWER: If we examine the net dipole moments (shown in red) for the bond moments (black),
we see that in *trans*-1,2-dichloroethene the bond moments cancel each other, whereas in *cis*-1,2-dichloroethene they aug-
ment each other.

Bond moments (black)
are in same general
direction. Resultant
dipole moment (red)
is large.

Bond moments
cancel each other.
Net dipole is zero.

cis-1,2-Dichloroethene
$\mu = 1.9$ D

trans-1,2-Dichloroethene
$\mu = 0$ D

● ● ●

PRACTICE PROBLEM 2.8   Indicate the direction of the important bond moments in each of the following compounds (neglect C—H bonds). You should also give the direction of the net dipole moment for the molecule. If there is no net dipole moment, state that $\mu = 0$ D.

(a) cis-CHF═CHF   **(b)** $trans$-CHF═CHF   **(c)** CH_2═CF_2 **(d)** CF_2═CF_2

● ● ●

PRACTICE PROBLEM 2.9   Write structural formulas for all of the alkenes with **(a)** the formula $C_2H_2Br_2$ and **(b)** the formula $C_2Br_2Cl_2$. In each instance designate compounds that are cis–trans isomers of each other. Predict the dipole moment of each one.

2.4 FUNCTIONAL GROUPS

● **Functional groups** are common and specific arrangements of atoms that impart predictable reactivity and properties to a molecule.

The functional group of an alkene, for example, is its carbon–carbon double bond. When we study the reactions of alkenes in greater detail in Chapter 8, we shall find that most of the chemical reactions of alkenes are the chemical reactions of the carbon–carbon double bond.

     The functional group of an alkyne is its carbon–carbon triple bond. Alkanes do not have a functional group. Their molecules have carbon–carbon single bonds and carbon–hydrogen bonds, but these bonds are present in molecules of almost all organic compounds, and C—C and C—H bonds are, in general, much less reactive than common functional groups. We shall introduce other common functional groups and their properties in Sections 2.5–2.11. Table 2.3 (Section 2.12) summarizes the most important functional groups. First, however, let us introduce some common alkyl groups, which are specific groups of carbon and hydrogen atoms that are not part of functional groups.

2.4A Alkyl Groups and the Symbol R

Alkyl groups are the groups that we identify for purposes of naming compounds. They are groups that would be obtained by removing a hydrogen atom from an alkane:

Alkane	Alkyl Group	Abbreviation	Bond-line	Model
CH_3—H	H_3C—⧵	Me-		
Methane	**Methyl**			
CH_3CH_2—H	CH_3CH_2—⧵	Et-		
Ethane	**Ethyl**			
$CH_3CH_2CH_2$—H	$CH_3CH_2CH_2$—⧵	Pr-	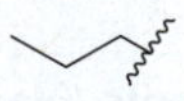	
Propane	**Propyl**			
$CH_3CH_2CH_2CH_2$—H	$CH_3CH_2CH_2CH_2$—⧵	Bu-		
Butane	**Butyl**			

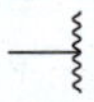

While only one alkyl group can be derived from methane or ethane (the **methyl** and **ethyl** groups, respectively), two groups can be derived from propane. Removal of a hydrogen from one of the end carbon atoms gives a group that is called the **propyl** group; removal of a hydrogen from the middle carbon atom gives a group that is called the **isopropyl** group. The names and structures of these groups are used so frequently in organic chemistry that you should learn them now. See Section 4.3C for names and structures of branched alkyl groups derived from butane and other hydrocarbons.

We can simplify much of our future discussion if, at this point, we introduce a symbol that is widely used in designating general structures of organic molecules: the symbol R. R *is used as a general symbol to represent any alkyl group*. For example, R might be a methyl group, an ethyl group, a propyl group, or an isopropyl group:

CH_3—   Methyl   ⎫
CH_3CH_2—   Ethyl   ⎪   These and
$CH_3CH_2CH_2$— Propyl   ⎬   others
CH_3CHCH_3 Isopropyl ⎪   can be
  | ⎭   designated by R.

Thus, the general formula for an alkane is R—H.

2.4B Phenyl and Benzyl Groups

When a benzene ring is attached to some other group of atoms in a molecule, it is called a **phenyl group**, and it is represented in several ways:

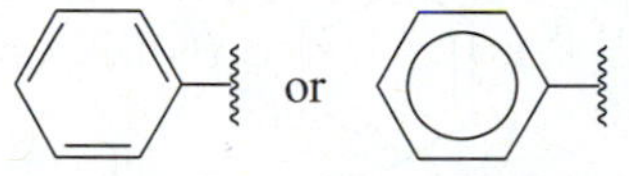

or   or C_6H_5— or Ph—

or φ— or Ar— (if ring substituents are present)

Ways of representing a phenyl group

The combination of a phenyl group and a **methylene group** ($-CH_2-$) is called a **benzyl group**:

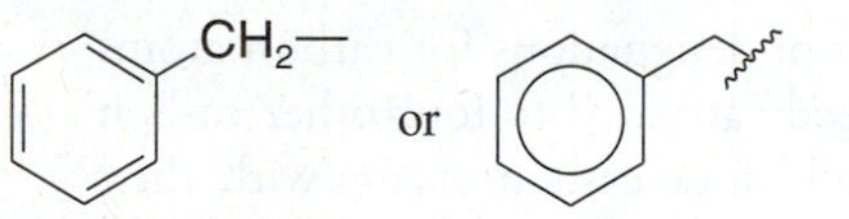

CH_2—

or

or $C_6H_5CH_2$— or Bn—

Ways of representing a benzyl group

2.5 ALKYL HALIDES OR HALOALKANES

Alkyl halides are compounds in which a halogen atom (fluorine, chlorine, bromine, or iodine) replaces a hydrogen atom of an alkane. For example, CH_3Cl and CH_3CH_2Br are alkyl halides. Alkyl halides are also called **haloalkanes**. The generic formula for an alkyl halide is R—$\ddot{X}$: where X = fluorine, chlorine, bromine, or iodine.

Alkyl halides are classified as being primary (1°), secondary (2°), or tertiary (3°). *This classification is based on the carbon atom to which the halogen is directly attached*. If the carbon atom that bears the halogen is directly attached to only one other carbon, the carbon atom is said to be a **primary carbon atom** and the alkyl halide is classified as a **primary alkyl halide**. If the carbon that bears the halogen is itself directly attached to two other carbon atoms, then the carbon is a **secondary carbon** and the alkyl halide is a **secondary alkyl halide**. If the carbon that bears the halogen is directly attached to three other carbon atoms, then the carbon is a **tertiary**

2-Chloropropane

Helpful Hint

Although we use the symbols 1°, 2°, 3°, we do not say first degree, second degree, and third degree; we say *primary*, *secondary*, and *tertiary*.

carbon and the alkyl halide is a **tertiary alkyl halide**. Examples of primary, secondary, and tertiary alkyl halides are the following:

1° Carbon

$$H-\underset{\underset{H}{|}}{\overset{\overset{H}{|}}{C}}-\underset{\underset{H}{|}}{\overset{\overset{H}{|}}{C}}-Cl \quad \text{or}$$

A 1° alkyl chloride

2° Carbon

$$H-\underset{\underset{H}{|}}{\overset{\overset{H}{|}}{C}}-\underset{\underset{Cl}{|}}{\overset{\overset{H}{|}}{C}}-\underset{\underset{H}{|}}{\overset{\overset{H}{|}}{C}}-H \quad \text{or}$$

A 2° alkyl chloride

3° Carbon

$$CH_3-\underset{\underset{CH_3}{|}}{\overset{\overset{CH_3}{|}}{C}}-Cl \quad \text{or}$$

A 3° alkyl chloride

An **alkenyl halide** is a compound with a halogen atom bonded to an alkene carbon. In older nomenclature such compounds were sometimes referred to as vinyl halides. An **aryl halide** is a compound with a halogen atom bonded to an aromatic ring such as a benzene ring.

An alkenyl chloride   **A phenyl bromide**

SOLVED PROBLEM 2.4

Write the structure of an alkane with the formula C_5H_{12} that has no secondary or tertiary carbon atoms. *Hint*: The compound has a quaternary (4°) carbon.

STRATEGY AND ANSWER: Following the pattern of designations for carbon atoms given above, a 4° carbon atom must be one that is directly attached to four other carbon atoms. If we start with this carbon atom, and then add four carbon atoms with their attached hydrogens, there is only one possible alkane. The other four carbons are all primary carbons; none is secondary or tertiary.

4° Carbon atom

$$\text{or} \quad CH_3-\underset{\underset{CH_3}{|}}{\overset{\overset{CH_3}{|}}{C}}-CH_3$$

●●●

PRACTICE PROBLEM 2.10 Write bond-line structural formulas for **(a)** two constitutionally isomeric primary alkyl bromides with the formula C_4H_9Br, **(b)** a secondary alkyl bromide, and **(c)** a tertiary alkyl bromide with the same formula. Build handheld molecular models for each structure and examine the differences in their connectivity.

●●●

PRACTICE PROBLEM 2.11 Although we shall discuss the naming of organic compounds later when we discuss the individual families in detail, one method of naming alkyl halides is so straightforward that it is worth describing here. We simply name the alkyl group attached to the halogen and add the word *fluoride, chloride, bromide*, or *iodide*. Write formulas for **(a)** ethyl fluoride and **(b)** isopropyl chloride.

What are the names for **(c)** 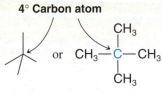 Br, **(d)** F , and **(e)** C_6H_5I?

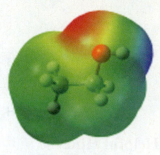

2.6 ALCOHOLS AND PHENOLS

Methyl alcohol (also called methanol) has the structural formula CH_3OH and is the simplest member of a family of organic compounds known as **alcohols**. The characteristic functional group of this family is the hydroxyl (—OH) group attached to an sp^3-hybridized carbon atom. Another example of an alcohol is ethyl alcohol, CH_3CH_2OH (also called ethanol).

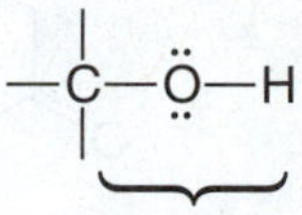

This is the functional group of an alcohol.

Ethanol

Alcohols may be viewed structurally in two ways: (1) as hydroxyl derivatives of alkanes and (2) as alkyl derivatives of water. Ethyl alcohol, for example, can be seen as an ethane molecule in which one hydrogen has been replaced by a hydroxyl group or as a water molecule in which one hydrogen has been replaced by an ethyl group:

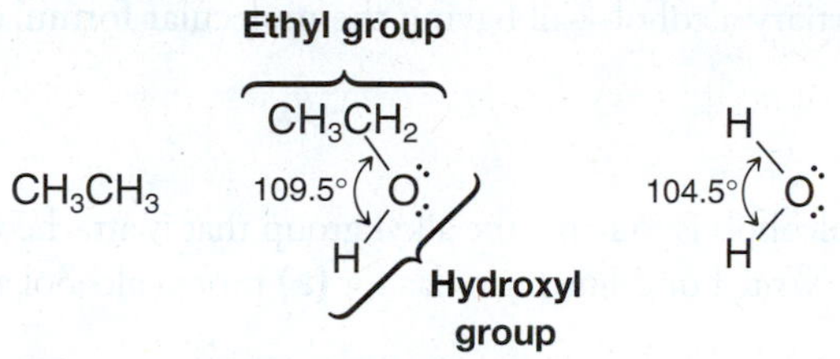

Ethyl group

CH_3CH_3 CH_3CH_2 109.5° $\ddot{O}$: H 104.5° $\ddot{O}$:

H **Hydroxyl group**

Ethane   **Ethyl alcohol (ethanol)**   **Water**

As with alkyl halides, alcohols are classified into three groups: primary (1°), secondary (2°), and tertiary (3°) alcohols. *This classification is based on the degree of substitution of the carbon to which the hydroxyl group is directly attached.* If the carbon has only one other carbon attached to it, the carbon is said to be a **primary carbon** and the alcohol is a **primary alcohol**:

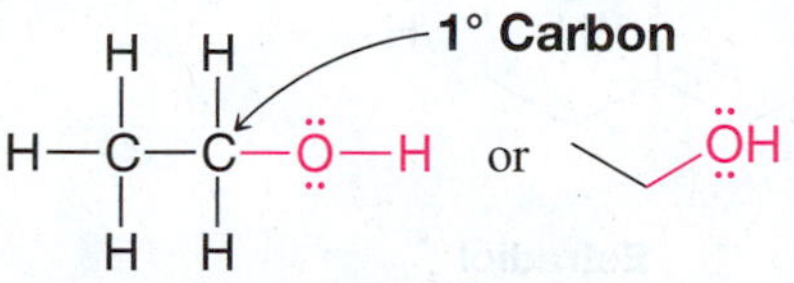

1° Carbon

Ethyl alcohol (a 1° alcohol) 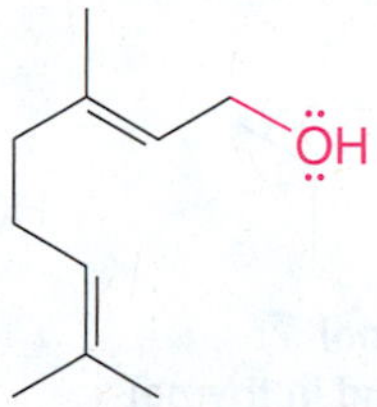   **Geraniol (a 1° alcohol)** 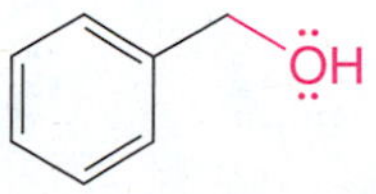   **Benzyl alcohol (a 1° alcohol)**

If the carbon atom that bears the hydroxyl group also has two other carbon atoms attached to it, this carbon is called a secondary carbon, and the alcohol is a secondary alcohol:

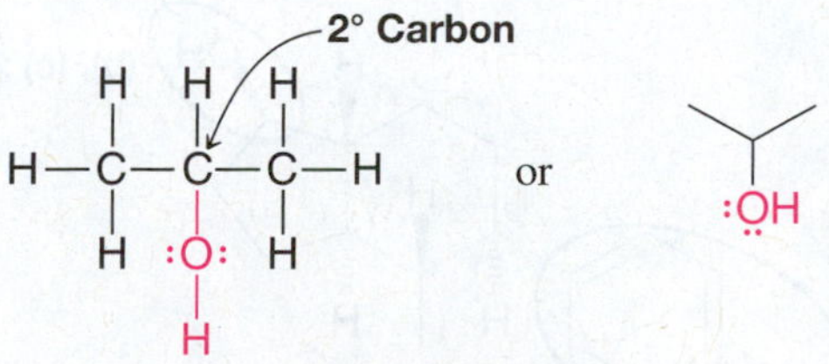

2° Carbon

Isopropyl alcohol (a 2° alcohol) 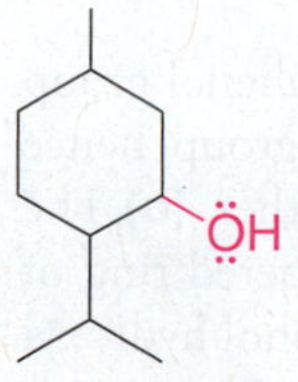   **Menthol (a 2° alcohol found in peppermint oil)**

If the carbon atom that bears the hydroxyl group has three other carbons attached to it, this carbon is called a tertiary carbon, and the alcohol is a tertiary alcohol:

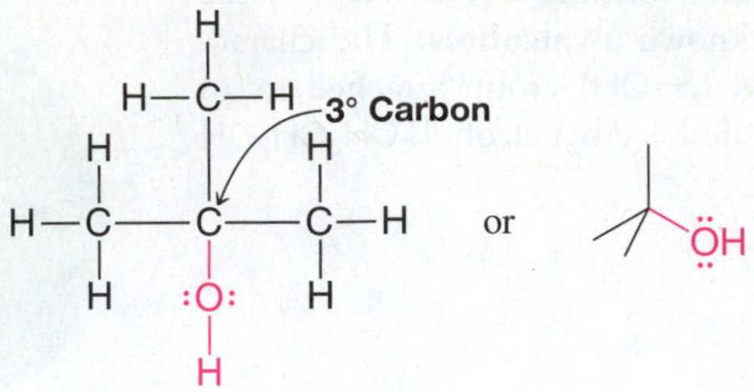

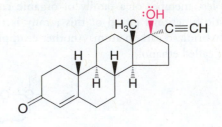

Helpful Hint

Practice with handheld molecular models by building models of as many of the compounds on this page as you can.

tert-Butyl alcohol
(a 3° alcohol)

Norethindrone
(an oral contraceptive that contains a 3° alcohol
group as well as a ketone group and
carbon–carbon double and triple bonds)

●●●
PRACTICE PROBLEM 2.12 Write bond-line structural formulas for **(a)** two primary alcohols, **(b)** a secondary alcohol, and **(c)** a tertiary alcohol—all having the molecular formula $C_4H_{10}O$.

●●●
PRACTICE PROBLEM 2.13 One way of naming alcohols is to name the alkyl group that is attached to the —OH and add the word *alcohol*. Write bond-line formulas for **(a)** propyl alcohol and **(b)** isopropyl alcohol.

When a hydroxyl group is bonded to a benzene ring the combination of the ring and the hydroxyl is called a phenol. Phenols differ significantly from alcohols in terms of their relative acidity, as we shall see in Chapter 3, and thus they are considered a distinct functional group.

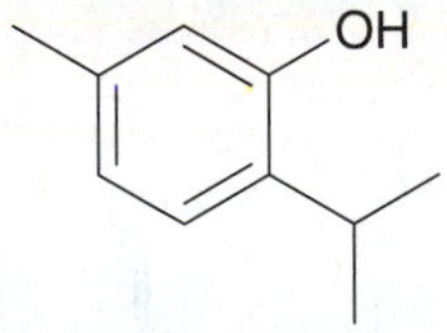

Thymol
(a phenol found in thyme)

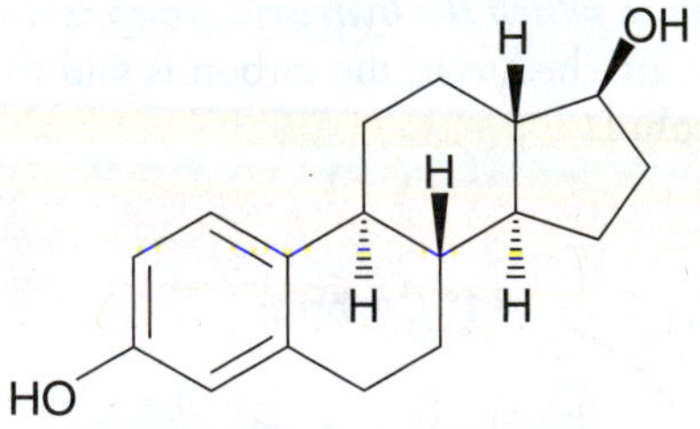

Estradiol
(a sex hormone that contains
both alcohol and phenol groups)

●●● **SOLVED PROBLEM 2.5**

Circle the atoms that comprise **(a)** the phenol and **(b)** the alcohol functional groups in estradiol. **(c)** What is the class of the alcohol?

STRATEGY AND ANSWER: **(a)** A phenol group consists of a benzene ring and a hydroxyl group, hence we circle these parts of the molecule together. **(b)** The alcohol group is found in the five-membered ring of estradiol. **(c)** The carbon bearing the alcohol hydroxyl group has two carbons directly bonded to it, thus it is a secondary alcohol.

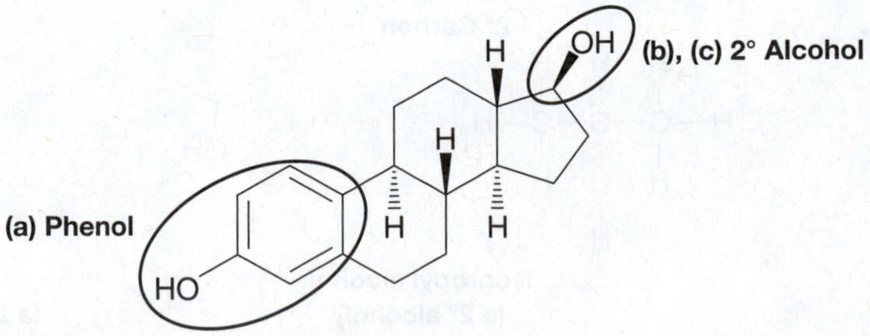

(b), (c) 2° Alcohol

(a) Phenol

2.7 ETHERS

Ethers have the general formula R—O—R or R—O—R′, where R′ may be an alkyl (or phenyl) group different from R. Ethers can be thought of as derivatives of water in which both hydrogen atoms have been replaced by alkyl groups. The bond angle at the oxygen atom of an ether is only slightly larger than that of water:

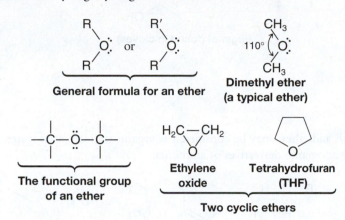

General formula for an ether

Dimethyl ether
(a typical ether)

The functional group
of an ether

Ethylene
oxide

Tetrahydrofuran
(THF)

Two cyclic ethers

Dimethyl ether

One way of naming ethers is to name the two alkyl groups attached to the oxygen atom in alphabetical order and add the word *ether*. If the two alkyl groups are the same, we use the prefix *di-*, for example, as in *dimethyl ether*. Write bond-line structural formulas for **(a)** diethyl ether, **(b)** ethyl propyl ether, and **(c)** ethyl isopropyl ether. What name would you give to **(d)** ⟋⟍⟋OMe **(e)** ⟋⟍O⟋⟍ and **(f)** $CH_3OC_6H_5$?

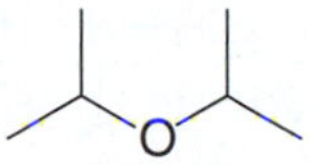

PRACTICE PROBLEM 2.14

THE CHEMISTRY OF... Ethers as General Anesthetics

Nitrous oxide (N_2O), also called laughing gas, was first used as an anesthetic in 1799, and it is still in use today, even though when used alone it does not produce deep anesthesia. The first use of an ether, diethyl ether, to produce deep anesthesia occurred in 1842. In the years that have passed since then, several different ethers, usually with halogen substituents, have replaced diethyl ether as anesthetics of choice. One reason: unlike diethyl ether, which is highly flammable, the halogenated ethers are not. Two halogenated ethers that are currently used for inhalation anesthesia are desflurane and sevoflurane.

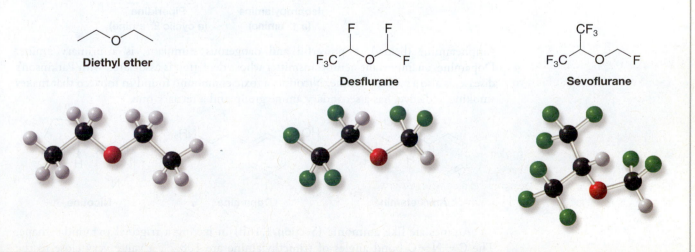

Diethyl ether

Desflurane

Sevoflurane

PRACTICE PROBLEM 2.15   Eugenol is the main constituent of the natural oil from cloves. Circle and label all of the functional groups in eugenol.

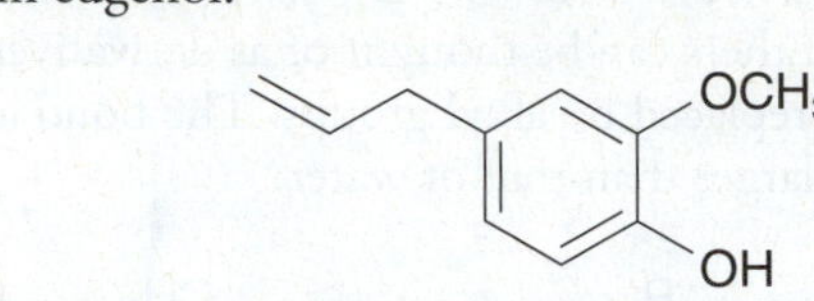

Eugenol (found in cloves)

2.8 AMINES

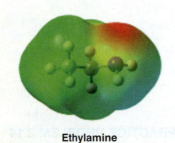

Ethylamine

Just as alcohols and ethers may be considered as organic derivatives of water, amines may be considered as organic derivatives of ammonia:

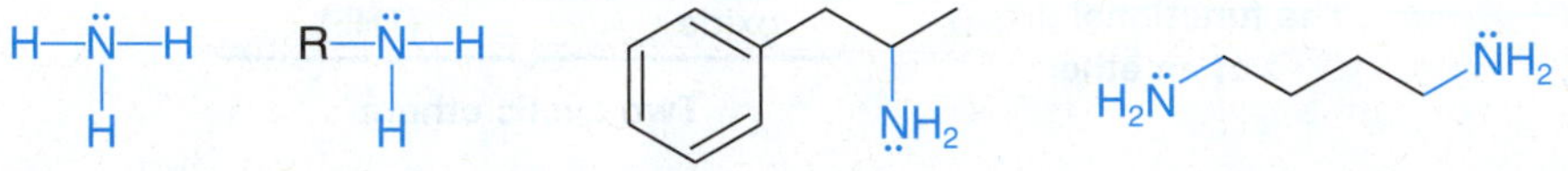

Ammonia   **An amine**   **Amphetamine**
(a dangerous stimulant)   **Putrescine**
(found in decaying meat)

Amines are classified as primary, secondary, or tertiary amines. **This classification is based on** *the number of organic groups that are attached to the nitrogen atom*:

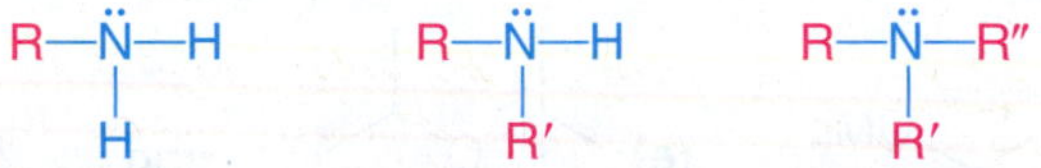

A primary (1°)   **A secondary (2°)**   **A tertiary (3°)**
amine   **amine**   **amine**

Notice that this is quite different from the way alcohols and alkyl halides are classified. Isopropylamine, for example, is a primary amine even though its —NH$_2$ group is attached to a secondary carbon atom. It is a primary amine because only one organic group is attached to the nitrogen atom:

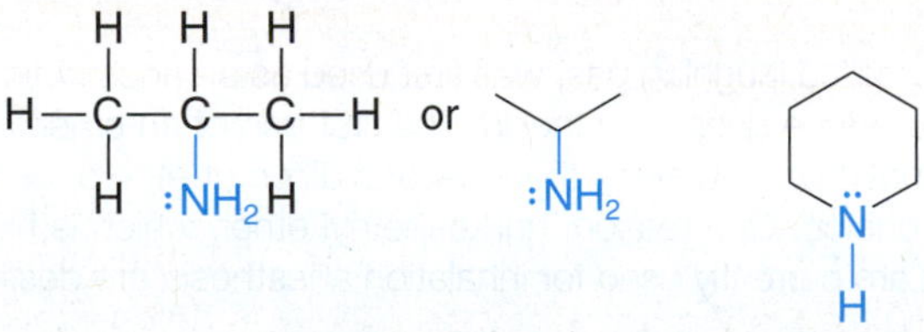

Isopropylamine   **Piperidine**
(a 1° amine)   **(a cyclic 2° amine)**

Amphetamine (below), a powerful and dangerous stimulant, is a primary amine. Dopamine, an important neurotransmitter whose depletion is associated with Parkinson's disease, is also a primary amine. Nicotine, a toxic compound found in tobacco that makes smoking addictive, has a secondary amine group and a tertiary one.

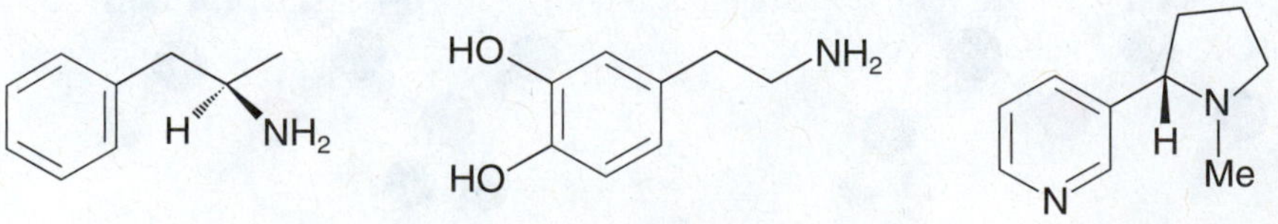

Amphetamine   **Dopamine**   **Nicotine**

Amines are like ammonia (Section 1.16B) in having a trigonal pyramidal shape. The C—N—C bond angles of trimethylamine are 108.7°, a value very close to the

H—C—H bond angles of methane. Thus, for all practical purposes, the nitrogen atom of an amine can be considered to be sp^3 hybridized with the unshared electron pair occupying one orbital (see below). This means that the unshared pair is relatively exposed, and as we shall see this is important because it is involved in almost all of the reactions of amines.

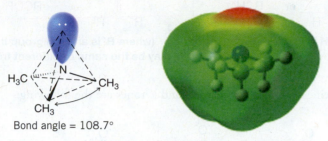

Bond angle = 108.7°

Trimethylamine

PRACTICE PROBLEM 2.16

One way of naming amines is to name in alphabetical order the alkyl groups attached to the nitrogen atom, using the prefixes *di-* and *tri-* if the groups are the same. An example is *isopropylamine*, whose formula is shown above. What are names for **(a)**, **(b)**, **(c)**, and **(d)**? Build hand-held molecular models for the compounds in parts **(a)**–**(d)**.

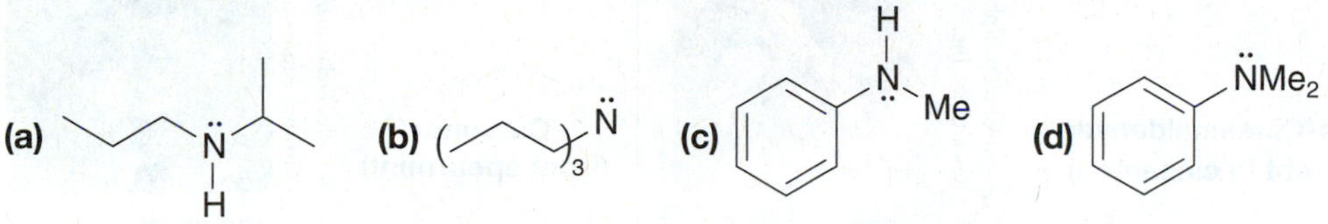

(a) **(b)** **(c)** **(d)**

Write bond-line formulas for **(e)** propylamine, **(f)** trimethylamine, and **(g)** ethylisopropylmethylamine.

PRACTICE PROBLEM 2.17

Which amines in Practice Problem 2.16 are **(a)** primary amines, **(b)** secondary amines, and **(c)** tertiary amines?

PRACTICE PROBLEM 2.18

Amines are like ammonia in being weak bases. They do this by using their unshared electron pair to accept a proton. **(a)** Show the reaction that would take place between trimethyl amine and HCl. **(b)** What hybridization state would you expect for the nitrogen atom in the product of this reaction?

2.9 ALDEHYDES AND KETONES

Aldehydes and ketones both contain the **carbonyl group**—a group in which a carbon atom has a double bond to oxygen:

The carbonyl group

The carbonyl group of an aldehyde is bonded to one hydrogen atom and one carbon atom (except for formaldehyde, which is the only aldehyde bearing two hydrogen atoms).

Acetaldehyde

The carbonyl group of a ketone is bonded to two carbon atoms. Using R, we can designate the general formulas for aldehydes and ketones as follows:

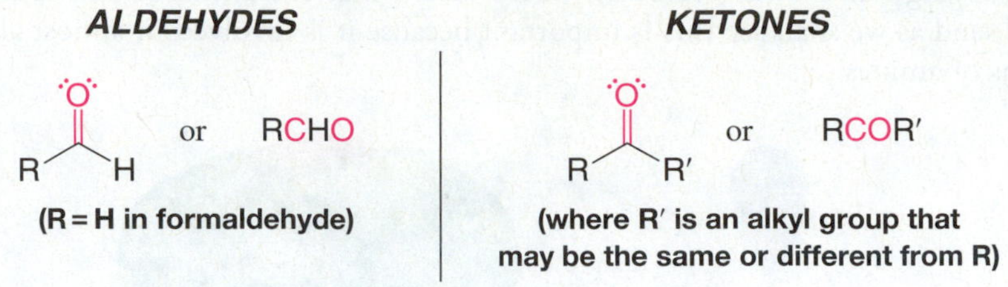

ALDEHYDES

or RCHO

(R = H in formaldehyde)

KETONES

or RCOR′

(where R′ is an alkyl group that may be the same or different from R)

Some specific examples of aldehydes and ketones are the following:

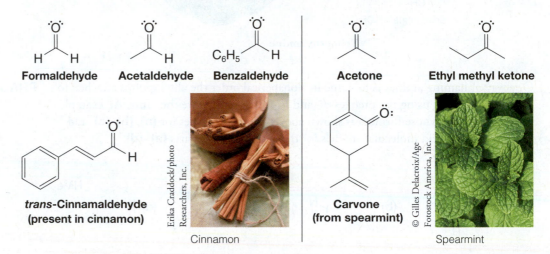

Formaldehyde   **Acetaldehyde**   **Benzaldehyde**   **Acetone**   **Ethyl methyl ketone**

trans-**Cinnamaldehyde**
(present in cinnamon)

Erika Craddock/photo Researchers, Inc.

Cinnamon

Carvone
(from spearmint)

© Gilles Delacroix/Age Fotostock America, Inc.

Spearmint

Helpful Hint

Computer molecular models can be found in the 3D Models section of the book's website for these and many other compounds we discuss in this book.

Aldehydes and ketones have a trigonal planar arrangement of groups around the carbonyl carbon atom. The carbon atom is sp^2 hybridized. In formaldehyde, for example, the bond angles are as follows:

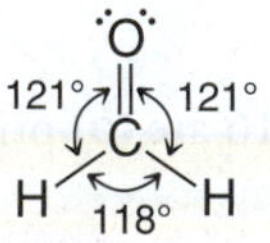

$$121° \quad 121°$$
$$118°$$

Retinal (below) is an aldehyde made from vitamin A that plays a vital role in vision. We discuss this further in Chapter 13.

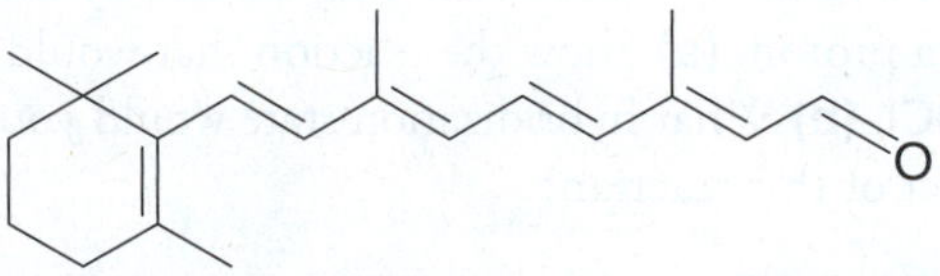

Retinal

PRACTICE PROBLEM 2.19 Write the resonance structure for carvone that results from moving the electrons as indicated. Include all formal charges.

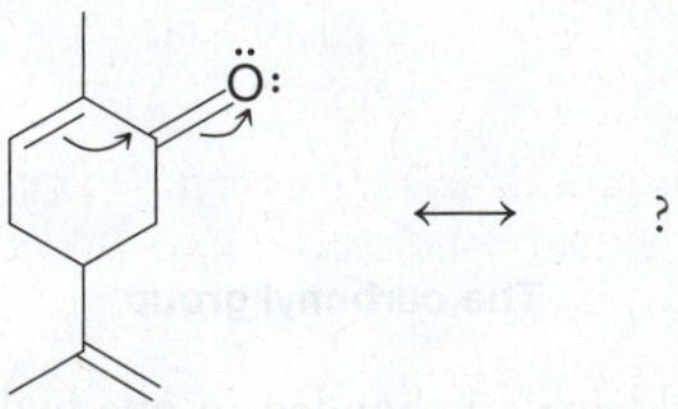

⟷   ?

Write bond-line formulas for **(a)** four aldehydes and **(b)** three ketones that have the formula $C_5H_{10}O$.

PRACTICE PROBLEM 2.20

2.10 CARBOXYLIC ACIDS, ESTERS, AND AMIDES

Carboxylic acids, esters, and amides all contain a carbonyl group that is bonded to an oxygen or nitrogen atom. As we shall learn in later chapters, all of these functional groups are interconvertible by appropriately chosen reactions.

2.10A Carboxylic Acids

Carboxylic acids have a carbonyl group bonded to a hydroxyl group, and they have the

general formula $\underset{R}{\overset{O}{\underset{\|}{C}}}_{O-H}$. The functional group, $\overset{O}{\underset{\|}{C}}_{O-H}$, is called the **carboxyl group** (**carbo**nyl + hydro**xyl**):

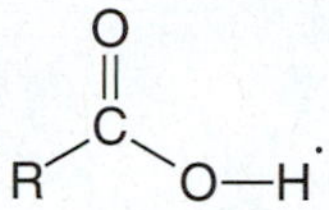

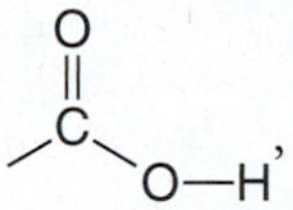

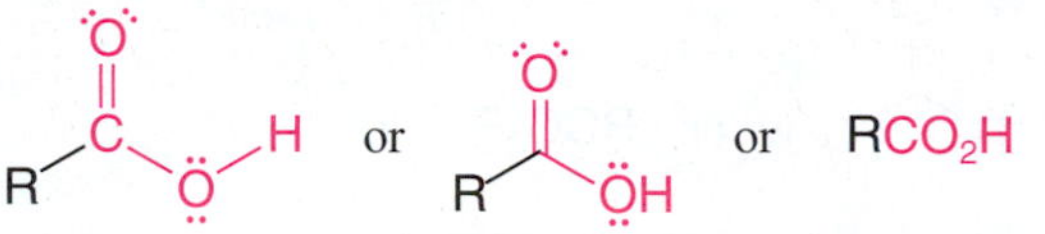

A carboxylic acid   or RCO_2H **The carboxyl group**   or $-CO_2H$ or $-COOH$

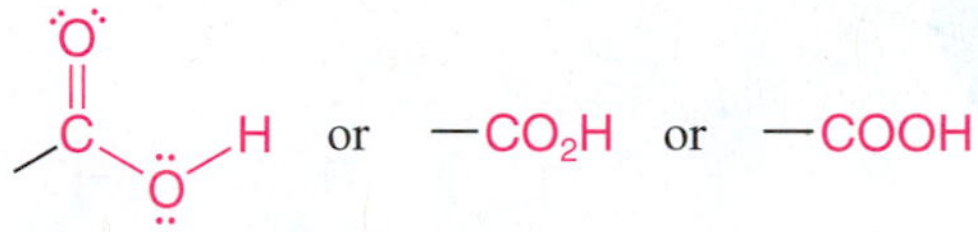

Examples of carboxylic acids are formic acid, acetic acid, and benzoic acid:

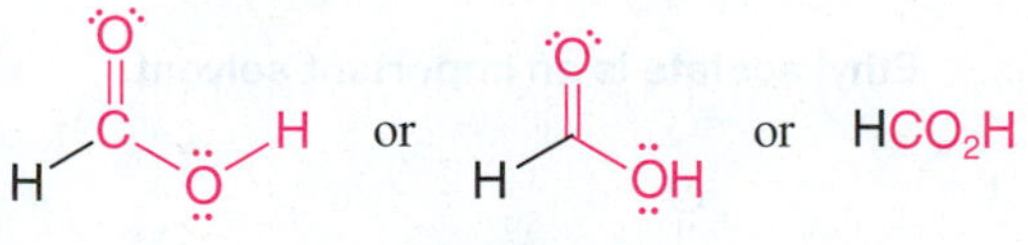

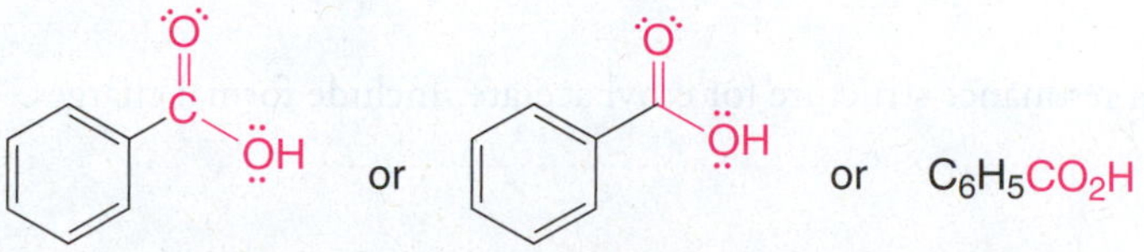

Formic acid   or HCO_2H

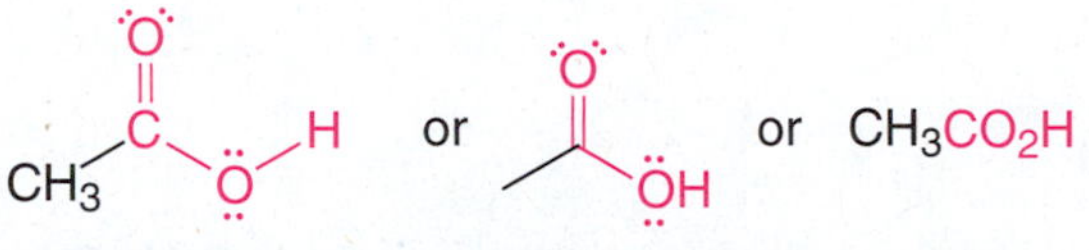

Acetic acid   or CH_3CO_2H

Benzoic acid   or $C_6H_5CO_2H$

Acetic acid

Formic acid is an irritating liquid produced by ants. (The sting of the ant is caused, in part, by formic acid being injected under the skin. *Formic* is the Latin word for ant.) Acetic acid, the substance responsible for the sour taste of vinegar, is produced when certain bacteria act on the ethyl alcohol of wine and cause the ethyl alcohol to be oxidized by air.

••• SOLVED PROBLEM 2.6

When formic acid (see above) donates a proton to a base, the result is the formation of a formate ion (HCO_2^-). **(a)** Write two resonance structures for the formate ion, and two resonance structures for formic acid. **(b)** Review the Rules for Resonance in Chapter 1, and identify which species, formate ion or formic acid, is most stabilized by resonance.

STRATEGY AND ANSWER: (a) We move the electron pairs as indicated below.

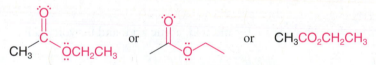

Formic acid   **Formate ion**

(b) The formate ion would be most stabilized because it does not have separated charges.

••• **PRACTICE PROBLEM 2.21** Write bond-line formulas for four carboxylic acids with the formula $C_5H_{10}O_2$.

2.10B Esters

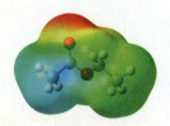

Ethyl acetate

Esters have the general formula RCO_2R' (or $RCOOR'$), where a carbonyl group is bonded to an alkoxyl (—OR) group:

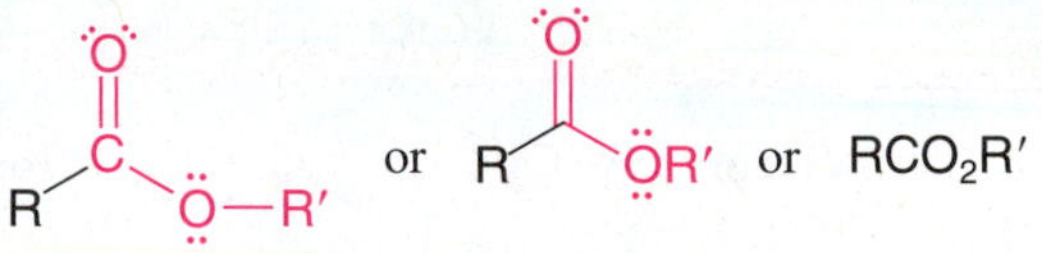

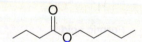

or R—CO$_2$R' or RCO_2R'

General formula for an ester

CH_3—CO—OCH$_2$CH$_3$ or or $CH_3CO_2CH_2CH_3$

Ethyl acetate is an important solvent.

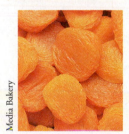

The ester pentyl butanoate has the odor of apricots and pears.

Pentyl butanoate has the odor of apricots and pears.

••• **PRACTICE PROBLEM 2.22** Write bond-line formulas for three esters with the formula $C_5H_{10}O_2$.

••• **PRACTICE PROBLEM 2.23** Write another resonance structure for ethyl acetate. Include formal charges.

Esters can be made from a carboxylic acid and an alcohol through the acid-catalyzed loss of a molecule of water. For example:

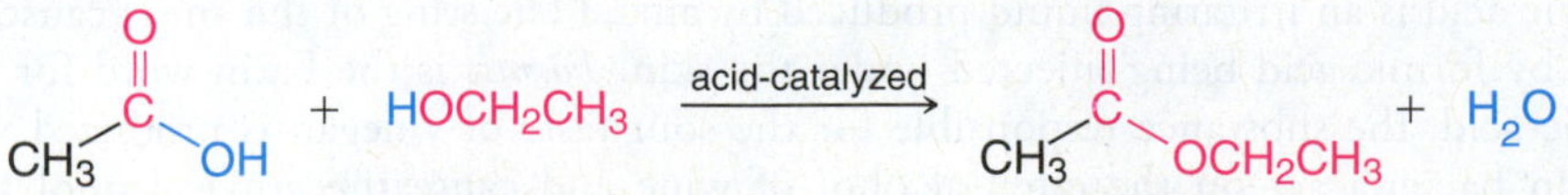

CH_3—CO—OH + HOCH$_2$CH$_3$ $\xrightarrow{\text{acid-catalyzed}}$ CH_3—CO—OCH$_2$CH$_3$ + H$_2$O

Acetic acid   **Ethyl alcohol**   **Ethyl acetate**

Your body makes esters from long-chain carboxylic acids called "fatty acids" by combining them with glycerol. We discuss their chemistry in detail in Chapter 23.

2.10C Amides

Amides have the formulas RCONH₂, RCONHR′, or RCONR′R″ where a carbonyl group is bonded to a nitrogen atom bearing hydrogen and/or alkyl groups. General formulas and some specific examples are shown below.

An unsubstituted amide   An N-substituted amide   An N,N-disubstituted amide

General formulas for amides

Nylon is a polymer comprised of regularly repeating amide groups.

Acetamide   N-Methylacetamide   N,N-Dimethylacetamide

Specific examples of amides

N- and *N,N*- indicate that the substituents are attached to the nitrogen atom.

Acetamide

Write another resonance structure for acetamide.

PRACTICE PROBLEM 2.24

2.11 NITRILES

A nitrile has the formula R—C≡N: (or R—CN). The carbon and the nitrogen of a nitrile are *sp* hybridized. In IUPAC systematic nomenclature, acyclic nitriles are named by adding the suffix -*nitrile* to the name of the corresponding hydrocarbon. The carbon atom of the —C≡N group is assigned number 1. The name acetonitrile is an acceptable common name for CH_3CN, and acrylonitrile is an acceptable common name for CH_2=CHCN:

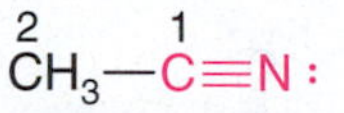

$$\overset{2}{CH_3}—\overset{1}{C}≡N: \qquad \overset{4}{CH_3}\overset{3}{CH_2}\overset{2}{CH_2}—\overset{1}{C}≡N:$$

Ethanenitrile   Butanenitrile
(acetonitrile)

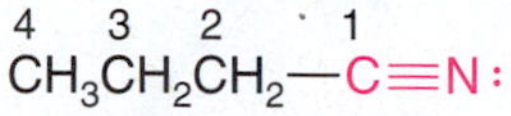

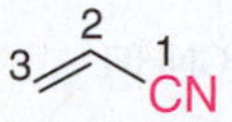

Propenenitrile   4-Pentenenitrile
(acrylonitrile)

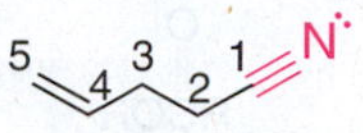

Acetonitrile

Cyclic nitriles are named by adding the suffix -*carbonitrile* to the name of the ring system to which the —CN group is attached. Benzonitrile is an acceptable common name for C_6H_5CN:

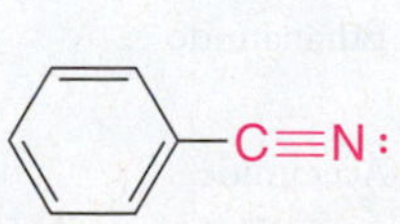

Benzenecarbonitrile   Cyclohexanecarbonitrile
(benzonitrile)

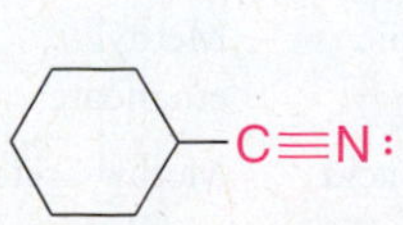

2.12 SUMMARY OF IMPORTANT FAMILIES OF ORGANIC COMPOUNDS

A summary of the important families of organic compounds is given in Table 2.3. You should learn to identify these common functional groups as they appear in other, more complicated molecules.

TABLE 2.3 IMPORTANT FAMILIES OF ORGANIC COMPOUNDS

	Alkane	Alkene	Alkyne	Aromatic	Haloalkane	Alcohol	Phenol	Ether
Functional group	$C-H$ and $C-C$ bonds	$C=C$	$-C\equiv C-$	Aromatic ring	$-C-\ddot{X}:$	$-C-\ddot{O}H$	OH 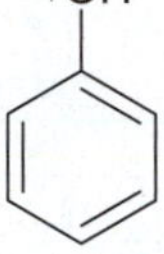 (phenol)	$-C-\ddot{O}-C-$
General formula	RH	$RCH=CH_2$, $RCH=CHR$, $R_2C=CHR$, $R_2C=CR_2$	$RC\equiv CH$, $RC\equiv CR$	ArH	RX	ROH	ArOH	ROR
Specific example	CH_3CH_3	$CH_2=CH_2$	$HC\equiv CH$	 (benzene ring)	CH_3CH_2Cl	CH_3CH_2OH	OH 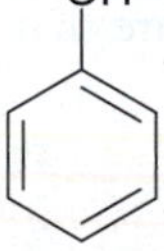 (phenol)	CH_3OCH_3
IUPAC name	Ethane	Ethene	Ethyne	Benzene	Chloroethane	Ethanol	Phenol	Methoxymethane
Common name[a]	Ethane	Ethylene	Acetylene	Benzene	Ethyl chloride	Ethyl alcohol	Phenol	Dimethyl ether

	Amine	Aldehyde	Ketone	Carboxylic Acid	Ester	Amide	Nitrile
Functional group	$-C-N$	$\overset{\ddot{O}}{C}-H$	$-C-\overset{\ddot{O}}{C}-C-$	$\overset{\ddot{O}}{C}-\ddot{O}H$	$\overset{\ddot{O}}{C}-\ddot{O}-C-$	$\overset{\ddot{O}}{C}-\ddot{N}$	$-C\equiv N:$
General formula	RNH_2, R_2NH, R_3N	$\overset{O}{RCH}$	$\overset{O}{RCH'}$	$\overset{O}{RCOH}$	$\overset{O}{RCOR'}$	$\overset{O}{RCNH_2}$, $\overset{O}{RCNHR'}$, $\overset{O}{RCNR'R''}$	RCN
Specific example	CH_3NH_2	$\overset{O}{CH_3CH}$	$\overset{O}{CH_3CCH_3}$	$\overset{O}{CH_3COH}$	$\overset{O}{CH_3COCH_3}$	$\overset{O}{CH_3CNH_2}$	$CH_3C\equiv N$
IUPAC name	Methanamine	Ethanal	Propanone	Ethanoic acid	Methyl ethanoate	Ethanamide	Ethanenitrile
Common name	Methylamine	Acetaldehyde	Acetone	Acetic acid	Methyl acetate	Acetamide	Acetonitrile

[a] These names are also accepted by the IUPAC.

2.12A Functional Groups in Biologically Important Compounds

Many of the functional groups we have listed in Table 2.3 are central to the compounds of living organisms. A typical sugar, for example, is glucose. Glucose contains several alcohol hydroxyl groups (—OH) and in one of its forms contains an aldehyde group. Fats and oils contain ester groups, and proteins contain amide groups. See if you can identify alcohol, aldehyde, ester, and amide groups in the following examples.

Glucose

A typical fat

Part of a protein

2.13 PHYSICAL PROPERTIES AND MOLECULAR STRUCTURE

So far, we have said little about one of the most obvious characteristics of organic compounds—that is, *their physical state or phase*. Whether a particular substance is a solid, or a liquid, or a gas would certainly be one of the first observations that we would note in any experimental work. The temperatures at which transitions occur between phases—that is, melting points (mp) and boiling points (bp)—are also among the more easily measured **physical properties**. Melting points and boiling points are also useful in identifying and isolating organic compounds.

Suppose, for example, we have just carried out the synthesis of an organic compound that is known to be a liquid at room temperature and 1 atm pressure. If we know the boiling point of our desired product and the boiling points of by-products and solvents that may be present in the reaction mixture, we can decide whether or not simple distillation will be a feasible method for isolating our product.

In another instance our product might be a solid. In this case, in order to isolate the substance by crystallization, we need to know its melting point and its solubility in different solvents.

The physical constants of known organic substances are easily found in handbooks and other reference books.* Table 2.4 lists the melting and boiling points of some of the compounds that we have discussed in this chapter.

Often in the course of research, however, the product of a synthesis is a new compound—one that has never been described before. In these instances, success in isolating the new compound depends on making reasonably accurate estimates of its melting point, boiling point, and solubilities. Estimations of these macroscopic physical properties are based on the most likely structure of the substance and on the forces that act between molecules and ions. The temperatures at which phase changes occur are an indication of the strength of these intermolecular forces.

Helpful Hint

Understanding how molecular structure influences physical properties is very useful in practical organic chemistry.

*Two useful handbooks are *Handbook of Chemistry*, Lange, N. A., Ed., McGraw-Hill: New York; and *CRC Handbook of Chemistry and Physics*, CRC: Boca Raton, FL.

TABLE 2.4 PHYSICAL PROPERTIES OF REPRESENTATIVE COMPOUNDS

Compound	Structure	mp (°C)	bp (°C) (1 atm)[a]
Methane	CH_4	−182.6	−162
Ethane	CH_3CH_3	−172	−88.2
Ethene	$CH_2{=}CH_2$	−169	−102
Ethyne	$HC{\equiv}CH$	−82	−84 subl
Chloromethane	CH_3Cl	−97	−23.7
Chloroethane	CH_3CH_2Cl	−138.7	13.1
Ethyl alcohol	CH_3CH_2OH	−114	78.5
Acetaldehyde	CH_3CHO	−121	20
Acetic acid	CH_3CO_2H	16.6	118
Sodium acetate	CH_3CO_2Na	324	dec
Ethylamine	$CH_3CH_2NH_2$	−80	17
Diethyl ether	$(CH_3CH_2)_2O$	−116	34.6
Ethyl acetate	$CH_3CO_2CH_2CH_3$	−84	77

[a]In this table dec = decomposes and subl = sublimes.

2.13A Ionic Compounds: Ion–Ion Forces

- The **melting point** of a substance is the temperature at which an equilibrium exists between the well-ordered crystalline state and the more random liquid state.

If the substance is an ionic compound, such as sodium acetate (Table 2.4), the **ion–ion forces** that hold the ions together in the crystalline state are the strong electrostatic lattice forces that act between the positive and negative ions in the orderly crystalline structure. In Fig. 2.6 each sodium ion is surrounded by negatively charged acetate ions, and each acetate ion is surrounded by positive sodium ions. A large amount of thermal energy is required to break up the orderly structure of the crystal into the disorderly open structure of a liquid. As a result, the temperature at which sodium acetate melts is quite high, 324 °C. The **boiling points** of ionic compounds are higher still, so high that most ionic organic compounds decompose (are changed by undesirable chemical reactions) before they boil. Sodium acetate shows this behavior.

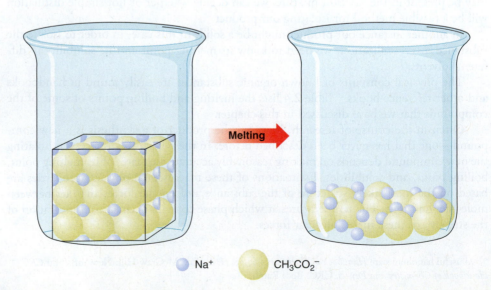

FIGURE 2.6 The melting of sodium acetate.

● Na^+ ● $CH_3CO_2{}^-$

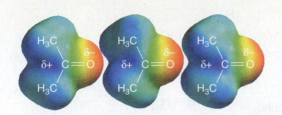

FIGURE 2.7 Electrostatic potential models for acetone molecules that show how acetone molecules might align according to attractions of their partially positive regions and partially negative regions (dipole–dipole interactions).

2.13B Intermolecular Forces (van der Waals Forces)

The forces that act between molecules are not as strong as those between ions, but they account for the fact that even completely nonpolar molecules can exist in liquid and solid states. These **intermolecular forces**, collectively called **van der Waals forces**, are all electrical in nature. We will focus our attention on three types:

1. Dipole–dipole forces
2. Hydrogen bonds
3. Dispersion forces

Dipole–Dipole Forces   Most organic molecules are not fully ionic but have instead a *permanent dipole moment* resulting from a nonuniform distribution of the bonding electrons (Section 2.3). Acetone and acetaldehyde are examples of molecules with permanent dipoles because the carbonyl group that they contain is highly polarized. In these compounds, the attractive forces between molecules are much easier to visualize. In the liquid or solid state, **dipole–dipole** attractions cause the molecules to orient themselves so that the positive end of one molecule is directed toward the negative end of another (Fig. 2.7).

Hydrogen Bonds

- Very strong dipole–dipole attractions occur between hydrogen atoms bonded to small, strongly electronegative atoms (O, N, or F) and nonbonding electron pairs on other such electronegative atoms. This type of intermolecular force is called a **hydrogen bond**.

Hydrogen bonds (bond dissociation energies of about 4–38 kJ mol^{-1}) are weaker than ordinary covalent bonds but much stronger than the dipole–dipole interactions that occur above, for example, in acetone.

Hydrogen bonding explains why water, ammonia, and hydrogen fluoride all have far higher boiling points than methane (bp −161.6 °C), even though all four compounds have similar molecular weights.

bp 100 °C   bp 19.5 °C   bp −33.4 °C

Hydrogen bonds are shown by the red dots.

One of the most important consequences of hydrogen bonding is that it causes water to be a liquid rather than a gas at 25 °C. Calculations indicate that in the absence of hydrogen bonding, water would have a boiling point near −80 °C and would not exist as a liquid unless the temperature were lower than that temperature. Had this been the case, it is highly unlikely that life, as we know it, could have developed on the planet Earth.

Hydrogen bonds hold the base pairs of double-stranded DNA together (see Section 25.4). Thymine hydrogen bonds with adenine. Cytosine hydrogen bonds with guanine.

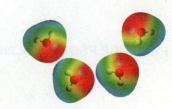

Water molecules associated by attraction of opposite partial charges.

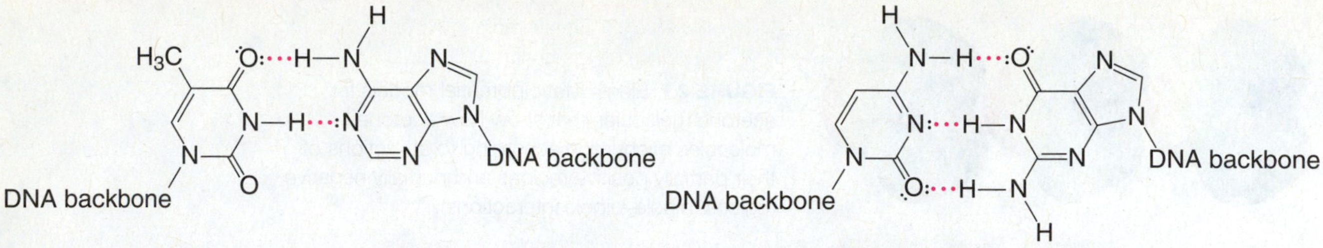

Thymine   **Adenine**   **Cytosine**   **Guanine**

Hydrogen bonding accounts for the fact that ethyl alcohol has a much higher boiling point (78.5 °C) than dimethyl ether (24.9 °C) even though the two compounds have the same molecular weight. Molecules of ethyl alcohol, because they have a hydrogen atom covalently bonded to an oxygen atom, can form strong hydrogen bonds to each other.

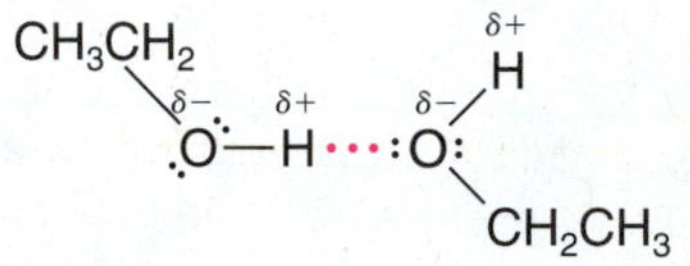

> **The red dots represent a hydrogen bond. Strong hydrogen bonding is limited to molecules having a hydrogen atom attached to an O, N, or F atom.**

Molecules of dimethyl ether, because they lack a hydrogen atom attached to a strongly electronegative atom, cannot form strong hydrogen bonds to each other. In dimethyl ether the intermolecular forces are weaker dipole–dipole interactions.

PRACTICE PROBLEM 2.25 The compounds in each part below have the same (or similar) molecular weights. Which compound in each part would you expect to have the higher boiling point? Explain your answers.

(a) 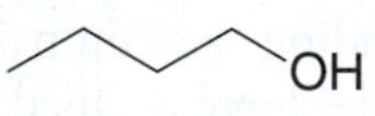 [structure] OH or [structure] O [structure] **(c)** 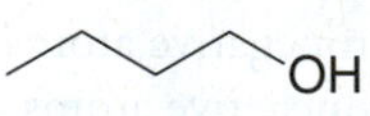 [structure] OH or HO [structure] OH

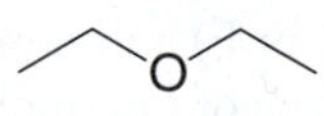

(b) (CH₃)₃N or [structure]

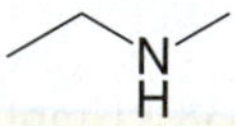

A factor (in addition to polarity and hydrogen bonding) that affects the *melting point* of many organic compounds is the compactness and rigidity of their individual molecules.

● Molecules that are symmetrical generally have abnormally high melting points. *tert*-Butyl alcohol, for example, has a much higher melting point than the other isomeric alcohols shown here:

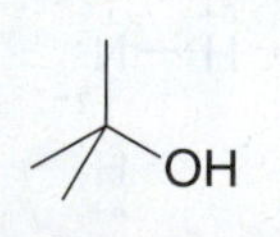

tert-Butyl alcohol   Butyl alcohol   Isobutyl alcohol   *sec*-Butyl alcohol
(mp 25 °C)   (mp −90 °C)   (mp −108 °C)   (mp −114 °C)

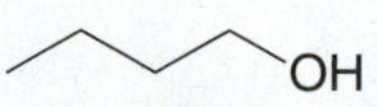

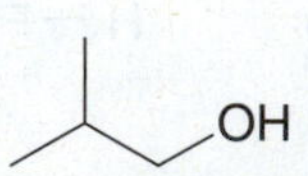

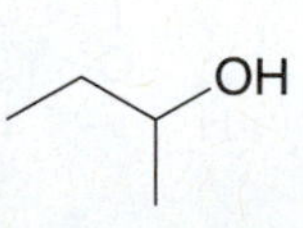

PRACTICE PROBLEM 2.26 Which compound would you expect to have the higher melting point, propane or cyclopropane? Explain your answer.

Dispersion Forces If we consider a substance like methane where the particles are non-polar molecules, we find that the melting point and boiling point are very low: −182.6 °C and −162 °C, respectively. Instead of asking, "Why does methane melt and boil at low

FIGURE 2.8 Temporary dipoles and induced dipoles in nonpolar molecules resulting from an uneven distribution of electrons at a given instant.

temperatures?" a more appropriate question might be "Why does methane, a nonionic, nonpolar substance, become a liquid or a solid at all?" The answer to this question can be given in terms of attractive intermolecular forces called **dispersion forces** or London forces.

An accurate account of the nature of dispersion forces requires the use of quantum mechanics. We can, however, visualize the origin of these forces in the following way. The average distribution of charge in a nonpolar molecule (such as methane) over a period of time is uniform. At any given instant, however, *because electrons move*, the electrons and therefore the charge may not be uniformly distributed. Electrons may, in one instant, be slightly accumulated on one part of the molecule, and, as a consequence, *a small temporary dipole will occur* (Fig. 2.8). This temporary dipole in one molecule can induce opposite (attractive) dipoles in surrounding molecules. It does this because the negative (or positive) charge in a portion of one molecule will distort the electron cloud of an adjacent portion of another molecule, causing an opposite charge to develop there. These temporary dipoles change constantly, but the net result of their existence is to produce attractive forces between nonpolar molecules and thus make possible the existence of their liquid and solid states.

Two important factors determine the magnitude of dispersion forces.

1. **The relative polarizability of electrons of the atoms involved**. *By polarizability we mean how easily the electrons respond to a changing electric field*. The electrons of large atoms such as iodine are loosely held and are easily polarized, while the electrons of small atoms such as fluorine are more tightly held and are much less polarizable.

CF_4 and CI_4 are both nonpolar molecules. But if we were to consider the intermolecular forces between two CI_4 molecules, which contain polarizable iodine atoms, we would find that the dispersion forces are much larger than between two CF_4 molecules, which contains fluorine atoms that are not very polarizable.

2. **The relative surface area of the molecules involved**. The larger the surface area, the larger is the overall attraction between molecules caused by dispersion forces. Molecules that are generally longer, flatter, or cylindrical have a greater surface area available for intermolecular interactions than more spherical molecules, and consequently have greater attractive forces between them than the tangential interactions between branched molecules. This is evident when comparing pentane, the unbranched C_5H_{12} hydrocarbon, with neopentane, the most highly branched C_5H_{12} isomer (in which one carbon bears four methyl groups). Pentane has a boiling point of 36.1 °C. Neopentane has a boiling point of 9.5 °C. The difference in their boiling points indicates that the attractive forces between pentane molecules are stronger than between neopentane molecules.

For large molecules, the cumulative effect of these small and rapidly changing dispersion forces can lead to a large net attraction.

2.13C Boiling Points

- The **boiling point** of a liquid is the temperature at which the vapor pressure of the liquid equals the pressure of the atmosphere above it.

The boiling points of liquids are *pressure dependent*, and boiling points are always reported as occurring at a particular pressure, at 1 atm (or at 760 torr), for example. A substance that boils at 150 °C at 1 atm pressure will boil at a substantially lower temperature if the pressure is reduced to, for example, 0.01 torr (a pressure easily obtained with a vacuum pump). The normal boiling point given for a liquid is its boiling point at 1 atm.

In passing from a liquid to a gaseous state, the individual molecules (or ions) of the substance must separate. Because of this, we can understand why ionic organic compounds often decompose before they boil. The thermal energy required to completely separate (volatilize) the ions is so great that chemical reactions (decompositions) occur first.

© RFcompany/Age Fotostock America, Inc.

Dispersion forces are what provides a gecko's grip to smooth surfaces.

THE CHEMISTRY OF... Fluorocarbons and Teflon

Fluorocarbons (compounds containing only carbon and fluorine) have extraordinarily low boiling points when compared to hydrocarbons of the same molecular weight. The fluorocarbon C_5F_{12} has a slightly lower boiling point than pentane (C_5H_{12}) even though it has a far higher molecular weight. The important factor in explaining this behavior is the very low polarizability of fluorine atoms that we mentioned earlier, resulting in very small dispersion forces.

The fluorocarbon called *Teflon* $[CF_2CF_2]_n$ (see Section 10.10) has self-lubricating properties that are exploited in making "nonstick" frying pans and lightweight bearings.

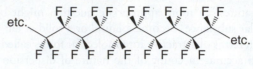

Teflon

Leonard Lessin/Photo Researchers, Inc.

Nonpolar compounds, where the intermolecular forces are very weak, usually boil at low temperatures even at 1 atm pressure. This is not always true, however, because of other factors that we have not yet mentioned: the effects of molecular weight and molecular shape and surface area. Heavier molecules require greater thermal energy in order to acquire velocities sufficiently great to escape the liquid phase, and because the surface areas of larger molecules can be much greater, intermolecular dispersion attractions can also be much larger. These factors explain why nonpolar ethane (bp −88.2 °C) boils higher than methane (bp −162 °C) at a pressure of 1 atm. It also explains why, at 1 atm, the even heavier and larger nonpolar molecule decane ($C_{10}H_{22}$) boils at 174 °C. The relationship between dispersion forces and surface area helps us understand why neopentane (2,2-dimethylpropane) has a lower boiling point (9.5 °C) than pentane (36.1 °C), even though they have the same molecular weight. The branched structure of neopentane allows less surface interaction between neopentane molecules, hence lower dispersion forces, than does the linear structure of pentane.

●●● SOLVED PROBLEM 2.7

Arrange the following compounds according to their expected boiling points, with the lowest boiling point first, and explain your answer. Notice that the compounds have similar molecular weights.

Diethyl ether   ***sec*-Butyl alcohol**   **Pentane**

STRATEGY AND ANSWER:

pentane < diethyl ether < *sec*-butyl alcohol

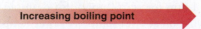

Increasing boiling point

Pentane has no polar groups and has only dispersion forces holding its molecules together. It would have the lowest boiling point. Diethyl ether has the polar ether group that provides dipole–dipole forces which are greater than dispersion forces, meaning it would have a higher boiling point than pentane. *sec*-Butyl alcohol has an —OH group that can form strong hydrogen bonds; therefore, it would have the highest boiling point.

PRACTICE PROBLEM 2.27

Arrange the following compounds in order of increasing boiling point. Explain your answer in terms of the intermolecular forces in each compound.

(a)   (b)   (c)   (d)

2.13D Solubilities

Intermolecular forces are of primary importance in explaining the **solubilities** of substances. Dissolution of a solid in a liquid is, in many respects, like the melting of a solid. The orderly crystal structure of the solid is destroyed, and the result is the formation of the more disorderly arrangement of the molecules (or ions) in solution. In the process of dissolving, too, the molecules or ions must be separated from each other, and energy must be supplied for both changes. The energy required to overcome lattice energies and intermolecular or interionic attractions comes from the formation of new attractive forces between solute and solvent.

Consider the dissolution of an ionic substance as an example. Here both the lattice energy and interionic attractions are large. We find that water and only a few other very polar solvents are capable of dissolving ionic compounds. These solvents dissolve ionic compounds by **hydrating** or **solvating** the ions (Fig. 2.9).

Water molecules, by virtue of their great polarity as well as their very small, compact shape, can very effectively surround the individual ions as they are freed from the crystal surface. Positive ions are surrounded by water molecules with the negative end of the water dipole pointed toward the positive ion; negative ions are solvated in exactly the opposite way. Because water is highly polar, and because water is capable of forming strong hydrogen bonds, the **ion–dipole forces** of attraction are also large. The energy supplied by the formation of these forces is great enough to overcome both the lattice energy and interionic attractions of the crystal.

A general rule for solubility is that "like dissolves like" in terms of comparable polarities.

- Polar and ionic solids are usually soluble in polar solvents.
- Polar liquids are usually miscible.
- Nonpolar solids are usually soluble in nonpolar solvents.
- Nonpolar liquids are usually miscible.
- Polar and nonpolar liquids, like oil and water, are usually not soluble to large extents.

Helpful Hint

Your ability to make qualitative predictions regarding solubility will prove very useful in the organic chemistry laboratory.

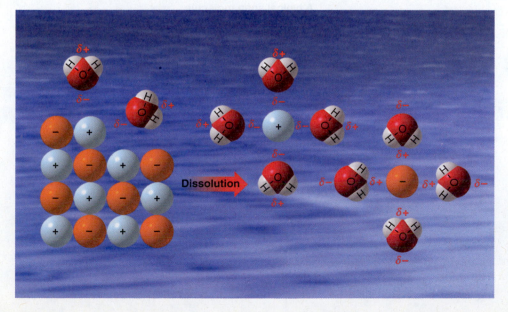

Dissolution

FIGURE 2.9 The dissolution of an ionic solid in water, showing the hydration of positive and negative ions by the very polar water molecules. The ions become surrounded by water molecules in all three dimensions, not just the two shown here.

Methanol and water are miscible in all proportions; so too are mixtures of ethanol and water and mixtures of both propyl alcohols and water. In these cases the alkyl groups of the alcohols are relatively small, and the molecules therefore resemble water more than they do an alkane. Another factor in understanding their solubility is that the molecules are capable of forming strong hydrogen bonds to each other:

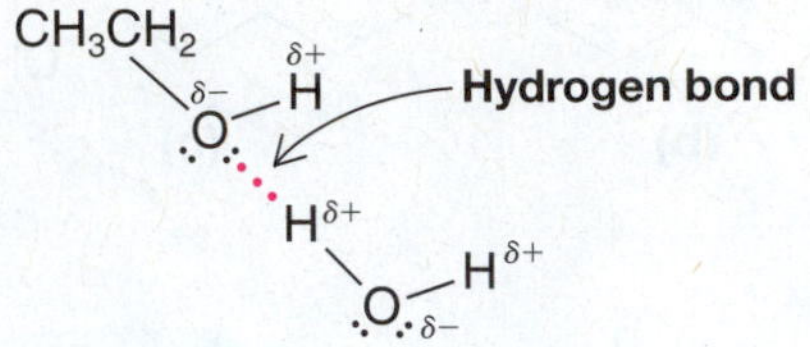

We often describe molecules or parts of molecules as being hydrophilic or hydrophobic. The alkyl groups of methanol, ethanol, and propanol are hydrophobic. Their hydroxyl groups are hydrophilic.

- **Hydrophobic** means incompatible with water (*hydro*, water; *phobic*, fearing or avoiding).
- **Hydrophilic** means compatible with water (*philic*, loving or seeking).

Decyl alcohol, with a chain of 10 carbon atoms, is a compound whose hydrophobic alkyl group overshadows its hydrophilic hydroxyl group in terms of water solubility.

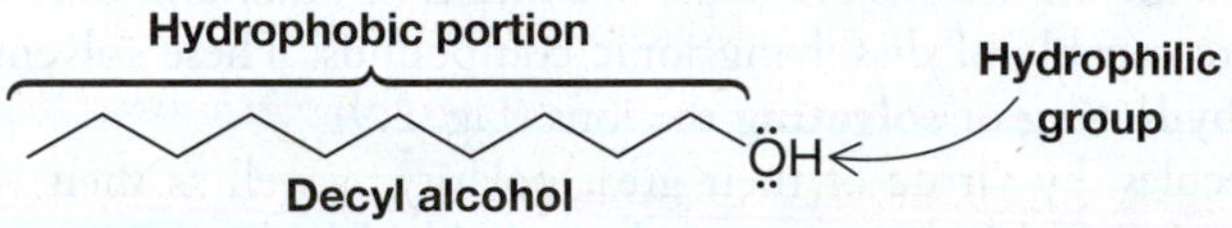

An explanation for why nonpolar groups such as long alkane chains avoid an aqueous environment—that is, for the so-called **hydrophobic effect**—is complex. The most important factor seems to involve an **unfavorable entropy change** in the water. Entropy changes (Section 3.10) have to do with changes from a relatively ordered state to a more disordered one or the reverse. Changes from order to disorder are favorable, whereas changes from disorder to order are unfavorable. For a nonpolar hydrocarbon chain to be accommodated by water, the water molecules have to form a more ordered structure around the chain, and for this, the entropy change is unfavorable.

We will see in Section 23.2C that the presence of a **hydrophobic group** and a hydrophilic group are essential components of soaps and detergents.

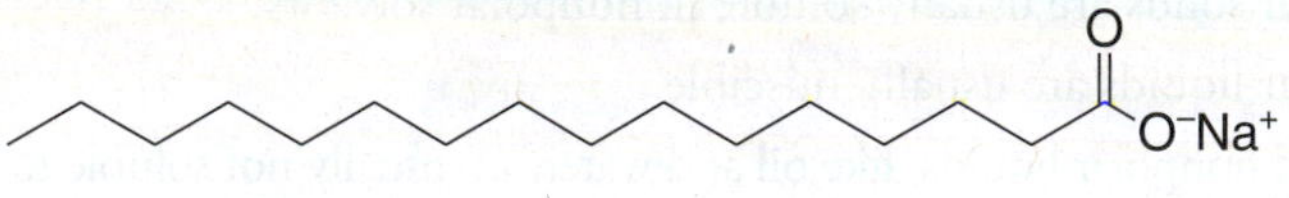

A typical soap molecule

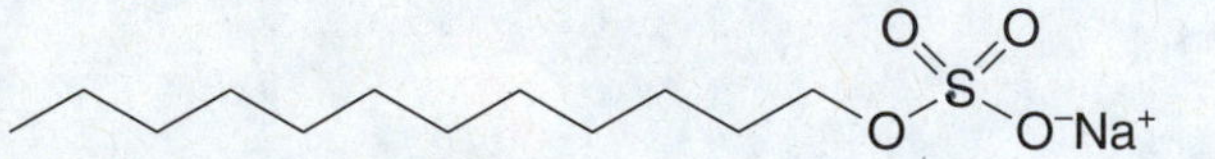

A typical detergent molecule

The hydrophobic long carbon chains of a soap or detergent embed themselves in the oily layer that typically surrounds the thing we want to wash away. The hydrophilic ionic groups at the ends of the chains are then left exposed on the surface and make the surface one that water molecules find attractive. Oil and water don't mix, but now the oily layer looks like something ionic and the water can take it "right down the drain."

2.13E Guidelines for Water Solubility

Organic chemists usually define a compound as water soluble if at least 3 g of the organic compound dissolves in 100 mL of water. We find that for compounds containing one hydrophilic group—and thus capable of forming strong **hydrogen bonds**—the following approximate guidelines hold: compounds with one to three carbon atoms are water soluble, compounds with four or five carbon atoms are borderline, and compounds with six carbon atoms or more are insoluble.

When a compound contains more than one hydrophilic group, these guidelines do not apply. Polysaccharides (Chapter 22), proteins (Chapter 24), and nucleic acids (Chapter 25) all contain thousands of carbon atoms *and many are water soluble*. They dissolve in water because they also contain thousands of hydrophilic groups.

2.13F Intermolecular Forces in Biochemistry

Later, after we have had a chance to examine in detail the properties of the molecules that make up living organisms, we shall see how **intermolecular forces** are extremely important in the functioning of cells. **Hydrogen bond** formation, the hydration of polar groups, and the tendency of nonpolar groups to avoid a polar environment all cause complex protein molecules to fold in precise ways—ways that allow them to function as biological catalysts of incredible efficiency. The same factors allow molecules of hemoglobin to assume the shape needed to transport oxygen. They allow proteins and molecules called lipids to function as cell membranes. Hydrogen bonding gives certain carbohydrates a globular shape that makes them highly efficient food reserves in animals. It gives molecules of other carbohydrates a rigid linear shape that makes them perfectly suited to be structural components in plants.

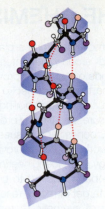

Hydrogen bonding (red dotted lines) in the α-helix structure of proteins

(Illustration, Irving Geis. Image from the Irving Geis Collection, HHMI. Rights owned by Howard Hughes Medical Institute. Not to be reproduced without permission.)

2.14 SUMMARY OF ATTRACTIVE ELECTRIC FORCES

The attractive forces occurring between molecules and ions that we have studied so far are summarized in Table 2.5.

TABLE 2.5 ATTRACTIVE ELECTRIC FORCES

Electric Force	Relative Strength	Type	Example
Cation–anion (in a crystal)	Very strong		Sodium chloride crystal lattice
Covalent bonds	Strong (140–523 kJ mol^{-1})	Shared electron pairs	H—H (436 kJ mol^{-1}) CH$_3$—CH$_3$ (378 kJ mol^{-1}) I—I (151 kJ mol^{-1})
Ion–dipole	Moderate		Na$^+$ in water (see Fig. 2.9)
Hydrogen bonds	Moderate to weak (4–38 kJ mol^{-1})	$-\overset{\delta-}{Z}:\cdots\overset{\delta+}{H}-$	
Dipole–dipole	Weak	$\overset{\delta+}{C}\overset{\delta-}{H_3Cl}\cdots\overset{\delta+}{C}\overset{\delta-}{H_3Cl}$	
Dispersion	Variable	Transient dipole	Interactions between methane molecules

THE CHEMISTRY OF... Organic Templates Engineered to Mimic Bone Growth

Intermolecular forces play a myriad of roles in life and in the world around us. Intermolecular forces hold together the strands of our DNA, provide structure to our cell membranes, cause the feet of gecko lizards to stick to walls and ceilings, keep water from boiling at room temperature and ordinary pressure, and literally provide the adhesive forces that hold our cells, bones, and tissues together. As these examples show, the world around us provides exquisite instruction in nanotechnology and bioengineering, and scientists throughout the ages have been inspired to create and innovate based on nature. One target of recent research in bioengineering is the development of synthetic materials that mimic nature's template for bone growth. A synthetic material with bone-promoting properties could be used to help repair broken bones, offset osteoporosis, and treat bone cancer.

Both natural bone growth and the synthetic system under development depend strongly on intermolecular forces. In living systems, bones grow by adhesion of specialized cells to a long fibrous natural template called collagen. Certain functional groups along the collagen promote the binding of bone-growing cells, while other functional groups facilitate calcium crystallization. Chemists at Northwestern University (led by S. I. Stupp) have engineered a molecule that can be made in the laboratory and that mimics this process. The molecule shown below spontaneously self-assembles into a long tubular aggregate, imitating the fibers of collagen. Dispersion forces between hydrophobic alkyl tails on the molecule cause self-assembly of the molecules into tubules. At the other end of the molecule, the researchers included functional groups that promote cell binding and still other functional groups that encourage calcium crystallization. Lastly, they included functional groups that allow one molecule to be covalently linked to its neighbors after the self-assembly process has occurred, thus adding further stabilization to the initially noncovalent structure. Designing all of these features into the molecular structure has paid off, because the self-assembled fiber promotes calcium crystallization along its axis, much like nature's collagen template. This example of molecular design is just one exciting development at the intersection of nanotechnology and bioengineering.

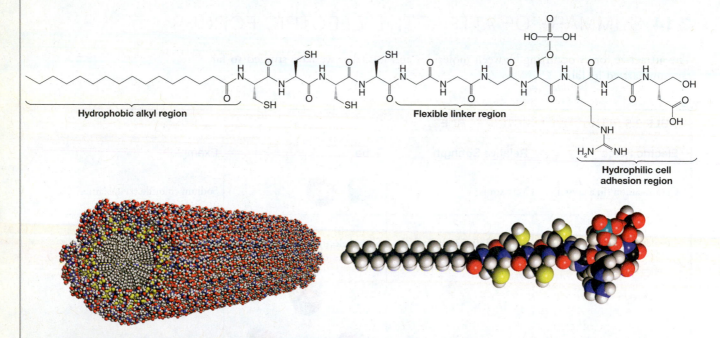

(From Hartgerink, J.D., Beniash, E. J, Stupp, S.I.:Self-assembly and Mineralization of Peptide-Amphiphile Nanofibers. SCIENCE 294:1684-1688, Figure 1 (2001). Reprinted with permission from AAAS.)

2.15 INFRARED SPECTROSCOPY: AN INSTRUMENTAL METHOD FOR DETECTING FUNCTIONAL GROUPS

Infrared (IR) spectroscopy is a simple, rapid, and nondestructive instrumental technique that can give evidence for the presence of various functional groups. If you had a sample of unknown identity, among the first things you would do is obtain an infrared spectrum, along with determining its solubility in common solvents and its melting and/or boiling point.

Infrared spectroscopy, as with all forms of spectroscopy, depends on the interaction of molecules or atoms with electromagnetic radiation. Infrared radiation causes atoms and groups of atoms of organic compounds to vibrate with increased amplitude about the

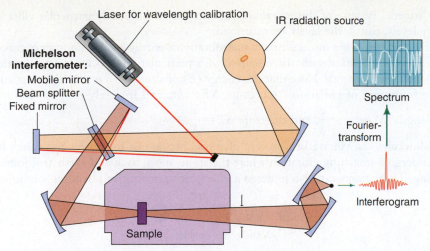

(Diagram adapted from the computer program IR Tutor, Columbia University.)

FIGURE 2.10 A diagram of a Fourier transform infrared (FTIR) spectrometer. FTIR spectrometers employ a Michelson interferometer, which splits the radiation beam from the IR source so that it reflects simultaneously from a moving mirror and a fixed mirror, leading to interference. After the beams recombine, they pass through the sample to the detector and are recorded as a plot of time versus signal intensity, called an interferogram. The overlapping wavelengths and the intensities of their respective absorptions are then converted to a spectrum by applying a mathematical operation called a Fourier transform.

The FTIR method eliminates the need to scan slowly over a range of wavelengths, as was the case with older types of instruments called dispersive IR spectrometers, and therefore FTIR spectra can be acquired very quickly. The FTIR method also allows greater throughput of IR energy. The combination of these factors gives FTIR spectra strong signals as compared to background noise (i.e., a high signal to noise ratio) because radiation throughput is high and rapid scanning allows multiple spectra to be averaged in a short period of time. The result is enhancement of real signals and cancellation of random noise.

covalent bonds that connect them. (Infrared radiation is not of sufficient energy to excite electrons, as is the case when some molecules interact with visible, ultraviolet, or higher energy forms of light.) Since the functional groups of organic molecules include specific arrangements of bonded atoms, absorption of IR radiation by an organic molecule will occur at specific frequencies characteristic of the types of bonds and atoms present in the specific functional groups of that molecule. These vibrations are *quantized*, and as they occur, the compounds absorb IR energy in particular regions of the IR portion of the spectrum.

An infrared spectrometer (Fig. 2.10) operates by passing a beam of IR radiation through a sample and comparing the radiation transmitted through the sample with that transmitted in the absence of the sample. Any frequencies absorbed by the sample will be apparent by the difference. The spectrometer plots the results as a graph showing absorbance versus frequency or wavelength.

- The position of an absorption band (peak) in an IR spectrum is specified in units of **wavenumbers ($\overline{v}$)**.

Wavenumbers are the reciprocal of wavelength when wavelength is expressed in centimeters (the unit is cm^{-1}), and therefore give the number of wave cycles per centimeter. The larger the wavenumber, the higher is the frequency of the wave, and correspondingly the higher is the frequency of the bond absorption. IR absorptions are sometimes, though less commonly, reported in terms of **wavelength (λ)**, in which case the units are micrometers (μm; old name micron, μ). Wavelength is the distance from crest to crest of a wave.

$$\overline{v} = \frac{1}{\lambda} \text{ (with } \lambda \text{ in cm)} \qquad \text{or} \qquad \overline{v} = \frac{10.000}{\lambda} \text{ (with } \lambda \text{ in } \mu\text{m)}$$

In their vibrations covalent bonds behave as if they were tiny springs connecting the atoms. When the atoms vibrate, they can do so only at certain frequencies, as if the bonds were "tuned." Because of this, covalently bonded atoms have only particular vibrational energy levels; that is, the levels are quantized.

The excitation of a molecule from one vibrational energy level to another occurs only when the compound absorbs IR radiation of a particular energy, meaning a particular wavelength or frequency. Note that the energy (E) of absorption is directly proportional to the **frequency** of radiation (v) because $\Delta E = hv$, and inversely proportional to the wavelength (λ) because $\dfrac{c}{\lambda}$, and therefore $\Delta E = \dfrac{hc}{\lambda}$.

Molecules can vibrate in a variety of ways. Two atoms joined by a covalent bond can undergo a stretching vibration where the atoms move back and forth as if joined by a spring. Three atoms can also undergo a variety of stretching and bending vibrations.

A stretching vibration

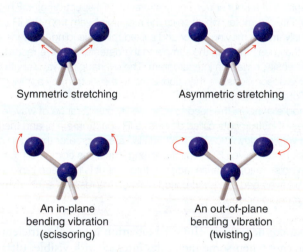

Symmetric stretching   Asymmetric stretching

An in-plane   An out-of-plane
bending vibration   bending vibration
(scissoring)   (twisting)

The *frequency* of a given stretching vibration *in an IR spectrum* can be related to two factors. These are *the masses of the bonded atoms*—light atoms vibrate at higher frequencies than heavier ones—*and the relative stiffness of the bond*. (These factors are accounted for in Hooke's law, a relationship you may study in introductory physics.) Triple bonds are stiffer (and vibrate at higher frequencies) than double bonds, and double bonds are stiffer (and vibrate at higher frequencies) than single bonds. We can see some of these effects in Table 2.6. Notice that stretching frequencies of groups involving hydrogen (a light atom) such as C—H, N—H, and O—H all occur at relatively high frequencies:

GROUP	BOND	FREQUENCY RANGE (cm^{-1})
Alkyl	C—H	2853–2962
Alcohol	O—H	3590–3650
Amine	N—H	3300–3500

Notice, too, that triple bonds vibrate at higher frequencies than double bonds:

GROUP	BOND	FREQUENCY RANGE (cm^{-1})
Alkyne	C≡C	2100–2260
Nitrile	C≡N	2220–2260
Alkene	C=C	1620–1680
Carbonyl	C=O	1630–1780

● Not all molecular vibrations result in the absorption of IR energy. *In order for a vibration to occur with the absorption of IR energy, the dipole moment of the molecule must change as the vibration occurs.*

TABLE 2.6 CHARACTERISTIC INFRARED ABSORPTIONS OF GROUPS

Group		Frequency Range (cm^{-1})	Intensity[a]
A. Alkyl			
C—H (stretching)		2853–2962	(m–s)
Isopropyl, —CH(CH$_3$)$_2$		1380–1385	(s)
	and	1365–1370	(s)
tert-Butyl, —C(CH$_3$)$_3$		1385–1395	(m)
	and	~1365	(s)
B. Alkenyl			
C—H (stretching)		3010–3095	(m)
C=C (stretching)		1620–1680	(v)
R—CH=CH$_2$		985–1000	(s)
(out-of-plane	and	905–920	(s)
C—H bendings)			
R$_2$C=CH$_2$		880–900	(s)
cis-RCH=CHR		675–730	(s)
trans-RCH=CHR		960–975	(s)
C. Alkynyl			
≡C—H (stretching)		~3300	(s)
C≡C (stretching)		2100–2260	(v)
D. Aromatic			
Ar—H (stretching)		~3030	(v)
C=C (stretching)		1450–1600	(m)
Aromatic substitution type			
(C—H out-of-plane bendings)			
Monosubstituted		690–710	(very s)
	and	730–770	(very s)
o-Disubstituted		735–770	(s)
m-Disubstituted		680–725	(s)
	and	750–810	(very s)
p-Disubstituted		800–860	(very s)
E. Alcohols, Phenols, and Carboxylic Acids			
O—H (stretching)			
Alcohols, phenols (dilute solutions)		3590–3650	(sharp, v)
Alcohols, phenols (hydrogen bonded)		3200–3550	(broad, s)
Carboxylic acids (hydrogen bonded)		2500–3000	(broad, v)
F. Ethers, Alcohols, and Esters			
C—O (stretching)		1020–1275	(s)
G. Aldehydes, Ketones, Esters, Carboxylic Acids, and Amides			
C=O (stretching)		1630–1780	(s)
Aldehydes		1690–1740	(s)
Ketones		1680–1750	(s)
Esters		1735–1750	(s)
Carboxylic acids		1710–1780	(s)
Amides		1630–1690	(s)
H. Amines			
N—H		3300–3500	(m)
I. Nitriles			
C≡N		2220–2260	(m)

[a]Abbreviations: s = strong, m = medium, w = weak, v = variable, ~ = approximately.

Thus, methane does not absorb IR energy for symmetric streching of the four C—H bonds; asymmetric stretching, on the other hand, does lead to an IR absorption. Symmetrical vibrations of the carbon–carbon double and triple bonds of ethene and ethyne do not result in the absorption of IR radiation, either.

●●● **SOLVED PROBLEM 2.8**

The infrared spectrum of l-hexyne shows a sharp absorption peak near 2100 cm^{-1} due to stretching of its triple bond. However, 3-hexyne shows no absorption in that region. Explain.

1-Hexyne   **3-Hexyne**

STRATEGY AND ANSWER: For an infrared absorption to occur there must be a change in the dipole moment of the molecule during the stretching process. Since 3-hexyne is symmetrical about its triple bond, there is no change in its dipole moment as stretching takes place, hence there is no IR absorption from the triple bond.

Vibrational absorption may occur outside the region measured by a particular IR spectrometer, and vibrational absorptions may occur so closely together that peaks fall on top of peaks.

Other factors bring about even more absorption peaks. Overtones (harmonics) of fundamental absorption bands may be seen in IR spectra even though these overtones occur with greatly reduced intensity. Bands called combination bands and difference bands also appear in IR spectra.

Because IR spectra of even relatively simple compounds contain so many peaks, the possibility that two different compounds will have the same IR spectrum is exceedingly small. It is because of this that an IR spectrum has been called the "fingerprint" of a molecule. Thus, with organic compounds, if two pure samples give different IR spectra, one can be certain that they are different compounds. If they give the same IR spectrum, then they are very likely to be the same compound.

2.16 INTERPRETING IR SPECTRA

IR spectra contain a wealth of information about the structures of compounds. We show some of the information that can be gathered from the spectra of octane and methylbenzene (commonly called toluene) in Figs. 2.11 and 2.12. In this section we shall learn how

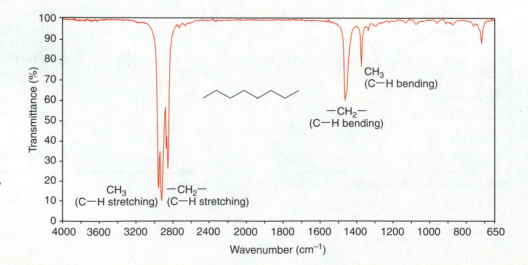

FIGURE 2.11 The IR spectrum of octane. (Notice that, in IR spectra, the peaks are usually measured in % transmittance. Thus, the peak at 2900 cm^{-1} has 10% transmittance—that is, an absorbance, A, of 0.90.)

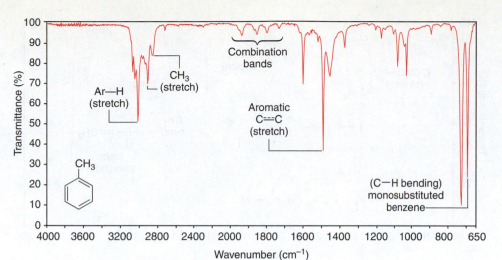

FIGURE 2.12 The IR spectrum of methylbenzene (toluene).

to recognize the presence of characteristic IR absorption peaks that result from vibrations of alkyl and functional groups. The data given in Table 2.6 will provide us with key information to use when correlating actual spectra with IR absorption frequencies that are typical for various groups.

2.16A Infrared Spectra of Hydrocarbons

- All hydrocarbons give absorption peaks in the 2800–3300-cm^{-1} region that are associated with carbon–hydrogen stretching vibrations.

We can use these peaks in interpreting IR spectra because the exact location of the peak depends on the strength (and stiffness) of the $C—H$ bond, which in turn depends on the hybridization state of the carbon that bears the hydrogen. The $C—H$ bonds involving sp-hybridized carbon are strongest and those involving sp^3-hybridized carbon are weakest. The order of bond strength is

$$sp > sp^2 > sp^3$$

This, too, is the order of the bond stiffness.

- The carbon–hydrogen stretching peaks of hydrogen atoms attached to sp-hybridized carbon atoms occur at highest frequencies, about 3300 cm^{-1}.

The carbon–hydrogen bond of a terminal alkyne ($\equiv C—H$) gives an absorption in the 3300-cm^{-1} region. We can see the absorption of the acetylenic (alkynyl) $C—H$ bond of 1-heptyne at 3320 cm^{-1} in Fig. 2.13.

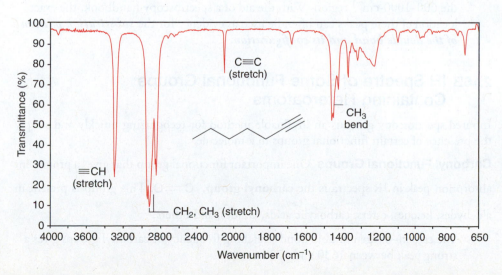

FIGURE 2.13 The IR spectrum of 1-heptyne.

FIGURE 2.14 The IR spectrum of 1-octene.

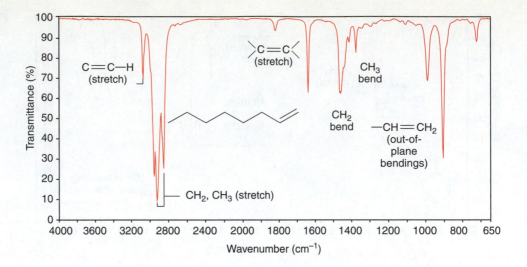

- The carbon–hydrogen stretching peaks of hydrogen atoms attached to sp^2-hybridized carbon atoms occur in the 3000–3100-cm^{-1} region.

Thus, alkenyl C—H bonds and the C—H groups of aromatic rings give absorption peaks in this region. We can see the alkenyl C—H absorption peak at 3080 cm^{-1} in the spectrum of 1-octene (Fig. 2.14), and we can see the C—H absorption of the aromatic hydrogen atoms at 3090 cm^{-1} in the spectrum of methylbenzene (Fig. 2.12).

- The carbon–hydrogen stretching bands of hydrogen atoms attached to sp^3-hybridized carbon atoms occur at lowest frequencies, in the 2800–3000-cm^{-1} region.

We can see methyl and methylene absorption peaks in the spectra of octane (Fig. 2.11), methylbenzene (Fig. 2.12), 1-heptyne (Fig. 2.13), and 1-octene (Fig. 2.14).

Hydrocarbons also give absorption peaks in their IR spectra that result from carbon–carbon bond stretchings. Carbon–carbon single bonds normally give rise to very weak peaks that are usually of little use in assigning structures. More useful peaks arise from carbon–carbon multiple bonds, however.

- Carbon–carbon double bonds give absorption peaks in the 1620–1680-cm^{-1} region, and carbon–carbon triple bonds give absorption peaks between 2100 and 2260 cm^{-1}.

These absorptions are not usually strong ones, and they are absent if the double or triple bond is symmetrically substituted. (No dipole moment change will be associated with the vibration.) The stretchings of the carbon–carbon bonds of benzene rings usually give a set of characteristic sharp peaks in the 1450–1600-cm^{-1} region.

- Absorptions arising from carbon–hydrogen bending vibrations of alkenes occur in the 600–1000-cm^{-1} region. With the aid of a spectroscopy handbook, the exact location of these peaks can often be used as evidence for the *substitution pattern of the double bond and its configuration*.

2.16B IR Spectra of Some Functional Groups Containing Heteroatoms

Helpful Hint

IR spectroscopy is an exceedingly useful tool for detecting functional groups.

Infrared spectroscopy gives us an invaluable method for recognizing quickly and simply the presence of certain functional groups in a molecule.

Carbonyl Functional Groups One important functional group that gives a prominent absorption peak in IR spectra is the **carbonyl group**, $C{=}O$. This group is present in aldehydes, ketones, esters, carboxylic acids, amides, and others.

- The carbon–oxygen double-bond stretching frequency of carbonyl groups gives a strong peak between 1630 and 1780 cm^{-1}.

The exact location of the absorption depends on whether it arises from an aldehyde, ketone, ester, and so forth.

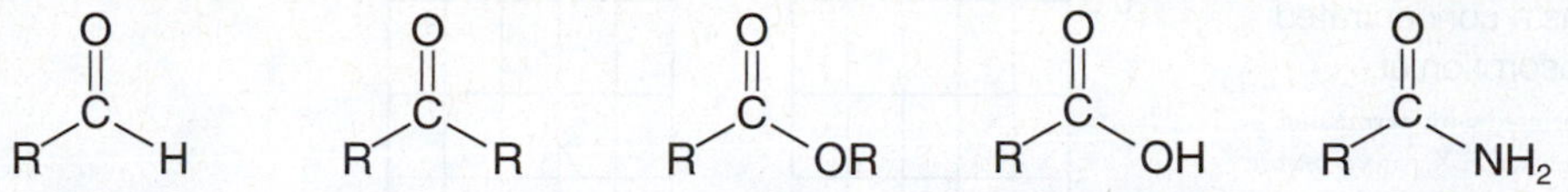

$\overset{O}{\underset{R\quad H}{\parallel}}$ C	$\overset{O}{\underset{R\quad R}{\parallel}}$ C	$\overset{O}{\underset{R\quad OR}{\parallel}}$ C	$\overset{O}{\underset{R\quad OH}{\parallel}}$ C	$\overset{O}{\underset{R\quad NH_2}{\parallel}}$ C
Aldehyde	**Ketone**	**Ester**	**Carboxylic acid**	**Amide**
1690–1740 cm^{-1}	1680–1750 cm^{-1}	1735–1750 cm^{-1}	1710–1780 cm^{-1}	1630–1690 cm^{-1}

●●● SOLVED PROBLEM 2.9

A compound with the molecular formula $C_4H_4O_2$ has a strong sharp absorbance near 3300 cm^{-1}, absorbances in the 2800–3000-cm^{-1} region, and a sharp absorbance peak near 2200 cm^{-1}. It also has a strong broad absorbance in the 2500–3600-cm^{-1} region and a strong peak in the 1710–1780-cm^{-1} region. Propose a possible structure for the compound.

STRATEGY AND ANSWER: The sharp peak near 3300 cm^{-1} is likely to arise from the stretching of a hydrogen attached to the sp-hybridized carbon of a triple bond. The sharp peak near 2200 cm^{-1}, where the triple bond of an alkyne stretches, is consistent with this. The peaks in the 2800–3000-cm^{-1} region suggest stretchings of the C—H bonds of alkyl groups, either CH_2 or CH_3 groups. The strong, broad absorbance in the 2500–3600-cm^{-1} region suggests a hydroxyl group arising from a carboxylic acid. The strong peak around 1710–1780 cm^{-1} is consistent with this since it could arise from the carbonyl group of a carboxylic acid. Putting all this together with the molecular formula suggests the compound is as shown at the right.

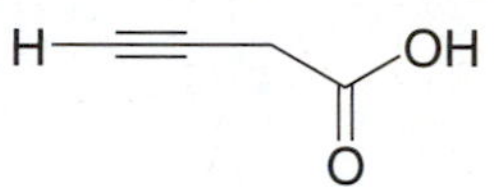

$$H-\!\!\!\equiv\!\!\!-\overset{\displaystyle OH}{\underset{\displaystyle O}{}}$$

●●● PRACTICE PROBLEM 2.28

Use arguments based on resonance and electronegativity effects to explain the trend in carbonyl IR stretching frequencies from higher frequency for esters and carboxylic acids to lower frequencies for amides. (*Hint:* Use the range of carbonyl stretching frequencies for aldehydes and ketones as the "base" frequency range of an unsubstituted carbonyl group and consider the influence of electronegative atoms on the carbonyl group and/or atoms that alter the resonance hybrid of the carbonyl.) What does this suggest about the way the nitrogen atom influences the distribution of electrons in an amide carbonyl group?

Alcohols and Phenols The **hydroxyl groups** of alcohols and phenols are also easy to recognize in IR spectra by their O—H stretching absorptions. These bonds also give us direct evidence for hydrogen bonding (Section 2.13B).

- The IR absorption of an alcohol or phenol O—H group is in the 3200–3550-cm^{-1} range, and most often it is broad.

The typical broadness of the peak is due to association of the molecules through hydrogen bonding (Section 2.13B), which causes a wider distribution of stretching frequencies for the O—H bond. If an alcohol or phenol is present as a very dilute solution in a solvent that cannot contribute to hydrogen bonding (e.g., CCl_4), O—H absorption occurs as a very sharp peak in the 3590–3650-cm^{-1} region. In very dilute solution in such a solvent or in the gas phase, formation of intermolecular hydrogen bonds does not take place because molecules of the analyte are too widely separated. A sharp peak in the 3590–3650-cm^{-1} region, therefore, is attributed to "free" (unassociated) hydroxyl groups. Increasing the concentration of the alcohol or phenol causes the sharp peak to be replaced by a broad band in the 3200–3550-cm^{-1} region. Hydroxyl absorptions in IR spectra of cyclohexylcarbinol (cyclohexylmethanol) run in dilute and concentrated solutions (Fig. 2.15) exemplify these effects.

FIGURE 2.15 (*a*) The IR spectrum of an alcohol (cyclohexyl-carbinol) in a dilute solution shows the sharp absorption of a "free" (non-hydrogen-bonded) hydroxyl group at 3600 cm^{-1}. (*b*) The IR spectrum of the same alcohol as a concentrated solution shows a broad hydroxyl group absorption at 3300 cm^{-1} due to hydrogen bonding. (Reprinted with permission of John Wiley & Sons, Inc. From Silverstein, R., and Webster, F. X., *Spectrometric Identification of Organic Compounds*, Sixth Edition, p. 89. Copyright 1998.)

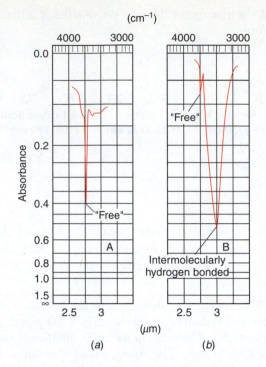

(a)   (b)

Carboxylic Acids

The **carboxylic acid group** can also be detected by IR spectroscopy. If both carbonyl and hydroxyl stretching absorptions are present in an IR spectrum, there is good evidence for a carboxylic acid functional group (although it is possible that isolated carbonyl and hydroxyl groups could be present in the molecule).

- The hydroxyl absorption of a carboxylic acid is often very broad, extending from 3600 cm^{-1} to 2500 cm^{-1}.

Figure 2.16 shows the IR spectrum of propanoic acid.

Amines

IR spectroscopy also gives evidence for N—H bonds (see Figure 2.17).

- Primary (1°) and secondary (2°) amines give absorptions of moderate strength in the 3300–3500-cm^{-1} region.
- Primary amines exhibit two peaks in this region due to symmetric and asymmetric stretching of the two N—H bonds.
- Secondary amines exhibit a single peak.
- Tertiary amines show no N—H absorption because they have no such bond.
- A basic pH is evidence for any class of amine.

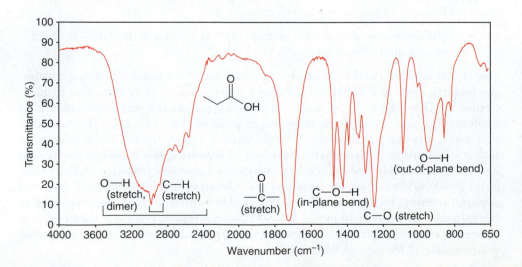

FIGURE 2.16 The IR spectrum of propanoic acid.

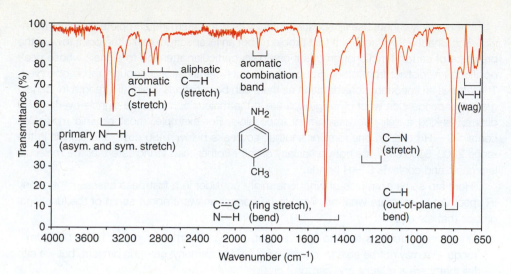

FIGURE 2.17 Annotated IR spectrum of 4-methylaniline.

RNH$_2$ (1° Amine)

Two peaks in 3300–3500-cm^{-1} region

Symmetric stretching

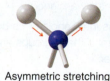

Asymmetric stretching

R$_2$NH (2° Amine)

One peak in 3300–3500-cm^{-1} region

Hydrogen bonding causes N—H stretching peaks of 1° and 2° amines to broaden. The NH groups of **amides** give similar absorption peaks and include a carbonyl absorption as well.

● ● ● SOLVED PROBLEM 2.10

What key peaks would you expect to find in the IR spectrum of the following compound?

STRATEGY AND ANSWER: The compound is an amide. We should expect a strong peak in the 1630–1690 cm^{-1} region arising from the carbonyl group and a single peak of moderate strength in the 3300–3500 cm^{-1} region for the N—H group.

HOW TO INTERPRET AN IR SPECTRUM WITHOUT ANY KNOWLEDGE OF THE STRUCTURE

IR spectroscopy is an incredibly powerful tool for functional group identification, as we have seen in the preceding sections. However, in introducing this technique, we have explored IR spectra from the perspective of compounds of known structure, explaining the peaks observed in reference to each critical grouping of atoms that we know to be present. In the real world, one often encounters brand new materials of unknown structure. How IR can help in this scenario is something that a forensics scientist or natural products isolation chemist might need to worry about on a daily basis.

We certainly cannot use IR spectroscopy by itself to determine complete structure (techniques in Chapter 9 will help with that problem), but an IR spectrum can often point toward the presence of certain functional groups if one pays particular attention to signals whose peak positions are distinct from other groups and is consistently strong enough to be observed. The latter is an important consideration as there can be variations in signal strength for certain groups dependent on what other groups are in the molecule, and some signals overlap with others, making a definitive assignment impossible. For example, most organic molecules contain C—H bonds in one form or another, so peaks below 1450 cm^{-1} and signals in the range 2800–3000 cm^{-1} are not particularly definitive other than to indicate that the molecule is organic and contains C—H bonds.

Here are some examples of what one might consider in a first-pass assessment of any IR spectrum to generate what are likely to be correct answers about some of the functional groups that are present:

● Only C=O stretches tend to have a tight, strong absorbance in the 1630–1780 cm^{-1} range. We may not be able to identify what kind of carbonyl group is present, but we can tell that there is at least one carbonyl group.

● Only the stretches of nitrile or alkyne bonds tend to appear between 2000 and 2300 cm^{-1}, so these can be fairly readily assigned.

● Only hydroxyl groups as in alcohols or carboxylic acids tend to create a large and broad signal at about 3300 cm^{-1}; these groups are easy to identify assuming the sample is not contaminated with water.

● Only amines tend to produce broad but smaller peaks than hydroxyl peaks around 3300 cm^{-1}. The number of those peaks can sometimes tell if there is one or two hydrogens attached to that nitrogen atom.

The examples below allow us to put these general principles into practice.

The IR spectrum of Unknown 1 (Fig. 2.18) has broad signals centered around 3300 cm^{-1} and a medium absorption at 2250 cm^{-1}. Based on the information above, we can surmise that the molecule likely contains a hydroxyl group and a group with a triple bond. Most likely the triply-bonded group is a nitrile since nitriles tend to appear at about 2250 cm^{-1}, whereas alkynes appear slightly lower at around 2000 cm^{-1}. We cannot be strictly sure that it is a nitrile, but that would be a good hypothesis in the absence of any other chemical evidence. Indeed, this turns out to be correct, as the molecule is 3-hydroxypropionitrile in this case.

In the IR spectrum of Unknown 2 (Fig. 2.19) there is a hydroxyl absorption once again centered around 3300 cm^{-1}, as well as a carbonyl peak at 1705 cm^{-1}. And, although we cannot always tell what kind of carbonyl is present, when the hydroxyl peak is extremely broad and has a ragged appearance (due to overlap of the C—H absorptions that extend below it, in contrast to the first spectrum where the hydroxyl was smooth, it is usually safe to assume that this hydroxyl group is attached to the

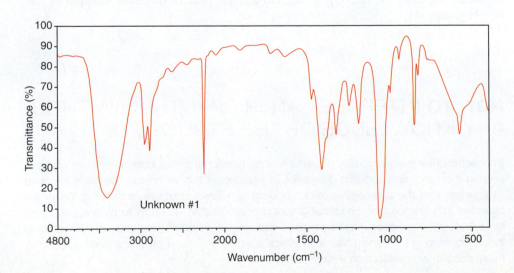

FIGURE 2.18 The IR Spectrum of Unknown 1. (SDBS, National Institute of Advanced Industrial Science and Technology)

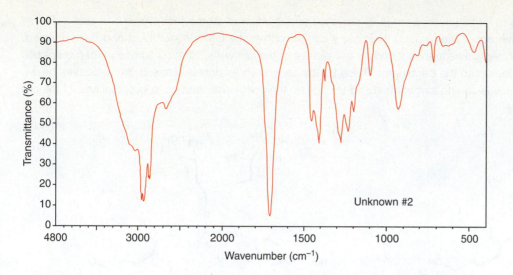

FIGURE 2.19 The IR Spectrum of Unknown 2. (SDBS, National Institute of Advanced Industrial Science and Technology)

carbonyl group; thus, these two groups are together part of a carboxylic acid functional group. Once again, we were able to identify the key functional group of the molecule since this is heptanoic acid.

2.17 APPLICATIONS OF BASIC PRINCIPLES

We now review how certain basic principles apply to phenomena that we have studied in this chapter.

Polar Bonds Are Caused by Electronegativity Differences We saw in Section 2.2 that when atoms with different electronegativities are covalently bonded, the more electronegative atom will be negatively charged and the less electronegative atom will be positively charged. The bond will be a *polar bond* and it will have a *dipole moment*.

Dipole moments are important in explaining physical properties of molecules (as we shall review below), and in explaining infrared spectra. For a vibration to occur with the absorption of IR energy, the dipole moment of the molecule must change during the course of the vibration.

Opposite Charges Attract This principle is central to understanding physical properties of organic compounds (Section 2.13). All of the forces that operate between individual molecules (and thereby affect boiling points, melting points, and solubilities) are between oppositely charged molecules (ions) or between oppositely charged portions of molecules. Examples are ion–ion forces (Section 2.13A) that exist between oppositely charged ions in crystals of ionic compounds, dipole–dipole forces (Section 2.13B) that exist between oppositely charged portions of polar molecules and that include the very strong dipole–dipole forces that we call *hydrogen bonds*, and the weak *dispersion* or *London forces* that exist between portions of molecules that bear small temporary opposite charges.

Molecular Structure Determines Properties We learned in Section 2.13 how physical properties are related to molecular structure.

[WHY Do These Topics Matter?

VANCOMYCIN AND ANTIBIOTIC RESISTANCE

Just as hydrogen bonds are critical in the pairing of nucleotides, they also play a major role in how one of the world's most powerful antibiotics kills bacteria. That antibiotic is vancomycin, a compound first isolated in 1956 by scientists at the Eli Lilly pharmaceutical company from the fermentation broth of a microbe found in the jungles of Borneo. Its name was derived from the verb "to vanquish," because it could kill every strain of gram-positive bacteria thrown at it, including the deadly strain known as MRSA (for methicillin-resistant *Staphylococcus aureus*), one of the so-called flesh-eating bacteria.

Vancomycin's success is due to its structure, a carefully designed arrangement of atoms that allows it to attack diverse bacterial strains. As bacteria move about their hosts, their cell walls are constantly being assembled and disassembled. Vancomycin targets one particular peptide sequence found on the surface of the cell walls, forming a network of five specific hydrogen bonds that allows it to lock onto the bacterium. These bonds are shown as dashed lines in the structures below. Once attached to vancomycin, bacteria can no longer build and strengthen their cell walls, leading to eventual lysis of the cell membrane and their death.

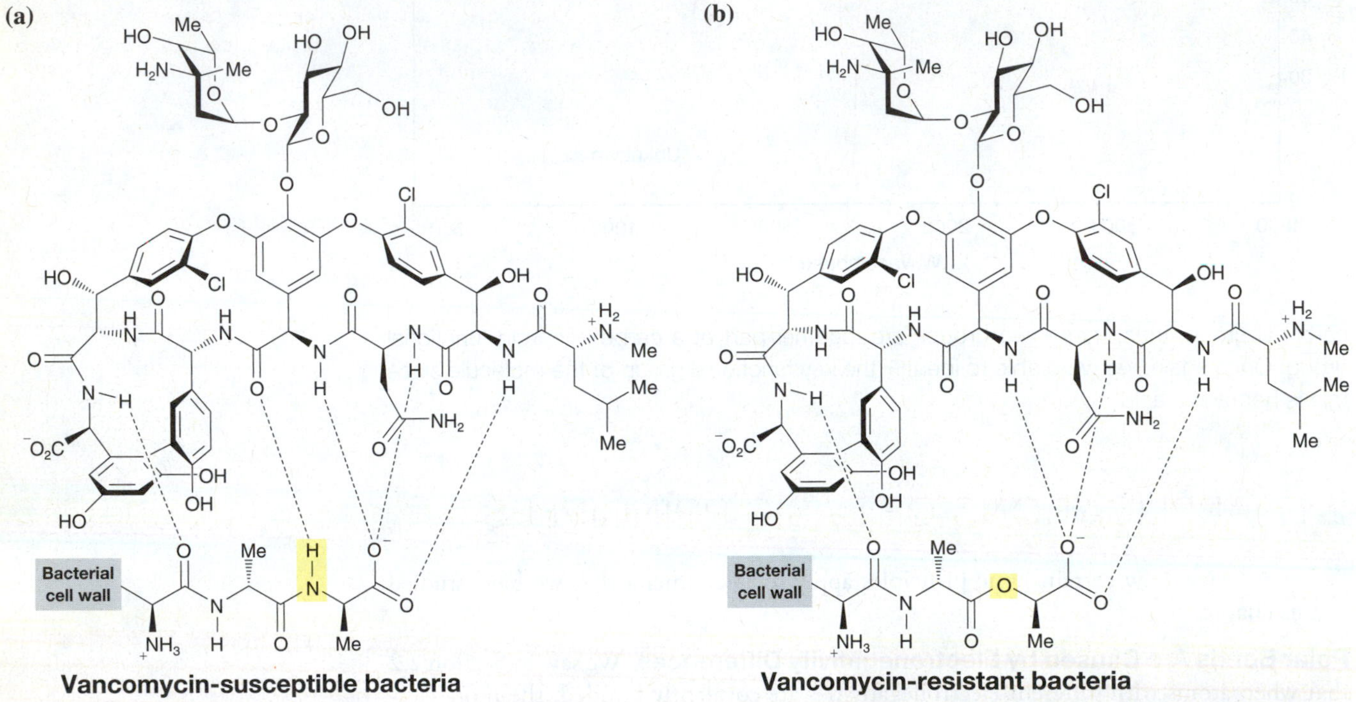

(a)

Vancomycin-susceptible bacteria

(b)

Vancomycin-resistant bacteria

Unfortunately, while vancomycin has proven effective for many decades in combating bacterial infections, in the past few years some bacteria have become resistant to it. These resistant bacteria have evolved a different set of peptides on their cell surface. The highlighted N—H group in (a) has been instead replaced with an O, as shown in (b). Although we will have much more to say about peptides and amino acids in Chapter 24, for now realize that this change has turned one hydrogen-bond donor (the N—H) into an atom that is a hydrogen-bond acceptor (O). As a result, vancomycin can form only four hydrogen bonds with the target. Although this constitutes a loss of just 20% of its hydrogen-bonding capacity, it turns out that its overall effectiveness in terms of its bacterial-killing ability is reduced by a factor of 1000. As a result, these bacteria are resistant to vancomycin, meaning that new chemical weapons are needed if patients infected with certain resistant gram-positive bacteria are to survive. Fortunately, there are several leads being explored in clinical trials, but given the ability of bacteria to constantly evolve and evade our therapies, we will need to keep developing new and better antibiotics.

© WorldIllustrated/Photoshot

Vancomycin was discovered in microbes from the jungles in Borneo.

To learn more about these topics, see:

1. Nicolaou, K. C.; Boddy, C. N. C., "Behind enemy lines" in *Scientific American*, May 2001, pp. 54–61.
2. Nicolaou, K. C.; Snyder, S. A. *Classics in Total Synthesis II*. Wiley-VCH: Weinheim, 2003, pp. 239–300.

SUMMARY AND REVIEW TOOLS

In Chapter 2 you learned about families of organic molecules, some of their physical properties, and how we can use an instrumental technique called infrared spectroscopy to study them.

You learned that functional groups define the families to which organic compounds belong. At this point you should be able to name functional groups when you see them in structural formulas, and, when given the name of a functional group, draw a general example of its structure.

You also built on your knowledge of how electronegativity influences charge distribution in a molecule and how, together with three-dimensional structure, charge distribution influences the overall polarity of a molecule. Based on polarity and three-dimensional structure, you should be able to predict the kind and relative strength of electrostatic forces between molecules. With this understanding you will be able to roughly estimate physical properties such as melting point, boiling point, and solubility.

Last, you learned to use IR spectroscopy as an indicator of the family to which an organic compound belongs. IR spectroscopy provides signatures (in the form of spectra) that suggest which functional groups are present in a molecule.

If you know the concepts in Chapters 1 and 2 well, you will be on your way to having the solid foundation you need for success in organic chemistry. Keep up the good work (including your diligent homework habits)!

The study aids for this chapter include key terms and concepts (which are hyperlinked to the glossary from the bold, blue terms in the *WileyPLUS* version of the book at wileyplus.com) and a Concept Map after the end-of-chapter problems.

KEY TERMS AND CONCEPTS PLUS

The key terms and concepts that are highlighted in bold, blue text within the chapter are defined in the glossary (at the back of the book) and have hyperlinked definitions in the accompanying *WileyPLUS* course (www.wileyplus.com).

PROBLEMS PLUS

Note to Instructors: Many of the homework problems are available for assignment via *WileyPLUS*, an online teaching and learning solution.

FUNCTIONAL GROUPS AND STRUCTURAL FORMULAS

2.29 Classify each of the following compounds as an alkane, alkene, alkyne, alcohol, aldehyde, amine, and so forth.

(a)

(b) $CH_3-C\equiv CH$

(c)

(d)

(e)
Obtained from oil of cloves

(f)
Sex attractant of the common housefly

2.30 Identify all of the functional groups in each of the following compounds:

(a)
Vitamin D₃

(b)
Aspartame

(c)
Amphetamine

(d)
Cholesterol

(e)
Demerol

(f)
A cockroach repellent found in cucumbers

(g)
A synthetic cockroach repellent

2.31 There are four alkyl bromides with the formula C_4H_9Br. Write their structural formulas and classify each as to whether it is a primary, secondary, or tertiary alkyl bromide.

2.32 There are seven isomeric compounds with the formula $C_4H_{10}O$. Write their structures and classify each compound according to its functional group.

2.33 Classify the following alcohols as primary, secondary, or tertiary:

(a) (b) (c) (d) (e)

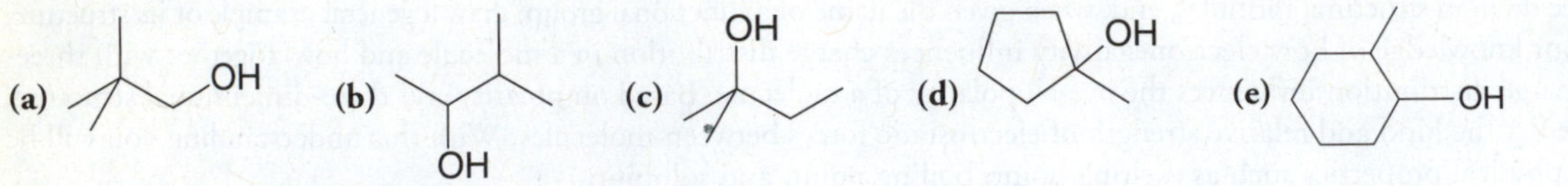

2.34 Classify the following amines as primary, secondary, or tertiary:

(a) (b) (c) (d) (e) (f)

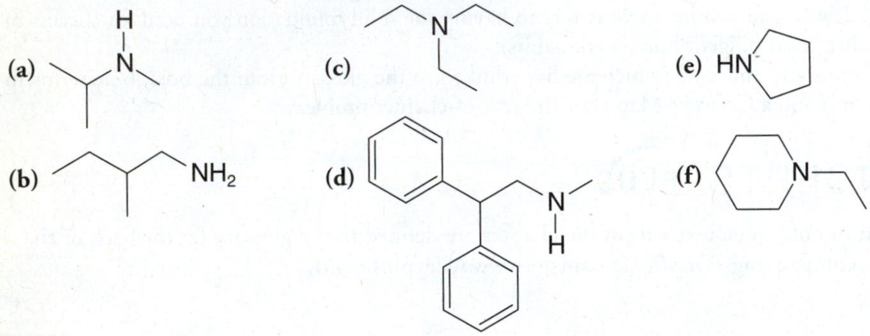

2.35 Write structural formulas for each of the following:

(a) Three ethers with the formula $C_4H_{10}O$.

(b) Three primary alcohols with the formula C_4H_8O.

(c) A secondary alcohol with the formula C_3H_6O.

(d) A tertiary alcohol with the formula C_4H_8O.

(e) Two esters with the formula $C_3H_6O_2$.

(f) Four primary alkyl halides with the formula $C_5H_{11}Br$.

(g) Three secondary alkyl halides with the formula $C_5H_{11}Br$.

(h) A tertiary alkyl halide with the formula $C_5H_{11}Br$.

(i) Three aldehydes with the formula $C_5H_{10}O$.

(j) Three ketones with the formula $C_5H_{10}O$.

(k) Two primary amines with the formula C_3H_9N.

(l) A secondary amine with the formula C_3H_9N.

(m) A tertiary amine with the formula C_3H_9N.

(n) Two amides with the formula C_2H_5NO.

2.36 Identify all of the functional groups in Crixivan, an important drug in the treatment of AIDS.

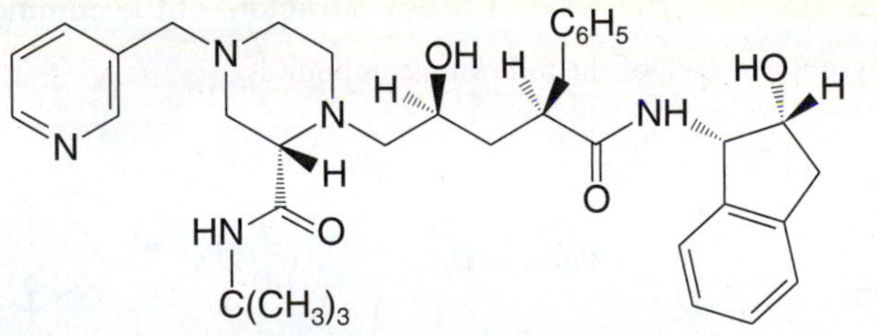

Crixivan (an HIV protease inhibitor)

2.37 Identify all of the functional groups in the following molecule.

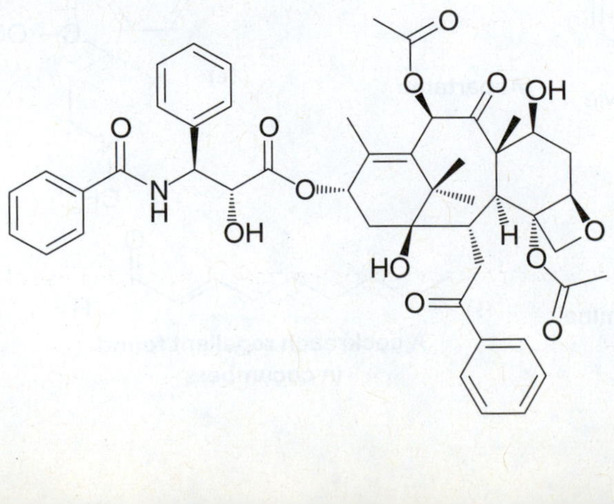

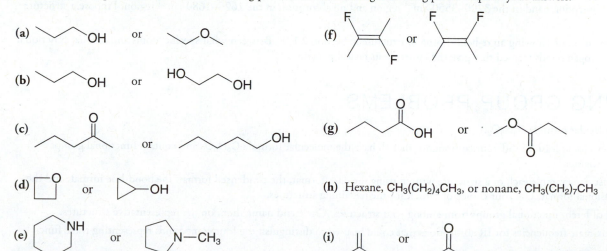

PHYSICAL PROPERTIES

2.38 (a) Indicate the hydrophobic and hydrophilic parts of vitamin A and comment on whether you would expect it to be soluble in water. (b) Do the same for vitamin B_3 (also called niacin).

Vitamin A   **Vitamin B_3 or niacin**

2.39 Hydrogen fluoride has a dipole moment of 1.83 D; its boiling point is 19.34 °C. Ethyl fluoride (CH_3CH_2F) has an almost identical dipole moment and has a larger molecular weight, yet its boiling point is −37.7 °C. Explain.

2.40 Why does one expect the cis isomer of an alkene to have a higher boiling point than the trans isomer?

2.41 Cetylethyldimethylammonium bromide is the common name for

Br^-

, a compound with antiseptic properties. Predict its solubility

behavior in water and in diethyl ether.

2.42 Which of the following solvents should be capable of dissolving ionic compounds?

(a) Liquid SO_2 (b) Liquid NH_3 (c) Benzene   (d) CCl_4

2.43 Write a three-dimensional formula for each of the following molecules using the wedge–dashed wedge–line formalism. If the molecule has a net dipole moment, indicate its direction with an arrow, ⟷. If the molecule has no net dipole moment, you should so state. (You may ignore the small polarity of C—H bonds in working this and similar problems.)

(a) CH_3F (c) CHF_3 (e) CH_2FCl (g) BeF_2 (i) CH_3OH

(b) CH_2F_2 (d) CF_4 (f) BCl_3 (h) CH_3OCH_3 (j) CH_2O

2.44 Consider each of the following molecules in turn: (a) dimethyl ether, $(CH_3)_2O$; (b) trimethylamine, $(CH_3)_3N$; (c) trimethylboron, $(CH_3)_3B$; and (d) dimethylberyllium, $(CH_3)_2Be$. Describe the hybridization state of the central atom (i.e., O, N, B, or Be) of each molecule, tell what bond angles you would expect at the central atom, and state whether the molecule would have a dipole moment.

2.45 Analyze the statement: For a molecule to be polar, the presence of polar bonds is necessary, but it is not a sufficient requirement.

2.46 Which compound in each of the following pairs would have the higher boiling point? Explain your answers.

(a)   OH   or   O

(b)   OH   or   HO   OH

(c)   (ketone structure)   or   OH

(d)   (oxetane)   or   (cyclopropanol) OH

(e)   NH   or   N—CH₃

(f)   (fluoroalkene)   or   (difluoroalkene)

(g)   OH   or   O (ester)

(h) Hexane, $CH_3(CH_2)_4CH_3$, or nonane, $CH_3(CH_2)_7CH_3$

(i)   (alkene)   or   (ketone)

IR SPECTROSCOPY

2.47 Predict the key IR absorption bands whose presence would allow each compound in pairs (a), (c), (d), (e), (g), and (i) from Problem 2.46 to be distinguished from each other.

2.48 The IR spectrum of propanoic acid (Fig. 2.17) indicates that the absorption for the O—H stretch of the carboxylic acid functional group is due to a hydrogen-bonded form. Draw the structure of two propanoic acid molecules showing how they could dimerize via hydrogen bonding.

2.49 In infrared spectra, the carbonyl group is usually indicated by a single strong and sharp absorption. However, in the case of carboxylic acid anhydrides, $R-\overset{\underset{||}{O}}{C}-O-\overset{\underset{||}{O}}{C}-R$, two peaks are observed even though the two carbonyl groups are chemically equivalent.

Explain this fact, considering what you know about the IR absorption of primary amines.

MULTICONCEPT PROBLEMS

2.50 Write structural formulas for four compounds with the formula C_3H_6O and classify each according to its functional group. Predict IR absorption frequencies for the functional groups you have drawn.

2.51 There are four amides with the formula C_3H_7NO. **(a)** Write their structures. **(b)** One of these amides has a melting and a boiling point that are substantially lower than those of the other three. Which amide is this? Explain your answer. **(c)** Explain how these amides could be differentiated on the basis of their IR spectra.

2.52 Write structures for all compounds with molecular formula C_4H_6O that would not be expected to exhibit infrared absorption in the $3200–3550\text{-cm}^{-1}$ and $1620–1780\text{-cm}^{-1}$ regions.

2.53 Cyclic compounds of the general type shown here are called lactones. What functional group does a lactone contain?

CHALLENGE PROBLEMS

2.54 Two constitutional isomers having molecular formula C_4H_6O are both symmetrical in structure. In their infrared spectra, neither isomer when in dilute solution in CCl_4 (used because it is nonpolar) has absorption in the 3600-cm^{-1} region. Isomer A has absorption bands at approximately 3080, 1620, and 700 cm^{-1}. Isomer B has bands in the 2900-cm^{-1} region and at 1780 cm^{-1}. Propose a structure for A and two possible structures for B.

2.55 When two substituents are on the same side of a ring skeleton, they are said to be cis, and when on opposite sides, trans (analogous to use of those terms with 1,2-disubstituted alkene isomers). Consider stereoisomeric forms of 1,2-cyclopentanediol (compounds having a five-membered ring and hydroxyl groups on two adjacent carbons that are cis in one isomer and trans in the other). At high dilution, both isomers have an infrared absorption band at approximately 3626 cm^{-1} but only one isomer has a band at 3572 cm^{-1}.
(a) Assume for now that the cyclopentane ring is coplanar (the interesting actuality will be studied later) and then draw and label the two isomers using the wedge–dashed wedge method of depicting the OH groups. **(b)** Designate which isomer will have the 3572-cm^{-1} band and explain its origin.

2.56 Compound C is asymmetric, has molecular formula $C_5H_{10}O$, and contains two methyl groups and a 3° functional group. It has a broad infrared absorption band in the $3200–3550\text{-cm}^{-1}$ region and no absorption in the $1620–1680\text{-cm}^{-1}$ region. Propose a structure for C.

2.57 Examine the diagram showing an α-helical protein structure in Section 2.13E. Between what specific atoms and of what functional groups are the hydrogen bonds formed that give the molecule its helical structure?

LEARNING GROUP PROBLEMS

Consider the molecular formula $C_4H_8O_2$.

1. Write structures for at least 15 different compounds that all have the molecular formula $C_4H_8O_2$ and contain functional groups presented in this chapter.

2. Provide at least one example each of a structure written using the dash format, the condensed format, the bond-line format, and the full three-dimensional format. Use your choice of format for the remaining structures.

3. Identify four different functional groups from among your structures. Circle and name them on the representative structures.

4. Predict approximate frequencies for IR absorptions that could be used to distinguish the four compounds representing these functional groups.

5. If any of the 15 structures you drew have atoms where the formal charge is other than zero, indicate the formal charge on the appropriate atom(s) and the overall charge for the molecule.

6. Identify which types of intermolecular forces would be possible in pure samples of all 15 compounds.

7. Pick five formulas you have drawn that represent a diversity of structures, and predict their order with respect to trend in increasing boiling point.

8. Explain your order of predicted boiling points on the basis of intermolecular forces and polarity.

[C O N C E P T   M A P]

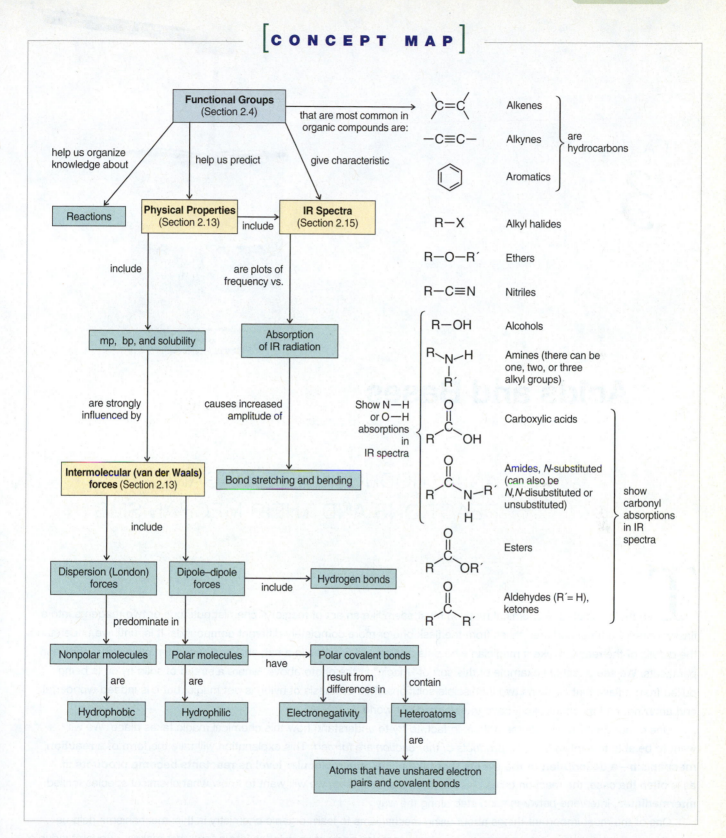

3

Acids and Bases

AN INTRODUCTION TO ORGANIC REACTIONS AND THEIR MECHANISMS

To the uninitiated, a chemical reaction must seem like an act of magic. A chemist puts one or two reagents into a flask, waits for a time, and then takes from the flask one or more completely different compounds. It is, until we understand the details of the reaction, like a magician who puts apples and oranges in a hat, shakes it, and then pulls out rabbits and parakeets. We see a real-life example of this sort of "magic" in the photo above, where a strand of solid nylon is being pulled from a flask that contains two immiscible solutions. This synthesis of nylon is not magic, but it is indeed wonderful and amazing, and reactions like it have transformed our world.

One of our goals in this course will be, in fact, to try to understand how this chemical magic takes place. We will want to be able to explain *how the products of the reaction are formed*. This explanation will take the form of a **reaction mechanism—a description of the events that take place on a molecular level as reactants become products**. If, as is often the case, the reaction takes place in more than one step, we will want to know what chemical species, called **intermediates**, intervene between each step along the way.

One of the most important things about using mechanisms to learn organic chemistry is this: mechanisms help us organize what otherwise might be an overwhelmingly complex body of knowledge into a form that makes it understandable. There are millions of organic compounds now known, and there are millions of reactions that these compounds undergo. If we had to learn them all by rote memorization, then we would soon give up. But, we don't have to do this. In the same way

PHOTO CREDITS: (making nylon) Charles D. WInters/Photo Researchers, Inc.; (magician's hand) © AndyL/iStockphoto

that functional groups help us organize compounds in a comprehensible way, mechanisms help us organize reactions. Fortunately, too, there are a relatively small number of basic mechanisms.

IN THIS CHAPTER WE WILL CONSIDER:

- rules that show how to classify reactive groups within molecules from the standpoints of acids and bases as well as from electron-rich and electron-poor domains
- the step-by-step processes of a chemical reaction and how to codify these processes into a few specific, easy-to-understand types

[**WHY** DO THESE TOPICS MATTER?]   At the end of the chapter, we will show a rare case where an important discovery that truly changed the world was made without any knowledge of these principles. However, the rare occurrence of such events argues for why real advances require a core understanding of the topics in this chapter.

3.1 ACID–BASE REACTIONS

We begin our study of chemical reactions and mechanisms by examining some of the basic principles of acid–base chemistry. There are several reasons for doing this:

- Many of the reactions that occur in organic chemistry are either acid–base reactions themselves or they involve an acid–base reaction at some stage.
- Acid–base reactions are simple fundamental reactions that will enable you to see how chemists use curved arrows to represent mechanisms of reactions and how they depict the processes of bond breaking and bond making that occur as molecules react.

3.1A Brønsted–Lowry Acids and Bases

Two classes of acid–base reactions are fundamental in organic chemistry: Brønsted–Lowry and Lewis acid–base reactions. We start our discussion with Brønsted–Lowry acid–base reactions.

- **Brønsted–Lowry acid–base** reactions involve the transfer of protons.
- A **Brønsted–Lowry acid** is a substance that can donate (or lose) a proton.
- A **Brønsted–Lowry base** is a substance that can accept (or remove) a proton.

Let us consider some examples.

Hydrogen chloride (HCl), in its pure form, is a gas. When HCl gas is bubbled into water, the following reaction occurs.

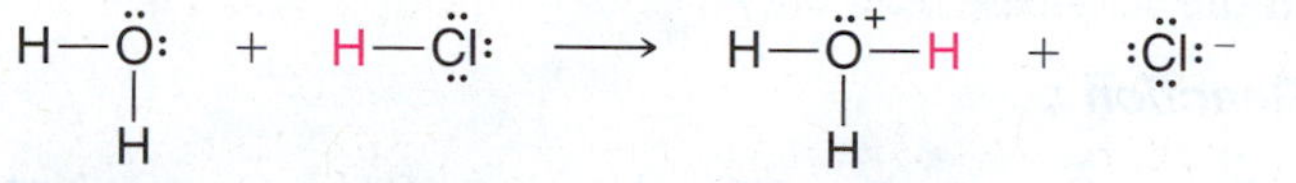

$$H-\overset{..}{\underset{|}{O}}:\ +\ H-\overset{..}{\underset{..}{Cl}}: \longrightarrow H-\overset{+}{\underset{|}{O}}-H\ +\ :\overset{..}{\underset{..}{Cl}}:^{-}$$

| Base (proton acceptor) | Acid (proton donor) | Conjugate acid of H_2O | Conjugate base of HCl |

In this reaction hydrogen chloride donates a proton; therefore it acts as a Brønsted–Lowry acid. Water accepts a proton from hydrogen chloride; thus water serves as a Brønsted–Lowry base. The products are a hydronium ion (H_3O^+) and chloride ion (Cl^-).

Just as we classified the reactants as either an acid or a base, we also classify the products in a specific way.

- The molecule or **ion** that forms when an acid loses its proton is called the **conjugate base** of that acid. In the above example, chloride ion is the conjugate base.
- The molecule or ion that forms when a base accepts a proton is called the **conjugate acid**. Hydronium ion is the conjugate acid of water.

The color of hydrangea flowers depends, in part, on the relative acidity of their soil.

Media Bakery

Hydrogen chloride is considered a strong acid because transfer of its proton in water proceeds essentially to completion. Other strong acids that completely transfer a proton when dissolved in water are hydrogen iodide, hydrogen bromide, and sulfuric acid.

$$HI + H_2O \longrightarrow H_3O^+ + I^-$$
$$HBr + H_2O \longrightarrow H_3O^+ + Br^-$$
$$H_2SO_4 + H_2O \longrightarrow H_3O^+ + HSO_4^-$$
$$HSO_4^- + H_2O \rightleftharpoons H_3O^+ + SO_4^{2-}$$

- The extent to which an acid transfers protons to a base, such as water, is a measure of its strength as an acid. Acid strength is therefore a measure of the percentage of ionization and *not* of concentration.

Sulfuric acid is called a diprotic acid because it can transfer two protons. Transfer of the first proton occurs completely, while the second is transferred only to the extent of about 10% (hence the equilibrium arrows in the equation for the second proton transfer).

3.1B Acids and Bases in Water

- Hydronium ion is the strongest acid that can exist in water to any significant extent. Any acid stronger than hydronium ion will simply transfer its proton to a water molecule to form hydronium ions.
- Hydroxide ion is the strongest base that can exist in water to any significant extent. Any base stronger than hydroxide will remove a proton from water to form hydroxide ions.

When an ionic compound dissolves in water the ions are solvated. With sodium hydroxide, for example, the positive sodium ions are stabilized by interaction with unshared electron pairs of water molecules, and the hydroxide ions are stabilized by hydrogen bonding of their unshared electron pairs with the partially positive hydrogens of water molecules.

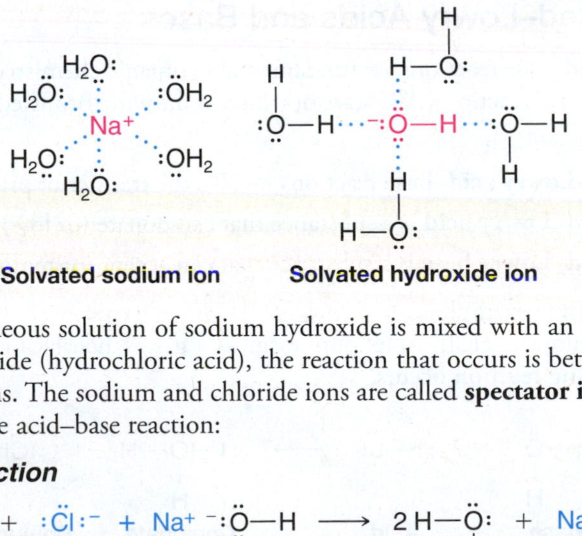

Solvated sodium ion   **Solvated hydroxide ion**

When an aqueous solution of sodium hydroxide is mixed with an aqueous solution of hydrogen chloride (hydrochloric acid), the reaction that occurs is between hydronium and hydroxide ions. The sodium and chloride ions are called **spectator ions** because they play no part in the acid–base reaction:

Total Ionic Reaction

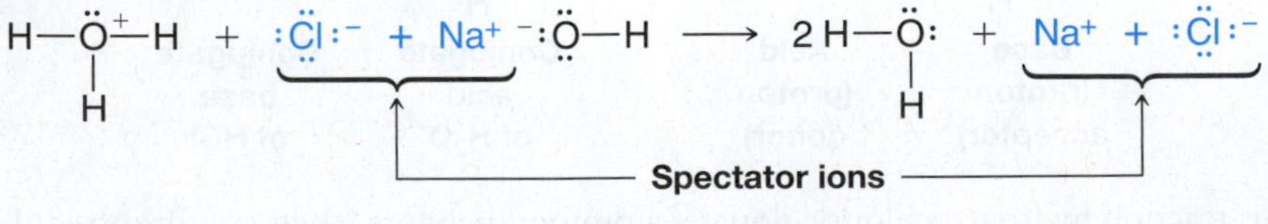

────── Spectator ions ──────

Net Reaction

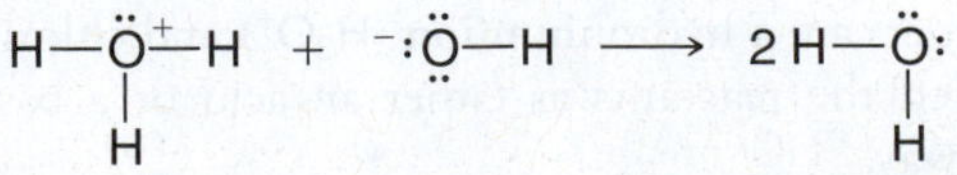

What we have just said about hydrochloric acid and aqueous sodium hydroxide is true when solutions of all aqueous strong acids and bases are mixed. The net **ionic reaction** is simply

$$H_3O^+ + HO^- \longrightarrow 2 H_2O$$

3.2 HOW TO USE CURVED ARROWS IN ILLUSTRATING REACTIONS

Up to this point we have not indicated how bonding changes occur in the reactions we have presented, but this can easily be done using curved-arrow notation. **Curved arrows** show the direction of electron flow in a reaction mechanism.

1. Draw the curved arrow so that it points from the source of an electron pair to the atom receiving the pair. (Curved arrows can also show the movement of single electrons. We shall discuss reactions of this type in a later chapter.)

2. Always show the flow of electrons from a site of higher electron density to a site of lower electron density.

3. **Never** use curved arrows to show the movement of atoms. Atoms are assumed to follow the flow of the electrons.

4. Make sure that the movement of electrons shown by the curved arrow does not violate the octet rule for elements in the second row of the periodic table.

The reaction of hydrogen chloride with water provides a simple example of how to use curved arrow notation. Here we invoke the first of many "A Mechanism for the Reaction" boxes, in which we show every key step in a mechanism using color-coded formulas accompanied by explanatory captions.

A MECHANISM FOR THE REACTION — Reaction of Water with Hydrogen Chloride: The Use of Curved Arrows

Reaction

$$H_2O + HCl \longrightarrow H_3O^+ + Cl^-$$

Mechanism

A water molecule uses one of the nonbonding electron pairs to form a bond to a proton of HCl. The bond between the hydrogen and chlorine breaks, and the electron pair goes to the chlorine atom.

This leads to the formation of a hydronium ion and a chloride ion.

Helpful Hint

Curved arrows point *from* electrons *to* the atom receiving the electrons.

The curved arrow begins with a covalent bond or unshared electron pair (a site of higher electron density) and points toward a site of electron deficiency. We see here that as the water molecule collides with a hydrogen chloride molecule, it uses one of its unshared electron pairs (shown in blue) to form a bond to the proton of HCl. This bond forms because the negatively charged electrons of the oxygen atom are attracted to the positively charged proton. As the bond between the oxygen and the proton forms, the hydrogen–chlorine bond of HCl breaks, and the chlorine of HCl departs with the electron pair that formerly bonded it to the proton. (If this did not happen, the proton would end up forming two covalent bonds, which, of course, a proton cannot do.) We, therefore, use a curved arrow to show the bond cleavage as well. By pointing from the bond to the chlorine, the arrow indicates that the bond breaks and the electron pair leaves with the chloride ion.

The following acid–base reactions give other examples of the use of the curved-arrow notation:

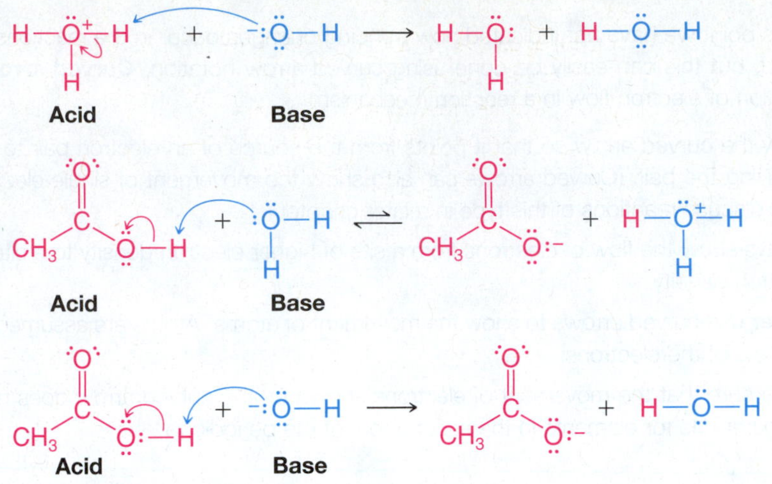

Acid   **Base**

Acid   **Base**

Acid   **Base**

●●● SOLVED PROBLEM 3.1

Add curved arrows to the following reactions to indicate the flow of electrons for all of the bond-forming and bond-breaking steps.

(a)

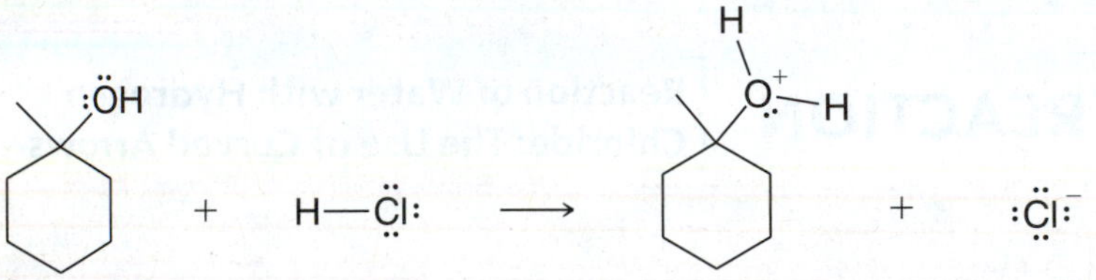

(b)

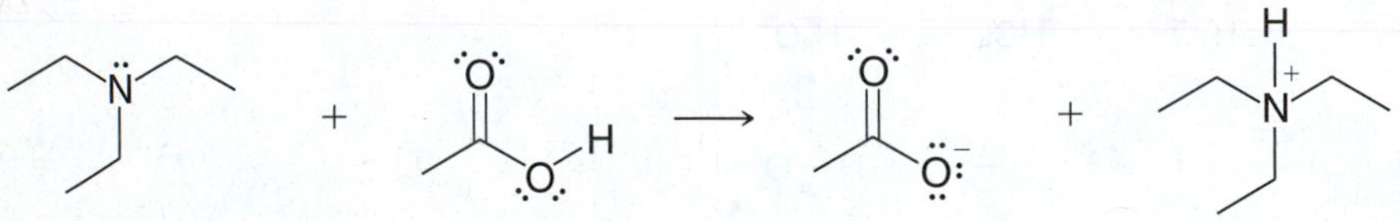

STRATEGY AND ANSWER: Recall the rules for use of curved arrows presented at the beginning of Section 3.2. Curved arrows point from the source of an electron pair to the atom receiving the pair, and always point from a site of higher electron density to a site of lower electron density. We must also not exceed two electrons for a hydrogen atom, or an octet of electrons for any elements in the second row of the periodic table. We must also account for the formal charges on atoms and write equations whose charges are balanced.

In (a), the hydrogen atom of HCl is partially positive (electrophilic) due to the electronegativity of the chlorine atom. The alcohol oxygen is a source of electrons (a Lewis base) that can be given to this partially positive proton. The proton must lose a pair of electrons as it gains a pair, however, and thus the chloride ion accepts a pair of electrons from the bond it had with the hydrogen atom as the hydrogen becomes bonded to the alcohol oxygen.

(a)

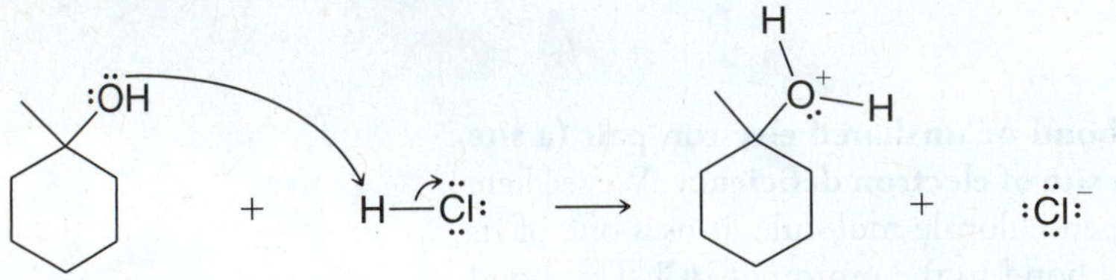

In (b), the carboxylic acid hydrogen is partially positive and therefore electrophilic, and the amine provides an unshared pair of electrons that forms a bond with the carboxylic acid hydrogen, causing departure of a carboxylate anion.

(b)

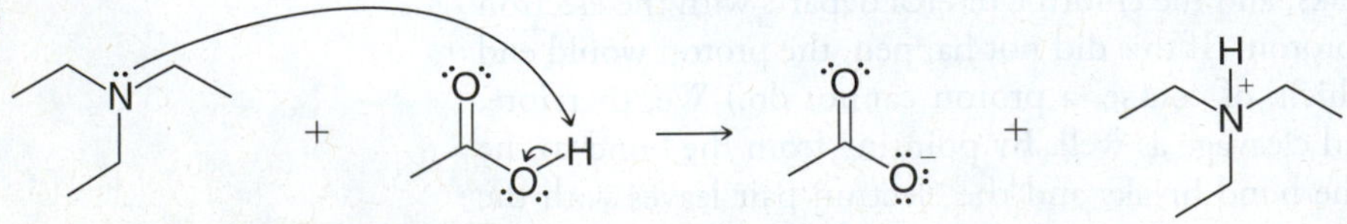

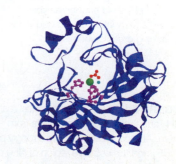

Add curved arrows to the following reactions to indicate the flow of electrons for all of the bond-forming and bond-breaking steps.

(a)

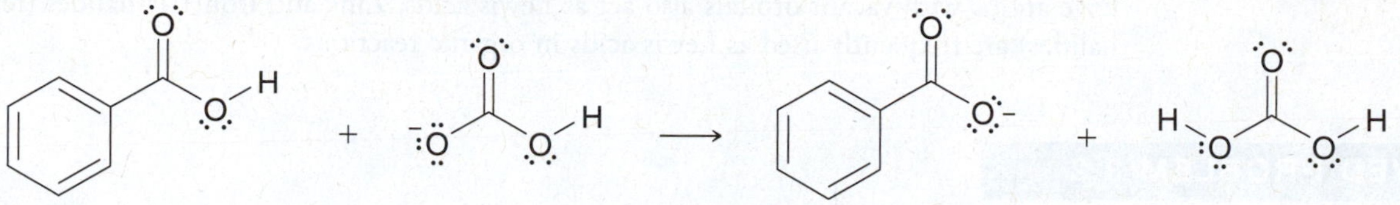

(b)

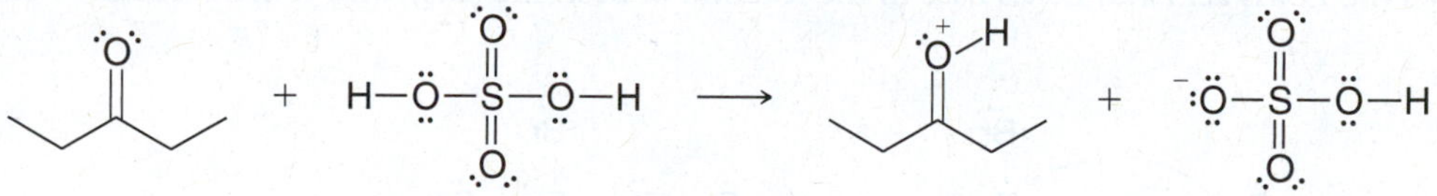

3.3 LEWIS ACIDS AND BASES

In 1923 G. N. Lewis proposed a theory that significantly broadened the understanding of acids and bases. As we go along we shall find that an understanding of **Lewis acid– base theory** is exceedingly helpful to understanding a variety of organic reactions. Lewis proposed the following definitions for acids and bases.

- Acids are electron pair acceptors.
- Bases are electron pair donors.

In Lewis acid–base theory, proton donors are not the only acids; many other species are acids as well. Aluminum chloride, for example, reacts with ammonia in the same way that a proton donor does. Using curved arrows to show the donation of the electron pair of ammonia (the Lewis base), we have the following examples:

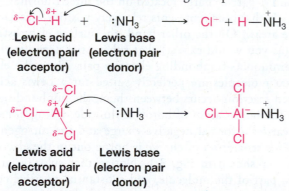

In the reaction with hydrogen chloride above, notice that the electron pair acceptor (the proton) must also lose an electron pair as the new bond is formed with nitrogen. This is necessary because the hydrogen atom had a full valence shell of electrons at the start. On the other hand, because the valence shell of the aluminum atom in aluminum chloride was not full at the beginning (it had only a sextet of valence electrons), it can accept an electron pair without breaking any bonds. The aluminum atom actually achieves an octet by accepting the pair from nitrogen, although it gains a formal negative charge. When it accepts the electron pair, aluminum chloride is, in the Lewis definition, *acting as an acid.*

Bases are much the same in the Lewis theory and in the Brønsted–Lowry theory, because in the Brønsted–Lowry theory a base must donate a pair of electrons in order to accept a proton.

- The Lewis theory, by virtue of its broader definition of acids, allows acid–base theory to include all of the Brønsted–Lowry reactions and, as we shall see, a great many others. Most of the reactions we shall study in organic chemistry involve Lewis acid–base interactions, and a sound understanding of Lewis acid–base chemistry will help greatly.

Helpful Hint

Verify for yourself that you can calculate the formal charges in these structures.

Carbonic anhydrase

A zinc ion acts as a Lewis acid in the mechanism of the enzyme carbonic anhydrase (Chapter 24).

Any *electron-deficient atom* can act as a Lewis acid. Many compounds containing group IIIA elements such as boron and aluminum are Lewis acids because group IIIA atoms have only a sextet of electrons in their outer shell. Many other compounds that have atoms with vacant orbitals also act as Lewis acids. Zinc and iron(III) halides (ferric halides) are frequently used as Lewis acids in organic reactions.

••• SOLVED PROBLEM 3.2

Write an equation that shows the Lewis acid and Lewis base in the reaction of bromine (Br_2) with ferric bromide ($FeBr_3$).

ANSWER:

$$:\ddot{Br}-\ddot{Br}: \quad + \quad \underset{\text{Lewis acid}}{\overset{:\ddot{Br}:}{\underset{:\ddot{Br}\quad\ddot{Br}:}{Fe}}} \quad \longrightarrow \quad :\ddot{Br}-\overset{+}{\ddot{Br}}-\underset{:\ddot{Br}:}{\overset{:\ddot{Br}:}{\overset{-}{Fe}}}-\ddot{Br}:$$

Lewis base   **Lewis acid**

3.3A Opposite Charges Attract

- In Lewis acid–base theory, as in many organic reactions, the attraction of oppositely charged species is fundamental to reactivity.

As one further example, we consider boron trifluoride, an even more powerful Lewis acid than aluminum chloride, and its reaction with ammonia. The calculated structure for boron trifluoride in Fig. 3.1 shows electrostatic potential at its van der Waals surface (like that in Section 2.2A for HCl). It is obvious from this figure (and you should be able to predict this) that BF_3 has substantial positive charge centered on the boron atom and negative charge located on the three fluorines. (The convention in these structures is that blue represents relatively positive areas and red represents relatively negative areas.) On the other hand, the surface electrostatic potential for ammonia shows (as you would expect) that substantial negative charge is localized in the region of ammonia's nonbonding electron pair. Thus, the electrostatic properties of these two molecules are perfectly suited for a Lewis acid–base reaction. When the expected reaction occurs between them, the nonbonding electron pair of ammonia attacks the boron atom of boron trifluoride, filling boron's valence shell. The boron now carries a formal negative charge and the nitrogen carries a formal positive charge. This separation of charge is borne out in the electrostatic potential map for the product shown in Fig. 3.1. Notice that substantial negative charge resides in the BF_3 part of the molecule, and substantial positive charge is localized near the nitrogen.

Although calculated electrostatic potential maps like these illustrate charge distribution and molecular shape well, it is important that you are able to draw the same conclusions based on what you would have predicted about the structures of BF_3 and NH_3 and their reaction product using orbital hybridization (Sections 1.13–1.15), VSEPR models (Section 1.17), consideration of formal charges (Section 1.5), and electronegativity (Sections 1.3A and 2.2).

Helpful Hint

The need for a firm understanding of structure, formal charges, and electronegativity can hardly be emphasized enough as you build a foundation of knowledge for learning organic chemistry.

FIGURE 3.1 Electrostatic potential maps for BF_3 and NH_3 and the product that results from reaction between them. Attraction between the strongly positive region of BF_3 and the negative region of NH_3 causes them to react. The electrostatic potential map for the product shows that the fluorine atoms draw in the electron density of the formal negative charge, and the nitrogen atom, with its hydrogens, carries the formal positive charge.

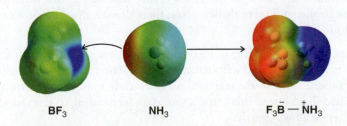

BF_3 NH_3 $F_3\bar{B}-\overset{+}{N}H_3$

PRACTICE PROBLEM 3.2

Write equations showing the Lewis acid–base reaction that takes place when:

(a) Methanol (CH_3OH) reacts with BF_3.

(b) Chloromethane (CH_3Cl) reacts with $AlCl_3$.

(c) Dimethyl ether (CH_3OCH_3) reacts with BF_3.

PRACTICE PROBLEM 3.3

Which of the following are potential Lewis acids and which are potential Lewis bases?

(a) $CH_3CH_2-\overset{\displaystyle CH_3}{\underset{\displaystyle CH_3}{\overset{|}{\underset{|}{N}}}}-CH_3$

(b) $H_3C-\overset{\displaystyle CH_3}{\underset{\displaystyle CH_3}{\overset{|}{\underset{|}{C^+}}}}$

(c) $(C_6H_5)_3P:$

(d) $:\ddot{Br}:^-$

(e) $(CH_3)_3B$

(f) $H:^-$

3.4 HETEROLYSIS OF BONDS TO CARBON: CARBOCATIONS AND CARBANIONS

Heterolysis of a bond to a carbon atom can lead to either of two ions: either to an ion with a positive charge on the carbon atom, called a **carbocation**, or to an ion with a negatively charged carbon atom, called a **carbanion**:

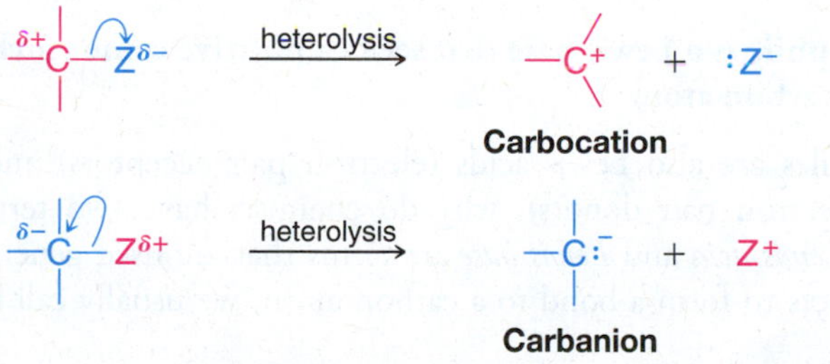

Carbocation

Carbanion

- Carbocations are electron deficient. They have only six electrons in their valence shell, and because of this, carbocations are Lewis acids.

In this way they are like BF_3 and $AlCl_3$. Most carbocations are also short-lived and highly reactive. They occur as intermediates in some organic reactions. Carbocations react rapidly with Lewis bases—with molecules or ions that can donate the electron pair that they need to achieve a stable octet of electrons (i.e., the electronic configuration of a noble gas):

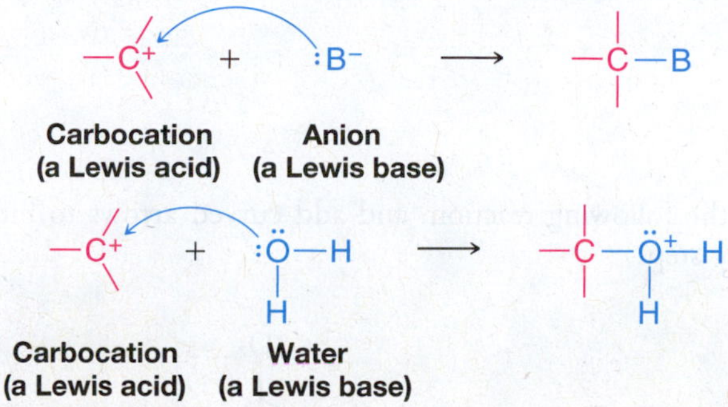

Carbocation   Anion
(a Lewis acid)   (a Lewis base)

Carbocation   Water
(a Lewis acid)   (a Lewis base)

- **Carbanions** are electron rich. They are anions and have an unshared electron pair. Carbanions, therefore, are **Lewis bases and react accordingly** (Section 3.3).

3.4A Electrophiles and Nucleophiles

Because carbocations are electron-seeking reagents chemists call them **electrophiles** (meaning electron-loving).

- **Electrophiles** are reagents that seek electrons so as to achieve a stable shell of electrons like that of a noble gas.
- **All Lewis acids are electrophiles**. By accepting an electron pair from a Lewis base, a carbocation fills its valence shell.

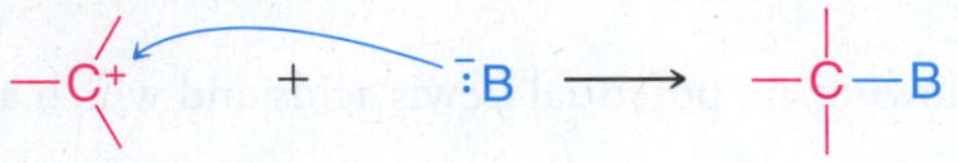

Carbocation
Lewis acid and
electrophile

Lewis
base

- **Carbon atoms that are electron poor because of bond polarity, but are not carbocations, can also be electrophiles**. They can react with the electron-rich centers of Lewis bases in reactions such as the following:

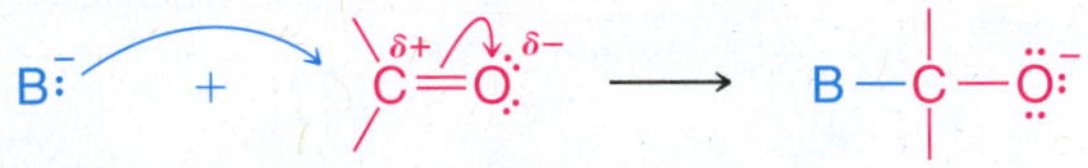

Lewis
base

Lewis acid
(electrophile)

Carbanions are Lewis bases. Carbanions seek a proton or some other positive center to which they can donate their electron pair and thereby neutralize their negative charge.

When a Lewis base *seeks a positive center other than a proton, especially that of a carbon atom*, chemists call it a **nucleophile** (meaning nucleus loving; the *nucleo-* part of the name comes from *nucleus*, the positive center of an atom).

- **A nucleophile** is a Lewis base that seeks a positive center such as a positively charged carbon atom.

Since electrophiles are also Lewis acids (electron pair acceptors) and nucleophiles are Lewis bases (electron pair donors), why do chemists have two terms for them? The answer is that *Lewis acid* and *Lewis base* are terms that are used generally, but when one or the other reacts to form a bond to a carbon atom, we usually call it an *electrophile* or a *nucleophile*.

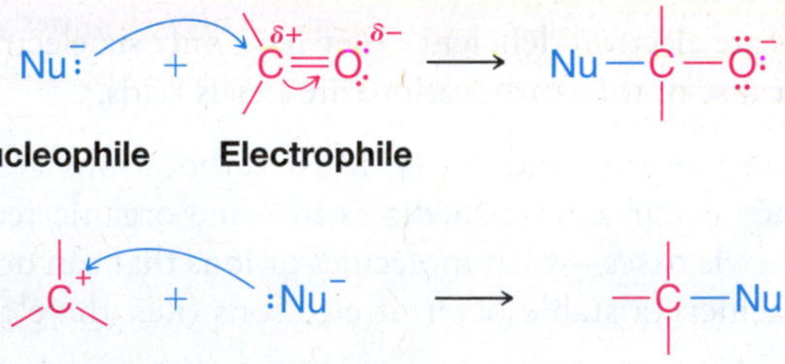

Nucleophile   Electrophile

Electrophile   Nucleophile

●●● **SOLVED PROBLEM 3.3**

Identify the electrophile and the nucleophile in the following reaction, and add curved arrows to indicate the flow of electrons for the bond-forming and bond-breaking steps.

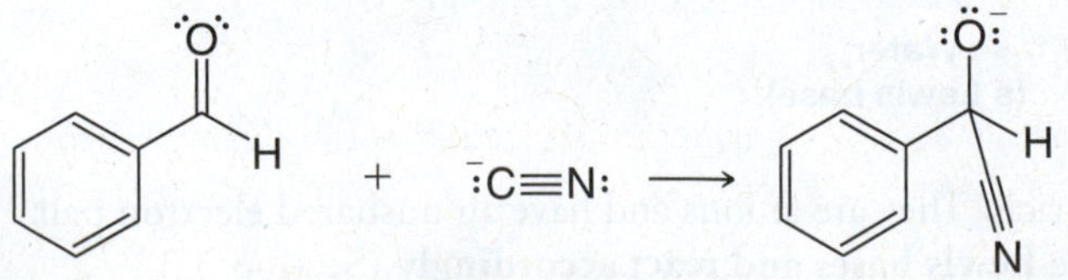

STRATEGY AND ANSWER: The aldehyde carbon is electrophilic due to the electronegativity of the carbonyl oxygen. The cyanide anion acts as a Lewis base and is the nucleophile, donating an electron pair to the carbonyl carbon, and causing an electron pair to shift to the oxygen so that no atom has more than an octet of electrons.

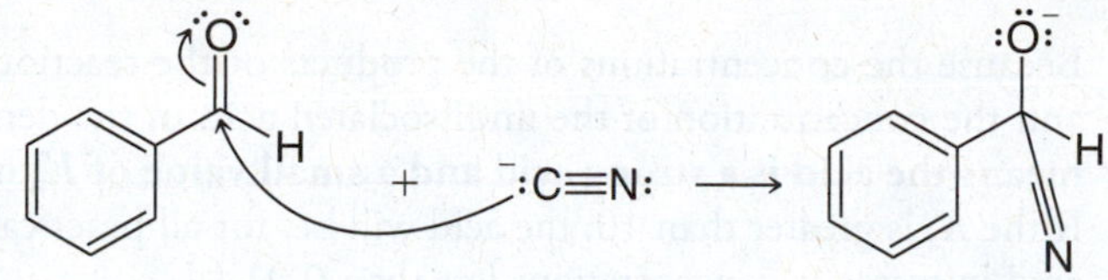

Use the curved-arrow notation to write the reaction that would take place between dimethylamine (CH$_3$)$_2$NH and boron trifluoride. Identify the Lewis acid, Lewis base, nucleophile, and electrophile and assign appropriate formal charges.

PRACTICE PROBLEM 3.4

3.5 THE STRENGTH OF BRØNSTED–LOWRY ACIDS AND BASES: K_a AND pK_a

Many organic reactions involve the transfer of a proton by an acid–base reaction. An important consideration, therefore, is the relative strengths of compounds that could potentially act as Brønsted–Lowry acids or bases in a reaction.

In contrast to the strong acids, such as HCl and H$_2$SO$_4$, acetic acid is a much weaker acid. When acetic acid dissolves in water, the following reaction does not proceed to completion:

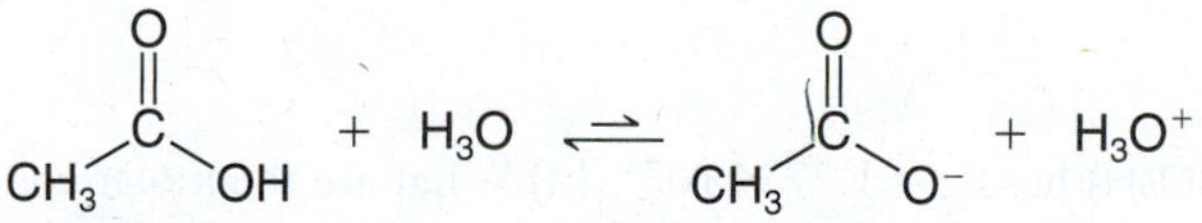

Experiments show that in a 0.1 M solution of acetic acid at 25 °C only about 1% of the acetic acid molecules ionize by transferring their protons to water. Therefore, acetic acid is a weak acid. As we shall see next, **acid strength** is characterized in terms of **acidity constant (K_a)** or pK_a values.

3.5A The Acidity Constant, K_a

Because the reaction that occurs in an aqueous solution of acetic acid is an equilibrium, we can describe it with an expression for the **equilibrium constant (K_{eq})**:

$$K_{eq} = \frac{[H_3O^+][CH_3CO_2^-]}{[CH_3CO_2H][H_2O]}$$

For dilute aqueous solutions, the concentration of water is essentially constant (~55.5 M), so we can rewrite the expression for the equilibrium constant in terms of a new constant (K_a) called the **acidity constant**:

$$K_a = K_{eq}[H_2O] = \frac{[H_3O^+][CH_3CO_2^-]}{[CH_3CO_2H]}$$

At 25 °C, the acidity constant for acetic acid is 1.76×10^{-5}.

We can write similar expressions for any weak acid dissolved in water. Using a generalized hypothetical acid (HA), the reaction in water is

$$HA \;+\; H_2O \;\rightleftharpoons\; H_3O^+ \;+\; A^-$$

and the expression for the acidity constant is

$$K_a = \frac{[H_3O^+][A^-]}{[HA]}$$

Because the concentrations of the products of the reaction are written in the numerator and the concentration of the undissociated acid in the denominator, **a large value of K_a means the acid is a strong acid and a small value of K_a means the acid is a weak acid**. If the K_a is greater than 10, the acid will be, for all practical purposes, completely dissociated in water at concentrations less than 0.01 M.

●●● SOLVED PROBLEM 3.4

Phenol (C_6H_5)OH has $K_a = 1.26 \times 10^{-10}$. **(a)** What is the molar concentration of hydronium ion in a 1.0 M solution of phenol? **(b)** What is the pH of the solution?

STRATEGY AND ANSWER: Use the equation for K_a for the equilibrium:

$$C_6H_5OH \ + \ H_2O \ \rightleftharpoons \ C_6H_5O^- \ + \ H_3O^+$$

Phenol   **Phenoxide**   **Hydronium**
                                  ion   **ion**

$$K_a = \frac{[H_3O]^+[C_6H_5O^-]}{[C_6H_5OH]} = 1.26 \times 10^{-10}$$

At equilibrium the concentration of hydronium ion will be the same as that of phenoxide ion, thus we can let them both equal x. Therefore

$$\frac{(x)(x)}{1.0} = \frac{x^2}{1.0} = 1.26 \times 10^{-10}$$

and

$$x = 1.1 \times 10^{-5}.$$

●●●

PRACTICE PROBLEM 3.5   Formic acid (HCO_2H) has $K_a = 1.77 \times 10^{-4}$. **(a)** What are the molar concentrations of the hydronium ion and formate ion (HCO_2^-) in a 0.1 M aqueous solution of formic acid? **(b)** What percentage of the formic acid is ionized?

3.5B Acidity and pK_a

Chemists usually express the acidity constant, K_a, as its negative logarithm, **pK_a**:

$$pK_a = -\log K_a$$

This is analogous to expressing the hydronium ion concentration as pH:

$$pH = -\log[H_3O^+]$$

For acetic acid the pK_a is 4.75:

$$pK_a = -\log(1.76 \times 10^{-5}) = -(-4.75) = 4.75$$

Notice that there is an inverse relationship between the magnitude of the pK_a and the strength of the acid.

Helpful Hint

K_a and pK_a are indicators of acid strengths.

• **The larger the value of the pK_a, the weaker is the acid.**

For example, acetic acid with p$K_a = 4.75$ is a weaker acid than trifluoroacetic acid with p$K_a = 0$ ($K_a = 1$). Hydrochloric acid with p$K_a = -7$ ($K_a = 10^7$) is a far stronger acid than trifluoroacetic acid. (It is understood that a positive pK_a is larger than a negative pK_a.)

$$CH_3CO_2H < CF_3CO_2H < HCl$$

$$pK_a = 4.75 \qquad pK_a = 0 \qquad pK_a = -7$$

Weak acid   **Very strong acid**

→ **Increasing acid strength**

Table 3.1 lists pK_a values for a selection of acids relative to water as the base.

TABLE 3.1 RELATIVE STRENGTH OF SELECTED ACIDS AND THEIR CONJUGATE BASES

	Acid	Approximate pK_a	Conjugate Base	
Strongest acid	$HSbF_6$	< -12	SbF_6^-	Weakest base
	HI	-10	I^-	
	H_2SO_4	-9	HSO_4^-	
	HBr	-9	Br^-	
	HCl	-7	Cl^-	
	$C_6H_5SO_3H$	-6.5	$C_6H_5SO_3^-$	
	$(CH_3)_2\overset{+}{O}H$	-3.8	$(CH_3)_2O$	
	$(CH_3)_2C=\overset{+}{O}H$	-2.9	$(CH_3)_2C=O$	
	$CH_3\overset{+}{O}H_2$	-2.5	CH_3OH	
	H_3O^+	-1.74	H_2O	
	HNO_3	-1.4	NO_3^-	
	CF_3CO_2H	0.18	$CF_3CO_2^-$	
	HF	3.2	F^-	
	$C_6H_5CO_2H$	4.21	$C_6H_5CO_2^-$	
	$C_6H_5NH_3^+$	4.63	$C_6H_5NH_2$	
	CH_3CO_2H	4.75	$CH_3CO_2^-$	
	H_2CO_3	6.35	HCO_3^-	
	$CH_3COCH_2COCH_3$	9.0	$CH_3CO\overset{-}{C}HCOCH_3$	
	NH_4^+	9.2	NH_3	
	C_6H_5OH	9.9	$C_6H_5O^-$	
	HCO_3^-	10.2	CO_3^{2-}	
	$CH_3NH_3^+$	10.6	CH_3NH_2	
	H_2O	15.7	HO^-	
	CH_3CH_2OH	16	$CH_3CH_2O^-$	
	$(CH_3)_3COH$	18	$(CH_3)_3CO^-$	
	CH_3COCH_3	19.2	$^-CH_2COCH_3$	
	$HC\equiv CH$	25	$HC\equiv C^-$	
	$C_6H_5NH_2$	31	$C_6H_5NH^-$	
	H_2	35	H^-	
	$(i\text{-}Pr)_2NH$	36	$(i\text{-}Pr)_2N^-$	
	NH_3	38	$^-NH_2$	
	$CH_2=CH_2$	44	$CH_2=CH^-$	
Weakest acid	CH_3CH_3	50	$CH_3CH_2^-$	Strongest base

↑ Increasing acid strength   ↓ Increasing base strength

The values in the middle pK_a range of Table 3.1 are the most accurate because they can be measured in aqueous solution. Special methods must be used to estimate the pK_a values for the very strong acids at the top of the table and for the very weak acids at the bottom.* The pK_a values for these very strong and weak acids are therefore approximate. All of the acids that we shall consider in this book will have strengths in between that of ethane (an extremely weak acid) and that of $HSbF_6$ (an acid that is so strong that it is called a "superacid"). As you examine Table 3.1, take care not to lose sight of the vast range of acidities that it represents (a factor of 10^{62}).

*Acids that are stronger than a hydronium ion and bases that are stronger than a hydroxide ion react completely with water (a phenomenon called the **leveling effect**; see Sections 3.1B and 3.14). Therefore, it is not possible to measure acidity constants for these acids in water. Other solvents and special techniques are used, but we do not have the space to describe those methods here.

⦁⦁⦁

PRACTICE PROBLEM 3.6   **(a)** An acid (HA) has $K_a = 10^{-7}$. What is its pK_a?   **(b)** Another acid (HB) has $K_a = 5$. What is its pK_a?   **(c)** Which is the stronger acid?

Water, itself, is a very weak acid and undergoes self-ionization even in the absence of acids and bases:

$$\text{H}-\overset{..}{\text{O}}: \; + \; \text{H}-\overset{..}{\underset{|}{\text{O}}}: \; \rightleftharpoons \; \text{H}-\overset{+}{\underset{|}{\text{O}}}-\text{H} \; + \; :\overset{..}{\underset{..}{\text{O}}}-\text{H}$$

In pure water at 25 °C, the concentrations of hydronium and hydroxide ions are equal to 10^{-7} M. Since the concentration of water in pure water is 55.5 M, we can calculate the K_a for water.

$$K_a = \frac{[\text{H}_3\text{O}^+][\text{OH}^-]}{[\text{H}_2\text{O}]} \qquad K_a = \frac{(10^{-7})(10^{-7})}{55.5} = 1.8 \times 10^{-16} \qquad pK_a = 15.7$$

⦁⦁⦁ **SOLVED PROBLEM 3.5**

Show calculations proving that the pK_a of the hydronium ion (H_3O^+) is -1.74 as given in Table 3.1.

STRATEGY AND ANSWER: When H_3O^+ acts as an acid in aqueous solution, the equilibrium is

$$\text{H}_3\text{O}^+ \; + \; \text{H}_2\text{O} \; \rightleftharpoons \; \text{H}_2\text{O} \; + \; \text{H}_3\text{O}^+$$

and K_a is equal to the molar concentration of water;

$$K_a = \frac{[\text{H}_2\text{O}][\text{H}_3\text{O}^+]}{[\text{H}_3\text{O}^+]} = [\text{H}_2\text{O}]$$

The molar concentration of H_2O in pure H_2O is equal to the number of moles of H_2O (MW = 18 g/mol) in 1000 g (one liter) of water. That is, $[\text{H}_2\text{O}] = 1000 \text{ g L}^{-1}/18 \text{ g/mole}^{-1} = 55.5$. Therefore, $K_a = 55.5$. The $pK = -\log 55.5 = -1.74$.

3.5C Predicting the Strength of Bases

In our discussion so far we have dealt only with the strengths of acids. Arising as a natural corollary to this is a principle that allows us to estimate the **base strength**. Simply stated, the principle is this:

⦁ **The stronger the acid, the weaker will be its conjugate base**.

We can, therefore, **relate the strength of a base to the pK_a of its conjugate acid**.

⦁ **The larger the pK_a of the conjugate acid, the stronger is the base**.

Consider the following as examples:

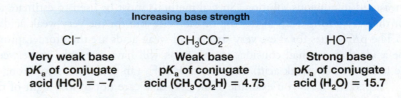

We see that the hydroxide ion is the strongest in this series of three bases because its conjugate acid, water, is the weakest acid. We know that water is the weakest acid because it has the largest pK_a.

Amines are like ammonia in that they are weak bases. Dissolving ammonia in water brings about the following equilibrium:

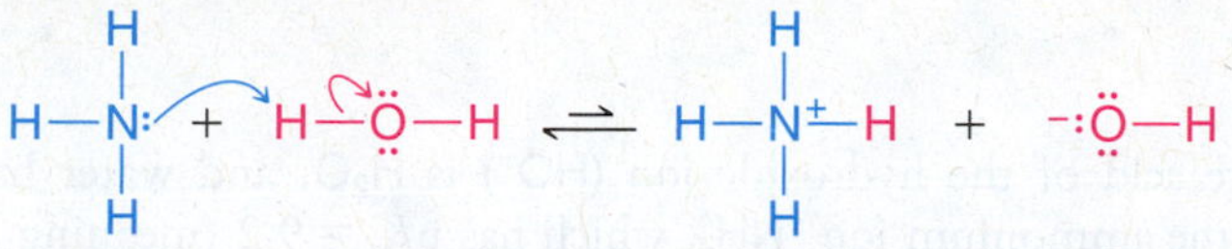

| Base | Acid | Conjugate acid | Conjugate base |

pK_a = 9.2

Dissolving methylamine in water causes the establishment of a similar equilibrium.

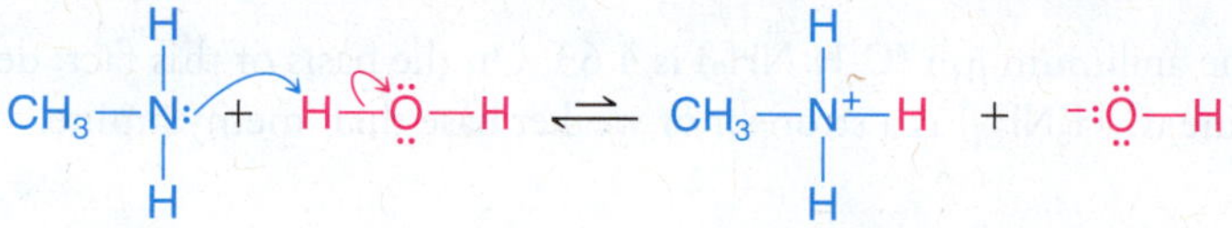

| Base | Acid | Conjugate acid | Conjugate base |

pK_a = 10.6

Again we can relate the basicity of these substances to the strength of their conjugate acids. The conjugate acid of ammonia is the ammonium ion, NH_4^+. The pK_a of the ammonium ion is 9.2. The conjugate acid of methylamine is the $CH_3NH_3^+$ ion. This ion, called the methylaminium ion, has pK_a = 10.6. Since the conjugate acid of methylamine is a weaker acid than the conjugate acid of ammonia, we can conclude that methylamine is a stronger base than ammonia.

••• SOLVED PROBLEM 3.6

Using the pK_a values in Table 3.1 decide which is the stronger base, CH_3OH or H_2O.

STRATEGY AND ANSWER: From Table 3.1, we find the pK_a values of the conjugate acids of water and methanol.

Weaker acid

pK_a = –1.74

H₃C—Ö⁺—H

Stronger acid

pK_a = –2.5

Because water is the conjugate base of the weaker acid, it is the stronger base.

Stronger base

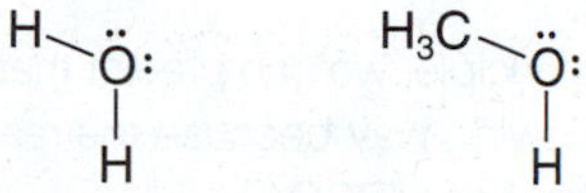

Weaker base

••• PRACTICE PROBLEM 3.7

Using the pK_a values of analogous compounds in Table 3.1, predict which would be the stronger base.

(a)

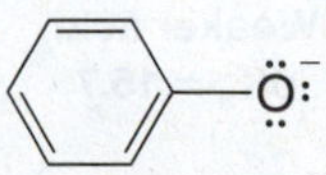

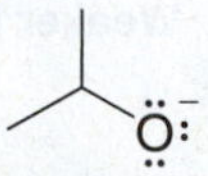

(b) $(CH_3)_3C\ddot{O}$:⁻   or

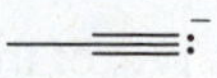

(c)

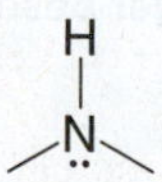

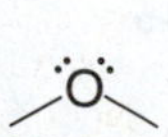

(d)

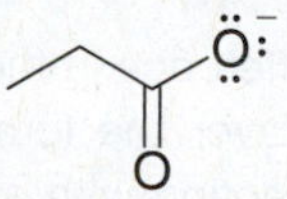

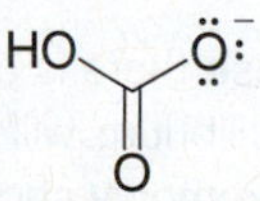

●●● **SOLVED PROBLEM 3.7**

Which would be the stronger base, HO^- or NH_3?

STRATEGY AND ANSWER: The conjugate acid of the hydroxide ion (HO^-) is H_2O, and water has $pK_a = 15.7$ (Table 3.1). The conjugate acid of ammonia is the ammonium ion $^+NH_4$, which has $pK_a = 9.2$ (meaning it is a stronger acid than water). Since ammonium ion is the stronger acid, its conjugate base NH_3 is the weaker base, and HO^-, the conjugate base of water (the weaker acid), is the stronger base.

●●● **PRACTICE PROBLEM 3.8**   The pK_a of the anilinium ion ($C_6H_5\overset{+}{N}H_3$) is 4.63. On the basis of this fact, decide whether aniline ($C_6H_5NH_2$) is a stronger or weaker base than methylamine.

3.6 HOW TO PREDICT THE OUTCOME OF ACID–BASE REACTIONS

Table 3.1 gives the approximate pK_a values for a range of representative compounds. While you probably will not be expected to memorize all of the pK_a values in Table 3.1, it is a good idea to begin to learn the general order of acidity and basicity for some of the common acids and bases. The examples given in Table 3.1 are representative of their class or functional group. For example, acetic acid has a $pK_a = 4.75$, and carboxylic acids generally have pK_a values near this value (in the range $pK_a = 3–5$). Ethyl alcohol is given as an example of an alcohol, and alcohols generally have pK_a values near that of ethyl alcohol (in the pK_a range 15–18), and so on. There are exceptions, of course, and we shall learn what these exceptions are as we go on.

By learning the relative scale of acidity of common acids now, you will be able to predict whether or not an acid–base reaction will occur as written.

- The general principle to apply is this: **acid–base reactions always favor the formation of the weaker acid and the weaker base**.

Helpful Hint

Formation of the weaker acid and base is an important general principle for predicting the outcome of acid–base reactions.

The reason for this is that the outcome of an acid–base reaction is determined by the position of an equilibrium. Acid–base reactions are said, therefore, to be **under equilibrium control**, and reactions under equilibrium control always favor the formation of the most stable (lowest potential energy) species. The weaker acid and weaker base are more stable (lower in potential energy) than the stronger acid and stronger base.

Using this principle, we can predict that a carboxylic acid (RCO_2H) will react with aqueous NaOH in the following way because the reaction will lead to the formation of the weaker acid (H_2O) and weaker base (RCO_2^-):

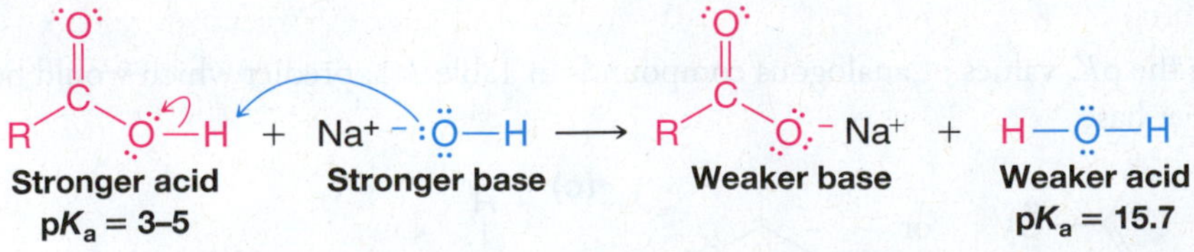

Because there is a large difference in the value of the pK_a of the two acids, the position of equilibrium will greatly favor the formation of the products. In instances like these we commonly show the reaction with a one-way arrow even though the reaction is an equilibrium.

Consider the mixing of an aqueous solution of phenol, C_6H_5OH (see Table 3.1), and NaOH. What acid–base reaction, if any, would take place?

STRATEGY: Consider the relative acidities of the reactant (phenol) and of the acid that might be formed (water) by a proton transfer to the base (the hydroxide ion).

ANSWER: The following reaction would take place because it would lead to the formation of a weaker acid (water) from the stronger acid (phenol). It would also lead to the formation of a weaker base, C_6H_5ONa, from the stronger base, NaOH.

$$C_6H_5-\ddot{O}-H \;+\; Na^+ \; {}^-{:}\ddot{O}-H \;\longrightarrow\; C_6H_5-\ddot{O}{:}^- \; Na^+ \;+\; H-\ddot{O}-H$$

Stronger acid	Stronger base	Weaker base	Weaker acid
$pK_a = 9.9$			$pK_a = 15.7$

Using Table 3.1, explain why the acid–base reaction that takes place between NaH (as source of $:H^-$ ions) and CH_3OH is

$$CH_3\ddot{O}H \;+\; :H^- \;\longrightarrow\; CH_3\ddot{O}{:}^- \;+\; H_2$$

rather than

$$CH_3\ddot{O}H \;+\; :H^- \;\xrightarrow{\;\;/\!\!\!\!\longrightarrow\;\;} \; {}^-{:}CH_2\ddot{O}H \;+\; H_2$$

ANSWER: A hydride ion is a very strong base, being the conjugate base of H_2 (a very weak acid, $pK_a = 35$). Hydride will remove the most acidic proton from CH_3OH. Although CH_3OH is not given in Table 3.1, we can compare it to CH_3CH_2OH, a similar alcohol whose hydroxyl group pK_a is 16, far more acidic than any proton attached to a carbon without a functional group (e.g., a proton of CH_3CH_3, which has $pK_a = 50$). Because the proton attached to the oxygen is much more acidic, it is removed preferentially.

Predict the outcome of the following reaction.

$$\text{(alkyne)} \;+\; {}^-NH_2 \;\longrightarrow$$

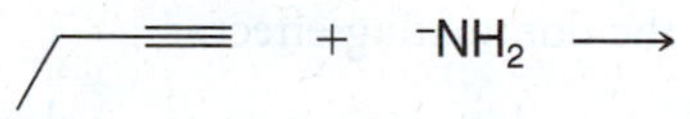

3.6A Water Solubility as the Result of Salt Formation

Although acetic acid and other carboxylic acids containing fewer than five carbon atoms are soluble in water, many other carboxylic acids of higher molecular weight are not appreciably soluble in water. Because of their acidity, however, *water-insoluble carboxylic acids dissolve in aqueous sodium hydroxide*; they do so by reacting to form water-soluble sodium salts:

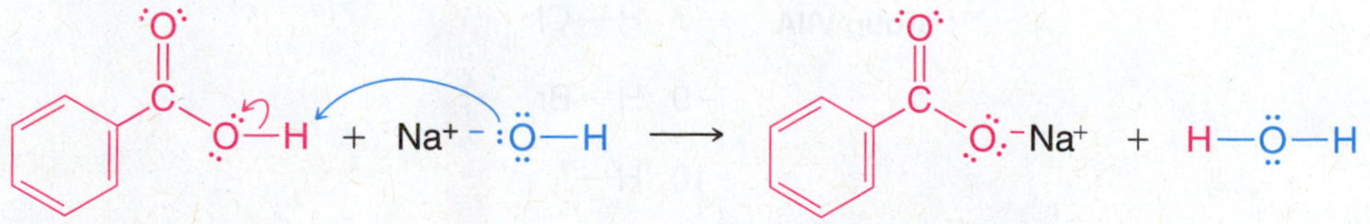

Insoluble in water	Soluble in water (due to its polarity as a salt)

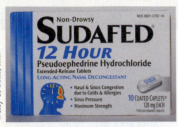

Pseudoephedrine is an amine that is sold as its hydrochloride salt.

We can also predict that an amine will react with aqueous hydrochloric acid in the following way:

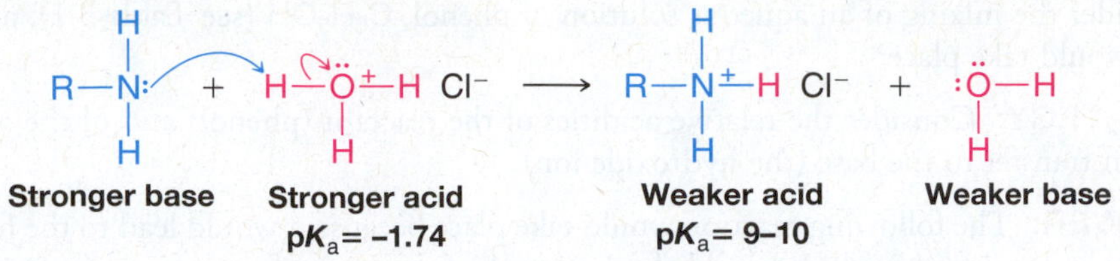

$$
\underset{\substack{\text{Stronger base}}}{R-\overset{\overset{\displaystyle H}{|}}{\underset{\underset{\displaystyle H}{|}}{N}}:} \;+\; \underset{\substack{\text{Stronger acid}\\ pK_a = -1.74}}{H-\overset{\overset{\displaystyle +}{\ddot{O}}}{\underset{\underset{\displaystyle H}{|}}{}}-H \;\; Cl^-} \longrightarrow \underset{\substack{\text{Weaker acid}\\ pK_a = 9-10}}{R-\overset{\overset{\displaystyle H}{|}}{\underset{\underset{\displaystyle H}{|}}{N}}\overset{+}{{}}-H \;\; Cl^-} \;+\; \underset{\substack{\text{Weaker base}}}{:\ddot{O}-H}
$$

While methylamine and most amines of low molecular weight are very soluble in water, amines with higher molecular weights, such as aniline ($C_6H_5NH_2$), have limited water solubility. However, these *water-insoluble amines dissolve readily in hydrochloric acid* because the acid–base reactions convert them into soluble salts:

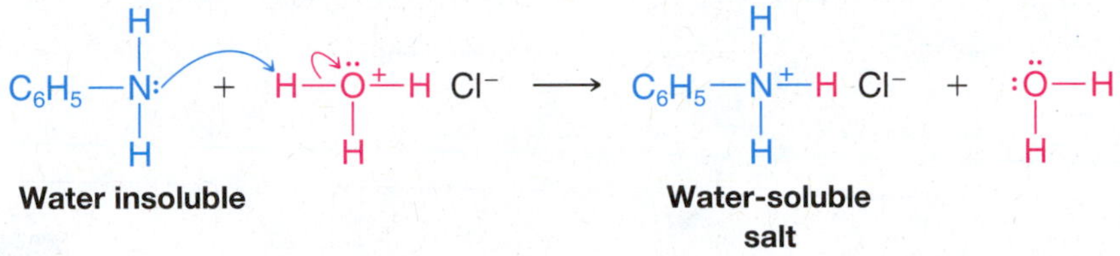

$$
\underset{\substack{\text{Water insoluble}}}{C_6H_5-\overset{\overset{\displaystyle H}{|}}{\underset{\underset{\displaystyle H}{|}}{N}}:} \;+\; H-\overset{\overset{\displaystyle +}{\ddot{O}}}{\underset{\underset{\displaystyle H}{|}}{}}-H \;\; Cl^- \longrightarrow \underset{\substack{\text{Water-soluble}\\ \text{salt}}}{C_6H_5-\overset{\overset{\displaystyle H}{|}}{\underset{\underset{\displaystyle H}{|}}{N}}\overset{+}{{}}-H \;\; Cl^-} \;+\; :\overset{\displaystyle }{\underset{\underset{\displaystyle H}{|}}{\ddot{O}}}-H
$$

●●●

PRACTICE PROBLEM 3.10 Most carboxylic acids dissolve in aqueous solutions of sodium bicarbonate ($NaHCO_3$) because, as carboxylate salts, they are more polar. Write curved arrows showing the reaction between a generic carboxylic acid and sodium bicarbonate to form a carboxylate salt and H_2CO_3. (Note that H_2CO_3 is unstable and decomposes to carbon dioxide and water. You do not need to show that process.)

3.7 RELATIONSHIPS BETWEEN STRUCTURE AND ACIDITY

The strength of a Brønsted–Lowry acid depends on the extent to which a proton can be separated from it and transferred to a base. Removing the proton involves breaking a bond to the proton, and it involves making the conjugate base more electrically negative.

When we compare compounds in a single column of the periodic table, the strength of the bond to the proton is the dominating effect.

- Bond strength to the proton decreases as we move down the column, increasing its acidity.

This phenomenon is mainly due to decreasing effectiveness of orbital overlap between the hydrogen $1s$ orbital and the orbitals of successively larger elements in the column. The less effective the orbital overlap, the weaker is the bond, and the stronger is the acid. The acidities of the hydrogen halides furnish an example:

Helpful Hint

Proton acidity increases as we descend a column in the periodic table due to decreasing bond strength to the proton.

pK_a		
3.2	H—F	
−7	H—Cl	**Group VIIA**
−9	H—Br	
−10	H—I	

Acidity increases ↓

Comparing the hydrogen halides with each other, H—F is the weakest acid and H—I is the strongest. This follows from the fact that the H—F bond is by far the strongest and the H—I bond is the weakest.

Because HI, HBr, and HCl are strong acids, their conjugate bases (I⁻, Br⁻, Cl⁻) are all weak bases. HF, however, is less acidic than the other hydrogen halides and fluoride ion is a stronger base. The fluoride anion is still not nearly as basic as other species we commonly think of as bases, such as the hydroxide anion, however.

- Acidity increases from left to right when we compare compounds in a given row of the periodic table.

Bond strengths vary somewhat, but the predominant factor becomes the electronegativity of the atom bonded to the hydrogen. The electronegativity of the atom in question affects acidity in two related ways: (1) it affects the polarity of the bond to the proton and (2) it affects the relative stability of the anion (conjugate base) that forms when the proton is lost.

We can see an example of this effect when we compare the acidities of the compounds CH_4, NH_3, H_2O, and HF. These compounds are all hydrides of first-row elements, and electronegativity increases across a row of the periodic table from left to right (see Table 1.2):

Electronegativity increases →

C   N   O   F

Because fluorine is the most electronegative, the bond in H—F is most polarized, and the proton in H—F is the most positive. Therefore, H—F loses a proton most readily and is the most acidic in this series:

Acidity increases →

$H_3C—H$ $\overset{\delta-}{H_2N}—\overset{\delta+}{H}$ $\overset{\delta-}{HO}—\overset{\delta+}{H}$ $\overset{\delta-}{F}—\overset{\delta+}{H}$

$pK_a = 48$ $pK_a = 38$ $pK_a = 15.7$ $pK_a = 3.2$

Electrostatic potential maps for these compounds directly illustrate this trend based on electronegativity and increasing polarization of the bonds to hydrogen (Fig. 3.2). Almost no positive charge (indicated by the extent of color trending toward blue) is evident at the hydrogens of methane. Very little positive charge is present at the hydrogens of ammonia. This is consistent with the weak electronegativity of both carbon and nitrogen and hence with the behavior of methane and ammonia as exceedingly weak acids (pK_a values of 48 and 38, respectively). Water shows significant positive charge at its hydrogens (pK_a more than 20 units lower than ammonia), and hydrogen fluoride clearly has the highest amount of positive charge at its hydrogen (pK_a of 3.2), resulting in strong acidity.

Because H—F is the strongest acid in this series, its conjugate base, the fluoride ion (F^-), will be the weakest base. Fluorine is the most electronegative atom and it accommodates the negative charge most readily:

Basicity increases ←

CH_3^- H_2N^- HO^- F^-

The methanide ion (CH_3^-) is the least stable anion of the four, because carbon being the least electronegative element is least able to accept the negative charge. The methanide ion, therefore, is the strongest base in this series. [The methanide ion, a **carbanion**, and the amide ion ($^-NH_2$) are exceedingly strong bases because they are the conjugate bases of extremely weak acids. We shall discuss some uses of these powerful bases in Section 3.14.]

Methane   Ammonia   Water   Hydrogen fluoride

FIGURE 3.2 The effect of increasing electronegativity among elements from left to right in the first row of the periodic table is evident in these maps of electrostatic potential for methane, ammonia, water, and hydrogen fluoride.

FIGURE 3.3 A summary of periodic trends in relative acidity. Acidity increases from left to right across a given row (electronegativity effect) and from top to bottom in a given column (bond strength effect) of the periodic table.

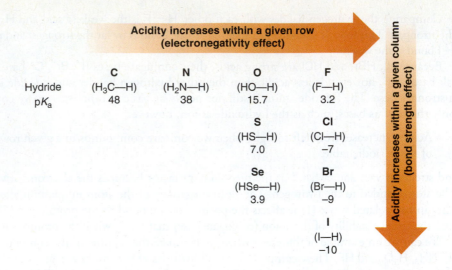

Trends in acidity within the periodic table are summarized in Fig. 3.3.

3.7A The Effect of Hybridization

- An alkyne hydrogen is weakly acid. Alkene and alkane hydrogens are essentially not acidic.

The pK_a values for ethyne, ethene, and ethane illustrate this trend.

$$H—C≡C—H$$

Ethyne
$pK_a = 25$

Ethene
$pK_a = 44$

Ethane
$pK_a = 50$

We can explain this order of acidities on the basis of the hybridization state of carbon in each compound. Electrons of $2s$ orbitals have lower energy than those of $2p$ orbitals because *electrons in 2s orbitals tend, on the average, to be much closer to the nucleus than electrons in 2p orbitals.* (Consider the shapes of the orbitals: $2s$ orbitals are spherical and centered on the nucleus; $2p$ orbitals have lobes on either side of the nucleus and are extended into space.)

- With hybrid orbitals, **having more *s* character means that the electrons of the anion will, on the average, be lower in energy, and the anion will be more stable.**

The sp orbitals of the C—H bonds of ethyne have 50% s character (because they arise from the combination of one s orbital and one p orbital), those of the sp^2 orbitals of ethene have 33.3% s character, while those of the sp^3 orbitals of ethane have only 25% s character. This means, in effect, that the sp carbon atoms of ethyne act as if they were more electronegative than the sp^2 carbon atoms of ethene and the sp^3 carbon atoms of ethane. (Remember: electronegativity measures an atom's ability to hold bonding electrons close to its nucleus, and having electrons closer to the nucleus makes it more stable.)

- An sp carbon atom is effectively more electronegative than an sp^2 carbon, which in turn is more electronegative than an sp^3 carbon.

The effect of hybridization and effective electronegativity on acidity is borne out in the calculated electrostatic potential maps for ethyne, ethene, and ethane shown in Fig. 3.4. Some positive charge (indicated by blue color) is clearly evident on the hydrogens of ethyne ($pK_a = 25$), but almost no positive charge is present on the hydrogens of ethene and ethane (both having pK_a values more than 20 units greater than ethyne).

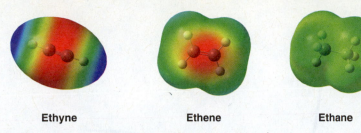

FIGURE 3.4 Electrostatic potential maps for ethyne, ethene, and ethane.

| Ethyne | Ethene | Ethane |

In summary, the order of relative acidities of ethyne, ethene, and ethane parallels the effective electronegativity of the carbon atom in each compound:

Relative Acidity of the Hydrocarbons

$$HC\equiv CH > H_2C=CH_2 > H_3C-CH_3$$

As expected based on the properties of acid–base conjugate pairs, an sp^3 carbanion is the strongest base in a series based on carbon hybridization, and an sp carbanion (an alkynide) is the weakest base. This trend is illustrated here with the conjugate bases of ethane, ethene, and ethyne.

Relative Basicity of the Carbanions

$$H_3C-CH_2{:}^- > H_2C=CH{:}^- > HC\equiv C{:}^-$$

3.7B Inductive Effects

The carbon–carbon bond of ethane is completely nonpolar because at each end of the bond there are two identical methyl groups:

$$CH_3-CH_3$$

Ethane

The C — C bond is nonpolar.

This is not the case with the carbon–carbon bond of ethyl fluoride, however:

$$\overset{\delta+}{CH_3}\rightarrow\overset{\delta+}{CH_2}\rightarrow\overset{\delta-}{F}$$
$$\quad 2 \qquad 1$$

One end of the bond, the one nearer the fluorine atom, is more negative than the other. This polarization of the carbon–carbon bond results from an intrinsic electron-attracting ability of the fluorine (because of its electronegativity) that is transmitted *through space* and *through the bonds of the molecule*. Chemists call this kind of effect an inductive effect.

- **Inductive effects** are electronic effects transmitted through bonds. The inductive effect of a group can be **electron donating** or **electron withdrawing**. Inductive effects weaken as the distance from the group increases.

In the case of ethyl fluoride, the positive charge that the fluorine imparts to C1 is greater than that imparted to C2 because the fluorine is closer to C1.

Figure 3.5 shows the dipole moment for ethyl fluoride (fluoroethane). The distribution of negative charge around the electronegative fluorine is plainly evident in the calculated electrostatic potential map.

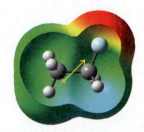

FIGURE 3.5 Ethyl fluoride, showing its dipole moment inside a cutaway view of the electrostatic potential at its van der Waals surface.

3.8 ENERGY CHANGES

Energy is defined as the capacity to do work. The two fundamental types of energy are **kinetic energy** and **potential energy**.

Kinetic energy is the energy an object has because of its motion; it equals one-half the object's mass multiplied by the square of its velocity (i.e., $\frac{1}{2}mv^2$).

Potential energy is stored energy. It exists only when an attractive or repulsive force exists between objects. Two balls attached to each other by a spring (an analogy

FIGURE 3.6 Potential energy exists between objects that either attract or repel each other. In the case of atoms joined by a covalent bond, or objects connected by a spring, the lowest potential energy state occurs when atoms are at their ideal internuclear distance (bond length), or when a spring between objects is relaxed. Lengthening or shortening the bond distance, or compressing or stretching a spring, raises the potential energy.

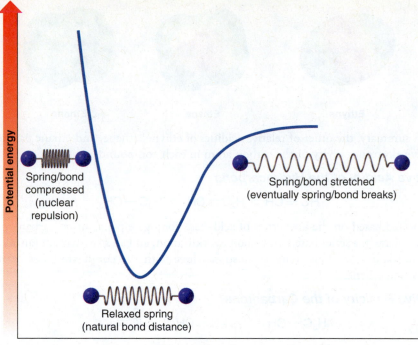

we used for covalent bonds when we discussed infrared spectroscopy in Section 2.15) can have their potential energy increased when the spring is stretched or compressed (Fig. 3.6). If the spring is stretched, an attractive force will exist between the balls. If it is compressed, a repulsive force will exist. In either instance releasing the balls will cause the potential energy (stored energy) of the balls to be converted into kinetic energy (energy of motion).

Chemical energy is a form of potential energy. It exists because attractive and repulsive electrical forces exist between different pieces of the molecules. Nuclei attract electrons, nuclei repel each other, and electrons repel each other.

It is usually impractical (and often impossible) to describe the *absolute* amount of potential energy contained by a substance. Thus we usually think in terms of its *relative potential energy*. We say that one system has *more* or *less* potential energy than another.

Another term that chemists frequently use in this context is the term **stability** or **relative stability**. *The relative stability of a system is inversely related to its relative potential energy.*

- **The *more* potential energy an object has, the *less stable* it is**.

Consider, as an example, the relative potential energy and the relative stability of snow when it lies high on a mountainside and when it lies serenely in the valley below. Because of the attractive force of gravity, the snow high on the mountain *has greater potential energy and is much less stable* than the snow in the valley. This greater potential energy of the snow on the mountainside can become converted into the enormous kinetic energy of an avalanche. By contrast, the snow in the valley, with its lower potential energy and with its greater stability, is incapable of releasing such energy.

3.8A Potential Energy and Covalent Bonds

Atoms and molecules possess potential energy—often called chemical energy—that can be released as heat when they react. Because heat is associated with molecular motion, this release of heat results from a change from potential energy to kinetic energy.

From the standpoint of covalent bonds, the state of greatest potential energy is the state of free atoms, the state in which the atoms are not bonded to each other at all. This is true because the formation of a chemical bond is always accompanied by the lowering of the potential energy of the atoms (cf. Fig. 1.8). Consider as an example the formation of hydrogen molecules from hydrogen atoms:

$$H\cdot + H\cdot \longrightarrow H—H \qquad \Delta H° = -436 \text{ kJ mol}^{-1}*$$

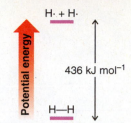

The potential energy of the atoms decreases by 436 kJ mol^{-1} as the covalent bond forms. This potential energy change is illustrated graphically in Fig. 3.7.

A convenient way to represent the relative potential energies of molecules is in terms of their relative **enthalpies**, or **heat contents**, H. (*Enthalpy* comes from *en + thalpein*, Greek: to heat.) The difference in relative enthalpies of reactants and products in a chemical change is called the **enthalpy change** and is symbolized by $\Delta H°$. [The Δ (delta) in front of a quantity usually means the difference, or change, in the quantity. The superscript ° indicates that the measurement is made under standard conditions.]

FIGURE 3.7 The relative potential energies of hydrogen atoms and a hydrogen molecule.

By convention, the sign of $\Delta H°$ for **exothermic reactions** (those evolving heat) is negative. **Endothermic reactions** (those that absorb heat) have a positive $\Delta H°$. The heat of reaction, $\Delta H°$, measures the change in enthalpy of the atoms of the reactants as they are converted to products. For an exothermic reaction, the atoms have a smaller enthalpy as products than they do as reactants. For endothermic reactions, the reverse is true.

3.9 THE RELATIONSHIP BETWEEN THE EQUILIBRIUM CONSTANT AND THE STANDARD FREE-ENERGY CHANGE, $\Delta G°$

An important **relationship exists between the equilibrium constant** (K_{eq}) **and the standard free-energy change** ($\Delta G°$). for a reaction.[†]

$$\Delta G° = -RT \ln K_{eq}$$

where R is the gas constant and equals 8.314 J K^{-1} mol^{-1} and T is the absolute temperature in kelvins (K).

This equation tells us the following:

- **For a reaction to favor the formation of products when equilibrium is reached it must have a negative value for $\Delta G°$.** Free energy must be *lost* as the reactants become products; that is, the reaction must go down an energy hill. For such a reaction the equilibrium constant will be greater than one. If $\Delta G°$ is more negative than 13 kJ mol^{-1} the equilibrium constant will be large enough for the reaction to *go to completion*, meaning that more than 99% of the reactants will be converted to products when equilibrium is reached.

- **For reactions with a positive $\Delta G°$, the formation of products at equilibrium is unfavorable.** The equilibrium constant for these reactions will be less than one.

The free-energy change ($\Delta G°$) has two components, the **enthalpy change** ($\Delta H°$) and the **entropy change** ($\Delta S°$). The relationship between these three thermodynamic quantities is

$$\Delta G° = \Delta H° - T\Delta S°$$

We have seen (Section 3.8) that $\Delta H°$ is associated with changes in bonding that occur in a reaction. If, collectively, stronger bonds are formed in the products than existed in the starting materials, then $\Delta H°$ will be negative (i.e., the reaction is *exothermic*). If the reverse is true, then $\Delta H°$ will be positive (the reaction is *endothermic*). **A negative value for $\Delta H°$, therefore, will contribute to making $\Delta G°$ negative and will consequently**

---

*The unit of energy in SI units is the joule, J, and 1 cal = 4.184 J. (Thus 1 kcal = 4.184 kJ.) A kilocalorie of energy (1000 cal) is the amount of energy in the form of heat required to raise by 1 °C the temperature of 1 kg (1000 g) of water at 15 °C.

[†]By standard free-energy change ($\Delta G°$), we mean that the products and reactants are taken as being in their standard states (1 atm of pressure for a gas and 1 M for a solution). The free-energy change is often called the **Gibbs free-energy change**, to honor the contributions to thermodynamics of J. Willard Gibbs, a professor of mathematical physics at Yale University in the latter part of the nineteenth century.

favor the formation of products. For the ionization of an acid, the less positive or more negative the value of $\Delta H°$, the stronger the acid will be.

Entropy changes have to do with *changes in the relative order of a system*. **The more random a system is, the greater is its entropy**. Therefore, a positive entropy change $(+\Delta S°)$ is always associated with a change from a more ordered system to a less ordered one. A negative entropy change $(-\Delta S°)$ accompanies the reverse process. In the equation $\Delta G° = \Delta H° - T \Delta S°$, the entropy change (multiplied by T) is preceded by a negative sign; this means that **a positive entropy change (from order to disorder) makes a negative contribution to $\Delta G°$ and is energetically favorable for the formation of products**.

For many reactions in which the number of molecules of products equals the number of molecules of reactants (e.g., when two molecules react to produce two molecules), the entropy change will be small. This means that except at high temperatures (where the term $T \Delta S°$ becomes large even if $\Delta S°$ is small) the value of $\Delta H°$ will largely determine whether or not the formation of products will be favored. If $\Delta H°$ is large and negative (if the reaction is exothermic), then the reaction will favor the formation of products at equilibrium. If $\Delta H°$ is positive (if the reaction is endothermic), then the formation of products will be unfavorable.

- - -

PRACTICE PROBLEM 3.11   State whether you would expect the entropy change, $\Delta S°$, to be positive, negative, or approximately zero for each of the following reactions. (Assume the reactions take place in the gas phase.)

(a) $A + B \longrightarrow C$ **(b)** $A + B \longrightarrow C + D$ **(c)** $A \longrightarrow B + C$

- - -

PRACTICE PROBLEM 3.12   **(a)** What is the value of $\Delta G°$ for a reaction where $K_{eq} = 1$?   **(b)** Where $K_{eq} = 10$? (The change in $\Delta G°$ required to produce a 10-fold increase in the equilibrium constant is a useful term to remember.)   **(c)** Assuming that the entropy change for this reaction is negligible (or zero), what change in $\Delta H°$ is required to produce a 10-fold increase in the equilibrium constant?

3.10 ACIDITY: CARBOXYLIC ACIDS VERSUS ALCOHOLS

Carboxylic acids are weak acids, typically having pK_a values in the range of 3–5. Alcohols, by comparison, have pK_a values in the range of 15–18, and essentially do not give up a proton unless exposed to a very strong base.

To understand the reasons for this difference, let's consider acetic acid and ethanol as representative examples of simple carboxylic acids and alcohols.

$$\underset{\substack{\textbf{Acetic acid} \\ \textbf{p}K_a = 4.75 \\ \Delta G° = 27 \text{ kJ mol}^{-1}}}{CH_3-\overset{\displaystyle O}{\overset{\displaystyle \|}{C}}-OH} \qquad\qquad \underset{\substack{\textbf{Ethanol} \\ \textbf{p}K_a = 16 \\ \Delta G° = 90.8 \text{ kJ mol}^{-1}}}{CH_3CH_2-OH}$$

($\Delta G°$ values are for OH proton ionization.)

Using the pK_a for acetic acid (4.75), one can calculate (Section 3.9) that the free-energy change ($\Delta G°$) for ionization of the carboxyl proton of acetic acid is $+27$ kJ mol^{-1}, a moderately endergonic (unfavorable) process, since the $\Delta G°$ value is positive. Using the pK_a of ethanol (16), one can calculate that the corresponding free-energy change for ionization of the hydroxyl proton of ethanol is $+90.8$ kJ mol^{-1}, a much more endergonic (and hence even less favorable) process. These calculations reflect the fact that ethanol

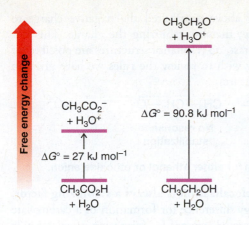

FIGURE 3.8 A diagram comparing the free-energy changes that accompany ionization of acetic acid and ethanol. Ethanol has a larger positive free-energy change and is a weaker acid because its ionization is more unfavorable.

is much less acidic than acetic acid. Figure 3.8 depicts the magnitude of these energy changes in a relative sense.

How do we explain the much greater acidity of carboxylic acids than alcohols? Consider first the structural changes that occur if both acetic acid and ethanol act as acids by donating a proton to water.

Acetic Acid Acting as an Acid

| Acetic acid | Water | | Acetate ion | Hydronium ion |

Ethanol Acting as an Acid

$$CH_3CH_2-\overset{..}{O}-H \ + \ :\overset{H}{\underset{H}{\overset{..}{O}}} \ \rightleftharpoons \ CH_3CH_2-\overset{..}{\underset{..}{O}}:^- \ + \ H-\overset{H}{\underset{H}{\overset{+}{O}}}:$$

| Ethanol | Water | Ethoxide ion | Hydronium ion |

What we need to focus on is the relative stability of the conjugate bases derived from a carboxylic acid and an alcohol. This is because the smaller free-energy change for ionization of a carboxylic acid (e.g., acetic acid) as compared to an alcohol (e.g., ethanol) has been attributed to greater stabilization of the negative charge in the carboxylate ion as compared to an alkoxide ion. Greater stabilization of the carboxylate ion appears to arise from two factors: (a) delocalization of charge (as depicted by resonance structures for the carboxylate ion, Section 3.10A), and (b) an inductive electron-withdrawing effect (Section 3.7B).

3.10A The Effect of Delocalization

Delocalization of the negative charge is possible in a carboxylate anion, but it is not possible in an alkoxide ion. We can show how delocalization is possible in carboxylate ions by writing **resonance** structures for the acetate ion.

Two Resonance Structures That Can Be Written for Acetate Anion

Resonance stabilization in acetate ion
(The structures are equivalent and there is no requirement for charge separation.)

The two resonance structures we drew above distributed the negative charge to both oxygen atoms of the carboxylate group, thereby stabilizing the charge. This is a **delocalization effect** (by resonance). In contrast, no resonance structures are possible for an alkoxide ion, such as ethoxide. (You may wish to review the rules we have given in Section 1.8 for writing proper resonance structures.)

$$CH_3-CH_2-\overset{..}{\underset{..}{O}}-H \ + \ H_2O \ \rightleftharpoons \ CH_3-CH_2-\overset{..}{\underset{..}{O}}:^- \ + \ H_3O^+$$

**No resonance   No resonance
stabilization   stabilization**

No resonance structures can be drawn for either ethanol or ethoxide anion.

A rule to keep in mind is that **charge delocalization is always a stabilizing factor**, and because of charge stabilization, the energy difference for formation of a carboxylate ion from a carboxylic acid is less than the energy difference for formation of an alkoxide ion from an alcohol. Since the energy difference for ionization of a carboxylic acid is less than for an alcohol, the carboxylic acid is a stronger acid.

3.10B The Inductive Effect

We have already shown how the negative charge in a carboxylate ion can be delocalized over two oxygen atoms by resonance. However, the electronegativity of these oxygen atoms further helps to stabilize the charge, by what is called an **inductive electron-withdrawing effect**. A carboxylate ion has two oxygen atoms whose combined electronegativity stabilizes the charge more than in an alkoxide ion, which has only a single electronegative oxygen atom. In turn, this lowers the energy barrier to forming the carboxylate ion, making a carboxylic acid a stronger acid than an alcohol. This effect is evident in electrostatic potential maps depicting approximately the bonding electron density for the two anions (Fig. 3.9). Negative charge in the acetate anion is evenly distributed over the two oxygen atoms, whereas in ethoxide the negative charge is localized on its sole oxygen atom (as indicated by red in the electrostatic potential map).

It is also reasonable to expect that a carboxylic acid would be a stronger acid than an alcohol when considering each as a neutral molecule (i.e., prior to loss of a proton), because both functional groups have a highly polarized O—H bond, which in turn weakens the bond to the hydrogen atom. However, the significant electron-withdrawing effect of the carbonyl group in acetic acid and the absence of an adjacent electron-withdrawing group in ethanol make the carboxylic acid hydrogen much more acidic than the alcohol hydrogen.

Acetate anion

Ethoxide anion

FIGURE 3.9 Calculated electrostatic potential maps at a surface approximating the bonding electron density for acetate anion and ethoxide anion. Although both molecules carry the same −1 net charge, acetate stabilizes the charge better by dispersing it over both oxygen atoms.

$$\underset{\substack{\textbf{Acetic acid}\\ \textbf{(stronger acid)}}}{CH_3-\overset{\overset{\textstyle O}{\|}}{C}-O\leftarrow H} \qquad \underset{\substack{\textbf{Ethanol}\\ \textbf{(weaker acid)}}}{CH_3-CH_2-O\leftarrow H}$$

Electrostatic potential maps at approximately the bond density surface for acetic acid and ethanol (Fig. 3.10) clearly show the positive charge at the carbonyl carbon of acetic acid, as compared to the CH_2 carbon of ethanol.

FIGURE 3.10 Maps of electrostatic potential at approximately the bond density surface for acetic acid and ethanol. The positive charge at the carbonyl carbon of acetic acid is evident in the blue color of the electrostatic potential map at that position, as compared to the hydroxyl carbon of ethanol. The inductive electron-withdrawing effect of the carbonyl group in carboxylic acids contributes to the acidity of this functional group.

 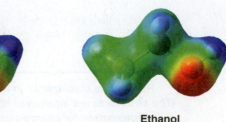

Acetic acid   **Ethanol**

3.10C Summary and a Comparison of Conjugate Acid–Base Strengths

In summary, the greater acidity of a carboxylic acid is predominantly due to the ability of its conjugate base (a carboxylate ion) to stabilize a negative charge better than an alkoxide ion, the conjugate base of an alcohol. In other words, the conjugate base of a carboxylic acid is a weaker base than the conjugate base of an alcohol. Therefore, since there is an inverse strength relationship between an acid and its conjugate base, a carboxylic acid is a stronger acid than an alcohol.

> **Helpful Hint**
>
> The more stable a conjugate base is, the stronger the corresponding acid.

●●● ....

PRACTICE PROBLEM 3.13

Draw contributing resonance structures and a hybrid resonance structure that explain two related facts: the carbon–oxygen bond distances in the acetate ion are the same, and the oxygens of the acetate ion bear equal negative charges.

3.10D Inductive Effects of Other Groups

The acid-strengthening effect of other electron-attracting groups (other than the carbonyl group) can be shown by comparing the acidities of acetic acid and chloroacetic acid:

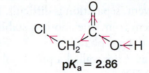

$$pK_a = 4.75 \qquad pK_a = 2.86$$

This is an example of a **substituent effect**. The greater acidity of chloroacetic acid can be attributed, in part, to the extra electron-attracting inductive effect of the electronegative chlorine atom. By adding its inductive effect to that of the carbonyl group and the oxygen, it makes the hydroxyl proton of chloroacetic acid even more positive than that of acetic acid. It also stabilizes the chloroacetate ion that is formed when the proton is lost *by dispersing its negative charge* (Fig. 3.11):

$$\text{Cl—CH}_2\text{—C(=O)—O—H} + H_2O \rightleftharpoons {}^{\delta-}\text{Cl—CH}_2\text{—C(=O}^{\delta-})\text{—O}^{\delta-} + H_3O^+$$

Dispersal of charge always makes a species more stable, and, as we have seen now in several instances, **any factor that stabilizes the conjugate base of an acid increases the strength of the acid**. (In Section 3.11, we shall see that entropy changes in the solvent are also important in explaining the increased acidity of chloroacetic acid.)

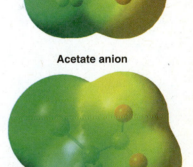

Acetate anion

Chloroacetate anion

FIGURE 3.11 The electrostatic potential maps for acetate and chloroacetate ions show the relatively greater ability of chloroacetate to disperse the negative charge.

●●● **SOLVED PROBLEM 3.10**

Which compound in each pair would be most acidic?

(a) [structure with F substituent] or [structure with Br substituent] **(b)** [structure with F near carbonyl] or [structure with F on adjacent carbon]

STRATEGY AND ANSWER: Decide what is similar in each pair and what is different. In pair (a), the difference is the halogen substituent on the carbon adjacent to the carboxyl group. In the first example it is fluorine; in the second it is bromine. Fluorine is much more electronegative (electron-attracting) than bromine (Table 1.2); therefore it will be able to disperse the negative charge of the anion formed when the proton is lost. Thus the first compound will be the stronger acid. In pair (b), the difference is the position of the fluorine substituents. In the second compound the fluorine is closer to the carboxyl group where it will be better able to disperse the negative charge in the anion formed when the proton is lost. The second compound will be the stronger acid.

•••

PRACTICE PROBLEM 3.14   Which would you expect to be the stronger acid? Explain your reasoning in each instance.

(a) CH_2ClCO_2H or $CHCl_2CO_2H$ (c) CH_2FCO_2H or CH_2BrCO_2H

(b) CCl_3CO_2H or $CHCl_2CO_2H$ (d) CH_2FCO_2H or $CH_2FCH_2CO_2H$

3.11 THE EFFECT OF THE SOLVENT ON ACIDITY

In the absence of a solvent (i.e., in the gas phase), most acids are far weaker than they are in solution. In the gas phase, for example, acetic acid is estimated to have a pK_a of about 130 (a K_a of $\sim 10^{-130}$)! The reason is this: when an acetic acid molecule donates a proton to a water molecule in the gas phase, the ions that are formed are oppositely charged particles and the particles must become separated:

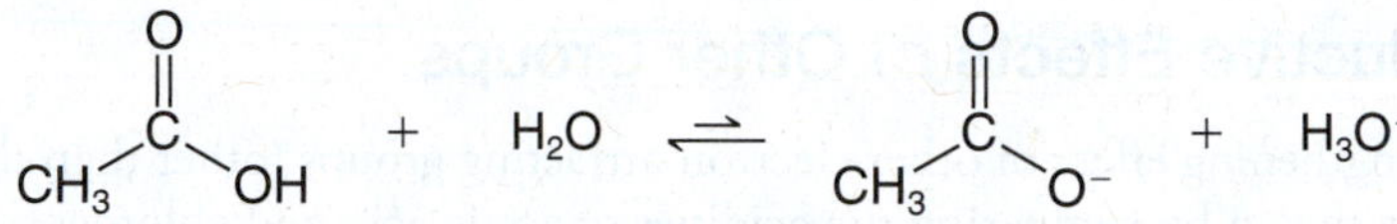

In the absence of a solvent, separation is difficult. In solution, solvent molecules surround the ions, insulating them from one another, stabilizing them, and making it far easier to separate them than in the gas phase.

In a solvent such as water, called a protic solvent, solvation by hydrogen bonding is important (Section 2.13D).

- A **protic solvent** is one that has a hydrogen atom attached to a strongly electronegative element such as oxygen or nitrogen.

A protic solvent, therefore, can form hydrogen bonds to the unshared electron pairs of an acid and its conjugate base, but they may not stabilize both equally.

- The stability of a conjugate base is enhanced if it is solvated to a greater extent than the corresponding acid.

Relative acidity cannot be predicted solely on the basis of solvation, however. Steric factors affecting solvation, and the relative order or disorder of the solvent molecules (entropic parameters), can enhance or decrease acidity.

3.12 ORGANIC COMPOUNDS AS BASES

If an organic compound contains an atom with an unshared electron pair, it is a potential base. We saw in Section 3.5C that compounds with an unshared electron pair on a nitrogen atom (i.e., amines) act as bases. Let us now consider several examples in which organic compounds having an unshared electron pair on an oxygen atom act in the same way.

Dissolving gaseous HCl in methanol brings about an acid–base reaction much like the one that occurs with water (Section 3.1A):

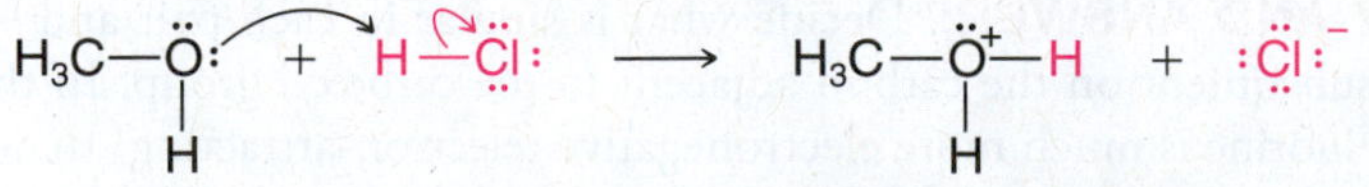

The conjugate acid of the alcohol is often called a **protonated alcohol**, although more formally it is called an **alkyloxonium ion** or simply an **oxonium ion**.

Alcohols, in general, undergo this same reaction when they are treated with solutions of strong acids such as HCl, HBr, HI, and H_2SO_4:

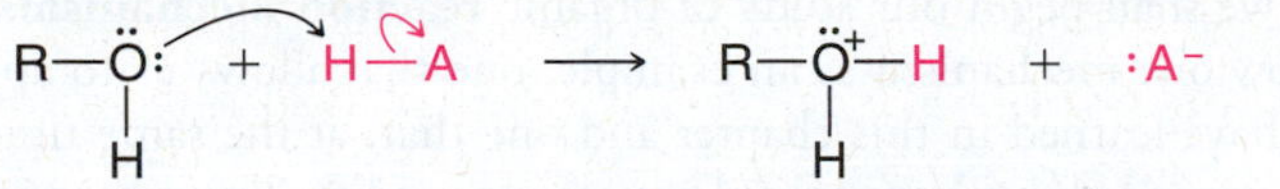

| Alcohol | Strong acid | Alkyloxonium ion | Weak base |

So, too, do ethers:

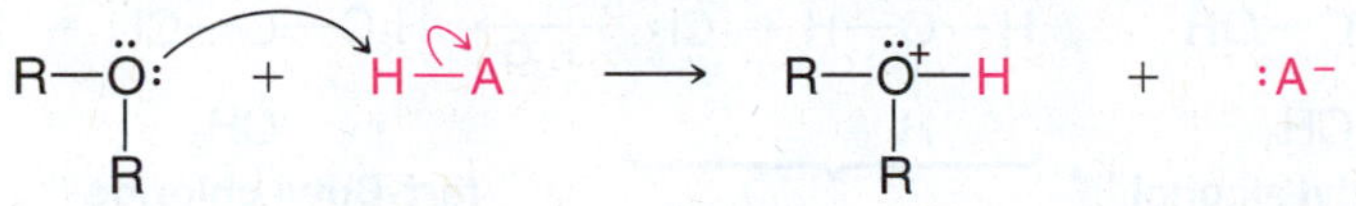

| Ether | Strong acid | Dialkyloxonium ion | Weak base |

Compounds containing a carbonyl group also act as bases in the presence of a strong acid:

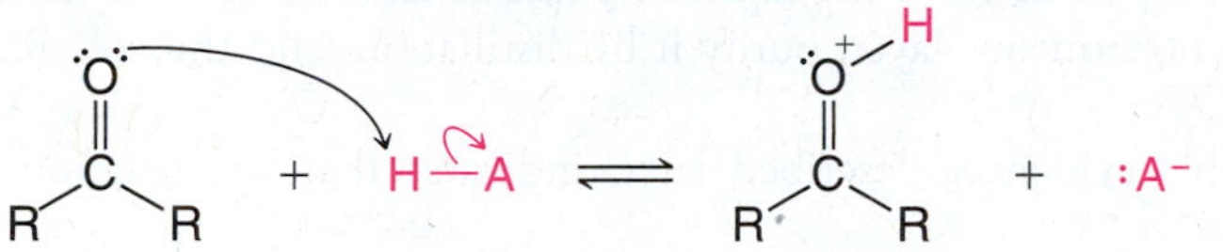

| Ketone | Strong acid | Protonated ketone | Weak base |

Proton transfer reactions like these are often the first step in many reactions that alcohols, ethers, aldehydes, ketones, esters, amides, and carboxylic acids undergo. The pK_a values for some of these protonated intermediates are given in Table 3.1.

An atom with an unshared electron pair is not the only locus that confers basicity on an organic compound. The π bond of an alkene can have the same effect. Later we shall study many reactions in which, as a first step, alkenes react with a strong acid by accepting a proton in the following way:

| Alkene | Strong acid | Carbocation | Weak base |

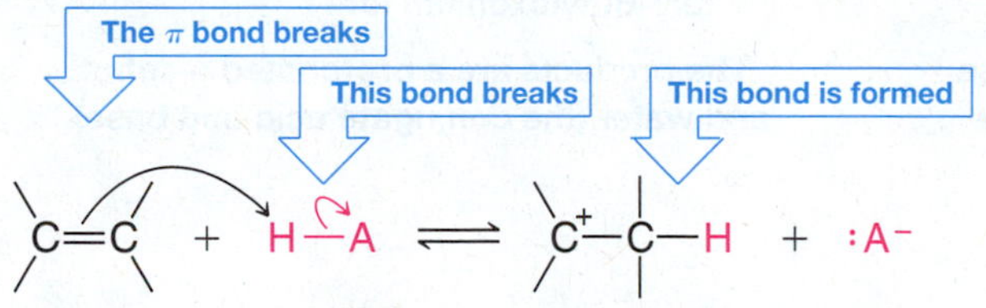

In this reaction the electron pair of the π bond of the alkene is used to form a bond between one carbon of the alkene and the proton donated by the strong acid. Notice that two bonds are broken in this process: the π bond of the double bond and the bond between the proton of the acid and its conjugate base. One new bond is formed: a bond between a carbon of the alkene and the proton. This process leaves the other carbon of the alkene trivalent, electron deficient, and with a formal positive charge. A compound containing a carbon of this type is called a **carbocation** (Section 3.4). As we shall see in later chapters, carbocations are unstable intermediates that react further to produce stable molecules.

> **Helpful Hint**
>
> Proton transfers are a common first step in many reactions we shall study.

● ● ●

PRACTICE PROBLEM 3.15

It is a general rule that any organic compound containing oxygen, nitrogen, or a multiple bond will dissolve in concentrated sulfuric acid. Explain the basis of this rule in terms of acid–base reactions and intermolecular forces.

3.13 A MECHANISM FOR AN ORGANIC REACTION

In Chapter 6 we shall begin our study of organic **reaction mechanisms** in earnest. Let us consider now one mechanism as an example, one that allows us to apply some of the chemistry we have learned in this chapter and one that, at the same time, will reinforce what we have learned about how curved arrows are used to illustrate mechanisms.

Dissolving *tert*-butyl alcohol in concentrated (concd) aqueous hydrochloric acid soon results in the formation of *tert*-butyl chloride. The reaction is a **substitution reaction**:

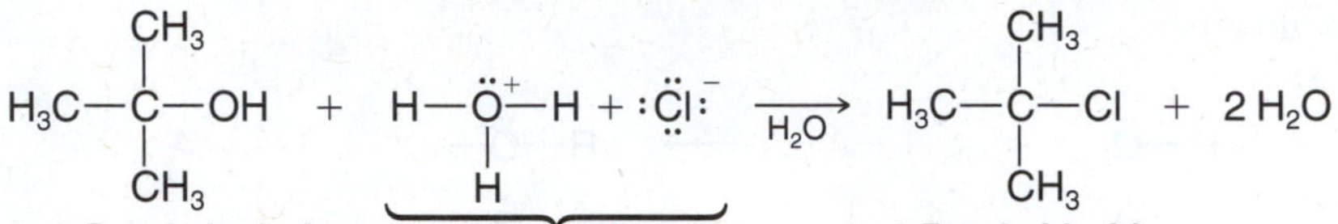

tert-**Butyl alcohol**
(soluble in H₂O)

Concd HCl

tert-**Butyl chloride**
(insoluble in H₂O)

That a reaction has taken place is obvious when one actually does the experiment. *tert*-Butyl alcohol is soluble in the aqueous medium; however, *tert*-butyl chloride is not, and consequently it separates from the aqueous phase as another layer in the flask. It is easy to remove this nonaqueous layer, purify it by distillation, and thus obtain the *tert*-butyl chloride.

Considerable evidence, described later, indicates that the reaction occurs in the following way.

⌈ A MECHANISM FOR THE REACTION — **Reaction of *tert*-Butyl Alcohol with Concentrated Aqueous HCl ⌉**

Step 1

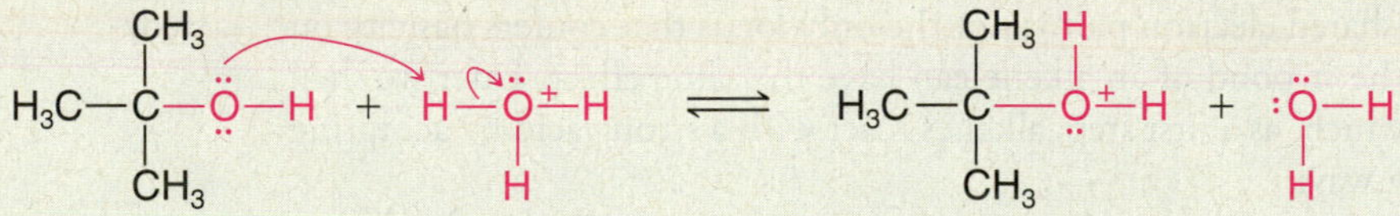

tert-**Butyloxonium ion**

tert-**Butyl alcohol acts as a base and accepts a proton from the hydronium ion. (Chloride anions are spectators in this step of the reaction.)**

The products are a protonated alcohol and water (the conjugate acid and base).

Step 2

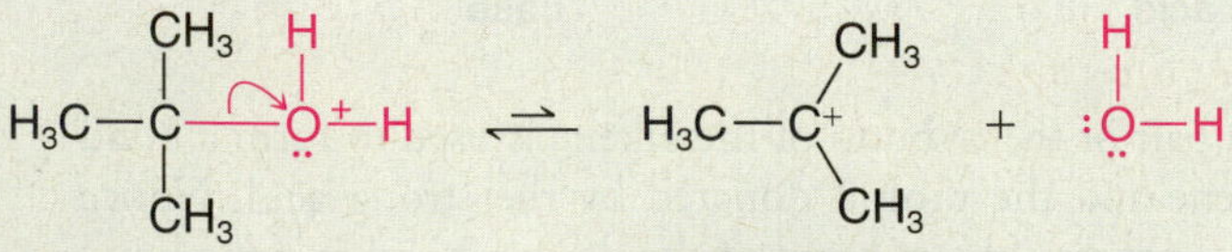

Carbocation

The bond between the carbon and oxygen of the *tert*-butyloxonium ion breaks heterolytically, leading to the formation of a carbocation and a molecule of water.

Step 3

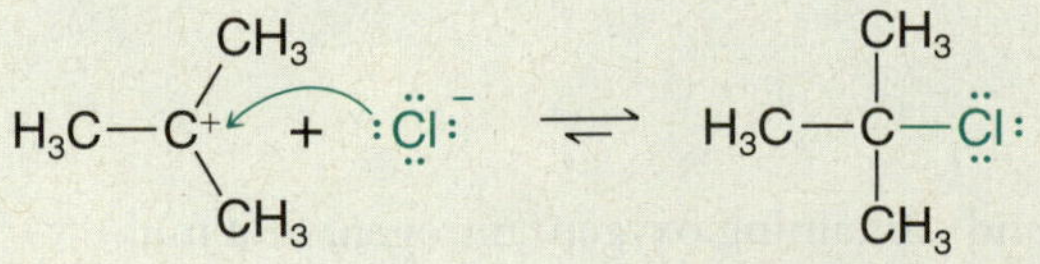

tert-**Butyl chloride**

The carbocation, acting as a Lewis acid, accepts an electron pair from a chloride ion to become the product.

Notice that **all of these steps involve acid–base reactions**. Step 1 is a straightforward Brønsted acid–base reaction in which the alcohol oxygen removes a proton from the hydronium ion. Step 2 is the reverse of a Lewis acid–base reaction. In it, the carbon–oxygen bond of the protonated alcohol breaks heterolytically as a water molecule departs with the electrons of the bond. This happens, in part, because the alcohol is protonated. The presence of a formal positive charge on the oxygen of the protonated alcohol weakens the carbon–oxygen bond by drawing the electrons in the direction of the positive oxygen. Step 3 is a Lewis acid–base reaction, in which a chloride anion (a Lewis base) reacts with the carbocation (a Lewis acid) to form the product.

A question might arise: why doesn't a molecule of water (also a Lewis base) instead of a chloride ion react with the carbocation? After all, there are many water molecules around, since water is the solvent. The answer is that this step does occur sometimes, but it is simply the reverse of step 2. That is to say, not all of the carbocations that form go on directly to become product. Some react with water to become protonated alcohols again. However, these will dissociate again to become carbocations (even if, before they do, they lose a proton to become the alcohol again). Eventually, however, most of them are converted to the product because, under the conditions of the reaction, the equilibrium of the last step lies far to the right, and the product separates from the reaction mixture as a second phase.

3.14 ACIDS AND BASES IN NONAQUEOUS SOLUTIONS

If you were to add sodium amide ($NaNH_2$) to water in an attempt to carry out a reaction using the amide ion ($^-NH_2$) as a very powerful base, the following reaction would take place immediately:

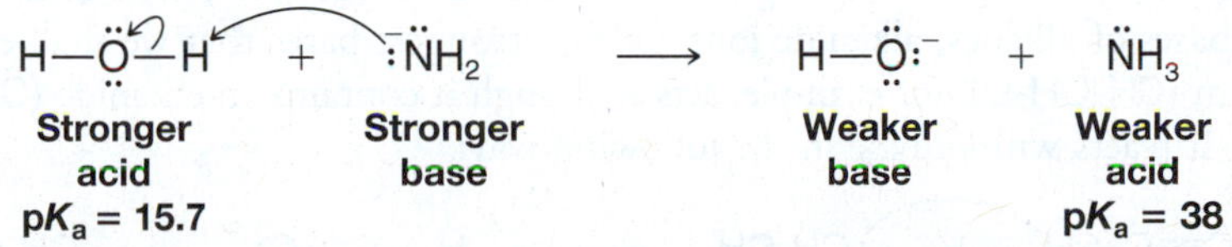

Stronger acid	**Stronger base**		**Weaker base**	**Weaker acid**
$pK_a = 15.7$				$pK_a = 38$

The amide ion would react with water to produce a solution containing hydroxide ion (a much weaker base) and ammonia. This example illustrates what is called the **leveling effect of the solvent**. *Water*, the solvent here, *donates a proton to any base stronger than a hydroxide ion*. Therefore, *it is not possible to use a base stronger than hydroxide ion in aqueous solution*.

We can use bases stronger than hydroxide ion, however, if we choose solvents that are weaker acids than water. We can use amide ion (e.g., from $NaNH_2$) in a solvent such as hexane, diethyl ether, or liquid NH_3 (the liquified gas, bp −33 °C, not the aqueous solution that you may have used in your general chemistry laboratory). All of these solvents are very weak acids (we generally don't think of them as acids), and therefore they will not donate a proton even to the strong base $^-NH_2$.

We can, for example, convert ethyne to its conjugate base, a carbanion, by treating it with sodium amide in liquid ammonia:

Stronger acid	**Stronger base**		**Weaker base**	**Weaker acid**
$pK_a = 25$	**(from $NaNH_2$)**			$pK_a = 38$

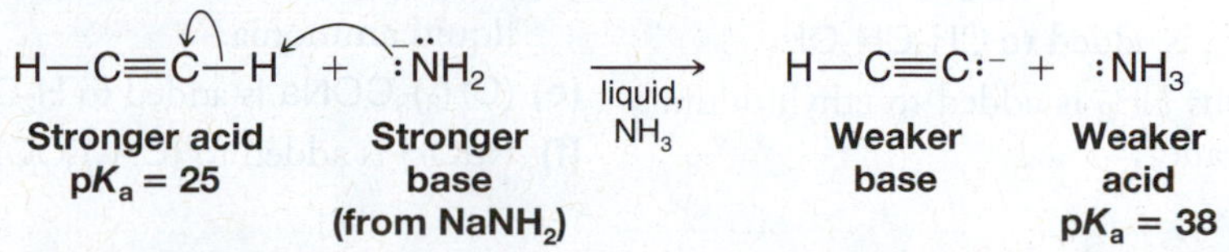

> **Helpful Hint**
>
> We shall use this reaction as part of our introduction to organic synthesis in Chapter 7.

Most **terminal alkynes** (alkynes with a proton attached to a triply bonded carbon) have pK_a values of about 25; therefore, all react with sodium amide in liquid ammonia in the same way that ethyne does. The general reaction is

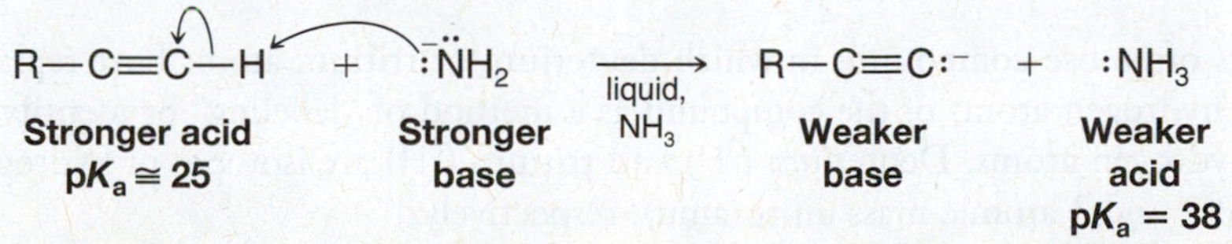

Stronger acid	**Stronger base**		**Weaker base**	**Weaker acid**
$pK_a \cong 25$				$pK_a = 38$

Alcohols are often used as solvents for organic reactions because, being somewhat less polar than water, they dissolve less polar organic compounds. Using alcohols as solvents also offers the advantage of using RO⁻ ions (called **alkoxide ions**) as bases. Alkoxide ions are somewhat stronger bases than hydroxide ions because alcohols are weaker acids than water. For example, we can create a solution of sodium ethoxide (CH_3CH_2ONa) in ethyl alcohol by adding sodium hydride (NaH) to ethyl alcohol. We use a large excess of ethyl alcohol because we want it to be the solvent. Being a very strong base, the hydride ion reacts readily with ethyl alcohol:

$$CH_3CH_2\ddot{O}-H \ + \ :H^- \xrightarrow{\text{ethyl alcohol}} CH_3CH_2\ddot{O}:^- \ + \ H_2$$

Stronger acid	Stronger base	Weaker base	Weaker acid
$pK_a = 16$	(from NaH)		$pK_a = 35$

The *tert*-butoxide ion, $(CH_3)_3CO^-$, in *tert*-butyl alcohol, $(CH_3)_3COH$, is a stronger base than the ethoxide ion in ethyl alcohol, and it can be prepared in a similar way:

$$(CH_3)_3C\ddot{O}-H \ + \ :H^- \xrightarrow{\text{tert-butyl alcohol}} (CH_3)_3C\ddot{O}:^- \ + \ H_2$$

Stronger acid	Stronger base	Weaker base	Weaker acid
$pK_a = 18$	(from NaH)		$pK_a = 35$

Although the carbon–lithium bond of an alkyllithium (RLi) has covalent character, it is polarized so as to make the carbon negative:

$$\overset{\delta-}{R}\leftarrow\overset{\delta+}{Li}$$

Alkyllithium reagents react as though they contain alkanide ($R:^-$) ions and, being the conjugate bases of alkanes, alkanide ions are the strongest bases that we shall encounter. Ethyllithium (CH_3CH_2Li), for example, acts as though it contains an ethanide ($CH_3CH_2:^-$) carbanion. It reacts with ethyne in the following way:

$$H-C\equiv C-H \ + \ ^-:CH_2CH_3 \xrightarrow{\text{hexane}} H-C\equiv C:^- \ + \ CH_3CH_3$$

Stronger acid	Stronger base	Weaker base	Weaker acid
$pK_a = 25$	(from CH_3CH_2Li)		$pK_a = 50$

Alkyllithiums can be easily prepared by allowing an alkyl bromide to react with lithium metal in an ether solvent (such as diethyl ether). See Section 12.6.

PRACTICE PROBLEM 3.16 Write equations for the acid–base reaction that would occur when each of the following compounds or solutions are mixed. In each case label the stronger acid and stronger base, and the weaker acid and weaker base, by using the appropriate pK_a values (Table 3.1). If no appreciable acid–base reaction would occur, you should indicate this.

(a) NaH is added to CH_3OH.
(b) $NaNH_3$ is added to CH_3CH_2OH.
(c) Gaseous NH_2 is added to ethyllithium in hexane.
(d) NH_4Cl is added to sodium amide in liquid ammonia.
(e) $(CH_3)_3CONa$ is added to H_2O.
(f) NaOH is added to $(CH_3)_3COH$.

3.15 ACID–BASE REACTIONS AND THE SYNTHESIS OF DEUTERIUM- AND TRITIUM-LABELED COMPOUNDS

Chemists often use compounds in which deuterium or tritium atoms have replaced one or more hydrogen atoms of the compound as a method of "labeling" or identifying particular hydrogen atoms. Deuterium (^2H) and tritium (^3H) are isotopes of hydrogen with masses of 2 and 3 atomic mass units (amu), respectively.

One way to introduce a deuterium or tritium atom into a specific location in a molecule is through the acid–base reaction that takes place when a very strong base is treated with D_2O or T_2O (water that has deuterium or tritium in place of its hydrogens). For example, treating a solution containing $(CH_3)_2CHLi$ (isopropyllithium) with D_2O results in the formation of propane labeled with deuterium at the central atom:

$$
\begin{array}{c}
CH_3 \\
| \\
CH_3CH:^- Li^+
\end{array}
+ \quad D_2O \quad \xrightarrow{\text{hexane}} \quad
\begin{array}{c}
CH_3 \\
| \\
CH_3CH-D
\end{array}
+ \quad DO^-
$$

Isopropyl-lithium	D₂O	2-Deuterio-propane	DO⁻
(stronger base)	*(stronger acid)*	*(weaker acid)*	*(weaker base)*

●●● **SOLVED PROBLEM 3.11**

Assuming you have available propyne, a solution of sodium amide in liquid ammonia, and T_2O, show how you would prepare the tritium-labeled compound $CH_3C{\equiv}CT$.

ANSWER: First add propyne to sodium amide in liquid ammonia. The following acid–base reaction will take place:

$$CH_3C{\equiv}CH + {}^-NH_2 \xrightarrow{\text{liq. ammonia}} CH_3C{\equiv}C:^- + NH_3$$

Stronger acid	Stronger base	Weaker base	Weaker acid

Then adding T_2O (a much stronger acid than NH_3) to the solution will produce $CH_3C{\equiv}CT$:

$$CH_3C{\equiv}C:^- + T_2O \xrightarrow{\text{liq. ammonia}} CH_3C{\equiv}CT + TO^-$$

Stronger base	Stronger acid	Weaker acid	Weaker base

●●● **PRACTICE PROBLEM 3.17**

Complete the following acid–base reactions:

(a) $HC{\equiv}CH + NaH \xrightarrow{\text{hexane}}$

(b) The solution obtained in **(a)** + $D_2O \longrightarrow$

(c) $CH_3CH_2Li + D_2O \xrightarrow{\text{hexane}}$

(d) $CH_3CH_2OH + NaH \xrightarrow{\text{hexane}}$

(e) The solution obtained in **(d)** + $T_2O \longrightarrow$

(f) $CH_3CH_2CH_2Li + D_2O \xrightarrow{\text{hexane}}$

3.16 APPLICATIONS OF BASIC PRINCIPLES

Again we review how certain basic principles apply to topics we have studied in this chapter.

Electronegativity Differences Polarize Bonds We saw how this principle applies to the heterolysis of bonds to carbon in Section 3.4 and in explaining the strength of acids in Sections 3.7 and 3.10B.

Polarized Bonds Underlie Inductive Effects In Section 3.10B we saw how polarized bonds explain effects that we call *inductive effects* and how these effects are part of the explanation for why carboxylic acids are more acidic than corresponding alcohols.

Opposite Charges Attract This principle is fundamental to understanding *Lewis acid–base theory* as we saw in Section 3.3A. Positively charged centers in molecules that are electron pair acceptors are attracted to negatively charged centers in electron pair donors. In Section 3.4 we saw this principle again in the reaction of carbocations (positively charged Lewis acids) with anions (which are negatively charged by definition) and other Lewis bases.

Nature Prefers States of Lower Potential Energy In Section 3.8A we saw how this principle explains the energy changes—called *enthalpy changes*—that take place when covalent bonds form, and in Section 3.9 we saw the role enthalpy changes play in explaining how large or how small the equilibrium constant for a reaction is. The lower the potential energy of the products, the larger is the equilibrium constant, and the more favored is the formation of the products when equilibrium is reached. This section also introduced a related principle: **Nature prefers disorder to order**—or, to put it another way, *a positive entropy change* for a reaction favors the formation of the products at equilibrium.

Resonance Effects Can Stabilize Molecules and Ions When a molecule or ion can be represented by two or more equivalent resonance structures, then the molecule or ion will be stabilized (will have its potential energy lowered) by delocalization of charge. In Section 3.10A we saw how this effect helps explain the greater acidity of carboxylic acids when compared to corresponding alcohols.

WHY Do These Topics Matter?]

THE RARITY OF CHEMICAL DISCOVERIES WITHOUT KNOWLEDGE OF THE MECHANISMS

From the time of its initial discovery in the 1630s until the middle of the twentieth century, the natural product quinine was the world's only real treatment for malaria. Yet, because it could only be obtained in small quantities from relatively remote places of the globe, it was a medicine that effectively was available to only a small number of very wealthy or well-connected individuals. In light of this issue, scientists began to wonder whether quinine could be synthesized in the laboratory, an idea that was first put to the test in 1856 by a graduate student in England named William Henry Perkin. Perkin's plan for synthesis was based on an idea posited in 1849 by his mentor, August Wilhelm von Hofmann, that quinine could be prepared from the constituents of coal tar. This notion was based on the balanced chemical equation shown below. The formulas were all that was known at the time, not the actual structures. We realize today that there was no chance for success in this endeavor simply because there is no mechanism by which these chemicals could react in the right way. Fortune, however, sometimes arrives in unexpected ways.

$$C_{10}H_{13}N \quad + \quad C_{10}H_{13}N \quad + \quad 3/2\ O_2 \quad \longrightarrow \quad C_{20}H_{24}N_2O_2 \quad + \quad H_2O$$

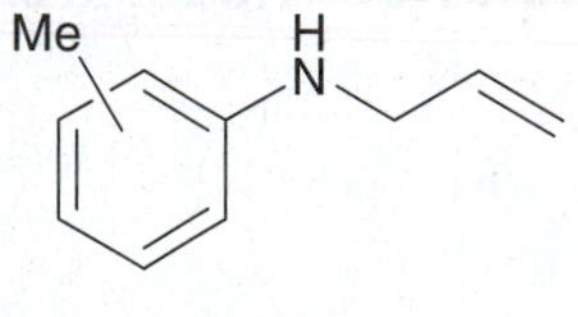

N-Allyltoluidine 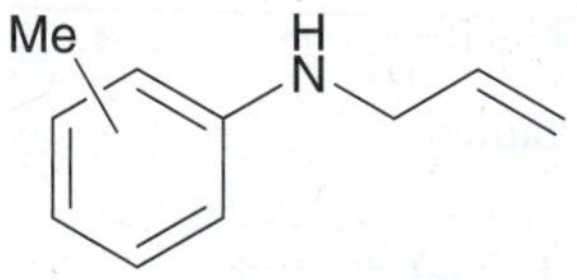   *N*-Allyltoluidine 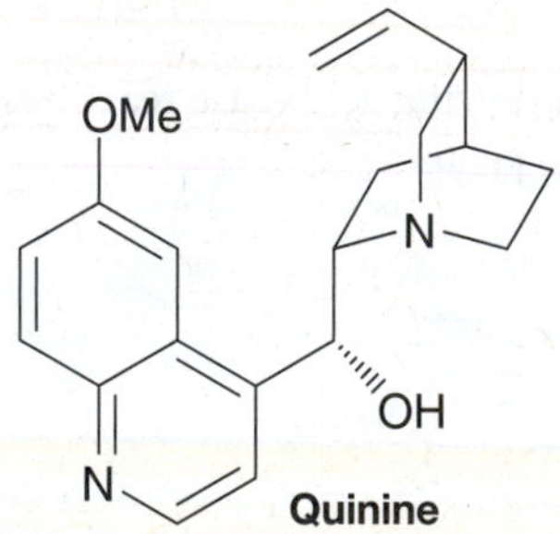   Quinine

Perkin did his most important experiment on this problem in a laboratory at his family home, an experiment in which he altered his supervisor's idea ever so slightly by using a different starting material (aniline, containing several different contaminants) and heating it in the presence of a strong oxidant (potassium dichromate). What resulted was a dark tar that looked a bit like asphalt. Although such products are often the result of reactions gone wrong, Perkin attempted to see if he could get anything in the tarry residue to dissolve by adding different solvents. Some did nothing, but when he added ethanol, a beautiful purple-colored solution was formed. This solution proved capable of turning any light-colored fabric the exact same purple shade. Although not quinine, what Perkin had discovered was the first synthetic dye, a way to color fabric a shade previously reserved for royalty. Indeed, before Perkin's discovery, the only way to obtain a purple-colored dye was by the tedious isolation of mucous secretions of select Mediterranean snails.

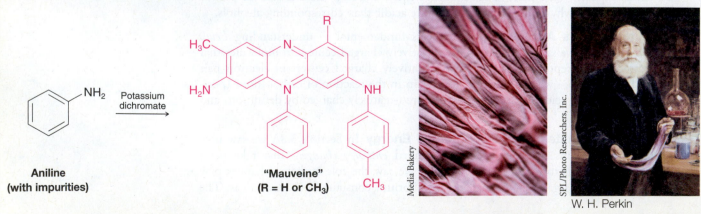

Aniline
(with impurities)

Potassium dichromate

"Mauveine"
(R = H or CH₃)

W. H. Perkin

Perkin ended up making a fortune from his discovery, a material he called mauveine, which is actually composed of two compounds. The more important outcome, however, was that it showed for the first time that organic chemistry could really change the world, launching an entire industry of other chemists looking to make ever more fanciful and wonderful colors not readily found in nature.

The key message, though, is that no matter how wonderful this story is, it is only one of a handful of cases in which there was such a significant outcome in the absence of any real chemical knowledge of mechanism. Major discoveries are much more likely when it is known what the given compounds might actually do when they react together! Otherwise, organic chemistry would be just alchemy. That might explain why it took nearly another century of work before quinine actually succumbed to laboratory synthesis.

To learn more about these topics, see:

1. Garfield, S. *Mauve: How One Man Invented a Color that Changed the world*. Faber and Faber, **2001**, p. 240.
2. Nicolaou, K. C.; Montagnon, T. *Molecules that Changed the World*. Wiley-VCH: Weinheim, **2008**, p. 366.
3. Meth-Cohn, O; Smith, M. "What did W. D. Perkin actually make when he oxidised aniline to obtain mauveine?", *J. Chem. Soc. Perkin Trans 1*, **1994**, 5–7.

SUMMARY AND REVIEW TOOLS

In Chapter 3 you studied acid–base chemistry, one of the most important topics needed to learn organic chemistry. If you master acid–base chemistry you will be able to understand most of the reactions that you study in organic chemistry, and by understanding how reactions work, you will be able to learn and remember them more easily.

You have reviewed the Brønsted–Lowry definition of acids and bases and the meanings of pH and pK_a. You have learned to identify the most acidic hydrogen atoms in a molecule based on a comparison of pK_a values. You will see in many cases that Brønsted–Lowry acid–base reactions either initiate or complete an organic reaction, or prepare an organic molecule for further reaction. The Lewis definition of acids and bases may have been new to you. However, you will see over and over again that Lewis acid–base reactions which involve either the donation of an electron pair to form a new covalent bond or the departure of an electron pair to break a covalent bond are central steps in many organic reactions. The vast majority of organic reactions you will study are either Brønsted–Lowry or Lewis acid–base reactions.

Your knowledge of organic structure and polarity from Chapters 1 and 2 has been crucial to your understanding of acid–base reactions. You have seen that stabilization of charge by delocalization is key to determining how readily an acid will give up a proton, or how readily a base will accept a proton. In addition, you have learned the essential skill of drawing curved arrows to accurately show the movement of electrons in these processes. With these concepts and skills you will be prepared to understand how organic reactions occur on a step-by-step basis—something organic chemists call "a mechanism for the reaction."

So, continue to work hard to master acid–base chemistry and other fundamentals. Your toolbox is quickly filling with the tools you need for overall success in organic chemistry!

The study aids for this chapter include key terms and concepts (which are hyperlinked to the glossary from the bold, blue terms in the *WileyPLUS* version of the book at wileyplus.com) and a Concept Map after the end-of-chapter problems.

PROBLEMS WILEY PLUS

Note to Instructors: Many of the homework problems are available for assignment via *WileyPLUS*, an online teaching and learning solution.

BRØNSTED–LOWRY ACIDS AND BASES

3.18 What is the conjugate base of each of the following acids?

(a) NH_3 (b) H_2O (c) H_2 (d) $HC\equiv CH$ (e) CH_3OH (f) H_3O^+

3.19 List the bases you gave as answers to Exercise 3.18 in order of decreasing basicity.

3.20 What is the conjugate acid of each of the following bases?

(a) HSO_4^- (b) H_2O (c) CH_3NH_2 (d) $^-NH_2$ (e) $CH_3\bar{C}H_2$ (f) $CH_3CO_2^-$

3.21 List the acids you gave as answers to Exercise 3.20 in order of decreasing acidity.

LEWIS ACIDS AND BASES

3.22 Designate the Lewis acid and Lewis base in each of the following reactions:

(a) 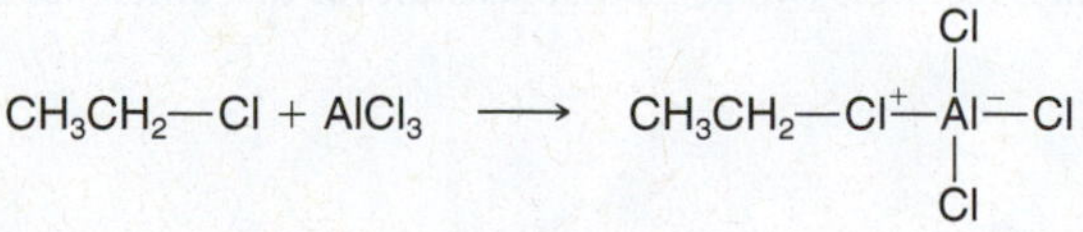 $CH_3CH_2{-}Cl + AlCl_3 \longrightarrow CH_3CH_2{-}Cl^+{-}\overset{Cl}{\underset{Cl}{Al}}{=}Cl$

(c) 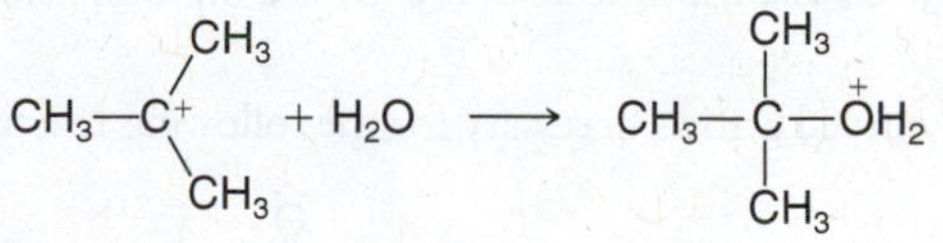 $CH_3{-}\overset{CH_3}{\underset{CH_3}{C^+}} + H_2O \longrightarrow CH_3{-}\overset{CH_3}{\underset{CH_3}{C}}{-}\overset{+}{O}H_2$

(b) 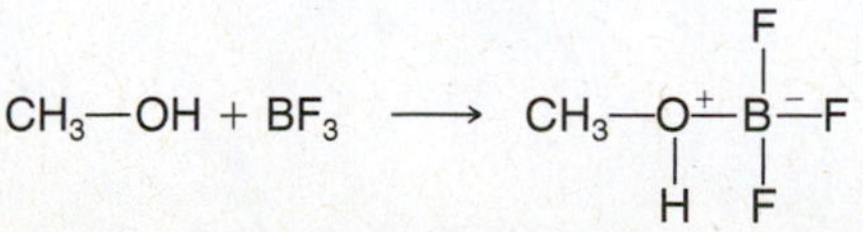 $CH_3{-}OH + BF_3 \longrightarrow CH_3{-}\overset{+}{\underset{H}{O}}{-}\overset{F}{\underset{F}{B}}{=}F$

CURVED-ARROW NOTATION

3.23 Rewrite each of the following reactions using curved arrows and show all nonbonding electron pairs:

(a) $CH_3OH + HI \longrightarrow CH_3\overset{+}{O}H_2 + I^-$

(b) $CH_3NH_2 + HCl \longrightarrow CH_3\overset{+}{N}H_3 + Cl^-$

(c)

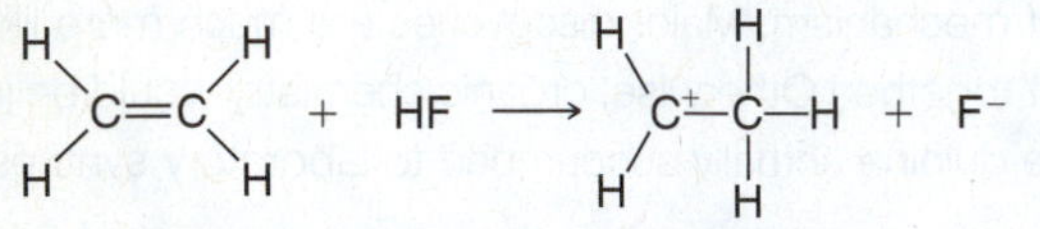

$$\begin{array}{c} H \\ C=C \\ H \end{array} \begin{array}{c} H \\ \\ H \end{array} + HF \longrightarrow \begin{array}{c} H \\ \overset{+}{C}-C-H \\ H \end{array} \begin{array}{c} H \\ \\ H \end{array} + F^-$$

3.24 Follow the curved arrows and write the products.

(a)

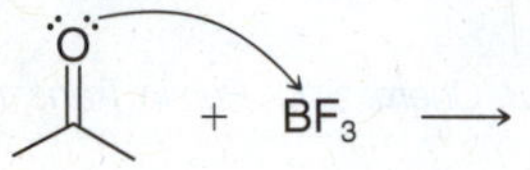

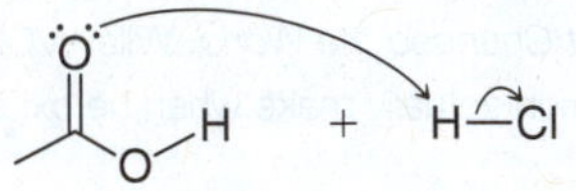

(b)

(c)

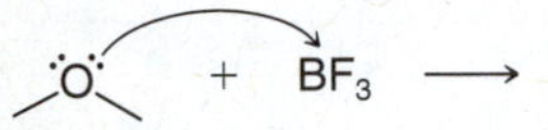

(d)

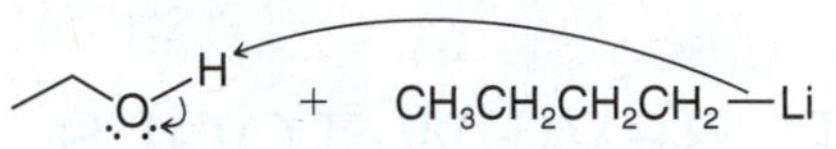

3.25 Write an equation, using the curved-arrow notation, for the acid–base reaction that will take place when each of the following are mixed. If no appreciable acid–base reaction takes place, because the equilibrium is unfavorable, you should so indicate.

(a) Aqueous NaOH and $CH_3CH_2CO_2H$

(b) Aqueous NaOH and $C_6H_5SO_3H$

(c) CH_3CH_2ONa in ethyl alcohol and ethyne

(d) CH_3CH_2Li in hexane and ethyne

(e) CH_3CH_2Li in hexane and ethyl alcohol

ACID–BASE STRENGTH AND EQUILIBRIA

3.26 When methyl alcohol is treated with NaH, the product is $CH_3O^-Na^+$ (and H_2) and not $Na^+ {}^-CH_2OH$ (and H_2). Explain why this is so.

3.27 What reaction will take place if ethyl alcohol is added to a solution of $HC\equiv C:^- Na^+$ in liquid ammonia?

3.28 (a) The K_a of formic acid (HCO_2H) is 1.77×10^{-4}. What is the pK_a?   **(b)** What is the K_a of an acid whose $pK_a = 13$?

3.29 Acid HA has $pK_a = 20$; acid HB has $pK_a = 10$.

(a) Which is the stronger acid?

(b) Will an acid–base reaction with an equilibrium lying to the right take place if Na^+A^- is added to HB? Explain your answer.

3.30 Starting with appropriate unlabeled organic compounds, show syntheses of each of the following:

(a) $C_6H_5-C\equiv C-T$ **(b)** $CH_3-\underset{\underset{CH_3}{|}}{CH}-O-D$ **(c)** $CH_3CH_2CH_2OD$

3.31 (a) Arrange the following compounds in order of decreasing acidity and explain your answer: $CH_3CH_2NH_2$, CH_3CH_2OH, and $CH_3CH_2CH_3$.   **(b)** Arrange the conjugate bases of the acids given in part (a) in order of increasing basicity and explain your answer.

3.32 Arrange the following compounds in order of decreasing acidity:

(a) $CH_3CH=CH_2$, $CH_3CH_2CH_3$, $CH_3C\equiv CH$ **(c)** CH_3CH_2OH, $CH_3CH_2\overset{+}{O}H_2$, CH_3OCH_3

(b) $CH_3CH_2CH_2OH$, $CH_3CH_2CO_2H$, $CH_3CHClCO_2H$

3.33 Arrange the following in order of increasing basicity:

(a) CH_3NH_2, $CH_3\overset{+}{N}H_3$, $CH_3\bar{N}H$ **(c)** $CH_3CH=\bar{C}H$, $CH_3CH_2\bar{C}H_2$, $CH_3C\equiv C^-$

(b) CH_3O^-, $CH_3\bar{N}H$, $CH_3\bar{C}H_2$

GENERAL PROBLEMS

3.34 Whereas H_3PO_4 is a triprotic acid, H_3PO_3 is a diprotic acid. Draw structures for these two acids that account for this difference in behavior.

3.35 Supply the curved arrows necessary for the following reactions:

(a)

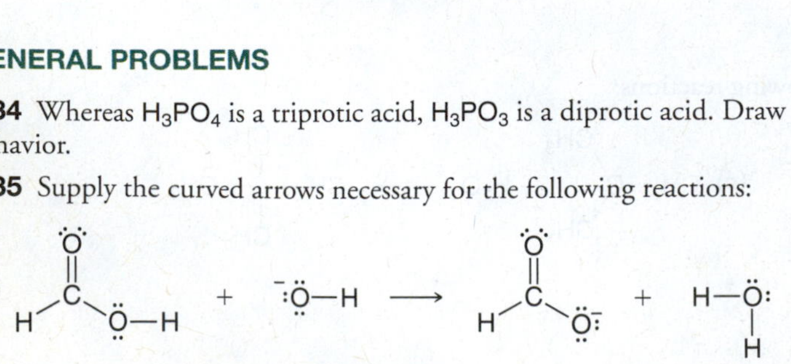

(b) [structure: methyl formate with carbonyl O, H—C—O—CH₃] + ⁻:Ö—H ⟶ [structure: H—C(—Ö⁻)(—Ö—H)(—Ö—CH₃)]

(c) [structure: H—C(—Ö⁻)(—O—CH₃)(—Ö—H) with :Ö—CH₃] ⟶ [structure: carbonyl, H—C(=O)—Ö—H] + ⁻:Ö—CH₃

(d) H—Ö:⁻ + CH₃—Ï: ⟶ H—Ö—CH₃ + :Ï:⁻

(e) H—Ö:⁻ + H—CH₂—C(CH₃)(CH₃)—Cl: ⟶ [structure: H₂C=C(CH₃)(CH₃) alkene] + :Cl:⁻ + H—Ö—H

3.36 Glycine is an amino acid that can be obtained from most proteins. In solution, glycine exists in equilibrium between two forms:

$$H_2NCH_2CO_2H \rightleftharpoons H_3\overset{+}{N}CH_2CO_2^-$$

(a) Consult Table 3.1 and state which form is favored at equilibrium.

(b) A handbook gives the melting point of glycine as 262 °C (with decomposition). Which of the structures given above best represents glycine?

3.37 Malonic acid, $HO_2CCH_2CO_2H$, is a diprotic acid. The pK_a for the loss of the first proton is 2.83; the pK_a for the loss of the second proton is 5.69.   **(a)** Explain why malonic acid is a stronger acid than acetic acid ($pK_a = 4.75$).   **(b)** Explain why the anion, $^-O_2CCH_2CO_2H$, is so much less acidic than malonic acid itself.

3.38 The free-energy change, $\Delta G°$, for the ionization of acid **HA** is 21 kJ mol^{-1}; for acid **HB** it is -21 kJ mol^{-1}. Which is the stronger acid?

3.39 At 25 °C the enthalpy change, $\Delta H°$, for the ionization of trichloroacetic acid is $+6.3$ kJ mol^{-1} and the entropy change, $\Delta S°$, is $+0.0084$ kJ mol^{-1} K^{-1}. What is the pK_a of trichloroacetic acid?

3.40 The compound at right has (for obvious reasons) been given the trivial name **squaric acid**. Squaric acid is a diprotic acid, with both protons being more acidic than acetic acid. In the dianion obtained after the loss of both protons, all of the carbon–carbon bonds are the same length as well as all of the carbon–oxygen bonds. Provide a resonance explanation for these observations.

Squaric acid

CHALLENGE PROBLEMS

3.41 CH₃CH₂SH + CH₃O⁻ ⟶ **A** (contains sulfur) + **B**

   A + CH₂—CH₂ (epoxide, O bridging) ⟶ **C** (which has the partial structure **A**—CH₂CH₂O)

   C + H₂O ⟶ **D** + **E** (which is inorganic)

(a) Given the above sequence of reactions, draw structures for **A** through **E**.

(b) Rewrite the reaction sequence, showing all nonbonding electron pairs and using curved arrows to show electron pair movements.

3.42 First, complete and balance each of the equations below. Then, choosing among ethanol, hexane, and liquid ammonia, state which (there may be more than one) might be suitable solvents for each of these reactions. Disregard the practical limitations that come from consideration of "like dissolves like" and base your answers only on relative acidities.

(a) CH₃(CH₂)₈OD + CH₃(CH₂)₈Li ⟶   **(c)** HCl + [benzene ring]—NH₂ ⟶

(b) NaNH₂ + CH₃C≡CH ⟶

   (The conjugate acid of this amine, aniline, has a pK_a of 4.63.)

3.43 Dimethylformamide (DMF), HCON(CH₃)₂, is an example of a polar aprotic solvent, aprotic meaning it has no hydrogen atoms attached to highly electronegative atoms.

(a) Draw its dash structural formula, showing unshared electron pairs.

(b) Draw what you predict to be its most important resonance forms [one is your answer to part (a)].

(c) DMF, when used as the reaction solvent, greatly enhances the reactivity of nucleophiles (e.g., ⁻CN from sodium cyanide) in reactions like this:

$$NaCN + CH_3CH_2Br \longrightarrow CH_3CH_2C \equiv N + NaBr$$

Suggest an explanation for this effect of DMF on the basis of Lewis acid–base considerations. (*Hint*: Although water or an alcohol solvates both cations and anions, DMF is only effective in solvating cations.)

3.44 As noted in Table 3.1, the pK_a of acetone, CH_3COCH_3, is 19.2.

(a) Draw the bond-line formula of acetone and of any other contributing resonance form.

(b) Predict and draw the structure of the conjugate base of acetone and of any other contributing resonance form.

(c) Write an equation for a reaction that could be used to synthesize CH_3COCH_2D.

3.45 Formamide ($HCONH_2$) has a pK_a of approximately 25. Predict, based on the map of electrostatic potential for formamide shown here, which hydrogen atom(s) has this pK_a value. Support your conclusion with arguments having to do with the electronic structure of formamide.

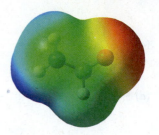

LEARNING GROUP PROBLEMS

Suppose you carried out the following synthesis of 3-methylbutyl ethanoate (isoamyl acetate):

| Ethanoic acid (excess) | + | 3-Methyl-1-butanol | H_2SO_4 (trace) | 3-Methylbutyl ethanoate | + | H_2O |

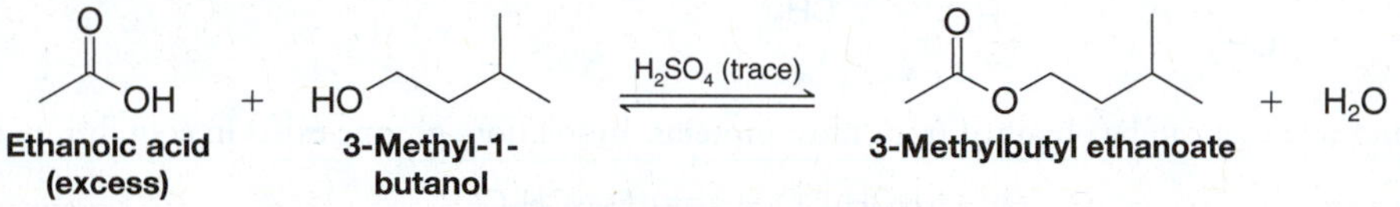

As the chemical equation shows, 3-methyl-1-butanol (also called isoamyl alcohol or isopentyl alcohol) was mixed with an excess of acetic acid (ethanoic acid by its systematic name) and a trace of sulfuric acid (which serves as a catalyst). This reaction is an equilibrium reaction, so it is expected that not all of the starting materials will be consumed. The equilibrium should lie quite far to the right due to the excess of acetic acid used, but not completely.

After an appropriate length of time, isolation of the desired product from the reaction mixture was begun by adding a volume of 5% aqueous sodium bicarbonate ($NaHCO_3$ has an effective pK_a of 7) roughly equal to the volume of the reaction mixture. Bubbling occurred and a mixture consisting of two layers resulted—a basic aqueous layer and an organic layer. The layers were separated and the aqueous layer was removed. The addition of aqueous sodium bicarbonate to the layer of organic materials and separation of the layers were repeated twice. Each time the predominantly aqueous layers were removed, they were combined in the same collection flask. The organic layer that remained after the three bicarbonate extractions was dried and then subjected to distillation in order to obtain a pure sample of 3-methylbutyl ethanoate (isoamyl acetate).

1. List all the chemical species likely to be present at the end of the reaction but before adding aqueous $NaHCO_3$. Note that the H_2SO_4 was not consumed (since it is a catalyst), and is thus still available to donate a proton to atoms that can be protonated.

2. Use a table of pK_a values, such as Table 3.1, to estimate pK_a values for any potentially acidic hydrogens in each of the species you listed in part 1 (or for the conjugate acid).

3. Write chemical equations for all the acid–base reactions you would predict to occur (based on the pK_a values you used) when the species you listed above encounter the aqueous sodium bicarbonate solution. (*Hint:* Consider whether each species might be an acid that could react with $NaHCO_3$.)

4. (a) Explain, on the basis of polarities and solubility, why separate layers formed when aqueous sodium bicarbonate was added to the reaction mixture. (*Hint:* Most sodium salts of organic acids are soluble in water, as are neutral oxygen-containing organic compounds of four carbons or less.)

(b) List the chemical species likely to be present after the reaction with $NaHCO_3$ in (i) the organic layer and (ii) the aqueous layer.

(c) Why was the aqueous sodium bicarbonate extraction step repeated three times?

[**C O N C E P T   M A P**]

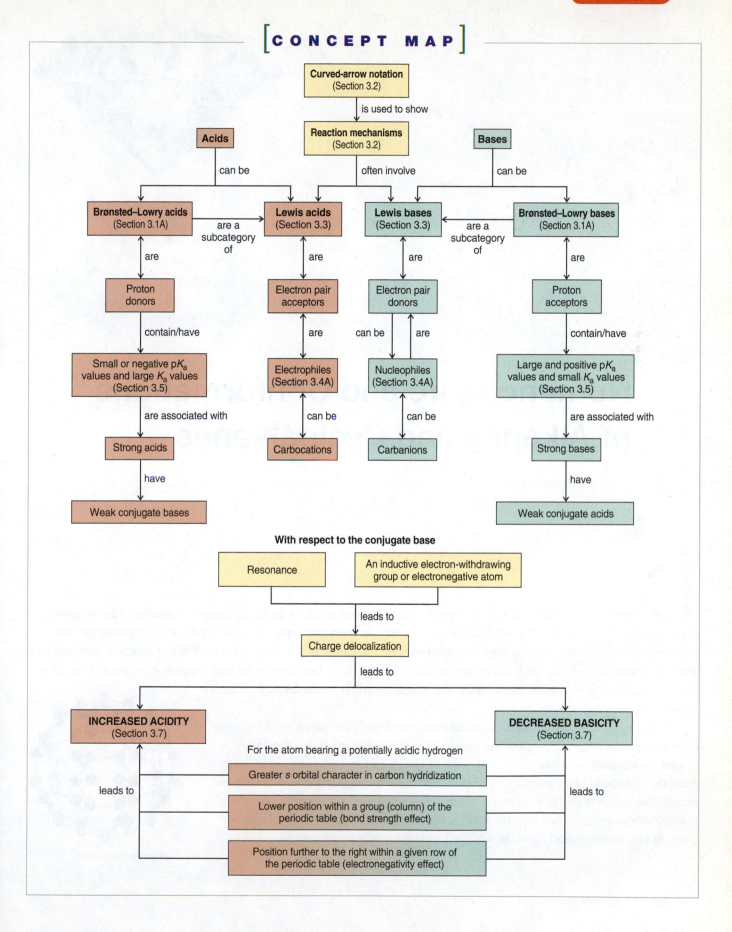

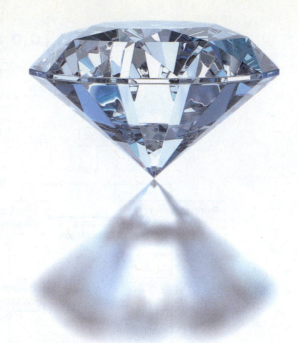

CHAPTER

4

Nomenclature and Conformations of Alkanes and Cycloalkanes

Diamond is an exceptionally hard material. One reason diamond is so strong is that it contains a rigid network of carbon-carbon bonds. Muscle, on the other hand, which also contains many carbon-carbon bonds, is strong yet has great flexibility. This remarkable contrast in properties, from the rigidity of diamond to the flexibility of muscles, depends on whether rotation is possible about individual carbon-carbon bonds. In this chapter we shall consider changes in molecular structure and energy that result from rotation about carbon-carbon bonds, using a process called conformational analysis.

We learned in Chapter 2 that our study of organic chemistry can be organized around functional groups. Now we consider the hydrocarbon framework to which functional groups are attached—the framework that consists of only carbon and hydrogen atoms. From the standpoint of an architect, hydrocarbon frameworks present a dream of limitless possibilities, which is part of what makes organic chemistry such a fascinating discipline. Buckminsterfullerene, named after the visionary architect Buckminster Fuller, is just one example of a carbon-based molecule with an intriguing molecular architecture.

Buckminsterfullerene

PHOTO CREDIT: © Evgeny Terentev/iStockphoto

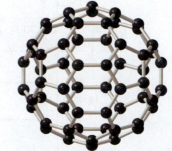

IN THIS CHAPTER WE WILL CONSIDER:

- how to name many simple organic molecules
- the flexible, three-dimensional nature of organic molecules
- an organic reaction that can convert alkenes and alkynes to alkanes

[WHY DO THESE TOPICS MATTER?]   At the end of the chapter, we will show how, using the same set of rules, both chemists and nature have created some unique arrangements of carbon and hydrogen atoms. Some of these structural arrangements were not expected to exist, one structural arrangement lets you write, and others are fueling advances in the area of materials research and nanotechnology.

4.1 INTRODUCTION TO ALKANES AND CYCLOALKANES

We noted earlier that the family of organic compounds called hydrocarbons can be divided into several groups on the basis of the type of bond that exists between the individual carbon atoms. Those hydrocarbons in which all of the carbon–carbon bonds are single bonds are called **alkanes**, those hydrocarbons that contain a carbon–carbon double bond are called **alkenes**, and those with a carbon–carbon triple bond are called **alkynes**.

Cycloalkanes are alkanes in which all or some of the carbon atoms are arranged in a ring. Alkanes have the general formula C_nH_{2n+2}; cycloalkanes containing a single ring have two fewer hydrogen atoms and thus have the general formula C_nH_{2n}.

Cyclohexane

4.1A Sources of Alkanes: Petroleum

The primary source of alkanes is petroleum. Petroleum is a complex mixture of organic compounds, most of which are alkanes and aromatic compounds (cf. Chapter 14). It also contains small amounts of oxygen-, nitrogen-, and sulfur-containing compounds.

Some of the molecules in petroleum are clearly of biological origin. Most scientists believe that petroleum originated with accumulation of dead microorganisms that settled to the bottom of the sea and that were entombed in sedimentary rock. These microbial remains eventually were transformed into oil by the heat radiating from Earth's core.

Hydrocarbons are also found in outer space. Asteroids and comets contain a variety of organic compounds. Methane and other hydrocarbons are found in the atmospheres of Jupiter, Saturn, and Uranus. Saturn's moon Titan has a solid form of methane–water ice at its surface and an atmosphere rich in methane. Whether of terrestrial or celestial origin, we need to understand the properties of alkanes. We begin with a consideration of their shapes and how we name them.

Tom McHugh/Photo Researchers, Inc.

Petroleum is a finite resource that likely originated with decay of primordial microbes. At the La Brea Tar Pits in Los Angeles, many prehistoric animals perished in a natural vat containing hydrocarbons.

THE CHEMISTRY OF... Petroleum Refining

The first step in refining petroleum is distillation; the object here is to separate the petroleum into fractions based on the volatility of its components. Complete separation into fractions containing individual compounds is economically impractical and virtually impossible technically. More than 500 different compounds are contained in the petroleum distillates boiling below 200 °C, and many have almost the same boiling points. Thus the fractions taken contain mixtures of alkanes of similar boiling points (see the table below). Mixtures of alkanes, fortunately, are perfectly suitable for uses as fuels, solvents, and lubricants, the primary uses of petroleum.

The demand for gasoline is much greater than that supplied by the gasoline fraction of petroleum. Important processes in the petroleum industry, therefore, are concerned with converting hydrocarbons from other fractions into gasoline. When a mixture of alkanes from the gas oil fraction (C_{12} and higher) is heated at very high temperatures (~500 °C) in the presence of a variety of catalysts, the molecules break apart and rearrange to smaller, more highly branched hydrocarbons containing 5–10 carbon atoms. This process is called *catalytic cracking*. **Cracking** can also be done in the absence of a catalyst—called ***thermal***

(continues on next page)

cracking—but in this process the products tend to have unbranched chains, and alkanes with unbranched chains have a very low "octane rating."

The highly branched compound 2,2,4-trimethylpentane (called isooctane in the petroleum industry) burns very smoothly (without knocking) in internal combustion engines and is used as one of the standards by which the octane rating of gasolines is established. According to this scale, 2,2,4-trimethylpentane has an octane rating of 100. Heptane, $CH_3(CH_2)_5CH_3$, a compound that produces much knocking when it is burned in an internal combustion engine, is given an octane rating of 0. Mixtures of 2,2,4-trimethylpentane and heptane are used as standards for octane ratings between 0 and 100. A gasoline, for example, that has the same characteristics in an engine as a mixture of 87% 2,2,4-trimethylpentane and 13% heptane would be rated as 87-octane gasoline.

2,2,4-Trimethylpentane ("isooctane")

TYPICAL FRACTIONS OBTAINED BY DISTILLATION OF PETROLEUM

Boiling Range of Fraction (°C)	Number of Carbon Atoms per Molecule	Use
Below 20	C_1–C_4	Natural gas, bottled gas, petrochemicals
20–60	C_5–C_6	Petroleum ether, solvents
60–100	C_6–C_7	Ligroin, solvents
40–200	C_5–C_{10}	Gasoline (straight-run gasoline)
175–325	C_{12}–C_{18}	Kerosene and jet fuel
250–400	C_{12} and higher	Gas oil, fuel oil, and diesel oil
Nonvolatile liquids	C_{20} and higher	Refined mineral oil, lubricating oil, and grease
Nonvolatile solids	C_{20} and higher	Paraffin wax, asphalt, and tar

Adapted with permission of John Wiley & Sons, Inc., from Holum, J. R., *General, Organic, and Biological Chemistry*, Ninth Edition, p. 213. Copyright 1995.

4.2 SHAPES OF ALKANES

A general tetrahedral orientation of groups—and thus sp^3 hybridization—is the rule for the carbon atoms of all alkanes and cycloalkanes. We can represent the shapes of **alkanes** as shown in Fig. 4.1.

Butane and pentane are examples of alkanes that are sometimes called "straight-chain" alkanes. One glance at three-dimensional models, however, shows that because of their tetrahedral carbon atoms the chains are zigzagged and not at all straight. Indeed,

FIGURE 4.1 Ball-and-stick models for three simple alkanes.

Propane	Butane	Pentane
$CH_3CH_2CH_3$	$CH_3CH_2CH_2CH_3$	$CH_3CH_2CH_2CH_2CH_3$
or	or	or

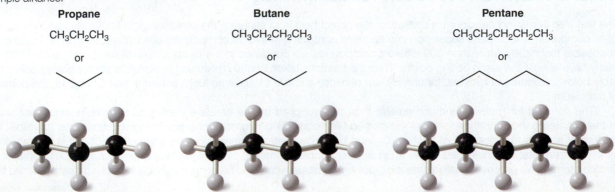

the structures that we have depicted in Fig. 4.1 are the straightest possible arrangements of the chains because rotations about the carbon–carbon single bonds produce arrangements that are even less straight. A better description is **unbranched**. This means that each carbon atom within the chain is bonded to no more than two other carbon atoms and that unbranched alkanes contain only primary and secondary carbon atoms. Primary, secondary, and tertiary carbon atoms were defined in Section 2.5.

Isobutane, isopentane, and neopentane (Fig. 4.2) are examples of branched-chain alkanes. In neopentane the central carbon atom is bonded to four carbon atoms.

> **Helpful Hint**
>
> You should build your own molecular models of the compounds in Figs. 4.1 and 4.2. View them from different perspectives and experiment with how their shapes change when you twist various bonds. Make drawings of your structures.

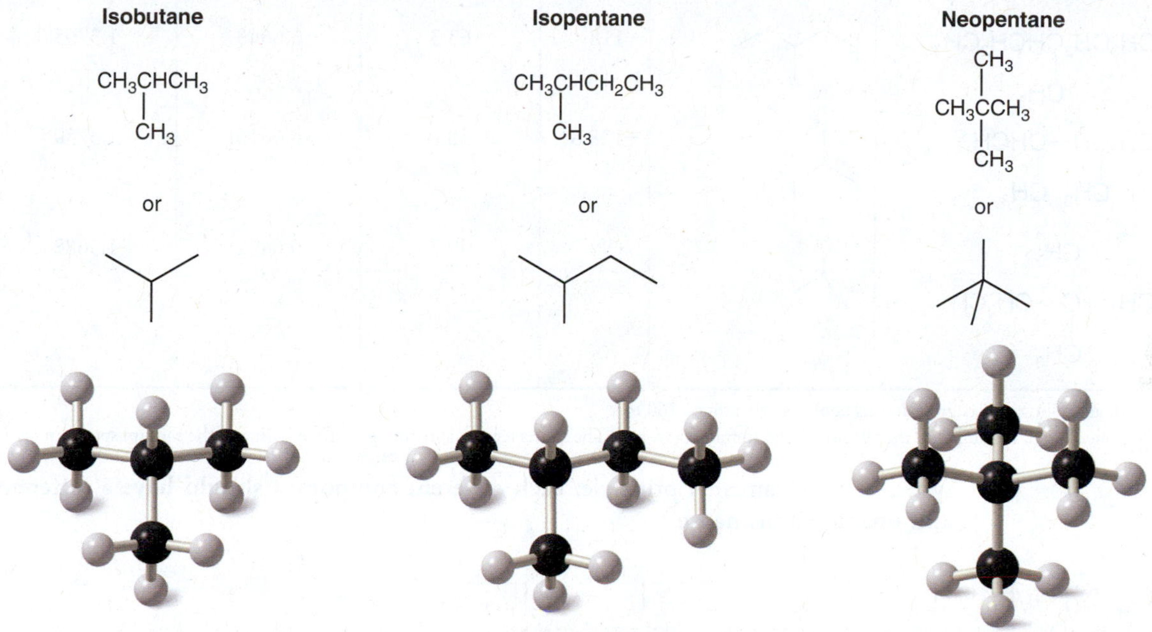

FIGURE 4.2 Ball-and-stick models for three branched-chain alkanes. In each of the compounds one carbon atom is attached to more than two other carbon atoms.

Butane and isobutane have the same molecular formula: C_4H_{10}. The two compounds have their atoms connected in a different order and are, therefore, **constitutional isomers** (Section 1.3). Pentane, isopentane, and neopentane are also constitutional isomers. They, too, have the same molecular formula (C_5H_{12}) but have different structures.

● ● ●

PRACTICE PROBLEM 4.1

Write condensed and bond-line structural formulas for all of the constitutional isomers with the molecular formula C_7H_{16}. (There are a total of nine constitutional isomers.)

Constitutional isomers, as stated earlier, have different physical properties. The differences may not always be large, but constitutional isomers are always found to have different melting points, boiling points, densities, indexes of refraction, and so forth. Table 4.1 gives some of the physical properties of the C_6H_{14} isomers, of which there are only five. Note that the number of constitutional isomers that is possible increases dramatically as the number of carbon atoms in the alkane increases.

Prior to the development near the end of the nineteenth century of a formal system for naming organic compounds, many organic compounds had already been discovered or synthesized. Early chemists named these compounds, often on the basis of the source of the compound. Acetic acid (systematically called ethanoic acid) is an example; it was obtained by distilling vinegar, and it got its name from the Latin word for vinegar, *acetum*. Formic acid (systematically called methanoic acid) had been obtained by the distillation of the bodies of ants, so it got the name from the Latin word for ants, *formicae*. Many of these older names for compounds, called common or trivial names, are still in wide use today.

Today, chemists use a systematic nomenclature developed and updated by the International Union of Pure and Applied Chemistry (IUPAC). Underlying the IUPAC

TABLE 4.1 PHYSICAL CONSTANTS OF THE HEXANE ISOMERS

Molecular Formula	Condensed Structural Formula	Bond-Line Formula	mp (°C)	bp (°C)[a] (1 atm)	Density (g mL^{-1}) at 20 °C	Index of Refraction[b] (n_D 20 °C)
C_6H_{14}	$CH_3CH_2CH_2CH_2CH_2CH_3$		−95	68.7	0.6594	1.3748
C_6H_{14}	$CH_3CHCH_2CH_2CH_3$ CH_3	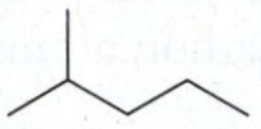	−153.7	60.3	0.6532	1.3714
C_6H_{14}	$CH_3CH_2CHCH_2CH_3$ CH_3	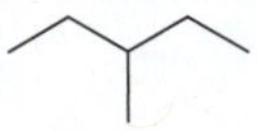	−118	63.3	0.6643	1.3765
C_6H_{14}	$CH_3CH-CHCH$ CH_3 CH_3	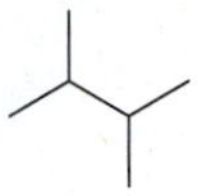	−128.8	58	0.6616	1.3750
C_6H_{14}	CH_3 CH_3-C-CH_2CH CH_3	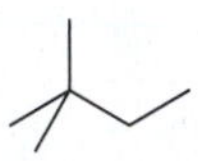	−98	49.7	0.6492	1.3688

[a]Unless otherwise indicated, all boiling points given in this book are at 1 atm or 760 torr.
[b]The index of refraction is a measure of the ability of the alkane to bend (refract) light rays. The values reported are for light of the D line of the sodium spectrum (n_D).

system is a fundamental principle: **each different compound should have a different and unambiguous name**.*

4.3 HOW TO NAME ALKANES, ALKYL HALIDES, AND ALCOHOLS: THE IUPAC SYSTEM

The **IUPAC system** for naming **alkanes** is not difficult to learn, and the principles involved are used in naming compounds in other families as well. For these reasons we begin our study of the IUPAC system with the rules for naming alkanes and then study the rules for alkyl halides and alcohols.

The names for several of the unbranched alkanes are listed in Table 4.2. The ending for all of the names of alkanes is -*ane*. The stems of the names of most of the alkanes (above C_4) are of Greek and Latin origin. Learning the stems is like learning to count in organic chemistry. Thus, one, two, three, four, and five become meth-, eth-, prop-, but-, and pent-.

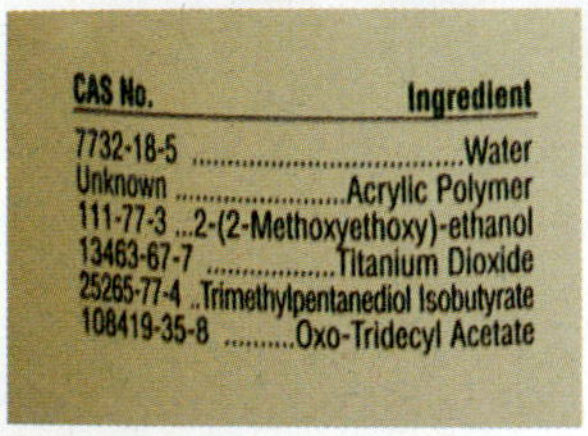

TABLE 4.2 THE UNBRANCHED ALKANES

Name	Number of Carbon Atoms	Structure	Name	Number of Carbon Atoms	Structure
Methane	1	CH_4	Undecane	11	$CH_3(CH_2)_9CH_3$
Ethane	2	CH_3CH_3	Dodecane	12	$CH_3(CH_2)_{10}CH_3$
Propane	3	$CH_3CH_2CH_3$	Tridecane	13	$CH_3(CH_2)_{11}CH_3$
Butane	4	$CH_3(CH_2)_2CH_3$	Tetradecane	14	$CH_3(CH_2)_{12}CH_3$
Pentane	5	$CH_3(CH_2)_3CH_3$	Pentadecane	15	$CH_3(CH_2)_{13}CH_3$
Hexane	6	$CH_3(CH_2)_4CH_3$	Hexadecane	16	$CH_3(CH_2)_{14}CH_3$
Heptane	7	$CH_3(CH_2)_5CH_3$	Heptadecane	17	$CH_3(CH_2)_{15}CH_3$
Octane	8	$CH_3(CH_2)_6CH_3$	Octadecane	18	$CH_3(CH_2)_{16}CH_3$
Nonane	9	$CH_3(CH_2)_7CH_3$	Nonadecane	19	$CH_3(CH_2)_{17}CH_3$
Decane	10	$CH_3(CH_2)_8CH_3$	Eicosane	20	$CH_3(CH_2)_{18}CH_3$

CAS No.	Ingredient
7732-18-5	Water
Unknown	Acrylic Polymer
111-77-3	2-(2-Methoxyethoxy)-ethanol
13463-67-7	Titanium Dioxide
25265-77-4	Trimethylpentanediol Isobutyrate
108419-35-8	Oxo-Tridecyl Acetate

Photo by Lisa Gee

The Chemical Abstracts Service assigns a CAS Registry Number to every compound. CAS numbers make it easy to find information about a compound in the chemical literature. The CAS numbers for ingredients in a can of latex paint are shown here.

*The complete IUPAC rules for nomenclature can be found through links at the IUPAC website.

4.3A **HOW TO** Name Unbranched Alkyl Groups

If we remove one hydrogen atom from an alkane, we obtain what is called an **alkyl group**. These alkyl groups have names that end in **-yl**. When the alkane is **unbranched**, and the hydrogen atom that is removed is a **terminal** hydrogen atom, the names are straightforward:

CH_3-H	CH_3CH_2-H	$CH_3CH_2CH_2-H$	$CH_3CH_2CH_2CH_2-H$
Methane	**Ethane**	**Propane**	**Butane**
CH_3-	CH_3CH_2-	$CH_3CH_2CH_2-$	$CH_3CH_2CH_2CH_2-$
Methyl	**Ethyl**	**Propyl**	**Butyl**
Me-	**Et-**	**Pr-**	**Bu-**

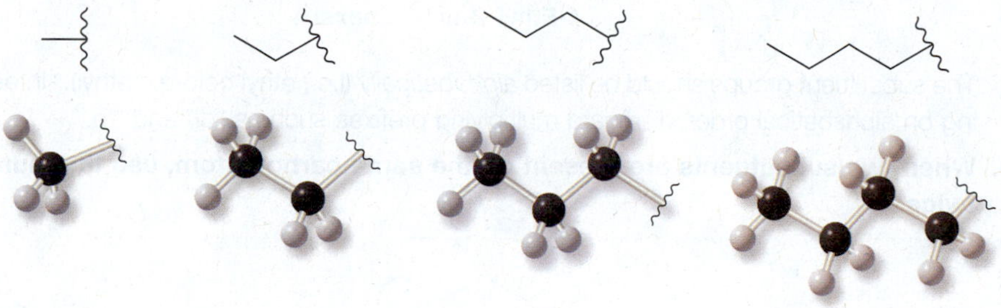

4.3B **HOW TO** Name Branched-Chain Alkanes

Branched-chain alkanes are named according to the following rules:

1. **Locate the longest continuous chain of carbon atoms; this chain determines the parent name for the alkane**. We designate the following compound, for example, as a *hexane* because the longest continuous chain contains six carbon atoms:

$$CH_3CH_2CH_2CH_2CHCH_3$$
$$| \quad CH_3$$

Longest chain

The longest continuous chain may not always be obvious from the way the formula is written. Notice, for example, that the following alkane is designated as a *heptane* because the longest chain contains seven carbon atoms:

2. **Number the longest chain beginning with the end of the chain nearer the substituent**. Applying this rule, we number the two alkanes that we illustrated previously in the following way:

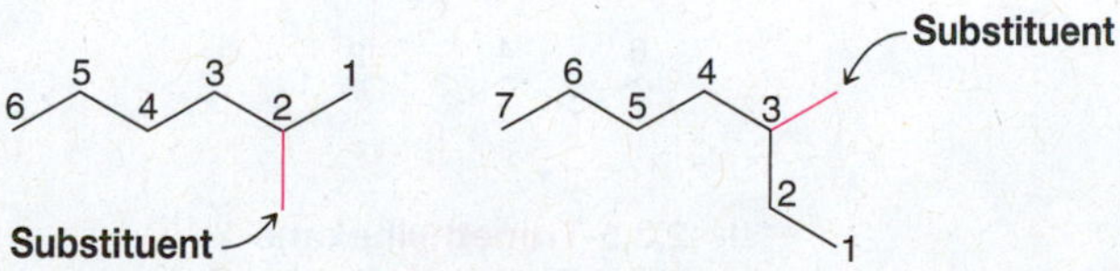

Substituent

3. **Use the numbers obtained by application of rule 2 to designate the location of the substituent group**. The parent name is placed last, and the substituent group, preceded by the number designating its location on the chain, is placed first. Numbers

are separated from words by a hyphen. Our two examples are 2-methylhexane and 3-methylheptane, respectively:

4. **When two or more substituents are present, give each substituent a number corresponding to its location on the longest chain.** For example, we designate the following compound as 4-ethyl-2-methylhexane:

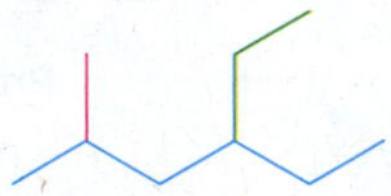

4-Ethyl-2-methylhexane

The substituent groups should be listed *alphabetically* (i.e., ethyl before methyl).* In deciding on alphabetical order, disregard multiplying prefixes such as "di" and "tri."

5. **When two substituents are present on the same carbon atom, use that number twice:**

3-Ethyl-3-methylhexane

6. **When two or more substituents are identical, indicate this by the use of the prefixes di-, tri-, tetra-,** and so on. Then make certain that each and every substituent has a number. Commas are used to separate numbers from each other:

2,3-Dimethylbutane   **2,3,4-Trimethylpentane**   **2,2,4,4-Tetramethylpentane**

Application of these six rules allows us to name most of the alkanes that we shall encounter. Two other rules, however, may be required occasionally:

7. **When two chains of equal length compete for selection as the parent chain, choose the chain with the greater number of substituents:**

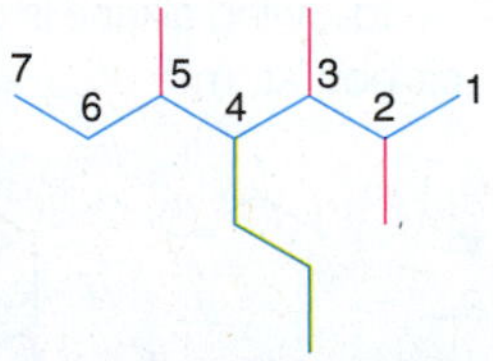

2,3,5-Trimethyl-4-propylheptane
(four substituents)

8. **When branching first occurs at an equal distance from either end of the longest chain, choose the name that gives the lower number at the first point of difference:**

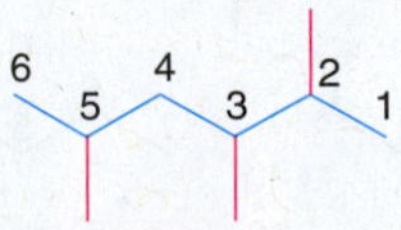

2,3,5-Trimethylhexane
(*not* 2,4,5-trimethylhexane)

---

*Some handbooks also list the groups in order of increasing size or complexity (i.e., methyl before ethyl). An alphabetical listing, however, is now by far the most widely used system.

●●● SOLVED PROBLEM 4.1

Provide an IUPAC name for the following alkane.

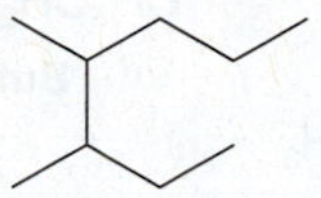

STRATEGY AND SOLUTION: We find the longest chain (shown in blue) to be seven carbons; therefore the parent name is heptane. There are two methyl substituents (shown in red). We number the chain so as to give the first methyl group the lower number. The correct name, therefore, is 3,4-dimethylheptane. Numbering the chain from the other end to give 4,5-dimethylheptane would have been incorrect.

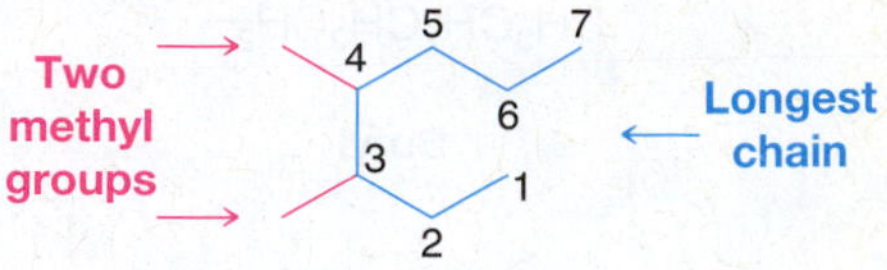

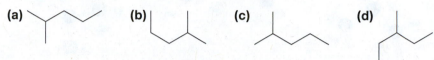

Two methyl groups

Longest chain

●●●

PRACTICE PROBLEM 4.2

Which structure does not represent 2-methylpentane?

(a)   **(b)**   **(c)**   **(d)**

●●●

PRACTICE PROBLEM 4.3

Write the structure and give the IUPAC name for an alkane with formula C_6H_{14} that has only primary and secondary carbon atoms.

●●●

PRACTICE PROBLEM 4.4

Draw bond-line formulas for all of the isomers of C_8H_{18} that have **(a)** methyl substituents, and **(b)** ethyl substituents.

4.3C HOW TO Name Branched Alkyl Groups

In Section 4.3A you learned the names for the unbranched alkyl groups such as methyl, ethyl, propyl, and butyl, groups derived by removing a terminal hydrogen from an alkane. For alkanes with more than two carbon atoms, more than one derived group is possible. Two groups can be derived from propane, for example; the **propyl group** is derived by removal of a terminal hydrogen, and the **1-methylethyl** or **isopropyl group** is derived by removal of a hydrogen from the central carbon:

Three-Carbon Groups

$$CH_3CH_2CH_2$$
Propane

$$CH_3CH_2CH_2—$$

Propyl

Pr-

$$CH_3CHCH_3$$

Isopropyl

i-Pr-

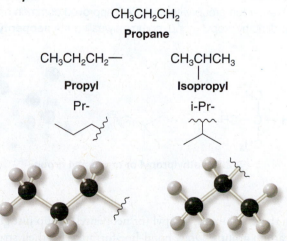

1-Methylethyl is the systematic name for this group; isopropyl is a common name. Systematic nomenclature for alkyl groups is similar to that for branched-chain alkanes, with the provision that *numbering always begins at the point where the group is attached to the main chain*. There are four C_4 groups.

Four-Carbon Groups

$$CH_3CH_2CH_2CH_3$$
Butane

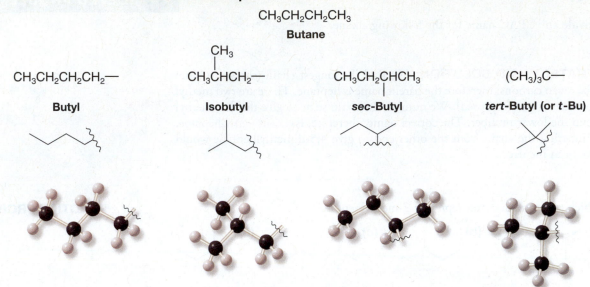

$CH_3CH_2CH_2CH_2-$

Butyl

$$\begin{array}{c} CH_3 \\ | \\ CH_3CHCH_2- \end{array}$$

Isobutyl

$$CH_3CH_2CHCH_3$$

sec-Butyl

$$(CH_3)_3C-$$

tert-Butyl (or t-Bu)

The following examples show how the names of these groups are employed:

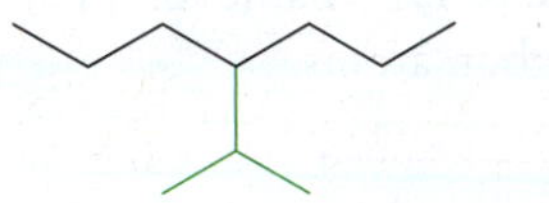

4-(1-Methylethyl)heptane or **4-isopropylheptane**

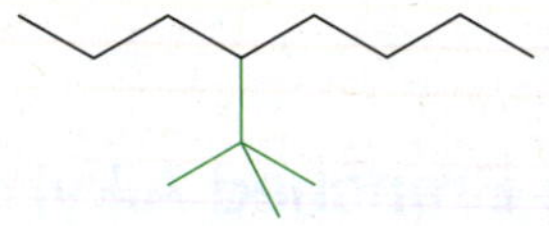

4-(1,1-Dimethylethyl)octane or **4-tert-butyloctane**

The common names **isopropyl, isobutyl, sec-butyl,** and **tert-butyl** are approved by the IUPAC for the unsubstituted groups, and they are still very frequently used. You should learn these groups so well that you can recognize them any way they are written. In deciding on alphabetical order for these groups you should disregard structure-defining prefixes that are written in italics and separated from the name by a hyphen. Thus *tert*-butyl precedes ethyl, but ethyl precedes isobutyl.*

There is one five-carbon group with an IUPAC approved common name that you should also know: the 2,2-dimethylpropyl group, commonly called the **neopentyl group**.

$$\begin{array}{c} CH_3 \\ | \\ CH_3-C-CH_2- \\ | \\ CH_3 \end{array}$$

2,2-Dimethylpropyl or neopentyl group

PRACTICE PROBLEM 4.5 **(a)** In addition to the 2,2-dimethylpropyl (or neopentyl) group just given, there are seven other five-carbon groups. Draw bond-line formulas for their structures and give each structure its systematic name. **(b)** Draw bond-line formulas and provide IUPAC names for all of the isomers of C_7H_{16}.

*The abbreviations *i-*, *s-*, and *t-* are sometimes used for iso-, *sec-*, and *tert-*, respectively.

4.3D **HOW TO** Classify Hydrogen Atoms

The hydrogen atoms of an alkane are classified on the basis of the carbon atom to which they are attached. A hydrogen atom attached to a primary carbon atom is a primary (1°) hydrogen atom, and so forth. The following compound, 2-methylbutane, has primary, secondary (2°), and tertiary (3°) hydrogen atoms:

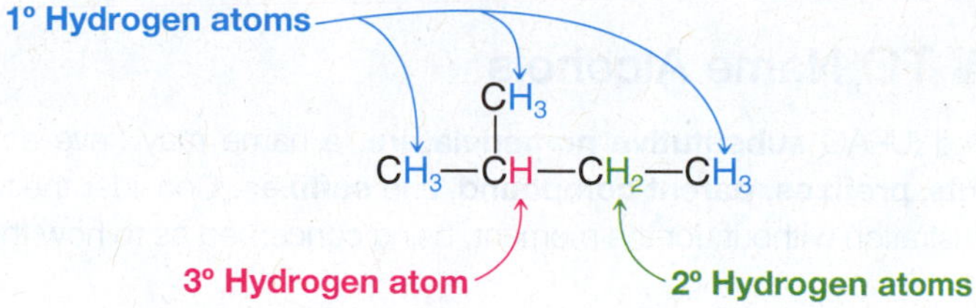

1° Hydrogen atoms

$$CH_3$$
$$CH_3-CH-CH_2-CH_3$$

3° Hydrogen atom   **2° Hydrogen atoms**

On the other hand, 2,2-dimethylpropane, a compound that is often called **neopentane**, has only primary hydrogen atoms:

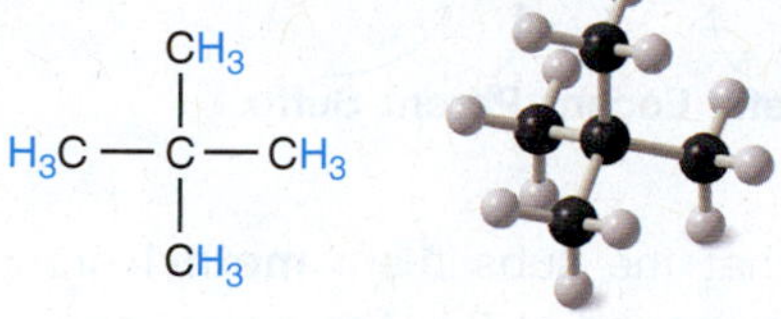

$$H_3C-\overset{\overset{\displaystyle CH_3}{|}}{\underset{\underset{\displaystyle CH_3}{|}}{C}}-CH_3$$

**2,2-Dimethylpropane
(neopentane)**

4.3E **HOW TO** Name Alkyl Halides

Alkanes bearing halogen substituents are named in the IUPAC substitutive system as haloalkanes:

CH_3CH_2Cl $CH_3CH_2CH_2F$ $CH_3CHBrCH_3$

Chloroethane   **1-Fluoropropane**   **2-Bromopropane**

- When the parent chain has both a halo and an alkyl substituent attached to it, number the chain from the end nearer the first substituent, regardless of whether it is halo or alkyl. If two substituents are at equal distance from the end of the chain, then number the chain from the end nearer the substituent that has alphabetical precedence:

2-Chloro-3-methylpentane   **2-Chloro-4-methylpentane**

Common names for many simple haloalkanes are still widely used, however. In this common nomenclature system, called **functional class nomenclature**, haloalkanes are named as alkyl halides. (The following names are also accepted by the IUPAC.)

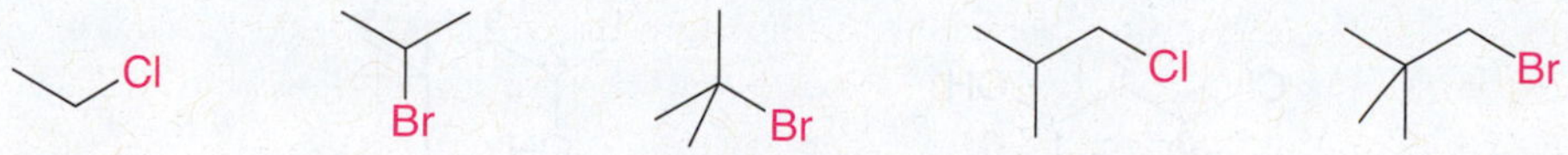

**Ethyl
chloride**   **Isopropyl
bromide**   **tert-Butyl
bromide**   **Isobutyl
chloride**   **Neopentyl
bromide**

• • •

PRACTICE PROBLEM 4.6 Draw bond-line formulas and give IUPAC substitutive names for all of the isomers of
(a) C_4H_9Cl and (b) $C_5H_{11}Br$.

4.3F HOW TO Name Alcohols

In what is called IUPAC **substitutive nomenclature**, a name may have as many as four
features: **locants, prefixes, parent compound**, and **suffixes**. Consider the following compound as an illustration without, for the moment, being concerned as to how the name arises:

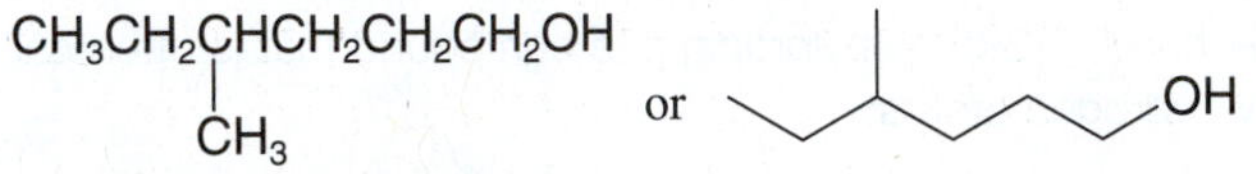

$$CH_3CH_2CHCH_2CH_2CH_2OH$$
$$| $$
$$CH_3$$

or [bond-line structure with OH]

4-Methyl-1-hexanol

Locant Prefix Locant Parent Suffix

The *locant* **4-** tells that the substituent **methyl** group, named as a *prefix*, is
attached to the *parent compound* at C4. The parent compound contains six carbon
atoms and no multiple bonds, hence the parent name **hexane**, and it is an alcohol;
therefore it has the *suffix* **-ol**. The locant **1-** tells that C1 bears the hydroxyl group.
**In general, numbering of the chain always begins at the end nearer the group
named as a suffix.**

The locant for a suffix (whether it is for an alcohol or another functional group) may be
placed before the parent name as in the above example or, according to a 1993 IUPAC revi-
sion of the rules, immediately before the suffix. Both methods are IUPAC approved. Therefore,
the above compound could also be named **4-methylhexan-1-ol**.

The following procedure should be followed in giving alcohols IUPAC substitutive
names:

1. Select the longest continuous carbon chain *to which the hydroxyl is directly attached*.
   Change the name of the alkane corresponding to this chain by dropping the final *-e* and
   adding the suffix *-ol*.

2. Number the longest continuous carbon chain so as to give the carbon atom bearing
   the hydroxyl group the lower number. Indicate the position of the hydroxyl group by
   using this number as a locant; indicate the positions of other substituents (as prefixes)
   by using the numbers corresponding to their positions along the carbon chain as
   locants.

The following examples show how these rules are applied:

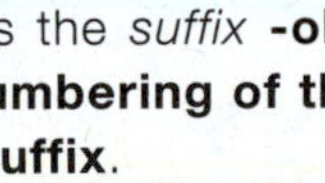

1-Propanol

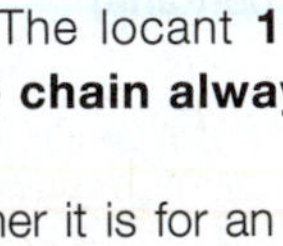

2-Butanol

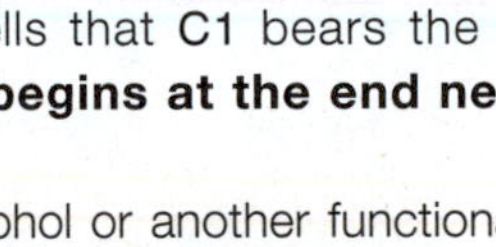

4-Methyl-1-pentanol
or 4-methylpentan-1-ol
(*not* 2-methyl-5-pentanol)

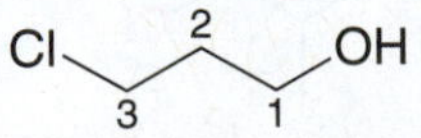

3-Chloro-1-propanol
or 3-chloropropan-1-ol

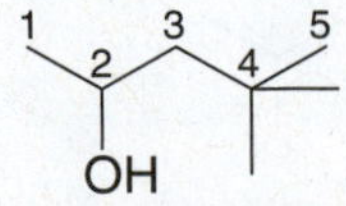

4,4-Dimethyl-2-pentanol
or 4,4-dimethylpentan-2-ol

••• SOLVED PROBLEM 4.2

Give an IUPAC name for the compound shown.

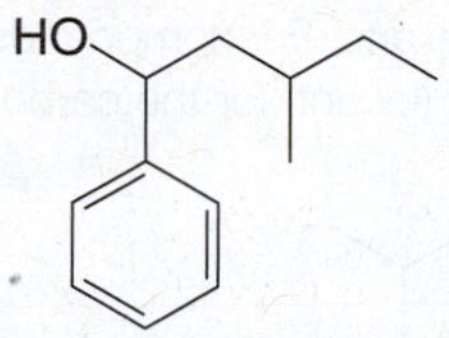

STRATEGY AND ANSWER: We find that the longest carbon chain (in red at right) has five carbons and it bears a hydroxyl group on the first carbon. So we name this part of the molecule as a 1-pentanol. There is a phenyl group on carbon-1 and a methyl group on carbon-3, so the full name is 3-methyl-1-phenyl-1-pentanol.

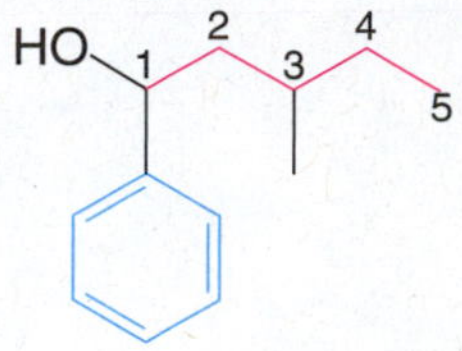

••• PRACTICE PROBLEM 4.7

Draw bond-line formulas and give IUPAC substitutive names for all of the isomeric alcohols with the formulas **(a)** $C_4H_{10}O$ and **(b)** $C_5H_{12}O$.

Simple alcohols are often called by *common* functional class names that are also approved by the IUPAC. We have seen several examples already (Section 2.6). In addition to *methyl alcohol, ethyl alcohol,* and *isopropyl alcohol*, there are several others, including the following:

Propyl alcohol	Butyl alcohol	*sec*-Butyl alcohol

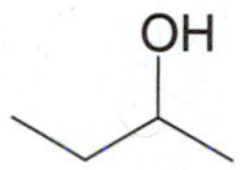

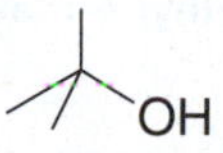

tert-Butyl alcohol	Isobutyl alcohol	Neopentyl alcohol

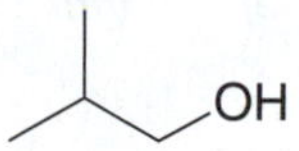

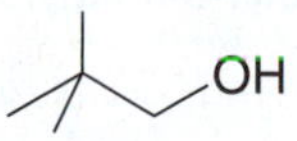

Alcohols containing two hydroxyl groups are commonly called **glycols**. In the IUPAC substitutive system they are named as **diols**:

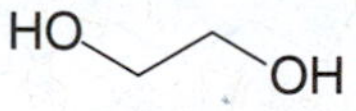

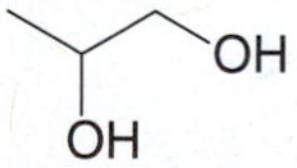

Substitutive	1,2-Ethanediol or ethane-1,2-diol	1,2-Propanediol or propane-1,2-diol	1,3-Propanediol or propane-1,3-diol
Common	Ethylene glycol	Propylene glycol	Trimethylene glycol

4.4 HOW TO NAME CYCLOALKANES

4.4A HOW TO Name Monocyclic Cycloalkanes

Cycloalkanes are named by adding "cyclo" before the parent name.

1. *Cycloalkanes with one ring and no substituents:* Count the number of carbon atoms in the ring, then add "cyclo" to the beginning of the name of the alkane with that number of carbons. For example, cyclopropane has three carbons and cyclopentane has five carbons.

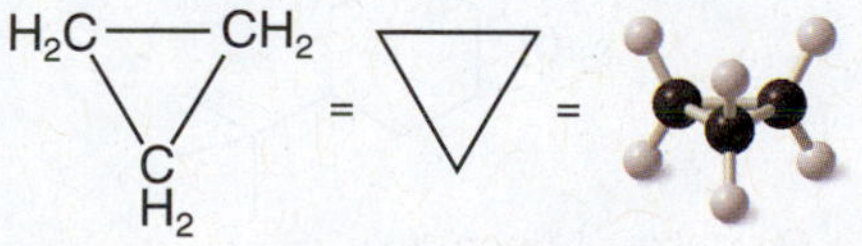

H_2C——CH_2
  \ /
   C
   H_2
= ▽ =

Cyclopropane

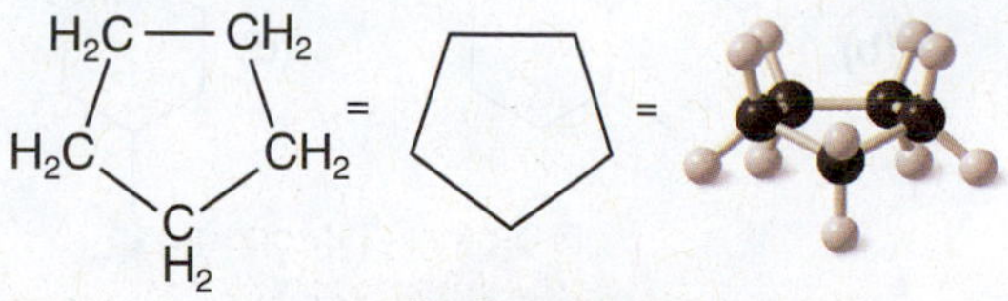

H_2C——CH_2
H_2C CH_2
     \ /
      C
      H_2
= ⬠ =

Cyclopentane

2. *Cycloalkanes with one ring and one substituent:* Add the name of the substituent to the beginning of the parent name. For example, cyclohexane with an attached isopropyl group is isopropylcyclohexane. For compounds with only one substituent, it is not necessary to specify a number (locant) for the carbon bearing the substituent.

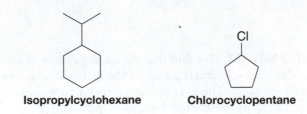

Isopropylcyclohexane   **Chlorocyclopentane**

3. *Cycloalkanes with one ring and two or more substituents:* For a ring with two substituents, begin by numbering the carbons in the ring, starting at the carbon with the substituent that is first in the alphabet and number in the direction that gives the next substituent the lower number possible. When there are three or more substituents, begin at the substituent that leads to the lowest set of numbers (locants). The substituents are listed in alphabetical order, not according to the number of their carbon atom.

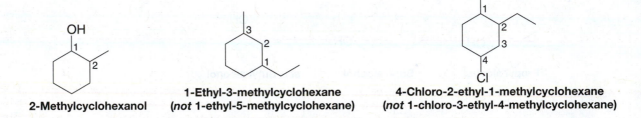

2-Methylcyclohexanol

1-Ethyl-3-methylcyclohexane
(*not* 1-ethyl-5-methylcyclohexane)

4-Chloro-2-ethyl-1-methylcyclohexane
(*not* 1-chloro-3-ethyl-4-methylcyclohexane)

4. When a single ring system is attached to a single chain with a greater number of carbon atoms, or when more than one ring system is attached to a single chain, then it is appropriate to name the compounds as *cycloalkylalkanes*. For example.

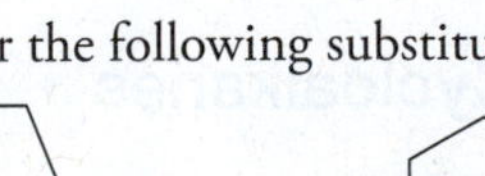

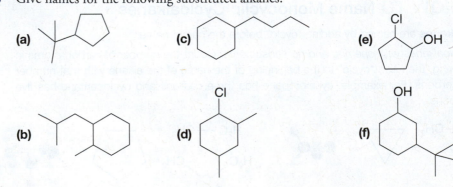

1-Cyclobutylpentane   **1,3-Dicyclohexylpropane**

⬤⬤⬤

PRACTICE PROBLEM 4.8 Give names for the following substituted alkanes:

(a)

(c)

(e)

(b)

(d)

(f)

4.4B HOW TO Name Bicyclic Cycloalkanes

1. We name compounds containing two fused or bridged rings as bicycloalkanes and we use the name of the alkane corresponding to the total number of carbon atoms in the rings as the parent name. The following compound, for example, contains seven carbon atoms and is, therefore, a bicycloheptane. The carbon atoms common to both rings are called bridgeheads, and each bond, or each chain of atoms connecting the bridgehead atoms, is called a bridge.

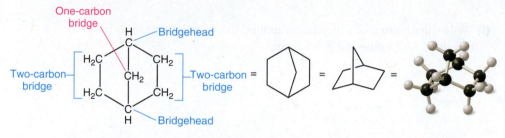

A bicycloheptane

> **Helpful Hint**
>
> Explore the structures of these **bicyclic compounds** by building handheld molecular models.

2. We then interpose an expression in brackets within the name that denotes the number of carbon atoms in each bridge (in order of decreasing length). Fused rings have zero carbons in their bridge. For example,

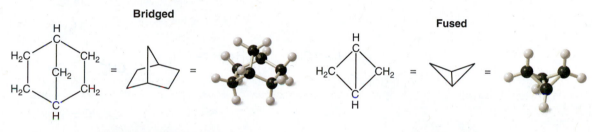

**Bicyclo[2.2.1]heptane
(also called *norbornane*)**

Bicyclo[1.1.0]butane

3. In bicycloalkanes with substituents, we number the bridged ring system beginning at one bridgehead, proceeding first along the longest bridge to the other bridgehead, then along the next longest bridge back to the first bridgehead; the shortest bridge is numbered last.

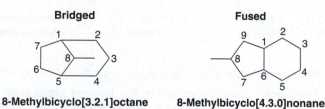

8-Methylbicyclo[3.2.1]octane   8-Methylbicyclo[4.3.0]nonane

●●● **SOLVED PROBLEM 4.3**

Write a structural formula for 7,7-dichlorobicyclo[2.2.1]heptane.

STRATEGY AND ANSWER: First we write a bicyclo[2.2.1]heptane ring and number it. Then we add the substituents (two chlorine atoms) to the proper carbon.

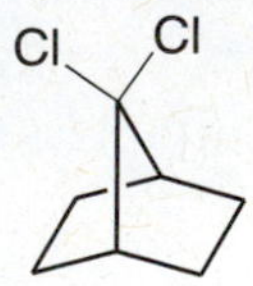

PRACTICE PROBLEM 4.9 Give names for each of the following bicyclic alkanes:

(a) 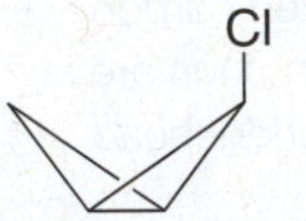   (b)    (c)    (d) 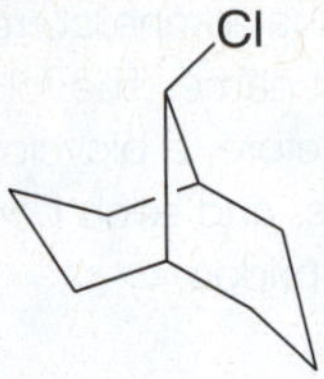   (e) 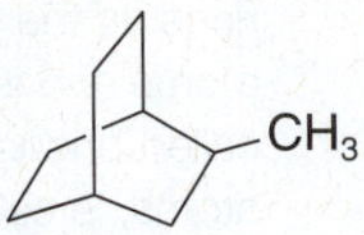 CH₃

(f) Write the structure of a bicyclic compound that is a constitutional isomer of
bicyclo[2.2.0]hexane and give its name.

4.5 HOW TO NAME ALKENES AND CYCLOALKENES

Many older names for **alkenes** are still in common use. Propene is often called propylene, and
2-methylpropene frequently bears the name isobutylene:

$$CH_2=CH_2 \qquad CH_3CH=CH_2 \qquad \begin{array}{c} CH_3 \\ | \\ CH_3-C=CH_2 \end{array}$$

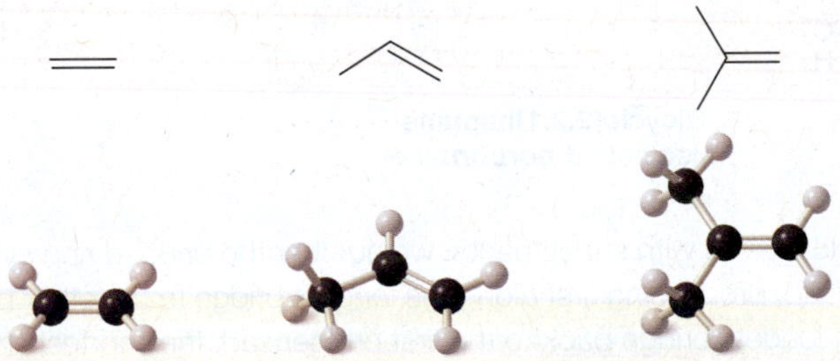

| IUPAC: | Ethene | Propene | 2-Methylpropene |
| Common: | Ethylene | Propylene | Isobutylene |

The IUPAC rules for naming alkenes are similar in many respects to those for naming alkanes:

1. **Determine the parent name by selecting the longest chain that contains the double bond and change the ending of the name of the alkane of identical length from -ane to -ene.** Thus, if the longest chain contains five carbon atoms, the parent name for the alkene is *pentene*; if it contains six carbon atoms, the parent name is *hexene*, and so on.

2. **Number the chain so as to include both carbon atoms of the double bond, and begin numbering at the end of the chain nearer the double bond. Designate the location of the double bond by using the number of the first atom of the double bond as a prefix. The locant for the alkene suffix may precede the parent name or be placed immediately before the suffix.** We will show examples of both styles:

$$\overset{1}{C}H_2=\overset{2}{C}H\overset{3}{C}H_2\overset{4}{C}H_3 \qquad CH_3CH=CHCH_2CH_2CH_3$$

1-Butene   **2-Hexene**
(*not* 3-butene)   (*not* 4-hexene)

3. **Indicate the locations of the substituent groups by the numbers of the carbon atoms to which they are attached:**

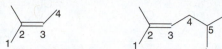

2-Methyl-2-butene
or 2-methylbut-2-ene

2,5-Dimethyl-2-hexene
or 2,5-dimethylhex-2-ene

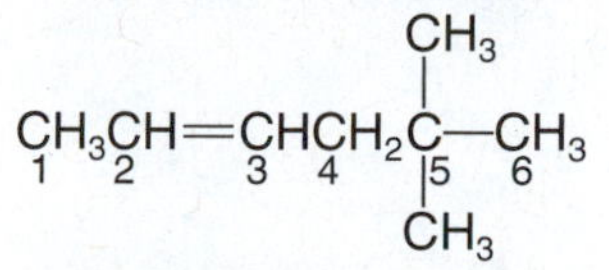

5,5-Dimethyl-2-hexene
or 5,5-dimethylhex-2-ene

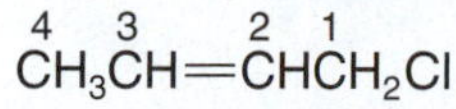

1-Chloro-2-butene
or 1-chlorobut-2-ene

4. **Number substituted cycloalkenes in the way that gives the carbon atoms of the double bond the 1 and 2 positions and that also gives the substituent groups the lower numbers at the first point of difference.** With substituted cycloalkenes it is not necessary to specify the position of the double bond since it will always begin with C1 and C2. The two examples shown here illustrate the application of these rules:

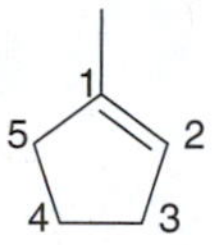

1-Methylcyclopentene
(*not* 2-methylcyclopentene)

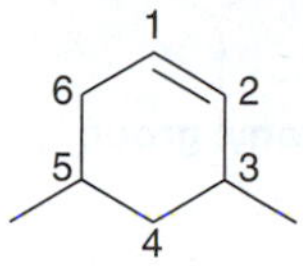

3,5-Dimethylcyclohexene
(*not* 4,6-dimethylcyclohexene)

5. Name compounds containing a double bond and an alcohol group as alkenols (or cyclo-alkenols) and give the alcohol carbon the lower number:

4-Methyl-3-penten-2-ol
or 4-methylpent-3-en-2-ol

2-Methyl-2-cyclohexen-1-ol
or 2-methylcyclohex-2-en-1-ol

6. Two frequently encountered alkenyl groups are the **vinyl group** and the **allyl group**:

The vinyl group 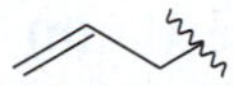   **The allyl group**

    Using substitutive nomenclature, the vinyl and allyl groups are called *ethenyl* and *prop-2-en-1-yl*, respectively. The following examples illustrate how these names are employed:

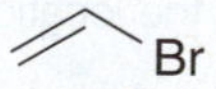

Bromoethene
or
vinyl bromide
(common)

Ethenylcyclopropane
or
vinylcyclopropane

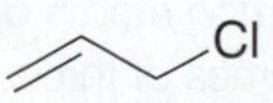

3-Chloropropene
or
allyl chloride
(common)

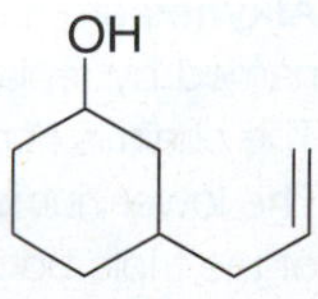

3-(Prop-2-en-1-yl)cyclohexan-1-ol
or
3-allylcyclohexanol

7. If two identical or substantial groups are on the same side of the double bond, the compound can be designated *cis*; if they are on opposite sides it can be designated *trans*:

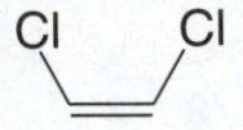

cis-1,2-Dichloroethene 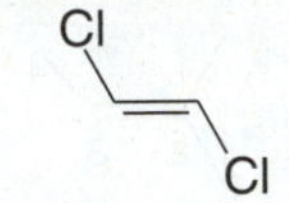   *trans*-1,2-Dichloroethene

(In Section 7.2 we shall see another method for designating the geometry of the double bond.)

●●● SOLVED PROBLEM 4.4

Give an IUPAC name for the molecule shown.

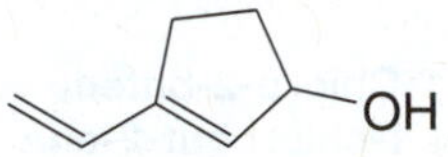

STRATEGY AND ANSWER: We number the ring as shown below starting with the hydroxyl group so as to give the double bond the lower possible number. We include in the name the substituent (an ethenyl group) and the double bond (*-ene-*), and the hydroxyl group (*-ol*) with numbers for their respective positions. Hence the IUPAC name is 3-ethenyl-2-cyclopenten-1-ol.

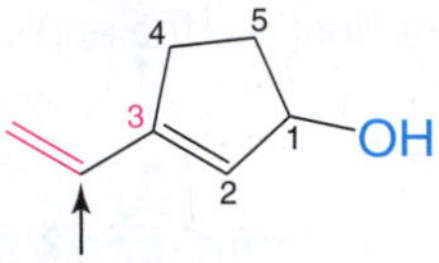

Ethenyl group

●●● PRACTICE PROBLEM 4.10   Give IUPAC names for the following alkenes:

(a)

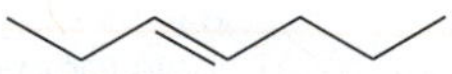

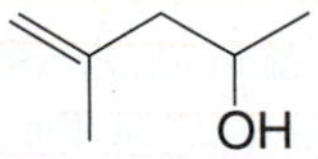

(b)

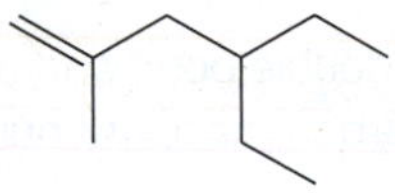

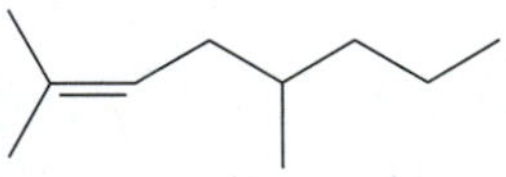

(c)

(d)

(e)

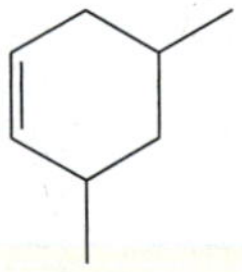

(f)

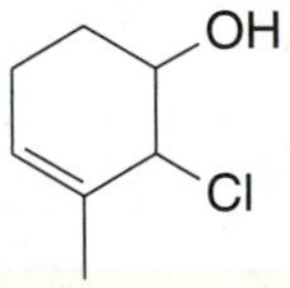

●●● PRACTICE PROBLEM 4.11   Write bond-line formulas for the following:

(a) *cis*-3-Octene
(b) *trans*-2-Hexene
(c) 2,4-Dimethyl-2-pentene
(d) *trans*-1-Chlorobut-2-ene
(e) 4,5-Dibromo-1-pentene
(f) 1-Bromo-2-methyl-1-(prop-2-en-1-yl)cyclopentane
(g) 3,4-Dimethylcyclopentene
(h) Vinylcyclopentane
(i) 1,2-Dichlorocyclohexane
(j) *trans*-1,4-Dichloro-2-pentene

4.6 HOW TO NAME ALKYNES

Alkynes are named in much the same way as alkenes. Unbranched alkynes, for example, are named by replacing the **-ane** of the name of the corresponding alkane with the ending **-yne.** The chain is numbered to give the carbon atoms of the triple bond the lower possible numbers. The lower number of the two carbon atoms of the triple bond is used to designate the location of the triple bond. The IUPAC names of three unbranched alkynes are shown here:

$H-C\equiv C-H$

Ethyne or acetylene*

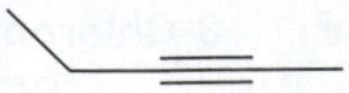

2-Pentyne

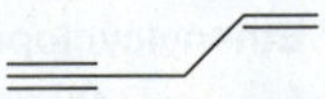

1-Penten-4-yne[†] or pent-1-en-4-yne

The locations of substituent groups of branched alkynes and substituted alkynes are also indicated with numbers. An —OH group has priority over the triple bond when numbering the chain of an alkynol:

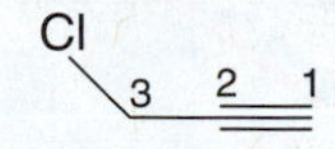

3-Chloropropyne

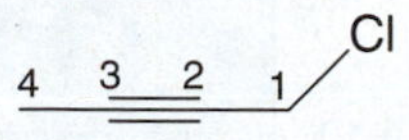

1-Chloro-2-butyne
or 1-chlorobut-2-yne

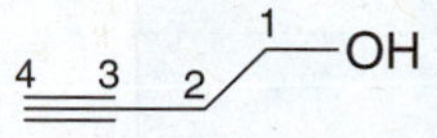

3-Butyn-1-ol
or but-3-yn-1-ol

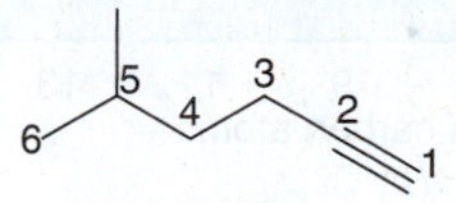

5-Methyl-1-hexyne
or 5-methylhex-1-yne

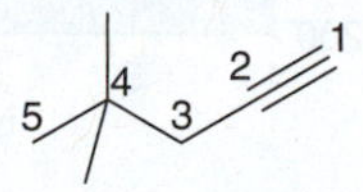

4,4-Dimethyl-1-pentyne
or 4,4-dimethylpent-1-yne

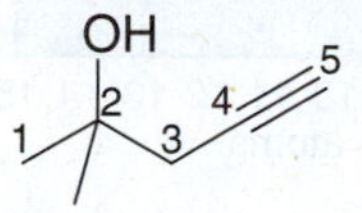

2-Methyl-4-pentyn-2-ol
or 2-methylpent-4-yn-2-ol

Give the structures and IUPAC names for all the alkynes with the formula C_6H_{10}.

PRACTICE PROBLEM 4.12

Monosubstituted acetylenes or 1-alkynes are called **terminal alkynes**, and the hydrogen attached to the carbon of the triple bond is called the **acetylenic hydrogen atom**:

Acetylenic hydrogen

$$R—C≡C—H$$
A terminal alkyne

When named as a substituent, the HC≡C— group is called the ethynyl group.

The anion obtained when the acetylenic hydrogen is removed is known as an *alkynide ion* or an *acetylide ion*. As we shall see in Section 7.11, these ions are useful in synthesis:

$$R—C≡C:^-$$ $$CH_3C≡C:^-$$
or   or
$$R—≡:^-$$ $$—≡:^-$$

An alkynide ion   **The propynide**
(an acetylide ion)   **ion**

4.7 PHYSICAL PROPERTIES OF ALKANES AND CYCLOALKANES

If we examine the unbranched **alkanes** in Table 4.2, we notice that each alkane differs from the preceding alkane by one —CH_2— group. Butane, for example, is $CH_3(CH_2)_2CH_3$ and pentane is $CH_3(CH_2)_3CH_3$. A series of compounds like this, where each member differs from the next member by a constant unit, is called a **homologous series**. Members of a homologous series are called **homologues.**

At room temperature (25 °C) and 1 atm pressure the first four members of the homologous series of unbranched alkanes are gases (Fig. 4.3), the C_5—C_{17} unbranched alkanes (pentane to heptadecane) are liquids, and the unbranched alkanes with 18 and more carbon atoms are solids.

*The name acetylene is retained by the IUPAC system for the compound HC≡CH and is used frequently.

†When double and triple bonds are present, the direction of numbering is chosen so as to give the lowest overall set of locants. In the face of equivalent options, then preference is given to assigning lowest numbers to the double bonds.

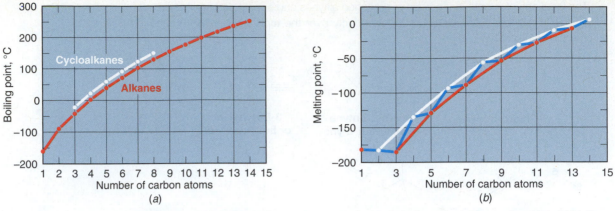

FIGURE 4.3 (*a*) Boiling points of unbranched alkanes (in red) and cycloalkanes (in white). (*b*) Melting points of unbranched alkanes.

Boiling Points   The boiling points of the unbranched alkanes show a regular increase with increasing molecular weight (Fig. 4.3*a*) in the homologous series of straight-chain alkanes. Branching of the alkane chain, however, lowers the boiling point. The hexane isomers in Table 4.1 examplify this trend.

Part of the explanation for these effects lies in the dispersion forces that we studied in Section 2.13B. With unbranched alkanes, as molecular weight increases, so too do molecular size and, even more importantly, molecular surface area. With increasing surface area, the dispersion forces between molecules increase; therefore, more energy (a higher temperature) is required to separate molecules from one another and produce boiling. Chain branching, on the other hand, makes a molecule more compact, reducing its surface area and with it the strength of the dispersion forces operating between it and adjacent molecules; this has the effect of lowering the boiling point. Figure 4.4 illustrates this for two **C₈** isomers.

FIGURE 4.4 Chain-branching decreases the contact surface area between molecules, as for the branched C8 isomer in (*b*), lessening the dispersion forces between them and leading to a lower boiling point than for the unbranched C8 isomer (*a*).

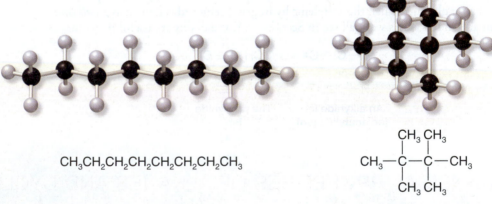

$$CH_3CH_2CH_2CH_2CH_2CH_2CH_2CH_3$$

(*a*) **Octane (bp 125.6 °C)**

(*b*) **2,2,3,3-Tetramethylbutane (bp 106.3 °C)**

Melting Points   The unbranched alkanes do not show the same smooth increase in melting points with increasing molecular weight (blue line in Fig. 4.3*b*) that they show in their boiling points. There is an alternation as one progresses from an unbranched alkane with an even number of carbon atoms to the next one with an odd number of carbon atoms. If, however, the even- and odd-numbered alkanes are plotted on *separate* curves (white and red lines in Fig. 4.3*b*), there *is* a smooth increase in melting point with increasing molecular weight.

X-ray diffraction studies, which provide information about molecular structure, have revealed the reason for this apparent anomaly. Alkane chains with an even number of carbon atoms pack more closely in the crystalline state. As a result, attractive forces between individual chains are greater and melting points are higher.

Cycloalkanes also have much higher melting points than their open-chain counterparts (Table 4.3).

TABLE 4.3 PHYSICAL CONSTANTS OF CYCLOALKANES

Number of Carbon Atoms	Name	bp (°C) (1 atm)	mp (°C)	Density at 20 °C (g mL^{-1})	Refractive Index (n_D^{20})
3	Cyclopropane	−33	−126.6	—	—
4	Cyclobutane	13	−90	—	1.4260
5	Cyclopentane	49	−94	0.751	1.4064
6	Cyclohexane	81	6.5	0.779	1.4266
7	Cycloheptane	118.5	−12	0.811	1.4449
8	Cyclooctane	149	13.5	0.834	—

Density As a class, the alkanes and cycloalkanes are the least dense of all groups of organic compounds. All alkanes and cycloalkanes have densities considerably less than 1.00 g mL^{-1} (the density of water at 4 °C). As a result, petroleum (a mixture of hydrocarbons rich in alkanes) floats on water.

Solubility Alkanes and cycloalkanes are almost totally insoluble in water because of their very low polarity and their inability to form hydrogen bonds. Liquid alkanes and cycloalkanes are soluble in one another, and they generally dissolve in solvents of low polarity. Good solvents for them are benzene, carbon tetrachloride, chloroform, and other hydrocarbons.

THE CHEMISTRY OF ... Pheromones: Communication by Means of Chemicals

Many animals communicate with other members of their species using a language based not on sounds or even visual signals but on the odors of chemicals called **pheromones** that these animals release. For insects, this appears to be the chief method of communication. Although pheromones are secreted by insects in extremely small amounts, they can cause profound and varied biological effects. Some insects use pheromones in courtship as sex attractants. Others use pheromones as warning substances, and still others secrete chemicals called "aggregation compounds" to cause members of their species to congregate. Often these pheromones are relatively simple compounds, and some are hydrocarbons. For example, a species of cockroach uses undecane as an aggregation pheromone:

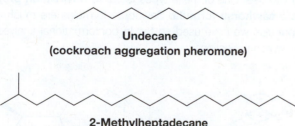

Undecane
(cockroach aggregation pheromone)

2-Methylheptadecane
(sex attractant of female tiger moth)

When a female tiger moth wants to mate, she secretes 2-methylheptadecane, a perfume that the male tiger moth apparently finds irresistible.

The sex attractant of the common housefly (*Musca domestica*) is a 23-carbon alkene with a cis double bond between atoms 9 and 10 called muscalure:

Muscalure
(sex attractant of common housefly)

(continues on next page)

Danilo Donadoni/Photoshot Holdings Ltd.

Many insect sex attractants have been synthesized and are used to lure insects into traps as a means of insect control, a much more environmentally sensitive method than the use of insecticides.

Research suggests there are roles for pheromones in the lives of humans as well. For example, studies have shown that the phenomenon of menstrual synchronization among women who live or work with each other is likely caused by pheromones. Olfactory sensitivity to musk, which includes steroids such as androsterone, large cyclic ketones, and lactones (cyclic esters), also varies cyclically in women, differs between the sexes, and may influence our behavior. Some of these compounds are used in perfumes, including civetone, a natural product isolated from glands of the civet cat, and pentalide, a synthetic musk.

Androsterone **Civetone** **Pentalide**

FANCY/Image Source

4.8 SIGMA BONDS AND BOND ROTATION

Two groups bonded by only a single bond can undergo rotation about that bond with respect to each other.

- The temporary molecular shapes that result from such a rotation are called **conformations** of the molecule.
- Each possible structure is called a **conformer**.
- An analysis of the energy changes that occur as a molecule undergoes rotations about single bonds is called a **conformational analysis**.

4.8A Newman Projections and HOW TO Draw Them

When we do conformational analysis, we will find that certain types of structural formulas are especially convenient to use. One of these types is called a **Newman projection formula** and another type is a **sawhorse formula**. Sawhorse formulas are much like dash–wedge three-dimensional formulas we have used so far. In conformational analyses, we will make substantial use of Newman projections.

Newman projection formula   **Sawhorse formula**

To write a Newman projection formula:

- We imagine ourselves taking a view from one atom (usually a carbon) directly along a selected bond axis to the next atom (also usually a carbon atom).
- The front carbon and its other bonds are represented as [Y shape].
- The back carbon and its bonds are represented as [Y circle shape].

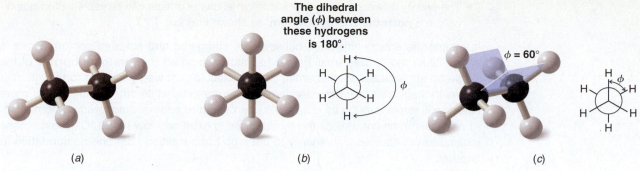

The dihedral angle (ϕ) between these hydrogens is 180°.

$\phi = 60°$

(a)   (b)   (c)

FIGURE 4.5 (*a*) The staggered conformation of ethane. (*b*) The Newman projection formula for the staggered conformation. (*c*) The dihedral angle between these hydrogen atoms is 60°.

In Figs. 4.5*a*,*b* we show ball-and-stick models and a Newman projection formula for the staggered conformation of ethane. The **staggered conformation** of a molecule is that conformation where the **dihedral angle** between the bonds at each of the carbon–carbon bonds is 180° and where atoms or groups bonded to carbons at each end of a carbon–carbon bond are as far apart as possible. The 180° dihedral angle in the staggered conformation of ethane is indicated in Fig. 4.5*b*.

The eclipsed conformation of ethane is shown in Fig. 4.6 using ball-and-stick models and a Newman projection. In an **eclipsed conformation** the atoms bonded to carbons at each end of a carbon–carbon bond are directly opposed to one another. The dihedral angle between them is 0°.

The dihedral angle (ϕ) between these hydrogens is 0°.

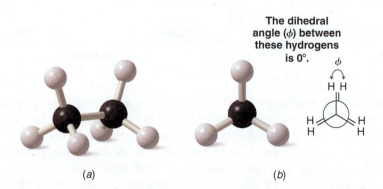

(a)   (b)

FIGURE 4.6 (*a*) The eclipsed conformation of ethane. (*b*) The Newman projection formula for the eclipsed conformation.

4.8B HOW TO Do a Conformational Analysis

Now let us consider a conformational analysis of ethane. Clearly, infinitesimally small changes in the dihedral angle between C—H bonds at each end of ethane could lead to an infinite number of conformations, including, of course, the staggered and eclipsed conformations. These different conformations are not all of equal stability, however, and it is known that the staggered conformation of ethane is the most stable conformation (i.e., it is the conformation of lowest potential energy). The explanation for greater stability of the staggered conformation relates mainly to steric repulsion between bonding pairs of electrons. In the eclipsed conformation electron clouds from the C—H bonds are closer and repel one another. The staggered conformation allows the maximum possible separation of the electron pairs in the C—H bonds. In addition, there is a phenomenon called **hyperconjugation** that involves favorable overlap between filled and unfilled sigma orbitals in the staggered conformation. Hyperconjugation helps to stabilize the staggered conformation. The more important factor, however, is the minimization of steric repulsions in the staggered form. In later chapters we shall explain hyperconjugation further and the role it plays in relative stability of reactive species called carbocations.

● The energy difference between the conformations of ethane can be represented graphically in a **potential energy diagram**, as shown in Figure 4.7.

In ethane the energy difference between the staggered and eclipsed conformations is about 12 kJ mol^{-1}. This small barrier to rotation is called the **torsional barrier** of the single bond. Because of this barrier, some molecules will wag back and forth with their atoms in staggered or nearly staggered conformations, while others with slightly more energy will rotate through an eclipsed conformation to another staggered conformation. At any given moment, unless the temperature is extremely low (−250 °C), most ethane molecules will have enough energy to undergo bond rotation from one conformation to another.

FIGURE 4.7 Potential energy changes that accompany rotation of groups about the carbon–carbon bond of ethane.

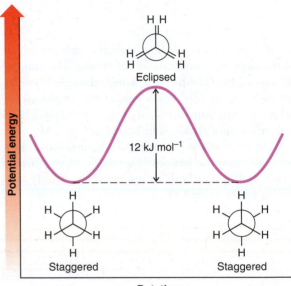

The idea that certain conformations of molecules are favored originates from the work of J.H. van't Hoff. He was also winner of the first Nobel Prize in Chemistry (1901) for his work in chemical kinetics.

What does all this mean about ethane? We can answer this question in two different ways. If we consider a single molecule of ethane, we can say, for example, that it will spend most of its time in the lowest energy, staggered conformation, or in a conformation very close to being staggered. Many times every second, however, it will acquire enough energy through collisions with other molecules to surmount the torsional barrier and it will rotate through an eclipsed conformation. If we speak in terms of a large number of ethane molecules (a more realistic situation), we can say that at any given moment most of the molecules will be in staggered or nearly staggered conformations.

4.9 CONFORMATIONAL ANALYSIS OF BUTANE

Now let us consider rotation about the **C2—C3** bond of butane. The barriers to rotation about the **C2—C3** bond in butane are larger than for rotation about the **C—C** bond in ethane, but still not large enough to prevent the rotations that lead to all possible butane conformers.

● The factors involved in barriers to bond rotation are together called **torsional strain** and include the repulsive interactions called **steric hindrance** between electron clouds of the bonded groups.

In butane, torsional strain results from steric hindrance between the terminal methyl groups and hydrogen atoms at **C-2** and **C-3** and from steric hindrance directly between

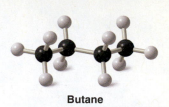

the two methyl groups. These interactions result in six important conformers of butane, shown as **I–VI** below.

CH₃	HCH₃	CH₃
I	**II**	**III**
An anti conformation	**An eclipsed conformation**	**A gauche conformation**

H₃C CH₃	CH₃	HCH₃
IV	**V**	**VI**
An eclipsed conformation	**A gauche conformation**	**An eclipsed conformation**

> ### Helpful Hint
>
> You should build a molecular model of butane and examine its various conformations as we discuss their relative potential energies.

Butane

The **anti conformation** (**I**) does not have torsional strain from steric hindrance because the groups are staggered and the methyl groups are far apart. The anti conformation is the most stable. The methyl groups in the **gauche conformations** **III** and **V** are close enough to each other that the dispersion forces between them are *repulsive*; the electron clouds of the two groups are so close that they repel each other. This repulsion causes the gauche conformations to have approximately 3.8 kJ mol⁻¹ more energy than the anti conformation.

The eclipsed conformations (**II**, **IV**, and **VI**) represent energy maxima in the potential energy diagram (Fig. 4.8). Eclipsed conformations **II** and **VI** have repulsive dispersion forces arising from the eclipsed methyl groups and hydrogen atoms. Eclipsed conformation **IV** has the greatest energy of all because of the added large repulsive dispersion forces between the eclipsed methyl groups as compared to **II** and **VI**.

Although the barriers to rotation in a butane molecule are larger than those of an ethane molecule (Section 4.8), they are still far too small to permit isolation of the gauche and anti conformations at normal temperatures. Only at extremely low temperatures would the molecules have insufficient energies to surmount these barriers.

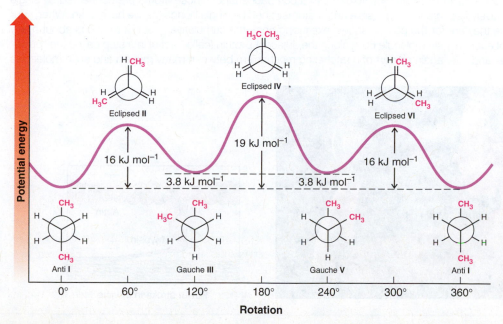

FIGURE 4.8 Energy changes that arise from rotation about the C2—C3 bond of butane.

We saw earlier (Section 2.16C) that dispersion forces can be *attractive*. Here, however, we find that they can also be *repulsive*, leading to steric hindrance. Whether dispersion interactions lead to attraction or to repulsion depends on the distance that separates the two groups. As two nonpolar groups are brought closer and closer together, the first effect is one in which a momentarily unsymmetrical distribution of electrons in one group induces an opposite polarity in the other. The opposite charges induced in those portions of the two groups that are in closest proximity lead to attraction between them. This attraction increases to a maximum as the internuclear distance of the two groups decreases. The internuclear distance at which the attractive force is at a maximum is equal to the sum of what are called the *van der Waals radii* of the two groups. The van der Waals radius of a group is, in effect, a measure of its size. If the two groups are brought still closer—closer than the sum of their van der Waals radii—their electron clouds begin to penetrate each other, and strong electron–electron repulsion occurs.

4.9A Stereoisomers and Conformational Stereoisomers

Gauche conformers **III** and **V** of butane are examples of stereoisomers.

- **Stereoisomers** have the same molecular formula and connectivity but different arrangements of atoms in three-dimensional space.
- **Conformational stereoisomers** are related to one another by bond rotations.

Conformational analysis is but one of the ways in which we will consider the three-dimensional shapes and stereochemistry of molecules. We shall see that there are other types of stereoisomers that cannot be interconverted simply by rotations about single bonds. Among these are cis–trans cycloalkane isomers (Section 4.13) and others that we shall consider in Chapter 5.

PRACTICE PROBLEM 4.13   Sketch a curve similar to that in Fig. 4.8 showing in general terms the energy changes that arise from rotation about the C2—C3 bond of 2-methylbutane. You need not concern yourself with the actual numerical values of the energy changes, but you should label all maxima and minima with the appropriate conformations.

THE CHEMISTRY OF ... Muscle Action

Muscle proteins are essentially very long linear molecules (folded into a compact shape) whose atoms are connected by single bonds in a chainlike fashion. Relatively free rotation is possible about atoms joined by single bonds, as we have seen. When your muscles contract to do work, like they are for the person shown exercising here, the cumulative effect of rotations about many single bonds is to move the tail of each myosin molecule 60 Å along the adjacent protein (called actin) in a step called the "power stroke." This process occurs over and over again as part of a ratcheting mechanism between many myosin and actin molecules for each muscle movement.

Blend/Image Source

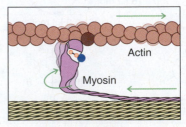

Power stroke in muscle

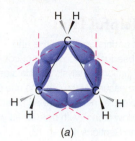

4.10 THE RELATIVE STABILITIES OF CYCLOALKANES: RING STRAIN

Cycloalkanes do not all have the same relative stability. Experiments have shown that cyclohexane is the most stable cycloalkane and that, in comparison, cyclopropane and cyclobutane are much less stable. This difference in relative stability is due to **ring strain**, which comprises **angle strain** and **torsional strain**.

- **Angle strain** is the result of deviation from ideal bond angles caused by inherent structural constraints (such as ring size).
- **Torsional strain** is the result of repulsive dispersion forces that cannot be relieved due to restricted conformational mobility.

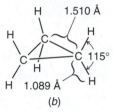

4.10A Cyclopropane

The carbon atoms of alkanes are sp^3 hybridized. The normal tetrahedral bond angle of an sp^3-hybridized atom is 109.5°. In cyclopropane (a molecule with the shape of a regular triangle), the internal angles must be 60° and therefore they must depart from this ideal value by a very large amount—by 49.5°:

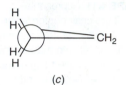

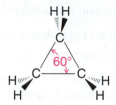

Angle strain exists in a cyclopropane ring because the sp^3 orbitals comprising the carbon–carbon σ bonds cannot overlap as effectively (Fig. 4.9a) as they do in alkanes (where perfect end-on overlap is possible). The carbon–carbon bonds of cyclopropane are often described as being "bent." Orbital overlap is less effective. (The orbitals used for these bonds are not purely sp^3; they contain more p character.) The carbon–carbon bonds of cyclopropane are weaker, and as a result the molecule has greater potential energy.

While angle strain accounts for most of the ring strain in cyclopropane, it does not account for it all. Because the ring is (of necessity) planar, the C—H bonds of the ring are all *eclipsed* (Figs. 4.9b,c), and the molecule has torsional strain from repulsive dispersion forces as well.

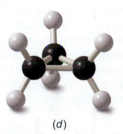

FIGURE 4.9 (a) Orbital overlap in the carbon–carbon bonds of cyclopropane cannot occur perfectly end-on. This leads to weaker "bent" bonds and to angle strain. (b) Bond distances and angles in cyclopropane. (c) A Newman projection formula as viewed along one carbon–carbon bond shows the eclipsed hydrogens. (Viewing along either of the other two bonds would show the same picture.) (d) Ball-and-stick model of cyclopropane.

4.10B Cyclobutane

Cyclobutane also has considerable angle strain. The internal angles are 88°—a departure of more than 21° from the normal tetrahedral bond angle. The cyclobutane ring is not planar but is slightly "folded" (Fig. 4.10a). If the cyclobutane ring were planar, the angle strain would be somewhat less (the internal angles would be 90° instead of 88°), but torsional strain would be considerably larger because all eight C—H bonds would be eclipsed. By folding or bending slightly the cyclobutane ring relieves more of its torsional strain than it gains in the slight increase in its angle strain.

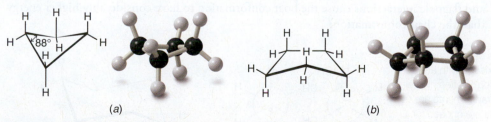

FIGURE 4.10 (a) The "folded" or "bent" conformation of cyclobutane. (b) The "bent" or "envelope" form of cyclopentane. In this structure the front carbon atom is bent upward. In actuality, the molecule is flexible and shifts conformations constantly.

4.10C Cyclopentane

The internal angles of a regular pentagon are 108°, a value very close to the normal tetrahedral bond angles of 109.5°. Therefore, if cyclopentane molecules were planar, they would have very little angle strain. Planarity, however, would introduce considerable torsional strain because all ten C—H bonds would be eclipsed. Consequently, like cyclobutane, cyclopentane assumes a slightly bent conformation in which one or two of the atoms of the ring are out of the plane of the others (Fig. 4.10b). This relieves some of the torsional strain. Slight twisting of carbon–carbon bonds can occur with little change in energy and causes the out-of-plane atoms to move into plane and causes others to move out. Therefore, the molecule is flexible and shifts rapidly from one conformation to another. With little torsional strain and angle strain, cyclopentane is almost as stable as cyclohexane.

4.11 CONFORMATIONS OF CYCLOHEXANE: THE CHAIR AND THE BOAT

Cyclohexane is more stable than the other **cycloalkanes** we have discussed, and it has several conformations that are important for us to consider.

- The most stable **conformation of cyclohexane** is the **chair conformation**.
- There is no angle or torsional strain in the chair form of cyclohexane.

In a chair conformation (Fig. 4.11), all of the carbon–carbon bond angles are 109.5°, and are thereby free of angle strain. The chair conformation is free of torsional strain, as well. When viewed along any carbon–carbon bond (viewing the structure from an end, Fig. 4.12), the bonds are seen to be perfectly staggered. Moreover, the hydrogen atoms at opposite corners of the cyclohexane ring are maximally separated.

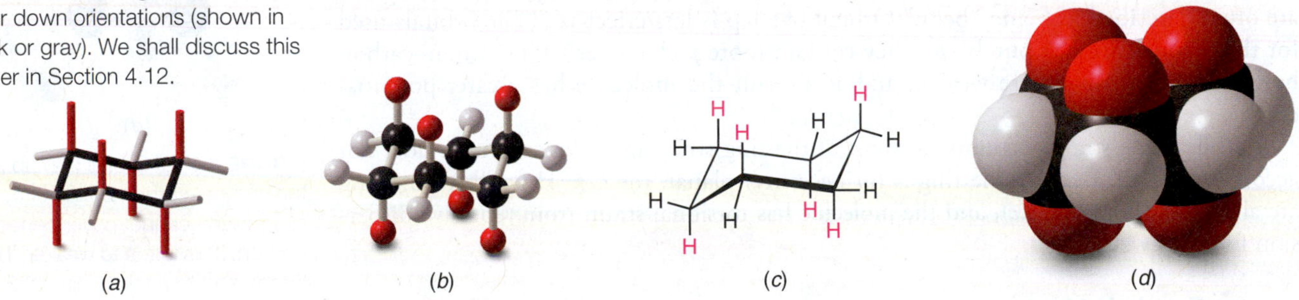

FIGURE 4.11 Representations of the chair conformation of cyclohexane: (a) tube format; (b) ball-and-stick format; (c) line drawing; (d) space-filling model of cyclohexane. Notice that there are two orientations for the hydrogen substituents—those that project obviously up or down (shown in red) and those that lie around the perimeter of the ring in more subtle up or down orientations (shown in black or gray). We shall discuss this further in Section 4.12.

(a)   (b)   (c)   (d)

- By partial rotations about the carbon–carbon single bonds of the ring, the chair conformation can assume another shape called the **boat conformation** (Fig. 4.13).
- The boat conformation has no angle strain, but it does have torsional strain.

When a model of the boat conformation is viewed down carbon–carbon bond axes along either side (Fig. 4.14a), the C—H bonds at those carbon atoms are found to be eclipsed, causing torsional strain. Additionally, two of the hydrogen atoms on C1 and C4 are close enough to each other to cause van der Waals repulsion (Fig. 4.14b). This latter effect has been called the "flagpole" interaction of the boat conformation. Torsional strain and flagpole interactions cause the boat conformation to have considerably higher energy than the chair conformation.

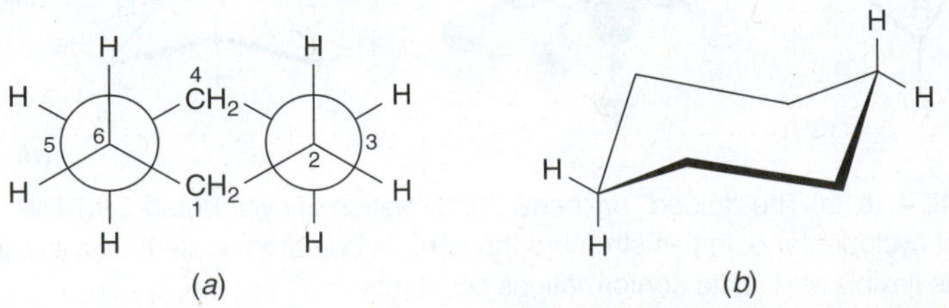

FIGURE 4.12 (a) A Newman projection of the chair conformation of cyclohexane. (Comparisons with an actual molecular model will make this formulation clearer and will show that similar staggered arrangements are seen when other carbon–carbon bonds are chosen for sighting.) (b) Illustration of large separation between hydrogen atoms at opposite corners of the ring (designated C1 and C4) when the ring is in the chair conformation.

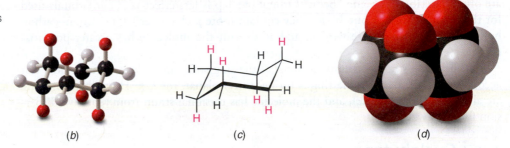

(a)   (b)

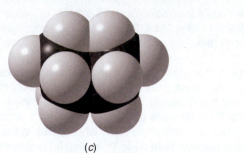

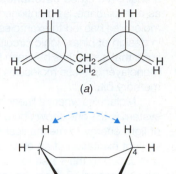

FIGURE 4.13 (*a*) The boat conformation of cyclohexane is formed by "flipping" one end of the chair form up (or down). This flip requires only rotations about carbon–carbon single bonds. (*b*) Ball-and-stick model of the boat conformation. (*c*) A space-filling model of the boat conformation.

FIGURE 4.14 (*a*) Illustration of the eclipsed conformation of the boat conformation of cyclohexane. (*b*) Flagpole interaction of the C1 and C4 hydrogen atoms of the boat conformation. The C1–C4 flagpole interaction is also readily apparent in Fig. 4.13*c*.

Although it is more stable, the chair conformation is much more rigid than the boat conformation. The boat conformation is quite flexible. By flexing to a new form—the twist conformation (Fig. 4.15)—the boat conformation can relieve some of its torsional strain and, at the same time, reduce the flagpole interactions.

- The twist boat conformation of cyclohexane has a lower energy than the pure boat conformation, but is not as stable as the chair conformation.

The stability gained by flexing is insufficient, however, to cause the twist conformation to be more stable than the chair conformation. The chair conformation is estimated to be lower in energy than the twist conformation by approximately 23 kJ mol^{-1}.

The energy barriers between the chair, boat, and twist conformations of cyclohexane are low enough (Fig. 4.16) to make separation of the conformers impossible at room temperature. At room temperature the thermal energies of the molecules are great enough to cause approximately 1 million interconversions to occur each second.

- *Because of the greater stability of the chair, more than 99% of the molecules are estimated to be in a chair conformation at any given moment.*

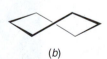

FIGURE 4.15 (*a*) Tube model and (*b*) line drawing of the twist conformation of cyclohexane.

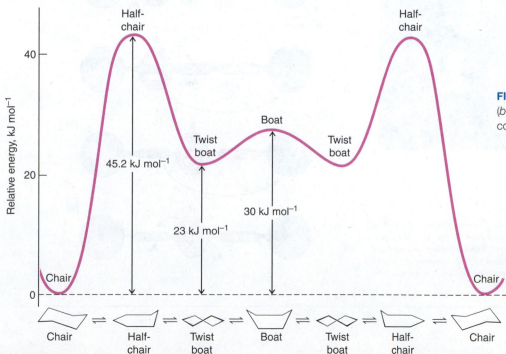

FIGURE 4.16 The relative energies of the various conformations of cyclohexane. The positions of maximum energy are conformations called half-chair conformations, in which the carbon atoms of one end of the ring have become coplanar.

THE CHEMISTRY OF ... Nanoscale Motors and Molecular Switches

Molecular rings that interlock with one another and compounds that are linear molecules threaded through rings are proving to have fascinating potential for the creation of molecular switches and motors. Molecules consisting of interlocking rings, like a chain, are called **catenanes**. The first catenanes were synthesized in the 1960s and have come to include examples such as olympiadane, as mentioned in Section 4.11A. Further research by J. F. Stoddart (UCLA) and collaborators on interlocking molecules has led to examples such as the catenane molecular switch shown here in (*i*). In an application that could be useful in design of binary logic circuits, one ring of this molecule can be made to circumrotate in controlled fashion about the other, such that it switches between two defined states. As a demonstration of its potential for application in electronics fabrication, a monolayer of these molecules has been "tiled" on a surface (*ii*) and shown to have characteristics like a conventional magnetic memory bit.

Molecules where a linear molecule is threaded through a ring are called **rotaxanes**. One captivating example of a rotaxane system is the one shown here in (*iii*), under development by V. Balzani (University of Bologna) and collaborators. By conversion of light energy to mechanical energy at the molecular level, this rotaxane behaves like a "four-stroke" shuttle engine. In step (*a*) light excitation of an electron in the **P** group leads to transfer of the electron to the initially +2 A_1 group, at which point A_1 is reduced to the +1 state. Ring **R**, which was attracted to A_1 when it was in the +2 state, now slides over to A_2 in step (*b*), which remains +2. Back transfer of the electron from A_1 to P^+ in step (*c*) restores the +2 state of **A1**, causing ring **R** to return to its original location in step (*d*). Modifications envisioned for this system include attaching binding sites to **R** such that some other molecular species could be transported from one location to another as **R** slides along the linear molecule, or linking **R** by a springlike tether to one end of the "piston rod" such that additional potential and mechanical energy can be incorporated in the system.

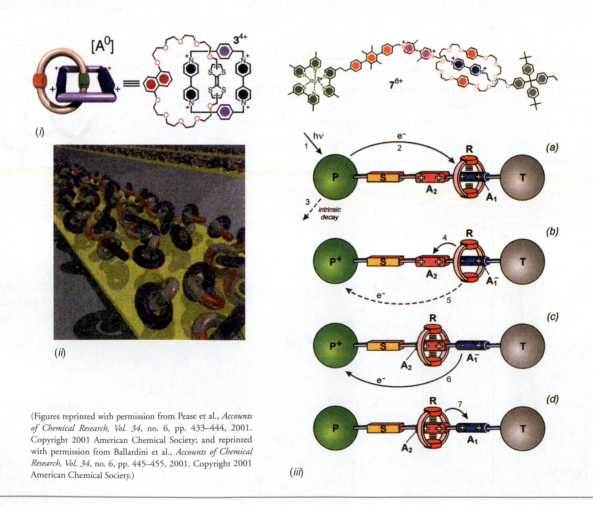

(*i*)

(*ii*)

(*iii*)

(*a*)

(*b*)

(*c*)

(*d*)

(Figures reprinted with permission from Pease et al., *Accounts of Chemical Research*, Vol. 34, no. 6, pp. 433–444, 2001. Copyright 2001 American Chemical Society; and reprinted with permission from Ballardini et al., *Accounts of Chemical Research*, Vol. 34, no. 6, pp. 445–455, 2001. Copyright 2001 American Chemical Society.)

4.11A Conformations of Higher Cycloalkanes

Cycloheptane, cyclooctane, and cyclononane and other higher cycloalkanes also exist in nonplanar conformations. The small instabilities of these higher cycloalkanes appear to be caused primarily by torsional strain and repulsive dispersion forces between hydrogen atoms across rings, called *transannular strain*. The nonplanar conformations of these rings, however, are essentially free of angle strain.

X-ray crystallographic studies of cyclodecane reveal that the most stable conformation has carbon–carbon–carbon bond angles of 117°. This indicates some angle strain. The wide bond angles apparently allow the molecules to expand and thereby minimize unfavorable repulsions between hydrogen atoms across the ring.

There is very little free space in the center of a cycloalkane unless the ring is quite large. Calculations indicate that cyclooctadecane, for example, is the smallest ring through which a $-CH_2CH_2CH_2-$ chain can be threaded. Molecules have been synthesized, however, that have large rings threaded on chains and that have large rings that are interlocked like links in a chain. These latter molecules are called **catenanes**:

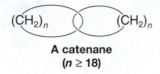

A catenane
($n \geq 18$)

In 1994 J. F. Stoddart and co-workers, then at the University of Birmingham (England), achieved a remarkable synthesis of a catenane containing a linear array of five interlocked rings. Because the rings are interlocked in the same way as those of the olympic symbol, they named the compound **olympiadane**.

Derek H. R. Barton (1918–1998) and Odd Hassel (1897–1981) shared the Nobel Prize in 1969 "for developing and applying the principles of conformation in chemistry." Their work led to fundamental understanding of not only the conformations of cyclohexane rings but also the structures of steroids (Section 23.4) and other compounds containing cyclohexane rings.

4.12 SUBSTITUTED CYCLOHEXANES: AXIAL AND EQUATORIAL HYDROGEN GROUPS

The six-membered ring is the most common ring found among nature's organic molecules. For this reason, we shall give it special attention. We have already seen that the chair conformation of cyclohexane is the most stable one and that it is the predominant conformation of the molecules in a sample of cyclohexane.

The chair conformation of a cyclohexane ring has two distinct orientations for the bonds that project from the ring. These positions are called axial and equatorial, as shown for cyclohexane in Fig. 4.17.

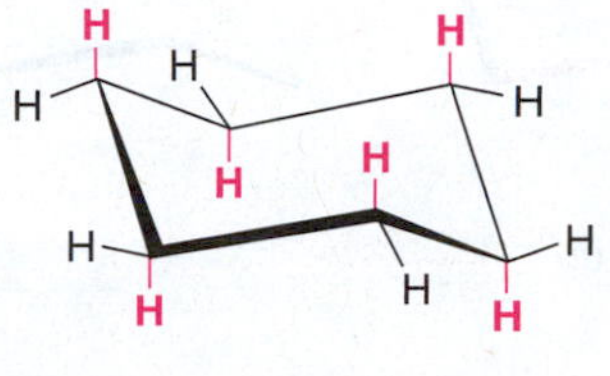

FIGURE 4.17 The chair conformation of cyclohexane. Axial hydrogen atoms are shown in red, equatorial hydrogens are shown in black.

- The **axial bonds** of cyclohexane are those that are perpendicular to the average plane of the ring. There are three axial bonds on each face of the cyclohexane ring, and their orientation (up or down) alternates from one carbon to the next.
- The **equatorial bonds** of cyclohexane are those that extend from the perimeter of the ring. The equatorial bonds alternate from slightly up to slightly down in their orientation from one carbon to the next.

● When a cyclohexane ring undergoes a chair–chair conformational change (a **ring flip**), all of the bonds that were axial become equatorial, and all bonds that were equatorial become axial.

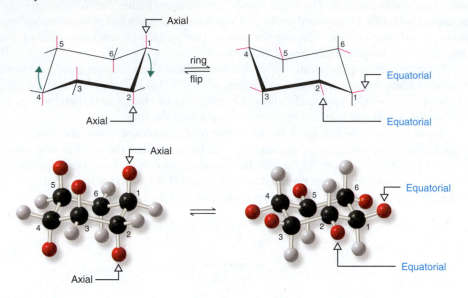

4.12A **HOW TO** Draw Chair Conformational Structures

A set of guidelines will help you draw chair conformational structures that are clear and that have unambiguous axial and equatorial bonds.

● Notice in Fig. 4.18a that sets of parallel lines define opposite sides of the chair. Notice, too, that equatorial bonds are parallel to ring bonds that are one bond away from them in either direction. When you draw chair conformational structures, try to make the corresponding bonds parallel in your drawings.

● When a chair formula is drawn as shown in Fig. 4.18, the axial bonds are all either up or down, in a vertical orientation (Fig. 4.18b). When a vertex of bonds in the ring points up, the axial bond at that position is also up, and the equatorial bond at the same carbon is angled slightly down. When a vertex of ring bonds is down, the axial bond at that position is also down, and the equatorial bond is angled slightly upward.

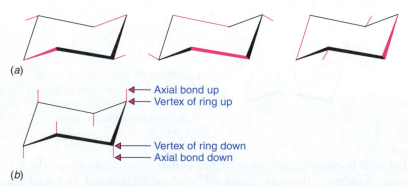

FIGURE 4.18 (a) Sets of parallel lines that constitute the ring and equatorial C—H bonds of the chair conformation. (b) The axial bonds are all vertical. When the vertex of the ring points up, the axial bond is up and vice versa.

Now, try to draw some chair conformational structures for yourself that include the axial and equatorial bonds. Then, compare your drawings with those here and with actual models. You will see that with a little practice your chair conformational structures can be perfect.

4.12B A Conformational Analysis of Methylcyclohexane

Now let us consider methylcyclohexane. Methylcyclohexane has two possible chair conformations (Fig. 4.19, **I** and **II**), and these are interconvertible through the bond rotations that constitute a ring flip. In conformation (Fig. 4.19a) the methyl group (with yellow hydrogens in the space-filling model) occupies an *axial* position, and in conformation **II** the methyl group occupies an *equatorial* position.

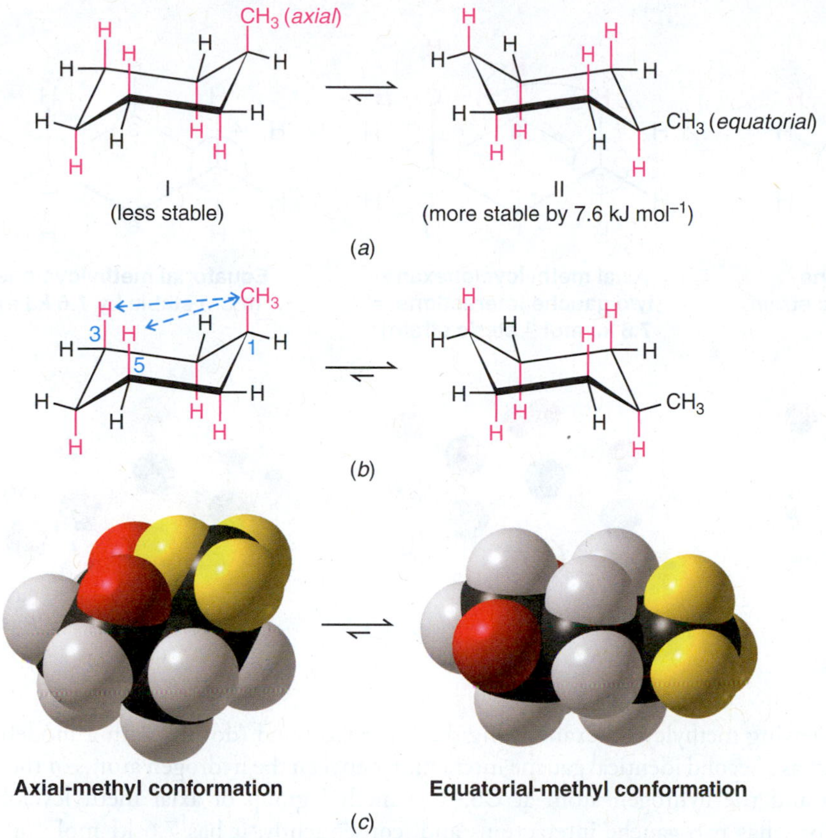

FIGURE 4.19 (a) The conformations of methylcyclohexane with the methyl group axial (**I**) and equatorial (**II**). (b) 1,3-Diaxial interactions between the two axial hydrogen atoms and the axial methyl group in the axial conformation of methylcyclohexane are shown with dashed arrows. Less crowding occurs in the equatorial conformation. (c) Space-filling molecular models for the axial–methyl and equatorial–methyl conformers of methylcyclohexane. In the axial–methyl conformer the methyl group (shown with yellow hydrogen atoms) is crowded by the 1,3-diaxial hydrogen atoms (red), as compared to the equatorial–methyl conformer, which has no 1,3-diaxial interactions with the methyl group.

- The most stable conformation for a monosubstituted cyclohexane ring (a cyclohexane ring where one carbon atom bears a group other than hydrogen) is the conformation where the substituent is equatorial.

Studies indicate that conformation **II** with the equatorial methyl group is more stable than conformation **I** with the axial methyl group by about 7.6 kJ mol^{-1}. Thus, in the equilibrium mixture, the conformation with the methyl group in the equatorial position is the predominant one, constituting about 95% of the equilibrium mixture.

The greater stability of methylcyclohexane with an equatorial methyl group can be understood through an inspection of the two forms as they are shown in Figs. 4.19a–c.

- Studies done with models of the two conformations show that when the methyl group is axial, it is so close to the two axial hydrogens on the same side of the ring (attached to the C3 and C5 atoms) that **the dispersion forces between them are repulsive**.
- This type of steric strain, because it arises from an interaction between an axial group on carbon atom 1 and an axial hydrogen on carbon atom 3 (or 5), is called a **1,3-diaxial interaction**.
- Studies with other substituents show that **there is generally less repulsion when any group larger than hydrogen is equatorial rather than axial**.

The strain caused by a 1,3-diaxial interaction in methylcyclohexane is the same as the strain caused by the close proximity of the hydrogen atoms of methyl groups in the gauche form of butane (Section 4.9). Recall that the interaction in *gauche*-butane (called, for convenience, a *gauche interaction*) causes *gauche*-butane to be less stable than *anti*-butane by 3.8 kJ mol⁻¹. The following Newman projections will help you to see that the two steric interactions are the same. In the second projection we view axial methyl-cyclohexane along the C1—C2 bond and see that what we call a 1,3-diaxial interaction is simply a gauche interaction between the hydrogen atoms of the methyl group and the hydrogen atom at C3:

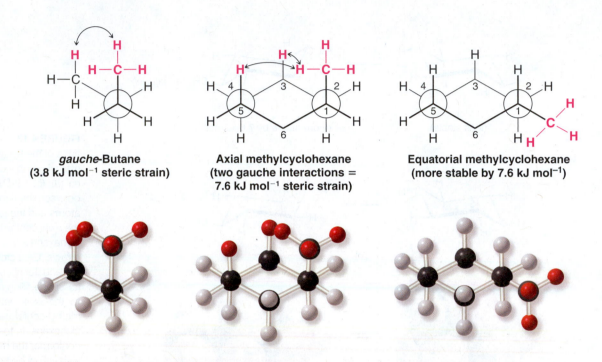

gauche-Butane
(3.8 kJ mol⁻¹ steric strain)

Axial methylcyclohexane
(two gauche interactions =
7.6 kJ mol⁻¹ steric strain)

Equatorial methylcyclohexane
(more stable by 7.6 kJ mol⁻¹)

Viewing methylcyclohexane along the C1—C6 bond (do this with a model) shows that it has a second identical gauche interaction between the hydrogen atoms of the methyl group and the hydrogen atom at C5. The methyl group of axial methylcyclohexane, therefore, has two gauche interactions and, consequently, it has 7.6 kJ mol⁻¹ of strain. The methyl group of equatorial methylcyclohexane does not have a gauche interaction because it is anti to C3 and C5.

PRACTICE PROBLEM 4.14 Show by a calculation (using the formula $\Delta G° = -RT \ln K_{eq}$) that a free-energy difference of 7.6 kJ mol⁻¹ between the axial and equatorial forms of methylcyclo-hexane at 25 °C (with the equatorial form being more stable) does correlate with an equilibrium mixture in which the concentration of the equatorial form is approximately 95%.

4.12C 1,3-Diaxial Interactions of a *tert*-Butyl Group

In cyclohexane derivatives with larger alkyl substituents, the strain caused by 1,3-diaxial interactions is even more pronounced. The conformation of *tert*-butylcyclohexane with the *tert*-butyl group equatorial is estimated to be approximately 21 kJ mol⁻¹ more stable than the axial form (Fig. 4.20). This large energy difference between the two conformations means that, at room temperature, 99.99% of the molecules of *tert*-butylcyclohexane have the *tert*-butyl group in the equatorial position. (The molecule is not conformationally "locked," however; it still flips from one chair conformation to the other.)

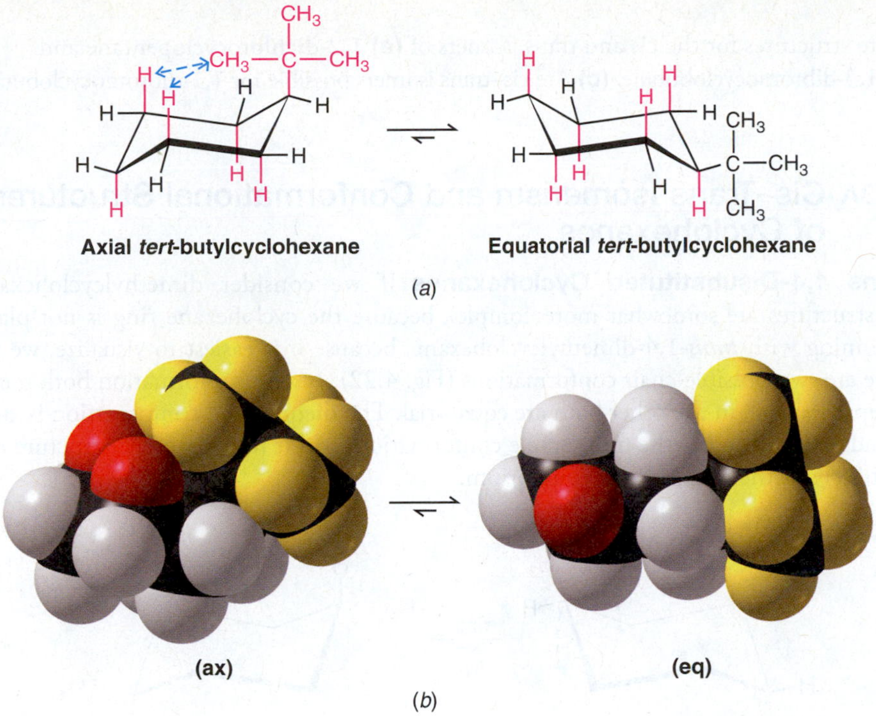

Axial *tert*-butylcyclohexane　　　　**Equatorial *tert*-butylcyclohexane**

(a)

(ax)　　　　　　　**(eq)**

(b)

FIGURE 4.20 (*a*) Diaxial interactions with the large *tert*-butyl group axial cause the conformation with the *tert*-butyl group equatorial to be the predominant one to the extent of 99.99%. (*b*) Space-filling molecular models of *tert*-butylcyclohexane in the axial (ax) and equatorial (eq) conformations, highlighting the position of the 1,3-hydrogens (red) and the *tert*-butyl group (shown with yellow hydrogen atoms).

4.13 DISUBSTITUTED CYCLOALKANES: CIS–TRANS ISOMERISM

The presence of two substituents on different carbons of a cycloalkane allows for the possibility of **cis–trans isomerism** similar to the kind we saw for alkenes in Section 1.14B. These cis–trans isomers are also **stereoisomers** because they differ from each other only in the arrangement of their atoms in space. Consider 1,2-dimethylcyclopropane (Fig. 4.21) as an example.

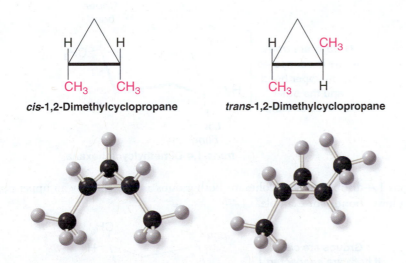

***cis*-1,2-Dimethylcyclopropane**　　***trans*-1,2-Dimethylcyclopropane**

FIGURE 4.21 The *cis*- and *trans*-1,2-dimethylcyclopropane isomers.

The planarity of the cyclopropane ring makes the cis–trans isomerism obvious. In the first structure the methyl groups are on the same side of the ring; therefore, they are cis. In the second structure, they are on opposite sides of the ring; they are trans.

Cis and trans isomers such as these cannot be interconverted without breaking carbon–carbon bonds. They will have different physical properties (boiling points, melting points, and so on). As a result, they can be separated, placed in separate bottles, and kept indefinitely.

PRACTICE PROBLEM 4.15   Write structures for the cis and trans isomers of **(a)** 1,2-dichlorocyclopentane and **(b)** 1,3-dibromocyclobutane. **(c)** Are cis–trans isomers possible for 1,1-dibromocyclobutane?

4.13A Cis–Trans Isomerism and Conformational Structures of Cyclohexanes

Trans 1,4-Disubstituted Cyclohexanes   If we consider dimethylcyclohexanes, the structures are somewhat more complex because the cyclohexane ring is not planar. Beginning with *trans*-1,4-dimethylcyclohexane, because it is easiest to visualize, we find there are two possible chair conformations (Fig. 4.22). In one conformation both methyl groups are axial; in the other both are equatorial. The diequatorial conformation is, as we would expect it to be, the more stable conformation, and it represents the structure of at least 99% of the molecules at equilibrium.

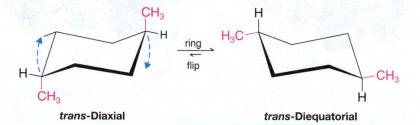

FIGURE 4.22 The two chair conformations of trans-1,4-dimethylcyclohexane: trans–diequatorial and trans–diaxial. The trans–diequatorial form is more stable by 15.2 kJ mol⁻¹.

trans-Diaxial   ⇌ ring flip ⇌   *trans*-Diequatorial

That the diaxial form of *trans*-1,4-dimethylcyclohexane is a trans isomer is easy to see; the two methyl groups are clearly on opposite sides of the ring. The trans relationship of the methyl groups in the diequatorial form is not as obvious, however.

How do we know two groups are cis or trans? A general way to recognize a trans-disubstituted cyclohexane is to notice that one group is attached by the *upper* bond (of the two to its carbon) and one by the *lower* bond:

Groups are *trans* if one is connected by an upper bond and the other by a lower bond

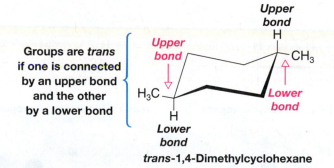

trans-1,4-Dimethylcyclohexane

In a cis 1,4-disubstituted cyclohexane both groups are attached by an upper bond or both by a lower bond. For example,

Groups are *cis* if both are connected by upper bonds or if both are connected by lower bonds

cis-1,4-Dimethylcyclohexane

Cis 1,4-Disubstituted Cyclohexanes   *cis*-1,4-Dimethylcyclohexane exists in two *equivalent* chair conformations (Fig. 4.23). In a cis 1,4-disubstituted cyclohexane, one group is axial and the other is equatorial in both of the possible chair conformations.

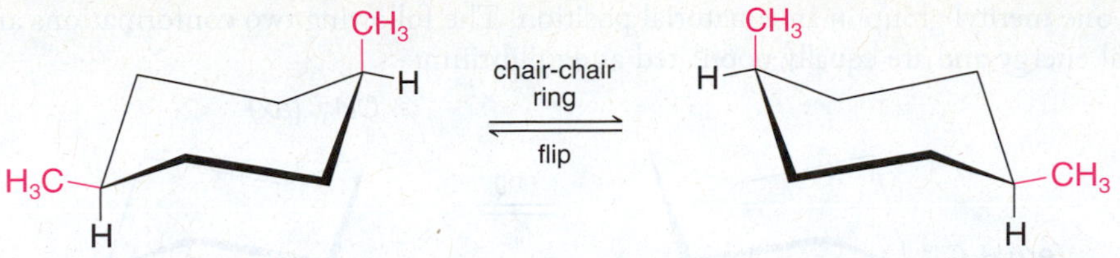

FIGURE 4.23 Equivalent conformations of *cis*-1,4-dimethylcyclohexane.

Equatorial–axial   Axial–equatorial

●●● **SOLVED PROBLEM 4.5**

Consider each of the following conformational structures and tell whether each is cis or trans:

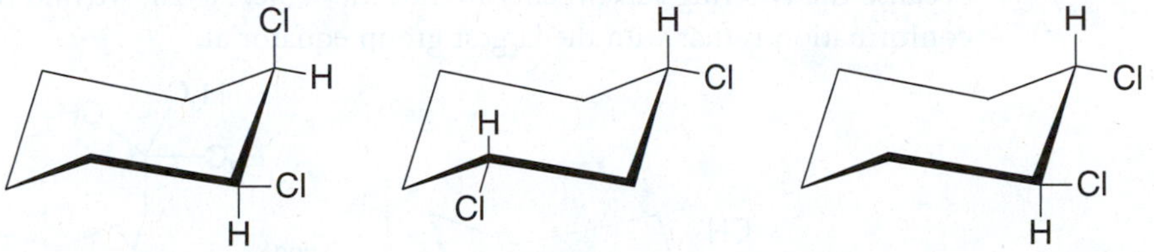

(a)   (b)   (c)

ANSWER: **(a)** Each chlorine is attached by the upper bond at its carbon; therefore, both chlorine atoms are on the same side of the molecule and this is a cis isomer. This is a *cis*-1,2-dichlorocyclohexane. **(b)** Here both chlorine atoms are attached by a lower bond; therefore, in this example, too, both chlorine atoms are on the same side of the molecule and this, too, is a cis isomer. It is *cis*-1,3-dichlorocyclohexane. **(c)** Here one chlorine atom is attached by a lower bond and one by an upper bond. The two chlorine atoms, therefore, are on opposite sides of the molecule, and this is a trans isomer. It is *trans*-1,2-dichlorocyclohexane. Verify these facts by building models.

The two conformations of cis 1,4-disubstituted cyclohexanes *are not equivalent* if one group is larger than the other. Consider *cis*-1-*tert*-butyl-4-methylcyclohexane:

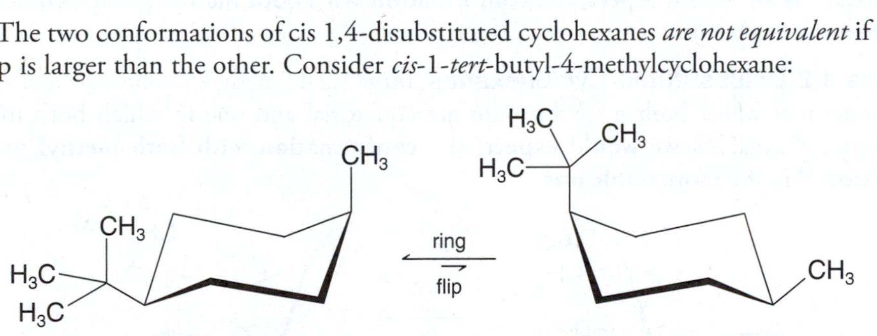

(More stable because large group is equatorial)   (Less stable because large group is axial)

cis-1-*tert*-**Butyl-4-methylcyclohexane**

Here the more stable conformation is the one with the larger group equatorial. This is a general principle:

- When one ring substituent group is larger than the other and they cannot both be equatorial, the conformation with the larger group equatorial will be more stable.

●●●

(a) Write structural formulas for the two chair conformations of *cis*-1-isopropyl-4-methylcyclohexane. **(b)** Are these two conformations equivalent? **(c)** If not, which would be more stable? **(d)** Which would be the preferred conformation at equilibrium?

PRACTICE PROBLEM 4.16

Trans 1,3-Disubstituted Cyclohexanes *trans*-1,3-Dimethylcyclohexane is like the cis 1,4 compound in that each conformation has one methyl group in an axial position and one methyl group in an equatorial position. The following two conformations are of equal energy and are equally populated at equilibrium:

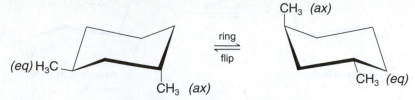

trans-1,3-Dimethylcyclohexane
Equal energy and equally populated conformations

The situation is different for *trans*-1-*tert*-butyl-3-methylcyclohexane (shown below) because the two ring substituents are not the same. Again, we find that the lower energy conformation is that with the largest group equatorial.

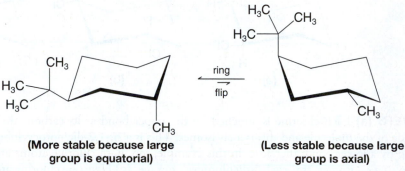

(More stable because large group is equatorial) (Less stable because large group is axial)

trans-1-*tert*-Butyl-3-methylcyclohexane

Cis 1,3-Disubstituted Cyclohexanes *cis*-1,3-Dimethylcyclohexane has a conformation in which both methyl groups are equatorial and one in which both methyl groups are axial. **As we would expect, the conformation with both methyl groups equatorial is the more stable one.**

Trans 1,2-Disubstituted Cyclohexanes *trans*-1,2-Dimethylcyclohexane has a conformation in which both methyl groups are equatorial and one in which both methyl groups are axial. **As we would expect, the conformation with both methyl groups equatorial is the more stable one.**

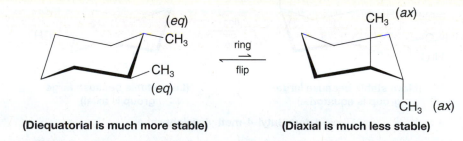

(Diequatorial is much more stable) (Diaxial is much less stable)

trans-1,2-Dimethylcyclohexane

Cis 1,2-Disubstituted Cyclohexanes *cis*-1,2-Dimethylcyclohexane has one methyl group that is axial and one methyl group that is equatorial in each of its chair conformations, thus its two conformations are of equal stability.

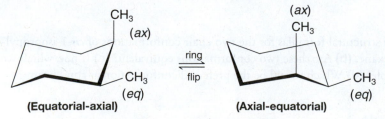

(Equatorial-axial) (Axial-equatorial)

cis-1,2-Dimethylcyclohexane
Equal energy and equally populated conformations

Write a conformational structure for 1,2,3-trimethylcyclohexane in which all the methyl groups are axial and then show its more stable conformation.

ANSWER: A ring flip gives a conformation in which all the groups are equatorial and, therefore, much more stable.

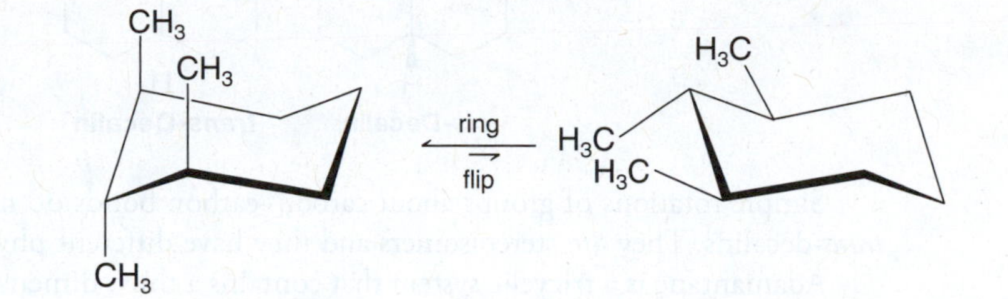

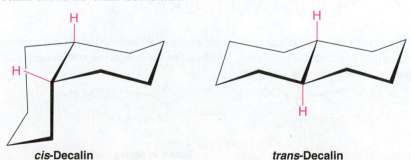

CH₃
CH₃
CH₃

ring flip

H₃C
H₃C
H₃C

All groups axial. Much less stable conformation.   **All groups are equatorial. Much more stable conformation.**

●●● **PRACTICE PROBLEM 4.17**

Write a conformational structure for 1-bromo-3-chloro-5-fluorocyclohexane in which all the substituents are equatorial. Then write its structure after a ring flip.

●●● **PRACTICE PROBLEM 4.18**

(a) Write the two conformations of *cis*-1-*tert*-butyl-2-methylcyclohexane. **(b)** Which conformer has the lowest potential energy?

4.14 BICYCLIC AND POLYCYCLIC ALKANES

Many of the molecules that we encounter in our study of organic chemistry contain more than one ring (Section 4.4B). One of the most important bicyclic systems is bicyclo [4.4.0]decane, a compound that is usually called by its common name, *decalin*:

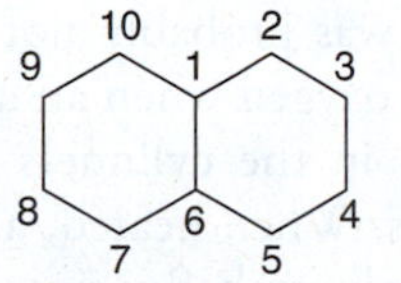

Decalin (bicyclo[4.4.0]decane)
(carbon atoms 1 and 6 are bridgehead carbon atoms)

Decalin shows cis–trans isomerism:

Helpful Hint

Chemical Abstracts Service (CAS) determines the number of rings by the formula $S - A + 1 = N$, where S is the number of single bonds in the ring system, A is the number of atoms in the ring system, and N is the calculated number of rings (see Problem 4.30).

cis-Decalin   *trans*-Decalin

In *cis*-decalin the two hydrogen atoms attached to the bridgehead atoms lie on the same side of the ring; in *trans*-decalin they are on opposite sides. We often indicate this by writing their structures in the following way:

cis-Decalin   *trans*-Decalin

Simple rotations of groups about carbon–carbon bonds do not interconvert *cis*- and *trans*-decalins. They are stereoisomers and they have different physical properties.

Adamantane is a tricyclic system that contains a three-dimensional array of cyclohexane rings, all of which are in the chair form.

Adamantane

In the chapter closer we shall see several examples of other unusual and highly strained, cyclic hydrocarbons.

4.15 CHEMICAL REACTIONS OF ALKANES

Alkanes, as a class, are characterized by a general inertness to many chemical reagents. Carbon–carbon and carbon–hydrogen bonds are quite strong; they do not break unless alkanes are heated to very high temperatures. Because carbon and hydrogen atoms have nearly the same electronegativity, the carbon–hydrogen bonds of alkanes are only slightly polarized. As a consequence, they are generally unaffected by most bases. Molecules of alkanes have no unshared electrons to offer as sites for attack by acids. This low reactivity of alkanes toward many reagents accounts for the fact that alkanes were originally called **paraffins** (*parum affinis*, Latin: little affinity).

The term paraffin, however, was probably not an appropriate one. We all know that alkanes react vigorously with oxygen when an appropriate mixture is ignited. This combustion occurs, for example, in the cylinders of automobiles, in furnaces, and, more gently, with paraffin candles. When heated, alkanes also react with chlorine and bromine, and they react explosively with fluorine. We shall study these reactions in Chapter 10.

4.16 SYNTHESIS OF ALKANES AND CYCLOALKANES

A chemical synthesis may require, at some point, the conversion of a carbon–carbon double or triple bond to a single bond. Synthesis of the following compound, used as an ingredient in some perfumes, is an example.

(used in some perfumes)

This conversion is easily accomplished by a reaction called **hydrogenation**. There are several reaction conditions that can be used to carry out hydrogenation, but among the common ways is use of hydrogen gas and a solid metal catalyst such as platinum, palladium, or nickel. Equations in the following section represent general examples for the hydrogenation of alkenes and alkynes.

4.16A Hydrogenation of Alkenes and Alkynes

- Alkenes and alkynes react with hydrogen in the presence of metal catalysts such as nickel, palladium, and platinum to produce **alkanes**.

The general reaction is one in which the atoms of the hydrogen molecule add to each atom of the carbon–carbon double or triple bond of the alkene or alkyne. This converts the alkene or alkyne to an alkane:

General Reaction

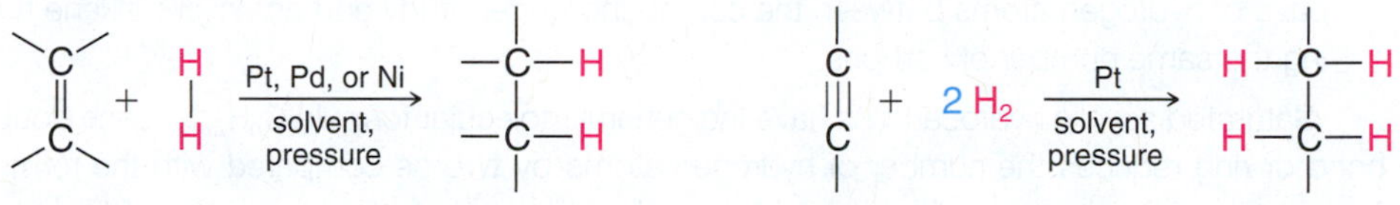

Alkene + H–H → (Pt, Pd, or Ni, solvent, pressure) → Alkane

Alkyne + 2 H₂ → (Pt, solvent, pressure) → Alkane

The reaction is usually carried out by dissolving the alkene or alkyne in a solvent such as ethyl alcohol (C_2H_5OH), adding the metal catalyst, and then exposing the mixture to hydrogen gas under pressure in a special apparatus. One molar equivalent of hydrogen is required to reduce an alkene to an alkane. Two molar equivalents are required to reduce an alkyne. (We shall discuss the mechanism of this reaction in Chapter 7.)

Specific Examples

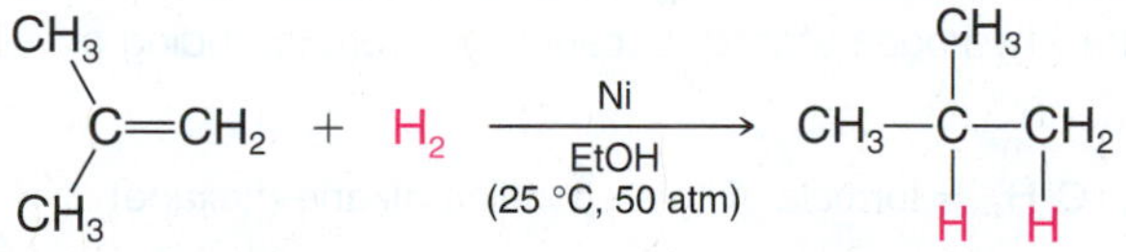

2-Methylpropene + H₂ → (Ni, EtOH, 25 °C, 50 atm) → Isobutane

Cyclohexene + H₂ → (Pd, EtOH, 25 °C, 1 atm) → Cyclohexane

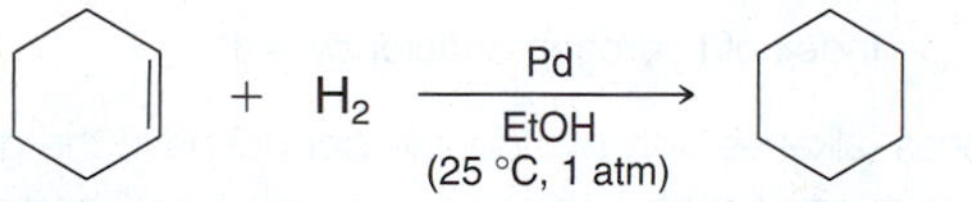

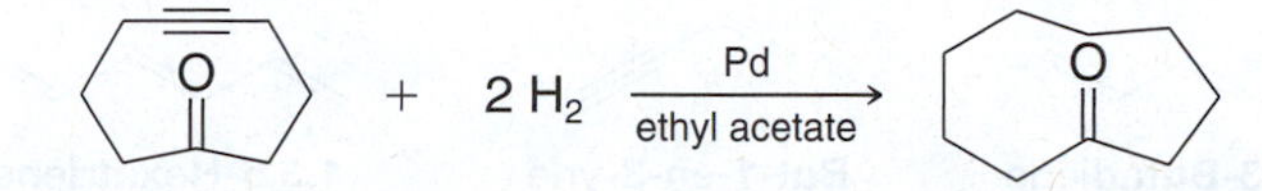

Cyclononyn-6-one + 2 H₂ → (Pd, ethyl acetate) → Cyclononanone

●●● SOLVED PROBLEM 4.7

Write the structures of three pentenes that would all yield pentane on hydrogenation.

ANSWER:

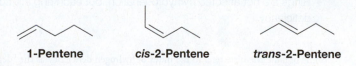

1-Pentene   *cis*-2-Pentene   *trans*-2-Pentene

PRACTICE PROBLEM 4.19 Show the reactions involved for hydrogenation of all the alkenes and alkynes that would yield 2-methylbutane.

4.17 HOW TO GAIN STRUCTURAL INFORMATION FROM MOLECULAR FORMULAS AND THE INDEX OF HYDROGEN DEFICIENCY

A chemist working with an unknown compound can obtain considerable information about its structure from the compound's molecular formula and its **index of hydrogen deficiency (IHD)**.

- The **index of hydrogen deficiency** (IHD)* is defined as the difference in the *number of pairs* of hydrogen atoms between the compound under study and an acyclic alkane having the same number of carbons.

  Saturated acyclic hydrocarbons have the general molecular formula C_nH_{2n+2}. Each double bond or ring reduces the number of hydrogen atoms by two as compared with the formula for a saturated compound. Thus each ring or double bond provides one unit of hydrogen deficiency. For example, 1-hexene and cyclohexane have the same molecular formula (C_6H_{12}) and they are constitutional isomers.

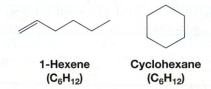

1-Hexene   **Cyclohexane**
(C_6H_{12})   **(C_6H_{12})**

Both 1-hexene and cyclohexane (C_6H_{12}) have an index of hydrogen deficiency equal to 1 (meaning one pair of hydrogen atoms), because the corresponding acyclic alkane is hexane (C_6H_{14}).

C_6H_{14} = formula of corresponding alkane (hexane)

$\underline{C_6H_{12}}$ = formula of compound (1-hexene or cyclohexane)

H_2 = difference = 1 pair of hydrogen atoms

Index of hydrogen deficiency = 1

Alkynes and alkadienes (alkenes with two double bonds) have the general formula C_nH_{2n-2}. Alkenynes (hydrocarbons with one double bond and one triple bond) and alkatrienes (alkenes with three double bonds) have the general formula C_nH_{2n-4}, and so forth.

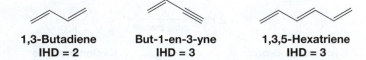

1,3-Butadiene   **But-1-en-3-yne**   **1,3,5-Hexatriene**
IHD = 2   **IHD = 3**   **IHD = 3**

The index of hydrogen deficiency is easily determined by comparing the molecular formula of a given compound with the formula for its hydrogenation product.

- Each double bond consumes one molar equivalent of hydrogen and counts for one unit of hydrogen deficiency.
- Each triple bond consumes two molar equivalents of hydrogen and counts for two units of hydrogen deficiency.
- Rings are not affected by hydrogenation, but each ring still counts for one unit of hydrogen deficiency.

*Some organic chemists refer to the index of hydrogen deficiency as the "degree of unsaturation" or "the number of double-bond equivalencies."

Hydrogenation, therefore, allows us to distinguish between rings and double or triple bonds. Consider again two compounds with the molecular formula C_6H_{12}: 1-hexene and cyclohexane. 1-Hexene reacts with one molar equivalent of hydrogen to yield hexane; under the same conditions cyclohexane does not react:

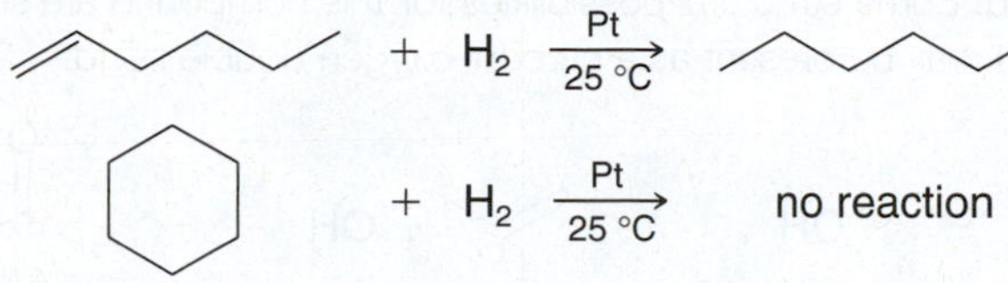

Or consider another example. Cyclohexene and 1,3-hexadiene have the same molecular formula (C_6H_{10}). Both compounds react with hydrogen in the presence of a catalyst, but cyclohexene, because it has a ring and only one double bond, reacts with only one molar equivalent. 1,3-Hexadiene adds two molar equivalents:

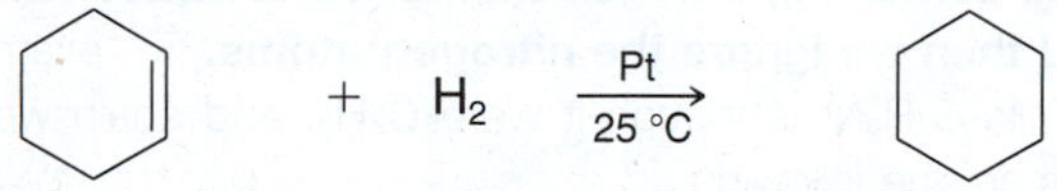

Cyclohexene

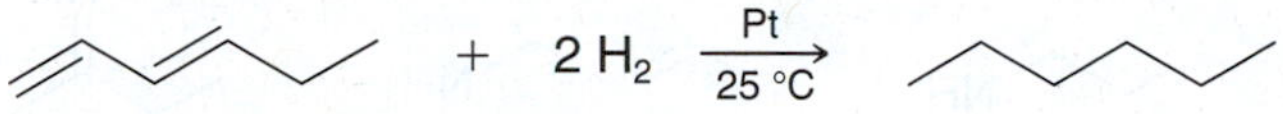

1,3-Hexadiene

PRACTICE PROBLEM 4.20

(a) What is the index of hydrogen deficiency of 2-hexene? **(b)** Of methylcyclopentane? **(c)** Does the index of hydrogen deficiency reveal anything about the location of the double bond in the chain? **(d)** About the size of the ring? **(e)** What is the index of hydrogen deficiency of 2-hexyne? **(f)** In general terms, what structural possibilities exist for a compound with the molecular formula $C_{10}H_{16}$?

PRACTICE PROBLEM 4.21

Zingiberene, a fragrant compound isolated from ginger, has the molecular formula $C_{15}H_{24}$ and is known not to contain any triple bonds. **(a)** What is the index of hydrogen deficiency of zingiberene? **(b)** When zingiberene is subjected to catalytic hydrogenation using an excess of hydrogen, 1 mol of zingiberene absorbs 3 mol of hydrogen and produces a compound with the formula $C_{15}H_{30}$. How many double bonds does a molecule of zingiberene have? **(c)** How many rings?

4.17A Compounds Containing Halogens, Oxygen, or Nitrogen

Calculating the index of hydrogen deficiency (IHD) for compounds other than hydrocarbons is relatively easy.

For compounds containing halogen atoms, we simply count the halogen atoms as though they were hydrogen atoms. Consider a compound with the formula $C_4H_6Cl_2$. To calculate the IHD, we change the two chlorine atoms to hydrogen atoms, considering the formula as though it were C_4H_8. This formula has two hydrogen atoms fewer than the formula for a saturated alkane (C_4H_{10}), and this tells us that the compound has IHD = 1. It could, therefore, have either one ring or one double bond. [We can tell which it has from a hydrogenation experiment: If the compound adds one molar equivalent of hydrogen (H_2) on catalytic hydrogenation at room temperature, then it must have a double bond; if it does not add hydrogen, then it must have a ring.]

For compounds containing oxygen, we simply ignore the oxygen atoms and calculate the IHD from the remainder of the formula. Consider as an example a compound with the formula C_4H_8O. For the purposes of our calculation we consider the compound to be simply C_4H_8 and we calculate IHD = 1. Again, this means that the compound contains either a ring or a double bond. Some structural possibilities for this compound are shown next. Notice that the double bond may be present as a carbon–oxygen double bond:

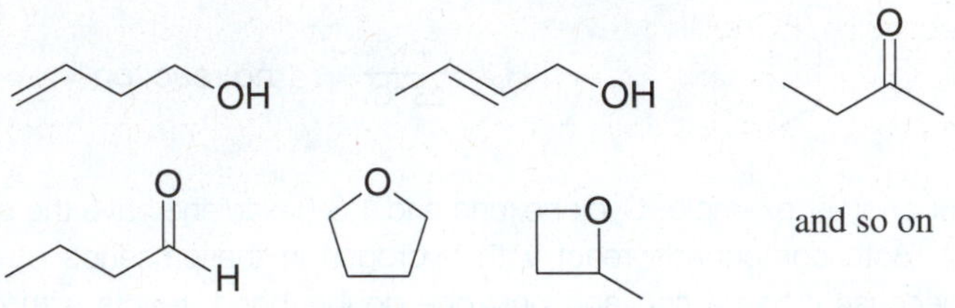

and so on

For compounds containing nitrogen atoms we subtract one hydrogen for each nitrogen atom, and then we ignore the nitrogen atoms. For example, we treat a compound with the formula C_4H_9N as though it were C_4H_8, and again we get IHD = 1. Some structural possibilities are the following:

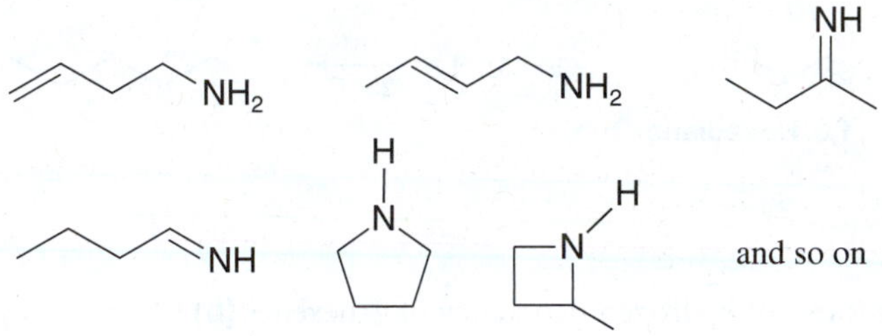

and so on

PRACTICE PROBLEM 4.22   Carbonyl groups also count for a unit of hydrogen deficiency. What are the indices of hydrogen deficiency for the reactant and for the product in the equation shown at the beginning of Section 4.16 for synthesis of a perfume ingredient?

4.18 APPLICATIONS OF BASIC PRINCIPLES

In this chapter we have seen repeated applications of one basic principle in particular:

Nature Prefers States of Lower Potential Energy This principle underlies our explanations of conformational analysis in Sections 4.8–4.13. The staggered conformation of ethane (Section 4.8) is preferred (more populated) in a sample of ethane because its potential energy is lowest. In the same way, the anti conformation of butane (Section 4.9) and the chair conformation of cyclohexane (Section 4.11) are the preferred conformations of these molecules because these conformations are of lowest potential energy. Methylcyclohexane (Section 4.12) exists mainly in the chair conformation with its methyl group equatorial for the same reason. Disubstituted cycloalkanes (Section 4.13) prefer a conformation with both substituents equatorial if this is possible, and, if not, they prefer a conformation with the larger group equatorial. The preferred conformation in each instance is the one of lowest potential energy.

Another effect that we encounter in this chapter, and one we shall see again and again, is how **steric factors** (spatial factors) can affect the stability and reactivity of molecules. Unfavorable spatial interactions between groups are central to explaining why certain conformations are higher in energy than others. But fundamentally this effect is derived itself from another familiar principle: **like charges repel.** Repulsive interactions between the electrons of groups that are in close proximity cause certain conformations to have higher potential energy than others. We call this kind of effect *steric hindrance.*

[WHY Do These Topics Matter?

PUSHING THE BOUNDARIES OF BONDING, ALL WITHIN THE RULES

In this chapter we have learned many of the rules of bond formation and of conformation. Although there are only a few kinds of bonds in organic molecules, they can be combined in an infinite number of ways, sometimes leading to molecules whose existence defies our expectations. For example, using just C—C and C—H bonds, chemists have been able to synthesize structures such as cubane, prismane, and bicyclo[1.1.0]butane, materials that have incredible strain built into their structures. Strained compounds are also found in nature, with one recent discovery being pentacycloanammoxic acid, a material isolated from a particular bacterial strain. This compound is also known as a ladderane because it has a connected set of five 4-membered rings that exist in three-dimensional space like a ladder, or staircase.

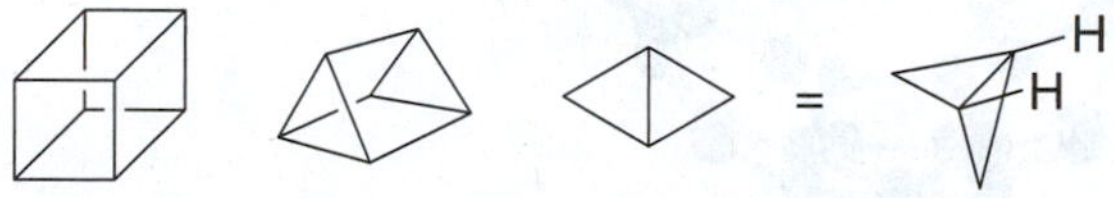

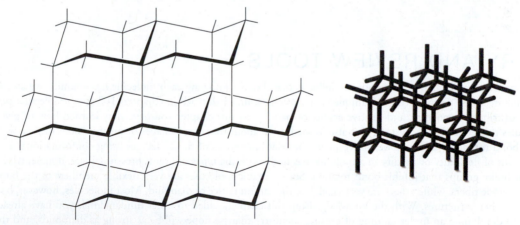

Cubane   Prismane   Bicyclo[1.1.0]butane

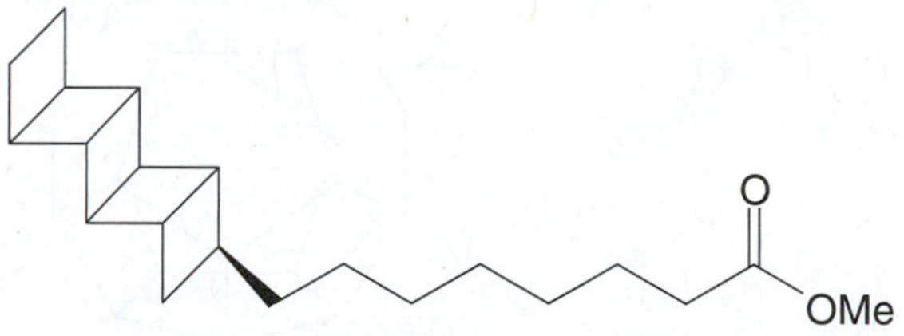

**Pentacycloanammoxic acid methyl ester
(a ladderane)**

An important thing to note is that these different bond combinations and resultant three-dimensional shapes lead to completely distinct physical properties that can be harnessed for unique practical applications. Perhaps one of the best illustrations of this concept is found in materials comprised only of carbon, materials also known as allotropes since they are formed solely from a single, pure element. For example, when carbon is bonded with itself through single bonds with sp^3 hybridization, the result is diamond, the hardest of all materials found in nature and a popular component of jewelry. When carbon is bonded with itself with sp^2 hybridization through a series of interconnecting C—C and C=C bonds, it forms flat, interconnected sheets of benzene-like rings. These sheets can stack with one another, forming graphite. This material is much softer than diamond and is the material that constitutes the "lead" of pencils. Graphene, which is just one of these sheets, can be wrapped through new bonds into tubes (also called nanotubes) that have impressive properties as thermal and electrical conductors.

A portion of the diamond structure

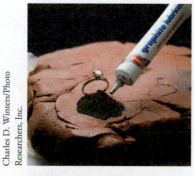

Charles D. Winters/Photo Researchers, Inc.

Carbon is shown here in its diamond and graphite forms

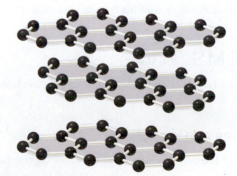

A portion of the structure of graphite

(continues on next page)

If the rings of graphite and graphene are combined together into discrete balls possessing a finite number of carbon atoms, then materials such as buckminsterfullerene (also known as buckyballs) result. The name of this material derives from its resemblance to the geodesic dome first designed by architect Buckminster Fuller. This particular compound, which is comprised of 60 carbon atoms, has bonds that look exactly like the seams of a soccer ball through its possession of 32 interlocking rings of which 20 are hexagons and 12 are pentagons. The center is large enough, in fact, to hold an atom of argon (such a compound has been made). Another variant of this type of structure is dodecahedrane, a compound composed of 20 carbon atoms and first synthesized in 1982 by scientists at the Ohio State University. Materials of this type collectively are believed to have potential in applications as diverse as armor, drug delivery, and superconductivity.

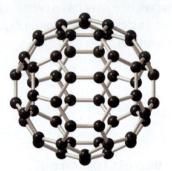

Dodecahedrane   **Buckminsterfullerene**

The key point is that the molecular variations are nearly endless, increasing exponentially as more and more atoms are added. This fact is one of the most beautiful elements of organic chemistry, because it means that we are largely limited in terms of possible structures by just two factors: our ability to imagine a molecule and having the tools necessary to forge it in the form of appropriate chemical reactions. This outcome is because the rules, the language, of organic chemistry are consistent.

To learn more about these topics, see:

Hopf, H. *Classics in Hydrocarbon Chemistry*. Wiley-VCH: Weinheim, **2000**, p. 560.

SUMMARY AND REVIEW TOOLS

One of the reasons we organic chemists love our discipline is that, besides knowing each molecule has a family, we also know that each one has its own architecture, "personality," and unique name. You have already learned in Chapters 1–3 about molecular personalities with regard to charge distribution, polarity, and relative acidity or basicity. In this chapter you have now learned how to give unique names to simple molecules using the IUPAC system. You also learned more about the overall shapes of organic molecules, how their shapes can change through bond rotations, and how we can compare the relative energies of those changes using conformational analysis. You now know that the extent of flexibility or rigidity in a molecule has to do with the types of bonds present (single, double, triple), and whether there are rings or bulky groups that inhibit bond rotation. Some organic molecules are very flexible members of the family, such as the molecules in our muscle fibers, while others are very rigid, like the carbon lattice of diamond. Most molecules, however, have both flexible and rigid aspects to their structures. With the knowledge from this chapter, added to other fundamentals you have already learned, you are on your way to developing an understanding of organic chemistry that we hope will be as strong as diamonds, and that you can flex like a muscle when you approach a problem. When you are finished with this chapter's homework, maybe you can even take a break by resting your mind on the chair conformation of cyclohexane.

PROBLEMS WILEY PLUS

Note to Instructors: Many of the homework problems are available for assignment via *WileyPLUS*, an online teaching and learning solution program.

NOMENCLATURE AND ISOMERISM

4.23 Write a bond-line formula for each of the following compounds:

(a) 1,4-Dichloropentane

(b) *sec*-Butyl bromide

(c) 4-Isopropylheptane

(d) 2,2,3-Trimethylpentane

(e) 3-Ethyl-2-methylhexane

(f) 1,1-Dichlorocyclopentane

(g) *cis*-1,2-Dimethylcyclopropane   **(j)** *trans*-4-Isobutylcyclohexanol   **(m)** Bicyclo[2.2.2]octane

(h) *trans*-1,2-Dimethylcyclopropane   **(k)** 1,4-Dicyclopropylhexane   **(n)** Bicyclo[3.1.1]heptane

(i) 4-Methyl-2-pentanol   **(l)** Neopentyl alcohol   **(o)** Cyclopentylcyclopentane

4.24 Give systematic IUPAC names for each of the following:

(a)

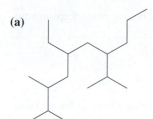

(c)

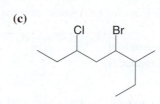

(e)

(g)

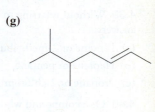

(b)

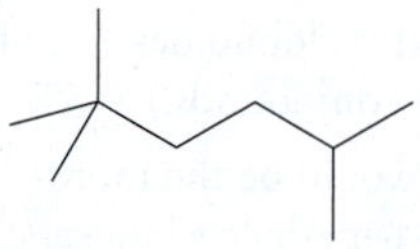

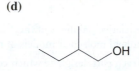

(d)

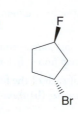

(f)

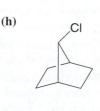

(h)

4.25 The name *sec*-butyl alcohol defines a specific structure but the name *sec*-pentyl alcohol is ambiguous. Explain.

4.26 Write the structure and give the IUPAC systematic name of an alkane or cycloalkane with the formulas **(a)** C_8H_{18} that has only primary hydrogen atoms, **(b)** C_6H_{12} that has only secondary hydrogen atoms, **(c)** C_6H_{12} that has only primary and secondary hydrogen atoms, and **(d)** C_8H_{14} that has 12 secondary and 2 tertiary hydrogen atoms.

4.27 Write the structure(s) of the simplest alkane(s), i.e., one(s) with the fewest number of carbon atoms, wherein each possesses primary, secondary, tertiary, and quaternary carbon atoms. (A quaternary carbon is one that is bonded to four other carbon atoms.) Assign an IUPAC name to each structure.

4.28 Ignoring compounds with double bonds, write structural formulas and give names for all of the isomers with the formula C_5H_{10}.

4.29 Write structures for the following bicyclic alkanes:

(a) Bicyclo[1.1.0]butane   **(c)** 2-Chlorobicyclo[3.2.0]heptane

(b) Bicyclo[2.1.0]pentane   **(d)** 7-Methylbicyclo[2.2.1]heptane

4.30 Use the $S - A + 1 = N$ method (Helpful Hint, Section 4.14) to determine the number of rings in cubane (Section 4.14).

4.31 A spiro ring junction is one where two rings that share no bonds originate from a single carbon atom. Alkanes containing such a ring junction are called spiranes.

(a) For the case of bicyclic spiranes of formula C_7H_{12}, write structures for all possibilities where all carbons are incorporated into rings.

(b) Write structures for other bicyclic molecules that fit this formula.

4.32 Tell what is meant by a homologous series and illustrate your answer by writing structures for a homologous series of alkyl halides.

HYDROGENATION

4.33 Four different cycloalkenes will all yield methylcyclopentane when subjected to catalytic hydrogenation. What are their structures? Show the reactions.

4.34 **(a)** Three different alkenes yield 2-methylbutane when they are hydrogenated in the presence of a metal catalyst. Give their structural formulas and write equations for the reactions involved. **(b)** One of these alkene isomers has characteristic absorptions at approximately 998 and 914 cm^{-1} in its IR spectrum. Which one is it?

4.35 An alkane with the formula C_6H_{14} can be prepared by hydrogenation of either of only two precursor alkenes having the formula C_6H_{12}. Write the structure of this alkane, give its IUPAC name, and show the reactions.

CONFORMATIONS AND STABILITY

4.36 Rank the following compounds in order of increasing stability based on relative ring strain.

4.37 Write the structures of two chair conformations of 1-*tert*-butyl-1-methylcyclohexane. Which conformation is more stable? Explain your answer.

4.38 Sketch curves similar to the one given in Fig. 4.8 showing the energy changes that arise from rotation about the C2—C3 bond of (a) 2,3-dimethylbutane and (b) 2,2,3,3-tetramethylbutane. You need not concern yourself with actual numerical values of the energy changes, but you should label all maxima and minima with the appropriate conformations.

4.39 Without referring to tables, decide which member of each of the following pairs would have the higher boiling point. Explain your answers.

(a) Pentane or 2-methylbutane

(b) Heptane or pentane

(c) Propane or 2-chloropropane

(d) Butane or 1-propanol

(e) Butane or CH_3COCH_3

4.40 One compound whose molecular formula is C_4H_6 is a bicyclic compound. Another compound with the same formula has an infrared absorption at roughly 2250 cm^{-1} (the bicyclic compound does not). Draw structures for each of these two compounds and explain how the IR absorption allows them to be differentiated.

4.41 Which compound would you expect to be the more stable: *cis*-1,2-dimethylcyclopropane or *trans*-1,2-dimethylcyclopropane? Explain your answer.

4.42 Consider that cyclobutane exhibits a puckered geometry. Judge the relative stabilities of the 1,2-disubstituted cyclobutanes and of the 1,3-disubstituted cyclobutanes. (You may find it helpful to build handheld molecular models of representative compounds.)

4.43 Write the two chair conformations of each of the following and in each part designate which conformation would be the more stable: (a) *cis*-1-*tert*-butyl-3-methylcyclohexane, (b) *trans*-1-*tert*-butyl-3-methylcyclohexane, (c) *trans*-1-*tert*-butyl-4-methylcyclohexane, (d) *cis*-1-*tert*-butyl-4-methylcyclohexane.

4.44 Provide an explanation for the surprising fact that all-*trans*-1,2,3,4,5,6-hexaisopropylcyclohexane is a stable molecule in which all isopropyl groups are axial. (You may find it helpful to build a handheld molecular model.)

4.45 *trans*-1,3-Dibromocyclobutane has a measurable dipole moment. Explain how this proves that the cyclobutane ring is not planar.

SYNTHESIS

4.46 Specify the missing compounds and/or reagents in each of the following syntheses:

(a) *trans*-5-Methyl-2-hexene $\xrightarrow{?}$ 2-methylhexane

(b)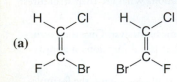

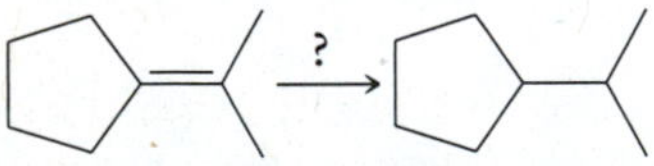

(c) Chemical reactions rarely yield products in such initially pure form that no trace can be found of the starting materials used to make them. What evidence in an IR spectrum of each of the crude (unpurified) products from the above reactions would indicate the presence of one of the organic reactants used to synthesize each target molecule? That is, predict one or two key IR absorptions for the reactants that would distinguish it/them from IR absorptions predicted for the product.

CHALLENGE PROBLEMS

4.47 Consider the cis and trans isomers of 1,3-di-*tert*-butylcyclohexane (build molecular models). What unusual feature accounts for the fact that one of these isomers apparently exists in a twist boat conformation rather than a chair conformation?

4.48 Using the rules found in this chapter, give systematic names for the following or indicate that more rules need to be provided:

(a)

(c)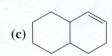

(b)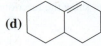

(d)

4.49 Open the energy-minimized 3D Molecular Models on the book's website for *trans*-1-*tert*-butyl-3-methylcyclohexane and *trans*-1,3-di-*tert*-butylcyclohexane. What conformations of cyclohexane do the rings in these two compounds resemble most closely? How can you account for the difference in ring conformations between them?

4.50 Open the 3D Molecular Models on the book's website for cyclopentane and vitamin B_{12}. Compare cyclopentane with the nitrogen-containing five-membered rings in vitamin B_{12}. Is the conformation of cyclopentane represented in the specified rings of vitamin B_{12}? What factor(s) account for any differences you observe?

4.51 Open the 3D Molecular Model on the book's website for buckminsterfullerene. What molecule has its type of ring represented 16 times in the surface of buckminsterfullerene?

LEARNING GROUP PROBLEMS

1. This is the predominant conformation for D-glucose:

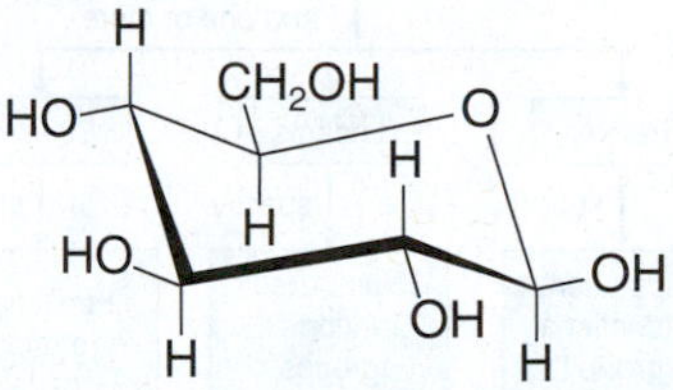

Why is it not surprising that D-glucose is the most commonly found sugar in nature? (*Hint*: Look up structures for sugars such as D-galactose and D-mannose, and compare these with D-glucose.)

2. Using Newman projections, depict the relative positions of the substituents on the bridgehead atoms of *cis*- and *trans*-decalin. Which of these isomers would be expected to be more stable, and why?

3. When 1,2-dimethylcyclohexene (below) is allowed to react with hydrogen in the presence of a platinum catalyst, the product of the reaction is a cycloalkane that has a melting point of $-50\,°C$ and a boiling point of $130\,°C$ (at 760 torr). **(a)** What is the structure of the product of this reaction? **(b)** Consult an appropriate resource (such as the web or a CRC handbook) and tell which stereoisomer it is. **(c)** What does this experiment suggest about the mode of addition of hydrogen to the double bond?

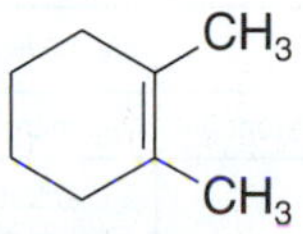

1,2-Dimethylcyclohexene

4. When cyclohexene is dissolved in an appropriate solvent and allowed to react with chlorine, the product of the reaction, $C_6H_{10}Cl_2$, has a melting point of $-7\,°C$ and a boiling point (at 16 torr) of $74\,°C$. **(a)** Which stereoisomer is this? **(b)** What does this experiment suggest about the mode of addition of chlorine to the double bond?

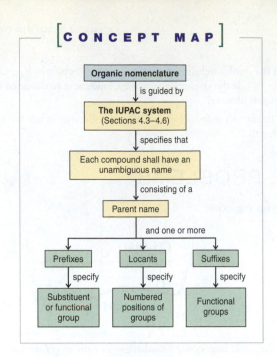

CONCEPT MAP

Organic nomenclature
is guided by
The IUPAC system (Sections 4.3–4.6)
specifies that
Each compound shall have an unambiguous name
consisting of a
Parent name
and one or more
Prefixes
Locants
Suffixes
specify
specify
specify
Substituent or functional group
Numbered positions of groups
Functional groups

CONCEPT MAP

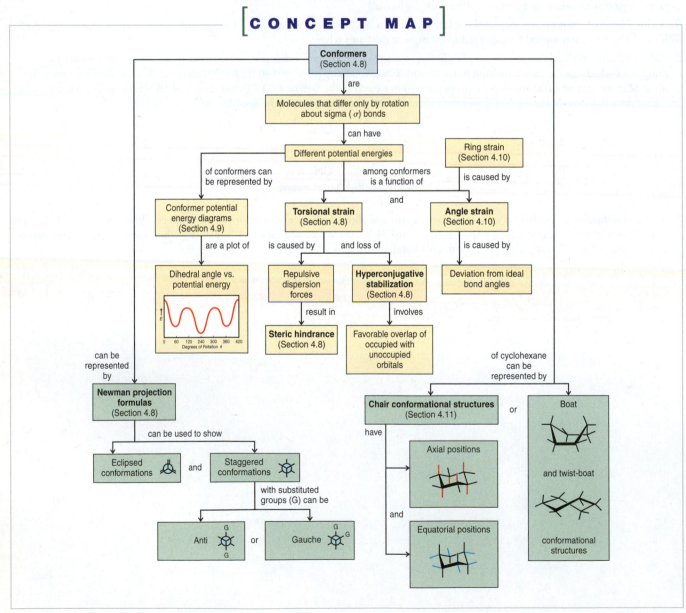

Conformers (Section 4.8)
are
Molecules that differ only by rotation about sigma (σ) bonds
can have
Different potential energies
of conformers can be represented by
among conformers is a function of
and
Ring strain (Section 4.10)
is caused by
Conformer potential energy diagrams (Section 4.9)
are a plot of
Dihedral angle vs. potential energy
E
0 60 120 240 300 360 420
Degrees of Rotation θ
Torsional strain (Section 4.8)
is caused by
and loss of
Angle strain (Section 4.10)
is caused by
Repulsive dispersion forces
result in
Steric hindrance (Section 4.8)
Hyperconjugative stabilization (Section 4.8)
involves
Favorable overlap of occupied with unoccupied orbitals
Deviation from ideal bond angles
can be represented by
Newman projection formulas (Section 4.8)
can be used to show
Eclipsed conformations
and
Staggered conformations
with substituted groups (G) can be
Anti
or
Gauche
G
G
of cyclohexane can be represented by
Chair conformational structures (Section 4.11)
or
Boat
and twist-boat
conformational structures
have
Axial positions
and
Equatorial positions

WILEY PLUS See Special Topic A in *WileyPLUS*

Stereochemistry

CHIRAL MOLECULES

We are all aware of the fact that certain everyday objects such as gloves and shoes possess the quality of "handedness". A right-handed glove only fits a right hand; a left-handed shoe only fits a left foot. Many other objects have the potential to exist in right- and left-handed forms, and those that do are said to be "chiral". For example, the screws shown above are chiral. One screw has a right-handed thread. A right-handed person would find using it to be quite comfortable. The other screw has a left-handed thread and would better suit a left-handed person. (Unfortunately, for left-handed persons, most screws are right-handed.) We shall now find that chirality also has important consequences for chemistry.

IN THIS CHAPTER WE WILL CONSIDER:

- how to identify, categorize, and name chiral molecules
- how chirality can affect the chemical and biochemical behavior of organic compounds

[WHY DO THESE TOPICS MATTER?] At the end of this chapter, we will explain what may have been the origin of chirality in the universe, and why many of the important molecules found in living organisms, such as peptides, DNA, and carbohydrates exist in only one chiral form when the other form seems equally likely.

PHOTO CREDITS: Nicholas Eveleigh/Stockbyte/Getty Images, Inc.

5.1 CHIRALITY AND STEREOCHEMISTRY

The glass and its mirror image are superposable.

Chirality is a phenomenon that pervades the universe. How can we know whether a particular object is **chiral** or **achiral** (not chiral)?

- We can tell if an object has **chirality** by examining the object and its mirror image.

Every object has a mirror image. Many objects are achiral. By this we mean that *the object and its mirror image are identical*—that is, the object and its mirror image are **superposable** one on the other.* Superposable means that one can, in one's mind's eye, place one object on the other so that all parts of each coincide. Simple geometrical objects such as a sphere or a cube are achiral. So is an object like a water glass.

- **A chiral object is one that cannot be superposed on its mirror image.**

FIGURE 5.1 The mirror image of a right hand is a left hand.

FIGURE 5.2 Left and right hands are not superposable.

Each of our hands is chiral. When you view your right hand in a mirror, the image that you see in the mirror *is a left hand* (Fig. 5.1). However, as we see in Fig. 5.2, your left hand and your right hand are not identical because *they are not superposable*. Your hands are chiral. In fact, the word chiral comes from the Greek word cheir meaning hand. An object such as a mug may or may not be chiral. If it has no markings on it, it is achiral. If the mug has a logo or image on one side, it is chiral.

This mug is chiral because it is not superposable on its mirror image.

*To be superposable is different than to be super*im*posable. Any two objects can be superimposed simply by putting one object on top of the other, whether or not the objects are the same. To *superpose* two objects (as in the property of superposition) means, on the other hand, that **all parts of each object must coincide**. The condition of superposability must be met for two things to be **identical**.

5.1A The Biological Significance of Chirality

The human body is structurally chiral, with the heart lying to the left of center and the liver to the right. Helical seashells are chiral and most are spiral, such as a right-handed screw. Many plants show chirality in the way they wind around supporting structures. Honeysuckle winds as a left-handed helix; bindweed winds in a right-handed way. DNA is a chiral molecule. The double helical form of DNA turns in a right-handed way.

Chirality in molecules, however, involves more than the fact that some molecules adopt left- or right-handed conformations. As we shall see in this chapter, it is the nature of groups bonded at specific atoms that can bestow chirality on a molecule. Indeed, all but one of the 20 amino acids that make up naturally occurring proteins are chiral, and all of these are classified as being left-handed. The molecules of natural sugars are almost all classified as being right-handed. In fact, most of the molecules of life are chiral, and most are found in only one mirror image form.*

Chirality has tremendous importance in our daily lives. Most pharmaceuticals are chiral. Usually only one mirror-image form of a drug provides the desired effect. The other mirror-image form is often inactive or, at best, less active. In some cases the other mirror-image form of a drug actually has severe side effects or toxicity (see Section 5.5 regarding thalidomide). Our senses of taste and smell also depend on chirality. As we shall see, one mirror-image form of a chiral molecule may have a certain odor or taste while its mirror image smells and tastes completely different. The food we eat is largely made of molecules of one mirror-image form. If we were to eat food that was somehow made of molecules with the unnatural mirror-image form, we would likely starve because the enzymes in our bodies are chiral and preferentially react with the natural mirror-image form of their substrates.

Let us now consider what causes some molecules to be chiral. To begin, we will return to aspects of isomerism.

Bindweed (top photo) (*Convolvulus sepium*) winds in a right-handed fashion, like the right-handed helix of DNA. (DNA spiral: Reprinted with permission of the McGraw-Hill Companies. From Neal, L.; *Chemistry and Biochemistry: A Comprehensive Introduction.* © 1971.)

Perennou Nuridsany/Photo Researchers, Inc.

5.2 ISOMERISM: CONSTITUTIONAL ISOMERS AND STEREOISOMERS

5.2A Constitutional Isomers

Isomers are different compounds that have the same molecular formula. In our study thus far, much of our attention has been directed toward isomers we have called constitutional isomers.

- **Constitutional isomers** have the same molecular formula but different connectivity, meaning that their atoms are connected in a different order. Examples of constitutional isomers are the following:

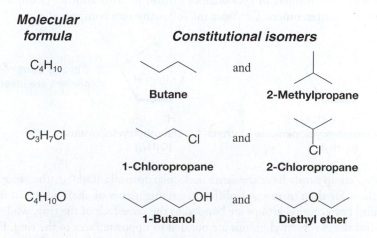

Molecular formula	Constitutional isomers	
C_4H_{10}	Butane	and 2-Methylpropane
C_3H_7Cl	1-Chloropropane	and 2-Chloropropane
$C_4H_{10}O$	1-Butanol	and Diethyl ether

*For interesting reading, see Hegstrum, R. A. and Kondepudi, D. K. The Handedness of the Universe. *Sci. Am.* **1990**, *262*, 98–105, and Horgan, J. The Sinister Cosmos. *Sci. Am.* **1997**, *276*, 18–19.

5.2B Stereoisomers

Stereoisomers are not constitutional isomers.

- **Stereoisomers** have their atoms connected in the same sequence (the same constitution), but they differ in the arrangement of their atoms in space. The consideration of such spatial aspects of molecular structure is called **stereochemistry**.

We have already seen examples of some types of stereoisomers. The cis and trans forms of alkenes are stereoisomers (Section 1.13B), as are the cis and trans forms of substituted cyclic molecules (Section 4.13).

5.2C Enantiomers and Diastereomers

Stereoisomers can be subdivided into two general categories: those that are enantiomers of each other, and those that are **diastereomers** of each other.

- **Enantiomers** are stereoisomers whose molecules are nonsuperposable mirror images of each other.

All other stereoisomers are diastereomers.

- **Diastereomers** are stereoisomers whose molecules are not mirror images of each other.

The alkene isomers *cis*- and *trans*-1,2-dichloroethene shown here are stereoisomers that are **diastereomers**.

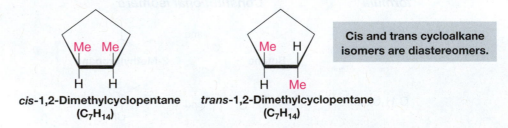

cis-1,2-Dichloroethene   *trans*-1,2-Dichloroethene
($C_2H_2Cl_2$)   ($C_2H_2Cl_2$)

Cis and trans alkene isomers are diastereomers.

By examining the structural formulas for *cis*- and *trans*-1,2-dichloroethene, we see that they have the same molecular formula ($C_2H_2Cl_2$) and the same connectivity (both compounds have two central carbon atoms joined by a double bond, and both compounds have one chlorine and one hydrogen atom attached to each carbon atom). But, their atoms have a different arrangement in space that is not interconvertible from one to another (due to the large barrier to rotation of the carbon–carbon double bond), making them stereoisomers. Furthermore, they are stereoisomers that are not mirror images of each other; therefore they are diastereomers and not enantiomers.

Cis and trans isomers of cycloalkanes furnish us with another example of stereoisomers that are diastereomers. Consider the following two compounds:

Me Me   Me H
H H   H Me
cis-1,2-Dimethylcyclopentane   *trans*-1,2-Dimethylcyclopentane
(C_7H_{14})   (C_7H_{14})

Cis and trans cycloalkane isomers are diastereomers.

These two compounds have the same molecular formula (C_7H_{14}), the same sequence of connections for their atoms, but different arrangements of their atoms in space. In one compound both methyl groups are bonded to the same face of the ring, while in the other compound the two methyl groups are bonded to opposite faces of the ring. Furthermore, the positions of the methyl groups cannot be interconverted by conformational changes. Therefore, these compounds are stereoisomers, and because they are stereoisomers that are not mirror images of each other, they can be further classified as diastereomers.

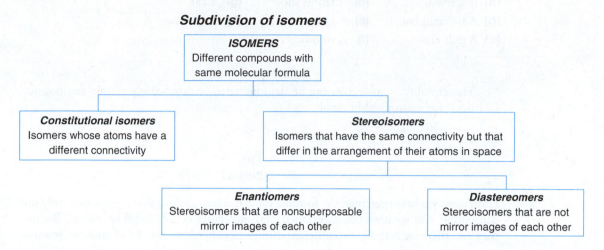

In Section 5.12 we shall study other molecules that can exist as diastereomers but are not cis and trans isomers of each other. First, however, we need to consider enantiomers further.

Subdivision of isomers

ISOMERS
Different compounds with same molecular formula

Constitutional isomers
Isomers whose atoms have a different connectivity

Stereoisomers
Isomers that have the same connectivity but that differ in the arrangement of their atoms in space

Enantiomers
Stereoisomers that are nonsuperposable mirror images of each other

Diastereomers
Stereoisomers that are not mirror images of each other

5.3 ENANTIOMERS AND CHIRAL MOLECULES

Enantiomers occur only with compounds whose molecules are chiral.

- A **chiral molecule** is one that is not superposable on its mirror image.

The trans isomer of 1,2-dimethylcyclopentane is **chiral** because it is **not superposable** on its mirror image, as the following formulas illustrate.

Mirror images of *trans*-1,2-dimethylcyclopentane
They are not superposable and therefore are enantiomers.

Enantiomers do not exist for achiral molecules.

- An **achiral molecule** is superposable on its mirror image.

The cis and trans isomers of 1,2-dichloroethene are both **achiral** because each isomer is **superposable** on its mirror image, as the following formulas illustrate.

cis-1,2-Dichloroethene mirror images   *trans*-1,2-Dichloroethene mirror images

The mirror images of the cis isomer are superposable on each other (try rotating one by 180° to see that it is identical to the other), and therefore the cis formulas both represent the same, achiral molecule. The same analysis is true for the trans isomer.

- Enantiomers only occur with compounds whose molecules are chiral.
- A chiral molecule and its mirror image are called a **pair of enantiomers**. The relationship between them is **enantiomeric**.

The universal test for chirality of a molecule, or any object, is the nonsuperposability of the molecule or object on its mirror image. We encounter chiral and achiral objects throughout our daily life. Shoes are chiral, for example, whereas most socks are achiral.

PRACTICE PROBLEM 5.1 Classify each of the following objects as to whether it is chiral or achiral:

(a) A screwdriver   **(d)** A tennis shoe   **(g)** A car

(b) A baseball bat   **(e)** An ear   **(h)** A hammer

(c) A golf club   **(f)** A woodscrew

The chirality of molecules can be demonstrated with relatively simple compounds. Consider, for example, 2-butanol:

2-Butanol

Until now, we have presented the formula for 2-butanol as though it represented only one compound and we have not mentioned that molecules of 2-butanol are chiral. Because they are, there are actually two different 2-butanols and these two 2-butanols are enantiomers. We can understand this if we examine the drawings and models in Fig. 5.3.

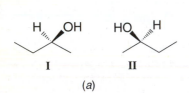

(a)

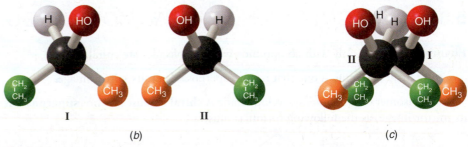

(b)   (c)

FIGURE 5.3 (a) Three-dimensional drawings of the 2-butanol enantiomers **I** and **II**. (b) Models of the 2-butanol enantiomers. (c) An unsuccessful attempt to superpose models of **I** and **II**.

Helpful Hint

Working with models is a helpful study technique whenever three-dimensional aspects of chemistry are involved.

If model **I** is held before a mirror, model **II** is seen in the mirror and vice versa. Models **I** and **II** are not superposable on each other; therefore, they represent different, but isomeric, molecules. *Because models I and II are nonsuperposable mirror images of each other, the molecules that they represent are enantiomers.*

PRACTICE PROBLEM 5.2 Construct handheld models of the 2-butanols represented in Fig. 5.3 and demonstrate for yourself that they are not mutually superposable. **(a)** Make similar models of 2-bromopropane. Are they superposable? **(b)** Is a molecule of 2-bromopropane chiral? **(c)** Would you expect to find enantiomeric forms of 2-bromopropane?

5.4 MOLECULES HAVING ONE CHIRALITY CENTER ARE CHIRAL

- A **chirality center** is a tetrahedral carbon atom that is bonded to four different groups.
- A molecule that contains **one chirality** center is chiral and can exist as a pair of enantiomers.

Molecules with more than one chirality center can also exist as enantiomers, but only if the molecule is not superposable on its mirror image. (We shall discuss that situation later in Section 5.12.) For now we will focus on molecules having a single chirality center.

Chirality centers are often designated with an asterisk (*). The chirality center in 2-butanol is **C2** (Figure 5.4). The four different groups attached to C2 are a hydroxyl group, a hydrogen atom, a methyl group, and an ethyl group. (It is important to note that

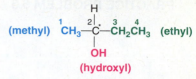

chirality is a property of a molecule as a whole, and that a chirality center is a structural feature that can cause a molecule to be chiral.)

An ability to find chirality centers in structural formulas will help us recognize molecules that are chiral, and that can exist as enantiomers.

- The presence of a single chirality center in a molecule guarantees that the molecule is chiral and that enantiomeric forms are possible.

Figure 5.5 demonstrates that enantiomeric compounds can exist whenever a molecule contains a single chirality center.

(hydrogen)

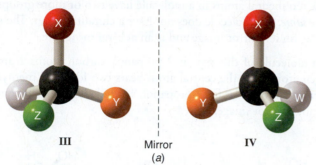

FIGURE 5.4 The tetrahedral carbon atom of 2-butanol that bears four different groups. [By convention, chirality centers are often designated with an asterisk (*).]

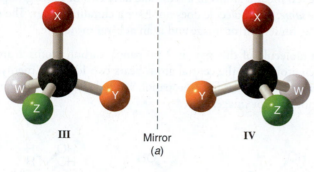

III   Mirror   IV

(a)

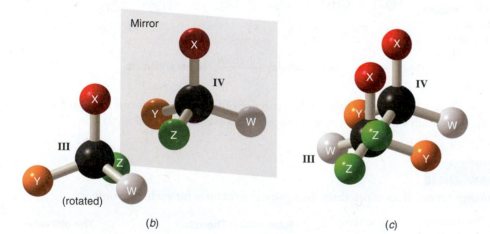

Mirror

III   (rotated)   IV

III

(b)   (c)

FIGURE 5.5 A demonstration of chirality of a generalized molecule containing one chirality center. (a) The four different groups around the carbon atom in **III** and **IV** are arbitrary. (b) **III** is rotated and placed in front of a mirror. **III** and **IV** are found to be related as an object and its mirror image. (c) **III** and **IV** are not superposable; therefore, the molecules that they represent are chiral and are enantiomers.

- An important property of enantiomers with a single chirality center is that *interchanging any two groups at the chirality center converts one enantiomer into the other*.

In Fig. 5.3*b* it is easy to see that interchanging the methyl and ethyl groups converts one enantiomer into the other. You should now convince yourself that interchanging any other two groups has the same result.

- Any atom at which an interchange of groups produces a stereoisomer is called a **stereogenic center**. (If the atom is a carbon atom it is usually called a **stereogenic carbon**.)

When we discuss interchanging groups like this, we must take care to notice that what we are describing is *something we do to a molecular model or something we do on paper*. An interchange of groups in a real molecule, if it can be done, requires breaking covalent bonds, and this is something that requires a large input of energy. This means that enantiomers such as the 2-butanol enantiomers ***do not interconvert*** spontaneously.

The *chirality center* of 2-butanol is one example of a *stereogenic center*, but there are stereogenic centers that are not chirality centers. The carbon atoms of *cis*-1,2-dichloroethene and of *trans*-1,2-dichloroethene (Section 5.2C) are stereogenic centers because an interchange of groups at either carbon atom produces the other stereoisomer. The carbon atoms of *cis*-1,2-dichloroethene and *trans*-1,2-dichloroethene are not chirality centers, however, because they do not have four different groups attached to them.

Helpful Hint

Interchanging two groups of a model or three-dimensional formula is a useful test for determining whether structures of two chiral molecules are the same or different.

●●●

PRACTICE PROBLEM 5.3   Demonstrate the validity of what we have represented in Fig. 5.5 by constructing models. Demonstrate for yourself that **III** and **IV** are related as an object and its mirror image and *that they are not superposable* (i.e., that **III** and **IV** are chiral molecules and are enantiomers). **(a)** Take **IV** and exchange the positions of any two groups. What is the new relationship between the molecules? **(b)** Now take either model and exchange the positions of any two groups. What is the relationship between the molecules now?

● If all of the tetrahedral atoms in a molecule have two or more groups attached that *are the same*, the molecule does not have a chirality center. The molecule is superposable on its mirror image and is an **achiral molecule**.

An example of a molecule of this type is 2-propanol; carbon atoms 1 and 3 bear three identical hydrogen atoms and the central atom bears two identical methyl groups. If we write three-dimensional formulas for 2-propanol, we find (Fig. 5.6) that one structure can be superposed on its mirror image.

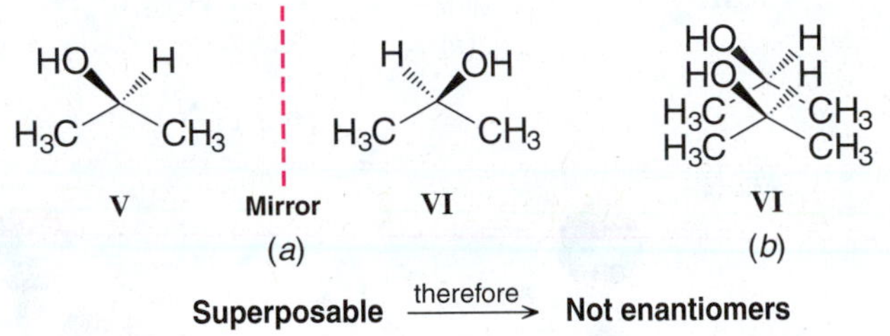

FIGURE 5.6 (a) 2-Propanol (**V**) and its mirror image (**VI**). (b) When either one is rotated, the two structures are superposable and so do not represent enantiomers. They represent two molecules of the same compound. 2-Propanol does not have a chirality center.

Thus, we would not predict the existence of enantiomeric forms of 2-propanol, and experimentally only one form of 2-propanol has ever been found.

SOLVED PROBLEM 5.1

Does 2-bromopentane have a chirality center? If so, write three-dimensional structures for each enantiomer.

STRATEGY AND ANSWER: First we write a structural formula for the molecule and look for a carbon atom that has four different groups attached to it. In this case, carbon 2 has four different groups: a hydrogen, a methyl group, a bromine, and a propyl group. Thus, carbon 2 is a **chirality center**.

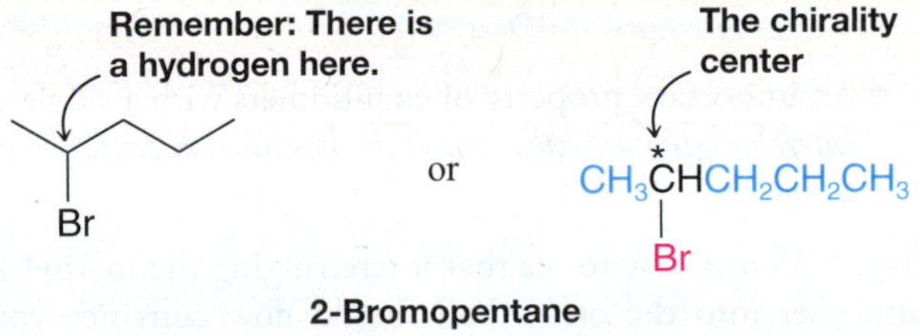

The enantiomers are

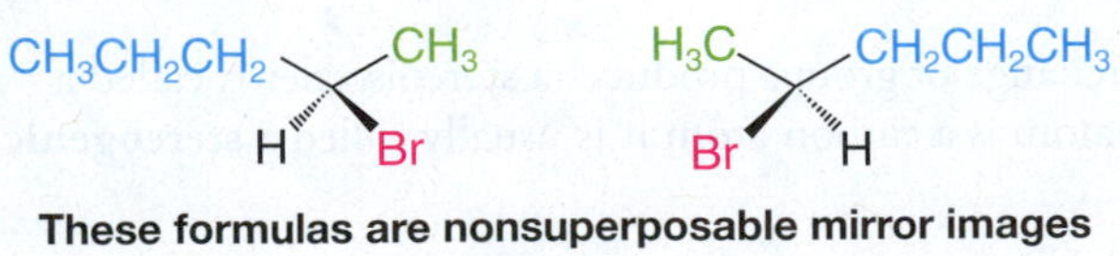

These formulas are nonsuperposable mirror images

●●●

PRACTICE PROBLEM 5.4   Some of the molecules listed here have a chirality center; some do not. Write three-dimensional formulas for both enantiomers of those molecules that do have a chirality center.

(a) 2-Fluoropropane
(b) 2-Methylbutane
(c) 2-Chlorobutane
(d) 2-Methyl-1-butanol

(e) *trans*-2-Butene
(f) 2-Bromopentane
(g) 3-Methylpentane
(h) 3-Methylhexane

(i) 2-Methyl-2-pentene
(j) 1-Chloro-2-methylbutane

5.4A Tetrahedral versus Trigonal Stereogenic Centers

It is important to clarify the difference between stereogenic centers, in general, and a chirality center, which is one type of stereogenic center. The chirality center in 2-butanol is a tetrahedral stereogenic center. The carbon atoms of *cis-* and *trans-*1,2-dichloroethene are also stereogenic centers, but they are trigonal stereogenic centers. They are *not* chirality centers. An interchange of groups at the alkene carbons of either 1,2-dichloroethene isomer produces a stereoisomer (a molecule with the same connectivity but a different arrangement of atoms in space), but it does not produce a nonsuperposable mirror image. A chirality center, on the other hand, is one that must have the possibility of nonsuperposable mirror images.

- Chirality centers are tetrahedral stereogenic centers.
- Cis and trans alkene isomers contain trigonal stereogenic centers.

5.5 MORE ABOUT THE BIOLOGICAL IMPORTANCE OF CHIRALITY

The origin of biological properties relating to chirality is often likened to the specificity of our hands for their respective gloves; the binding specificity for a chiral molecule (like a hand) at a chiral receptor site (a glove) is only favorable in one way. If either the molecule or the biological receptor site had the wrong handedness, the natural physiological response (e.g., neural impulse, reaction catalysis) would not occur. A diagram showing how only one amino acid in a pair of enantiomers can interact in an optimal way with a hypothetical binding site (e.g., in an enzyme) is shown in Fig. 5.7. Because of the chirality center of the amino acid, three-point binding can occur with proper alignment for only one of the two enantiomers.

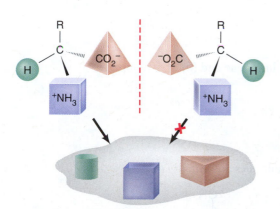

FIGURE 5.7 Only one of the two amino acid enantiomers shown (the left-hand one) can achieve three-point binding with the hypothetical binding site (e.g., in an enzyme).

Chiral molecules can show their handedness in many ways, including the way they affect human beings. One enantiomeric form of a compound called limonene (Section 23.3) is primarily responsible for the odor of oranges and the other enantiomer for the odor of lemons.

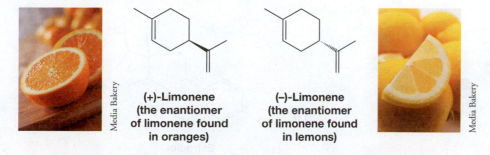

(+)-Limonene
(the enantiomer
of limonene found
in oranges)

(−)-Limonene
(the enantiomer
of limonene found
in lemons)

One enantiomer of a compound called carvone (Practice Problem 5.14) is the essence of caraway, and the other is the essence of spearmint.

The activity of drugs containing chirality centers can similarly vary between enantiomers, sometimes with serious or even tragic consequences. For several years before 1963 the drug thalidomide was used to alleviate the symptoms of morning sickness in pregnant women. In 1963 it was discovered that thalidomide was the cause of horrible birth defects in many children born subsequent to the use of the drug.

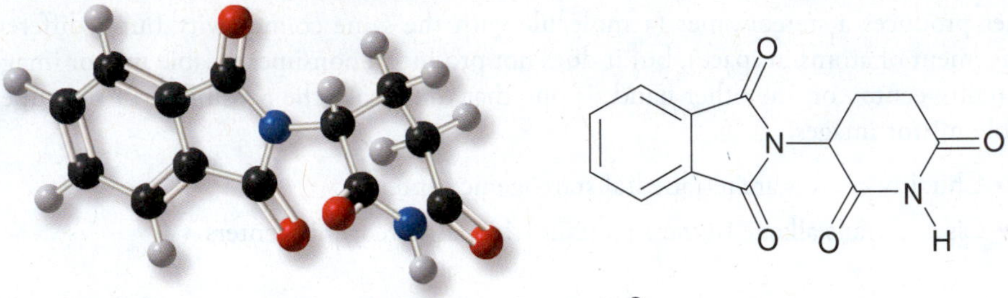

Thalidomide (Thalomid®)

Even later, evidence began to appear indicating that whereas one of the thalidomide enantiomers (the right-handed molecule) has the intended effect of curing morning sickness, the other enantiomer, which was also present in the drug (in an equal amount), may be the cause of the birth defects. The evidence regarding the effects of the two enantiomers is complicated by the fact that, under physiological conditions, the two enantiomers are interconverted. Now, however, thalidomide is approved under highly strict regulations for treatment of some forms of cancer and a serious complication associated with leprosy. Its potential for use against other conditions including AIDS and rheumatoid arthritis is also under investigation. We shall consider other aspects of chiral drugs in Section 5.11.

● ● ●

PRACTICE PROBLEM 5.5   Which atom is the chirality center of **(a)** limonene and **(b)** of thalidomide?

● ● ●

PRACTICE PROBLEM 5.6   Which atoms in each of the following molecules are chirality centers?

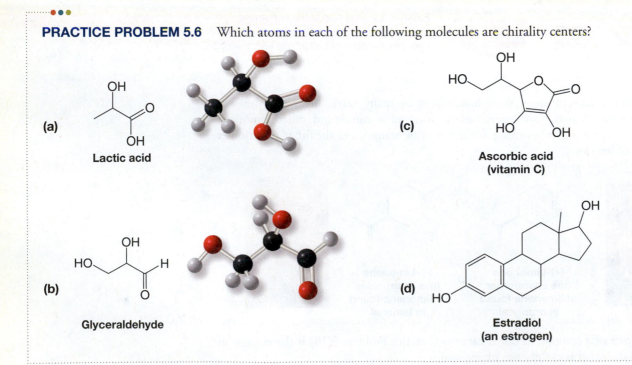

(a)

Lactic acid

(b)

Glyceraldehyde

(c)

**Ascorbic acid
(vitamin C)**

(d)

**Estradiol
(an estrogen)**

5.6 HOW TO TEST FOR CHIRALITY: PLANES OF SYMMETRY

The ultimate way to test for molecular **chirality** is to construct models of the molecule and its mirror image and then determine whether they are superposable. If the two models are superposable, the molecule that they represent is achiral. If the models are not superposable, then the molecules that they represent are chiral. We can apply this test with actual models, as we have just described, or we can apply it by drawing three-dimensional structures and attempting to superpose them in our minds.

There are other aids, however, that will assist us in recognizing chiral molecules. We have mentioned one already: **the presence of a *single* chirality center**. Other aids are based on the absence of certain symmetry elements in the molecule.

- A molecule will not be chiral if it possesses a plane of symmetry.
- A **plane of symmetry** (also called a mirror plane) is defined as an imaginary plane that bisects a molecule in such a way that the two halves of the molecule are mirror images of each other.

The plane may pass through atoms, between atoms, or both. For example, 2-chloropropane has a plane of symmetry (Fig. 5.8a), whereas 2-chlorobutane does not (Fig. 5.8b).

- All molecules with a plane of symmetry in their most symmetric conformation are achiral.

FIGURE 5.8 (a) 2-Chloropropane has a plane of symmetry and is achiral. (b) 2-Chlorobutane does not possess a plane of symmetry and is chiral.

Plane of symmetry

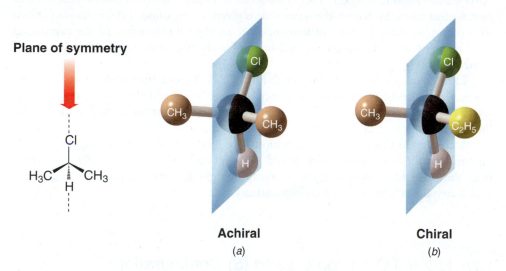

Achiral
(a)

Chiral
(b)

Glycerol, $CH_2OHCHOHCH_2OH$, is an important constituent in the biological synthesis of fats, as we shall see in Chapter 23. **(a)** Does glycerol have a plane of symmetry? If so, write a three-dimensional structure for glycerol and indicate where it is. **(b)** Is glycerol chiral?

STRATEGY AND ANSWER: **(a)** Yes, glycerol has a plane symmetry. Notice that we have to choose the proper conformation and orientation of the molecule to see the plane of symmetry. **(b)** No, glycerol is achiral because it has a conformation containing a plane of symmetry.

Plane of symmetry

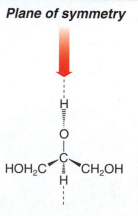

●●●

Which of the objects listed in Practice Problem 5.1 possess a plane of symmetry and are, therefore, achiral?

PRACTICE PROBLEM 5.7

PRACTICE PROBLEM 5.8 Write three-dimensional formulas and designate a plane of symmetry for all of the achiral molecules in Practice Problem 5.4. (In order to be able to designate a plane of symmetry you may need to write the molecule in an appropriate conformation.)

5.7 NAMING ENANTIOMERS: THE *R,S*-SYSTEM

The two enantiomers of 2-butanol are the following:

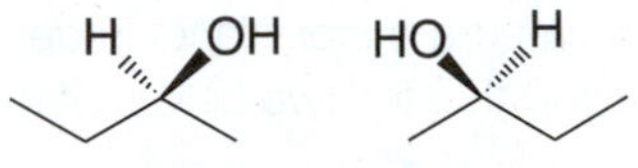

If we name these two **enantiomers** using only the IUPAC system of nomenclature that we have learned so far, both enantiomers will have the same name: 2-butanol (or *sec*-butyl alcohol) (Section 4.3F). This is undesirable because *each compound must have its own distinct name*. Moreover, the name that is given a compound should allow a chemist who is familiar with the rules of nomenclature to write the structure of the compound from its name alone. Given the name 2-butanol, a chemist could write either structure **I** or structure **II**.

Three chemists, R. S. Cahn (England), C. K. Ingold (England), and V. Prelog (Switzerland), devised a system of nomenclature that, when added to the IUPAC system, solves both of these problems. This system, called the ***R,S*-system** or the Cahn–Ingold–Prelog system, is part of the IUPAC rules.

According to this system, one enantiomer of 2-butanol should be designated (*R*)-2-butanol and the other enantiomer should be designated (*S*)-2-butanol. [(*R*) and (*S*) are from the Latin words *rectus* and *sinister*, meaning right and left, respectively.] These molecules are said to have opposite **configurations** at C2.

5.7A HOW TO Assign (*R*) and (*S*) Configurations

We assign (*R*) and (*S*) configurations on the basis of the following procedure.

1. Each of the four groups attached to the chirality center is assigned a **priority** or **preference** *a*, *b*, *c*, or *d*. Priority is first assigned on the basis of the **atomic number** of the atom that is directly attached to the chirality center. The group with the lowest atomic number is given the lowest priority, *d*; the group with next higher atomic number is given the next higher priority, *c*; and so on. (In the case of isotopes, the isotope of greatest atomic mass has highest priority.)

We can illustrate the application of the rule with the following 2-butanol enantiomer:

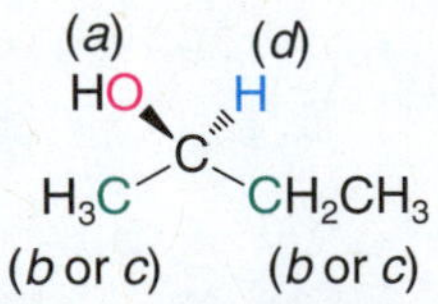

One of the 2-butanol enantiomers

Oxygen has the highest atomic number of the four atoms attached to the chirality center and is assigned the highest priority, *a*. Hydrogen has the lowest atomic number and is assigned the lowest priority, *d*. A priority cannot be assigned for the methyl group and the ethyl group by this approach because the atom that is directly attached to the chirality center is a carbon atom in both groups.

2. When a priority cannot be assigned on the basis of the atomic number of the atoms that are directly attached to the chirality center, then the next set of atoms in the unassigned groups is examined. This process is continued until a decision can be made. *We assign a priority at the first point of difference.**

When we examine the methyl group of the 2-butanol enantiomer above, we find that the next set of atoms bonded to the carbon consists of three hydrogen atoms (H, H, H). In the ethyl group the next set of atoms bonded to the carbon consists of one carbon atom and two hydrogen atoms (C, H, H). Carbon has a higher atomic number than hydrogen, so we assign the ethyl group the higher priority, *b*, and the methyl group the lower priority, *c*, since (C, H, H) > (H, H, H):

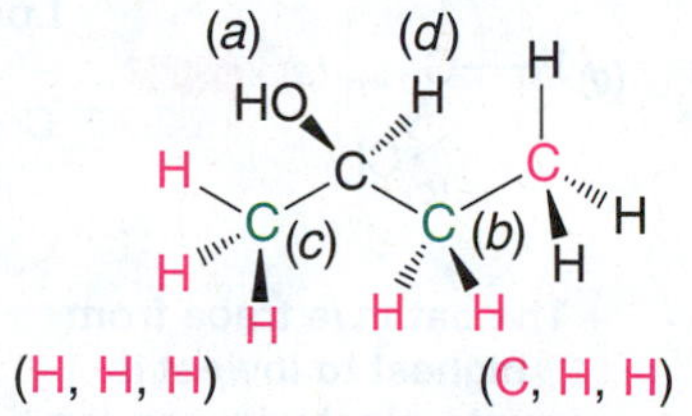

3. We now rotate the formula (or model) so that the group with lowest priority (*d*) is directed away from us:

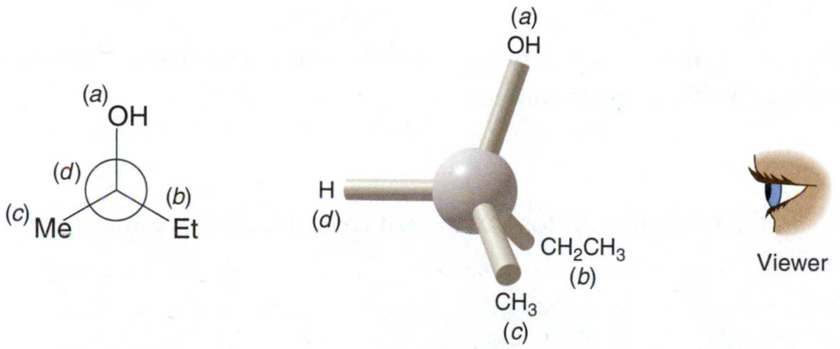

Newman projection	**One of the 2-butanol enantiomers**

Then we trace a path from *a* to *b* to *c*. If, as we do this, the direction of our finger (or pencil) is *clockwise*, the enantiomer is designated (*R*). If the direction is *counterclockwise*, the enantiomer is designated (*S*).

On this basis the 2-butanol enantiomer **II** is (*R*)-2-butanol:

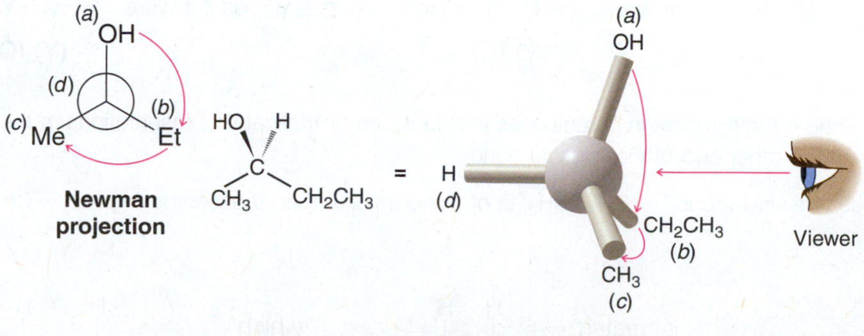

Arrows are clockwise.

(*R*)-2-Butanol

---

*The rules for a branched chain require that we follow the chain with the highest priority atoms.

••• SOLVED PROBLEM 5.3

Shown here is an enantiomer of bromochlorofluoroiodomethane. Is it (*R*) or (*S*)?

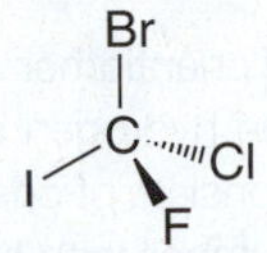

STRATEGY AND ANSWER:

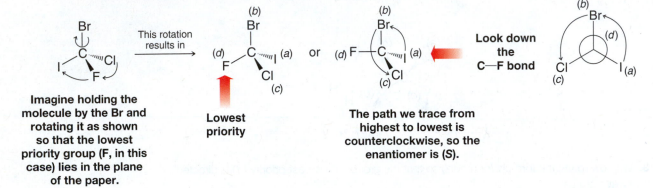

Imagine holding the molecule by the Br and rotating it as shown so that the lowest priority group (F, in this case) lies in the plane of the paper.

Lowest priority

The path we trace from highest to lowest is counterclockwise, so the enantiomer is (S).

Look down the C—F bond

•••

PRACTICE PROBLEM 5.9 Write the enantiomeric forms of bromochlorofluoromethane and assign each enantiomer its correct (*R*) or (*S*) designation.

•••

PRACTICE PROBLEM 5.10 Give (*R*) and (*S*) designations for each pair of enantiomers given as answers to Practice Problem 5.4.

The first three rules of the Cahn–Ingold–Prelog system allow us to make an (*R*) or (*S*) designation for most compounds containing single bonds. For compounds containing multiple bonds one other rule is necessary:

4. Groups containing double or triple bonds are assigned priorities as if both atoms were duplicated or triplicated—that is,

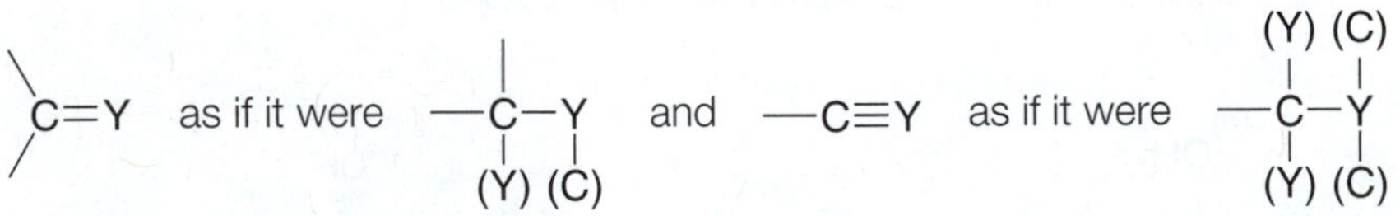

$$\overset{}{\underset{}{\diagdown}}C{=}Y \quad \text{as if it were} \quad -\overset{|}{\underset{(Y)\ (C)}{C}}-Y \quad \text{and} \quad -C{\equiv}Y \quad \text{as if it were} \quad -\overset{(Y)\ (C)}{\underset{(Y)\ (C)}{C}}-Y$$

where the symbols in parentheses are duplicate or triplicate representations of the atoms at the other end of the multiple bond.

Thus, the vinyl group, —CH=CH$_2$, is of higher priority than the isopropyl group, —CH(CH$_3$)$_2$. That is,

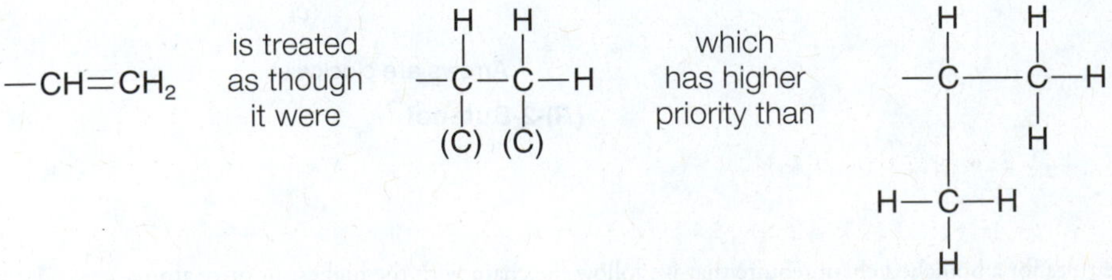

—CH=CH$_2$ is treated as though it were

$$-\overset{H}{\underset{(C)}{C}}-\overset{H}{\underset{(C)}{C}}-H$$

which has higher priority than

$$-\overset{H}{\underset{\underset{\underset{H}{|}}{\overset{|}{\underset{H-C-H}{|}}}}{C}}-\overset{H}{\underset{H}{C}}-H$$

because at the second set of atoms out, the vinyl group (see the following structure) is C, H, H, whereas the isopropyl group along either branch is H, H, H. (At the first set of atoms both groups are the same: C, C, H.)

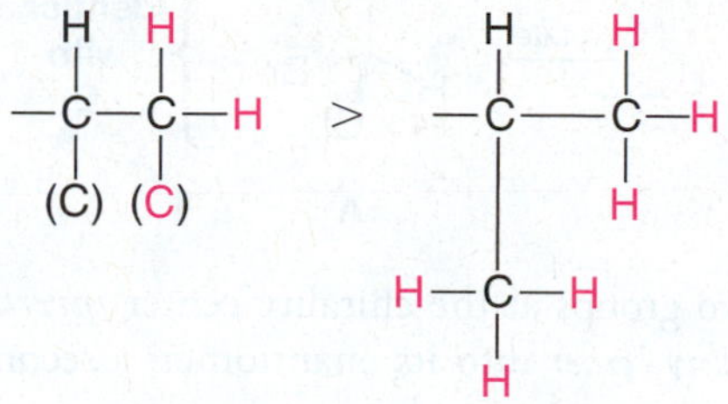

$$
\begin{array}{c}
\overset{\displaystyle H \quad H}{-\overset{|}{C}-\overset{|}{C}-H} \\
\underset{\text{(C) (C)}}{}
\end{array}
\quad > \quad
\begin{array}{c}
\overset{\displaystyle H \quad H}{-\overset{|}{C}-\overset{|}{C}-H} \\
\overset{|}{H} \\
H-\overset{|}{\underset{|}{C}}-H \\
H
\end{array}
$$

C, H, H	>	**H, H, H**
Vinyl group		**Isopropyl group**

Other rules exist for more complicated structures, but we shall not study them here.*

PRACTICE PROBLEM 5.11

List the substituents in each of the following sets in order of priority, from highest to lowest:

(a) —Cl, —OH, —SH, —H

(b) —CH₃, —CH₂Br, —CH₂Cl, —CH₂OH

(c) —H, —OH, —CHO, —CH₃

(d) —CH(CH₃)₂, —C(CH₃)₃, —H, —CH=CH₂

(e) —H, —N(CH₃)₂, —OCH₃, —CH₃

(f) —OH, —OPO₃H₂, —H, —CHO

PRACTICE PROBLEM 5.12

Assign (*R*) or (*S*) designations to each of the following compounds:

(a)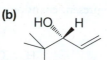

(b)

(c)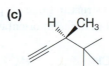

(d)

D-Glyceraldehyde-3-phosphate (a glycolysis intermediate)

●●● SOLVED PROBLEM 5.4

Consider the following pair of structures and tell whether they represent enantiomers or two molecules of the same compound in different orientations:

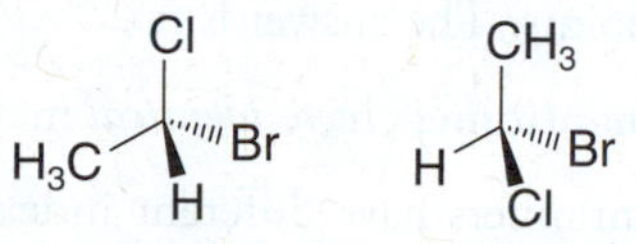

Cl	CH₃
H₃C — C ⋯Br	H — C ⋯Br
H	Cl
A	**B**

STRATEGY: One way to approach this kind of problem is to take one structure and, in your mind, hold it by one group. Then rotate the other groups until at least one group is in the same place as it is in the other structure. (Until you can do this easily in your mind, practice with models.) By a series of rotations like this you will be able to convert the structure you are manipulating into one that is either identical with or the mirror image of the other. For example, take **A**, hold it

(continues on next page)

---

*Further information can be found in the Chemical Abstracts Service *Index Guide*.

by the Cl atom and then rotate the other groups about the C*—Cl bond until the hydrogen occupies the same position as in **B**. Then hold it by the H and rotate the other groups about the C*—H bond. This will make **B** identical with **A**:

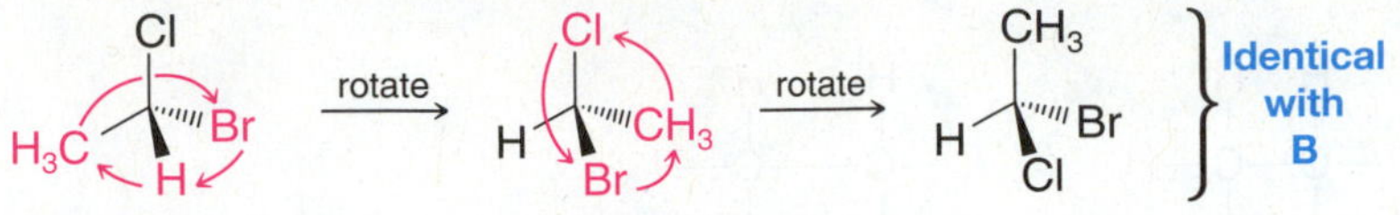

Another approach is to recognize that exchanging two groups at the chirality center *inverts the configuration of* that carbon atom and converts a structure *with only one chirality center* into its enantiomer; a second exchange recreates the original molecule. So we proceed this way, keeping track of how many exchanges are required to convert **A** into **B**. In this instance we find that two exchanges are required, and, again, we conclude that **A** and **B** are the same:

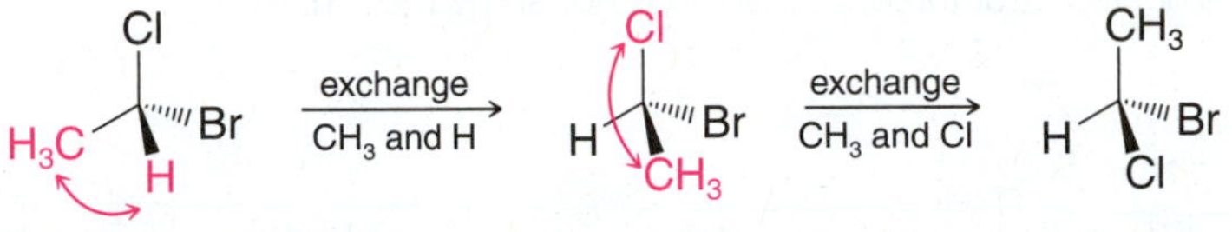

A useful check is to name each compound including its (R,S) designation. If the names are the same, then the structures are the same. In this instance both structures are (R)-1-bromo-1-chloroethane.

Another method for assigning (R) and (S) configurations using one's hands as chiral templates has been described (Huheey, J. E., *J. Chem. Educ.* **1986**, *63*, 598–600). Groups at a chirality center are correlated from lowest to highest priority with one's wrist, thumb, index finger, and second finger, respectively. With the ring and little finger closed against the palm and viewing one's hand with the wrist away, if the correlation between the chirality center is with the left hand, the configuration is (S), and if with the right hand, (R).

ANSWER: **A** and **B** are two molecules of the same compound oriented differently.

PRACTICE PROBLEM 5.13   Tell whether the two structures in each pair represent enantiomers or two molecules of the same compound in different orientations.

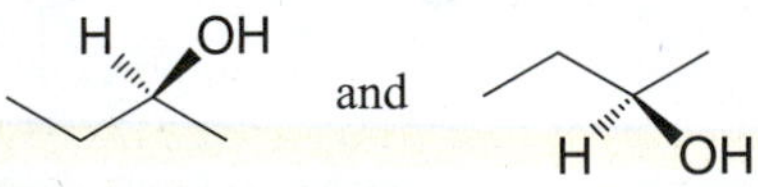

5.8 PROPERTIES OF ENANTIOMERS: OPTICAL ACTIVITY

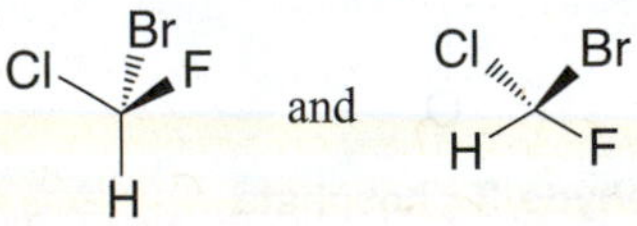

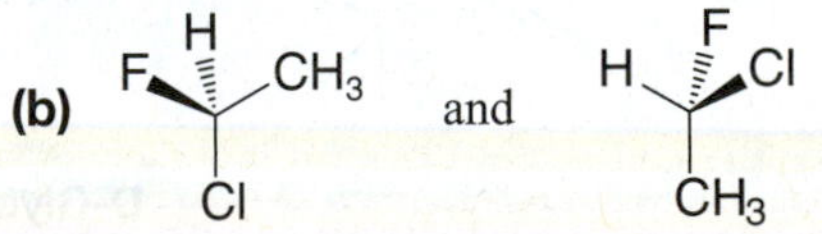

The molecules of enantiomers are not superposable and, on this basis alone, we have concluded that enantiomers are different compounds. How are they different? Do enantiomers resemble constitutional isomers and diastereomers in having different melting and boiling points? The answer is *no*.

- Pure **enantiomers** have *identical* melting and boiling points.

Do pure enantiomers have different indexes of refraction, different solubilities in common solvents, different infrared spectra, and different rates of reaction with achiral reagents? The answer to each of these questions is also *no*.

Many of these properties (e.g., boiling points, melting points, and solubilities) are dependent on the magnitude of the intermolecular forces operating between the molecules (Section 2.13), and for molecules that are mirror images of each other these forces will be identical. We can see an example of this if we examine Table 5.1, where boiling points of the 2-butanol enantiomers are listed.

Mixtures of the enantiomers of a compound have different properties than pure samples of each, however. The data in Table 5.1 illustrate this for tartaric acid. The natural isomer, (+)-tartaric acid, has a melting point of 168–170°C, as does its unnatural

TABLE 5.1 PHYSICAL PROPERTIES OF 2-BUTANOL AND TARTARIC ACID ENANTIOMERS	
Compound	Boiling Point (bp) or Melting Point (mp)
(R)-2-Butanol	99.5 °C (bp)
(S)-2-Butanol	99.5 °C (bp)
(+)-(R,R)-Tartaric acid	168–170 °C (mp)
(−)-(S,S)-Tartaric acid	168–170 °C (mp)
(+/−)-Tartaric acid	210–212 °C (mp)

enantiomer, (−)-tartaric acid. An equal mixture tartaric acid enantiomers, (+/−)-tartaric acid, has a melting point of 210–212 °C, however.

- Enantiomers show different behavior only when they interact with other chiral substances, including their own enantiomer.

This is evident in the melting point data above. Enantiomers also show different rates of reaction toward other chiral molecules—that is, toward reagents that consist of a single enantiomer or an excess of a single enantiomer. And, enantiomers show different solubilities in solvents that consist of a single enantiomer or an excess of a single enantiomer.

One easily observable way in which enantiomers differ is in *their behavior toward plane-polarized light.*

- When a beam of plane-polarized light passes through an enantiomer, the plane of polarization **rotates.**
- Separate enantiomers rotate the plane of plane-polarized light equal amounts *but in opposite directions.*
- Separate enantiomers are said to be **optically active compounds**. Because of their effect on plane-polarized light.

In order to understand this behavior of enantiomers, we need to understand the nature of plane-polarized light. We also need to understand how an instrument called a polarimeter operates.

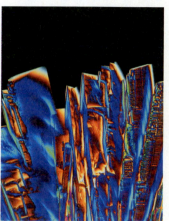

Tartaric acid is found naturally in grapes and many other plants. Crystals of tartaric acid can sometimes be found in wine.

5.8A Plane-Polarized Light

Light is an electromagnetic phenomenon. A beam of light consists of two mutually perpendicular oscillating fields: an oscillating electric field and an oscillating magnetic field (Fig. 5.9).

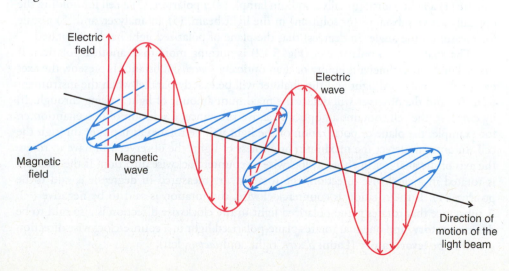

FIGURE 5.9 The oscillating electric and magnetic fields of a beam of ordinary light in one plane. The waves depicted here occur in all possible planes in ordinary light.

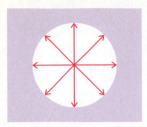

FIGURE 5.10 Oscillation of the electric field of ordinary light occurs in all possible planes perpendicular to the direction of propagation.

If we were to view a beam of ordinary light from one end, and if we could actually see the planes in which the electrical oscillations were occurring, we would find that oscillations of the electric field were occurring in all possible planes perpendicular to the direction of propagation (Fig. 5.10). (The same would be true of the magnetic field.)

When ordinary light is passed through a polarizer, the polarizer interacts with the electric field so that the electric field of the light that emerges from the polarizer (and the magnetic field perpendicular to it) is oscillating only in one plane. Such light is called **plane-polarized light** (Fig. 5.11a). If the plane-polarized beam encounters a filter with perpendicular polarization, the light is blocked (Fig. 5.11b). This phenomenon can readily be demonstrated with lenses from a pair of polarizing sunglasses or a sheet of polarizing film (Fig. 5.11c).

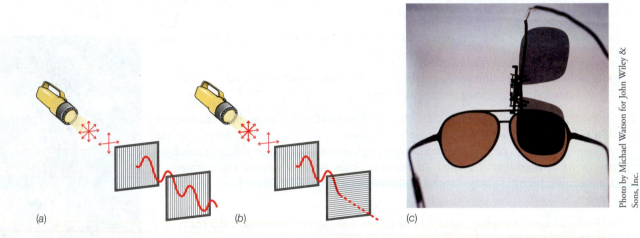

(a)   (b)   (c)

Photo by Michael Watson for John Wiley & Sons, Inc.

FIGURE 5.11 (a) Ordinary light passing through the first polarizing filter emerges with an electric wave oscillating in only one plane (and a perpendicular magnetic wave plane not shown). When the second filter is aligned with its polarizing direction the same as the first filter, as shown, the plane-polarized light can pass through. (b) If the second filter is turned 90°, the plane-polarized light is blocked. (c) Two polarizing sunglass lenses oriented perpendicular to each other block the light beam.

5.8B The Polarimeter

- The device that is used for measuring the effect of optically active compounds on plane-polarized light is a **polarimeter**.

A sketch of a polarimeter is shown in Fig. 5.12. The principal working parts of a polarimeter are (1) a light source (usually a sodium lamp), (2) a polarizer, (3) a cell for holding the optically active substance (or solution) in the light beam, (4) an analyzer, and (5) a scale for measuring the angle (in degrees) that the plane of polarized light has been rotated.

The analyzer of a polarimeter (Fig. 5.12) is nothing more than another polarizer. If the cell of the polarimeter is empty or if an optically *inactive* substance is present, the axes of the plane-polarized light and the analyzer will be exactly parallel when the instrument reads 0°, and the observer will detect the maximum amount of light passing through. If, by contrast, the cell contains an optically active substance, a solution of one enantiomer, for example, the plane of polarization of the light will be rotated as it passes through the cell. In order to detect the maximum brightness of light, the observer will have to rotate the axis of the analyzer in either a clockwise or counterclockwise direction. If the analyzer is rotated in a clockwise direction, the rotation, α (measured in degrees), is said to be positive (+). If the rotation is counterclockwise, the rotation is said to be negative (−). A substance that rotates plane-polarized light in the clockwise direction is also said to be **dextrorotatory**, and one that rotates plane-polarized light in a counterclockwise direction is said to be **levorotatory** (Latin: *dexter*, right, and *laevus*, left).

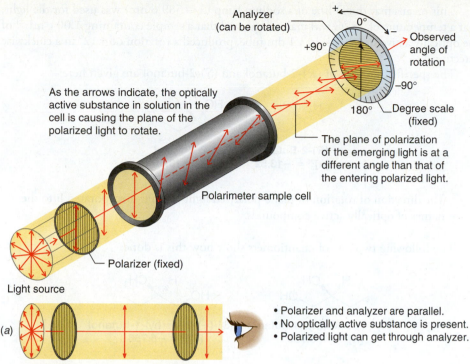

Analyzer
(can be rotated)

Observed
angle of
rotation

Degree scale
(fixed)

The plane of polarization
of the emerging light is at a
different angle than that of
the entering polarized light.

As the arrows indicate, the optically
active substance in solution in the
cell is causing the plane of the
polarized light to rotate.

Polarimeter sample cell

Polarizer (fixed)

Light source

(a)
• Polarizer and analyzer are parallel.
• No optically active substance is present.
• Polarized light can get through analyzer.

(b)
• Polarizer and analyzer are perpendicular.
• No optically active substance is present.
• No polarized light can emerge from analyzer.

(c)
• Substance in cell between polarizer and analyzer is optically active.
• Analyzer has been rotated to the left (from observer's point of view) to permit rotated polarized light through (substance is levorotatory in this example).

Polarizer   Analyzer Observer

FIGURE 5.12 The principal working parts of a polarimeter and the measurement of optical rotation. (Reprinted with permission of John Wiley & Sons, Inc. from Holum, J. R., *Organic Chemistry: A Brief Course*, p. 316. Copyright 1975.)

5.8C Specific Rotation

● The number of degrees that the plane of polarization is rotated as the light passes through a solution of an enantiomer depends on the number of chiral molecules that it encounters.

To normalize optical rotation data relative to experimental variables such as tube length and the concentration of the enantiomer, chemists calculate a quantity called the **specific rotation**, $[\alpha]$, by the following equation:

$$[\alpha] = \frac{\alpha}{c \cdot l}$$

where $[\alpha]$ = the specific rotation

α = the observed rotation

c = the concentration of the solution in grams per milliliter of solution (or density in g mL^{-1} for neat liquids)

l = the length of the cell in decimeters (1 dm = 10 cm)

The specific rotation also depends on the temperature and the wavelength of light that is employed. Specific rotations are reported so as to incorporate these quantities as well. A specific rotation might be given as follows:

$$[\alpha]_D^{25} = +3.12$$

This means that the D line of a sodium lamp ($\lambda = 589.6$ nm) was used for the light, that a temperature of 25 °C was maintained, and that a sample containing 1.00 g mL^{-1} of the optically active substance, in a 1 dm tube, produced a rotation of 3.12° in a clockwise direction.*

The specific rotations of (R)-2-butanol and (S)-2-butanol are given here:

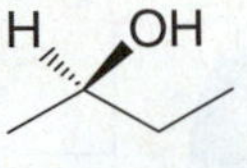

(R)-2-Butanol
$[\alpha]_D^{25} = -13.52$

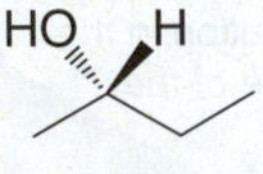

(S)-2-Butanol
$[\alpha]_D^{25} = +13.52$

- The direction of rotation of plane-polarized light is often incorporated into the names of optically active compounds.

The following two sets of enantiomers show how this is done:

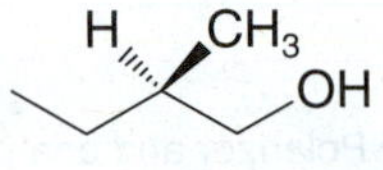

(R)-(+)-2-Methyl-1-butanol
$[\alpha]_D^{25} = +5.756$

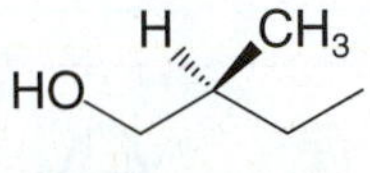

(S)-(−)-2-Methyl-1-butanol
$[\alpha]_D^{25} = -5.756$

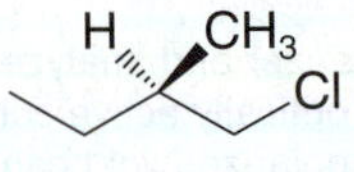

(R)-(−)-1-Chloro-2-methylbutane
$[\alpha]_D^{25} = -1.64$

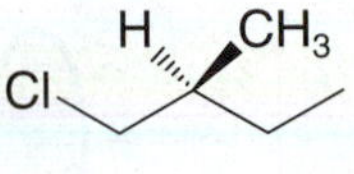

(S)-(+)-1-Chloro-2-methylbutane
$[\alpha]_D^{25} = +1.64$

The previous compounds also illustrate an important principle:

- No obvious correlation exists between the (R) and (S) configurations of enantiomers and the direction [(+) or (−)] in which they rotate plane-polarized light.

(R)-(+)-2-Methyl-1-butanol and (R)-(−)-1-chloro-2-methylbutane have the same *configuration*; that is, they have the same general arrangement of their atoms in space. They have, however, an opposite effect on the direction of rotation of the plane of plane-polarized light:

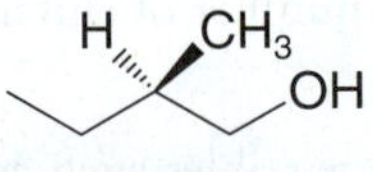

Same configuration

(R)-(+)-2-Methyl-1-butanol

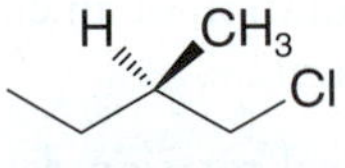

(R)-(−)-1-Chloro-2-methylbutane

These same compounds also illustrate a second important principle:

- No necessary correlation exists between the (R) and (S) designation and the direction of rotation of plane-polarized light.

(R)-2-Methyl-1-butanol is dextrorotatory (+), and (R)-1-chloro-2-methylbutane is levorotatory (−).

A method based on the measurement of optical rotation at many different wavelengths, called optical rotatory dispersion, has been used to correlate configurations of chiral molecules. A discussion of the technique of optical rotatory dispersion, however, is beyond the scope of this text.

*The magnitude of rotation is dependent on the solvent used when solutions are measured. This is the reason the solvent is specified when a rotation is reported in the chemical literature.

PRACTICE PROBLEM 5.14

Shown is the configuration of (+)-carvone. (+)-Carvone is the principal component of caraway seed oil and is responsible for its characteristic odor. (−)-Carvone, its enantiomer, is the main component of spearmint oil and gives it its characteristic odor. The fact that the carvone enantiomers do not smell the same suggests that the receptor sites in the nose for these compounds are chiral, and that only the correct enantiomer binds well to its particular site (just as a hand requires a glove of the correct chirality for a proper fit). Give the correct (*R*) and (*S*) designations for (+)- and (−)-carvone.

(+)-Carvone

5.9 THE ORIGIN OF OPTICAL ACTIVITY

Optical activity is measured by the degree of rotation of plane-polarized light passing through a chiral medium. The theoretical explanation for the origin of optical activity requires consideration of *circularly*-polarized light, however, and its interaction with chiral molecules. While it is not possible to provide a full theoretical explanation for the origin of optical activity here, the following explanation will suffice. A beam of plane-polarized light (Fig. 5.13*a*) can be described in terms of circularly-polarized light. A beam of

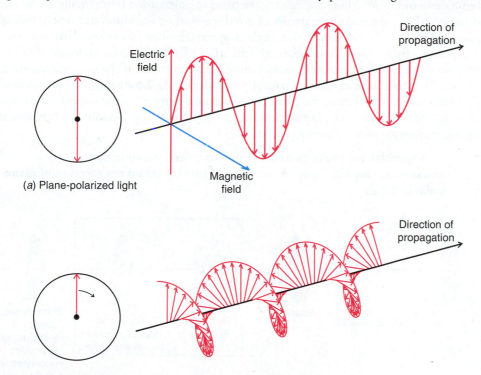

(*a*) Plane-polarized light

(*b*) Circularly-polarized light

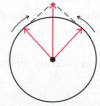

(*c*) Two circularly-polarized beams counter-rotating at the same velocity (in phase), and their vector sum. The net result is like (*a*).

(*d*) Two circularly-polarized beams counter-rotating at different velocities, such as after interaction with a chiral molecule, and their vector sum. The net result is like (*b*).

FIGURE 5.13 (*a*) Plane-polarized light. (*b*) Circularly-polarized light. (*c*) Two circularly-polarized beams counterrotating at the same velocity (in phase) and their vector sum. The net result is like (*a*). (*d*) Two circularly-polarized beams counter-rotating at different velocities, such as after interaction with a chiral molecule, and their vector sum. The net result is like (*b*).
(From ADAMSON. *A TEXTBOOK OF PHYSICAL CHEMISTRY*, 3E. © 1986 Brooks/Cole, a part of Cengage Learning, Inc. Reproduced by permission. *www.cengage.com/ permissions*.)

circularly-polarized light rotating in one direction is shown in Fig. 5.13*b*. The vector sum of *two* counterrotating in-phase circularly-polarized beams is a beam of plane-polarized light (Fig. 5.13*c*). The optical activity of chiral molecules results from the fact that the two *counterrotating circularly-polarized beams travel with different velocities through the chiral medium*. As the difference between the two circularly-polarized beams propagates through the sample, their vector sum describes a plane that is progressively rotated (Fig. 5.13*d*). What we measure when light emerges from the sample is the net rotation of the plane-polarized light caused by differences in velocity of the circularly-polarized beam components. The origin of the differing velocities has ultimately to do with interactions between electrons in the chiral molecule and light.

Achiral molecules in solution cause no difference in velocity of the two circularly-polarized beams; hence there is no rotation of the plane of polarized light described by their vector sum. Randomly-oriented achiral molecules, therefore, are not optically active. (However, oriented achiral molecules and crystals having specific symmetric characteristics can rotate plane-polarized light.)

5.9A Racemic Forms

A sample that consists exclusively or predominantly of one enantiomer causes a net rotation of plane-polarized light. Figure 5.14*a* depicts a plane of polarized light as it encounters a molecule of (*R*)-2-butanol, causing the plane of polarization to rotate slightly in one direction. (For the remaining purposes of our discussion we shall limit our description of polarized light to the resultant plane, neglecting consideration of the circularly-polarized components from which plane-polarized light arises.) Each additional molecule of (*R*)-2-butanol that the beam encounters would cause further rotation in the same direction. If, on the other hand, the mixture contained molecules of (*S*)-2-butanol, each molecule of that enantiomer would cause the plane of polarization to rotate in the opposite direction (Fig. 5.14*b*). If the (*R*) and (*S*) enantiomers were present in equal amounts, there would be no net rotation of the plane of polarized light.

- An equimolar mixture of two enantiomers is called a **racemic mixture** (or **racemate** or **racemic form**). **A racemic mixture causes no net rotation of plane-polarized light**.

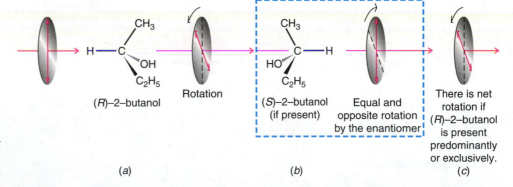

FIGURE 5.14 (*a*) A beam of plane-polarized light encounters a molecule of (*R*)-2-butanol, a chiral molecule. This encounter produces a slight rotation of the plane of polarization. (*b*) Exact cancellation of this rotation occurs if a molecule of (*S*)-2-butanol is encountered. (*c*) Net rotation of the plane of polarization occurs if (*R*)-2-butanol is present predominantly or exclusively.

(*R*)–2–butanol   Rotation   (*S*)–2–butanol (if present)   Equal and opposite rotation by the enantiomer   There is net rotation if (*R*)–2–butanol is present predominantly or exclusively.

(*a*)   (*b*)   (*c*)

In a racemic mixture the effect of each molecule of one enantiomer on the circularly-polarized beam cancels the effect of molecules of the other enantiomer, resulting in no net optical activity.

The racemic form of a sample is often designated as being (±). A racemic mixture of (*R*)-(−)-2-butanol and (*S*)-(+)-2-butanol might be indicated as

$$(\pm)\text{-2-butanol} \quad \text{or} \quad (\pm)\text{-CH}_3\text{CH}_2\text{CHOHCH}_3$$

5.9B Racemic Forms and Enantiomeric Excess

A sample of an optically active substance that consists of a single enantiomer is said to be **enantiomerically pure** or to have an **enantiomeric excess** of 100%. An enantiomerically pure sample of (S)-(+)-2-butanol shows a specific rotation of +13.52 ($[\alpha]_D^{25} = +13.52$). On the other hand, a sample of (S)-(+)-2-butanol that contains less than an equimolar amount of (R)-(−)-2-butanol will show a specific rotation that is less than +13.52 but greater than zero. Such a sample is said to have an *enantiomeric excess* less than 100%. The **enantiomeric excess (ee)**, also known as the **optical purity**, is defined as follows:

$$\% \text{ Enantiomeric excess} = \frac{\text{moles of one enantiomer} - \text{moles of other enantiomer}}{\text{total moles of both enantiomers}} \times 100$$

The enantiomeric excess can be calculated from optical rotations:

$$\% \text{ Enantiomeric excess}^* = \frac{\text{observed specific rotation}}{\text{specific rotation of the pure enantiomer}} \times 100$$

Let us suppose, for example, that a mixture of the 2-butanol enantiomers showed a specific rotation of +6.76. We would then say that the enantiomeric excess of the (S)-(+)-2-butanol is 50%:

$$\text{Enantiomeric excess} = \frac{+6.76}{+13.52} \times 100 = 50\%$$

When we say that the enantiomeric excess of this mixture is 50%, we mean that 50% of the mixture consists of the (+) enantiomer (the excess) and the other 50% consists of the racemic form. Since for the 50% that is racemic the optical rotations cancel one another out, only the 50% of the mixture that consists of the (+) enantiomer contributes to the observed optical rotation. The observed rotation is, therefore, 50% (or one-half) of what it would have been if the mixture had consisted only of the (+) enantiomer.

●●● SOLVED PROBLEM 5.5

What is the actual stereoisomeric composition of the mixture referred to above?

ANSWER: Of the total mixture, 50% consists of the racemic form, which contains equal numbers of the two enantiomers. Therefore, half of this 50%, or 25%, is the (−) enantiomer and 25% is the (+) enantiomer. The other 50% of the mixture (the excess) is also the (+) enantiomer. Consequently, the mixture is 75% (+) enantiomer and 25% (−) enantiomer.

PRACTICE PROBLEM 5.15

A sample of 2-methyl-1-butanol (see Section 5.8C) has a specific rotation, $[\alpha]_D^{25}$, equal to +1.151. **(a)** What is the percent enantiomeric excess of the sample? **(b)** Which enantiomer is in excess, the (R) or the (S)?

5.10 THE SYNTHESIS OF CHIRAL MOLECULES

5.10A Racemic Forms

Reactions carried out with achiral reactants can often lead to *chiral* products. In the absence of any chiral influence from a catalyst, reagent, or solvent, the outcome of such a reaction is a racemic mixture. In other words, the chiral product is obtained as a 50:50 mixture of enantiomers.

*This calculation should be applied to a single enantiomer or to mixtures of enantiomers only. It should not be applied to mixtures in which some other compound is present.

An example is the synthesis of 2-butanol by the nickel-catalyzed hydrogenation of butanone. In this reaction the hydrogen molecule adds across the carbon–oxygen double bond in much the same way that it adds to a carbon–carbon double bond.

$$\underset{\underset{\text{(achiral}}{\text{Butanone}}}{CH_3CH_2CCH_3} + \underset{\underset{\text{(achiral}}{\text{Hydrogen}}}{H{-}H} \xrightarrow{\text{Ni}} \underset{\underset{\text{(±)-2-Butanol}}{}}{(\pm)\text{-}CH_3CH_2\overset{*}{C}HCH_3}$$

Butanone	**Hydrogen**
(achiral	(achiral
molecules)	molecules)

(±)-2-Butanol
[chiral molecules
but 50:50 mixture (*R*) and (*S*)]

Figure 5.15 illustrates the stereochemical aspects of this reaction. Because butanone is achiral, there is no difference in presentation of either face of the molecule to the surface of the metal catalyst. The two faces of the trigonal planar carbonyl group interact with the metal surface with equal probability. Transfer of the hydrogen atoms from the metal to the carbonyl group produces a chirality center at carbon 2. Since there has been no chiral influence in the reaction pathway, the product is obtained as a racemic mixture of the two enantiomers, (*R*)-(−)-2-butanol and (*S*)-(+)-2-butanol.

We shall see that when reactions like this are carried out in the presence of a chiral influence, such as an enzyme or chiral catalyst, the result is usually not a racemic mixture.

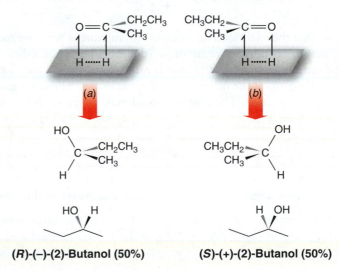

FIGURE 5.15 The reaction of buta-none with hydrogen in the presence of a nickel catalyst. The reaction rate by path (*a*) is equal to that by path (*b*). (*R*-15)-(−)-2-Butanol and (*S*)-(+)-2-butanol are produced in equal amounts, as a racemate.

(*R*)-(−)-(2)-Butanol (50%)   **(*S*)-(+)-(2)-Butanol (50%)**

5.10B Stereoselective Syntheses

Stereoselective reactions are reactions that lead to a preferential formation of one stereo-isomer over other stereoisomers that could possibly be formed.

- If a reaction produces preferentially one enantiomer over its mirror image, the reaction is said to be an **enantioselective reaction**.
- If a reaction leads preferentially to one diastereomer over others that are possible, the reaction is said to be an **diastereoselective reaction**.

For a reaction to be either enantioselective or diastereoselective, a chiral reagent, catalyst, or solvent must assert an influence on the course of the reaction.

In nature, where most reactions are stereoselective, the chiral influences come from protein molecules called **enzymes**. Enzymes are biological catalysts of extraordinary efficiency. Not only do they have the ability to cause reactions to take place much more rapidly than they would otherwise, they also have the ability to assert a *dramatic chiral influence* on a reaction. Enzymes do this because they, too, are chiral, and they possess an active site where the reactant molecules are momentarily bound while the reaction takes place. The active site is chiral (see Fig. 5.7), and only one enantiomer of a chiral reactant fits it properly and is able to undergo the reaction.

Many enzymes have found use in the organic chemistry laboratory, where organic chemists take advantage of their properties to bring about stereoselective reactions. One of these is an enzyme called **lipase**. Lipase catalyzes a reaction called **hydrolysis**, whereby an ester (Section 2.10B) reacts with a molecule of water to produce a carboxylic acid and an alcohol.

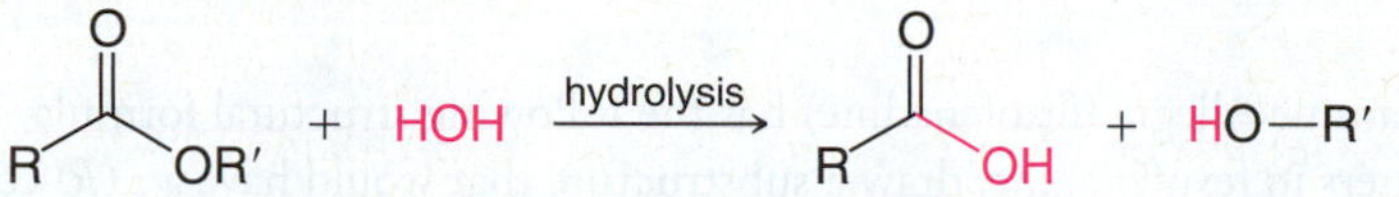

| Ester | Water | | Carboxylic acid | Alcohol |

If the starting ester is chiral and present as a mixture of its enantiomers, the lipase enzyme reacts selectively with one enantiomer to release the corresponding chiral carboxylic acid and an alcohol, while the other ester enantiomer remains unchanged or reacts much more slowly. The result is a mixture that consists predominantly of one stereoisomer of the reactant and one stereoisomer of the product, which can usually be separated easily on the basis of their different physical properties. Such a process is called a **kinetic resolution**, where the rate of a reaction with one enantiomer is different than with the other, leading to a preponderance of one product stereoisomer. We shall say more about the resolution of enantiomers in Section 5.16. The following hydrolysis is an example of a kinetic resolution using lipase:

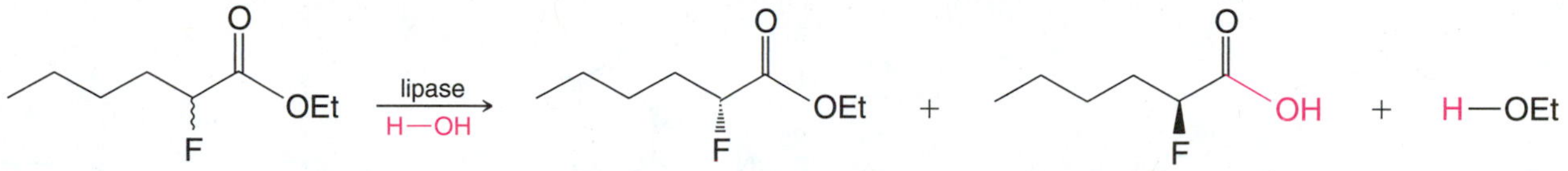

Ethyl (±)-2-fluorohexanoate [an ester that is a racemate of (R) and (S) forms]

Ethyl (R)-(+)-2-fluorohexanoate (>99% enantiomeric excess)

(S)-(−)-2-Fluorohexanoic acid (>69% enantiomeric excess)

Other enzymes called hydrogenases have been used to effect enantioselective versions of carbonyl reductions like that in Section 5.10A. We shall have more to say about the stereoselectivity of enzymes in Chapter 12.

5.11 CHIRAL DRUGS

The U.S. Food and Drug Administration and the pharmaceutical industry are very interested in the production of chiral drugs—that is, drugs that contain a single enantiomer rather than a racemate. The antihypertensive drug **methyldopa** (Aldomet), for example, owes its effect exclusively to the (S) isomer. In the case of **penicillamine**, the (S) isomer is a highly potent therapeutic agent for primary chronic arthritis, while the (R) isomer has no therapeutic action and is highly toxic. The anti-inflammatory agent **ibuprofen** (Advil, Motrin, Nuprin) is marketed as a racemate even though only the (S) enantiomer is the active agent. The (R) isomer of ibuprofen has no anti-inflammatory action and is slowly converted to the (S) isomer in the body. A formulation of ibuprofen based on solely the (S) isomer, however, would be more effective than the racemate.

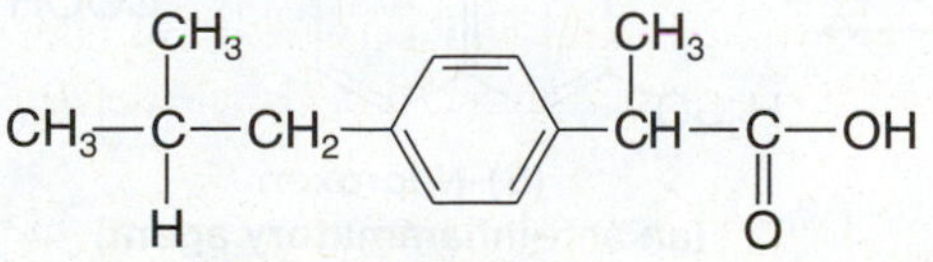

Ibuprofen 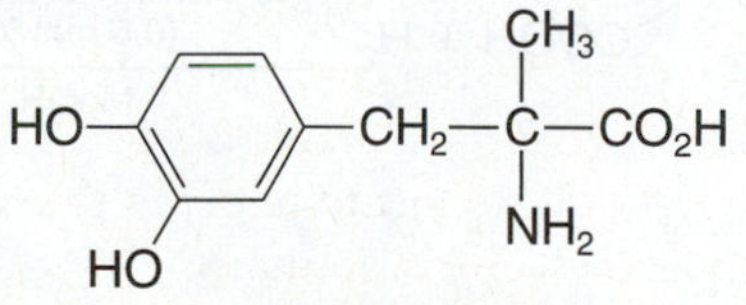 **Methyldopa** 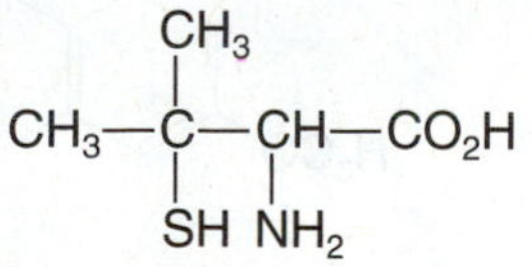 **Penicillamine**

PRACTICE PROBLEM 5.16 Write three-dimensional formulas for the (S) isomers of **(a)** methyldopa, **(b)** penicillamine, and **(c)** ibuprofen.

PRACTICE PROBLEM 5.17 The antihistamine Allegra (fexofenadine) has the following structural formula. For any chirality centers in fexofenadine, draw a substructure that would have an (R) configuration.

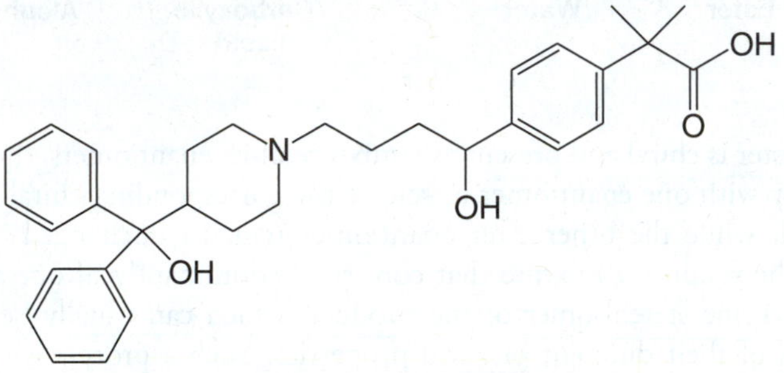

Fexofenadine (Allegra)

PRACTICE PROBLEM 5.18 Assign the (R,S) configuration at each chirality center in Darvon (dextropropoxyphene).

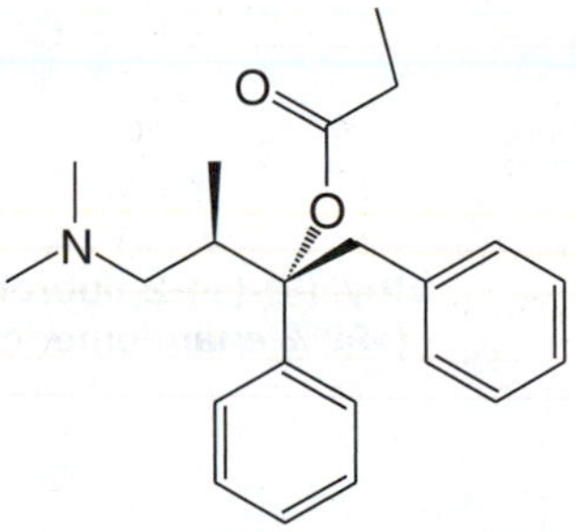

Darvon

WILLIAM KNOWLES, RYOJI NOYORI, AND BARRY SHARPLESS were awarded the 2001 Nobel Prize in Chemistry for catalytic stereoselective reactions.

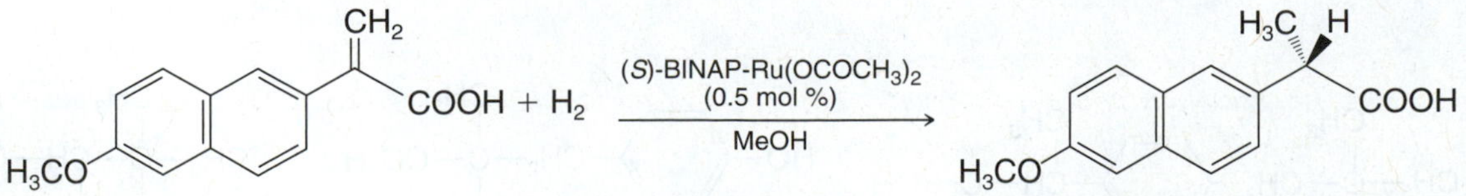

There are many other examples of drugs like these, including drugs where the enantiomers have distinctly different effects. The preparation of enantiomerically pure drugs, therefore, is one factor that makes stereoselective synthesis (Section 5.10B) and the resolution of racemic drugs (separation into pure enantiomers, Section 5.16) major areas of research today.

Underscoring the importance of stereoselective synthesis is the fact that the 2001 Nobel Prize in Chemistry was given to researchers who developed reaction catalysts that are now widely used in industry and academia. William Knowles (Monsanto Company, deceased 2012) and Ryoji Noyori (Nagoya University) were awarded half of the prize for their development of reagents used for catalytic stereoselective hydrogenation reactions. The other half of the prize was awarded to Barry Sharpless (Scripps Research Institute) for development of catalytic stereoselective oxidation reactions (see Chapter 11). An important example resulting from the work of Noyori and based on earlier work by Knowles is a synthesis of the anti-inflammatory agent **naproxen**, involving a stereoselective catalytic hydrogenation reaction:

$$CH_2 \quad \text{---COOH} + H_2 \xrightarrow[\text{MeOH}]{\substack{(S)\text{-BINAP-Ru(OCOCH}_3)_2 \\ (0.5 \text{ mol \%})}} \quad H_3C \quad H \quad \text{---COOH}$$

(S)-Naproxen
(an anti-inflammatory agent)
(92% yield, 97% ee)

The hydrogenation catalyst in this reaction is an organometallic complex formed from ruthenium and a chiral organic ligand called (S)-BINAP. The reaction itself is truly remarkable because it proceeds with excellent enantiomeric excess (97%) and in very high yield (92%). We will have more to say about BINAP ligands and the origin of their chirality in Section 5.18.

THE CHEMISTRY OF... Selective Binding of Drug Enantiomers to Left- and Right-Handed Coiled DNA

Would you like left- or right-handed DNA with your drug? That's a question that can now be answered due to the recent discovery that each enantiomer of the drug daunorubicin selectively binds DNA coiled with opposite handedness. (+)-Daunorubicin binds selectively to DNA coiled in the typical right-handed conformation (B-DNA). (−)-Daunorubicin binds selectively to DNA coiled in the left-handed conformation (Z-DNA). Furthermore, daunorubicin is capable of inducing conformational changes in DNA from one coiling direction to the other, depending on which coiling form is favored when a given daunorubicin enantiomer binds to the DNA. It has long been known that DNA adopts a number of secondary and tertiary structures, and it is presumed that some of these conformations are involved in turning on or off transcription of a given section of DNA. The discovery of specific interactions between each daunorubicin enantiomer and the left- and right-handed coil forms of DNA will likely assist in design and discovery of new drugs with anticancer or other activities.

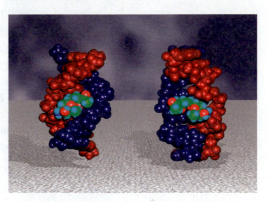

Enantiomeric forms of daunorubicin bind with DNA and cause it to coil with opposite handedness.
(Graphic courtesy John O. Trent, Brown Cancer Center, Department of Medicine, University of Louisville, KY. Based on work from Qu, X., Trent, J.O., Fokt, I., Priebe, W., and Chaires, J.B., *Allosteric, Chiral-Selective Drug Binding to DNA*, Proc. Natl. Acad. Sci. U.S.A., 2000 (Oct 24): 97(22), 12032–12037.)

5.12 MOLECULES WITH MORE THAN ONE CHIRALITY CENTER

So far we have mainly considered **chiral molecules** that contain only one chirality center. Many organic molecules, especially those important in biology, contain more than one chirality center. Cholesterol (Section 23.4B), for example, contains eight chirality centers. (Can you locate them?) We can begin, however, with simpler molecules. Let us consider 2,3-dibromopentane, shown here in a two-dimensional bond-line formula. 2,3-Dibromopentane has two chirality centers:

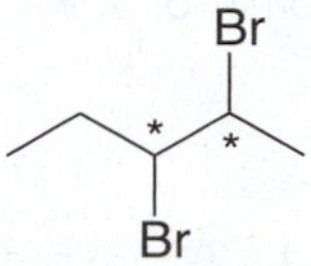

2,3-Dibromopentane

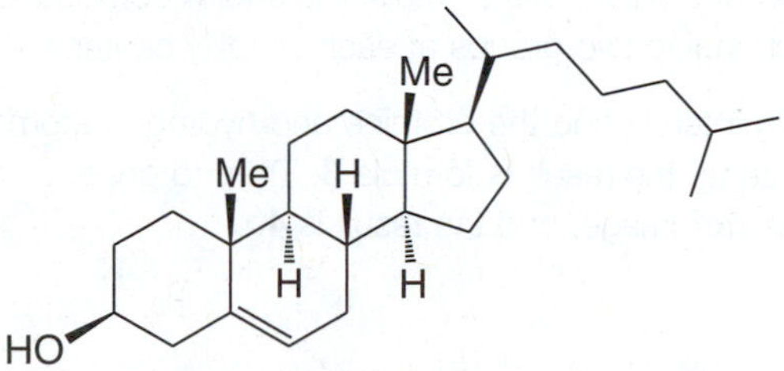

Cholesterol

A useful rule gives the maximum number of stereoisomers:

- In compounds whose stereoisomerism is due to chirality centers, *the total number of stereoisomers will not exceed 2^n, where n is equal to the number of chirality centers*.

For 2,3-dibromopentane we should not expect more than four stereoisomers ($2^2 = 4$). Our next task is to write three-dimensional bond-line formulas for the possible stereoisomers.

Helpful Hint

Cholesterol, having eight chirality centers, hypothetically could exist in 2^8 (256) stereoisomeric forms, yet biosynthesis via enzymes produces only *one* stereoisomer.

5.12A HOW TO Draw Stereoisomers for Molecules Having More Than One Chirality Center

Using 2,3-dibromopentane as an example, the following sequence explains how we can draw all of the possible isomers for a molecule that contains more than one chirality center. Remember that in the case of 2,3-dibromopentane we expect a maximum of four possible isomers because there are two chirality centers (2^n, where n is the number of chirality centers).

1. Start by drawing the portion of the carbon skeleton that contains the chirality centers in such a way that as many of the chirality centers are placed in the plane of the paper as possible, and as symmetrically as possible. In the case of 2,3-dibromopentane, we simply begin by drawing the bond between C2 and C3, since these are the only chirality centers.

2. Next we add the remaining groups that are bonded at the chirality centers in such a way as to maximize the symmetry between the chirality centers. In this case we start by drawing the two bromine atoms so that they project either both outward or both inward relative to the plane of the paper, and we add the hydrogen atoms at each chirality center. Drawing the bromine atoms outward results in formula **1**, shown below. Even though there are eclipsing interactions in this conformation, and it is almost certainly not the most stable conformation for the molecule, we draw it this way so as to maximize the possibility of finding symmetry in the molecule.

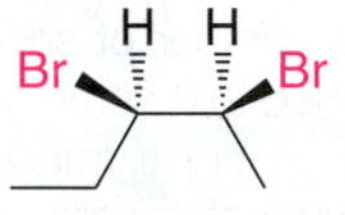

3. To draw the enantiomer of the first stereoisomer, we simply draw its mirror image, either side-by-side or top and bottom, by imagining a mirror between them. The result is formula **2**.

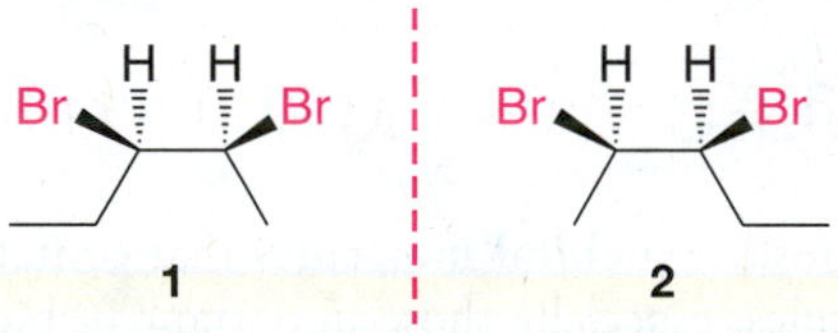

4. To draw another stereoisomer, we interchange two groups at any one of the chirality centers. By doing so we invert the *R,S* configuration at that chirality center.

   • All of the possible stereoisomers for a compound can be drawn by successively interchanging two groups at each chirality center.

   If we interchange the bromine and hydrogen atoms at C2 in formula **1** for 2,3-dibromopentane, the result is formula **3**. Then to generate the enantiomer of **3**, we simply draw its mirror image, and the result is **4**.

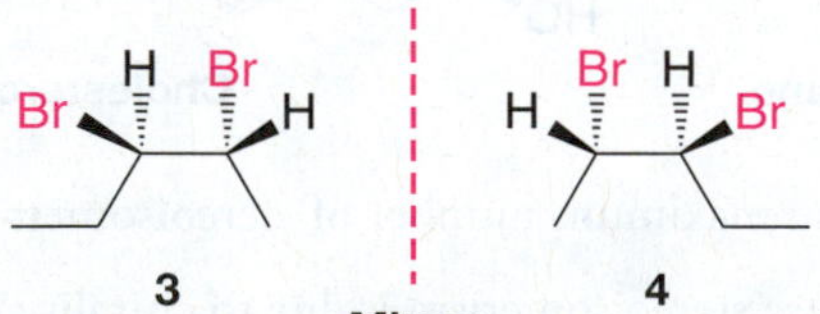

5. Next we examine the relationship between all of the possible pairings of formulas to determine which are pairs of enantiomers, which are diastereomers, and, for special cases like we shall see in Section 5.12B, which formulas are actually identical due to an internal plane of symmetry.

Since structures **1** and **2** are not superposable, they represent different compounds. Since structures **1** and **2** differ only in the arrangement of their atoms in space, they represent stereoisomers. Structures **1** and **2** are also mirror images of each other; thus **1** and **2** represent a pair of enantiomers. Structures **3** and **4** correspond to another pair of enantiomers. Structures **1–4** are all different, so there are, in total, four stereoisomers of 2,3-dibromopentane.

At this point you should convince yourself that there are no other stereoisomers by writing other structural formulas. You will find that rotation about the single bonds, or of the entire structure, or of any other arrangement of the atoms will cause the structure to become superposable with one of the structures that we have written here. Better yet, using different colored balls, make molecular models as you work this out.

The compounds represented by structures **1–4** are all optically active compounds. Any one of them, if placed separately in a polarimeter, would show optical activity.

The compounds represented by structures **1** and **2** are enantiomers. The compounds represented by structures **3** and **4** are also enantiomers. But what is the isomeric relation between the compounds represented by **1** and **3**?

We can answer this question by observing that **1** and **3** *are stereoisomers* and that they *are not mirror images of each other*. They are, therefore, *diastereomers*.

- Diastereomers have different physical properties—different melting points and boiling points, different solubilities, and so forth.

PRACTICE PROBLEM 5.19

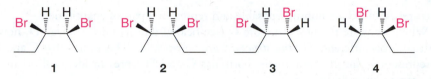

| 1 | 2 | 3 | 4 |

(a) If **3** and **4** are enantiomers, what are **1** and **4**? **(b)** What are **2** and **3**, and **2** and **4**? **(c)** Would you expect **1** and **3** to have the same melting point? **(d)** The same boiling point? **(e)** The same vapor pressure?

●●● SOLVED PROBLEM 5.6

Draw all possible stereoisomers for 2-bromo-4-chloropentane.

STRATEGY AND ANSWER: C2 and C4 are chirality centers in 2-bromo-4-chloropentane. We begin by drawing the carbon chain with as many carbons depicted in the plane of the paper as possible, and in a way that maximizes the symmetry between C2 and C4. In this case, an ordinary zig-zag bond-line formula provides symmetry between C2 and C4. Then we add the bromine and chlorine atoms at C2 and C4, respectively, as well as the hydrogen atoms at these carbons, resulting in formula I. To draw its enantiomer (II), we imagine a mirror and draw a reflection of the molecule.

To draw another stereoisomer we invert the configuration at one chirality center by interchanging two groups at one chirality center, as shown for C2 in III. Then we draw the enantiomer of III by imagining its mirror reflection.

(continues on next page)

Last, we check that none of these formulas is identical to another by testing the superposability of each one with the others. We should not expect any to be identical because none of the formulas has an internal plane of symmetry. The case would have been different for 2,4-dibromopentane, however, in which case there would have been one meso stereoisomer (a type of stereoisomer that we shall study in the next section).

5.12B Meso Compounds

A structure with two chirality centers does not always have four possible stereoisomers. Sometimes there are only *three*. As we shall see:

- Some molecules are achiral even though they contain chirality centers.

To understand this, let us write stereochemical formulas for 2,3-dibromobutane. We begin in the same way as we did before. We write formulas for one stereoisomer and for its mirror image:

Structures **A** and **B** are nonsuperposable and represent a pair of enantiomers.

When we write the new structure **C** (see below) and its mirror image **D**, however, the situation is different. *The two structures are superposable.* This means that **C** and **D** do not represent a pair of enantiomers. Formulas **C** and **D** represent identical orientations of the same compound:

The molecule represented by structure **C** (or **D**) is not chiral even though it contains two chirality centers.

- If a molecule has an internal plane of symmetry it is achiral.
- A **meso compound** is an achiral molecule that contains chirality centers and has an internal plane of symmetry. Meso compounds are not optically active.

Another test for molecular chirality is to construct a model (or write the structure) of the molecule and then test whether or not the model (or structure) is superposable on its mirror image. If it is, the molecule is achiral. If it *is not*, the molecule is chiral.

We have already carried out this test with structure **C** and found that it is achiral. We can also demonstrate that **C** is achiral in another way. Figure 5.16 shows that structure **C** *has an internal plane of symmetry* (Section 5.6).

The following two problems relate to compounds **A–D** in the preceding paragraphs.

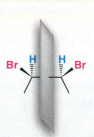

FIGURE 5.16 The plane of symmetry of meso-2,3-dibromobutane. This plane divides the molecule into halves that are mirror images of each other.

PRACTICE PROBLEM 5.20   Which of the following would be optically active?

(a) A pure sample of **A**

(b) A pure sample of **B**

(c) A pure sample of **C**

(d) An equimolar mixture of **A** and **B**

PRACTICE PROBLEM 5.21

The following are formulas for three compounds, written in noneclipsed conformations. In each instance tell which compound (**A**, **B**, or **C** above) each formula represents.

(a) 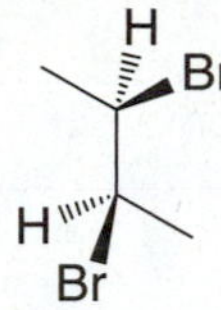 (b) 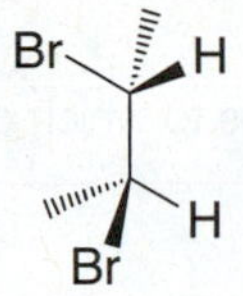 (c)

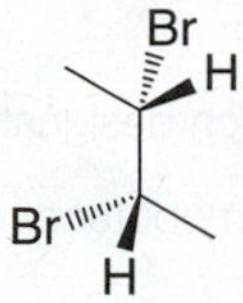

●●● **SOLVED PROBLEM 5.7**

Which of the following (**X**, **Y**, or **Z**) is a meso compound?

STRATEGY AND ANSWER: In each molecule, rotating the groups joined by the C_2—C_3 bond by 180° brings the two methyl groups into comparable position. In the case of compound Z, a plane of symmetry results, and therefore, Z is a meso compound. No plane of symmetry is possible in X and Y.

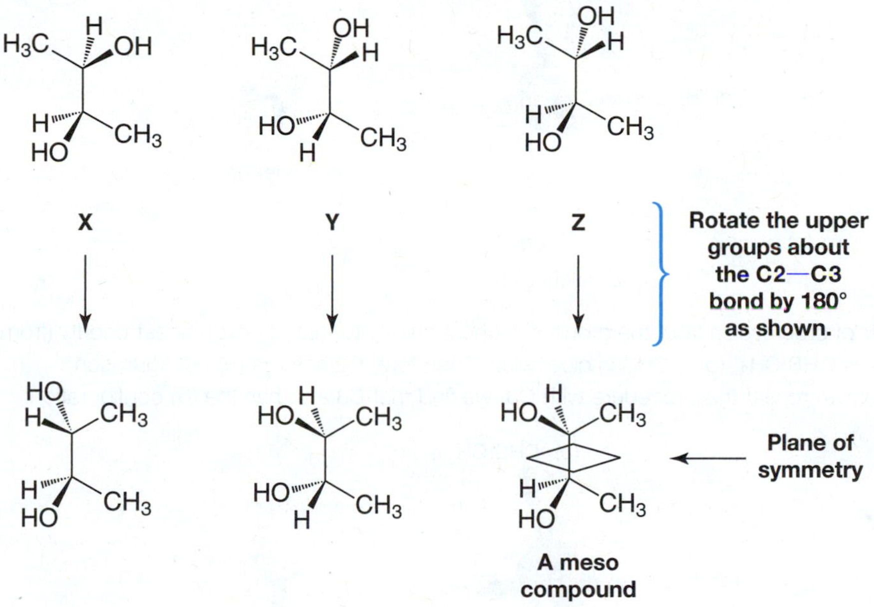

X   Y   Z

Rotate the upper groups about the C2—C3 bond by 180° as shown.

Plane of symmetry

A meso compound

PRACTICE PROBLEM 5.22

Write three-dimensional formulas for all of the stereoisomers of each of the following compounds. Label pairs of enantiomers and label meso compounds.

(a) 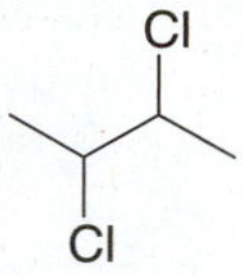 (b) 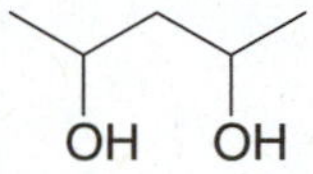 (c)

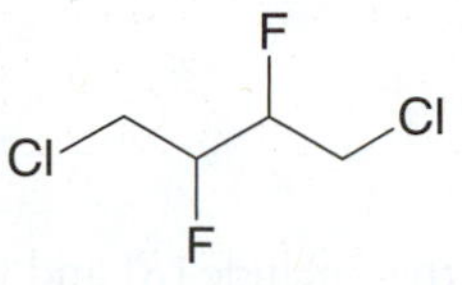

(d) 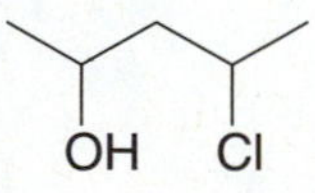 (e) 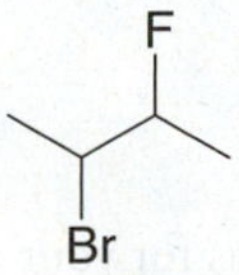 (f)

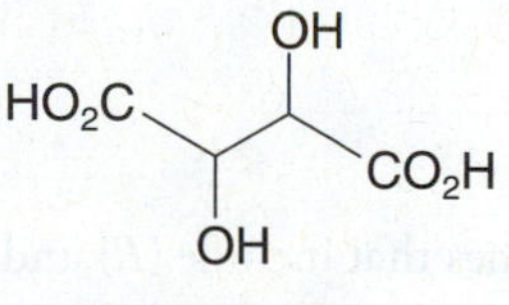

Tartaric acid

5.12C **HOW TO** Name Compounds with More Than One Chirality Center

1. If a compound has more than one chirality center, we analyze each center separately and decide whether it is (R) or (S).

2. Then, using numbers, we tell which designation refers to which carbon atom.

Consider stereoisomer **A** of 2,3-dibromobutane:

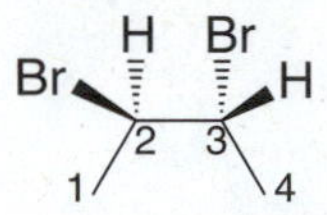

H   Br
Br↙‖‖↘H
  /2 3\
 1/   \4

A

2,3-Dibromobutane

When this formula is rotated so that the group of lowest priority attached to C2 is directed away from the viewer, it resembles the following:

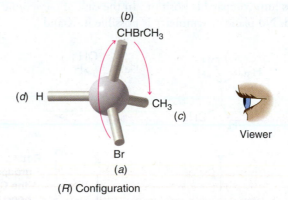

(R) Configuration

The order of progression from the group of highest priority to that of next highest priority (from —Br, to —CHBrCH₃, to —CH₃) is clockwise. Therefore, C2 has the (R) configuration.

When we repeat this procedure with C3, we find that C3 also has the (R) configuration:

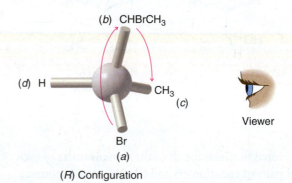

(R) Configuration

Compound **A**, therefore, is (2R,3R)-2,3-dibromobutane.

• • •

PRACTICE PROBLEM 5.23   Give names that include (R) and (S) designations for compounds **B** and **C** in Section 5.12B.

• • •

PRACTICE PROBLEM 5.24   Give names that include (R) and (S) designations for your answers to Practice Problem 5.22.

Chloramphenicol (at right) is a potent antibiotic, isolated from *Streptomyces venezuelae*, that is particularly effective against typhoid fever. It was the first naturally occurring substance shown to contain a nitro ($-NO_2$) group attached to an aromatic ring. Both chirality centers in chloramphenicol are known to have the (R) configuration. Identify the two chirality centers and write a three-dimensional formula for chloramphenicol.

PRACTICE PROBLEM 5.25

NO_2

HO—C—H

H—C—NHCOCHCl$_2$

CH$_2$OH

Chloramphenicol

5.13 FISCHER PROJECTION FORMULAS

So far in writing structures for chiral molecules we have only used formulas that show three dimensions with solid and dashed wedges, and we shall largely continue to do so until we study carbohydrates in Chapter 22. The reason is that formulas with solid and dashed wedges unambiguously show three dimensions, and they can be manipulated on paper in any way that we wish so long as we do not break bonds. Their use, moreover, teaches us to see molecules (in our mind's eye) in three dimensions, and this ability will serve us well.

Chemists, however, sometimes use formulas called **Fischer projections** to show three dimensions in chiral molecules such as acyclic carbohydrates. Fischer projection formulas are useful in cases where there are chirality centers at several adjacent carbon atoms, as is often the case in carbohydrates. Use of Fischer projection formulas requires rigid adherence to certain conventions, however. **Used carelessly, these projection formulas can easily lead to incorrect conclusions**.

5.13A HOW TO Draw and Use Fischer Projections

Let us consider how we would relate a three-dimensional formula for 2,3-dibromobutane using solid and dashed wedges to the corresponding Fischer projection formula.

1. The carbon chain in a Fischer projection is always drawn from top to bottom, rather than side to side as is often the case with bond-line formulas. We consider the molecule in a conformation that has eclipsing interactions between the groups at each carbon.

    For 2,3-dibromobutane we turn the bond-line formula so that the carbon chain runs up and down and we orient it so that groups attached to the main carbon chain project out of the plane like a bow tie. The carbon–carbon bonds of the chain, therefore, either lie in the plane of the paper or project behind it.

$$A \qquad = \qquad A$$

2. From the vertical formula with the groups at each carbon eclipsed we "project" all of the bonds onto the paper, replacing all solid and dashed wedges with ordinary lines. The vertical line of the formula now represents the carbon chain, each point of intersection between the vertical line and a horizontal line represents a carbon atom in the chain, and we interpret the horizontal lines as bonds that project out toward us.

Doing this with the vertical, eclipsed form of 2,3-dibromobutane leads to the Fischer projection shown here.

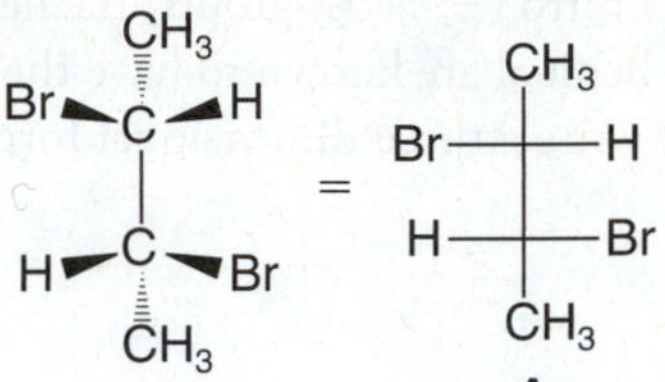

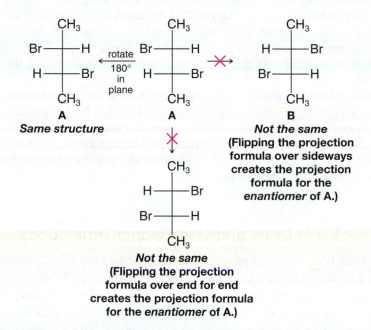

Fischer projection formula

3. To test the superposability of two structures represented by Fischer projections we are allowed to rotate them in the plane of the paper by 180°, *but by no other angle*. We must always keep the Fischer projection formulas in the plane of the paper, and **we are not allowed to flip them over**. If we flip a Fischer projection over, the horizontal bonds project behind the plane instead of in front, and every configuration would be *misrepresented* as the opposite of what was intended.

Because Fischer projections must be used with such care, we introduce them now only so that you can understand Fischer projections when you see them in the context of other courses. Our emphasis for most of this book will be on the use of solid and dashed wedges to represent three-dimensional formulas (or chair conformational structures in the case of cyclohexane derivatives), except in Chapter 22 when we will use Fischer projections again in our discussion of carbohydrates. If your instructor wishes to utilize Fischer projections further, you will be so advised.

PRACTICE PROBLEM 5.26   **(a)** Give the (R,S) designations for each chirality center in compound **A** and for compound **B**. **(b)** Write the Fischer projection formula for a compound **C** that is the diastereomer of **A** and **B**. **(c)** Would **C** be optically active?

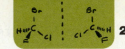

5.14 STEREOISOMERISM OF CYCLIC COMPOUNDS

Cyclopentane derivatives offer a convenient starting point for a discussion of the stereoisomerism of cyclic compounds. For example, 1,2-dimethylcyclopentane has two chirality centers and exists in three stereoisomeric forms **5**, **6**, and **7**:

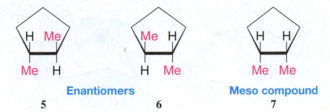

The trans compound exists as a pair of enantiomers **5** and **6**. *cis*-1,2-Dimethylcyclopentane (**7**) is a meso compound. It has a plane of symmetry that is perpendicular to the plane of the ring:

PRACTICE PROBLEM 5.27

(a) Is *trans*-1,2-dimethylcyclopentane (**5**) superposable on its mirror image (i.e., on compound **6**)? **(b)** Is *cis*-1,2-dimethylcyclopentane (**7**) superposable on its mirror image? **(c)** Is *cis*-1,2-dimethylcyclopentane a chiral molecule? **(d)** Would *cis*-1,2-dimethylcyclopentane show optical activity? **(e)** What is the stereoisomeric relationship between **5** and **7**? **(f)** Between **6** and **7**?

PRACTICE PROBLEM 5.28

Write structural formulas for all of the stereoisomers of 1,3-dimethylcyclopentane. Label pairs of enantiomers and meso compounds if they exist.

5.14A Cyclohexane Derivatives

1,4-Dimethylcyclohexanes If we examine a formula of 1,4-dimethylcyclohexane, we find that it does not contain any chirality centers. However, it does have two **stereogenic centers**. As we learned in Section 4.13, 1,4-dimethylcyclohexane can exist as cis–trans isomers. The cis and trans forms (Fig. 5.17) are *diastereomers*. Neither compound is chiral and, therefore, neither is optically active. Notice that both the cis and trans forms of 1,4-dimethylcyclohexane have a plane of symmetry.

Helpful Hint

Build handheld molecular models of the 1,4-, 1,3-, and 1,2-dimethylcyclohexane isomers discussed here and examine their stereochemical properties. Experiment with flipping the chairs and also switching between cis and trans isomers.

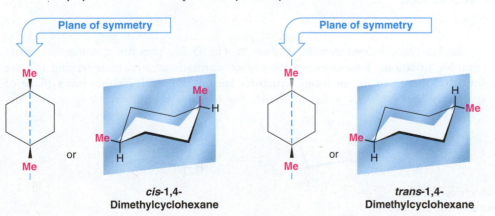

FIGURE 5.17 The cis and trans forms of 1,4-dimethylcyclohexane are diastereomers of each other. Both compounds are achiral, as the internal plane of symmetry (blue) shows for each.

1,3-Dimethylcyclohexanes 1,3-Dimethylcyclohexane has two chirality centers; we can, therefore, expect as many as four stereoisomers ($2^2 = 4$). In reality there are only three. *cis*-1,3-Dimethylcyclohexane has a plane of symmetry (Fig. 5.18) and is achiral.

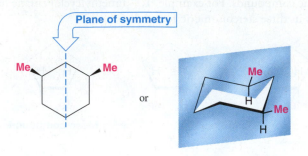

FIGURE 5.18 *cis*-1,3-Dimethylcyclohexane has a plane of symmetry, shown in blue, and is therefore achiral.

trans-1,3-Dimethylcyclohexane does not have a plane of symmetry and exists as a pair of enantiomers (Fig. 5.19). You may want to make models of the *trans*-1,3-dimethylcyclo-hexane enantiomers. Having done so, convince yourself that they cannot be superposed as they stand and that they cannot be superposed after one enantiomer has undergone a ring flip.

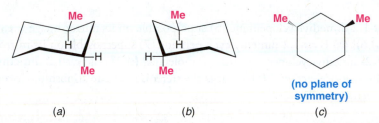

FIGURE 5.19 *trans*-1,3-Dimethylcyclohexane does not have a plane of symmetry and exists as a pair of enantiomers. The two structures (*a* and *b*) shown here are not superposable as they stand, and flipping the ring of either structure does not make it superposable on the other. (*c*) A simplified representation of (*b*).

1,2-Dimethylcyclohexanes 1,2-Dimethylcyclohexane also has two chirality centers, and again we might expect as many as four stereoisomers. Indeed there are four, but we find that we can *isolate* only *three* stereoisomers. *trans*-1,2-Dimethylcyclohexane (Fig. 5.20) exists as a pair of enantiomers. Its molecules do not have a plane of symmetry.

FIGURE 5.20 *trans*-1,2-Dimethylcyclohexane has no plane of symmetry and exists as a pair of enantiomers (*a* and *b*). [Notice that we have written the most stable confor-mations for (*a*) and (*b*). A ring flip of either (*a*) or (*b*) would cause both methyl groups to become axial.]

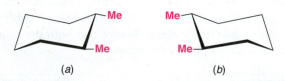

cis-1,2-Dimethylcyclohexane, shown in Fig. 5.21, presents a somewhat more complex situation. If we consider the two conformational structures (*c*) and (*d*), we find that these two mirror-image structures are not identical. Neither has a plane of symmetry and each is a chiral molecule, *but they are interconvertible by a ring flip*.

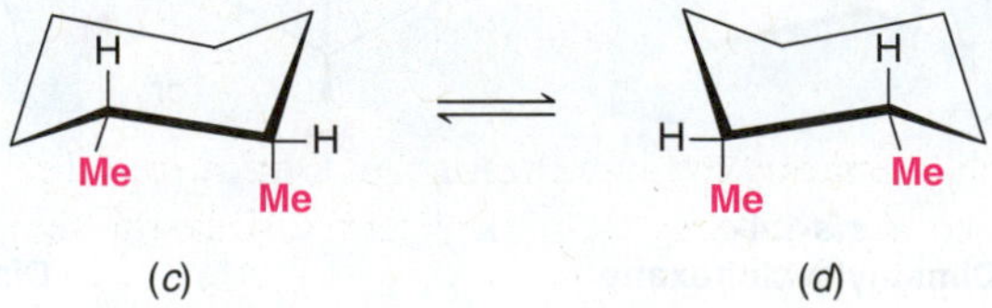

FIGURE 5.21 *cis*-1,2-Dimethylcyclohexane exists as two rapidly interconverting chair conformations (*c*) and (*d*).

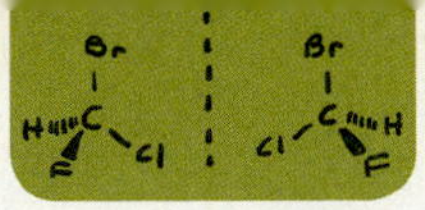

Therefore, although the two structures represent enantiomers, *they cannot be separated* because they rapidly interconvert even at low temperature. They simply represent *different conformations of the same compound.* Therefore, structures (c) and (d) are not configurational stereoisomers; they are **conformational stereoisomers** (see Section 4.9A). This means that at normal temperatures there are only three *isolable stereoisomers* of 1,2-dimethylcyclohexane.

As we shall see later, there are some compounds whose conformational stereoisomers *can* be isolated in enantiomeric forms. Isomers of this type are called atropisomers (Section 5.18).

●●● ·······

Write formulas for all of the isomers of each of the following. Designate pairs of enantiomers and achiral compounds where they exist.

PRACTICE PROBLEM 5.29

(a) 1-Bromo-2-chlorocyclohexane   **(c)** 1-Bromo-4-chlorocyclohexane
(b) 1-Bromo-3-chlorocyclohexane

●●● ·······

Give the (*R*,*S*) designation for each compound given as an answer to Practice Problem 5.29.

PRACTICE PROBLEM 5.30

5.15 RELATING CONFIGURATIONS THROUGH REACTIONS IN WHICH NO BONDS TO THE CHIRALITY CENTER ARE BROKEN

- A reaction is said to proceed with retention of **configuration** at a chirality center if no bonds to the chirality center are broken. This is true even if the *R*,*S* designation for the chirality center changes because the relative priorities of groups around it changes as a result of the reaction.

First consider an example that occurs with retention of configuration and that also retains the same *R*,*S* designation in the product as in the reactant. Such is the case when (*S*)-(−)-2-methyl-l-butanol reacts with hydrochloric acid to form (*S*)-(+)-l-chloro-2-methylbutanol. Note that none of the bonds at the chirality center are broken (we shall study how this reaction takes place in Section 11.8A).

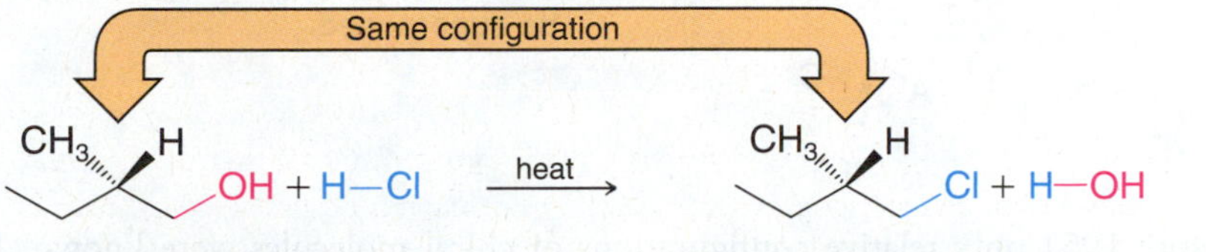

Same configuration

CH$_3$,,,, H
OH + H—Cl   →(heat)   CH$_3$,,,, H   Cl + H—OH

(*S*)-(−)-2-Methyl-1-butanol
$[\alpha]_D^{25} = -5.756$

(*S*)-(+)-1-Chloro-2-methylbutane
$[\alpha]_D^{25} = +1.64$

This example also reminds us that the sign of optical rotation is not directly correlated with the *R*,*S* configuration of a chirality center, since the sign of rotation changes but the *R*,*S* configuration does not.

Next consider the reaction of (*R*)-1-bromo-2-butanol with zinc and acid to form (*S*)-2-butanol. At this point we do not need to know how this reaction takes place, except to observe that none of the bonds to the chirality center are broken.

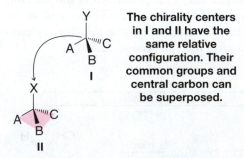

This reaction takes place with retention of configuration because no bonds to the chirality center are broken, but the *R,S* configuration changes because the relative priorities of groups bonded at the chirality center changes due to substitution of hydrogen for bromine.

●●● **SOLVED PROBLEM 5.8**

When (*R*)-1-bromo-2-butanol reacts with **KI** in acetone the product is 1-iodo-2-butanol. Would the product be (*R*) or (*S*)?

STRATEGY AND ANSWER: No bonds to the chirality center would be broken, so we can reason that the product would be the following.

The configuration of the product would still be (*R*) because replacing the bromine at **C1** with an iodine atom does not change the relative priority of **C1**.

5.15A Relative and Absolute Configurations

Reactions in which no bonds to the chirality center are broken are useful in relating configurations of chiral molecules. That is, they allow us to demonstrate that certain compounds have the same relative configuration. In each of the examples that we have just cited, the products of the reactions have the same *relative configurations* as the reactants.

- Chirality centers in different molecules have the same **relative configuration** if they share three groups in common and if these groups **with** the central carbon can be superposed in a pyramidal arrangement.

The chirality centers in I and II have the same relative configuration. Their common groups and central carbon can be superposed.

Before 1951 only relative configurations of chiral molecules were known. No one prior to that time had been able to demonstrate with certainty what the actual spatial arrangement of groups was in any chiral molecule. To say this another way, no one had been able to determine the absolute configuration of an optically active compound.

- The **absolute configuration** of a chirality center is its (*R*) or (*S*) designation, which can only be specified by knowledge of the actual arrangement of groups in space at the chirality center.

Prior to any known absolute configurations, the configurations of chiral molecules were related to each other *through reactions of known stereochemistry*. Attempts were also

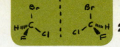

made to relate all configurations to a single compound that had been chosen arbitrarily to be the standard. This standard compound was glyceraldehyde:

OH
HO
*
O
H
Glyceraldehyde

Glyceraldehyde has one chirality center; therefore, glyceraldehyde exists as a pair of enantiomers:

H   OH
HO   O
H
(R)-Glyceraldehyde

(also known as D-glyceraldehyde)

and

HO   H
HO   O
H
(S)-Glyceraldehyde

(also known as L-glyceraldehyde)

In one system for designating configurations, (R)-glyceraldehyde is called D-glyceraldehyde and (S)-glyceraldehyde is called L-glyceraldehyde. This system of nomenclature is used with a specialized meaning in the nomenclature of carbohydrates. (See Section 22.2B.)

One glyceraldehyde enantiomer is dextrorotatory (+) and the other, of course, is levorotatory (−). Before 1951 no one was sure, however, which configuration belonged to which enantiomer. Chemists decided arbitrarily to assign the (R) configuration to the (+)-enantiomer. Then, configurations of other molecules were related to one glyceraldehyde enantiomer or the other through reactions of known stereochemistry.

For example, the configuration of (−)-lactic acid can be related to (+)-glyceraldehyde through the following sequence of reactions in which no bond to the chirality center is broken:

H   OH
HO   C   O
H
(+)-Glyceraldehyde   **This bond is broken.**

$\xrightarrow{\text{HgO}}$ (oxidation)

H   OH
HO   C   O
OH
(−)-Glyceric acid

$\xleftarrow[\text{H}_2\text{O}]{\text{HNO}_2}$

H   OH
H₂N   C   O
OH
This bond is broken.   **(+)-Isoserine**

$\xrightarrow[\text{HBr}]{\text{HNO}_2}$

Note that no bonds directly to the chirality center are broken.

This bond is broken.

H   OH
Br   C   O
OH
(−)-3-Bromo-2-hydroxy-propanoic acid

$\xrightarrow{\text{Zn, H}_3\text{O}^+}$

H   OH
H₃C   C   O
OH
(−)-Lactic acid

The stereochemistry of all of these reactions is known. Because none of the bonds to the chirality center (shown in red) has been broken during the sequence, its original configuration is retained. If the assumption is made that (+)-glyceraldehyde is the (R) stereoisomer, and therefore has the following configuration,

H   OH
HO   C   O
H
(R)-(+)-Glyceraldehyde

then (−)-lactic acid is also an (R) stereoisomer and its configuration is

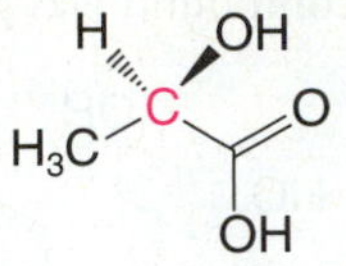

(R)-(−)-Lactic acid

●●●

PRACTICE PROBLEM 5.31 Write bond-line three-dimensional formulas for the starting compound, the product, and all of the intermediates in a synthesis similar to the one just given that relates the configuration of (−)-glyceraldehyde with (+)-lactic acid. Label each compound with its proper (R) or (S) and (+) or (−) designation.

The configuration of (−)-glyceraldehyde was also related through reactions of known stereochemistry to (+)-tartaric acid:

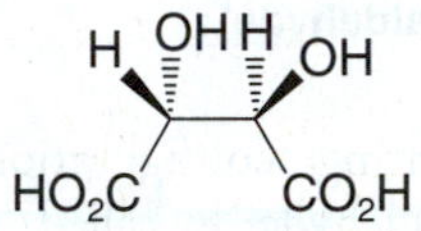

(+)-Tartaric acid

In 1951 J. M. Bijvoet, the director of the van't Hoff Laboratory of the University of Utrecht in the Netherlands, using a special technique of X-ray diffraction, was able to show conclusively that (+)-tartaric acid had the absolute configuration shown above. This meant that the original arbitrary assignment of the configurations of (+)- and (−)-glyceraldehyde was also correct. It also meant that the configurations of all of the compounds that had been related to one glyceraldehyde enantiomer or the other were now known with certainty and were now **absolute configurations**.

●●●

PRACTICE PROBLEM 5.32 Fischer projection formulas are often used to depict compounds such as glyceraldehyde, lactic acid, and tartaric acid. Draw Fischer projections for both enantiomers of **(a)** glyceraldehyde, **(b)** tartaric acid, and **(c)** lactic acid, and specify the (R) or (S) configuration at each chirality center. [Note that in Fischer projection formulas the terminal carbon that is most highly oxidized is placed at the top of the formula (an aldehyde or carboxylic acid group in the specific examples here).]

●●● **SOLVED PROBLEM 5.9**

Write a Fischer projection formula for a tartaric acid isomer that is not chiral.

STRATEGY AND ANSWER: We reason that because tartaric acid has two chirality centers, the achiral isomer must have a plane of symmetry and be a meso compound.

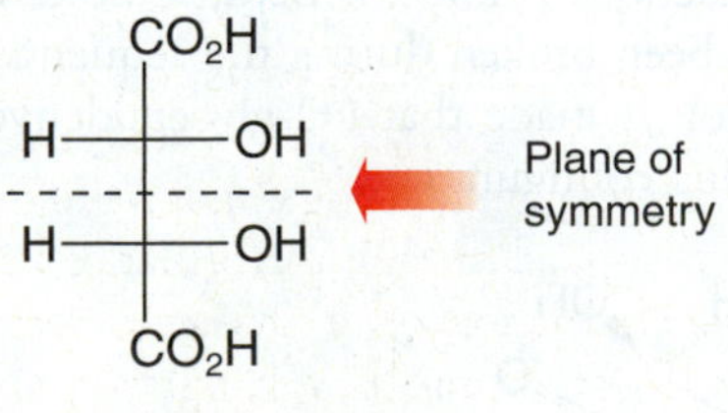

meso-Tartaric acid
(achiral)

5.16 SEPARATION OF ENANTIOMERS: RESOLUTION **231**

5.16 SEPARATION OF ENANTIOMERS: RESOLUTION

So far we have left unanswered an important question about optically active compounds and racemic forms: How are enantiomers separated? **Enantiomers** have identical solubilities in ordinary solvents, and they have identical boiling points. Consequently, the conventional methods for separating organic compounds, such as crystallization and distillation, fail when applied to a racemic form.

5.16A Pasteur's Method for Separating Enantiomers

It was, in fact, Louis Pasteur's separation of a racemic form of a salt of tartaric acid in 1848 that led to the discovery of the phenomenon called enantiomerism. Pasteur, consequently, is often considered to be the founder of the field of stereochemistry.

(+)-Tartaric acid is one of the by-products of wine making (nature usually only synthesizes one enantiomer of a chiral molecule). Pasteur had obtained a sample of racemic tartaric acid from the owner of a chemical plant. In the course of his investigation Pasteur began examining the crystal structure of the sodium ammonium salt of racemic tartaric acid. He noticed that two types of crystals were present. One was identical with crystals of the sodium ammonium salt of (+)-tartaric acid that had been discovered earlier and had been shown to be dextrorotatory. Crystals of the other type were *non*superposable mirror images of the first kind. The two types of crystals were actually chiral. Using tweezers and a magnifying glass, Pasteur separated the two kinds of crystals, dissolved them in water, and placed the solutions in a polarimeter. The solution of crystals of the first type was dextrorotatory, and the crystals themselves proved to be identical with the sodium ammonium salt of (+)-tartaric acid that was already known. The solution of crystals of the second type was levorotatory; it rotated plane-polarized light in the opposite direction and by an equal amount. The crystals of the second type were of the sodium ammonium salt of (−)-tartaric acid. The chirality of the crystals themselves disappeared, of course, as the crystals dissolved into their solutions, *but the optical activity* remained. Pasteur reasoned, therefore, that the molecules themselves must be chiral.

Pasteur's discovery of enantiomerism and his demonstration that the optical activity of the two forms of tartaric acid was a property of the molecules themselves led, in 1874, to the proposal of the tetrahedral structure of carbon by van't Hoff and Le Bel.

Unfortunately, few organic compounds give chiral crystals as do the (+)- and (−)-tartaric acid salts. Few organic compounds crystallize into separate crystals (containing separate enantiomers) that are visibly chiral like the crystals of the sodium ammonium salt of tartaric acid. Pasteur's method, therefore, is not generally applicable to the separation of enantiomers.

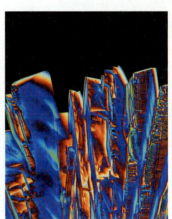

Tartaric acid crystals.

5.16B Modern Methods for Resolution of Enantiomers

One of the most useful procedures for separating enantiomers is based on the following:

- When a racemic mixture reacts with a single enantiomer of another compound, a mixture of diastereomers results, and diastereomers, because they have different melting points, boiling points, and solubilities, can be separated by conventional means.

Diastereomeric recrystallization is one such process. We shall see how this is done in Section 20.3F. Another method is **resolution** by enzymes, whereby an enzyme selectively converts one enantiomer in a racemic mixture to another compound, after which the unreacted enantiomer and the new compound are separated. The reaction by lipase in Section 5.10B is an example of this type of resolution. Chromatography using chiral media is also widely used to resolve enantiomers. This approach is applied in high-performance liquid chromatography (HPLC) as well as in other forms of chromatography. Diastereomeric interactions between molecules of the racemic mixture and the chiral chromatography medium cause enantiomers of the racemate to move through the chromatography apparatus at different rates. The enantiomers are then collected separately as they elute from the chromatography device. (See "*The Chemistry of*... HPLC Resolution of Enantiomers," Section 20.3.)

5.17 COMPOUNDS WITH CHIRALITY CENTERS OTHER THAN CARBON

Any tetrahedral atom with four different groups attached to it is a **chirality center**. Shown here are general formulas of compounds whose molecules contain chirality centers other than carbon. Silicon and germanium are in the same group of the periodic table as carbon. They form tetrahedral compounds as carbon does. When four different groups are situated around the central atom in silicon, germanium, and nitrogen compounds, the molecules are chiral and the enantiomers can, in principle, be separated. Sulfoxides, like certain examples of other functional groups where one of the four groups is a nonbonding electron pair, are also chiral. This is not the case for amines, however (Section 20.2B):

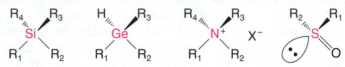

5.18 CHIRAL MOLECULES THAT DO NOT POSSESS A CHIRALITY CENTER

A molecule is chiral if it is not superposable on its mirror image. The presence of a tetrahedral atom with four different groups is only one type of chirality center, however. While most of the chiral molecules we shall encounter have chirality centers, there are other structural attributes that can confer chirality on a molecule. For example, there are compounds that have such large rotational barriers between conformers that individual conformational isomers can be separated and purified, and some of these conformational isomers are stereoisomers.

Conformational isomers that are stable, isolable compounds are called **atropisomers**. The existence of chiral atropisomers has been exploited to great effect in the development of chiral catalysts for stereoselective reactions. An example is BINAP, shown below in its enantiomeric forms:

(S)-BINAP   *(R)*-BINAP

The origin of chirality in BINAP is the restricted rotation about the bond between the two nearly perpendicular naphthalene rings. This torsional barrier leads to two resolvable enantiomeric conformers, (*S*)- and (*R*)-BINAP. When each enantiomer is used as a ligand for metals such as ruthenium and rhodium (bound by unshared electron pairs on the phosphorus atoms), chiral organometallic complexes result that are capable of catalyzing stereoselective hydrogenation and other important industrial reactions. The significance of chiral ligands is highlighted by the industrial synthesis each year of approximately 3500 *tons* of (−)-menthol using an isomerization reaction involving a rhodium (*S*)-BINAP catalyst.

Allenes are compounds that also exhibit stereoisomerism. Allenes are molecules that contain the following double-bond sequence:

$$\text{C=C=C}$$

The planes of the π bonds of allenes are perpendicular to each other:

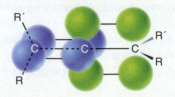

This geometry of the π bonds causes the groups attached to the end carbon atoms to lie in perpendicular planes, and, because of this, allenes with different substituents on the end carbon atoms are chiral (Fig. 5.22). (Allenes do not show cis–trans isomerism.)

Mirror

FIGURE 5.22 Enantiomeric forms of 1,3-dichloroallene. These two molecules are nonsuperposable mirror images of each other and are therefore chiral. They do not possess a tetrahedral atom with four different groups, however.

[WHY Do These Topics Matter?

THE POTENTIAL ORIGIN OF CHIRALITY

In the opening chapter of the book, we described the ground-breaking 1952 experiment by two chemists at the University of Chicago, Harold Urey and Stanley Miller, who showed how many of the amino acids found in living things were made spontaneously under simple, primordial-like conditions with simple chemicals. What we did not mention, however, was the proof that these amino acids had actually been synthesized during the experiment and were not the product of some contaminant within the apparatus itself. Urey and Miler's proof was that all of the amino acids were produced as racemates. As this chapter has shown, any amino acid produced by a life form on Earth exists as a single enantiomer. The question we are left with, then, is why do the molecules of life (such as amino acids) exist as single enantiomers? In other words, what is the origin of chirality on our planet? Potential answers to this question are more recent in origin, though it is a question that has interested scientists for well over a century.

In 1969, a large meteorite landed near the town of Murchison, Australia. Chemical analysis of its organic molecules showed it possessed over 100 amino acids, including dozens not found on Earth. Some of the amino acids possessed enantiomeric excess (e.e.) to the extent of 2–15%, all in favor of the L-amino acids, the same enantiomers found in all of

Earth's life forms. Careful analytical work proved that this optical activity was not the result of some Earth-based contaminant. In the past decade experiments have shown that with only the small amount of enantiomeric excess that these amino acids possess, some of them, such as the two shown below which have a fully substituted chirality center and cannot racemize, can effect a resolution of racemic amino acids through relatively simple processes such as crystallization. These events leave behind aqueous solutions of L-amino acids in high enantiomeric excess. Moreover, once these chiral L-amino acid solutions are generated, they can catalyze the enantiocontrolled synthesis of D-carbohydrates, which is what we all possess as well. As such, it is conceivable that the origin of chirality may well have come from outer space.

15.2% e.e   **2.8% e.e**

But what generated that initial enantiomeric excess? No one is quite sure, but some scientists speculate that electromagnetic radiation emitted in a corkscrew fashion from the poles of spinning neutron stars could lead to the bias of one mirror-image isomer over another when those molecules were formed in interstellar space. If that is true, then it is possible that on the other side of the galaxy there is a world that is the chiral opposite of Earth, where there are life forms with D-amino acids and L-sugars. Ronald Breslow of Columbia University, a leading researcher in this area, has said of such a possibility: "Since such life forms could well be advanced versions of dinosaurs, assuming that mammals did not have the good fortune to have the dinosaurs wiped out by an asteroidal collision as on earth, we may be better off not finding out."

Argonne National Laboratory/Photo Researchers, Inc.

To learn more about these topics, see:
Breslow, R. "The origin of homochirality in amino acids and sugars on prebiotic earth" *Tetrahedron Lett.* **2011**, *52*, 2028–2032 and references therein.

SUMMARY AND REVIEW TOOLS

In this chapter you learned that the handedness of life begins at the molecular level. Molecular recognition, signaling, and chemical reactions in living systems often hinge on the handedness of chiral molecules. Molecules that bear four different groups at a tetrahedral carbon atom are chiral if they are nonsuperposable with their mirror image. The atoms bearing four different groups are called chirality centers.

Mirror planes of symmetry have been very important to our discussion. If we want to draw the enantiomer of a molecule, one way to do so is to draw the molecule as if it were reflected in a mirror. If a mirror plane of symmetry exists *within* a molecule, then it is achiral (not chiral), even if it contains chirality centers. Using mirror planes to test for symmetry is an important technique.

In this chapter you learned how to give unique names to chiral molecules using the Cahn–Ingold–Prelog *R,S* system. You have also exercised your mind's eye in visualizing molecular structures in three dimensions, and you have refined your skill at drawing three-dimensional molecular formulas. You learned that pairs of enantiomers have identical physical properties except for the equal and opposite rotation of plane-polarized light, whereas diastereomers have different physical properties from one another. Interactions between each enantiomer of a chiral molecule and any other chiral material lead to diastereomeric interactions, which lead to different physical properties that can allow the separation of enantiomers.

Chemistry happens in three dimensions. Now, with the information from this chapter building on fundamentals you have learned about molecular shape and polarity in earlier chapters, you are ready to embark on your study of the reactions of organic molecules. Use the Key Terms and Concepts (which are hyperlinked to the Glossary from the bold, blue terms in the *WileyPLUS* version of the book at wileyplus.com) and the Concept Map that follows to help you reveiw and see the relationships between topics. Practice drawing molecules that show three dimensions at chirality centers, practice naming molecules, and label their regions of partial positive and negative charge. Paying attention to these things will help you learn about the reactivity of molecules in succeeding chapters. Most important of all, do your homework!

PROBLEMS

Note to Instructors: Many of the homework problems are available for assignment via *WileyPLUS*, an online teaching and learning solution.

CHIRALITY AND STEREOISOMERISM

5.33 Which of the following are chiral and, therefore, capable of existing as enantiomers?

(a) 1,3-Dichlorobutane

(b) 1,2-Dibromopropane

(c) 1,5-Dichloropentane

(d) 3-Ethylpentane

(e) 2-Bromobicyclo[1.1.0]butane

(f) 2-Fluorobicyclo[2.2.2]octane

(g) 2-Chlorobicyclo[2.1.1]hexane

(h) 5-Chlorobicyclo[2.1.1]hexane

5.34 (a) How many carbon atoms does an alkane (not a cycloalkane) need before it is capable of existing in enantiomeric forms? (b) Give correct names for two sets of enantiomers with this minimum number of carbon atoms.

5.35 Designate the (*R*) or (*S*) configuration at each chirality center in the following molecules.

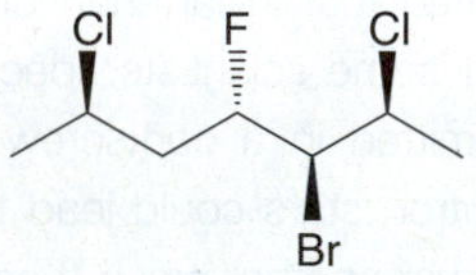

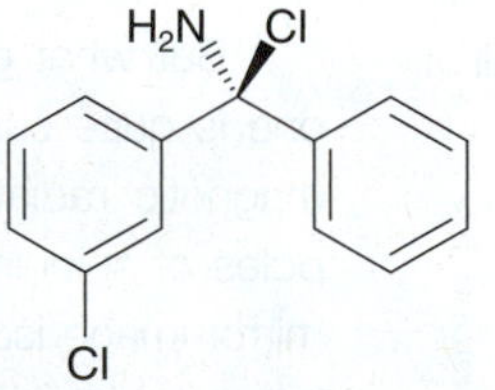

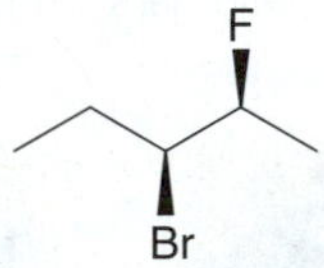

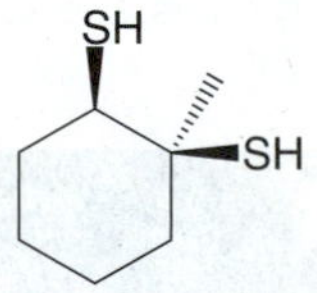

5.36 Albuterol, shown here, is a commonly prescribed asthma medication. For either enantiomer of albuterol, draw a three-dimensional formula using dashes and wedges for bonds that are not in the plane of the paper. Choose a perspective that allows as many carbon atoms as possible to be in the plane of the paper, and show all unshared electron pairs and hydrogen atoms (except those on the methyl groups labeled Me). Specify the (*R,S*) configuration of the enantiomer you drew.

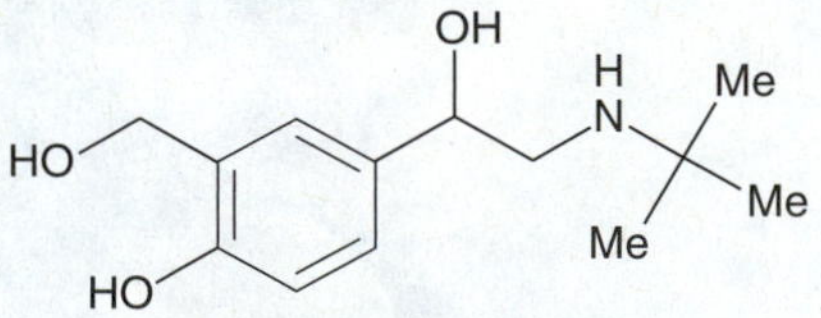

Albuterol

5.37 (a) Write the structure of 2,2-dichlorobicyclo[2.2.1]heptane. (b) How many chirality centers does it contain? (c) How many stereoisomers are predicted by the 2^n rule? (d) Only one pair of enantiomers is possible for 2,2-dichlorobicyclo[2.2.1]heptane. Explain.

5.38 Shown below are Newman projection formulas for (*R,R*)-, (*S,S*)-, and (*R,S*)-2,3-dichlorobutane. **(a)** Which is which? **(b)** Which formula is a meso compound?

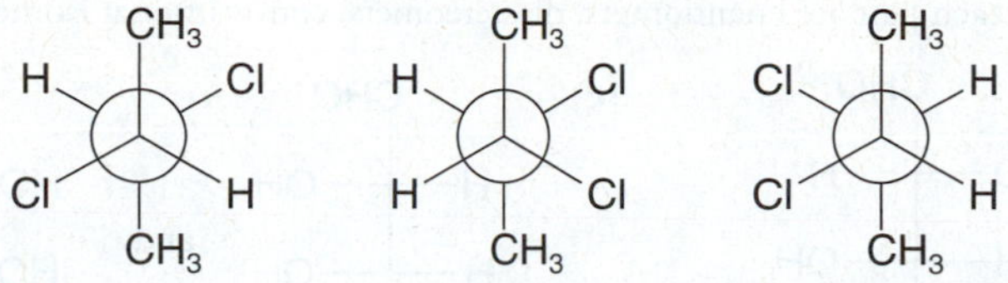

A B C

5.39 Write appropriate structural formulas for **(a)** a cyclic molecule that is a constitutional isomer of cyclohexane, **(b)** molecules with the formula C_6H_{12} that contain one ring and that are enantiomers of each other, **(c)** molecules with the formula C_6H_{12} that contain one ring and that are diastereomers of each other, **(d)** molecules with the formula C_6H_{12} that contain no ring and that are enantiomers of each other, and **(e)** molecules with the formula C_6H_{12} that contain no ring and that are diastereomers of each other.

5.40 Consider the following pairs of structures. Designate each chirality center as (*R*) or (*S*) and identify the relationship between them by describing them as representing enantiomers, diastereomers, constitutional isomers, or two molecules of the same compound. Use handheld molecular models to check your answers.

(a) 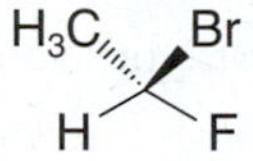 and 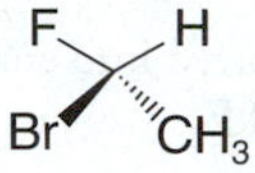

(b) 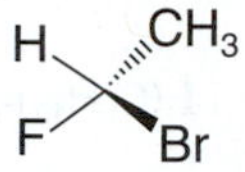 and 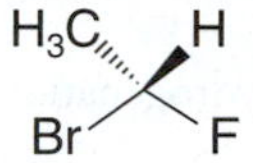

(c) 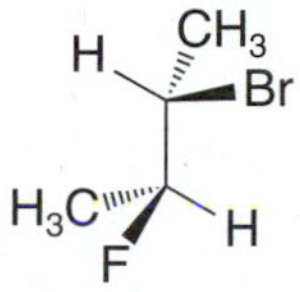 and 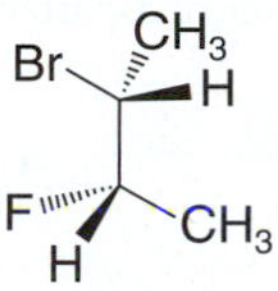

(d) 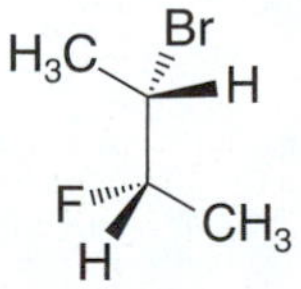 and 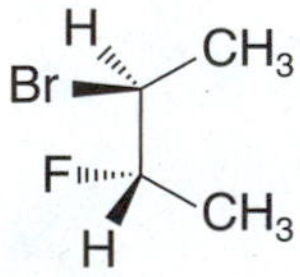

(e) 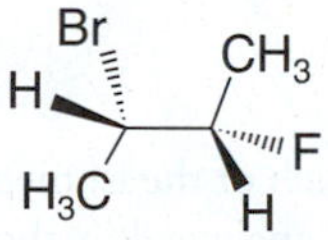 and 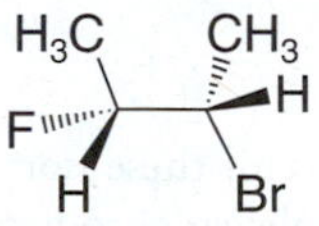

(f) 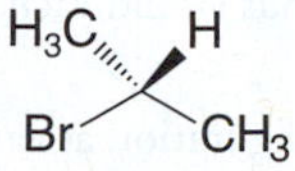 and 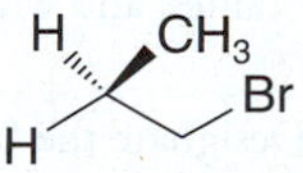

(g) 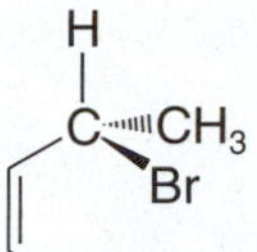 and 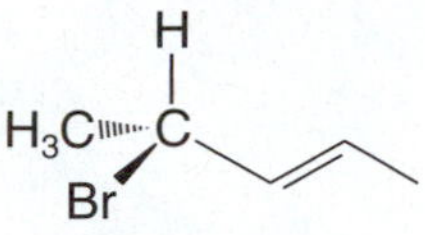

(h) 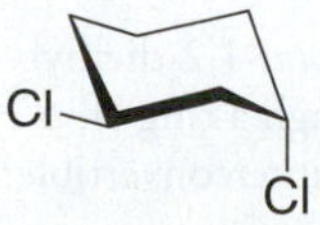 and 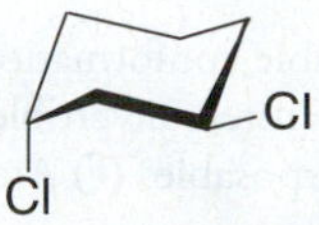

(i) 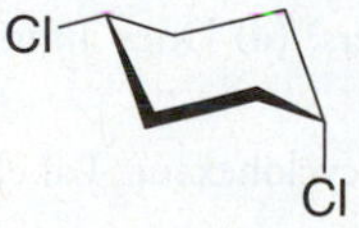 and 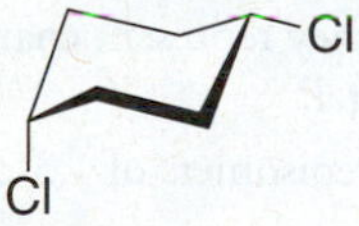

(j) 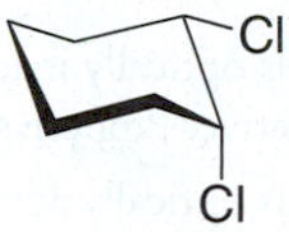 and 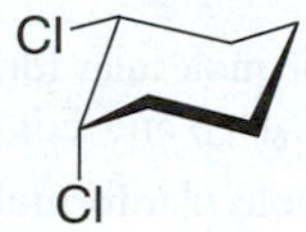

(k) 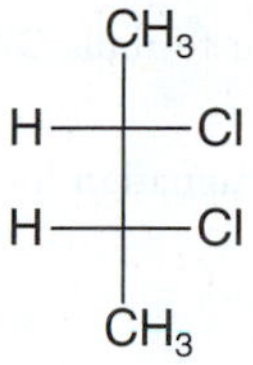 and 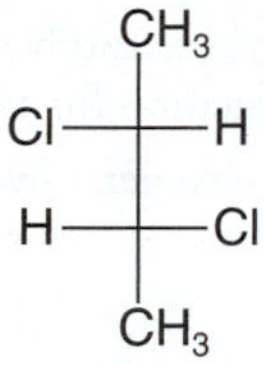

(l) 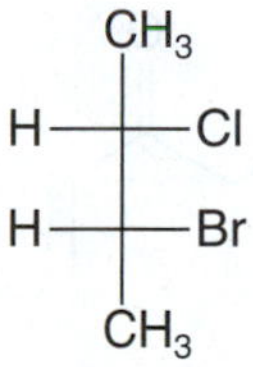 and 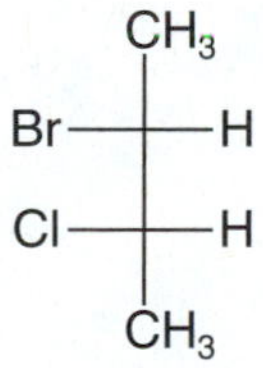

(m) 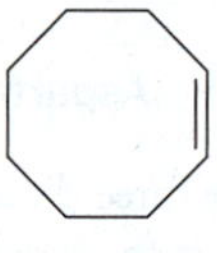 and 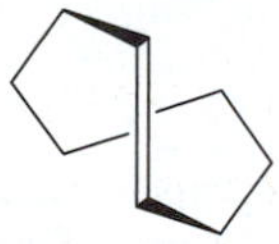

(n) 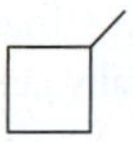 and 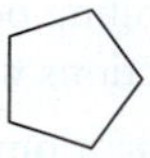

(o) 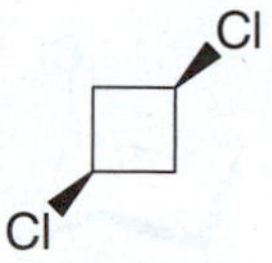 and 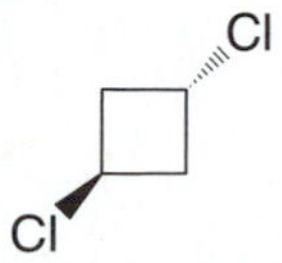

(p) 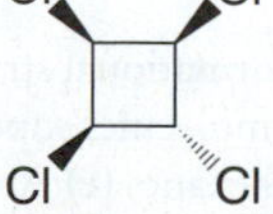 and 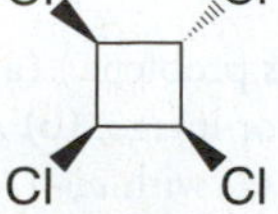

(q) 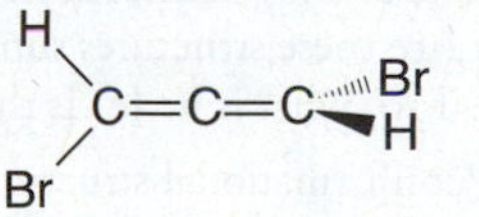 and 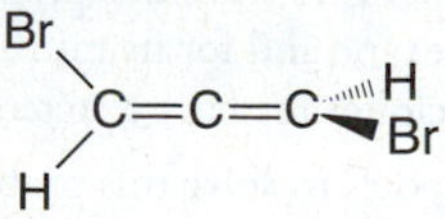

5.41 Discuss the anticipated stereochemistry of each of the following compounds.
(a) $ClCH=C=C=CHCl$ (b) $CH_2=C=C=CHCl$ (c) $ClCH=C=C=CCl_2$

5.42 Tell whether the compounds of each pair are enantiomers, diastereomers, constitutional isomers, or not isomeric.

(a)

```
      CHO                CHO
  H——OH            HO——H
            and
  H——OH             H——OH
     CH2OH              CH2OH
```

(b)

```
      CHO                CHO
 HO——H             H——OH
            and
  H——OH            HO——H
     CH2OH              CH2OH
```

(c)

```
      CHO                CHO
  H——OH            HO——H
            and
  H——OH            HO——H
     CH2OH              CH2OH
```

(d)

```
     CH2OH               CHO
  H——OH            HO——H
            and
  H——OH            HO——H
      CHO               CH2OH
```

5.43 A compound **D** with the molecular formula C_6H_{12} is optically inactive but can be resolved into enantiomers. On catalytic hydrogenation, **D** is converted to **E** (C_6H_{14}) and **E** is optically inactive. Propose structures for **D** and **E**.

5.44 Compound **F** has the molecular formula C_5H_8 and is optically active. On catalytic hydrogenation **F** yields **G** (C_5H_{12}) and **G** is optically inactive. Propose structures for **F** and **G**.

5.45 Compound **H** is optically active and has the molecular formula C_6H_{10}. On catalytic hydrogenation **H** is converted to **I** (C_6H_{12}) and **I** is optically inactive. Propose structures for **H** and **I**.

5.46 Aspartame is an artificial sweetener. Give the (R,S) designation for each chirality center of aspartame.

Aspartame

5.47 There are four dimethylcyclopropane isomers. (a) Write three-dimensional formulas for these isomers. (b) Which of the isomers are chiral? (c) If a mixture consisting of 1 mol of each of these isomers were subjected to simple gas chromatography (an analytical method that can separate compounds according to boiling point), how many fractions would be obtained and which compounds would each fraction contain? (d) How many of these fractions would be optically active?

5.48 For the following molecule, draw its enantiomer as well as one of its diastereomers. Designate the (R) or (S) configuration at each chirality center.

5.49 (Use models to solve this problem.) (a) Write a conformational structure for the most stable conformation of *trans*-1,2-diethyl-cyclohexane and write its mirror image. (b) Are these two molecules superposable? (c) Are they interconvertible through a ring "flip"? (d) Repeat the process in part (a) with *cis*-1,2-diethylcyclohexane. (e) Are these structures superposable? (f) Are they interconvertible?

5.50 (Use models to solve this problem.) (a) Write a conformational structure for the most stable conformation of *trans*-1,4-diethylcyclohexane and for its mirror image. (b) Are these structures superposable? (c) Do they represent enantiomers? (d) Does *trans*-1,4-diethylcyclohexane have a stereoisomer, and if so, what is it? (e) Is this stereoisomer chiral?

5.51 (Use models to solve this problem.) Write conformational structures for all of the stereoisomers of 1,3-diethylcyclohexane. Label pairs of enantiomers and meso compounds if they exist.

CHALLENGE PROBLEMS

5.52 Tartaric acid [$HO_2CCH(OH)CH(OH)CO_2H$] was an important compound in the history of stereochemistry. Two naturally occurring forms of tartaric acid are optically inactive. One optically inactive form has a melting point of 210–212°C, the other a melting point of 140°C. The inactive tartaric acid with a melting point of 210–212°C can be separated into two optically active forms of tartaric acid with the same melting point (168–170°C). One optically active tartaric acid has $[\alpha]_D^{25} = +12$, and the other, $[\alpha]_D^{25} = -12$. All attempts to separate the other inactive tartaric acid (melting point 140°C) into optically active compounds fail. **(a)** Write the three-dimensional structure of the tartaric acid with melting point 140°C. **(b)** Write structures for the optically active tartaric acids with melting points of 168–170°C. **(c)** Can you determine from the formulas which tartaric acid in (b) has a positive rotation and which has a negative rotation? **(d)** What is the nature of the form of tartaric acid with a melting point of 210–212°C?

5.53 (a) An aqueous solution of pure stereoisomer X of concentration 0.10 g mL^{-1} had an observed rotation of $-30°$ in a 1.0-dm tube at 589.6 nm (the sodium D line) and 25°C. What do you calculate its $[\alpha]_D$ to be at this temperature?

(b) Under identical conditions but with concentration 0.050 g mL^{-1}, a solution of X had an observed rotation of $+165°$. Rationalize how this could be and recalculate $[\alpha]_D$ for stereoisomer X.

(c) If the optical rotation of a substance studied at only one concentration is zero, can it definitely be concluded to be achiral? Racemic?

5.54 If a sample of a pure substance that has two or more chirality centers has an observed rotation of zero, it could be a racemate. Could it possibly be a pure stereoisomer? Could it possibly be a pure enantiomer?

5.55 Unknown Y has a molecular formula of $C_3H_6O_2$. It contains one functional group that absorbs infrared radiation in the 3200–3550-cm^{-1} region (when studied as a pure liquid; i.e., "neat"), and it has no absorption in the 1620–1780-cm^{-1} region. No carbon atom in the structure of Y has more than one oxygen atom bonded to it, and Y can exist in two (and only two) stereoisomeric forms. What are the structures of these forms of Y?

LEARNING GROUP PROBLEMS

1. Streptomycin is an antibiotic that is especially useful against penicillin-resistant bacteria. The structure of streptomycin is shown in Section 22.17. **(a)** Identify all of the chirality centers in the structure of streptomycin. **(b)** Assign the appropriate (R) or (S) designation for the configuration of each chirality center in streptomycin.

2. D-Galactitol is one of the toxic compounds produced by the disease galactosemia. Accumulation of high levels of D-galactitol causes the formation of cataracts. A Fischer projection for D-galactitol is shown at right:

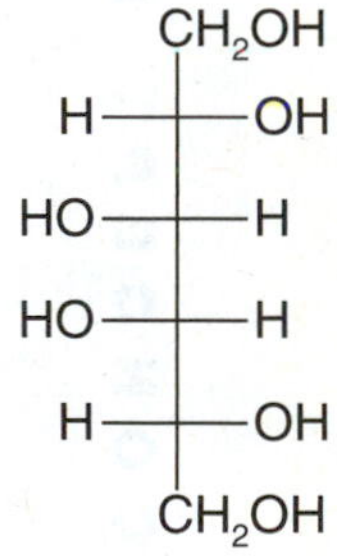

(a) Draw a three-dimensional structure for D-galactitol.

(b) Draw the mirror image of D-galactitol and write its Fischer projection formula.

(c) What is the stereochemical relationship between D-galactitol and its mirror image?

3. Cortisone is a natural steroid that can be isolated from the adrenal cortex. It has anti-inflammatory properties and is used to treat a variety of disorders (e.g., as a topical application for common skin diseases). The structure of cortisone is shown in Section 23.4D. **(a)** Identify all of the chirality centers in cortisone. **(b)** Assign the appropriate (R) or (S) designation for the configuration of each chirality center in cortisone.

[**CONCEPT MAP**]

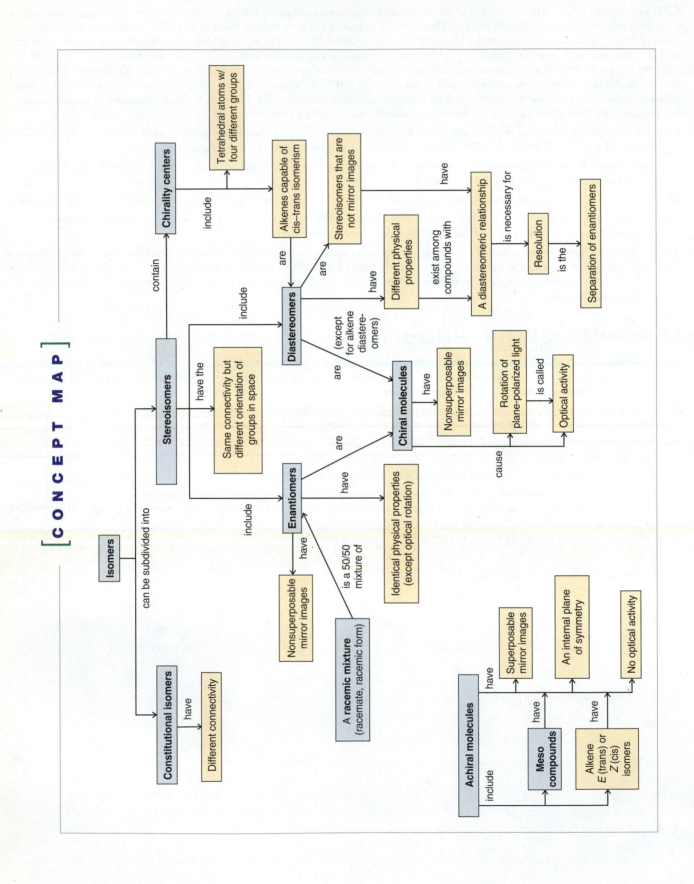

Ionic Reactions

NUCLEOPHILIC SUBSTITUTION AND ELIMINATION REACTIONS OF ALKYL HALIDES

Not all substitutions are a good thing; for instance, we wouldn't want to accidentally use salt in place of the needed amount of sugar in a batch of chocolate chip cookies. But with some substitutions, we get something even better. In organic chemistry that is often the case, since nucleophilic substitution reactions (which we will learn about in this chapter) allow the conversion of functional groups within a given molecule into entirely different functional groups, leading to new compounds with distinct properties. Moreover, nature utilizes a number of specific substitution reactions that are required for life.

IN THIS CHAPTER WE WILL CONSIDER:

- what groups can be replaced (i.e., substituted) or eliminated
- the various mechanisms by which such processes occur
- the conditions that can promote such reactions

[WHY DO THESE TOPICS MATTER?] At the end of the chapter, we will show an example where just a few substitution reactions can convert table sugar into a sweetener that has no calories—a sugar substitute that is not salty, but is in fact 600 times sweeter than sugar itself!

6.1 ALKYL HALIDES

- An **alkyl halide** has a halogen atom bonded to an sp^3-hybridized (tetrahedral) carbon atom.

- The carbon–halogen bond in an alkyl halide is polarized because the halogen is more electronegative than carbon. Therefore, the carbon atom has a partial positive charge ($\delta+$) and the halogen has a partial negative charge ($\delta-$).

- Alkyl halides are classified as primary (1°), secondary (2°), or tertiary (3°) according to the number of carbon groups (R) directly bonded to the carbon bearing the halogen atom (Section 2.5).

Halogen atom size increases as we go down the periodic table: fluorine atoms are the smallest and iodine atoms the largest. Consequently, the carbon–halogen *bond length increases* and carbon–halogen *bond strength decreases* as we go down the periodic table (Table 6.1). Maps of electrostatic potential (see Table 6.1) at the van der Waals surface for the four methyl halides, with ball-and-stick models inside, illustrate the trend in polarity, C—X bond length, and halogen atom size as one progresses from fluorine to iodine substitution. Fluoromethane is highly polar and has the shortest C—X bond length and the strongest C—X bond. Iodomethane is much less polar and has the longest C—X bond length and the weakest C—X bond.

In the laboratory and in industry, alkyl halides are used as solvents for relatively nonpolar compounds, and they are used as the starting materials for the synthesis of many compounds. As we shall learn in this chapter, the halogen atom of an alkyl halide can be easily replaced by other groups, and the presence of a halogen atom on a carbon chain also affords us the possibility of introducing a multiple bond.

Compounds in which a halogen atom is bonded to an alkene carbon are called **alkenyl halides**. In older nomenclature these were called vinylic halides. Compounds having a halogen bonded to an aromatic ring are called **aryl halides**. When the aromatic

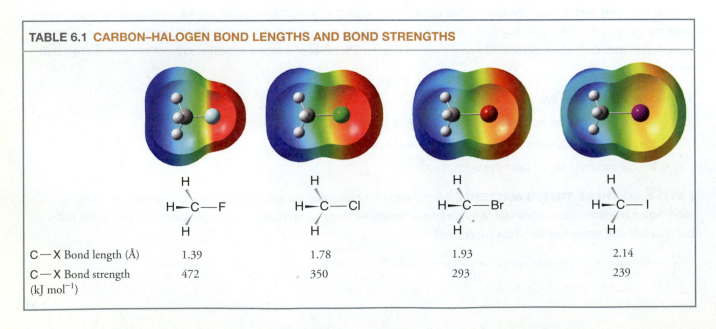

TABLE 6.1 CARBON–HALOGEN BOND LENGTHS AND BOND STRENGTHS

	H—C—F	H—C—Cl	H—C—Br	H—C—I
C—X Bond length (Å)	1.39	1.78	1.93	2.14
C—X Bond strength (kJ mol^{-1})	472	350	293	239

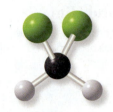

ring is specifically a benzene ring these compounds are called **phenyl halides**. The reactivity of compounds in which a halogen is bonded to an sp^2 carbon, as in alkenyl, aryl, and phenyl halides, is markedly different than when a halogen is bonded to an sp^3 carbon, as in an alkyl halide. The reactions that we discuss in this chapter will pertain to alkyl halides, not alkenyl, aryl, or phenyl halides.

An alkenyl halide   A phenyl halide or aryl halide

6.1A Physical Properties of Alkyl Halides

Most alkyl halides have very low solubilities in water, but as we might expect, they are miscible with each other and with other relatively nonpolar solvents. Dichloromethane (CH_2Cl_2, also called *methylene chloride*), trichloromethane ($CHCl_3$, also called *chloroform*), and tetrachloromethane (CCl_4, also called *carbon tetrachloride*) are sometimes used as solvents for nonpolar and moderately polar compounds. Many chloroalkanes, including CH_2Cl_2, $CHCl_3$, and CCl_4, have cumulative toxicity and are carcinogenic, however, and should therefore be used only in fume hoods and with great care.

Dichloromethane (CH_2Cl_2), a common laboratory solvent

PRACTICE PROBLEM 6.1

Give IUPAC names for each of the following.

(a)   (b)   (c)

PRACTICE PROBLEM 6.2

Classify each of the following organic halides as primary, secondary, tertiary, alkenyl, or aryl.

(a)   (b)   (c)   (d)   (e)

6.2 NUCLEOPHILIC SUBSTITUTION REACTIONS

Nucleophilic substitution reactions are among the most fundamental types of organic reactions. In a **nucleophilic substitution reaction** a nucleophile (Nu:) displaces a leaving group (LG) in the molecule that undergoes the substitution (the substrate).

- The **nucleophile** is always a Lewis base, and it may be negatively charged or neutral.
- The **leaving group** is always a species that takes a pair of electrons with it when it departs.

Often the **substrate** is an alkyl halide (R—Ẍ:) and the leaving group is a halide anion (:Ẍ:⁻). The following equations include a generic nucleophilic substitution reaction and some specific examples.

$$Nu:^- \quad + \quad R-LG \quad \longrightarrow \quad R-Nu \quad + \quad {}^-:LG$$

The nucleophile is a Lewis base that donates an electron pair to the substrate.	The bond between the carbon and the leaving group breaks, giving both electrons from the bond to the leaving group.	The nucleophile uses its electron pair to form a new covalent bond with the substrate carbon.	The leaving group gains the pair of electrons that originally bonded it in the substrate.

Helpful Hint

In color-coded reactions of this chapter, we will use red to indicate a nucleophile and blue to indicate a leaving group.

$$HO:^- \quad + \quad CH_3-\ddot{I}: \quad \longrightarrow \quad CH_3-\ddot{O}H \quad + \quad :\ddot{I}:^-$$

Helpful Hint

In Section 6.14 we shall see examples of biological nucleophilic substitution.

$$CH_3\ddot{O}:^- \quad + \quad CH_3CH_2-\ddot{B}r: \quad \longrightarrow \quad CH_3CH_2-\ddot{O}CH_3 \quad + \quad :\ddot{B}r:^-$$

$$:\ddot{I}:^- \quad + \quad \wedge\!\!\!\wedge\,\ddot{C}l: \quad \longrightarrow \quad \wedge\!\!\!\wedge\,\ddot{I}: \quad + \quad :\ddot{C}l:^-$$

$$\begin{array}{c} R \\ | \\ R-N: \\ | \\ R \end{array} \quad + \quad H_3C-\ddot{I}: \quad \longrightarrow \quad \begin{array}{c} R \\ | \\ R-\overset{+}{N}-CH_3 \\ | \\ R \end{array} \quad + \quad :\ddot{I}:^-$$

In nucleophilic substitution reactions the bond between the substrate carbon and the leaving group undergoes *heterolytic* bond cleavage. The unshared electron pair of the nucleophile forms the new bond to the carbon atom.

A key question we shall want to address later in this chapter is this: when does the bond between the leaving group and the carbon break? Does it break at the same time that the new bond between the nucleophile and carbon forms, as shown below?

$$Nu:^- \quad + \quad R-\ddot{X}: \quad \longrightarrow \quad \overset{\delta-}{Nu}\text{---}R\text{---}\overset{\delta-}{\ddot{X}}: \quad \longrightarrow \quad Nu-R \quad + \quad :\ddot{X}:^-$$

Or, does the bond to the leaving group break first?

$$R-\ddot{X}: \quad \longrightarrow \quad R^+ \quad + \quad :\ddot{X}:^-$$

Followed by

$$Nu:^- \quad + \quad R^+ \quad \longrightarrow \quad Nu-R$$

We shall find in Sections 6.9 and 6.14A that the answer depends greatly on the structure of the substrate.

●●● SOLVED PROBLEM 6.1

(a) A solution containing methoxide ions, CH_3O^- ions (as $NaOCH_3$), in methanol can be prepared by adding sodium hydride (NaH) to methanol (CH_3OH). A flammable gas is the other product. Write the acid–base reaction that takes place.
(b) Write the nucleophilic substitution that takes place when CH_3I is added and the resulting solution is heated.

STRATEGY AND ANSWER:

(a) We recall from Section 3.15 that sodium hydride consists of Na^+ ions and hydride ions ($H:^-$ ions), and that the hydride ion is a very strong base. [It is the conjugate base of H_2, a very weak acid ($pK_a = 35$, see Table 3.1).] The acid–base reaction that takes place is

$$CH_3\ddot{O}-H \quad + \quad Na^+:H^- \quad \longrightarrow \quad H_3C-\ddot{O}:^- Na^+ \quad + \quad H:H$$

Methanol (stronger acid)	Sodium hydride (stronger base)	Sodium methoxide (weaker base)	Hydrogen (weaker acid)

(b) The methoxide ion reacts with the alkyl halide (CH_3I) in a nucleophilic substitution:

$$CH_3-\ddot{O}:^- Na^+ \quad + \quad CH_3-\ddot{I}: \quad \xrightarrow[CH_3OH]{} \quad H_3C-\ddot{O}-CH_3 \quad + \quad Na^+ \quad + \quad :\ddot{I}:^-$$

6.3 NUCLEOPHILES

A nucleophile is a reagent that seeks a positive center.

- Any negative ion or uncharged molecule with an unshared electron pair is a potential nucleophile.

When a nucleophile reacts with an alkyl halide, the carbon atom bearing the halogen atom is the positive center that attracts the nucleophile. This carbon carries a partial positive charge because the electronegative halogen pulls the electrons of the carbon–halogen bond in its direction.

Helpful Hint

You may wish to review Section 3.3A, "Opposite Charges Attract."

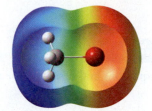

This is the positive center that the nucleophile seeks.

The electronegative halogen polarizes the C—X bond.

Let us look at two examples, one in which the nucleophile is a negatively charged Lewis base, and one in which the nucleophile is a neutral Lewis base.

1. Use of a **negatively charged nucleophile** (hydroxide in this case) **results in a neutral product** (an alcohol in this case). Formation of the covalent bond between the negative nucleophile and the substrate neutralizes the formal charge of the nucleophile.

Nucleophilic Substitution by a Negatively Charged Nucleophile Results Directly in a Neutral Product

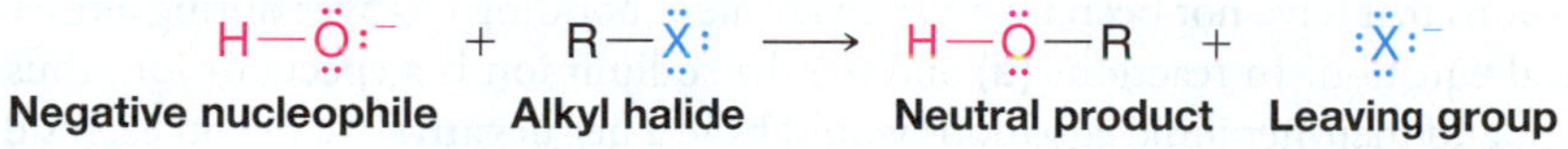

$$H\!-\!\overset{..}{\underset{..}{O}}\!:^{-} \quad + \quad R\!-\!\overset{..}{\underset{..}{X}}\!: \quad \longrightarrow \quad H\!-\!\overset{..}{\underset{..}{O}}\!-\!R \quad + \quad :\overset{..}{\underset{..}{X}}\!:^{-}$$

Negative nucleophile **Alkyl halide**   **Neutral product**   **Leaving group**

2. Use of a **neutral nucleophile** (water in this case) **results initially in a positively charged product**. The neutral nucleophile gains a positive formal charge through formation of the covalent bond with the substrate. A neutral product results only after a proton is removed from the atom with the formal positive charge in the initial product.

Nucleophilic Substitution by a Neutral Nucleophile Results Initially in a Positively Charged Product

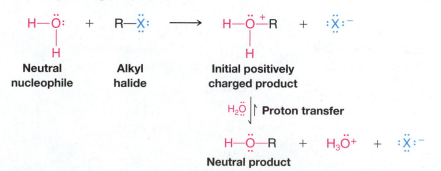

Helpful Hint

A deprotonation step is always required to complete the reaction when the nucleophile was a neutral atom that bore a proton.

In a reaction like this the nucleophile is a solvent molecule (as is often the case when neutral nucleophiles are involved). Since solvent molecules are present in great excess, the equilibrium favors transfer of a proton from the alkyloxonium ion to a water molecule. This type of reaction is an example of solvolysis, which we shall discuss further in Section 6.12B.

The reaction of ammonia (NH_3) with an alkyl halide, as shown below, provides another example where the nucleophile is uncharged. An excess of ammonia favors equilibrium removal of a proton from the alkylaminium ion to form the neutral amine.

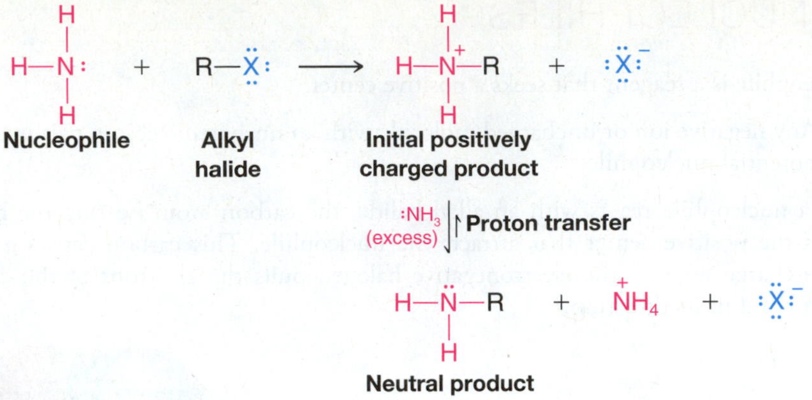

Nucleophile   **Alkyl halide**   **Initial positively charged product**

:NH₃ (excess) ↑ **Proton transfer**

Neutral product

••• SOLVED PROBLEM 6.2

Write the following as net ionic equations and designate the nucleophile, substrate, and leaving group in each case.

(a)

(b)

(c)

STRATEGY: A net ionic equation does not include spectator ions but is still balanced in terms of charges and the remaining species. Spectator ions are those ions that have not been involved in covalent bonding changes during a reaction, and that appear on both sides of a chemical equation. In reactions **(a)** and **(b)** the sodium ion is a spectator ion, thus the net ionic equation would not include them, and their net ionic equations would have a net negative charge on each side of the arrow. Equation **(c)** has no ions present among the reactants, and thus the ions found with the products are not spectator ions—they have resulted from covalent bonding changes. Equation **(c)** cannot be simplified to a net ionic equation.

Nucleophiles use a pair of electrons to form a covalent bond that is present in a product molecule. In all of the above reactions we can identify a species that used a pair of electrons in this way. These are the nucleophiles. **Leaving groups** depart from one of the reactant molecules and take a pair of electrons with them. In each reaction above we can identify such a species. Finally, the reactants to which the nucleophiles became bonded and from which the leaving groups departed are the **substrates**.

ANSWER: The net ionic equations are as follows for **(a)** and **(b)**, and there is no abbreviated equation possible for **(c)**. Nucleophiles, substrates, and leaving groups are labeled accordingly.

(a)

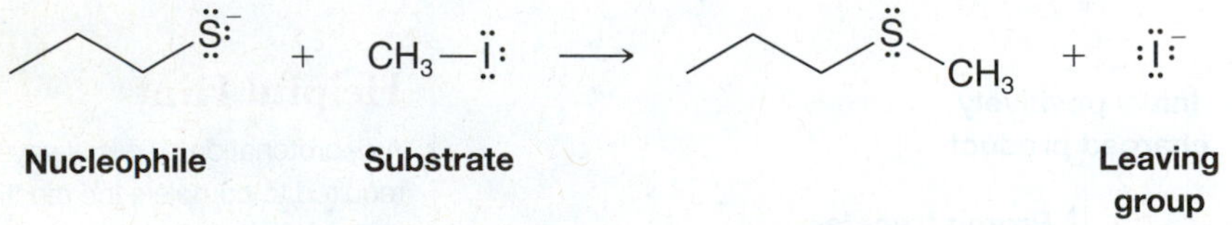

Nucleophile   Substrate   Leaving group

(b)

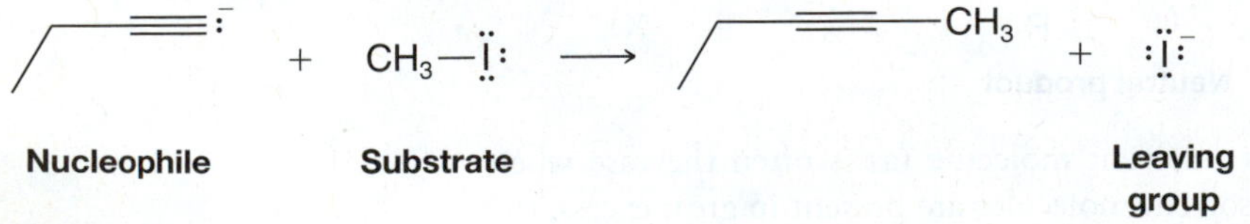

Nucleophile   Substrate   Leaving group

(c)

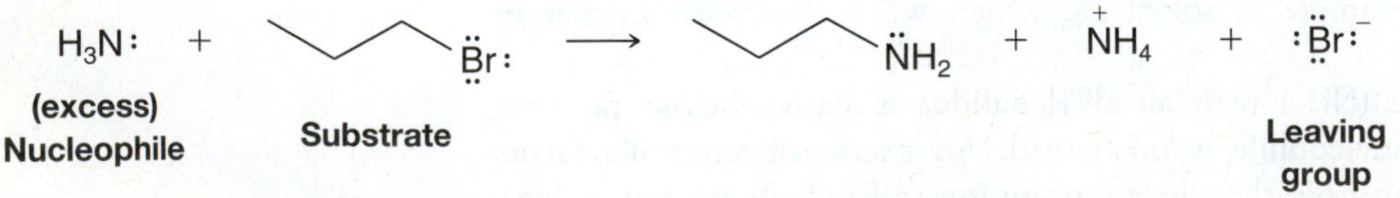

(excess)
Nucleophile   Substrate   Leaving group

Write the following as net ionic equations and designate the nucleophile, substrate, and leaving group in each reaction:

(a) CH$_3$I + CH$_3$CH$_2$ONa $\longrightarrow$ CH$_3$OCH$_2$CH$_3$ + NaI

(b) NaI + CH$_3$CH$_2$Br $\longrightarrow$ CH$_3$CH$_2$I + NaBr

(c) 2 CH$_3$OH + (CH$_3$)$_3$CCl $\longrightarrow$ (CH$_3$)$_3$COCH$_3$ + CH$_3$$\overset{+}{O}H_2$ + Cl$^-$

(d) /\/Br + NaCN $\longrightarrow$ /\/CN + NaBr

(e) (benzyl)Br + 2 NH$_3$ $\longrightarrow$ (benzyl)NH$_2$ + NH$_4$Br

6.4 LEAVING GROUPS

To act as the substrate in a nucleophilic substitution reaction, a molecule must have a good leaving group.

- A good **leaving group** is a substituent that can leave as a relatively stable, weakly basic molecule or ion.

In the examples shown above (Sections 6.2 and 6.3) the leaving group has been a halogen. Halide anions are weak bases (they are the conjugate bases of strong acids, HX), and therefore halogens are good leaving groups.

Some leaving groups depart as neutral molecules, such as a molecule of water or an alcohol. For this to be possible, the leaving group must have a formal positive charge while it is bonded to the substrate. When this group departs with a pair of electrons the leaving group's formal charge goes to zero. The following is an example where the leaving group departs as a water molecule.

$$CH_3-\overset{..}{\underset{|}{O}}: \;+\; CH_3-\overset{+}{\underset{|}{O}}-H \;\longrightarrow\; CH_3-\overset{+}{\underset{|}{O}}-CH_3 \;+\; :\overset{}{\underset{|}{O}}-H$$
$$\quad\;\; H \qquad\qquad\quad H \qquad\qquad\qquad H \qquad\qquad\quad H$$

> **Helpful Hint**
>
> Note that the net charge is the same on each side of a properly written chemical equation.

As we shall see later, the positive charge on a leaving group (like that above) usually results from protonation of the substrate by an acid. However, use of an acid to protonate the substrate and make a positively charged leaving group is feasible only when the nucleophile itself is not strongly basic, and when the nucleophile is present in abundance (such as in solvolysis).

Let us now begin to consider the mechanisms of nucleophilic substitution reactions. How does the nucleophile replace the leaving group? Does the reaction take place in one step or is more than one step involved? If more than one step is involved, what kinds of intermediates are formed? Which steps are fast and which are slow? In order to answer these questions, we need to know something about the rates of chemical reactions.

6.5 KINETICS OF A NUCLEOPHILIC SUBSTITUTION REACTION: AN S$_N$2 REACTION

To understand how the rate of a reaction (**kinetics**) might be measured, let us consider an actual example: the reaction that takes place between chloromethane and hydroxide ion in aqueous solution:

$$CH_3-Cl \;+\; {}^-OH \;\xrightarrow[H_2O]{60°C}\; CH_3-OH \;+\; Cl^-$$

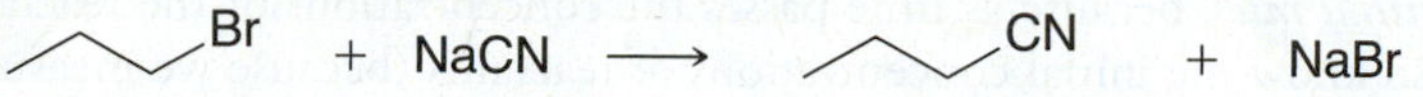

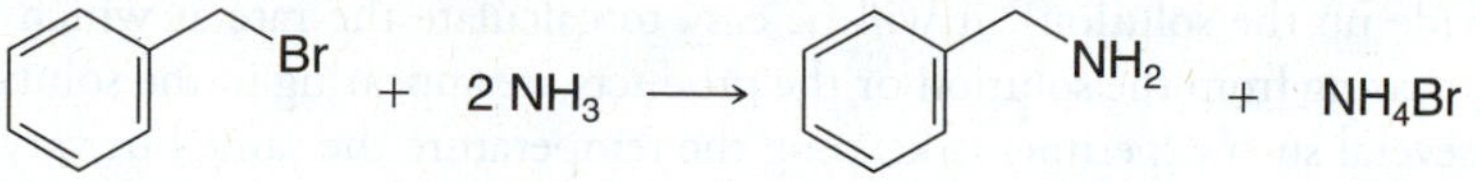

Although chloromethane is not highly soluble in water, it is soluble enough to carry out our kinetic study in an aqueous solution of sodium hydroxide. Because reaction rates are known to be temperature dependent (Section 6.7), we carry out the reaction at a constant temperature.

6.5A How Do We Measure the Rate of This Reaction?

The rate of the reaction can be determined experimentally by measuring the rate at which chloromethane or hydroxide ion *disappears* from the solution or the rate at which methanol or chloride ion *appears* in the solution. We can make any of these measurements by withdrawing a small sample from the reaction mixture soon after the reaction begins and analyzing it for the concentrations of CH_3Cl or HO^- and CH_3OH or Cl^-. We are interested in what are called *initial rates*, because as time passes the concentrations of the reactants change. Since we also know the initial concentrations of reactants (because we measured them when we made up the solution), it will be easy to calculate the rate at which the reactants are disappearing from the solution or the products are appearing in the solution.

We perform several such experiments keeping the temperature the same but varying the initial concentrations of the reactants. The results that we might get are shown in Table 6.2.

TABLE 6.2 RATE STUDY OF REACTION OF CH₃Cl WITH HO⁻ AT 60 °C

Experiment Number	Initial [CH₃Cl]	Initial [HO⁻]	Initial Rate (mol L⁻¹ s⁻¹)
1	0.0010	1.0	4.9×10^{-7}
2	0.0020	1.0	9.8×10^{-7}
3	0.0010	2.0	9.8×10^{-7}
4	0.0020	2.0	19.6×10^{-7}

Notice that the experiments show that the rate depends on the concentration of chloromethane *and* on the concentration of hydroxide ion. When we doubled the concentration of chloromethane in experiment 2, the rate *doubled*. When we doubled the concentration of hydroxide ion in experiment 3, the rate *doubled*. When we doubled both concentrations in experiment 4, the rate increased by a factor of *four*.

We can express these results as a proportionality,

$$\text{Rate} \propto [CH_3Cl][HO^-]$$

and this proportionality can be expressed as an equation through the introduction of a proportionality constant (k) called the rate constant:

$$\text{Rate} = k[CH_3Cl][HO^-]$$

For this reaction at this temperature we find that $k = 4.9 \times 10^{-4}$ L mol⁻¹ s⁻¹. (Verify this for yourself by doing the calculation.)

6.5B What Is the Order of This Reaction?

This reaction is said to be **second order overall**.* It is reasonable to conclude, therefore, that *for the reaction to take place a hydroxide ion and a chloromethane molecule must collide.* We also say that the reaction is **bimolecular**. (By *bimolecular* we mean that two species are involved in the step whose rate is being measured. In general the number of species involved in a reaction step is called the **molecularity** of the reaction.) We call this kind of reaction an **S_N2 reaction**, meaning **substitution, nucleophilic, bimolecular**.

6.6 A MECHANISM FOR THE S_N2 REACTION

A schematic representation of orbitals involved in an S_N2 reaction—based on ideas proposed by Edward D. Hughes and Sir Christopher Ingold in 1937—is outlined below.

*In general, the overall order of a reaction is equal to the sum of the exponents a and b in the rate equation Rate = $k[A]^a [B]^b$. If in some other reaction, for example, we found that Rate = $k[A]^2 [B]$, then we would say that the reaction is second order with respect to [A], first order with respect to [B], and third order overall.

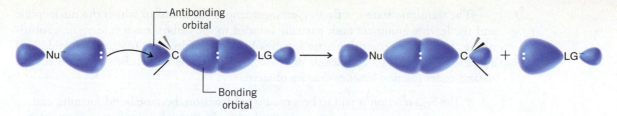

According to this mechanism:

- The nucleophile approaches the carbon bearing the leaving group from the **back side**, that is, from the side directly opposite the leaving group.

The orbital that contains the electron pair of the nucleophile (its highest occupied molecular orbital, or HOMO) begins to overlap with an empty orbital (the lowest unoccupied molecular orbital, or LUMO) of the carbon atom bearing the leaving group. As the reaction progresses, the bond between the nucleophile and the carbon atom strengthens, and the bond between the carbon atom and the leaving group weakens.

- As the nucleophile forms a bond and the leaving group departs, the substrate carbon atom undergoes **inversion***—its tetrahedral bonding configuration is turned inside out.

The formation of the bond between the nucleophile and the carbon atom provides most of the energy necessary to break the bond between the carbon atom and the leaving group. We can represent this mechanism with chloromethane and hydroxide ion as shown in the "Mechanism for the S$_N$2 Reaction" box below.

- The S$_N$2 reaction proceeds in a single step (without any intermediates) through an unstable arrangement of atoms called the **transition state**.

A MECHANISM FOR THE REACTION — Mechanism for the S$_N$2 Reaction

Reaction

$$HO^- + CH_3Cl \longrightarrow CH_3OH + Cl^-$$

Mechanism

| The negative hydroxide ion brings a pair of electrons to the partially positive carbon from the back side with respect to the leaving group. The chlorine begins to move away with the pair of electrons that bonded it to the carbon. | In the transition state, a bond between oxygen and carbon is partially formed and the bond between carbon and chlorine is partially broken. The configuration of the carbon atom begins to invert. | Now the bond between the oxygen and carbon has formed and the chloride ion has departed. The configuration of the carbon has inverted. |

*Considerable evidence had appeared in the years prior to Hughes and Ingold's 1937 publication indicating that in reactions like this an inversion of configuration of the carbon bearing the leaving group takes place. The first observation of such an inversion was made by the Latvian chemist Paul Walden in 1896, and such inversions are called **Walden inversions** in his honor. We shall study this aspect of the S$_N$2 reaction further in Section 6.8.

The transition state is a fleeting arrangement of the atoms in which the nucleophile and the leaving group are both partially bonded to the carbon atom undergoing substitution. Because the transition state involves both the nucleophile (e.g., a hydroxide ion) and the substrate (e.g., a molecule of chloromethane), this mechanism accounts for the second-order reaction kinetics that we observe.

- The S_N2 reaction is said to be a **concerted reaction**, because bond forming and bond breaking occur in concert (*simultaneously*) through a single transition state.

The transition state has an extremely brief existence. It lasts only as long as the time required for one molecular vibration, about 10^{-12} s. The structure and energy of the transition state are highly important aspects of any chemical reaction. We shall, therefore, examine this subject further in Section 6.7.

6.7 TRANSITION STATE THEORY: FREE-ENERGY DIAGRAMS

- A reaction that proceeds with a negative free-energy change (releases energy to its surroundings) is said to be **exergonic**; one that proceeds with a positive free-energy change (absorbs energy from its surroundings) is said to be **endergonic**.

The reaction between chloromethane and hydroxide ion in aqueous solution is highly exergonic; at 60°C (333 K), $\Delta G° = -100$ kJ mol^{-1}. (The reaction is also exothermic, $\Delta H° = -75$ kJ mol^{-1}.)

$$CH_3-Cl + {}^-OH \longrightarrow CH_3-OH + Cl^- \qquad \Delta G° = -100 \text{ kJ mol}^{-1}$$

The equilibrium constant for the reaction is extremely large, as we show by the following calculation:

$$\Delta G° = -RT \ln K_{eq}$$

$$\ln K_{eq} = \frac{-\Delta G°}{RT}$$

$$\ln K_{eq} = \frac{-(-100 \text{ kJ mol}^{-1})}{0.00831 \text{ kJ K}^{-1} \text{mol}^{-1} \times 333 \text{ K}}$$

$$\ln K_{eq} = 36.1$$

$$K_{eq} = 5.0 \times 10^{15}$$

An equilibrium constant as large as this means that the reaction goes to completion.

Because the free-energy change is negative, we can say that in energy terms the reaction goes **downhill**. The products of the reaction are at a lower level of free energy than the reactants. However, if covalent bonds are broken in a reaction, the reactants must go up an energy hill first, before they can go downhill. This will be true even if the reaction is exergonic.

We can represent the energy changes in a reaction using a graph called a **free-energy diagram**, where we plot the free energy of the reacting particles (y-axis) against the reaction coordinate (x-axis). Figure 6.1 is an example for a generalized S_N2 reaction.

- The **reaction coordinate** indicates the progress of the reaction, in terms of the conversion of reactants to products.
- The top of the energy curve corresponds to the **transition state** for the reaction.
- The **free energy of activation** ($\Delta G^{\ddagger}$) for the reaction is the difference in energy between the reactants and the transition state.
- The **free energy change for the reaction** ($\Delta G°$) is the difference in energy between the reactants and the products.

The top of the energy hill corresponds to the transition state. *The difference in free energy between the reactants and the transition state is the free energy of activation, $\Delta G^{\ddagger}$. The difference in free energy between the reactants and products is the free-energy change for the reaction, $\Delta G°$. For our example in Fig. 6.1, the free-energy level of the products is lower*

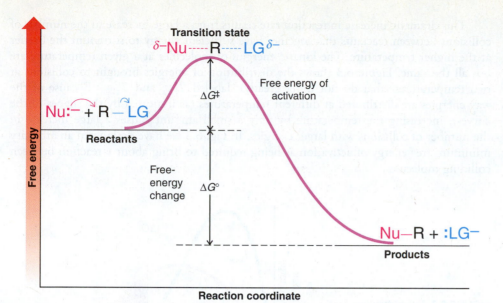

FIGURE 6.1 A free-energy diagram for a hypothetical exergonic S_N2 reaction (i.e., that takes place with a negative $\Delta G°$, releasing energy to the surroundings).

than that of the reactants. In terms of our analogy, we can say that the reactants in one energy valley must surmount an energy hill (the transition state) in order to reach the lower energy valley of the products.

If a reaction in which covalent bonds are broken proceeds with a positive free-energy change (Fig. 6.2), there will still be a free energy of activation. That is, if the products have greater free energy than reactants, the free energy of activation will be even higher. ($\Delta G^{\ddagger}$ will be larger than $\Delta G°$.) In other words, in the **uphill** (endergonic) reaction an even larger energy hill lies between the reactants in one valley and the products in a higher one.

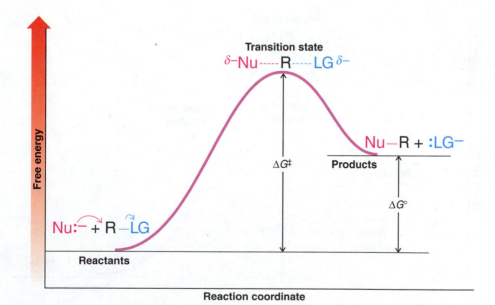

FIGURE 6.2 A free-energy diagram for a hypothetical endergonic S_N2 reaction (i.e., that takes place with a positive $\Delta G°$, absorbing energy from the surroundings).

6.7A Temperature and Reaction Rate

Most chemical reactions occur much more rapidly at higher temperatures. The increase in reaction rate for S_N2 reactions relates to the fact that at higher temperatures the number of collisions between reactants with sufficient energy to surmount the activation energy ($\Delta G^{\ddagger}$) increases significantly (see Fig. 6.3).

- A 10 °C increase in temperature will cause the reaction rate to double for many reactions taking place near room temperature.

This dramatic increase in reaction rate results from a large increase in the number of collisions between reactants that together have sufficient energy to surmount the barrier at the higher temperature. The kinetic energies of molecules at a given temperature are not all the same. Figure 6.3 shows the distribution of energies brought to collisions at two temperatures (that do not differ greatly), labeled T_{Low} and T_{High}. Because of the way energies are distributed at different temperatures (as indicated by the shapes of the curves), increasing the temperature by only a small amount causes a large increase in the number of collisions with larger energies. In Fig. 6.3 we have designated an arbitrary minimum free energy of activation as being required to bring about a reaction between colliding molecules.

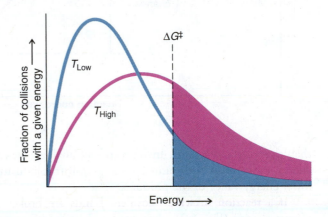

FIGURE 6.3 The distribution of energies at two different temperatures, T_{Low} and T_{High}. The number of collisions with energies greater than the free energy of activation is indicated by the corresponding shaded area under each curve.

A free-energy diagram for the reaction of chloromethane with hydroxide ion is shown in Fig. 6.4. At 60°C, $\Delta G^{\ddagger} = 103$ kJ mol^{-1}, which means that at this temperature the reaction reaches completion in a matter of a few hours.

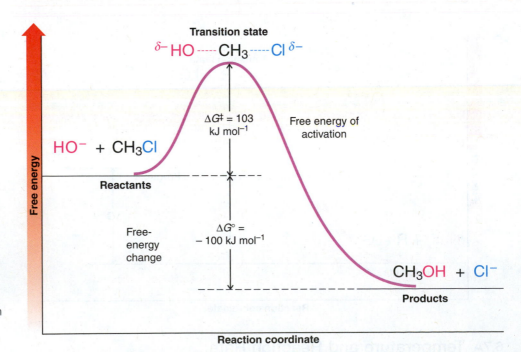

FIGURE 6.4 A free-energy diagram for the reaction of chloromethane with hydroxide ion at 60°C.

●●●

PRACTICE PROBLEM 6.4 Draw a hypothetical free-energy diagram for the S$_N$2 reaction of iodide anion with 1-chlorobutane. Label the diagram as in Fig. 6.4, and assume it is exergonic but without specific values for $\Delta G^{\ddagger}$ and $\Delta G°$.

6.8 THE STEREOCHEMISTRY OF S$_N$2 REACTIONS

The **stereochemistry** of S$_N$2 reactions is directly related to key features of the mechanism that we learned earlier:

- **The nucleophile approaches the substrate carbon from the back side with respect to the leaving group.** In other words, the bond to the nucleophile that is forming is opposite (at 180°) to the bond to the leaving group that is breaking.
- Nucleophilic displacement of the leaving group in an S$_N$2 reaction causes **inversion of configuration** at the substrate carbon.

We depict the inversion process as follows. It is much like the way an umbrella is inverted in a strong wind.

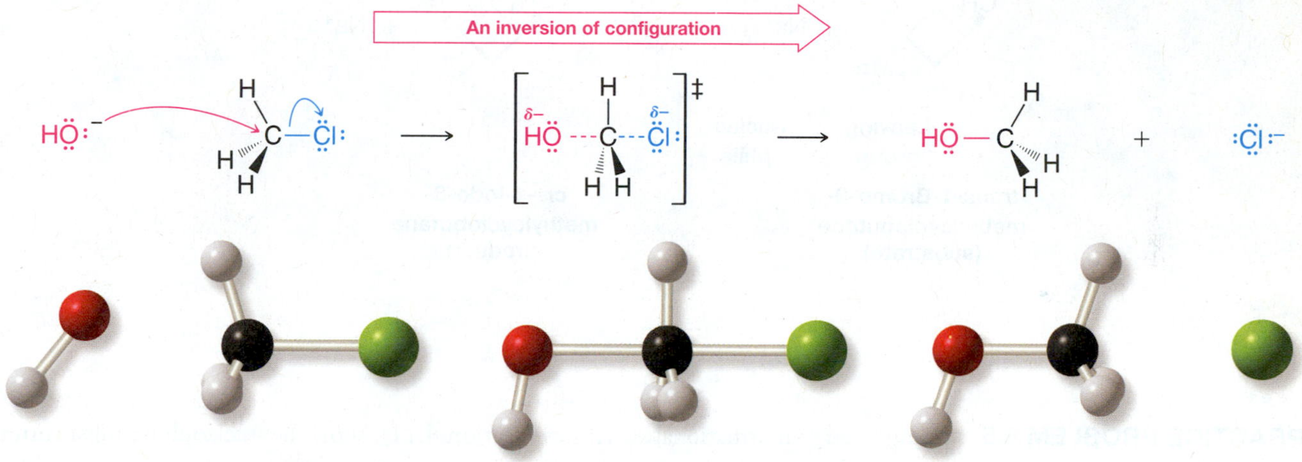

Transition state for an S$_N$2 reaction

An inversion of configuration

With a molecule such as chloromethane, however, there is no way to prove that attack by the nucleophile has involved inversion of configuration of the carbon atom because one form of methyl chloride is identical to its inverted form. With a molecule containing chirality centers such as *cis*-1-chloro-3-methylcyclopentane, however, we can observe the results of an inversion of configuration by the change in stereochemistry that occurs. When *cis*-1-chloro-3-methylcyclopentane reacts with hydroxide ion in an S$_N$2 reaction, the product is *trans*-3-methylcyclopentanol. *The hydroxyl group ends up bonded on the opposite side of the ring from the chlorine it replaces:*

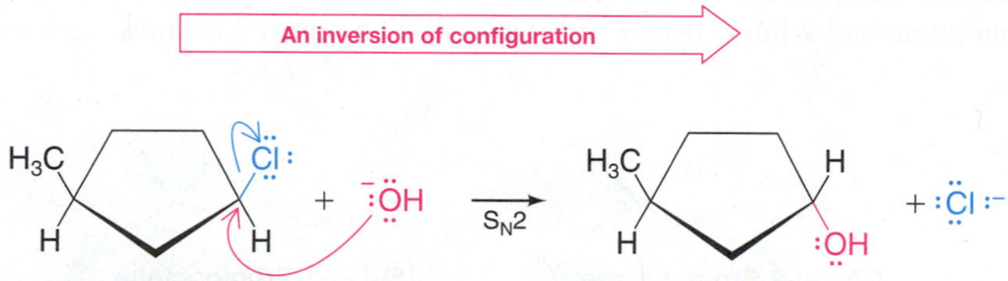

An inversion of configuration

***cis*-1-Chloro-3-methylcyclopentane**   ***trans*-3-Methylcyclopentanol**

Presumably, the transition state for this reaction is like that shown here.

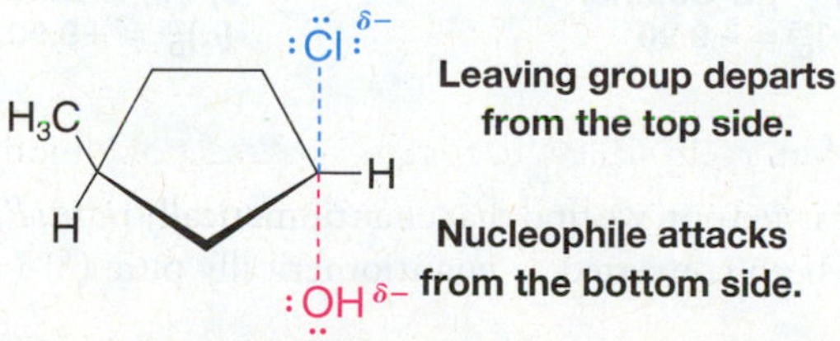

Leaving group departs from the top side.

Nucleophile attacks from the bottom side.

SOLVED PROBLEM 6.3

Give the structure of the product that would be formed when *trans*-1-bromo-3-methylcyclobutane undergoes an S_N2 reaction with NaI.

STRATEGY AND ANSWER: First, write the formulas for the reactants and identify the nucleophile, the substrate, and the leaving group. Then, recognizing that the nucleophile will attack the back side of the substrate carbon atom that bears the leaving group, causing an inversion of configuration at that carbon, write the structure of the product.

Leaving group **Nucleophile**

***trans*-1-Bromo-3-methylcyclobutane (substrate)**

***cis*-1-Iodo-3-methylcyclobutane (product)**

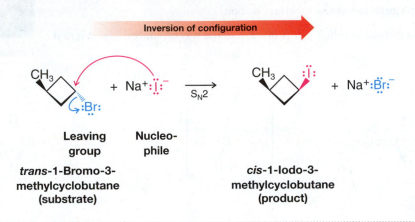

PRACTICE PROBLEM 6.5

Using chair conformational structures (Section 4.11), show the nucleophilic substitution reaction that would take place when *trans*-1-bromo-4-*tert*-butylcyclohexane reacts with iodide ion. (Show the most stable conformation of the reactant and the product.)

- **S_N2 reactions always occur with inversion of configuration.**

We can also observe inversion of configuration when an S_N2 reaction occurs at a chirality center in an acyclic molecule. The reaction of (*R*)-(−)-2-bromooctane with sodium hydroxide provides an example. We can determine whether or not inversion of configuration occurs in this reaction because the configurations and optical rotations for both enantiomers of 2-bromooctane and the expected product, 2-octanol, are known.

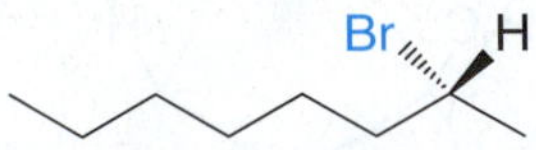

(*R*)-(−)-2-Bromooctane
$[\alpha]_D^{25} = -34.25$

(*S*)-(+)-2-Bromooctane
$[\alpha]_D^{25} = +34.25$

(*R*)-(−)-2-Octanol
$[\alpha]_D^{25} = -9.90$

(*S*)-(+)-2-Octanol
$[\alpha]_D^{25} = +9.90$

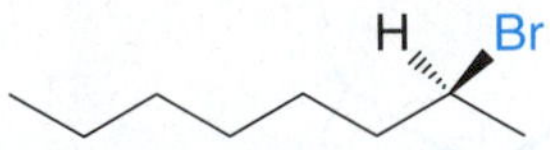

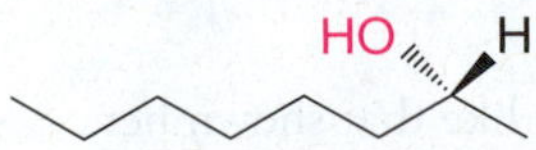

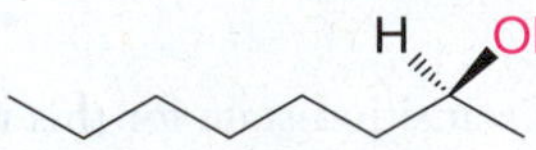

When the reaction is carried out, we find that enantiomerically pure (*R*)-(−)-2-bromooctane ($[\alpha]_D^{25} = -34.25$) has been converted to enantiomerically pure (*S*)-(+)-2-octanol ($[\alpha]_D^{25} = +9.90$).

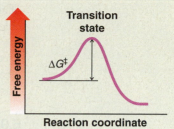

A MECHANISM FOR THE REACTION — The Stereochemistry of an S$_N$2 Reaction

The reaction of (*R*)-(−)-2-bromooctane with hydroxide is an S$_N$2 reaction and takes place with *inversion of configuration*:

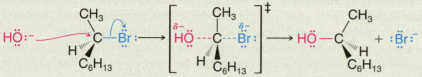

An inversion of configuration

Transition state

(*R*)-(−)-2-Bromooctane	(*S*)-(1)-2-Octanol	
$[\alpha]_D^{25} = -34.25°$	$[\alpha]_D^{25} = +9.90°$	An S$_N$2 reaction has one
Enantiomeric purity = 100%	**Enantiomeric purity = 100%**	transition state.

S$_N$2 reactions that involve breaking a bond to a chirality center can be used to relate configurations of molecules because the *stereochemistry* of the reaction is known.

PRACTICE PROBLEM 6.6

(a) Illustrate how this is true by assigning *R,S* configurations to the 2-chlorobutane enantiomers based on the following data. [The configuration of (−)-2-butanol is given in Section 5.8C.]

$$\text{(+)-2-Chlorobutane} \xrightarrow[\text{S}_N2]{\text{HO}^-} \text{(−)-2-Butanol}$$

$[\alpha]_D^{25} = +36.00$ $[\alpha]_D^{25} = -13.52$

Enantiomerically pure   **Enantiomerically pure**

(b) When optically pure (+)-2-chlorobutane is allowed to react with potassium iodide in acetone in an S$_N$2 reaction, the 2-iodobutane that is produced has a minus rotation. What is the configuration of (−)-2-iodobutane? Of (+)-2-iodobutane?

6.9 THE REACTION OF *TERT*-BUTYL CHLORIDE WITH WATER: AN S$_N$1 REACTION

Let us now consider another mechanism for nucleophilic substitution: the S$_N$1 reaction. When Hughes (Section 6.6) and co-workers studied the reaction of *tert*-butyl chloride with water they found the kinetics leading to formation of *tert*-butyl alcohol to be quite different than for other substitution reactions that they had studied.

$$\underset{\textbf{\textit{tert}-Butyl chloride}}{CH_3-\overset{\overset{\displaystyle CH_3}{|}}{\underset{\underset{\displaystyle CH_3}{|}}{C}}-Cl} + H_2O \longrightarrow \underset{\textbf{\textit{tert}-Butyl alcohol}}{CH_3-\overset{\overset{\displaystyle CH_3}{|}}{\underset{\underset{\displaystyle CH_3}{|}}{C}}-OH} + HCl$$

Hughes found that the rate of *tert*-butyl chloride substitution was the same whether the reaction was run at pH 7, where the hydroxide ion concentration is 10^{-7} M and the predominant nucleophile is water, or in 0.05 M hydroxide, where the more powerful hydroxide nucleophile is present in roughly 500,000 times higher concentration. These results suggest that neither water nor hydroxide ions are involved in the rate-determining step of the reaction. Instead, the rate of substitution is dependent only on the concentration of *tert*-butyl chloride. The reaction is thus first order in *tert*-butyl chloride, and first order overall.

Rate = $k[(CH_3)_3CCl]$

The reaction rate is first order in *tert*-butyl chloride and first order overall.

Furthermore, these results indicate that the transition state governing the rate of reaction involves only molecules of *tert*-butyl chloride, and not water or hydroxide ions. The reaction is said to be **unimolecular** (first-order) in the rate-determining step, and we call it an **S_N1 reaction** (**substitution, nucleophilic, unimolecular**). In Section 6.15 we shall see that elimination reactions can compete with S_N1 reactions, leading to the formation of alkenes, but in the case of the conditions used above for the experiments with *tert*-butyl chloride (moderate temperature and dilute base), S_N1 is the dominant process.

How can we explain an S_N1 reaction in terms of a mechanism? To do so, we shall need to consider the possibility that the mechanism involves more than one step. But what kind of kinetic results should we expect from a multistep reaction? Let us consider this point further.

6.9A Multistep Reactions and the Rate-Determining Step

- If a reaction takes place in a series of steps, and if one step is intrinsically slower than all the others, then the rate of the overall reaction will be essentially the same as the rate of this slow step. This slow step, consequently, is called the **rate-limiting step** or the **rate-determining step**.

Consider a multistep reaction such as the following:

$$\text{Reactant} \xrightarrow[\text{(slow)}]{k_1} \text{intermediate 1} \xrightarrow[\text{(fast)}]{k_2} \text{intermediate 2} \xrightarrow[\text{(fast)}]{k_3} \text{product}$$

$$\textit{Step 1} \qquad\qquad \textit{Step 2} \qquad\qquad \textit{Step 3}$$

When we say that the first step in this example is intrinsically slow, we mean that the rate constant for step 1 is very much smaller than the rate constant for step 2 or for step 3. That is, $k_1 << k_2$ or k_3. When we say that steps 2 and 3 are *fast*, we mean that because their rate constants are larger, they could (in theory) take place rapidly if the concentrations of the two intermediates ever became high. In actuality, the concentrations of the intermediates are always very small because of the slowness of step 1.

As an analogy, imagine an hourglass modified in the way shown in Fig. 6.5. The opening between the top chamber and the one just below is considerably smaller than the other two. The overall rate at which sand falls from the top to the bottom of the hourglass is limited by the rate at which sand passes through the small orifice. This step, in the passage of sand, is analogous to the rate-determining step of the multistep reaction.

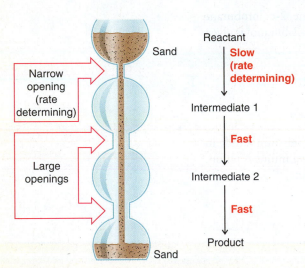

Reactant

Slow (rate determining)

↓

Intermediate 1

Fast

↓

Intermediate 2

Fast

↓

Product

FIGURE 6.5 A modified hourglass that serves as an analogy for a multistep reaction. The overall rate is limited by the rate of the slow step.

6.10 A MECHANISM FOR THE S_N1 REACTION

The mechanism for the reaction of *tert*-butyl chloride with water (Section 6.9) can be described in three steps. See the box "Mechanism for the S_N1 Reaction" below, with a schematic free-energy diagram highlighted for each step. Two distinct **intermediates** are formed. The first step is the slow step—it is the rate-determining step. In it a molecule of *tert*-butyl chloride ionizes and becomes a *tert*-butyl cation and a chloride ion. In the transition state for this step the carbon–chlorine bond of *tert*-butyl chloride is largely broken and ions are beginning to develop:

$$CH_3 - \underset{\underset{CH_3}{|}}{\overset{\overset{CH_3}{|}}{C}}{}^{\delta+} ---- Cl^{\delta-}$$

The solvent (water) stabilizes these developing ions by solvation. Carbocation formation, in general, takes place slowly because it is usually a highly endothermic process and is uphill in terms of free energy.

The first step requires heterolytic cleavage of the carbon–chlorine bond. Because no other bonds are formed in this step, it should be highly endothermic and it should have a high free energy of activation, as we see in the free-energy diagram. **That departure of the halide takes place at all is largely because of the ionizing ability of the solvent, water**. Experiments indicate that in the gas phase (i.e., in the absence of a solvent), the free energy of activation is about 630 kJ mol^{-1}! In aqueous solution, however, the free energy of activation is much lower—about 84 kJ mol^{-1}. **Water molecules surround and stabilize the cation and anion that are produced** (cf. Section 2.13D).

In the second step the intermediate *tert*-butyl cation reacts rapidly with water to produce a *tert*-butyloxonium ion, (CH$_3$)$_3$COH$_2$$^+$, which in the third step, rapidly transfers a proton to a molecule of water producing *tert*-butyl alcohol.

[A MECHANISM FOR THE REACTION — Mechanism for the S$_N$1 Reaction]

Reaction

$$CH_3-\underset{\underset{CH_3}{|}}{\overset{\overset{CH_3}{|}}{C}}-\ddot{C}l: \ + \ 2\,H_2O \longrightarrow CH_3-\underset{\underset{CH_3}{|}}{\overset{\overset{CH_3}{|}}{C}}-\ddot{O}H \ + \ H_3O^+ \ + \ :\ddot{C}l:^-$$

Mechanism

Step 1

$$CH_3-\underset{\underset{CH_3}{|}}{\overset{\overset{CH_3}{|}}{C}}-\ddot{C}l: \ \underset{H_2O}{\overset{\text{slow}}{\rightleftharpoons}} \ CH_3-\underset{\underset{CH_3}{|}}{\overset{\overset{CH_3}{|}}{C}}{}^+ \ + \ :\ddot{C}l:^-$$

Aided by the polar solvent, a chlorine departs with the electron pair that bonded it to the carbon.

This slow step produces the 3° carbocation intermediate and a chloride ion. Although not shown here, the ions are solvated (and stabilized) by water molecules.

Step 1
Transition state 1
Free energy
$\Delta G^{\ddagger}{}_{(1)}$
Reaction coordinate

$\Delta G^{\ddagger}{}_{(1)}$ is much larger than $\Delta G^{\ddagger}{}_{(2)}$ or $\Delta G^{\ddagger}{}_{(3)}$, hence this is the slowest step

Step 2

$$CH_3-\underset{\underset{CH_3}{|}}{\overset{\overset{CH_3}{|}}{C}}{}^+ \ + \ :\underset{\underset{H}{|}}{\ddot{O}}-H \ \overset{\text{fast}}{\rightleftharpoons} \ CH_3-\underset{\underset{CH_3}{|}}{\overset{\overset{CH_3}{|}}{C}}-\underset{\underset{H}{|}}{\overset{+}{\ddot{O}}}-H$$

A water molecule acting as a Lewis base donates an electron pair to the carbocation (a Lewis acid). This gives the cationic carbon eight electrons.

The product is a *tert*-butyloxonium ion (or protonated *tert*-butyl alcohol).

Step 2
Transition state 2
Free energy
$\Delta G^{\ddagger}{}_{(2)}$
Reaction coordinate

Step 3

$$CH_3-\underset{\underset{CH_3}{|}}{\overset{\overset{CH_3}{|}}{C}}-\underset{\underset{H}{|}}{\overset{+}{\ddot{O}}}-H \ + \ :\underset{\underset{H}{|}}{\ddot{O}}-H \ \overset{\text{fast}}{\rightleftharpoons} \ CH_3-\underset{\underset{CH_3}{|}}{\overset{\overset{CH_3}{|}}{C}}-\underset{\underset{H}{|}}{\ddot{O}}: \ + \ H-\underset{\underset{H}{|}}{\overset{+}{\ddot{O}}}-H$$

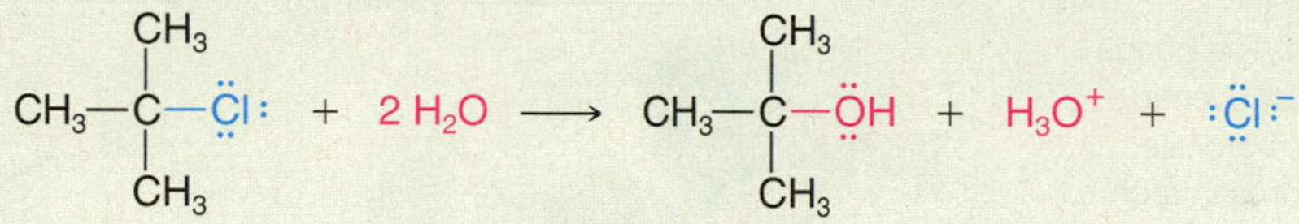

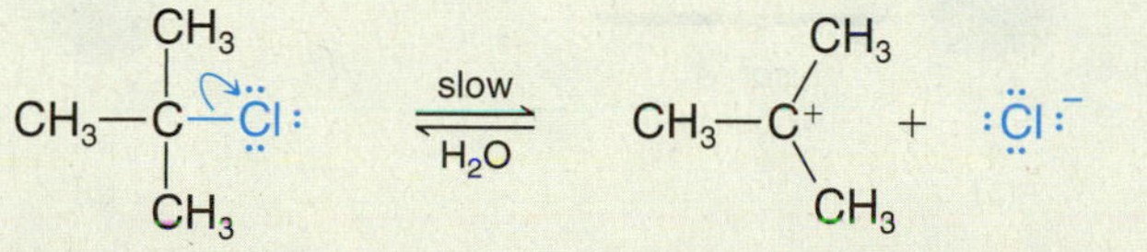

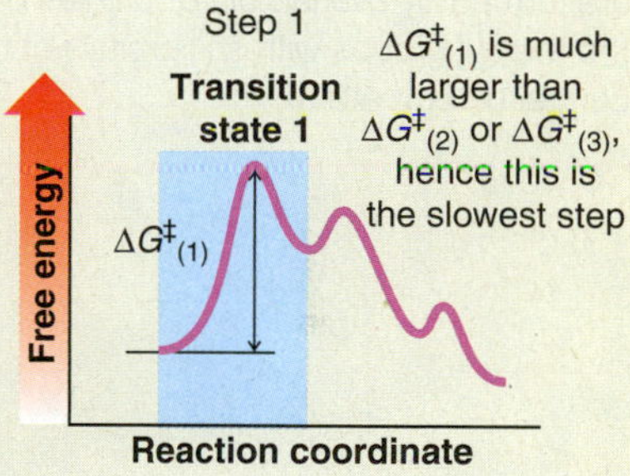

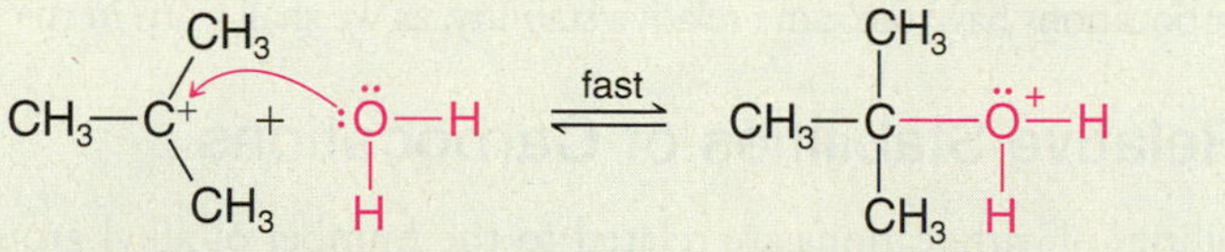

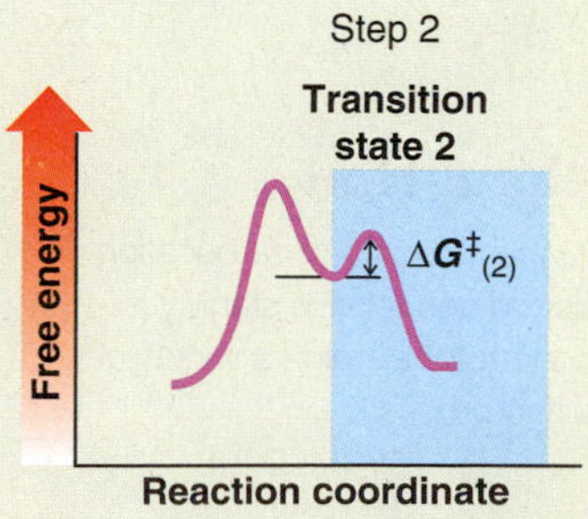

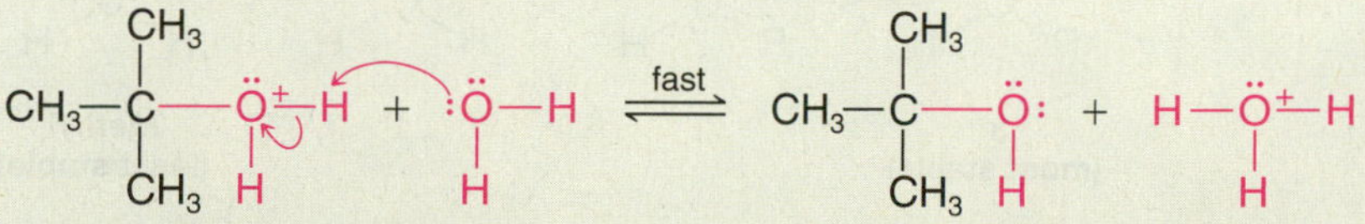

A water molecule acting as a Brønsted base accepts a proton from the *tert*-butyloxonium ion.

The products are *tert*-butyl alcohol and a hydronium ion.

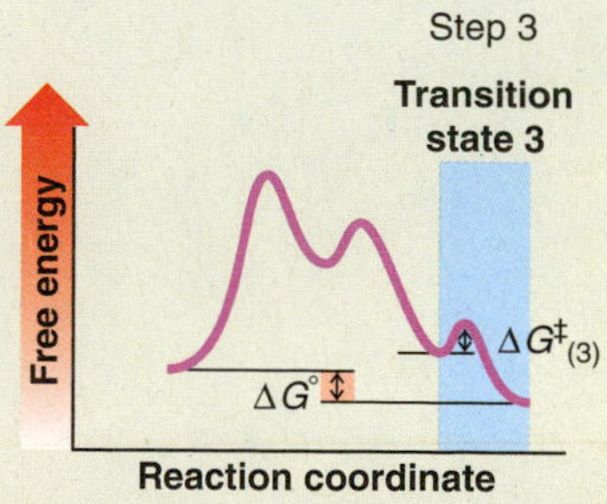

Step 3
Transition state 3
Free energy
$\Delta G^{\ddagger}{}_{(3)}$
$\Delta G°$
Reaction coordinate

6.11 CARBOCATIONS

 GEORGE A. OLAH was awarded the 1994 Nobel Prize in Chemistry.

Beginning in the 1920s much evidence began to accumulate implicating simple alkyl cations as intermediates in a variety of ionic reactions. However, because alkyl cations are highly unstable and highly reactive, they were, in all instances studied before 1962, very short-lived, transient species that could not be observed directly.* However, in 1962 George A. Olah (University of Southern California) and co-workers published the first of a series of papers describing experiments in which alkyl cations were prepared in an environment in which they were reasonably stable and in which they could be observed by a number of spectroscopic techniques.

6.11A The Structure of Carbocations

- Carbocations are trigonal planar.

Just as the trigonal planar structure of BF_3 (Section 1.16D) can be accounted for on the basis of sp^2 hybridization, so, too (Fig. 6.6), can the trigonal planar structure of carbocations.

FIGURE 6.6 *(a)* A stylized orbital structure of the methyl cation. The bonds are sigma (σ) bonds formed by overlap of the carbon atom's three sp^2 orbitals with the 1s orbitals of the hydrogen atoms. The p orbital is vacant. *(b)* A dashed line–wedge representation of the *tert*-butyl cation. The bonds between carbon atoms are formed by overlap of sp^3 orbitals of the methyl groups with sp^2 orbitals of the central carbon atom.

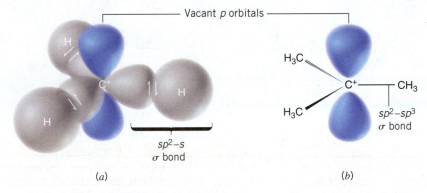

- The central carbon atom in a carbocation is electron deficient; it has only six electrons in its valence shell.

In our model (Fig. 6.6) these six electrons are used to form three sigma (σ) covalent bonds to hydrogen atoms or alkyl groups.

- The p orbital of a carbocation contains no electrons, but it can accept an electron pair when the carbocation undergoes further reaction.

Not all types of carbocations have the same relative stability, as we shall learn in the next section.

6.11B The Relative Stabilities of Carbocations

The relative stabilities of carbocations are related to the number of alkyl groups attached to the positively charged trivalent carbon.

- Tertiary carbocations are the most stable, and the methyl carbocation is the least stable.
- The overall order of stability is as follows:

$$\underset{\substack{3^\circ \\ \text{(most stable)}}}{R\overset{R}{\underset{R}{C^+}}} > \underset{2^\circ}{R\overset{R}{\underset{H}{C^+}}} > \underset{1^\circ}{R\overset{H}{\underset{H}{C^+}}} > \underset{\substack{\text{Methyl} \\ \text{(least stable)}}}{H\overset{H}{\underset{H}{C^+}}}$$

This order of carbocation stability can be explained on the basis of hyperconjugation.

- **Hyperconjugation** involves electron delocalization (via partial orbital overlap) from a filled bonding orbital to an adjacent unfilled orbital (Section 4.8).

---

*As we shall learn later, carbocations bearing aromatic groups can be much more stable; one of these had been studied as early as 1901.

Helpful Hint

An understanding of carbocation structure and relative stability is important for learning a variety of reaction processes.

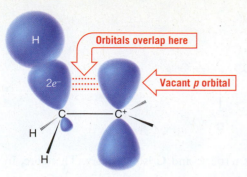

FIGURE 6.7 How an adjacent sigma bond helps stabilize the positive charge of a carbocation. Electron density from one of the carbon–hydrogen sigma bonds of the methyl group flows into the vacant *p* orbital of the carbocation because the orbitals can partly overlap. Shifting electron density in this way makes the sp^2-hybridized carbon of the carbocation somewhat less positive, and the hydrogens of the methyl group assume some of the positive charge. Delocalization (dispersal) of the charge in this way leads to greater stability. This interaction of a bond orbital with a *p* orbital is called hyperconjugation.

In the case of a carbocation, the unfilled orbital is the vacant *p* orbital of the carbocation, and the filled orbitals are C—H or C—C sigma bonds at the carbons *adjacent* to the *p* orbital of the carbocation. Sharing of electron density from adjacent C—H or C—C sigma bonds with the carbocation *p* orbital delocalizes the positive charge.

- Any time a charge can be dispersed or delocalized by hyperconjugation, inductive effects, or resonance, a system will be stabilized.

Figure 6.7 shows a stylized representation of hyperconjugation between a sigma bonding orbital and an adjacent carbocation *p* orbital.

Tertiary carbocations have three carbons with C—H bonds (or, depending on the specific example, C—C bonds instead of C—H) adjacent to the carbocation that can overlap partially with the vacant *p* orbital. Secondary carbocations have only two adjacent carbons with C—H or C—C bonds to overlap with the carbocation; hence, the possibility for hyperconjugation is less and the secondary carbocation is less stable. Primary carbocations have only one adjacent carbon from which to derive hyperconjugative stabilization, and so they are even less stable. A methyl carbocation has no possibility for hyperconjugation, and it is the least stable of all in this series. The following are specific examples:

tert-Butyl cation (3°) (most stable) is more stable than **Isopropyl cation (2°)** is more stable than **Ethyl cation (1°)** is more stable than **Methyl cation (least stable)**

In summary:

- **The relative stability of carbocations is $3° > 2° > 1° > $ methyl**.

This trend is also readily seen in electrostatic potential maps for these carbocations (Fig. 6.8).

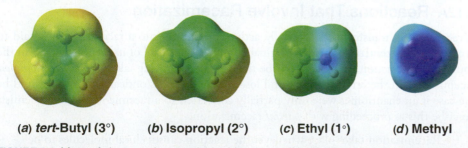

(a) tert-Butyl (3°)   **(b) Isopropyl (2°)**   **(c) Ethyl (1°)**   **(d) Methyl**

FIGURE 6.8 Maps of electrostatic potential for (a) tert-butyl (3°), (b) isopropyl (2°), (c) ethyl (1°), and (d) methyl carbocations show the trend from greater to lesser delocalization (stabilization) of the positive charge in these structures. Less blue color indicates greater delocalization of the positive charge. (The structures are mapped on the same scale of electrostatic potential to allow direct comparison.)

●●● **SOLVED PROBLEM 6.4**

Rank the following carbocations in order of increasing stability:

A   **B**   **C**

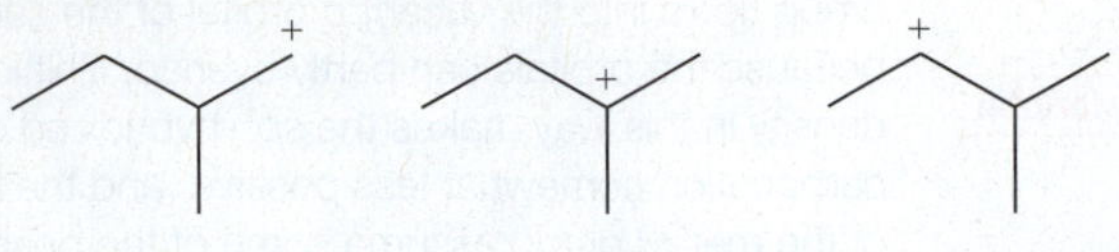

STRATEGY AND ANSWER: Structure **A** is a primary carbocation, **B** is tertiary, and **C** is secondary. Therefore, in order of increasing stability, **A < C < B**.

●●●

PRACTICE PROBLEM 6.7   Rank the following carbocations in order of increasing stability:

(a)   **(b)**   **(c)**

6.12 THE STEREOCHEMISTRY OF S_N1 REACTIONS

Because the carbocation formed in the first step of an S_N1 reaction has a trigonal planar structure (Section 6.11A), when it reacts with a nucleophile, it may do so from either the front side or the back side (see below). With the *tert*-butyl cation this makes no difference; since the *tert*-butyl group is not a chirality center, the same product is formed by either mode of attack. (Convince yourself of this result by examining models.)

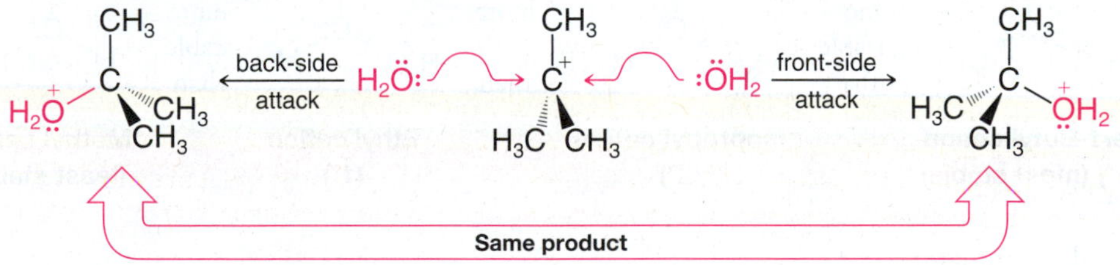

With some cations, however, stereoisomeric products arise from the two reaction possibilities. We shall study this point next.

6.12A Reactions That Involve Racemization

A reaction that transforms an optically active compound into a racemic form is said to proceed with **racemization**. If the original compound loses all of its optical activity in the course of the reaction, chemists describe the reaction as having taken place with *complete* racemization. If the original compound loses only part of its optical activity, as would be the case if an enantiomer were only partially converted to a racemic form, then chemists describe this as proceeding with *partial* racemization.

● Racemization takes place whenever the reaction causes chiral molecules to be converted to an achiral intermediate.

Examples of this type of reaction are S_N1 reactions in which the leaving group departs from a chirality center. These reactions almost always result in extensive and sometimes complete racemization. For example, heating optically active (S)-3-bromo-3-methylhexane with aqueous acetone results in the formation of 3-methyl-3-hexanol as a mixture of 50% (*R*) and 50% (*S*).

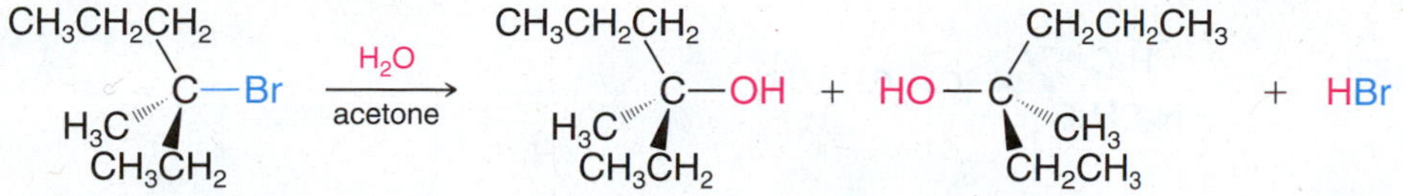

(S)-3-Bromo-3-methylhexane
(optically active)

50%
(S)-3-Methyl-3-hexanol

50%
(R)-3-Methyl-3-hexanol
(optically inactive, a racemic form)

The reason: the S$_N$1 reaction proceeds through the formation of an intermediate carbocation and the carbocation, because of its trigonal planar configuration, *is achiral*. It reacts with water at equal rates from either side to form the enantiomers of 3-methyl-3-hexanol in equal amounts.

[A MECHANISM FOR THE REACTION — The Stereochemistry of an S$_N$1 Reaction]

Reaction

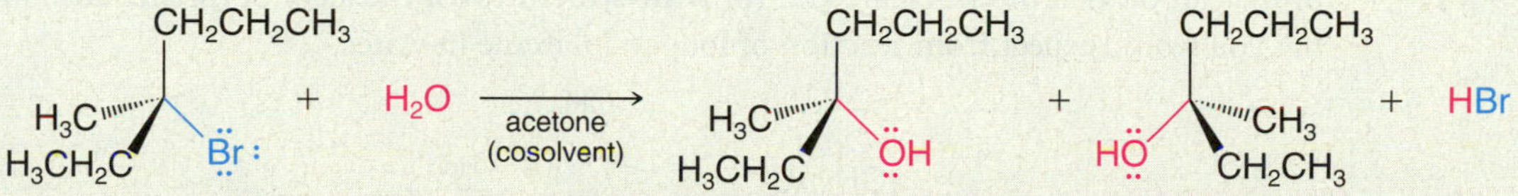

Mechanism

Step 1

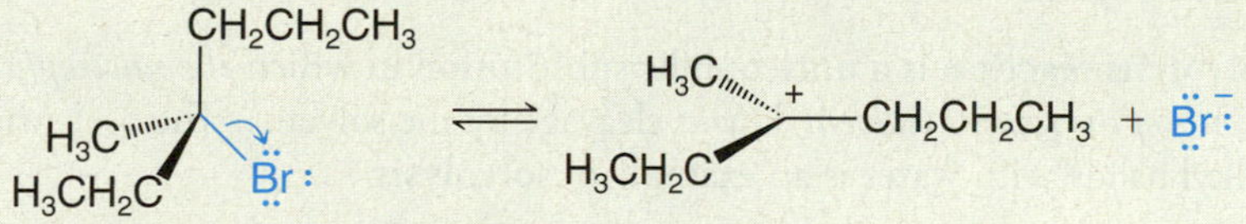

Departure of the leaving group (assisted by hydrogen bonding with water) leads to the carbocation.

Step 2

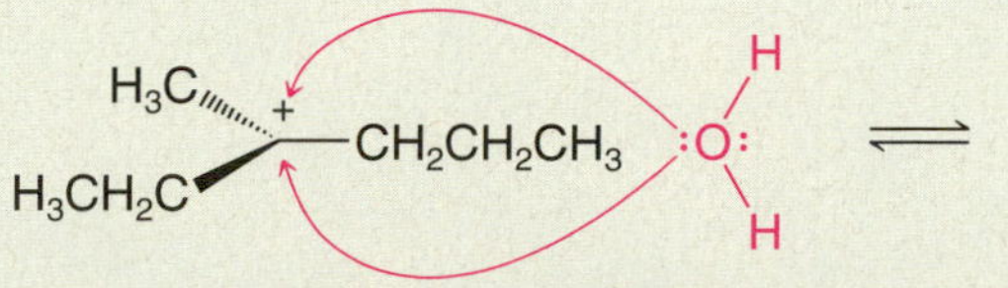

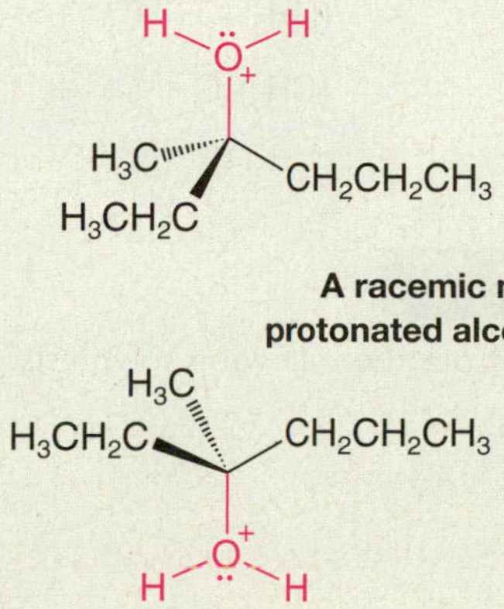

Attack at either face:
The carbocation is an achiral intermediate.
Because both faces of the carbocation are the same, the nucleophile can bond with either face to form a mixture of stereoisomers.

A racemic mixture of protonated alcohols results.

(mechanism continues on the next page)

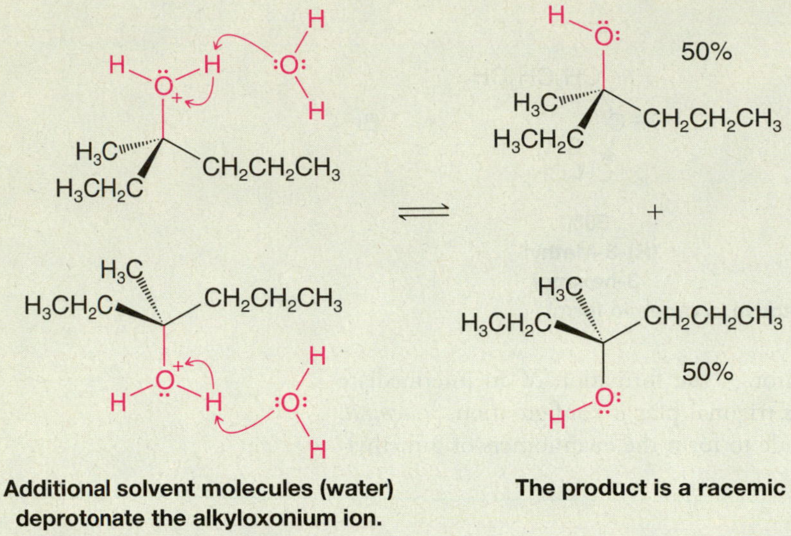

Additional solvent molecules (water) deprotonate the alkyloxonium ion.

The product is a racemic mixture.

The S$_N$1 reaction of (S)-3-bromo-3-methylhexane proceeds with racemization because the intermediate carbocation is achiral and attack by the nucleophile can occur from either side.

PRACTICE PROBLEM 6.8 Keeping in mind that carbocations have a trigonal planar structure, (a) write a structure for the carbocation intermediate and (b) write structures for the alcohol (or alcohols) that you would expect from reaction of iodocyclohexane in water:

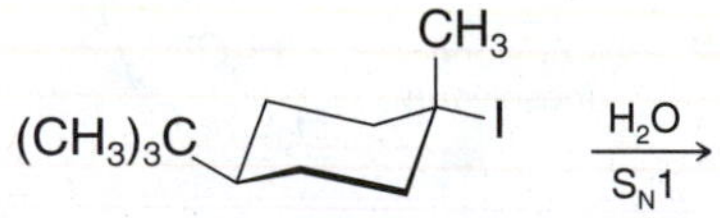

6.12B Solvolysis

- A **solvolysis reaction** is a nucleophilic substitution in which *the nucleophile is a molecule of the solvent* (*solvent* + *lysis*: cleavage by the solvent). The S$_N$1 reaction of an alkyl halide with water is an example of **solvolysis**.

When the solvent is water, we could also call the reaction a **hydrolysis**. If the reaction had taken place in methanol, we would call it a **methanolysis**.

Examples of Solvolysis

$$(CH_3)_3C—Br + H_2O \longrightarrow (CH_3)_3C—OH + HBr$$

$$(CH_3)_3C—Cl + CH_3OH \longrightarrow (CH_3)_3C—OCH_3 + HCl$$

SOLVED PROBLEM 6.5

What product(s) would you expect from the following solvolysis?

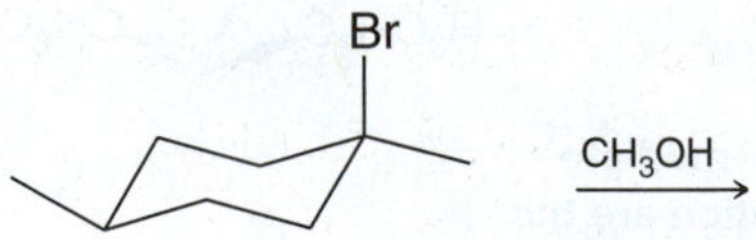

STRATEGY AND ANSWER: We observe that this cyclohexyl bromide is tertiary, and therefore in methanol it should lose a bromide ion to form a tertiary carbocation. Because the carbocation is trigonal planar at the positive carbon, it can react with a solvent molecule (methanol) to form two products.

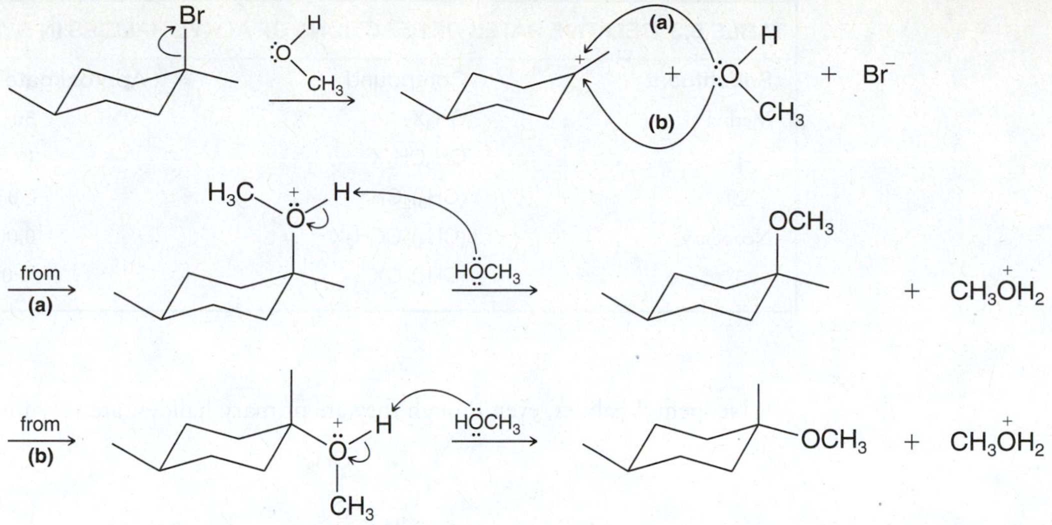

What product(s) would you expect from the methanolysis of the iodocyclohexane derivative given as the reactant in Practice Problem 6.8?

PRACTICE PROBLEM 6.9

6.13 FACTORS AFFECTING THE RATES OF S$_N$1 AND S$_N$2 REACTIONS

Now that we have an understanding of the mechanisms of S$_N$2 and S$_N$1 reactions, our next task is to explain why chloromethane reacts by an S$_N$2 mechanism and *tert*-butyl chloride by an S$_N$1 mechanism. We would also like to be able to predict which pathway—S$_N$1 or S$_N$2—would be followed by the reaction of any alkyl halide with any nucleophile under varying conditions.

The answer to this kind of question is to be found in the *relative rates of the reactions that occur*. If a given alkyl halide and nucleophile react *rapidly* by an S$_N$2 mechanism but *slowly* by an S$_N$1 mechanism under a given set of conditions, then an S$_N$2 pathway will be followed by most of the molecules. On the other hand, another alkyl halide and another nucleophile may react very slowly (or not at all) by an S$_N$2 pathway. If they react rapidly by an S$_N$1 mechanism, then the reactants will follow an S$_N$1 pathway.

- A number of factors affect the relative rates of S$_N$1 and S$_N$2 reactions. The most important factors are:

  1. the structure of the substrate,

  2. the concentration and reactivity of the nucleophile (for S$_N$2 reactions only),

  3. the effect of the solvent, and

  4. the nature of the leaving group.

6.13A The Effect of the Structure of the Substrate

S$_N$2 Reactions Simple alkyl halides show the following general order of reactivity in S$_N$2 reactions:

Methyl > primary > secondary >> (tertiary—unreactive)

Methyl halides react most rapidly, and tertiary halides react so slowly as to be unreactive by the S$_N$2 mechanism. Table 6.3 gives the relative rates of typical S$_N$2 reactions.

TABLE 6.3 RELATIVE RATES OF REACTIONS OF ALKYL HALIDES IN S_N2 REACTIONS

Substituent	Compound	Approximate Relative Rate
Methyl	CH_3X	30
1°	CH_3CH_2X	1
2°	$(CH_3)_2CHX$	0.03
Neopentyl	$(CH_3)_3CCH_2X$	0.00001
3°	$(CH_3)_3CX$	~0

Neopentyl halides, even though they are primary halides, are very unreactive:

A neopentyl halide

The important factor behind this order of reactivity is a steric effect, and in this case, steric hindrance.

- A **steric effect** is an effect on the relative rates caused by the space-filling properties of those parts of a molecule attached at or near the reacting site.
- **Steric hindrance** is when the spatial arrangement of atoms or groups at or near a reacting site of a molecule hinders or retards a reaction.

For particles (molecules and ions) to react, their reactive centers must be able to come within bonding distance of each other. Although most molecules are reasonably flexible, very large and bulky groups can often hinder the formation of the required transition state. In some cases they can prevent its formation altogether.

An S_N2 reaction requires an approach by the nucleophile to a distance within the bonding range of the carbon atom bearing the leaving group. Because of this, bulky substituents on *or near* that carbon atom have a dramatic inhibiting effect (Fig. 6.9). Nearby bulky groups cause the free energy of an S_N2 transition state to be higher and, consequently, the free energy of activation for the reaction is larger, and the rate of reaction is slower. Of the simple alkyl halides, methyl halides react most rapidly in S_N2 reactions because only three small hydrogen atoms interfere with the approaching nucleophile. Neopentyl and tertiary halides are the least reactive because bulky groups present a strong hindrance to the approaching nucleophile. (Tertiary substrates, for all practical purposes, do not react by an S_N2 mechanism.)

Helpful Hint

You can best appreciate the steric effects in these structures by building models.

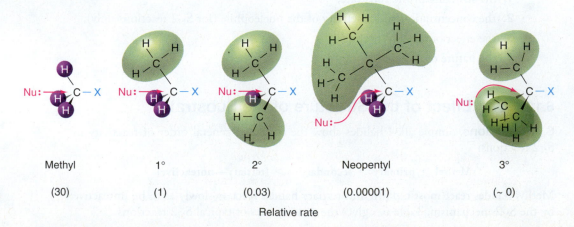

Methyl	1°	2°	Neopentyl	3°
(30)	(1)	(0.03)	(0.00001)	(~ 0)

Relative rate

FIGURE 6.9 Steric effects and relative rates in the S_N2 reaction.

SOLVED PROBLEM 6.6

Rank the following alkyl bromides in order of decreasing reactivity (from fastest to slowest) as a substrate in an S_N2 reaction.

A   B   C   D

STRATEGY AND ANSWER: We examine the carbon bearing the leaving group in each instance to assess the steric hindrance to an S_N2 reaction at that carbon. In **C** it is 3°; therefore, three groups would hinder the approach of a nucleophile, so this alkyl bromide would react most slowly. In **D** the carbon bearing the leaving group is 2° (two groups hinder the approach of the nucleophile), while in both **A** and **B** it is 1° (one group hinders the nucleophile's approach). Therefore, **D** would react faster than **C**, but slower than either **A** or **B**. But, what about **A** and **B**? They are both 1° alkyl bromides, but **B** has a methyl group on the carbon adjacent to the one bearing the bromine, which would provide steric hindrance to the approaching nucleophile that would not be present in **A**. The order of reactivity, therefore, is **A > B > D ≫ C**.

S_N1 Reactions

- The primary factor that determines the reactivity of organic substrates in an S_N1 reaction is the relative stability of the carbocation that is formed.

Of the simple alkyl halides that we have studied so far, this means (for all practical purposes) that only tertiary halides react by an S_N1 mechanism. (Later we shall see that certain organic halides, called *allylic halides* and *benzylic halides*, can also react by an S_N1 mechanism because they can form relatively stable carbocations; see Sections 13.4 and 15.15.)

Tertiary carbocations are stabilized because sigma bonds at three adjacent carbons contribute electron density to the carbocation *p* orbital by hyperconjugation (Section 6.11B). Secondary and primary carbocations have less stabilization by hyperconjugation. A methyl carbocation has no stabilization. Formation of a relatively stable carbocation is important in an S_N1 reaction because it means that the free energy of activation for the slow step of the reaction (e.g., $R—L \longrightarrow R^+ + L^-$) will be low enough for the reaction to take place at a reasonable rate.

PRACTICE PROBLEM 6.10

Which of the following alkyl halides is most likely to undergo substitution by an S_N1 mechanism?

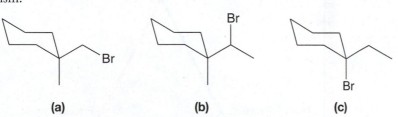

(a)   (b)   (c)

The Hammond–Leffler Postulate If you review the free-energy diagrams that accompany the mechanism for the S_N1 reaction of *tert*-butyl chloride and water (Section 6.10), you will see that step 1, the ionization of the leaving group to form the carbocation, is *uphill in terms of free energy* ($\Delta G°$ for this step is positive). It is also uphill in terms of enthalpy ($\Delta H°$ is also positive), and, therefore, this step is *endothermic*.

- According to the **Hammond–Leffler postulate**, **the transition-state structure for a step that is uphill in energy should show a strong resemblance to the structure of the product of that step.**

Since the product of this step (actually an intermediate in the overall reaction) is a carbocation, any factor that stabilizes the carbocation—such as dispersal of the positive charge by electron-releasing groups—should also stabilize the transition state in which the positive charge is developing.

Ionization of the Leaving Group

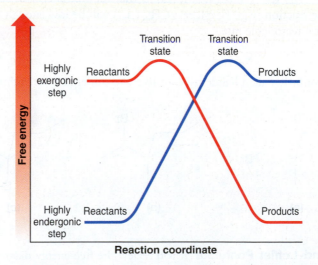

A methyl, primary, or secondary alkyl halide would have to ionize to form a methyl, primary, or secondary carbocation to react by an S_N1 mechanism. These carbocations, however, are much higher in energy than a tertiary carbocation, and the transition states leading to these carbocations are even higher in energy.

- The activation energy for an S_N1 reaction of a simple methyl, primary, or secondary halide is so large (therefore the reaction is so slow) that, for all practical purposes, an S_N1 reaction with a methyl, primary, or secondary halide does not compete with the corresponding S_N2 reaction.

The Hammond–Leffler postulate is quite general and can be better understood through consideration of Fig. 6.10. One way that the postulate can be stated is to say that *the structure of a transition state resembles the stable species that is nearest it in free energy.* For example, in a highly **endergonic** step (blue curve) the transition state lies close to the products in free energy, and we assume, therefore, that **it resembles the products of that step in structure**. Conversely, in a highly exergonic step (red curve) the transition state lies close to the reactants in free energy, and we assume **it resembles the reactants in structure** as well. The great value of the Hammond–Leffler postulate is that it gives us an intuitive way of visualizing those important, but fleeting, species that we call transition states. We shall make use of it in many future discussions.

FIGURE 6.10 The transition state for a highly exergonic step (red curve) lies close to and resembles the reactants. The transition state for an endergonic step (blue curve) lies close to and resembles the products of a reaction. (Reprinted with permission of the McGraw-Hill Companies from Pryor, W., *Free Radicals*, p. 156, copyright 1966.)

PRACTICE PROBLEM 6.11   The relative rates of ethanolysis of four primary alkyl halides are as follows: CH_3CH_2Br, 1.0; $CH_3CH_2CH_2Br$, 0.28; $(CH_3)_2CHCH_2Br$, 0.030; $(CH_3)_3CCH_2Br$, 0.00000042.

(a) Is each of these reactions likely to be S_N1 or S_N2?

(b) Provide an explanation for the relative reactivities that are observed.

6.13B The Effect of the Concentration and Strength of the Nucleophile

- The rate of an S_N1 reaction is unaffected by either the concentration or the identity of the nucleophile, because the nucleophile does not participate in the rate-determining step of an S_N1 reaction.

- The rate of an S_N2 reaction depends on *both* the concentration *and* the identity of the attacking nucleophile.

We saw in Section 6.5 how increasing the concentration of the nucleophile increases the rate of an S_N2 reaction. We can now examine how the rate of an S_N2 reaction depends on the identity of the nucleophile.

- The relative strength of a nucleophile (its **nucleophilicity**) is measured in terms of the relative rate of its S_N2 reaction with a given substrate.

A good nucleophile is one that reacts rapidly in an S_N2 reaction with a given substrate. A poor nucleophile is one that reacts slowly in an S_N2 reaction with the same substrate under comparable reaction conditions. (As mentioned above, we cannot compare nucleophilicities with regard to S_N1 reactions because the nucleophile does not participate in the rate-determining step of an S_N1 reaction.)

Methoxide anion, for example, is a good nucleophile for a substitution reaction with iodomethane. It reacts rapidly by an S_N2 mechanism to form dimethyl ether:

$$CH_3O^- + CH_3I \xrightarrow{\text{rapid}} CH_3OCH_3 + I^-$$

Methanol, on the other hand, is a poor nucleophile for reaction with iodomethane. Under comparable conditions it reacts very slowly. It is not a sufficiently powerful Lewis base (i.e., nucleophile) to cause displacement of the iodide leaving group at a significant rate:

$$CH_3OH + CH_3I \xrightarrow{\text{very slow}} CH_3\overset{+}{\underset{|}{O}}CH_3 + I^-$$
$$\qquad\qquad\qquad\qquad\qquad\;\; H$$

- The relative strengths of nucleophiles can be correlated with three structural features:

  1. **A negatively charged nucleophile is always a more reactive nucleophile than its conjugate acid**. Thus HO^- is a better nucleophile than H_2O and RO^- is better than ROH.

  2. **In a group of nucleophiles in which the nucleophilic atom is the same, nucleophilicities parallel basicities**. Oxygen compounds, for example, show the following order of reactivity:

     $$RO^- > HO^- \gg RCO_2^- > ROH > H_2O$$

     This is also their order of basicity. An alkoxide ion (RO^-) is a slightly stronger base than a hydroxide ion (HO^-), a hydroxide ion is a much stronger base than a carboxylate ion (RCO_2^-), and so on.

  3. **When the nucleophilic atoms are different, nucleophilicities may not parallel basicities**. For example, in protic solvents HS^-, $N\equiv C^-$, and I^- are all weaker bases than HO^-, yet they are **stronger nucleophiles** than HO^-.

     $$HS^- > N\equiv C^- > I^- > HO^-$$

Nucleophilicity versus Basicity While nucleophilicity and basicity are related, they are not measured in the same way.

- Basicity, as expressed by pK_a, is measured *by the position of an equilibrium* in an acid–base reaction.

- Nucleophilicity is measured *by the relative rates of substitution reactions*.

For example, the hydroxide ion (HO⁻) is a stronger base than a cyanide ion (N≡C⁻); at equilibrium it has the greater affinity for a proton (the pK_a of H_2O is ∼16, while the pK_a of HCN is ∼10). Nevertheless, cyanide ion is a stronger nucleophile; it reacts more rapidly with a carbon bearing a leaving group than does hydroxide ion.

PRACTICE PROBLEM 6.12 Rank the following in terms of *decreasing* nucleophilicity:

$$CH_3CO_2^- \qquad CH_3OH \qquad CH_3O^- \qquad CH_3CO_2H \qquad N≡C^-$$

6.13C Solvent Effects in S_N2 and S_N1 Reactions

- S_N2 reactions are favored by **polar aprotic solvents** (e.g., acetone, DMF, DMSO).
- S_N1 reactions are favored by **polar protic solvents** (e.g., EtOH, MeOH, H_2O).

Important reasons for these **solvent effects** have to do with (a) minimizing the solvent's interaction with the nucleophile in S_N2 reactions, and (b) facilitating ionization of the leaving group and stabilizing ionic intermediates by solvents in S_N1 reactions. In the following subsections we will explain these factors in further detail.

Polar Aprotic Solvents Favor S_N2 Reactions

- An aprotic solvent does not have hydrogen atoms that are capable of hydrogen bonding.
- Polar, aprotic solvents such as acetone, DMF, DMSO, and HMPA are often used alone or as co-solvents for S_N2 reactions.

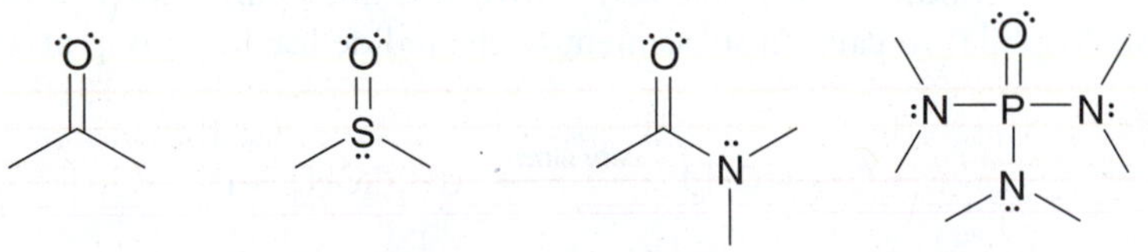

| Acetone | DMSO (Dimethylsulfoxide) | DMF (N,N-Dimethyl-formamide) | HMPA (Hexamethyl phosphoramide) |

- The rates of S_N2 reactions generally are vastly increased when they are carried out in polar aprotic solvents. The increase in rate can be as large as a millionfold.

Polar aprotic solvents solubilize cations well using their unshared electron pairs, but do not interact as strongly with anions because they cannot hydrogen bond with them and because the positive regions of the solvent are shielded by steric effects from the anion. This differential solvation leaves anions more free to act as nucleophiles because they are less encumbered by the cation and solvent, thus enhancing the rate of S_N2 reaction. Sodium ions of sodium iodide, for example, can be solvated by DMSO as shown here, leaving the iodide anion more free to act as a nucleophile.

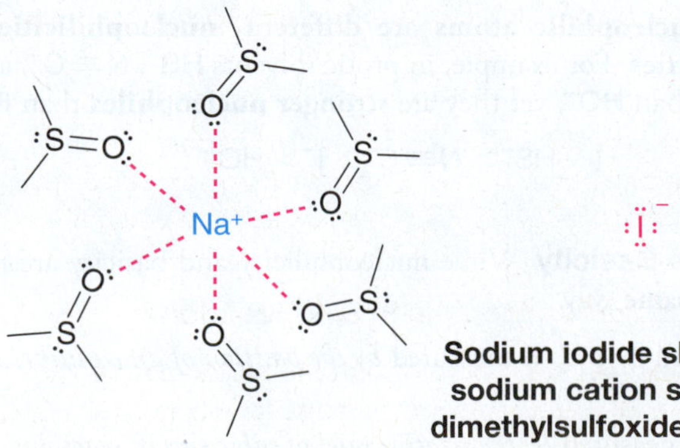

Sodium iodide showing the sodium cation solvated by dimethylsulfoxide molecules.

"Naked" anions in polar aprotic solvents are also more reactive as bases, as well as in their capacity as nucleophiles. In DMSO, for example, the relative order of halide ion basicity is the same as their relative order of nucleophilicity. Halide basicity is opposite to nucleophilicity in protic solvents, however, as we shall explain shortly.

$$F^- \; > \; Cl^- \; > \; Br^- \; > \; I^-$$

Halide nucleophilicity in aprotic solvents

Polar Protic Solvents Favor S$_N$1 Reactions

- A protic solvent has at least one hydrogen atom capable of participating in a hydrogen bond.
- Protic solvents, such as water, EtOH, and MeOH, facilitate formation of a carbocation by forming hydrogen bonds with the leaving group as it departs, thereby lowering the energy of the transition state leading to a carbocation.

> **Helpful Hint**
> Polar aprotic solvents increase S$_N$2 rates.

| **Hydrogen bonding with the substrate** | **Departure of the leaving group is assisted by hydrogen bonding in the transition state** | **Carbocation intermediate** | **Solvated leaving group** |

A rough indication of a solvent's polarity is a quantity called the **dielectric constant**. The dielectric constant is a measure of the solvent's ability to insulate opposite charges (or separate ions) from each other. Electrostatic attractions and repulsions between ions are smaller in solvents with higher dielectric constants. Table 6.4 gives the dielectric constants of some common solvents.

TABLE 6.4 DIELECTRIC CONSTANTS OF COMMON SOLVENTS

	Solvent	Formula	Dielectric Constant
	Water	H_2O	80
	Formic acid	HCO_2H	59
	Dimethyl sulfoxide (DMSO)	CH_3SOCH_3	49
Increasing solvent polarity	N,N-Dimethylformamide (DMF)	$HCON(CH_3)_2$	37
	Acetonitrile	$CH_3C{\equiv}N$	36
	Methanol	CH_3OH	33
	Hexamethylphosphoramide (HMPA)	$[(CH_3)_2N]_3P{=}O$	30
	Ethanol	CH_3CH_2OH	24
	Acetone	CH_3COCH_3	21
	Acetic acid	CH_3CO_2H	6

Water is the most effective solvent for promoting ionization, but most organic compounds do not dissolve appreciably in water. They usually dissolve, however, in alcohols, and quite often mixed solvents are used. Methanol—water and ethanol—water are common mixed solvents for nucleophilic substitution reactions.

Protic Solvents Hinder the Nucleophile in S$_N$2 Reactions

A solvated nucleophile must shed some of its solvent molecules to react with the substrate. In a polar aprotic solvent, the nucleophile is less unencumbered by solvent molecules because hydrogen bonding between the solvent and the nucleophile is not possible.

- Hydrogen bonding with a protic solvent such as water, EtOH, or MeOH, encumbers a nucleophile and hinders its reactivity in a nucleophilic substitution reaction.

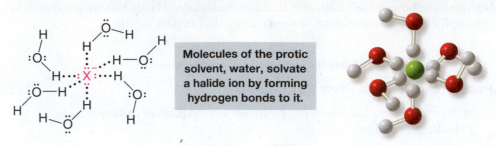

> Molecules of the protic solvent, water, solvate a halide ion by forming hydrogen bonds to it.

- The extent of hydrogen bonding with the nucleophile varies with the identity of the nucleophile. Hydrogen bonding with a small nucleophilic atom is stronger than to a larger nucleophilic atom among elements in the same group (column) of the periodic table.

For example, fluoride anion is more strongly solvated than the other halides because it is the smallest halide anion and its charge is the most concentrated. Hence, in a protic solvent fluoride is not as effective a nucleophile as the other halide anions. Iodide is the largest halide anion and it is the most weakly solvated in a protic solvent; hence, it is the strongest nucleophile among the halide anions.

- In a protic solvent, the general trend in *nucleophilicity* among the halide anions is as follows:

$$I^- > Br^- > Cl^- > F^-$$

Halide nucleophilicity in protic solvents

The same effect holds true when we compare sulfur nucleophiles with oxygen nucleophiles. Sulfur atoms are larger than oxygen atoms and hence they are not solvated as strongly in a protic solvent. Thus, thiols (R—SH) are stronger nucleophiles than alcohols, and RS$^-$ anions are better nucleophiles than RO$^-$ anions.

The greater reactivity of nucleophiles with large nucleophilic atoms is not entirely related to solvation. Larger atoms have greater **polarizability** (their electron clouds are more easily distorted); therefore, a larger nucleophilic atom can donate a greater degree of electron density to the substrate than a smaller nucleophile whose electrons are more tightly held.

The relative nucleophilicities of some common nucleophiles in protic solvents are as follows:

$$HS^- > N{\equiv}C^- > I^- > HO^- > N_3^- > Br^- > CH_3CO_2^- > Cl^- > F^- > H_2O$$

Relative nucleophilicity in protic solvents

• • •

PRACTICE PROBLEM 6.13   Rank the following in terms of decreasing nucleophilicity in a protic solvent.

$$CH_3CO_2^- \quad CH_3O^- \quad CH_3S^- \quad CH_3SH \quad CH_3OH$$

• • •

PRACTICE PROBLEM 6.14   Classify the following solvents as being protic or aprotic: formic acid, HCO_2H; acetone, CH_3COCH_3; acetonitrile, $CH_3C{\equiv}N$; formamide, $HCONH_2$; sulfur dioxide, SO_2; ammonia, NH_3; trimethylamine, $N(CH_3)_3$; ethylene glycol, $HOCH_2CH_2OH$.

• • •

PRACTICE PROBLEM 6.15   Would you expect the reaction of propyl bromide with sodium cyanide (NaCN), that is,

$$CH_3CH_2CH_2Br + NaCN \longrightarrow CH_3CH_2CH_2CN + NaBr$$

to occur faster in DMF or in ethanol? Explain your answer.

PRACTICE PROBLEM 6.16

Which would you expect to be the stronger nucleophile in a polar aprotic solvent?
(a) $CH_3CO_2^-$ or CH_3O^- **(b)** H_2O or H_2S **(c)** $(CH_3)_3P$ or $(CH_3)_3N$

PRACTICE PROBLEM 6.17

When *tert*-butyl bromide undergoes solvolysis in a mixture of methanol and water, the rate of solvolysis (measured by the rate at which bromide ions form in the mixture) *increases* when the percentage of water in the mixture is increased. **(a)** Explain this occurrence. **(b)** Provide an explanation for the observation that the rate of the S$_N$2 reaction of ethyl chloride with potassium iodide in methanol and water *decreases* when the percentage of water in the mixture is increased.

6.13D The Nature of the Leaving Group

- Leaving groups depart with the electron pair that was used to bond them to the substrate.
- The best leaving groups are those that become either a relatively stable anion or a neutral molecule when they depart.

> **Helpful Hint**
> Good leaving groups are weak bases.

First, let us consider leaving groups that become anions when they separate from the substrate. Because weak bases stabilize a negative charge effectively, leaving groups that become weak bases are good leaving groups.

The reason that stabilization of the negative charge is important can be understood by considering the structure of the transition states. In either an S$_N$1 or S$_N$2 reaction the leaving group begins to acquire a negative charge as the transition state is reached:

S$_N$1 Reaction (Rate-Limiting Step)

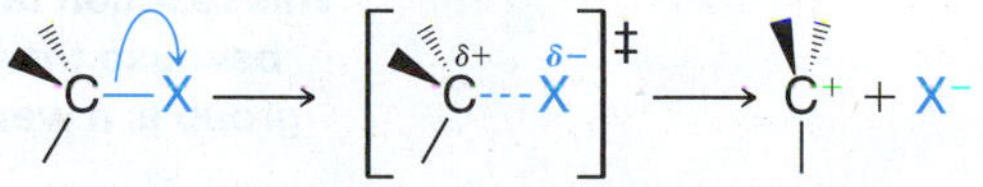

Transition state

S$_N$2 Reaction

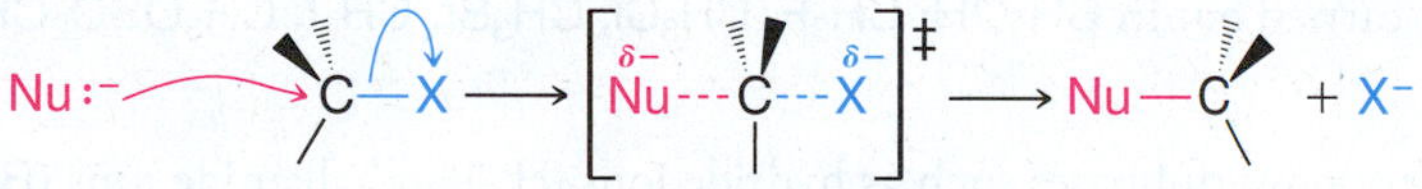

Transition state

Stabilization of this developing negative charge at the leaving group stabilizes the transition state (lowers its free energy); this lowers the free energy of activation and thereby increases the rate of the reaction.

- Among the halogens, an iodide ion is the best leaving group and a fluoride ion is the poorest:

$$I^- > Br^- > Cl^- \gg F^-$$

The order is the opposite of the basicity in an aprotic solvent:

$$F^- \gg Cl^- > Br^- > I^-$$

Other weak bases that are good leaving groups, which we shall study later, are alkanesulfonate ions, alkyl sulfate ions, and the *p*-toluenesulfonate ion:

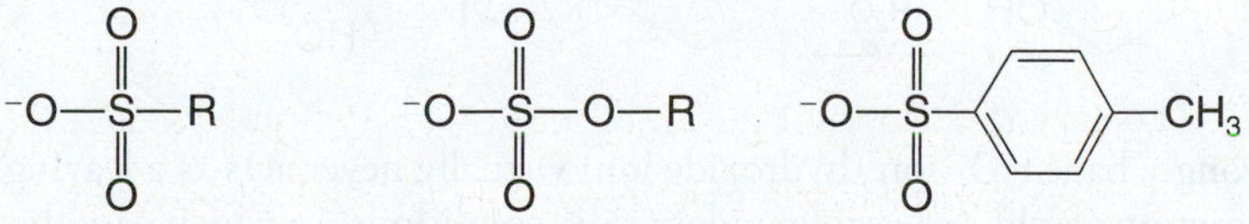

An alkanesulfonate ion   **An alkyl sulfate ion**   **p-Toluenesulfonate ion**

These anions are all the conjugate bases of very strong acids.

The trifluoromethanesulfonate ion ($CF_3SO_3^-$, commonly called the **triflate ion**) is one of the best leaving groups known to chemists. It is the conjugate base of CF_3SO_3H, an exceedingly strong acid ($pK_a \sim -5$ to -6):

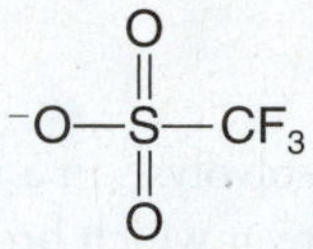

$$^-O-\overset{\overset{\displaystyle O}{\|}}{\underset{\underset{\displaystyle O}{\|}}{S}}-CF_3$$

Triflate ion
(a "super" leaving group)

- Strongly basic ions rarely act as leaving groups.

The hydroxide ion, for example, is a strong base and thus reactions like the following do not take place:

$$Nu:^- \longrightarrow R\overset{\frown}{-}\overset{..}{O}-H \quad \times \longrightarrow \quad R-Nu + :\overset{..}{O}-H$$

This reaction does not take place because the leaving group is a strongly basic hydroxide ion.

However, when an alcohol is dissolved in a strong acid, it can undergo substitution by a nucleophile. Because the acid protonates the —OH group of the alcohol, the leaving group no longer needs to be a hydroxide ion; it is now a molecule of water—a much weaker base than a hydroxide ion and a good leaving group:

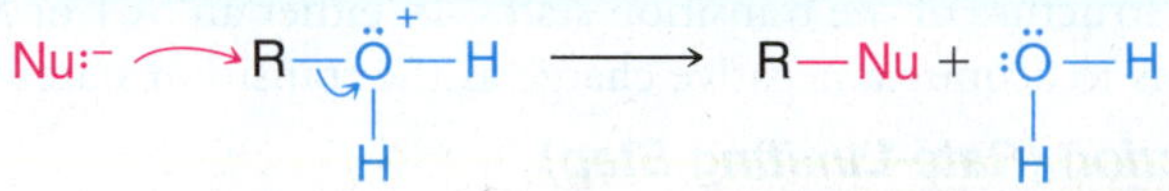

$$Nu:^- \longrightarrow R\overset{\frown}{-}\overset{+}{\underset{\underset{\displaystyle H}{|}}{\overset{..}{O}}}-H \longrightarrow R-Nu + :\overset{..}{\underset{\underset{\displaystyle H}{|}}{O}}-H$$

This reaction takes place because the leaving group is a weak base.

•••

PRACTICE PROBLEM 6.18   List the following compounds in order of decreasing reactivity toward CH_3O^- in an S_N2 reaction carried out in CH_3OH: CH_3F, CH_3Cl, CH_3Br, CH_3I, $CH_3OSO_2CF_3$, $^{14}CH_3OH$.

- Very powerful bases such as hydride ions ($H:^-$) and alkanide ions ($R:^-$) virtually never act as leaving groups.

Therefore, **reactions such as the following are not feasible**:

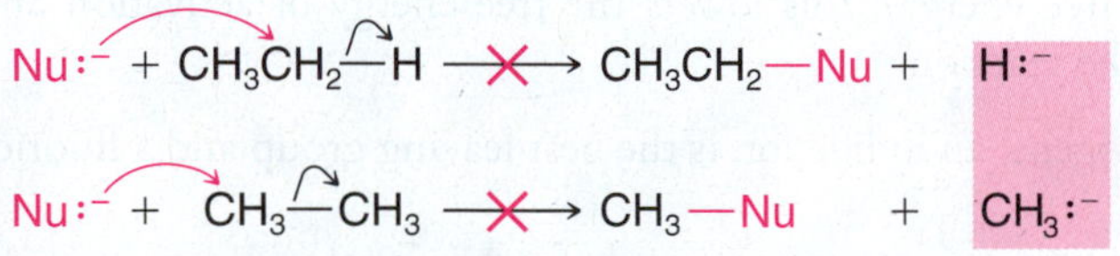

$$Nu:^- + CH_3CH_2\overset{\frown}{-}H \quad \times \longrightarrow \quad CH_3CH_2-Nu + \boxed{H:^-}$$

$$Nu:^- + CH_3\overset{\frown}{-}CH_3 \quad \times \longrightarrow \quad CH_3-Nu \quad + \quad \boxed{CH_3:^-}$$

These are not leaving groups.

Remember: The best leaving groups are weak bases after they depart.

••• **SOLVED PROBLEM 6.7**

Explain why the following reaction is not feasible as a synthesis of butyl iodide.

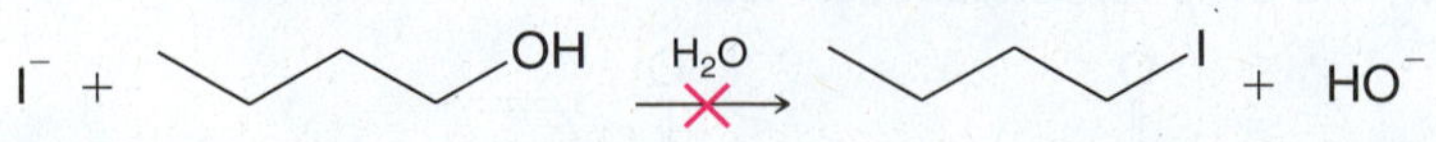

$$I^- + \qquad \diagdown\diagup\diagdown\diagup OH \quad \xrightarrow[\times]{H_2O} \quad \diagdown\diagup\diagdown\diagup I \quad + \quad HO^-$$

STRATEGY AND ANSWER: The strongly basic HO^- ion (hydroxide ion) virtually never acts as a leaving group, something this reaction would require. This reaction would be feasible under acidic conditions, in which case the leaving group would be a water molecule.

SUMMARY OF S$_N$1 VERSUS S$_N$2 REACTIONS

Helpful Hint

S$_N$1 versus S$_N$2.

S$_N$1: The Following Conditions Favor an S$_N$1 Reaction

1. A substrate that can form a relatively stable carbocation (such as a substrate with a leaving group at a tertiary position)
2. A relatively weak nucleophile
3. A polar, protic solvent such as EtOH, MeOH, or H$_2$O

The S$_N$1 mechanism is, therefore, important in solvolysis reactions of tertiary alkyl halides, especially when the solvent is highly polar. In a solvolysis reaction the nucleophile is weak because it is a neutral molecule (of the polar protic solvent) rather than an anion.

S$_N$2: The Following Conditions Favor an S$_N$2 Reaction

1. A substrate with a relatively unhindered leaving group (such as a methyl, primary, or secondary alkyl halide). The order of reactivity is

$$CH_3-X \; > \; R-CH_2-X \; > \; R-\overset{\displaystyle R}{\underset{\textstyle }{CH}}-X$$

$$\textbf{Methyl} \; > \; \textbf{1°} \; > \; \textbf{2°}$$

Tertiary halides do not react by an S$_N$2 mechanism.

2. A strong nucleophile (usually negatively charged)
3. High concentration of the nucleophile
4. A polar, aprotic solvent

The trend in reaction rate for a halogen as the leaving group is the same in S$_N$1 and S$_N$2 reactions:

$$R-I > R-Br > R-Cl \qquad S_N1 \text{ or } S_N2$$

Because alkyl fluorides react so slowly, they are seldom used in nucleophilic substitution reactions.

These factors are summarized in Table 6.5.

TABLE 6.5 FACTORS FAVORING S$_N$1 VERSUS S$_N$2 REACTIONS

Factor	S$_N$1	S$_N$2
Substrate	3° (requires formation of a relatively stable carbocation)	Methyl > 1° > 2° (requires unhindered substrate)
Nucleophile	Weak Lewis base, neutral molecule, nucleophile may be the solvent (solvolysis)	Strong Lewis base, rate increased by high concentration of nucleophile
Solvent	Polar protic (e.g., alcohols, water)	Polar aprotic (e.g., DMF, DMSO)
Leaving group	I > Br > Cl > F for both S$_N$1 and S$_N$2 (the weaker the base after the group departs, the better the leaving group)	

6.14 ORGANIC SYNTHESIS: FUNCTIONAL GROUP TRANSFORMATIONS USING S$_N$2 REACTIONS

S$_N$2 reactions are highly useful in organic synthesis because they enable us to convert one functional group into another—a process that is called a **functional group transformation** or a **functional group interconversion**. With the S$_N$2 reactions shown in Fig. 6.11, methyl, primary, or secondary alkyl halides can be transformed into alcohols, ethers, thiols, thioethers, nitriles, esters, and so on. (*Note*: The use of the prefix *thio-* in a name means that a sulfur atom has replaced an oxygen atom in the compound.)

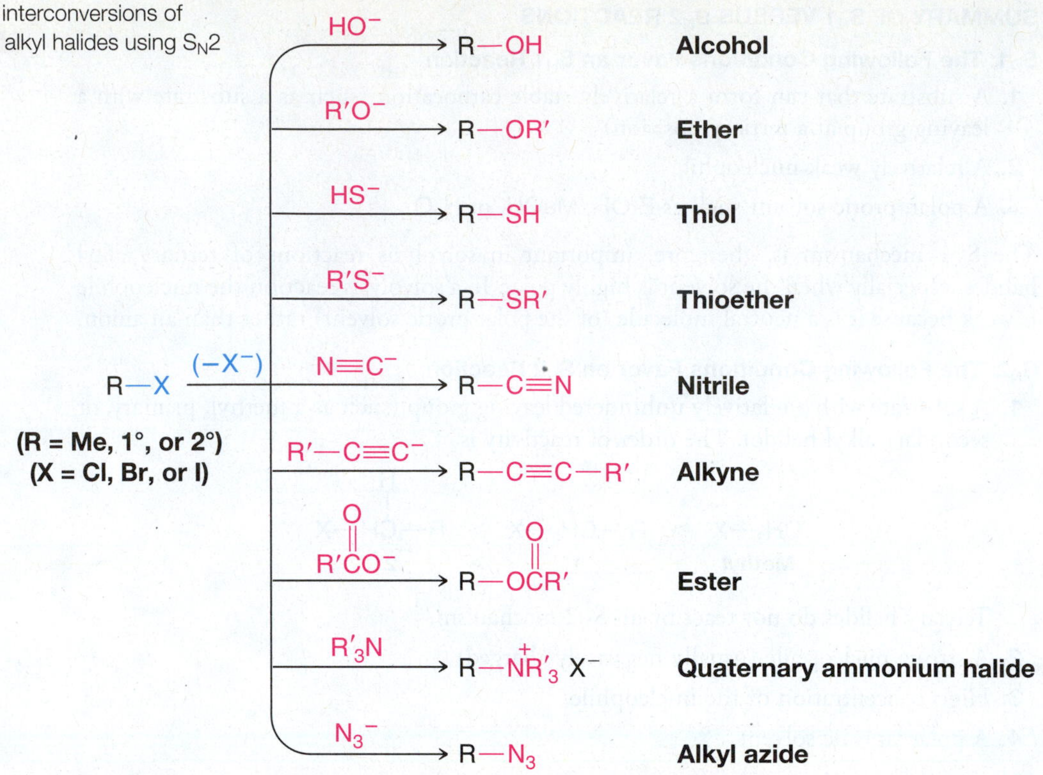
R—X
(−X⁻)
(R = Me, 1°, or 2°)
(X = Cl, Br, or I)
HO⁻
R—OH
Alcohol
R′O⁻
R—OR′
Ether
HS⁻
R—SH
Thiol
R′S⁻
R—SR′
Thioether
N≡C⁻
R—C≡N
Nitrile
R′—C≡C⁻
R—C≡C—R′
Alkyne
R′CO⁻
R—OCR′
Ester
R′3N
R—NR′3 X⁻
Quaternary ammonium halide
N3⁻
R—N3
Alkyl azide

FIGURE 6.11 Functional group interconversions of methyl, primary, and secondary alkyl halides using S_N2 reactions.

Alkyl chlorides and bromides are also easily converted to alkyl iodides by nucleophilic substitution reactions.

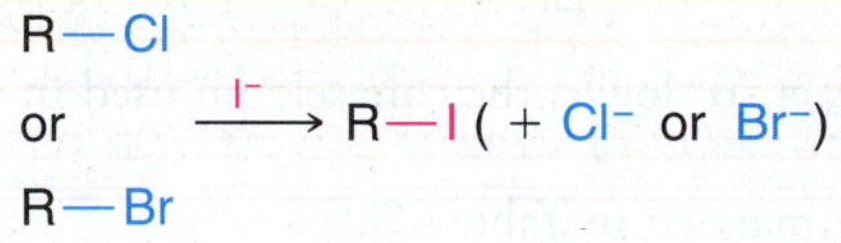
R—Cl
or
R—Br
I⁻
R—I (+ Cl⁻ or Br⁻)

One other aspect of the S_N2 reaction that is of great importance is **stereochemistry** (Section 6.8). S_N2 reactions always occur with **inversion of configuration** at the atom that bears the leaving group. This means that when we use S_N2 reactions in syntheses we can be sure of the configuration of our product if we know the configuration of our reactant. For example, suppose we need a sample of the following nitrile with the (*S*) configuration:

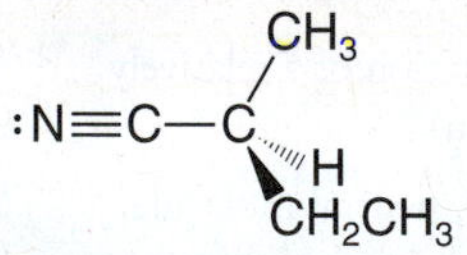

(*S*)-2-Methylbutanenitrile

If we have available (*R*)-2-bromobutane, we can carry out the following synthesis:

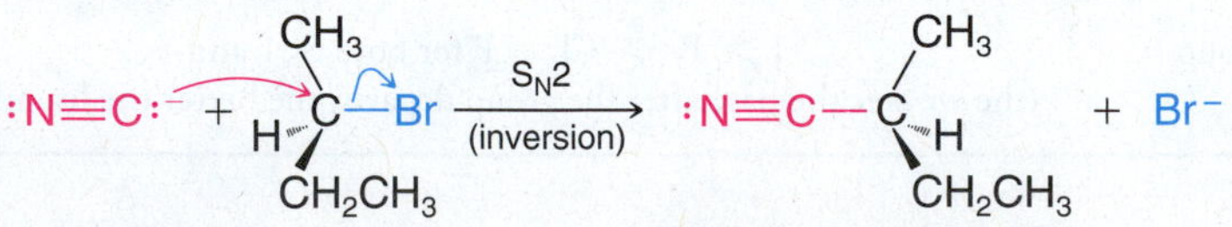
SN2
(inversion)
+ Br⁻

(*R*)-2-Bromobutane **(*S*)-2-Methylbutanenitrile**

PRACTICE PROBLEM 6.19 Starting with (*S*)-2-bromobutane, outline syntheses of each of the following compounds:

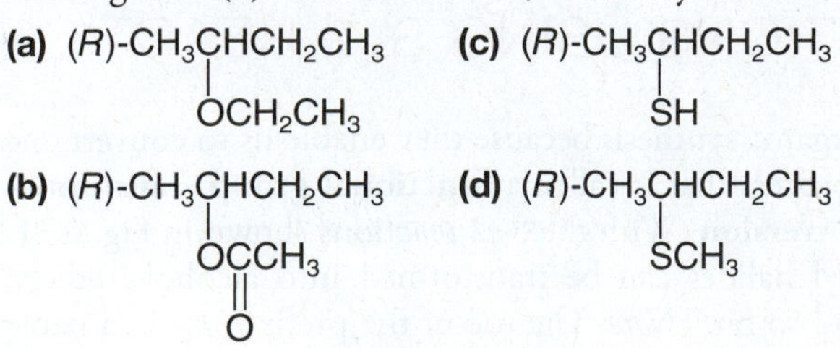
(a) (R)-CH3CHCH2CH3
OCH2CH3
(b) (R)-CH3CHCH2CH3
OCCH3
O
(c) (R)-CH3CHCH2CH3
SH
(d) (R)-CH3CHCH2CH3
SCH3

THE CHEMISTRY OF...

Biological Methylation: A Biological Nucleophilic Substitution Reaction

The cells of living organisms synthesize many of the compounds they need from smaller molecules. Often these biosyntheses resemble the syntheses organic chemists carry out in their laboratories. Let us examine one example now.

Many reactions taking place in the cells of plants and animals involve the transfer of a methyl group from an amino acid called methionine to some other compound. That this transfer takes place can be demonstrated experimentally by feeding a plant or animal methionine containing an isotopically labeled carbon atom (e.g., ^{13}C or ^{14}C) in its methyl group. Later, other compounds containing the "labeled" methyl group can be isolated from the organism. Some of the compounds that get their methyl groups from methionine are the following. The isotopically labeled carbon atom is shown in green.

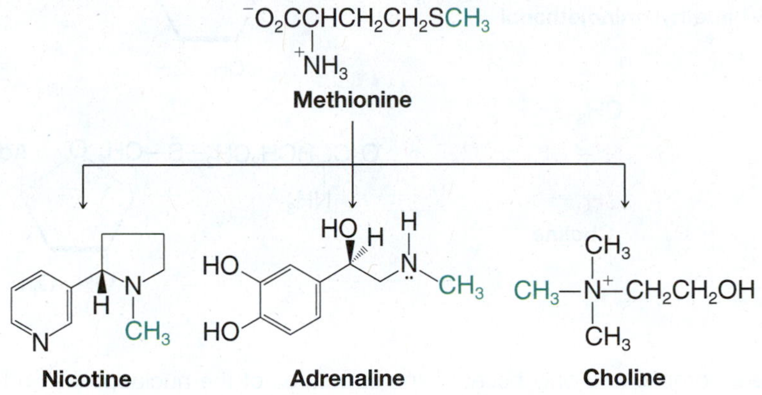

$$^-O_2CCHCH_2CH_2SCH_3$$
$$\overset{|}{\overset{+}{N}H_3}$$

Methionine

Nicotine

Adrenaline

Choline

Choline is important in the transmission of nerve impulses, adrenaline causes blood pressure to increase, and nicotine is the compound contained in tobacco that makes smoking tobacco addictive. (In large doses nicotine is poisonous.)

The transfer of the methyl group from methionine to these other compounds does not take place directly. The actual methylating agent is not methionine; it is *S*-adenosylmethionine,* a compound that results when methionine reacts with adenosine triphosphate (ATP):

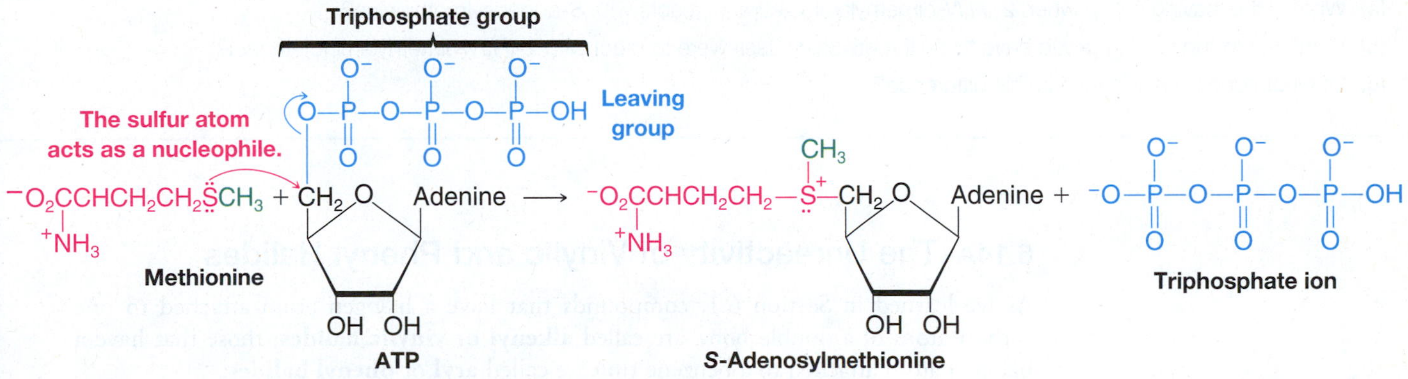

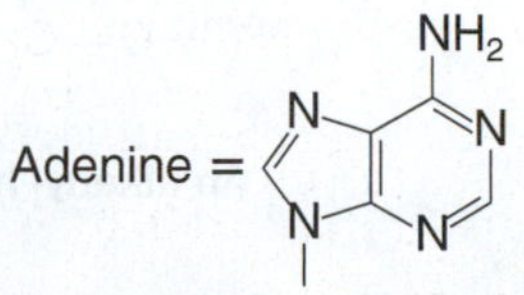

Triphosphate group

The sulfur atom acts as a nucleophile.

Leaving group

$$^-O_2CCHCH_2CH_2\overset{..}{\underset{\overset{|}{\overset{+}{N}H_3}}{S}}CH_3 + CH_2O\text{—Adenine} \longrightarrow {}^-O_2CCHCH_2CH_2\overset{CH_3}{\underset{\overset{|}{\overset{+}{N}H_3}}{\overset{|}{\overset{+}{S}}}}CH_2O\text{—Adenine} + {}^-O\text{—P—O—P—O—P—OH}$$

Methionine

ATP

OH OH

S-Adenosylmethionine

OH OH

Triphosphate ion

Adenine =

---

*The prefix *S* is a locant meaning "on the sulfur atom" and should not be confused with the (*S*) used to define absolute configuration. Another example of this kind of locant is *N*, meaning "on the nitrogen atom."

(continues on next page)

This reaction is a nucleophilic substitution reaction. The nucleophilic atom is the sulfur atom of methionine. The leaving group is the weakly basic triphosphate group of ATP. The product, *S*-adenosylmethionine, contains a methyl-sulfonium group,

$$CH_3-\overset{|}{\underset{|}{S}}{}^+-.$$

S-Adenosylmethionine then acts as the substrate for other nucleophilic substitution reactions. In the biosynthesis of choline, for example, it transfers its methyl group to a nucleophilic nitrogen atom of 2-(*N,N*-dimethylamino)ethanol:

2-(*N,N*-Dimethylamino)ethanol

Choline

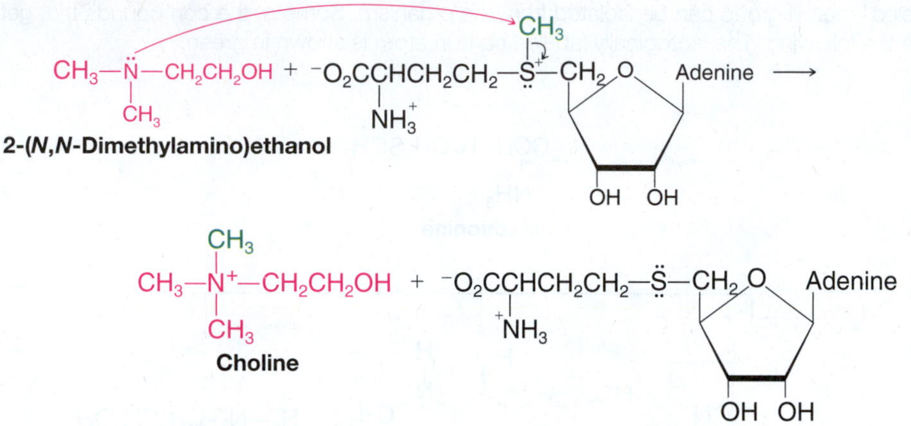

These reactions appear complicated only because the structures of the nucleophiles and substrates are complex. Yet conceptually they are simple, and they illustrate many of the principles we have encountered thus far in Chapter 6. In them we see how nature makes use of the high nucleophilicity of sulfur atoms. We also see how a weakly basic group (e.g., the triphosphate group of ATP) functions as a leaving group. In the reaction of 2-(*N,N*-dimethylamino) ethanol we see that the more basic $(CH_3)_2N-$ group acts as the nucleophile rather than the less basic $-OH$ group. And when a nucleophile attacks *S*-adenosylmethionine, we see that the attack takes place at the less hindered CH_3- group rather than at one of the more hindered $-CH_2-$ groups.

Study Problem

(a) What is the leaving group when 2-(*N,N*-dimethylamino)ethanol reacts with *S*-adenosylmethionine?

(b) What would the leaving group have to be if methionine itself were to react with 2-(*N,N*-dimethylamino)ethanol?

(c) Of what special significance is this difference?

6.14A The Unreactivity of Vinylic and Phenyl Halides

As we learned in Section 6.1, compounds that have a halogen atom attached to one carbon atom of a double bond are called **alkenyl** or **vinylic halides**; those that have a halogen atom attached to a benzene ring are called **aryl** or **phenyl halides**:

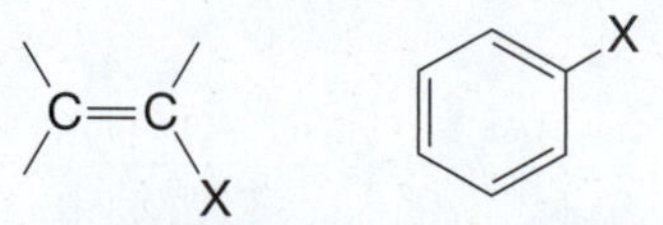

An alkenyl halide **A phenyl halide**

- Alkenyl and phenyl halides are generally unreactive in S_N1 or S_N2 reactions.

They are unreactive in S_N1 reactions because alkenyl and phenyl cations are relatively unstable and do not form readily. They are unreactive in S_N2 reactions because the carbon–halogen bond of an alkenyl or phenyl halide is stronger than that of an alkyl halide (we shall see why later), and the electrons of the double bond or benzene ring repel the approach of a nucleophile from the back side.

6.15 ELIMINATION REACTIONS OF ALKYL HALIDES

Elimination reactions of alkyl halides are important reactions that compete with substitution reactions. In an **elimination reaction** the fragments of some molecule (YZ) are removed (eliminated) from adjacent atoms of the reactant. This elimination leads to the creation of a multiple bond:

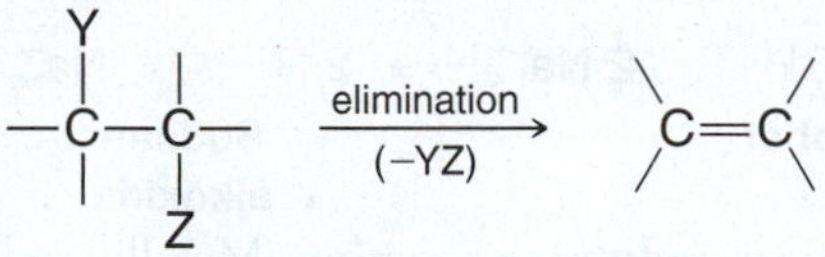
Y
Z
elimination
(−YZ)
C=C

6.15A Dehydrohalogenation

A widely used method for synthesizing alkenes is the elimination of HX from adjacent atoms of an alkyl halide. Heating the alkyl halide with a strong base causes the reaction to take place. The following are two examples:

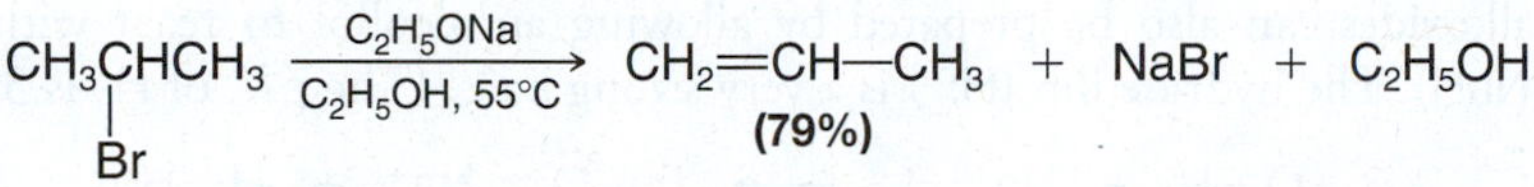
CH3CHCH3
Br
C2H5ONa
C2H5OH, 55°C
CH2=CH—CH3 + NaBr + C2H5OH
(79%)

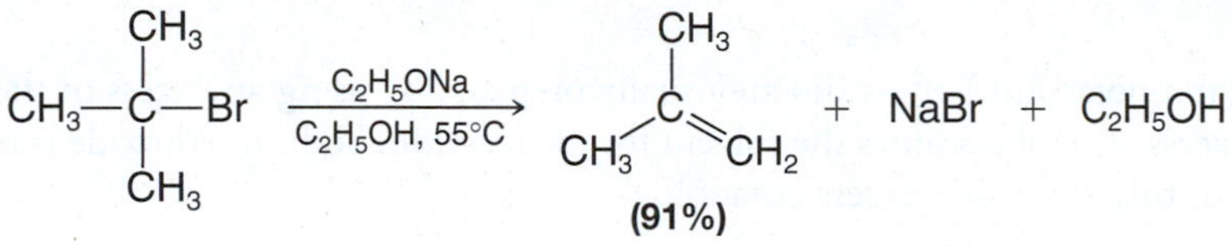
CH3
CH3—C—Br
CH3
C2H5ONa
C2H5OH, 55°C
CH3
C
CH3
CH2
+ NaBr + C2H5OH
(91%)

Reactions like these are not limited to the elimination of hydrogen bromide. Chloroalkanes also undergo the elimination of hydrogen chloride, iodoalkanes undergo the elimination of hydrogen iodide, and, in all cases, alkenes are produced. When the elements of a hydrogen halide are eliminated from a haloalkane in this way, the reaction is often called **dehydrohalogenation**:

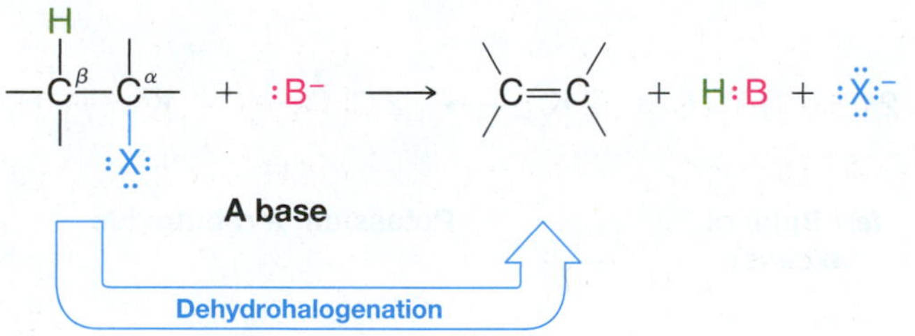
H
β
α
:X:
+ :B⁻
C=C
+ H:B + :X:⁻
A base
Dehydrohalogenation

In these eliminations, as in S_N1 and S_N2 reactions, there is a leaving group and an attacking Lewis base that possesses an electron pair.

Chemists often call the carbon atom that bears the leaving group (e.g., the halogen atom in the previous reaction) the **alpha (*α*) carbon atom** and any carbon atom adjacent to it a **beta (*β*) carbon atom**. A hydrogen atom attached to the *β* carbon atom is called a ***β* hydrogen atom**. Since the hydrogen atom that is eliminated in dehydrohalogenation is from the *β* carbon atom, these reactions are often called ***β* eliminations**. They are also often referred to as **1,2 eliminations**.

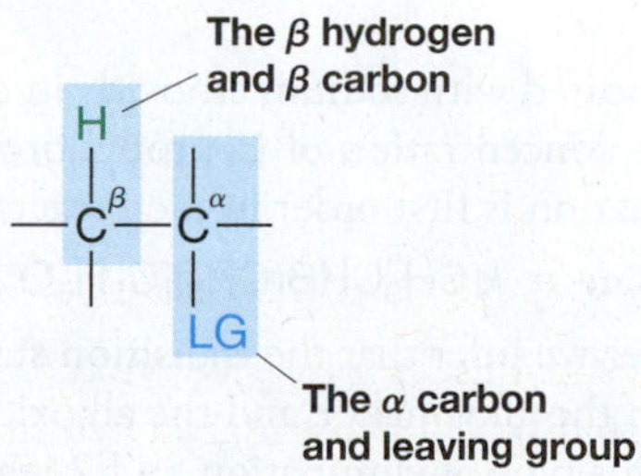
The β hydrogen
and β carbon
H
β
α
LG
The α carbon
and leaving group

We shall have more to say about dehydrohalogenation in Chapter 7, but we can examine several important aspects here.

6.15B Bases Used in Dehydrohalogenation

Various strong bases have been used for dehydrohalogenations. Potassium hydroxide dissolved in ethanol (KOH/EtOH) is a reagent sometimes used, but the conjugate bases of alcohols, such as sodium ethoxide (EtONa), often offer distinct advantages.

The conjugate base of an alcohol (an alkoxide) can be prepared by treating an alcohol with an alkali metal. For example:

$$2\ R\text{—}\ddot{O}H\ +\ 2\ Na\ \longrightarrow\ 2\ R\text{—}\ddot{O}{:}^-\ Na^+\ +\ H_2$$
Alcohol   **Sodium alkoxide**

This reaction is an **oxidation–reduction reaction**. Metallic sodium reacts with hydrogen atoms that are bonded to oxygen atoms to generate hydrogen gas, sodium cations, and the alkoxide anion. The reaction with water is vigorous and at times explosive.

$$2\ H\ddot{O}H\ +\ 2\ Na\ \longrightarrow\ 2\ H\ddot{O}{:}^-\ Na^+\ +\ H_2$$
Sodium hydroxide

Sodium alkoxides can also be prepared by allowing an alcohol to react with sodium hydride (NaH). The hydride ion (H:$^-$) is a very strong base. (The pK_a of H_2 is 35.)

$$R\text{—}\ddot{O}\text{—}H\ +\ Na^+{:}H^-\ \longrightarrow\ R\text{—}\ddot{O}{:}^-\ Na^+\ +\ H\text{—}H$$

Sodium (and potassium) alkoxides are usually prepared by using an excess of the alcohol, and the excess alcohol becomes the solvent for the reaction. Sodium ethoxide is frequently prepared in this way using excess ethanol.

Helpful Hint

EtONa/EtOH is a common abbreviation for sodium ethoxide dissolved in ethanol.

$$2\ CH_3CH_2\ddot{O}H\ +\ 2\ Na\ \longrightarrow\ 2\ CH_3CH_2\ddot{O}{:}^-\ Na^+\ +\ H_2$$
Ethanol (excess)   **Sodium ethoxide dissolved in excess ethanol**

Potassium *tert*-butoxide (*t*-BuOK) is another highly effective dehydrohalogenating reagent. It can be made by the reaction below, or purchased as a solid.

Helpful Hint

t-BuOK/*t*-BuOH represents potassium *tert*-butoxide dissolved in *tert*-butanol.

$$2\ CH_3\underset{\underset{CH_3}{|}}{\overset{\overset{CH_3}{|}}{C}}\text{—}\ddot{O}H\ +\ 2\ K\ \longrightarrow\ 2\ CH_3\underset{\underset{CH_3}{|}}{\overset{\overset{CH_3}{|}}{C}}\text{—}\ddot{O}{:}^-\ K^+\ +\ H_2$$
tert-Butanol (excess)   **Potassium tert-butoxide**

6.15C Mechanisms of Dehydrohalogenations

Elimination reactions occur by a variety of mechanisms. With alkyl halides, two mechanisms are especially important because they are closely related to the S_N2 and S_N1 reactions that we have just studied. One mechanism, called the **E2 reaction**, is bimolecular in the rate-determining step; the other mechanism is the **E1 reaction**, which is unimolecular in the rate-determining step.

6.16 THE E2 REACTION

When isopropyl bromide is heated with sodium ethoxide in ethanol to form propene, the reaction rate depends on the concentration of isopropyl bromide and the concentration of ethoxide ion. The rate equation is first order in each reactant and second order overall:

$$\text{Rate} = k[CH_3CHBrCH_3][C_2H_5O^-]$$

- From the reaction order we infer that the transition state for the rate-determining step must involve both the alkyl halide and the alkoxide ion: The reaction must be bimolecular. We call this type of elimination an E2 reaction.

Considerable experimental evidence indicates that an E2 reaction takes place in the following way:

A MECHANISM FOR THE REACTION — Mechanism for the E2 Reaction

Reaction

$$C_2H_5O^- \ + \ CH_3CHBrCH_3 \longrightarrow CH_2{=}CHCH_3 \ + \ C_2H_5OH \ + \ Br^-$$

Mechanism

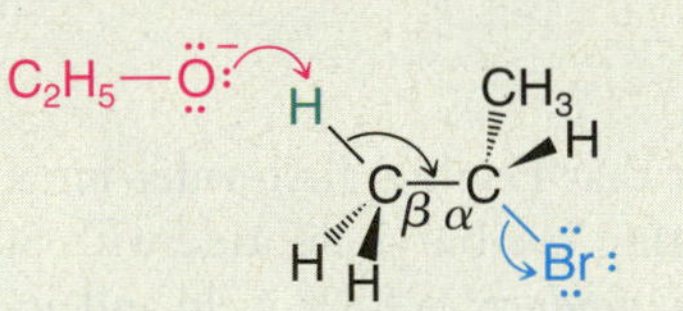

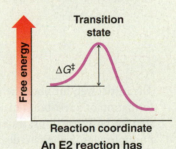

Transition state

The basic ethoxide ion begins to remove a proton from the β carbon using its electron pair to form a bond to it. At the same time, the electron pair of the β C—H bond begins to move in to become the π bond of a double bond, and the bromine begins to depart with the electrons that bonded it to the α carbon.

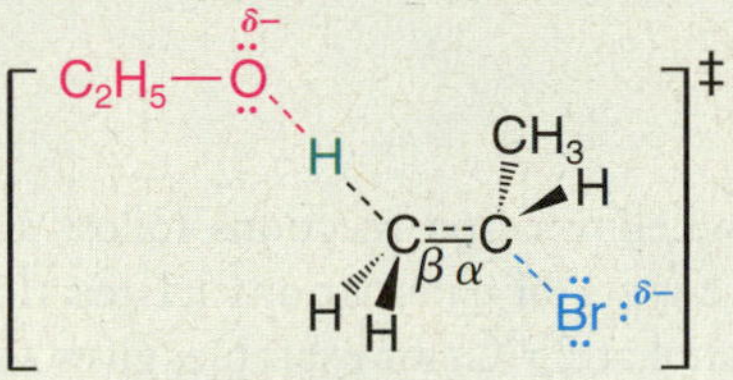

Partial bonds in the transition state extend from the oxygen atom that is removing the β hydrogen, through the carbon skeleton of the developing double bond, to the departing leaving group. The flow of electron density is from the base toward the leaving group as an electron pair fills the π bonding orbital of the alkene.

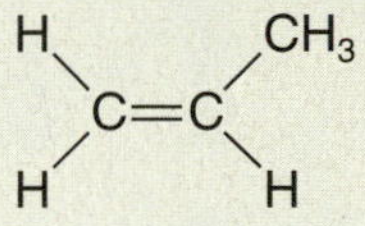

At completion of the reaction, the double bond is fully formed and the alkene has a trigonal planar geometry at each carbon atom. The other products are a molecule of ethanol and a bromide ion.

Transition state

An E2 reaction has one transition state

When we study the E2 reaction further in Section 7.6D, we shall find that the orientations of the hydrogen atom being removed and the leaving group are not arbitrary and that an orientation where they are all in the same plane, like that shown above and in the example that follows, is required.

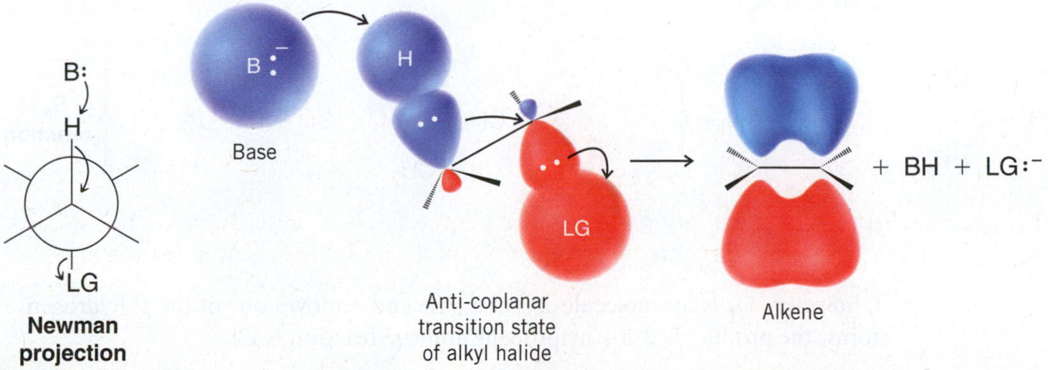

B:

H

Newman projection

Base

Anti-coplanar transition state of alkyl halide

Alkene

$+ \ BH \ + \ LG{:}^-$

Notice that the geometry required here is similar to that of the S_N2 reaction. In the S_N2 reaction (Section 6.6) the nucleophile must push out the leaving group from the **opposite side**. In the E2 reaction the **electron pair of the C—H** bond pushes the leaving group away from the **opposite side** as the base removes the hydrogen. (We shall also find in Section 7.7C that a syn-coplanar E2 transition state is possible, though not as favorable.)

6.17 THE E1 REACTION

Some elimination reactions follow a pathway that exhibits first-order kinetics. We call these types of eliminations E1 reactions. Treating *tert*-butyl chloride with 80% aqueous ethanol at 25°C, for example, gives *substitution products* in 83% yield and an elimination product (2-methylpropene) in 17% yield:

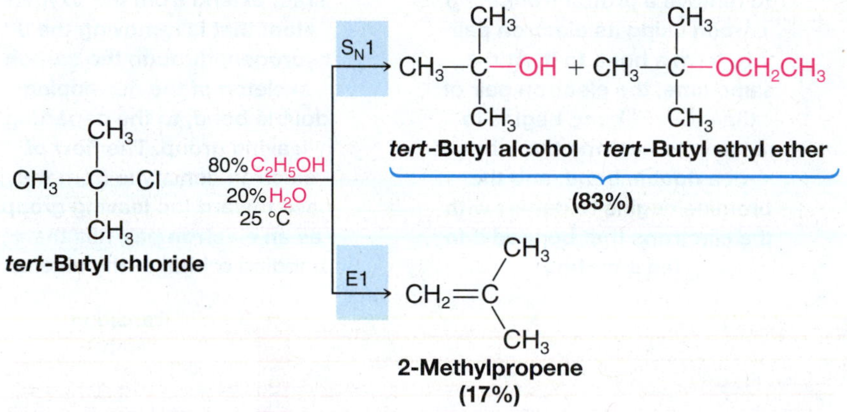

- The initial step for both the substitution and the elimination pathways is the formation of a *tert*-butyl cation as a common intermediate. This is also the slowest step for both reactions; thus both reactions exhibit first-order kinetics and are unimolecular in the rate-determining step.

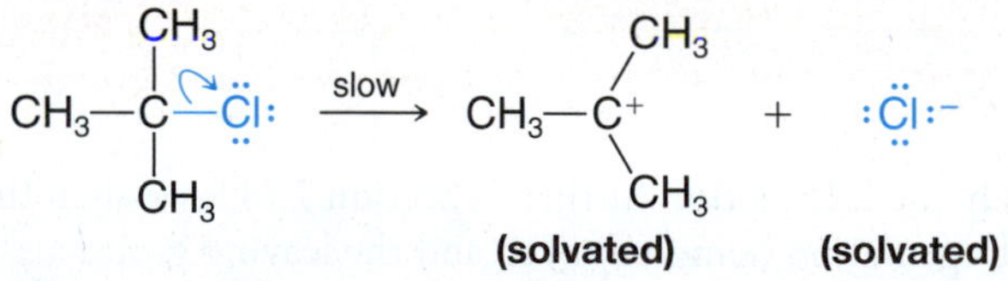

Whether substitution or elimination takes place depends on the next step (the fast step).

- If a solvent molecule reacts as a nucleophile at the positive carbon atom of the *tert*-butyl cation, the product is *tert*-butyl alcohol or *tert*-butyl ethyl ether and the reaction is S_N1:

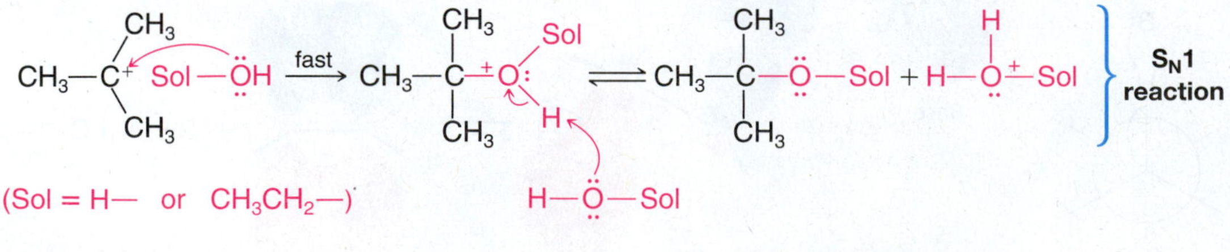

- If, however, a solvent molecule acts as a base and removes one of the β hydrogen atoms, the product is 2-methylpropene and the reaction is E1.

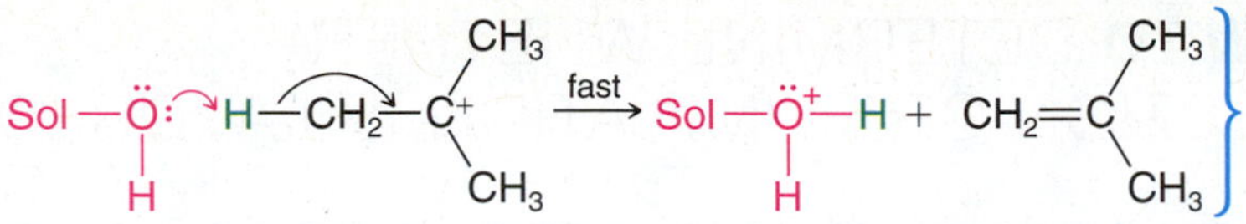

$$\text{Sol}\!-\!\overset{..}{\underset{H}{O}}\!:\!\overset{\curvearrowleft}{H}\!-\!CH_2\!-\!\overset{CH_3}{\underset{CH_3}{\overset{|}{C}}}{}^{+}\;\xrightarrow{\text{fast}}\;\text{Sol}\!-\!\overset{+}{\underset{H}{\overset{..}{O}}}\!-\!H\;+\;CH_2\!=\!\overset{CH_3}{\underset{CH_3}{\overset{|}{C}}}\left.\begin{array}{c}\\\\\end{array}\right\} \begin{array}{c}\textbf{E1}\\\textbf{reaction}\end{array}$$

2-Methylpropene

E1 reactions almost always accompany S_N1 reactions.

[A MECHANISM FOR THE REACTION ──| Mechanism for the E1 Reaction]

Reaction

$$(CH_3)_3CCl \;+\; H_2O \;\longrightarrow\; CH_2\!=\!C(CH_3)_2 \;+\; H_3O^+ \;+\; Cl^-$$

Mechanism

Step 1

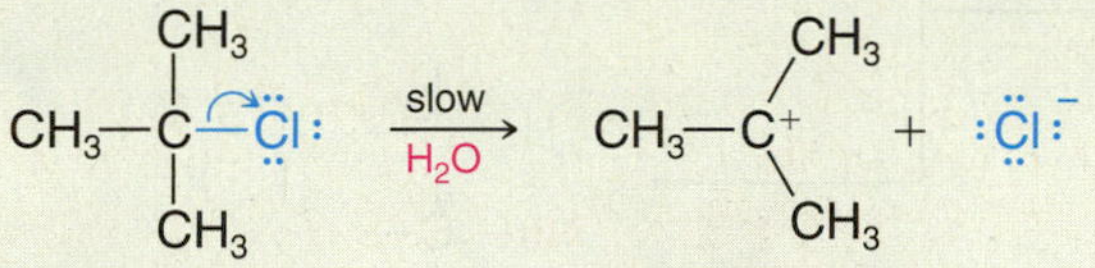

$$CH_3\!-\!\overset{CH_3}{\underset{CH_3}{\overset{|}{C}}}\!-\!\overset{\frown}{\overset{..}{Cl}}\!:\;\xrightarrow[\text{H}_2\text{O}]{\text{slow}}\;CH_3\!-\!\overset{CH_3}{\underset{CH_3}{\overset{|}{C}}}{}^{+}\;+\;:\!\overset{..}{\underset{..}{Cl}}\!:^{-}$$

Aided by the polar solvent, a chlorine departs with the electron pair that bonded it to the carbon.

This slow step produces the relatively stable 3° carbocation and a chloride ion. The ions are solvated (and stabilized) by surrounding water molecules.

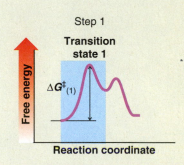

Step 1

Transition state 1

Free energy

$\Delta G^{\ddagger}_{(1)}$

Reaction coordinate

Step 2

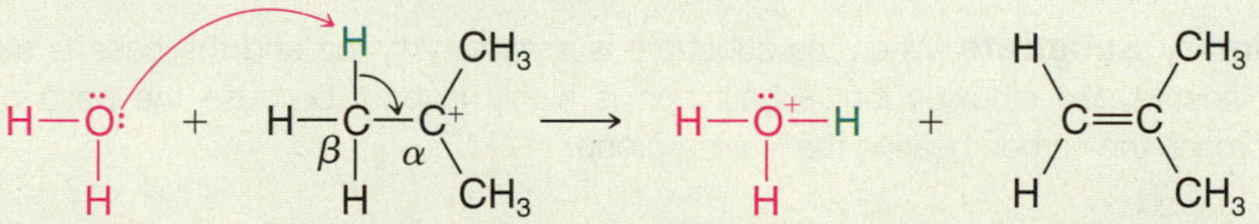

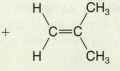

$$H\!-\!\overset{..}{\underset{H}{O}}\!: \;+\; H\!-\!\underset{\beta}{\overset{|}{C}}\!-\!\underset{\alpha}{\overset{+}{\overset{|}{C}}}\underset{CH_3}{\overset{CH_3}{<}}\;\longrightarrow\; H\!-\!\overset{+}{\underset{H}{\overset{..}{O}}}\!-\!H \;+\; \overset{H}{\underset{H}{>}}C\!=\!C\overset{CH_3}{\underset{CH_3}{<}}$$

A molecule of water removes one of the hydrogens from the β carbon of the carbocation. These hydrogens are acidic due to the adjacent positive charge. At the same time an electron pair moves in to form a double bond between the α and β carbon atoms.

This step produces the alkene and a hydronium ion.

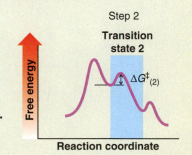

Step 2

Transition state 2

Free energy

$\Delta G^{\ddagger}_{(2)}$

Reaction coordinate

6.18 HOW TO DETERMINE WHETHER SUBSTITUTION OR ELIMINATION IS FAVORED

All nucleophiles are potential bases and all bases are potential nucleophiles. This is because the reactive part of both nucleophiles and bases is an unshared electron pair. It should not be surprising, then, that nucleophilic substitution reactions and elimination reactions often compete with each other. We shall now summarize factors that influence which type of reaction is favored, and provide some examples.

Helpful Hint

This section draws together the various factors that influence the competition between substitution and elimination.

6.18A S_N2 versus E2

S_N2 and E2 reactions are both favored by a high concentration of a strong nucleophile or base. When the nucleophile (base) attacks a β hydrogen atom, elimination occurs. When the nucleophile attacks the carbon atom bearing the leaving group, substitution results:

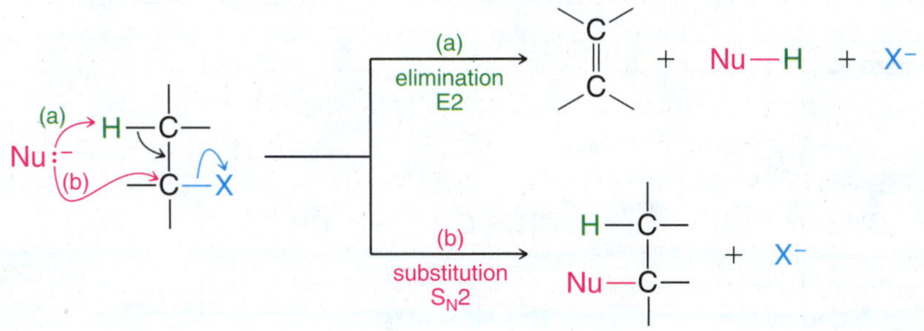

The following examples illustrate the effects of several parameters on substitution and elimination: relative steric hindrance in the substrate (class of alkyl halide), temperature, size of the base/nucleophile (**EtONa** versus **t-BuOK**), and the effects of basicity and polarizability. In these examples we also illustrate a very common way of writing organic reactions, where reagents are written over the reaction arrow, solvents and temperatures are written under the arrow, and only the substrate and major organic products are written to the left and right of the reaction arrow. We also employ typical shorthand notations of organic chemists, such as exclusive use of bond-line formulas and use of commonly accepted abbreviations for some reagents and solvents.

Primary Substrate When the substrate is a *primary* halide and the base is strong and unhindered, like ethoxide ion, substitution is highly favored because the base can easily approach the carbon bearing the leaving group:

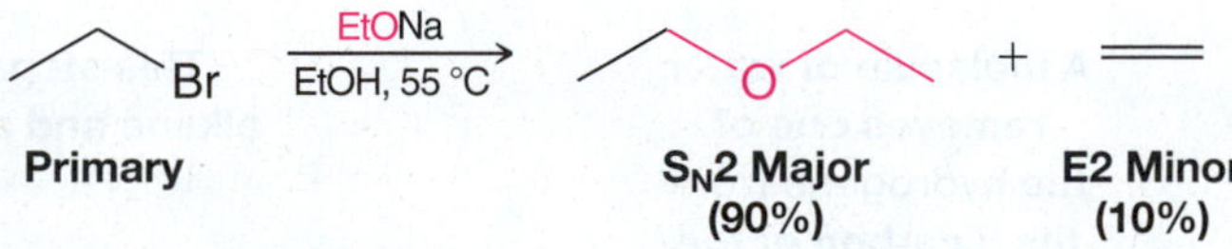

Secondary Substrate With *secondary* halides, however, a strong base favors elimination because steric hindrance in the substrate makes substitution more difficult:

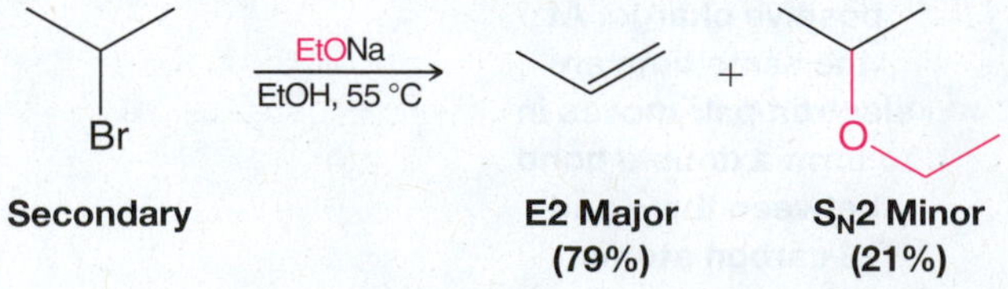

Tertiary Substrate With *tertiary* halides, steric hindrance in the substrate is severe and an S_N2 reaction cannot take place. Elimination is highly favored, especially when the reaction is carried out at higher temperatures. Any substitution that occurs must take place through an S_N1 mechanism:

Without Heating

Tertiary   EtONa, EtOH, 25 °C (room temp.) → E2 Major (91%) + SN1 Minor (9%)

With Heating

Tertiary   EtONa, EtOH, 55 °C → E2 + E1 Only (100%)

Temperature Increasing the reaction temperature favors elimination (E1 and E2) over substitution. Elimination reactions have greater free energies of activation than substitution reactions because more bonding changes occur during elimination. When higher temperature is used, the proportion of molecules able to surmount the energy of activation barrier for elimination increases more than the proportion of molecules able to undergo substitution, although the rate of both substitution and elimination will be increased. Furthermore, elimination reactions are entropically favored over substitution because the products of an elimination reaction are greater in number than the reactants. Additionally, because temperature is the coefficient of the entropy term in the Gibbs free-energy equation $\Delta G° = \Delta H° - T\Delta S°$, an increase in temperature further enhances the entropy effect.

Size of the Base/Nucleophile Increasing the reaction temperature is one way of favorably influencing an elimination reaction of an alkyl halide. Another way is to use a *strong sterically hindered base* such as the *tert*-butoxide ion. The bulky methyl groups of the *tert*-butoxide ion inhibit its reaction by substitution, allowing elimination reactions to take precedence. We can see an example of this effect in the following two reactions. The relatively unhindered methoxide ion reacts with octadecyl bromide primarily by *substitution*, whereas the bulky *tert*-butoxide ion gives mainly *elimination*.

Unhindered (Small) Base/Nucleophile

CH₃ONa, CH₃OH, 65 °C → E2 (1%) + SN2 (99%)

Hindered Base/Nucleophile

t-BuOK, t-BuOK, 40 °C → E2 (85%) + SN2 (15%)

Basicity and Polarizability Another factor that affects the relative rates of E2 and SN2 reactions is the relative basicity and polarizability of the base/nucleophile. Use of a strong, slightly polarizable base such as hydroxide ion, amide ion (NH_2^-), or alkoxide ion (especially a hindered one) tends to increase the likelihood of elimination (E2). Use of a weakly basic ion such as a chloride ion (Cl^-) or an acetate ion ($CH_3CO_2^-$) or a weakly basic and highly polarizable one such as Br^-, I^-, or RS^- increases the likelihood of substitution (SN2). Acetate ion, for example, reacts with isopropyl bromide almost exclusively by the SN2 path:

SN2 (~100%) → + Br⁻

The more strongly basic ethoxide ion (Section 6.15B) reacts with the same compound mainly by an E2 mechanism.

6.18B Tertiary Halides: S$_N$1 versus E1

Because E1 and S$_N$1 reactions proceed through the formation of a common intermediate, the two types respond in similar ways to factors affecting reactivities. E1 reactions are favored with substrates that can form stable carbocations (i.e., tertiary halides); they are also favored by the use of poor nucleophiles (weak bases) and they are generally favored by the use of polar solvents.

It is usually difficult to influence the relative partition between S$_N$1 and E1 products because the free energy of activation for either reaction proceeding from the carbocation (loss of a proton or combination with a molecule of the solvent) is very small.

In most unimolecular reactions the S$_N$1 reaction is favored over the E1 reaction, especially at lower temperatures. *In general, however, substitution reactions of tertiary halides do not find wide use as synthetic methods. Such halides undergo eliminations much too easily.*

Increasing the temperature of the reaction favors reaction by the E1 mechanism at the expense of the S$_N$1 mechanism.

- **If an elimination product is desired from a tertiary substrate, it is advisable to use a strong base so as to encourage an E2 mechanism over the competing E1 and S$_N$1 mechanisms.**

6.19 OVERALL SUMMARY

The most important reaction pathways for the substitution and elimination reactions of simple alkyl halides are summarized in Table 6.6.

Helpful Hint

Use this table as an overall summary.

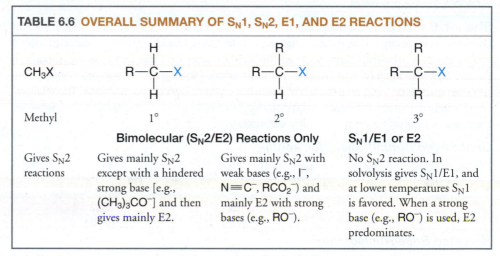

TABLE 6.6 OVERALL SUMMARY OF S$_N$1, S$_N$2, E1, AND E2 REACTIONS

CH$_3$X	R—C(H)(H)—X	R—C(R)(H)—X	R—C(R)(R)—X
Methyl	1°	2°	3°
	Bimolecular (S$_N$2/E2) Reactions Only		S$_N$1/E1 or E2
Gives S$_N$2 reactions	Gives mainly S$_N$2 except with a hindered strong base [e.g., (CH$_3$)$_3$CO$^-$] and then gives mainly E2.	Gives mainly S$_N$2 with weak bases (e.g., I$^-$, N≡C$^-$, RCO$_2$$^-$) and mainly E2 with strong bases (e.g., RO$^-$).	No S$_N$2 reaction. In solvolysis gives S$_N$1/E1, and at lower temperatures S$_N$1 is favored. When a strong base (e.g., RO$^-$) is used, E2 predominates.

Let us examine several sample exercises that will illustrate how the information in Table 6.6 can be used.

●●● **SOLVED PROBLEM 6.8**

Give the product (or products) that you would expect to be formed in each of the following reactions. In each case give the mechanism (S$_N$1, S$_N$2, E1, or E2) by which the product is formed and predict the relative amount of each (i.e., would the product be the only product, the major product, or a minor product?).

(a)   ⟍⟋⟍Br $\xrightarrow[\text{CH}_3\text{OH, 50 °C}]{\text{CH}_3\text{O}^-}$

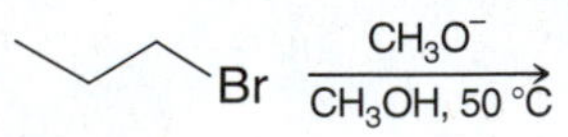

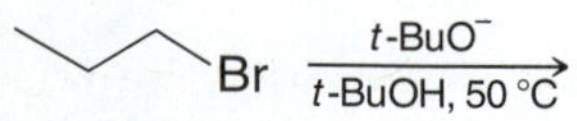

(b)   ⟍⟋⟍Br $\xrightarrow[\text{t-BuOH, 50 °C}]{\text{t-BuO}^-}$

(c)   H‚‚‚⟍⟋Br $\xrightarrow[\text{CH}_3\text{OH, 50 °C}]{\text{HS}^-}$

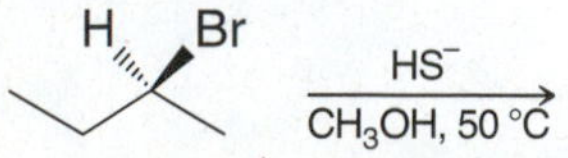

(d)   (Br)   ⟍⟋C⟍⟋ $\xrightarrow[\text{CH}_3\text{OH, 50 °C}]{\text{HO}^-}$

(e)   (Br)   ⟍⟋C⟍⟋ $\xrightarrow[\text{CH}_3\text{OH, 25 °C}]{}$

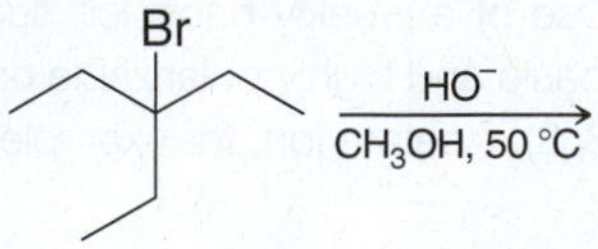

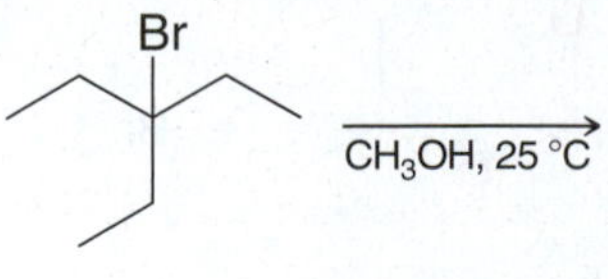

STRATEGY AND ANSWER:

(a) The substrate is a 1° halide. The base/nucleophile is CH₃O⁻, a strong base (but not a hindered one) and a good nucleophile. According to Table 6.6, we should expect an S$_N$2 reaction mainly, and the major product should be

OCH₃. A minor product might be 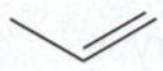   by an E2 pathway.

(b) Again the substrate is a 1° halide, but the base/nucleophile, *t*-BuO⁻, is a strong hindered base. We should expect, therefore, the major product to be 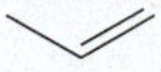   by an E2 pathway and a minor product to be    O-*t*-Bu by an S$_N$2 pathway.

(c) The reactant is (S)-2-bromobutane, a 2° halide and one in which the leaving group is attached to a chirality center. The base/nucleophile is HS⁻, a strong nucleophile but a weak base. We should expect mainly an S$_N$2 reaction, causing an inversion of configuration at the chirality center and producing the (R) stereoisomer:

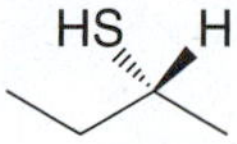

(d) The base/nucleophile is HO⁻, a strong base and a strong nucleophile. The substrate is a 3° halide; therefore, we should not expect an S$_N$2 reaction. The major product should be 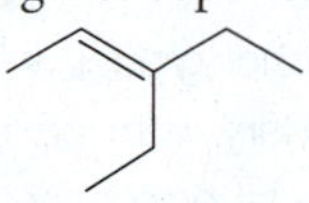   via an E2 reaction. At this higher temperature and in

the presence of a strong base, we should not expect an appreciable amount of the S$_N$1 solvolysis, product,   OCH₃ .

(e) This is solvolysis; the only base/nucleophile is the solvent, CH₃OH, which is a weak base (therefore, no E2 reaction) and a poor nucleophile. The substrate is tertiary (therefore, no S$_N$2 reaction). At this lower temperature we should expect mainly an S$_N$1 pathway leading to   OCH₃ . A minor product, by an E1 pathway, would be 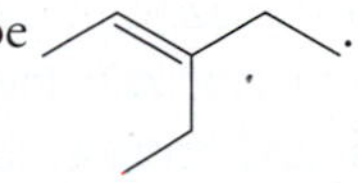   .

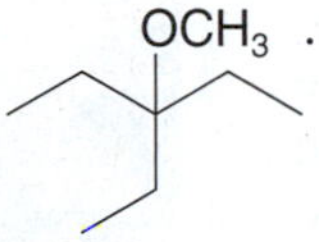

[WHY Do These Topics Matter?

SUBSTITUTING THE CALORIES OF TABLE SUGAR

As we shall see in more detail in Chapter 24, simple carbohydrates, or monosaccharides, can exist in the form of a six-membered ring system with a chair conformation. The name carbohydrate derives from "hydrated carbon" since most carbon atoms have an H and OH attached. In the examples below, the structural differences of the monosaccharides glucose, mannose, and galactose are based on the change of one or more chirality centers through what we could formally consider to be an inversion reaction. As such, all of these carbohydrates are diastereomers of each other. Based on what you already know about torsional strain from Chapter 4, it should come as no surprise that D-glucose is the most common monosaccharide: D-glucose has the least strain because all of its substituents are in equatorial positions. All other six-carbon sugars have at least one axial group, and thus possess some 1,3-diaxial strain. Standard table sugar, or sucrose, is a disaccharide, since it combines a molecule of D-glucose with the slightly less common carbohydrate called D-fructose.

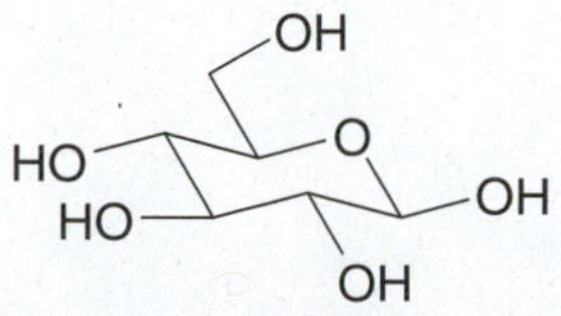

D-glucose 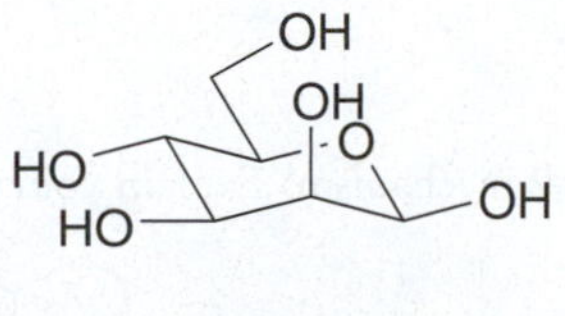   **D-mannose** 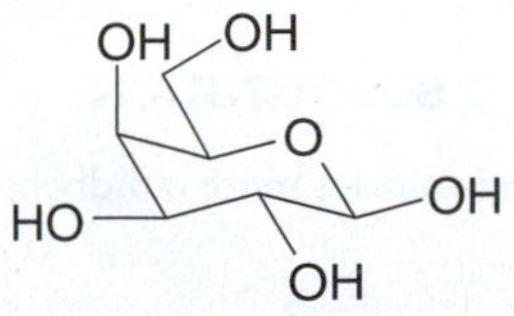   **D-galactose** 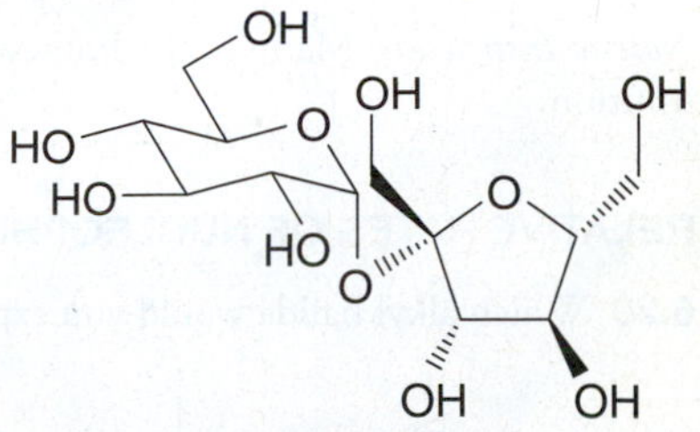   **Sucrose**

All carbohydrates taste sweet, though not equally so. D-Fructose, for example, tastes approximately 1.5 times sweeter than the same amount of simple table sugar, while D-glucose is only about 0.75 times as sweet. Irrespective of their individual degrees of sweetness, however, it is the fact that they are all sweet that lets us perceive their presence in foods whether they are found naturally or have been added (often from corn syrup or cane sugar) to create a more unique flavor profile. Either way, their

(continues on next page)

sweet taste always comes at a cost: calories that can be converted into fat in our bodies. At present, it is estimated by some that Americans consume well over 100 pounds of sugar per person per year from sources both natural and unnatural. That amounts to a lot of calories! What is amazing is that organic chemistry can come to the rescue and knock those calories out. Shown below is the structure of a popular artificial (or synthetic) sweetener known as sucralose, or Splenda. It is the product of some of the chemistry that you have learned in this chapter. Can you guess what that chemistry is?

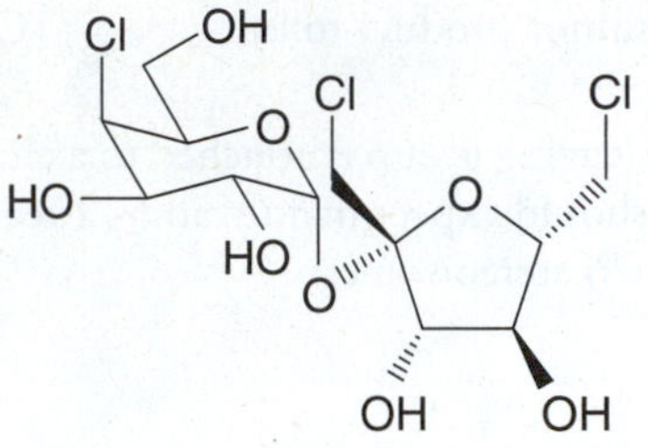

Sucralose

Absolutely right—it is the replacement of three alcohol groups within sucrose, two of them primary and one of them secondary, with chloride through an inversion reaction. Achieving these events in a laboratory setting is quite difficult, since it means selective reaction of only certain hydroxyls in the presence of many others, but it is possible over several steps under the right conditions, including solvent, temperature, and time. What results is a compound that, when ingested, is sensed by our taste receptors as being sweet like table sugar—in fact, 600 times as sweet! What is perhaps even more amazing, however, is that sucralose has, in effect, no calories. We have metabolic pathways that can, in principle, carry out the reverse reactions and replace those chlorines with alcohols through inversion chemistry, thereby re-creating table sugar and leading to calories. But those replacements do not happen fast enough physiologically. As a result, sucralose leaves our bodies before it can be converted into energy and/or stored as fat. Pretty amazing what just a few substitutions can do!

Media Bakery

SUMMARY AND REVIEW TOOLS

The study aids for this chapter include Key Terms and Concepts (which are highlighted in bold, blue text within the chapter and defined in the glossary (at the back of the book) and have hyperlinked definitions in the accompanying *WileyPLUS* course (www.wileyplus.com), and a Mechanism Review regarding substitution and elimination reactions.

PROBLEMS   WILEY PLUS

Note to Instructors: Many of the homework problems are available for assignment via *WileyPlus*, an online teaching and learning solution.

RELATIVE RATES OF NUCLEOPHILIC SUBSTITUTION

6.20 Which alkyl halide would you expect to react more rapidly by an S_N2 mechanism? Explain your answer.

(a) 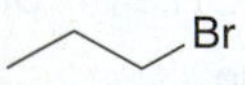   Br   or   Br

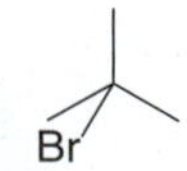

(b)    Cl or   I

(c) 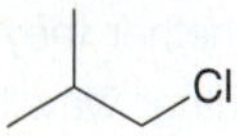   Cl   or   Cl

(d) 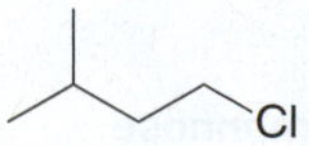   Cl   or   Cl

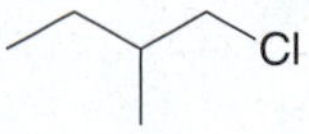

(e) 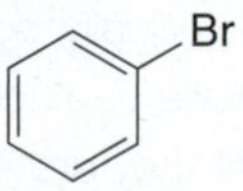   Br   or   Cl

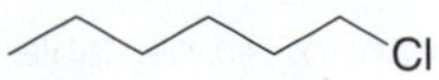

6.21 Which S$_N$2 reaction of each pair would you expect to take place more rapidly in a protic solvent?

(a) 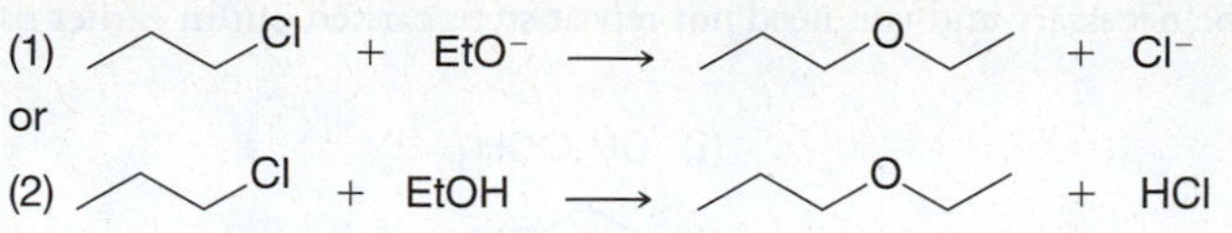 (1) $\diagup$Cl + EtO$^-$ $\longrightarrow$ $\diagup$O$\diagdown$ + Cl$^-$

or

(2) $\diagup$Cl + EtOH $\longrightarrow$ $\diagup$O$\diagdown$ + HCl

(b) 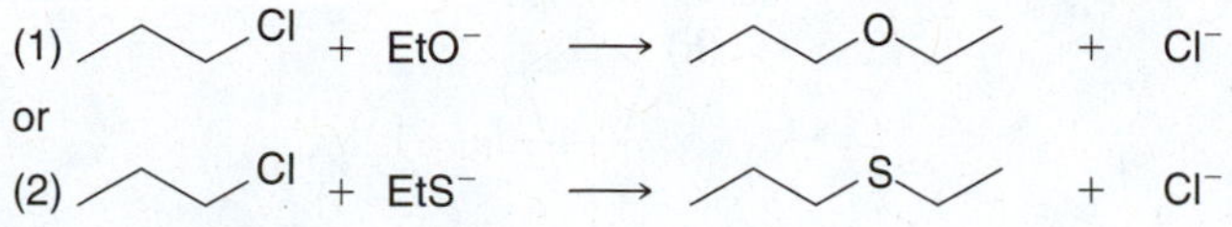 (1) $\diagup$Cl + EtO$^-$ $\longrightarrow$ $\diagup$O$\diagdown$ + Cl$^-$

or

(2) $\diagup$Cl + EtS$^-$ $\longrightarrow$ $\diagup$S$\diagdown$ + Cl$^-$

(c) 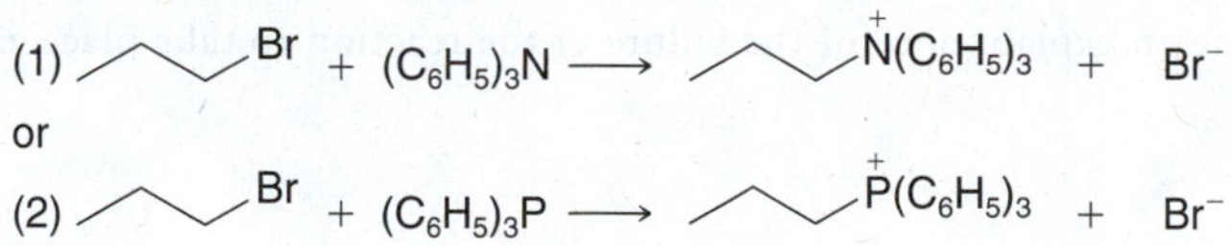 (1) $\diagup$Br + (C$_6$H$_5$)$_3$N $\longrightarrow$ $\diagup\overset{+}{N}$(C$_6$H$_5$)$_3$ + Br$^-$

or

(2) $\diagup$Br + (C$_6$H$_5$)$_3$P $\longrightarrow$ $\diagup\overset{+}{P}$(C$_6$H$_5$)$_3$ + Br$^-$

(d) 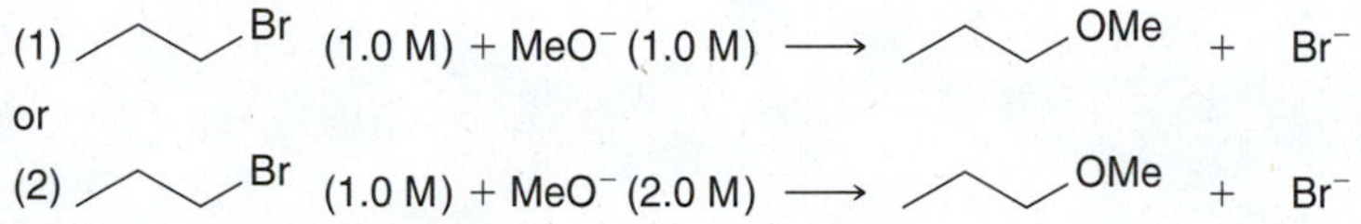 (1) $\diagup$Br (1.0 M) + MeO$^-$ (1.0 M) $\longrightarrow$ $\diagup$OMe + Br$^-$

or

(2) $\diagup$Br (1.0 M) + MeO$^-$ (2.0 M) $\longrightarrow$ $\diagup$OMe + Br$^-$

6.22 Which S$_N$1 reaction of each pair would you expect to take place more rapidly? Explain your answer.

(a) 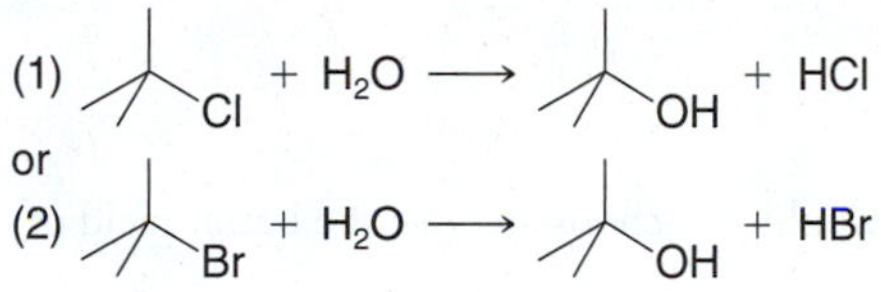 (1) $\overset{}{\diagup}$Cl + H$_2$O $\longrightarrow$ $\diagup$OH + HCl

or

(2) $\diagup$Br + H$_2$O $\longrightarrow$ $\diagup$OH + HBr

(b) 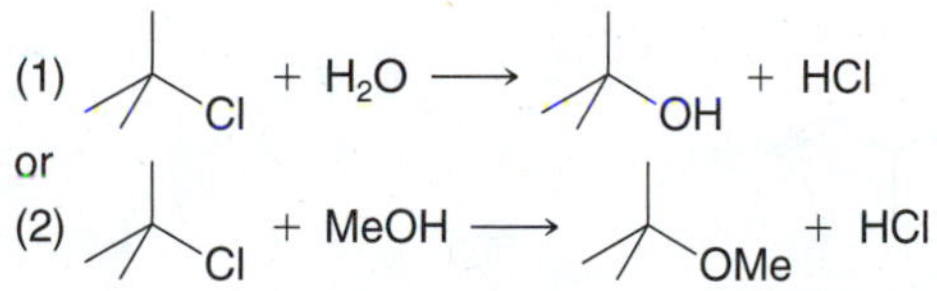 (1) $\diagup$Cl + H$_2$O $\longrightarrow$ $\diagup$OH + HCl

or

(2) $\diagup$Cl + MeOH $\longrightarrow$ $\diagup$OMe + HCl

(c) 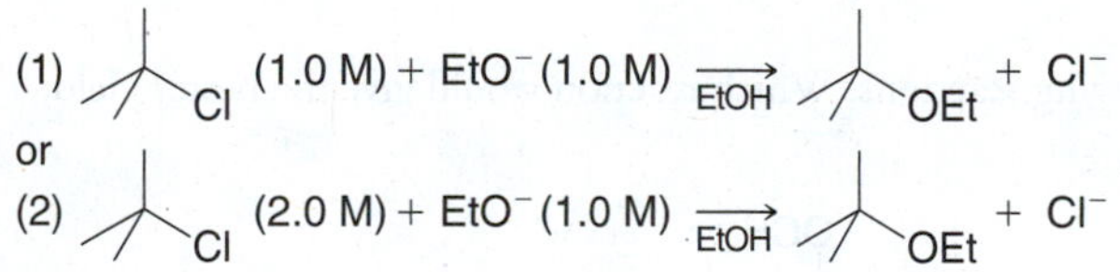 (1) $\diagup$Cl (1.0 M) + EtO$^-$ (1.0 M) $\xrightarrow{\text{EtOH}}$ $\diagup$OEt + Cl$^-$

or

(2) $\diagup$Cl (2.0 M) + EtO$^-$ (1.0 M) $\xrightarrow{\text{EtOH}}$ $\diagup$OEt + Cl$^-$

(d) 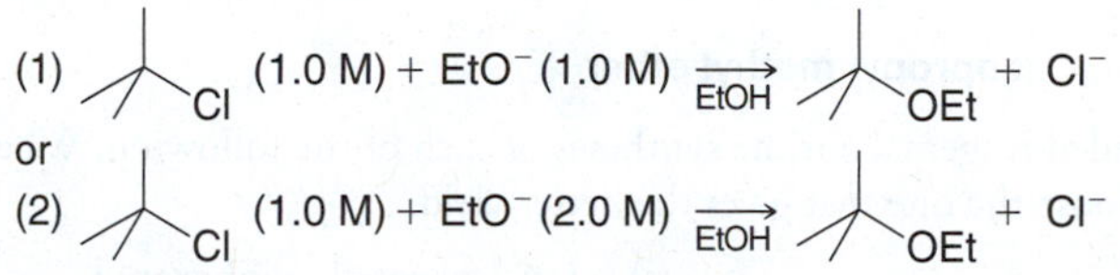 (1) $\diagup$Cl (1.0 M) + EtO$^-$ (1.0 M) $\xrightarrow{\text{EtOH}}$ $\diagup$OEt + Cl$^-$

or

(2) $\diagup$Cl (1.0 M) + EtO$^-$ (2.0 M) $\xrightarrow{\text{EtOH}}$ $\diagup$OEt + Cl$^-$

(e) 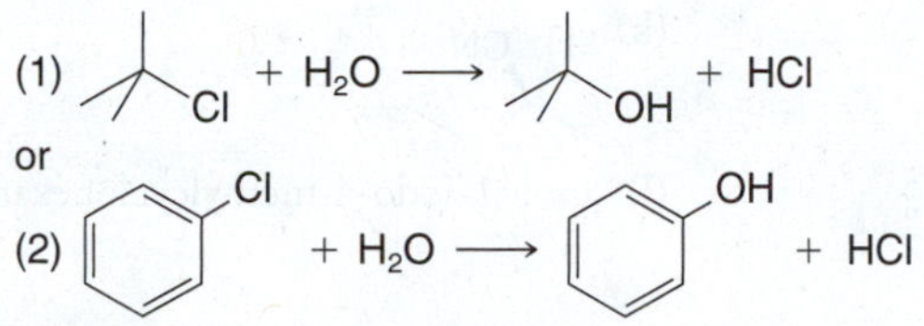 (1) $\diagup$Cl + H$_2$O $\longrightarrow$ $\diagup$OH + HCl

or

(2) (C$_6$H$_5$)Cl + H$_2$O $\longrightarrow$ (C$_6$H$_5$)OH + HCl

SYNTHESIS

6.23 Show how you might use a nucleophilic substitution reaction of 1-bromopropane to synthesize each of the following compounds. (You may use any other compounds that are necessary.)

(a)  $\diagup\diagdown$OH

(b) 1-Iodopropane

(c) 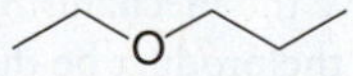 $\diagup$O$\diagdown$

(d) CH$_3$CH$_2$CH$_2$—S—CH$_3$

(e) 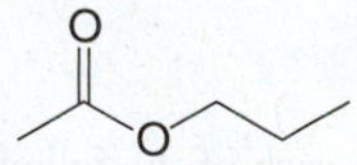 $\diagup$O—C(=O)—$\diagdown$ (propyl ester)

(f)  $\diagup\diagdown$N$_3$

(g) 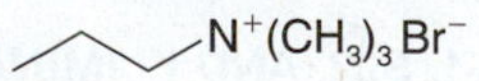 $\diagup\diagdown$N$^+$(CH$_3$)$_3$ Br$^-$

(h) 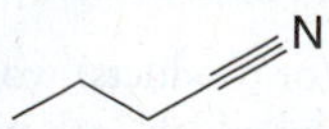 $\diagup\diagdown$C$\equiv$N

(i)  $\diagup\diagdown$SH

6.24 With methyl, ethyl, or cyclopentyl halides as your organic starting materials and using any needed solvents or inorganic reagents, outline syntheses of each of the following. More than one step may be necessary and you need not repeat steps carried out in earlier parts of this problem.

(a) CH_3I

(b)  ⌢⌢I

(c) CH_3OH

(d) 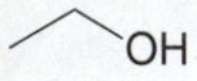 ⌢⌢OH

(e) CH_3SH

(f) 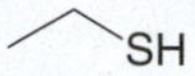 ⌢⌢SH

(g) CH_3CN

(h) 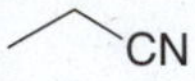 ⌢⌢CN

(i) CH_3OCH_3

(j) 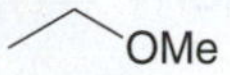 ⌢⌢OMe

(k) 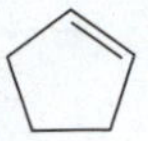 ⬠

6.25 Listed below are several hypothetical nucleophilic substitution reactions. None is synthetically useful because the product indicated is not formed at an appreciable rate. In each case provide an explanation for the failure of the reaction to take place as indicated.

(a)  ⌢⌢ + HO⁻ ⤫→ ⌢⌢OH + ⁻CH₃

(b)  ⌢⌢ + HO⁻ ⤫→ ⌢⌢OH + H⁻

(c)  ☐ + HO⁻ ⤫→ ⁻⌢⌢⌢OH

(d) 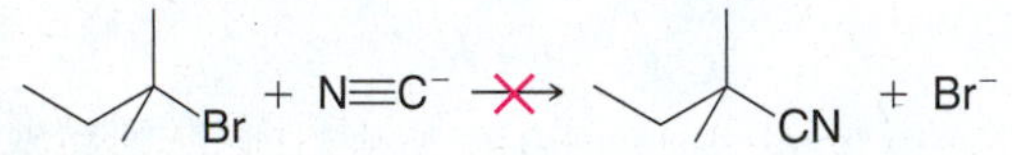 ⌁Br + N≡C⁻ ⤫→ ⌁CN + Br⁻

(e) $NH_3 + CH_3OCH_3$ ⤫→ $CH_3NH_2 + CH_3OH$

(f) $NH_3 + CH_3\overset{+}{O}H_2$ ⤫→ $CH_3\overset{+}{N}H_3 + H_2O$

6.26 Your task is to prepare styrene by one of the following reactions. Which reaction would you choose to give the better yield of styrene? Explain your answer.

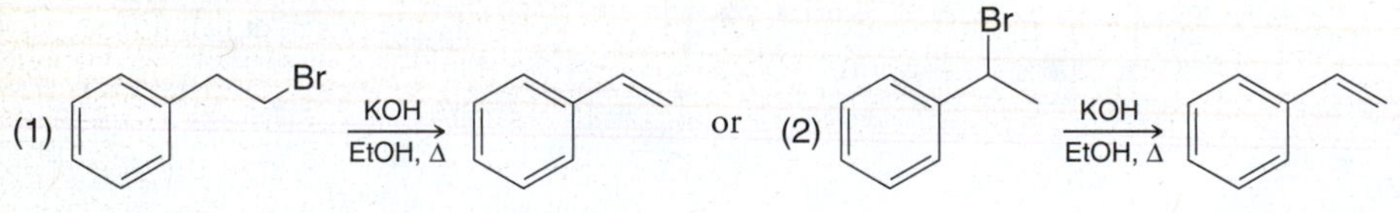

(1) [benzene-CH₂CH₂Br] $\xrightarrow[\text{EtOH, }\Delta]{\text{KOH}}$ [styrene]   or   (2) [benzene-CHBr-CH₃] $\xrightarrow[\text{EtOH, }\Delta]{\text{KOH}}$ [styrene]

Styrene   **Styrene**

6.27 Your task is to prepare isopropyl methyl ether by one of the following reactions. Which reaction would give the better yield? Explain your answer.

(1) [isopropyl-I] + CH_3ONa ⟶ [isopropyl-OCH₃]   or   (2) [isopropyl-ONa] + CH_3I ⟶ [isopropyl-OCH₃]

Isopropyl methyl ether   **Isopropyl methyl ether**

6.28 Starting with an appropriate alkyl halide and using any other needed reagents, outline syntheses of each of the following. When alternative possibilities exist for a synthesis, you should be careful to choose the one that gives the better yield.

(a) Butyl *sec*-butyl ether

(b) 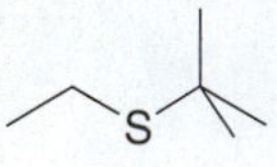 ⌢⌢S⧖

(c) Methyl neopentyl ether

(d) Methyl phenyl ether

(e) 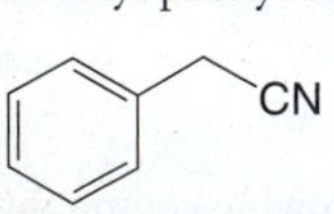 [benzene-CH₂CN]

(f) 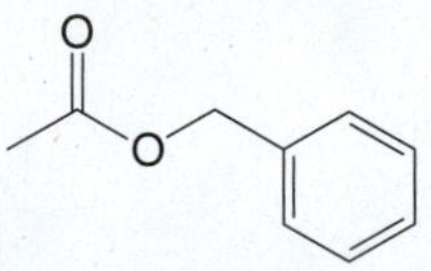 [CH₃-C(=O)-O-CH₂-phenyl]

(g) (S)-2-Pentanol

(h) (R)-2-Iodo-4-methylpentane

(i) 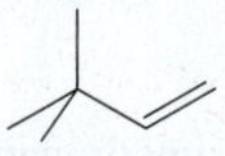 ⧖⌣

(j) *cis*-4-Isopropylcyclohexanol

(k) 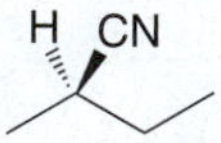 [H⟍ CN structure]

(l) *trans*-1-Iodo-4-methylcyclohexane

GENERAL S_N1, S_N2, AND ELIMINATION

6.29 Which product (or products) would you expect to obtain from each of the following reactions? In each part give the mechanism (S_N1, S_N2, E1, or E2) by which each product is formed and predict the relative amount of each product (i.e., would the product be the only product, the major product, a minor product, etc.?).

(a) 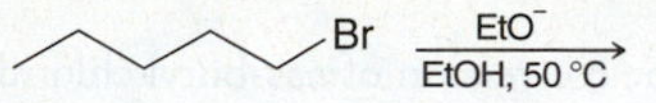 $\xrightarrow[\text{EtOH, 50 °C}]{\text{EtO}^-}$

(b) 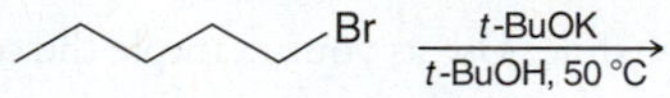 $\xrightarrow[\text{t-BuOH, 50 °C}]{\text{t-BuOK}}$

(c) $\xrightarrow[\text{MeOH, 50 °C}]{\text{MeO}^-}$

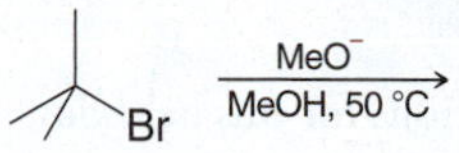

(d) 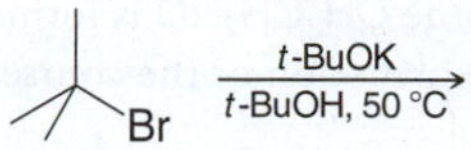 $\xrightarrow[\text{t-BuOH, 50 °C}]{\text{t-BuOK}}$

(e) 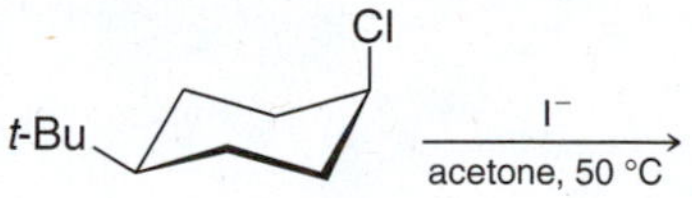 $\xrightarrow[\text{acetone, 50 °C}]{\text{I}^-}$

(f) 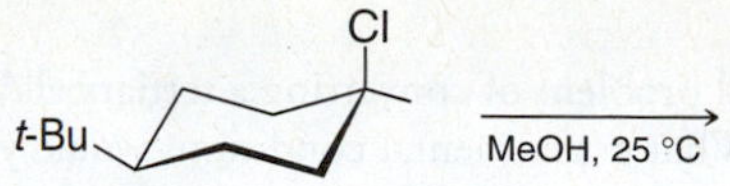 $\xrightarrow[\text{MeOH, 25 °C}]{}$

(g) 3-Chloropentane $\xrightarrow[\text{MeOH, 50 °C}]{\text{MeO}^-}$

(h) 3-Chloropentane $\xrightarrow[\text{CH}_3\text{CO}_2\text{H, 50 °C}]{\text{CH}_3\text{CO}_2^-}$

(i) (R)-2-bromobutane $\xrightarrow[\text{25 °C}]{\text{HO}^-}$

(j) (S)-3-Bromo-3-methylhexane $\xrightarrow[\text{MeOH}]{\text{25 °C}}$

(k) (S)-2-Bromooctane $\xrightarrow[\text{MeOH, 50 °C}]{\text{I}^-}$

6.30 Write conformational structures for the substitution products of the following deuterium-labeled compounds:

(a) 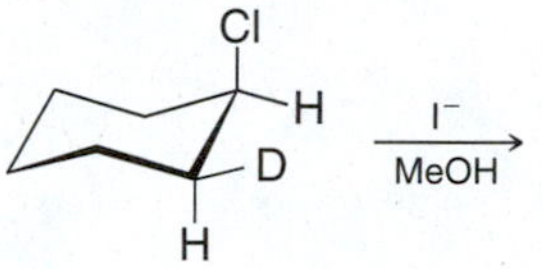 $\xrightarrow[\text{MeOH}]{\text{I}^-}$

(c) 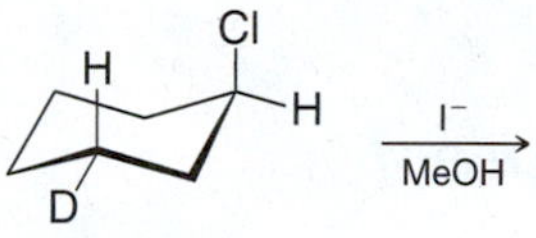 $\xrightarrow[\text{MeOH}]{\text{I}^-}$

(b) 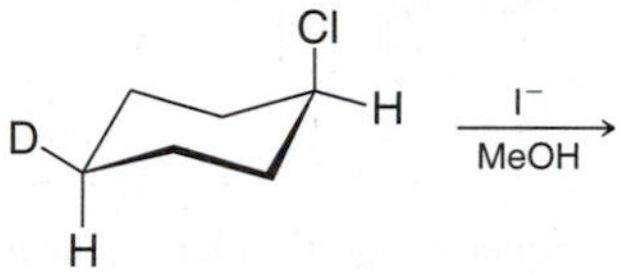 $\xrightarrow[\text{MeOH}]{\text{I}^-}$

(d) 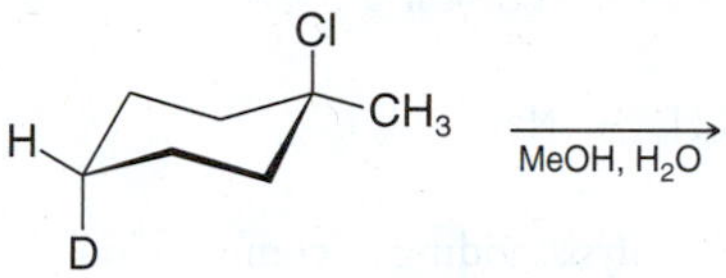 $\xrightarrow[\text{MeOH, H}_2\text{O}]{}$

6.31 Although ethyl bromide and isobutyl bromide are both primary halides, ethyl bromide undergoes S_N2 reactions more than 10 times faster than isobutyl bromide does. When each compound is treated with a strong base/nucleophile (EtO$^-$), isobutyl bromide gives a greater yield of elimination products than substitution products, whereas with ethyl bromide this behavior is reversed. What factor accounts for these results?

6.32 Consider the reaction of I$^-$ with CH$_3$CH$_2$Cl.
(a) Would you expect the reaction to be S_N1 or S_N2? The rate constant for the reaction at 60 °C is 5×10^{-5} L mol^{-1} s^{-1}.
(b) What is the reaction rate if [I$^-$] = 0.1 mol L^{-1} and [CH$_3$CH$_2$Cl] = 0.1 mol L^{-1}?
(c) If [I$^-$] = 0.1 mol L^{-1} and [CH$_3$CH$_2$Cl] = 0.2 mol L^{-1}?
(d) If [I$^-$] = 0.2 mol L^{-1} and [CH$_3$CH$_2$Cl] = 0.1 mol L^{-1}?
(e) If [I$^-$] = 0.2 mol L^{-1} and [CH$_3$CH$_2$Cl] = 0.2 mol L^{-1}?

6.33 Which reagent in each pair listed here would be the more reactive nucleophile in a polar aprotic solvent?
(a) CH$_3\bar{\text{N}}$H   or   CH$_3$NH$_2$
(b) CH$_3$O$^-$ or   CH$_3$CO$_2^-$ ($^-$OAc)
(c) CH$_3$SH   or   CH$_3$OH
(d) (C$_6$H$_5$)$_3$N   or   (C$_6$H$_5$)$_3$P
(e) H$_2$O   or   H$_3$O$^+$
(f) NH$_3$ or $^+$NH$_4$
(g) H$_2$S   or   HS$^-$
(h) CH$_3$CO$_2^-$ ($^-$OAc)   or   HO$^-$

6.34 Write mechanisms that account for the products of the following reactions:

(a) 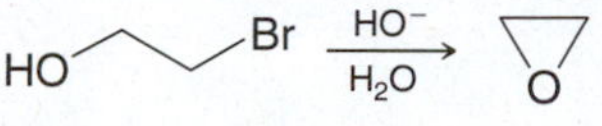 $\xrightarrow[\text{H}_2\text{O}]{\text{HO}^-}$

(b) 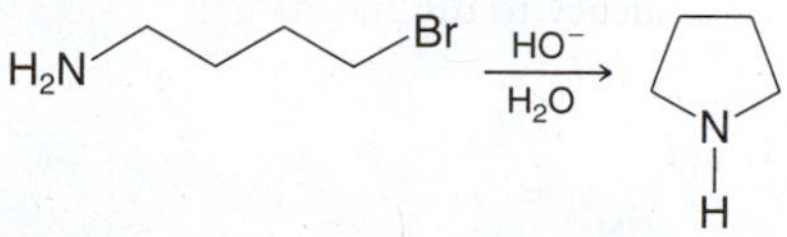 $\xrightarrow[\text{H}_2\text{O}]{\text{HO}^-}$

6.35 Draw a three-dimensional representation for the transition state structure in the S_N2 reaction of N≡C:$^-$ (cyanide anion) with bromoethane, showing all nonbonding electron pairs and full or partial charges.

6.36 Many S_N2 reactions of alkyl chlorides and alkyl bromides are catalyzed by the addition of sodium or potassium iodide. For example, the hydrolysis of methyl bromide takes place much faster in the presence of sodium iodide. Explain.

6.37 Explain the following observations: When *tert*-butyl bromide is treated with sodium methoxide in a mixture of methanol and water, the rate of formation of *tert*-butyl alcohol and *tert*-butyl methyl ether does not change appreciably as the concentration of sodium methoxide is increased. However, increasing the concentration of sodium methoxide causes a marked increase in the rate at which *tert*-butyl bromide disappears from the mixture.

6.38

(a) Consider the general problem of converting a tertiary alkyl halide to an alkene, for example, the conversion of *tert*-butyl chloride to 2-methylpropene. What experimental conditions would you choose to ensure that elimination is favored over substitution?

(b) Consider the opposite problem, that of carrying out a substitution reaction on a tertiary alkyl halide. Use as your example the conversion of *tert*-butyl chloride to *tert*-butyl ethyl ether. What experimental conditions would you employ to ensure the highest possible yield of the ether?

6.39 1-Bromobicyclo[2.2.1]heptane is extremely unreactive in either S_N2 or S_N1 reactions. Provide explanations for this behavior.

6.40 When ethyl bromide reacts with potassium cyanide in methanol, the major product is CH_3CH_2CN. Some CH_3CH_2NC is formed as well, however. Write Lewis structures for the cyanide ion and for both products and provide a mechanistic explanation of the course of the reaction.

6.41 Give structures for the products of each of the following reactions:

(a) 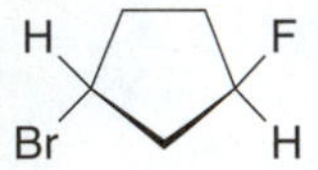
$\xrightarrow[\text{acetone}]{\text{NaI (1 mol)}}$ C_5H_8FI + NaBr

(b) 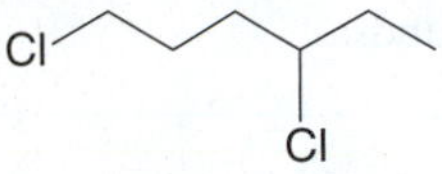 Cl—（structure）（1 mol） $\xrightarrow[\text{acetone}]{\text{NaI (1 mol)}}$ $C_6H_{12}CII$ + NaCl

(c) 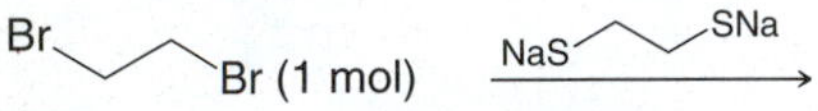 Br—（structure）—Br (1 mol) $\xrightarrow{\text{NaS}\sim\text{SNa}}$ $C_4H_8S_2$ + 2 NaBr

(d)  Cl—（structure）—OH $\xrightarrow[\text{Et}_2\text{O}]{\text{NaH (}-\text{H}_2\text{)}}$ C_4H_8ClONa $\xrightarrow{\text{Et}_2\text{O, heat}}$ C_4H_8O + NaCl

(e) 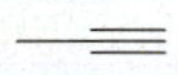 —≡— $\xrightarrow[\text{liq. NH}_3]{\text{NaNH}_2\text{ (}-\text{NH}_3\text{)}}$ C_3H_3Na $\xrightarrow{\text{CH}_3\text{I}}$ C_4H_6 + NaI

6.42 When *tert*-butyl bromide undergoes S_N1 hydrolysis, adding a "common ion" (e.g., NaBr) to the aqueous solution has no effect on the rate. On the other hand, when $(C_6H_5)_2CHBr$ undergoes S_N1 hydrolysis, adding NaBr retards the reaction. Given that the $(C_6H_5)_2CH^+$ cation is known to be much more stable than the $(CH_3)_3C^+$ cation (and we shall see why in Section 15.12A), provide an explanation for the different behavior of the two compounds.

6.43 When the alkyl bromides (listed here) were subjected to hydrolysis in a mixture of ethanol and water (80% EtOH/20% H_2O) at 55 °C, the rates of the reaction showed the following order:

$$(CH_3)_3CBr > CH_3Br > CH_3CH_2Br > (CH_3)_2CHBr$$

Provide an explanation for this order of reactivity.

6.44 The reaction of 1° alkyl halides with nitrite salts produces both RNO_2 and RONO. Account for this behavior.

6.45 What would be the effect of increasing solvent polarity on the rate of each of the following nucleophilic substitution reactions?

(a) Nu: + R—L ⟶ R—Nu$^+$ + :L$^-$ **(b)** R—L$^+$ ⟶ R$^+$ + :L

6.46 Competition experiments are those in which two reactants at the same concentration (or one reactant with two reactive sites) compete for a reagent. Predict the major product resulting from each of the following competition experiments:

(a) 
$\xrightarrow[\text{DMF}]{\text{I}^-}$

(b) 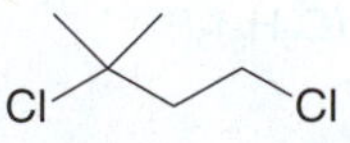
$\xrightarrow[\text{acetone}]{\text{H}_2\text{O}}$

6.47 In contrast to S_N2 reactions, S_N1 reactions show relatively little nucleophile selectivity. That is, when more than one nucleophile is present in the reaction medium, S_N1 reactions show only a slight tendency to discriminate between weak nucleophiles and strong nucleophiles, whereas S_N2 reactions show a marked tendency to discriminate.

(a) Provide an explanation for this behavior.

(b) Show how your answer accounts for the following:

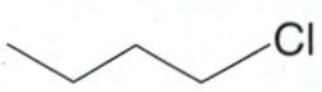

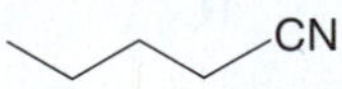

Major product

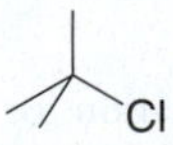

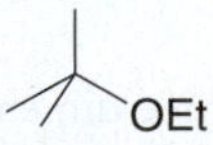

Major product

CHALLENGE PROBLEMS

6.48 The reaction of chloroethane with water *in the gas phase* to produce ethanol and hydrogen chloride has $\triangle H° = +26.6$ kJ mol^{-1} and $\triangle S° = +4.81$ J K^{-1} mol^{-1} at 25 °C.

(a) Which of these terms, if either, favors the reaction going to completion?

(b) Calculate $\triangle G°$ for the reaction. What can you now say about whether the reaction will proceed to completion?

(c) Calculate the equilibrium constant for the reaction.

(d) In aqueous solution the equilibrium constant is very much larger than the one you just calculated. How can you account for this fact?

6.49 When (S)-2-bromopropanoic acid [(S)-CH₃CHBrCO₂H] reacts with concentrated sodium hydroxide, the product formed (after acidification) is (R)-2-hydroxypropanoic acid [(R)-CH₃CHOHCO₂H, commonly known as (R)-lactic acid]. This is, of course, the normal stereochemical result for an S_N2 reaction. However, when the same reaction is carried out with a low concentration of hydroxide ion in the presence of Ag₂O (where Ag⁺ acts as a Lewis acid), it takes place with overall *retention of configuration* to produce (S)-2-hydroxypropanoic acid. The mechanism of this reaction involves a phenomenon called **neighboring-group participation**. Write a detailed mechanism for this reaction that accounts for the net retention of configuration when Ag⁺ and a low concentration of hydroxide are used.

6.50 The phenomenon of configuration inversion in a chemical reaction was discovered in 1896 by Paul Walden (Section 6.6). Walden's proof of configuration inversion was based on the following cycle:

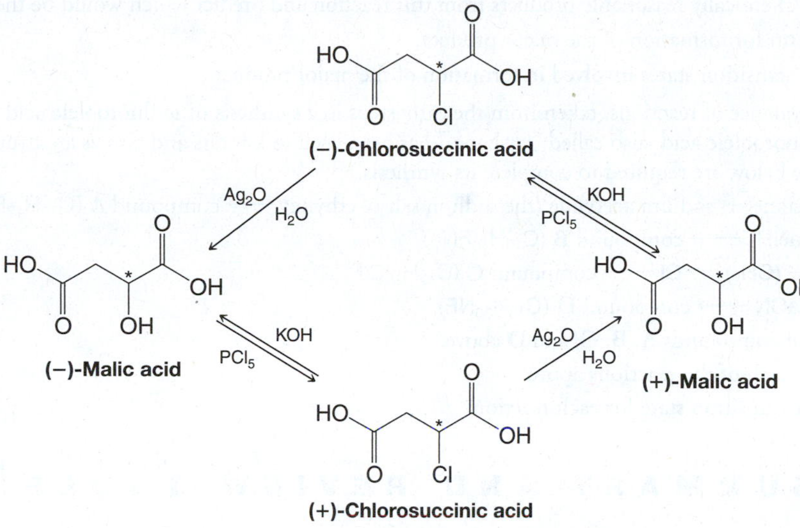

The Walden Cycle

(a) Basing your answer on the preceding problem, which reactions of the Walden cycle are likely to take place with overall inversion of configuration and which are likely to occur with overall retention of configuration?

(b) Malic acid with a negative optical rotation is now known to have the (S) configuration. What are the configurations of the other compounds in the Walden cycle?

(c) Walden also found that when (+)-malic acid is treated with thionyl chloride (rather than PCl₅), the product of the reaction is (+)-chlorosuccinic acid. How can you explain this result?

(d) Assuming that the reaction of (−)-malic acid and thionyl chloride has the same stereochemistry, outline a Walden cycle based on the use of thionyl chloride instead of PCl₅.

6.51 (R)-(3-Chloro-2-methylpropyl) methyl ether (**A**) on reaction with azide ion (N₃⁻) in aqueous ethanol gives (S)-(3-azido-2-methylpropyl) methyl ether (**B**). Compound **A** has the structure ClCH₂CH(CH₃)CH₂OCH₃.

(a) Draw wedge–dashed wedge–line formulas of both **A** and **B**.

(b) Is there a change of configuration during this reaction?

6.52 Predict the structure of the product of this reaction:

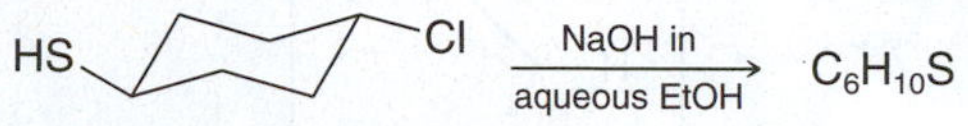

The product has no infrared absorption in the 1620–1680-cm⁻¹ region.

6.53 *cis*-4-Bromocyclohexanol $\xrightarrow[\text{t-BuOH}]{\text{t-BuO}^-}$ racemic C₆H₁₀O (compound **C**)

Compound **C** has infrared absorption in the 1620–1680-cm⁻¹ and in the 3590–3650-cm⁻¹ regions. Draw and label the (R) and (S) enantiomers of product **C**.

6.54 1-Bromo[2.2.1]bicycloheptane is unreactive toward both S_N2 and S_N1 reactions. Open the computer molecular model at the book's website titled "1-Bromo[2.2.1]bicycloheptane" and examine the structure. What barriers are there to substitution of 1-bromo[2.2.1]bicycloheptane by both S_N2 and S_N1 reaction mechanisms?

6.55 Open the computer molecular model titled "1-Bromo[2.2.1]bicycloheptane LUMO" at the book's website for the lowest unoccupied molecular orbital (LUMO) of this compound. Where is the lobe of the LUMO with which the HOMO of a nucleophile would interact in an S_N2 reaction?

6.56 In the previous problem and the associated molecular model at the book's website, you considered the role of HOMOs and LUMOs in an S_N2 reaction.

(a) What is the LUMO in an S_N1 reaction and in what reactant and species is it found?

(b) Open the molecular model at the book's website titled "Isopropyl Methyl Ether Carbocation LUMO." Identify the lobe of the LUMO in this carbocation model with which a nucleophile would interact.

(c) Open the model titled "Isopropyl Methyl Ether Carbocation HOMO." Why is there a large orbital lobe between the oxygen and the carbon of the carbocation?

LEARNING GROUP PROBLEMS

1. Consider the solvolysis reaction of (1S,2R)-1-bromo-1,2-dimethylcyclohexane in 80% H_2O/20% CH_3CH_2OH at room temperature.

(a) Write the structure of all chemically reasonable products from this reaction and predict which would be the major one.

(b) Write a detailed mechanism for formation of the major product.

(c) Write the structure of all transition states involved in formation of the major product.

2. Consider the following sequence of reactions, taken from the early steps in a synthesis of ω-fluorooleic acid, a toxic natural compound from an African shrub. (ω-Fluorooleic acid, also called "ratsbane," has been used to kill rats and also as an arrow poison in tribal warfare. Two more steps beyond those below are required to complete its synthesis.)

    (i)   1-Bromo-8-fluorooctane + sodium acetylide (the sodium salt of ethyne) $\longrightarrow$ compound **A** ($C_{10}H_{17}F$)

    (ii) Compound **A** + $NaNH_2$ $\longrightarrow$ compound **B** ($C_{10}H_{16}FNa$)

    (iii) Compound **B** + I—$(CH_2)_7$—Cl $\longrightarrow$ compound **C** ($C_{17}H_{30}ClF$)

    (iv) Compound **C** + NaCN $\longrightarrow$ compound **D** ($C_{18}H_{30}NF$)

(a) Elucidate the structures of compounds **A**, **B**, **C**, and **D** above.

(b) Write the mechanism for each of the reactions above.

(c) Write the structure of the transition state for each reaction.

[S U M M A R Y   A N D   R E V I E W   T O O L S]

Mechanism Review: Substitution versus Elimination

S_N2

Primary substrate
Back side attack of Nu: with respect to LG
Strong/polarizable unhindered nucleophile

Bimolecular in rate-determining step
Concerted bond forming/bond breaking
Inversion of stereochemistry
Favored by polar aprotic solvent

S_N1 and E1

Tertiary substrate
Carbocation intermediate
Weak nucleophile/base (e.g., solvent)

Unimolecular in rate-determining step
Racemization if S_N1
Removal of β-hydrogen if E1
Protic solvent assists ionization of LG
Low temperature (S_N1) / high temperature (E1)

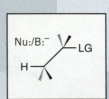

S_N2 and E2

Secondary or primary substrate
Strong unhindered base/nucleophile
  leads to S_N2
Strong hindered base/nucleophile leads to E2
Low temperature (S_N2) / high
  temperature (E2)

E2

Tertiary or secondary substrate
Concerted anti-coplanar transition state

Bimolecular in rate-determining step
Strong hindered base
High temperature

Alkenes and Alkynes I

PROPERTIES AND SYNTHESIS. ELIMINATION
REACTIONS OF ALKYL HALIDES

Despite being a world of seven billion people spread over seven continents, a popular but unproven theory claims that there are only six degrees of separation between each of us and every other person. In other words, we are all a friend of a friend, and so on. As strange as it might sound, organic molecules are not much different, with alkenes and alkynes being the key connectors to numerous other functional groups as well as to C—C bond-formation processes that can rapidly create molecular complexity. In truth, it rarely takes six steps to find where an alkene or alkyne may have played a role in the synthesis of a molecule; more typically, it takes only one or two steps.

IN THIS CHAPTER WE WILL CONSIDER:

- the properties of alkenes and alkynes and how they are named
- how alkenes and alkynes can be transformed into alkanes
- how to plan the synthesis of any organic molecule

[**WHY** DO THESE TOPICS MATTER?] At the end of the chapter, we will show how simple changes in the placement of alkene functional groups can lead to distinct properties, from the strength of the rubber in our tires to our ability to see.

PHOTO CREDIT: Media Bakery

7.1 INTRODUCTION

Alkenes are hydrocarbons whose molecules contain a carbon–carbon double bond. An old name for this family of compounds that is still often used is the name **olefins**. Ethene (ethylene), the simplest olefin (alkene), was called olefiant gas (Latin: *oleum,* oil + *facere,* to make) because gaseous ethene (C_2H_4) reacts with chlorine to form $C_2H_4Cl_2$, a liquid (oil).

Hydrocarbons whose molecules contain the carbon–carbon triple bond are called alkynes. The common name for this family is **acetylenes,** after the simplest member, $HC \equiv CH$:

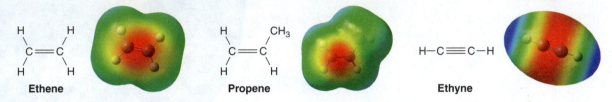

7.1A Physical Properties of Alkenes and Alkynes

Alkenes and alkynes have physical properties similar to those of corresponding alkanes. Alkenes and alkynes up to four carbons (except 2-butyne) are gases at room temperature. Being relatively nonpolar themselves, alkenes and alkynes dissolve in nonpolar solvents or in solvents of low polarity. Alkenes and alkynes are only *very slightly soluble* in water (with alkynes being slightly more soluble than alkenes). The densities of alkenes and alkynes are lower than that of water.

7.2 THE (*E*)–(*Z*) SYSTEM FOR DESIGNATING ALKENE DIASTEREOMERS

In Section 4.5 we learned to use the terms cis and trans to designate the stereochemistry of alkene diastereomers (**cis–trans isomers**). These terms are unambiguous, however, only when applied to disubstituted alkenes. If the alkene is trisubstituted or tetrasubstituted, the terms cis and trans are either ambiguous or do not apply at all. Consider the following alkene as an example:

$$
\begin{array}{ccc}
\text{Br} & & \text{Cl} \\
& \text{C}=\text{C} & \\
\text{H} & & \text{F} \\
& \textbf{A} &
\end{array}
$$

It is impossible to decide whether **A** is cis or trans since no two groups are the same.

A system that works in all cases is based on the priorities of groups in the Cahn–Ingold–Prelog convention (Section 5.7). This system, called the **(*E*)–(*Z*) system**, applies to alkene diastereomers of all types.

7.2A HOW TO Use The (*E*)–(*Z*) System

1. Examine the two groups attached to one carbon atom of the double bond and decide which has higher priority.

2. Repeat that operation at the other carbon atom:

Higher priority → Cl, F

Higher priority → Br, H

Higher priority → F, Cl ← Higher priority

Higher priority → Br, H

Cl > F

Br > H

(*Z*)-2-Bromo-1-chloro-1-fluoroethene

(*E*)-2-Bromo-1-chloro-1-fluoroethene

3. Compare the group of higher priority on one carbon atom with the group of higher priority on the other carbon atom. If the two groups of higher priority are on the same side of the double bond, the alkene is designated (Z) (from the German word *zusammen,* meaning together). If the two groups of higher priority are on opposite sides of the double bond, the alkene is designated (E) (from the German word *entgegen,* meaning opposite). The following isomers provide another example.

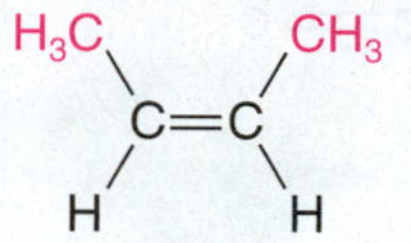

$CH_3 > H$

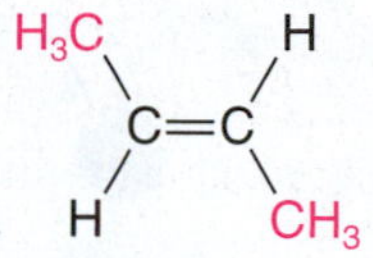

(Z)-2-Butene or (Z)-but-2-ene
(cis-2-butene)

(E)-2-Butene or (E)-but-2-ene
(trans-2-butene)

••• SOLVED PROBLEM 7.1

The two stereoisomers of 1-bromo-1,2-dichloroethene cannot be designated as cis and trans in the normal way because the double bond is trisubstituted. They can, however, be given (E) and (Z) designations. Write a structural formula for each isomer and give each the proper designation.

STRATEGY AND ANSWER: We write the structures (below), then note that chlorine has a higher priority than hydrogen, and bromine has a higher priority than chlorine. The group with higher priority on C1 is bromine and the group with higher priority at C2 is chlorine. In the first structure the higher priority chlorine and bromine atoms are on opposite sides of the double bond, and therefore this isomer is (E). In the second structure those chlorine and bromine atoms are on the same side, so the latter isomer is (Z).

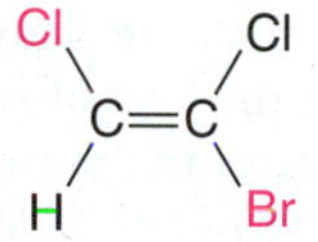

$Cl > H$
$Br > Cl$

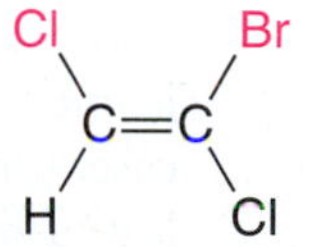

(E)-1-Bromo-1,2-dichloroethene

(Z)-1-Bromo-1,2-dichloroethene

••• PRACTICE PROBLEM 7.1

Using the (E)–(Z) designation [and in parts (e) and (f) the (R)–(S) designation as well] give IUPAC names for each of the following:

(a) 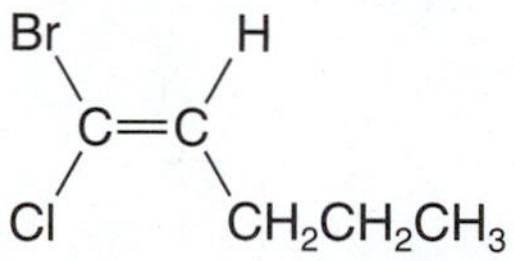 (b) 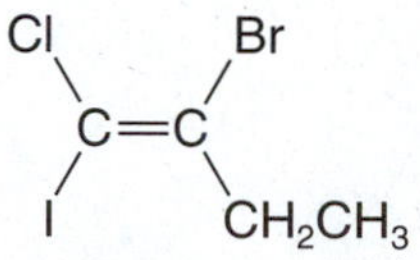 (c) (d) 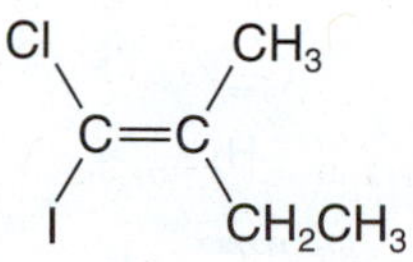 (e) (f)

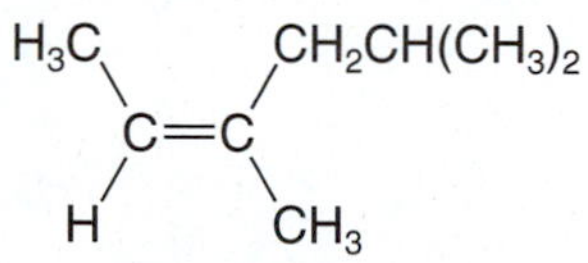

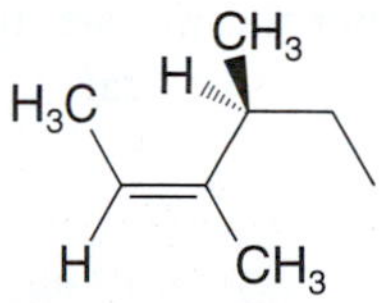

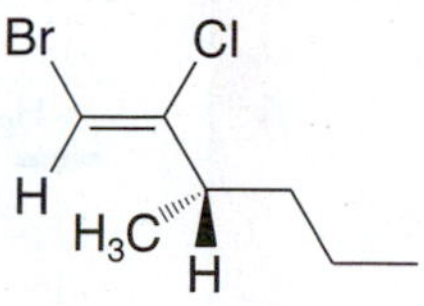

7.3 RELATIVE STABILITIES OF ALKENES

Cis and trans isomers of alkenes do not have the same stability.

• Strain caused by crowding of two alkyl groups on the same side of a double bond makes cis isomers generally less stable than trans isomers (Fig. 7.1).

This effect can be measured quantitatively by comparing thermodynamic data from experiments involving alkenes with related structures, as we shall see later.

FIGURE 7.1 Cis and trans alkene isomers. The cis isomer is less stable due to greater strain from crowding by the adjacent alkyl groups.

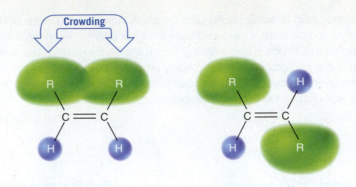

7.3A Heat of Reaction

The addition of hydrogen to an alkene (**hydrogenation**, Sections 4.16A and 7.13) is an exothermic reaction; the enthalpy change involved is called the **heat of reaction** or, in this specific case, the **heat of hydrogenation**.

$$\underset{\diagdown}{\overset{\diagup}{C}}=\underset{\diagup}{\overset{\diagdown}{C}} + H{-}H \xrightarrow{Pt} \overset{|}{\underset{\underset{H}{|}}{C}}{-}\overset{|}{\underset{\underset{H}{|}}{C}}{-} \qquad \Delta H^\circ \cong -120 \text{ kJ mol}^{-1}$$

We can gain a quantitative measure of relative alkene stabilities by comparing the heats of hydrogenation for a family of alkenes that all become the same alkane product on hydrogenation. The results of such an experiment involving platinum-catalyzed hydrogenation of three butene isomers are shown in Fig. 7.2. All three isomers yield the same product—butane—but the heat of reaction is different in each case. On conversion to butane, 1-butene liberates the most heat (127 kJ mol^{-1}), followed by *cis*-2-butene (120 kJ mol^{-1}), with *trans*-2-butene producing the least heat (115 kJ mol^{-1}). These data indicate that the trans isomer is more stable than the cis isomer, since less energy is released when the trans isomer is converted to butane. Furthermore, it shows that the terminal alkene, 1-butene, is less stable than either of the disubstituted alkenes, since its reaction is the most exothermic. Of course, alkenes that do not yield the same hydrogenation products cannot be compared on the basis of their respective heats of hydrogenation. In such cases it is necessary to compare other thermochemical data, such as heats of combustion, although we will not go into analyses of that type here.

FIGURE 7.2 An energy diagram for platinum-catalyzed hydrogenation of the three butene isomers. The order of stability based on the differences in their heats of hydrogenation is *trans*-2-butene > *cis*-2-butene > 1-butene.

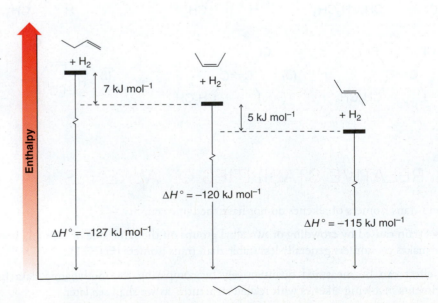

7.3B Overall Relative Stabilities of Alkenes

Studies of numerous alkenes reveal a pattern of stabilities that is related to the number of alkyl groups attached to the carbon atoms of the double bond.

- The greater the number of attached alkyl groups (i.e., the more highly substituted the carbon atoms of the double bond), the greater is the alkene's stability.

This order of stabilities can be given in general terms as follows:*

Relative Stabilities of Alkenes

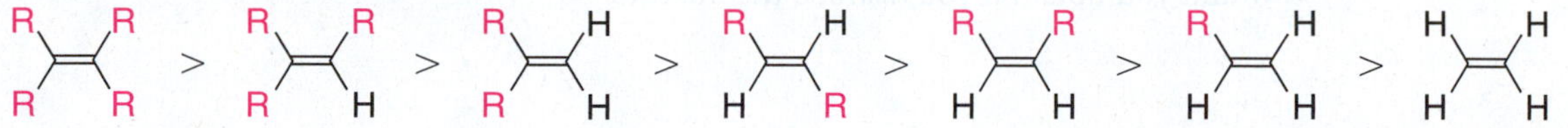

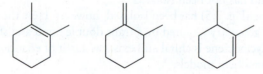

Tetrasubstituted   Trisubstituted   ⟵———————— Disubstituted ————————⟶   Monosubstituted   Unsubstituted

●●● **SOLVED PROBLEM 7.2**

Consider the two alkenes 2-methyl-1-pentene and 2-methyl-2-pentene and decide which would be most stable.

STRATEGY AND ANSWER: First write the structures of the two alkenes, then decide how many substituents the double bond of each has.

2-Methyl-1-pentene
(disubstituted, less stable)

2-Methyl-2-pentene
(trisubstituted, more stable)

2-Methyl-2-pentene has three substituents on its double bond, whereas 2-methyl-1-pentene has two, and therefore 2-methyl-2-pentene is the more stable.

●●●
PRACTICE PROBLEM 7.2

Rank the following cycloalkenes in order of increasing stability.

●●●
PRACTICE PROBLEM 7.3

Heats of hydrogenation of three alkenes are as follows:

2-methyl-1-butene (-119 kJ mol^{-1})

3-methyl-1-butene (-127 kJ mol^{-1})

2-methyl-2-butene (-113 kJ mol^{-1})

(a) Write the structure of each alkene and classify it as to whether its doubly bonded atoms are monosubstituted, disubstituted, trisubstituted, or tetrasubstituted. **(b)** Write the structure of the product formed when each alkene is hydrogenated. **(c)** Can heats of hydrogenation be used to relate the relative stabilities of these three alkenes? **(d)** If so, what is the predicted order of stability? If not, why not? **(e)** What other alkene isomers are possible for these alkenes? Write their structures. **(f)** What are the relative stabilities among just these isomers?

*This order of stabilities may seem contradictory when compared with the explanation given for the relative stabilities of cis and trans isomers. Although a detailed explanation of the trend given here is beyond our scope, the relative stabilities of substituted alkenes can be rationalized. Part of the explanation can be given in terms of the electron-releasing effect of alkyl groups (Section 6.11B), an effect that satisfies the electron-withdrawing properties of the sp^2-hybridized carbon atoms of the double bond.

● ● ●

PRACTICE PROBLEM 7.4   Predict the more stable alkene of each pair: **(a)** 2-methyl-2-pentene or 2,3-dimethyl-2-butene; **(b)** *cis*-3-hexene or *trans*-3-hexene; **(c)** 1-hexene or *cis*-3-hexene; **(d)** *trans*-2-hexene or 2-methyl-2-pentene.

● ● ●

PRACTICE PROBLEM 7.5   How many stereoisomers are possible for 4-methyl-2-hexene, and how many fractions would you obtain if you distilled the mixture?

7.4 CYCLOALKENES

The rings of cycloalkenes containing five carbon atoms or fewer exist only in the cis form (Fig. 7.3). The introduction of a trans double bond into rings this small would, if it were possible, introduce greater strain than the bonds of the ring atoms could accommodate.

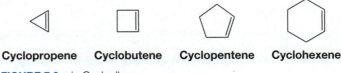

Cyclopropene   **Cyclobutene**   **Cyclopentene**   **Cyclohexene**

FIGURE 7.3 *cis*-Cycloalkenes.

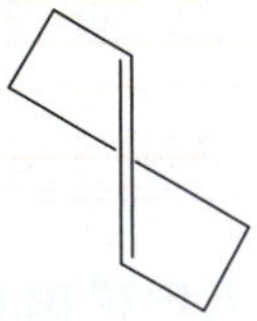

FIGURE 7.4 Hypothetical *trans*-cyclohexene. This molecule is apparently too strained to exist at room temperature.

(Verify this with handheld molecular models.) *trans*-Cyclohexene might resemble the structure shown in Fig. 7.4. There is evidence that it can be formed as a very reactive short-lived intermediate in some chemical reactions, but it is not isolable as a stable molecule.

    trans-Cycloheptene has been observed spectroscopically, but it is a substance with a very short lifetime and has not been isolated.

    trans-Cyclooctene (Fig. 7.5) has been isolated, however. Here the ring is large enough to accommodate the geometry required by a trans double bond and still be stable at room temperature. *trans*-Cyclooctene is chiral and exists as a pair of enantiomers. You may wish to verify this using handheld models.

Helpful Hint

Exploring all of these cycloalkenes with handheld molecular models, including both enantiomers of *trans*-cyclooctene, will help illustrate their structural differences.

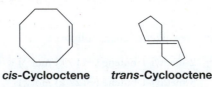

***cis*-Cyclooctene**   ***trans*-Cyclooctene**

FIGURE 7.5 The cis and trans forms of cyclooctene.

7.5 SYNTHESIS OF ALKENES VIA ELIMINATION REACTIONS

Elimination reactions are the most important means for synthesizing alkenes. In this chapter we shall study two methods for alkene synthesis based on elimination reactions: dehydrohalogenation of alkyl halides and dehydration of alcohols.

Dehydrohalogenation of Alkyl Halides (Sections 6.15, 6.16, and 7.6)

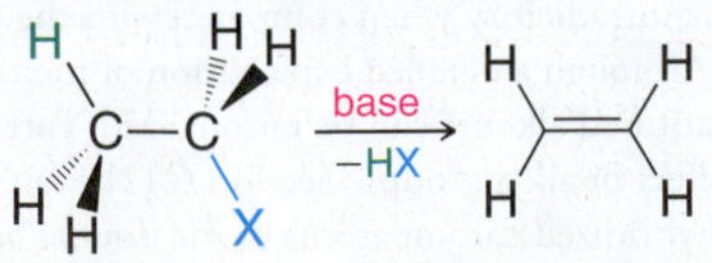

Dehydration of Alcohols (Sections 7.7 and 7.8)

$$\text{CH}_3\text{CH}_2\text{OH} \xrightarrow[-\text{HOH}]{\text{H}^+,\ \text{heat}} \text{CH}_2{=}\text{CH}_2$$

7.6 DEHYDROHALOGENATION OF ALKYL HALIDES

- The best reaction conditions to use when synthesizing an alkene by **dehydrohalogenation** are those that promote an E2 mechanism.

In an E2 mechanism, a base removes a β hydrogen from the β carbon, as the double bond forms and a leaving group departs from the α carbon.

$$\overset{\beta}{C}\!-\!\overset{\alpha}{C}\,X \xrightarrow{\text{E2}} \ \ C{=}C \ + \ B{:}H \ + \ {:}X^-$$

Reaction conditions that favor elimination by an E1 mechanism should be avoided because the results can be too variable. The carbocation intermediate that accompanies an E1 reaction can undergo rearrangement of the carbon skeleton, as we shall see in Section 7.8, and it can also undergo substitution by an S_N1 mechanism, which competes strongly with formation of products by an E1 path.

7.6A HOW TO Favor an E2 Mechanism

1. **Use a secondary or tertiary alkyl halide if possible.**
   Why? Because steric hindrance in the substrate will inhibit substitution.

2. **When a synthesis must begin with a primary alkyl halide, use a bulky base.**
   Why? Because the steric bulk of the base will inhibit substitution.

3. **Use a high concentration of a strong and nonpolarizable base such as an alkoxide.**
   Why? Because a weak and polarizable base would not drive the reaction toward a bimolecular reaction, thereby allowing unimolecular processes (such as S_N1 or E1 reactions) to compete.

4. **Sodium ethoxide in ethanol (EtONa/EtOH) and potassium *tert*-butoxide in *tert*-butyl alcohol (*t*-BuOK/*t*-BuOH) are bases typically used to promote E2 reactions.**
   Why? Because they meet criterion 3 above. Note that in each case the alkoxide base is dissolved in its corresponding alcohol. (Potassium hydroxide dissolved in ethanol or *tert*-butyl alcohol is also sometimes used, in which case the active base includes both the alkoxide and hydroxide species present at equilibrium.)

5. **Use elevated temperature because heat generally favors elimination over substitution.**
   Why? Because elimination reactions are entropically favored over substitution reactions (because the products are greater in number than the reactants). Hence $\Delta S°$ in the Gibbs free-energy equation, $\Delta G° = \Delta H° - T\Delta S°$ is significant, and $\Delta S°$ will be increased by higher temperature since T is a coefficient, leading to a more negative (favorable) $\Delta G°$.

7.6B Zaitsev's Rule: Formation of the More Substituted Alkene Is Favored with a Small Base

We showed examples in Sections 6.15–6.17 of dehydrohalogenations where only a single elimination product was possible. For example:

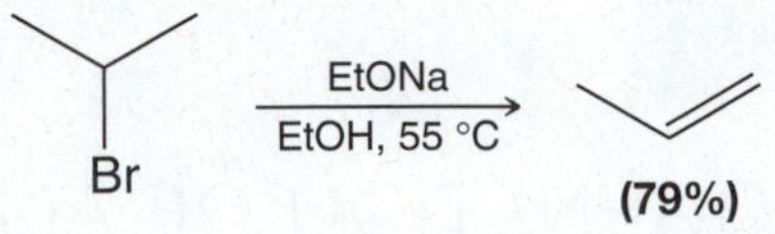

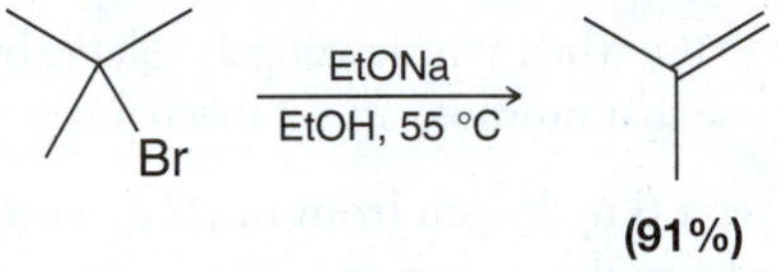

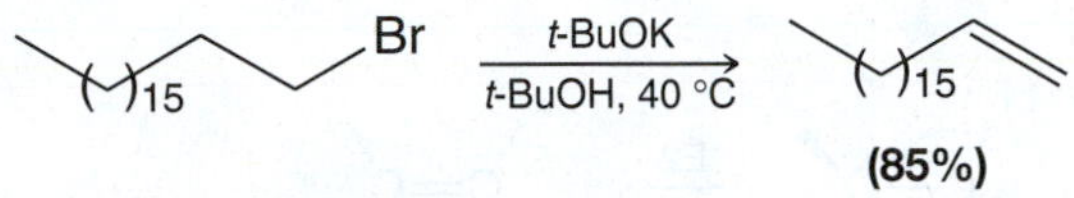

Dehydrohalogenation of many alkyl halides, however, yields more than one product. For example, dehydrohalogenation of 2-bromo-2-methylbutane can yield two products: 2-methyl-2-butene and 2-methyl-1-butene, as shown here by pathways (a) and (b), respectively:

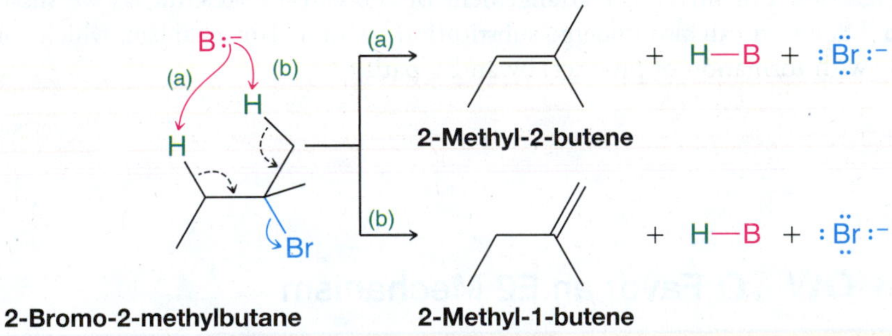

- If we use a small base such as ethoxide or hydroxide, the major product of the reaction will be the more highly substituted alkene (which is also the more stable alkene).

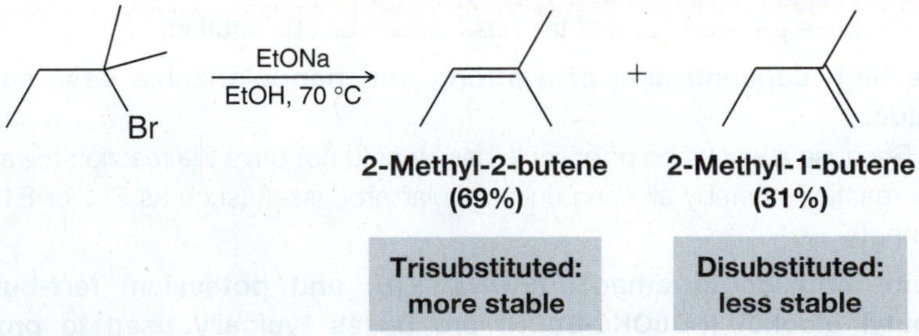

2-Methyl-2-butene is a trisubstituted alkene (three methyl groups are attached to carbon atoms of the double bond), whereas 2-methyl-1-butene is only disubstituted. 2-Methyl-2-butene is the major product.

- Whenever an elimination occurs to give the more stable, more highly substituted alkene, chemists say that the elimination follows **Zaitsev's rule**, named for the nineteenth-century Russian chemist A. N. Zaitsev (1841–1910) who formulated it. (Zaitsev's name is also transliterated as Zaitzev, Saytzeff, Saytseff, or Saytzev.)

The reason for this behavior is related to the double-bond character that develops in the transition state (cf. Section 6.16) for each reaction:

Helpful Hint

The Zaitsev product is that which is the more stable product.

The β hydrogen and leaving group are anti coplanar.

Transition state for an E2 reaction

The carbon–carbon bond has the developing character of a double bond.

The transition state for the reaction leading to 2-methyl-2-butene (Fig. 7.6) has the developing character of the double bond in a trisubstituted alkene. The transition state for the reaction leading to 2-methyl-1-butene has the developing character of a double bond in a disubstituted alkene. Because the transition state leading to 2-methyl-2-butene resembles a more stable alkene, this transition state is more stable (recall the Hammond–Leffler postulate, Fig. 6.10). Because this transition state is more stable (occurs at lower free energy), the free energy of activation for this reaction is lower and 2-methyl-2-butene is formed faster. This explains why 2-methyl-2-butene is the major product.

- In general, the preferential formation of one product because the free energy of activation leading to its formation is lower than that for another product, and therefore the rate of its formation faster, is called **kinetic control** of product formation. (See also Section 13.10A.)

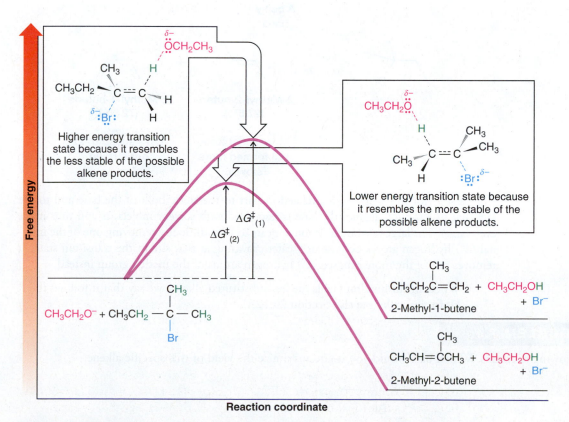

FIGURE 7.6 Reaction (2) leading to the more stable alkene occurs faster than reaction (1) leading to the less stable alkene; $\Delta G^{\ddagger}_{(2)}$ is less than $\Delta G^{\ddagger}_{(1)}$.

●●● SOLVED PROBLEM 7.3

Using Zaitsev's rule, predict which would be the major product of the following reaction:

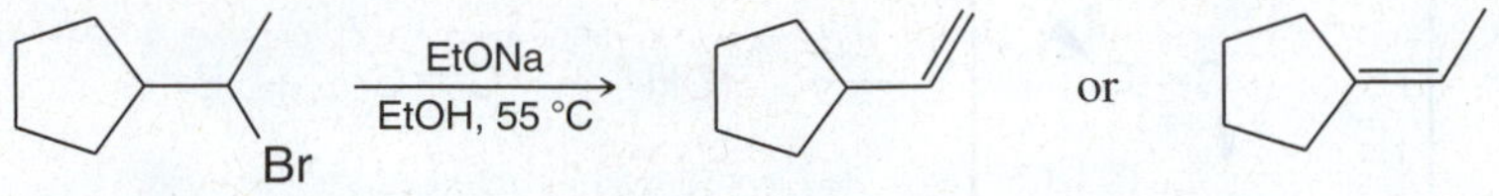

A   **B**

STRATEGY AND ANSWER: Alkene **B** has a trisubstituted double bond whereas the double bond of **A** is only mono-substituted. Therefore, **B** is more stable and, according to Zaitsev's rule, would be the major product.

●●●

PRACTICE PROBLEM 7.6   Predict the major product formed when 2-bromobutane is subjected to dehydrobromi-nation using sodium ethoxide in ethanol at 55 °C.

●●●

PRACTICE PROBLEM 7.7   List the alkenes that would be formed when each of the following alkyl halides is subjected to dehydrohalogenation with potassium ethoxide in ethanol and use Zaitsev's rule to predict the major product of each reaction: **(a)** 2-bromo-3-methylbutane and **(b)** 2-bromo-2,3-dimethylbutane.

7.6C Formation of the Less Substituted Alkene Using a Bulky Base

- Carrying out dehydrohalogenations with a bulky base such as potassium *tert*-butoxide (*t*-BuOK) in *tert*-butyl alcohol (*t*-BuOH) favors the formation of the **less substituted alkene:**

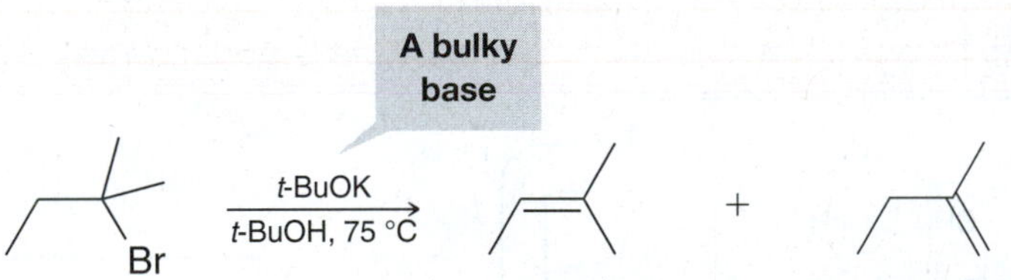

2-Methyl-2-butene   **2-Methyl-1-butene**
(27.5%)   **(72.5%)**

| More substituted but formed more slowly. | Less substituted but formed faster. |

The reasons for this behavior are related in part to the steric bulk of the base and to the fact that in *tert*-butyl alcohol the base is associated with solvent molecules and thus made even larger. The large *tert*-butoxide ion appears to have difficulty removing one of the internal (2°) hydrogen atoms because of greater crowding at that site in the transition state. It removes one of the more exposed (1°) hydrogen atoms of the methyl group instead.

- When an elimination yields the less substituted alkene, we say that it follows the **Hofmann rule** (see also Section 20.12A).

●●● SOLVED PROBLEM 7.4

Your task is the following synthesis. Which base would you use to maximize the yield of this specific alkene?

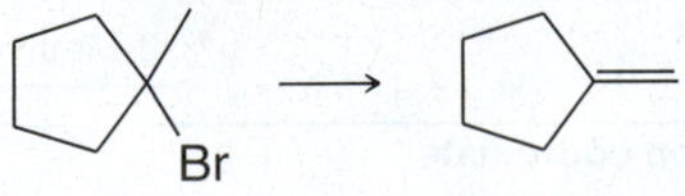

STRATEGY AND ANSWER: Here you want the Hofmann rule to apply (you want the less substituted alkene to be formed). Therefore, use a bulky base such as potassium *tert*-butoxide in *tert*-butyl alcohol.

Examine Solved Problem 7.3. Your task is to prepare **A** in the highest possible yield by dehydrobromination. Which base would you use?

7.6D The Stereochemistry of E2 Reactions: The Orientation of Groups in the Transition State

- The five atoms involved in the transition state of an E2 reaction (including the base) must be **coplanar**, i.e., lie in the same plane.

The requirement for coplanarity of the H—C—C—LG unit arises from a need for proper overlap of orbitals in the developing π bond of the alkene that is being formed (see Section 6.16). There are two ways that this can happen:

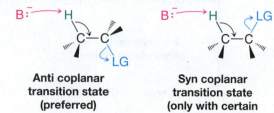

Anti coplanar transition state (preferred)   **Syn coplanar transition state (only with certain rigid molecules)**

- The **anti coplanar** conformation is the preferred transition state geometry.

The **syn coplanar** transition state occurs only with rigid molecules that are unable to assume the anti arrangement. The reason: the anti coplanar transition state is staggered (and therefore of lower energy), while the syn coplanar transition state is eclipsed. Practice Problem 7.9 will help to illustrate this difference.

Helpful Hint

Be able to draw a three-dimensional representation of an anti coplanar E2 transition state.

PRACTICE PROBLEM 7.9

Consider a simple molecule such as ethyl bromide and show with Newman projection formulas how the anti coplanar transition state would be favored over the syn coplanar one.

Part of the evidence for the preferred anti coplanar arrangement of groups comes from experiments done with cyclic molecules. Two groups axially oriented on adjacent carbons in a chair conformation of cyclohexane are anti coplanar. If one of these groups is a hydrogen and the other a leaving group, the geometric requirements for an anti coplanar E2 transition state are met. Neither an axial–equatorial nor an equatorial–equatorial orientation of the groups allows formation of an anti coplanar transition state. (Note that there are no syn coplanar groups in a chair conformation, either.)

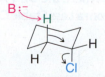

Here the β hydrogen and the chlorine are both axial. This allows an anti coplanar transition state.   **A Newman projection formula shows that the β hydrogen and the chlorine are anti coplanar when they are both axial.**

As examples, let us consider the different behavior in E2 reactions shown by two compounds containing cyclohexane rings, neomenthyl chloride and menthyl chloride:

Neomenthyl chloride   **Menthyl chloride**

Helpful Hint

Examine the conformations of neomenthyl chloride using handheld models.

In the more stable conformation of neomenthyl chloride (see the following mechanism), the alkyl groups are both equatorial and the chlorine is axial. There are also axial hydrogen atoms on both C1 and C3. The base can attack either of these hydrogen atoms and achieve an anti coplanar transition state for an E2 reaction. Products corresponding to each of these transition states (2-menthene and 1-menthene) are formed rapidly. In accordance with Zaitsev's rule, 1-menthene (with the more highly substituted double bond) is the major product.

[**A MECHANISM FOR THE REACTION** — **E2 Elimination Where There Are Two Axial β Hydrogens**]

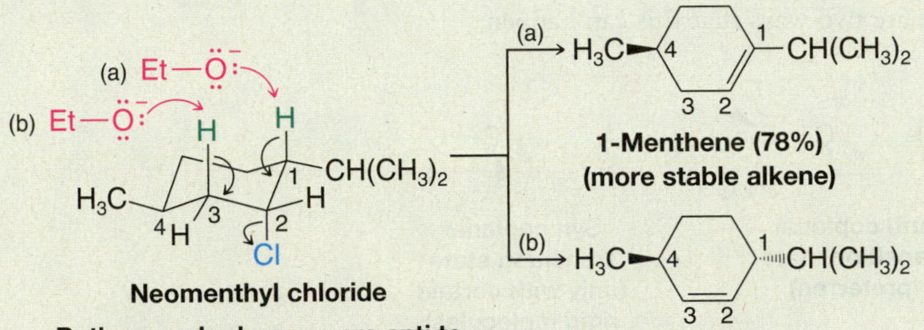

Neomenthyl chloride

Both green hydrogens are anti to the chlorine in this, the more stable conformation. Elimination by path (a) leads to 1-menthene; by path (b) to 2-menthene.

1-Menthene (78%)
(more stable alkene)

2-Menthene (22%)
(less stable alkene)

On the other hand, the more stable conformation of menthyl chloride has all three groups (including the chlorine) equatorial. For the chlorine to become axial, menthyl chloride has to assume a conformation in which the large isopropyl group and the methyl group are also axial. This conformation is of much higher energy, and the free energy of activation for the reaction is large because it includes the energy necessary for the conformational change. Consequently, menthyl chloride undergoes an E2 reaction very slowly, and the product is entirely 2-menthene because the hydrogen atom at C1 cannot be anti to the chlorine. This product (or any resulting from an elimination to yield the less substituted alkene) is sometimes called the *Hofmann product* (Sections 7.6C and 20.12A).

[**A MECHANISM FOR THE REACTION** — **E2 Elimination Where the Only Axial β Hydrogen Is from a Less Stable Conformer**]

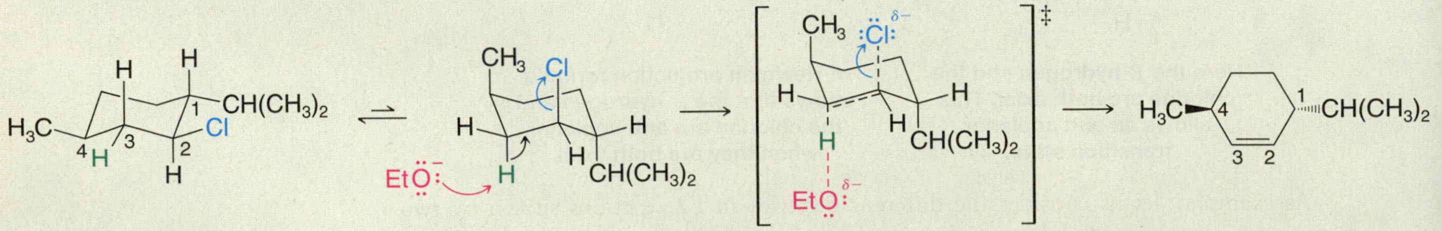

Menthyl chloride
(more stable conformation)
Elimination is not possible for this conformation because no hydrogen is anti to the leaving group.

Menthyl chloride
(less stable conformation)
Elimination is possible from this conformation because the green hydrogen is anti to the chlorine.

The transition state for the E2 elimination is anti coplanar.

2-Menthene (100%)

Predict the major product formed when the following compound is subjected to dehydrochlorination with sodium ethoxide in ethanol.

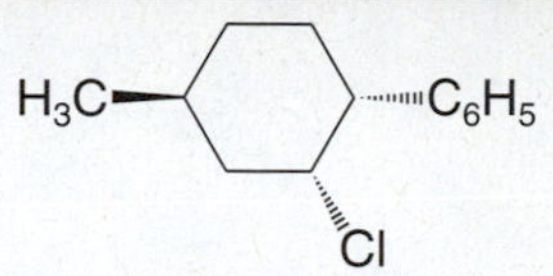

STRATEGY AND ANSWER: We know that for an E2 dehydrochlorination to take place the chlorine will have to be
axial. The following conformation has the chlorine axial and has two hydrogen atoms that are anti coplanar to the chlorine. Two products will be formed but **(B)** being more stable should be the major product.

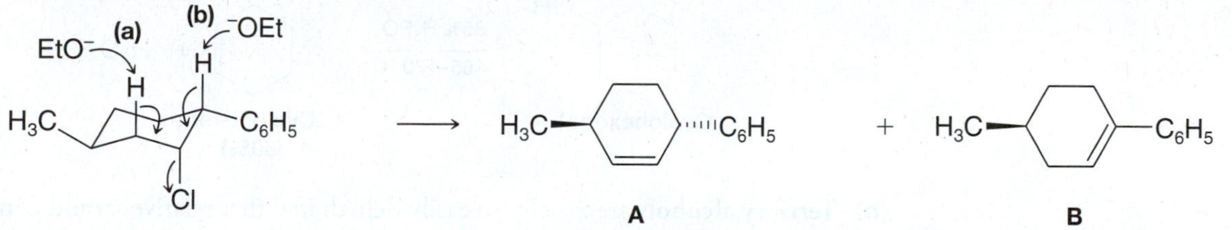

A

Disubstituted, less stable
(minor product)

B

Trisubstituted, more stable
(major product)

When *cis*-1-bromo-4-*tert*-butylcyclohexane is treated with sodium ethoxide in ethanol, it
reacts rapidly; the product is 4-*tert*-butylcyclohexene. Under the same conditions, *trans*-
1-bromo-4-*tert*-butylcyclohexane reacts very slowly. Write conformational structures and
explain the difference in reactivity of these cis–trans isomers.

PRACTICE PROBLEM 7.10

(a) When *cis*-1-bromo-2-methylcyclohexane undergoes an E2 reaction, two products
(cycloalkenes) are formed. What are these two cycloalkenes, and which would you expect
to be the major product? Write conformational structures showing how each is formed.
(b) When *trans*-1-bromo-2-methylcyclohexane reacts in an E2 reaction, only one cycloalkene is formed. What is this product? Write conformational structures showing why it
is the only product.

PRACTICE PROBLEM 7.11

7.7 ACID-CATALYZED DEHYDRATION OF ALCOHOLS

- Most alcohols undergo **dehydration** (lose a molecule of water) to form an alkene
  when heated with a strong acid.

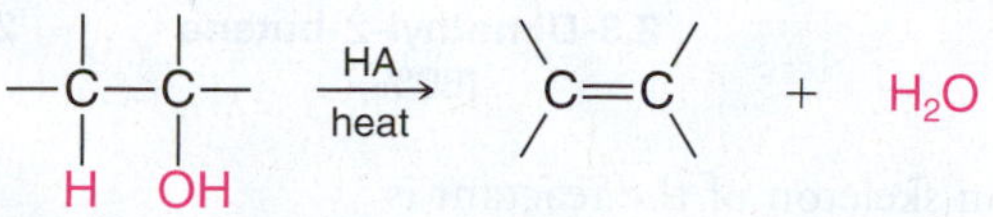

$$\underset{\underset{H}{|} \quad \underset{OH}{|}}{-C-C-} \quad \xrightarrow[\text{heat}]{HA} \quad \overset{}{\underset{}{C}}=\overset{}{\underset{}{C}} \quad + \quad H_2O$$

The reaction is an **elimination** and is favored at higher temperatures (Section 6.18A).
The most commonly used acids in the laboratory are Brønsted acids—proton donors
such as sulfuric acid and phosphoric acid. Lewis acids such as alumina (Al_2O_3) are often
used in industrial, gas-phase dehydrations.

1. **The temperature and concentration of acid required to dehydrate an alcohol
   depend on the structure of the alcohol substrate.**

   (a) Primary alcohols are the most difficult to dehydrate. Dehydration of ethanol,
   for example, requires concentrated sulfuric acid and a temperature of 180 °C:

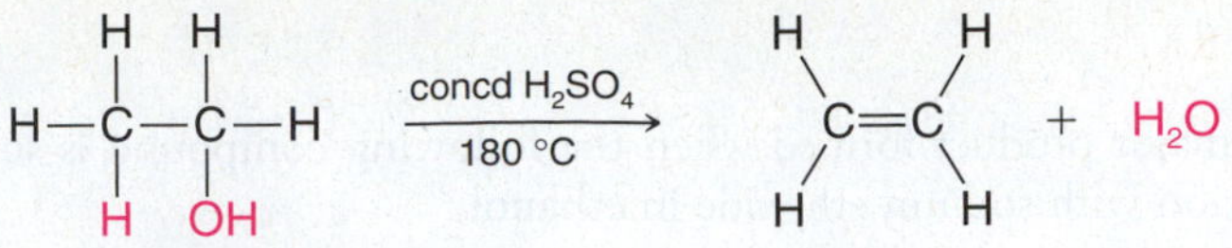

$$\underset{\substack{\text{Ethanol}\\\text{(a 1° alcohol)}}}{\overset{\displaystyle H-\overset{\displaystyle H}{\underset{\displaystyle H}{C}}-\overset{\displaystyle H}{\underset{\displaystyle OH}{C}}-H}{}} \xrightarrow[\text{180 °C}]{\text{concd H}_2\text{SO}_4} \underset{\text{Ethene}}{\overset{\displaystyle \overset{H}{\underset{H}{C}}=\overset{H}{\underset{H}{C}}}{}} + \;{\color{red}\text{H}_2\text{O}}$$

(b) Secondary alcohols usually dehydrate under milder conditions. Cyclohexanol, for example, dehydrates in 85% phosphoric acid at 165–170 °C:

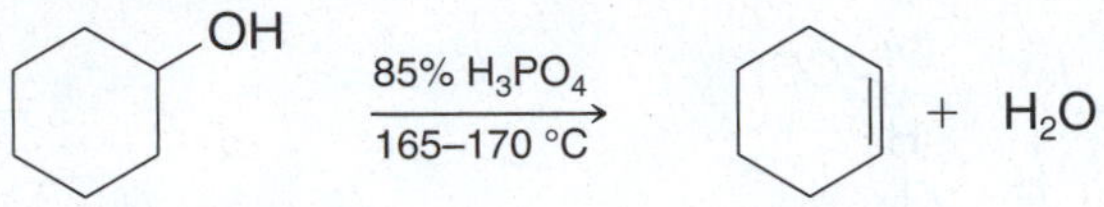

Cyclohexanol $\xrightarrow[\text{165–170 °C}]{85\% \text{ H}_3\text{PO}_4}$ Cyclohexene (80%) $+\;\text{H}_2\text{O}$

(c) Tertiary alcohols are usually so easily dehydrated that relatively mild conditions can be used. *tert*-Butyl alcohol, for example, dehydrates in 20% aqueous sulfuric acid at a temperature of 85 °C:

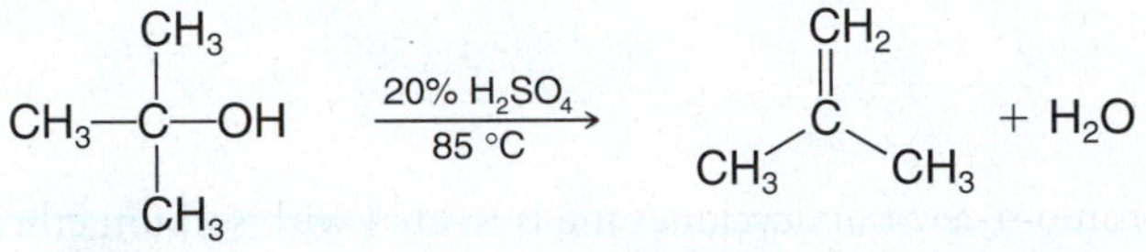

$$\underset{\substack{\textit{tert}\text{-Butyl}\\\text{alcohol}}}{\overset{\displaystyle CH_3-\overset{\displaystyle CH_3}{\underset{\displaystyle CH_3}{C}}-OH}{}} \xrightarrow[\text{85 °C}]{20\% \text{ H}_2\text{SO}_4} \underset{\substack{\text{2-Methylpropene}\\\text{(84%)}}}{\overset{\displaystyle \overset{CH_2}{\underset{CH_3\quad CH_3}{C}}}{}} + \;\text{H}_2\text{O}$$

- The relative ease with which alcohols undergo dehydration is 3° > 2° > 1°.

$$\underset{\text{3° Alcohol}}{\overset{\displaystyle R-\overset{\displaystyle R}{\underset{\displaystyle R}{C}}-OH}{}} \;>\; \underset{\text{2° Alcohol}}{\overset{\displaystyle R-\overset{\displaystyle R}{\underset{\displaystyle H}{C}}-OH}{}} \;>\; \underset{\text{1° Alcohol}}{\overset{\displaystyle R-\overset{\displaystyle H}{\underset{\displaystyle H}{C}}-OH}{}}$$

This behavior, as we shall see in Section 7.7B, is related to the relative stabilities of carbocations.

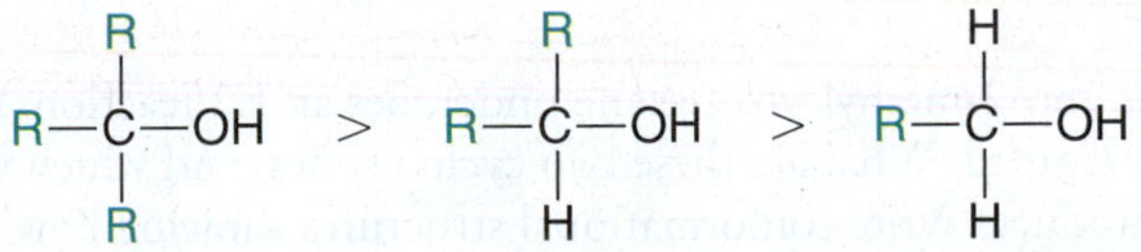

2. **Some primary and secondary alcohols also undergo rearrangements of their carbon skeletons during dehydration.** Such a rearrangement occurs in the dehydration of 3,3-dimethyl-2-butanol:

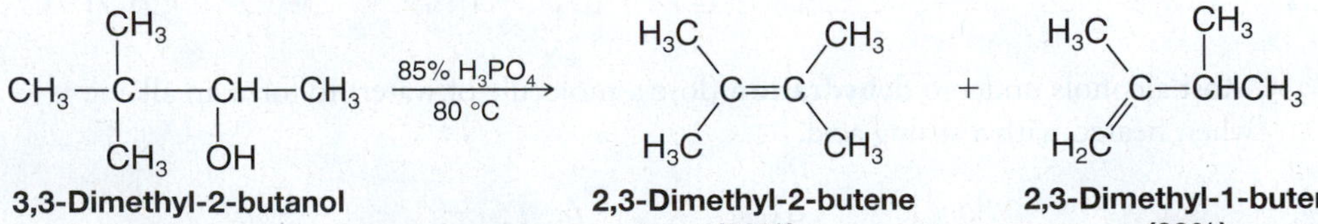

$$\underset{\text{3,3-Dimethyl-2-butanol}}{\overset{\displaystyle CH_3-\overset{\displaystyle CH_3}{\underset{\displaystyle CH_3}{C}}-\overset{\displaystyle}{\underset{\displaystyle OH}{C}H}-CH_3}{}} \xrightarrow[\text{80 °C}]{85\% \text{ H}_3\text{PO}_4} \underset{\substack{\text{2,3-Dimethyl-2-butene}\\\text{(80%)}}}{\overset{\displaystyle \overset{H_3C\quad CH_3}{\underset{H_3C\quad CH_3}{C=C}}}{}} + \underset{\substack{\text{2,3-Dimethyl-1-butene}\\\text{(20%)}}}{\overset{\displaystyle \overset{H_3C\quad CH_3}{\underset{H_2C}{C-CHCH_3}}}{}}$$

Notice that the carbon skeleton of the reactant is

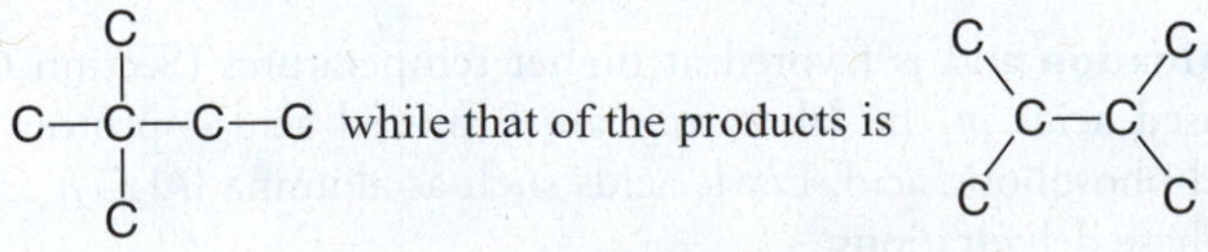

$$\overset{\displaystyle C}{\underset{\displaystyle C}{C-C-C-C}} \text{ while that of the products is } \overset{\displaystyle C\quad\quad C}{\underset{\displaystyle C\quad\quad C}{C-C}}$$

| The carbon skeleton has rearranged |

We shall see in Section 7.8 that this reaction involves the migration of a methyl group from one carbon to the next so as to form a more stable carbocation. (Rearrangements to carbocations of approximately equal energy may also be possible with some substrates.)

7.7A Mechanism for Dehydration of Secondary and Tertiary Alcohols: An E1 Reaction

Explanations for these observations can be based on a stepwise mechanism originally proposed by F. Whitmore (of Pennsylvania State University).

The mechanism is an E1 reaction in which the substrate is a protonated alcohol. Consider the dehydration of *tert*-butyl alcohol as an example:

Step 1

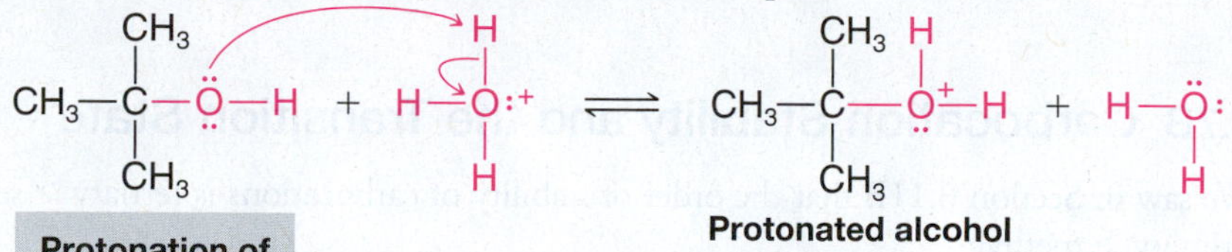

Protonation of the alcohol

Protonated alcohol

In this step, an acid–base reaction, a proton is rapidly transferred from the acid to one of the unshared electron pairs of the alcohol. In dilute sulfuric acid the acid is a hydronium ion; in concentrated sulfuric acid the initial proton donor is sulfuric acid itself. This step is characteristic of all reactions of an alcohol with a strong acid.

The presence of the positive charge on the oxygen of the protonated alcohol weakens all bonds to oxygen, including the carbon–oxygen bond, and in step 2 the carbon–oxygen bond breaks. The leaving group is a molecule of water:

Step 2

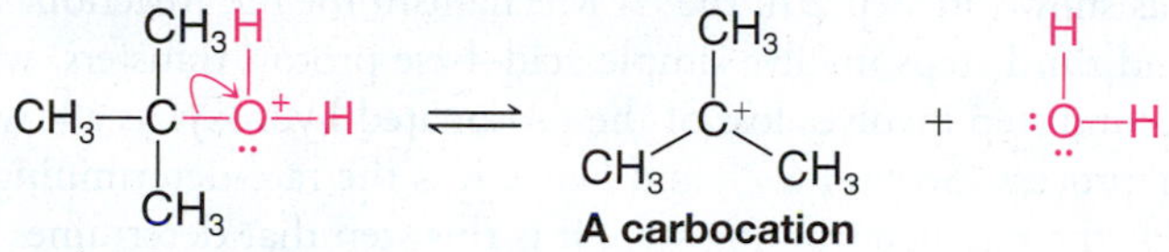

A carbocation

Departure of a water molecule

The carbon–oxygen bond breaks **heterolytically.** The bonding electrons depart with the water molecule and leave behind a carbocation. The carbocation is, of course, highly reactive because the central carbon atom has only six electrons in its valence level, not eight.

Finally, in step 3, a water molecule removes a proton from the β carbon of the carbocation by the process shown below. The result is the formation of a hydronium ion and an alkene:

Step 3

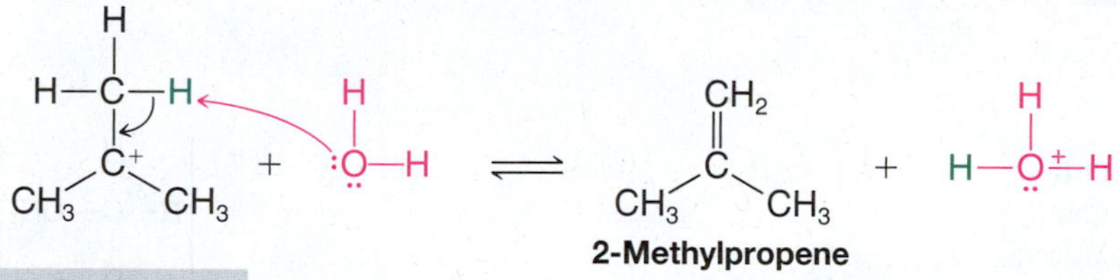

2-Methylpropene

Removal of a β hydrogen

In step 3, also an acid–base reaction, any one of the nine protons available at the three methyl groups can be transferred to a molecule of water. The electron pair left behind when a proton is removed becomes the second bond of the double bond of the alkene. Notice that this step restores an octet of electrons to the central carbon atom. An orbital representation of this process, with the transition state, is as follows.

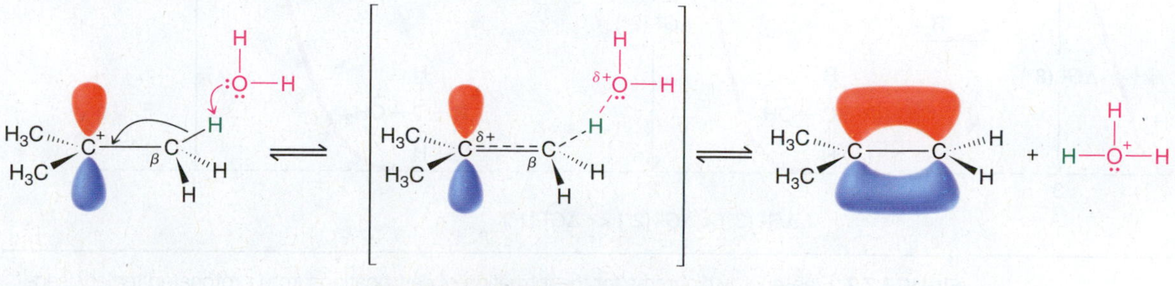

Transition state for removal of a proton from the β carbon of the carbocation

• • •

PRACTICE PROBLEM 7.12 Dehydration of 2-propanol occurs in 14 M H_2SO_4 at 100 °C. **(a)** Using curved arrows, write all steps in a mechanism for the dehydration. **(b)** Explain the essential role performed in alcohol dehydrations by the acid catalyst. [*Hint:* Consider what would have to happen if no acid were present.]

7.7B Carbocation Stability and the Transition State

We saw in Section 6.11B that the order of stability of carbocations is tertiary > secondary > primary > methyl:

$$R-\overset{R}{\underset{R}{\overset{|}{C}}}{}^{+} \quad \cdot > \quad R-\overset{H}{\underset{R}{\overset{|}{C}}}{}^{+} \quad > \quad R-\overset{H}{\underset{H}{\overset{|}{C}}}{}^{+} \quad > \quad H-\overset{H}{\underset{H}{\overset{|}{C}}}{}^{+}$$

3°	>	2°	>	1°	>	Methyl
(most stable)						**(least stable)**

In the dehydration of secondary and tertiary alcohols the slowest step is formation of the carbocation as shown in step 2 of the "A Mechanism for the Reaction" box in this section. The first and third steps involve simple acid–base proton transfers, which occur very rapidly. The second step involves loss of the protonated hydroxyl as a leaving group, a highly endergonic process (Section 6.7), and hence it is the rate-determining step.

Because step 2 is the rate-determining step, it is this step that determines the overall reactivity of alcohols toward dehydration. With that in mind, we can now understand why tertiary alcohols are the most easily dehydrated. The formation of a tertiary carbocation is easiest because the free energy of activation for step 2 of a reaction leading to a tertiary carbocation is lowest (see Fig. 7.7). Secondary alcohols are not so easily dehydrated because the free energy of activation for their dehydration is higher—a secondary carbocation is less stable. The free energy of activation for dehydration of primary alcohols via a carbocation is so high that they undergo dehydration by another mechanism (Section 7.7C).

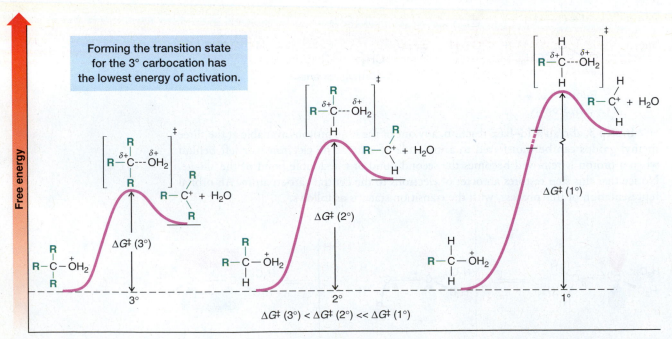

FIGURE 7.7 Free-energy diagrams for the formation of carbocations from protonated tertiary, secondary, and primary alcohols. The relative free energies of activation are tertiary < secondary ≪ primary.

[A MECHANISM FOR THE REACTION ̶ Acid-Catalyzed Dehydration of Secondary or Tertiary Alcohols: An E1 Reaction]

Step 1

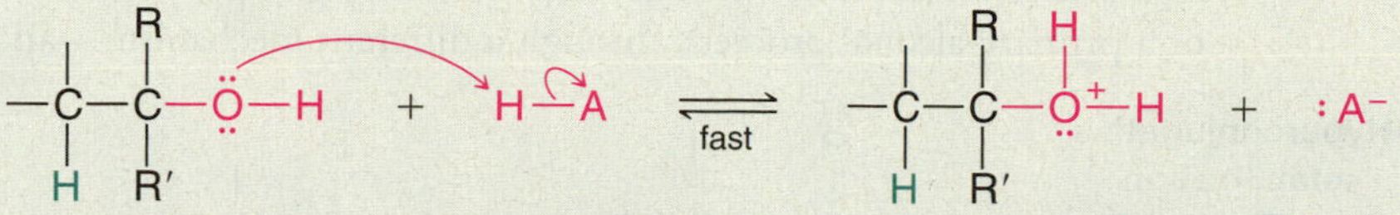

| **2° or 3° Alcohol (R′ may be H)** | **Acid catalyst (typically sulfuric or phosphoric acid)** | | **Protonated alcohol** | **Conjugate base** |

The alcohol accepts a proton from the acid in a fast step.

Step 2

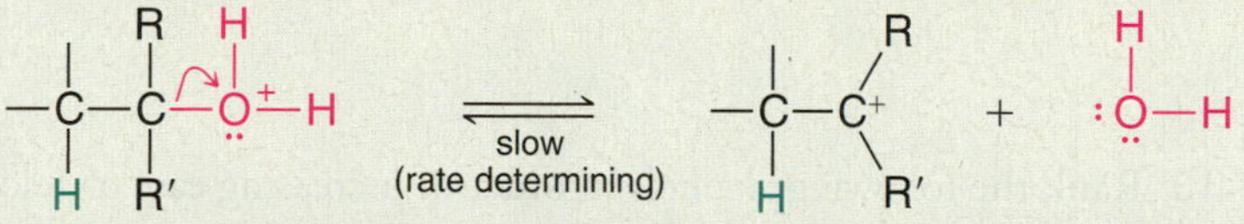

The protonated alcohol loses a molecule of water to become a carbocation. This step is slow and rate determining.

Step 3

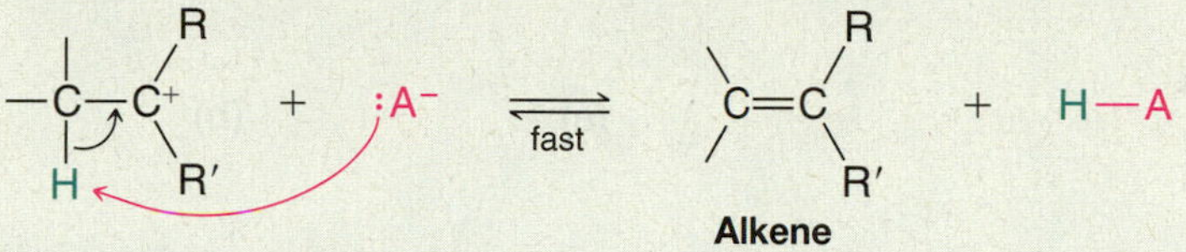

Alkene

The carbocation loses a proton to a base. In this step, the base may be another molecule of the alcohol, water, or the conjugate base of the acid. The proton transfer results in the formation of the alkene. Note that the overall role of the acid is catalytic (it is used in the reaction and regenerated).

The reactions by which carbocations are formed from protonated alcohols are all highly *endergonic*. Based on the Hammond–Leffler postulate (Section 6.13A), there should be a strong resemblance between the transition state and the carbocation in each case.

- *The transition state that leads to the tertiary carbocation is lowest in free energy because it resembles the carbocation that is lowest in energy.*

By contrast, the transition state that leads to the primary carbocation occurs at highest free energy because it resembles the carbocation that is highest in energy. In each instance, moreover, the same factor stabilizes the transition state that stabilizes the carbocation itself: **delocalization of the charge**. We can understand this if we examine the process by which the transition state is formed:

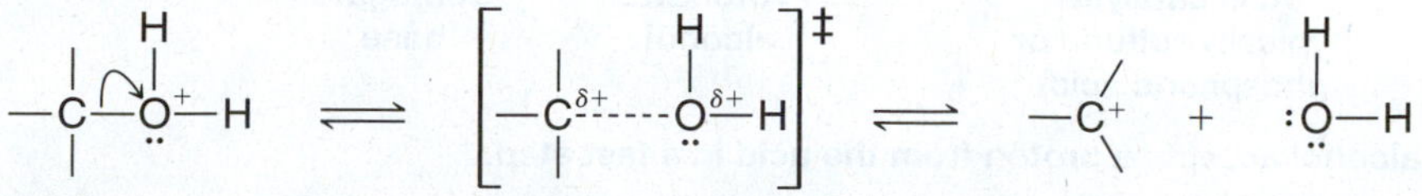

| **Protonated alcohol** | **Transition state** | **Carbocation** |

The oxygen atom of the protonated alcohol bears a full positive charge. As the transition state develops, this oxygen atom begins to separate from the carbon atom to which it is attached. The carbon atom begins to develop a partial positive charge because it is losing the electrons that bonded it to the oxygen atom. This developing positive charge *is most effectively delocalized in the transition state leading to a tertiary carbocation because three alkyl groups are present to contribute electron density by hyperconjugation (Section 6.11B) to the developing carbocation.* The positive charge is less effectively delocalized in the transition state leading to a secondary carbocation (*two* electron-releasing groups) and is least effectively delocalized in the transition state leading to a primary carbocation (*one* electron-releasing group). For this reason the dehydration of a primary alcohol proceeds through a different mechanism—an E2 mechanism.

Hyperconjugative stabilization (see Figure 6.7) is greatest for a tertiary carbocation.	Transition state leading to 3° carbocation (most stable)	Transition state leading to 2° carbocation	Transition state leading to 1° carbocation (least stable)

PRACTICE PROBLEM 7.13 Rank the following alcohols in order of increasing ease of acid-catalyzed dehydration.

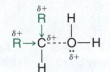

(a)   (b)   (c)

7.7C A Mechanism for Dehydration of Primary Alcohols: An E2 Reaction

Dehydration of primary alcohols apparently proceeds through an E2 mechanism because the primary carbocation required for dehydration by an E1 mechanism is relatively unstable. The first step in dehydration of a primary alcohol is protonation, just as in the E1 mechanism. Then, with the protonated hydroxyl as a good leaving group, a Lewis base in the reaction mixture removes a β hydrogen simultaneously with formation of the alkene double bond and departure of the protonated hydroxyl group (water).

[A MECHANISM FOR THE REACTION ── Dehydration of a Primary Alcohol: An E2 Reaction]

Primary alcohol	Acid catalyst (typically sulfuric or phosphoric acid)	Protonated alcohol	Conjugate base

The alcohol accepts a proton from the acid in a fast step.

Alkene

A base removes a hydrogen from the β carbon as the double bond forms and the protonated hydroxyl group departs. The base may be another molecule of the alcohol or the conjugate base of the acid.

7.8 CARBOCATION STABILITY AND THE OCCURRENCE OF MOLECULAR REARRANGEMENTS

With an understanding of carbocation stability and its effect on transition states, we can now proceed to explain the rearrangements of carbon skeletons that occur in some alcohol **dehydrations**.

7.8A Rearrangements During Dehydration of Secondary Alcohols

Consider again the **rearrangement** that occurs when 3,3-dimethyl-2-butanol is dehydrated:

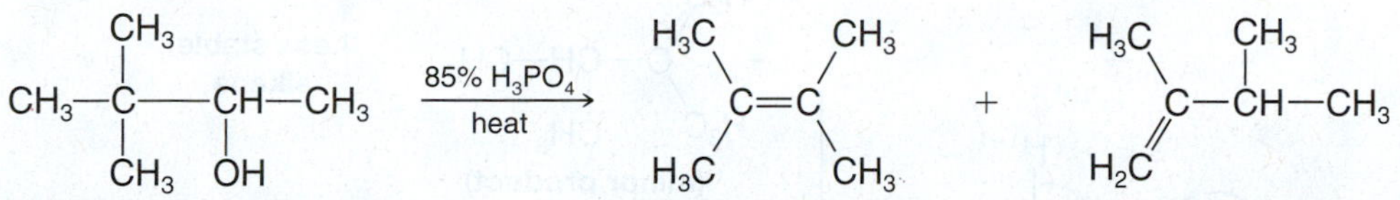

3,3-Dimethyl-2-butanol → **2,3-Dimethyl-2-butene (major product)** + **2,3-Dimethyl-1-butene (minor product)**

(85% H_3PO_4, heat)

The first step of this dehydration is the formation of the protonated alcohol in the usual way:

Step 1

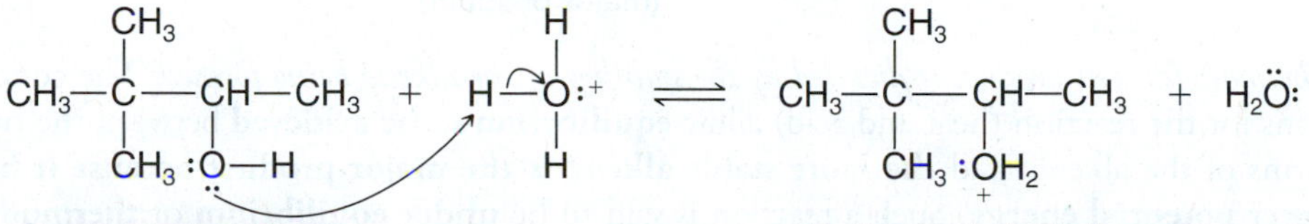

Protonation of the alcohol

Protonated alcohol

In the second step the protonated alcohol loses water and a secondary carbocation forms:

Step 2

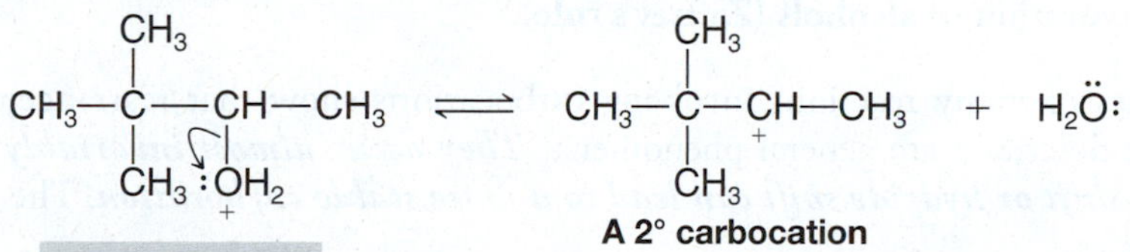

Departure of a water molecule

A 2° carbocation

Now the rearrangement occurs. *The less stable, secondary carbocation rearranges to a more stable tertiary carbocation:*

Step 3

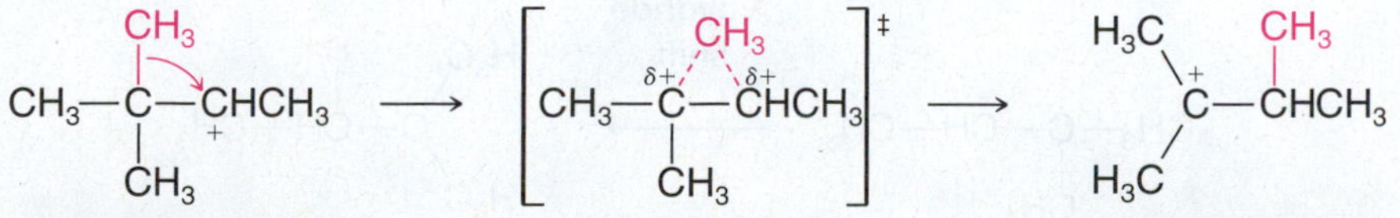

2° Carbocation (less stable) → **Transition state** → **3° Carbocation (more stable)**

Rearrangement by migration of a methyl group

The rearrangement occurs through the migration of an alkyl group (methyl) from the carbon atom adjacent to the one with the positive charge. The methyl group migrates **with its pair of electrons** (called a **methanide** shift). In the transition state the shifting methyl is partially bonded to both carbon atoms by the pair of electrons with which it migrates. It never leaves the carbon skeleton. After the migration is complete, the carbon atom that the methyl anion left has become a carbocation, and the positive charge on the carbon atom to which it migrated has been neutralized. Because a group migrates from one carbon to an adjacent one, this kind of rearrangement is also called a **1,2 shift**.

The final step of the reaction is the removal of a proton from the new carbocation (by a Lewis base in the reaction mixture) and the formation of an alkene. This step, however, can occur in two ways:

Step 4

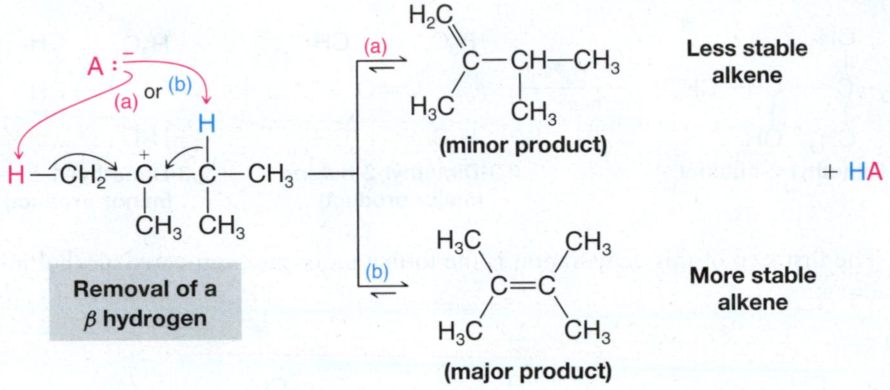

The more favored product is dictated by the stability of the alkene being formed. The conditions for the reaction (heat and acid) allow **equilibrium to be achieved** between the two forms of the alkene, and **the more stable alkene is the major product because it has lower potential energy.** Such a reaction is said to be **under equilibrium** or **thermodynamic control**. Path (b) leads to the highly stable tetrasubstituted alkene and this is the path followed by most of the carbocations. Path (a), on the other hand, leads to a less stable, disubstituted alkene, and because its potential energy is higher, it is the minor product of the reaction.

Helpful Hint

Alcohol dehydration follows Zaitsev's rule.

- **Formation of the more stable alkene is the general rule in acid-catalyzed dehydration of alcohols (Zaitsev's rule).**

Studies of many reactions involving carbocations show that rearrangements like those just described are general phenomena. *They occur almost invariably when an alkanide shift or hydride shift can lead to a more stable carbocation.* The following are examples:

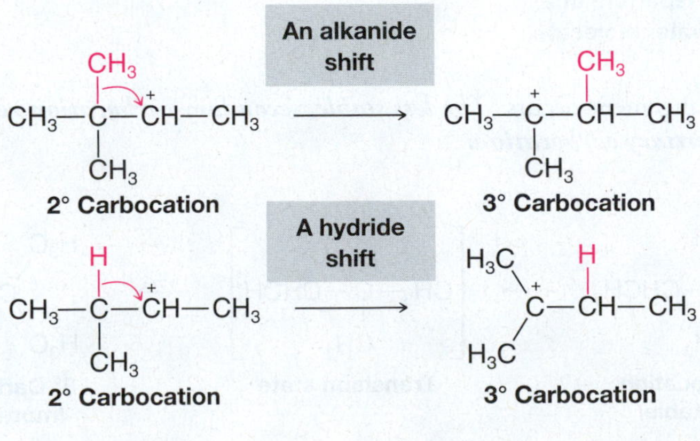

We shall see biological examples of alkanide (specifically methanide) and hydride migrations in "The Chemistry of … Cholesterol Biosynthesis" (online in *WileyPLUS* for Chapter 8).

Rearrangements of carbocations can also lead to a change in ring size, as the following example shows:

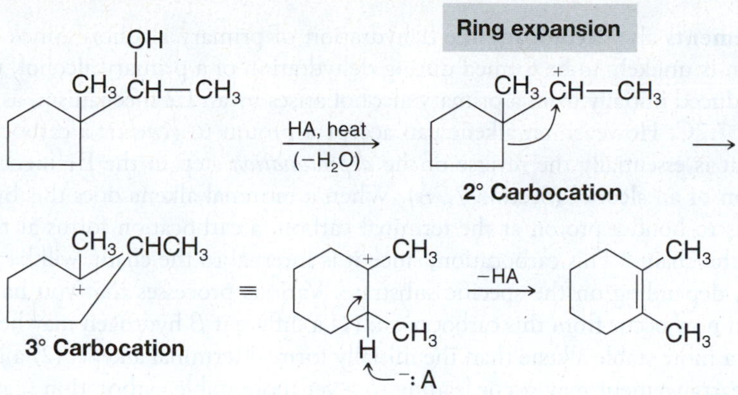

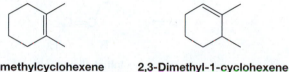

Ring expansion by migration is especially favorable if relief in ring strain occurs.

It is important to note that rearrangements to carbocations having approximately equal energy are also possible (e.g., from one secondary carbocation to another), and this can complicate the mixture of products that might be obtained from a reaction.

●●● **SOLVED PROBLEM 7.6**

Explain why the major product of the dehydration above is 1,2-dimethylcyclohexene (as shown) and not 2,3-dimethyl-1-cyclohexene.

1,2-Dimethylcyclohexene
(major product)

2,3-Dimethyl-1-cyclohexene
(minor product)

STRATEGY AND ANSWER: We have just learned that dehydration leads mainly to the more stable alkene (when two are possible). We also know that the stability of an alkene is related to the number of alkyl groups that are attached to the carbons of the double bond. 1,2-Dimethylcyclohexene has a tetrasubstituted double bond (and is more stable), while in 2,3-dimethylcyclohexene the double bond is only trisubstituted.

●●●

Acid-catalyzed dehydration of neopentyl alcohol, $(CH_3)_3CCH_2OH$, yields 2-methyl-2-butene as the major product. Outline a mechanism showing all steps in its formation.

PRACTICE PROBLEM 7.14

●●●

Acid-catalyzed dehydration of either 2-methyl-1-butanol or 3-methyl-1-butanol gives 2-methyl-2-butene as the major product. Write plausible mechanisms that explain these results.

PRACTICE PROBLEM 7.15

●●●

When the compound called *isoborneol* is heated with 9 M sulfuric acid, the product of the reaction is the compound called camphene and not bornylene, as one might expect. Using models to assist you, write a step-by-step mechanism showing how camphene is formed.

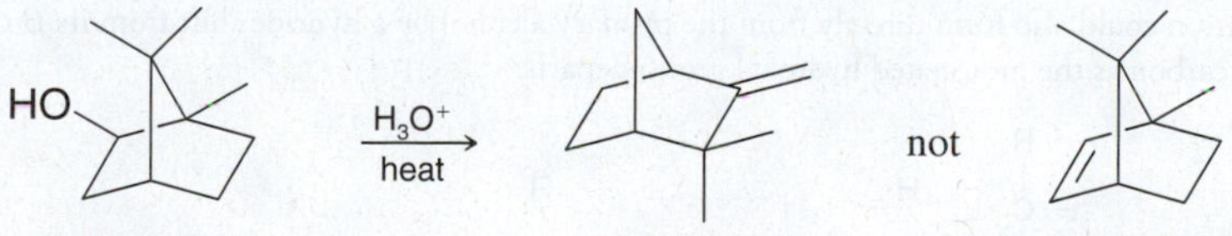

PRACTICE PROBLEM 7.16

Isoborneol $\xrightarrow[\text{heat}]{H_3O^+}$ **Camphene**   not   **Bornylene**

7.8B Rearrangement After Dehydration of a Primary Alcohol

Rearrangements also accompany the dehydration of primary alcohols. Since a primary carbocation is unlikely to be formed during dehydration of a primary alcohol, the alkene that is produced initially from a primary alcohol arises by an E2 mechanism, as described in Section 7.7C. However, an alkene can accept a proton to *generate* a carbocation in a process that is essentially the reverse of the *deprotonation* step in the E1 mechanism for dehydration of an alcohol (Section 7.7A). When a terminal alkene does this by using its π electrons to bond a proton at the terminal carbon, a carbocation forms at the second carbon of the chain.* This carbocation, since it is internal to the chain, will be secondary or tertiary, depending on the specific substrate. Various processes that you have already learned can now occur from this carbocation: (1) a different β hydrogen may be removed, leading to a more stable alkene than the initially formed terminal alkene; (2) a hydride or alkanide rearrangement may occur leading to a yet more stable carbocation (e.g., moving from a 2° to a 3° carbocation) or to a carbocation of approximately equal stability, after which the elimination may be completed; or (3) a nucleophile may attack any of these carbocations to form a substitution product. Under the high-temperature conditions for alcohol dehydration the principal products will be alkenes rather than substitution products.

[A MECHANISM FOR THE REACTION ── Formation of a Rearranged Alkene During Dehydration of a Primary Alcohol]

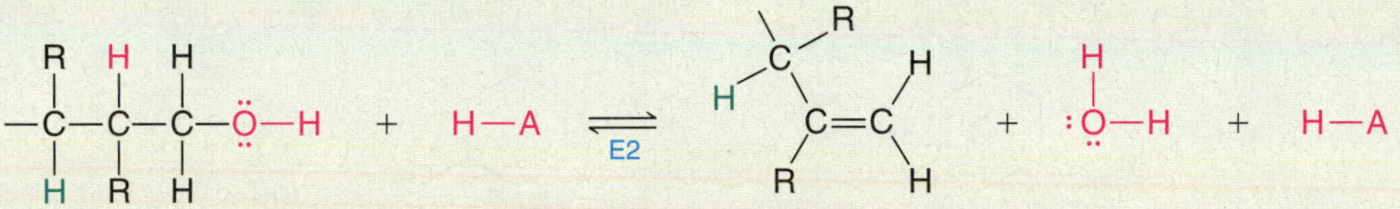

The primary alcohol initially undergoes acid-catalyzed dehydration by an E2 mechanism (Section 7.7C).

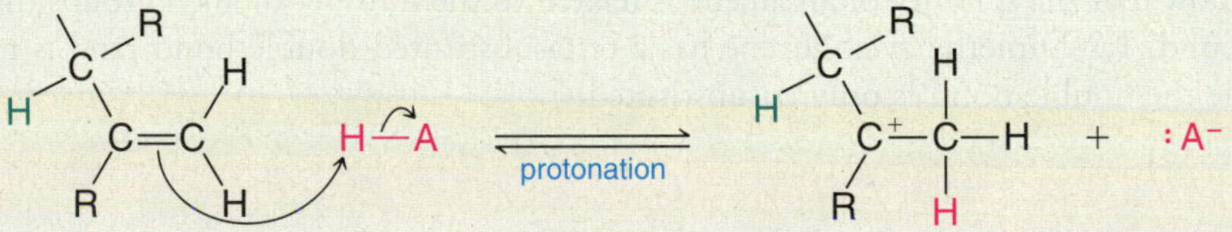

The π electrons of the initial alkene can then be used to form a bond with a proton at the terminal carbon, forming a secondary or tertiary carbocation.*

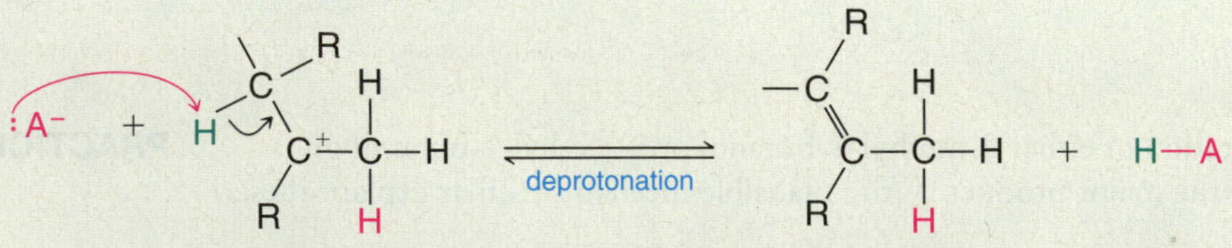

A different β hydrogen can be removed from the carbocation, so as to form a more highly substituted alkene than the initial alkene. This deprotonation step is the same as the usual completion of an E1 elimination. (This carbocation could experience other fates, such as further rearrangement before elimination or substitution by an S$_N$1 process.)

*The carbocation could also form directly from the primary alcohol by a hydride shift from its β carbon to the terminal carbon as the protonated hydroxyl group departs:

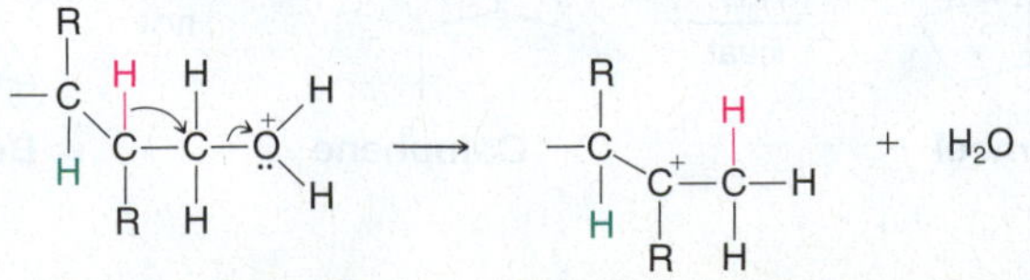

7.9 THE ACIDITY OF TERMINAL ALKYNES

The hydrogen bonded to the carbon of a terminal alkyne, called an **acetylenic hydrogen atom**, is considerably more acidic than those bonded to carbons of an alkene or alkane (see Section 3.8A). The pK_a values for ethyne, ethene, and ethane illustrate this point:

A terminal alkyne is ~10^{20} times more acidic than an alkene or alkane.

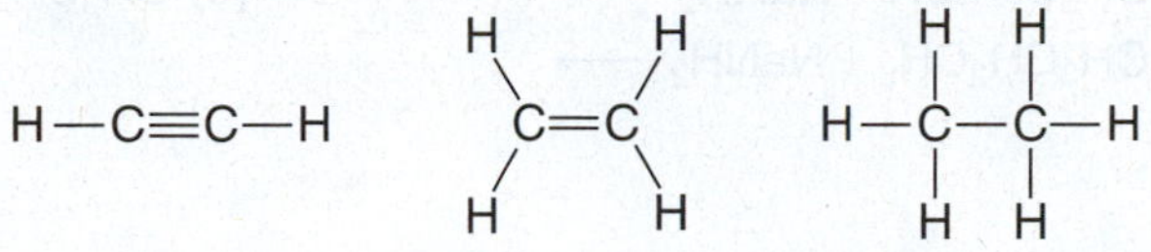

$$H-C\equiv C-H \qquad \underset{H}{\overset{H}{C}}{=}\underset{H}{\overset{H}{C}} \qquad H-\underset{H}{\overset{H}{C}}-\underset{H}{\overset{H}{C}}-H$$

$$pK_a = 25 \qquad\qquad pK_a = 44 \qquad\qquad pK_a = 50$$

The order of basicity of their anions is opposite that of their relative acidity:

Relative Basicity

$$CH_3CH_2{:}^- > CH_2{=}CH{:}^- > HC\equiv C{:}^-$$

If we include in our comparison hydrogen compounds of other first-row elements of the periodic table, we can write the following orders of relative acidities and basicities. This comparison is useful as we consider what bases and solvents to use with terminal alkynes.

Relative Acidity

Most acidic				Least acidic	

$$H{-}\ddot{O}H > H{-}\ddot{O}R > H{-}C\equiv CR > H{-}\ddot{N}H_2 > H{-}CH{=}CH_2 > H{-}CH_2CH_3$$

pK_a 15.7 16–17 25 38 44 50

Relative Basicity

Least basic		Most basic	

$${}^-\!{:}\ddot{O}H < {}^-\!{:}\ddot{O}R < {}^-\!{:}C\equiv CR < {}^-\!{:}\ddot{N}H_2 < {}^-\!{:}CH{=}CH_2 < {}^-\!{:}CH_2CH_3$$

We see from the order just given that while terminal alkynes are more acidic than ammonia, they are less acidic than alcohols and are less acidic than water.

●●● **SOLVED PROBLEM 7.7**

As we shall soon see, sodium amide ($NaNH_2$) is useful, especially when a reaction requires a very strong base. Explain why a solvent such as methanol cannot be used to carry out a reaction in which you might want to use sodium amide as a base.

STRATEGY AND ANSWER: An alcohol has $pK_a = 16$–17, and ammonia has $pK_a = 38$. This means that methanol is a significantly stronger acid than ammonia, and the conjugate base of ammonia (the ${}^-NH_2$ ion) is a significantly stronger base than an alkoxide ion. Therefore, the following acid–base reaction would take place as soon as sodium amide is added to methanol.

$$\underset{\substack{\text{Stronger}\\\text{acid}}}{CH_3OH} \;+\; \underset{\substack{\text{Stronger}\\\text{base}}}{NaNH_2} \;\xrightarrow{\;CH_3OH\;}\; \underset{\substack{\text{Weaker}\\\text{base}}}{CH_3ONa} \;+\; \underset{\substack{\text{Weaker}\\\text{acid}}}{NH_3}$$

With a pK_a difference this large, the sodium amide would convert all of the methanol to sodium methoxide, a much weaker base than sodium amide. (This is an example of what is called the leveling effect of a solvent.)

● ● ●

PRACTICE PROBLEM 7.17   Predict the products of the following acid–base reactions. If the equilibrium would not result in the formation of appreciable amounts of products, you should so indicate. In each case label the stronger acid, the stronger base, the weaker acid, and the weaker base:

(a) $CH_3CH{=}CH_2 + NaNH_2 \longrightarrow$

(b) $CH_3C{\equiv}CH + NaNH_2 \longrightarrow$

(c) $CH_3CH_2CH_3 + NaNH_2 \longrightarrow$

(d) $CH_3C{\equiv}C{:}^- + CH_3CH_2OH \longrightarrow$

(e) $CH_3C{\equiv}C{:}^- + NH_4Cl \longrightarrow$

7.10 SYNTHESIS OF ALKYNES BY ELIMINATION REACTIONS

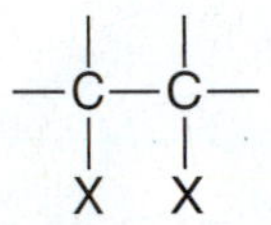

A vic–dihalide

● Alkynes can be synthesized from alkenes via compounds called vicinal dihalides.

A vicinal dihalide (abbreviated **vic-dihalide**) is a compound bearing the halogens on adjacent carbons (*vicinus*, Latin: adjacent). Vicinal dihalides are also called 1,2-dihalides. A vicinal dibromide, for example, can be synthesized by addition of bromine to an alkene (Section 8.1). The *vic*-dibromide can then be subjected to a double dehydrohalogenation reaction with a strong base to yield an alkyne.

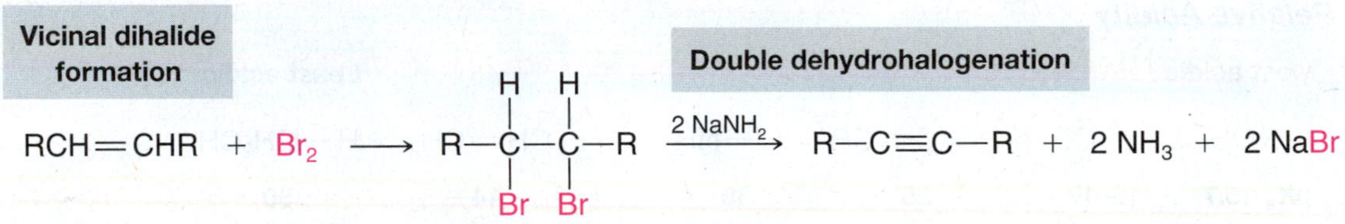

The dehydrohalogenations occur in two steps, the first yielding a bromoalkene, and the second, the alkyne.

7.10A Laboratory Application of This Alkyne Synthesis

The two dehydrohalogenations may be carried out as separate reactions, or they may be carried out consecutively in a single mixture. Sodium amide ($NaNH_2$), a very strong base, can be used to cause both reactions in a single mixture. At least two molar equivalents of sodium amide per mole of the dihalide must be used. For example, adding bromine to 1,2-diphenylethene provides the vicinal dihalide needed for a synthesis of 1,2-diphenylethyne:

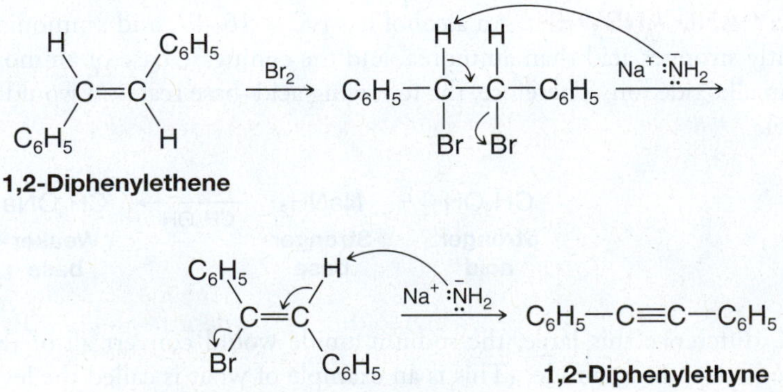

1,2-Diphenylethene

1,2-Diphenylethyne

[A MECHANISM FOR THE REACTION — Dehydrohalogenation of *vic*-Dibromides to Form Alkynes]

Reaction

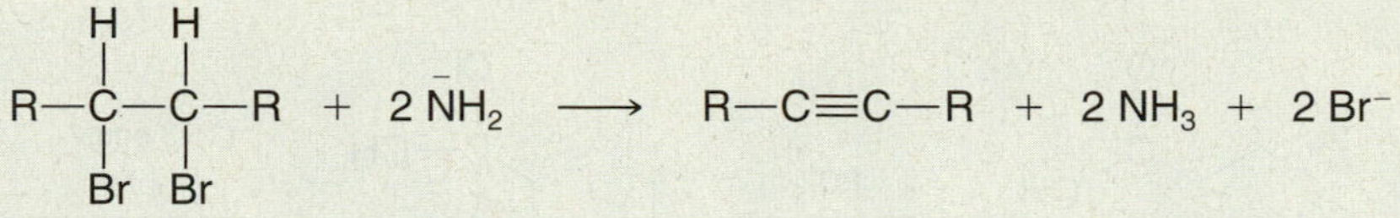

$$
\underset{\substack{\text{H H} \\ \text{R—C—C—R} \\ \text{Br Br}}}{} + 2\ \overset{-}{N}H_2 \longrightarrow R—C\!\equiv\!C—R + 2\ NH_3 + 2\ Br^-
$$

Mechanism

Step 1

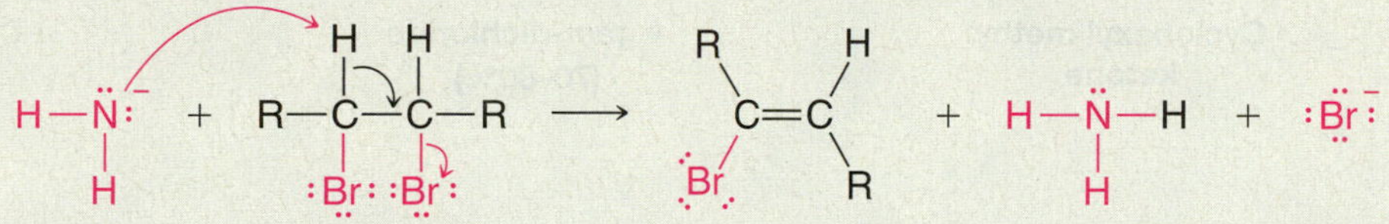

Amide ion	*vic*-Dibromide	Bromoalkene	Ammonia	Bromide ion

The strongly basic amide ion brings about an E2 reaction.

Step 2

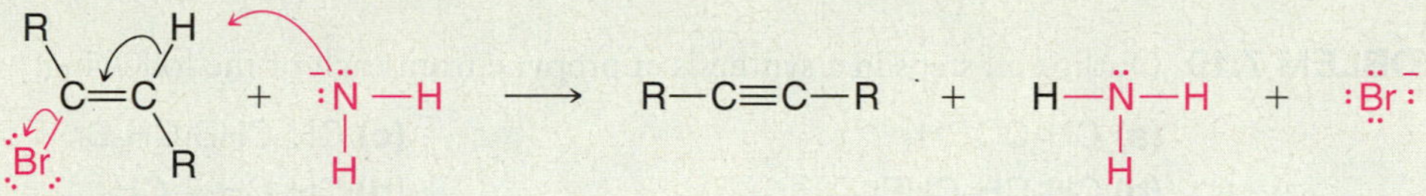

$$ \longrightarrow R—C\!\equiv\!C—R + \ \ + \ \ : Br : ^- $$

Bromoalkene	Amide ion	Alkyne	Ammonia	Bromide ion

A second E2 reaction produces the alkyne.

- If the product is to be an alkyne with a triple bond at the end of the chain (a terminal alkyne) as we show in the example below, then three molar equivalents of sodium amide are required.

Initial dehydrohalogenation of the *vic*-dihalide produces a mixture of two bromoalkenes that are not isolated but that undergo a second dehydrohalogenation. The terminal alkyne that results from this step is deprotonated (because of its acidity) by the third mole of sodium amide (see Section 7.9). To complete the process, addition of ammonium chloride converts the sodium alkynide to the desired product, 1-butyne.

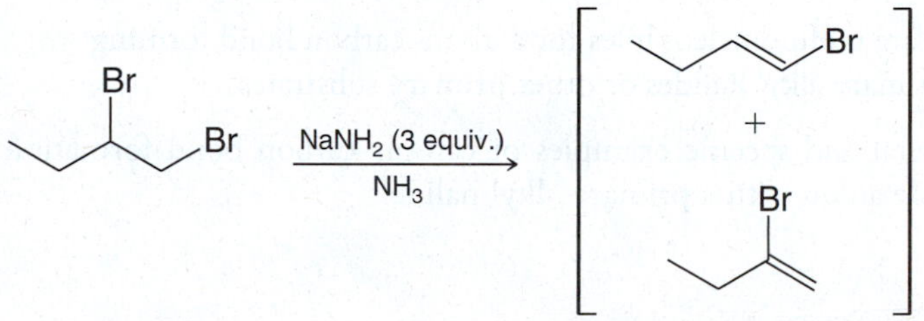

Result of the first dehydrohalogenation

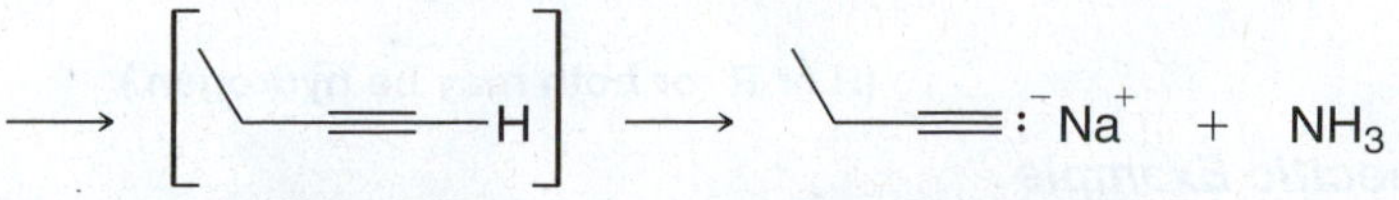

Result of the second dehydrohalogenation

The initial alkyne is deprotonated by the third equivalent of base.

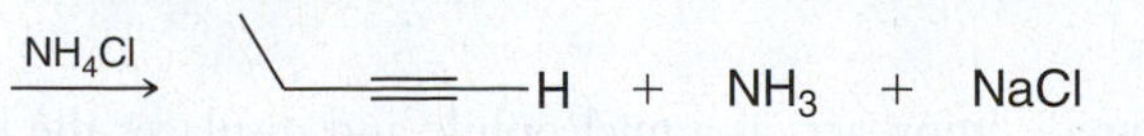

1-Butyne

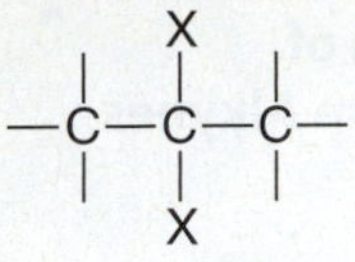

A gem–dihalide

- **Geminal** dihalides can also be converted to alkynes by dehydrohalogenation.

A geminal dihalide (abbreviated **gem-dihalide**) has two halogen atoms bonded to the same carbon (*geminus*, Latin: twins). Ketones can be converted to *gem*-dichlorides by reaction with phosphorus pentachloride, and the *gem*-dichlorides can be used to synthesize alkynes.

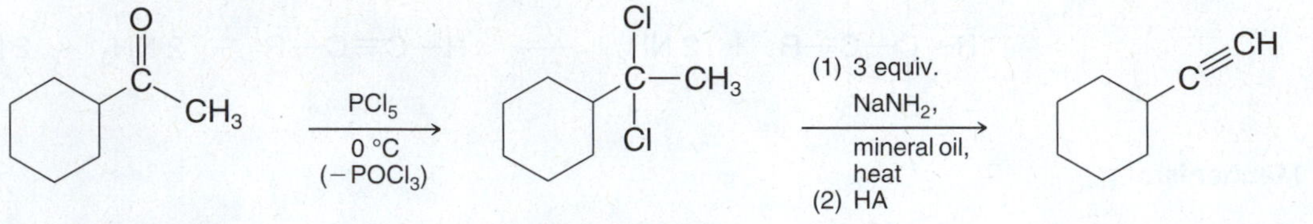

Cyclohexyl methyl ketone → $\xrightarrow[\substack{0\,°C \\ (-\,POCl_3)}]{PCl_5}$ → **A gem-dichloride (70–80%)** → $\xrightarrow[\substack{\text{mineral oil,} \\ \text{heat} \\ (2)\ HA}]{(1)\ 3\ \text{equiv.} \\ NaNH_2,}$ → **Cyclohexylacetylene (46%)**

PRACTICE PROBLEM 7.18 Show how you might synthesize ethynylbenzene from methyl phenyl ketone.

PRACTICE PROBLEM 7.19 Outline all steps in a synthesis of propyne from each of the following:

(a) CH_3COCH_3

(b) $CH_3CH_2CHBr_2$

(c) $CH_3CHBrCH_2Br$

(d) $CH_3CH=CH_2$

7.11 TERMINAL ALKYNES CAN BE CONVERTED TO NUCLEOPHILES FOR CARBON–CARBON BOND FORMATION

- The acetylenic proton of ethyne or any terminal alkyne (pK_a 25) can be removed with a strong base such as sodium amide ($NaNH_2$). The result is an alkynide anion.

$$H-C\equiv C-H + NaNH_2 \xrightarrow{\text{liq. } NH_3} H-C\equiv C:^- Na^+ + NH_3$$

$$CH_3C\equiv C-H + NaNH_2 \xrightarrow{\text{liq. } NH_3} CH_3C\equiv C:^- Na^+ + NH_3$$

- Alkynide anions are useful nucleophiles for carbon–carbon bond forming reactions with primary alkyl halides or other primary substrates.

The following are general and specific examples of carbon–carbon bond formation by alkylation of an alkynide anion with a primary alkyl halide.

General Example

$$R-C\equiv C:^- Na^+ + R'CH_2-Br \longrightarrow R-C\equiv C-CH_2R' + NaBr$$

| Sodium alkynide | Primary alkyl halide | Mono- or disubstituted acetylene |

(R or R′ or both may be hydrogen.)

Specific Example

$$CH_3CH_2C\equiv C:^- Na^+ + CH_3CH_2-Br \xrightarrow[\text{6 h}]{\text{liq. } NH_3} CH_3CH_2C\equiv CCH_2CH_3 + NaBr$$

3-Hexyne (75%)

The alkynide anion acts as a nucleophile and displaces the halide ion from the primary alkyl halide. We now recognize this as an S_N2 reaction (Section 6.5).

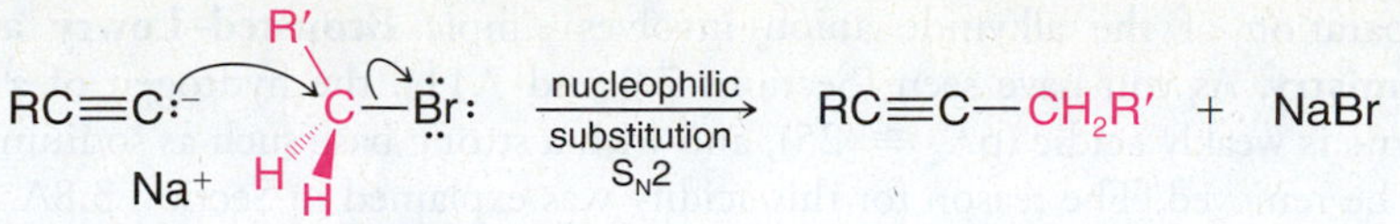

Sodium alkynide | 1° Alkyl halide

$$RC\equiv C{:}^- \cdot Na^+ + R'CH_2Br \xrightarrow[\text{substitution } S_N2]{\text{nucleophilic}} RC\equiv C{-}CH_2R' + NaBr$$

- Primary alkyl halides should be used in the alkylation of alkynide anions, so as to avoid competition by elimination.

Use of a secondary or tertiary substrate causes E2 elimination instead of substitution because the alkynide anion is a strong base as well as a good nucleophile.

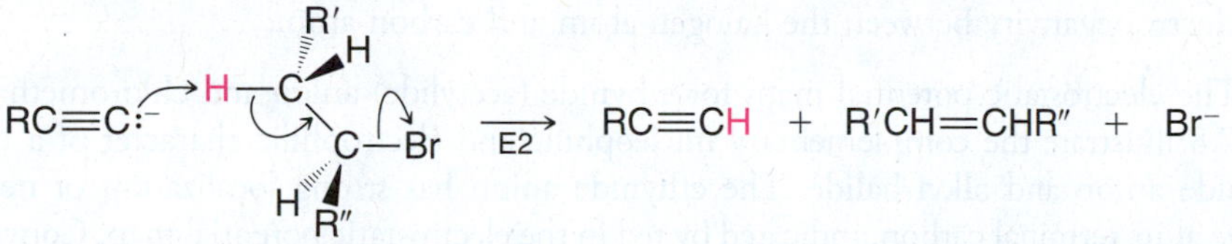

2° Alkyl halide

$$RC\equiv CH + R'CH{=}CHR'' + Br^-$$

Outline a synthesis of 4-phenyl-2-butyne from 1-propyne.

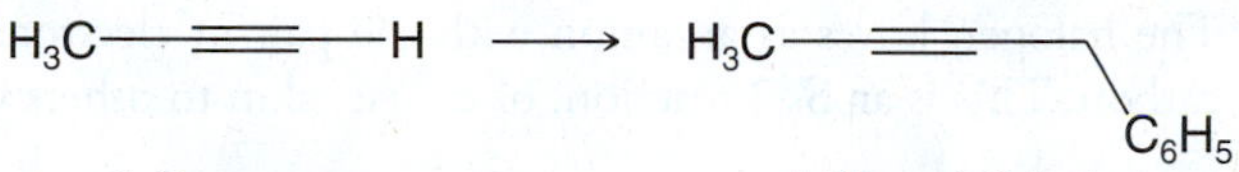

1-Propyne → 4-Phenyl-2-butyne

STRATEGY AND ANSWER: Take advantage of the acidity of the acetylenic hydrogen of propyne and convert it to an alkynide anion using sodium amide, a base that is strong enough to remove the acetylenic hydrogen. Then use the akynide ion as a nucleophile in an S_N2 reaction with benzyl bromide.

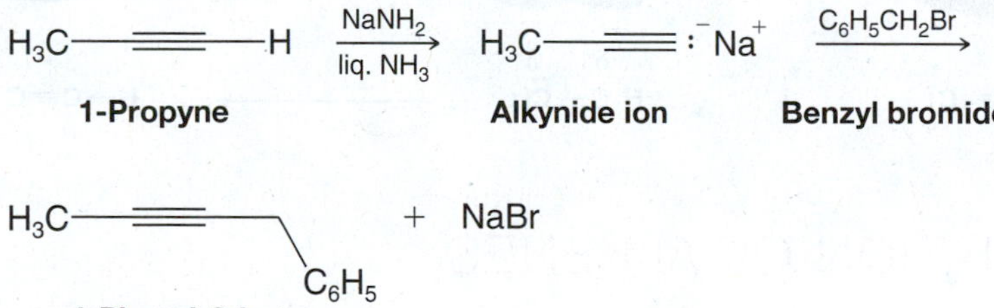

1-Propyne → (NaNH₂, liq. NH₃) → Alkynide ion → (C₆H₅CH₂Br) → Benzyl bromide

4-Phenyl-2-butyne + NaBr

●●● **PRACTICE PROBLEM 7.20**

Your goal is to synthesize 4,4-dimethyl-2-pentyne. You have a choice of beginning with any of the following reagents:

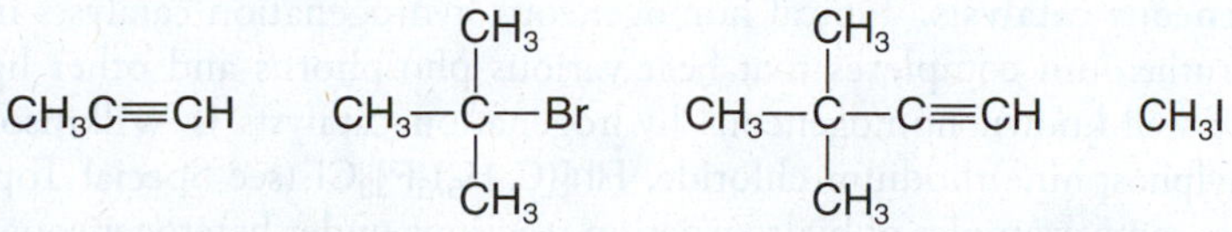

$$CH_3C\equiv CH \qquad CH_3{-}\overset{\displaystyle CH_3}{\underset{\displaystyle CH_3}{\overset{|}{\underset{|}{C}}}}{-}Br \qquad CH_3{-}\overset{\displaystyle CH_3}{\underset{\displaystyle CH_3}{\overset{|}{\underset{|}{C}}}}{-}C\equiv CH \qquad CH_3I$$

Assume that you also have available sodium amide and liquid ammonia. Outline the best synthesis of the required compound.

7.11A General Principles of Structure and Reactivity Illustrated by the Alkylation of Alkynide Anions

The **alkylation** of alkynide anions illustrates several essential aspects of structure and reactivity that have been important to our study of organic chemistry thus far.

1. Preparation of the alkynide anion involves simple **Brønsted–Lowry acid–base chemistry.** As you have seen (Sections 7.9 and 7.11), the hydrogen of a terminal alkyne is weakly acidic (p$K_a \cong 25$), and with a strong base such as sodium amide it can be removed. The reason for this acidity was explained in Section 3.8A.

2. Once formed, the alkynide anion is a **Lewis base** (Section 3.3) with which the alkyl halide reacts as an electron pair acceptor (a **Lewis acid**). The alkynide anion can thus be called a *nucleophile* (Sections 3.4 and 6.3) because of the negative charge concentrated at its terminal carbon—it is a reagent that seeks positive charge.

3. The alkyl halide can be called an *electrophile* (Sections 3.4 and 8.1) because of the partial positive charge at the carbon bearing the halogen—it is a reagent that seeks negative charge. Polarity in the alkyl halide is the direct result of the difference in electronegativity between the halogen atom and carbon atom.

The electrostatic potential maps for ethynide (acetylide) anion and chloromethane in Fig. 7.8 illustrate the complementary nucleophilic and electrophilic character of a typical alkynide anion and alkyl halide. The ethynide anion has strong localization of negative charge at its terminal carbon, indicated by red in the electrostatic potential map. Conversely, chloromethane has partial positive charge at the carbon bonded to the electronegative chlorine atom. (The dipole moment for chloromethane is aligned directly along the carbon–chlorine bond.) Thus, acting as a Lewis base, the alkynide anion is attracted to the partially positive carbon of the alkyl halide. Assuming a collision between the two occurs with the proper orientation and sufficient kinetic energy, as the alkynide anion brings two electrons to the alkyl halide to form a new bond, it will displace the halogen from the alkyl halide. The halogen leaves as an anion with the pair of electrons that formerly bonded it to the carbon. This is an S_N2 reaction, of course, akin to others we discussed in Chapter 6.

FIGURE 7.8 The reaction of ethynide (acetylide) anion and chloromethane. Electrostatic potential maps illustrate the complementary nucleophilic and electrophilic character of the alkynide anion and the alkyl halide. The dipole moment of chloromethane is shown by the red arrow.

$$H-C\equiv\bar{C}: \quad + \quad \overset{\delta^+ \quad \delta^-}{H_3C-\overset{..}{\underset{..}{Cl}}:} \quad \longrightarrow \quad H-C\equiv C-CH_3 \quad + \quad :\overset{..}{\underset{..}{Cl}}:^-$$

7.12 HYDROGENATION OF ALKENES

- Alkenes react with hydrogen in the presence of a variety of metal catalysts to add one hydrogen atom to each carbon atom of the double bond (Sections 4.16A, 5.10A).

Hydrogenation reactions that involve *insoluble* platinum, palladium, or nickel catalysts (Section 4.16A) proceed by **heterogeneous catalysis** because the catalyst is not soluble in the reaction mixture. Hydrogenation reactions that involve soluble catalysts occur by **homogeneous catalysis.** Typical homogeneous hydrogenation catalysts include rhodium and ruthenium complexes that bear various phosphorus and other ligands. One of the most well-known homogeneous hydrogenation catalysts is Wilkinson's catalyst, tris(triphenylphosphine)rhodium chloride, $Rh[(C_6H_5)_3P]_3Cl$ (see Special Topic G.) The following are some examples of hydrogenation reactions under heterogeneous and homogeneous catalysis:

Catalytic hydrogenation reactions, like those shown above, are a type of **addition reaction** (versus substitution or elimination), and they are also a type of reduction. This leads to a distinction between compounds that are saturated versus those that are unsaturated.

- Compounds containing only carbon–carbon single bonds (alkanes and others) are said to be **saturated compounds** because they contain the maximum number of hydrogen atoms that a given formula can possess.
- Compounds containing carbon–carbon multiple bonds (alkenes, alkynes, and aromatic compounds) are said to be **unsaturated compounds** because they contain fewer than the maximum number of hydrogen atoms possible for a given formula.

Unsaturated compounds can be **reduced** to saturated compounds by **catalytic hydrogenation**. The following example shows conversion of an unsaturated triglyceride to a saturated triglyceride (both are fats), in a catalytic hydrogenation reaction as might be done in the food industry to change the physical properties of a fat.

H_2 (excess) and Pt, Pd or Ni catalyst

> An unsaturated fat can be hydrogenated to form a saturated fat.

Molecules of a natural unsaturated fat can align less evenly with each other than can molecules of saturated fats due to the "kinks" from the cis double bonds in unsaturated fats. Hence intermolecular forces between unsaturated fat molecules are weaker and they have lower melting points than saturated fats. See "The Chemistry of … Hydrogenation in the Food Industry."

THE CHEMISTRY OF... Hydrogenation in the Food Industry

The food industry makes use of catalytic hydrogenation to convert liquid vegetable oils to semisolid fats in making margarine and solid cooking fats. Examine the labels of many prepared foods and you will find that they contain "partially hydrogenated vegetable oils." There are several reasons why foods contain these oils, but one is that partially hydrogenated vegetable oils have a longer shelf life.

Fats and oils (Section 23.2) are glyceryl esters of carboxylic acids with long carbon chains, called "fatty acids." Fatty acids are saturated (no double bonds), monounsaturated (one double bond), or polyunsaturated (more than one double bond). Oils typically contain a higher proportion of fatty acids with one or more double bonds than fats do. Partial hydrogenation of an oil converts some of its double bonds to single bonds, and this conversion has the effect of producing a fat with the consistency of margarine or a semisolid cooking fat.

One potential problem that arises from using catalytic hydrogenation to produce partially hydrogenated vegetable oils is that

the catalysts used for hydrogenation cause isomerization of some of the double bonds of the fatty acids (some of those that do not absorb hydrogen). In most natural fats and oils, the double bonds of the fatty acids have the cis configuration. The catalysts used for hydrogenation convert some of these cis double bonds to the unnatural trans configuration. The health effects of trans fatty acids are still under study, but experiments thus far indicate that they cause an increase in serum levels of cholesterol and triacylglycerols, which in turn increases the risk of cardiovascular disease.

A product used in baking that contains oils and mono- and diacylglycerols that are partially hydrogenated.

Photo by Lisa Gee

No (or zero%) trans fatty acids.

© Jonathan Vasata/iStock photo

7.13   HYDROGENATION: THE FUNCTION OF THE CATALYST

Hydrogenation of an alkene is an exothermic reaction ($\Delta H° \cong -120$ kJ mol^{-1}):

$$R-CH=CH-R + H_2 \xrightarrow{\text{hydrogenation}} R-CH_2-CH_2-R + \text{heat}$$

Although the process is exothermic, there is usually a high free energy of activation for uncatalyzed alkene hydrogenation, and therefore, the uncatalyzed reaction does not take place at room temperature. However, hydrogenation will take place readily at room temperature in the presence of a catalyst because the catalyst provides a new pathway for the reaction that involves lower free energy of activation (Fig. 7.9).

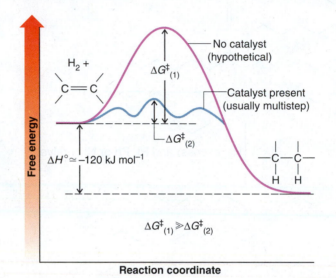

FIGURE 7.9 Free-energy diagram for the hydrogenation of an alkene in the presence of a catalyst and the hypothetical reaction in the absence of a catalyst. The free energy of activation for the uncatalyzed reaction ($\Delta G^{\ddagger}_{(1)}$) is very much larger than the largest free energy of activation for the catalyzed reaction ($\Delta G^{\ddagger}_{(2)}$). The uncatalyzed hydrogenation reaction does not occur.

Heterogeneous hydrogenation catalysts typically involve finely divided platinum, palladium, nickel, or rhodium deposited on the surface of powdered carbon (charcoal). Hydrogen gas introduced into the atmosphere of the reaction vessel adsorbs to the metal by a chemical reaction where unpaired electrons on the surface of the metal *pair* with the electrons of hydrogen (Fig. 7.10*a*) and bind the hydrogen to the surface. The collision of an alkene with the surface bearing adsorbed hydrogen causes adsorption of the alkene as well (Fig. 7.10*b*). A stepwise transfer of hydrogen atoms takes place, and this produces an alkane before the organic molecule leaves the catalyst surface (Figs. 7.10*c, d*). As a consequence, *both hydrogen atoms usually add from the same side of the molecule.* This mode of addition is called a **syn** addition (Section 7.14A):

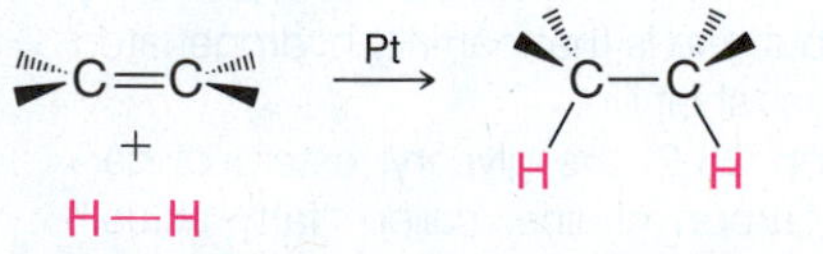

Catalytic hydrogenation is a syn addition.

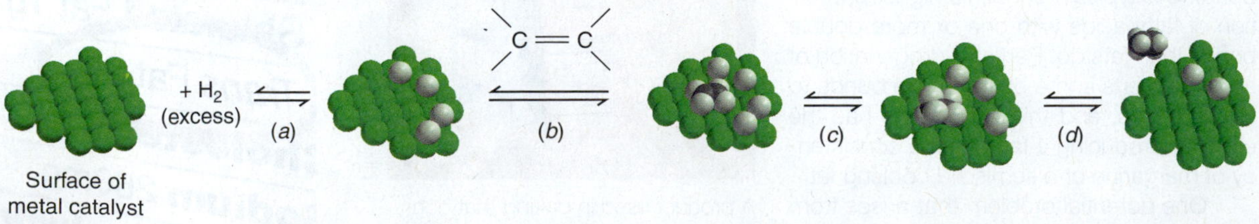

FIGURE 7.10 The mechanism for the hydrogenation of an alkene as catalyzed by finely divided platinum metal: (*a*) hydrogen adsorption; (*b*) adsorption of the alkene; (*c, d*) stepwise transfer of both hydrogen atoms to the same face of the alkene (syn addition).

7.13A Syn and Anti Additions

An addition that places the parts of the adding reagent on the same side (or face) of the reactant is called **syn addition**. We have just seen that the platinum-catalyzed addition of hydrogen (X = Y = H) is a syn addition:

$$\overset{\text{\tiny{}}}{\underset{\text{\tiny{}}}{C}}=\overset{\text{\tiny{}}}{\underset{\text{\tiny{}}}{C} \quad + \quad X—Y \quad \longrightarrow \quad \begin{Bmatrix} C—C \\ | \quad | \\ X \quad Y \end{Bmatrix} \begin{array}{l} \textbf{Syn} \\ \textbf{addition} \end{array}$$

The opposite of a syn addition is an **anti addition**. An anti addition places the parts of the adding reagent on opposite faces of the reactant.

$$\overset{\text{\tiny{}}}{\underset{\text{\tiny{}}}{C}}=\overset{\text{\tiny{}}}{\underset{\text{\tiny{}}}{C} \quad + \quad X—Y \quad \longrightarrow \quad \begin{Bmatrix} C—C \\ | \quad | \\ X \quad Y \end{Bmatrix} \begin{array}{l} \textbf{Anti} \\ \textbf{addition} \end{array}$$

In Chapter 8 we shall study a number of important syn and anti additions.

7.14 HYDROGENATION OF ALKYNES

Depending on the conditions and the catalyst employed, one or two molar equivalents of hydrogen will add to a carbon–carbon triple bond. When a platinum catalyst is used, the alkyne generally reacts with two molar equivalents of hydrogen to give an alkane:

$$CH_3C\equiv CCH_3 \xrightarrow{Pt,\ H_2} [CH_3CH=CHCH_3] \xrightarrow{Pt,\ H_2} CH_3CH_2CH_2CH_3$$

However, **hydrogenation** of an alkyne to an alkene can be accomplished through the use of special catalysts or reagents. Moreover, these special methods allow the preparation of either (*E*)- or (*Z*)-alkenes from disubstituted alkynes.

7.14A Syn Addition of Hydrogen: Synthesis of *cis*-Alkenes

A **heterogeneous catalyst** that permits hydrogenation of an alkyne to an alkene is the nickel boride compound called P-2 catalyst. The P-2 catalyst can be prepared by the reduction of nickel acetate with sodium borohydride:

$$Ni\begin{pmatrix} O \\ \| \\ OCCH_3 \end{pmatrix}_2 \xrightarrow[\text{EtOH}]{NaBH_4} \underset{\textbf{P-2}}{Ni_2B}$$

- Hydrogenation of alkynes in the presence of P-2 catalyst causes **syn addition of hydrogen**. The alkene formed from an internal alkyne has the (*Z*) or cis configuration.

The hydrogenation of 3-hexyne illustrates this method. The reaction takes place on the surface of the catalyst (Section 7.14), accounting for the syn addition:

Syn addition of hydrogen to an alkyne		
(structure) 3-Hexyne	$\xrightarrow[\text{(syn addition)}]{H_2/Ni_2B\ (P-2)}$	(structure) (**Z**)-3-Hexene (*cis*-3-hexene) (97%)

Other specially conditioned catalysts can be used to prepare *cis*-alkenes from disubstituted alkynes. Metallic palladium deposited on calcium carbonate can be used in this way after it has been conditioned with lead acetate and quinoline (an amine, see Section 20.1B). This special catalyst is known as **Lindlar's catalyst:**

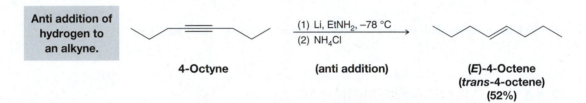

7.14B Anti Addition of Hydrogen: Synthesis of *trans*-Alkenes

- **Anti addition** of hydrogen to the triple bond of alkynes occurs when they are treated with lithium or sodium metal in ammonia or ethylamine at low temperatures.

This reaction, called a **dissolving metal reduction,** takes place in solution and produces an (*E*)- or *trans*-alkene. The mechanism involves radicals, which are molecules that have unpaired electrons (see Chapter 10).

Anti addition of hydrogen to an alkyne.

4-Octyne — (1) Li, EtNH₂, −78 °C (2) NH₄Cl — (anti addition) — (*E*)-4-Octene (*trans*-4-octene) (52%)

[A MECHANISM FOR THE REACTION — The Dissolving Metal Reduction of an Alkyne]

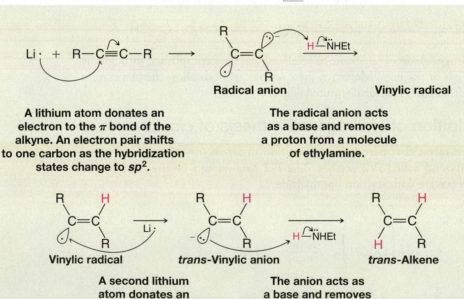

A lithium atom donates an electron to the π bond of the alkyne. An electron pair shifts to one carbon as the hybridization states change to *sp²*.

The radical anion acts as a base and removes a proton from a molecule of ethylamine.

A second lithium atom donates an electron to the vinylic radical.

The anion acts as a base and removes a proton from a second molecule of ethylamine.

The mechanism for this reduction, shown in the preceding box, involves successive electron transfers from lithium (or sodium) atoms and proton transfers from amines (or ammonia). In the first step, a lithium atom transfers an electron to the alkyne to produce an intermediate that bears a negative charge and has an unpaired electron, called a **radical anion.** In the second step, an amine transfers a proton to produce a **vinylic radical.** Then, transfer of another electron gives a **vinylic anion.** It is this step that determines the stereochemistry of the reaction. The *trans*-vinylic anion is formed preferentially because it is more stable; the bulky alkyl groups are farther apart. Protonation of the *trans*-vinylic anion leads to the *trans*-alkene.

Write the structure of compound **A**, used in this synthesis of the perfume ingredient (*Z*)-jasmone.

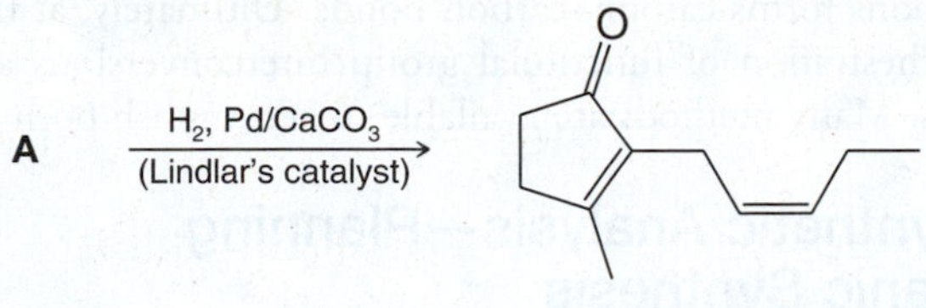

PRACTICE PROBLEM 7.21

$$\textbf{A} \xrightarrow[\text{(Lindlar's catalyst)}]{\text{H}_2,\ \text{Pd/CaCO}_3}$$

(Z)-Jasmone

How would you convert 2-nonyne into (*E*)-2-nonene?

PRACTICE PROBLEM 7.22

7.15 AN INTRODUCTION TO ORGANIC SYNTHESIS

You have learned quite a few tools that are useful for organic synthesis, including nucleophilic substitution reactions, elimination reactions, and the hydrogenation reactions covered in Sections 7.12–7.14. Now we will consider the logic of organic synthesis and the important process of retrosynthetic analysis. Then we will apply nucleophilic substitution (in the specific case of alkylation of alkynide anions) and **hydrogenation** reactions to the synthesis of some simple target molecules.

7.15A Why Do Organic Synthesis?

Organic synthesis is the process of building organic molecules from simpler precursors. Syntheses of organic compounds are carried out for many reasons. Chemists who develop new drugs carry out organic syntheses in order to discover molecules with structural attributes that enhance certain medicinal effects or reduce undesired side effects. Crixivan, whose structure is shown below, was designed by small-scale synthesis in a research laboratory and then quickly moved to large-scale synthesis after its approval as a drug. In other situations, organic synthesis may be needed to test a hypothesis about a reaction mechanism or about how a certain organism metabolizes a compound. In cases like these we often will need to synthesize a particular compound "labeled" at a certain position (e.g., with deuterium, tritium, or an isotope of carbon).

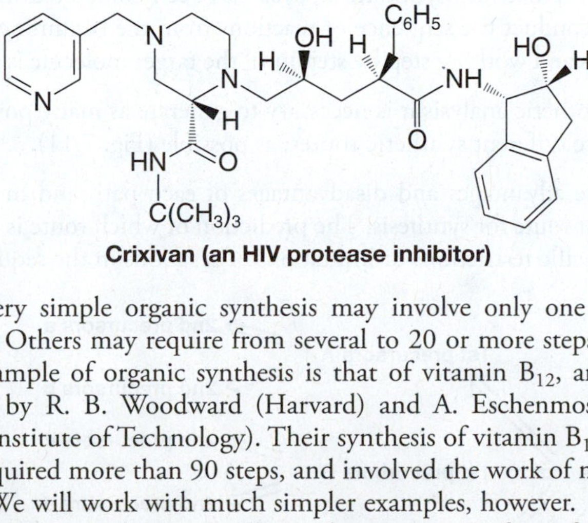

Crixivan (an HIV protease inhibitor)

A carbon–cobalt σ bond

Vitamin B$_{12}$

A very simple organic synthesis may involve only one chemical reaction. Others may require from several to 20 or more steps. A landmark example of organic synthesis is that of vitamin B$_{12}$, announced in 1972 by R. B. Woodward (Harvard) and A. Eschenmoser (Swiss Federal Institute of Technology). Their synthesis of vitamin B$_{12}$ took 11 years, required more than 90 steps, and involved the work of nearly 100 people. We will work with much simpler examples, however.

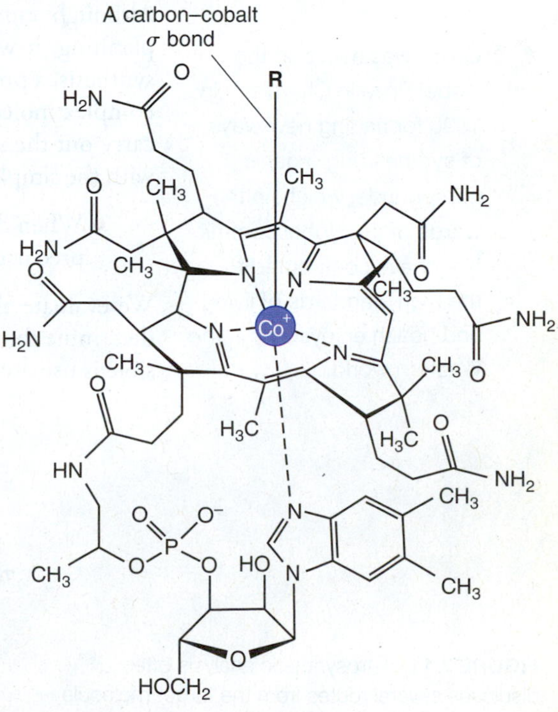

An organic synthesis typically involves two types of transformations:

1. reactions that convert functional groups from one to another
2. reactions that create new carbon–carbon bonds.

You have studied examples of both types of reactions already. For example, hydrogenation transforms the carbon–carbon double- or triple-bond functional groups in alkenes and alkynes to single bonds (actually removing a functional group in this case), and alkylation of alkynide anions forms carbon–carbon bonds. Ultimately, at the heart of organic synthesis is the orchestration of functional group interconversions and carbon–carbon bond-forming steps. Many methods are available to accomplish both of these things.

7.15B Retrosynthetic Analysis—Planning an Organic Synthesis

Sometimes it is possible to visualize from the start all the steps necessary to synthesize a desired (target) molecule from obvious precursors. Often, however, the sequence of transformations that would lead to the desired compound is too complex for us to "see" a path from the beginning to the end. In this case, since we know where we want to finish (the target molecule) but not where to start, we envision the sequence of steps that is required in a backward fashion, one step at a time. We begin by identifying immediate precursors that could react to make the target molecule. Once these have been chosen, they in turn become new intermediate target molecules, and we identify the next set of precursors that could react to form them, and so on, and so on. This process is repeated until we have worked backward to compounds that are sufficiently simple that they are readily available in a typical laboratory:

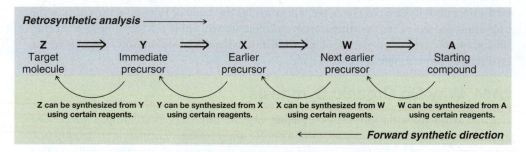

- The process we have just described is called **retrosynthetic analysis**.
- The open arrow is called a **retrosynthesis arrow,** and means that a molecule can be synthesized from its most immediate precursor by some chemical reaction.

Although some of the earliest organic syntheses likely required some type of analytical planning, it was E. J. Corey who first formalized a set of global principles for chemical synthesis, a process which he termed retrosynthetic analysis, that enabled anyone to plan a complex molecule synthesis. Once retrosynthetic analysis has been completed, to actually carry out the synthesis we conduct the sequence of reactions from the beginning, starting with the simplest precursors and working step by step until the target molecule is achieved.

- When doing retrosynthetic analysis it is necessary to generate as many possible precursors, and hence different synthetic routes, as possible (Fig. 7.11).

We evaluate all the possible advantages and disadvantages of each path and in so doing determine the most efficient route for synthesis. The prediction of which route is most feasible is usually based on specific restrictions or limitations of reactions in the sequence, the

 COREY was awarded the Nobel Prize in Chemistry in 1990 for finding new ways of synthesizing organic compounds, which, in the words of the Nobel committee, "have contributed to the high standards of living and health enjoyed … in the Western world."

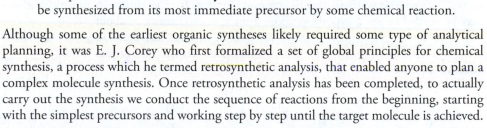

FIGURE 7.11 Retrosynthetic analysis often discloses several routes from the target molecule back to varied precursors.

availability of materials, or other factors. We shall see an example of this in Section 7.16C. In actuality more than one route may work well. In other cases it may be necessary to try several approaches in the laboratory in order to find the most efficient or successful route.

7.15C Identifying Precursors

In the case of functional groups we need to have a toolbox of reactions from which to choose those we know can convert one given functional group into another. You will develop such a toolbox of reactions as you proceed through your study of organic chemistry. Similarly, with regard to making carbon–carbon bonds in synthesis, you will develop a repertoire of reactions for that purpose. In order to choose the appropriate reaction for either purpose, you will inevitably consider basic principles of structure and reactivity.

As we stated in Sections 3.3A and 7.12:

> **Helpful Hint**
>
> Over time you will add to your toolbox reactions for two major categories of synthetic operations: carbon–carbon bond formation and functional group interconversion.

- Many organic reactions depend on the interaction of molecules that have complementary full or partial charges.

One very important aspect of retrosynthetic analysis is being able to identify those atoms in a target molecule that could have had complementary (opposite) charges in synthetic precursors. Consider, for example, the synthesis of 1-cyclohexyl-1-butyne. On the basis of reactions learned in this chapter, you might envision an alkynide anion and an alkyl halide as precursors having complementary polarities that when allowed to react together would lead to this molecule:

Retrosynthetic Analysis

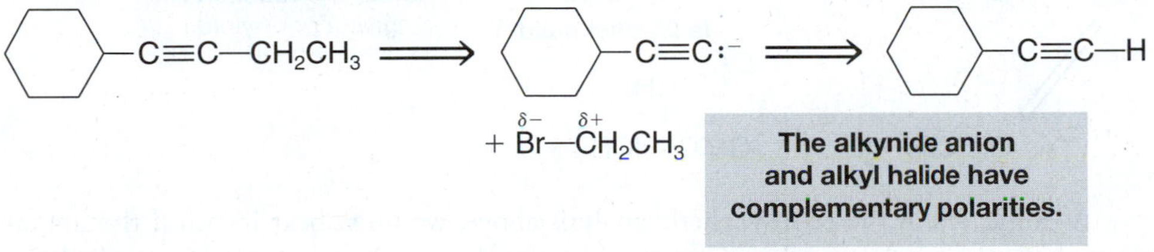

$$\text{Cy}-C\equiv C-CH_2CH_3 \Longrightarrow \text{Cy}-C\equiv C:^- \Longrightarrow \text{Cy}-C\equiv C-H$$

$$+\ \overset{\delta-}{Br}-\overset{\delta+}{CH_2CH_3}$$

The alkynide anion and alkyl halide have complementary polarities.

Synthesis

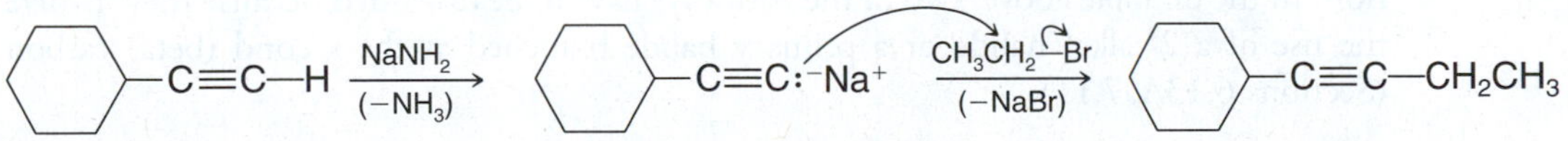

$$\text{Cy}-C\equiv C-H \xrightarrow[(-NH_3)]{NaNH_2} \text{Cy}-C\equiv C:^-Na^+ \xrightarrow[(-NaBr)]{CH_3CH_2-Br} \text{Cy}-C\equiv C-CH_2CH_3$$

Sometimes, however, it will not at first be obvious where the retrosynthetic bond disconnections are in a target molecule that would lead to oppositely charged or complementary precursors. The synthesis of an alkane would be such an example. An alkane does not contain carbon atoms that could directly have had opposite charges in precursor molecules. However, if one supposes that certain carbon–carbon single bonds in the alkane could have arisen by hydrogenation of a corresponding alkyne (a functional group interconversion), then, in turn, two atoms of the alkyne could have been joined from precursor molecules that had complementary charges (i.e., an alkynide anion and an alkyl halide).

THE CHEMISTRY OF... From the Inorganic to the Organic

In 1862, Friedrich Wöhler discovered calcium carbide (CaC_2) by heating carbon with an alloy of zinc and calcium. He then synthesized acetylene by allowing the calcium carbide to react with water:

$$C \xrightarrow{\text{zinc-calcium alloy, heat}} CaC_2 \xrightarrow{2H_2O} HC\equiv CH + Ca(OH)_2$$

Acetylene produced this way burned in lamps of some lighthouses and in old-time miners' headlamps. From the standpoint of organic synthesis, it is theoretically possible to synthesize anything using reactions of alkynes to form carbon–carbon bonds and to prepare other functional groups. Thus, while Wöhler's 1828 conversion of ammonium cyanate to urea was the first synthesis of an organic compound from an inorganic precursor (Section 1.1A), his discovery of calcium carbide and its reaction with water to form acetylene gives us a formal link from inorganic materials to the entire realm of organic synthesis.

Consider the following retrosynthetic analysis for **2-methylhexane**:

Retrosynthetic Analysis

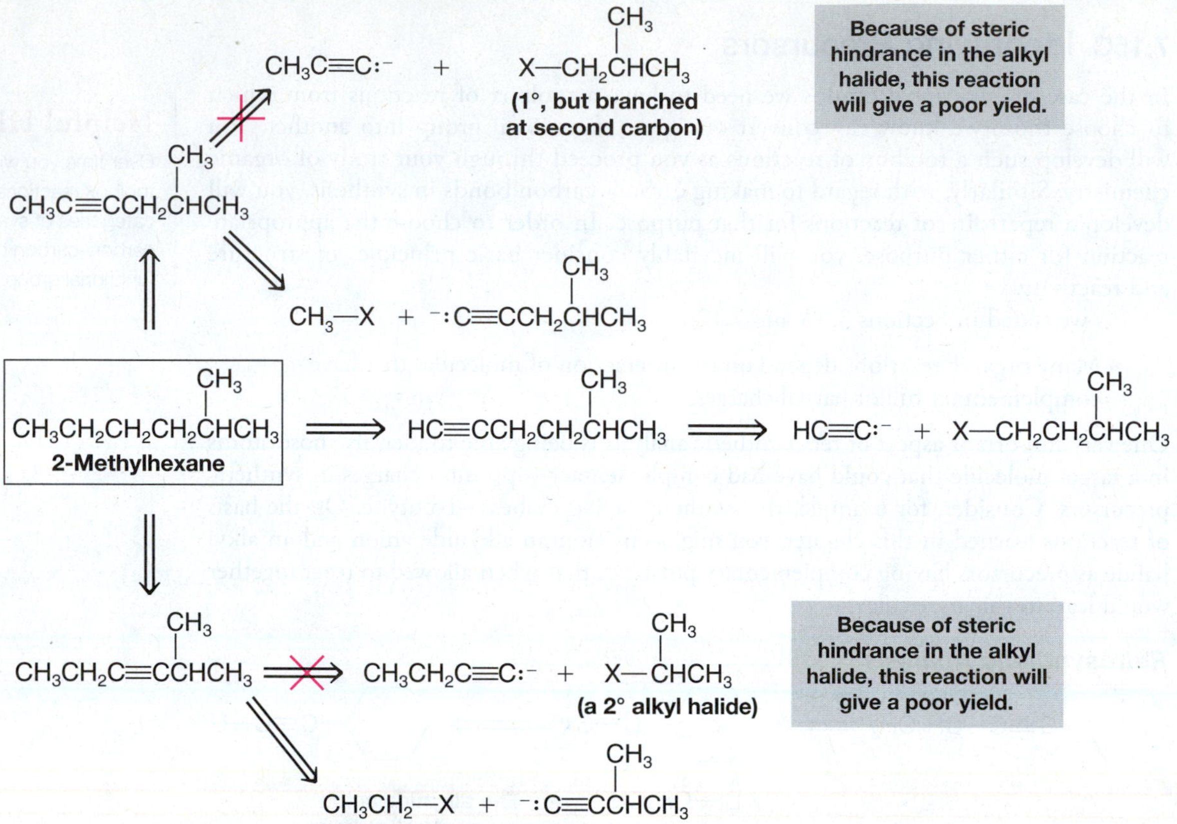

As indicated in the retrosynthetic analysis above, we must bear in mind the limitations that exist for the reactions that would be applied in the synthetic (forward) direction. In the example above, two of the pathways have to be discarded because they involve the use of a 2° alkyl halide or a primary halide branched at the second (beta) carbon (Sections 6.13A, 7.11).

••• SOLVED PROBLEM 7.9

Outline a retrosynthetic pathway that leads from 'muscalure', the sex attractant pheromone of the common housefly back to the simplest alkyne, ethyne (acetylene). Then show the synthesis. You may use any inorganic compounds, or solvents, you need and alkyl halides of any length necessary.

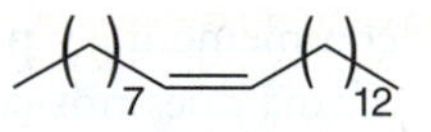

Muscalure

STRATEGY AND ANSWER: We make use of two reactions that we have just studied in this chapter: syn addition of hydrogen to an alkyne, and alkylation of alkynide ions.

Retrosynthetic Analysis

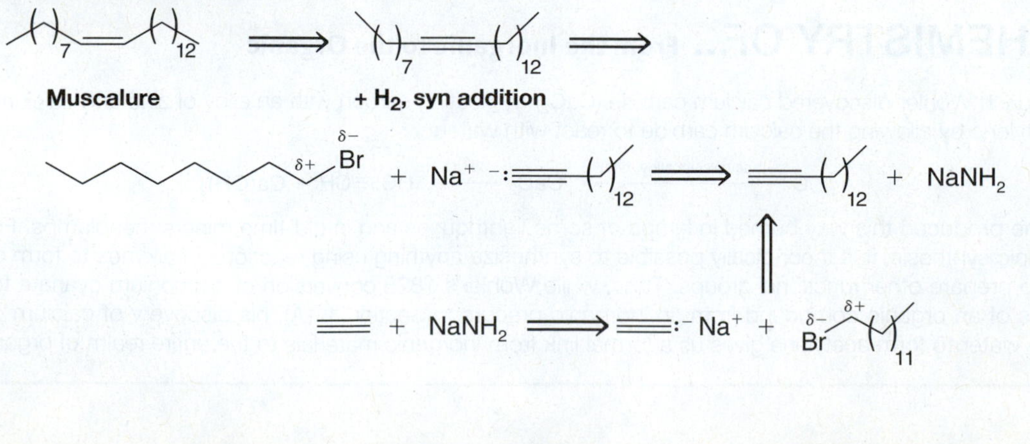

Synthesis

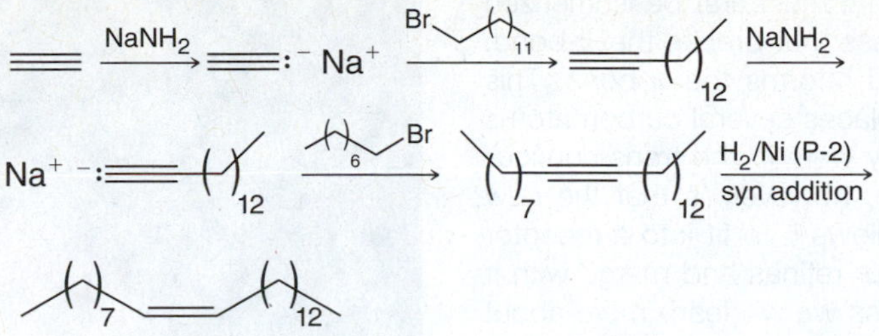

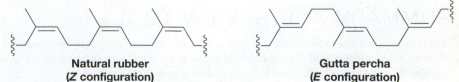

Referring to the retrosynthetic analysis for 2-methylhexane in this section, write reactions for those synthesis routes that are feasible.

PRACTICE PROBLEM 7.23

(a) Devise retrosynthetic schemes for all conceivable alkynide anion alkylation syntheses of the insect pheromones undecane and 2-methylheptadecane (see "The Chemistry of … Pheromones" box in Chapter 4). **(b)** Write reactions for two feasible syntheses of each pheromone.

PRACTICE PROBLEM 7.24

7.15D Raison d'Etre

Solving synthetic puzzles by application of retrosynthetic analysis is one of the joys of learning organic chemistry. As you might imagine, there is skill and some artistry involved. Over the years many chemists have set their minds to organic synthesis, and because of this we have all prospered from the fruits of their endeavors.

[WHY Do These Topics Matter?

ALKENE GEOMETRY, RUBBER, AND THE CHEMISTRY OF VISION

The (*E*) or (*Z*) configurations of substituted double bonds is not just a matter for exercise sets and exams. In the real world, they define the properties of many compounds. For example, natural rubber, which can be obtained from the sap of certain trees, has all (*Z*) configurations about its trisubstituted double bonds. Some other trees make the all (*E*) version, a compound known as gutta percha. While gutta percha is also a latex-like material, the change in stereochemistry actually makes it inelastic so that it does not have the same useful properties as natural rubber.

Björn Svensson/Photo Researchers, Inc.

Natural rubber
(Z configuration)

Gutta percha
(E configuration)

Alkene stereochemistry is also critical to our ability to see. In our eyes, the key molecule is a compound called *trans*-retinal, a material that can be synthesized in our bodies and that comes from our diets through foods like carrots.

(continues on next page)

In order for retinal to participate in the vision process, one particular double bond within it must first be isomerized from trans to cis through a process that breaks the π-bond, rotates about a single bond, and reforms the π-bond. This new stereochemical orientation places several carbon atoms in a different orientation than they were in the trans configuration. The value of that change, however, is that the new spatial orientation of *cis*-retinal allows it to fit into a receptor in a protein known as opsin in our retinas and merge with it through a general reaction process we will learn more about in Chapter 16; this step generates a new complex known as rhodopsin. The key thing to understand for now, however, is that when rhodopsin is exposed to light of a certain wavelength (of which we will learn more in Chapter 13), the cis double bond isomerizes back to the more stable all trans configuration through a series of steps to become metarhodopsin II as shown below. It is believed that the repositioning of the cyclohexene ring within the retinal portion of metarhodopsin II following this isomerization induces some further conformational changes within the protein. These changes ultimately lead to a nerve impulse that is interpreted by our brains as vision. The picture shown here is but a small part in the overall process, but these initial, critical steps are based solely on alkene stereochemistry. As such, it really does matter whether a double bond is cis or trans, (*E*) or (*Z*)!

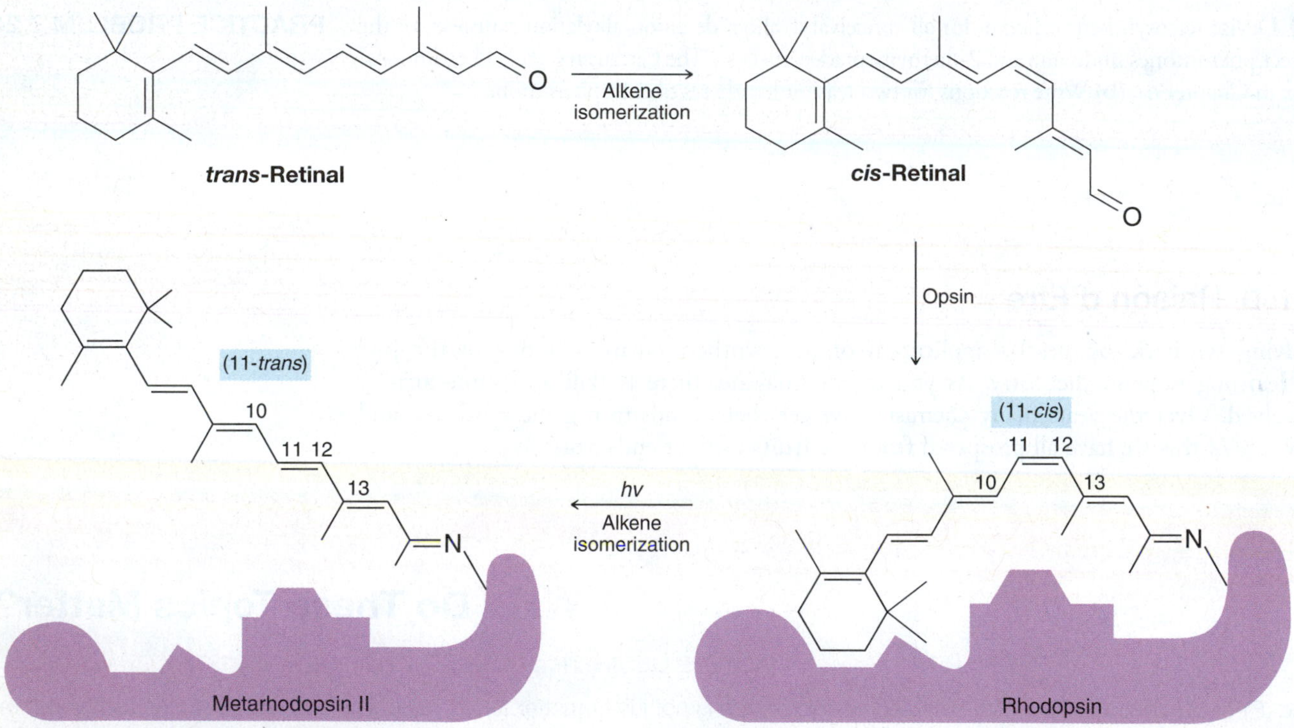

Image Source

trans-Retinal → Alkene isomerization → **cis-Retinal**

Opsin

(11-*trans*) 10 11 12 13

Metarhodopsin II ← *hv* Alkene isomerization ← (11-*cis*) 11 12 10 13

Rhodopsin

SUMMARY AND REVIEW TOOLS

In this chapter we described methods for the synthesis of alkenes using dehydrohalogenation, dehydration of alcohols, and reduction of alkynes. We also introduced the alkylation of alkynide anions as a method for forming new carbon–carbon bonds, and we introduced retrosynthetic analysis as a means of logically planning an organic synthesis.

The study aids for this chapter include the list of methods below that we have mentioned for synthesizing alkenes, as well as key terms and concepts (which are hyperlinked to the Glossary from the bold, blue terms in the *WileyPLUS* version of the book at wileyplus.com). Following the end of chapter problems you will find graphical overviews of the mechanisms for E2 and E1 reactions, a Synthetic Connections scheme for alkynes, alkenes, alkyl halides, and alcohols, and a Concept Map regarding organic synthesis involving alkenes.

SUMMARY OF METHODS FOR THE PREPARATION OF ALKENES AND ALKYNES

1. **Dehydrohalogenation of alkyl halides (Section 7.6):**

General Reaction

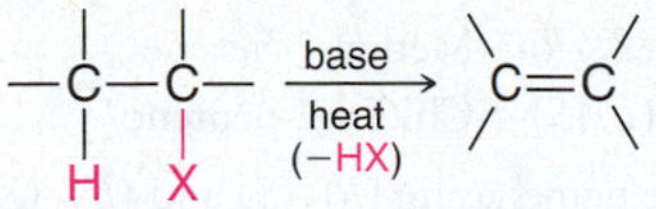

Specific Examples

CH$_3$CH$_2$CHCH$_3$ $\xrightarrow[\text{EtOH}]{\text{EtONa}}$ CH$_3$CH=CHCH$_3$ + CH$_3$CH$_2$CH=CH$_2$
$\quad\quad$ |
$\quad\quad$ Br $\quad\quad\quad\quad\quad\quad\quad\quad$ **(cis and trans, 81%)** $\quad$ **(19%)**

CH$_3$CH$_2$CHCH$_3$ $\xrightarrow[\substack{t\text{-BuOH}\\70\,°C}]{t\text{-BuOK}}$ CH$_3$CH=CHCH$_3$ + CH$_3$CH$_2$CH=CH$_2$
$\quad\quad$ |
$\quad\quad$ Br $\quad\quad\quad\quad\quad\quad\quad$ **Disubstituted alkenes** $\quad$ **Monosubstituted alkene**
$\quad\quad\quad\quad\quad\quad\quad\quad\quad\quad\quad\quad$ **(cis and trans, 47%)** $\quad\quad\quad$ **(53%)**

2. **Dehydration of alcohols (Sections 7.7 and 7.8):**

General Reaction

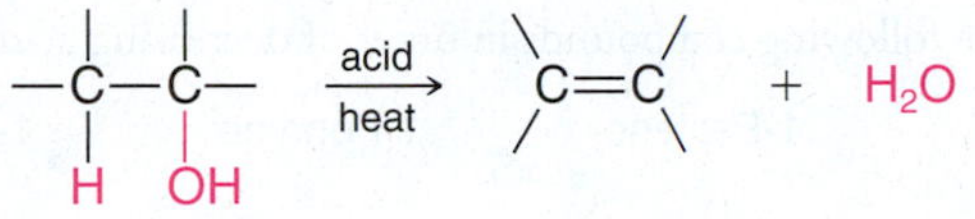

Specific Examples

CH$_3$CH$_2$OH $\xrightarrow[\text{180 °C}]{\text{concd H}_2\text{SO}_4}$ CH$_2$=CH$_2$ + H$_2$O

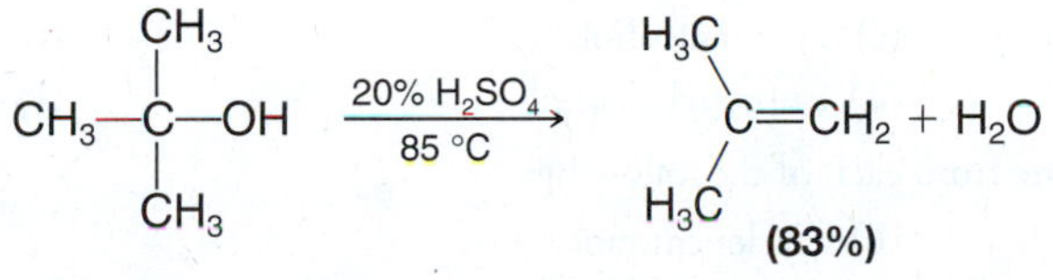

3. **Hydrogenation of alkynes (Section 7.15):**

General Reaction

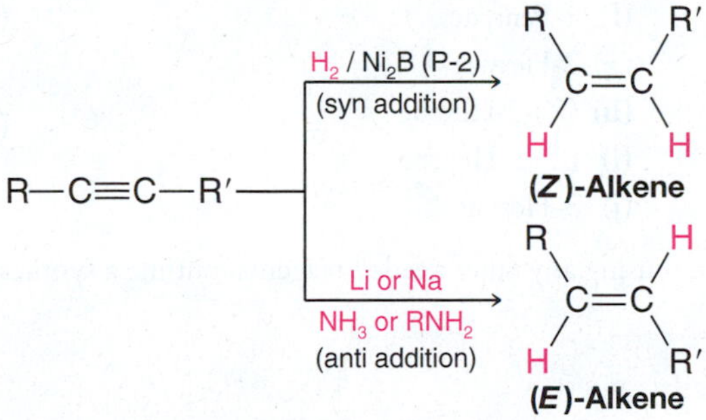

In subsequent chapters we shall see a number of other methods for alkene synthesis.

PROBLEMS WILEY PLUS

Note to Instructors: Many of the homework problems are available for assignment via *WileyPLUS*, an online teaching and learning solution.

STRUCTURE AND NOMENCLATURE

7.25 Each of the following names is incorrect. Give the correct name and explain your reasoning.

(a) *trans*-3-Pentene

(b) 1,1-Dimethylethene

(c) 2-Methylcyclohexene

(d) 4-Methylcyclobutene

(e) (*Z*)-3-Chloro-2-butene

(f) 5,6-Dichlorocyclohexene

7.26 Write a structural formula for each of the following:

(a) 3-Methylcyclobutene
(b) 1-Methylcyclopentene
(c) 2,3-Dimethyl-2-pentene
(d) (Z)-3-Hexene

(e) (E)-2-Pentene
(f) 3,3,3-Tribromopropene
(g) (Z,4R)-4-Methyl-2-hexene
(h) (E,4S)-4-Chloro-2-pentene

(i) (Z)-1-Cyclopropyl-1-pentene
(j) 5-Cyclobutyl-1-pentene
(k) (R)-4-Chloro-2-pentyne
(l) (E)-4-Methylhex-4-en-1-yne

7.27 Write three-dimensional formulas for and give names using (R)–(S) and (E)–(Z) designations for the isomers of:

(a) 4-Bromo-2-hexene
(b) 3-Chloro-1,4-hexadiene

(c) 2,4-Dichloro-2-pentene
(d) 2-Bromo-4-chlorohex-2-en-5-yne

7.28 Give the IUPAC names for each of the following:

(a) 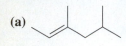   (c)    (e)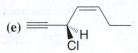

(b) 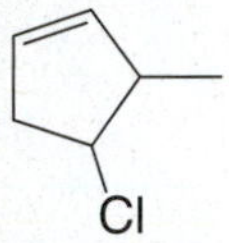   (d) 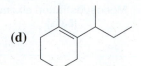   (f)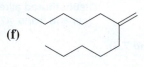

7.29 Without consulting tables, arrange the following compounds in order of decreasing acidity:

Pentane   1-Pentene   1-Pentyne   1-Pentanol

SYNTHESIS

7.30 Outline a synthesis of propene from each of the following:

(a) Propyl chloride
(b) Isopropyl chloride

(c) Propyl alcohol
(d) Isopropyl alcohol

(e) 1,2-Dibromopropane
(f) Propyne

7.31 Outline a synthesis of cyclopentene from each of the following:

(a) Bromocyclopentane

(b) Cyclopentanol

7.32 Starting with ethyne, outline syntheses of each of the following. You may use any other needed reagents, and you need not show the synthesis of compounds prepared in earlier parts of this problem.

(a) Propyne
(b) 1-Butyne
(c) 2-Butyne
(d) cis-2-Butene
(e) trans-2-Butene

(f) 1-Pentyne
(g) 2-Hexyne
(h) (Z)-2-Hexene
(i) (E)-2-Hexene
(j) 3-Hexyne

(k) $CH_3CH_2C\equiv CD$

(l)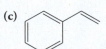

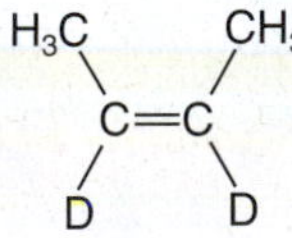

7.33 Starting with 1-methylcyclohexene and using any other needed reagents, outline a synthesis of the following deuterium-labeled compound:

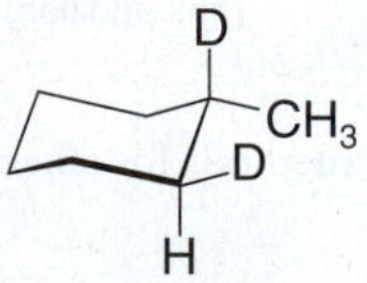

7.34 Outline a synthesis of phenylethyne from each of the following:

(a) 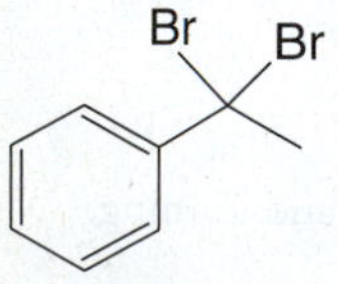   (b) 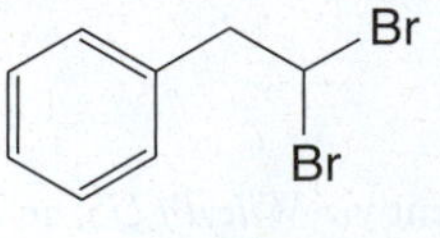   (c)   (d)

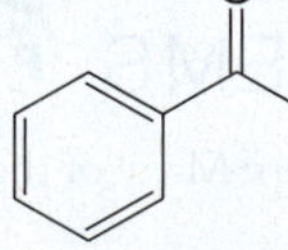

DEHYDROHALOGENATION AND DEHYDRATION

7.35 Write a three-dimensional representation for the transition state structure leading to formation of 2-methyl-2-butene from reaction of 2-bromo-2-methylbutane with sodium ethoxide.

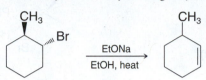

7.36 When *trans*-2-methylcyclohexanol (see the following reaction) is subjected to acid-catalyzed dehydration, the major product is 1-methylcyclohexene:

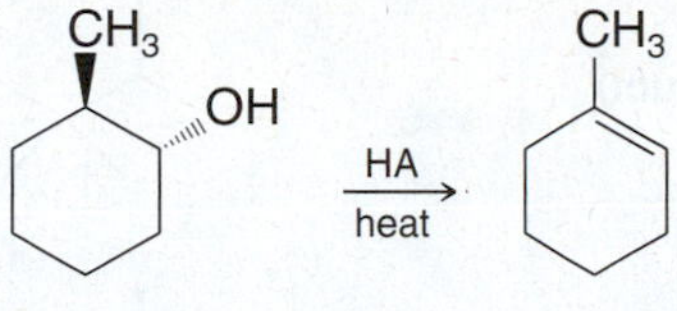

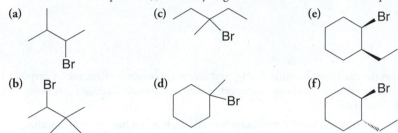

However, when *trans*-1-bromo-2-methylcyclohexane is subjected to dehydrohalogenation, the major product is 3-methylcyclohexene:

Account for the different products of these two reactions.

7.37 Write structural formulas for all the products that would be obtained when each of the following alkyl halides is heated with sodium ethoxide in ethanol. When more than one product results, you should indicate which would be the major product and which would be the minor product(s). You may neglect cis–trans isomerism of the products when answering this question.

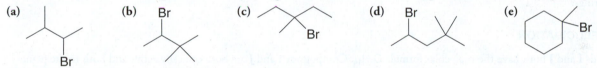

7.38 Write structural formulas for all the products that would be obtained when each of the following alkyl halides is heated with potassium *tert*-butoxide in *tert*-butyl alcohol. When more than one product results, you should indicate which would be the major product and which would be the minor product(s). You may neglect cis–trans isomerism of the products when answering this question.

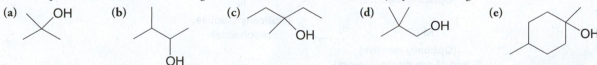

7.39 Starting with an appropriate alkyl halide and base, outline syntheses that would yield each of the following alkenes as the major (or only) product:

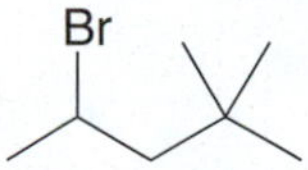

(a) 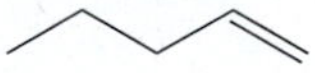 (b) 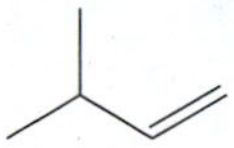 (c) 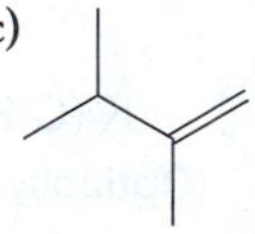 (d) 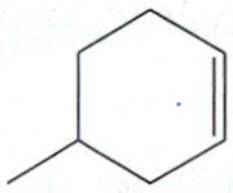 (e)

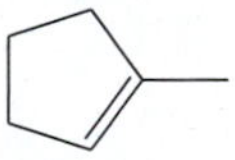

7.40 Arrange the following alcohols in order of their reactivity toward acid-catalyzed dehydration (with the most reactive first):

1-Pentanol   2-Methyl-2-butanol   3-Methyl-2-butanol

7.41 Give the products that would be formed when each of the following alcohols is subjected to acid-catalyzed dehydration. If more than one product would be formed, designate the alkene that would be the major product. (Neglect cis–trans isomerism.)

(a) (b) (c) (d) (e)

7.42 1-Bromobicyclo[2.2.1]heptane does not undergo elimination (below) when heated with a base. Explain this failure to react. (Construction of molecular models may help.)

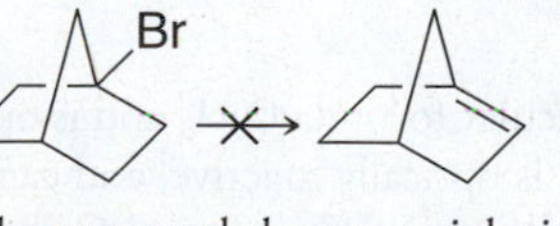

7.43 When the deuterium-labeled compound shown at right is subjected to dehydrohalogenation using sodium ethoxide in ethanol, the only alkene product is 3-methylcyclohexene. (The product contains no deuterium.) Provide an explanation for this result.

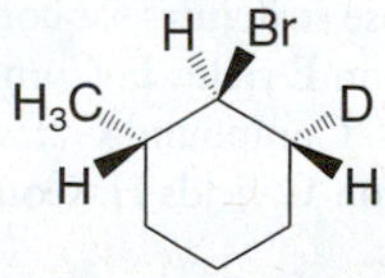

7.44 Provide a mechanistic explanation for each of the following reactions:

(a)

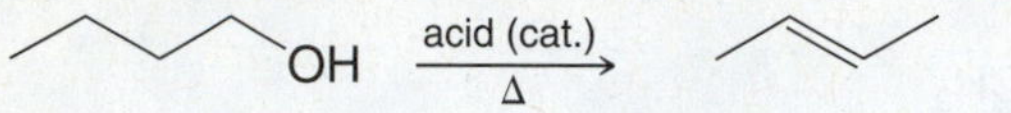

(major product)

(b)

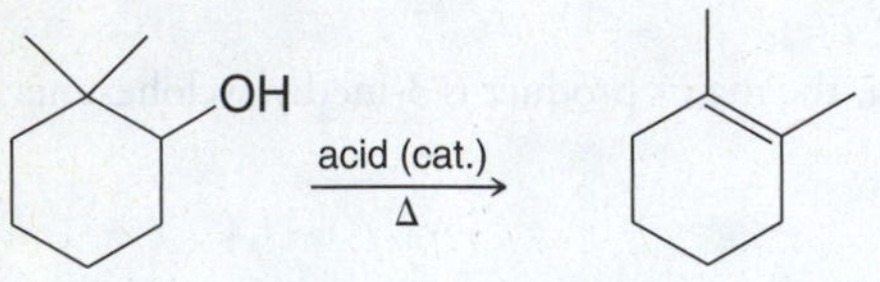

(major product)

(c)

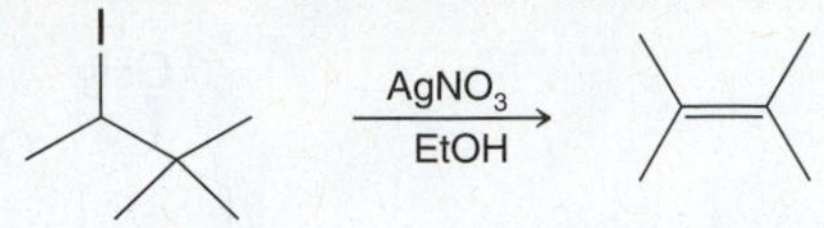

(major product)

(d)

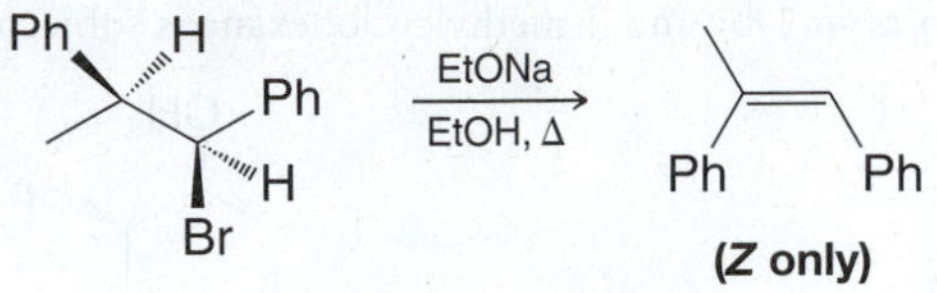

(Z only)

INDEX OF HYDROGEN DEFICIENCY

7.45 What is the index of hydrogen deficiency (IHD) (degree of unsaturation) for each of the following compounds?

(a)

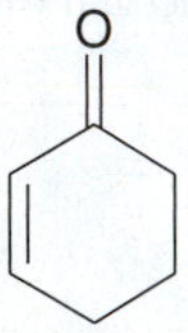

(b) $C_6H_8Br_4$

7.46 Caryophyllene, a compound found in oil of cloves, has the molecular formula $C_{15}H_{24}$ and has no triple bonds. Reaction of caryophyllene with an excess of hydrogen in the presence of a platinum catalyst produces a compound with the formula $C_{15}H_{28}$. How many **(a)** double bonds and **(b)** rings does a molecule of caryophyllene have?

7.47 Squalene, an important intermediate in the biosynthesis of steroids, has the molecular formula $C_{30}H_{50}$ and has no triple bonds.

(a) What is the index of hydrogen deficiency of squalene?

(b) Squalene undergoes catalytic hydrogenation to yield a compound with the molecular formula $C_{30}H_{62}$. How many double bonds does a molecule of squalene have?

(c) How many rings?

STRUCTURE ELUCIDATION

7.48 Compounds **I** and **J** both have the molecular formula C_7H_{14}. Compounds **I** and **J** are both optically active and both rotate plane-polarized light in the same direction. On catalytic hydrogenation **I** and **J** yield the same compound **K** (C_7H_{16}). Compound **K** is optically active. Propose possible structures for **I**, **J**, and **K**.

$$(+)-\text{I } (C_7H_{14}) \xrightarrow{\text{H}_2, \text{ Pt}}$$
$$\text{K } (C_7H_{16})$$
$$\textbf{(Optically active)}$$
$$(+)-\text{J } (C_7H_{14}) \xrightarrow{\text{H}_2, \text{ Pt}}$$

7.49 Compounds **L** and **M** have the molecular formula C_7H_{14}. Compounds **L** and **M** are optically inactive, are nonresolvable, and are diastereomers of each other. Catalytic hydrogenation of either **L** or **M** yields **N**. Compound **N** is optically inactive but can be resolved into separate enantiomers. Propose possible structures for **L**, **M**, and **N**.

$$\text{L } (C_7H_{14}) \xrightarrow{\text{H}_2, \text{ Pt}}$$
$$\textbf{(Optically inactive)}$$
$$\textbf{N}$$
$$\textbf{(Optically inactive, resolvable)}$$
$$\text{M } (C_7H_{14}) \xrightarrow{\text{H}_2, \text{ Pt}}$$
$$\textbf{(Optically inactive)}$$
$$\textbf{(L and M are diastereomers)}$$

CHALLENGE PROBLEMS

7.50 Propose structures for compounds **E–H**. Compound **E** has the molecular formula C_5H_8 and is optically active. On catalytic hydrogenation **E** yields **F**. Compound **F** has the molecular formula C_5H_{10}, is optically inactive, and cannot be resolved into separate enantiomers. Compound **G** has the molecular formula C_6H_{10} and is optically active. Compound **G** contains no triple bonds. On catalytic hydrogenation **G** yields **H**. Compound **H** has the molecular formula C_6H_{14}, is optically inactive, and cannot be resolved into separate enantiomers.

7.51 Consider the interconversion of *cis*-2-butene and *trans*-2-butene.
(a) What is the value of $\Delta H°$ for the reaction *cis*-2-butene → *trans*-2-butene (see Section 7.3A)?
(b) Assume $\Delta H° \cong \Delta G°$. What minimum value of $\Delta G^\ddagger$ would you expect for this reaction (see Section 1.13A)?
(c) Sketch a free-energy diagram for the reaction and label $\Delta G°$ and $\Delta G^\ddagger$.

7.52 (a) Partial dehydrohalogenation of either (1*R*,2*R*)-1,2-dibromo-1,2-diphenylethane or (1*S*,2*S*)-1,2-diphenylethane enantiomers (or a racemate of the two) produces (*Z*)-1-bromo-1,2-diphenylethene as the product, whereas **(b)** partial dehydrohalogenation of (1*R*,2*S*)-1,2-dibromo-1,2-diphenylethane (the meso compound) gives only (*E*)-1-bromo-1,2-diphenylethene. **(c)** Treating (1*R*,2*S*)-1,2-dibromo-1,2-diphenylethane with sodium iodide in acetone produces only (*E*)-1,2-diphenylethene. Explain these results.

7.53 (a) Using reactions studied in this chapter, show steps by which this alkyne could be converted to the seven-membered ring homolog of the product obtained in Problem 7.44(b).

(b) Could the homologous products obtained in these two cases be relied upon to show infrared absorption in the 1620–1680-cm^{-1} region?

7.54 Predict the structures of compounds **A**, **B**, and **C**:

A is an unbranched C_6 alkyne that is also a primary alcohol.

B is obtained from **A** by use of hydrogen and nickel boride catalyst or dissolving metal reduction.

C is formed from **B** on treatment with aqueous acid at room temperature. Compound **C** has no infrared absorption in either the 1620–1680-cm^{-1} or the 3590–3650-cm^{-1} region. It has an index of hydrogen deficiency of 1 and has one chirality center but forms as the racemate.

7.55 What is the index of hydrogen deficiency for **(a)** $C_7H_{10}O_2$ and **(b)** $C_5H_4N_4$?

LEARNING GROUP PROBLEMS

1. Write the structure(s) of the major product(s) obtained when 2-chloro-2,3-dimethylbutane (either enantiomer) reacts with **(a)** sodium ethoxide (EtONa) in ethanol (EtOH) at 80 °C or (in a separate reaction) with **(b)** potassium *tert*-butoxide (*t*-BuOK) in *tert*-butyl alcohol (*t*-BuOH) at 80 °C. If more than one product is formed, indicate which one would be expected to be the major product. **(c)** Provide a detailed mechanism for formation of the major product from each reaction, including a drawing of the transition state structures.

2. Explain using mechanistic arguments involving Newman projections or other three-dimensional formulas why the reaction of 2-bromo-1,2-diphenylpropane (either enantiomer) with sodium ethoxide (EtONa) in ethanol (EtOH) at 80 °C produces mainly (*E*)-1,2-diphenylpropene [little of the (*Z*) diastereomer is formed].

(major)   **(minor)**

3. (a) Write the structure of the product(s) formed when 1-methylcyclohexanol reacts with 85% (concd) H_3PO_4 at 150 °C. **(b)** Write a detailed mechanism for the reaction.

4. Consider the reaction of 1-cyclobutylethanol (1-hydroxyethylcyclobutane) with concentrated H_2SO_4 at 120 °C. Write structures of all reasonable organic products. Assuming that methylcyclopentene is one product, write a mechanism that accounts for its formation. Write mechanisms that account for formation of all other products as well.

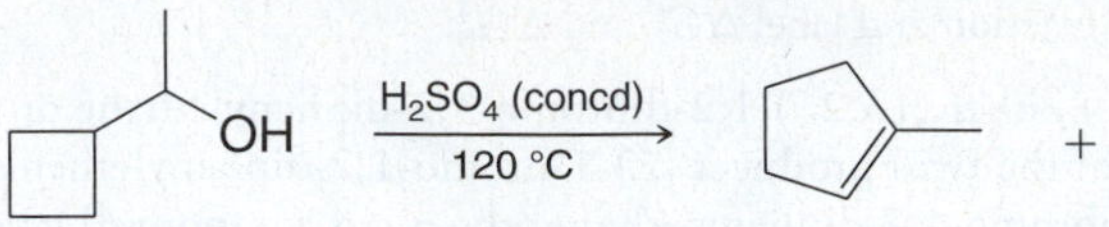

5. Consider the following compound:

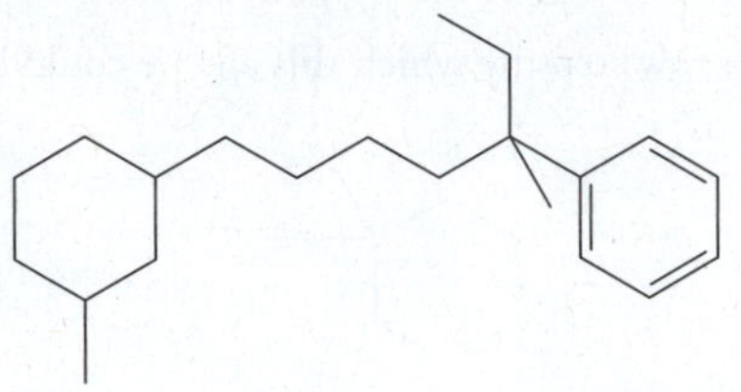

(a) Develop all reasonable retrosynthetic analyses for this compound (any diastereomer) that, at some point, involve carbon–carbon bond formation by alkylation of an alkynide ion.

(b) Write reactions, including reagents and conditions, for syntheses of this compound that correspond to the retrosynthetic analyses you developed above.

(c) Infrared spectroscopy could be used to show the presence of certain impurities in your final product that would result from leftover intermediates in your syntheses. Which of your synthetic intermediates would show IR absorptions that are distinct from those in the final product, and in what regions of the IR spectrum would the absorptions occur?

(d) Draw a three-dimensional structure for either the cis or trans form of the target molecule. Use dashed and solid wedges where appropriate in the alkyl side chain and use a chair conformational structure for the ring. (*Hint:* Draw the structure so that the carbon chain of the most complicated substituent on the cyclohexane ring and the ring carbon where it is attached are all in the plane of the paper. In general, for three-dimensional structures choose an orientation that allows as many carbon atoms as possible to be in the plane of the paper.)

SUMMARY AND REVIEW TOOLS

Reaction Summary of E2 and E1 elimination

E2 via small base

- Strong **unhindered base**, e.g., CH_3CH_2ONa (EtONa), HO^-
- Predominant formation of **most substituted alkene** (Zaitsev product)
- Anti coplanar transition state
- Bimolecular in rate-determining step

E2 via bulky base

- Strong **hindered base**, e.g. $(CH_3)_3COK$ (*t*-BuOK)
- Predominant formation of **least substituted alkene** (Hofmann product)
- Anti coplanar transition state
- Bimolecular in rate-determining step

E1 (including Alcohol Dehydration)

- Absence of strong base (solvent is often the base)
- Alcohols require **strong acid catalyst**
- Carbocation formation is unimolecular rate-determining step
- **Carbocation may rearrange**
- Predominant formation of most substituted alkene (Zaitsev product)
- Leaving group for alcohols is — $\overset{+}{O}H_2$
- 1° Alcohols react by an E2-type mechanism

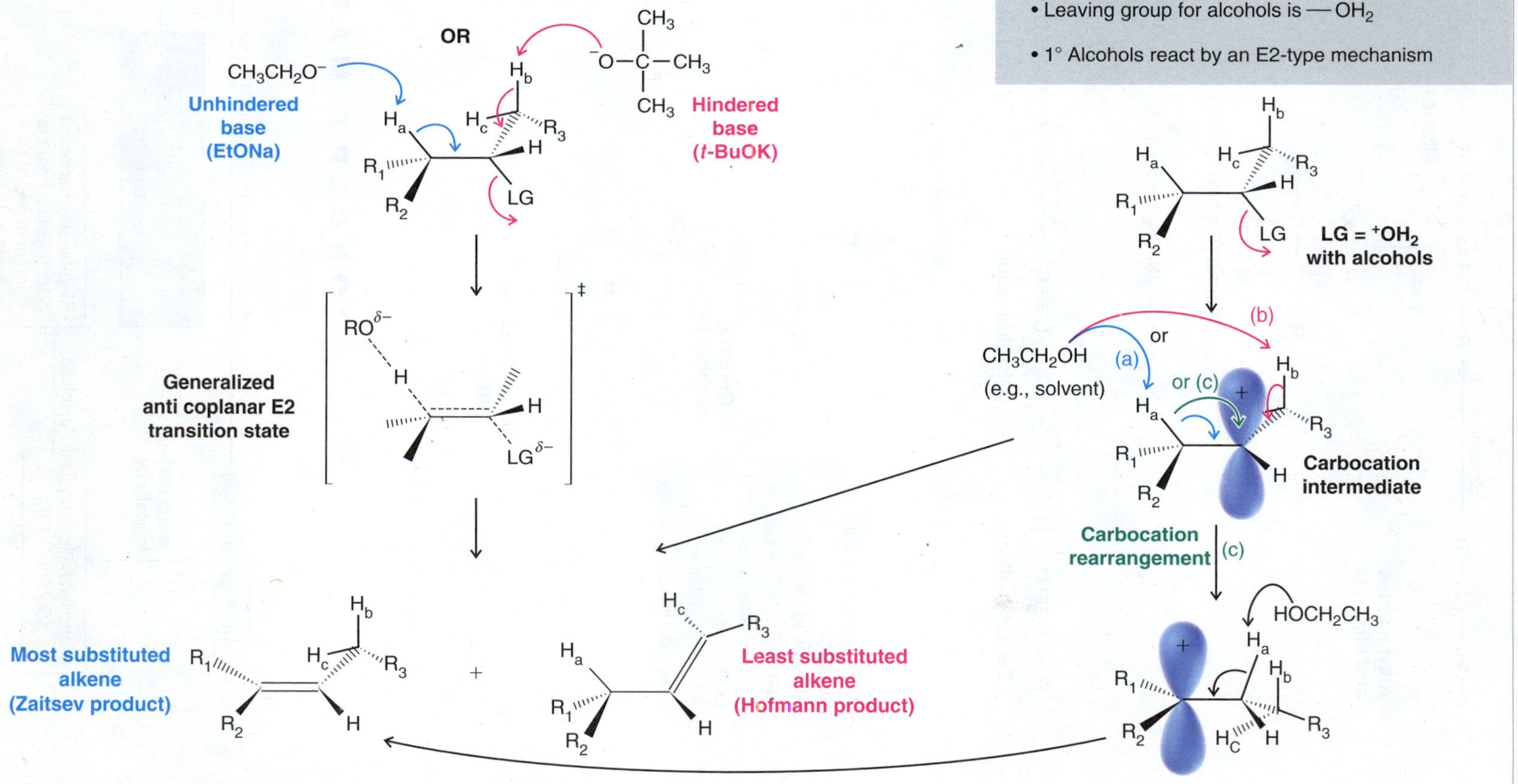

$LG = {}^+OH_2$ with alcohols

Carbocation intermediate

Carbocation rearrangement

Least substituted alkene (Hofmann product)

Most substituted alkene (Zaitsev product)

Hindered base (*t*-BuOK)

Unhindered base (EtONa)

Generalized anti coplanar E2 transition state

[S U M M A R Y   A N D   R E V I E W   T O O L S]

Synthetic Connections of Alkynes, Alkenes, Alkyl Halides, and Alcohols

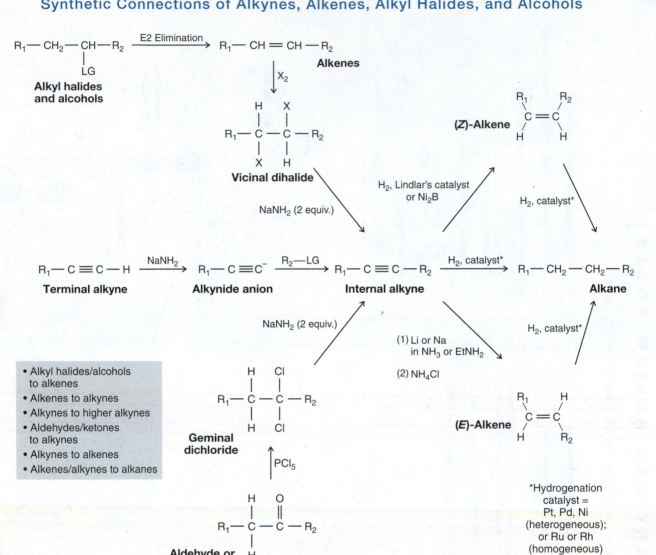

- Alkyl halides/alcohols to alkenes
- Alkenes to alkynes
- Alkynes to higher alkynes
- Aldehydes/ketones to alkynes
- Alkynes to alkenes
- Alkenes/alkynes to alkanes

*Hydrogenation catalyst = Pt, Pd, Ni (heterogeneous); or Ru or Rh (homogeneous)

[C O N C E P T   M A P]

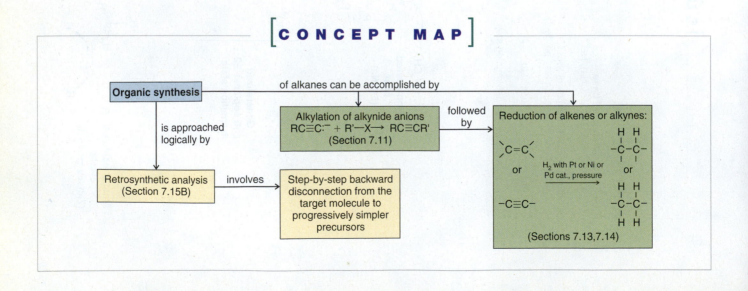

CHAPTER

8

Alkenes and Alkynes II

ADDITION REACTIONS

In recent chapters we have discussed mechanisms that involve electron pairs in bond-forming and bond-breaking steps of substitution and elimination reactions. Nucleophiles and bases served as electron pair donors in these reactions. In this chapter we discuss reactions of **alkenes** and **alkynes** in which a double or triple bond acts as the electron pair donor for bond formation. These reactions are called **addition reactions**.

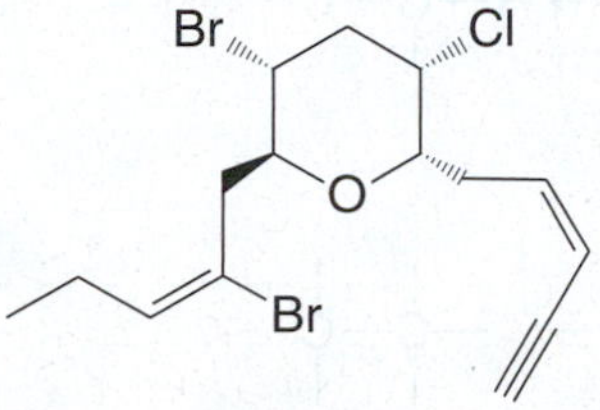

Dactylyne

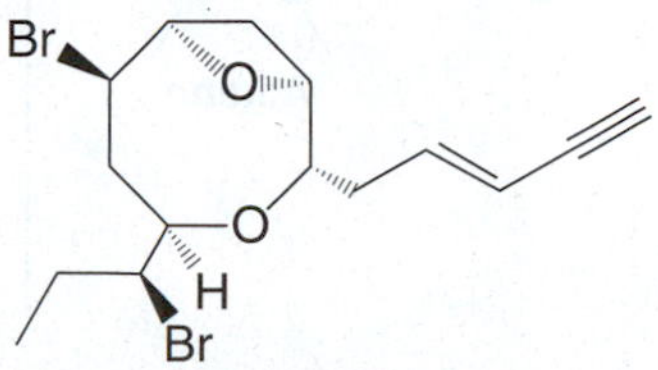

(3*E*)-Laureatin

Alkenes and alkynes are very common in nature, both on land and in the sea. Examples from the sea include dactylyne and (3*E*)-laureatin, whose formulas are shown here. These compounds include halogens in their structures, as is the case for many other natural marine compounds. Certain marine organisms may produce compounds like these for the purpose of self-defense, since a number of them have cytotoxic properties. Interestingly, the halogens in these marine compounds are

PHOTO CREDITS: (coral) © Mehmet Torlak/iStockphoto; (discus fish) © cynoclub/iStockphoto incorporated by biological reactions similar to those we shall study in this chapter (Section 8.11). Not only, therefore, do compounds like dactylyne and (3E)-laureatin have intriguing structures and properties, and arise in the beautiful environment of the sea, but they also have fascinating chemistry behind them.

IN THIS CHAPTER WE WILL CONSIDER:

- the regio- and stereochemistry of addition reactions of alkenes
- processes that can add molecules of water, halogens, carbon, and other functionalities across alkenes
- events that cleave double bonds and provide more highly oxidized compounds
- alkyne reactions that are analogous to alkene reactions

[WHY DO THESE TOPICS MATTER?]   At the end of the chapter, we will show how, in nature, a special class of alkenes is involved in the creation of tens of thousands of bioactive molecules, all through processes that mirror the core reactions discussed in this chapter.

8.1 ADDITION REACTIONS OF ALKENES

We have already studied one addition reaction of alkenes—hydrogenation—in which a hydrogen atom is added at each end of a double (or triple) bond. In this chapter we shall study other alkene addition reactions that do not involve the same mechanism as hydrogenation. We can depict this type of reaction generally, using **E** for an electrophilic portion of a reagent and **Nu** for a nucleophilic portion, as follows.

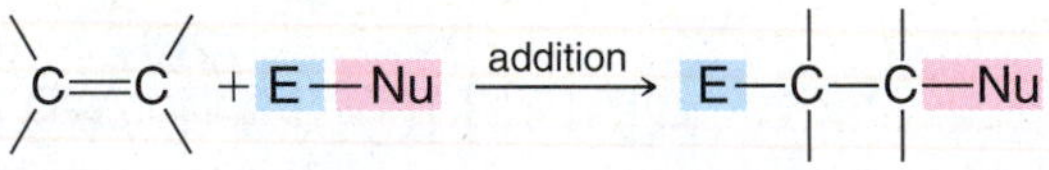

$$\text{C=C} + \text{E-Nu} \xrightarrow{\text{addition}} \text{E-C-C-Nu}$$

Some specific reactions of this type that we shall study in this chapter include addition of hydrogen halides, sulfuric acid, water (in the presence of an acid catalyst), and halogens. Later we shall also study some specialized reagents that undergo addition reactions with alkenes.

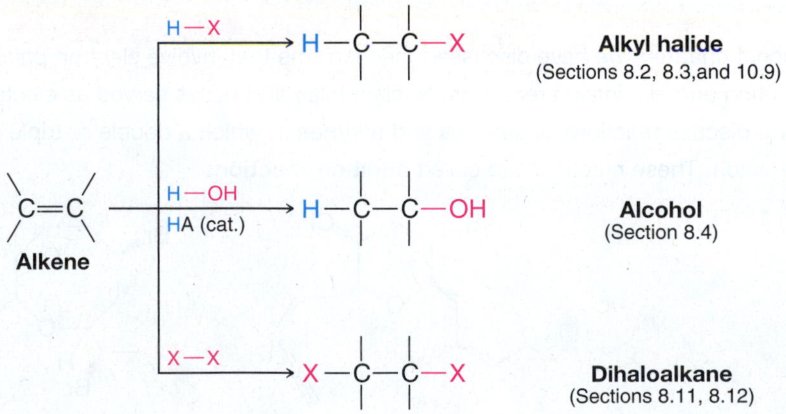

8.1A HOW TO Understand Additions to Alkenes

Two characteristics of the double bond help us understand why these addition reactions occur:

1. An addition reaction results in the conversion of one π bond and one σ bond (Sections 1.12 and 1.13) into two σ bonds. The result of this change is usually energetically favorable. The energy released in making two σ bonds exceeds that needed to break

one σ bond and one π bond (because π bonds are weaker), and, therefore, addition reactions are usually exothermic:

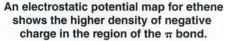

2. The electrons of the π bond are exposed. Because the π bond results from overlapping p orbitals, the π electrons lie above and below the plane of the double bond:

 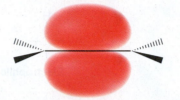

An electrostatic potential map for ethene shows the higher density of negative charge in the region of the π bond.

The electron pair of the π bond is distributed throughout both lobes of the π molecular orbital.

Electrophilic Addition

- Electrons in the π bond of alkenes react with electrophiles.

- **Electrophiles** are electron-seeking reagents. They have the property of being **electrophilic**.

Electrophiles include proton donors such as Brønsted–Lowry acids, neutral reagents such as bromine (because it can be polarized so that one end is positive), and Lewis acids such as BH_3, BF_3, and $AlCl_3$. Metal ions that contain vacant orbitals—the silver ion (Ag^+), the mercuric ion (Hg^{2+}), and the platinum ion (Pt^{2+}), for example—also act as electrophiles.

Hydrogen halides, for example, react with alkenes by accepting a pair of electrons from the π bond to form a σ bond between the hydrogen and one of the carbon atoms, with loss of the halide ion. This leaves a vacant p orbital and a + charge on the other carbon. The overall result is the formation of a carbocation and a halide ion from the alkene and **HX**:

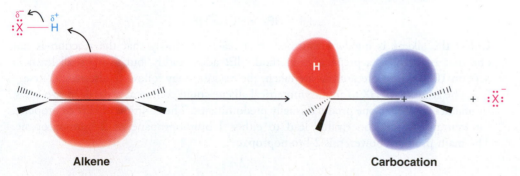

Alkene   Carbocation

Being highly reactive, the carbocation may then combine with the halide ion by accepting one of its electron pairs:

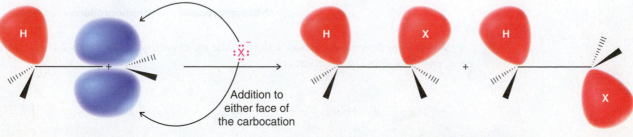

Carbocation   Addition to either face of the carbocation   Addition products

Electrophiles Are Lewis Acids Electrophiles are molecules or ions that can accept an electron pair. Nucleophiles are molecules or ions that can furnish an electron pair (i.e., Lewis bases). Any reaction of an electrophile also involves a nucleophile. In the protonation of an alkene the electrophile is the proton donated by an acid; the nucleophile is the alkene:

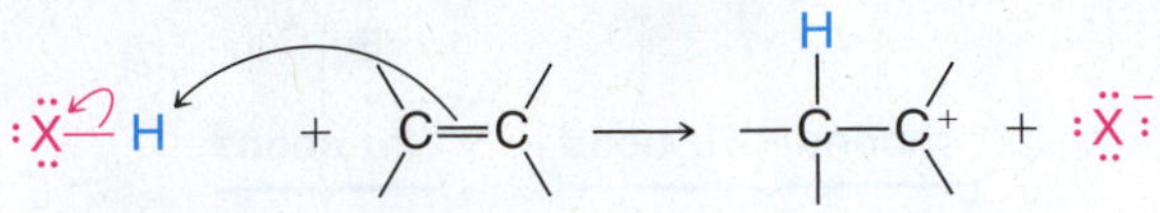

Electrophile   Nucleophile

In the next step, the reaction of the carbocation with a halide ion, the carbocation is the electrophile and the halide ion is the nucleophile:

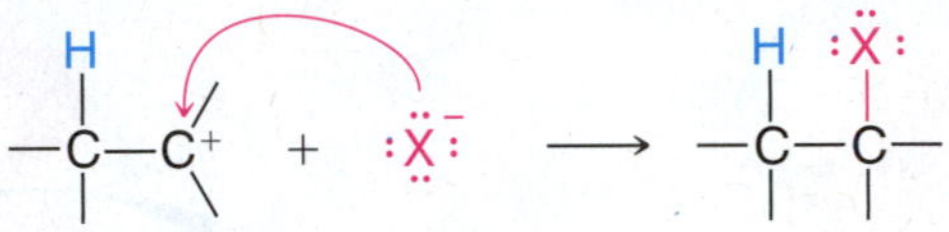

Electrophile   Nucleophile

8.2 ELECTROPHILIC ADDITION OF HYDROGEN HALIDES TO ALKENES: MECHANISM AND MARKOVNIKOV'S RULE

Hydrogen halides (HI, HBr, HCl, and HF) add to the double bond of alkenes:

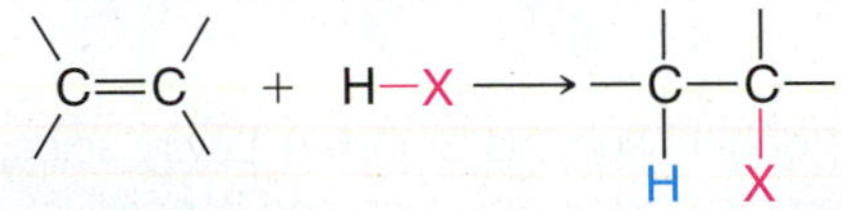

These additions are sometimes carried out by dissolving the hydrogen halide in a solvent, such as acetic acid or CH_2Cl_2, or by bubbling the gaseous hydrogen halide directly into the alkene and using the alkene itself as the solvent. **HF** is prepared as polyhydrogen fluoride in pyridine.

- The order of reactivity of the hydrogen halides in alkene addition is

$$HI > HBr > HCl > HF$$

Unless the alkene is highly substituted, HCl reacts so slowly that the reaction is not one that is useful as a preparative method. HBr adds readily, but as we shall learn in Section 10.9, unless precautions are taken, the reaction may follow an alternate course.

The addition of **HX** to an unsymmetrical alkene could conceivably occur in two ways. In practice, however, one product usually predominates. The addition of **HBr** to propene, for example, could conceivably lead to either 1-bromopropane or 2-bromopropane. The main product, however, is 2-bromopropane:

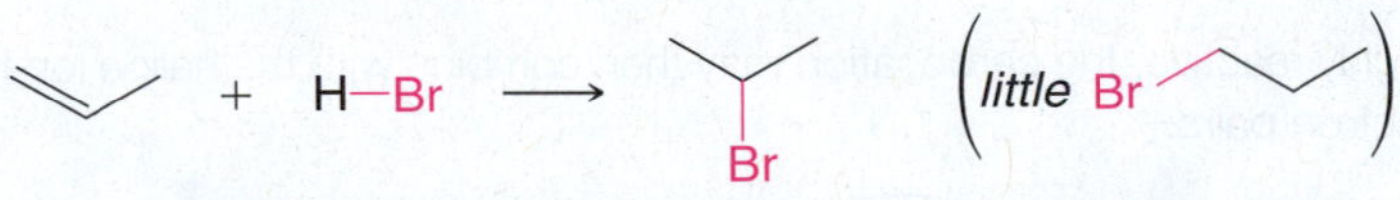

2-Bromopropane   1-Bromopropane

When 2-methylpropene reacts with **HBr**, the main product is 2-bromo-2-methylpropane, not 1-bromo-2-methylpropane:

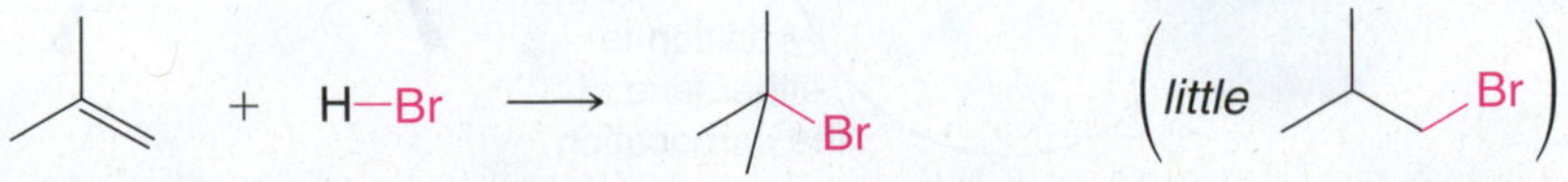

2-Methylpropene   2-Bromo-2-methylpropane   1-Bromo-2-methylpropane

Consideration of many examples like this led the Russian chemist Vladimir Markovnikov in 1870 to formulate what is now known as Markovnikov's rule.

- One way to state **Markovnikov's rule** is to say that *in the addition of HX to an alkene, the hydrogen atom adds to the carbon atom of the double bond that already has the greater number of hydrogen atoms.*[*]

The addition of HBr to propene is an illustration:

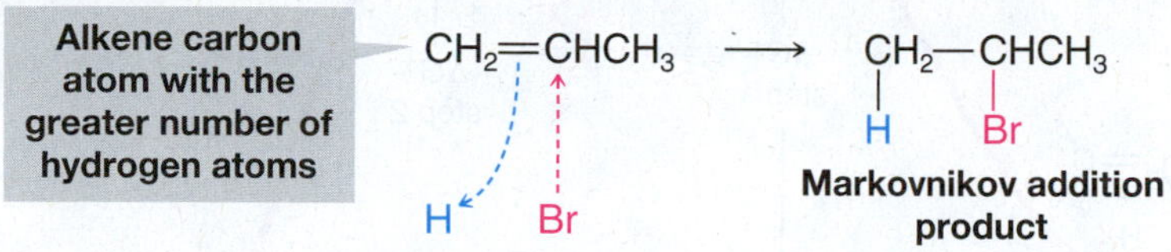

Reactions that illustrate Markovnikov's rule are said to be *Markovnikov additions*.

A mechanism for addition of a hydrogen halide to an alkene involves the following two steps:

A MECHANISM FOR THE REACTION — Addition of a Hydrogen Halide to an Alkene

Step 1

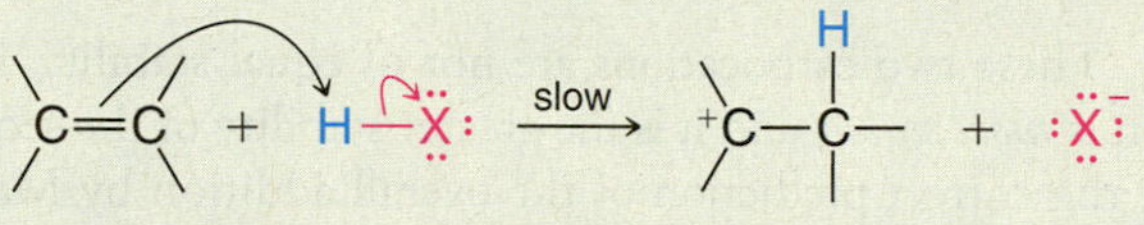

The π electrons of the alkene form a bond with a proton from **HX** to form a carbocation and a halide ion.

Step 2

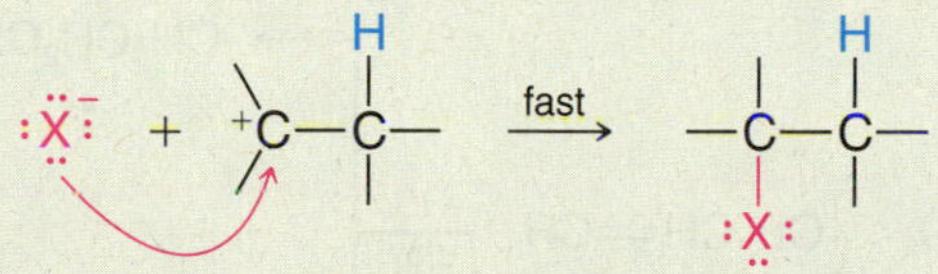

The halide ion reacts with the carbocation by donating an electron pair; the result is an alkyl halide.

The important step—because it is the **rate-determining step**—is step 1. In step 1 the alkene donates a pair of electrons to the proton of the hydrogen halide and forms a carbocation. This step (Fig. 8.1) is highly endergonic and has a high free energy of activation. Consequently, it takes place slowly. In step 2 the highly reactive carbocation stabilizes itself by combining with a halide ion. This exergonic step has a very low free energy of activation and takes place very rapidly.

8.2A Theoretical Explanation of Markovnikov's Rule

If the alkene that undergoes addition of a hydrogen halide is an unsymmetrical alkene such as propene, then step 1 could conceivably lead to two different carbocations:

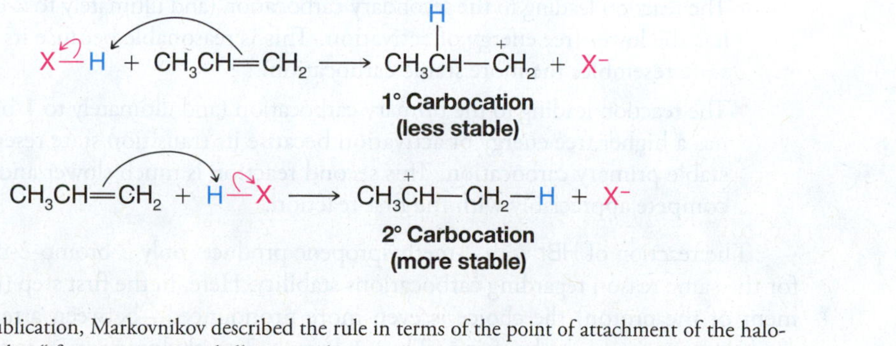

1° Carbocation
(less stable)

2° Carbocation
(more stable)

---

[*]In his original publication, Markovnikov described the rule in terms of the point of attachment of the halogen atom, stating that "if an unsymmetrical alkene combines with a hydrogen halide, the halide ion adds to the carbon atom with the fewer hydrogen atoms."

FIGURE 8.1 Free-energy diagram for the addition of **HX** to an alkene. The free energy of activation for step 1 is much larger than that for step 2.

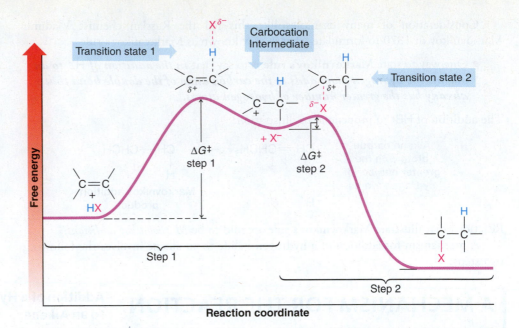

These two carbocations are not of equal stability, however. The secondary carbocation is *more stable*, and it is the greater stability of the secondary carbocation that accounts for the correct prediction of the overall addition by Markovnikov's rule. In the addition of **HBr** to propene, for example, the reaction takes the following course:

$$
CH_3CH=CH_2 \xrightarrow[\text{slow}]{HBr}
\begin{cases}
\xrightarrow{\times} CH_3CH_2\overset{+}{C}H_2 \xrightarrow{Br^-} CH_3CH_2CH_2Br \\
\qquad\qquad\quad 1° \qquad\qquad\qquad \textbf{1-Bromopropane} \\
\qquad\qquad\qquad\qquad\qquad\qquad (\textit{little} \text{ formed}) \\
\\
\longrightarrow CH_3\overset{+}{C}HCH_3 \xrightarrow[\text{fast}]{Br^-} CH_3\underset{\underset{Br}{|}}{C}HCH_3 \\
\qquad\qquad\quad 2° \\
\qquad\qquad\qquad\qquad\qquad \textbf{2-Bromopropane} \\
\qquad\qquad\qquad\qquad\qquad (\textbf{main product})
\end{cases}
$$

|— Step 1 ————————|————————— Step 2 —|

The chief product of the reaction is 2-bromopropane because the more stable secondary carbocation is formed preferentially in the first step.

- The more stable carbocation predominates because it is formed faster.

We can understand why this is true if we examine the free-energy diagrams in Fig. 8.2.

- The reaction leading to the secondary carbocation (and ultimately to 2-bromopropane) has the lower free energy of activation. This is reasonable because its transition state resembles the more stable carbocation.

- The reaction leading to the primary carbocation (and ultimately to 1-bromopropane) has a higher free energy of activation because its transition state resembles a less stable primary carbocation. This second reaction is much slower and does not compete appreciably with the first reaction.

The reaction of **HBr** with 2-methylpropene produces only 2-bromo-2-methylpropane, for the same reason regarding carbocations stability. Here, in the first step (i.e., the attachment of the proton) the choice is even more pronounced—between a tertiary carbocation and a primary carbocation. Thus, 1-bromo-2-methylpropane is *not* obtained as a product of the reaction because its formation would require the formation of a primary

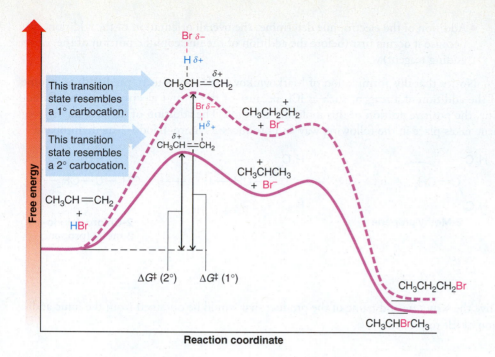

FIGURE 8.2 Free-energy diagrams for the addition of HBr to propene. $\Delta G^{\ddagger}(2°)$ is less than $\Delta G^{\ddagger}(1°)$.

carbocation. Such a reaction would have a much higher free energy of activation than that leading to a tertiary carbocation.

- Rearrangements invariably occur when the carbocation initially formed by addition of **HX** to an alkene can rearrange to a more stable one (see Section 7.8 and Practice Problem 8.3).

[A MECHANISM FOR THE REACTION — Addition of HBr to 2-Methylpropene]

This reaction takes place:

3° Carbocation
(more stable carbocation)

2-Bromo-2-methylpropane

CH$_3$—C—CH$_3$ **Major product**

This reaction *does not* occur to any appreciable extent:

1° Carbocation
(less stable carbocation)

1-Bromo-2-methylpropane

Little formed

8.2B General Statement of Markovnikov's Rule

With this understanding of the mechanism for the ionic addition of hydrogen halides to alkenes, we can now generalize about how electrophiles add to alkenes.

- General statement of **Markovnikov's rule**: In the ionic addition of an unsymmetrical reagent to a double bond, the positive portion of the adding reagent attaches itself to a carbon atom of the double bond so as to yield the more stable carbocation as an intermediate.

- Addition of the electrophile determines the overall orientation of the addition, because it occurs first (before the addition of the nucleophilic portion of the adding reagent).

Notice that this formulation of Markovnikov's rule allows us to predict the outcome of the addition of a reagent such as ICl. Because of the greater electronegativity of chlorine, the positive portion of this molecule is iodine. The addition of ICl to 2-methylpropene takes place in the following way and produces 2-chloro-1-iodo-2-methylpropane:

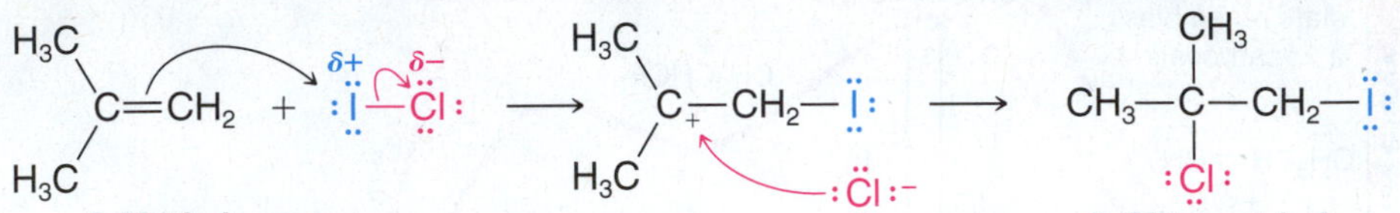

2-Methylpropene   **2-Chloro-1-iodo-2-methylpropane**

PRACTICE PROBLEM 8.1   Give the structure and name of the product that would be obtained from the ionic addition of IBr to propene.

PRACTICE PROBLEM 8.2   Outline mechanisms for the following addition reactions:

(a) 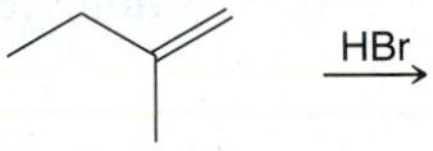 $\xrightarrow{\text{HBr}}$ (b) 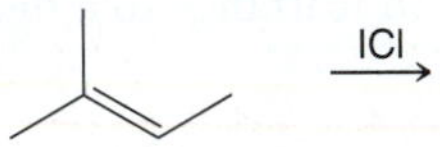 $\xrightarrow{\text{ICl}}$ (c) 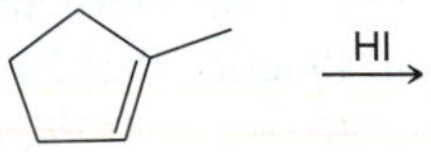 $\xrightarrow{\text{HI}}$

PRACTICE PROBLEM 8.3   Provide mechanistic explanations for the following observations:

(a)

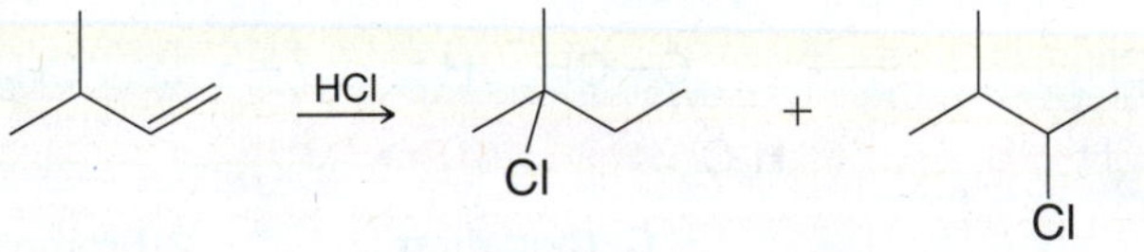

(b)

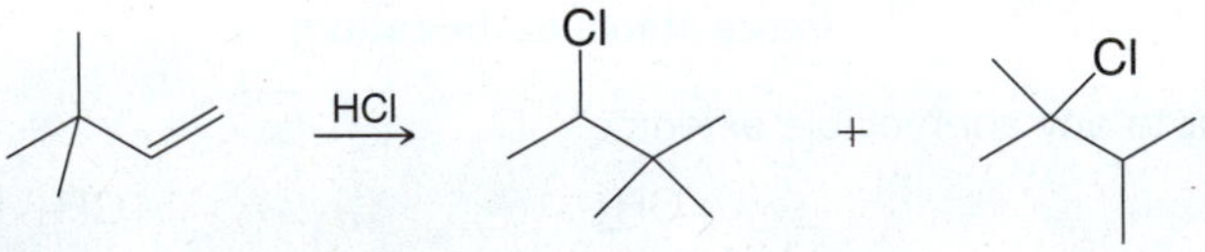

8.2C Regioselective Reactions

Chemists describe reactions like the Markovnikov additions of hydrogen halides to alkenes as being **regioselective**. *Regio* comes from the Latin word *regionem* meaning direction.

- When a reaction that can potentially yield two or more constitutional isomers actually produces only one (or a predominance of one), the reaction is said to be a **regioselective reaction**.

The addition of HX to an unsymmetrical alkene such as propene could conceivably yield two constitutional isomers, for example. As we have seen, however, the reaction yields only one, and therefore it is regioselective.

8.2D An Exception to Markovnikov's Rule

In Section 10.9 we shall study an exception to Markovnikov's rule. This exception concerns the addition of HBr to alkenes **when the addition is carried out in the presence of peroxides** (i.e., compounds with the general formula ROOR).

- When alkenes are treated with HBr in the presence of peroxides, an **anti-Markovnikov addition** occurs in the sense that the hydrogen atom becomes attached to the carbon atom with the fewer hydrogen atoms.

With propene, for example, the addition takes place as follows:

$$CH_3CH{=}CH_2 + HBr \xrightarrow{ROOR} CH_3CH_2CH_2Br$$

In Section 10.10 we shall find that this addition occurs by *a radical mechanism*, and not by the ionic mechanism given at the beginning of Section 8.2.

- This anti-Markovnikov addition occurs **only when HBr is used in the presence of peroxides** and does not occur significantly with HF, HCl, and HI even when peroxides are present.

8.3 STEREOCHEMISTRY OF THE IONIC ADDITION TO AN ALKENE

Consider the following addition of HX to 1-butene and notice that the reaction leads to the formation of a 2-halobutane that contains a chirality center:

$$CH_3CH_2CH{=}CH_2 \ + \ HX \longrightarrow CH_3CH_2\overset{*}{C}HCH_3$$
$$| \atop X$$

The product, therefore, can exist as a pair of enantiomers. The question now arises as to how these enantiomers are formed. Is one enantiomer formed in greater amount than the other? The answer is *no*; the carbocation that is formed in the first step of the addition (see the following scheme) is trigonal planar and is **achiral** (a model will show that it has a plane of symmetry). When the halide ion reacts with this achiral carbocation in the second step, **reaction is equally likely at either face**. The reactions leading to the two enantiomers occur at the same rate, and the enantiomers, therefore, are produced in equal amounts **as a racemic mixture**.

[**THE STEREOCHEMISTRY OF THE REACTION...** ─ **Ionic Addition to an Alkene**]

Achiral, trigonal planar carbocation

(S)-2-Halobutane
(50%)

(R)-2-Halobutane
(50%)

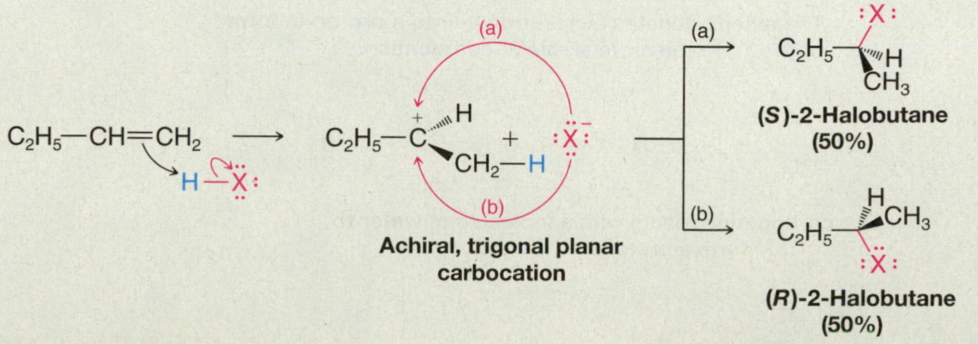

1-Butene donates a pair of electrons to the proton of HX to form an achiral carbocation.

The carbocation reacts with the halide ion at equal rates by path (a) or (b) to form the enantiomers as a racemate.

8.4 ADDITION OF WATER TO ALKENES: ACID-CATALYZED HYDRATION

The acid-catalyzed addition of water to the double bond of an alkene (**hydration** of an alkene) is a method for the preparation of low-molecular-weight alcohols. This reaction has its greatest utility in large-scale industrial processes. The acids most commonly used to catalyze the hydration of alkenes are dilute aqueous solutions of sulfuric acid and phosphoric acid. These reactions, too, are usually regioselective, and the addition of water to the double bond follows Markovnikov's rule. In general, the reaction takes the form that follows:

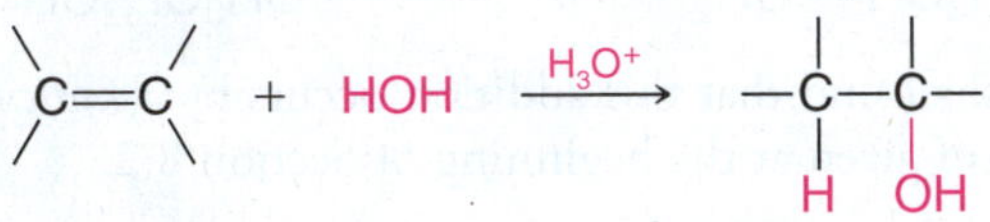

An example is the hydration of 2-methylpropene:

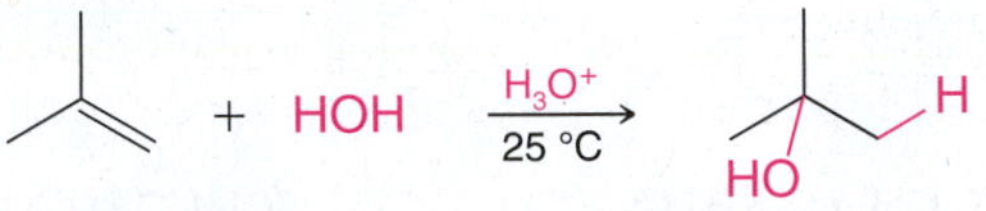

2-Methylpropene (isobutylene)

2-Methyl-2-propanol (*tert*-butyl alcohol)

Because the reactions follow Markovnikov's rule, acid-catalyzed hydrations of alkenes do not yield primary alcohols except in the special case of the hydration of ethene:

$$CH_2=CH_2 + HOH \xrightarrow[300\ °C]{H_3PO_4} CH_3CH_2OH$$

8.4A Mechanism

The mechanism for the hydration of an alkene is simply the reverse of the mechanism for the dehydration of an alcohol. We can illustrate this by giving the mechanism for the **hydration** of 2-methylpropene and by comparing it with the mechanism for the **dehydration** of 2-methyl-2-propanol given in Section 7.7A.

$$\boxed{\text{A MECHANISM FOR THE REACTION} - \begin{array}{l}\text{Acid-Catalyzed Hydration}\\ \text{of an Alkene}\end{array}}$$

Step 1

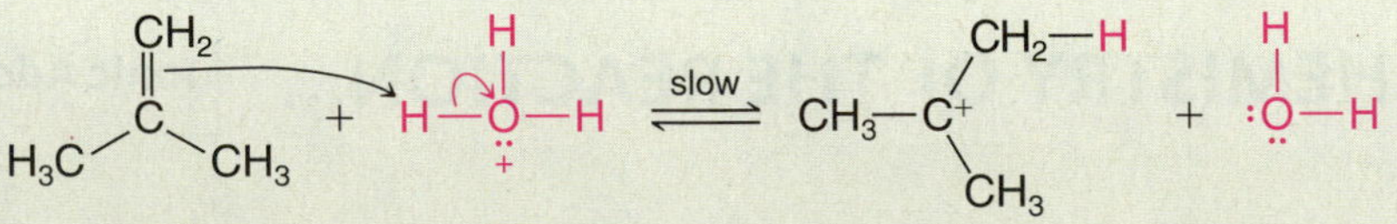

The alkene donates an electron pair to a proton to form the more stable 3° carbocation.

Step 2

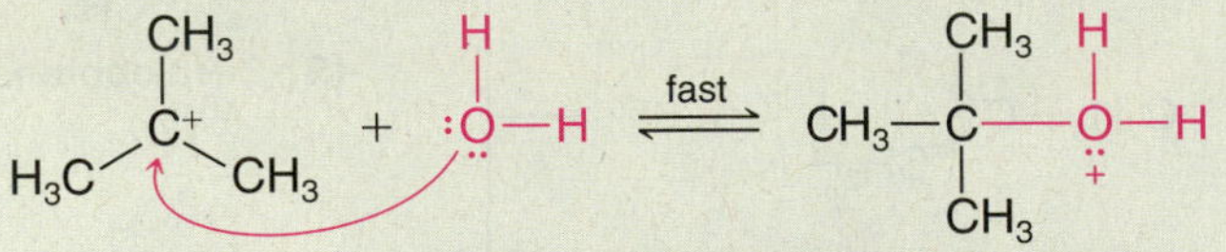

The carbocation reacts with a molecule of water to form a protonated alcohol.

Step 3

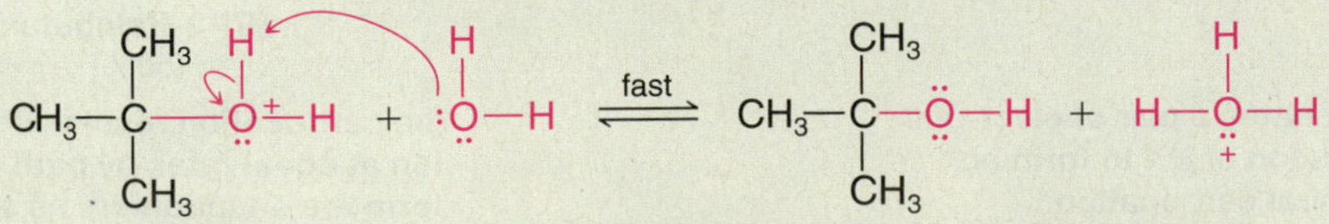

A transfer of a proton to a molecule of water leads to the product.

The rate-determining step in the *hydration* mechanism is step 1: the formation of the carbocation. It is this step, too, that accounts for the Markovnikov addition of water to the double bond. The reaction produces 2-methyl-2-propanol because step 1 leads to the formation of the more stable tertiary (3°) cation rather than the much less stable primary (1°) cation:

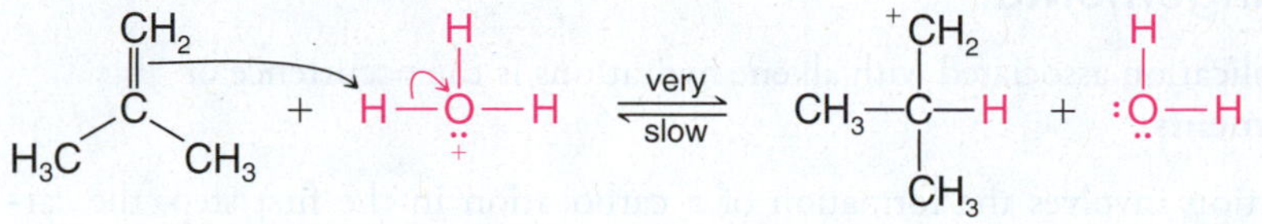

> For all practical purposes this reaction does not take place because it produces a 1° carbocation.

The reactions whereby *alkenes are hydrated or alcohols are dehydrated* are reactions in which the ultimate product is governed by the position of an equilibrium. Therefore, in the *dehydration of an alcohol* it is best to use a concentrated acid so that the concentration of water is low. (The water can be removed as it is formed, and it helps to use a high temperature.) In the *hydration of an alkene* it is best to use dilute acid so that the concentration of water is high. It also usually helps to use a lower temperature.

●●● **SOLVED PROBLEM 8.1**

Write a mechanism that explains the following reaction.

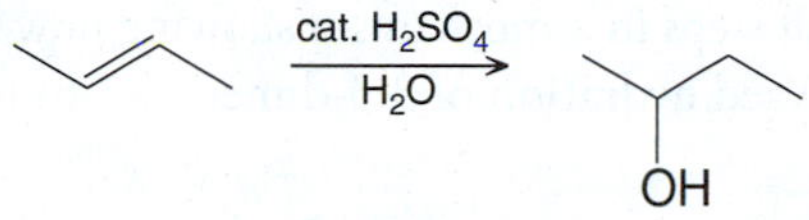

STRATEGY AND ANSWER: We know that a hydronium ion, formed from sulfuric acid and water, can donate a proton to an alkene to form a carbocation. The carbocation can then accept an electon pair from a molecule of water to form a protonated alcohol. The protonated alcohol can donate a proton to water to become an alcohol.

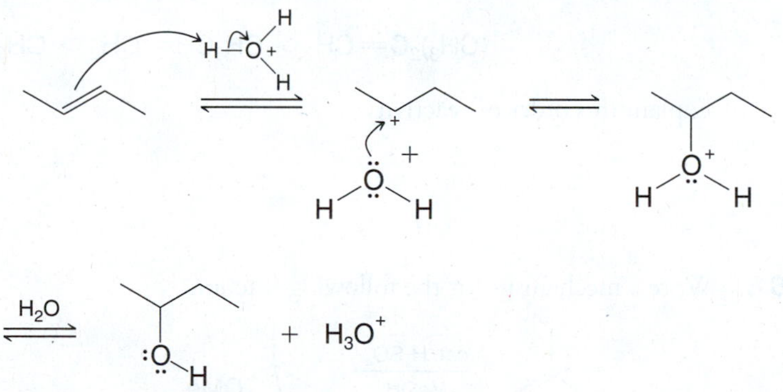

●●●

(a) Write a mechanism for the following reaction.

PRACTICE PROBLEM 8.4

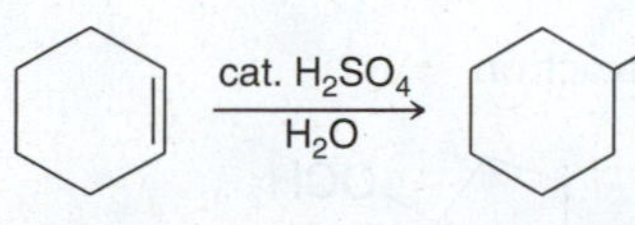

(b) What general conditions would you use to ensure a good yield of the product?

(c) What general conditions would you use to carry out the reverse reaction, i.e., the dehydration of cyclohexanol to produce cyclohexene?

(d) What product would you expect to obtain from the acid-catalyzed hydration of 1-methylcyclohexene? Explain your answer.

●●●
PRACTICE PROBLEM 8.5   In one industrial synthesis of ethanol, ethene is first dissolved in 95% sulfuric acid. In a second step water is added and the mixture is heated. Outline the reactions involved.

8.4B Rearrangements

• One complication associated with alkene hydrations is the occurrence of **rearrangements**.

Because the reaction involves the formation of a carbocation in the first step, the carbocation formed initially invariably rearranges to a more stable one (or possibly to an isoenergetic one) if such a rearrangement is possible. An illustration is the formation of 2,3-dimethyl-2-butanol as the major product when 3,3-dimethyl-1-butene is hydrated:

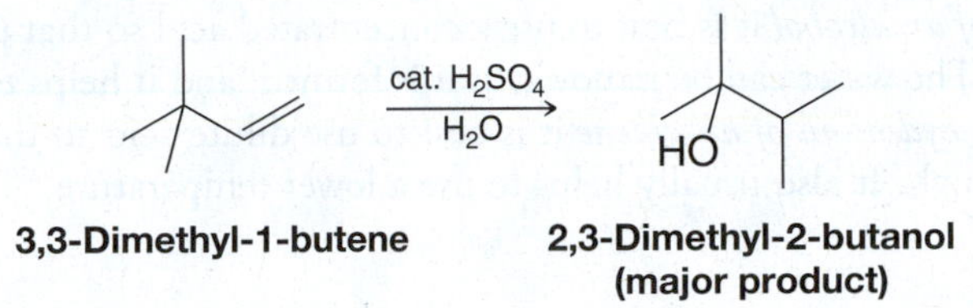

3,3-Dimethyl-1-butene   **2,3-Dimethyl-2-butanol**
                                              (major product)

●●●
PRACTICE PROBLEM 8.6   Outline all steps in a mechanism showing how 2,3-dimethyl-2-butanol is formed in the acid-catalyzed hydration of 3,3-dimethyl-1-butene.

●●●
PRACTICE PROBLEM 8.7   The following order of reactivity is observed when the following alkenes are subjected to acid-catalyzed hydration:

$$(CH_3)_2C=CH_2 > CH_3CH=CH_2 > CH_2=CH_2$$

Explain this order of reactivity.

●●●
PRACTICE PROBLEM 8.8   Write a mechanism for the following reaction.

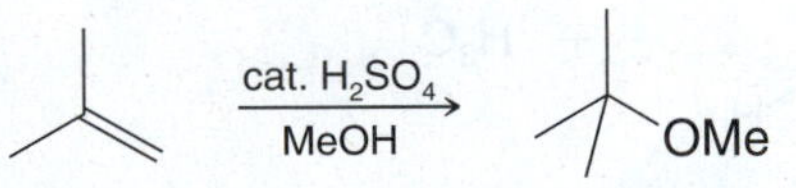

●●● SOLVED PROBLEM 8.2

Write a mechanism that will explain the course of the following reaction

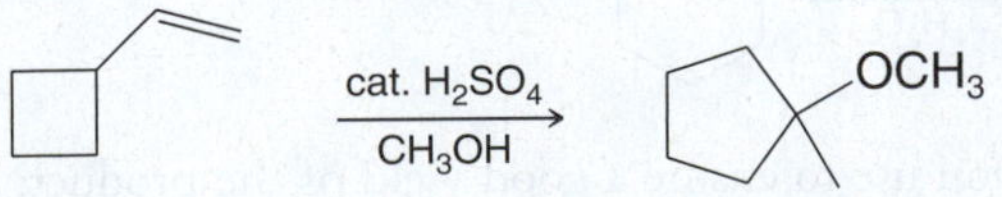

STRATEGY AND ANSWER: As we have learned, in a strongly acidic medium such as methanol containing catalytic sulfuric acid, an alkene can accept a proton to become a carbocation. In the reaction above, the 2° carbocation formed initially can undergo an alkanide shift and a hydride shift as shown below to become a 3° carbocation, which can then react with the solvent (methanol) to form an ether.

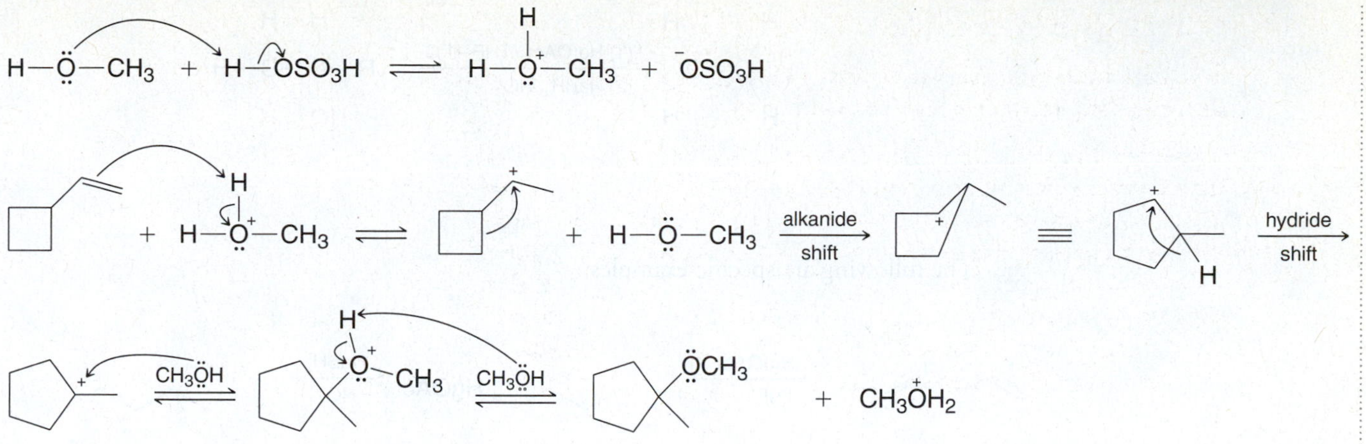

8.5 ALCOHOLS FROM ALKENES THROUGH OXYMERCURATION– DEMERCURATION: MARKOVNIKOV ADDITION

A useful laboratory procedure for synthesizing alcohols from alkenes that avoids rearrangement is a two-step method called **oxymercuration–demercuration**.

- Alkenes react with mercuric acetate in a mixture of tetrahydrofuran (THF) and water to produce (hydroxyalkyl)mercury compounds. These (hydroxyalkyl) mercury compounds can be reduced to alcohols with sodium borohydride.

Step 1: Oxymercuration

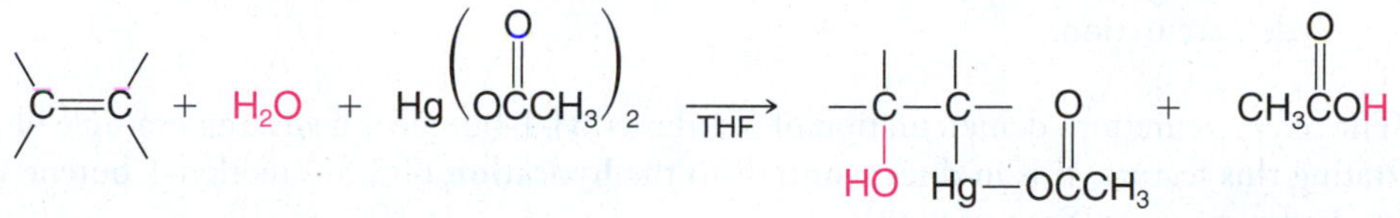

Step 2: Demercuration

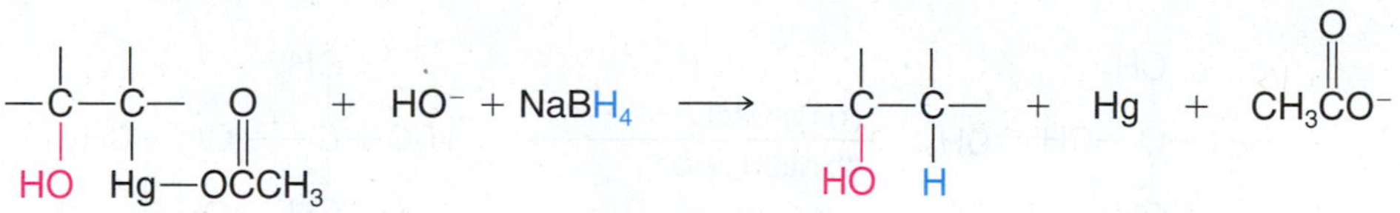

- In the first step, **oxymercuration**, water and mercuric acetate add to the double bond.
- In the second step, **demercuration**, sodium borohydride reduces the acetoxymercury group and replaces it with hydrogen. (The acetate group is often abbreviated —OAc.)

Both steps can be carried out in the same vessel, and both reactions take place very rapidly at room temperature or below. The first step—oxymercuration—usually goes to completion within a period of seconds to minutes. The second step—demercuration—normally requires less than an hour. The overall reaction gives alcohols in very high yields, usually greater than 90%.

8.5A Regioselectivity of Oxymercuration–Demercuration

Oxymercuration–demercuration is also highly regioselective.

- In oxymercuration–demercuration, the net orientation of the addition of the elements of water, H— and —OH, *is in accordance with Markovnikov's rule*. The H— becomes attached to the carbon atom of the double bond with the greater number of hydrogen atoms.

Mercury compounds are extremely hazardous. Before you carry out a reaction involving mercury or its compounds, you should familiarize yourself with current procedures for its use and containment. There are no satisfactory methods for disposal of mercury.

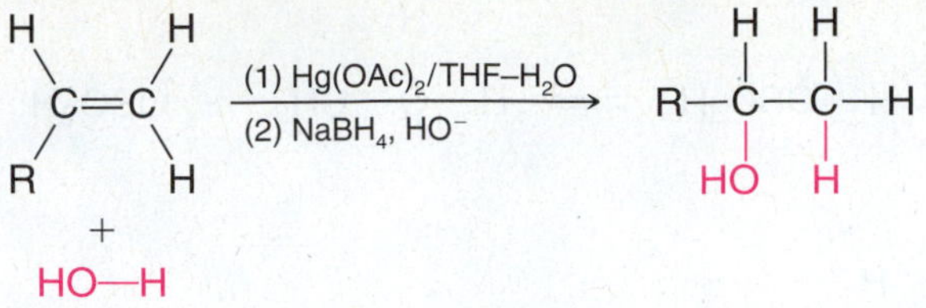

$$
\underset{R}{\overset{H}{\underset{\displaystyle}{}}}C=C\overset{H}{\underset{H}{}}
\quad
\xrightarrow[\text{(2) NaBH}_4,\ \text{HO}^-]{\text{(1) Hg(OAc)}_2/\text{THF–H}_2\text{O}}
\quad
R-\overset{H}{\underset{HO}{C}}-\overset{H}{\underset{H}{C}}-H
$$

$$
+\\
HO-H
$$

The following are specific examples:

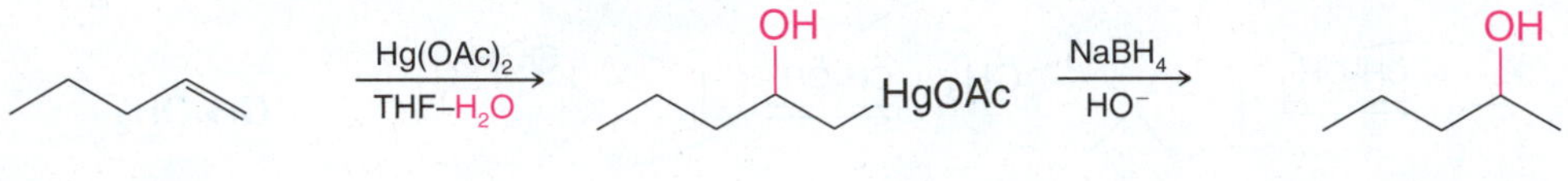

1-Pentene → (Hg(OAc)₂, THF–H₂O) → (OH, HgOAc) → (NaBH₄, HO⁻) → **2-Pentanol (93%)**

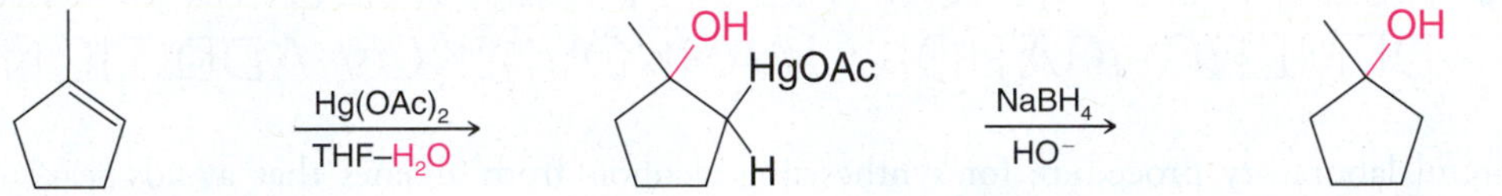

1-Methylcyclopentene → (Hg(OAc)₂, THF–H₂O) → (OH, HgOAc, H) → (NaBH₄, HO⁻) → **1-Methylcyclopentanol**

8.5B Rearrangements Seldom Occur in Oxymercuration–Demercuration

- Rearrangements of the carbon skeleton seldom occur in oxymercuration–demercuration.

Helpful Hint

Oxymercuration–demercuration is not prone to hydride or alkanide rearrangements.

The oxymercuration–demercuration of 3,3-dimethyl-1-butene is a striking example illustrating this feature. It is in direct contrast to the **hydration** of 3,3-dimethyl-1-butene we studied previously (Section 8.4B).

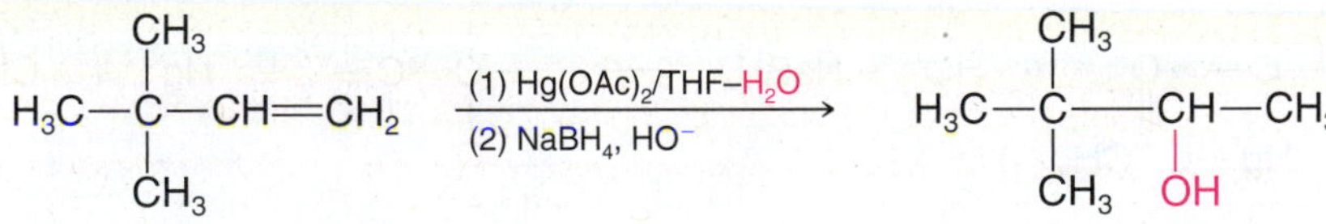

$$
\underset{\underset{CH_3}{|}}{\overset{\overset{CH_3}{|}}{H_3C-C}}-CH=CH_2
\quad
\xrightarrow[\text{(2) NaBH}_4,\ \text{HO}^-]{\text{(1) Hg(OAc)}_2/\text{THF–H}_2\text{O}}
\quad
\underset{\underset{CH_3}{|}}{\overset{\overset{CH_3}{|}}{H_3C-C}}-\underset{\underset{OH}{|}}{CH}-CH_3
$$

3,3-Dimethyl-1-butene   **3,3-Dimethyl-2-butanol (94%)**

Analysis of the mixture of products by gas chromatography failed to reveal the presence of any 2,3-dimethyl-2-butanol. The acid-catalyzed hydration of 3,3-dimethyl-1-butene, by contrast, gives 2,3-dimethyl-2-butanol as the major product.

8.5C Mechanism of Oxymercuration

A mechanism that accounts for the orientation of addition in the oxymercuration stage, and one that also explains the general lack of accompanying rearrangements, is shown below.

- Central to this mechanism is an electrophilic attack by the mercury species, $\overset{+}{H}gOAc$, at the less substituted carbon of the double bond (i.e., at the carbon atom that bears the greater number of hydrogen atoms), and the formation of a bridged intermediate.

We illustrate the mechanism using 3,3-dimethyl-1-butene as the example:

[A MECHANISM FOR THE REACTION ─ Oxymercuration]

Step 1

$$Hg(OAc)_2 \rightleftharpoons \overset{+}{H}gOAc + AcO^-$$

Mercuric acetate dissociates to form a $\overset{+}{H}gOAc$ cation and an acetate anion.

Step 2

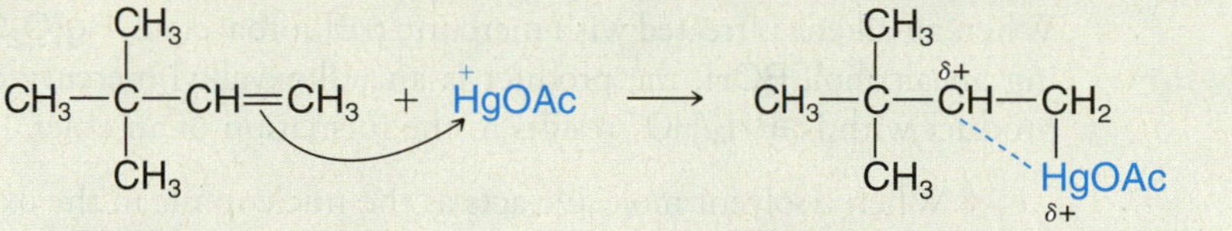

3,3-Dimethyl-1-butene   **Mercury-bridged carbocation**

The alkene donates a pair of electrons to the electrophilic $\overset{+}{H}gOAc$ cation to form a mercury-bridged carbocation. In this carbocation, the positive charge is shared between the 2° (more substituted) carbon atom and the mercury atom. The charge on the carbon atom is large enough to account for the Markovnikov orientation of the addition, but not large enough for a rearrangement to occur.

Step 3

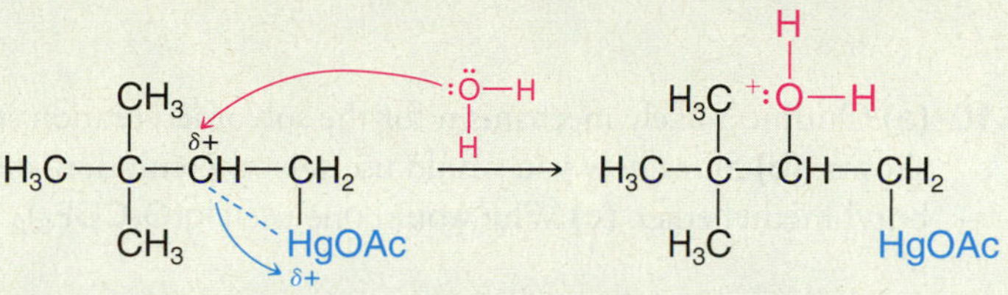

A water molecule attacks the carbon of the bridged mercurinium ion that is better able to bear the partial positive charge.

Step 4

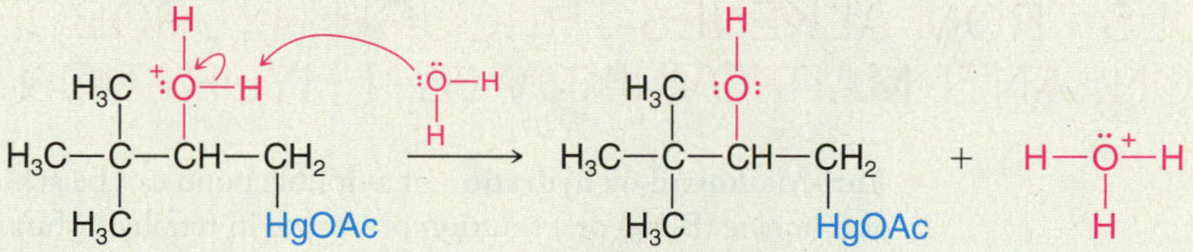

(Hydroxyalkyl)mercury compound

An acid–base reaction transfers a proton to another water molecule (or to an acetate ion). This step produces the (hydroxyalkyl)mercury compound.

Calculations indicate that mercury-bridged carbocations (termed mercurinium ions) such as those formed in this reaction retain much of the positive charge on the mercury moiety. Only a small portion of the positive charge resides on the more substituted carbon atom. The charge is large enough to account for the observed Markovnikov addition, but it is too small to allow the usual rapid carbon skeleton rearrangements that take place with more fully developed carbocations.

Although attack by water on the bridged mercurinium ion leads to anti addition of the hydroxyl and mercury groups, the reaction that replaces mercury with hydrogen is not stereocontrolled (it likely involves radicals; see Chapter 10). This step scrambles the overall stereochemistry.

- The net result of oxymercuration–demercuration is a mixture of syn and anti addition of —H and —OH to the alkene.

- As already noted, oxymercuration–demercuration takes place with Markovnikov regiochemistry.

PRACTICE PROBLEM 8.9   Write the structure of the appropriate alkene and specify the reagents needed for synthesis of the following alcohols by oxymercuration–demercuration:

(a) 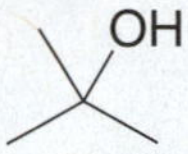 (structure with OH)   **(b)** 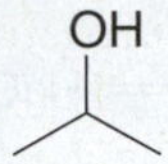 (structure with OH)   **(c)** 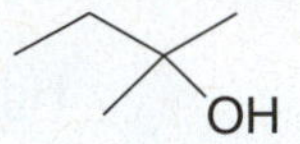 (structure with OH)

When an alkene is treated with mercuric trifluoroacetate, $Hg(O_2CCF_3)_2$, in THF containing an alcohol, ROH, the product is an (alkoxyalkyl)mercury compound. Treating this product with $NaBH_4/HO^-$ results in the formation of an ether.

- When a solvent molecule acts as the nucleophile in the oxymercuration step the overall process is called *solvomercuration–demercuration*:

$$\text{C=C} \xrightarrow[\text{solvomercuration}]{Hg(O_2CCF_3)_2/\text{THF–ROH}} \text{(alkoxyalkyl)mercuric trifluoroacetate} \xrightarrow[\text{demercuration}]{NaBH_4,\ HO^-} \text{Ether}$$

Alkene   **(Alkoxyalkyl)mercuric trifluoroacetate**   **Ether**

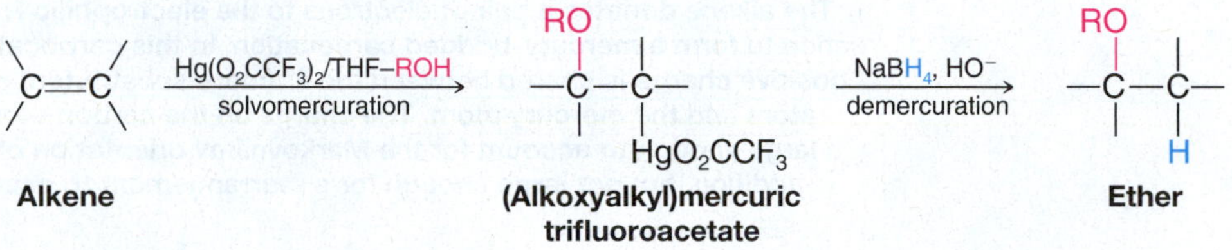

PRACTICE PROBLEM 8.10   **(a)** Outline a likely mechanism for the solvomercuration step of the ether synthesis just shown. **(b)** Show how you would use solvomercuration–demercuration to prepare *tert*-butyl methyl ether. **(c)** Why would one use $Hg(O_2CCF_3)_2$ instead of $Hg(OAc)_2$?

8.6 ALCOHOLS FROM ALKENES THROUGH HYDROBORATION–OXIDATION: ANTI-MARKOVNIKOV SYN HYDRATION

- **Anti-Markovnikov hydration** of a double bond can be achieved through the use of diborane (B_2H_6) or a solution of borane in tetrahydrofuran (BH_3:THF).

The addition of water is indirect in this process, and two reactions are involved. The first is the addition of a boron atom and hydrogen atom to the double bond, called **hydroboration**; the second is **oxidation** and hydrolysis of the alkylborane intermediate to an alcohol and boric acid. The anti-Markovnikov regiochemistry of the addition is illustrated by the hydroboration–oxidation of propene:

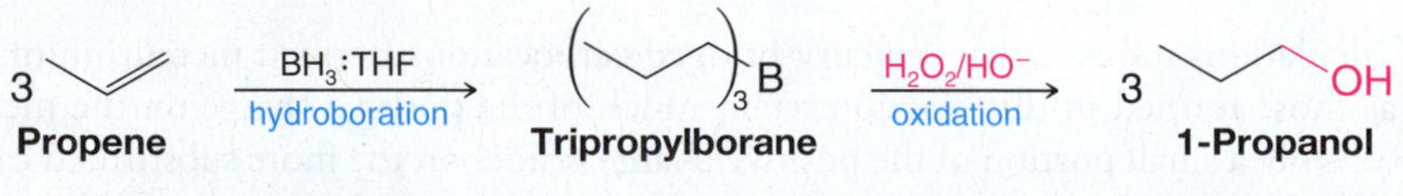

$$3\ \text{Propene} \xrightarrow[\text{hydroboration}]{BH_3:THF} \text{Tripropylborane} \xrightarrow[\text{oxidation}]{H_2O_2/HO^-} 3\ \text{1-Propanol}$$

- Hydroboration–oxidation takes place with **syn** stereochemistry, as well as anti-Markovnikov regiochemistry.

This can be seen in the following example with 1-methylcyclopentene:

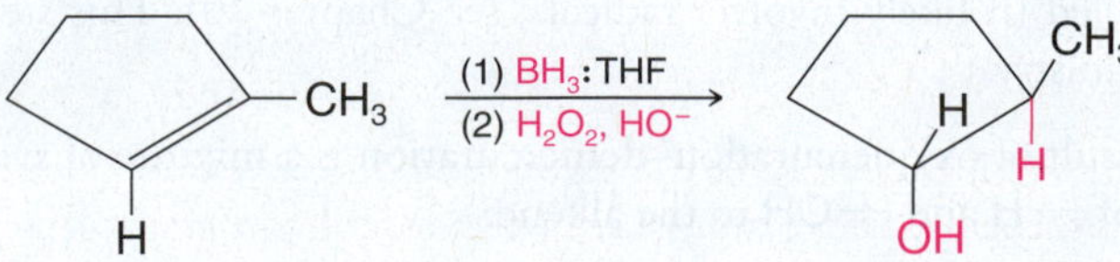

(reaction scheme: 1-methylcyclopentene → (1) BH_3:THF (2) H_2O_2, HO^- → product)

In the following sections we shall consider details of the mechanism that lead to the anti-Markovnikov regiochemistry and syn stereochemistry of hydroboration–oxidation.

8.7 HYDROBORATION: SYNTHESIS OF ALKYLBORANES

Hydroboration of an alkene is the starting point for a number of useful synthetic procedures, including the anti-Markovnikov syn **hydration** procedure we have just mentioned. Hydroboration was discovered by Herbert C. Brown (Purdue University), and it can be represented in its simplest terms as follows:

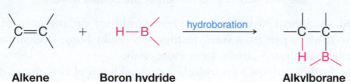

| Alkene | Boron hydride | Alkylborane |

Hydroboration can be accomplished with diborane (B_2H_6), which is a gaseous dimer of borane (BH_3), or more conveniently with a reagent prepared by dissolving diborane in THF. When diborane is introduced to THF, it reacts to form a Lewis acid–base complex of borane (the Lewis acid) and THF. The complex is represented as BH_3:THF.

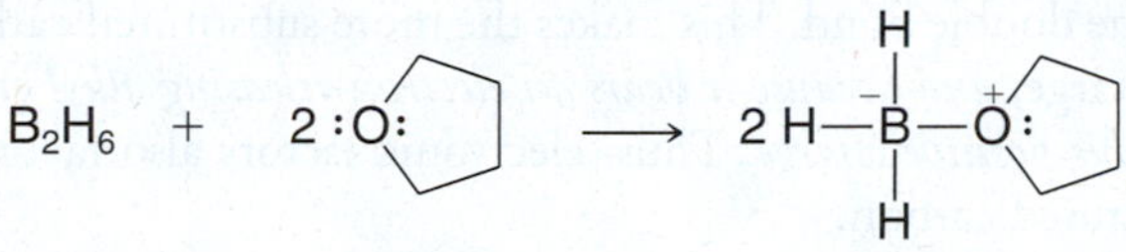

$$B_2H_6 \quad + \quad 2 \text{ :O} \quad \longrightarrow \quad 2 \text{ H—B—O}$$

| Diborane | THF (tetrahydrofuran) | | BH_3:THF |

Solutions containing the BH_3:THF complex can be obtained commercially. Hydroboration reactions are usually carried out in ethers: either in diethyl ether, $(CH_3CH_2)_2O$, or in some higher molecular weight ether such as "diglyme" $[(CH_3OCH_2CH_2)_2O$, *di*ethylene *gly*col *di*methyl ether]. Great care must be used in handling diborane and alkylboranes because they ignite spontaneously in air (with a green flame). The solution of BH_3:THF must be used in an inert atmosphere (e.g., argon or nitrogen) and with care.

8.7A Mechanism of Hydroboration

When a terminal alkene such as propene is treated with a solution containing BH_3:THF, the boron hydride adds successively to the double bonds of three molecules of the alkene to form a trialkylborane:

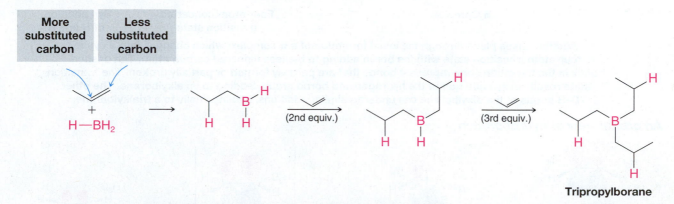

Tripropylborane

- In each addition step *the boron atom becomes attached to the less substituted carbon atom of the double bond*, and a hydrogen atom is transferred from the boron atom to the other carbon atom of the double bond.
- Hydroboration is **regioselective** and it is **anti-Markovnikov** (the hydrogen atom becomes attached to the carbon atom with fewer hydrogen atoms).

Other examples that illustrate the tendency for the boron atom to become attached to the less substituted carbon atom are shown here. The percentages designate where the boron atom becomes attached.

BROWN'S discovery of hydroboration led to his being named a co-winner of the 1979 Nobel Prize in Chemistry.

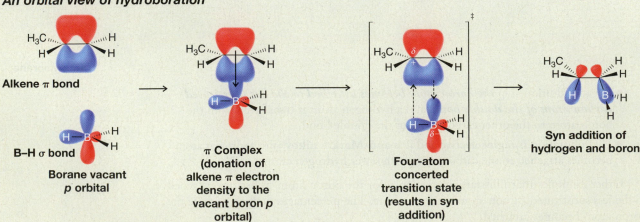

These percentages, indicating where boron becomes attached in reactions using these starting materials, illustrate the tendency for boron to bond at the less substituted carbon of the double bond.

This observed attachment of boron to the less substituted carbon atom of the double bond seems to result in part from **steric factors**—the bulky boron-containing group can approach the less substituted carbon atom more easily.

In the mechanism proposed for hydroboration, addition of BH_3 to the double bond begins with a donation of π electrons from the double bond to the vacant p orbital of BH_3 (see the following mechanism). In the next step this complex becomes the addition product by passing through a four-atom transition state in which the boron atom is partially bonded to the less substituted carbon atom of the double bond and one hydrogen atom is partially bonded to the other carbon atom. As this transition state is approached, electrons shift in the direction of the boron atom and away from the more substituted carbon atom of the double bond. This makes the more substituted carbon atom develop a partial positive charge, *and because it bears an electron-releasing alkyl group, it is better able to accommodate this positive charge.* Thus, electronic factors also favor addition of boron at the least substituted carbon.

- Overall, both *electronic* and *steric factors* account for the anti-Markovnikov orientation of the addition.

A MECHANISM FOR THE REACTION ┤ Hydroboration

Hydroboration

Addition takes place through the initial formation of a π complex, which changes into a cyclic four-atom transition state with the boron adding to the less hindered carbon atom. The dashed bonds in the transition state represent bonds that are partially formed or partially broken. The transition state results in syn addition of the hydrogen and boron group, leading to an alkylborane. The other B–H bonds of the alkylborane can undergo similar additions, leading finally to a trialkylborane.

An orbital view of hydroboration

8.7B Stereochemistry of Hydroboration

- The transition state for hydroboration requires that the boron atom and the hydrogen atom add to the same face of the double bond:

Stereochemistry of Hydroboration

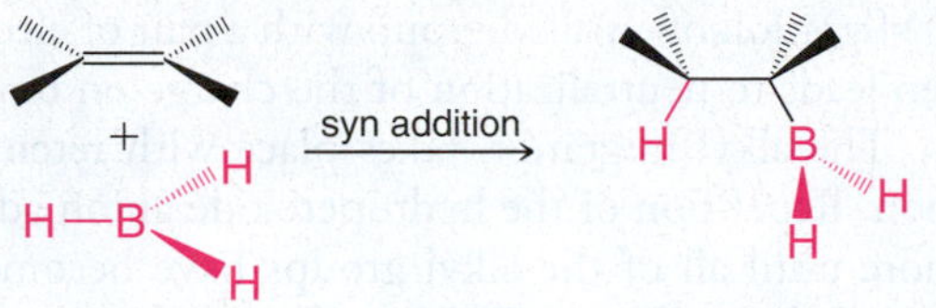

We can see the results of a syn addition in our example involving the hydroboration of 1-methylcyclopentene. Formation of the enantiomer, which is equally likely, results when the boron hydride adds to the top face of the 1-methylcyclopentene ring:

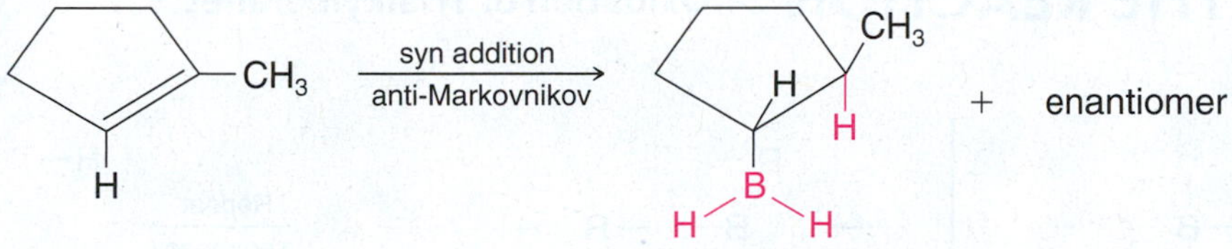

Specify the alkene needed for synthesis of each of the following alkylboranes by hydroboration:

PRACTICE PROBLEM 8.11

(a)

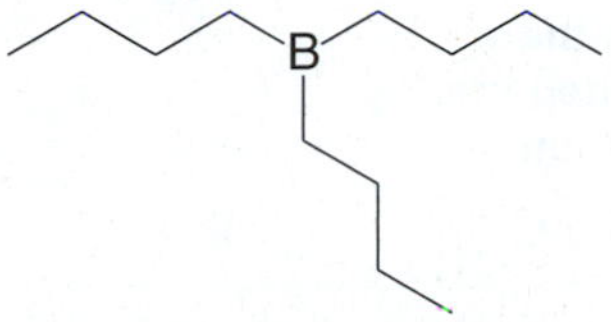

(b)

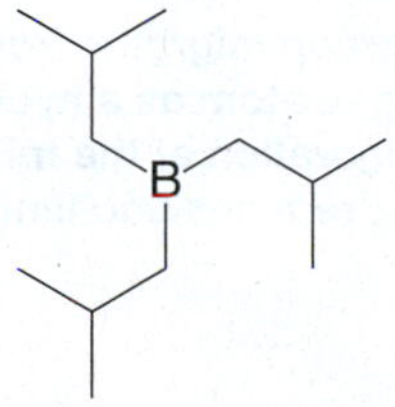

(c)

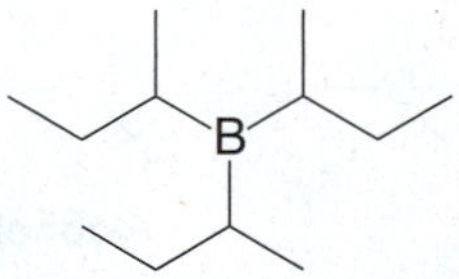

(d) Show the stereochemistry involved in the hydroboration of 1-methylcyclohexene.

PRACTICE PROBLEM 8.12

Treating a hindered alkene such as 2-methyl-2-butene with BH_3:THF leads to the formation of a dialkylborane instead of a trialkylborane. When 2 mol of 2-methyl-2-butene is added to 1 mol of BH_3, the product formed is bis(3-methyl-2-butyl)borane, nicknamed "disiamylborane." Write its structure. Bis(3-methyl-2-butyl)borane is a useful reagent in certain syntheses that require a sterically hindered borane. (The name "disiamyl" comes from "*di*secondary-*iso*-amyl," a completely unsystematic and unacceptable name. The name "amyl" is an old common name for a five-carbon alkyl group.)

8.8 OXIDATION AND HYDROLYSIS OF ALKYLBORANES

The alkylboranes produced in the hydroboration step are usually not isolated. They are oxidized and hydrolyzed to alcohols in the same reaction vessel by the addition of hydrogen peroxide in an aqueous base:

$$R_3B \xrightarrow[\text{Oxidation and hydrolysis}]{H_2O_2,\ aq.\ NaOH,\ 25\ ^\circ C} 3R\!-\!OH + B(ONa)_3$$

- **The oxidation and hydrolysis steps take place with retention of configuration** at the carbon initially bearing boron and ultimately bearing the hydroxyl group.

We shall see how this occurs by considering the mechanisms of oxidation and hydrolysis. Alkylborane oxidation begins with addition of a hydroperoxide anion (HOO^-) to the trivalent boron atom. An unstable intermediate is formed that has a formal negative charge on the boron. Migration of an alkyl group with a pair of electrons from the boron to the adjacent oxygen leads to neutralization of the charge on boron and displacement of a hydroxide anion. The alkyl migration takes place with retention of configuration at the migrating carbon. Repetition of the hydroperoxide anion addition and migration steps occurs twice more until all of the alkyl groups have become attached to oxygen atoms, resulting in a trialkyl borate ester, $B(OR)_3$. The borate ester then undergoes basic hydrolysis to produce three molecules of the alcohol and an inorganic borate anion.

[A MECHANISM FOR THE REACTION ─ Oxidation of Trialkylboranes]

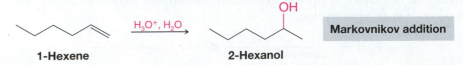

Trialkyl-   **Hydroperoxide**   **Unstable intermediate**   **Borate ester**
borane   **ion**

The boron atom accepts an electron pair from the hydroperoxide ion to form an unstable intermediate.

An alkyl group migrates from boron to the adjacent oxygen atom as a hydroxide ion departs. The configuration at the migrating carbon remains unchanged.

Hydrolysis of the Borate Ester

Trialkyl borate ester   **Alcohol**

Hydroxide anion attacks the boron atom of the borate ester.

An alkoxide anion departs from the borate anion, reducing the formal change on boron to zero.

Proton transfer completes the formation of one alcohol molecule. The sequence repeats until all three alkoxy groups are released as alcohols and inorganic borate remains.

8.8A Regiochemistry and Stereochemistry of Alkylborane Oxidation and Hydrolysis

- Hydroboration–oxidation reactions are **regioselective**; the net result of hydroboration–oxidation is **anti-Markovnikov** addition of water to an alkene.

- As a consequence, hydroboration–oxidation gives us a method for the preparation of alcohols that cannot normally be obtained through the acid-catalyzed hydration of alkenes or by oxymercuration–demercuration.

For example, the acid-catalyzed hydration (or oxymercuration–demercuration) of 1-hexene yields 2-hexanol, the Markovnikov addition product.

1-Hexene $\xrightarrow{H_3O^+,\ H_2O}$ **2-Hexanol**   **Markovnikov addition**

In contrast, hydroboration–oxidation of 1-hexene yields 1-hexanol, the anti-Markovnikov product.

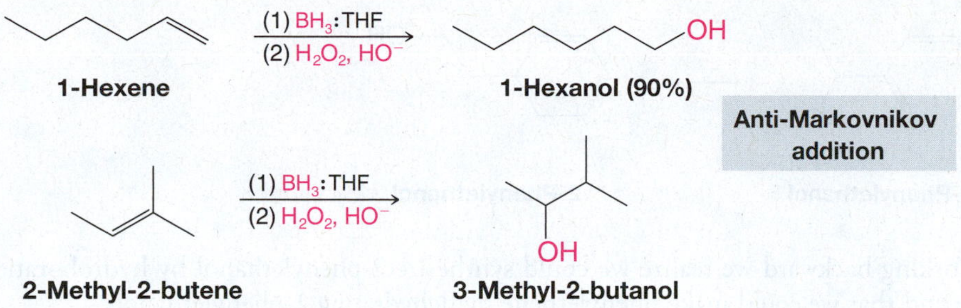

1-Hexene $\xrightarrow{\text{(1) BH}_3\text{:THF}}_{\text{(2) H}_2\text{O}_2,\ \text{HO}^-}$ **1-Hexanol (90%)**   Anti-Markovnikov addition

2-Methyl-2-butene $\xrightarrow{\text{(1) BH}_3\text{:THF}}_{\text{(2) H}_2\text{O}_2,\ \text{HO}^-}$ **3-Methyl-2-butanol**

- Hydroboration–oxidation reactions are **stereospecific**; the net addition of —H and —OH is **syn**, and if chirality centers are formed, their configuration depends on the stereochemistry of the starting alkene.

Because the oxidation step in the hydroboration–oxidation synthesis of alcohols takes place with retention of configuration, **the hydroxyl group replaces the boron atom where it stands in the alkylboron compound**. The net result of the two steps (hydroboration and oxidation) is the syn addition of —H and —OH. We can review the anti-Markovnikov and syn aspects of hydroboration–oxidation by considering the hydration of 1-methylcyclopentene, as shown in Fig. 8.3.

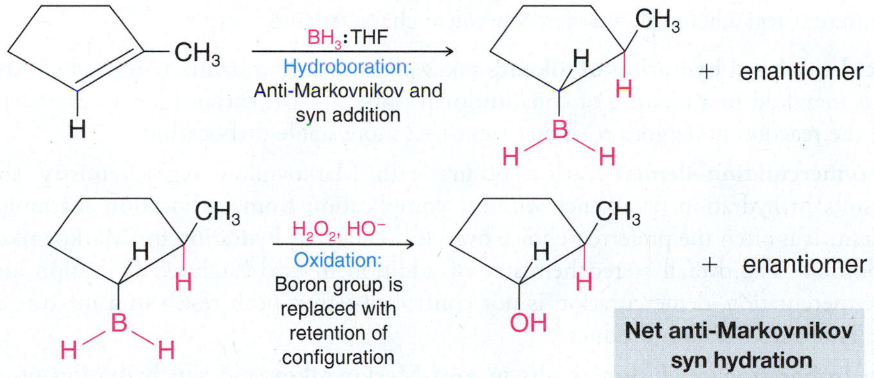

$\xrightarrow[\text{syn addition}]{\substack{\text{BH}_3\text{:THF}\\ \text{Hydroboration:}\\ \text{Anti-Markovnikov and}}}$ + enantiomer

$\xrightarrow[\text{configuration}]{\substack{\text{H}_2\text{O}_2,\ \text{HO}^-\\ \text{Oxidation:}\\ \text{Boron group is}\\ \text{replaced with}\\ \text{retention of}}}$ + enantiomer   Net anti-Markovnikov syn hydration

FIGURE 8.3 The hydroboration–oxidation of 1-methylcyclopentene. The first reaction is a syn addition of borane. In this illustration we have shown the boron and hydrogen entering from the bottom side of 1-methylcyclopentene. The reaction also takes place from the top side at an equal rate to produce the enantiomer. In the second reaction the boron atom is replaced by a hydroxyl group with retention of configuration. The product is *trans*-2-methylcyclopentanol, and the overall result is the syn addition of —H and —OH.

Specify the appropriate alkene and reagents for synthesis of each of the following alcohols by hydroboration–oxidation.

PRACTICE PROBLEM 8.13

(a)

(b)

(c)

(d)

(e)

(f)

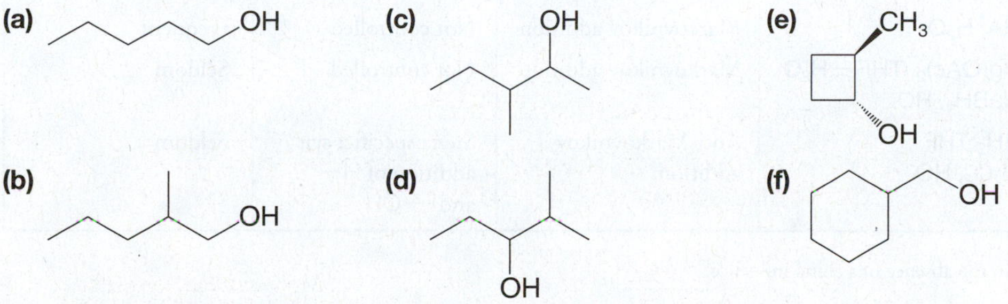

● ● ●   **SOLVED PROBLEM 8.3**

Outline a method for carrying out the following conversion.

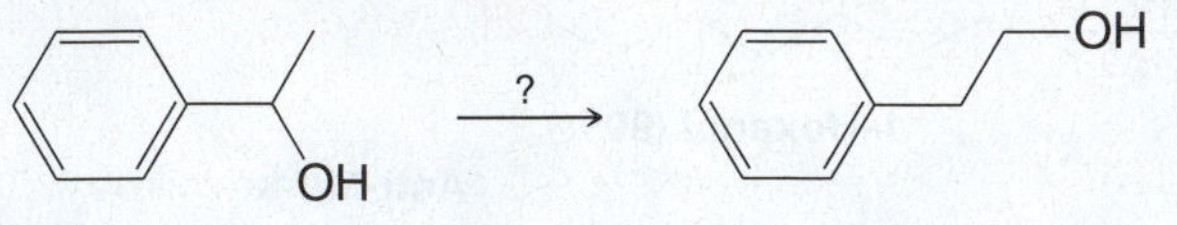

1-Phenylethanol   **2-Phenylethanol**

STRATEGY AND ANSWER: Working backward we realize we could synthesize 2-phenylethanol by hydroboration–oxidation of phenylethene (styrene), and that we could make phenylethene by dehydrating 1-phenylethanol.

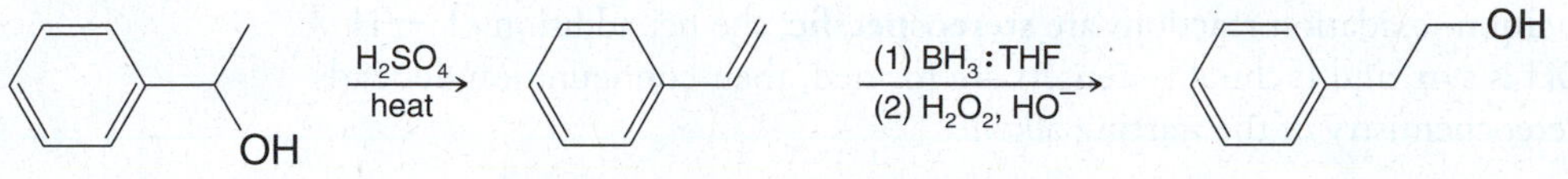

Phenylethene

8.9 SUMMARY OF ALKENE HYDRATION METHODS

The three methods we have studied for alcohol synthesis by addition reactions to alkenes have different regiochemical and stereochemical characteristics.

1. **Acid-catalyzed hydration of alkenes** takes place with Markovnikov regiochemistry but may lead to a mixture of constitutional isomers if the carbocation intermediate in the reaction undergoes rearrangement to a more stable carbocation.

2. **Oxymercuration–demercuration** occurs with Markovnikov regiochemistry and results in hydration of alkenes without complication from carbocation rearrangement. It is often the preferred choice over acid-catalyzed hydration for Markovnikov addition. The overall stereochemistry of addition in acid-catalyzed hydration and oxymercuration–demercuration is not controlled—they both result in a mixture of cis and trans addition products.

3. **Hydroboration–oxidation** results in **anti-Markovnikov** and syn hydration of an alkene.

The complementary regiochemical and stereochemical aspects of these methods provide useful alternatives when we desire to synthesize a specific alcohol by hydration of an alkene. We summarize them here in Table 8.1.

TABLE 8.1 SUMMARY OF METHODS FOR CONVERTING AN ALKENE TO AN ALCOHOL

Reaction	Conditions	Regiochemistry	Stereochemistry[a]	Occurrence of Rearrangements
Acid-catalyzed hydration	cat. HA, H_2O	Markovnikov addition	Not controlled	Frequent
Oxymercuration–demercuration	(1) $Hg(OAc)_2$, THF — H_2O (2) $NaBH_4$, HO^-	Markovnikov addition	Not controlled	Seldom
Hydroboration–oxidation	(1) BH_3:THF (2) H_2O_2, HO^-	Anti-Markovnikov addition	Stereospecific: syn addition of H— and —OH	Seldom

[a]All of these methods produce racemic mixtures in the absence of a chiral influence.

8.10 PROTONOLYSIS OF ALKYLBORANES

Heating an alkylborane with acetic acid causes cleavage of the carbon–boron bond and replacement with hydrogen:

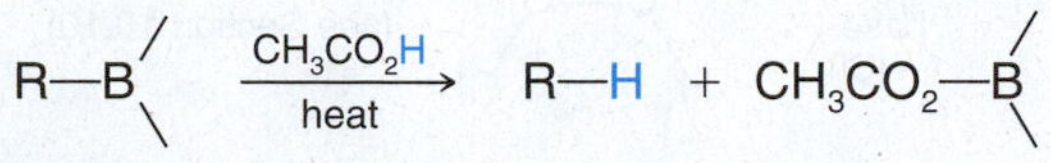

$$R—B \xrightarrow[\text{heat}]{CH_3CO_2H} R—H + CH_3CO_2—B$$

Alkylborane   **Alkane**

- Protonolysis of an alkylborane takes place with retention of configuration; hydrogen replaces boron **where it stands** in the alkylborane.
- The overall stereochemistry of hydroboration–protonolysis, therefore, is **syn** (like that of the oxidation of alkylboranes).

Hydroboration followed by protonolysis of the resulting alkylborane can be used as an alternative method for hydrogenation of alkenes, although catalytic hydrogenation (Section 7.12) is the more common procedure. Reaction of alkylboranes with deuterated or tritiated acetic acid also provides a very useful way to introduce these isotopes into a compound in a specific way.

● ● ●
PRACTICE PROBLEM 8.14

Starting with any needed alkene (or cycloalkene) and assuming you have deuterioacetic acid (CH_3CO_2D) available, outline syntheses of the following deuterium-labeled compounds.

(a) $(CH_3)_2CHCH_2CH_2D$ (b) $(CH_3)_2CHCHDCH_3$ (c)

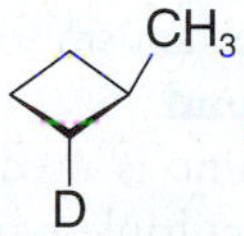

(+ enantiomer)

(d) Assuming you also have available BD_3:THF and CH_3CO_2T, can you suggest a synthesis of the following?

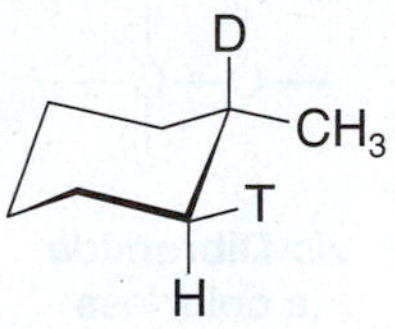

(+ enantiomer)

8.11 ELECTROPHILIC ADDITION OF BROMINE AND CHLORINE TO ALKENES

Alkenes react rapidly with bromine and chlorine in nonnucleophilic solvents to form **vicinal dihalides**. An example is the addition of chlorine to ethene.

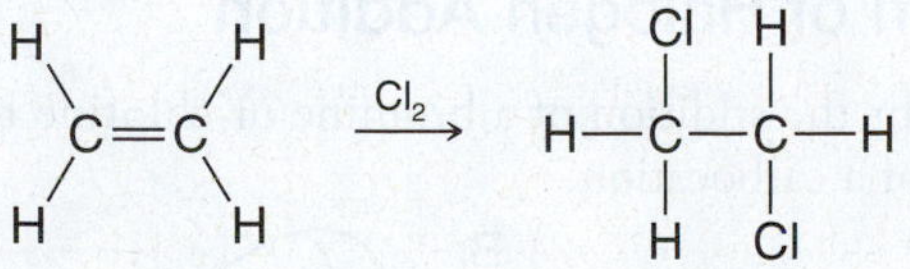

Ethene   **1,2-Dichloroethane**

This addition is a useful industrial process because 1,2-dichloroethane can be used as a solvent and can be used to make vinyl chloride, the starting material for poly(vinyl chloride).

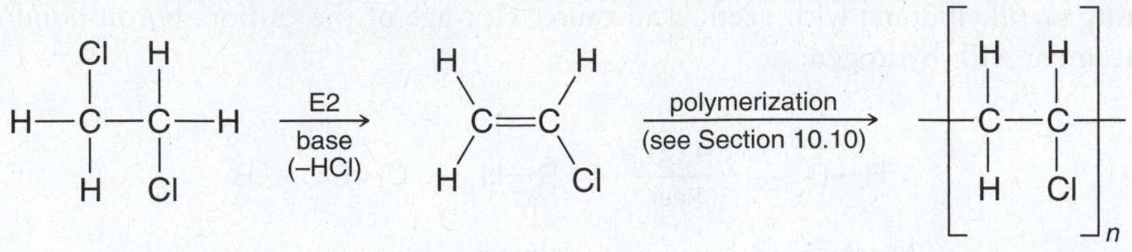

1,2-Dichloroethane → **Vinyl chloride** → **Poly(vinyl chloride)**

Other examples of the addition of halogens to a double bond are the following:

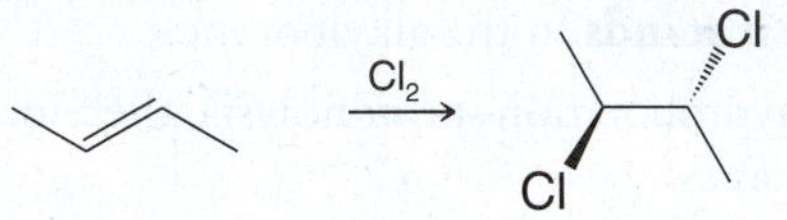

trans-2-Butene   *meso*-1,2-Dichlorobutane

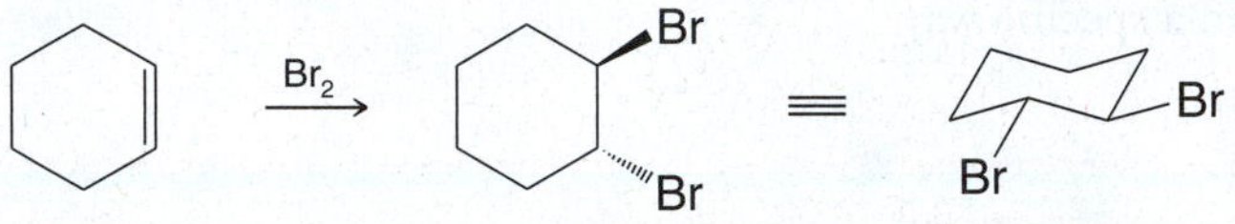

Cyclohexene   *trans*-1,2-Dibromocyclohexane (racemic)

These two examples show an aspect of these additions that we shall address later when we examine a mechanism for the reaction: **the addition of halogens is an anti addition to the double bond**.

When bromine is used for this reaction, it can serve as a test for the presence of carbon–carbon multiple bonds. If we add bromine to an alkene (or alkyne, see Section 8.17), the red-brown color of the bromine disappears almost instantly as long as the alkene (or alkyne) is present in excess:

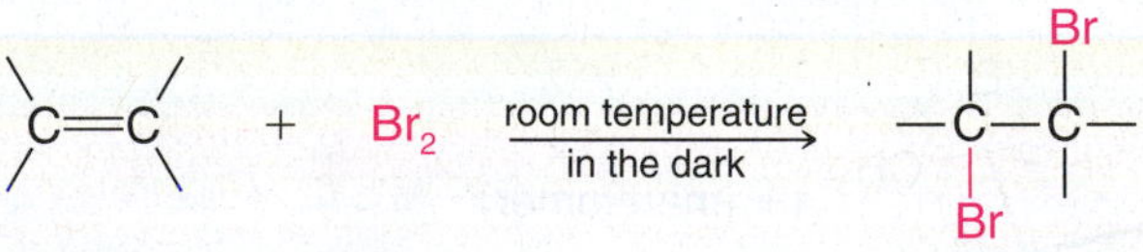

An alkene (colorless) + **Bromine (red-brown)** → (room temperature in the dark) → **vic-Dibromide (a colorless compound)**

Rapid decolorization of Br_2 is a positive test for alkenes and alkynes.

This behavior contrasts markedly with that of **alkanes**. Alkanes do not react appreciably with bromine or chlorine at room temperature and in the absence of light. When alkanes *do* react under those conditions, however, it is by substitution rather than addition and by a mechanism involving radicals that we shall discuss in Chapter 10:

R—H + Br_2 → (room temperature in the dark) → **No appreciable reaction**
Alkane (colorless)   Bromine (red-brown)

8.11A Mechanism of Halogen Addition

A possible mechanism for the addition of a bromine or chlorine to an alkene is one that involves the formation of a carbocation.

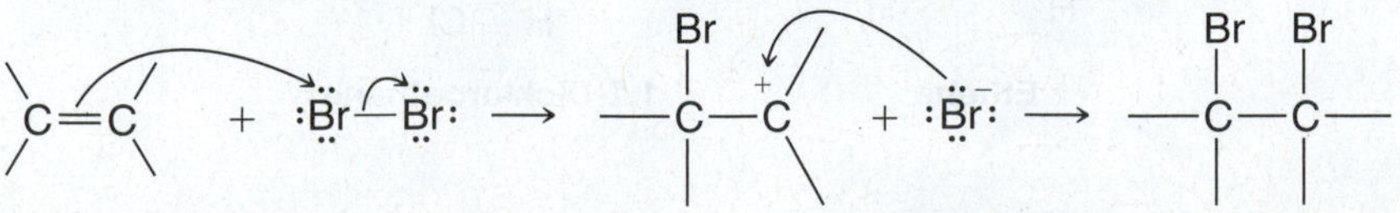

Although this mechanism is similar to ones we have studied earlier, such as the addition of H—X to an alkene, it does not explain an important fact. As we have just seen (in Section 8.11) the addition of bromine or chlorine to an alkene is an **anti addition**.

The addition of bromine to cyclopentene, for example, produces *trans*-1,2-dibromocyclopentane, not *cis*-1,2-dibromocyclopentane.

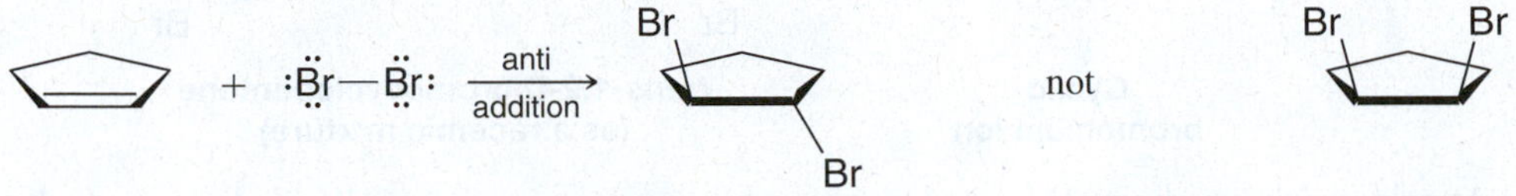

trans-1,2-Dibromocyclopentane
(as a racemic mixture)

not

cis-1,2-Dibromocyclohexane
(a meso compound)

A mechanism that explains anti addition is one in which a bromine molecule transfers a bromine atom to the alkene to form a cyclic **bromonium ion** and a bromide ion, as shown in step 1 of "A Mechanism for the Reaction" that follows. The cyclic bromonium ion causes net anti addition, as follows.

In step 2, a bromide ion attacks the back side of either carbon 1 or carbon 2 of the bromonium ion (an S_N2 process) to open the ring and produce the *trans*-1,2-dibromide. Attack occurs from the side **opposite the bromine of the bromonium ion** because attack from this direction is unhindered. Attack at the other carbon of the cyclic bromonium ion produces the enantiomer.

A MECHANISM FOR THE REACTION — Addition of Bromine to an Alkene

Step 1

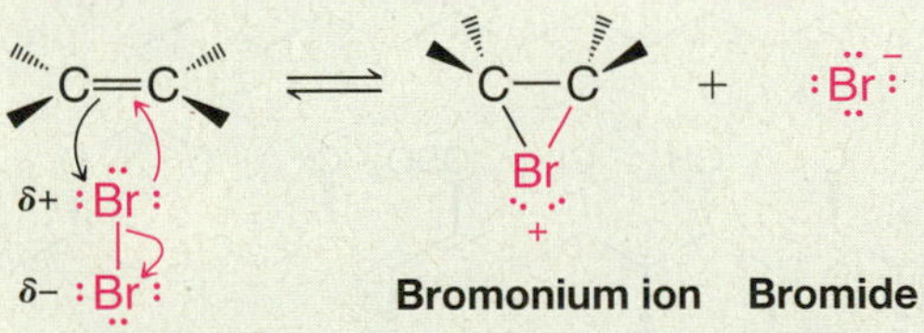

Bromonium ion   Bromide ion

As a bromine molecule approaches an alkene, the electron density of the alkene π bond repels electron density in the closer bromine, polarizing the bromine molecule and making the closer bromine atom electrophilic. The alkene donates a pair of electrons to the closer bromine, causing displacement of the distant bromine atom. As this occurs, the newly bonded bromine atom, due to its size and polarizability, donates an electron pair to the carbon that would otherwise be a carbocation, thereby stabilizing the positive charge by delocalization. The result is a bridged bromonium ion intermediate.

Step 2

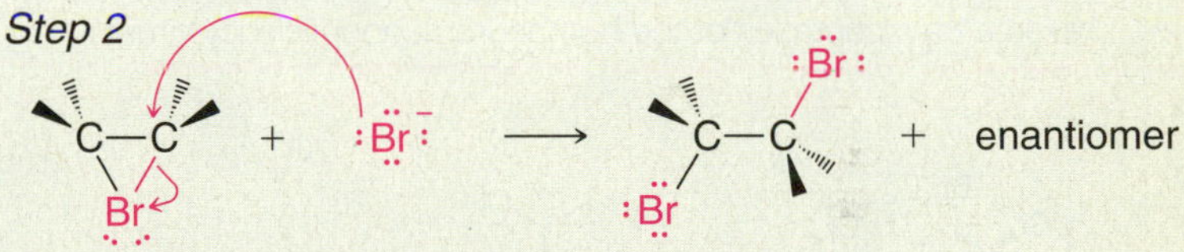

Bromonium ion   Bromide ion   vic-Dibromide

A bromide anion attacks at the back side of one carbon (or the other) of the bromonium ion in an S_N2 reaction, causing the ring to open and resulting in the formation of a *vic*-dibromide.

This process is shown for the addition of bromine to cyclopentene below.

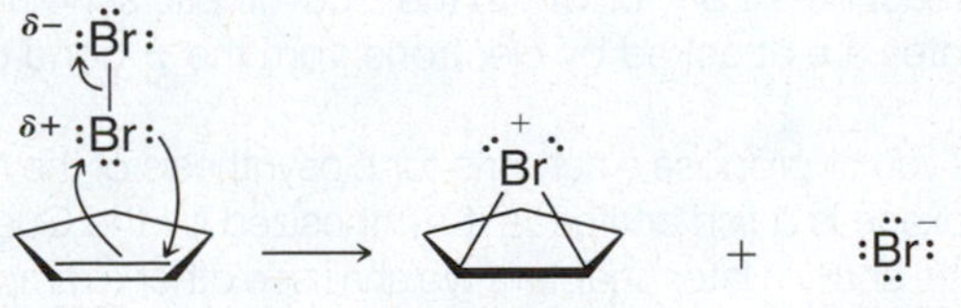

Cyclic   **Bromide**
bromonium ion   **ion**

Plane of symmetry

Enantiomers

Cyclic bromonium ion

***trans*-1,2-Dibromocyclopentane (as a racemic mixture)**

Attack at either carbon of the cyclopentene bromonium ion is equally likely because the cyclic bromonium ion is symmetric. It has a vertical plane of symmetry passing through the bromine atom and halfway between carbons 1 and 2. The *trans*-dibromide, therefore, is formed as a racemic mixture.

◆ THE CHEMISTRY OF... The Sea: A Treasury of Biologically Active Natural Products

The world's oceans are a vast storehouse of dissolved halide ions. The concentration of halides in the ocean is approximately 0.5 M in chloride, 1 mM in bromide, and 1 μM in iodide ions. Perhaps it is not surprising, then, that marine organisms have incorporated halogen atoms into the structures of many of their metabolites. Among these are such intriguing polyhalogenated compounds as halomon, dactylyne, tetrachloromertensene, (3*E*)-laureatin, peyssonol A, azamerone, and a structurally complex member of the polychlorinated sulfolipid family of natural products. Just the sheer number of halogen atoms in these metabolites is cause for wonder. For the organisms that make them, some of these molecules are part of defense mechanisms that serve to promote the species' survival by deterring predators or inhibiting the growth of competing organisms. For humans, the vast resource of marine natural products shows ever-greater potential as a source of new therapeutic agents. Halomon, for example, is in preclinical evaluation as a cytotoxic agent against certain tumor cell types, dactylyne is an inhibitor of pentobarbital metabolism, and peyssonol A is a modest allosteric inhibitor of the reverse transcriptases of the human immunodeficiency virus.

Image Source

Halomon   **Tetrachloromertensene**

A polychlorinated sulfolipid

Dactylyne   **(3*E*)-Laureatin**   **Peyssonol A**   **Azamerone**

The biosynthesis of certain halogenated marine natural products is intriguing. Some of their halogens appear to have been introduced as *electrophiles* rather than as Lewis bases or nucleophiles, which is their character when they are solutes in seawater. But how do marine organisms transform nucleophilic halide anions into *electrophilic* species for incorporation into their metabolites? It happens that many marine organisms have enzymes called haloperoxidases that convert nucleophilic iodide, bromide, or chloride anions into electrophilic species that react like I^+, Br^+, or Cl^+. In the biosynthetic schemes proposed for some halogenated natural products, positive halogen intermediates are attacked by electrons from the π bond of an alkene or alkyne in an addition reaction.

The final Learning Group Problem for this chapter asks you to propose a scheme for biosynthesis of the marine natural product kumepaloxane by electrophilic halogen addition. Kumepaloxane is a fish antifeedant synthesized by the Guam bubble snail *Haminoea cymbalum*, presumably as a defense mechanism for the snail. In later chapters we shall see other examples of truly remarkable marine natural products, such as brevetoxin B, associated with deadly "red tides," and eleutherobin, a promising anticancer agent.

The mechanisms for addition of Cl_2 and I_2 to alkenes are similar to that for Br_2, involving formation and ring opening of their respective **halonium ions**.

As with bridged mercurinium ions, the bromonium ion does not necessarily have symmetrical charge distribution at its two carbon atoms. If one carbon of the bromonium ion is more highly substituted than the other, and therefore able to stabilize positive charge better, it may bear a greater fraction of positive charge than the other carbon (i.e., the positively charged bromine draws electron density from the two carbon atoms of the ring, but not equally if they are of different substitution). Consequently, the more positively charged carbon may be attacked by the reaction nucleophile more often than the other carbon. However, in reactions with symmetrical reagents (e.g., Br_2, Cl_2, and I_2) there is no observed difference. We shall discuss this point further in Section 8.13, where we will study a reaction where we can discern regioselectivity of attack on a halonium ion by the nucleophile.

8.12 STEREOSPECIFIC REACTIONS

The anti addition of a halogen to an alkene provides us with an example of what is called a **stereospecific reaction**.

- A **stereospecific reaction** is one where a particular stereoisomer of the starting material yields a specific stereoisomeric form of the product.

Consider the reactions of *cis-* and *trans-*2-butene with bromine shown below. When *trans-*2-butene adds bromine, the product is the meso compound, (2R,3S)-2,3-dibromobutane. When *cis-*2-butene adds bromine, the product is a *racemic mixture* of (2R,3R)-2,3-dibromobutane and (2S,3S)-2,3-dibromobutane:

Reaction 1

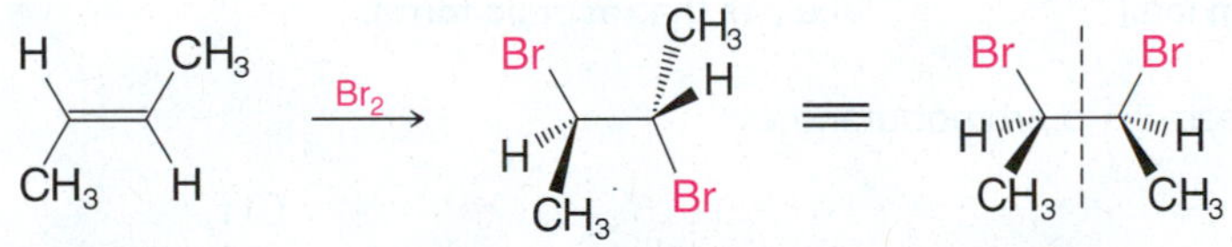

trans-2-Butene   (2R,3S)-2,3-Dibromobutane
                         (a meso compound)

Reaction 2

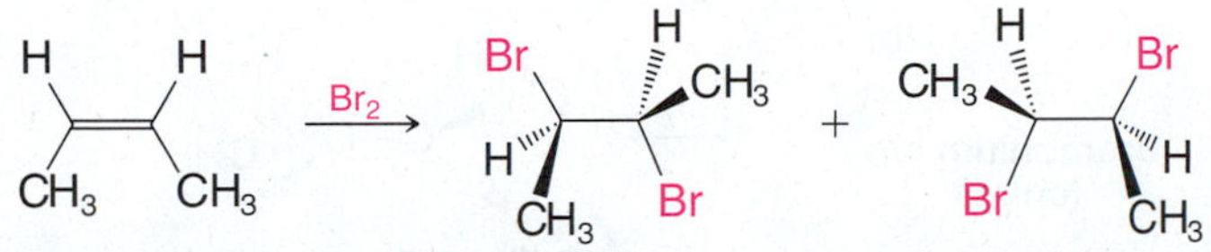

cis-2-Butene   (2R,3R)   (2S,3S)
              (a pair of enantiomers)

The reactants *cis-*2-butene and *trans-*2-butene are stereoisomers; they are *diastereomers*. The product of reaction 1, (2R,3S)-2,3-dibromobutane, is a meso compound, and it is a stereoisomer of both of the products of reaction 2 (the enantiomeric 2,3-dibromobutanes). Thus, by definition, both reactions are stereospecific. One stereoisomeric form of the reactant (e.g., *trans-*2-butene) gives one product (the meso compound), whereas the other stereoisomeric form of the reactant (*cis-*2-butene) gives a stereoisomerically different product (the enantiomers).

We can better understand the results of these two reactions if we examine their mechanisms. The first mechanism in the following box shows how *cis-*2-butene adds bromine to yield intermediate bromonium ions that are achiral. (The bromonium ion has a plane of symmetry.) These bromonium ions can then react with bromide ions by either path (a) or path (b). Reaction by path (a) yields one 2,3-dibromobutane enantiomer; reaction by path (b) yields the other enantiomer. The reaction occurs at the same rate by either path; therefore, the two enantiomers are produced in equal amounts (as a racemic form).

The second mechanism in the box shows how *trans*-2-butene reacts at the bottom face to yield an intermediate bromonium ion that is chiral. (Reaction at the other face would produce the enantiomeric bromonium ion.) Reaction of this chiral bromonium ion (or its enantiomer) with a bromide ion either by path (a) or by path (b) yields the same achiral product, *meso*-2,3-dibromobutane.

THE STEREOCHEMISTRY OF THE REACTION... ⎤ Addition of Bromine to *cis*- and *trans*-2-Butene

cis-2-Butene reacts with bromine to yield the enantiomeric 2,3-dibromobutanes by the following mechanism:

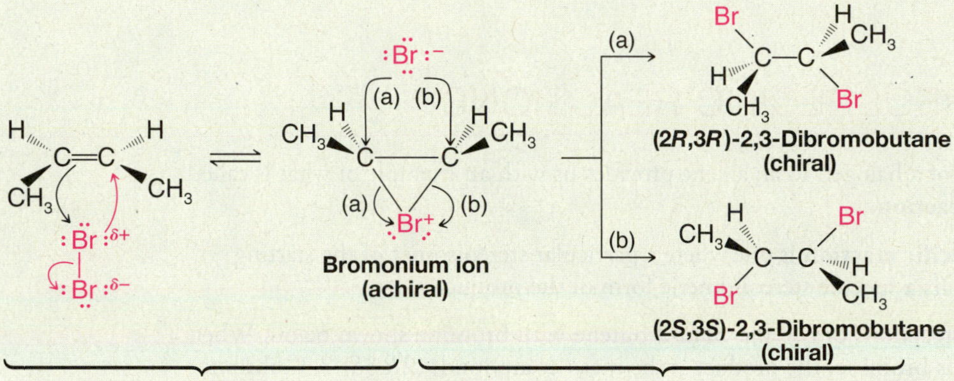

cis-2-Butene reacts with bromine to yield an achiral bromonium ion and a bromide ion. [Reaction at the other face of the alkene (top) would yield the same bromonium ion.]

The bromonium ion reacts with the bromide ions at equal rates by paths (a) and (b) to yield the two enantiomers in equal amounts (i.e., as the racemic form).

trans-2-Butene reacts with bromine to yield *meso*-2,3-dibromobutane.

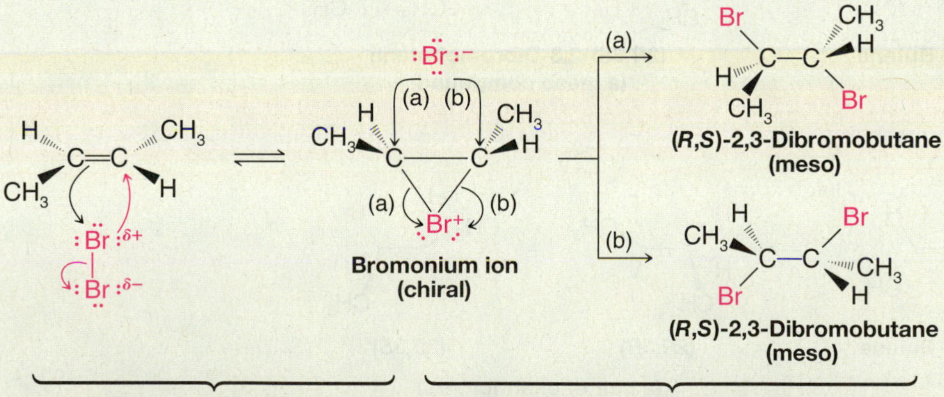

trans-2-Butene reacts with bromine to yield chiral bromonium ions and bromide ions. [Reaction at the other face (top) would yield the enantiomer of the bromonium ion as shown here.]

When the bromonium ions react by either path (a) or path (b), they yield the *same* achiral meso compound. (Reaction of the enantiomer of the intermediate bromonium ion would produce the same result.)

8.13 HALOHYDRIN FORMATION

- When the halogenation of an alkene is carried out in aqueous solution, rather than in a non-nucleophilic solvent, the major product is a **halohydrin** (also called a halo alcohol) instead of a *vic*-dihalide.

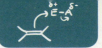

Molecules of water react with the halonium ion intermediate as the predominant nucleophile because they are in high concentration (as the solvent). The result is formation of a halohydrin as the major product. If the halogen is bromine, it is called a **bromohydrin**, and if chlorine, a **chlorohydrin**.

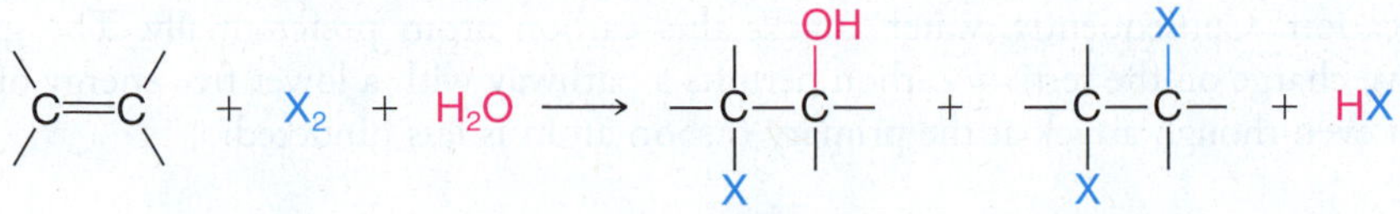

$$\underset{X = Cl \text{ or } Br}{C=C} + X_2 + H_2O \longrightarrow \underset{\substack{\text{Halohydrin} \\ \text{(major)}}}{-\overset{OH}{\underset{X}{C}}-\overset{}{\underset{}{C}}-} + \underset{\substack{vic\text{-Dihalide} \\ \text{(minor)}}}{-\overset{X}{\underset{X}{C}}-\overset{}{\underset{}{C}}-} + HX$$

Halohydrin formation can be described by the following mechanism.

A MECHANISM FOR THE REACTION — Halohydrin Formation from an Alkene

Step 1

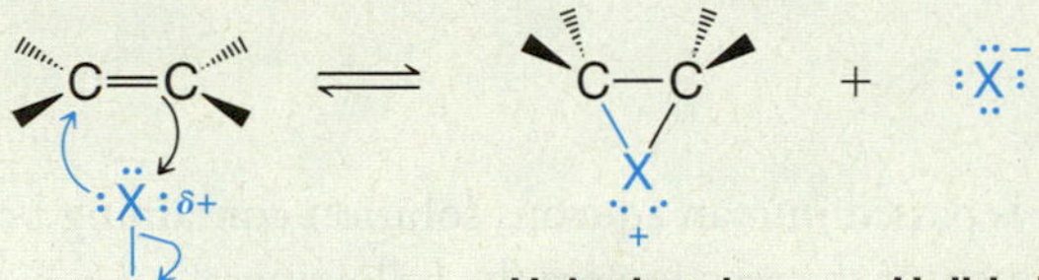

> Halonium ion   Halide ion
>
> **This step is the same as for halogen addition to an alkene (see Section 8.11A).**

Steps 2 and 3

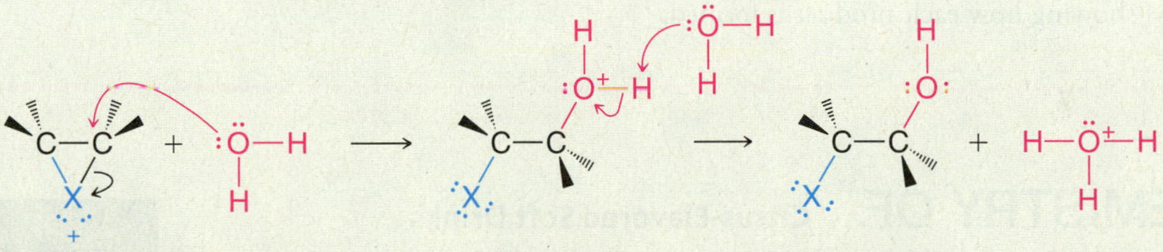

> Halonium ion   Protonated halohydrin   Halohydrin
>
> **Here, however, a water molecule acts as the nucleophile and attacks a carbon of the ring, causing the formation of a protonated halohydrin.**
>
> **The protonated halohydrin loses a proton (it is transferred to a molecule of water). This step produces the halohydrin and hydronium ion.**

The first step is the same as that for halogen addition. In the second step, however, the two mechanisms differ. In halohydrin formation, water acts as the nucleophile and attacks one carbon atom of the halonium ion. The three-membered ring opens, and a protonated halohydrin is produced. Loss of a proton then leads to the formation of the halohydrin itself.

Write a mechanism to explain the following reaction.

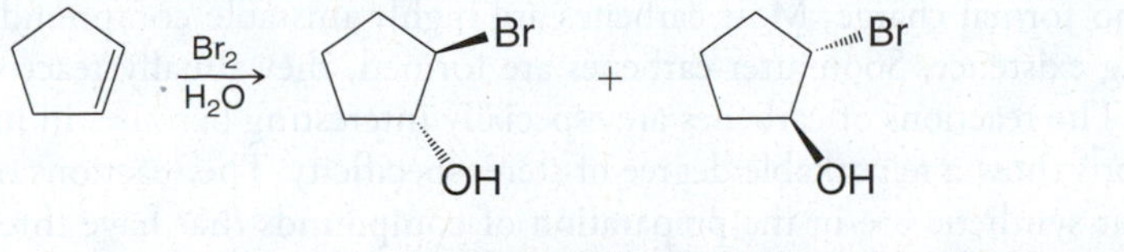

PRACTICE PROBLEM 8.15

(as a racemic mixture)

• If the alkene is unsymmetrical, the halogen ends up on the carbon atom with the greater number of hydrogen atoms.

Bonding in the intermediate bromonium ion is *unsymmetrical*. The more highly substituted carbon atom bears the greater positive charge because it resembles the more stable carbocation. Consequently, water attacks this carbon atom preferentially. The greater positive charge on the tertiary carbon permits a pathway with a lower free energy of activation even though attack at the primary carbon atom is less hindered:

> **This bromonium ion is bridged asymmetrically because the 3° carbon can accommodate more positive charge than the 1° carbon.**

(73%)

• • •

PRACTICE PROBLEM 8.16 When ethene gas is passed into an aqueous solution containing bromine and sodium chloride, the products of the reaction are the following:

Write mechanisms showing how each product is formed.

THE CHEMISTRY OF... Citrus-Flavored Soft Drinks

In Chapter 7 we discussed how double bonds within unsaturated fats could be hydrogenated to change their physical properties to convert materials like butter into margarine. It turns out that similar unsaturated fats can be used in the food industry in other ways, as well! For example, the properties of some unsaturated emulsifying agents can be enhanced if just a small percentage of their double bonds are brominated using Br_2, via the chemistry in this chapter. The increased density of these emulsifying agents, due to the bromine atoms, helps match the density of water more closely and creates a more stable, cloudy, colloidal-like mixture. The real value of this process, however, lies in what can now occur: other lipid-soluble molecules, such as many citrus flavors, can be used in aqueous foods due to the solubilizing action of these higher-density emulsifiers. The results are seen in beverages such as Mountain Dew, Squirt, or Fresca, soft drinks that all take advantage of this chemistry and that can be identified by the presence of "brominated vegetable oil" in the listing of ingredients.

© Donald Erickson/iStockphoto

8.14 DIVALENT CARBON COMPOUNDS: CARBENES

There are compounds in which a carbon has an unshared electron pair and only *two bonds*. These divalent carbon compounds are called **carbenes**. Carbenes are neutral and have no formal charge. Most carbenes are highly unstable compounds that have only a fleeting existence. Soon after carbenes are formed, they usually react with another molecule. The reactions of carbenes are especially interesting because, in many instances, the reactions show a remarkable degree of stereospecificity. The reactions of carbenes are also of great synthetic use in the preparation of compounds that have three-membered rings such as bicyclo[4.1.0]heptane, shown on the next page.

8.14A Structure and Reactions of Methylene

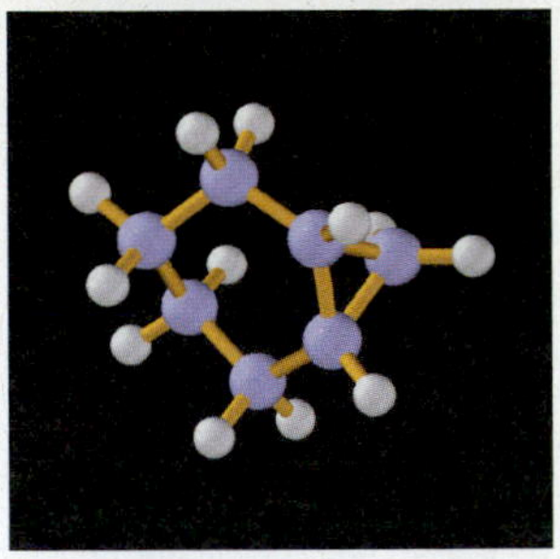

The simplest carbene is the compound called **methylene** ($:CH_2$). Methylene can be prepared by the decomposition of diazomethane (CH_2N_2), a very poisonous yellow gas. This decomposition can be accomplished by heating diazomethane (thermolysis) or by irradiating it with light of a wavelength that it can absorb (photolysis):

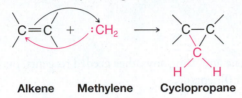

Diazomethane   **Methylene   Nitrogen**

The structure of diazomethane is actually a resonance hybrid of three structures:

$$:\bar{C}H_2 - \overset{+}{N} \equiv N: \longleftrightarrow CH_2 = \overset{+}{N} = \bar{N}: \longleftrightarrow :\bar{C}H_2 - \ddot{N} = \overset{+}{N}:$$

I   II   III

We have chosen resonance structure **I** to illustrate the decomposition of diazomethane because with **I** it is readily apparent that heterolytic cleavage of the carbon–nitrogen bond results in the formation of methylene and molecular nitrogen.

Methylene reacts with alkenes by adding to the double bond to form cyclopropanes:

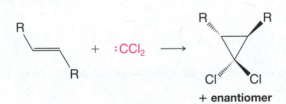

Alkene   Methylene   Cyclopropane

8.14B Reactions of Other Carbenes: Dihalocarbenes

Dihalocarbenes are also frequently employed in the synthesis of cyclopropane derivatives from alkenes. Most reactions of dihalocarbenes are **stereospecific**:

> The addition of $:CX_2$ is stereospecific. If the R groups of the alkene are trans, they will be trans in the product. (If the R groups were initially cis, they would be cis in the product.)

+ enantiomer

Dichlorocarbene can be synthesized by an **α elimination** of hydrogen chloride from chloroform. [The hydrogen of chloroform is mildly acidic ($pK_a \approx 24$) due to the inductive effect of the chlorine atoms.] This reaction resembles the β-elimination reactions by which alkenes are synthesized from alkyl halides (Section 6.15), except that the leaving group is on the same carbon as the proton that is removed.

$$R - \ddot{O}:^- K^+ + H:CCl_3 \rightleftharpoons R - \ddot{O}:H + \ ^-:CCl_3 + K^+ \xrightarrow{slow} :CCl_2 + :\ddot{C}l:^-$$

Dichlorocarbene

Compounds *with a β hydrogen* react by β elimination preferentially. Compounds with no β hydrogen but with an α hydrogen (such as chloroform) react by α elimination.

A variety of cyclopropane derivatives have been prepared by generating dichlorocarbene in the presence of alkenes. Cyclohexene, for example, reacts with dichlorocarbene generated by treating chloroform with potassium *tert*-butoxide to give a bicyclic product:

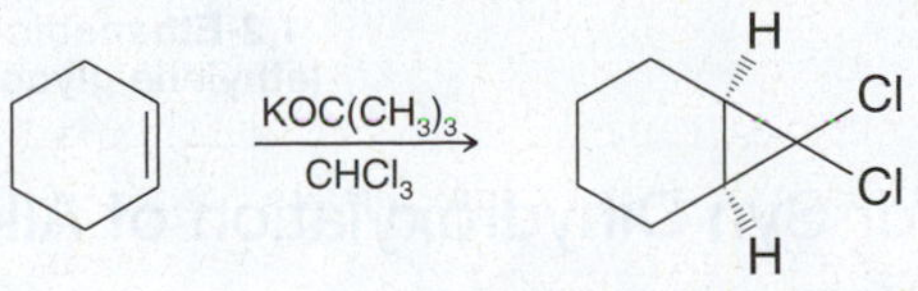

7,7-Dichlorobicyclo[4.1.0]heptane
(59%)

Bicyclo[4.1.0]heptane

8.14C Carbenoids: The Simmons–Smith Cyclopropane Synthesis

A useful cyclopropane synthesis was developed by H. E. Simmons and R. D. Smith of the DuPont Company. In this synthesis diiodomethane and a zinc–copper couple are stirred together with an alkene. The diiodomethane and zinc react to produce a carbene-like species called a **carbenoid**:

$$CH_2I_2 + Zn(Cu) \longrightarrow ICH_2ZnI$$
A carbenoid

The carbenoid then brings about the stereospecific addition of a CH_2 group directly to the double bond.

PRACTICE PROBLEM 8.17 What products would you expect from each of the following reactions?

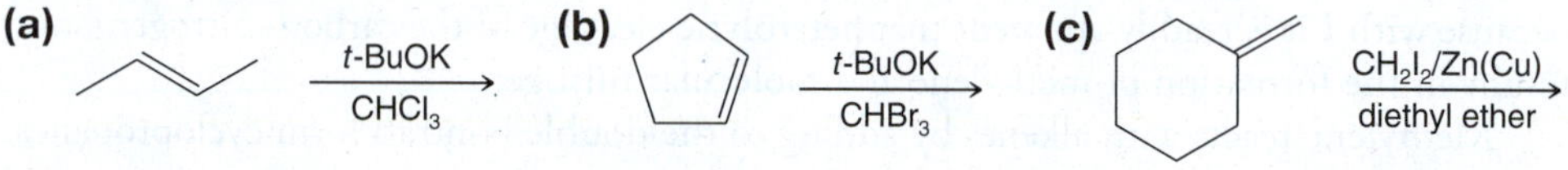

(a)

$$\xrightarrow[\text{CHCl}_3]{t\text{-BuOK}}$$

(b)

$$\xrightarrow[\text{CHBr}_3]{t\text{-BuOK}}$$

(c)

$$\xrightarrow[\text{diethyl ether}]{\text{CH}_2\text{I}_2/\text{Zn(Cu)}}$$

PRACTICE PROBLEM 8.18 Starting with cyclohexene and using any other needed reagents, outline a synthesis of 7,7-dibromobicyclo[4.1.0]heptane.

PRACTICE PROBLEM 8.19 Treating cyclohexene with 1,1-diiodoethane and a zinc–copper couple leads to two isomeric products. What are their structures?

8.15 OXIDATION OF ALKENES: SYN 1,2-DIHYDROXYLATION

Alkenes undergo a number of reactions in which the carbon–carbon double bond is oxidized.

- **1,2-Dihydroxylation** is an important oxidative addition reaction of alkenes.

Osmium tetroxide is widely used to synthesize **1,2-diols** (the products of 1,2-**dihydroxylation**, sometimes also called **glycols**). Potassium permanganate can also be used, although because it is a stronger oxidizing agent it is prone to cleave the diol through further oxidation (Section 8.15).

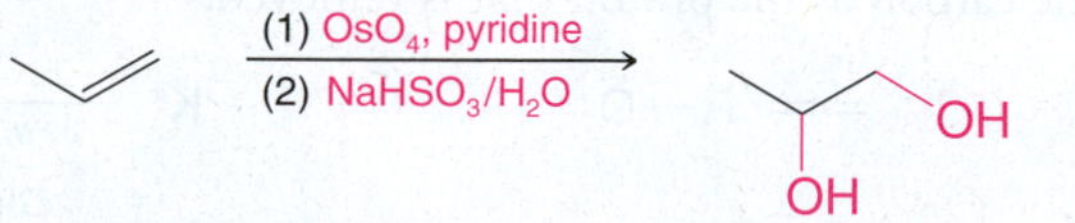

$$\xrightarrow[\text{(2) NaHSO}_3/\text{H}_2\text{O}]{\text{(1) OsO}_4, \text{ pyridine}}$$

Propene   **1,2-Propanediol (propylene glycol)**

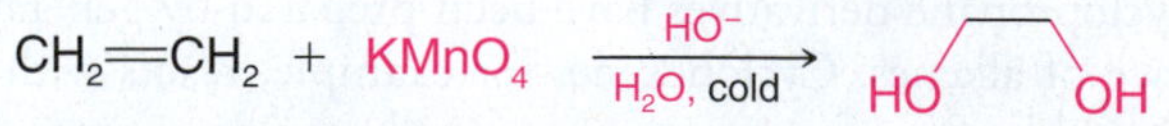

$$CH_2{=}CH_2 + KMnO_4 \xrightarrow[\text{H}_2\text{O, cold}]{\text{HO}^-}$$

Ethene   **1,2-Ethanediol (ethylene glycol)**

8.15A Mechanism for Syn Dihydroxylation of Alkenes

- The mechanism for the formation of a 1,2-diol by osmium tetroxide involves a cyclic intermediate that results in **syn addition** of the oxygen atoms (see below).

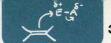

After formation of the cyclic intermediate with osmium, cleavage at the oxygen–metal bonds takes place without altering the stereochemistry of the two new C—O bonds.

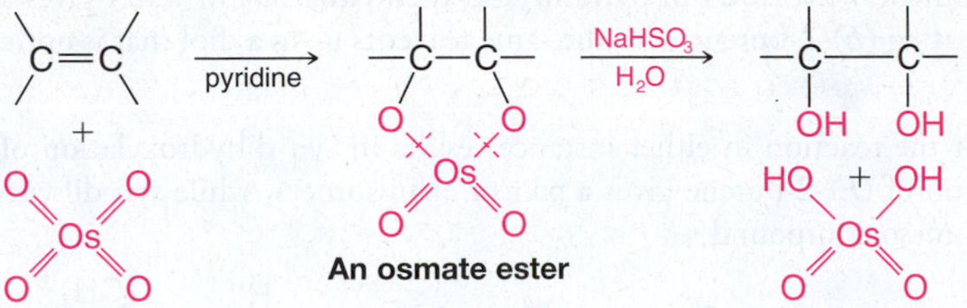

The syn stereochemistry of this dihydroxylation can readily be observed by the reaction of cyclopentene with osmium tetroxide. The product is *cis*-1,2-cyclopentanediol.

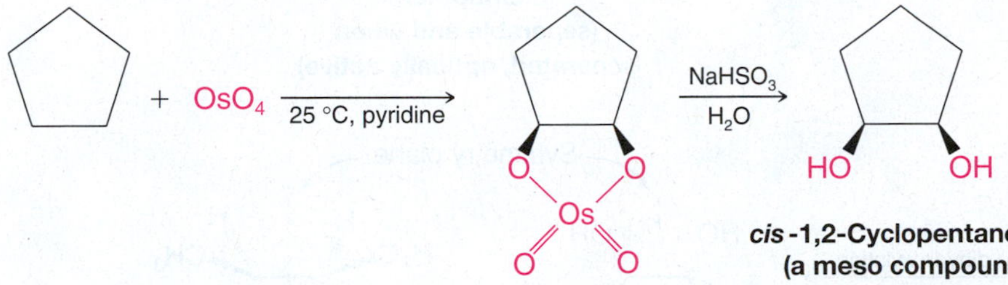

cis-1,2-Cyclopentanediol
(a meso compound)

Osmium tetroxide is highly toxic, volatile, and very expensive. For these reasons, methods have been developed that permit OsO₄ to be used *catalytically* in conjunction with a co-oxidant.* A very small molar percentage of OsO₄ is placed in the reaction mixture to do the dihydroxylation step, while a stoichiometric amount of co-oxidant reoxidizes the OsO₄ as it is used in each cycle, allowing oxidation of the alkene to continue until all has been converted to the diol. *N*-Methylmorpholine *N*-oxide (NMO) is one of the most commonly used co-oxidants with catalytic OsO₄. The method was discovered at Upjohn Corporation in the context of reactions for synthesis of a prostaglandin† (Section 23.5):

Catalytic OsO₄ 1,2-Dihydroxylation

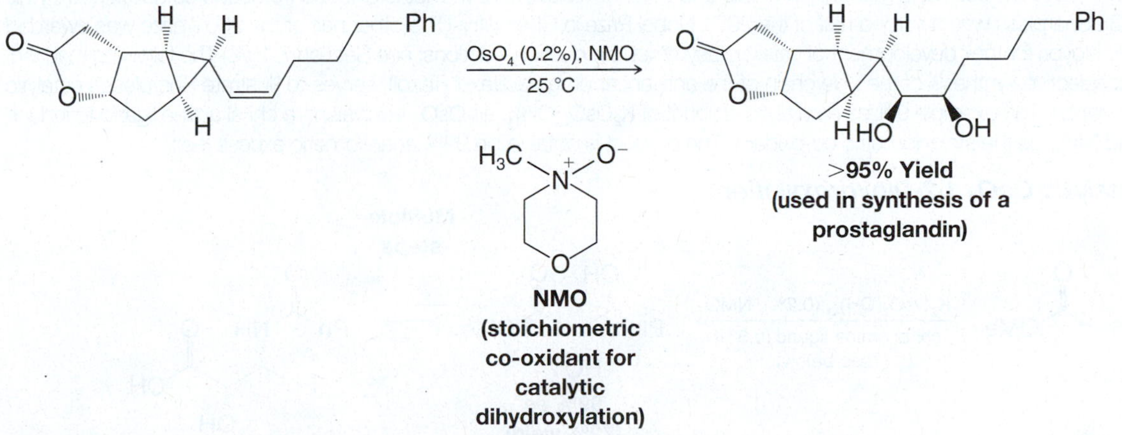

>95% Yield
(used in synthesis of a
prostaglandin)

NMO
(stoichiometric
co-oxidant for
catalytic
dihydroxylation)

PRACTICE PROBLEM 8.20

Specify the alkene and reagents needed to synthesize each of the following diols.

(a)

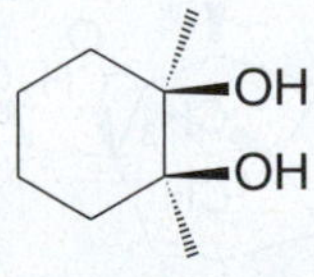

(b)

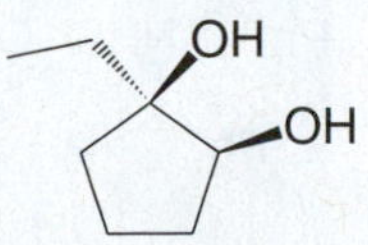

(racemic)

(c)

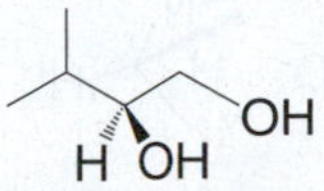

(racemic)

---

*See Nelson, D. W., et al., *J. Am. Chem. Soc.* **1997**, *119*, 1840–1858; and Corey, E. J., et al., *J. Am. Chem. Soc.* **1996**, *118*, 319–329.

†Van Rheenan, V., Kelley, R. C., and Cha, D. Y., *Tetrahedron Lett.* **1976**, *25*, 1973.

●●● SOLVED PROBLEM 8.4

Explain the following facts: Treating (Z)-2-butene with OsO_4 in pyridine and then $NaHSO_3$ in water gives a diol that is optically inactive and cannot be resolved. Treating (E)-2-butene with the same reagents gives a diol that is optically inactive but can be resolved into enantiomers.

STRATEGY AND ANSWER: Recall that the reaction in either instance results in syn dihydroxylation of the double bond of each compound. Syn dihydroxylation of (E)-2-butene gives a pair of enantiomers, while syn dihydroxylation of (Z)-2-butene gives a single product that is a meso compound.

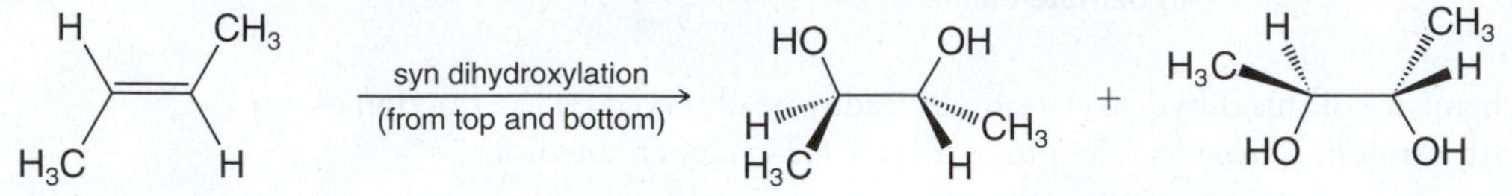

(E)-2-Butene → syn dihydroxylation (from top and bottom) →

Enantiomers (separable and when separated, optically active)

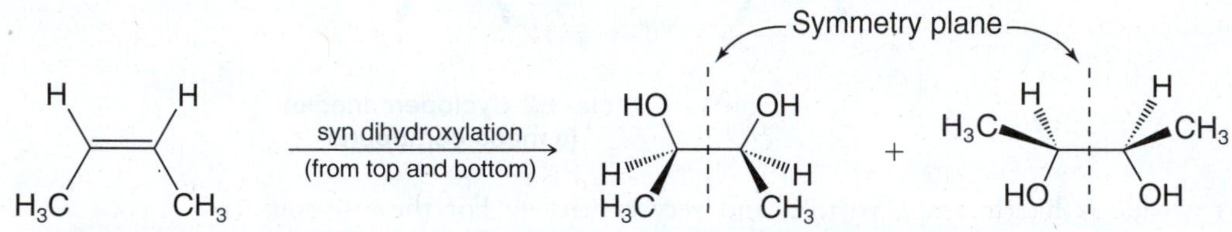

(Z)-2-Butene → syn dihydroxylation (from top and bottom) →

Identical (a meso compound) — Symmetry plane

THE CHEMISTRY OF... Catalytic Asymmetric Dihydroxylation

SHARPLESS shared the 2001 Nobel Prize in Chemistry for his development of asymmetric oxidation methods.

Methods for catalytic *asymmetric* **syn dihydroxylation** have been developed that significantly extend the synthetic utility of dihydroxylation. K. B. Sharpless (The Scripps Research Institute) and co-workers discovered that addition of a chiral amine to the oxidizing mixture leads to enantioselective catalytic syn dihydroxylation. Asymmetric dihydroxylation has become an important and widely used tool in the synthesis of complex organic molecules. In recognition of this and other advances in asymmetric oxidation procedures developed by his group (Section 11.13), Sharpless was awarded half of the 2001 Nobel Prize in Chemistry. (The other half of the 2001 prize was awarded to W. Knowles and R. Noyori for their development of catalytic asymmetric reduction reactions; see Section 7.13A.) The following reaction, involved in an enantioselective synthesis of the side chain of the anticancer drug paclitaxel (Taxol), serves to illustrate Sharpless's catalytic asymmetric dihydroxylation. The example utilizes a catalytic amount of $K_2OsO_2(OH)_4$, an OsO_4 equivalent, a chiral amine ligand to induce enantioselectivity, and NMO as the stoichiometric co-oxidant. The product is obtained in 99% enantiomeric excess (ee):

*Asymmetric Catalytic OsO₄ 1,2-Dihydroxylation**

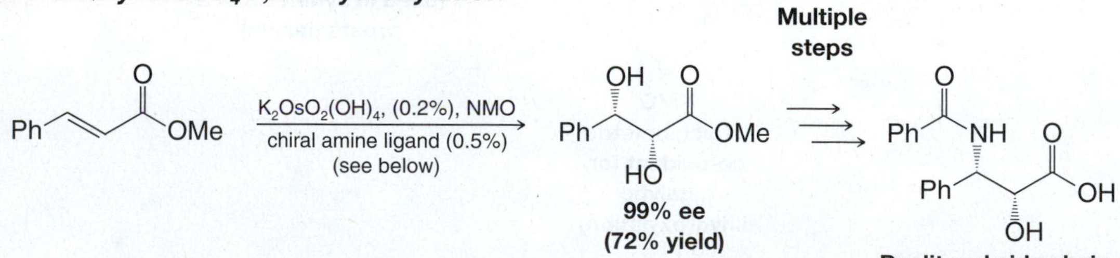

$K_2OsO_2(OH)_4$, (0.2%), NMO / chiral amine ligand (0.5%) (see below)

Multiple steps

99% ee (72% yield)

Paclitaxel side chain

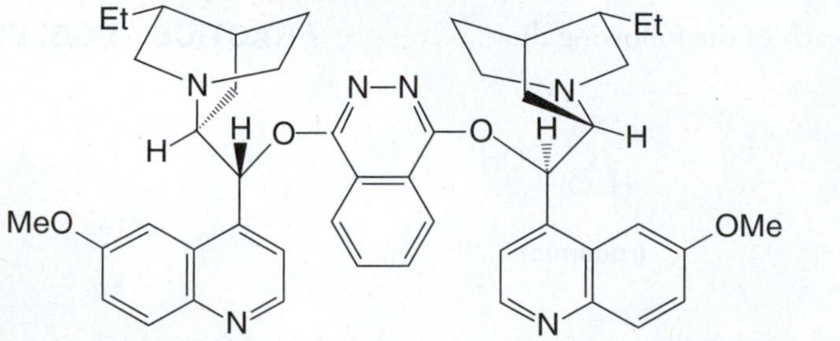

A chiral amine ligand used in catalytic asymmetric dihydroxylation

Paclitaxel

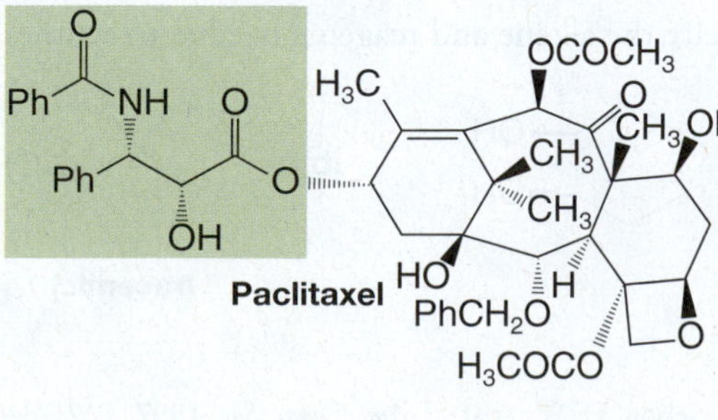

(*Adapted with permission from Sharpless et al., *The Journal of Organic Chemistry*, Vol. 59, p. 5104, 1994. Copyright 1994 American Chemical Society.)

8.16 OXIDATIVE CLEAVAGE OF ALKENES

Alkenes can be **oxidatively cleaved** using potassium permanganate or ozone (as well as by other reagents). Potassium permanganate ($KMnO_4$) is used when strong oxidation is needed. Ozone (O_3) is used when mild oxidation is desired. [Alkynes and aromatic rings are also oxidized by $KMnO_4$ and O_3 (Sections 8.19 and 15.13D).]

8.16A Cleavage with Hot Basic Potassium Permanganate

- Treatment with hot basic potassium permanganate oxidatively cleaves the double bond of an alkene.

Cleavage is believed to occur via a cyclic intermediate similar to the one formed with osmium tetroxide (Section 8.15A) and intermediate formation of a 1,2-diol.

- Alkenes with monosubstituted carbon atoms are oxidatively cleaved to salts of carboxylic acids.
- Disubstituted alkene carbons are oxidatively cleaved to ketones.
- Unsubstituted alkene carbons are oxidized to carbon dioxide.

The following examples illustrate the results of potassium permanganate cleavage of alkenes with different substitution patterns. In the case where the product is a carboxylate salt, an acidification step is required to obtain the carboxylic acid.

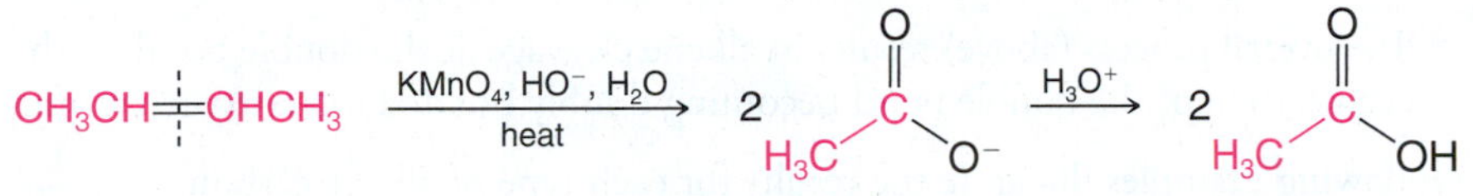

(cis or trans)   **Acetate ion**   **Acetic acid**

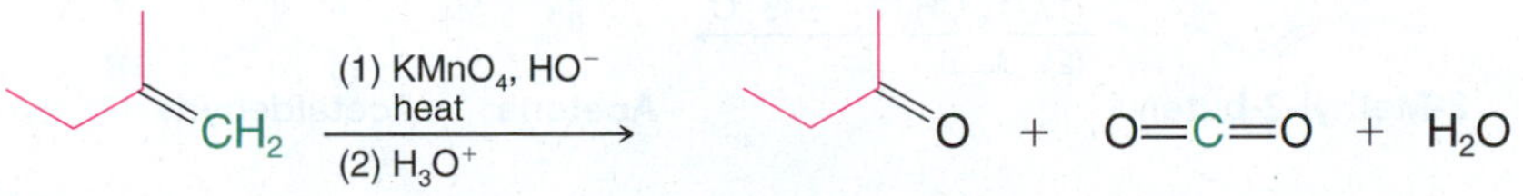

One of the uses of potassium permanganate, other than for oxidative cleavage, is as a **chemical test for unsaturation** in an unknown compound.

- If an alkene is present (or an alkyne, Section 8.19), the purple color of a potassium permanganate solution is discharged and a brown precipitate of manganese dioxide (MnO_2) forms as the oxidation takes place.

The oxidative cleavage of alkenes has also been used to establish the location of the double bond in an alkene chain or ring. The reasoning process requires us to think backward much as we do with retrosynthetic analysis. Here we are required to work backward from the products to the reactant that might have led to those products. We can see how this might be done with the following example.

●●● **SOLVED PROBLEM 8.5**

An unknown alkene with the formula C_8H_{16} was found, on oxidation with hot basic permanganate, to yield a three-carbon carboxylic acid (propanoic acid) and a five-carbon carboxylic acid (pentanoic acid). What was the structure of this alkene?

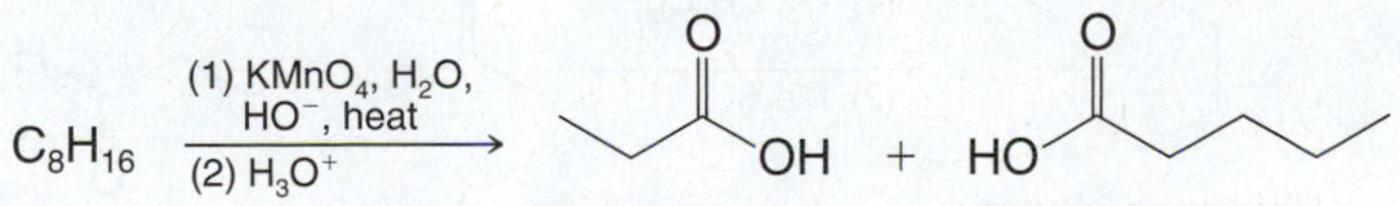

Propanoic acid   **Pentanoic acid**

(continues on next page)

STRATEGY AND ANSWER: The carbonyl groups in the products are the key to seeing where the oxidative cleavage occurred. Therefore, oxidative cleavage must have occurred as follows, and the unknown alkene must have been *cis-* or *trans-*3-octene, which is consistent with the molecular formula given.

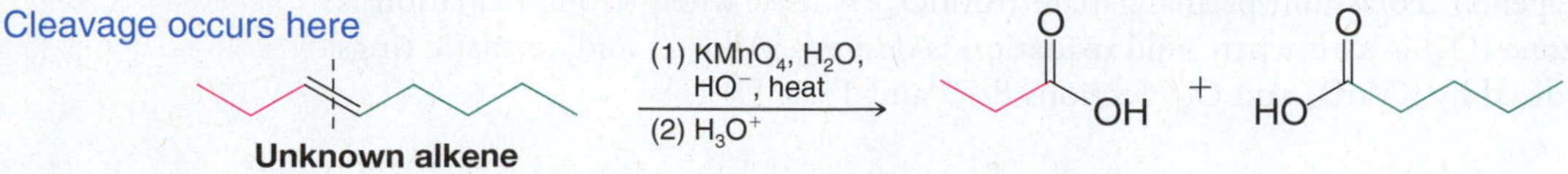

Cleavage occurs here

$$\xrightarrow[\text{(2) } H_3O^+]{\text{(1) } KMnO_4, H_2O, \ HO^-, \text{ heat}}$$

Unknown alkene
(either *cis-* or *trans-*3-octene)

8.16B Cleavage with Ozone

- The most useful method for cleaving alkenes is to use ozone (O_3).

Ozonolysis consists of bubbling ozone into a very cold ($-78°C$) solution of the alkene in CH_2Cl_2, followed by treatment of the solution with dimethyl sulfide (or zinc and acetic acid). The overall result is as follows:

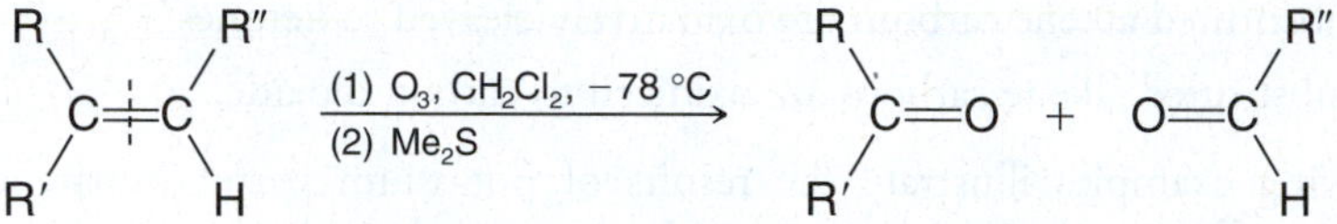

$$\underset{R'}{\overset{R}{}}C{=}C\underset{H}{\overset{R''}{}} \xrightarrow[\text{(2) } Me_2S]{\text{(1) } O_3, CH_2Cl_2, -78\ °C} \underset{R'}{\overset{R}{}}C{=}O \ + \ O{=}C\underset{H}{\overset{R''}{}}$$

The reaction is useful as a synthetic tool, as well as a method for determining the location of a double bond in an alkene by reasoning backward from the structures of the products.

- The overall process (above) results in alkene cleavage at the double bond, with each carbon of the double bond becoming doubly bonded to an oxygen atom.

The following examples illustrate the results for each type of alkene carbon.

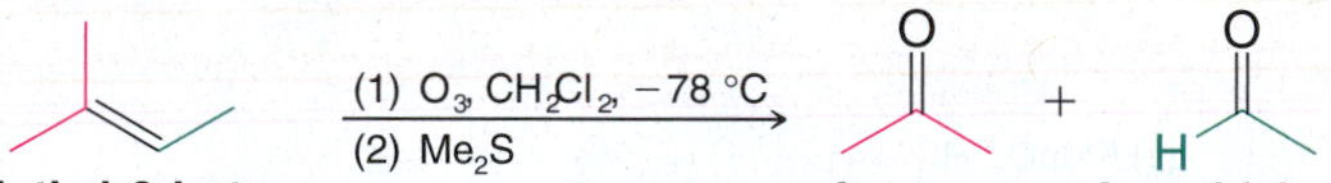

$$\xrightarrow[\text{(2) } Me_2S]{\text{(1) } O_3, CH_2Cl_2, -78\ °C}$$

2-Methyl-2-butene **Acetone** **Acetaldehyde**

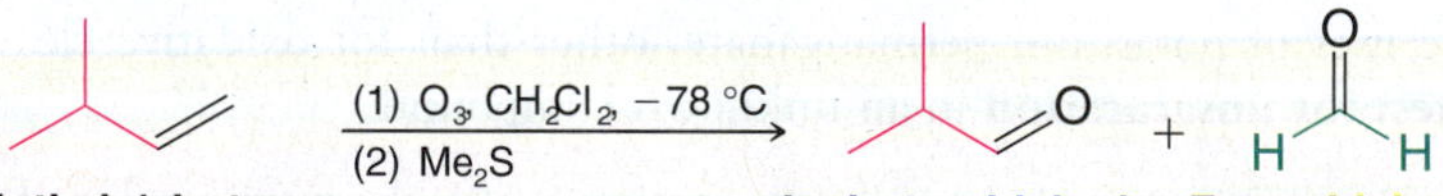

$$\xrightarrow[\text{(2) } Me_2S]{\text{(1) } O_3, CH_2Cl_2, -78\ °C}$$

3-Methyl-1-butene **Isobutyraldehyde Formaldehyde**

••• SOLVED PROBLEM 8.6

Give the structure of an unknown alkene with the formula C_7H_{12} that undergoes ozonolysis to yield, after acidification, *only the following product*:

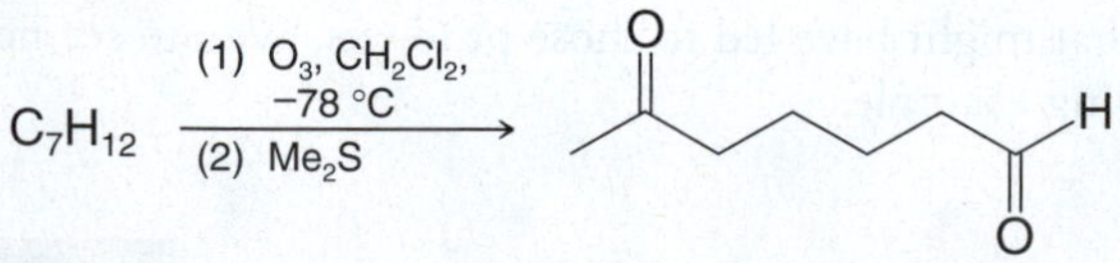

$$C_7H_{12} \xrightarrow[\text{(2) } Me_2S]{\text{(1) } O_3, CH_2Cl_2, \ -78\ °C}$$

STRATEGY AND ANSWER: Since there is only a single product containing the same number of carbon atoms as the reactant, the only reasonable explanation is that the reactant has a double bond contained in a ring. Ozonolysis of the double bond opens the ring:

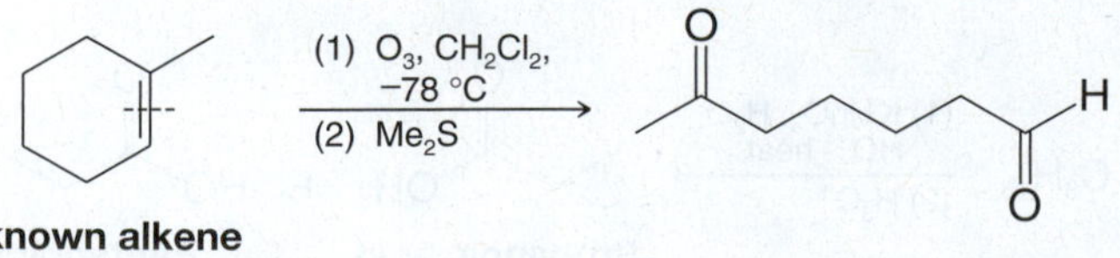

$$\xrightarrow[\text{(2) } Me_2S]{\text{(1) } O_3, CH_2Cl_2, \ -78\ °C}$$

Unknown alkene
(1-methylcyclohexene)

Predict the products of the following ozonolysis reactions.

(a)

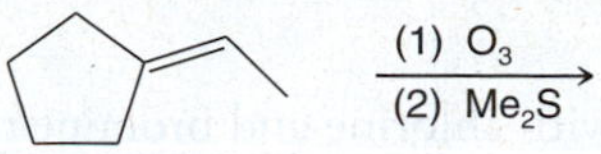

$$\xrightarrow[\text{(2) Me}_2\text{S}]{\text{(1) O}_3}$$

(c)

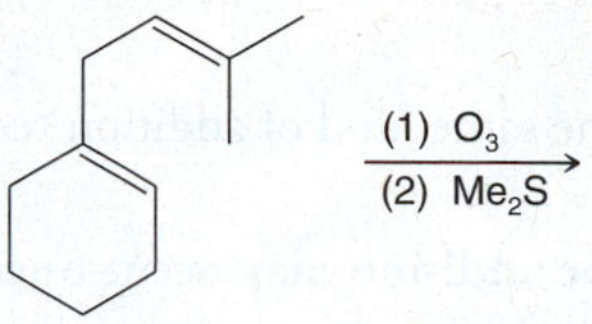

$$\xrightarrow[\text{(2) Me}_2\text{S}]{\text{(1) O}_3}$$

(b)

$$\xrightarrow[\text{(2) Me}_2\text{S}]{\text{(1) O}_3}$$

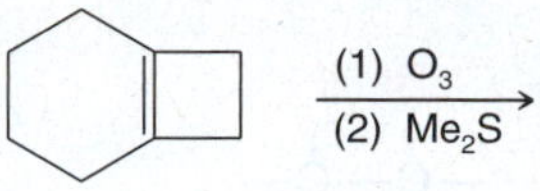

The mechanism of ozone addition to alkenes begins with formation of unstable compounds called *initial ozonides* (sometimes called molozonides). The process occurs vigorously and leads to spontaneous (and sometimes noisy) rearrangement to compounds known as **ozonides.** The rearrangement is believed to occur with dissociation of the initial ozonide into reactive fragments that recombine to yield the ozonide. Ozonides are very unstable compounds, and low-molecular-weight ozonides often explode violently.

[A MECHANISM FOR THE REACTION ─ Ozonolysis of an Alkene]

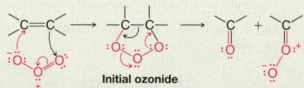

Ozone adds to the alkene to form an initial ozonide.

Initial ozonide

The initial ozonide fragments.

Ozonide
The fragments recombine to form the ozonide.

Aldehydes and/or ketones   **Dimethyl sulfoxide**

Write the structures of the alkenes that would yield the following carbonyl compounds when treated with ozone and then with dimethyl sulfide.

(a)

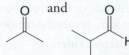 and

(c)

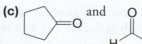 and

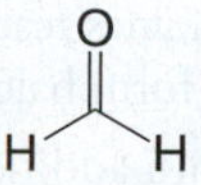

(b)

 (2 mol is produced from 1 mol of alkene)

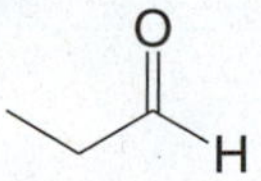

8.17 ELECTROPHILIC ADDITION OF BROMINE AND CHLORINE TO ALKYNES

- Alkynes show the same kind of addition reactions with chlorine and bromine that alkenes do.
- With alkynes **the addition may occur once or twice**, depending on the number of molar equivalents of halogen we employ:

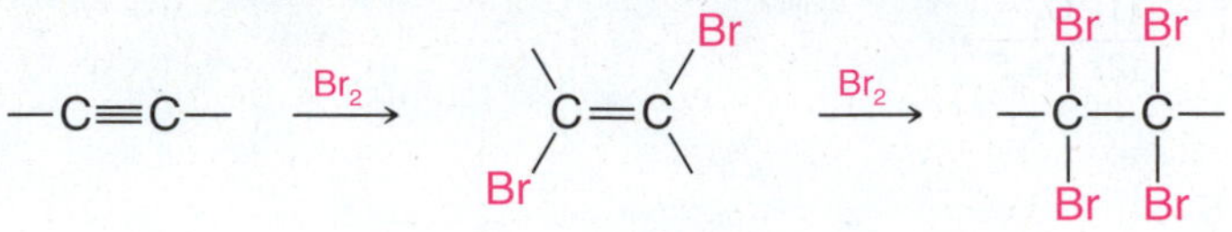

Dibromoalkene   **Tetrabromoalkane**

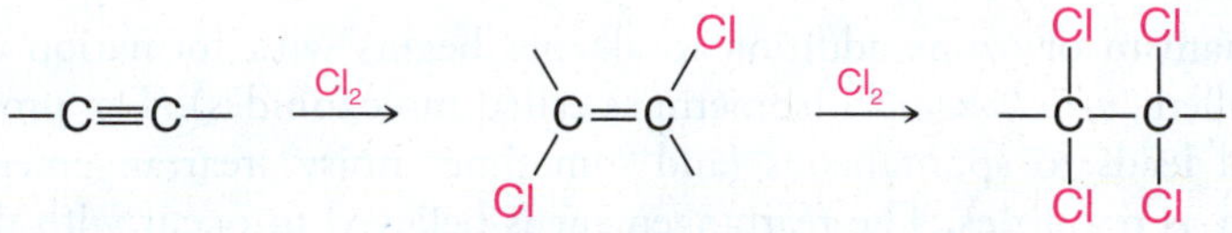

Dichloroalkene   **Tetrachloroalkane**

It is usually possible to prepare a dihaloalkene by simply adding one molar equivalent of the halogen:

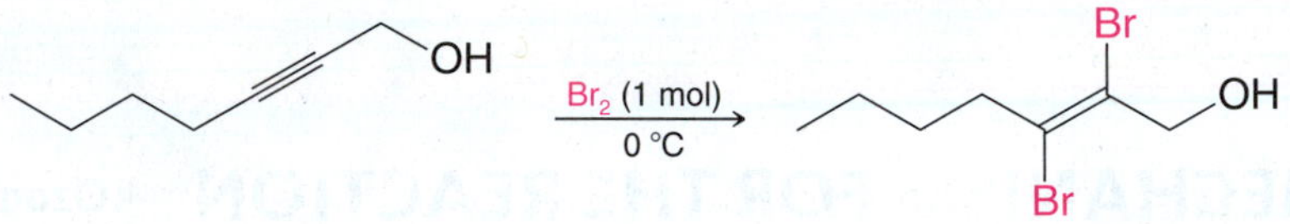

$$\xrightarrow[0\,°C]{Br_2\ (1\ mol)}$$

- Addition of one molar equivalent of chlorine or bromine to an alkyne generally results in anti addition and yields a *trans*-dihaloalkene.

Addition of bromine to acetylenedicarboxylic acid, for example, gives the trans isomer in 70% yield:

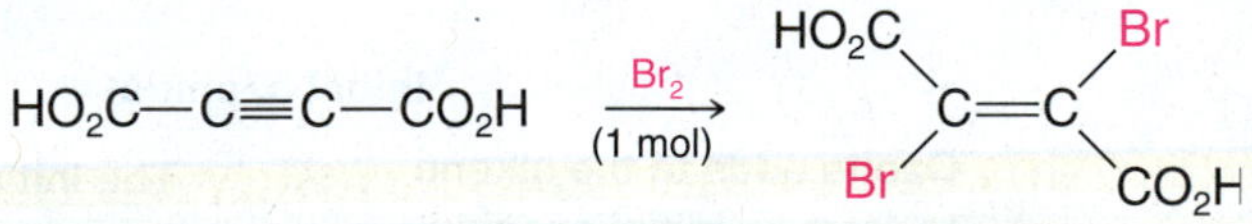

$$HO_2C-C\equiv C-CO_2H \xrightarrow[(1\ mol)]{Br_2}$$

Acetylenedicarboxylic acid   **(70%)**

PRACTICE PROBLEM 8.23   Alkenes are more reactive than alkynes toward addition of electrophilic reagents (i.e., Br_2, Cl_2, or HCl). Yet when alkynes are treated with one molar equivalent of these same electrophilic reagents, it is easy to stop the addition at the "alkene stage." This appears to be a paradox and yet it is not. Explain.

8.18 ADDITION OF HYDROGEN HALIDES TO ALKYNES

- Alkynes react with one molar equivalent of hydrogen chloride or hydrogen bromide to form haloalkenes, and with two molar equivalents to form geminal dihalides.
- Both additions are **regioselective** and follow **Markovnikov's rule**:

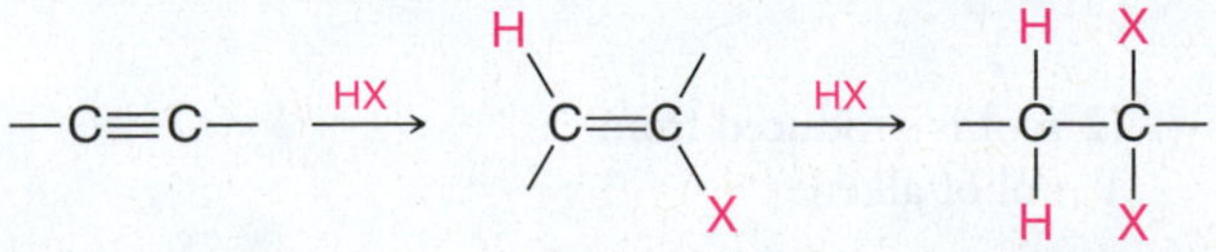

Haloalkene   *gem*-**Dihalide**

The hydrogen atom of the hydrogen halide becomes attached to the carbon atom that has the greater number of hydrogen atoms. 1-Hexyne, for example, reacts slowly with one molar equivalent of hydrogen bromide to yield 2-bromo-1-hexene and with two molar equivalents to yield 2,2-dibromohexane:

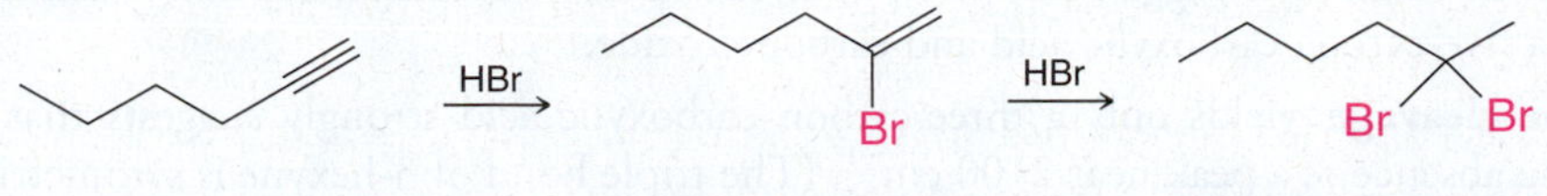

2-Bromo-1-hexene   **2,2-Dibromohexane**

The addition of HBr to an alkyne can be facilitated by using acetyl bromide (CH_3COBr) and alumina instead of aqueous HBr. Acetyl bromide acts as an HBr precursor by reacting with the alumina to generate HBr. For example, 1-heptyne can be converted to 2-bromo-1-heptene in good yield using this method:

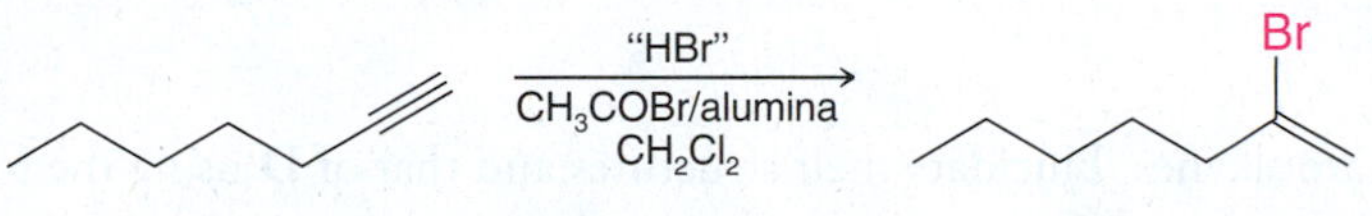

(82%)

Anti-Markovnikov addition of hydrogen bromide to alkynes occurs when peroxides are present in the reaction mixture. These reactions take place through a free-radical mechanism (Section 10.10):

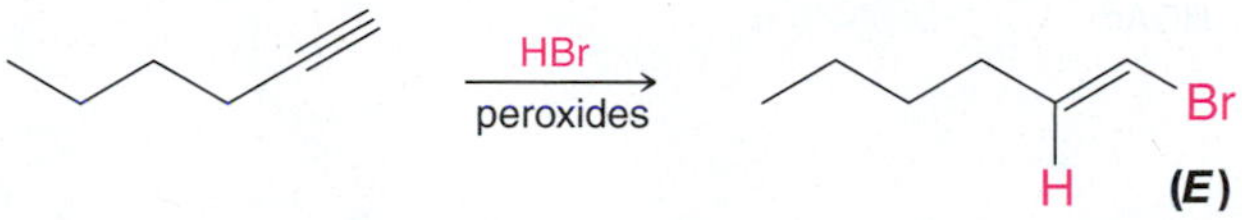

(74%)   (*E*) and (*Z*)

8.19 OXIDATIVE CLEAVAGE OF ALKYNES

Treating alkynes with **ozone** followed by acetic acid, or with basic potassium permanganate followed by acid, leads to cleavage at the carbon–carbon triple bond. The products are carboxylic acids:

$$R-C\equiv C-R' \xrightarrow[\text{(2) HOAc}]{\text{(1) O}_3} RCO_2H + R'CO_2H$$

or

$$R-C\equiv C-R' \xrightarrow[\text{(2) H}_3O^+]{\text{(1) KMnO}_4,\ HO^-} RCO_2H + R'CO_2H$$

●●● **SOLVED PROBLEM 8.7**

Three alkynes, **X**, **Y**, and **Z**, each have the formula C_6H_{10}. When allowed to react with excess hydrogen in the presence of a platinum catalyst each alkyne yields only hexane as a product.

(1) The IR spectrum of compound **X** shows, among others, a peak near 3320 cm^{-1}, several peaks in the 2800–3000-cm^{-1} region, and a peak near 2100 cm^{-1}. On oxidation with hot, basic potassium permanganate followed by acidification, **X** produces a five-carbon carboxylic acid and a gas.

(2) Compound **Y** has no IR peak in the 3300-cm^{-1} region and when oxidized with hot, basic $KMnO_4$ produces on acidification a three-carbon carboxylic acid only. Compound **Y** has peaks in the 2800–3000-cm^{-1} region, but no peak near 2100 cm^{-1}.

(3) On treatment with hot basic $KMnO_4$ followed by acid, **Z** produces a two-carbon carboxylic acid and a four-carbon one. In its IR spectrum, **Z** has peaks in the 2800–3000-cm^{-1} region and a peak near 2100 cm^{-1}, but no peaks near the 3300-cm^{-1} region. Consult Section 2.16A and propose structures for each alkyne.

(continues on next page)

STRATEGY AND ANSWER: That all three alkynes yield hexane on catalytic hydrogenation shows that they are all unbranched hexynes.

(1) That compound **X** has a peak near 3200 cm^{-1} indicates that it has a terminal triple bond. The peak near 2100 cm^{-1} is also associated with that triple bond. These facts suggest that compound **X** is 1-hexyne, something that is confirmed by the results of its oxidation to a five-carbon carboxylic acid and carbon dioxide.

(2) That compound **Y**, on **oxidative cleavage**, yields only a three-carbon carboxylic acid strongly suggests that it is 3-hexyne; this is confirmed by the absence of a peak near 2100 cm^{-1}. (The triple bond of 3-hexyne is symmetrically substituted and, therefore, the absence of an IR peak in this region is consistent with there being no dipole moment change associated with its vibration.)

(3) That compound **Z** has a peak near 2100 cm^{-1} indicates the presence of an unsymmetrically substituted triple bond, and this is consistent with the formation of two different carboxylic acids (one with two carbons and one with four) when it is oxidized. **Z**, therefore, is 2-hexyne.

• • •

PRACTICE PROBLEM 8.24 **A**, **B**, and **C** are alkynes. Elucidate their structures and that of **D** using the following reaction roadmap.

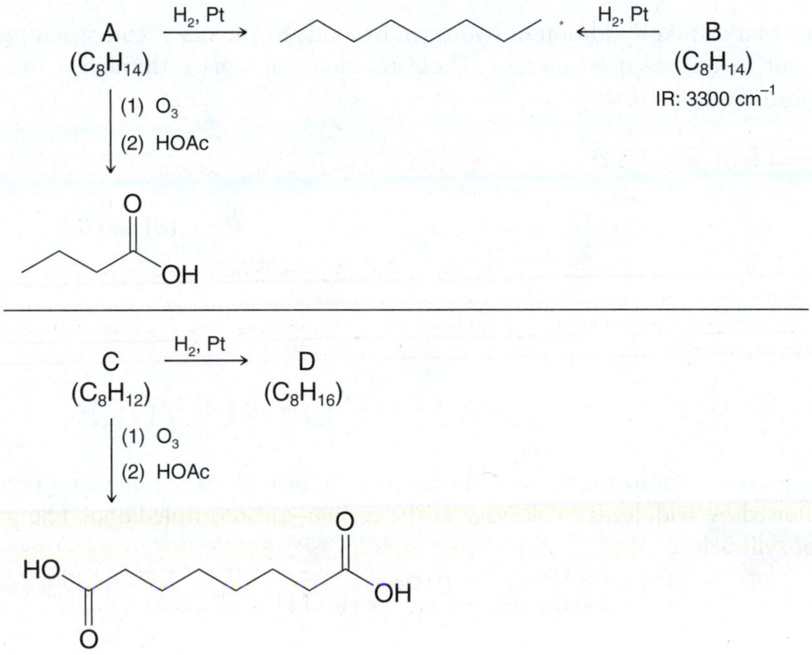

8.20 HOW TO PLAN A SYNTHESIS: SOME APPROACHES AND EXAMPLES

In planning a synthesis we often have to consider four interrelated aspects:

1. construction of the carbon skeleton,
2. functional group interconversions,
3. control of regiochemistry, and
4. control of stereochemistry.

You have had some experience with certain aspects of synthetic strategies in earlier sections.

- In Section 7.15B you learned about *retrosynthetic analysis* and how this kind of thinking could be applied to the construction of carbon skeletons of alkanes and cycloalkanes.
- In Section 6.14 you learned the meaning of a *functional group interconversion* and how nucleophilic substitution reactions could be used for this purpose.

In other sections, perhaps without realizing it, you have begun adding to your basic store of methods for construction of carbon skeletons and for making functional group interconversions. This is the time to begin keeping a card file for all the reactions that you have learned, noting especially their applications to synthesis. This file will become your **Tool Kit for Organic Synthesis**. Now is also the time to look at some new examples and to see how we integrate all four aspects of synthesis into our planning.

8.20A Retrosynthetic Analysis

Consider a problem in which we are asked to outline a synthesis of 2-bromobutane from compounds of two carbon atoms or fewer. This synthesis, as we shall see, involves construction of the carbon skeleton, functional group interconversion, and control of regiochemistry.

HOW TO Apply Retrosynthetic Analysis to 2-Bromobutane

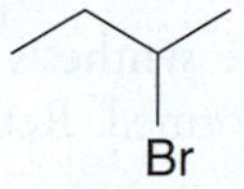

We begin by thinking backward. The final target, 2-bromobutane, can be made in one step from 1-butene by addition of hydrogen bromide. The regiochemistry of this functional group interconversion must be Markovnikov addition:

Retrosynthetic Analysis

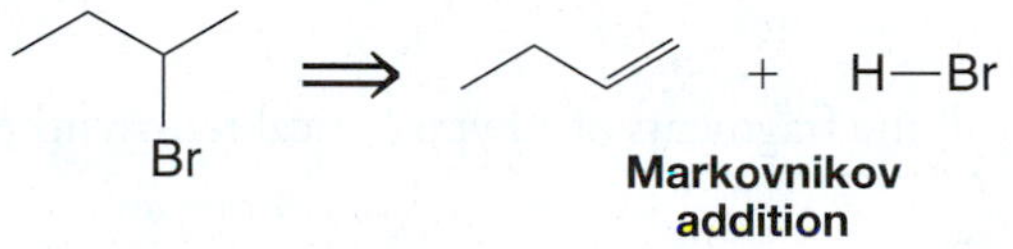

Synthesis

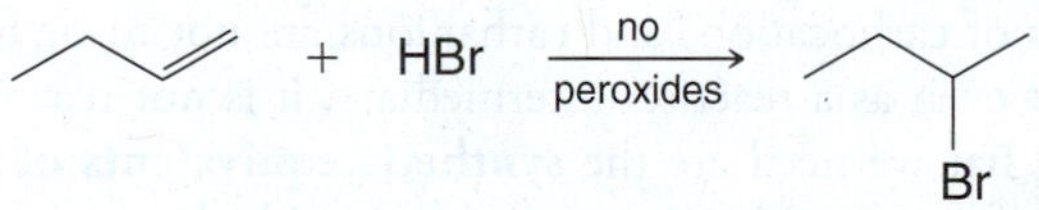

Remember: The open arrow is a symbol used to show a retrosynthetic process that relates the target molecule to its precursors:

$$\text{Target molecule} \Longrightarrow \text{precursors}$$

Continuing to work backward one hypothetical reaction at a time, we realize that a synthetic precursor of 1-butene is 1-butyne. Addition of 1 mol of hydrogen to 1-butyne would lead to 1-butene. With 1-butyne as our new target, and bearing in mind that we are told that we have to construct the carbon skeleton from compounds with two carbons or fewer, we realize that 1-butyne can be formed in one step from ethyl bromide and acetylene by an alkynide anion alkylation.

- The **key to retrosynthetic analysis** is to think of how to synthesize each target molecule in one reaction from an immediate precursor, considering first the ultimate target molecule and working backward.

Retrosynthetic Analysis

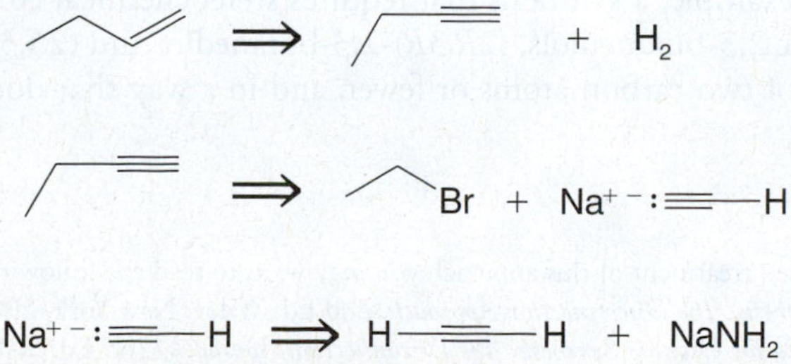

Synthesis

$$H—\!\!\!\equiv\!\!\!—H \ + \ Na^+ \ ^-NH_2 \xrightarrow[\text{liq. } NH_3, \ -33 \ ^\circ C]{} Na^+ \ ^-:\!\!\!\equiv\!\!\!—H$$

$$\diagdown\!\!\diagup\!Br \ + \ Na^+ \ ^-:\!\!\!\equiv\!\!\!—H \xrightarrow[\text{liq. } NH_3, \ -33 \ ^\circ C]{} \diagup\!\!\!\equiv\!\!\!\diagdown$$

$$\diagup\!\!\!\equiv\!\!\!\diagdown \ + \ H_2 \xrightarrow{Ni_2B \ (P\text{-}2)} \diagdown\!\!\diagup\!\!\diagdown$$

8.20B Disconnections, Synthons, and Synthetic Equivalents

- One approach to retrosynthetic analysis is to consider a retrosynthetic step as a "disconnection" of one of the bonds (Section 7.15).*

For example, an important step in the synthesis that we have just given is the one in which a new carbon–carbon bond is formed. Retrosynthetically, it can be shown in the following way:

$$\diagup\!\!\!\equiv \ \Longrightarrow \ \diagup^+ \ + \ ^-:\!\!\!\equiv\!\!\!—H$$

The hypothetical fragments of this disconnection are an ethyl cation and an ethynide anion.

- In general, we call the fragments of a hypothetical retrosynthetic disconnection **synthons**.

Seeing the synthons above may help us to reason that we could, in theory, synthesize a molecule of 1-butyne by combining an ethyl cation with an ethynide anion. We know, however, that bottles of carbocations and carbanions are not to be found on our laboratory shelves and that even as a reaction intermediate, it is not reasonable to consider an ethyl carbocation. What we need are the **synthetic equivalents** of these synthons. The synthetic equivalent of an ethynide ion is sodium ethynide, because sodium ethynide contains an ethynide ion (and a sodium cation). The synthetic equivalent of an ethyl cation is ethyl bromide. To understand how this is true, we reason as follows: if ethyl bromide were to react by an S_N1 reaction, it would produce an ethyl cation and a bromide ion. However, we know that, being a primary halide, ethyl bromide is unlikely to react by an S_N1 reaction. Ethyl bromide, however, will react readily with a strong nucleophile such as sodium ethynide by an S_N2 reaction, and when it reacts, the product that is obtained is the same as the product that would have been obtained from the reaction of an ethyl cation with sodium ethynide. Thus, ethyl bromide, in this reaction, functions as the synthetic equivalent of an ethyl cation.

2-Bromobutane could also be synthesized from compounds of two carbons or fewer by a route in which (E)- or (Z)-2-butene is an intermediate. You may wish to work out the details of that synthesis for yourself.

8.20C Stereochemical Considerations

Consider another example, a synthesis that requires stereochemical control: the synthesis of the enantiomeric 2,3-butanediols, (2R,3R)-2,3-butanediol and (2S,3S)-2,3-butanediol, from compounds of two carbon atoms or fewer, and in a way that does not produce the meso stereoisomer.

---

*For an excellent detailed treatment of this approach you may want to read the following: Warren, S., and Wyatt, P., *Organic Synthesis, The Disconnection Approach*, 2nd Ed. Wiley: New York, 2009; and Warren, S., and Wyatt, P., *Workbook for Organic Synthesis, The Disconnection Approach*, 2nd Ed. Wiley: New York, 2009.

HOW TO Apply Stereochemical Considerations in Planning a Synthesis of 2,3-Butanediol Enantiomers

Here we see that a possible final step to the enantiomers is syn dihydroxylation of *trans*-2-butene. This reaction is stereospecific and produces the desired enantiomeric 2,3-butanediols as a racemic form. Here we have made the key choice **not** to use *cis*-2-butene. Had we chosen *cis*-2-butene, our product would have been the meso 2,3-butanediol stereoisomer.

Retrosynthetic Analysis

Enantiomeric 2,3-butanediols

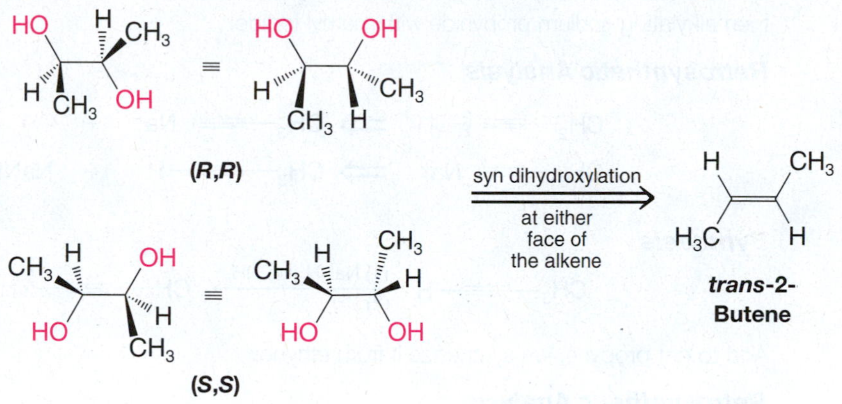

(R,R)

(S,S)

syn dihydroxylation at either face of the alkene

***trans*-2-Butene**

Synthesis

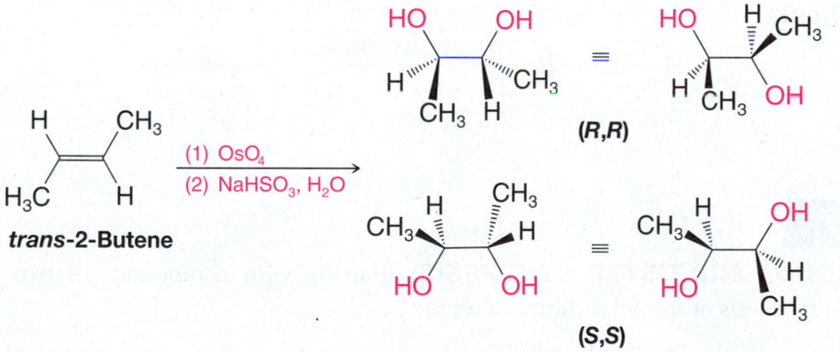

***trans*-2-Butene**

(1) OsO_4
(2) $NaHSO_3$, H_2O

(R,R)

(S,S)

Enantiomeric 2,3-butanediols

Synthesis of *trans*-2-butene can be accomplished by treating 2-butyne with lithium in liquid ammonia. The anti addition of hydrogen by this reaction gives us the trans product that we need.

Retrosynthetic Analysis

***trans*-2-Butene**

anti addition

CH_3—≡—CH_3 + H_2

2-Butyne

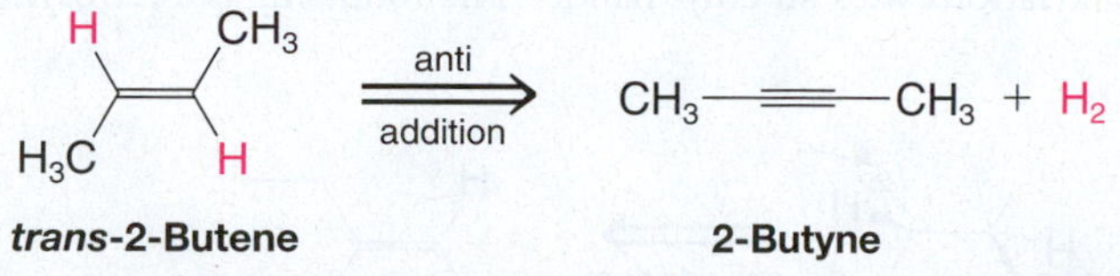

Synthesis

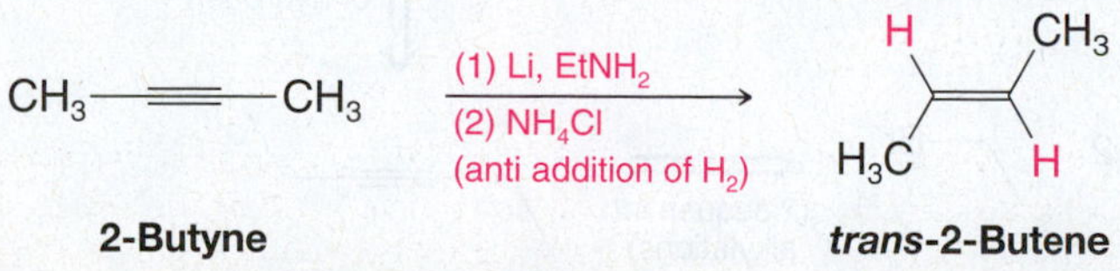

CH_3—≡—CH_3

2-Butyne

(1) Li, $EtNH_2$
(2) NH_4Cl
(anti addition of H_2)

***trans*-2-Butene**

- The reaction above is an example of a **stereoselective reaction**. A **stereoselective reaction** is one in which the reactant is not necessarily chiral (as in the case of an alkyne) but in which the reaction produces predominantly or exclusively one stereoisomeric form of the product (or a certain subset of stereoisomers from among all those that are possible).

- Note the difference between stereoselective and stereospecific. A stereospecific reaction is one that produces predominantly or exclusively one stereoisomer of the product when a specific stereoisomeric form of the reactant is used. (All stereospecific reactions are stereoselective, but the reverse is not necessarily true.)

We can synthesize 2-butyne from propyne by first converting it to sodium propynide and then alkylating sodium propynide with methyl iodide:

Retrosynthetic Analysis

$$CH_3 \!\!\equiv\!\!\text{-}CH_3 \implies CH_3\!\!\equiv\!\!:^-Na^+ \; + \; CH_3\!\!-\!\!I$$

$$CH_3\!\!\equiv\!\!:^-Na^+ \implies CH_3\!\!\equiv\!\!-H \; + \; NaNH_2$$

Synthesis

$$CH_3\!\!\equiv\!\!-H \xrightarrow[\text{(2) } CH_3I]{\text{(1) } NaNH_2/\text{liq. } NH_3} CH_3\!\!\equiv\!\!-CH_3$$

And to get propyne, we synthesize it from ethyne:

Retrosynthetic Analysis

$$H\!\!\equiv\!\!-CH_3 \implies H\!\!\equiv\!\!:^-Na^+ \; + \; CH_3\!\!-\!\!I$$

Synthesis

$$H\!\!\equiv\!\!-H \xrightarrow[\text{(2) } CH_3I]{\text{(1) } NaNH_2/\text{liq. } NH_3} CH_3\!\!\equiv\!\!-H$$

●●● SOLVED PROBLEM 8.8

ILLUSTRATING A STEREOSPECIFIC MULTISTEP SYNTHESIS: Starting with compounds of two carbon atoms or fewer, outline a stereospecific synthesis of *meso*-3,4-dibromohexane.

STRATEGY AND ANSWER: We begin by working backward from the target molecule. Since the target molecule is a meso compound, it is convenient to start by drawing a formula that illustrates its internal plane of symmetry, as shown below. But since we also know that a vicinal dibromide can be formed by anti addition of bromine to an alkene, we redraw the target molecule formula in a conformation that shows the bromine atoms anti to each other, as they would be after addition to an alkene. Then, retaining the relative spatial relationship of the alkyl groups, we draw the alkene precursor to the 1,2-dibromide, and find that this compound is (*E*)-3-hexene. Knowing that an (*E*) alkene can be formed by anti addition of hydrogen to an alkyne using lithium in ethylamine or ammonia (Section 7.14B), we see that 3-hexyne is a suitable synthetic precursor to (*E*)-3-hexene. Last, because we know it is possible to alkylate terminal alkynes, we recognize that 3-hexyne could be synthesized from acetylene by two successive alkylations with an ethyl halide. The following is a retrosynthetic analysis.

Retrosynthetic Analysis

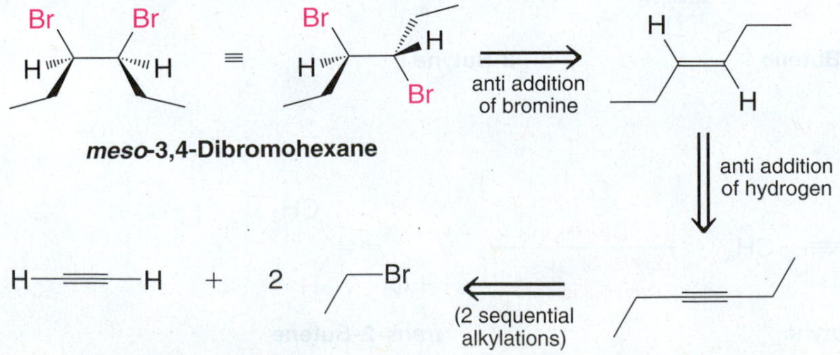

meso-3,4-Dibromohexane

The synthesis could be written as follows:

$$H—\!\!\equiv\!\!—H \xrightarrow[\text{(2) CH}_3\text{CH}_2\text{Br}]{\text{(1) NaNH}_2\text{, liq. NH}_3} \quad \text{(propyne)} \xrightarrow[\text{(2) CH}_3\text{CH}_2\text{Br}]{\text{(1) NaNH}_2\text{, liq. NH}_3}$$

then

$$\xrightarrow[\text{(2) NH}_4\text{Cl}]{\text{(1) Li, EtNH}_2}$$

and finally

meso-3,4-Dibromohexane

⋯⋯⋯ ● ● ●

How would you modify the procedure given in Solved Problem 8.8 so as to synthesize a racemic form of (3R,4R)- and (3S,4S)-3,4-dibromohexane?   **PRACTICE PROBLEM 8.25**

[**WHY** Do These Topics Matter?

ALKENES IN NATURAL CHEMICAL SYNTHESES

As illustrated in Chapters 7 and 8, unsaturation within molecules provides numerous possibilities for the addition of functional groups and the creation of C—C bonds. Thus, it should probably come as no surprise that the synthesis of complex molecules in nature also involves sites of unsaturation. Alkenes, not alkynes, are the main players in such processes, often in the form of isoprene building blocks. The five-carbon isoprene unit is easily recognized as an unsaturated four carbon chain with a methyl branch. In nature, several isoprene units combine to make long carbon chains that terminate with a reactive pyrophosphate group, such as geranylgeranyl pyrophosphate (GGPP). Such compounds take part in highly controlled reaction processes that generate tens of thousands of distinct natural products—compounds that serve as critical hormones and signaling molecules, among a myriad of other functions.

A eucalyptus tree, the source of eucalyptol.

A Pacific yew tree, the source of Taxol.

Corbis RF/Image Source

© NHPA/Photoshot Holdings Ltd.

Isoprene building block   **Geranylgeranyl pyrophosphate (GGPP)**

$$PP = -\overset{\displaystyle O}{\underset{\displaystyle O_-}{\overset{\displaystyle \|}{P}}}-O-\overset{\displaystyle O}{\underset{\displaystyle O_-}{\overset{\displaystyle \|}{P}}}-O^-$$

This chemistry begins with an enzyme that folds the isoprene-containing building block into a distinct conformation, one meant to trigger specific C—C bond formations where the OPP group will serve as a departing group in S_N2 or S_N1 processes. In some

(continues on next page)

cases, the leaving group initially helps to reposition the site of C—C bond formation through chemistry such as that shown below for the synthesis of eucalyptol. After that key step, though, it is all chemistry of the type you have seen in this chapter where alkenes are attacking electrophilic species, with the **OPP** group serving to shuttle around protons (try filling in the missing proton transfer steps for yourself).

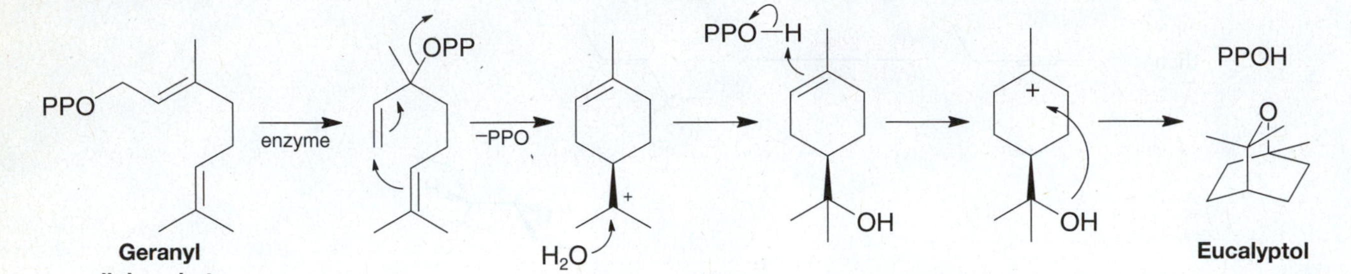

Geranyl diphosphate   **Eucalyptol**

In other cases, the OPP group is replaced directly. For example, following enzymatic organization of GGPP as shown below, removal of the indicated proton by a base can cause the neighboring alkene to displace the OPP group, leading to a molecule known as cembrene A. Then, through a series of further alkene-based C—C bond formation reactions (using more of the standard principles of nucleophiles and electrophiles as we have discussed), and oxidations, and again controlled by specific enzymes, this carbon core can be converted into materials like kempene-2. For termites, this and related molecules serve as critical protective agents against invading species.

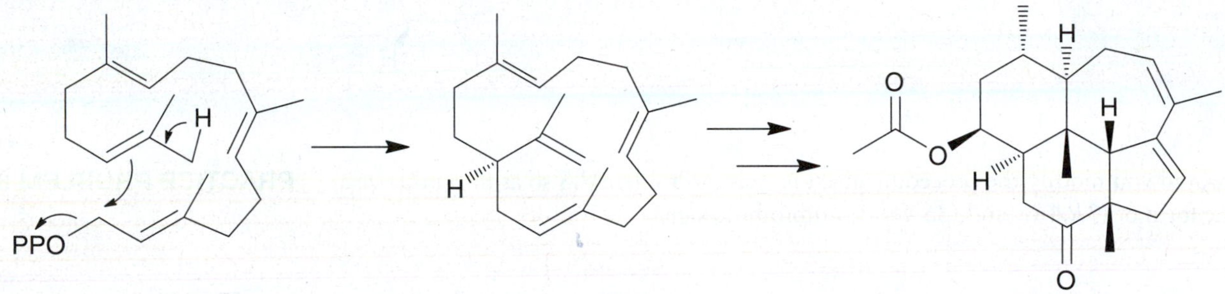

GGPP   **Cembrene A**   **Kempene-2**

What is amazing, however, is that other organisms can take the same starting piece and make completely different molecules through exactly the same processes (folding and oxidation). In the Pacific yew tree, for example, GGPP is converted into Taxol, a compound that is currently one of the world's leading cancer therapies. Several other plant species and fungi, by contrast, turn GGPP into a signaling molecule known as gibberellic acid 3. There is a lot of complex organic chemistry going on in these processes, but the key take-home message is both simple and elegant: from one single set of starting materials a large number of diverse compounds can be synthesized, all through the power of alkenes, one additional reactive group, and some very specialized and highly evolved enzymes.

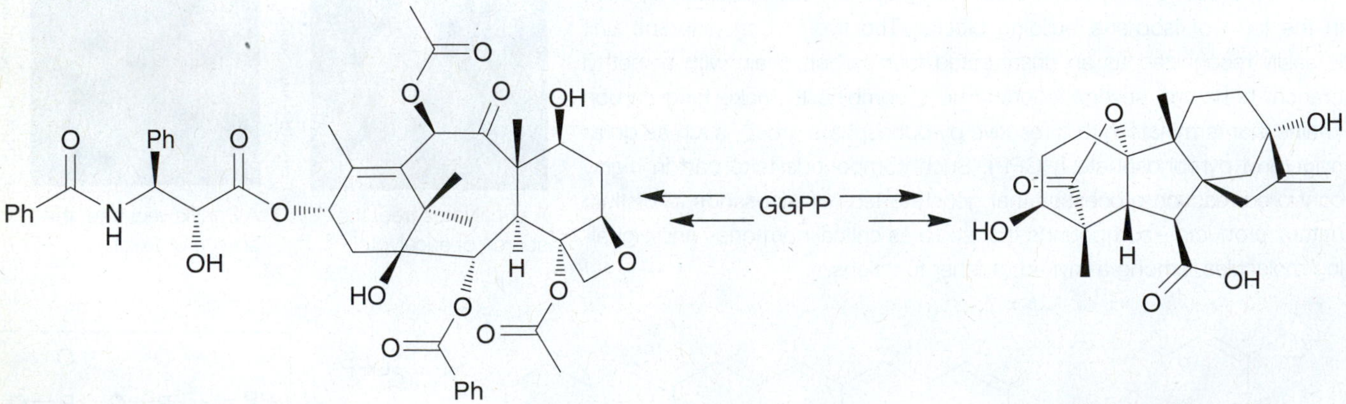

Taxol   **Gibberellic acid 3**

To learn more about these topics, see:

1. Fischbach, M. A.; Clardy, J. "One pathway, many products" in *Nature: Chem. Bio.* **2007**, *3*, 353–355.
2. Ishihara, Y.; Baran, P.S. "Two-Phase Terpene Total Synthesis: Historical Perspective and Application to the Taxol® Problem" in *Synlett* **2010**, *12*, 1733–1745.

SUMMARY AND REVIEW TOOLS

The study aids for this chapter include key terms and concepts (which are hyperlinked to the Glossary from the bold, blue terms in the *WileyPLUS* version of the book at wileyplus.com), a Mechanism Review of Alkene Addition Reactions, and a Synthetic Connections roadmap involving alkenes and alkynes.

PROBLEMS WILEY PLUS

Note to Instructors: Many of the homework problems are available for assignment via *WileyPLUS*, an online teaching and learning solution.

ALKENES AND ALKYNES REACTION TOOLKIT

8.26 Write structural formulas for the products that form when 1-butene reacts with each of the following reagents:

(a) HI

(b) H_2, Pt

(c) Dilute H_2SO_4, warm

(d) Cold concentrated H_2SO_4

(e) Cold concentrated H_2SO_4, then H_2O and heat

(f) HBr

(g) Br_2 in CCl_4

(h) Br_2 in H_2O

(i) HCl

(j) O_3, then Me_2S

(k) OsO_4, then $NaHSO_3/H_2O$

(l) $KMnO_4$, HO^-, heat, then H_3O^+

(m) $Hg(OAc)_2$ in THF and H_2O, then $NaBH_4$, HO^-

(n) BH_3:THF, then H_2O_2, HO^-

8.27 Repeat Exercise 8.26 using 1-methylcyclopentene instead of 1-butene.

8.28 Write structures for the major organic products from the following reactions. Show stereoisomers where applicable.

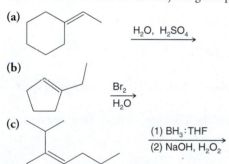

(a) $\xrightarrow{\text{H}_2\text{O, H}_2\text{SO}_4}$

(b) $\xrightarrow[\text{H}_2\text{O}]{\text{Br}_2}$

(c) $\xrightarrow[\text{(2) NaOH, H}_2\text{O}_2]{\text{(1) BH}_3\text{:THF}}$

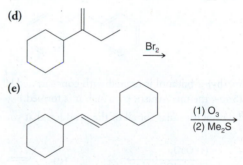

(d) $\xrightarrow{\text{Br}_2}$

(e) $\xrightarrow[\text{(2) Me}_2\text{S}]{\text{(1) O}_3}$

8.29 Give the structure of the products that you would expect from the reaction of 1-butyne with:

(a) One molar equivalent of Br_2

(b) One molar equivalent of HBr

(c) Two molar equivalents of HBr

(d) H_2 (in excess)/Pt

(e) H_2, Ni_2B (P-2)

(f) $NaNH_2$ in liquid NH_3, then CH_3I

(g) $NaNH_2$ in liquid NH_3, then $(CH_3)_3CBr$

8.30 Give the structure of the products you would expect from the reaction (if any) of 2-butyne with:

(a) One molar equivalent of HBr

(b) Two molar equivalents of HBr

(c) One molar equivalent of Br_2

(d) Two molar equivalents of Br_2

(e) H_2, Ni_2B (P-2)

(f) One molar equivalent of HCl

(g) Li/liquid NH_3

(h) H_2 (in excess), Pt

(i) Two molar equivalents of H_2, Pt

(j) Hot $KMnO_4$, HO^-, then H_3O^+

(k) O_3, then HOAc

(l) $NaNH_2$, liquid NH_3

8.31 Write structures for the major organic products from the following reactions. Show stereoisomers where applicable.

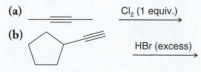

(a) $\xrightarrow{\text{Cl}_2 \text{ (1 equiv.)}}$

(b) $\xrightarrow{\text{HBr (excess)}}$

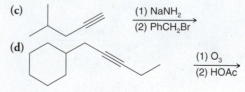

(c) $\xrightarrow[\text{(2) PhCH}_2\text{Br}]{\text{(1) NaNH}_2}$

(d) $\xrightarrow[\text{(2) HOAc}]{\text{(1) O}_3}$

8.32 Show how 1-butyne could be synthesized from each of the following:

(a) 1-Butene   (b) 1-Chlorobutane   (c) 1-Chloro-1-butene   (d) 1,1-Dichlorobutane   (e) Ethyne and ethyl bromide

8.33 Starting with 2-methylpropene (isobutylene) and using any other needed reagents, outline a synthesis of each of the following:

(a)

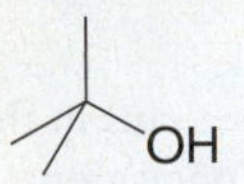

(b)

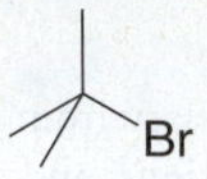

(c)

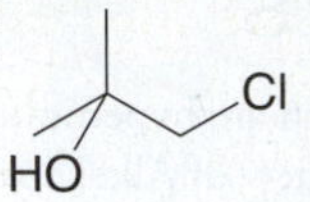

(d)

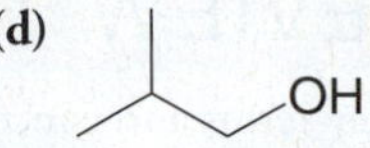

MECHANISMS

8.34 Write a three-dimensional formula for the product formed when 1-methylcyclohexene is treated with each of the following reagents. In each case, designate the location of deuterium or tritium atoms.

(a) (1) BH_3:THF, (2) CH_3CO_2T

(c) (1) BD_3:THF, (2) NaOH, H_2O_2, H_2O

(b) (1) BD_3:THF, (2) CH_3CO_2D

8.35 Write a mechanism that accounts for the formation of ethyl isopropyl ether in the following reaction.

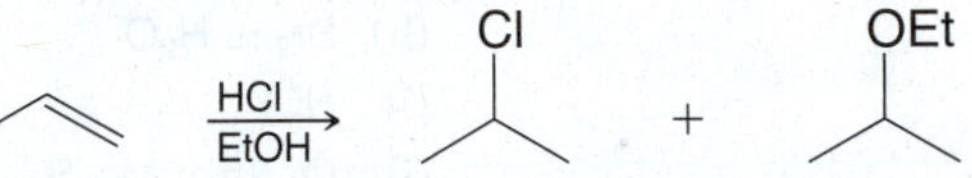

8.36 When, in separate reactions, 2-methylpropene, propene, and ethene are allowed to react with HI under the same conditions (i.e., identical concentration and temperature), 2-methylpropene is found to react fastest and ethene slowest. Provide an explanation for these relative rates.

8.37 Propose a mechanism that accounts for the following reaction.

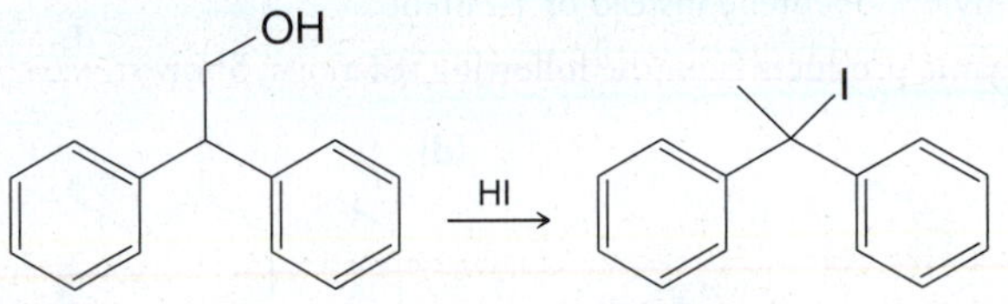

8.38 When 3,3-dimethyl-2-butanol is treated with concentrated HI, a rearrangement takes place. Which alkyl iodide would you expect from the reaction? (Show the mechanism by which it is formed.)

8.39 Write stereochemical formulas for all of the products that you would expect from each of the following reactions. (You may find models helpful.)

(a) 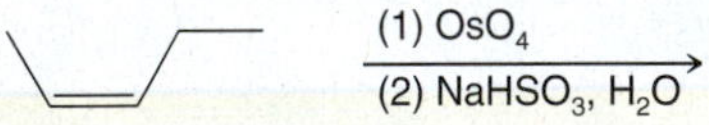 $\xrightarrow{\text{(1) OsO}_4}{\text{(2) NaHSO}_3, H_2O}$

(c) $\xrightarrow{\text{Br}_2}$

(b) $\xrightarrow{\text{(1) OsO}_4}{\text{(2) NaHSO}_3, H_2O}$

(d) 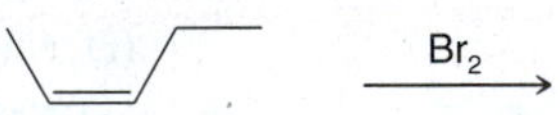 $\xrightarrow{\text{Br}_2}$

8.40 Give (R, S) designations for each different compound given as an answer to Problem 8.39.

8.41 The double bond of tetrachloroethene is undetectable in the bromine test for unsaturation. Give a plausible explanation for this behavior.

8.42 The reaction of bromine with cyclohexene involves anti addition, which generates, initially, the diaxial conformation of the addition product that then undergoes a ring flip to the diequatorial conformation of *trans*-1,2-dibromo-cyclohexane. However, when the unsaturated bicyclic compound **I** is the alkene, instead of cyclohexene, the addition product is exclusively in a stable diaxial conformation. Account for this. (You may find it helpful to build handheld molecular models.)

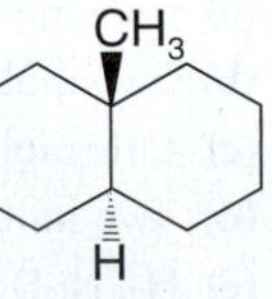

8.43 Propose a mechanism that explains formation of the products from the following reaction, including the distribution of the products as major and minor.

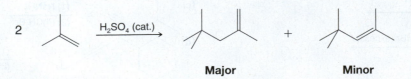

Major   **Minor**

8.44 Internal alkynes can be isomerized to terminal alkynes on treatment with $NaNH_2$. The process is much less successful when NaOH is used. Why is there this difference?

8.45 Write a mechanism that explains the following reaction.

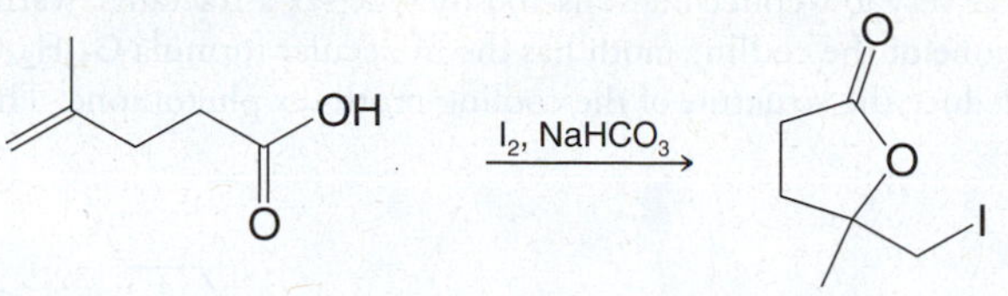

8.46 Write a mechanism for the following reaction.

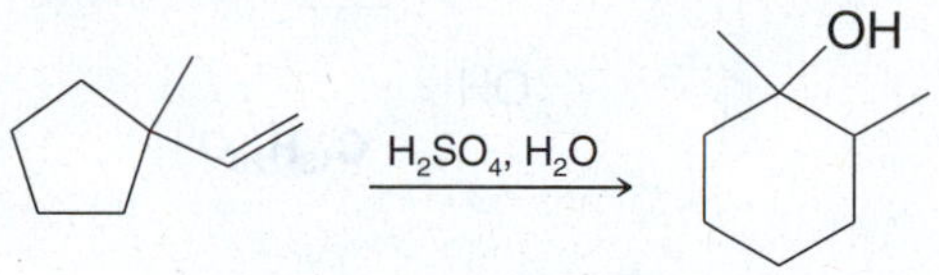

8.47 Write a mechanism that explains formation of the products shown in the following reaction.

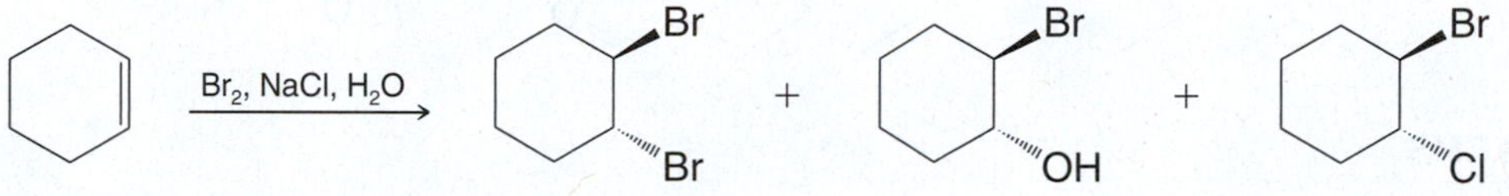

STRUCTURE ELUCIDATION

8.48 Myrcene, a fragrant compound found in bayberry wax, has the formula $C_{10}H_{16}$ and is known not to contain any triple bonds.

(a) What is the index of hydrogen deficiency of myrcene? When treated with excess hydrogen and a platinum catalyst, myrcene is converted to a compound **(A)** with the formula $C_{10}H_{22}$.

(b) How many rings does myrcene contain?

(c) How many double bonds? Compound **A** can be identified as 2,6-dimethyloctane. Ozonolysis of myrcene followed by treatment with dimethyl sulfide yields 2 mol of formaldehyde (HCHO), 1 mol of acetone (CH_3COCH_3), and a third compound **(B)** with the formula $C_5H_6O_3$.

(d) What is the structure of compound **B**?

(e) What is the structure of myrcene?

8.49 Farnesene (below) is a compound found in the waxy coating of apples. **(a)** Give the structure and IUPAC name of the product formed when farnesene is allowed to react with excess hydrogen in the presence of a platinum catalyst. **(b)** How many stereoisomers of the product are possible?

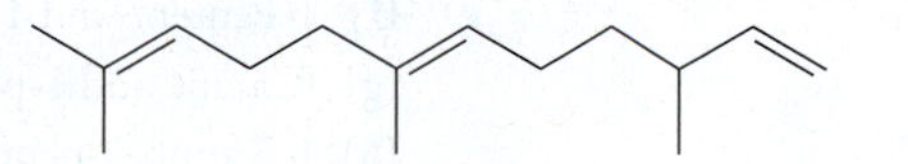

Farnesene

8.50 Write structural formulas for the products that would be formed when geranial, a component of lemongrass oil, is treated with ozone and then with dimethyl sulfide (Me_2S).

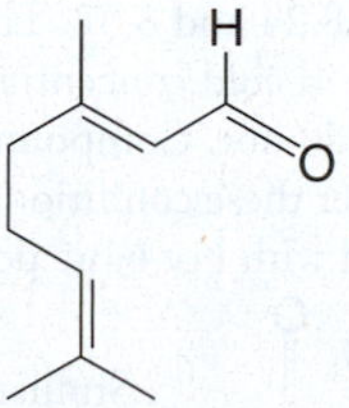

Geranial

8.51 Limonene is a compound found in orange oil and lemon oil. When limonene is treated with excess hydrogen and a platinum catalyst, the product of the reaction is 1-isopropyl-4-methylcyclohexane. When limonene is treated with ozone and then with dimethyl sulfide (Me_2S), the products of the reaction are formaldehyde (HCHO) and the following compound. Write a structural formula for limonene.

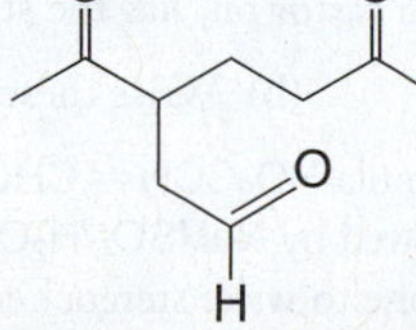

8.52 Pheromones (Section 4.7) are substances secreted by animals that produce a specific behavioral response in other members of the same species. Pheromones are effective at very low concentrations and include sex attractants, warning substances, and "aggregation" compounds. The sex attractant pheromone of the codling moth has the molecular formula $C_{13}H_{24}O$. Using information you can glean from the following reaction diagram, deduce the structure of the codling moth sex pheromone. The double bonds are known (on the basis of other evidence) to be (2Z,6E).

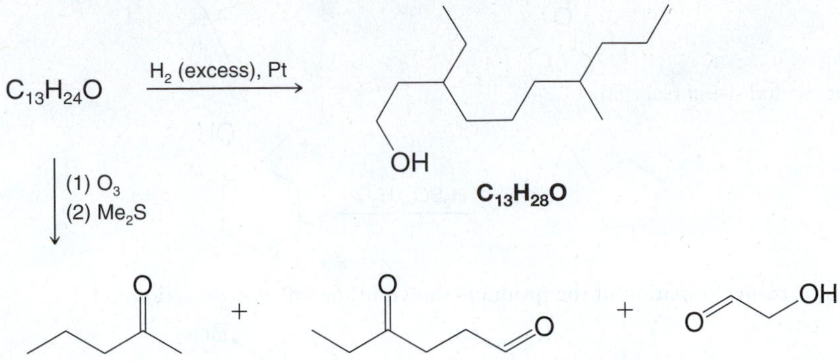

GENERAL PROBLEMS

8.53 Synthesize the following compound starting with ethyne and 1-bromopentane as your only organic reagents (except for solvents) and using any needed inorganic compounds.

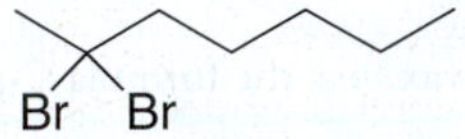

8.54 Shown below is the final step in a synthesis of an important perfume constituent, *cis*-jasmone. Which reagents would you choose to carry out this last step?

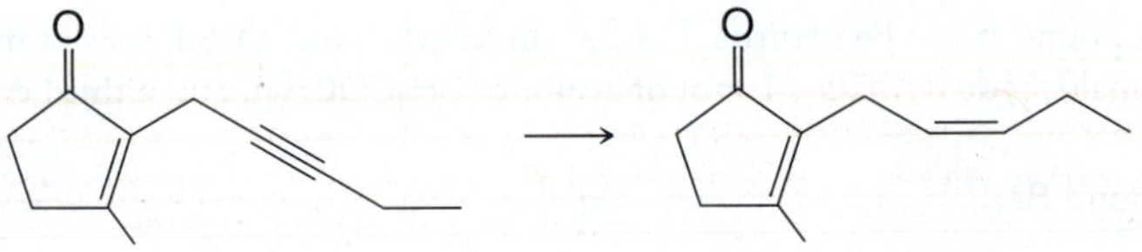

cis-Jasmone

8.55 Predict features of their IR spectra that you could use to distinguish between the members of the following pairs of compounds. You may find the IR chart in the endpapers of the book and Table 2.1 useful.

(a) Pentane and 1-pentyne

(b) Pentane and 1-pentene

(c) 1-Pentene and 1-pentyne

(d) Pentane and 1-bromopentane

(e) 2-Pentyne and 1-pentyne

(f) 1-Pentene and 1-pentanol

(g) Pentane and 1-pentanol

(h) 1-Bromo-2-pentene and 1-bromopentane

(i) 1-Pentanol and 2-penten-1-ol

8.56 Deduce the structures of compounds **A**, **B**, and **C**, which all have the formula C_6H_{10}. As you read the information that follows, draw reaction flowcharts (roadmaps) like those in Problems 8.24 and 8.52. This approach will help you solve the problem. All three compounds rapidly decolorize bromine; all three are soluble in cold concentrated sulfuric acid. Compound **A** has an absorption in its IR spectrum at about 3300 cm^{-1}, but compounds **B** and **C** do not. Compounds **A** and **B** both yield hexane when they are treated with excess hydrogen in the presence of a platinum catalyst. Under these conditions **C** absorbs only one molar equivalent of hydrogen and gives a product with the formula C_6H_{12}. When **A** is oxidized with hot basic potassium permanganate and the resulting solution acidified,

the only organic product that can be isolated is 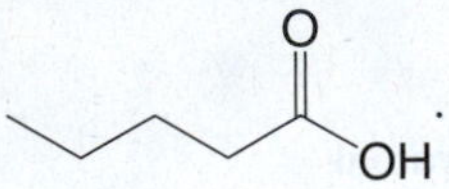 . Similar oxidation of **B** gives only 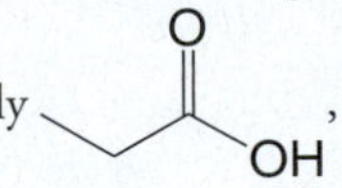 , and similar

treatment of **C** gives only

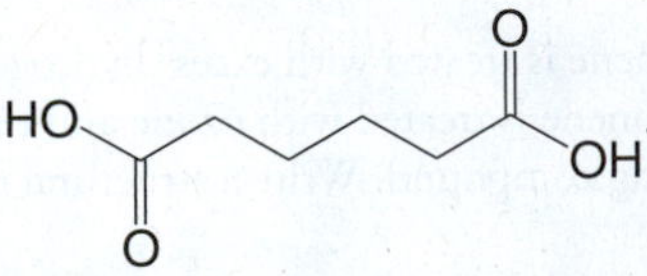

8.57 Ricinoleic acid, a compound that can be isolated from castor oil, has the structure $CH_3(CH_2)_5CHOHCH_2CH{=}CH(CH_2)_7CO_2H$.

(a) How many stereoisomers of this structure are possible?   (b) Write these structures.

8.58 There are two dicarboxylic acids with the general formula $HO_2CCH{=}CHCO_2H$. One dicarboxylic acid is called maleic acid; the other is called fumaric acid. When treated with OsO_4, followed by $NaHSO_3/H_2O$, maleic acid yields *meso*-tartaric acid and fumaric acid yields ($\pm$)-tartaric acid. Show how this information allows one to write stereochemical formulas for maleic acid and fumaric acid.

8.59 Use your answers to the preceding problem to predict the stereochemical outcome of the addition of bromine to maleic acid and to fumaric acid. **(a)** Which dicarboxylic acid would add bromine to yield a meso compound? **(b)** Which would yield a racemic form?

8.60 Alkyl halides add to alkenes in the presence of AlCl₃; yields are the highest when tertiary halides are used. Predict the outcome of the reaction of *tert*-pentyl chloride (1-chloro-2,2-dimethylpropane) with propene and specify the mechanistic steps.

8.61 Explain the stereochemical results observed in this catalytic hydrogenation. (You may find it helpful to build hand-held molecular models.)

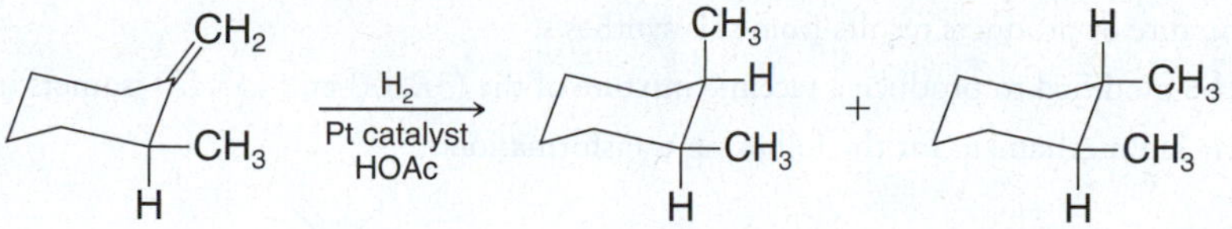

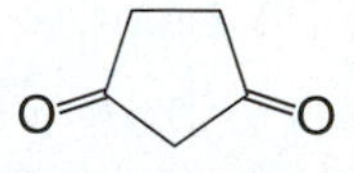

1,3-Cyclopentanedione

8.62 Make a reaction flowchart (roadmap diagram), as in previous problems, to organize the information provided to solve this problem. An optically active compound **A** (assume that it is dextrorotatory) has the molecular formula C₇H₁₁Br. **A** reacts with hydrogen bromide, in the absence of peroxides, to yield isomeric products, **B** and **C**, with the molecular formula C₇H₁₂Br₂. Compound **B** is optically active; **C** is not. Treating **B** with 1 mol of potassium *tert*-butoxide yields (+)-**A**. Treating **C** with 1 mol of potassium *tert*-butoxide yields (±)-**A**. Treating **A** with potassium *tert*-butoxide yields **D** (C₇H₁₀). Subjecting 1 mol of **D** to ozonolysis followed by treatment with dimethyl sulfide (Me₂S) yields 2 mol of formaldehyde and 1 mol of 1,3-cyclopentanedione. Propose stereochemical formulas for **A**, **B**, **C**, and **D** and outline the reactions involved in these transformations.

8.63 A naturally occurring antibiotic called mycomycin has the structure shown here. Mycomycin is optically active. Explain this by writing structures for the enantiomeric forms of mycomycin.

HC≡C—C≡C—CH=C=CH—(CH=CH)₂CH₂CO₂H

Mycomycin

8.64 An optically active compound **D** has the molecular formula C₆H₁₀ and shows a peak at about 3300 cm⁻¹ in its IR spectrum. On catalytic hydrogenation **D** yields **E** (C₆H₁₄). Compound **E** is optically inactive and cannot be resolved. Propose structures for **D** and **E**.

8.65 (a) Based on the following information, draw three-dimensional formulas for **A**, **B**, and **C**.

Reaction of cyclopentene with bromine in water gives **A**.

Reaction of **A** with aqueous NaOH (1 equivalent, cold) gives **B**, C₅H₈O (no 3590–3650-cm⁻¹ infrared absorption). (See the squalene cyclization discussion in "The Chemistry of... Cholesterol Biosynthesis" in *WileyPLUS* for a hint.)

Heating of **B** in methanol containing a catalytic amount of strong acid gives **C**, C₆H₁₂O₂, which does show 3590–3650-cm⁻¹ infrared absorption.

(b) Specify the (R) or (S) configuration of the chirality centers in your predicted structures for **C**. Would **C** be formed as a single stereoisomer or as a racemate?

(c) How could you experimentally confirm your predictions about the stereochemistry of **C**?

CHALLENGE PROBLEMS

8.66 Propose a mechanism that explains the following transformation. (Note its similarity to the cyclization of squalene oxide to lanosterol, as shown in "The Chemistry of... Cholesterol Biosynthesis." in *WileyPLUS*.)

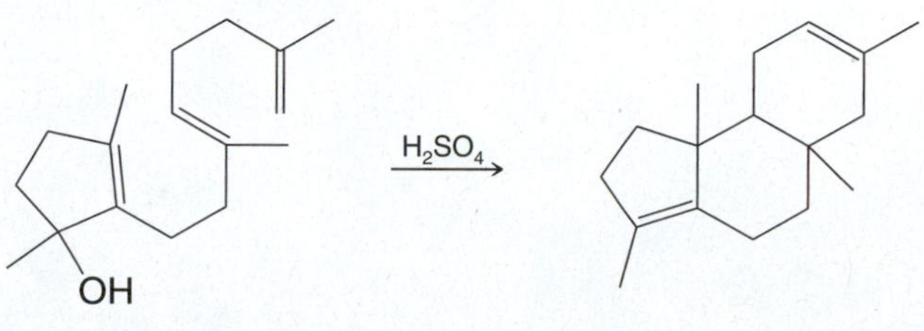

8.67 Triethylamine, (C₂H₅)₃N, like all amines, has a nitrogen atom with an unshared pair of electrons. Dichlorocarbene also has an unshared pair of electrons. Both can be represented as shown below. Draw the structures of compounds **D**, **E**, and **F**.

(C₂H₅)₃N: + :CCl₂ ⟶ **D**   (an unstable adduct)

D ⟶ **E** + C₂H₄   (by an intramolecular E2 reaction)

E →[H₂O] **F**   (Water effects a replacement that is the reverse of that used to make *gem*-dichlorides.)

LEARNING GROUP PROBLEMS

1. (a) Synthesize (3S,4R)-3,4-dibromo-1-cyclohexylpentane (and its enantiomer, since a racemic mixture will be formed) from ethyne, 1-chloro-2-cyclohexylethane, bromomethane, and any other reagents necessary. (Use ethyne, 1-chloro-2-cyclohexylethane, and bromomethane as the sole sources of carbon atoms.) Start the problem by showing a retrosynthetic analysis. In the process, decide which atoms of the target molecule will come from which atoms of the starting reagents. Also, bear in mind how the stereospecificity of the reactions you employ can be used to achieve the required stereochemical form of the final product.

(b) Explain why a racemic mixture of products results from this synthesis.

(c) How could the synthesis be modified to produce a racemic mixture of the (3R,4R) and (3S,4S) isomers instead?

2. Write a reasonable and detailed mechanism for the following transformation:

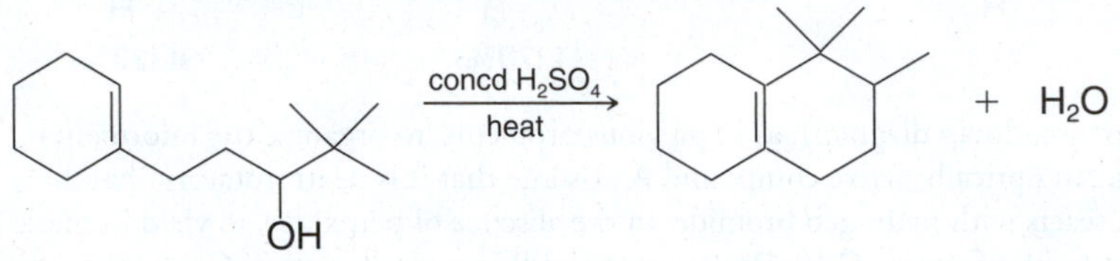

3. Deduce the structures of compounds **A–D**. Draw structures that show stereochemistry where appropriate:

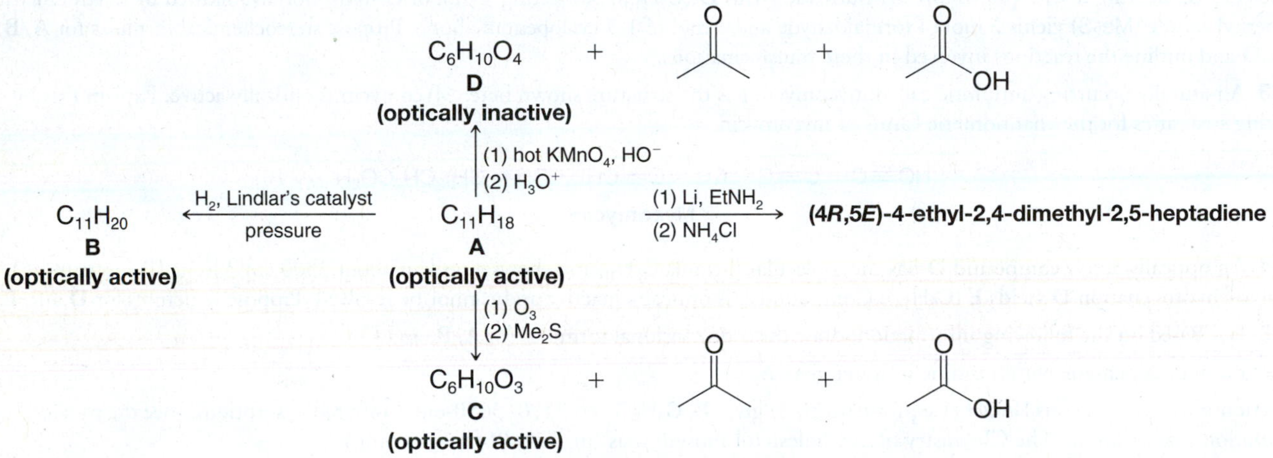

4. The Guam bubble snail (*Haminoea cymbalum*) contains kumepaloxane (shown below), a chemical signal agent discharged when this mollusk is disturbed by predatory carnivorous fish. The biosynthesis of bromoethers like kumepaloxane is thought to occur via the enzymatic intermediacy of a "Br⁺" agent. Draw the structure of a possible biosynthetic precursor (*hint*: an alkene alcohol) to kumepaloxane and write a plausible and detailed mechanism by which it could be converted to kumepaloxane using Br^+ and some generic proton acceptor Y^-.

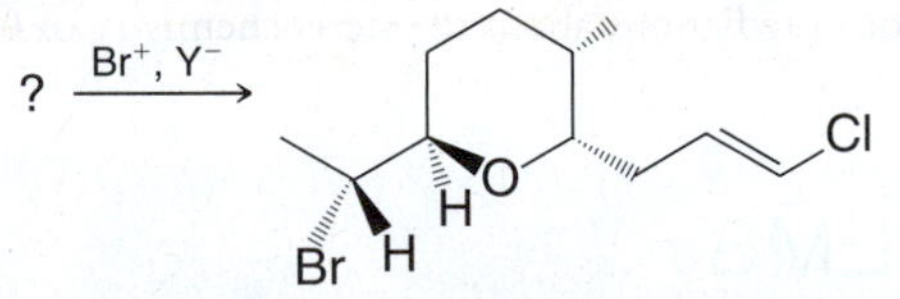

Kumepaloxane

[SUMMARY AND REVIEW TOOLS]

Summary of Alkene Addition Reactions

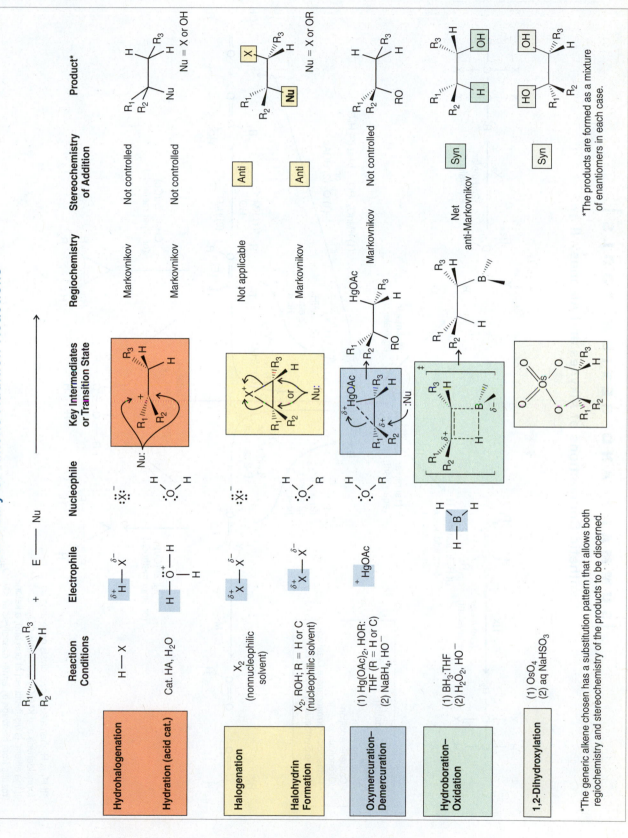

*The products are formed as a mixture of enantiomers in each case.

*The generic alkene chosen has a substitution pattern that allows both regiochemistry and stereochemistry of the products to be discerned.

SUMMARY AND REVIEW TOOLS

Synthetic Connections of Alkynes and Alkenes: II

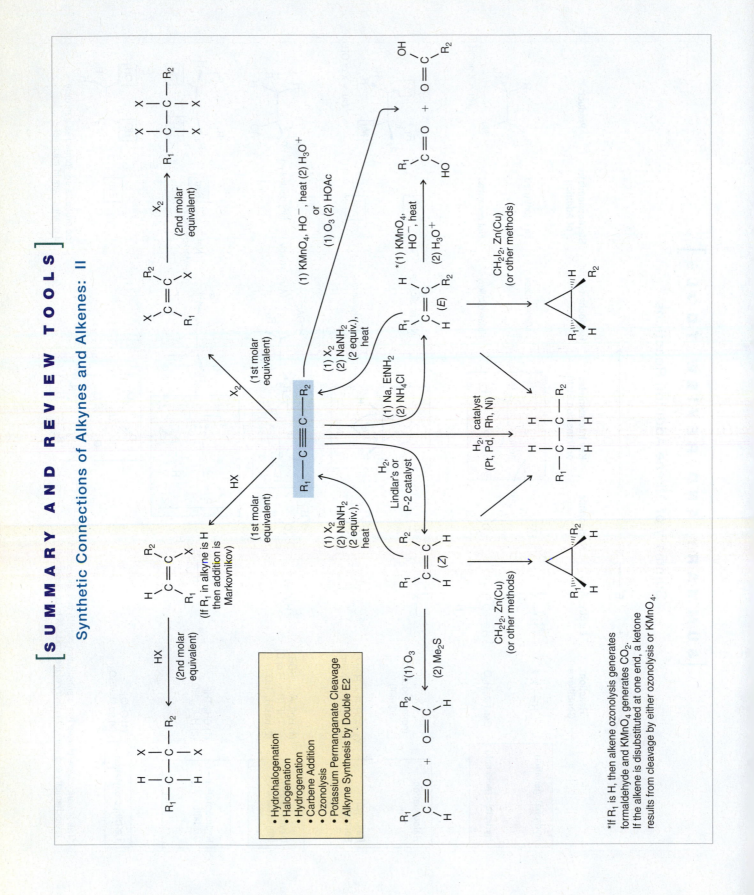

- Hydrohalogenation
- Halogenation
- Hydrogenation
- Carbene Addition
- Ozonolysis
- Potassium Permanganate Cleavage
- Alkyne Synthesis by Double E2

*If R_1 is H, then alkene ozonolysis generates formaldehyde and $KMnO_4$ generates CO_2. If the alkene is disubstituted at one end, a ketone results from cleavage by either ozonolysis or $KMnO_4$.

CHAPTER
9

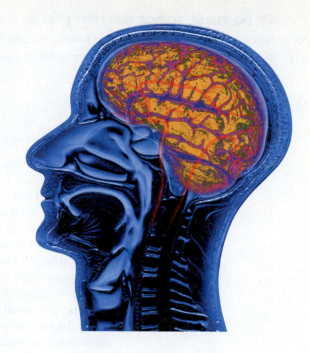

Nuclear Magnetic Resonance and Mass Spectrometry
TOOLS FOR STRUCTURE DETERMINATION

Have you known someone who needed an MRI (magnetic resonance imaging) scan for a medical condition, or have you needed one yourself? Have you ever observed someone in an airport security line having their belongings wiped down with a pad that was then placed in some kind of analytical instrument? Have you wondered how scientists determine the structures of compounds found in nature, or have you known a fellow student in a laboratory class who extracted bark, leaves, or fruit to isolate and identify some natural compounds? Or have you wondered how forensic evidence is analyzed in criminal cases, or how pesticides are identified in food samples?

If you have wondered about any of these things, then some of your curiosity will be satisfied by learning about spectroscopic methods such as nuclear magnetic resonance (NMR) spectrometry, which involves the same physical principles as MRI imaging, and MS (mass spectrometry), which is used in some airport screening processes as well as many forensic applications. NMR and MS are workhorse techniques for the study of both biological and nonbiological molecular structure.

IN THIS CHAPTER WE WILL CONSIDER:

- nuclear magnetic resonance (NMR), a form of spectroscopy that is one of the most powerful tools for the identification of functional groups and for the determination of connections between the atoms in molecules

- mass spectroscopy (MS), which allows the determination of exact molecular formulas of molecules both large and small

[**WHY** DO THESE TOPICS MATTER?] At the end of the chapter, we will show just how critical these techniques are for determining the structure of organic molecules. Indeed, before spectroscopy, structure determination could take years or even decades, sometimes providing challenges that stymied future chemistry Nobel laureates!

9.1 INTRODUCTION

- **Spectroscopy** is the study of the interaction of energy with matter.

When energy is applied to matter, it can be absorbed, emitted, cause a chemical change, or be transmitted. In this chapter we shall see how detailed information about molecular structure can be obtained by interpreting results from the interaction of energy with molecules. In our study of nuclear magnetic resonance (NMR) spectroscopy we shall focus our attention on energy absorption by molecules that have been placed in a strong magnetic field. When we study mass spectrometry (MS), we shall learn how molecular structure can be probed by bombarding molecules with a beam of high-energy electrons. These two techniques (NMR and MS) are a powerful combination for elucidating the structures of organic molecules. Together with infrared (IR) spectroscopy (Section 2.15), these methods comprise the typical array of spectroscopic tools used by organic chemists. Later, we shall briefly discuss how gas chromatography (GC) is linked with mass spectrometry in GC/MS instruments to obtain mass spectrometric data from individual components of a mixture.

We begin our study with a discussion of nuclear magnetic resonance spectroscopy.

9.2 NUCLEAR MAGNETIC RESONANCE (NMR) SPECTROSCOPY

The 1952 Nobel Prize in Physics was awarded to Felix Bloch (Stanford) and Edward M. Purcell (Harvard) for their discoveries relating to nuclear magnetic resonance.

The nuclei of certain elements, including ^1H nuclei (protons) and ^{13}C (carbon-13) nuclei, behave as though they were magnets spinning about an axis. When a compound containing protons or carbon-13 nuclei is placed in a very strong magnetic field and simultaneously irradiated with electromagnetic energy of the appropriate frequency, nuclei of the compound absorb energy through a process called magnetic resonance. The absorption of energy is quantized.

- A graph that shows the characteristic energy absorption frequencies and intensities for a sample in a magnetic field is called a **nuclear magnetic resonance (NMR) spectrum**.

As a typical example, the proton (^1H) NMR spectrum of bromoethane is shown in Fig. 9.1.

We can use NMR spectra to provide valuable information about the structure of any molecule we might be studying. In the following sections we shall explain how four features of a molecule's proton NMR spectrum can help us arrive at its structure.

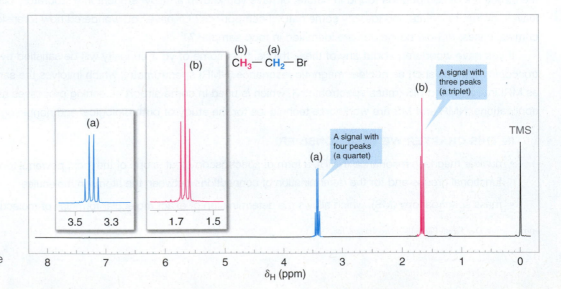

FIGURE 9.1 The 300-MHz ^1H NMR spectrum of bromoethane (ethyl bromide). Zoomed-in expansions of the signals are shown in the offset plots.

1. **The number of signals in the spectrum** tells us **how many different sets of protons there are in the molecule**. In the spectrum for bromoethane (Fig. 9.1) there are *two signals arising from two different sets of protons*. One signal (consisting of four peaks) is shown in blue and labeled (a). The other signal (consisting of three peaks) is in red and is labeled (b). These signals are shown twice in the figure, at a smaller scale on the baseline spectrum, and expanded and moved to the left above the base spectrum. [Don't worry now about the signal at the far right of the spectrum (labeled TMS); it comes from a compound (tetramethylsilane) that was added to the bromoethane so as to calibrate the positions of the other signals.]

2. **The position of the signals in the spectrum along the *x*-axis** tells us about the magnetic environment of each set of protons arising largely from the electron density in their environment. We'll learn more about this in Section 9.2A.

3. **The area under the signal** tells us about **how many protons there are in the set being measured**. We'll learn how this is done in Section 9.2B.

4. **The multiplicity (or splitting pattern) of each signal** tells us about **the number of protons on atoms adjacent to the one whose signal is being measured**. In bromo-ethane, signal (a) is split into a **quartet** of peaks by the three protons of set (b), and signal (b) is split into a *triplet* of peaks by the two protons of set (a). We'll explain splitting patterns in Section 9.2C.

9.2A Chemical Shift

- The position of a signal along the *x*-axis of an NMR spectrum is called its **chemical shift**.
- **The chemical shift of each signal gives information about the structural environment of the nuclei producing that signal.**
- Counting the number of signals in a ^{1}H NMR spectrum indicates, at a first approximation, the number of distinct proton environments in a molecule.

Tables and charts have been developed that allow us to correlate chemical shifts of NMR signals with likely structural environments for the nuclei producing the signals. Table 9.1

TABLE 9.1 APPROXIMATE PROTON CHEMICAL SHIFTS

Type of Proton	Chemical Shift (δ, ppm)	Type of Proton	Chemical Shift (δ, ppm)
1° Alkyl, RCH$_3$	0.8–1.2	Alkyl bromide, RCH$_2$Br	3.4–3.6
2° Alkyl, RCH$_2$R	1.2–1.5	Alkyl chloride, RCH$_2$Cl	3.6–3.8
3° Alkyl, R$_3$CH	1.4–1.8	Vinylic, R$_2$C=CH$_2$	4.6–5.0
Allylic, R$_2$C=C—CH$_3$ (R)	1.6–1.9	Vinylic, R$_2$C=CH (R)	5.2–5.7
Ketone, RCCH$_3$ (O)	2.1–2.6	Aromatic, ArH	6.0–8.5
Benzylic, ArCH$_3$	2.2–2.5	Aldehyde, RCH (O)	9.5–10.5
Acetylenic, RC≡CH	2.5–3.1	Alcohol hydroxyl, ROH	0.5–6.0[a]
Alkyl iodide, RCH$_2$I	3.1–3.3	Amino, R—NH$_2$	1.0–5.0[a]
Ether, ROCH$_2$R	3.3–3.9	Phenolic, ArOH	4.5–7.7[a]
Alcohol, HOCH$_2$R	3.3–4.0	Carboxylic, RCOH (O)	10–13[a]

[a]The chemical shifts of these protons vary in different solvents and with temperature and concentration.

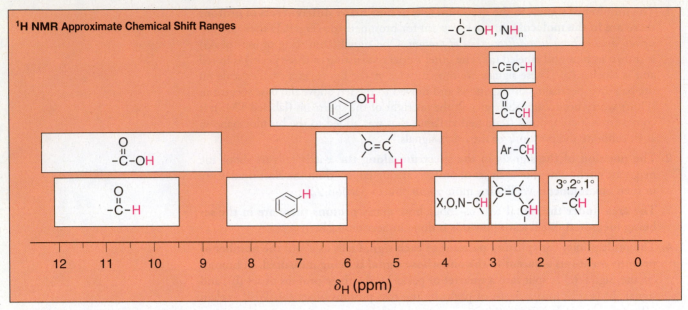

FIGURE 9.2 Approximate proton chemical shifts.

and Fig. 9.2, for example, are useful for this purpose. ^{1}H NMR chemical shifts generally fall in the range of 13–0 ppm (δ).

The chemical shift of a signal in an NMR spectrum depends on the local magnetic environment of the nucleus producing the signal. The local magnetic environment of a nucleus is influenced by electron density and other factors we shall discuss shortly. The physical meaning of chemical shift values relates to the actual frequency of the NMR signals produced by the nuclei. The *practical* importance of chemical shift information is that it gives important clues about molecular structure. Each NMR signal indicates the presence of nuclei in a different magnetic environment.

Chemical shifts are measured along the spectrum axis using a delta (δ) scale, in units of parts per million (ppm). When comparing one signal with another:

- A signal with a chemical shift further to the left in the spectrum than another signal (i.e., at a larger δ or ppm value) has a higher frequency.

- A signal to the right of another signal has a lower frequency.

●●● SOLVED PROBLEM 9.1

Consider the spectrum of bromoethane (Fig. 9.1). What is the chemical shift of the signal in blue for the CH$_2$ group?

STRATEGY AND ANSWER: The signal for the CH$_2$ group of bromoethane appears as a symmetrical pattern of four peaks. For a signal with multiple peaks, such as a quartet, the chemical shift is reported as the midpoint of the peaks in the signal. Estimating as well as possible from the zoomed-in offset expansion in Fig. 9.1, the chemical shift of the bromoethane quartet is 3.4 ppm.

The ^{1}H NMR spectrum of 1,4-dimethylbenzene (*p*-xylene), shown in Fig. 9.3, is a simple example that we can use to learn how to interpret chemical shifts. First, note that there is a signal at δ 0. The signal at δ 0 is *not* from 1,4-dimethylbenzene, but from tetramethylsilane (TMS), a compound that is sometimes added to samples as an internal standard to calibrate the chemical shift scale. If the signal from TMS appears at zero ppm, the chemical shift axis is calibrated correctly.

Next we observe that there are only two signals in the ^{1}H NMR spectrum of 1,4-dimethylbenzene other than for TMS, at approximately δ 7.0 and δ 2.3. The existence of just two signals implies that there are only two distinct proton environments in 1,4-dimethylbenzene, a fact we can easily verify for ourselves by examining its structure.

We say, then, that there are "two types" of hydrogen atoms in 1,4-dimethylbenzene, and these are the hydrogen atoms of the methyl groups and the hydrogen atoms of the

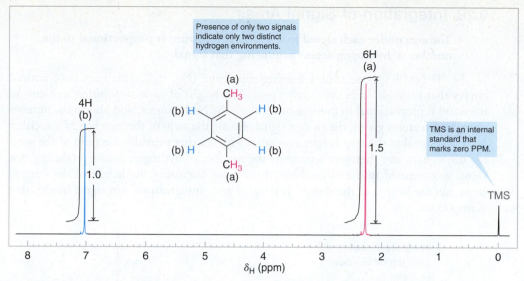

FIGURE 9.3 The 300-MHz ^1H NMR spectrum of 1,4-dimethylbenzene.

benzene ring. The two methyl groups produce only one signal because they are equivalent by virtue of the plane of symmetry between them. Furthermore, the three hydrogen atoms of each methyl group are equivalent due to free rotation about the bond between the methyl carbon and the ring. The benzene ring hydrogen atoms also produce only one signal because they are equivalent to each other by symmetry.

Referring to Table 9.1 or Fig. 9.2, we can see that ^1H NMR signals for hydrogen atoms bonded to a benzene ring typically occur between δ 6 and 8.5, and that signals for hydrogen atoms on an sp^3 carbon bonded to a benzene ring (benzylic hydrogens) typically occur between δ 2 and 3. Thus, chemical shifts for the signals from 1,4-dimethylbenzene occur where we would expect them to according to NMR spectral correlation charts.

In the case of this example, the structure of the compound under consideration was known from the outset. Had we not known its structure in advance, however, we would have used chemical shift correlation tables to infer likely structural environments for the hydrogen atoms. We would also have considered the relative area of the signals and signal multiplicity, factors we shall discuss in the following sections.

●●● **SOLVED PROBLEM 9.2**

Based on the information in Table 9.1, in what ppm range would you expect to find the protons of **(a)** acetone (CH_3COCH_3) and **(b)** ethanol?

STRATEGY AND ANSWER: We use a chemical shift correlation table, such as Table 9.1, to find the closest match between the compound of interest and the partial structures shown in the table.

(a) Acetone is a ketone bearing hydrogen atoms on the carbons adjacent to its carbonyl group. Ketones are listed in Table 9.1 as a representative substructure whose protons have a chemical shift range of 2.1–2.6 ppm. Thus, we expect the proton NMR signal from acetone to appear in the 2.1–2.6 ppm range. There will be one signal for all of the hydrogen atoms in acetone because, due to free rotation, they can occupy equivalent magnetic environments at any given instant.

(b) Ethanol is expected to exhibit three proton NMR signals, one for each of its three distinct hydrogen environments. Ethanol contains an alcohol hydroxyl proton, which Table 9.1 lists in the range of 0.5–6.0 ppm; two protons on the carbon bearing the hydroxyl group, which according to Table 9.1 we expect in the 3.3–4.0 ppm range; and a methyl group bonded to no functional groups, which, as a 1° alkyl group, should appear in the 0.8–1.2 ppm range.

●●● **PRACTICE PROBLEM 9.1**

In what chemical shift ranges would you expect to find the proton NMR signals of ethyl acetate ($CH_3CO_2CH_2CH_3$)?

9.2B Integration of Signal Areas

- **The area under each signal in a ¹H NMR spectrum is proportional to the number of hydrogen atoms producing that signal.**

In the ¹H NMR spectrum of 1,4-dimethylbenzene (Fig. 9.3), you may have noticed curves that resemble steps over each signal. The height of each step (using any unit of measure) is proportional to the area underneath the NMR signal, and also to the number of hydrogen atoms giving rise to that signal. Taking the ratio of the step height associated with one signal to the step height associated with another provides the ratio of the areas for the signals, and therefore represents the number of hydrogen atoms producing one signal as compared to the other. Note that we are discussing the height of the integral steps, not the heights of the signals. It is signal area (**integration**), not signal height, that is important.

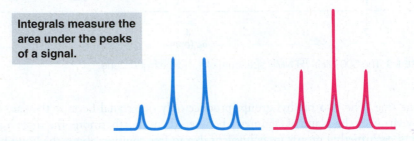

Integrals measure the area under the peaks of a signal.

The area under each signal (shown with blue shading above) is measured (integrated) and taken as a ratio against the area of other signals to compare the relative numbers of hydrogen atoms producing each signal in an NMR spectrum.

In Fig. 9.3 we have indicated the relative integral step heights as 1.0 and 1.5 (in dimensionless units). Had these values not been given, we would have measured the step heights with a ruler and taken their ratio. Since the actual number of hydrogen atoms giving rise to the signals is not likely to be 1 and 1.5 (we cannot have a fraction of an atom), we can surmise that the true number of hydrogens producing the signals is probably 2 and 3, or 4 and 6, etc. For 1,4-dimethylbenzene the actual values are, of course, 4 and 6.

Whether NMR data are provided as in Fig. 9.3 with an integral step over each signal, or simply with numbers that represent each signal's relative area, the process of interpreting the data is the same because the area of each signal is proportional to the number of hydrogen atoms producing that signal. It is important to note that in ¹³C NMR spectroscopy signal area is not relevant in routine analyses.

●●● SOLVED PROBLEM 9.3

What integral values (as whole number ratios) would you expect for signals in the proton NMR spectrum of 3-methyl-2-butanone?

STRATEGY AND ANSWER: There are three distinct proton environments in 3-methyl-2-butanone: the methyl at C1, the methine hydrogen at C3, and the two methyl groups bonded to C3, which are equivalent. The ratio of these signals, in the order just listed, would be 3 : 1 : 6.

9.2C Coupling (Signal Splitting)

Coupling, also referred to as **signal splitting** or signal multiplicity, is a third feature of ¹H NMR spectra that provides very useful information about the structure of a compound.

- Coupling is caused by the magnetic effect of nonequivalent hydrogen atoms that are within 2 or 3 bonds of the hydrogens producing the signal.

The effect of the nearby hydrogens is to split (or couple with) the energy levels of the hydrogens whose signal is being observed, and the result is a signal with multiple peaks.

(Notice that we have been careful to differentiate use of the words signal and peak. A group of equivalent hydrogen atoms produces one *signal* that may be split into multiple *peaks*.) We shall explain the physical origin of coupling further in Section 9.9; however, **the importance of coupling is that it is predictable, and it gives us specific information about the constitution of the molecule under study**.

The typical coupling we observe is from nonequivalent, **vicinal** hydrogens, that is, from hydrogens on adjacent carbons, separated by three bonds from the hydrogens producing the signal. Coupling can also occur between nonequivalent **geminal** hydrogens (hydrogens bonded to the same carbon) if the geminal hydrogens are in a chiral or conformationally restricted molecule. (We shall discuss cases of chiral and conformationally restricted molecules in Section 9.8.)

- A simple rule exists for predicting the number of peaks expected from coupling in ^{1}H NMR:

Number of peaks $= n + 1$ in a ^{1}H NMR signal	Where n is the number of vicinal and geminal hydrogen atoms that are nonequivalent to the hydrogens producing the signal

This rule is applicable in general to achiral molecules without conformational barriers.

The ^{1}H NMR spectrum of 1,4-dimethylbenzene (Fig. 9.3) is an example where $n = 0$ (in the above equation) regarding the hydrogen atoms producing the signals at δ 7.0 and at δ 2.3. There are no hydrogen atoms on the carbons adjacent to the methyl groups; hence $n = 0$ for the signal at δ 2.3, and the signal is a singlet (signals with only one peak are called **singlets**). And, since all of the hydrogen atoms on the ring are equivalent by symmetry and there are no adjacent nonequivalent hydrogen atoms, $n = 0$ for the hydrogens on the ring producing the signal at δ 7.0, and hence this signal is a singlet as well.

The ^{1}H NMR spectrum of 1,1,2-trichloroethane, shown in Fig. 9.4, provides an example where n is not equal to zero, and coupling is therefore evident. In the spectrum of 1,1,2-trichloroethane we see two signals: one with three peaks and one with two peaks. These signals are called, respectively, a **triplet** and a **doublet**. The signal for the —CHCl$_2$ hydrogen is a triplet because there are two hydrogen atoms on the adjacent carbon ($n = 2$). The signal for the —CH$_2$Cl hydrogens is a doublet because there is one hydrogen on the adjacent carbon ($n = 1$). We shall consider why this is so in Section 9.9.

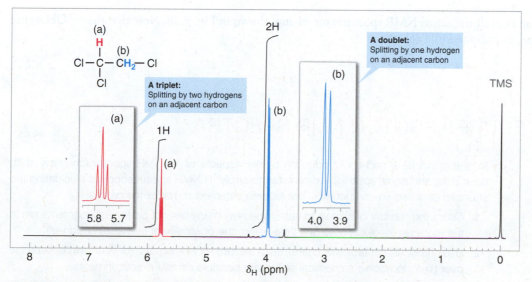

FIGURE 9.4 The 300-MHz ^{1}H NMR spectrum of 1,1,2-trichloroethane. Zoomed-in expansions of the signals are shown in the offset plots.

● ● ● **SOLVED PROBLEM 9.4**

Sketch a predicted proton NMR spectrum for ethanol, showing signals in the expected chemical shift ranges (based on Table 9.1) and with the appropriate number of peaks in each. (Note one important fact: hydrogen atoms bonded to oxygen and nitrogen do not usually show coupling, but often exhibit a single broad peak instead. We shall explain why later in Section 9.10.)

STRATEGY AND ANSWER: There are four things to pay attention to: (1) the number of signals, (2) the chemical shifts of the signals, (3) the coupling patterns (signal splitting) in the signals, and (4) the relative signal areas. We have already predicted the first two of these in Solved Problem 9.2, part b.

1. In ethanol there are protons in three distinct environments; thus, we expect three signals.

2. The predicted chemical shifts are 3.3–4.0 ppm for the two protons on the alcohol carbon, 0.8–1.2 ppm for the three methyl protons, and 0.5–6.0 ppm for the hydroxyl proton (showing it anywhere in this broad range is acceptable—we shall explain why the range is broad in Section 9.10).

3. Regarding coupling patterns, the alcohol hydrogen does not couple, as we stated earlier. The alcohol —CH_2— group has three vicinal protons (the methyl group); following the $n + 1$ rule these should appear as a quartet. The methyl group has two vicinal protons (the alcohol —CH_2— group), thus it should be a triplet.

4. The relative signal areas are 1 : 2 : 3, according to the number of protons producing each signal, which we indicate as 1H, 2H, and 3H in our sketch.

Last, it is helpful to use letters to assign the protons in a formula to signals associated with them in a spectrum, and we shall do that here.

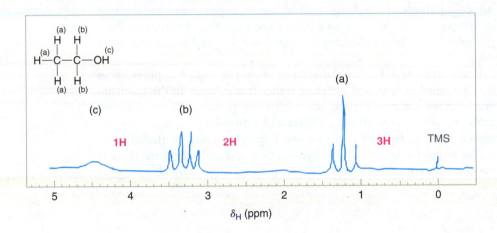

To verify our sketch we can consult the actual NMR spectrum for ethanol shown in Fig. 9.28. Note that the —OH signal can appear in a wide range, as indicated in Table 9.1.

9.3 HOW TO INTERPRET PROTON NMR SPECTRA

Now that we have had an introduction to key aspects of ¹H NMR spectra (chemical shift, peak area, and signal splitting), we can start to apply ¹H NMR spectroscopy to elucidating the structure of unknown compounds. The following steps summarize the process:

1. Count the number of signals to determine how many distinct proton environments are in the molecule (neglecting, for the time being, the possibility of overlapping signals).

2. Use chemical shift tables or charts, such as Table 9.1 or Fig. 9.2 (or your own experience over time), to correlate chemical shifts with possible structural environments.

3. Determine the relative area of each signal, as compared with the area of other signals, as an indication of the relative number of protons producing the signal.

4. Interpret the splitting pattern for each signal to determine how many hydrogen atoms are present on carbon atoms adjacent to those producing the signal and sketch possible molecular fragments.

5. Join the fragments to make a molecule in a fashion that is consistent with the data.

As a beginning example, let's interpret the ^1H NMR spectrum in Fig. 9.5 for a compound with the molecular formula C_3H_7Br.

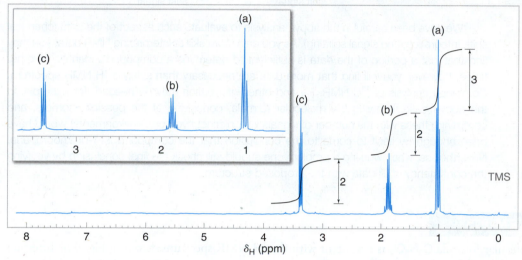

FIGURE 9.5 The 300-MHz ^1H NMR spectrum of a compound with the formula C_3H_7Br. Expansions of the signals are shown in the offset plots. (Spectra adapted from Sigma-Aldrich Co. © Sigma-Aldrich Co.)

1. First, we observe that there are three distinct signals, with chemical shifts of approximately δ 3.4, 1.8, and 1.1. One of these signals (δ 3.4) has a noticeably higher frequency chemical shift from the others, indicating hydrogen atoms that are likely to be near an electronegative group. This is not surprising given the presence of bromine in the formula. The presence of three distinct signals suggests that there are only three distinct proton environments in the molecule. For this example, this information alone makes it possible to reach a conclusion about the structure of the compound, since its molecular formula is as simple as C_3H_7Br. (Do you know what the compound is? Even if you do, you should still demonstrate that all of the information in the spectrum is consistent with the structure you propose.)

2. Next, we measure (or estimate) the step heights of the integral curves and reduce them to whole number ratios. Doing so, we find that the ratio is 2 : 2 : 3 (for the signals at δ 3.4, 1.8, and 1.1, respectively). Given a molecular formula that contains seven hydrogen atoms, we infer that these signals likely arise from two CH_2 groups and one CH_3 group, respectively. One of the CH_2 groups must bear the bromine. (Although you almost certainly know the structure of the compound at this point, let's continue with the analysis.) At this point we can begin to sketch molecular fragments, if we wish.

3. Next we evaluate the multiplicity of the signals. The signal at δ 3.4 is a triplet, indicating that there are two hydrogen atoms on the adjacent carbon. Since this signal is **downfield** (vs. **upfied**) and has an integral value that suggests two hydrogens, we conclude that this signal is from the CH_2Br group, and that it is next to a CH_2 group. The signal at δ 1.8 is a sextet, indicating five hydrogen atoms on adjacent carbons. The presence of five neighboring hydrogen atoms ($n = 5$, producing six peaks) is consistent with a CH_2 group on one side and a CH_3 group on the other. Last, the signal at δ 1.1 is a triplet, indicating two adjacent hydrogen atoms. Joining these molecular pieces together on paper or in our mind gives us $BrCH_2CH_2CH_3$ for the structural formula.

1-Bromopropane

triplet   sextet   triplet
($n = 2$) ($n = 5$) ($n = 2$)

Electro-
negative   Br←CH$_2$—CH$_2$—CH$_3$
atom   (c)   (b)   (a)

Highest   Lowest
chemical   chemical
shift signal   shift signal

We have been careful in the above analysis to evaluate each aspect of the data (chemical shift, integration, and signal splitting). As you gain more skill at interpreting NMR data, you may find that just a portion of the data is sufficient to determine a compound's identity. At other times, however, you will find that more data are necessary than solely a ^{1}H NMR spectrum. Combined analysis of ^{13}C NMR, IR, and other information may be needed, for example. In the above case, knowing the molecular formula, conceiving of the possible isomers, and comparing these with the number of signals (i.e., distinct hydrogen environments) would have been enough by itself to come to the conclusion that the compound is 1-bromopropane. Nevertheless, when working a problem one should still check the final conclusion by verifying the consistency of all data with the proposed structure.

●●● **SOLVED PROBLEM 9.5**

What compound with molecular formula $C_3H_6Cl_2$ is consistent with the ^{1}H NMR spectrum shown in Fig. 9.6? Interpret the data by assigning each aspect of the spectrum to the structure you propose.

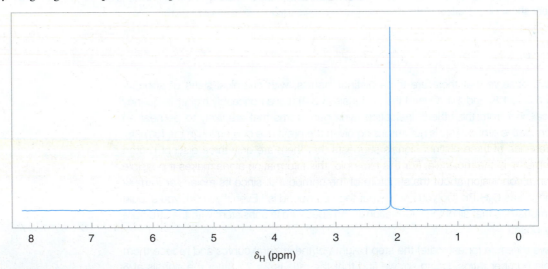

FIGURE 9.6 The 300-MHz ^{1}H NMR spectrum of the compound in Solved Problem 9.5 with molecular formula $C_3H_6Cl_2$. (Spectra adapted from Sigma-Aldrich © Sigma-Aldrich Co.)

STRATEGY AND ANSWER: The spectrum shown in Fig. 9.6 shows only one signal (therefore its integral is irrelevant and not shown). This must mean that the six hydrogen atoms in the formula $C_3H_6Cl_2$ all exist in the same magnetic environment. The presence of two equivalent methyl groups is a likely scenario for six equivalent hydrogen atoms. The only way to have two identical methyl groups with the formula $C_3H_6Cl_2$ is for both chlorine atoms to be bonded at C2 resulting in the structure shown to the right.

●●●

PRACTICE PROBLEM 9.2   What compound with molecular formula $C_3H_6Cl_2$ is consistent with the ^{1}H NMR spectrum shown in Fig. 9.7? Interpret the data by assigning each aspect of the spectrum to the structure you propose. (In other words, explain how the chemical shifts, signal areas, and splitting patterns support your conclusion.)

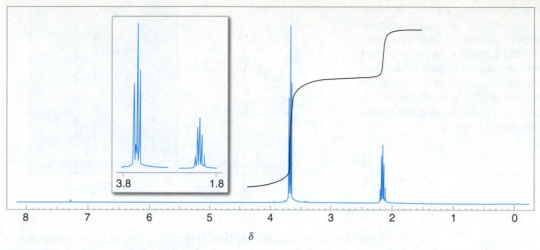

FIGURE 9.7 The 300-MHz ^{1}H NMR spectrum of the compound in Practice Problem 9.2 with formula $C_3H_6Cl_2$. Expansions of the signals are shown in the offset plots. (Spectra adapted from Sigma-Aldrich © Sigma-Aldrich Co.)

Now that we have had an introduction to interpreting NMR spectra, let us briefly explain the physical origin of NMR signals and how NMR spectrometers work, before returning to important information about factors that influence chemical shift and signal splitting.

9.4 NUCLEAR SPIN: THE ORIGIN OF THE SIGNAL

The nuclei of certain isotopes possess the quality of spin, and therefore these nuclei have spin quantum numbers, designated I. The nucleus of ordinary hydrogen, ^{1}H, has a spin quantum number of $\frac{1}{2}$, and it can assume either of two spin states: $+\frac{1}{2}$ or $-\frac{1}{2}$. These correspond to the magnetic moments (m) allowed for $I = \frac{1}{2}$, which are $m = +\frac{1}{2}$ or $m = -\frac{1}{2}$. Other nuclei with spin quantum numbers $I = \frac{1}{2}$ are ^{13}C, ^{19}F, and ^{31}P. Some nuclei, such as ^{12}C, ^{16}O, and ^{32}S, have no spin ($I = 0$), and these nuclei do not give an NMR spectrum. Other nuclei have spin quantum numbers greater than $\frac{1}{2}$. In our treatment here, however, we are concerned primarily with the spectra that arise from ^{1}H and from ^{13}C, both of which have $I = \frac{1}{2}$.

Since the proton is electrically charged, the spinning charge generates a tiny magnetic moment—one that coincides with the axis of spin (Fig. 9.8). This tiny magnetic moment gives the spinning proton properties analogous to those of a tiny bar magnet.

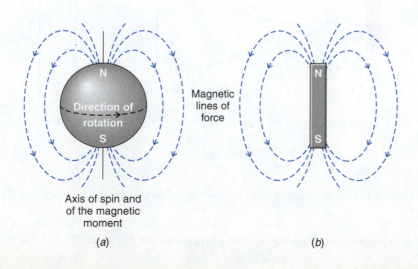

Axis of spin and of the magnetic moment

(a)

(b)

FIGURE 9.8 (a) The magnetic field associated with a spinning proton. (b) The spinning proton resembles a tiny bar magnet.

FIGURE 9.9 (a) In the absence of a magnetic field the magnetic moments of protons (represented by arrows) are randomly oriented. (b) When an external magnetic field (**B₀**) is applied, the protons orient themselves. Some are aligned with the applied field (α spin state) and some against it (β spin state). The difference in the number of protons aligned with and against the applied field is very small, but is observable with an NMR spectrometer.

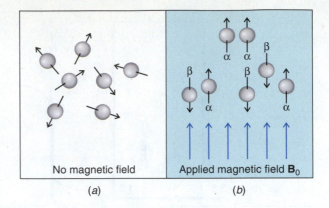

In the absence of a magnetic field (Fig. 9.9*a*), the magnetic moments of the protons of a given sample are randomly oriented. When a compound containing hydrogen (and thus protons) is placed in an applied external magnetic field, however, the magnetic moment of the protons may assume one of two possible orientations with respect to the external magnetic field (other orientations are disallowed on the basis of quantum mechanics). The magnetic moment of the proton may be aligned "with" the external field or "against" it (Fig. 9.9*b*). These alignments correspond to the two spin states mentioned earlier.

- The two alignments of the proton's magnetic moment in an external field are not of equal energy. When the proton's magnetic moment is aligned with the magnetic field, its energy is lower than when it is aligned against the magnetic field. The lower energy state is slightly more populated in the ground state.

- Energy is required to "flip" the proton's magnetic moment from its lower energy state (with the field) to its higher energy state (against the field). In an NMR spectrometer this energy is supplied by electromagnetic radiation in the RF (radio frequency) region. When this energy absorption occurs, the nuclei are said to be *in resonance* with the electromagnetic radiation.

The energy required to excite the proton is proportional to the strength of the magnetic field (Fig. 9.10). One can show by relatively simple calculations that, in a magnetic field of approximately 7.04 tesla, for example, electromagnetic radiation of 300×10^6 cycles per second (300 MHz) supplies the correct amount of energy for protons.* The proton NMR spectra shown in this chapter are 300-MHz spectra.

FIGURE 9.10 The energy difference between the two spin states of a proton depends on the strength of the applied external magnetic field, **B₀**. (a) If there is no applied field (**B₀** = 0), there is no energy difference between the two states. (b) If **B₀** ≅ 1.41 tesla, the energy difference corresponds to that of electromagnetic radiation of 60×10^6 Hz (60 MHz). (c) In a magnetic field of approximately 7.04 tesla, the energy difference corresponds to electromagnetic radiation of 300×10^6 Hz (300 MHz). Instruments are available that operate at these and even higher frequencies (as high as 800 MHz to 1 GHz).

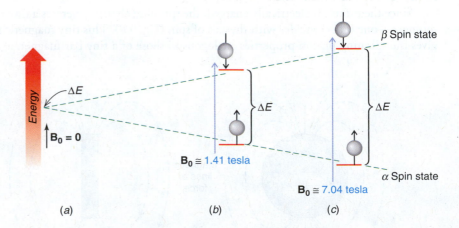

*The relationship between the frequency of the radiation (ν) and the strength of the magnetic field (**B₀**) is

$$\nu = \frac{\gamma \, \mathbf{B_0}}{2\pi}$$

where γ is the magnetogyric (or gyromagnetic) ratio. For a proton, $\gamma = 26.753$ rad s^{-1} tesla^{-1}.

Let us now consider how the signal from nuclei that are in resonance is detected by NMR spectrometers, and how it is converted to an NMR spectrum.

9.5 DETECTING THE SIGNAL: FOURIER TRANSFORM NMR SPECTROMETERS

Most NMR spectrometers use superconducting magnets that have very high magnetic field strengths. Superconducting magnets operate in a bath of liquid helium at 4.3 degrees above absolute zero, and they have magnetic field strengths more than 100,000 times as strong as Earth's magnetic field.

The stronger the magnet is in a spectrometer, the more sensitive the instrument. Figure 9.11 shows a diagram of a **Fourier transform NMR** spectrometer.

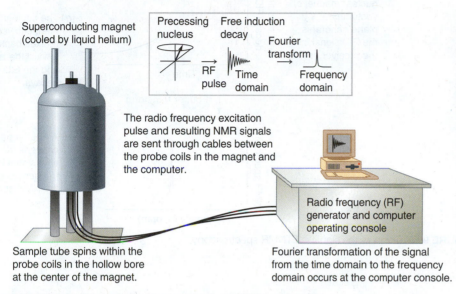

Superconducting magnet (cooled by liquid helium)

Precessing nucleus — RF pulse → Free induction decay — Time domain → Fourier transform → Frequency domain

The radio frequency excitation pulse and resulting NMR signals are sent through cables between the probe coils in the magnet and the computer.

Radio frequency (RF) generator and computer operating console

Sample tube spins within the probe coils in the hollow bore at the center of the magnet.

Fourier transformation of the signal from the time domain to the frequency domain occurs at the computer console.

FIGURE 9.11 Diagram of a Fourier transform NMR spectrometer.

The superconducting magnet of an FTNMR spectrometer.

Photo by Craig Fryhle

As we discussed in the previous section, certain nuclei in the presence of a magnetic field behave as though they were tiny bar magnets that align themselves with or against the applied magnetic field. The nuclei spin (precess) about the spectrometer's magnetic field axis (the "applied" magnetic field), much the same way that a spinning top gyrates around the axis of gravity. The precessional frequency of each nucleus is directly related to its chemical shift. We can illustrate a nuclear magnetic moment precessing about the axis of an applied magnetic field ($\mathbf{B}_0$) using a vector representation, as shown in Fig. 9.12a.

Applying a pulse of radio frequency energy that matches the precessional frequency of the nuclear magnetic moment causes the magnetic vector of the nucleus to tip away from the applied magnetic field axis (the z-axis) toward the x–y plane (Fig. 9.12b). The nuclear magnetic vector still precesses about the z-axis, but it lies in the x–y plane. From the perspective of a tiny coil of wire (called a receiver coil) situated next to the x–y plane, rotation of this vector around the z-axis but in the x–y plane presents an oscillating magnetic field to the receiver coil. And just as with large-scale electrical generators, this oscillating magnetic field induces an oscillating electric current in the coil (Fig. 9.12c). This current is the signal detected by the NMR spectrometer. Let us briefly explain the properties of this signal further.

Tipping the nuclear magnetic vector away from the axis of the applied magnetic field requires absorption of radio frequency energy by the nucleus. This energy comes from a radio frequency pulse generated by the NMR spectrometer. In a matter of seconds or less, however, the nucleus releases the energy it absorbed back to the sample environment, returning the nucleus to its ground state energy as it moves back toward the z-axis. As it does so, the vector component of magnetization in the x–y axis diminishes, and the observed electrical signal decays (Fig. 9.12d). The oscillating electrical signal produced by

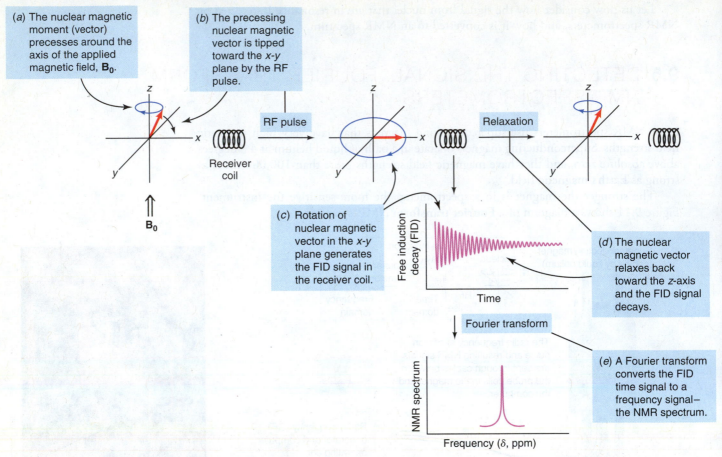

(a) The nuclear magnetic moment (vector) precesses around the axis of the applied magnetic field, **B₀**.

(b) The precessing nuclear magnetic vector is tipped toward the x-y plane by the RF pulse.

(c) Rotation of nuclear magnetic vector in the x-y plane generates the FID signal in the receiver coil.

(d) The nuclear magnetic vector relaxes back toward the z-axis and the FID signal decays.

(e) A Fourier transform converts the FID time signal to a frequency signal— the NMR spectrum.

FIGURE 9.12 Origin of the signal in FTNMR spectroscopy.

the excited nucleus is therefore not a signal of steady amplitude, but one that dies away exponentially. This signal is called a free induction decay (FID). The NMR computer applies a mathematical operation called a Fourier transform to convert the signal from a time versus amplitude signal (the FID) to a frequency versus amplitude signal (the NMR spectrum that we interpret, Fig. 9.12e).

There is much more that could be said about the origin of NMR signals and how NMR spectrometers work. The interested student is referred to advanced texts on spectroscopy for further information. However, let us conclude with a few final points.

As we mentioned, the chemical shift of an NMR signal is directly related to its precessional frequency. Since most compounds have nuclei in a variety of environments, they have nuclei that precess at a variety of frequencies, and therefore exhibit signals at a variety of chemical shifts. The FID signal detected by the NMR spectrometer is an aggregate of all of these frequencies. A powerful aspect of the Fourier transform (FT), as a mathematical process, is that it extracts these combined frequencies from the FID and converts them to discrete signals that we can interpret in an NMR spectrum.

Another great advantage to Fourier transform spectrometers is that the FT process allows computerized signal averaging of many data scans, which cancels out random electronic noise and enhances the actual NMR signals. This is especially important for samples that produce weak signals. Furthermore, acquisition of the data from each scan is very fast. The radio-frequency pulse used to excite the sample is typically on the order of only 10^{-5} s, and pulses can be repeated within a few seconds or less. Thus, many data scans can be acquired over just a short time, so as to maximize signal averaging and enhance the clarity of the data.

With this introduction to the origin of NMR signals and how spectrometers work, we return to consider further aspects of chemical shift, shielding and deshielding, and signal splitting.

9.6 THE CHEMICAL SHIFT

All protons do not absorb at the same chemical shift (δ) in an external magnetic field. The chemical shift of a given proton is dependent on its chemical environment, as we shall discuss in Section 9.7. The δ value that we report for a proton's chemical shift is actually a measure of its NMR absorption frequency, which is proportional to the external magnetic field strength (Section 9.6A).

- Smaller chemical shift (δ) values correspond with lower absorption frequency.
- Larger chemical shift (δ) values correspond with higher absorption frequency.

Chemical shifts are most often reported in reference to the absorption of the protons of TMS (tetramethylsilane), which is defined as zero on the δ scale. A small amount of TMS is either included as an internal standard in the solvent for a sample, or the NMR spectrometer itself is calibrated electronically to a chemical shift standard.

$$Si(CH_3)_4$$
Tetramethylsilane (TMS)

- The signal from TMS defines zero ppm on the chemical shift (δ) scale.

Tetramethylsilane was chosen as a reference compound for several reasons. It has 12 equivalent hydrogen atoms, and, therefore, a very small amount of TMS gives a relatively large signal. Because the hydrogen atoms are all equivalent, they give a *single signal*. Since silicon is less electronegative than carbon, the protons of TMS are in regions of high electron density. They are, as a result, highly shielded, and the signal from TMS occurs in a region of the spectrum where few other hydrogen atoms absorb. Thus, their signal seldom interferes with the signals from other hydrogen atoms. Tetramethylsilane, like an alkane, is relatively inert. It is also volatile, having a boiling point of 27 °C. After the spectrum has been determined, the TMS can be removed from the sample easily by evaporation.

9.6A PPM and the δ Scale

The chemical shift of a proton, when expressed in **hertz (Hz)**, is proportional to the strength of the external magnetic field. Since spectrometers with different magnetic field strengths are commonly used, it is desirable to express chemical shifts in a form that is independent of the strength of the external field. This can be done easily by dividing the chemical shift by the frequency of the spectrometer, with both numerator and denominator of the fraction expressed in frequency units (hertz). Since chemical shifts are always very small (typically <5000 Hz) compared with the total field strength (commonly the equivalent of 60, 300, or 600 *million* hertz), it is convenient to express these fractions in units of *parts per million* (ppm). This is the origin of the delta scale for the expression of chemical shifts relative to TMS:

$$\delta = \frac{\text{(observed shift from TMS in hertz)} \times 10^6}{\text{(operating frequency of the instrument in hertz)}}$$

For example, the chemical shift for benzene protons is 2181 Hz when the instrument is operating at 300 MHz. Therefore,

$$\delta = \frac{2181 \text{ Hz} \times 10^6}{300 \times 10^6 \text{ Hz}} = 7.27$$

The chemical shift of benzene protons in a 60-MHz instrument is 436 Hz:

$$\delta = \frac{436 \text{ Hz} \times 10^6}{60 \times 10^6 \text{ Hz}} = 7.27$$

Thus, the chemical shift expressed in ppm is the same whether measured with an instrument operating at 300 or 60 MHz (or any other field strength).

Figure 9.2 (Section 9.2A) gives the *approximate* values of proton chemical shifts for some common hydrogen-containing groups.

9.7 SHIELDING AND DESHIELDING OF PROTONS

- Protons absorb at different NMR frequencies depending on the electron density around them and the effects of local induced magnetic fields.

Protons of Hydrogen Atoms in Alkyl C—H Groups

The external (applied) magnetic field of an NMR spectrometer causes the σ electrons in an alkyl C—H bond to circulate in a way that generates an induced local magnetic field at the proton that is **opposite** to the applied magnetic field (Fig. 9.13). The hydrogen of an alkyl C—H group thus experiences a net smaller magnetic field than the applied field, causing its proton to resonate at a lower frequency (smaller chemical shift). The proton is said to be **shielded** from the applied magnetic field by the circulating σ electrons.

- The chemical shift for hydrogens of unsubstituted alkanes is typically in the range of δ 0.8–1.8.

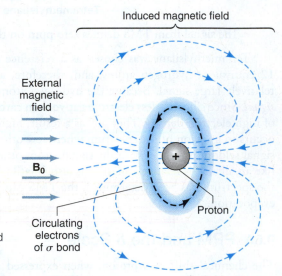

FIGURE 9.13 Circulations of the electrons of a C—H bond under the influence of an external magnetic field. The electron circulations generate a small magnetic field (an induced field) that shields the proton from the external field.

Protons of Hydrogens Near Electronegative Groups

Electronegative groups withdraw electron density from nearby hydrogen atoms, diminishing shielding of their protons by circulating σ electrons. Protons of hydrogen atoms near an electronegative group are said to be **deshielded** from the applied magnetic field, and they resonate at a higher frequency (larger chemical shift) than more shielded protons.

- The chemical shift of hydrogens bonded to a carbon bearing oxygen or a halogen is typically in the range of δ 3.1–4.0.

Protons of Hydrogen Atoms Near π Electrons

The π electrons in alkenes, alkynes, benzene, and others π-bonded groups also circulate so as to generate an induced local magnetic field in the presence of an external magnetic field. Whether shielding or deshielding occurs depends on the location of the protons in the induced magnetic field.

In the case of benzene, where the π-electron system is a closed loop (we shall discuss this in detail in Chapter 14), the external magnetic field induces a local magnetic field with flux lines that add to the external magnetic field in the region of the hydrogens (Fig. 9.14). The result is a deshielding effect on the protons and resonance at a higher frequency (larger chemical shift) than for protons of an alkyl C—H group.

- The hydrogens of benzene absorb at δ 7.27. Hydrogens bonded to substituted benzene rings have chemical shifts in the range of δ 6.0–8.5, depending on the electron-donating or electron-withdrawing effect of the substituents.

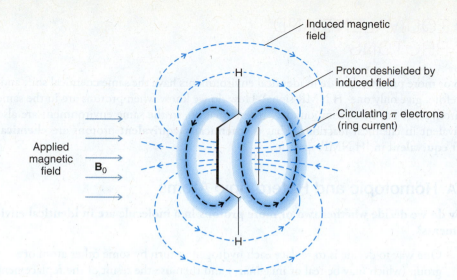

FIGURE 9.14 The induced magnetic field of the π electrons of benzene deshields the benzene protons. Deshielding occurs because at the location of the protons the induced field is in the same direction as the applied field.

The π electrons of an alkene circulate at the π bond itself to also generate an induced local magnetic field that adds to the applied magnetic field in the region of the alkene hydrogens (Fig. 9.15), though not as substantially as in benzene.

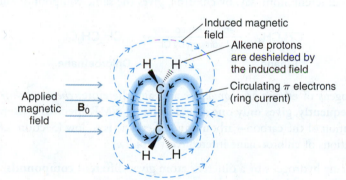

FIGURE 9.15 The induced magnetic field of the π electrons in an alkene deshield the hydrogens of an alkene. Their chemical shift of alkene hydrogens is approximately δ 6.0–4.0.

- The chemical shift of alkene hydrogens is typically in the range of δ 4.0–6.0.

The π electrons of an alkyne also circulate with respect to its π bonds, but in a way that generates an induced magnetic field that is opposite to the applied magnetic field near a terminal alkyne (acetylenic) hydrogen (Fig. 9.16).

- The chemical shift of an alkyne hydrogen is typically in the range of δ 2.5–3.1.

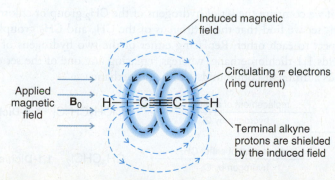

FIGURE 9.16 The induced magnetic field of the π electrons in a triple bond shield terminal alkyne hydrogens. Their chemical shift is approximately δ 3.0–2.1.

9.8 CHEMICAL SHIFT EQUIVALENT AND NONEQUIVALENT PROTONS

Two or more protons that are in identical environments have the same chemical shift and, therefore, give only one ^1H NMR signal. How do we know when protons are in the same environment? For most compounds, protons that are in the same environment are also equivalent in chemical reactions. That is, **chemically equivalent** protons are **chemical shift equivalent** in ^1H NMR spectra.

9.8A Homotopic and Heterotopic Atoms

How do we decide whether two or more protons in a molecule are in identical environments?

- One way to decide is to replace each hydrogen in turn by some other atom or group (which may be real or imaginary) and then use the result of the replacement to make our decision.

If replacing the hydrogens by a different atom gives the same compound, the hydrogens are said to be **homotopic**.

- Homotopic hydrogens have identical environments and will have the same chemical shift. They are said to be **chemical shift equivalent**.

Consider the hydrogens of ethane as an example. Replacing any one of the six hydrogens of ethane by a different atom, say, by chlorine, gives the same compound: chloroethane.

$$CH_3CH_3 \xrightarrow[\text{hydrogen by Cl}]{\text{replacement of any}} CH_3CH_2Cl$$

Ethane   **Chloroethane**

The six hydrogens of ethane are *homotopic* and are, therefore, *chemical shift equivalent*. **Ethane, consequently, gives only one signal in its ^1H NMR spectrum**. [Remember, the barrier to rotation of the carbon–carbon bond of ethane is so low (Section 4.8), the various conformations of chloroethane interconvert rapidly.]

- If replacing hydrogens by a different atom gives **different compounds**, the hydrogens are said to be **heterotopic**.
- **Heterotopic atoms have different chemical shifts and are not chemical shift equivalent**.

Consider the set of methyl hydrogens at C2 of chloroethane. Replacing any one of the three hydrogens of the CH_3 group of chloroethane with chlorine yields the same compound, 1,2-dichloroethane. The three protons of the CH_3 group are **homotopic** with respect to each other, and the CH_3 group gives only one ^1H NMR signal.

$$CH_3CH_2Cl \xrightarrow[\text{hydrogen by Cl}]{\text{replacement of } CH_3} ClCH_2CH_2Cl$$

Chloroethane   **1,2-Dichloroethane**

However, if we compare the set of hydrogens of the CH_2 group of chloroethane with those of its CH_3 set we find that the hydrogens of the CH_3 and CH_2 groups are **heterotopic** with respect to each other. Replacing either of the two hydrogens of the CH_2 set by chlorine yields 1,1-dichloroethane, whereas replacing any one of the set of three CH_3 hydrogens yields a different compound, 1,2-dichloroethane.

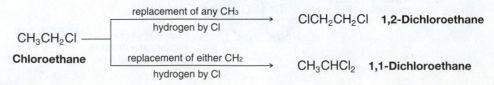

Chloroethane, therefore, has two sets of hydrogens that are heterotopic with respect to each other, the CH_3 hydrogens and the CH_2 hydrogens. The hydrogens of these two sets are not chemical shift equivalent, and chloroethane gives two ^1H NMR signals.

Consider 2-methylpropene as a further example:

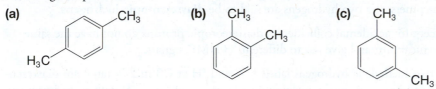

3-Chloro-2-methylpropene   1-Chloro-2-methylpropene

The six methyl hydrogens (b) are one set of homotopic hydrogens; replacing any one of them with chlorine, for example, leads to the same compound, 3-chloro-2-methylpropene. The two vinyl hydrogens (a) are another set of homotopic hydrogens; replacing either of these leads to 1-chloro-2-methylpropene. 2-Methylpropene, therefore, gives two ^1H NMR signals.

PRACTICE PROBLEM 9.3

Using the method of Section 9.8A, determine the number of expected signals for the following compounds.

(a) CH_3, H_3C

(b) CH_3, CH_3

(c) CH_3, CH_3

PRACTICE PROBLEM 9.4

How many signals would each compound give in its ^1H NMR spectrum?

(a) CH_3OCH_3 **(c)** OCH_3 **(e)** (*Z*)-2-Butene

(b)   **(d)** 2,3-Dimethyl-2-butene   **(f)** (*E*)-2-Butene

Application to ^{13}C NMR Spectroscopy As a preview of what is to come later in this chapter when we study ^{13}C NMR spectroscopy, let us look briefly at the carbon atoms of ethane to see whether we can use a similar method to decide whether they are homotopic or heterotopic, and whether ethane would give one or two ^{13}C signals. Here we can make our imaginary replacements using a silicon atom.

$$CH_3CH_3 \xrightarrow[\text{an Si atom}]{\text{replacement of either carbon atom by}} SiH_3CH_3$$

Ethane

Only one product is possible; therefore, the carbons of ethane are **homotopic**, and **ethane would give only one signal in its ^{13}C spectrum**.

On the other hand, if we consider chloroethane, replacement of a carbon atom by a silicon atom gives two possibilities:

$$CH_3CH_2Cl
\begin{cases}
\xrightarrow[\text{carbon by an Si atom}]{\text{replacement of the } CH_3} SiH_3CH_2Cl \\
\xrightarrow[\text{carbon by an Si atom}]{\text{replacement of the } CH_2} CH_3SiH_2Cl
\end{cases}$$

We do not get the same compounds from each replacement. We can conclude, therefore, that the two carbon atoms of chloroethane are **heterotopic**. They are not chemical shift equivalent, and each carbon atom of chloroethane would give a ^{13}C signal at a different chemical shift. **Chloroethane gives two ^{13}C NMR signals.**

9.8B Enantiotopic and Diastereotopic Hydrogen Atoms

If replacement of each of two hydrogen atoms by the same group yields compounds that are enantiomers, the two hydrogen atoms are said to be **enantiotopic hydrogens**.

- Enantiotopic hydrogen atoms have the same chemical shift and give only one ^{1}H NMR signal:*

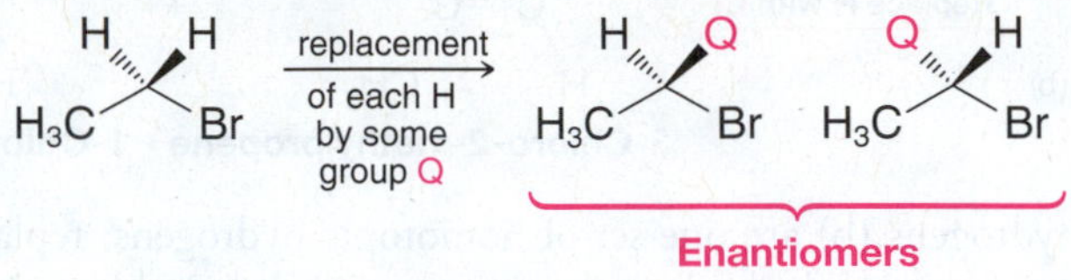

The two hydrogen atoms of the —CH_2Br group of bromoethane are enantiotopic. Bromoethane, then, gives two signals in its ^{1}H NMR spectrum. The three equivalent protons of the —CH_3 group give one signal; the two enantiotopic protons of the —CH_2Br group give the other signal. [The ^{1}H NMR spectrum of bromoethane, as we shall see, actually consists of seven peaks (three in one signal, four in the other). This is a result of signal splitting, which is explained in Section 9.9.]

If replacement of each of two hydrogen atoms by a group, Q, gives compounds that are diastereomers, the two hydrogens are said to be **diastereotopic hydrogens**.

- Except for accidental coincidence, diastereotopic protons do not have the same chemical shift and give rise to different ^{1}H NMR signals.

The two methylene hydrogens labeled ^{a}H and ^{b}H at C3 in 2-butanol are **diastereotopic**. We can illustrate this by imagining replacement of ^{a}H or ^{b}H with some imaginary group Q. The result is a pair of diastereomers. As diastereomers, they have different physical properties, including chemical shifts, especially for those protons near the chirality center.

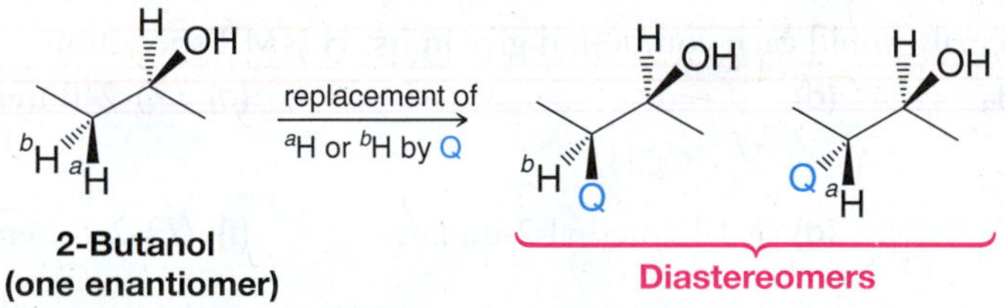

The diastereotopic nature of ^{a}H and ^{b}H at C3 in 2-butanol can also be appreciated by viewing Newman projections. In the conformations shown below (Fig. 9.17), as is the case for every possible conformation of 2-butanol, ^{a}H and ^{b}H experience different environments because of the asymmetry from the chirality center at C2. That is, the "molecular landscape" of 2-butanol appears different to each of these diastereotopic hydrogens. ^{a}H and ^{b}H

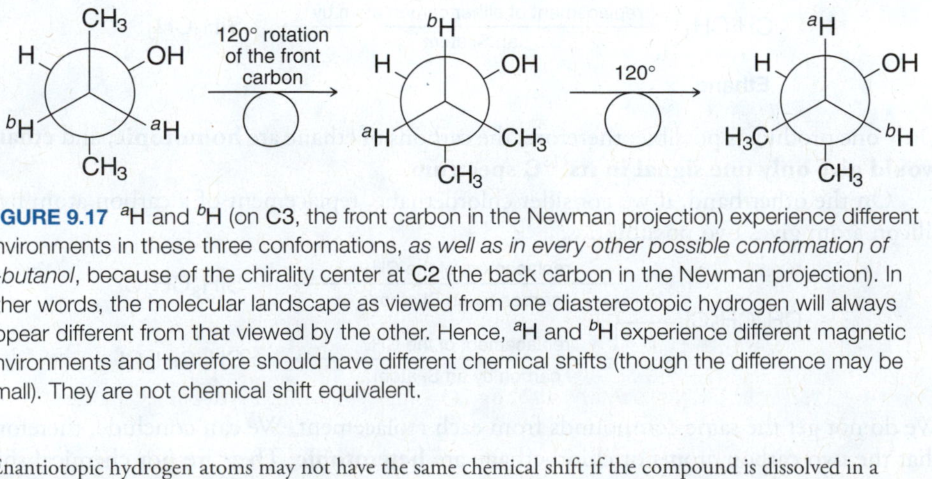

FIGURE 9.17 ^{a}H and ^{b}H (on C3, the front carbon in the Newman projection) experience different environments in these three conformations, *as well as in every other possible conformation of 2-butanol*, because of the chirality center at C2 (the back carbon in the Newman projection). In other words, the molecular landscape as viewed from one diastereotopic hydrogen will always appear different from that viewed by the other. Hence, ^{a}H and ^{b}H experience different magnetic environments and therefore should have different chemical shifts (though the difference may be small). They are not chemical shift equivalent.

*Enantiotopic hydrogen atoms may not have the same chemical shift if the compound is dissolved in a chiral solvent. However, most ^{1}H NMR spectra are determined using achiral solvents, and in this situation enantiotopic protons have the same chemical shift.

experience different magnetic environments, and are therefore not chemical shift equivalent. This is true in general: **diastereotopic hydrogens are not chemical shift equivalent**.

Alkene hydrogens can also be diastereotopic. The two protons of the $=CH_2$ group of chloroethene are diastereotopic:

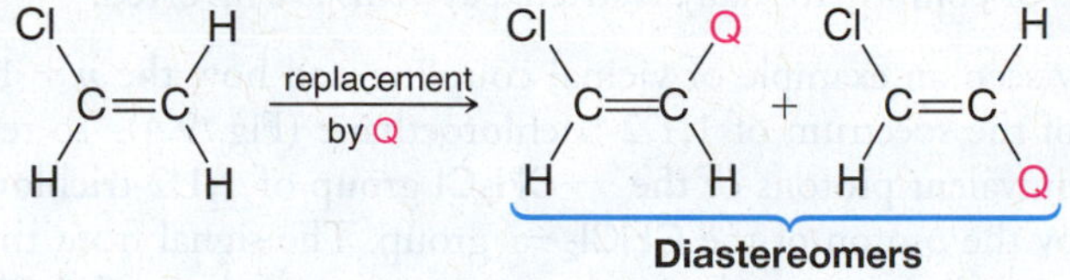

Diastereomers

Chloroethene, then, should give signals from each of its three nonequivalent protons: one for the proton of the $ClCH=$ group, and one for each of the diastereotopic protons of the $=CH_2$ group.

PRACTICE PROBLEM 9.5

(a) Show that replacing each of the CH_2 protons by some group Q in the (S) enantiomer of 2-butanol leads to a pair of diastereomers, as it does for the (R) enantiomer.

(b) How many chemically nonequivalent sets of protons are there in 2-butanol?

(c) How many ^1H NMR signals would you expect to find in the spectrum of 2-butanol?

PRACTICE PROBLEM 9.6

How many ^1H NMR signals would you expect from each of the following compounds?

(a)

(b) 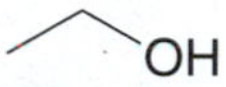 OH

(c)

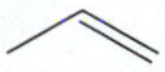

(d)

(e) 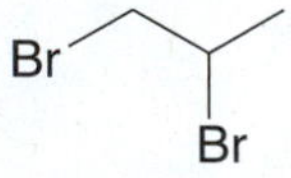 Br Br

(f)

(g)

(h)

(i)

(j) 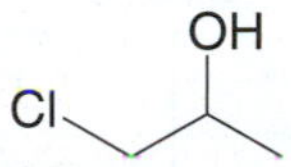 OH / Cl

(k)

(l)

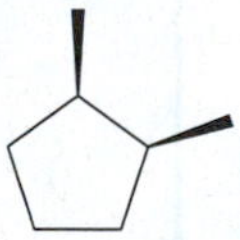

(m)

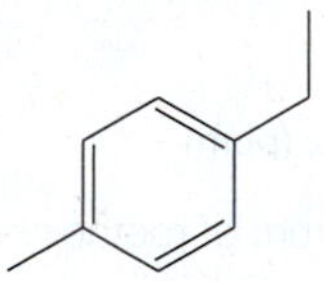

(n) 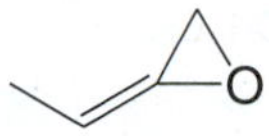 O

(o)

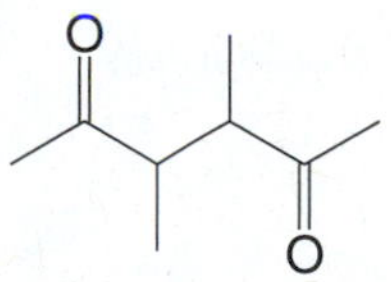

(p)

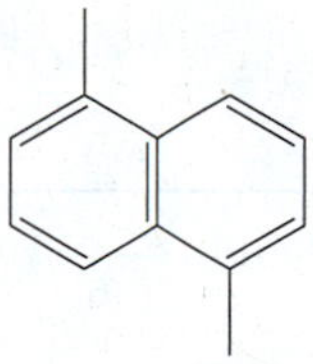

9.9 SIGNAL SPLITTING: SPIN–SPIN COUPLING

Signal splitting arises from a phenomenon known as spin–spin coupling. Spin–spin coupling effects are transferred primarily through bonding electrons and lead to **spin–spin splitting**.

• **Vicinal coupling** is coupling between hydrogen atoms on adjacent carbons (vicinal hydrogens), where separation between the hydrogens is by three σ bonds.

The most common occurrence of coupling is vicinal coupling. Hydrogens bonded to the same carbon (geminal hydrogens) can also couple, but only if they are diastereotopic. Long-range coupling can be observed over more than three bond lengths in very rigid molecules such as bicyclic compounds, and in systems where π bonds are involved. We shall limit our discussion to vicinal coupling, however.

9.9A Vicinal Coupling

- Vicinal coupling between heterotopic protons generally follows the $n + 1$ rule (Section 9.2C). Exceptions to the $n + 1$ rule can occur when diastereotopic hydrogens or conformationally restricted systems are involved.

We have already seen an example of vicinal coupling and how the $n + 1$ rule applies in our discussion of the spectrum of 1,1,2-trichloroethane (Fig. 9.4). To review, the signal from the two equivalent protons of the $-CH_2Cl$ group of 1,1,2-trichloroethane is split into a doublet by the proton of the $CHCl_2-$ group. The signal from the proton of the $CHCl_2-$ group is split into a triplet by the two protons of the $-CH_2Cl$ group.

Before we explain the origin of signal splitting, however, let us also consider two examples where signal splitting would *not* be observed. Part of understanding signal splitting is recognizing when you would not observe it. Consider ethane and methoxy-acetonitrile. All of the hydrogen atoms in ethane are equivalent, and therefore they have the same chemical shift and do not split each other. The ^1H NMR spectrum of ethane consists of one signal that is a singlet. The spectrum of methoxyacetonitrile is shown in Fig. 9.18. While there are two signals in the spectrum of methoxyacetonitrile, no coupling is observed and therefore both signals are singlets because (1) the hydrogens labeled (a) and (b) are more than three single bonds apart, and (2) the hydrogens labeled (a) are homotopic and those labeled (b) are enantiotopic.

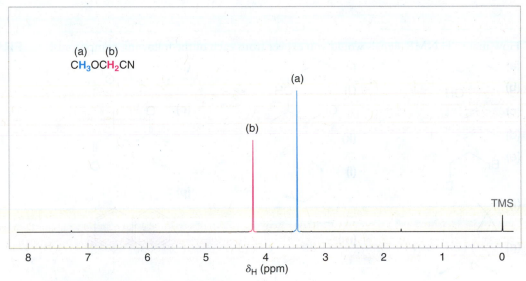

FIGURE 9.18 The 300-MHz ^1H NMR spectrum of methoxyacetonitrile. The signal of the enantiotopic protons (b) is not split.

- **Signal splitting is not observed for protons that are homotopic (chemical shift equivalent) or enantiotopic.**

Let us now explain how signal splitting arises from coupled sets of protons that are not homotopic.

9.9B Splitting Tree Diagrams and the Origin of Signal Splitting

Signal splitting is caused by the magnetic effect of protons that are nearby and nonequivalent to those protons producing a given signal. Nearby protons have magnetic moments that can either add to or subtract from the magnetic field around the proton being observed. This effect splits the energy levels of the protons whose signal is being observed into a signal with multiple peaks.

We can illustrate the origin of signal splitting using **splitting tree diagrams** and by showing the possible combinations of magnetic moment alignments for the adjacent protons (Figs. 9.19, 9.20).

Splitting Analysis for a Doublet Figure 9.19 shows a splitting tree diagram for a doublet. The signal from the observed hydrogen (^{a}H) is split into two peaks of **1 : 1 intensity** by the additive and subtractive effects of the magnetic field from a single adjacent hydrogen (^{b}H) on the applied magnetic field, B_0. The two possible magnetic orientations for the adjacent hydrogen (^{b}H) that align either against or with the applied magnetic field are shown underneath the splitting tree using arrows. J_{ab}, the spacing between the peaks (measured in hertz), is called the coupling constant. (We shall have more to say about coupling constants later.)

Splitting Analysis for a Triplet Figure 9.20 shows a splitting tree diagram for a triplet. The signal from the observed hydrogen (^{a}H) is split into three peaks of **1 : 2 : 1 intensity** by the magnetic effects of two adjacent equivalent hydrogens (^{b}H). The upper level in the diagram represents splitting from one of the adjacent ^{b}H hydrogens, leading initially to two legs that appear like the diagram for a doublet. Each of these legs is split by the second ^{b}H hydrogen, as shown at the next level. The center legs at this level overlap, however, because J_{ab} is the same for coupling of both of the ^{b}H hydrogens with ^{a}H. This overlap of the two center legs reflects the observed 1 : 2 : 1 ratio of intensities in a spectrum, as shown in the simulated triplet in Fig. 9.20. (Note that in any splitting tree diagram, the lowermost level schematically represents the peaks we observe in the actual spectrum.)

The possible magnetic orientations of the two ^{b}H hydrogens that cause the triplet are shown under the splitting diagram with arrows. The arrows indicate that both of the adjacent hydrogens may be aligned with the applied field, or one may be aligned with and the other against (in two equal energy combinations, causing a doubling of intensity), or both may be aligned against the applied field. Diagraming the possible combinations for the nuclear magnetic moments is another way (in addition to the splitting tree diagram) to show the origin of the 1 : 2 : 1 peak intensities that we observe in a triplet.

Splitting Analysis for a Quartet The NMR signal for hydrogens split by three equivalent vicinal hydrogens appears as a quartet with peak intensities in a 1 : 3 : 3 : 1 ratio. The NMR spectrum of bromoethane (Fig. 9.1), for example, exhibits a quartet for the hydrogens at C1 because they are split by the three equivalent hydrogens of the methyl

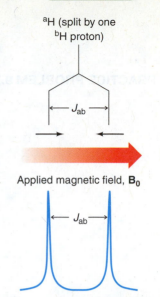

FIGURE 9.19 Splitting tree diagram for a doublet. The signal from the observed hydrogen (^{a}H) is split into two peaks of 1 : 1 intensity by the additive and subtractive effects of the magnetic field from one adjacent hydrogen (^{b}H) on B_0 (the applied field). J_{ab}, the spacing between the peaks (measured in hertz), is called the coupling constant.

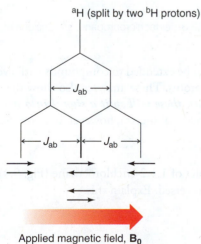

FIGURE 9.20 Splitting tree diagram for a triplet. The signal from the observed hydrogen (^{a}H) is split into three peaks of 1 : 2 : 1 intensity by two adjacent equivalent hydrogens (^{b}H). The upper level of splitting in the diagram represents splitting from one of the adjacent ^{b}H hydrogens, producing a doublet shown as two legs. The second ^{b}H hydrogen splits each of these legs again, as shown at the next level. The center legs at this level overlap however, because J_{ab} is the same* for the coupling of both ^{b}H hydrogens with ^{a}H. This analysis accounts for the observed 1 : 2 : 1 ratio of intensities in a spectrum (simulated in blue). In any splitting tree diagram, the lowermost level most closely represents what we observe in the actual spectrum. The possible magnetic orientations of the two ^{b}H hydrogens are shown under the tree diagram with arrows indicating that both of the adjacent hydrogens may be aligned with the applied field, or one may be aligned with and the other against (in two equal energy combinations, hence twice the intensity), or both may be aligned against the applied field.

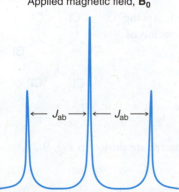

*In this example, J_{ab} is the same for both ^{b}H hydrogens because we assume them to be homotopic or enantiotopic (chemical shift equivalent). If they were diastereotopic or otherwise chemical shift non-equivalent, each may have had a different coupling constant with ^{a}H, and the splitting pattern would not have been a pure triplet (or not even a triplet at all). For example, if the two coupling constants had been significantly different, the pattern would have been a doublet of doublets instead of a triplet. Diastereotopic geminal hydrogens that couple with a vicinal hydrogen typically produce a doublet of doublets, because geminal coupling constants are often larger than vicinal coupling constants.

group at **C2**. A splitting tree analysis for a quartet would be generated following the same path of analysis as for doublets and triplets, but carried to one further level of splitting.

● ● ●
PRACTICE PROBLEM 9.7   Draw a splitting tree diagram for a quartet by adding one more level to the diagram shown in Fig 9.20 for a triplet. Underneath your quartet splitting tree show, using arrows as in Fig. 9.20, the combinations of magnetic orientations that are possible for the three vicinal hydrogens and that lead to the observed 1 : 3 : 3 : 1 ratio of intensities.

Let us conclude this section with two last examples. The spectrum of 1,1,2,3,3-pentachloropropane (Fig. 9.21) is similar to that of 1,1,2-trichloroethane in that it also consists of a 1:2:1 triplet and a 1:1 doublet. The two hydrogen atoms ^{b}H of 1,1,2,3,3-pentachloropropane are equivalent even though they are on separate carbon atoms.

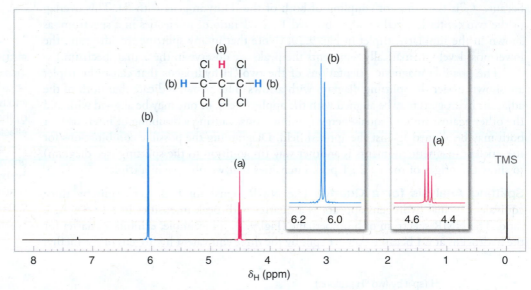

FIGURE 9.21 The 300-MHz ^{1}H NMR spectrum of 1,1,2,3,3-pentachloropropane. Expansions of the signals are shown in the offset plots.

The kind of analysis that we have just given can be extended to compounds with even larger numbers of equivalent protons on adjacent atoms. These analyses also show that *if there are n equivalent protons on adjacent atoms, these will split a signal into n + 1 peaks*. (We may not always see all of these peaks in actual spectra, however, because some of them may be very small.)

● ● ●
PRACTICE PROBLEM 9.8   The relative chemical shifts of the doublet and triplet of 1,1,2-trichloroethane (Fig. 9.4) and 1,1,2,3,3-pentachloropropane (Fig. 9.21) are reversed. Explain this.

● ● ●
PRACTICE PROBLEM 9.9   Sketch the ^{1}H NMR spectrum you would expect for the following compound, showing the splitting patterns and relative position of each signal.

● ● ●
PRACTICE PROBLEM 9.10   Propose a structure for each of the compounds whose spectra are shown in Fig. 9.22, and account for the splitting pattern of each signal.

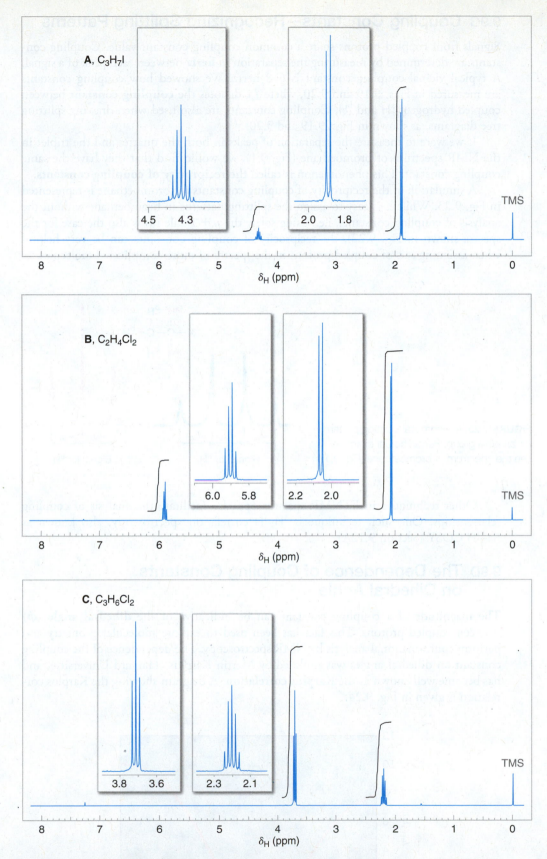

FIGURE 9.22 The 300-MHz ¹H NMR spectra for compounds **A**, **B**, and **C** in Practice Problem 9.10. Expansions are shown in the offset plots.

9.9C Coupling Constants—Recognizing Splitting Patterns

Signals from coupled protons share a common coupling constant value. Coupling constants are determined by measuring the separation in **hertz** between each peak of a signal. A typical vicinal coupling constant is 6–8 hertz. We showed how coupling constants are measured in Figs. 9.19 and 9.20, where J_{ab} denotes the **coupling constant** between coupled hydrogens ^aH and ^bH. Coupling constants are also used when drawing splitting tree diagrams, as shown in Figs. 9.19 and 9.20.

If we were to measure the separation of peaks in both the quartet and the triplet in the NMR spectrum of bromoethane (Fig. 9.1), we would find that they have the same coupling constant. This phenomenon is called **the reciprocity of coupling constants**.

A simulation of the reciprocity of coupling constants for bromoethane is represented in Fig. 9.23. While it is easy to assign the splitting patterns in bromoethane without the analysis of coupling constants, i.e., using solely the $n + 1$ rule (as is also the case for the spectra shown in Fig. 9.22), the reciprocity of coupling constants can be very helpful when assigning sets of coupled protons in the spectra of more complicated molecules.

Helpful Hint

Observing the reciprocity of coupling constants can help assign sets of coupled protons.

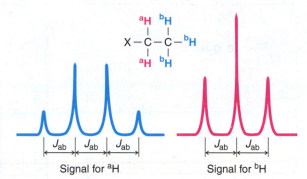

FIGURE 9.23 A theoretical splitting pattern for an ethyl group. For an actual example, see the spectrum of bromoethane (Fig. 9.1).

Signal for ^aH Signal for ^bH

Other techniques in FTNMR spectroscopy also facilitate the analysis of coupling relationships. One such technique is $^1H–^1H$ **co**rrelation **s**pectroscop**y**, also known as $^1H–^1H$ COSY (Section 9.12A).

9.9D The Dependence of Coupling Constants on Dihedral Angle

The magnitude of a coupling constant can be indicative of the **dihedral angle (ϕ)** between coupled protons. This fact has been used to explore molecular geometry and perform conformational analysis by NMR spectroscopy. The dependence of the coupling constant on dihedral angles was explored by Martin Karplus (Harvard University), and has become well known as the **Karplus correlation**. A diagram showing the Karplus correlation is given in Fig. 9.24.

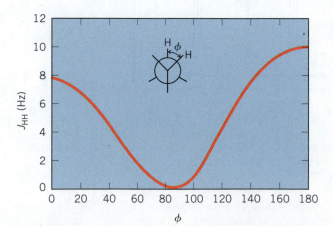

FIGURE 9.24 The Karplus correlation defines a relationship between dihedral angle (ϕ) and coupling constant for vicinal protons. (Reprinted with permission of John Wiley & Sons, Inc. From Silverstein, R., and Webster, F. X., *Spectrometric Identification of Organic Compounds, Sixth Edition*, p. 186. Copyright 1998.)

The influence of dihedral angles on coupling constants is often evident in the ^{1}H NMR spectra of substituted cyclohexanes. The coupling constant between vicinal axial protons ($J_{ax,ax}$) is typically 8–10 Hz, which is larger than the coupling constant between vicinal axial and equatorial protons ($J_{ax,eq}$), which is typically 2–3 Hz. Measuring coupling constants in the NMR spectrum of a substituted cyclohexane can therefore provide information about low energy conformations available to the compound.

PRACTICE PROBLEM 9.11

What is the dihedral angle and expected coupling constant between the labeled protons in each of the following molecules?

(a)

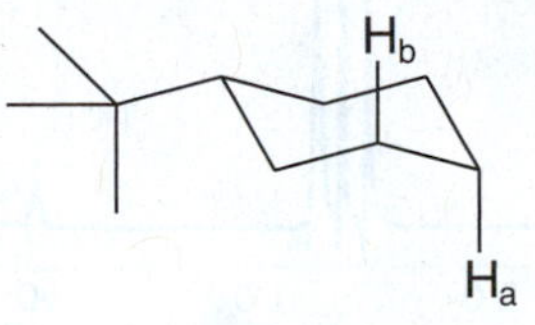

H$_b$

H$_a$

(b)

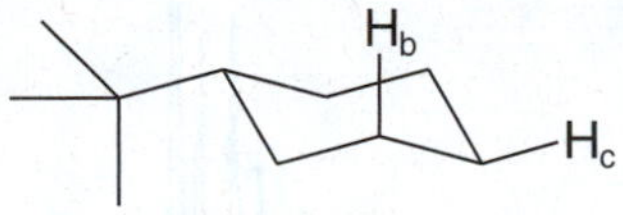

H$_b$

H$_c$

PRACTICE PROBLEM 9.12

Draw the most stable chair conformation of 1-bromo-2-chlorocyclohexane, if the coupling constant between hydrogens on C1 and C2 was found to be 7.8 Hz ($J_{1,2}$ = 7.8 Hz).

PRACTICE PROBLEM 9.13

Explain how you could distinguish between the following two compounds using NMR coupling constants. (These compounds are derived from glucose, by a reaction we shall study in Chapters 16 and 22.)

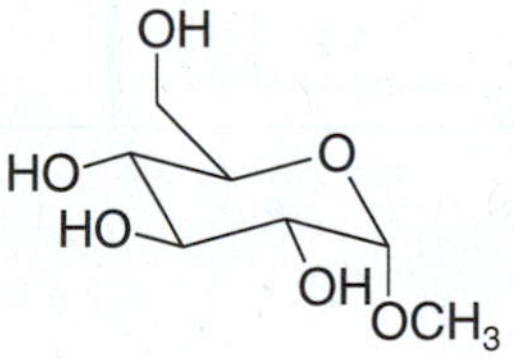

HO
HO

OH

O

OH
OCH$_3$

A

HO
HO

OH

O

OCH$_3$
OH

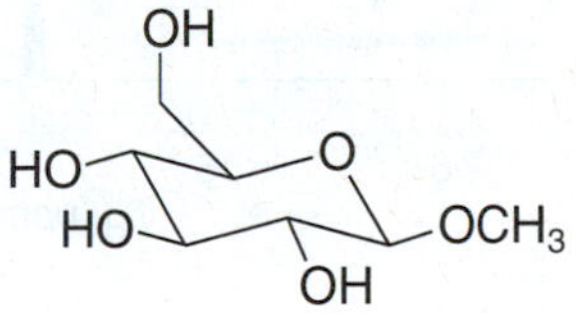

B

9.9E Complicating Features

Proton NMR spectra may have features that complicate the analysis when we try to determine the structure of a compound. For example:

1. Signals may overlap. This happens when the chemical shifts of the signals are very nearly the same. In the 60-MHz spectrum of ethyl chloroacetate (Fig. 9.25, top) we see that the singlet of the —CH$_2$Cl group falls directly on top of one of the outermost peaks of the ethyl quartet. Using NMR spectrometers with higher magnetic field strength (corresponding to ^{1}H resonance frequencies of 300–900 MHz) often allows separation of signals that would overlap at lower magnetic field strengths (Fig. 9.25, bottom).

2. Spin–spin couplings between the protons of nonadjacent atoms may occur. This long-range coupling happens frequently in compounds when π-bonded atoms intervene between the atoms bearing the coupled protons, and in some cyclic molecules that are rigid.

FIGURE 9.25 (*Top*) The 60-MHz ¹H NMR spectrum of ethyl chloroacetate. Note the overlapping signals at δ 4. (*Bottom*) The 300-MHz ¹H NMR spectrum of ethyl chloroacetate, showing resolution at higher magnetic field strength of the signals that overlapped at 60 MHz. Expansions of the signals are shown in the offset plots.

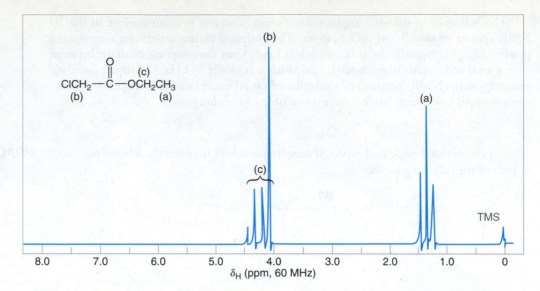

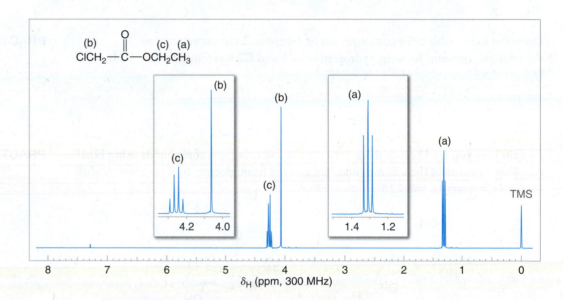

3. The splitting patterns of aromatic groups can be difficult to analyze. A monosubstituted benzene ring (a phenyl group) has three different kinds of protons:

The chemical shifts of these protons may be so similar that the phenyl group gives a signal that resembles a singlet. Or the chemical shifts may be different and, because of long-range couplings, the phenyl group signal may appear as a very complicated multiplet.

9.9F Analysis of Complex Interactions

In all of the ¹H NMR spectra that we have considered so far, we have restricted our attention to signal splittings arising from interactions of only two sets of equivalent protons on adjacent atoms. What kinds of patterns should we expect from compounds in which more than two sets of equivalent protons are interacting? We cannot answer this question completely because of limitations of space, but we can give an example that illustrates the kind of analysis that is involved. Let us consider a 1-substituted propane:

$$\overset{\text{(a)}}{CH_3}-\overset{\text{(b)}}{CH_2}-\overset{\text{(c)}}{CH_2}-Z$$

Here, there are three sets of equivalent protons. We have no problem in deciding what kind of signal splitting to expect from the protons of the CH_3- group or the $-CH_2Z$ group. The methyl group is spin–spin coupled only to the two protons of the central $-CH_2-$ group. Therefore, the methyl group should appear as a triplet. The protons of the $-CH_2Z$ group are similarly coupled only to the two protons of the central $-CH_2-$ group. Thus, the protons of the $-CH_2Z$ group should also appear as a triplet.

But what about the protons of the central $-CH_2-$ group (b)? They are spin–spin coupled with the three protons at (a) and with two protons at (c). The protons at (a) and (c), moreover, are not equivalent. If the coupling constants J_{ab} and J_{bc} have quite different values, then the protons at (b) could be split into a quartet by the three protons at (a) and each line of the quartet could be split into a triplet by the two protons at (c), resulting in 12 peaks (Fig. 9.26).

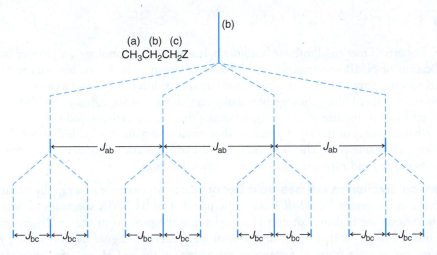

FIGURE 9.26 The splitting pattern that would occur for the (b) protons of $CH_3CH_2CH_2Z$ if J_{ab} were much larger than J_{bc}. Here $J_{ab} = 3J_{bc}$.

It is unlikely, however, that we would observe as many as 12 peaks in an actual spectrum because the coupling constants are such that peaks usually fall on top of peaks. The ^1H NMR spectrum of 1-nitropropane (Fig. 9.27) is typical of 1-substituted propane compounds, in that the central $-CH_2-$ group "sees" five approximately equivalent adjacent protons. Hence, by the $n + 1$ rule, we see that the (b) protons are split into six major peaks.

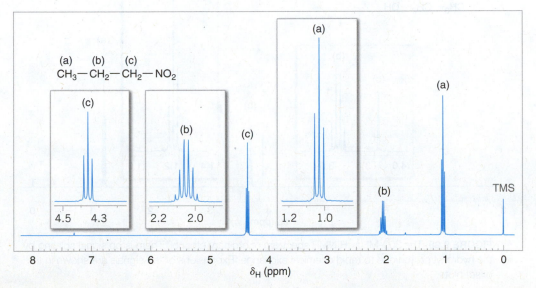

FIGURE 9.27 The 300-MHz ^1H NMR spectrum of 1-nitropropane. Expansions of the signals are shown in the offset plots.

● ● ●

PRACTICE PROBLEM 9.14   Carry out an analysis like that shown in Fig. 9.26 and show how many peaks the signal from (b) would be split into if $J_{ab} = 2J_{bc}$ and if $J_{ab} = J_{bc}$. (*Hint*: In both cases peaks will fall on top of peaks so that the total number of peaks in the signal is fewer than 12.)

The presentation we have given here applies only to what are called *first-order spectra*. In first-order spectra, the distance in hertz ($\Delta\nu$) that separates the coupled signals is very much larger than the coupling constant, J. That is, $\Delta\nu \gg J$. In *second-order spectra* (which we have not discussed), $\Delta\nu$ approaches J in magnitude and the situation becomes much more complex. The number of peaks increases and the intensities are not those that might be expected from first-order considerations.

9.10 PROTON NMR SPECTRA AND RATE PROCESSES

J. D. Roberts (Emeritus Professor, California Institute of Technology), a pioneer in the application of NMR spectroscopy to problems of organic chemistry, has compared the NMR spectrometer to a camera with a relatively slow shutter speed. Just as a camera with a slow shutter speed blurs photographs of objects that are moving rapidly, the NMR spectrometer blurs its picture of molecular processes that are occurring rapidly.

What are some of the rapid processes that occur in organic molecules? Two processes that we shall mention are chemical exchange of hydrogen atoms bonded to heteroatoms (such as oxygen and nitrogen), and conformational changes.

Chemical Exchange Causes Spin Decoupling   An example of a rapidly occurring process can be seen in ^{1}H NMR spectra of ethanol. The ^{1}H NMR spectrum of ordinary ethanol shows the hydroxyl proton as a singlet and the protons of the $-CH_2-$ group as a quartet (Fig. 9.28). In ordinary ethanol we observe *no signal splitting arising from coupling between the hydroxyl proton and the protons of the $-CH_2-$ group even though they are on adjacent atoms.*

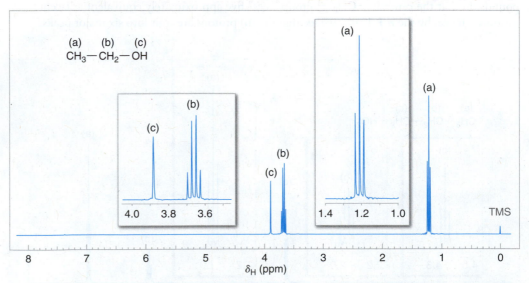

FIGURE 9.28 The 300-MHz ^{1}H NMR spectrum of ordinary ethanol. There is no signal splitting by the hydroxyl proton due to rapid chemical exchange. Expansions of the signals are shown in the offset plots.

If we were to examine a ^{1}H NMR spectrum of *very pure* ethanol, however, we would find that the signal from the hydroxyl proton was split into a triplet and that the signal from the protons of the —CH_2— group was split into a multiplet of eight peaks. Clearly, in very pure ethanol the spin of the proton of the hydroxyl group is coupled with the spins of the protons of the —CH_2— groups.

Whether coupling occurs between the hydroxyl protons and the methylene protons depends on the length of time the proton spends on a particular ethanol molecule.

- Protons attached to electronegative atoms with lone pairs such as oxygen (or nitrogen) can undergo rapid **chemical exchange**. That is, they can be transferred rapidly from one molecule to another and are therefore called **exchangeable protons**.

The chemical exchange in very pure ethanol is slow and, as a consequence, we see the signal splitting of and by the hydroxyl proton in the spectrum. In ordinary ethanol, acidic and basic impurities catalyze the chemical exchange; the exchange occurs so rapidly that the hydroxyl proton gives an unsplit signal and that of the methylene protons is split only by coupling with the protons of the methyl group.

- Rapid exchange causes **spin decoupling**.
- Spin decoupling is found in the ^{1}H NMR spectra of alcohols, amines, and carboxylic acids. The signals of OH and NH protons are normally unsplit and broad.
- Protons that undergo rapid chemical exchange (i.e., those attached to oxygen or nitrogen) can be easily detected by placing the compound in D_2O. The protons are rapidly replaced by deuterons, and the proton signal disappears from the spectrum.

Conformational Changes   At temperatures near room temperature, groups connected by carbon–carbon single bonds rotate very rapidly (unless rotation is prevented by some structural constraint, e.g., a rigid ring system). Because of this, when we determine spectra of compounds with single bonds that allow rotation, the spectra that we obtain often reflect the individual hydrogen atoms in their average environment—that is, in an environment that is an average of all the environments that the protons have as a result of conformational changes.

To see an example of this effect, let us consider the spectrum of bromoethane again. The most stable conformation is the one in which the groups are perfectly staggered. In this staggered conformation one hydrogen of the methyl group (in red in the following structure) is in a different environment from that of the other two methyl hydrogen atoms. If the NMR spectrometer were to detect this specific conformation of bromoethane, it would show the protons of the methyl group at *different chemical shifts*. We know, however, that in the spectrum of bromoethane (Fig. 9.1), the three protons of the methyl group give *one signal* (a signal that is split into a triplet by spin–spin coupling with the two protons of the adjacent carbon).

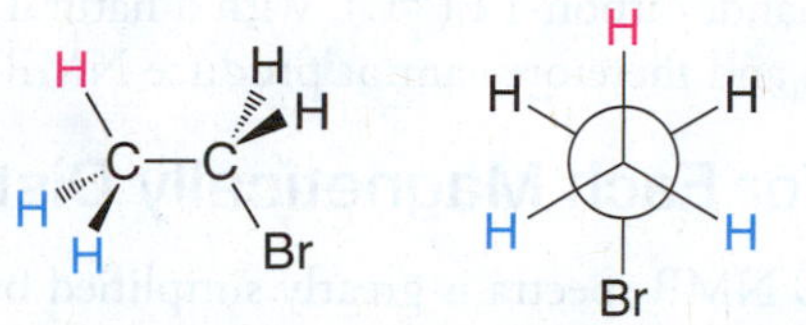

The methyl protons of bromoethane give a single signal because at room temperature the groups connected by the carbon–carbon single bond rotate approximately 1 million times each second. The "shutter speed" of the NMR spectrometer is too slow to "photograph" this rapid rotation; instead, it photographs the methyl hydrogen atoms in their average environments, and in this sense, it gives us a blurred picture of the methyl group.

Rotations about single bonds slow down as the temperature of the compound is lowered. Sometimes, this slowing of rotations allows us to "see" the different conformations of a molecule when we determine the spectrum at a sufficiently low temperature.

An example of this phenomenon, and one that also shows the usefulness of deuterium labeling, can be seen in the low-temperature ^{1}H NMR spectra of cyclohexane and of undecadeuteriocyclohexane. (These experiments originated with F. A. L. Anet, Emeritus Professor, University of California, Los Angeles, another pioneer in the applications of NMR spectroscopy to organic chemistry, especially to conformational analysis.)

Undecadeuteriocyclohexane

At room temperature, ordinary cyclohexane gives one signal because interconversion of chair forms occurs very rapidly. At low temperatures, however, ordinary cyclohexane gives a very complex ^{1}H NMR spectrum. At low temperatures interconversions are slow; the chemical shifts of the axial and equatorial protons are resolved, and complex spin–spin couplings occur.

At $-100\,^\circ$C, however, undecadeuteriocyclohexane gives only two signals of equal intensity. These signals correspond to the axial and equatorial hydrogen atoms of the following two chair conformations. Interconversions between these conformations occur at this low temperature, but they happen slowly enough for the NMR spectrometer to detect the individual conformations. (The nucleus of a deuterium atom (a deuteron) has a much smaller magnetic moment than a proton, and signals from deuteron absorption do not occur in ^{1}H NMR spectra.)

PRACTICE PROBLEM 9.15 How many signals would you expect to obtain in the ^{1}H NMR spectrum of undecadeuteriocyclohexane at room temperature?

9.11 CARBON-13 NMR SPECTROSCOPY

9.11A Interpretation of ^{13}C NMR Spectra

Helpful Hint

You may also wish to refer to ^{13}C NMR Spectroscopy: A Practical Introduction, Special Topic A in *WileyPLUS*.

We begin our study of ^{13}C NMR spectroscopy with a brief examination of some special features of spectra arising from carbon-13 nuclei. Although ^{13}C accounts for only 1.1% of naturally occurring carbon, the fact that ^{13}C can produce an NMR signal has profound importance for the analysis of organic compounds. In some important ways ^{13}C spectra are usually less complex and easier to interpret than ^{1}H NMR spectra. The major isotope of carbon, on the other hand, carbon-12 (^{12}C), with a natural abundance of about 99%, has no net magnetic spin and therefore cannot produce NMR signals.

9.11B One Peak for Each Magnetically Distinct Carbon Atom

The interpretation of ^{13}C NMR spectra is greatly simplified by the following facts:

- Each distinct carbon produces one signal in a ^{13}C NMR spectrum.

- Splitting of ^{13}C signals into multiple peaks is not observed in routine ^{13}C NMR spectra.

Recall that in ^{1}H NMR spectra, hydrogen nuclei that are near each other (within a few bonds) couple with each other and cause the signal for each hydrogen to be split into a multiplet of peaks. Coupling is not observed for adjacent carbons because only one out of every 100 carbon atoms is a ^{13}C nucleus (1.1% natural abundance). Therefore, the probability of there being two ^{13}C atoms adjacent to each other in a molecule is only about 1 in 10,000 (1.1% $\times$ 1.1%), essentially eliminating the possibility of two neighboring carbon atoms splitting each other's signal into a multiplet of peaks. The low natural abundance of ^{13}C nuclei and its inherently low sensitivity also have another effect: carbon-13 NMR spectra can be obtained only on pulse FTNMR spectrometers, where signal averaging is possible.

Whereas carbon–carbon signal splitting does not occur in ^{13}C NMR spectra, hydrogen atoms attached to carbon can split ^{13}C NMR signals into multiple peaks. However, it is useful to simplify the appearance of ^{13}C NMR spectra by initially eliminating signal splitting for ^1H–^{13}C coupling. This can be done by choosing instrumental parameters that decouple the proton–carbon interactions, and such a spectrum is said to be **broadband (BB) proton decoupled**.

- In a broadband **proton-decoupled** ^{13}C NMR spectrum, each carbon atom in a distinct environment gives a signal consisting of only one peak.

Most ^{13}C NMR spectra are obtained in the simplified broadband decoupled mode first and then in modes that provide information from the ^1H–^{13}C couplings (Sections 9.11D and 9.11E).

9.11C ^{13}C Chemical Shifts

As we found with ^1H spectra, the **chemical shift** of a given nucleus depends on the relative electron density around that atom.

- Decreased electron density around an atom **deshields** the atom from the magnetic field and causes its signal to occur further to the left in the NMR spectrum at a larger chemical shift (δ) value and higher frequency.
- Relatively higher electron density around an atom **shields** the atom from the magnetic field and causes the signal to occur further to the right in the NMR spectrum at a smaller chemical shift (δ) value and lower frequency.

For example, carbon atoms that are attached only to other carbon and hydrogen atoms are relatively shielded from the magnetic field by the density of electrons around them, and, as a consequence, alkyl carbons produce peaks that are further to the right in ^{13}C NMR spectra. On the other hand, carbon atoms bearing electronegative groups are deshielded from the magnetic field by the electron-withdrawing effects of these groups and, therefore, produce peaks that are further to the left in the NMR spectrum.

- Electronegative groups such as halogens, hydroxyl groups, and other electron-withdrawing functional groups deshield the carbons to which they are attached, causing their ^{13}C NMR peaks to occur at greater chemical shift (δ) values than those of unsubstituted carbon atoms.

Reference tables of approximate chemical shift ranges for carbons bearing different substituents are available. Figure 9.29 and Table 9.2 are examples. [The reference

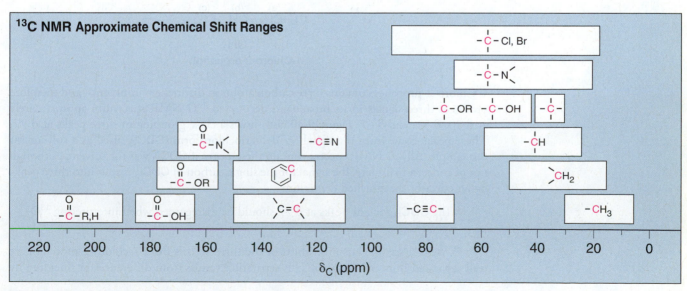

FIGURE 9.29 Approximate ^{13}C chemical shifts.

TABLE 9.2 APPROXIMATE CARBON-13 CHEMICAL SHIFTS

Type of Carbon Atom	Chemical Shift (δ, ppm)
1° Alkyl, RCH_3	0–40
2° Alkyl, RCH_2R	10–50
3° Alkyl, $RCHR_2$	15–50
Alkyl halide or amine, $-\overset{\mid}{\underset{\mid}{C}}-X\left(X = Cl, Br, \text{ or } N-\right)$	10–65
Alcohol or ether, $-\overset{\mid}{\underset{\mid}{C}}-O-$	50–90
Alkyne, $-C\equiv$	60–90
Alkene, $\overset{\diagdown}{C}=$	100–170
Aryl, ⟨benzene ring⟩$C-$	100–170
Nitrile, $-C\equiv N$	120–130
Amide, $-\overset{O}{\overset{\|}{C}}-\overset{\mid}{N}-$	150–180
Carboxylic acid or ester, $-\overset{O}{\overset{\|}{C}}-O-$	160–185
Aldehyde or ketone, $-\overset{O}{\overset{\|}{C}}-$	182–215

standard assigned as zero ppm in ^{13}C NMR spectra is also tetramethylsilane (TMS), $Si(CH_3)_4$.]

As a first example of the interpretation of ^{13}C NMR spectra, let us consider the ^{13}C spectrum of 1-chloro-2-propanol (Fig. 9.30*a*):

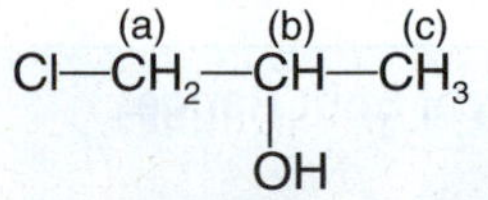

$$\overset{(a)}{Cl-CH_2}-\overset{(b)}{\underset{\underset{OH}{\mid}}{CH}}-\overset{(c)}{CH_3}$$

1-Chloro-2-propanol

1-Chloro-2-propanol contains three carbon atoms in distinct environments, and therefore produces three signals in its broadband decoupled ^{13}C NMR spectrum: approximately at δ 20, δ 51, and δ 67. Figure 9.30 also shows a close grouping of three peaks at δ 77. These peaks come from the signal for deuteriochloroform ($CDCl_3$) used as a solvent for the sample. All ^{13}C NMR spectra contain these peaks if $CDCl_3$ was the solvent. Although not of concern to us here, the signal for the single carbon of $CDCl_3$ is split into three peaks by an effect of the attached deuterium.

• The $CDCl_3$ solvent peaks at δ 77 should be disregarded when interpreting ^{13}C spectra.

As we can see, the chemical shifts of the three signals from 1-chloro-2-propanol are well separated from one another. This separation results from differences in shielding by

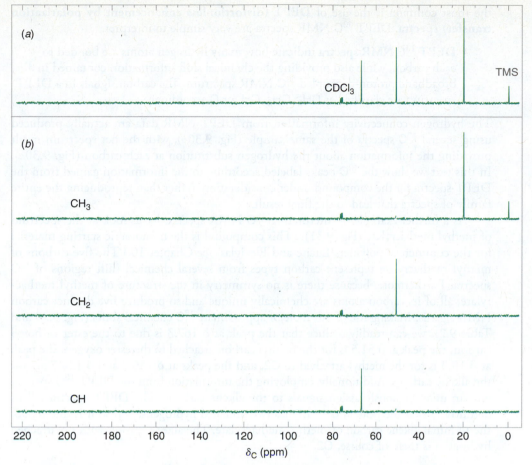

FIGURE 9.30 (*a*) The broadband proton-decoupled ^{13}C NMR spectrum of 1-chloro-2-propanol. (*b*) These three spectra show the DEPT ^{13}C NMR data from 1-chloro-2-propanol (see Section 9.11D). (This will be the only full display of a DEPT spectrum in the text. Other ^{13}C NMR figures will show the full broadband proton-decoupled spectrum but with information from the DEPT ^{13}C NMR spectra indicated near each peak as **C**, **CH**, **CH₂**, or **CH₃**.)

circulating electrons in the local environment of each carbon. Remember: the lower the electron density in the vicinity of a given carbon, the less that carbon will be shielded, and the further to the left will be the signal for that carbon. The oxygen of the hydroxyl group is the most electronegative atom; it withdraws electrons most effectively. Therefore, the carbon bearing the —OH group is the most *deshielded* carbon, and so this carbon gives the signal at δ 67. Chlorine is less electronegative than oxygen, causing the peak for the carbon to which it is attached to occur at δ 51. The methyl group carbon has no electronegative groups directly attached to it, so it has the smallest chemical shift, appearing at δ 20. Using tables of typical chemical shift values (such as Fig. 9.29 and Table 9.2), one can usually assign ^{13}C NMR signals to each carbon in a molecule, on the basis of the groups attached to each carbon.

9.11D DEPT ^{13}C Spectra

At times, more information than a predicted chemical shift is needed to assign an NMR signal to a specific carbon atom of a molecule. Fortunately, NMR spectrometers can differentiate among carbon atoms on the basis of the number of hydrogen atoms that are attached to each carbon. Several methods to accomplish this are available. One of the most common is the use of **DEPT (distortionless enhancement by polarization transfer) spectra**. DEPT ^{13}C NMR spectra are very simple to interpret.

- **DEPT ^{13}C NMR spectra** indicate how many hydrogen atoms are bonded to each carbon, while also providing the chemical shift information contained in a broadband proton-decoupled ^{13}C NMR spectrum. The carbon signals in a DEPT spectrum are classified as CH_3, CH_2, CH, or C accordingly.

The hydrogen connectivity information from DEPT NMR data are actually produced using several ^{13}C spectra of the same sample (Fig. 9.30*b*), with the net spectrum result providing the information about the hydrogen substitution at each carbon (Fig. 9.30*a*). In this text we show the ^{13}C peaks labeled according to the information gained from the DEPT spectra for the compound under consideration, rather than reproducing the entire family of spectra that lead to the final result.

As a further example of interpreting ^{13}C NMR spectra, let us look at the spectrum of methyl methacrylate (Fig. 9.31). (This compound is the monomeric starting material for the commercial polymers Lucite and Plexiglas, see Chapter 10.) The five carbons of methyl methacrylate represent carbon types from several chemical shift regions of ^{13}C spectra. Furthermore, because there is no symmetry in the structure of methyl methacrylate, all of its carbon atoms are chemically unique and so produce five distinct carbon NMR signals. Making use of our table of approximate ^{13}C chemical shifts (Fig. 9.29 and Table 9.2), we can readily deduce that the peak at δ 167.3 is due to the ester carbonyl carbon, the peak at δ 51.5 is for the methyl carbon attached to the ester oxygen, the peak at δ 18.3 is for the methyl attached to C2, and the peaks at δ 136.9 and δ 124.7 are for the alkene carbons. Additionally, employing the information from the DEPT ^{13}C spectra, we can unambiguously assign signals to the alkene carbons. The DEPT spectra tell us definitively that the peak at δ 124.7 has two attached hydrogens, and so it is due to C3, the terminal alkene carbon of methyl methacrylate. The alkene carbon with no attached hydrogens is then, of course, C2.

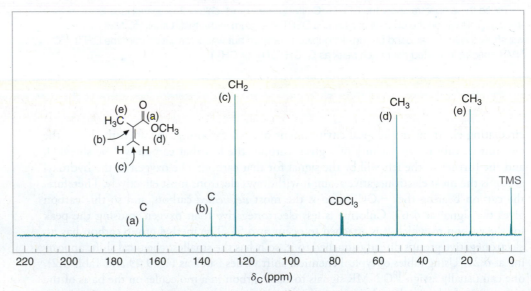

FIGURE 9.31 The broadband proton-decoupled ^{13}C NMR spectrum of methyl methacrylate. Information from the DEPT ^{13}C NMR spectra is given above the peaks.

PRACTICE PROBLEM 9.16 Compounds **A**, **B**, and **C** are isomers with the formula $C_5H_{11}Br$. Their broadband proton-decoupled ^{13}C NMR spectra are given in Fig. 9.32. Information from the DEPT ^{13}C NMR spectra is given near each peak. Give structures for **A**, **B**, and **C**.

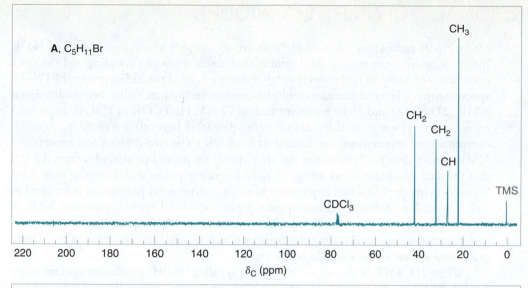

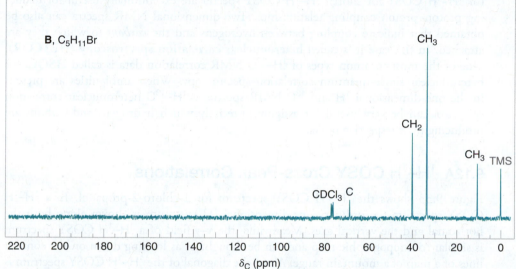

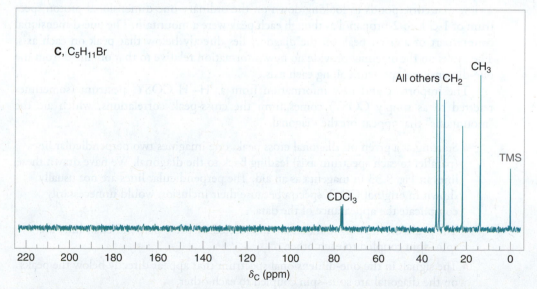

FIGURE 9.32 The broadband proton-decoupled ^{13}C NMR spectra of compounds **A**, **B**, and **C**, Practice Problem 9.16. Information from the DEPT ^{13}C NMR spectra is given above the peaks.

9.12 TWO-DIMENSIONAL (2D) NMR TECHNIQUES

Many NMR techniques are available that greatly simplify the interpretation of NMR spectra. Chemists can readily glean information about spin–spin coupling and the exact *connectivity* of atoms in molecules through techniques called **multidimensional FTNMR spectroscopy**. The most common multidimensional techniques utilize **two-dimensional NMR** (**2D NMR**) and go by acronyms such as **COSY**, HETCOR or **HSQC**, and a variety of others. [Even three-dimensional techniques (and beyond) are possible, although computational requirements can limit their feasibility.] The two-dimensional sense of 2D NMR spectra does not refer to the way they appear on paper but instead reflects the fact that the data are accumulated using two radio frequency pulses with a varying time delay between them. Sophisticated application of other instrumental parameters is involved as well. Discussion of these parameters and the physics behind multidimensional NMR is beyond the scope of this text. The result, however, is an NMR spectrum with the usual one-dimensional spectrum along the horizontal and vertical axes and a set of correlation peaks that appear in the *x–y* field of the graph.

When 2D NMR is applied to ^{1}H NMR it is called **^{1}H–^{1}H correlation spectroscopy** (or **^{1}H–^{1}H COSY** for short). ^{1}H–^{1}H COSY spectra are exceptionally useful for deducing proton–proton coupling relationships. Two-dimensional NMR spectra can also be obtained that indicate coupling between hydrogens and the *carbons* to which they are attached. In this case it is called **heteronuclear correlation spectroscopy** (**HETCOR**). One of the most common types of ^{1}H–^{13}C NMR correlation data is called **HSQC**, for heteronuclear single-quantum correlation spectroscopy. When ambiguities are present in the one-dimensional ^{1}H and ^{13}C NMR spectra, a ^{1}H–^{13}C heteronuclear correlation spectrum can be very useful for assigning precisely which hydrogens and carbons are producing their respective peaks.

9.12A ^{1}H–^{1}H COSY Cross-Peak Correlations

Figure 9.33 shows the ^{1}H–^{1}H COSY spectrum for 1-chloro-2-propanol. In a ^{1}H–^{1}H COSY spectrum the ordinary one-dimensional ^{1}H spectrum is shown along both the horizontal and the vertical axes. Meanwhile, the *x–y* field of a ^{1}H–^{1}H COSY spectrum is similar to a topographic map and can be thought of as looking down on the contour lines of a map of a mountain range. Along the diagonal of the ^{1}H–^{1}H COSY spectrum is a view that corresponds to looking down on the ordinary one-dimensional proton spectrum of 1-chloro-2-propanol as though each peak were a mountain. The one-dimensional counterpart of a given peak on the diagonal lies directly below that peak on each axis. The peaks on the diagonal provide no new information relative to that obtained from the one-dimensional spectrum along each axis.

The important and new information from a ^{1}H–^{1}H COSY spectrum (sometimes referred to as simply COSY) comes from the cross-peak correlations, which are the "mountains" that appear off the diagonal.

- Starting at a given off-diagonal cross peak, one imagines two perpendicular lines (parallel to each spectrum axis) leading back to the diagonal. We have drawn these lines in Fig. 9.33 in magenta as an aid. The perpendicular lines are not usually shown in original COSY spectra because their inclusion would unnecessarily complicate the appearance of the data.

- The peaks on the diagonal that are intersected by these perpendicular lines are spin–spin coupled to each other.

- The signals in the one-dimensional spectrum that appear directly below the peaks on the diagonal are spin–spin coupled to each other.

The cross peaks above the diagonal are mirror reflections of those below the diagonal; thus the information is redundant and only cross peaks on one side of the diagonal need to be interpreted.

Let's trace the coupling relationships in 1-chloro-2-propanol made evident in its COSY spectrum (Fig. 9.33). (Even though coupling relationships from the ordinary

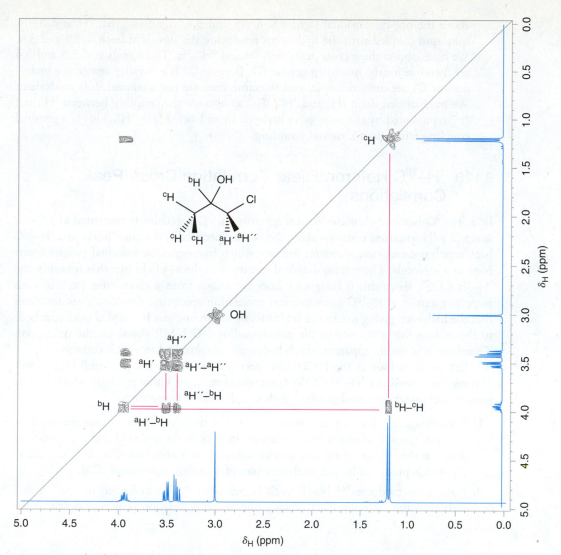

FIGURE 9.33 ^1H–^1H COSY spectrum of 1-chloro-2-propanol.

one-dimensional spectrum for 1-chloro-2-propanol are fairly readily interpreted, this compound makes a good beginning example for interpretation of COSY spectra.)

1. First we choose a starting point on a one-dimensional axis of the COSY spectrum from which to begin tracing the coupling relationships. It is best to choose a signal whose proton assignment is relatively obvious. For 1-chloro-2-propanol, the doublet from the methyl group at δ 1.2 is readily assigned.

2. Next we find the peak on the diagonal that corresponds to the methyl doublet. This peak is on the diagonal directly above the doublet in the one-dimensional spectrum, and we have labeled it ^cH in Fig. 9.33.

3. Now we determine which signal is coupled to the methyl doublet by looking for a cross peak that is correlated with it. We draw (or imagine) a line parallel to either axis that starts from the peak for the methyl doublet on the diagonal. Doing so we find that this line leads to a cross peak labeled ^bH–^cH in Fig. 9.33.

4. Taking a perpendicular line from cross peak ^bH–^cH back to the diagonal leads to the peak labeled ^bH on the diagonal. The signal at δ 3.9 in the one-dimensional spectrum below the ^bH diagonal peak is the one that is spin–spin coupled to the methyl doublet, and it is now clear that this signal is from the hydrogen on the alcohol carbon in 1-chloro-2-propanol.

5. To continue the sequence of correlations, we can now take a line from diagonal peak ^bH back to other cross peaks to which it is correlated. Doing so, we find cross peaks

above the one-dimensional signals at δ 3.5 and 3.4, signifying that ^bH at δ 3.9 is spin–spin coupled with the hydrogens producing the signals at both δ 3.5 and 3.4. We have labeled these cross peaks $^aH'$–^bH and $^aH''$–^bH. The signals at δ 3.5 and 3.4 are therefore for the two hydrogens at C1. Because C2 is a chirality center, the hydrogens at C1 are diastereotopic, and therefore they are not chemical shift equivalent. We have labeled them $^aH'$ and $^aH''$. We can also see the coupling between $^aH'$ and $^aH''$ represented by the cross peak between them labeled $^aH'$–$^aH''$. This is a geminal coupling rather than a **vicinal coupling**.

9.12B ^1H–^{13}C Heteronuclear Correlation Cross-Peak Correlations

In a ^1H–^{13}C heteronuclear correlation spectrum a ^{13}C spectrum is presented along one axis and a ^1H spectrum is shown along the other. Specifically, the cross peaks in a ^1H–^{13}C heteronuclear correlation spectrum indicate which hydrogens are attached to which carbons in a molecule. There is no diagonal spectrum in the x–y field like that found in the ^1H–^1H COSY spectrum. If imaginary lines are drawn from a given cross peak to each respective axis in a ^1H–^{13}C heteronuclear correlation spectrum, the cross peak indicates that the hydrogen giving rise to the ^1H NMR signal on one axis is coupled (and attached) to the carbon that gives rise to the corresponding ^{13}C NMR signal on the other axis. Therefore, it is readily apparent which hydrogens are attached to which carbons.

Let us take a look at the HETCOR spectrum for 1-chloro-2-propanol (Fig. 9.34). Having interpreted the ^1H–^1H COSY spectrum already, we know precisely which hydrogens of 1-chloro-2-propanol produce each signal in the ^1H spectrum.

1. If an imaginary line is taken from the methyl doublet of the proton spectrum at 1.2 ppm (vertical axis) out to the correlation peak in the x–y field and then dropped down to the ^{13}C spectrum axis (horizontal axis), it is apparent that the ^{13}C peak at 20 ppm is produced by the methyl carbon of 1-chloro-2-propanol (C3).

2. Having assigned the ^1H NMR peak at 3.9 ppm to the hydrogen on the alcohol carbon of the molecule (C2), tracing out to the correlation peak and down to the

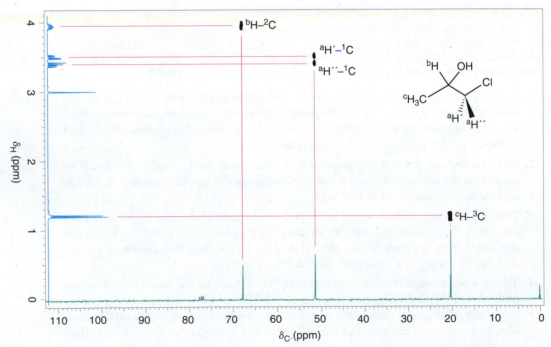

FIGURE 9.34 ^1H–^{13}C heteronuclear correlation NMR spectrum of 1-chloro-2-propanol. The ^1H NMR spectrum is shown in blue and the ^{13}C NMR spectrum is shown in green. Correlations of the ^1H–^{13}C cross peaks with the one-dimensional spectra are indicated by red lines.

THE CHEMISTRY OF... Magnetic Resonance Imaging in Medicine

An important application of ¹H NMR spectroscopy in medicine today is a technique called **magnetic resonance imaging**, or **MRI**. One great advantage of MRI is that, unlike X-rays, it does not use dangerous ionizing radiation, and it does not require the injection of potentially harmful chemicals in order to produce contrasts in the image. In MRI, a portion of the patient's body is placed in a powerful magnetic field and irradiated with RF energy.

A typical MRI image is shown at the right. The instruments used in producing images like this one use the pulse method (Section 9.5) to excite the protons in the tissue under observation and use a Fourier transformation to translate the information into an image. The brightness of various regions of the image is related to two things.

The first factor is the number of protons in the tissue at that particular place. The second factor arises from what are called the **relaxation times** of the protons. When protons are excited to a higher energy state by the pulse of RF energy, they absorb energy. They must lose this energy to return to the lower energy spin state before they can be excited again by a second pulse. The process by which the nuclei lose this energy is called **relaxation**, and the time it takes to occur is the relaxation time.

There are two basic modes of relaxation available to protons. In one, called *spin–lattice relaxation*, the extra energy is transferred to neighboring molecules in the surroundings (or lattice). The time required for this to happen is called T_1 and is characteristic of the time required for the spin system to return to thermal equilibrium with its surroundings. In solids, T_1 can be hours long. For protons in pure liquid water, T_1 is only a few seconds. In the other type of relaxation, called spin–spin relaxation, the extra energy is dissipated by transfer to nuclei of nearby atoms. The time required for this is called T_2. In liquids

An image obtained by magnetic resonance imaging.

the magnitude of T_2 is approximately equal to T_1. In solids, however, the T_1 is very much longer.

Various techniques based on the time between pulses of RF radiation have been developed to utilize the differences in relaxation times in order to produce contrasts between different regions in soft tissues. The soft tissue contrast is inherently higher than that produced with X-ray techniques. Magnetic resonance imaging is being used to great effect in locating tumors, lesions, and edemas. Improvements in this technique are occurring rapidly, and the method is not restricted to observation of proton signals.

One important area of medical research is based on the observation of signals from ³¹P. Compounds that contain phosphorus as phosphate esters (Section 11.10) such as adenosine triphosphate (ATP) and adenosine diphosphate (ADP), are involved in most metabolic processes. Using techniques based on NMR, researchers now have a non-invasive way to follow cellular metabolism.

¹³C spectrum indicates that the ¹³C NMR signal at 67 ppm arises from the alcohol carbon (C2).

3. Finally, from the ¹H NMR peaks at 3.4–3.5 ppm for the two hydrogens on the carbon bearing the chlorine, our interpretation leads us out to the cross peak and down to the ¹³C peak at 51 ppm (C1).

Thus, by a combination of ¹H–¹H COSY and ¹H–¹³C heteronuclear correlation spectra, all ¹³C and ¹H peaks can be unambiguously assigned to their respective carbon and hydrogen atoms in 1-chloro-2-propanol. In this simple example using 1-chloro-2-propanol, we could have arrived at complete assignment of these spectra without COSY and heteronuclear correlation data. For many compounds, however, the assignments are quite difficult to make without the aid of these 2D NMR techniques.

9.13 AN INTRODUCTION TO MASS SPECTROMETRY

Mass spectrometry (MS) involves formation of ions in a mass spectrometer followed by separation and detection of the ions according to mass and charge. A mass spectrum is a graph that on the *x*-axis represents the formula weights of the detected ions, and on the *y*-axis represents the abundance of each detected ion. The *x*-axis is labeled *m/z*, where m = mass and z = charge. In examples we shall consider, z equals +1, and hence the *x*-axis effectively represents the formula weight of each detected ion. The *y*-axis expresses relative ion abundance, usually as a percentage of the tallest peak or directly as the number of detected ions. The tallest peak is called the **base peak**. As a typical example, the mass spectrum of propane is shown in Fig. 9.35.

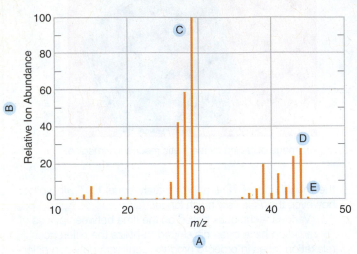

A The x-axis, in units of m/z, represents the formula weight of the detected ions. **m/z** is the mass (m) to charge (z) ratio. Because z is typically +1, m/z represents the formula weight of each ion.

B The y-axis represents the relative abundance of each detected ion.

C The most abundant ion (tallest peak) is called the **base peak**. The base peak is usually an easily formed fragment of the original compound. In this case it is an ethyl fragment ($C_2H_5^+$, m/z 29).

D One of the higher value m/z peaks may or may not represent the **molecular ion** (the ion with the formula weight of the original compound). When present, the molecular ion (m/z 44 in the case of propane) is usually not the base peak, because ions from the original molecule tend to fragment, resulting in the other m/z peaks in the spectrum.

E Small peaks having m/z values 1 or 2 higher than the formula weight of the compound are due to ^{13}C and other isotopes (Section 9.17).

FIGURE 9.35 A mass spectrum of propane. (NIST Mass Spec Data Center, S. E. Stein, director, "Mass Spectra" in NIST Chemistry WebBook, NIST Standard Reference Database Number 69, Eds. P. J. Linstrom and W. G. Mallard, June 2005, National Institute of Standards and Technology, Gaithersburg, MD, 20899 http://webbook.nist.gov.)

9.14 FORMATION OF IONS: ELECTRON IMPACT IONIZATION

The ions in mass spectrometry may be formed in a variety of ways. One method for converting molecules to ions (**ionization**) in a mass spectrometer is to place a sample under high vacuum and bombard it with a beam of high-energy electrons ($\sim$70 eV, or $\sim6.7 \times 10^3$ kJ mol^{-1}). This method is called **electron impact (EI)** ionization mass spectrometry. The impact of the electron beam dislodges a valence electron from the gas-phase molecules, leaving them with a +1 charge and an unshared electron. This species is called the **molecular ion** (M$^{+\cdot}$). We can represent this process as follows:

> **Note the $\overset{+}{\cdot}$, representing an unshared electron and the net +1 charge.**

$$M \quad + \quad e^- \quad \longrightarrow \quad M^{+\cdot} \quad + \quad 2\,e^-$$

Molecule **High-energy electron** **Molecular ion**

The molecular ion is a **radical cation** because it contains both an unshared electron and a positive charge. Using propane as an example, we can write the following equation to represent formation of its molecular ion by electron impact ionization:

> **A radical cation**

$$CH_3CH_2CH_3 \quad + \quad e^- \quad \longrightarrow \quad [CH_3CH_2CH_3]^{+\cdot} \quad + \quad 2e^-$$

9.15 DEPICTING THE MOLECULAR ION

Notice that we have written the above formula for the propane radical cation in brackets. This is because we do not know precisely from where the electron was lost in propane. We only know that one valence electron in propane was dislodged by the electron impact process. However, depicting the **molecular ion** with a localized charge and odd electron is sometimes useful (as we shall discuss in Section 9.16 when considering fragmentation reactions). One possible formula representing the molecular ion from propane with a localized charge and an odd electron is the following:

$$CH_3CH_2{}^{\overset{+}{\cdot}}CH_3$$

In many cases, the choice of just where to localize the odd electron and charge is arbitrary, however. This is especially true if there are only carbon–carbon and carbon–hydrogen single bonds, as in propane. When possible, though, we write the structure showing the molecular ion that would result from the removal of one of the most loosely held valence electrons of the original molecule. Just which valence electrons are most loosely held can usually be estimated from ionization potentials (Table 9.3). The ionization potential of a molecule is the amount of energy (in electron volts) required to remove a valence electron from the molecule.

As we might expect, ionization potentials indicate that in formation of radical cations the nonbonding electrons of nitrogen, oxygen, and halogen atoms, and the π electrons of alkenes and aromatic molecules, are held more loosely than the electrons of carbon–carbon and carbon–hydrogen σ bonds. Therefore we have the following general rule.

- When a molecule contains oxygen, nitrogen, or a π bond, we place the odd electron and charge at a nitrogen, oxygen, halogen, or π bond. If resonance is possible, the radical cation may be delocalized.

The following are examples of these cases.

TABLE 9.3 IONIZATION POTENTIALS OF SELECTED MOLECULES

Compound	Ionization Potential (eV)
$CH_3(CH_2)_3NH_2$	8.7
C_6H_6 (benzene)	9.2
C_2H_4	10.5
CH_3OH	10.8
C_2H_6	11.5
CH_4	12.7

Radical cations from ionization of nonbonding or π electrons.

Methanol	Trimethylamine	1-Butene
$CH_3-\overset{+}{\underset{..}{O}}H$	$CH_3-\overset{+}{\underset{CH_3}{N}}-CH_3$	$CH_2{}^{\overset{+}{\cdot}}CHCH_2CH_3$

9.16 FRAGMENTATION

Molecular ions formed by EI mass spectrometry are highly energetic species, and in the case of complex molecules, a great many things can happen to them. A molecular ion can break apart in a variety of ways, the fragments that are produced can undergo further **fragmentation**, and so on. We cannot go into all of the processes that are possible, but we can examine a few of the more important ones.

As we begin, let us keep three important principles in mind:

1. The reactions that take place in a mass spectrometer are unimolecular, that is, they do not involve collisions between molecules or ions. This is true because the pressure is kept so low (10^{-6} torr) that reactions involving bimolecular collisions do not occur.

2. We use single-barbed arrows to depict mechanisms involving single electron movements (see Section 3.1A).

3. The relative ion abundances, as indicated by peak intensities, are very important. We shall see that the appearance of certain prominent peaks in the spectrum gives us key information about the structures of the fragments produced and about their original locations in the molecule.

9.16A Fragmentation by Cleavage at a Single Bond

One important type of fragmentation is the simple cleavage of a single bond. With a radical cation this cleavage can take place in at least two ways; each way produces a *cation* and a *radical*. Only the cations are detected in a positive ion mass spectrometer. (The radicals, because they are not charged, are not detected.) With the molecular ion obtained from propane by loss of one carbon–carbon σ bonding electron, for example, two possible modes of cleavage are

$$[CH_3CH_2CH_3]^{\ddagger} \longrightarrow \begin{array}{l} CH_3CH_2^+ \ + \ \cdot CH_3 \\ CH_3CH_2^{\cdot} \ + \ {}^+CH_3 \end{array}$$

These two modes of cleavage do not take place at equal rates, however. Although the relative abundance of cations produced by such a cleavage is influenced by the stability of both the carbocation and the radical, the *carbocation's stability is more important.** In the spectrum of propane shown earlier (Fig. 9.35), the peak at *m/z* 29 ($CH_3CH_2^+$) is the most intense peak; the peak at *m/z* 15 (CH_3^+) has a relative abundance of only 5.6%. This reflects the greater stability of $CH_3CH_2^+$ as compared to CH_3^+.

When drawing mechanism arrows to show cleavage reactions it is convenient to choose a localized representation of the radical cation, as we have done above for propane. (When showing only an equation for the cleavage and not a mechanism, however, we would use the convention of brackets around the formula with the odd electron and charge shown outside.) Fragmentation equations for propane would be written in the following way (note the use of single-barbed arrows):

Helpful Hint

Recall that we use single-barbed arrows to show the movement of single electrons, as in the case of these homolytic bond cleavages and other processes involving radicals (see Section 3.1A).

$$CH_3CH_2CH_3 \xrightarrow{-e^-} \text{or}$$

$$CH_3CH_2^+CH_3 \longrightarrow CH_3CH_2^+ \ + \ \cdot CH_3 \\ \textbf{\textit{m/z}} \ \textbf{29}$$

$$CH_3CH_2^+CH_3 \longrightarrow CH_3CH_2^{\cdot} \ + \ {}^+CH_3 \\ \textbf{\textit{m/z}} \ \textbf{15}$$

••• SOLVED PROBLEM 9.6

The mass spectrum of CH_3F is given in Fig. 9.36. **(a)** Draw a likely structure for the molecular ion (*m/z* 34). **(b)** Assign structural formulas to the two other high abundance peaks (*m/z* 33 and *m/z* 15) in the spectrum. **(c)** Propose an explanation for the low abundance of the peak at *m/z* 19.

STRATEGY AND ANSWER:

(a) Nonbonding electrons have lower ionization energies than bonding electrons, so we can expect that the molecular ion for CH_3F was formed by loss of an electron from the fluorine atom.

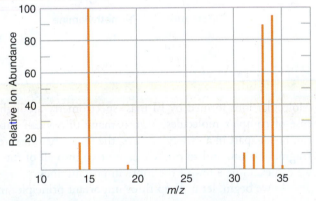

FIGURE 9.36 Mass spectrum for Solved Problem 9.6.

$$e^- \ + \ CH_3-\ddot{\underset{\cdot\cdot}{F}}: \xrightarrow[\text{(EI)}]{\begin{array}{c}\text{electron}\\\text{impact}\\\text{ionization}\end{array}} CH_3-\ddot{\underset{\cdot\cdot}{F}}:^{\cdot+} \ + \ 2\,e^-$$

(b) The ion with *m/z* 33 differs from the molecular ion by one atomic mass unit, thus a hydrogen atom must have been lost. Cleavage with loss of a hydrogen atom could occur as follows, leaving both the carbon and fluorine with full valence electron shells, but as a cationic species overall.

$$\underset{\substack{|\\H}}{\overset{\substack{H\\|}}{H-C}}-\ddot{\underset{\cdot\cdot}{F}}:^{\cdot+} \xrightarrow{-H\cdot} \underset{\substack{|\\H}}{\overset{\substack{H\\|}}{\cdot C}}-\ddot{\underset{\cdot\cdot}{F}}:^{+} \longleftrightarrow \underset{H}{\overset{H}{\diagup}}C=\ddot{\underset{\cdot\cdot}{F}}:^{+}$$

---

*This can be demonstrated through thermochemical calculations that we cannot go into here. The interested student is referred to McLafferty, F. W., *Interpretation of Mass Spectra*, 2nd ed.; Benjamin: Reading, MA, 1973; pp. 41, 210–211.

The ion with *m/z* 15 must be a methyl carbocation formed by loss of a fluorine atom, as shown below. The fleeting existence of a methyl carbocation is possible in electron impact ionization (EI) mass spectrometry (MS) because electrons with high kinetic energy bombard the species undergoing analysis, allowing higher energy pathways to be followed than occur with reactions that take place in solution.

$$H-\overset{\overset{\displaystyle H}{|}}{\underset{\underset{\displaystyle H}{|}}{C}} - \overset{..}{\overset{+}{\underset{..}{F}}}: \longrightarrow \quad \overset{H}{\underset{H}{}}\overset{}{C^{+}} \quad + \quad :\overset{.}{\underset{..}{F}}:$$

(c) The very small *m/z* 19 peak in this spectrum would have to be a fluorine cation. The presence of only 6 valence electrons in an F^+ ion and the strong electronegativity of fluorine would create a very high energy barrier to formation of F^+ and hence, cause it to be formed in very low abundance relative to other ionization and cleavage pathways for CH_3F^+.

9.16B Fragmentation of Longer Chain and Branched Alkanes

The mass spectrum of hexane shown in Fig. 9.37 illustrates the kind of fragmentation that a longer chain alkane can undergo. Here we see a reasonably abundant molecular ion at *m/z* 86 accompanied by a small $M^{+} + 1$ peak. There is also a smaller peak at *m/z* 71 ($M^{+} - 15$) corresponding to the loss of $\cdot CH_3$, and the base peak is at *m/z* 57 ($M^{+} - 29$) corresponding to the loss of $\cdot CH_2CH_3$. The other prominent peaks are at *m/z* 43 ($M^{+} - 43$) and *m/z* 29 ($M^{+} - 57$), corresponding to the loss of $\cdot CH_2CH_2CH_3$ and $\cdot CH_2CH_2CH_2CH_3$, respectively. The important fragmentations are just the ones we would expect:

$$[CH_3CH_2CH_2CH_2CH_2CH_3]^{+} \longrightarrow$$

$$\longrightarrow CH_3CH_2CH_2CH_2CH_2^{+} + \cdot CH_3$$
$$\textbf{\textit{m/z} 71}$$

$$\longrightarrow CH_3CH_2CH_2CH_2^{+} + \cdot CH_2CH_3$$
$$\textbf{\textit{m/z} 57}$$

$$\longrightarrow CH_3CH_2CH_2^{+} + \cdot CH_2CH_2CH_3$$
$$\textbf{\textit{m/z} 43}$$

$$\longrightarrow CH_3CH_2^{+} + \cdot CH_2CH_2CH_2CH_3$$
$$\textbf{\textit{m/z} 29}$$

Chain branching increases the likelihood of cleavage at a branch point because a more stable carbocation can result. When we compare the mass spectrum of 2-methylbutane (Fig. 9.38) with the spectrum of hexane, we see a much more intense peak at $M^{+} - 15$.

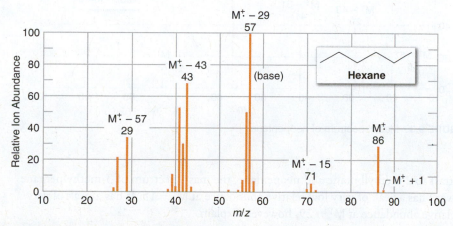

FIGURE 9.37 Mass spectrum of hexane.

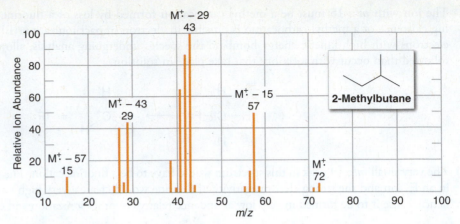

FIGURE 9.38 Mass spectrum of 2-methylbutane.

Loss of a methyl radical from the molecular ion of 2-methylbutane can give a secondary carbocation:

$$\begin{bmatrix} CH_3 \\ | \\ CH_3CHCH_2CH_3 \end{bmatrix}^{+\cdot} \longrightarrow CH_3\overset{+}{C}HCH_2CH_3 + \cdot CH_3$$

$$\begin{array}{ccc} \textit{m/z 72} & & \textit{m/z 57} \\ \textbf{M}^{+\cdot} & & \textbf{M}^{+\cdot}-\textbf{15} \end{array}$$

whereas with hexane loss of a methyl radical can yield only a primary carbocation.

With neopentane (Fig. 9.39), this effect is even more dramatic. Loss of a methyl radical by the molecular ion produces a *tertiary* carbocation, and this reaction takes place so readily that virtually none of the molecular ions survive long enough to be detected:

$$\begin{bmatrix} CH_3 \\ | \\ CH_3-C-CH_3 \\ | \\ CH_3 \end{bmatrix}^{+\cdot} \longrightarrow CH_3-\overset{+}{\underset{\underset{CH_3}{|}}{C}}-CH_3 \; + \; \cdot CH_3$$

$$\begin{array}{ccc} \textit{m/z 72} & & \textit{m/z 57} \\ \textbf{M}^{+\cdot} & & \textbf{M}^{+\cdot}-\textbf{15} \end{array}$$

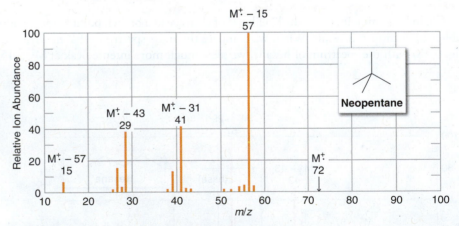

FIGURE 9.39 Mass spectrum of neopentane.

• • •

PRACTICE PROBLEM 9.17   In contrast to 2-methylbutane and neopentane, the mass spectrum of 3-methylpentane (not given) has a peak of very low relative abundance at $M^{+\cdot} - 15$. It has a peak of very high relative abundance at $M^{+\cdot} - 29$, however. Explain.

9.16C Fragmentation to Form Resonance-Stabilized Cations

Carbocations stabilized by resonance are usually prominent in mass spectra. Several ways that resonance-stabilized carbocations can be produced are outlined in the following list. These examples begin by illustrating the likely sites for initial ionization (π and nonbonding electrons), as well.

1. Alkenes ionize and frequently undergo fragmentations that yield resonance-stabilized allylic cations:

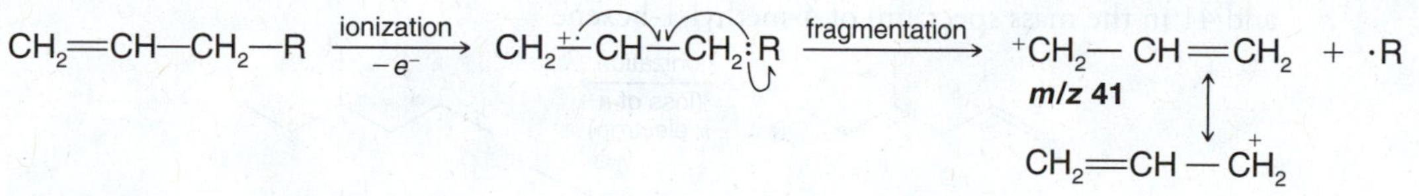

2. Carbon–carbon bonds next to an atom with an unshared electron pair usually break readily because the resulting carbocation is resonance stabilized:

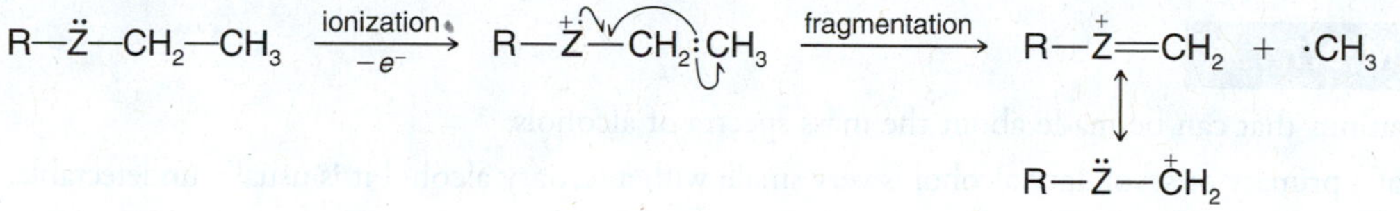

where Z = N, O, or S; R may also be H.

3. Carbon–carbon bonds next to the carbonyl group of an aldehyde or ketone break readily because resonance-stabilized ions called **acylium ions** are produced:

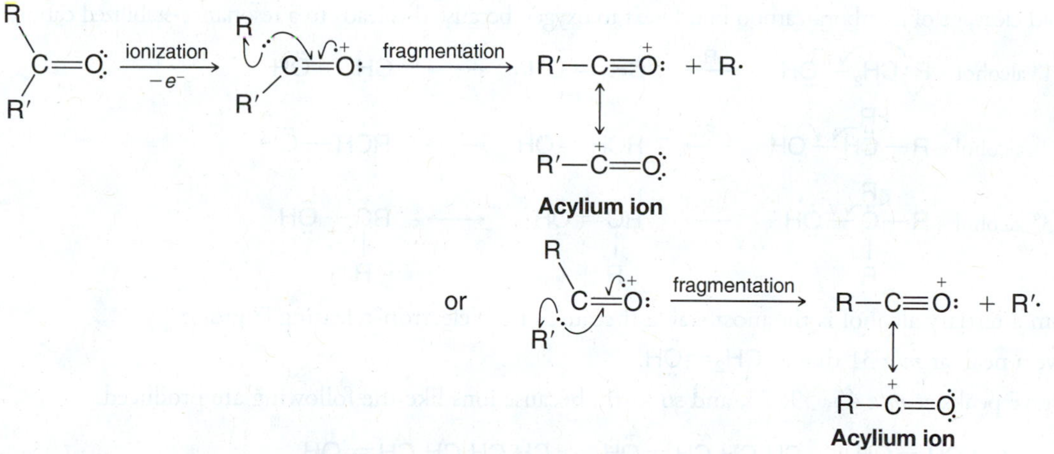

Acylium ion

or

Acylium ion

4. Alkyl-substituted benzenes ionize by loss of a π electron and undergo loss of a hydrogen atom or methyl group to yield the relatively stable tropylium ion (see Section 14.7C). This fragmentation gives a prominent peak (sometimes the base peak) at *m/z* 91:

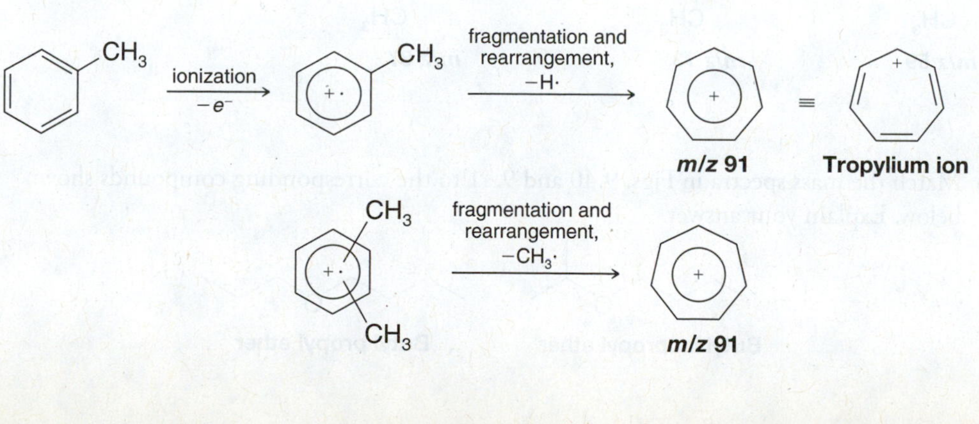

m/z 91   **Tropylium ion**

m/z 91

5. Monosubstituted benzenes with other than alkyl groups also ionize by loss of a π electron and then lose their substituent to yield a phenyl cation with m/z 77:

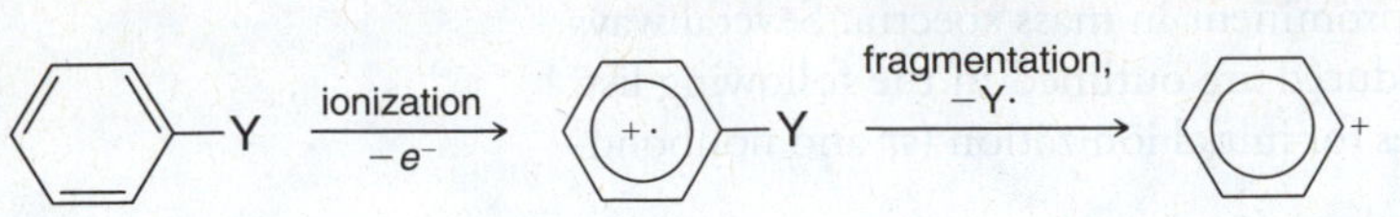

$$ Y = \text{halogen}, -NO_2, -\overset{\overset{\displaystyle O}{\|}}{C}R, -R, \text{ and so on} $$

m/z 77

PRACTICE PROBLEM 9.18 Propose structures and fragmentation mechanisms corresponding to ions with m/z 57 and 41 in the mass spectrum of 4-methyl-1-hexene.

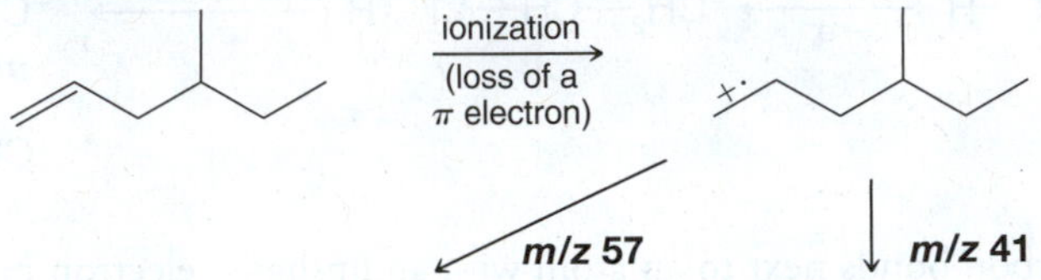

ionization
(loss of a
π electron)

m/z 57   **m/z 41**

••• SOLVED PROBLEM 9.7

Explain the following observations that can be made about the mass spectra of alcohols:

(a) The molecular ion peak of a primary or secondary alcohol is very small; with a tertiary alcohol it is usually undetectable.

(b) Primary alcohols show a prominent peak at m/z 31.

(c) Secondary alcohols usually give prominent peaks at m/z 45, 59, 73, and so on.

(d) Tertiary alcohols have prominent peaks at m/z 59, 73, 87, and so on.

STRATEGY AND ANSWER:

(a) Alcohols undergo rapid cleavage of a carbon–carbon bond next to oxygen because this leads to a resonance-stabilized cation.

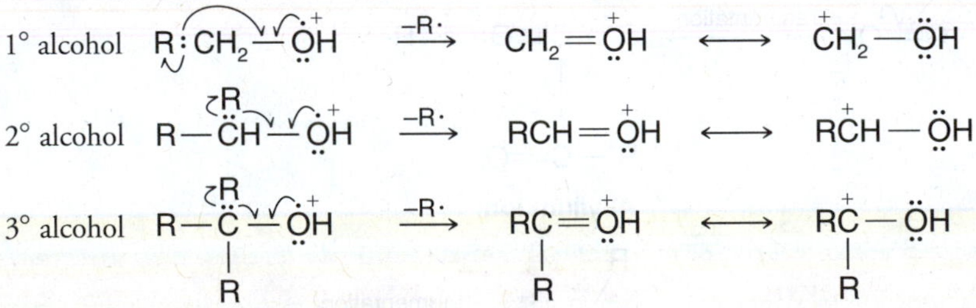

The cation obtained from a tertiary alcohol is the most stable (because of the electron-releasing R groups).

(b) Primary alcohols give a peak at m/z 31 due to $CH_2 \overset{+}{=} \ddot{O}H$.

(c) Secondary alcohols give peaks at m/z 45, 59, 73, and so forth, because ions like the following are produced.

$$ CH_3CH \overset{+}{=} \ddot{O}H \qquad CH_3CH_2CH \overset{+}{=} \ddot{O}H \qquad CH_3CH_2CH_2CH \overset{+}{=} \ddot{O}H $$

m/z 45   **m/z 59**   **m/z 73**

(d) Tertiary alcohols give peaks at m/z 59, 73, 87, and so forth, because ions like the following are produced.

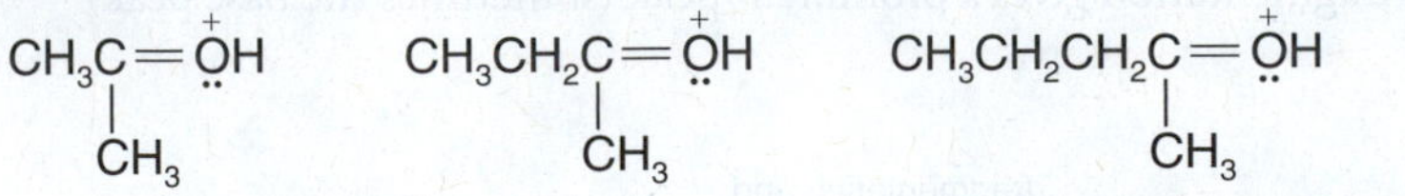

m/z 59   **m/z 73**   **m/z 87**

PRACTICE PROBLEM 9.19 Match the mass spectra in Figs. 9.40 and 9.41 to the corresponding compounds shown below. Explain your answer.

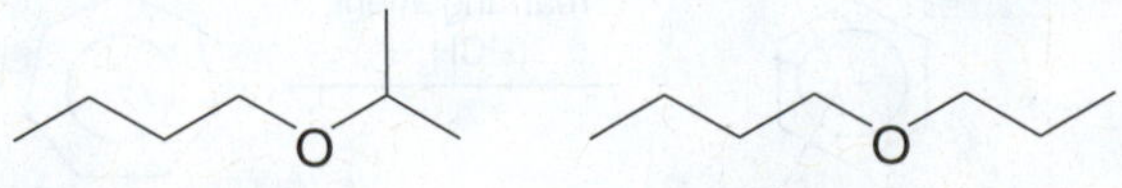

Butyl isopropyl ether   **Butyl propyl ether**

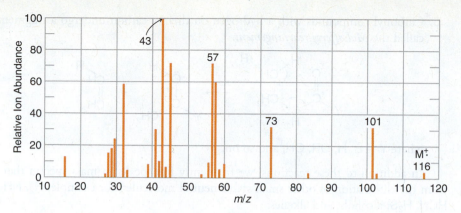

FIGURE 9.40 Mass spectrum for Practice Problem 9.19.

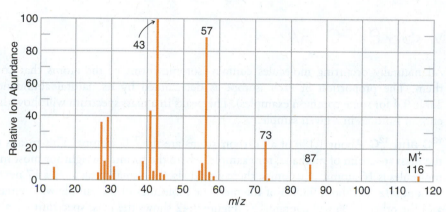

FIGURE 9.41 Mass spectrum for Practice Problem 9.19.

9.16D Fragmentation by Cleavage of Two Bonds

Many peaks in mass spectra can be explained by fragmentation reactions that involve the breaking of two covalent bonds. When a radical cation undergoes this type of fragmentation, the products are *a new radical cation* and *a neutral molecule*. Some important examples, starting from the initial radical cation, are the following:

1. Alcohols frequently show a prominent peak at $M^{\ddagger} - 18$. This corresponds to the loss of a molecule of water (See Solved Problem 9.7):

$$R-\overset{..}{C}H-CH_2 \longrightarrow R-CH \overset{.}{\overset{+}{-}} CH_2 + H-\overset{..}{\overset{..}{O}}-H$$
$$\textbf{M}^{\ddagger} \qquad\qquad \textbf{M}^{\ddagger} - \textbf{18}$$

which can also be written as

$$[R-CH_2-CH_2-OH]^{\ddagger} \longrightarrow [R-CH=CH_2]^{\ddagger} + H_2O$$
$$\textbf{M}^{\ddagger} \qquad\qquad\qquad \textbf{M}^{\ddagger} - \textbf{18}$$

2. Cycloalkenes can undergo a retro-Diels–Alder reaction (Section 13.11) that produces an alkene and an alkadienyl radical cation:

$$\left[\bigcirc \right]^{\ddagger} \longrightarrow \left[\bigcirc \right]^{\ddagger} + \begin{array}{c} CH_2 \\ \| \\ CH_2 \end{array}$$

which can also be written as

$$\left[\bigcirc \right] \longrightarrow \left[\bigcirc \right] + \|$$

3. Carbonyl compounds with a hydrogen on their γ carbon undergo a fragmentation called the *McLafferty rearrangement.*

where Y = R, H, OR, OH, and so on.

In addition to these reactions, we frequently find peaks in mass spectra that result from the elimination of other small stable neutral molecules, for example, H_2, NH_3, CO, HCN, H_2S, alcohols, and alkenes.

9.17 ISOTOPES IN MASS SPECTRA

All naturally occurring molecules contain isotopic forms of the atoms that comprise them. The proportion of each isotope is determined by its natural abundance (see Table 9.4 for some common examples). The peaks in a mass spectrum will show the presence of isotopes in a given sample.

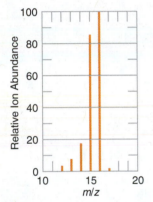

FIGURE 9.42 Mass spectrum for methane.

^{13}C and ^{12}C About 1.1% of all carbon atoms are the ^{13}C isotope. This means that in the mass spectrum of methane, for example, where the formula weight for most methane molecules is 16 atomic mass units, there would also be a small peak at m/z 17 next to the peak at m/z 16. About 98.9% of the methane molecules in the sample will contain ^{12}C, and the other 1.1% will contain ^{13}C. Figure 9.42 shows the mass spectrum for a sample of methane, in which a small m/z 17 peak can be seen for the $M^{+\cdot} + 1$ ion.

For molecules with more than one carbon, the intensity of the $M^{+\cdot} + 1$ peak taken in proportion to the $M^{+\cdot}$ peak can be used as an approximation of the number of carbon atoms in the molecule. This is because there is a 1.1% chance that each carbon in the molecule could be a ^{13}C isotope. However, in large molecules the $(M^{+\cdot} + 1)/M^{+\cdot}$ ratio is altered by the existence of other isotopes with a nominal mass one unit larger than their most abundant form, such as for ^{2}H, ^{17}O, and so on. Therefore, for large molecules the $(M^{+\cdot} + 1)/M^{+\cdot}$ ratio cannot be used reliably as an indication of the number of carbons.

^{35}Cl and ^{37}Cl; ^{79}Br and ^{81}Br Some elements that are common in organic molecules have isotopes that differ by two atomic mass units. These include ^{16}O and ^{18}O, ^{32}S and ^{34}S, ^{35}Cl and

	Most Common Isotope	%	Natural Abundance of Other Isotopes			
Element				%		%
Carbon	^{12}C	98.93	^{13}C	1.07		
Hydrogen	^{1}H	99.99	^{2}H	0.011		
Nitrogen	^{14}N	99.63	^{15}N	0.368		
Oxygen	^{16}O	99.76	^{17}O	0.038	^{18}O	0.205
Fluorine	^{19}F	100				
Silicon	^{28}Si	92.23	^{29}Si	4.68	^{30}Si	3.09
Phosphorus	^{31}P	100				
Sulfur	^{32}S	94.93	^{33}S	0.76	^{34}S	4.29
Chlorine	^{35}Cl	75.78	^{37}Cl	24.22		
Bromine	^{79}Br	50.69	^{81}Br	49.31		
Iodine	^{127}I	100				

TABLE 9.4 PRINCIPAL STABLE ISOTOPES OF COMMON ELEMENTS[a]

[a]Data based on the 1997 Technical Report of the International Union of Pure and Applied Chemistry (IUPAC), Rosman, K. J. R., Taylor, P. D. P. *Pure and Applied Chemistry,* **1998,** Vol. 70, No. 1, 217–235.

^{37}Cl, and ^{79}Br and ^{81}Br. It is particularly easy to identify the presence of chlorine or bromine using mass spectrometry because the isotopes of chlorine and bromine are relatively abundant.

- The natural abundance of ^{35}Cl is 75.5% and that of ^{37}Cl is 24.5%.
- In the mass spectrum for a sample containing chlorine, we would expect to find peaks separated by two mass units, in an approximately 3:1 (75.5%:24.5%) ratio for the **molecular ion** or any fragments that contain chlorine.

Figure 9.43*a* shows the mass spectrum of chlorobenzene. The peaks at *m/z* 112 and *m/z* 114 in approximately a 3:1 intensity ratio are a clear indication that chlorine atoms are present. An *m/z* 77 peak for the phenyl cation fragment is evident, as well.

- The natural abundance of ^{79}Br is 51.5%, and that of ^{81}Br is 49.5%.
- In the mass spectrum for a sample containing bromine we would expect to find peaks separated by two mass units in an approximately 1:1 ratio (49.5%:51.5%).

Figure 9.43*b* shows the mass spectrum of bromomethylbenzene. The peaks at *m/z* 170 and *m/z* 172 in an approximately 1:1 intensity ratio are a clear indication that bromine is present. Note that the base peak is *m/z* 91, most likely representing a tropylium cation (Section 9.16C).

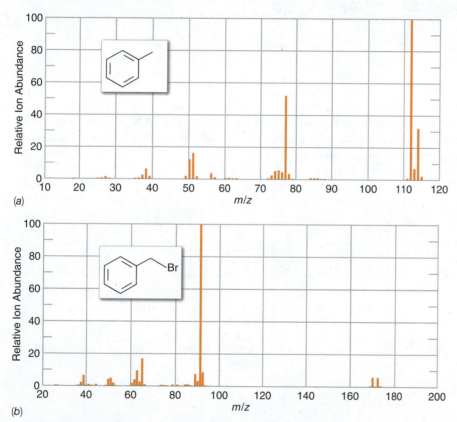

(a)

(b)

FIGURE 9.43 (*a*) The mass spectrum of chlorobenzene. Note the approximately 3:1 intensity ratio of peaks at *m/z* 112 and 114 due to the presence of ^{35}Cl and ^{37}Cl. (*b*) The mass spectrum of bromomethylbenzene. Note the 1:1 intensity ratio of peaks at *m/z* 170 and 172 due to the presence of ^{79}Br and ^{81}Br. (a,b: P. J. Linstrom and W. G. Mallard, Eds., NIST Chemistry WebBook, NIST Standard Reference Database Number 69, National Institute of Standards and Technology, Gaithersburg, MD 20899, http://webbook.nist.gov.)

If two bromine or chlorine atoms are present in a molecule, then an M$^{+\cdot}$ + 4 peak will appear in addition to the M$^{+\cdot}$ + 2 and M$^{+\cdot}$ peaks. In a molecule containing two bromine atoms, for example, the intensity of the M$^{+\cdot}$ + 4 peak due to two ^{81}Br atoms present in one molecule is the same as the M$^{+\cdot}$ peak for two ^{79}Br atoms in one molecule. But the probability of having one ^{79}Br and one ^{81}Br is double when two bromines are present (because either bromine atom could be either isotope). Thus, the ratio of intensities for the M$^{+\cdot}$, M$^{+\cdot}$ + 2, and M$^{+\cdot}$ + 4 peaks will be 1:2:1 when two bromine atoms are present.

●●● **SOLVED PROBLEM 9.8**

(a) What approximate intensities would you expect for the M$^{+\cdot}$ and M$^{+\cdot}$ + 2 peaks of CH$_3$Cl?

(b) For the M$^{+\cdot}$ and M$^{+\cdot}$ + 2 peaks of CH$_3$Br?

(c) An organic compound gives an M$^{+\cdot}$ peak at *m/z* 122 and a peak of nearly equal intensity at *m/z* 124. What is a likely molecular formula for the compound?

STRATEGY AND ANSWER:

(a) The $M^{\ddagger} + 2$ peak due to CH_3—^{37}Cl (at m/z 52) should be almost one-third as large as the $M^{\ddagger}$ peak at m/z 50 because of the relative natural abundances of ^{35}Cl and ^{37}Cl.

(b) The peaks due to CH_3—^{79}Br and CH_3—^{81}Br (at m/z 94 and m/z 96, respectively) should be of nearly equal intensity due to the relative natural abundances of ^{79}Br and ^{81}Br.

(c) That the $M^{\ddagger}$ and $M^{\ddagger} + 2$ peaks are of nearly equal intensity tells us that the compound contains bromine. C_3H_7Br is therefore a likely molecular formula.

$$
\begin{array}{ll}
C_3 = 36 & C_3 = 36 \\
H_7 = 7 & H_7 = 7 \\
{}^{79}Br = \underline{79} & {}^{81}Br = \underline{81} \\
m/z = 122 & m/z = 124
\end{array}
$$

PRACTICE PROBLEM 9.20   What are the expected ratios of the $M^{\ddagger}$, $M^{\ddagger} + 2$, and $M^{\ddagger} + 4$ peaks for the following compounds?

(a)

(b)

PRACTICE PROBLEM 9.21   Given the mass spectrum in Figure 9.44 and the fact that the ^1H NMR spectrum for this compound consists of only a large doublet and a small septet, what is the structure of the compound? Explain your reasoning.

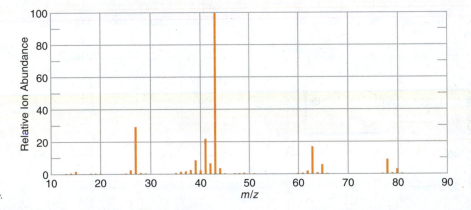

FIGURE 9.44 Mass spectrum for Practice Problem 9.21. P. J. Linstrom and W. G. Mallard, Eds., NIST Chemistry WebBook, NIST Standard Reference Database Number 69, National Institute of Standards and Technology, Gaithersburg, MD 20899, http://webbook.nist.gov.

9.17A High-Resolution Mass Spectrometry

All of the spectra that we have described so far have been determined on what are called "low-resolution" mass spectrometers. These spectrometers, as we noted earlier, measure m/z values to the nearest whole-number mass unit. Many laboratories are equipped with this type of mass spectrometer.

Some laboratories, however, are equipped with the more expensive "high-resolution" mass spectrometers. These spectrometers can measure m/z values to three or four decimal places and thus provide an extremely accurate method for determining molecular weights. And because molecular weights can be measured so accurately, these spectrometers also allow us to determine molecular formulas.

The determination of a molecular formula by an accurate measurement of a molecular weight is possible because the actual masses of atomic particles (nuclides) are not integers (see Table 9.5). Consider, as examples, the three molecules O_2, N_2H_4, and CH_3OH.

The actual atomic masses of the molecules are all different (though nominally they all have atomic mass of 32):

$$O_2 = 2(15.9949) = 31.9898$$
$$N_2H_4 = 2(14.0031) + 4(1.00783) = 32.0375$$
$$CH_4O = 12.00000 + 4(1.00783) + 15.9949 = 32.0262$$

High-resolution mass spectrometers are capable of measuring mass with an accuracy of 1 part in 40,000 or better. Thus, such a spectrometer can easily distinguish among these three molecules and, in effect, tell us the molecular formula.

The ability of high-resolution instruments to measure exact masses has been put to great use in the analysis of biomolecules such as proteins and nucleic acids. For example, one method that has been used to determine the amino acid sequence in oligopeptides is to measure the exact mass of fragments derived from an original oligopeptide, where the mixture of fragments includes oligopeptides differing in length by one amino acid residue. The exact mass difference between each fragment uniquely indicates the amino acid residue that occupies that position in the intact oligopeptide (see Section 24.5E). Another application of exact mass determinations is the identification of peptides in mixtures by comparison of mass spectral data with a database of exact masses for known peptides. This technique has become increasingly important in the field of proteomics (Section 24.14).

9.18 GC/MS ANALYSIS

Gas chromatography is often coupled with mass spectrometry in a technique called **GC/MS analysis**. The gas chromatograph separates components of a mixture, while the mass spectrometer then gives structural information about each one (Fig. 9.45). GC/MS can also provide quantitative data when standards of known concentration are used with the unknown.

In GC analysis, a minute amount of a mixture to be analyzed, typically 0.001 mL (1.0 μL) or less of a dilute solution containing the sample, is injected by syringe into a heated port of the gas chromatograph. The sample is vaporized in the injector port and swept by a flow of helium into a capillary column. The capillary column is a thin tube usually 10–30 meters long and 0.1–0.5 mm in diameter. It is contained in a chamber (the "oven") whose temperature can be varied according to the volatility of the samples being analyzed. The inside of the capillary column is typically coated with a "stationary phase" of low polarity (essentially a high-boiling and very viscous liquid that is often a nonpolar silicon-based polymer). As molecules of the mixture are swept through the column by the helium, they travel at different rates according to their boiling points and the degree of affinity for the stationary phase. Materials with higher boiling points or stronger affinity for the stationary phase take longer to pass through the column. Low-boiling and nonpolar materials pass through very quickly. The length of time each component takes to travel through the column is called the retention time. Retention times typically range from 1 to about 30 minutes, depending on the sample and the specific type of column used.

As each component of the mixture exits the GC column it travels into a mass spectrometer. Here, molecules of the sample are bombarded by electrons; ions and fragments of the molecule are formed, and a mass spectrum results similar to those we have studied earlier in this chapter. The important thing, however, is that mass spectra are obtained for *each* component of the original mixture that is separated. This ability of GC/MS to separate mixtures and give information about the structure of each component makes it a virtually indispensable tool in analytical, forensic, and organic synthesis laboratories.

TABLE 9.5 EXACT MASSES OF NUCLIDES	
Isotope	Mass
^{1}H	1.00783
^{2}H	2.01410
^{12}C	12.00000 (std)
^{13}C	13.00336
^{14}N	14.0031
^{15}N	15.0001
^{16}O	15.9949
^{17}O	16.9991
^{18}O	17.9992
^{19}F	18.9984
^{32}S	31.9721
^{33}S	32.9715
^{34}S	33.9679
^{35}Cl	34.9689
^{37}Cl	36.9659
^{79}Br	78.9183
^{81}Br	80.9163
^{127}I	126.9045

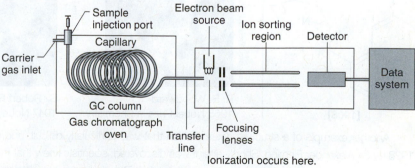

FIGURE 9.45 Schematic of a typical capillary gas chromatograph/mass spectrometer (GC/MS).

9.19 MASS SPECTROMETRY OF BIOMOLECULES

Advances in mass spectrometry have made it a tool of exceptional power for analysis of large biomolecules. **Electrospray ionization**, **MALDI (matrix-assisted laser desorption ionization)**, and other "soft ionization" techniques for nonvolatile compounds and macromolecules make possible analyses of proteins, nucleic acids, and other biologically relevant compounds with molecular weights up to and in excess of 100,000 daltons. Electrospray ionization with quadrupole mass analysis is now routine for biomolecule analysis, as is analysis using MALDI–TOF (time of flight) instruments. Extremely high resolution can be achieved using Fourier transform–ion cyclotron resonance (FT ICR, or FTMS). We shall discuss ESI and MALDI applications of mass spectrometry to protein sequencing and analysis in Sections 24.5E, 24.13B, and 24.14.

WHY Do These Topics Matter?]

STRUCTURE DETERMINATION WITHOUT NMR

With the advent of the techniques you have learned about in this chapter, chemists can determine the complete structures of most organic molecules with only a few milligrams of material and in a relatively short time. However, prior to the introduction of spectroscopy in the late 1950s and 1960s, it was a much different story. In that era, chemists needed grams of a compound and, in some cases, decades of time to determine a compound's structure.

Without spectroscopy, structure determination began with combustion analysis, a method that could be used to determine the molecular formula of the sample by literally burning it and measuring the relative amounts of the combustion products, such as water and carbon dioxide. Then, through painstaking detective work, chemists would perform different chemical reactions to degrade the compound into smaller components. They would then attempt to rebuild those materials into the original compound to determine how those atoms were combined. With unusual or particularly complex compounds, structural determination could be a very slow process; for example, establishing the connectivities of the atoms of quinine, the world's first antimalarial drug, took 54 years.

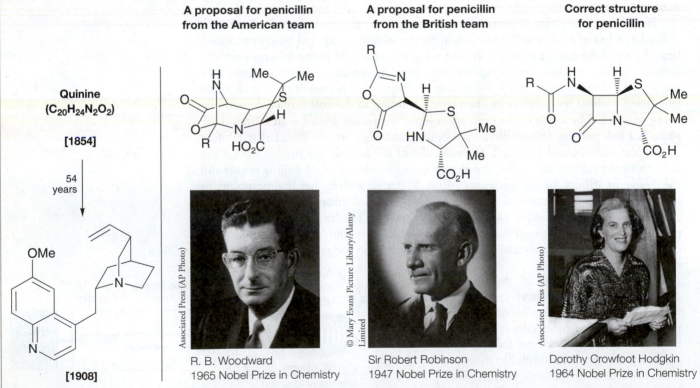

Quinine
($C_{20}H_{24}N_2O_2$)

[1854]

54 years

OMe

[1908]

A proposal for penicillin from the American team

A proposal for penicillin from the British team

Correct structure for penicillin

R. B. Woodward
1965 Nobel Prize in Chemistry

Sir Robert Robinson
1947 Nobel Prize in Chemistry

Dorothy Crowfoot Hodgkin
1964 Nobel Prize in Chemistry

Another example of a structure determination that was particularly difficult, and costly, involved the antibiotic penicillin, first isolated in 1928 by Sir Alexander Fleming. Shortly after it was discovered, scientists knew that this molecule would have tremendous value in combating infections that were previously viewed as untreatable. The challenge was obtaining sufficient supplies to treat everyone who needed it, an issue that became particularly salient during World War II when tens of thousands of soldiers suffered wounds. In response, the American and British governments began an extensive project involving hundreds of scientists on both sides of the Atlantic seeking to make penicillin in the laboratory through organic synthesis. The problem was that the structure of penicillin had not been established, and the American and

British teams, each of which included a future Nobel laureate, generally had different, and incorrect, theories for its connections. As a result, no real quantities of penicillin were synthesized during the war, with fermentation from mold being the main supply for those in need.

Ultimately, it would take a different future Nobel laureate and a different technique to solve the problem—namely, Dorothy Crowfoot Hodgkin and X-ray crystallography. In this method, if a material can be solidified into a regular crystalline form, light can be shone on it, and based on the resultant diffraction pattern due to interactions of the light with the atoms in the crystal, the connections of every non-hydrogen atom can be determined. With X-ray crystallography, penicillin was shown to possess a four-membered ring, a motif not expected to exist because of strain, as we have discussed previously. That ring, in fact, would be a major challenge for its eventual chemical synthesis as we will discuss in Chapter 17, and the problem would take another decade to solve once the structure of penicillin was established.

To learn more about these topics, see:

1. Sheehan, J. C. *The Enchanted Ring: The Untold Story of Penicillin*. MIT Press: Cambridge, **1984**, p. 224.
2. Nicolaou, K. C.; Montagnon, T. *Molecules that Changed the World*. Wiley-VCH: Weinheim, **2008**, p. 366.

SUMMARY AND REVIEW TOOLS

The study aids for this chapter include key terms and concepts (which are hyperlinked to the Glossary from the bold, blue terms in the *WileyPLUS* version of the book at wileyplus.com), a Concept Map, and NMR chemical shift correlation charts.

PROBLEMS PLUS

Note to Instructors: Many of the homework problems are available for assignment via *WileyPLUS*, an online teaching and learning solution.

NMR SPECTROSCOPY

The following are some abbreviations used to report spectroscopic data:

^{1}H **NMR:** s = singlet, d = doublet, t = triplet, q = quartet, bs = broad singlet, m = multiplet

IR absorptions: s = strong, m = moderate, br = broad

9.22 How many ^{1}H NMR signals (not peaks) would you predict for each of the following compounds? (Consider all protons that would be chemical shift nonequivalent.)

9.23 How many ^{13}C NMR signals would you predict for each of the compounds shown in Problem 9.22?

9.24 Propose a structure for an alcohol with molecular formula $C_5H_{12}O$ that has the ^{1}H NMR spectrum given in Fig. 9.46. Assign the chemical shifts and splitting patterns to specific aspects of the structure you propose.

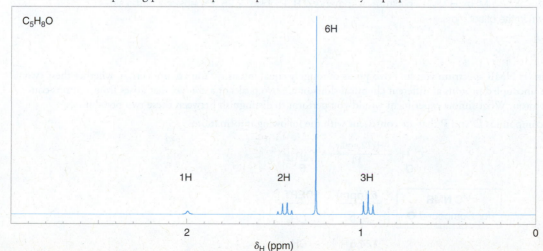

FIGURE 9.46 The ^{1}H NMR spectrum (simulated) of alcohol C_5H_8O, Problem 9.24.

9.25 Propose structures for the compounds **G** and **H** whose ¹H NMR spectra are shown in Figs. 9.47 and 9.48.

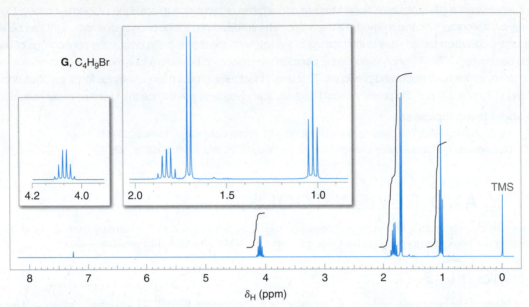

FIGURE 9.47 The 300-MHz ¹H NMR spectrum of compound **G**, Problem 9.25. Expansions of the signals are shown in the offset plots.

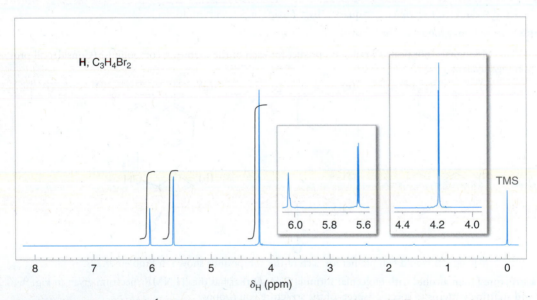

FIGURE 9.48 The 300-MHz ¹H NMR spectrum of compound **H**, Problem 9.25. Expansions of the signals are shown in the offset plots.

9.26 Assume that in a certain ¹H NMR spectrum you find two peaks of roughly equal intensity. You are not certain whether these two peaks are *singlets* arising from uncoupled protons at different chemical shifts or are two peaks of a *doublet* that arises from protons coupling with a single adjacent proton. What simple experiment would you perform to distinguish between these two possibilities?

9.27 Propose structures for compounds **O** and **P** that are consistent with the following information.

$$C_6H_8 \xrightarrow[\text{Pt}]{\text{H}_2 \text{ (2 equiv.)}} C_6H_{12}$$
$$\textbf{O} \qquad\qquad\qquad \textbf{P}$$

¹³C NMR for Compound **O**	δ (ppm)	DEPT
	26.0	CH$_2$
	124.5	CH

9.28 Compound **Q** has the molecular formula C_7H_8. The broad-band proton decoupled ^{13}C spectrum of **Q** has signals at δ 50 (CH), 85 (CH$_2$), and 144 (CH). On catalytic hydrogenation **Q** is converted to **R** (C_7H_{12}). Propose structures for **Q** and **R**.

9.29 Explain in detail how you would distinguish between the following sets of compounds using the indicated method of spectroscopy.

(a) ^1H NMR

(c) ^{13}C NMR

(b) ^{13}C and ^1H NMR

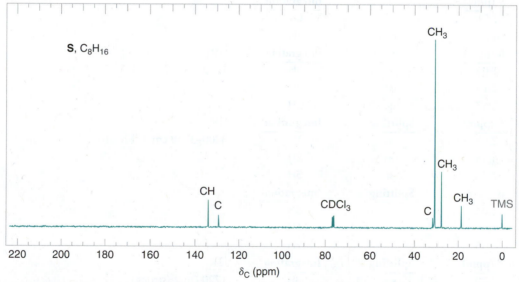

9.30 Compound **S** (C_8H_{16}) reacts with one mole of bromine to form a compound with molecular formula $C_8H_{16}Br_2$. The broadband proton-decoupled ^{13}C spectrum of **S** is given in Fig. 9.49. Propose a structure for **S**.

FIGURE 9.49 The broadband proton-decoupled ^{13}C NMR spectrum of compound **S**, Problem 9.30. Information from the DEPT ^{13}C NMR spectra is given above each peak.

MASS SPECTROMETRY

9.31 A compound with molecular formula C_4H_8O has a strong IR absorption at 1730 cm^{-1}. Its mass spectrum includes key peaks at m/z 44 (the base peak) and m/z 29. Propose a structure for the compound and write fragmentation equations showing how peaks having these m/z values arise.

9.32 In the mass spectrum of 2,6-dimethyl-4-heptanol there are prominent peaks at m/z 87, 111, and 126. Propose reasonable structures for these fragment ions.

9.33 In the mass spectrum of 4-methyl-2-pentanone a McLafferty rearrangement and two other major fragmentation pathways occur. Propose reasonable structures for these fragment ions and specify the m/z value for each.

9.34 What are the masses and structures of the ions produced in the following cleavage pathways?

(a) α-cleavage of 2-methyl-3-hexanone (two pathways)

(b) dehydration of cyclopentanol

(c) McLafferty rearrangement of 4-methyl-2-octanone (two pathways)

9.35 Predict the masses and relative intensities of the peaks in the molecular ion region for the following compound.

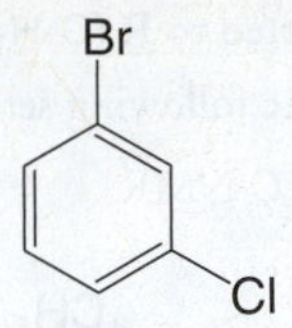

9.36 Ethyl bromide and methoxybenzene (shown below) have the same nominal molecular weights, displaying a significant peak at m/z 108. Regarding their molecular ions, what other features would allow the two compounds to be distinguished on the basis of their mass spectra?

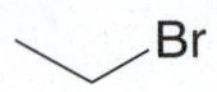

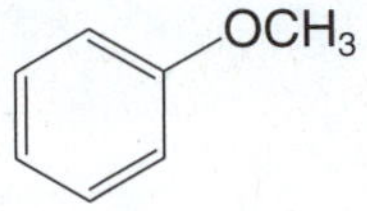

9.37 The homologous series of primary amines, $CH_3(CH_2)_nNH_2$, from CH_3NH_2 to $CH_3(CH_2)_{13}NH_2$ all have their base (largest) peak at m/z 30. What ion does this peak represent, and how is it formed?

INTEGRATED STRUCTURE ELUCIDATION

9.38 Propose a structure that is consistent with each set of ^1H **NMR** data. **IR** data is provided for some compounds.

(a) $C_4H_{10}O$

δ (ppm)	Splitting	Integration
1.28	s	9H
1.35	s	1H

(b) C_3H_7Br

δ (ppm)	Splitting	Integration
1.71	d	6H
4.32	septet	1H

(c) C_4H_8O

δ (ppm)	Splitting	Integration	**IR**
1.05	t	3H	1720 cm^{-1} (strong)
2.13	s	3H	
2.47	q	2H	

(d) C_7H_8O

δ (ppm)	Splitting	Integration	**IR**
2.43	s	1H	3200–3550 cm^{-1} (broad)
4.58	s	2H	
7.28	m	5H	

(e) C_4H_9Cl

δ (ppm)	Splitting	Integration
1.04	d	6H
1.95	m	1H
3.35	d	2H

(f) $C_{15}H_{14}O$

δ (ppm)	Splitting	Integration	**IR**
2.20	s	3H	1720 cm^{-1} (strong)
5.08	s	1H	
7.25	m	10H	

(g) $C_4H_7BrO_2$

δ (ppm)	Splitting	Integration	**IR**
1.08	t	3H	2500–3500 cm^{-1} (broad)
2.07	m	2H	1715 cm^{-1} (strong)
4.23	t	1H	
10.97	s	1H	

(h) C_8H_{10}

δ (ppm)	Splitting	Integration
1.25	t	3H
2.68	q	2H
7.23	m	5H

(i) $C_4H_8O_3$

δ (ppm)	Splitting	Integration	**IR**
1.27	t	3H	2500–3550 cm^{-1} (broad)
3.66	q	2H	1715 cm^{-1} (strong)
4.13	s	2H	
10.95	s	1H	

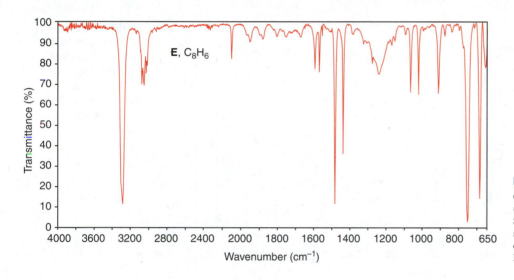

(j) $C_3H_7NO_2$

δ (ppm)	Splitting	Integration
1.55	d	6H
4.67	septet	1H

(k) $C_4H_{10}O_2$

δ (ppm)	Splitting	Integration
3.25	s	6H
3.45	s	4H

(l) $C_5H_{10}O$

δ (ppm)	Splitting	Integration	IR
1.10	d	6H	1720 cm^{-1} (strong)
2.10	s	3H	
2.50	septet	1H	

(m) C_8H_9Br

δ (ppm)	Splitting	Integration
2.0	d	3H
5.15	q	1H
7.35	m	5H

9.39 Propose structures for compounds **E** and **F**. Compound **E** (C_8H_6) reacts with 2 molar equivalents of bromine to form **F** ($C_8H_6Br_4$). **E** has the IR spectrum shown in Fig. 9.50. What are the structures of **E** and **F**?

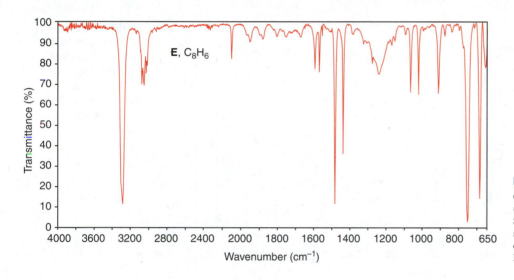

FIGURE 9.50 The IR spectrum of compound **E**, Problem 9.39.

9.40 Regarding compound **J**, $C_2H_xCl_y$, use the ^1H NMR and IR data below to propose a stereochemical formula that is consistent with the data.

^1H NMR	δ (ppm)	Splitting	Integration
	6.3	s	—
IR	3125 cm^{-1}		
	1625 cm^{-1}		
	1280 cm^{-1}		
	820 cm^{-1}		
	695 cm^{-1}		

9.41 When dissolved in $CDCl_3$, a compound (**K**) with the molecular formula $C_4H_8O_2$ gives a ^1H NMR spectrum that consists of a doublet at δ 1.35, a singlet at δ 2.15, a broad singlet at δ 3.75 (1H), and a quartet at δ 4.25 (1H). When dissolved in D_2O, the compound gives a similar ^1H NMR spectrum, with the exception that the signal at δ 3.75 has disappeared. The IR spectrum of the compound shows a strong absorption peak near 1720 cm^{-1}.

(a) Propose a structure for compound **K**.

(b) Explain why the NMR signal at δ 3.75 disappears when D_2O is used as the solvent.

9.42 Compound **T** (C_5H_8O) has a strong IR absorption band at 1745 cm^{-1}. The broad-band proton decoupled ^{13}C spectrum of **T** shows three signals: at δ 220 (C), 23 (CH_2), and 38 (CH_2). Propose a structure for **T**.

9.43 Deduce the structure of the compound that gives the following ^1H, ^{13}C, and IR spectra (Figs. 9.51–9.53). Assign all aspects of the ^1H, and ^{13}C spectra to the structure you propose. Use letters to correlate protons with signals in the ^1H NMR spectrum, and numbers to correlate carbons with signals in the ^{13}C spectrum. The mass spectrum of this compound shows the molecular ion at m/z 96.

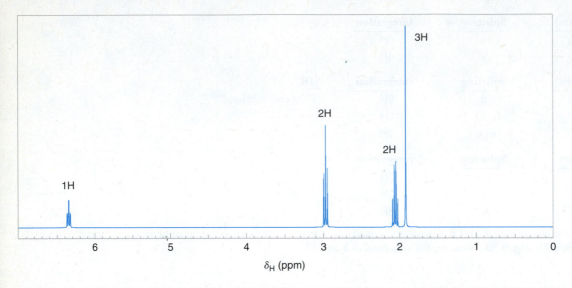

FIGURE 9.51 The ^1H NMR spectrum (simulated) for Problem 9.43.

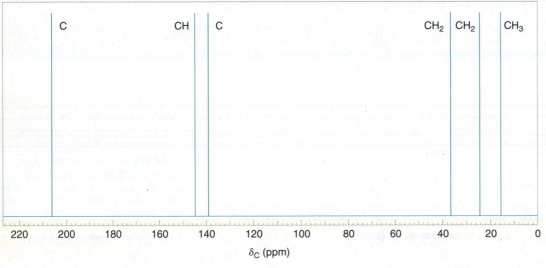

FIGURE 9.52 A simulated broadband proton-decoupled ^{13}C NMR spectrum for Problem 9.43. Information from the DEPT ^{13}C spectra is given above each peak.

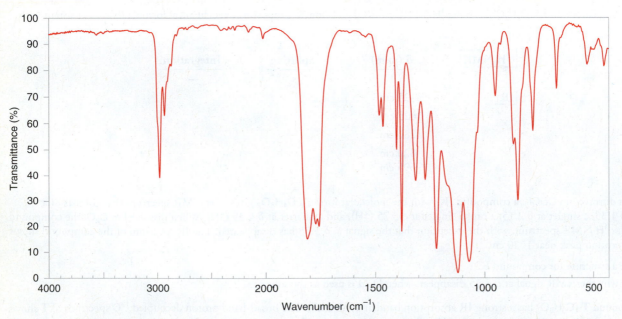

FIGURE 9.53 The IR spectrum for Problem 9.43 (Spectra adapted from Sigma-Aldrich Co. © Sigma-Aldrich Co.)

9.44 Deduce the structure of the compound that gives the following ^{1}H, ^{13}C, and IR spectra (Figs. 9.54–9.56). Assign all aspects of the ^{1}H and ^{13}C spectra to the structure you propose. Use letters to correlate protons with the signals in the ^{1}H NMR spectrum, and numbers to correlate carbons with the signals in the ^{13}C spectrum. The mass spectrum of this compound shows the molecular ion at m/z 148.

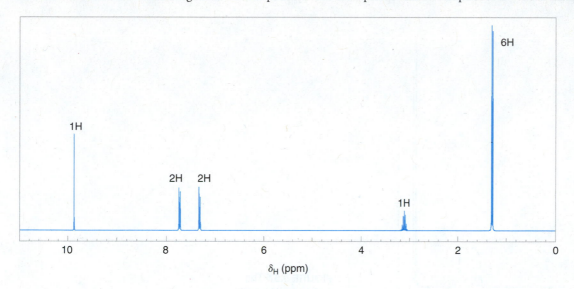

FIGURE 9.54 The 300-MHz ^{1}H NMR spectrum (simulated) for Problem 9.44.

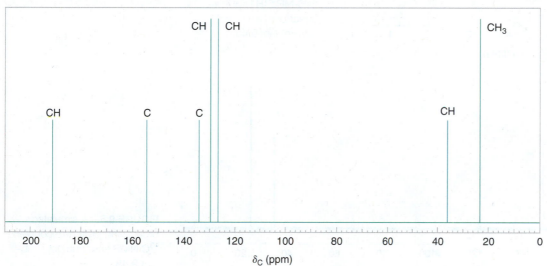

FIGURE 9.55 A simulated broadband proton-decoupled ^{13}C NMR spectrum for Problem 9.44. Information from the DEPT ^{13}C spectra is given above each peak.

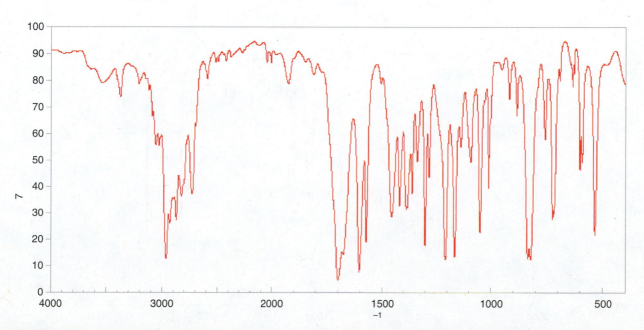

FIGURE 9.56 The IR spectrum for Problem 9.44. (SDBS, National Institute of Advanced Industrial Science and Technology)

9.45 Deduce the structure of the compound that gives the following ^1H, ^{13}C, and IR spectra (Figs. 9.57–9.59). Assign all aspects of the ^1H and ^{13}C spectra to the structure you propose. Use letters to correlate protons with signals in the ^1H NMR spectrum, and numbers to correlate carbons with signals in the ^{13}C spectrum. The mass spectrum of this compound shows the molecular ion at m/z 204.

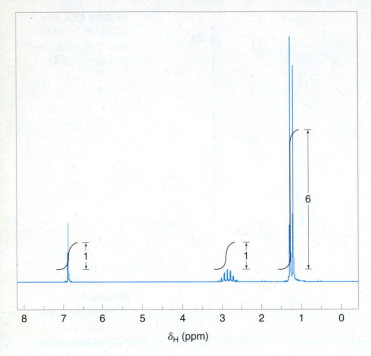

FIGURE 9.57 The 300-MHz ^1H NMR spectrum (simulated) for Problem 9.45.

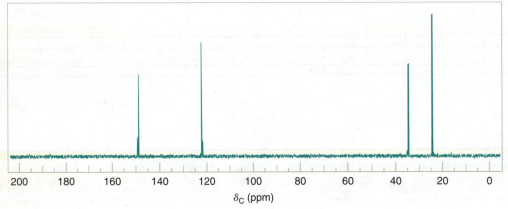

FIGURE 9.58 A simulated broadband proton-decoupled ^{13}C NMR spectrum for Problem 9.45.

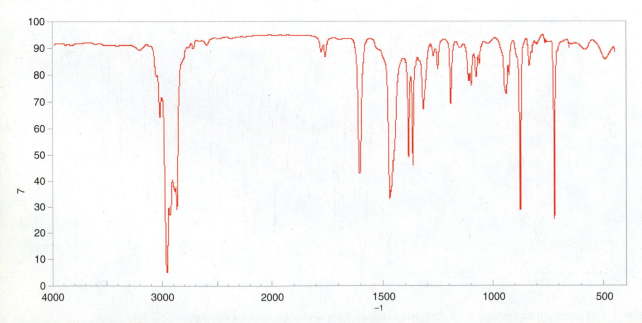

FIGURE 9.59 The IR spectrum for Problem 9.45. (SDBS, National Institute of Advanced Industrial Science and Technology)

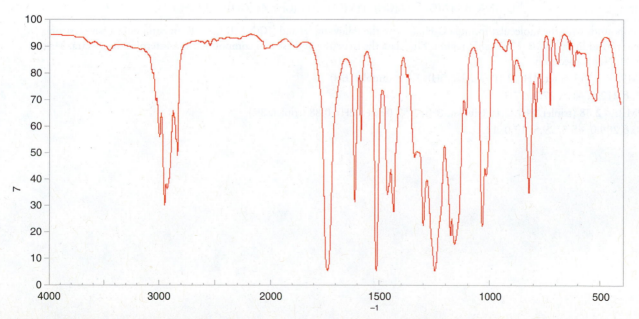

9.46 Deduce the structure of the compound ($C_5H_{10}O_3$) that gives the following ^1H, ^{13}C, and IR spectra (Figs. 9.60–9.62), Assign all aspects of the ^1H and ^{13}C spectra to the structure you propose. Use letters to correlate protons with signals in the ^1H NMR spectrum, and numbers to correlate carbons with signals in the ^{13}C spectrum.

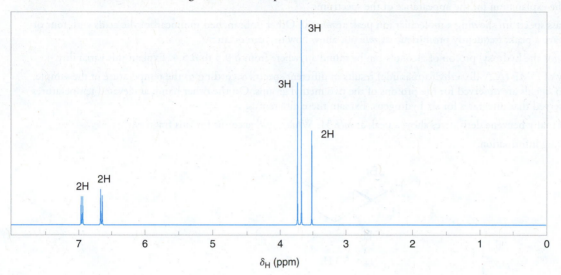

FIGURE 9.60 The 300-MHz ^1H NMR spectrum (simulated) for Problem 9.46.

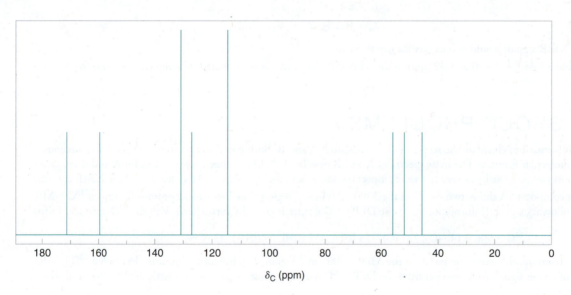

FIGURE 9.61 A simulated broadband proton-decoupled ^{13}C NMR spectrum for Problem 9.46.

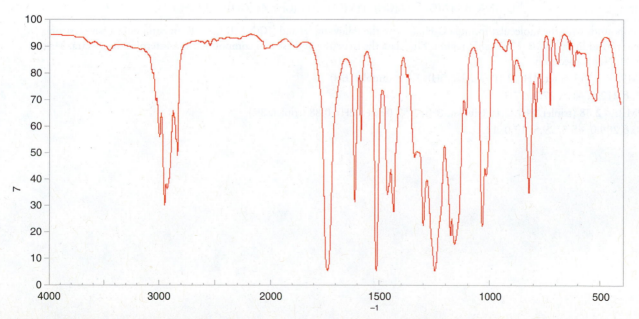

FIGURE 9.62 The IR spectrum for Problem 9.46. (SDBS, National Institute of Advanced Industrial Science and Technology)

CHALLENGE PROBLEMS

9.47 The ^{1}H NMR spectrum of a solution of 1,3-dimethylcyclopentadiene in concentrated sulfuric acid shows three peaks with relative areas of 6:4:1. What is the explanation for the appearance of the spectrum?

9.48 Acetic acid has a mass spectrum showing a molecular ion peak at m/z 60. Other unbranched monocarboxylic acids with four or more carbon atoms also have a peak, frequently prominent, at m/z 60. Show how this can occur.

9.49 The ^{1}H NMR peak for the hydroxyl proton of alcohols can be found anywhere from δ 0.5 to δ 5.4. Explain this variability.

9.50 The ^{1}H NMR study of DMF (*N,N*-dimethylformamide) results in different spectra according to the temperature of the sample. At room temperature, two signals are observed for the protons of the two methyl groups. On the other hand, at elevated temperatures (>130 °C) a singlet is observed that integrates for six hydrogens. Explain these differences.

9.51 The mass spectra of many benzene derivatives show a peak at m/z 51. What could account for this fragment?

9.52 Consider the following information.

$$J_{ab} = 5.3 \text{ Hz}$$
$$J_{ac} = 8.2 \text{ Hz}$$
$$J_{bc} = 10.7 \text{ Hz}$$

(a) How many total ^{1}H NMR signals would you expect for the above molecule?

(b) H_a appears as a doublet of doublets (dd) at 1.32 ppm in the ^{1}H NMR spectrum. Draw a labeled splitting tree diagram for H_a using the coupling constant values given above.

LEARNING GROUP PROBLEMS

1. Given the following information, elucidate the structures of compounds **A** and **B**. Both compounds are soluble in dilute aqueous HCl, and both have the same molecular formula. The mass spectra of **A** and **B** have M^{+} 149. Other spectroscopic data for **A** and **B** are given below. Justify the structures you propose by assigning specific aspects of the data to the structures. Make sketches of the NMR spectra.

(a) The IR spectrum for compound **A** shows two bands in the 3300–3500-cm^{-1} region. The broadband proton-decoupled ^{13}C NMR spectrum displayed the following signals (information from the DEPT ^{13}C spectra is given in parentheses with the ^{13}C chemical shifts):

^{13}C NMR: δ 140 (C), 127 (C), 125 (CH), 118 (CH), 24 (CH$_2$), 13 (CH$_3$)

(b) The IR spectrum for compound **B** shows no bands in the 3300–3500-cm^{-1} region. The broadband proton-decoupled ^{13}C NMR spectrum displayed the following signals (information from the DEPT ^{13}C spectra is given in parentheses with the ^{13}C chemical shifts):

^{13}C NMR: δ 147 (C), 129 (CH), 115 (CH), 111 (CH), 44 (CH$_2$), 13 (CH$_3$)

2. Two compounds with the molecular formula $C_5H_{10}O$ have the following ^{1}H and ^{13}C NMR data. Both compounds have a strong IR absorption band in the 1710–1740-cm^{-1} region. Elucidate the structure of these two compounds and interpret the spectra. Make a sketch of each NMR spectrum.

(a) ^{1}H NMR: δ 2.55 (septet, 1H), 2.10 (singlet, 3H), 1.05 (doublet, 6H)
^{13}C NMR: δ 212.6, 41.5, 27.2, 17.8

(b) ^{1}H NMR: δ 2.38 (triplet, 2H), 2.10 (singlet, 3H), 1.57 (sextet, 2H), 0.88 (triplet, 3H)
^{13}C NMR: δ 209.0, 45.5, 29.5, 17.0, 13.2

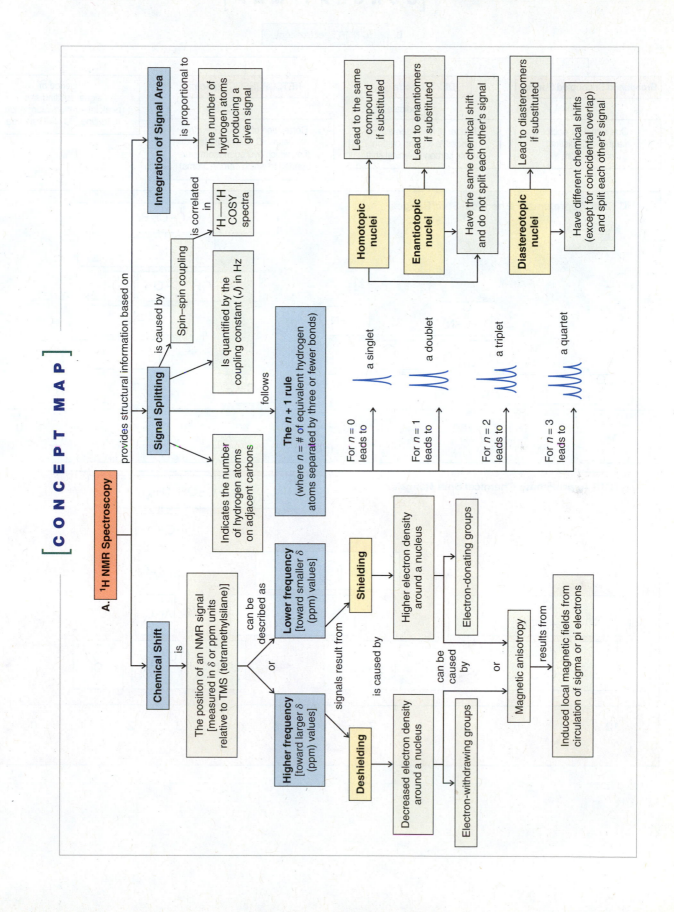

[CONCEPT MAP]

A. ¹H NMR Spectroscopy

provides structural information based on

Chemical Shift

is

The position of an NMR signal [measured in δ or ppm units relative to TMS (tetramethylsilane)]

can be described as

or

Higher frequency [toward larger δ (ppm) values]

Lower frequency [toward smaller δ (ppm) values]

signals result from

is caused by

Deshielding

Decreased electron density around a nucleus

Electron-withdrawing groups

Shielding

Higher electron density around a nucleus

Electron-donating groups

can be caused by

or

Magnetic anisotropy

results from

Induced local magnetic fields from circulation of sigma or pi electrons

Signal Splitting

is caused by

Spin–spin coupling

is correlated in

¹H—¹H COSY spectra

Is quantified by the coupling constant (*J*) in Hz

Indicates the number of hydrogen atoms on adjacent carbons

follows

The *n* + 1 rule (where *n* = # of equivalent hydrogen atoms separated by three or fewer bonds)

For *n* = 0 leads to → a singlet

For *n* = 1 leads to → a doublet

For *n* = 2 leads to → a triplet

For *n* = 3 leads to → a quartet

Integration of Signal Area

is proportional to

The number of hydrogen atoms producing a given signal

Homotopic nuclei

Lead to the same compound if substituted

Enantiotopic nuclei

Lead to enantiomers if substituted

Have the same chemical shift and do not split each other's signal

Diastereotopic nuclei

Lead to diastereomers if substituted

Have different chemical shifts (except for coincidental overlap) and split each other's signal

[C O N C E P T   M A P]

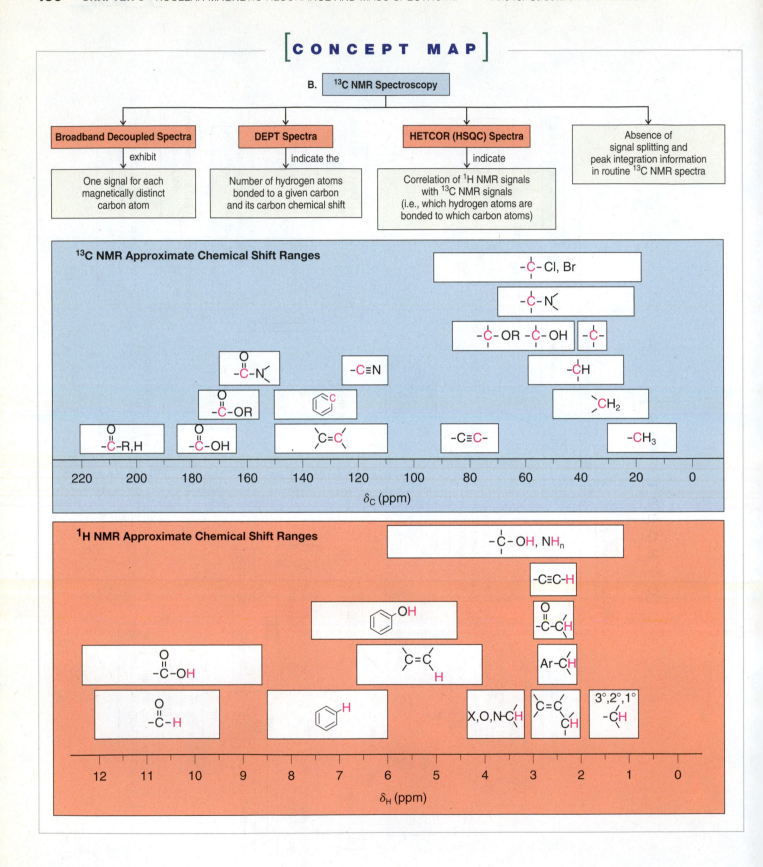

CHAPTER

10

Radical Reactions

Unpaired electrons lead to many burning questions about radical types of reactivity. In fact, species with unpaired electrons are called radicals, and they are involved in the chemistry of burning, aging, disease, as well as in reactions related to destruction of the ozone layer and the synthesis of products that enhance our everyday lives. For example, polyethylene, which can have a molecular weight from the thousands to the millions, and practical uses ranging from plastic films and wraps to water bottles, bulletproof vests, and hip and knee replacements, is made by a reaction involving radicals. Oxygen that we breathe and nitric oxide that serves as a chemical signaling agent for some fundamental biological processes are both molecules with unpaired electrons. Highly colored natural compounds like those found in blueberries and carrots react with radicals and may protect us from undesirable biological radical reactions. Large portions of the economy hinge on radicals, as well, from reactions used to make polymers like polyethylene, to the target action of pharmaceuticals like Cialis, Levitra, and Viagra, which act on a nitric oxide biological signaling pathway.

IN THIS CHAPTER WE WILL CONSIDER:

- the properties of radicals, their formation, and their reactivity
- significant radical-based reactions in nature

[**WHY** DO THESE TOPICS MATTER?]   At the end of the chapter, we will show that there is a natural molecule that combines radical chemistry and molecular shape in a way that can cause cell death. Chemists have used this knowledge to fashion a few anticancer drugs.

10.1 INTRODUCTION: HOW RADICALS FORM AND HOW THEY REACT

So far almost all of the reactions whose mechanisms we have studied have been **ionic reactions**. Ionic reactions are those in which covalent bonds break **heterolytically** and in which ions are involved as reactants, intermediates, or products.

Another broad category of reactions has mechanisms that involve **homolysis** of covalent bonds with the production of intermediates possessing unpaired electrons called **radicals** (or **free radicals**):

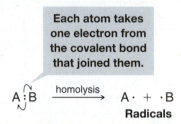

Each atom takes one electron from the covalent bond that joined them.

$$A\!:\!B \xrightarrow{\text{homolysis}} A\cdot + \cdot B$$
Radicals

Helpful Hint

A single-barbed curved arrow shows movement of one electron.

This simple example illustrates the way we use **single-barbed curved arrows** to show the movement of **a single electron** (not of an electron pair as we have done earlier). In this instance, each group, A and B, comes away with one of the electrons of the covalent bond that joined them.

10.1A Production of Radicals

- Energy in the form of heat or light must be supplied to cause homolysis of covalent bonds (Section 10.2).

For example, compounds with an oxygen–oxygen single bond, called **peroxides**, undergo homolysis readily when heated, because the oxygen–oxygen bond is weak. The products are two radicals, called alkoxyl radicals:

$$R—\overset{..}{\underset{..}{O}}\!:\!\overset{..}{\underset{..}{O}}—R \xrightarrow{\text{heat}} 2\,R—\overset{..}{\underset{..}{O}}\cdot$$

Homolysis of a dialkyl peroxide.

Dialkyl peroxide   **Alkoxyl radicals**

Halogen molecules (X_2) also contain a relatively weak bond. As we shall soon see, halogens undergo homolysis readily when heated or when irradiated with light of a wavelength that can be absorbed by the halogen molecule:

$$:\!\overset{..}{\underset{..}{X}}\!:\!\overset{..}{\underset{..}{X}}\!: \xrightarrow[\substack{\text{heat} \\ \text{or light } (h\nu)}]{\text{homolysis}} 2\,:\!\overset{..}{\underset{..}{X}}\cdot$$

Homolysis of a halogen molecule.

The products of this homolysis are halogen atoms, and because halogen atoms contain an unpaired electron, they are radicals.

10.1B Reactions of Radicals

- Almost all small radicals are short-lived, highly reactive species.

When radicals collide with other molecules, they tend to react in a way that leads to pairing of their unpaired electron. One way they can do this is by abstracting an atom from another molecule. To abstract an atom means to remove an atom by homolytic bond cleavage as the atom forms a bond with another radical. For example, a halogen atom may abstract a hydrogen atom from an alkane. This **hydrogen abstraction** gives the halogen atom an electron (from the hydrogen atom) to pair with its unpaired electron. Notice, however, that the other product of this abstraction *is another radical intermediate*, in this case, an alkyl radical, R·, which goes on to react further, as we shall see in this chapter.

[A MECHANISM FOR THE REACTION ─ Hydrogen Atom Abstraction]

General Reaction

$$:\ddot{X}\cdot \; + \; H\!:\!R \longrightarrow \; :\ddot{X}\!:\!H \; + \; R\cdot$$

Reactive	Alkane		Alkyl radical
radical			intermediate
intermediate			(reacts further)

Specific Example

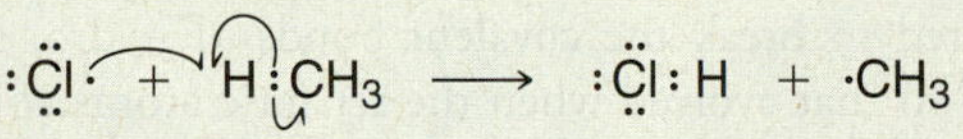

$$:\ddot{Cl}\cdot \; + \; H\!:\!CH_3 \longrightarrow \; :\ddot{Cl}\!:\!H \; + \; \cdot CH_3$$

Chlorine	Methane		Methyl radical
atom			intermediate
(a radical)			(reacts further)

This behavior is characteristic of **radical reactions**. Consider another example, one that shows another way in which radicals can react: they can combine with a compound containing a multiple bond to produce a new radical, which goes on to react further. (We shall study reactions of this type in Section 10.10.)

[A MECHANISM FOR THE REACTION ─ Radical Addition to a π Bond]

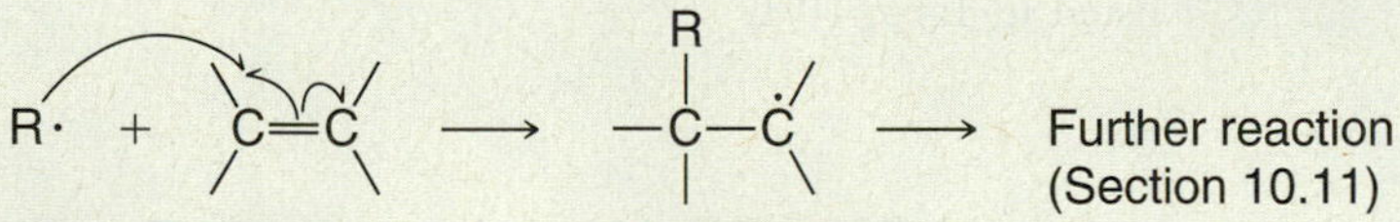

$$R\cdot \; + \; C\!=\!C \longrightarrow \; -\overset{R}{\underset{|}{C}}-\dot{C} \longrightarrow \; \text{Further reaction (Section 10.11)}$$

Reactive	Alkene	New radical
alkyl radical		intermediate
intermediate		

THE CHEMISTRY OF... Acne Medications

It turns out that although certain peroxides are great at initiating radical reactions, peroxy radicals also have many valuable uses on their own. For example, benzoyl peroxide is an active ingredient typically found in many acne medications that breaks apart and forms radicals through the warmth of our skin and exposure to light. These radicals can then kill the bacteria that cause break-outs.

The same compound is also used as a whitening and bleaching agent. As we will see in Chapter 13, many colored compounds possess double bonds in conjugation; the benzoyl peroxide radicals can add to those bonds, break their conjugation, and remove their color to leave behind new white materials. If you have ever wiped your face after using an acne medication with a colored towel, you may already have seen these effects!

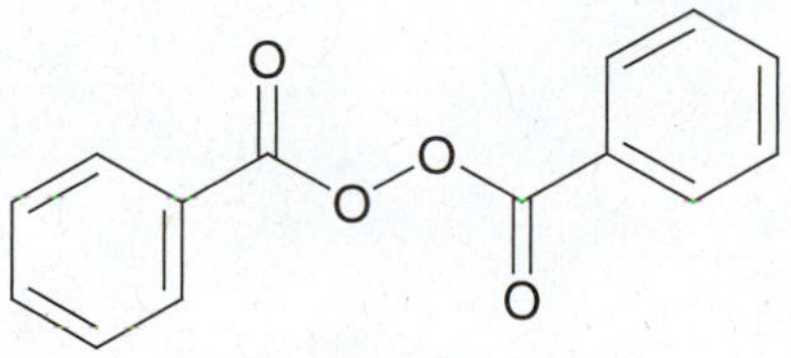

Benzoyl peroxide

10.2 HOMOLYTIC BOND DISSOCIATION ENERGIES ($DH°$)

When atoms combine to form molecules, energy is released as covalent bonds form. The molecules of the products have lower enthalpy than the separate atoms. When hydrogen atoms combine to form hydrogen molecules, for example, the reaction is *exothermic*; it evolves 436 kJ of heat for every mole of hydrogen that is produced. Similarly, when chlorine atoms combine to form chlorine molecules, the reaction evolves 243 kJ mol^{-1} of chlorine produced:

$$H\cdot + H\cdot \longrightarrow H{-}H \qquad \Delta H° = -436 \text{ kJ mol}^{-1}$$
$$Cl\cdot + Cl\cdot \longrightarrow Cl{-}Cl \qquad \Delta H° = -243 \text{ kJ mol}^{-1}$$

> **Bond formation is an exothermic process: $\Delta H°$ is negative.**

Reactions in which only bond breaking occurs are always endothermic. The energy required to break the covalent bonds of hydrogen or chlorine homolytically is exactly equal to that evolved when the separate atoms combine to form molecules. In the bond cleavage reaction, however, $\Delta H°$ is positive:

$$H{-}H \longrightarrow H\cdot + H\cdot \qquad \Delta H° = +436 \text{ kJ mol}^{-1}$$
$$Cl{-}Cl \longrightarrow Cl\cdot + Cl\cdot \qquad \Delta H° = +243 \text{ kJ mol}^{-1}$$

> **Bond breaking is an endothermic process: $\Delta H°$ is positive.**

- Energy must be supplied to break covalent bonds.
- The energies required to break covalent bonds homolytically are called **homolytic bond dissociation energies**, and they are usually abbreviated by the symbol **$DH°$**.

The homolytic bond dissociation energies of hydrogen and chlorine, for example, can be written in the following way:

$$H{-}H \qquad\qquad Cl{-}Cl$$
$$(DH° = 436 \text{ kJ mol}^{-1}) \qquad (DH° = 243 \text{ kJ mol}^{-1})$$

The homolytic bond dissociation energies of a variety of covalent bonds have been determined experimentally or calculated from related data. Some of these $DH°$ values are listed in Table 10.1.

10.2A HOW TO Use Homolytic Bond Dissociation Energies to Determine the Relative Stabilities of Radicals

Homolytic bond dissociation energies also provide us with a convenient way to estimate the relative stabilities of radicals. If we examine the data given in Table 10.1, we find the following values of $DH°$ for the primary and secondary C—H bonds of propane:

$$(DH° = 423 \text{ kJ mol}^{-1}) \qquad (DH° = 413 \text{ kJ mol}^{-1})$$

This means that for the reaction in which the designated C—H bonds are broken homolytically, the values of $\Delta H°$ are those given here.

$$\text{(propane)} {-}H \longrightarrow \text{(propyl radical)} + H\cdot \qquad \Delta H° = +423 \text{ kJ mol}^{-1}$$

**Propyl radical
(a 1° radical)**

$$\text{(propane)} {-}H \longrightarrow \text{(isopropyl radical)} + H\cdot \qquad \Delta H° = +413 \text{ kJ mol}^{-1}$$

**Isopropyl radical
(a 2° radical)**

These reactions resemble each other in two respects: they both begin with the same alkane (propane), and they both produce an alkyl radical and a hydrogen atom. They differ, however, in the amount of energy required and in the type of carbon radical produced. These two differences are related to each other.

- Alkyl radicals are classified as being 1°, 2°, or 3° based on the carbon atom that has the unpaired electron, the same way that we classify carbocations based on the carbon atom with the positive charge.

More energy must be supplied to produce a primary alkyl radical (the propyl radical) from propane than is required to produce a secondary carbon radical (the isopropyl radical) from the same compound. This must mean that the primary radical has absorbed more energy and thus has greater *potential energy*. Because the relative stability of a chemical species is inversely related to its potential energy, the secondary radical must be the *more stable* radical (Fig. 10.1a). In fact, the secondary isopropyl radical is more stable than the primary propyl radical by 10 kJ mol^{-1}.

TABLE 10.1 SINGLE-BOND HOMOLYTIC DISSOCIATION ENERGIES (*DH°*) AT 25 °C^{a}

$$A{:}B \longrightarrow A{\cdot} + B{\cdot}$$

Bond Broken (shown in red)	kJ mol^{-1}	Bond Broken (shown in red)	kJ mol^{-1}	Bond Broken (shown in red)	kJ mol^{-1}
H—H	436	CH_3CH_2—OCH_3	352	CH_2=$CHCH_2$—H	369
D—D	443	$CH_3CH_2CH_2$—H	423	CH_2=CH—H	465
F—F	159	$CH_3CH_2CH_2$—F	444	C_6H_5—H	474
Cl—Cl	243	$CH_3CH_2CH_2$—Cl	354	HC≡C—H	547
Br—Br	193	$CH_3CH_2CH_2$—Br	294	CH_3—CH_3	378
I—I	151	$CH_3CH_2CH_2$—I	239	CH_3CH_2—CH_3	371
H—F	570	$CH_3CH_2CH_2$—OH	395	$CH_3CH_2CH_2$—CH_3	374
H—Cl	432	$CH_3CH_2CH_2$—OCH_3	355	CH_3CH_2—CH_2CH_3	343
H—Br	366	$(CH_3)_2CH$—H	413	$(CH_3)_2CH$—CH_3	371
H—I	298	$(CH_3)_2CH$—F	439	$(CH_3)_3C$—CH_3	363
CH_3—H	440	$(CH_3)_2CH$—Cl	355	HO—H	499
CH_3—F	461	$(CH_3)_2CH$—Br	298	HOO—H	356
CH_3—Cl	352	$(CH_3)_2CH$—I	222	HO—OH	214
CH_3—Br	293	$(CH_3)_2CH$—OH	402	$(CH_3)_3CO$—$OC(CH_3)_3$	157
CH_3—I	240	$(CH_3)_2CH$—OCH_3	359		
CH_3—OH	387	$(CH_3)_2CHCH_2$—H	422	$C_6H_5\overset{\displaystyle O}{\overset{\|}{C}}O$—$\overset{\displaystyle O}{\overset{\|}{O}}CC_6H_5$	139
CH_3—OCH_3	348	$(CH_3)_3C$—H	400	CH_3CH_2O—OCH_3	184
CH_3CH_2—H	421	$(CH_3)_3C$—Cl	349	CH_3CH_2O—H	431
CH_3CH_2—F	444	$(CH_3)_3C$—Br	292		
CH_3CH_2—Cl	353	$(CH_3)_3C$—I	227	$CH_3\overset{\displaystyle O}{\overset{\|}{C}}$—H	364
CH_3CH_2—Br	295	$(CH_3)_3C$—OH	400		
CH_3CH_2—I	233	$(CH_3)_3C$—OCH_3	348		
CH_3CH_2—OH	393	$C_6H_5CH_2$—H	375		

^{a}Data compiled from the National Institute of Standards (NIST) Standard Reference Database Number 69, July 2001 Release, Accessed via NIST Chemistry WebBook (http://webbook.nist.gov/chemistry/) Copyright 2000. Data from CRC Handbook of Chemistry and Physics, Updated 3rd Electronic Edition; Lide, David R., ed. DH° values were obtained directly or calculated from heat of formation (H$_f$) data using the equation DH° [A-B]= H$_f$ [A.] + H$_f$ [B.] - H$_f$ [A-B].

We can use the data in Table 10.1 to make a similar comparison of the *tert*-butyl radical (a 3° radical) and the isobutyl radical (a 1° radical) relative to isobutane:

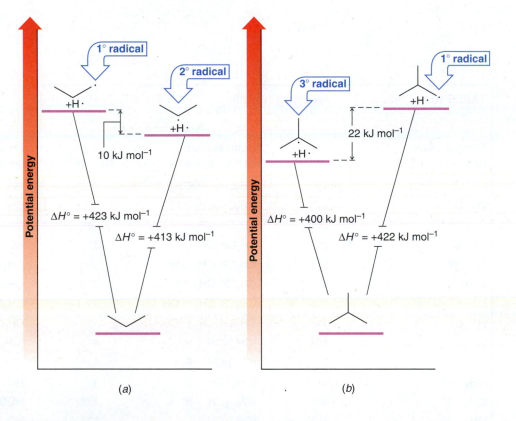

tert-**Butyl radical**
(a 3° radical)   + H· $\Delta H° = +400$ kJ mol^{-1}

Isobutyl radical
(a 1° radical)   + H· $\Delta H° = +422$ kJ mol^{-1}

Here we find (Fig. 10.1*b*) that the difference in stability of the two radicals is even larger. The tertiary radical is more stable than the primary radical by 22 kJ mol^{-1}.

FIGURE 10.1 (*a*) Comparison of the potential energies of the propyl radical (+H·) and the isopropyl radical (+H·) relative to propane. The isopropyl radical (a 2° radical) is more stable than the 1° radical by 10 kJ mol^{-1}. (*b*) Comparison of the potential energies of the *tert*-butyl radical (+H·) and the isobutyl radical (+H·) relative to isobutane. The 3° radical is more stable than the 1° radical by 22 kJ mol^{-1}.

(a)   *(b)*

The kind of pattern that we find in these examples is found with alkyl radicals generally.

- **Overall, the relative stabilities of radicals are 3° > 2° > 1° > methyl.**

Tertiary (3°)	>	Secondary (2°)	>	Primary (1°)	>	Methyl

$$\underset{\overset{|}{C}}{\overset{C}{C-C\cdot}} \quad > \quad \underset{\overset{|}{H}}{\overset{C}{C-C\cdot}} \quad > \quad \underset{\overset{|}{H}}{\overset{H}{C-C\cdot}} \quad > \quad \underset{\overset{|}{H}}{\overset{H}{H-C\cdot}}$$

- **The order of stability of alkyl radicals is the same as for carbocations** (Section 6.11B).

Although alkyl radicals are uncharged, the carbon that bears the odd electron is *electron deficient*. Therefore, alkyl groups attached to this carbon provide a stabilizing effect through hyperconjugation, and the more alkyl groups bonded to it, the more stable the radical is. Thus, the reasons for the relative stabilities of radicals and carbocations are similar.

Helpful Hint

Knowing the relative stability of radicals is important for predicting reaction pathways.

Classify each of the following radicals as being 1°, 2°, or 3°, and rank them in order of decreasing stability.

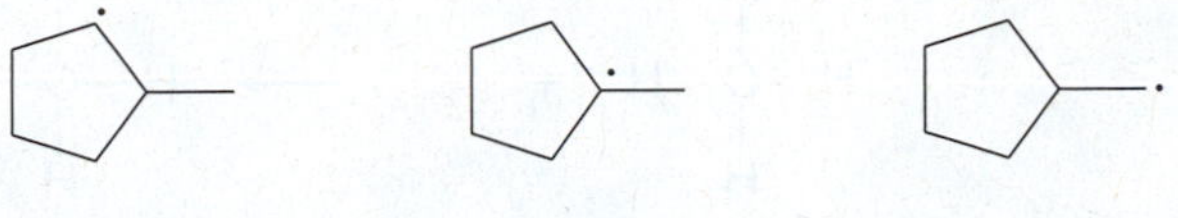

A   **B**   **C**

STRATEGY AND ANSWER: We examine the carbon bearing the unpaired electron in each radical to classify the radical as to its type. **B** is a tertiary radical (the carbon bearing the unpaired electron is tertiary) and is, therefore, most stable. **C** is a primary radical and is least stable. **A**, being a secondary radical, falls in between. The order of stability is **B > A > C**.

••• ⋯⋯

PRACTICE PROBLEM 10.1

List the following radicals in order of decreasing stability:

·CH₃

10.3 REACTIONS OF ALKANES WITH HALOGENS

- Alkanes react with molecular halogens to produce alkyl halides by a **substitution reaction** called **radical halogenation**.

A general reaction showing formation of a monohaloalkane by radical halogenation is shown below. It is called radical halogenation because, as we shall see, the mechanism involves species with unpaired electrons called radicals. This reaction is not a nucleophilic substitution reaction.

$$R-H + X_2 \longrightarrow R-X + HX$$

- A halogen atom replaces one or more of the hydrogen atoms of the alkane, and the corresponding hydrogen halide is formed as a by-product.

Only fluorine, chlorine, and bromine react this way with alkanes. Iodine is essentially unreactive due to unfavorable reaction energetics.

10.3A Multiple Halogen Substitution

One complicating factor of alkane halogenations is that multiple substitutions almost always occur unless we use an excess of the alkane (see Solved Problem 10.2). The following example illustrates this phenomenon. If we mix an equimolar ratio of methane and chlorine (both substances are gases at room temperature) and then either heat the mixture or irradiate it with light of the appropriate wavelength, a reaction begins to occur vigorously and ultimately produces the following mixture of products:

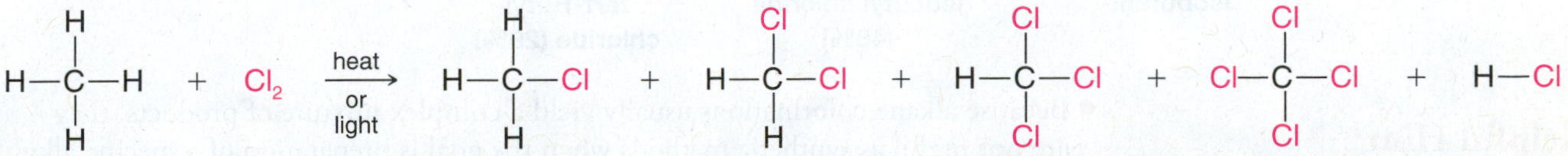

H		H	Cl	Cl	Cl
H—C—H + Cl₂ →(heat or light) H—C—Cl	+	H—C—Cl	+ H—C—Cl	+ Cl—C—Cl	+ H—Cl
H		H	H	Cl	Cl
Methane Chlorine		Chloromethane	Dichloromethane	Trichloromethane Tetrachloromethane	Hydrogen chloride

(The sum of the number of moles of each chlorinated methane produced equals the number of moles of methane that reacted.)

To understand the formation of this mixture, we need to consider how the concentration of reactants and products changes as the reaction proceeds. At the outset, the only compounds that are present in the mixture are chlorine and methane, and the only reaction that can take place is one that produces chloromethane and hydrogen chloride:

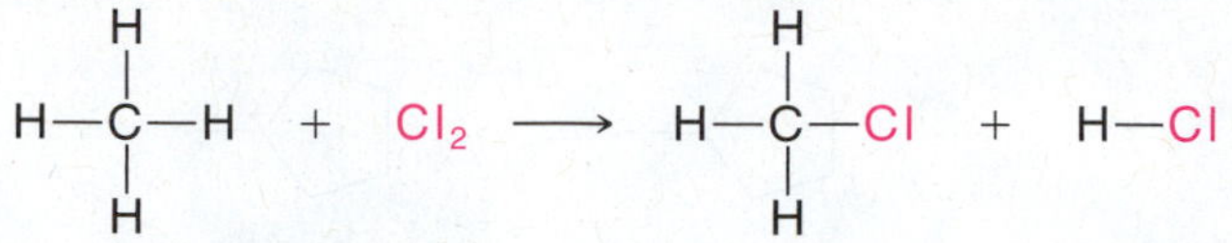

As the reaction progresses, however, the concentration of chloromethane in the mixture increases, and a second substitution reaction begins to occur. Chloromethane reacts with chlorine to produce dichloromethane:

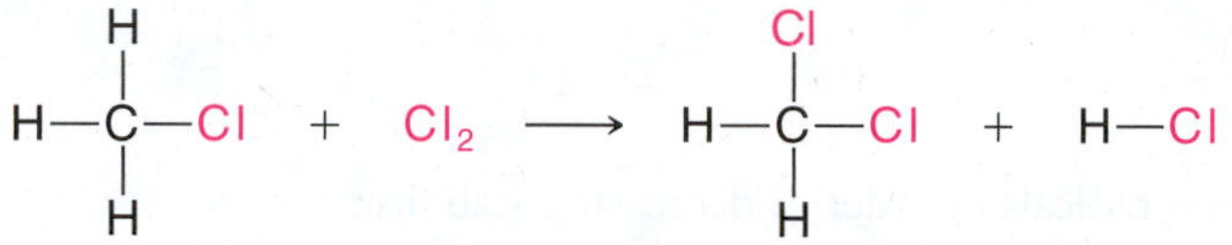

The dichloromethane produced can then react to form trichloromethane, and trichloromethane, as it accumulates in the mixture, can react with chlorine to produce tetrachloromethane. Each time a substitution of —Cl for —H takes place, a molecule of H—Cl is produced.

●●● SOLVED PROBLEM 10.2

If the goal of a synthesis is to prepare chloromethane (CH_3Cl), its formation can be maximized and the formation of CH_2Cl_2, $CHCl_3$, and CCl_4 minimized by using a large excess of methane in the reaction mixture. Explain why this is possible.

ANSWER: The use of a large excess of methane maximizes the probability that chlorine will attack methane molecules because the concentration of methane in the mixture will always be relatively large. It also minimizes the probability that chlorine will attack molecules of CH_3Cl, CH_2Cl_2, and $CHCl_3$, because their concentrations will always be relatively small. After the reaction is over, the unreacted excess methane can be recovered and recycled.

10.3B Lack of Chlorine Selectivity

Chlorination of most higher alkanes gives a mixture of isomeric monochlorinated products as well as more highly halogenated compounds.

- Chlorine is relatively *unselective*; it does not discriminate greatly among the different types of hydrogen atoms (primary, secondary, and tertiary) in an alkane.

An example is the light-promoted chlorination of isobutane:

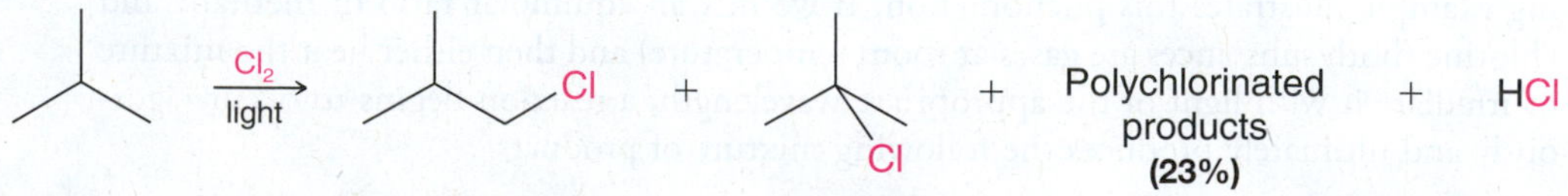

Isobutane   Isobutyl chloride   *tert*-Butyl   Polychlorinated   + HCl
                  (48%)   chloride (29%)   products
                                                            (23%)

- Because alkane chlorinations usually yield a complex mixture of products, they are not useful as synthetic methods when the goal is preparation of a specific alkyl chloride.
- An exception is the halogenation of an alkane (or cycloalkane) whose hydrogen atoms *are all equivalent* (i.e., homotopic). [Homotopic hydrogen atoms are defined as those that on replacement by some other group (e.g., chlorine) yield the same compound (Section 9.8).]

Neopentane, for example, can form only one monohalogenation product, and the use of a large excess of neopentane minimizes polychlorination:

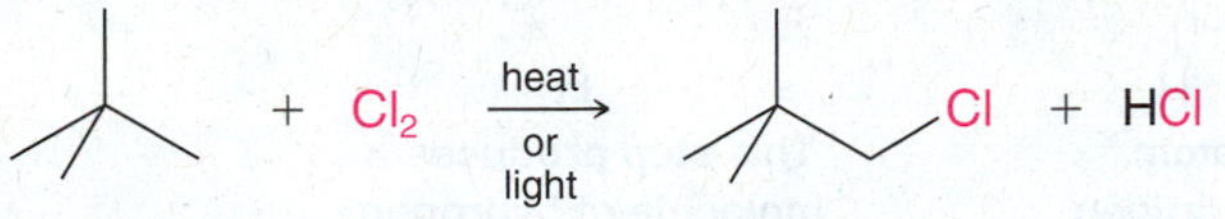

Neopentane   **Neopentyl chloride**
(excess)

- Bromine is generally less reactive toward alkanes than chlorine, and bromine is *more selective* in the site of attack when it does react.

We shall examine the selectivity of bromination further in Section 10.5A.

10.4 CHLORINATION OF METHANE: MECHANISM OF REACTION

The reaction of methane with chlorine (in the gas phase) provides a good example for studying the mechanism of radical **halogenation**.

$$CH_4 + Cl_2 \longrightarrow CH_3Cl + HCl \ (+ \ CH_2Cl_2, \ CHCl_3, \ and \ CCl_4)$$

Several experimental observations help in understanding the mechanism of this reaction:

1. **The reaction is promoted by heat or light**. At room temperature methane and chlorine do not react at a perceptible rate as long as the mixture is kept away from light. Methane and chlorine do react, however, at room temperature if the gaseous reaction mixture is irradiated with UV light at a wavelength absorbed by Cl_2, and they react in the dark if the gaseous mixture is heated to temperatures greater than 100 °C.

2. **The light-promoted reaction is highly efficient**. A relatively small number of light photons permits the formation of relatively large amounts of chlorinated product.

A mechanism that is consistent with these observations has several steps, shown below. The first step involves the dissociation of a chlorine molecule, by heat or light, into two chlorine atoms. The second step involves hydrogen abstraction by a chlorine atom.

[**A MECHANISM FOR THE REACTION** — **Radical Chlorination of Methane**]

Reaction

$$CH_4 + Cl_2 \xrightarrow[\text{or light}]{\text{heat}} CH_3Cl + HCl$$

Mechanism

Chain Initiation

Step 1: Halogen dissociation

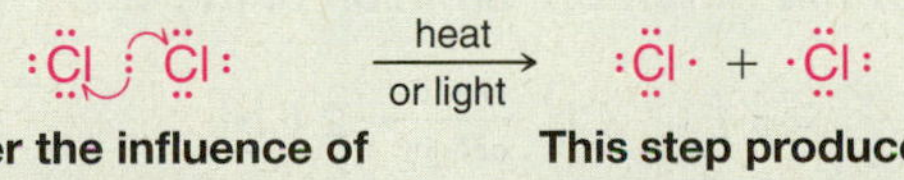

$$:\ddot{C}l \frown \ddot{C}l: \xrightarrow[\text{or light}]{\text{heat}} :\ddot{C}l\cdot \ + \ \cdot\ddot{C}l:$$

Under the influence of heat or light a molecule of chlorine dissociates; each atom takes one of the bonding electrons.

This step produces two highly reactive chlorine atoms.

(mechanism continues on the next page)

Chain Propagation

Step 2: Hydrogen abstraction

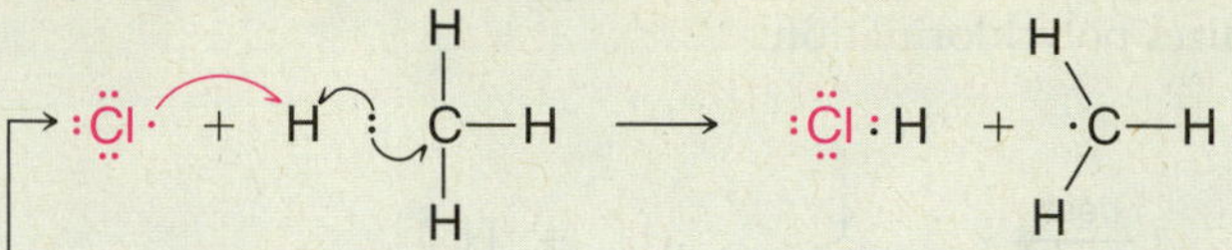

A chlorine atom abstracts a hydrogen atom from a methane molecule.

This step produces a molecule of hydrogen chloride and a methyl radical.

Step 3: Halogen abstraction

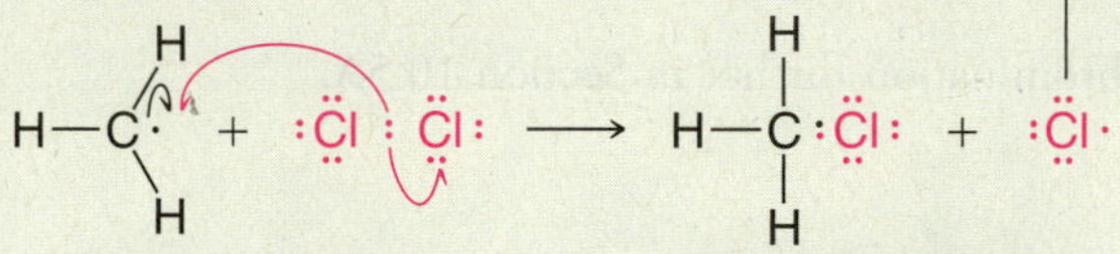

A methyl radical abstracts a chlorine atom from a chlorine molecule.

This step produces a molecule of chloromethane and a chlorine atom. The chlorine atom can now cause a repetition of step 2.

Chain Termination

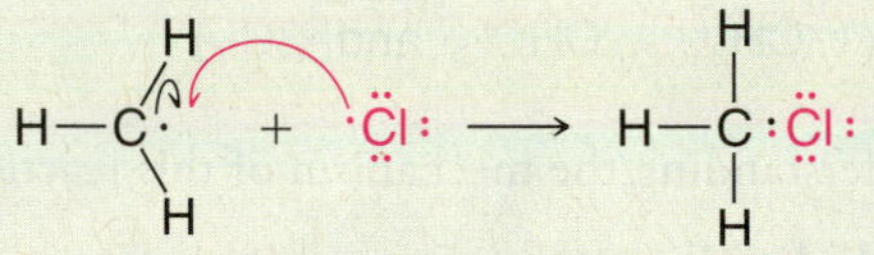

Coupling of any two radicals depletes the supply of reactive intermediates and terminates the chain. Several pairings are possible for radical coupling termination steps (see the text).

> **Helpful Hint**
>
> *Remember:* These conventions are used in illustrating reaction mechanisms in this text.
>
> 1. Curved arrows ⌢ or ⌢ always show the direction of movement of electrons.
> 2. Single-barbed arrows ⌢ show the attack (or movement) of an unpaired electron.
> 3. Double-barbed arrows ⌢ show the attack (or movement) of an electron pair.

In step 3 the highly reactive methyl radical reacts with a chlorine molecule by abstracting a chlorine atom. This results in the formation of a molecule of chloromethane (one of the ultimate products of the reaction) and a *chlorine atom*. The latter product is particularly significant, for the chlorine atom formed in step 3 can attack another methane molecule and cause a repetition of step 2. Then, step 3 is repeated, and so forth, for hundreds or thousands of times. (With each repetition of step 3 a molecule of chloromethane is produced.)

- This type of sequential, stepwise mechanism, in which each step generates the reactive intermediate that causes the next cycle of the reaction to occur, is called a **chain reaction**.

Step 1 is called the **chain-initiating step**. In the chain-initiating step *radicals are created*. Steps 2 and 3 are called **chain-propagating steps**. In chain-propagating steps *one radical generates another*.

Chain Initiation: *creation of radicals*

Step 1 $Cl_2 \xrightarrow[\text{or light}]{\text{heat}} 2\,Cl\cdot$

Chain Propagation: *reaction and regeneration of radicals*

Step 2 $CH_4 + Cl\cdot \longrightarrow \cdot CH_3 + H{-}Cl$

Step 3 $\cdot CH_3 + Cl_2 \longrightarrow CH_3Cl + Cl\cdot$

The chain nature of the reaction accounts for the observation that the light-promoted reaction is highly efficient. The presence of a relatively few atoms of chlorine at any given moment is all that is needed to cause the formation of many thousands of molecules of chloromethane.

What causes the chain reaction to terminate? Why does one photon of light not promote the **chlorination** of all of the methane molecules present? We know that this does not happen because we find that, at low temperatures, continuous irradiation is required or the reaction slows and stops. The answer to these questions is the existence of ***chain-terminating steps***: steps that happen infrequently but occur often enough to *use up one or both of the reactive intermediates*. The continuous replacement of intermediates used up by chain-terminating steps requires continuous irradiation. Plausible chain-terminating steps are as follows.

Chain Termination: *consumption of radicals (e.g., by coupling)*

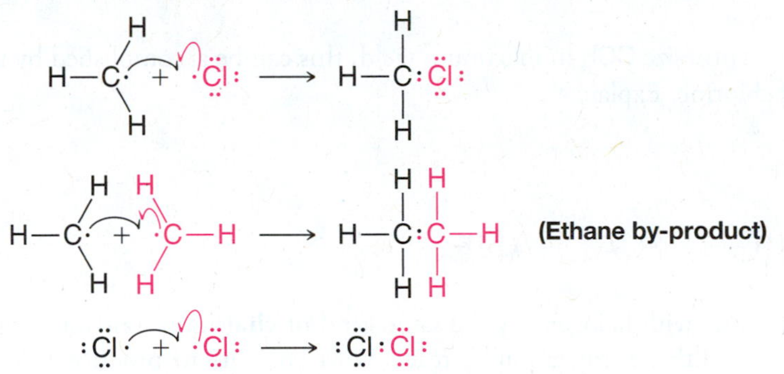

(Ethane by-product)

Our radical mechanism also explains how the reaction of methane with chlorine produces the more highly halogenated products, CH_2Cl_2, $CHCl_3$, and CCl_4 (as well as additional HCl). As the reaction progresses, chloromethane (CH_3Cl) accumulates in the mixture and its hydrogen atoms, too, are susceptible to abstraction by chlorine. Thus chloromethyl radicals are produced that lead to dichloromethane (CH_2Cl_2).

Side Reactions: *multihalogenated by-product formation*

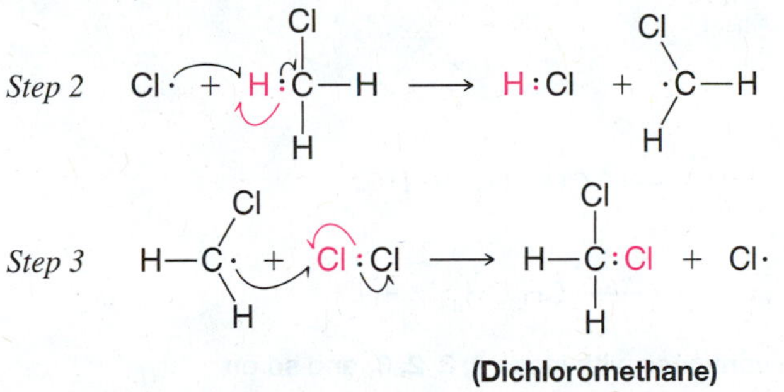

Step 2

Step 3

(Dichloromethane)

Then step 2 is repeated, then step 3 is repeated, and so on. Each repetition of step 2 yields a molecule of HCl, and each repetition of step 3 yields a molecule of CH_2Cl_2.

●●● **SOLVED PROBLEM 10.3**

When methane is chlorinated, among the products found are traces of chloroethane. How is it formed? Of what significance is its formation?

STRATEGY AND ANSWER: A small amount of ethane is formed by the combination of two methyl radicals:

$$2 \cdot CH_3 \longrightarrow CH_3 : CH_3$$

The ethane byproduct formed by coupling then reacts with chlorine in a radical halogenation reaction (see Section 10.5) to form chloroethane.

The significance of this observation is that it is evidence for the proposal that the combination of two methyl radicals is one of the chain-terminating steps in the chlorination of methane.

• • •

PRACTICE PROBLEM 10.2 Suggest a method for separating and isolating the CH_3Cl, CH_2Cl_2, $CHCl_3$, and CCl_4 that may be formed as a mixture when methane is chlorinated. (You may want to consult a handbook.) What analytical method could be used to separate this mixture and give structural information about each component?

• • •

PRACTICE PROBLEM 10.3 How would the molecular ion peaks in the respective mass spectra of CH_3Cl, CH_2Cl_2, $CHCl_3$, and CCl_4 differ on the basis of the number of chlorines? (Remember that chlorine has isotopes ^{35}Cl and ^{37}Cl found in a $3:1$ ratio.)

• • •

PRACTICE PROBLEM 10.4 If the goal is to synthesize CCl_4 in maximum yield, this can be accomplished by using a large excess of chlorine. Explain.

10.5 HALOGENATION OF HIGHER ALKANES

Higher alkanes react with halogens by the same kind of **chain mechanism** as those that we have just seen. Ethane, for example, reacts with chlorine to produce chloroethane (ethyl chloride). The mechanism is as follows:

[A MECHANISM FOR THE REACTION ── Radical Halogenation of Ethane]

Chain Initiation

Step 1

$$Cl_2 \xrightarrow[\text{or heat}]{\text{light}} 2\ Cl\cdot$$

Chain Propagation

Step 2

$$CH_3CH_2{:}H\ +\ \cdot Cl \longrightarrow CH_3CH_2{\cdot}\ +\ H{:}Cl$$

Step 3

$$CH_3CH_2{\cdot}\ +\ Cl{:}Cl \longrightarrow CH_3CH_2{:}Cl\ +\ Cl\cdot$$

Chain propagation continues with steps 2, 3, 2, 3, and so on.

Chain Termination

$$CH_3CH_2{\cdot}\ +\ \cdot Cl \longrightarrow CH_3CH_2{:}Cl$$

$$CH_3CH_2{\cdot}\ +\ \cdot CH_2CH_3 \longrightarrow CH_3CH_2{:}CH_2CH_3$$

$$Cl\cdot\ +\ \cdot Cl \longrightarrow Cl{:}Cl$$

• • •

PRACTICE PROBLEM 10.5 When ethane is chlorinated, 1,1-dichloroethane and 1,2-dichloroethane, as well as more highly chlorinated ethanes, are formed in the mixture (see Section 10.3A). Write chain reaction mechanisms accounting for the formation of 1,1-dichloroethane and 1,2-dichloroethane.

Chlorination of most alkanes whose molecules contain more than two carbon atoms gives a mixture of isomeric monochloro products (as well as more highly chlorinated compounds). Several examples follow. The percentages given are based on the total amount of monochloro products formed in each reaction.

These examples show the nonselectivity of chlorination.

$$\text{Propane} \xrightarrow[\text{light, 25 °C}]{\text{Cl}_2} \text{1-Chloropropane (45\%)} + \text{2-Chloropropane (55\%)}$$

Propane → **1-Chloropropane (45%)** + **2-Chloropropane (55%)**

2-Methylpropane $\xrightarrow[\text{25 °C}]{\underset{\text{light}}{\text{Cl}_2}}$ **1-Chloro-2-methyl propane (63%)** + **2-Chloro-2-methyl propane (37%)**

2-Methylbutane $\xrightarrow[\text{300 °C}]{\text{Cl}_2}$ **1-Chloro-2-methylbutane (30%)** + **2-Chloro-2-methyl butane (22%)**

+ **2-Chloro-3-methyl-butane (33%)** + **1-Chloro-3-methyl butane (15%)**

> **Helpful Hint**
>
> Chlorination is not selective.

The ratios of products that we obtain from chlorination reactions of higher alkanes are not identical to what we would expect if all the hydrogen atoms of the alkane were equally reactive. We find that there is a correlation between reactivity of different hydrogen atoms and the type of hydrogen atom (1°, 2°, or 3°) being replaced. The tertiary hydrogen atoms of an alkane are most reactive, secondary hydrogen atoms are next most reactive, and primary hydrogen atoms are the least reactive (see Practice Problem 10.6).

PRACTICE PROBLEM 10.6

(a) What percentages of 1-chloropropane and 2-chloropropane would you expect to obtain from the chlorination of propane if 1° and 2° hydrogen atoms were equally reactive?

$$\xrightarrow[h\nu]{\text{Cl}_2} \quad \text{Cl} \quad + \quad \text{Cl}$$

(b) What percentages of 1-chloro-2-methylpropane and 2-chloro-2-methylpropane would you expect from the chlorination of 2-methylpropane if the 1° and 3° hydrogen atoms were equally reactive?

$$\xrightarrow[h\nu]{\text{Cl}_2} \quad \text{Cl} \quad + \quad \text{Cl}$$

(c) Compare these calculated answers with the results actually obtained (above in Section 10.5) and justify the assertion that the order of reactivity of the hydrogen atoms is 3° > 2° > 1°.

We can account for the relative reactivities of the primary, secondary, and tertiary hydrogen atoms in a chlorination reaction on the basis of the homolytic bond dissociation energies we saw earlier (Table 10.1). Of the three types, breaking a tertiary C—H bond requires the least energy, and breaking a primary C—H bond requires the most. Since the step in which the C—H bond is broken (i.e., the hydrogen atom–abstraction step) determines the location or orientation of the chlorination, we would expect the E_{act} for abstracting a tertiary hydrogen atom to be least and the E_{act} for abstracting a primary hydrogen atom to be greatest. Thus tertiary hydrogen atoms should be most reactive, secondary hydrogen atoms should be the next most reactive, and primary hydrogen atoms should be the least reactive.

The differences in the rates with which primary, secondary, and tertiary hydrogen atoms are replaced by chlorine are not large, however.

- Chlorine does not discriminate among the different types of hydrogen atoms in a way that makes chlorination of higher alkanes a generally useful laboratory synthesis.

●●● SOLVED PROBLEM 10.4

An alkane with the formula C_5H_{12} undergoes chlorination to give only one product with the formula $C_5H_{11}Cl$. What is the structure of this alkane?

STRATEGY AND ANSWER: The hydrogen atoms of the alkane must all be equivalent (homotopic), so that replacing any one of them leads to the same product. The only five-carbon alkane for which this is true is neopentane.

$$H_3C-\underset{\underset{CH_3}{|}}{\overset{\overset{CH_3}{|}}{C}}-CH_3$$

●●●

PRACTICE PROBLEM 10.7 Chlorination reactions of certain alkanes can be used for laboratory preparations. Examples are the preparation of chlorocyclopropane from cyclopropane and chlorocyclobutane from cyclobutane.

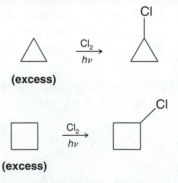

What structural feature of these molecules makes this possible?

●●●

PRACTICE PROBLEM 10.8 Each of the following alkanes reacts with chlorine to give a single monochloro substitution product. On the basis of this information, deduce the structure of each alkane.

(a) C_5H_{10} **(b)** C_8H_{18}

10.5A Selectivity of Bromine

Bromine shows a much greater ability to discriminate among the different types of hydrogen atoms.

- Bromine is less reactive than chlorine toward alkanes in general but bromine is more *selective* in the site of attack.
- Bromination is selective for substitution where the most stable radical intermediate can be formed.

The reaction of 2-methylpropane and bromine, for example, gives almost exclusive replacement of the tertiary hydrogen atom:

$$\underset{}{} \xrightarrow[h\nu,\ 127\ °C]{Br_2} \underset{Br}{} + \underset{}{Br}$$

(>99%) (trace)

A very different result is obtained when 2-methylpropane reacts with chlorine:

$$\underset{}{} \xrightarrow[h\nu,\ 25\ °C]{Cl_2} \underset{Cl}{} + \underset{}{Cl}$$

(37%) (63%)

Fluorine, being much more reactive than chlorine, *is even less selective than chlorine.* Because the **energy of activation** for the abstraction of any type of hydrogen by a fluorine atom is low, there is very little difference in the rate at which a 1°, 2°, or 3° hydrogen reacts with fluorine. Reactions of alkanes with fluorine give (almost) the distribution of products that we would expect if all of the hydrogens of the alkane were equally reactive.

(a)

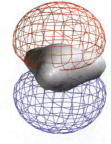

(b)

FIGURE 10.2 (a) Drawing of a methyl radical showing the sp^2-hybridized carbon atom at the center, the unpaired electron in the half-filled p orbital, and the three pairs of electrons involved in covalent bonding. The unpaired electron could be shown in either lobe. (b) Calculated structure for the methyl radical showing the highest occupied molecular orbital, where the unpaired electron resides, in red and blue. The region of bonding electron density around the carbons and hydrogens is in gray.

10.6 THE GEOMETRY OF ALKYL RADICALS

Experimental evidence indicates that the geometric structure of most alkyl **radicals** is trigonal planar at the carbon having the unpaired electron. This structure can be accommodated by an sp^2-hybridized central carbon. In an alkyl radical, the p orbital contains the unpaired electron (Fig. 10.2).

10.7 REACTIONS THAT GENERATE TETRAHEDRAL CHIRALITY CENTERS

- When **achiral molecules** react to produce a compound with a single tetrahedral **chirality center**, the product will be a **racemic form**.

This will always be true in the absence of any chiral influence on the reaction such as an enzyme or the use of a chiral reagent or solvent.

Let us examine a reaction that illustrates this principle, the **radical** chlorination of pentane:

$$\underset{\substack{\textbf{Pentane} \\ \textbf{(achiral)}}}{} \xrightarrow[\text{(achiral)}]{Cl_2} \underset{\substack{\textbf{1-Chloropentane} \\ \textbf{(achiral)}}}{\qquad Cl} + \underset{\substack{\textbf{(\pm)-2-Chloropentane} \\ \textbf{(a racemic form)}}}{} + \underset{\substack{\textbf{3-Chloropentane} \\ \textbf{(achiral)}}}{}$$

Pentane
(achiral)

1-Chloropentane
(achiral)

(±)-2-Chloropentane
(a racemic form)

3-Chloropentane
(achiral)

The reaction will lead to the products shown here, as well as more highly chlorinated products. (We can use an excess of pentane to minimize multiple chlorinations.) Neither 1-chloropentane nor 3-chloropentane contains a chirality center, but 2-chloropentane does, and it is *obtained as a racemic form*. If we examine the mechanism we shall see why.

[A MECHANISM FOR THE REACTION — The Stereochemistry of Chlorination at C2 of Pentane]

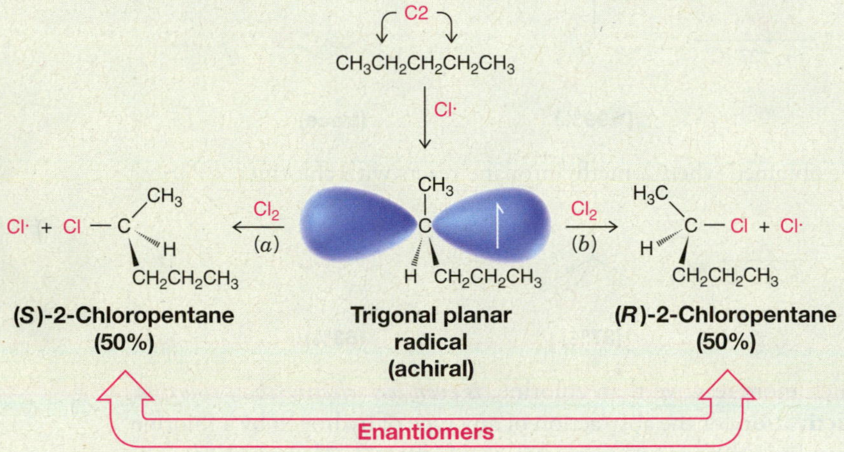

Abstraction of a hydrogen atom from C2 produces a trigonal planar radical that is achiral. This radical then reacts with chlorine at either face [by path (a) or path (b)]. Because the radical is achiral, the probability of reaction by either path is the same; therefore, the two enantiomers are produced in equal amounts, and a racemic form of 2-chloropentane results.

We can also say that the C2 hydrogens of pentane are **enantiotopic** because enantiomers are formed by reaction at each C2 hydrogen.

10.7A Generation of a Second Chirality Center in a Radical Halogenation

Let us now examine what happens when a chiral molecule (containing one chirality center) reacts so as to yield a product with a second chirality center. As an example consider what happens when (S)-2-chloropentane undergoes chlorination at C3 (other products are formed, of course, by chlorination at other carbon atoms). The results of chlorination at C3 are shown in the box below.

The products of the reactions are (2S,3S)-2,3-dichloropentane and (2S,3R)-2,3-dichloropentane. These two compounds are **diastereomers**. (They are stereoisomers but they are not mirror images of each other.) They each resulted by substitution of one of the **diastereotopic** hydrogens at C3. The two diastereomers are *not* produced in equal amounts. Because the intermediate radical itself is chiral, reactions at the two faces are not equally likely. The radical reacts with chlorine to a greater extent at one face than the other (although we cannot easily predict which). That is, the presence of a chirality center in the radical (at C2) influences the reaction that introduces the new chirality center (at C3).

Both of the 2,3-dichloropentane diastereomers are chiral and, therefore, each exhibits optical activity. Moreover, because the two compounds are *diastereomers*, they have different physical properties (e.g., different melting points and boiling points) and are separable by conventional means (by gas chromatography or by careful fractional distillation).

A MECHANISM FOR THE REACTION — The Stereochemistry of Chlorination at C3 of (S)-2-Chloropentane

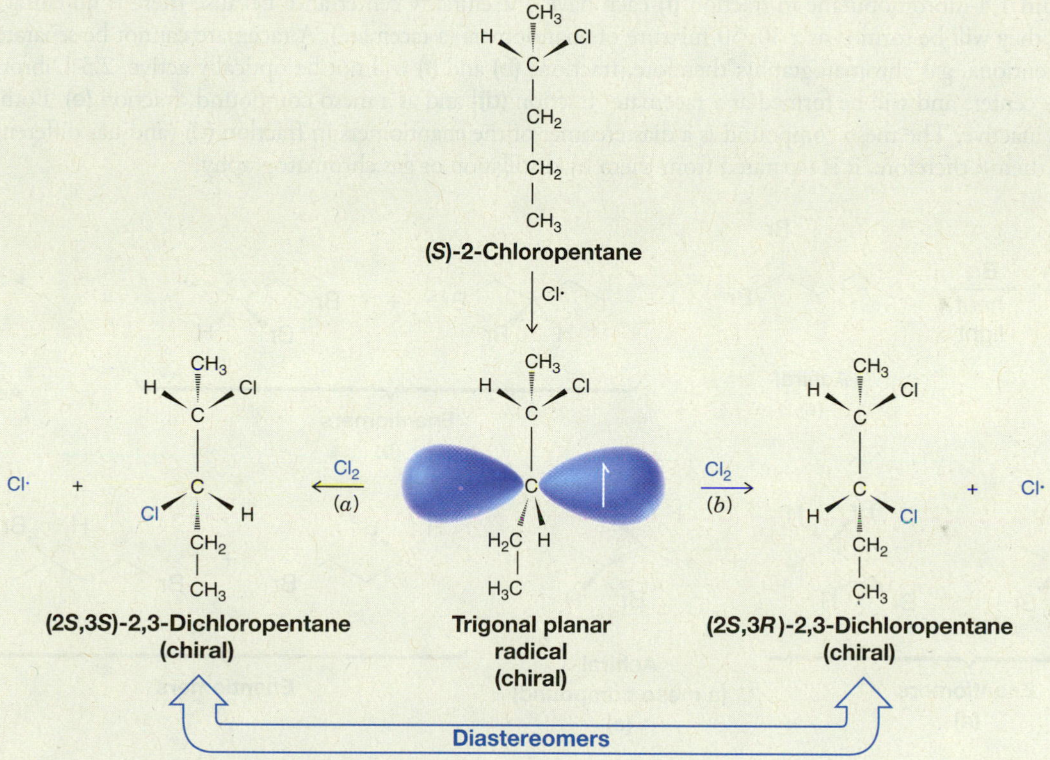

Abstraction of a hydrogen atom from C3 of (S)-2-chloropentane produces a radical that is chiral (it contains a chirality center at C2). This chiral radical can then react with chlorine at one face [path (a)] to produce (2S, 3S)-2,3-dichloropentane and the other face [path (b)] to yield (2S, 3R)-2,3-dichloropentane. These two compounds are diastereomers, and they are not produced in equal amounts. Each product is chiral, and each alone would be optically active.

Consider the chlorination of (S)-2-chloropentane at C4. **(a)** Write structural formulas for the products, showing three dimensions at all chirality centers. Give each its proper (R,S) designation. **(b)** What is the stereoisomeric relationship between these products? **(c)** Are both products chiral? **(d)** Are both optically active? **(e)** Could the products be separated by conventional means? **(f)** What other dichloropentanes would be obtained by chlorination of (S)-2-chloropentane? **(g)** Which of these are optically active?

PRACTICE PROBLEM 10.9

●●● SOLVED PROBLEM 10.5

Consider the bromination of butane using sufficient bromine to cause dibromination. After the reaction is over, you separate all the dibromobutane isomers by gas chromatography or by fractional distillation. How many fractions would you obtain, and what compounds would the individual fractions contain? Which if any of the fractions would be optically active?

STRATEGY AND ANSWER: The construction of handheld models will help in solving this problem. First, decide how many constitutional isomers are possible by replacing two hydrogens of butane with two bromine atoms. There are six: 1,1-dibromobutane, 1,2-dibromobutane, 2,2-dibromobutane, 2,3-dibromobutane, 1,3-dibromobutane, and 1,4-dibromobutane. Then recall that constitutional isomers have different physical properties (i.e., boiling points and retention times in a gas chromatograph), so there should be at least six fractions. In actuality there are seven. See fractions **(a)**–**(g)** below. We soon see why there are seven fractions if we examine each constitutional isomer looking for chirality centers and stereoisomers. Isomers **(a)**, **(c)**, and **(g)** have no chirality centers and are, therefore, achiral and are optically inactive. 1,2-Dibromobutane in fraction **(b)** and 1,4-dibromobutane in fraction **(f)** each have one chirality center and, because there is no chiral influence on the reaction, they will be formed as a 50:50 mixture of enantiomers (a racemate). A racemate cannot be separated by distillation or conventional gas chromatography; therefore, fractions **(b)** and **(f)** will not be optically active. 2,3-Dibromobutane has two chirality centers and will be formed as a racemate [fraction **(d)**] and as a meso compound, fraction **(e)**. Both fractions will be optically inactive. The meso compound is a diastereomer of the enantiomers in fraction **(d)** (and has different physical properties from them); therefore, it is separated from them by distillation or gas chromatography.

●●●

PRACTICE PROBLEM 10.10 Consider the monochlorination of 2-methylbutane.

(a) Assuming that the product mixture was subjected to fractional distillation, which fractions, if any, would show optical activity? **(b)** Could any of these fractions be resolved, theoretically, into enantiomers? **(c)** Could the components of each fraction from the distillation be identified on the basis of ^{1}H NMR spectroscopy? What specific characteristics in a ^{1}H NMR spectrum of each fraction would indicate the identity of the component(s) in that fraction?

10.8 ALLYLIC SUBSTITUTION AND ALLYLIC RADICALS

- An atom or group that is bonded to an sp^3-hybridized carbon adjacent to an alkene double bond is called an **allylic group**. The group is said to be bonded at the **allylic position**.

The following are some examples.

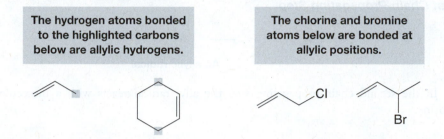

| The hydrogen atoms bonded to the highlighted carbons below are allylic hydrogens. | The chlorine and bromine atoms below are bonded at allylic positions. |

Allylic hydrogens are especially reactive in radical substitution reactions. We can synthesize allylic halides by substitution of allylic hydrogens. For example, when propene reacts with bromine or chlorine at high temperatures or under radical conditions where the concentration of the halogen is small, the result is **allylic substitution**.

$$\text{Propene} + X_2 \xrightarrow[\text{and low concentration of } X_2]{\text{high temperature}} \diagdown\!\diagup\!\diagdown^{X} + HX$$

Propene

> **At high temperature (or in the presence of a radical initiator) and low concentration of X_2 a substitution reaction occurs.**

On the other hand, when propene reacts with bromine or chlorine at low temperatures, an addition reaction of the type we studied in Chapter 8 occurs.

$$\diagup\!\diagdown + X_2 \xrightarrow[\text{CCl}_4]{\text{low temperature}} \text{(product)}$$

> **At low temperature an addition reaction occurs.**

To bias the reaction toward allylic substitution we need to use reaction conditions that favor formation of radicals and that provide a low but steady concentration of halogen.

10.8A Allylic Chlorination (High Temperature)

Propene undergoes allylic chlorination when propene and chlorine react in the gas phase at 400 °C.

$$\text{Propene} + Cl_2 \xrightarrow[\text{gas phase}]{400\ °C} \diagdown\!\diagup\!\diagdown^{Cl} + HCl$$

Propene   **3-Chloropropene (allyl chloride)**

The mechanism for allylic substitution is the same as the chain mechanism for alkane **halogenations** that we saw earlier in the chapter. In the chain-initiating step, the chlorine molecule dissociates into chlorine atoms.

Chain-Initiating Step

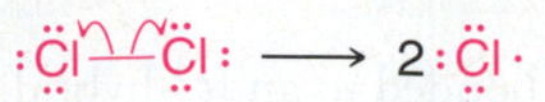

$$:\ddot{Cl} \frown \ddot{Cl}: \longrightarrow 2:\ddot{Cl}\cdot$$

In the first chain-propagating step the chlorine atom abstracts one of the allylic hydrogen atoms. The radical that is produced in this step is called an **allylic radical**.

First Chain-Propagating Step

An allylic radical

In the second chain-propagating step the allyl radical reacts with a molecule of chlorine.

Second Chain-Propagating Step

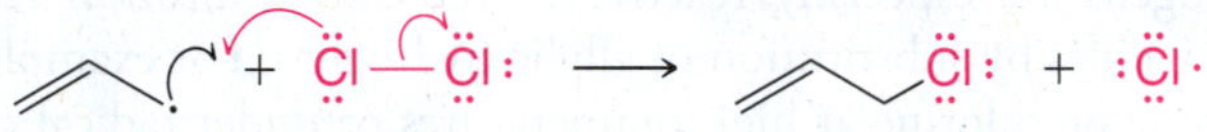

Allyl chloride

This step results in the formation of a molecule of allyl chloride (2-chloro-1-propene) and a chlorine atom. The chlorine atom then brings about a repetition of the first chain-propagating step. The chain reaction continues until the usual chain-terminating steps (see Section 10.4) consume the radicals.

10.8B Allylic Bromination with *N*-Bromosuccinimide (Low Concentration of Br₂)

Propene undergoes allylic bromination when it is treated with *N*-bromosuccinimide (NBS) in the presence of peroxides or light:

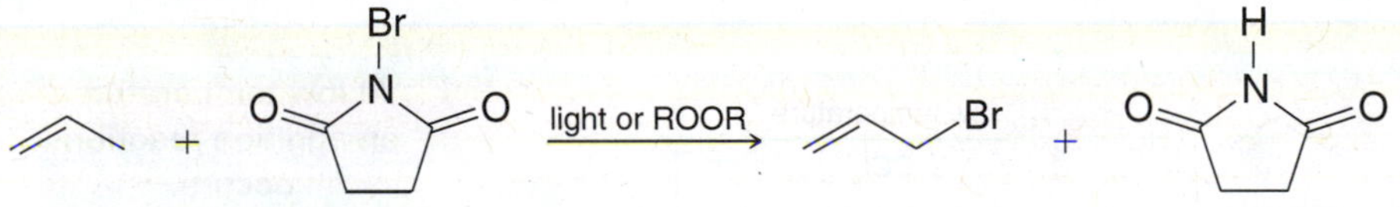

| *N*-Bromosuccinimide (NBS) | 3-Bromopropene (allyl bromide) | Succinimide |

The reaction is initiated by the formation of a small amount of Br· (possibly formed by dissociation of the N—Br bond of the NBS). The main propagation steps for this reaction are the same as for allylic chlorination (Section 10.2A):

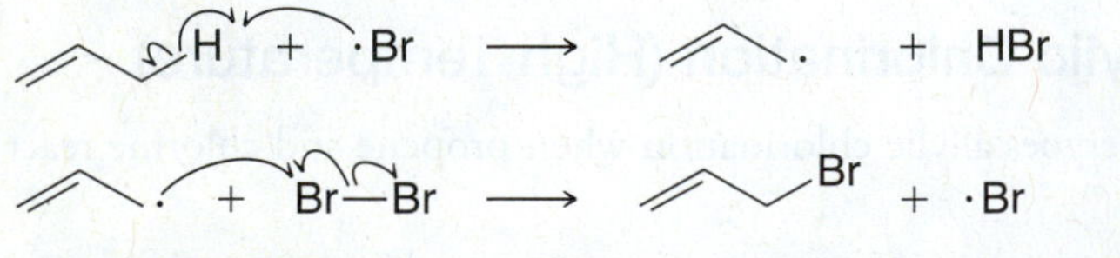

N-Bromosuccinimide is a solid that provides a constant but very low concentration of bromine in the reaction mixture. It does this by reacting very rapidly with the HBr formed in the substitution reaction. Each molecule of HBr is replaced by one molecule of Br₂.

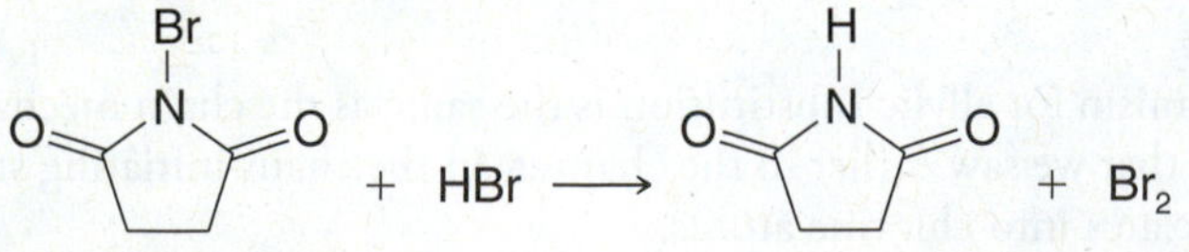

Under these conditions, that is, *in a nonpolar solvent and with a very low concentration of bromine*, very little bromine adds to the double bond; it reacts by substitution and replaces an allylic hydrogen atom instead.

The following reaction with cyclohexene is another example of allylic bromination with NBS:

82–87%

- In general, NBS is a good reagent to use for allylic bromination.

10.8C Allylic Radicals Are Stabilized by Electron Delocalization

Let us examine the bond dissociation energy of an allylic carbon–hydrogen bond and compare it with the bond dissociation energies of other carbon–hydrogen bonds.

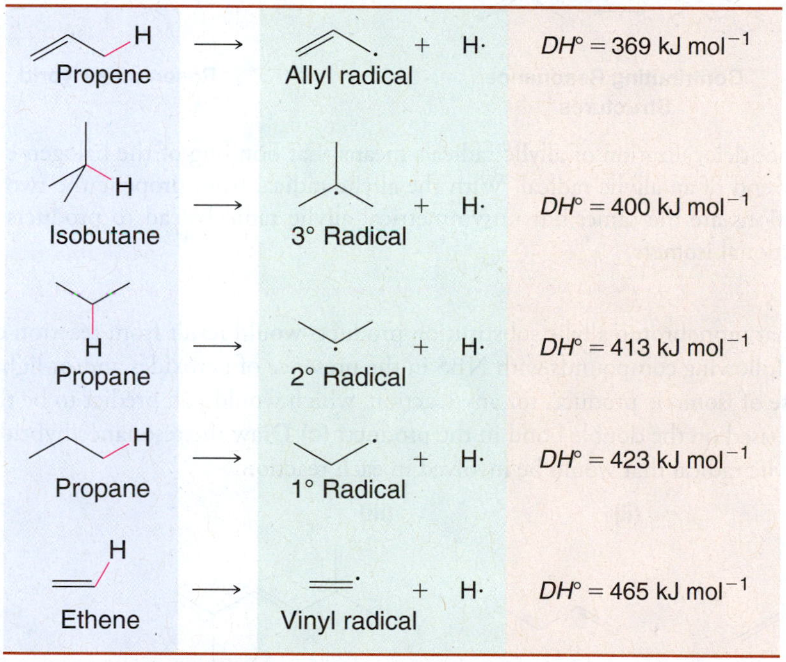

Propene → Allyl radical + H·	$DH° = 369$ kJ mol^{-1}
Isobutane → 3° Radical + H·	$DH° = 400$ kJ mol^{-1}
Propane → 2° Radical + H·	$DH° = 413$ kJ mol^{-1}
Propane → 1° Radical + H·	$DH° = 423$ kJ mol^{-1}
Ethene → Vinyl radical + H·	$DH° = 465$ kJ mol^{-1}

See Table 10.1 for a list of additional bond dissociation energies.

We see that an allylic carbon–hydrogen bond of propene is broken with greater ease than even the tertiary carbon–hydrogen bond of isobutane and with far greater ease than a vinylic carbon–hydrogen bond:

Allyl radical   +   HX E_{act} is low.

Vinylic radical   +   HX E_{act} is high.

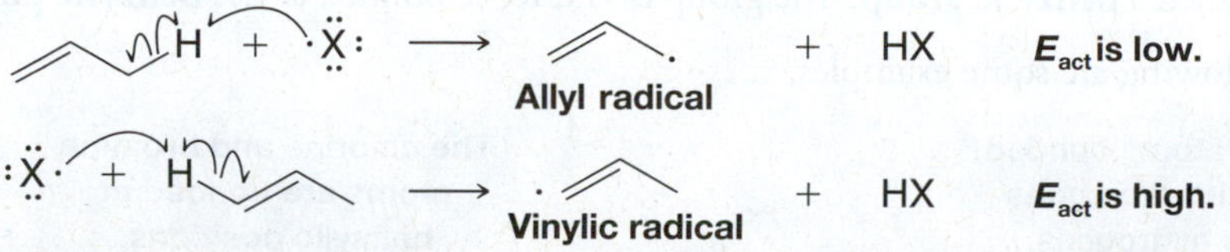

- The ease with which an allylic carbon–hydrogen bond is broken means that relative to primary, secondary, tertiary, and vinylic free radicals an allylic radical is the *most stable* (Fig. 10.3):

Relative stability: allylic or allyl > 3° > 2° > 1° > vinyl or vinylic

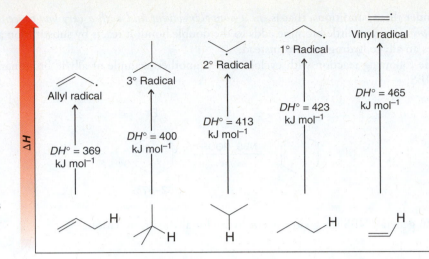

FIGURE 10.3 The relative stability of the allyl radical compared to 1°, 2°, 3°, and vinyl radicals. (The stabilities of the radicals are relative to the hydrocarbon from which each was formed, and the overall order of stability is allyl > 3° > 2° > 1° > vinyl.)

The reason that allylic radicals are more stable than alkyl radicals is due to electron delocalization. For example, we can draw the following contributing resonance structures and the corresponding resonance hybrid for the allylic radical from propene.

Contributing Resonance Structures   **Resonance Hybrid**

Resonance delocalization of allylic radicals means that bonding of the halogen can occur at either end of an allylic radical. With the allylic radical from propene the two possible substitutions are the same, but unsymmetrical allylic radicals lead to products that are constitutional isomers.

PRACTICE PROBLEM 10.11 **(a)** What monobromo allylic substitution products would result from reaction of each of the following compounds with NBS in the presence of peroxides and/or light? **(b)** In the case of isomeric products for any reaction, which would you predict to be the most stable based on the double bond in the product? **(c)** Draw the resonance hybrid(s) for the allylic radical that would be involved in each reaction.

(i)   **(ii)**   **(iii)**

10.9 BENZYLIC SUBSTITUTION AND BENZYLIC RADICALS

- An atom or group bonded to an sp^3-hybridized carbon adjacent to a benzene ring is called a **benzylic group**. The group is said to be bonded at the **benzylic position**.

The following are some examples.

The hydrogen atoms bonded at the highlighted carbons are benzylic hydrogens.

The chlorine and bromine atoms are bonded at benzylic positions.

Benzylic hydrogens are even more reactive than allylic hydrogens in radical substitution reactions due to the additional delocalization that is possible for a benzylic radical intermediate (see Practice Problem 10.12).

When methylbenzene (toluene) reacts with *N*-bromosuccinimide (NBS) in the presence of light, for example, the major product is benzyl bromide. *N*-Bromosuccinimide furnishes a low concentration of Br_2, and the reaction is analogous to that for allylic bromination that we studied in Section 10.8B.

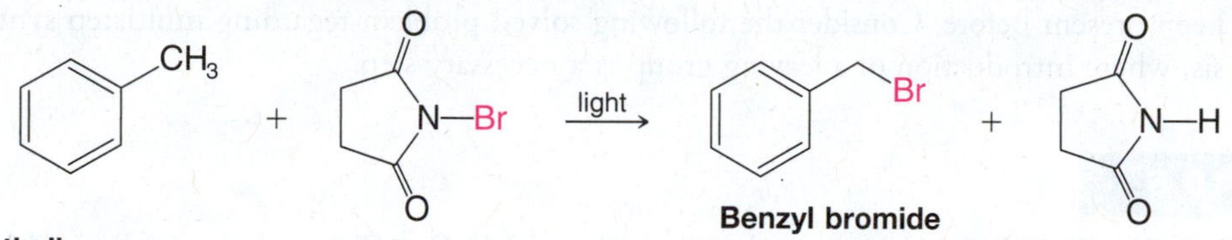

Methylbenzene   **NBS**   **Benzyl bromide**
(Toluene)   **(α-bromotoluene)**
                                                (64%)

Benzylic chlorination of methylbenzene takes place in the gas phase at 400–600 °C or in the presence of UV light. When an excess of chlorine is used, multiple chlorinations of the side chain occur:

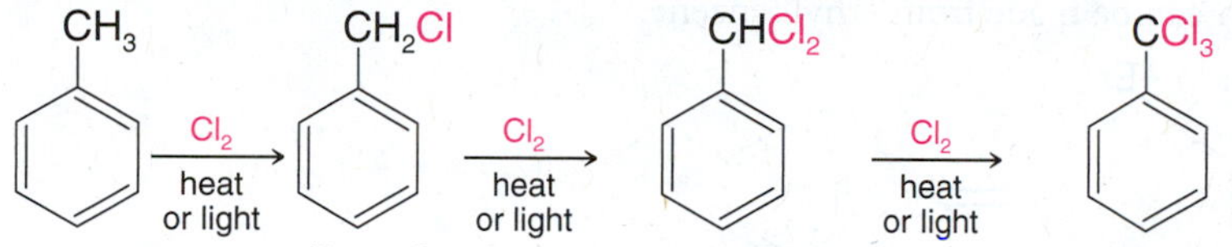

                        Benzyl   **Dichloromethyl**   **Trichloromethyl**
                        chloride   **benzene**   **benzene**

These halogenations take place through the same radical mechanism we saw for alkanes in Section 10.4. The halogens dissociate to produce halogen atoms and then the halogen atoms initiate chain reactions by abstracting hydrogens of the methyl group.

PRACTICE PROBLEM 10.12

Benzylic radicals, due to the adjacent benzene ring, have even greater possibility for delocalization than allylic radicals. Draw contributing resonance structures that show this delocalization for the benzylic radical derived from methylbenzene. (*Hint:* There are four contributing resonance structures for this benzylic radical.)

The greater stability of benzylic radicals accounts for the fact that when ethylbenzene is halogenated, the major product is the 1-halo-1-phenylethane. The benzylic radical is formed much faster than the 1° radical:

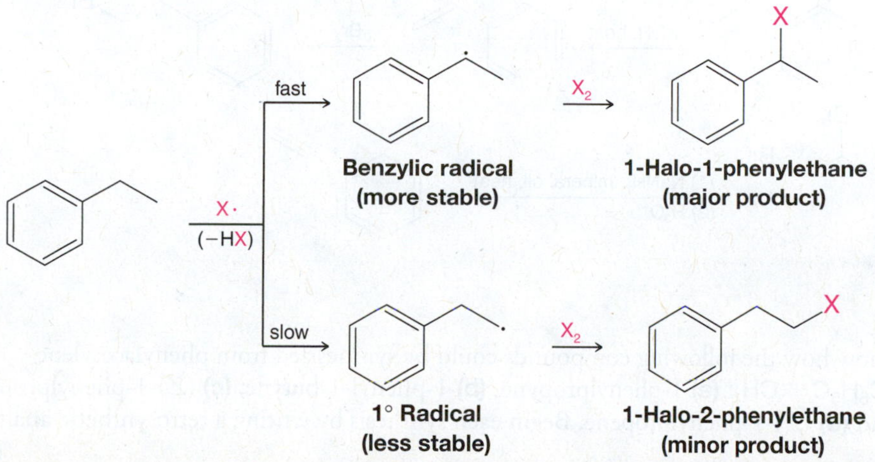

                        Benzylic radical   **1-Halo-1-phenylethane**
                        (more stable)   **(major product)**

                        1° Radical   **1-Halo-2-phenylethane**
                        (less stable)   **(minor product)**

PRACTICE PROBLEM 10.13 When propylbenzene reacts with chlorine in the presence of UV radiation, the major product is 1-chloro-1-phenylpropane. Both 2-chloro-1-phenylpropane and 3-chloro-1-phenylpropane are minor products. Write the structure of the radical leading to each product and account for the fact that 1-chloro-1-phenylpropane is the major product.

Benzylic halogenation is useful for introducing a leaving group where none may have been present before. Consider the following solved problem regarding multistep synthesis, where introduction of a leaving group is a necessary step.

SOLVED PROBLEM 10.6

ILLUSTRATING A MULTISTEP SYNTHESIS: Show how phenylacetylene (C₆H₅C≡CH) could be synthesized from ethylbenzene (phenylethane). Begin by writing a retrosynthetic analysis, and then write reactions needed for the synthesis.

ANSWER: Working backward, that is, using *retrosynthetic analysis*, we find that we can easily envision two syntheses of phenylacetylene. We can make phenylacetylene by dehydrohalogenation of 1,1-dibromo-1-phenylethane, which could have been prepared by allowing ethylbenzene (phenylethane) to react with 2 mol of NBS. Alternatively, we can prepare phenylacetylene from 1,2-dibromo-1-phenylethane, which could be prepared from styrene (phenylethene). Styrene can be made from 1-bromo-1-phenylethane, which can be made from ethylbenzene.

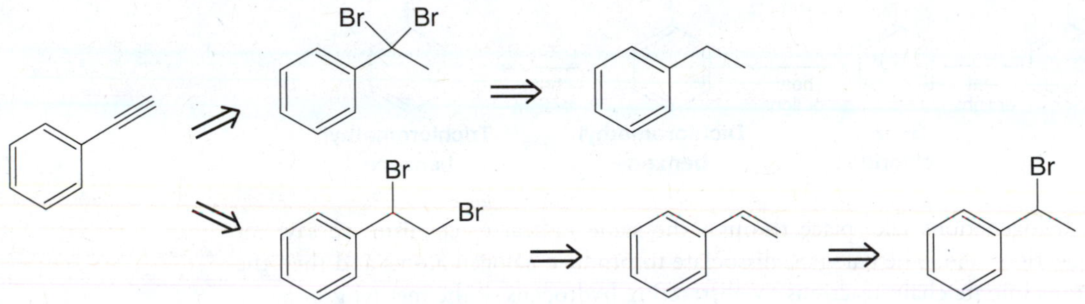

Following are the synthetic reactions we need for the two retrosynthetic analyses above:

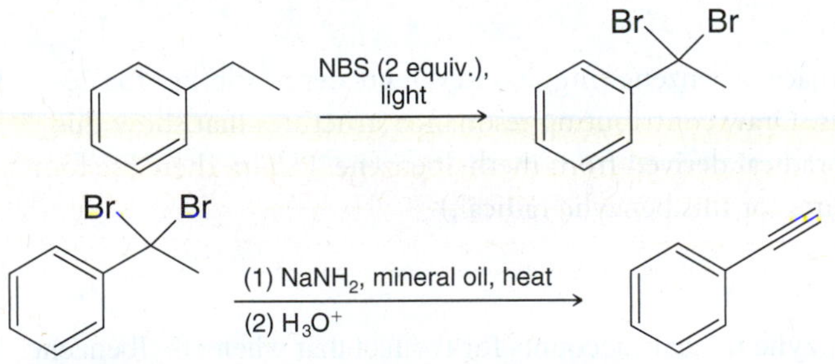

or

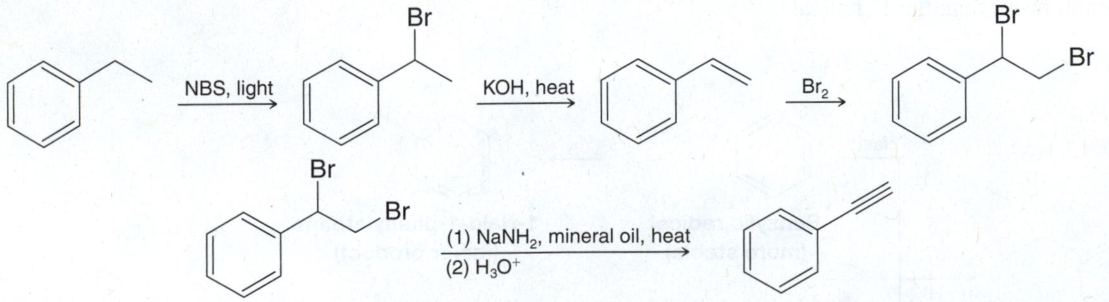

PRACTICE PROBLEM 10.14 Show how the following compounds could be synthesized from phenylacetylene (C₆H₅C≡CH): **(a)** 1-phenylpropyne, **(b)** 1-phenyl-1-butyne, **(c)** (*Z*)-1-phenylpropene, and **(d)** (*E*)-1-phenylpropene. Begin each synthesis by writing a retrosynthetic analysis.

10.10 RADICAL ADDITION TO ALKENES: THE ANTI-MARKOVNIKOV ADDITION OF HYDROGEN BROMIDE

Before 1933, the orientation of the addition of hydrogen bromide to alkenes was the subject of much confusion. At times addition occurred in accordance with Markovnikov's rule; at other times it occurred in just the opposite manner. Many instances were reported where, under what seemed to be the same experimental conditions, Markovnikov additions were obtained in one laboratory and anti-Markovnikov additions in another. At times even the same chemist would obtain different results using the same conditions but on different occasions.

The mystery was solved in 1933 by the research of M. S. Kharasch and F. R. Mayo (of the University of Chicago). The explanatory factor turned out to be organic peroxides present in the alkenes—peroxides that were formed by the action of atmospheric oxygen on the alkenes (Section 10.12D).

$$R—\ddot{O}—\ddot{O}—R \qquad\qquad R—\ddot{O}—\ddot{O}—H$$

An organic peroxide   **An organic hydroperoxide**

- When alkenes containing peroxides or hydroperoxides react with hydrogen bromide, anti-Markovnikov addition of HBr occurs.

For example, in the *presence* of peroxides propene yields 1-bromopropane. In the *absence* of peroxides, or in the presence of compounds that "trap" radicals, normal Markovnikov addition occurs.

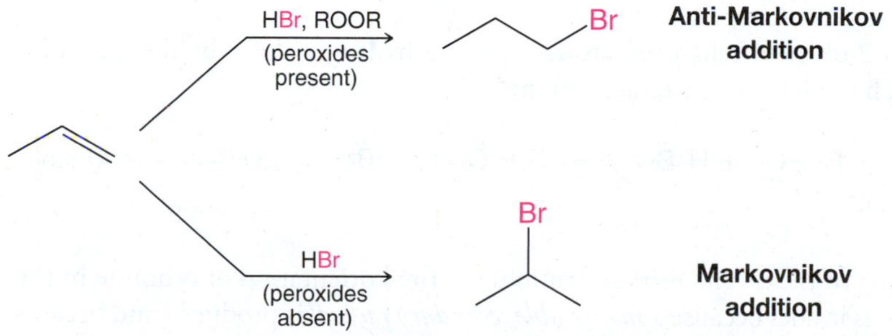

- Hydrogen bromide is the only hydrogen halide that gives anti-Markovnikov addition when peroxides are present.

Hydrogen fluoride, hydrogen chloride, and hydrogen iodide *do not* give anti-Markovnikov addition even when peroxides are present.

The mechanism for **anti-Markovnikov addition of hydrogen bromide** is a **radical chain reaction** initiated by peroxides.

[A MECHANISM FOR THE REACTION] — Anti-Markovnikov Addition of HBr

Chain Initiation

Step 1

$$R—\ddot{O}:\ddot{O}—R \xrightarrow{heat} 2\,R—\ddot{O}\cdot$$

Heat brings about homolytic cleavage of the weak oxygen–oxygen bond.

Step 2

$$R—\ddot{O}\cdot + H:\ddot{Br}: \longrightarrow R—\ddot{O}:H + :\ddot{Br}\cdot$$

The alkoxyl radical abstracts a hydrogen atom from HBr, producing a bromine radical.

(mechanism continues on the next page)

Chain Propagation

Step 3

$$:\ddot{Br}\cdot \; + \; H_2C{=}CH{-}CH_3 \longrightarrow :\ddot{Br}:CH_2{-}\dot{CH}{-}CH_3$$

2° Radical

**A bromine radical adds to the double bond
to produce the more stable 2° alkyl radical.**

Step 4

$$:\ddot{Br}{-}CH_2{-}\dot{CH}{-}CH_3 \; + \; H{:}\ddot{Br}: \longrightarrow :\ddot{Br}{-}CH_2{-}\underset{H}{CH}{-}CH_3 \; + \; \cdot\ddot{Br}:$$

1-Bromopropane

**The alkyl radical abstracts a hydrogen atom from HBr.
This leads to the product and regenerates a bromine radical.
Then repetitions of steps 3 and 4 lead to a chain reaction.**

Step 1 is the simple homolytic cleavage of the peroxide molecule to produce two alkoxyl radicals. The oxygen–oxygen bond of peroxides is weak, and such reactions are known to occur readily:

$$R{-}\ddot{O}{:}\ddot{O}{-}R \longrightarrow 2\,R{-}\ddot{O}\cdot \qquad \Delta H^\circ \cong +150 \text{ kJ mol}^{-1}$$

Peroxide   **Alkoxyl radical**

Step 2 of the mechanism, abstraction of a hydrogen atom by the radical, is exothermic and has a low energy of activation:

$$R{-}\ddot{O}\cdot \; + \; H{:}\ddot{Br}: \longrightarrow R{-}\ddot{O}{:}H \; + \; :\ddot{Br}\cdot \qquad \Delta H^\circ \cong -96 \text{ kJ mol}^{-1}$$

$$E_{act} \text{ is low}$$

Step 3 of the mechanism determines the final orientation of bromine in the product. It occurs as it does because a *more stable secondary radical* is produced and because *attack at the primary carbon atom is less hindered*. Had the bromine attacked propene at the secondary carbon atom, a less stable, primary radical would have been the result,

$$Br\cdot \; + \; CH_2{=}CHCH_3 \; \xrightarrow{\;\; \times \;\;} \; \cdot CH_2\underset{Br}{CHCH_3}$$

**1° Radical
(less stable)**

and attack at the secondary carbon atom would have been more hindered.

Step 4 of the mechanism is simply the abstraction of a hydrogen atom from hydrogen bromide by the radical produced in step 3. This hydrogen atom abstraction produces a bromine atom (which, of course, is a radical due to its unpaired electron) that can bring about step 3 again; then step 4 occurs again—a chain reaction.

10.10A Summary of Markovnikov versus Anti-Markovnikov Addition of HBr to Alkenes

We can now see the contrast between the two ways that HBr can add to an alkene. In the *absence* of peroxides, the reagent that attacks the double bond first is a proton. Because a proton is small, steric effects are unimportant. It attaches itself to a carbon atom by an

Helpful Hint

How to achieve regioselective alkyl halide synthesis through alkene addition.

ionic mechanism so as to form the more stable carbocation. The result is Markovnikov addition. Polar, protic solvents favor this process.

Ionic Addition

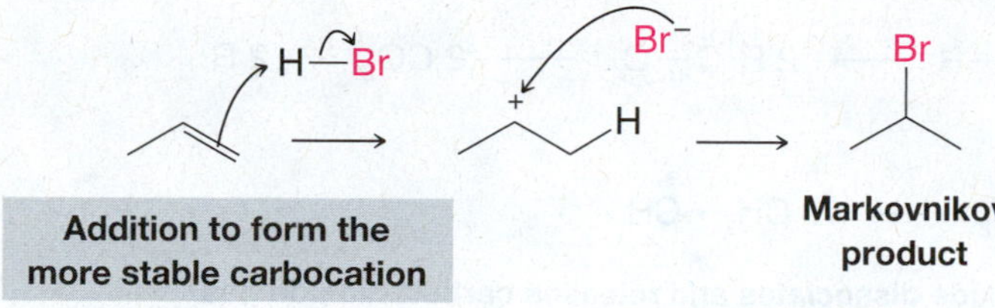

Addition to form the more stable carbocation

Markovnikov product

In the *presence* of peroxides, the reagent that attacks the double bond first is the larger bromine atom. It attaches itself to the less hindered carbon atom by a radical mechanism, so as to form the more stable radical intermediate. The result is anti-Markovnikov addition. Nonpolar solvents are preferable for reactions involving radicals.

Radical Addition

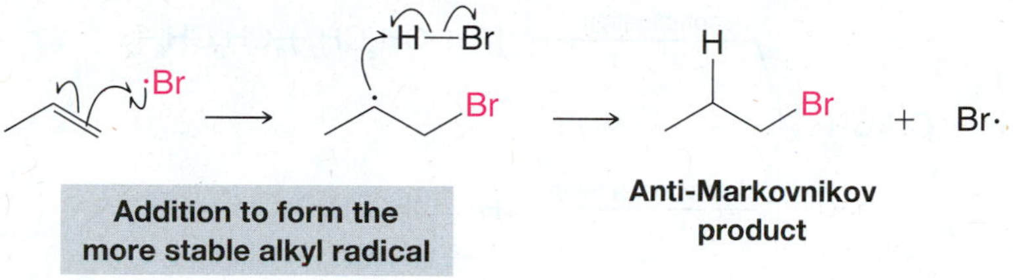

Addition to form the more stable alkyl radical

Anti-Markovnikov product + Br·

10.11 RADICAL POLYMERIZATION OF ALKENES: CHAIN-GROWTH POLYMERS

Polymers are substances that consist of very large molecules called **macromolecules** that are made up of many repeating subunits. The molecular subunits that are used to synthesize polymers are called **monomers**, and the reactions by which monomers are joined together are called **polymerizations**. Many polymerizations can be initiated by radicals.

Ethylene (ethene), for example, is the monomer that is used to synthesize the familiar polymer called *polyethylene*.

Monomeric units

$$m\, CH_2{=}CH_2 \xrightarrow{\text{polymerization}} -CH_2CH_2\!\left(CH_2CH_2\right)_{\!n}\!CH_2CH_2-$$

Ethylene monomer   **Polyethylene polymer**

(*m* and *n* are large numbers)

Because polymers such as polyethylene are made by addition reactions, they are often called **chain-growth polymers** or **addition polymers**. Let us now examine in some detail how polyethylene is made.

Ethene (ethylene) polymerizes by a radical mechanism when it is heated at a pressure of 1000 atm with a small amount of an organic peroxide (called a diacyl peroxide).

A MECHANISM FOR THE REACTION — Radical Polymerization of Ethene (Ethylene)

Chain Initiation

Step 1

$$R-\overset{O}{\underset{\|}{C}}-\overset{..}{\underset{..}{O}}:\overset{..}{\underset{..}{O}}-\overset{O}{\underset{\|}{C}}-R \longrightarrow 2\ R:\overset{O}{\underset{\|}{C}}-\overset{..}{\underset{..}{O}}\cdot \longrightarrow 2\ CO_2\ +\ 2\ R\cdot$$

Diacyl peroxide

Step 2

$$R\cdot\ +\ CH_2{=}CH_2 \longrightarrow R:CH_2{-}CH_2\cdot$$

The diacyl peroxide dissociates and releases carbon dioxide gas. Alkyl radicals are produced, which in turn initiate chains.

Chain Propagation

Step 3

$$R{-}CH_2CH_2\cdot\ +\ n\,CH_2{=}CH_2 \longrightarrow R{\left(CH_2CH_2\right)}_n CH_2CH_2\cdot$$

Chains propagate by adding successive ethylene units, until their growth is stopped by combination or disproportionation.

Chain Termination

Step 4

$$2\ R{\left(CH_2CH_2\right)}_n CH_2CH_2\cdot$$

combination $\longrightarrow [R{\left(CH_2CH_2\right)}_n CH_2CH_2{\tfrac{}{}}]_2$

disproportionation $\longrightarrow R{\left(CH_2CH_2\right)}_n CH{=}CH_2\ +$
$\qquad\qquad\qquad\qquad R{\left(CH_2CH_2\right)}_n CH_2CH_3$

The radical at the end of the growing polymer chain can also abstract a hydrogen atom from itself by what is called "black biting." This leads to chain branching.

Chain Branching

$$R{-}CH_2\overset{..}{C}H\underset{(CH_2CH_2)_n}{}\overset{\overset{\cdot CH_2}{|}}{\underset{CH_2}{}} \longrightarrow RCH_2\overset{\cdot}{C}H{\left(CH_2CH_2\right)}_n CH_2CH_2{-}H$$

$$\downarrow CH_2{=}CH_2$$

$$RCH_2\overset{}{C}H{\left(CH_2CH_2\right)}_n CH_2CH_3$$
$$\underset{\overset{|}{\underset{\cdot}{CH_2}}}{\overset{|}{CH_2}}$$

$$\downarrow \text{etc.}$$

The polyethylene produced by radical polymerization is not generally useful unless it has a molecular weight of nearly 1,000,000. Very high molecular weight polyethylene can be obtained by using a low concentration of the initiator. This initiates the growth of only a few chains and ensures that each chain will have a large excess of the monomer available. More initiator may be added as chains terminate during the polymerization, and, in this way, new chains are begun.

Polyethylene has been produced commercially since 1943. It is used in manufacturing flexible bottles, films, sheets, and insulation for electric wires. Polyethylene produced by radical polymerization has a softening point of about 110 °C.

Polyethylene can be produced in a different way using (see Special Topic B in *WileyPLUS*) catalysts called **Ziegler–Natta catalysts** that are organometallic complexes

The 1963 Nobel Prize was awarded to Karl Ziegler and Guilio Natta for their research in polymers.

of transition metals. In this process no radicals are produced, no back biting occurs, and, consequently, there is no chain branching. The polyethylene that is produced is of higher density, has a higher melting point, and has greater strength.

Another familiar polymer is *polystyrene*. The monomer used in making polystyrene is phenylethene, a compound commonly known as *styrene*.

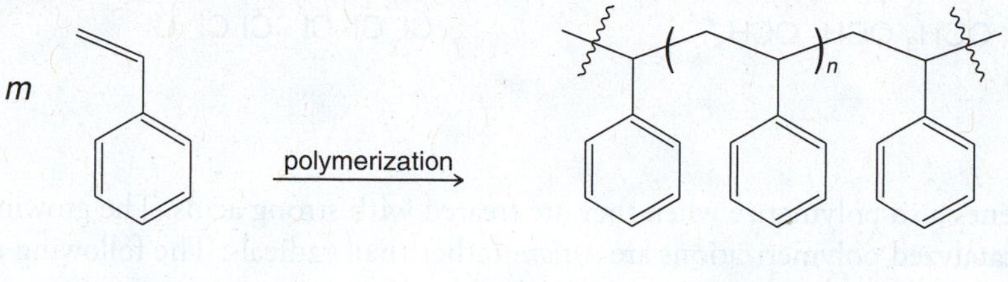

Styrene   **Polystyrene**

Table 10.2 lists several other common chain-growth polymers. Further information on each is provided in Special Topic B.

TABLE 10.2 OTHER COMMON CHAIN-GROWTH POLYMERS

Monomer	Polymer	Names
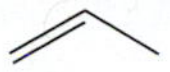	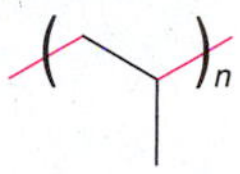	Polypropylene
	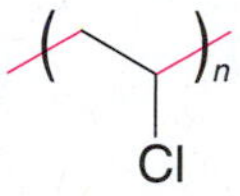	Poly(vinyl chloride), PVC
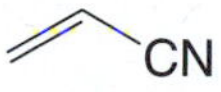	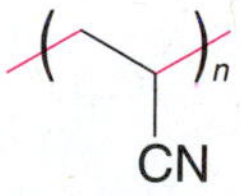	Polyacrylonitrile, Orlon
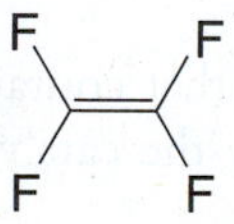	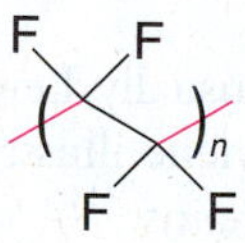	Poly(tetrafluoroethene), Teflon
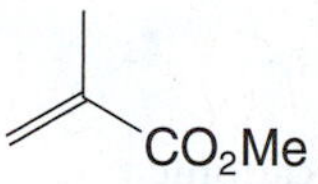	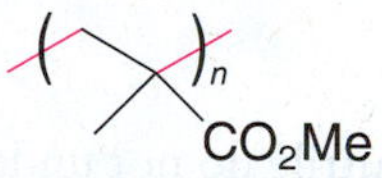	Poly(methyl methacrylate), Lucite, Plexiglas, Perspex

PRACTICE PROBLEM 10.15

Can you suggest an explanation that accounts for the fact that the radical polymerization of styrene ($C_6H_5CH\!=\!CH_2$) to produce polystyrene occurs in a head-to-tail fashion,

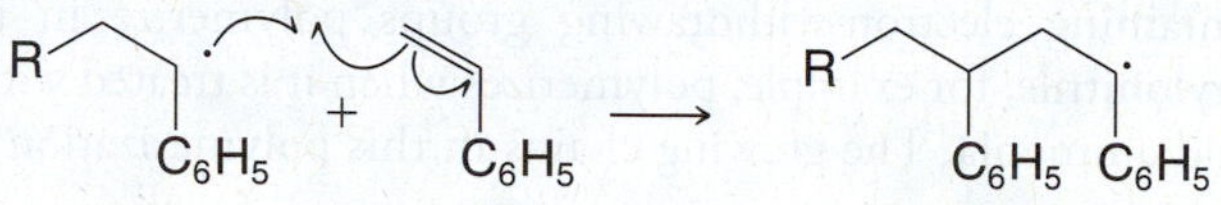

Polystyrene

rather than the head-to-head manner shown here?

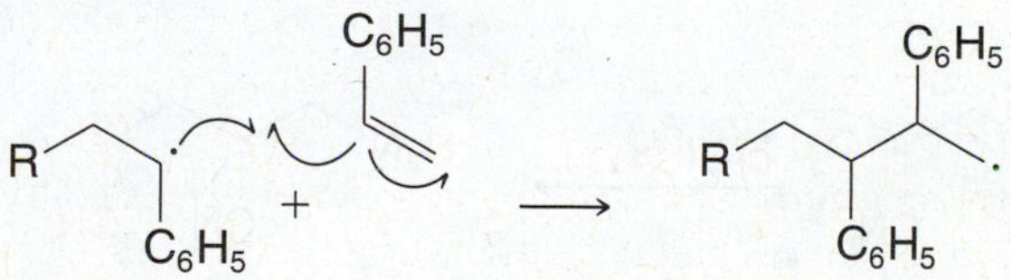

• • •

PRACTICE PROBLEM 10.16 Outline a general method for the synthesis of each of the following polymers by radical polymerization. Show the monomers that you would use.

(a)

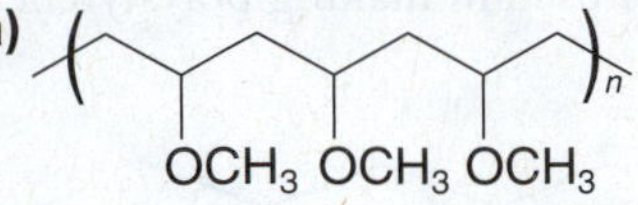

$$\left(\text{CH}_2\text{-CH-CH}_2\text{-CH-CH}_2\text{-CH} \right)_n$$

OCH₃ OCH₃ OCH₃

(b)

$$\left(\text{CH}_2\text{-C-CH}_2\text{-C-CH}_2\text{-C} \right)_n$$

Cl Cl Cl Cl Cl Cl

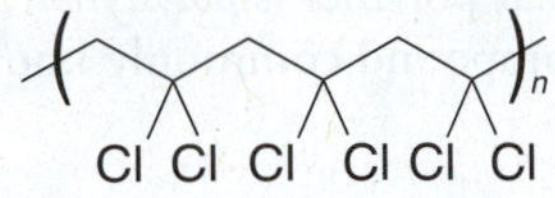

Alkenes also polymerize when they are treated with strong acids. The growing chains in acid-catalyzed polymerizations are *cations* rather than radicals. The following reactions illustrate the cationic polymerization of isobutylene:

Step 1 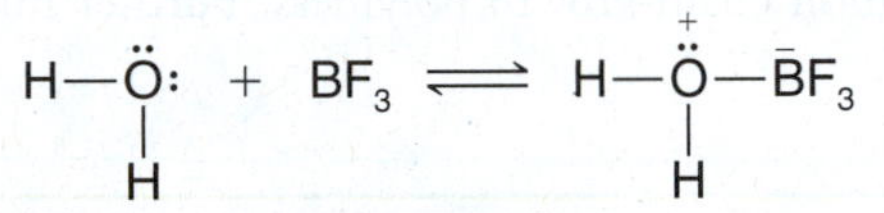   H—O: + BF₃ ⇌ H—O—BF₃
              | |
              H   H

Step 2 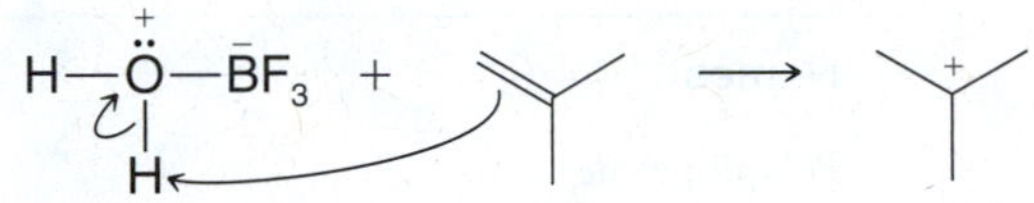   H—O—BF₃ + [alkene] ⟶ [cation]
              |
              H

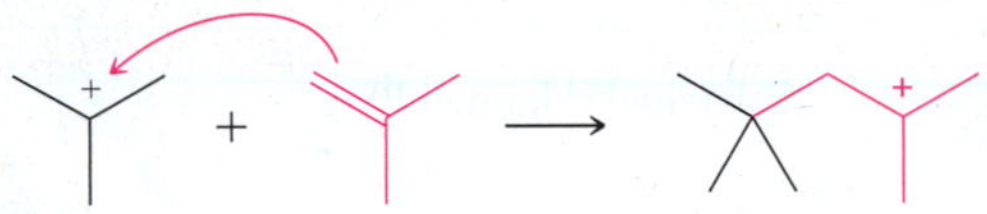

Step 3   [cation] + [alkene] ⟶ [chain cation]

Step 4   [chain cation] + [alkene] ⟶ [extended chain cation] →^etc.

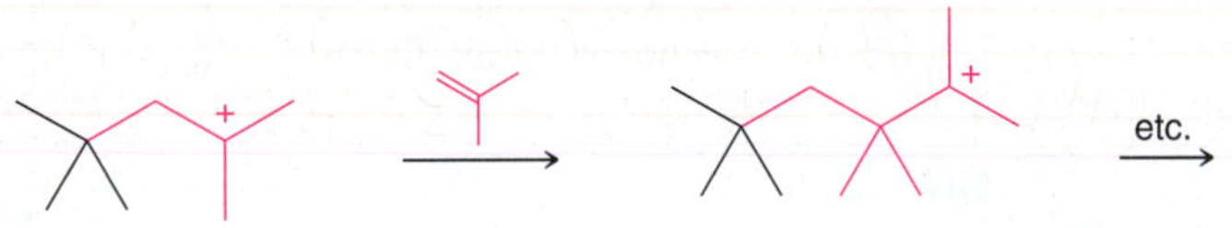

The catalysts used for cationic polymerizations are usually Lewis acids that contain a small amount of water. The polymerization of isobutylene illustrates how the catalyst (BF₃ and H_2O) functions to produce growing cationic chains.

• • •

PRACTICE PROBLEM 10.17 Alkenes such as ethene, vinyl chloride, and acrylonitrile do not undergo cationic polymerization very readily. On the other hand, isobutylene undergoes cationic polymerization rapidly. Provide an explanation for this behavior.

Alkenes containing electron-withdrawing groups polymerize in the presence of strong bases. Acrylonitrile, for example, polymerizes when it is treated with sodium amide (NaNH₂) in liquid ammonia. The growing chains in this polymerization are anions:

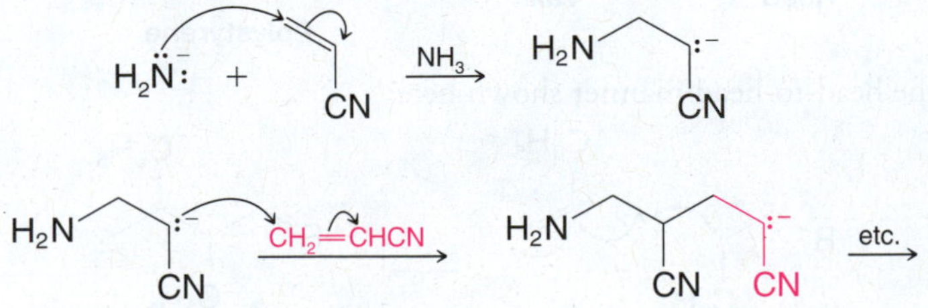

$$H_2\ddot{N}:^- + \text{[CH}_2\text{=CHCN]} \xrightarrow{NH_3} H_2N\text{-CH}_2\text{-CH:}^-$$
                                              |
                                              CN

$$H_2N\text{-CH}_2\text{-CH:}^- \xrightarrow{CH_2=CHCN} H_2N\text{-CH}_2\text{-CH-CH}_2\text{-CH:}^- \xrightarrow{etc.}$$
                    | | |
                    CN   CN   CN

Anionic polymerization of acrylonitrile is less important in commercial production than the radical process illustrated in Special Topic B.

The remarkable adhesive called "superglue" is a result of anionic polymerization. Superglue is a solution containing methyl cyanoacrylate:

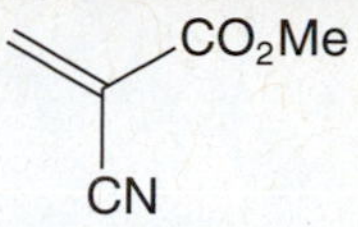

CO₂Me

CN

Methyl cyanoacrylate

Methyl cyanoacrylate can be polymerized by anions such as hydroxide ion, but it is even polymerized by traces of water found on the surfaces of the two objects being glued together. (These two objects, unfortunately, have often been two fingers of the person doing the gluing.) Show how methyl cyanoacrylate would undergo anionic polymerization.

STRATEGY AND ANSWER:

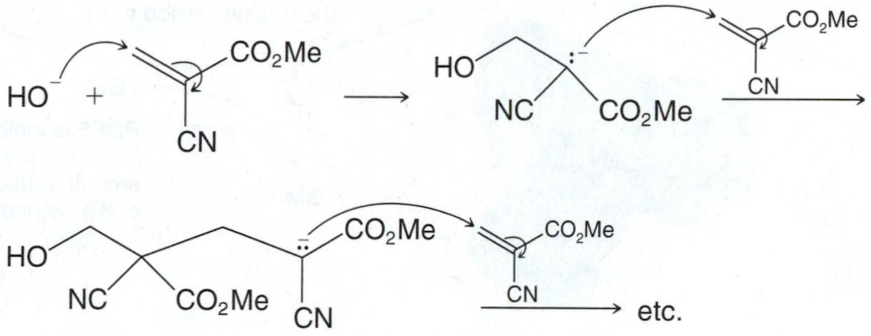

HO⁻ + [structure with CO₂Me and CN] ⟶ HO—[structure with NC, CO₂Me, CN] ⟶ [structure with CO₂Me, CN] ⟶

HO—[structure with NC, CO₂Me, CN, CO₂Me] [structure with CO₂Me, CN] ⟶ etc.

10.12 OTHER IMPORTANT RADICAL REACTIONS

Radical mechanisms are important in understanding many other organic reactions. We shall see other examples in later chapters, but let us examine a few important radicals and radical reactions here: oxygen and superoxide, the combustion of alkanes, DNA cleavage, autoxidation, antioxidants, and some reactions of chlorofluoromethanes that have threatened the protective layer of ozone in the stratosphere.

10.12A Molecular Oxygen and Superoxide

One of the most important radicals (and one that we encounter every moment of our lives) is molecular oxygen. Molecular oxygen in the ground state is a diradical with one unpaired electron on each oxygen. As a radical, oxygen can abstract hydrogen atoms just like other radicals we have seen. This is one way oxygen is involved in autoxidation (Section 10.12C) and combustion reactions (Section 10.12D). In biological systems, oxygen is an electron acceptor. When molecular oxygen accepts one electron, it becomes a radical anion called superoxide (O_2^{-}). Superoxide is involved in both positive and negative physiological roles. The immune system uses superoxide in its defense against pathogens, yet superoxide is also suspected of being involved in degenerative disease processes associated with aging and oxidative damage to healthy cells. The enzyme superoxide dismutase regulates the level of superoxide by catalyzing conversion of superoxide to hydrogen peroxide and molecular oxygen. Hydrogen peroxide, however, is also harmful because it can produce hydroxyl ($HO\cdot$) radicals. The enzyme catalase helps to prevent release of hydroxyl radicals by converting hydrogen peroxide to water and oxygen:

 The 1998 Nobel Prize in Physiology or Medicine was awarded to R. F. Furchgott, L. J. Ignarro, and F. Murad for their discovery that NO is an important signaling molecule.

$$2\,O_2^{-} + 2\,H^{+} \xrightarrow{\text{superoxide dismutase}} H_2O_2 + O_2$$

$$2\,H_2O_2 \xrightarrow{\text{catalase}} 2\,H_2O + O_2$$

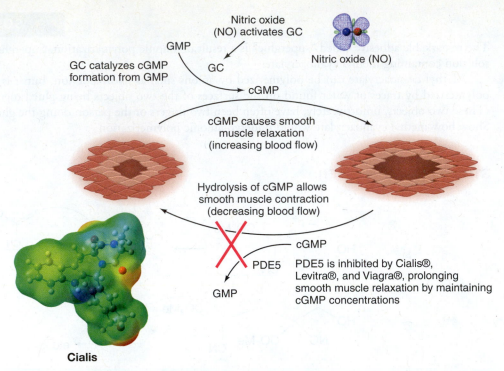

Cialis

FIGURE 10.4 Nitric oxide (NO) activates guanylate cyclase (GC), leading to production of cyclic guanosine monophosphate (cGMP). cGMP signals processes that cause smooth muscle relaxation, ultimately resulting in increased blood flow to certain tissues. Phosphodiesterase V (PDE5) degrades cGMP, leading to smooth muscle contraction and a reduction of blood flow. Cialis, Levitra, and Viagra take their effect by inhibiting PDE5, thus maintaining concentrations of cGMP and sustaining smooth muscle relaxation and tissue engorgement. (Reprinted with permission from Christianson, *Accounts of Chemical Research, 38*, p197, Figure 6b 2005. Copyright 2005 American Chemical Society.)

10.12B Nitric Oxide

Nitric oxide, synthesized in the body from the amino acid arginine, serves as a chemical messenger in a variety of biological processes, including blood pressure regulation and the immune response. Its role in relaxation of smooth muscle in vascular tissues is shown in Fig. 10.4.

10.12C Autoxidation

Linoleic acid is an example of a *polyunsaturated fatty acid*, the kind of polyunsaturated acid that occurs as an ester in **polyunsaturated fats** (Section 7.13, "The Chemistry of . . . Hydrogenation in the Food Industry," and Chapter 23). By polyunsaturated, we mean that the compound contains two or more double bonds:

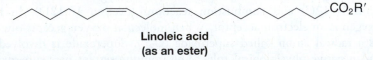

**Linoleic acid
(as an ester)**

Polyunsaturated fats occur widely in the fats and oils that are components of our diets. They are also widespread in the tissues of the body where they perform numerous vital functions.

The hydrogen atoms of the —CH_2— group located between the two double bonds of linoleic ester (Lin—H) are especially susceptible to abstraction by radicals (we shall see why in Chapter 13). Abstraction of one of these hydrogen atoms produces a new radical (Lin·) that can react with oxygen in a chain reaction that belongs to a general type of reaction called **autoxidation** (Fig. 10.5). The result of autoxidation is the formation of a hydroperoxide. Autoxidation is a process that occurs in many substances; for example, autoxidation is responsible for the development of the rancidity that occurs when fats and oils spoil and for the spontaneous combustion of oily rags left open to the air. Autoxidation also occurs in the body, and here it may cause irreversible damage.

Chain Initiation
Step 1

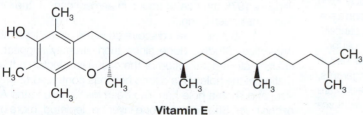

Chain Propagation
Step 2

Another radical

Step 3

**Hydrogen abstraction from
another molecule of
the linoleic ester**

A hydroperoxide   + Lin·

FIGURE 10.5 Autoxidation of a linoleic acid ester. In step 1 the reaction is initiated by the attack of a radical on one of the hydrogen atoms of the —CH_2— group between the two double bonds; this hydrogen abstraction produces a radical that is a resonance hybrid. In step 2 this radical reacts with oxygen in the first of two chain-propagating steps to produce an oxygen-containing radical, which in step 3 can abstract a hydrogen from another molecule of the linoleic ester (**Lin**—**H**). The result of this second chain-propagating step is the formation of a hydroperoxide and a radical (**Lin**·) that can bring about a repetition of step 2.

THE CHEMISTRY OF... Antioxidants

If you want to stop the ability of radicals to generate more of themselves, especially in scenarios where they could be damaging like autoxidation, one needs to find a suitable trapping reagent. Such materials, known as antioxidants, succeed when they can lead to a new, and more stable, radical species that terminates the chain by no longer reacting, or by further consuming reactive radicals to generate additional nonradical species. Two such compounds are vitamin E (also known as α-tocopherol) and BHT (butylated hydroxytoluene), shown further below:

**Vitamin E
(α-tocopherol)**

Vitamin E is found in vegetable oils.

In both cases, reaction with a radical species initially leads to a phenoxy radical, shown below with BHT in its reaction with a peroxy radical (ROO·). This event is the key for antioxidant behavior, in that it turns a highly reactive radical species into a fully covalent molecule that is less reactive (here a hydroperoxide, **ROOH**), with the newly formed phenoxy radical stabilized by the neighboring aromatic ring and attendant steric bulk of the *tert*-butyl groups.

ROO· →   + ROOH

**BHT
(Butylated hydroxytoluene)**   **BHT radical**

Worth noting is that vitamin E could be considered a natural antioxidant, since it is found in many foods and may work in our bodies to scavenge potentially damaging radical species, while BHT is a synthetic material that is added to many foods as a preservative.

10.12D Combustion of Alkanes

When alkanes react with oxygen (e.g., in oil and gas furnaces and in internal combustion engines) a complex series of reactions takes place, ultimately converting the alkane to carbon dioxide and water. Although our understanding of the detailed mechanism of combustion is incomplete, we do know that the important reactions occur by radical chain mechanisms with chain-initiating and chain-propagating steps such as the following reactions:

$$RH + O_2 \longrightarrow R\cdot + \cdot OOH \qquad \textbf{Initiating}$$

$$R\cdot + O_2 \longrightarrow R{-}OO\cdot$$
$$R{-}OO\cdot + R{-}H \longrightarrow R{-}OOH + R\cdot \quad \Bigg\} \textbf{Propagating}$$

One product of the second chain-propagating step is R—OOH, called an alkyl hydroperoxide. The oxygen–oxygen bond of an alkyl hydroperoxide is quite weak, and it can break and produce radicals that can initiate other chains:

$$RO{-}OH \longrightarrow RO\cdot + \cdot OH$$

THE CHEMISTRY OF... Ozone Depletion and Chlorofluorocarbons (CFCs)

In the stratosphere at altitudes of about 25 km, very high-energy (very short wavelength) UV light converts diatomic oxygen (O_2) into ozone (O_3). The reactions that take place may be represented as follows:

Step 1 $O_2 + h\nu \longrightarrow O + O$
Step 2 $O + O_2 + M \longrightarrow O_3 + M + heat$

where M is some other particle that can absorb some of the energy released in the second step.

The ozone produced in step 2 can also interact with high-energy UV light in the following way:

Step 3 $O_3 + h\nu \longrightarrow O_2 + O + heat$

The oxygen atom formed in step 3 can cause a repetition of step 2, and so forth. The net result of these steps is to convert highly energetic UV light into heat. This is important because the existence of this cycle shields Earth from radiation that is destructive to living organisms. This shield makes life possible on Earth's surface. Even a relatively small increase in high-energy UV radiation at Earth's surface would cause a large increase in the incidence of skin cancers.

Production of chlorofluoromethanes (and of chlorofluoroethanes) called chlorofluorocarbons (CFCs) or **freons** began in 1930. These compounds have been used as refrigerants, solvents, and propellants in aerosol cans. Typical freons are trichlorofluoromethane, $CFCl_3$ (called Freon-11), and dichlorodifluoromethane, CF_2Cl_2 (called Freon-12).

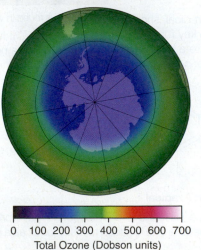

0   100 200 300 400 500 600 700
Total Ozone (Dobson units)

Chain Initiation

Step 1 $CF_2Cl_2 + h\nu \longrightarrow CF_2Cl\cdot + Cl\cdot$

Chain Propagation

Step 2 $Cl\cdot + O_3 \longrightarrow ClO\cdot + O_2$

Step 3 $ClO\cdot + O \longrightarrow O_2 + Cl\cdot$

In the chain-initiating step, UV light causes homolytic cleavage of one C—Cl bond of the freon. The chlorine atom thus produced is the real villain; it can set off a chain reaction that destroys thousands of molecules of ozone before it diffuses out of the stratosphere or reacts with some other substance.

In 1975 a study by the National Academy of Sciences supported the predictions of Rowland and Molina, and since January 1978 the use of freons in aerosol cans in the United States has been banned.

In 1985 a hole was discovered in the ozone layer above Antarctica. Studies done since then strongly suggest that chlorine atom destruction of the ozone is a factor in the formation of the hole. This ozone hole has continued to grow in size, and such a hole has also been discovered in the Arctic ozone layer. Should the ozone layer be depleted, more of the sun's damaging rays would penetrate to the surface of Earth.

 By 1974 world freon production was about 2 billion pounds annually. Most freon, even that used in refrigeration, eventually makes its way into the atmosphere where it diffuses unchanged into the stratosphere. In June 1974 F. S. Rowland and M. J. Molina published an article indicating, for the first time, that in the stratosphere freon is able to initiate radical chain reactions that can upset the natural ozone balance. The 1995 Nobel Prize in Chemistry was awarded to P. J. Crutzen, M. J. Molina, and F. S. Rowland for their combined work in this area. The reactions that take place are the following. (Freon-12 is used as an example.)

Recognizing the global nature of the problem, the "Montreal Protocol" was initiated in 1987. This treaty required the signing nations to reduce their production and consumption of chlorofluorocarbons. Accordingly, the industrialized nations of the world ceased production of chlorofluorocarbons as of 1996, and over 120 nations have signed the Montreal Protocol. Increased worldwide understanding of stratospheric ozone depletion, in general, has accelerated the phasing out of chlorofluorocarbons.

[WHY Do These Topics Matter?

RADICALS FROM THE BERGMAN CYCLOAROMATIZATION REACTION

In 1972, chemists at the University of California at Berkeley under the direction of Robert Bergman discovered a new chemical reaction that could be used to synthesize a benzene ring from starting materials that contained two alkyne triple bonds connected by a cis double bond, an array of atoms also known as an enediyne. This process, known as a cycloaromatization since it makes a ring that is aromatic, involves radicals as shown below both in making new bonds as well as in adding the final hydrogen atoms needed to make a benzene ring system.

© Joel Carillet/iStockphoto

Equally interesting, it was discovered on further exploration that the temperature needed to make the reaction occur was directly correlated to the distance between the termini of the two alkynes. For most molecules, that distance is greater than 3.6 Å and temperatures in excess of 200 °C are required to initiate the event. However, if that length can be made shorter, for example by placing the alkynes within a constrained ring, then these processes can occur at lower temperatures. Typically, distances between 3.2 to 3.3 Å allow for the cycloaromatization at ambient temperature (i.e., 37 °C), while even shorter distances, such as 3.0 Å, allow the reaction to proceed at room temperature (i.e., 25 °C).

Globally, such studies highlight the ability of chemists to discover new reactivity and understand molecular processes at a very sophisticated level. In this case, however, it turns out that the process occurs in nature, a fact that chemists just did not know until the natural product calicheamicin γ_1^I was isolated from a bacterial strain!

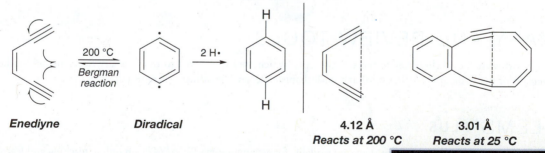

Enediyne   **Diradical**   **4.12 Å**   **3.01 Å**
Reacts at 200 °C   Reacts at 25 °C

As you can see from its structure below, calicheamicin γ_1^I is highly complex. It has an enediyne motif, one that is stable at 37 °C (body temperature) since the distance between the ends of the two alkyne units in its natural form is calculated to be just longer than 3.3 Å. However, when this compound is brought into a cell's nucleus, the unique trisulfide portion of the molecule can be converted into a sulfide nucleophile. Once unveiled, this reactive group can then attack a neighboring group of atoms through a chemical reaction we will learn more about in Chapter 19. What is important to know for now, however, is that this event changes the conformation of the entire right-hand half of the molecule, bringing the ends of the two alkynes closer together, to a distance of ~3.2 Å. As a result, a Bergman cycloaromatization can now occur at 37 °C, immediately generating a diradical that can abstract hydrogen from DNA, creating new DNA radicals that lead to cell death.

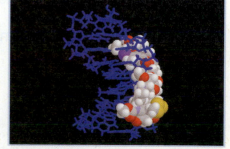

Calicheamicin bound to DNA. (PDB ID: 2PIK. Kumar, R. A.; Ikemoto, N., and Patel, D. J., Solution structure of the calicheamicin γ_1^I–DNA complex, J. Mol. Biol. 1997, 265, 187.) [Calicheamicin γ_1^I structure from Chemistry and Biology, 1994, 1(1). Nicolaou, K.C., Pitsinos, E.N., Theodorakis, A., Saimoto, H., and Wrasidio, W., Chemistry and Biology of the Calicheamicins, pp. 25-30. Copyright Elsevier 1994.

(continues on next page)

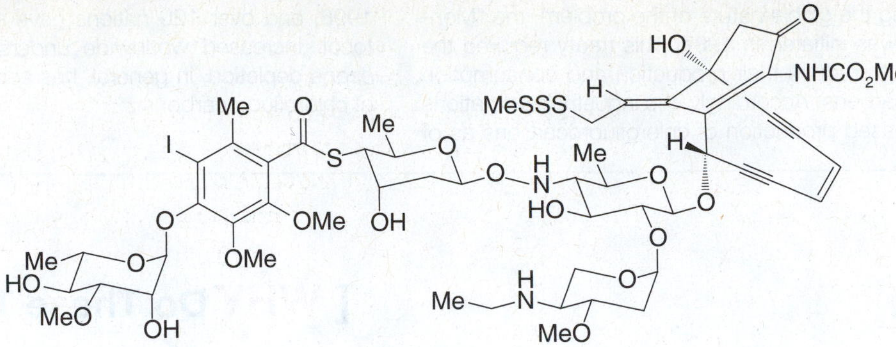

Calicheamicin γ₁¹

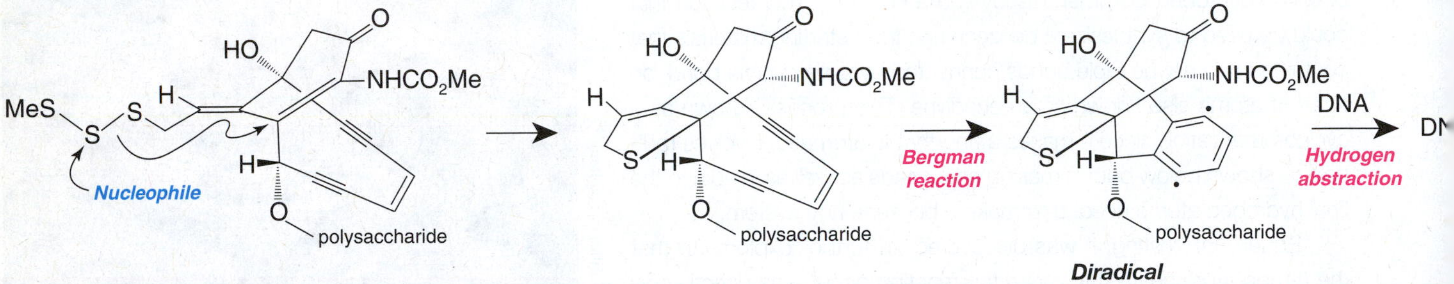

This chemistry shows that the Bergman reaction can occur naturally and that it functions in the formation of a molecule with a special triggering system that takes advantage of the differences in reactivity between different enediynes. Scientists in the pharmaceutical industry have since used this trigger system, as well as those of other related enediyne molecules from Nature, to create new drugs that have undergone clinical testing targeting a number of cancers such as acute forms of leukemia.

To learn more about these topics, see:

1. Bergman, R. G. "Reactive 1,4-Dehydroaromatics" in *Acc. Chem. Res.* **1973**, *6*, 25–31.
2. Nicolaou, K. C.; Smith, A. L.; Yue, E. W. "Chemistry and biology of natural and designed enediynes" in *Proc. Natl. Acad. Sci. USA* **1993**, *90*, 5881–5888 and references therein.

SUMMARY AND REVIEW TOOLS

The study aids for this chapter include key terms and concepts (which are hyperlinked to the Glossary from the bold, blue terms in the *WileyPLUS* version of the book at wileyplus.com) and a Mechanism Review regarding radical reactions.

PROBLEMS WILEY PLUS

Note to Instructors: Many of the homework problems are available for assignment via *WileyPLUS*, an online teaching and learning solution.

RADICAL MECHANISMS AND PROPERTIES

10.18 Write a mechanism for the following radical halogenation reaction.

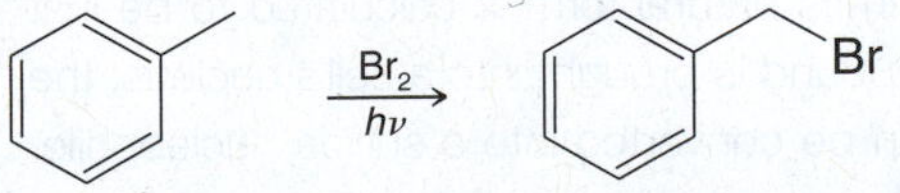

10.19 Explain the relative distribution of products below using reaction energy diagrams for the hydrogen abstraction step that leads to each product. (The rate-determining step in radical halogenation is the hydrogen abstraction step.) In energy diagrams for the two pathways, show the relative energies of the transition states and of the alkyl radical intermediate that results in each case.

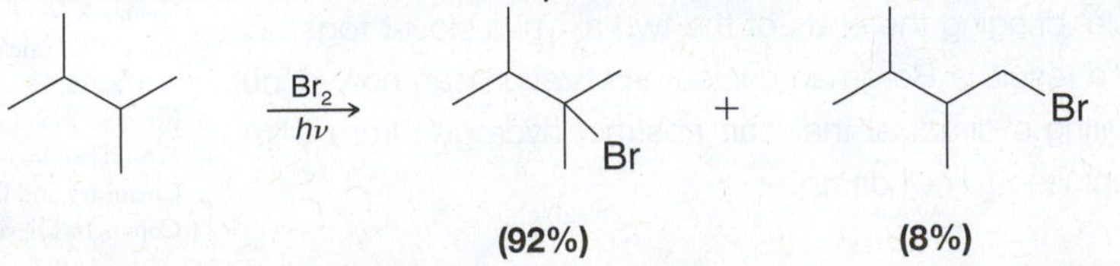

10.20 Which of the following compounds can be prepared by radical halogenation with little complication by formation of isomeric by-products?

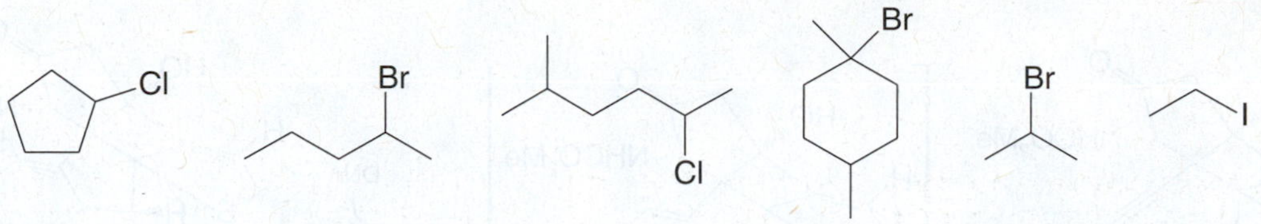

10.21 The radical reaction of propane with chlorine yields (in addition to more highly halogenated compounds) 1-chloropropane and 2-chloropropane.

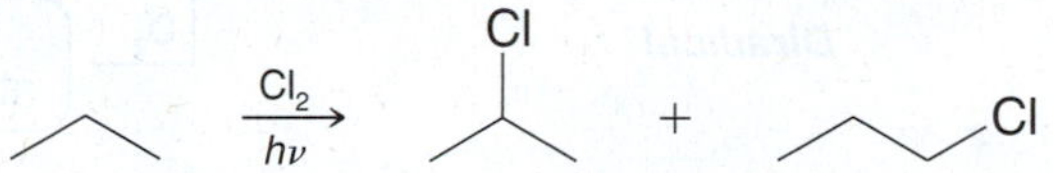

Write chain-initiating and chain-propagating steps showing how each of the products above is formed.

10.22 In addition to more highly chlorinated products, chlorination of butane yields a mixture of compounds with the formula C_4H_9Cl.

$$\text{butane} \xrightarrow[h\nu]{Cl_2} C_4H_9Cl$$

(multiple isomers)

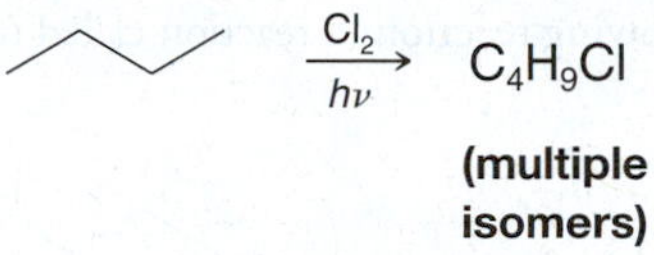

(a) Taking stereochemistry into account, how many different isomers with the formula C_4H_9Cl would you expect?

(b) If the mixture of C_4H_9Cl isomers were subjected to fractional distillation (or gas chromatography), how many fractions (or peaks) would you expect?

(c) Which fractions would be optically *inactive*?

(d) Which fractions could theoretically be resolved into enantiomers?

(e) Predict features in the ^1H and ^{13}C DEPT NMR spectra for each that would differentiate among the isomers separated by distillation or GC.

(f) How could fragmentation in their mass spectra be used to differentiate the isomers?

10.23 Chlorination of (*R*)-2-chlorobutane yields a mixture of dichloro isomers.

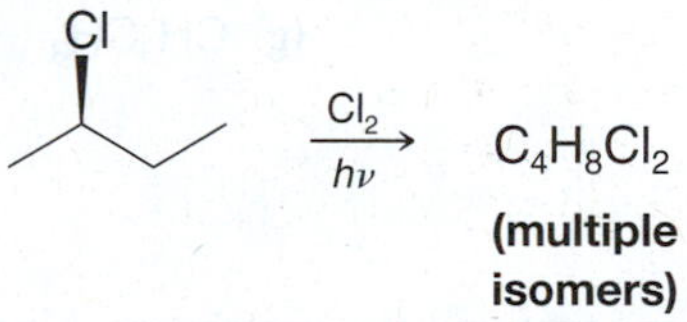

$$\xrightarrow[h\nu]{Cl_2} C_4H_8Cl_2$$

(multiple isomers)

(a) Taking into account stereochemistry, how many different isomers would you expect? Write their structures.

(b) How many fractions would be obtained upon fractional distillation?

(c) Which of these fractions would be optically active?

10.24 Peroxides are often used to initiate radical chain reactions such as in the following radical halogenation.

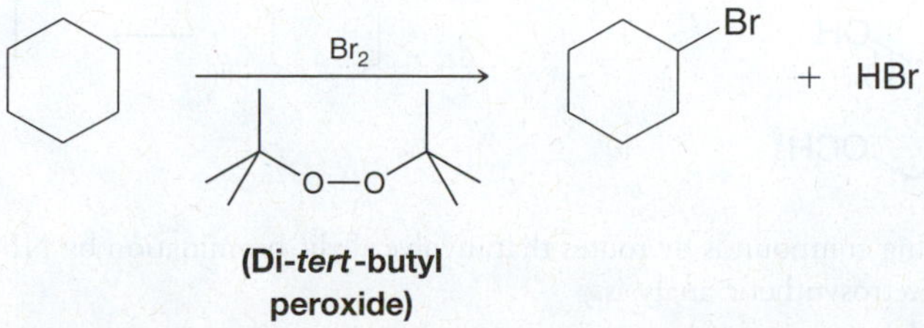

$$\xrightarrow{Br_2} \quad + \quad HBr$$

(Di-*tert*-butyl peroxide)

(a) Using bond dissociation energies in Table 10.1, explain why peroxides are especially effective as radical initiators.

(b) Write a mechanism for the reaction above showing how it could be initiated by di-*tert*-butyl peroxide.

10.25 List in order of decreasing stability all of the radicals that can be obtained by abstraction of a hydrogen atom from 2-methylbutane.

10.26 Draw mechanism arrows to show electron movements in the Bergman cycloaromatization reaction that leads to the diradical believed responsible for the DNA-cleaving action of the antitumor agent calicheamicin (see "Why Do These Topics Matter?" after Section 10.12).

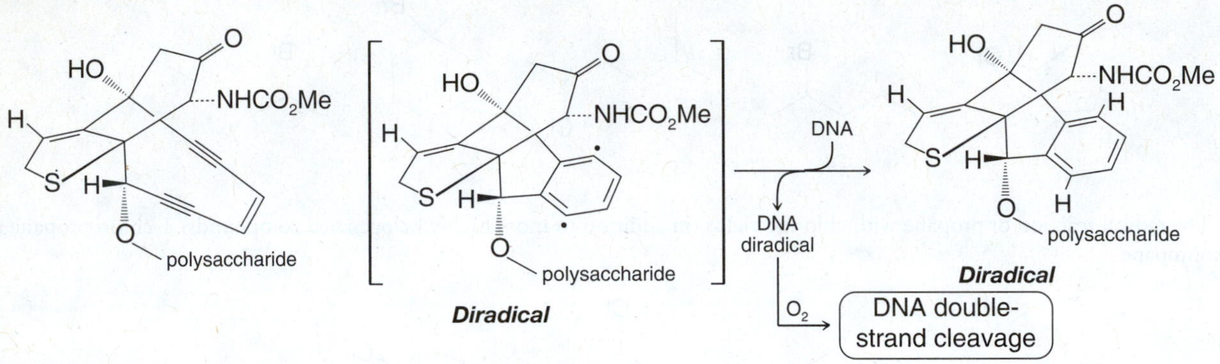

10.27 Find examples of C—H bond dissociation energies in Table 10.1 that are as closely related as possible to the bonds to H_a, H_b, and H_c in the molecule at right. Use these values to answer the questions below.

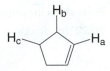

(a) What can you conclude about the relative ease of radical halogenation at H_a?

(b) Comparing H_b and H_c, which would more readily undergo substitution by radical halogenation?

10.28 Write a radical chain mechanism for the following reaction (a reaction called the Hunsdiecker reaction).

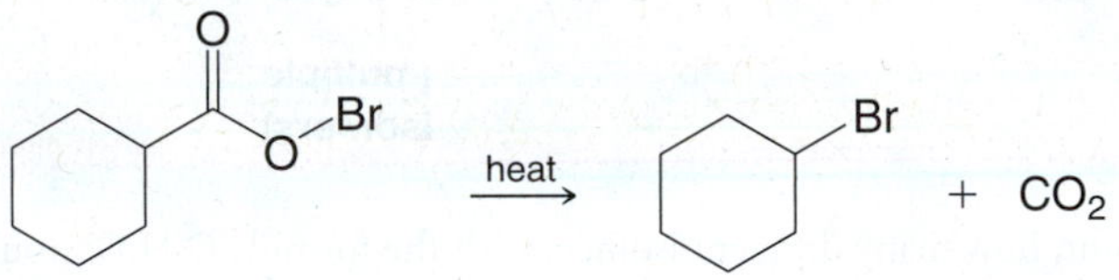

SYNTHESIS

10.29 Starting with the compound or compounds indicated in each part and using any other needed reagents, outline syntheses of each of the following compounds. (You need not repeat steps carried out in earlier parts of this problem.)

(a) $CH_3CH_3 \longrightarrow$

(b) $CH_3CH_3 \longrightarrow$

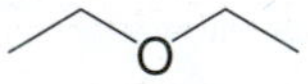

(c)

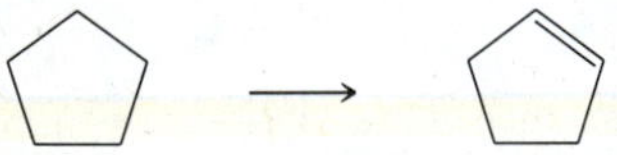

(d)

(e) $CH_4 + HC≡CH \longrightarrow$

(f) $CH_3CH_3 + HC≡CH \longrightarrow$

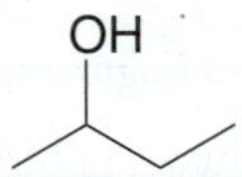

(g) $CH_3CH_3 \longrightarrow$

10.30 Provide the reagents necessary for the following synthetic transformations. More than one step may be required.

(a)

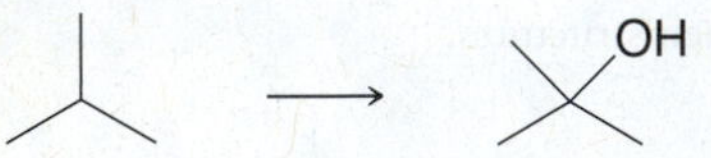

(b)

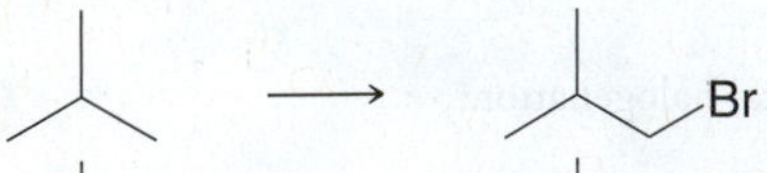

(c)

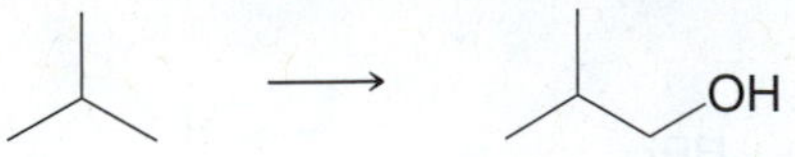

(d)

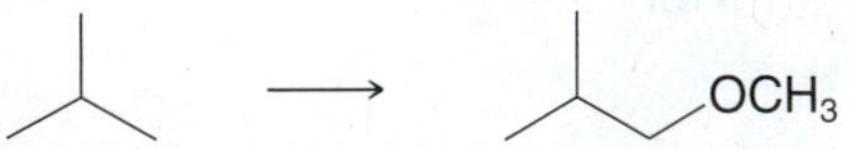

(e)

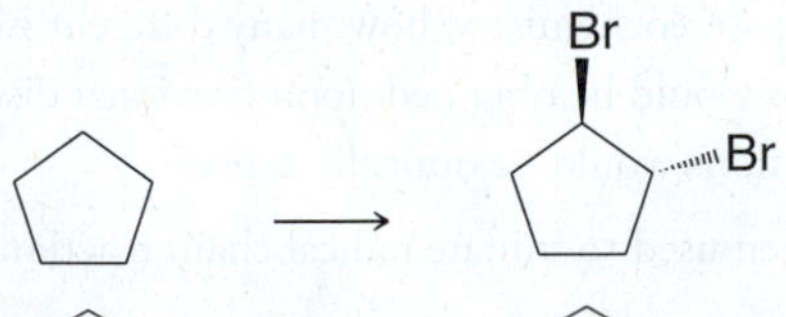

(f)

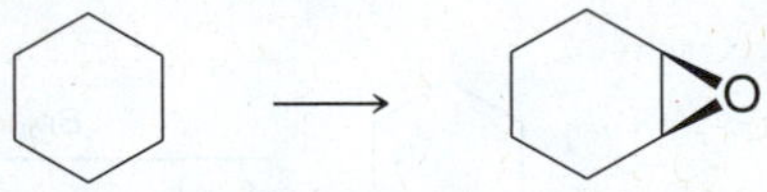

10.31 Synthesize each of the following compounds by routes that involve allylic bromination by NBS. Use starting materials having four carbons or fewer. Begin by writing a retrosynthetic analysis.

(a)

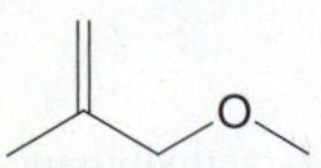

(b)

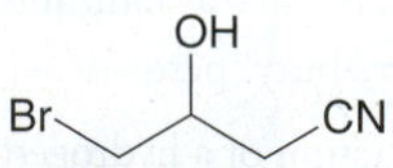

(c)

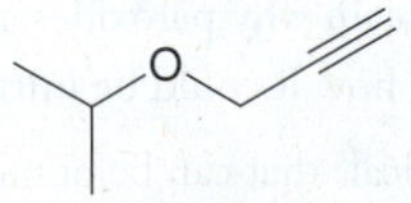

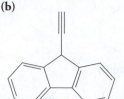

10.32 Synthesize each of the following compounds by routes that involve benzylic bromination by NBS and any other synthetic steps necessary. Begin by writing a retrosynthetic analysis.

(a)

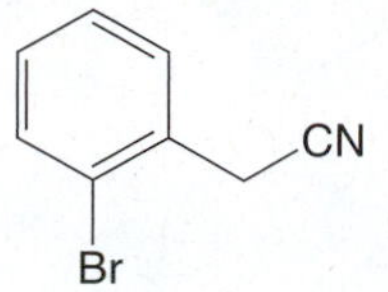

(b)

(c)

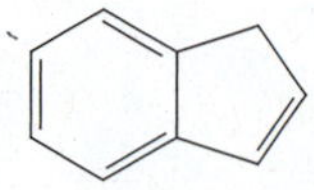

10.33 Synthesize each of the following compounds by routes that involve both allylic and benzylic bromination by NBS.

(a)

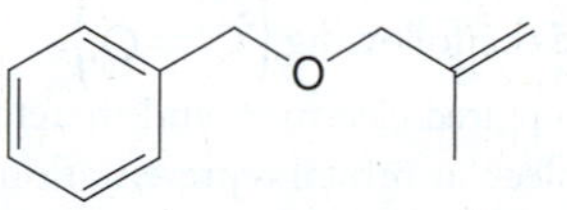

(b)

CHALLENGE PROBLEMS

10.34 The following reaction is the first step in the industrial synthesis of acetone and phenol (C_6H_5OH). AIBN (2,2′-azobisisobutyronitrile) initiates radical reactions by breaking down upon heating to form two isobutyronitrile radicals and nitrogen gas. Using an isobutyronitrile radical to initiate the reaction, write a mechanism for the following process.

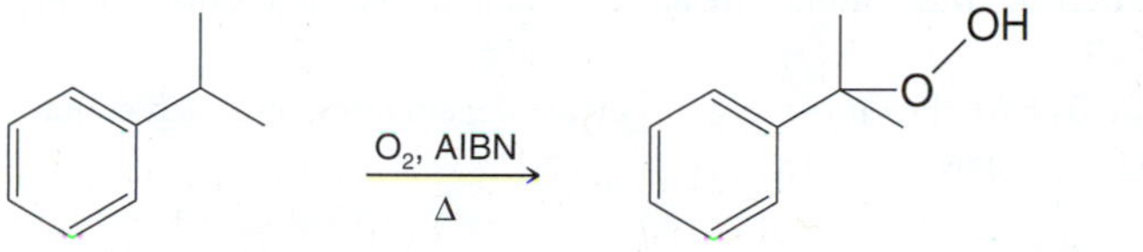

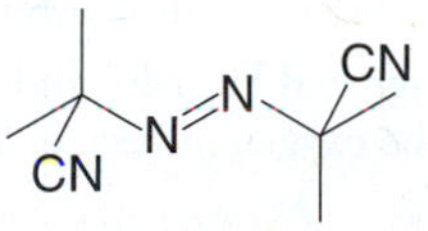

AIBN

10.35 In the radical chlorination of 2,2-dimethylhexane, chlorine substitution occurs much more rapidly at C5 than it does at a typical secondary carbon (e.g., C2 in butane). Consider the mechanism of radical polymerization and then suggest an explanation for the enhanced rate of substitution at C5 in 2,2-dimethylhexane.

10.36 Write a mechanism for the following reaction.

$$\xrightarrow{(PhCO_2)_2,\ heat}\quad +\ CO$$

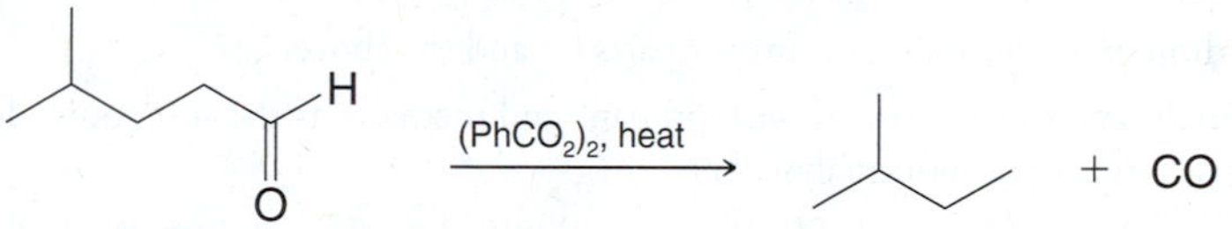

10.37 Hydrogen peroxide and ferrous sulfate react to produce hydroxyl radical (HO·), as reported in 1894 by English chemist H. J. H. Fenton. When *tert*-butyl alcohol is treated with HO· generated this way, it affords a crystalline reaction product **X**, mp 92 °C, which has these spectral properties:

MS: heaviest mass peak is at *m/z* 131
IR: 3620, 3350 (broad), 2980, 2940, 1385, 1370 cm^{-1}
^{1}H NMR: sharp singlets at δ 1.22, 1.58, and 2.95 (6:2:1 area ratio)
^{13}C NMR: δ 28 (CH_3), 35 (CH_2), 68 (C)

Draw the structure of **X** and write a mechanism for its formation.

10.38 The halogen atom of an alkyl halide can be replaced by the hydrogen atom bonded to tin in tributyltin hydride (Bu_3SnH). The process, called dehalogenation, is a radical reaction, and it can be initiated by AIBN (2,2′-azobisisobutyronitrile). AIBN decomposes to form nitrogen gas and two isobutyronitrile radicals, which initiate the reaction. Write a mechanism for the reaction.

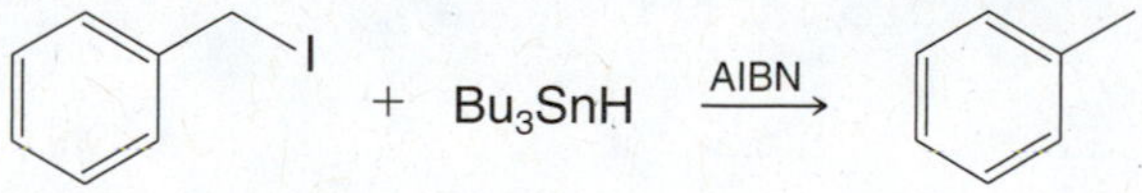

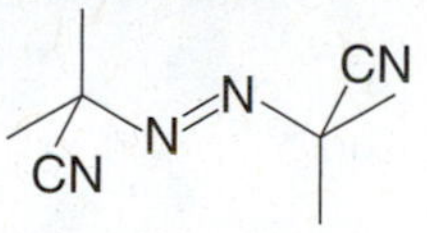

$$+\ Bu_3SnH\ \xrightarrow{AIBN}$$

AIBN

10.39 Write a mechanism that accounts for the following reaction. Note that the hydrogen atom bonded to tin in tributyltin hydride is readily transferred in radical mechanisms.

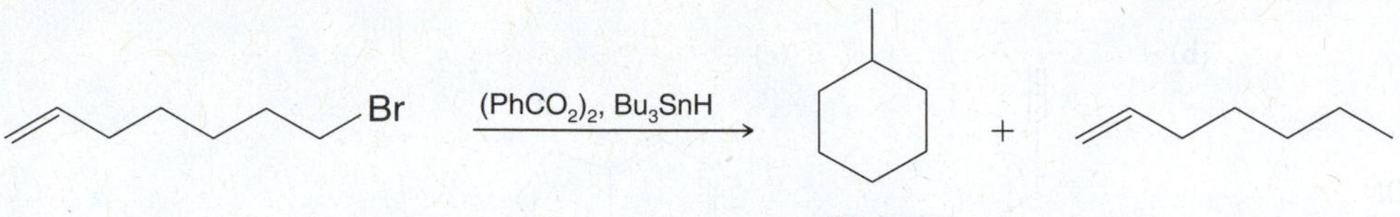

(Major)

10.40 Molecular orbital calculations can be used to model the location of electron density from unpaired electrons in a radical. Open the molecular models on the book's website for the methyl, ethyl, and *tert*-butyl radicals. The gray wire mesh surfaces in these models represent volumes enclosing electron density from unpaired electrons. What do you notice about the distribution of unpaired electron density in the ethyl radical and *tert*-butyl radical, as compared to the methyl radical? What bearing does this have on the relative stabilities of the radicals in this series?

10.41 If one were to try to draw the simplest Lewis structure for molecular oxygen, the result might be the following $\left(:\!\overset{..}{O}\!=\!\overset{..}{O}:\right)$.

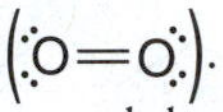

However, it is known from the properties of molecular oxygen and experiments that O_2 contains two unpaired electrons, and therefore, the Lewis structure above is incorrect. To understand the structure of O_2, it is necessary to employ a molecular orbital representation. To do so, we will need to recall (1) the shapes of bonding and antibonding σ and π molecular orbitals, (2) that each orbital can contain a maximum of two electrons, (3) that molecular oxygen has 16 electrons in total, and (4) that the two unpaired electrons in oxygen occupy separate degenerate (equal-energy) orbitals. Now, open the molecular model on the book's website for oxygen and examine its molecular orbitals in sequence from the HOMO-7 orbital to the LUMO. [HOMO-7 means the seventh orbital in energy below the highest occupied molecular orbital (HOMO), HOMO-6 means the sixth below the HOMO, and so forth.] Orbitals HOMO-7 through HOMO-4 represent the $\sigma 1s$, $\sigma 1s^*$, $\sigma 2s$, and $\sigma 2s^*$ orbitals, respectively, each containing a pair of electrons.

(a) What type of orbital is represented by HOMO-3 and HOMO-2? (*Hint*: What types of orbitals are possible for second-row elements like oxygen, and which orbitals have already been used?)

(b) What type of orbital is HOMO-1? [*Hint*: The $\sigma 2s$ and $\sigma 2s^*$ orbitals are already filled, as are the HOMO-3 and HOMO-2 orbitals identified in part (a). What bonding orbital remains?]

(c) The orbitals designated HOMO and LUMO in O_2 have the same energy (they are degenerate), and each contains one of the unpaired electrons of the oxygen molecule. What type of orbitals are these?

LEARNING GROUP PROBLEMS

1. (a) Draw structures for all organic products that would result when an *excess* of *cis*-1,3-dimethylcyclohexane reacts with Br_2 in the presence of heat and light. Use three-dimensional formulas to show stereochemistry.

(b) Draw structures for all organic products that would result when an *excess* of *cis*-1,3-dimethylcyclohexane reacts with Cl_2 in the presence of heat and light. Use three-dimensional formulas to show stereochemistry.

(c) As an alternative, use *cis*-1,2-dimethylcyclohexane to answer parts (a) and (b) above.

2. (a) Propose a synthesis of 2-methoxypropene starting with propane and methane as the sole source for carbon atoms. You may use any other reagents necessary. Devise a retrosynthetic analysis first.

(b) 2-Methoxypropene will form a polymer when treated with a radical initiator. Write the structure of this polymer and a mechanism for the polymerization reaction assuming a radical mechanism initiated by a diacyl peroxide.

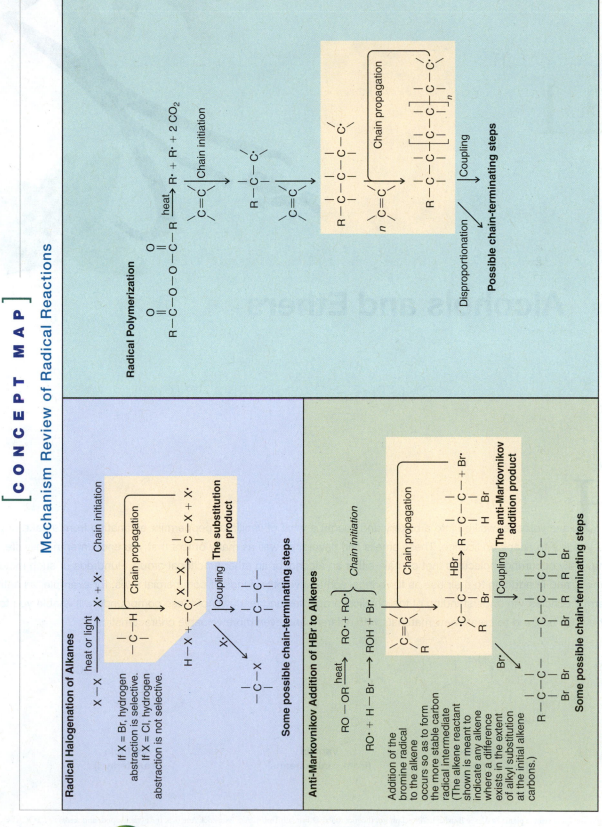

[CONCEPT MAP]

Mechanism Review of Radical Reactions

Radical Polymerization

Radical Halogenation of Alkanes

If X = Br, hydrogen abstraction is selective.
If X = Cl, hydrogen abstraction is not selective.

The substitution product

Some possible chain-terminating steps

Anti-Markovnikov Addition of HBr to Alkenes

Addition of the bromine radical to the alkene occurs so as to form the more stable carbon radical intermediate (The alkene reactant shown is meant to indicate any alkene where a difference exists in the extent of alkyl substitution at the initial alkene carbons.)

The anti-Markovnikov addition product

Some possible chain-terminating steps

WILEY PLUS See Special Topic B in *WileyPLUS*

11

Alcohols and Ethers

SYNTHESIS AND REACTIONS

Have you ever walked into a bakery and caught a whiff of vanilla or peppermint emanating from a cake or pastry? Maybe you like to snack on licorice. These smells and flavors, as well as many others that you encounter in daily life, arise from naturally occurring molecules that contain either an alcohol or an ether functional group. Hundreds of such molecules are known, and in addition to their use as flavorings, some have other kinds of commercial roles, for example, as antifreezes or pharmaceuticals. An understanding of the physical properties and reactivity of these compounds will enable you to see how they can be used to create new materials with different and even more valuable characteristics.

(–)-Menthol
(from peppermint)

Vanillin
(from vanilla beans)

Anethole
(from fennel)

PHOTO CREDITS: (peppermint plant) © Alexey Ilyashenko/iStockphoto; (licorice roots) © Fabrizio Troiani/Age Fotostock America, Inc.; (vanilla pods and seeds) © STOCKFOOD LBRF/Age Fotostock America, Inc.

IN THIS CHAPTER WE WILL CONSIDER:

- the structures, properties, and nomenclature of common alcohols and ethers
- key molecules that contain such groups
- the reactivity of alcohols, ethers, and a special group of ethers known as epoxides

[WHY DO THESE TOPICS MATTER?]   At the end of the chapter, we will see how the reactivity of epoxides can not only make highly complex molecules containing dozens of rings from acyclic precursors in a single step, but also help detoxify cancer-causing compounds from grilled meat, cigarettes, and peanuts.

11.1 STRUCTURE AND NOMENCLATURE

Alcohols have a hydroxyl (—OH) group bonded to a *saturated* carbon atom. The alcohol carbon atom may be part of a simple alkyl group, as in some of the following examples, or it may be part of a more complex molecule, such as cholesterol. Alcohols are also classified as 1°, 2°, or 3°, depending on the number of carbons bonded to the alcohol carbon.

CH₃OH

**Methanol
(methyl alcohol)**

1°

**Ethanol
(ethyl alcohol),
a 1° alcohol**

2°

**2-Propanol
(isopropyl alcohol),
a 2° alcohol**

3°

**2-Methyl-2-propanol
(*tert*-butyl alcohol),
a 3° alcohol**

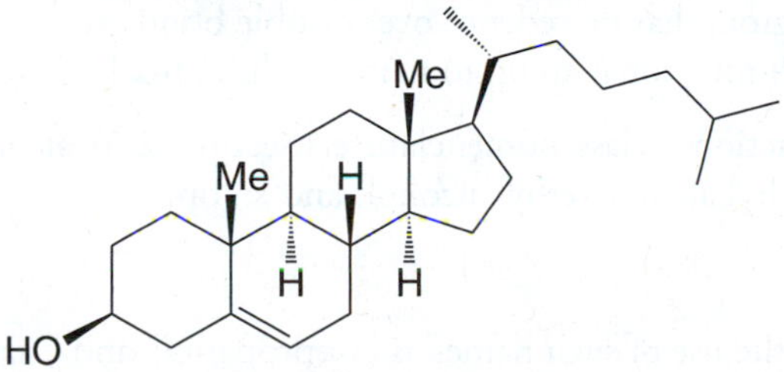

Cholesterol

The alcohol carbon atom may also be a saturated carbon atom adjacent to an alkenyl or alkynyl group, or the carbon atom may be a saturated carbon atom that is attached to a benzene ring:

allylic, 1° position

benzylic, 1° position

**2-Propenol
(or prop-2-en-1-ol,
or allyl alcohol),
an allylic alcohol**

**2-Propynol
(or prop-2-yn-1-ol,
or propargyl alcohol)**

**Benzyl alcohol
(a benzylic alcohol)**

Compounds that have a hydroxyl group attached *directly* to a benzene ring are called **phenols**. (Phenols are discussed in detail in Chapter 21.)

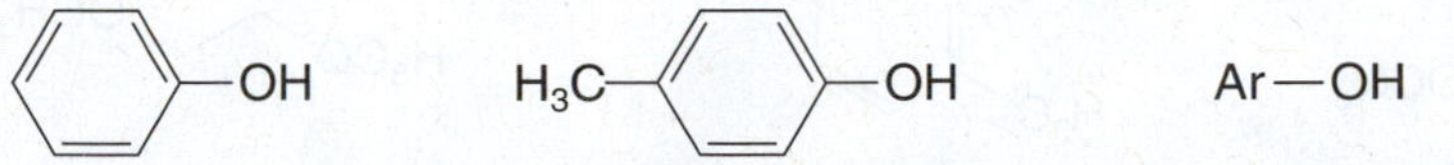

—OH

Phenol

H₃C—⟨⟩—OH

**p-Methylphenol,
a substituted phenol**

Ar—OH

**General formula
for a phenol**

Ethers differ from alcohols in that the oxygen atom of an ether is bonded to two carbon atoms. The hydrocarbon groups may be alkyl, alkenyl, vinyl, alkynyl, or aryl. Several examples are shown here:

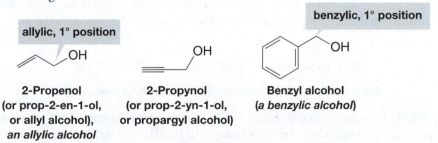

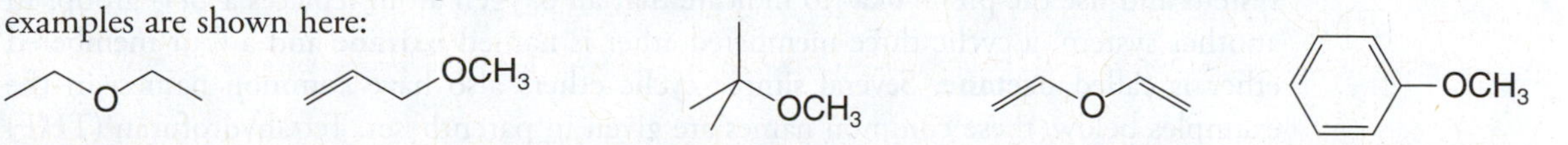

Diethyl ether   **Allyl methyl ether**   ***tert*-Butyl methyl ether**   **Divinyl ether**   **Methyl phenyl ether**

11.1A Nomenclature of Alcohols

We studied the IUPAC system of nomenclature for alcohols in Sections 2.6 and 4.3F. As a review consider the following problem.

●●● SOLVED PROBLEM 11.1

Give IUPAC substitutive names for the following alcohols:

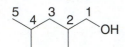

ANSWER: The longest chain *to which the hydroxyl group is attached* gives us the *base name*. The ending is *-ol*. We then number *the longest chain from the end that gives the carbon bearing the hydroxyl group the lower number*. Thus, the names, in both of the accepted IUPAC formats, are

(a)

5 4 3 2 1
OH

2,4-Dimethyl-1-pentanol
(or 2,4-dimethylpentan-1-ol)

(b)

1 2 3 4 5
OH C₆H₅

4-Phenyl-2-pentanol
(or 4-phenylpentan-2-ol)

(c)

1 2 3 4 5
OH

4-Penten-2-ol
(or pent-4-en-2-ol)

- The hydroxyl group has precedence over double bonds and triple bonds in deciding which functional group to name as the suffix [see example (c) above].

In common functional class nomenclature (Section 2.6) alcohols are called **alkyl alcohols** such as methyl alcohol, ethyl alcohol, and so on.

●●●

PRACTICE PROBLEM 11.1 What is wrong with the use of such names as "isopropanol" and "*tert*-butanol"?

11.1B Nomenclature of Ethers

Simple ethers are frequently given common functional class names. One simply lists (in alphabetical order) both groups that are attached to the oxygen atom and adds the word *ether*:

OCH₃   O   OC₆H₅

Ethyl methyl ether   **Diethyl ether**   *tert*-**Butyl phenyl ether**

IUPAC substitutive names should be used for complicated ethers, however, and for compounds with more than one ether linkage. In this IUPAC style, ethers are named as alkoxyalkanes, alkoxyalkenes, and alkoxyarenes. The RO— group is an **alkoxy** group.

OCH₃   OCH₃
                H₃C   H₃CO

2-Methoxypentane   1-Ethoxy-4-methylbenzene   1,2-Dimethoxyethane (DME)

Cyclic ethers can be named in several ways. One simple way is to use **replacement nomenclature**, in which we relate the cyclic ether to the corresponding hydrocarbon ring system and use the prefix **oxa-** to indicate that an oxygen atom replaces a CH₂ group. In another system, a cyclic three-membered ether is named **oxirane** and a four-membered ether is called **oxetane.** Several simple cyclic ethers also have common names; in the examples below, these common names are given in parentheses. Tetrahydrofuran (THF) and 1,4-dioxane are useful solvents:

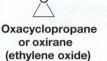

Oxacyclopropane or oxirane (ethylene oxide)

Oxacyclobutane or oxetane

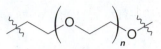

Oxacyclopentane (tetrahydrofuran or THF)

1,4-Dioxacyclohexane (1,4-dioxane)

Polyethylene oxide (PEO) (a water-soluble polymer made from ethylene oxide)

Polyethylene oxide is used in some skin creams.

Ethylene oxide is the starting material for polyethylene oxide (PEO, also called polyethylene glycol, PEG). Polyethylene oxide has many practical uses, including covalent attachment to therapeutic proteins such as interferon, a use that has been found to increase the circulatory lifetime of the drug. PEO is also used in some skin creams, and as a laxative prior to digestive tract procedures.

●●● **SOLVED PROBLEM 11.2**

Albuterol (used in some commonly prescribed respiratory medications) and vanillin (from vanilla beans) each contain several functional groups. Name the functional groups in albuterol and vanillin and, if appropriate for a given group, classify them as primary (1°), secondary (2°), or tertiary (3°).

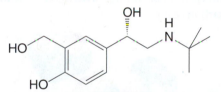

Albuterol (an asthma medication)   **Vanillin (from vanilla beans)**

Albuterol is used in some respiratory medications.

STRATEGY AND ANSWER: Albuterol has the following functional groups: 1° alcohol, 2° alcohol, phenol, and 2° amine. Vanillin has aldehyde, ether, and phenol functional groups. See Chapter 2 for a review of how to classify alcohol and amine functional groups as 1°, 2°, or 3°.

●●●

Give bond-line formulas and appropriate names for all of the alcohols and ethers with the formulas **(a)** C_3H_8O and **(b)** $C_4H_{10}O$.

PRACTICE PROBLEM 11.2

11.2 PHYSICAL PROPERTIES OF ALCOHOLS AND ETHERS

The physical properties of a number of alcohols and ethers are given in Tables 11.1 and 11.2.

- Ethers have boiling points that are roughly comparable with those of hydrocarbons of the same molecular weight (MW).

For example, the boiling point of diethyl ether (MW = 74) is 34.6 °C; that of pentane (MW = 72) is 36 °C.

- Alcohols have much higher boiling points than comparable ethers or hydrocarbons.

The boiling point of butyl alcohol (MW = 74) is 117.7 °C. We learned the reason for this behavior in Section 2.13C.

TABLE 11.1 PHYSICAL PROPERTIES OF SOME COMMON ALCOHOLS

Name	Formula	mp (°C)	bp (°C) (1 atm)	Water Solubility (g/100 mL H₂O)
Monohydroxy Alcohols				
Methanol	CH_3OH	−97	64.7	∞
Ethanol	CH_3CH_2OH	−117	78.3	∞
Propyl alcohol	$CH_3CH_2CH_2OH$	−126	97.2	∞
Isopropyl alcohol	$CH_3CH(OH)CH_3$	−88	82.3	∞
Butyl alcohol	$CH_3CH_2CH_2CH_2OH$	−90	117.7	8.3
Isobutyl alcohol	$CH_3CH(CH_3)CH_2OH$	−108	108.0	10.0
sec-Butyl alcohol	$CH_3CH_2CH(OH)CH_3$	−114	99.5	26.0
tert-Butyl alcohol	$(CH_3)_3COH$	25	82.5	∞
Diols and Triols				
Ethylene glycol	CH_2OHCH_2OH	−12.6	197	∞
Propylene glycol	$CH_3CHOHCH_2OH$	−59	187	∞
Trimethylene glycol	$CH_2OHCH_2CH_2OH$	−30	215	∞
Glycerol	$CH_2OHCHOHCH_2OH$	18	290	∞

Photo courtesy of AMSOIL, INC.

Propylene glycol (1,2-propanediol) is used as an environmentally friendly engine coolant because it is biodegradable, has a high boiling point, and is miscible with water.

- Alcohol molecules can associate with each other through **hydrogen bonding**, whereas those of ethers and hydrocarbons cannot.

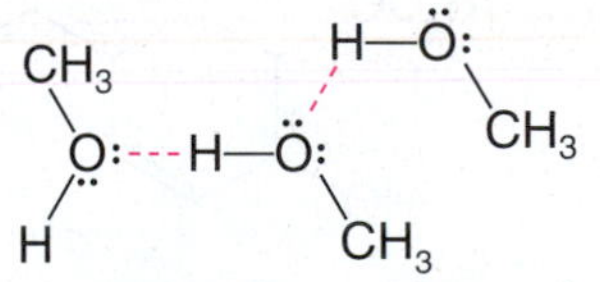

Hydrogen bonding between molecules of methanol

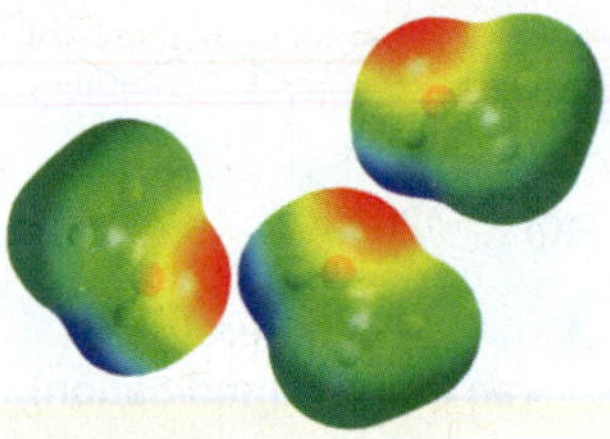

Ethers, however, *are* able to form hydrogen bonds with compounds such as water. Ethers, therefore, have solubilities in water that are similar to those of alcohols of the same molecular weight and that are very different from those of hydrocarbons.

TABLE 11.2 PHYSICAL PROPERTIES OF SOME COMMON ETHERS

Name	Formula	mp (°C)	bp (°C) (1 atm)
Dimethyl ether	CH_3OCH_3	−138	−24.9
Ethyl methyl ether	$CH_3OCH_2CH_3$		10.8
Diethyl ether	$CH_3CH_2OCH_2CH_3$	−116	34.6
1,2-Dimethoxyethane (DME)	$CH_3OCH_2CH_2OCH_3$	−68	83
Oxirane		−112	12
Tetrahydrofuran (THF)		−108	65.4
1,4-Dioxane	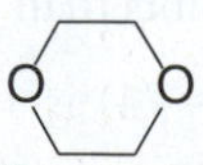	11	101

Diethyl ether and 1-butanol, for example, have the same solubility in water, approximately 8 g per 100 mL at room temperature. Pentane, by contrast, is virtually insoluble in water.

Methanol, ethanol, both propyl alcohols, and *tert*-butyl alcohol are completely miscible with water (Table 11.1). The solubility of alcohols in water gradually decreases as the hydrocarbon portion of the molecule lengthens; long-chain alcohols are more "alkane-like" and are, therefore, less like water.

••• **SOLVED PROBLEM 11.3**

1,2-Propanediol (propylene glycol) and 1,3-propanediol (trimethylene glycol) have higher boiling points than any of the butyl alcohols (see Table 11.1), even though they all have roughly the same molecular weight. Propose an explanation.

STRATEGY AND ANSWER: The presence of two hydroxyl groups in each of the diols allows their molecules to form more hydrogen bonds than the butyl alcohols. Greater hydrogen-bond formation means that the molecules of 1,2-propanediol and 1,3-propanediol are more highly associated and, consequently, their boiling points are higher.

11.3 IMPORTANT ALCOHOLS AND ETHERS

11.3A Methanol

At one time, most methanol was produced by the destructive distillation of wood (i.e., heating wood to a high temperature in the absence of air). It was because of this method of preparation that methanol came to be called "wood alcohol." Today, most methanol is prepared by the catalytic hydrogenation of carbon monoxide. This reaction takes place under high pressure and at a temperature of 300–400 °C:

$$CO \ + \ 2\,H_2 \xrightarrow[\substack{200\text{–}300\ atm \\ ZnO\text{–}Cr_2O_3}]{300\text{–}400\ °C} CH_3OH$$

Methanol is highly toxic. Ingestion of even small quantities of methanol can cause blindness; large quantities cause death. Methanol poisoning can also occur by inhalation of the vapors or by prolonged exposure to the skin.

11.3B Ethanol

Ethanol can be made by the fermentation of sugars, and it is the alcohol of all alcoholic beverages. The synthesis of ethanol in the form of wine by the fermentation of the sugars of fruit juices was among our first accomplishments in the field of organic synthesis. Sugars from a wide variety of sources can be used in the preparation of alcoholic beverages. Often, these sugars are from grains, and it is this derivation that accounts for ethanol having the synonym "grain alcohol."

Fermentation is usually carried out by adding yeast to a mixture of sugars and water. Yeast contains enzymes that promote a long series of reactions that ultimately convert a simple sugar ($C_6H_{12}O_6$) to ethanol and carbon dioxide:

$$C_6H_{12}O_6 \xrightarrow{\text{yeast}} 2\,CH_3CH_2OH \ + \ 2\,CO_2$$
$$\textbf{(~95\% yield)}$$

Fermentation alone does not produce beverages with an ethanol content greater than 12–15% because the enzymes of the yeast are deactivated at higher concentrations. To produce beverages of higher alcohol content, the aqueous solution must be distilled.

Ethanol is an important industrial chemical. Most ethanol for industrial purposes is produced by the acid-catalyzed hydration of ethene:

$$\diagup\diagup \ + \ H_2O \xrightarrow{\text{acid}} \diagup\diagdown_{OH}$$

About 5% of the world's ethanol supply is produced this way.

Vineyard grapes for use in fermentation.

THE CHEMISTRY OF... Ethanol as a Biofuel

Ethanol is said to be a renewable energy source because it can be made by fermentation of grains and other agricultural sources such as switchgrass and sugarcane. The crops themselves grow, of course, by converting light energy from the sun to chemical energy through photosynthesis. Once obtained, the ethanol can be combined with gasoline in varying proportions and used in internal combustion engines. During the year 2007, the United States led the world in ethanol production with 6.5 billion U.S. gallons, followed closely by Brazil with 5 billion gallons.

When used as a replacement for gasoline, ethanol has a lower energy content, by about 34% per unit volume. This, and other factors, such as costs in energy required to produce the agricultural feedstock, especially corn, have created doubts about the wisdom of an ethanol-based program as a renewable energy source. Production of ethanol from corn is 5 to 6 times less efficient than producing it from sugarcane, and it also diverts production of a food crop into an energy source. World food shortages may be a result.

Media Bakery

Ethanol is a *hypnotic* (sleep producer). It depresses activity in the upper brain even though it gives the illusion of being a stimulant. Ethanol is also toxic, but it is much less toxic than methanol. In rats the lethal dose of ethanol is 13.7 g kg^{-1} of body weight.

11.3C Ethylene and Propylene Glycols

Ethylene glycol ($HOCH_2CH_2OH$) has a low molecular weight, a high boiling point, and is miscible with water (Table 11.1). These properties made ethylene glycol a good automobile antifreeze. Unfortunately, however, ethylene glycol is toxic. Propylene glycol (1,2-propanediol) is now widely used as a low-toxicity, environmentally friendly alternative to ethylene glycol.

11.3D Diethyl Ether

Diethyl ether is a very low boiling, highly flammable liquid. Care should always be taken when diethyl ether is used in the laboratory, because open flames or sparks from light switches can cause explosive combustion of mixtures of diethyl ether and air.

Most ethers react slowly with oxygen by a radical process called **autoxidation** (see Section 10.12D) to form hydroperoxides and peroxides:

Step 1

$$\cdot \ddot{O}-\ddot{O}\cdot \; + \; \overset{H}{\underset{|}{\overset{|}{C}}}-OR' \; \longrightarrow \; \cdot\ddot{O}-\ddot{O}-H \; + \; -\overset{OR'}{\underset{\diagdown}{C}}$$

> **Hydrogen abstraction adjacent to the ether oxygen occurs readily.**

Step 2

$$-\overset{OR'}{\underset{|}{C}} \; + \; O_2 \; \longrightarrow \; -\overset{OO\cdot}{\underset{|}{C}}-OR'$$

Step 3a

$$-\overset{OO\cdot}{\underset{|}{C}}-OR' \; + \; \overset{H}{\underset{|}{\overset{|}{C}}}-OR' \; \longrightarrow \; -\overset{OOH}{\underset{|}{C}}-OR' \; + \; -\overset{OR'}{\underset{\diagdown}{C}}$$

A hydroperoxide

or

Step 3b

$$-\overset{OO\cdot}{\underset{|}{C}}-OR' \; + \; -\overset{OR'}{\underset{\diagup}{C}} \; \longrightarrow \; R'O-\overset{}{\underset{|}{C}}-OO-\overset{}{\underset{|}{C}}-OR'$$

> **Hydroperoxides and peroxides can be explosive.**

A peroxide

THE CHEMISTRY OF... Cholesterol and Heart Disease

Cholesterol (below, see also Section 23.4B) is an alcohol that is a precursor of steroid hormones and a vital constituent of cell membranes. It is essential to life. On the other hand, deposition of cholesterol in arteries is a cause of heart disease and atherosclerosis, two leading causes of death in humans. For an organism to remain healthy, there has to be a delicate balance between the biosynthesis of cholesterol and its utilization, so that arterial deposition is kept at a minimum.

a drug called a *statin*, a drug designed to interfere with the biosynthesis of cholesterol.

In the body, the biosynthesis of cholesterol takes place through a series of steps, one of which is catalyzed by the enzyme *HMG-CoA reductase* and which uses mevalonate ion as a substrate. The statin interferes with this step and thereby reduces blood cholesterol levels.

Cholesterol

Lovastatin   **Mevalonate ion**

For some individuals with high blood levels of cholesterol, the remedy is as simple as following a diet low in cholesterol and fat. For those who suffer from elevated blood cholesterol levels for genetic reasons, other means of cholesterol reduction are required. One remedy involves taking

Lovastatin, a compound isolated from the fungus *Apergillus terreus*, was the first statin to be marketed. Now many others are in use.

Lovastatin, because a part of its structure resembles mevalonate ion, can apparently bind at the active site of HMGA-CoA-reductase and act as a competitive inhibitor of this enzyme and thereby reduce cholesterol biosynthesis.

These hydroperoxides and peroxides, which often accumulate in ethers that have been stored for months or longer in contact with air (the air in the top of the bottle is enough), are dangerously explosive. They often detonate without warning when ether solutions are distilled to near dryness. Since ethers are used frequently in extractions, one should take care to test for and decompose any peroxides present in the ether before a distillation is carried out. (Consult a laboratory manual for instructions.)

Diethyl ether was at one time used as a surgical anesthetic. The most popular modern anesthetic is halothane ($CF_3CHBrCl$). Unlike diethyl ether, halothane is not flammable. (See "The Chemistry of... Ethers as General Anesthetics," Section 2.7, for more information.)

11.4 SYNTHESIS OF ALCOHOLS FROM ALKENES

We have already studied the acid-catalyzed **hydration** of alkenes, **oxymercuration**–demercuration, and **hydroboration**–oxidation as methods for the synthesis of alcohols from alkenes (see Sections 8.4, 8.5, and 8.6, respectively). Below, we briefly summarize these methods.

1. **Acid-Catalyzed Hydration of Alkenes** Alkenes add water in the presence of an acid catalyst to yield alcohols (Section 8.5). The addition takes place with **Markovnikov regioselectivity**. The reaction is reversible, and the mechanism for the acid-catalyzed hydration of an alkene is simply the reverse of that for the dehydration of an alcohol (Section 7.7).

Alkene   **Alcohol**

Acid-catalyzed hydration of alkenes has limited synthetic utility, however, because the carbocation intermediate may rearrange if a more stable or isoenergetic carbocation is possible by hydride or alkanide migration. Thus, a mixture of isomeric alcohol products may result.

2. **Oxymercuration–Demercuration** Alkenes react with mercuric acetate in a mixture of water and tetrahydrofuran (THF) to produce (hydroxyalkyl)mercury compounds. These can be reduced to alcohols with sodium borohydride and water (Section 8.5).

Oxymercuration

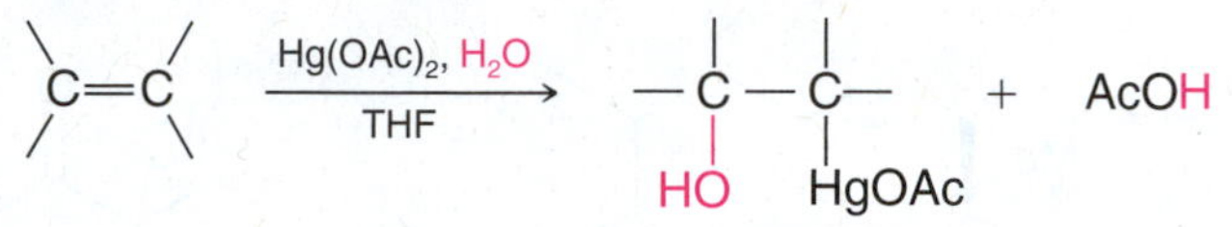

Demercuration

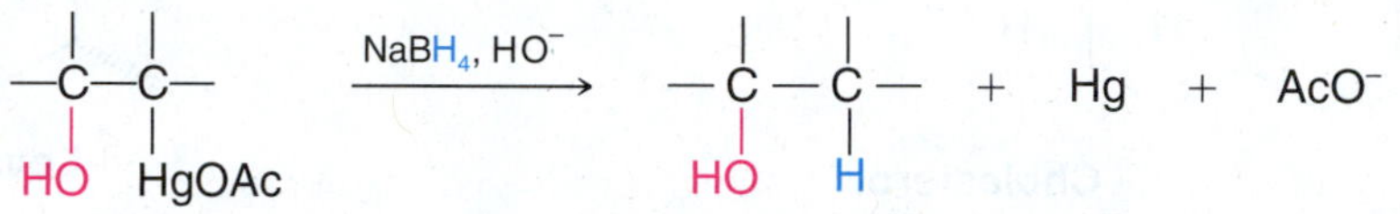

In the oxymercuration step, water and mercuric acetate add to the double bond; in the demercuration step, sodium borohydride reduces the acetoxymercury group and replaces it with hydrogen. The net addition of H— and —OH takes place with **Markovnikov regioselectivity** and **generally takes place without the complication of rearrangements**, as sometimes occurs with acid-catalyzed hydration of alkenes. The overall alkene hydration is not stereoselective because even though the oxymercuration step occurs with anti addition, the demercuration step is not stereoselective (radicals are thought to be involved), and hence a mixture of syn and anti products results.

3. **Hydroboration–Oxidation** An alkene reacts with BH_3:THF or diborane to produce an alkylborane. Oxidation and hydrolysis of the alkylborane with hydrogen peroxide and base yield an alcohol (Section 8.6).

Hydroboration

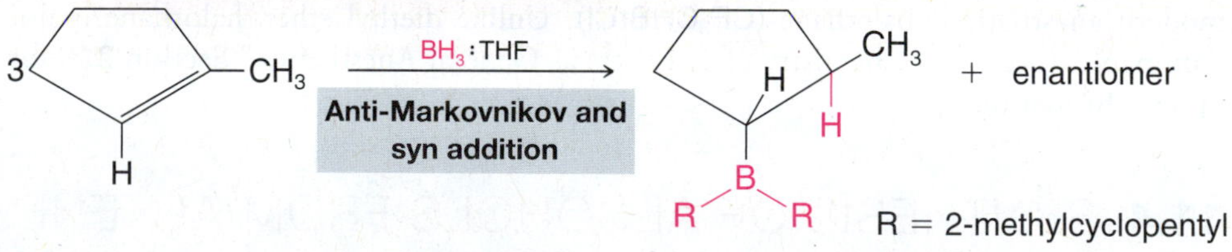

Oxidation

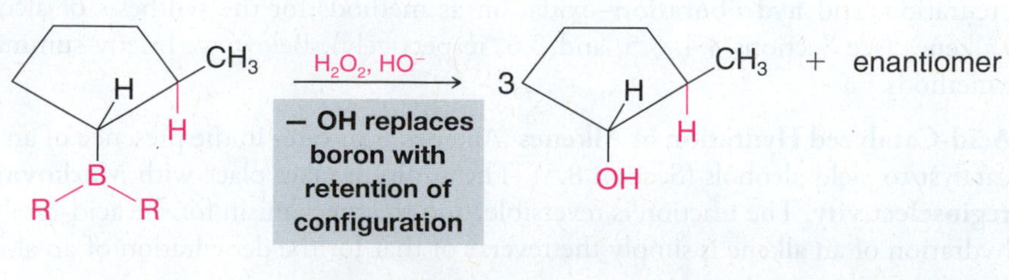

In the first step, boron and hydrogen undergo syn addition to the alkene; in the second step, treatment with hydrogen peroxide and base replaces the boron with —OH with retention of configuration. The net addition of —H and —OH occurs with **anti-Markovnikov regioselectivity** and **syn stereoselectivity.** Hydroboration–oxidation, therefore, serves as a useful regiochemical complement to oxymercuration–demercuration.

Mercury compounds are hazardous. Before you carry out a reaction involving mercury or its compounds, you should familiarize yourself with current procedures for its use and disposal.

Helpful Hint

Oxymercuration–demercuration and hydroboration–oxidation have complementary regioselectivity.

What conditions would you use for each reaction?

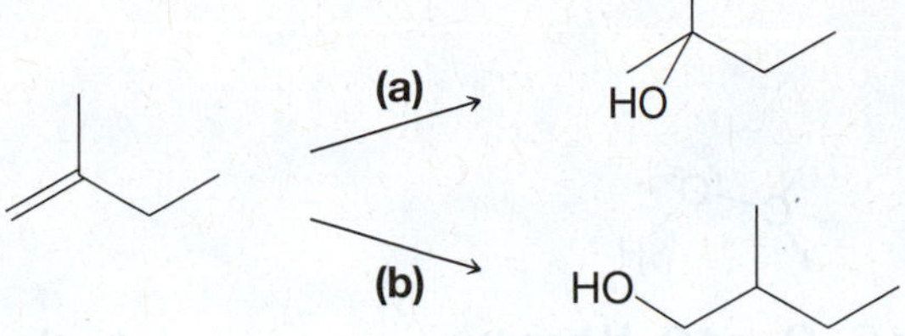

STRATEGY AND ANSWER: We recognize that synthesis by path (a) would require a Markovnikov addition of water to the alkene. So, we could use either acid-catalyzed hydration or oxymercuration–demercuration.

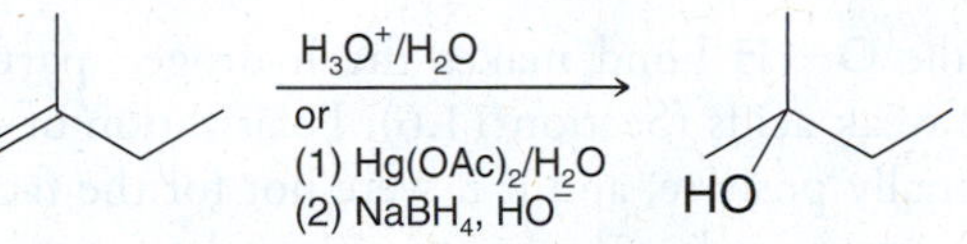

H_3O^+/H_2O
or
(1) $Hg(OAc)_2/H_2O$
(2) $NaBH_4$, HO^-

Markovnikov addition of H— and —OH

Synthesis by path (b) requires an anti-Markovnikov addition, so we would choose hydroboration–oxidation.

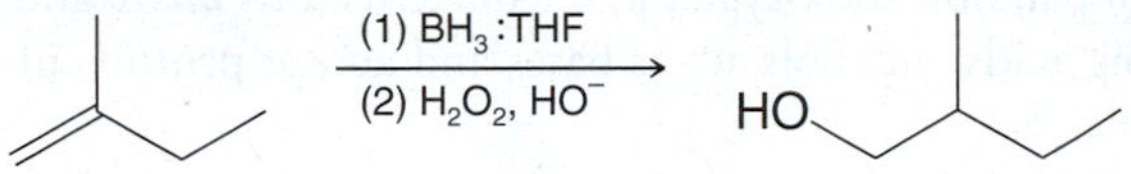

(1) BH_3:THF
(2) H_2O_2, HO^-

anti-Markovnikov addition of H— and —OH

Predict the major products of the following reactions:

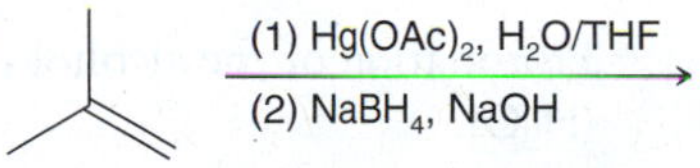

●●●
PRACTICE PROBLEM 11.3

(a)   cat. H_2SO_4
H_2O

(c)   (1) $Hg(OAc)_2$, H_2O/THF
(2) $NaBH_4$, $NaOH$

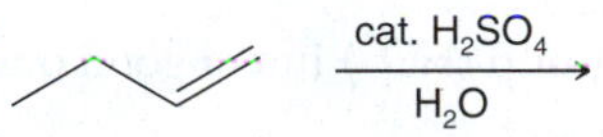

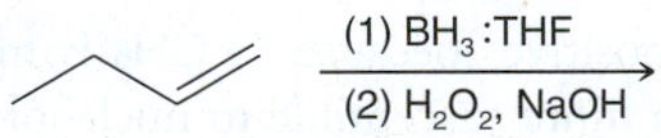

(b)   (1) BH_3:THF
(2) H_2O_2, $NaOH$

The following reaction does not produce the product shown.

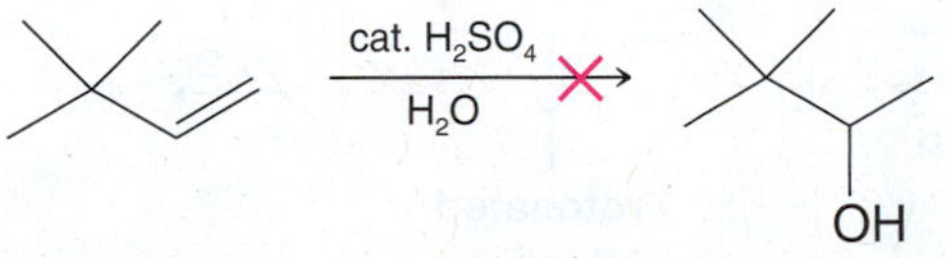

●●●
PRACTICE PROBLEM 11.4

cat. H_2SO_4 ✗
H_2O

OH

(a) Predict the major product from the conditions shown above, and write a detailed mechanism for its formation.

(b) What reaction conditions would you use to successfully synthesize the product shown above (3,3-dimethyl-2-butanol).

11.5 REACTIONS OF ALCOHOLS

The reactions of alcohols have mainly to do with the following:

- The oxygen atom of the hydroxyl group is nucleophilic and weakly basic.
- The hydrogen atom of the hydroxyl group is weakly acidic.
- The hydroxyl group can be converted to a leaving group so as to allow substitution or elimination reactions.

Our understanding of the reactions of alcohols will be aided by an initial examination of the electron distribution in the alcohol functional group and of how this distribution affects its reactivity. The oxygen atom of an alcohol polarizes both the C—O bond and the O—H bond of an alcohol:

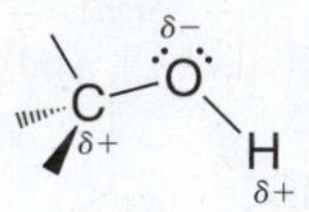

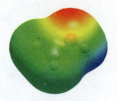

The C—O and O—H bonds of an alcohol are polarized

An electrostatic potential map for methanol shows partial negative charge at the oxygen and partial positive charge at the hydroxyl proton.

Polarization of the O—H bond makes the hydrogen partially positive and explains why alcohols are weak acids (Section 11.6). Polarization of the C—O bond makes the carbon atom partially positive, and if it were not for the fact that HO^- is a strong base and, therefore, a very poor leaving group, this carbon would be susceptible to nucleophilic attack.

The electron pairs on the oxygen atom make it both **basic** and **nucleophilic**. In the presence of strong acids, alcohols act as bases and accept protons in the following way:

Protonation of an alcohol

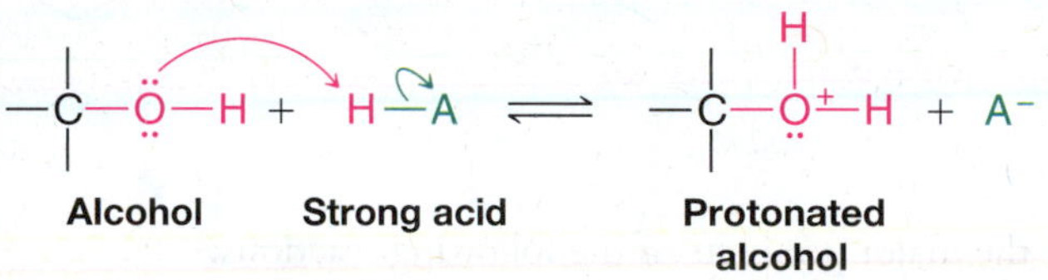

$$-\overset{|}{\underset{|}{C}}-\overset{..}{\underset{..}{O}}-H \ + \ H-A \ \rightleftharpoons \ -\overset{|}{\underset{|}{C}}-\overset{H}{\underset{..}{\overset{+}{O}}}-H \ + \ A^-$$

Alcohol   Strong acid   Protonated alcohol

- Protonation of the alcohol converts a poor leaving group (HO^-) into a good one (H_2O).

Protonation also makes the carbon atom even more positive (because $-\overset{+}{O}H_2$ is more electron withdrawing than $-OH$) and, therefore, even more susceptible to nucleophilic attack.

- Once the alcohol is protonated substitution reactions become possible (S_N2 or S_N1, depending on the class of alcohol, Section 11.8).

The protonated hydroxyl group is a good leaving group (H_2O)

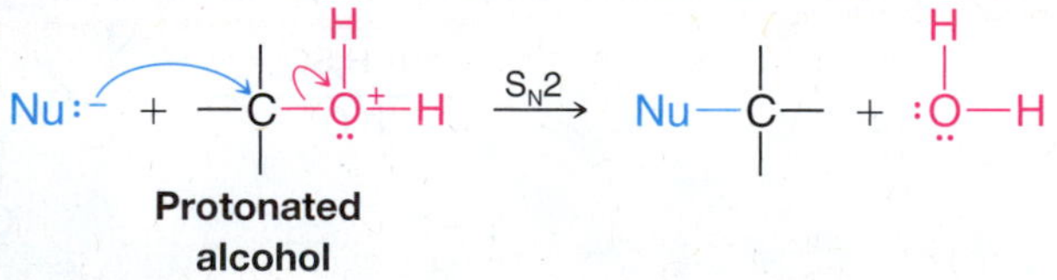

$$Nu:^- \ + \ -\overset{|}{\underset{|}{C}}-\overset{H}{\underset{..}{\overset{+}{O}}}-H \ \xrightarrow{S_N2} \ Nu-\overset{|}{\underset{|}{C}}- \ + \ :\overset{H}{\underset{..}{O}}-H$$

Protonated alcohol

Because alcohols are nucleophiles, they, too, can react with protonated alcohols. This, as we shall see in Section 11.11A, is an important step in one synthesis of ethers:

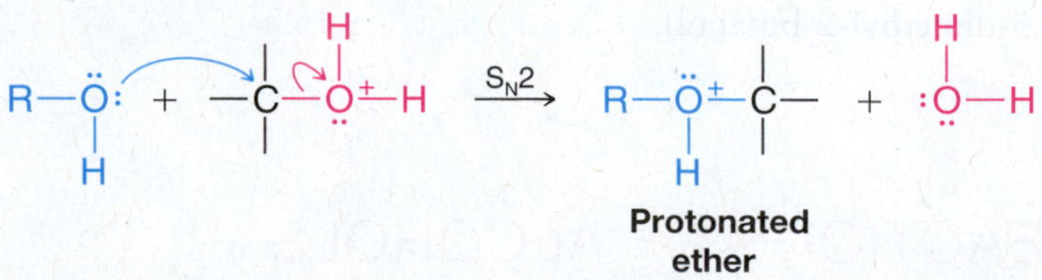

$$R-\overset{..}{\underset{|}{O}}:\ + \ -\overset{|}{\underset{|}{C}}-\overset{H}{\underset{..}{\overset{+}{O}}}-H \ \xrightarrow{S_N2} \ R-\overset{+}{\underset{|}{O}}-\overset{|}{\underset{|}{C}}- \ + \ :\overset{H}{\underset{..}{O}}-H$$
$$H H$$

Protonated ether

At a high enough temperature and in the absence of a good nucleophile, protonated alcohols are capable of undergoing E1 or E2 reactions. This is what happens in alcohol dehydrations (Section 7.7).

Alcohols also react with PBr_3 and $SOCl_2$ to yield alkyl bromides and alkyl chlorides. These reactions, as we shall see in Section 11.9, are initiated by the alcohol using its unshared electron pairs to act as a nucleophile.

11.6 ALCOHOLS AS ACIDS

TABLE 11.3 pK$_a$ VALUES FOR SOME WEAK ACIDS

Acid	pK$_a$
CH_3OH	15.5
H_2O	15.74
CH_3CH_2OH	15.9
$(CH_3)_3COH$	18.0

- Alcohols have acidities similar to that of water.

Methanol is a slightly stronger acid than water (pK_a = 15.7) but most alcohols are somewhat weaker acids. Values of pK_a for several alcohols are listed in Table 11.3.

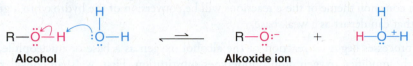

Alcohol   **Alkoxide ion**

(If R is bulky, there is less stabilization of the alkoxide by solvation and greater destabilization due to inductive effects. Consequently, the equilibrium lies even further toward the alcohol.)

- Sterically hindered alcohols such as *tert*-butyl alcohol are less acidic, and hence their conjugate bases more basic, than unhindered alcohols such as ethanol or methanol.

One reason for this difference in acidity has to do with the effect of solvation. With an unhindered alcohol, water molecules can easily surround, solvate, and hence stabilize the alkoxide anion that would form by loss of the alcohol proton to a base. As a consequence of this stabilization, formation of the alcohol's conjugate base is easier, and therefore its acidity is increased. If the R group of the alcohol is bulky, solvation of the alkoxide anion is hindered. Stabilization of the conjugate base is not as effective, and consequently the hindered alcohol is a weaker acid. Another reason that hindered alcohols are less acidic has to do with the inductive electron-donating effect of alkyl groups. The alkyl groups of a hindered alcohol donate electron density, making formation of an alkoxide anion more difficult than with a less hindered alcohol.

- All alcohols are much stronger acids than terminal alkynes, and they are very much stronger acids than hydrogen, ammonia, and alkanes (see Table 3.1).

Helpful Hint

Remember: Any factor that stabilizes the conjugate base of an acid increases its acidity.

Relative Acidity

Water and alcohols are the strongest acids in this series.	$H_2O > ROH > RC\equiv CH > H_2 > NH_3 > RH$

Sodium and potassium alkoxides can be prepared by treating alcohols with sodium or potassium metal or with the metal hydride (Section 6.15B). Because most alcohols are weaker acids than water, most alkoxide ions are stronger bases than the hydroxide ion.

- Conjugate bases of compounds with higher pK_a values than an alcohol will deprotonate an alcohol.

Relative Basicity

$R^- > H_2N^- > H^- > RC\equiv C^- > RO^- > HO^-$

Hydroxide is the weakest base in this series.

PRACTICE PROBLEM 11.5

Write equations for the acid–base reactions that would occur (if any) if ethanol were added to solutions of each of the following compounds. In each reaction, label the stronger acid, the stronger base, and so forth (consult Table 3.1).

(a) NaNH$_2$ (b) H————:$^-$Na$^+$ (c)
$$\underset{ONa}{\overset{O}{\parallel}}$$
     (d) NaOH

Sodium and potassium alkoxides are often used as bases in organic syntheses (Section 6.15B). We use alkoxides, such as ethoxide and *tert*-butoxide, when we carry out reactions that require stronger bases than hydroxide ion but do not require exceptionally powerful bases, such as the amide ion or the anion of an alkane. We also use alkoxide ions when, for reasons of solubility, we need to carry out a reaction in an alcohol solvent rather than in water.

11.7 CONVERSION OF ALCOHOLS INTO ALKYL HALIDES

In this and several following sections we will be concerned with reactions that involve substitution of the alcohol hydroxyl group.

- A hydroxyl group is such a poor leaving group (it would depart as hydroxide) that a common theme of these reactions will be conversion of the hydroxyl to a group that can depart as a weak base.

These processes begin by reaction of the alcohol oxygen as a base or nucleophile, after which the modified oxygen group undergoes substitution. First, we shall consider reactions that convert alcohols to alkyl halides.

The most commonly used reagents for conversion of alcohols to alkyl halides are the following:

- Hydrogen halides (HCl, HBr, HI)
- Phosphorus tribromide (PBr_3)
- Thionyl chloride ($SOCl_2$)

Examples of the use of these reagents are the following. All of these reactions result in cleavage of the C—O bond of the alcohol. In each case, the hydroxyl group is first converted to a suitable leaving group. We will see how this is accomplished when we study each type of reaction.

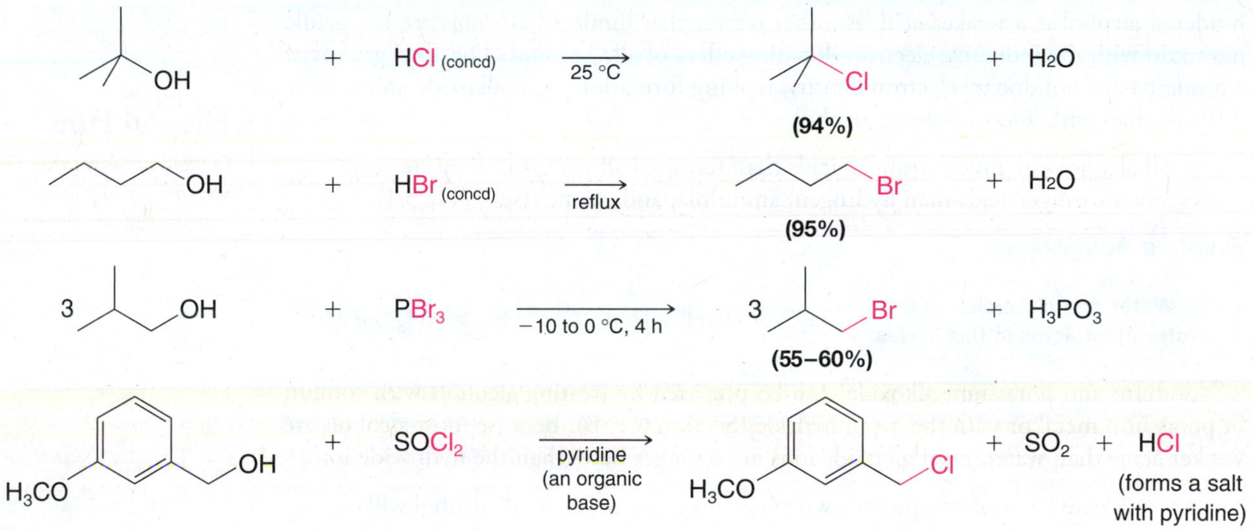

11.8 ALKYL HALIDES FROM THE REACTION OF ALCOHOLS WITH HYDROGEN HALIDES

When alcohols react with a hydrogen halide, a substitution takes place producing an alkyl halide and water:

$$R\text{—}OH + HX \longrightarrow R\text{—}X + H_2O$$

- The order of reactivity of alcohols is $3° > 2° > 1° < $ methyl.
- The order of reactivity of the hydrogen halides is HI > HBr > HCl (HF is generally unreactive).

The reaction is *acid catalyzed*. Alcohols react with the strongly acidic hydrogen halides HCl, HBr, and HI, but they do not react with nonacidic NaCl, NaBr, or NaI. Primary and secondary alcohols can be converted to alkyl chlorides and bromides by allowing them to react with a mixture of a sodium halide and sulfuric acid:

$$ROH + NaX \xrightarrow{H_2SO_4} RX + NaHSO_4 + H_2O$$

11.8A Mechanisms of the Reactions of Alcohols with HX

- Secondary, tertiary, allylic, and benzylic alcohols appear to react by a mechanism that involves the formation of a carbocation—a mechanism that we first saw in Section 3.13 and that you should now recognize *as an S_N1 reaction with the protonated alcohol acting as the substrate.*

We again illustrate this mechanism with the reaction of *tert*-butyl alcohol and aqueous hydrochloric acid (H_3O^+, Cl^-).

The first two steps in this S_N1 substitution mechanism are the same as in the mechanism for the dehydration of an alcohol (Section 7.7).

Step 1   **The alcohol accepts a proton.**

Step 2   **The protonated hydroxyl group departs as a leaving group to form a carbocation and water.**

In step 3 the mechanisms for the dehydration of an alcohol and the formation of an alkyl halide differ. In dehydration reactions the carbocation loses a proton in an E1 reaction to form an alkene. In the formation of an alkyl halide, the carbocation reacts with a nucleophile (a halide ion) in an S_N1 reaction.

Step 3   **A halide anion reacts with the carbocation.**

How can we account for S_N1 substitution in this case versus elimination in others?

When we dehydrate alcohols, we usually carry out the reaction in concentrated sulfuric acid and at high temperature. The hydrogen sulfate (HSO_4^-) present after protonation of the alcohol is a weak nucleophile, and at high temperature the highly reactive carbocation forms a more stable species by losing a proton and becoming an alkene. Furthermore, the alkene is usually volatile and distills from the reaction mixture as it is formed, thus drawing the equilibrium toward alkene formation. The net result is *an E1 reaction.*

In the reverse reaction, that is, the hydration of an alkene (Section 8.5), the carbocation *does* react with a nucleophile. It reacts with water. Alkene hydrations are carried out in dilute sulfuric acid, where the water concentration is high. In some instances, too, carbocations may react with HSO_4^- ions or with sulfuric acid, itself. When they do, they form alkyl hydrogen sulfates ($R-OSO_2OH$).

When we convert an alcohol to an alkyl halide, we carry out the reaction in the presence of acid and *in the presence of halide ions*, and not at elevated temperature. Halide ions are good nucleophiles (they are much stronger nucleophiles than water), and since halide ions are present in high concentration, most of the carbocations react with an electron pair of a halide ion to form a more stable species, the alkyl halide product. The overall result is an S_N1 reaction.

These two reactions, dehydration and the formation of an alkyl halide, also furnish another example of the competition between nucleophilic substitution and elimination (see Section 6.18). Very often, in conversions of alcohols to alkyl halides, we find that the reaction is accompanied by the formation of some alkene (i.e., by elimination). The free energies of activation for these two reactions of carbocations are not very different from one another. Thus, not all of the carbocations become stable products by reacting with nucleophiles; some lose a β proton to form an alkene.

Primary Alcohols Not all acid-catalyzed conversions of alcohols to alkyl halides proceed through the formation of carbocations.

- Primary alcohols and methanol react to form alkyl halides under acidic conditions by an S$_N$2 mechanism.

In these reactions the function of the acid is to produce *a protonated alcohol*. The halide ion then displaces a molecule of water (a good leaving group) from carbon; this produces an alkyl halide:

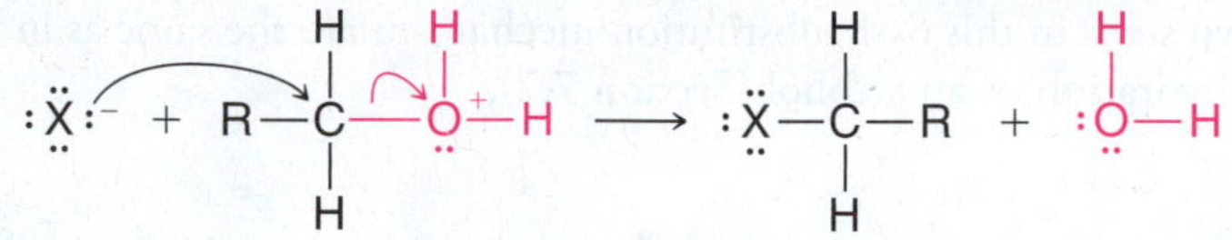

(Protonated 1° alcohol   (A good
or methanol)   leaving group)

Acid Is Required Although halide ions (particularly iodide and bromide ions) are strong nucleophiles, they are not strong enough to carry out substitution reactions with alcohols themselves.

- Reactions like the following do not occur because the leaving group would have to be a strongly basic hydroxide ion:

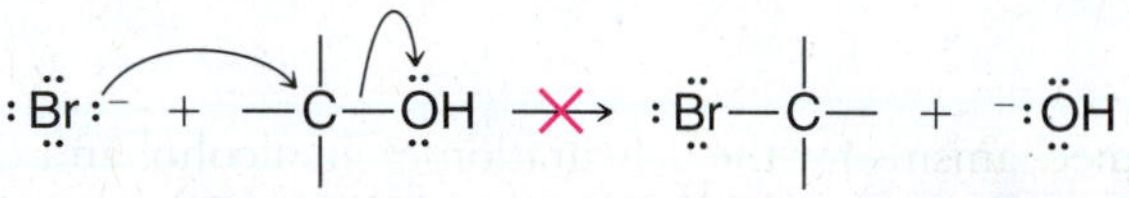

Helpful Hint

The reverse reaction, that is, the reaction of an alkyl halide with hydroxide ion, does occur and is a method for the synthesis of alcohols. We saw this reaction in Chapter 6.

We can see now why the reactions of alcohols with hydrogen halides are acid-promoted.

- Acid protonates the alcohol hydroxyl group, making it a good leaving group.

Because the chloride ion is a weaker nucleophile than bromide or iodide ions, hydrogen chloride does not react with primary or secondary alcohols unless zinc chloride or some similar Lewis acid is added to the reaction mixture as well. Zinc chloride, a good Lewis acid, forms a complex with the alcohol through association with an unshared pair of electrons on the oxygen atom. This enhances the hydroxyl group's leaving potential sufficiently that chloride can displace it.

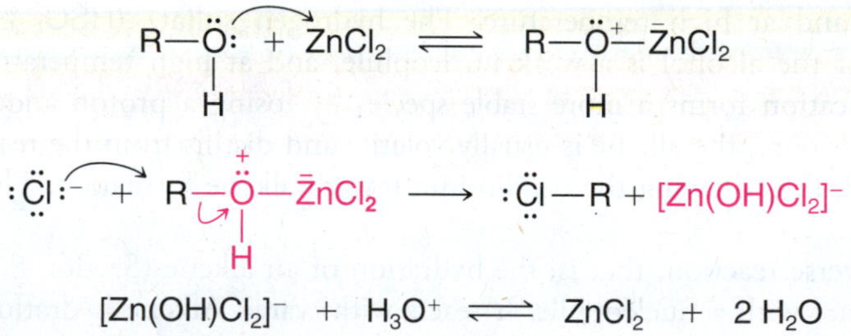

- As we might expect, many reactions of alcohols with hydrogen halides, particularly those in which carbocations are formed, *are accompanied by rearrangements.*

How do we know that rearrangements can occur when secondary alcohols are treated with a hydrogen halide? Results like that in Solved Problem 11.5 indicate this to be the case.

●●● **SOLVED PROBLEM 11.5**

Treating 3-methyl-2-butanol (see the following reaction) yields 2-bromo-2-methylbutane as the sole product. Propose a mechanism that explains the course of the reaction.

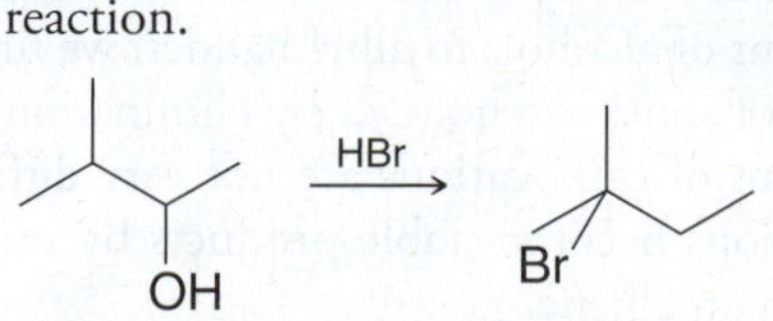

STRATEGY AND ANSWER: The reaction must involve a rearrangement by a hydride shift from the initially formed carbocation.

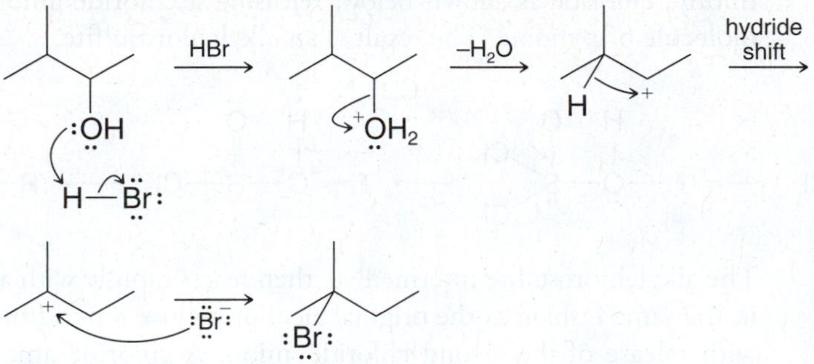

Write a detailed mechanism for the following reaction.

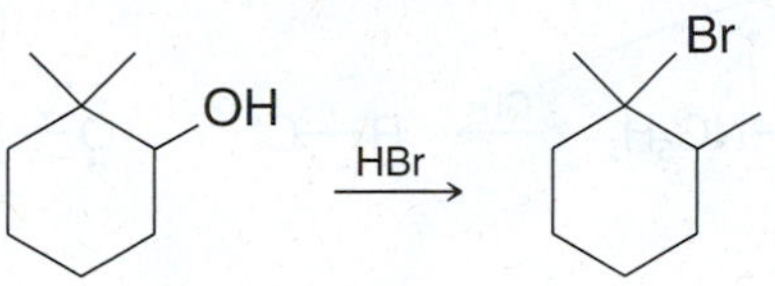

PRACTICE PROBLEM 11.6

(a) What factor explains the observation that tertiary alcohols react with HX faster than secondary alcohols? **(b)** What factor explains the observation that methanol reacts with HX faster than a primary alcohol?

PRACTICE PROBLEM 11.7

Since rearrangements can occur when some alcohols are treated with hydrogen halides, how can we successfully convert a secondary alcohol to an alkyl halide without rearrangement? The answer to this question comes in the next section, where we discuss the use of reagents such as thionyl chloride (SOCl₂) and phosphorus tribromide (PBr₃).

11.9 ALKYL HALIDES FROM THE REACTION OF ALCOHOLS WITH PBr₃ OR SOCl₂

Primary and secondary alcohols react with phosphorus tribromide to yield alkyl bromides.

$$3\ \text{R}\!-\!\text{OH} + \text{PBr}_3 \longrightarrow 3\ \text{R}\!-\!\text{Br} + \text{H}_3\text{PO}_3$$

(1° or 2°)

> **Helpful Hint**
>
> PBr₃: A reagent for synthesizing 1° and 2° alkyl bromides.

- The reaction of an alcohol with PBr₃ does not involve the formation of a carbocation and *usually occurs without rearrangement* of the carbon skeleton (especially if the temperature is kept below 0 °C).

- Phosphorus tribromide is often preferred as a reagent for the transformation of an alcohol to the corresponding alkyl bromide.

The mechanism for the reaction involves sequential replacement of the bromine atom in PBr₃ by three molecules of the alcohol to form a trialkylphosphite, P(OR)₃, and three molecules of HBr.

$$\text{ROH} + \text{PBr}_3 \longrightarrow \text{P(OR)}_3 + 3\ \text{HBr}$$

The trialkylphosphite goes on to react with three molecules of HBr to form three molecules of the alkyl bromide and a molecule of phosphonic acid.

$$\text{P(OR)}_3 + 3\ \text{HBr} \longrightarrow 3\ \text{RBr} + \text{H}_3\text{PO}_3$$

Helpful Hint

SOCl₂: A reagent for synthesizing 1° and 2° alkyl chlorides.

Thionyl chloride (SOCl₂) converts primary and secondary alcohols to alkyl chlorides. Pyridine (C_5H_5N) is often included to promote the reaction. The alcohol substrate attacks thionyl chloride as shown below, releasing a chloride anion and losing its proton to a molecule of pyridine. The result is an alkylchlorosulfite.

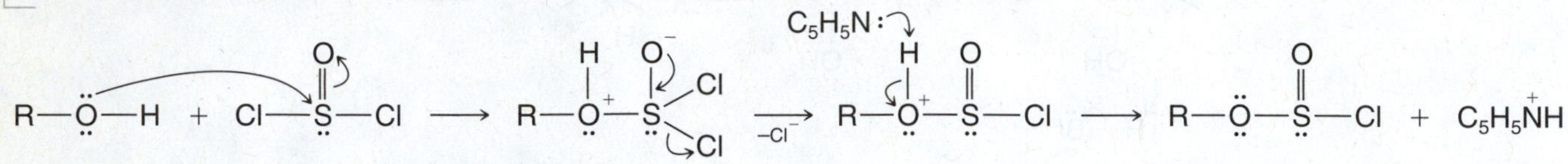

The alkylchlorosulfite intermediate then reacts rapidly with another molecule of pyridine, in the same fashion as the original alcohol, to give a pyridinium alkylsulfite intermediate, with release of the second chloride anion. A chloride anion then attacks the substrate carbon, displacing the sulfite leaving group, which in turn decomposes to release gaseous SO_2 and pyridine. (In the absence of pyridine the reaction occurs with retention of configuration. See Problem 11.55.)

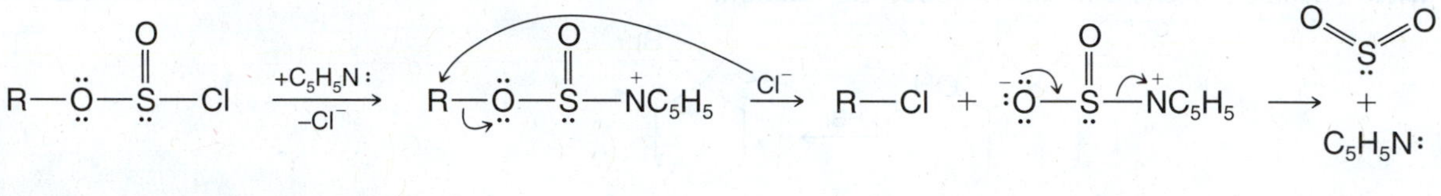

●●● SOLVED PROBLEM 11.6

Starting with alcohols, outline a synthesis of each of the following: **(a)** benzyl bromide, **(b)** cyclohexyl chloride, and **(c)** butyl bromide.

POSSIBLE ANSWERS:

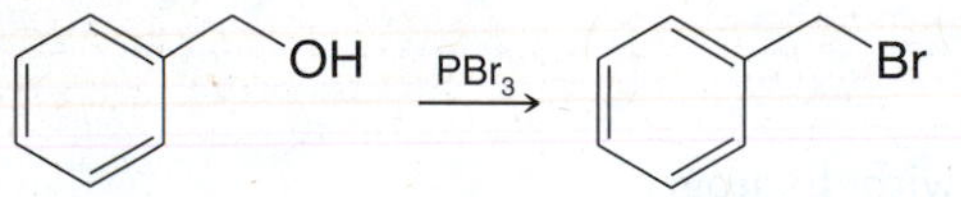

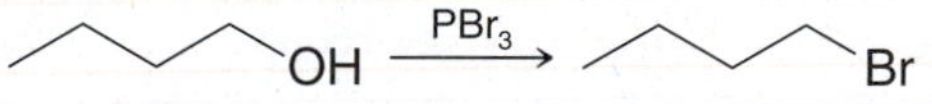

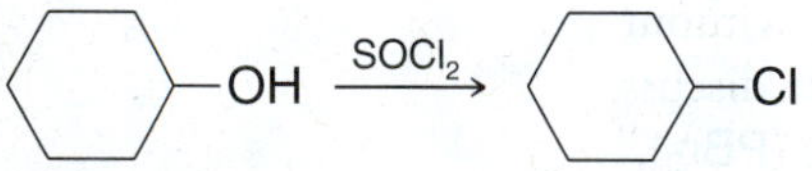

11.10 TOSYLATES, MESYLATES, AND TRIFLATES: LEAVING GROUP DERIVATIVES OF ALCOHOLS

The hydroxyl group of an alcohol can be converted to a good leaving group by conversion to a **sulfonate ester** derivative. The most common sulfonate esters used for this purpose are methanesulfonate esters ("**mesylates**"), *p*-toluenesulfonate esters ("**tosylates**"), and trifluoromethanesulfonates ("**triflates**").

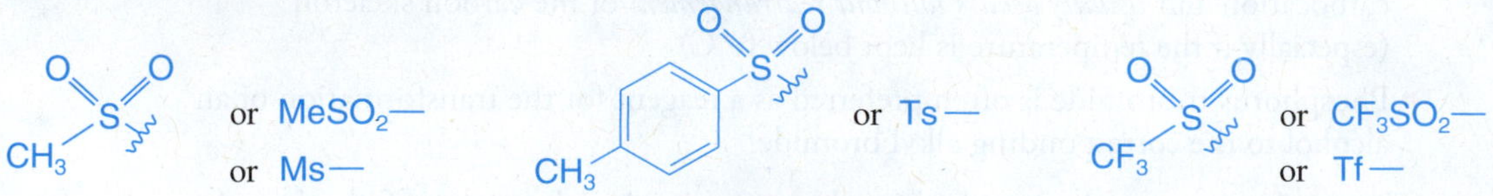

The mesyl group   **The tosyl group**   **The trifyl group**

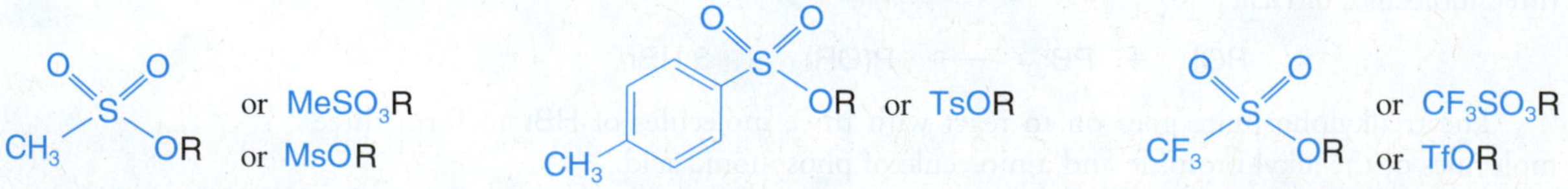

An alkyl mesylate   **An alkyl tosylate**   **An alkyl triflate**

The desired sulfonate ester is usually prepared by reaction of the alcohol in pyridine with the appropriate sulfonyl chloride, that is, methanesulfonyl chloride (mesyl chloride) for a mesylate, *p*-toluenesulfonyl chloride (tosyl chloride) for a tosylate, or trifluoromethane-sulfonyl chloride [or trifluoromethanesulfonic anhydride (triflic anhydride)] for a triflate. Pyridine (C_5H_5N, pyr) serves as the solvent and to neutralize the HCl formed. Ethanol, for example, reacts with methanesulfonyl chloride to form ethyl methanesulfonate and with *p*-toluenesulfonyl chloride to form ethyl *p*-toluenesulfonate:

Helpful Hint

A method for making an alcohol hydroxyl group into a leaving group.

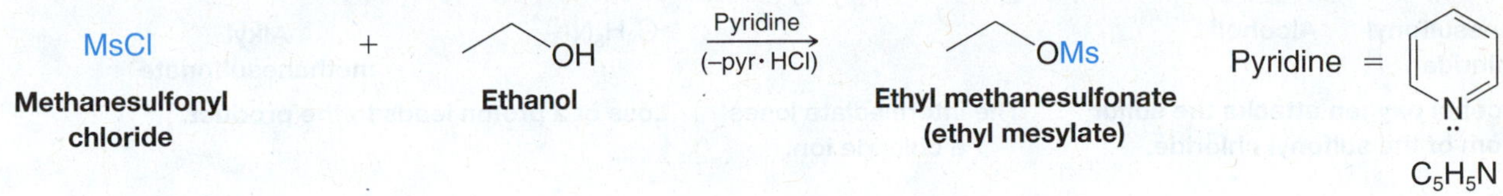

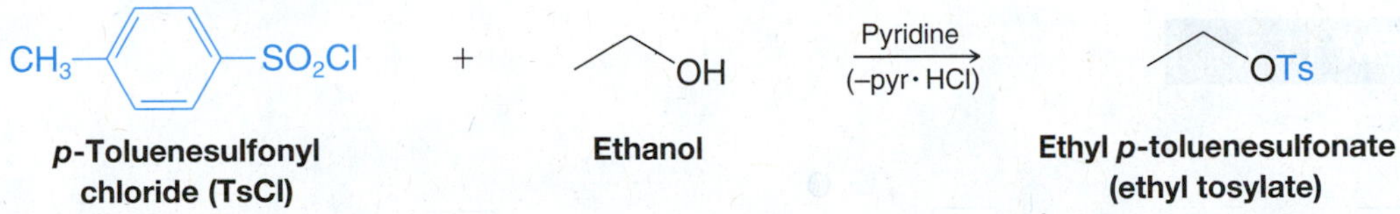

MsCl	+	⌒OH	Pyridine (−pyr·HCl) →	⌒OMs
Methanesulfonyl chloride		**Ethanol**		**Ethyl methanesulfonate (ethyl mesylate)**

Pyridine = (pyridine structure) C_5H_5N

CH_3—⬡—SO_2Cl	+	⌒OH	Pyridine (−pyr·HCl) →	⌒OTs
p-Toluenesulfonyl chloride (TsCl)		**Ethanol**		**Ethyl p-toluenesulfonate (ethyl tosylate)**

It is important to note that formation of the sulfonate ester does not affect the stereochemistry of the alcohol carbon, because the C—O bond is not involved in this step. Thus, if the alcohol carbon is a chirality center, no change in configuration occurs on making the sulfonate ester—the reaction proceeds with **retention of configuration.** On reaction of the sulfonate ester with a nucleophile, the usual parameters of nucleophilic substitution reactions become involved.

Substrates for Nucleophilic Substitution Mesylates, tosylates, and triflates, because they are good leaving groups, are frequently used as substrates for nucleophilic substitution reactions. They are good leaving groups because the sulfonate anions they become when they depart are very weak bases. The triflate anion is the weakest base in this series, and is thus the best leaving group among them.

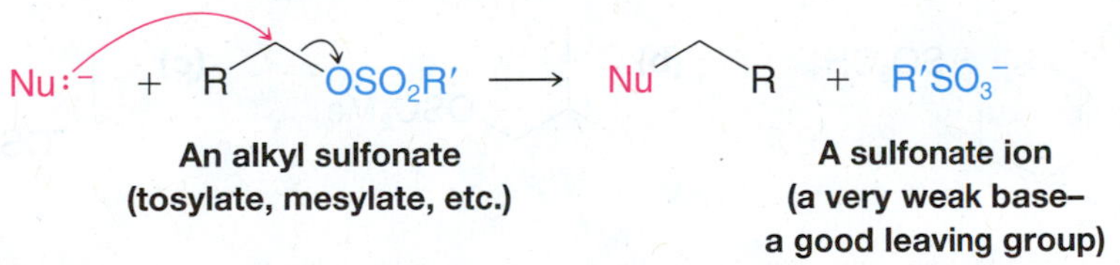

$$Nu{:}^- \; + \; R\text{—}OSO_2R' \longrightarrow Nu\text{—}R \; + \; R'SO_3^-$$

An alkyl sulfonate (tosylate, mesylate, etc.)		**A sulfonate ion (a very weak base– a good leaving group)**

- To carry out a nucleophilic substitution on an alcohol, we first convert the alcohol to an alkyl sulfonate and then, in a second reaction, allow it to react with a nucleophile.
- If the mechanism is S_N2, as shown in the second reaction of the following example, **inversion of configuration** takes place at the carbon that originally bore the alcohol hydroxyl group:

Step 1 (structure with R, H, R', OH) + **TsCl** $\xrightarrow[(-\text{pyr·HCl})]{\text{retention}}$ (structure with R, H, R', OTs)

Step 2 **Nu:**⁻ + (structure with R, H, R', OTs) $\xrightarrow{\text{inversion} \atop S_N2}$ (structure with R, Nu, H, R') + **TsO⁻**

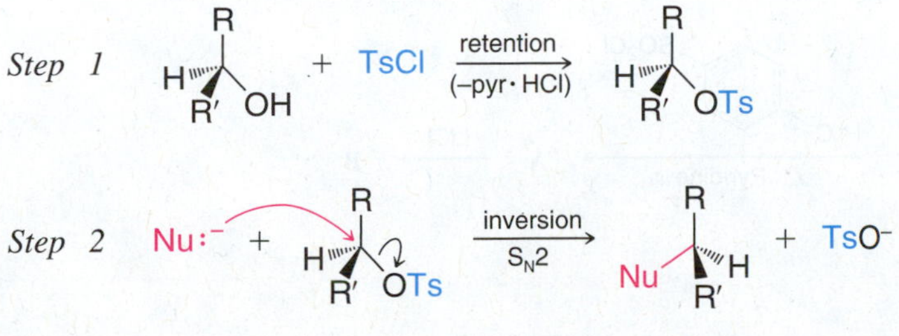

The fact that the C—O bond of the alcohol does not break during formation of the sulfonate ester is accounted for by the following mechanism. Methanesulfonyl chloride is used in the example.

A MECHANISM FOR THE REACTION — Conversion of an Alcohol into a Mesylate (an Alkyl Methanesulfonate)

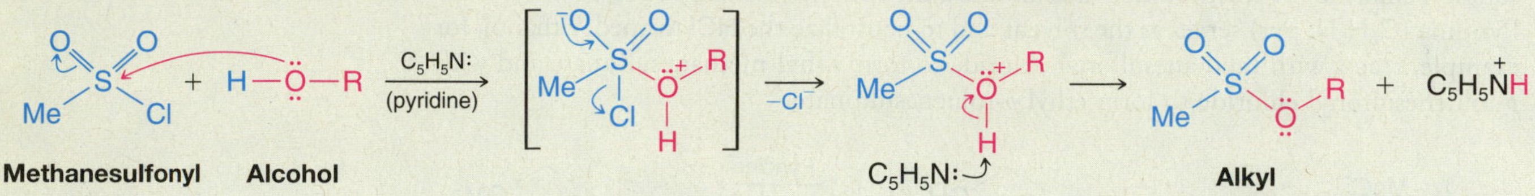

Methanesulfonyl chloride   **Alcohol**

The alcohol oxygen attacks the sulfur atom of the sulfonyl chloride.

The intermediate loses a chloride ion.

Loss of a proton leads to the product.

Alkyl methanesulfonate

●●● SOLVED PROBLEM 11.7

Supply the missing reagents.

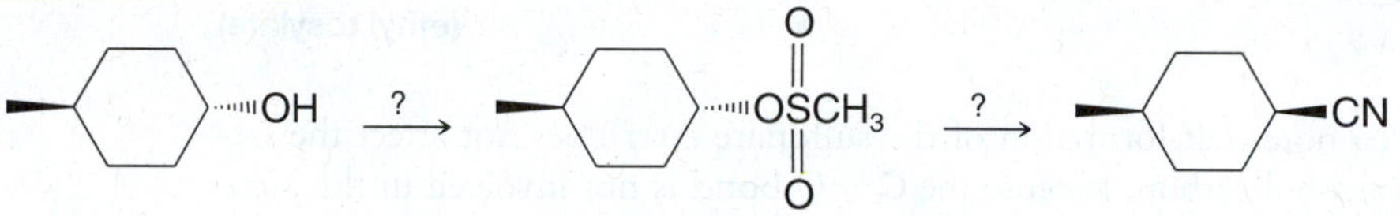

STRATEGY AND ANSWER: The overall transformation over two steps involves replacing an alcohol hydroxyl group by a cyano group with inversion of configuration. To accomplish this, we need to convert the alcohol hydroxyl to a good leaving group in the first step, which we do by making it a methanesulfonate ester (a mesylate) using methanesulfonyl chloride in pyridine. The second step is an S_N2 substitution of the methanesulfonate (mesyl) group, which we do using potassium or sodium cyanide in a polar aprotic solvent such as dimethylformamide (DMF).

●●●

PRACTICE PROBLEM 11.8   Show how you would prepare the following compounds from the appropriate sulfonyl chlorides.

(a) 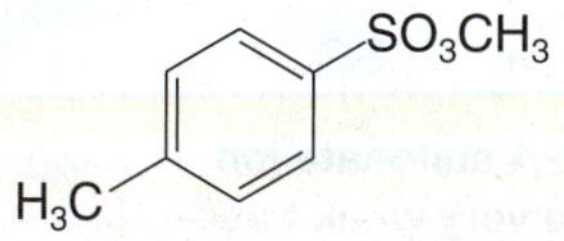 SO_3CH_3 (b) 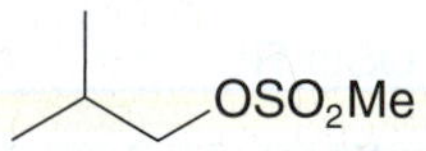 OSO_2Me (c) 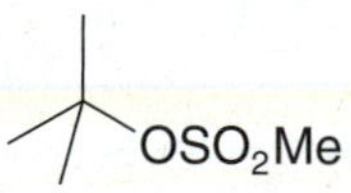 OSO_2Me

H_3C

●●●

PRACTICE PROBLEM 11.9   Write structures for products **X, Y, A,** and **B,** showing stereochemistry.

(a) 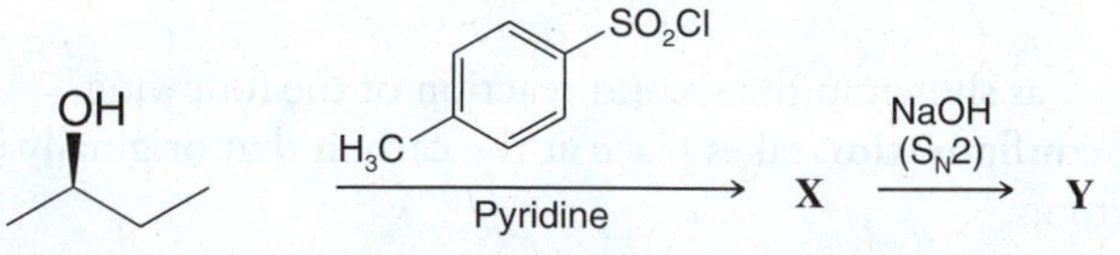   OH

(*R*)-2-Butanol

SO_2Cl

H_3C

Pyridine → **X** → NaOH (S_N2) → **Y**

(b)   OH

SO_2Cl

H_3C

Pyridine → **A** → LiCl → **B**

●●●

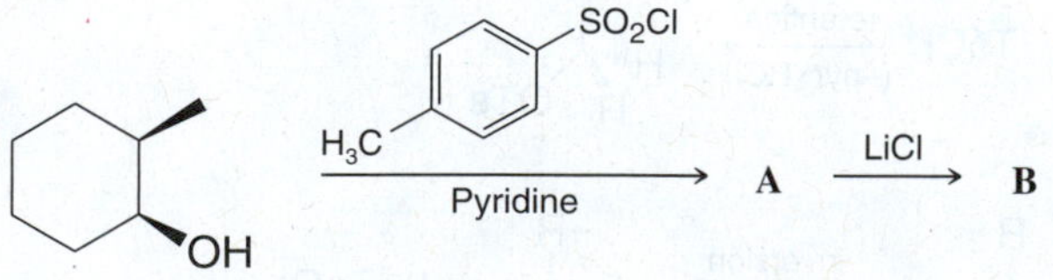

PRACTICE PROBLEM 11.10   Suggest an experiment using an isotopically labeled alcohol that would prove that the formation of an alkyl sulfonate does not cause cleavage at the C—O bond of the alcohol.

11.11 SYNTHESIS OF ETHERS

11.11A Ethers by Intermolecular Dehydration of Alcohols

Two alcohol molecules can form an ether by loss of water through an acid-catalyzed substitution reaction.

$$R\!-\!OH + HO\!-\!R \xrightarrow[-H_2O]{HA} R\!-\!O\!-\!R$$

This reaction competes with the formation of alkenes by acid-catalyzed alcohol dehydration (Sections 7.7 and 7.8). Intermolecular dehydration of alcohols usually takes place at lower temperature than dehydration to an alkene, and dehydration to the ether can be aided by distilling the ether as it is formed. For example, diethyl ether is made commercially by dehydration of ethanol. Diethyl ether is the predominant product at 140 °C; ethene is the predominant product at 180 °C.

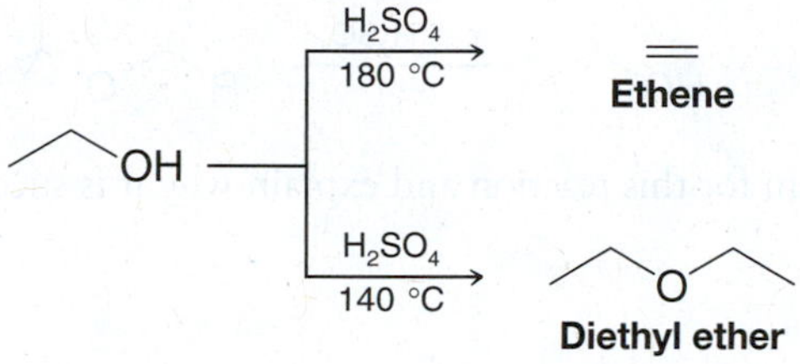

Ethene

H_2SO_4, 180 °C

H_2SO_4, 140 °C

Diethyl ether

The formation of the ether occurs by an S_N2 mechanism with one molecule of the alcohol acting as the nucleophile and another protonated molecule of the alcohol acting as the substrate (see Section 11.5).

[A MECHANISM FOR THE REACTION — Intermolecular Dehydration of Alcohols to Form an Ether]

Step 1

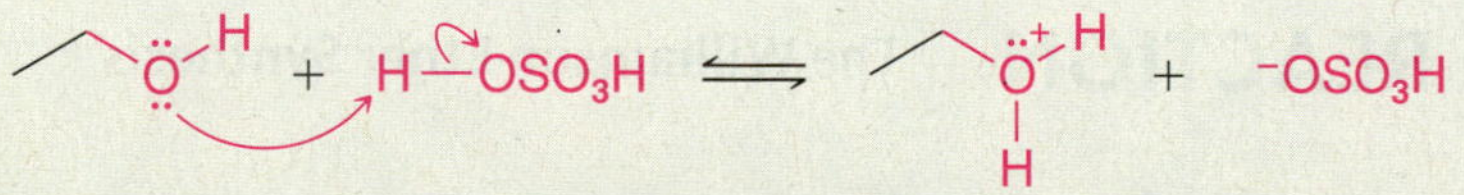

This is an acid–base reaction in which the alcohol accepts a proton from the sulfuric acid.

Step 2

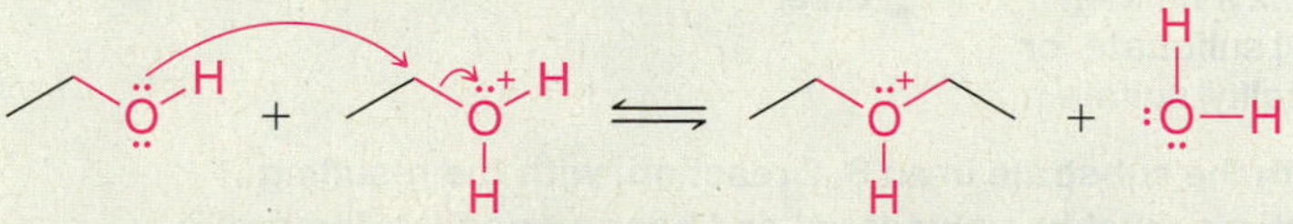

Another molecule of the alcohol acts as a nucleophile and attacks the protonated alcohol in an S_N2 reaction.

Step 3

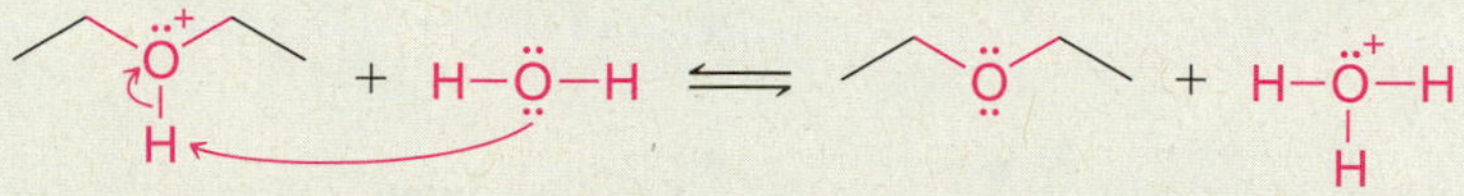

Another acid–base reaction converts the protonated ether to an ether by transferring a proton to a molecule of water (or to another molecule of the alcohol).

Complications of Intermolecular Dehydration The method of synthesizing ethers by intermolecular dehydration has some important limitations.

- Attempts to synthesize ethers by intermolecular dehydration of secondary alcohols are usually unsuccessful because alkenes form too easily.
- Attempts to make ethers with tertiary alkyl groups lead predominantly to alkenes.

● Intermolecular dehydration is not useful for the preparation of unsymmetrical ethers from primary alcohols because the reaction leads to a mixture of products:

$$\underbrace{ROH \ + \ R'OH}_{1° \ \textbf{Alcohols}} \quad \xrightleftharpoons{H_2SO_4} \quad \begin{array}{c} ROR \\ + \\ ROR' \ + \ H_2O \\ + \\ R'OR' \end{array}$$

PRACTICE PROBLEM 11.11 An exception to what we have just said has to do with syntheses of unsymmetrical ethers in which one alkyl group is a *tert*-butyl group and the other group is primary. For example, this synthesis can be accomplished by adding *tert*-butyl alcohol to a mixture of the primary alcohol and H_2SO_4 at room temperature.

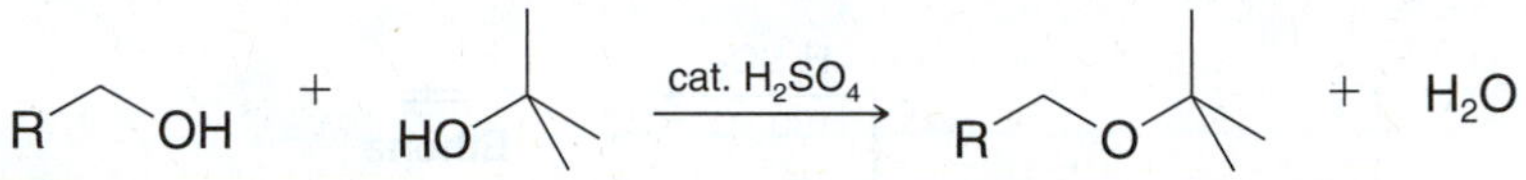

$$R\diagup OH \quad + \quad HO\diagdown\diagup \quad \xrightarrow{\text{cat. } H_2SO_4} \quad R\diagup O\diagdown\diagup \quad + \quad H_2O$$

Give a likely mechanism for this reaction and explain why it is successful.

11.11B The Williamson Ether Synthesis

An important route to unsymmetrical ethers is a nucleophilic substitution reaction known as the **Williamson ether synthesis**.

● The Williamson ether synthesis consists of an S_N2 reaction of a sodium alkoxide with an alkyl halide, alkyl sulfonate, or alkyl sulfate.

[A MECHANISM FOR THE REACTION ┤ The Williamson Ether Synthesis]

$$\underset{\substack{\textbf{Sodium} \\ \textbf{(or potassium)} \\ \textbf{alkoxide}}}{R-\ddot{O}\text{:} \ Na^+} \quad + \quad \underset{\substack{\textbf{Alkyl halide,} \\ \textbf{alkyl sulfonate, or} \\ \textbf{dialkyl sulfate}}}{R'-LG} \quad \longrightarrow \quad \underset{\textbf{Ether}}{R-\ddot{O}-R'} \ + \ Na^+ \ \text{:}LG^-$$

The alkoxide ion reacts with the substrate in an S_N2 reaction, with the resulting formation of an ether. The substrate must be unhindered and bear a good leaving group. Typical substrates are 1° or 2° alkyl halides, alkyl sulfonates, and dialkyl sulfates:

$$-LG \ = \ -\ddot{B}r\text{:}, \ -\ddot{I}\text{:}, \ -OSO_2R'', \text{ or } -OSO_2OR''$$

The following reaction is a specific example of the Williamson ether synthesis. The sodium alkoxide can be prepared by allowing an alcohol to react with NaH:

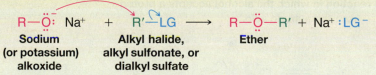

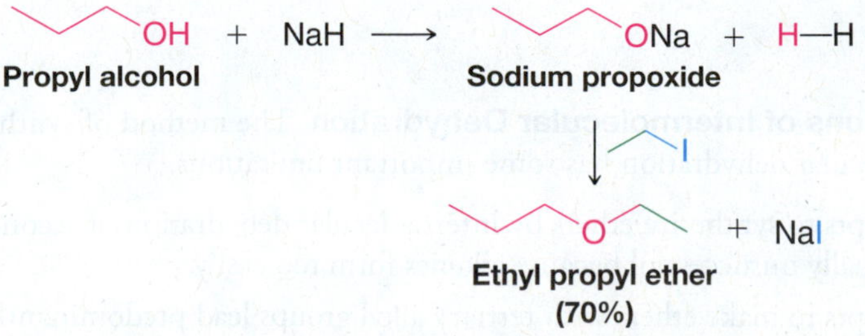

$$\underset{\textbf{Propyl alcohol}}{\diagup\diagdown OH} \ + \ NaH \ \longrightarrow \ \underset{\textbf{Sodium propoxide}}{\diagup\diagdown ONa} \ + \ H-H$$

$$\downarrow \diagup\diagdown I$$

$$\underset{\substack{\textbf{Ethyl propyl ether} \\ \textbf{(70\%)}}}{\diagup\diagdown O\diagup} \ + \ Na I$$

The usual limitations of S_N2 reactions apply here.

- Best results are obtained when the alkyl halide, sulfonate, or sulfate is primary (or methyl). **If the substrate is tertiary, elimination is the exclusive result.** Substitution is also favored over elimination at lower temperatures.

> **Helpful Hint**
>
> Conditions that favor a Williamson ether synthesis.

●●●

PRACTICE PROBLEM 11.12

(a) Outline two methods for preparing isopropyl methyl ether by a Williamson ether synthesis.

(b) One method gives a much better yield of the ether than the other. Explain which is the better method and why.

●●● **SOLVED PROBLEM 11.8**

The cyclic ether tetrahydrofuran (THF) can be synthesized by treating 4-chloro-1-butanol with aqueous sodium hydroxide (see below). Propose a mechanism for this reaction.

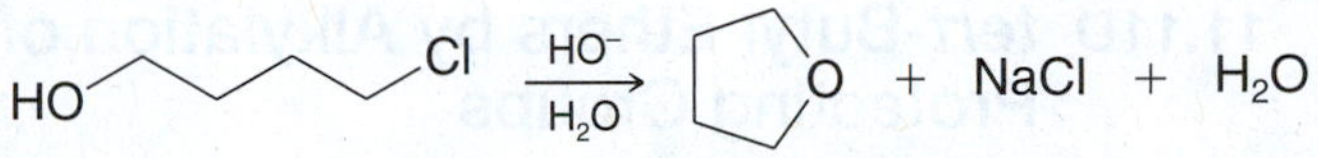

Tetrahydrofuran

STRATEGY AND ANSWER: Removal of a proton from the hydroxyl group of 4-chloro-1-butanol gives an alkoxide ion that can then react with itself in an intramolecular S_N2 reaction to form a ring.

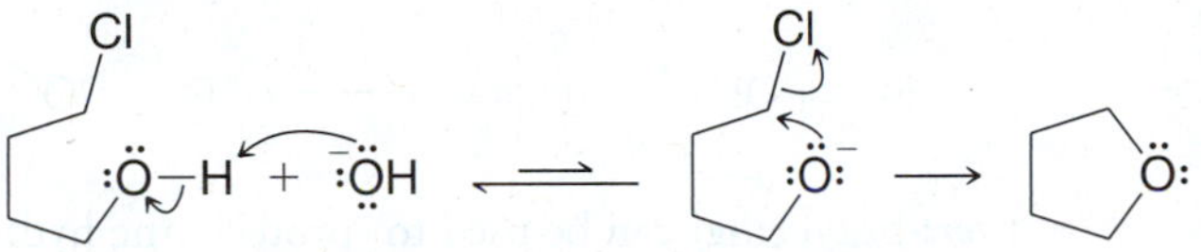

Even though treatment of the alcohol with hydroxide does not favor a large equilibrium concentration of the alkoxide, the alkoxide anions that are present react rapidly by the intramolecular S_N2 reaction. As alkoxide anions are consumed by the substitution reaction, their equilibrium concentration is replenished by deprotonation of additional alcohol molecules, and the reaction is drawn to completion.

●●●

PRACTICE PROBLEM 11.13

Epoxides can be synthesized by treating halohydrins with aqueous base. Propose a mechanism for reactions **(a)** and **(b)**, and explain why no epoxide formation is observed in **(c)**.

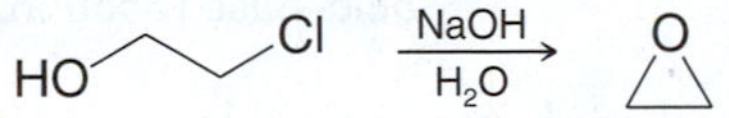

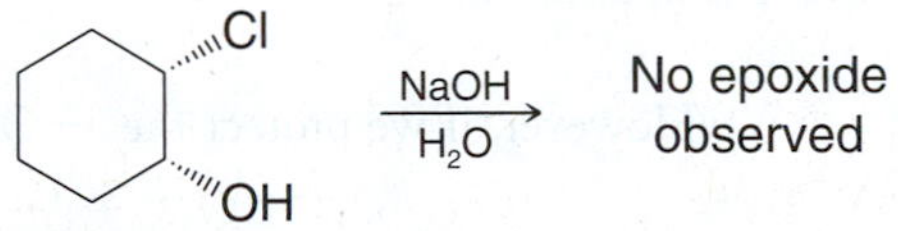

●●●

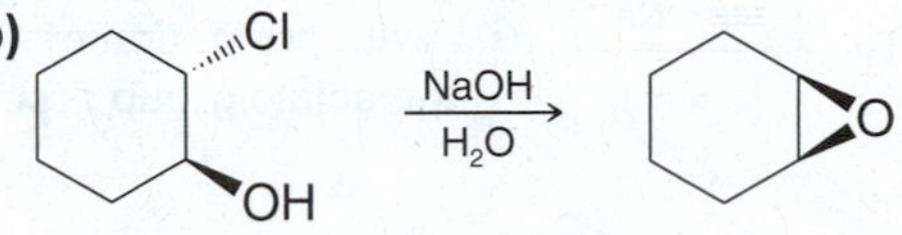

PRACTICE PROBLEM 11.14

Write structures for products **A, B, C,** and **D**, showing stereochemistry. (*Hint:* **B** and **D** are stereoisomers.)

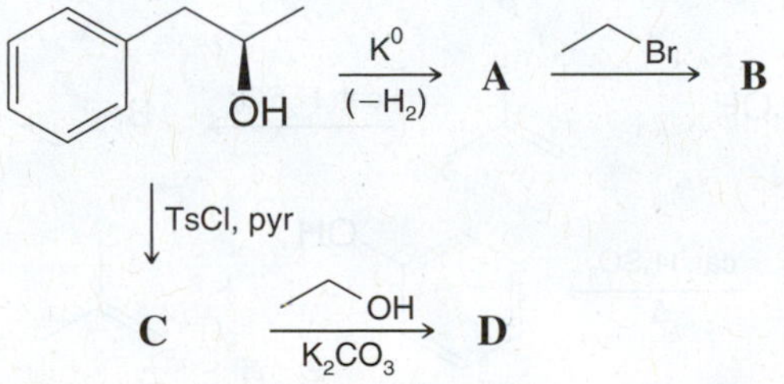

11.11C Synthesis of Ethers by Alkoxymercuration–Demercuration

Alkoxymercuration–demercuration is another method for synthesizing ethers.

- The reaction of an alkene with an alcohol in the presence of a mercury salt such as mercuric acetate or trifluoroacetate leads to an alkoxymercury intermediate, which on reaction with sodium borohydride yields an ether.

When the alcohol reactant is also the solvent, the method is called solvomercuration–demercuration. This method directly parallels hydration by oxymercuration–demercuration (Section 8.5):

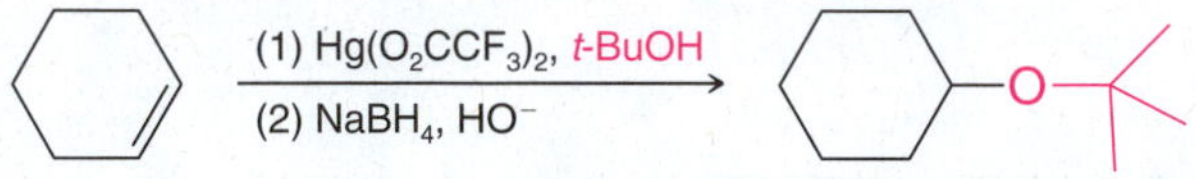

(98% Yield)

11.11D *tert*-Butyl Ethers by Alkylation of Alcohols: Protecting Groups

Primary alcohols can be converted to *tert*-butyl ethers by dissolving them in a strong acid such as sulfuric acid and then adding isobutylene to the mixture. (This procedure minimizes dimerization and polymerization of the isobutylene.)

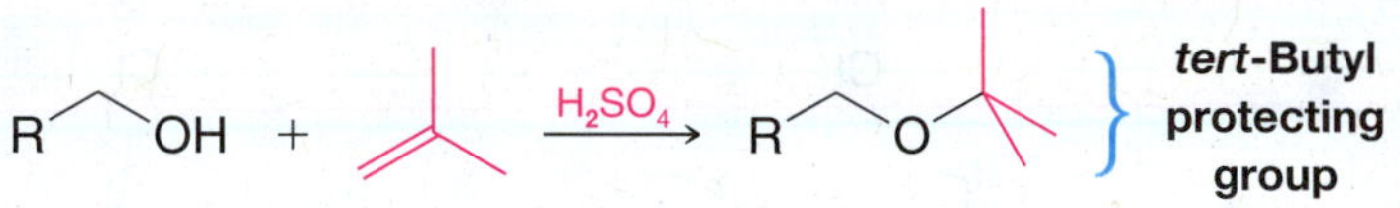

- A *tert*-butyl ether can be used to "protect" the hydroxyl group of a primary alcohol while another reaction is carried out on some other part of the molecule.
- A *tert*-butyl **protecting group** can be removed easily by treating the ether with dilute aqueous acid.

Suppose, for example, we wanted to prepare 4-pentyn-1-ol from 3-bromo-1-propanol and sodium acetylide. If we allow them to react directly, the strongly basic sodium acetylide will react first with the hydroxyl group, making the alkylation unsuccessful:

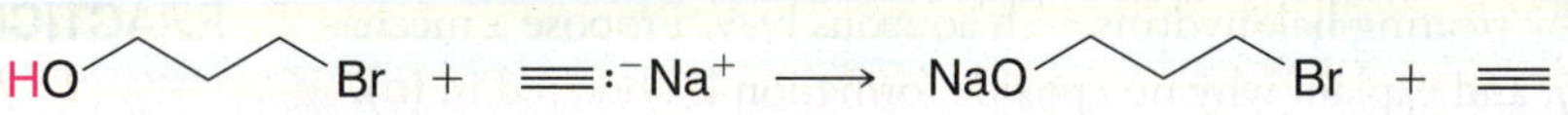

3-Bromo-1-propanol

Unsuccessful alkylation due to competing acid–base reaction.

However, if we protect the —OH group first, the synthesis becomes feasible:

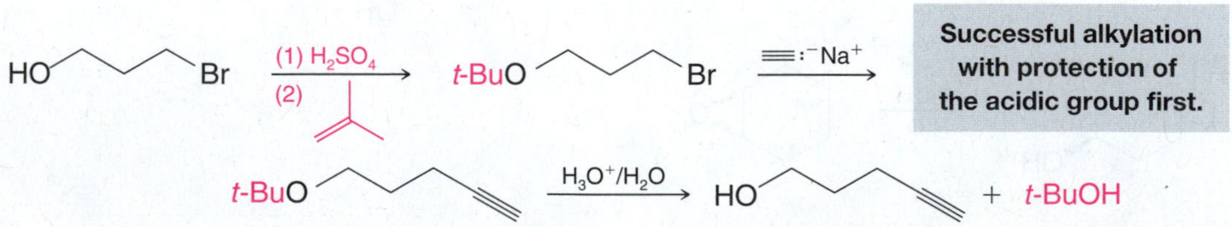

Successful alkylation with protection of the acidic group first.

4-Pentyn-1-ol

PRACTICE PROBLEM 11.15 Propose mechanisms for the following reactions.

(a)

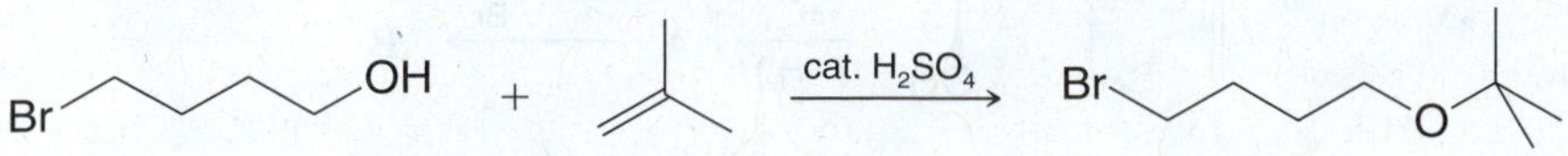

(b)

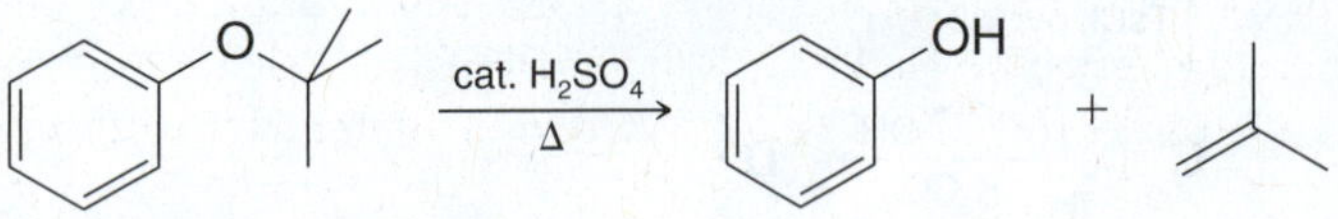

11.11E Silyl Ether Protecting Groups

- A hydroxyl group can be protected from acid–base reactions by converting it to a **silyl ether** group.

One of the most common silyl ether **protecting groups** is the *tert*-butyldimethylsilyl ether group [*tert*-butyl $(Me)_2Si$—O—R, or TBS—O—R], although triethylsilyl, triisopropylsilyl, *tert*-butyldiphenylsilyl, and others can be used. The *tert*-butyldimethylsilyl ether is stable over a pH range of roughly 4–12. A TBS group can be added by allowing the alcohol to react with *tert*-butyldimethylsilyl chloride in the presence of an aromatic amine (a base) such as imidazole or pyridine:

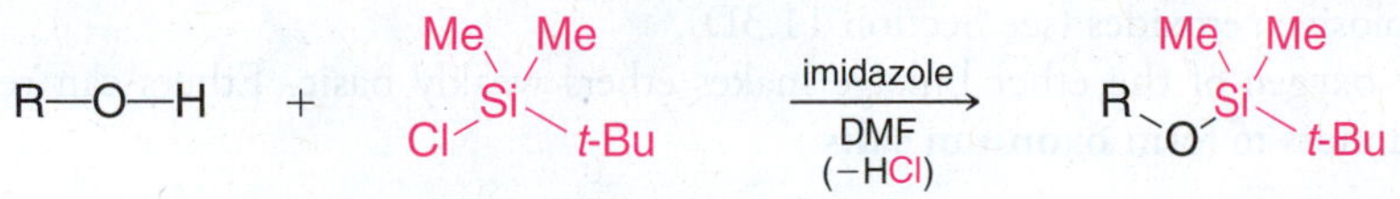

tert-Butylchlorodimethylsilane
(TBSCl)

(R—O—TBS)

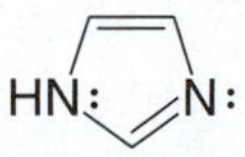

Imidazole

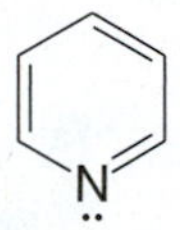

Pyridine

- The TBS group can be removed by treatment with fluoride ion (tetrabutylammonium fluoride or aqueous HF is frequently used). These conditions tend not to affect other functional groups, which is why TBS ethers are such good protecting groups.

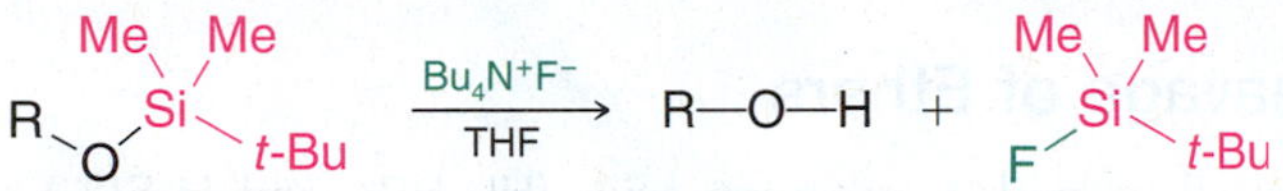

(R—O—TBS)

Converting an alcohol to a silyl ether also makes it much more volatile. This increased volatility makes the alcohol (as a silyl ether) much more amenable to analysis by gas chromatography. Trimethylsilyl ethers are often used for this purpose. The trimethylsilyl ether group is too labile to use as a protecting group in most reactions, however.

●●● **SOLVED PROBLEM 11.9**

Supply the missing reagents and intermediates **A–E**.

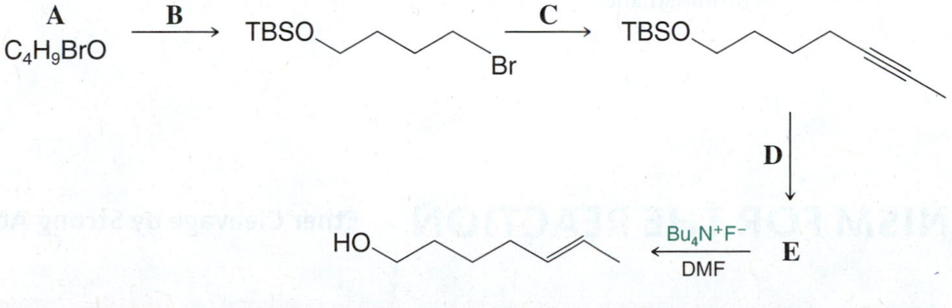

STRATEGY AND ANSWER: We start by noticing several things: a TBS (*tert*-butyldimethylsilyl) protecting group is involved, the carbon chain increases from four carbons in **A** to seven in the final product, and an alkyne is reduced to a *trans* alkene. **A** does not contain any silicon atoms, whereas the product after the reaction under conditions **B** does. Therefore, **A** must be an alcohol that is protected as a TBS ether by conditions specified as **B**. **A** is therefore 4-bromo-1-butanol, and conditions **B** are TBSCl (*tert*-butyldimethylsilyl chloride) with imidazole in DMF. Conditions **C** involve loss of the bromine and chain extension by three carbons with incorporation of an alkyne. Thus, the reaction conditions for **C** must involve sodium propynide, which would come from deprotonation of propyne using an appropriate base, such as $NaNH_2$ or CH_3MgBr. The conditions leading from **E** to the final product are those for removal of a TBS group, and not those for converting an alkyne to a *trans* alkene; thus, **E** must still contain the TBS ether but already contain the *trans* alkene. Conditions **D**, therefore, must be (1) Li, Et_2NH, (2) NH_4Cl, which are those required for converting the alkyne to a *trans* alkene. **E**, therefore, must be the TBS ether of 5-heptyn-1-ol (which can also be named 1-*tert*-butyldimethylsiloxy-5-heptynol).

11.12 REACTIONS OF ETHERS

Dialkyl ethers react with very few reagents other than acids. The only reactive sites that molecules of a dialkyl ether present to another reactive substance are the C—H bonds of the alkyl groups and the —Ö— group of the ether linkage. Ethers resist attack by nucleophiles (why?) and by bases. This lack of reactivity coupled with the ability of ethers to solvate cations (by donating an electron pair from their oxygen atom) makes ethers especially useful as solvents for many reactions.

Ethers are like alkanes in that they undergo halogenation reactions (Chapter 10), but these reactions are of little synthetic importance. They also undergo slow autoxidation to form explosive peroxides (see Section 11.3D).

The oxygen of the ether linkage makes ethers weakly basic. Ethers can react with proton donors to form **oxonium salts**:

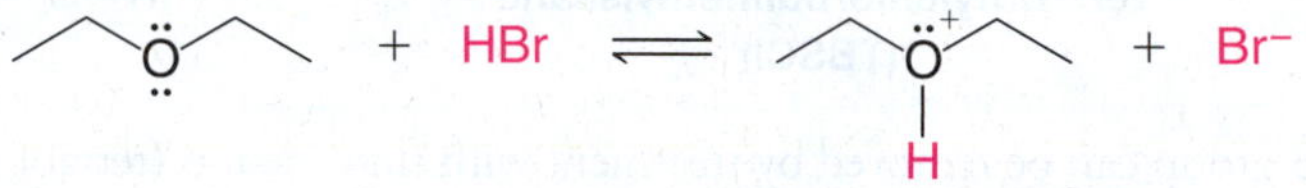

An oxonium salt

11.12A Cleavage of Ethers

Heating dialkyl ethers with very strong acids (HI, HBr, and H_2SO_4) causes them to undergo reactions in which the carbon–oxygen bond breaks. Diethyl ether, for example, reacts with hot concentrated hydrobromic acid to give two molecular equivalents of bromoethane:

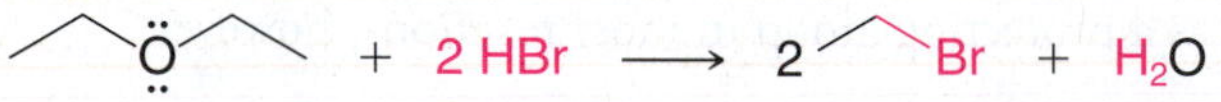

Cleavage of an ether

The mechanism for this reaction begins with formation of an oxonium cation. Then, an S_N2 reaction with a bromide ion acting as the nucleophile produces ethanol and bromoethane. Excess HBr reacts with the ethanol produced to form the second molar equivalent of bromoethane.

[**A MECHANISM FOR THE REACTION** — **Ether Cleavage by Strong Acids**]

Step 1

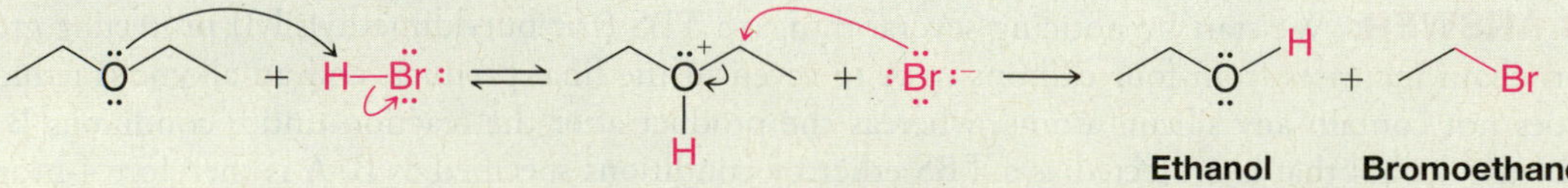

Ethanol   Bromoethane

Step 2   **In step 2 the ethanol (just formed) reacts with HBr (present in excess) to form a second molar equivalent of bromoethane**

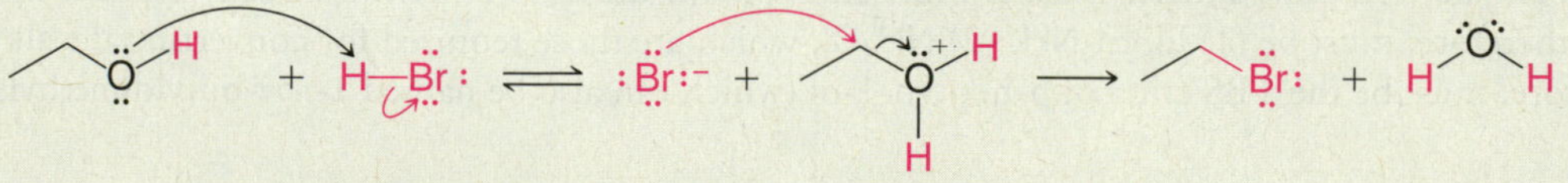

● ● ●
PRACTICE PROBLEM 11.16

When an ether is treated with *cold* concentrated HI, cleavage occurs as follows:

$$R—O—R + HI \longrightarrow ROH + RI$$

When mixed ethers are used, the alcohol and alkyl iodide that form depend on the nature of the alkyl groups. Use mechanisms to explain the following observations:

(a) OMe $\xrightarrow{\text{HI}}$ OH + MeI

(b) OMe $\xrightarrow{\text{HI}}$ I + MeOH

● ● ●
PRACTICE PROBLEM 11.17

Write a detailed mechanism for the following reaction.

$$\xrightarrow[\Delta]{\text{HBr (excess)}} 2 \quad \text{Br}$$

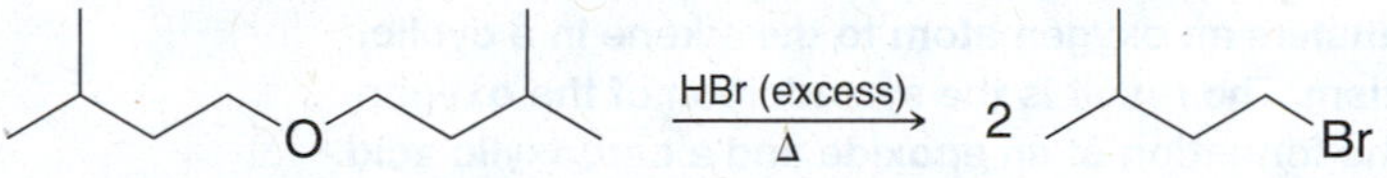

● ● ●
PRACTICE PROBLEM 11.18

Provide a mechanism for the following reaction.

OCH₃ $\xrightarrow{\text{HCl}}$ Cl

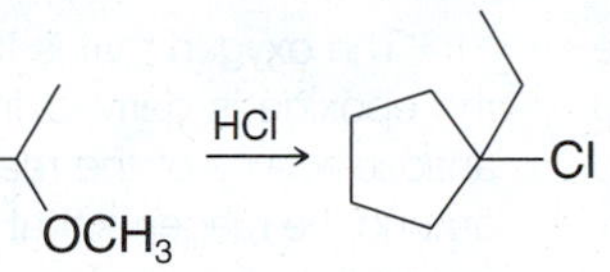

11.13 EPOXIDES

Epoxides are cyclic ethers with three-membered rings. In IUPAC nomenclature epoxides are called **oxiranes**. The simplest epoxide has the common name ethylene oxide:

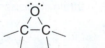

An epoxide

IUPAC name: oxirane
Common name: ethylene oxide

11.13A Synthesis of Epoxides: Epoxidation

Epoxides can be synthesized by the reaction of an alkene with an organic **peroxy acid** (RCO₃H—sometimes called simply a **peracid**), a process that is called **epoxidation**. *meta*-Chloroperoxybenzoic acid (mCPBA) is one peroxy acid reagent commonly used for epoxidation. The following reaction is an example.

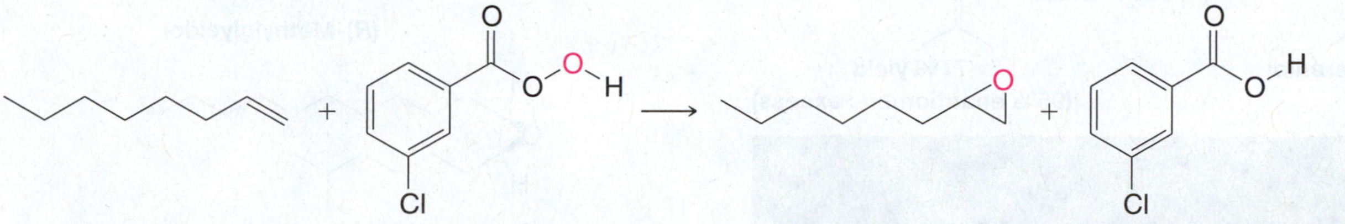

1-Octene   **mCPBA**   **(81%)**   ***meta*-Chlorobenzoic acid**

meta-Chlorobenzoic acid is a by-product of the reaction. Often it is not written in the chemical equation, as the following example illustrates.

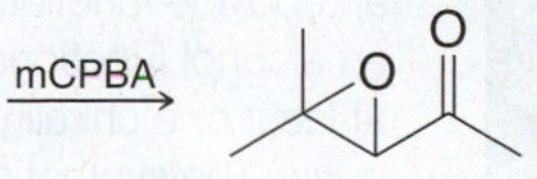

$$\xrightarrow{\text{mCPBA}}$$

(77%)

As the first example illustrates, the peroxy acid transfers an oxygen atom to the alkene. The following mechanism has been proposed.

[A MECHANISM FOR THE REACTION — **Alkene Epoxidation**]

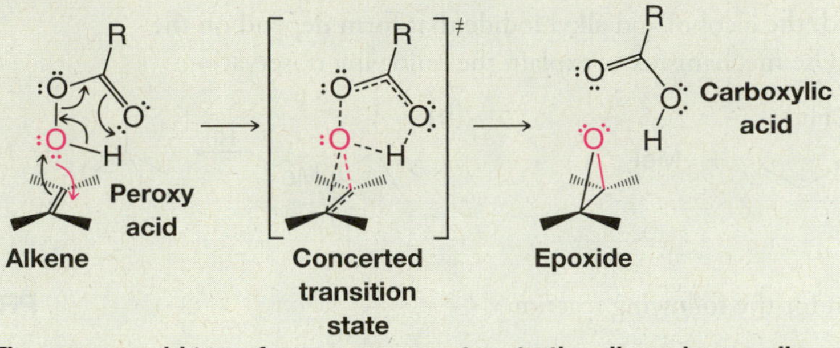

Peroxy acid

Alkene

Concerted transition state

Epoxide

Carboxylic acid

The peroxy acid transfers an oxygen atom to the alkene in a cyclic, single-step mechanism. The result is the syn addition of the oxygen to the alkene, with the formation of an epoxide and a carboxylic acid.

THE CHEMISTRY OF...   The Sharpless Asymmetric Epoxidation

In 1980, K. B. Sharpless (then at the Massachusetts Institute of Technology, presently at The Scripps Research Institute) and co-workers reported a method that has since become one of the most valuable tools for chiral synthesis. The Sharpless asymmetric epoxidation is a method for converting allylic alcohols (Section 11.1) to chiral epoxy alcohols with very high enantioselectivity (i.e., with preference for one enantiomer rather than formation of a racemic mixture). In recognition of this and other work in asymmetric oxidation methods (see Section 8.16A), Sharpless received half of the 2001 Nobel Prize in Chemistry (the other half was awarded to W. S. Knowles and R. Noyori; see Section 7.14). The Sharpless asymmetric epoxidation involves treating the allylic alcohol with *tert*-butyl hydroperoxide, titanium(IV) tetraisopropoxide [Ti(O—*i*-Pr)$_4$], and a specific stereoisomer of a tartrate ester. (The tartrate stereoisomer that is chosen depends on the specific enantiomer of the epoxide desired). The following is an example:

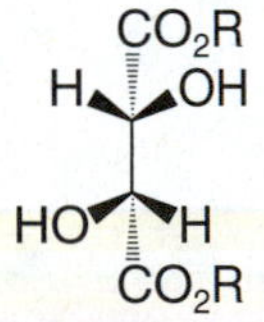

A (+)-dialkyl tartrate ester

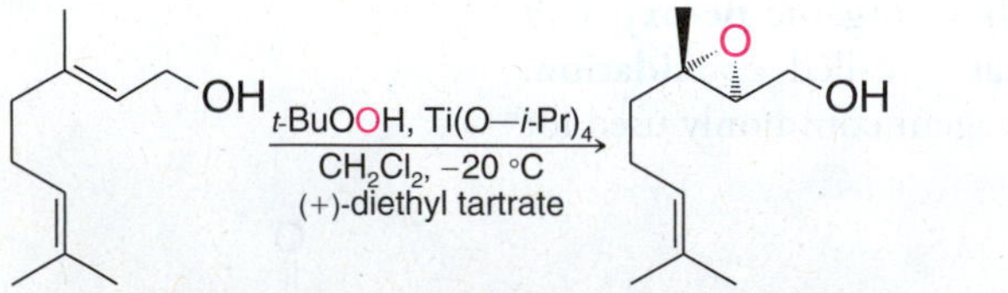

The oxygen that is transferred to the allylic alcohol to form the epoxide is derived from *tert*-butyl hydroperoxide. The enantioselectivity of the reaction results from a titanium complex among the reagents that includes the enantiomerically pure tartrate ester as one of the ligands. The choice of whether to use the (+)- or (−)-tartrate ester for stereochemical control depends on which enantiomer of the epoxide is desired. [The (+)- and (−)-tartrates are either diethyl or diisopropyl esters.] The stereochemical preferences of the reaction have been well studied, such that it is possible to prepare either enantiomer of a chiral epoxide in high enantiomeric excess, simply by choosing the appropriate (+)- or (−)-tartrate stereoisomer as the chiral ligand:

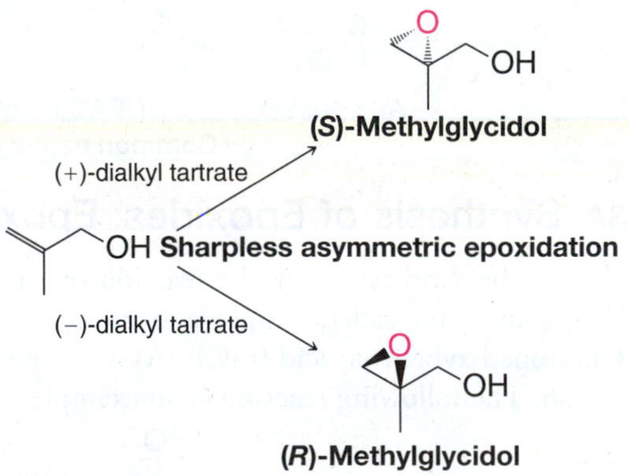

(S)-Methylglycidol

(+)-dialkyl tartrate

Sharpless asymmetric epoxidation

(−)-dialkyl tartrate

(R)-Methylglycidol

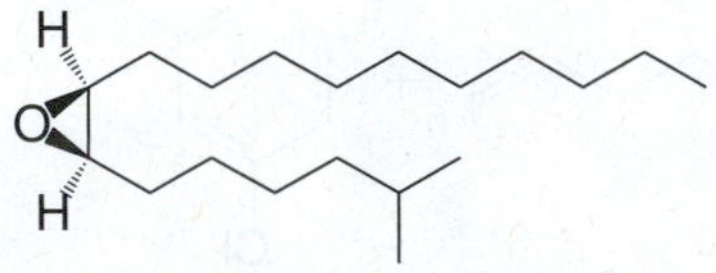

(7R,8S)-Disparlure

Geraniol

t-BuOOH, Ti(O—*i*-Pr)$_4$
CH$_2$Cl$_2$, −20 °C
(+)-diethyl tartrate

**77% yield
(95% enantiomeric excess)**

Compounds of this general structure are extremely useful and versatile synthons because combined in one molecule are an epoxide functional group (a highly reactive electrophilic site), an alcohol functional group (a potentially nucleophilic site), and at least one chirality center that is present in high enantiomeric purity. The synthetic utility of chiral epoxy alcohol synthons produced by the Sharpless asymmetric epoxidation has been demonstrated over and over in enantioselective syntheses of many important compounds. Some examples include the synthesis

© Universal Images Group/SuperStock

of the polyether antibiotic X-206 by E. J. Corey (Harvard), the J. T. Baker commercial synthesis of the gypsy moth pheromone (7R,8S)-disparlure, and synthesis by K. C. Nicolaou (University of California San Diego and Scripps Research Institute) of

zaragozic acid A (which is also called squalestatin S1 and has been shown to lower serum cholesterol levels in test animals by inhibition of squalene biosynthesis; see "The Chemistry of... Cholesterol Biosynthesis," online in *WileyPLUS* after Chapter 8).

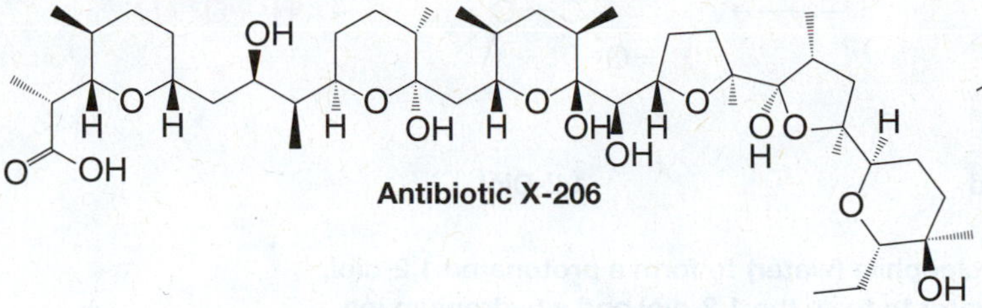

Antibiotic X-206

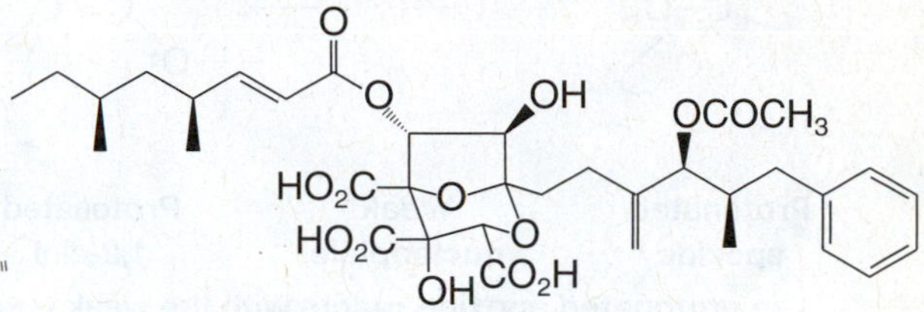

Zaragozic acid A (squalestatin S1)

11.13B Stereochemistry of Epoxidation

- The reaction of alkenes with peroxy acids is, of necessity, a **syn** addition, and it is **stereospecific**. Furthermore, the oxygen atom can add to either face of the alkene.

For example, *trans*-2-butene yields racemic *trans*-2,3-dimethyloxirane, because addition of oxygen to each face of the alkene generates an enantiomer. *cis*-2-Butene, on the other hand, yields only *cis*-2,3-dimethyloxirane, no matter which face of the alkene accepts the oxygen atom, due to the plane of symmetry in both the reactant and the product. If additional chirality centers are present in a substrate, then diastereomers would result.

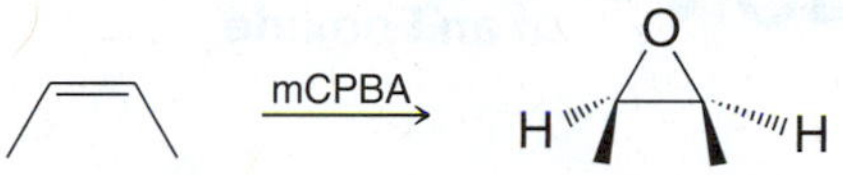

cis-2-Butene → (mCPBA) → **cis-2,3-Dimethyloxirane (a meso compound)**

trans-2-Butene → (mCPBA) → **Enantiomeric trans-2,3-dimethyloxiranes**

In Special Topic D (Section D.3, in *WileyPLUS*) we present a method for synthesizing epoxides from aldehydes and ketones.

11.14 REACTIONS OF EPOXIDES

- The highly strained three-membered ring of epoxides makes them much more reactive toward nucleophilic substitution than other ethers.

Acid catalysis assists epoxide ring opening by providing a better leaving group (an alcohol) at the carbon atom undergoing nucleophilic attack. This catalysis is especially important if the nucleophile is a weak nucleophile such as water or an alcohol. An example is the acid-catalyzed hydrolysis of an epoxide.

[A MECHANISM FOR THE REACTION — Acid-Catalyzed Ring Opening of an Epoxide]

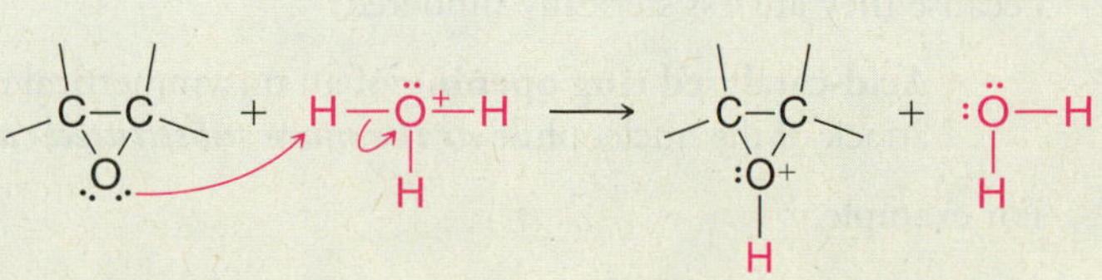

Epoxide   **Protonated epoxide**
The acid reacts with the epoxide to produce a protonated epoxide.

(mechanism continues on the next page)

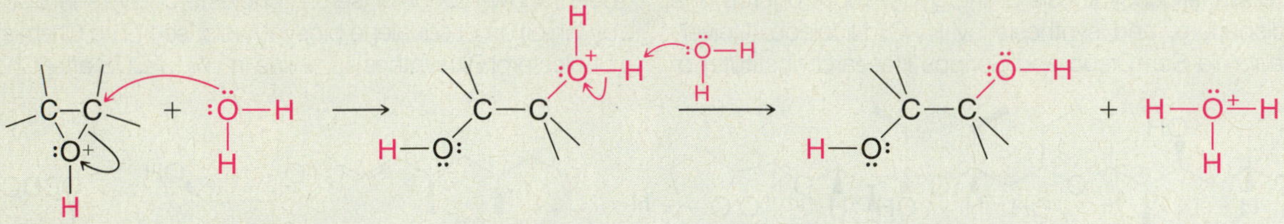

Protonated   **Weak**   **Protonated**   **1,2-Diol**
epoxide   **nucleophile**   **1,2-diol**

The protonated epoxide reacts with the weak nucleophile (water) to form a protonated 1,2-diol, which then transfers a proton to a molecule of water to form the 1,2-diol and a hydronium ion.

Epoxides can also undergo base-catalyzed ring opening. Such reactions do not occur with other ethers, but they are possible with epoxides (because of ring strain), provided that the attacking nucleophile is also a strong base such as an alkoxide ion or hydroxide ion.

[**A MECHANISM FOR THE REACTION** — **Base-Catalyzed Ring Opening of an Epoxide**]

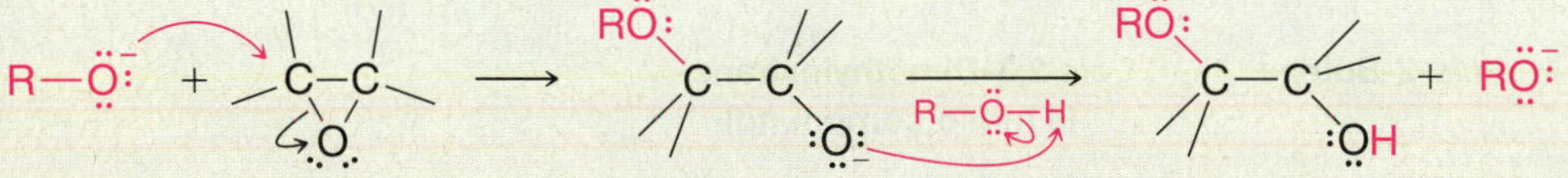

Strong   **Epoxide**   **An alkoxide ion**
nucleophile

A strong nucleophile such as an alkoxide ion or a hydroxide ion is able to open the strained epoxide ring in a direct S$_N$2 reaction.

Helpful Hint

Regioselectivity in the opening of epoxides.

- **Base-catalyzed ring opening** of an unsymmetrical epoxide occurs primarily by attack of the nucleophile *at the less substituted carbon atom*.

For example, methyloxirane reacts with an alkoxide ion mainly at its primary carbon atom:

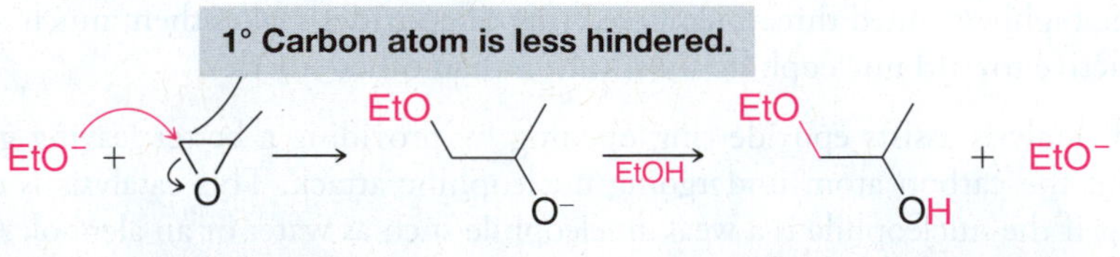

1° Carbon atom is less hindered.

Methyloxirane   **1-Ethoxy-2-propanol**

This is just what we should expect: the reaction is, after all, an S$_N$2 reaction, and, as we learned earlier (Section 6.13A), primary substrates react more rapidly in S$_N$2 reactions because they are less sterically hindered.

- **Acid-catalyzed ring opening** of an unsymmetrical epoxide occurs primarily by attack of the nucleophile *at the more substituted carbon atom*.

For example,

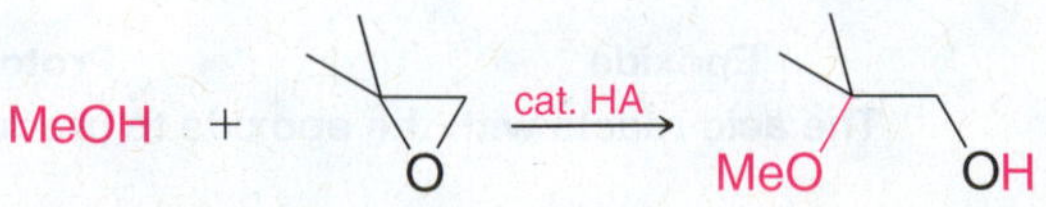

The reason: bonding in the protonated epoxide (see the following reaction) is unsymmetrical, with the more highly substituted carbon atom bearing a considerable positive charge; the reaction is S_N1 like. The nucleophile, therefore, attacks this carbon atom even though it is more highly substituted:

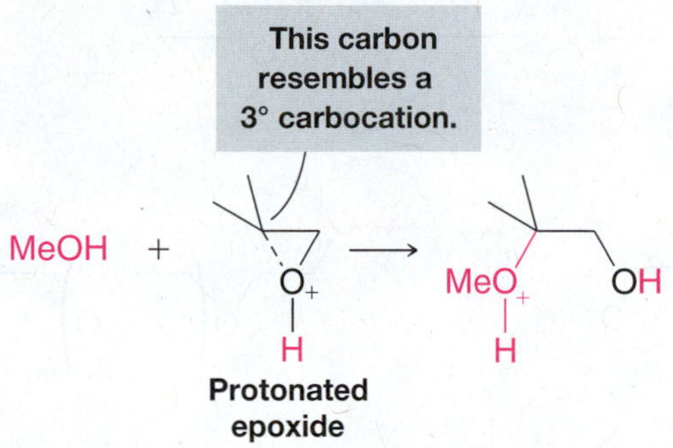

This carbon
resembles a
3° carbocation.

MeOH +

Protonated
epoxide

The more highly substituted carbon atom bears a greater positive charge because it resembles a more stable tertiary carbocation. [Notice how this reaction (and its explanation) resembles that given for halohydrin formation from unsymmetrical alkenes in Section 8.14 and attack on mercurinium ions.]

PRACTICE PROBLEM 11.19

Propose structures for each of the following products derived from oxirane (ethylene oxide):

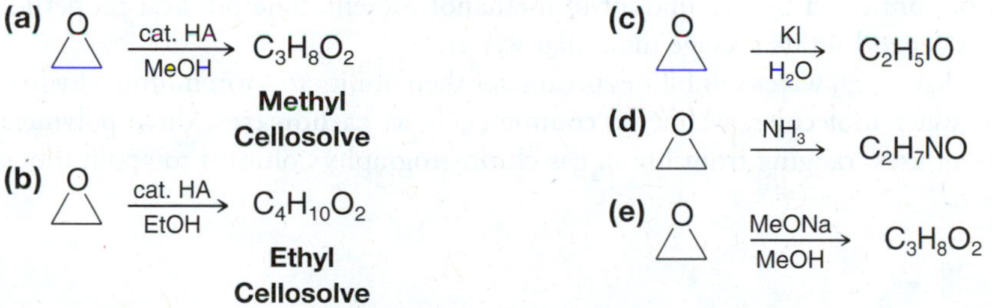

(a)
$\xrightarrow[\text{MeOH}]{\text{cat. HA}}$ $C_3H_8O_2$
Methyl Cellosolve

(b)
$\xrightarrow[\text{EtOH}]{\text{cat. HA}}$ $C_4H_{10}O_2$
Ethyl Cellosolve

(c)
$\xrightarrow[\text{H}_2\text{O}]{\text{KI}}$ C_2H_5IO

(d)
$\xrightarrow{\text{NH}_3}$ C_2H_7NO

(e)
$\xrightarrow[\text{MeOH}]{\text{MeONa}}$ $C_3H_8O_2$

PRACTICE PROBLEM 11.20

Provide a mechanistic explanation for the following observation.

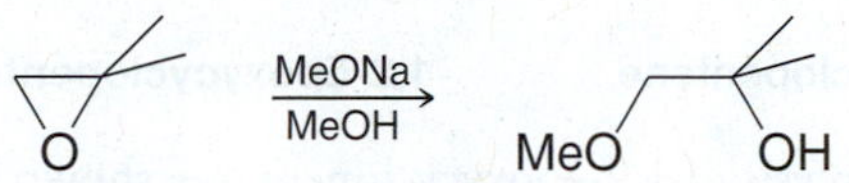

$\xrightarrow[\text{MeOH}]{\text{MeONa}}$

PRACTICE PROBLEM 11.21

When sodium ethoxide reacts with 1-(chloromethyl)oxirane (also called epichlorohydrin), labeled with ^{14}C as shown by the asterisk in **I**, the major product is **II**. Provide a mechanistic explanation for this result.

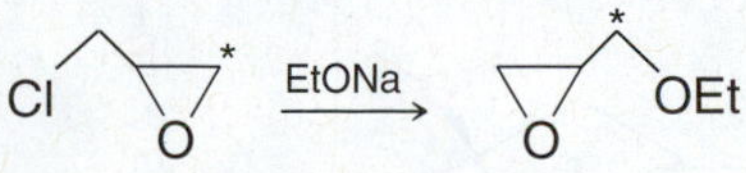

$\xrightarrow{\text{EtONa}}$

I

II

**1-(Chloromethyl)oxirane
(epichlorohydrin)**

11.14A Polyethers from Epoxides

Treating ethylene oxide with sodium methoxide (in the presence of a small amount of methanol) can result in the formation of a **polyether**:

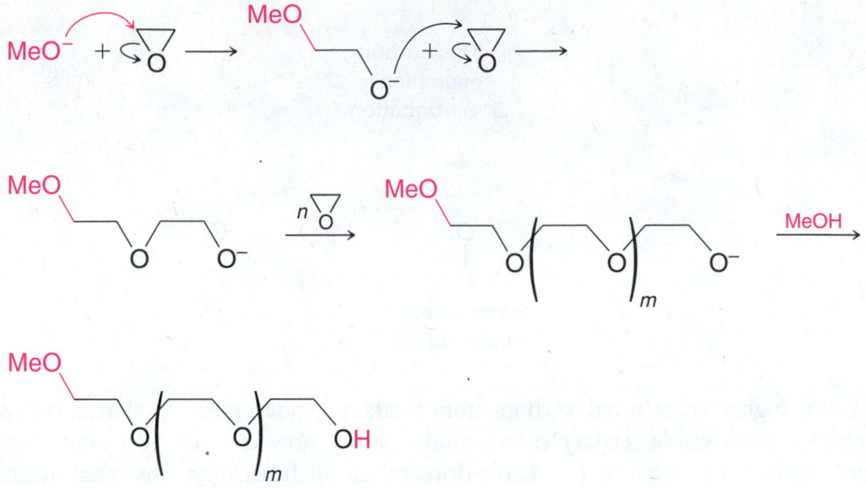

Poly(ethylene glycol)
(a polyether)

This is an example of **anionic polymerization** (Section 10.11). The polymer chains continue to grow until methanol protonates the alkoxide group at the end of the chain. The average length of the growing chains and, therefore, the average molecular weight of the polymer can be controlled by the amount of methanol present. The physical properties of the polymer depend on its average molecular weight.

Polyethers have high water solubilities because of their ability to form multiple hydrogen bonds to water molecules. Marketed commercially as **carbowaxes**, these polymers have a variety of uses, ranging from use in gas chromatography columns to applications in cosmetics.

11.15 ANTI 1,2-DIHYDROXYLATION OF ALKENES VIA EPOXIDES

Epoxidation of cyclopentene with a peroxycarboxylic acid produces 1,2-epoxycyclopentane:

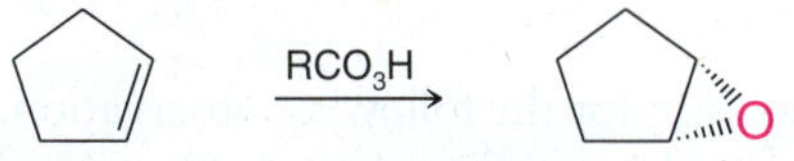

Cyclopentene   **1,2-Epoxycyclopentane**

Helpful Hint

A synthetic method for anti 1,2-dihydroxylation.

Acid-catalyzed hydrolysis of 1,2-epoxycyclopentane, shown below, yields a trans diol, *trans*-1,2-cyclopentanediol. Water acting as a nucleophile attacks the protonated epoxide from the side opposite the epoxide group. The carbon atom being attacked undergoes an inversion of configuration. We show here only one carbon atom being attacked. Attack at the other carbon atom of this symmetrical system is equally likely and produces the enantiomeric form of *trans*-1,2-cyclopentanediol:

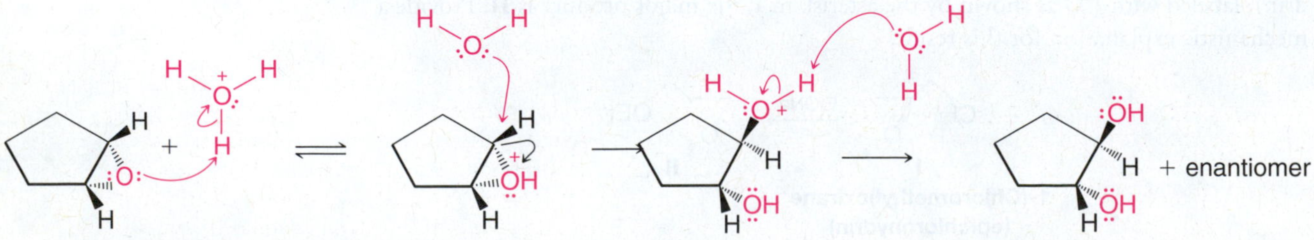

trans-1,2-Cyclopentanediol

Epoxidation followed by acid-catalyzed hydrolysis gives us, therefore, a method for **anti 1,2-dihydroxylation** of a double bond (as opposed to syn 1,2-dihydroxylation, Section 8.16). The stereochemistry of this technique parallels closely the stereochemistry of the bromination of cyclopentene given earlier (Section 8.13).

●●●

PRACTICE PROBLEM 11.22

Outline a mechanism similar to the one just given that shows how the enantiomeric form of *trans*-1,2-cyclopentanediol is produced.

●●● **SOLVED PROBLEM 11.10**

In Section 11.13B we showed the epoxidation of *cis*-2-butene to yield *cis*-2,3-dimethyloxirane and epoxidation of *trans*-2-butene to yield *trans*-2,3-dimethyloxirane. Now consider acid-catalyzed hydrolysis of these two epoxides and show what product or products would result from each. Are these reactions stereospecific?

ANSWER: (a) The meso compound, *cis*-2,3-dimethyloxirane (Fig. 11.1), yields on hydrolysis (2R,3R)-2,3-butanediol and (2S,3S)-2,3-butanediol. These products are enantiomers. Since the attack by water at either carbon [path **(a)** or path **(b)** in Fig. 11.1] occurs at the same rate, the product is obtained in a racemic form.

When either of the *trans*-2,3-dimethyloxirane enantiomers undergoes acid-catalyzed hydrolysis, the only product that is obtained is the meso compound, (2R,3S)-2,3-butanediol. The hydrolysis of one enantiomer is shown in Fig. 11.2. (You might construct a similar diagram showing the hydrolysis of the other enantiomer to convince yourself that it, too, yields the same product.)

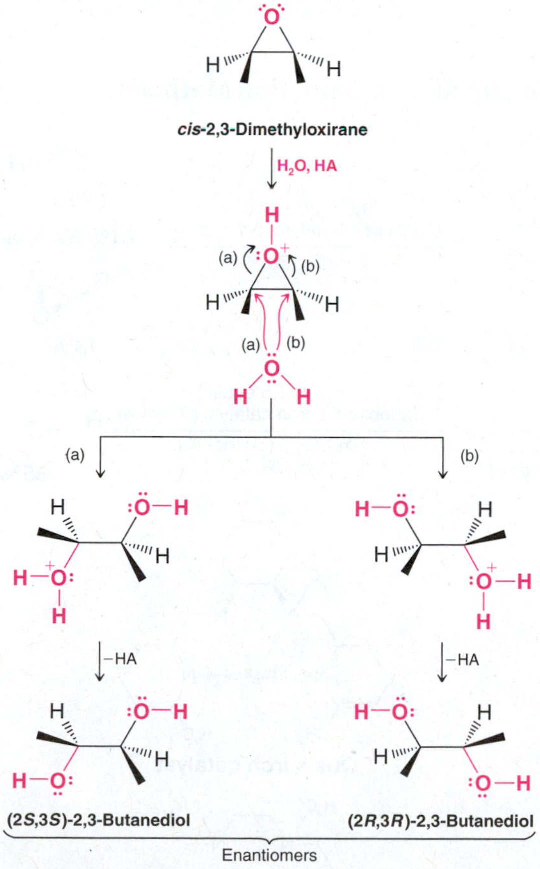

FIGURE 11.1 Acid-catalyzed hydrolysis of *cis*-2,3-dimethyloxirane yields (2S,3S)-2,3-butanediol by path (a) and (2R,3R)-2,3-butanediol by path (b). (Use models to convince yourself.)

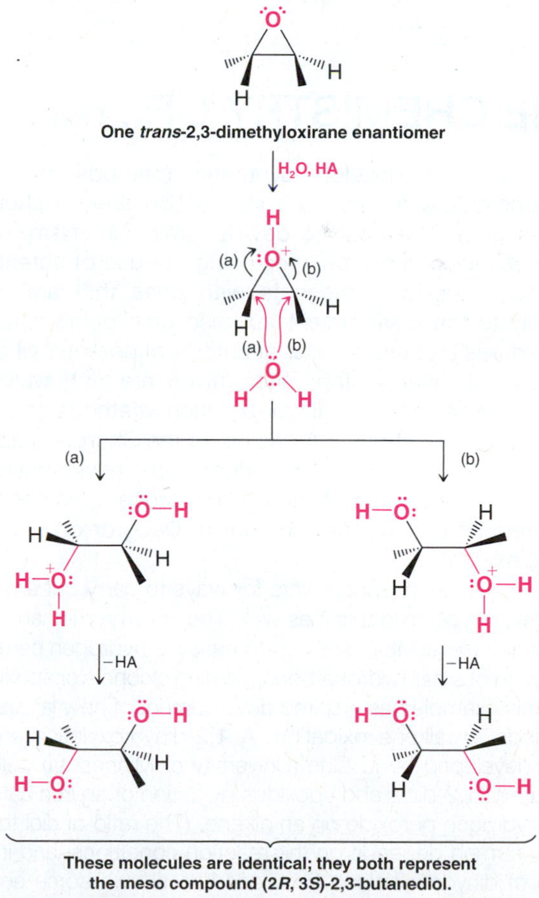

FIGURE 11.2 The acid-catalyzed hydrolysis of one *trans*-2,3-dimethyloxirane enantiomer produces the meso compound, (2R,3S)-2,3-butanediol, by path (a) or by path (b). Hydrolysis of the other enantiomer (or the racemic modification) would yield the same product. (You should use models to convince yourself that the two structures given for the products do represent the same compound.)

(continues on next page)

(b) Since both steps in this method for the conversion of an alkene to a 1,2-diol (glycol) are stereospecific (i.e., both the epoxidation step and the acid-catalyzed hydrolysis), the net result is a stereospecific anti 1,2-dihydroxylation of the double bond (Fig. 11.3).

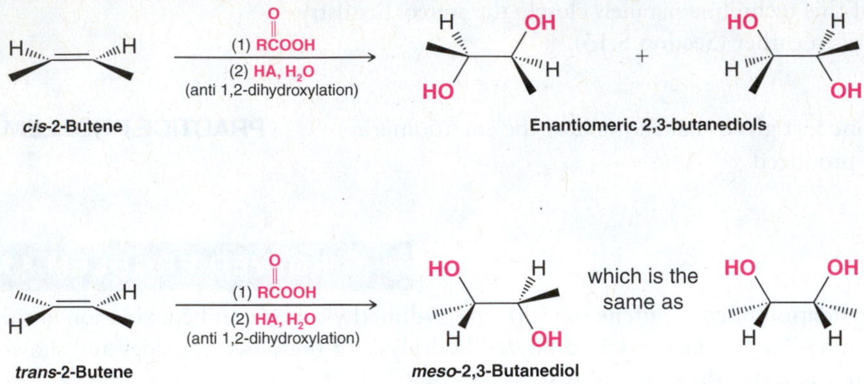

Enantiomeric 2,3-butanediols

cis-2-Butene

(1) RCOOH
(2) HA, H$_2$O
(anti 1,2-dihydroxylation)

trans-2-Butene

(1) RCOOH
(2) HA, H$_2$O
(anti 1,2-dihydroxylation)

which is the same as

meso-2,3-Butanediol

FIGURE 11.3 The overall result of epoxidation followed by acid-catalyzed hydrolysis is a stereospecific anti 1,2-dihydroxylation of the double bond. *cis*-2-Butene yields the enantiomeric 2,3-butanediols; *trans*-2-butene yields the meso compound.

• • •

PRACTICE PROBLEM 11.23 Supply the missing reagents and intermediates **A–E**.

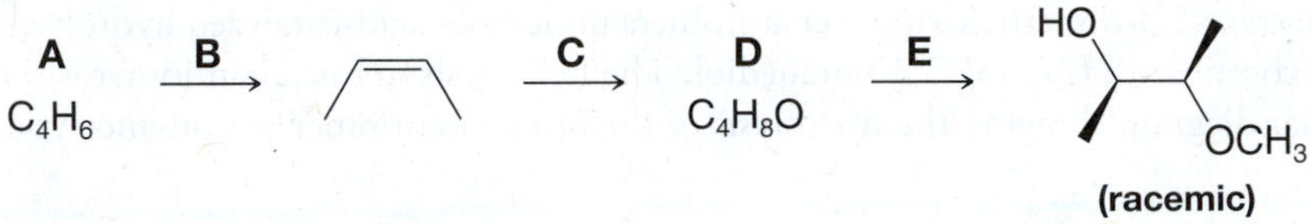

A
C$_4$H$_6$

B

C

D
C$_4$H$_8$O

E

(racemic)

THE CHEMISTRY OF... Environmentally Friendly Alkene Oxidation Methods

The effort to develop synthetic methods that are environmentally friendly is a very active area of chemistry research. The push to devise "green chemistry" procedures includes not only replacing the use of potentially hazardous or toxic reagents with ones that are more friendly to the environment but also developing catalytic procedures that use smaller quantities of potentially harmful reagents when other alternatives are not available. The catalytic syn 1,2-dihydroxylation methods that we described in Section 8.16 (including the Sharpless asymmetric dihydroxylation procedure) are environmentally friendly modifications of the original procedures because they require only a small amount of OsO$_4$ or other heavy metal oxidant.

Nature has provided hints for ways to carry out environmentally sound oxidations as well. The enzyme methane monooxygenase (MMO) uses iron to catalyze hydrogen peroxide oxidation of small hydrocarbons, yielding alcohols or epoxides, and this example has inspired development of new laboratory methods for alkene oxidation. A 1,2-dihydroxylation procedure developed by L. Que (University of Minnesota) yields a mixture of 1,2-diols and epoxides by action of an iron catalyst and hydrogen peroxide on an alkene. (The ratio of diol to epoxide formed depends on the reaction conditions, and in the case of dihydroxylation, the procedure shows some enantioselectivity.) Another green reaction is the epoxidation method developed by E. Jacobsen (Harvard University). Jacobsen's procedure uses hydrogen peroxide and a similar iron catalyst to epoxidize alkenes (without the complication of diol formation). Que's and Jacobsen's methods are environmentally friendly because their procedures employ catalysts containing a nontoxic metal, and an inexpensive, relatively safe oxidizing reagent is used that is converted to water in the course of the reaction.

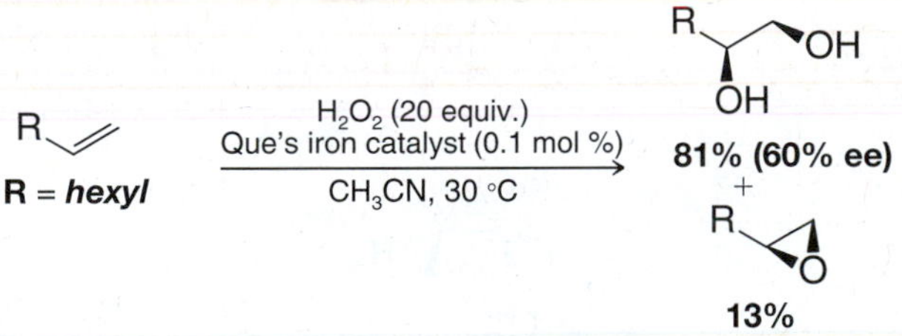

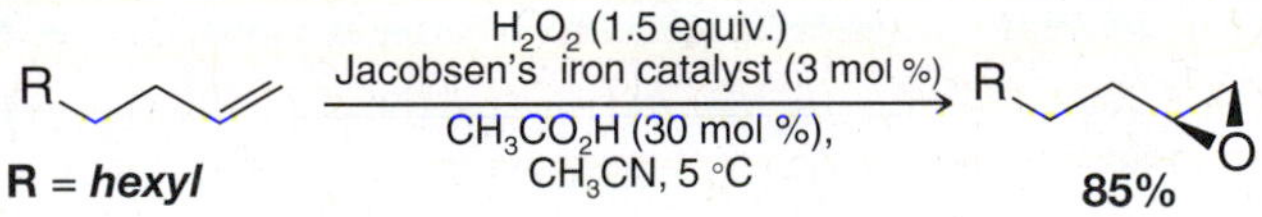

R
R = *hexyl*

H$_2$O$_2$ (20 equiv.)
Que's iron catalyst (0.1 mol %)
CH$_3$CN, 30 °C

81% (60% ee)
+

13%

R
R = *hexyl*

H$_2$O$_2$ (1.5 equiv.)
Jacobsen's iron catalyst (3 mol %)
CH$_3$CO$_2$H (30 mol %),
CH$_3$CN, 5 °C

85%

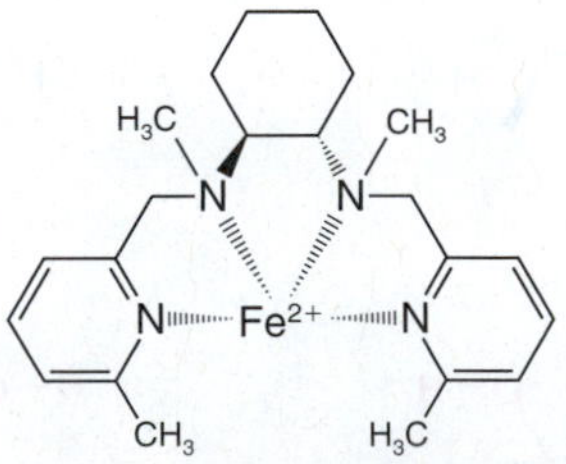

Que's iron catalyst

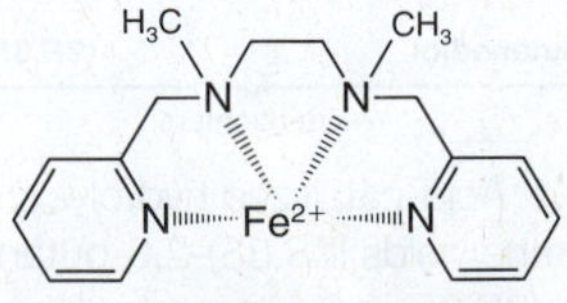

Jacobsen's iron catalyst

The quest for more methods in green chemistry, with benign reagents and by-products, catalytic cycles, and high yields, will no doubt drive further research by present and future chemists. In coming chapters we shall see more examples of green chemistry in use or under development.

11.16 CROWN ETHERS

Crown ethers are compounds having structures like that of 18-crown-6, below. 18-Crown-6 is a cyclic oligomer of ethylene glycol. Crown ethers are named as *x*-crown-*y*, where *x* is the total number of atoms in the ring and *y* is the number of oxygen atoms. A key property of crown ethers is that they are able to bind cations, as shown below for 18-crown-6 and a potassium ion.

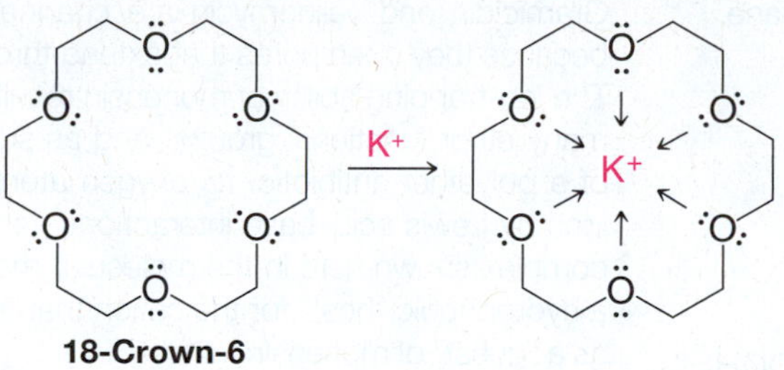

18-Crown-6

Crown ethers render many salts soluble in nonpolar solvents. For this reason they are called **phase transfer catalysts**. When a crown ether coordinates with a metal cation it masks the ion with a hydrocarbon-like exterior. 18-Crown-6 coordinates very effectively with potassium ions because the cavity size is correct and because the six oxygen atoms are ideally situated to donate their electron pairs to the central ion in a Lewis acid–base complex.

- The relationship between a crown ether and the ion it binds is called a **host–guest relationship**.

Salts such as KF, KCN, and potassium acetate can be transferred into aprotic solvents using catalytic amounts of 18-crown-6. Use of a crown ether with a nonpolar solvent can be very favorable for an S_N2 reaction because the nucleophile (such as F^-, CN^-, or acetate from the compounds just listed) is unencumbered by solvent in an aprotic solvent, while at the same time the cation is prevented by the crown ether from associating with the nucleophile. Dicyclohexano-18-crown-6 is another example of a phase transfer catalyst. It is even more soluble in nonpolar solvents than 18-crown-6 due to its additional hydrocarbon groups. Phase transfer catalysts can also be used for reactions such as oxidations. (There are phase transfer catalysts that are not crown ethers, as well.)

Dicyclohexano-18-crown-6

The development of crown ethers and other molecules "with structure specific interactions of high selectivity" led to awarding of the 1987 Nobel Prize in Chemistry to Charles J. Pedersen (DuPont Company, deceased), Donald J. Cram (University of California, Los Angeles, deceased), and Jean-Marie Lehn (Louis Pasteur University, Strasbourg, France). Their contributions to our understanding of what is now called "molecular recognition" have implications for how enzymes recognize their substrates, how hormones cause their effects, how antibodies recognize antigens, how neurotransmitters propagate their signals, and many other aspects of biochemistry.

 The 1987 Nobel Prize in Chemistry was awarded to Pedersen, Cram, and Lehn for their work relating to crown ethers.

Write structures for **(a)** 15-crown-5 and **(b)** 12-crown-4.

PRACTICE PROBLEM 11.24

THE CHEMISTRY OF... Transport Antibiotics and Crown Ethers

There are several antibiotics called ionophores. Some notable examples are monensin, nonactin, gramicidin, and valinomycin. The structures of monensin and nonactin are shown below. Ionophore antibiotics like monensin and nonactin coordinate with metal cations in a manner similar to crown ethers. Their mode of action has to do with disrupting the natural gradient of ions on each side of the cell membrane.

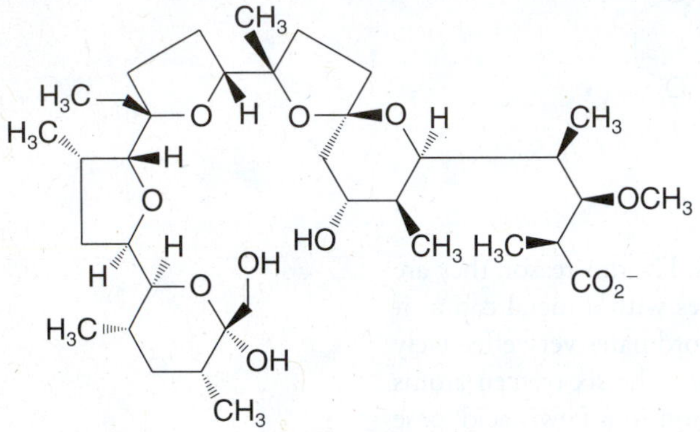

Monensin

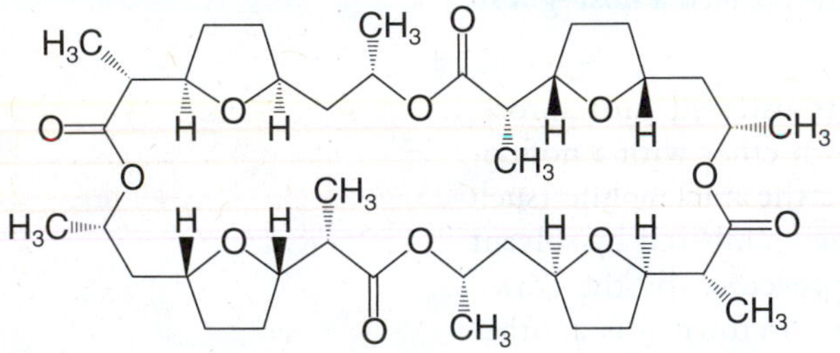

Nonactin

through the cell membrane is slow, and requires an expenditure of energy by the cell. (The 1997 Nobel Prize in Chemistry was awarded in part for work regarding sodium and potassium cell membrane transport.*)

Monensin is called a carrier ionophore because it binds with sodium ions and carries them across the cell membrane. Gramicidin and valinomycin are channel-forming antibiotics because they open pores that extend through the membrane. The ion-trapping ability of monensin results principally from its many ether functional groups, and as such, it is an example of a polyether antibiotic. Its oxygen atoms bind with sodium ions by Lewis acid–base interactions, forming the octahedral complex shown here in the molecular model. The complex is a hydrophobic "host" for the cation that allows it to be carried as a "guest" of monensin from one side of the cell membrane to the other. The transport process destroys the critical sodium concentration gradient needed for cell function. Nonactin is another ionophore that upsets the concentration gradient by binding strongly to potassium ions, allowing the membrane to be permeable to potassium ions, also destroying the essential concentration gradient.

The ionophore antibiotic monensin complexed with a sodium cation.

The cell membrane, in its interior, is like a hydrocarbon because it consists in this region primarily of the hydrocarbon portions of lipids (Chapter 23). Normally, cells must maintain a gradient between the concentrations of sodium and potassium ions inside and outside the cell membrane. Potassium ions are "pumped" in, and sodium ions are pumped out. This gradient is essential to the functions of nerves, transport of nutrients into the cell, and maintenance of proper cell volume. The biochemical transport of sodium and potassium ions

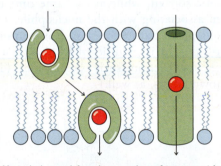

Carrier (left) and channel-forming modes of transport ionophores. (Reprinted with permission of John Wiley & Sons, Inc. from Voet, D. and Voet, J. G. *Biochemistry*, Second Edition. Copyright 1995 Voet, D., and Voet, J. G.)

*Discovery and characterization of the actual molecular pump that establishes the sodium and potassium concentration gradient (Na^+, K^+ -ATPase) earned Jens Skou (Aarhus University, Denmark) half of the 1997 Nobel Prize in Chemistry. The other half went to Paul D. Boyer (UCLA) and John E. Walker (Cambridge) for elucidating the enzymatic mechanism of ATP synthesis.

11.17 SUMMARY OF REACTIONS OF ALKENES, ALCOHOLS, AND ETHERS

Helpful Hint

Some tools for synthesis.

We have studied reactions in this chapter and in Chapter 8 that can be extremely useful in designing syntheses. Most of these reactions involving alcohols and ethers are summarized in the Summary Review Tools at the end of the chapter.

- We can use alcohols to make alkyl halides, sulfonate esters, ethers, and alkenes.
- We can oxidize alkenes to make epoxides, diols, aldehydes, ketones, and carboxylic acids (depending on the specific alkene and conditions).
- We can use alkenes to make alkanes, alcohols, and alkyl halides.
- If we have a terminal alkyne, such as could be made from an appropriate vicinal dihalide, we can use the alkynide anion derived from it to form carbon–carbon bonds by nucleophilic substitution.

All together, we have a repertoire of reactions that can be used to directly or indirectly interconvert almost all of the functional groups we have studied so far. In Section 11.17A we summarize some reactions of alkenes.

11.17A HOW TO Use Alkenes in Synthesis

- Alkenes are an entry point to virtually all of the other functional groups that we have studied.

For this reason, and because many of the reactions afford us some degree of control over the regiochemical and/or stereochemical form of the products, alkenes are versatile intermediates for synthesis.

- We have two methods to **hydrate a double bond in a Markovnikov orientation**: (1) *oxymercuration–demercuration* (Section 8.5), and (2) *acid-catalyzed hydration* (Section 8.4).

Of these methods oxymercuration–demercuration is the most useful in the laboratory because it is easy to carry out and *is not accompanied by rearrangements*.

- We can **hydrate a double bond in an anti-Markovnikov orientation** by *hydroboration–oxidation* (Section 8.6). With hydroboration–oxidation we can also achieve a *syn addition of the* H— *and* —OH *groups*.

Remember, too, **the boron group of an organoborane can be replaced by hydrogen, deuterium, or tritium** (Section 8.11), and that hydroboration, itself, involves a *syn addition of* H— *and* —B—.

- We can **add HX to a double bond in a Markovnikov sense** (Section 8.2) using HF, HCl, HBr, or HI.
- We can **add HBr in an anti-Markovnikov orientation** (Section 10.9), by treating an alkene with HBr *and a peroxide*. (The other hydrogen halides do not undergo anti-Markovnikov addition when peroxides are present.)
- We can **add bromine or chlorine to a double bond** (Section 8.12) and the addition is an *anti addition* (Section 8.13).
- We can also **add X— and —OH to a double bond** (i.e., synthesize a halohydrin) by carrying out a bromination or chlorination in water (Section 8.14). This addition, too, is an *anti addition*.
- We can carry out a **syn 1,2-dihydroxylation of a double bond** using either $KMnO_4$ in cold, dilute, and basic solution or OsO_4 followed by $NaHSO_3$ (Section 8.16). Of these two methods, the latter is preferable because of the tendency of $KMnO_4$ to overoxidize the alkene and cause cleavage at the double bond.
- We can carry out **anti 1,2-dihydroxylation of a double bond** by converting the alkene to an *epoxide* and then carrying out an acid-catalyzed hydrolysis (Section 11.15).

Equations for most of these reactions are given in the Synthetic Connections reviews for Chapters 7 and 8 and this chapter.

WHY Do These Topics Matter?]

IMPORTANT, BUT HIDDEN, EPOXIDES

Because of the strain and reactivity of epoxides, it is quite rare to isolate a compound from nature that actually contains an epoxide ring. That does not mean that this functional group does not serve diverse purposes. In fact, there are many instances where epoxides appear to play a critical role in the formation of new bonds within complex natural products. For example, if a long chain of alkenes such as that shown below could be epoxidized at every double bond in a stereocontrolled way (likely using enzymes), then subsequent activation of the terminal epoxide with a proton could potentially initiate a cascade, or domino-like, set of cyclizations leading to many new ring systems with complete stereocontrol. This process is shown here specifically for gymnocin B, one member of a large class of marine-based natural products known as cyclic polyethers. These compounds are potent neurotoxins.

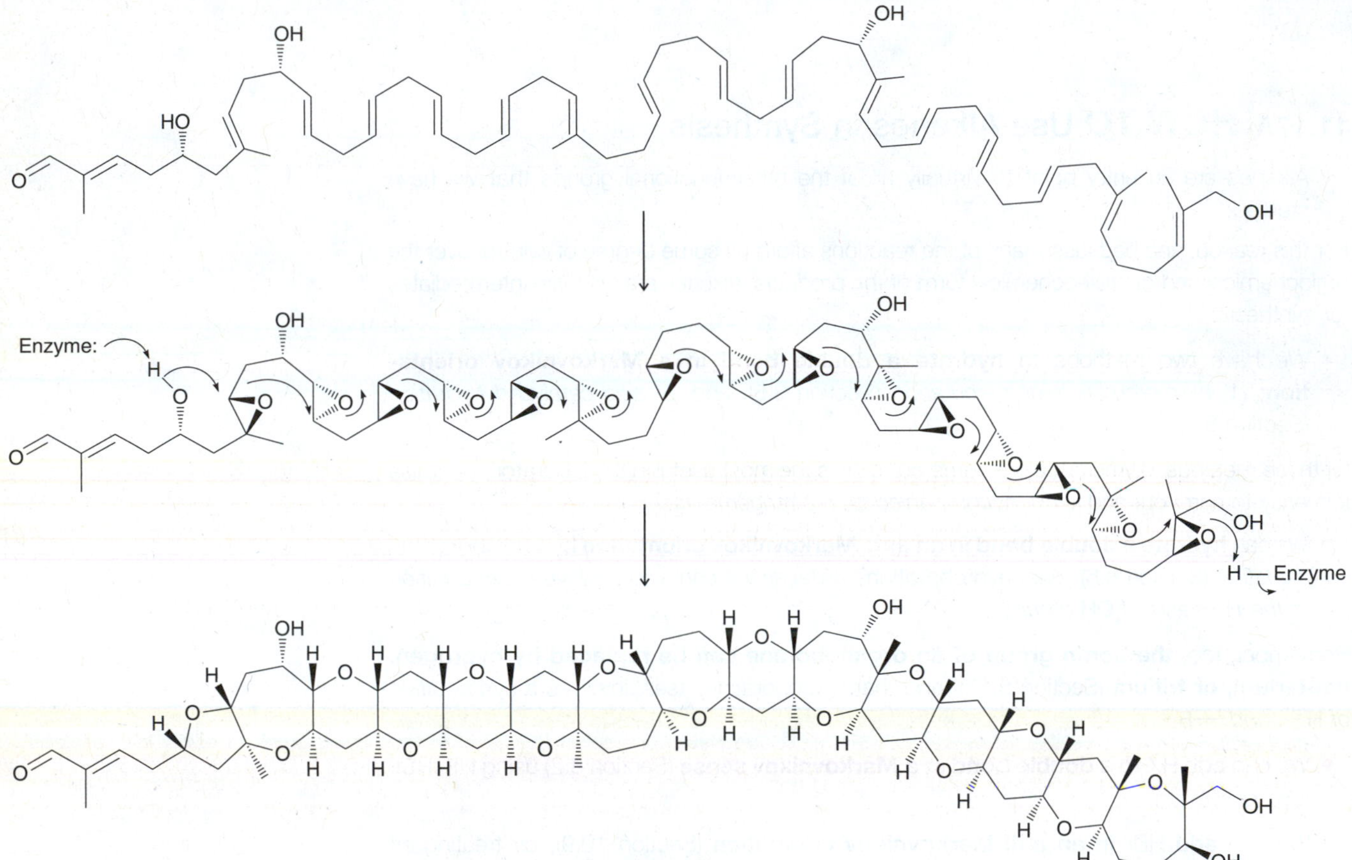

Gymnocin B

Above structure from Vilotijevic, I.; Jamison, T.F.: Epoxide-Opening Cascades Promoted by Water. SCIENCE 317:1189 (2007). Reprinted with permission from AAAS.

Epoxides also play a critical role in eliminating some dangerous molecules we might ingest, a role that again is hidden if we only look at starting materials and products. We will consider two compounds. The first is aflatoxin B_1, a compound that can contaminate peanuts and some cereal grains depending on the soil conditions where the crop was grown. The second is benzo[a]pyrene, a substance found in cigarette smoke and grilled meat (it is a component of the char marks). Aflatoxin B_1 is a carcinogen and benzo[a]pyrene can intercalate with DNA and prevent gene transcription (a topic we will discuss in more detail in Chapter 25).

Michael Patrick O'Neill/Photo Researchers, Inc.

Image Source

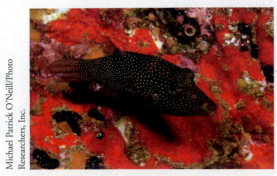

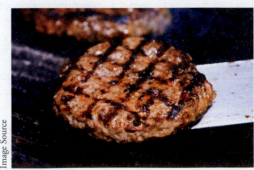

The body's system to eliminate these toxic chemicals begins by oxidizing their carbon frameworks using enzymes known as cytochrome P450s; these enzymes are found in the liver and intestines. For both aflatoxin B$_1$ and benzo[a]pyrene, at least one of their double bonds can be converted into an epoxide, as shown below. The next step is for a highly polar nucleophile, such as glutathione, to add to that reactive ring system and make the resulting molecule water soluble so it can be excreted quickly. However, these reactions are risky because other nucleophiles can attack as well. For example, nucleotide bases within DNA can also react with these epoxides. If that happens, as shown for the epoxidized form of benzo[a]pyrene, cancer can result. Thus, the epoxide in these instances is a two-edged sword—it serves as a way to remove a potentially toxic molecule while also creating a species that is sometimes even more dangerous and reactive than the original material. As a challenge question with this closing essay, why do you think the two nucleophile additions shown below occur only at the indicated positions?

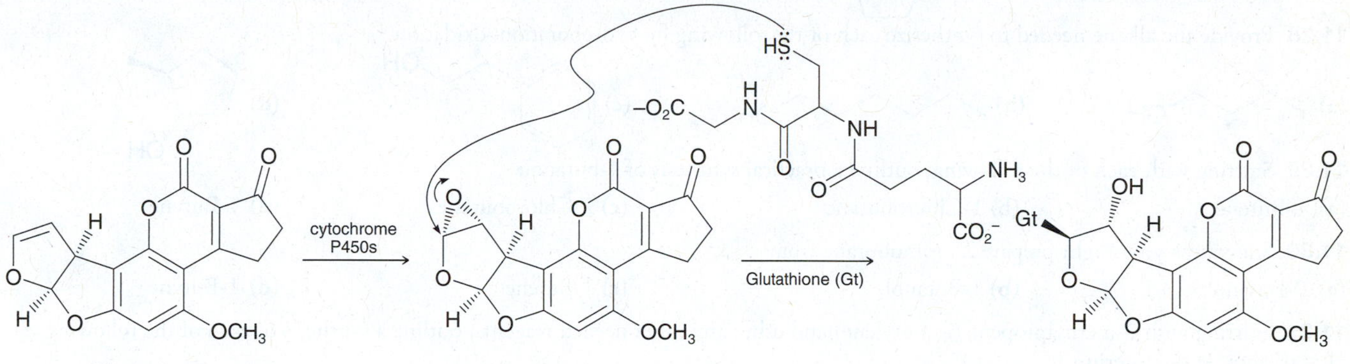

Aflatoxin B$_1$

[can be excreted]

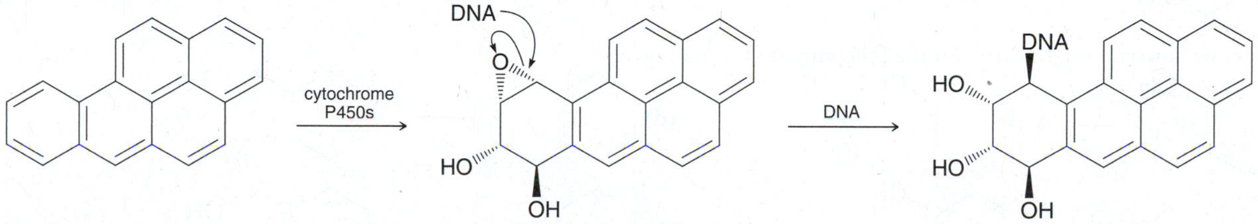

Benzo[a]pyrene

[carcinogenic]

To learn more about these topics, see:

1. Vilotijevic, I.; Jamison, T. F. "Epoxide–Opening Cascades Promoted by Water" in *Science* **2007**, *317*, 1189 and references therein.
2. Nakanishi, K. "The Chemistry of Brevetoxins: A Review" in *Toxicon* **1985**, *23*, 473–479.

SUMMARY AND REVIEW TOOLS

In addition to Section 11.17, which summarizes many of the reactions of alkenes, alcohols, and ethers, the study aids for this chapter also include key terms and concepts (which are hyperlinked to the Glossary from the bold, blue terms in the *WileyPLUS* version of the book at wileyplus.com) and a Synthetic Connections chart.

PROBLEMS WILEY PLUS+

Note to Instructors: Many of the homework problems are available for assignment via *WileyPLUS*, an online teaching and learning solution.

NOMENCLATURE

11.25 Give an IUPAC substitutive name for each of the following alcohols:

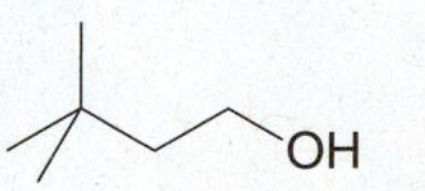

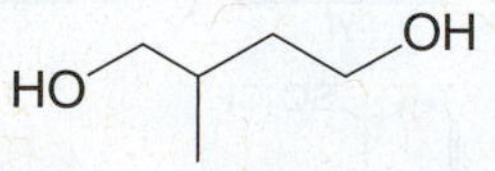

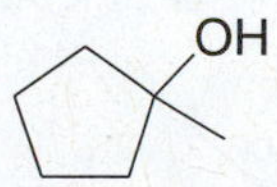

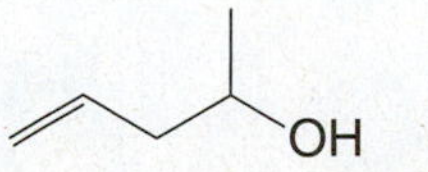

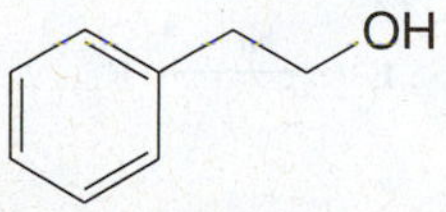

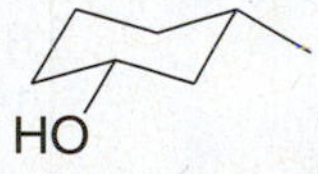

11.26 Write structural formulas for each of the following:

(a) (Z)-But-2-en-1-ol
(b) (R)-Butane-1,2,4-triol
(c) (1R,2R)-Cyclopentane-1,2-diol

(d) 1-Ethylcyclobutanol
(e) 2-Chlorohex-3-yn-1-ol
(f) Tetrahydrofuran

(g) 2-Ethoxypentane
(h) Ethyl phenyl ether
(i) Diisopropyl ether

(j) 2-Ethoxyethanol

REACTIONS AND SYNTHESIS

11.27 Provide the alkene needed to synthesize each of the following by oxymercuration–demercuration.

(a)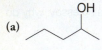
(b)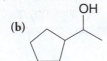
(c)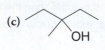
(d)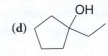

11.28 Provide the alkene needed to synthesize each of the following by hydroboration–oxidation.

(a)
(b)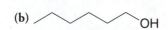
(c)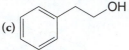
(d)

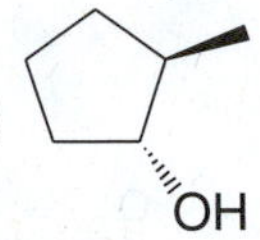

11.29 Starting with each of the following, outline a practical synthesis of 1-butanol:

(a) 1-Butene
(b) 1-Chlorobutane
(c) 2-Chlorobutane
(d) 1-Butyne

11.30 Show how you might prepare 2-bromobutane from

(a) 2-Butanol
(b) 1-Butanol
(c) 1-Butene
(d) 1-Butyne

11.31 Starting with 2-methylpropene (isobutylene) and using any other needed reagents, outline a synthesis of each of the following (T = tritium, D = deuterium):

(a)
(b)
(c)
(d)

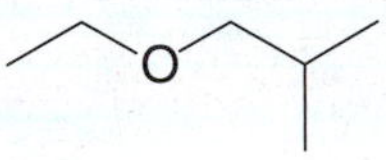

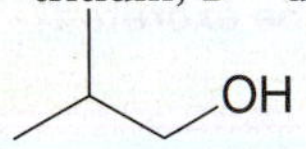

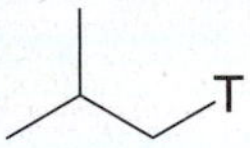

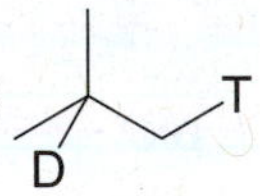

11.32 Show how you might carry out the following transformations:

(a) 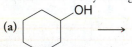 $\longrightarrow$

(b) 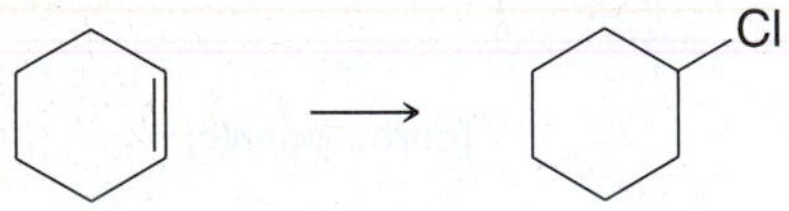 $\longrightarrow$

(c) 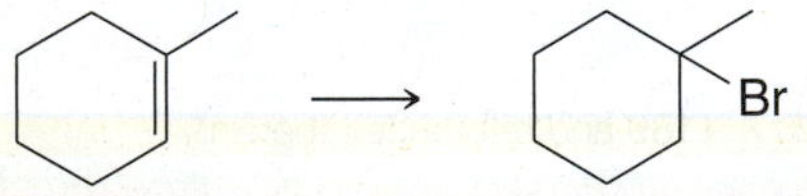 $\longrightarrow$

(d) 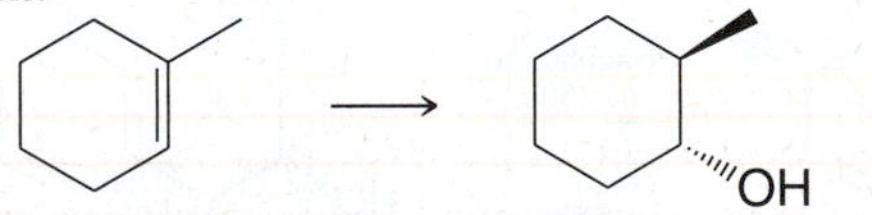 $\longrightarrow$

(e) 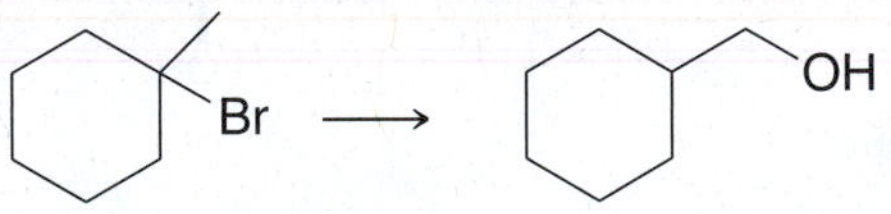 $\longrightarrow$

11.33 What compounds would you expect to be formed when each of the following ethers is refluxed with excess concentrated hydrobromic acid?

(a) 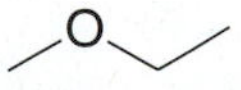
(b)
(c) 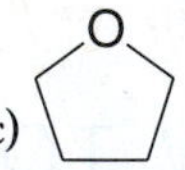 **(THF)**
(d) 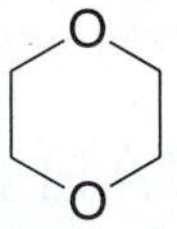 **(1,4-dioxane)**

11.34 Considering **A–L** to represent the major products formed in each of the following reactions, provide a structure for each of **A** through **L**. If more than one product can reasonably be conceived from a given reaction, include those as well.

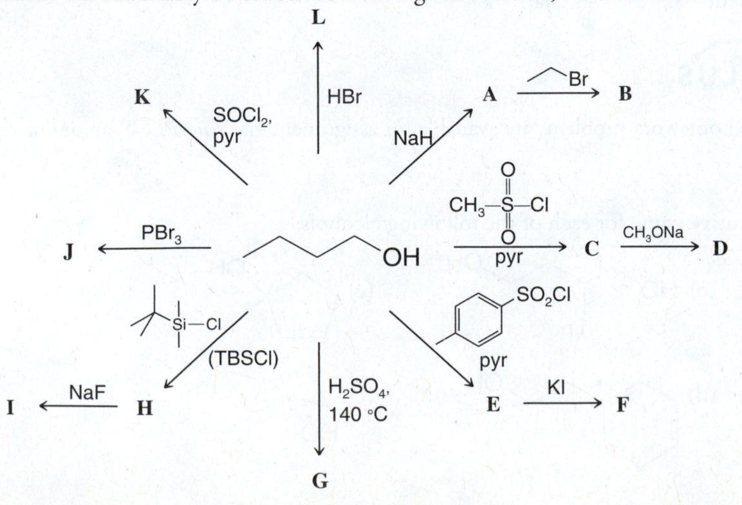

11.35 Write structures for the products that would be formed under the conditions in Problem 11.34 if cyclopentanol had been used as the starting material. If more than one product can reasonably be conceived from a given reaction, include those as well.

11.36 Starting with isobutane, show how each of the following could be synthesized. (You need not repeat the synthesis of a compound prepared in an earlier part of this problem.)

(a) *tert*-Butyl bromide
(b) 2-Methylpropene
(c) Isobutyl bromide
(d) Isobutyl iodide
(e) Isobutyl alcohol (two ways)
(f) *tert*-Butyl bromide

(g) Isobutyl methyl ether

(h)

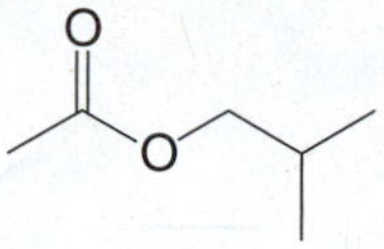

(i)

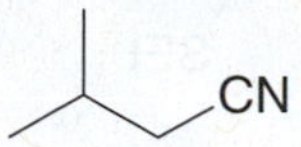

(j) 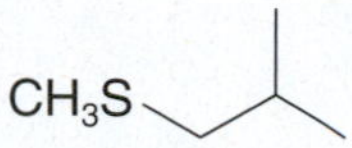 CH_3S

(k)

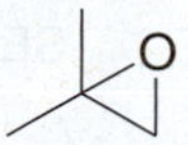

(l)

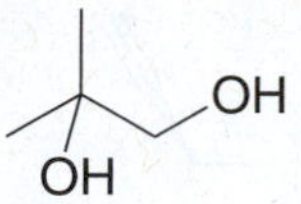

(m)

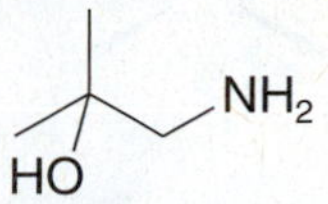

(n)

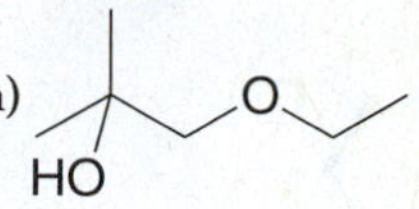

11.37 Outlined below is a synthesis of the gypsy moth sex attractant disparlure (a pheromone). Give the structure of disparlure and intermediates **A–D**.

$HC\equiv CNa$ $\xrightarrow[\text{liq. NH}_3]{\text{1-bromo-5-methylhexane}}$ **A** (C_9H_{16}) $\xrightarrow[\text{liq. NH}_3]{\text{NaNH}_2}$ **B** $(C_9H_{15}Na)$

$\xrightarrow{\text{1–bromodecane}}$ **C** $(C_{19}H_{36})$ $\xrightarrow[\text{Ni}_2\text{B(P-2)}]{\text{H}_2}$ **D** $(C_{19}H_{38})$ $\xrightarrow{\text{mCPBA}}$ **Disparlure** $(C_{19}H_{38}O)$

11.38 Provide the reagents necessary for the following syntheses. More than one step may be required.

(a)

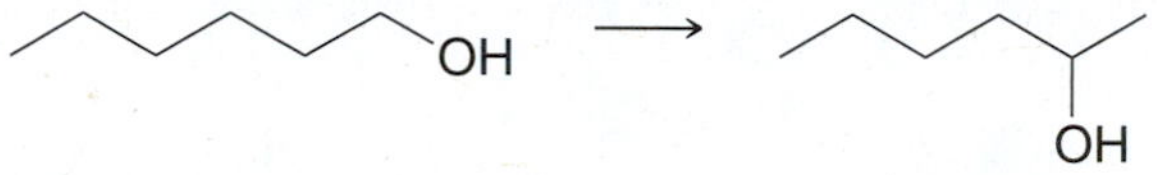

(b)

(c)

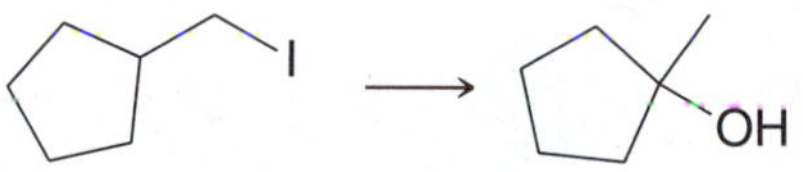

(d)

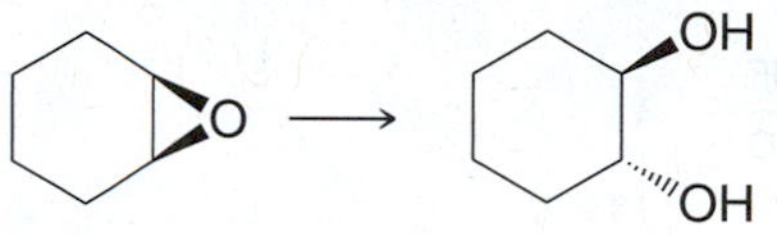

(e)

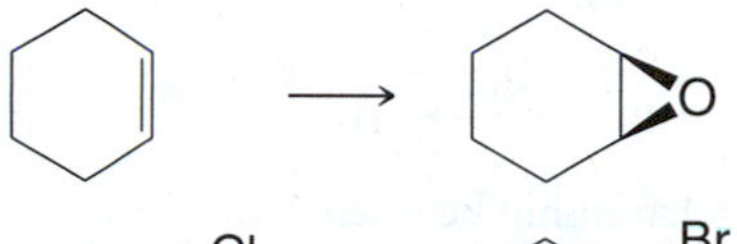

(f)

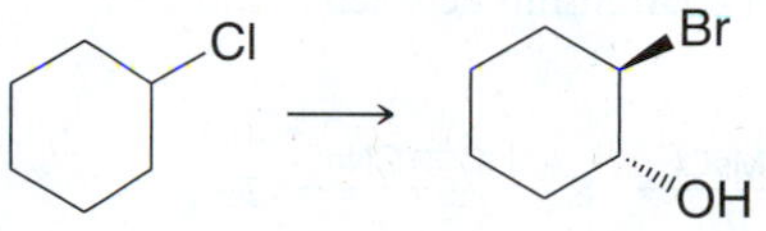

11.39 Predict the major product from each of the following reactions.

(a) 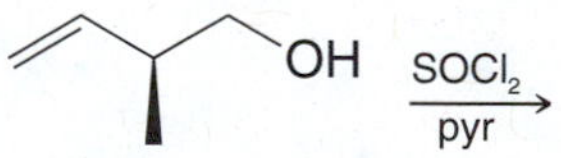 $\xrightarrow[\text{pyr}]{\text{SOCl}_2}$

(b) 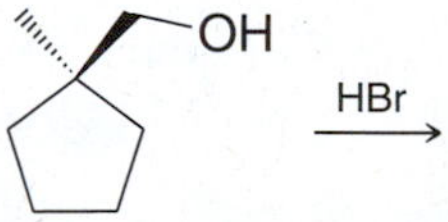 $\xrightarrow{\text{HBr}}$

(c) 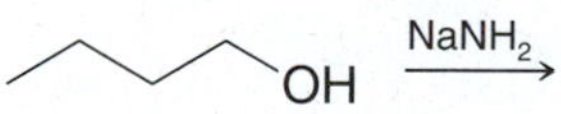 $\xrightarrow{\text{NaNH}_2}$

(d) 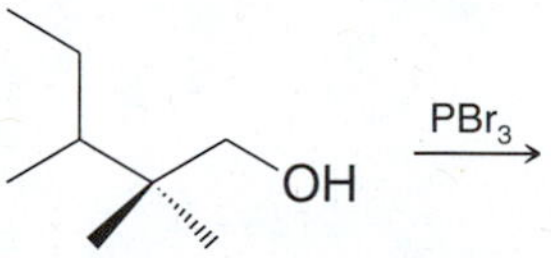 $\xrightarrow{\text{PBr}_3}$

(e) 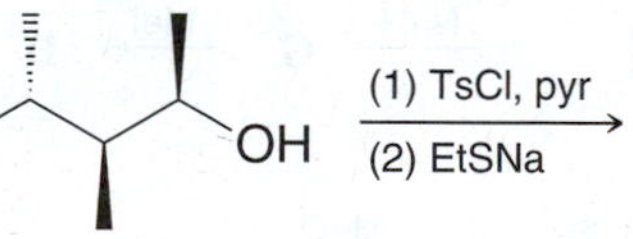 $\xrightarrow[\text{(2) EtSNa}]{\text{(1) TsCl, pyr}}$

(f)  $\xrightarrow{\text{NaI, H}_2\text{SO}_4}$

11.40 Predict the products from each of the following reactions.

(a) 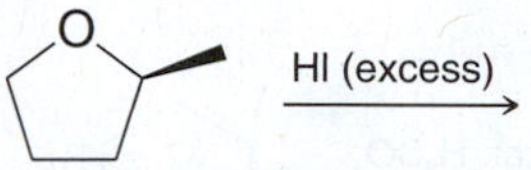 $\xrightarrow{\text{HI (excess)}}$

(b) 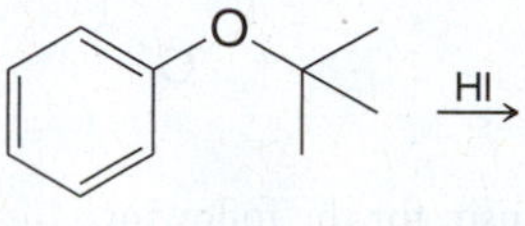 $\xrightarrow{\text{HI}}$

(c) 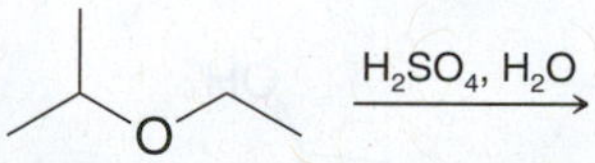 $\xrightarrow{\text{H}_2\text{SO}_4, \text{H}_2\text{O}}$

(d) 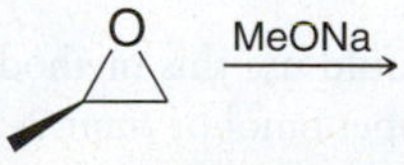 $\xrightarrow{\text{MeONa}}$

(e) 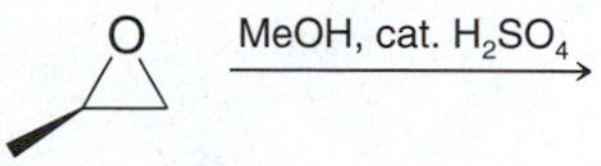 $\xrightarrow{\text{MeOH, cat. H}_2\text{SO}_4}$

(f) 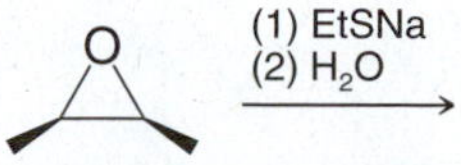 $\xrightarrow[\text{(2) H}_2\text{O}]{\text{(1) EtSNa}}$

(g) 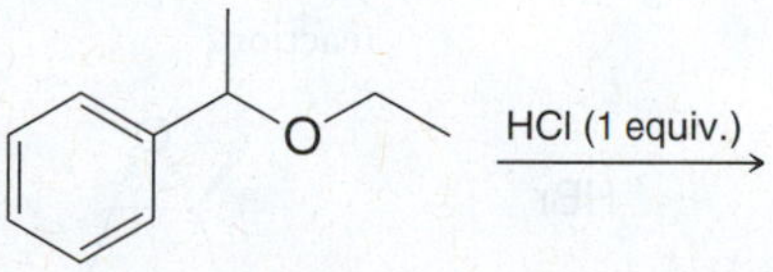 $\xrightarrow{\text{HCl (1 equiv.)}}$

(h) $\xrightarrow{\text{MeONa}}$

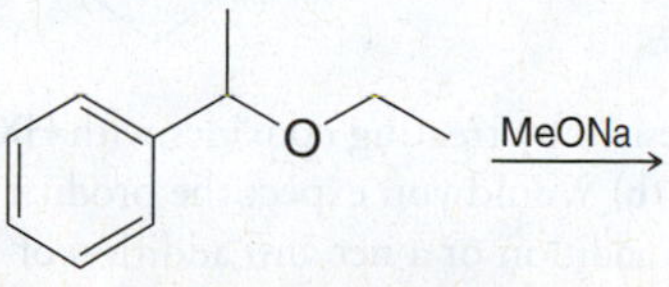

(i) 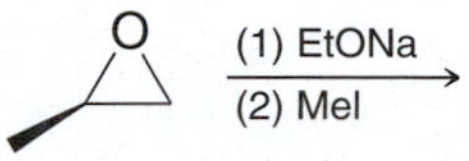 $\xrightarrow[\text{(2) MeI}]{\text{(1) EtONa}}$

(j) 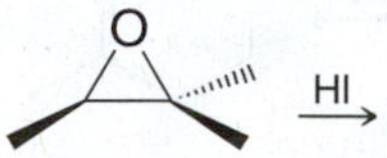 $\xrightarrow{\text{HI}}$

11.41 Provide the reagents necessary to accomplish the following syntheses.

(a) 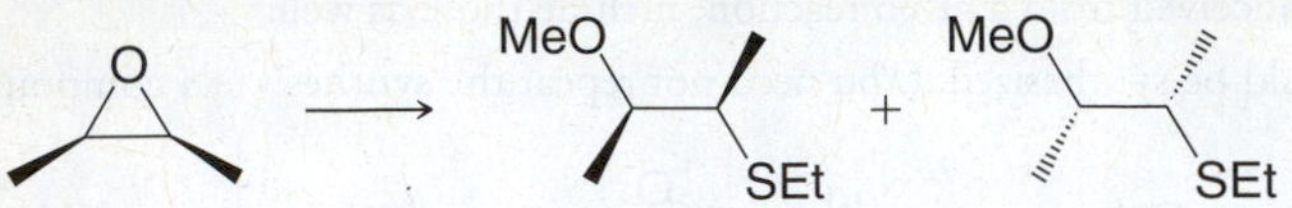
O
MeO
SEt
+
MeO
SEt

(c) 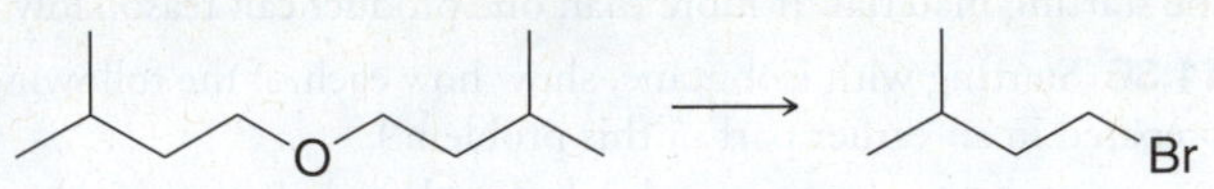
O
Br

(b) 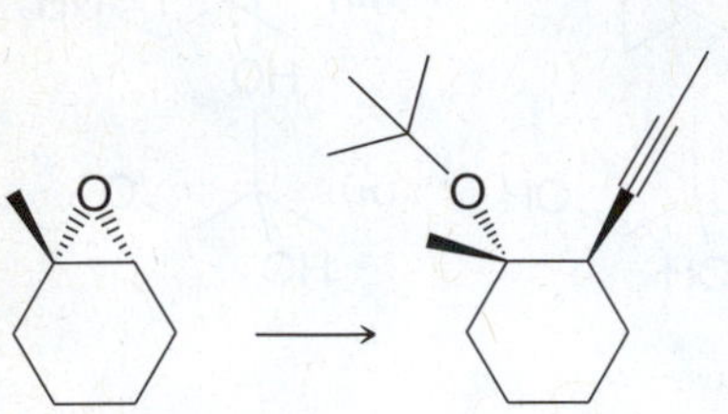
O
O

(d) 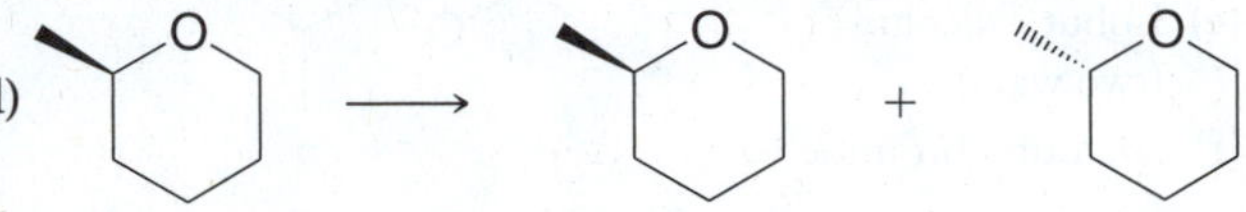
O
O
+
O

11.42 Provide reagents that would accomplish the follwing syntheses.

(a) 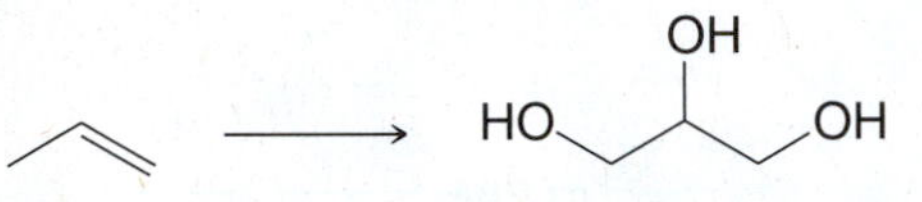
OH
HO
OH

Glycerol

(b) 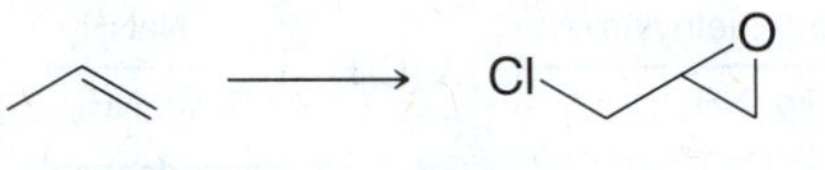
Cl
O

Epichlorohydrin

11.43 Write structures for compounds **A–J** showing stereochemistry where appropriate.

(a) 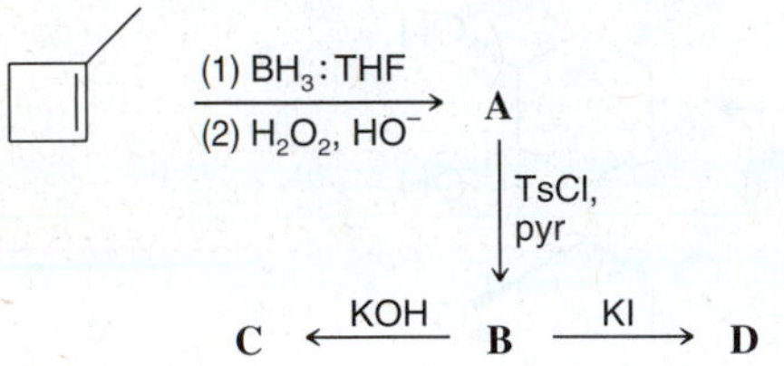
(1) BH3:THF
(2) H2O2, HO−
A
TsCl,
pyr
C
KOH
B
KI
D

What is the stereochemical relationship between **A** and **C**?

(b) 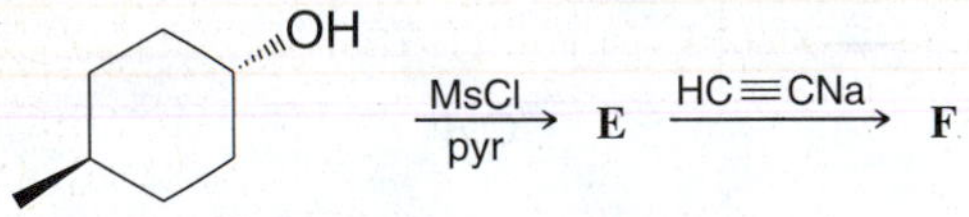
OH
MsCl
pyr
E
HC≡CNa
F

(c) 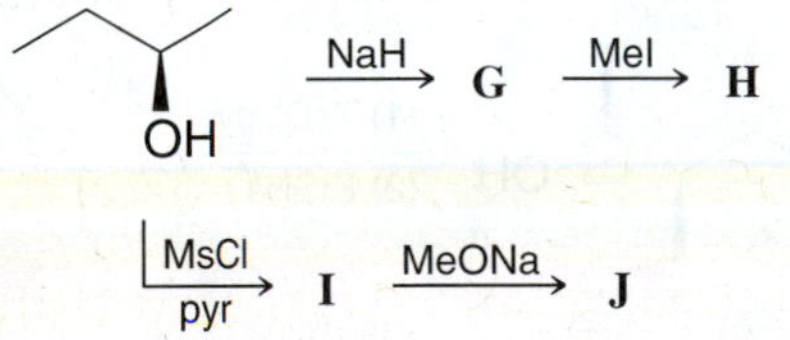
OH
NaH
G
MeI
H
MsCl
pyr
I
MeONa
J

What is the stereochemical relationship between **H** and **J**?

MECHANISMS

11.44 Write a mechanism that accounts for the following reaction:

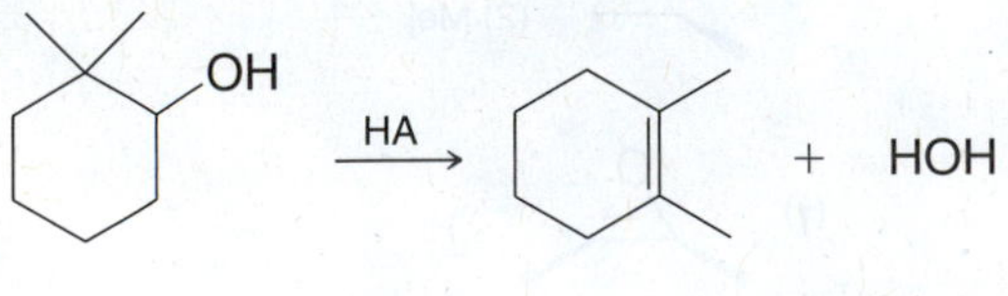
OH
HA
+ HOH

11.45 Propose a reasonable mechanism for the following reaction.

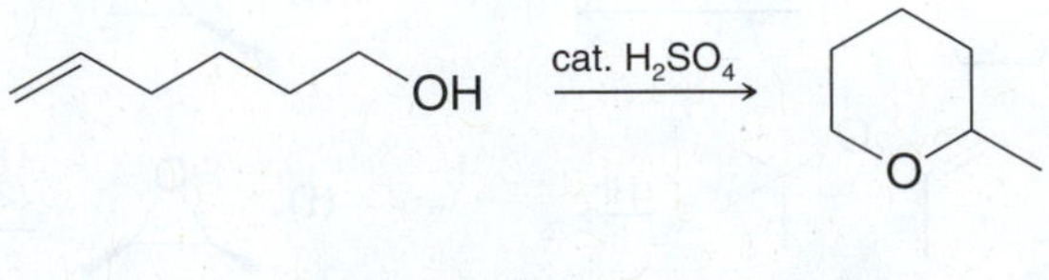
OH
cat. H2SO4
O

11.46 Propose a reasonable mechanism for the following reaction.

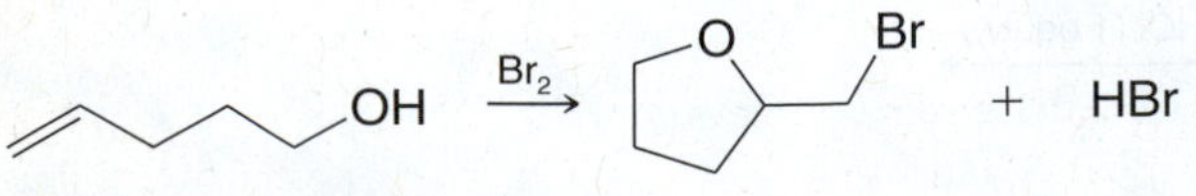
OH
Br2
O
Br
+ HBr

11.47 Propose a reasonable mechanism for the following reaction.

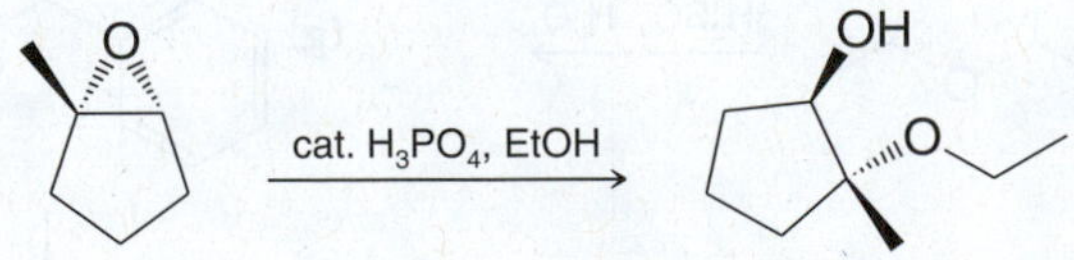
O
cat. H3PO4, EtOH
OH
O

11.48 Vicinal halo alcohols (halohydrins) can be synthesized by treating epoxides with HX. (**a**) Show how you would use this method to synthesize 2-chlorocyclopentanol from cyclopentene. (**b**) Would you expect the product to be *cis*-2-chlorocyclopentanol or *trans*-2-chlorocyclopentanol; that is, would you expect a net syn addition or a net anti addition of —Cl and —OH? Explain.

11.49 Base-catalyzed hydrolysis of the 1,2-chlorohydrin **1** is found to give a chiral glycol **2** with retention of configuration. Propose a reasonable mechanism that would account for this transformation. Include all formal charges and arrows showing the movement of electrons.

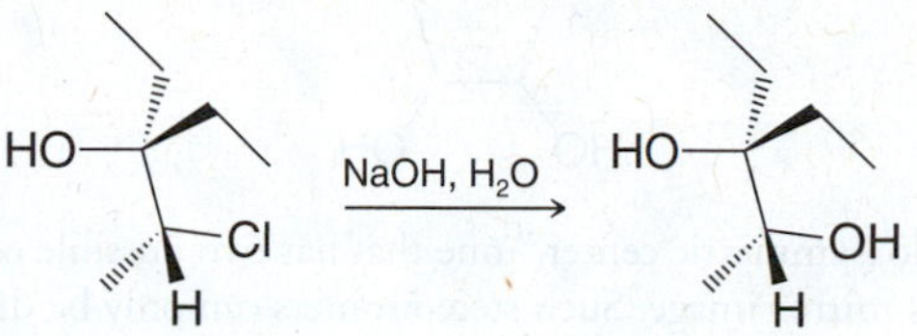

1 **2**

11.50 Compounds of the type 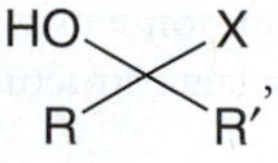

, called α-haloalcohols, are unstable and cannot be isolated. Propose a mechanistic explanation for why this is so.

11.51 While simple alcohols yield alkenes on reaction with dehydrating acids, diols form carbonyl compounds. Rationalize mechanistically the outcome of the following reaction:

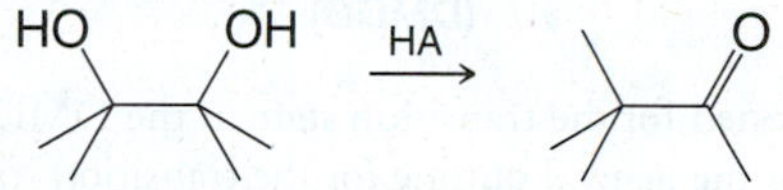

11.52 When the bicyclic alkene **I**, a *trans*-decalin derivative, reacts with a peroxy acid, **II** is the major product. What factor favors the formation of **II** in preference to **III**? (You may find it helpful to build a handheld molecular model.)

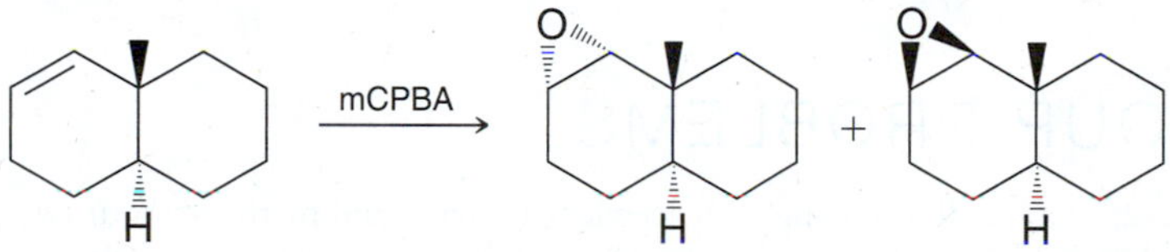

I **II (major)** **III (minor)**

11.53 Use Newman projection formulas for ethylene glycol (1,2-ethanediol) and butane to explain why the gauche conformer of ethylene glycol is expected to contribute more to its ensemble of conformers than would the gauche conformer of butane to its respective set of conformers.

CHALLENGE PROBLEMS

11.54 When the 3-bromo-2-butanol with the stereochemical structure **A** is treated with concentrated HBr, it yields *meso*-2,3-dibromobutane; a similar reaction of the 3-bromo-2-butanol **B** yields (±)-2,3-dibromobutane. This classic experiment performed in 1939 by S. Winstein and H. J. Lucas was the starting point for a series of investigations of what are called *neighboring group effects*. Propose mechanisms that will account for the stereochemistry of these reactions.

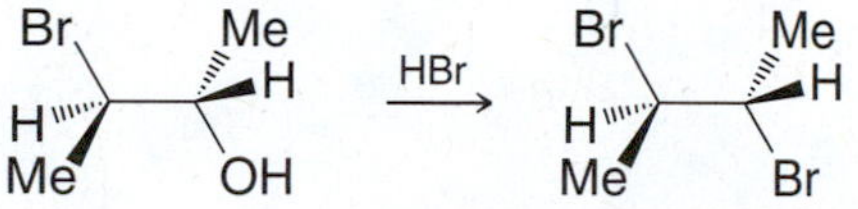

A **2,3-Dibromobutane (meso)**

versus

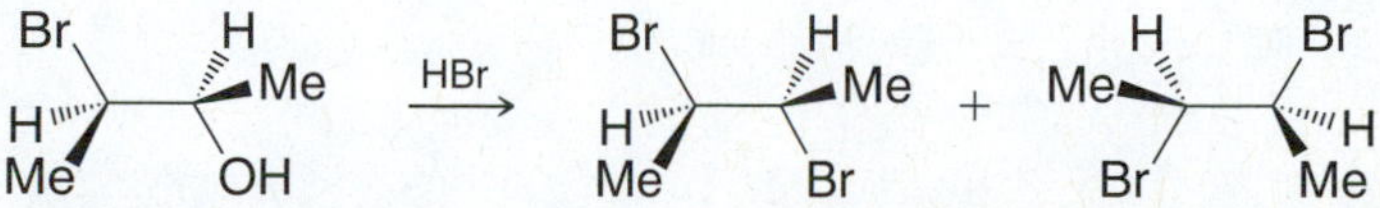

B **2,3-Dibromobutane (racemic)**

11.55 Reaction of an alcohol with thionyl chloride in the presence of a tertiary amine (e.g., pyridine) affords replacement of the OH group by Cl *with inversion of configuration* (Section 11.9). However, if the amine is omitted, the result is usually replacement with retention of configuration. The same chlorosulfite intermediate is involved in both cases. Suggest a mechanism by which this intermediate can give the chlorinated product without inversion.

11.56 Draw all of the stereoisomers that are possible for 1,2,3-cyclopentanetriol. Label their chirality centers and say which are enantiomers and which are diastereomers.

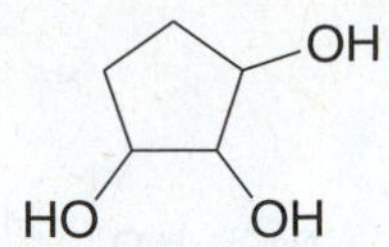

(*Hint*: Some of the isomers contain a "pseudoasymmetric center," one that has two possible configurations, each affording a different stereoisomer, each of which is identical to its mirror image. Such stereoisomers can only be distinguished by the order of attachment of **R** versus **S** groups at the pseudoasymmetric center. Of these the **R** group is given higher priority than the **S**, and this permits assignment of configuration as **r** or **s**, lowercase letters being used to designate the pseudoasymmetry.)

11.57 Dimethyldioxirane (DMDO), whose structure is shown below, is another reagent commonly used for alkene epoxidation. Write a mechanism for the epoxidation of (*Z*)-2-butene by DMDO, including a possible transition state structure. What is the by-product of a DMDO epoxidation?

Dimethyldioxirane
(DMDO)

11.58 Two configurations can actually be envisioned for the transition state in the DMDO epoxidation of (*Z*)-2-butene, based on analogy with geometric possibilities fitting within the general outline for the transition state in a peroxycarboxylic acid epoxidation of (*Z*)-2-butene. Draw these geometries for the DMDO epoxidation of (*Z*)-2-butene. Then, open the molecular models on the book's website for these two possible transition state geometries in the DMDO epoxidation of (*Z*)-2-butene and speculate as to which transition state would be lower in energy.

LEARNING GROUP PROBLEMS

1. Devise two syntheses for *meso*-2,3-butanediol starting with acetylene (ethyne) and methane. Your two pathways should take different approaches during the course of the reactions for controlling the origin of the stereochemistry required in the product.

2. (a) Write as many chemically reasonable syntheses as you can think of for ethyl 2-methylpropyl ether (ethyl isobutyl ether). Be sure that at some point in one or more of your syntheses you utilize the following reagents (not all in the same synthesis, however): PBr$_3$, SOCl$_2$, *p*-toluenesulfonyl chloride (tosyl chloride), NaH, ethanol, 2-methyl-1-propanol (isobutyl alcohol), concentrated H$_2$SO$_4$, Hg(OAc)$_2$, ethene (ethylene).
(b) Evaluate the relative merits of your syntheses on the basis of selectivity and efficiency. (Decide which ones could be argued to be the "best" syntheses and which might be "poorer" syntheses.)

3. Synthesize the compound shown below from methylcyclopentane and 2-methylpropane using those compounds as the source of the carbon atoms and any other reagents necessary. Synthetic tools you might need include Markovnikov or anti-Markovnikov hydration, Markovnikov or anti-Markovnikov hydrobromination, radical halogenation, elimination, and nucleophilic substitution reactions.

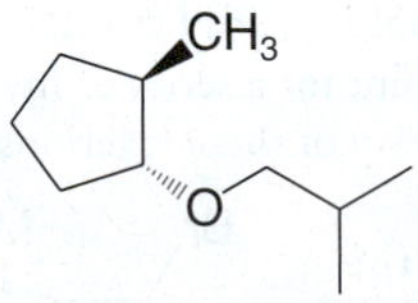

[SUMMARY AND REVIEW TOOLS]

Some Synthetic Connections of Alkenes, Alkynes, Alcohols, Alkyl Halides, and Ethers

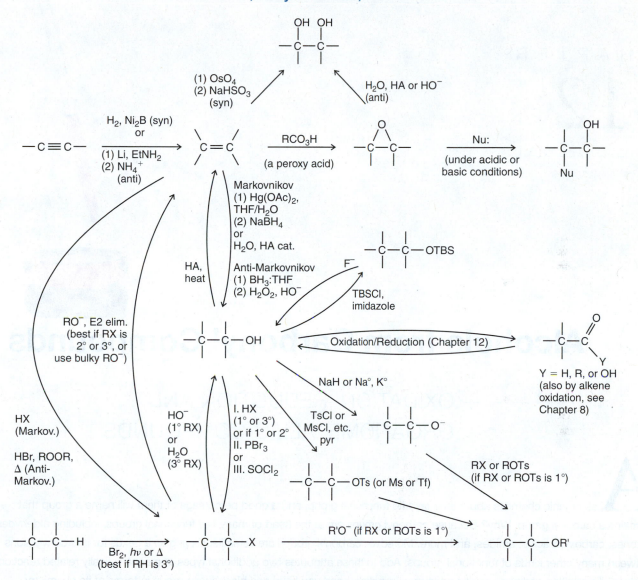

- Alkynes to alkenes
- Alkenes and alcohols
- Alcohols and alkyl halides
- Alkenes and alkyl halides
- Alcohols and ethers
- Alkenes, epoxides, and 1,2-diols
- Alkanes to alkyl halides
- Alcohol silyl protecting group
- Alcohols to carbonyl compounds

CHAPTER

12

Alcohols from Carbonyl Compounds

OXIDATION–REDUCTION AND
ORGANOMETALLIC COMPOUNDS

Ask organic chemists about their favorite functional group, and a good percentage of them will name a group that contains a carbonyl group. Why? Because carbonyl groups are at the heart of many key functional groups, including aldehydes, ketones, carboxylic acids, amides, and more. Moreover, carbonyl groups are versatile, serving as a nexus for interconversions between many other kinds of functional groups. Add to these attributes two additional types of mechanistically related reactions— nucleophilic addition and nucleophilic addition-elimination—and you have one blockbuster group in terms of its chemistry.

As we have seen in earlier chapters, carbonyl groups are essential components of many natural compounds, they are intrinsic to some important synthetic materials, such as nylon, and they are central to the organic chemistry of life, whether in the form of carbohydrates or DNA, or in key biochemical processes.

IN THIS CHAPTER WE WILL CONSIDER:

- the structure and reactivity of carbonyl compounds
- the interconversion of carbonyl functional groups and alcohols through oxidation-reduction reactions
- the formation of new C—C bonds by reaction of certain carbonyl groups with organometallic reagents

[**WHY** DO THESE TOPICS MATTER?]   At the end of the chapter, we will see how the simple change from an alcohol to a ketone and back can fundamentally change the properties and uses of a molecule by looking at a few cases where such reactions occur in nature.

PHOTO CREDIT: FSTOP/Image Source Limited

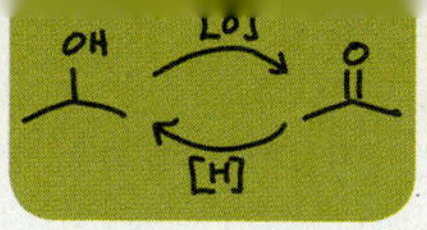

12.1 STRUCTURE OF THE CARBONYL GROUP

Carbonyl compounds are a broad group of compounds that includes aldehydes, ketones, carboxylic acids, esters, and amides.

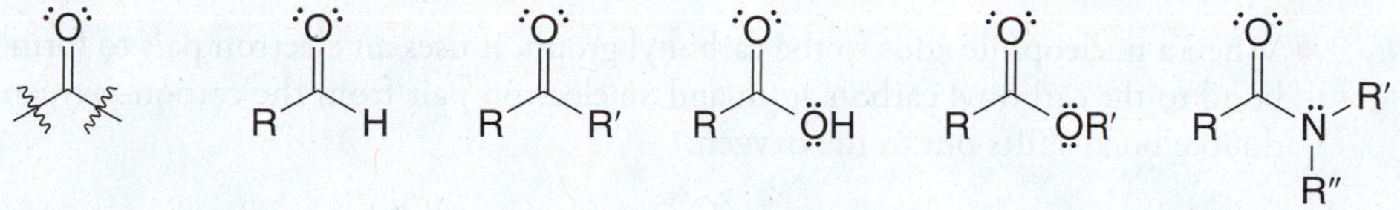

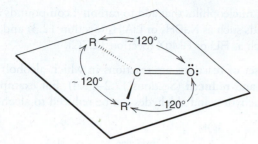

The carbonyl group   **An aldehyde**   **A ketone**   **A carboxylic acid**   **An ester**   **An amide**

The carbonyl carbon atom is sp^2 hybridized; thus, it and the three atoms attached to it lie in the same plane. The bond angles between the three attached atoms are what we would expect of a trigonal planar structure; they are approximately 120°:

The carbon–oxygen double bond consists of two electrons in a σ bond and two electrons in a π bond. The π bond is formed by overlap of the carbon p orbital with a p orbital from the oxygen atom. The electron pair in the π bond occupies both lobes (above and below the plane of the σ bonds).

Formaldehyde

The π bonding molecular orbital of formaldehyde. The electron pair of the π bond occupies both lobes.

- The more electronegative oxygen atom strongly attracts the electrons of both the σ bond and the π bond, causing the carbonyl group to be highly polarized; the carbon atom bears a substantial positive charge and the oxygen atom bears a substantial negative charge.

Polarization of the π bond can be represented by the following resonance structures for the carbonyl group:

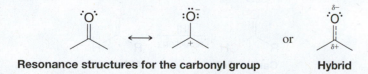

Resonance structures for the carbonyl group   or   **Hybrid**

Evidence for the polarity of the carbon–oxygen bond can be found in the rather large dipole moments associated with carbonyl compounds.

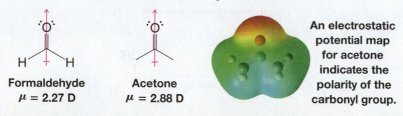

Formaldehyde $\mu = 2.27$ D   **Acetone** $\mu = 2.88$ D   An electrostatic potential map for acetone indicates the polarity of the carbonyl group.

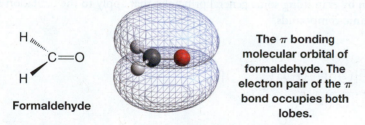

12.1A Reactions of Carbonyl Compounds with Nucleophiles

One of the most important reactions of carbonyl compounds is **nucleophilic addition to the carbonyl group**. The carbonyl group is susceptible to nucleophilic attack because, as we have just seen, the carbonyl carbon bears a partial positive charge.

- When a nucleophile adds to the carbonyl group, it uses an electron pair to form a bond to the carbonyl carbon atom and an electron pair from the carbon–oxygen double bond shifts out to the oxygen:

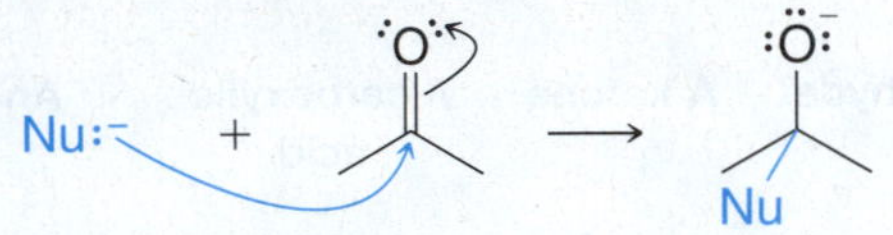

As the reaction takes place, the carbon atom undergoes a change from trigonal planar geometry and sp^2 hybridization to tetrahedral geometry and sp^3 hybridization.

- Two important nucleophiles that add to carbonyl compounds are **hydride ions** from compounds such as $NaBH_4$ or $LiAlH_4$ (Section 12.3) and **carbanions** from compounds such as RLi or RMgX (Section 12.7C).

Another related set of reactions are reactions in which alcohols and carbonyl compounds are **oxidized** and **reduced** (Sections 12.2–12.4). For example, primary alcohols can be oxidized to aldehydes, and aldehydes can be reduced to alcohols:

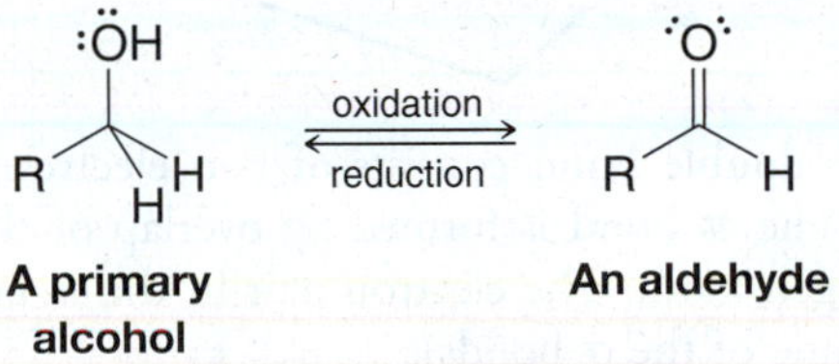

A primary alcohol ⇌ (oxidation / reduction) **An aldehyde**

Let us begin by examining some general principles that apply to the oxidation and reduction of organic compounds.

12.2 OXIDATION–REDUCTION REACTIONS IN ORGANIC CHEMISTRY

- **Reduction** of an organic molecule usually corresponds to increasing its hydrogen content or to decreasing its oxygen content.

For example, converting a carboxylic acid to an aldehyde is a reduction because the oxygen content is decreased:

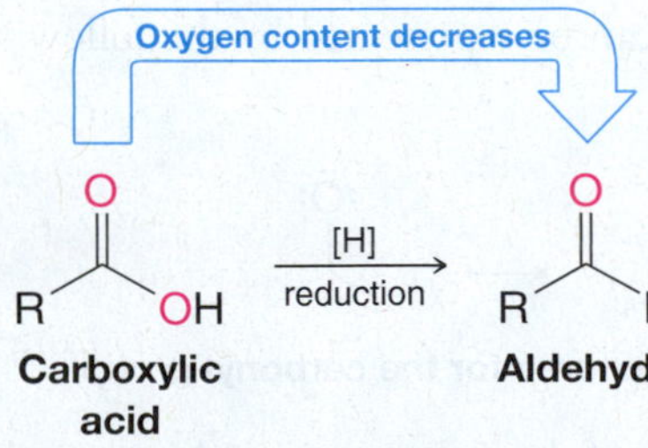

Oxygen content decreases

Carboxylic acid —[H] reduction→ **Aldehyde**

Converting an aldehyde to an alcohol is a reduction:

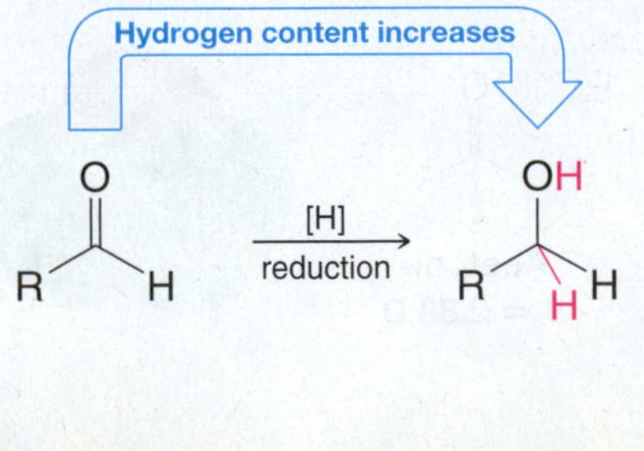

Hydrogen content increases

Converting an alcohol to an alkane is also a reduction:

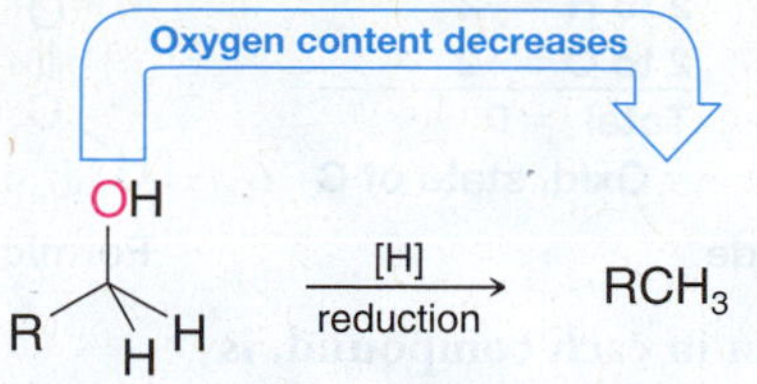

Oxygen content decreases

In these examples we have used the symbol [H] to indicate that a reduction of the organic compound has taken place. We do this when we want to write a general equation without specifying what the reducing agent is.

- The opposite of reduction is **oxidation**. Increasing the oxygen content of an organic molecule or decreasing its hydrogen content is an **oxidation**.

The reverse of each reaction that we have just given is an oxidation of the organic molecule, and we can summarize these oxidation–reduction reactions as shown below. We use the symbol [O] to indicate in a general way that the organic molecule has been oxidized.

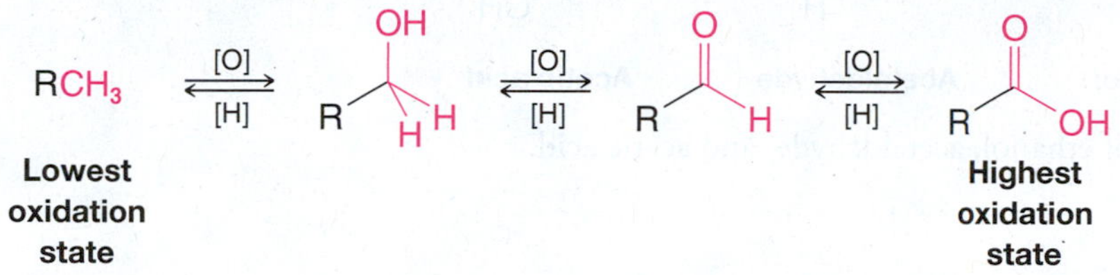

Lowest oxidation state   **Highest oxidation state**

- Oxidation of an organic compound may be more broadly defined as a reaction that increases its content of any element more electronegative than carbon.

For example, replacing hydrogen atoms by chlorine atoms is an oxidation:

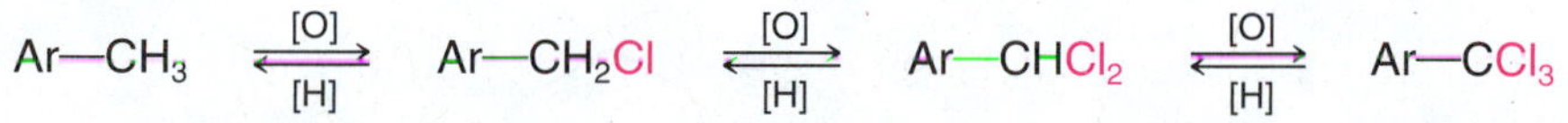

$$Ar\text{—}CH_3 \underset{[H]}{\overset{[O]}{\rightleftharpoons}} Ar\text{—}CH_2Cl \underset{[H]}{\overset{[O]}{\rightleftharpoons}} Ar\text{—}CHCl_2 \underset{[H]}{\overset{[O]}{\rightleftharpoons}} Ar\text{—}CCl_3$$

Of course, when an organic compound is reduced, something else—the **reducing agent**—must be oxidized. And when an organic compound is oxidized, something else—the **oxidizing agent**—is reduced. These oxidizing and reducing agents are often inorganic compounds, and in the next two sections we shall see what some of them are.

12.2A Oxidation States in Organic Chemistry

One method for assigning oxidation states in organic compounds is similar to the method we used for assigning formal charges (Section 1.7). We base the assignment on **the groups attached to the carbon (or carbons) whose oxidation state undergoes change in the reaction we are considering**. Recall that with formal charges we assumed that electrons in covalent bonds are shared equally. *When assigning oxidation states to carbon atoms we assign electrons to the more electronegative element* (see Section 1.4A and Table 1.2). For example, a bond to hydrogen (or to any atom less electronegative than carbon) makes that carbon negative by one unit (−1), and a bond to oxygen, nitrogen, or a halogen (F, Cl, and Br) makes the carbon positive by one unit (+1). A bond to another carbon does not change its oxidation state.

Using this method the carbon atom of methane, for example, is assigned an oxidation state of −4, and that of carbon dioxide, +4.

Helpful Hint

Note the general interpretation of oxidation–reduction regarding organic compounds.

Helpful Hint

A method for balancing organic oxidation–reduction reactions is described in the Study Guide that accompanies this text.

●●● **SOLVED PROBLEM 12.1**

Using the method just described, assign oxidation states to the carbon atoms of methanol (CH_3OH), formaldehyde (HCHO), and formic acid (HCO_2H) and arrange these compounds along with carbon dioxide and methane (see above) in order of increasing oxidation state.

STRATEGY AND ANSWER: We calculate the oxidation state of each carbon based on the number of bonds it is forming to atoms more (or less) electronegative than carbon.

(continues on next page)

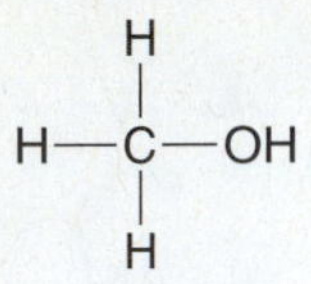

Methanol	Formaldehyde	Formic acid
H—C—OH (with H above, H below)	O=C with H, H	O=C with H, OH
Bonds to C	**Bonds to C**	**Bonds to C**
3 to H = −3	2 to H = −2	1 to H = −1
1 to O = +1	2 to O = +2	3 to O = +3
Total = −2	Total = 0	Total = +2
= Oxid. state of C	= Oxid. state of C	= Oxid. state of C

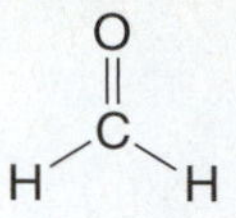

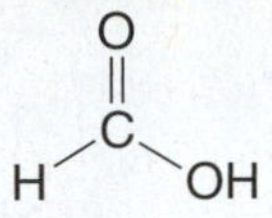

The order overall, based on the oxidation state of carbon in each compound, is

$$CH_4 = -4 < CH_3OH = -2 < HCHO = 0 < HCOOH = +2 < CO_2 = +4$$

Lowest
Oxid. state
of carbon

Highest
Oxid. state
of carbon

● ● ●

PRACTICE PROBLEM 12.1

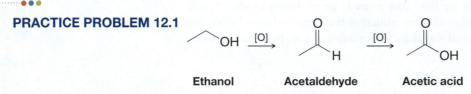

Ethanol   [O]→   **Acetaldehyde**   [O]→   **Acetic acid**

Assign oxidation states to each carbon of ethanol, acetaldehyde, and acetic acid.

● ● ●

PRACTICE PROBLEM 12.2 **(a)** Although we have described the hydrogenation of an alkene as an addition reaction, organic chemists often refer to it as a "reduction." Use the method described in Section 12.2A for assigning oxidation states to explain this usage of the word "reduction."

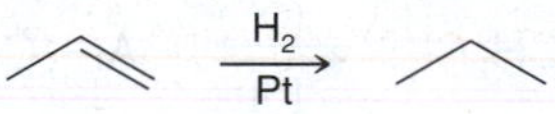

$$\overset{\text{H}_2}{\underset{\text{Pt}}{\longrightarrow}}$$

(b) Provide a similar analysis for this reaction:

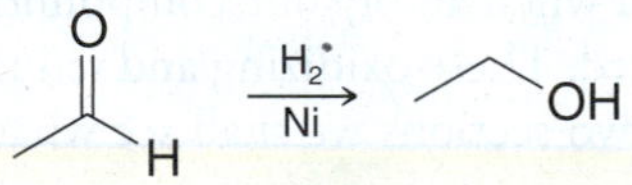

$$\overset{\text{H}_2^*}{\underset{\text{Ni}}{\longrightarrow}} \quad \text{OH}$$

12.3 ALCOHOLS BY REDUCTION OF CARBONYL COMPOUNDS

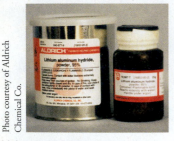

Unless special precautions are taken, lithium aluminum hydride reductions can be very dangerous. You should consult an appropriate laboratory manual before attempting such a reduction, and the reaction should be carried out on a small scale.

Primary and secondary alcohols can be synthesized by the **reduction** of a variety of compounds that contain the carbonyl group. Several general examples are shown here:

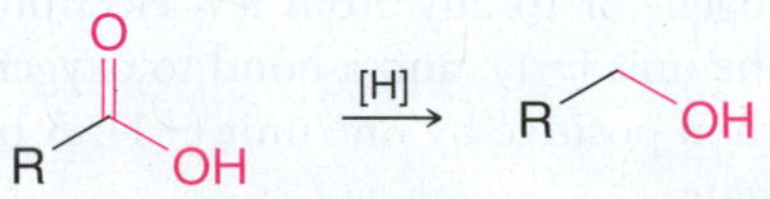

$$\underset{\textbf{Carboxylic acid}}{R-\overset{O}{\overset{\|}{C}}-OH} \xrightarrow{[H]} \underset{\textbf{1° Alcohol}}{R-OH}$$

$$\underset{\textbf{Ester}}{R-\overset{O}{\overset{\|}{C}}-OR'} \xrightarrow{[H]} \underset{\textbf{1° Alcohol}}{R-OH} \quad (+\ R'OH)$$

$$\underset{\textbf{Aldehyde}}{R-\overset{O}{\overset{\|}{C}}-H} \xrightarrow{[H]} \underset{\textbf{1° Alcohol}}{R-OH}$$

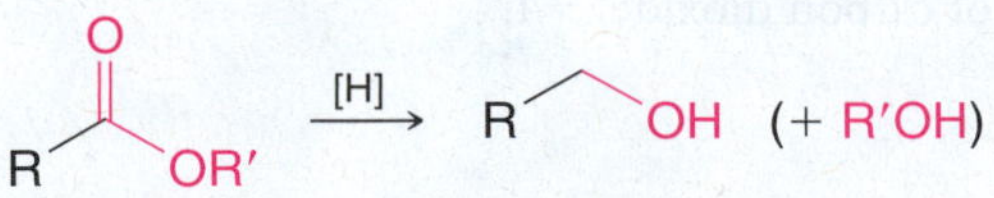

$$\underset{\textbf{Ketone}}{R-\overset{O}{\overset{\|}{C}}-R'} \xrightarrow{[H]} \underset{\textbf{2° Alcohol}}{R-\overset{OH}{\underset{}{C}}-R'}$$

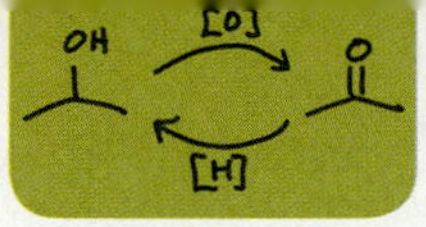

12.3A Lithium Aluminum Hydride

- Lithium aluminum hydride (LiAlH$_4$, abbreviated LAH) reduces carboxylic acids and esters to primary alcohols.

An example of lithium aluminum hydride reduction is the conversion of 2,2-dimethylpropanoic acid to 2,2-dimethylpropanol (neopentyl alcohol).

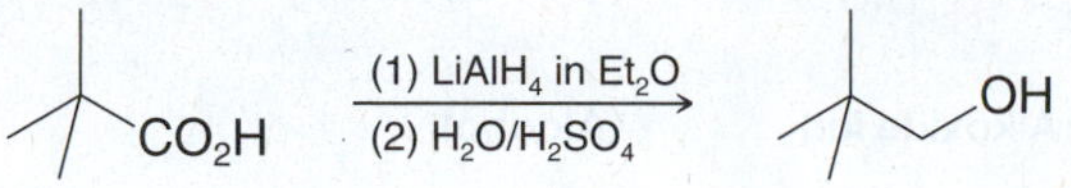

2,2-Dimethylpropanoic acid → **Neopentyl alcohol (92%)**

LAH reduction of a carboxylic acid

LAH reduction of an ester yields two alcohols, one derived from the carbonyl part of the ester group, and the other from the alkoxyl part of the ester.

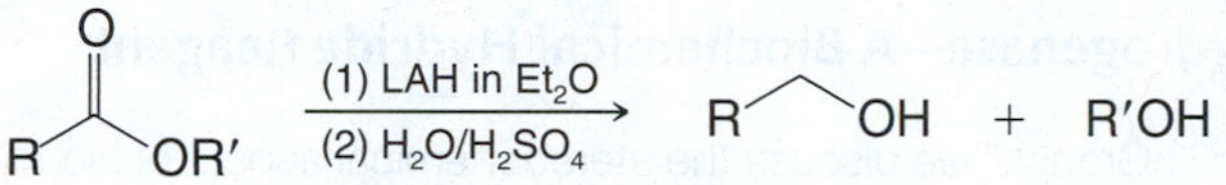

LAH reduction of an ester

Carboxylic acids and esters are more difficult to reduce than aldehydes and ketones. LAH, however, is a strong enough **reducing agent** to accomplish this transformation. Sodium borohydride (NaBH$_4$), which we shall discuss shortly, is commonly used to reduce aldehydes and ketones, but it is not strong enough to reduce carboxylic acids and esters.

Great care must be taken when using LAH to avoid the presence of water or any other weakly acidic solvent (e.g., alcohols). **LAH reacts violently with proton donors to release hydrogen gas.** Anhydrous diethyl ether (Et$_2$O) is a commonly used solvent for LAH reductions. After all of the LAH has been consumed by the reduction step of the reaction, however, water and acid are added to neutralize the resulting salts and facilitate isolation of the alcohol products.

12.3B Sodium Borohydride

- Aldehydes and ketones are easily reduced by sodium borohydride (NaBH$_4$).

Sodium borohydride is usually preferred over LAH for the reduction of aldehydes and ketones. Sodium borohydride can be used safely and effectively in water as well as alcohol solvents, whereas special precautions are required when using LAH.

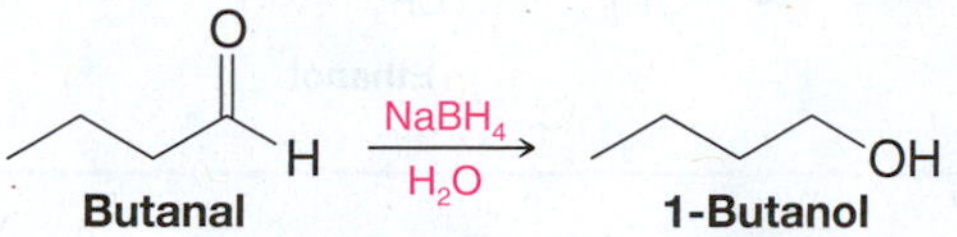

Butanal → **1-Butanol (85%)**

NaBH$_4$ reduction of a aldehyde

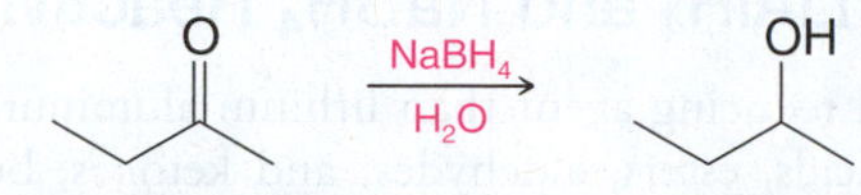

Butanone → **2-Butanol (87%)**

NaBH$_4$ reduction of an ketone

Aldehydes and ketones can be reduced using hydrogen and a metal catalyst, as well, and by sodium metal in an alcohol solvent.

The key step in the reduction of a carbonyl compound by either lithium aluminum hydride or sodium borohydride is the transfer of a **hydride ion** from the metal to the carbonyl carbon. In this transfer the hydride ion acts as a *nucleophile*. The mechanism for the reduction of a ketone by sodium borohydride is illustrated here.

[A MECHANISM FOR THE REACTION — Reduction of Aldehydes and Ketones by Hydride Transfer]

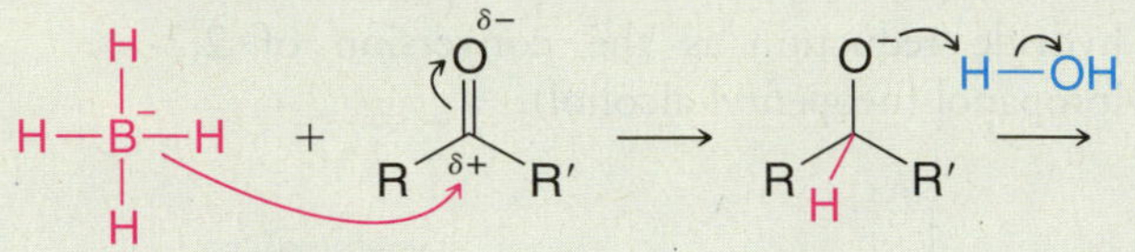

Hydride transfer   **Alkoxide ion**   **Alcohol**

These steps are repeated until all hydrogen atoms attached to boron have been transferred.

THE CHEMISTRY OF... Alcohol Dehydrogenase—A Biochemical Hydride Reagent

When the enzyme alcohol dehydrogenase converts acetaldehyde to ethanol, NADH acts as a reducing agent by transferring a hydride from **C4** of the nicotinamide ring to the carbonyl group of acetaldehyde. The nitrogen of the nicotinamide ring facilitates this process by contributing its non-bonding electron pair to the ring, which together with loss of the hydride converts the ring to the energetically more stable ring found in NAD^+ (we shall see why it is more stable in Chapter 14). The ethoxide anion resulting from hydride transfer to acetaldehyde is then protonated by the enzyme to form ethanol.

Although the carbonyl carbon of acetaldehyde that accepts the hydride is inherently electrophilic because of its electronegative oxygen, the enzyme enhances this property by providing a zinc ion as a Lewis acid to coordinate with the carbonyl oxygen. The Lewis acid stabilizes the negative charge that develops on the oxygen in the transition state. The role of the enzyme's protein scaffold, then, is to hold the zinc ion, coenzyme, and substrate in the three-dimensional array required to lower the energy of the transition state. The reaction is entirely reversible, of course, and when the relative concentration of ethanol is high, alcohol dehydrogenase carries out the oxidation of ethanol by removal of a hydride. This role of alcohol dehydrogenase is important in detoxification. In "The Chemistry of... Stereoselective Reductions of Carbonyl

Groups" we discuss the stereochemical aspect of alcohol dehydrogenase reactions.

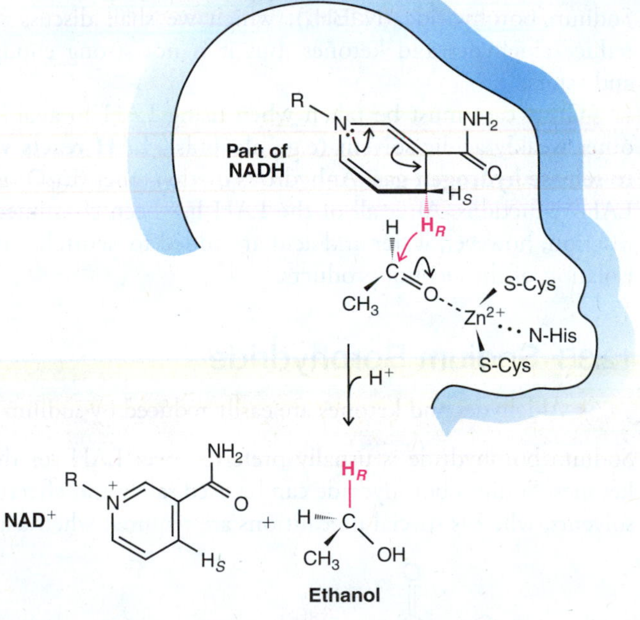

Ethanol

12.3C Overall Summary of LiAlH₄ and NaBH₄ Reactivity

Sodium borohydride is a less powerful reducing agent than lithium aluminum hydride. Lithium aluminum hydride reduces acids, esters, aldehydes, and ketones, but sodium borohydride reduces only aldehydes and ketones:

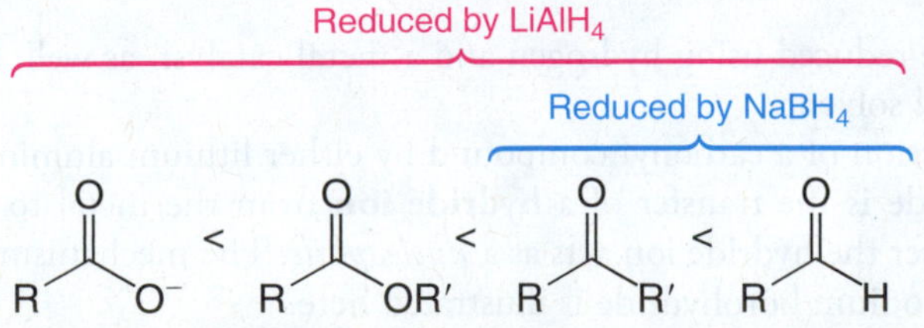

Reduced by LiAlH₄

Reduced by NaBH₄

The products from all of these reductions are alcohols

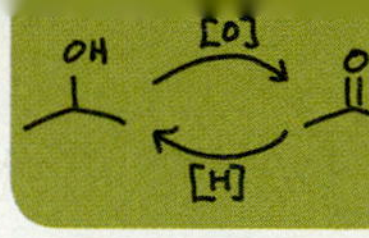

Lithium aluminum hydride reacts violently with water, and therefore reductions with lithium aluminum hydride must be carried out in anhydrous solutions, usually in anhydrous ether. (Ethyl acetate is added cautiously after the reaction is over to decompose excess $LiAlH_4$; then water is added to decompose the aluminum complex.) Sodium borohydride reductions, by contrast, can be carried out in water or alcohol solutions.

12.3D Reduction of Alkyl Halides to Hydrocarbons: $RX \longrightarrow RH$

Replacement of the halogen atom of an alkyl halide by hydrogen can be accomplished by treating the alkyl halide with lithium aluminum hydride (Section 12.3A). Because a halogen atom with a higher oxidation state is replaced by a hydrogen atom with a lower oxidation state, this reaction is a reduction. Almost all types of alkyl halides (primary, secondary, tertiary) can be reduced by $LiAlH_4$. $LiAlD_4$ can be used to replace the halogen atom with a deuterium atom.

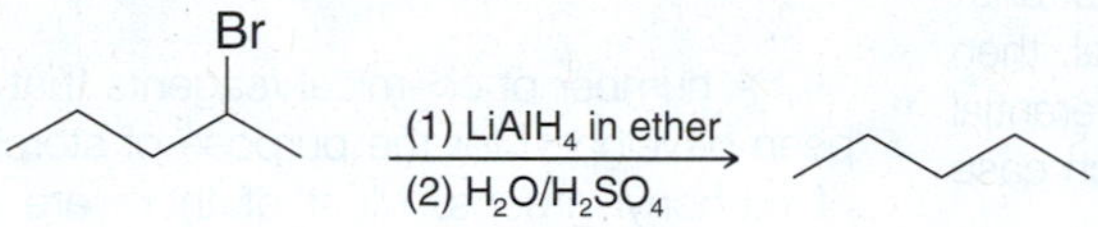

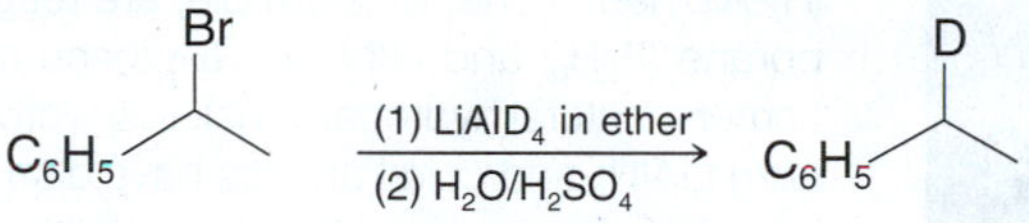

Which reducing agent, $LiAlH_4$ or $NaBH_4$, would you use to carry out the following transformations?

PRACTICE PROBLEM 12.3

(a)

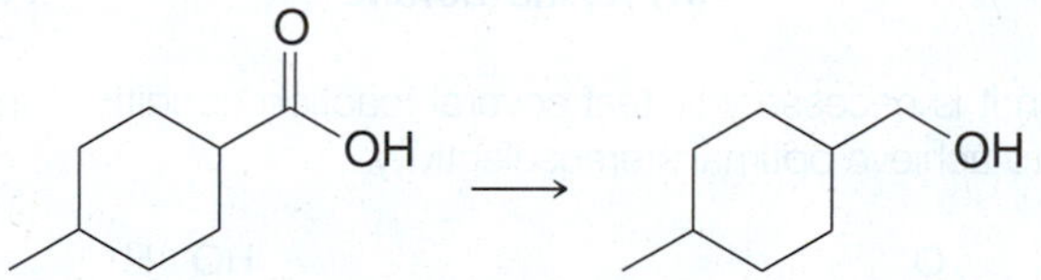

(d)

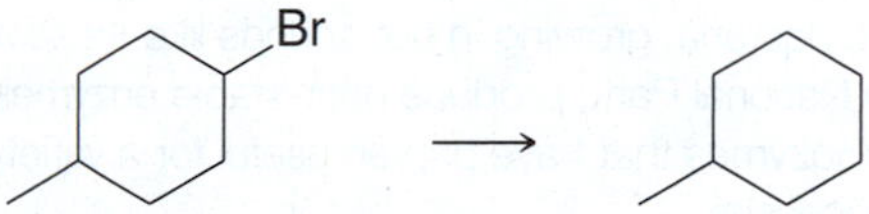

(b)

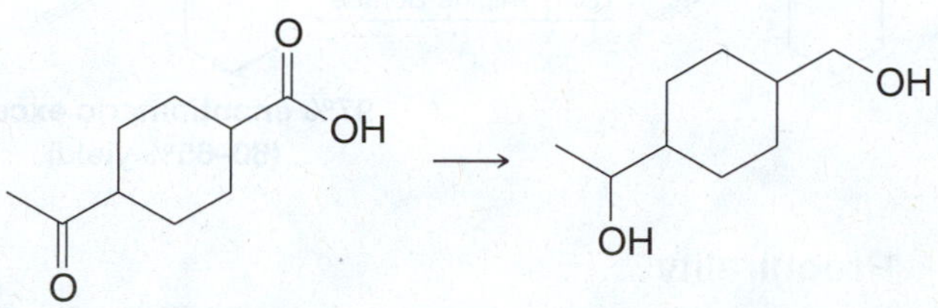

(e)

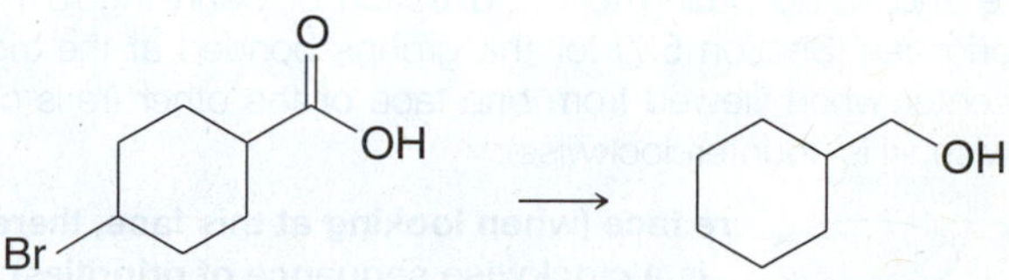

(c)

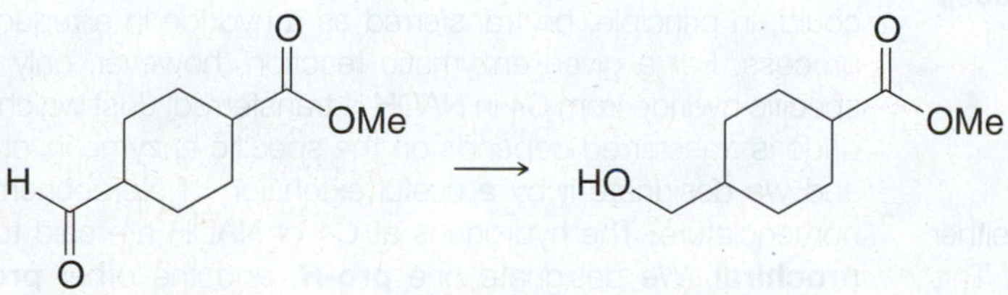

◆ THE CHEMISTRY OF... Stereoselective Reductions of Carbonyl Groups

Enantioselectivity

The possibility of **stereoselective** reduction of a carbonyl group is an important consideration in many syntheses. Depending on the structure about the carbonyl group that is being reduced, the tetrahedral carbon that is formed by transfer of a hydride could be a new chirality center. Achiral reagents, like $NaBH_4$ and $LiAlH_4$, react with equal rates at either face of an achiral trigonal planar substrate, leading to a racemic form of the product. But enzymes, for example, are chiral, and reactions involving a chiral reactant typically lead to a predominance of one enantiomeric form of a chiral product. Such a reaction is said to be **enantioselective**. Thus, when enzymes like alcohol dehydrogenase reduce carbonyl groups using the coenzyme NADH (see "The Chemistry of... Alcohol Dehydrogenase"), they discriminate between the two faces of the trigonal planar carbonyl substrate, such that a predominance of one of the two possible stereoisomeric forms of the tetrahedral product results. (If the original reactant was chiral, then formation of the new chirality center may result in preferential formation of one diastereomer of the product, in which case the reaction is said to be **diastereoselective**.)

Thermophilic bacteria, growing in hot springs like these at Yellowstone National Park, produce heat-stable enzymes called extremozymes that have proven useful for a variety of chemical processes.

The two faces of a trigonal planar center are designated re and si, according to the direction of Cahn–Ingold–Prelog priorities (Section 5.7) for the groups bonded at the trigonal center when viewed from one face or the other (re is clockwise, si is counterclockwise):

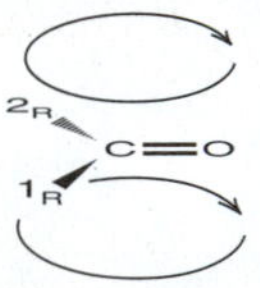

re face (when looking at this face, there is a clockwise sequence of priorities)

si face (when looking at this face, there is a counterclockwise sequence of priorities)

The **re** and **si** faces of a carbonyl group (where O > ¹R > ²R in terms of Cahn–Ingold–Prelog priorities)

The preference of many NADH-dependent enzymes for either the re or si face of their respective substrates is known. This

knowledge has allowed some of these enzymes to become exceptionally useful stereoselective reagents for synthesis. One of the most widely used is yeast alcohol dehydrogenase. Others that have become important are enzymes from thermophilic bacteria (bacteria that grow at elevated temperatures). Use of heat-stable enzymes (called **extremozymes**) allows reactions to be completed faster due to the rate-enhancing factor of elevated temperature (over 100 °C in some cases), although greater enantioselectivity is achieved at lower temperatures.

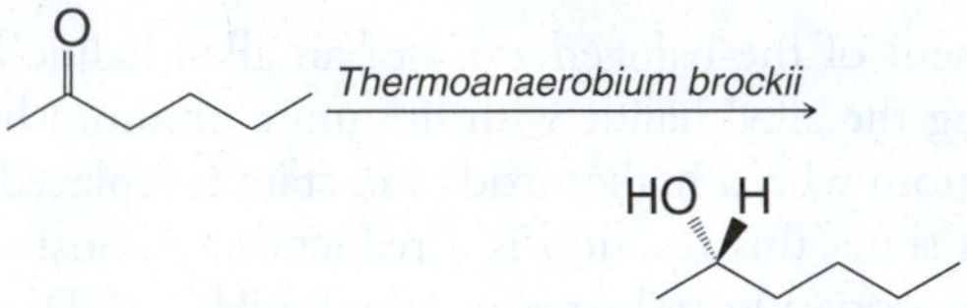

Thermoanaerobium brockii

96% enantiomeric excess (85% yield)

A number of chemical reagents that are chiral have also been developed for the purpose of stereoselective reduction of carbonyl groups. Most of them are derivatives of standard aluminum or boron hydride reducing agents that involve one or more chiral organic ligands. (S)-Alpine-Borane and (R)-Alpine-Borane, for example, are reagents derived from diborane (B_2H_6) and either (−)-α-pinene or (+)-α-pinene (enantiomeric natural hydrocarbons), respectively. Reagents derived from $LiAlH_4$ and chiral amines have also been developed. The extent of stereoselectivity achieved either by enzymatic reduction or reduction by a chiral reducing agent depends on the specific structure of the substrate.

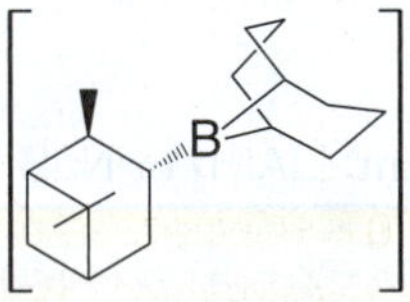

(R)-Alpine-Borane

Often it is necessary to test several reaction conditions in order to achieve optimal stereoselectivity.

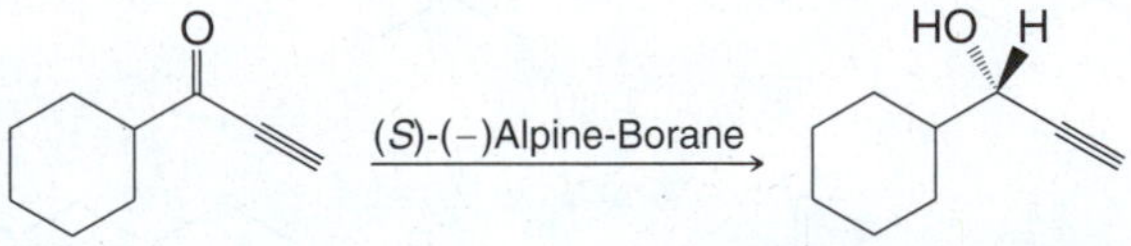

(S)-(−)Alpine-Borane

97% enantiomeric excess (60–65% yield)

Prochirality

A second aspect of the stereochemistry of NADH reactions results from NADH having two hydrogens at C4, either of which could, in principle, be transferred as a hydride in a reduction process. For a given enzymatic reaction, however, only one specific hydride from C4 in NADH is transferred. Just which hydride is transferred depends on the specific enzyme involved, and we designate it by a useful extension of stereochemical nomenclature. The hydrogens at C4 of NADH are said to be **prochiral**. We designate one **pro-R**, and the other **pro-S**,

depending on whether the configuration would be R or S when, in our imagination, each is replaced by a group of higher priority than hydrogen. If this exercise produces the R configuration, the hydrogen "replaced" is pro-R, and if it produces the S configuration it is pro-S. In general, a **prochiral center** is one for which addition of a group to a trigonal planar atom (as in reduction of a ketone) or replacement of one of two identical groups at a tetrahedral atom leads to a new chirality center.

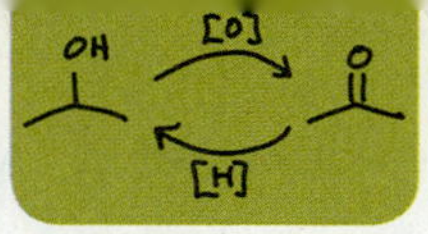

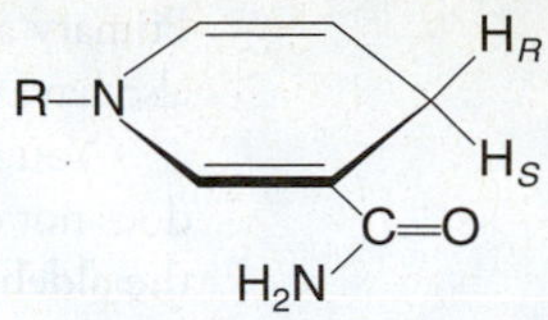

Nicotinamide ring of NADH, showing the pro-R and pro-S hydrogens

12.4 OXIDATION OF ALCOHOLS

Primary alcohols can be oxidized to aldehydes, and aldehydes can be oxidized to carboxylic acids:

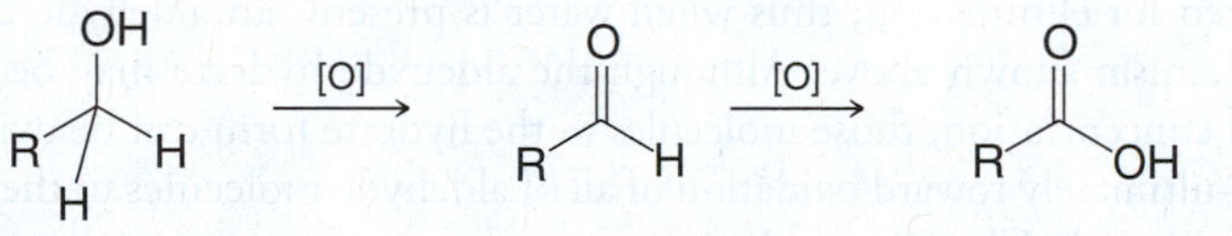

1° Alcohol → **Aldehyde** → **Carboxylic acid**

Secondary alcohols can be oxidized to ketones:

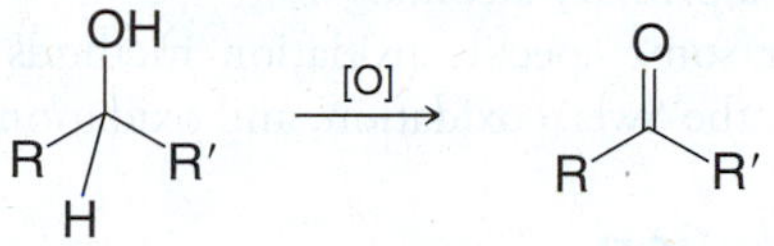

2° Alcohol → **Ketone**

Tertiary alcohols cannot be oxidized to carbonyl compounds.

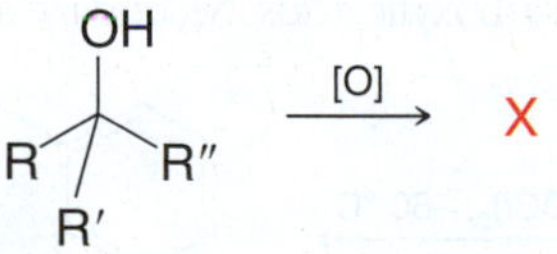

3° Alcohol $\xrightarrow{[O]}$ **X**

These examples have one aspect in common: when **oxidation** takes place, a hydrogen atom is lost from the alcohol or aldehyde carbon. A tertiary alcohol has no hydrogen on the alcohol carbon, and thus it cannot be oxidized in this way.

12.4A A Common Mechanistic Theme

Oxidations of primary and secondary alcohols, like those above, follow a common mechanistic path when certain reagents are used. These reagents, some of which we will discuss below, temporarily install a leaving group on the hydroxyl oxygen during the reaction. Loss of a hydrogen from the hydroxyl carbon and departure of the leaving group from the oxygen result in an elimination that forms the C=O π bond. Formation of the carbonyl double bond essentially occurs in a fashion analogous to formation of an alkene double bond by an elimination reaction. The general pathway is shown here.

Alcohol Oxidation by Elimination

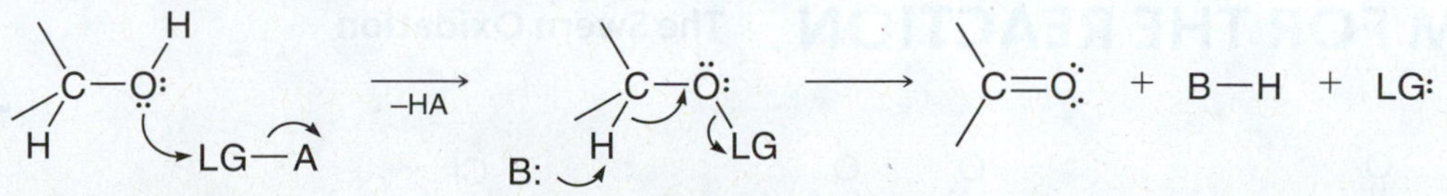

A 1° or 2° alcohol reacts with a reagent that installs a leaving group (LG) on the alcohol oxygen atom.

In an elimination step, a base removes a hydrogen from the alcohol carbon, the C=O π bond forms, and the leaving group departs, resulting in the oxidized product.

Primary and secondary alcohols have the required hydrogen atom at the alcohol carbon. They also have the hydroxyl hydrogen that is lost when the leaving group is installed, as shown above.

You might ask how an aldehyde can be oxidized by this mechanism, since an aldehyde does not contain a hydroxyl group to participate as shown above. The answer lies in whether the aldehyde reaction mixture includes water or not. In the presence of water, an aldehyde can form an aldehyde hydrate (by an **addition reaction** that we shall study in Chapter 16).

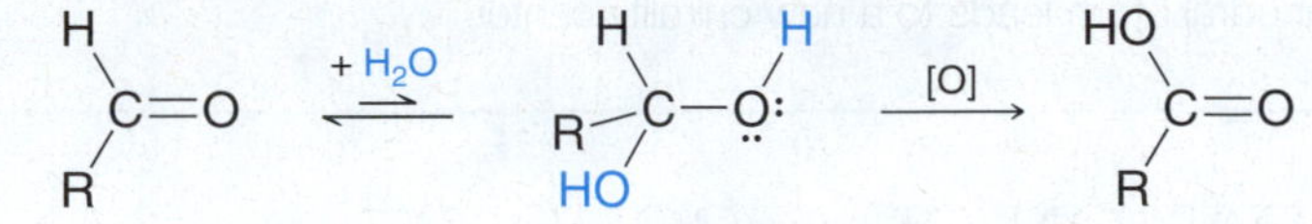

The carbon of an aldehyde hydrate has both a hydroxyl group and the hydrogen atom required for elimination; thus when water is present, an aldehyde can be oxidized by the mechanism shown above. Although the aldehyde hydrate may be present in low equilibrium concentration, those molecules in the hydrate form can be oxidized, drawing the reaction ultimately toward oxidation of all of aldehyde molecules to the corresponding carboxylic acid via LeChatelier's principle.

Aldehydes cannot be oxidized by the general mechanism above when water is absent. This fact proves to be useful when choosing conditions leading to specifically an aldehyde or a carboxylic acid from a primary alcohol.

Now let us consider some specific oxidation methods that hinge on the general mechanism shown above: the Swern oxidation, and oxidations involving chromate esters.

12.4B Swern Oxidation

The Swern oxidation is broadly useful for the synthesis of aldehydes and ketones from primary and secondary alcohols, respectively. The reaction is conducted in the absence of water, thus primary alcohols form aldehydes and not carboxylic acids. Secondary alcohols are oxidized to ketones.

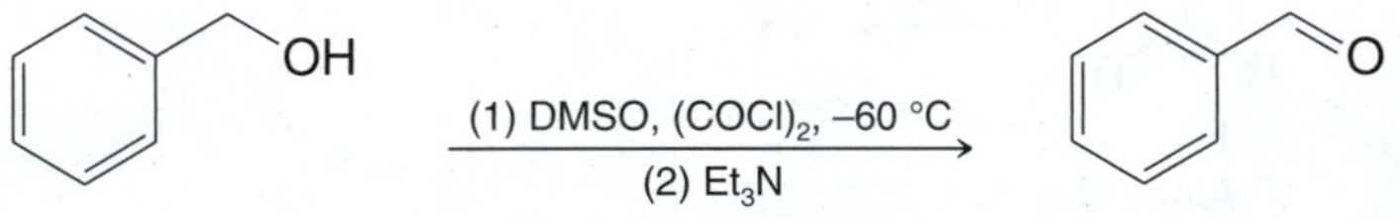

$$\xrightarrow[\text{(2) Et}_3\text{N}]{\text{(1) DMSO, (COCl)}_2, -60\,°\text{C}}$$

Swern oxidation of a 1° alcohol to an aldehyde

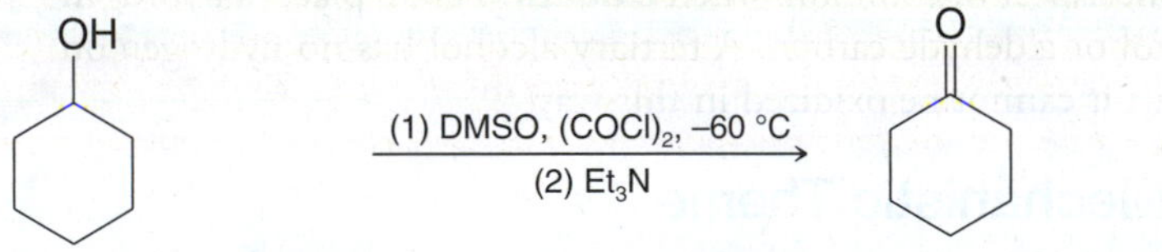

$$\xrightarrow[\text{(2) Et}_3\text{N}]{\text{(1) DMSO, (COCl)}_2, -60\,°\text{C}}$$

Swern oxidation of a 2° alcohol to a ketone

The reaction is carried out in sequential operations. First, oxalyl chloride (ClCOCOCl) is added to dimethyl sulfoxide (DMSO), usually at low temperature, to generate a chloro-dimethylsulfonium salt (as well as CO_2, CO, and HCl by-products). The alcohol substrate is then added to the chlorodimethylsulfonium salt, during which time a dimethylsulfonium group is installed as a leaving group on the hydroxyl oxygen. Third, an amine is added as a base to promote the elimination reaction.

[A MECHANISM FOR THE REACTION — The Swern Oxidation]

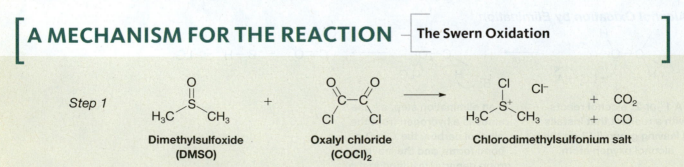

Step 1

Dimethylsulfoxide (DMSO) + Oxalyl chloride (COCl)₂ → Chlorodimethylsulfonium salt + CO_2 + CO

DMSO and oxalyl chloride react to form a chlorodimethylsulfonium salt.

Step 2

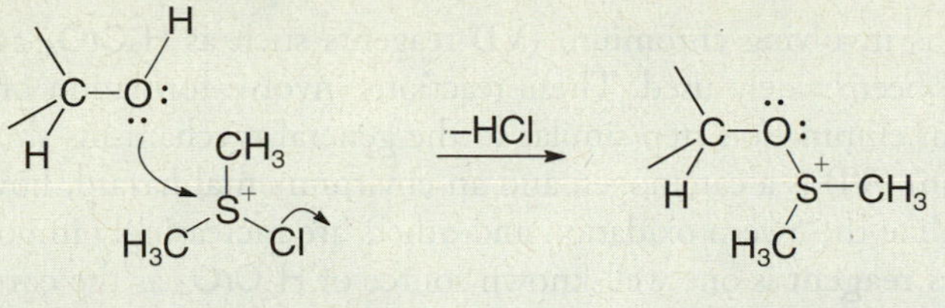

The 1° or 2° alcohol reacts with the sulfonium salt, installing a leaving group on the alcohol oxygen atom, along with loss of the hydroxyl proton.

The oxygen now bears a leaving group that can be lost in an elimination reaction.

Step 3

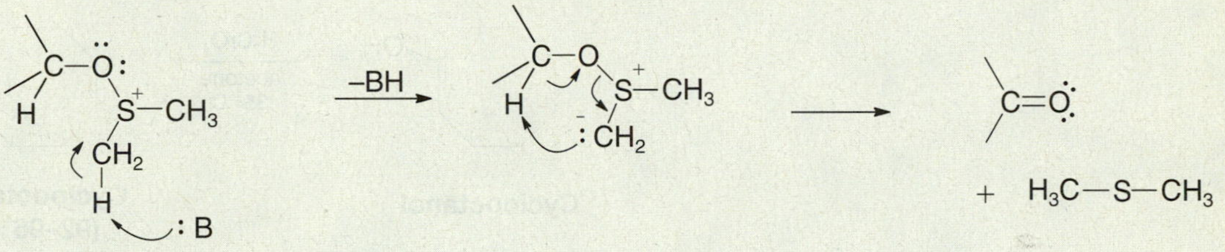

$+ \; H_3C-S-CH_3$

A base (usually triethylamine or diisopropylamine) removes a hydrogen from a methyl group adjacent to the positively–charged sulfur.

The anionic methyl group removes a proton from the alcohol carbon, forming the C=O π–bond. Dimethylsulfide departs as a leaving group, resulting in the oxidized product.

Reagents other than oxalyl chloride, including trifluoroacetic anhydride, have been used to generate various dimethylsulfonium salts for reaction with the alcohol.

What oxidation product would result from each of the following reactions?

PRACTICE PROBLEM 12.4

(a)

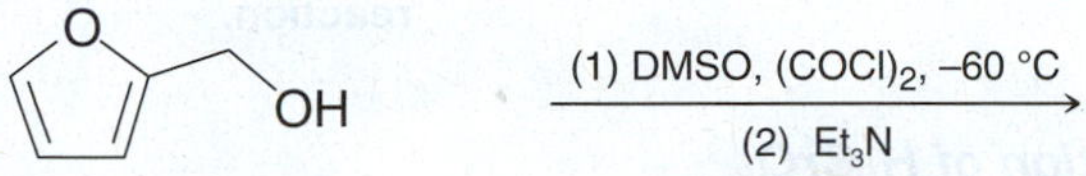

$$\xrightarrow[\text{(2) Et}_3\text{N}]{\text{(1) DMSO, (COCl)}_2, \; -60 \; °C}$$

(b)

$$\xrightarrow[\text{(2) Et}_3\text{N}]{\text{(1) DMSO, (COCl)}_2, \; -60 \; °C}$$

(c)

$$\xrightarrow[\text{(2) Et}_3\text{N (excess)}]{\text{(1) DMSO, (COCl)}_2 \; 2 \; \text{equiv.,} \; -60 \; °C}$$

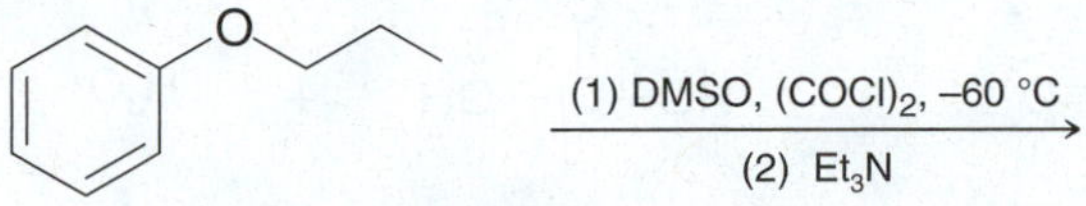

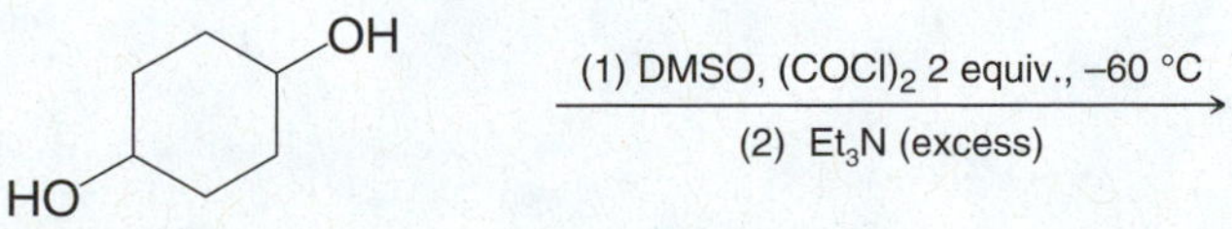

12.4C Chromic Acid (H_2CrO_4) Oxidation

Oxidations involving chromium (VI) reagents such as H_2CrO_4 are simple to carry out and have been widely used. These reactions involve formation of chromate esters, and include an elimination step similar to the general mechanisms shown in Section 12.4A. Chromium (VI) is a carcinogen and an environmental hazard, however. For this reason, methods like the Swern oxidation and others are increasingly important.

Jones reagent is one well-known source of H_2CrO_4 as the chromium (VI) oxidizing species. It can be prepared by adding CrO_3 or Na_2CrO_4 to aqueous sulfuric acid. Jones reagent is typically used by addition to solutions of an alcohol or aldehyde in acetone or acetic acid (solvents that cannot be oxidized). Primary alcohols are oxidized to carboxylic acids, via the aldehyde hydrate mentioned above. Secondary alcohols are oxidized to ketones. The following is an example of an oxidation using Jones reagent.

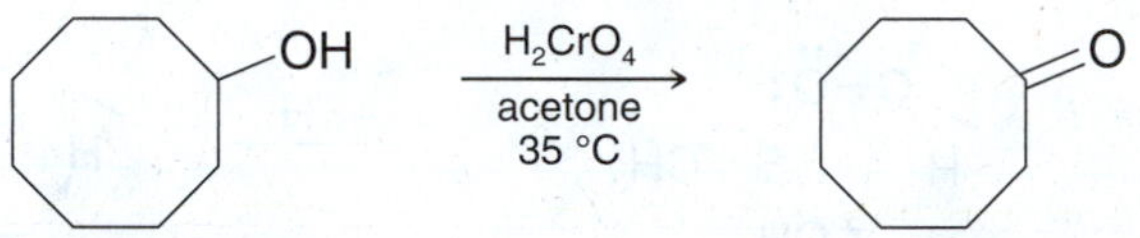

Cyclooctanol

$\xrightarrow[\substack{\text{acetone} \\ 35\,°C}]{H_2CrO_4}$

**Cyclooctanol
(92–96%)**

As mentioned earlier, the chromate oxidation mechanism first involves formation of a chromate ester with the alcohol. Then a molecule of H_2CrO_3 serves as a leaving group during the elimination step that generates the C=O bond of the carbonyl compound.

A MECHANISM FOR THE REACTION — Chromic Acid Oxidation

Formation of the Chromate Ester

Step 1

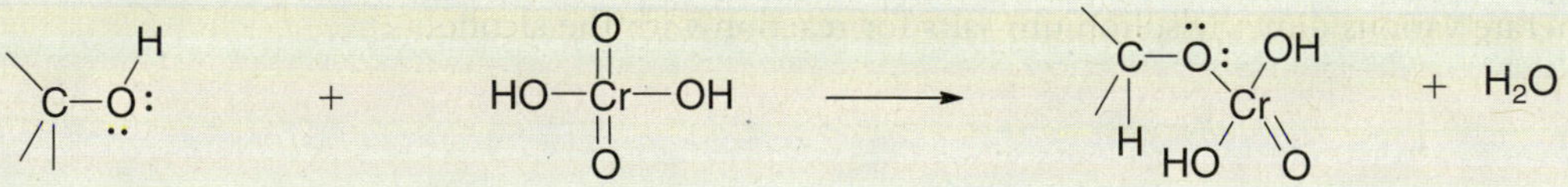

The 1° or 2° alcohol reacts with chromic acid to form a chromate ester with loss of water, installing a leaving group on the alcohol oxygen.

The oxygen now bears a leaving group that can be lost in an elimination reaction.

Oxidation by Elimination of H_2CrO_3

Step 2

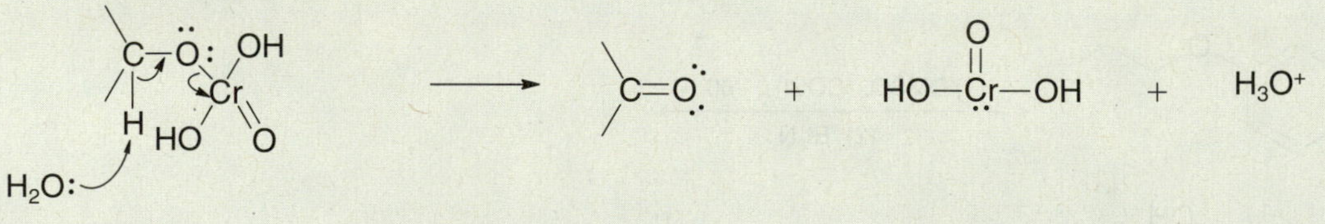

A water molecule removes a proton from the alcohol carbon, forming the C=O π–bond. The chromium atom is reduced as H_2CrO_3 departs, resulting in the oxidized product.

Chromic acid solutions are orange-red in color, and the product mixture, containing Cr(III), is a greenish blue. Thus, reagents like Jones reagent can serve as a color-based functional group test. Primary or secondary alcohols and aldehydes are rapidly oxidized by Jones reagent, turning the solution an opaque greenish blue within a few seconds. If none of these groups are present, the solution remains orange-red until side reactions eventually change the color. This color change is the basis for the original **breathalyzer alcohol test**.

$$H_2CrO_4 \xrightarrow[\text{or aldehyde}]{\text{Add 1° or 2° alcohol}} H_2CrO_3 + \text{Oxidation products}$$

Clear orange-red solution   **Opaque green-blue solution**

12.4D Pyridinium Chlorochromate (PCC)

Pyridinium chlorochromate (PCC) is a Cr(VI) salt formed between pyridine (C_6H_5N), HCl, and CrO_3. PCC is soluble in dichloromethane, thus it can be used under conditions that exclude water, allowing for the oxidization of primary alcohols to aldehydes because the aldehyde hydrate is not present under anhydrous conditions. Jones reagent, on the other hand, oxidizes primary alcohols to carboxylic acids because it is an aqueous reagent. The following are some general examples of PCC oxidations.

Pyridinium chlorochromate (PCC)

PCC oxidation of a 1° alcohol to an aldehyde

2-Ethyl-2-methyl-1-butanol   **2-Ethyl-2-methylbutanal**

PCC oxidation of a 2° alcohol to a ketone

12.4E Potassium Permanganate ($KMnO_4$)

Primary alcohols and aldehydes can be oxidized by potassium permanganate ($KMnO_4$) to the corresponding carboxylic acids. Secondary alcohols can be oxidized to ketones. These reactions do not proceed by the type of mechanism described above (and we shall not discuss the mechanism here). The reaction is usually carried out in basic aqueous solution, from which MnO_2 precipitates as the oxidation takes place. After the oxidation is complete, filtration allows removal of the MnO_2 and acidification of the filtrate gives the carboxylic acid.

●●● **SOLVED PROBLEM 12.2**

Which reagents would you use to accomplish the following transformations?

STRATEGY AND ANSWER:

(a) To oxidize a primary alcohol to a carboxylic acid, use (1) potassium permanganate in aqueous base, followed by (2) H_3O^+, or use chromic acid (H_2CrO_4).

(b) To reduce a carboxylic acid to a primary alcohol, use $LiAlH_4$.

(c) To oxidize a primary alcohol to an aldehyde, use the Swern oxidation or pyridinium chlorochromate (PCC).

(d) To reduce an aldehyde to a primary alcohol, use $NaBH_4$ (preferably) or $LiAlH_4$.

PRACTICE PROBLEM 12.5 Show how each of the following transformations could be accomplished:

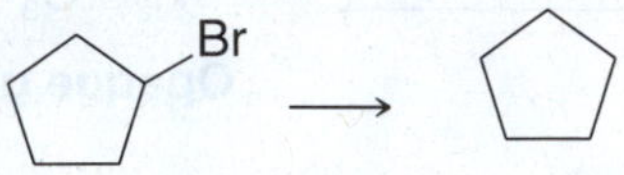

12.4F Spectroscopic Evidence for Alcohols

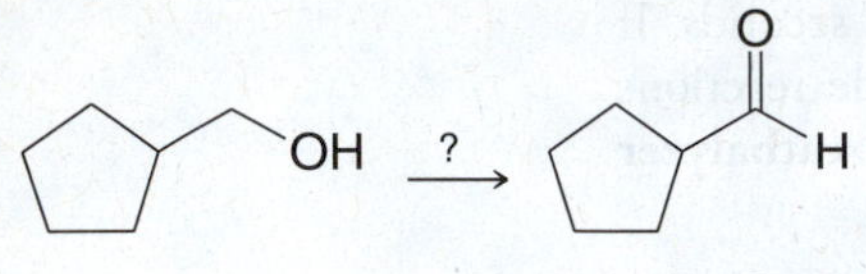

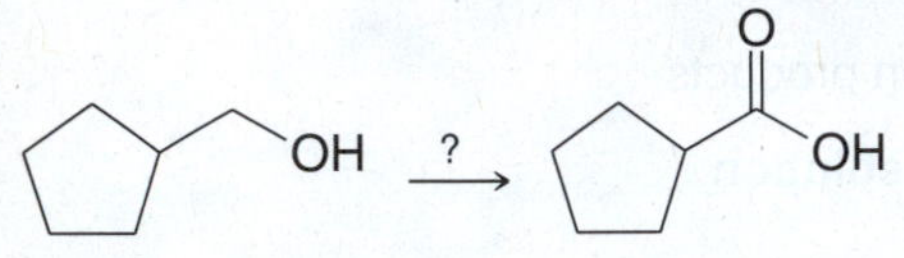

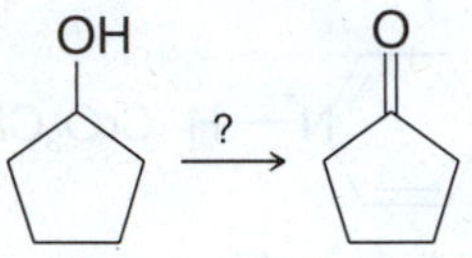

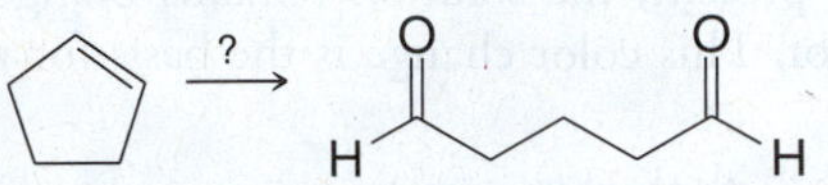

- Alcohols give rise to broad O—H stretching absorptions from 3200 to 3600 cm^{-1} in infrared spectra.
- The alcohol hydroxyl hydrogen typically produces a broad ^{1}H NMR signal of variable chemical shift which can be eliminated by exchange with deuterium from D_2O (see Table 9.1).
- Hydrogen atoms on the carbon of a primary or secondary alcohol produce a signal in the ^{1}H NMR spectrum between δ 3.3 and δ 4.0 (see Table 9.1) that integrates for 2 and 1 hydrogens, respectively.
- The ^{13}C NMR spectrum of an alcohol shows a signal between δ 50 and δ 90 for the alcohol carbon (see Table 9.2).

12.5 ORGANOMETALLIC COMPOUNDS

- Compounds that contain carbon–metal bonds are called **organometallic compounds**.

The nature of the carbon–metal bond varies widely, ranging from bonds that are essentially ionic to those that are primarily covalent. Whereas the structure of the organic portion of the organometallic compound has some effect on the nature of the carbon–metal bond, the identity of the metal itself is of far greater importance. Carbon–sodium and carbon–potassium bonds are largely ionic in character; carbon–lead, carbon–tin, carbon–thallium, and carbon–mercury bonds are essentially covalent. Carbon–lithium and carbon–magnesium bonds lie between these extremes.

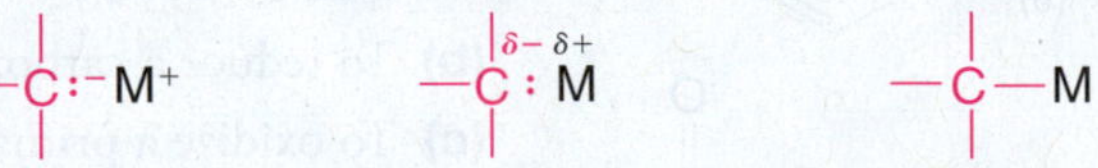

| Primarily ionic (M = Na⁺ or K⁺) | (M = Mg or Li) | Primarily covalent (M = Pb, Sn, Hg, or Tl) |

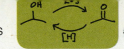

The reactivity of organometallic compounds increases with the percent ionic character of the carbon–metal bond. Alkylsodium and alkylpotassium compounds are highly reactive and are among the most powerful of bases. They react explosively with water and burst into flame when exposed to air. Organomercury and organolead compounds are much less reactive; they are often volatile and are stable in air. They are all poisonous. They are generally soluble in nonpolar solvents. Tetraethyllead, for example, was once used as an "antiknock" compound in gasoline, but because of the lead pollution it contributed to the environment it has been replaced by other antiknock agents. *tert*-Butyl methyl ether is another antiknock additive, though there are concerns about its presence in the environment, as well.

Organometallic compounds of lithium and magnesium are of great importance in organic synthesis. They are relatively stable in ether solutions, but their carbon–metal bonds have considerable ionic character. Because of this ionic nature, the carbon atom that is bonded to the metal atom of an organolithium or organomagnesium compound is a strong base and powerful nucleophile. We shall soon see reactions that illustrate both of these properties.

Helpful Hint

A number of organometallic reagents are very useful for carbon–carbon bond forming reactions (see Section 12.8 and Special Topic G).

12.6 PREPARATION OF ORGANOLITHIUM AND ORGANOMAGNESIUM COMPOUNDS

12.6A Organolithium Compounds

Organolithium compounds are often prepared by the reduction of organic halides with lithium metal. These reductions are usually carried out in ether solvents, and since organolithium compounds are strong bases, care must be taken to exclude moisture. (Why?) The ethers most commonly used as solvents are diethyl ether and tetrahydrofuran. (Tetrahydrofuran is a cyclic ether.)

Diethyl ether (Et₂O)   **Tetrahydrofuran (THF)**

• Organolithium compounds are prepared in this general way:

$$R{-}X + 2\,Li \xrightarrow{Et_2O} RLi + LiX$$
(or Ar—X)   **(or ArLi)**

The order of reactivity of halides is RI > RBr > RCl. (Alkyl and aryl fluorides are seldom used in the preparation of organolithium compounds.)

For example, butyl bromide reacts with lithium metal in diethyl ether to give a solution of butyllithium:

$$\text{Br} + 2\,Li \xrightarrow[-10\,°C]{Et_2O,} \text{Li} + LiBr$$
Butyl bromide   **Butyllithium (80–90%)**

Several alkyl- and aryllithium reagents are commercially available in hexane and other hydrocarbon solvents.

12.6B Grignard Reagents

Organomagnesium halides were discovered by the French chemist Victor Grignard in 1900. Grignard received the Nobel Prize for his discovery in 1912, and organomagnesium halides are now called **Grignard reagents** in his honor. Grignard reagents have great use in organic synthesis.

- Grignard reagents are prepared by the reaction of an organic halide with magnesium metal in an anhydrous ether solvent:

$$\text{RX} + \text{Mg} \xrightarrow{\text{Et}_2\text{O}} \text{RMgX}$$

$$\text{ArX} + \text{Mg} \xrightarrow{\text{Et}_2\text{O}} \text{ArMgX}$$

$\left.\begin{array}{c}\\\\\end{array}\right\}$ **Grignard reagents**

The order of reactivity of halides with magnesium is also RI > RBr > RCl. Very few organomagnesium fluorides have been prepared. Aryl Grignard reagents are more easily prepared from aryl bromides and aryl iodides than from aryl chlorides, which react very sluggishly. Once prepared, a Grignard reagent is usually used directly in a subsequent reaction.

The actual structures of Grignard reagents are more complex than the general formula RMgX indicates. Experiments have established that for most Grignard reagents there is an equilibrium between an alkylmagnesium halide and a dialkylmagnesium.

$$2\ \text{RMgX} \rightleftharpoons \text{R}_2\text{Mg} + \text{MgX}_2$$

Alkylmagnesium halide   **Dialkylmagnesium**

For convenience in this text, however, we shall write the formula for the Grignard reagent as though it were simply RMgX.

A Grignard reagent forms a complex with its ether solvent; the structure of the complex can be represented as follows:

$$
\begin{array}{c}
\text{R} \quad \ddot{\text{O}} \quad \text{R} \\
\\
\text{R}-\text{Mg}-\text{X} \\
\\
\text{R} \quad \ddot{\text{O}} \quad \text{R}
\end{array}
$$

Complex formation with molecules of ether is an important factor in the formation and stability of Grignard reagents.

The mechanism by which Grignard reagents form is complicated and has been a matter of debate. There seems to be general agreement that radicals are involved and that a mechanism similar to the following is likely:

$$\text{R}-\text{X} + :\text{Mg} \longrightarrow \text{R}\cdot + \cdot\text{MgX}$$

$$\text{R}\cdot + \cdot\text{MgX} \longrightarrow \text{RMgX}$$

12.7 REACTIONS OF ORGANOLITHIUM AND ORGANOMAGNESIUM COMPOUNDS

12.7A Reactions with Compounds Containing Acidic Hydrogen Atoms

- Grignard reagents and organolithium compounds are very strong bases. They react with any compound that has a hydrogen atom attached to an electronegative atom such as oxygen, nitrogen, or sulfur.

We can understand how these reactions occur if we represent the Grignard reagent and organolithium compounds in the following ways:

$$\overset{\delta-}{\text{R}}:\overset{\delta+}{\text{MgX}} \quad \text{and} \quad \overset{\delta-}{\text{R}}:\overset{\delta+}{\text{Li}}$$

When we do this, we can see that the reactions of Grignard reagents with water and alcohols are nothing more than acid–base reactions; they lead to the formation of the weaker conjugate acid and weaker conjugate base.

- A Grignard reagent behaves as if it contained the anion of an alkane, *as if it contained a carbanion*:

$$\overset{\delta-}{R}-\overset{\delta+}{Mg}X \ + \ \overset{\delta+}{H}-\overset{\delta-}{\ddot{O}}-H \ \longrightarrow \ R-H \ + \ H\ddot{O}:^- \ + \ Mg^{2+} \ + \ X^-$$

| Stronger base | Stronger acid (pK_a 15.7) | Weaker acid (pK_a 40–50) | Weaker base |

$$\overset{\delta-}{R}-\overset{\delta+}{Mg}X \ + \ \overset{\delta+}{H}-\overset{\delta-}{\ddot{O}}-R \ \longrightarrow \ R-H \ + \ R\ddot{O}:^- \ + \ Mg^{2+} \ + \ X^-$$

| Stronger base | Stronger acid (pK_a 15–18) | Weaker acid (pK_a 40–50) | Weaker base |

●●● **SOLVED PROBLEM 12.3**

Write an equation for the reaction that would take place when phenyllithium is treated with water. Designate the stronger acid and stronger base.

STRATEGY AND ANSWER: Recognizing that phenyllithium, like a Grignard reagent, acts as though it contains a carbanion, a very powerful base ($pK_a = 40$–50), we conclude that the following acid–base reaction would occur.

$$\overset{\delta-}{Ar}-\overset{\delta+}{Li} \ + \ H:\ddot{O}H \ \longrightarrow \ Ar-H \ + \ H\ddot{O}:^- \ + \ Li^+$$

| Stronger base | Stronger acid | Weaker acid | Weaker base |

●●● **PRACTICE PROBLEM 12.6**

Predict the products of the following acid–base reactions. Using pK_a values, indicate which side of each equilibrium reaction is favored, and label the species representing the stronger acid and stronger base in each case.

(a)

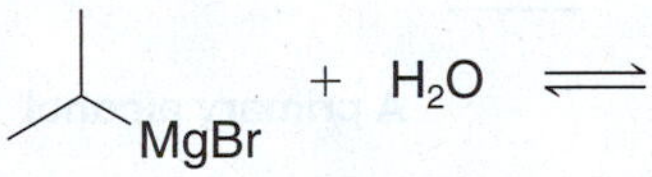

(isopropyl)MgBr + H₂O ⇌

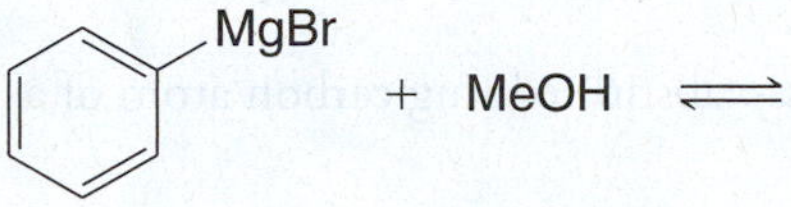

(c)

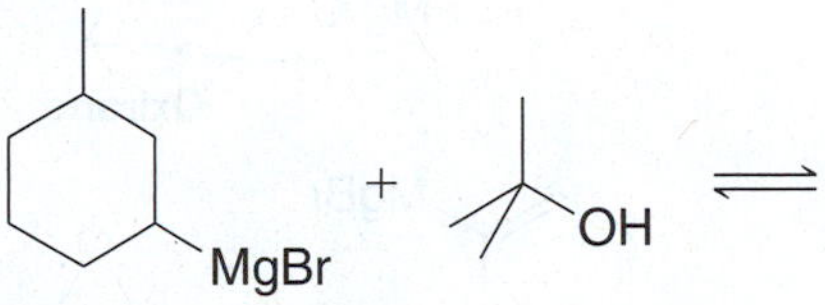

(methylcyclohexyl)MgBr + (tert-butanol)OH ⇌

(b)

(phenyl)MgBr + MeOH ⇌

(d)

(butyl)Li + (tert-butanol)OH ⇌

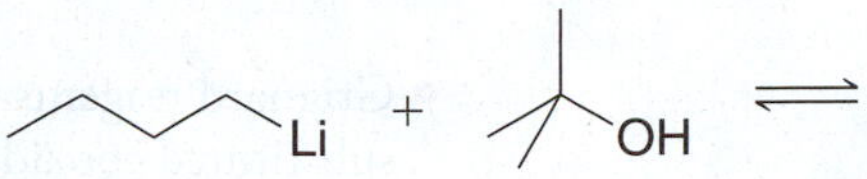

●●● **PRACTICE PROBLEM 12.7**

Provide the reagents necessary to achieve the following transformations.

(a)

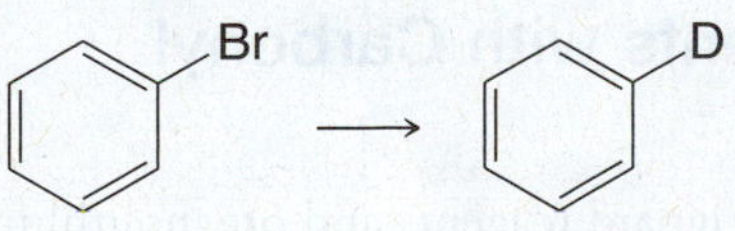

(bromobenzene) Br → (deuterated benzene) D

(b)

(isobutane) → (deuterated isobutane) D

(D = deuterium)

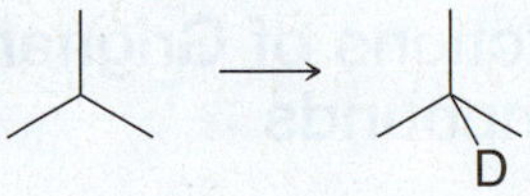

Grignard reagents and organolithium compounds remove protons that are much less acidic than those of water and alcohols.

- Grignard reagents react with the terminal hydrogen atoms of 1-alkynes by an acid–base reaction, and this is a useful method for the preparation of alkynylmagnesium halides and alkynyllithiums.

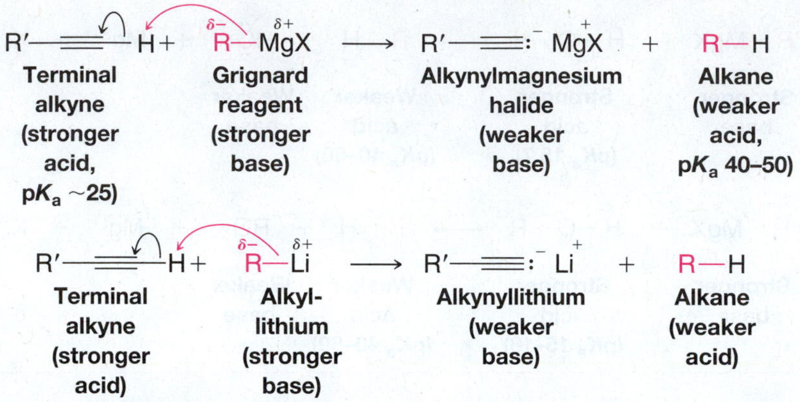

The fact that these reactions go to completion is not surprising when we recall that alkanes have pK_a values of 40–50, whereas those of terminal alkynes are ~25 (Table 3.1). Not only are Grignard reagents strong bases, they are also *powerful nucleophiles*.

- Reactions in which Grignard reagents act as nucleophiles are by far the most important and we shall consider these next.

12.7B Reactions of Grignard Reagents with Epoxides (Oxiranes)

- Grignard reagents react as nucleophiles with epoxides (oxiranes), providing convenient synthesis of alcohols.

The nucleophilic alkyl group of the Grignard reagent attacks the partially positive carbon of the epoxide ring. Because it is highly strained, the ring opens, and the reaction leads to the alkoxide salt of an alcohol. Subsequent acidification produces the alcohol. (Compare this reaction with the base-catalyzed ring opening we studied in Section 11.14.) The following are examples with oxirane.

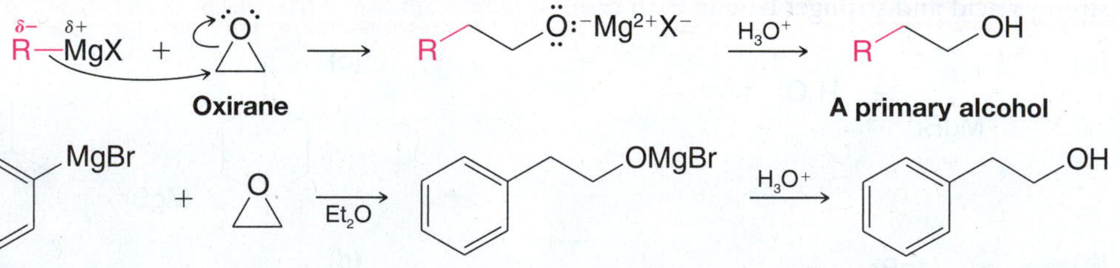

- Grignard reagents react primarily at the less-substituted ring carbon atom of a substituted epoxide.

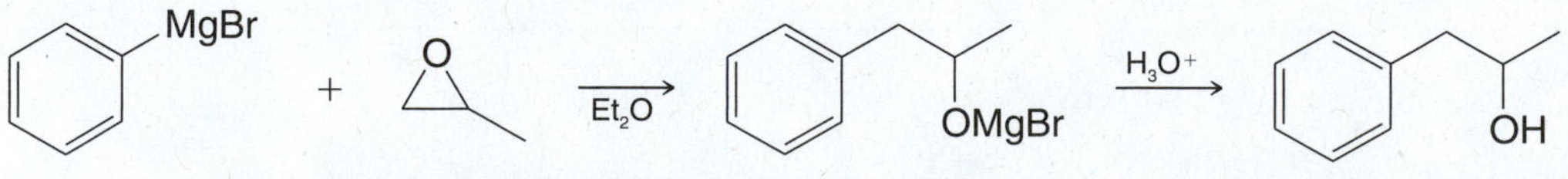

12.7C Reactions of Grignard Reagents with Carbonyl Compounds

- The most important synthetic reactions of Grignard reagents and organolithium compounds are those in which they react as nucleophiles and attack an unsaturated carbon—*especially the carbon of a carbonyl group.*

We saw in Section 12.1A that carbonyl compounds are highly susceptible to nucleophilic attack. Grignard reagents react with carbonyl compounds (aldehydes and ketones) in the following way:

[A MECHANISM FOR THE REACTION — The Grignard Reaction]

Reaction

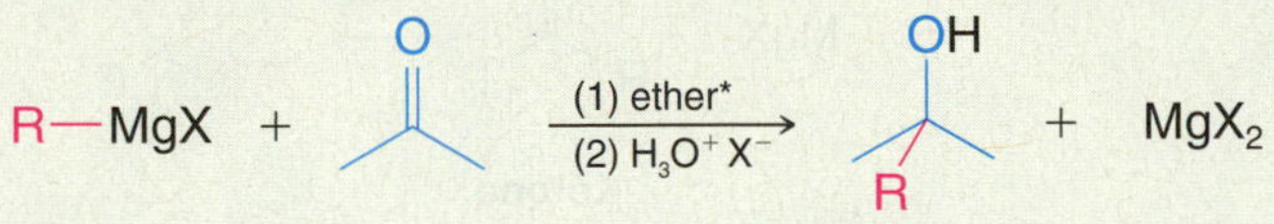

$$R-MgX \quad + \quad \text{(ketone)} \quad \xrightarrow[\text{(2) } H_3O^+ X^-]{\text{(1) ether*}} \quad \text{(alcohol)} \quad + \quad MgX_2$$

Mechanism

Step 1

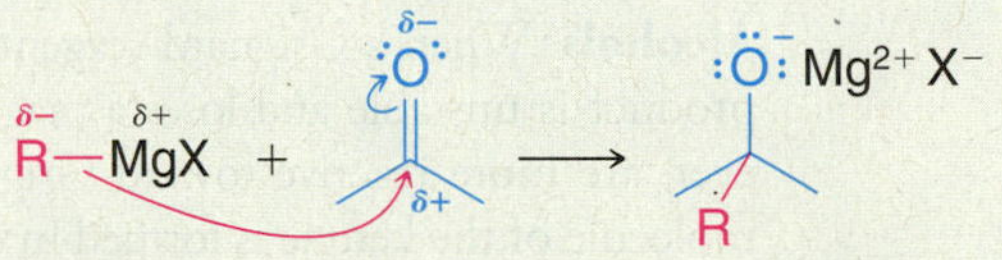

Grignard reagent   **Carbonyl compound**   **Halomagnesium alkoxide**

The strongly nucleophilic Grignard reagent uses its electron pair to form a bond to the carbon atom. One electron pair of the carbonyl group shifts out to the oxygen. This reaction is a nucleophilic addition to the carbonyl group, and it results in the formation of an alkoxide ion associated with Mg^{2+} and X^-.

Step 2

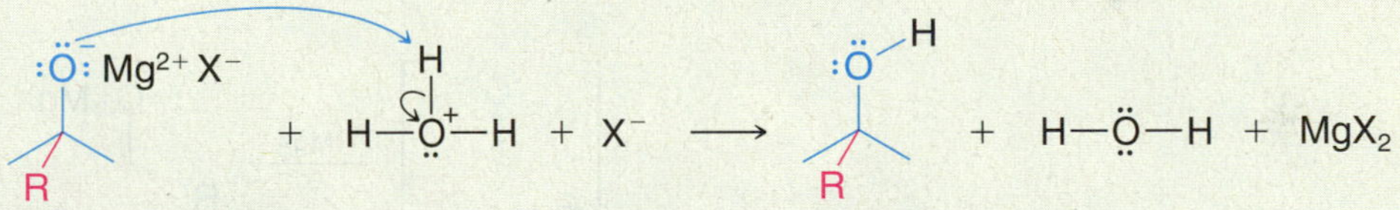

Halomagnesium alkoxide   **Alcohol**

In the second step, the addition of aqueous HX causes protonation of the alkoxide ion; this leads to the formation of the alcohol and MgX_2.

*By writing "(1) ether" over the arrow and "(2) $H_3O^+ X^-$" under the arrow, we mean that in the first laboratory step the Grignard reagent and the carbonyl compound are allowed to react in an ether solvent. Then in a second step, after the reaction of the Grignard reagent and the carbonyl compound is over, we add aqueous acid (e.g., dilute HX) to convert the salt of the alcohol (ROMgX) to the alcohol itself. If the alcohol is tertiary, it will be susceptible to acid-catalyzed dehydration. In this case, a solution of NH_4Cl in water is often used because it is acidic enough to convert ROMgX to ROH while not allowing acid-catalyzed reactions of the resulting tertiary alcohol.

12.8 ALCOHOLS FROM GRIGNARD REAGENTS

Grignard additions to carbonyl compounds are especially useful because they can be used to prepare primary, secondary, or tertiary alcohols:

1. Grignard Reagents React with Formaldehyde to Give a Primary Alcohol

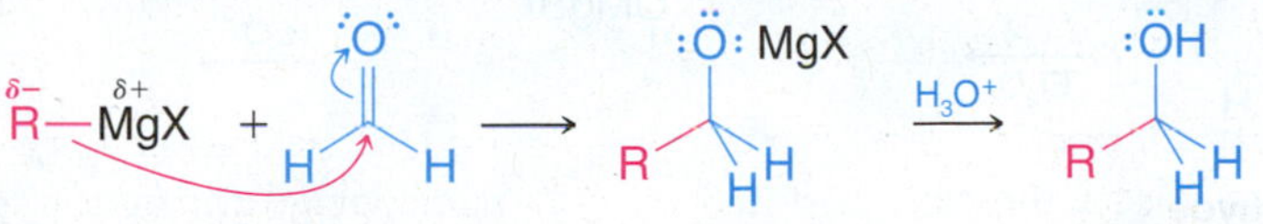

Formaldehyde   **1° Alcohol**

2. Grignard Reagents React with All Other Aldehydes to Give Secondary Alcohols

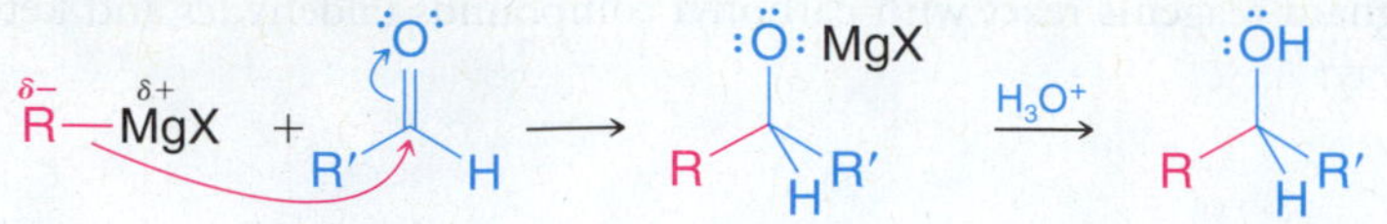

Higher aldehyde → **2° Alcohol**

3. Grignard Reagents React with Ketones to Give Tertiary Alcohols

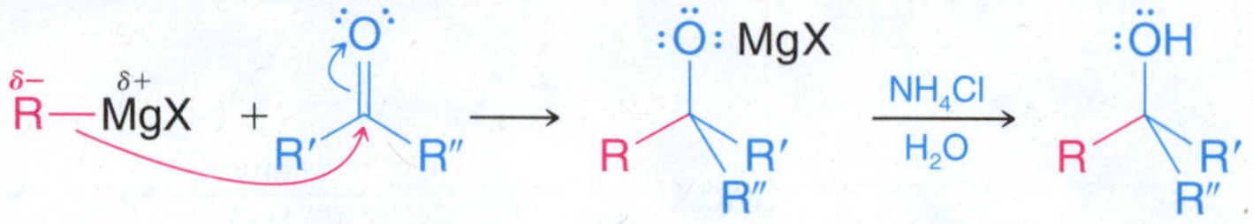

Ketone → **3° Alcohol**

4. Esters React with Two Molar Equivalents of a Grignard Reagent to Form Tertiary Alcohols

When a Grignard reagent adds to the carbonyl group of an ester, the initial product is unstable and loses a magnesium alkoxide to form a ketone. Ketones, however, are more reactive toward Grignard reagents than esters. Therefore, as soon as a molecule of the ketone is formed in the mixture, it reacts with a second molecule of the Grignard reagent. After hydrolysis, **the product is a tertiary alcohol with two identical alkyl groups**, groups that correspond to the alkyl portion of the Grignard reagent:

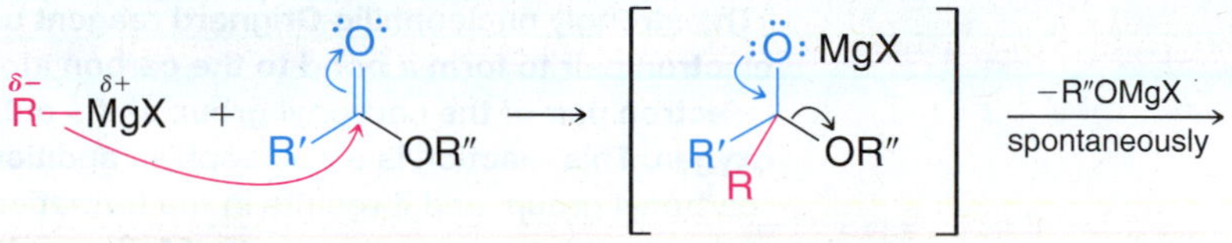

Ester → **Initial product (unstable)**

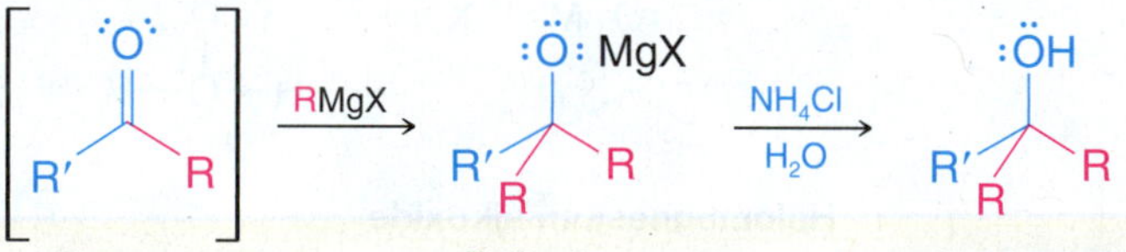

Ketone → **Salt of an alcohol (not isolated)** → **3° Alcohol (with two or three identical R groups)**

Specific examples of these reactions are shown here.

Grignard Reagent	Carbonyl Reactant			Final Product

Reaction with Formaldehyde

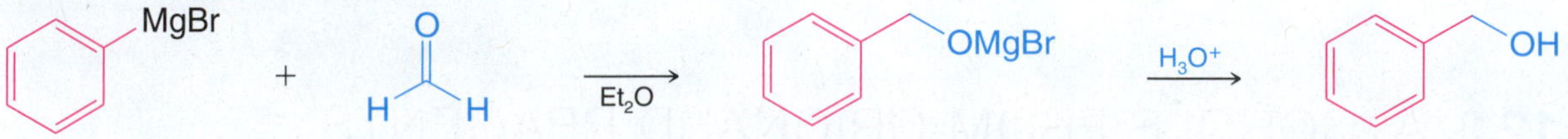

Phenylmagnesium bromide + **Formaldehyde** → **Benzyl alcohol (90%)**

Reaction with a Higher Aldehyde

Ethylmagnesium bromide + **Acetaldehyde** → **2-Butanol (80%)**

Reaction with a Ketone

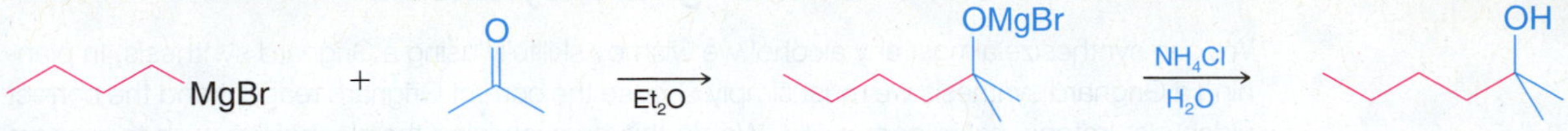

Butylmagnesium bromide   **Acetone**

2-Methyl-2-hexanol (92%)

Reaction with an Ester

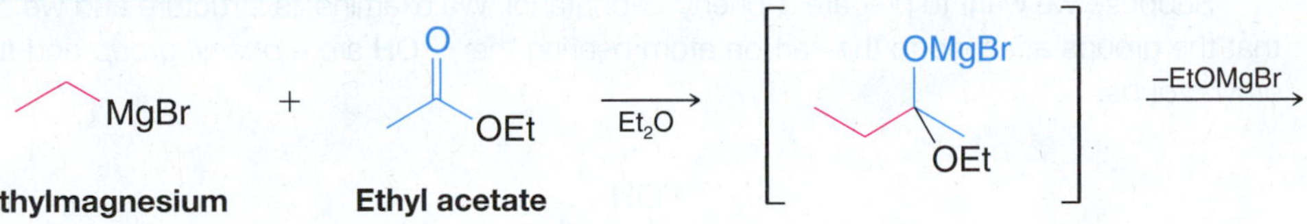

Ethylmagnesium bromide   **Ethyl acetate**

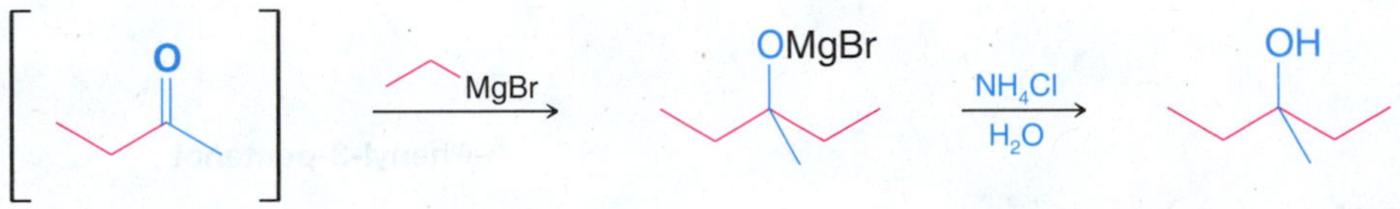

3-Methyl-3-pentanol (67%)

●●● **SOLVED PROBLEM 12.4**

How would you carry out the following synthesis?

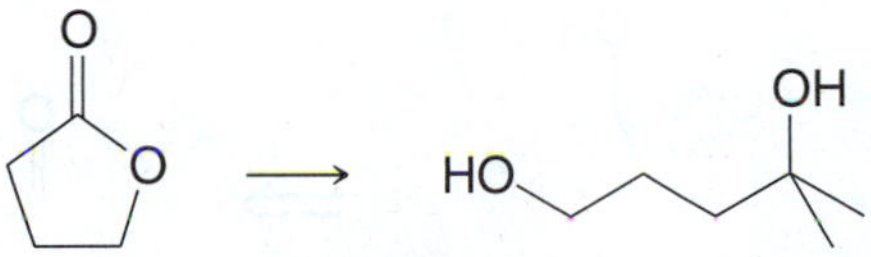

STRATEGY AND ANSWER: Here we are converting an ester (a cyclic ester) to **a tertiary alcohol with two identical alkyl groups** (methyl groups). So, we should use two molar equivalents of the Grignard reagent that contains the required alkyl groups, in this case, methyl magnesium iodide.

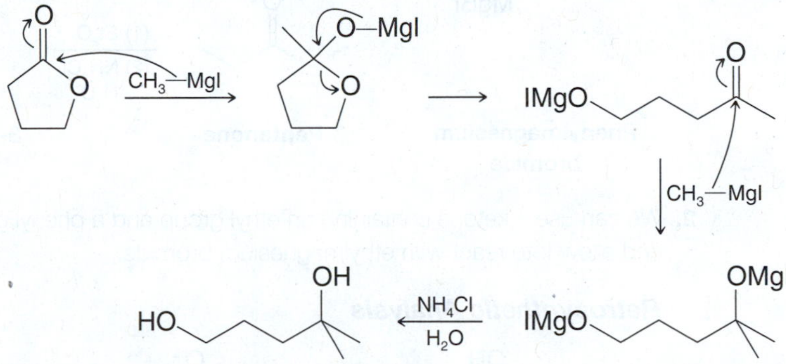

●●● **PRACTICE PROBLEM 12.8**

Provide a mechanism for the following reaction, based on your knowledge of the reaction of esters with Grignard reagents.

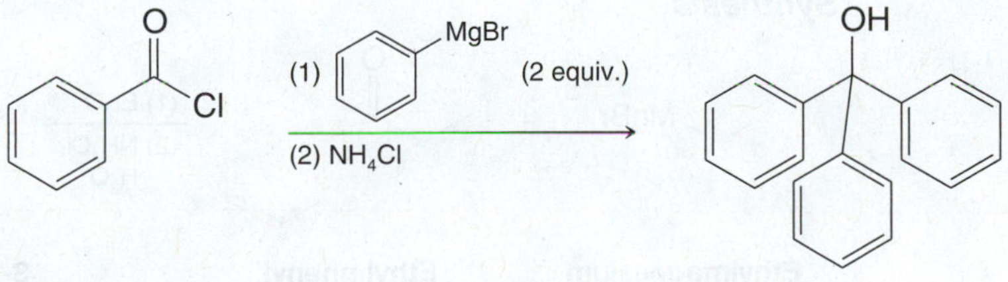

12.8A HOW TO Plan a Grignard Synthesis

We can synthesize almost any alcohol we wish by skillfully using a Grignard synthesis. In planning a Grignard synthesis we must simply choose the correct Grignard reagent and the correct aldehyde, ketone, ester, or epoxide. We do this by examining the alcohol we wish to prepare and by paying special attention to the groups attached to the carbon atom bearing the —OH group. Many times there may be more than one way of carrying out the synthesis. In these cases our final choice will probably be dictated by the availability of starting compounds. Let us consider an example.

Suppose we want to prepare 3-phenyl-3-pentanol. We examine its structure and we see that the groups attached to the carbon atom bearing the —OH are a *phenyl group* and *two ethyl groups*:

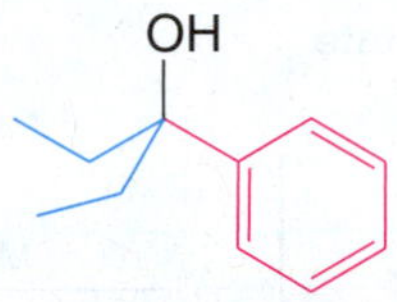

3-Phenyl-3-pentanol

This means that we can synthesize this compound in several different ways:

1. We can use a ketone with two ethyl groups (3-pentanone) and allow it to react with phenylmagnesium bromide:

Retrosynthetic Analysis

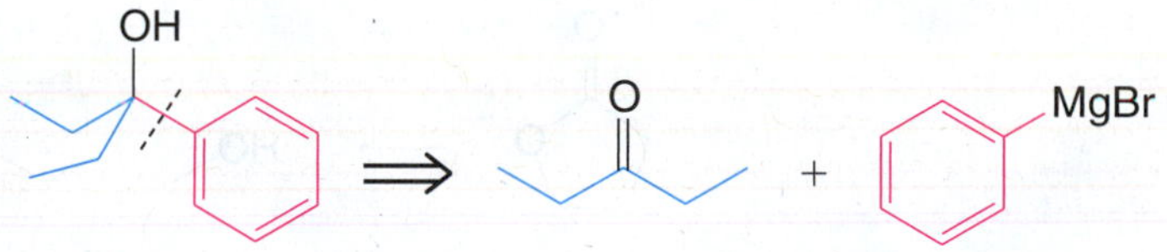

Synthesis

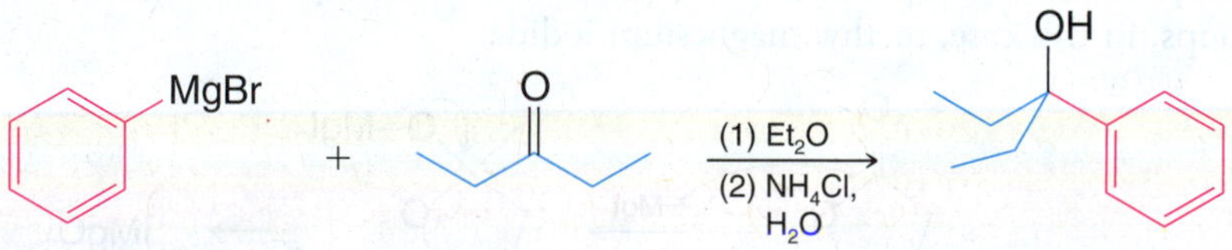

| **Phenylmagnesium bromide** | **3-Pentanone** | **3-Phenyl-3-pentanol** |

2. We can use a ketone containing an ethyl group and a phenyl group (ethyl phenyl ketone) and allow it to react with ethylmagnesium bromide:

Retrosynthetic Analysis

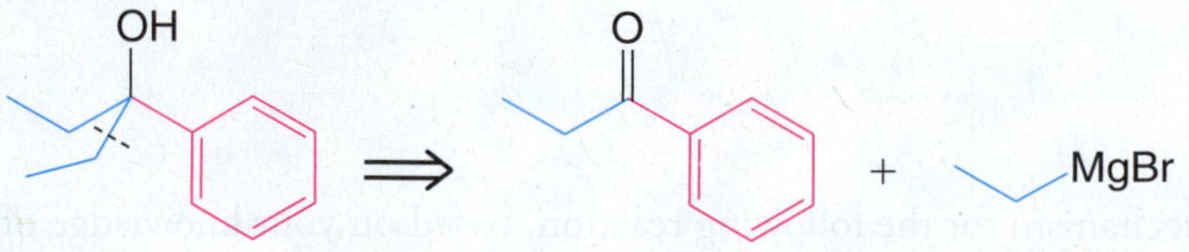

Synthesis

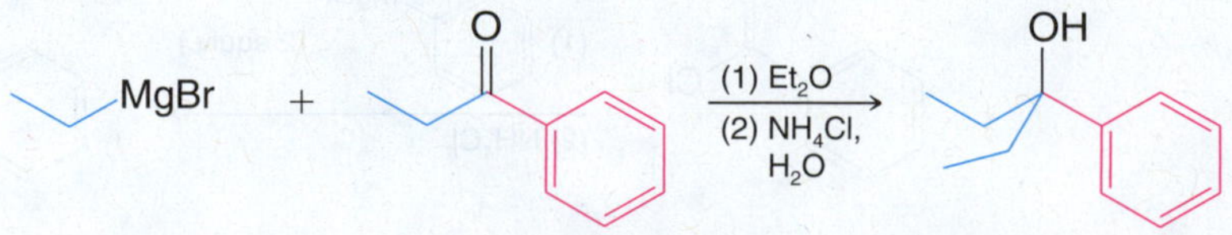

| **Ethylmagnesium bromide** | **Ethyl phenyl ketone** | **3-Phenyl-3-pentanol** |

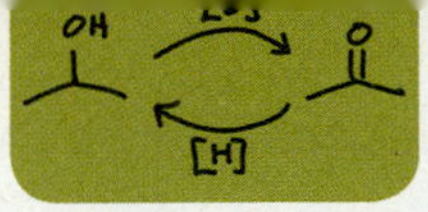

3. We can use an ester of benzoic acid and allow it to react with two molar equivalents of ethylmagnesium bromide:

Retrosynthetic Analysis

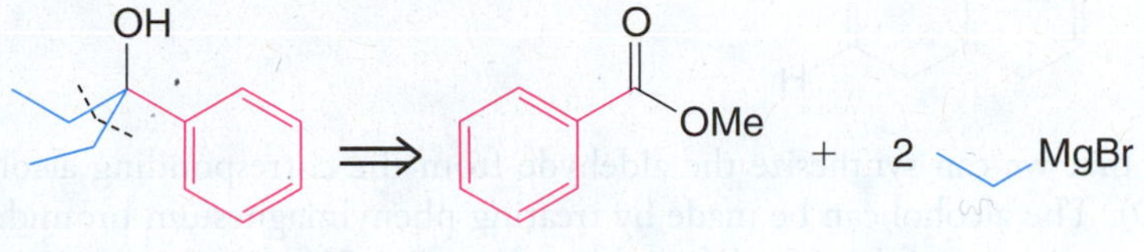

Synthesis

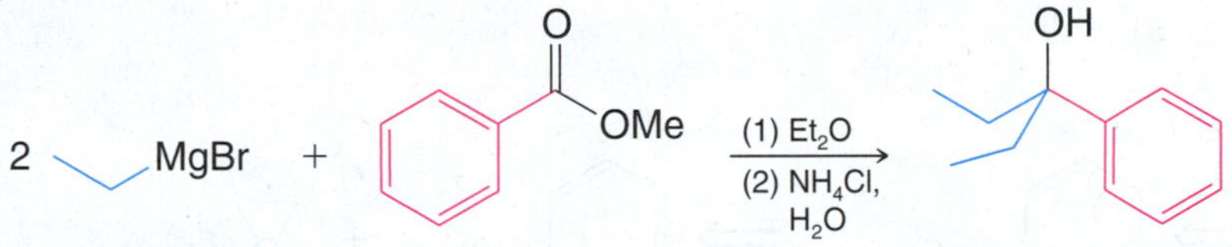

| Ethylmagnesium bromide | Methyl benzoate | 3-Phenyl-3-pentanol |

All of these methods will likely give us our desired compound in high yields.

●●● **SOLVED PROBLEM 12.5**

ILLUSTRATING A MULTISTEP SYNTHESIS: Using an alcohol of no more than four carbon atoms as your only organic starting material, outline a synthesis of **A**:

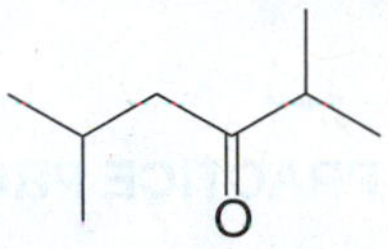

ANSWER: We can construct the carbon skeleton from two four-carbon compounds using a Grignard reaction. Then oxidation of the alcohol produced will yield the desired ketone.

Retrosynthetic Analysis

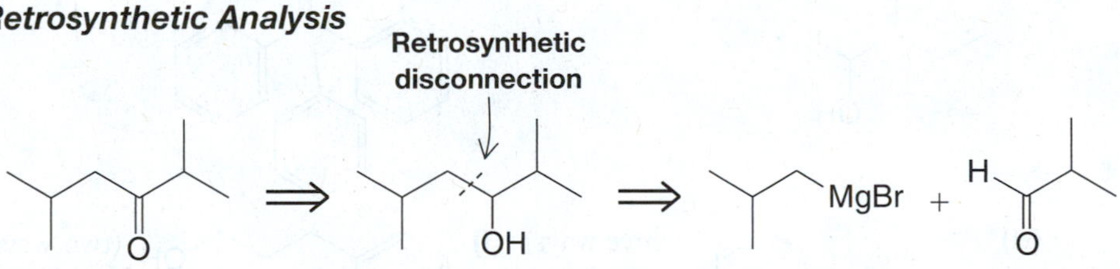

Retrosynthetic disconnection

Synthesis

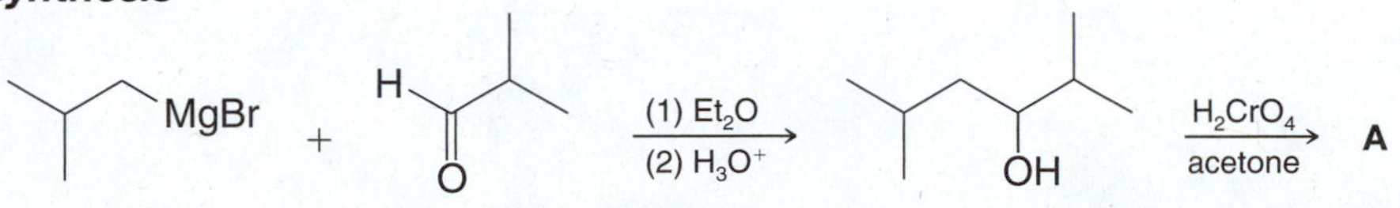

We can synthesize the Grignard reagent (**B**) and the aldehyde (**C**) from isobutyl alcohol:

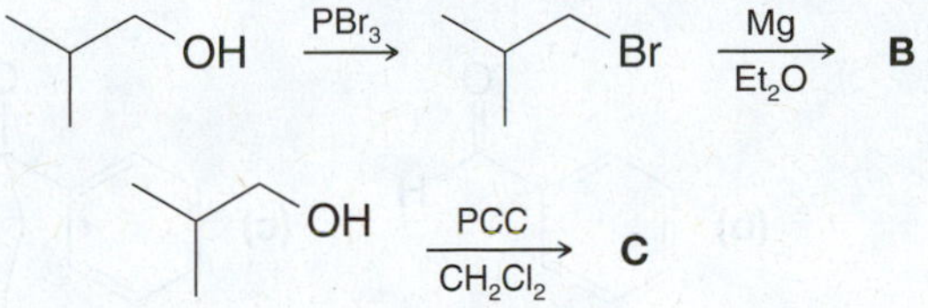

••• SOLVED PROBLEM 12.6

ILLUSTRATING A MULTISTEP SYNTHESIS: Starting with bromobenzene and any other needed reagents, outline a synthesis of the following aldehyde:

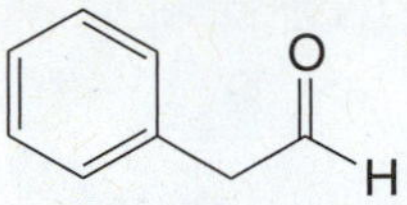

ANSWER: Working backward, we remember that we can synthesize the aldehyde from the corresponding alcohol by the Swern oxidation or PCC (Sections 12.4B, D). The alcohol can be made by treating phenylmagnesium bromide with oxirane. [Adding oxirane to a Grignard reagent is a very useful method for adding a —CH$_2$CH$_2$OH unit to an organic group (Section 12.7B).] Phenylmagnesium bromide can be made in the usual way, by treating bromobenzene with magnesium in an ether solvent.

Retrosynthetic Analysis

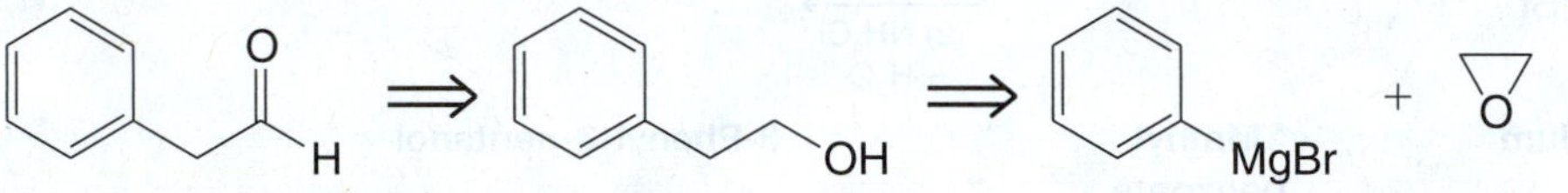

Synthesis

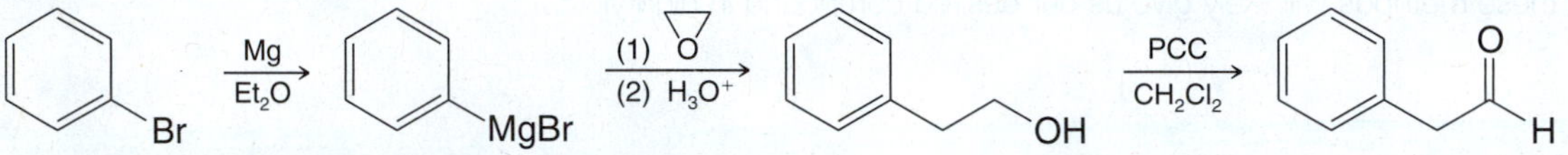

•••

PRACTICE PROBLEM 12.9 Provide retrosynthetic analyses and syntheses for each of the following alcohols, starting with appropriate alkyl or aryl halides.

(a)  OH (three ways)

(c) 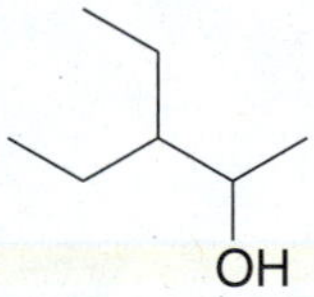 (two ways)

(e) 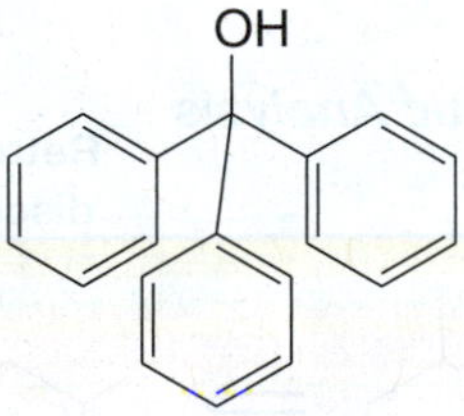 OH (two ways)

(b) 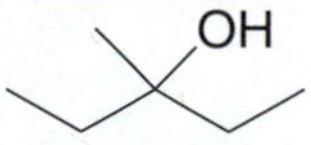 OH (three ways)

(d) 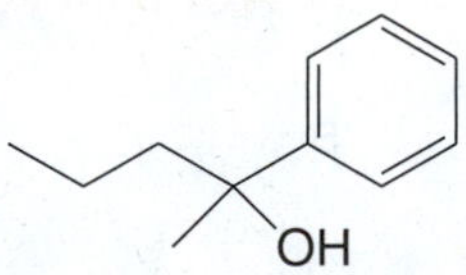 (three ways)

(f) 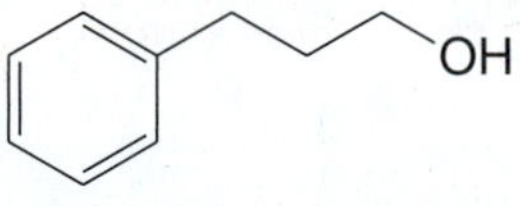 OH (two ways)

•••

PRACTICE PROBLEM 12.10 Provide a retrosynthetic analysis and synthesis for each of the following compounds. Permitted starting materials are phenylmagnesium bromide, oxirane, formaldehyde, and alcohols or esters of four carbon atoms or fewer. You may use any inorganic reagents and oxidizing conditions such as Swern oxidation or pyridinium chlorochromate (PCC).

(a) 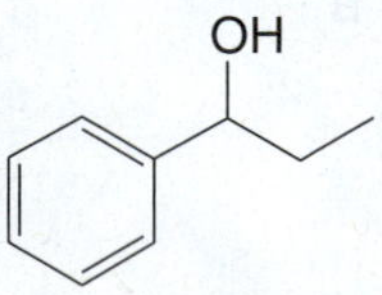 OH

(b) 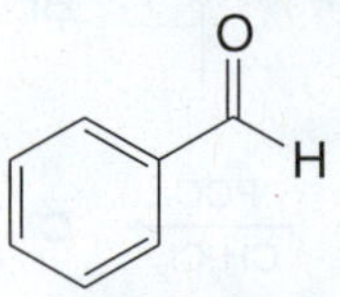 O H

(c) 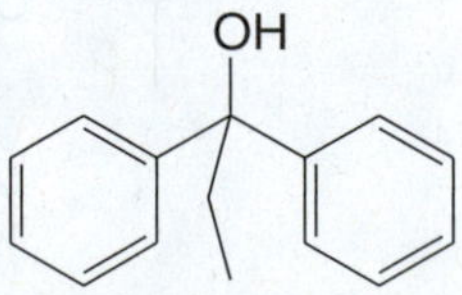 OH

(d) 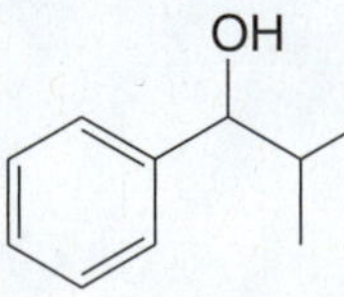 OH

12.8B Restrictions on the Use of Grignard Reagents

Although the Grignard synthesis is one of the most versatile of all general synthetic procedures, it is not without its limitations. Most of these limitations arise from the very feature of the Grignard reagent that makes it so useful—its *extraordinary reactivity as a nucleophile and a base.*

The Grignard reagent is a very powerful base; in effect it contains a carbanion.

- It is not possible to prepare a Grignard reagent from a compound that contains any hydrogen more acidic than the hydrogen atoms of an alkane or alkene.

We cannot, for example, prepare a Grignard reagent from a compound containing an —OH group, an —NH— group, an —SH group, a —CO₂H group, or an —SO₃H group. If we were to attempt to prepare a Grignard reagent from an organic halide containing any of these groups, the formation of the Grignard reagent would simply fail to take place. (Even if a Grignard reagent were to form, it would immediately be neutralized by the acidic group.)

- Since Grignard reagents are powerful nucleophiles, we cannot prepare a Grignard reagent from any organic halide that contains a carbonyl, epoxy, nitro, or cyano (—CN) group.

If we were to attempt to carry out this kind of reaction, any Grignard reagent that formed would only react with the unreacted starting material:

$-OH, -NH_2, -NHR, -CO_2H, -SO_3H, -SH, -C{\equiv}C-H$

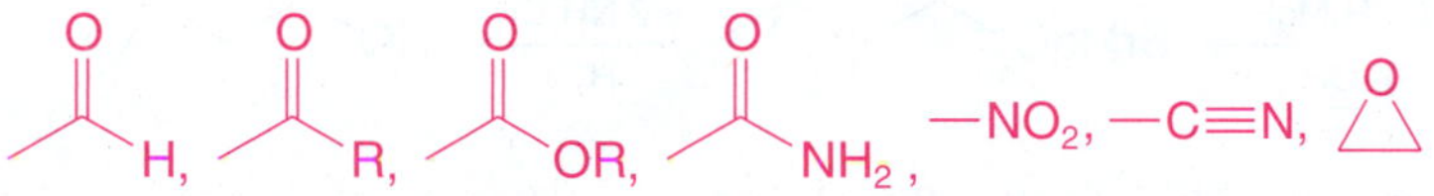

Grignard reagents cannot be prepared in the presence of these groups because they will react with them.

Helpful Hint

A protecting group can sometimes be used to mask the reactivity of an incompatible group (see Sections 11.11D, 11.11E, and 12.9).

This means that when we prepare Grignard reagents, we are effectively limited to alkyl halides or to analogous organic halides containing carbon–carbon double bonds, internal triple bonds, ether linkages, and —NR₂ groups.

Grignard reactions are so sensitive to acidic compounds that when we prepare a Grignard reagent we must take special care to exclude moisture from our apparatus, and we must use an anhydrous ether as our solvent.

As we saw earlier, acetylenic hydrogens are acidic enough to react with Grignard reagents. This is a limitation that we can use, however.

- We can make acetylenic Grignard reagents by allowing terminal alkynes to react with alkyl Grignard reagents (cf. Section 12.7A).

We can then use these acetylenic Grignard reagents to carry out other syntheses. For example,

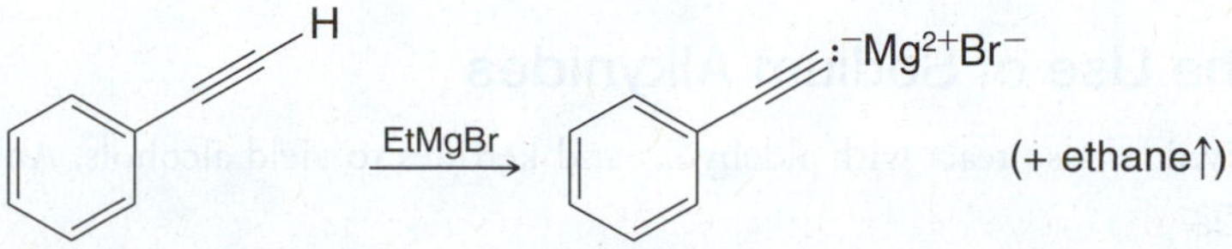

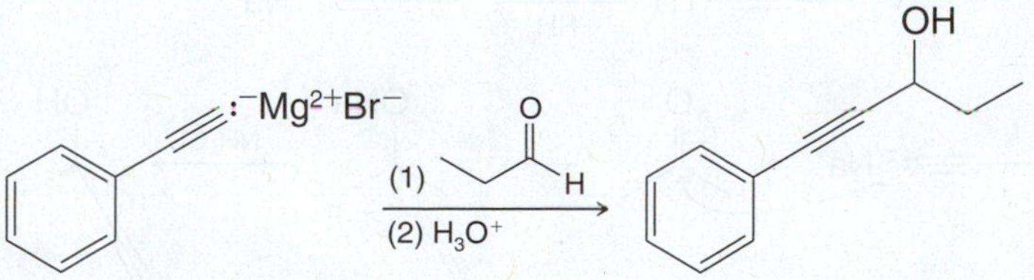

(52%)

• When we plan a Grignard synthesis, we must also take care that any aldehyde, ketone, epoxide, or ester that we use as a substrate does not also contain an acidic group (other than when we deliberately let it react with a terminal alkyne).

If we were to do this, the Grignard reagent would simply react as a base with the acidic hydrogen rather than reacting at the carbonyl or epoxide carbon as a nucleophile. If we were to treat 4-hydroxy-2-butanone with methylmagnesium bromide, for example, the reaction that would take place is

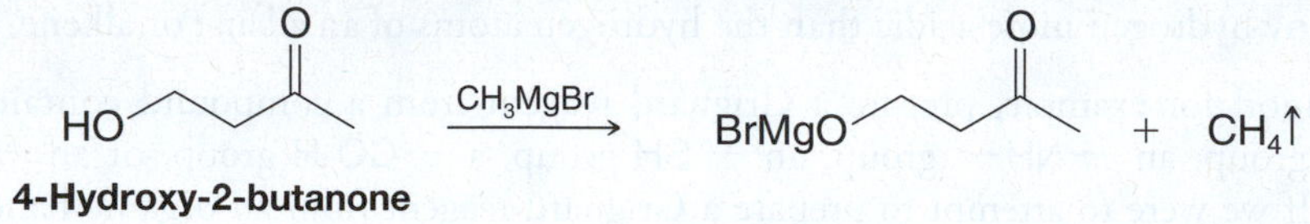

4-Hydroxy-2-butanone

rather than

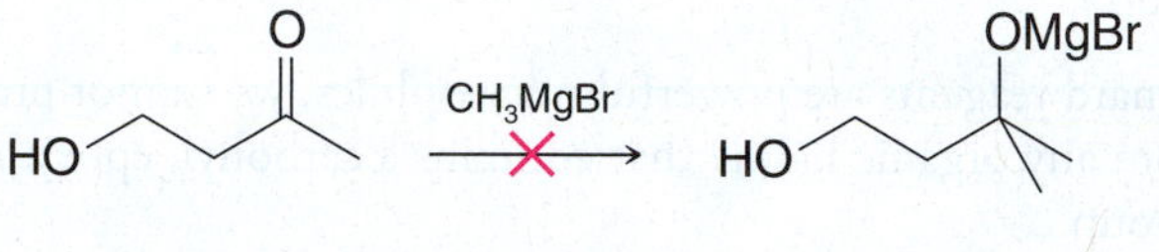

If we were prepared to waste one molar equivalent of the Grignard reagent, we can treat 4-hydroxy-2-butanone with two molar equivalents of the Grignard reagent and thereby get addition to the carbonyl group:

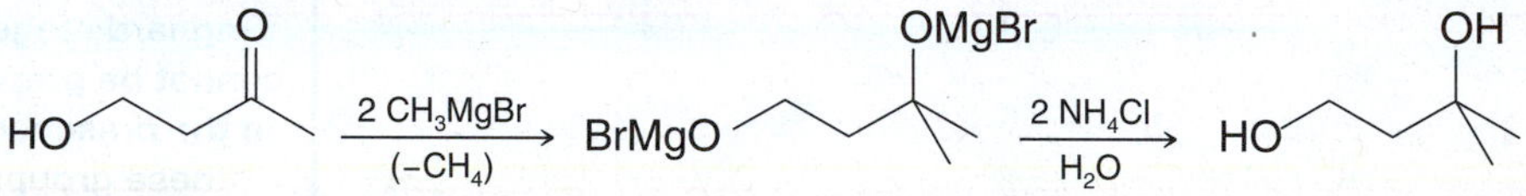

This technique is sometimes employed in small-scale reactions when the Grignard reagent is inexpensive and the other reagent is expensive.

12.8C The Use of Lithium Reagents

Organolithium reagents (RLi) react with carbonyl compounds in the same way as Grignard reagents and thus provide an alternative method for preparing alcohols.

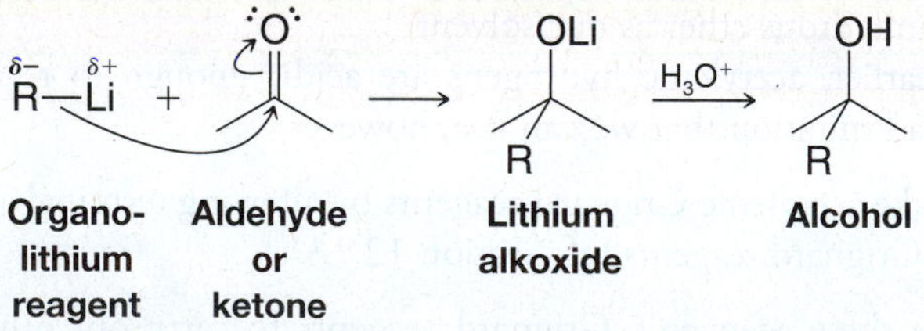

| Organo- lithium reagent | Aldehyde or ketone | Lithium alkoxide | Alcohol |

Organolithium reagents have the advantage of being somewhat more reactive than Grignard reagents although they are more difficult to prepare and handle.

12.8D The Use of Sodium Alkynides

Sodium alkynides also react with aldehydes and ketones to yield alcohols. An example is the following:

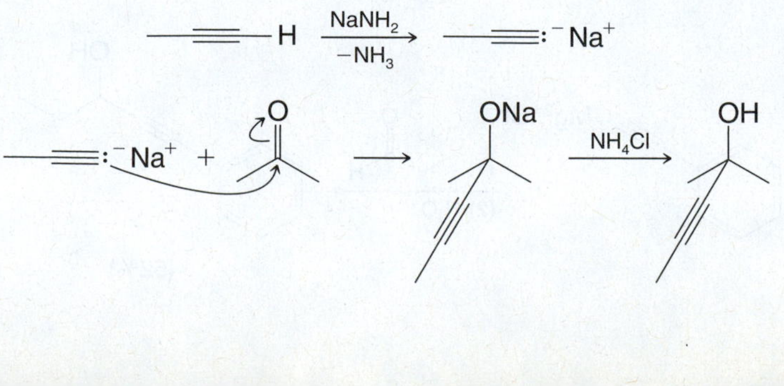

SOLVED PROBLEM 12.7

ILLUSTRATING MULTISTEP SYNTHESES: For the following compounds, write a retrosynthetic scheme and then synthetic reactions that could be used to prepare each one. Use hydrocarbons, organic halides, alcohols, aldehydes, ketones, or esters containing six carbon atoms or fewer and any other needed reagents.

(a) 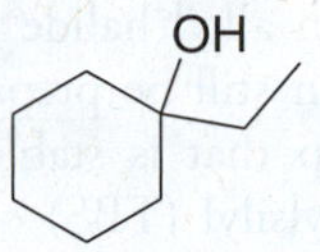

(b) 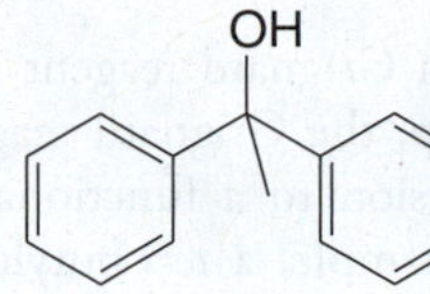

(c) 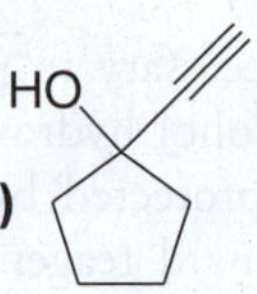

ANSWERS:

(a)

Retrosynthetic Analysis

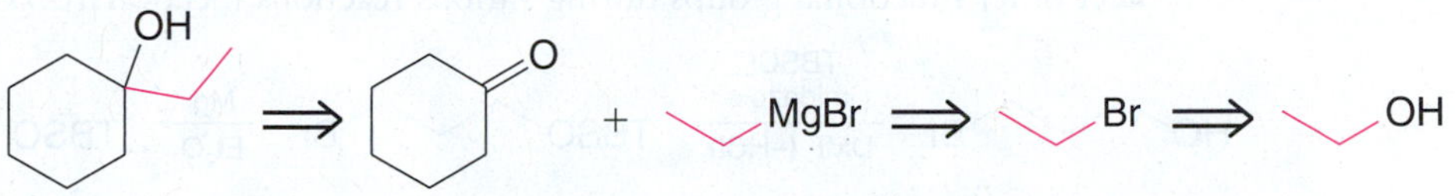

Synthesis

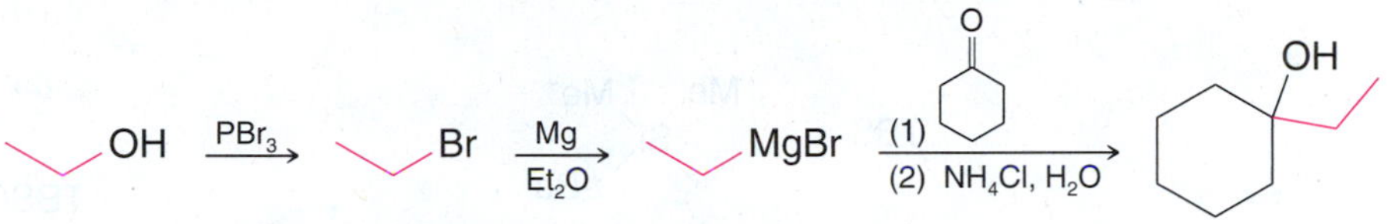

(b)

Retrosynthetic Analysis

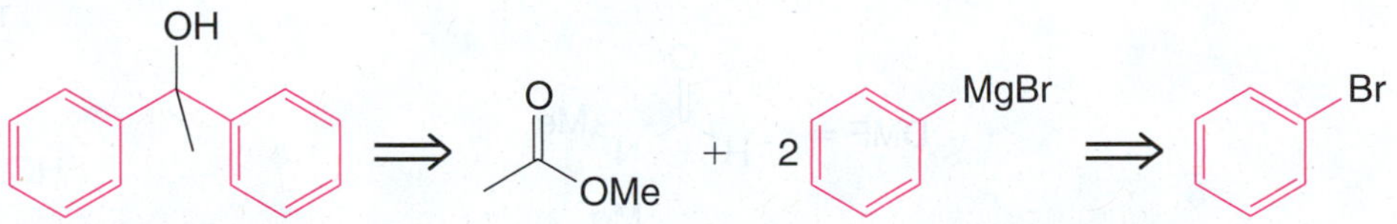

Synthesis

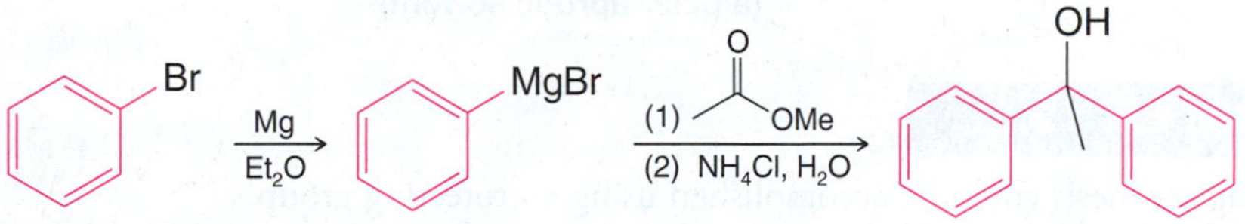

(c)

Retrosynthetic Analysis

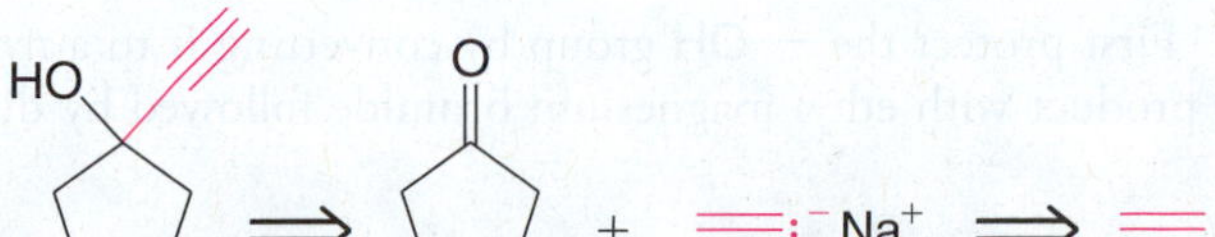

Synthesis

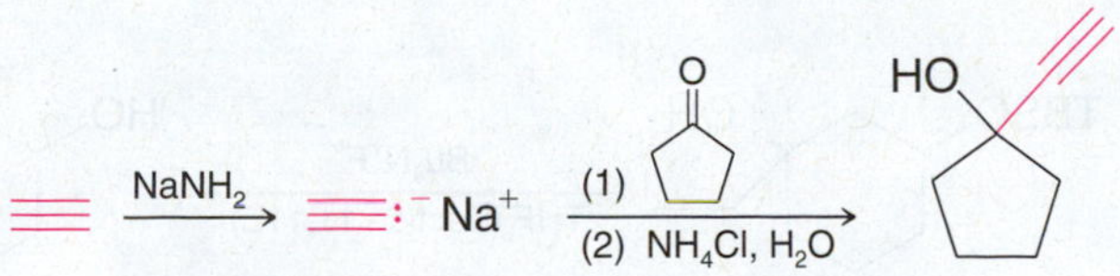

12.9 PROTECTING GROUPS

- A **protecting group** can be used in some cases where a reactant contains a group that is incompatible with the reaction conditions necessary for a given transformation.

For example, if it is necessary to prepare a Grignard reagent from an alkyl halide that already contains an alcohol hydroxyl group, the Grignard reagent can still be prepared if the alcohol is first protected by conversion to a functional group that is stable in the presence of a Grignard reagent, for example, a *tert*-butyldimethylsilyl (TBS) ether (Section 11.11E). The Grignard reaction can be conducted, and then the original alcohol group can be liberated by cleavage of the silyl ether with fluoride ion (see Problem 12.36). An example is the following synthesis of 1,4-pentanediol. This same strategy can be used when an organolithium reagent or alkynide anion must be prepared in the presence of an incompatible group. In later chapters we will encounter strategies that can be used to protect other functional groups during various reactions (Section 16.7C).

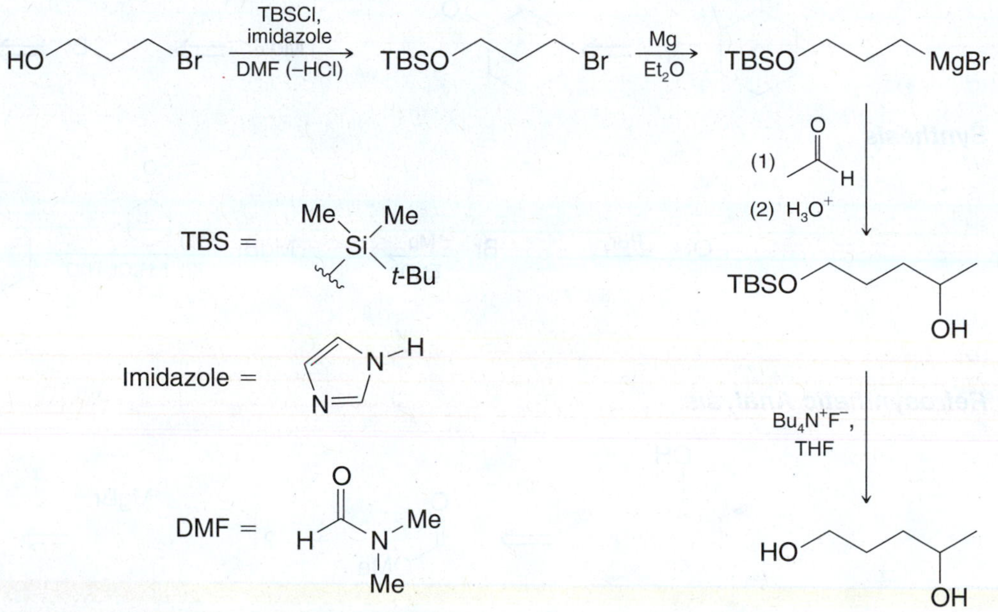

Dimethyl formamide (a polar aprotic solvent)

1,4-Pentanediol

●●● **SOLVED PROBLEM 12.8**

Show how the following synthesis could be accomplished using a protecting group.

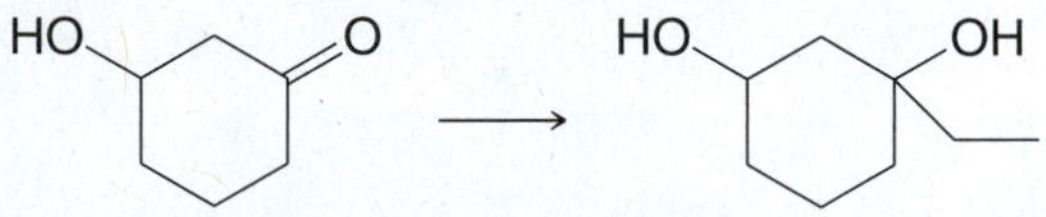

STRATEGY AND ANSWER: First protect the —OH group by converting it to a *tert*-butyldimethylsilyl (TBS) ether (Section 11.11E), then treat the product with ethyl magnesium bromide followed by dilute acid. Then remove the protecting group.

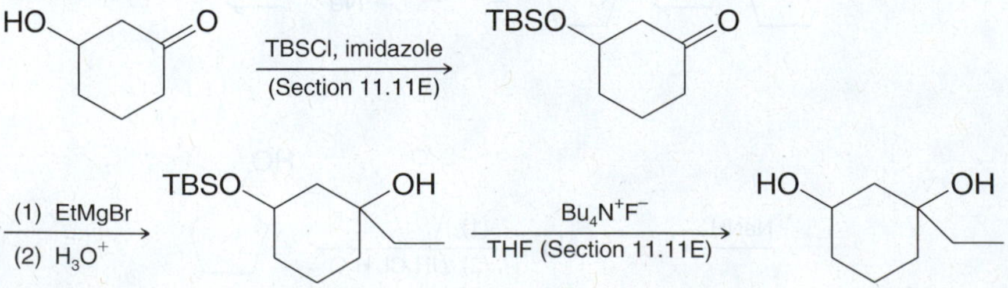

[WHY Do These Topics Matter?

CHANGING PROPERTIES BY CHANGING OXIDATION STATE

Although you have now learned several tools to interconvert primary and secondary alcohols into aldehydes, ketones, and carboxylic acids, what you may not have fully realized is how those operations can alter a compound's properties. Specifically, we mean changes other than the standard ones of melting or boiling points, polarity, and physical appearance (i.e., solid versus a liquid) that are true of any functional group alteration. Indeed, moving from a hydroxyl group to a carbonyl group or vice versa causes many molecules to have completely different biochemical profiles as well, something that occurs frequently in nature. Here we will consider just a few examples.

Codeine, a natural compound found in opium poppies, is currently prescribed as a medication to treat mild or moderate pain (i.e., as an analgesic). If its secondary alcohol is oxidized to a ketone, however, a compound known as codeinone results. While it also can serve as a pain medication, it is only 33% as effective as codeine. Similarly, pregnenolone is a steroid used in the body largely as a key synthetic precursor to progesterone. In the needed oxidation event, it turns out that not only is the alcohol oxidized, but the neighboring double bond moves into conjugation as well, a phenomenon that will make more sense once

© NHPA/Photoshot Holdings Ltd.

you have read Chapter 13. For now, however, what is important to note is that the new molecule that is created plays a critical role in the menstrual cycle and in pregnancy. In fact, progesterone is currently prescribed in many different forms, particularly to support a woman's effort to become pregnant during procedures such as *in vitro* fertilization (IVF). Pregnenolone itself does not appear to have such important properties, though intriguingly at least one derivative of its alcohol function can promote the generation of neurons in the hippocampus, a region of the brain that is affected by Alzheimer's disease.

Codeine   [O] →   **Codeinone**   **Pregnenolone**   [O] →   **Progesterone**

As a final example, consider the structure of borneol, a compound found in several plant species and used in some traditional Chinese medicines. This compound is a component of several essential oils and is a natural insect repellant. When it is oxidized, a new natural product results—camphor. Camphor has many additional uses, from serving as a plasticizer, to being a flavorant in several foods, as well as being an active ingredient in products such as Vicks VapoRub®. Interestingly, an attempt to reduce the alcohol within camphor with a simple reagent such as **NaBH₄** creates isoborneol rather than borneol because the steric bulk of the methyl groups on the upper carbon bridge ensures that hydride adds from the bottom face. Isoborneol, in fact, is quite similar to camphor in its properties. Overall, it is pretty amazing what some small adjustments in oxidation state can do!

Borneol   [O] →   **Camphor**   NaBH₄ →   **Isoborneol**

SUMMARY AND REVIEW TOOLS

The study aids for this chapter include key terms and concepts (which are hyperlinked to the Glossary from the bold, blue terms in the *WileyPLUS* version of the book at wileyplus.com) and Synthetic Connections summaries of oxidation, reduction, and carbon–carbon bond-forming reactions related to alcohol and carbonyl compounds.

PROBLEMS *WILEY PLUS*

Note to Instructors: Many of the homework problems are available for assignment via *WileyPLUS*, an online teaching and learning solution.

REAGENTS AND REACTIONS

12.11 What products would you expect from the reaction of ethylmagnesium bromide (CH_3CH_2MgBr) with each of the following reagents?

(a) H_2O

(b) D_2O

(c) 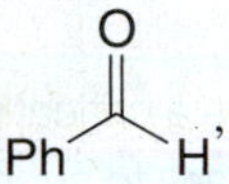 Ph–CHO, then H_3O^+

(d) 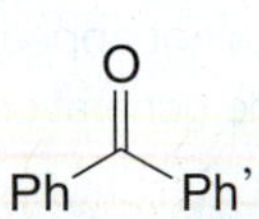 Ph–CO–Ph, then NH_4Cl, H_2O

(e) 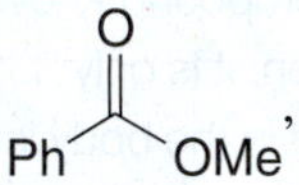 Ph–CO–OMe, then NH_4Cl, H_2O

(f) 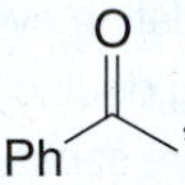 Ph–CHO, then NH_4Cl, H_2O

(g) alkyne–H, then acetaldehyde, then H_3O^+

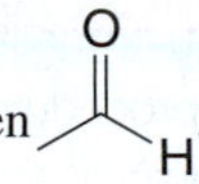

12.12 What products would you expect from the reaction of propyllithium ($CH_3CH_2CH_2Li$) with each of the following reagents?

(a) 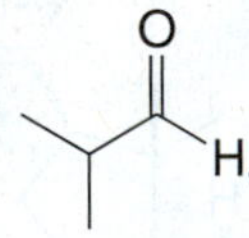 isobutyraldehyde–H, then H_3O^+

(b) 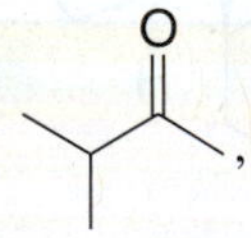 methyl isopropyl ketone, then NH_4Cl, H_2O

(c) 1-Pentyne, then acetone, then NH_4Cl, H_2O

(d) Ethanol

(e) 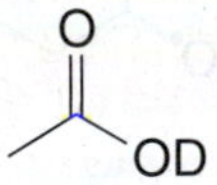 ester–OD

12.13 What product (or products) would be formed from the reaction of 1-bromo-2-methylpropane (isobutyl bromide) under each of the following conditions?

(a) HO^-, H_2O
(b) NC^-, ethanol
(c) *t*-BuOK, *t*-BuOH
(d) MeONa, MeOH
(e) (1) Li, Et_2O; (2) 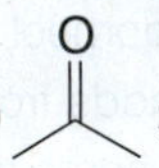 ketone ; (3) NH_4Cl, H_2O
(f) Mg, Et_2O, then CH_3CH=O, then H_3O^+

(g) (1) Mg, Et_2O; (2) 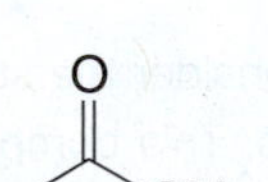 ester–OMe ; (3) NH_4Cl, H_2O
(h) (1) Mg, Et_2O; (2)  epoxide ; (3) H_3O^+
(i) (1) Mg, Et_2O; (2) 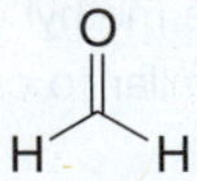 H–CO–H ; (3) NH_4Cl, H_2O
(j) Li, Et_2O; (2) MeOH
(k) Li, Et_2O; (2) H–C≡C–H

12.14 Which oxidizing or reducing agent would you use to carry out the following transformations?

(a) 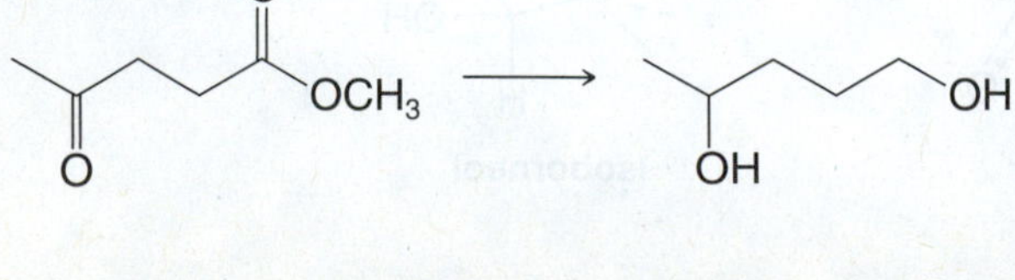 ketodiester–OCH_3 → diol–OH, OH

(b) 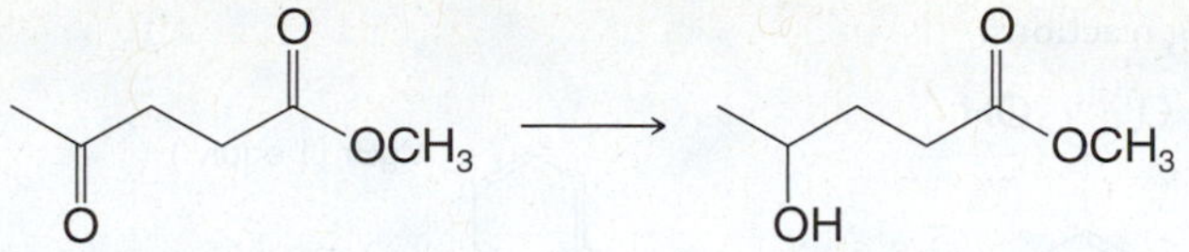
O
OCH3
O
O
OCH3
OH

(c) 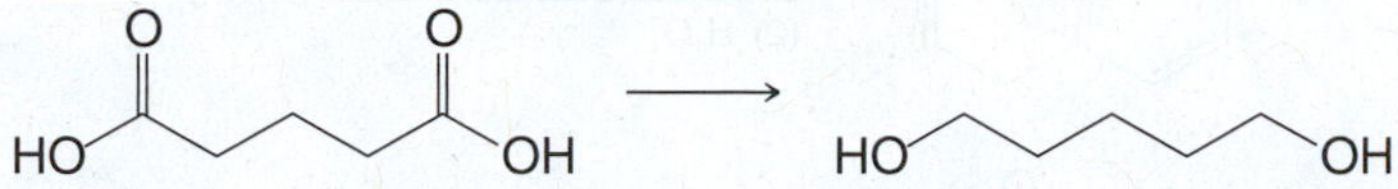
O O
HO OH
HO OH

(d) 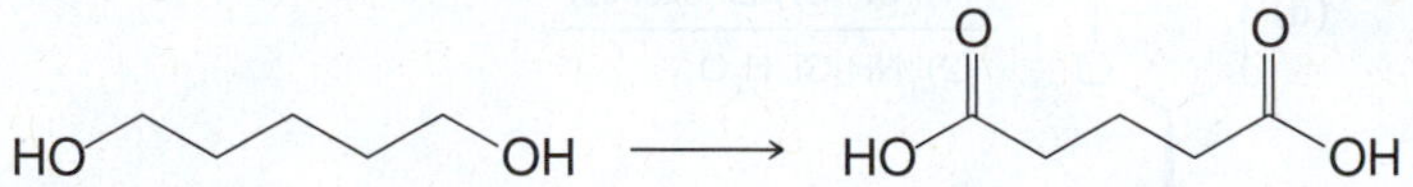
HO OH
O O
HO OH

(e) 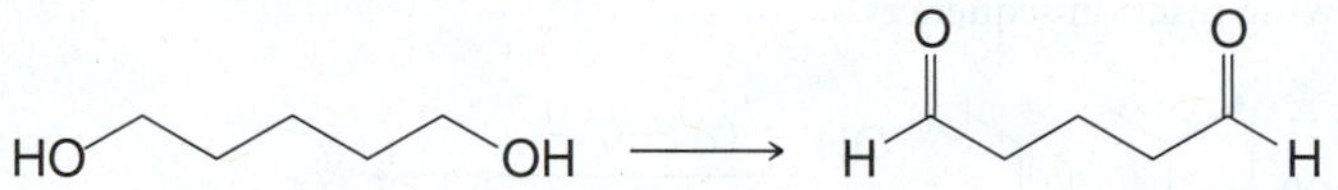
HO OH
O O
H H

12.15 Write reaction conditions and the product from Swern oxidation of the following compounds.

(a) 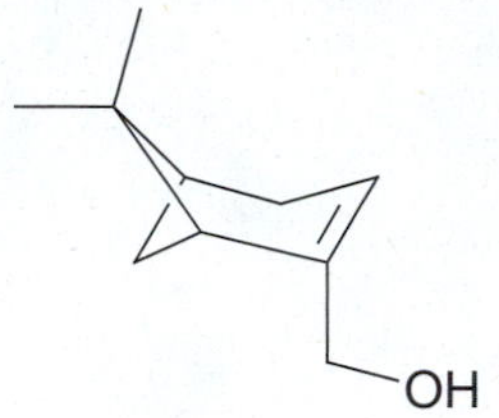
OH

Myrtenol

(b) 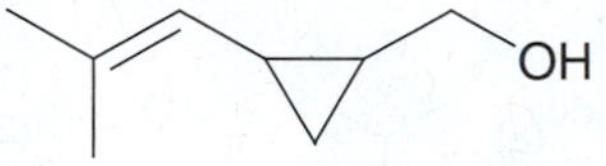
OH

Chrysanthemyl alcohol

12.16 Predict the products of the following reactions.

(a) 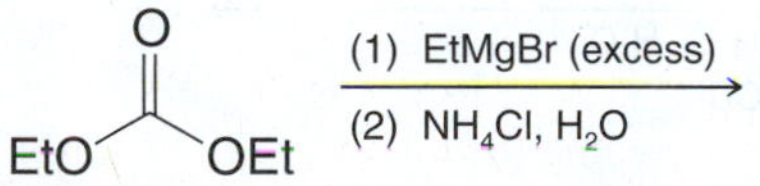
O
EtO OEt
(1) EtMgBr (excess)
(2) NH4Cl, H2O

(b) 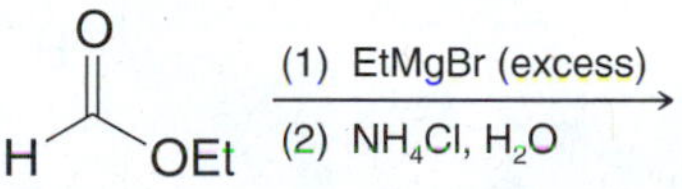
O
H OEt
(1) EtMgBr (excess)
(2) NH4Cl, H2O

12.17 Predict the organic product from each of the following reduction reactions.

(a) 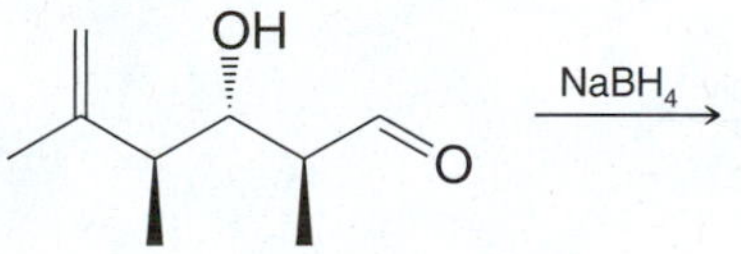
OH
O
NaBH4

(b) 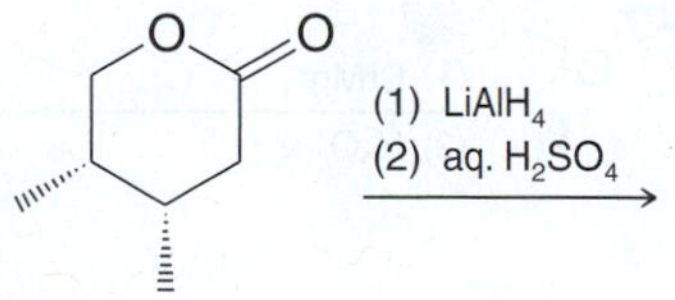
O O
(1) LiAlH4
(2) aq. H2SO4

(c) 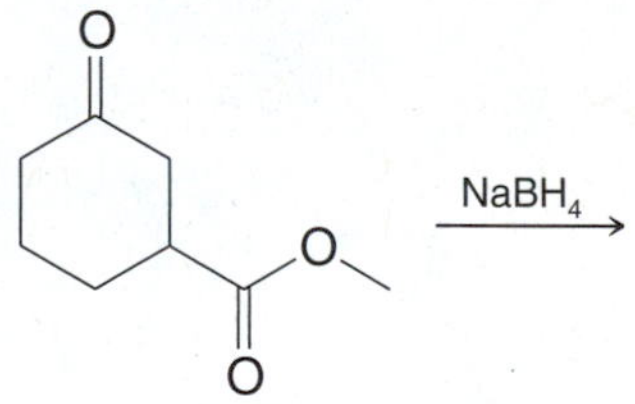
O
O
O
NaBH4

12.18 Predict the organic product from each of the following oxidation reactions.

(a) 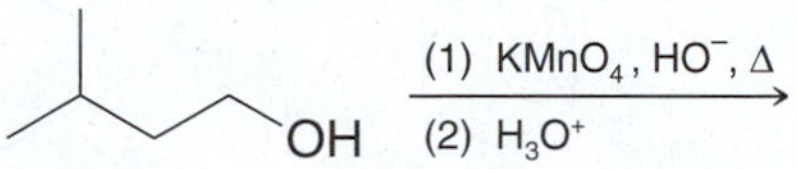
OH
(1) KMnO4, HO⁻, Δ
(2) H3O+

(c) 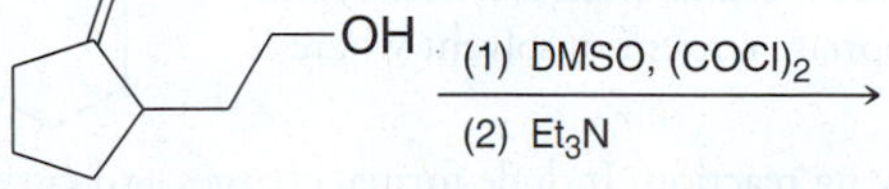
OH
(1) DMSO, (COCl)2
(2) Et3N

(e) 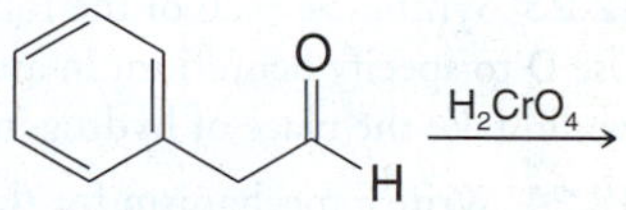
O
H
H2CrO4

(b) 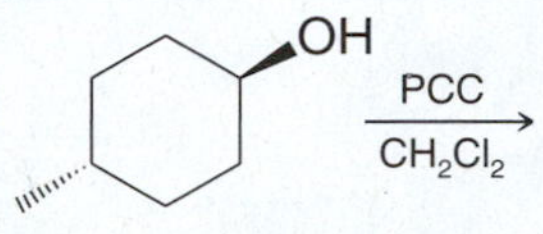
OH
PCC
CH2Cl2

(d) 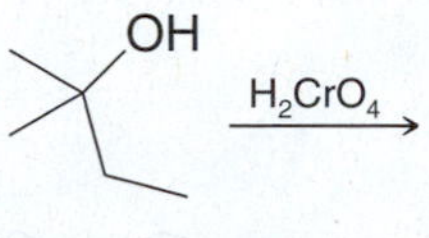
OH
H2CrO4

12.19 Predict the organic product from each of the following oxidation and reduction reactions.

(a) 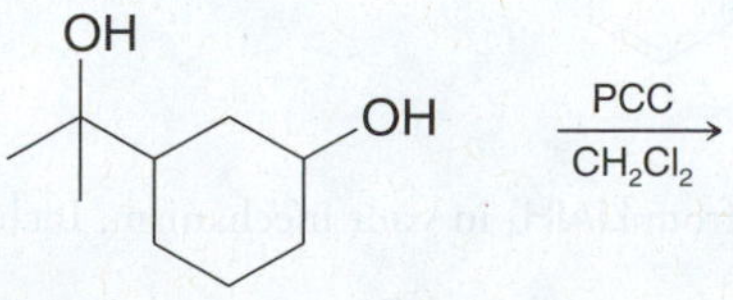
OH
OH
PCC
CH2Cl2

(c) 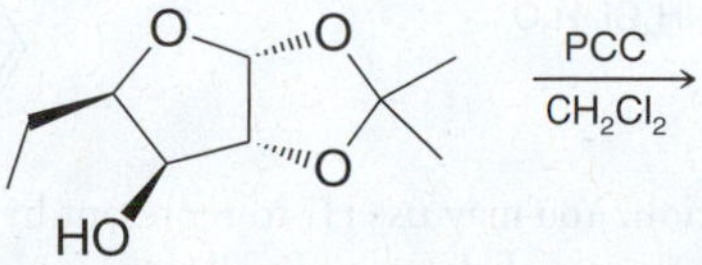
O
O
O
HO
PCC
CH2Cl2

(e) 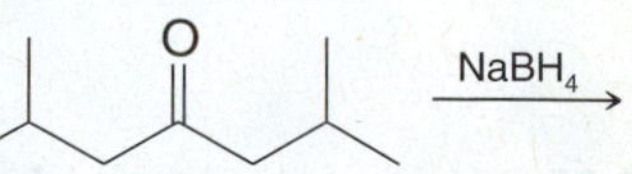
O
NaBH4

(b) 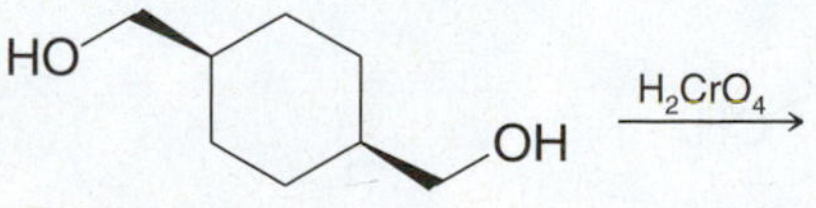
HO
OH
H2CrO4

(d) 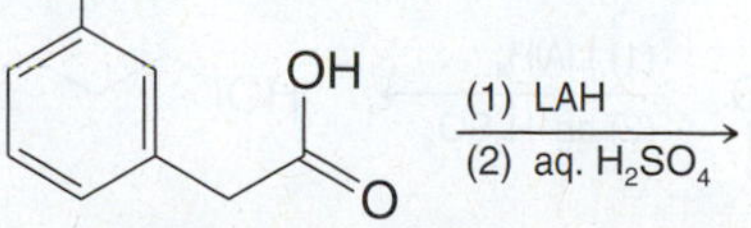
OH
O
(1) LAH
(2) aq. H2SO4

12.20 Predict the major organic product from each of the following reactions.

(a)
(1) CH₃MgBr
(2) H₃O⁺

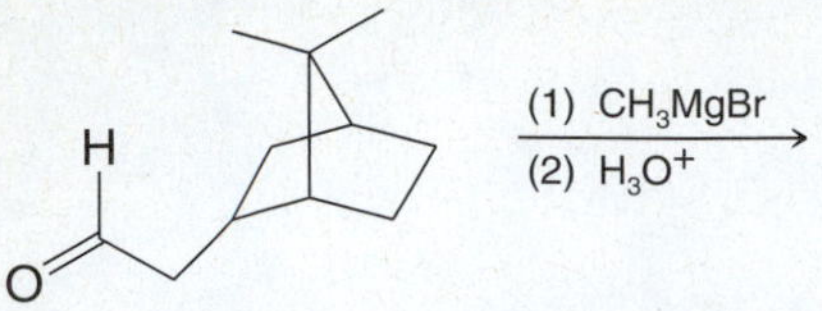

(c)
(1) [4-methoxyphenyl]MgBr (1 equiv.)
(2) H₃O⁺

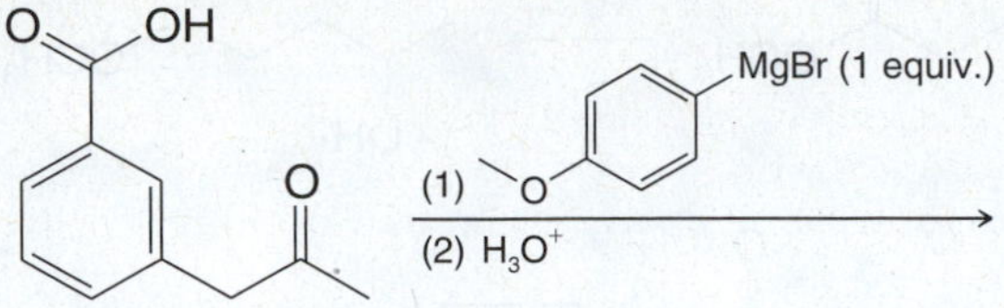

(b)
(1) [allyl]MgBr
(2) NH₄Cl, H₂O

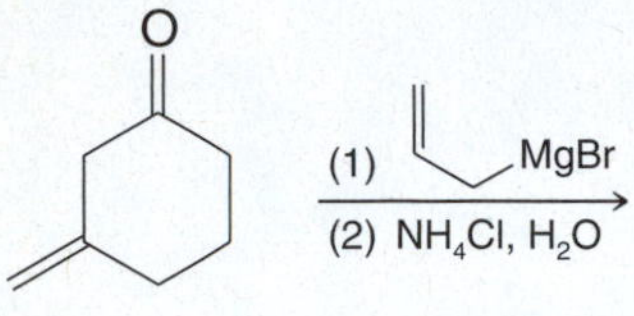

(d)
(1) CH₃CH₂Li (excess)
(2) NH₄Cl, H₂O

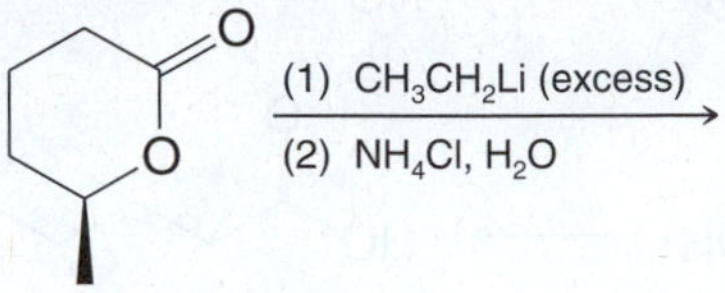

12.21 Predict the major organic product from each of the following reaction sequences.

(a)
(1) MeMgBr (excess)
(2) NH₄Cl, H₂O

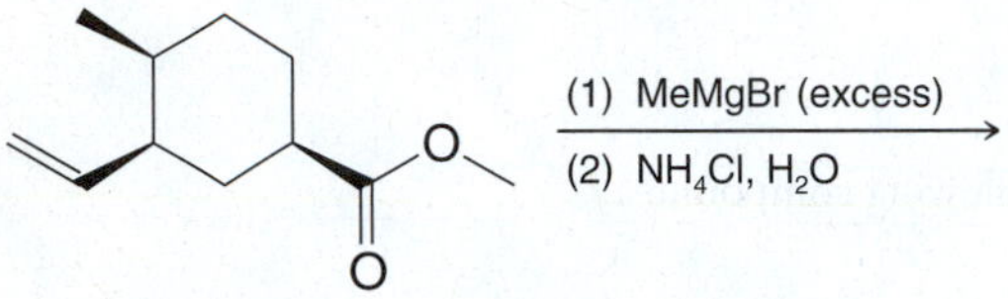

(d)
(1) PCC
(2) [propyl]MgBr
(3) H₃O⁺

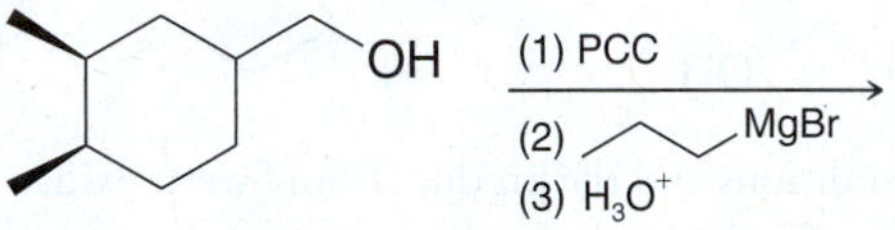

(b)
(1) Mg
(2) H–CHO
(3) H₃O⁺

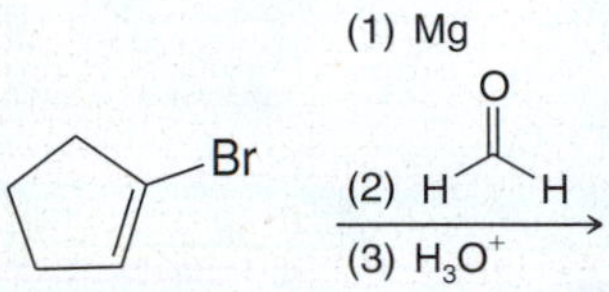

(e)
(1) EtMgBr
(2) H₃O⁺
(3) NaH
(4) CH₃Br

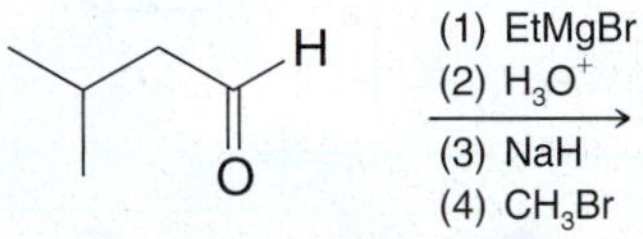

(c)
(1) PBr₃
(2) Mg
(3) H₃O⁺

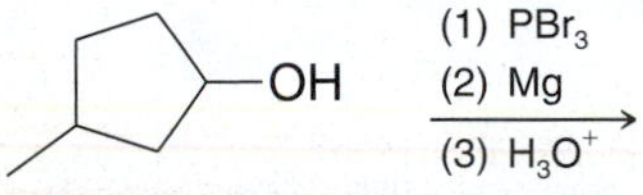

(f)
(1) [reagent] MgBr (excess)
(2) NH₄Cl, H₂O
(3) PCC

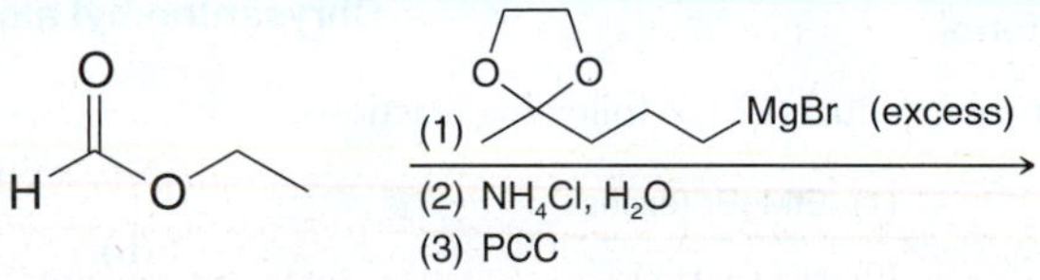

12.22 Predict the product of the following reaction.

(1) BrMg~~~~~~~~~MgBr (1 equiv.)
(2) H₃O⁺

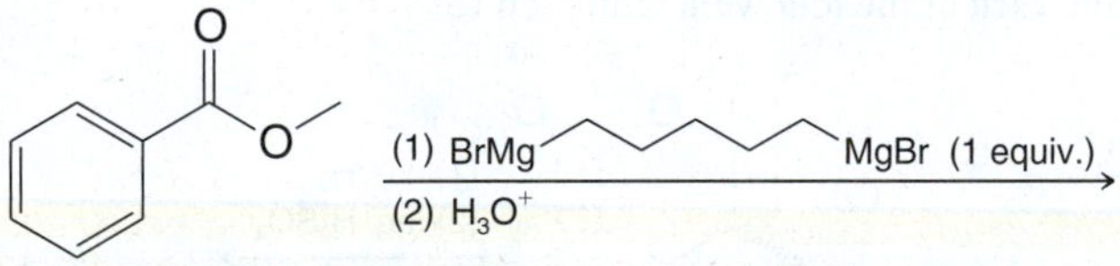

MECHANISMS

12.23 Synthesize each of the following compounds from cyclohexanone. Use D to specify deuterium in any appropriate reagent or solvent where it would take the place of hydrogen.

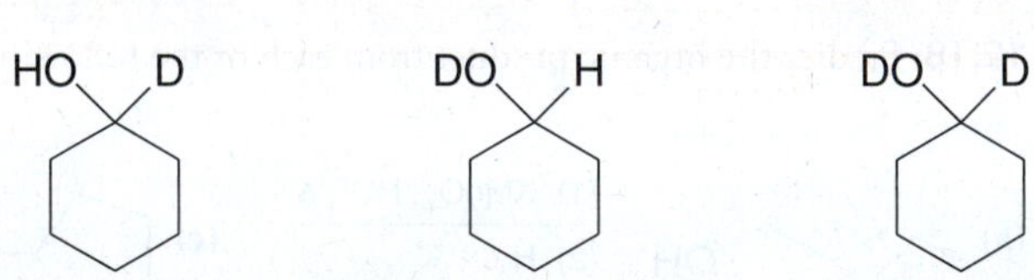

12.24 Write a mechanism for the following reaction. Include formal charges and curved arrows to show the movement of electrons in all steps.

(1) [phenyl]MgBr (excess)
(2) NH₄Cl, H₂O

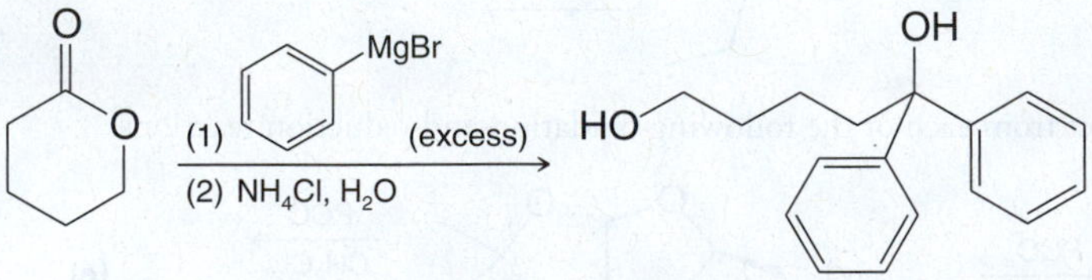

12.25 Write a mechanism for the following reaction. You may use H⁻ to represent hydride ions from LiAlH₄ in your mechanism. Include formal charges and curved arrows to show the movement of electrons in all steps.

(1) LiAlH₄
(2) aq. H₂SO₄

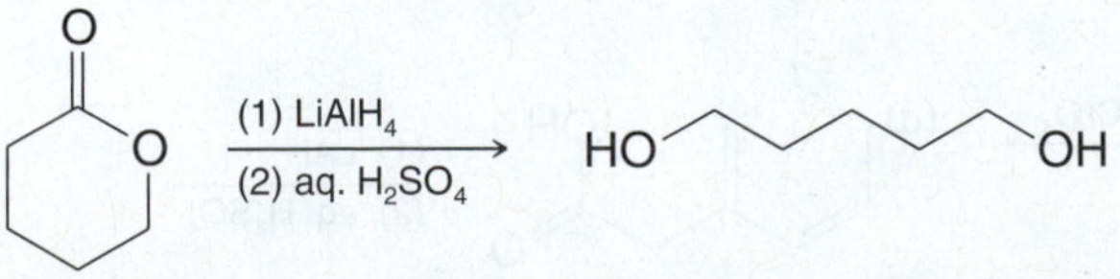

12.26 Although oxirane (oxacyclopropane) and oxetane (oxacyclobutane) react with Grignard and organolithium reagents to form alcohols, tetrahydrofuran (oxacyclopentane) is so unreactive that it can be used as the solvent in which these organometallic compounds are prepared. Explain the difference in reactivity of these oxygen heterocycles.

12.27 Studies suggest that attack by a Grignard reagent at a carbonyl group is facilitated by involvement of a second molecule of the Grignard reagent that participates in an overall cyclic ternary complex. The second molecule of Grignard reagent assists as a Lewis acid. Propose a structure for the ternary complex and write all of the products that result from it.

SYNTHESIS

12.28 What organic products **A–H** would you expect from each of the following reactions?

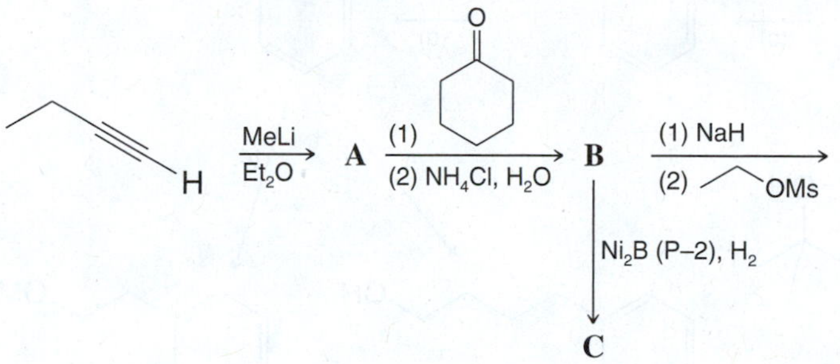

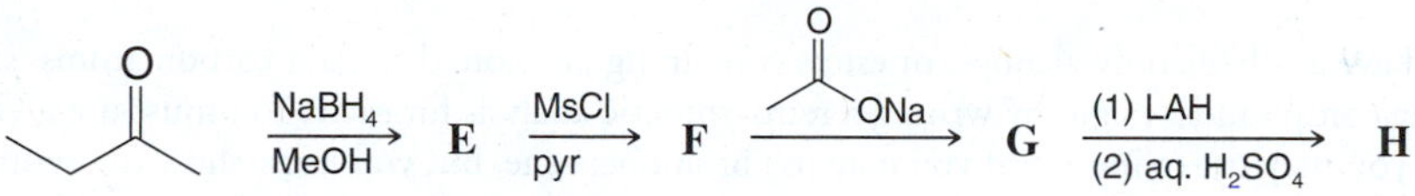

12.29 Outline all steps in a synthesis that would transform 2-propanol (isopropyl alcohol) into each of the following:

(a)

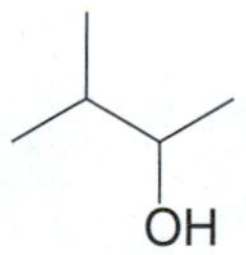

(d)

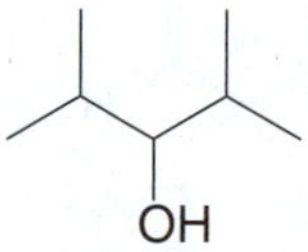

(b)

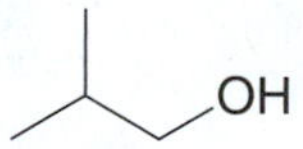

(e)

(c)

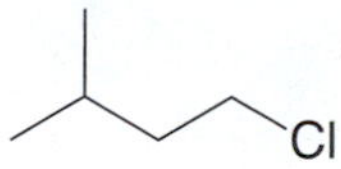

12.30 Show how 1-pentanol could be transformed into each of the following compounds. (You may use any needed inorganic reagents and you need not show the synthesis of a particular compound more than once.)

(a) 1-Bromopentane
(b) 1-Pentene
(c) 2-Pentanol
(d) Pentane
(e) 2-Bromopentane
(f) 1-Hexanol
(g) 1-Heptanol

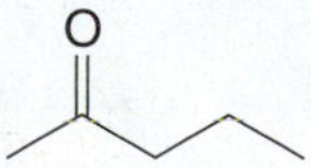

(j) Pentanoic acid,

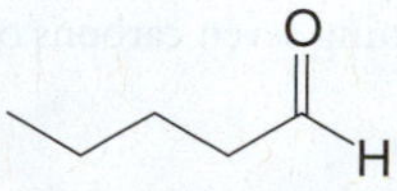

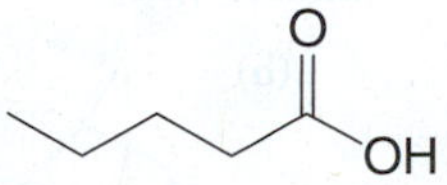

(k) Dipentyl ether (two ways)
(l) 1-Pentyne
(m) 2-Bromo-1-pentene
(n) Pentyllithium
(o) 4-Methyl-4-nonanol

(h) Pentanal,

(i) 2-Pentanone,

12.31 Provide the reagents needed to accomplish transformations (a)–(g). More than one step may be necessary.

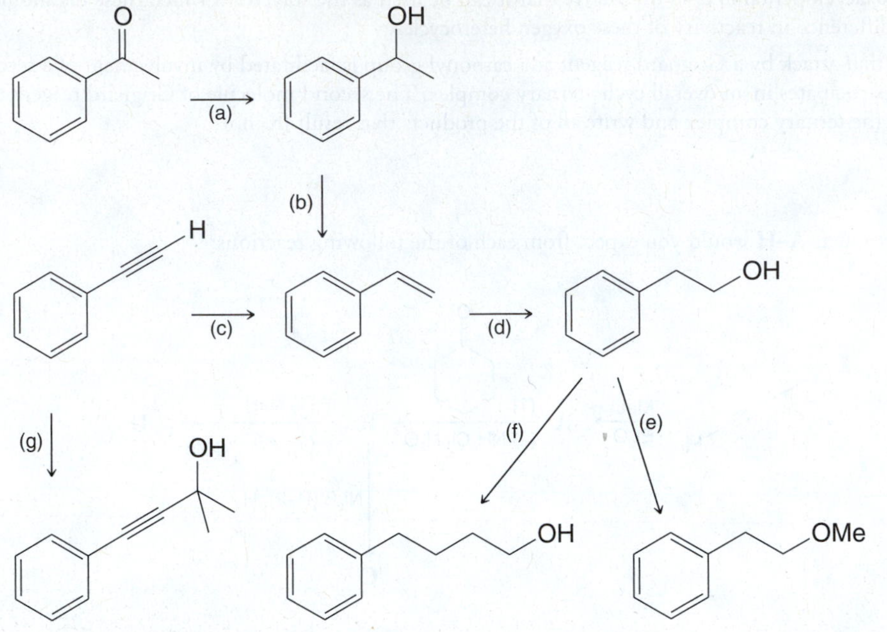

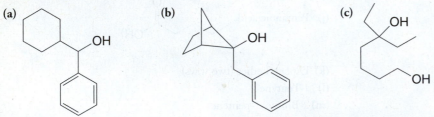

12.32 Assuming that you have available only alcohols or esters containing no more than four carbon atoms, show how you might synthesize each of the following compounds. Begin by writing a retrosynthetic analysis for each. You must use a Grignard reagent at one step in the synthesis. If needed, you may use oxirane and you may use bromobenzene, but you must show the synthesis of any other required organic compounds. Assume you have available any solvents and any inorganic compounds, including oxidizing and reducing agents, that you require.

(a)

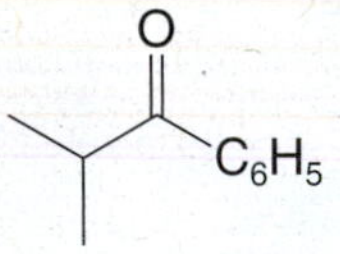

(b)

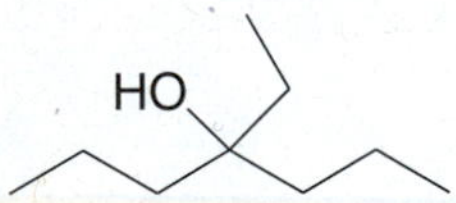

(c)

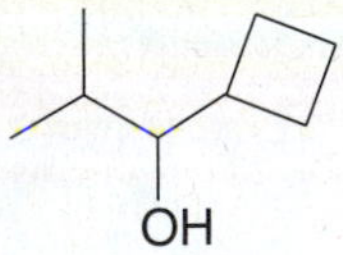

(d)

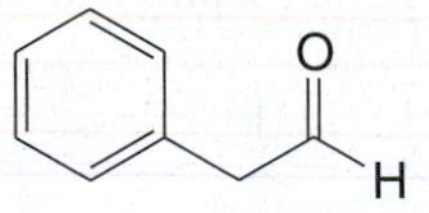

(e)

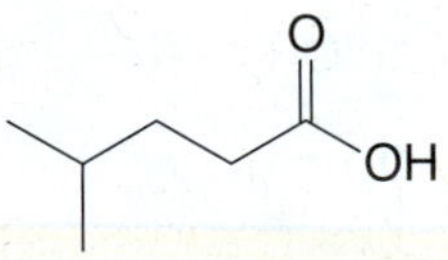

(f)

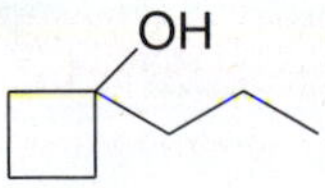

(g)

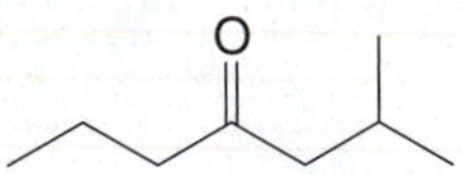

(h)

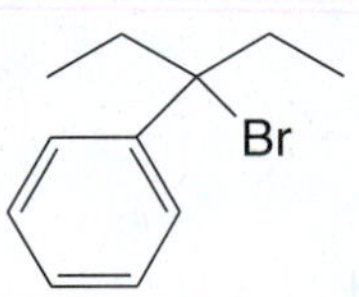

12.33 For each of the following alcohols, write a retrosynthetic analysis and synthesis that involves an appropriate organometallic reagent (either a Grignard or alkyllithium reagent).

(a)

(b)

(c)

12.34 Synthesize each of the following compounds starting from primary or secondary alcohols containing seven carbons or less and, if appropriate, bromobenzene.

(a)

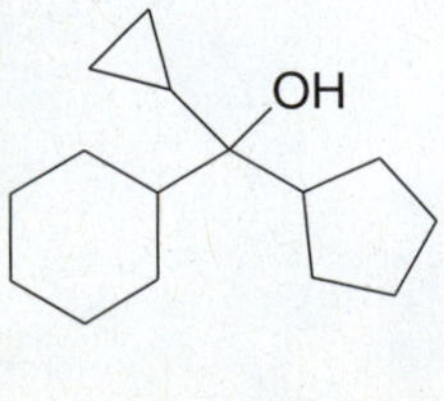

(b)

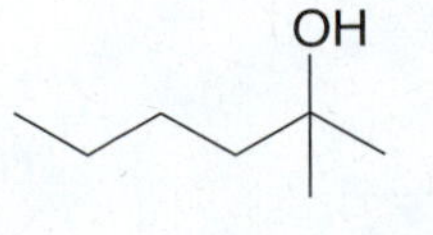

(c)

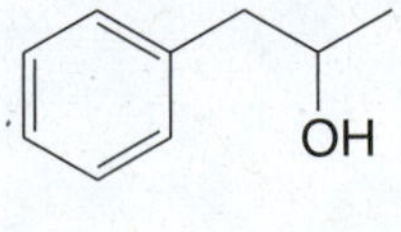

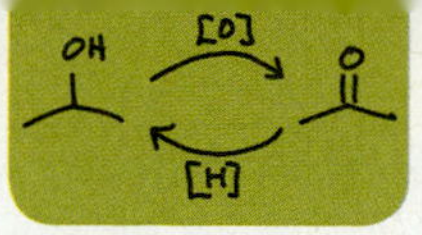

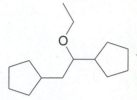

12.35 The alcohol shown here is used in making perfumes. Write a retrosynthetic analysis and then synthetic reactions that could be used to prepare this alcohol from bromobenzene and 1-butene.

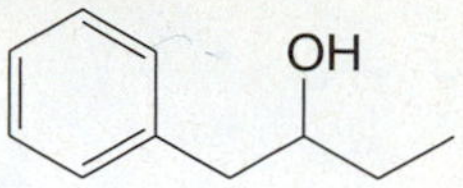

12.36 Write a retrosynthetic analysis and then synthetic reactions that could be used to prepare racemic Meparfynol, a mild hypnotic (sleep-inducing compound), starting with compounds of four carbon atoms or fewer.

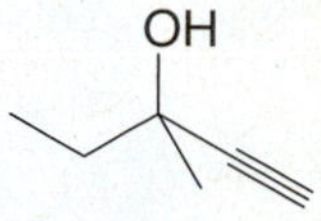

Meparfynol

12.37 Write a retrosynthetic analysis and synthesis for the following transformation.

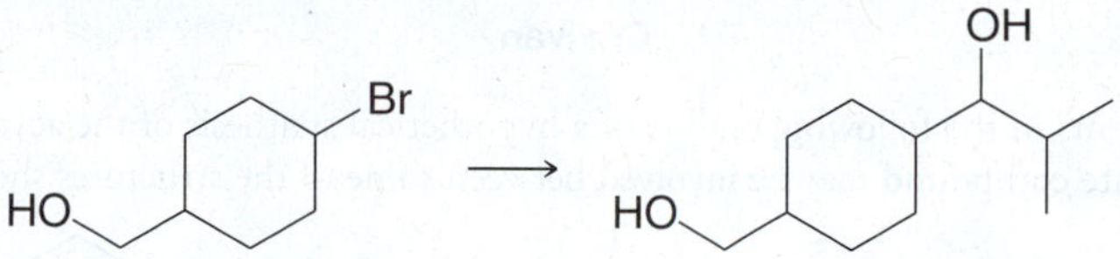

12.38 Synthesize the following compound using cyclopentane and ethyne (acetylene) as the sole source of carbon atoms.

CHALLENGE PROBLEMS

12.39 Explain how ^{1}H NMR, ^{13}C NMR, and IR spectroscopy could be used to differentiate among the following compounds.

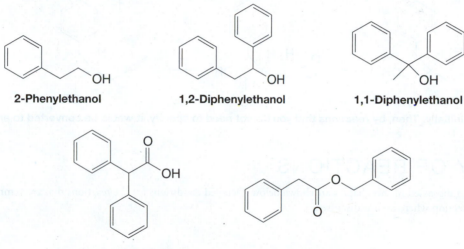

2-Phenylethanol   **1,2-Diphenylethanol**   **1,1-Diphenylethanol**

2,2-Diphenylethanoic acid   **Benzyl 2-phenylethanoate**

12.40 When sucrose (common table sugar) is treated with aqueous acid, it is cleaved and yields simpler sugars of these types:

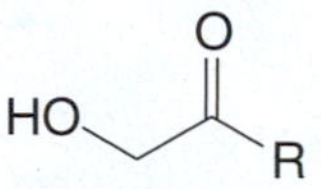

and

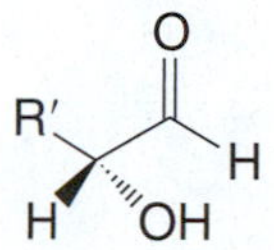

For reasons to be studied later, in the use of this procedure for the identification of the sugars incorporated in a saccharide like sucrose, the product mixtures are often treated with sodium borohydride before analysis. What limitation(s) does this put on identification of the sugar building blocks of the starting saccharide?

12.41 An unknown **X** shows a broad absorption band in the infrared at 3200–3550 cm^{-1} but none in the 1620–1780 cm^{-1} region. It contains only C, H, and O. A 116-mg sample was treated with an excess of methylmagnesium bromide, producing 48.7 mL of methane gas collected over mercury at 20 °C and 750 mm Hg. The mass spectrum of **X** has its molecular ion (barely detectable) at 116 *m/z* and a fragment peak at 98. What does this information tell you about the structure of **X**?

LEARNING GROUP PROBLEMS

The problem below is directed toward devising a hypothetical pathway for the synthesis of the acyclic central portion of Crixivan (Merck and Company's HIV protease inhibitor). Note that your synthesis might not adequately control the stereochemistry during each step, but for this particular exercise that is not expected.

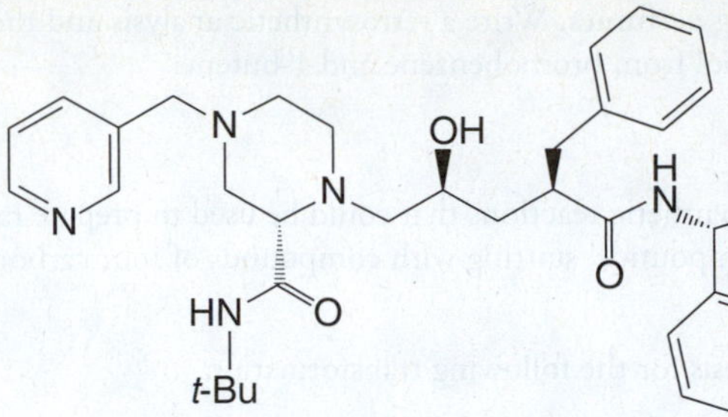

Crixivan

Fill in missing compounds and reagents in the following outline of a hypothetical synthesis of the acyclic central portion of Crixivan. Note that more than one intermediate compound may be involved between some of the structures shown below.

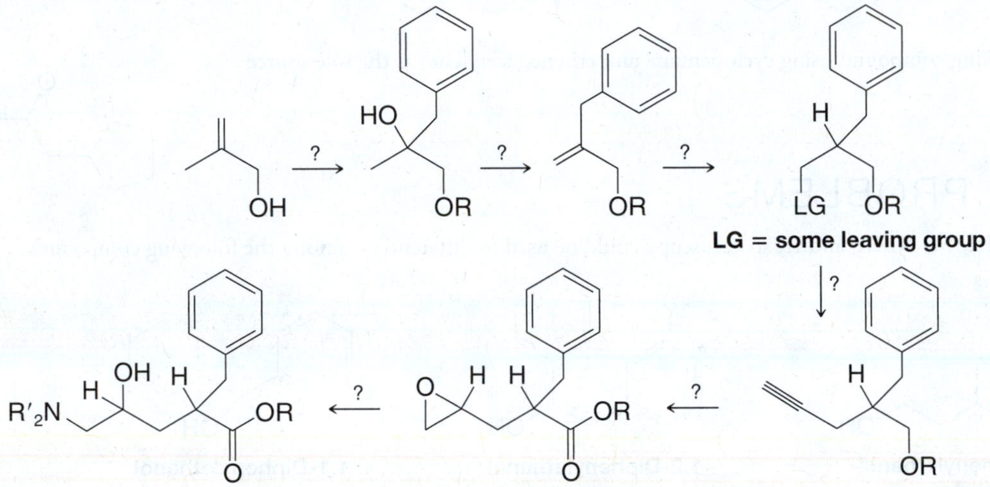

LG = some leaving group

(R would be H initially. Then, by reactions that you do not need to specify, it would be converted to an alkyl group.)

SUMMARY OF REACTIONS

Summaries of reactions discussed in this chapter are shown below. Detailed conditions for the reactions that are summarized can be found in the chapter section where each is discussed.

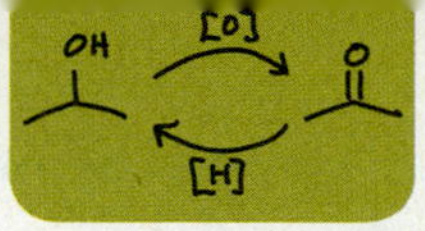

[SUMMARY AND REVIEW TOOLS]

Synthetic Connections of Alcohols and Carbonyl Compounds

1. Carbonyl Reduction Reactions

- Aldehydes to primary alcohols
- Ketones to secondary alcohols
- Esters to alcohols
- Carboxylic acids to primary alcohols

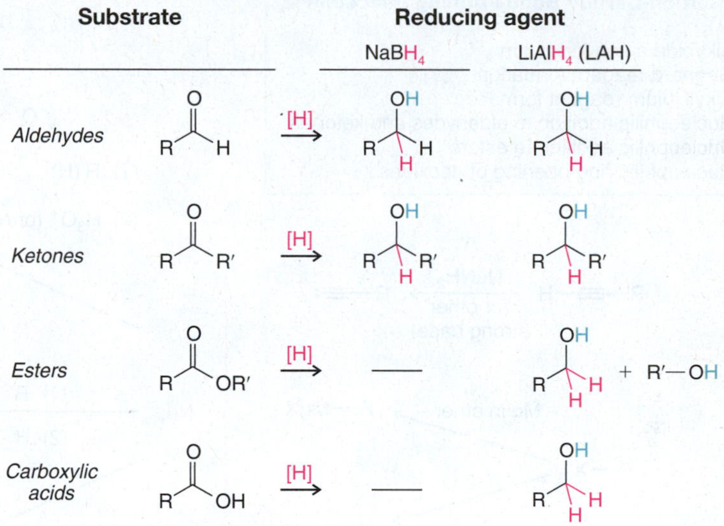

Substrate		Reducing agent	
		NaBH$_4$	LiAlH$_4$ (LAH)
Aldehydes	R–CHO	R–CH(OH)H	R–CH(OH)H
Ketones	R–CO–R'	R–CH(OH)R'	R–CH(OH)R'
Esters	R–CO–OR'	—	R–CH(OH)H + R'—OH
Carboxylic acids	R–CO–OH	—	R–CH(OH)H

(Hydrogen atoms in blue are added during the reaction workup by water or aqueous acid.)

2. Alcohol Oxidation Reactions

- Primary alcohols to aldehydes
- Primary alcohols to carboxylic acids
- Secondary alcohols to ketones

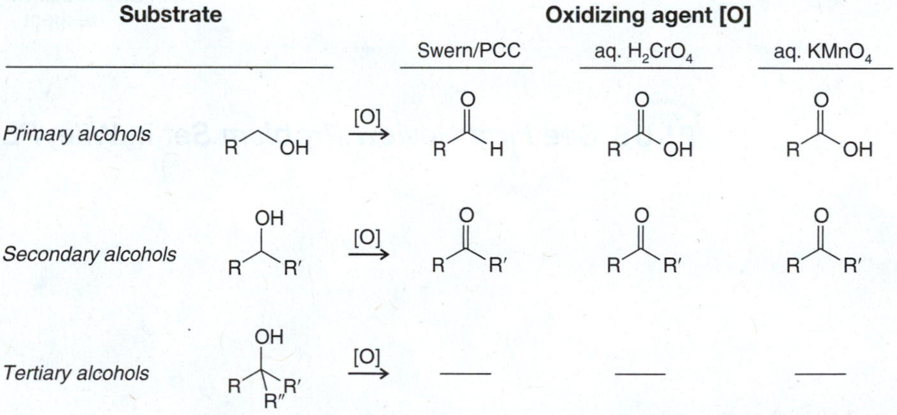

Substrate		Oxidizing agent [O]		
		Swern/PCC	aq. H$_2$CrO$_4$	aq. KMnO$_4$
Primary alcohols	R–OH	R–CHO	R–CO–OH	R–CO–OH
Secondary alcohols	R–CH(OH)–R'	R–CO–R'	R–CO–R'	R–CO–R'
Tertiary alcohols	R–C(OH)(R')–R''	—	—	—

[**S U M M A R Y   A N D   R E V I E W   T O O L S**]

Synthetic Connections of Alcohols and Carbonyl Compounds

3. Carbon-Carbon Bond Forming Reactions

• Alkynide anion formation
• Grignard reagent formation
• Alkyllithium reagent formation
• Nucleophilic addition to aldehydes and ketones
• Nucleophilic addition to esters
• Nucleophilic ring-opening of epoxides

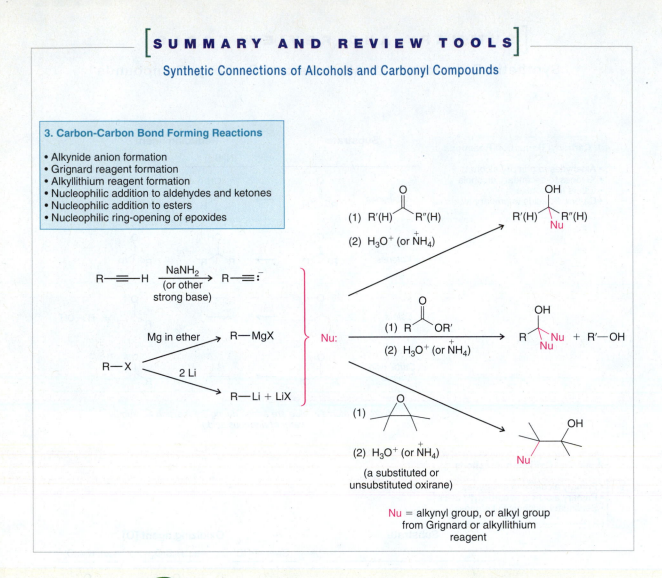

PLUS See *First Review Problem Set* in *WileyPLUS*

CHAPTER
13

Conjugated Unsaturated Systems

What do the colors of a plant, a carrot, your favorite pair of blue jeans, and our ability to see all have in common? They all result from molecules that have a sequence of alternating double and single bonds. This bonding pattern leads to a phenomenon known as conjugation. Conjugated compounds include β-carotene, which produces the orange color of carrots, chlorophyll a, which is the green pigment that carries on photosynthesis, and indigo, the pigment that gives your blue jeans their distinctive color. It is the conjugation of their double bonds and the way these compounds interact with light (both visible and ultraviolet) that produces their colors. Moreover, such molecules also have unique aspects of reactivity because the anions, cations, and radical species formed from them possess greater than normal stabilization. As a result, they can participate in a large and unique spectrum of organic chemistry.

IN THIS CHAPTER WE WILL CONSIDER:

- conjugation and resonance structures based on radicals, cations, and anions
- the unique physical properties of conjugated systems, especially as observed by UV–Vis spectroscopy
- a special transformation, named the Diels–Alder reaction, that can combine conjugated molecules known as 1,3-dienes with certain partners to create six-membered rings containing up to four new chirality centers

[WHY DO THESE TOPICS MATTER?]   At the end of the chapter, we will see just how Otto Diels and Kurt Alder first discovered the reaction that would later bear their names and garner them the Nobel Prize, noting that it was nearly discovered by a number of other people who had done the same chemistry, but could not figure out what they had made!

13.1 INTRODUCTION

At its essence, a **conjugated system** involves at least one atom with a *p* orbital adjacent to at least one π bond. The adjacent atom with the *p* orbital can be part of another π bond, as in 1,3-butadiene, or a radical, cationic, or anionic reaction intermediate. If an example derives specifically from a propenyl group, the common name for this group is **allyl**. In general when we are considering a radical, cation, or anion that is adjacent to one or more π bonds in a molecule other than propene, the adjacent position is called **allylic**. Below we show the formula for butadiene, resonance hybrids for the allyl radical and an **allylic carbocation**, and molecular orbital representations for each one.

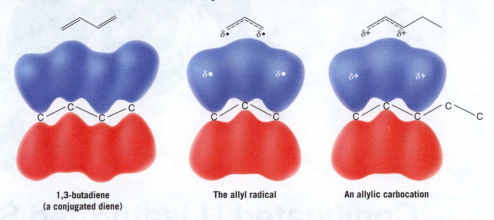

| 1,3-butadiene (a conjugated diene) | The allyl radical | An allylic carbocation |

Radical substitution at an allylic position, as we saw in Chapter 10, is especially favorable precisely because the intermediate radical is part of a conjugated system.

● ● ● SOLVED PROBLEM 13.1

Identify all of the positions bearing allylic hydrogen atoms in cryptoxanthin (a natural pigment).

STRATEGY AND ANSWER: Allylic hydrogen atoms in cryptoxanthin are found on the sp^3-hybridized carbons adjacent to π bonds. These positions are labeled by boxes in the formula below.

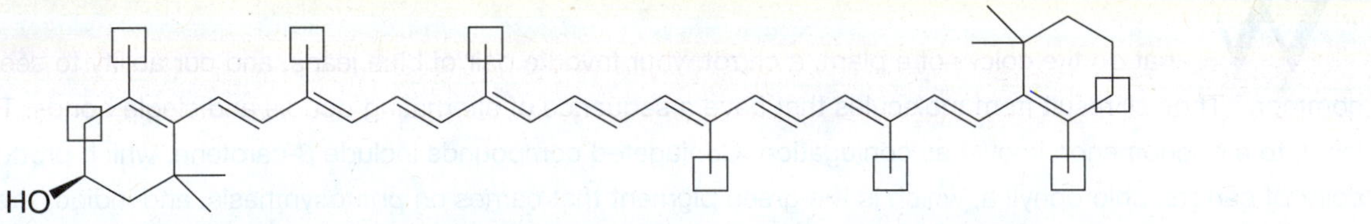

Cryptoxanthin (with positions bearing allylic hydrogen atoms shown in boxes)

13.2 THE STABILITY OF THE ALLYL RADICAL

An explanation of the stability of the allyl radical can be approached in two ways: in terms of molecular orbital theory and in terms of resonance theory (Section 1.8). As we shall see soon, both approaches give us equivalent descriptions of the allyl radical. The molecular orbital approach is easier to visualize, so we shall begin with it. (As preparation for this section, it may help the reader to review the molecular orbital theory given in Sections 1.11 and 1.13.)

13.2A Molecular Orbital Description of the Allyl Radical

As an allylic hydrogen atom is abstracted from propene (see the following diagram), the sp^3-hybridized carbon atom of the methyl group changes its hybridization state to sp^2 (see

Section 10.6). The *p* orbital of this new *sp²*-hybridized carbon atom overlaps with the *p* orbital of the central carbon atom.

- In the allyl radical three *p* orbitals overlap to form a set of *π* molecular orbitals that encompass all three carbon atoms.

- The new *p* orbital of the allyl radical is said to be *conjugated* with those of the double bond, and the allyl radical is said to be a *conjugated unsaturated system*.

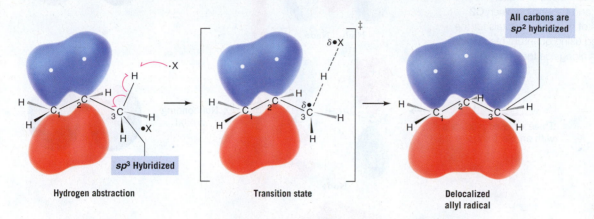

Hydrogen abstraction   **Transition state**   **Delocalized allyl radical**

- The unpaired electron of the allyl radical and the two electrons of the *π* bond are **delocalized** over all three carbon atoms.

Delocalization of the unpaired electron accounts for the greater stability of the allyl radical when compared to primary, secondary, and tertiary radicals. Although some delocalization occurs in primary, secondary, and tertiary radicals, delocalization is not as effective because it occurs only through hyperconjugation (Section 6.11B) with *σ* bonds.

The diagram in Fig. 13.1 illustrates how the three *p* orbitals of the allyl radical combine to form three *π* molecular orbitals. (*Remember*: The number of molecular orbitals that results always equals the number of atomic orbitals that combine; see Section 1.11.) The bonding *π* molecular orbital is of lowest energy; it encompasses all three carbon atoms and is occupied by two spin-paired electrons. This bonding *π* orbital is the result of having *p* orbitals with lobes of the same sign overlap between adjacent carbon atoms. This type of overlap, as we recall, increases the *π*-electron density in the regions between the atoms where it is needed for bonding. The nonbonding *π* orbital is occupied by one unpaired electron, and it has a node at the central carbon atom. This node means that the unpaired electron is located in the vicinity of carbon atoms 1 and 3 only. The antibonding *π* molecular orbital results when orbital lobes of opposite sign overlap between adjacent carbon atoms. Such overlap means that in the antibonding *π* orbital there is a node between each pair of carbon atoms. This antibonding orbital of the allyl radical is of highest energy and is empty in the ground state of the radical.

We can illustrate the picture of the allyl radical given by molecular orbital theory with the following structure:

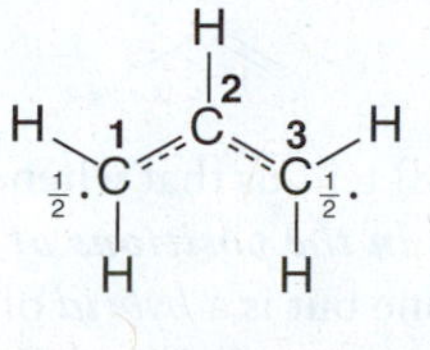

$$H \underset{H}{\overset{H}{\underset{\frac{1}{2}}{\cdot C}}} \overset{H}{\underset{2}{C}} \underset{H}{\overset{3}{\underset{\frac{1}{2}}{C \cdot}}} H$$

We indicate with dashed lines that both carbon–carbon bonds are partial double bonds. This accommodates one of the things that molecular orbital theory tells us: *that there is a π bond encompassing all three atoms.* We also place the symbol $\frac{1}{2}$ beside the C1 and C3 atoms. This presentation denotes a second thing molecular orbital theory tells us: *that electron density from the unpaired electron is equal in the vicinity of C1 and C3.* Finally, implicit in the molecular orbital picture of the allyl radical is this: the two ends of the allyl radical are *equivalent.* This aspect of the molecular orbital description is also implicit in the formula just given.

FIGURE 13.1 Combination of three atomic *p* orbitals to form three *π* molecular orbitals in the allyl radical. The bonding *π* molecular orbital is formed by the combination of the three *p* orbitals with lobes of the same sign overlapping above and below the plane of the atoms. The nonbonding *π* molecular orbital has a node at C2. The antibonding *π* molecular orbital has two nodes: between C1 and C2 and between C2 and C3. The shapes of molecular orbitals for the allyl radical calculated using quantum mechanical principles are shown alongside the schematic orbitals.

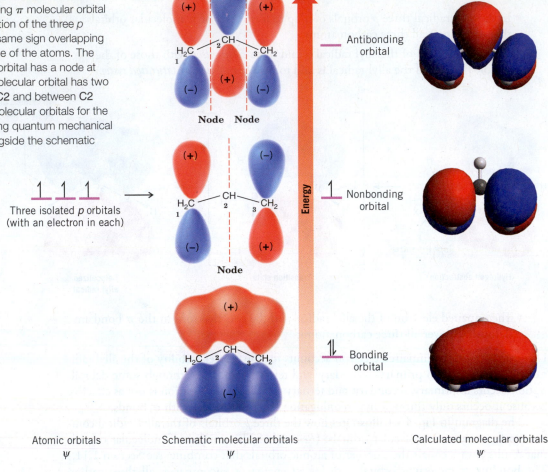

Atomic orbitals ψ Schematic molecular orbitals ψ Calculated molecular orbitals ψ

13.2B Resonance Description of the Allyl Radical

One structure that we can write for the allyl radical is **A**:

However, we might just as well have written the equivalent structure, **B**:

In writing structure **B**, we do not mean to imply that we have simply taken structure **A** and turned it over. We have not moved the nuclei. What we have done is move the electrons in the following way:

Resonance theory (Section 1.8) tells us that whenever we can write two structures for a chemical entity *that differ only in the positions of the electrons*, the entity cannot be represented by either structure alone but is a *hybrid* of both. We can represent the hybrid in two ways. We can write both structures **A** and **B** and connect them with a double-headed arrow, the special arrow we use to indicate that they are resonance structures:

Or we can write a single structure, **C**, that blends the features of both resonance structures:

We see, then, that resonance theory gives us exactly the same picture of the allyl radical that we obtained from molecular orbital theory. Structure **C** describes the carbon–carbon bonds of the allyl radical as partial double bonds. The resonance structures **A** and **B** also tell us that the unpaired electron is associated only with the C1 and C3 atoms. We indicate this in structure **C** by placing a δ beside C1 and C3. Because resonance structures **A** and **B** are equivalent, the electron density from the unpaired electron is shared equally by C1 and C3.

Another rule in resonance theory is the following:

- Whenever equivalent **resonance structures** can be written for a chemical species, the chemical species is much more stable than any single resonance structure would indicate.

If we were to examine either **A** or **B** alone, we might decide incorrectly that it resembled a primary radical. Thus, we might estimate the stability of the allyl radical as approximately that of a primary radical. In doing so, we would greatly underestimate the stability of the allyl radical. Resonance theory tells us, however, that since **A** and **B** are *equivalent resonance structures*, the allyl radical should be much more stable than either, that is, much more stable than a primary radical. This correlates with what experiments have shown to be true: **the allyl radical is even more stable than a tertiary radical**.

We should point out that a structure that would indicate an unpaired electron on the central carbon of the allyl system, as shown here, is not a proper resonance structure because resonance theory dictates that all resonance structures must have the same number of unpaired electrons (Section 13.4A). This structure shows three unpaired electrons, whereas the other resonance structures for the allyl radical have only one unpaired electron.

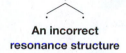

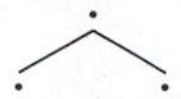

**An incorrect
resonance structure**

●●● SOLVED PROBLEM 13.2

Subjecting propene labeled with ^{13}C at carbon 1 to allylic chlorination (see below) leads to a 50:50 mixture of 1-chloropropene labeled at C1 and at C3. Write a mechanism that explains this result. (An asterisk * next to a carbon atom indicates that the carbon atom is ^{13}C.)

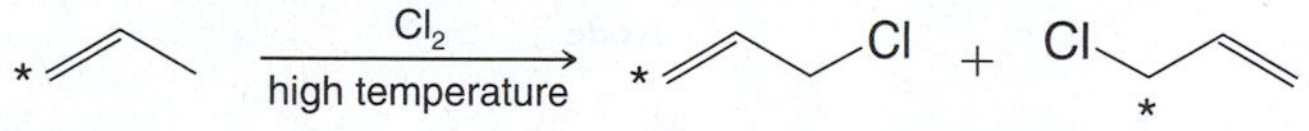

$$ * \diagup\!\!\!\diagdown \xrightarrow[\text{high temperature}]{Cl_2} *\diagup\!\!\!\diagdown\!\!\diagup Cl \; + \; Cl\diagdown\!\!\diagup\!\!\!\diagdown * $$

STRATEGY AND ANSWER: We recall (Section 10.8A) that the mechanism for allylic chlorination involves the formation of a resonance-stabilized radical created by having a chlorine atom abstract an allylic hydrogen atom. Because the radical formed in this case is a hybrid of two structures (which are equivalent except for the position of the label), it can react with Cl_2 at either end to give a 50 : 50 mixture of the differently labeled products.

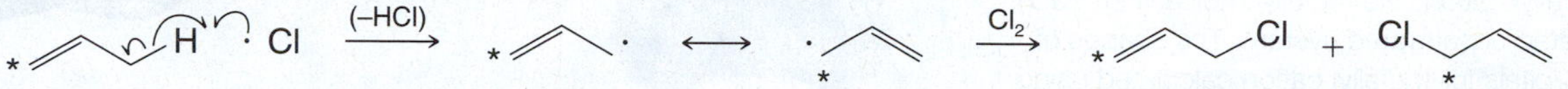

$$ *\diagup\!\!\!\diagdown\!\!\!\curvearrowright H \cdot Cl \xrightarrow{(-HCl)} *\diagup\!\!\!\diagdown\!\!\!\diagdown\cdot \longleftrightarrow \cdot\diagdown\!\!\!\diagup\!\!\!\diagdown * \xrightarrow{Cl_2} *\diagup\!\!\!\diagdown\!\!\!\diagup Cl \; + \; Cl\diagdown\!\!\!\diagup\!\!\!\diagdown * $$

                                                                           50%   50%

●●● ⋯⋯⋯⋯⋯

Consider the allylic bromination of cyclohexene labeled at C3 with ^{13}C. Neglecting stereoisomers, what products would you expect from this reaction?

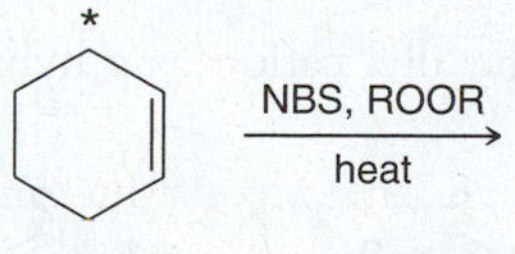

PRACTICE PROBLEM 13.1

$$ \underset{}{\overset{*}{\hexagon}} \xrightarrow[\text{heat}]{NBS, ROOR} $$

(* = ^{13}C-labeled position)

13.3 THE ALLYL CATION

Carbocations can be allylic as well.

- The **allyl (propenyl) cation** 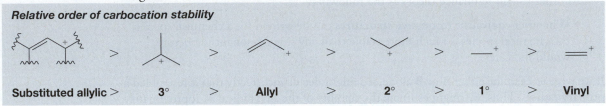 is even more stable than a secondary carbocation and is almost as stable as a tertiary carbocation.

In general terms, the relative order of stabilities of carbocations is that given here.

Relative order of carbocation stability

| Substituted allylic > | 3° | > | Allyl | > | 2° | > | 1° | > | Vinyl |

The molecular orbital description of the allyl cation is shown in Fig. 13.2.

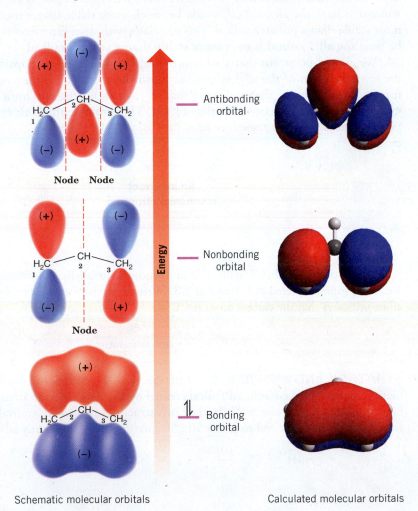

FIGURE 13.2 The π molecular orbitals of the allyl cation. The allyl cation, like the allyl radical (Fig. 13.1), is a conjugated unsaturated system. The shapes of molecular orbitals for the allyl cation calculated using quantum mechanical principles are shown alongside the schematic orbitals.

Schematic molecular orbitals   Calculated molecular orbitals

The bonding π molecular orbital of the allyl cation, like that of the **allyl radical** (Fig. 13.1), contains two spin-paired electrons. The nonbonding π molecular orbital of the allyl cation, however, is empty.

Resonance theory depicts the allyl cation as a hybrid of structures **D** and **E** represented here:

D   E

Because **D** and **E** are *equivalent* resonance structures, resonance theory predicts that the allyl cation should be unusually stable. Since the positive charge is located on C3 in **D** and on C1 in **E**, resonance theory also tells us that the positive charge should be delocalized over both carbon atoms. Carbon atom 2 carries none of the positive charge. The hybrid structure **F** includes charge and bond features of both **D** and **E**:

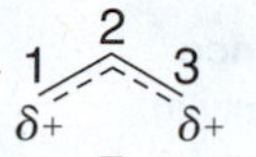

●●● SOLVED PROBLEM 13.3

Allyl bromide (3-bromo-1-propene) forms a carbocation readily. For example, it undergoes S_N1 reactions. Explain this observation.

STRATEGY AND ANSWER: Ionization of allyl bromide (at right) produces an allyl cation that is unusually stable (far more stable than a simple primary carbocation) because it is resonance stabilized.

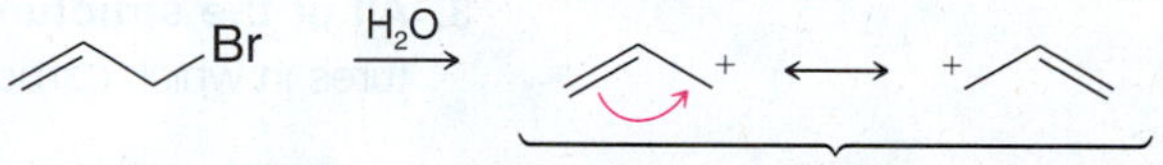

Resonance-stabilized carbocation

●●●·········· **PRACTICE PROBLEM 13.2**

(a) Draw resonance structures for the carbocation that could be formed from (*E*)-2-butenyl trifluoromethanesulfonate.

(b) One of the resonance structures for this carbocation should be a more important contributor to the resonance hybrid than the other. Which resonance structure would be the greater contributor?

(c) What products would you expect if this carbocation reacted with a chloride ion?

13.4 RESONANCE THEORY REVISITED

We have already used **resonance** theory in earlier chapters, and we have been using it extensively in this chapter because we are describing radicals and ions with delocalized electrons (and charges) in π bonds. Resonance theory is especially useful with systems like this, and we shall use it again and again in the chapters that follow. In Section 1.8 we had an introduction to resonance theory and an initial presentation of some rules for writing resonance structures. It should now be helpful, in light of our previous discussions of relative carbocation stability and radicals, and our growing understanding of conjugated systems, to review and expand on those rules as well as those for the ways in which we estimate the relative contribution a given structure will make to the overall hybrid.

Helpful Hint

Resonance is an important tool we use frequently when discussing structure and reactivity.

13.4A HOW TO Write Proper Resonance Structures

1. **Resonance structures exist only on paper.** Although they have no real existence of their own, **resonance structures** are useful because they allow us to describe molecules, radicals, and ions for which a single Lewis structure is inadequate. Instead, we write two or more Lewis structures, calling them resonance structures or resonance contributors. We connect these structures by double-headed arrows (⟷), and we say that the hybrid of all of them represents the real molecule, radical, or ion.

2. **In writing resonance structures, we are only allowed to move electrons.** The positions of the nuclei of the atoms must remain the same in all of the structures. Structure **3**

is not a resonance structure for the allylic cation, for example, because in order to form it we would have to move a hydrogen atom and this is not permitted:

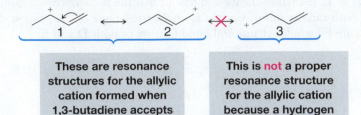

> **These are resonance structures for the allylic cation formed when 1,3-butadiene accepts a proton.**

> **This is not a proper resonance structure for the allylic cation because a hydrogen atom has been moved.**

Generally speaking, when we move electrons we move only those of π bonds (as in the example above) and those of lone pairs.

3. **All of the structures must be proper Lewis structures.** We should not write structures in which carbon has five bonds, for example:

> **This shift would give a carbon 10 electrons**

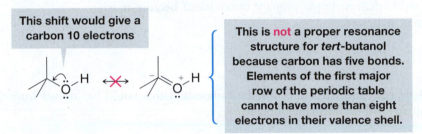

> **This is not a proper resonance structure for *tert*-butanol because carbon has five bonds. Elements of the first major row of the periodic table cannot have more than eight electrons in their valence shell.**

4. **All resonance structures must have the same number of unpaired electrons.** The structure on the right is not a proper resonance structure for the allyl radical because it contains three unpaired electrons whereas the allyl radical contains only one:

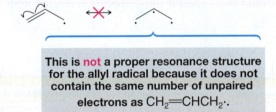

> **This is not a proper resonance structure for the allyl radical because it does not contain the same number of unpaired electrons as $CH_2=CHCH_2 \cdot$.**

5. **All atoms that are part of the delocalized π-electron system must lie in a plane or be nearly planar.** For example, 2,3-di-*tert*-butyl-1,3-butadiene behaves like a *nonconjugated* diene because the large *tert*-butyl groups twist the structure and prevent the double bonds from lying in the same plane. Because they are not in the same plane, the p orbitals at C2 and C3 do not overlap and delocalization (and therefore resonance) is prevented:

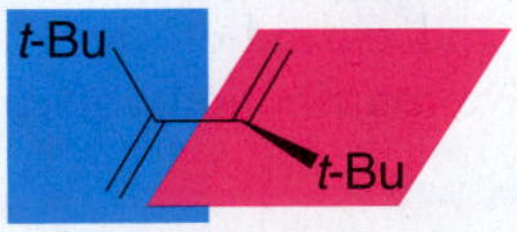

2,3-Di-tert-butyl-1,3-butadiene

6. **The energy of the actual molecule is lower than the energy that might be estimated for any contributing structure.** The actual allyl cation, for example, is more stable than either resonance structure **4** or **5** taken separately would indicate. Structures **4** and **5** resemble primary carbocations and yet the allyl cation is more stable (has lower energy) than a secondary carbocation. Chemists often call this kind of stabilization *resonance stabilization*:

4   5

In Chapter 14 we shall find that benzene is highly resonance stabilized because it is a hybrid of the two equivalent forms that follow:

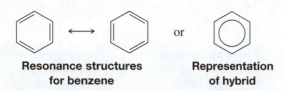

Resonance structures   or   **Representation**
for benzene   **of hybrid**

7. **Equivalent resonance structures make equal contributions to the hybrid, and a system described by them has a large resonance stabilization.** Structures **4** and **5** above make equal contributions to the allylic cation because they are equivalent. They also make a large stabilizing contribution and account for allylic cations being unusually stable. The same can be said about the contributions made by the equivalent structures **A** and **B** (Section 13.2B) for the allyl radical.

8. **The more stable a structure is (when taken by itself), the greater is its contribution to the hybrid.** Structures that are not equivalent do not make equal contributions. For example, the following cation is a hybrid of structures **6** and **7**. Structure **6** makes a greater contribution than **7** because structure **6** is a more stable tertiary carbocation while structure **7** is a primary cation:

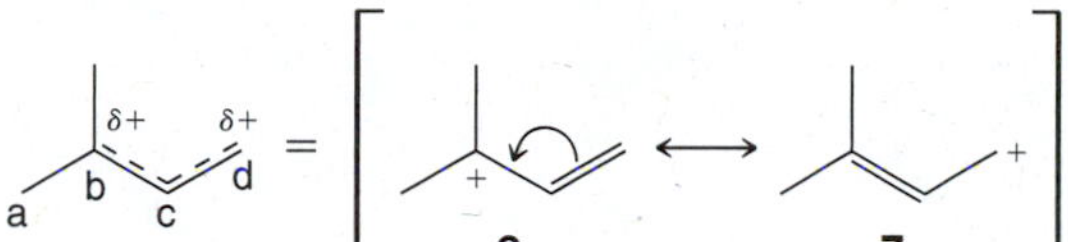

That **6** makes a larger contribution means that the partial positive charge on carbon b of the hybrid will be larger than the partial positive charge on carbon d. It also means that the bond between carbon atoms c and d will be more like a double bond than the bond between carbon atoms b and c.

13.4B HOW TO Estimate the Relative Stability of Contributing Resonance Structures

The following rules will help us in making decisions about the relative stabilities of resonance structures.

a. **The more covalent bonds a structure has, the more stable it is.** This is exactly what we would expect because we know that forming a covalent bond lowers the energy of atoms. This means that of the following structures for 1,3-butadiene, **8** is by far the most stable and makes by far the largest contribution because it contains one more bond. (It is also more stable for the reason given under rule **c**.)

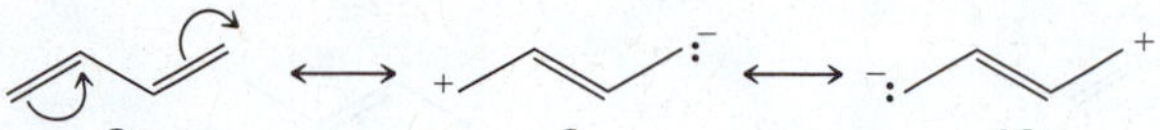

This structure is the most stable because it contains more covalent bonds.

b. **Structures in which all of the atoms have a complete valence shell of electrons (i.e., the noble gas structure) are especially stable and make large contributions to the hybrid.** Again, this is what we would expect from what we know about bonding. This means, for example, that **12** makes a larger stabilizing contribution to the cation below than **11** because all of the atoms of **12** have a complete valence shell. (Notice too that **12** has more covalent bonds than **11**; see rule **a**.)

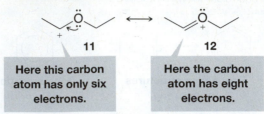

11	**12**
Here this carbon atom has only six electrons.	Here the carbon atom has eight electrons.

c. **Charge separation decreases stability.** Separating opposite charges requires energy. Therefore, structures in which opposite charges are separated have greater energy (lower stability) than those that have no charge separation. This means that of the following two structures for vinyl chloride, structure **13** makes a larger contribution because it does not have separated charges. (This does not mean that structure **14** does not contribute to the hybrid; it just means that the contribution made by **14** is smaller.)

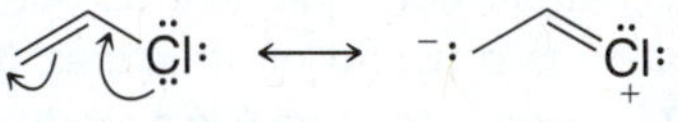

13 ⟷ **14**

●●● SOLVED PROBLEM 13.4

Write resonance structures for ⟍⟋⟍Ö. and tell which structure would be the greater contributor to the resonance hybrid.

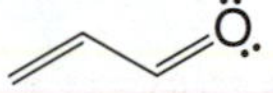

STRATEGY AND ANSWER: We write the structure for the molecule and then move the electron pairs as shown below. Then we examine the two structures. The second structure contains separated charges (which would make it less stable) and the first structure contains more bonds (which would make it more stable). Both factors tell us that the first structure is more stable. Consequently, it should be the greater contributor to the hybrid.

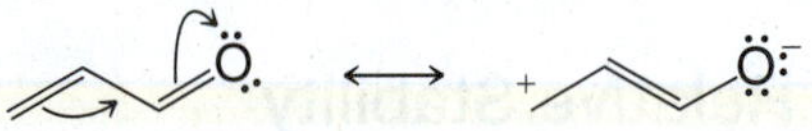

More bonds   **Separated charges**

········●●●

PRACTICE PROBLEM 13.3 Write the important resonance structures for each of the following:

(a)

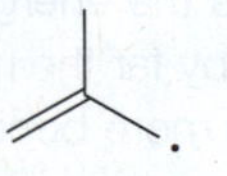

(b)

(c)

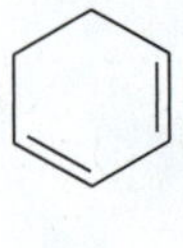

(d)

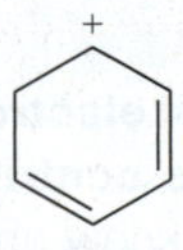

(e)

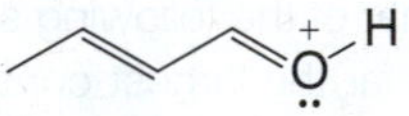

(f) $CH_2=CH-Br$

(g)

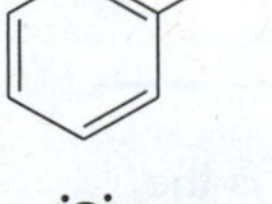

(h)

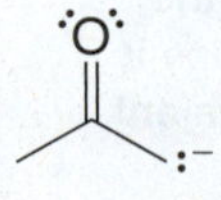

(i)

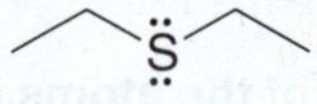

(j)

From each set of resonance structures that follow, designate the one that would contribute most to the hybrid and explain your choice:

(a)

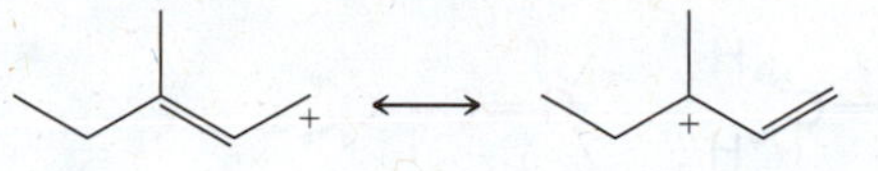

(b)

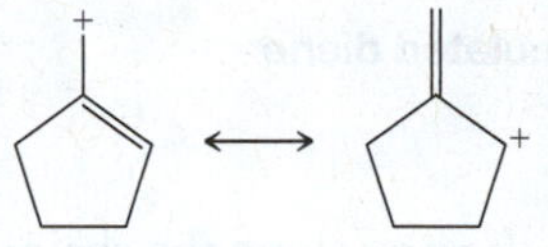

(c)

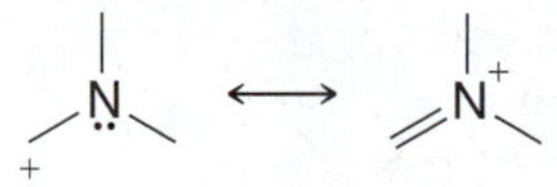

(d)

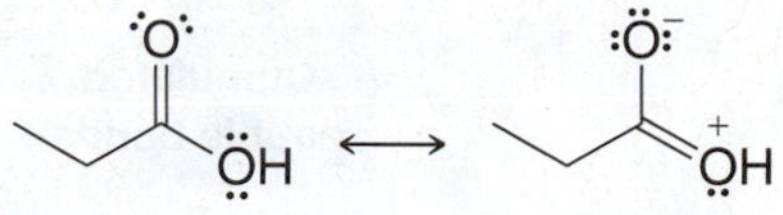

(e)

(f) $:NH_2$—C≡N: ⟷ $\overset{+}{NH_2}$=C=N̈:⁻

The following enol (an alk*ene*-alcoh*ol*) and keto (a *keto*ne) forms of C_2H_4O differ in the positions for their electrons, but they are not resonance structures. Explain why they are not.

C_2H_4O

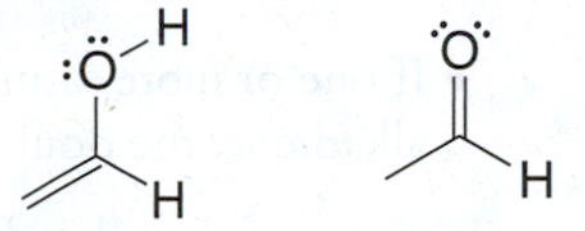

Enol form   Keto form

13.5 ALKADIENES AND POLYUNSATURATED HYDROCARBONS

Many hydrocarbons are known that contain more than one double or triple bond. A hydrocarbon that contains two double bonds is called an **alkadiene**; one that contains three double bonds is called an **alkatriene**, and so on. Colloquially, these compounds are often referred to simply as dienes or trienes. A hydrocarbon with two triple bonds is called an **alkadiyne**, and a hydrocarbon with a double and triple bond is called an **alkenyne**.

The following examples of polyunsaturated hydrocarbons illustrate how specific compounds are named. Recall from IUPAC rules (Sections 4.5 and 4.6) that the numerical locants for double and triple bonds can be placed at the beginning of the name or immediately preceding the respective suffix. We provide examples of both styles.

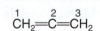

$$\overset{1}{C}H_2=\overset{2}{C}=\overset{3}{C}H_2$$
1,2-Propadiene
(allene, or propa-1,2-diene)

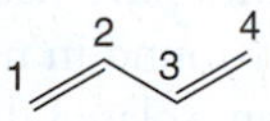

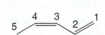

1,3-Butadiene
(buta-1,3-diene)

(3Z)-Penta-1,3-diene
(*cis*-penta-1,3-diene)

Pent-1-en-4-yne

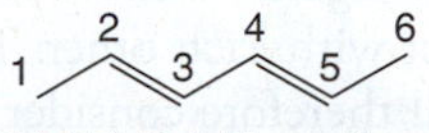

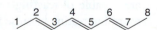

(2E,4E)-2,4-Hexadiene
(*trans,trans*-2,4-hexadiene)

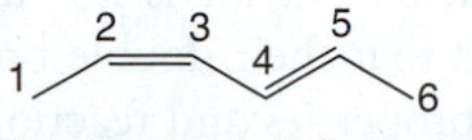

(2Z,4E)-Hexa-2,4-diene
(*cis,trans*-hexa-2,4-diene)

(2E,4E,6E)-Octa-2,4,6-triene
(*trans,trans,trans*-octa-2,4,6-triene)

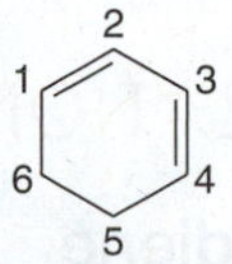

1,3-Cyclohexadiene 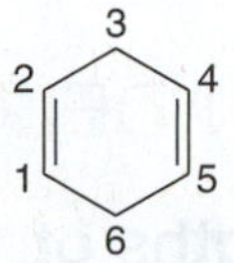   1,4-Cyclohexadiene

The multiple bonds of polyunsaturated compounds are classified as being **cumulated, conjugated,** or **isolated.**

- The double bonds of a 1,2-diene (such as 1,2-propadiene, also called allene) are said to be **cumulated** because one carbon (the central carbon) participates in two double bonds.

Hydrocarbons whose molecules have cumulated double bonds are called **cumulenes**. The name **allene** (Section 5.18) is also used as a class name for molecules with two cumulated double bonds:

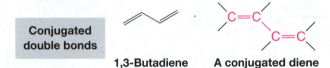

Cumulated double bonds

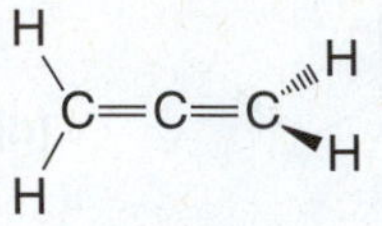

Allene 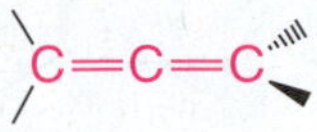   A cumulated diene

An example of a conjugated diene is 1,3-butadiene.

- In **conjugated** polyenes the double and single bonds *alternate* along the chain:

Conjugated double bonds

1,3-Butadiene   A conjugated diene

(2*E*,4*E*,6*E*)-Octa-2,4,6-triene is an example of a conjugated alkatriene.

- If one or more saturated carbon atoms intervene between the double bonds of an alkadiene, the double bonds are said to be **isolated**.

An example of an isolated diene is 1,4-pentadiene:

Isolated double bonds

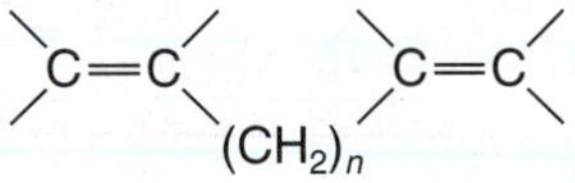

An isolated diene
(*n* ≠ 0)    1,4-Pentadiene

---

PRACTICE PROBLEM 13.6   **(a)** Which other compounds in Section 13.5 are conjugated dienes?
(b) Which other compounds are isolated dienes?
(c) Which compound is an isolated enyne?

In Chapter 5 we saw that appropriately substituted cumulated dienes (allenes) give rise to chiral molecules. Cumulated dienes have had some commercial importance, and cumulated double bonds are occasionally found in naturally occurring molecules. In general, cumulated dienes are less stable than isolated dienes.

The double bonds of isolated dienes behave just as their name suggests—as isolated "enes." They undergo all of the reactions of alkenes, and, except for the fact that they are capable of reacting twice, their behavior is not unusual. Conjugated dienes are far more interesting because we find that their double bonds interact with each other. This interaction leads to unexpected properties and reactions. We shall therefore consider the chemistry of conjugated dienes in detail.

13.6 1,3-BUTADIENE: ELECTRON DELOCALIZATION

13.6A Bond Lengths of 1,3-Butadiene

The carbon–carbon bond lengths of 1,3-butadiene have been determined and are shown here:

Double bonds are shorter then single bonds.

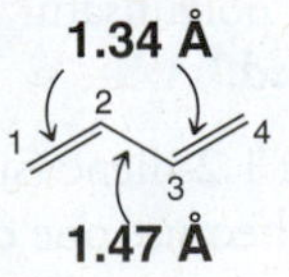

1.34 Å

1.47 Å

The C1—C2 bond and the C3—C4 bond are (within experimental error) the same length as the carbon–carbon double bond of ethene. The central bond of 1,3-butadiene (1.47 Å), however, is considerably shorter than the single bond of ethane (1.54 Å).

This should not be surprising. All of the carbon atoms of 1,3-butadiene are sp^2 hybridized and, as a result, the central bond of butadiene results from overlapping sp^2 orbitals. And, as we know, a sigma bond that is sp^3–sp^3 is *longer*. There is, in fact, a steady decrease in bond length of carbon–carbon single bonds as the hybridization state of the bonded atoms changes from sp^3 to sp (Table 13.1).

TABLE 13.1	CARBON–CARBON SINGLE-BOND LENGTHS AND HYBRIDIZATION STATE	
Compound	Hybridization State	Bond Length (Å)
$H_3C—CH_3$	sp^3–sp^3	1.54
$CH_2=CH—CH_3$	sp^2–sp^3	1.50
$CH_2=CH—CH=CH_2$	sp^2–sp^2	1.47
$HC≡C—CH_3$	sp–sp^3	1.46
$HC≡C—CH=CH_2$	sp–sp^2	1.43
$HC≡C—C≡CH$	sp–sp	1.37

13.6B Conformations of 1,3-Butadiene

There are two possible planar conformations of 1,3-butadiene: the s-cis and the s-trans conformations.

s-cis Conformation
of 1,3-butadiene

rotate about C2—C3

s-trans Conformation
of 1,3-butadiene

These are not true cis and trans forms since the s-cis and s-trans conformations of 1,3-butadiene can be interconverted through rotation about the single bond (hence the prefix s). The s-trans conformation is the predominant one at room temperature. We shall see that the s-cis conformation of 1,3-butadiene and other 1,3-conjugated alkenes is necessary for the Diels–Alder reaction (Section 13.10).

●●● **SOLVED PROBLEM 13.5**

Provide an explanation for the fact that many more molecules are in the s-trans conformation of 1,3-butadiene at equilibrium.

STRATEGY AND ANSWER: The s-cis conformation of 1,3-butadiene is less stable, and therefore less populated, than the s-trans conformer because it has steric interference between the hydrogen atoms at carbons 1 and 4. Interference of this kind does not exist in the s-trans conformation, and therefore, the s-trans conformation is more stable and more populated at equilibrium.

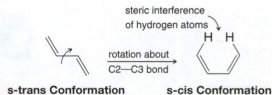

steric interference of hydrogen atoms

rotation about C2—C3 bond

s-trans Conformation
is more stable

s-cis Conformation
is less stable

13.6C Molecular Orbitals of 1,3-Butadiene

The central carbon atoms of 1,3-butadiene (Fig. 13.3) are close enough for overlap to occur between the p orbitals of C2 and C3. This overlap is not as great as that between the orbitals of C1 and C2 (or those of C3 and C4). The C2–C3 orbital overlap, however,

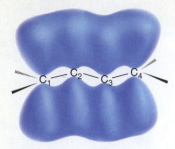

FIGURE 13.3 The p orbitals of 1,3-butadiene. (See Fig. 13.4 for the shapes of calculated molecular orbitals for 1,3-butadiene.)

gives the central bond partial double-bond character and allows the four π electrons of 1,3-butadiene to be delocalized over all four atoms.

Figure 13.4 shows how the four p orbitals of 1,3-butadiene combine to form a set of four π molecular orbitals.

- Two of the π molecular orbitals of 1,3-butadiene are bonding molecular orbitals. In the ground state these orbitals hold the four π electrons with two spin-paired electrons in each.

- The other two π molecular orbitals are antibonding molecular orbitals. In the ground state these orbitals are unoccupied.

An electron can be excited from the highest occupied molecular orbital (HOMO) to the lowest unoccupied molecular orbital (LUMO) when 1,3-butadiene absorbs light with a wavelength of 217 nm. (We shall study the absorption of light by unsaturated molecules in Section 13.8.)

- The delocalized bonding that we have just described for 1,3-butadiene is characteristic of all conjugated polyenes.

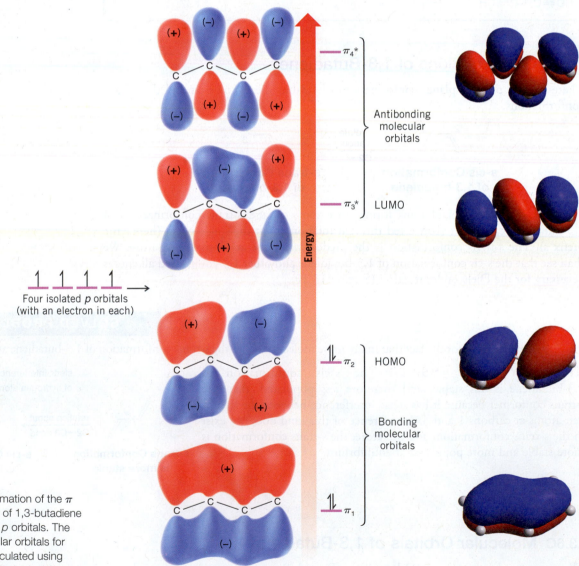

FIGURE 13.4 Formation of the π molecular orbitals of 1,3-butadiene from four isolated p orbitals. The shapes of molecular orbitals for 1,3-butadiene calculated using quantum mechanical principles are shown alongside the schematic orbitals.

Schematic molecular orbitals

Calculated molecular orbitals

13.7 THE STABILITY OF CONJUGATED DIENES

- Conjugated alkadienes are thermodynamically more stable than isomeric isolated alkadienes.

Two examples of this extra stability of conjugated dienes can be seen in an analysis of the heats of hydrogenation given in Table 13.2.

TABLE 13.2 HEATS OF HYDROGENATION OF ALKENES AND ALKADIENES

Compound	H_2 (mol)	$\Delta H°$ (kJ mol^{-1})
1-Butene	1	−127
1-Pentene	1	−126
trans-2-Pentene	1	−115
1,3-Butadiene	2	−239
trans-1,3-Pentadiene	2	−226
1,4-Pentadiene	2	−254
1,5-Hexadiene	2	−253

In itself, 1,3-butadiene cannot be compared directly with an isolated diene of the same chain length. However, a comparison can be made between the heat of hydrogenation of 1,3-butadiene and that obtained when two molar equivalents of 1-butene are hydrogenated:

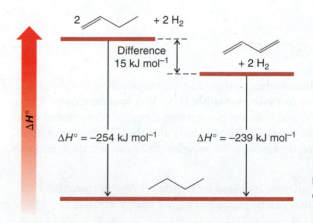

$$\Delta H° \text{ (kJ mol}^{-1})$$

2 /\\=/ $+ 2 H_2 \longrightarrow 2$ /\\/ $\qquad 2 \times (-127) = -254$
1-Butene

/=\\/ $+ 2 H_2 \longrightarrow$ /\\/ $\qquad\qquad\qquad = -239$
1,3-Butadiene $\qquad\qquad\qquad$ Difference $\qquad$ 15 kJ mol^{-1}

Because 1-butene has a monosubstituted double bond like those in 1,3-butadiene, we might expect hydrogenation of 1,3-butadiene to liberate the same amount of heat (254 kJ mol^{-1}) as two molar equivalents of 1-butene. We find, however, that 1,3-butadiene liberates only 239 kJ mol^{-1}, 15 kJ mol^{-1} *less* than expected. We conclude, therefore, that conjugation imparts some extra stability to the conjugated system (Fig. 13.5).

$\Delta H°$

2 /\\=/ $+ 2 H_2$

Difference
15 kJ mol^{-1}

/=\\/ $+ 2 H_2$

$\Delta H° = -254$ kJ mol^{-1} $\qquad$ $\Delta H° = -239$ kJ mol^{-1}

FIGURE 13.5 Heats of hydrogenation of 2 mol of 1-butene and 1 mol of 1,3-butadiene.

An assessment of the stabilization that conjugation provides *trans*-1,3-pentadiene can be made by comparing the heat of hydrogenation of *trans*-1,3-pentadiene to the sum of the heats of hydrogenation of 1-pentene and *trans*-2-pentene. This way we are comparing double bonds of comparable types:

1-Pentene $\Delta H° = -126$ kJ mol^{-1}

trans-2-Pentene $\Delta H° = -115$ kJ mol^{-1}
                                    Sum $= -241$ kJ mol^{-1}

trans-1,3-Pentadiene $\Delta H° = -226$ kJ mol^{-1}
                                    Difference $= \ \ \ 15$ kJ mol^{-1}

We see from these calculations that conjugation affords *trans*-1,3-pentadiene an extra stability of 15 kJ mol^{-1}, a value that is equivalent, to two significant figures, to the one we obtained for 1,3-butadiene (15 kJ mol^{-1}).

When calculations like these are carried out for other conjugated dienes, similar results are obtained; *conjugated dienes are found to be more stable than isolated dienes*. The question, then, is this: what is the source of the extra stability associated with conjugated dienes? There are two factors that contribute. The extra stability of conjugated dienes arises in part from the stronger central bond that they contain and, in part, from the additional delocalization of the π electrons that occurs in conjugated dienes.

●●● SOLVED PROBLEM 13.6

Which diene would you expect to be more stable: 1,3-cyclohexadiene or 1,4-cyclohexadiene? Why? What experiment could you carry out to confirm your answer?

STRATEGY AND ANSWER: 1,3-Cyclohexadiene is conjugated, and on that basis we would expect it to be more stable. We could determine the heats of hydrogenation of the two compounds, and since on hydrogenation each compound yields the same product, the diene with the smaller heat of hydrogenation would be the more stable one.

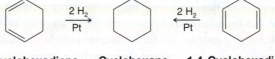

1,3-Cyclohexadiene   **Cyclohexane**   **1,4-Cyclohexadiene**
(a conjugated diene)   **(an isolated diene)**

13.8 ULTRAVIOLET–VISIBLE SPECTROSCOPY

The extra stability of conjugated dienes when compared to corresponding unconjugated dienes can also be seen in data from **ultraviolet–visible (UV–Vis) spectroscopy**. When electromagnetic radiation in the UV and visible regions passes through a compound containing multiple bonds, a portion of the radiation is usually absorbed by the compound. Just how much radiation is absorbed depends on the wavelength of the radiation and the structure of the compound.

- The absorption of UV–Vis radiation is caused by transfer of energy from the radiation beam to electrons that can be excited to higher energy orbitals.

In Section 13.8C we shall return to discuss specifically how data from UV–Vis spectroscopy demonstrate the additional stability of conjugated dienes. First, in Section 13.8A we briefly review the properties of electromagnetic radiation, and then in Section 13.8B we look at how data from a UV–Vis spectrophotometer are obtained.

13.8A The Electromagnetic Spectrum

According to quantum mechanics, electromagnetic radiation has a dual and seemingly contradictory nature.

- Electromagnetic radiation can be described as a wave occurring simultaneously in electrical and magnetic fields. It can also be described as if it consisted of particles called quanta or photons.

Different experiments disclose these two different aspects of electromagnetic radiation. They are not seen together in the same experiment.

- A wave is usually described in terms of its **wavelength** (λ) or its **frequency** (ν).

A simple wave is shown in Fig. 13.6. The distance between consecutive crests (or troughs) is the wavelength. The number of full cycles of the wave that pass a given point each second, as the wave moves through space, is called the *frequency* and is measured in cycles per second (cps), or **hertz (Hz)**.*

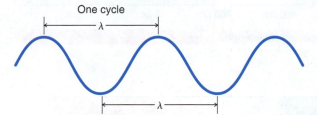

FIGURE 13.6 A simple wave and its wavelength, λ.

All electromagnetic radiation travels through a vacuum at the same velocity. This velocity (c), called the velocity of light, is 2.99792458×10^8 m s^{-1} and relates to wavelength and frequency as $c = \lambda\nu$. The wavelengths of electromagnetic radiation are expressed either in meters (m), millimeters (1 mm $= 10^{-3}$ m), micrometers (1 μm $= 10^{-6}$ m), or nanometers (1 nm $= 10^{-9}$ m). [An older term for micrometer is *micron* (abbreviated μ) and an older term for nanometer is *millimicron*.]

The energy of a quantum of electromagnetic energy is directly related to its frequency:

$$E = h\nu$$

where h = Planck's constant, 6.63×10^{-34} J s
ν = frequency (Hz)

> **The higher the frequency (ν) of radiation, the greater is its energy.**

X-rays, for example, are much more energetic than rays of visible light. The frequencies of X-rays are on the order of 10^{19} Hz, while those of visible light are on the order of 10^{15} Hz.

Since $\nu = c/\lambda$, the energy of electromagnetic radiation is inversely proportional to its wavelength:

$$E = \frac{hc}{\lambda}$$

where c = velocity of light

> **The shorter the wavelength (λ) of radiation, the greater is its energy.**

X-rays have wavelengths on the order of 0.1 nm and are very energetic, whereas visible light has wavelengths between 400 and 750 nm and is, therefore, of lower energy than X-rays.[†]

*The term hertz (after the German physicist H. R. Hertz), abbreviated Hz, is used in place of the older term *cycles per second* (cps). Frequency of electromagnetic radiation is also sometimes expressed in *wavenumbers*— that is, the number of waves per centimeter.

[†]A convenient formula that relates wavelength (in nm) to the energy of electromagnetic radiation is the following:

$$E \text{ (in kJ mol}^{-1}) = \frac{1.20 \times 10^{-9} \text{ kJ mol}^{-1}}{\text{wavelength in nanometers}}$$

It may be helpful to point out, too, that for visible light, wavelengths (and thus frequencies) are related to what we perceive as colors. The light that we call red light has a wavelength of approximately 650 nm. The light we call violet light has a wavelength of approximately 400 nm. All of the other colors of the visible spectrum (the rainbow) lie in between these wavelengths.

The different regions of the **electromagnetic spectrum** are shown in Fig. 13.7. Nearly every portion of the electromagnetic spectrum from the region of X-rays to that of microwaves and radio waves has been used in elucidating structures of atoms and molecules. Although techniques differ according to the portion of the electromagnetic spectrum in which we are working, there is a consistency and unity of basic principles.

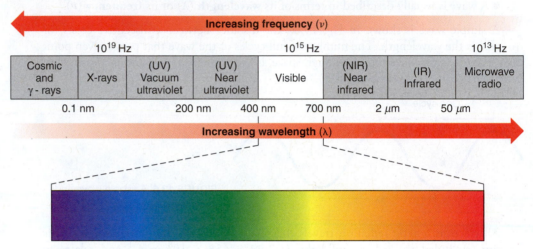

FIGURE 13.7 The electromagnetic spectrum.

13.8B UV–Vis Spectrophotometers

- A UV–Vis spectrophotometer (Fig. 13.8) measures the amount of light absorbed by a sample at each wavelength of the UV and visible regions of the electromagnetic spectrum.

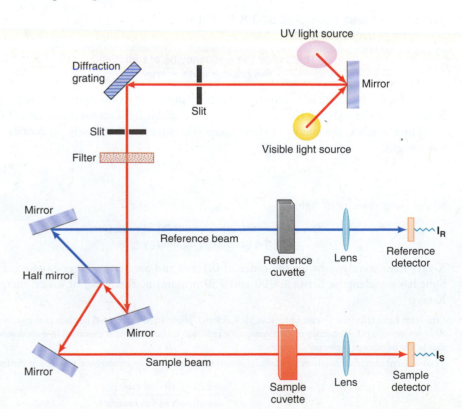

FIGURE 13.8 A diagram of a UV–Vis spectrophotometer.
(Courtesy William Reusch, http://www2.chemistry.msu.edu/faculty/reusch/VirtTxtJml/Spectrpy/UV-Vis/uvspec.htm#uv1. © 1999)

UV and visible radiation are of higher energy (shorter wavelength) than infrared radiation (used in IR spectroscopy) and radio frequency radiation (used in NMR) but not as energetic as X-radiation (Fig. 13.7).

In a standard UV–Vis spectrophotometer (Fig. 13.8) a beam of light is split; one half of the beam (the sample beam) is directed through a transparent cell containing a solution of the compound being analyzed, and one half (the reference beam) is directed through an identical cell that does not contain the compound but contains the solvent. Solvents are chosen to be transparent in the region of the spectrum being used for analysis. The instrument is designed so that it can make a comparison of the intensities of the two beams as it scans over the desired region of wavelengths. If the compound absorbs light at a particular wavelength, the intensity of the sample beam (I_S) will be less than that of the reference beam (I_R). The absorbance at a particular wavelength is defined by the equation $A_\lambda = \log(I_R/I_S)$.

- Data from a UV–Vis spectrophotometer are presented as an **absorption spectrum**, which is a graph of wavelength (λ) versus sample absorbance (A) at each wavelength in the spectral region of interest.

In diode-array UV–Vis spectrophotometers the absorption of all wavelengths of light in the region of analysis is measured simultaneously by an array of photodiodes. The absorption of the solvent is measured over all wavelengths of interest first, and then the absorption of the sample is recorded over the same range. Data from the solvent are electronically subtracted from the data for the sample. The difference is then displayed as the absorption spectrum for the sample.

A typical UV absorption spectrum, that of 2,5-dimethyl-2,4-hexadiene, is given in Fig. 13.9. It shows a broad absorption band in the region between 210 and 260 nm, with the maximum absorption at 242.5 nm.

- The wavelength of maximum absorption in a given spectrum is usually reported in the chemical literature as λ_{max}.

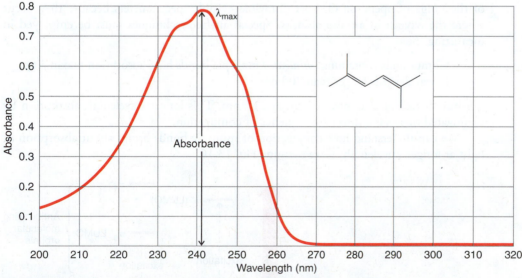

FIGURE 13.9 The UV absorption spectrum of 2,5-dimethyl-2,4-hexadiene in methanol at a concentration of 5.95×10^{-5} M in a 1.00-cm cell.

In addition to reporting the wavelength of maximum absorption (λ_{max}), chemists often report another quantity called the molar absorptivity, ε. In older literature, the molar absorptivity, ε, is often referred to as the molar extinction coefficient.

- The **molar absorptivity** (ε, in units of $M^{-1}cm^{-1}$) indicates the intensity of the absorbance for a sample at a given wavelength. It is a proportionality constant that relates absorbance to molar concentration of the sample (M) and the path length (l, in cm) of light through the sample.

- The equation that relates absorbance (A) to concentration (C) and path length (l) via molar absorptivity (ε) is called **Beer's law**.

$$A = \varepsilon \times C \times l \quad \text{or} \quad \varepsilon = \frac{A}{C \times l} \quad \boxed{\text{Beer's law}}$$

For 2,5-dimethyl-2,4-hexadiene dissolved in methanol the molar absorptivity at the wavelength of maximum absorbance (242.5 nm) is 13,100 M^{-1} cm^{-1}. In the chemical literature this would be reported as

$$\text{2,5-Dimethyl-2,4-hexadiene, } \lambda_{max}^{methanol} \text{ 242.5 nm} \quad (\varepsilon = 13,100)$$

13.8C Absorption Maxima for Nonconjugated and Conjugated Dienes

As we noted earlier, when compounds absorb light in the UV and visible regions, electrons are excited from lower electronic energy levels to higher ones. For this reason, visible and UV spectra are often called **electronic spectra**. The absorption spectrum of 2,5-dimethyl-2,4-hexadiene is a typical electronic spectrum because the absorption band (or peak) is very broad. Most absorption bands in the visible and UV region are broad because each electronic energy level has associated with it vibrational and rotational levels. Thus, electron transitions may occur from any of several vibrational and rotational states of one electronic level to any of several vibrational and rotational states of a higher level.

- Alkenes and nonconjugated dienes usually have absorption maxima (λ_{max}) below 200 nm.

Ethene, for example, gives an absorption maximum at 171 nm; 1,4-pentadiene gives an absorption maximum at 178 nm. These absorptions occur at wavelengths that are out of the range of operation of most ultraviolet–visible spectrometers because they occur where the oxygen in air also absorbs. Special air-free techniques must be employed in measuring them.

- Compounds containing *conjugated* multiple bonds have absorption maxima (λ_{max}) at wavelengths longer than 200 nm.

1,3-Butadiene, for example, absorbs at 217 nm. This longer wavelength absorption by conjugated dienes is a direct consequence of conjugation.

We can understand how conjugation of multiple bonds brings about absorption of light at longer wavelengths if we examine Fig. 13.10.

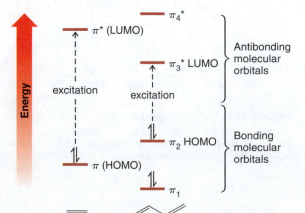

FIGURE 13.10 The relative energies of the π molecular orbitals of ethene and 1,3-butadiene (Section 13.6C).

- When a molecule absorbs light at its longest wavelength, an electron is excited from its **highest occupied molecular orbital (HOMO)** to the **lowest unoccupied molecular orbital (LUMO)**.
- For most alkenes and alkadienes the HOMO is a bonding π orbital and the LUMO is an antibonding π^* orbital.

The wavelength of the absorption maximum is determined by the difference in energy between these two levels. The energy gap between the HOMO and LUMO of ethene is greater than that between the corresponding orbitals of 1,3-butadiene. Thus, the $\pi \longrightarrow \pi^*$ electron excitation of ethene requires absorption of light of greater energy (shorter wavelength) than the corresponding $\pi_2 \longrightarrow \pi_3^*$ excitation in 1,3-butadiene. The energy difference between the HOMOs and the LUMOs of the two compounds is reflected in their absorption spectra. Ethene has its λ_{max} at 171 nm; 1,3-butadiene has a λ_{max} at 217 nm.

The narrower gap between the HOMO and the LUMO in 1,3-butadiene results from the conjugation of the double bonds. Molecular orbital calculations indicate that a much larger gap should occur in isolated alkadienes. This is borne out experimentally. Isolated alkadienes give absorption spectra similar to those of alkenes. Their λ_{max} are at shorter wavelengths, usually below 200 nm. As we mentioned, 1,4-pentadiene has its λ_{max} at 178 nm.

Conjugated alkatrienes absorb at longer wavelengths than conjugated alkadienes, and this too can be accounted for in molecular orbital calculations. The energy gap between the HOMO and the LUMO of an alkatriene is even smaller than that of an alkadiene.

- In general, *the greater the number of conjugated multiple bonds in a molecule, the longer will be its λ_{max}*.

Polyenes with eight or more conjugated double bonds absorb light in the visible region of the spectrum. For example, β-carotene, a precursor of vitamin A and a compound that imparts its orange color to carrots, has 11 conjugated double bonds; β-carotene has an absorption maximum at 497 nm, well into the visible region. Light of 497 nm has a blue-green color; this is the light that is absorbed by β-carotene. We perceive the complementary color of blue-green, which is red-orange.

β-Carotene

Lycopene, a compound partly responsible for the red color of tomatoes, also has 11 conjugated double bonds. Lycopene has an absorption maximum at 505 nm where it absorbs intensely. (Approximately 0.02 g of lycopene can be isolated from 1 kg of fresh, ripe tomatoes.)

Lycopene

Table 13.3 gives the values of λ_{max} for a number of unsaturated compounds.

TABLE 13.3 LONG-WAVELENGTH ABSORPTION MAXIMA OF UNSATURATED HYDROCARBONS

Compound	Structure	λ_{max} (nm)	ε_{max} (M^{-1} cm^{-1})
Ethene	$CH_2 = CH_2$	171	15,530
trans-3-Hexene		184	10,000
Cyclohexene	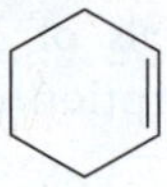	182	7,600
1-Octene	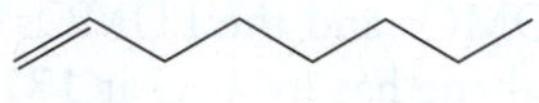	177	12,600
1-Octyne	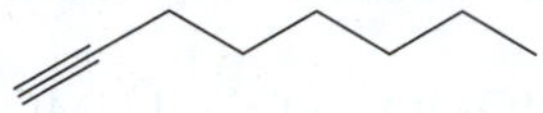	185	2,000
1,3-Butadiene	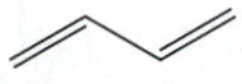	217	21,000
cis-1,3-Pentadiene	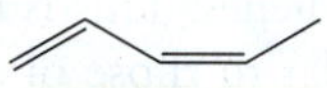	223	22,600
trans-1,3-Pentadiene	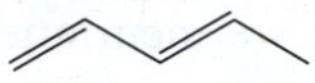	223.5	23,000
But-1-en-3-yne	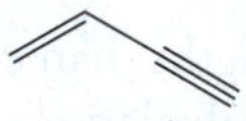	228	7,800
1,4-Pentadiene		178	17,000
1,3-Cyclopentadiene		239	3,400
1,3-Cyclohexadiene	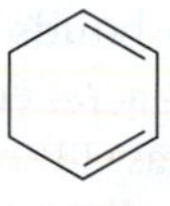	256	8,000
trans-1,3,5-Hexatriene		274	50,000

Compounds with carbon–oxygen double bonds also absorb light in the UV region. Acetone, for example, has a broad absorption peak at 280 nm that corresponds to the excitation of an electron from one of the unshared pairs (a nonbonding or "n" electron) to the π^* orbital of the carbon–oxygen double bond:

Acetone

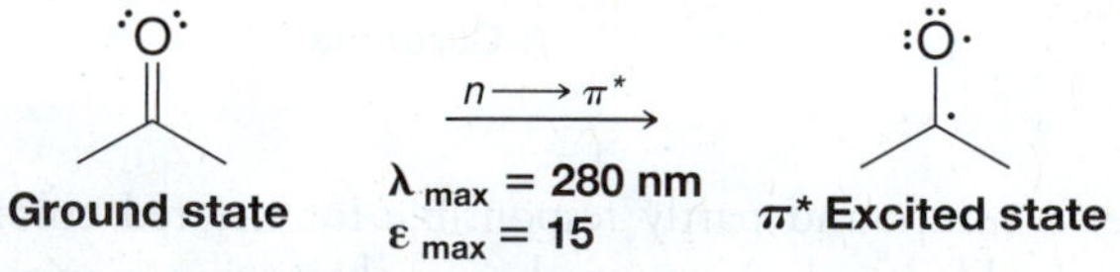

$$n \longrightarrow \pi^*$$
$$\lambda_{max} = 280 \text{ nm}$$
$$\varepsilon_{max} = 15$$

Ground state π^* **Excited state**

Compounds in which the carbon–oxygen double bond is conjugated with a carbon–carbon double bond have absorption maxima corresponding to $n \longrightarrow \pi^*$ excitations and $\pi \longrightarrow \pi^*$ excitations. The $n \longrightarrow \pi^*$ absorption maxima occur at longer wavelengths but are much weaker (i.e., have smaller molar absorptivity (ε) values):

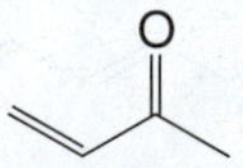

$$n \longrightarrow \pi^* \quad \lambda_{max} = 324 \text{ nm}, \ \varepsilon_{max} = 24$$
$$\pi \longrightarrow \pi^* \quad \lambda_{max} = 219 \text{ nm}, \ \varepsilon_{max} = 3600$$

13.8D Analytical Uses of UV–Vis Spectroscopy

UV–Vis spectroscopy can be used in the structure elucidation of organic molecules to indicate whether conjugation is present in a given sample. Although conjugation in a molecule may be indicated by data from IR, NMR, or mass spectrometry, UV–Vis analysis can provide corroborating information.

A more widespread use of UV–Vis spectroscopy, however, has to do with determining the concentration of an unknown sample. As mentioned in Section 13.8B, the relationship $A = \varepsilon Cl$ indicates that the amount of absorption by a sample at a certain wavelength is dependent on its concentration. This relationship is usually linear over a range of concentrations suitable for analysis. To determine the unknown concentration of a sample, a graph of absorbance versus concentration is made for a set of standards of known concentrations. The wavelength used for analysis is usually the λ_{max} of the sample. The concentration of the sample is obtained by measuring its absorbance and determining the corresponding value of concentration from the graph of known concentrations. Quantitative analysis using UV–Vis spectroscopy is routinely used in biochemical studies to measure the rates of enzymatic reactions. The concentration of a species involved in the reaction (as related to its UV–Vis absorbance) is plotted versus time to determine the rate of reaction. UV–Vis spectroscopy is also used in environmental chemistry to determine the concentration of various metal ions (sometimes involving absorption spectra for organic complexes with the metal) and as a detection method in high-performance liquid chromatography (HPLC).

••• SOLVED PROBLEM 13.7

Two isomeric compounds, **A** and **B**, have the molecular formula C_7H_{12}. Compound **A** shows no absorption in the UV–visible region. The ^{13}C NMR spectrum of **A** shows only three signals. Compound **B** shows a UV–visible peak in the region of 180 nm, its ^{13}C NMR spectrum shows four signals, and its DEPT ^{13}C NMR data show that none of its carbon atoms is a methyl group. On catalytic hydrogenation with excess hydrogen, **B** is converted to heptane. Propose structures for **A** and **B**.

STRATEGY AND ANSWER: On the basis of their molecular formulas, both compounds have an index of hydrogen deficiency (Section 4.17) equal to 2. Therefore on this basis alone, each could contain two double bonds, one ring and one double bond, two rings, or a triple bond. Consider **A** first. The fact that **A** does not absorb in the UV–visible region suggests that it does not have any double bonds; therefore, it must contain two rings. A compound with two rings that would give only three signals in its ^{13}C spectrum is bicyclo[2.2.1]heptane (because it has only three distinct types of carbon atoms).

Now consider **B**. The fact that **B** is converted to heptane on catalytic hydrogenation suggests that **B** is a heptadiene or a heptyne with an unbranched chain. UV–visible absorption in the 180-nm region suggests that **B** does not contain conjugated π bonds. Given that the DEPT ^{13}C data for **B** shows the absence of any methyl groups, and only four ^{13}C signals in total, **B** must be 1,6-hexadiene.

A
Bicyclo[2.2.1]heptane
(three ^{13}C signals)

B
1,6-Hexadiene
(four ^{13}C signals,
none comes from a methyl group)

••• **PRACTICE PROBLEM 13.7**

Two compounds, **A** and **B**, have the same molecular formula, C_6H_8. Both **A** and **B** react with two molar equivalents of hydrogen in the presence of platinum to yield cyclohexane. Compound **A** shows three signals in its broadband decoupled ^{13}C NMR spectrum. Compound **B** shows only two ^{13}C NMR signals. Compound **A** shows an absorption maximum at 256 nm, whereas **B** shows no absorption maximum at wavelengths longer than 200 nm. What are the structures of **A** and **B**?

C_6H_8
A
^{13}C NMR: 3 signals
UV: λ_{max} ~256 nm

H_2, Pt

H_2, Pt

C_6H_8
B
^{13}C NMR: 2 signals
UV: λ_{max} < 200 nm

PRACTICE PROBLEM 13.8 Propose structures for **D, E,** and **F**.

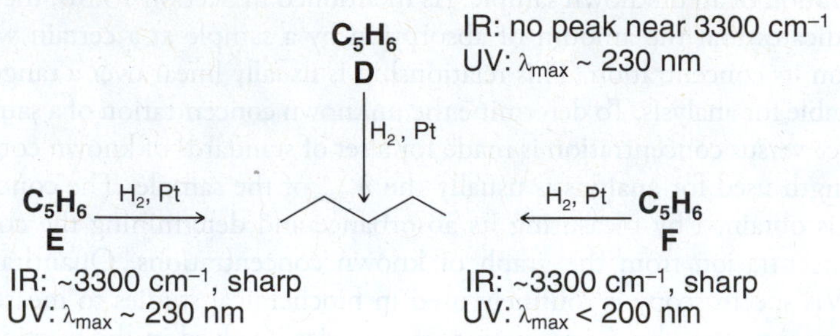

$$C_5H_6$$
D
IR: no peak near 3300 cm⁻¹
UV: λ_{max} ~ 230 nm

H_2, Pt

$C_5H_6 \xrightarrow{H_2, Pt}$
E
IR: ~3300 cm⁻¹, sharp
UV: λ_{max} ~ 230 nm

$\xleftarrow{H_2, Pt} C_5H_6$
F
IR: ~3300 cm⁻¹, sharp
UV: λ_{max} < 200 nm

13.9 ELECTROPHILIC ATTACK ON CONJUGATED DIENES: 1,4-ADDITION

Not only are conjugated dienes somewhat more stable than nonconjugated dienes, they also display special behavior when they react with electrophilic reagents.

- Conjugated dienes undergo both 1,2- and 1,4-addition through an allylic intermediate that is common to both.

For example, 1,3-butadiene reacts with one molar equivalent of hydrogen chloride to produce two products, 3-chloro-1-butene and 1-chloro-2-butene:

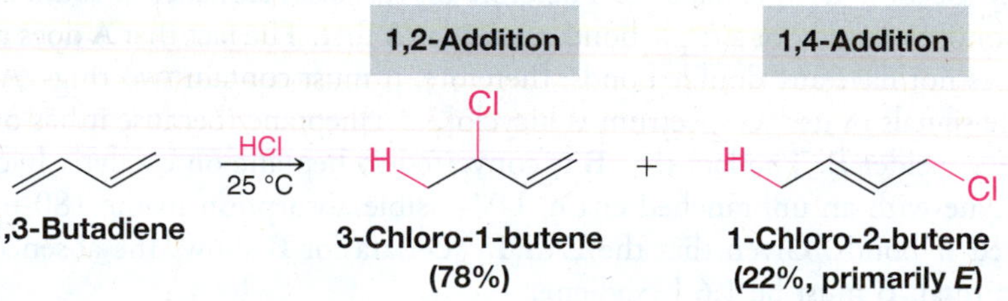

1,2-Addition	**1,4-Addition**

1,3-Butadiene $\xrightarrow[25\ °C]{HCl}$ **3-Chloro-1-butene** (78%) + **1-Chloro-2-butene** (22%, primarily *E*)

If only the first product (3-chloro-1-butene) were formed, we would not be particularly surprised. We would conclude that hydrogen chloride had added to one double bond of 1,3-butadiene in the usual way. It is the second product, 1-chloro-2-butene, that is initially surprising. Its double bond is between the central atoms, and the elements of hydrogen chloride have added to the C1 and C4 atoms.

To understand how both 1,2- and 1,4-addition products result from reaction of 1,3-butadiene with HCl, consider the following mechanism.

Step 1

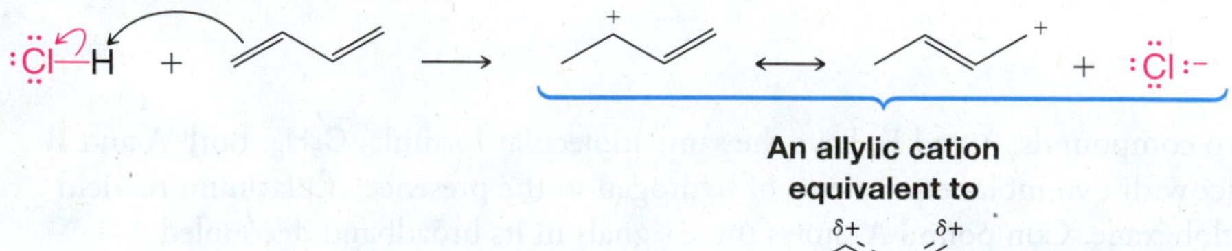

An allylic cation equivalent to

Step 2

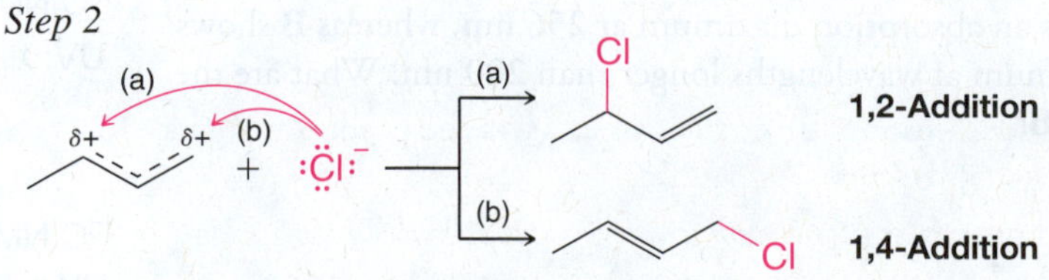

(a) **1,2-Addition**

(b) **1,4-Addition**

In step 1 a proton adds to one of the terminal carbon atoms of 1,3-butadiene to form, as usual, the more stable carbocation, in this case a resonance-stabilized allylic cation.

Addition to one of the inner carbon atoms would have produced a much less stable primary cation, one that could not be stabilized by resonance:

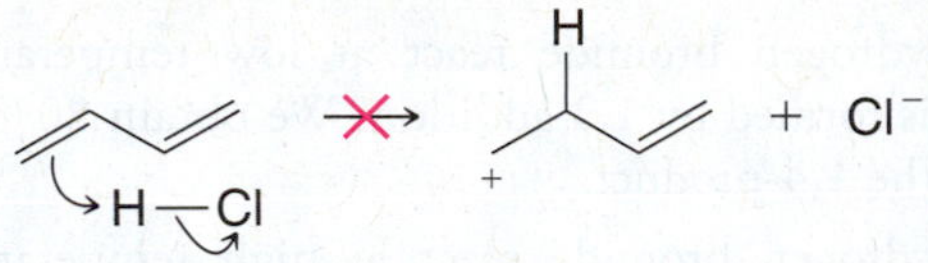

> **Addition of the electrophile in this fashion does not lead to an allylic (resonance-stabilized) carbocation.**

In step 2 a chloride ion forms a bond to one of the carbon atoms of the allylic cation that bears a partial positive charge. Reaction at one carbon atom results in the 1,2-addition product; reaction at the other gives the 1,4-addition product.

Note that the designations 1,2 and 1,4 only coincidentally relate to the IUPAC numbering of carbon atoms in this example.

• Chemists typically use 1,2 and 1,4 to refer to modes of addition to any conjugated diene system, regardless of where the conjugated double bonds are in the overall molecule.

Thus, addition reactions of 2,4-hexadiene would still involve references to 1,2 and 1,4 modes of addition.

Predict the products of the following reactions.

PRACTICE PROBLEM 13.9

(a) 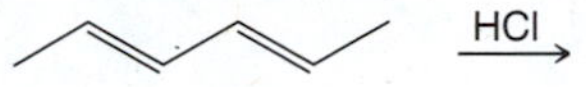   (b)

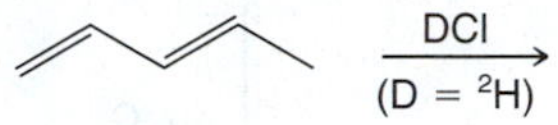

1,3-Butadiene shows 1,4-addition reactions with electrophilic reagents other than hydrogen chloride. Two examples are shown here, the addition of hydrogen bromide (in the absence of peroxides) and the addition of bromine:

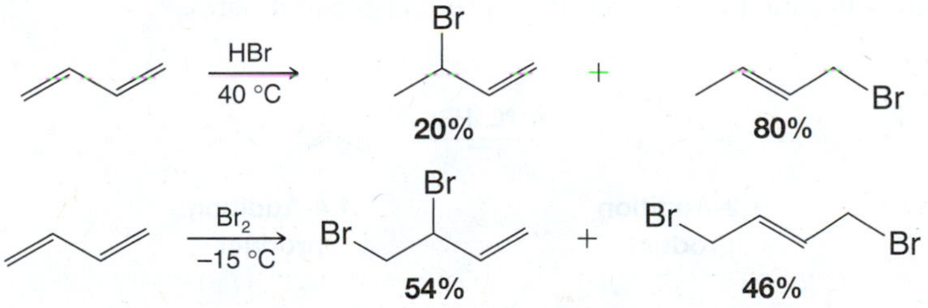

Reactions of this type are quite general with other conjugated dienes. Conjugated trienes often show 1,6-addition. An example is the 1,6-addition of bromine to 1,3,5-cyclooctatriene:

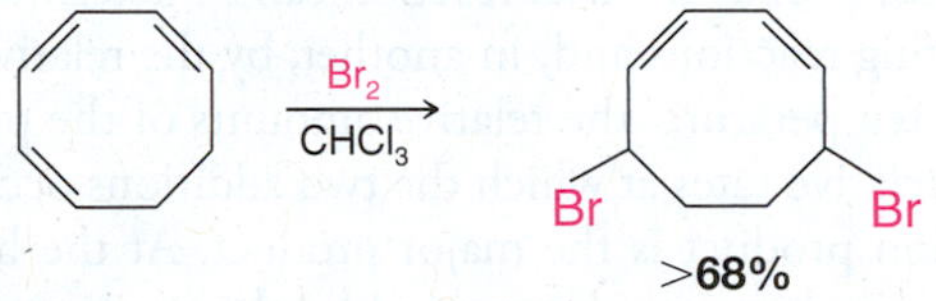

13.9A Kinetic Control versus Thermodynamic Control of a Chemical Reaction

The addition of hydrogen bromide to 1,3-butadiene allows the illustration of another important aspect of reactivity—the way temperature affects product distribution in a reaction that can take multiple paths. In general:

• The favored products in a reaction at *lower temperature* are those formed by the pathway having the smallest energy of activation barrier. In this case the reaction is said to be under **kinetic (or rate) control**, and the predominant products are called the **kinetic products**.

• The favored products at *higher temperature* in a *reversible* reaction are those that are most stable. In this case the reaction is said to be under **thermodynamic (or equilibrium) control**, and the predominant products are called the **thermodynamic (or equilibrium) products**.

Let's consider specific reaction conditions for the ionic addition of hydrogen bromide to 1,3-butadiene.

Case 1. When 1,3-butadiene and hydrogen bromide react at low temperature (−80 °C), the major product is formed by 1,2-addition. We obtain 80% of the 1,2-product and 20% of the 1,4-product.

Case 2. When 1,3-butadiene and hydrogen bromide react at high temperature (40 °C), the major product is formed by 1,4-addition. We obtain about 20% of the 1,2-product and about 80% of the 1,4-product.

Case 3. When the product mixture from the low temperature reaction is warmed to the higher temperature, the product distribution becomes the same as when the reaction was carried out at high temperature—that is, the 1,4-product predominates.

We summarize these scenarios here:

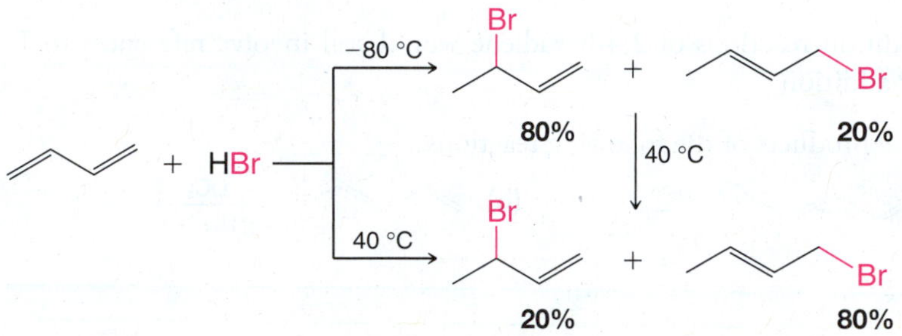

Furthermore, when a pure sample of 3-bromo-1-butene (the predominant product at low temperature) is subjected to the high temperature reaction conditions, an equilibrium mixture results in which the 1,4-addition product predominates.

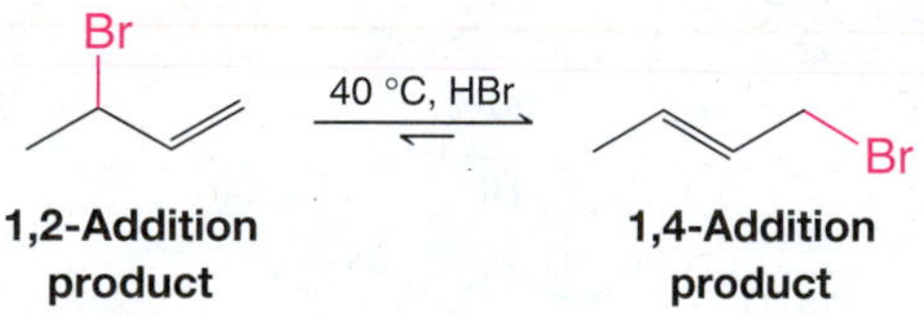

1,2-Addition product   **1,4-Addition product**

Because this equilibrium favors the 1,4-addition product, *that product must be more stable*.

The reactions of hydrogen bromide with 1,3-butadiene serve as a striking illustration of the way that the outcome of a chemical reaction can be determined, in one instance, by relative rates of competing reactions and, in another, by the relative stabilities of the final products. At the lower temperature, the relative amounts of the products of the addition are determined by the relative rates at which the two additions occur; 1,2-addition occurs faster so the 1,2-addition product is the major product. At the higher temperature, the relative amounts of the products are determined by the position of an equilibrium. The 1,4-addition product is the more stable, so it is the major product.

This behavior of 1,3-butadiene and hydrogen bromide can be more fully understood if we examine the diagram shown in Fig. 13.11.

- The step that determines the overall outcome of this reaction is the step in which the hybrid allylic cation combines with a bromide ion.

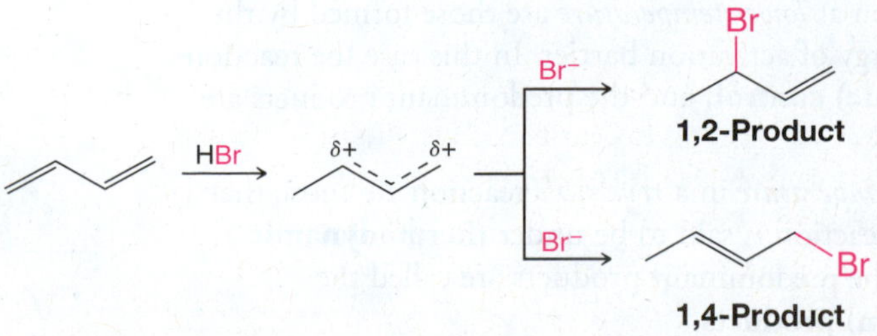

The reaction of the bromide anion with the allylic cation determines the regioselectivity of the reaction.

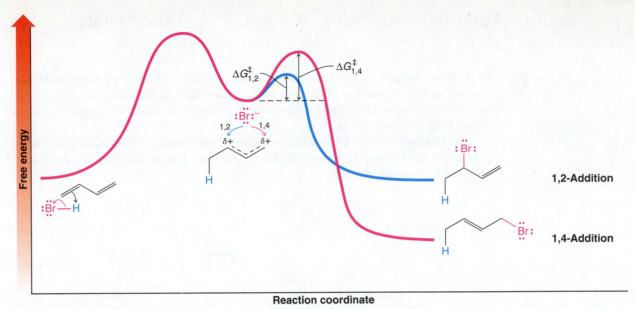

FIGURE 13.11 A schematic free-energy versus reaction coordinate diagram for the 1,2- and 1,4-addition of HBr to 1,3-butadiene. An allylic carbocation is common to both pathways. The energy barrier for attack of bromide ion on the allylic cation to form the 1,2-addition product is less than that to form the 1,4-addition product. The 1,2-addition product is kinetically favored. The 1,4-addition product is more stable, and so it is the thermodynamically favored product.

We see in Fig. 13.11 that the free energy of activation leading to the 1,2-addition product is less than the free energy of activation leading to the 1,4-addition product, even though the 1,4-product is more stable.

- At **low temperature**, the fraction of collisions capable of surmounting the higher energy barrier leading to formation of the 1,4-product is smaller than the fraction that can cross the barrier leading to the 1,2-product.

- At low temperature, formation of the 1,2- and 1,4-products is essentially *irreversible* because there is not enough energy for either product to cross back over the barrier to reform the allylic cation. Thus, the 1,2-product predominates at lower temperature because it is formed faster and it is not formed reversibly. It is the **kinetic product** of this reaction.

- At **higher temperature**, collisions between the intermediate ions are sufficiently energetic to allow rapid formation of *both* the 1,2- and 1,4-products. *But*, there is also sufficient energy for both products to revert to the **allylic carbocation**.

- Because the 1,2-product has a smaller energy barrier for conversion back to the allylic cation than does the 1,4-product, more of the 1,2-product reverts to the allylic cation than does the 1,4-product. But since both the 1,4- and the 1,2-products readily form from the allylic cation at high temperature, eventually this equilibrium leads to a preponderance of the 1,4-product because it is more stable. The 1,4-product is the **thermodynamic** or **equilibrium product** of this reaction.

Before we leave this subject, one final point should be made. This example clearly demonstrates that predictions of relative reaction rates made on the basis of product stabilities alone can be wrong. This is not always the case, however. For many reactions in which a common intermediate leads to two or more products, the most stable product is formed fastest.

● ● ●

(a) Suggest a structural explanation for the fact that the 1,2-addition reaction of **PRACTICE PROBLEM 13.10** 1,3-butadiene and hydrogen bromide occurs faster than 1,4-addition? (*Hint*: Consider the relative contributions that the two forms ⌒⌒⁺ and ⌒⌒⁺ make to the resonance hybrid of the allylic cation.)

(b) How can you account for the fact that the 1,4-addition product is more stable?

13.10 THE DIELS–ALDER REACTION: A 1,4-CYCLOADDITION REACTION OF DIENES

In 1928 two German chemists, Otto Diels and Kurt Alder, developed a **1,4-cycloaddition** reaction of dienes that has since come to bear their names. The reaction proved to be one of such great versatility and synthetic utility that Diels and Alder were awarded the Nobel Prize in Chemistry in 1950.

An example of the Diels–Alder reaction is the reaction that takes place when 1,3-butadiene and maleic anhydride are heated together at 100 °C. The product is obtained in quantitative yield:

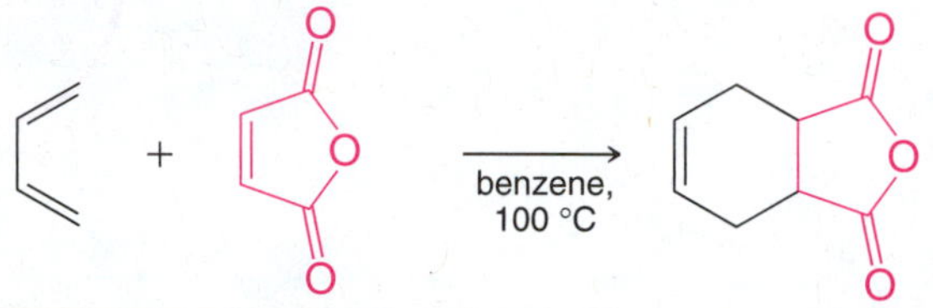

1,3-Butadiene (diene)	**Maleic anhydride** (dienophile)	benzene, 100 °C	**Adduct** (100%)	

- In general terms, the **Diels–Alder reaction** is one between a conjugated **diene** (a 4π-electron system) and a compound containing a double bond (a 2π-electron system) called a **dienophile** (diene + *philia*, Greek: to love). The product of a Diels–Alder reaction is often called an **adduct**.

In the Diels–Alder reaction, two new σ bonds are formed at the expense of two π bonds of the diene and dienophile. The adduct contains a new six-membered ring with a double bond. Since σ bonds are usually stronger than π bonds, formation of the adduct is usually favored energetically, *but most Diels–Alder reactions are reversible.*

We can account for all of the bond changes in a Diels–Alder reaction like that above by using curved arrows in the following way:

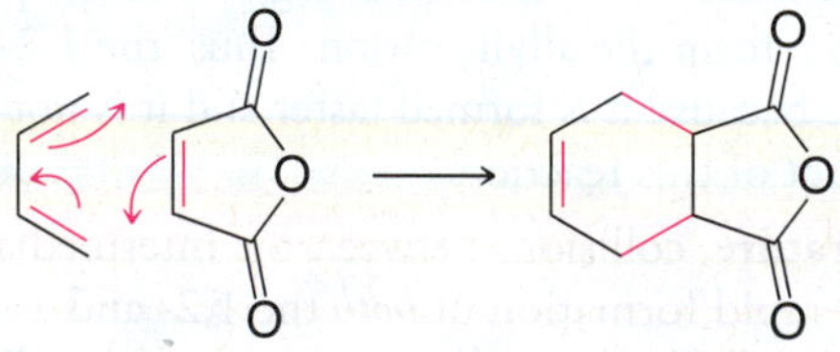

Diene   Dienophile   Adduct

The Diels–Alder reaction is an example of a **pericyclic reaction** (see Special Topic H in *WileyPLUS*). **Pericyclic reactions are concerted reactions that take place in one step through a cyclic transition state in which symmetry characteristics of molecular orbitals control the course of the reaction.**

We will discuss the mechanism of the Diels–Alder reaction in terms of molecular orbitals in Special Topic H in *WileyPLUS*. For the moment we will continue to describe the Diels–Alder reaction using bonds and curved arrows to describe the movement of electrons that take place.

The simplest example of a Diels–Alder reaction is the one that takes place between 1,3-butadiene and ethene. This reaction, however, takes place much more slowly than the reaction of butadiene with maleic anhydride and also must be carried out under pressure:

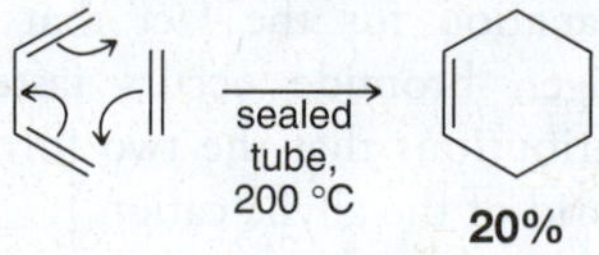

sealed tube, 200 °C   **20%**

Helpful Hint

The Diels–Alder reaction is a very useful synthetic tool for preparing cyclohexene rings.

Another example is the preparation of an intermediate in the synthesis of the anticancer drug Taxol (paclitaxel) by K. C. Nicolaou (The Scripps Research Institute and the University of California, San Diego):

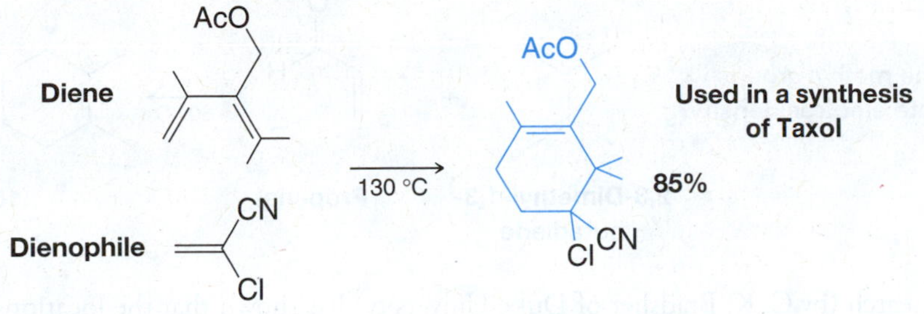

Diene

Dienophile

130 °C

85%

Used in a synthesis of Taxol

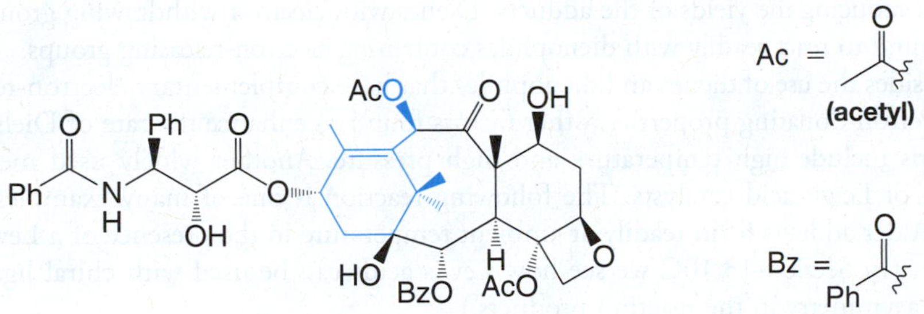

Ac = (acetyl)

Bz = Ph (benzoyl)

Taxol
(blue atoms derived from the Diels–Alder adduct above)

In general, the dienophile reacts with a conjugated diene by 1,4-addition to form a six-membered ring. The process is called a [4+2] **cycloaddition**, named according to the number of atoms from each reactant that join to form the ring, and it is brought about by heat (a thermal reaction). Any position on either side of the diene or dienophile can be substituted. Some representative electron-withdrawing groups that can be part of the dienophile are shown below as Z and Z'.

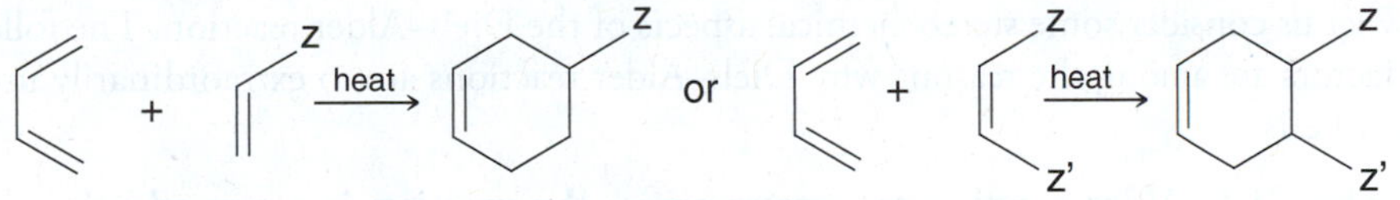

heat or heat

Where Z and Z′ can be CHO, COR, CO₂H, CO₂R, CN, Ar, CO-O-CO,
or halogen, as well as others.

Pericyclic reactions in which two alkenes combine in the following way are also known.

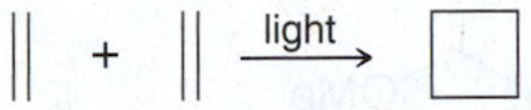

|| + || light →

These are called [2 + 2] cycloadditions and require light energy (they are photochemical reactions). We discuss them in Special Topic H in *WileyPLUS*.

13.10A Factors Favoring the Diels–Alder Reaction

Alder originally stated that the Diels–Alder reaction is favored by the presence of electron-withdrawing groups in the dienophile and by electron-releasing groups in the diene. Maleic anhydride, a very potent dienophile, has two electron-withdrawing carbonyl groups on carbon atoms adjacent to the double bond.

The helpful effect of electron-releasing groups in the diene can also be demonstrated; 2,3-dimethyl-1,3-butadiene, for example, is nearly five times as reactive in Diels–Alder reactions as is 1,3-butadiene. The methyl groups inductively release electron density, just as alkyl groups do when stabilizing a carbocation (though no carbocations are involved here). When 2,3-dimethyl-1,3-butadiene reacts with propenal (acrolein) at only 30 °C, the adduct is obtained in quantitative yield:

The methyl groups donate electron density.				
	2,3-Dimethyl-1,3-butadiene	Propenal		100%

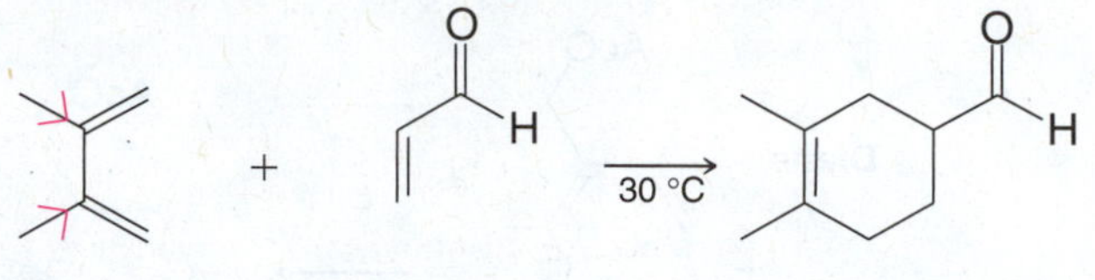

Research (by C. K. Bradsher of Duke University) has shown that the locations of electron-withdrawing and electron-releasing groups in the dienophile and diene can be reversed without reducing the yields of the adducts. Dienes with electron-withdrawing groups have been found to react readily with dienophiles containing electron-releasing groups.

Besides the use of dienes and dienophiles that have complementary electron-releasing and electron-donating properties, other factors found to enhance the rate of Diels–Alder reactions include high temperature and high pressure. Another widely used method is the use of Lewis acid catalysts. The following reaction is one of many examples where Diels–Alder adducts form readily at ambient temperature in the presence of a Lewis acid catalyst. (In Section 13.10C we see how Lewis acids can be used with chiral ligands to induce asymmetry in the reaction products.)

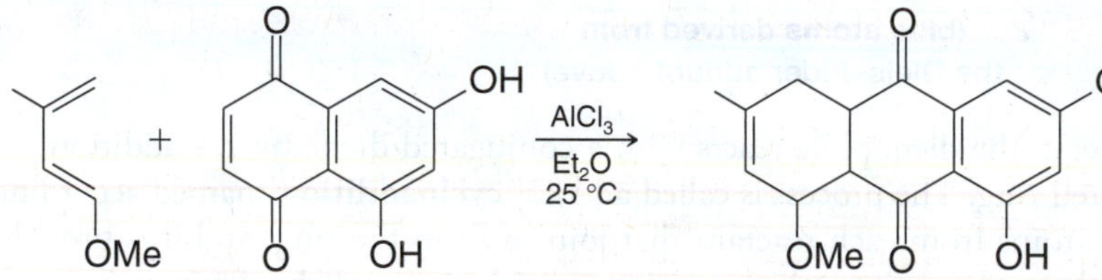

13.10B Stereochemistry of the Diels–Alder Reaction

Now let us consider some stereochemical aspects of the Diels–Alder reaction. The following factors are among the reasons why Diels–Alder reactions are so extraordinarily useful in synthesis.

1. **The Diels–Alder reaction is stereospecific: the reaction is a syn addition, and the configuration of the dienophile is *retained* in the product.** Two examples that illustrate this aspect of the reaction are shown here:

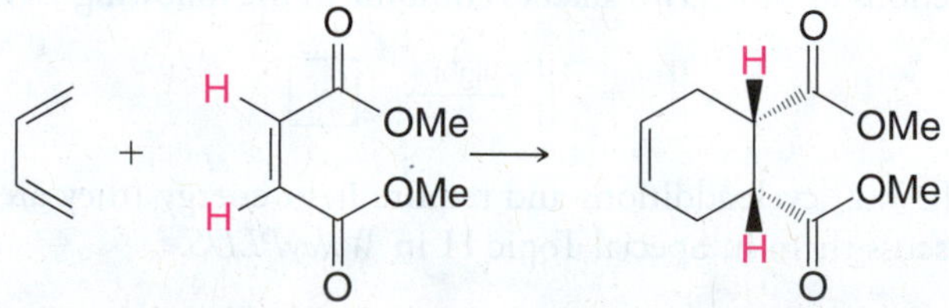

Dimethyl maleate (a *cis*-dienophile)	Dimethyl cyclohex-4-ene-*cis*-1,2-dicarboxylate

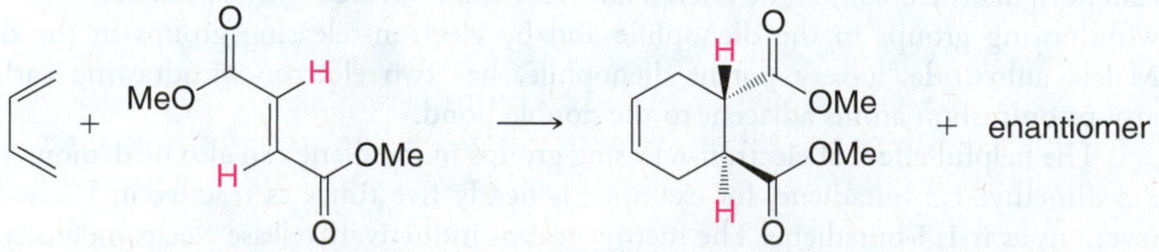

Dimethyl fumarate (a *trans*-dienophile)	Dimethyl cyclohex-4-ene-*trans*-1,2-dicarboxylate

In the first example, a dienophile with cis ester groups reacts with 1,3-butadiene to give an adduct with cis ester groups. In the second example just the reverse is true. A *trans*-dienophile gives a trans adduct.

2. **The diene, of necessity, reacts in the s-cis rather than in the s-trans conformation:**

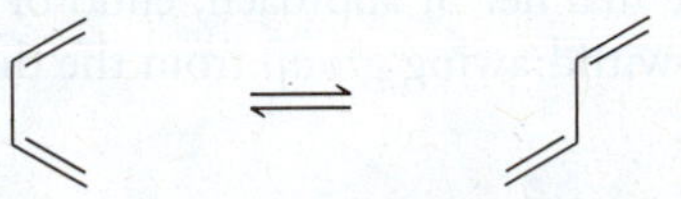

s-cis Conformation   **s-trans Conformation**

Reaction in the s-trans conformation would, if it occurred, produce a six-membered ring with a highly strained trans double bond. This course of the Diels–Alder reaction has never been observed.

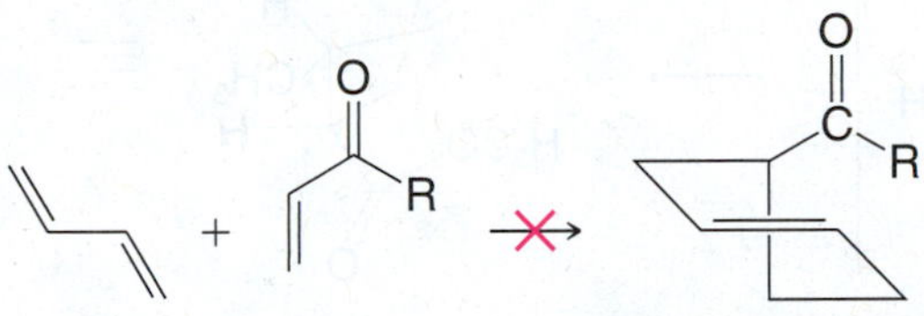

Highly strained

> **Helpful Hint**
>
> Use handheld molecular models to investigate the strained nature of hypothetical *trans*-cyclohexene.

Cyclic dienes in which the double bonds are held in the s-cis conformation are usually highly reactive in the Diels–Alder reaction. Cyclopentadiene, for example, reacts with maleic anhydride at room temperature to give the following adduct in quantitative yield:

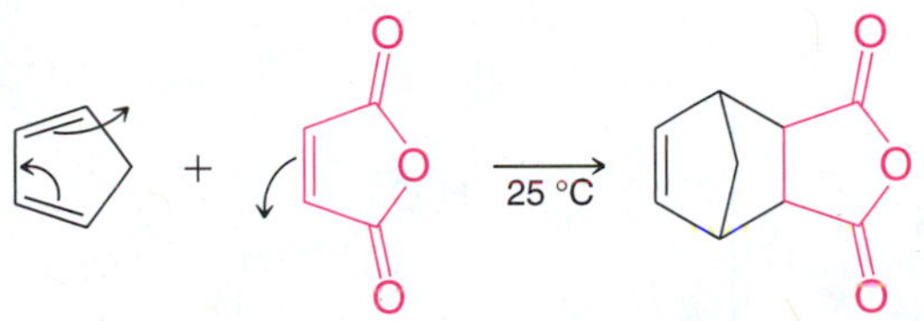

Cyclopentadiene is so reactive that on standing at room temperature it slowly undergoes a Diels–Alder reaction with itself:

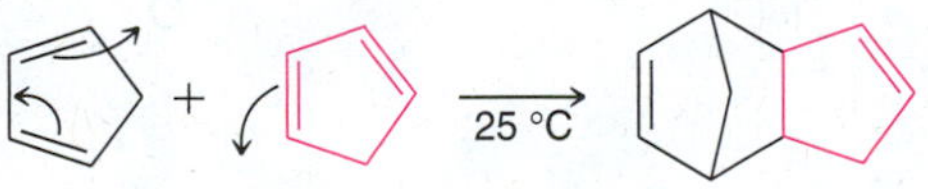

Dicyclopentadiene

The reaction is reversible, however. When dicyclopentadiene is distilled, it dissociates (is "cracked") into two molar equivalents of cyclopentadiene.

The reactions of cyclopentadiene illustrate a third stereochemical characteristic of the Diels–Alder reaction.

3. **The Diels–Alder reaction occurs primarily in an endo rather than an exo fashion when the reaction is kinetically controlled.** A dienophile often contains an electron-withdrawing group, such as a carbonyl or other electronegative group with π electrons, as in the example below and in all of the examples we have shown thus far.

- In Diels–Alder reactions, endo and exo refer to the orientation of the dienophile and its electron-withdrawing group when it reacts with the diene.
- When the dienophile reacts such that the π electron orbitals of its electron-withdrawing group align under (or above) the π electron orbitals of the diene, the orientation of approach is called **endo**.
- When the electron-withdrawing group of the dienophile is aligned away from the π electrons of the diene in the transition state, the orientation of approach is called **exo**.
- When a Diels–Alder reaction takes place, products from both the endo and exo transition states can be formed, but the endo product typically predominates because the endo transition state is usually of lower energy.

Although we shall not discuss the details here, the reason that the endo approach is generally favored has to do with orbital overlap that lowers the transition state energy in the endo orientation. For this reason, the endo product is formed faster (it is the kinetic product).

Consider the example in Figure 13.12 involving (2E,4E)-hexa-2,4-diene and methyl propenoate. Note that the manner of approach, endo or exo, affects the product stereochemistry of the electron-withdrawing group from the dienophile.

(a)

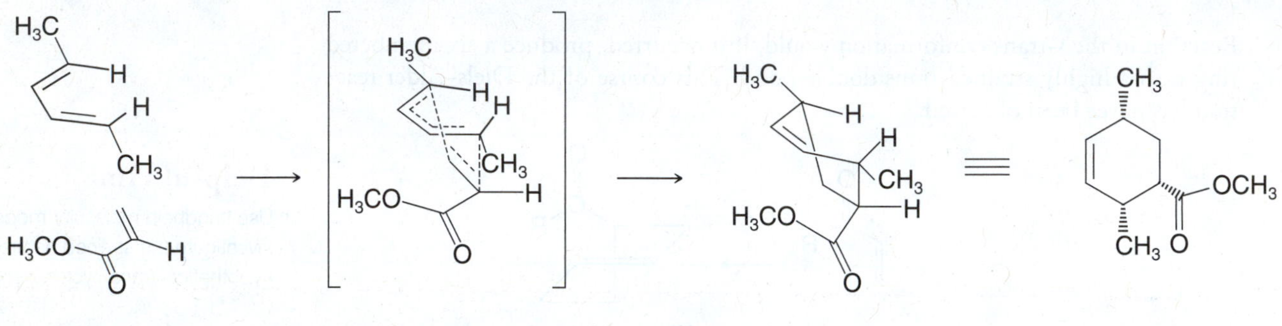

(plus enantiomer)

Endo approach: the dienophile aligns with its electron-withdrawing substituent underneath the π system, as shown, or directly above it (not shown). The two possible endo approaches lead to enantiomers.

(b)

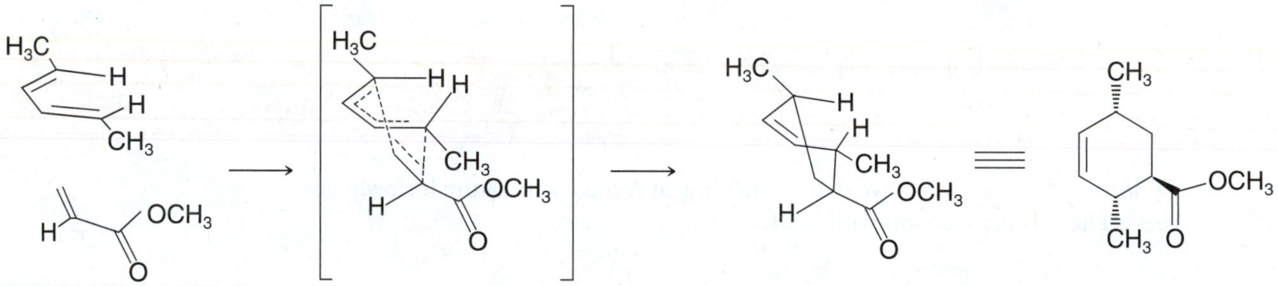

(plus enantiomer)

Exo approach: the dienophile aligns with its electron-withdrawing substituent below and away from the diene π system, as shown, or above it and away (not shown). The two possible exo approaches lead to enantiomers.

FIGURE 13.12 The Diels-Alder reaction of (2E,4E)-hexa-2,4-diene and methyl propenoate showing (a) endo and (b) exo approaches. The endo transition state is favored over exo because the π electron orbitals of the electron-withdrawing group in the dienophile align closer to the π electron orbitals of the diene.

Although in both the endo and exo examples that we have shown in Figure 13.12 the dienophile approaches the diene from below, the dienophile can also approach the diene from above, which is why the enantiomer of the product that we have drawn is formed in each case. In general, Diels–Alder reactions result in the formation of enantiomers unless one or both of the reactants is chiral or there is an influence from a chiral catalyst.

PRACTICE PROBLEM 13.11 Draw transition states for the endo and exo Diels–Alder reaction of maleic anhydride with cyclopentadiene. The transition state leading to only one enantiomer need be shown in each case.

If we were to write a chemical equation for the reaction of (2*E*,4*E*)-hexa-2,4-diene and methyl propenoate we might write it as shown below. **(a)** Explain why we can predict that the endo and exo products will each be formed as a pair of enantiomers. **(b)** What is the stereochemical relationship between either one of the endo enantiomers and either of the exo enantiomers?

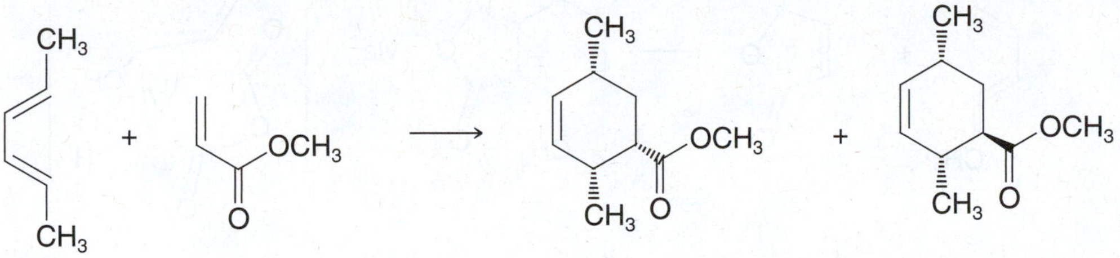

Major–by endo addition
(plus enantiomer)

Major–by exo addition
(plus enantiomer)

The terms endo and exo are also used to designate the orientation of substituents at tetrahedral carbons in the bridges of bicyclic ring systems.

- In a bicyclic system, a substituent is exo if the bond attaching it to the bridge is oriented away from the longest of the two remaining bridges in the system, and endo if the bond to the substituent is oriented toward the longest of the two remaining bridges.

The following illustration shows an R group in an exo position and an endo position.

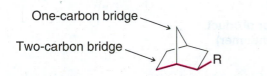

One-carbon bridge

Two-carbon bridge

R is exo
(oriented away from the longest bridge)

R is endo
(oriented closer to the longest bridge)

Consider the Diels–Alder reaction of cyclopentadiene with methyl propenoate shown below. The ester group in the major product is endo because the ester substituent is oriented closer to the two-carbon (the longer) bridge than to the one-carbon (shorter) bridge. In the minor product the ester substituent is exo because it is oriented away from the longest bridge.

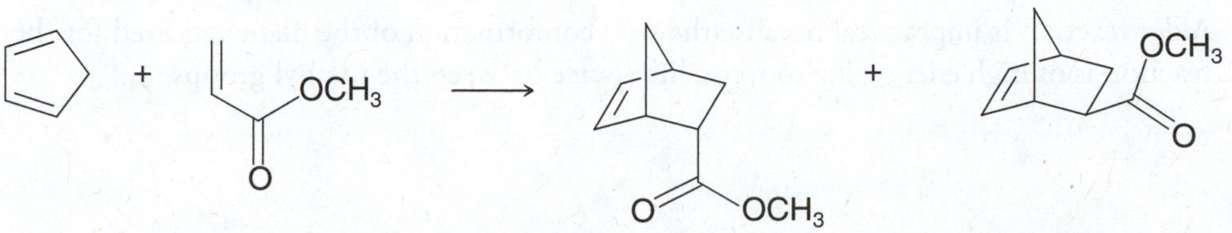

Endo product–major
(plus enantiomer)

Exo product–major
(plus enantiomer)

4. **The configuration of the diene is retained in a Diels–Alder reaction.** Just as the configuration of the dienophile is retained in the adduct of a Diels–Alder reaction (see 1 above), the configuration of the diene is also retained. What we mean is that the *E*,*Z* alkene stereochemistry of both the diene and the dienophile are transferred

to new tetrahedral chirality centers in the Diels–Alder adduct. Consider the reactions below of maleic anhydride with (2E,4E)-hexa-2,4-diene, and with the alkene diastereomer and (2E,4Z)-hexa-2,4-diene, showing the endo (major) product in each case.

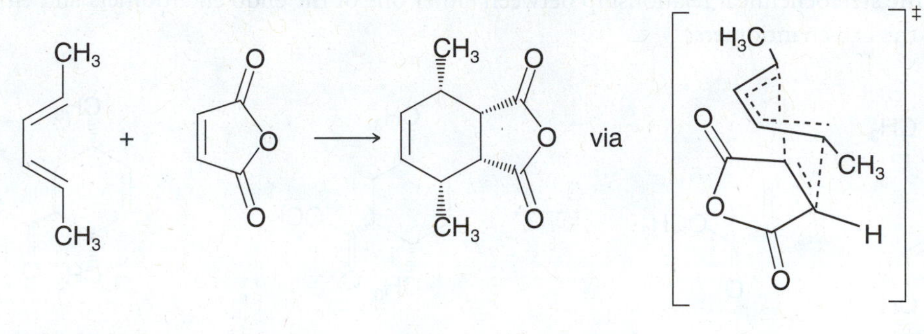

(2E,4E)–Hexa-   **Maleic**   **Endo–major product**
2,4–diene   **anhydride**   **(plus enantiomer)**

When the diene stereochemistry is changed, the stereochemistry of the adduct differs accordingly, as shown below.

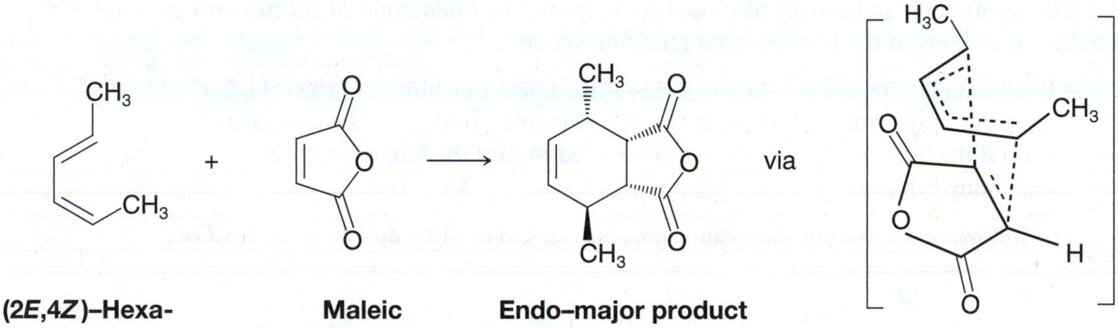

(2E,4Z)–Hexa-   **Maleic**   **Endo–major product**
2,4–diene   **anhydride**   **(plus enantiomer)**

If the diene substituents are both on the outside of the diene component when the diene is in the s-cis conformation the substituents will be cis.

Consideration of the transition states for each reaction, as given above, helps to show how the stereochemistry is transferred from the diene to the adduct in each case.

PRACTICE PROBLEM 13.13 If (2Z,4Z)-hexa-2,4-diene were able to undergo a Diels–Alder reaction with methyl propenoate, what would be the products? (*Hint:* There are four products comprised of two pairs of enantiomers. One enantiomer pair would predominate.) In reality, this Diels–Alder reaction is impractical because the s-cis conformation of the diene required for the reaction is of high energy due to steric hindrance between the methyl groups.

13.10C HOW TO Predict the Products of a Diels–Alder Reaction

Problem: Predict the product of the Diels–Alder reaction between the following compounds.

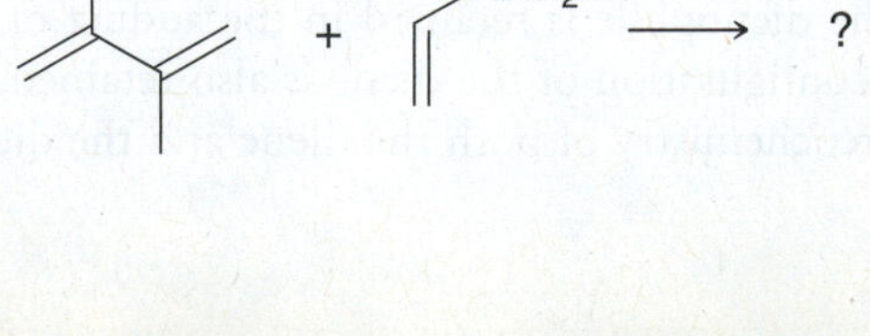

Strategy and Solution: Draw the diene component in the s-cis conformation so that the ends of both double bonds are near the double bond of the dienophile. Then show the movement of electron pairs that will convert the two molecules into one cyclic molecule.

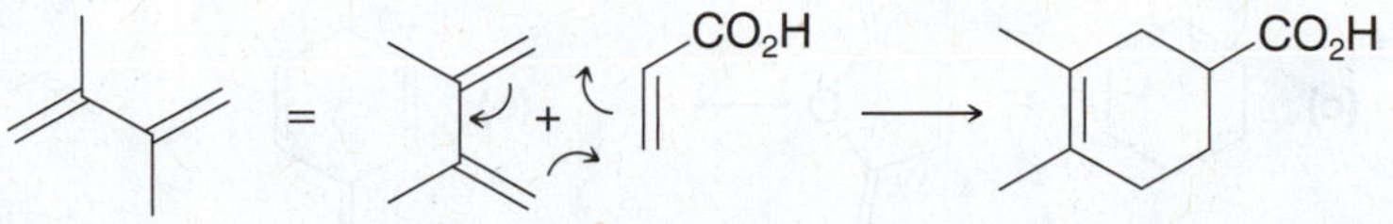

13.10D **HOW TO** Use a Diels–Alder Reaction in a Retrosynthetic Analysis

Problem: Outline a stererospecific synthesis of the all-cis stereoisomer of 1,2,3,4-tetramethycyclohexane (that is, where the methyl groups are all on the same side of the ring).

Retrosynthetic Analysis and Solution: Devising this synthesis will help you see how a Diels–Alder reaction can be used to impart specific stereochemistry in a synthesis. It will also help you review some other reactions you have just learned. Here is our retrosynthetic scheme:

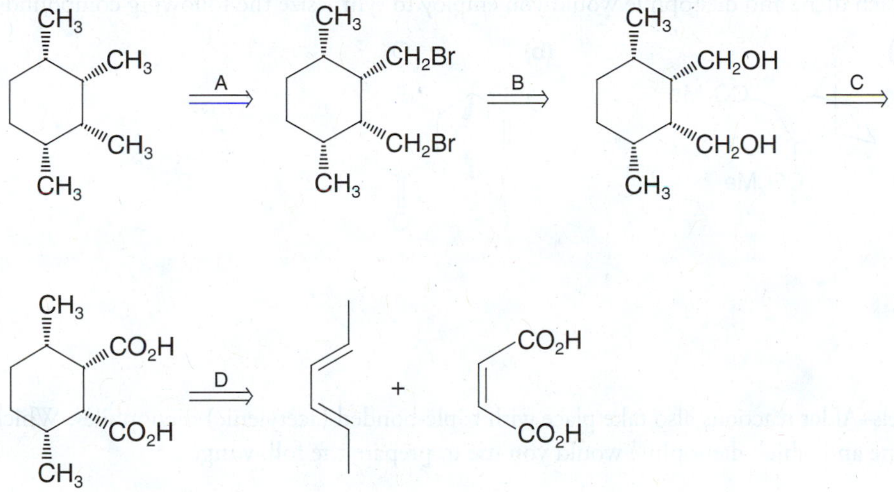

Here is our synthesis. We use a Diels–Alder reaction to create a six-membered ring with the all-cis stereochemistry from the endo product. Then we convert the two $-CO_2H$ groups to $-CH_2OH$ groups, then to $-CH_2Br$ groups, and finally to $-CH_3$ groups.

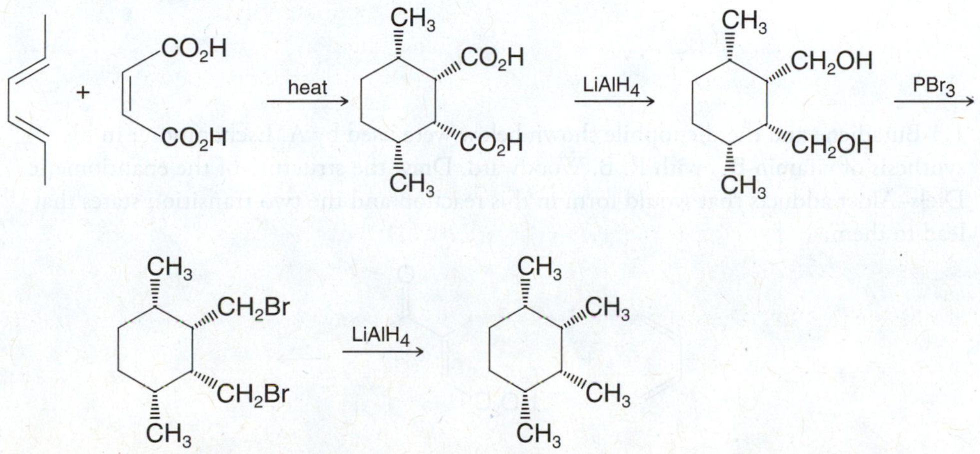

PRACTICE PROBLEM 13.14 What products would you expect from the following reactions?

(a)

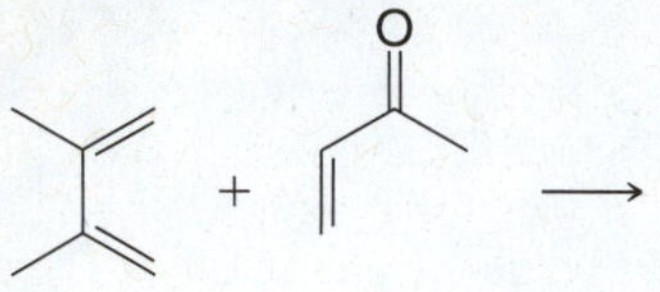

(b)

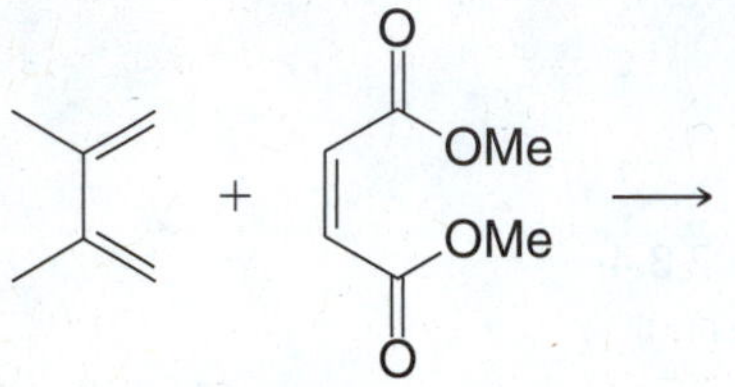

(c)

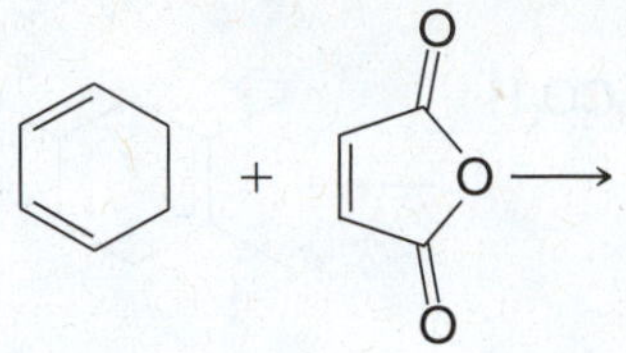

(d)

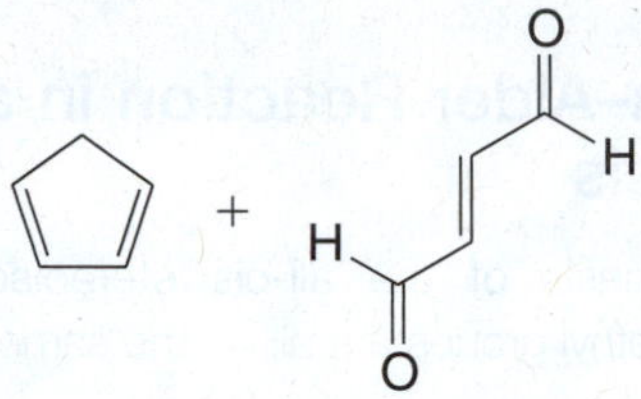

(e)

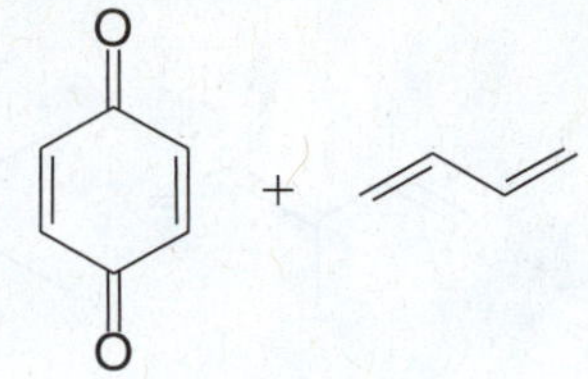

PRACTICE PROBLEM 13.15 Which diene and dienophile would you employ to synthesize the following compounds?

(a)

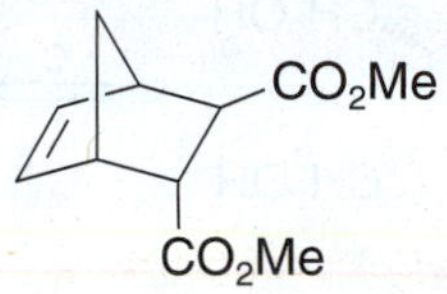

(b)

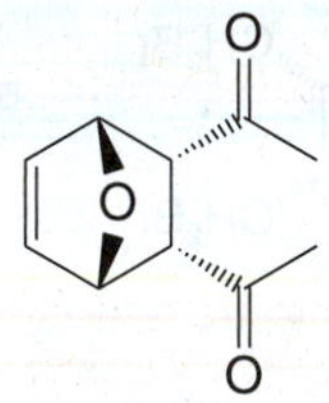

PRACTICE PROBLEM 13.16 Diels–Alder reactions also take place with triple-bonded (acetylenic) dienophiles. Which diene and which dienophile would you use to prepare the following?

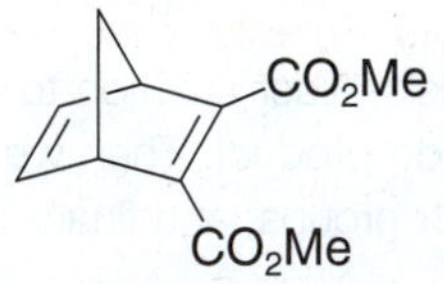

PRACTICE PROBLEM 13.17 1,3-Butadiene and the dienophile shown below were used by A. Eschenmoser in his synthesis of vitamin B_{12} with R. B. Woodward. Draw the structure of the enantiomeric Diels–Alder adducts that would form in this reaction and the two transition states that lead to them.

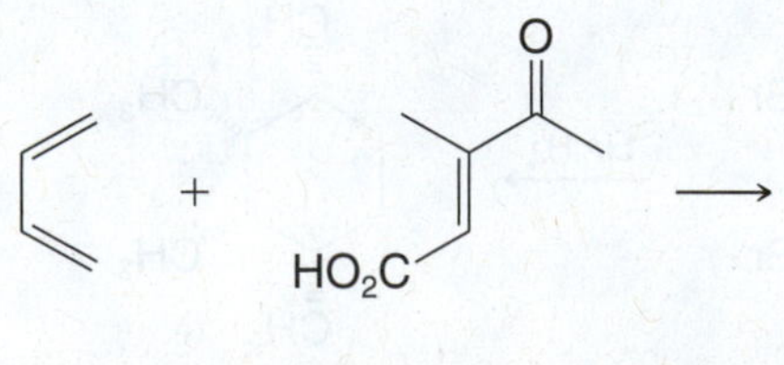

THE CHEMISTRY OF... Molecules with the Nobel Prize in Their Synthetic Lineage

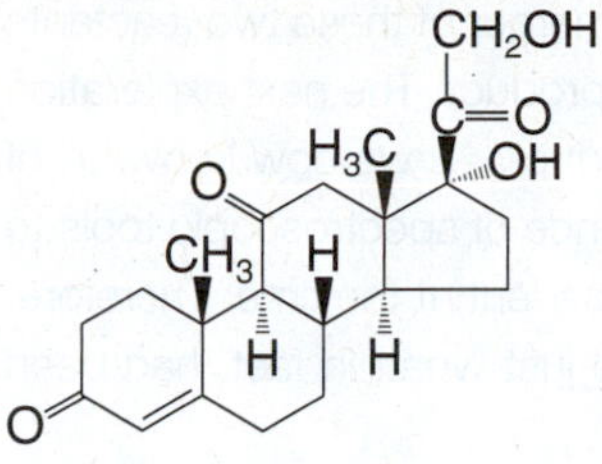

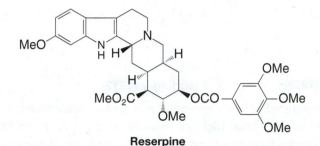

Cortisone

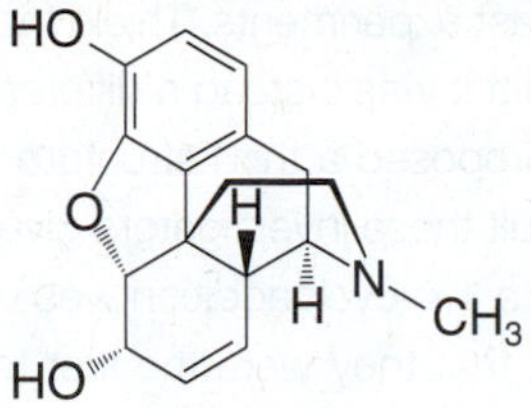

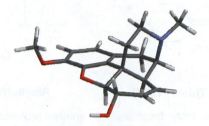

Morphine

Many organic molecules from among the great targets for synthesis have the Diels–Alder reaction in their synthetic lineage. As we have learned, from acyclic precursors the Diels–Alder reaction can form a six-membered ring, with as many as four new chirality centers created in a single stereospecific step. It also produces a double bond that can be used to introduce other functionalities. The great utility of the Diels–Alder reaction earned Otto Diels and Kurt Alder the Nobel Prize in Chemistry in 1950 for developing the reaction that bears their names.

Reserpine

Morphine, the synthesis of which involved the Diels–Alder reaction.

Molecules that have been synthesized using the Diels–Alder reaction (and the chemists who led the work) include morphine (above, and shown as a model), the hypnotic sedative used after many surgical procedures (M. Gates); reserpine (above), a clinically used antihypertensive agent (R. B. Woodward); cholesterol, precursor of all steroids in the body, and cortisone (also above), the antiinflammatory agent (both by R. B. Woodward); prostaglandins $F_{2\alpha}$ and E_2 (Section 13.10C), members of a family of hormones that mediate blood pressure, smooth muscle contraction, and inflammation (E. J. Corey); vitamin B_{12} (Section 7.16A), used in the production of blood and nerve cells (A. Eschenmoser and R. B. Woodward); and Taxol (chemical name paclitaxel, Section 13.10), a potent cancer chemotherapy agent (K. C. Nicolaou). This list alone is a veritable litany of monumental synthetic accomplishments, yet there are many other molecules that have also succumbed to synthesis using the Diels–Alder reaction. It could be said that all of these molecules have a certain sense of "Nobel-ity" in their heritage.

[WHY Do These Topics Matter?

A REACTION THAT COULD HAVE HAD A DIFFERENT NAME

The specific reaction that Otto Diels (the professor) and Kurt Alder (his graduate student) reported in their famous 1928 paper was the merger of two molecules of cyclopentadiene with benzoquinone, as shown below. We have already highlighted the importance and significance of this general reaction process in facilitating the synthesis of complex molecules, something Diels and Alder themselves recognized at the time they made their discovery: *"Thus it appears to us that the possibility of synthesis of complex compounds related to or identical with natural products such as terpenes, sesquiterpenes, perhaps even alkaloids, has been moved to the near prospect."* In fact, it is because of its broad applicability and utility that this reaction won them the Nobel Prize in 1950.

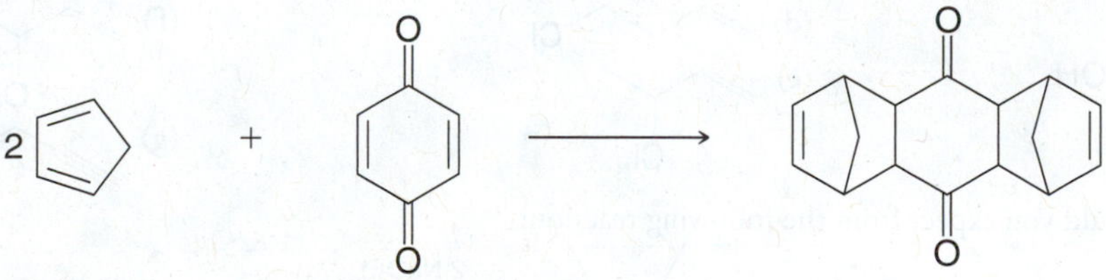

Cyclopentadiene   ***p*-Benzoquinone**

(continues on next page)

However, what you may not know is that Diels and Alder were not the first to explore the addition of cyclopentadiene to benzoquinone. Several others, in fact, had performed the same reaction before them. The first was Johannes Thiele and his graduate student Walther Albrecht, who did their work in 1906; based on past experiments, Thiele thought that merger of these two reactants would yield the molecule shown below, while Albrecht believed that it was instead a different addition product. The next exploration of this reaction came from Hermann Staudinger, who, in 1912, proposed a third structure for the product. As we now know, all of these earlier proposals were wrong. Still, it would be difficult to fault these investigators given the absence of spectroscopic tools to aid in structure determination and the fact that reactivity involving a 1,4-cycloaddition was without precedent at the time. Therefore, a major part of the true genius of Diels and Alder lies in the fact that they were the first to recognize just what, in fact, had been formed from the experiment and to suggest that the reaction involved could be a general process.

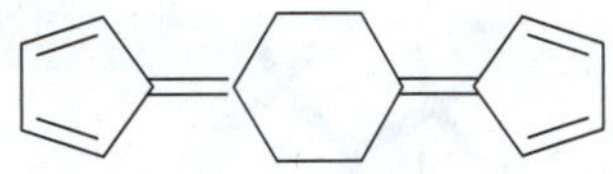

Thiele (1906) 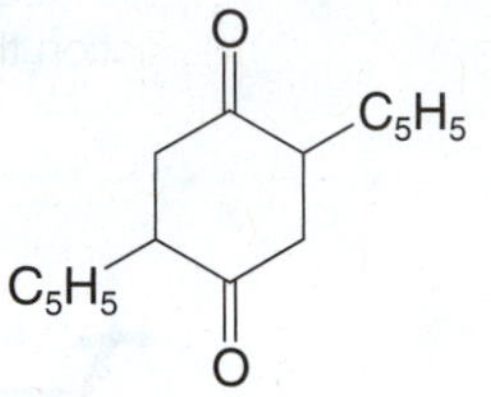   **Albrecht (1906)** 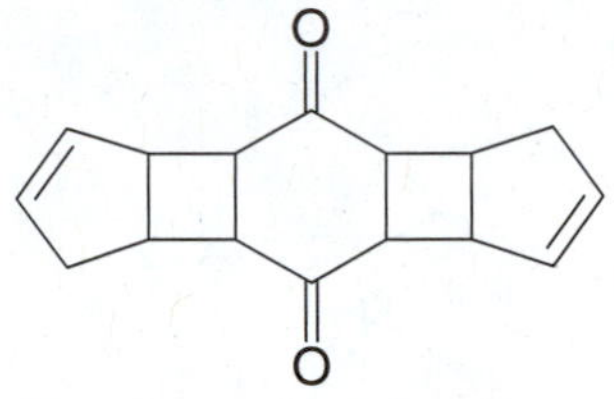   **Staudinger (1912)**

As a final twist to the story, there was yet another professor-and-student team who had run a similar tranformation, and correctly predicted a 1,4-cycloaddition product eight years before Diels and Alder. That work was by Hans von Euler and Karl Josephson as shown below. However, they were tentative in their structural assignment, and though they promised to perform a follow-up study to prove their proposal in that paper, for whatever reason that work never appeared. Diels and Alder, by contrast, did much work with the process and expanded it dramatically, hence the reaction bears their names.

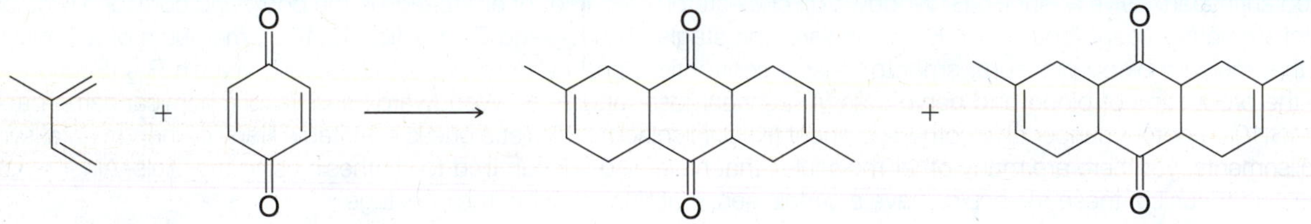

Above structure adapted with permission from Berson, J., Tetrahedron 1992, 48, 3–17, Wiley-VCH and Berson, J., *Chemical Creativity: Ideas from the Work of Woodward, Hückel, Meerwein, and Others*, © **1999**, Wiley-VCH.

To learn more about these topics, see:

1. Berson, J. "Discoveries missed, discoveries made: creativity, influence, and fame in chemistry" in *Tetrahedron* **1992**, *48*, 3–17.
2. Berson, J. *Chemical Creativity: Ideas from the Work of Woodward, Hückel, Meerwein, and Others*. Wiley-VCH: Weinheim, **1999**, p.198.

SUMMARY AND REVIEW TOOLS

The study aids for this chapter include key terms and concepts (which are hyperlinked to the Glossary from the bold, blue terms in the *WileyPLUS* version of the book at wileyplus.com) and a Concept Map relating to properties and reactivity of conjugated systems and the Diels–Alder reaction.

PROBLEMS PLUS

Note to Instructors: Many of the homework problems are available for assignment via *WileyPLUS*, an online teaching and learning solution.

CONJUGATED SYSTEMS

13.18 Provide the reagents needed to synthesize 1,3-butadiene starting from

(a) 1,4-Dibromobutane

(b)

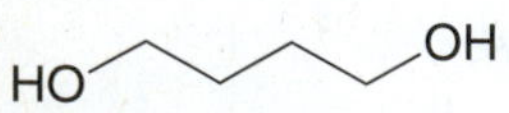

(c)

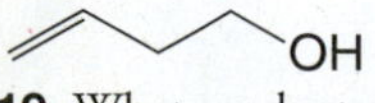

(d)

(e)

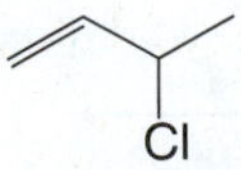

(f)

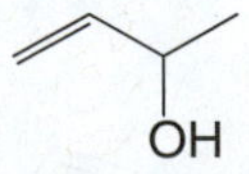

(g)

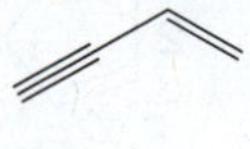

13.19 What product would you expect from the following reaction?

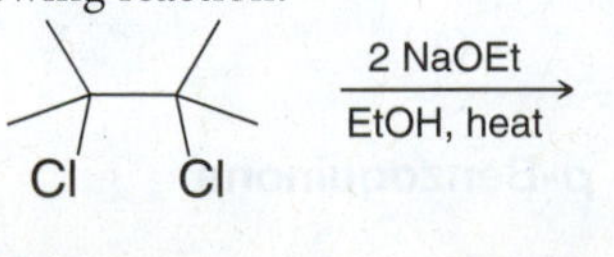

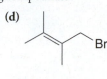

13.20 What products would you expect from the reaction of 1 mol of 1,3-butadiene and each of the following reagents? (If no reaction would occur, you should indicate that as well.)

(a) 1 mol of Cl_2

(b) 2 mol of Cl_2

(c) 2 mol of Br_2

(d) 2 mol of H_2, Ni

(e) 1 mol of Cl_2 in H_2O

(f) Hot $KMnO_4$ (excess)

(g) H_2O, cat. H_2SO_4

13.21 Provide the reagents necessary to transform 2,3-dimethyl-1,3-butadiene into each of the following compounds.

(a) 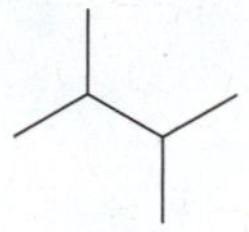   **(b)** 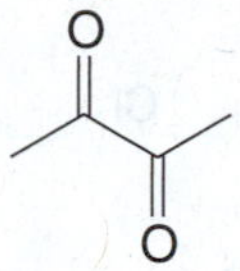   **(c)**   **(d)**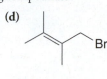

13.22 Provide the reagents necessary for each of the following transformations. In some cases several steps may be necessary.

(a) 1-Butene ⟶ 1,3-butadiene

(b) 1-Pentene ⟶ 1,3-pentadiene

(c)

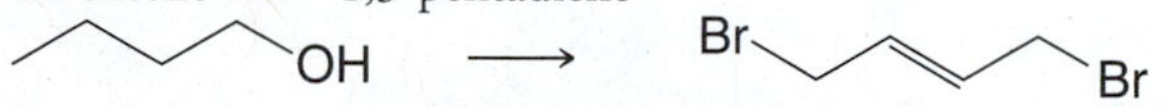

(d)

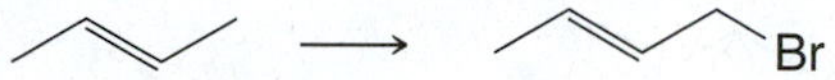

(e)

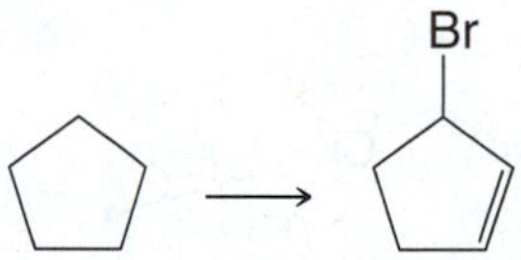

(f)

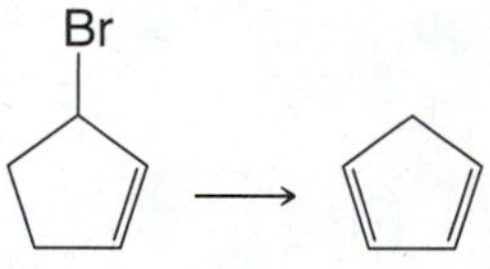

13.23 Conjugated dienes react with radicals by both 1,2- and 1,4-addition. Write a detailed mechanism to account for this fact using the peroxide-promoted addition of one molar equivalent of HBr to 1,3-butadiene as an illustration.

13.24 UV–Vis, IR, NMR, and mass spectrometry are spectroscopic tools we use to obtain structural information about compounds. For each pair of compounds below, describe at least one aspect from each of two spectroscopic methods (UV–Vis, IR, NMR, or mass spectrometry) that would distinguish one compound in a pair from the other.

(a) 1,3-Butadiene and 1-butyne

(b) 1,3-Butadiene and butane

(c) Butane and

(d) 1,3-Butadiene and

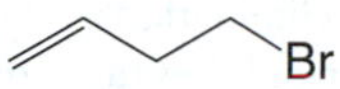

(e) 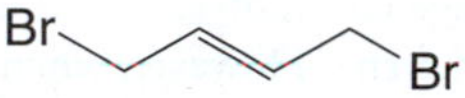   and

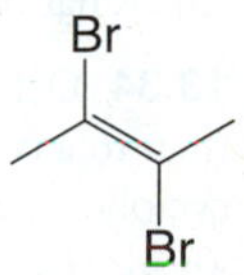

13.25 When 2-methyl-1,3-butadiene (isoprene) undergoes a 1,4-addition of hydrogen chloride, the major product that is formed is 1-chloro-3-methyl-2-butene. Little or no 1-chloro-2-methyl-2-butene is formed. How can you explain this?

13.26 When 1-pentene reacts with *N*-bromosuccinimide (NBS), two products with the formula C_5H_9Br are obtained. What are these products and how are they formed?

13.27 **(a)** The hydrogen atoms attached to C3 of 1,4-pentadiene are unusually susceptible to abstraction by radicals. How can you account for this? **(b)** Can you provide an explanation for the fact that the protons attached to C3 of 1,4-pentadiene are more acidic than the methyl hydrogen atoms of propene?

13.28 Provide a mechanism that explains formation of the following products. Include all intermediates, formal charges, and arrows showing electron flow.

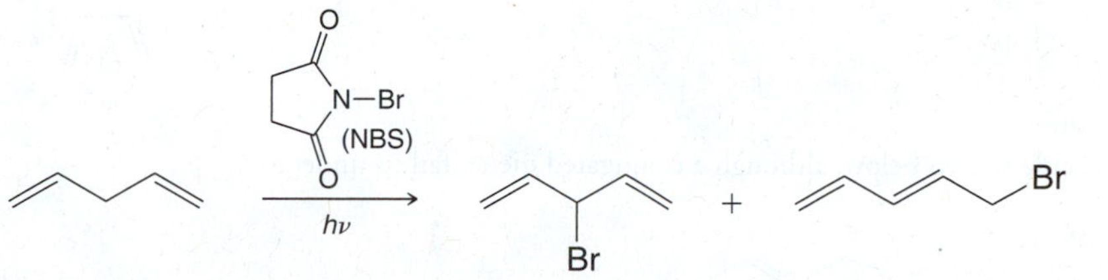

13.29 Provide a mechanism for the following reaction. Draw a reaction energy coordinate diagram that illustrates the kinetic and thermodynamic pathways for this reaction.

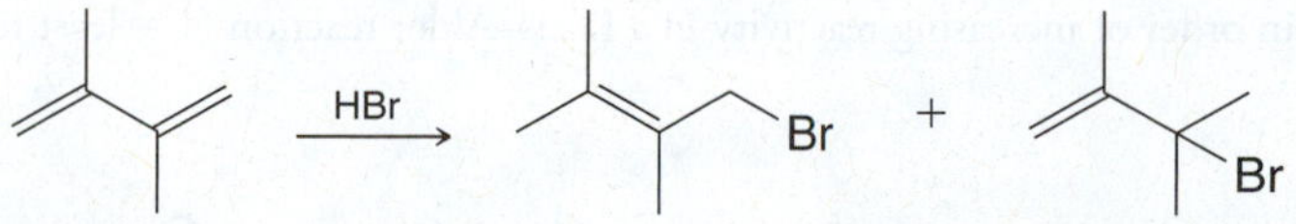

13.30 Predict the products of the following reactions.

(a) 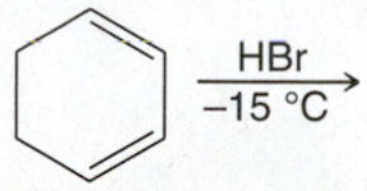 $\xrightarrow[-15\,°C]{HBr}$

(b) $\xrightarrow[40\,°C]{HBr}$

(c) $\xrightarrow[hv,\ heat]{NBS}$

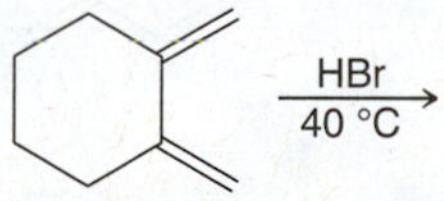

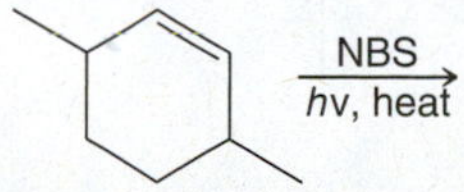

13.31 Provide a mechanism that explains formation of the following products.

$$\text{(structure)} \xrightarrow[\text{}]{\text{HCl (concd)}} \text{(structures)}$$

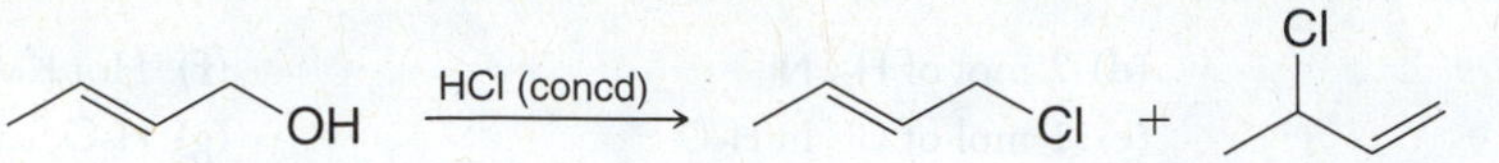

13.32 Provide a mechanism that explains formation of the following products.

$$\text{(structure)} \xrightarrow[\text{MeOH}]{\text{Cl}_2} \text{(structures)}$$

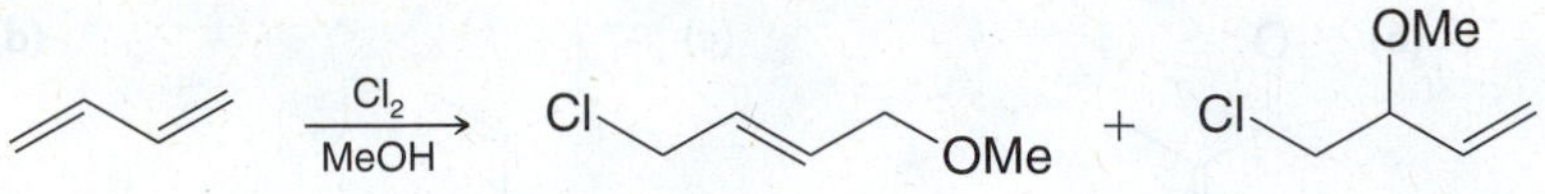

13.33 Treating either 1-chloro-3-methyl-2-butene or 3-chloro-3-methyl-1-butene with Ag_2O in water gives (in addition to AgCl) the following mixture of alcohol products.

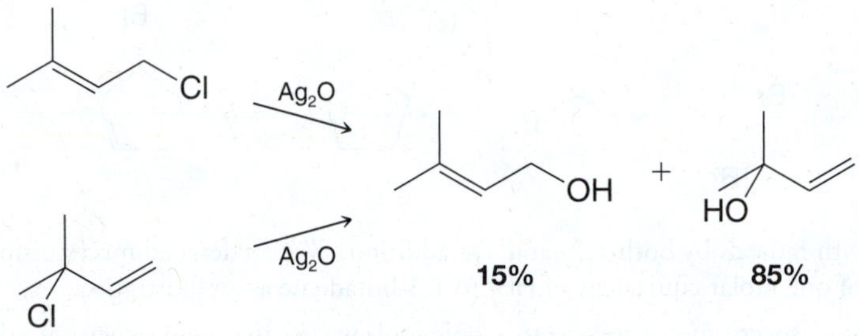

$$\text{(structures with Ag}_2\text{O)} \quad 15\% \quad + \quad 85\%$$

(a) Write a mechanism that accounts for the formation of these products.

(b) What might explain the relative proportions of the two alkenes that are formed?

13.34 Dehydrohalogenation of 1,2-dihalides (with the elimination of two molar equivalents of HX) normally leads to an alkyne rather than to a conjugated diene. However, when 1,2-dibromocyclohexane is dehydrohalogenated, 1,3-cyclohexadiene is produced and not cyclohexyne. What factor accounts for this?

13.35 The heat of hydrogenation of allene is 298 kJ mol^{-1}, whereas that of propyne is 290 kJ mol^{-1}. **(a)** Which compound is more stable? **(b)** Treating allene with a strong base causes it to isomerize to propyne. Explain.

13.36 Although both 1-bromobutane and 4-bromo-1-butene are primary halides, the latter undergoes elimination more rapidly. How can this behavior be explained?

DIELS–ALDER REACTIONS

13.37 Complete the molecular orbital description for the ground state of cyclopentadiene shown at right. Shade the appropriate lobes to indicate phase signs in each molecular orbital according to increasing energy of the molecular orbitals. Label the HOMO and LUMO orbitals, and place the appropriate number of electrons in each level, using a straight single-barbed arrow to represent each electron.

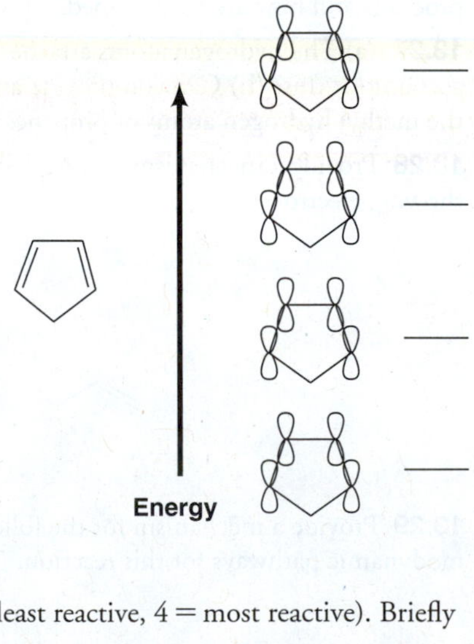

13.38 Why does the molecule shown below, although a conjugated diene, fail to undergo a Diels–Alder reaction?

13.39 Rank the following dienes in order of increasing reactivity in a Diels–Alder reaction (1 = least reactive, 4 = most reactive). Briefly explain your ranking.

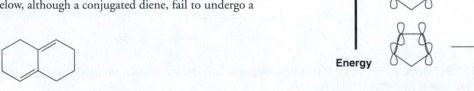

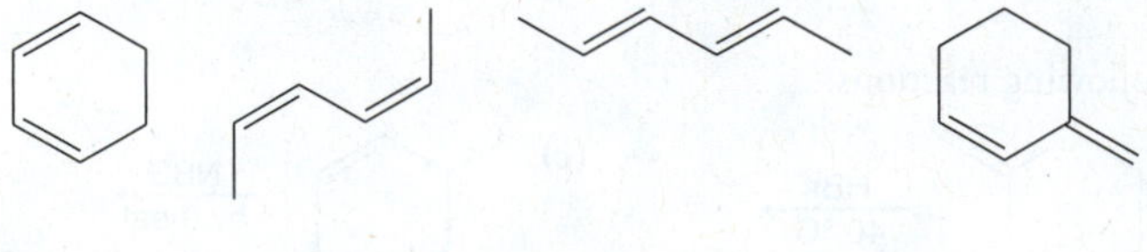

13.40 Give the structures of the products that would be formed when 1,3-butadiene reacts with each of the following:

(a)

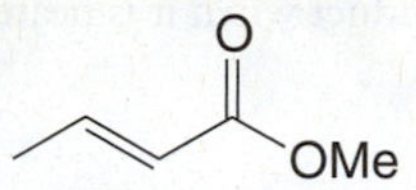

O
‖
/\=/_C_OMe

(b)

O
‖
_/=_C_OMe

(c)

/\=/_CN

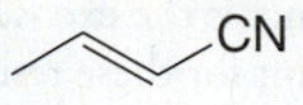

(d)

_/=_CN

13.41 Cyclopentadiene undergoes a Diels–Alder reaction with ethene at 160–180 °C. Write the structure of the product of this reaction.

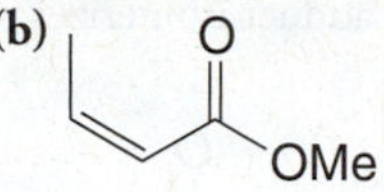

13.42 Acetylenic compounds may be used as dienophiles in the Diels–Alder reaction (see Practice Problem 13.16). Write structures for the adducts that you expect from the reaction of 1,3-butadiene with

(a)

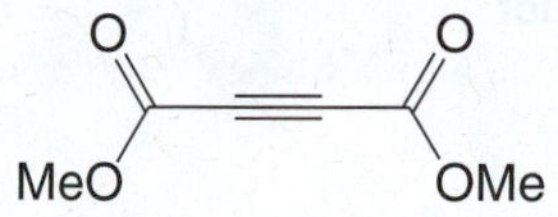

O   O
‖   ‖
MeO_C≡C_OMe

(dimethyl acetylenedicarboxylate)

(b)

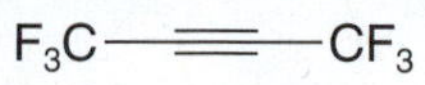

F_3C—≡—CF_3

(hexafluoro-2-butyne)

13.43 Predict the products of the following reactions.

(a)

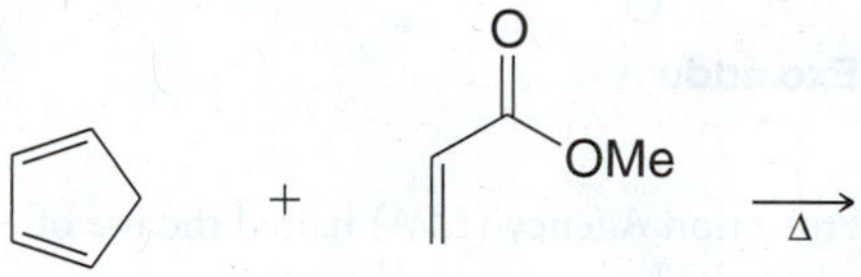

⬡(cyclopentadiene) + CH₂=CH–CO–OMe —Δ→

(e)

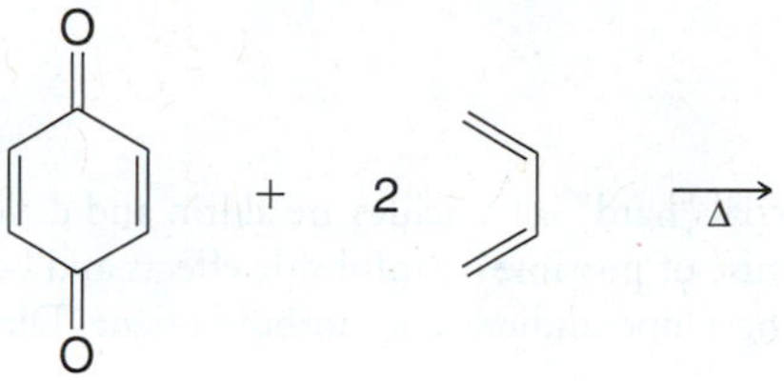

(p-benzoquinone) + 2 (1,3-butadiene) —Δ→

(b)

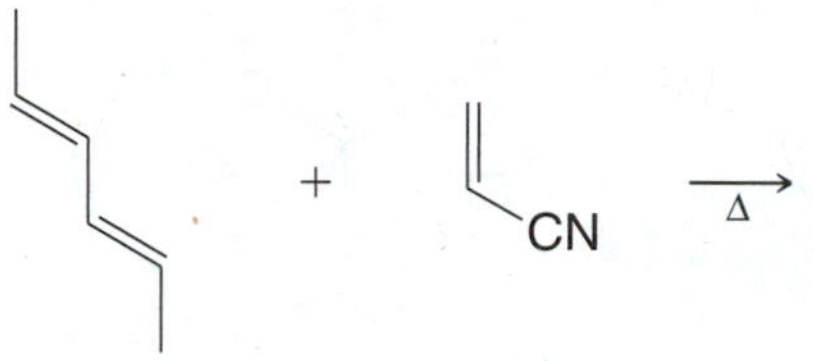

(2,4-hexadiene) + CH₂=CH–CN —Δ→

(f)

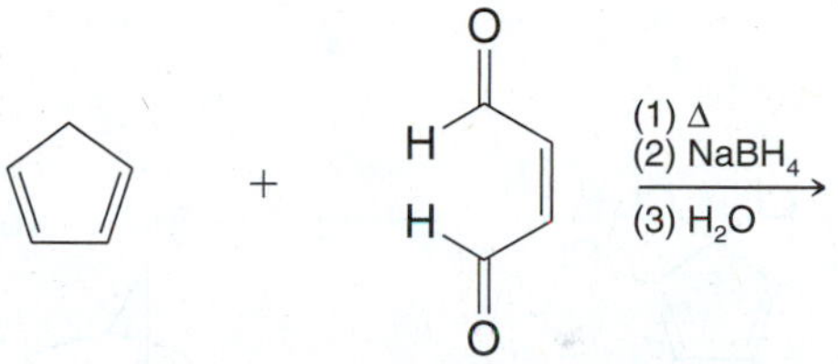

(cyclopentadiene) + maleic dialdehyde (1) Δ / (2) NaBH₄ / (3) H₂O →

(c)

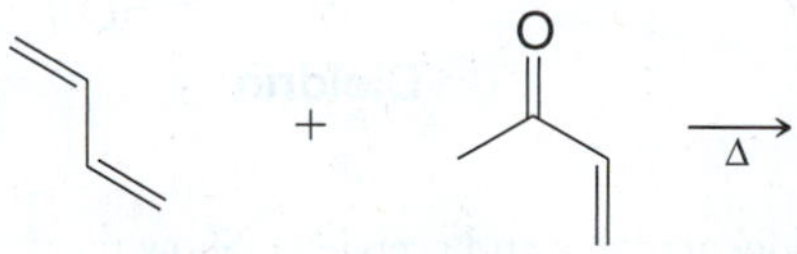

(1,3-butadiene) + CH₃–CO–CH=CH₂ —Δ→

(g)

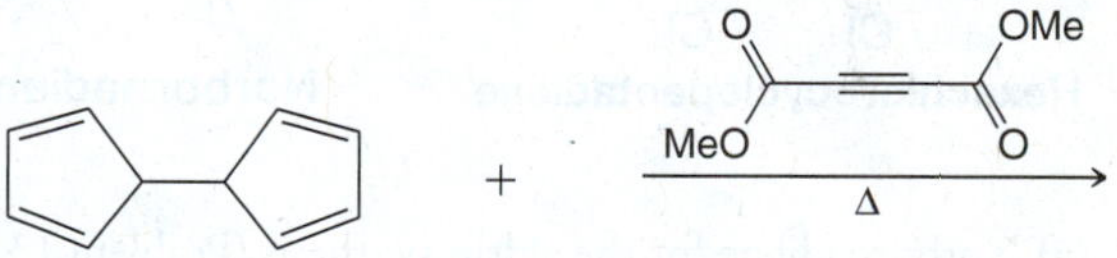

(dicyclopentadienyl compound) + MeO–CO–C≡C–CO–OMe —Δ→

(d)

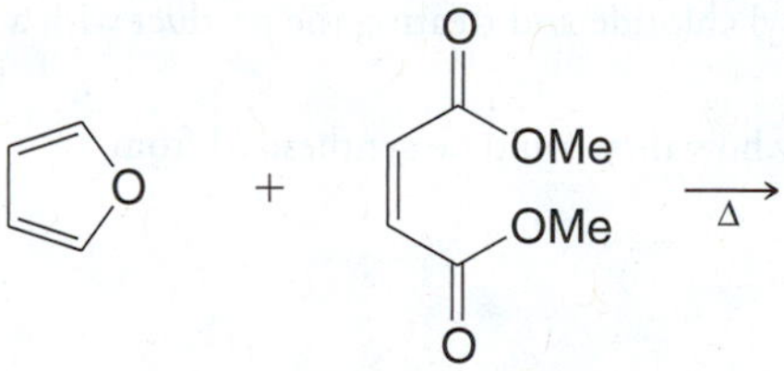

(furan) + dimethyl maleate (OMe, OMe) —Δ→

(h)

(cyclohexene) + maleic dinitrile (CN, CN) —Δ→

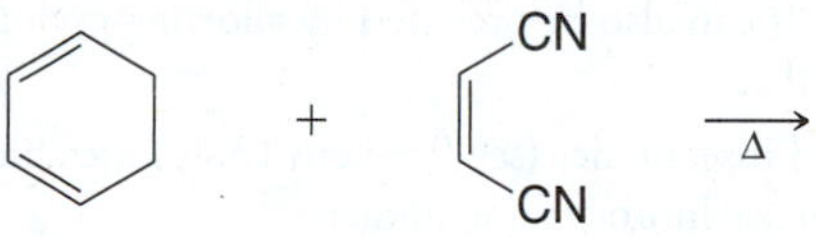

13.44 Which diene and dienophile would you employ in a synthesis of each of the following?

(a)

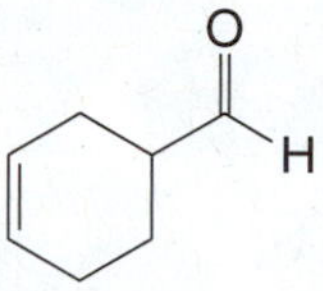

O
‖
cyclohexene–C–H

(c)

O
‖
dimethyl ester of substituted cyclohexene (OMe, OMe)

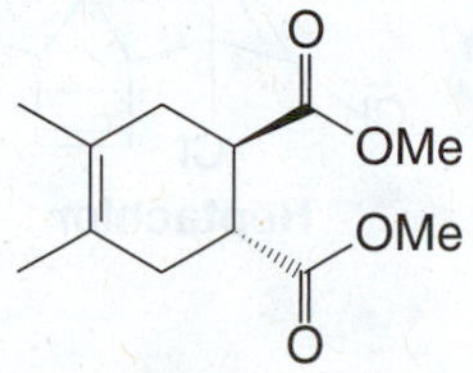

(e)

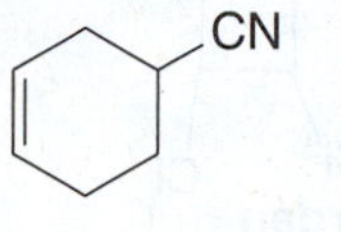

CN
cyclohexene–CN

(g)

O
‖
substituted cyclohexane aldehyde (–H)

(b)

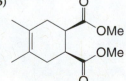

(d)

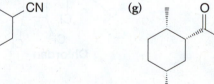

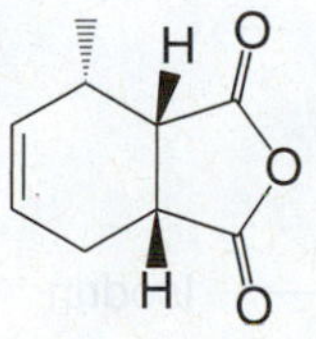

(f)

bicyclic compound with CO₂Me

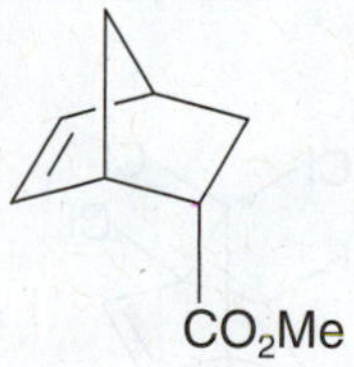

13.45 When furan and maleimide undergo a Diels–Alder reaction at 25 °C, the major product is the endo adduct **G**. When the reaction is carried out at 90 °C, however, the major product is the exo isomer **H**. The endo adduct isomerizes to the exo adduct when it is heated to 90 °C. Propose an explanation that will account for these results.

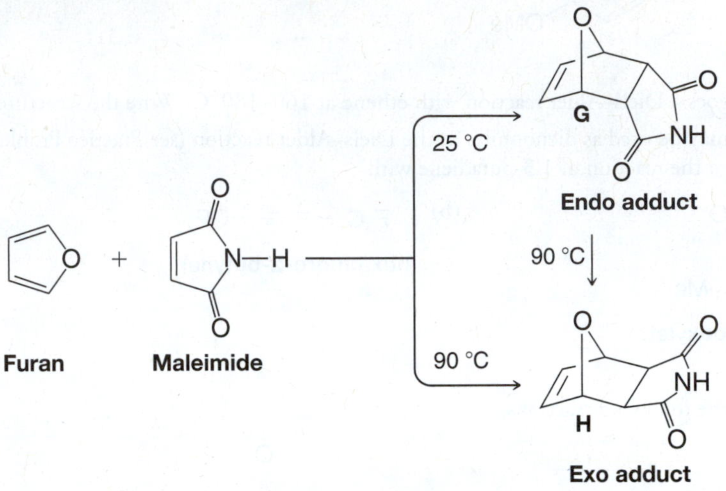

13.46 Two controversial "hard" insecticides are aldrin and dieldrin. [The Environmental Protection Agency (EPA) halted the use of these insecticides because of possible harmful side effects and because they are not biodegradable.] The commercial synthesis of aldrin began with hexachlorocyclopentadiene and norbornadiene. Dieldrin was synthesized from aldrin. Show how these syntheses might have been carried out.

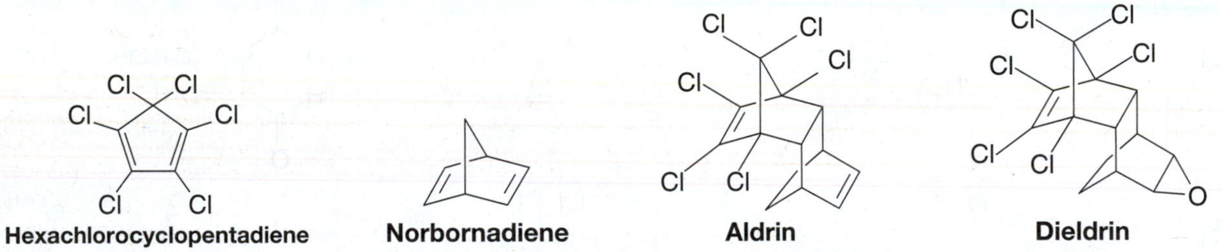

13.47 **(a)** Norbornadiene for the aldrin synthesis (Problem 13.46) can be prepared from cyclopentadiene and acetylene. Show the reaction involved. **(b)** It can also be prepared by allowing cyclopentadiene to react with vinyl chloride and treating the product with a base. Outline this synthesis.

13.48 Two other hard insecticides (see Problem 13.46) are chlordan and heptachlor. Show how they could be synthesized from cyclopentadiene and hexachlorocyclopentadiene.

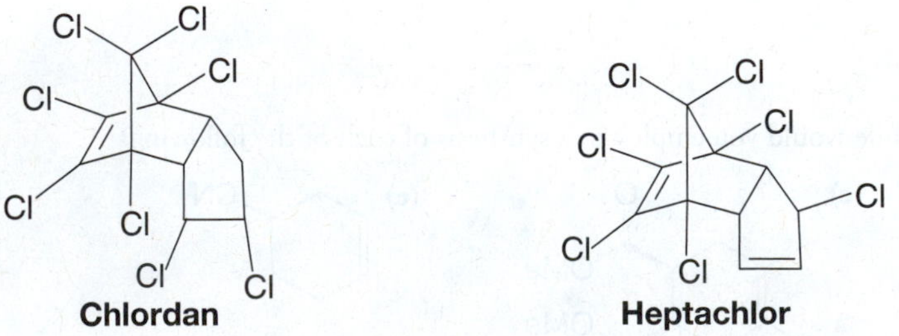

13.49 Isodrin, an isomer of aldrin, is obtained when cyclopentadiene reacts with the hexachloronorbornadiene, shown here. Propose a structure for isodrin.

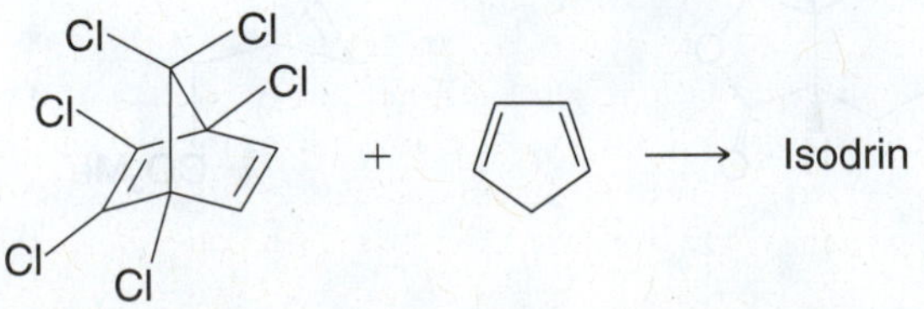

13.50 Provide the reagents necessary to achieve the following synthetic transformations. More than one step may be required.

(a)

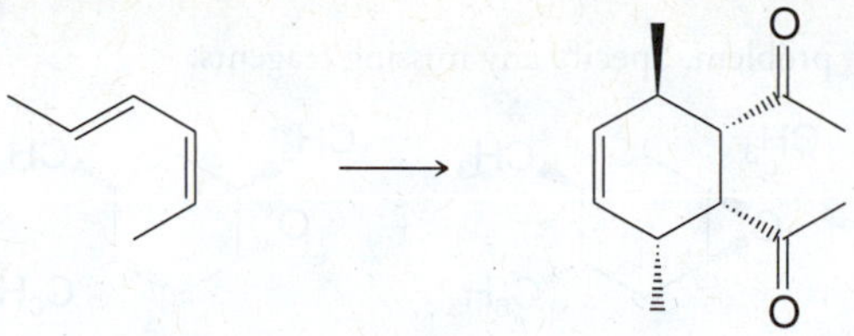

(d)

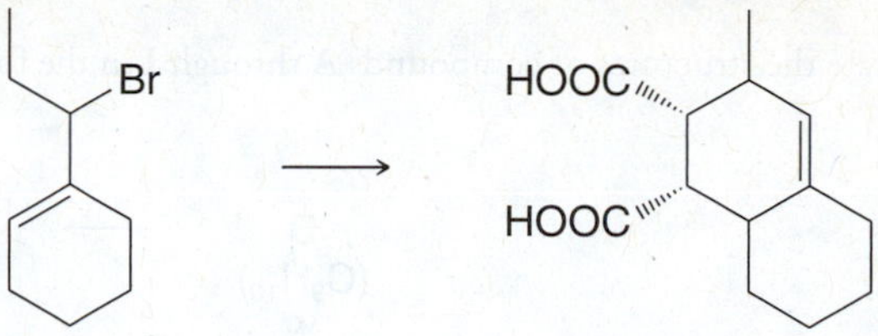

(b)

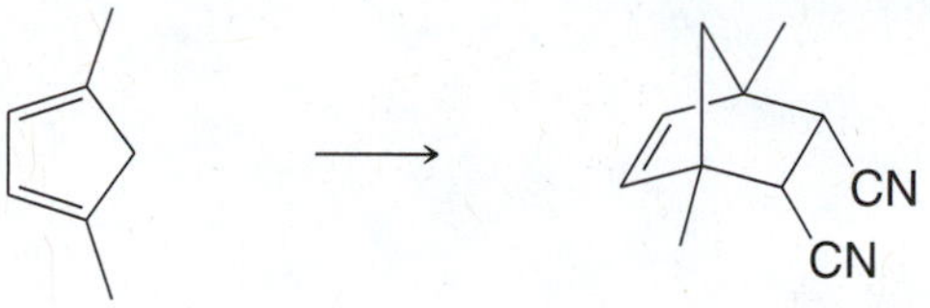

(e)

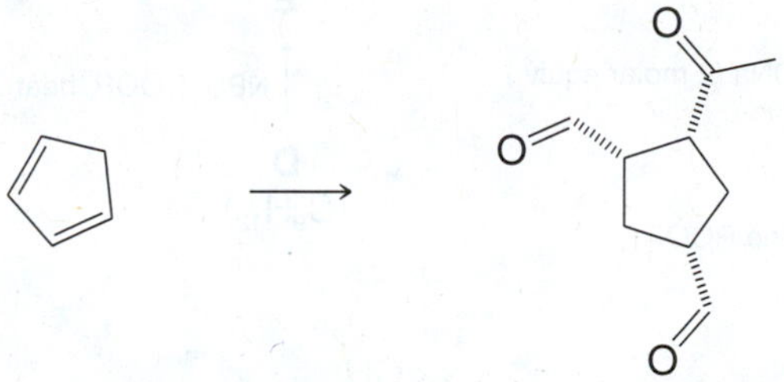

(c)

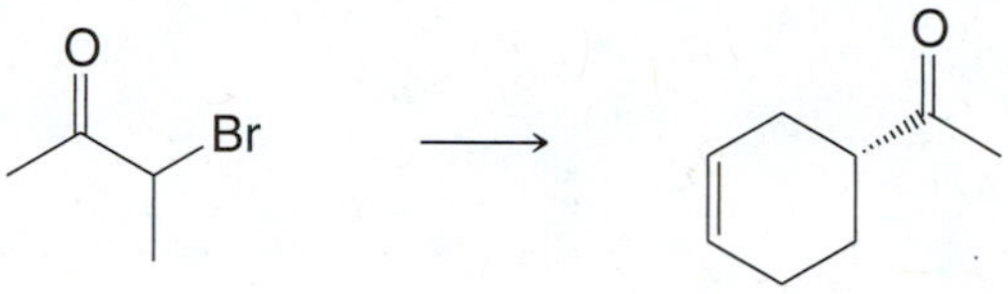

CHALLENGE PROBLEMS

13.51 Explain the product distribution below based on the polarity of the diene and dienophile, as predicted by contributing resonance structures for each.

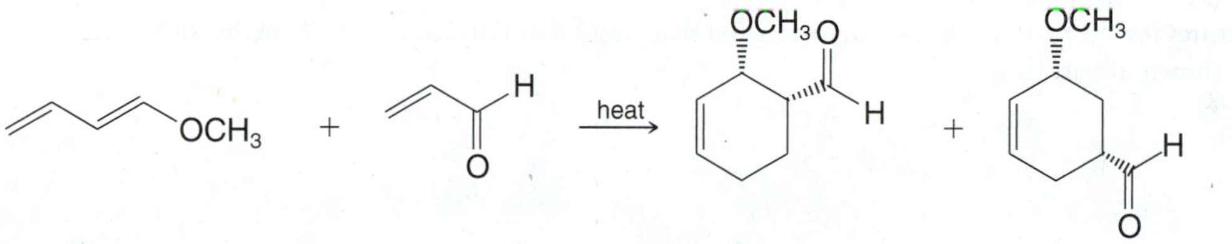

Major
(plus enantiomer)

Minor
(plus enantiomer)

13.52 Mixing furan (Problem 13.45) with maleic anhydride in diethyl ether yields a crystalline solid with a melting point of 125 °C. When melting of this compound takes place, however, one can notice that the melt evolves a gas. If the melt is allowed to resolidify, one finds that it no longer melts at 125 °C but instead it melts at 56 °C. Consult an appropriate chemistry handbook and provide an explanation for what is taking place.

13.53 Draw the structure of the product from the following reaction (formed during a synthesis of one of the endiandric acids by K. C. Nicolaou):

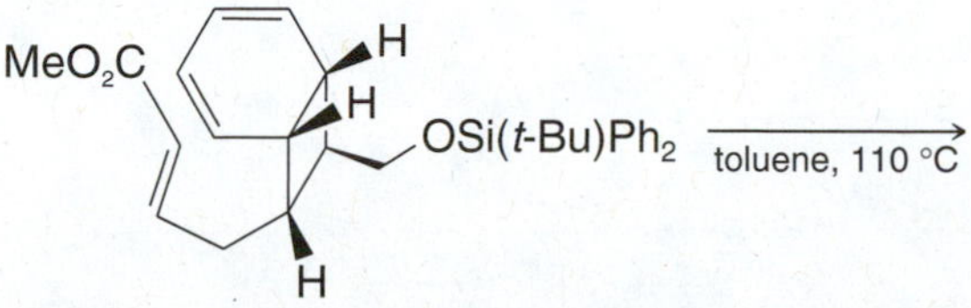

13.54 Draw all of the contributing resonance structures and the resonance hybrid for the carbocation that would result from ionization of bromine from 5-bromo-1,3-pentadiene. Open the computer molecular model at the book's website depicting a map of electrostatic potential for the pentadienyl carbocation. Based on the model, which is the most important contributing resonance structure for this cation? Is this consistent with what you would have predicted based on your knowledge of relative carbocation stabilities? Why or why not?

LEARNING GROUP PROBLEMS

1. Elucidate the structures of compounds **A** through **I** in the following "road map" problem. Specify any missing reagents.

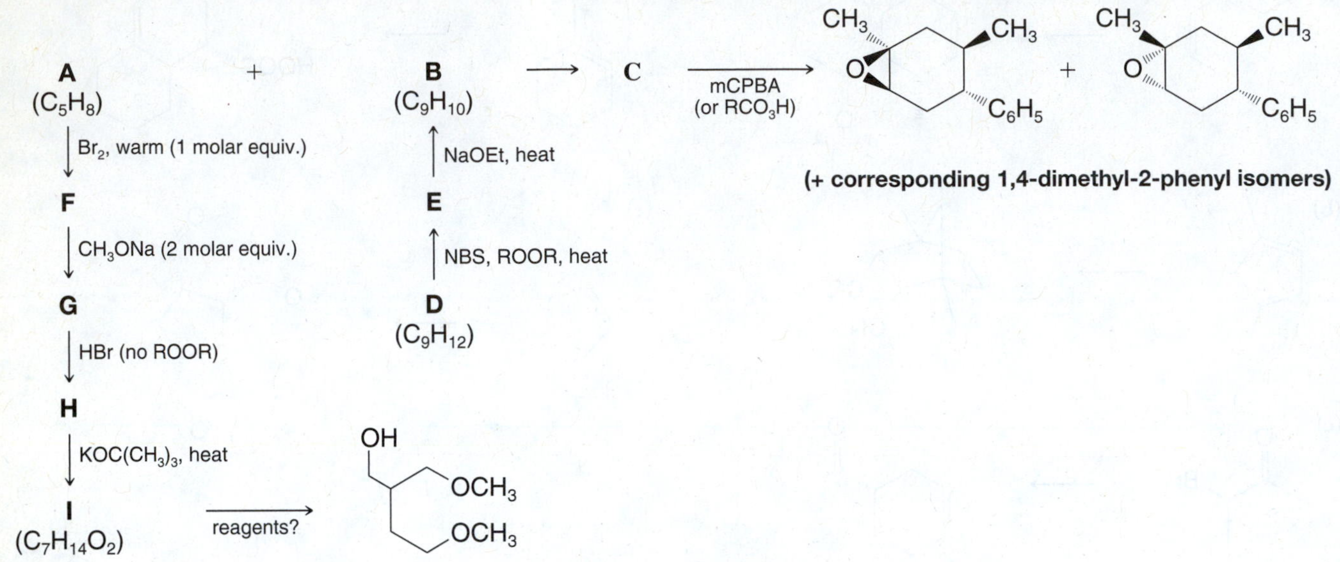

A
(C_5H_8) +   **B**
(C_9H_{10}) $\longrightarrow$ **C** $\xrightarrow[\text{(or RCO}_3\text{H)}]{\text{mCPBA}}$

(+ corresponding 1,4-dimethyl-2-phenyl isomers)

A $\downarrow$ Br$_2$, warm (1 molar equiv.)

F

$\downarrow$ CH$_3$ONa (2 molar equiv.)

G

$\downarrow$ HBr (no ROOR)

H

$\downarrow$ KOC(CH$_3$)$_3$, heat

I
$(C_7H_{14}O_2)$ $\xrightarrow{\text{reagents?}}$

B (C_9H_{10}) $\uparrow$ NaOEt, heat

E $\uparrow$ NBS, ROOR, heat

D (C_9H_{12})

2. (a) Write reactions to show how you could convert 2-methyl-2-butene into 2-methyl-1,3-butadiene.

(b) Write reactions to show how you could convert ethylbenzene into the following compound:

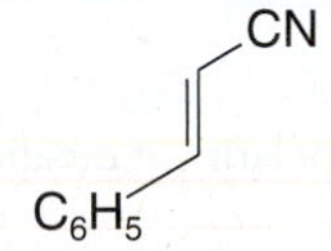

(c) Write structures for the various Diels–Alder adduct(s) that could result on reaction of 2-methyl-1,3-butadiene with the compound shown in part (b).

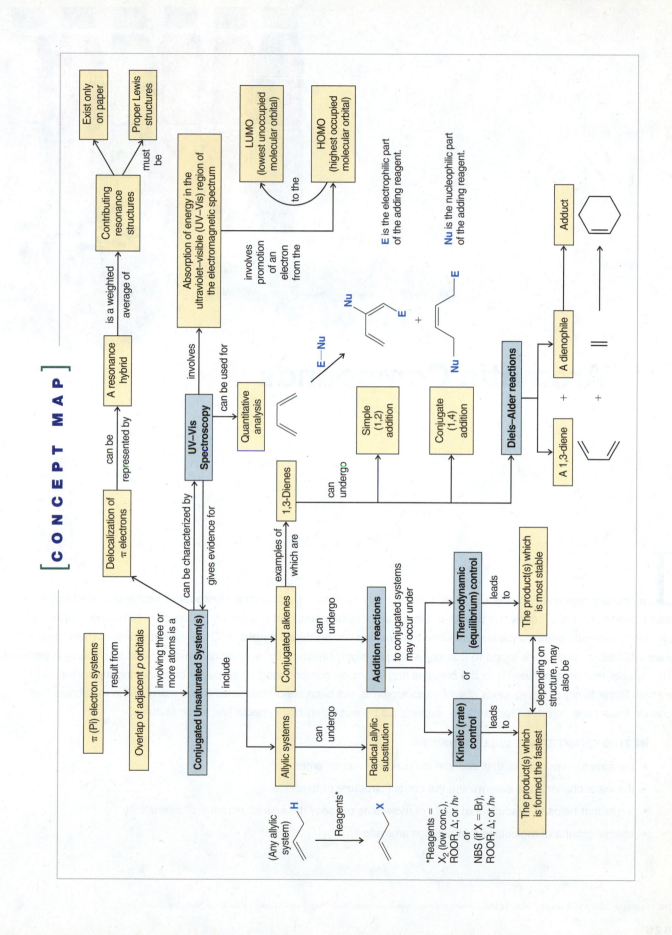

[CONCEPT MAP]

Contributing resonance structures
- Exist only on paper
- must be → Proper Lewis structures

is a weighted average of ← **A resonance hybrid**

can be represented by ← **Delocalization of π electrons**

result from ← **π (Pi) electron systems**

Overlap of adjacent p orbitals
involving three or more atoms is a

Conjugated Unsaturated System(s)

can be characterized by / gives evidence for ← **UV–Vis Spectroscopy**

involves → **Absorption of energy in the ultraviolet–visible (UV–Vis) region of the electromagnetic spectrum**

involves promotion of an electron from the → **LUMO (lowest unoccupied molecular orbital)** / **HOMO (highest occupied molecular orbital)**

to the

can be used for → **Quantitative analysis**

E is the electrophilic part of the adding reagent.

Nu is the nucleophilic part of the adding reagent.

E—Nu

include:
- **Conjugated alkenes**
- **Allylic systems**

examples of which are → **1,3-Dienes**

1,3-Dienes can undergo:
- **Simple (1,2) addition**
- **Conjugate (1,4) addition**
- **Diels–Alder reactions**

Diels–Alder reactions →
- **A dienophile**
- **A 1,3-diene**

→ **Adduct**

+ +

Conjugated alkenes can undergo → **Addition reactions**

to conjugated systems may occur under:
- **Kinetic (rate) control** → leads to → The product(s) which is formed the fastest
- **Thermodynamic (equilibrium) control** → leads to → The product(s) which is most stable

or

depending on structure, may also be

Allylic systems can undergo → **Radical allylic substitution**

(Any allylic system)
H

Reagents* →

X

*Reagents =
X₂ (low conc.),
ROOR, Δ; or hν
or
NBS (if X = Br),
ROOR, Δ; or hν

Aromatic Compounds

In ordinary conversation, the word "aromatic" conjures pleasant associations—the odor of freshly prepared coffee, a warm cinnamon bun, a freshly cut pine tree. Similar associations occurred in the early history of organic chemistry when pleasantly aromatic compounds were isolated from natural oils produced by plants. Once the structures of these materials were elucidated, many were found to possess a unique, highly unsaturated, six-carbon structural unit also found in benzene. This special ring became known as the benzene ring. Aromatic compounds that contain a benzene ring are now part of a much larger family of compounds classified as aromatic, not because of their smell (since many of the molecules that contain them have no odor—for example, aspirin), but because they have special electronic features.

IN THIS CHAPTER WE WILL CONSIDER:

- the structural principles that underlie the use of the term "aromatic"
- the initial challenge of determining the correct structure of benzene
- a rule that helps to predict what kinds of molecules possess the special property of aromaticity
- special groups of molecules that are also aromatic

[**WHY** DO THESE TOPICS MATTER?]   At the end of the chapter, we will explore the question of just how large the rings of these molecules can be and still be aromatic, noting that chemists have been able to make aromatic rings far larger in size than those of molecules obtained from nature, but largely using design clues derived from those natural molecules.

14.1 THE DISCOVERY OF BENZENE

The following are a few examples of aromatic compounds, including benzene itself. In these formulas we foreshadow our discussion of the special properties of the benzene ring by using a circle in a hexagon to depict the six π electrons and six-membered ring of these compounds, whereas up to now we have shown benzene rings only as indicated in the left-hand formula for benzene below.

Benzene

Benzaldehyde
(in oil of almonds)

Methyl salicylate
(in oil of wintergreen)

Eugenol
(in oil of cloves)

Cinnamaldehyde
(in oil of cinnamon)

Vanillin
(in oil of vanilla)

Acetylsalicylic acid
(aspirin)

The study of the class of compounds that organic chemists call aromatic compounds (Section 2.1D) began with the discovery in 1825 of a new hydrocarbon by the English chemist Michael Faraday (Royal Institution). Faraday called this new hydrocarbon "bicarburet of hydrogen"; we now call it benzene. Faraday isolated benzene from a compressed illuminating gas that had been made by pyrolyzing whale oil.

In 1834 the German chemist Eilhardt Mitscherlich (University of Berlin) synthesized benzene by heating benzoic acid with calcium oxide. Using vapor density measurements, Mitscherlich further showed that benzene has the molecular formula C_6H_6:

$$C_6H_5CO_2H \ + \ CaO \ \xrightarrow{\text{heat}} \ C_6H_6 \ + \ CaCO_3$$
Benzoic acid   **Benzene**

The molecular formula itself was surprising. Benzene has *only as many hydrogen atoms as it has carbon atoms*. Most compounds that were known then had a far greater proportion of hydrogen atoms, usually twice as many. Benzene, having the formula of C_6H_6, should be a highly unsaturated compound because it has an index of hydrogen deficiency equal to 4. Eventually, chemists began to recognize that benzene was a member of a new class of organic compounds with unusual and interesting properties. **As we shall see in Section 14.3, benzene does not show the behavior expected of a highly unsaturated compound**.

During the latter part of the nineteenth century the Kekulé–Couper–Butlerov theory of valence was systematically applied to all known organic compounds. One result of this effort was the placing of organic compounds in either of two broad categories; compounds were classified as being either **aliphatic** or **aromatic**. To be classified as aliphatic meant then that the chemical behavior of a compound was "fatlike." (Now it means that the compound reacts like an alkane, an alkene, an alkyne, or one of their derivatives.) To be classified as aromatic meant then that the compound had a low hydrogen-to-carbon

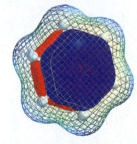

One of the π molecular orbitals of benzene, seen through a mesh representation of its electrostatic potential at its van der Waals surface.

ratio and that it was "fragrant." Most of the early aromatic compounds were obtained from balsams, resins, or essential oils.

Kekulé was the first to recognize that these early aromatic compounds all contain a six-carbon unit and that they retain this six-carbon unit through most chemical transformations and degradations. Benzene was eventually recognized as being the parent compound of this new series. It was not until the development of quantum mechanics in the 1920s, however, that a reasonably clear understanding of its structure emerged.

14.2 NOMENCLATURE OF BENZENE DERIVATIVES

Two systems are used in naming monosubstituted benzenes.

- In many simple compounds, *benzene* is the parent name and the substituent is simply indicated by a prefix.

We have, for example,

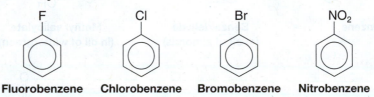

Fluorobenzene   **Chlorobenzene**   **Bromobenzene**   **Nitrobenzene**

- For other simple and common compounds, the substituent and the benzene ring taken together may form a commonly accepted parent name.

Methylbenzene is usually called *toluene*, hydroxybenzene is almost always called *phenol*, and aminobenzene is almost always called *aniline*. These and other examples are indicated here:

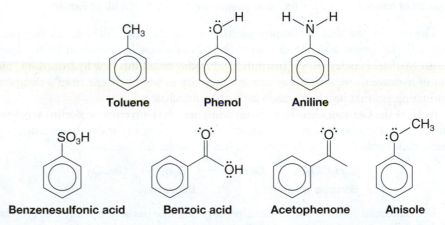

Toluene   **Phenol**   **Aniline**

Benzenesulfonic acid   **Benzoic acid**   **Acetophenone**   **Anisole**

- When two substituents are present, their relative positions are indicated by the prefixes **ortho-, meta-,** and **para-** (abbreviated **o-, m-,** and **p-**) or by the use of numbers.

For the dibromobenzenes we have

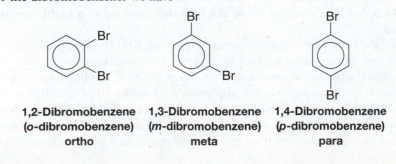

1,2-Dibromobenzene   **1,3-Dibromobenzene**   **1,4-Dibromobenzene**
(o-dibromobenzene)   **(m-dibromobenzene)**   **(p-dibromobenzene)**
ortho   meta   para

and for the nitrobenzoic acids

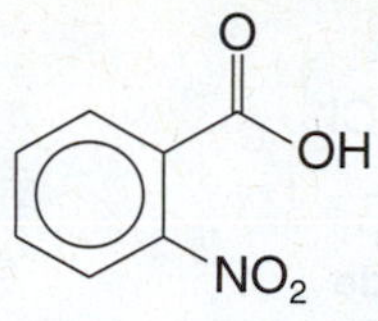

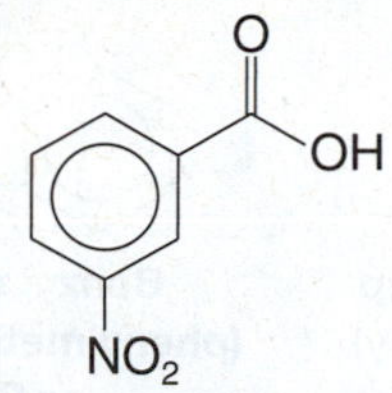

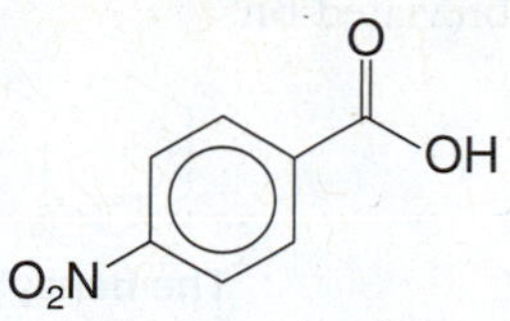

2-Nitrobenzoic acid
(o-nitrobenzoic acid)

3-Nitrobenzoic acid
(m-nitrobenzoic acid)

4-Nitrobenzoic acid
(p-nitrobenzoic acid)

The dimethylbenzenes are often called *xylenes*:

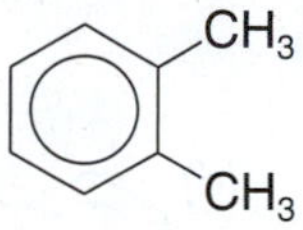

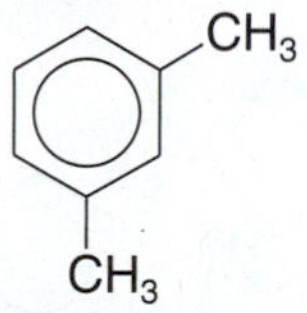

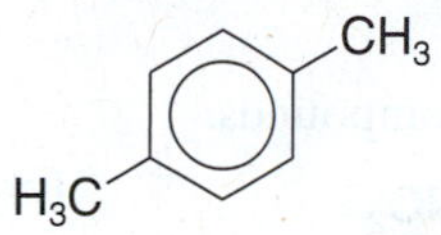

1,2-Dimethylbenzene
(o-xylene)

1,3-Dimethylbenzene
(m-xylene)

1,4-Dimethylbenzene
(p-xylene)

- If more than two groups are present on the benzene ring, their positions must be indicated by the use of *numbers*.

As examples, consider the following two compounds:

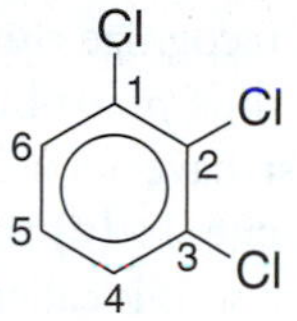

1,2,3-Trichlorobenzene

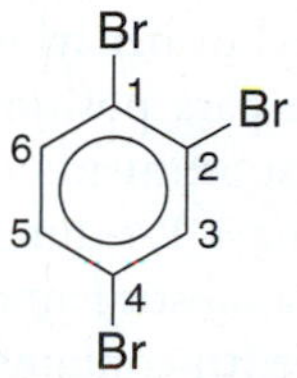

1,2,4-Tribromobenzene
(*not* 1,3,4-tribromobenzene)

- The benzene ring is numbered so as to give **the lowest possible numbers to the substituents**.
- When more than two substituents are present and the substituents are different, they are listed in alphabetical order.
- When a substituent is one that together with the benzene ring gives a new base name, that substituent is assumed to be in position 1 and the new parent name is used.

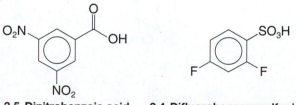

3,5-Dinitrobenzoic acid   **2,4-Difluorobenzenesulfonic acid**

- When the C_6H_5— group is named as a substituent, it is called a **phenyl** group. The phenyl group is often abbreviated as C_6H_5—, Ph—, or φ—.

A hydrocarbon composed of one saturated chain and one benzene ring is usually named as a derivative of the larger structural unit. However, if the chain is unsaturated, the compound may be named as a derivative of that chain, regardless of ring size. The following are examples:

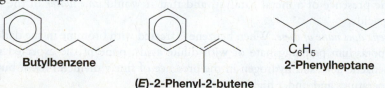

Butylbenzene

(E)-2-Phenyl-2-butene

2-Phenylheptane

Helpful Hint

Note the abbreviations for common aromatic groups.

• **Benzyl** is an alternative name for the phenylmethyl group. It is sometimes abbreviated Bn.

**The benzyl group
(the phenylmethyl
group)**

**Benzyl chloride
(phenylmethyl chloride
or BnCl)**

●●● SOLVED PROBLEM 14.1

Provide a name for each of the following compounds.

(a)

(b) NO₂

(c)

(d)

STRATEGY AND ANSWER: In each compound we look first to see if a commonly named unit containing a benzene ring is present. If not, we consider whether the compound can be named as a simple derivative of benzene, or if the compound incorporates the benzene ring as a phenyl or benzyl group. In **(a)** we recognize the common structural unit of acetophenone, and find a *tert*-butyl group in the para position. The name is thus *p-tert*-butylacetophenone or 4-*tert*-butylacetophenone. Compound **(b)**, having three substituents on the ring, must have its substituents named in alphabetical order and their positions numbered. The name is 1,4-dimethyl-2-nitrobenzene. In **(c)** there would appear to be a benzyl group, but the benzene ring can be considered a substituent on the alkyl chain, so it is called phenyl in this case. The name is 2-chloro-2-methyl-1-phenylpentane. Because **(d)** contains an ether functional group, we name it according to the groups bonded to the ether oxygen. The name is benzyl ethyl ether, or ethyl phenylmethyl ether.

●●● PRACTICE PROBLEM 14.1 Provide a name for each of the following compounds.

(a)

O

OH

Br

(b)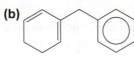

(c)

Cl

(d)

O

14.3 REACTIONS OF BENZENE

In the mid-nineteenth century, benzene presented chemists with a real puzzle. They knew from its formula (Section 14.1) that benzene was highly unsaturated, and they expected it to react accordingly. They expected it to react like an alkene by decolorizing bromine through *addition*. They expected that it would change the color of aqueous potassium permanganate by being *oxidized*, that it would *add hydrogen* rapidly in the presence of a metal catalyst, and that it would *add water* in the presence of strong acids.

Benzene does none of these. When benzene is treated with bromine in the dark or with aqueous potassium permanganate or with dilute acids, none of the expected reactions occurs. Benzene does add hydrogen in the presence of finely divided nickel, but only at high temperatures and under high pressures:

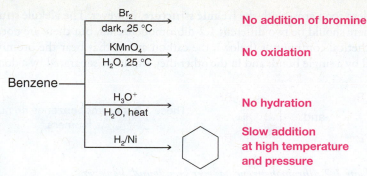

Benzene

- Br_2, dark, 25 °C → **No addition of bromine**
- $KMnO_4$, H_2O, 25 °C → **No oxidation**
- H_3O^+, H_2O, heat → **No hydration**
- H_2/Ni → **Slow addition at high temperature and pressure**

Benzene *does* react with bromine but only in the presence of a Lewis acid catalyst such as ferric bromide. Most surprisingly, however, it reacts not by addition but by *substitution*—**benzene substitution**.

Substitution

$$C_6H_6 + Br_2 \xrightarrow{FeBr_3} C_6H_5Br + HBr$$ **Substitution is observed.**

Addition

$$C_6H_6 + Br_2 \not\longrightarrow C_6H_6Br_2 + C_6H_6Br_4 + C_6H_6Br_6$$ **Addition is not observed.**

When benzene reacts with bromine, *only one monobromobenzene* is formed. That is, only one compound with the formula C_6H_5Br is found among the products. Similarly, when benzene is chlorinated, *only one monochlorobenzene* results.

Two possible explanations can be given for these observations. The first is that only one of the six hydrogen atoms in benzene is reactive toward these reagents. The second is that all six hydrogen atoms in benzene are equivalent, and replacing any one of them with a substituent results in the same product. As we shall see, the second explanation is correct.

Listed below are four compounds that have the molecular formula C_6H_6. Which of these compounds would yield only one monosubstitution product, if, for example, one hydrogen were replaced by bromine?

PRACTICE PROBLEM 14.2

(a) 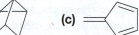   (b) [structure]   (c) [structure]   (d) [structure]

14.4 THE KEKULÉ STRUCTURE FOR BENZENE

In 1865, August Kekulé, the originator of the structural theory (Section 1.3), proposed the first definite structure for benzene,* a structure that is still used today (although as we shall soon see, we give it a meaning different from the meaning Kekulé gave it). Kekulé suggested that the carbon atoms of benzene are in a ring, that they are bonded to each other by alternating single and double bonds, and that one hydrogen atom is attached to each carbon atom. This structure satisfied the requirements of the structural theory that carbon atoms form four bonds and that all the hydrogen atoms of benzene are equivalent:

[Kekulé structure of benzene]   or   [hexagon with inscribed circle/double bonds]

The Kekulé formula for benzene

*In 1861 the Austrian chemist Johann Josef Loschmidt represented the benzene ring with a circle, but he made no attempt to indicate how the carbon atoms were actually arranged in the ring.

A problem soon arose with the **Kekulé structure**, however. The Kekulé structure predicts that there should be two different 1,2-dibromobenzenes, but there are not. In one of these hypothetical compounds (below), the carbon atoms that bear the bromines would be separated by a single bond, and in the other they would be separated by a double bond.

These 1,2-dibromobenzenes do not exist as isomers.

- *Only one 1,2-dibromobenzene has ever been found, however.*

To accommodate this objection, Kekulé proposed that the two forms of benzene (and of benzene derivatives) are in a state of equilibrium and that this equilibrium is so rapidly established that it prevents isolation of the separate compounds. Thus, the two 1,2-dibromobenzenes would also be rapidly equilibrated, and this would explain why chemists had not been able to isolate the two forms:

There is no such equilibrium between benzene ring bond isomers.

- We now know that this proposal was also incorrect and that *no such equilibrium exists.*

Nonetheless, the Kekulé formulation of benzene's structure was an important step forward and, for very practical reasons, it is still used today. Now we understand its meaning differently, however.

The tendency of benzene to react by substitution rather than addition gave rise to another concept of **aromaticity**. For a compound to be called aromatic meant, experimentally, that it gave substitution reactions rather than addition reactions even though it was highly unsaturated.

Before 1900, chemists assumed that the ring of alternating single and double bonds was the structural feature that gave rise to the aromatic properties. Since benzene and benzene derivatives (i.e., compounds with six-membered rings) were the only aromatic compounds known, chemists naturally sought other examples. The compound cyclooctatetraene seemed to be a likely candidate:

Cyclooctatetraene

In 1911, Richard Willstätter succeeded in synthesizing cyclooctatetraene. Willstätter found, however, that it is not at all like benzene. Cyclooctatetraene reacts with bromine by addition, it adds hydrogen readily, it is oxidized by solutions of potassium permanganate, and thus it is clearly *not aromatic*. While these findings must have been a keen disappointment to Willstätter, they were very significant for what they did not prove. Chemists, as a result, had to look deeper to discover the origin of benzene's aromaticity.

14.5 THE THERMODYNAMIC STABILITY OF BENZENE

We have seen that benzene shows unusual behavior by undergoing substitution reactions when, on the basis of its Kekulé structure, we should expect it to undergo addition. Benzene is unusual in another sense: it is *more stable thermodynamically* than the Kekulé structure suggests. To see how, consider the following thermochemical results.

Cyclohexene, a six-membered ring containing one double bond, can be hydrogenated easily to cyclohexane. When the $\Delta H°$ for this reaction is measured, it is found to be -120 kJ mol^{-1}, very much like that of any similarly substituted alkene:

$$\text{Cyclohexene} \quad + \quad H_2 \quad \xrightarrow{Pt} \quad \text{Cyclohexane} \qquad \Delta H° = -120 \text{ kJ mol}^{-1}$$

Cyclohexene   **Cyclohexane**

We would expect that hydrogenation of 1,3-cyclohexadiene would liberate roughly twice as much heat and thus have a $\Delta H°$ equal to about -240 kJ mol^{-1}. When this experiment is done, the result is $\Delta H° = -232$ kJ mol^{-1}. This result is quite close to what we calculated, and the difference can be explained by taking into account the fact that compounds containing conjugated double bonds are usually somewhat more stable than those that contain isolated double bonds (Section 13.8):

$$\text{1,3-Cyclohexadiene} \quad + \quad 2\,H_2 \quad \xrightarrow{Pt} \quad \text{Cyclohexane}$$

1,3-Cyclohexadiene   **Cyclohexane**

Calculated
$\Delta H° = 2 \times (-120) = -240$ kJ mol^{-1}
Observed
$\Delta H° = -232$ kJ mol^{-1}

If we extend this kind of thinking, and if benzene is simply 1,3,5-cyclohexatriene, we would predict benzene to liberate approximately 360 kJ mol^{-1} [$3 \times (-120)$] when it is hydrogenated. When the experiment is actually done, the result is surprisingly different. The reaction is exothermic, but only by 208 kJ mol^{-1}:

$$\text{Benzene} \quad + \quad 3\,H_2 \quad \xrightarrow{Pt} \quad \text{Cyclohexane}$$

Benzene   **Cyclohexane**

Calculated
$\Delta H° = 3 \times (-120) = -360$ kJ mol^{-1}

Observed	$\Delta H° = -208$ kJ mol^{-1}
Difference	$= 152$ kJ mol^{-1}

When these results are represented as in Fig. 14.1, it becomes clear that benzene is much more stable than we calculated it to be. Indeed, it is more stable than the hypothetical 1,3,5-cyclohexatriene by 152 kJ mol^{-1}. This difference between the amount of heat actually released and that calculated on the basis of the Kekulé structure is now called the **resonance energy** of the compound.

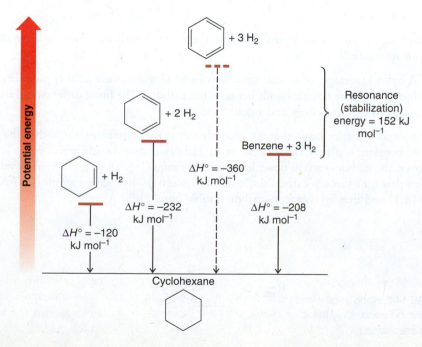

FIGURE 14.1 Relative stabilities of cyclohexene, 1,3-cyclohexadiene, 1,3,5-cyclohexatriene (hypothetical), and benzene.

14.6 MODERN THEORIES OF THE STRUCTURE OF BENZENE

It was not until the development of quantum mechanics in the 1920s that the unusual behavior and stability of benzene began to be understood. Quantum mechanics, as we have seen, produced two ways of viewing bonds in molecules: resonance theory and molecular orbital theory. We now look at both of these as they apply to benzene.

14.6A The Resonance Explanation of the Structure of Benzene

A basic postulate of resonance theory (Sections 1.8 and 13.4) is that whenever two or more Lewis structures can be written for a molecule that *differ only in the positions of their electrons*, none of the structures will be in complete accord with the compound's chemical and physical properties. If we recognize this, we can now understand the true nature of the two Kekulé structures (**I** and **II**) for benzene.

- Kekulé structures **I** and **II** below differ only in the positions of their electrons; they do not represent two separate molecules in equilibrium as Kekulé had proposed.

Instead, structures **I** and **II** are the closest we can get to a structure for benzene within the limitations of its molecular formula, the classic rules of valence, and the fact that the six hydrogen atoms are chemically equivalent. The problem with the Kekulé structures is that they are Lewis structures, and Lewis structures portray electrons in localized distributions. (With benzene, as we shall see, the electrons are delocalized.) Resonance theory, fortunately, does not stop with telling us when to expect this kind of trouble; it also gives us a way out.

- According to resonance theory, we consider Kekulé structures **I** and **II** below as *resonance contributors* to the real structure of benzene, and we relate them to each other with one double-headed, double-barbed arrow (not two separate arrows, which we reserve for equilibria).

Resonance contributors, we emphasize again, are not in equilibrium. They are not structures of real molecules. They are the closest we can get if we are bound by simple rules of valence, but they are very useful in helping us visualize the actual molecule as a hybrid:

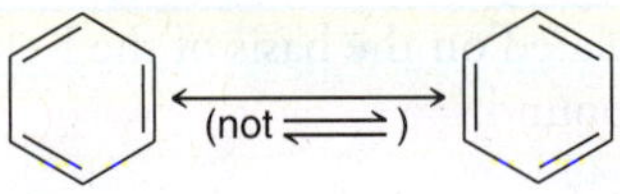

Look at the structures carefully. All of the single bonds in structure **I** are double bonds in structure **II**.

- A hybrid (average) of Kekulé structures **I** and **II** would have neither pure single bonds nor pure double bonds between the carbons. The bond order would be between that of a single and a double bond.

Experimental evidence bears this out. Spectroscopic measurements show that the molecule of benzene is planar and that all of its carbon–carbon bonds are of equal length. Moreover, the carbon–carbon bond lengths in benzene (Fig. 14.2) are 1.39 Å, a value in between that for a carbon–carbon single bond between sp^2-hybridized atoms (1.47 Å) (see Table 13.1) and that for a carbon–carbon double bond (1.34 Å).

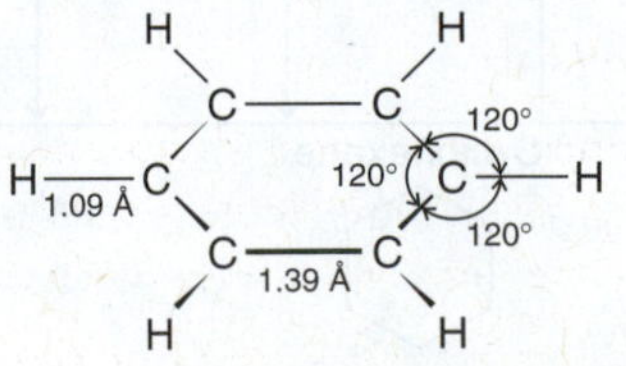

FIGURE 14.2 Bond lengths and angles in benzene. (Only the σ bonds are shown.)

- The hybrid structure of benzene is represented by inscribing a circle inside the hexagon as shown in formula **III** below.

III

There are times when an accounting of the π electron pairs must be made, however, and for these purposes we use either Kekulé structure **I** or **II**. We do this simply because the electron pairs and total π electron count is obvious in a Kekulé structure, whereas the number of π electron pairs represented by a circle can be ambiguous. As we shall see later in this chapter, there are systems having different ring sizes and different numbers of delocalized π electrons that can also be represented by a circle. In benzene, however, the circle is understood to represent six π electrons that are delocalized around the six carbons of the ring.

- An actual molecule of benzene (depicted by the resonance hybrid **III**) is more stable than either contributing resonance structure because more than one equivalent resonance structure can be drawn for benzene (**I** and **II** above).

The difference in energy between hypothetical 1,3,5-cyclohexatriene (which if it existed would have higher energy) and benzene is called *resonance energy*, and it is an indication of the extra stability of benzene due to electron delocalization.

● ● ●

PRACTICE PROBLEM 14.3

If benzene were 1,3,5-cyclohexatriene, the carbon–carbon bonds would be alternately long and short as indicated in the following structures. However, to consider the structures here as resonance contributors (or to connect them by a double-headed arrow) violates a basic principle of resonance theory. Explain.

14.6B The Molecular Orbital Explanation of the Structure of Benzene

The fact that the bond angles of the carbon atoms in the benzene ring are all 120° strongly suggests that the carbon atoms are sp^2 hybridized. If we accept this suggestion and construct a planar six-membered ring from sp^2 carbon atoms, representations like those shown in Figs. 14.3a and b emerge. In these models, each carbon is sp^2 hybridized and has a p orbital available for overlap with p orbitals of its neighboring carbons. If we consider favorable overlap of these p orbitals all around the ring, the result is the model shown in Fig. 14.3c.

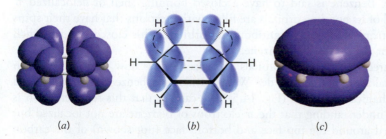

(a)   (b)   (c)

FIGURE 14.3 (a) Six sp^2-hybridized carbon atoms joined in a ring (each carbon also bears a hydrogen atom). Each carbon has a p orbital with lobes above and below the plane of the ring. (b) A stylized depiction of the p orbitals in (a). (c) Overlap of the p orbitals around the ring results in a molecular orbital encompassing the top and bottom faces of the ring. (Differences in the mathematical phase of the orbital lobes are not shown in these representations.)

- As we recall from the principles of quantum mechanics (Section 1.11), the number of molecular orbitals in a molecule is the same as the number of atomic orbitals from which they are derived, and each orbital can accommodate a maximum of two electrons if their spins are opposed.

If we consider only the p atomic orbitals contributed by the carbon atoms of benzene, there should be six π molecular orbitals. These orbitals are shown in Fig. 14.4.

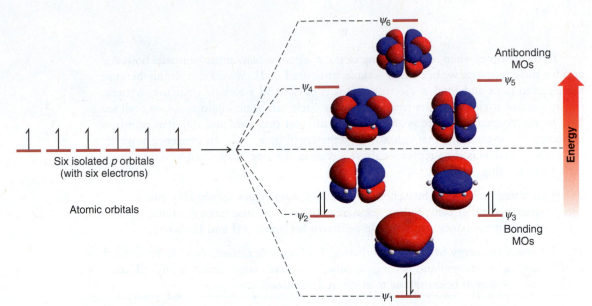

FIGURE 14.4 How six p atomic orbitals (one from each carbon of the benzene ring) combine to form six π molecular orbitals. Three of the molecular orbitals have energies lower than that of an isolated p orbital; these are the bonding molecular orbitals. Three of the molecular orbitals have energies higher than that of an isolated p orbital; these are the antibonding molecular orbitals. Orbitals ψ_2 and ψ_3 have the same energy and are said to be degenerate; the same is true of orbitals ψ_4 and ψ_5.

The electronic configuration of the ground state of benzene is obtained by adding the six π electrons to the π molecular orbitals shown in Fig. 14.4, starting with the orbitals of lowest energy. The lowest energy π molecular orbital in benzene has overlap of p orbitals with the same mathematical phase sign all around the top and bottom faces of the ring. In this orbital there are no nodal planes (changes in orbital phase sign) perpendicular to the atoms of the ring. The orbitals of next higher energy each have one nodal plane. (In general, each set of higher energy π molecular orbitals has an additional nodal plane.) Each of these orbitals is filled with a pair of electrons, as well. These orbitals are of equal energy (degenerate) because they both have one nodal plane. Together, these three orbitals comprise the bonding π molecular orbitals of benzene. The next higher energy set of π molecular orbitals each has two nodal planes, and the highest energy π molecular orbital of benzene has three nodal planes. These three orbitals are the antibonding π molecular orbitals of benzene, and they are unoccupied in the ground state. Benzene is said to have a closed bonding shell of delocalized π electrons because all of its bonding orbitals are filled with electrons that have their spins paired, and no electrons are found in antibonding orbitals. This closed bonding shell accounts, in part, for the stability of benzene.

Having considered the molecular orbitals of benzene, it is now useful to view an electrostatic potential map of the van der Waals surface for benzene, also calculated from quantum mechanical principles (Fig. 14.5). We can see that this representation is consistent with our understanding that the π electrons of benzene are not localized but are evenly distributed around the top face and bottom face (not shown) of the carbon ring in benzene.

It is interesting to note the recent discovery that crystalline benzene involves perpendicular interactions between benzene rings, so that the relatively positive periphery of one molecule associates with the relatively negative faces of the benzene molecules aligned above and below it.

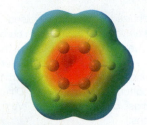

FIGURE 14.5 Electrostatic potential map of benzene.

14.7 HÜCKEL'S RULE: THE $4n + 2$ π ELECTRON RULE

In 1931 the German physicist Erich Hückel carried out a series of mathematical calculations based on the kind of theory that we have just described. **Hückel's rule** is concerned with compounds containing **one planar ring in which each atom has a p orbital** as in benzene. His calculations showed that planar monocyclic rings containing $4n + 2$ π electrons, where $n = 0, 1, 2, 3$, and so on (i.e., rings containing 2, 6, 10, 14,..., etc., π electrons), have closed shells of delocalized electrons like benzene and should have substantial resonance energies.

- In other words, Hückel's rule states that **planar monocyclic rings with 2, 6, 10, 14,..., delocalized electrons should be aromatic**.

14.7A HOW TO Diagram the Relative Energies of π Molecular Orbitals in Monocyclic Systems Based on Hückel's Rule

There is a simple way to make a diagram of the relative energies of orbitals in monocyclic conjugated systems based on Hückel's calculations. To do so, we use the following procedure.

1. We start by drawing a polygon corresponding to the number of carbons in the ring, *placing a corner of the polygon at the bottom*.

2. Next we surround the polygon with a circle that touches each corner of the polygon.

3. At the points where the polygon touches the circle, we draw short horizontal lines outside the circle. The height of each line represents the relative energy of each π molecular orbital.

4. Next we draw a dashed horizontal line across and halfway up the circle. The energies of bonding π molecular orbitals are below this line. The energies of antibonding π molecular orbitals are above, and those for nonbonding orbitals are at the level of the dashed line.

5. Based on the number of π electrons in the ring, we then place electron arrows on the lines corresponding to the respective orbitals, beginning at the lowest energy level and working upward. In doing so, we fill degenerate orbitals each with one electron first, then add to each unpaired electron another with opposite spin if it is available.

Applying this method to benzene, for example (Fig. 14.6), furnishes the same energy levels that we saw earlier in Fig. 14.4, energy levels that were based on quantum mechanical calculations.

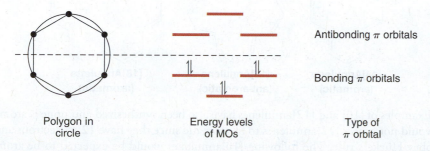

| Polygon in circle | Energy levels of MOs | Type of π orbital |

Antibonding π orbitals

Bonding π orbitals

FIGURE 14.6 The polygon-and-circle method for deriving the relative energies of the π molecular orbitals of benzene. A horizontal line halfway up the circle divides the bonding orbitals from the antibonding orbitals. If an orbital falls on this line, it is a nonbonding orbital. This method was developed by C. A. Coulson (of Oxford University).

We can now understand why cyclooctatetraene is not aromatic. Cyclooctatetraene has a total of eight π electrons. Eight is not a Hückel number; it is a *4n number*, not a *4n + 2 number*. Using the polygon-and-circle method (Fig. 14.7), we find that cyclooctatetraene, if it were planar, *would not* have a closed shell of π electrons like benzene; it would have an unpaired electron in each of two nonbonding orbitals. Molecules with unpaired electrons (radicals) are *not* unusually stable; they are typically highly reactive and unstable. A planar form of cyclooctatetraene, therefore, should not be at all like benzene and should not be aromatic.

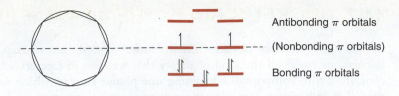

FIGURE 14.7 The π molecular orbitals that cyclooctatetraene would have if it were planar. Notice that, unlike benzene, this molecule is predicted to have two nonbonding orbitals, and because it has eight π electrons, it would have an unpaired electron in each of the two nonbonding orbitals (Hund's rule, Section 1.11). Such a system would not be expected to be aromatic.

Because cyclooctatetraene does not gain stability by becoming planar, it assumes the tub shape shown below. (In Section 14.7E we shall see that cyclooctatetraene would actually lose stability by becoming planar.) The bonds of cyclooctatetraene are known to be alternately long and short; X-ray studies indicate that they are 1.48 and 1.34 Å, respectively.

14.7B The Annulenes

The word **annulene** is incorporated into the class name for monocyclic compounds that can be represented by structures having alternating single and double bonds. The ring size of an annulene is indicated by a number in brackets. Thus, benzene is [6]annulene and cyclooctatetraene is [8]annulene.

- Hückel's rule predicts that annulenes will be aromatic if their molecules have $4n + 2$ π electrons and have a planar carbon skeleton:

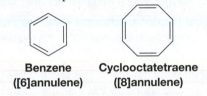

Benzene
([6]annulene)

Cyclooctatetraene
([8]annulene)

Before 1960 the only annulenes that were available to test Hückel's predictions were benzene and cyclooctatetraene. During the 1960s, and largely as a result of research by F. Sondheimer, a number of large-ring annulenes were synthesized, and the predictions of Hückel's rule were verified.

Consider the [14], [16], [18], [20], [22], and [24]annulenes as examples. Of these, *as Hückel's rule predicts*, the [14], [18], and [22]annulenes ($4n + 2$ when $n = 3, 4, 5$, respectively) have been found to be aromatic. The [16]annulene and the [24]annulene are not aromatic; they are *antiaromatic* (see Section 14.7E). They are $4n$ compounds, not $4n + 2$ compounds:

[18]Annulene.

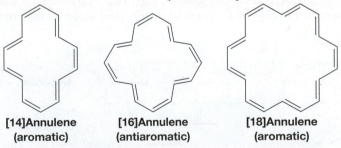

[14]Annulene
(aromatic)

[16]Annulene
(antiaromatic)

[18]Annulene
(aromatic)

Examples of [10] and [12]annulenes have also been synthesized and none is aromatic. We would not expect [12]annulenes to be aromatic since they have 12 π electrons and do not obey Hückel's rule. The following [10]annulenes would be expected to be aromatic on the basis of electron count, but their rings are not planar.

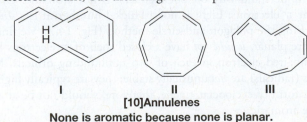

I
II
III
[10]Annulenes
None is aromatic because none is planar.

The [10]annulene **I** has two trans double bonds. Its bond angles are approximately 120°; therefore, it has no appreciable angle strain. The carbon atoms of its ring, however, are prevented from becoming coplanar because the two hydrogen atoms in the center of the ring interfere with each other. Because the ring is not planar, the *p* orbitals of the carbon atoms are not parallel and, therefore, cannot overlap effectively around the ring to form the π molecular orbitals of an aromatic system.

The [10]annulene with all cis double bonds (**II**) would, if it was planar, have considerable angle strain because the internal bond angles would be 144°. Consequently, any stability this isomer gained by becoming planar in order to become aromatic would be more than offset by the destabilizing effect of the increased angle strain. A similar problem of a large angle strain associated with a planar form prevents molecules of the [10]annulene isomer with one trans double bond (**III**) from being aromatic.

After many unsuccessful attempts over many years, in 1965 [4]annulene (or cyclobutadiene) was synthesized by R. Pettit and co-workers at the University of Texas, Austin. Cyclobutadiene is a $4n$ molecule, not a $4n + 2$ molecule, and, as we would expect, it is a highly unstable compound and *it is antiaromatic* (see Section 14.7E):

Cyclobutadiene
or [4]annulene
(antiaromatic)

•●● **SOLVED PROBLEM 14.2**

Using the polygon-and-circle method to outline the molecular orbitals of cyclobutadiene, explain why cyclobutadiene is not aromatic.

STRATEGY AND ANSWER: We inscribe a square inside a circle with one corner at the bottom.

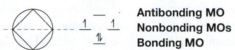

We see that cyclobutadiene, according to this model, would have an unpaired electron in each of its two nonbonding molecular orbitals. We would, therefore, not expect cyclobutadiene to be aromatic.

14.7C NMR Spectroscopy: Evidence for Electron Delocalization in Aromatic Compounds

The ^1H NMR spectrum of benzene consists of a single unsplit signal at δ 7.27. That only a single unsplit signal is observed is further proof that all of the hydrogens of benzene are equivalent. That the signal occurs at relatively high frequency is, as we shall see, compelling evidence for the assertion that the π electrons of benzene are delocalized.

We learned in Section 9.6 that circulations of σ electrons of C—H bonds cause the protons of alkanes to be *shielded* from the applied magnetic field of an NMR spectrometer and, consequently, these protons absorb at lower frequency. We shall now explain the high-frequency absorption of benzene protons on the basis of *deshielding caused by circulation of the π electrons of benzene*, and this explanation, as you will see, requires that the π electrons be delocalized.

When benzene molecules are placed in the powerful magnetic field of the NMR spectrometer, electrons circulate in the direction shown in Fig. 14.8; by doing so, they generate a **ring current**. (If you have studied physics, you will understand why the electrons circulate in this way.)

- The circulation of π electrons in benzene creates an induced magnetic field that, *at the position of the protons, reinforces the applied magnetic field.* This reinforcement causes the protons to be strongly *deshielded* and to have a relatively high frequency ($\delta \sim 7$) absorption.

By "deshielded" we mean that the protons sense the sum of the two fields, and, therefore, the net magnetic field strength is greater than it would have been in the absence of the induced field. This strong deshielding, which we attribute to a ring current created by the *delocalized* π electrons, explains why aromatic protons absorb at relatively high frequency.

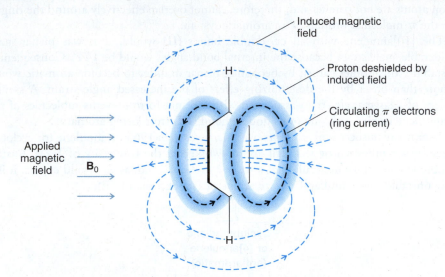

FIGURE 14.8 The induced magnetic field of the π electrons of benzene deshields the benzene protons. Deshielding occurs because at the location of the protons the induced field is in the same direction as the applied field.

The deshielding of external aromatic protons that results from the ring current is one of the best pieces of physical evidence that we have for π-electron delocalization in aromatic rings. In fact, relatively high frequency proton absorption is often used as a criterion for assessing aromaticity in newly synthesized conjugated cyclic compounds.

Not all aromatic protons have high frequency absorptions, however. The internal protons of large-ring aromatic compounds that have hydrogens in the center of the ring (in the π-electron cavity) absorb at unusually low frequency because they are highly shielded by the opposing induced magnetic field in the center of the ring (see Fig. 14.8). An example is [18]annulene (Fig. 14.9). The internal protons of [18]annulene absorb far upfield at δ −3.0, above the signal for tetramethylsilane (TMS); the external protons, on the other hand, absorb far downfield at δ 9.3. Considering that [18]annulene has $4n + 2$ π electrons, this evidence provides strong support for π-electron delocalization as a criterion for aromaticity and for the predictive power of Hückel's rule.

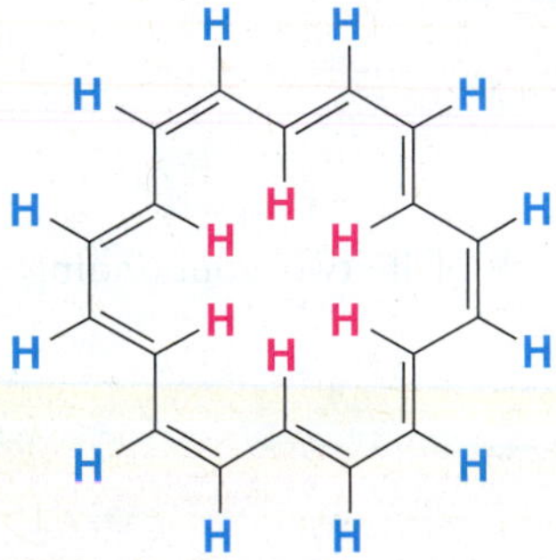

FIGURE 14.9 [18]Annulene. The internal protons (red) are highly shielded and absorb at δ −3.0. The external protons (blue) are highly deshielded and absorb at δ 9.3.

14.7D Aromatic Ions

In addition to the neutral molecules that we have discussed so far, there are a number of monocyclic species that bear either a positive or a negative charge. Some of these ions show unexpected stabilities that suggest that they are **aromatic ions**. Hückel's rule is helpful in accounting for the properties of these ions as well. We shall consider two examples: the cyclopentadienyl anion and the cycloheptatrienyl cation.

Cyclopentadiene is not aromatic; however, it is unusually acidic for a hydrocarbon. (The pK_a for cyclopentadiene is 16 and, by contrast, the pK_a for cycloheptatriene is 36.) Because of its acidity, cyclopentadiene can be converted to its anion by treatment with moderately strong bases. The cyclopentadienyl anion, moreover, is unusually stable, and NMR spectroscopy shows that all five hydrogen atoms in the cyclopentadienyl anion are equivalent and absorb downfield.

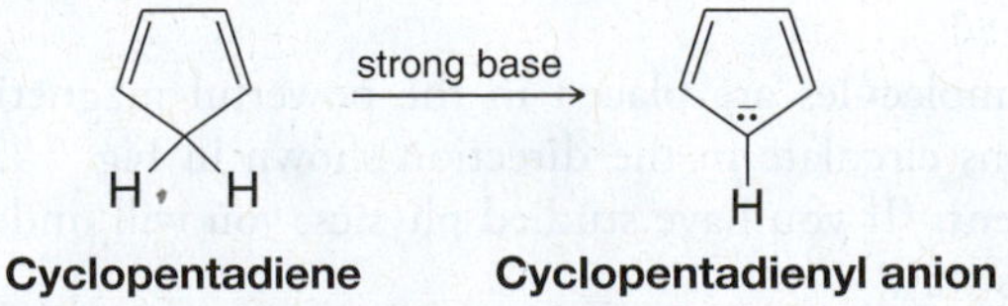

$$\text{Cyclopentadiene} \xrightarrow{\text{strong base}} \text{Cyclopentadienyl anion}$$

The orbital structure of cyclopentadiene (Fig. 14.10) shows why cyclopentadiene, itself, is not aromatic. Not only does it not have the proper number of π electrons, but the π electrons cannot be delocalized about the entire ring because of the intervening sp^3-hybridized —CH$_2$— group with no available p orbital.

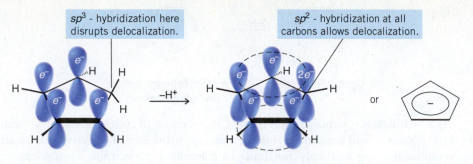

Cyclopentadiene   **Cyclopentadienyl anion**

FIGURE 14.10 Cyclopentadiene is not aromatic because it has only four π electrons and the sp^3-hybridized carbon prevents complete delocalization around the ring. Removal of a proton produces the cyclopentadienyl anion, which is aromatic because it has 6 π electrons and all of its carbon atoms have a p orbital.

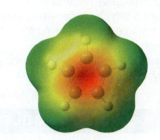

FIGURE 14.11 An electrostatic potential map of the cyclopentadienyl anion. The ion is negatively charged overall, of course, but regions with greatest negative potential are shown in red, and regions with least negative potential are in blue. The concentration of negative potential in the center of the top face and bottom face (not shown) indicates that the extra electron of the ion is involved in the aromatic π-electron system.

On the other hand, if the —CH_2— carbon atom becomes sp^2 hybridized after it loses a proton (Fig. 14.10), the two electrons left behind can occupy the new p orbital that is produced. Moreover, this new p orbital can overlap with the p orbitals on either side of it and give rise to a ring with *six* delocalized π electrons. Because the electrons are delocalized, all of the hydrogen atoms are equivalent, and this agrees with what NMR spectroscopy tells us. A calculated electrostatic potential map for cyclopentadienyl anion (Fig. 14.11) also shows the symmetrical distribution of negative charge within the ring, and the overall symmetry of the ring structure.

Six, the number of π electrons in the cyclopentadienyl anion is, of course, a Hückel number ($4n + 2$, where $n = 1$).

- The cyclopentadienyl anion is, therefore, an **aromatic anion**, and the unusual acidity of cyclopentadiene is a result of the unusual stability of its anion.

Cycloheptatriene (Fig. 14.12) (a compound with the common name tropylidene) has six π electrons. However, the six π electrons of cycloheptatriene cannot be fully delocalized because of the presence of the —CH_2— group, a group that does not have an available p orbital (Fig. 14.12).

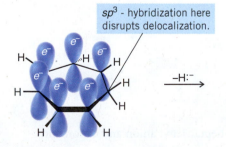

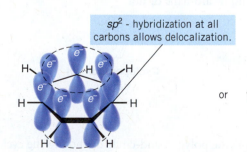

Cycloheptatriene   **Cycloheptatrienyl cation**

FIGURE 14.12 Cycloheptatriene is not aromatic, even though it has six π electrons, because it has an sp^3-hybridized carbon that prevents delocalization around the ring. Removal of a hydride (H:⁻) produces the cycloheptatrienyl cation, which is aromatic because all of its carbon atoms now have a p orbital, and it still has 6 π electrons.

When cycloheptatriene is treated with a reagent that can abstract a hydride ion, it is converted to the cycloheptatrienyl (or tropylium) cation. The loss of a hydride ion from cycloheptatriene occurs with unexpected ease, and the cycloheptatrienyl cation is found to be unusually stable. The NMR spectrum of the cycloheptatrienyl cation indicates that all seven hydrogen atoms are equivalent. If we look closely at Fig. 14.12, we see how we can account for these observations.

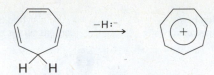

Cycloheptatriene   **Cycloheptatrienyl cation**
                            (tropylium cation)

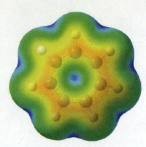

FIGURE 14.13 An electrostatic potential map of the tropylium cation. The ion is positive overall, of course, but a region of relatively greater negative electrostatic potential can clearly be seen around the top face (and bottom face, though not shown) of the ring where electrons are involved in the π system of the aromatic ring.

As a hydride ion is removed from the —CH_2— group of cycloheptatriene, a vacant p orbital is created, and the carbon atom becomes sp^2 hybridized. The cation that results has seven overlapping p orbitals containing *six* delocalized π electrons. The cycloheptatrienyl cation is, therefore, an aromatic cation, and all of its hydrogen atoms should be equivalent; again, this is exactly what we find experimentally.

The calculated electrostatic potential map for cycloheptatrienyl (tropylium) cation (Fig. 14.13) also shows the symmetry of this ion. Electrostatic potential from the π electrons involved in the aromatic system is indicated by the yellow-orange color that is evenly distributed around the top face (and bottom face, though not shown) of the carbon framework. The entire ion is positive, of course, and the region of greatest positive potential is indicated by blue around the periphery of the ion.

●●● SOLVED PROBLEM 14.3

Apply the polygon-and-circle method to explain why the cyclopentadienyl anion is aromatic.

STRATEGY AND ANSWER: We inscribe a pentagon inside a circle with one corner at the bottom and find that the energy levels of the molecular orbitals are such that three molecular orbitals are bonding and two are antibonding:

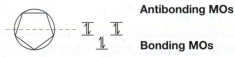

Antibonding MOs

Bonding MOs

Cyclopentadienyl anion has six π electrons, which is a Hückel number, and they fill all the bonding orbitals. There are no unpaired electrons and no electrons in antibonding orbitals. This is what we would expect of an aromatic ion.

●●●

PRACTICE PROBLEM 14.4 Apply the polygon-and-circle method to cyclopentadienyl cation and explain whether it would be aromatic or not.

Cyclopentadienyl cation

●●●

PRACTICE PROBLEM 14.5 Apply the polygon-and-circle method to the cycloheptatrienyl anion and cation and explain whether each would be aromatic or not.

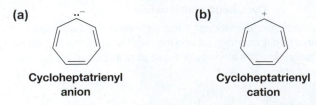

(a) **Cycloheptatrienyl anion**   **(b)** **Cycloheptatrienyl cation**

●●●

PRACTICE PROBLEM 14.6 1,3,5-Cycloheptatriene is even less acidic than 1,3,5-heptatriene. Explain how this experimental observation might help to confirm your answer to part (b) of the previous problem.

When 1,3,5-cycloheptatriene reacts with one molar equivalent of bromine at 0 °C, it undergoes 1,6 addition. **(a)** Write the structure of this product. **(b)** On heating, this 1,6-addition product loses HBr readily to form a compound with the molecular formula C_7H_7Br, called *tropylium bromide*. Tropylium bromide is insoluble in nonpolar solvents but is soluble in water; it has an unexpectedly high melting point (mp 203 °C), and when treated with silver nitrate, an aqueous solution of tropylium bromide gives a precipitate of AgBr. What do these experimental results suggest about the bonding in tropylium bromide?

14.7E Aromatic, Antiaromatic, and Nonaromatic Compounds

- An aromatic compound has its π electrons *delocalized* over the entire ring and it is *stabilized* by the π-electron delocalization.

As we have seen, a good way to determine whether the π electrons of a cyclic system are delocalized is through the use of NMR spectroscopy. It provides direct physical evidence of whether or not the π electrons are delocalized.

But what do we mean by saying that a compound is stabilized by π-electron delocalization? We have an idea of what this means from our comparison of the heat of hydrogenation of benzene and that calculated for the hypothetical 1,3,5-cyclohexatriene. We saw that benzene—in which the π electrons are delocalized—is much more stable than 1,3,5-cyclohexatriene (a model in which the π electrons are not delocalized). We call the energy difference between them the resonance energy (delocalization energy) or stabilization energy.

In order to make similar comparisons for other aromatic compounds, we need to choose proper models. But what should these models be?

One way to evaluate whether a cyclic compound is stabilized by delocalization of π electrons through its ring is to compare it with an open-chain compound having the same number of π electrons. This approach is particularly useful because it furnishes us with models not only for annulenes but for aromatic cations and anions, as well. (Corrections need to be made, of course, when the cyclic system is strained.)

To use this approach we do the following:

1. We take as our model a linear chain of sp^2-hybridized atoms having the same number of π electrons as our cyclic compound.

2. Then we imagine removing a hydrogen atom from each end of the chain and joining the ends to form a ring.

   - If, based on sound calculations or experiments, the ring has *lower* π-electron energy, then the ring is aromatic.
   - If the ring and the chain have the *same* π-electron energy, then the ring is nonaromatic.
   - If the ring has *greater* π-electron energy than the open chain, then the ring is **antiaromatic**.

The actual calculations and experiments used in determining π-electron energies are beyond our scope, but we can study four examples that illustrate how this approach has been used.

Cyclobutadiene For cyclobutadiene we consider the change in π-electron energy for the following *hypothetical* transformation:

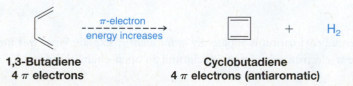

1,3-Butadiene	Cyclobutadiene
4 π electrons	4 π electrons (antiaromatic)

Calculations indicate and experiments appear to confirm that the π-electron energy of cyclobutadiene is higher than that of its open-chain counterpart. Thus cyclobutadiene is classified as antiaromatic.

Benzene Here our comparison is based on the following hypothetical transformation:

1,3,5-Hexatriene
6 π electrons

Benzene
6 π electrons (aromatic)

Calculations indicate and experiments confirm that benzene has a much lower π-electron energy than 1,3,5-hexatriene. Benzene is classified as being aromatic on the basis of this comparison as well.

Cyclopentadienyl Anion Here we use a linear anion for our hypothetical transformation:

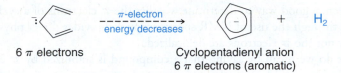

6 π electrons

Cyclopentadienyl anion
6 π electrons (aromatic)

Both calculations and experiments confirm that the cyclic anion has a lower π-electron energy than its open-chain counterpart. Therefore the cyclopentadienyl anion is classified as aromatic.

Cyclooctatetraene For cyclooctatetraene we consider the following hypothetical transformation:

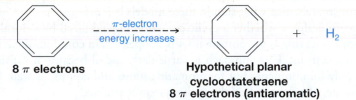

8 π electrons

Hypothetical planar cyclooctatetraene
8 π electrons (antiaromatic)

Here calculations and experiments indicate that a planar cyclooctatetraene would have higher π-electron energy than the open-chain octatetraene. Therefore, a planar form of cyclooctatetraene would, if it existed, be *antiaromatic*. As we saw earlier, cyclooctatetraene is not planar and behaves like a simple cyclic polyene.

●●● SOLVED PROBLEM 14.4

Calculations indicate that the π-electron energy decreases for the hypothetical transformation from the allyl cation to the cyclopropenyl cation below. What does this indicate about the possible aromaticity of the cyclopropenyl cation

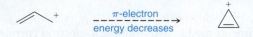

STRATEGY AND ANSWER: Because the π-electron energy of the cyclic cation is less than that of the allyl cation, we can conclude that the cyclopropenyl cation would be aromatic. (See Practice Problem 14.9 for more information on this cation.)

●●●

PRACTICE PROBLEM 14.8 The cyclopentadienyl cation is apparently *antiaromatic*. Explain what this means in terms of the π-electron energies of a cyclic and an open-chain compound.

In 1967 R. Breslow (of Columbia University) and co-workers showed that adding $SbCl_5$ to a solution of 3-chlorocyclopropene in CH_2Cl_2 caused the precipitation of a white solid with the composition $C_3H_3^+SbCl_6^-$. NMR spectroscopy of a solution of this salt showed that all of its hydrogen atoms were equivalent. **(a)** What new aromatic ion had these researchers prepared? **(b)** How many ^{13}C NMR signals would you predict for this ion?

14.8 OTHER AROMATIC COMPOUNDS

14.8A Benzenoid Aromatic Compounds

In addition to those that we have seen so far, there are many other examples of aromatic compounds. Representatives of one broad class of **benzenoid aromatic compounds**, called **polycyclic aromatic hydrocarbons (PAH)**, are illustrated in Fig. 14.14.

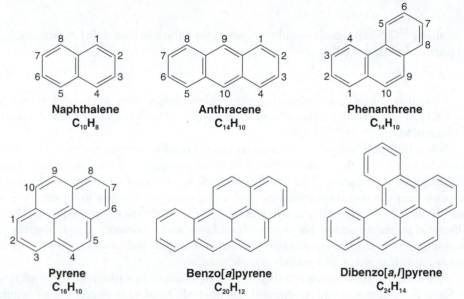

Naphthalene
$C_{10}H_8$

Anthracene
$C_{14}H_{10}$

Phenanthrene
$C_{14}H_{10}$

Pyrene
$C_{16}H_{10}$

Benzo[a]pyrene
$C_{20}H_{12}$

Dibenzo[a,l]pyrene
$C_{24}H_{14}$

FIGURE 14.14 Benzenoid aromatic hydrocarbons. Some polycyclic aromatic hydrocarbons (PAHs), such as dibenzo[a,l]pyrene, are carcinogenic. (See "Important, but hidden, epoxides" at the end of Chapter 11.)

- Benzenoid polycyclic **aromatic** hydrocarbons consist of molecules having two or more benzene rings *fused* together.

A close look at one example, naphthalene, will illustrate what we mean by this.

According to resonance theory, a molecule of naphthalene can be considered to be a hybrid of three Kekulé structures. One of these Kekulé structures, the most important one, is shown in Fig. 14.15. There are two carbon atoms in naphthalene (C4a and C8a) that are common to both rings. These two atoms are said to be at the points of *ring fusion*. They direct all of their bonds toward other carbon atoms and do not bear hydrogen atoms.

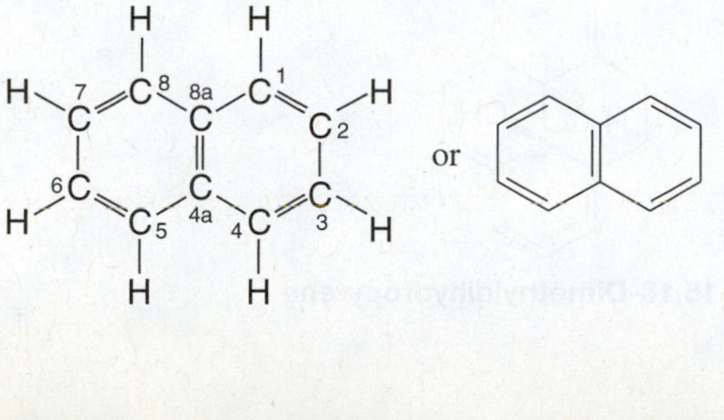

or

FIGURE 14.15 One Kekulé structure for naphthalene.

Acenaphthylene

●●● SOLVED PROBLEM 14.5

How many ^{13}C NMR signals would you expect for acenaphthylene?

STRATEGY AND ANSWER: Acenaphthylene has a plane of symmetry which makes the five carbon atoms on the left (a–e, at right) equivalent to those on the right. Carbon atoms f and g are unique. Consequently, acenaphthylene should give seven ^{13}C NMR signals.

Acenaphthylene

●●●

PRACTICE PROBLEM 14.10 How many ^{13}C NMR signals would you predict for **(a)** naphthalene, **(b)** anthracene, **(c)** phenanthrene, and **(d)** pyrene?

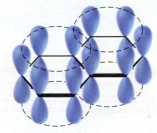

FIGURE 14.16 The stylized *p* orbitals of naphthalene.

Molecular orbital calculations for naphthalene begin with the model shown in Fig. 14.16. The *p* orbitals overlap around the periphery of both rings and across the points of ring fusion.

When molecular orbital calculations are carried out for naphthalene using the model shown in Fig. 14.16, the results of the calculations correlate well with our experimental knowledge of naphthalene. The calculations indicate that delocalization of the 10 π electrons over the two rings produces a structure with considerably lower energy than that calculated for any individual Kekulé structure. Naphthalene, consequently, has a substantial resonance energy. Based on what we know about benzene, moreover, naphthalene's tendency to react by substitution rather than addition and to show other properties associated with aromatic compounds is understandable.

Anthracene and phenanthrene (Fig. 14.14) are isomers. In anthracene the three rings are fused in a linear way, and in phenanthrene they are fused so as to produce an angular molecule. Both of these molecules also show large resonance energies and chemical properties typical of aromatic compounds.

Pyrene (Fig. 14.17) is also aromatic. Pyrene itself has been known for a long time; a pyrene derivative, however, has been the object of research that shows another interesting application of Hückel's rule.

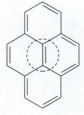

FIGURE 14.17 One Kekulé structure for pyrene. The internal double bond is enclosed in a dotted circle for emphasis.

To understand this particular research, we need to pay special attention to the Kekulé structure for pyrene (Fig. 14.17). The total number of π electrons in pyrene is 16 (8 double bonds = 16 π electrons). Sixteen is a non-Hückel number, but **Hückel's rule is intended to be applied only to monocyclic compounds** and pyrene is clearly tetracyclic. If we disregard the internal double bond of pyrene, however, and look only at the periphery, we see that the periphery is a planar ring with 14 π electrons. The periphery is, in fact, very much like that of [14]annulene. Fourteen *is* a Hückel number ($4n + 2$, where $n = 3$), and one might then predict that the periphery of pyrene would be aromatic by itself, in the absence of the internal double bond.

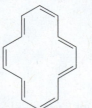

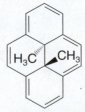

[14]Annulene   ***trans*-15,16-Dimethyldihydropyrene**

This prediction was confirmed when V. Boekelheide (University of Oregon) synthe-sized *trans*-15,16-dimethyldihydropyrene and showed that it is aromatic.

In addition to a signal downfield, the ¹H NMR spectrum of *trans*-15,16-dimethyldihydro-pyrene has a signal far upfield at δ −4.2. Account for the presence of this upfield signal.

PRACTICE PROBLEM 14.11

14.8B Nonbenzenoid Aromatic Compounds

Naphthalene, phenanthrene, and anthracene are examples of *benzenoid* aromatic com-pounds. On the other hand, the cyclopentadienyl anion, the cycloheptatrienyl cation, *trans*-15,16-dimethyldihydropyrene, and the aromatic annulenes (except for [6]annulene) are classified as **nonbenzenoid aromatic compounds**.

Another example of a *nonbenzenoid* aromatic hydrocarbon is the compound azulene. Azulene has a resonance energy of 205 kJ mol⁻¹. There is substantial separation of charge between the rings in azulene, as is indicated by the electrostatic potential map for azulene shown in Fig. 14.18. Factors related to aromaticity account for this property of azulene (see Practice Problem 14.12).

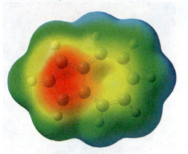

Azulene

FIGURE 14.18 A calculated electrostatic potential map for azulene. (Red areas are more negative and blue areas are less negative.)

Azulene has an appreciable dipole moment. Write resonance structures for azulene that explain this dipole moment and that help explain its aromaticity.

PRACTICE PROBLEM 14.12

14.8C Fullerenes

In 1990 W. Krätschmer (Max Planck Institute, Heidelberg), D. Huffman (University of Arizona), and their co-workers described the first practical synthesis of C₆₀, a molecule shaped like a soccer ball and called buckminsterfullerene. Formed by the resistive heating of graphite in an inert atmosphere, C₆₀ is a member of an exciting new group of aromatic compounds called **fullerenes**. Fullerenes are cagelike molecules with the geometry of a truncated icosahedron or geodesic dome, named after the architect Buckminster Fuller, renowned for his development of structures with geodesic domes. The structure of C₆₀ and its existence had been established five years earlier, by H. W. Kroto (University of Sussex), R. E. Smalley and R. F. Curl (Rice University), and their co-workers. Kroto, Curl, and Smalley had found both C₆₀ and C₇₀ (Fig. 14.19) as highly stable components

The Nobel Prize in Chemistry was awarded in 1996 to Professors Curl, Kroto, and Smalley for their discovery of fullerenes.

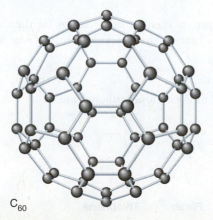

C₆₀

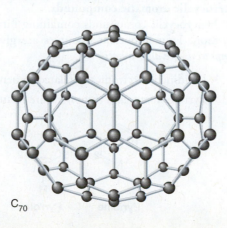

C₇₀

FIGURE 14.19 The structures of C₆₀ and C₇₀. (Reprinted with permission from Diederich, F., and Whetten, R. L. *Accounts of Chemical Research, Vol. 25,* pp. 119–126. Copyright 1992 American Chemical Society.)

of a mixture of carbon clusters formed by laser-vaporizing graphite. Since 1990 chemists have synthesized many other higher and lower fullerenes and have begun exploring their interesting chemistry.

Like a geodesic dome, a fullerene is composed of a network of pentagons and hexagons. To close into a spheroid, a fullerene must have exactly 12 five-membered faces, but the number of six-membered faces can vary widely. The structure of C_{60} has 20 hexagonal faces; C_{70} has 25. Each carbon of a fullerene is sp^2 hybridized and forms σ bonds to three other carbon atoms. The remaining electron at each carbon is delocalized into a system of molecular orbitals that gives the whole molecule aromatic character.

The chemistry of fullerenes is proving to be even more fascinating than their synthesis. Fullerenes have a high electron affinity and readily accept electrons from alkali metals to produce a new metallic phase—a "buckide" salt. One such salt, K_3C_{60}, is a stable metallic crystal consisting of a face-centered-cubic structure of "buckyballs" with a potassium ion in between; it becomes a superconductor when cooled below 18 K. Fullerenes have even been synthesized that have metal atoms in the interior of the carbon atom cage.

THE CHEMISTRY OF... Nanotubes

Nanotubes are a relatively new class of carbon-based materials related to buckminsterfullerenes. A **nanotube** is a structure that looks as though it were formed by rolling a sheet of graphite-like carbon (a flat network of fused benzene rings resembling chicken wire) into the shape of a tube and capping each end with half of a buckyball. Nanotubes are very tough—about 100 times as strong as steel. Besides their potential as strengtheners for new composite materials, some nanotubes have been shown to act as electrical conductors or semiconductors depending on their precise form. They are also being used as probe tips for analysis of DNA and proteins by atomic force microscopy (AFM). Many other applications have been envisioned for them as well, including use as molecular-size test tubes or capsules for drug delivery.

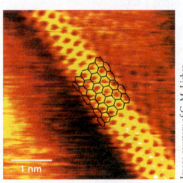

Image courtesy of C. M. Lieber, Harvard University

A network of benzene rings, highlighted in black on this scanning tunneling microscopy (STM) image, comprise the wall of a nanotube.

14.9 HETEROCYCLIC AROMATIC COMPOUNDS

Almost all of the cyclic molecules that we have discussed so far have had rings composed solely of carbon atoms. However, in many cyclic compounds an element other than carbon is present in the ring.

- Cyclic compounds that include an element other than carbon are called **heterocyclic compounds**.

Heterocyclic molecules are quite commonly encountered in nature. For this reason, and because some of these molecules are aromatic, we shall now describe a few examples of **heterocyclic aromatic compounds**.

Heterocyclic compounds containing nitrogen, oxygen, or sulfur are by far the most common. Four important examples are given here in their Kekulé forms. *These four compounds are all aromatic:*

- Pyridine is electronically related to benzene.
- Pyrrole, furan, and thiophene are related to the cyclopentadienyl anion.

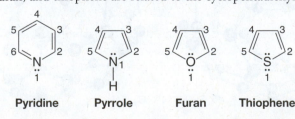

Pyridine   Pyrrole   Furan   Thiophene

The nitrogen atoms in molecules of both pyridine and pyrrole are sp^2 hybridized. In pyridine (Fig. 14.20) the sp^2-hybridized nitrogen donates one bonding electron to the π system. This electron, together with one from each of the five carbon atoms, gives pyridine a sextet of electrons like benzene. The two unshared electrons of the nitrogen of pyridine are in an sp^2 orbital that lies in the same plane as the atoms of the ring. This sp^2 orbital does not overlap with the p orbitals of the ring (it is, therefore, said to be *orthogonal* to the p orbitals). The unshared pair on nitrogen is not a part of the π system, and these electrons confer on pyridine the properties of a weak base.

In pyrrole (Fig. 14.21) the electrons are arranged differently. Because only four π electrons are contributed by the carbon atoms of the pyrrole ring, the sp^2-hybridized nitrogen must contribute two electrons to give an aromatic sextet. Because these electrons are a part of the aromatic sextet, they are not available for donation to a proton. Thus, in aqueous solution, pyrrole is not appreciably basic.

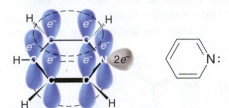

FIGURE 14.20 Pyridine is aromatic and weakly basic. Its nitrogen atom has an unshared electron pair in an sp^2 orbital (shown in gray) that is not part of the aromatic system.

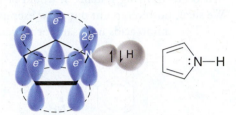

FIGURE 14.21 Pyrrole is aromatic but not basic. It does not have any unshared electron pairs. The electron pair on nitrogen is part of the aromatic system.

●●● **SOLVED PROBLEM 14.6**

Imidazole (at right) has two nitrogens. N3 is relatively basic (like the nitrogen of pyridine). N1 is relatively nonbasic (like the nitrogen of pyrrole). Explain the different basicities of these two nitrogens.

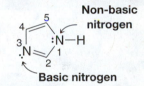

STRATEGY AND ANSWER: When imidazole accepts a proton at N3 the electron pair that accepts the proton is not a part of the π system of six electrons that makes imidazole aromatic. Consequently, the conjugate base that is formed is still aromatic (it is an aromatic cation) and retains its resonance energy of stabilization.

On the other hand, if imidazole were to accept a proton at N1 the resulting ion (which is not formed) would **not** be aromatic and would have much greater potential energy (its resonance stabilization would be lost). Hence, N1 is not appreciably basic.

Furan and thiophene are structurally quite similar to pyrrole. The oxygen atom in furan and the sulfur atom in thiophene are sp^2 hybridized. In both compounds the p orbital of the heteroatom donates two electrons to the π system. The oxygen and sulfur atoms of furan and thiophene carry an unshared pair of electrons in an sp^2 orbital (Fig. 14.22) that is orthogonal to the π system.

FIGURE 14.22 Furan and thiophene are aromatic. In each case, the heteroatom provides a pair of electrons to the aromatic system, but each also has an unshared electron pair in an sp^2 orbital that is not part of the aromatic system.

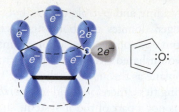

Furan

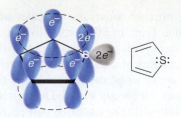

Thiophene

14.10 AROMATIC COMPOUNDS IN BIOCHEMISTRY

Compounds with aromatic rings occupy numerous and important positions in reactions that occur in living systems. It would be impossible to describe them all in this chapter. We shall, however, point out a few examples now and we shall see others later.

Two amino acids necessary for protein synthesis contain the benzene ring:

Phenylalanine   **Tyrosine**

A third aromatic amino acid, tryptophan, contains a benzene ring fused to a pyrrole ring. (This aromatic ring system is called an indole system; see Section 20.1B.)

Tryptophan   **Indole**

It appears that humans, because of the course of evolution, do not have the biochemical ability to synthesize the benzene ring. As a result, phenylalanine and tryptophan derivatives are essential in the human diet. Because tyrosine can be synthesized from phenylalanine in a reaction catalyzed by an enzyme known as *phenylalanine hydroxylase*, it is not essential in the diet as long as phenylalanine is present.

Heterocyclic aromatic compounds are also present in many biochemical systems. Derivatives of purine and pyrimidine are essential parts of DNA and RNA:

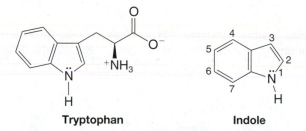

Purine   **Pyrimidine**

DNA is the molecule responsible for the storage of genetic information, and RNA is prominently involved in the synthesis of enzymes and other proteins (Chapter 25).

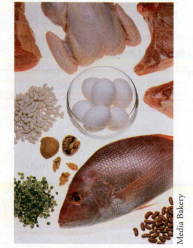

Dairy products, beans, fish, meat, and poultry are dietary sources of the essential amino acids.

Media Bakery

● ● ●

PRACTICE PROBLEM 14.13   (a) The —SH group is sometimes called the *mercapto group*. 6-Mercaptopurine is used in the treatment of acute leukemia. Write its structure. (b) Allopurinol, a compound used to treat gout, is 6-hydroxypurine. Write its structure.

Nicotinamide adenine dinucleotide, one of the most important coenzymes (Section 24.9) in biological oxidations and reductions, includes both a pyridine derivative (nicotinamide) and a purine derivative (adenine) in its structure. Its formula is shown in Fig. 14.23 as NAD^+, the oxidized form that contains the pyridinium aromatic ring. The reduced form of the coenzyme is NADH, in which the pyridine ring is no longer aromatic due to presence of an additional hydrogen and two electrons in the ring.

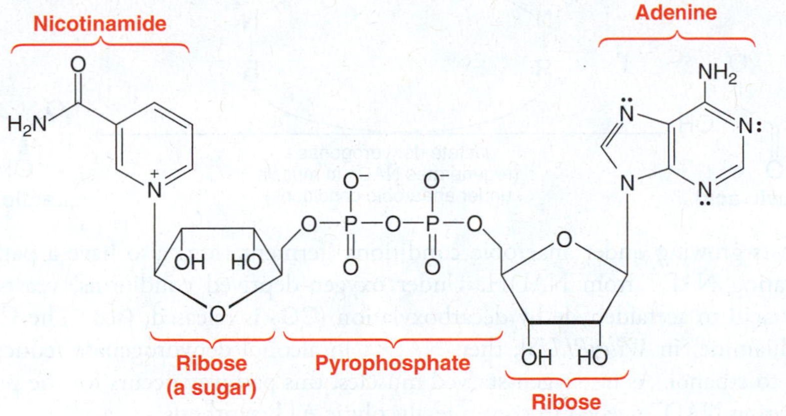

FIGURE 14.23 Nicotinamide adenine dinucleotide (NAD^+).

A key role of NAD^+ in metabolism is to serve as a coenzyme for glyceraldehyde-3-phosphate dehydrogenase (GAPDH) in glycolysis, the pathway by which glucose is broken down for energy production. In the reaction catalyzed by GAPDH (Fig. 14.24), the aldehyde group of glyceraldehyde-3-phosphate (GAP) is oxidized to a carboxyl group (incorporated as a phosphoric anhydride) in 1,3-bisphosphoglycerate (1,3-BPG). Concurrently, the aromatic pyridinium ring of NAD^+ is reduced to its higher energy form, NADH. One of the ways the chemical energy stored in the nonaromatic ring of NADH is used is in the mitochondria for the production of ATP, where cytochrome electron transport and oxidative phosphorylation take place. There, release of chemical energy from NADH by oxidation to the more stable aromatic form NAD^+ (and a proton) is coupled with the pumping of protons across the inner mitochondrial membrane. An electrochemical gradient is created across the mitochondrial membrane, which drives the synthesis of ATP by the enzyme ATP synthase.

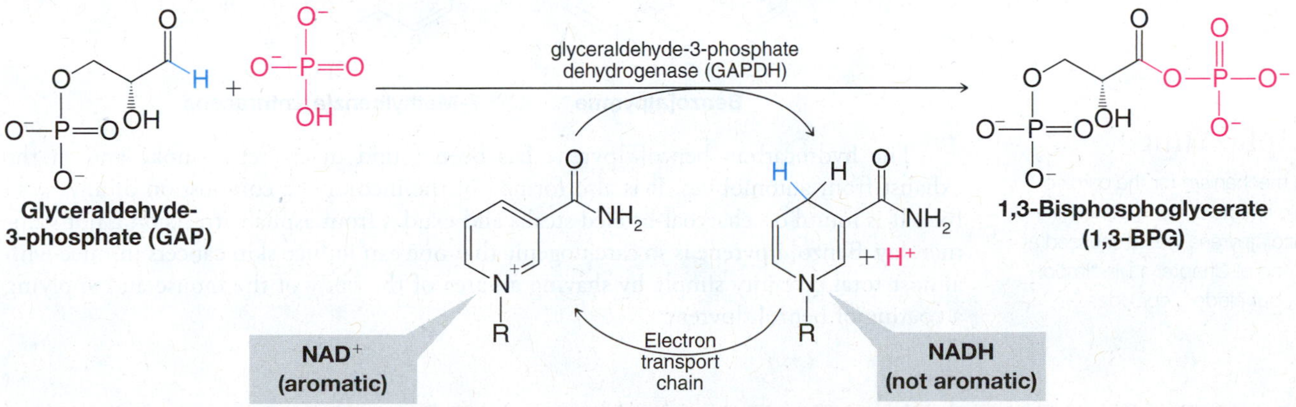

FIGURE 14.24 NAD^+, as the coenzyme in glyceraldehyde-3-phosphate dehydrogenase (GAPDH), is used to oxidize glyceraldehyde-3-phosphate (GAP) to 1,3-bisphosphoglycerate during the degradation of glucose in glycolysis. One of the ways that NADH can be reoxidized to NAD^+ is by the electron transport chain in mitochondria, where, under aerobic conditions, rearomatization of NADH helps to drive ATP synthesis.

The chemical energy stored in NADH is used to bring about many other essential biochemical reactions as well. NADH is part of an enzyme called lactate dehydrogenase that reduces the ketone group of pyruvic acid to the alcohol group of lactic acid. Here, the nonaromatic ring of NADH is converted to the aromatic ring of NAD^+. This process is important in muscles operating under oxygen-depleted conditions (anaerobic metabolism), where reduction of pyruvic acid to lactic acid by NADH serves to regenerate NAD^+ that is needed to continue glycolytic synthesis of ATP:

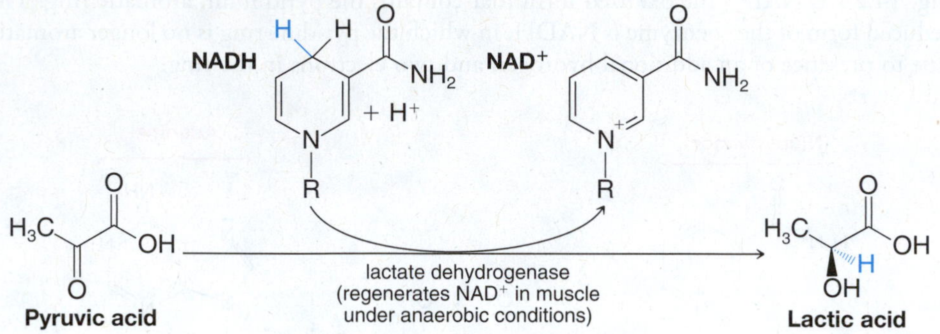

Pyruvic acid   lactate dehydrogenase (regenerates NAD^+ in muscle under anaerobic conditions)   **Lactic acid**

Yeasts growing under anaerobic conditions (fermentation) also have a pathway for regenerating NAD^+ from NADH. Under oxygen-deprived conditions, yeasts convert pyruvic acid to acetaldehyde by decarboxylation (CO_2 is released, (see "The Chemistry of... Thiamine" in *WileyPLUS*); then NADH in alcohol dehydrogenase reduces acetaldehyde to ethanol. As in oxygen-starved muscles, this pathway occurs for the purpose of regenerating NAD^+ needed to continue glycolytic ATP synthesis:

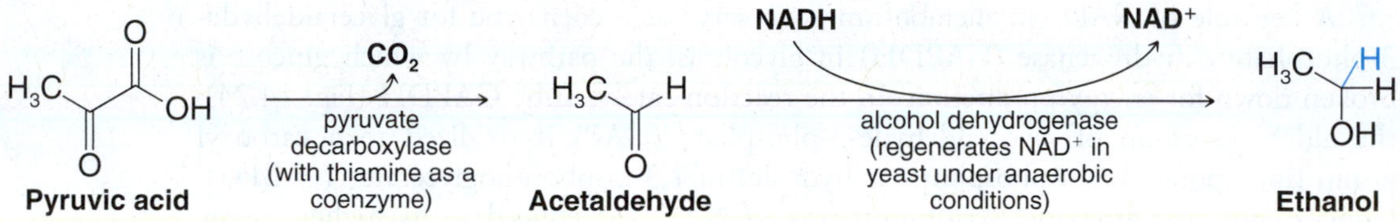

Pyruvic acid   pyruvate decarboxylase (with thiamine as a coenzyme)   **Acetaldehyde**   alcohol dehydrogenase (regenerates NAD^+ in yeast under anaerobic conditions)   **Ethanol**

Although many aromatic compounds are essential to life, others are hazardous. Many are quite toxic, and several benzenoid compounds, including benzene itself, are **carcinogenic.** Two other examples are benzo[*a*]pyrene and 7-methylbenz[*a*]anthracene:

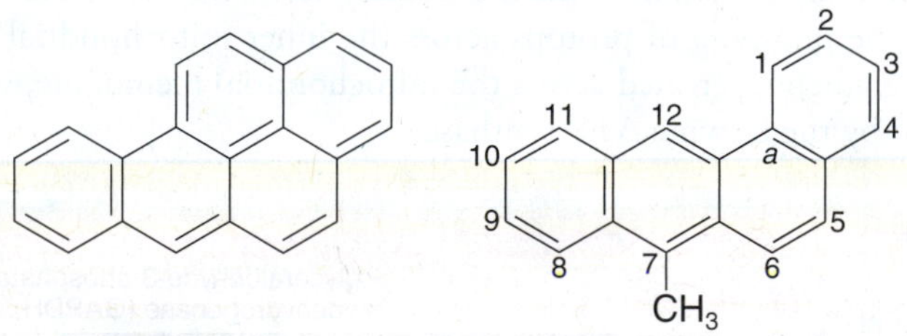

Benzo[a]pyrene   **7-Methylbenz[a]anthracene**

Helpful Hint

The mechanism for the carcinogenic effects of compounds like benzo[a]pyrene was discussed at the end of Chapter 11 in "Important, but hidden, epoxides."

The hydrocarbon benzo[*a*]pyrene has been found in cigarette smoke and in the exhaust from automobiles. It is also formed in the incomplete combustion of any fossil fuel. It is found on charcoal-broiled steaks and exudes from asphalt streets on a hot summer day. Benzo[*a*]pyrene is so carcinogenic that one can induce skin cancers in mice with almost total certainty simply by shaving an area of the body of the mouse and applying a coating of benzo[*a*]pyrene.

14.11 SPECTROSCOPY OF AROMATIC COMPOUNDS

14.11A ^{1}H NMR Spectra

- The ring hydrogens of benzene derivatives absorb downfield in the region between δ 6.0 and δ 9.5.

In Section 14.7C we found that absorption takes place far downfield because a ring current generated in the benzene ring creates a magnetic field, called "the induced field," which reinforces the applied magnetic field at the position of the protons of the ring. This reinforcement causes the protons of benzene to be highly deshielded.

We also learned in Section 14.7C that internal hydrogens of large-ring **aromatic** compounds such as [18]annulene, because of their position, are highly shielded by this induced field. They therefore absorb at unusually low frequency, often at negative delta values.

14.11B ^{13}C NMR Spectra

- The carbon atoms of benzene rings generally absorb in the δ 100–170 region of ^{13}C NMR spectra.

Figure 14.25 gives the broadband proton-decoupled ^{13}C NMR spectrum of 4-N,N-diethylaminobenzaldehyde and permits an exercise in making ^{13}C assignments of a compound with both aromatic and aliphatic carbon atoms.

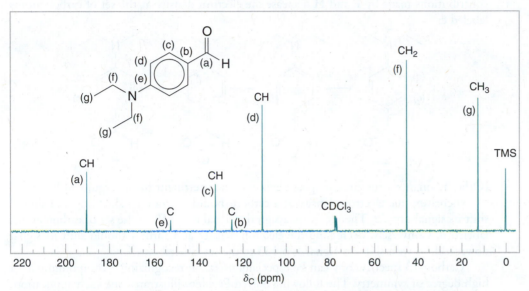

FIGURE 14.25 The broadband proton-decoupled ^{13}C NMR spectrum of 4-N,N-diethylaminobenz-aldehyde. DEPT information and carbon assignments are shown by each peak.

The DEPT spectra (not given to save space) show that the signal at δ 45 arises from a CH_2 group and the one at δ 13 arises from a CH_3 group. This allows us to assign these two signals immediately to the two carbons of the equivalent ethyl groups.

The signals at δ 126 and δ 153 appear in the DEPT spectra as carbon atoms that do not bear hydrogen atoms and are assigned to carbons b and e (see Fig. 14.25). The greater electronegativity of nitrogen (when compared to carbon) causes the signal from e to be further downfield (at δ 153). The signal at δ 190 appears as a CH group in the DEPT spectra and arises from the carbon of the aldehyde group. Its chemical shift is the most downfield of all the peaks because of the great electronegativity of its oxygen and because the second resonance structure below contributes to the hybrid. Both factors cause the electron density at this carbon to be very low, and, therefore, this carbon is strongly deshielded.

Resonance contributors for an aldehyde group

This leaves the signals at δ 112 and δ 133 and the two sets of carbon atoms of the benzene ring labeled c and d to be accounted for. Both signals are indicated as CH groups in the DEPT spectra. But which signal belongs to which set of carbon atoms? Here we find another interesting application of resonance theory.

If we write resonance structures **A–D** involving the unshared electron pair of the amino group, we see that contributions made by **B** and **D** increase the electron density at the set of carbon atoms labeled d:

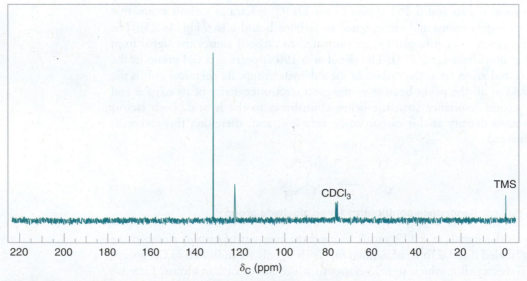

On the other hand, writing structures **E–H** involving the aldehyde group shows us that contributions made by **F** and **H** decrease the electron density at the set of carbon atoms labeled c:

(Other resonance structures are possible but are not pertinent to the argument here.)

Increasing the electron density at a carbon should increase its shielding and should shift its signal upfield. Therefore, we assign the signal at δ 112 to the set of carbon atoms labeled d. Conversely, decreasing the electron density at a carbon should shift its signal downfield, so we assign the signal at δ 133 to the set labeled c.

Carbon-13 spectroscopy can be especially useful in recognizing a compound with a high degree of symmetry. The following Solved Problem illustrates one such application.

●●● **SOLVED PROBLEM 14.7**

The broadband proton-decoupled ^{13}C spectrum given in Fig. 14.26 is of a tribromobenzene ($C_6H_3Br_3$). Which tribromobenzene is it?

FIGURE 14.26 The broadband proton-decoupled ^{13}C NMR spectrum of a tribromobenzene.

ANSWER: There are three possible tribromobenzenes:

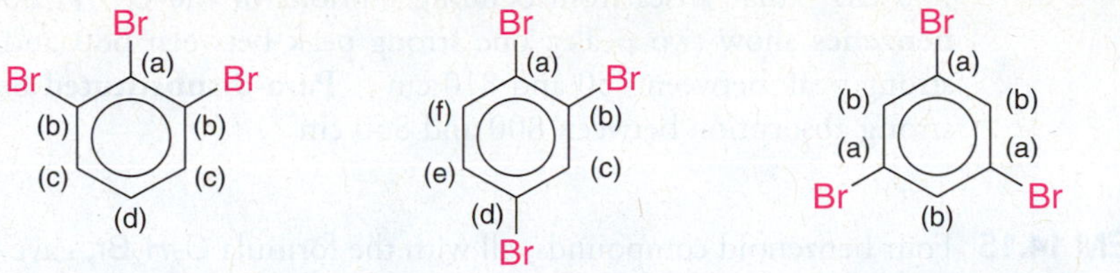

1,2,3-Tribromobenzene   **1,2,4-Tribromobenzene**   **1,3,5-Tribromobenzene**

Our spectrum (Fig. 14.26) consists of only two signals, indicating that only two different types of carbon atoms are present in the compound. Only 1,3,5-tribromobenzene has a degree of symmetry such that it would give only two signals, and, therefore, it is the correct answer. 1,2,3-Tribromobenzene would give four ^{13}C signals and 1,2,4-tribromobenzene would give six.

Explain how ^{13}C NMR spectroscopy could be used to distinguish the *ortho-*, *meta-*, and *para-*dibromobenzene isomers one from another.

PRACTICE PROBLEM 14.14

14.11C Infrared Spectra of Substituted Benzenes

Benzene derivatives give characteristic C—H stretching peaks near 3030 cm^{-1} (Table 2.7). Stretching motions of the benzene ring can give as many as four bands in the 1450–1600-cm^{-1} region, with two peaks near 1500 and 1600 cm^{-1} being stronger.

Absorption peaks in the 680–860-cm^{-1} region from out-of-plane C—H bending can often (but not always) be used to characterize the substitution patterns of benzene compounds (Table 14.1). **Monosubstituted benzenes** give two very strong peaks, between 690 and 710 cm^{-1} and between 730 and 770 cm^{-1}.

TABLE 14.1 INFRARED ABSORPTIONS IN THE 680–860-CM^{-1} REGION^a

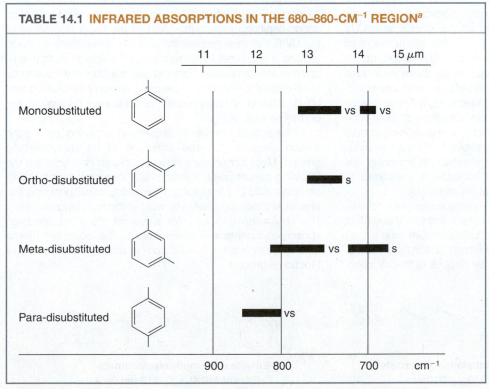

^{a}s, strong; vs, very strong.

Ortho-disubstituted benzenes show a strong absorption peak between 735 and 770 cm^{-1} that arises from bending motions of the C—H bonds. **Meta-disubstituted benzenes** show two peaks: one strong peak between 680 and 725 cm^{-1} and one very strong peak between 750 and 810 cm^{-1}. **Para-disubstituted benzenes** give a single very strong absorption between 800 and 860 cm^{-1}.

● ● ●

PRACTICE PROBLEM 14.15 Four benzenoid compounds, all with the formula C$_7$H$_7$Br, gave the following IR peaks in the 680–860-cm^{-1} region:

A, 740 cm^{-1} (strong)   **C**, 680 cm^{-1} (strong) and 760 cm^{-1} (very strong)

B, 800 cm^{-1} (very strong)   **D**, 693 cm^{-1} (very strong) and 765 cm^{-1} (very strong)

Propose structures for **A**, **B**, **C**, and **D**.

14.11D Ultraviolet–Visible Spectra of Aromatic Compounds

The conjugated π electrons of a benzene ring give characteristic ultraviolet absorptions that indicate the presence of a benzene ring in an unknown compound. One absorption band of moderate intensity occurs near 205 nm and another, less intense band appears in the 250–275-nm range. Conjugation outside the benzene ring leads to absorptions at other wavelengths.

THE CHEMISTRY OF... Sunscreens (Catching the Sun's Rays and What Happens to Them)

The use of sunscreens in recent years has increased due to heightened concern over the risk of skin cancer and other conditions caused by exposure to UV radiation. In DNA, for example, UV radiation can cause adjacent thymine bases to form mutagenic dimers. Sunscreens afford protection from UV radiation because they contain aromatic molecules that absorb energy in the UV region of the electromagnetic spectrum. Absorption of UV radiation by these molecules promotes π and nonbonding electrons to higher energy levels (Section 13.9C), after which the energy is dissipated by relaxation through molecular vibration. In essence, the UV radiation is converted to heat (IR radiation).

Sunscreens are classified according to the portion of the UV spectrum where their maximum absorption occurs. Three regions of the UV spec-

trum are typically discussed. The region from 320 to 400 nm is called UV-A, the region from 280 to 320 nm is called UV-B, and the region from 100 to 280 nm is called UV-C. The UV-C region is potentially the most dangerous because it encompasses the shortest UV wavelengths and is therefore of the highest energy. However, ozone and other components in Earth's atmosphere absorb UV-C wavelengths, and thus we are protected from radiation in this part of the spectrum so long as Earth's atmosphere is not compromised further by ozone-depleting pollutants. Most of the UV-A and some of the UV-B radiation passes through the atmosphere to reach us, and it is against these regions of the spectrum that sunscreens are formulated. Tanning and sunburn are caused by UV-B radiation. Risk of skin cancer is primarily associated with UV-B radiation, although some UV-A wavelengths may be important as well.

The specific range of protection provided by a sunscreen depends on the structure of its UV-absorbing groups. Most sunscreens have structures derived from the following parent compounds: *p*-aminobenzoic acid (PABA), cinnamic acid (3-phenylpropenoic acid), benzophenone (diphenyl ketone), and salicylic acid (*o*-hydroxybenzoic acid). The structures and λ_{max} for a few of the most common sunscreen agents are given below. The common theme among them is an aromatic core in conjugation with other functional groups.

A UV-A and UV-B sunscreen product whose active ingredients are octyl 4-methoxycinnamate and 2-hydroxy-4-methoxy-benzophenone (oxybenzone).

**Octyl 4-*N*,*N*-dimethylaminobenzoate
(Padimate O),** λ_{max} **310 nm**

**2-Ethylhexyl 4-methoxycinnamate
(Parsol MCX),** λ_{max} **310 nm**

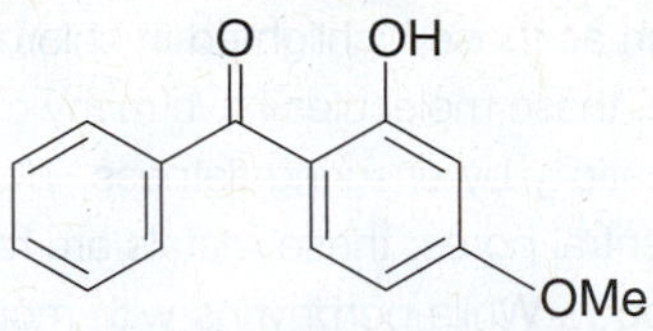

**2-Hydroxy-4-methoxybenzophenone
(Oxybenzone),** λ_{max} **288 and 325 nm**

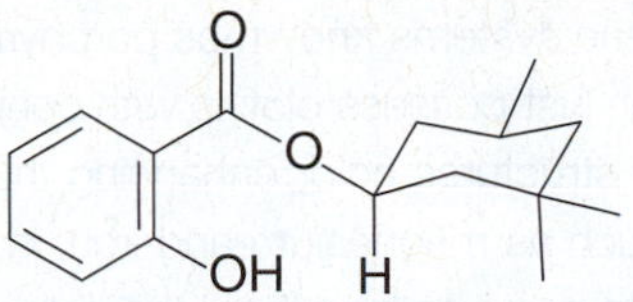

**Homomenthyl salicylate
(Homosalate),** λ_{max} **309 nm**

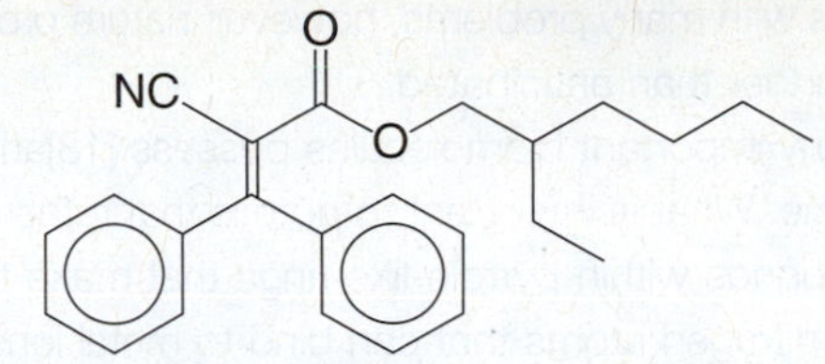

**2-Ethylhexyl 2-cyano-3,3-diphenylacrylate
(Octocrylene),** λ_{max} **310 nm**

14.11E Mass Spectra of Aromatic Compounds

The major ion in the mass spectrum of an alkyl-substituted benzene is often m/z 91 ($C_6H_5CH_2^+$), resulting from cleavage between the first and second carbons of the alkyl chain attached to the ring. The ion presumably originates as a benzylic cation that rearranges to a tropylium cation ($C_7H_7^+$, Section 14.7D). Another ion frequently seen in mass spectra of monoalkylbenzene compounds is m/z 77, corresponding to $C_6H_5^+$.

[WHY Do These Topics Matter?

MAKING EVEN LARGER AROMATIC MOLECULES

Although Hückel's rule can readily be used to suggest whether or not a molecule might be aromatic, we know that it is not always accurate because it cannot predict molecular shape. For example, as we discussed previously, [10]annulene is not aromatic, because two hydrogen atoms would have to occupy the same space for the system to be flat; instead, the molecule twists out of conjugation to allow space for the hydrogens. While larger molecules, such as [18]annulene, can overcome that problem and are aromatic as a result ($4n + 2$, $n = 4$), for a long time it was expected that there might be an upper limit to how large a molecule could be and still have aromaticity as predicted by Hückel's rule. In fact, until the late 1980s, a system with 22 π electrons appeared to be the maximum that could still retain aromaticity ($4n + 2$, $n = 5$) based on both experimental and theoretical calculations. Larger rings were thought to be too flexible, rendering them unable to maintain a structure that consistently placed all of their π systems in a flat, planar, conjugated

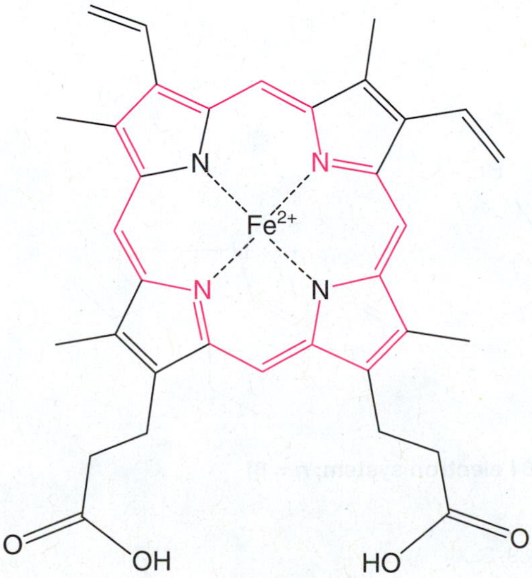

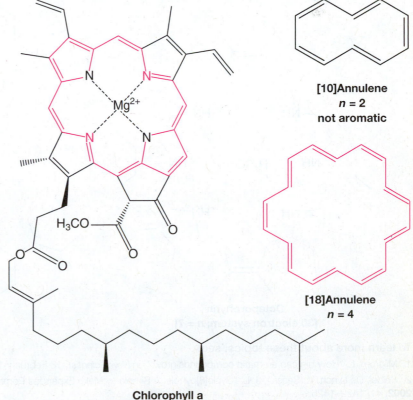

**[10]Annulene
$n = 2$
not aromatic**

**[18]Annulene
$n = 4$**

**Heme
[18 electron system; $n = 4$]**

**Chlorophyll a
[18 electron system; $n = 4$]**

(continues on the next page)

array. As with many problems, however, nature provided chemists with the inspiration for how to push the boundaries of aromaticity much further than anticipated.

Many important biomolecules possess [18]annulene systems known as porphyrins, such as those highlighted in chlorophyll a and heme. What is important to note is that rather than just possess olefins with conjugation, these molecules have many of those double bonds within pyrrole-like rings that make their structures rigid, enhancing their aromaticity by ensuring flatness. They also provide nitrogen atoms that can bind to metal ions, such as magnesium and iron, in their central pores; these metals are required to perform the redox chemistry of photosynthesis or to carry oxygen to all the cells of the body. While porphyrins with more than 18 electrons have not yet been found in nature, the idea of using rings to help enhance ridigity is what inspired chemists to design artificial porphyrins that are not only larger, but also break the hypothesized 22-electron aromaticity barrier.

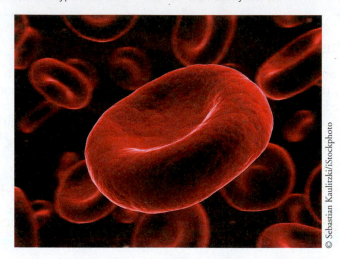

One example, shown below, is from molecules synthesized in 2001 that contain 8 pyrrole rings. This system has 30 electrons based on the highlighted atoms, showing that n can equal 7 in the Hückel paradigm of $4n + 2$. Another example comes from chemists in Germany who made a molecule containing 34 π electrons; ^{1}H NMR spectroscopy revealed that it was aromatic based on ring current, meaning that Hückel's rule can apply to systems as large as $n = 8$. Its central cavity is so big, in fact, that a number of molecules can fit within it. Where the upper limit to Hückel's rule is, no one knows, but clearly even large molecules have the potential to show the unique properties of aromaticity.

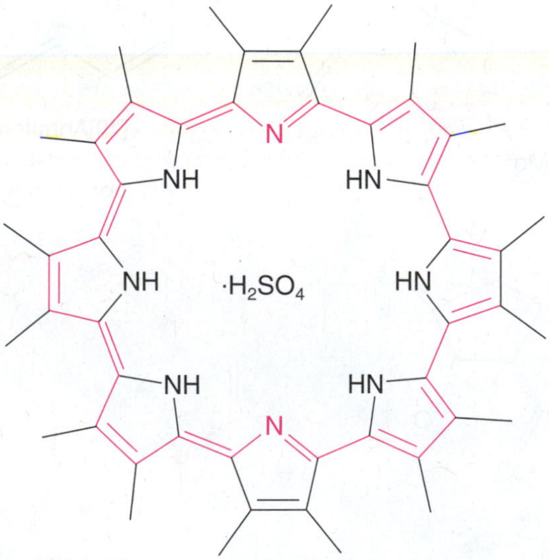

Octaporphyrin
[30 electron system; $n = 7$]

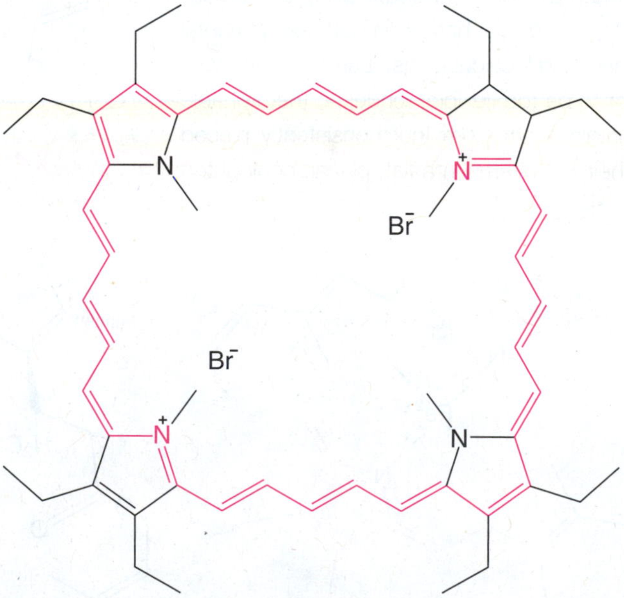

[34 electron system; $n = 8$]

To learn more about these topics, see:

1. Milgrom, L. "How big can aromatic compounds grow?" in *New Scientist*, 18 February 1989, 32.

2. Seidel, D.; Lynch, V.; Sessler, J. L. "Cyclo[8]pyrrole: A Simple-to-Make Expanded Porphyrin with No Meso Bridges" in *Angew. Chem. Int. Ed.* **2002**, *41*, 1422–1425.

3. Knubel, G.; Franck, B. Biomimetic Synthesis of an Octavinylogous Porphyrin with an Aromatic [34]Annulene System in *Angew. Chem. Int. Ed. Engl.* **1988**, *27*, 1170–1172.

SUMMARY AND REVIEW TOOLS

The study aids for this chapter include key terms and concepts (which are hyperlinked to the Glossary from the bold, blue terms in the *WileyPLUS* version of the book at wileyplus.com) and a Concept Map relating to properties and reactivity of aromatic compounds.

PROBLEMS WILEY PLUS

Note to Instructors: Many of the homework problems are available for assignment via *WileyPLUS*, an online teaching and learning solution.

NOMENCLATURE

14.16 Write structural formulas for each of the following:

(a) 3-Nitrobenzoic acid
(b) *p*-Bromotoluene
(c) *o*-Dibromobenzene
(d) *m*-Dinitrobenzene
(e) 3,5-Dinitrophenol
(f) *p*-Nitrobenzoic acid

(g) 3-Chloro-1-ethoxybenzene
(h) *p*-Chlorobenzenesulfonic acid
(i) Methyl *p*-toluenesulfonate
(j) Benzyl bromide
(k) *p*-Nitroaniline
(l) *o*-Xylene

(m) *tert*-Butylbenzene
(n) *p*-Methylphenol
(o) *p*-Bromoacetophenone
(p) 3-Phenylcyclohexanol
(q) 2-Methyl-3-phenyl-1-butanol
(r) *o*-Chloroanisole

14.17 Write structural formulas and give acceptable names for all representatives of the following:

(a) Tribromobenzenes
(b) Dichlorophenols

(c) Nitroanilines
(d) Methylbenzenesulfonic acids

(e) Isomers of C_6H_5—C_4H_9

AROMATICITY

14.18 Which of the following molecules would you expect to be aromatic?

(a) 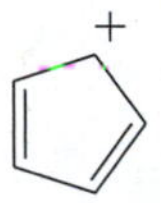   (c) 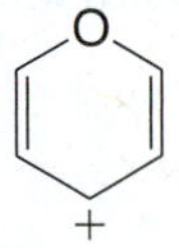   (e) 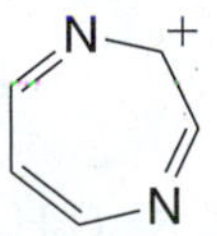   (g) 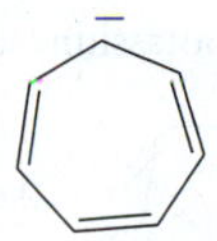   (i) 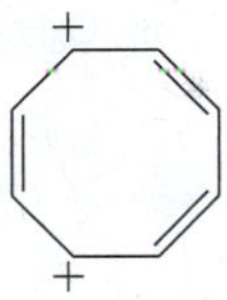   (k)

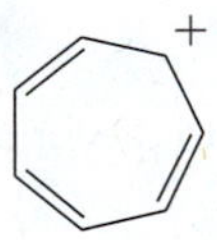

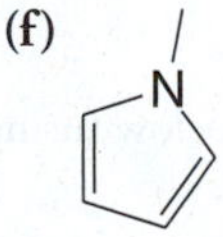

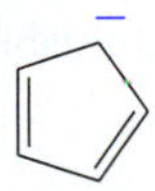

(b)   (d)   (f)   (h)   (j)   (l)

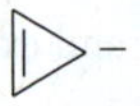

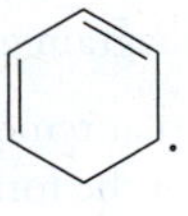

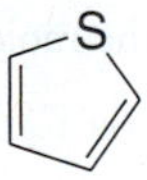

14.19 Use the polygon-and-circle method to draw an orbital diagram for each of the following compounds.

(a) 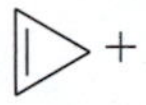 + (b)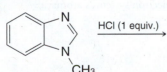

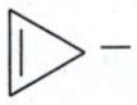

14.20 Write the structure of the product formed when each of the following compounds reacts with one molar equivalent of HCl.

(a) 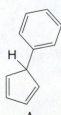 (b)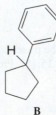

14.21 Which of the hydrogen atoms shown below is more acidic? Explain your answer.

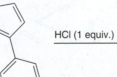

A   B

14.22 The rings below are joined by a double bond that undergoes cis–trans isomerization much more readily than the bond of a typical alkene. Provide an explanation.

14.23 Although Hückel's rule (Section 14.7) strictly applies only to monocyclic compounds, it does appear to have application to certain bicyclic compounds, if one assumes use of resonance structures involving only the perimeter double bonds, as shown with one resonance contributor for naphthalene below.

Both naphthalene (Section 14.8A) and azulene (Section 14.8B) have 10 π electrons and are aromatic. Pentalene (below) is apparently antiaromatic and is unstable even at −100 °C. Heptalene has been made but it adds bromine, it reacts with acids, and it is not planar. Is Hückel's rule applicable to these compounds? If so, explain their lack of aromaticity.

Pentalene   **Heptalene**

14.24

(a) In 1960 T. Katz (Columbia University) showed that cyclooctatetraene adds two electrons when treated with potassium metal and forms a stable, planar dianion, $C_8H_8^{2-}$ (as the dipotassium salt):

$$\xrightarrow[\text{THF}]{\text{2 equiv. K}} \quad 2\ K^+\ C_8H_8^{2-}$$

Use the molecular orbital diagram given in Fig. 14.7 and explain this result.

(b) In 1964 Katz also showed that removing two protons from the compound below (using butyllithium as the base) leads to the formation of a stable dianion with the formula $C_8H_6^{2-}$ (as the dilithium salt).

$$\xrightarrow{\text{2 BuLi}} \quad 2\ Li\ +\ C_8H_6^{2-}$$

Propose a reasonable structure for the product and explain why it is stable.

14.25 Although none of the [10]annulenes given in Section 14.7B is aromatic, the following 10 π-electron system is aromatic:

What factor makes this possible?

14.26 Cycloheptatrienone (**I**) is very stable. Cyclopentadienone (**II**) by contrast is quite unstable and rapidly undergoes a Diels–Alder reaction with itself.

(a) Propose an explanation for the different stabilities of these two compounds.

(b) Write the structure of the Diels–Alder adduct of cyclopentadienone.

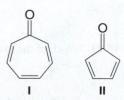

I   II

14.27 5-Chloro-1,3-cyclopentadiene (below) undergoes S_N1 solvolysis in the presence of silver ion extremely slowly even though the chlorine is doubly allylic and allylic halides normally ionize readily (Section 15.15). Provide an explanation for this behavior.

14.28 Explain the following: (**a**) Cyclononatetraenyl anion is planar (in spite of the angle strain involved) and appears to be aromatic. (**b**) Although [16]annulene is not aromatic, it adds two electrons readily to form an aromatic dianion.

14.29 Furan possesses less aromatic character than benzene as measured by their resonance energies (96 kJ mol^{-1} for furan; 151 kJ mol^{-1} for benzene). What reaction have we studied earlier that shows that furan is less aromatic than benzene and can react in a way characteristic of some dienes?

SPECTROSCOPY AND STRUCTURE ELUCIDATION

14.30 For each of the pairs below, predict specific aspects in their ^{1}H NMR spectra that would allow you to distinguish one compound from the other.

(a)

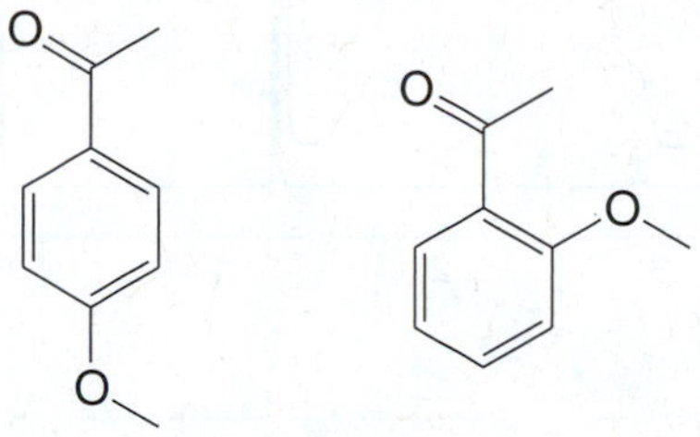

(c)

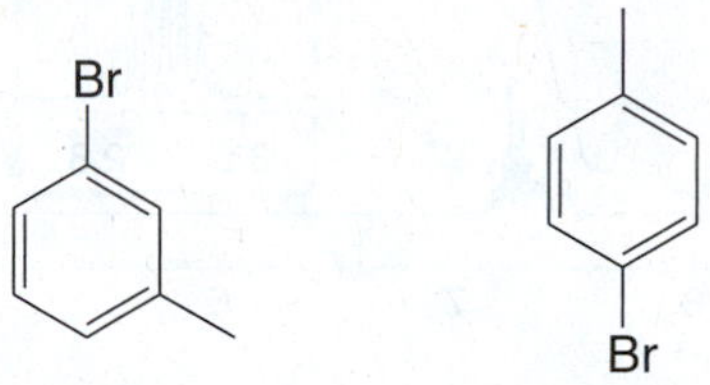

(b)

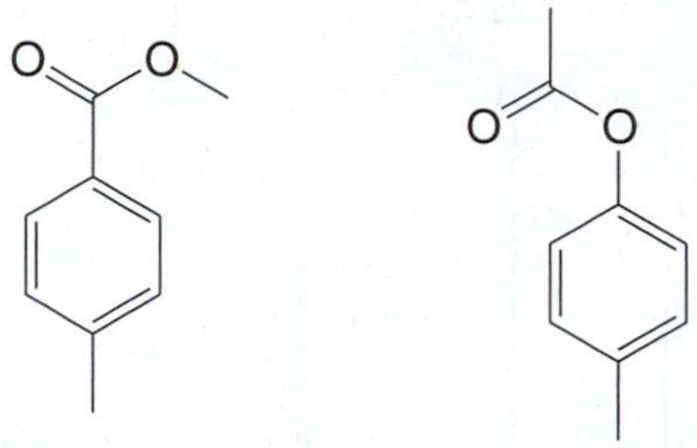

14.31 Assign structures to each of the compounds **A**, **B**, and **C** whose ^{1}H NMR spectra are shown in Fig. 14.27.

14.32 The ^{1}H NMR spectrum of cyclooctatetraene consists of a single line located at δ 5.78. What does the location of this signal suggest about electron delocalization in cyclooctatetraene?

14.33 Give a structure for compound **F** that is consistent with the ^{1}H NMR spectrum in Fig. 14.28 and IR absorptions at 3020, 2965, 2940, 2870, 1517, 1463, and 818 cm^{-1}.

14.34 A compound (**L**) with the molecular formula C_9H_{10} reacts with bromine and gives an IR absorption spectrum that includes the following absorption peaks: 3035 cm^{-1}(m), 3020 cm^{-1}(m), 2925 cm^{-1}(m), 2853 cm^{-1}(w), 1640 cm^{-1}(m), 990 cm^{-1}(s), 915 cm^{-1}(s), 740 cm^{-1}(s), 695 cm^{-1}(s). The ^{1}H NMR spectrum of **L** consists of:

Doublet δ 3.1 (2H) Multiplet δ 5.1 Multiplet δ 7.1 (5H)
Multiplet δ 4.8 Multiplet δ 5.8

The UV spectrum shows a maximum at 255 nm. Propose a structure for compound **L** and make assignments for each of the IR peaks.

14.35 Compound **M** has the molecular formula C_9H_{12}. The ^{1}H NMR spectrum of **M** is given in Fig. 14.29 and the IR spectrum in Fig. 14.30. Propose a structure for **M**.

14.36 A compound (**N**) with the molecular formula $C_9H_{10}O$ reacts with osmium tetroxide. The ^{1}H NMR spectrum of **N** is shown in Fig. 14.31 and the IR spectrum of **N** is shown in Fig. 14.32. Propose a structure for **N**.

14.37 The IR and ^{1}H NMR spectra for compound **X** (C_8H_{10}) are given in Fig. 14.33. Propose a structure for compound **X**.

14.38 The IR and ^{1}H NMR spectra of compound **Y** ($C_9H_{12}O$) are given in Fig. 14.34. Propose a structure for **Y**.

14.39

(a) How many signals would you expect to find in the ^{1}H NMR spectrum of caffeine?

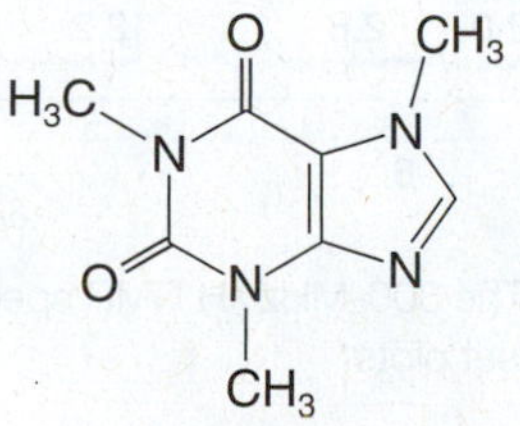

Caffeine

(b) What characteristic peaks would you expect to find in the IR spectrum of caffeine?

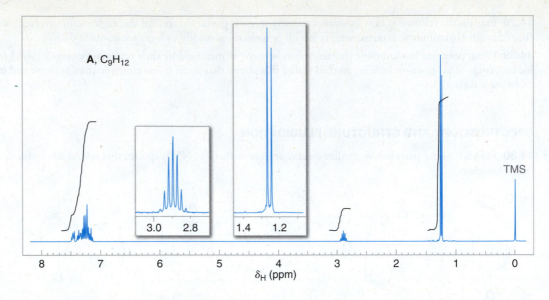

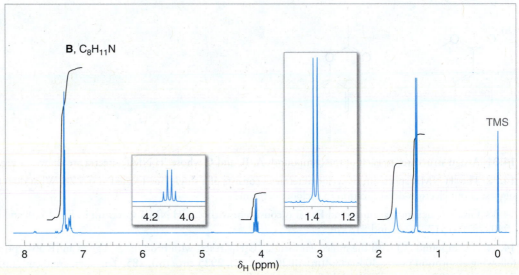

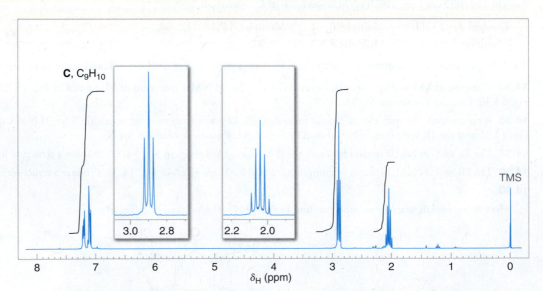

FIGURE 14.27 The 300-MHz ^{1}H NMR spectra for Problem 14.31. Expansions of the signals are shown in the offset plots.

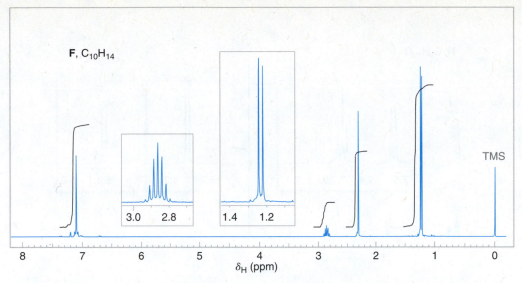

FIGURE 14.28 The 300-MHz ^{1}H NMR spectrum of compound **F**, Problem 14.33. Expansions of the signals are shown in the offset plots.

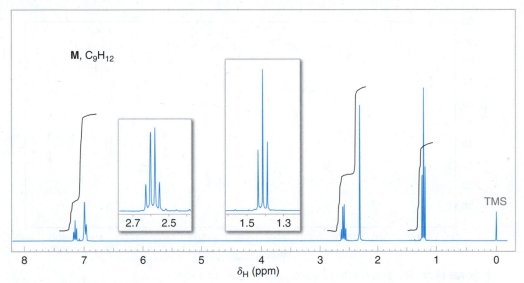

FIGURE 14.29 The 300-MHz ^{1}H NMR spectrum of compound **M**, Problem 14.35. Expansions of the signals are shown in the offset plots.

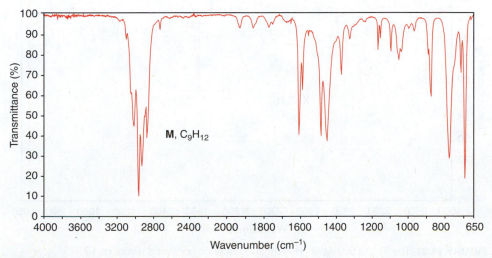

FIGURE 14.30 The IR spectrum of compound **M**, Problem 14.35.

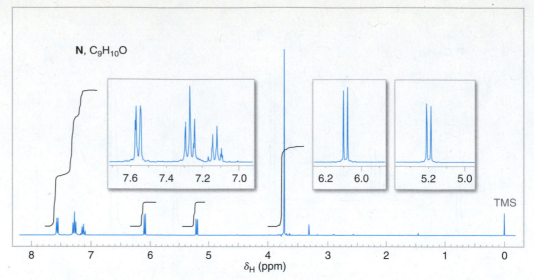

FIGURE 14.31 The 300-MHz ^{1}H NMR spectrum of compound **N**, Problem 14.36. Expansions of the signals are shown in the offset plots.

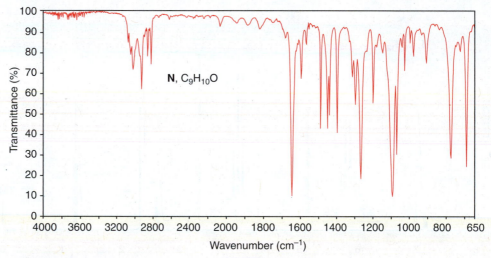

FIGURE 14.32 The IR spectrum of compound **N**, Problem 14.36.

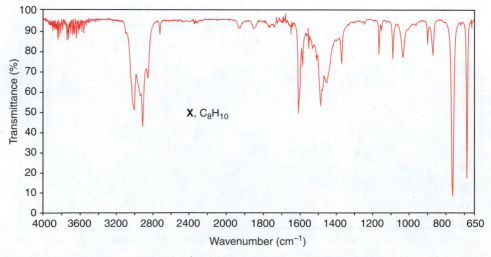

FIGURE 14.33 The IR and 300-MHz ^{1}H NMR spectra of compound **X**, Problem 14.37. Expansions of the signals are shown in the offset plots.

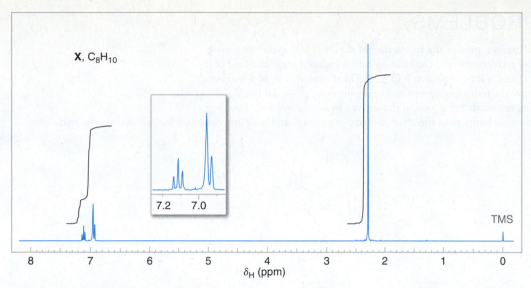

FIGURE 14.33 *(Continued)*

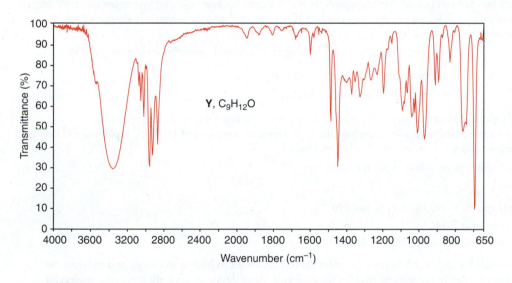

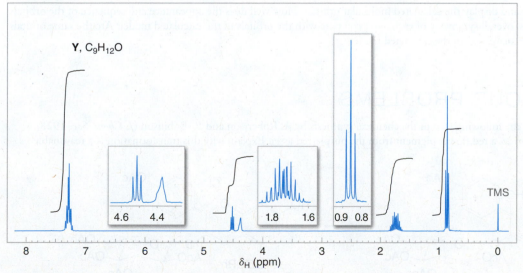

FIGURE 14.34 The IR and 300-MHz ^{1}H NMR spectra (next page) of compound **Y**, Problem 14.38. Expansions of the signals are shown in the offset plots.

CHALLENGE PROBLEMS

14.40 Given the following information, predict the appearance of the ^{1}H NMR spectrum arising from the vinyl hydrogen atoms of *p*-chlorostyrene. Deshielding by the induced magnetic field of the ring is greatest at proton c (δ 6.7) and is least at proton b (δ 5.3). The chemical shift of a is about δ 5.7. The coupling constants have the following approximate magnitudes: $J_{ac} \cong 18$ Hz, $J_{bc} \cong 11$ Hz, and $J_{ab} \cong 2$ Hz. (These coupling constants are typical of those given by vinylic systems: coupling constants for trans hydrogen atoms are larger than those for cis hydrogen atoms, and coupling constants for geminal vinylic hydrogen atoms are very small.)

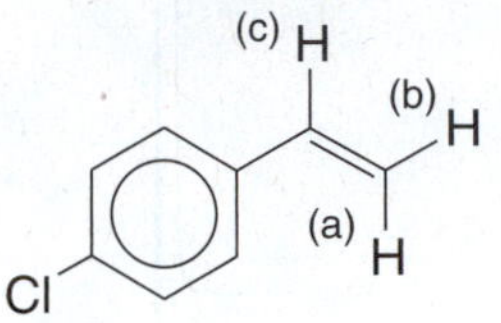

14.41 Consider these reactions:

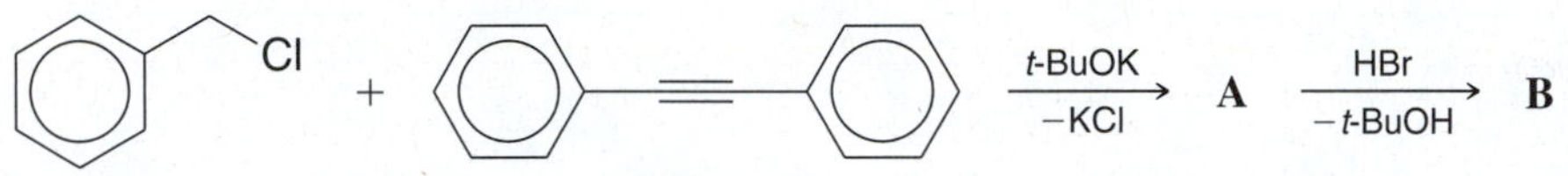

The intermediate **A** is a covalently bonded compound that has typical ^{1}H NMR signals for aromatic ring hydrogens and only one additional signal at δ 1.21, with an area ratio of 5:3, respectively. Final product **B** is ionic and has only aromatic hydrogen signals.
What are the structures of **A** and **B**?

14.42 The final product of this sequence, **D**, is an orange, crystalline solid melting at 174 °C and having molecular weight 186:

$$\text{Cyclopentadiene} + \text{Na} \longrightarrow \textbf{C} + \text{H}_2$$

$$2\,\textbf{C} + \text{FeCl}_2 \longrightarrow \textbf{D} + 2\,\text{NaCl}$$

In its ^{1}H and ^{13}C NMR spectra, product **D** shows only one kind of hydrogen and only one kind of carbon, respectively.
Draw the structure of **C** and make a structural suggestion as to how the high degree of symmetry of **D** can be explained. (**D** belongs to a group of compounds named after something you might get at a deli for lunch.)

14.43 Compound **E** has the spectral features given below. What is its structure?

MS (*m/z*):   M$^{+\cdot}$ 202
IR (cm^{-1}): 3030–3080, 2150 (very weak), 1600, 1490, 760, and 690
^{1}**H NMR** (δ): narrow multiplet centered at 7.34
UV (nm): 287 ($\epsilon = 25{,}000$), 305 ($\epsilon = 36{,}000$), and 326 ($\epsilon = 33{,}000$)

14.44 Draw all of the π molecular orbitals for (3*E*)-1,3,5-hexatriene, order them from lowest to highest in energy, and indicate the number of electrons that would be found in each in the ground state for the molecule. After doing so, open the computer molecular model for (3*E*)-1,3,5-hexatriene and display the calculated molecular orbitals. How well does the appearance and sequence of the orbitals you drew (e.g., number of nodes, overall symmetry of each, etc.) compare with the orbitals in the calculated model? Are the same orbitals populated with electrons in your analysis as in the calculated model?

LEARNING GROUP PROBLEMS

1. Write mechanism arrows for the following step in the chemical synthesis by A. Robertson and R. Robinson (*J. Chem. Soc.* **1928**, 1455–1472) of callistephin chloride, a red flower pigment from the purple-red aster. Explain why this transformation is a reasonable process.

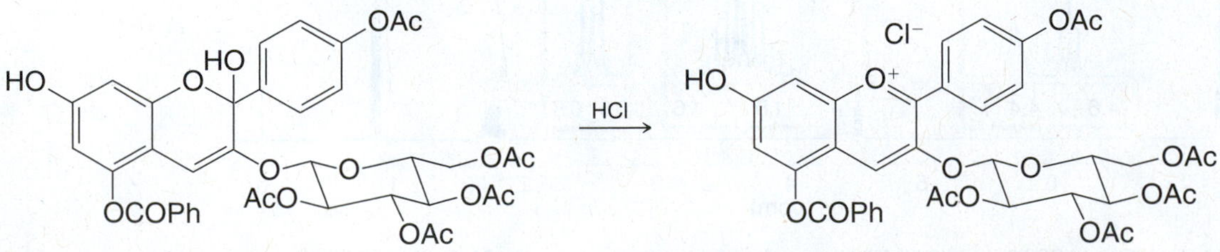

Callistephin chloride

2. The following reaction sequence was used by E. J. Corey (*J. Am. Chem. Soc.* **1969**, *91*, 5675–5677) at the beginning of a synthesis of prostaglandin $F_{2\alpha}$ and prostaglandin E_2. Explain what is involved in this reaction and why it is a reasonable process.

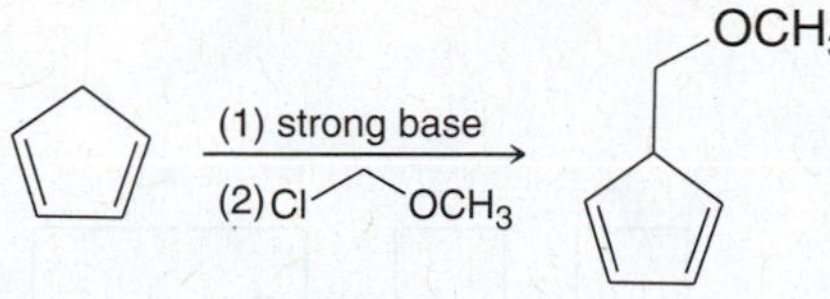

3. The ^1H NMR signals for the aromatic hydrogens of methyl *p*-hydroxybenzoate appear as two doublets at approximately 7.05 and 8.04 ppm (δ). Assign these two doublets to the respective hydrogens that produce each signal. Justify your assignments using arguments of relative electron density based on contributing resonance structures.

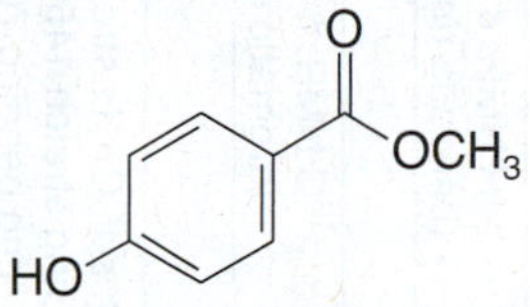

4. Draw the structure of adenine, a heterocyclic aromatic compound incorporated in the structure of DNA. Identify the nonbonding electron pairs that are *not* part of the aromatic system in the rings of adenine. Which nitrogen atoms in the rings would you expect to be more basic and which should be less basic?

5. Draw structures of the nicotinamide ring in NADH and NAD$^+$. In the transformation of NADH to NAD$^+$, in what form must a hydrogen be transferred in order to produce the aromatic pyridinium ion in NAD$^+$?

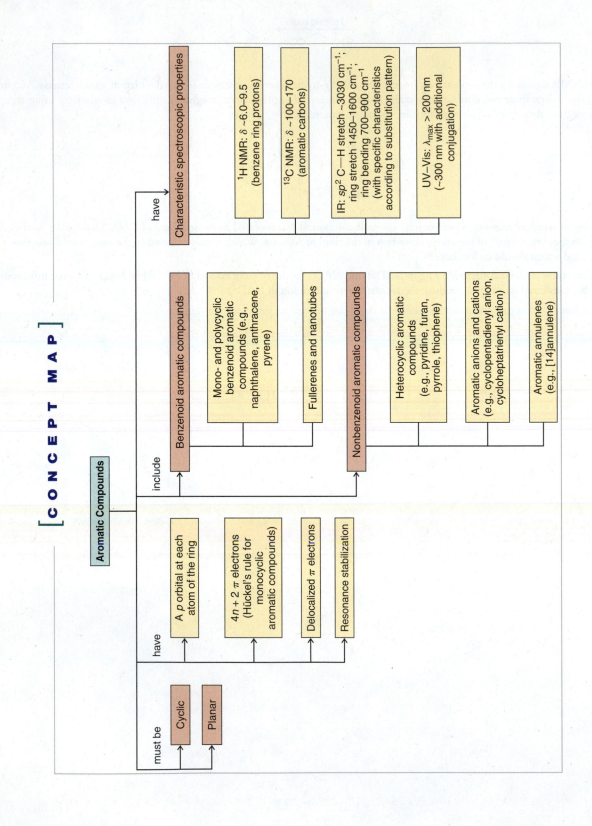

CHAPTER
15

Reactions of Aromatic Compounds

Although aromatic molecules have special electronic properties that render them inert to many standard reaction conditions, there are a number of ways to change the atoms that are attached to such systems by a process called electrophilic aromatic substitution. For instance, the six hydrogen atoms on benzene can all be replaced with different groups—for example, with halogens, carbonyl groups, or aliphatic chains. Such processes can convert benzene, a material that is a liquid at room temperature and serves as a solvent, into thousands of different molecules, including drugs like aspirin and explosives like trinitrotoluene (TNT). In biosynthesis, similar reactions produce biological molecules like thyroxine, a key hormone involved in metabolism, and pallidol, a compound produced by grapes. We will learn about these processes later in the chapter. The synthetic possibilities are nearly endless, but key to unlocking that potential is an understanding of the concepts, logic, and rules that determine how these reactions can be achieved.

IN THIS CHAPTER WE WILL CONSIDER:

- the general parameters that allow for substitution reactions of benzene
- how substituents on a benzene ring can impact reactivity and the ability to undergo additional substitutions
- reactions that can convert a given substituent into new functional groups

[WHY DO THESE TOPICS MATTER?]   At the end of the chapter, we will explore a special group of molecules that undergo different versions of the same reactions, both in nature and in the laboratory, to produce a diverse array of structures from similar starting materials.

PHOTO CREDIT: Image Source

15.1 ELECTROPHILIC AROMATIC SUBSTITUTION REACTIONS

Some of the most important reactions of aromatic compounds are those in which an electrophile replaces one of the hydrogen atoms of the ring.

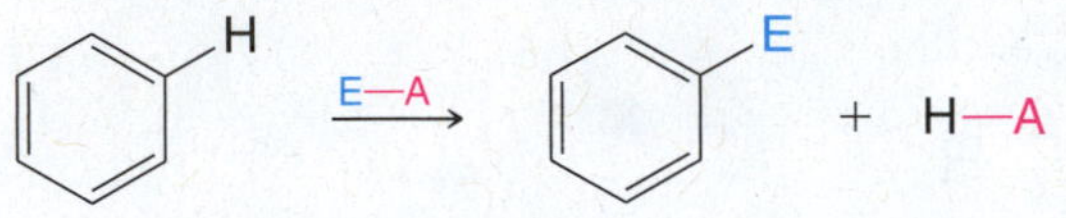

(E—A is an electrophilic reactant)

These reactions, called **electrophilic aromatic substitutions** (**EAS**), allow the direct introduction of groups onto aromatic rings such as benzene, and they provide synthetic routes to many important compounds. Figure 15.1 outlines five different types of electrophilic aromatic substitutions that we will study in this chapter, including carbon–carbon bond-forming reactions and halogenations.

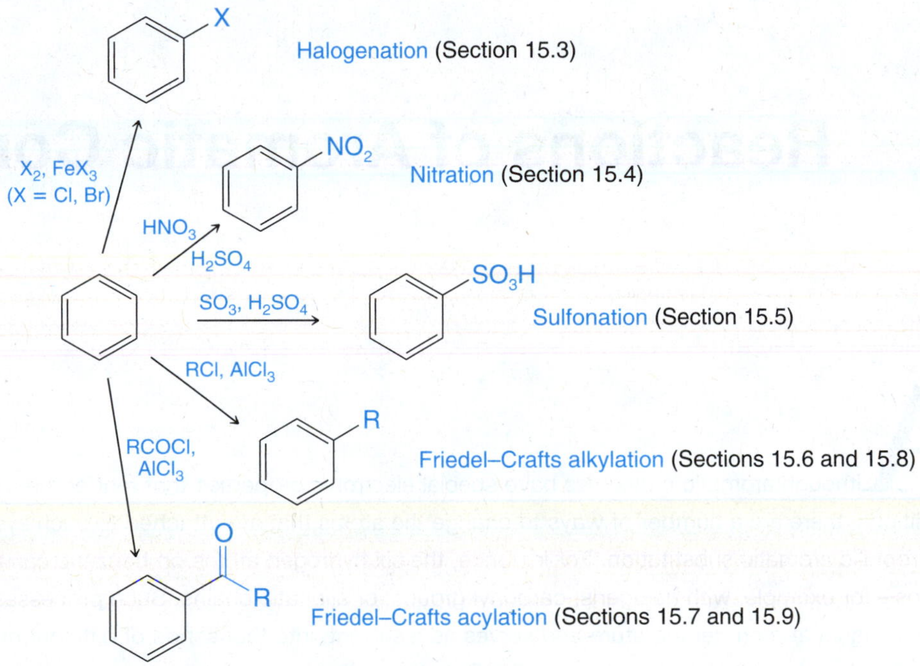

FIGURE 15.1 Electrophilic aromatic substitution reactions.

A noteworthy example of electrophilic aromatic substitution in nature, as mentioned in the introduction, is biosynthesis of the thyroid hormone thyroxine, where iodine is incorporated into benzene rings that are derived from tyrosine.

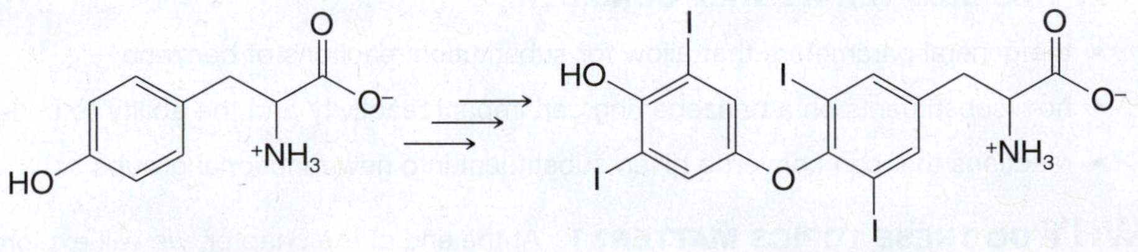

Tyrosine
(a dietary amino acid)

Thyroxine
(a thyroid hormone)

In the next section we shall learn the general mechanism for the way an electrophile reacts with a benzene ring. Then, in Sections 15.3–15.7, we shall see specific examples of electrophiles and how each is formed in a reaction mixture.

15.2 A GENERAL MECHANISM FOR ELECTROPHILIC AROMATIC SUBSTITUTION

The π electrons of benzene react with strong electrophiles. In this respect, benzene has something in common with alkenes. When an alkene reacts with an electrophile, as in the addition of HBr (Section 8.2), electrons from the alkene π bond react with the electrophile, leading to a carbocation intermediate.

Alkene   **Electrophile**   **Carbocation**

The carbocation formed from the alkene then reacts with the nucleophilic bromide ion to form the addition product.

Carbocation   **Addition product**

The similarity of benzene reactivity with that of an alkene ends, however, at the carbocation stage, prior to nucleophilic attack. As we saw in Chapter 14, benzene's closed shell of six π electrons give it special stability.

- Although benzene is susceptible to electrophilic attack, it undergoes *substitution reactions* rather than *addition reactions*.

Substitution reactions allow the aromatic sextet of π electrons in benzene to be regenerated after attack by the electrophile. We can see how this happens if we examine a general mechanism for **electrophilic aromatic substitution**.

Experimental evidence indicates that electrophiles attack the π system of benzene to form a **nonaromatic cyclohexadienyl carbocation** known as an **arenium ion**. In showing this step, it is convenient to use Kekulé structures, because these make it much easier to keep track of the π electrons:

> **Helpful Hint**
>
> Resonance structures (like those used here for the arenium ion) will be important for our study of electrophilic aromatic substitution.

Step 1

Arenium ion
(a delocalized cyclohexadienyl cation)

- In step 1 the electrophile takes two electrons of the six-electron π system to form a σ bond to one carbon atom of the benzene ring.

Formation of this bond interrupts the cyclic system of π electrons, because in the formation of the arenium ion the carbon that forms a bond to the electrophile becomes sp^3 hybridized and, therefore, no longer has an available p orbital. Now only five carbon atoms of the ring are sp^2 hybridized and still have p orbitals. The four π electrons of the arenium ion are delocalized through these five p orbitals. A calculated electrostatic potential map for the arenium ion formed by electrophilic addition of bromine to benzene indicates that positive charge is distributed in the arenium ion ring (Fig. 15.2), just as was shown in the contributing resonance structures.

FIGURE 15.2 A calculated structure for the arenium ion intermediate formed by electrophilic addition of bromine to benzene (Section 15.3). The electrostatic potential map for the principal location of bonding electrons (indicated by the solid surface) shows that positive charge (blue) resides primarily at the ortho and para carbons relative to the carbon where the electrophile has bonded. This distribution of charge is consistent with the resonance model for an arenium ion. (The van der Waals surface is indicated by the wire mesh.)

• In step 2 a proton is removed from the carbon atom of the arenium ion that bears the electrophile, restoring aromaticity to the ring.

Step 2

The two electrons that bonded the proton to the ring become a part of the π system. The carbon atom that bears the electrophile becomes sp^2 hybridized again, and a benzene derivative with six fully delocalized π electrons is formed. The proton is removed by any of the bases present, for example, by the anion derived from the electrophile.

● ● ●

PRACTICE PROBLEM 15.1   Show how loss of a proton can be represented using each of the three resonance structures for the arenium ion and show how each representation leads to the formation of a benzene ring with three alternating double bonds (i.e., six fully delocalized π electrons).

Kekulé structures are more appropriate for writing mechanisms such as electrophilic aromatic substitution because they permit the use of resonance theory, which, as we shall soon see, is invaluable as an aid to our understanding. If, for brevity, however, we wish to show the mechanism using the hybrid formula for benzene we can do it in the following way. We draw the arenium ion as a delocalized cyclohexadienyl cation:

Helpful Hint

In our color scheme for chemical formulas, blue generally indicates groups that are electrophilic or have electron-withdrawing character. Red indicates groups that are or become Lewis bases, or have electron-donating character.

Step 1

Arenium ion

Step 2

There is firm experimental evidence that the arenium ion is a true *intermediate* in electrophilic substitution reactions. It is not a transition state. This means that in a free-energy diagram (Fig. 15.3) the arenium ion lies in an energy valley between two transition states.

The free energy of activation for step 1, $\Delta G_{(1)}^{\ddagger}$, has been shown to be much greater than the free energy of activation for step 2, $\Delta G_{(2)}^{\ddagger}$, as depicted in Figure 15.3. This is

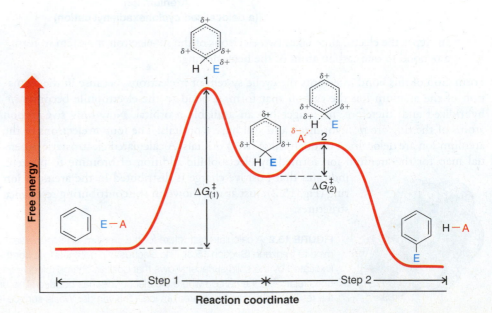

FIGURE 15.3 The free-energy diagram for an electrophilic aromatic substitution reaction. The arenium ion is a true intermediate lying between transition states 1 and 2. In transition state 1 the bond between the electrophile and one carbon atom of the benzene ring is only partially formed. In transition state 2 the bond between the same benzene carbon atom and its hydrogen atom is partially broken. The bond between the hydrogen atom and the conjugate base is partially formed.

Reaction coordinate

consistent with what we would expect. The reaction leading from benzene and an electrophile to the arenium ion is highly endothermic, because the aromatic stability of the benzene ring is lost. The reaction leading from the arenium ion to the substituted benzene, by contrast, is highly exothermic because it restores aromaticity to the system.

Of the following two steps, step 1 (the formation of the arenium ion) is usually the rate-determining step in electrophilic aromatic substitution because of its higher free energy of activation:

Step 1   + E—A ⟶   + :A⁻   **Slow, rate determining**

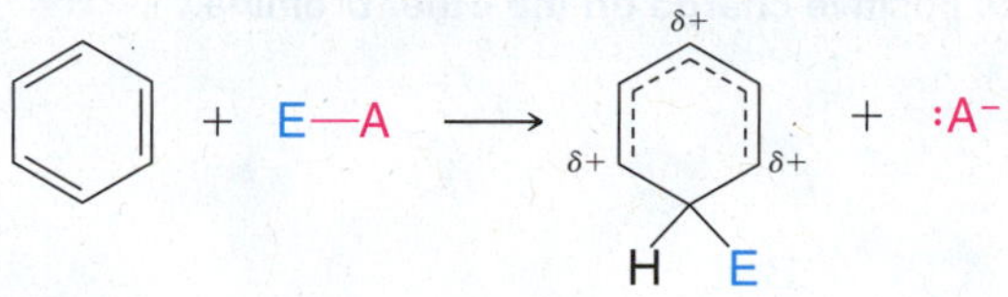

Step 2   ⟶   + H—A   **Fast**

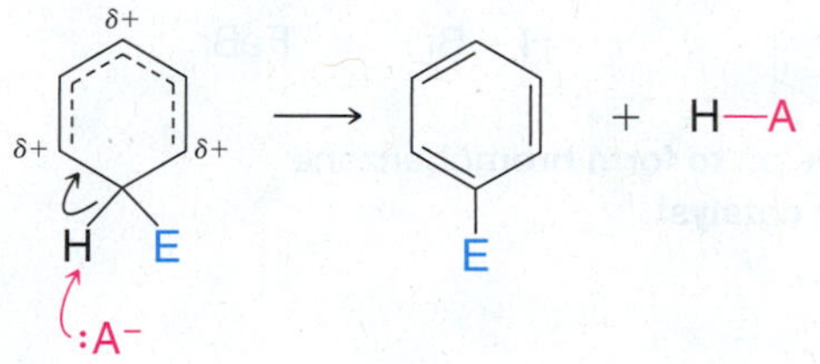

Step 2, the removal of a proton, occurs rapidly relative to step 1 and has no effect on the overall rate of reaction.

15.3 HALOGENATION OF BENZENE

Benzene reacts with bromine and chlorine in the presence of Lewis acids to give halogenated substitution products in good yield.

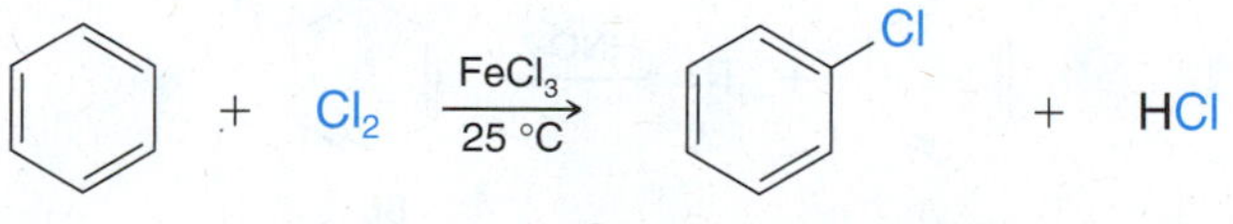

    + Cl_2 $\xrightarrow{\text{FeCl}_3 \atop 25\,°C}$ + HCl

Chlorobenzene (90%)

    + Br_2 $\xrightarrow{\text{FeBr}_3 \atop \text{heat}}$ + HBr

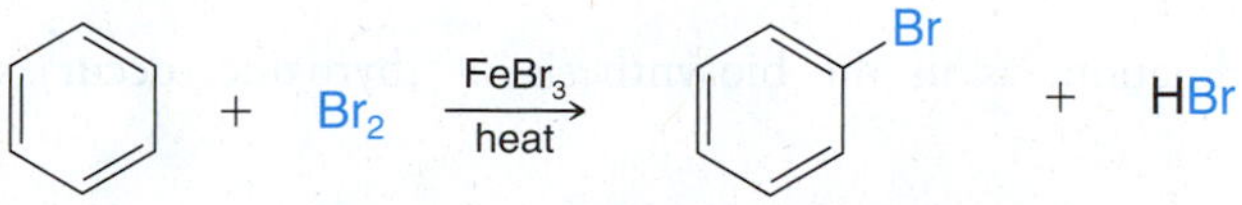

Bromobenzene (75%)

The Lewis acids typically used are aluminum chloride ($AlCl_3$) and iron chloride ($FeCl_3$) for chlorination, and iron bromide ($FeBr_3$) for bromination. The purpose of the Lewis acid is to make the halogen a stronger electrophile. A mechanism for electrophilic aromatic bromination is shown here.

[A MECHANISM FOR THE REACTION — Electrophilic Aromatic Bromination]

Step 1 $:\!\ddot{Br}\!-\!\ddot{Br}\!:$ + $FeBr_3$ ⇌ $:\!\ddot{Br}\!-\!\overset{+}{\ddot{Br}}\!-\!\bar{F}eBr_3$

Bromine combines with FeBr₃ to form a complex.

(mechanism continues on the next page)

Step 2

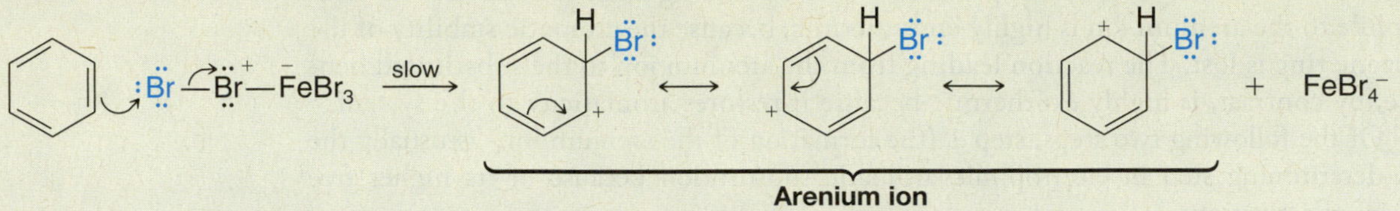

Arenium ion

Helpful Hint

An electrostatic potential map for this arenium ion is shown in Fig. 15.2.

The benzene ring donates an electron pair to the terminal bromine, forming the arenium ion and neutralizing the formal positive charge on the other bromine.

Step 3

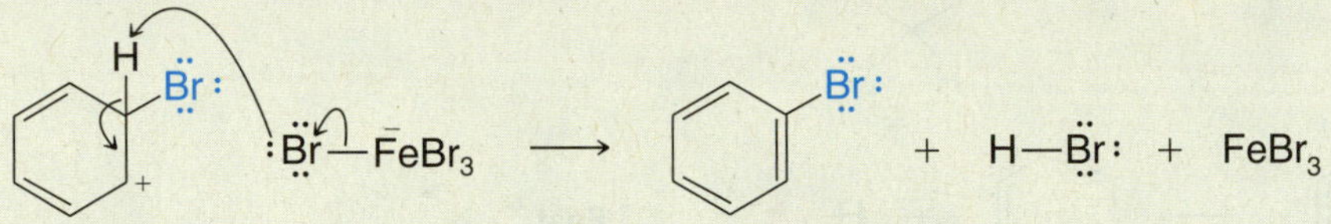

A proton is removed from the arenium ion to form bromobenzene and regenerate the catalyst.

The mechanism of the chlorination of benzene in the presence of ferric chloride is analogous to the one for bromination.

Fluorine reacts so rapidly with benzene that aromatic fluorination requires special conditions and special types of apparatus. Even then, it is difficult to limit the reaction to monofluorination. Fluorobenzene can be made, however, by an indirect method that we shall see in Section 20.7D.

Iodine, on the other hand, is so unreactive that a special technique has to be used to effect direct iodination; the reaction has to be carried out in the presence of an oxidizing agent such as nitric acid:

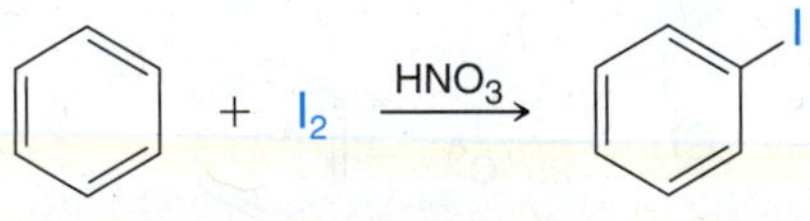

86%

Biochemical iodination, as in the biosynthesis of thyroxine, occurs with enzymatic catalysis.

15.4 NITRATION OF BENZENE

Benzene undergoes nitration on reaction with a mixture of concentrated nitric acid and concentrated sulfuric acid.

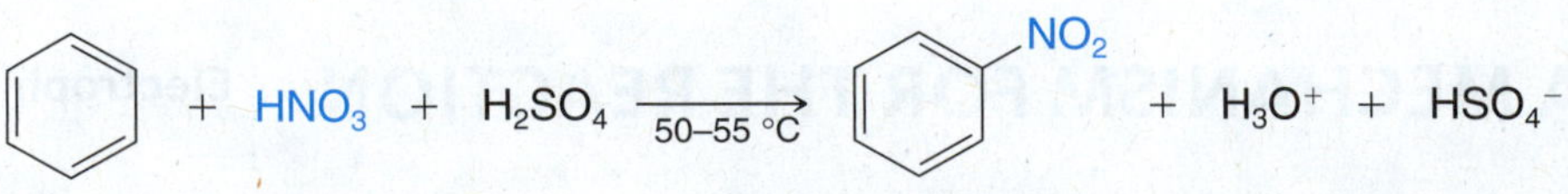

85%

Concentrated sulfuric acid increases the rate of the reaction by increasing the concentration of the electrophile, the nitronium ion (NO_2^+), as shown in the first two steps of the following mechanism.

A MECHANISM FOR THE REACTION — Nitration of Benzene

Step 1

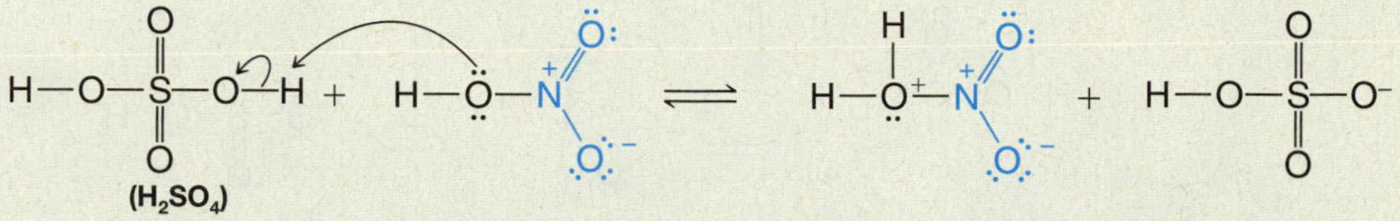

In this step nitric acid accepts a proton from the stronger acid, sulfuric acid.

Step 2

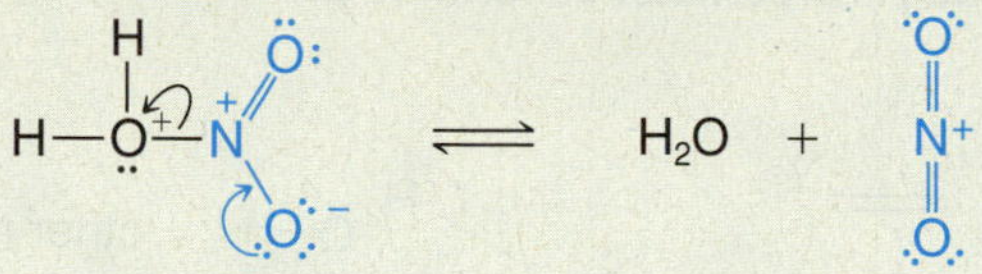

Nitronium ion

Now that it is protonated, nitric acid can dissociate to form a nitronium ion.

Step 3

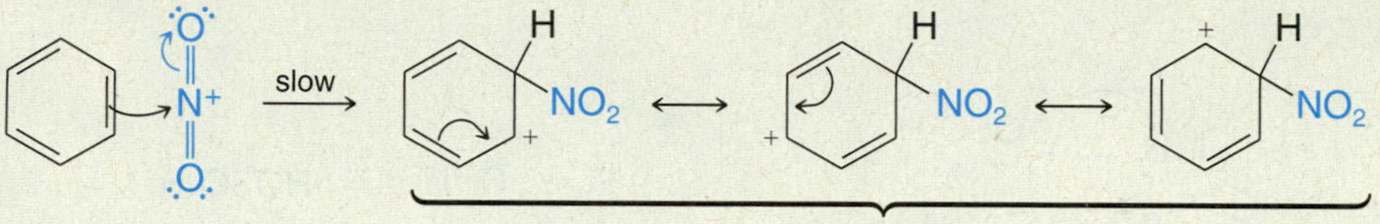

Arenium ion

The nitronium ion is the electrophile in nitration; it reacts with benzene to form a resonance-stabilized arenium ion.

Step 4

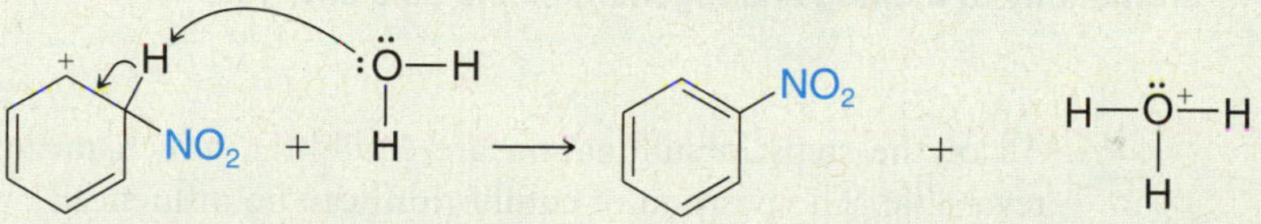

The arenium ion then loses a proton to a Lewis base and becomes nitrobenzene.

PRACTICE PROBLEM 15.2

Given that the pK_a of H_2SO_4 is −9 and that of HNO_3 is −1.4, explain why nitration occurs more rapidly in a mixture of concentrated nitric and sulfuric acids than in concentrated nitric acid alone.

15.5 SULFONATION OF BENZENE

Benzene reacts with fuming sulfuric acid at room temperature to produce benzenesulfonic acid. Fuming sulfuric acid is sulfuric acid that contains added sulfur trioxide (SO_3). Sulfonation also takes place in concentrated sulfuric acid alone, but more slowly. Under either condition, the electrophile appears to be sulfur trioxide.

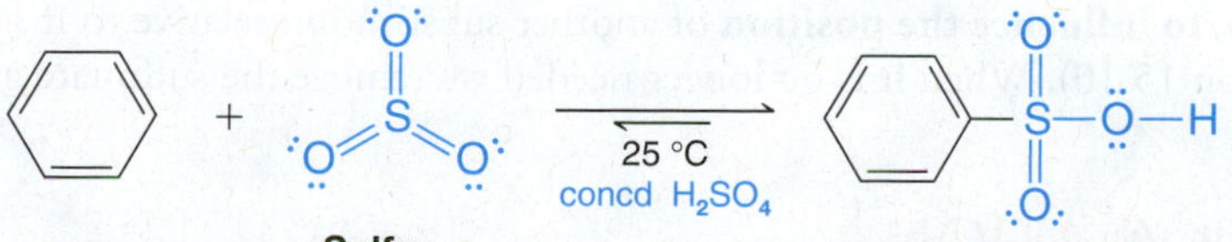

Sulfur trioxide

25 °C
concd H_2SO_4

Benzenesulfonic acid
(56%)

In concentrated sulfuric acid, sulfur trioxide is produced in an equilibrium in which H_2SO_4 acts as both an acid and a base (see step 1 of the following mechanism).

[A MECHANISM FOR THE REACTION ─ Sulfonation of Benzene]

Step 1

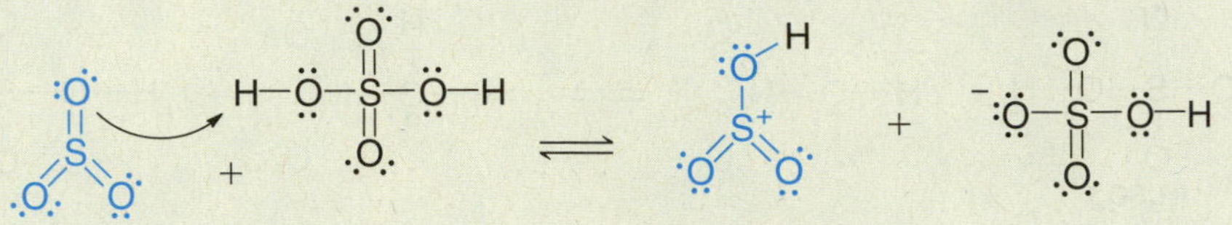

SO$_3$ is protonated to form SO$_3$H$^+$

Step 2

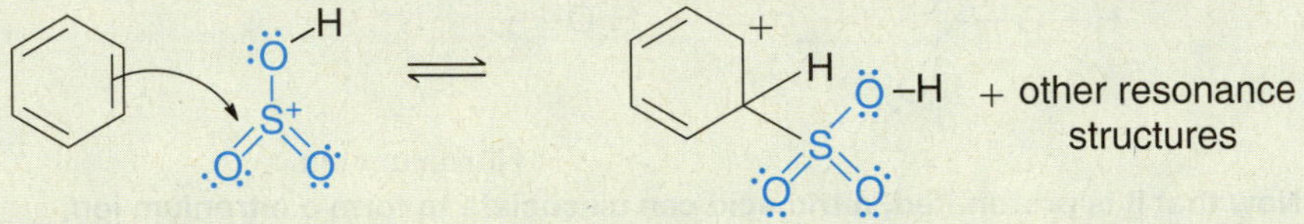

+ other resonance structures

SO$_3$H$^+$ reacts as an electrophile with the benzene ring to form an arenium ion.

Step 3

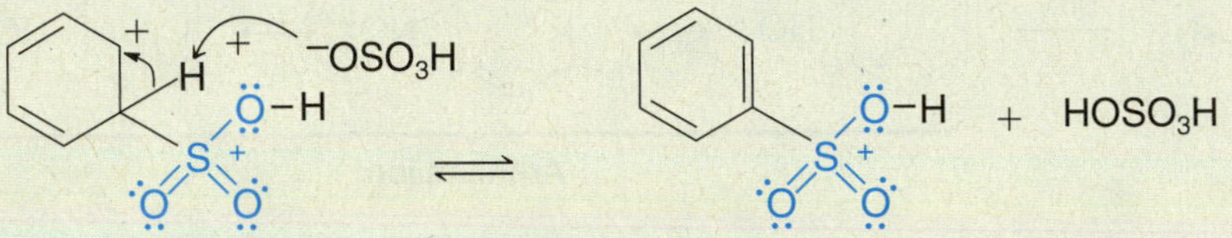

+ HOSO$_3$H

Loss of a proton from the arenium ion restores aromaticity to the ring and regenerates the acid catalyst.

All of the steps in sulfonation are equilibria, which means that the overall reaction is reversible. The position of equilibrium can be influenced by the conditions we employ.

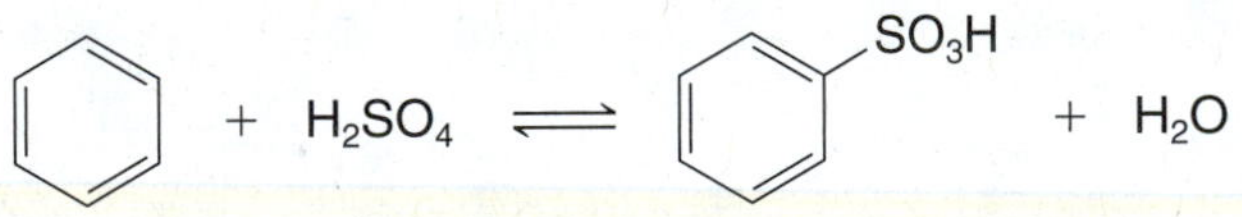

- If we want to sulfonate the ring (install a sulfonic acid group), we use concentrated sulfuric acid or—better yet—fuming sulfuric acid. Under these conditions the position of equilibrium lies appreciably to the right, and we obtain benzenesulfonic acid in good yield.

- If we want to desulfonate the ring (**remove** a sulfonic acid group), we employ dilute sulfuric acid and usually pass steam through the mixture. Under these conditions—with a high concentration of water—the equilibrium lies appreciably to the left and desulfonation occurs.

We shall see later that sulfonation and desulfonation reactions are often used in synthetic work.

- We sometimes install a sulfonate group **as a protecting group**, to temporarily block its position from electrophilic aromatic substitution, or **as a directing group, to influence the position** of another substitution relative to it (Section 15.10). When it is no longer needed we remove the sulfonate group.

Helpful Hint

Sulfonation–desulfonation is a useful tool in syntheses involving electrophilic aromatic substitution.

15.6 FRIEDEL–CRAFTS ALKYLATION

Charles Friedel, a French chemist, and his American collaborator, James M. Crafts, discovered new methods for the preparation of alkylbenzenes (ArR) and acylbenzenes (ArCOR) in 1877. These reactions are now called the Friedel–Crafts **alkylation** and acylation reactions. We shall study the Friedel–Crafts alkylation reaction here and take up the **Friedel–Crafts acylation** reaction in Section 15.7.

- The following is a general equation for a **Friedel–Crafts alkylation** reaction:

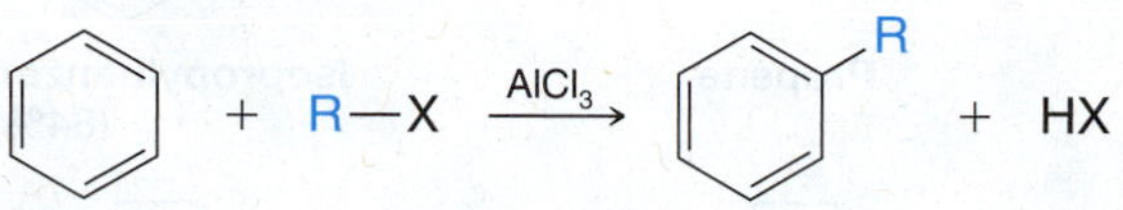

- The mechanism for the reaction starts with the formation of a carbocation.
- The carbocation then acts as an electrophile and attacks the benzene ring to form an arenium ion.
- The arenium ion then loses a proton.

This mechanism is illustrated below using 2-chloropropane and benzene.

[A MECHANISM FOR THE REACTION — Friedel–Crafts Alkylation]

Step 1

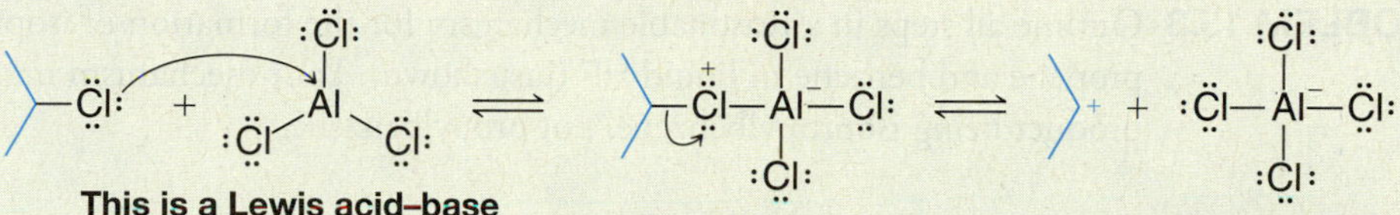

This is a Lewis acid–base reaction (see Section 15.3). **The complex dissociates to form a carbocation and AlCl$_4^-$.**

Step 2

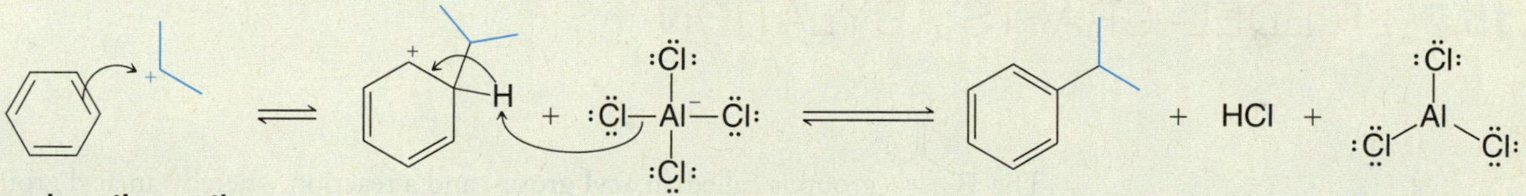

The carbocation, acting as an electrophile, reacts with benzene to produce an arenium ion.

A proton is removed from the arenium ion to form isopropylbenzene. This step also regenerates the AlCl$_3$ and liberates HCl.

- When R—X is a primary halide, a simple carbocation probably does not form. Instead, the aluminum chloride forms a complex with the alkyl halide, and this complex acts as the electrophile.

The complex is one in which the carbon–halogen bond is nearly broken—and one in which the carbon atom has a considerable positive charge:

$$\overset{\delta+}{RCH_2}\text{----}\overset{\delta-}{\ddot{C}l}:AlCl_3$$

Even though this complex is not a simple carbocation, it acts as if it were and it transfers a positive alkyl group to the aromatic ring.

- These complexes react so much like carbocations that they also undergo typical carbocation rearrangements (Section 15.8).
- Friedel–Crafts alkylations are not restricted to the use of alkyl halides and aluminum chloride. Other pairs of reagents that form carbocations (or species like carbocations) may be used in Friedel–Crafts alkylations as well.

These possibilities include the use of a mixture of an alkene and an acid:

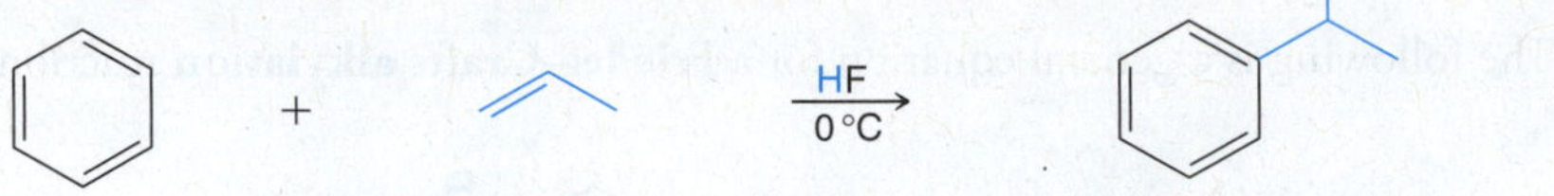

Propene $\xrightarrow[0\,°C]{HF}$ **Isopropylbenzene (cumene)**
(84%)

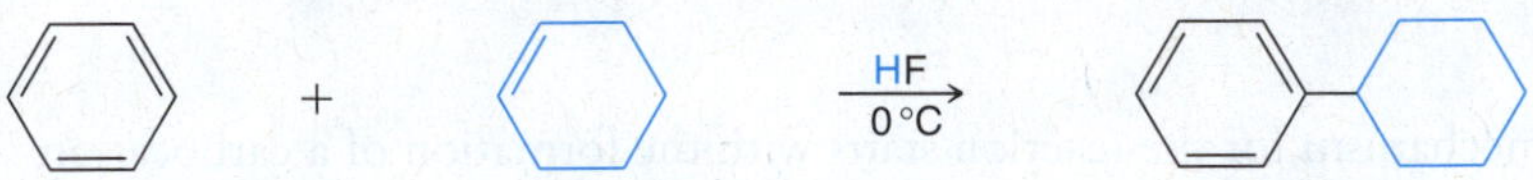

Cyclohexene $\xrightarrow[0\,°C]{HF}$ **Cyclohexylbenzene**
(62%)

A mixture of an alcohol and an acid may also be used:

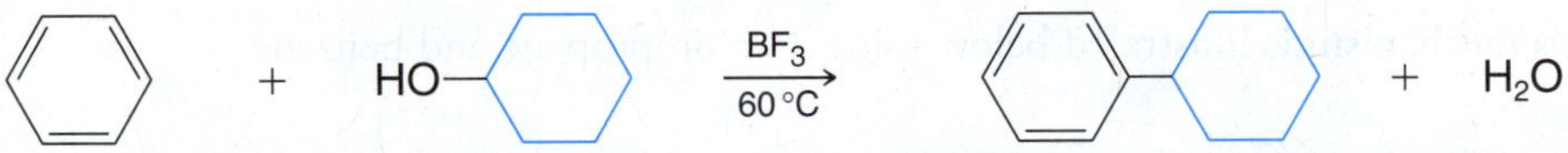

Cyclohexanol $\xrightarrow[60\,°C]{BF_3}$ **Cyclohexylbenzene** $+$ H_2O
(56%)

There are several important limitations of the Friedel–Crafts reaction. These are discussed in Section 15.8.

PRACTICE PROBLEM 15.3 Outline all steps in a reasonable mechanism for the formation of isopropylbenzene from propene and benzene in liquid HF (just shown). Your mechanism must account for the product being isopropylbenzene, not propylbenzene.

15.7 FRIEDEL–CRAFTS ACYLATION

The $R\overset{O}{\underset{}{\overset{\|}{C}}}$ group is called an **acyl group**, and a reaction whereby an acyl group is introduced into a compound is called an **acylation** reaction. Two common acyl groups are the acetyl group and the benzoyl group. (The benzoyl group should not be confused with the benzyl group, $-CH_2C_6H_5$; see Section 14.2.)

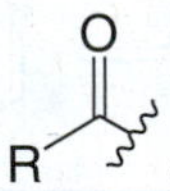

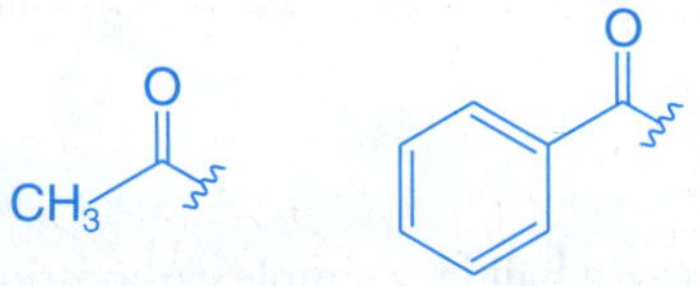

Acetyl group   **Benzoyl**
(ethanoyl group)   **group**

The **Friedel–Crafts acylation** reaction is often carried out by treating the aromatic compound with an **acyl halide** (often an acyl chloride). Unless the aromatic compound is one that is highly reactive, the reaction requires the addition of at least one equivalent of a Lewis acid (such as $AlCl_3$) as well. The product of the reaction is an aryl ketone:

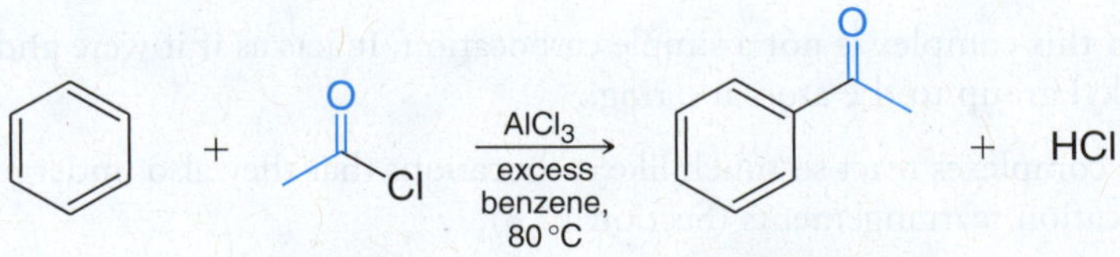

Acetyl $\xrightarrow[\substack{excess \\ benzene, \\ 80\,°C}]{AlCl_3}$ **Acetophenone** $+$ HCl
chloride   **(methyl phenyl ketone)**
            (97%)

Acyl chlorides, also called **acid chlorides**, are easily prepared (Section 18.5) by treating carboxylic acids with thionyl chloride ($SOCl_2$) or phosphorus pentachloride (PCl_5):

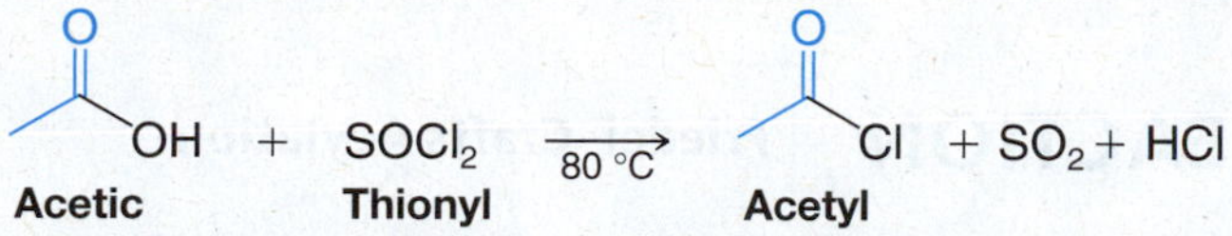

Acetic acid | Thionyl chloride | Acetyl chloride (80–90%)

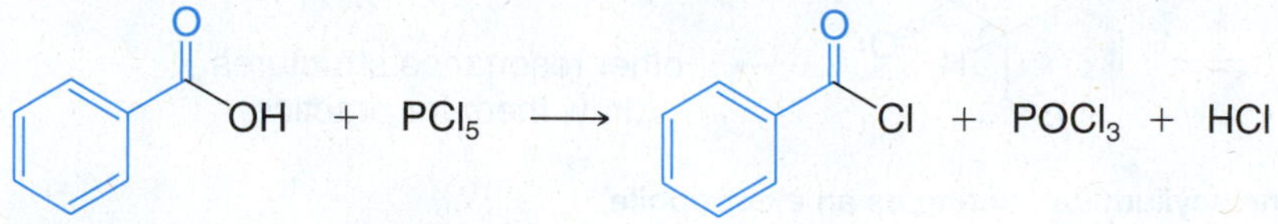

Benzoic acid | Phosphorus pentachloride | Benzoyl chloride (90%)

Friedel–Crafts acylations can also be carried out using carboxylic acid anhydrides. For example,

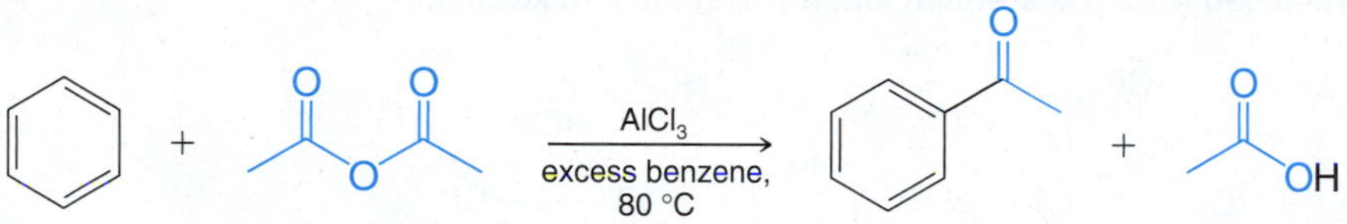

Acetic anhydride (a carboxylic acid anhydride) | Acetophenone (82–85%)

In most Friedel–Crafts acylations the electrophile appears to be an **acylium ion** formed from an acyl halide in the following way:

Step 1

Step 2

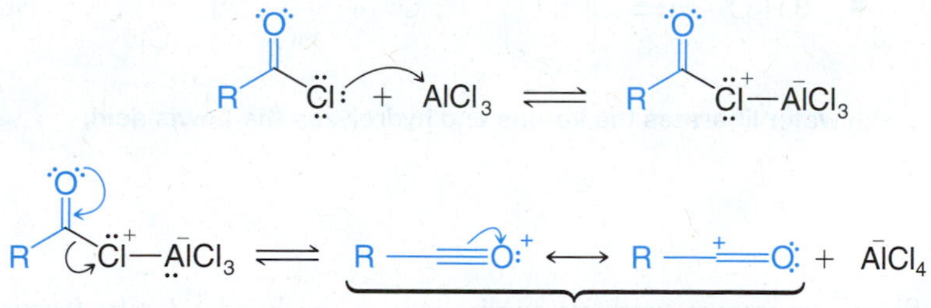

An acylium ion
(a resonance hybrid)

SOLVED PROBLEM 15.1

Show how an acylium ion could be formed from acetic anhydride in the presence of $AlCl_3$.

STRATEGY AND ANSWER: We recognize that $AlCl_3$ is a Lewis acid and that an acid anhydride, because it has multiple unshared electron pairs, is a Lewis base. A reasonable mechanism starts with a Lewis acid–base reaction and proceeds to form an acylium ion in the following way.

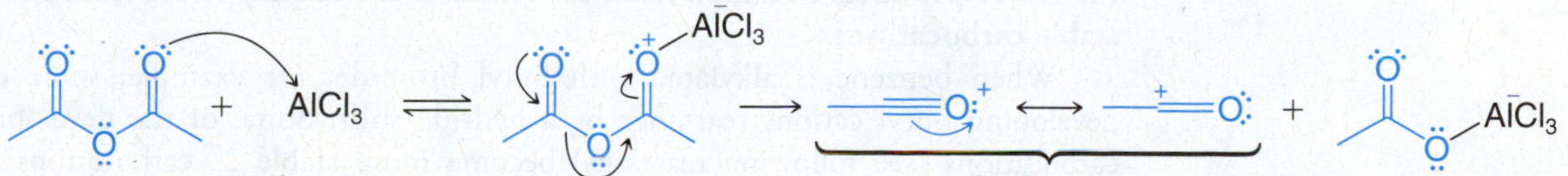

Acylium ion

The remaining steps in the Friedel–Crafts acylation of benzene are the following:

[A MECHANISM FOR THE REACTION — Friedel–Crafts Acylation]

Step 3

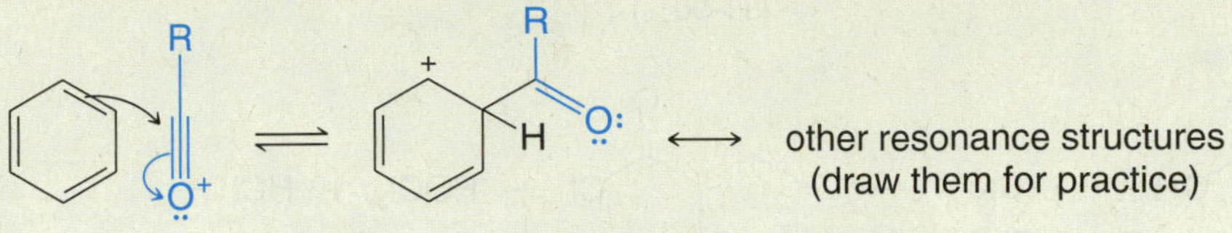

The acylium ion, acting as an electrophile,
reacts with benzene to form the arenium ion.

Step 4

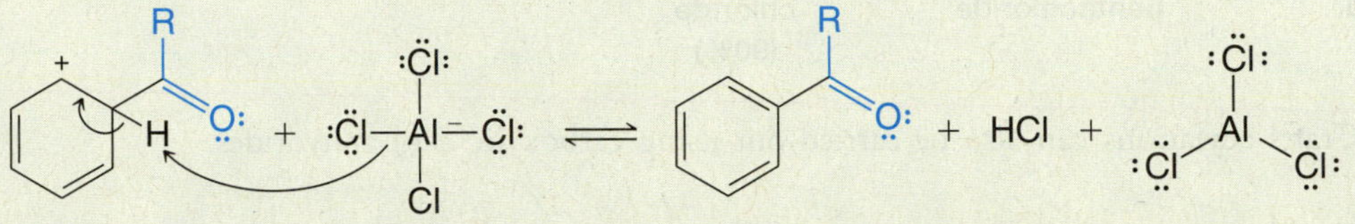

A proton is removed from the arenium ion, forming the aryl ketone.

Step 5

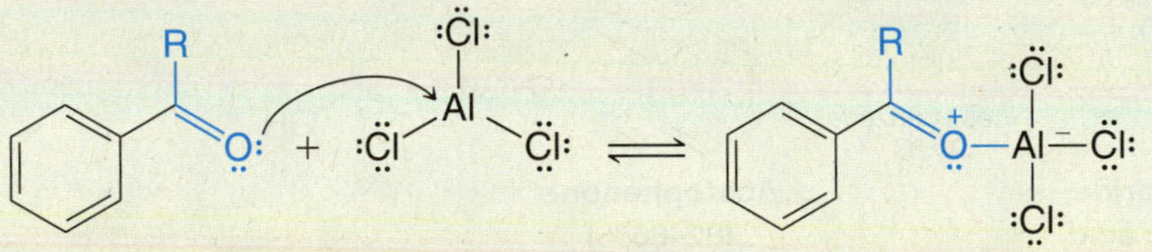

The ketone, acting as a Lewis base, reacts with
aluminum chloride (a Lewis acid) to form a complex.

Step 6

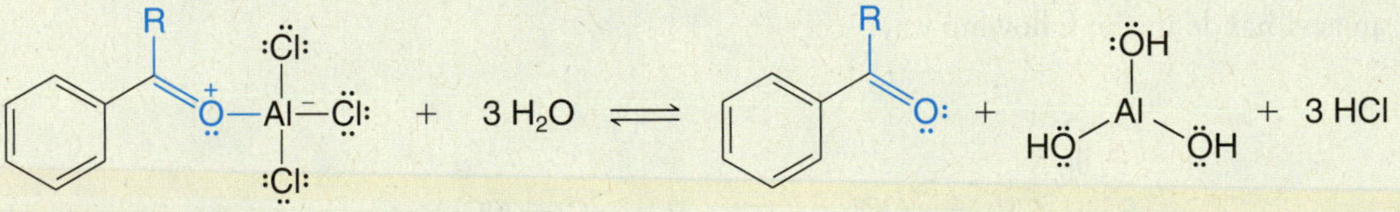

Treating the complex with water liberates the ketone and hydrolyzes the Lewis acid.

Several important synthetic applications of the Friedel–Crafts reaction are given in Section 15.9.

15.8 LIMITATIONS OF FRIEDEL–CRAFTS REACTIONS

Several restrictions limit the usefulness of Friedel–Crafts reactions:

1. **When the carbocation formed from an alkyl halide, alkene, or alcohol can rearrange to one or more carbocations that are more stable, it usually does so, and the major products obtained from the reaction are usually those from the more stable carbocations.**

   When benzene is alkylated with butyl bromide, for example, some of the developing butyl cations rearrange by a hydride shift. Some of the developing 1° carbocations (see following reactions) become more stable 2° carbocations. Then benzene reacts with both kinds of carbocations to form both butylbenzene and *sec*-butylbenzene:

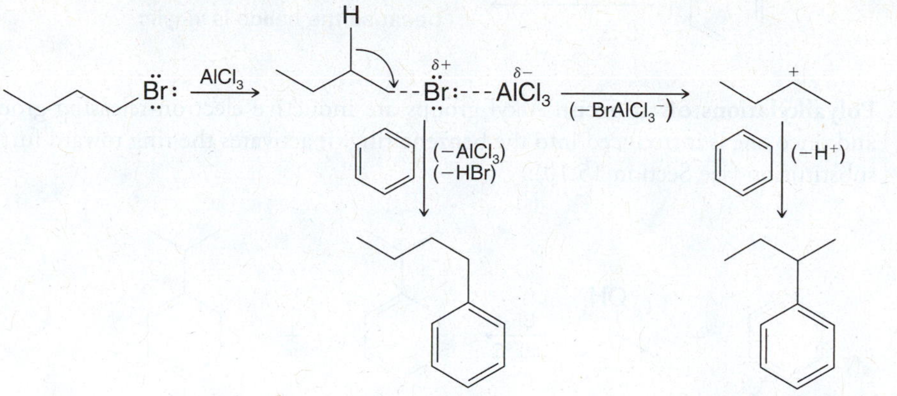

Butylbenzene
(32–36% of mixture)

sec-**Butylbenzene**
(64–68% of mixture)

2. **Friedel–Crafts reactions usually give poor yields when powerful electron-withdrawing groups** (Section 15.11) **are present on the aromatic ring or when the ring bears an** —NH_2, —NHR, **or** —NR_2 **group.** This applies to both alkylations and acylations.

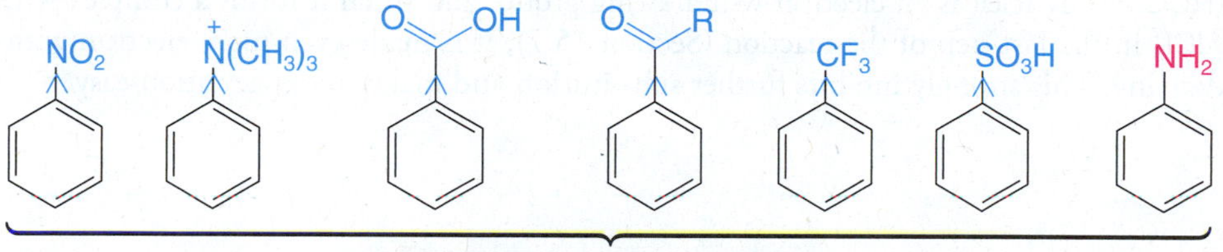

These usually give poor yields in Friedel-Crafts reactions.

We shall learn in Section 15.10 that groups present on an aromatic ring can have a large effect on the reactivity of the ring toward electrophilic aromatic substitution. **Electron-withdrawing groups make the ring less reactive by making it electron deficient.** Any substituent more electron withdrawing (or deactivating) than a halogen, that is, **any meta-directing group** (Section 15.11C), **makes an aromatic ring too electron deficient to undergo a Friedel–Crafts reaction.** The amino groups, —NH_2, —NHR, and —NR_2, are changed into powerful electron-withdrawing groups by the Lewis acids used to catalyze Friedel–Crafts reactions. For example,

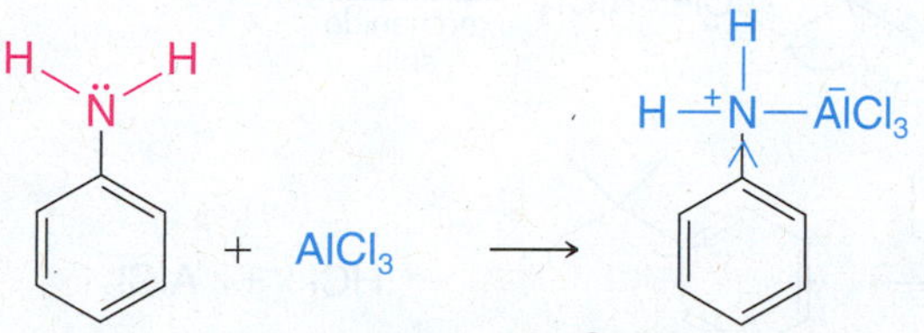

Does not undergo a Friedel–Crafts reaction

3. **Aryl and vinylic halides cannot be used as the halide component because they do not form carbocations readily** (see Section 6.14A):

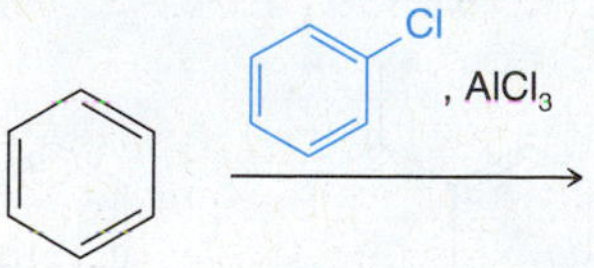

No Friedel–Crafts reaction because the halide is aryl

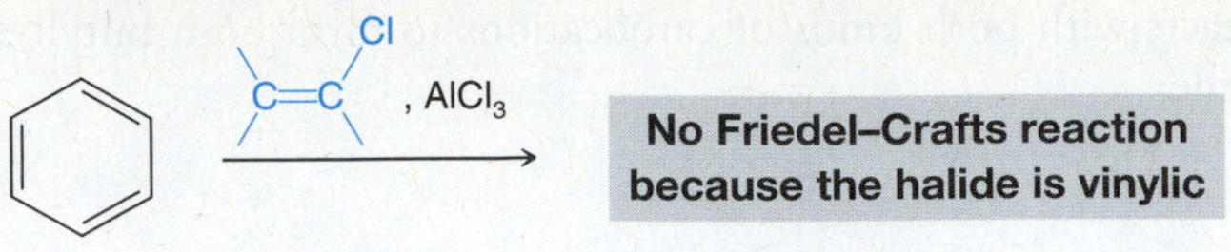

No Friedel–Crafts reaction because the halide is vinylic

4. **Polyalkylations often occur.** Alkyl groups are inductive electron-releasing groups, and once one is introduced into the benzene ring, it activates the ring toward further substitution (see Section 15.10):

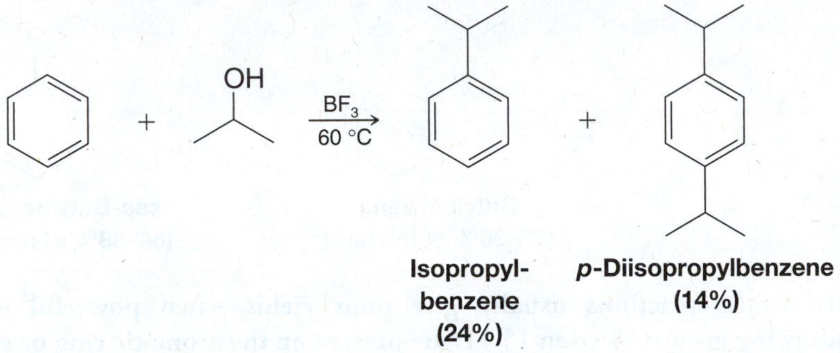

Isopropyl-benzene (24%) **p-Diisopropylbenzene (14%)**

Polyacylations are not a problem in Friedel–Crafts acylations. The acyl group (RCO—) by itself is an electron-withdrawing group, and when it forms a complex with $AlCl_3$ in the last step of the reaction (Section 15.7), it is made even more electron withdrawing. This strongly inhibits further substitution and makes monoacylation easy.

●●● **SOLVED PROBLEM 15.2**

When benzene reacts with 1-chloro-2,2-dimethylpropane (neopentyl chloride) in the presence of aluminum chloride, the major product is 2-methyl-2-phenylbutane, not 2,2-dimethyl-1-phenylpropane (neopentylbenzene). Explain this result.

STRATEGY AND ANSWER: The carbocation formed by direct reaction of $AlCl_3$ with 1-chloro-2,2-dimethylpropane would be a primary carbocation; however, it rearranges to the more stable tertiary carbocation before it can react with the benzene ring.

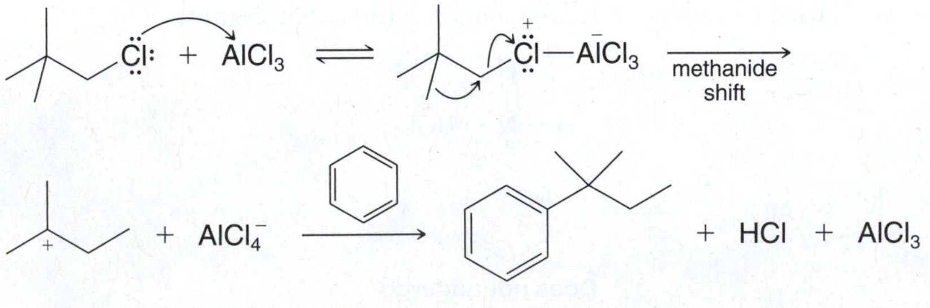

●●●

PRACTICE PROBLEM 15.4 Provide a mechanism that accounts for the following result.

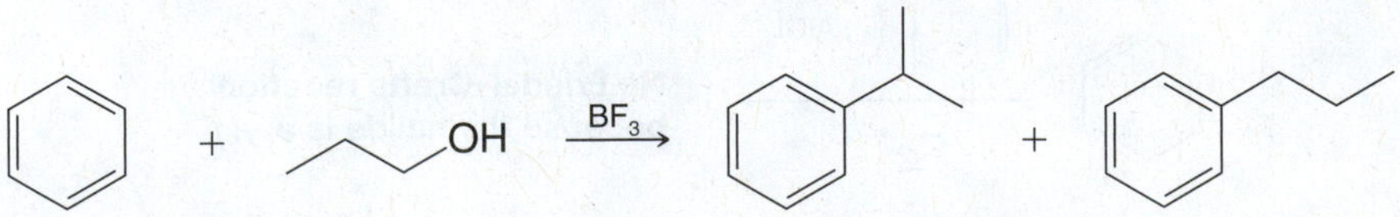

15.9 SYNTHETIC APPLICATIONS OF FRIEDEL–CRAFTS ACYLATIONS: THE CLEMMENSEN AND WOLFF–KISHNER REDUCTIONS

- Rearrangements of the carbon chain do not occur in Friedel–Crafts acylations.

The acylium ion, because it is stabilized by resonance, is more stable than most other carbocations. Thus, there is no driving force for a rearrangement. Because rearrangements do not occur, Friedel–Crafts acylations followed by reduction of the carbonyl group to a CH_2 group often give us much better routes to unbranched alkylbenzenes than do Friedel–Crafts alkylations.

- The carbonyl group of an aryl ketone can be reduced to a CH_2 group.

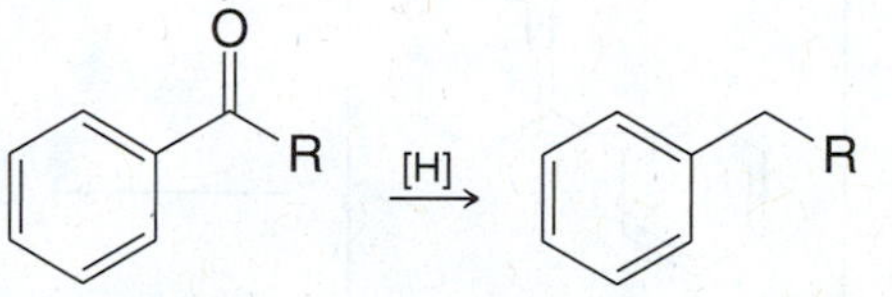

As an example, let us consider the problem of synthesizing propylbenzene. If we attempt this synthesis through a Friedel–Crafts alkylation, a rearrangement occurs and the major product is isopropylbenzene (see also Practice Problem 15.4):

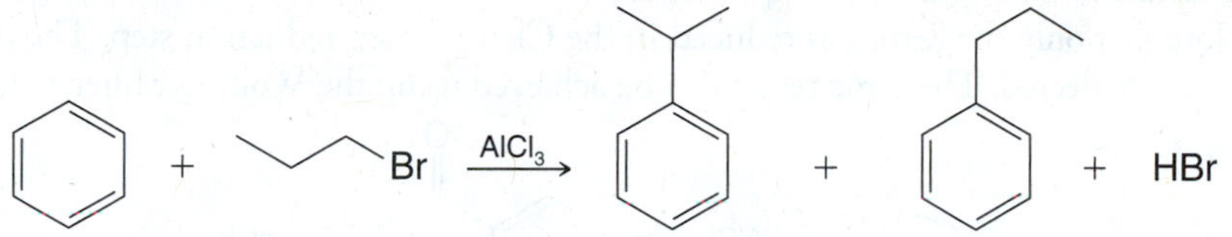

Isopropylbenzene   **Propylbenzene**
(major product)   **(minor product)**

By contrast, the Friedel–Crafts acylation of benzene with propanoyl chloride produces a ketone with an unrearranged carbon chain in excellent yield:

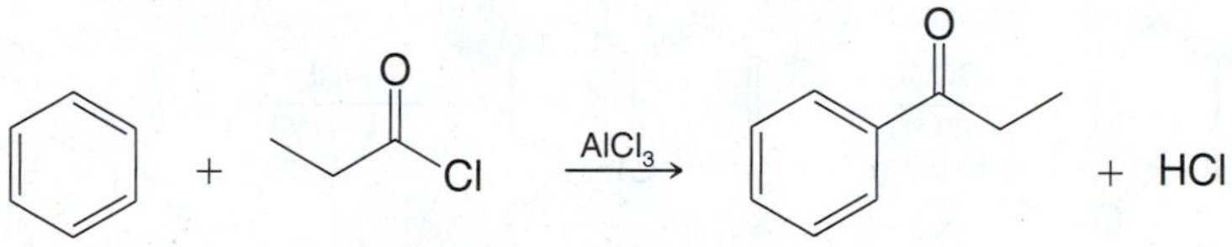

Propanoyl   **Ethyl phenyl**
chloride   **ketone (90%)**

This ketone can then be reduced to propylbenzene by several methods.

15.9A The Clemmensen Reduction

One general method for reducing a ketone to a methylene group—called the **Clemmensen reduction**—consists of refluxing the ketone with hydrochloric acid containing amalgamated zinc. [*Caution*: As we shall discuss later (Section 20.4B), zinc and hydrochloric acid will also reduce nitro groups to amino groups.]

A Clemmensen Reduction

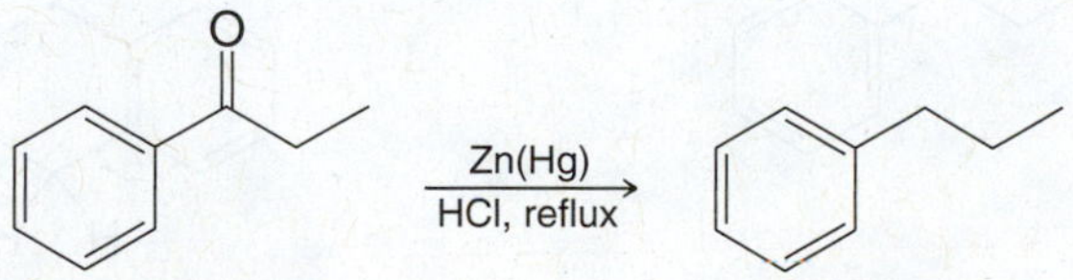

Ethyl phenyl   **Propylbenzene**
ketone   **(80%)**

Helpful Hint

Friedel–Crafts acylation followed by ketone reduction is the synthetic equivalent of Friedel–Crafts alkylation.

In general,

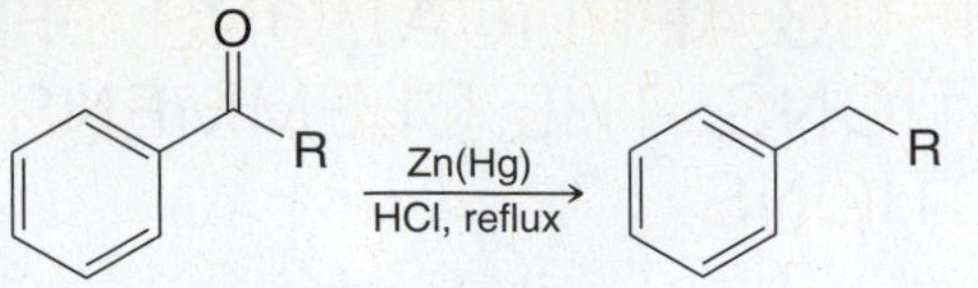

15.9B The Wolff–Kishner Reduction

Another method for reducing a ketone to a methylene group is the Wolff–Kishner reduction, which involves heating the ketone with hydrazine and base. The Wolff–Kishner reduction complements the Clemmensen reduction in that it is conducted under basic conditions, whereas the Clemmensen reduction involves acidic conditions. The Wolff–Kishner reduction proceeds via a hydrazone intermediate (Section 16.8B) that is not isolated during the reaction. Ethyl phenyl ketone can be reduced to propylbenzene by the Wolff–Kishner reduction as follows, for example.

A Wolff–Kishner Reduction

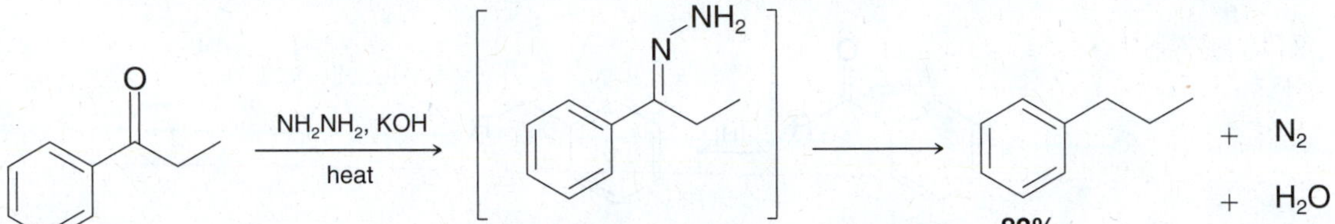

Hydrazone intermediate (see Section 16.8B)

When cyclic anhydrides are used as one component, the Friedel–Crafts acylation provides a means of adding a new ring to an aromatic compound. One illustration is shown here. Note that only the ketone is reduced in the Clemmensen reduction step. The carboxylic acid is unaffected. The same result can be achieved using the Wolff–Kishner reduction.

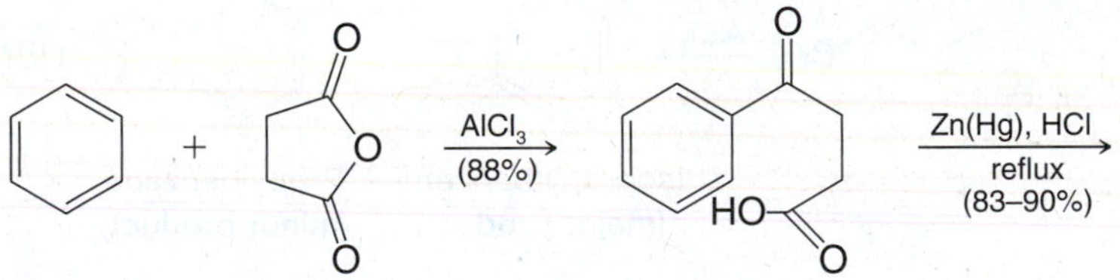

Benzene (excess)   **Succinic anhydride**   **3-Benzoylpropanoic acid**

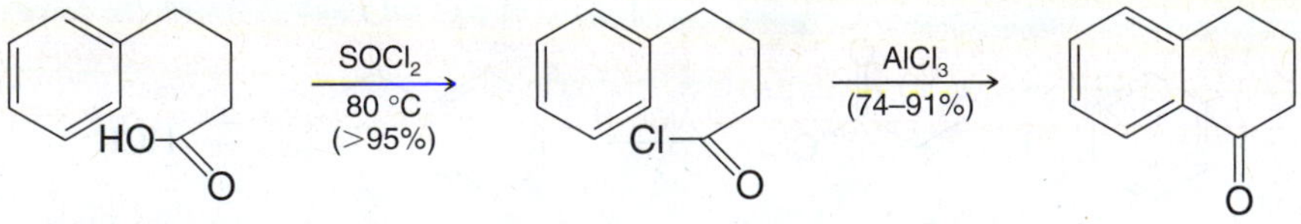

4-Phenylbutanoic acid   **4-Phenylbutanoyl chloride**   **α-Tetralone**

PRACTICE PROBLEM 15.5 Starting with benzene and the appropriate acyl chloride or acid anhydride, outline a synthesis of each of the following:

(a) Butylbenzene

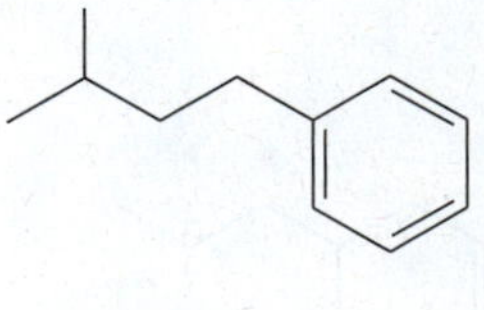

(c)

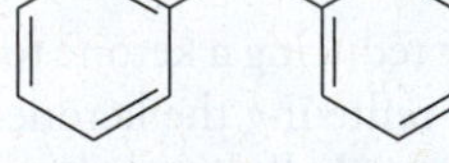

Diphenylmethane

(b)

(d)

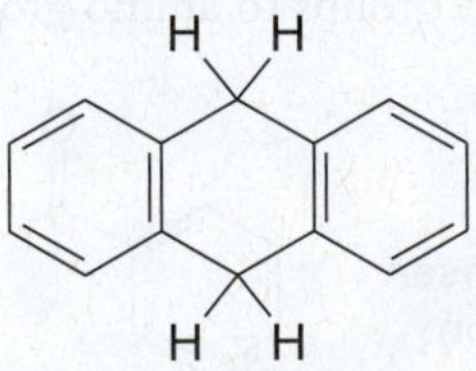

9,10-Dihydroanthracene

15.10 SUBSTITUENTS CAN AFFECT BOTH THE REACTIVITY OF THE RING AND THE ORIENTATION OF THE INCOMING GROUP

A **substituent group** already present on a benzene ring can affect both the **reactivity** of the ring toward electrophilic substitution and the **orientation** that the incoming group takes on the ring.

- A substituent can make the ring **more reactive** than benzene (i.e., it can make the compound react faster than benzene reacts). Such a group is called an **activating group**.
- A substituent can make the ring **less reactive** than benzene (i.e., it can make the compound react more slowly than benzene reacts). Such groups are called **deactivating groups**.

15.10A How Do Substituents Affect Reactivity?

Recall from Fig. 15.3 and Section 15.2 that the slow step in electrophilic aromatic substitution, the step that determines the overall rate of reaction, is the first step. In this step an electron-seeking reagent reacts by accepting an electron pair from the benzene ring.

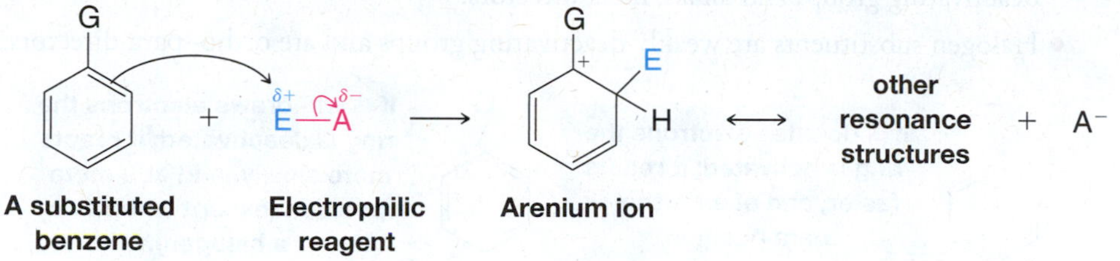

A substituted benzene   **Electrophilic reagent**   **Arenium ion**   other resonance structures   + A⁻

If a substituent that is already present on the ring makes the ring more electron rich by donating electrons to it, then the ring will be more nucleophilic, more reactive toward the electrophile, and the reaction will take place faster.

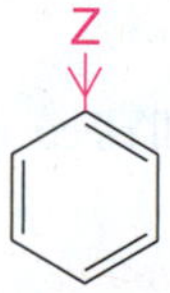

If Z donates electrons the ring is more electron rich and it reacts faster with an electrophile.

On the other hand, if the substituent on the ring withdraws electrons, the ring will be electron poor and an electrophile will react with the ring more slowly.

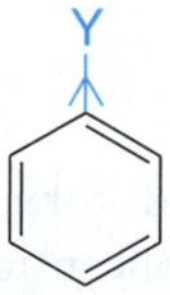

If Y withdraws electrons the ring is electron poor and it reacts more slowly with an electrophile.

15.10B Ortho–Para-Directing Groups and Meta-Directing Groups

A substituent on the ring can also affect the **orientation** that the incoming group takes when it replaces a hydrogen atom on the ring. Substituents fall into two general classes:

- **Ortho–para directors** predominantly direct the incoming group to a position ortho or para to itself.

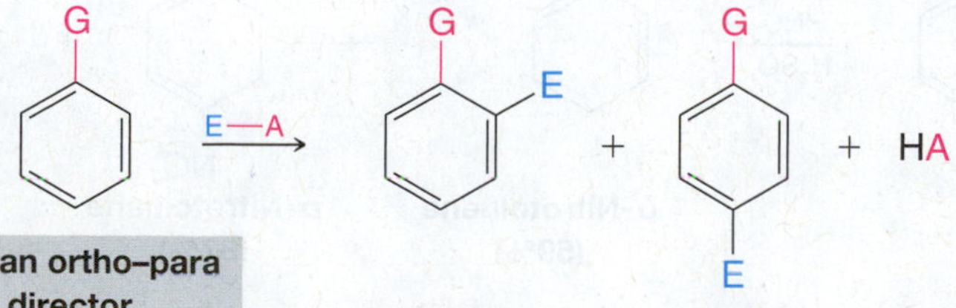

G is an ortho–para director.

Ortho product   Para product   + HA

• **Meta directors** predominantly direct the incoming group to a position meta to itself.

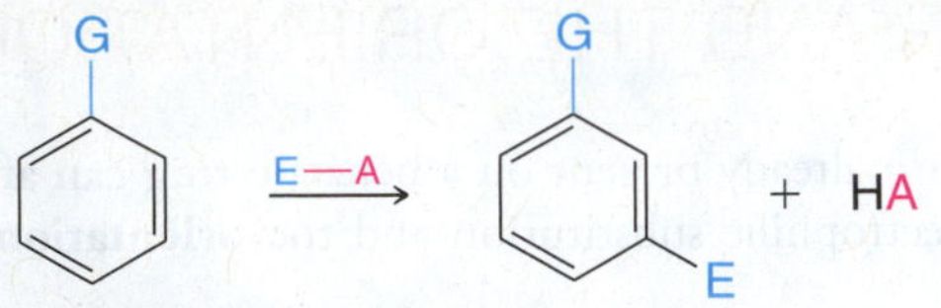

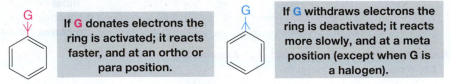

G is a meta director.   **Meta product**

15.10C Electron-Donating and Electron-Withdrawing Substituents

Whether a substituent is an activating group or a deactivating group, and whether it is an ortho–para director or a meta director, depends largely on whether the substituent donates electrons to the ring or whether it withdraws electrons.

• All electron-donating groups are activating groups and all are ortho–para directors.
• With the exception of halogen substituents, all electron-withdrawing groups are deactivating groups and all are meta directors.
• Halogen substituents are weakly deactivating groups and are ortho–para directors.

If G donates electrons the ring is activated; it reacts faster, and at an ortho or para position.

If G withdraws electrons the ring is deactivated; it reacts more slowly, and at a meta position (except when G is a halogen).

15.10D Groups: Ortho–Para Directors

• Alkyl substituents are electron-donating groups and they are **activating groups**. They are also **ortho–para directors**.

Toluene, for example, reacts considerably faster than benzene in all electrophilic substitutions:

CH_3 ← **An activating group**

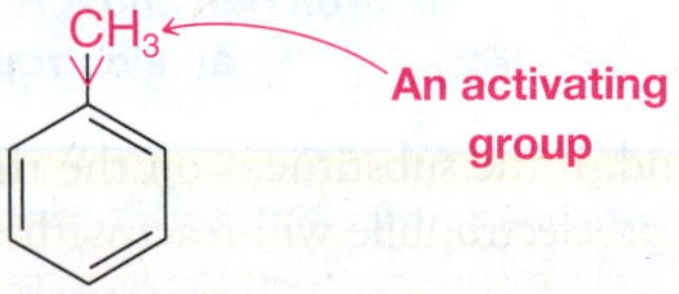

Toluene is more reactive than benzene toward electrophilic substitution.

We observe the greater reactivity of toluene in several ways. We find, for example, that with toluene, milder conditions—lower temperatures and lower concentrations of the electrophile—can be used in electrophilic substitutions than with benzene. We also find that under the same conditions toluene reacts faster than benzene. In nitration, for example, toluene reacts 25 times as fast as benzene.

We find, moreover, that when toluene undergoes electrophilic substitution, most of the substitution takes place at its ortho and para positions. When we nitrate toluene with nitric and sulfuric acids, we get mononitrotoluenes in the following relative proportions:

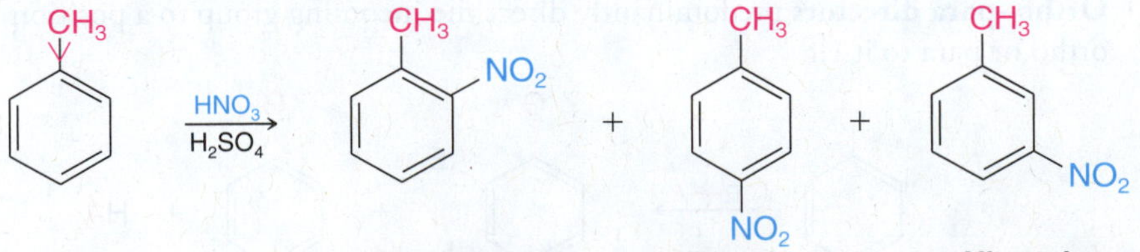

CH_3 $\xrightarrow[\text{H}_2\text{SO}_4]{\text{HNO}_3}$ CH_3 NO_2 + CH_3 + CH_3
                                                                                      NO_2 NO_2

o-Nitrotoluene   **p-Nitrotoluene**   **m-Nitrotoluene**
(59%)   **(37%)**   **(4%)**

Of the mononitrotoluenes obtained from the reaction, 96% (59% + 37%) have the nitro group in an ortho or para position. Only 4% have the nitro group in a meta position.

PRACTICE PROBLEM 15.6

Explain how the percentages just given show that the methyl group exerts an ortho–para directive effect by considering the percentages that would be obtained if the methyl group had no effect on the orientation of the incoming electrophile.

Predominant substitution of toluene at the ortho and para positions is not restricted to nitration reactions. The same behavior is observed in halogenation, sulfonation, and so forth.

- Groups that have an unshared electron pair on the atom attached to the aromatic ring, such as amino, hydroxyl, alkoxyl, and amides or esters with the oxygen or nitrogen directly bonded to the ring, are powerful activating groups and are strong ortho–para directors.

Phenol and aniline react with bromine in water (no catalyst is required) at room temperature to produce compounds in which both of the ortho positions and the para position become substituted.

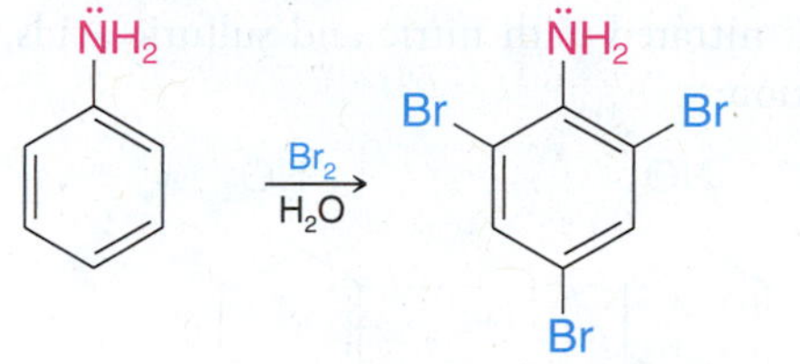

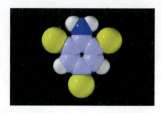

2,4,6-Tribromoaniline
(~100%)

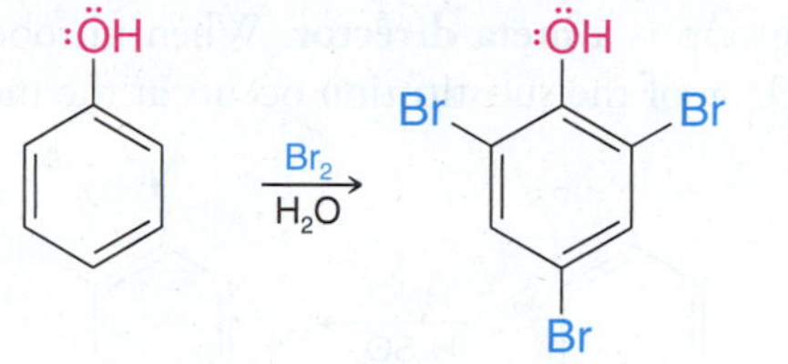

2,4,6-Tribromophenol
(~100%)

2,4,6-Tribromoaniline

- In general, substituent groups with unshared electron pairs on the atom adjacent to the benzene ring (e.g., hydroxyl, amino) are stronger activating groups than groups without unshared electron pairs (i.e., alkyl groups).
- Contribution of electron density to the benzene ring through resonance is generally stronger than through an inductive effect.

As a corollary, even though amides and esters have an unshared electron pair on the atom adjacent to the ring, their activating effect is diminished because the carbonyl group provides a resonance structure where electron density is directed away from the benzene ring. This makes amides and esters less activating than groups where the only resonance possibilities involve donation of electron density toward the benzene ring.

Examples of stabilization of an arenium ion by resonance and inductive effects

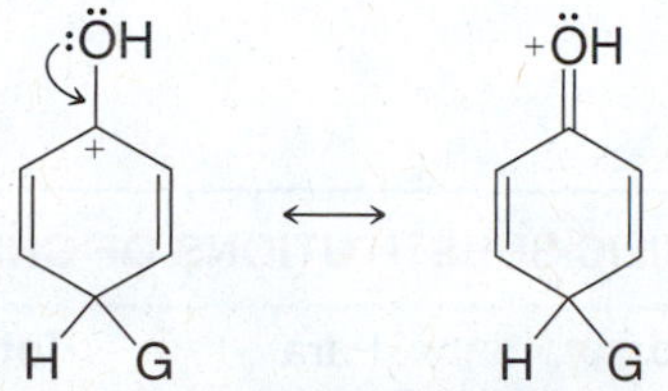

Electron donation through resonance

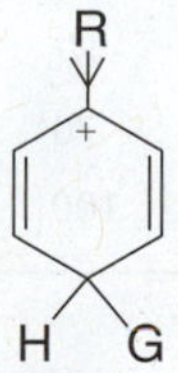

Electron donation through the inductive effect

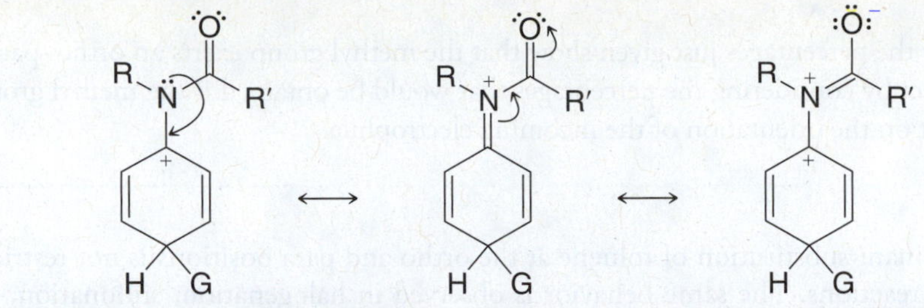

Electron donation to the ring by resonance is reduced when there is an alternative resonance pathway away from the ring.

15.10E Deactivating Groups: Meta Directors

- The nitro group is a very strong **deactivating group** and, because of the combined electronegativities of the nitrogen and oxygen atoms, it is a powerful electron-withdrawing group.

Nitrobenzene undergoes nitration at a rate only 10^{-4} times that of benzene. The nitro group is a **meta director**. When nitrobenzene is nitrated with nitric and sulfuric acids, 93% of the substitution occurs at the meta position:

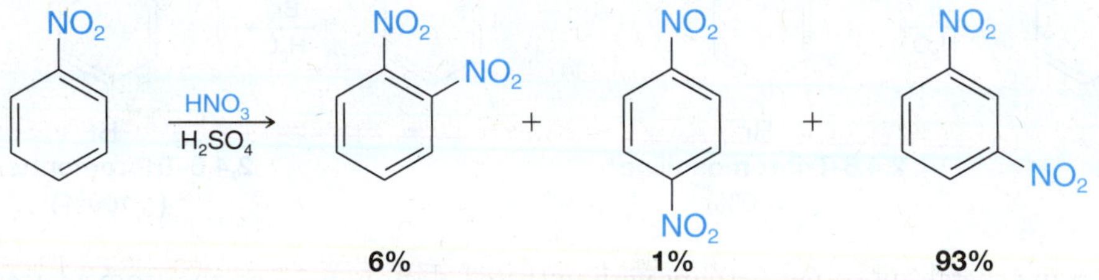

- The carboxyl group ($-CO_2H$), the sulfonic acid group ($-SO_3H$), and the trifluoromethyl group ($-CF_3$) are also deactivating groups; they are also meta directors.

15.10F Halo Substituents: Deactivating Ortho–Para Directors

- The chloro, bromo, and iodo groups are ortho–para directors. However, even though they contain unshared electron pairs, they are deactivating toward electrophilic aromatic substitution because of the electronegative effect of the halogens.

Chlorobenzene and bromobenzene, for example, undergo nitration at a rate approximately 30 times slower than benzene. The relative percentages of monosubstituted products that are obtained when chlorobenzene is chlorinated, brominated, nitrated, or sulfonated are shown in Table 15.1.

TABLE 15.1 ELECTROPHILIC SUBSTITUTIONS OF CHLOROBENZENE				
Reaction	Ortho Product (%)	Para Product (%)	Total Ortho and Para (%)	Meta Product (%)
Chlorination	39	55	94	6
Bromination	11	87	98	2
Nitration	30	70	100	
Sulfonation		100	100	

Similar results are obtained from electrophilic substitutions of bromobenzene.

15.10G Classification of Substituents

A summary of the effects of some substituents on reactivity and orientation is provided in Table 15.2.

TABLE 15.2 EFFECT OF SUBSTITUENTS ON ELECTROPHILIC AROMATIC SUBSTITUTION	
Ortho–Para Directors	**Meta Directors**
Strongly Activating	**Moderately Deactivating**
—N̈H₂, —N̈HR, —N̈R₂	—C≡N
—ÖH, —Ö:⁻	—SO₃H
Moderately Activating	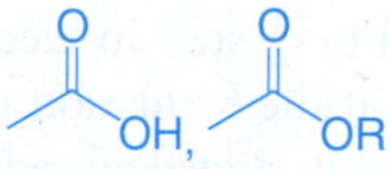
Weakly Activating	**Strongly Deactivating**
—R (alkyl)	—NO₂
—C₆H₅ (phenyl)	—NR₃⁺
Weakly Deactivating	—CF₃, CCl₃
—F̈:, —C̈l:, —B̈r:, —Ï:	

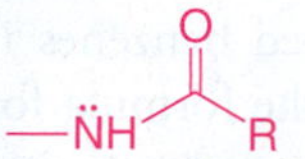

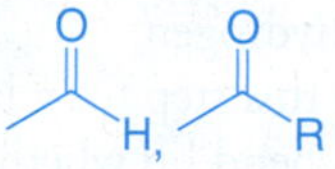

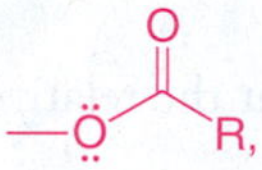

● ● ● **SOLVED PROBLEM 15.3**

Label each of the following aromatic rings as activated or deactivated based on the substituent attached, and state whether the group is an ortho–para or meta director.

(a) 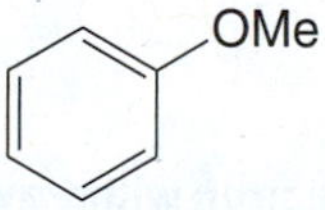 OMe   **(c)** O–C(=O)   **(e)** Cl

(b) C(=O)OMe   **(d)** 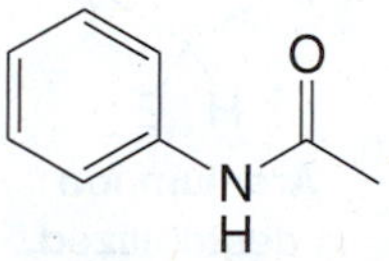 N–H–C(=O)   **(f)** 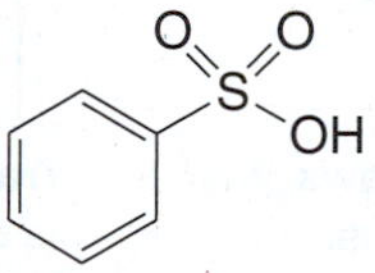 S(=O)(=O)OH

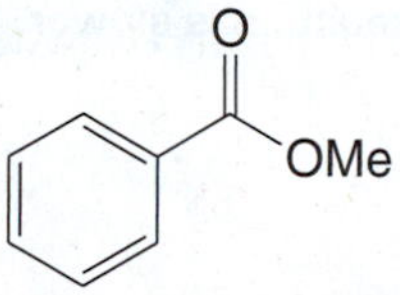

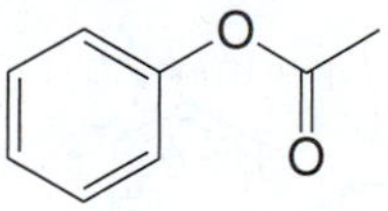

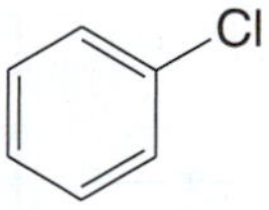

STRATEGY AND ANSWER: If a substituent donates electron density it will activate the ring and cause ortho and para substitution. If a substituent withdraws electron density it will deactivate the ring and cause meta substitution (except for halogens, which are electron withdrawing but cause ortho–para substitution). **(a)** Activated; an ether is an ortho–para director; **(b)** deactivated; the ester carbonyl is a meta director; **(c)** activated; the single-bonded oxygen of the ester is directly bonded to the ring, and therefore it is an ortho–para director; **(d)** activated; the amide nitrogen is an ortho–para director; **(e)** deactivated; however, the halogen is ortho–para director through resonance; **(f)** deactivated; the sulfonate group is a meta director.

● ● ● ....

PRACTICE PROBLEM 15.7

Predict the major products formed when:

(a) Toluene is sulfonated.   **(c)** Nitrobenzene is brominated.

(b) Benzoic acid is nitrated.   **(d)** Isopropylbenzene reacts with acetyl chloride and AlCl₃.

If the major products would be a mixture of ortho and para isomers, you should so state.

15.11 HOW SUBSTITUENTS AFFECT ELECTROPHILIC AROMATIC SUBSTITUTION: A CLOSER LOOK

15.11A Reactivity: The Effect of Electron-Releasing and Electron-Withdrawing Groups

We can account for relative reaction rates by examining the transition state for the rate-determining steps.

- Any factor that reduces the energy of the transition state relative to that of the reactants lowers the free energy of activation and increases the relative rate of the reaction.

- Any factor that raises the energy of the transition state relative to that of the reactants increases the free energy of activation and decreases the relative rate of the reaction.

The rate-determining step in electrophilic substitutions of substituted benzenes is the step that results in the formation of the arenium ion. We can write the formula for a substituted benzene in a generalized way if we use the letter G to represent any ring substituent, including hydrogen.

When we examine this step for a large number of reactions, we find that the relative rates of the reactions depend on whether G **releases** or **withdraws** electrons.

- If G is an electron-releasing group (relative to hydrogen), the reaction occurs faster than the corresponding reaction of benzene.

- If G is an electron-withdrawing group, the reaction is slower than that of benzene:

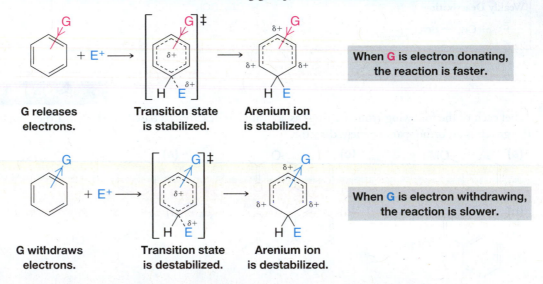

It appears, then, that the substituent (G) must affect the stability of the transition state relative to that of the reactants.

- Electron-releasing groups make the transition state more stable, whereas electron-withdrawing groups make it less stable.

That this is so is entirely reasonable, because the transition state resembles the arenium ion, and the arenium ion is a delocalized *carbocation*.

This effect illustrates another application of the Hammond–Leffler postulate (Section 6.13A). The arenium ion is a high-energy intermediate, and the step that leads to it is a *highly endothermic step*. Thus, according to the Hammond–Leffler postulate, there should be a strong resemblance between the arenium ion itself and the transition state leading to it.

Since the arenium ion is positively charged, we would expect an electron-releasing group to stabilize the arenium ion *and the transition state leading to it,* for the transition state is a developing delocalized carbocation. We can make the same kind of arguments about the effect of electron-withdrawing groups. An electron-withdrawing group should make the arenium ion *less stable*, and in a corresponding way it should make the transition state leading to the arenium ion *less stable*.

Figure 15.4 shows how the electron-withdrawing and electron-releasing abilities of substituents affect the relative free energies of activation of electrophilic aromatic substitution reactions.

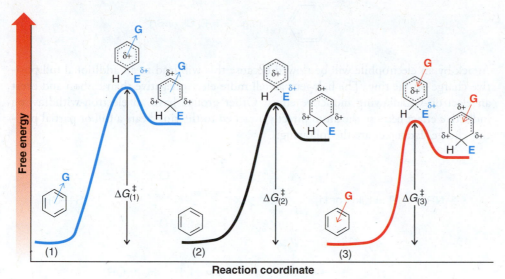

FIGURE 15.4 A comparison of free-energy profiles for arenium ion formation in a ring with an electron-withdrawing substituent (➤ G), no substituent, and an electron-donating substituent (◄ G). In (1) (blue energy profile), the electron-withdrawing group G raises the transition state energy. The energy of activation barrier is the highest, and therefore the reaction is the slowest. Reaction (2), with no substituent, serves as a reference for comparison. In (3) (red energy profile), an electron-donating group G stabilizes the transition state. The energy of activation barrier is the lowest, and therefore the reaction is the fastest.

Calculated electrostatic potential maps for two arenium ions comparing the charge-stabilizing effect of an electron-donating methyl group with the charge-destabilizing effect of an electron-withdrawing trifluoromethyl group are shown in Fig. 15.5. The arenium ion at the left (Fig. 15.5*a*) is that from electrophilic addition of bromine to methylbenzene (toluene) at the para position. The arenium ion at the right (Fig. 15.5*b*) is that from electrophilic addition of bromine to trifluoromethylbenzene at the meta position. Notice that the atoms of the ring in Fig. 15.5*a* have much less blue color associated with them, showing that they are much less positive and that the ring is stabilized.

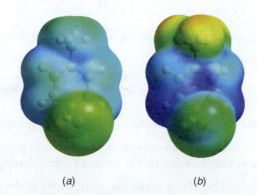

(a)   (b)

FIGURE 15.5 Calculated electrostatic potential maps for the arenium ions from electrophilic addition of bromine to (a) methylbenzene (toluene) and (b) trifluoromethylbenzene. The positive charge in the arenium ion ring of methylbenzene (a) is delocalized by the electron-releasing ability of the methyl group, whereas the positive charge in the arenium ion of trifluoromethylbenzene (b) is enhanced by the electron-withdrawing effect of the trifluoromethyl group. (The electrostatic potential maps for the two structures use the same color scale with respect to potential so that they can be directly compared.)

15.11B Inductive and Resonance Effects: Theory of Orientation

We can account for the electron-withdrawing and electron-releasing properties of groups on the basis of two factors: *inductive effects* and *resonance effects*. We shall also see that these two factors determine orientation in aromatic substitution reactions.

Inductive Effects The **inductive effect** of a substituent G arises from the electrostatic interaction of the polarized bond to G with the developing positive charge in the ring as it is attacked by an electrophile. If, for example, G is a more electronegative atom (or group) than carbon, then the ring will be at the positive end of the dipole:

$$\overset{\delta-}{G} \overset{\delta+}{\longleftarrow} \hspace{1em} \text{(e.g., G = F, Cl, or Br)}$$

Attack by an electrophile will be slowed because this will lead to an additional full positive charge on the ring. The halogens are all more electronegative than carbon and exert an electron-withdrawing inductive effect. Other groups have an electron-withdrawing inductive effect because the atom directly attached to the ring bears a full or partial positive charge. Examples are the following:

$$\rightarrow \overset{+}{N}R_3 \quad \text{(R = alkyl or H)} \qquad \overset{X^{\delta-}}{\underset{X^{\delta-}}{\rightarrow \overset{\uparrow}{\underset{\downarrow}{C^{\delta+}}} \rightarrow X^{\delta-}}} \qquad \rightarrow \overset{O}{\underset{O^-}{N^+}} \qquad \rightarrow \overset{O}{\underset{O}{S^{+}}}-OH$$

$$\rightarrow \overset{\overset{\cdot\cdot}{O}}{\underset{G}{\overset{\|}{C}}} \longleftrightarrow \rightarrow \overset{:\overset{\cdot\cdot}{O}:^-}{\underset{G}{C^+}} \qquad \text{(G = H, R, OH, or OR)}$$

> **Electron-withdrawing groups with a full or partial charge on the atom attached to the ring**

Resonance Effects The **resonance** effect of a substituent G refers to the possibility that the presence of G may increase or decrease the resonance stabilization of the intermediate arenium ion. The G substituent may, for example, cause one of the three contributors to the resonance hybrid for the arenium ion to be better or worse than the case when G is hydrogen. Moreover, when G is an atom bearing one or more nonbonding electron pairs, it may lend extra stability to the arenium ion by providing a *fourth* resonance contributor in which the positive charge resides on G:

$$E \overset{+}{\underset{H}{\bigcirc}}\overset{\ddot{G}}{} \longleftrightarrow E \overset{}{\underset{H}{\bigcirc}}G^+$$

This electron-donating resonance effect applies with decreasing strength in the following order:

Most electron donating $\overset{\cdot\cdot}{-NH_2}, \; -\overset{\cdot\cdot}{N}R_2 \; > \; -\overset{\cdot\cdot}{O}H, \; -\overset{\cdot\cdot}{O}R \; > \; -\overset{\cdot\cdot}{\underset{\cdot\cdot}{X}}:$ **Least electron donating**

This order also indicates the relative activating ability of these groups.

- Amino groups are highly activating, hydroxyl and alkoxyl groups are somewhat less activating, and halogen substituents are weakly deactivating.

When X = F, this order can be related to the electronegativity of the atoms with the nonbonding pair. The more electronegative the atom is, the less able it is to accept the positive charge (fluorine is the most electronegative, nitrogen the least). When X = Cl, Br, or I, the relatively poor electron-donating ability of the halogens by resonance is understandable on a different basis. These atoms (Cl, Br, and I) are all larger than carbon, and, therefore, the orbitals that contain the nonbonding pairs are further from the nucleus and do not overlap well with the $2p$ orbital of carbon. this is a general phenomenon: resonance effects are not transmitted well between atoms of different rows in the periodic table.

15.11C Meta-Directing Groups

- All meta-directing groups have either a partial positive charge or a full positive charge on the atom directly attached to the ring.

As a typical example let us consider the trifluoromethyl group. The trifluoromethyl group, because of the three highly electronegative fluorine atoms, is strongly electron withdrawing. It is a strong deactivating group and a powerful meta director in electrophilic aromatic substitution reactions. We can account for both of these characteristics of the trifluoromethyl group in the following way.

The trifluoromethyl group affects the rate of reaction by causing the transition state leading to the arenium ion to be highly unstable. It does this by withdrawing electrons from the developing carbocation, thus increasing the positive charge on the ring:

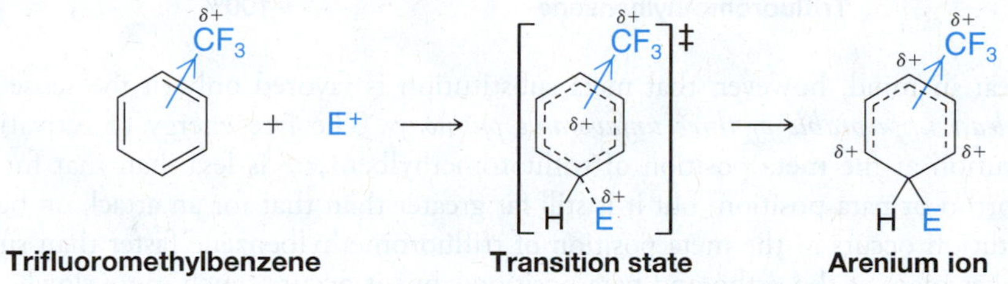

Trifluoromethylbenzene   **Transition state**   **Arenium ion**

We can understand how the trifluoromethyl group affects *orientation* in electrophilic aromatic substitution if we examine the resonance structures for the arenium ion that would be formed when an electrophile attacks the ortho, meta, and para positions of trifluoromethylbenzene.

Ortho Attack

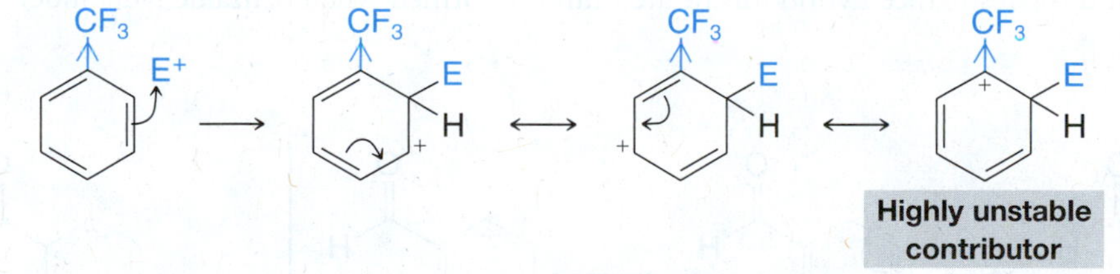

Highly unstable contributor

Meta Attack

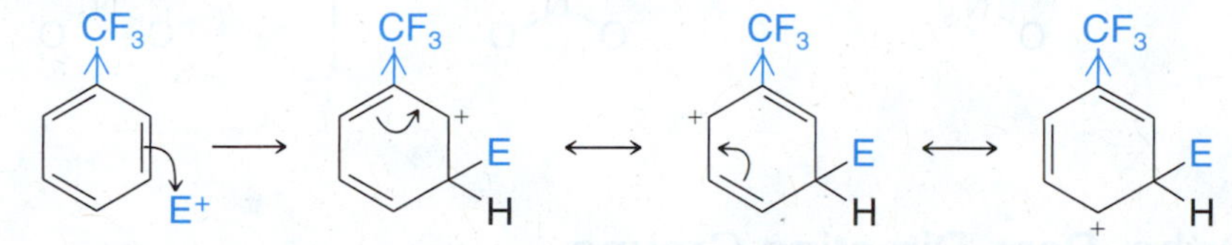

Para Attack

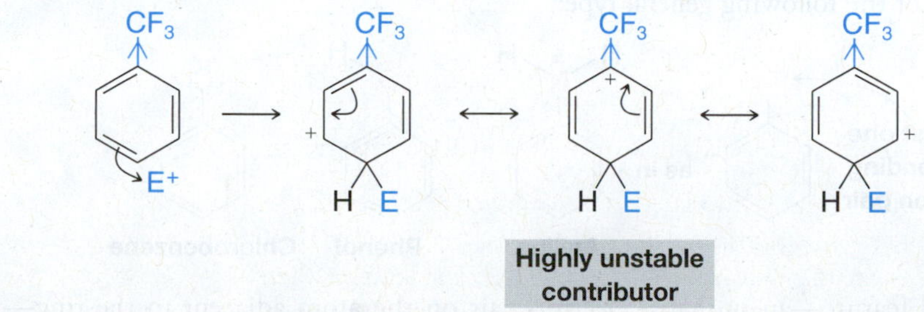

Highly unstable contributor

- The arenium ion arising from ortho and para attack each has *one contributing structure that is highly unstable relative to all the others because the positive charge is located on the ring carbon that bears the electron-withdrawing group.*

- The arenium ion arising from meta attack has *no* such highly unstable resonance structure.
- By the usual reasoning we would also expect the transition state leading to the meta-substituted arenium ion to be the least unstable and, therefore, that meta attack would be favored.

This is exactly what we find experimentally. The trifluoromethyl group is a powerful meta director:

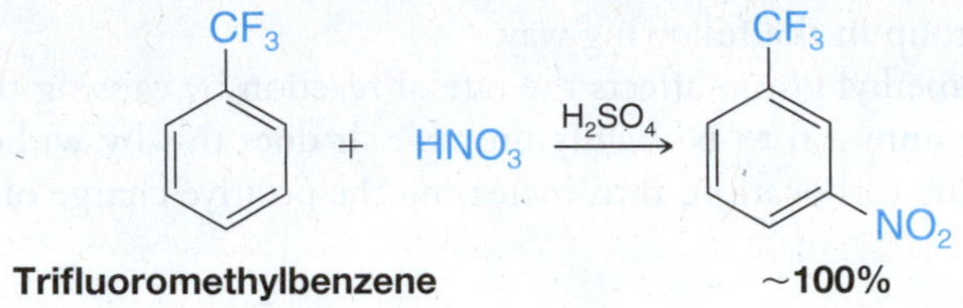

Trifluoromethylbenzene

~100%

Bear in mind, however, that meta substitution is favored only in the sense that *it is the least unfavorable of three unfavorable pathways*. The free energy of activation for substitution at the meta position of trifluoromethylbenzene is less than that for attack at an ortho or para position, but it is still far greater than that for an attack on benzene. Substitution occurs at the meta position of trifluoromethylbenzene faster than substitution takes place at the ortho and para positions, but it occurs much more slowly than it does with benzene.

- The nitro group, the carboxyl group, and other meta-directing groups (see Table 15.2) are all powerful electron-withdrawing groups and act in a similar way.

●●● SOLVED PROBLEM 15.4

Write contributing resonance structures and the resonance hybrid for the arenium ion formed when benzaldehyde undergoes nitration at the meta position.

STRATEGY AND ANSWER:

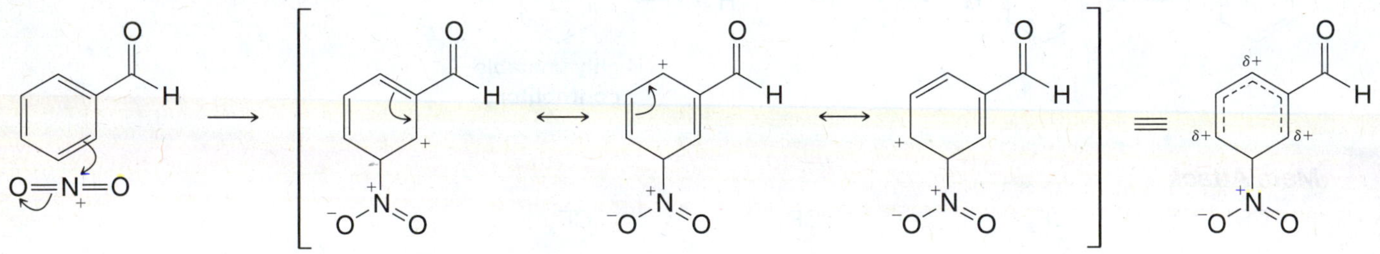

15.11D Ortho–Para-Directing Groups

Except for the alkyl and phenyl substituents, all of the ortho–para-directing groups in Table 15.2 are of the following general type:

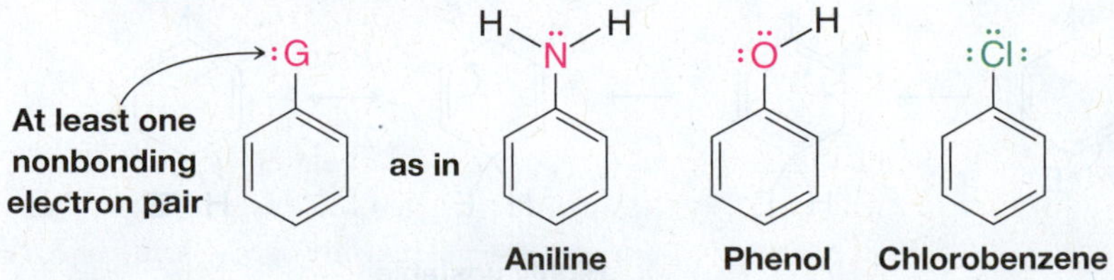

At least one nonbonding electron pair

as in

Aniline **Phenol** **Chlorobenzene**

This structural feature—an unshared electron pair on the atom adjacent to the ring—determines the orientation and influences reactivity in electrophilic substitution reactions.

The *directive effect* of groups with an unshared pair is predominantly caused by an electron-releasing resonance effect. The resonance effect, moreover, operates primarily in the arenium ion and, consequently, in the transition state leading to it.

Except for the halogens, the primary effect of these groups on relative reactivity of the benzene ring is also caused by an electron-releasing resonance effect. And, again, this effect operates primarily in the transition state leading to the arenium ion.

In order to understand these resonance effects, let us begin by recalling the effect of the amino group on electrophilic aromatic substitution reactions. The amino group is not only a powerful activating group, it is also a powerful ortho–para director. We saw earlier (Section 15.10D) that aniline reacts with bromine in aqueous solution at room temperature and in the absence of a catalyst to yield a product in which both ortho positions and the para position are substituted.

The inductive effect of the amino group makes it slightly electron withdrawing. Nitrogen, as we know, is more electronegative than carbon. The difference between the electronegativities of nitrogen and carbon in aniline is not large, however, because the carbon of the benzene ring is sp^2 hybridized and so it is somewhat more electronegative than it would be if it were sp^3 hybridized.

- The resonance effect of the amino group is far more important than its inductive effect in electrophilic aromatic substitution, and this resonance effect makes the amino group electron releasing.

We can understand this effect if we write the resonance structures for the arenium ions that would arise from ortho, meta, and para attack on aniline:

Ortho Attack

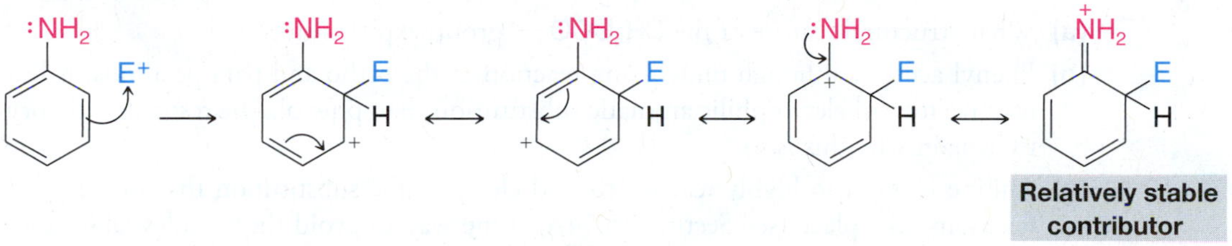

Relatively stable contributor

Meta Attack

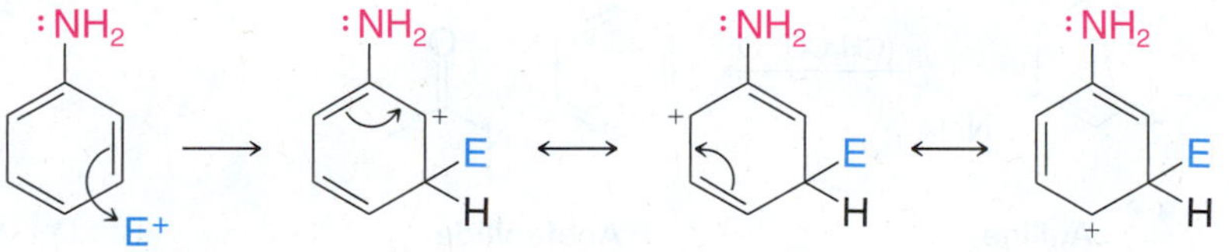

Para Attack

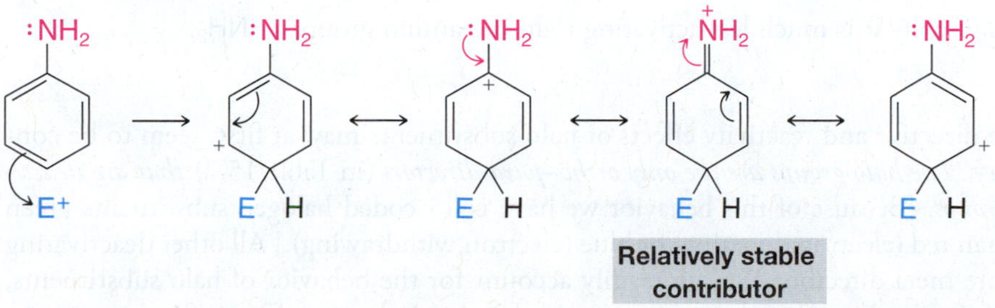

Relatively stable contributor

Four reasonable resonance structures can be written for the arenium ions resulting from ortho and para attack, whereas only three can be written for the arenium ion that results from meta attack. This observation, in itself, suggests that the ortho- and para-substituted arenium ions should be more stable. Of greater importance, however, are the relatively stable structures that contribute to the hybrid for the ortho- and para-substituted arenium ions. In these structures, nonbonding pairs of electrons from nitrogen form an additional covalent bond to the carbon of the ring. This extra bond—and the fact that every atom in each of these structures has a complete outer octet of electrons—makes these structures the most stable of all of the contributors. Because these structures are unusually stable, they make a large—*and stabilizing*—contribution to the hybrid. This means, of course, that the ortho- and para-substituted arenium ions themselves are considerably more stable than the arenium ion that results from the meta attack. The transition states leading to the ortho- and para-substituted arenium ions occur at unusually low free energies. As a result, electrophiles react at the ortho and para positions very rapidly.

PRACTICE PROBLEM 15.8 Use resonance theory to explain why the hydroxyl group of phenol is an activating group and an ortho–para director. Illustrate your explanation by showing the arenium ions formed when phenol reacts with a Br^+ ion at the ortho, meta, and para positions.

PRACTICE PROBLEM 15.9 Phenol reacts with acetic anhydride in the presence of sodium acetate to produce the ester phenyl acetate:

$$\text{Phenol} \xrightarrow[\text{CH}_3\text{CO}_2\text{Na}]{(\text{CH}_3\text{CO})_2\text{O}} \text{Phenyl acetate}$$

The CH_3COO- group of phenyl acetate, like the $-OH$ group of phenol (Practice Problem 15.8), is an ortho–para director.

(a) What structural feature of the CH_3COO- group explains this?

(b) Phenyl acetate, although undergoing reaction at the ortho and para positions, is less reactive toward electrophilic aromatic substitution than phenol. Use resonance theory to explain why this is so.

(c) Aniline is often so highly reactive toward electrophilic substitution that undesirable reactions take place (see Section 15.14A). One way to avoid these undesirable reactions is to convert aniline to acetanilide (below) by treating aniline with acetyl chloride or acetic anhydride:

$$\text{Aniline} \xrightarrow{(\text{CH}_3\text{CO})_2\text{O}} \text{Acetanilide}$$

What kind of directive effect would you expect the acetamido group (CH_3CONH-) to have?

(d) Explain why it is much less activating than the amino group, $-NH_2$.

The directive and reactivity effects of halo substituents may, at first, seem to be contradictory. *The halo groups are the only ortho–para directors* (in Table 15.2) *that are deactivating groups.* [Because of this behavior we have color coded halogen substituents green rather than red (electron donating) or blue (electron withdrawing).] All other deactivating groups are meta directors. We can readily account for the behavior of halo substituents, however, if we assume that their electron-withdrawing inductive effect influences *reactivity* and their electron-donating resonance effect governs *orientation*.

Let us apply these assumptions specifically to chlorobenzene. The chlorine atom is highly electronegative. Thus, we would expect a chlorine atom to withdraw electrons from the benzene ring and thereby deactivate it:

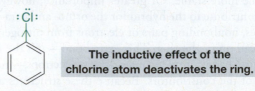

The inductive effect of the
chlorine atom deactivates the ring.

On the other hand, when electrophilic attack does take place, the chlorine atom stabilizes the arenium ions resulting from ortho and para attack relative to that from meta attack. The chlorine atom does this in the same way as amino groups and hydroxyl groups do—*by donating an unshared pair of electrons*. These electrons give rise to relatively stable resonance structures contributing to the hybrids for the ortho- and para-substituted arenium ions.

Ortho Attack

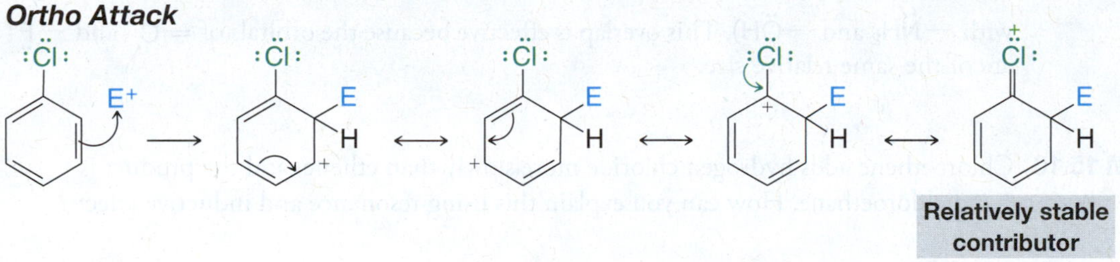

Relatively stable contributor

Meta Attack

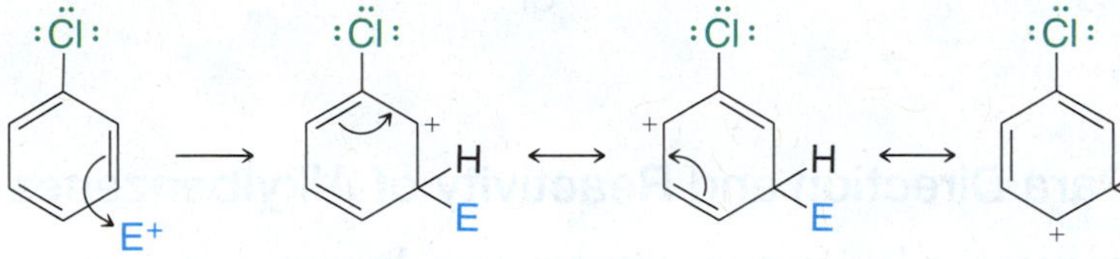

Para Attack

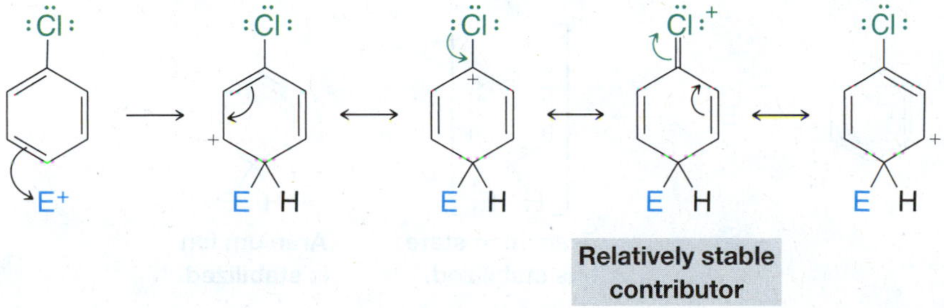

Relatively stable contributor

What we have said about chlorobenzene is also true of bromobenzene.

We can summarize the inductive and resonance effects of halo substituents in the following way.

- Through their electron-withdrawing inductive effect, halo groups make the ring more electron deficient than that of benzene. This causes the free energy of activation for any electrophilic aromatic substitution reaction to be greater than that for benzene, and, therefore, halo groups are deactivating.

- Through their electron-donating resonance effect, however, halo substituents cause the free energies of activation leading to ortho and para substitution to be lower than the free energy of activation leading to meta substitution. This makes halo substituents ortho–para directors.

You may have noticed an apparent contradiction between the rationale offered for the unusual effects of the halogens and that offered earlier for amino or hydroxyl groups. That is, oxygen is *more* electronegative than chlorine or bromine (and especially iodine). Yet the hydroxyl group is an activating group, whereas halogens are deactivating groups. An explanation for this can be obtained if we consider the relative stabilizing contributions made to the transition state leading to the arenium ion by resonance structures involving a group $-\ddot{G}(-\ddot{G} = -\ddot{N}H_2, -\ddot{O}-H, -\ddot{F}:, -\ddot{C}l:, -\ddot{B}r:, -\ddot{I}:)$ that is directly attached to the benzene ring in which G donates an electron pair. If $-\ddot{G}$ is $-\ddot{O}H$, or $-\ddot{N}H_2$ these resonance structures arise because of the overlap of a $2p$ orbital of carbon with that of oxygen or nitrogen. Such overlap is favorable because the atoms are almost the same size. With chlorine, however, donation of an electron pair to the benzene ring requires overlap of a carbon $2p$ orbital with a chlorine $3p$ orbital. Such overlap is less effective; the chlorine atom is much larger and its $3p$ orbital is much further from its nucleus. With bromine and iodine, overlap is even less effective. Justification for this explanation can be found in the observation that fluorobenzene (G $=-\ddot{\text{F}}\!:$) is the most reactive halobenzene in spite of the high electronegativity of fluorine and the fact that $-\ddot{\text{F}}\!:$ is the most powerful ortho–para director of the halogens. With fluorine, donation of an electron pair arises from overlap of a $2p$ orbital of fluorine with a $2p$ orbital of carbon (as with $-\ddot{\text{N}}\text{H}_2$ and $-\ddot{\text{O}}\text{H}$). This overlap is effective because the orbitals of $=\text{C}$ and $-\ddot{\text{F}}\!:$ are of the same relative size.

PRACTICE PROBLEM 15.10 Chloroethene adds hydrogen chloride more slowly than ethene, and the product is 1,1-dichloroethane. How can you explain this using resonance and inductive effects?

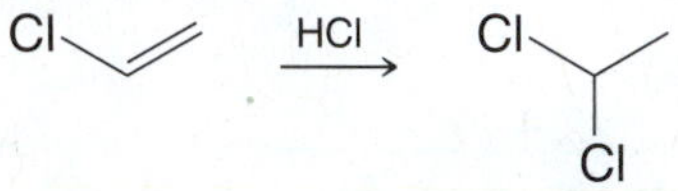

15.11E Ortho–Para Direction and Reactivity of Alkylbenzenes

Alkyl groups are better electron-releasing groups than hydrogen. Because of this, they can activate a benzene ring toward electrophilic substitution by stabilizing the transition state leading to the arenium ion:

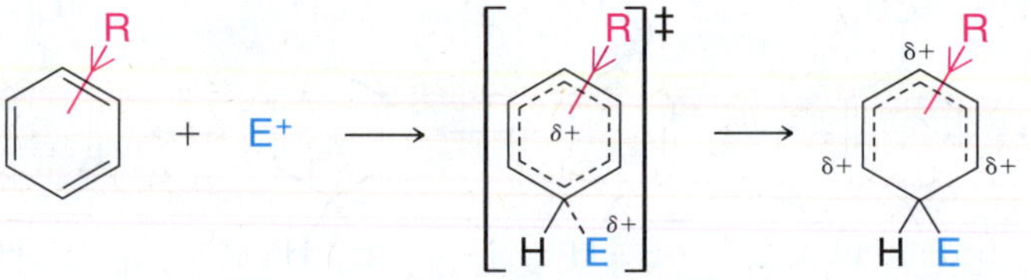

Transition state is stabilized.   **Arenium ion is stabilized.**

For an alkylbenzene the free energy of activation of the step leading to the arenium ion (just shown) is lower than that for benzene, and alkylbenzenes react faster.

Alkyl groups are ortho–para directors. We can also account for this property of alkyl groups on the basis of their ability to release electrons—an effect that is particularly important when the alkyl group is attached directly to a carbon that bears a positive charge. (Recall the ability of alkyl groups to stabilize carbocations that we discussed in Section 6.11 and in Fig. 6.8.)

If, for example, we write resonance structures for the arenium ions formed when toluene undergoes electrophilic substitution, we get the results shown below:

Ortho Attack

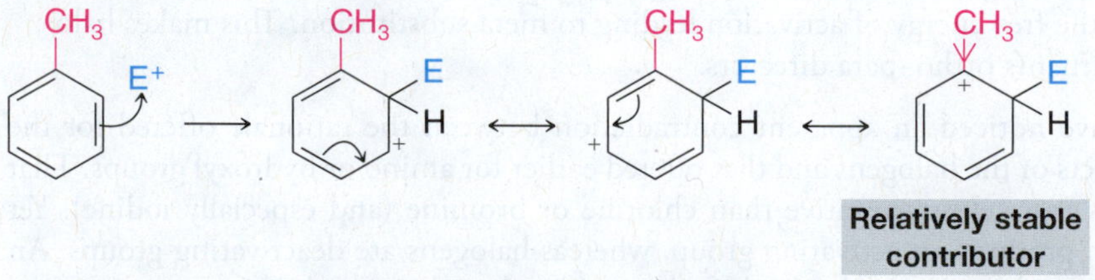

Relatively stable contributor

Meta Attack

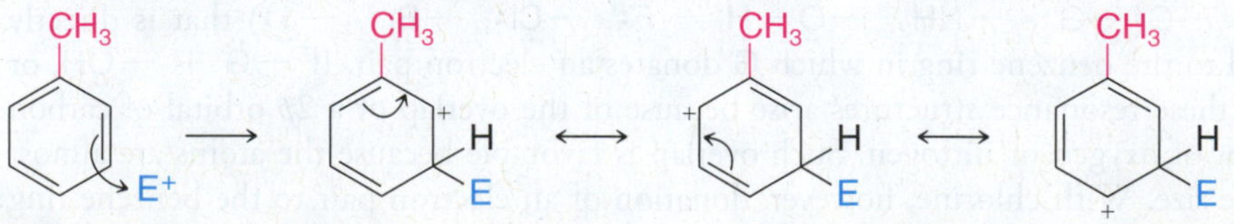

Para Attack

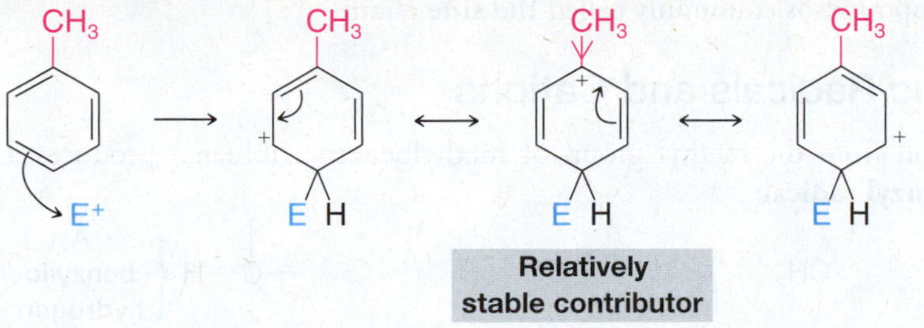

Relatively
stable contributor

In ortho attack and para attack we find that we can write resonance structures in which the methyl group is directly attached to a positively charged carbon of the ring. These structures are more *stable* relative to any of the others because in them the stabilizing influence of the methyl group (by inductive electron release) is most effective. These structures, therefore, make a large (stabilizing) contribution to the overall hybrid for ortho- and para-substituted arenium ions. No such relatively stable structure contributes to the hybrid for the meta-substituted arenium ion, and as a result it is less stable than the ortho- or para-substituted arenium ions. Since the ortho- and para-substituted arenium ions are more stable, the transition states leading to them occur at lower energy and ortho and para substitutions take place most rapidly.

Write resonance structures for the arenium ions formed when ethylbenzene reacts with a Br^+ ion (as formed from $Br_2/FeBr_3$) to produce the following ortho and para products.

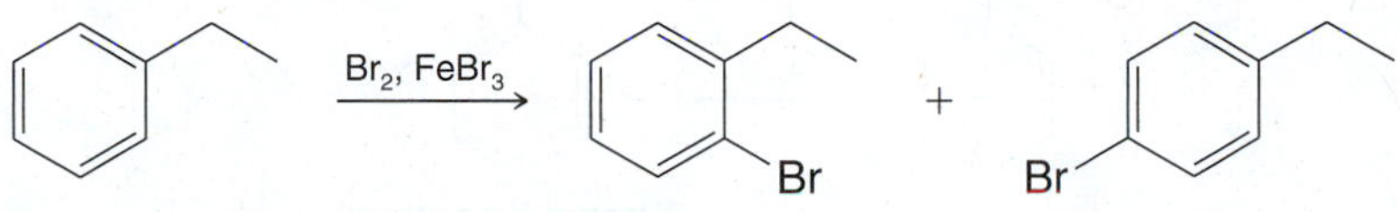

PRACTICE PROBLEM 15.11

Provide a mechanism for the following reaction and explain why it occurs faster than nitration of benzene.

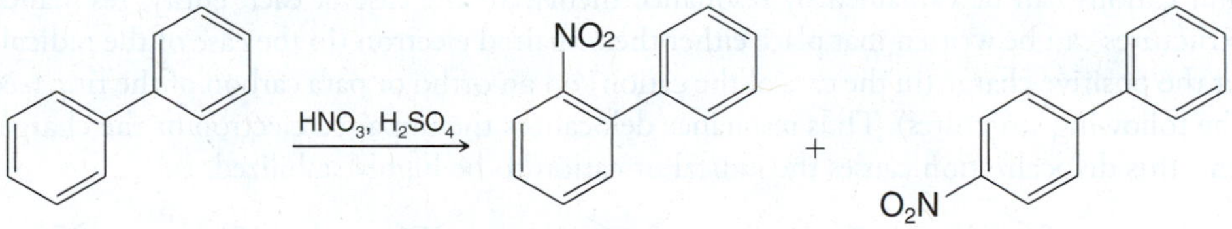

PRACTICE PROBLEM 15.12

15.11F Summary of Substituent Effects on Orientation and Reactivity

With a theoretical understanding now in hand of **substituent effects** on orientation and reactivity, we refer you back to Table 15.2 for a summary of specific groups and their effects.

15.12 REACTIONS OF THE SIDE CHAIN OF ALKYLBENZENES

Hydrocarbons that consist of both aliphatic and aromatic groups are also known as **arenes**. Toluene, ethylbenzene, and isopropylbenzene are **alkylbenzenes**:

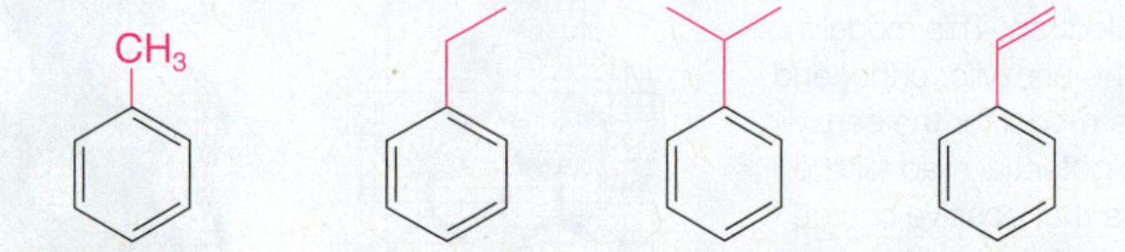

Methylbenzene (toluene)   Ethylbenzene   Isopropyl-benzene (cumene)   Phenylethene (styrene or vinylbenzene)

Phenylethene, usually called styrene, is an example of an **alkenylbenzene**. The aliphatic portion of these compounds is commonly called the **side chain**.

15.12A Benzylic Radicals and Cations

Hydrogen abstraction from the methyl group of methylbenzene (toluene) produces a radical called the **benzyl radical**:

Methylbenzene
(toluene)

−RH →

The benzyl
radical

A benzylic
radical

Benzylic
carbon

A
benzylic
hydrogen

Benzylic
carbon

The name benzyl radical is used as a specific name for the radical produced in this reaction. The general name **benzylic radical** applies to all radicals that have an unpaired electron on the side-chain carbon atom that is directly attached to the benzene ring (Section 10.9). The hydrogen atoms of the carbon atom directly attached to the benzene ring are called **benzylic hydrogen atoms**. A group bonded at a benzylic position is called a **benzylic substituent**.

Departure of a leaving group (**LG**) from a benzylic position produces a **benzylic cation**:

−LG⁻ →

A benzylic cation

Benzylic radicals and benzylic cations are *conjugated unsaturated systems* and *both are unusually stable*. They have approximately the same stabilities as allylic radicals (Section 10.8) and allylic cations (Section 13.4). This exceptional stability of benzylic radicals and cations can be explained by resonance theory. In the case of each entity, resonance structures can be written that place either the unpaired electron (in the case of the radical) or the positive charge (in the case of the cation) on an ortho or para carbon of the ring (see the following structures). Thus resonance delocalizes the unpaired electron or the charge, and this delocalization causes the radical or cation to be highly stabilized.

Benzylic radicals are stabilized by resonance.

Benzylic cations are stabilized by resonance.

Calculated structures for the benzyl radical and benzyl cation are presented in Fig. 15.6. These structures show the presence at their ortho and para carbons of unpaired electron density in the radical and positive charge in the cation, consistent with the resonance structures above.

FIGURE 15.6 The gray lobes in the calculated structure for the benzyl radical (left) show the location of density from the unpaired electron. This model indicates that the unpaired electron resides primarily at the benzylic, ortho, and para carbons, which is consistent with the resonance model for the benzylic radical discussed earlier. The calculated electrostatic potential map for the bonding electrons in the benzyl cation (right) indicates that positive charge (blue regions) resides primarily at the benzylic, ortho, and para carbons, which is consistent with the resonance model for the benzylic cation. The van der Waals surface of both structures is represented by the wire mesh.

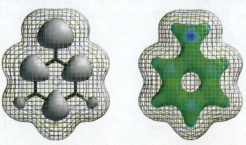

THE CHEMISTRY OF... Industrial Styrene Synthesis

Styrene is one of the most important industrial chemicals—more than 11 billion pounds is produced each year. The starting material for a major commercial synthesis of styrene is ethylbenzene, produced by Friedel–Crafts alkylation of benzene:

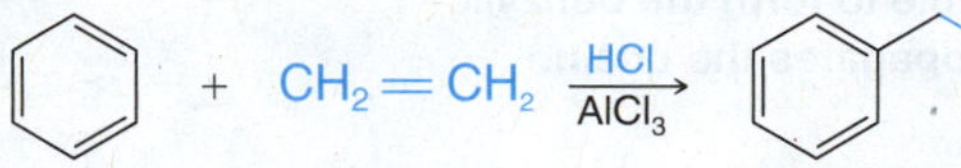

$$C_6H_5-CH_2CH_3 \xrightarrow[630\,°C]{catalyst} C_6H_5-CH=CH_2 + H_2$$

Styrene
(90–92% yield)

$$C_6H_5 + CH_2=CH_2 \xrightarrow[AlCl_3]{HCl} \text{Ethylbenzene}$$

Ethylbenzene

Ethylbenzene is then dehydrogenated in the presence of a catalyst (zinc oxide or chromium oxide) to produce styrene.

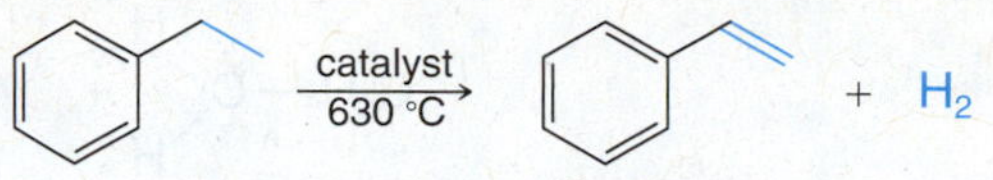

Most styrene is polymerized (Special Topic A) to the familiar plastic, polystyrene:

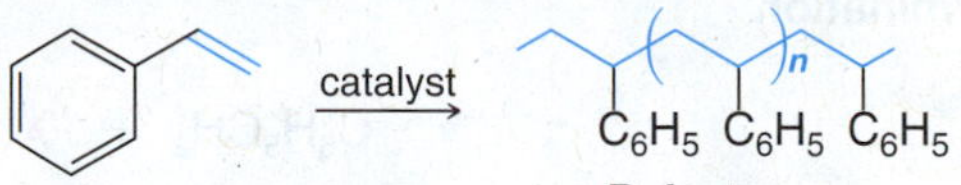

$$styrene \xrightarrow{catalyst} \overbrace{(CH_2CH)_n}^{C_6H_5\ C_6H_5\ C_6H_5}$$

Polystyrene

15.12B Benzylic Halogenation of the Side Chain

We have already seen in this chapter that we can substitute bromine and chlorine for hydrogen atoms on the benzene ring of toluene and other alkylaromatic compounds using electrophilic aromatic substitution reactions. We can also substitute bromine and chlorine for hydrogen atoms on the **benzylic** carbons of alkyl side chains by radical reactions in the presence of heat, light, or a radical initiator like a peroxide, as we first saw in Chapter 10, (Section 10.9). This is made possible by the special stability of the benzylic radical intermediate (Section 15.12A). For example, benzylic chlorination of toluene takes place in the gas phase at 400–600 °C or in the presence of UV light, as shown here. Multiple substitutions occur with an excess of chlorine.

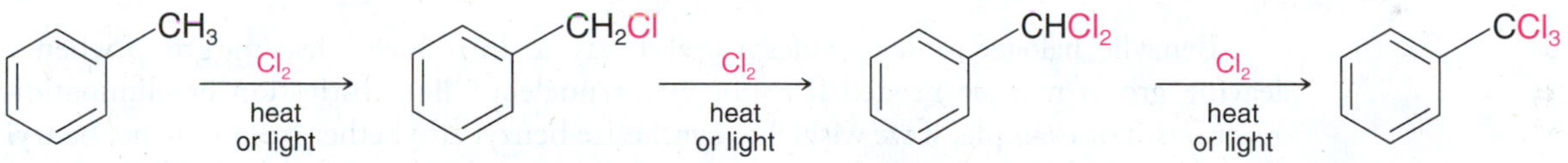

$$C_6H_5CH_3 \xrightarrow[\text{heat or light}]{Cl_2} C_6H_5CH_2Cl \xrightarrow[\text{heat or light}]{Cl_2} C_6H_5CHCl_2 \xrightarrow[\text{heat or light}]{Cl_2} C_6H_5CCl_3$$

Benzyl chloride   **Dichloromethyl-benzene**   **Trichloromethyl-benzene**

[A MECHANISM FOR THE REACTION — Benzylic Halogenation]

Chain Initiation

Step 1

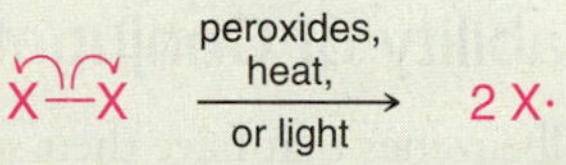

$$X-X \xrightarrow[\text{or light}]{\text{peroxides, heat,}} 2\,X\cdot$$

Peroxides, heat, or light cause halogen molecules to cleave into redicals.

Chain Propagation

Step 2

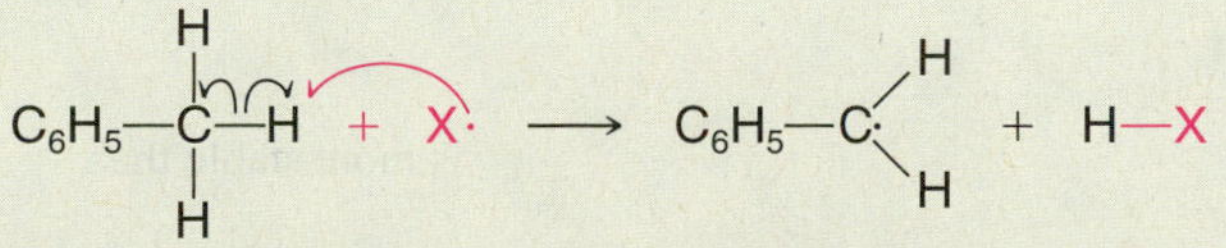

$$C_6H_5-\underset{H}{\overset{H}{C}}-H + X\cdot \longrightarrow C_6H_5-\overset{H}{\underset{H}{C}}\cdot + H-X$$

Benzyl radical

A halogen radical abstracts a benzylic hydrogen atom, forming a benzylic radical and a molecule of the hydrogen halide.

(mechanism continues on the next page)

Step 3

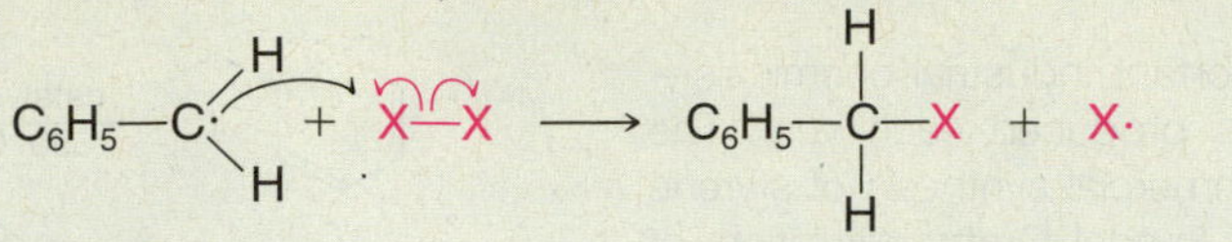

Benzyl radical   **Benzyl halide**

**The benzylic radical reacts with a halogen molecule to form the benzylic
halide product and a halogen radical that propagates the chain.**

Chain Termination

Step 4

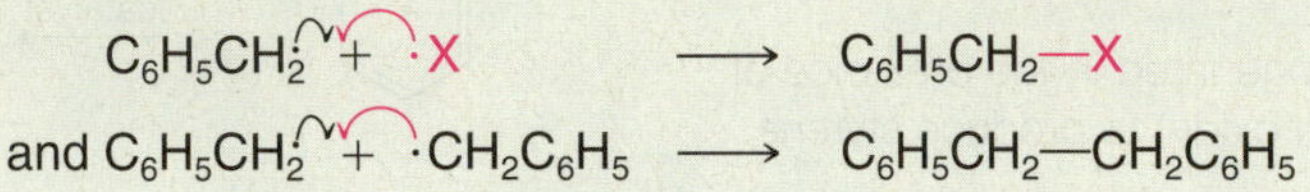

$$C_6H_5CH_2 + \cdot X \longrightarrow C_6H_5CH_2{-}X$$

$$\text{and } C_6H_5CH_2 + \cdot CH_2C_6H_5 \longrightarrow C_6H_5CH_2{-}CH_2C_6H_5$$

Various radical coupling reactions terminate the chain.

NBS (*N*-bromosuccinimide, Section 10.9) is often used in benzylic brominations
because it provides a stoichiometric amount of bromine in low concentration.

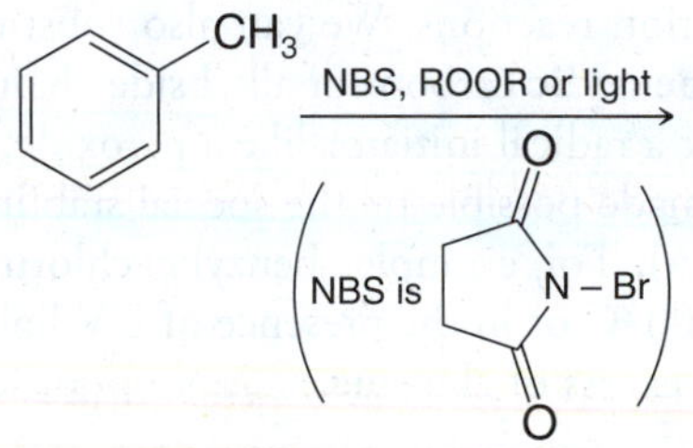

**Benzyl bromide
(α–bromotoluene)
(64%)**

Benzylic halogenation provides a useful way to introduce a leaving group when a
leaving group may be needed for subsequent nucleophilic substitution or elimination
reactions. For example, if we wished to synthesize benzyl ethyl ether from toluene, benzyl
bromide could be prepared from toluene as above, and then benzyl bromide could be
allowed to react with sodium ethoxide as follows.

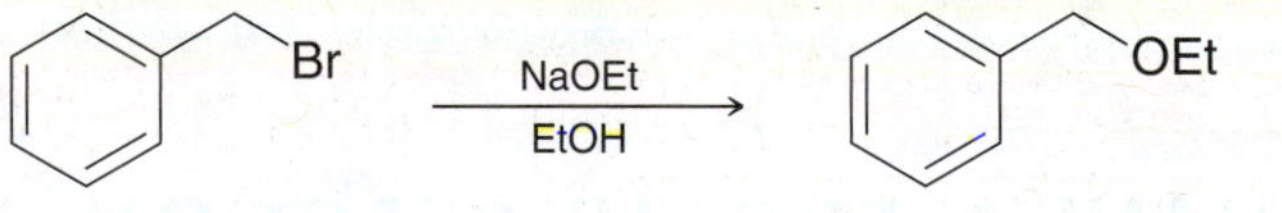

Benzyl ethyl ether

15.13 ALKENYLBENZENES

15.13A Stability of Conjugated Alkenylbenzenes

- Alkenylbenzenes that have their side-chain double bond conjugated with the
  benzene ring are more stable than those that do not:

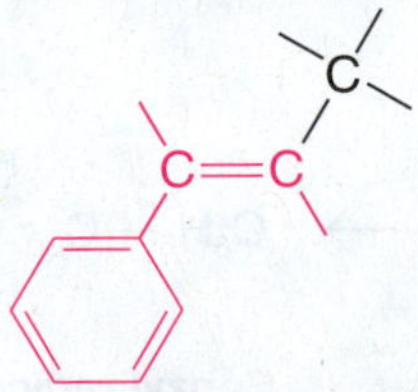

Conjugated system   is more stable than   **Nonconjugated system**

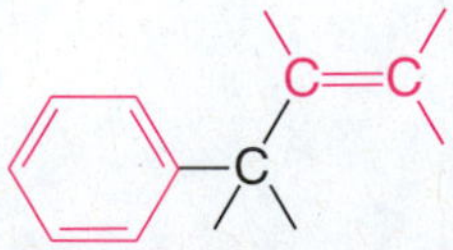

Part of the evidence for this comes from acid-catalyzed alcohol dehydrations, which are
known to yield the most stable alkene (Section 7.8A). For example, dehydration of an
alcohol such as the one that follows yields exclusively the conjugated system:

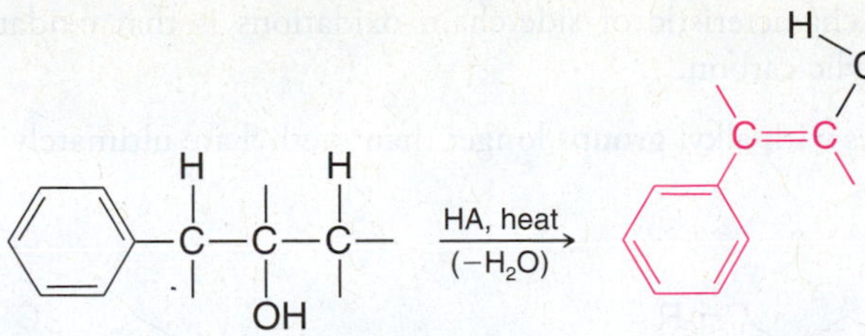

Because conjugation always lowers the energy of an unsaturated system by allowing the π electrons to be delocalized, this behavior is just what we would expect.

15.13B Additions to the Double Bond of Alkenylbenzenes

In the presence of peroxides, HBr adds to the double bond of 1-phenylpropene to give 2-bromo-1-phenylpropane as the major product:

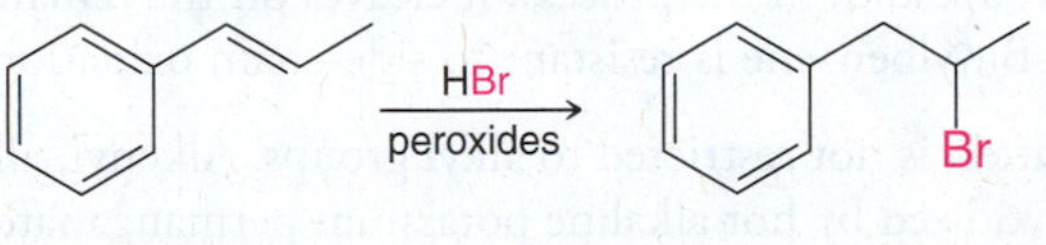

1-Phenylpropene   **2-Bromo-1-phenylpropane**

In the absence of peroxides, HBr adds in just the opposite way:

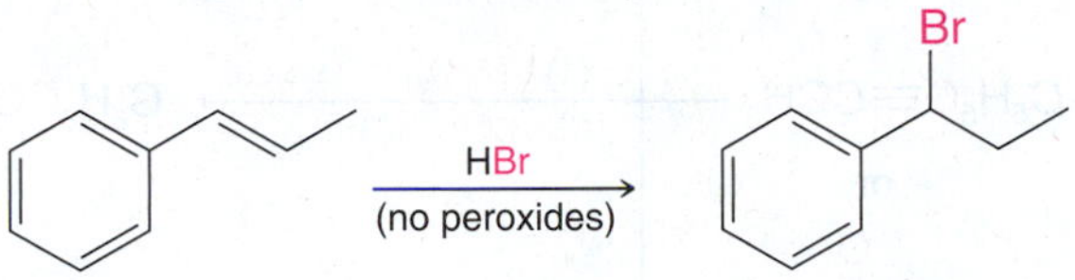

1-Phenylpropene   **1-Bromo-1-phenylpropane**

The addition of hydrogen bromide to 1-phenylpropene proceeds through a benzylic radical in the presence of peroxides and through a benzylic cation in their absence (see Practice Problem 15.15 and Section 10.9).

● ● ●

Write mechanisms for the reactions whereby HBr adds to 1-phenylpropene **(a)** in the presence of peroxides and **(b)** in the absence of peroxides. In each case account for the regiochemistry of the addition (i.e., explain why the major product is 2-bromo-1-phenylpropane when peroxides are present and why it is 1-bromo-1-phenylpropane when peroxides are absent).

PRACTICE PROBLEM 15.13

● ● ●

(a) What would you expect to be the major product when 1-phenylpropene reacts with HCl?
(b) What product would you expect when it is subjected to oxymercuration–demercuration?

PRACTICE PROBLEM 15.14

15.13C Oxidation of the Side Chain

Strong oxidizing agents oxidize toluene to benzoic acid. The oxidation can be carried out by the action of hot alkaline potassium permanganate. This method gives benzoic acid in almost quantitative yield:

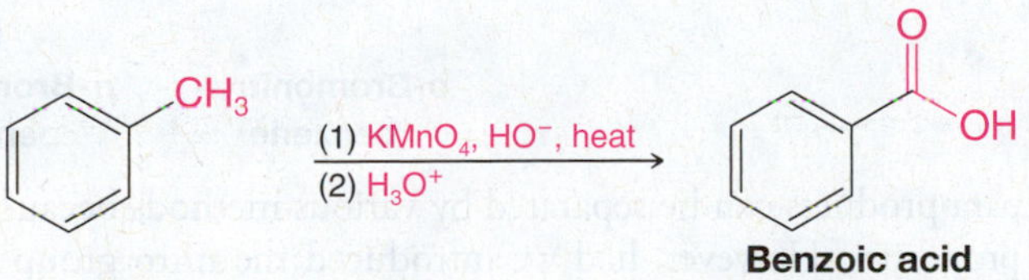

Benzoic acid
(~100%)

An important characteristic of side-chain oxidations is that oxidation takes place initially at the benzylic carbon.

- Alkylbenzenes with alkyl groups longer than methyl are ultimately degraded to benzoic acids:

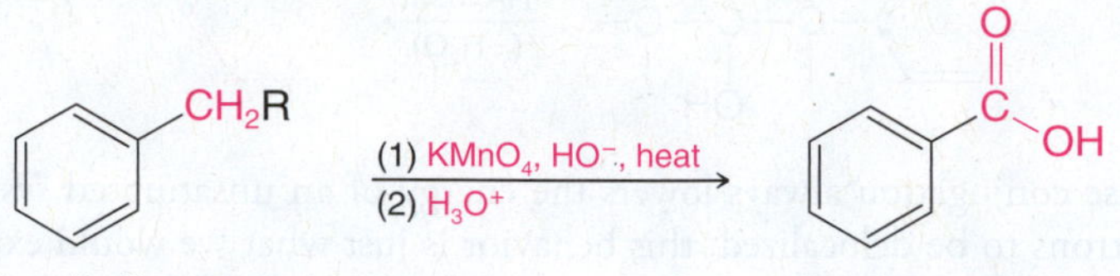

An alkylbenzene $\xrightarrow[\text{(2) } H_3O^+]{\text{(1) } KMnO_4, HO^-, \text{ heat}}$ **Benzoic acid**

Side-chain oxidations are similar to benzylic halogenations, because in the first step the oxidizing agent abstracts a benzylic hydrogen. Once oxidation is begun at the benzylic carbon, it continues at that site. Ultimately, the oxidizing agent oxidizes the benzylic carbon to a carboxyl group, and, in the process, it cleaves off the remaining carbon atoms of the side chain. (*tert*-Butylbenzene is resistant to side-chain oxidation. Why?)

- Side-chain oxidation is not restricted to alkyl groups. Alkenyl, alkynyl, and acyl groups are also oxidized by hot alkaline potassium permanganate.

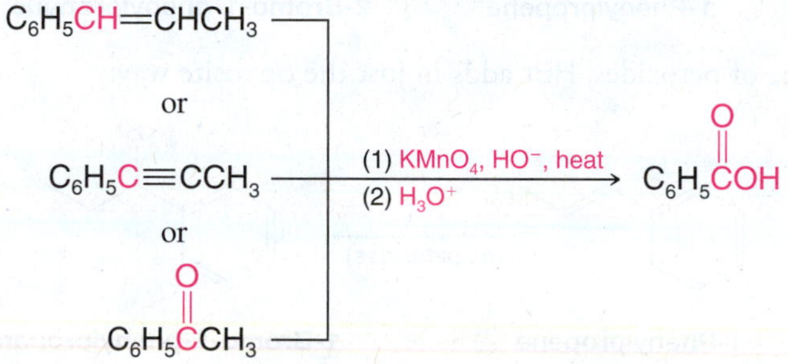

$$C_6H_5CH\!=\!CHCH_3$$

or

$$C_6H_5C\!\equiv\!CCH_3 \xrightarrow[\text{(2) } H_3O^+]{\text{(1) } KMnO_4, HO^-, \text{ heat}} C_6H_5\overset{\displaystyle O}{\overset{\|}{C}}OH$$

or

$$C_6H_5\overset{\displaystyle O}{\overset{\|}{C}}CH_3$$

15.13D Oxidation of the Benzene Ring

The benzene ring carbon where an alkyl group is bonded can be converted to a carboxyl group by ozonolysis, followed by treatment with hydrogen peroxide.

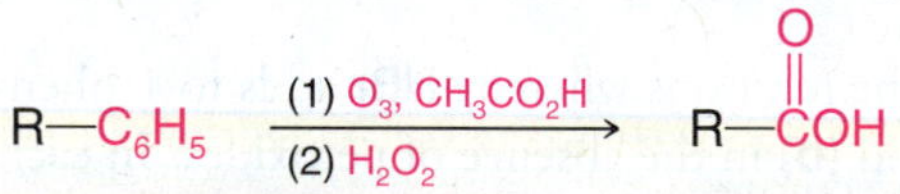

$$R\!-\!C_6H_5 \xrightarrow[\text{(2) } H_2O_2]{\text{(1) } O_3, CH_3CO_2H} R\!-\!\overset{\displaystyle O}{\overset{\|}{C}}OH$$

15.14 SYNTHETIC APPLICATIONS

The substitution reactions of aromatic rings and the reactions of the side chains of alkyl- and alkenylbenzenes, when taken together, offer us a powerful set of reactions for organic synthesis. By using these reactions skillfully, we shall be able to synthesize a large number of benzene derivatives.

- Part of the skill in planning a synthesis is deciding in what order to carry out the reactions.

Let us suppose, for example, that we want to synthesize *o*-bromonitrobenzene. We can see very quickly that we should introduce the bromine into the ring first because it is an ortho–para director:

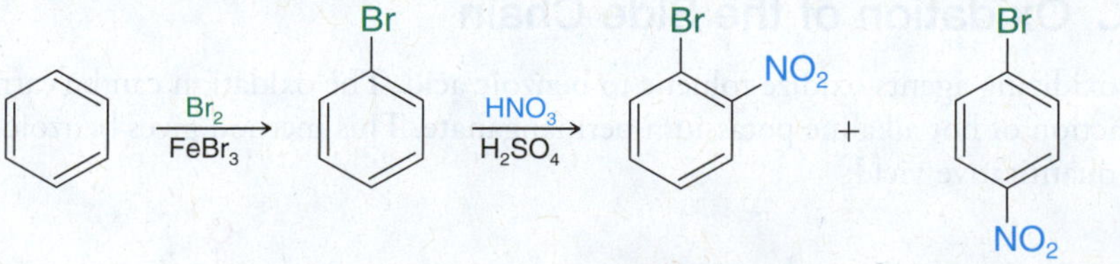

o-Bromonitro-   *p*-Bromonitro-
benzene   benzene

The ortho and para products can be separated by various methods because they have different physical properties. However, had we introduced the nitro group first, we would have obtained *m*-bromonitrobenzene as the major product.

Other examples in which choosing the proper order for the reactions is important are the syntheses of the *ortho*-, *meta*-, and *para*-nitrobenzoic acids. Because the methyl group of toluene is an electron-donating group (shown in red below), we can synthesize the *ortho*- and *para*-nitrobenzoic acids from toluene by nitrating it, separating the *ortho*- and *para*-nitrotoluenes, and then oxidizing the methyl groups to carboxyl groups:

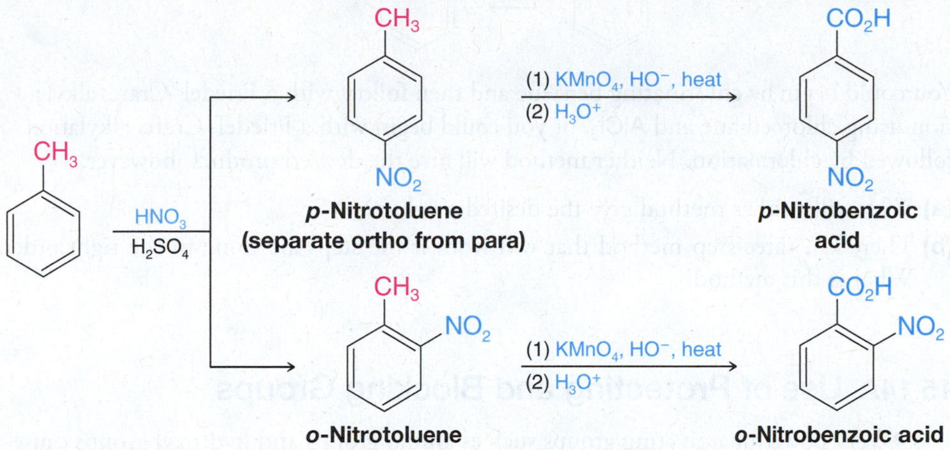

We can synthesize *m*-nitrobenzoic acid by reversing the order of the reactions. We oxidize the methyl group to a carboxylic acid, then use the carboxyl as an electron-withdrawing group (shown in blue) to direct nitration to the meta position.

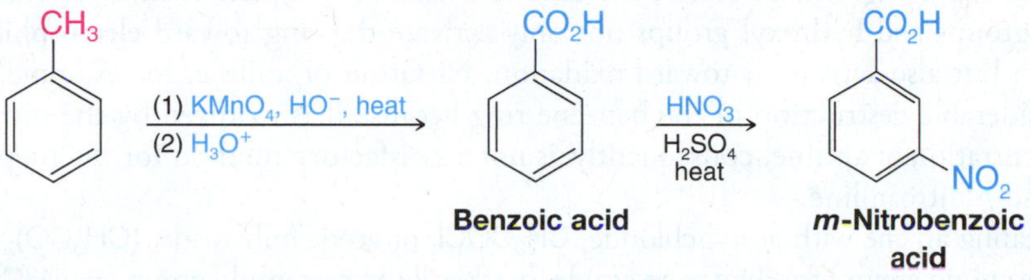

●●● **SOLVED PROBLEM 15.5**

Starting with toluene, outline a synthesis of **(a)** 1-bromo-2-trichloromethylbenzene, **(b)** 1-bromo-3-trichloromethylbenzene, and **(c)** 1-bromo-4-trichloromethylbenzene.

ANSWER: Compounds **(a)** and **(c)** can be obtained by ring bromination of toluene followed by benzylic radical chlorination of the side chain using three molar equivalents of chlorine:

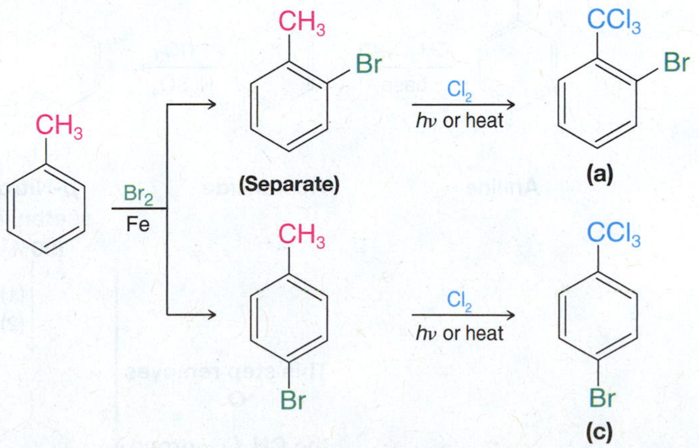

To make compound **(b)**, we reverse the order of the reactions. By converting the side chain to a —CCl$_3$ group first, we create a meta director, which causes the bromine to enter the desired position:

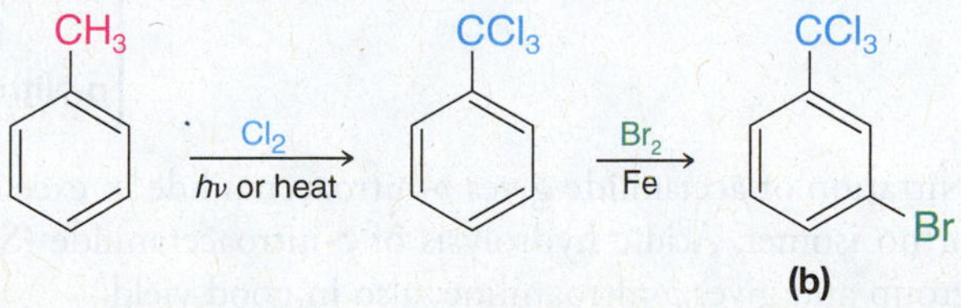

PRACTICE PROBLEM 15.15 Suppose you needed to synthesize *m*-chloroethylbenzene from benzene.

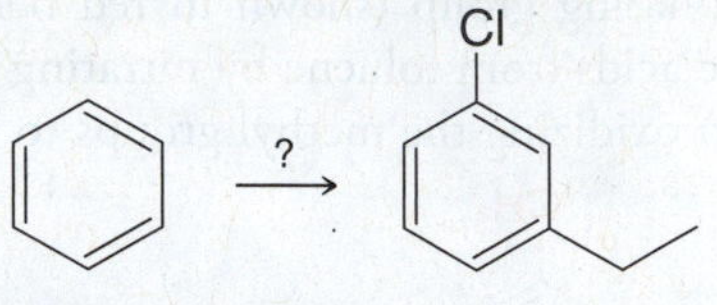

You could begin by chlorinating benzene and then follow with a Friedel–Crafts alkylation using chloroethane and $AlCl_3$, or you could begin with a Friedel–Crafts alkylation followed by chlorination. Neither method will give the desired product, however.

(a) Why will neither method give the desired product?

(b) There is a three-step method that will work if the steps are done in the right order. What is this method?

15.14A Use of Protecting and Blocking Groups

- Very powerful activating groups such as amino groups and hydroxyl groups cause the benzene ring to be so reactive that undesirable reactions may take place.

Some reagents used for electrophilic substitution reactions, such as nitric acid, are also strong *oxidizing agents*. Both electrophiles and oxidizing agents seek electrons. Thus, amino groups and hydroxyl groups not only activate the ring toward electrophilic substitution but also activate it toward oxidation. Nitration of aniline, for example, results in considerable destruction of the benzene ring because it is oxidized by the nitric acid. Direct nitration of aniline, consequently, is not a satisfactory method for the preparation of *o*- and *p*-nitroaniline.

Treating aniline with acetyl chloride, CH_3COCl, or acetic anhydride, $(CH_3CO)_2O$, converts the amino group of aniline to an amide (specifically an acetamido group, $—NHCOCH_3$), forming acetanilide. An amide group is only moderately activating, and it does not make the ring highly susceptible to oxidation during nitration (see Practice Problem 15.9). Thus, with the amino group of aniline blocked in acetanilide, direct nitration becomes possible:

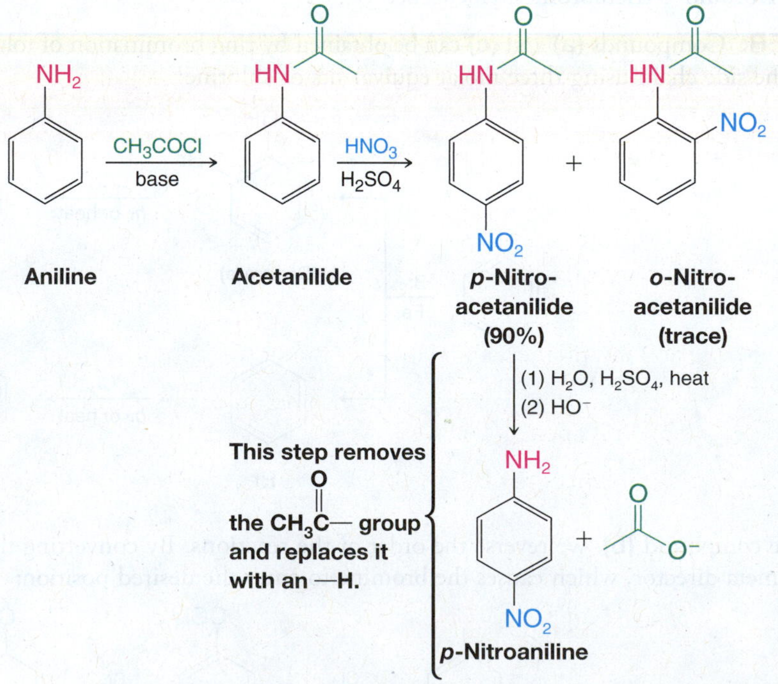

Nitration of acetanilide gives *p*-nitroacetanilide in excellent yield with only a trace of the ortho isomer. Acidic hydrolysis of *p*-nitroacetanilide (Section 18.8F) removes the acetyl group and gives *p*-nitroaniline, also in good yield.

Suppose, however, that we need *o*-nitroaniline. The synthesis that we just outlined would obviously not be a satisfactory method, for only a trace of *o*-nitroacetanilide is obtained in the nitration reaction. (The acetamido group is purely a para director in many reactions. Bromination of acetanilide, for example, gives *p*-bromoacetanilide almost exclusively.)

We can synthesize *o*-nitroaniline, however, through the reactions that follow:

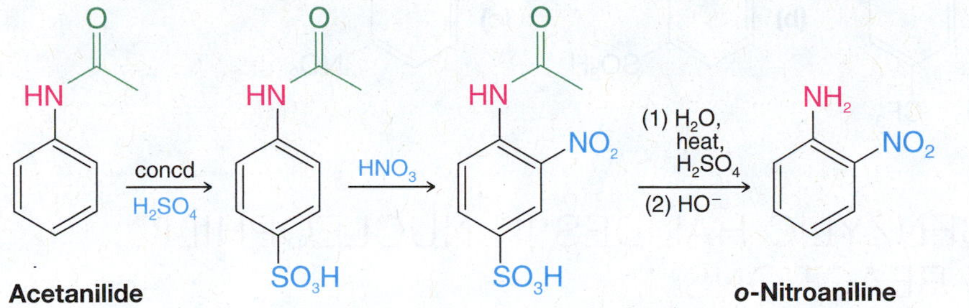

Acetanilide **o-Nitroaniline (56%)**

Here we see how a sulfonic acid group can be used as a "blocking group." We can remove the sulfonic acid group by desulfonation at a later stage. In this example, the reagent used for desulfonation (dilute H_2SO_4) also conveniently removes the acetyl group that we employed to "protect" the benzene ring from oxidation by nitric acid.

15.14B Orientation in Disubstituted Benzenes

- When two different groups are present on a benzene ring, the more powerful activating group (Table 15.2) generally determines the outcome of the reaction.

Let us consider, as an example, the orientation of electrophilic substitution of *p*-methylacetanilide. The amide group is a much stronger activating group than the methyl group. The following example shows that the amide group determines the outcome of the reaction. Substitution occurs primarily at the position ortho to the amide group:

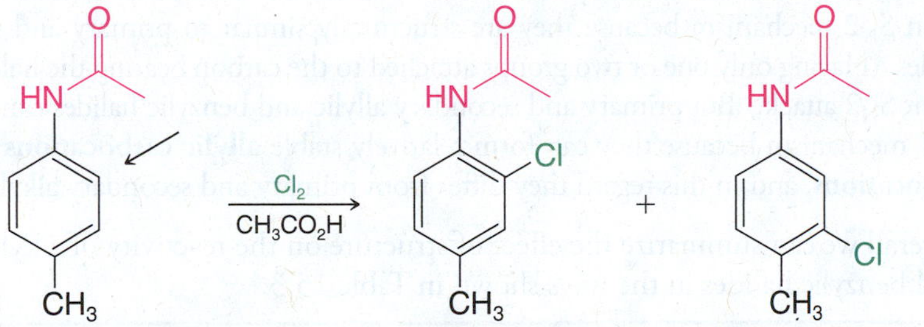

Major product **Minor product**

- An ortho–para director takes precedence over a meta director in determining the position of substitution because all ortho–para-directing groups are more activating than meta directors.

Steric effects are also important in aromatic substitutions.

- Substitution does not occur to an appreciable extent between meta substituents if another position is open.

A good example of this effect can be seen in the nitration of *m*-bromochlorobenzene:

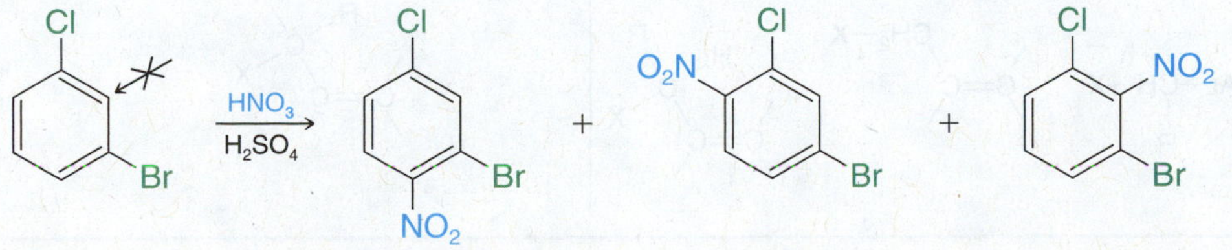

62% **37%** **1%**

Only 1% of the mononitro product has the nitro group between the bromine and chlorine.

PRACTICE PROBLEM 15.16 Predict the major product (or products) that would be obtained when each of the following compounds is nitrated:

(a) (b) (c)

OH — CF$_3$ (para)   CN — SO$_3$H   OCH$_3$ — NO$_2$

15.15 ALLYLIC AND BENZYLIC HALIDES IN NUCLEOPHILIC SUBSTITUTION REACTIONS

Allylic and benzylic halides can be classified in the same way that we have classified other organic halides:

1° Allylic   2° Allylic   3° Allylic   1° Benzylic   2° Benzylic   3° Benzylic

All of these compounds undergo nucleophilic substitution reactions. As with other tertiary halides (Section 6.13A), the steric hindrance associated with having three bulky groups on the carbon bearing the halogen prevents tertiary allylic and tertiary benzylic halides from reacting by an S$_N$2 mechanism. They react with nucleophiles only by an S$_N$1 mechanism.

Primary and secondary allylic and benzylic halides can react either by an S$_N$2 mechanism or by an S$_N$1 mechanism in ordinary nonacidic solvents. We would expect these halides to react by an S$_N$2 mechanism because they are structurally similar to primary and secondary alkyl halides. (Having only one or two groups attached to the carbon bearing the halogen does not prevent S$_N$2 attack.) But primary and secondary allylic and benzylic halides can also react by an S$_N$1 mechanism because they can form relatively stable **allylic carbocations** and **benzylic carbocations**, and in this regard they differ from primary and secondary alkyl halides.*

- Overall we can summarize the effect of structure on the reactivity of alkyl, allylic, and benzylic halides in the ways shown in Table 15.3.

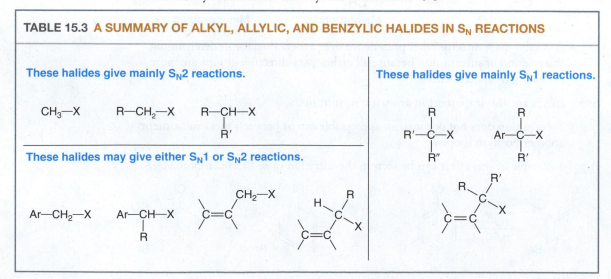

TABLE 15.3 A SUMMARY OF ALKYL, ALLYLIC, AND BENZYLIC HALIDES IN S$_N$ REACTIONS

*There is some dispute as to whether 2° alkyl halides react by an S$_N$1 mechanism to any appreciable extent in ordinary nonacidic solvents such as mixtures of water and alcohol or acetone, but it is clear that reaction by an S$_N$2 mechanism is, for all practical purposes, the more important pathway.

••• SOLVED PROBLEM 15.6

When either enantiomer of 3-chloro-1-butene [(*R*) or (*S*)] is subjected to hydrolysis, the products of the reaction are optically inactive. Explain these results.

ANSWER: The solvolysis reaction is S_N1. The intermediate allylic cation is achiral and therefore reacts with water to give the enantiomeric 3-buten-2-ols in equal amounts and to give some of the achiral 2-buten-1-ol:

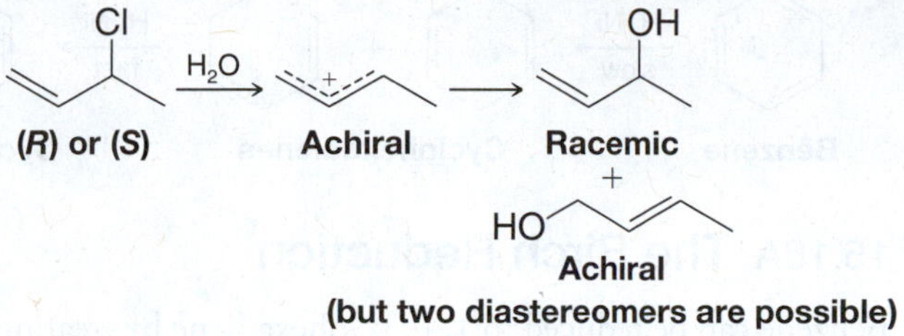

(*R*) or (*S*)   Achiral   Racemic
                                            +
                                         Achiral
                        (but two diastereomers are possible)

••• PRACTICE PROBLEM 15.17

Account for the following observations with mechanistic explanations.

(a) Cl⌒⌒⌒   EtONa (in high concentration) / EtOH →   EtO⌒⌒⌒

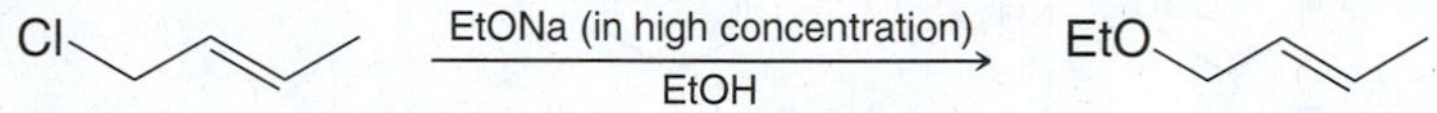

At high concentration of ethoxide, the rate depends on both the allylic halide and ethoxide concentrations.

(b) Cl⌒⌒⌒   EtONa (in low concentration) →   EtO⌒⌒⌒   +   OEt⌒⌒

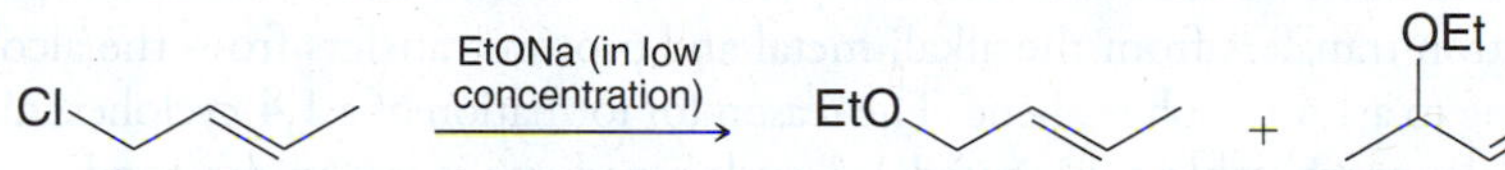

At low concentration of ethoxide, the rate depends only on the allylic halide concentration.

••• PRACTICE PROBLEM 15.18

1-Chloro-3-methyl-2-butene undergoes hydrolysis in a mixture of water and dioxane at a rate that is more than a thousand times that of 1-chloro-2-butene. **(a)** What factor accounts for the difference in reactivity? **(b)** What products would you expect to obtain? [Dioxane is a cyclic ether (below) that is miscible with water in all proportions and is a useful cosolvent for conducting reactions like these. Dioxane is carcinogenic (i.e., cancer causing), however, and like most ethers, it tends to form peroxides.]

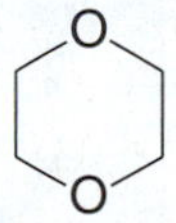

Dioxane

••• PRACTICE PROBLEM 15.19

Primary halides of the type $ROCH_2X$ apparently undergo S_N1-type reactions, whereas most primary halides do not. Can you propose a resonance explanation for the ability of halides of the type $ROCH_2X$ to undergo S_N1 reactions?

••• PRACTICE PROBLEM 15.20

The following chlorides (Ph = phenyl) undergo solvolysis in ethanol at the relative rates given in parentheses. How can you explain these results?

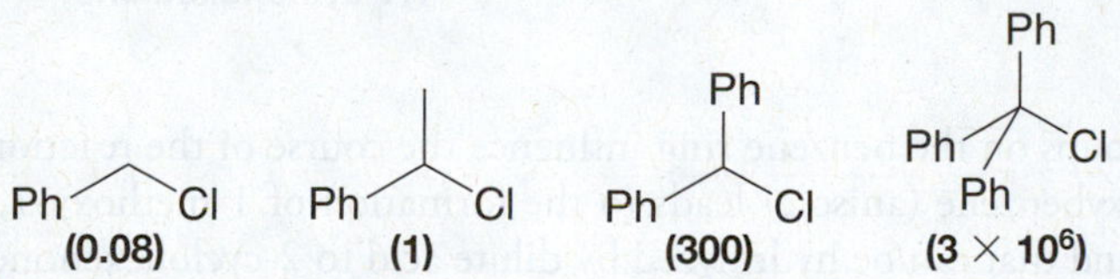

Ph⌒Cl   Ph⌒Cl   Ph⌒Cl   Ph₃C—Cl
(0.08)   **(1)**   **(300)**   **(3 × 10⁶)**

15.16 REDUCTION OF AROMATIC COMPOUNDS

Hydrogenation of benzene under pressure using a metal catalyst such as nickel results in the addition of three molar equivalents of hydrogen and the formation of cyclohexane (Section 14.3). The intermediate cyclohexadienes and cyclohexene cannot be isolated because these undergo catalytic hydrogenation faster than benzene does.

15.16A The Birch Reduction

Benzene can be reduced to 1,4-cyclohexadiene by treating it with an alkali metal (sodium, lithium, or potassium) in a mixture of liquid ammonia and an alcohol. This reaction is called the **Birch reduction**, after A. J. Birch, the Australian chemist who developed it.

The Birch reduction is a dissolving metal reduction, and the mechanism for it resembles the mechanism for the reduction of alkynes that we studied in Section 7.15B. A sequence of electron transfers from the alkali metal and proton transfers from the alcohol takes place, leading to a 1,4-cyclohexadiene. The reason for formation of a 1,4-cyclohexadiene in preference to the more stable conjugated 1,3-cyclohexadiene is not understood.

[A MECHANISM FOR THE REACTION — Birch Reduction]

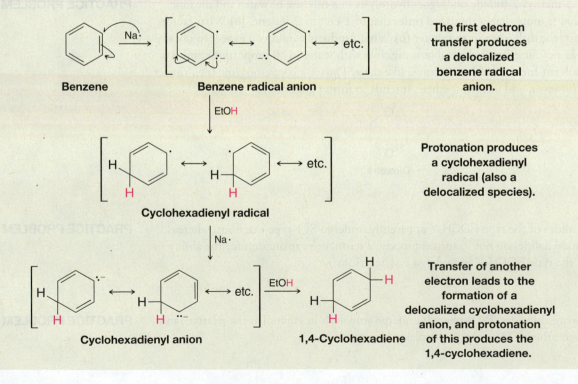

The first electron transfer produces a delocalized benzene radical anion.

Protonation produces a cyclohexadienyl radical (also a delocalized species).

Transfer of another electron leads to the formation of a delocalized cyclohexadienyl anion, and protonation of this produces the 1,4-cyclohexadiene.

Substituent groups on the benzene ring influence the course of the reaction. Birch reduction of methoxybenzene (anisole) leads to the formation of 1-methoxy-1,4-cyclohexadiene, a compound that can be hydrolyzed by dilute acid to 2-cyclohexenone. This method provides a useful synthesis of 2-cyclohexenones:

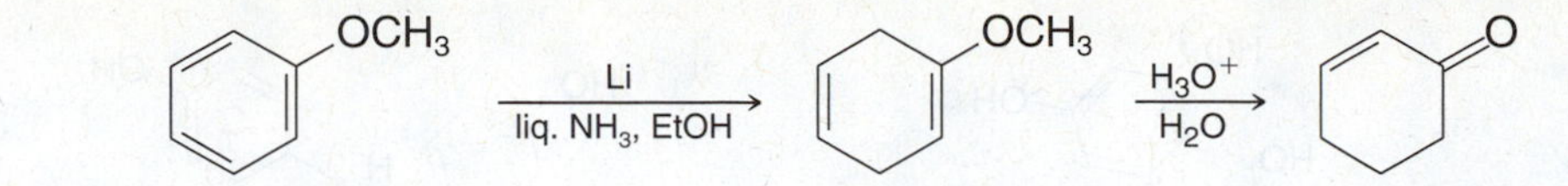

Methoxybenzene
(anisole)

$\xrightarrow[\text{liq. NH}_3, \text{EtOH}]{\text{Li}}$

1-Methoxy-1,4-
cyclohexadiene
(84%)

$\xrightarrow[\text{H}_2\text{O}]{\text{H}_3\text{O}^+}$

2-Cyclohexenone

PRACTICE PROBLEM 15.21

Birch reduction of toluene leads to a product with the molecular formula C_7H_{10}. On ozonolysis followed by reduction with dimethyl sulfide, the product is transformed into

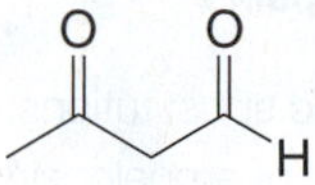

and

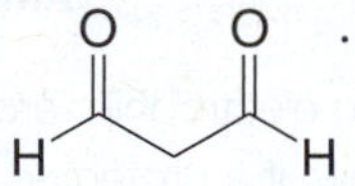

. What is the structure of the Birch reduction product?

[WHY Do These Topics Matter?

SYNTHESIZING ARCHITECTURALLY UNIQUE NATURAL PRODUCTS

When certain plants like grapevines are attacked by foreign pathogens, such as bacteria and fungi, they use a compound called resveratrol and like Lego building blocks combine it with other resveratrol molecules in different ways to create dozens of new, and larger, molecules. The goal is to synthesize at least one compound with the biological activity required to eradicate, or at least slow, the pathogen so the plant has a chance to survive. A few examples of these compounds are shown below, illustrating just a small portion of the architectural diversity that the family possesses. What is particularly fascinating is that the synthesis of these molecules likely involves two major types of bond formation—radical chemistry and electrophilic aromatic substitutions. Here we focus on the latter.

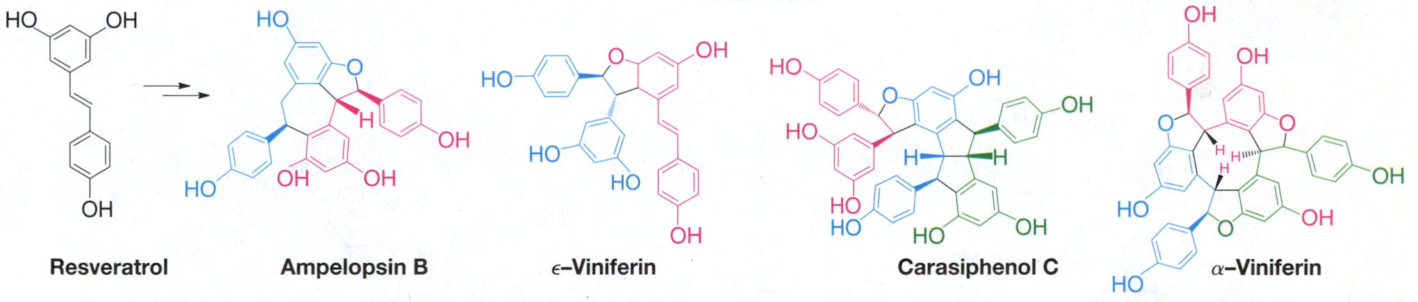

© Konstantin Suryagin/iStockphoto

Resveratrol   **Ampelopsin B**   **ε–Viniferin**   **Carasiphenol C**   **α–Viniferin**

If the resveratrol dimers quadrangularin A and ampelopsin D are exposed to an appropriate acid, it is reasonable to believe that proton activation of their alkenes would create new carbocations, shown here in their resonance-stabilized forms by shifting electrons from the neighboring *para*-phenoxy ring system. Attack by the neighboring electron-rich 3,5-diphenoxy ring system through Friedel–Crafts reactions, as shown, would then forge new C—C bonds (highlighted in red), leading to completely different structures in the form of pallidol and ampelopsin F. All of the benzene rings are labeled in each case so that you can see where they end up. Not only are new structures formed, but new and different biological properties are gained as well. Quadrangularin A is a good radical scavenger, while pallidol possesses potential cancer-fighting properties.

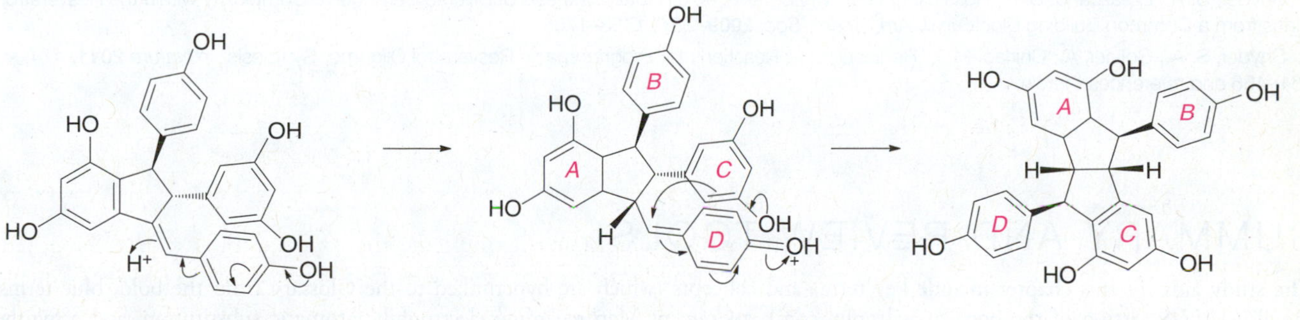

Quadrangularin A   **Pallidol**

(continues on next page)

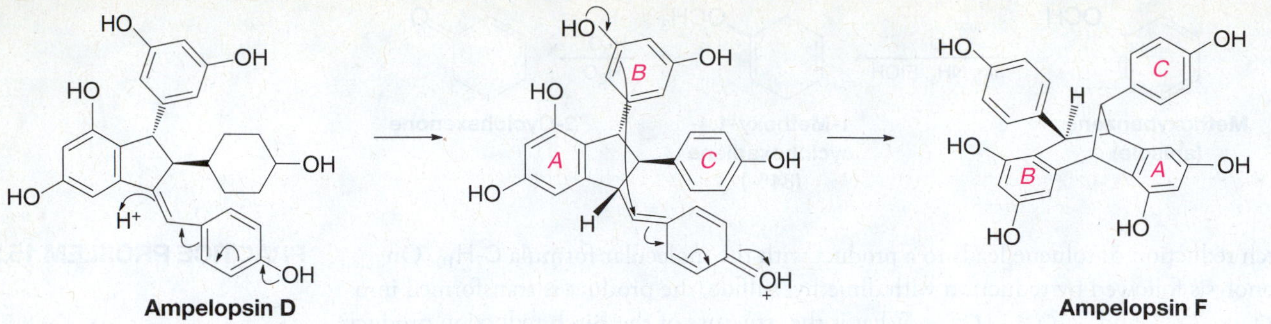

Ampelopsin D

Ampelopsin F

Chemists have also found creative ways to make these compounds in the laboratory using electrophilic aromatic substitutions. One of the key tools has been substituting aryl hydrogen atoms with bromine. As shown below, if a protected form of ampelopsin F is exposed to an electrophilic bromine source, such as *N*-bromosuccinimide (NBS), an electrophilic aromatic substitution reaction can happen directly without a Lewis acid due to the electron-rich nature of some of the aromatic rings. What is surprising, however, is that while there are many places where that addition can occur, the indicated monobrominated compound is formed selectively. Most other bromonium sources behave in the same way. However, there is one that substitutes a different site first, and that is Et$_2$SBr•SbCl$_5$Br (BDSB). Because of the ability to achieve this different and specific tailoring, the atoms needed to complete syntheses of the resveratrol trimers carasiphenol B and ampelopsin G could then be added in some additional operations.

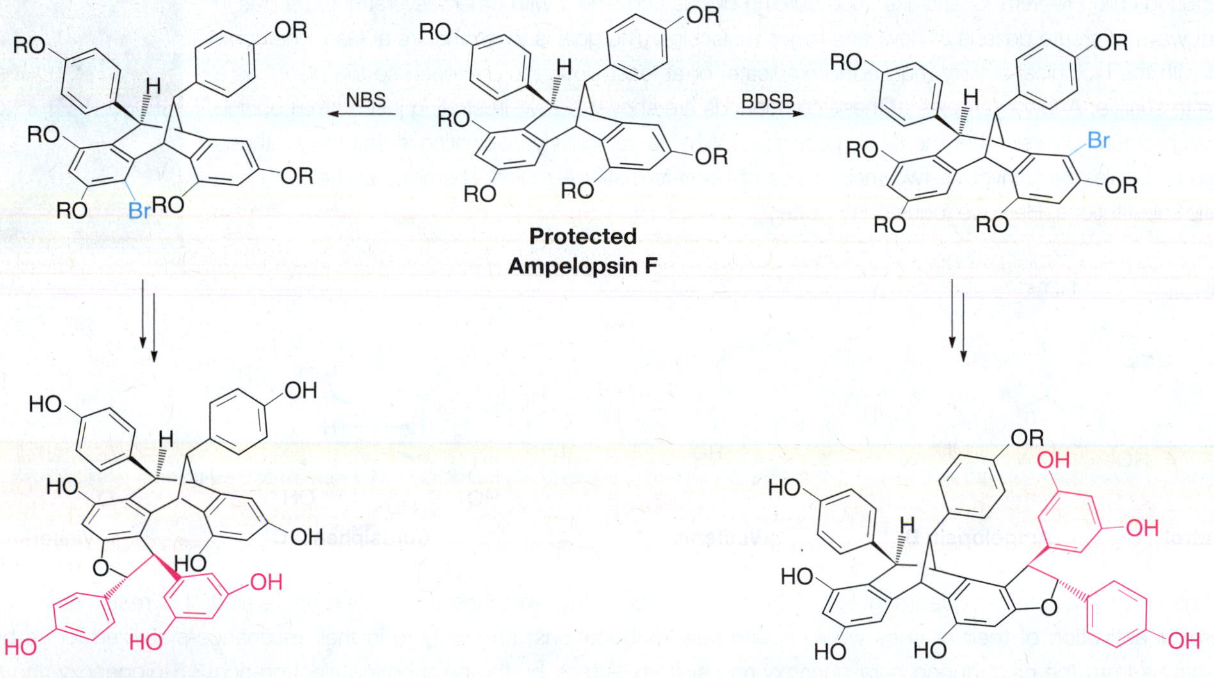

NBS

Protected Ampelopsin F

BDSB

Carasiphenol B

Ampelopsin G

To learn more about these topics, see:

1. Snyder, S. A.; Zografos, A. L.; Lin, Y. "Total Synthesis of Resveratrol-based Natural Products: A Chemoselective Solution" in *Angew. Chem. Int. Ed.* **2007**, *46*, 8186–8191.

2. Snyder, S. A.; Breazzano, S. P.; Ross, A. G.; Lin, Y.; Zografos, A. L. "Total Synthesis of Diverse Carbogenic Complexity within the Resveratrol Class from a Common Building Block" in *J. Am. Chem. Soc.* **2009**, *131*, 1753–1765.

3. Snyder, S. A.; Gollner, A.; Chiriac, M. I. "Regioselective Reactions for Programmable Resveratrol Oligomer Synthesis" in *Nature* **2011**, *474*, 461–466 and references therein.

SUMMARY AND REVIEW TOOLS

The study aids for this chapter include key terms and concepts (which are hyperlinked to the Glossary from the bold, blue terms in the *WileyPLUS* version of the book at wileyplus.com), a Concept Map regarding electrophilic aromatic substitution, and a Synthetic Connections scheme for reactions that link benzene and a variety of aryl derivatives.

PROBLEMS WILEY PLUS

Note to Instructors: Many of the homework problems are available for assignment via *WileyPLUS*, an online teaching and learning solution.

MECHANISMS

15.22 Provide a detailed mechanism for each of the following reactions. Include contributing resonance structures and the resonance hybrid for the arenium ion intermediates.

(a) [toluene] $\xrightarrow{HNO_3, H_2SO_4}$ [p-nitrotoluene, NO₂]

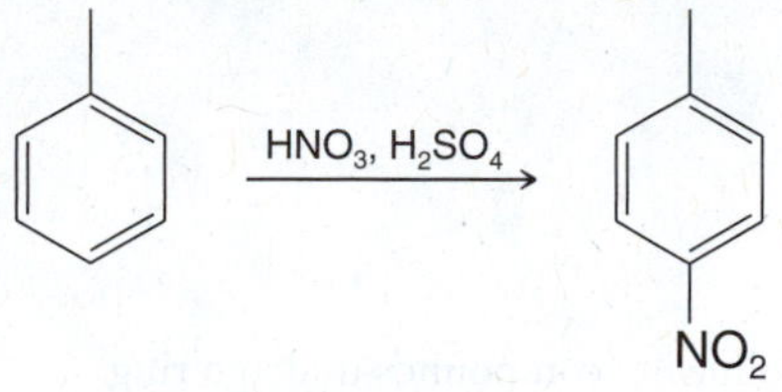

(b) [p-xylene] $\xrightarrow{Br_2, FeBr_3}$ [bromo-p-xylene, Br]

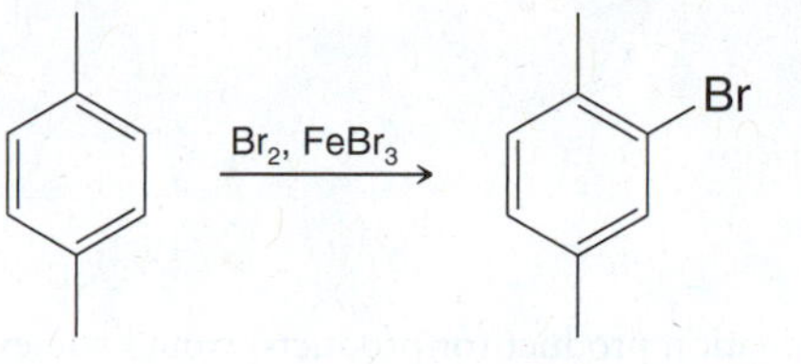

(c) [benzene] + [cyclobutylmethyl bromide, Br] $\xrightarrow{AlBr_3}$ [cyclopentylbenzene]

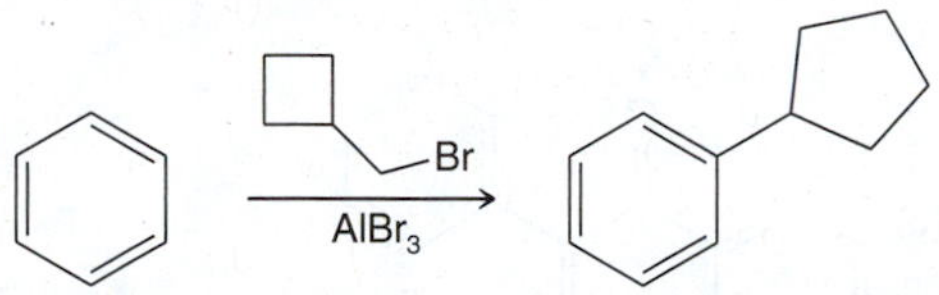

15.23 Provide a detailed mechanism for the following reaction.

[benzene] + [2,2,5,5-tetramethyltetrahydrofuran] $\xrightarrow{H_2SO_4}$ [1,1,4,4-tetramethyltetrahydronaphthalene] + H_2O

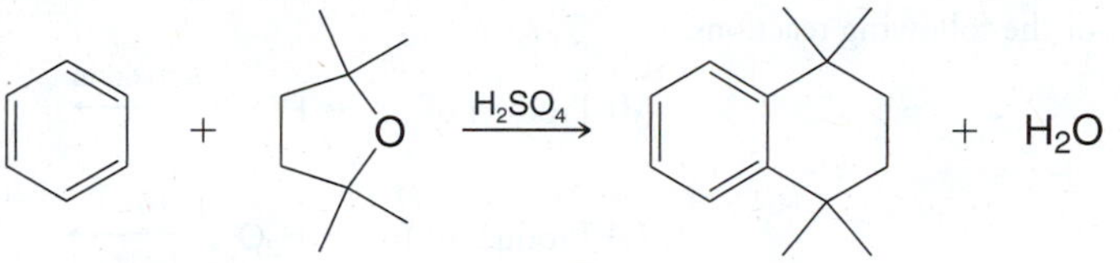

15.24 One ring of phenyl benzoate undergoes electrophilic aromatic substitution much more readily than the other. (a) Which one is it? (b) Explain your answer.

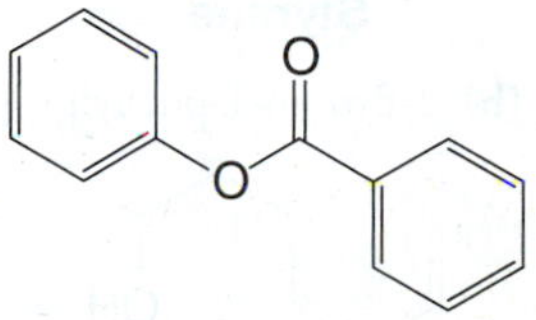

15.25 Many polycyclic aromatic compounds have been synthesized by a cyclization reaction known as the **Bradsher reaction** or **aromatic cyclodehydration**. This method can be illustrated by the following synthesis of 9-methylphenanthrene:

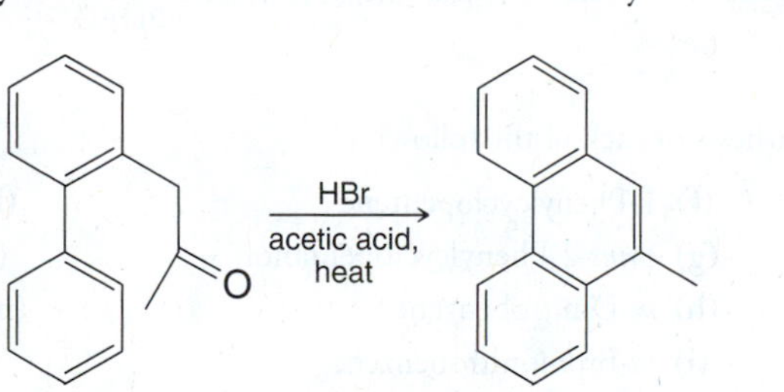

[substrate] $\xrightarrow[\text{heat}]{\substack{HBr \\ \text{acetic acid,}}}$ [9-methylphenanthrene]

9-Methylphenanthrene

An arenium ion is an intermediate in this reaction, and the last step involves the dehydration of an alcohol. Propose a plausible mechanism for this example of the Bradsher reaction.

15.26 Write mechanisms that account for the products of the following reactions:

(a) [fluorenyl methanol, OH] $\xrightarrow[(-H_2O)]{HA}$ [phenanthrene]

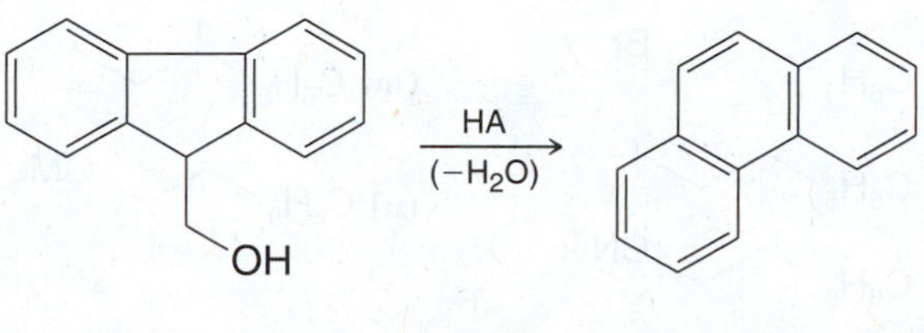

(b) [isopropenylbenzene] $\xrightarrow{HA}$ [1-phenyl-1,3,3-trimethylindane]

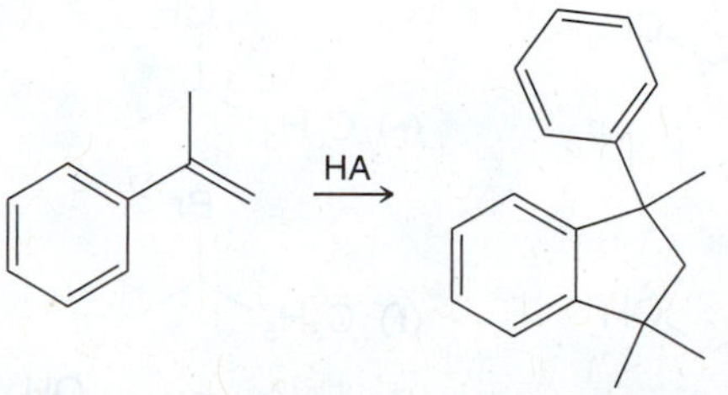

15.27 The addition of a hydrogen halide (hydrogen bromide or hydrogen chloride) to 1-phenyl-1,3-butadiene produces (only) 1-phenyl-3-halo-1-butene. (a) Write a mechanism that accounts for the formation of this product. (b) Is this 1,4 addition or 1,2 addition to the butadiene system? (c) Is the product of the reaction consistent with the formation of the most stable intermediate carbocation? (d) Does the reaction appear to be under kinetic control or equilibrium control? Explain.

REACTIONS AND SYNTHESIS

15.28 Predict the major product (or products) formed when each of the following reacts with Cl_2 and $FeCl_3$:

(a) Ethylbenzene

(b) Anisole (methoxybenzene)

(c) Fluorobenzene

(d) Benzoic acid

(e) Nitrobenzene

(f) Chlorobenzene

(g) Biphenyl (C_6H_5—C_6H_5)

(h) Ethyl phenyl ether

15.29 Predict the major product (or products) formed when each of the following reacts with a mixture of concentrated HNO_3 and H_2SO_4.

(a)

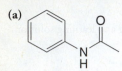

Acetanilide

(b)

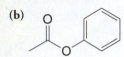

Phenyl acetate

(c) 4-Chlorobenzoic acid
(d) 3-Chlorobenzoic acid

(e)

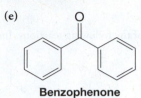

Benzophenone

15.30 What monobromination product (or products) would you expect to obtain when the following compounds undergo ring bromination with Br_2 and $FeBr_3$?

(a) **(b)** 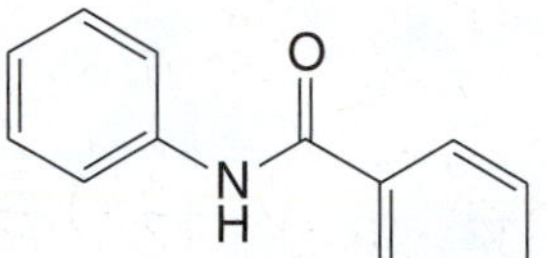 **(c)**

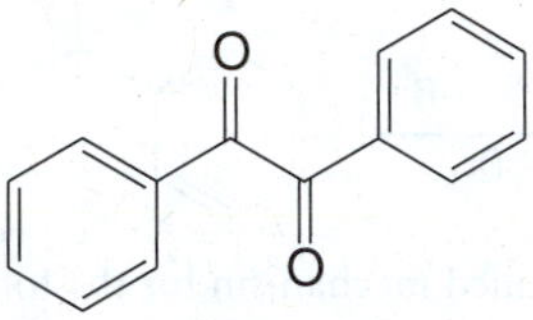

15.31 Predict the major products of the following reactions:

(a)

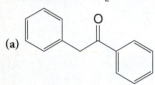
$\xrightarrow{\text{HCl}}$

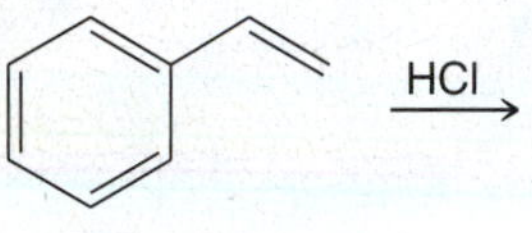

Styrene

(b) 2-Bromo-1-phenylpropane $\xrightarrow{\text{EtONa}}$

(c)

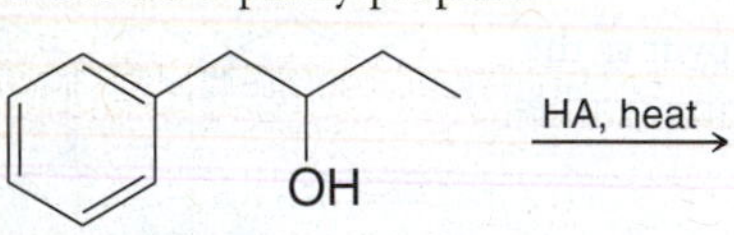

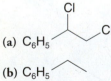
$\xrightarrow{\text{HA, heat}}$

(d) Product of (c) + HBr $\xrightarrow{\text{peroxides}}$

(e) Product of (c) + H_2O $\xrightarrow[\text{heat}]{\text{HA}}$

(f) Product of (c) + H_2 (1 molar equivalent) $\xrightarrow[25\,°C]{\text{Pt}}$

(g) Product of (f) $\xrightarrow[\text{(2) } H_3O^+]{\text{(1) } KMnO_4,\ HO^-,\ heat}$

15.32 Starting with benzene, outline a synthesis of each of the following:

(a) Isopropylbenzene
(b) *tert*-Butylbenzene
(c) Propylbenzene
(d) Butylbenzene
(e) 1-*tert*-Butyl-4-chlorobenzene

(f) 1-Phenylcyclopentene
(g) *trans*-2-Phenylcyclopentanol
(h) *m*-Dinitrobenzene
(i) *m*-Bromonitrobenzene
(j) *p*-Bromonitrobenzene

(k) *p*-Chlorobenzenesulfonic acid
(l) *o*-Chloronitrobenzene
(m) *m*-Nitrobenzenesulfonic acid

15.33 Starting with styrene, outline a synthesis of each of the following:

(a) C_6H_5 (with Cl, Cl)
(b) C_6H_5
(c) C_6H_5 (with OH, OH)
(d) 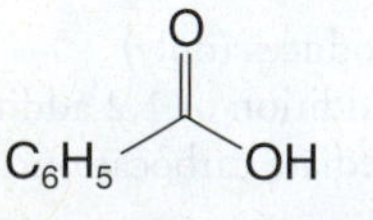 C_6H_5 (with =O, OH)

(e) 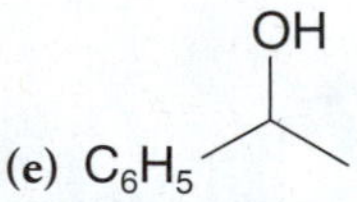 C_6H_5 (with OH)
(f) 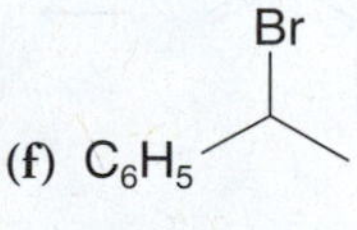 C_6H_5 (with Br)
(g) 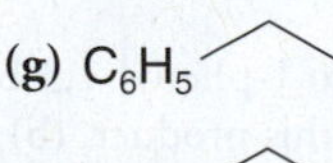 C_6H_5 (with OH)
(h) 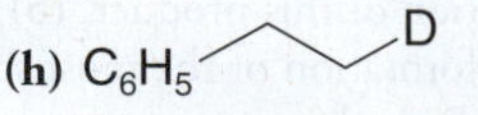 C_6H_5 (with D)

(i) 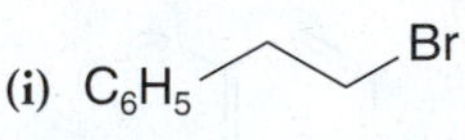 C_6H_5 (with Br)
(j) 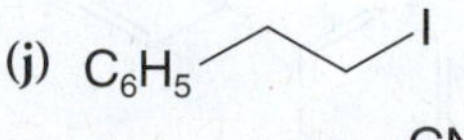 C_6H_5 (with I)
(k) 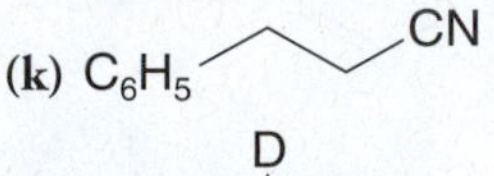 C_6H_5 (with CN)
(l) 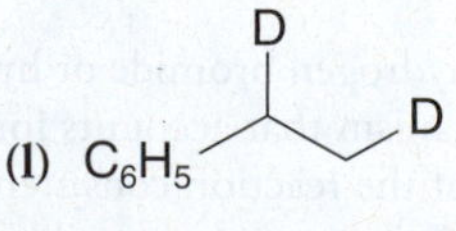 C_6H_5 (with D, D)

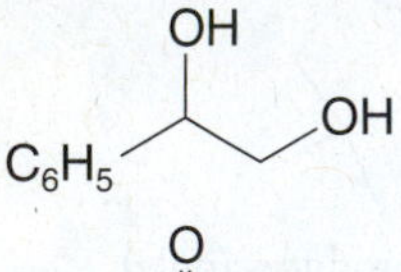

(m) 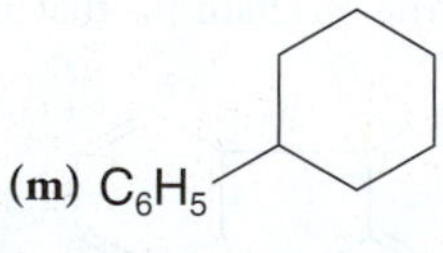 C_6H_5
(n) 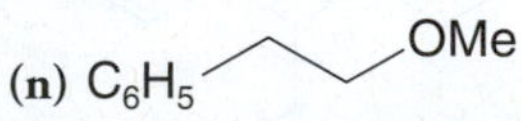 C_6H_5 (with OMe)

15.34 Starting with toluene, outline a synthesis of each of the following:

(a) *m*-Chlorobenzoic acid
(b) *p*-Methylacetophenone
(c) 2-Bromo-4-nitrotoluene
(d) *p*-Bromobenzoic acid

(e) 1-Chloro-3-trichloromethylbenzene
(f) *p*-Isopropyltoluene (*p*-cymene)
(g) 1-Cyclohexyl-4-methylbenzene
(h) 2,4,6-Trinitrotoluene (TNT)

(i) 4-Chloro-2-nitrobenzoic acid
(j) 1-Butyl-4-methylbenzene

15.35 Starting with aniline, outline a synthesis of each of the following:

(a) *p*-Bromoaniline

(b) *o*-Bromoaniline

(c) 2-Bromo-4-nitroaniline

(d) 4-Bromo-2-nitroaniline

(e) 2,4,6-Tribromoaniline

15.36 Both of the following syntheses will fail. Explain what is wrong with each one.

(a)

(1) HNO₃/H₂SO₄
(2) CH₃COCl/AlCl₃
(3) NH₂NH₂, HO⁻, heat

(b)

(1) NBS, light
(2) NaOEt, EtOH, heat
(3) Br₂, FeBr₃

15.37 Propose structures for compounds **G–I**:

$$\underset{\text{60–65 °C}}{\xrightarrow{\text{concd H}_2\text{SO}_4}} \textbf{G (C}_6\text{H}_6\text{S}_2\text{O}_8) \underset{\text{concd H}_2\text{SO}_4}{\xrightarrow{\text{concd HNO}_3}} \textbf{H (C}_6\text{H}_5\text{NS}_2\text{O}_{10}) \underset{\text{heat}}{\xrightarrow{\text{H}_3\text{O}^+, \text{H}_2\text{O}}} \textbf{I (C}_6\text{H}_5\text{NO}_4)$$

15.38 2,6-Dichlorophenol has been isolated from the females of two species of ticks (*Amblyomma americanum* and *A. maculatum*), where it apparently serves as a sex attractant. Each female tick yields about 5 ng of 2,6-dichlorophenol. Assume that you need larger quantities than this and outline a synthesis of 2,6-dichlorophenol from phenol. (*Hint*: When phenol is sulfonated at 100 °C, the product is chiefly *p*-hydroxybenzenesulfonic acid.)

15.39 2-Methylnaphthalene can be synthesized from toluene through the following sequence of reactions. Write the structure of each intermediate.

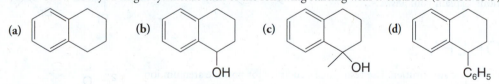

$$\text{Toluene} + \underset{\text{AlCl}_3}{\xrightarrow{\qquad}} \textbf{A (C}_{11}\text{H}_{12}\text{O}_3) \underset{\text{heat}}{\xrightarrow{\text{NH}_2\text{NH}_2, \text{KOH}}} \textbf{B (C}_{11}\text{H}_{14}\text{O}_2)$$

$$\xrightarrow{\text{SOCl}_2} \textbf{C (C}_{11}\text{H}_{13}\text{ClO}) \xrightarrow{\text{AlCl}_3} \textbf{D (C}_{11}\text{H}_{12}\text{O}) \xrightarrow{\text{NaBH}_4} \textbf{E (C}_{11}\text{H}_{14}\text{O})$$

$$\underset{\text{heat}}{\xrightarrow{\text{H}_2\text{SO}_4}} \textbf{F (C}_{11}\text{H}_{12}) \underset{\text{light}}{\xrightarrow{\text{NBS}}} \textbf{G (C}_{11}\text{H}_{12}\text{Br}) \underset{\substack{\text{EtOH,} \\ \text{heat}}}{\xrightarrow{\text{NaOEt}}}$$

15.40 Show how you might synthesize each of the following starting with α-tetralone (Section 15.9):

(a)

(b)

(c)

(d)

15.41 Give structures (including stereochemistry where appropriate) for compounds **A–G**:

(a) Benzene +

$$\underset{\text{(Section 7.10)}}{\xrightarrow{\text{AlCl}_3}} \textbf{A} \underset{\text{0 °C}}{\xrightarrow{\text{PCl}_5}} \textbf{B (C}_9\text{H}_{10}\text{Cl}_2) \underset{\substack{\text{mineral oil,} \\ \text{heat}}}{\xrightarrow{\text{2 NaNH}_2}} \textbf{C (C}_9\text{H}_8) \xrightarrow{\text{H}_2, \text{Ni}_2\text{B (P-2)}} \textbf{D (C}_9\text{H}_{10})$$

(*Hint*: The ¹H NMR spectrum of compound **C** consists of a multiplet at δ 7.20 (5H) and a singlet at δ 2.0 (3H).)

(b) **C** $\underset{\text{(2) NH}_4\text{Cl (Section 7.15B)}}{\xrightarrow{\text{(1) Li, EtNH}_2}}$ **E (C₉H₁₀)**

(d) **E** $\xrightarrow[\text{2-5 °C}]{\text{Br}_2}$ **G** + enantiomer (major products)

(c) **D** $\xrightarrow[\text{2-5 °C}]{\text{Br}_2}$ **F** + enantiomer (major products)

GENERAL PROBLEMS

15.42 Show how you might synthesize each of the following compounds starting with either benzyl bromide or allyl bromide:

(a) C₆H₅～CN

(b) C₆H₅～OMe

(c) C₆H₅～O～(C=O)～

(d) C₆H₅～I

(e) ～～N₃

(f) ～～O～

15.43 Provide structures for compounds **A** and **B**:

$$\text{Benzene} \xrightarrow[\text{liq. } NH_3, \text{ EtOH}]{\text{Na}} \textbf{A } (C_6H_8) \xrightarrow{\text{NBS}} \textbf{B } (C_6H_7Br)$$

15.44 Ring nitration of a dimethylbenzene (a xylene) results in the formation of only one dimethylnitrobenzene. Which dimethylbenzene isomer was the reactant?

15.45 The compound phenylbenzene (C_6H_5—C_6H_5) is called *biphenyl*, and the ring carbons are numbered in the following manner:

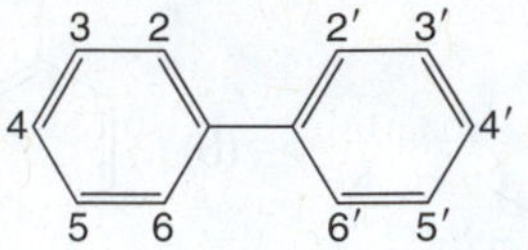

Use models to answer the following questions about substituted biphenyls. **(a)** When certain large groups occupy three or four of the *ortho* positions (e.g., 2, 6, 2′, and 6′), the substituted biphenyl may exist in enantiomeric forms. An example of a biphenyl that exists in enantiomeric forms is the compound in which the following substituents are present: 2-NO_2, 6-CO_2H, 2′-NO_2, 6′-CO_2H. What factors account for this? **(b)** Would you expect a biphenyl with 2-Br, 6-CO_2H, 2′-CO_2H, 6′-H to exist in enantiomeric forms? **(c)** The biphenyl with 2-NO_2, 6-NO_2, 2′-CO_2H, 6′-Br cannot be resolved into enantiomeric forms. Explain.

15.46 Treating cyclohexene with acetyl chloride and $AlCl_3$ leads to the formation of a product with the molecular formula $C_8H_{13}ClO$. Treating this product with a base leads to the formation of 1-acetylcyclohexene. Propose mechanisms for both steps of this sequence of reactions.

15.47 The *tert*-butyl group can be used as a blocking group in certain syntheses of aromatic compounds. **(a)** How would you introduce a *tert*-butyl group? **(b)** How would you remove it? **(c)** What advantage might a *tert*-butyl group have over a —SO_3H group as a blocking group?

15.48 When toluene is sulfonated (concentrated H_2SO_4) at room temperature, predominantly (about 95% of the total) ortho and para substitution occurs. If elevated temperatures (150–200 °C) and longer reaction times are employed, meta (chiefly) and para substitution account for some 95% of the products. Account for these differences in terms of kinetic and thermodynamic pathways. (*Hint: m*-Toluenesulfonic acid is the most stable isomer.)

15.49 A C—D bond is harder to break than a C—H bond, and, consequently, reactions in which C—D bonds are broken proceed more slowly than reactions in which C—H bonds are broken. What mechanistic information comes from the observation that perdeuterated benzene, C_6D_6, is nitrated at the same rate as normal benzene, C_6H_6?

15.50 Heating 1,1,1-triphenylmethanol with ethanol containing a trace of a strong acid causes the formation of 1-ethoxy-1,1,1-triphenylmethane. Write a plausible mechanism that accounts for the formation of this product.

15.51
(a) Which of the following halides would you expect to be most reactive in an S_N2 reaction? **(b)** In an S_N1 reaction? Explain your answers.

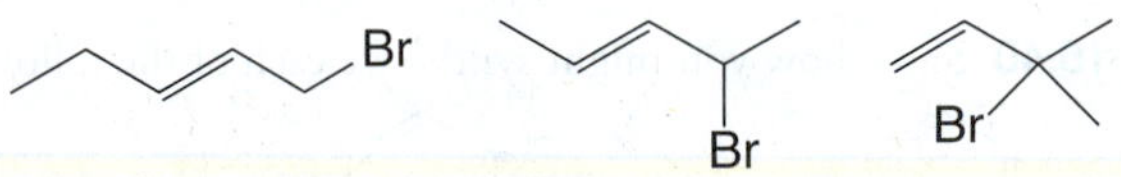

CHALLENGE PROBLEMS

15.52 Furan undergoes electrophilic aromatic substitution. Use resonance structures for possible arenium ion intermediates to predict whether furan is likely to undergo bromination more rapidly at C2 or at C3.

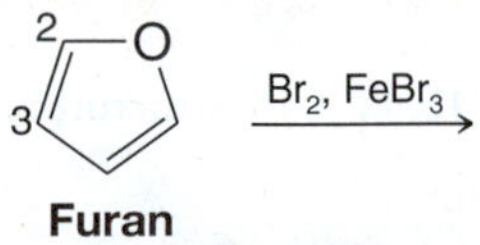

Furan

$$\xrightarrow{Br_2, \text{ FeBr}_3}$$

15.53 Acetanilide was subjected to the following sequence of reactions: (1) concd H_2SO_4; (2) HNO_3, heat; (3) H_2O, H_2SO_4, heat, then HO^-. The ^{13}C NMR spectrum of the final product gives six signals. Write the structure of the final product.

15.54 The lignins are macromolecules that are major components of the many types of wood, where they bind cellulose fibers together in these natural composites. The lignins are built up out of a variety of small molecules (most having phenylpropane skeletons). These precursor molecules are covalently connected in varying ways, and this gives the lignins great complexity. To explain the formation of compound **B** below as one of many products obtained when lignins are ozonized, lignin model compound **A** was treated as shown. Use the following information to determine the structure of **B**.

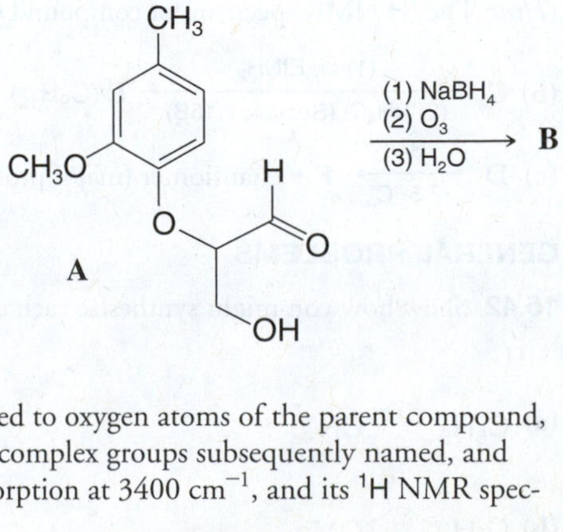

$$\xrightarrow[\text{(3) } H_2O]{\begin{array}{c}\text{(1) NaBH}_4 \\ \text{(2) } O_3\end{array}} \textbf{B}$$

A

To make **B** volatile enough for GC/MS (gas chromatography–mass spectrometry, Section 9.19), it was first converted to its tris(*O*-trimethylsilyl) derivative, which had M.$^+$ 308 *m/z*. ["Tris" means that three of the indicated complex groups named (e.g., trimethylsilyl groups here) are present. The capital, italicized *O* means these are attached to oxygen atoms of the parent compound, taking the place of hydrogen atoms. Similarly, the prefix "bis" indicates the presence of two complex groups subsequently named, and "tetrakis" (used in the problem below), means four.] The IR spectrum of **B** had a broad absorption at 3400 cm^{-1}, and its ^1H NMR spectrum showed a single multiplet at δ 3.6. What is the structure of **B**?

15.55 When compound **C**, which is often used to model a more frequently occurring unit in lignins, was ozonized, product **D** was obtained. In a variety of ways it has been established that the stereochemistry of the three-carbon side chain of such lignin units remains largely if not completely unchanged during oxidations like this.

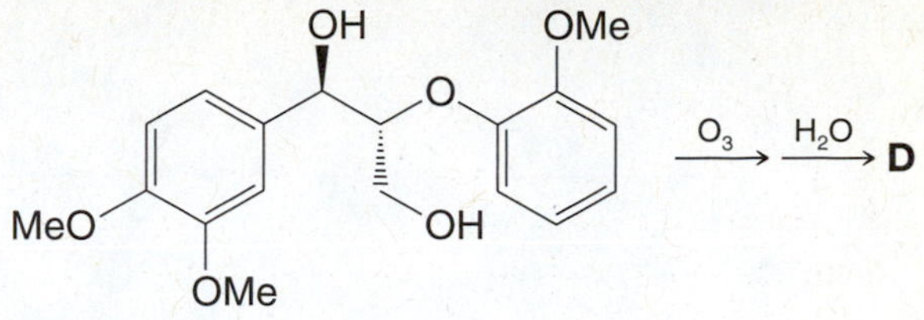

For GC/MS, **D** was converted to its tetrakis(*O*-trimethylsilyl) derivative, which had M$^{+\cdot}$ 424 *m/z*. The IR spectrum of **D** had bands at 3000 cm^{-1} (broad, strong) and 1710 cm^{-1} (strong). Its ^{1}H NMR spectrum had peaks at δ 3.7 (multiplet, 3H) and δ 4.2 (doublet, 1H) after treatment with D$_2$O. Its DEPT ^{13}C NMR spectra had peaks at δ 64 (CH$_2$), δ 75 (CH), δ 82 (CH), and δ 177 (C). What is the structure of **D**, including its stereochemistry?

LEARNING GROUP PROBLEMS

1. The structure of thyroxine, a thyroid hormone that helps to regulate metabolic rate, was determined in part by comparison with a synthetic compound believed to have the same structure as natural thyroxine. The final step in the laboratory synthesis of thyroxine by Harington and Barger, shown below, involves an electrophilic aromatic substitution. Draw a detailed mechanism for this step and explain why the iodine substitutions occur ortho to the phenolic hydroxyl and not ortho to the oxygen of the aryl ether. [One reason iodine is required in our diet (e.g., in iodized salt) is for the biosynthesis of thyroxine.]

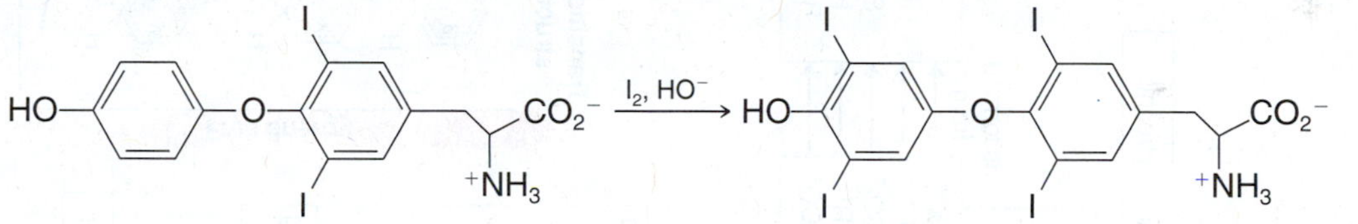

2. Synthesize 2-chloro-4-nitrobenzoic acid from toluene and any other reagents necessary. Begin by writing a retrosynthetic analysis.

3. Deduce the structures of compounds **E–L** in the roadmap below.

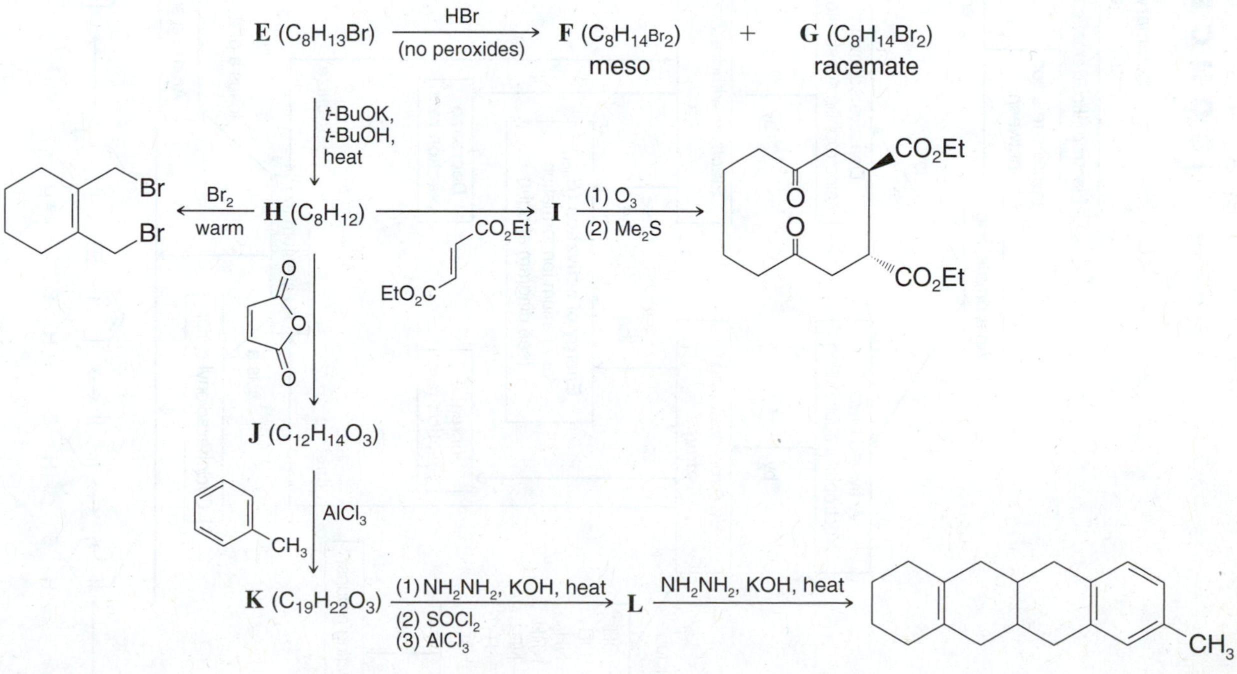

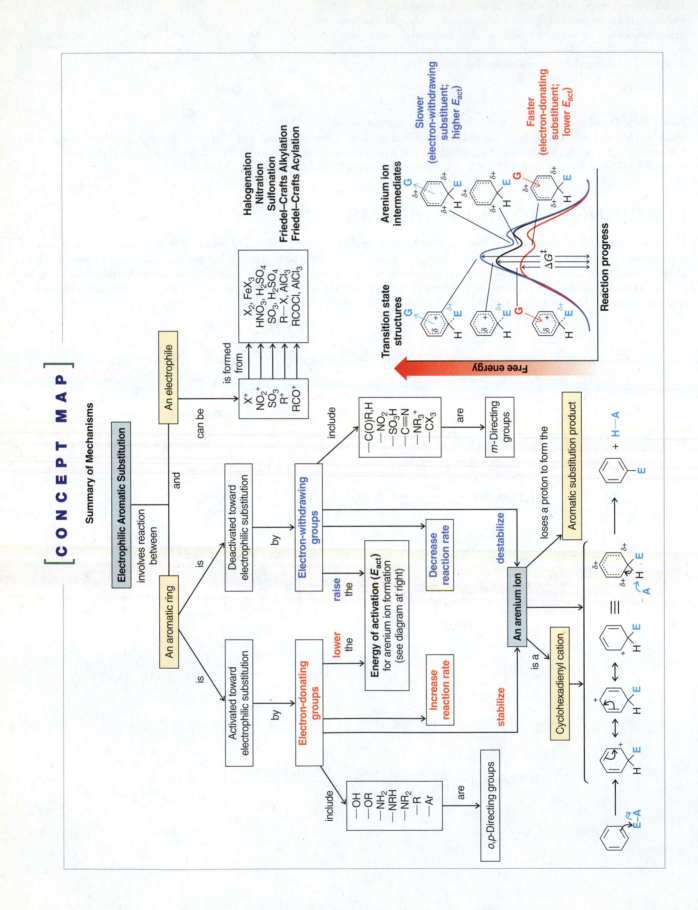

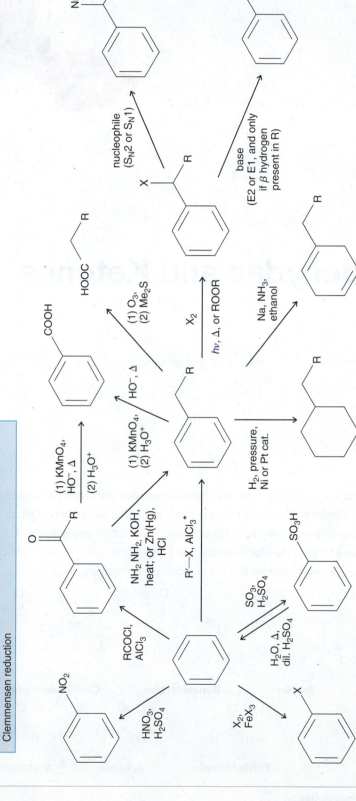

[C O N C E P T   M A P]

Some Synthetic Connections of Benzene and Aryl Derivatives

- Nitration
- Halogenation
- Sulfonation/desulfonation
- Friedel–Crafts alkylation
- Friedel–Crafts acylation
- Wolff–Kishner or
  Clemmensen reduction

- Side-chain oxidation
- Ring oxidation
- Catalytic hydrogenation of ring
- Birch reduction
- Benzylic radical halogenation
- Benzylic substitution/elimination

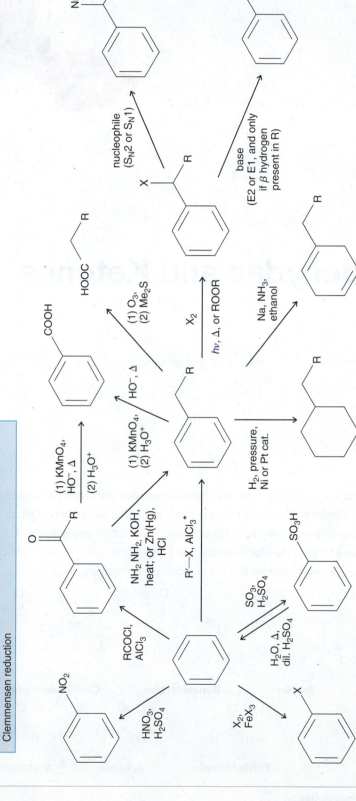

*In the Friedel–Crafts alkylation example shown here, R' is a primary alkyl halide.
If carbocation rearrangements are likely, then Friedel–Crafts acylation followed
by Wolff–Kishner or Clemmensen reduction should be used to incorporate a primary alkyl group.

CHAPTER

16

Aldehydes and Ketones

NUCLEOPHILIC ADDITION TO THE CARBONYL GROUP

Y ou may not know it, but you already have experience with aldehydes and ketones based on things you have likely smelled and tasted. Vanillin is responsible for the smell of vanilla, while almond flavor results from benzaldehyde, cinnamon from cinnamaldehyde, and spearmint from (R)-carvone. Other odors and sensations that are far less pleasant can also be caused by aldehydes and ketones—for example, the pungent odor of formaldehyde or acetone, or the hangover caused by acetaldehyde that results from drinking too many alcoholic beverages.

Vanillin   **Benzaldehyde**   **Cinnamaldehyde**   **(R)–carvone**

Formaldehyde   **Acetone**   **Acetaldehyde**

IN THIS CHAPTER WE WILL CONSIDER:

- the structure and reactivity of aldehydes and ketones
- methods for their synthesis from other functional groups
- unique functional groups that can arise from aldehydes and ketones that have special reactivity of their own

[**WHY** DO THESE TOPICS MATTER?] At the end of the chapter, we will show how some of the functional groups that can be obtained from aldehydes and ketones provide a triggering device that sea sponges use in molecules meant to kill or injure predators. Amazingly, these same molecules and functional groups provide a potential treatment for various forms of human cancer.

16.1 INTRODUCTION

- Aldehydes have a **carbonyl group** bonded to a carbon atom on one side and a hydrogen atom on the other side. (Formaldehyde is an exception because it has hydrogen atoms on both sides.)
- Ketones have a carbonyl group bonded to carbon atoms on both sides.

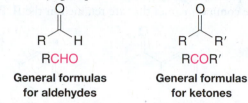

General formulas for aldehydes

General formulas for ketones

Acetone

Although earlier chapters have given us some insight into the chemistry of carbonyl compounds, we shall now consider their chemistry in detail. The reason: the chemistry of the carbonyl group is central to the chemistry of most of the chapters that follow.

In this chapter we focus our attention on the preparation of aldehydes and ketones, their physical properties, and especially *nucleophilic addition reactions at their carbonyl groups.*

16.2 NOMENCLATURE OF ALDEHYDES AND KETONES

- Aliphatic aldehydes are named substitutively in the IUPAC system by replacing the final **-e** of the name of the corresponding alkane with **-al**.

Since the aldehyde group must be at an end of the carbon chain, there is no need to indicate its position. When other substituents are present the carbonyl group carbon is assigned position 1. Many aldehydes also have common names; these are given below in parentheses. These common names are derived from the common names for the corresponding carboxylic acids (Section 17.2A), and some of them are retained by the IUPAC as acceptable names.

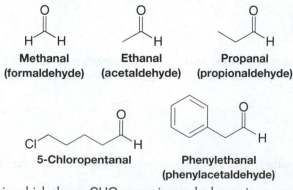

Methanal (formaldehyde) **Ethanal** (acetaldehyde) **Propanal** (propionaldehyde)

5-Chloropentanal **Phenylethanal** (phenylacetaldehyde)

- Aldehydes in which the —CHO group is attached to a ring system are named substitutively by adding the suffix **carbaldehyde**. Several examples follow:

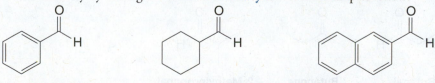

Benzenecarbaldehyde (benzaldehyde) **Cyclohexanecarbaldehyde** **2-Naphthalenecarbaldehyde**

The common name *benzaldehyde* is far more frequently used than benzenecarbalde-hyde for C_6H_5CHO, and it is the name we shall use in this text.

- Aliphatic ketones are named substitutively by replacing the final **-e** of the name of the corresponding alkane with **-one**.

The chain is then numbered in the way that gives the carbonyl carbon atom the lower possible number, and this number is used to designate its position.

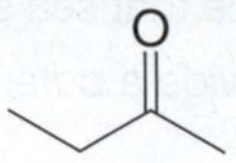

Butanone
(ethyl methyl ketone)

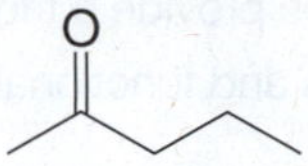

2-Pentanone
(methyl propyl ketone)

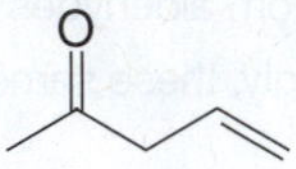

Pent-4-en-2-one
(not 1-penten-4-one)
(allyl methyl ketone)

Common functional group names for ketones (in parentheses above) are obtained simply by separately naming the two groups attached to the carbonyl group and adding the word **ketone** as a separate word.

Some ketones have common names that are retained in the IUPAC system:

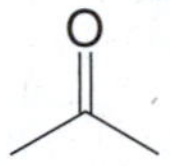

Acetone
(propanone)

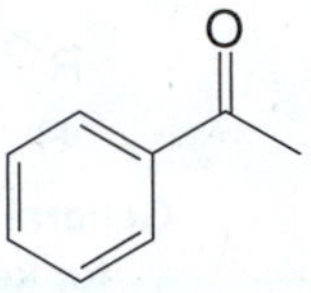

Acetophenone
(1-phenylethanone
or methyl phenyl ketone)

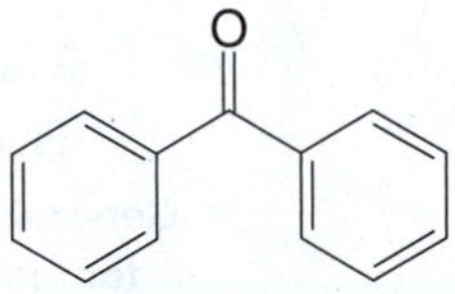

Benzophenone
(diphenylmethanone
or diphenyl ketone)

When it is necessary to name the $\overset{O}{\underset{}{\parallel}}$ H group as a prefix, it is the **methanoyl** or **formyl group**. The $\overset{O}{\underset{}{\parallel}}$ group is called the **ethanoyl** or **acetyl group** (often abbreviated as Ac). When $\overset{O}{\underset{}{\parallel}}$ R groups are named as substituents, they are called **alkanoyl** or **acyl groups**.

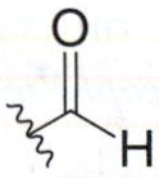

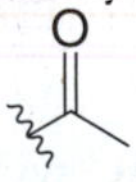

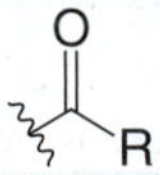

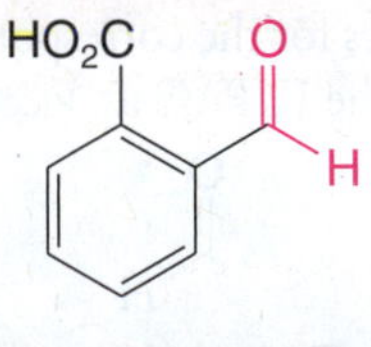

2-Methanoylbenzoic acid
(*o*-formylbenzoic acid)

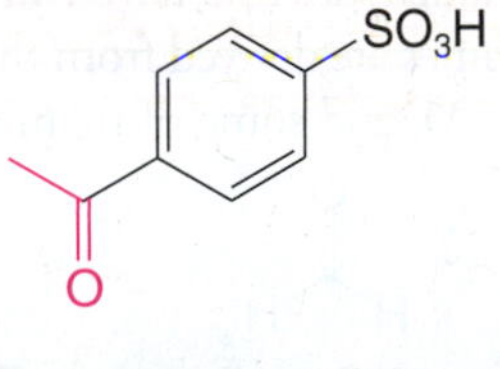

4-Ethanoylbenzenesulfonic acid
(*p*-acetylbenzenesulfonic acid)

●●● SOLVED PROBLEM 16.1

Write bond-line formulas for three isomeric compounds that contain a carbonyl group and have the molecular formula C_4H_8O. Then give their IUPAC names.

STRATEGY AND ANSWER: Write the formulas and then name the compounds.

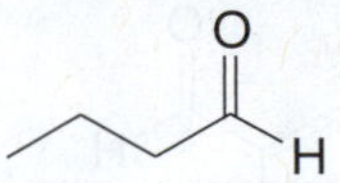

Butanal 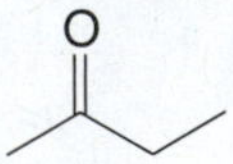   **Butanone** 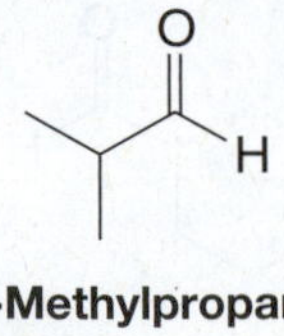   **2-Methylpropanal**

(a) Give IUPAC substitutive names for the seven isomeric aldehydes and ketones with the formula $C_5H_{10}O$. **(b)** Give structures and names (common or IUPAC substitutive names) for all the aldehydes and ketones that contain a benzene ring and have the formula C_8H_8O.

PRACTICE PROBLEM 16.1

16.3 PHYSICAL PROPERTIES

The carbonyl group is a polar group; therefore, aldehydes and ketones have higher boiling points than hydrocarbons of the same molecular weight. However, since aldehydes and ketones cannot have strong hydrogen bonds *between their molecules*, they have lower boiling points than the corresponding alcohols. The following compounds that have similar molecular weights exemplify this trend:

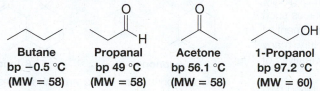

Butane	**Propanal**	**Acetone**	**1-Propanol**
bp −0.5 °C	bp 49 °C	bp 56.1 °C	bp 97.2 °C
(MW = 58)	(MW = 58)	(MW = 58)	(MW = 60)

A map of electrostatic potential for acetone shows the polarity of the carbonyl $C{=}O$ bond.

Which compound in each of the following pairs has the higher boiling point? (Answer this problem without consulting tables.)

(a) Pentanal or 1-pentanol

(b) 2-Pentanone or 2-pentanol

(c) Pentane or pentanal

(d) Acetophenone or 2-phenylethanol

(e) Benzaldehyde or benzyl alcohol

PRACTICE PROBLEM 16.2

The carbonyl oxygen atom allows molecules of aldehydes and ketones to form strong hydrogen bonds to molecules of water. As a result, low-molecular-weight aldehydes and ketones show appreciable solubilities in water. Acetone and acetaldehyde are soluble in water in all proportions.

Hydrogen bonding (shown in red) between water molecules and acetone

TABLE 16.1 PHYSICAL PROPERTIES OF ALDEHYDES AND KETONES

Formula	Name	mp (°C)	bp (°C)	Solubility in Water
HCHO	Formaldehyde	−92	−21	Very soluble
CH_3CHO	Acetaldehyde	−125	21	∞
CH_3CH_2CHO	Propanal	−81	49	Very soluble
$CH_3(CH_2)_2CHO$	Butanal	−99	76	Soluble
$CH_3(CH_2)_3CHO$	Pentanal	−92	102	Slightly soluble
$CH_3(CH_2)_4CHO$	Hexanal	−51	131	Slightly soluble
C_6H_5CHO	Benzaldehyde	−26	178	Slightly soluble
$C_6H_5CH_2CHO$	Phenylacetaldehyde	33	193	Slightly soluble
CH_3COCH_3	Acetone	−95	56.1	∞
$CH_3COCH_2CH_3$	Butanone	−86	80	Very soluble
$CH_3COCH_2CH_2CH_3$	2-Pentanone	−78	102	Soluble
$CH_3CH_2COCH_2CH_3$	3-Pentanone	−39	102	Soluble
$C_6H_5COCH_3$	Acetophenone	21	202	Insoluble
$C_6H_5COC_6H_5$	Benzophenone	48	306	Insoluble

Table 16.1 lists the physical properties of a number of common aldehydes and ketones. Some aldehydes obtained from natural sources have very pleasant fragrances. The following are some in addition to those we mentioned at the beginning of this chapter.

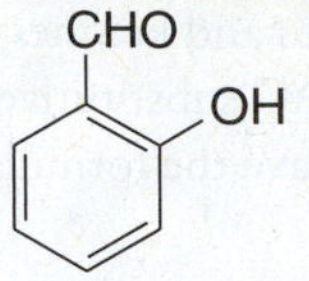

Salicylaldehyde
(from meadowsweet)

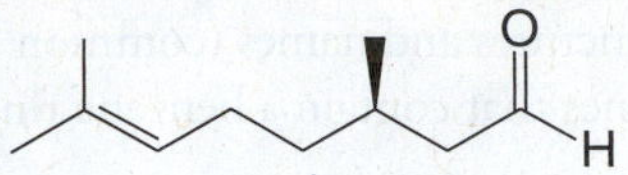

Citronellal
(the scent of lemon in certain plants)

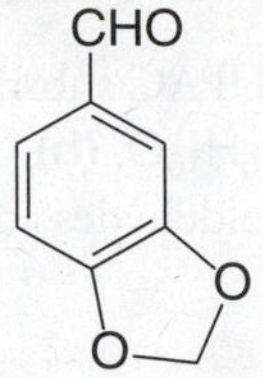

Piperonal
(made from safrole; odor of heliotrope)

THE CHEMISTRY OF... Aldehydes and Ketones in Perfumes

Many aldehydes and ketones have pleasant fragrances and, because of this, they have found use in perfumes. Originally, the ingredients for perfumes came from natural sources such as essential oils (Section 23.3), but with the development of synthetic organic chemistry in the nineteenth century, many ingredients now used in perfumes result from the creativity of laboratory chemists.

Practitioners of the perfumer's art, those who blend perfumes, talk of their ingredients in a language derived from music. The cabinet that holds the bottles containing the compounds that the perfumer blends is called an "organ." The ingredients themselves are described as having certain "notes." For example, highly volatile substances are said to display "head notes," those less volatile and usually associated with flowers are said to have "heart notes," and the least volatile ingredients, usually with woody, balsamic, or musklike aromas, are described as "base notes."*

(Z)-Jasmone (with the odor of jasmine) and α-damascone (odor of roses) have "heart notes," as do the ionones (with the odor of violets). All of these ketones can be obtained from natural sources.

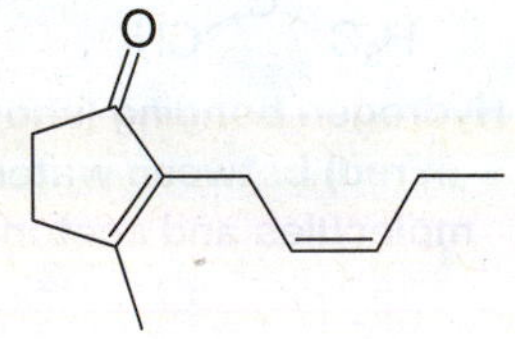

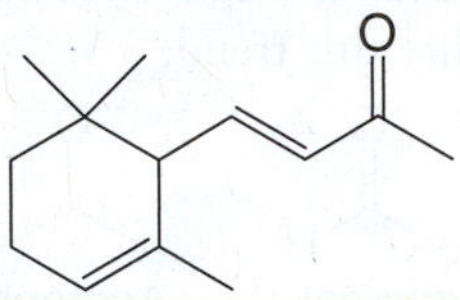

α-Ionone   **β-Ionone**

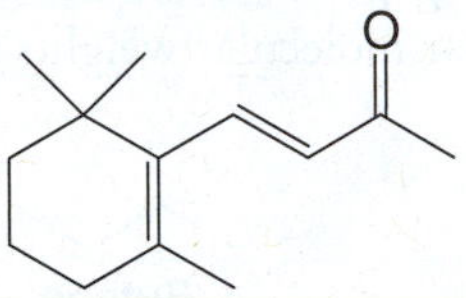

Two ketones from exotic natural sources are muscone (from the Himalayan musk deer) and civetone (from the African civet cat).

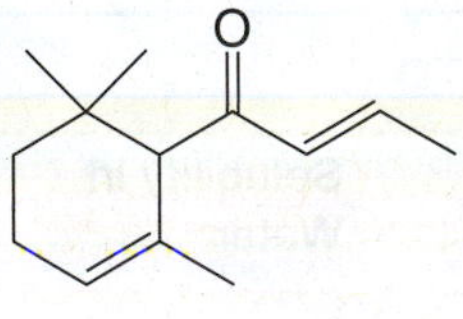

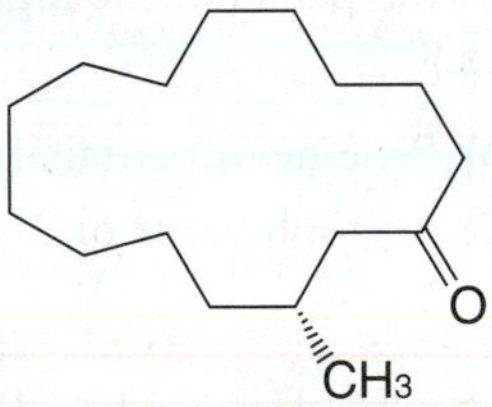

Muscone 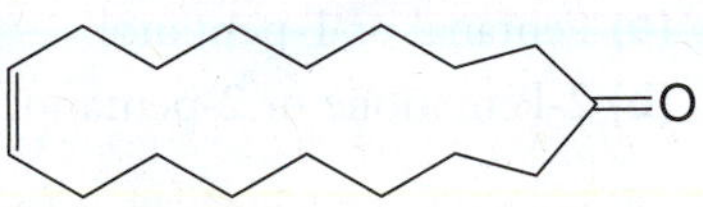   **Civetone**

Stereochemistry has a marked influence on odors. For example, the (R)-enantiomer of muscone (depicted above) is described as having a "rich and powerful musk," whereas the (S)-enantiomer is described as being "poor and less strong." The (R)-enantiomer of α-damascone has a rose petal odor with more apple and fruitier notes than the (S)-enantiomer.

Z-Jasmone   **α-Damascone**

*For an in-depth discussion of the perfume industry, see Fortineau, A.-D. "Chemistry Perfumes Your Daily Life," *J. Chem. Educ.*, **2004**, *81*, 45–50.

16.4 SYNTHESIS OF ALDEHYDES

16.4A Aldehydes by Oxidation of 1° Alcohols

- The oxidation state of an aldehyde lies between that of a 1° alcohol and a carboxylic acid (Section 12.4A).

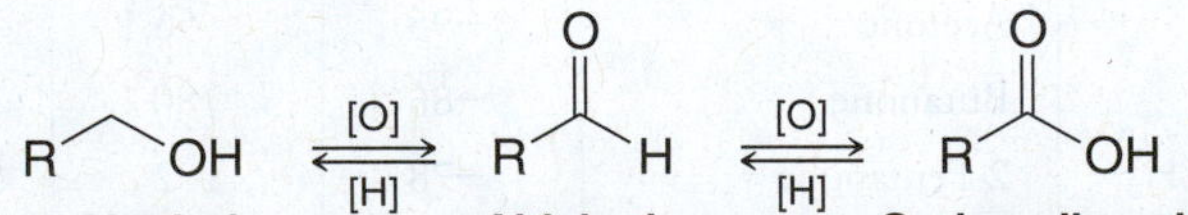

$$R \overset{[O]}{\underset{[H]}{\rightleftharpoons}} \quad R \overset{[O]}{\underset{[H]}{\rightleftharpoons}} \quad R$$

1° Alcohol   **Aldehyde**   **Carboxylic acid**

Aldehydes can be prepared from 1° alcohols by the Swern oxidation (Section 16.4B) and oxidation with pyridinium chlorochromate ($C_5H_5\overset{+}{N}HCrO_3Cl^-$, or PCC, Section 12.4D):

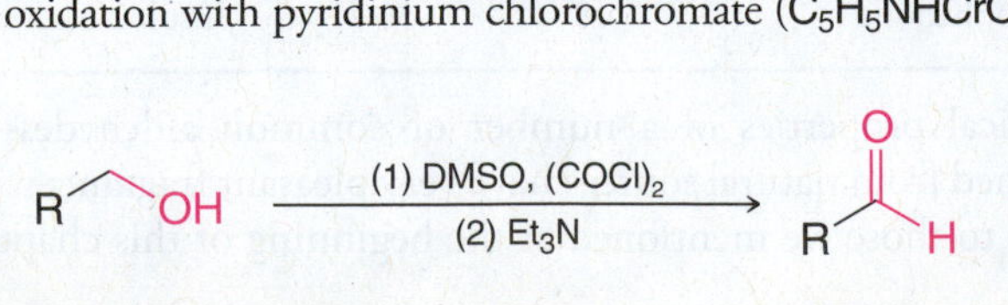

$$R \overset{OH}{\quad} \xrightarrow[\text{(2) Et}_3\text{N}]{\text{(1) DMSO, (COCl)}_2} \quad R \overset{O}{\underset{H}{\quad}}$$

Swern oxidation of a 1° alcohol to an aldehyde.

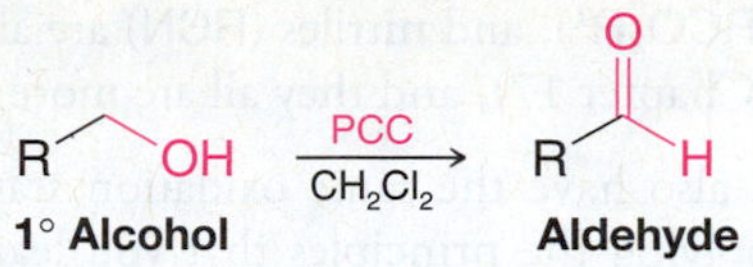

PCC oxidation of a 1° alcohol to an aldehyde

$$\text{1° Alcohol} \xrightarrow[\text{CH}_2\text{Cl}_2]{\text{PCC}} \text{Aldehyde}$$

The following are examples of the use of the Swern oxidation and PCC in the synthesis of aldehydes.

Benzyl alcohol → (1) DMSO, (COCl)$_2$ / (2) Et$_3$N → **Benzaldehyde**

1-Heptanol → PCC / CH$_2$Cl$_2$ → **Heptanal**

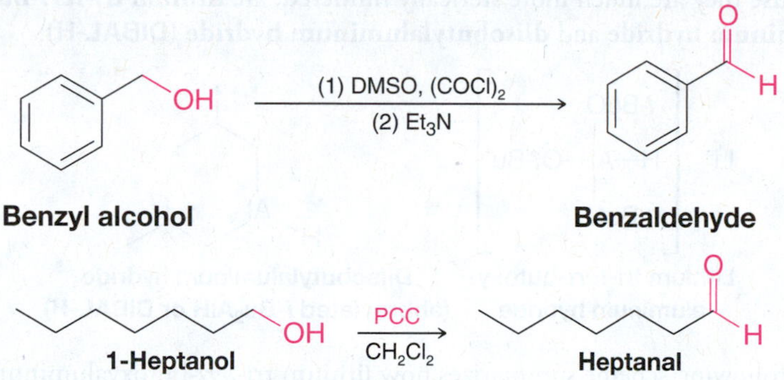

16.4B Aldehydes by Ozonolysis of Alkenes

- Alkenes can be cleaved by ozonolysis of their double bond (Section 8.17B). The products are aldehydes and ketones.

In Chapter 8 we also saw how this procedure has utility in structure determination. The following examples illustrate the synthesis of aldehydes by ozonolysis of alkenes.

(1) O$_3$, CH$_2$Cl$_2$, −78 °C / (2) Me$_2$S

(1) O$_3$, CH$_2$Cl$_2$, −78 °C / (2) Me$_2$S

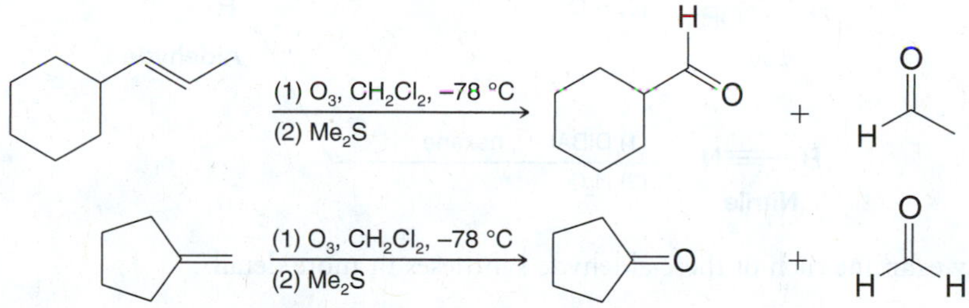

16.4C Aldehydes by Reduction of Acyl Chlorides, Esters, and Nitriles

Theoretically, it ought to be possible to prepare aldehydes by reduction of carboxylic acids. In practice, this is not possible with the reagent normally used to reduce a carboxylic acid, lithium aluminum hydride (LiAlH$_4$ or LAH).

- When any carboxylic acid is treated with LAH, it is reduced all the way to the 1° alcohol.
- This happens because LAH is a very powerful reducing agent and aldehydes are very easily reduced.

Any aldehyde that might be formed in the reaction mixture is immediately reduced by LAH to the 1° alcohol. (It does not help to use a stoichiometric amount of LAH, because as soon as the first few molecules of aldehyde are formed in the mixture, there will still be much unreacted LAH present and it will reduce the aldehyde.)

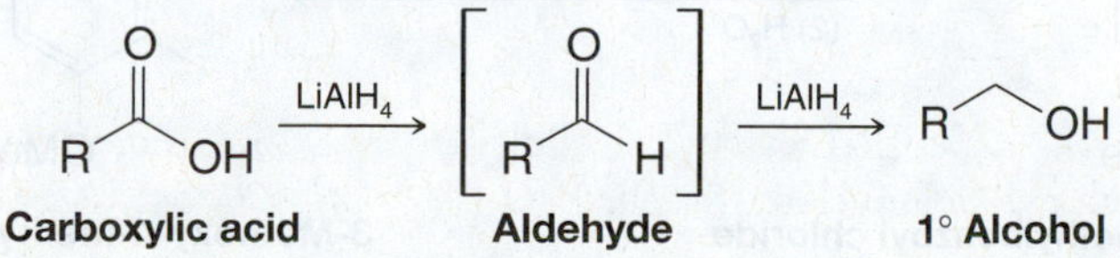

Carboxylic acid → LiAlH$_4$ → [**Aldehyde**] → LiAlH$_4$ → **1° Alcohol**

The secret to success here is not to use a carboxylic acid itself, but to use a derivative of a carboxylic acid that is more easily reduced, and an aluminum hydride derivative that is less reactive than LAH.

- Acyl chlorides (RCOCl), esters (RCO$_2$R′), and nitriles (RCN) are all easily prepared from carboxylic acids (Chapter 17), and they all are more easily reduced.

(Acyl chlorides, esters, and nitriles all also have the same oxidation state as carboxylic acids. Convince yourself of this by applying the principles that you learned in Section 12.2A).

- Two derivatives of aluminum hydride that are less reactive than LAH, in part because they are much more sterically hindered, are **lithium tri-*tert*-butoxy-aluminum hydride** and **diisobutylaluminum hydride (DIBAL-H)**:

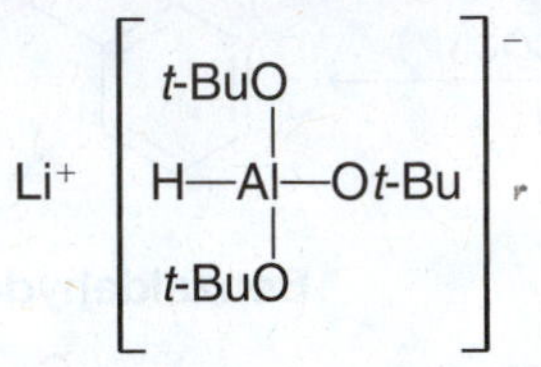

Lithium tri-*tert*-butoxy-aluminum hydride

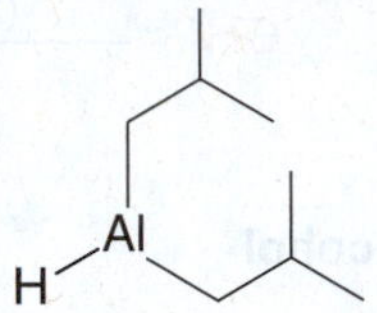

Diisobutylaluminum hydride (abbreviated *i*-Bu$_2$AlH or DIBAL-H)

- The following scheme summarizes how lithium tri-*tert*-butoxyaluminum hydride and DIBAL-H can be used to synthesize aldehydes from acid derivatives:

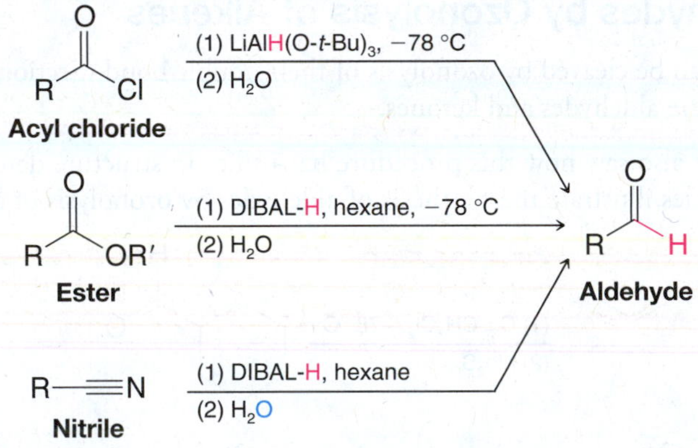

We now examine each of these aldehyde syntheses in more detail.

Aldehydes from Acyl Chlorides: RCOCl → RCHO

- Acyl chlorides can be reduced to aldehydes by treating them with LiAlH[OC(CH$_3$)$_3$]$_3$, lithium tri-*tert*-butoxyaluminum hydride, at −78 °C.
- Carboxylic acids can be converted to acyl chlorides by using SOCl$_2$ (see Section 15.7).

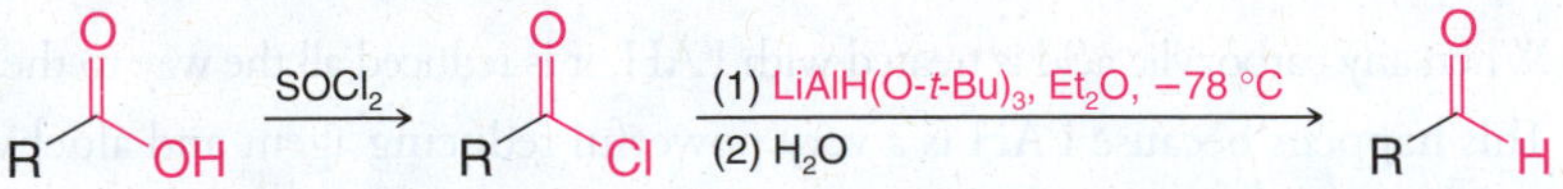

The following is a specific example:

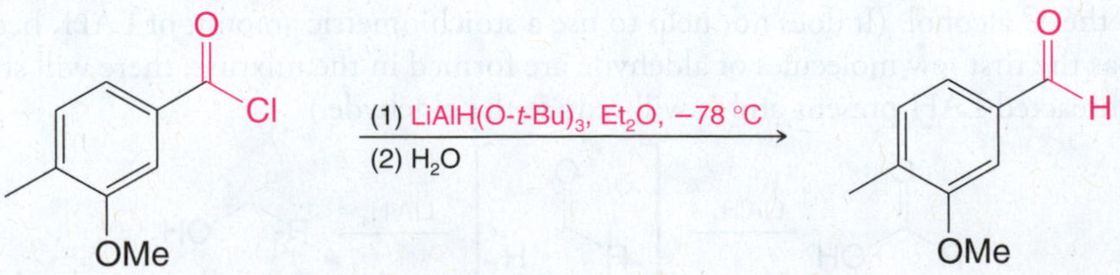

3-Methoxy-4-methylbenzoyl chloride   **3-Methoxy-4-methylbenzaldehyde**

Mechanistically, the reduction is brought about by the transfer of a hydride ion from the aluminum atom to the carbonyl carbon of the acyl chloride (see Section 12.3). Subsequent hydrolysis frees the aldehyde.

$$\underset{\text{R}}{\overset{\text{O}}{\parallel}}\text{C}-\text{R'(H)}$$

[A MECHANISM FOR THE REACTION — Reduction of an Acyl Chloride to an Aldehyde]

Transfer of a hydride ion to the carbonyl carbon brings about the reduction.

Acting as a Lewis acid, the aluminum atom accepts an electron pair from oxygen.

This intermediate loses a chloride ion as an electron pair from the oxygen assists.

The addition of water causes hydrolysis of this aluminum complex to take place, producing the aldehyde. (Several steps are involved.)

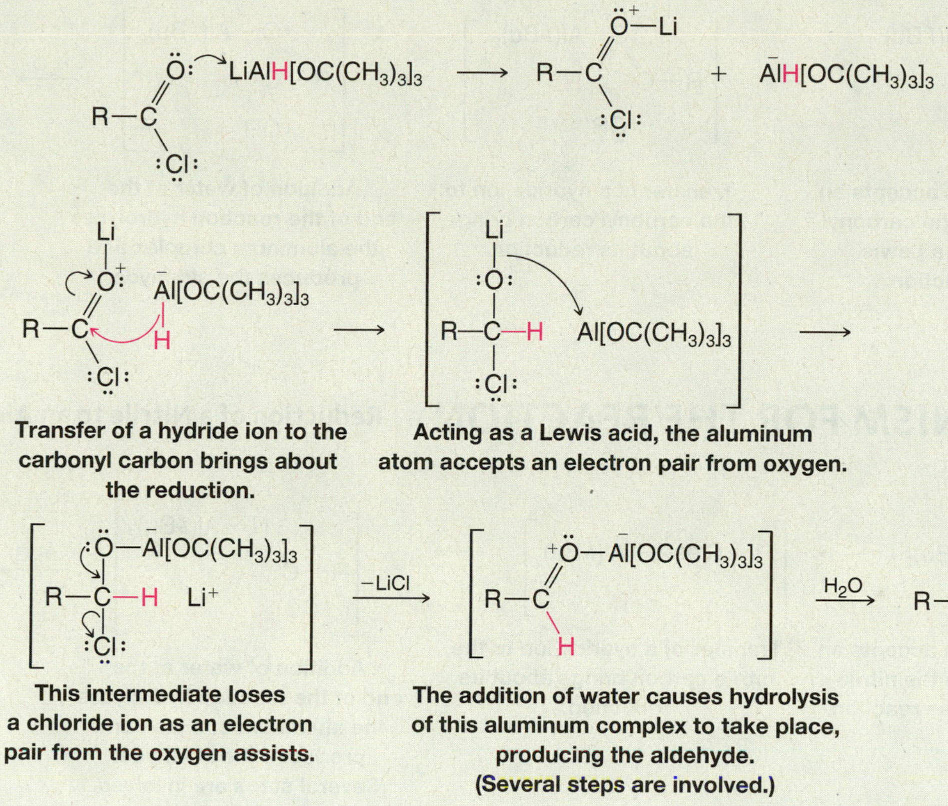

●●● **SOLVED PROBLEM 16.2**

Provide the reagents for transformations (1), (2), and (3).

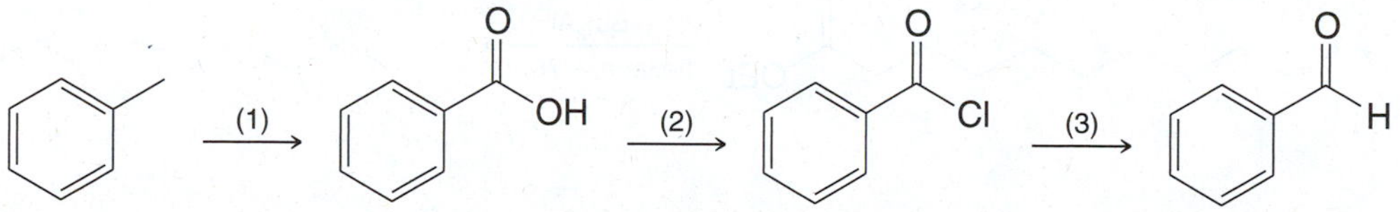

STRATEGY AND ANSWER: In (1), we must oxidize methylbenzene to benzoic acid. To do this we use hot potassium permanganate in a basic solution followed by an acidic workup (see Section 15.13C). For (2), we must convert a carboxylic acid to an acid chloride. For this transformation we use thionyl chloride or phosphorus pentachloride (see Section 15.7). For (3), we must reduce an acid chloride to an aldehyde. For this we use lithium tri-*tert*-butoxyaluminum hydride (see above).

Aldehydes from Esters and Nitriles: $RCO_2R' \rightarrow RCHO$ and $RC{\equiv}N \rightarrow RCHO$

- Both esters and nitriles can be reduced to aldehydes by DIBAL-H.

Carefully controlled amounts of DIBAL-H must be used to avoid overreduction, and the ester reduction must be carried out at low temperatures. Both reductions result in the formation of a relatively stable intermediate by the addition of a hydride ion to the carbonyl carbon of the ester or to the carbon of the $-C{\equiv}N$ group of the nitrile. Hydrolysis of the intermediate liberates the aldehyde. Schematically, the reactions can be viewed in the following way:

A MECHANISM FOR THE REACTION — Reduction of an Ester to an Aldehyde

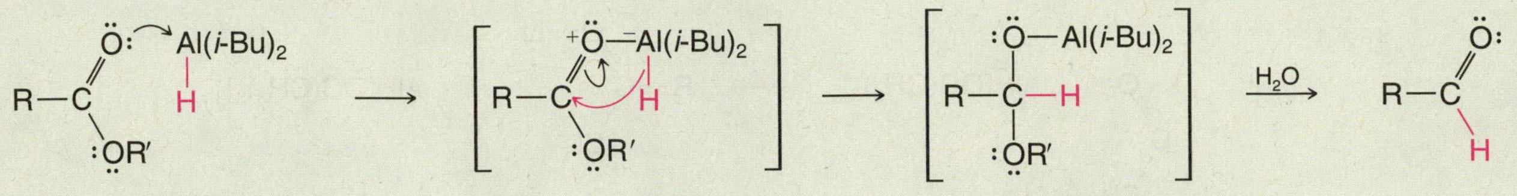

| The aluminum atom accepts an electron pair from the carbonyl oxygen atom in a Lewis acid-base reaction. | Transfer of a hydride ion to the carbonyl carbon brings about its reduction. | Addition of water at the end of the reaction hydrolyzes the aluminum complex and produces the aldehyde. |

A MECHANISM FOR THE REACTION — Reduction of a Nitrile to an Aldehyde

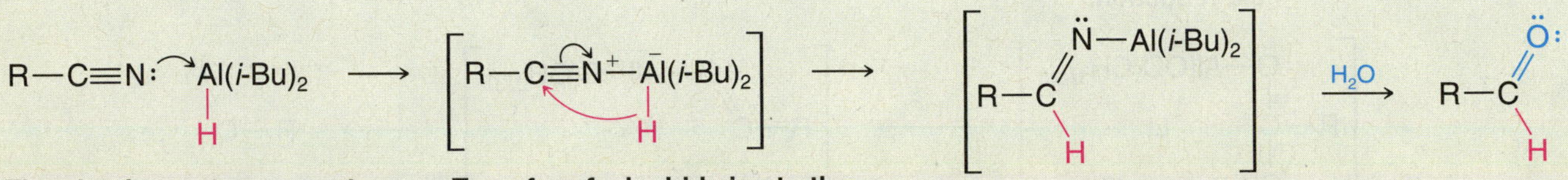

| The aluminum atom accepts an electron pair from the nitrile in a Lewis acid–base reaction. | Transfer of a hydride ion to the nitrile carbon brings about its reduction. | Addition of water at the end of the reaction hydrolyzes the aluminum complex and produces the aldehyde. (Several steps are involved. See Section 16.8 relating to imines.) |

The following specific examples illustrate these syntheses:

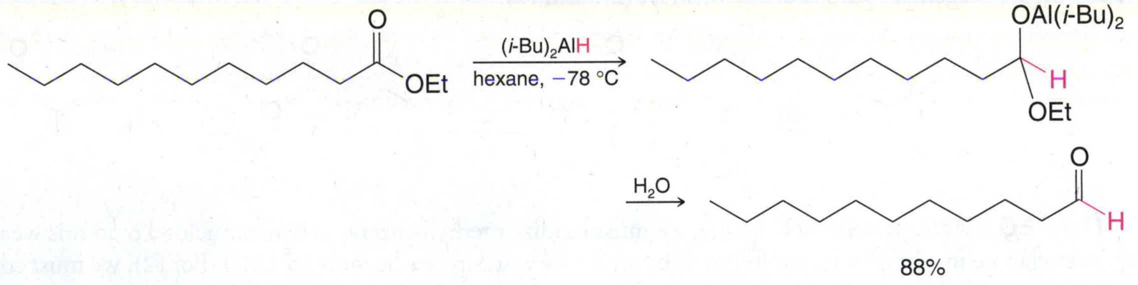

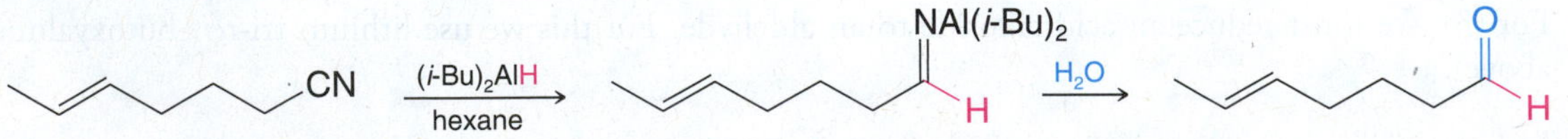

88%

●●● SOLVED PROBLEM 16.3

What is the product of the following reaction?

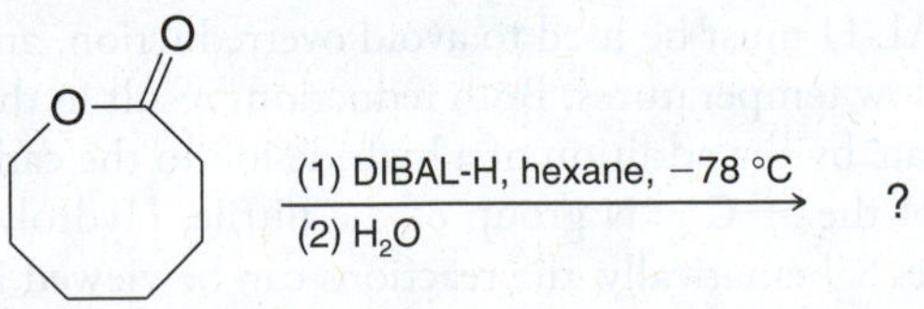

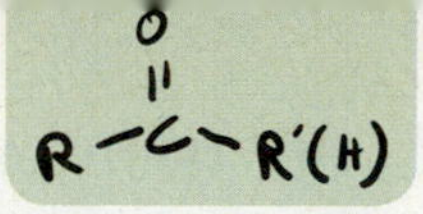

STRATEGY AND ANSWER: The starting compound is a cyclic ester, so the product would be an aldehyde that also contains an alcohol hydroxyl group.

SOLVED PROBLEM 16.4

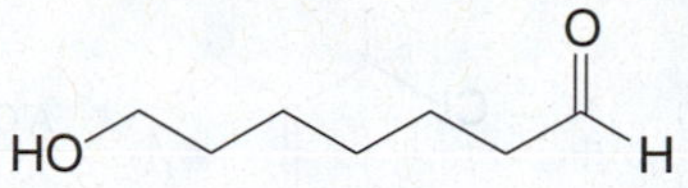

Starting with benzyl alcohol, outline a synthesis of phenylethanal.

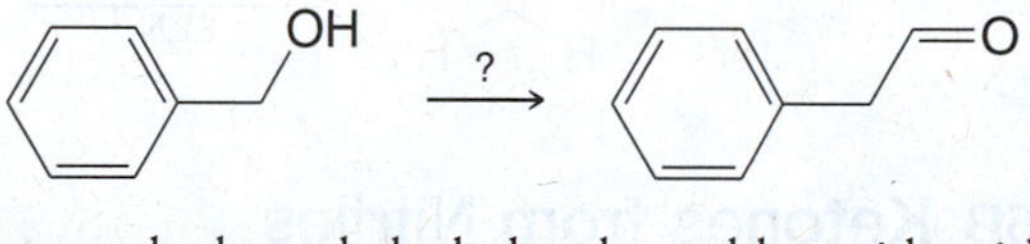

STRATEGY AND ANSWER: Convert the benzyl alcohol to benzyl bromide with PBr_3, then replace the bromine by cyanide in an S_N2 reaction. Lastly, reduce the nitrile to phenylethanal.

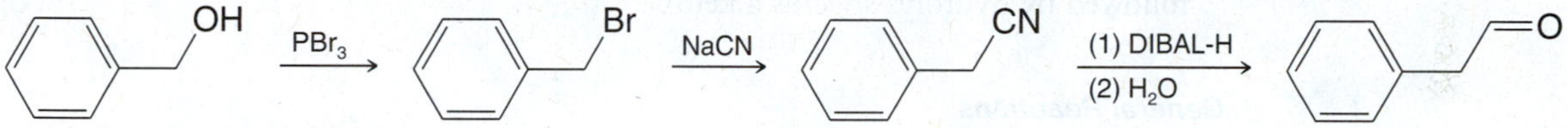

PRACTICE PROBLEM 16.3

Show how you would synthesize propanal from each of the following: **(a)** 1-propanol and **(b)** propanoic acid ($CH_3CH_2CO_2H$).

16.5 SYNTHESIS OF KETONES

16.5A Ketones from Alkenes, Arenes, and 2° Alcohols

We have seen three laboratory methods for the preparation of ketones in earlier chapters:

1. Ketones (and aldehydes) by ozonolysis of alkenes (discussed in Section 8.17B).

Ketone **Aldehyde**

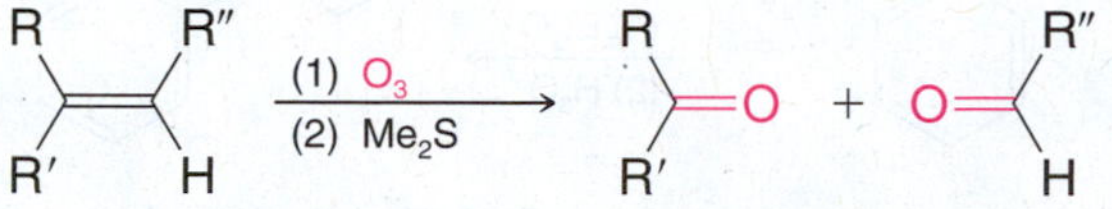

2. Ketones from arenes by Friedel–Crafts acylations (discussed in Section 15.7). For example:

An alkyl aryl ketone

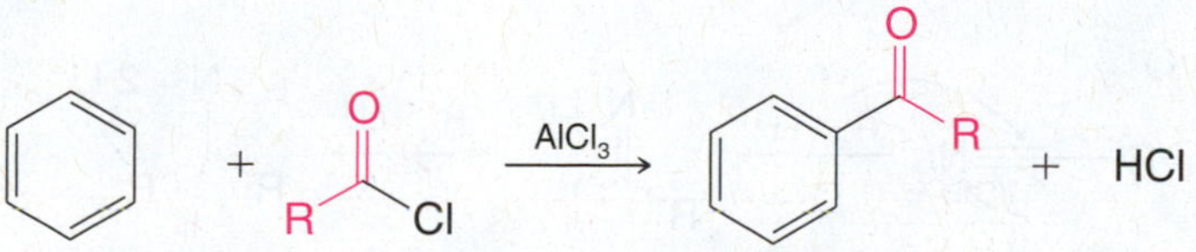

Alternatively,

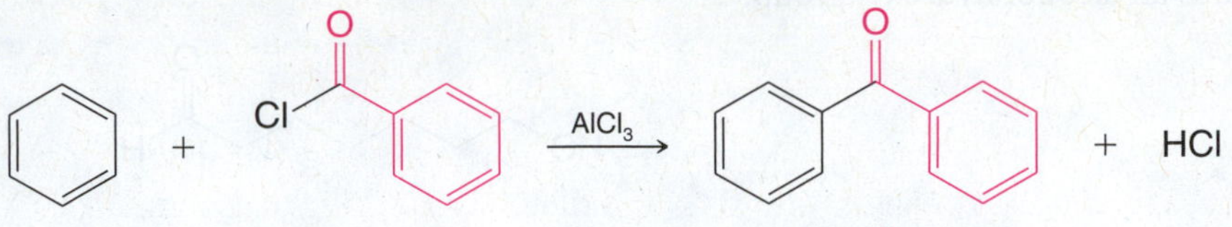

A diaryl ketone

3. Ketones from secondary alcohols by Swern oxidation and other methods (discussed in Section 12.4):

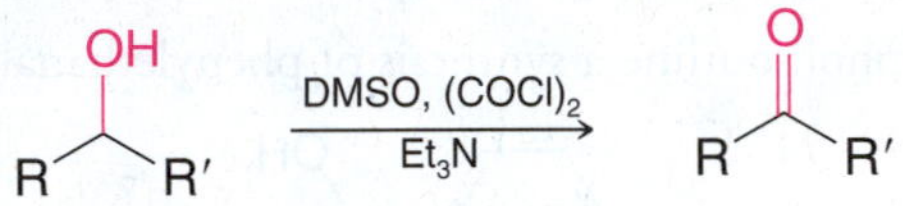

16.5B Ketones from Nitriles

Treating a nitrile (R—C≡N) with either a Grignard reagent or an organolithium reagent followed by hydrolysis yields a ketone.

General Reactions

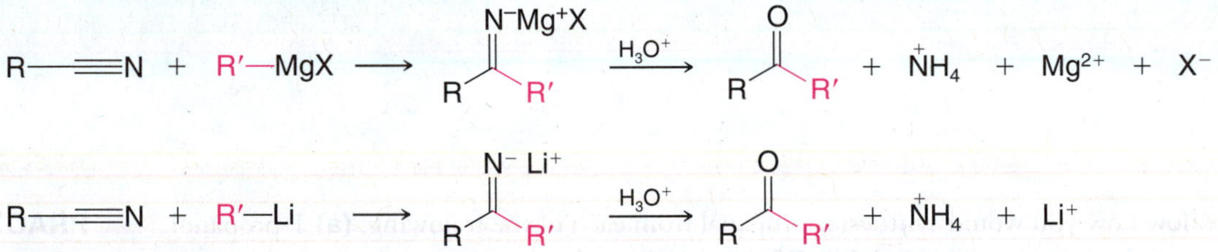

The mechanism for the acidic hydrolysis step is the reverse of one that we shall study for imine formation in Section 16.8A.

Specific Examples

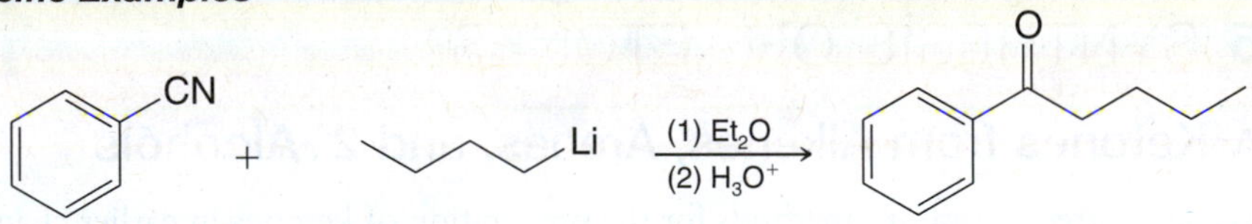

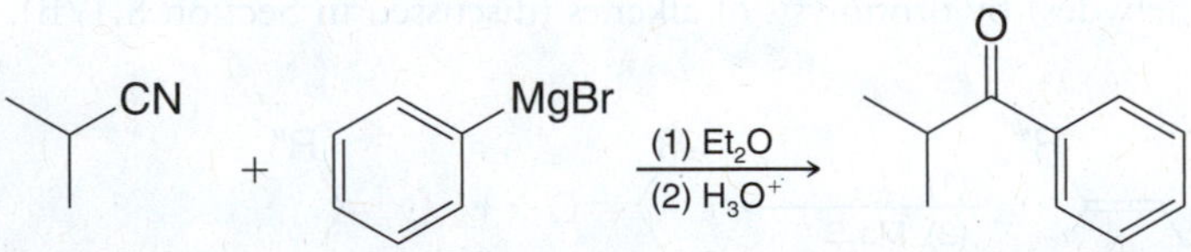

2-Cyanopropane **2-Methyl-1-phenylpropanone (isopropyl phenyl ketone)**

Even though a nitrile has a triple bond, addition of the Grignard or lithium reagent takes place only once. The reason: if addition took place twice, this would place a double negative charge on the nitrogen.

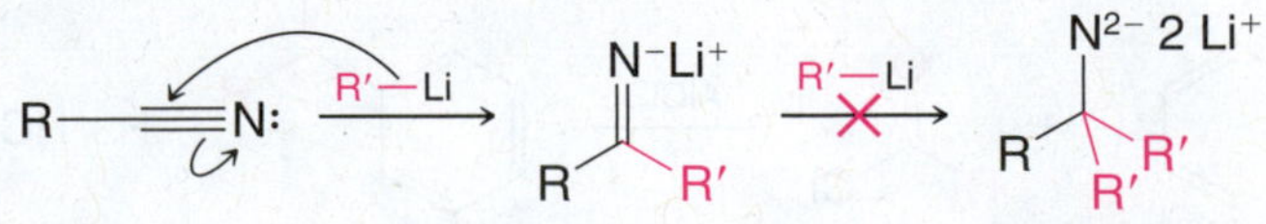

(The dianion does not form.)

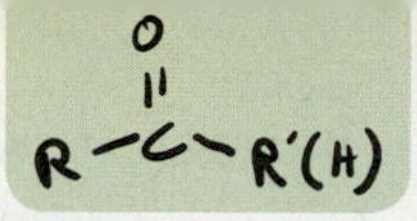

ILLUSTRATING A MULTISTEP SYNTHESIS: With 1-butanol as your only organic starting compound, devise a synthesis of 5-nonanone. Begin by writing a retrosynthetic analysis.

ANSWER: Retrosynthetic disconnection of 5-nonanone suggests butylmagnesium bromide and pentanenitrile as immediate precursors. Butylmagnesium bromide can, in turn, be synthesized from 1-bromobutane. Pentanenitrile can also be synthesized from 1-bromobutane, via S_N2 reaction of 1-bromobutane with cyanide. To begin the synthesis, 1-bromobutane can be prepared from 1-butanol by reaction with phosphorus tribromide.

Retrosynthetic Analysis

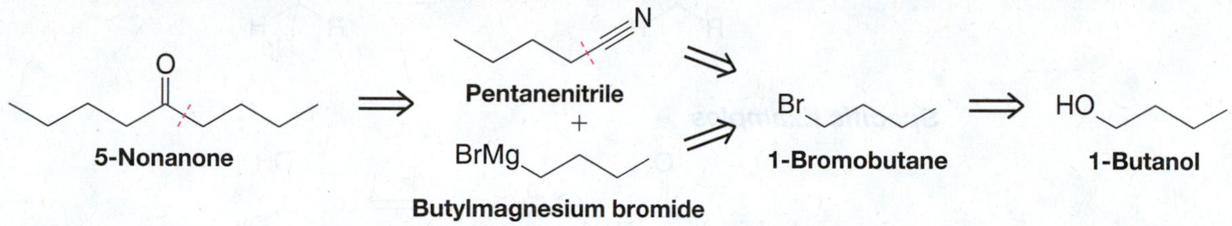

5-Nonanone ⟹ **Pentanenitrile** + **Butylmagnesium bromide** ⟹ **1-Bromobutane** ⟹ **1-Butanol**

Synthesis

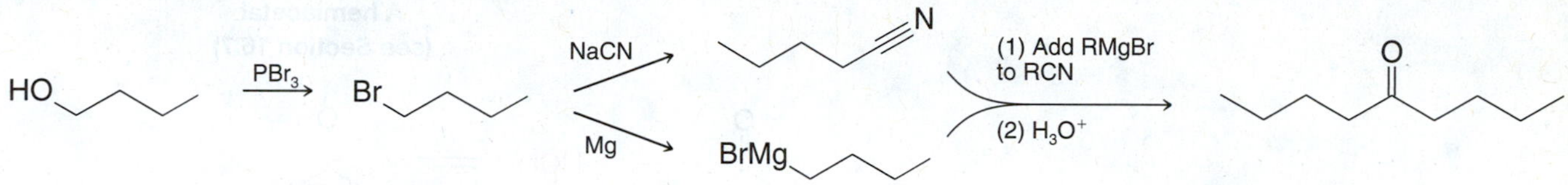

HO⌒⌒ →(PBr₃) Br⌒⌒ →(NaCN) / →(Mg) →BrMg⌒⌒ →(1) Add RMgBr to RCN (2) H₃O⁺→

Provide the reagents and indicated intermediates in each of the following syntheses.

(a)

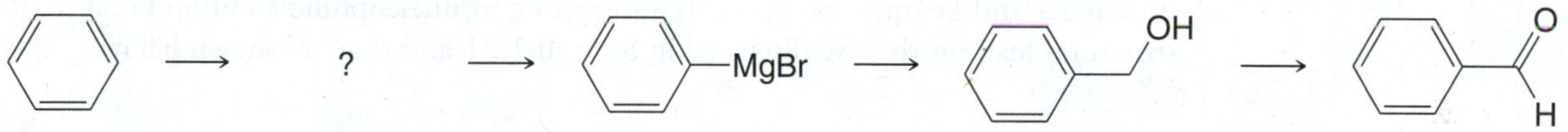

(b)

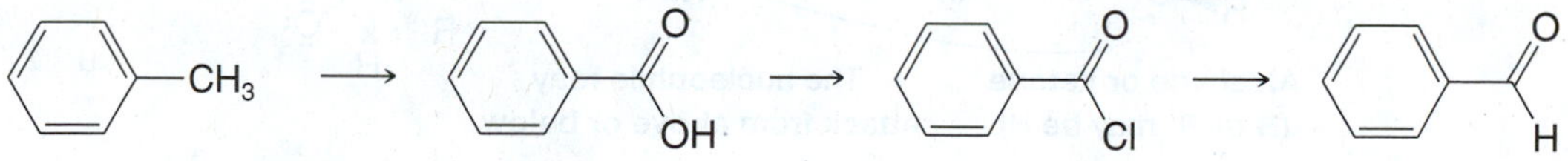

(c)

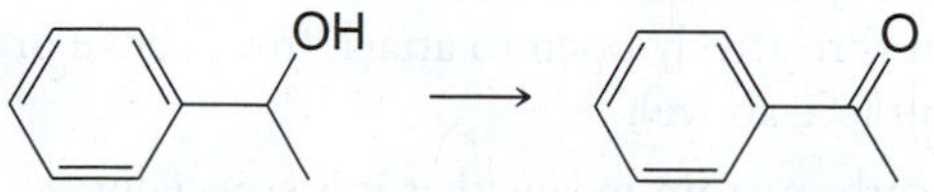

(d)

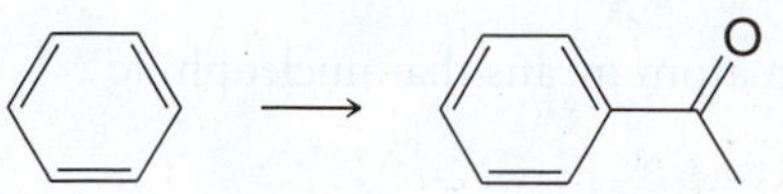

(e)

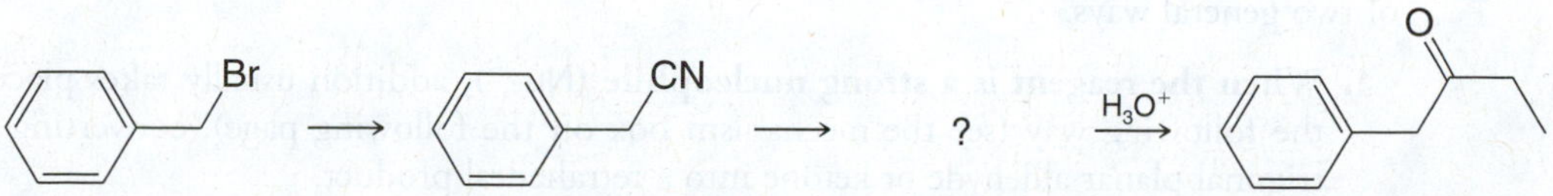

(f)

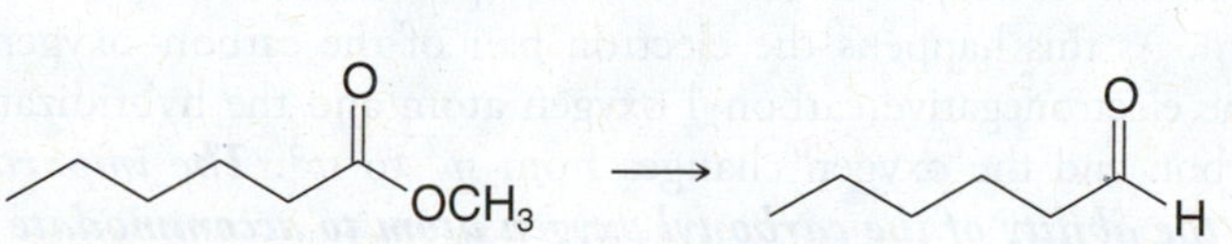

16.6 NUCLEOPHILIC ADDITION TO THE CARBON–OXYGEN DOUBLE BOND

- The most characteristic reaction of aldehydes and ketones is *nucleophilic addition* to the carbon–oxygen double bond.

General Reaction

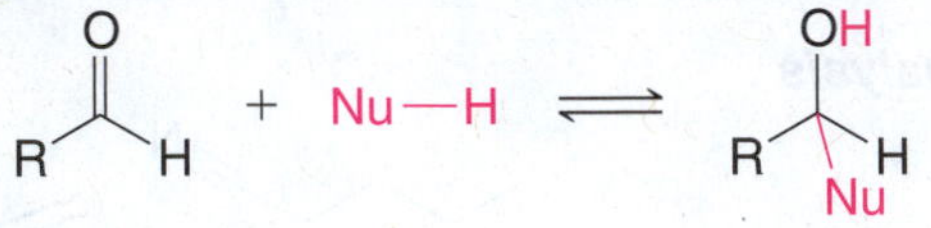

Specific Examples

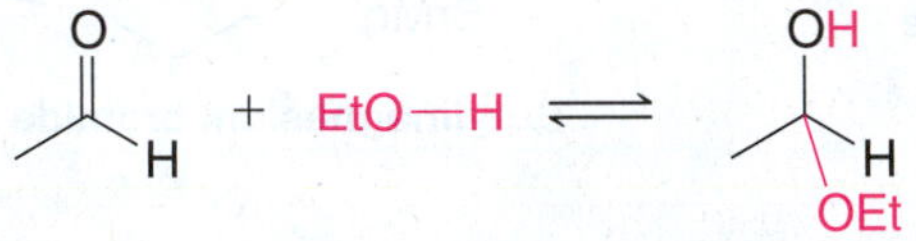

A hemiacetal
(see Section 16.7)

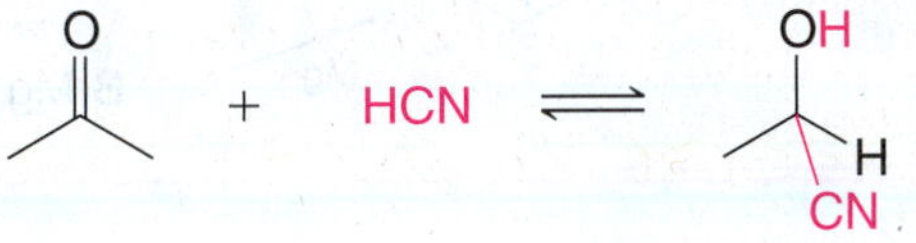

A cyanohydrin
(see Section 16.9)

Aldehydes and ketones are especially susceptible to nucleophilic addition because of the structural features that we discussed in Section 12.1 and that are shown below.

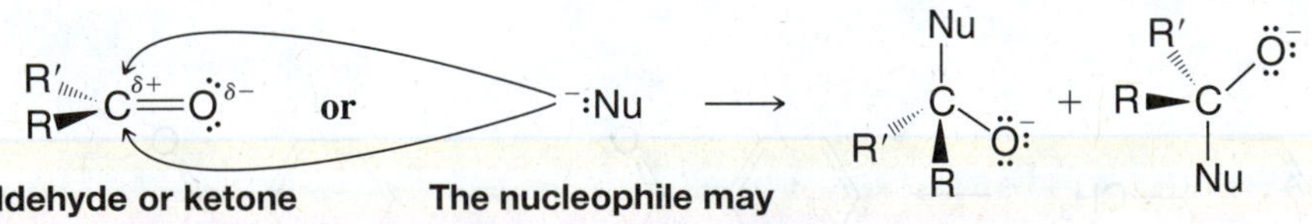

Aldehyde or ketone   **The nucleophile may**
(R or R′ may be H)   **attack from above or below.**

- The trigonal planar arrangement of groups around the carbonyl carbon atom means that the carbonyl carbon atom is relatively open to attack from above or below the plane of the carbonyl group (see above).
- The positive charge on the carbonyl carbon atom means that it is especially susceptible to attack by a nucleophile.
- The negative charge on the carbonyl oxygen atom means that nucleophilic addition is susceptible to acid catalysis.

Nucleophilic addition to the carbon–oxygen double bond occurs, therefore, in either of two general ways.

1. **When the reagent is a strong nucleophile (Nu:⁻),** addition usually takes place in the following way (see the mechanism box on the following page), converting the trigonal planar aldehyde or ketone into a tetrahedral product.

In this type of addition the nucleophile uses its electron pair to form a bond to the carbonyl carbon atom. As this happens the electron pair of the carbon–oxygen π bond shifts out to the electronegative carbonyl oxygen atom and the hybridization state of both the carbon and the oxygen changes from sp^2 to sp^3. *The important aspect of this step is the ability of the carbonyl oxygen atom to accommodate the electron pair of the carbon–oxygen double bond.*

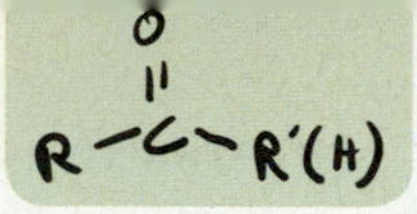

A MECHANISM FOR THE REACTION — Addition of a Strong Nucleophile to an Aldehyde or Ketone

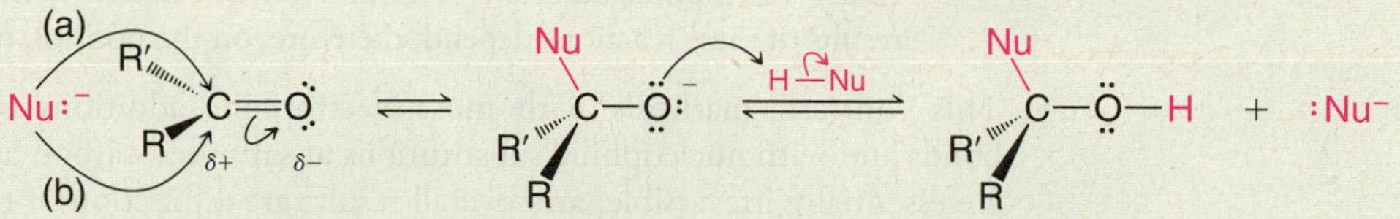

Trigonal planar   **Tetrahedral intermediate (plus enantiomer)**   **Tetrahedral product (plus enantiomer)**

In this step the nucleophile forms a bond to the carbon by donating an electron pair to the top or bottom face of the carbonyl group [path (a) or (b)]. An electron pair shifts out to the oxygen.

In the second step the alkoxide oxygen, because it is strongly basic, removes a proton from H—Nu or some other acid.

In the second step the oxygen atom accepts a proton. This happens because the oxygen atom is now much more basic; it carries a full negative charge as an alkoxide anion.

2. **When an acid catalyst is present and the nucleophile is weak**, reaction of the carbonyl oxygen with the acid enhances electrophilicity of the carbonyl group.

A MECHANISM FOR THE REACTION — Acid-Catalyzed Nucleophilic Addition to an Aldehyde or Ketone

Step 1

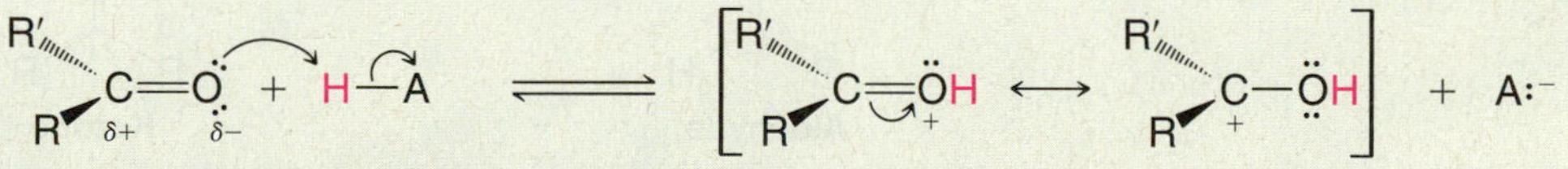

(or a Lewis acid)

In this step an electron pair of the carbonyl oxygen accepts a proton from the acid (or associates with a Lewis acid), producing an oxonium cation. The carbon of the oxonium cation is more susceptible to nucleophilic attack than the carbonyl of the starting ketone.

Step 2

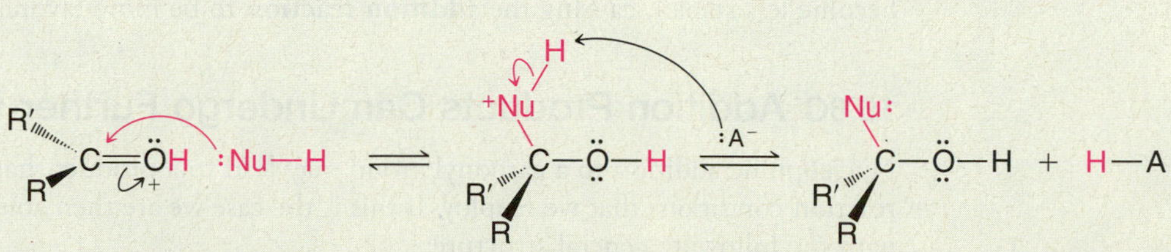

In the first of these two steps, the oxonium cation accepts the electron pair of the nucleophile. In the second step, a base removes a proton from the positively charged atom, regenerating the acid.

This mechanism operates when carbonyl compounds are treated with *strong acids* in the presence of *weak nucleophiles*. In the first step the acid donates a proton to an electron pair of the carbonyl oxygen atom. The resulting protonated carbonyl compound, an **oxonium cation**, is highly reactive toward nucleophilic attack at the carbonyl carbon atom because the carbonyl carbon atom carries more positive charge than it does in the unprotonated compound.

Helpful Hint

Any compound containing a positively charged oxygen atom that forms three covalent bonds is an *oxonium cation*.

16.6A Reversibility of Nucleophilic Additions to the Carbon–Oxygen Double Bond

- Many nucleophilic additions to carbon–oxygen double bonds are reversible; the overall results of these reactions depend, therefore, on the position of an equilibrium.

This contrasts markedly with most electrophilic additions to carbon–carbon double bonds and with nucleophilic substitutions at saturated carbon atoms. The latter reactions are essentially irreversible, and overall results are a function of relative reaction rates.

16.6B Relative Reactivity: Aldehydes versus Ketones

- In general, aldehydes are more reactive in nucleophilic additions than are ketones. Both steric and electronic factors favor aldehydes.

Steric Factors In aldehydes, where one group is a hydrogen atom, the central carbon of the tetrahedral product formed from the aldehyde is less crowded and the product is more stable. Formation of the product, therefore, is favored at equilibrium. With ketones, the two alkyl substituents at the carbonyl carbon cause greater steric crowding in the tetrahedral product and make it less stable. Therefore, a smaller concentration of the product is present at equilibrium.

Electronic Factors Because alkyl groups are electron releasing, aldehydes are more reactive on electronic grounds as well. Aldehydes have only one electron-releasing group to partially neutralize, and thereby stabilize, the positive charge at their carbonyl carbon atom. Ketones have two electron-releasing groups and are stabilized more. Greater stabilization of the ketone (the reactant) relative to its product means that the equilibrium constant for the formation of the tetrahedral product from a ketone is smaller and the reaction is less favorable:

Aldehyde   **Ketone**

The ketone carbonyl carbon is less positive because it has two electron-releasing alkyl groups.

On the other hand, electron-withdrawing substituents (e.g., $-CF_3$ or $-CCl_3$ groups) cause the carbonyl carbon to be more positive (and the starting compound to become less stable), causing the **addition reaction** to be more favorable.

16.6C Addition Products Can Undergo Further Reactions

Nucleophilic addition to a carbonyl group may lead to a product that is stable under the reaction conditions that we employ. If this is the case we are then able to isolate products with the following general structure:

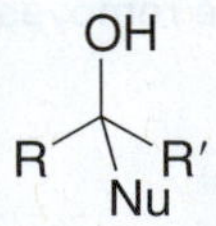

In other reactions the product formed initially may be unstable and may spontaneously undergo subsequent reactions. One common subsequent reaction is an *elimination reaction*, especially *dehydration*. Even if the initial addition product is stable, we may deliberately bring about a subsequent reaction by our choice of reaction conditions.

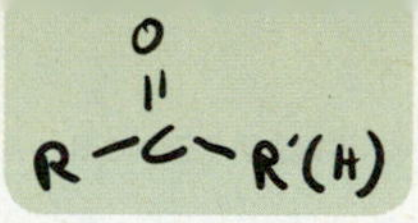

PRACTICE PROBLEM 16.5

The reaction of an aldehyde or ketone with a Grignard reagent (Section 12.8) is a nucleophilic addition to the carbon–oxygen double bond. **(a)** What is the nucleophile? **(b)** The magnesium portion of the Grignard reagent plays an important part in this reaction. What is its function? **(c)** What product is formed initially? **(d)** What product forms when water is added?

PRACTICE PROBLEM 16.6

The reactions of aldehydes and ketones with $LiAlH_4$ and $NaBH_4$ (Section 12.3) are nucleophilic additions to the carbonyl group. What is the nucleophile in these reactions?

16.7 THE ADDITION OF ALCOHOLS: HEMIACETALS AND ACETALS

- Aldehydes and ketones react with alcohols to form **hemiacetals** and **acetals** by an equilibrium reaction.

16.7A Hemiacetals

- The essential structural features of a **hemiacetal** are an —OH and an —OR group attached to the same carbon atom.

The hemiacetal results by nucleophilic addition of an alcohol oxygen to the carbonyl carbon of an aldehyde or ketone.

[A MECHANISM FOR THE REACTION — Hemiacetal Formation]

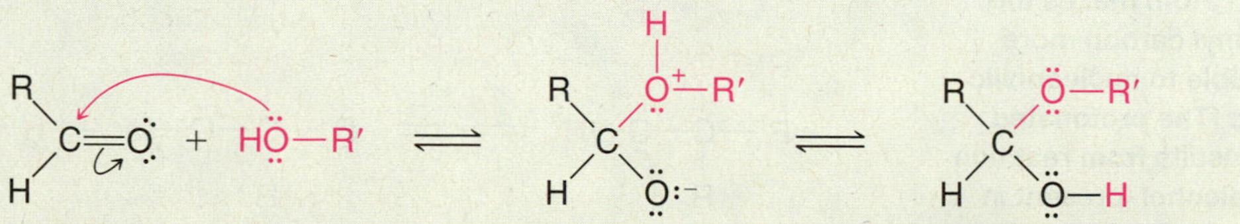

| Aldehyde (or ketone) | Alcohol | | Hemiacetal (usually too unstable to isolate) |

In this step the alcohol attacks the carbonyl carbon.

In two intermolecular steps, a proton is removed from the positive oxygen and a proton is gained at the negative oxygen.

- Most open-chain hemiacetals are not sufficiently stable to allow their isolation. Cyclic hemiacetals with five- or six-membered rings, however, are usually much more stable.

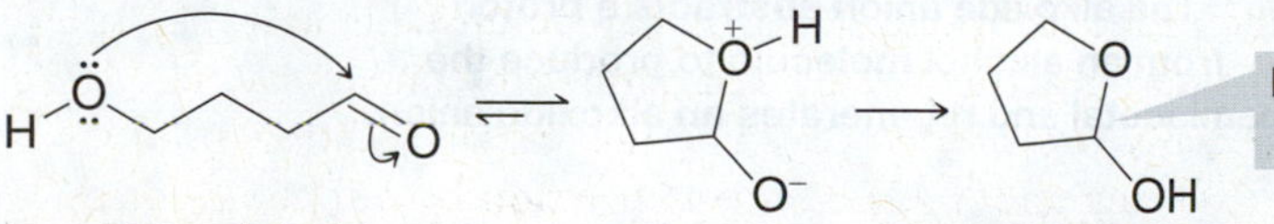

Butanal-4-ol

A cyclic hemiacetal

Hemiacetal: OH and OR groups bonded to the same carbon

Most simple sugars (Chapter 22) exist primarily in a cyclic hemiacetal form. Glucose is an example:

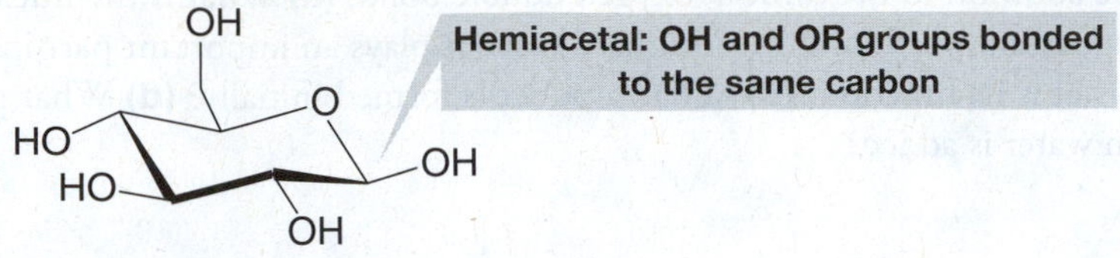

| Hemiacetal: OH and OR groups bonded to the same carbon |

(+)-Glucose
(a cyclic hemiacetal)

Whether the carbonyl reactant is an aldehyde or a ketone, the product with an —OH and an —OR group is called a **hemiacetal**.

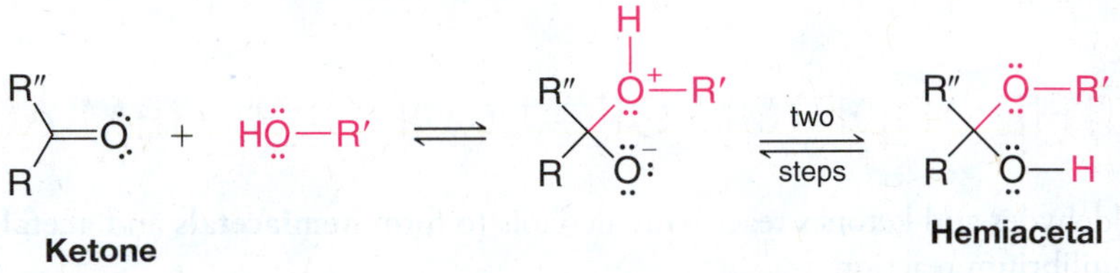

Ketone   two steps   **Hemiacetal**

- The formation of hemiacetals is catalyzed by acids and bases.

[A MECHANISM FOR THE REACTION — Acid-Catalyzed Hemiacetal Formation]

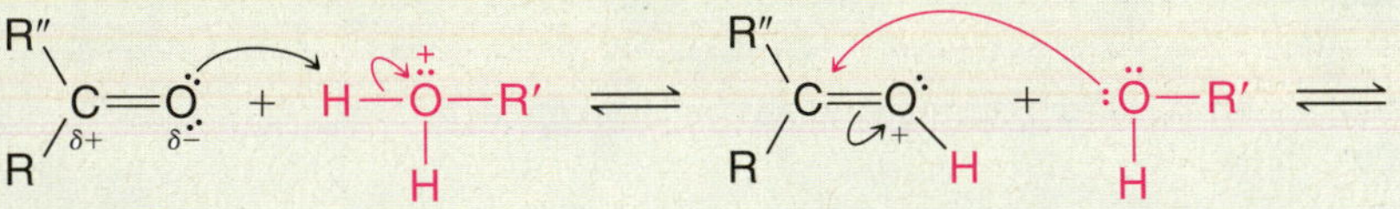

(R″ may be H)
Protonation of the aldehyde or ketone oxygen atom makes the carbonyl carbon more susceptible to nucleophilic attack. [The protonated alcohol results from reaction of the alcohol (present in excess) with the acid catalyst, e.g., gaseous (anhydrous) HCl.]

An alcohol molecule adds to the carbon of the oxonium cation.

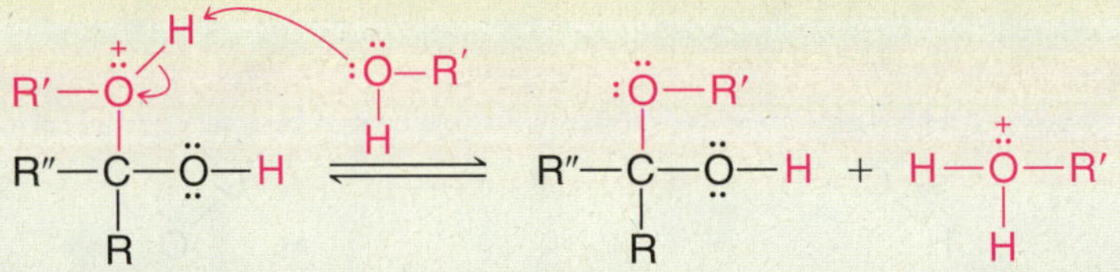

The transfer of a proton from the positive oxygen to another molecule of the alcohol leads to the hemiacetal.

[A MECHANISM FOR THE REACTION — Base-Catalyzed Hemiacetal Formation]

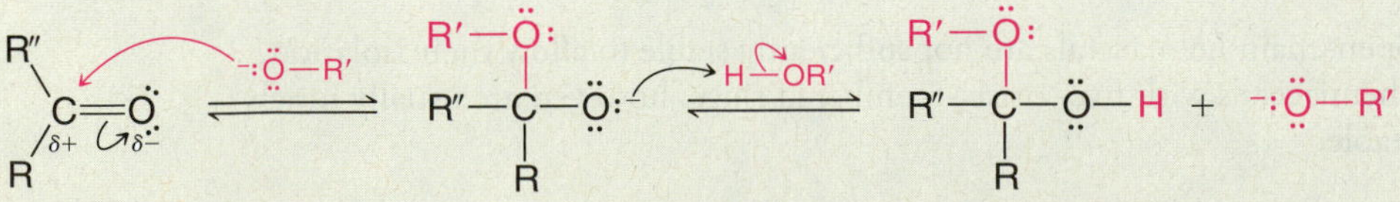

(R″ may be H)
An alkoxide anion acting as a nucleophile attacks the carbonyl carbon atom. An electron pair shifts onto the oxygen atom, producing a new alkoxide anion.

The alkoxide anion abstracts a proton from an alcohol molecule to produce the hemiacetal and regenerates an alkoxide anion.

Aldehyde Hydrates: *gem*-Diols Dissolving an aldehyde such as acetaldehyde in water causes the establishment of an equilibrium between the aldehyde and its **hydrate**. This hydrate is in actuality a 1,1-diol, called a geminal diol (or simply a *gem*-diol).

Acetaldehyde $\ce{C=O}$ + H_2O ⇌   **Hydrate** (a *gem*-diol)

The *gem*-diol results from a nucleophilic addition of water to the carbonyl group of the aldehyde.

A MECHANISM FOR THE REACTION — Hydrate Formation

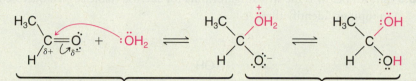

In this step water attacks the carbonyl carbon atom.

In two intermolecular steps a proton is lost from the positive oxygen atom and a proton is gained at the negative oxygen atom.

The equilibrium for the addition of water to most ketones is unfavorable, whereas some aldehydes (e.g., formaldehyde) exist primarily as the *gem*-diol in aqueous solution.

It is not possible to isolate most *gem*-diols from the aqueous solutions in which they are formed. Evaporation of the water, for example, simply displaces the overall equilibrium to the right and the *gem*-diol (or hydrate) reverts to the carbonyl compound:

$$\underset{\text{R}\quad\text{H}}{\text{HO}\quad\text{OH}} \xrightarrow{\text{distillation}} \underset{\text{R}\quad\text{H}}{\overset{\text{O}}{\|}} + H_2O$$

Compounds with strong electron-withdrawing groups attached to the carbonyl group can form stable *gem*-diols. An example is the compound called chloral hydrate:

$$\underset{\text{Cl}_3\text{C}\qquad\text{H}}{\text{HO}\qquad\text{OH}}$$

Chloral hydrate

• • •

PRACTICE PROBLEM 16.7

Dissolving formaldehyde in water leads to a solution containing primarily the *gem*-diol $CH_2(OH)_2$. Show the steps in its formation from formaldehyde.

• • •

PRACTICE PROBLEM 16.8

When acetone is dissolved in water containing ^{18}O instead of ordinary ^{16}O (i.e., $H_2{}^{18}O$ instead of $H_2{}^{16}O$), the acetone soon begins to acquire ^{18}O and becomes $CH_3\overset{\overset{\displaystyle ^{18}O}{\|}}{C}CH_3$. The formation of this oxygen-labeled acetone is catalyzed by traces of strong acids and by strong bases (e.g., HO^-). Show the steps that explain both the acid-catalyzed reaction and the base-catalyzed reaction.

16.7B Acetals

- An **acetal** has two —OR groups attached to the same carbon atom.

If we take an alcohol solution of an aldehyde (or ketone) and pass into it a small amount of gaseous HCl, a hemiacetal forms, and then the hemiacetal reacts with a second molar equivalent of the alcohol to produce an **acetal**.

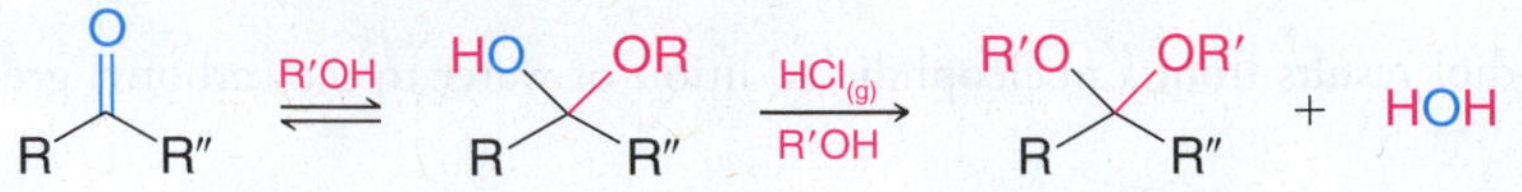

A hemiacetal
(R″ may be H)

An acetal
(R″ may be H)

PRACTICE PROBLEM 16.9 Shown below is the structural formula for sucrose (table sugar). Sucrose has two acetal groupings. Identify these.

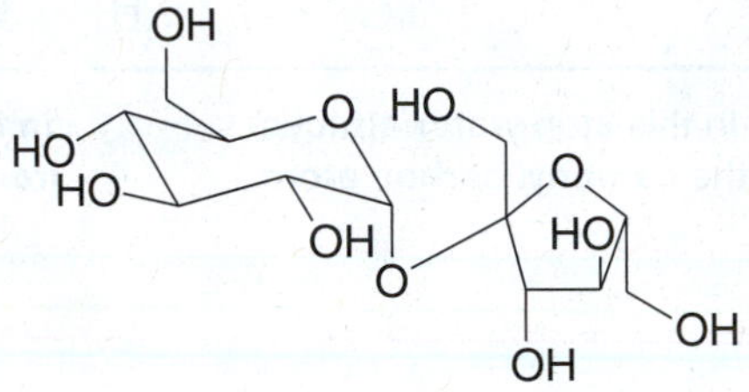

- The mechanism for acetal formation involves acid-catalyzed formation of the hemiacetal, then an acid-catalyzed elimination of water, followed by a second *addition* of the alcohol and loss of a proton.
- All steps in the formation of an acetal from an aldehyde are reversible.

If we dissolve an aldehyde in a large excess of an anhydrous alcohol and add a small amount of an anhydrous acid (e.g., gaseous HCl or concentrated H_2SO_4), the equilibrium will strongly favor the formation of an acetal. After the equilibrium is established, we can isolate the acetal by neutralizing the acid and evaporating the excess alcohol.

If we then place the acetal in water and add a catalytic amount of acid, all of the steps reverse. Under these conditions (an excess of water), the equilibrium favors the formation of the aldehyde. The acetal undergoes *hydrolysis*:

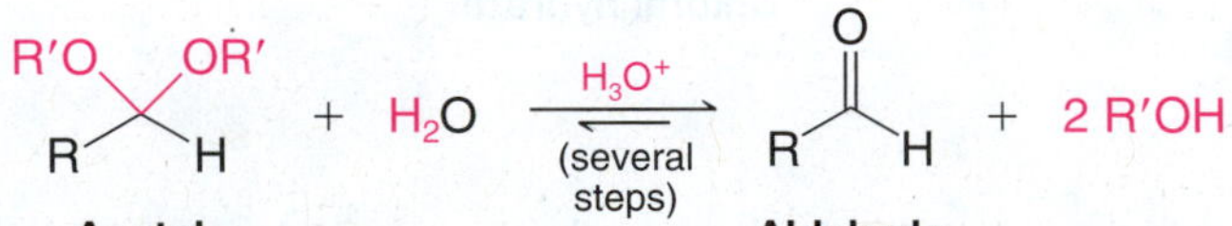

Helpful Hint

Equilibrium conditions govern the formation and hydrolysis of hemiacetals and acetals.

Acetal

Aldehyde

A MECHANISM FOR THE REACTION — Acid-Catalyzed Acetal Formation

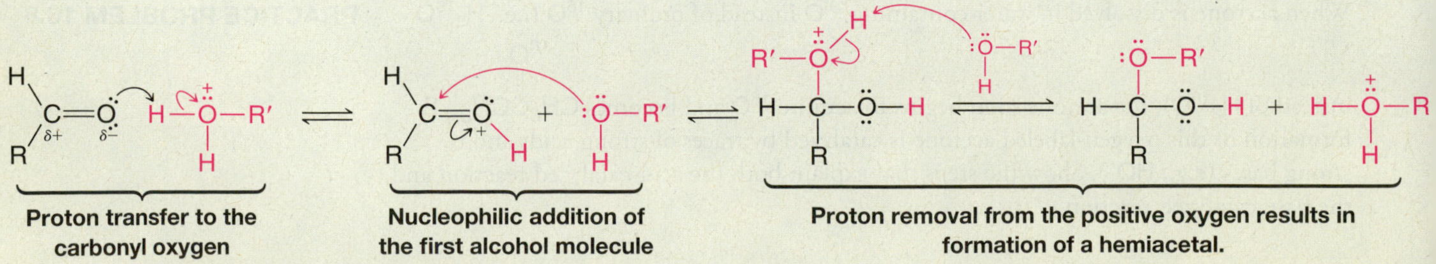

Proton transfer to the carbonyl oxygen

Nucleophilic addition of the first alcohol molecule

Proton removal from the positive oxygen results in formation of a hemiacetal.

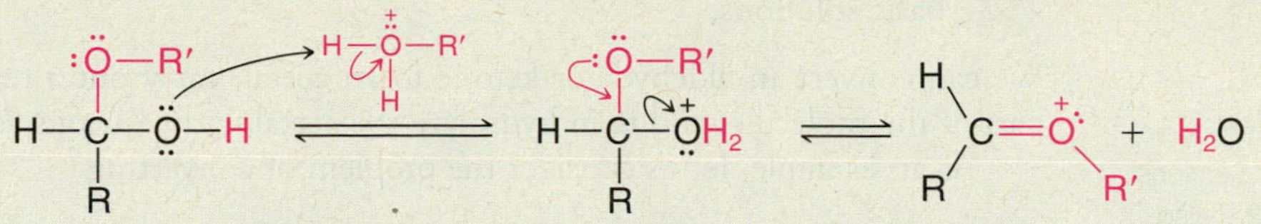

Protonation of the hydroxyl group leads to elimination of water and formation of a highly reactive oxonium cation.

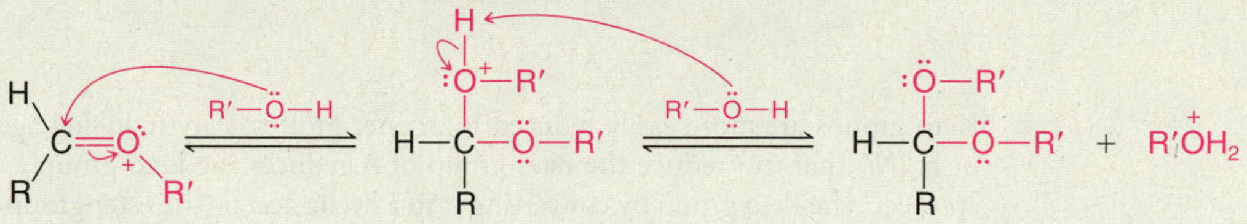

Attack on the carbon of the oxonium ion by a second molecule of the alcohol, followed by removal of a proton, leads to the acetal.

PRACTICE PROBLEM 16.10

Write a detailed mechanism for the formation of an acetal from benzaldehyde and methanol in the presence of an acid catalyst.

Cyclic Acetals

- Cyclic acetal formation is favored when a ketone or an aldehyde is treated with an excess of a 1,2-diol and a trace of acid:

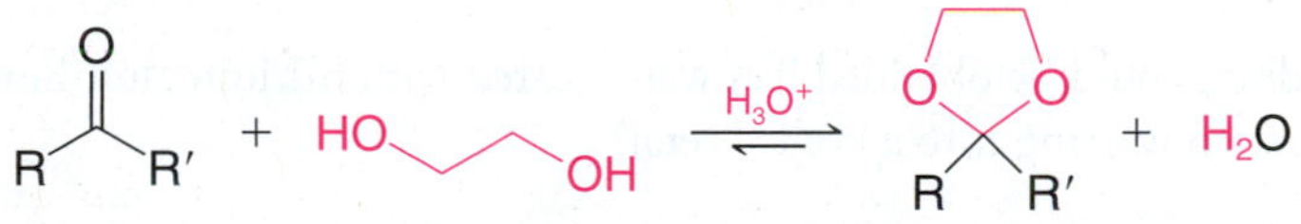

Ketone **(excess)** **Cyclic acetal**

This reaction, too, can be reversed by treating the acetal with aqueous acid:

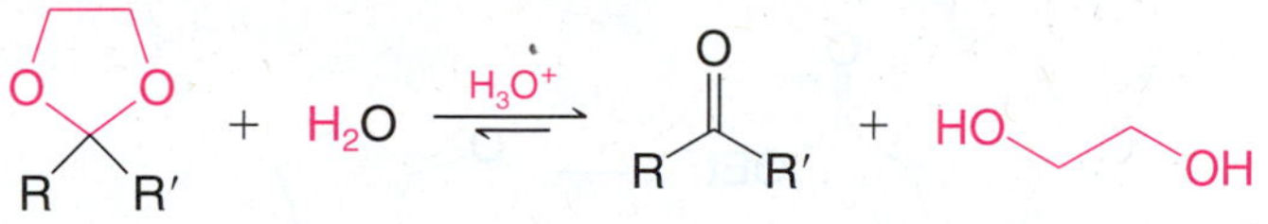

Acetal formation is not favored when ketones are treated with simple alcohols and gaseous HCl.

PRACTICE PROBLEM 16.11

Outline all steps in the mechanism for the formation of a cyclic acetal from acetone and ethylene glycol (1,2-ethanediol) in the presence of gaseous HCl.

16.7C Acetals Are Used as Protecting Groups

- Although acetals are hydrolyzed to aldehydes and ketones in aqueous acid, ***acetals are stable in basic solutions***:

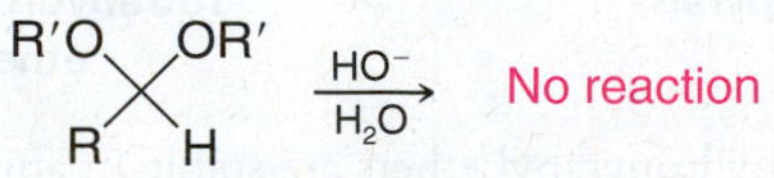

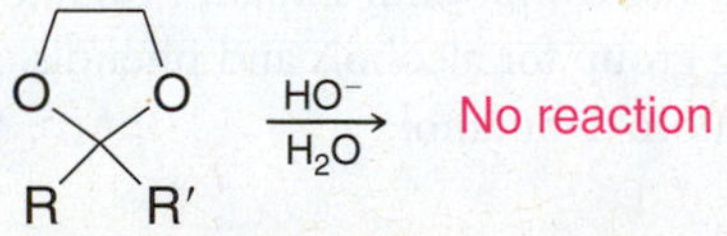

- Acetals are used to protect aldehydes and ketones from undesired reactions in basic solutions.

We can convert an aldehyde or ketone to an acetal, carry out a reaction on some other part of the molecule, and then hydrolyze the acetal with aqueous acid.

As an example, let us consider the problem of converting

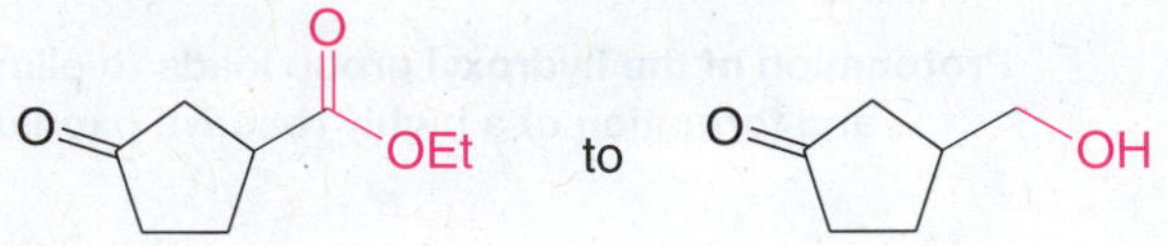

Helpful Hint

Protecting groups are strategic tools for synthesis. See Sections 11.11D, 11.11E, and 12.9 also.

A   to   **B**

Keto groups are more easily reduced than ester groups. Any reducing agent (e.g., LiAlH$_4$ or H$_2$/Ni) that can reduce the ester group of **A** reduces the keto group as well. But if we "protect" the keto group by converting it to a cyclic acetal (the ester group does not react), we can reduce the ester group in basic solution without affecting the cyclic acetal. After we finish the ester reduction, we can hydrolyze the cyclic acetal and obtain our desired product, **B**:

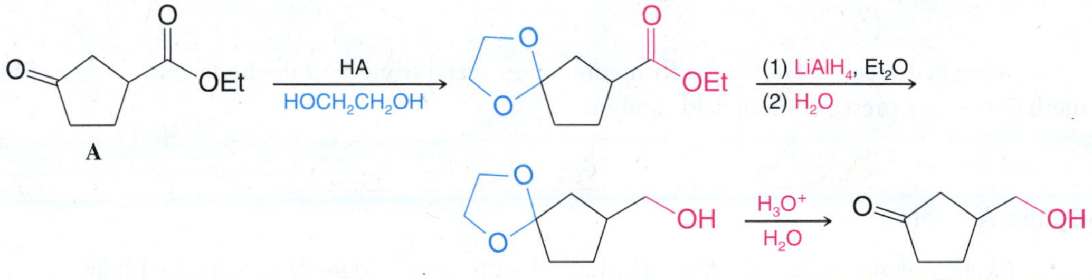

●●●

PRACTICE PROBLEM 16.12   What product would be obtained if **A** were treated with lithium aluminum hydride without first converting it to a cyclic acetal?

●●●

PRACTICE PROBLEM 16.13   **(a)** Show how you might use a cyclic acetal in carrying out the following transformation:

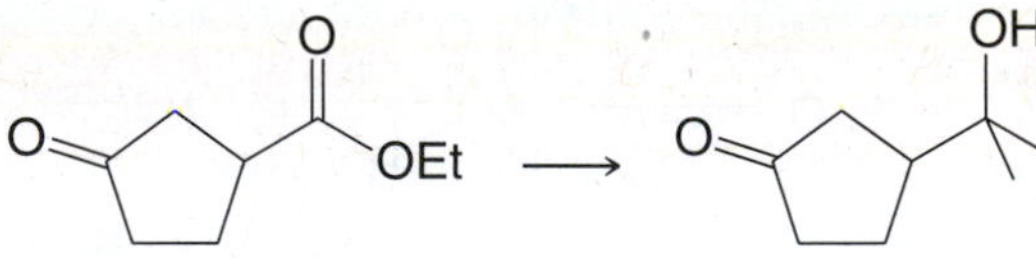

A   **B**

(b) Why would a direct addition of methylmagnesium bromide to **A** fail to give **B**?

●●●

PRACTICE PROBLEM 16.14   Dihydropyran reacts readily with an alcohol in the presence of a trace of anhydrous HCl or H$_2$SO$_4$ to form a tetrahydropyranyl (THP) ether:

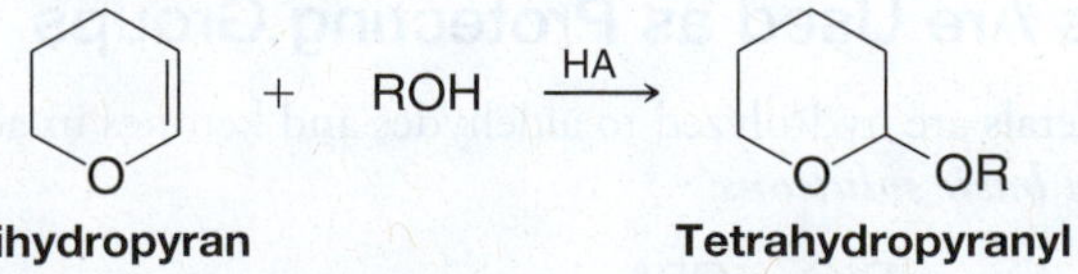

Dihydropyran   **Tetrahydropyranyl ether**

(a) Write a plausible mechanism for this reaction. **(b)** Tetrahydropyranyl ethers are stable in aqueous base but hydrolyze rapidly in aqueous acid to yield the original alcohol and another compound. Explain. (What is the other compound?) **(c)** The tetrahydropyranyl group can be used as a protecting group for alcohols and phenols. Show how you might use it in a synthesis of 5-methyl-1,5-hexanediol starting with 4-chloro-1-butanol.

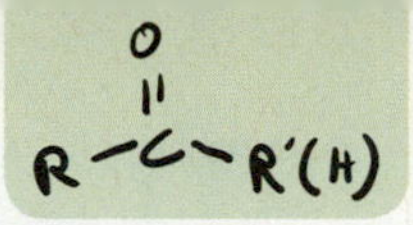

16.7D Thioacetals

- Aldehydes and ketones react with thiols to form *thioacetals*:

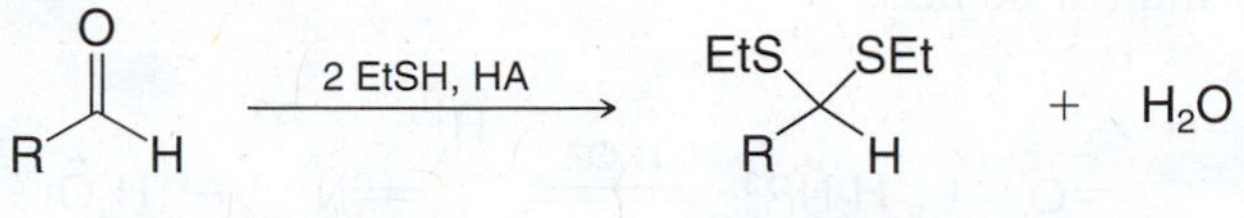

Thioacetal

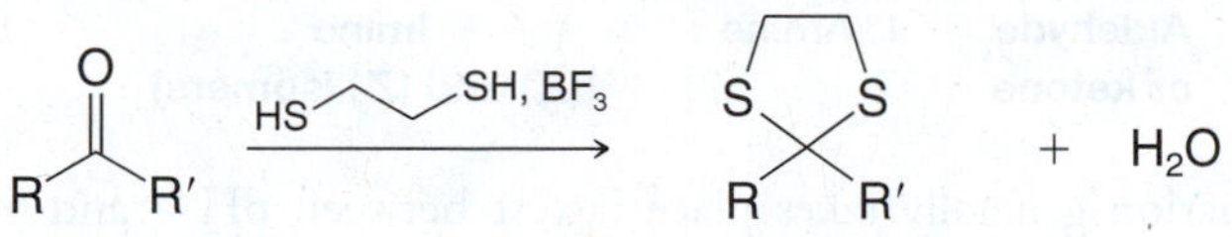

Cyclic thioacetal

Thioacetals are important in organic synthesis because they react with hydrogen and Raney nickel to yield hydrocarbons. Raney nickel is a special nickel catalyst that contains adsorbed hydrogen.

- Thioacetal formation with subsequent "desulfurization" with hydrogen and Raney nickel gives us an additional method for converting carbonyl groups of aldehydes and ketones to $-CH_2-$ groups:

> **Helpful Hint**
>
> A method for reducing the carbonyl group of aldehydes and ketones to $-CH_2-$ groups.

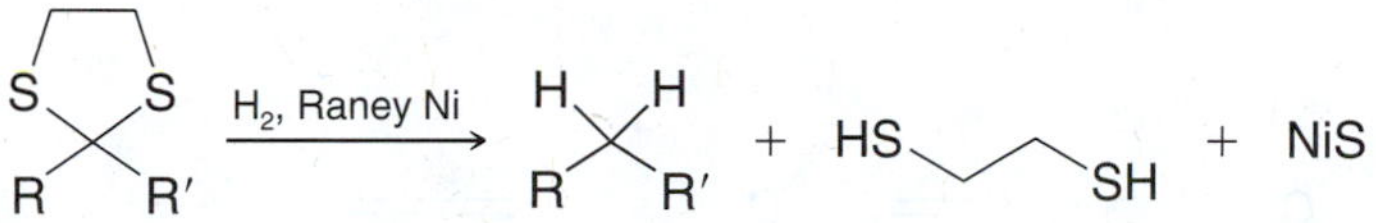

The other methods we have studied are the **Clemmensen reduction** (Section 15.9A) and the **Wolff–Kishner reduction** (Section 15.9B). In Section 16.8C we will discuss the mechanism of the Wolff–Kishner reduction.

● ● ●

Show how you might use thioacetal formation and Raney nickel desulfurization to convert: **(a)** cyclohexanone to cyclohexane and **(b)** benzaldehyde to toluene.

PRACTICE PROBLEM 16.15

16.8 THE ADDITION OF PRIMARY AND SECONDARY AMINES

- Aldehydes and ketones react with primary amines to form **imines** and with secondary amines to form **enamines**.

Imines have a carbon–nitrogen double bond. Enamines have an amino group joined to a carbon–carbon double bond (they are alk*ene*amines).

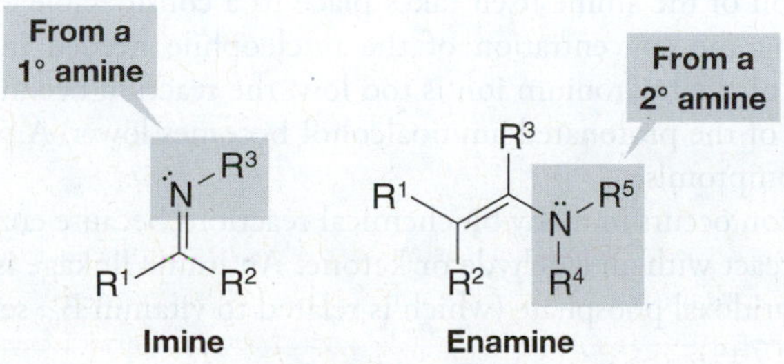

From a 1° amine

From a 2° amine

Imine

Enamine

$R^1, R^2, R^3 = C$ or H;
$R^4, R^5 = C$

16.8A Imines

A general equation for the formation of an imine from a primary amine and an aldehyde or ketone is shown here. Imine formation is acid catalyzed, and the product can form as a mixture of (E) and (Z) isomers:

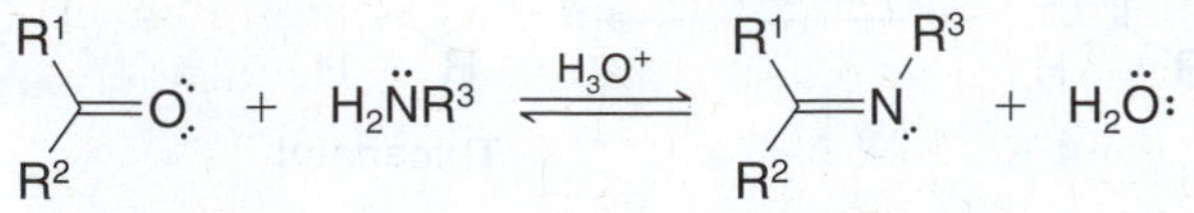

$$
\overset{R^1}{\underset{R^2}{\diagdown}}C{=}\ddot{O}: \ + \ H_2\ddot{N}R^3 \ \underset{}{\overset{H_3O^+}{\rightleftharpoons}} \ \overset{R^1}{\underset{R^2}{\diagdown}}C{=}\underset{\displaystyle \cdot}{N}\diagup^{R^3} \ + \ H_2\ddot{O}:
$$

Aldehyde   **1° Amine**   **Imine**
or ketone   **[(E) and (Z) isomers]**

Imine formation generally takes place fastest between pH 4 and 5 and is slow at very low or very high pH. We can understand why an acid catalyst is necessary if we consider the mechanism that has been proposed for imine formation. The important step is the step in which the protonated aminoalcohol loses a molecule of water to become an iminium ion. By protonating the alcohol group, the acid converts a poor leaving group (an —OH group) into a good one (an —$^+$OH$_2$ group).

[A MECHANISM FOR THE REACTION — Imine Formation]

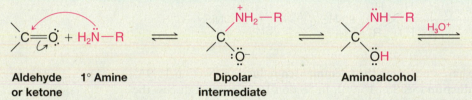

Aldehyde   **1° Amine**   **Dipolar**   **Aminoalcohol**
or ketone   **intermediate**

The amine adds to the carbonyl group   **Intermolecular proton transfer from nitrogen**
to form a dipolar tetrahedral intermediate.   **to oxygen produces an aminoalcohol.**

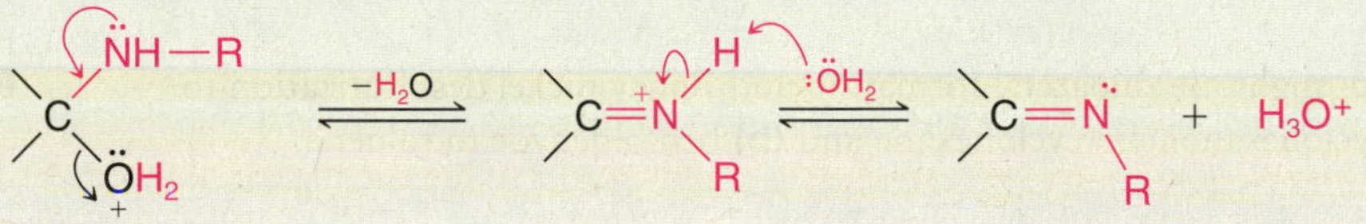

Protonated   **Iminium ion**   **Imine**
aminoalcohol   **[(E) and (Z) isomers]**

Protonation of the oxygen   **Transfer of a proton to water**
produces a good leaving group.   **produces the imine and regenerates**
Loss of a molecule of water   **the catalytic hydronium ion.**
yields an iminium ion.

The reaction proceeds more slowly if the hydronium ion concentration is too high, because protonation of the amine itself takes place to a considerable extent; this has the effect of decreasing the concentration of the nucleophile needed in the first step. If the concentration of the hydronium ion is too low, the reaction becomes slower because the concentration of the protonated aminoalcohol becomes lower. A pH between 4 and 5 is an effective compromise.

Imine formation occurs in many biochemical reactions because enzymes often use an —NH$_2$ group to react with an aldehyde or ketone. An imine linkage is important in the biochemistry of pyridoxal phosphate (which is related to vitamin B$_6$; see "The Chemistry of ..." box on page 744).

Imines are also formed as intermediates in a useful laboratory synthesis of amines that we shall study in Section 20.4.

Helpful Hint

See "The Chemistry of ... A Very Versatile Vitamin, Pyridoxine (Vitamin B$_6$)" on page 744, and "The Chemistry of... Pyridoxal Phosphate" in *WileyPLUS*.

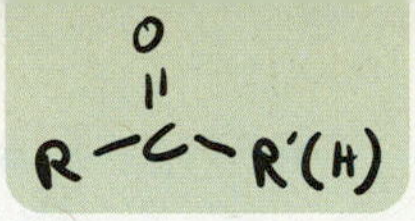

16.8B Oximes and Hydrazones

- Compounds such as hydroxylamine (NH_2OH), hydrazine (NH_2NH_2), and substituted hydrazines such as phenylhydrazine ($C_6H_5NHNH_2$) and 2,4-dinitrophenylhydrazine form $C=N$ derivatives of aldehydes and ketones.

These derivatives are called oximes, hydrazones, phenylhydrazones, and 2,4-dinitrophenylhydrazones, respectively. The mechanisms by which these $C=N$ derivatives form are similar to the mechanism for imine formation from a primary amine. As with imines, the formation of (E) and (Z) isomers is possible.

Oximes and the various **hydrazone** derivatives of aldehydes and ketones are sometimes used to identify unknown aldehydes and ketones. These derivatives are usually relatively insoluble solids that have sharp, characteristic melting points. The melting point of the derivative of an unknown compound can be compared with the melting point for the same derivative of a known compound or with data found in a reference table, and on this basis one can propose an identity for the unknown compound. Most laboratory textbooks for organic chemistry include extensive tables of derivative melting points. The method of comparing melting points is only useful, however, for compounds that have derivative melting points previously reported in the literature. Spectroscopic methods (especially IR, NMR, and mass spectrometry) are more generally applicable to identification of unknown compounds (Section 16.13).

Another important use of hydrazones is the Wolff–Kishner reduction, first mentioned in Section 15.9B, and whose mechanism we shall present below now that we have discussed hydrazones.

16.8C The Wolff–Kishner Reduction

A ketone can be reduced to a methylene group using the Wolff–Kishner reduction, which involves heating the ketone with hydrazine and base. The mechanism for the Wolff–Kishner reduction involves initial formation of a hydrazone followed by a tautomerization and elimination of nitrogen. The Wolff–Kishner reduction is complementary to the Clemmensen reduction (Section 15.9A), which involves acid, and to the reduction of dithioacetals (Section 16.7D), which involves catalytic hydrogenation.

Benzophenone can be reduced to diphenylmethane by the Wolff–Kishner reduction, for example.

A Wolff–Kishner reduction

$$\text{Benzophenone} \xrightarrow[\text{heat}]{\substack{NH_2NH_2 \ (85\%) \\ KOH}} \text{Diphenylmethane} + N_2 + H_2O$$

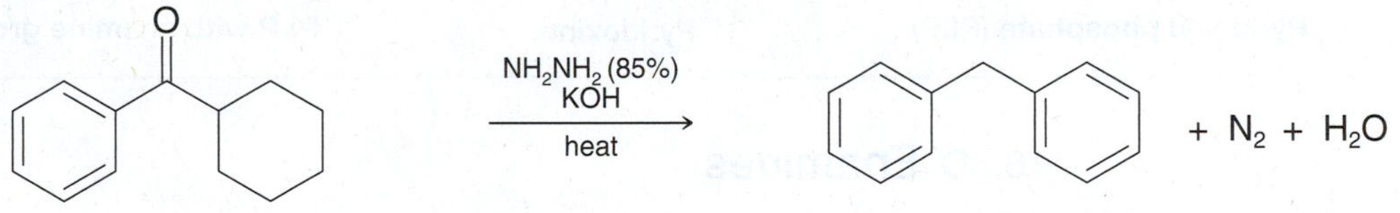

Benzophenone

Diphenylmethane (83%)

The mechanism for the Wolff–Kishner reaction is as follows:

[A MECHANISM FOR THE REACTION — **The Wolff–Kishner Reduction**]

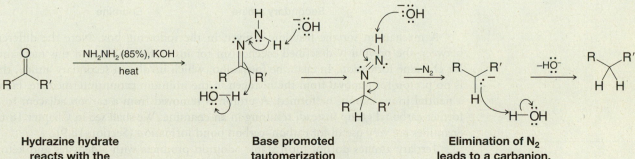

Hydrazine hydrate reacts with the carbonyl compound to form a hydrazone.

Base promoted tautomerization transfers a proton from nitrogen to carbon.

Elimination of N_2 leads to a carbanion, which is then protonated.

THE CHEMISTRY OF... A Very Versatile Vitamin, Pyridoxine (Vitamin B₆)

Pyridoxal phosphate (PLP) is at the heart of chemistry conducted by a number of enzymes. Many of us know the coenzyme pyridoxal phosphate by the closely related vitamin from which it is derived in our diet—pyridoxine, or vitamin B_6. Wheat is a good dietary source of vitamin B_6. Although pyridoxal phosphate (see below and the model) is a member of the aldehyde family, when it is involved in biological chemistry, it often contains the closely related functional group with a carbon–nitrogen double bond, the imine group.

Some enzymatic reactions that involve PLP include *transaminations*, which convert amino acids to ketones for use in the citric acid cycle and other pathways; *decarboxylation* of amino acids for biosynthesis of neurotransmitters such as histamine, dopamine, and serotonin; and *inversion* of amino acid chirality centers, such as required for the biosynthesis of cell walls in bacteria.

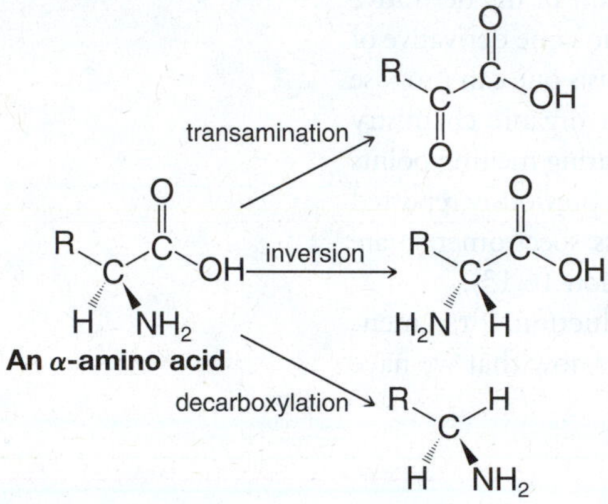

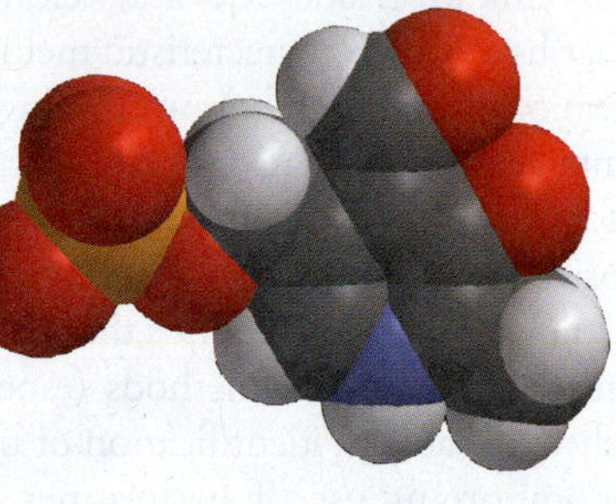

An α-amino acid
transamination / inversion / decarboxylation

Pyridoxal phosphate (vitamin B_6)

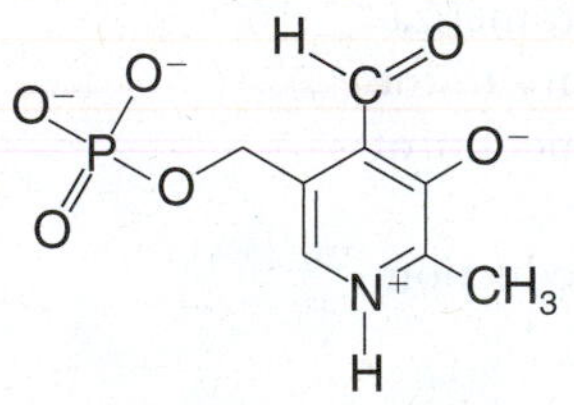

Pyridoxal phosphate (PLP) 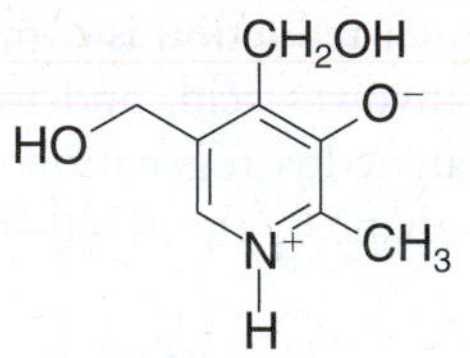   **Pyridoxine** 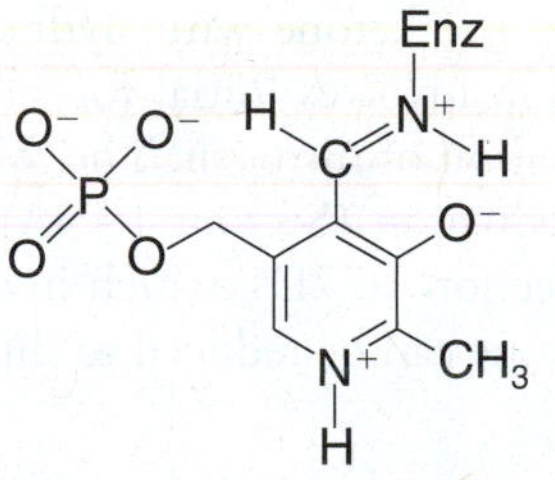   **PLP with an imine group**

16.8D Enamines

Aldehydes and ketones react with secondary amines to form enamines. The following is a general equation for enamine formation:

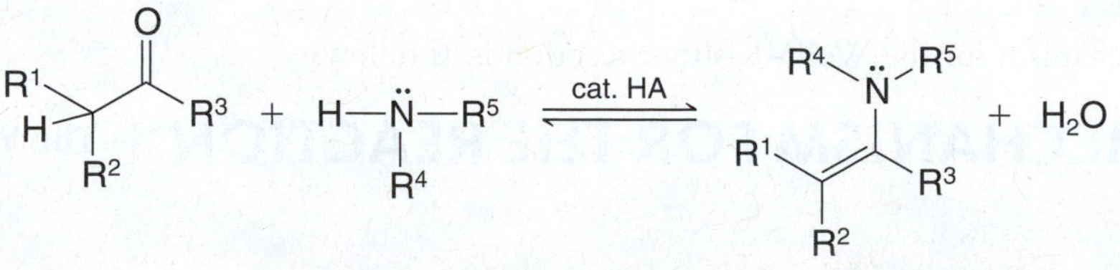

Secondary amine   Enamine

A mechanism for the reaction is given in the following box. Note the difference between the previously described mechanism for imine formation and this mechanism for enamine formation. In enamine formation, which involves a secondary amine, there is no proton for removal from the nitrogen in the iminium cation intermediate. Hence, a neutral imine cannot be formed. A proton is removed from a carbon adjacent to the former carbonyl group instead, resulting in an enamine. We shall see in Chapter 18 that enamines are very useful for carbon–carbon bond formation (Section 18.9).

Tertiary amines do not form stable addition products with aldehydes and ketones because, on forming the tetrahedral intermediate, the resulting formal positive charge cannot be neutralized by loss of a proton.

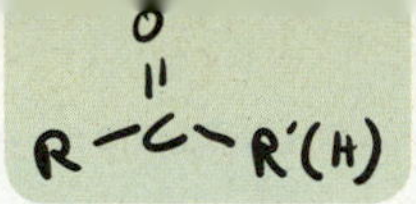

[A MECHANISM FOR THE REACTION — Enamine Formation]

Step 1

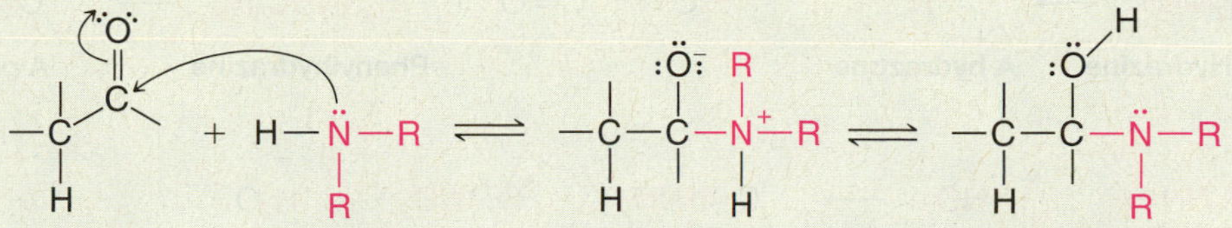

Aldehyde or ketone   **Secondary amine**   **Aminoalcohol intermediate**

The amine adds to the ketone or aldehyde carbonyl
to form a tetrahedral adduct. Intermolecular proton
transfer leads to the aminoalcohol intermediate.

Step 2

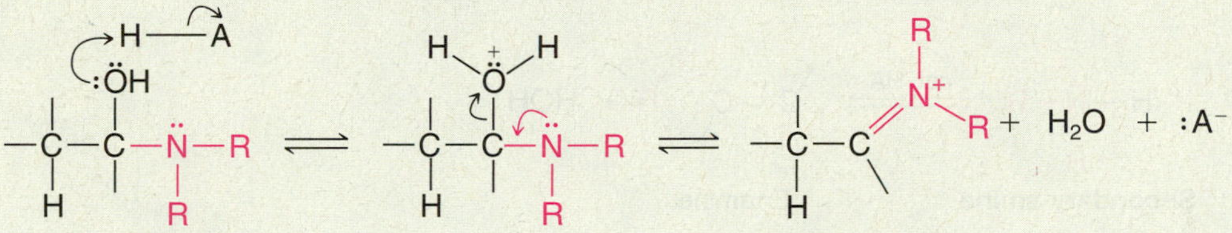

Aminoalcohol intermediate   **Iminium ion intermediate**

The aminoalcohol intermediate is protonated by the catalytic acid.
Contribution of an unshared electron pair from the nitrogen atom and
departure of a water molecule lead to an iminium cation intermediate.

Step 3

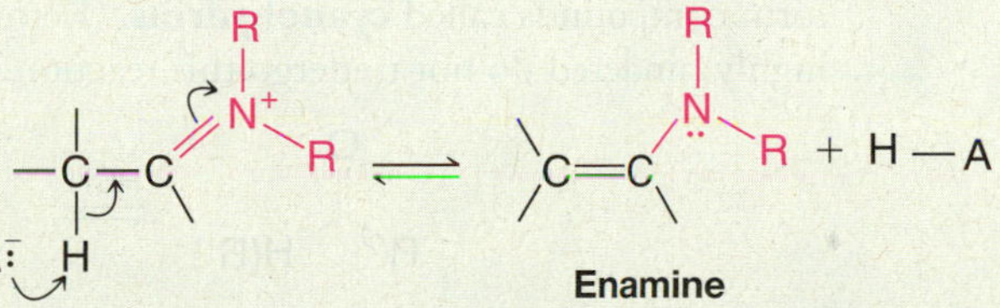

Enamine

A proton is removed from the carbon adjacent to the iminium group. Proton removal occurs
from the carbon because there is no proton to remove from the nitrogen of the iminium cation
(as there would have been if a primary amine had been used). This step forms the enamine,
neutralizes the formal charge, and regenerates the catalytic acid. (If there had been a proton to
remove from the nitrogen of the iminium cation, the final product would have been an imine.)

Table 16.2 summarizes the reactions of aldehydes and ketones with derivatives of
ammonia.

TABLE 16.2 REACTIONS OF ALDEHYDES AND KETONES WITH DERIVATIVES OF AMMONIA

1. **Imine formation—reaction with a primary amine**

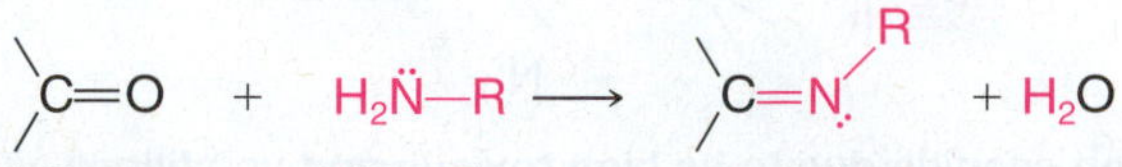

Aldehyde or ketone   **A 1° amine**   **An imine**   [(E) and (Z) isomers]

2. **Oxime formation—reaction with hydroxylamine**

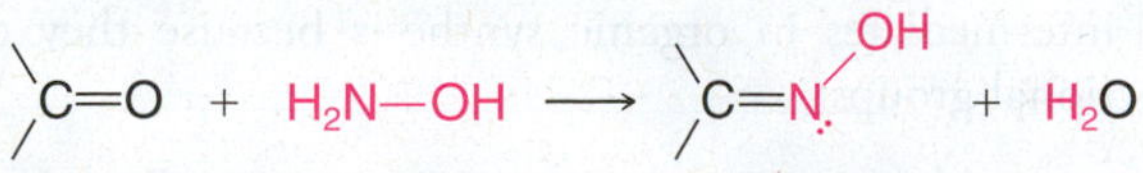

Aldehyde or ketone   **Hydroxylamine**   **An oxime**   [(E) and (Z) isomers]

(Table continues on next page)

3. **Hydrazone and substituted hydrazone formation—reactions with hydrazine, phenylhydrazine, and 2,4-dinitrophenylhydrazine [each derivative can form as an (E) or (Z) isomer]**

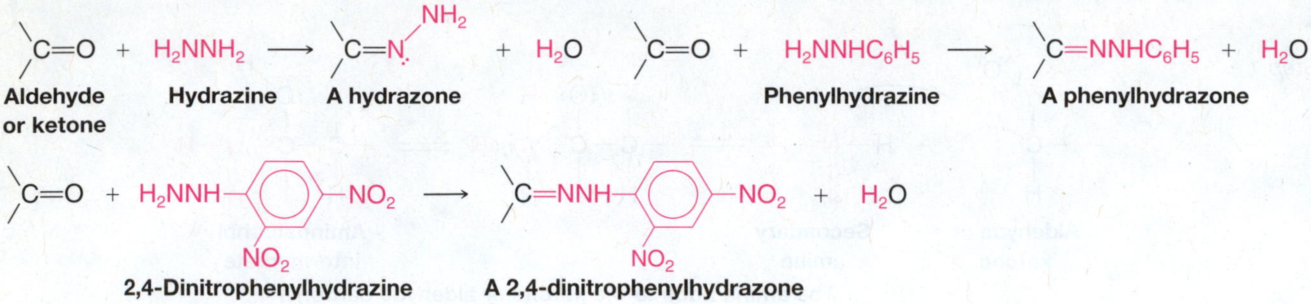

Aldehyde or ketone + Hydrazine → A hydrazone + H_2O

Phenylhydrazine → A phenylhydrazone + H_2O

2,4-Dinitrophenylhydrazine → A 2,4-dinitrophenylhydrazone + H_2O

4. **Enamine formation—reaction with a secondary amine**

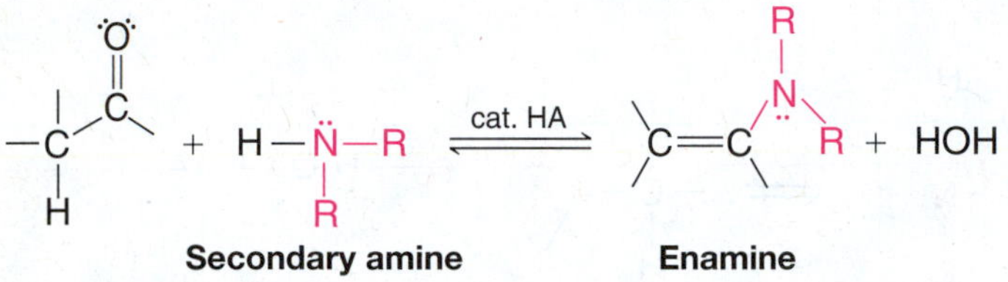

Secondary amine → Enamine + HOH

16.9 THE ADDITION OF HYDROGEN CYANIDE: CYANOHYDRINS

- Hydrogen cyanide adds to the carbonyl groups of aldehydes and most ketones to form compounds called **cyanohydrins**. (Ketones in which the carbonyl group is highly hindered do not undergo this reaction.)

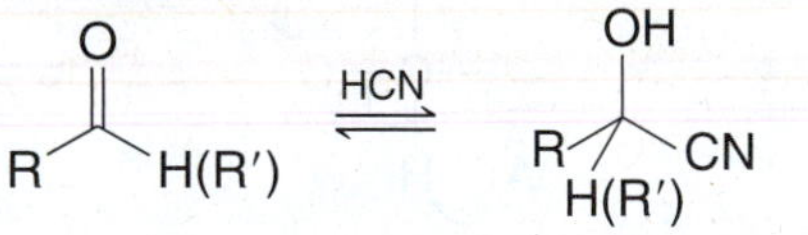

A cyanohydrin

Cyanohydrins form fastest under conditions where cyanide anions are present to act as the nucleophile. Use of potassium cyanide, or any base that can generate cyanide anions from HCN, increases the reaction rate as compared to the use of HCN alone. The addition of hydrogen cyanide itself to a carbonyl group is slow because the weak acidity of HCN ($pK_a \sim 9$) provides only a small concentration of the nucleophilic cyanide anion. The following is a mechanism for formation of a cyanohydrin.

[A MECHANISM FOR THE REACTION — Cyanohydrin Formation]

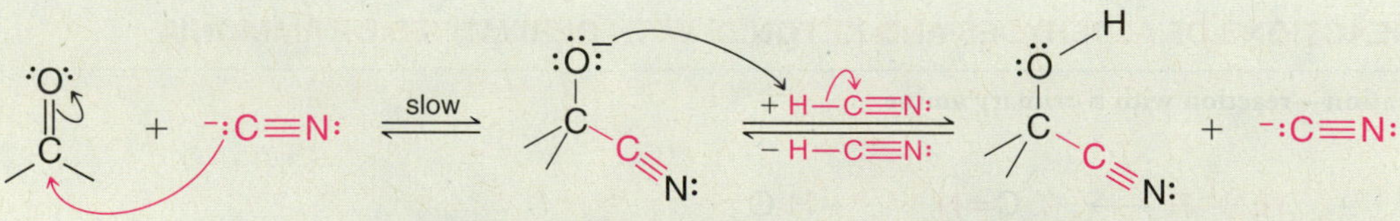

Great care must be taken when working with hydrogen cyanide due to its high toxicity and volatility. Reactions involving HCN must be conducted in an efficient fume hood.

Cyanohydrins are useful intermediates in organic synthesis because they can be converted to several other functional groups.

- Acidic hydrolysis converts cyanohydrins to α-hydroxy acids or to α,β-unsaturated acids.

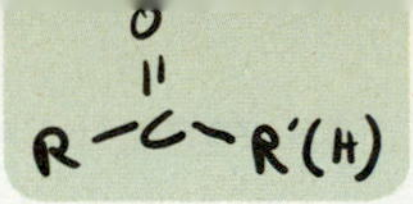

The mechanism for this hydrolysis is discussed in Section 17.8H. The preparation of α-hydroxy acids from cyanohydrins is part of the Kiliani–Fischer synthesis of simple sugars (Section 22.9A):

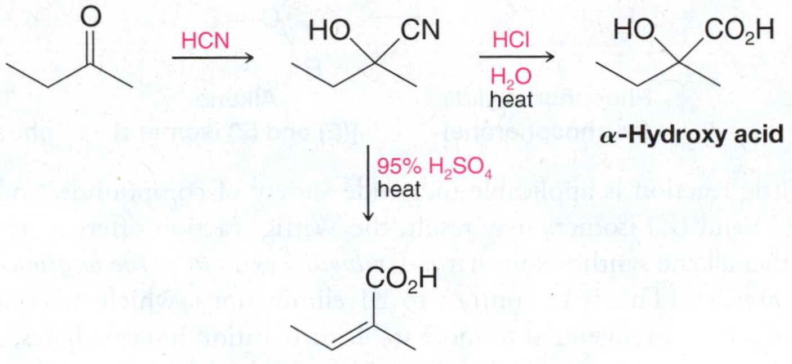

α-Hydroxy acid

α,β-Unsaturated acid

- Reduction of a cyanohydrin with lithium aluminum hydride gives a β-aminoalcohol:

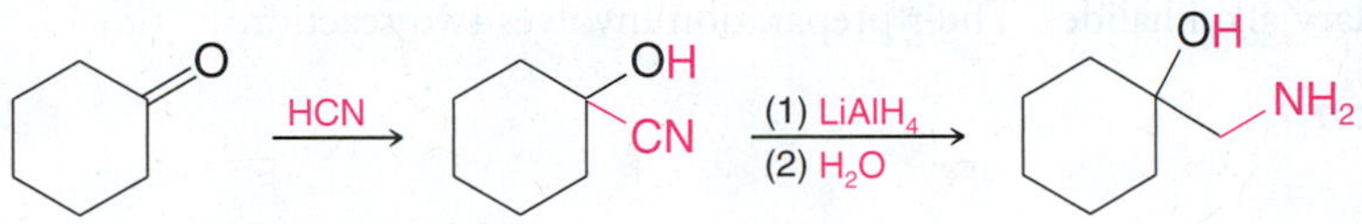

●●● **SOLVED PROBLEM 16.6**

Provide the missing reagents and intermediate in the following synthesis.

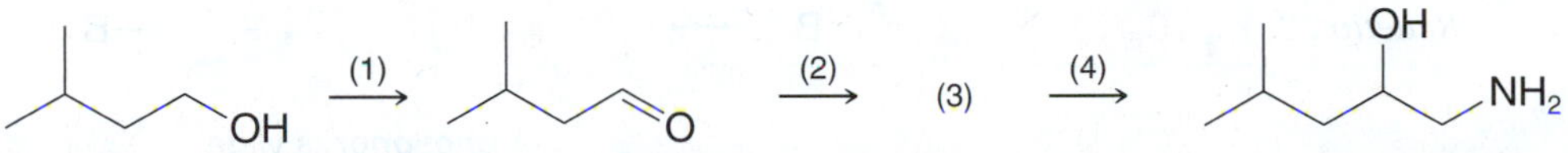

STRATEGY AND ANSWER: Step (1) requires oxidation of a primary alcohol to an aldehyde; use PCC (Section 12.4). To reach the final product from the aldehyde we need to add a carbon atom to the chain and introduce a primary amine. This combination suggests use of a nitrile, which we know can be reduced to a primary amine. Thus, adding HCN to the aldehyde in step (2) forms the cyanohydrin (3), shown below. This step also affords the alcohol group present in the final product. In step (4) we reduce the nitrile to a primary amine using LiAlH$_4$.

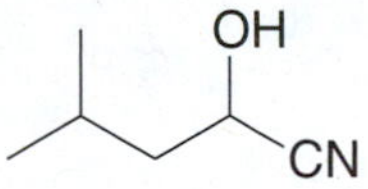

●●●

PRACTICE PROBLEM 16.16

(a) Show how you might prepare lactic acid from acetaldehyde through a cyanohydrin intermediate. **(b)** What stereoisomeric form of lactic acid would you expect to obtain?

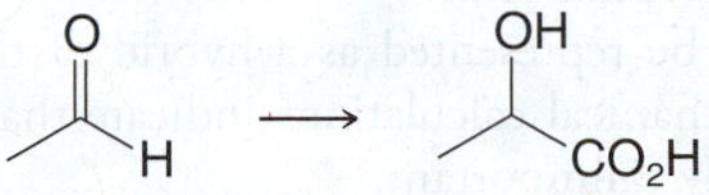

16.10 THE ADDITION OF YLIDES: THE WITTIG REACTION

- Aldehydes and ketones react with phosphorus ylides to yield alkenes and triphenylphosphine oxide (a by-product). This reaction is known as the **Wittig reaction**.

The Wittig reaction has proved to be a valuable method for synthesizing alkenes. The **ylide** required for the reaction is a molecule with no net charge but which has a negative carbon atom adjacent to a positive heteroatom, which in the Wittig reaction is a phosphorus atom. Phosphorus ylides are also called phosphoranes.

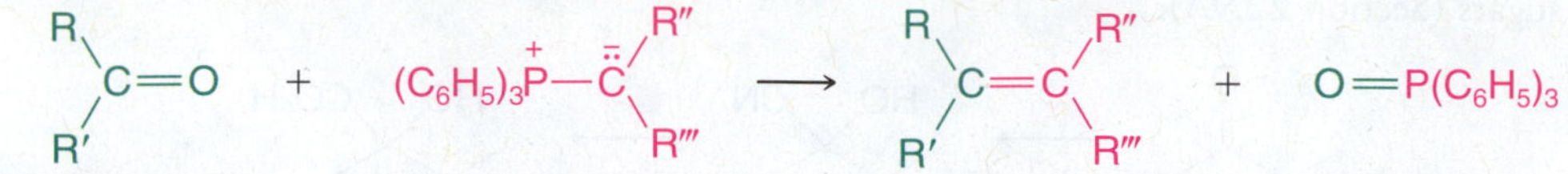

$$
\underset{\substack{\text{Aldehyde or}\\\text{ketone}}}{\overset{R}{\underset{R'}{\diagup}}C=O} \;+\; \underset{\substack{\text{Phosphorus ylide}\\\text{(or phosphorane)}}}{(C_6H_5)_3\overset{+}{P}-\overset{\ddots}{\underset{R'''}{\overset{R''}{C}}}} \longrightarrow \underset{\substack{\text{Alkene}\\\text{[(}E\text{) and (}Z\text{) isomers]}}}{\overset{R}{\underset{R'}{\diagup}}C=C\overset{R''}{\underset{R'''}{\diagdown}}} \;+\; \underset{\substack{\text{Triphenyl-}\\\text{phosphine oxide}}}{O=P(C_6H_5)_3}
$$

The Wittig reaction is applicable to a wide variety of compounds, and although a mixture of (E) and (Z) isomers may result, the Wittig reaction offers a great advantage over most other alkene syntheses in that *no ambiguity exists as to the location of the double bond in the product.* (This is in contrast to E1 eliminations, which may yield multiple alkene products by rearrangement to more stable carbocation intermediates, and both E1 and E2 elimination reactions, which may produce multiple products when different β hydrogens are available for removal.)

Phosphorus ylides are easily prepared from triphenylphosphine and primary or secondary alkyl halides. Their preparation involves two reactions:

General Reaction

Reaction 1 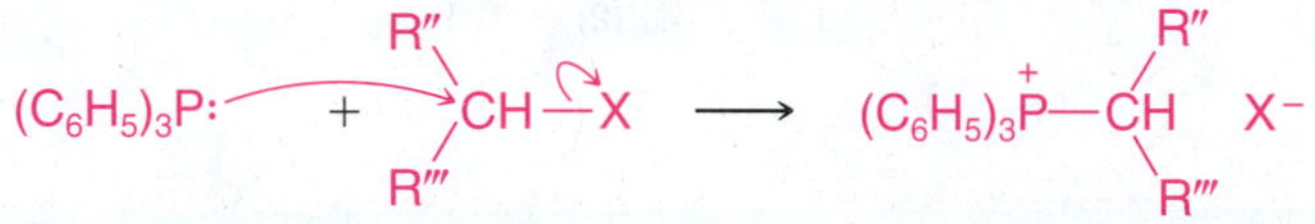
$$
(C_6H_5)_3P: \;+\; \overset{R''}{\underset{R'''}{\diagup}}CH-X \longrightarrow (C_6H_5)_3\overset{+}{P}-\overset{R''}{\underset{R'''}{\overset{\mid}{C}H}} \quad X^-
$$
                           Triphenylphosphine   **An alkyltriphenylphosphonium halide**

Reaction 2 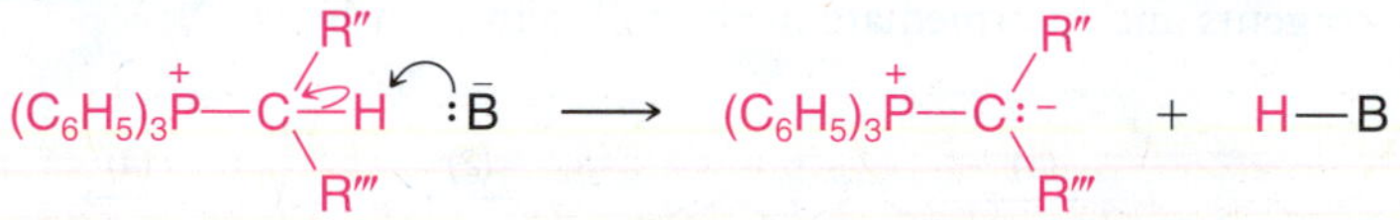
$$
(C_6H_5)_3\overset{+}{P}-\overset{R''}{\underset{R'''}{\overset{\mid}{C}}}H \quad :\bar{B} \longrightarrow (C_6H_5)_3\overset{+}{P}-\overset{R''}{\underset{R'''}{\overset{\mid}{C}}}:^- \;+\; H-B
$$
                                                                A phosphorus ylide

Specific Example

Reaction 1 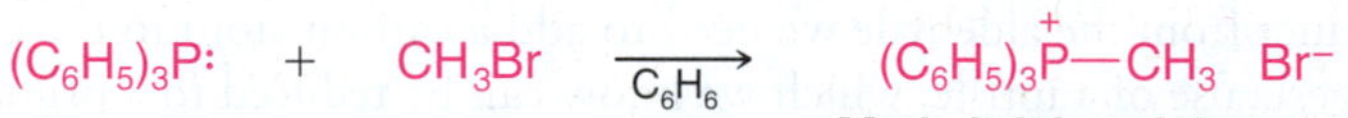
$$
(C_6H_5)_3P: \;+\; CH_3Br \xrightarrow{\;C_6H_6\;} (C_6H_5)_3\overset{+}{P}-CH_3 \quad Br^-
$$
                                                       Methyltriphenylphosphonium bromide (89%)

Reaction 2
$$
(C_6H_5)_3\overset{+}{P}-CH_3 \;+\; C_6H_5Li \longrightarrow (C_6H_5)_3\overset{+}{P}-CH_2:^- \;+\; C_6H_6 \;+\; LiBr
$$
$$
\underset{Br^-}{}
$$

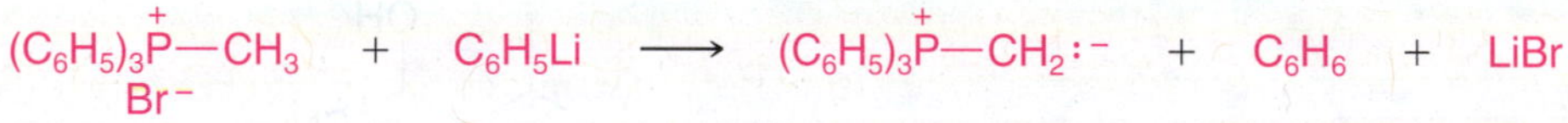

The first reaction is a nucleophilic substitution reaction. Triphenylphosphine is an excellent nucleophile and a weak base. It reacts readily with 1° and 2° alkyl halides by an S_N2 mechanism to displace a halide ion from the alkyl halide to give an alkyltriphenylphosphonium salt. **The second reaction is an acid–base reaction.** A strong base (usually an alkyllithium or phenyllithium) removes a proton from the carbon that is attached to phosphorus to give the ylide.

Phosphorus ylides can be represented as a hybrid of the two resonance structures shown here. Quantum mechanical calculations indicate that the contribution made by the first structure is relatively unimportant.

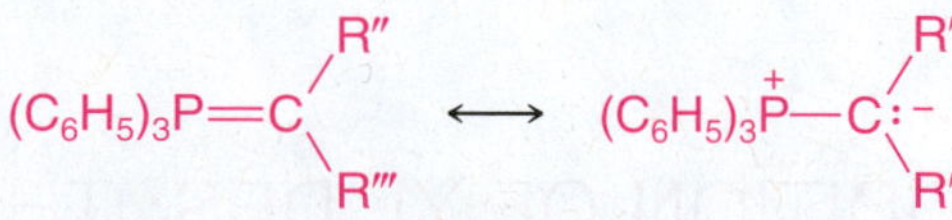

$$
(C_6H_5)_3P=C\overset{R''}{\underset{R'''}{\diagdown}} \longleftrightarrow (C_6H_5)_3\overset{+}{P}-C:^-\overset{R''}{\underset{R'''}{\diagdown}}
$$

Studies by E. Vedejs (University of Michigan) indicate that the Wittig reaction takes place in two steps. In the first step (below), the aldehyde or ketone combines with the ylide in a cycloaddition reaction to form the four-membered ring of an oxaphosphetane. Then in a second step, the oxaphosphetane decomposes to form the alkene and triphenylphosphine oxide. The driving force for the reaction is the formation of the very strong ($DH° = 540$ kJ mol^{-1}) phosphorus–oxygen bond in triphenylphosphine oxide.

A MECHANISM FOR THE REACTION — The Wittig Reaction

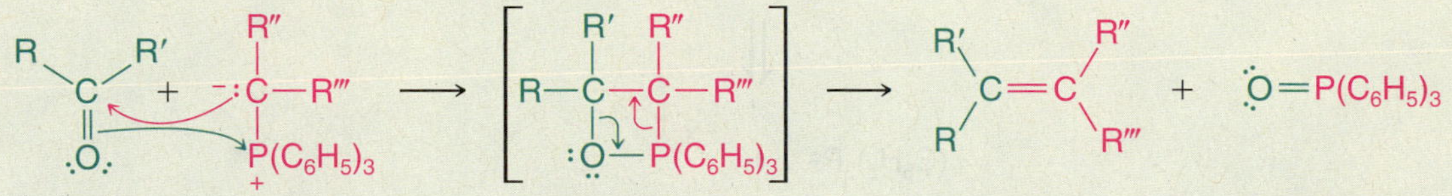

| Aldehyde or ketone | Ylide | Oxaphosphetane | Alkene (+ diastereomer) | Triphenylphosphine oxide |

Specific Example

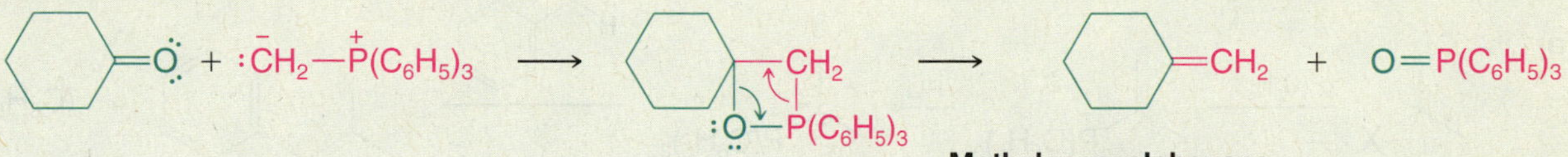

Methylenecyclohexane (86%)

While Wittig syntheses may appear to be complicated, in practice they are easy to carry out. Most of the steps can be carried out in the same reaction vessel, and the entire synthesis can be accomplished in a matter of hours. The overall result of a Wittig synthesis is:

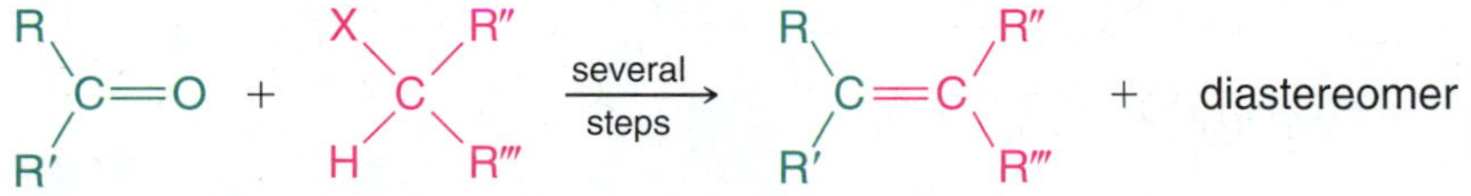

16.10A HOW TO Plan a Wittig Synthesis

Planning a Wittig synthesis begins with recognizing in the desired alkene what can be the aldehyde or ketone component and what can be the halide component. Any or all of the R groups may be hydrogen, although yields are generally better when at least one group is hydrogen. The halide component must be a primary, secondary, or methyl halide.

●●● **SOLVED PROBLEM 16.7**

Synthesize 2-methyl-1-phenylprop-1-ene using a Wittig reaction. Begin by writing a retrosynthetic analysis.

STRATEGY AND ANSWER: We examine the structure of the compound, paying attention to the groups on each side of the double bond:

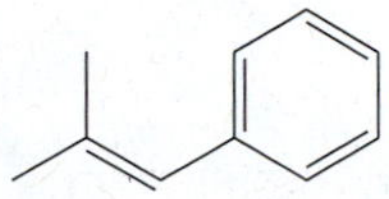

2–Methyl–1–phenylprop–1–ene

We see that two retrosynthetic analyses are possible.

Retrosynthetic Analysis

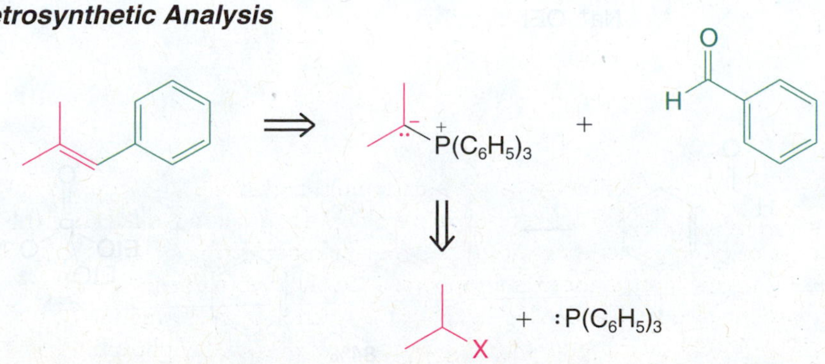

(continues on next page)

(b)

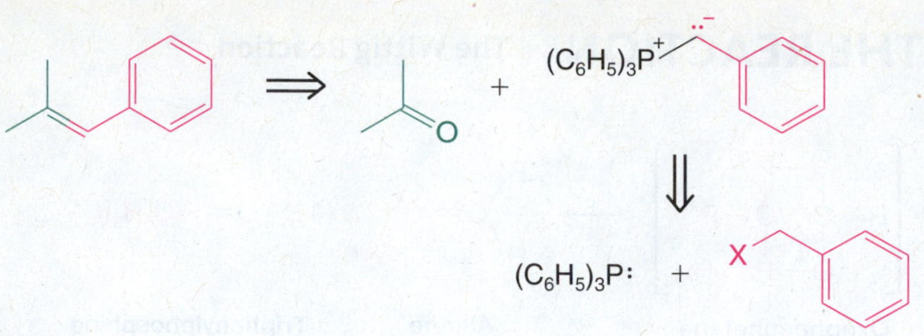

Synthesis

Following retrosynthetic analysis (a), we begin by making the ylide from a 2-halopropane and then allow the ylide to react with benzaldehyde:

(a)

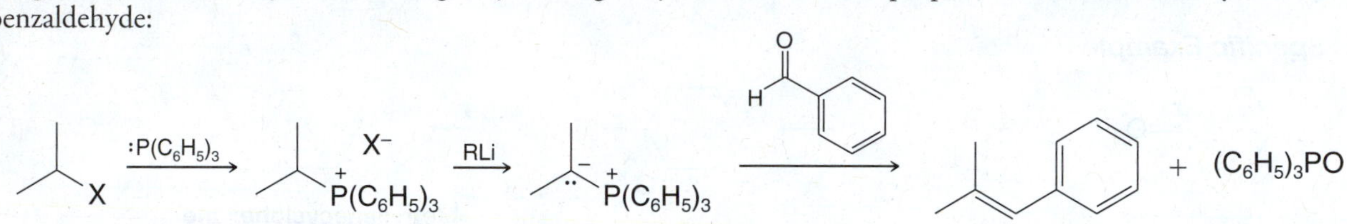

Following retrosynthetic analysis (b), we make the ylide from a benzyl halide and allow it to react with acetone:

(b)

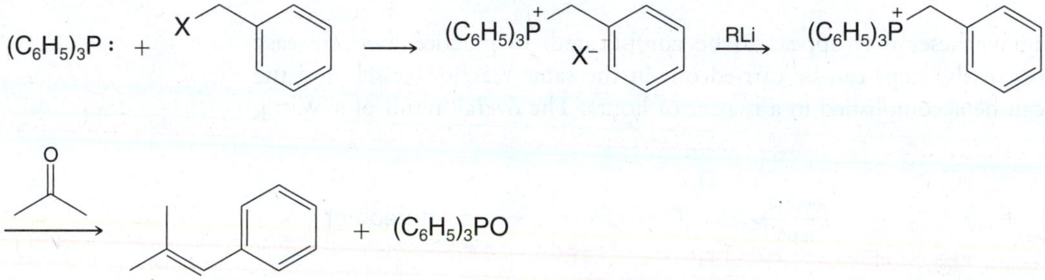

16.10B The Horner–Wadsworth–Emmons Reaction: A Modification of the Wittig Reaction

A widely used variation of the Wittig reaction is the **Horner–Wadsworth–Emmons** modification.

- The Horner–Wadsworth–Emmons reaction involves use of a phosphonate ester instead of a triphenylphosphonium salt. The major product is usually the (*E*)-alkene isomer.

Some bases that are typically used to form the phosphonate ester carbanion include sodium hydride, potassium *tert*-butoxide, and butyllithium. The following reaction sequence is an example:

Step 1

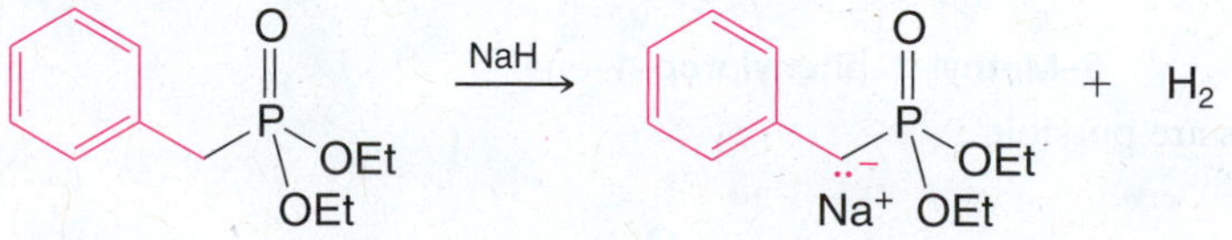

A phosphonate ester

Step 2

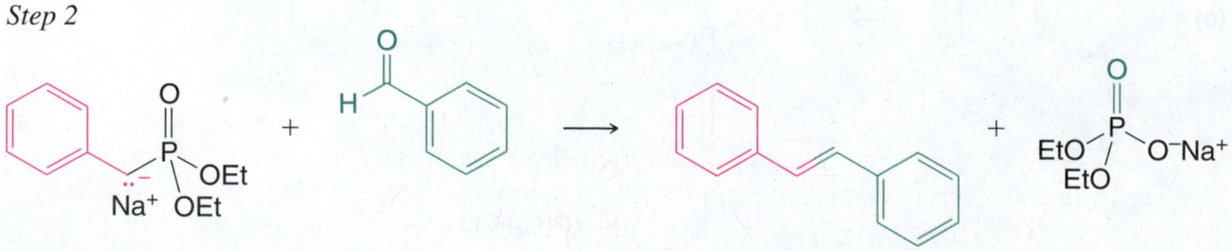

84%

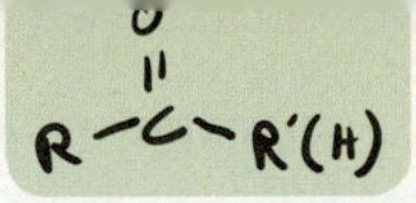

The phosphonate ester is prepared by reaction of a trialkyl phosphite [(RO)₃P] with an appropriate halide (a process called the Arbuzov reaction). The following is an example:

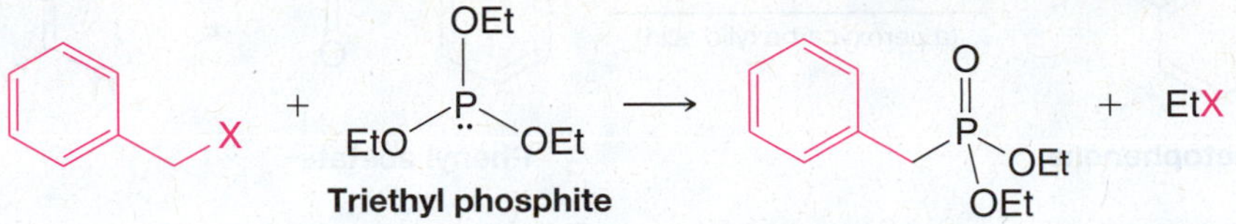

Triethyl phosphite

PRACTICE PROBLEM 16.17

In addition to triphenylphosphine, assume that you have available as starting materials any necessary aldehydes, ketones, and organic halides. Show how you might synthesize each of the following alkenes using the Wittig reaction:

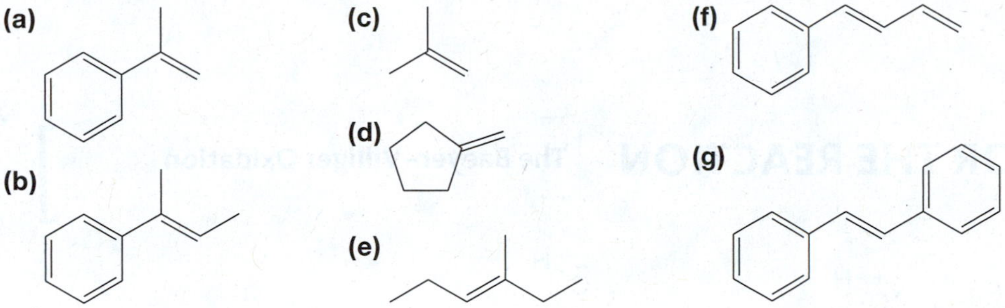

(a)

(c)

(f)

(b)

(d)

(g)

(e)

PRACTICE PROBLEM 16.18

Triphenylphosphine can be used to convert epoxides to alkenes—for example,

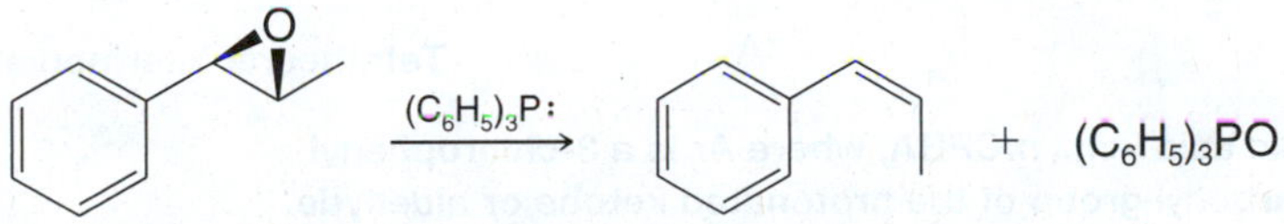

Propose a likely mechanism for this reaction.

16.11 OXIDATION OF ALDEHYDES

Aldehydes are much more easily oxidized than ketones. Aldehydes are readily oxidized by strong oxidizing agents such as potassium permanganate, and they are also oxidized by such mild oxidizing agents as silver oxide:

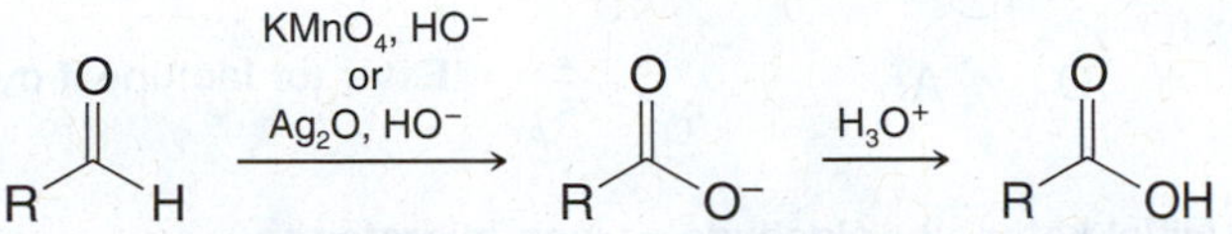

Notice that in these oxidations aldehydes lose the hydrogen that is attached to the carbonyl carbon atom. Because ketones lack this hydrogen, they are more resistant to oxidation. Aldehydes undergo slow oxidation by oxygen in the air, and thus stored samples of aldehydes often contain the corresponding carboxylic acid as an impurity.

16.12 THE BAEYER–VILLIGER OXIDATION

The Baeyer-Villiger oxidation is a useful method for conversion of aldehydes or ketones to esters by the insertion of an oxygen atom from a peroxycarboxylic acid (RCO_3H). For example, treating acetophenone with a peroxycarboxylic acid converts it to the ester, phenyl acetate.

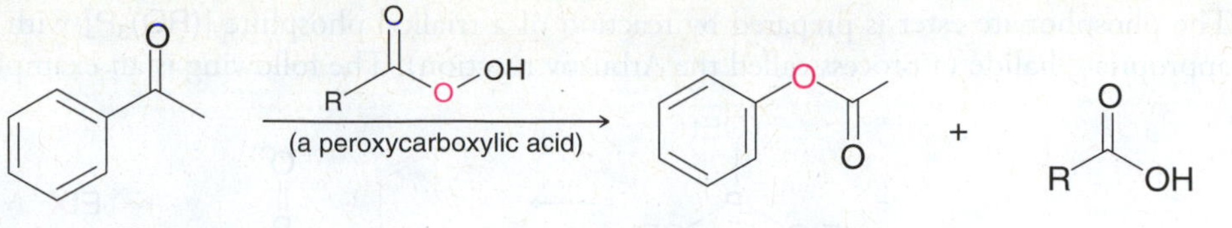

Acetophenone   **Phenyl acetate**

(a peroxycarboxylic acid)

The Baeyer–Villiger oxidation is also widely used for synthesizing lactones (cyclic esters) from cyclic ketones. A common reagent used to carry out the Baeyer–Villiger oxidation is *meta*-chloroperoxybenzoic acid (*m*CPBA). Certain other peroxycarboxylic acids can be used as well. The following is a mechanism for Baeyer–Villiger oxidation.

A MECHANISM FOR THE REACTION — The Baeyer–Villiger Oxidation

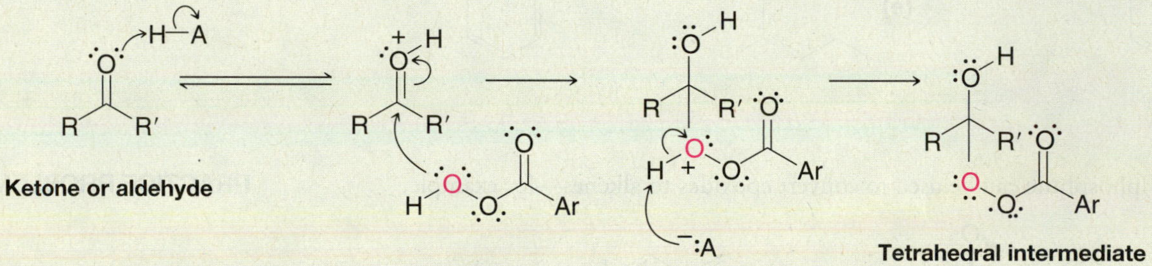

Ketone or aldehyde   **Tetrahedral intermediate**

The peroxycarboxylic acid (e.g., *m*CPBA, where Ar is a 3-chlorophenyl group) attacks the carbonyl group of the protonated ketone or aldehyde, leading to a tetrahedral intermediate.

Tetrahedral intermediate   **Ester (or lactone if cyclic)**

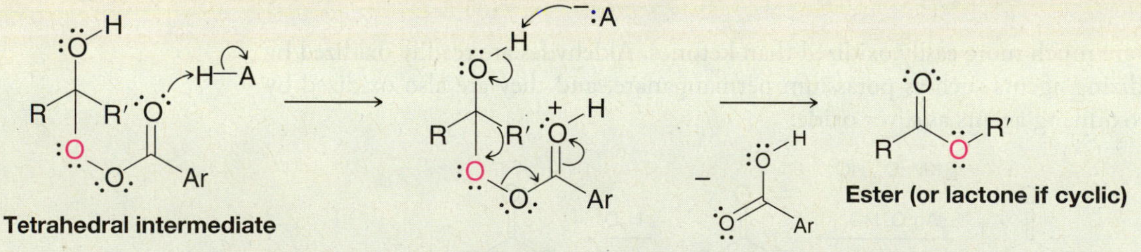

A group bonded to the initial ketone or aldehyde carbon migrates to oxygen, producing the ester (or lactone) as the peroxycarboxylic acid is released as a carboxylic acid.

The group that migrates from the original ketone or aldehyde carbon to the oxygen of the peroxycarboxylic acid is a function of "migratory aptitude." Studies have shown the migratory aptitude of groups is H > phenyl > 3° alkyl > 2° alkyl > 1° alkyl > methyl.

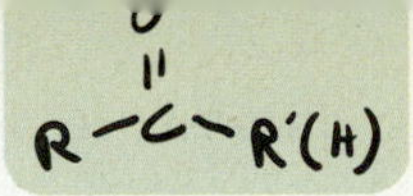

PRACTICE PROBLEM 16.19

When benzaldehyde reacts with a peroxy acid, the product is benzoic acid. The mechanism for this reaction is analogous to the one just given for the oxidation of acetophenone, and the outcome illustrates the greater migratory aptitude of a hydrogen atom compared to phenyl. Outline all the steps involved.

PRACTICE PROBLEM 16.20

Give the structure of the product that would result from a Baeyer–Villiger oxidation of cyclopentanone.

PRACTICE PROBLEM 16.21

What would be the major product formed in the Baeyer-Villiger oxidation of 3-methyl-2-butanone?

16.13 CHEMICAL ANALYSES FOR ALDEHYDES AND KETONES

16.13A Derivatives of Aldehydes and Ketones

Aldehydes and ketones can be differentiated from noncarbonyl compounds through their reactions with derivatives of ammonia (Section 16.8B). 2,4-Dinitrophenylhydrazine and hydroxylamine react with aldehydes and ketones to form precipitates. Oximes are usually colorless, whereas 2,4-dinitrophenylhydrazones are usually orange. The melting points of these derivatives can also be used in identifying specific aldehydes and ketones.

16.13B Tollens' Test (Silver Mirror Test)

The ease with which aldehydes undergo oxidation differentiates them from most ketones. Mixing aqueous silver nitrate with aqueous ammonia produces a solution known as Tollens' reagent. The reagent contains the diaminosilver(I) ion, $Ag(NH_3)_2^+$. Although this ion is a very weak oxidizing agent, it oxidizes aldehydes to carboxylate anions. As it does this, silver is reduced from the +1 oxidation state [of $Ag(NH_3)_2^+$] to metallic silver. If the rate of reaction is slow and the walls of the vessel are clean, metallic silver deposits on the walls of the test tube as a mirror; if not, it deposits as a gray-to-black precipitate. Tollens' reagent gives a negative result with all ketones except α-hydroxy ketones:

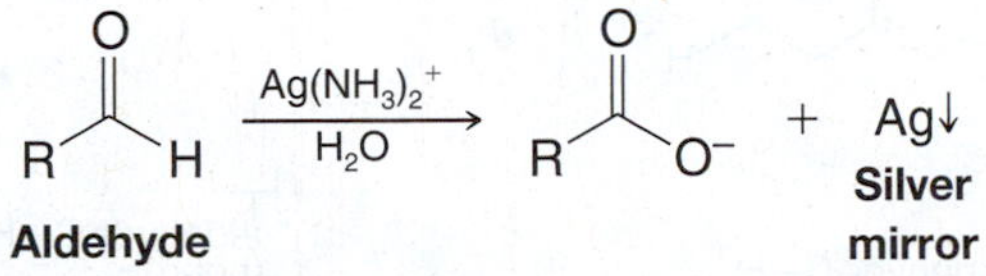

16.14 SPECTROSCOPIC PROPERTIES OF ALDEHYDES AND KETONES

16.14A IR Spectra of Aldehydes and Ketones

- Carbonyl groups of aldehydes and ketones give rise to very strong C=O stretching absorption bands in the 1665–1780-cm^{-1} region.

The exact location of the carbonyl IR absorption (Table 16.3) depends on the structure of the aldehyde or ketone and is one of the most useful and characteristic absorptions in the IR spectrum.

- Saturated acyclic aldehydes typically absorb near 1730 cm^{-1}; similar ketones absorb near 1715 cm^{-1}.
- Conjugation of the carbonyl group with a double bond or a benzene ring shifts the C=O absorption to lower frequencies by about 40 cm^{-1}.

TABLE 16.3 IR CARBONYL STRETCHING BANDS OF ALDEHYDES AND KETONES

C=O Stretching Frequencies			
Compound	Range (cm^{-1})	Compound	Range (cm^{-1})
R—CHO	1720–1740	RCOR	1705–1720
Ar—CHO	1695–1715	ArCOR	1680–1700
C=C, CHO	1680–1690	C=C, COR	1665–1680
		Cyclohexanone	1715
		Cyclopentanone	1751
		Cyclobutanone	1785

This shift to lower frequencies occurs because the carbonyl double bond of a conjugated compound has more single-bond character (see the resonance structures below), and single bonds are easier to stretch than double bonds.

←——— **Single bond**

The location of the carbonyl absorption of cyclic ketones depends on the size of the ring (compare the cyclic compounds in Table 16.3). *As the ring grows smaller, the* C=O *stretching peak is shifted to higher frequencies.*

Vibrations of the C—H bond of the CHO group of aldehydes also give two weak bands in the 2700–2775-cm^{-1} and 2820–2900-cm^{-1} regions that are easily identified.

Figure 16.1 shows the IR spectrum of phenylethanal.

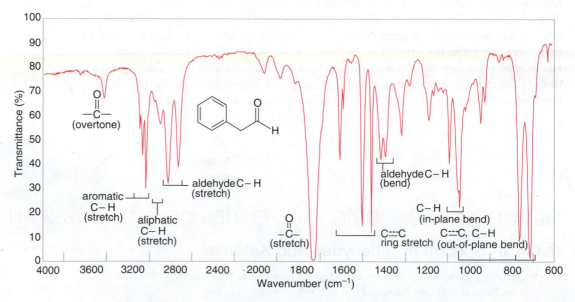

FIGURE 16.1
The infrared spectrum of phenylethanal.

16.14B NMR Spectra of Aldehydes and Ketones

^{13}C NMR Spectra

- The carbonyl carbon of an aldehyde or ketone gives characteristic NMR signals in the δ 180–220 region of ^{13}C spectra.

Since almost no other signals occur in this region, *the presence of a signal in this region (near δ 200) strongly suggests the presence of a carbonyl group.*

¹H NMR Spectra

- An aldehyde proton gives a distinct ¹H NMR signal downfield in the δ 9–12 region where almost no other protons absorb; therefore, it is easily identified.

The aldehyde proton of an aliphatic aldehyde shows spin–spin coupling with protons on the adjacent α carbon, and the splitting pattern reveals the degree of substitution of the α carbon. For example, in acetaldehyde (CH_3CHO) the aldehyde proton signal is split into a quartet by the three methyl protons, and the methyl proton signal is split into a doublet by the aldehyde proton. The coupling constant is small, however (about 3 Hz, as compared with typical vicinal splitting of about 7 Hz).

- Protons on the α carbon are deshielded by the carbonyl group, and their signals generally appear in the δ 2.0–2.3 region.
- Methyl ketones show a characteristic (3H) singlet near δ 2.1.

Figures 16.2 and 16.3 show annotated ¹H and ¹³C spectra of phenylethanal.

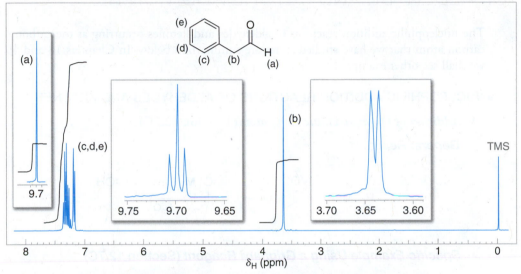

FIGURE 16.2 The 300-MHz ¹H NMR spectrum of phenylethanal. The small coupling between the aldehyde and methylene protons (2.6 Hz) is shown in the expanded offset plots.

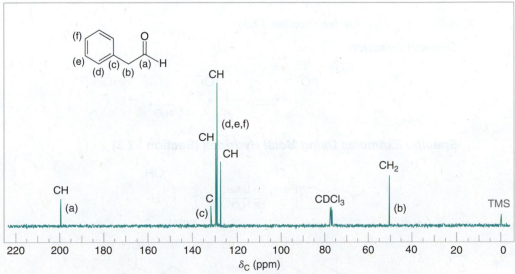

FIGURE 16.3 The broadband proton-decoupled ¹³C NMR spectrum of phenylethanal. DEPT ¹³C NMR information and carbon assignments are shown near each peak.

16.14C Mass Spectra of Aldehydes and Ketones

The mass spectra of ketones usually show a peak corresponding to the molecular ion. Aldehydes typically produce a prominent $M^{\cdot+}-1$ peak in their mass spectra from cleavage of the aldehyde hydrogen. Ketones usually undergo cleavage on either side of the carbonyl group to produce acylium ions, $RC{\equiv}O{:}^+$, where R can be the alkyl group from either side of the ketone carbonyl. Cleavage via the McLafferty rearrangement (Section 9.16D) is also possible in many aldehydes and ketones.

16.14D UV Spectra

The carbonyl groups of saturated aldehydes and ketones give a weak absorption band in the UV region between 270 and 300 nm. This band is shifted to longer wavelengths (300–350 nm) when the carbonyl group is conjugated with a double bond.

16.15 SUMMARY OF ALDEHYDE AND KETONE ADDITION REACTIONS

The nucleophilic addition reactions of aldehydes and ketones occurring at the carbonyl carbon atom that we have studied so far are summarized below. In Chapters 18 and 19 we shall see other examples.

NUCLEOPHILIC ADDITION REACTIONS OF ALDEHYDES AND KETONES

1. *Addition of Organometallic Compounds* (Section 12.7C)

General Reaction

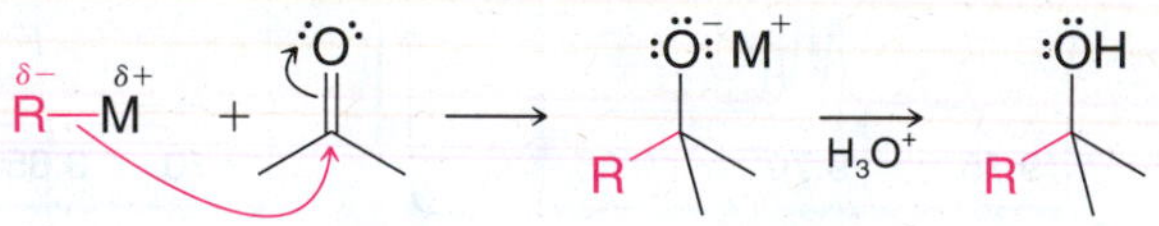

Specific Example Using a Grignard Reagent (Section 12.7C)

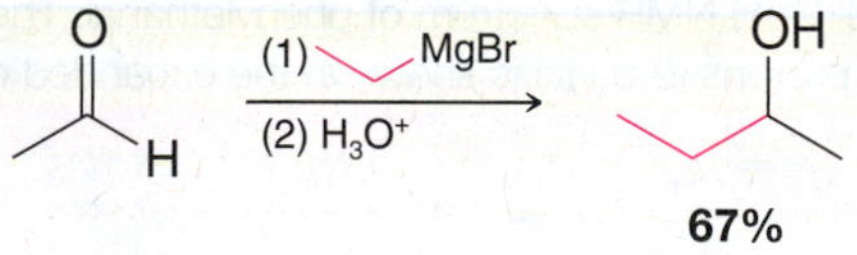

67%

2. *Addition of Hydride Ion* (Section 12.3)

General Reaction

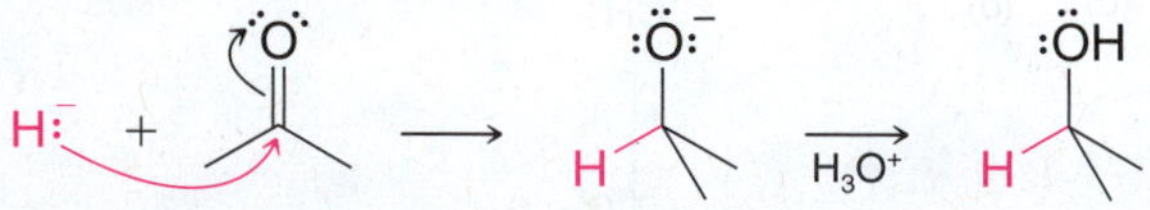

Specific Examples Using Metal Hydrides (Section 12.3)

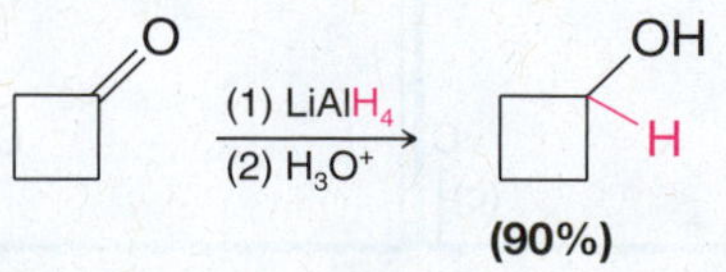

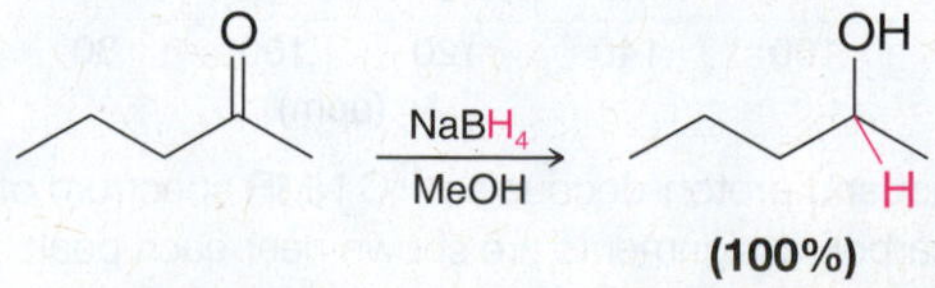

(90%)

(100%)

3. *Addition of Hydrogen Cyanide* (Section 16.9)
General Reaction

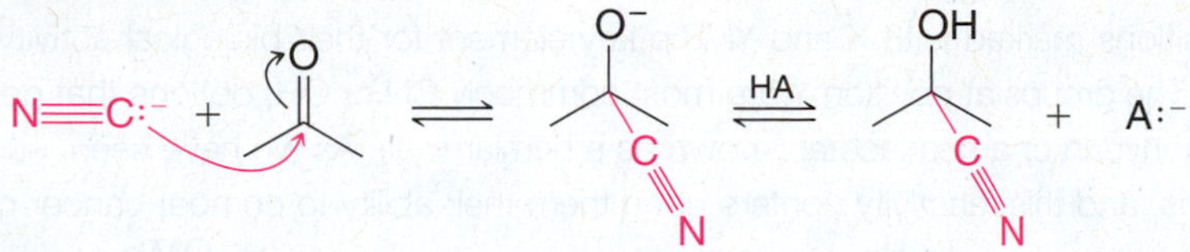

Specific Example

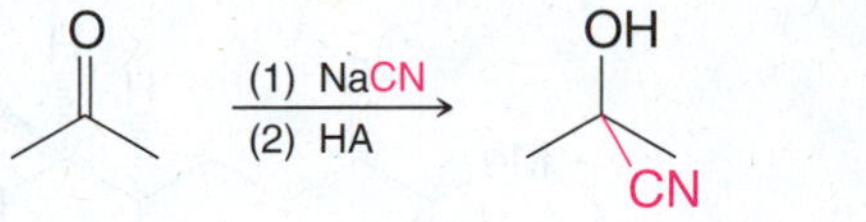

Acetone cyanohydrin
(78%)

4. *Addition of Ylides* (Section 16.10)
The Wittig Reaction

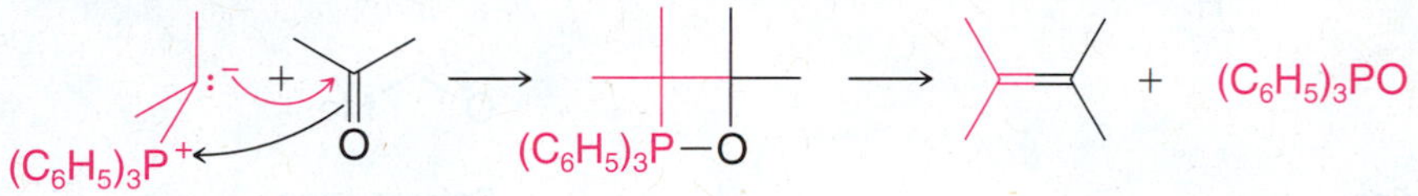

5. *Addition of Alcohols* (Section 16.7)
General Reaction

Hemiacetal   **Acetal**

Specific Example

6. *Addition of Derivatives of Ammonia* (Section 16.8)
Imines

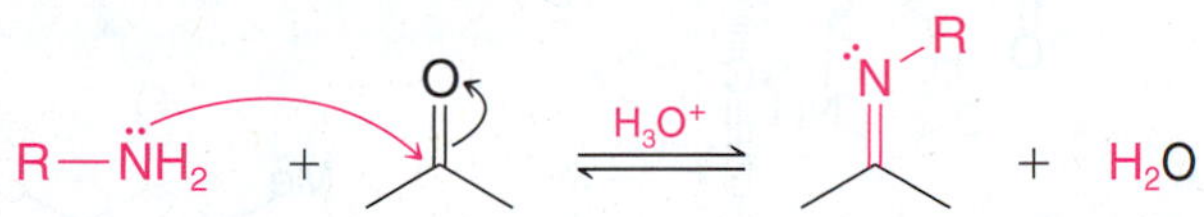

| 1° Amine | Aldehyde or ketone | Imine [(E) and (Z) isomers] |

Enamines

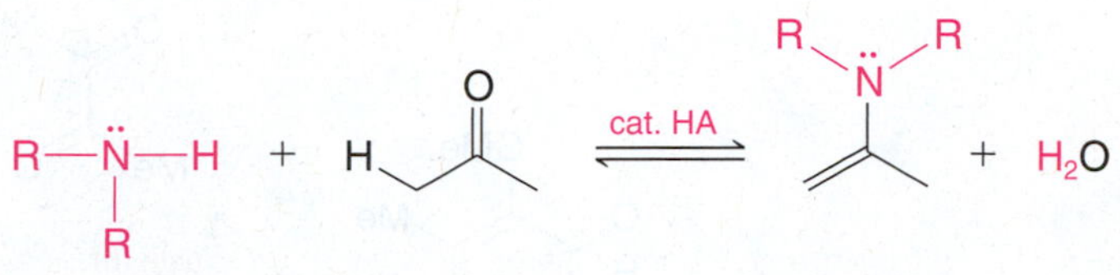

2° Amine   **Enamine**

[WHY Do These Topics Matter?

TRIGGERS FOR BIOCHEMICAL REACTIVITY

One of the things that makes sea sponges and other organisms that comprise coral reefs so beautiful is their bright and varied colors. However, this same feature, coupled with their inability to move, renders them easy targets for predators. Yet they survive because they have a chemical defense system that uses small, highly toxic molecules to ward off, injure, or even kill sea-based organism that might consume them. What is perhaps even more amazing is that many of these compounds have a far different

(continues on next page)

effect in humans: the ability to treat cancer by attacking cells that are replicating aberrantly. Moreover, the way in which this happens sometimes takes advantage of functional groups that you have seen in this chapter!

The ecteinascidins and the saframycins are two such groups of compounds. There are several variants of these compounds based on the identity of the atoms at the positions marked with X and Y. The key element for their biological activity is the highlighted configuration of atoms that includes Y. The groups at position Y are most commonly CN or OH, options that generate either the nitrogen functional group analog of a cyanohydrin or a hemiacetal (known as a hemiaminal). As we have seen, such functional groups can participate in a number of reactions, and this reactivity confers upon them their ability to combat cancer cells.

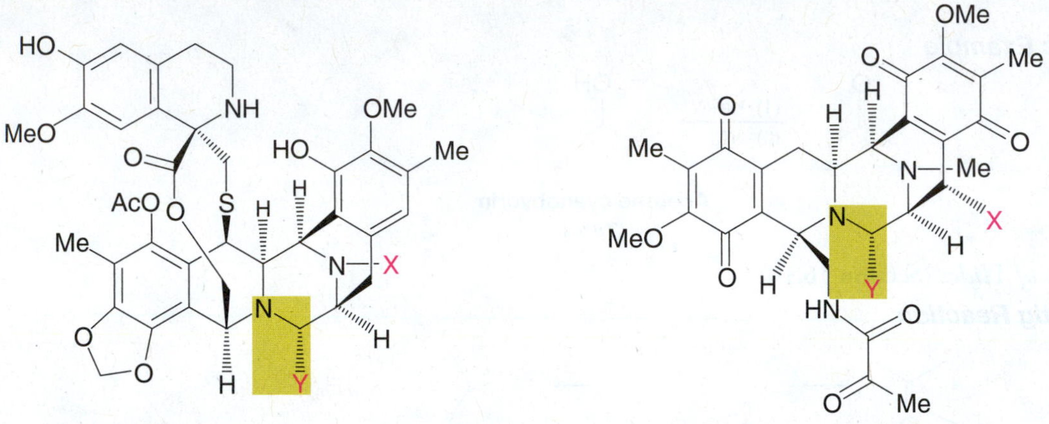

Ecteinascidins   **Saframycins**

In saframycin A, shown below, proton activation of the nitrile functional group creates a better leaving group, one that can lead to the formation of an iminium ion through the participation of the neighboring nitrogen atom. This reactive iminium species can then either be trapped reversibly by water (path a) to generate a hemiaminal, or if formed in the nucleus of a cell, it can be attacked by the nucleophilic free amine of a guanine residue from DNA (path b). If the latter happens, the other aromatic rings within saframycin A can then convert molecular oxygen into new reactive radical species that can damage the DNA and lead to cell death (we will learn this chemistry in Chapter 21).

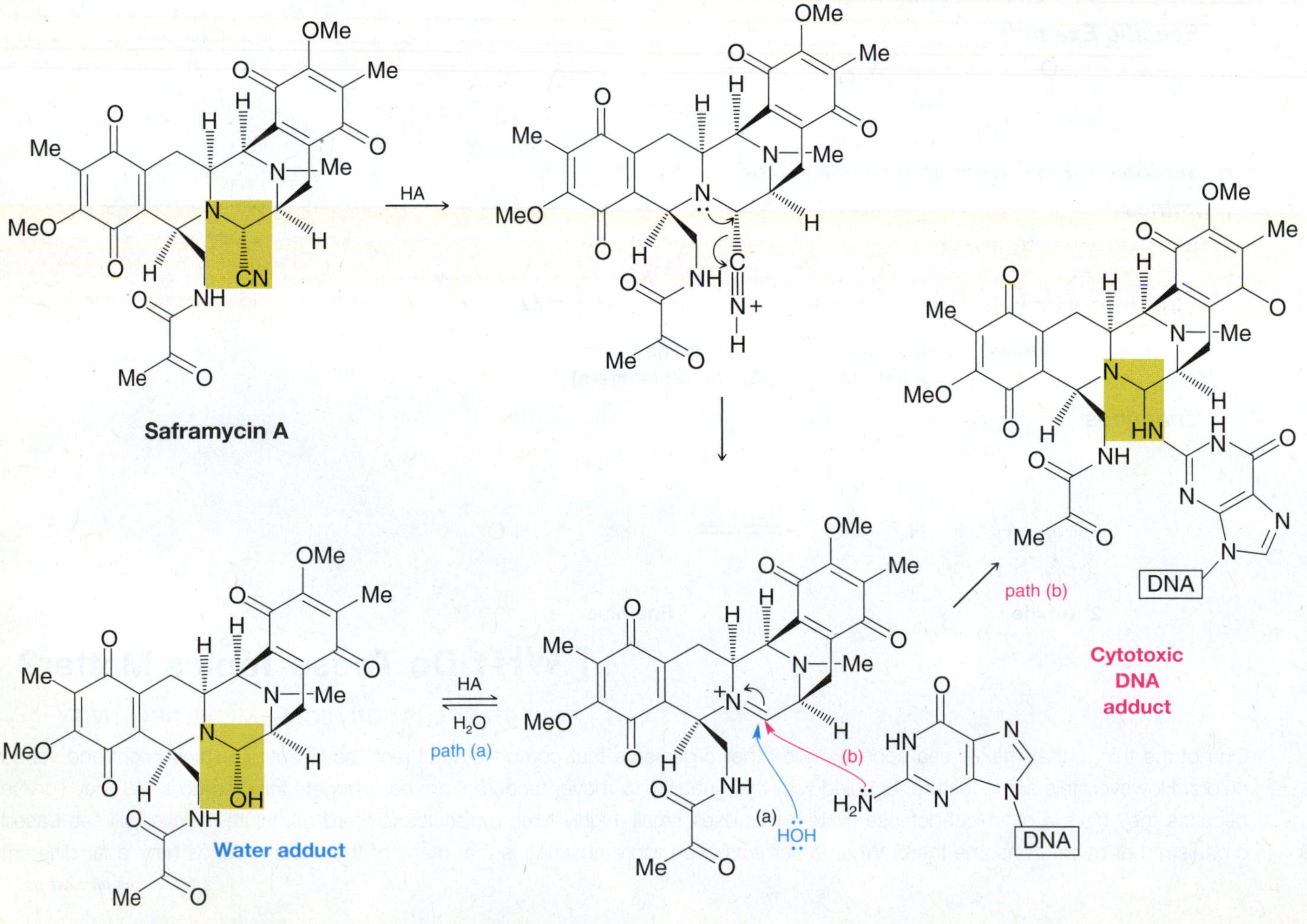

Saframycin A

Water adduct

path (a)

Cytotoxic DNA adduct

path (b)

What can be appreciated for now is a beautifully engineered triggering system for activity based entirely on some of the functional groups that can arise from carbonyl groups. To put the power of that design into some perspective, these compounds are among the most potent antitumor agents that have ever been identified from marine sources. In fact, some studies have estimated that a 5-mg dose of some compounds would be more than sufficient to eradicate several forms of human cancer. Clinical trials are currently evaluating that potential.

To learn more about these topics, see:

1. Lown, J. W.; Joshua, A. V.; Lee, J. S. "Molecular mechanisms of binding and single-strand scission of deoxyribonucleic acid by the antitumor antibiotics saframycins A and C" in *Biochemistry* **1982**, *21*, 419–428.
2. Nicolaou, K. C.; Snyder, S. A. *Classics in Total Synthesis II.* Wiley-VCH: Weinheim, **2003**, pp. 109–136 and references therein.

SUMMARY AND REVIEW TOOLS

The study aids for this chapter include key terms and concepts (which are hyperlinked to the Glossary from the bold, blue terms in the *WileyPLUS* version of the book at wileyplus.com), Mechanism Summaries regarding reactions of aldehydes and ketones with amines as well as with other nucleophiles, and a Synthetic Connections scheme regarding transformations of aldehydes and ketones.

PROBLEMS WILEY PLUS

Note to Instructors: Many of the homework problems are available for assignment via *WileyPLUS*, an online teaching and learning solution.

REACTIONS AND NOMENCLATURE

16.22 Give a structural formula and another acceptable name for each of the following compounds:

(a) Formaldehyde
(b) Acetaldehyde
(c) Phenylacetaldehyde
(d) Acetone
(e) Ethyl methyl ketone

(f) Acetophenone
(g) Benzophenone
(h) Salicylaldehyde
(i) Vanillin
(j) Diethyl ketone

(k) Ethyl isopropyl ketone
(l) Diisopropyl ketone
(m) Dibutyl ketone
(n) Dipropyl ketone
(o) Cinnamaldehyde

16.23 Write structural formulas for the products formed when propanal reacts with each of the following reagents:

(a) $NaBH_4$ in aqueous NaOH
(b) C_6H_5MgBr, then H_3O^+
(c) $LiAlH_4$, then H_2O
(d) Ag_2O, HO^-
(e) $(C_6H_5)_3\overset{+}{P}-\overset{..}{C}H_2$
(f) H_2 and Pt
(g) $HO\diagup\diagdown OH$ and HA
(h) $CH_3\overset{..}{C}-\overset{+}{P}(C_6H_5)_3$

(i) $Ag(NH_3)_2^+$
(j) Hydroxylamine
(k) Phenylhydrazine
(l) Cold dilute $KMnO_4$
(m) $HS\diagup\diagdown\diagup^{SH}$, HA
(n) $HS\diagup\diagdown\diagup^{SH}$, HA, then Raney nickel
(o) *m*CPBA

16.24 Give structural formulas for the products formed (if any) from the reaction of acetone with each reagent in Exercise 16.23.

16.25 What products would be obtained when acetophenone reacts under each of the following conditions?

Acetophenone

(a) $\xrightarrow[H_2SO_4]{HNO_3}$
(b) $\xrightarrow{C_6H_5NHNH_2,\ HA}$
(c) $\xrightarrow{H_2\overset{..}{C}-\overset{+}{P}(C_6H_5)_3}$
(d) $\xrightarrow[CH_3OH]{NaBH_4}$

(e) $\xrightarrow[(2)\ NH_4Cl]{(1)\ C_6H_5MgBr}$
(f) $\xrightarrow[\Delta]{NH_2NH_2,\ HO^-}$
(g) $\xrightarrow{mCPBA}$

16.26 Predict the major organic product from each of the following reactions.

(a)
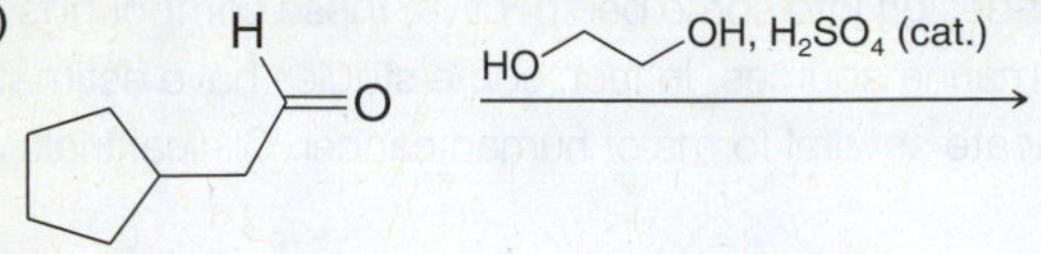
H
O
HO
OH, H_2SO_4 (cat.)

(b)
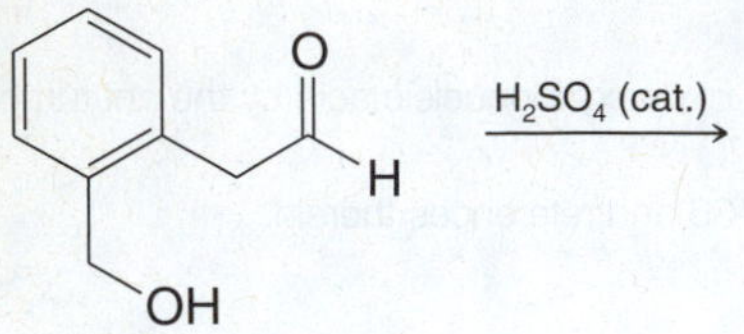
O
H
OH
H_2SO_4 (cat.)

(c)
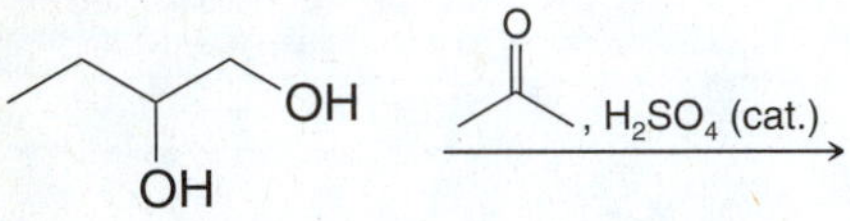
OH
OH
O
, H_2SO_4 (cat.)

(d)
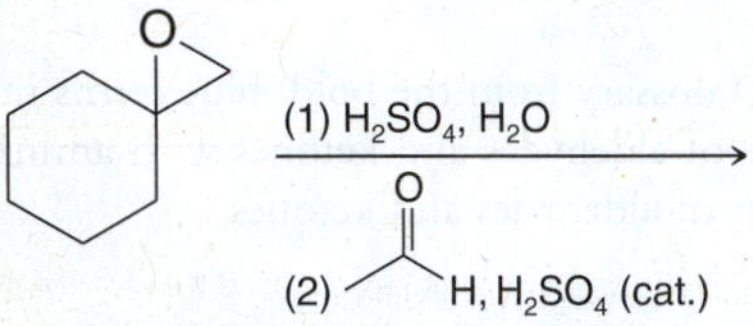
O
(1) H_2SO_4, H_2O
(2)
O
H, H_2SO_4 (cat.)

(e)
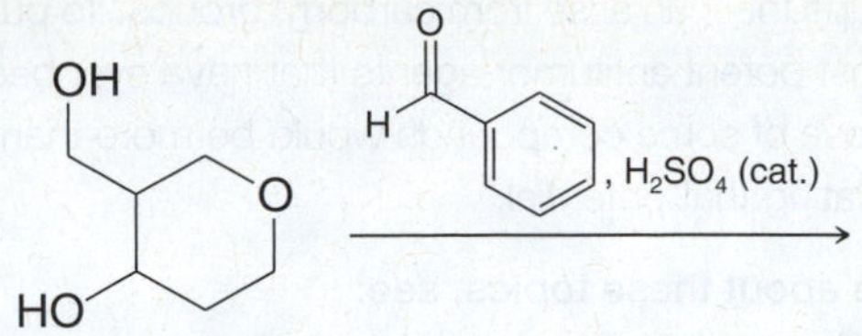
OH
O
HO
O
H
, H_2SO_4 (cat.)

(f)
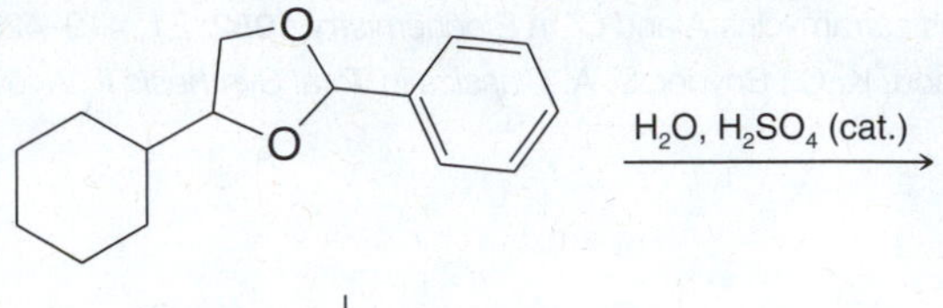
O
O
H_2O, H_2SO_4 (cat.)

(g)
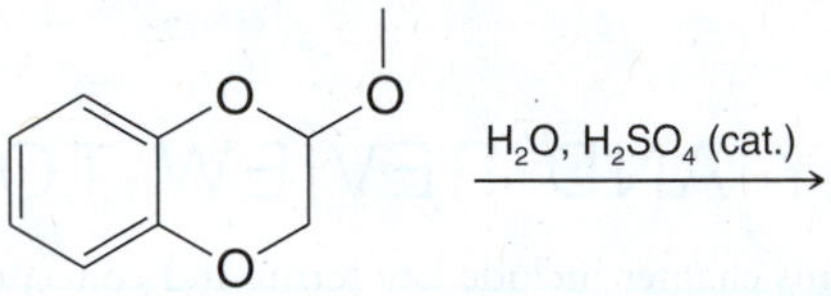
O
O
O
H_2O, H_2SO_4 (cat.)

16.27 Predict the major product from each of the following reactions.

(a)
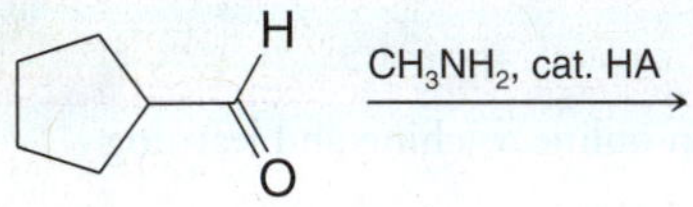
H
O
CH_3NH_2, cat. HA

(b)
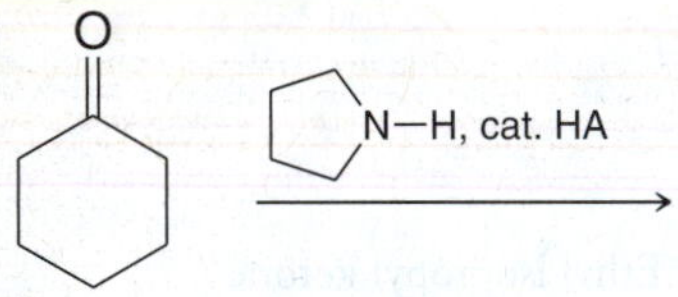
O
N–H, cat. HA

(c)
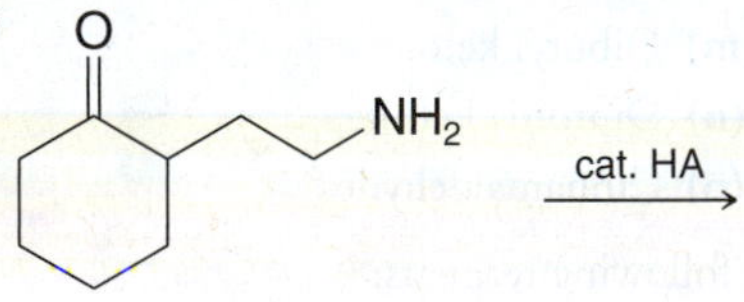
O
NH_2
cat. HA

(d)
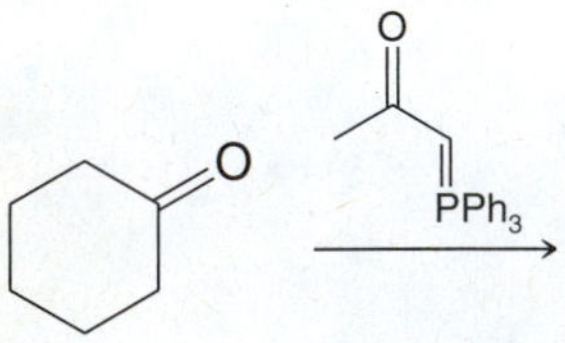
O
O
PPh_3

(e)
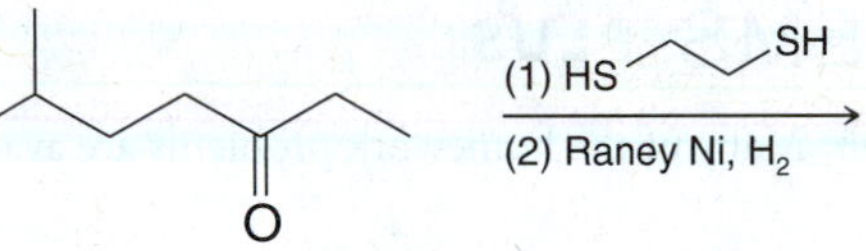
O
(1) HS
SH
(2) Raney Ni, H_2

(f)
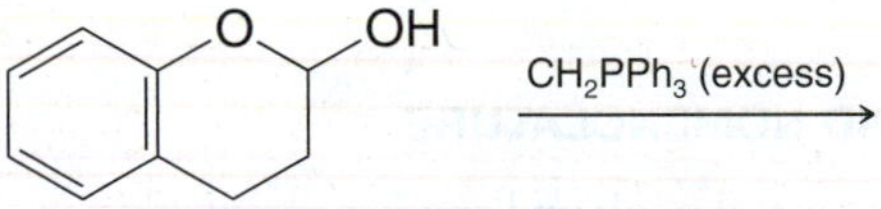
O
OH
CH_2PPh_3 (excess)

(g)
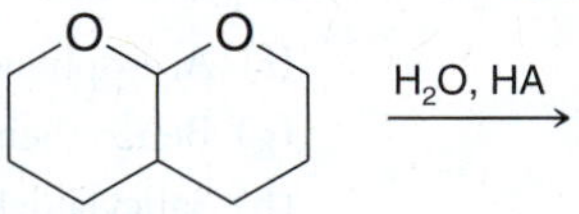
O
O
H_2O, HA

(h)
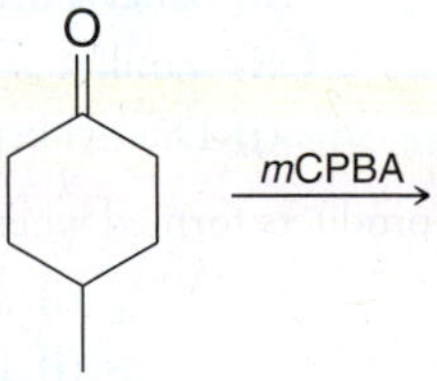
O
*m*CPBA

(i)
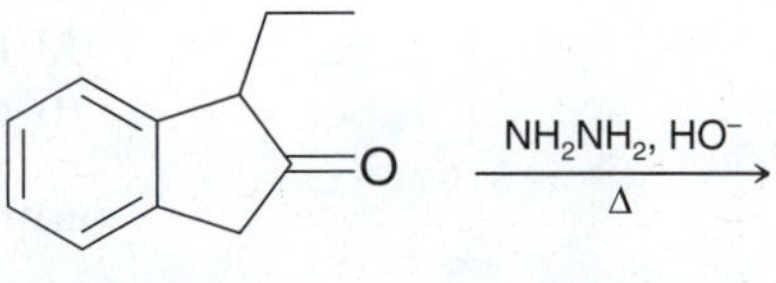
O
NH_2NH_2, HO^-
Δ

16.28 Predict the major product from each of the following reactions.

(a)
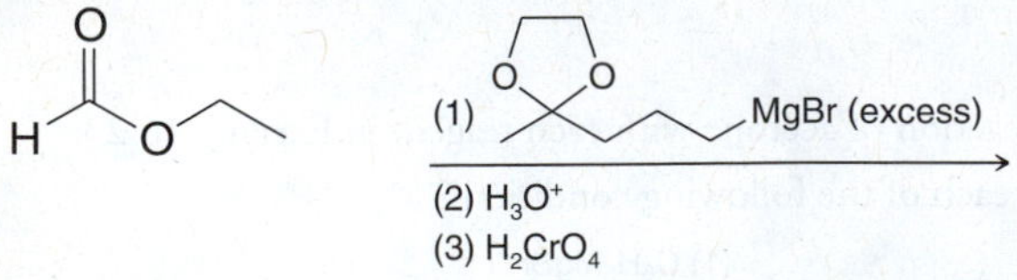
O
H
O
(1)
O
O
MgBr (excess)
(2) H_3O^+
(3) H_2CrO_4

(b)
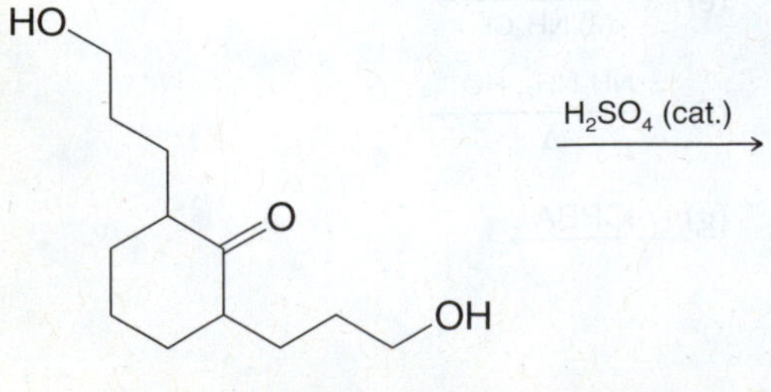
HO
O
OH
H_2SO_4 (cat.)

(c)
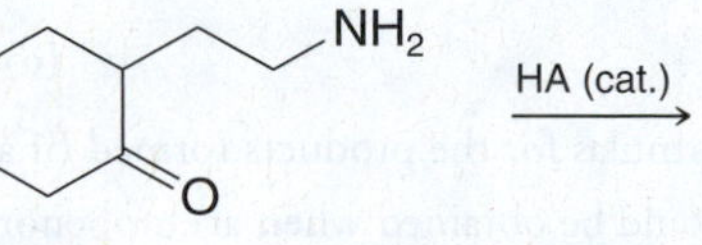
NH_2
O
HA (cat.)

(d)
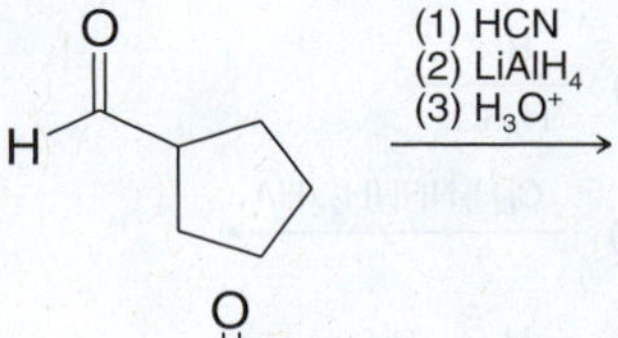
O
H
(1) HCN
(2) $LiAlH_4$
(3) H_3O^+

(e)
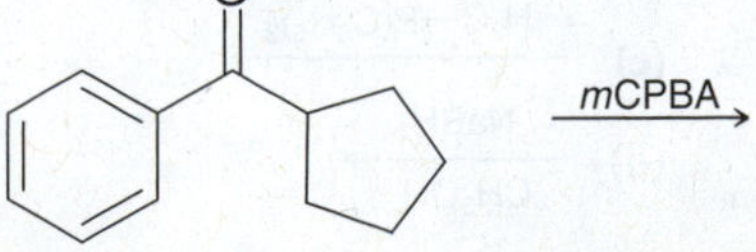
O
*m*CPBA

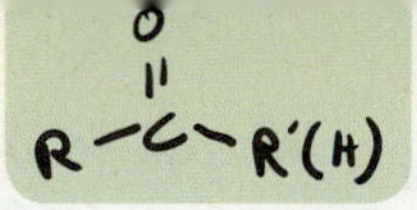

16.29 Provide the reagents needed to accomplish each of the following transformations.

(a)

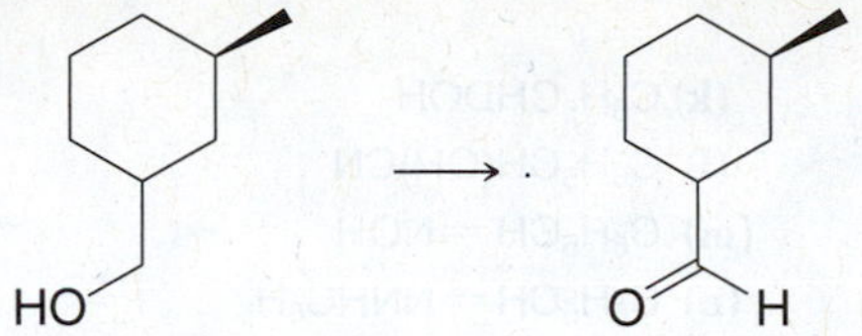

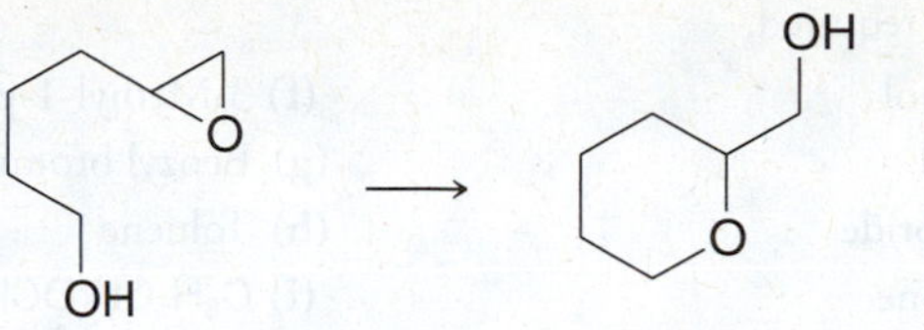

(b)

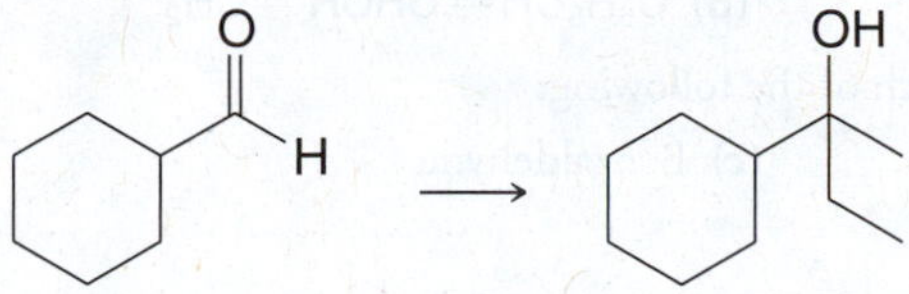

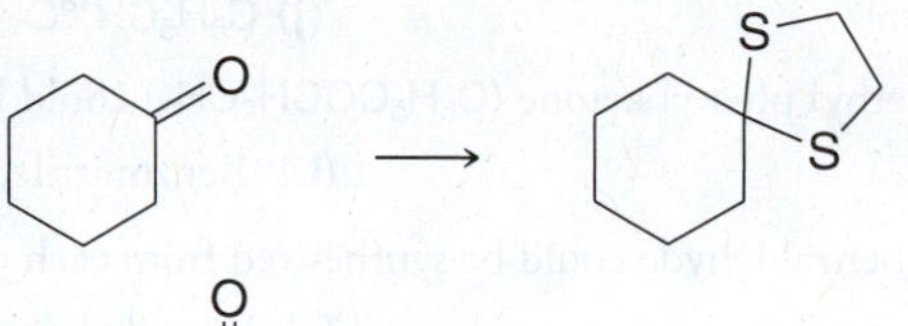

(c)

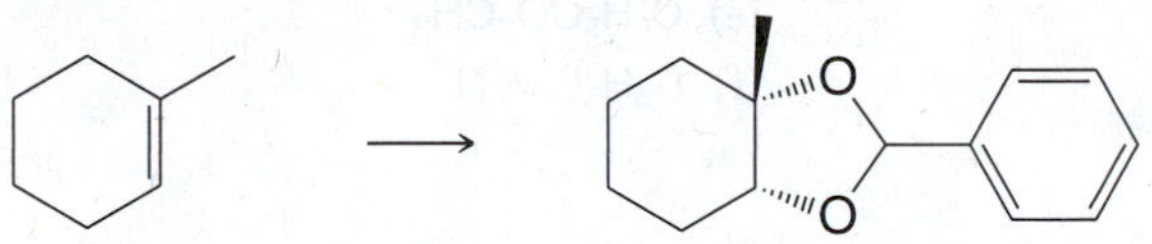

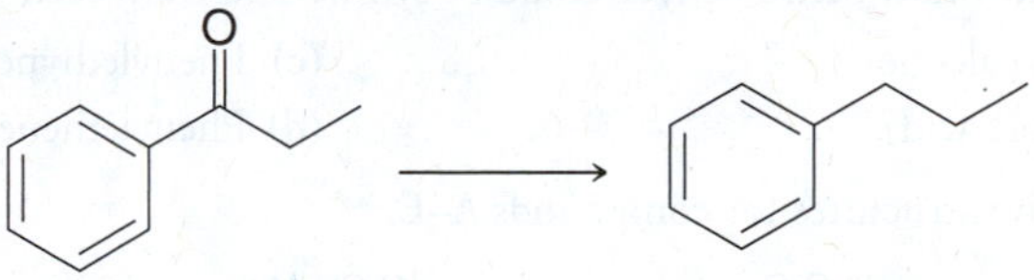

(d)

(e)

(f)

(g)

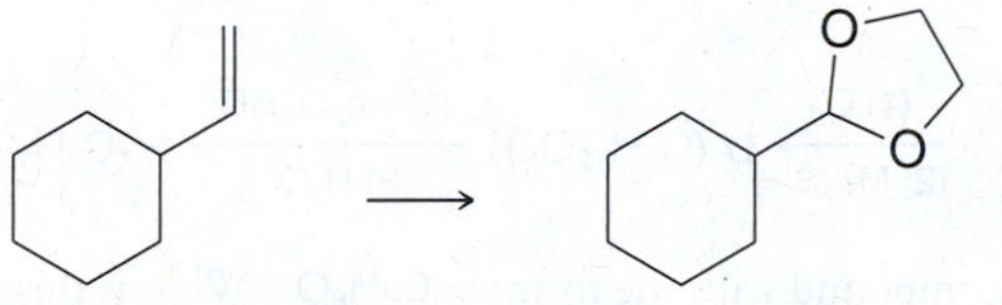

(h)

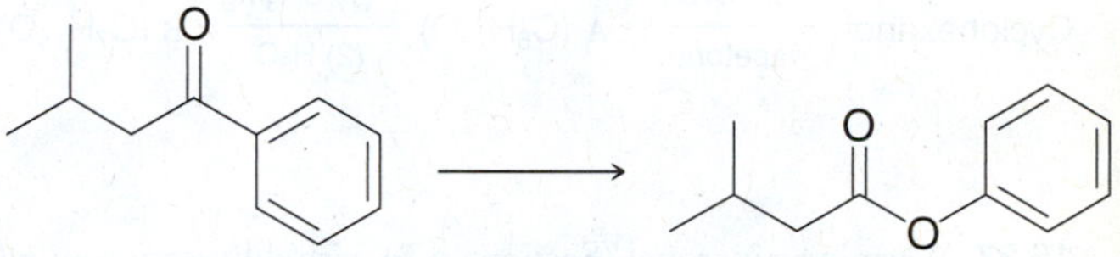

16.30 Write detailed mechanisms for each of the following reactions.

(a)

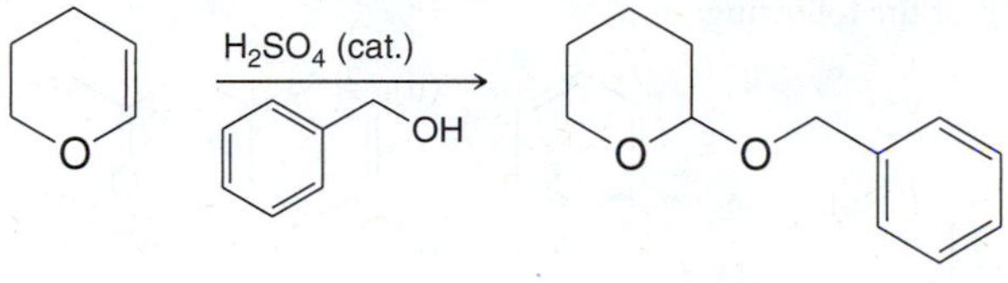

H_2SO_4 (cat.)

(b)

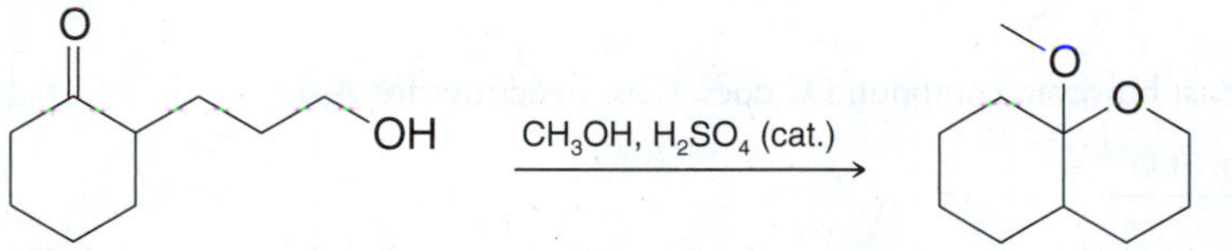

CH_3OH, H_2SO_4 (cat.)

(c)

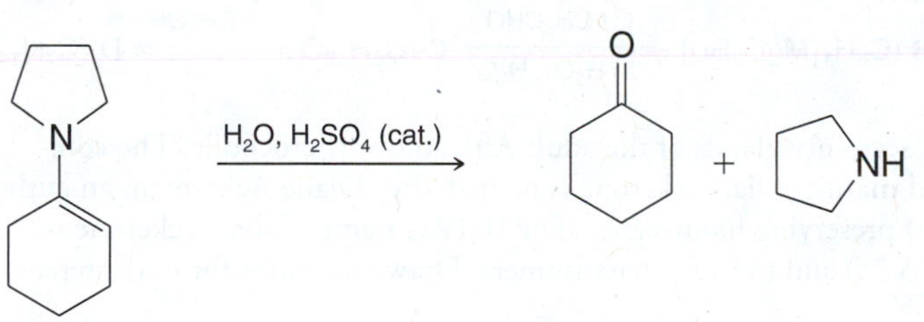

H_2O, H_2SO_4 (cat.)

(d)

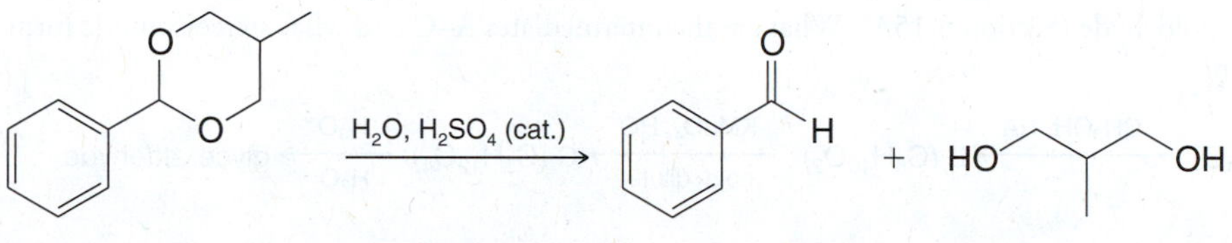

H_2O, H_2SO_4 (cat.)

16.31 Provide the reagents necessary for the following synthesis.

Start with

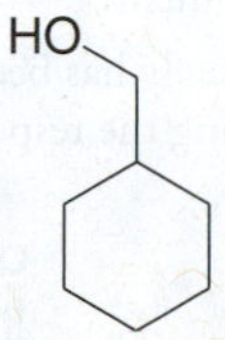

and make

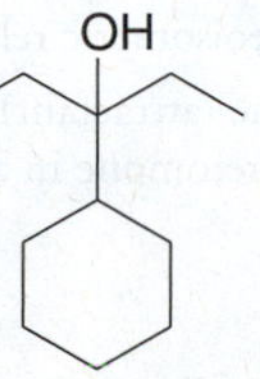

SYNTHESIS

16.32

(a) Synthesize phenyl propyl ketone from benzene and any other needed reagents.

(b) Give two methods for transforming phenyl propyl ketone into butylbenzene.

16.33 Show how you would convert benzaldehyde into each of the following. You may use any other needed reagents, and more than one step may be required.

(a) Benzyl alcohol

(b) Benzoic acid

(c) Benzoyl chloride

(d) Benzophenone

(e) 1-Phenylethanol

(f) 3-Methyl-1-phenyl-1-butanol

(g) Benzyl bromide

(h) Toluene

(i) $C_6H_5CH(OCH_3)_2$

(j) $C_6H_5CH^{18}O$

(k) C_6H_5CHDOH

(l) $C_6H_5CH(OH)CN$

(m) $C_6H_5CH=NOH$

(n) $C_6H_5CH=NNHC_6H_5$

(o) $C_6H_5CH=CHCH=CH_2$

16.34 Show how ethyl phenyl ketone ($C_6H_5COCH_2CH_3$) could be synthesized from each of the following:

(a) Benzene

(b) Benzonitrile, C_6H_5CN

(c) Benzaldehyde

16.35 Show how benzaldehyde could be synthesized from each of the following:

(a) Benzyl alcohol

(b) Benzoic acid

(c) Phenylethyne

(d) Phenylethene (styrene)

(e) $C_6H_5CO_2CH_3$

(f) $C_6H_5C\equiv N$

16.36 Give structures for compounds **A–E**.

Cyclohexanol $\xrightarrow[\text{acetone}]{H_2CrO_4}$ **A** ($C_6H_{10}O$) $\xrightarrow[\text{(2) }H_3O^+]{\text{(1) }CH_3MgI}$ **B** ($C_7H_{14}O$) $\xrightarrow[\text{heat}]{HA}$

C (C_7H_{12}) $\xrightarrow[\text{(2) }Me_2S]{\text{(1) }O_3}$ **D** ($C_7H_{12}O_2$) $\xrightarrow[\text{(2) }H_3O^+]{\text{(1) }Ag_2O,\ HO^-}$ **E** ($C_7H_{12}O_3$)

16.37 Warming piperonal (Section 16.3) with dilute aqueous HCl converts it to a compound with the formula $C_7H_6O_3$. What is this compound, and what type of reaction is involved?

16.38 Starting with benzyl bromide, show how you would synthesize each of the following:

(a)

(b)

(c)

(d)

16.39 Compounds **A** and **D** do not give positive Tollens' tests; however, compound **C** does. Give structures for **A–D**.

4-Bromobutanal $\xrightarrow[]{HOCH_2CH_2OH,\ HA}$ **A** ($C_6H_{11}O_2Br$) $\xrightarrow[]{Mg,\ Et_2O}$

[**B** ($C_6H_{11}MgO_2Br$)] $\xrightarrow[\text{(2) }H_3O^+,\ H_2O]{\text{(1) }CH_3CHO}$ **C** ($C_6H_{12}O_2$) $\xrightarrow[\text{HA}]{CH_3OH}$ **D** ($C_7H_{14}O_2$)

16.40 Dianeackerone is a volatile natural product isolated from secretory glands of the adult African dwarf crocodile. The compound is believed to be a pheromone associated with nesting and mating. Dianeackerone is named after Diane Ackerman, an author in the field of natural history and champion of the importance of preserving biodiversity. The IUPAC name of dianeackerone is 3,7-diethyl-9-phenylnonan-2-one, and it is found as both the (3*S*,7*S*) and (3*S*,7*R*) stereoisomers. Draw structures for both stereoisomers of dianeackerone.

16.41 Outlined here is a synthesis of glyceraldehyde (Section 5.15A). What are the intermediates **A–C** and what stereoisomeric form of glyceraldehyde would you expect to obtain?

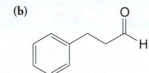 $\xrightarrow[CH_2Cl_2]{PCC}$ **A** (C_3H_4O) $\xrightarrow[]{CH_3OH,\ HA}$ **B** ($C_5H_{10}O_2$) $\xrightarrow[\text{cold, dilute}]{KMnO_4,\ HO^-}$ **C** ($C_5H_{12}O_4$) $\xrightarrow[H_2O]{H_3O^+}$ glyceraldehyde

16.42 Consider the reduction of (*R*)-3-phenyl-2-pentanone by sodium borohydride. After the reduction is complete, the mixture is separated by chromatography into two fractions. These fractions contain isomeric compounds, and each isomer is optically active. What are these two isomers and what is the stereoisomeric relationship between them?

16.43 The structure of the sex pheromone (attractant) of the female tsetse fly has been confirmed by the following synthesis. Compound **C** appears to be identical to the natural pheromone in all respects (including the response of the male tsetse fly). Provide structures for **A, B,** and **C**.

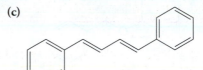 $\xrightarrow[\text{(2) 2 RLi}]{\text{(1) 2 }(C_6H_5)_3P}$ **A** ($C_{45}H_{46}P_2$) 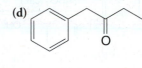 **B** ($C_{37}H_{72}$) $\xrightarrow[]{H_2,\ Pt}$ **C** ($C_{37}H_{76}$)

16.44 Provide reagents that would accomplish each of the following syntheses. Begin by writing a retrosynthetic analysis.

(a)

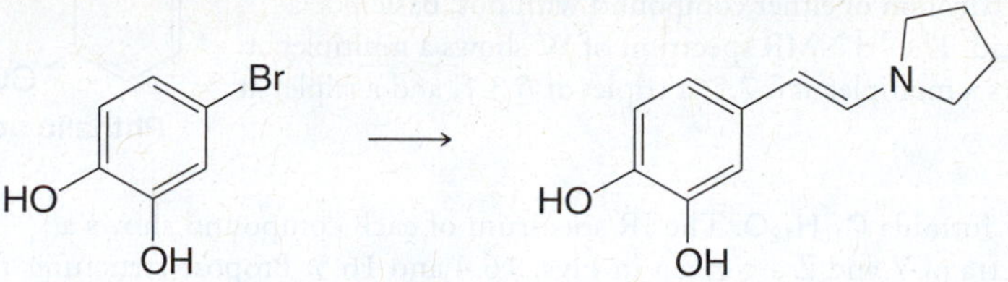

(b)

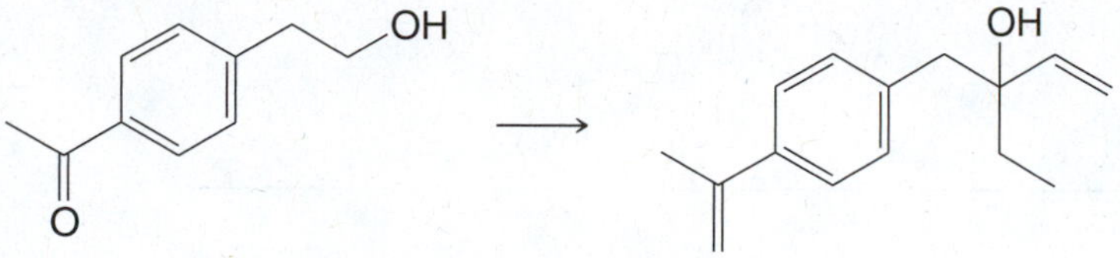

MECHANISMS AND STRUCTURE ELUCIDATION

16.45 Write a detailed mechanism for the following reaction.

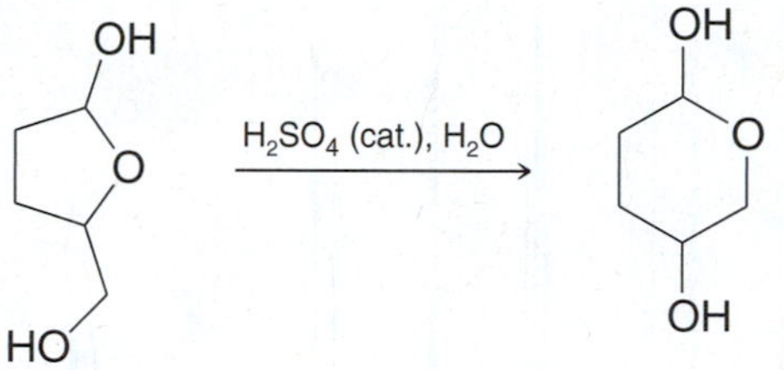

16.46 When H_2N—(C=O)—$NHNH_2$ (semicarbazide) reacts with a ketone (or an aldehyde) to form a derivative known as a semicarbazone, only one nitrogen atom of semicarbazide acts as a nucleophile and attacks the carbonyl carbon atom of the ketone. The product of

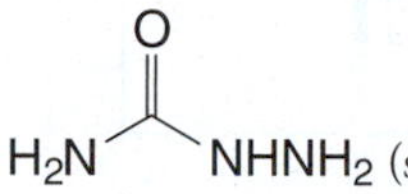

the reaction, consequently, is 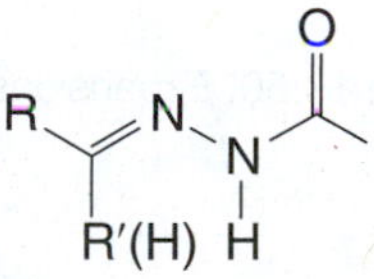 rather than 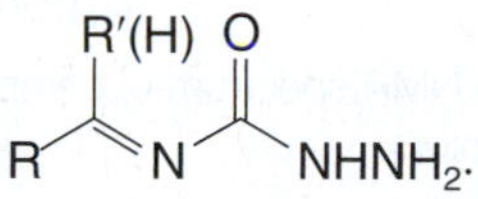 . What factor accounts for the fact that two

nitrogen atoms of semicarbazide are relatively non-nucleophilic?

16.47 Dutch elm disease is caused by a fungus transmitted to elm trees by the elm bark beetle. The female beetle, when she has located an attractive elm tree, releases several pheromones, including multistriatin, below. These pheromones attract male beetles, which bring with them the deadly fungus.

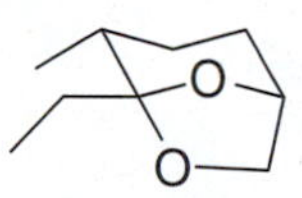

Multistriatin

Treating multistriatin with dilute aqueous acid at room temperature leads to the formation of a product, $C_{10}H_{20}O_3$, which shows a strong infrared peak near 1715 cm^{-1}. Propose a structure for this product.

16.48 The following structure is an intermediate in a synthesis of prostaglandins $F_{2\alpha}$ and E_2 by E. J. Corey (Harvard University). A Horner–Wadsworth–Emmons reaction was used to form the (E)-alkene. Write structures for the phosphonate ester and carbonyl reactant that were used in this process. (*Note*: The carbonyl component of the reaction included the cyclopentyl group.)

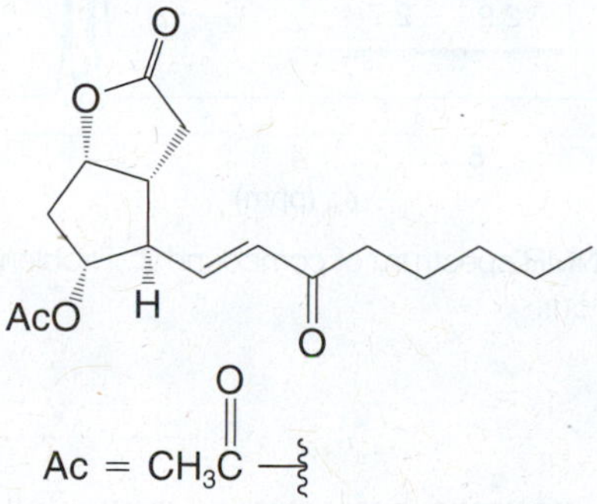

$$Ac = CH_3\overset{O}{\underset{||}{C}} —\xi$$

16.49 Compounds **W** and **X** are isomers; they have the molecular formula C_9H_8O. The IR spectrum of each compound shows a strong absorption band near 1715 cm^{-1}. Oxidation of either compound with hot, basic potassium permanganate followed by acidification yields phthalic acid. The ^1H NMR spectrum of **W** shows a multiplet at δ 7.3 and a singlet at δ 3.4. The ^1H NMR spectrum of **X** shows a multiplet at δ 7.5, a triplet at δ 3.1, and a triplet at δ 2.5. Propose structures for **W** and **X**.

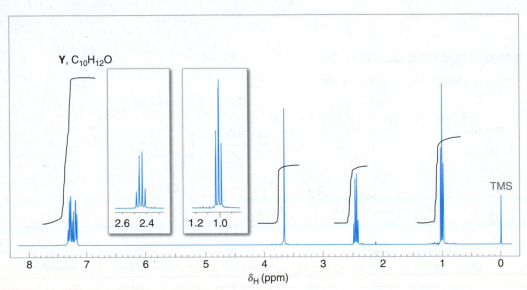

Phthalic acid

16.50 Compounds **Y** and **Z** are isomers with the molecular formula $C_{10}H_{12}O$. The IR spectrum of each compound shows a strong absorption band near 1710 cm^{-1}. The ^1H NMR spectra of **Y** and **Z** are given in Figs. 16.4 and 16.5. Propose structures for **Y** and **Z**.

FIGURE 16.4 The 300-MHz ^1H NMR spectrum of compound **Y**, Problem 16.50. Expansions of the signals are shown in the offset plots.

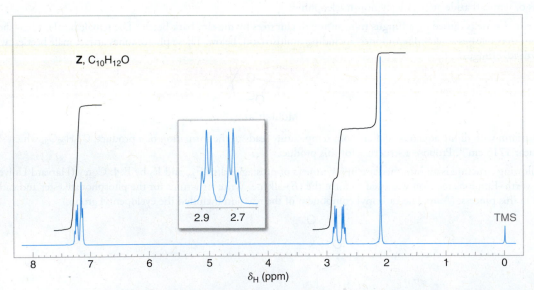

FIGURE 16.5 The 300-MHz ^1H NMR spectrum of compound **Z**, Problem 16.50. Expansions of the signals are shown in the offset plots.

16.51 Compound **A** ($C_9H_{18}O$) forms a phenylhydrazone, but it gives a negative Tollens' test. The IR spectrum of **A** has a strong band near 1710 cm^{-1}. The broadband proton-decoupled ^{13}C NMR spectrum of **A** is given in Fig. 16.6. Propose a structure for **A**.

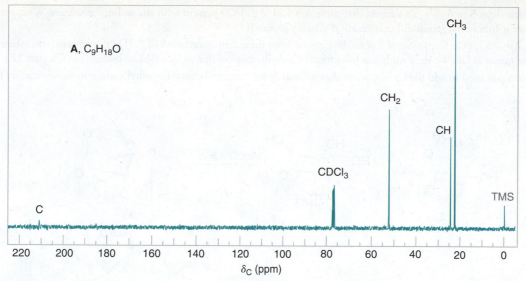

FIGURE 16.6 The broadband proton-decoupled ^{13}C NMR spectrum of compound **A**, Problem 16.51. Information from the DEPT ^{13}C NMR spectra is given above the peaks.

16.52 Compound **B** ($C_8H_{12}O_2$) shows a strong carbonyl absorption in its IR spectrum. The broadband proton-decoupled ^{13}C NMR spectrum of **B** has only three signals, at δ 19 (CH$_3$), 71 (C), and 216 (C). Propose a structure for **B**.

CHALLENGE PROBLEMS

16.53 **(a)** What would be the frequencies of the two absorption bands expected to be most prominent in the infrared spectrum of 4-hydroxycycloheptanone (**C**)?
(b) In reality, the lower frequency band of these two is very weak. Draw the structure of an isomer that would exist in equilibrium with **C** and that explains this observation.

16.54 One of the important reactions of benzylic alcohols, ethers, and esters is the ease of cleavage of the benzyl–oxygen bond during hydrogenation. This is another example of "hydrogenolysis," the cleavage of a bond by hydrogen. It is facilitated by the presence of acid. Hydrogenolysis can also occur with strained-ring compounds.

On hydrogenation of compound **D** (see below) using Raney nickel catalyst in a dilute solution of hydrogen chloride in dioxane and water, most products have a 3,4-dimethoxyphenyl group attached to a side chain. Among these, an interesting product is **E**, whose formation illustrates not only hydrogenolysis but also the migratory aptitude of phenyl groups. For product **E**, these are key spectral data:

> **MS** (*m/z*): 196.1084 (M^{+}, at high resolution), 178
>
> **IR** (cm^{-1}): 3400 (broad), 3050, 2850 (CH$_3$—O stretch)
>
> ^{1}**H NMR** (δ, in CDCl$_3$): 1.21 (d, 3H, $J=7$ Hz), 2.25 (s, 1H), 2.83 (m, 1H), 3.58 (d, 2H, $J=7$ Hz), 3.82 (s, 6H), 6.70 (s, 3H).
>
> What is the structure of compound **E**?

LEARNING GROUP PROBLEMS

A synthesis of ascorbic acid (vitamin C, **1**) starting from D-(+)-galactose (**2**) is shown below (Haworth, W. N., et al., *J. Chem. Soc.*, **1933**, 1419–1423). Consider the following questions about the design and reactions used in this synthesis:

(a) Why did Haworth and co-workers introduce the acetal functional groups in **3**?
(b) Write a mechanism for the formation of one of the acetals.
(c) Write a mechanism for the hydrolysis of one of the acetals (**4** to **5**). Assume that water was present in the reaction mixture.

(d) In the reaction from **5** to **6** you can assume that there was acid (e.g., HCl) present with the sodium amalgam. What reaction occurred here and from what functional group did that reaction actually proceed?

(e) Write a mechanism for the formation of a phenylhydrazone from the aldehyde carbonyl of **7**. [Do not be concerned about the phenylhydrazone group at C2. We shall study the formation of bishydrazones of this type (called an osazone) in Chapter 22.]

(f) What reaction was used to add the carbon atom that ultimately became the lactone carbonyl carbon in ascorbic acid (**1**)?

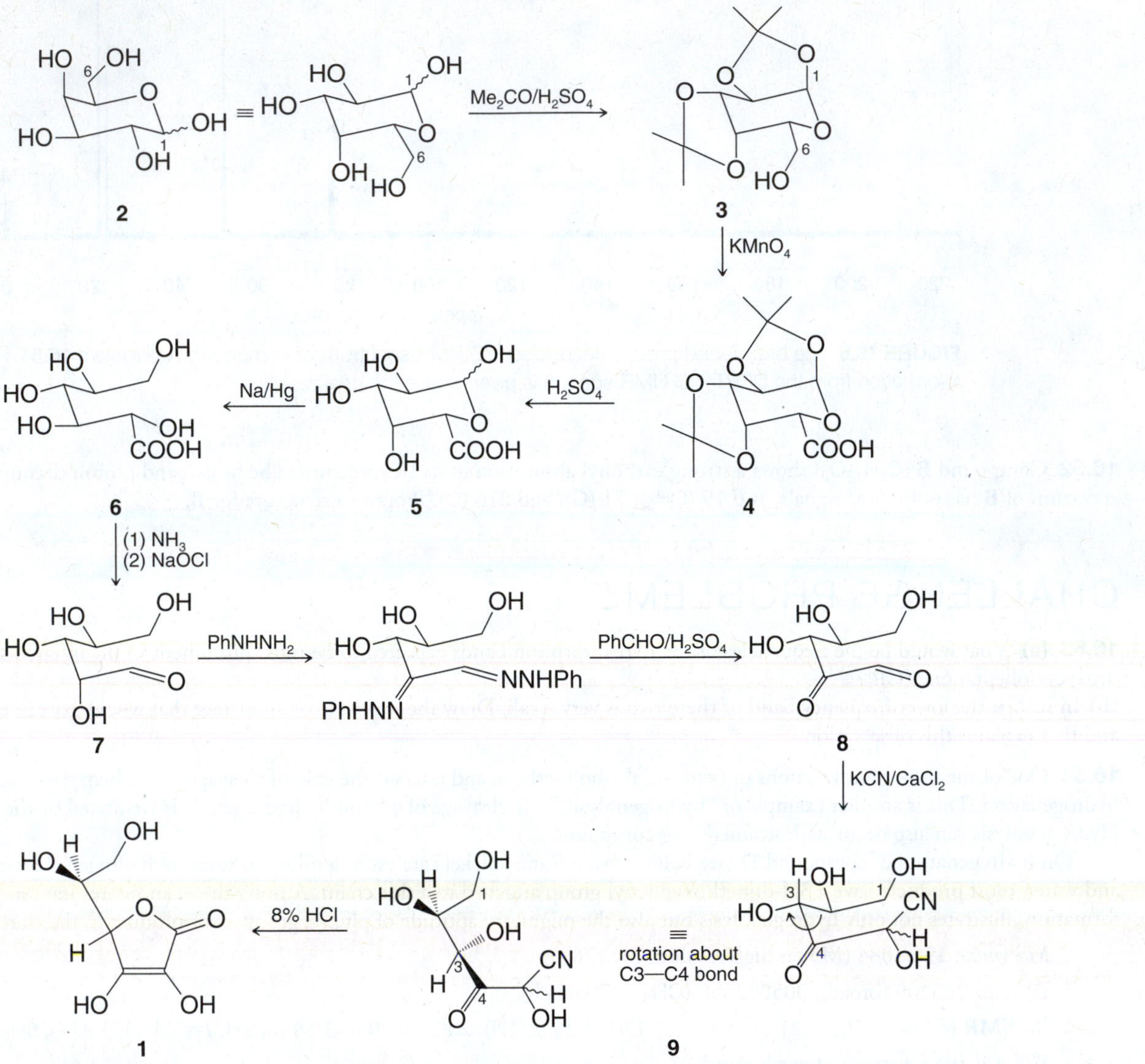

[SUMMARY OF MECHANISMS]

Acetals, Imines, and Enamines: Common Mechanistic Themes in Their Acid-catalyzed Formation from Aldehydes and Ketones

Many steps are nearly the same in acid-catalyzed reactions of aldehydes and ketones with alcohols and amines. Compare the mechanisms vertically to see the similarities and differences. Note differences in completion of the mechanism for each type of product.

I. Hemiacetal and acetal formation: reaction with alcohols

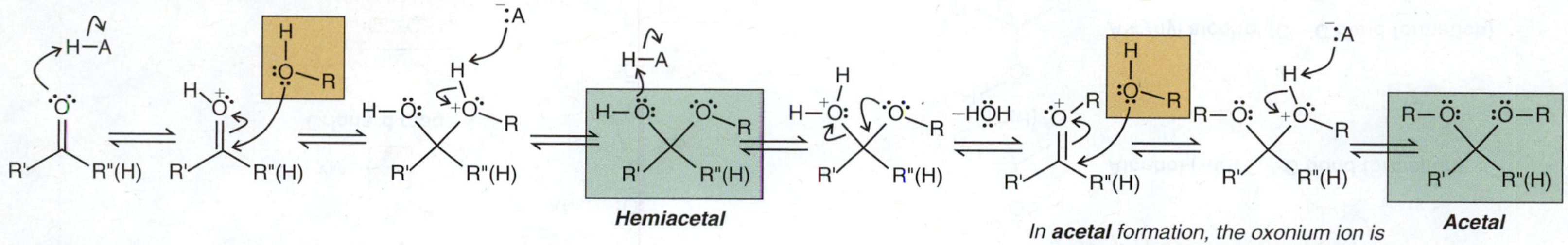

Acetal

Hemiacetal

In **acetal** formation, the oxonium ion is attacked by a second alcohol molecule.

II. Imine formation: reaction with primary amines

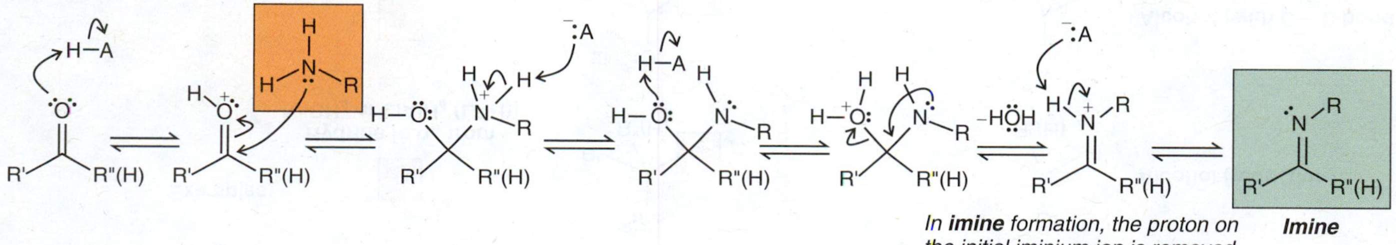

Imine

In **imine** formation, the proton on the initial iminium ion is removed, leading to the stable imine product.

III. Enamine formation: reaction with secondary amines

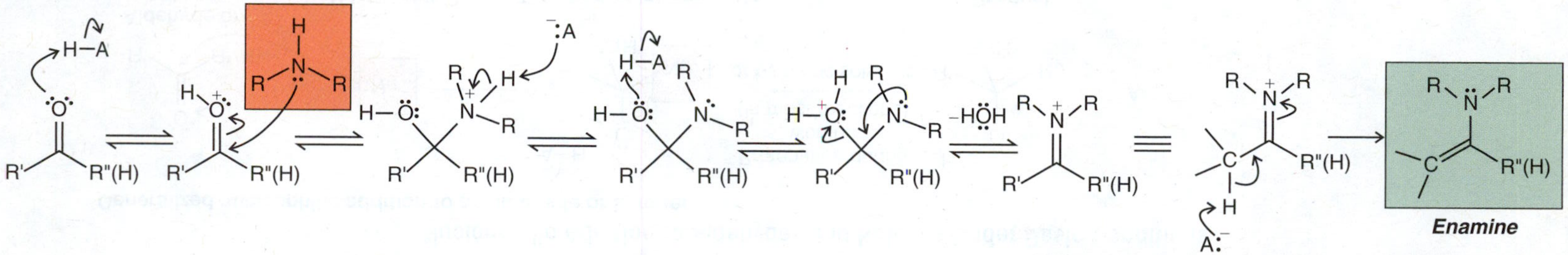

Enamine

In **enamine** formation, a proton is removed from a carbon adjacent to the iminium carbon (because no proton is available for removal from the nitrogen).

SUMMARY OF MECHANISMS

Nucleophilic Addition to Aldehydes and Ketones Under Basic Conditions

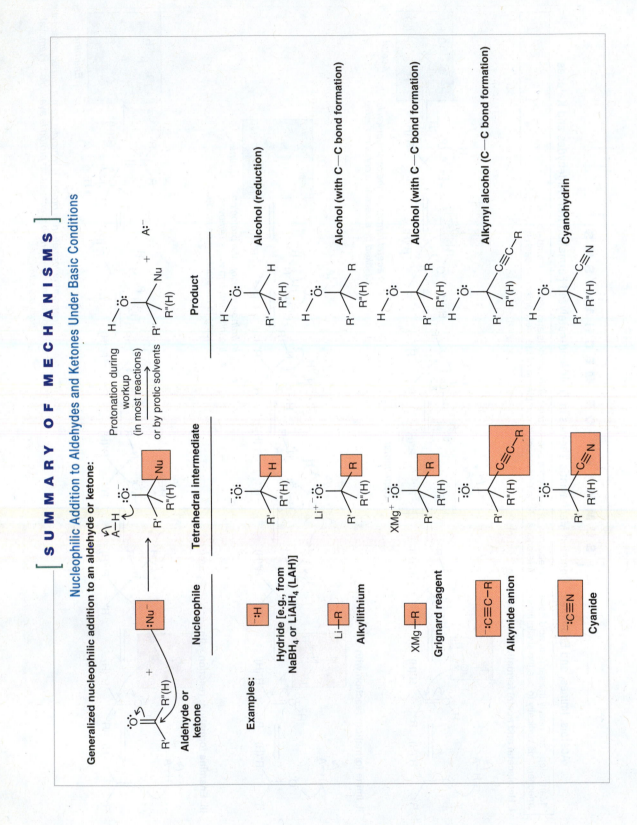

Generalized nucleophilic addition to an aldehyde or ketone:

Protonation during workup (in most reactions) or by protic solvents

Nucleophile	Tetrahedral intermediate	Product

Examples:

Hydride [e.g., from NaBH₄ or LiAlH₄ (LAH)]

Alkyllithium

Grignard reagent

Alkynide anion

Cyanide

Alcohol (reduction)

Alcohol (with C—C bond formation)

Alcohol (with C—C bond formation)

Alkynyl alcohol (C—C bond formation)

Cyanohydrin

SUMMARY OF MECHANISMS

Nucleophilic Addition to Aldehydes and Ketones Under Basic Conditions

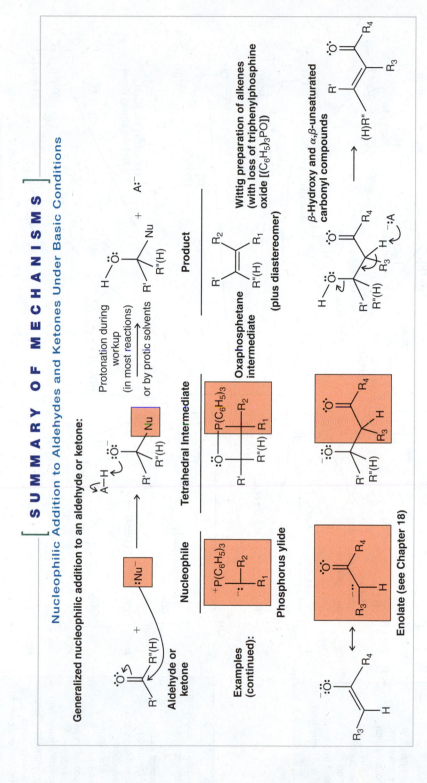

Generalized nucleophilic addition to an aldehyde or ketone:

Aldehyde or ketone + Nucleophile → Tetrahedral Intermediate → Product

Protonation during workup (in most reactions) or by protic solvents

Examples (continued):

Phosphorus ylide → Oxaphosphetane intermediate → Product (plus diastereomer)

Wittig preparation of alkenes (with loss of triphenylphosphine oxide [$(C_6H_5)_3PO$])

Enolate (see Chapter 18)

β-Hydroxy and α,β-unsaturated carbonyl compounds

[**S Y N T H E T I C C O N N E C T I O N S**]

Some Synthetic Connections of Aldehydes, Ketones, and Other Functional Groups

ROH, cat. HA $\rightleftharpoons$ H₂O, cat. HA

$\xrightarrow{\text{H}_2, \text{Raney Ni}}$

$\begin{array}{c}\text{ROH,}\\\text{cat. HA}\end{array}$

$\begin{array}{c}\text{H}_2\text{O,}\\\text{cat. HA}\end{array}$

RSH (2 equiv.), cat. HA

$\begin{array}{c}\text{H}_2\text{O,}\\\text{cat. HA}\end{array}$

NH₂NH₂, HO⁻ Δ

(1) RC≡C⁻
(2) H₃O⁺

(1) NC⁻
(2) H₃O⁺

(1) CHR³R⁴Br, (C₆H₅)₃P
(2) RLi (as strong base)

(E) and (Z)

(when R′ bore a hydrogen for removal)

R′—C≡N

(1) R″MgBr or R″Li
(2) H₃O⁺

R₂NH, cat. HA

RNH₂, cat. HA

RCO₃H

(1) ArH, AlCl₃;
(2) HOH
(leads to R″ = Ar)

(1) O₃
(2) Me₂S
(products depend on R groups)

H₂CrO₄ or KMnO₄

NaBH₄; or (1) LAH; (2) H₂O

PCC

(1) LiAlH(O-t-Bu)₃
(2) H₂O

(1) DIBAL-H
(2) H₂O

SOCl₂

(1) DIBAL-H
(2) H₂O

R′—C≡N

Clockwise from center, bottom:

I. Preparation of aldehydes and ketones:

- Nitrile, ester, acyl halide reduction
- Alcohol oxidation
- Ozonolysis
- Friedel–Crafts acylation
- Grignard with nitrile
- Acetal and hemiacetal hydrolysis

II. Reactions of aldehydes and ketones:

- Hemiacetal and acetal formation
- Thioacetal formation and reduction
- Wolff–Kishner reduction
- Alkynide anion addition
- Nitrile addition (cyanohydrin formation)
- Wittig synthesis of alkenes
- Enamine synthesis
- Baeyer–Villiger oxidation
- Imine synthesis
- Reduction to alcohols (left, center)

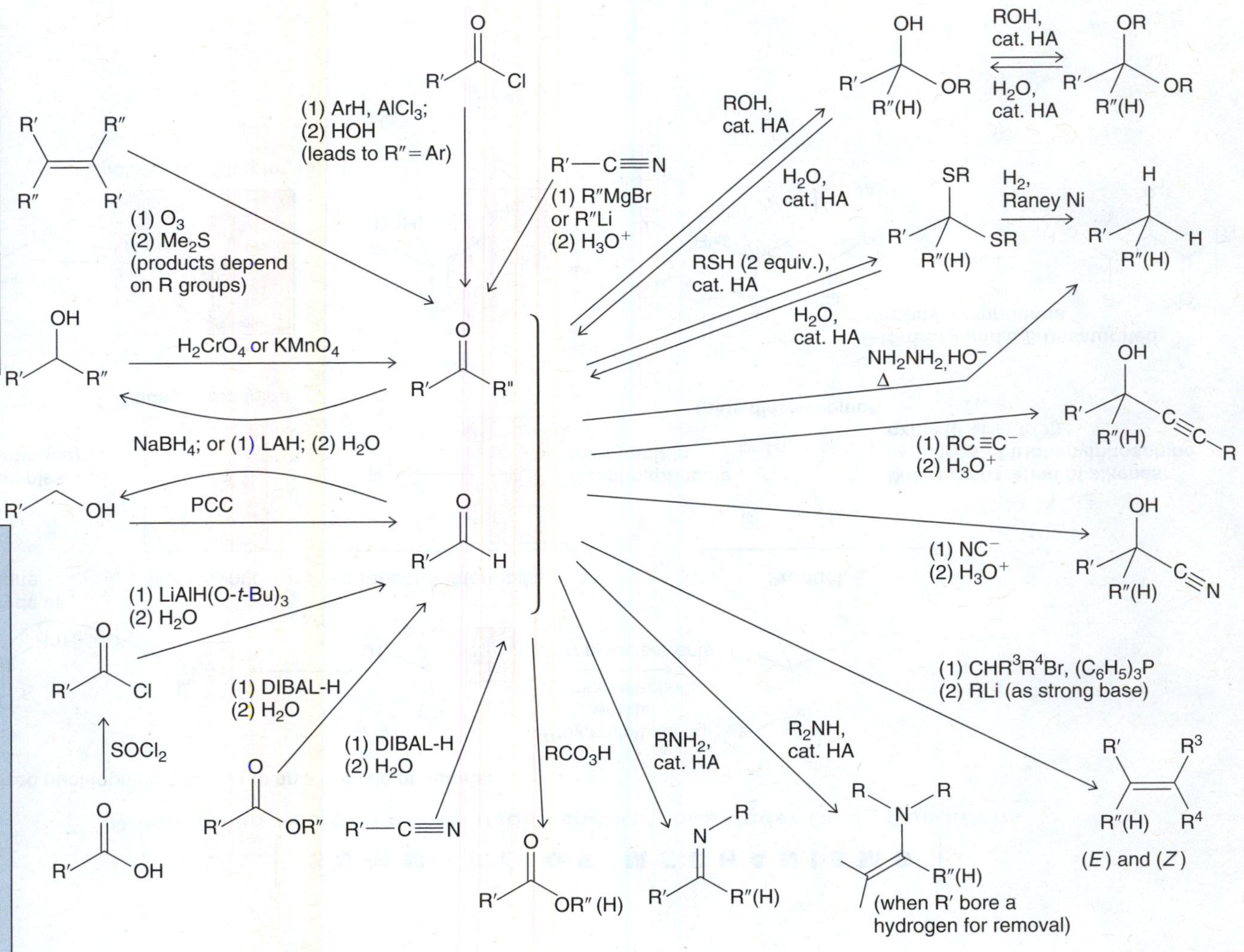

Carboxylic Acids and Their Derivatives

NUCLEOPHILIC ADDITION–ELIMINATION AT THE ACYL CARBON

Although there are many different derivatives of carboxylic acids, variations that can account for millions of distinct organic molecules, the vast majority can arise via a common and mechanistically consistent bond-formation process. This event is known as nucleophilic acyl substitution, and it involves the creation of a new bond by a nucleophilic addition and elimination at a carbonyl group. This process is utilized industrially in the synthesis of complex polymers, such as nylon and polyesters (see Special Topic C in *WileyPLUS*). It also occurs in metabolism, in the synthesis of proteins, fats, and steroid precursors, as well as in the breakdown of food for energy and for other biosynthetic raw materials (see Special Topic E in *WileyPLUS*). Its versatility is truly amazing.

IN THIS CHAPTER WE WILL CONSIDER:

- the structure and reactivity of various carboxylic acid derivatives
- many different examples of nucleophilic acyl substitutions, all of which proceed by a similar mechanism though they lead to different products
- methods for the preparation of carboxylic acid derivatives from other functional groups, such as nitriles

[WHY DO THESE TOPICS MATTER?]   At the end of the chapter, we will show you how a key problem in chemical synthesis requiring a nucleophilic acyl substitution—the laboratory preparation of the penicillins—served as inspiration for the development of a powerful class of reagents that has enabled the facile synthesis of amide bonds in many contexts.

PHOTO CREDIT: © design56/iStockphoto

17.1 INTRODUCTION

The carboxyl group, [structure: O=C–OH] (abbreviated —CO_2H or —COOH), is one of the most widely occurring functional groups in chemistry and biochemistry. Not only are carboxylic acids themselves important, but the carboxyl group is the parent group of a large family of related compounds called **acyl compounds** or **carboxylic acid derivatives**, shown in Table 17.1.

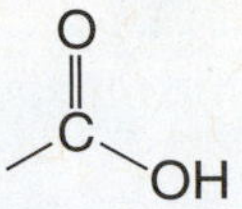

TABLE 17.1 CARBOXYLIC ACID DERIVATIVES

Structure	Name	Structure	Name
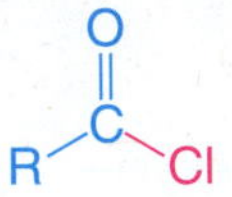 R–C(=O)–Cl	Acyl (or acid) chloride	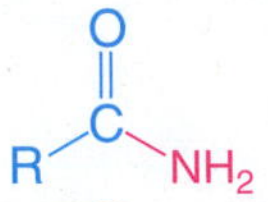 R–C(=O)–NH₂	Amide
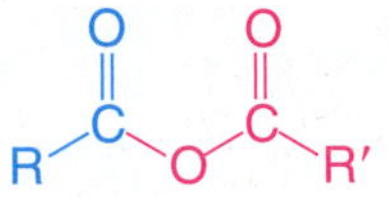 R–C(=O)–O–C(=O)–R′	Acid anhydride	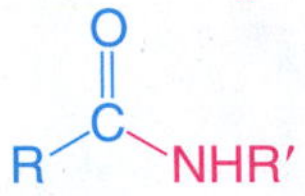 R–C(=O)–NHR′	
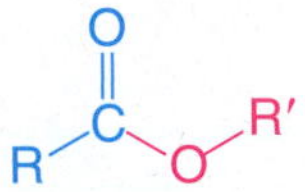 R–C(=O)–O–R′	Ester	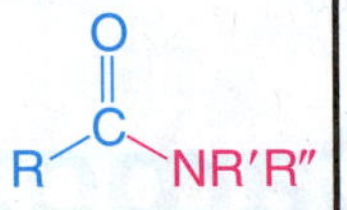 R–C(=O)–NR′R″	
R–C≡N	Nitrile		

17.2 NOMENCLATURE AND PHYSICAL PROPERTIES

17.2A Carboxylic Acids

- Systematic or substitutive names for carboxylic acids are obtained by dropping the final -*e* of the name of the alkane corresponding to the longest chain in the acid and by adding -*oic acid*. The carboxyl carbon atom is assigned number 1.

The following examples show how this is done:

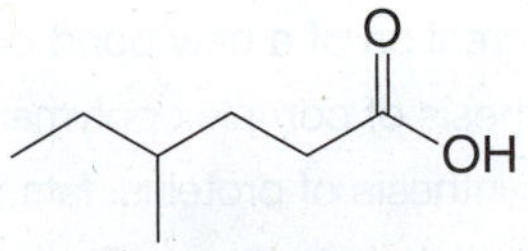

4-Methylhexanoic acid

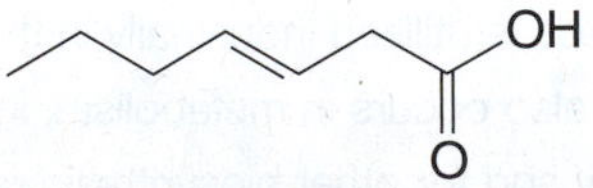

(*E*)-3-Heptenoic acid
[or (*E*)-hept-3-enoic acid]

Many carboxylic acids have common names that are derived from Latin or Greek words that indicate one of their natural sources. Methanoic acid is called formic acid (*formica*, Latin: ant). Ethanoic acid is called acetic acid (*acetum*, Latin: vinegar). Butanoic acid is one compound responsible for the odor of rancid butter, so its common name is butyric acid (*butyrum*, Latin: butter). Pentanoic acid, as a result of its occurrence in valerian, a perennial herb, is named valeric acid. Hexanoic acid is one compound associated with the odor of goats, hence its common name, caproic acid (*caper*, Latin: goat). Octadecanoic acid takes its common name, stearic acid, from the Greek word *stear*, for tallow.

Most of these common names have been used for a long time and some are likely to remain in common usage, so it is helpful to be familiar with them. In this text we shall refer to methanoic acid and ethanoic acid as formic acid and acetic acid, respectively. However, in almost all other instances we shall use IUPAC systematic or substitutive names.

Valerian is a source of valeric acid.

Carboxylic acids are polar substances. Their molecules can form strong hydrogen bonds with each other and with water. As a result, carboxylic acids generally have high boiling points, and low-molecular-weight carboxylic acids show appreciable solubility in water. As the length of the carbon chain increases, water solubility declines.

17.2B Carboxylate Salts

Salts of carboxylic acids are named as -ates; in both common and systematic names, -ate replaces -ic acid. The name of the cation precedes that of the carboxylate anion. Thus, CH_3CO_2Na is sodium acetate or sodium ethanoate.

Sodium and potassium salts of most carboxylic acids are readily soluble in water. This is true even of the long-chain carboxylic acids. Sodium or potassium salts of long-chain carboxylic acids are the major ingredients of soap (see Section 23.2C).

●●● **SOLVED PROBLEM 17.1**

Give an IUPAC systematic name for the following compound.

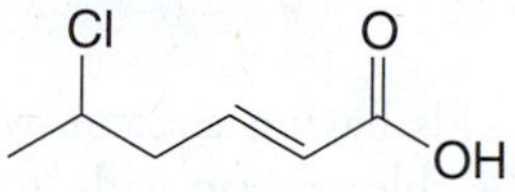

STRATEGY AND ANSWER: First we number the chain beginning with the carbon of the carboxylic acid group.

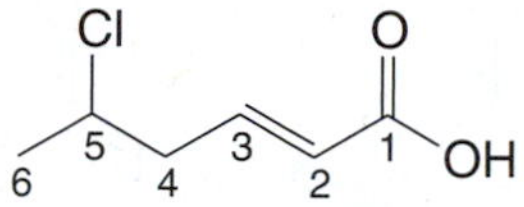

This chain contains six carbons with one double bond, so the base name is **hexenoic acid**. Then we give the position of the double bond and its stereochemistry, and the position and name of the substituent. The name, therefore, is (*E*)-5-chloro-2-hexenoic acid.

●●● **PRACTICE PROBLEM 17.1**

Give an IUPAC systematic name for each of the following:

(a)

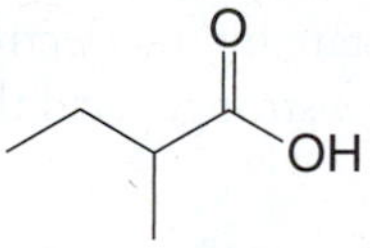

(b)

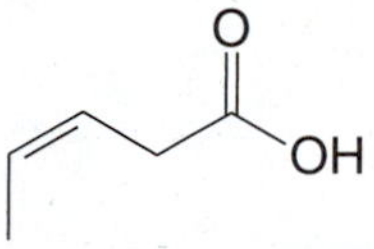

(c)

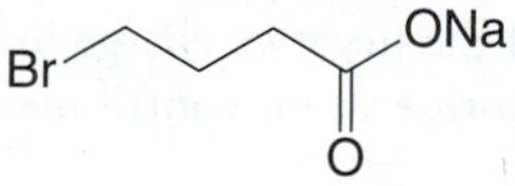

(d)

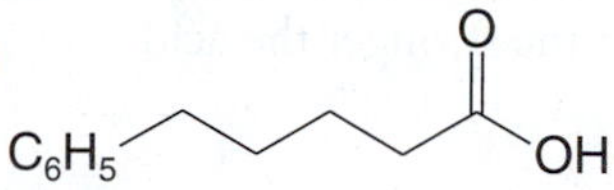

(e)

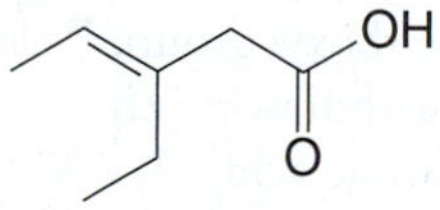

●●● **PRACTICE PROBLEM 17.2**

Experiments show that the molecular weight of acetic acid in the vapor state (just above its boiling point) is approximately 120. Explain the discrepancy between this experimental value and the true value of approximately 60.

17.2C Acidity of Carboxylic Acids

Most unsubstituted carboxylic acids have K_a values in the range of 10^{-4}–10^{-5} ($pK_a = 4$–5). The pK_a of water is about 16, and the apparent pK_a of H_2CO_3 is about 7. These relative acidities mean that carboxylic acids react readily with aqueous solutions of sodium hydroxide and sodium bicarbonate to form soluble sodium salts. We can use solubility tests, therefore, to distinguish water-insoluble carboxylic acids from water-insoluble phenols (Chapter 21) and alcohols.

Helpful Hint

Solubility tests such as these are rapid and useful ways to classify unknown compounds.

• Water-insoluble carboxylic acids dissolve in either aqueous sodium hydroxide or aqueous sodium bicarbonate.

Benzoic acid
water insoluble

$\xrightarrow{\text{aq. NaOH or}}$
$\xrightarrow{\text{aq. NaHCO}_3}$

Sodium benzoate
water soluble

• Water-insoluble phenols (Section 21.5) dissolve in aqueous sodium hydroxide but (except for some nitrophenols) do not dissolve in aqueous sodium bicarbonate.
• Water-insoluble alcohols do not dissolve in either aqueous sodium hydroxide or sodium bicarbonate.

Carboxylic acids having electron-withdrawing groups are more acidic than unsubstituted acids. The chloroacetic acids, for example, show the following order of acidities:

pK_a 0.70   1.48   2.86   4.76

As we saw in Section 3.11, this acid-strengthening effect of electron-withdrawing groups arises from a combination of inductive effects and entropy effects. We can visualize inductive charge delocalization when we compare the electrostatic potential maps for carboxylate anions of acetic acid and trichloroacetetic acid in Fig. 17.1. The maps show more negative charge localized near the acetate carboxyl group than the trichloroacetate carboxyl group. Delocalization of the negative charge in trichloroacetate by the electron-withdrawing effect of its three chlorine atoms contributes to its being a stronger acid than acetic acid.

• In general, the more delocalization of charge in the conjugate base, the more stable is the anion, and the stronger the acid.

FIGURE 17.1 Electrostatic potential maps for the carboxylate anions of (a) acetic acid and (b) trichloroacetic acid. There is greater delocalization of negative charge in trichloroacetate than acetate due to the inductive electron-withdrawing effect of the three chlorine atoms in trichloroacetate.

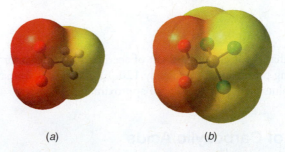

(a)   (b)

Since inductive effects are not transmitted very effectively through covalent bonds, the acid-strengthening effect decreases as the distance between the electron-withdrawing group and the carboxyl group increases. Of the chlorobutanoic acids that follow, the strongest acid is 2-chlorobutanoic acid:

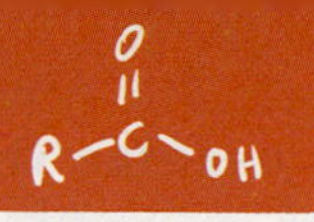

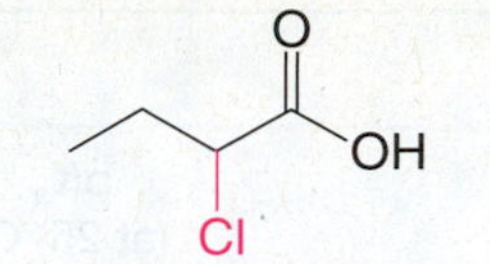

2-Chlorobutanoic acid
$(pK_a = 2.85)$

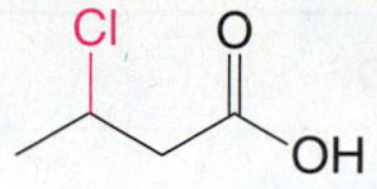

3-Chlorobutanoic acid
$(pK_a = 4.05)$

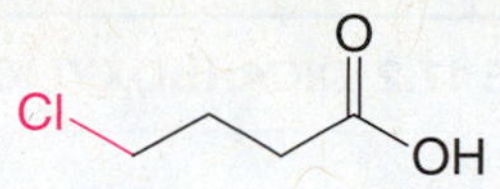

4-Chlorobutanoic acid
$(pK_a = 4.50)$

●●● **SOLVED PROBLEM 17.2**

Which carboxylic acid would you expect to be stronger, **A** or **B**?

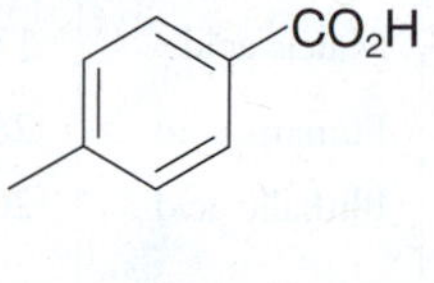

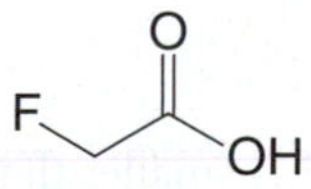

.CO₂H

or

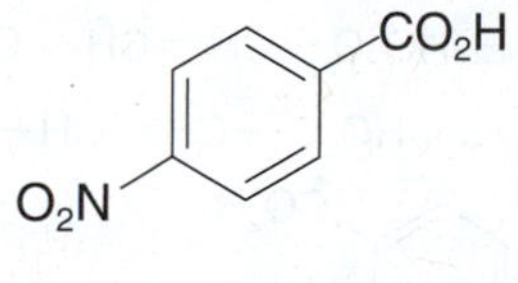

.CO₂H

O_2N

A   **B**

STRATEGY AND ANSWER: The electron-withdrawing effect of the nitro group would help stabilize the conjugate base of **B**, whereas the electron-donating effect of the methyl group in **A** would destabilize its conjugate base. Therefore, **B** is expected to be the stronger acid.

●●●

PRACTICE PROBLEM 17.3

Which acid of each pair shown here would you expect to be stronger?

(a)

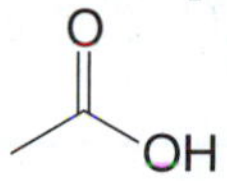

OH   or   F   OH

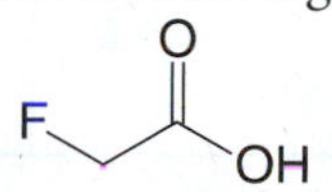

(b)

F   OH   or   Cl   OH

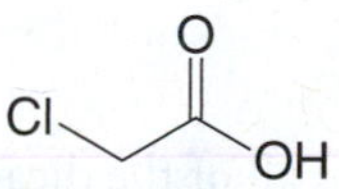

(c)

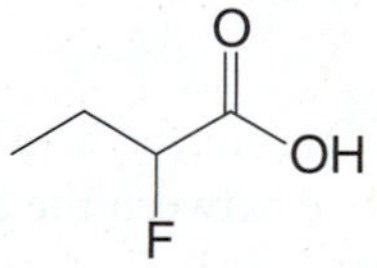

OH   or   F   OH

F

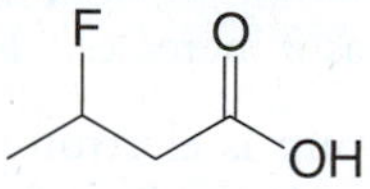

(d)

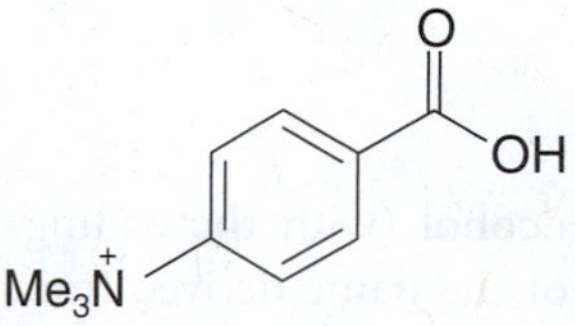

OH   or   OH

$Me_3\overset{+}{N}$

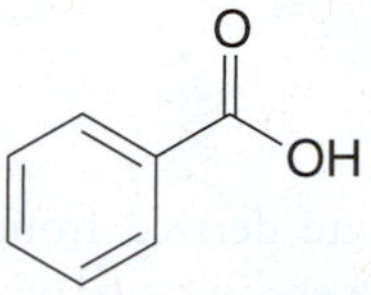

(e)

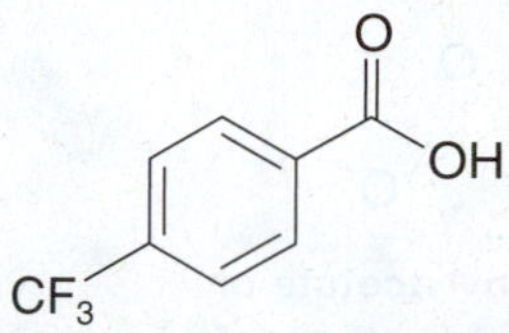

OH   or   OH

CF_3

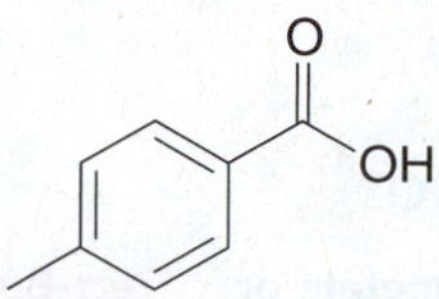

17.2D Dicarboxylic Acids

Dicarboxylic acids are named as **alkanedioic acids** in the IUPAC systematic or substitutive system. Most simple dicarboxylic acids have common names (Table 17.2).

TABLE 17.2 DICARBOXYLIC ACIDS

Structure	Common Name	mp (°C)	pK_a (at 25 °C)	
			pK_{a1}	pK_{a2}
HO_2C-CO_2H	Oxalic acid	189 dec	1.2	4.2
$HO_2CCH_2CO_2H$	Malonic acid	136	2.9	5.7
$HO_2C(CH_2)_2CO_2H$	Succinic acid	187	4.2	5.6
$HO_2C(CH_2)_3CO_2H$	Glutaric acid	98	4.3	5.4
$HO_2C(CH_2)_4CO_2H$	Adipic acid	153	4.4	5.6
cis-$HO_2C-CH=CH-CO_2H$	Maleic acid	131	1.9	6.1
$trans$-$HO_2C-CH=CH-CO_2H$	Fumaric acid	287	3.0	4.4
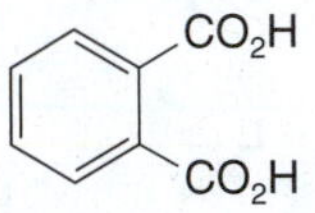 (structure: benzene with two ortho CO₂H groups)	Phthalic acid	206–208 dec	2.9	5.4
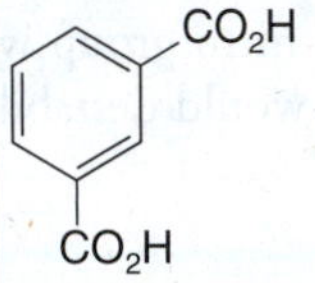 (structure: benzene with two meta CO₂H groups)	Isophthalic acid	345–348	3.5	4.6
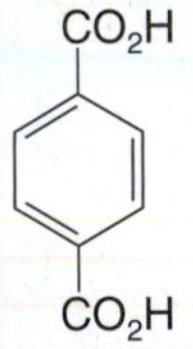 (structure: benzene with two para CO₂H groups)	Terephthalic acid	Sublimes	3.5	4.8

Succinic and fumaric acids are key metabolites in the citric acid pathway. Adipic acid is used in the synthesis of nylon. The isomers of phthalic acid are used in making polyesters. See Special Topic C in *WileyPLUS* for further information on polymers.

••• **SOLVED PROBLEM 17.3**

Suggest explanations for the following. **(a)** The pK_{a1} for all of the dicarboxylic acids in Table 17.2 is smaller than the pK_a for a monocarboxylic acid with the same number of carbon atoms. **(b)** The difference between pK_{a1} and pK_{a2} for dicarboxylic acids of the type $HO_2C(CH_2)_nCO_2H$ decreases as n increases.

STRATEGY AND ANSWER: (a) The carboxyl group is electron-withdrawing; thus, in a dicarboxylic acid such as those in Table 17.2, one carboxylic acid group increases the acidity of the other. **(b)** As the distance between the carboxyl groups increases, the acid-strengthening, inductive effect decreases.

17.2E Esters

The names of esters are derived from the names of the alcohol (with the ending **-yl**) and the acid (with the ending **-ate** or **-oate**). The portion of the name derived from the alcohol comes first:

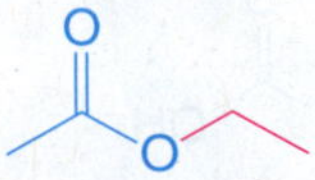

Ethyl acetate or ethyl ethanoate 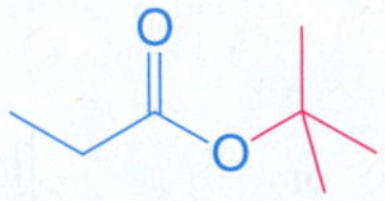   ***tert*-Butyl propanoate** 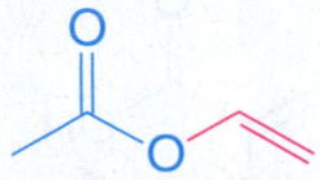   **Vinyl acetate or ethenyl ethanoate**

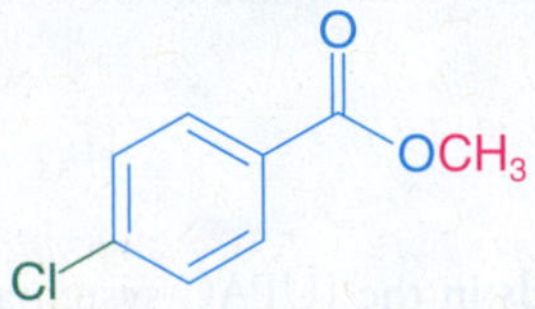

Methyl *p*-chlorobenzoate 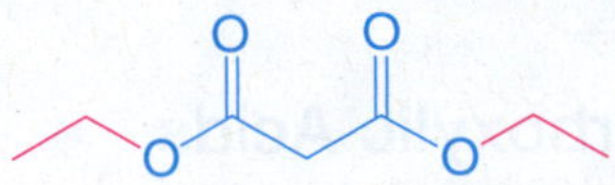   **Diethyl malonate**

Esters are polar compounds, but, lacking a hydrogen attached to oxygen, their molecules cannot form strong hydrogen bonds to each other. As a result, esters have boiling points that are lower than those of acids and alcohols of comparable molecular weight. The boiling points of esters are about the same as those of comparable aldehydes and ketones.

Unlike the low-molecular-weight acids, esters usually have pleasant odors, some resembling those of fruits, and these are used in the manufacture of synthetic flavors:

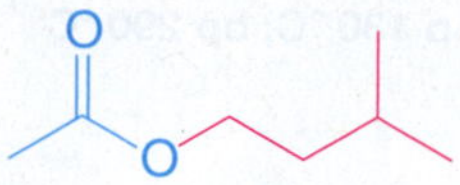

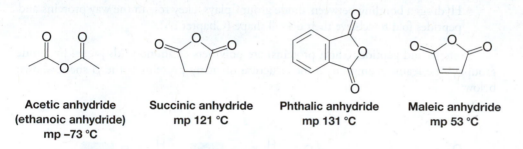

Isopentyl acetate
(used in synthetic banana flavor)

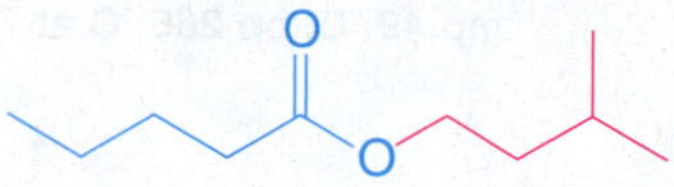

Isopentyl pentanoate
(used in synthetic apple flavor)

17.2F Carboxylic Anhydrides

Most anhydrides are named by dropping the word **acid** from the name of the carboxylic acid and then adding the word **anhydride**:

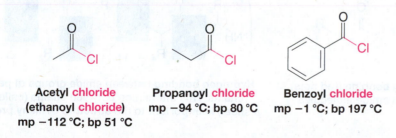

Acetic anhydride
(ethanoic anhydride)
mp −73 °C

Succinic anhydride
mp 121 °C

Phthalic anhydride
mp 131 °C

Maleic anhydride
mp 53 °C

17.2G Acyl Chlorides

Acyl chlorides are also called **acid chlorides**. They are named by dropping -**ic acid** from the name of the acid and then adding -**yl chloride**. Examples are

Acetyl chloride
(ethanoyl chloride)
mp −112 °C; bp 51 °C

Propanoyl chloride
mp −94 °C; bp 80 °C

Benzoyl chloride
mp −1 °C; bp 197 °C

Acyl chlorides and carboxylic anhydrides have boiling points in the same range as esters of comparable molecular weight.

17.2H Amides

Amides that have no substituent on nitrogen are named by dropping -**ic acid** from the common name of the acid (or -**oic acid** from the substitutive name) and then adding -**amide**. Alkyl groups on the nitrogen atom of amides are named as substituents, and the named substituent is prefaced by *N*- or *N,N*-. Examples are

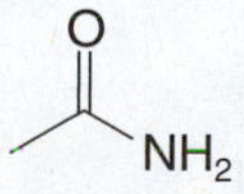

Acetamide
(ethanamide)
mp 82 °C; bp 221 °C

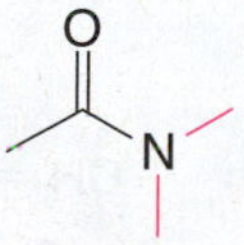

***N,N*-Dimethyl**acetamide
mp −20 °C; bp 166 °C

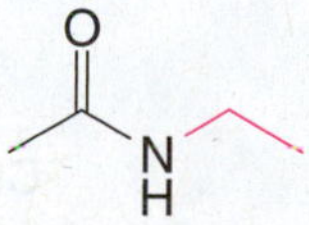

***N*-Ethyl**acetamide
bp 205 °C

N-Phenyl-N-propylacetamide
mp 49 °C; bp 266 °C at 712 torr

Benzamide
mp 130 °C; bp 290 °C

- Amides with nitrogen atoms bearing one or two hydrogen atoms are able to form strong hydrogen bonds to each other.

Such amides have high melting points and boiling points. On the other hand, molecules of *N,N*-disubstituted amides cannot form strong hydrogen bonds to each other, and they have lower melting points and boiling points. The melting and boiling data given above illustrate this trend.

- Hydrogen bonding between amide groups plays a key role in the way proteins and peptides fold to achieve their overall shape (Chapter 24).

Proteins and peptides (short proteins) are polymers of amino acids joined by amide groups. One feature common to the structure of many proteins is the β sheet, shown below:

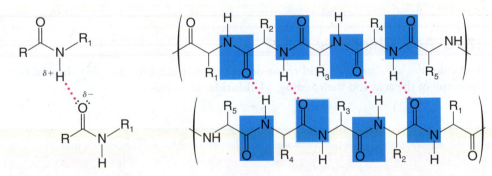

Hydrogen bonding (red dots) between amide molecules

Hydrogen bonding between amide groups of peptide chains. This interaction between chains (called a β sheet) is important to the structure of many proteins.

17.21 Nitriles

Carboxylic acids can be converted to nitriles and vice versa. In IUPAC substitutive nomenclature, acyclic nitriles are named by adding the suffix *-nitrile* to the name of the corresponding hydrocarbon. The carbon atom of the —C≡N group is assigned number 1. Additional examples of nitriles were presented in Section 2.11 with other functional groups of organic molecules. The name acetonitrile is an acceptable common name for CH_3CN, and acrylonitrile is an acceptable common name for $CH_2=CHCN$:

$$\overset{2}{C}H_3-\overset{1}{C}\equiv N\colon \qquad \overset{3}{C}H_2=\overset{2}{C}H-\overset{1}{C}\equiv N\colon$$

Ethanenitrile
(acetonitrile)

Propenenitrile
(acrylonitrile)

N,N-Diethyl-3-methylbenzamide (also called *N,N*-diethyl-*m*-toluamide, or DEET) is used in many insect repellants. Write its structure.

ANSWER:

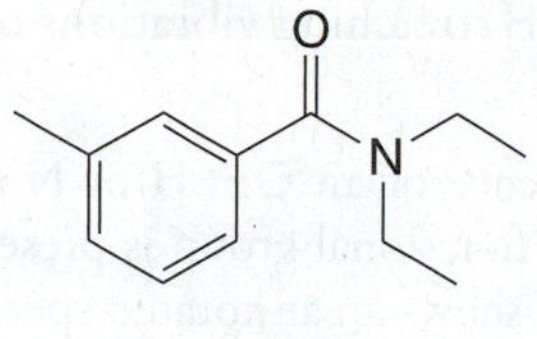

DEET

Write structural formulas for the following:

(a) Methyl propanoate

(b) Ethyl *p*-nitrobenzoate

(c) Dimethyl malonate

(d) *N,N*-Dimethylbenzamide

(e) Pentanenitrile

(f) Dimethyl phthalate

(g) Dipropyl maleate

(h) *N,N*-Dimethylformamide

(i) 2-Bromopropanoyl bromide

(j) Diethyl succinate

17.2J Spectroscopic Properties of Acyl Compounds

IR Spectra Infrared spectroscopy is of considerable importance in identifying carboxylic acids and their derivatives. The C=O stretching band is one of the most prominent in their IR spectra since it is always a strong band. Figure 17.2 gives the location of this band for most acyl compounds.

> **Helpful Hint**
>
> Infrared spectroscopy is useful for classifying acyl compounds.

- The C=O stretching band occurs at different frequencies for acids, esters, and amides, and its precise location is often helpful in structure determination.
- Conjugation and electron-donating groups bonded to the carbonyl shift the location of the C=O absorption to lower frequencies.

Functional Group	Approximate Frequency Range (cm^{-1})	
Acid chloride	1815–1785 1800–1770 (conj.)	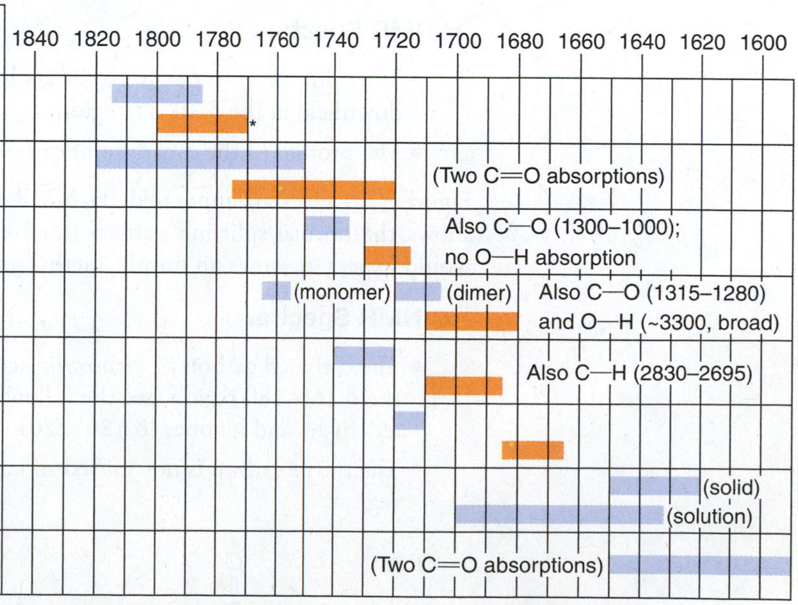 1840 1820 1800 1780 1760 1740 1720 1700 1680 1660 1640 1620 1600
Acid anhydride	1820–1750 1775–1720 (conj.)	(Two C=O absorptions)
Ester/lactone	1750–1735 1730–1715 (conj.)	Also C—O (1300–1000); no O—H absorption
Carboxylic acid	~1760 or 1720–1705 1710–1680 (conj.)	(monomer) (dimer) Also C—O (1315–1280) and O—H (~3300, broad)
Aldehyde	1740–1720 1710–1685 (conj.)	Also C—H (2830–2695)
Ketone	1720–1710 1685–1665 (conj.)	
Amide/lactam	1700–1620	(solid) (solution)
Carboxylate salt	1650–1550	(Two C=O absorptions)

*Orange bars represent absorption ranges for conjugated species.

FIGURE 17.2 Approximate carbonyl IR absorption frequencies. (Frequency ranges based on Silverstein and Webster, reprinted with permission of John Wiley & Sons, Inc. from Silverstein, R. and Webster, F. X., *Spectrometric Identification of Organic Compounds*, Sixth Edition. Copyright 1998.)

- Electron-withdrawing groups bonded to the carbonyl shift the $C{=}O$ absorption to higher frequencies.
- The hydroxyl groups of carboxylic acids also give rise to a broad peak in the $2500{-}3100\text{-cm}^{-1}$ region arising from $O{-}H$ stretching vibrations.
- The $N{-}H$ stretching vibrations of amides absorb between 3140 and 3500 cm^{-1}.

Presence or absence of an $O{-}H$ or $N{-}H$ absorption can be an important clue as to which carbonyl functional group is present in an unknown compound.

Figure 17.3 shows an annotated spectrum of propanoic acid. Nitriles show an intense and characteristic infrared absorption band near 2250 cm^{-1} that arises from stretching of the carbon–nitrogen triple bond.

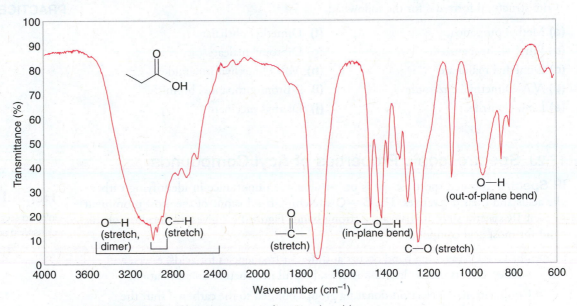

FIGURE 17.3 The infrared spectrum of propanoic acid.

¹H NMR Spectra

- The acidic protons of carboxylic acids are highly deshielded and absorb far downfield in the δ 10–12 region.
- The protons of the α carbon of carboxylic acids absorb in the δ 2.0–2.5 region.

Figure 17.4 gives an annotated ¹H NMR spectrum of an ester, methyl propanoate; it shows the normal splitting pattern (quartet and triplet) of an ethyl group, and, as we would expect, it shows an unsplit methyl group.

¹³C NMR Spectra

- The carbonyl carbon of carboxylic acids and their derivatives occurs downfield in the δ 160–180 region (see the following examples), but not as far downfield as for aldehydes and ketones (δ 180–220).
- The nitrile carbon is not shifted so far downfield and absorbs in the δ 115–120 region.

δ 177.2 δ 170.7 δ 170.3 δ 172.6 δ 117.4

¹³C NMR chemical shifts for the carbonyl or nitrile carbon atom

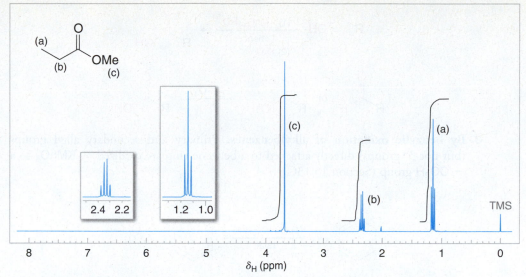

FIGURE 17.4 The 300-MHz ^{1}H NMR spectrum of methyl propanoate. Expansions of the signals are shown in the offset plots.

The carbon atoms of the alkyl groups of carboxylic acids and their derivatives have ^{13}C chemical shifts much further upfield. The chemical shifts for each carbon of pentanoic acid are as follows:

$$13.5 \quad 27.0 \quad 179.7$$
$$22.0 \quad 34.1$$

^{13}C NMR chemical shifts (δ)

17.3 PREPARATION OF CARBOXYLIC ACIDS

Most of the methods for the preparation of carboxylic acids have been presented previously:

1. **By oxidation of alkenes.** We learned in Section 8.17A that alkenes can be oxidized to carboxylic acids with hot alkaline $KMnO_4$:

$$R\text{—CH}=\text{CH—}R' \xrightarrow[\text{(2) } H_3O^+]{\substack{\text{(1) } KMnO_4,\ HO^- \\ \text{heat}}} R\text{—COOH} + R'\text{—COOH}$$

E or *Z*

Alternatively, ozonides (Section 8.17B) can be subjected to an oxidative workup that yields carboxylic acids:

$$R\text{—CH}=\text{CH—}R' \xrightarrow[\text{(2) } H_2O_2]{\text{(1) } O_3} R\text{—COOH} + R'\text{—COOH}$$

E or *Z*

2. **By oxidation of aldehydes and primary alcohols.** Aldehydes can be oxidized to carboxylic acids with mild oxidizing agents such as $Ag(NH_3)_2{}^+HO^-$ (Section 16.12B). Primary alcohols can be oxidized with $KMnO_4$. Aldehydes and primary alcohols are oxidized to carboxylic acids with chromic acid (H_2CrO_4) in aqueous acetone (the Jones oxidation; Section 12.4C).

$$R\text{—CHO} \xrightarrow[\text{(2) } H_3O^+]{\text{(1) } Ag_2O \text{ or } Ag(NH_3)_2{}^+HO^-} R\text{—COOH}$$

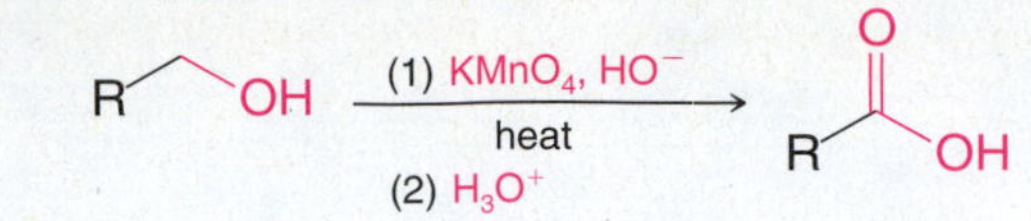

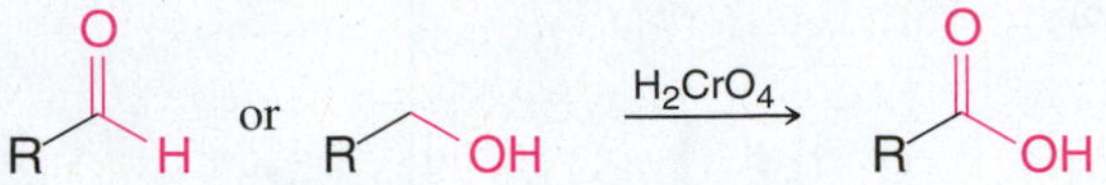

$$R \diagdown OH \xrightarrow[\text{(2) } H_3O^+]{\substack{\text{(1) KMnO}_4, \text{ HO}^- \\ \text{heat}}} R \diagup C(=O) OH$$

$$R \diagup C(=O) H \text{ or } R \diagdown OH \xrightarrow{H_2CrO_4} R \diagup C(=O) OH$$

3. By benzylic oxidation of alkylbenzenes. Primary and secondary alkyl groups (but not 3° groups) directly attached to a benzene ring are oxidized by **KMnO₄** to a —CO₂H group (Section 15.13C):

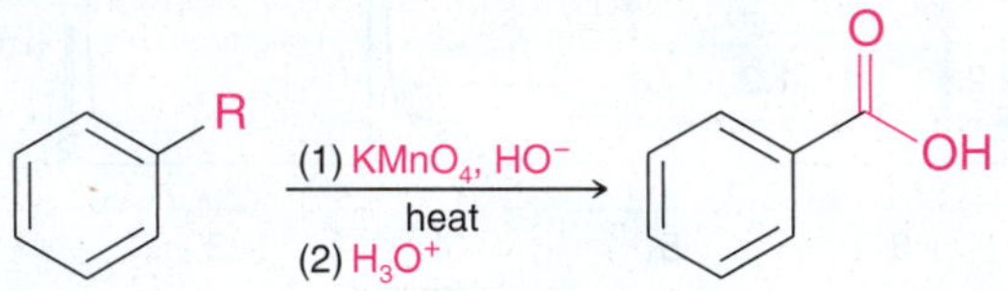

$$C_6H_5{-}R \xrightarrow[\text{(2) } H_3O^+]{\substack{\text{(1) KMnO}_4, \text{ HO}^- \\ \text{heat}}} C_6H_5{-}C(=O)OH$$

4. By oxidation of the benzene ring. The benzene ring of an alkylbenzene can be converted to a carboxyl group by ozonolysis, followed by treatment with hydrogen peroxide (Section 15.13D):

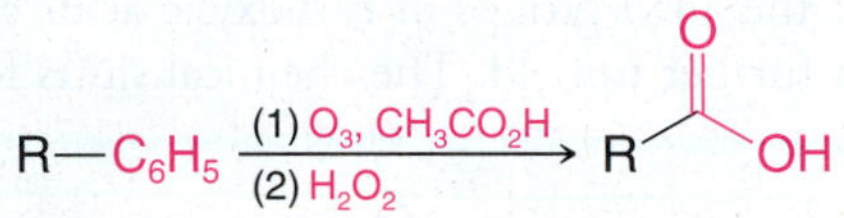

$$R{-}C_6H_5 \xrightarrow[\text{(2) } H_2O_2]{\text{(1) } O_3, \text{ CH}_3CO_2H} R{-}C(=O)OH$$

5. By hydrolysis of cyanohydrins and other nitriles. We saw, in Section 16.9, that aldehydes and ketones can be converted to **cyanohydrins** and that these can be hydrolyzed to α-hydroxy acids. In the hydrolysis the —CN group is converted to a —CO₂H group. The mechanism of nitrile hydrolysis is discussed in Section 17.8H:

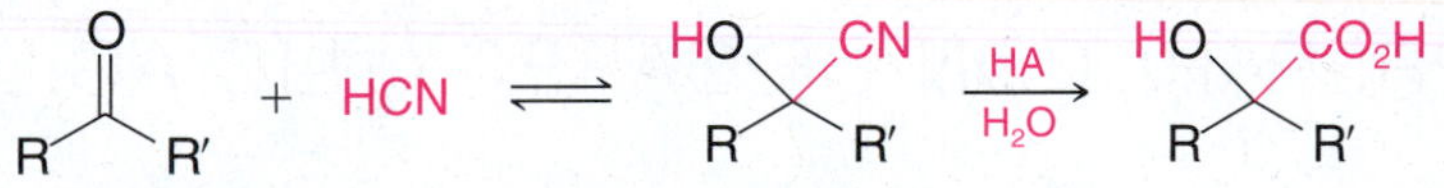

$$R{-}C(=O){-}R' + HCN \rightleftharpoons \underset{R,R'}{\overset{HO,CN}{C}} \xrightarrow[H_2O]{HA} \underset{R,R'}{\overset{HO,CO_2H}{C}}$$

Nitriles can also be prepared by nucleophilic substitution reactions of alkyl halides with sodium cyanide. Hydrolysis of the nitrile yields a carboxylic acid *with a chain one carbon atom longer* than the original alkyl halide:

General Reaction

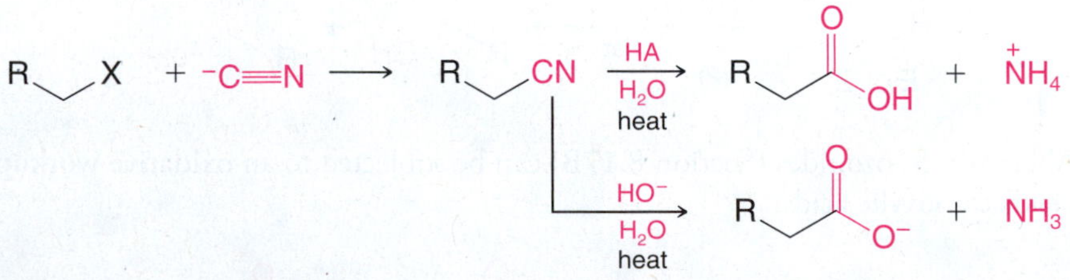

$$R{-}X + {}^-C{\equiv}N \longrightarrow R{-}CN \nearrow \xrightarrow[\substack{H_2O \\ \text{heat}}]{HA} R{-}CH_2C(=O)OH + {}^+NH_4$$

$$\searrow \xrightarrow[\substack{H_2O \\ \text{heat}}]{HO^-} R{-}CH_2C(=O)O^- + NH_3$$

Specific Examples

$$HO{-}\diagdown{-}Cl \xrightarrow[\text{(80\%)}]{\text{NaCN}} HO{-}\diagdown{-}CN \xrightarrow[\substack{\text{(2) } H_3O^+ \\ \text{(75–80\%)}}]{\text{(1) HO}^-, H_2O} HO{-}\diagdown{-}C(=O)OH$$

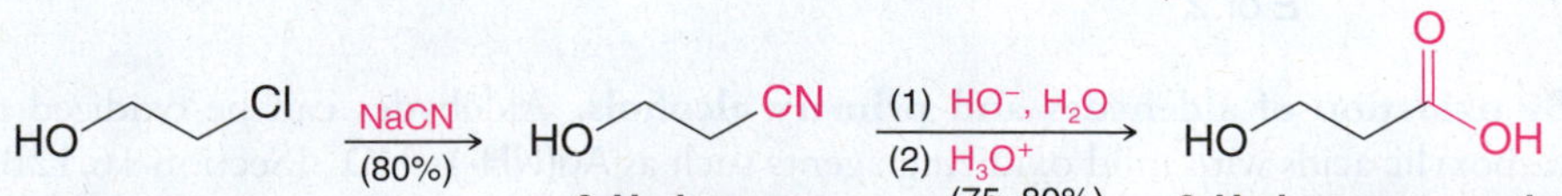

3-Hydroxy-propanenitrile   **3-Hydroxypropanoic acid**

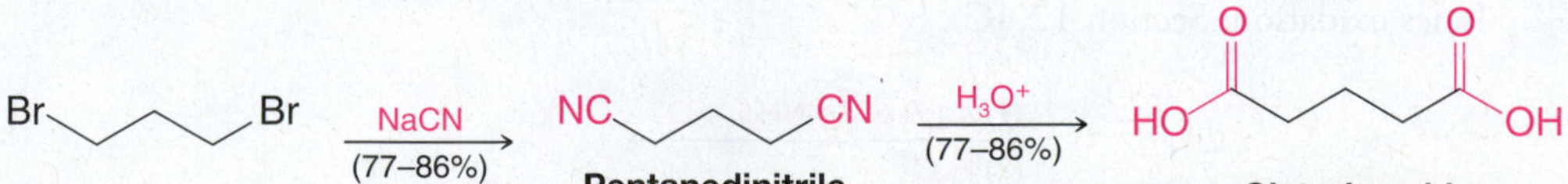

$$Br{\diagdown}Br \xrightarrow[\text{(77–86\%)}]{\text{NaCN}} NC{\diagdown}CN \xrightarrow[\text{(77–86\%)}]{H_3O^+} HO{-}C(=O){\diagdown}C(=O){-}OH$$

Pentanedinitrile   **Glutaric acid**

This synthetic method is generally limited to the use of *primary alkyl halides*. The cyanide ion is a relatively strong base, and the use of a secondary or tertiary alkyl halide leads primarily to an alkene (through E2 elimination) rather than to a nitrile (through S_N2 substitution). Aryl halides (except for those with ortho and para nitro groups) do not react with sodium cyanide.

6. **By carbonation of Grignard reagents.** Grignard reagents react with carbon dioxide to yield magnesium carboxylates. Acidification produces carboxylic acids:

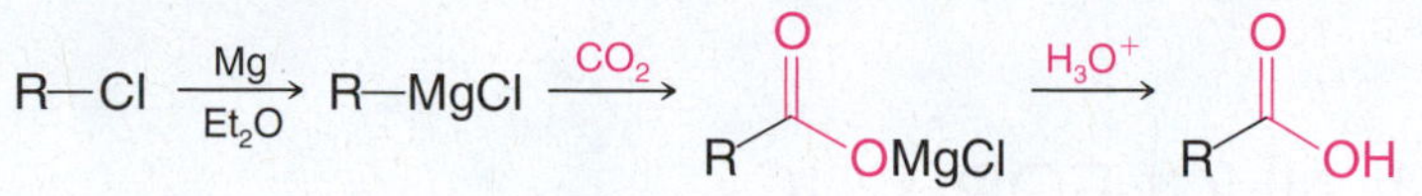

$$R-Cl \xrightarrow[Et_2O]{Mg} R-MgCl \xrightarrow{CO_2} \underset{R}{\overset{O}{\|}}OMgCl \xrightarrow{H_3O^+} \underset{R}{\overset{O}{\|}}OH$$

or

$$Ar-Br \xrightarrow[Et_2O]{Mg} Ar-MgBr \xrightarrow{CO_2} \underset{Ar}{\overset{O}{\|}}OMgBr \xrightarrow{H_3O^+} \underset{Ar}{\overset{O}{\|}}OH$$

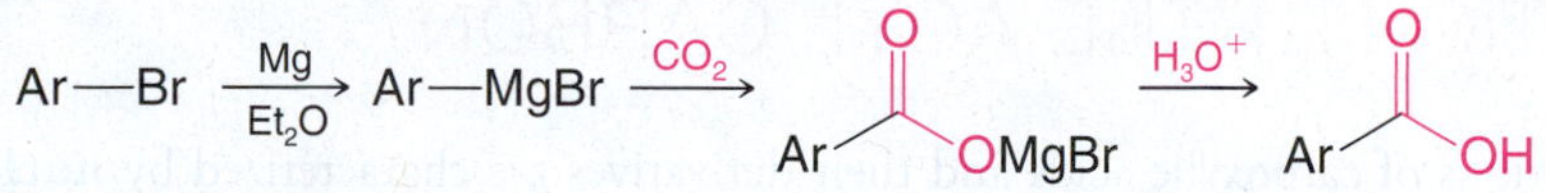

This synthesis of carboxylic acids is applicable to primary, secondary, tertiary, allyl, benzyl, and aryl halides, provided they have no groups incompatible with a Grignard reaction (see Section 12.8B):

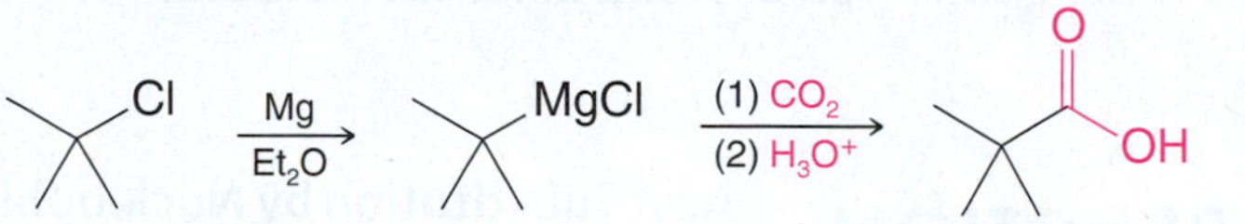

tert-Butyl chloride → 2,2-Dimethylpropanoic acid (79–80% overall)

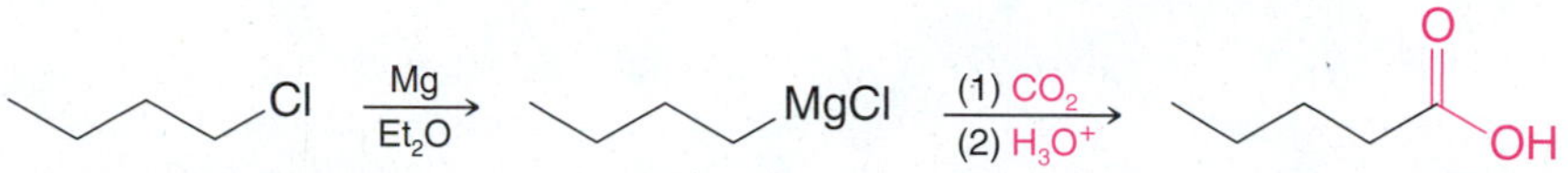

Butyl chloride → Pentanoic acid (80% overall)

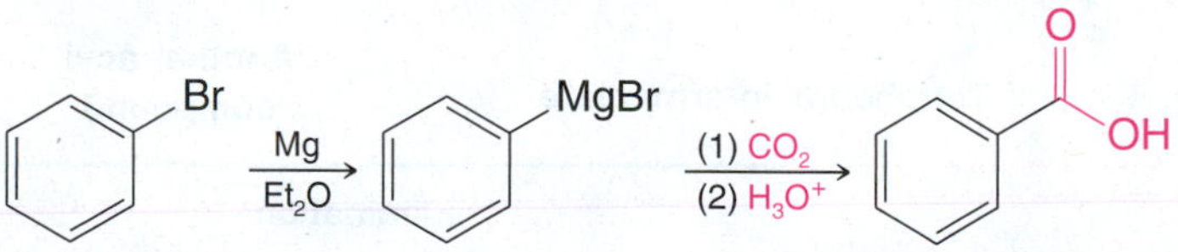

Bromobenzene → Benzoic acid (85%)

Show how each of the following compounds could be converted to benzoic acid:

(a) 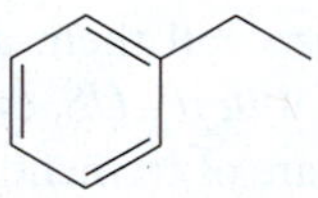 [structure: ethylbenzene]

(b) 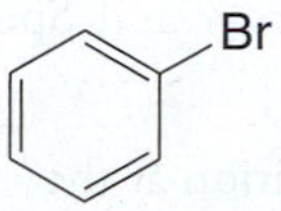 [structure: bromobenzene, Br]

(c) 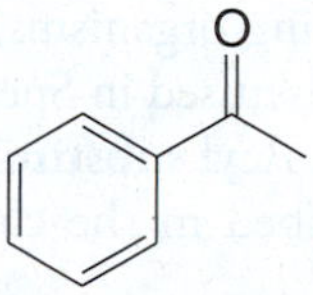 [structure: acetophenone, O]

(d) 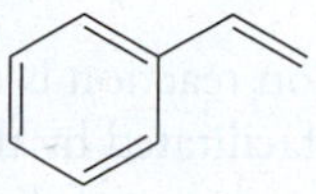 [structure: styrene]

(e) Benzyl alcohol

(f) Benzaldehyde

PRACTICE PROBLEM 17.5

Show how you would prepare each of the following carboxylic acids through a Grignard synthesis:

(a) [structure: phenylacetic acid, OH, O]

(b)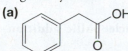

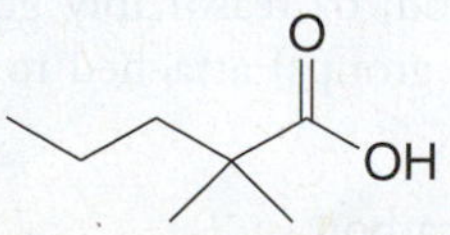

(c) 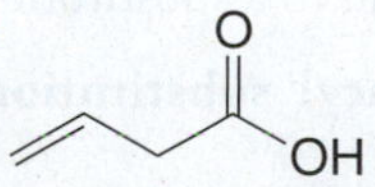 [structure: but-3-enoic acid, O, OH]

(d) 4-Methylbenzoic acid

(e) Hexanoic acid

PRACTICE PROBLEM 17.6

PRACTICE PROBLEM 17.7 **(a)** Which of the carboxylic acids in Practice Problem 17.6 could be prepared by a nitrile synthesis as well? **(b)** Which synthesis, Grignard or nitrile, would you choose to prepare

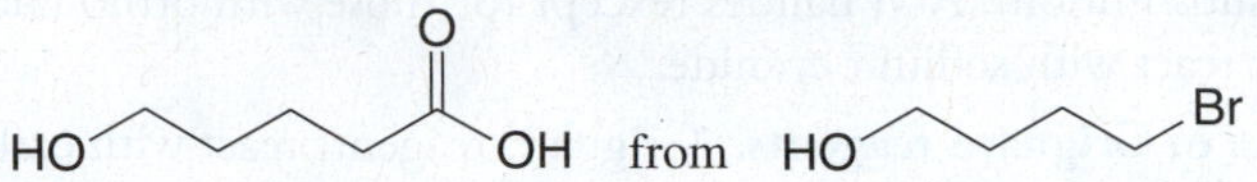

? Why?

17.4 ACYL SUBSTITUTION: NUCLEOPHILIC ADDITION–ELIMINATION AT THE ACYL CARBON

The reactions of carboxylic acids and their derivatives are characterized by **nucleophilic addition–elimination** at their acyl (carbonyl) carbon atoms. The result is a substitution at the acyl carbon. Key to this mechanism is formation of a **tetrahedral intermediate** that returns to a carbonyl group after the elimination of a leaving group. We shall encounter many reactions of this general type, as shown in the following box.

[**A MECHANISM FOR THE REACTION** ─ **Acyl Substitution by Nucleophilic Addition–Elimination**]

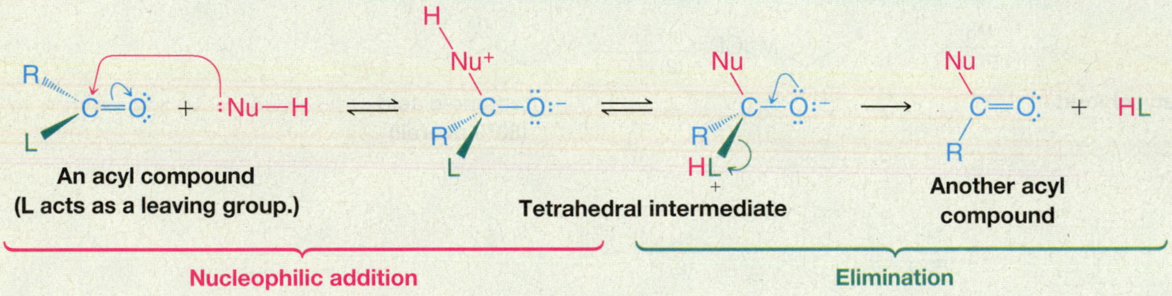

Helpful Hint

If you bear in mind the general mechanism for acyl substitution, you will see the common theme among reactions in this chapter.

Many reactions like this occur in living organisms, and biochemists call them **acyl transfer reactions**. Acetyl-coenzyme A, discussed in Special Topic E in *WileyPLUS*, often serves as a biochemical acyl transfer agent. Acyl **substitution reactions** are of tremendous importance in industry as well, as described in the chapter opening essay and Special Topic C in *WileyPLUS*.

- The initial step in an acyl substitution reaction is nucleophilic addition at the carbonyl carbon atom. This step is facilitated by the relative steric openness of the carbonyl carbon atom and the ability of the carbonyl oxygen atom to accommodate an electron pair of the carbon–oxygen double bond.

- In the second step the tetrahedral intermediate eliminates a leaving group (**L** in the mechanism above); this **elimination** leads to regeneration of the carbon–oxygen double bond and to a substitution product.

The overall process, therefore, is **acyl substitution** by a **nucleophilic addition–elimination** mechanism.

Acyl compounds react as they do because they all have good, or reasonably good, leaving groups (or they can be protonated to form good leaving groups) attached to the carbonyl carbon atom.

- Acyl substitution requires a leaving group at the carbonyl carbon.

An acyl chloride, for example, generally reacts by losing a *chloride ion*—a very weak base and thus a very good leaving group. The reaction of an acyl chloride with water is an example.

Specific Example

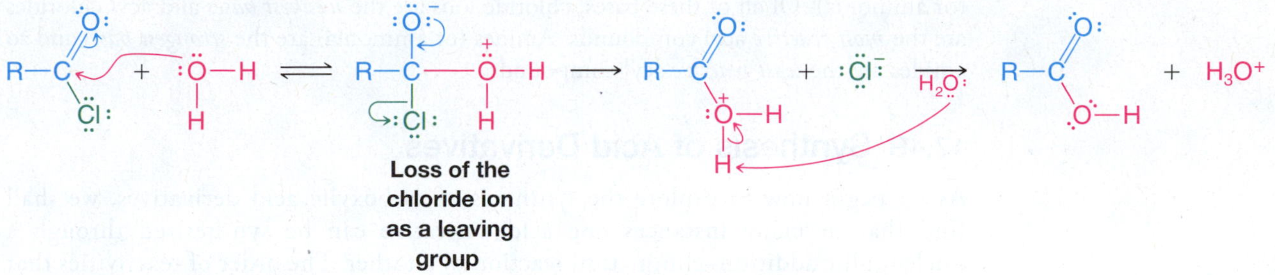

Loss of the chloride ion as a leaving group

An acid anhydride generally reacts by losing *a carboxylate anion* or a molecule of a *carboxylic acid*—both are weak bases and good leaving groups.

As we shall see later, esters generally undergo nucleophilic addition–elimination by losing a molecule of an *alcohol* (Section 17.7B), acids react by losing a molecule of *water* (Section 17.7A), and amides react by losing a molecule of *ammonia* or of an *amine* (Section 17.8F). All of the molecules lost in these reactions are weak bases and are reasonably good leaving groups.

For an aldehyde or ketone to react by nucleophilic addition–elimination, the tetrahedral intermediate would need to eject a hydride ion (H:⁻) or an alkanide ion (R:⁻). Both are *very powerful bases*, and both are therefore *very poor leaving groups*:

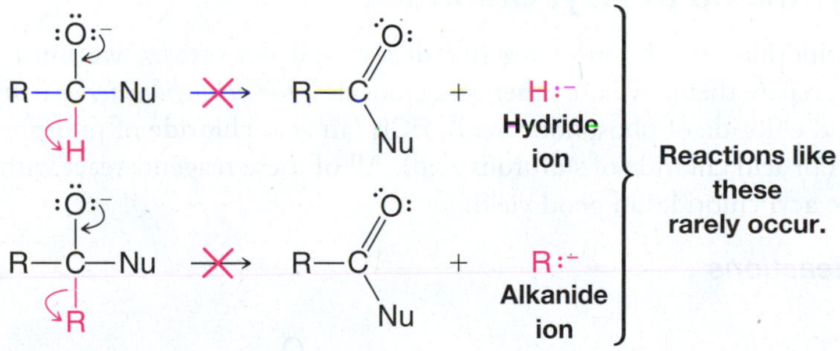

Reactions like these rarely occur.

[The haloform reaction (Section 18.3C) is one of the rare instances in which an alkanide anion can act as a leaving group, but then only, as we shall see, because the leaving group is a weakly basic trihalomethyl anion.]

17.4A Relative Reactivity of Acyl Compounds

Of the acid derivatives that we study in this chapter, acyl chlorides are the most reactive toward nucleophilic addition–elimination, and amides are the least reactive. In general, the overall order of reactivity is

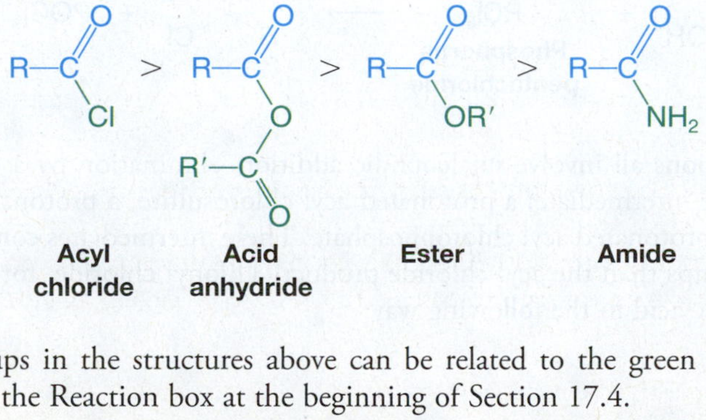

| Acyl chloride | Acid anhydride | Ester | Amide |

The green groups in the structures above can be related to the green **L** group in the Mechanism for the Reaction box at the beginning of Section 17.4.

- The general order of reactivity of acid derivatives can be explained by taking into account the basicity of the leaving groups.

When acyl chlorides react, the leaving group is a *chloride ion*. When acid anhydrides react, the leaving group is a carboxylic acid or a carboxylate ion. When esters react, the leaving group is an alcohol, and when amides react, the leaving group is an amine (or ammonia). Of all of these bases, chloride ions are the *weakest bases* and acyl chlorides are the *most reactive* acyl compounds. Amines (or ammonia) are the *strongest bases* and so amides are the *least reactive* acyl compounds.

17.4B Synthesis of Acid Derivatives

As we begin now to explore the syntheses of carboxylic acid derivatives, we shall find that in many instances one acid derivative can be synthesized through a nucleophilic addition–elimination reaction of another. The order of reactivities that we have presented gives us a clue as to which syntheses are practical and which are not. In general, *less reactive acyl compounds can be synthesized from more reactive ones, but the reverse is usually difficult and, when possible, requires special reagents.*

- Synthesis of acid derivatives by acyl substitution requires that the reactant have a better leaving group at the acyl carbon than the product.

17.5 ACYL CHLORIDES

17.5A Synthesis of Acyl Chlorides

Since acyl chlorides are the most reactive of the acid derivatives, we must use special reagents to prepare them. We use other acid chlorides, *the acid chlorides of inorganic acids*: PCl_5 (an acid chloride of phosphoric acid), PCl_3 (an acid chloride of phosphorous acid), and $SOCl_2$ (an acid chloride of sulfurous acid). All of these reagents react with carboxylic acids to give acyl chlorides in good yield:

General Reactions

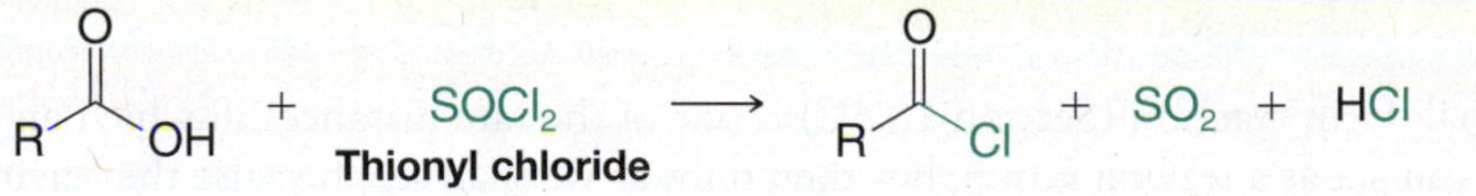

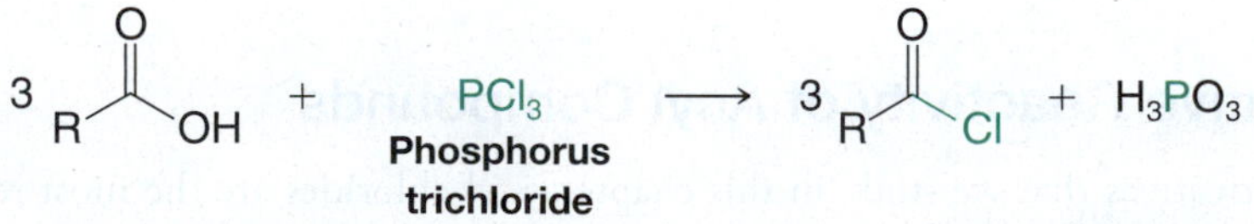

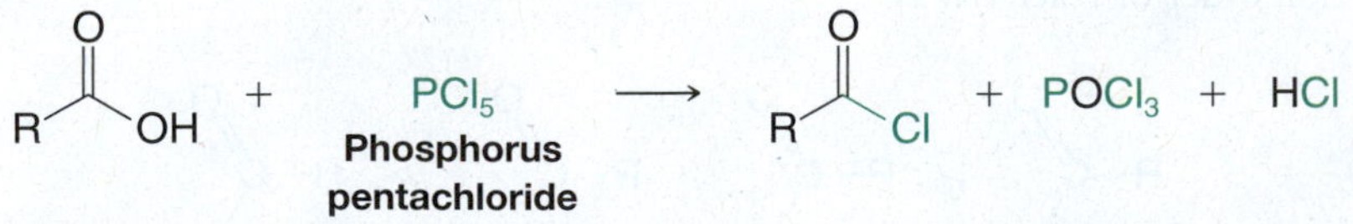

These reactions all involve nucleophilic addition–elimination by a chloride ion on a highly reactive intermediate: a protonated acyl chlorosulfite, a protonated acyl chlorophosphite, or a protonated acyl chlorophosphate. These intermediates contain even better acyl leaving groups than the acyl chloride product. Thionyl chloride, for example, reacts with a carboxylic acid in the following way:

A MECHANISM FOR THE REACTION — Synthesis of Acyl Chlorides Using Thionyl Chloride

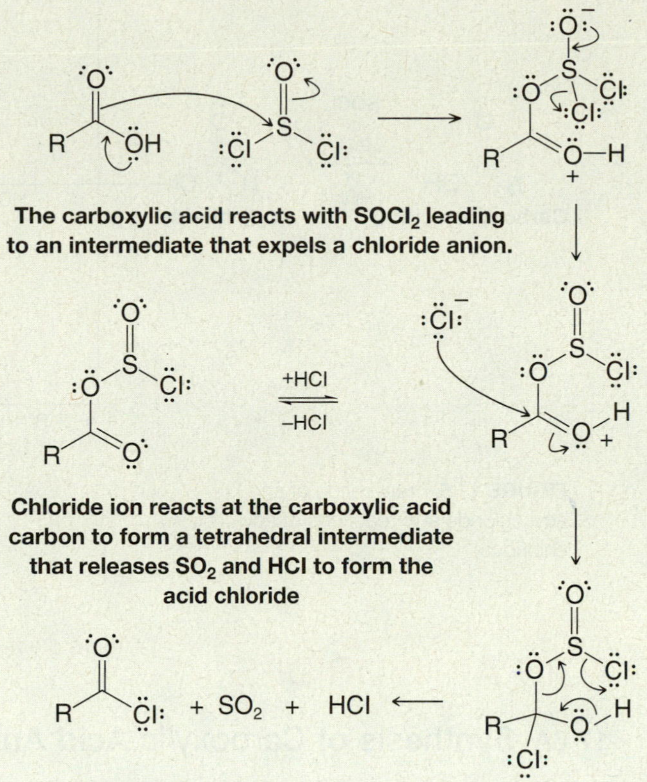

The carboxylic acid reacts with SOCl$_2$ leading to an intermediate that expels a chloride anion.

+HCl
−HCl

Chloride ion reacts at the carboxylic acid carbon to form a tetrahedral intermediate that releases SO$_2$ and HCl to form the acid chloride

R—C(O)—Cl + SO$_2$ + HCl

17.5B Reactions of Acyl Chlorides

Because acyl chlorides are the most reactive of the acyl derivatives, they are easily converted to less reactive ones.

- Often the best synthetic route to an anhydride, an ester, or an amide is synthesis of an acyl chloride from the carboxylic acid and then conversion of the acyl chloride to the desired acyl derivative.

The scheme given in Fig. 17.5 illustrates how this can be done. We examine these reactions in detail in Sections 17.6–17.8.

Acyl chlorides also react with water and (even more rapidly) with aqueous base, but these reactions are usually not carried out deliberately because they destroy the useful acyl chloride reactant by regenerating either the carboxylic acid or its salt:

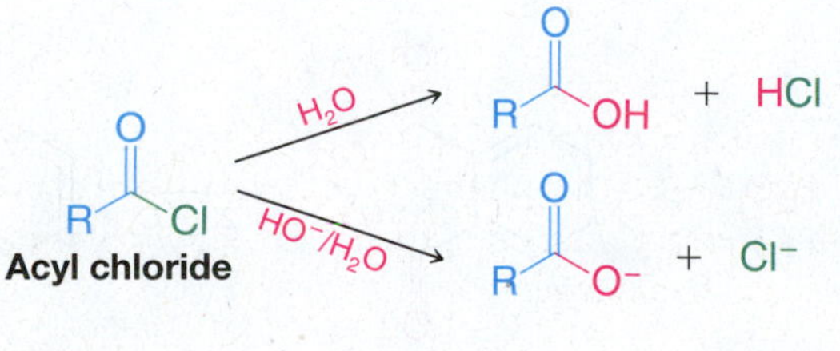

H$_2$O → R—C(O)—OH + HCl

HO$^-$/H$_2$O → R—C(O)—O$^-$ + Cl$^-$

Acyl chloride

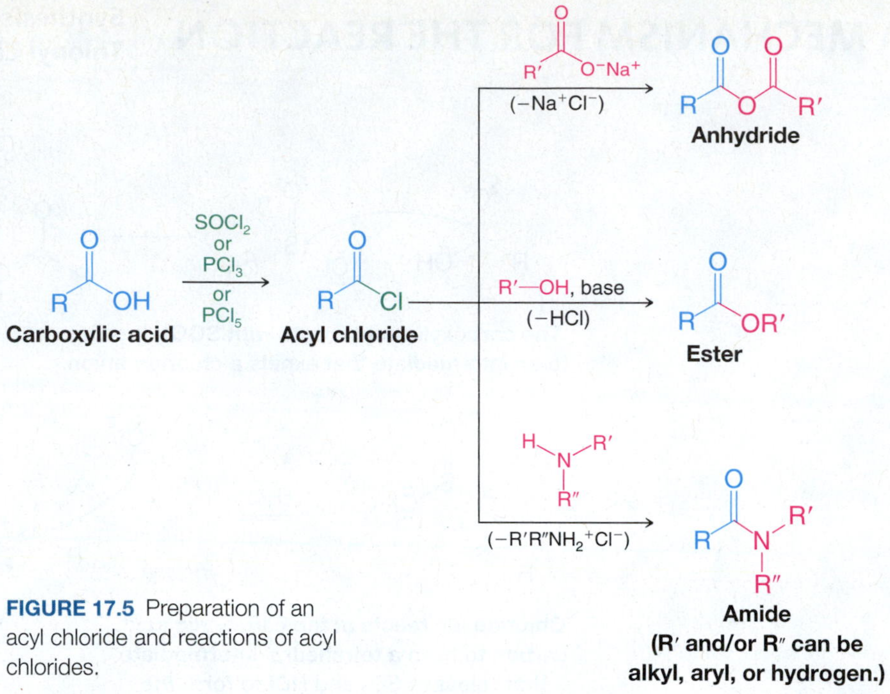

FIGURE 17.5 Preparation of an acyl chloride and reactions of acyl chlorides.

17.6 CARBOXYLIC ACID ANHYDRIDES

17.6A Synthesis of Carboxylic Acid Anhydrides

Carboxylic acids react with acyl chlorides in the presence of pyridine to give carboxylic acid anhydrides. Pyridine deprotonates the carboxylic acid, enhancing its nucleophilicity.

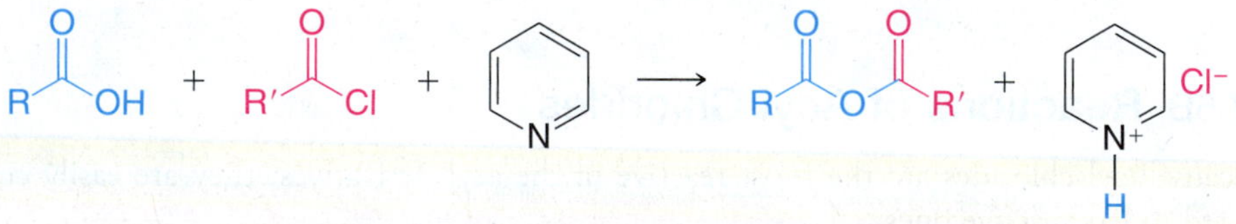

This method is frequently used in the laboratory for the preparation of anhydrides. The method is quite general and can be used to prepare mixed anhydrides (R ≠ R') or symmetric anhydrides (R = R').

Sodium salts of carboxylic acids also react with acyl chlorides to give anhydrides:

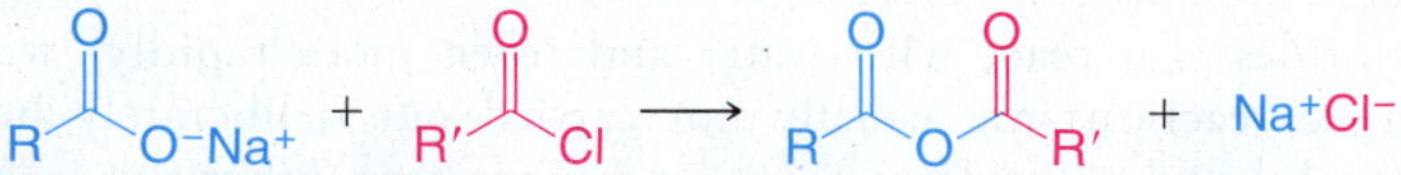

Cyclic anhydrides can sometimes be prepared simply by heating the appropriate dicarboxylic acid. This method succeeds, however, only when anhydride formation leads to a five- or six-membered ring:

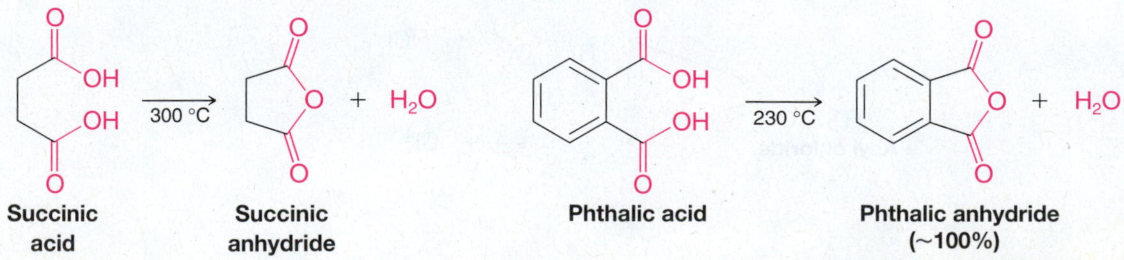

Succinic acid → Succinic anhydride + H$_2$O (300 °C)

Phthalic acid → Phthalic anhydride (~100%) + H$_2$O (230 °C)

PRACTICE PROBLEM 17.8

When maleic acid is heated to 200 °C, it loses water and becomes maleic anhydride. Fumaric acid, a diastereomer of maleic acid, requires a much higher temperature before it dehydrates; when it does, it also yields maleic anhydride. Provide an explanation for these observations.

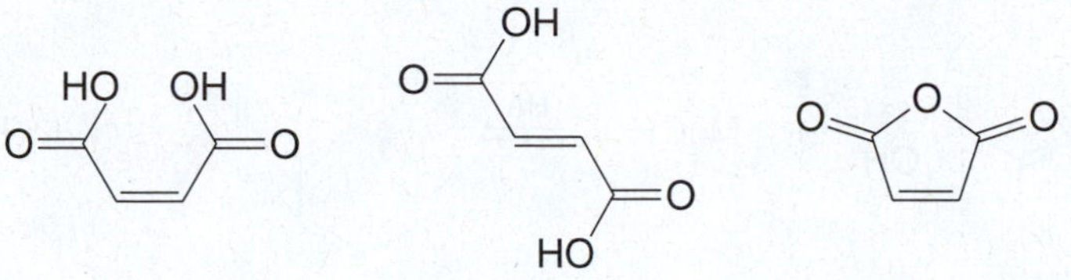

Maleic acid **Fumaric acid** **Maleic anhydride**

17.6B Reactions of Carboxylic Acid Anhydrides

Because carboxylic acid anhydrides are highly reactive, they can be used to prepare esters and amides (Fig. 17.6). We study these reactions in detail in Sections 17.7 and 17.8.

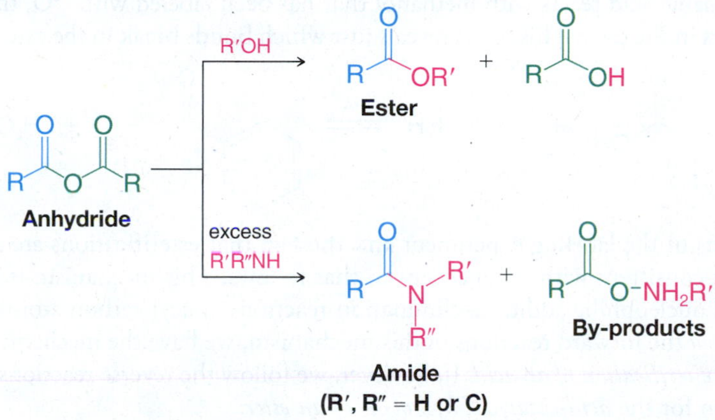

FIGURE 17.6 Reactions of carboxylic acid anhydrides.

Carboxylic acid anhydrides also undergo hydrolysis:

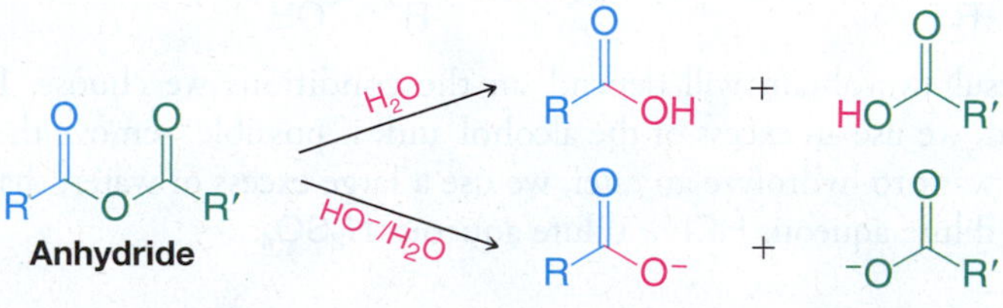

17.7 ESTERS

17.7A Synthesis of Esters: Esterification

Carboxylic acids react with alcohols to form esters through a condensation reaction known as **esterification**:

General Reaction

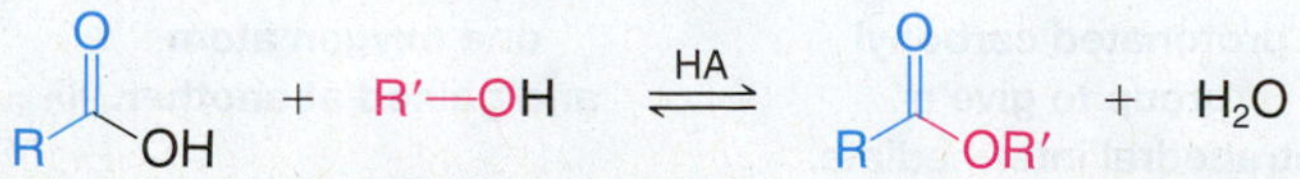

Specific Examples

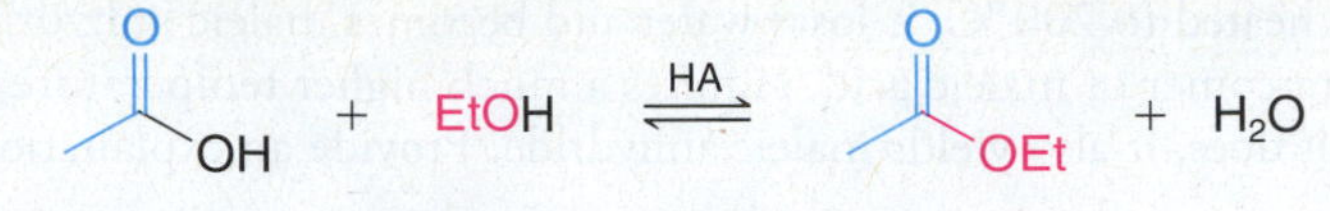

Acetic acid   **Ethanol**   **Ethyl acetate**

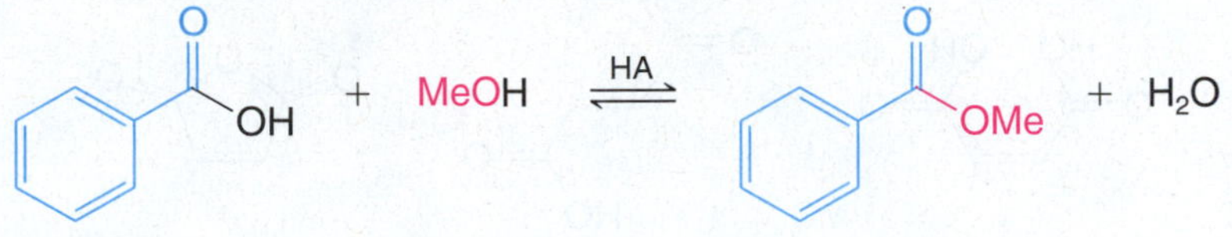

Benzoic acid   **Methanol**   **Methyl benzoate**

- Acid-catalyzed esterifications, such as these examples, are called Fischer esterifications.

Fischer esterifications proceed very slowly in the absence of strong acids, but they reach equilibrium within a matter of a few hours when an acid and an alcohol are refluxed with a small amount of concentrated sulfuric acid or hydrogen chloride. Since the position of equilibrium controls the amount of the ester formed, the use of an excess of either the carboxylic acid or the alcohol increases the yield based on the limiting reagent. Just which component we choose to use in excess will depend on its availability and cost. The yield of an esterification reaction can also be increased by removing water from the reaction mixture as it is formed.

When benzoic acid reacts with methanol that has been labeled with ^{18}O, the labeled oxygen appears in the ester. This result reveals just which bonds break in the esterification:

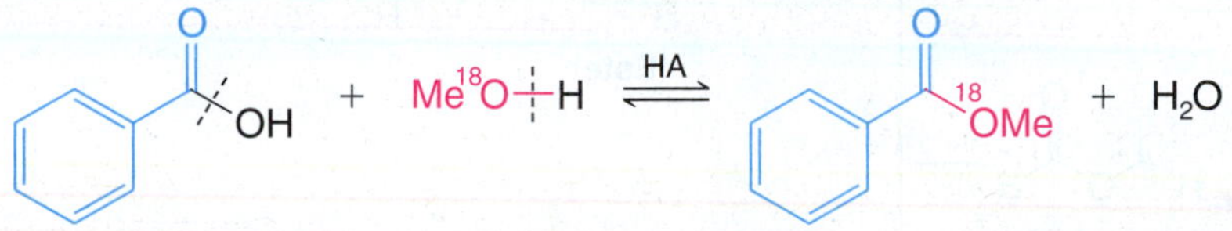

The results of the labeling experiment and the fact that esterifications are acid catalyzed are both consistent with the mechanism that follows. This mechanism is typical of acid-catalyzed nucleophilic addition–elimination reactions at acyl carbon atoms.

If we follow the forward reactions in this mechanism, we have the mechanism for the *acid-catalyzed esterification of an acid*. If, however, we follow the reverse reactions, we have the mechanism for the *acid-catalyzed hydrolysis of an ester*:

Acid-Catalyzed Ester Hydrolysis

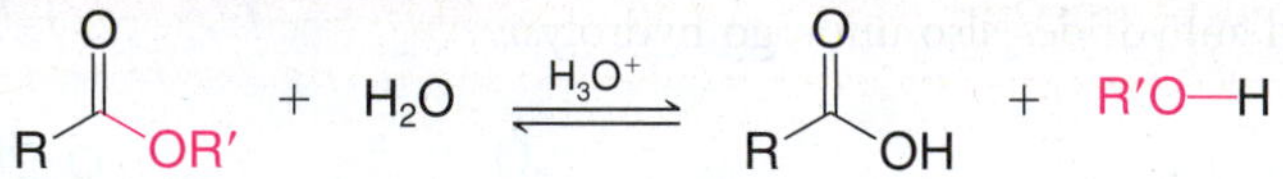

Which result we obtain will depend on the conditions we choose. If we want to esterify an acid, we use an excess of the alcohol and, if possible, remove the water as it is formed. If we want to hydrolyze an ester, we use a large excess of water; that is, we reflux the ester with dilute aqueous HCl or dilute aqueous H_2SO_4.

[A MECHANISM FOR THE REACTION — Acid-Catalyzed Esterification]

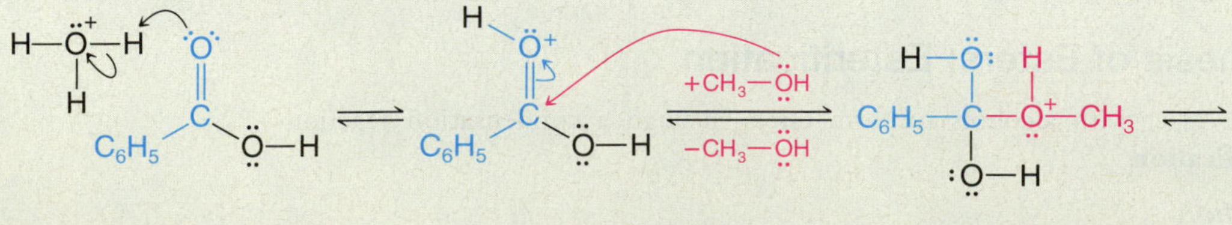

The carboxylic acid accepts a proton from the strong acid catalyst.

The alcohol attacks the protonated carbonyl group to give a tetrahedral intermediate.

A proton is lost at one oxygen atom and gained at another.

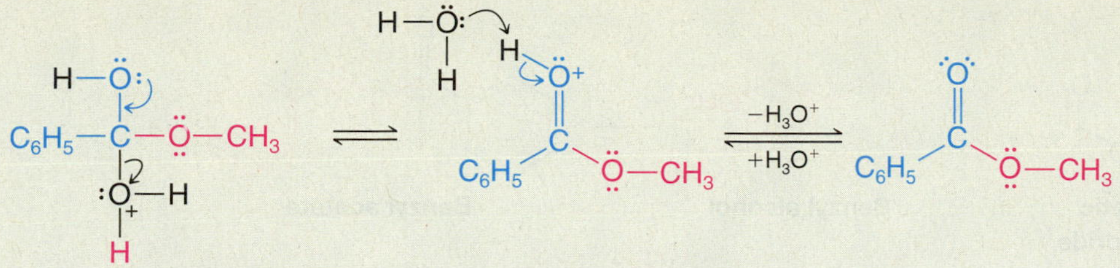

Loss of a molecule of water gives a protonated ester.

Transfer of a proton to a base leads to the ester.

Where would you expect to find the labeled oxygen if you carried out an acid-catalyzed hydrolysis of methyl benzoate in ^{18}O-labeled water? Write a detailed mechanism to support your answer.

PRACTICE PROBLEM 17.9

Steric factors strongly affect the rates of acid-catalyzed hydrolyses of esters. Large groups near the reaction site, whether in the alcohol component or the acid component, slow both reactions markedly. Tertiary alcohols, for example, react so slowly in acid-catalyzed esterifications that they usually undergo elimination instead. However, they can be converted to esters safely through the use of acyl chlorides and anhydrides in the ways that follow.

Esters from Acyl Chlorides

- The reaction of acyl chlorides with alcohols is one of the best ways to synthesize an ester.

The reaction of an acyl chloride with an alcohol to form an ester occurs rapidly and does not require an acid catalyst. Pyridine is often added to the reaction mixture to react with the HCl that forms. (Pyridine may also react with the acyl chloride to form an acylpyridinium ion, an intermediate that is even more reactive toward the nucleophile than the acyl chloride.)

General Reaction

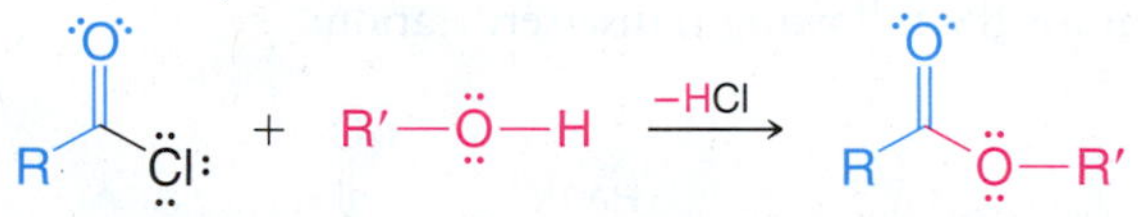

Specific Example

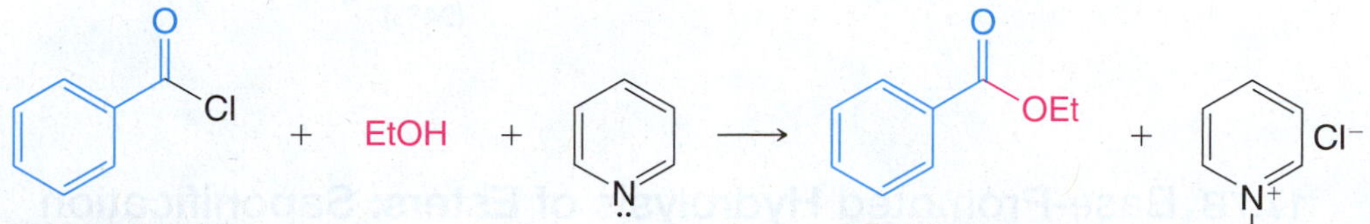

Benzoyl chloride

Ethyl benzoate (80%)

Esters from Carboxylic Acid Anhydrides

Carboxylic acid anhydrides also react with alcohols to form esters in the absence of an acid catalyst.

General Reaction

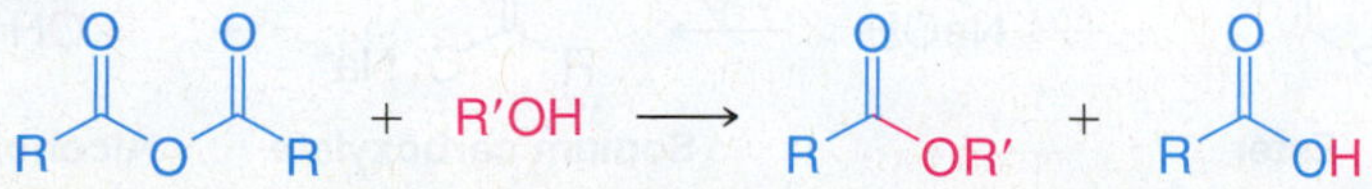

Specific Example

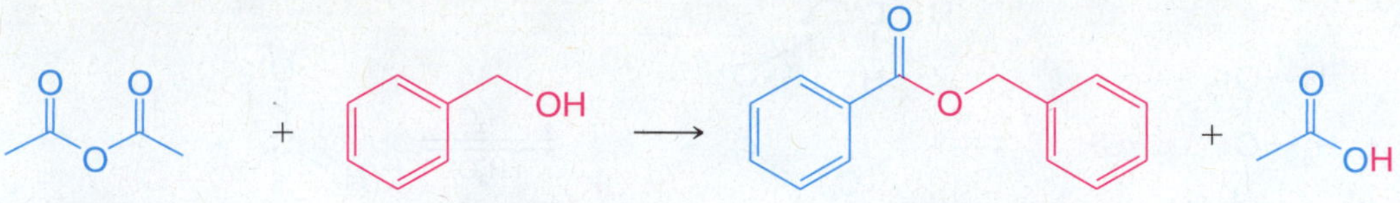

| **Acetic anhydride** | | **Benzyl alcohol** | | **Benzyl acetate** | |

- Ester synthesis is often accomplished best by the reaction of an alcohol with an acyl chloride or anhydride. These reagents avoid the use of a strong acid, as is needed for acid-catalyzed esterification. A strong acid may cause side reactions depending on what other functional groups are present.

Cyclic anhydrides react with one molar equivalent of an alcohol to form compounds that are *both esters and acids*:

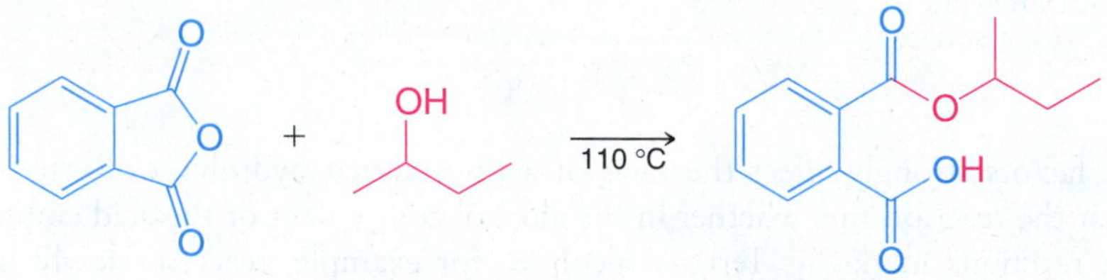

| **Phthalic anhydride** | | *sec*-**Butyl alcohol** | | *sec*-**Butyl hydrogen phthalate** (97%) |

PRACTICE PROBLEM 17.10 Esters can also be synthesized by **transesterification**:

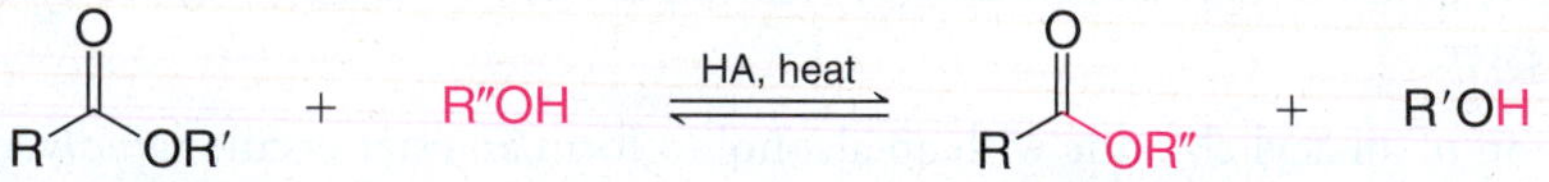

| **High-boiling ester** | | **High-boiling alcohol** | | **Higher boiling ester** | | **Lower boiling alcohol** |

In this procedure we shift the equilibrium to the right by allowing the low-boiling alcohol to distill from the reaction mixture. The mechanism for transesterification is similar to that for an acid-catalyzed esterification (or an acid-catalyzed ester hydrolysis). Write a detailed mechanism for the following transesterification:

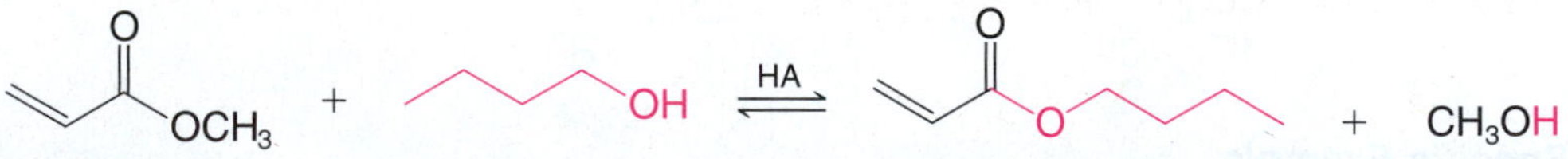

| **Methyl acrylate** | | **Butyl alcohol** | | **Butyl acrylate** (94%) | | **Methanol** |

17.7B Base-Promoted Hydrolysis of Esters: Saponification

- Esters undergo *base-promoted hydrolysis* as well as acid hydrolysis.

Base-promoted hydrolysis is called **saponification**, from the Latin word *sapo*, for soap (see Section 23.2C). Refluxing an ester with aqueous sodium hydroxide, for example, produces an alcohol and the sodium salt of the acid:

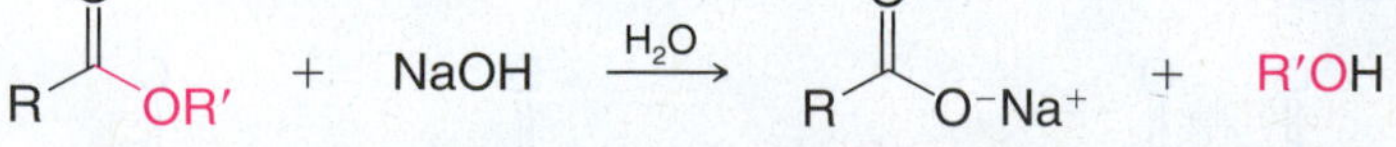

| **Ester** | | | | **Sodium carboxylate** | | **Alcohol** |

The carboxylate ion is very unreactive toward nucleophilic substitution because it is negatively charged. Base-promoted hydrolysis of an ester, as a result, is an essentially irreversible reaction.

The mechanism for base-promoted hydrolysis of an ester also involves a nucleophilic addition–elimination at the acyl carbon.

[A MECHANISM FOR THE REACTION — Base-Promoted Hydrolysis of an Ester]

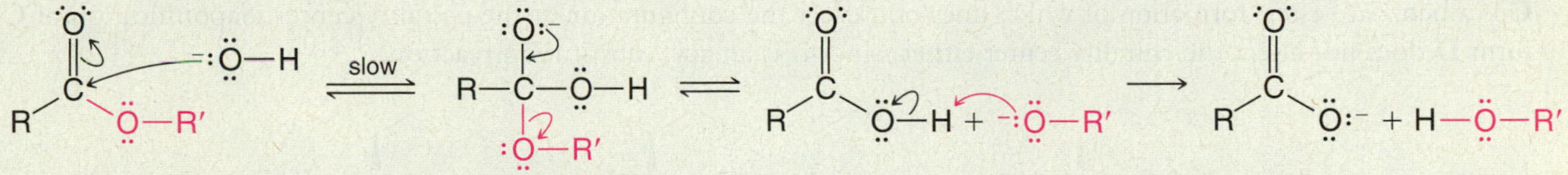

| A hydroxide ion attacks the carbonyl carbon atom. | The tetrahedral intermediate expels an alkoxide ion. | Transfer of a proton leads to the products of the reaction. |

Evidence for this mechanism comes from studies done with isotopically labeled esters. When ethyl propanoate labeled with ^{18}O in the ether-type oxygen of the ester (below) is subjected to hydrolysis with aqueous NaOH, all of the ^{18}O shows up in the ethanol that is produced. None of the ^{18}O appears in the propanoate ion:

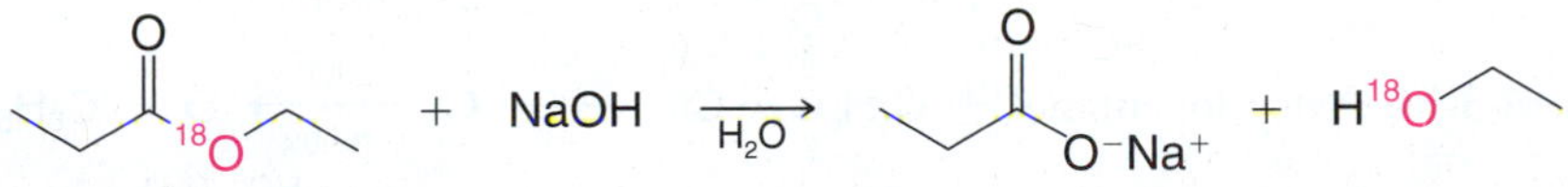

This labeling result is completely consistent with the mechanism given above (outline the steps for yourself and follow the labeled oxygen through to the products). If the hydroxide ion had attacked the alkyl carbon instead of the acyl carbon, the alcohol obtained would not have been labeled. Attack at the alkyl carbon is almost never observed. (For one exception see Practice Problem 17.12.)

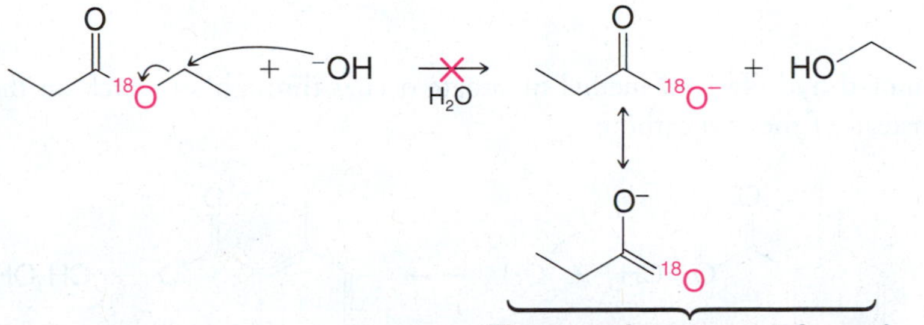

These products are not formed.

Although nucleophilic attack at the alkyl carbon seldom occurs with esters of carboxylic acids, it is the preferred mode of attack with esters of sulfonic acids (e.g., tosylates, mesylates, and triflates; Section 11.10).

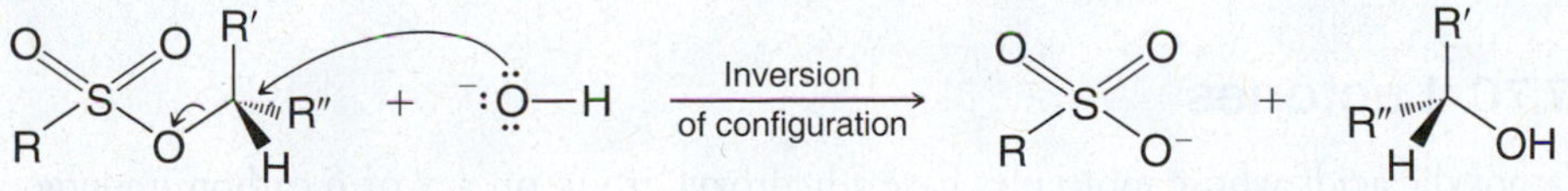

An alkyl sulfonate

This mechanism is preferred with alkyl sulfonates.

●●● **SOLVED PROBLEM 17.5**

Give stereochemical formulas for **A–D**. [*Hint*: **B** and **D** are enantiomers of each other.]

$$(R)\text{-2-Butanol} \xrightarrow{C_6H_5SO_2Cl} A \xrightarrow{HO^-/H_2O} B + C_6H_5SO_3^-$$

$$\xrightarrow{C_6H_5COCl} C \xrightarrow{HO^-/H_2O} D + C_6H_5CO_2^-$$

STRATEGY AND ANSWER: Compound **A** is a benzenesulfonate ester, which forms with retention of configuration from (*R*)-2-butanol. **B** is the S_N2 product formed by reaction with hydroxide, which occurs with **inversion** of configuration. **C** is a benzoate ester, formation of which does not affect the configuration at the chirality center. Saponification of **C** to form **D** does not affect the chirality center either, since it is an acyl substitution reaction.

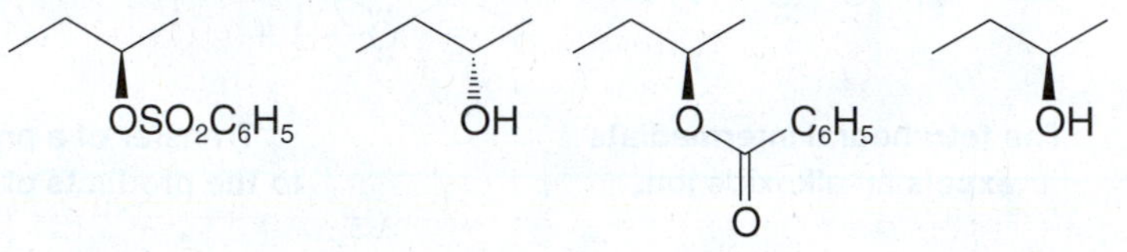

| A | B | C | D |

●●●

PRACTICE PROBLEM 17.11   (a) Write stereochemical formulas for compounds **A–F**:

1. *cis*-3-Methylcyclopentanol + $C_6H_5SO_2Cl \longrightarrow A \xrightarrow[\text{heat}]{HO^-} B + C_6H_5SO_3^-$

2. *cis*-3-Methylcyclopentanol + $C_6H_5\overset{\overset{O}{\|}}{C}-Cl \longrightarrow C \xrightarrow[\text{reflux}]{HO^-} D + C_6H_5CO_2^-$

3. (*R*)-2-Bromooctane + $CH_3CO_2^- Na^+ \longrightarrow E + NaBr \xrightarrow[\text{(reflux)}]{HO^-, H_2O} F$

4. (*R*)-2-Bromooctane + $HO^- \xrightarrow{\text{acetone}} F + Br^-$

(b) Which of the last two methods, **3** or **4**, would you expect to give a higher yield of **F**? Why?

●●●

PRACTICE PROBLEM 17.12   Base-promoted hydrolysis of methyl mesitoate occurs through an attack on the alcohol carbon instead of the acyl carbon:

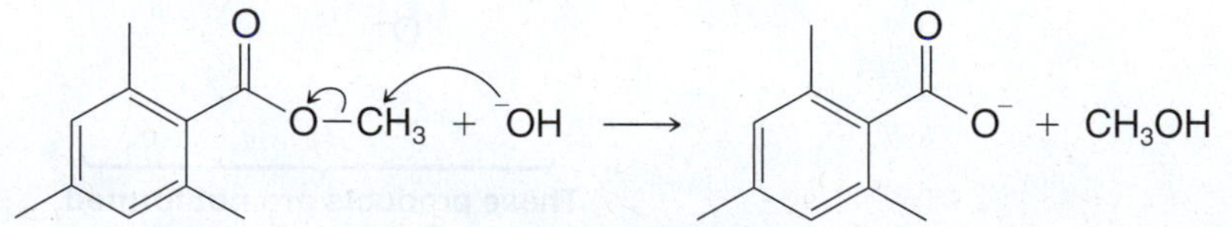

Methyl mesitoate

(a) Can you suggest a reason that accounts for this unusual behavior? (b) Suggest an experiment with labeled compounds that would confirm this mode of attack.

17.7C Lactones

Carboxylic acids whose molecules have a hydroxyl group on a γ or δ carbon undergo an intramolecular esterification to give cyclic esters known as γ- or δ-*lactones*. The reaction is acid catalyzed:

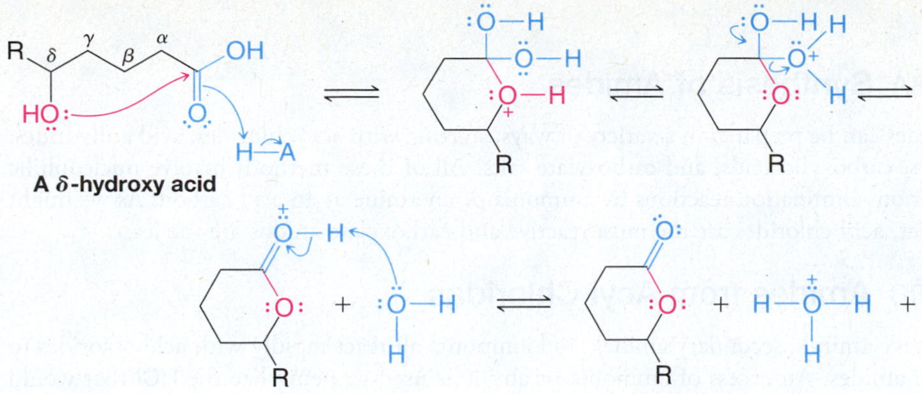

A δ-hydroxy acid

A δ-lactone

Lactones are hydrolyzed by aqueous base just as other esters are. Acidification of the sodium salt, however, may lead spontaneously back to the γ- or δ-lactone, particularly if excess acid is used:

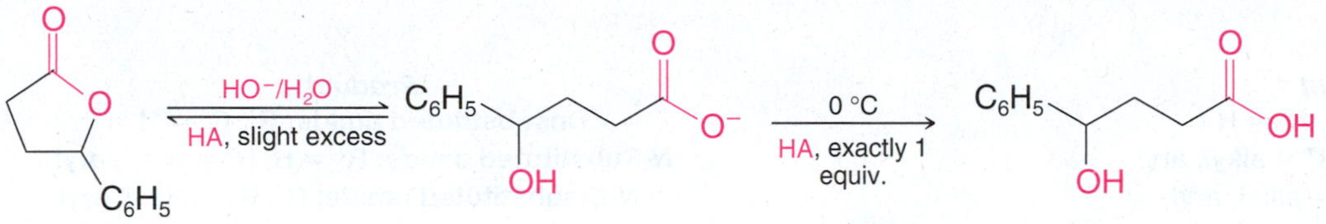

Many lactones occur in nature. Vitamin C (below), for example, is a γ-lactone. Some antibiotics, such as erythromycin and nonactin (Section 11.16), are lactones with very large rings (called macrocyclic lactones), but most naturally occurring lactones are γ- or δ-lactones; that is, most contain five- or six-membered rings.

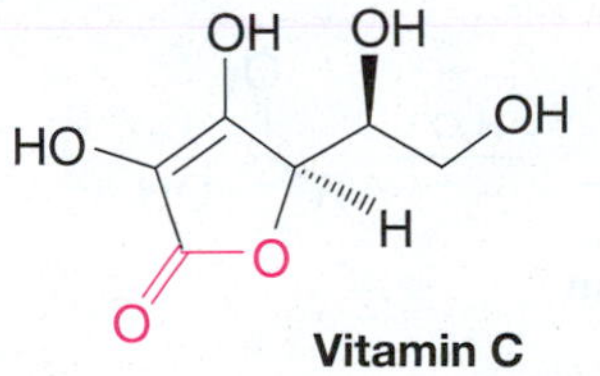

Vitamin C
(ascorbic acid)

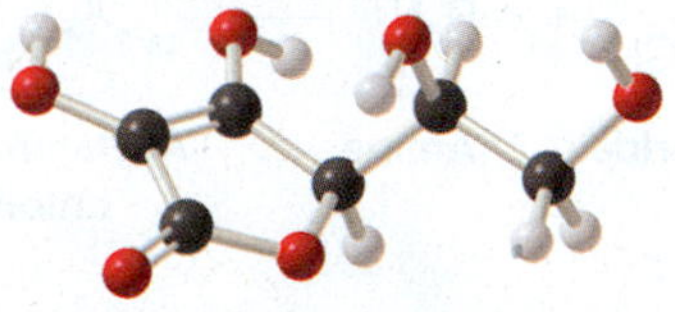

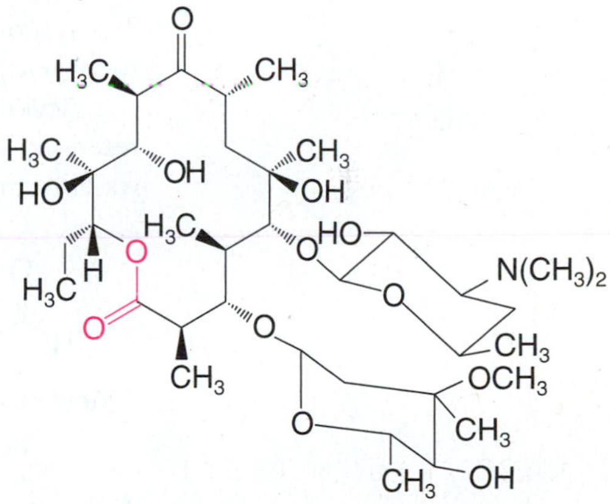

Erythromycin A

β-Lactones (lactones with four-membered rings) have been detected as intermediates in some reactions, and several have been isolated. They are highly reactive, however. If one attempts to prepare a β-lactone from a β-hydroxy acid, β elimination usually occurs instead:

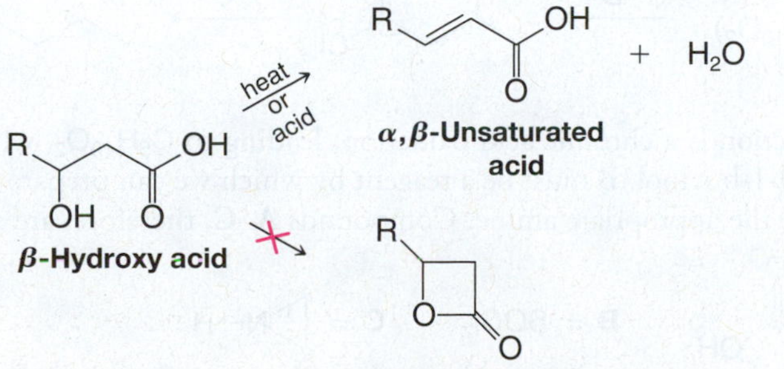

α,β-Unsaturated
acid

β-Hydroxy acid

β-Lactone
(does not form)

17.8 AMIDES

17.8A Synthesis of Amides

Amides can be prepared in a variety of ways, starting with acyl chlorides, acid anhydrides, esters, carboxylic acids, and carboxylate salts. All of these methods involve nucleophilic addition–elimination reactions by ammonia or an amine at an acyl carbon. As we might expect, acid chlorides are the most reactive and carboxylate anions are the least.

17.8B Amides from Acyl Chlorides

Primary amines, secondary amines, and ammonia all react rapidly with acid chlorides to form amides. An excess of ammonia or amine is used to neutralize the HCl that would be formed otherwise:

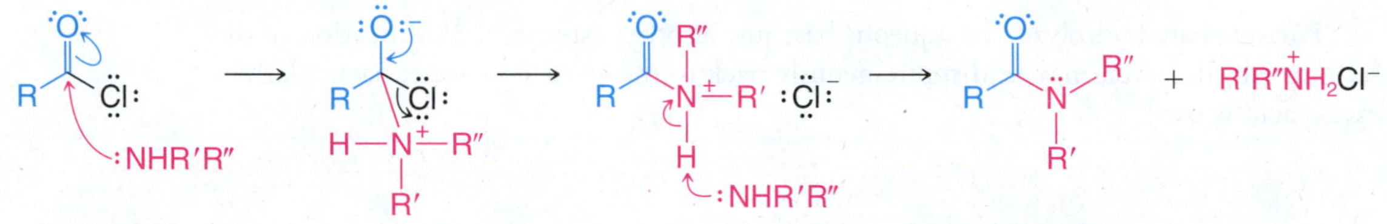

Reactant	**Product**
Ammonia; R′, R″ = H	Unsubstituted amide; R′, R″ = H
1° Amine; R′ = H, R″ = alkyl, aryl	N-Substituted amide; R′ = H, R″ = alkyl, aryl
2° Amine; R′, R″ = alkyl, aryl	N,N-Disubstituted amide; R′, R″ = alkyl, aryl

- The reaction of an amine with an acyl chloride is one of the most widely used laboratory methods for the synthesis of amides, because acyl chlorides are themselves easily prepared from carboxylic acids.

The reaction between an acyl chloride and an amine (or ammonia) usually takes place at room temperature (or below) and produces the amide in high yield.

Acyl chlorides also react with tertiary amines by a nucleophilic addition–elimination reaction. The acylammonium ion that forms, however, is not stable in the presence of water or any hydroxylic solvent:

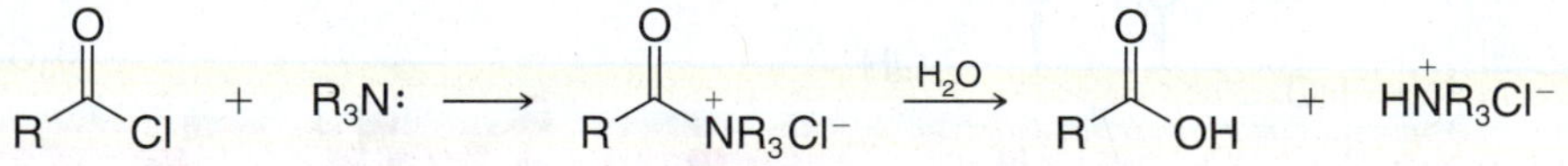

| Acyl chloride | 3° Amine | Acylammonium chloride |

Acylpyridinium ions are probably involved as intermediates in those reactions of acyl chlorides that are carried out in the presence of pyridine.

●●● SOLVED PROBLEM 17.6

Provide the missing compounds, **A–C**, in the following synthesis.

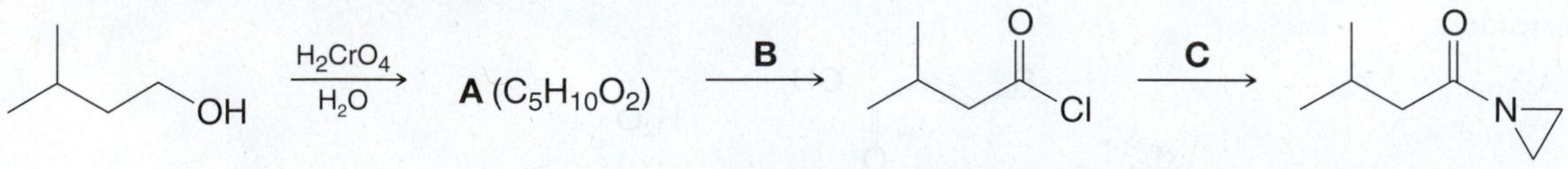

STRATEGY AND ANSWER: The first reaction is a chromic acid oxidation, leading to $C_5H_{10}O_2$, which is consistent with the carboxylic acid derived from 3-methyl-1-butanol. **B** must be a reagent by which we can prepare an acid chloride. The final product is an amide, thus **C** must be the appropriate amine. Compounds **A–C**, therefore, are as follows:

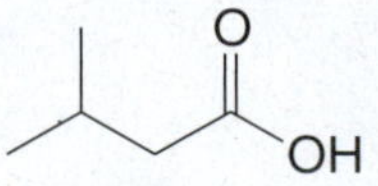

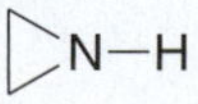

17.8C Amides from Carboxylic Anhydrides

Acid anhydrides react with ammonia and with primary and secondary amines to form amides through reactions that are analogous to those of acyl chlorides:

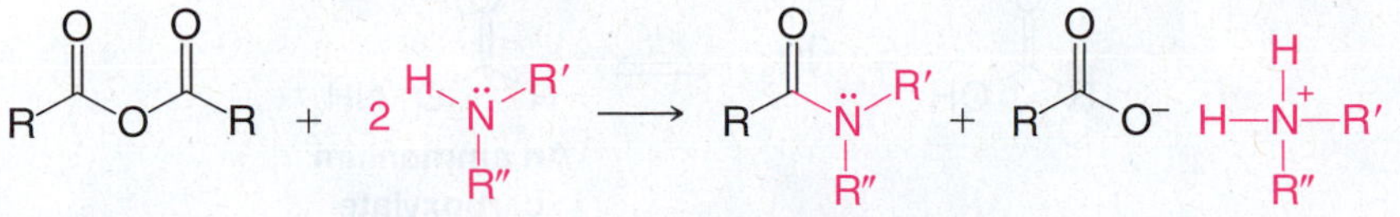

R', R" can be H, alkyl, or aryl.

Cyclic anhydrides react with ammonia or an amine in the same general way as acyclic anhydrides; however, the reaction yields a product that is both an amide and an ammonium salt. Acidifying the ammonium salt gives a compound that is both an amide and an acid:

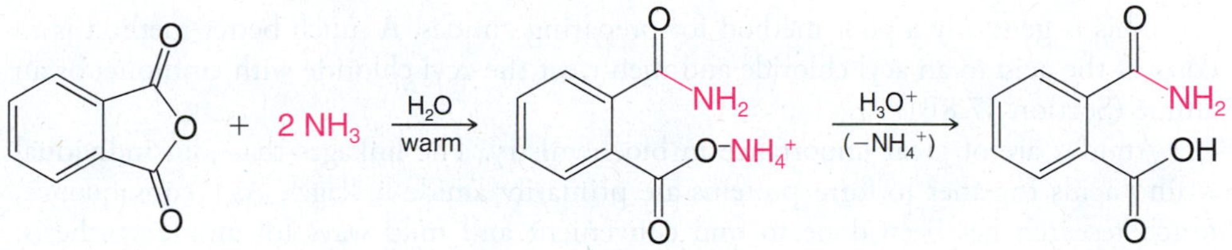

| Phthalic anhydride | Ammonium phthalamate (94%) | Phthalamic acid (81%) |

Heating the amide acid causes dehydration to occur and gives an *imide*. Imides contain the linkage

$$-\overset{O}{\underset{\|}{C}}-NH-\overset{O}{\underset{\|}{C}}-.$$

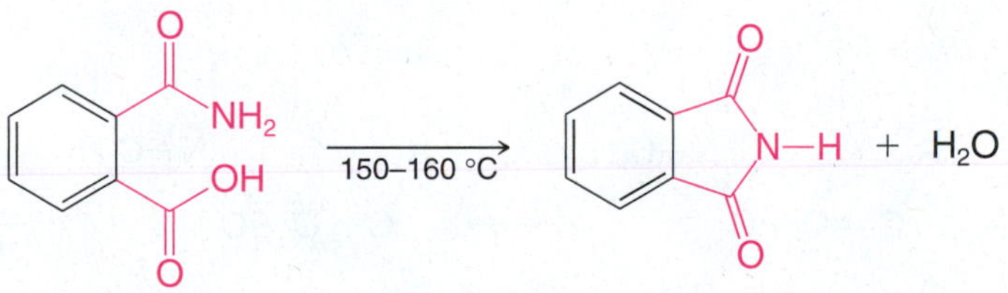

Phthalamic acid   150–160 °C   Phthalimide (~100%) + H₂O

17.8D Amides from Esters

Esters undergo nucleophilic addition–elimination at their acyl carbon atoms when they are treated with ammonia (called *ammonolysis*) or with primary and secondary amines. These reactions take place much more slowly than those of acyl chlorides and anhydrides, but they can still be synthetically useful:

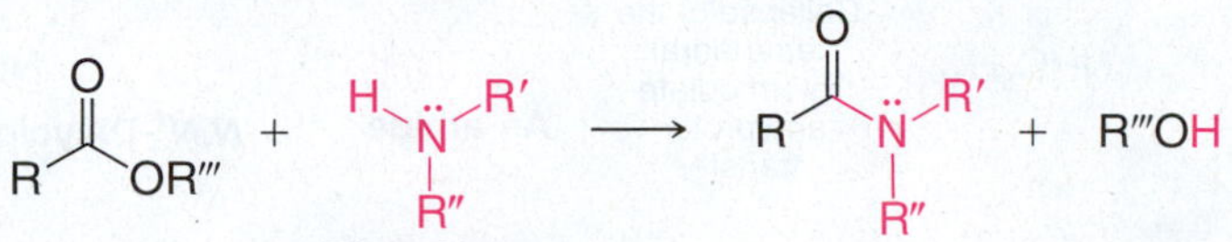

R' and/or R"may be H.

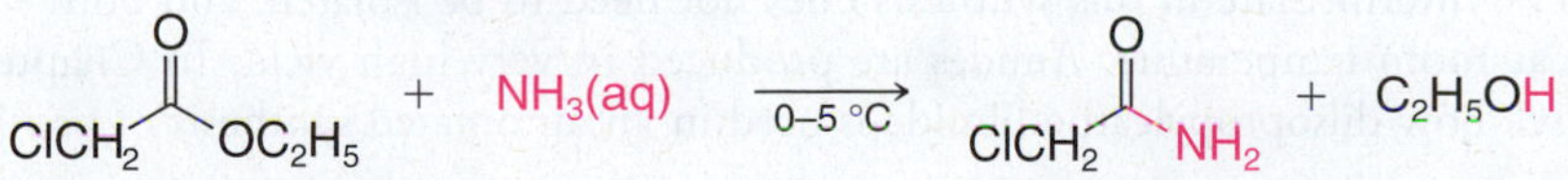

Ethyl chloroacetate   + NH₃(aq) 0–5 °C   Chloroacetamide (62–87%) + C₂H₅OH

17.8E Amides from Carboxylic Acids and Ammonium Carboxylates

Carboxylic acids react with aqueous ammonia to form ammonium salts:

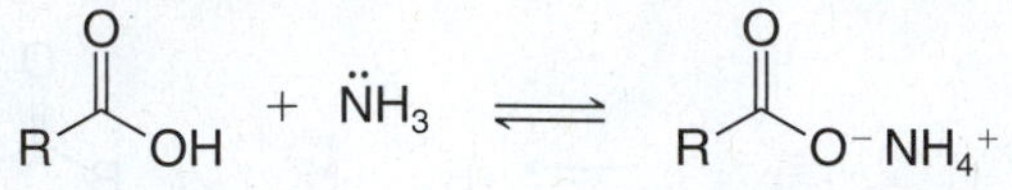

An ammonium carboxylate

Because of the low reactivity of the carboxylate ion toward nucleophilic addition–elimination, further reaction does not usually take place in aqueous solution. However, if we evaporate the water and subsequently heat the dry salt, dehydration produces an amide:

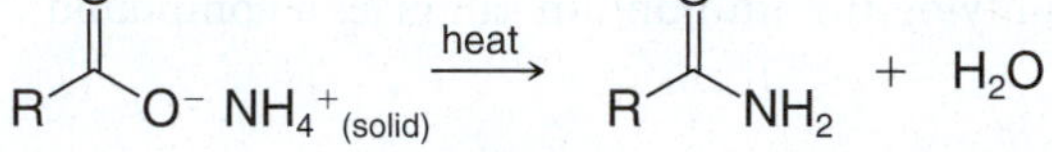

This is generally a poor method for preparing amides. A much better method is to convert the acid to an acyl chloride and then treat the acyl chloride with ammonia or an amine (Section 17.8B).

Amides are of great importance in biochemistry. The linkages that join individual amino acids together to form proteins are primarily amide linkages. As a consequence, much research has been done to find convenient and mild ways for amide synthesis. Dialkylcarbodiimides (R—N=C=N—R), such as diisopropylcarbodiimide and dicyclohexylcarbodiimide (DCC), are especially useful reagents for amide synthesis. Dialkylcarbodiimides promote amide formation by reacting with the carboxyl group of an acid and activating it toward nucleophilic addition–elimination.

[A MECHANISM FOR THE REACTION — DCC-Promoted Amide Synthesis]

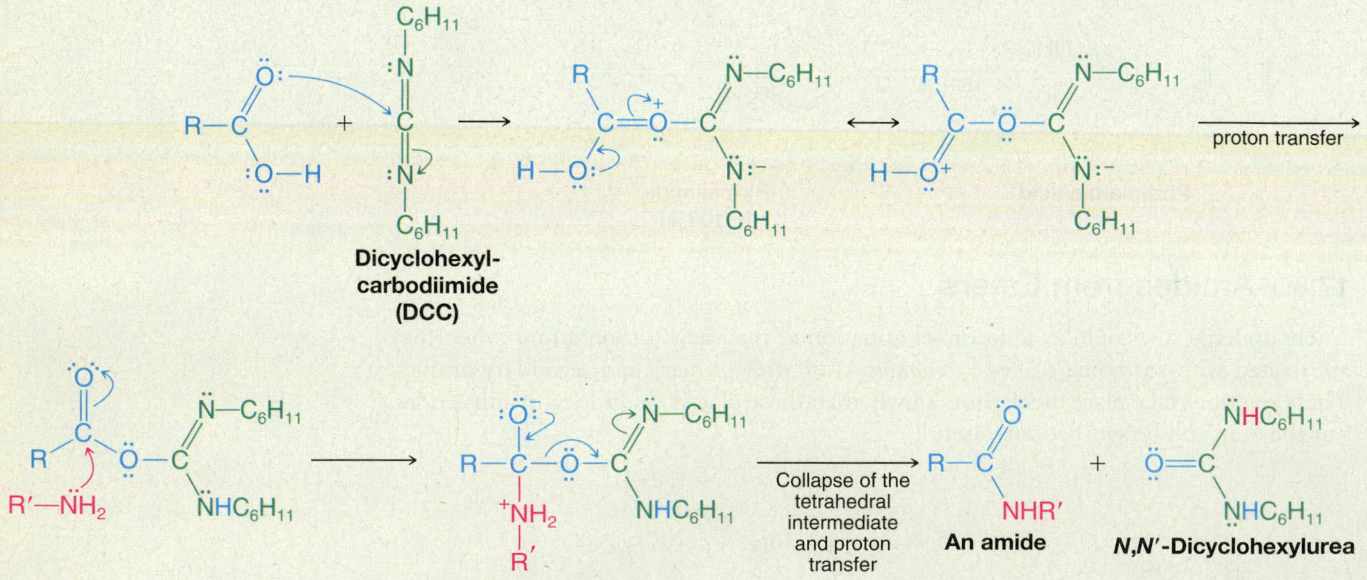

The intermediate in this synthesis does not need to be isolated, and both steps take place at room temperature. Amides are produced in very high yield. In Chapter 24 we shall see how diisopropylcarbodiimide is used in an automated synthesis of peptides.

17.8F Hydrolysis of Amides

• Amides undergo hydrolysis when they are heated with aqueous acid or aqueous base.

Acidic Hydrolysis

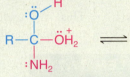

Basic Hydrolysis

N-Substituted amides and *N,N*-disubstituted amides also undergo hydrolysis in aqueous acid or base. Amide hydrolysis by either method takes place more slowly than the corresponding hydrolysis of an ester. Thus, amide hydrolyses generally require the forcing conditions of heat and strong acid or base.

The mechanism for acid hydrolysis of an amide is similar to that given in Section 17.7A for the acid hydrolysis of an ester. Water acts as a nucleophile and attacks the protonated amide. The leaving group in the acidic hydrolysis of an amide is ammonia (or an amine).

[A MECHANISM FOR THE REACTION — Acidic Hydrolysis of an Amide]

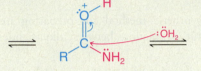

| The amide carbonyl accepts a proton from the aqueous acid. | A water molecule attacks the protonated carbonyl to give a tetrahedral intermediate. | A proton is lost at one oxygen and gained at the nitrogen. |

Loss of a molecule of ammonia gives a protonated carboxylic acid.

Transfer of a proton to ammonia leads to the carboxylic acid and an ammonium ion.

There is evidence that in basic hydrolyses of amides, hydroxide ions act both as nucleophiles and as bases.

[A MECHANISM FOR THE REACTION — Basic Hydrolysis of an Amide]

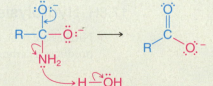

A hydroxide ion attacks the acyl carbon of the amide.

A hydroxide ion removes a proton to give a dianion.

The dianion loses a molecule of ammonia (or an amine); this step is synchronized with a proton transfer from water due to the basicity of NH_2^-.

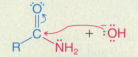
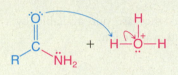

Hydrolysis of amides by enzymes is central to the digestion of proteins. The mechanism for protein hydrolysis by the enzyme chymotrypsin is presented in Section 24.11.

PRACTICE PROBLEM 17.13   What products would you obtain from acidic and basic hydrolysis of each of the following amides?

(a) *N,N*-Diethylbenzamide

(b)

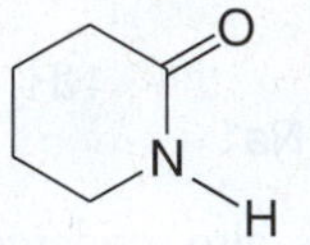

(c) 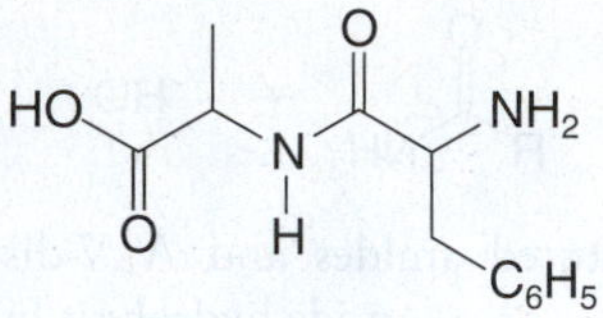 (a dipeptide)

17.8G Nitriles from the Dehydration of Amides

Amides react with P_4O_{10} (a compound that is often called phosphorus pentoxide and written P_2O_5) or with boiling acetic anhydride to form nitriles:

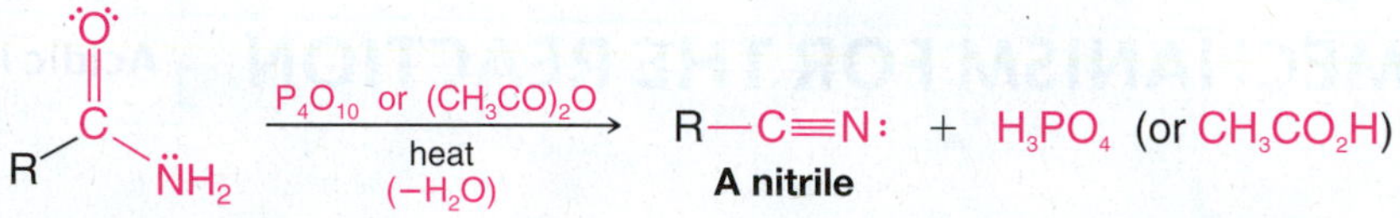

$$ \underset{R}{\overset{\ddot{O}}{\parallel}}\overset{}{C}-\ddot{N}H_2 \xrightarrow[\text{(−H}_2\text{O)}]{\overset{P_4O_{10} \text{ or } (CH_3CO)_2O}{\text{heat}}} R-C\equiv N\colon\ +\ H_3PO_4 \text{ (or } CH_3CO_2H) $$

A nitrile

This is a useful synthetic method for preparing nitriles that are not available by nucleophilic substitution reactions between alkyl halides and cyanide ion.

●●● SOLVED PROBLEM 17.7

At first glance the conversion of bromobenzene to benzenenitrile looks simple—just carry out a nucleophilic substitution using cyanide ion as the nucleophile. Then we remember that bromobenzene does not undergo either an S_N1 or an S_N2 reaction (Section 6.14A). The conversion can be accomplished, however, though it involves several steps. Outline possible steps.

ANSWER:

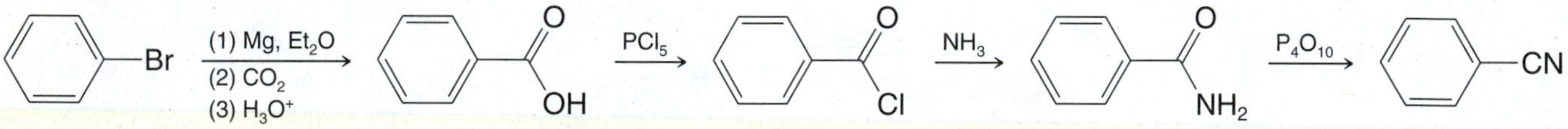

C₆H₅—Br $\xrightarrow[\text{(3) H}_3\text{O}^+]{\overset{\text{(1) Mg, Et}_2\text{O}}{\text{(2) CO}_2}}$ C₆H₅COOH $\xrightarrow{PCl_5}$ C₆H₅COCl $\xrightarrow{NH_3}$ C₆H₅CONH₂ $\xrightarrow{P_4O_{10}}$ C₆H₅—CN

●●● PRACTICE PROBLEM 17.14

(a) Provide the reagents required to accomplish the following transformation.

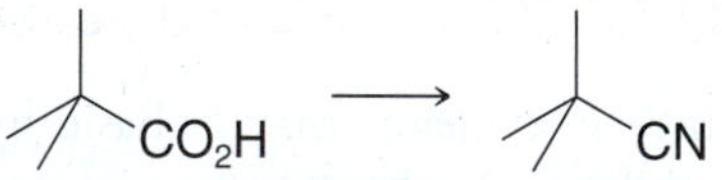

$(CH_3)_3C{-}CO_2H \longrightarrow (CH_3)_3C{-}CN$

(b) What product would you likely obtain if you attempted to synthesize the nitrile above by the following method?

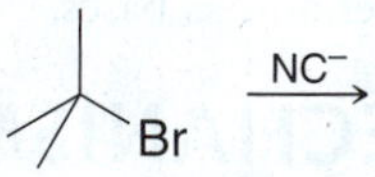

$(CH_3)_3C{-}Br \xrightarrow{NC^-}$

17.8H Hydrolysis of Nitriles

- Nitriles are related to carboxylic acids because complete hydrolysis of a nitrile produces a carboxylic acid or a carboxylate anion (Sections 16.9 and 17.3):

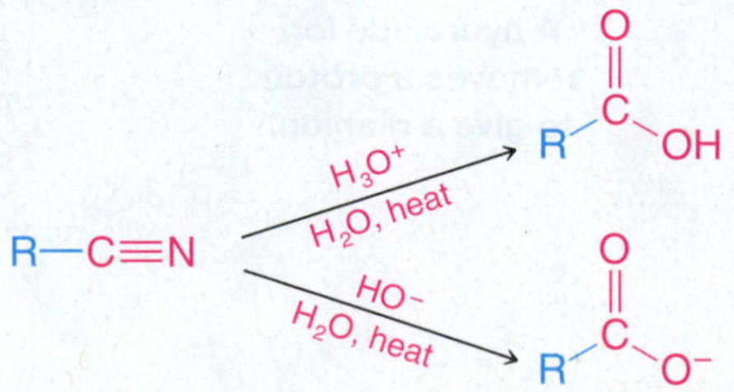

$$ R-C\equiv N \quad \begin{cases} \xrightarrow[\text{H}_2\text{O, heat}]{\text{H}_3\text{O}^+} & R-\overset{O}{\overset{\parallel}{C}}-OH \\[2ex] \xrightarrow[\text{H}_2\text{O, heat}]{\text{HO}^-} & R-\overset{O}{\overset{\parallel}{C}}-O^- \end{cases} $$

The mechanisms for these hydrolyses are related to those for the acidic and basic hydrolyses of amides.

In **acidic hydrolysis** of a nitrile the first step is protonation of the nitrogen atom. This protonation (in the following sequence) enhances polarization of the nitrile group and makes the carbon atom more susceptible to nucleophilic attack by the weak nucleophile, water. The loss of a proton from the oxygen atom then produces a tautomeric form of an amide. Gain of a proton at the nitrogen atom gives a **protonated amide**, and from this point on the steps are the same as those given for the acidic hydrolysis of an amide in Section 17.8F. (In concentrated H_2SO_4 the reaction stops at the protonated amide, and this is a useful way of making amides from nitriles.)

A MECHANISM FOR THE REACTION — Acidic Hydrolysis of a Nitrile

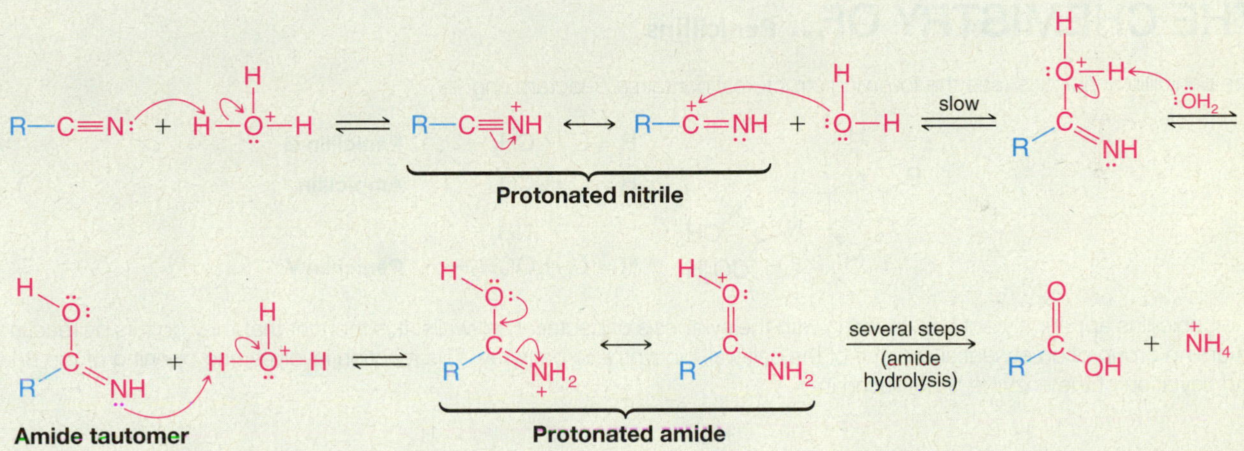

In **basic hydrolysis**, a hydroxide ion attacks the nitrile carbon atom, and subsequent protonation leads to the amide tautomer. Further attack by the hydroxide ion leads to hydrolysis in a manner analogous to that for the basic hydrolysis of an amide (Section 17.8F). (Under the appropriate conditions, amides can be isolated when nitriles are hydrolyzed.)

A MECHANISM FOR THE REACTION — Basic Hydrolysis of a Nitrile

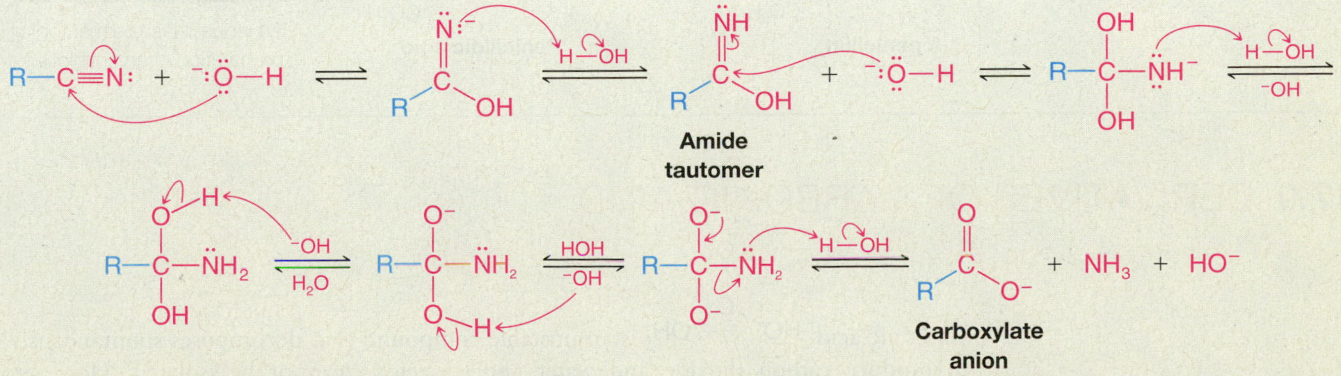

17.8I Lactams

Cyclic amides are called **lactams**. The size of the lactam ring is designated by Greek letters in a way that is analogous to lactone nomenclature (Section 17.7C):

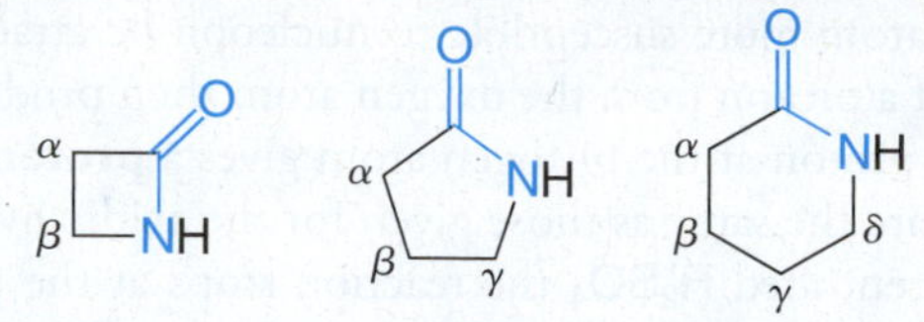

A β-lactam   **A γ-lactam**   **A δ-lactam**

γ-Lactams and δ-lactams often form spontaneously from γ- and δ-amino acids. β-Lactams, however, are highly reactive; their strained four-membered rings open easily in the presence of nucleophilic reagents.

THE CHEMISTRY OF... Penicillins

The penicillin antibiotics (see the following structures) contain a β-lactam ring:

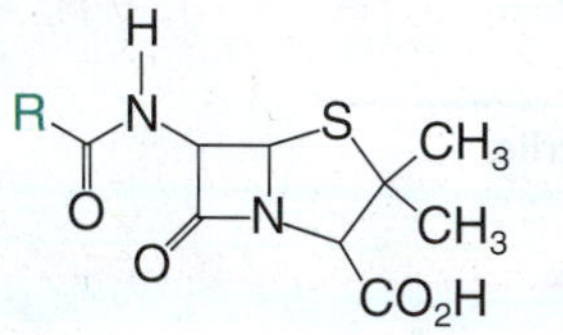

$R = C_6H_5CH_2-$ **Penicillin G**

$R = C_6H_5CH-$ **Ampicillin**
         |
        NH_2

$R = C_6H_5OCH_2-$ **Penicillin V**

The penicillins apparently act by interfering with the synthesis of bacterial cell walls. It is thought that they do this by reacting with an amino group of an essential enzyme of the cell wall biosynthetic pathway. This reaction involves ring opening of the β-lactam and acylation of the enzyme, inactivating it.

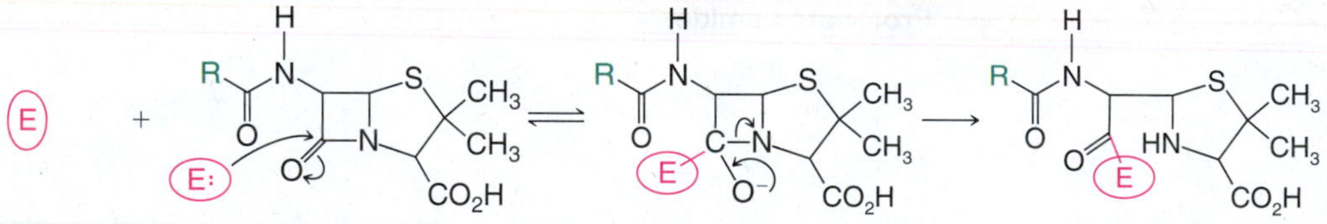

Active enzyme   **A penicillin**   **Inactive enzyme**   **Inactive enzyme**

Bacterial resistance to the penicillin antibiotics is a serious problem for the treatment of infections. Bacteria that have developed resistance to penicillin produce an enzyme called penicillinase. Penicillinase hydrolyzes the β-lactam ring of penicillin, resulting in penicilloic acid. Because penicilloic acid cannot act as an acylating agent, it is incapable of blocking bacterial cell wall synthesis by the mechanism shown above.

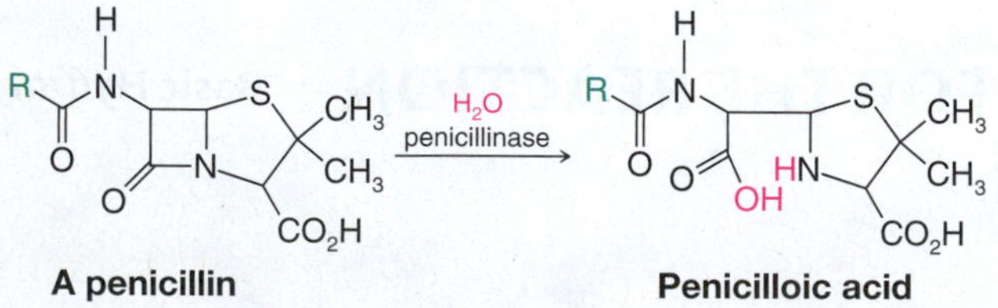

A penicillin   **Penicilloic acid**

An industrial-scale reactor for preparation of an antibiotic.

Michael Rosenfield/Stone/Getty Images

17.9 DERIVATIVES OF CARBONIC ACID

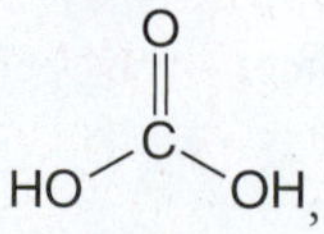

Carbonic acid, $HO-\overset{\displaystyle O}{\overset{\displaystyle \|}{C}}-OH$, is an unstable compound that decomposes spontaneously to produce carbon dioxide and water and, therefore, cannot be isolated. However, many acyl chlorides, esters, and amides that are derived from carbonic acid are stable compounds that have important applications.

Carbonyl dichloride (ClCOCl), a highly toxic compound that is also called *phosgene*, can be thought of as the diacyl chloride of carbonic acid. Carbonyl dichloride reacts by nucleophilic addition–elimination with two molar equivalents of an alcohol to yield a **dialkyl carbonate**:

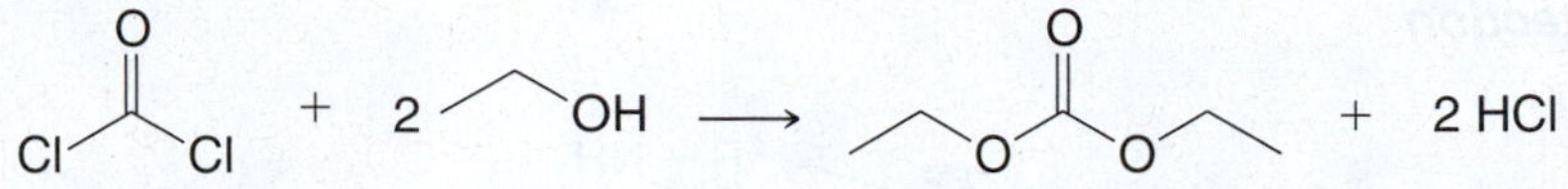

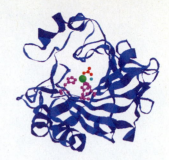

Carbonic anhydrase

Carbonic anhydrase is an enzyme that interconverts water and carbon dioxide with carbonic acid. A carbonate dianion is shown in red within the structure of carbonic anhydrase above.

Carbonyl dichloride + 2 ⟍OH ⟶ **Diethyl carbonate** + 2 HCl

A tertiary amine is usually added to the reaction to neutralize the hydrogen chloride that is produced.

Carbonyl dichloride reacts with ammonia to yield **urea** (Section 1.1A):

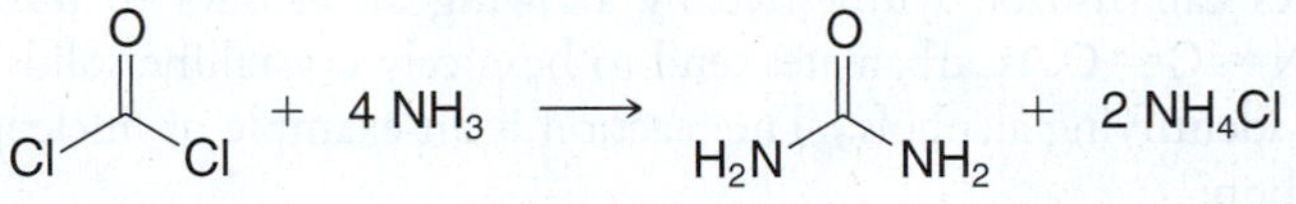

+ 4 NH_3 ⟶ (urea) + 2 NH_4Cl

Urea is the end product of the metabolism of nitrogen-containing compounds in most mammals and is excreted in the urine.

17.9A Alkyl Chloroformates and Carbamates (Urethanes)

Treating carbonyl dichloride with one molar equivalent of an alcohol leads to the formation of an alkyl chloroformate:

ROH + (carbonyl dichloride) ⟶ **Alkyl chloroformate** + HCl

Specific Example

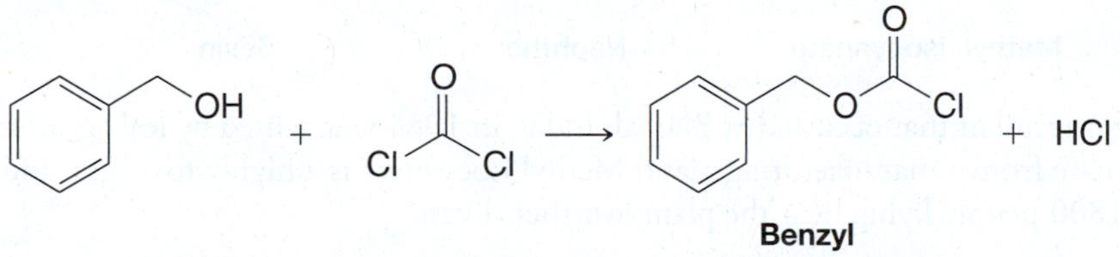

(benzyl alcohol) + (carbonyl dichloride) ⟶ **Benzyl chloroformate** + HCl

Alkyl chloroformates react with ammonia or amines to yield compounds called *carbamates* or *urethanes*:

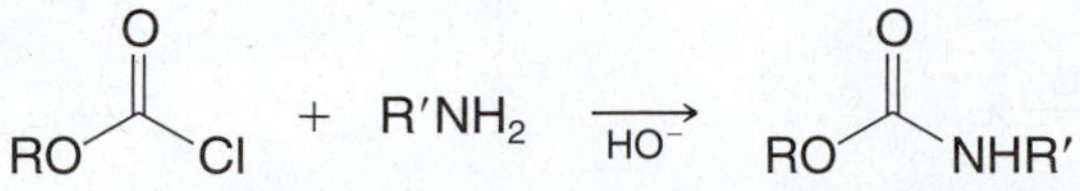

(alkyl chloroformate) + R′NH_2 →[HO$^-$] **A carbamate (or urethane)**

Benzyl chloroformate is used to install an amino protecting (blocking) group called the benzyloxycarbonyl group. We shall see in Section 24.7A how this protecting group is used in the synthesis of peptides and proteins. One advantage of the benzyloxycarbonyl group is that it can be removed under mild conditions. Treating the benzyloxycarbonyl derivative with hydrogen and a catalyst or with cold HBr in acetic acid removes the protecting group:

Protection

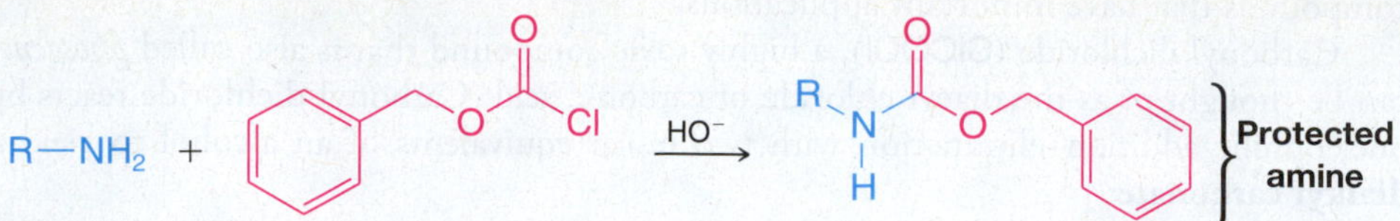

$$R-NH_2 + \text{[benzyl chloroformate]} \xrightarrow{HO^-} \text{[protected amine structure]} \left.\begin{array}{c}\end{array}\right\} \text{Protected amine}$$

Deprotection

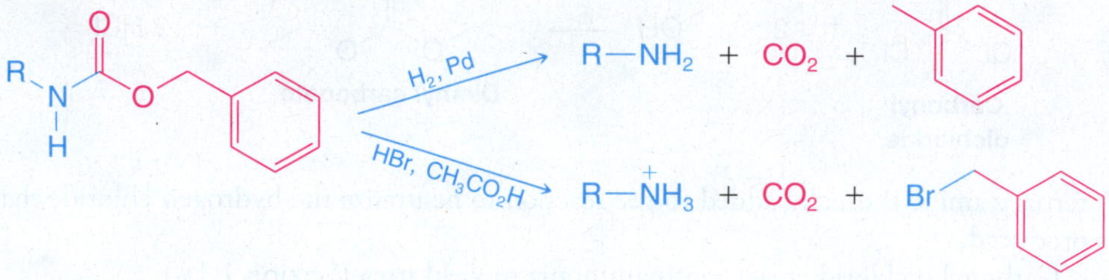

$$\text{[carbamate structure]} \xrightarrow{H_2,\ Pd} R-NH_2 + CO_2 + \text{[toluene]}$$

$$\xrightarrow{HBr,\ CH_3CO_2H} R-\overset{+}{N}H_3 + CO_2 + \text{Br}\text{[benzyl bromide]}$$

Carbamates can also be synthesized by allowing an alcohol to react with an isocyanate, $R-N=C=O$. (Carbamates tend to be nicely crystalline solids and are useful derivatives for identifying alcohols.) The reaction is an example of nucleophilic addition to the acyl carbon:

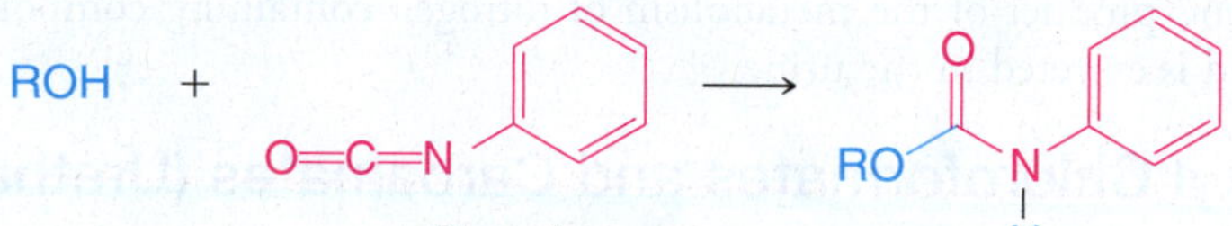

$$ROH + \underset{\substack{\textbf{Phenyl} \\ \textbf{isocyanate}}}{O=C=N\text{[phenyl]}} \longrightarrow RO\text{[carbamate structure]}$$

The insecticide called *Sevin* is a carbamate made by allowing 1-naphthol to react with methyl isocyanate:

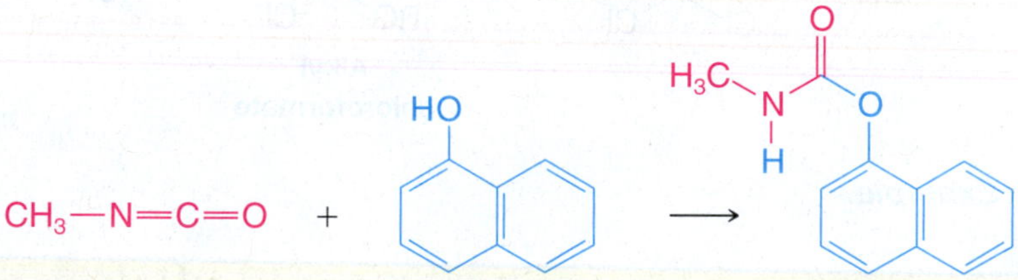

$$\underset{\textbf{Methyl isocyanate}}{CH_3-N=C=O} + \underset{\textbf{1-Naphthol}}{\text{[1-naphthol structure]}} \longrightarrow \underset{\textbf{Sevin}}{\text{[Sevin structure]}}$$

A tragic accident that occurred at Bhopal, India, in 1984 was caused by leakage of methyl isocyanate from a manufacturing plant. Methyl isocyanate is a highly toxic gas, and more than 1800 people living near the plant lost their lives.

● ● ●

PRACTICE PROBLEM 17.15 Write structures for the products of the following reactions:

(a) $C_6H_5CH_2OH + C_6H_5N=C=O \longrightarrow$

(b) $ClCOCl + \text{excess } CH_3NH_2 \longrightarrow$

(c) Glycine $(H_3\overset{+}{N}CH_2CO_2^-) + C_6H_5CH_2OCOCl \xrightarrow{HO^-}$

(d) Product of (c) $+ H_2, Pd \longrightarrow$

(e) Product of (c) $+ \text{cold HBr, } CH_3CO_2H \longrightarrow$

(f) Urea $+ HO^-, H_2O, \text{heat}$

Although alkyl chloroformates ($ROCOCl$), dialkyl carbonates ($ROCOOR$), and carbamates ($ROCONH_2$, $ROCONHR$, etc.) are stable, chloroformic acid ($HOCOCl$), alkyl hydrogen carbonates ($ROCOOH$), and carbamic acid ($HOCONH_2$) are not. These latter compounds decompose spontaneously to liberate carbon dioxide:

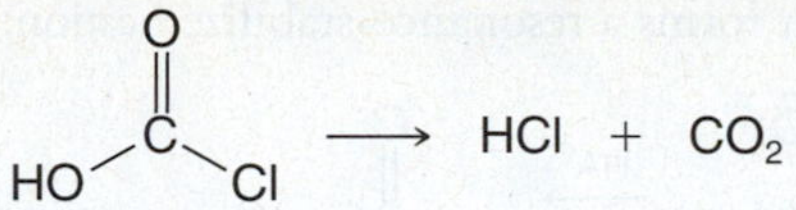

Chloroformic acid
(unstable)

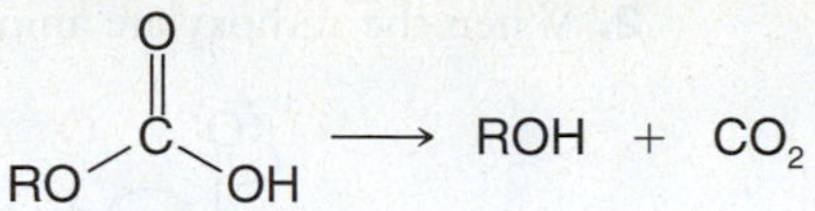

An alkyl hydrogen carbonate
(unstable)

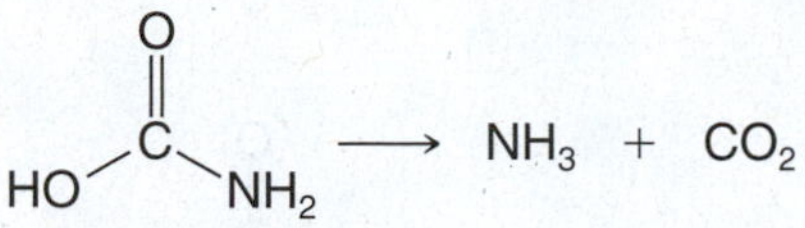

A carbamic acid
(unstable)

This instability is a characteristic that these compounds share with their functional parent, carbonic acid:

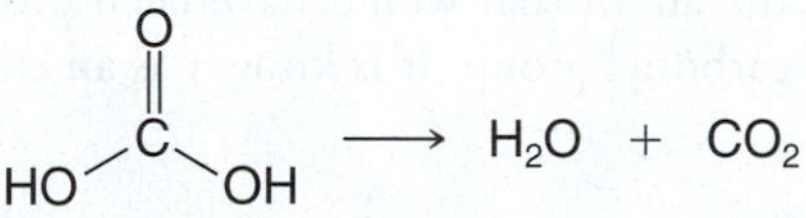

Carbonic acid
(unstable)

17.10 DECARBOXYLATION OF CARBOXYLIC ACIDS

The reaction whereby a carboxylic acid loses CO_2 is called a **decarboxylation**:

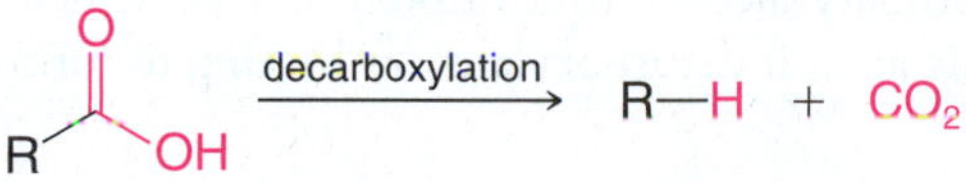

Although the unusual stability of carbon dioxide means that decarboxylation of most acids is exothermic, in practice the reaction is not always easy to carry out because the reaction is very slow. Special groups usually have to be present in the molecule for decarboxylation to be rapid enough to be synthetically useful.

- Carboxylic acids that have a carbonyl group one carbon removed from the carboxylic acid group, called β-**keto acids**, decarboxylate readily when they are heated to 100–150 °C. Some β-keto acids even decarboxylate slowly at room temperature.

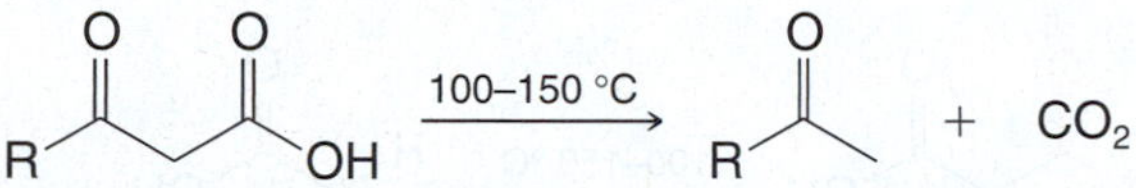

A β-keto acid

There are two reasons for this ease of decarboxylation:

1. When the acid itself decarboxylates, it can do so through a six-membered cyclic transition state:

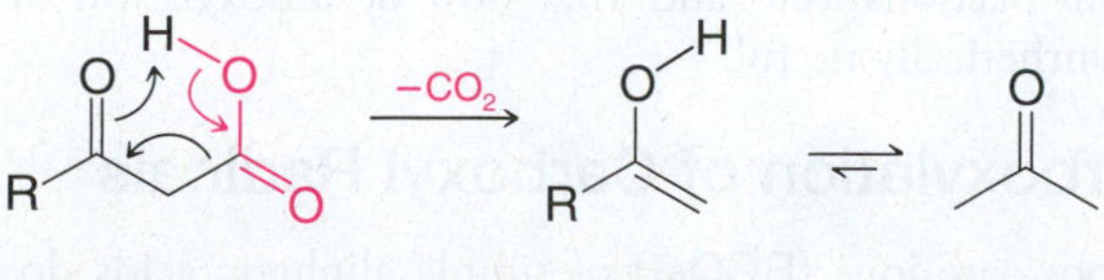

β-Keto acid   **Enol**   **Ketone**

This reaction produces an enol (alk**en**e-alco**hol**) directly and avoids an anionic intermediate. The enol then tautomerizes to a methyl ketone.

2. When the carboxylate anion decarboxylates, it forms a resonance-stabilized anion:

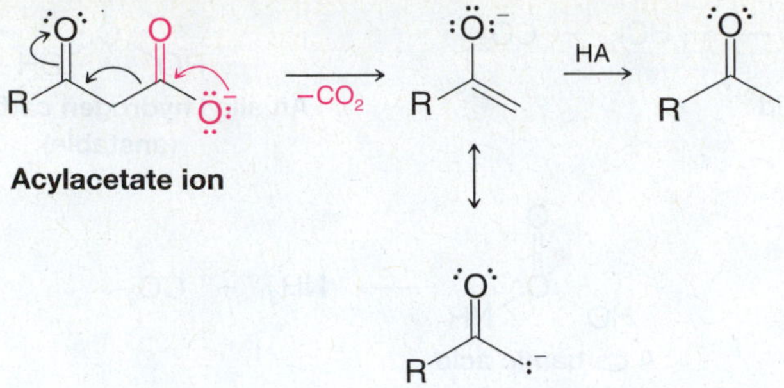

Acylacetate ion

Resonance-stabilized anion

This type of anion, which we shall study further in Chapter 19, is much more stable than simply RCH_2:$^-$, the anion that would have been produced by decarboxylation in the absence of a β-carbonyl group. It is known as an enolate.

●●● **SOLVED PROBLEM 17.8**

Provide structures for **A** and **B**.

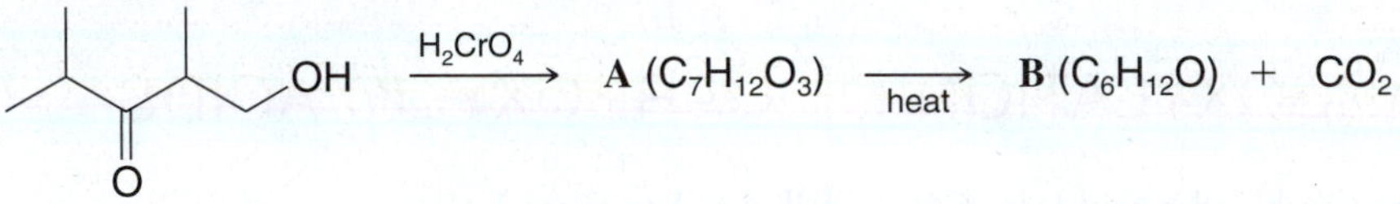

$$\text{A } (C_7H_{12}O_3) \xrightarrow{\text{heat}} \text{B } (C_6H_{12}O) + CO_2$$

STRATEGY AND ANSWER: H_2CrO_4 oxidizes a primary alcohol to a carboxylic acid, which is consistent with the formula provided for **A**. Because **A** is a β-ketocarboxylic acid, it decarboxylates on heating to form **B**.

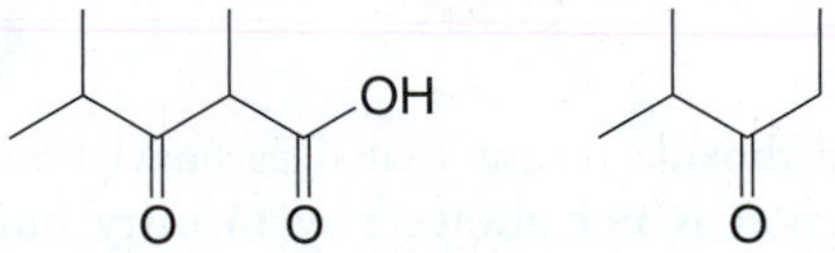

β-Dicarboxylic acids (1,3-dicarboxylic acids, also called malonic acids) decarboxylate readily for reasons similar to β-keto acids.

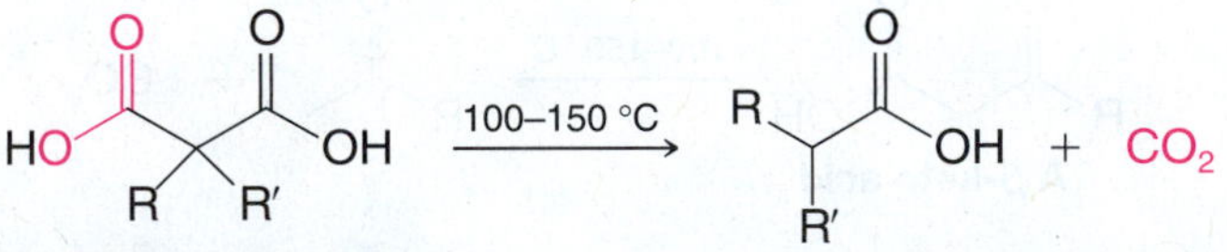

A β-dicarboxylic acid

β-Dicarboxylic acids undergo decarboxylation so readily that they do not form cyclic anhydrides (Section 17.6A).

We shall see in Sections 18.6 and 18.7 how decarboxylation of β-keto acids and malonic acids is synthetically useful.

17.10A Decarboxylation of Carboxyl Radicals

Although the carboxylate ions (RCO_2^-) of simple aliphatic acids do not decarboxylate readily, carboxyl radicals ($RCO_2\cdot$) do. They decarboxylate by losing CO_2 and producing alkyl radicals:

$$RCO_2\cdot \longrightarrow R\cdot + CO_2$$

Using decarboxylation reactions, outline a synthesis of each of the following from appropriate starting materials:

(a) 2-Hexanone

(b) 2-Methylbutanoic acid

(c) Cyclohexanone

(d) Pentanoic acid

Diacyl peroxides, R—C(=O)—O—O—C(=O)—R, decompose readily when heated.

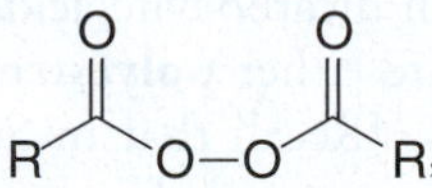

(a) What factor accounts for this instability?

(b) The decomposition of a diacyl peroxide produces CO_2. How is it formed?

(c) Diacyl peroxides are often used to initiate radical reactions, for example, the polymerization of an alkene:

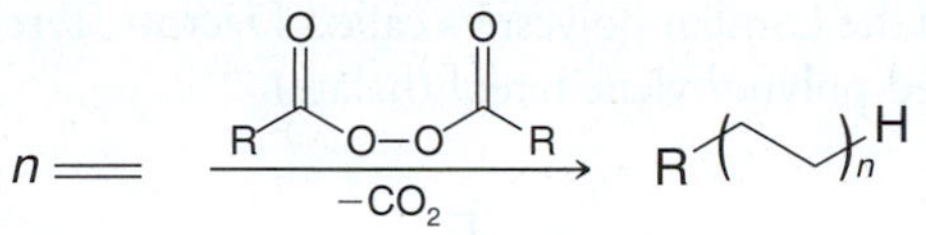

$$n \longrightarrow \quad \xrightarrow[-CO_2]{} \quad R\text{---}(\ )_n\text{---}H$$

Show the steps involved.

17.11 CHEMICAL TESTS FOR ACYL COMPOUNDS

Carboxylic acids are weak acids, and their acidity helps us to detect them.

- Aqueous solutions of water-soluble carboxylic acids give an acid test with blue litmus paper.
- Water-insoluble carboxylic acids dissolve in aqueous sodium hydroxide and aqueous sodium bicarbonate (see Section 17.2C).
- Sodium bicarbonate helps us distinguish carboxylic acids from most phenols. Except for the di- and trinitrophenols, phenols do not dissolve in aqueous sodium bicarbonate. When carboxylic acids dissolve in aqueous sodium bicarbonate, they also cause the evolution of carbon dioxide.

17.12 POLYESTERS AND POLYAMIDES: STEP-GROWTH POLYMERS

We have seen in Section 17.7A that carboxylic acids react with alcohols to form esters.

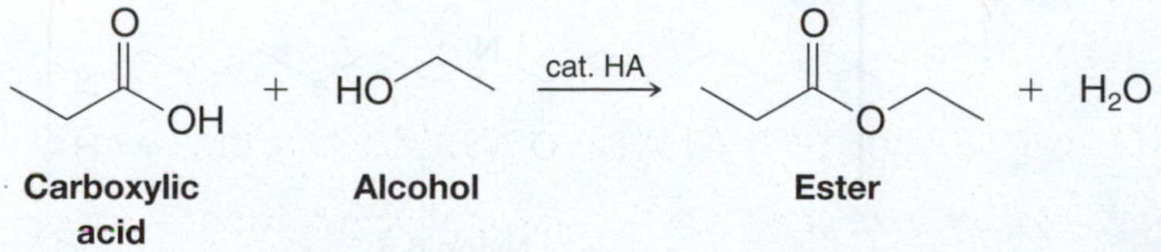

$$\text{Carboxylic acid} \quad + \quad \text{Alcohol} \quad \xrightarrow{\text{cat. HA}} \quad \text{Ester} \quad + \quad H_2O$$

In a similar way carboxylic acid derivatives (L is a leaving group) react with amines (Section 17.8) to form amides.

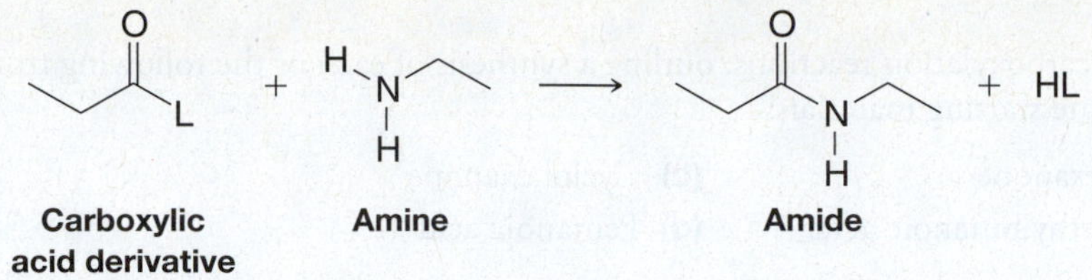

Carboxylic   Amine   Amide
acid derivative

In each reaction the two reactants become joined and a small molecule is lost. Such reactions are often called **condensation reactions**.

Similar condensation reactions beginning with dicarboxylic acids and either diols or diamines can be used to form polymers that are either **polyesters** or **polyamides**. These polymers are called **step-growth polymers**. [Recall that in Section 10.10 and Special Topic B in *WileyPLUS*, we studied another group of polymers called *chain-growth polymers* (also called *addition polymers*), which are formed by radicals undergoing chain-reactions.]

- **Polyesters.** When a dicarboxylic acid reacts with a diol under the appropriate conditions, the product is a polyester. For example, the reaction of 1,4-benzenedicarboxylic acid (terephthalic acid) with 1,2-ethanediol leads to the formation of the familiar polyesters called Dacron, Terelene or Mylar, and systemically called poly(ethylene terephthalate).

Terephthalic   1,2-Ethanediol   Poly(ethylene terephthalate)
acid   (a polyester)

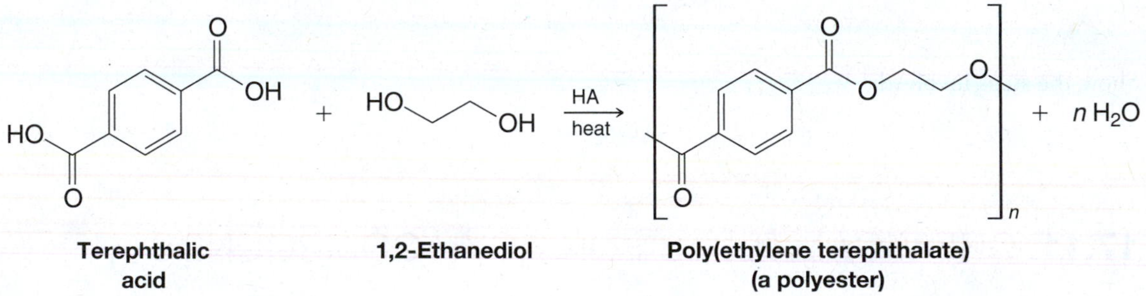

- **Polyamides.** When a dicarboxylic acid or acid chloride or anhydride reacts with a diamine under the appropriate conditions, the product is a polyamide. For example, 1,6-hexanedioic acid (adipic acid) can react with 1,6-hexanediamine with heat in an industrial process to form a familiar polyamide called Nylon. This example of nylon is called nylon 6,6 because both components of the polymer have six carbon atoms. Other nylons can be made in a similar way.

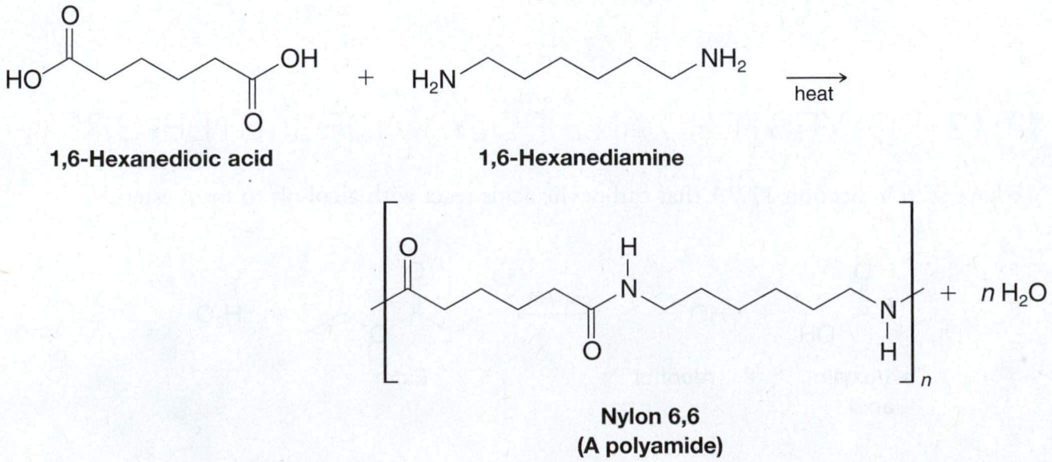

1,6-Hexanedioic acid   1,6-Hexanediamine

Nylon 6,6
(A polyamide)

Special Topic C in *WileyPLUS* continues our discussion of Step-Growth Polymers.

17.13 SUMMARY OF THE REACTIONS OF CARBOXYLIC ACIDS AND THEIR DERIVATIVES

The reactions of carboxylic acids and their derivatives are summarized here. Many (but not all) of the reactions in this summary are acyl substitution reactions (they are principally the reactions referenced to Sections 17.5 and beyond). As you use this summary, you will find it helpful to also review Section 17.4, which presents the general nucleophilic addition–elimination mechanism for acyl substitution. It is instructive to relate aspects of the specific acyl substitution reactions below to this general mechanism. In some cases proton transfer steps are also involved, such as to make a leaving group more suitable by prior protonation or to transfer a proton to a stronger base at some point in a reaction, but in all acyl substitution the essential nucleophilic addition–elimination steps are identifiable.

REACTIONS OF CARBOXYLIC ACIDS

1. As acids (discussed in Sections 3.11 and 17.2C):

2. Reduction (discussed in Section 12.3):

3. Conversion to acyl chlorides (discussed in Section 17.5):

4. Conversion to esters (Fischer esterification) or lactones (discussed in Section 17.7A):

5. Conversion to amides (discussed in Section 17.8E, but a very limited method):

6. Decarboxylation (discussed in Section 17.10):

REACTIONS OF ACYL CHLORIDES

1. Conversion (hydrolysis) to acids (discussed in Section 17.5B):

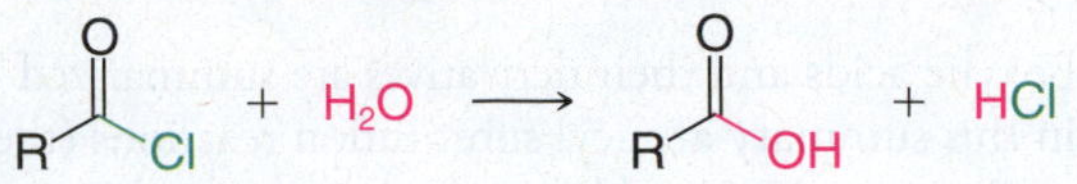

$$\text{R–CO–Cl} + \text{H}_2\text{O} \longrightarrow \text{R–CO–OH} + \text{HCl}$$

2. Conversion to anhydrides (discussed in Section 17.6A):

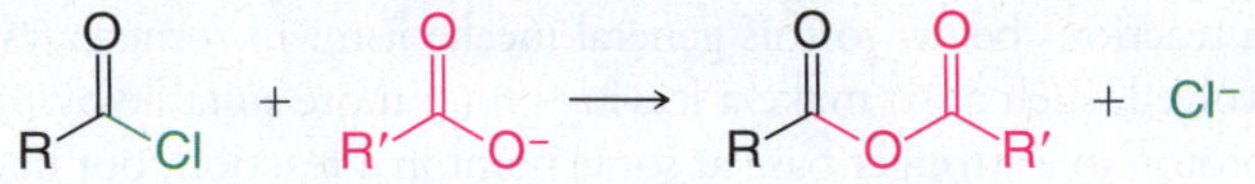

$$\text{R–CO–Cl} + \text{R'–CO–O}^- \longrightarrow \text{R–CO–O–CO–R'} + \text{Cl}^-$$

3. Conversion to esters (discussed in Section 17.7A):

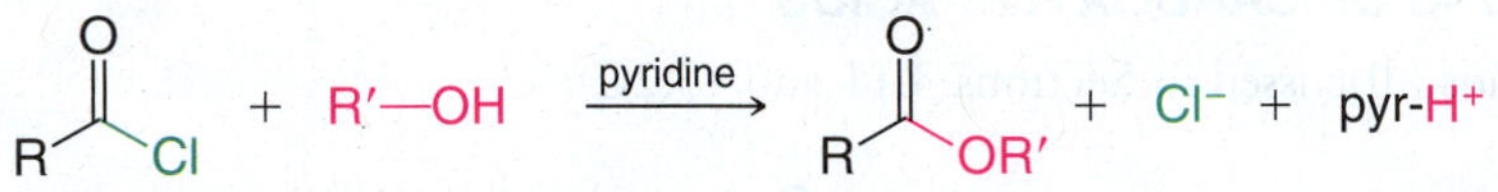

$$\text{R–CO–Cl} + \text{R'–OH} \xrightarrow{\text{pyridine}} \text{R–CO–OR'} + \text{Cl}^- + \text{pyr-H}^+$$

4. Conversion to amides (discussed in Section 17.8B):

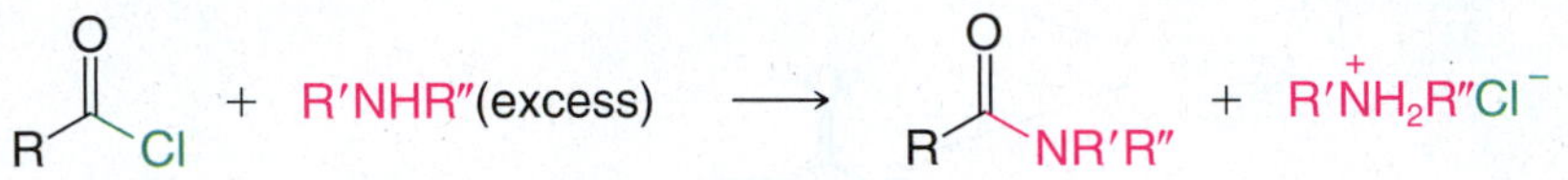

$$\text{R–CO–Cl} + \text{R'NHR''(excess)} \longrightarrow \text{R–CO–NR'R''} + \text{R'}\overset{+}{\text{N}}\text{H}_2\text{R''Cl}^-$$

R' and/or R'' may be H.

5. Conversion to ketones (Friedel–Crafts acylation, Sections 15.7–15.9):

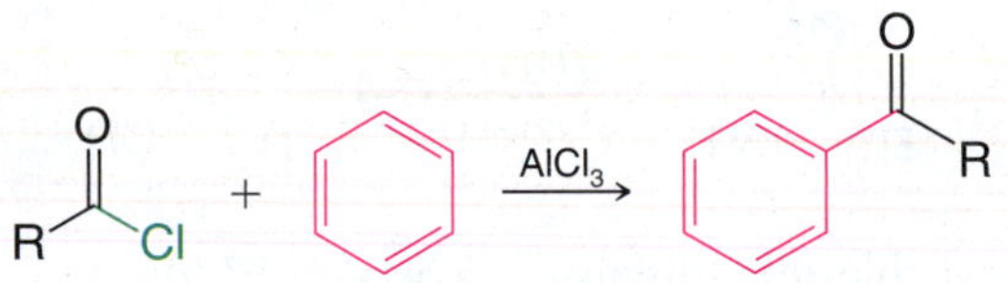

$$\text{R–CO–Cl} + \text{C}_6\text{H}_6 \xrightarrow{\text{AlCl}_3} \text{C}_6\text{H}_5\text{–CO–R}$$

6. Conversion to aldehydes (discussed in Section 16.4C):

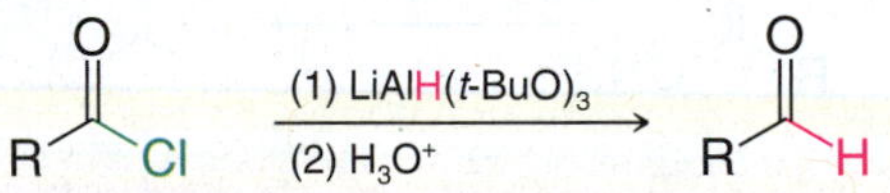

$$\text{R–CO–Cl} \xrightarrow[\text{(2) H}_3\text{O}^+]{\text{(1) LiAlH}(t\text{-BuO})_3} \text{R–CHO}$$

REACTIONS OF ACID ANHYDRIDES

1. Conversion (hydrolysis) to acids (discussed in Section 17.6B):

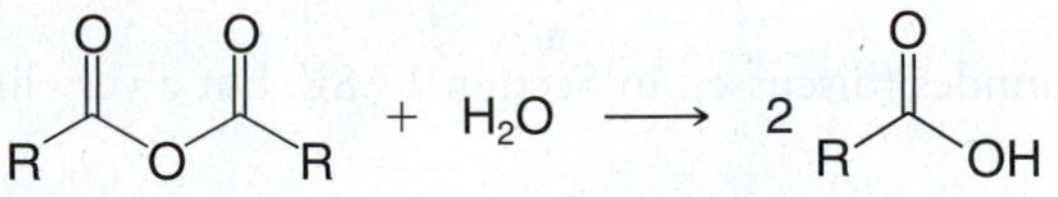

$$\text{R–CO–O–CO–R} + \text{H}_2\text{O} \longrightarrow 2\ \text{R–CO–OH}$$

2. Conversion to esters (discussed in Sections 17.6B and 17.7A):

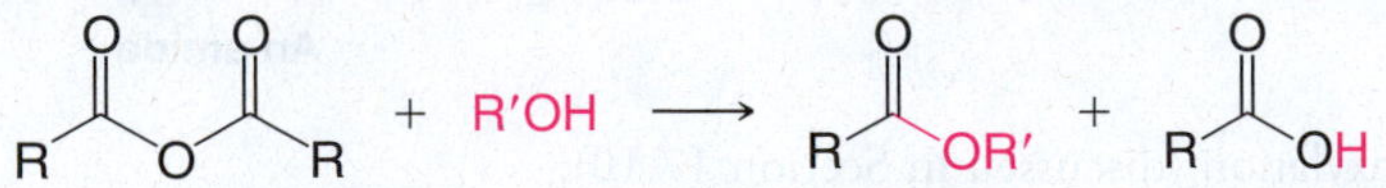

$$\text{R–CO–O–CO–R} + \text{R'OH} \longrightarrow \text{R–CO–OR'} + \text{R–CO–OH}$$

3. Conversion to amides (discussed in Section 17.8C):

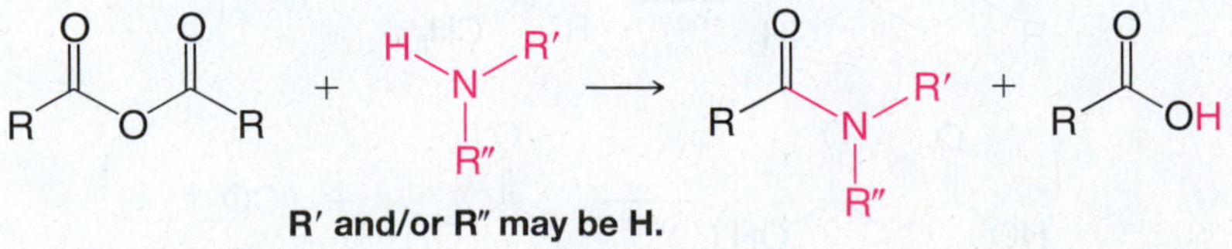

$$\text{R–CO–O–CO–R} + \text{HN}(\text{R'})(\text{R''}) \longrightarrow \text{R–CO–N}(\text{R'})(\text{R''}) + \text{R–CO–OH}$$

R' and/or R'' may be H.

4. Conversion to aryl ketones (Friedel–Crafts acylation, Sections 15.7–15.9):

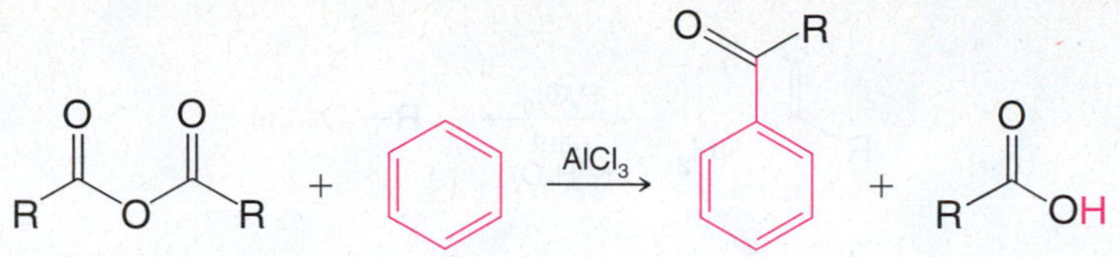

REACTIONS OF ESTERS

1. Hydrolysis (discussed in Section 17.7B):

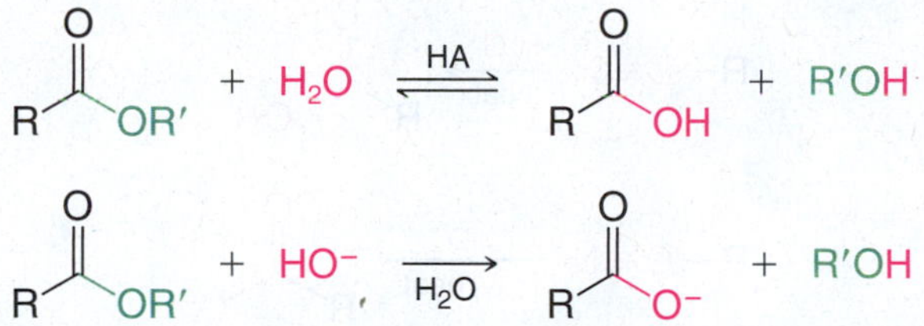

2. Conversion to other esters: transesterification (discussed in Practice Problem 17.10):

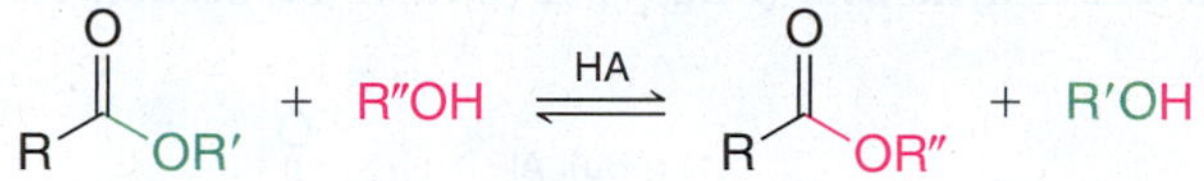

3. Conversion to amides (discussed in Section 17.8D):

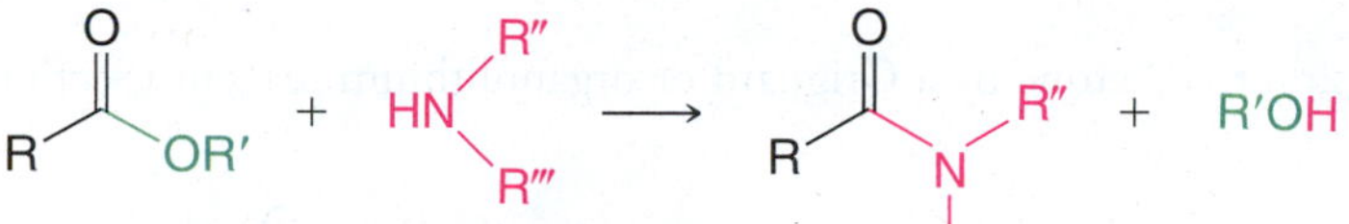

R″ and/or R‴ may be H.

4. Reaction with Grignard reagents (discussed in Section 12.8):

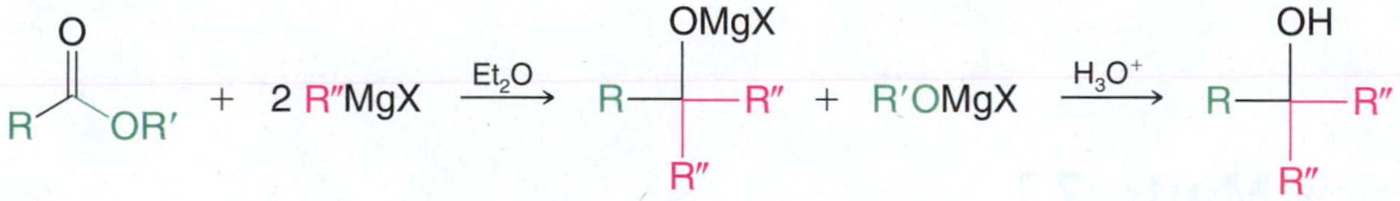

5. Reduction (discussed in Section 12.3):

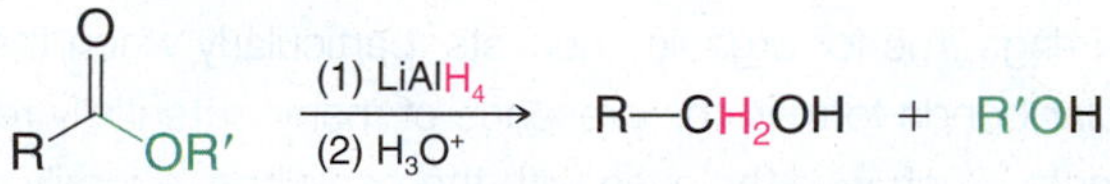

REACTIONS OF AMIDES

1. Hydrolysis (discussed in Section 17.8F):

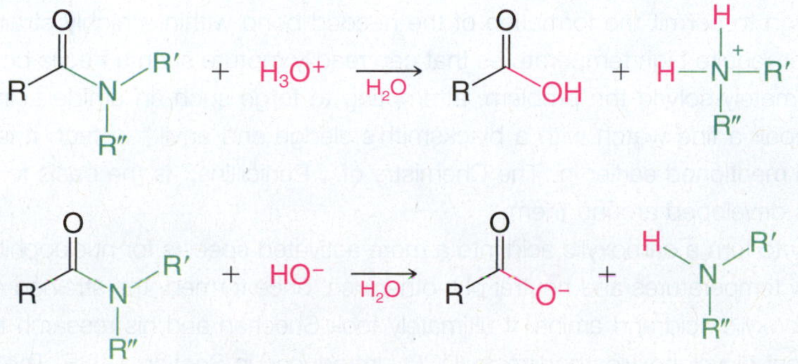

R, R′, and/or R″ may be H.

2. Conversion to nitriles: dehydration (discussed in Section 17.8G):

$$R-\overset{\overset{\displaystyle O}{\|}}{C}-NH_2 \xrightarrow[\substack{\text{heat} \\ (-H_2O)}]{P_4O_{10}} R-C\equiv N$$

REACTIONS OF NITRILES

1. Hydrolysis to a carboxylic acid or carboxylate anion (Section 17.8H):

$$R-C\equiv N \xrightarrow[\text{heat}]{H_3O^+} R-\overset{\overset{\displaystyle O}{\|}}{C}-OH$$

$$R-C\equiv N \xrightarrow[H_2O,\ \text{heat}]{HO^-} R-\overset{\overset{\displaystyle O}{\|}}{C}-O^-$$

2. Reduction to an aldehyde with $(i\text{-Bu})_2\text{AlH}$ (DIBAL-H, Section 16.4C):

$$R-C\equiv N \xrightarrow[\text{(2) } H_2O]{\text{(1) } (i\text{-Bu})_2\text{AlH}} R-\overset{\overset{\displaystyle O}{\|}}{C}-H$$

3. Conversion to a ketone by a Grignard or organolithium reagent (Section 16.5B):

$$R-C\equiv N \ + \xrightarrow[\text{(2) } H_3O^+]{\text{(1) } R'\text{MgBr or } R'\text{Li}} R-\overset{\overset{\displaystyle O}{\|}}{C}-R'$$

[WHY Do These Topics Matter?]

FORGING THE UNIQUE, STRAINED AMIDE OF THE PENICILLINS

The saying that necessity is the mother of invention often rings true for organic chemists, particularly when they are trying to synthesize unique structures. Indeed, efforts to make particular bonds form in the presence of many potentially reactive groups often require the development of new and more selective reagents. Such was the case with the penicillins, a family of molecules whose structural determination was discussed at the end of Chapter 9 and whose unique lactam structures were presented earlier in this chapter.

In 1945, the year when their structures were finally established, chemists knew several ways to make amides, including the use of acid halides and amines through nucleophilic acyl substitution reactions that you learned about in this chapter. None of the known processes, however, was mild enough to permit the formation of the needed bond within a highly strained system because they produce acidic by-products and/or require high temperatures that can readily rupture such a fragile bond. Indeed, as noted by John C. Sheehan of MIT, who ultimately solved the problem, attempting to forge such an amide using methods available at the time was "like attempting to repair a fine watch with a blacksmith's sledge and anvil." In fact, it is that same lability and strain in their lactam rings which, as mentioned earlier in "The Chemistry of... Penicillins," is the basis for how these antibiotics act and how bacterial resistance has developed around them.

What was needed chemically was the ability to turn a carboxylic acid into a more activated species for nucleophilic acyl substitution and effect its merger with an amine at low temperatures and neutral pH; otherwise, once formed, the strained amide bond would simply hydrolyze back to the starting carboxylic acid and amine. It ultimately took Sheehan and his research team over a decade to find a solution in the form of the reagent dicyclohexylcarbodiimide (DCC), introduced in Section 17.8E. The importance of this discovery cannot be understated. Not only did it allow for the production of penicillins in greater quantities, it also allowed chemists to create new penicillin analogs such as methicillin that have superior and/or distinct properties from the original structures found in nature. It also provided access to other classes of antibiotics that possess strained lactam rings such as the cephalosporins and the carbapenems. And, it provided insights into how to create even milder and more powerful amide-bond-forming reagents, tools that have now led to the automated synthesis of peptides (see Chapter 24) as well as a number of peptide-based drugs used by hundreds of thousands of patients, including several treatments for the human immunodeficiency virus.

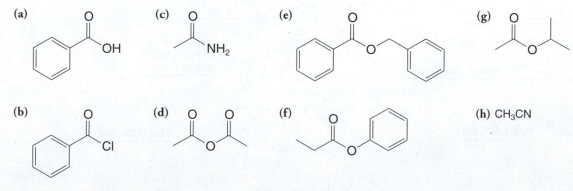

Penicillins
[R = variable group]

Methicillin

Cephalosporins
[R = variable group]

Carbapenems
[R = variable group]

To learn more about these topics, see:

1. Sheehan, J. C. *The Enchanted Ring: The Untold Story of Penicillin*. MIT Press: Cambridge, **1984**, p. 224.
2. "Penicillin Synthesis" in *Time* magazine. March 18, 1957.
3. Nicolaou, K. C.; Montagnon, T. *Molecules that Changed the World*. Wiley-VCH: Weinheim, **2008**, pp. 97–106..

SUMMARY AND REVIEW TOOLS

The study aids for this chapter include key terms and concepts (which are hyperlinked to the Glossary from the bold, blue terms in the *WileyPLUS* version of the book at wileyplus.com) and the Summary of Reactions of Carboxylic Acids and Their Derivatives found in Section 17.13.

PROBLEMS

Note to Instructors: Many of the homework problems are available for assignment via *WileyPLUS*, an online teaching and learning solution.

STRUCTURE AND NOMENCLATURE

17.18 Write a structural formula for each of the following compounds:

(a) Octanoic acid

(b) Propanamide

(c) *N,N*-Diethylhexanamide

(d) 2-Methyl-4-hexenoic acid

(e) Butanedioic acid

(f) 1,2-Benzenedioic acid (phthalic acid)

(g) 1,4-Benzenedioic acid (terephthalic acid)

(h) Acetic anhydride

(i) Isobutyl propanoate

(j) Benzyl acetate

(k) Ethanoyl chloride (acetyl chloride)

(l) 2-Methylpropanenitrile

(m) Ethyl 3-oxobutanoate (ethyl acetoacetate)

(n) Diethyl propanedioate (diethyl malonate)

17.19 Give an IUPAC systematic or common name for each of the following compounds:

(h) CH_3CN

17.20 Amides are weaker bases than corresponding amines. For example, most water-insoluble amines (RNH_2) will dissolve in dilute aqueous acids (aqueous HCl, H_2SO_4, etc.) by forming water-soluble alkylaminium salts ($RNH_3{}^+X^-$). Corresponding amides ($RCONH_2$) *do not dissolve in dilute aqueous acids*, however. Propose an explanation for the much lower basicity of amides when compared to amines.

17.21 While amides are much less basic than amines, they are much stronger acids. Amides have pK_a values in the range 14–16, whereas for amines, $pK_a = 33$–35.

(a) What factor accounts for the much greater acidity of amides?

(b) Imides, that is, compounds with the structure

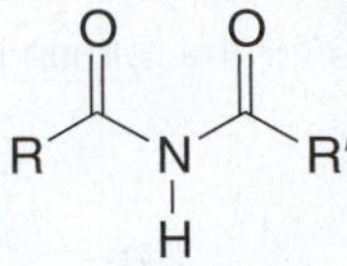

, are even stronger acids than amides.

For imides, $pK_a = 9$–10, and as a consequence, water-insoluble imides dissolve in aqueous NaOH by forming soluble sodium salts. What extra factor accounts for the greater acidity of imides?

FUNCTIONAL GROUP TRANSFORMATIONS

17.22 What major organic product would you expect to obtain when acetyl chloride reacts with each of the following?

(a) H_2O

(b) BuLi (excess)

(c)

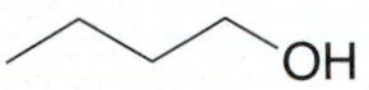

and pyridine

(d) NH_3 (excess)

(e)

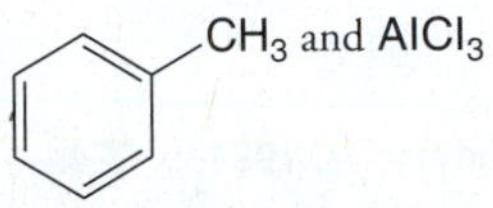

CH_3 and $AlCl_3$

(f) LiAlH(t-BuO)$_3$

(g) NaOH/H_2O

(h) CH_3NH_2 (excess)

(i) $(CH_3)_2NH$ (excess)

(j) EtOH and pyridine

(k) $CH_3CO_2^-Na^+$

(l) CH_3CO_2H and pyridine

17.23 What major organic product would you expect to obtain when acetic anhydride reacts with each of the following?

(a) NH_3 (excess)

(b) H_2O

(c) $CH_3CH_2CH_2OH$

(d) $C_6H_6 + AlCl_3$

(e) $CH_3CH_2NH_2$ (excess)

(f) $(CH_3CH_2)_2NH$ (excess)

17.24 What major organic product would you expect to obtain when succinic anhydride reacts with each of the reagents given in Problem 17.23?

17.25 What products would you expect to obtain when ethyl propanoate reacts with each of the following?

(a) H_3O^+, H_2O

(b) HO^-, H_2O

(c) 1-Octanol, HCl

(d) CH_3NH_2

(e) LiAlH$_4$, then H_2O

(f) Excess C_6H_5MgBr, then H_2O, NH_4Cl

17.26 What products would you expect to obtain when propanamide reacts with each of the following?

(a) H_3O^+, H_2O

(b) HO^-, H_2O

(c) P_4O_{10} and heat

17.27 What products would you expect to obtain when each of the following compounds is heated?

(a) 4-Hydroxybutanoic acid

(b) 3-Hydroxybutanoic acid

(c) 2-Hydroxybutanoic acid

(d) Glutaric acid

(e)

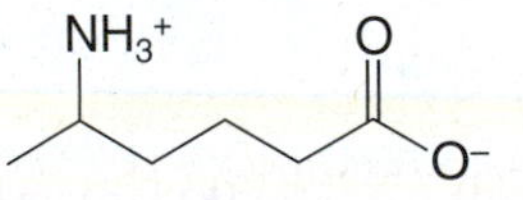

(f)

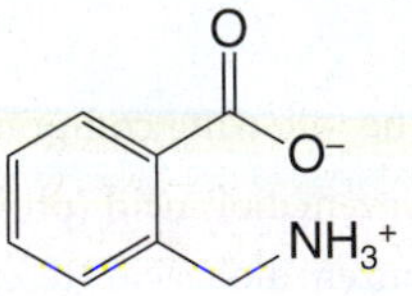

GENERAL PROBLEMS

17.28 Write structural formulas for the major organic products from each of the following reactions.

(a)

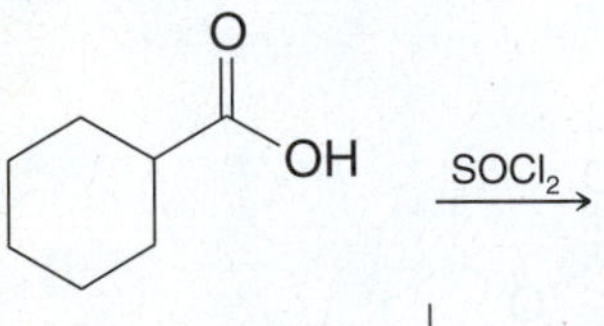

SOCl$_2$

(b)

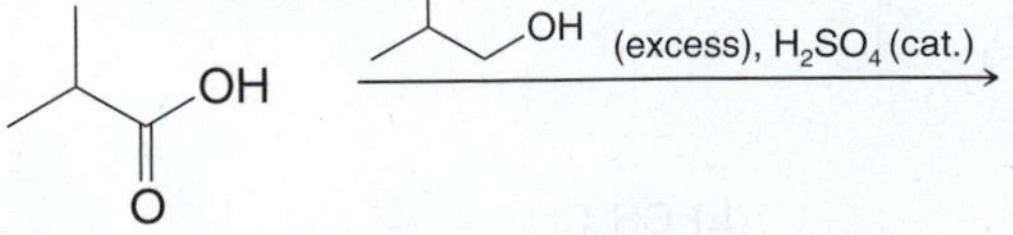

(excess), H_2SO_4 (cat.)

(c)

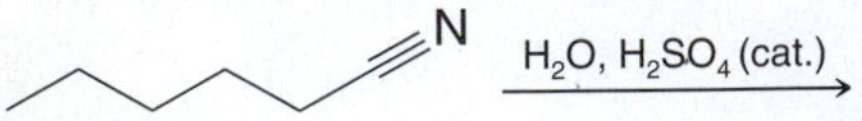

H_2O, H_2SO_4 (cat.)

(d)

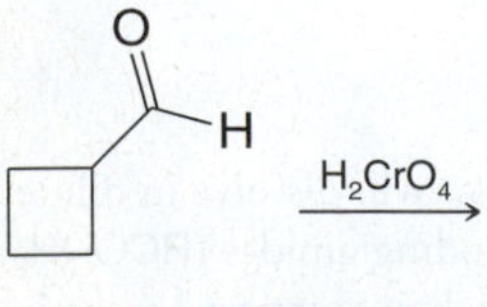

H_2CrO_4

(e)

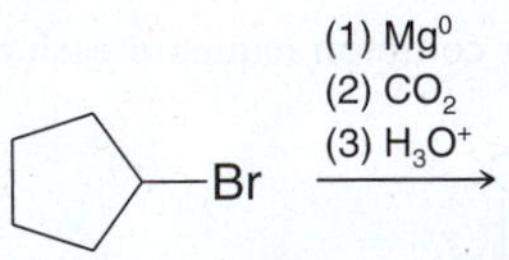

(1) Mg^0
(2) CO_2
(3) H_3O^+

(f)

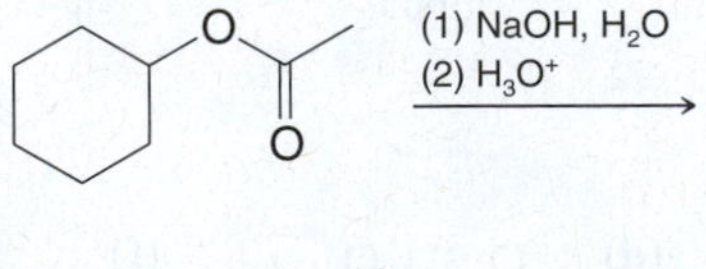

(1) NaOH, H_2O
(2) H_3O^+

(g)

H_2O, H_2SO_4 (cat.)

(h)

(1) H_2CrO_4
(2) SOCl$_2$

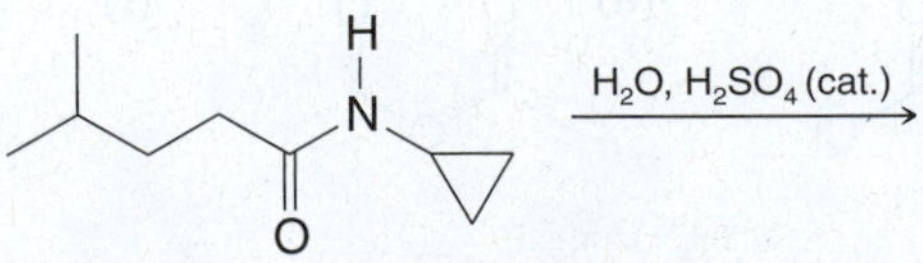

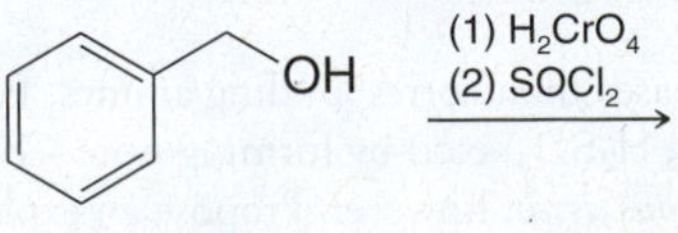

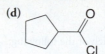

17.29 Indicate reagents that would accomplish each of following transformations. More than one reaction may be necessary in some cases.

(a)

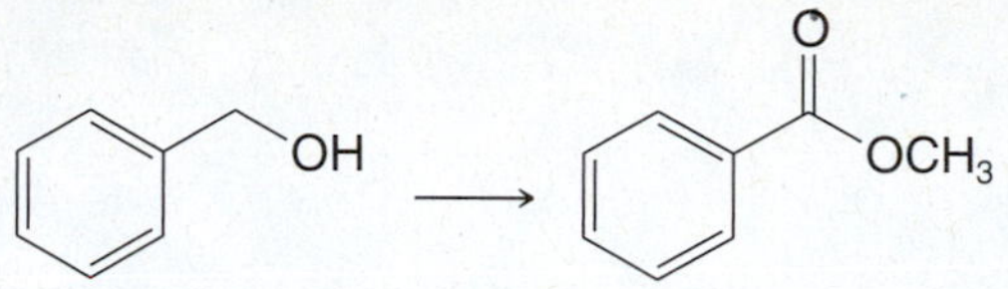

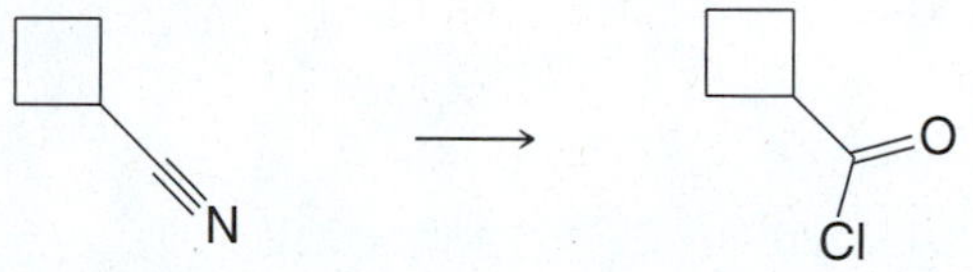

(d)

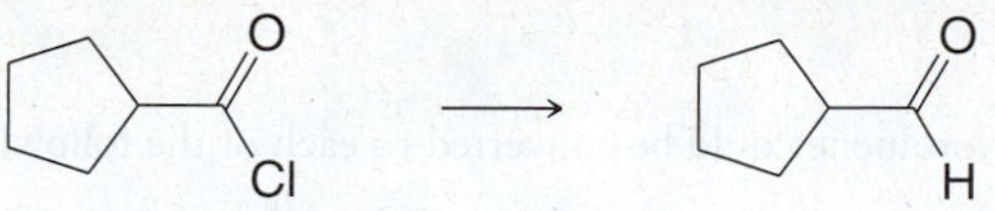

(b)

(e)

(c)

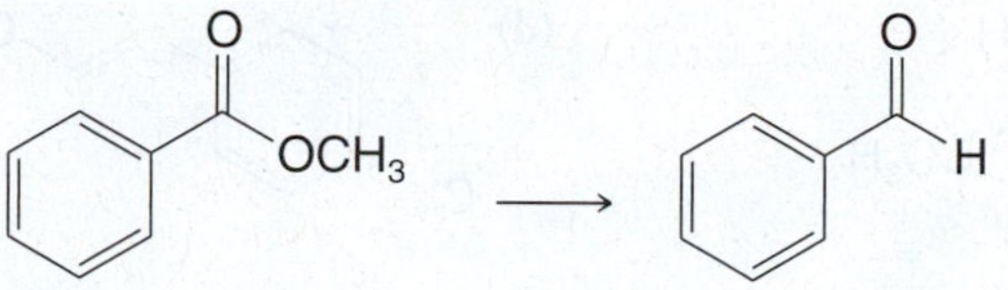

(f)

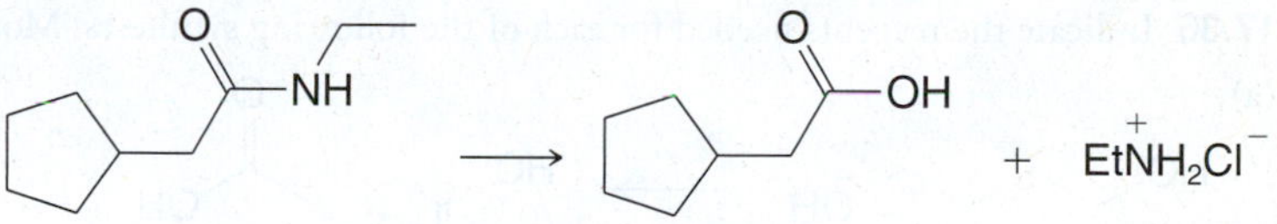

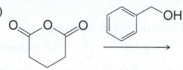

17.30 Write structural formulas for the major organic products from each of the following reactions.

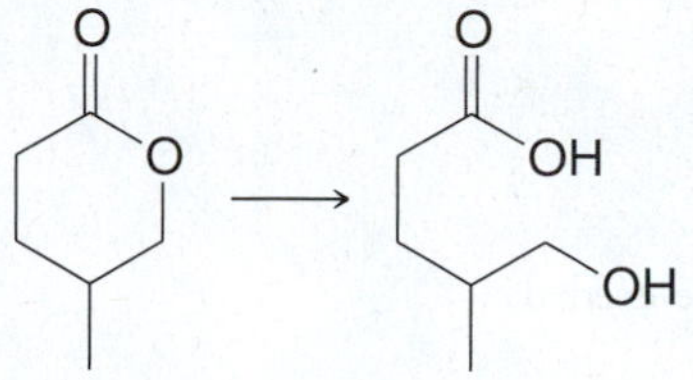

(a)

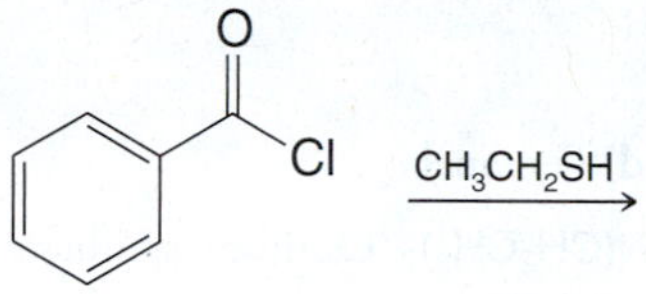

CH₃CH₂SH

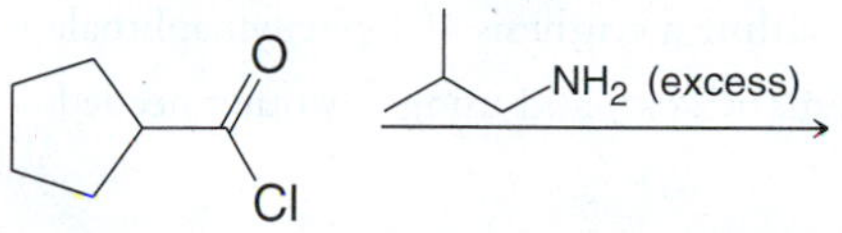

(c)

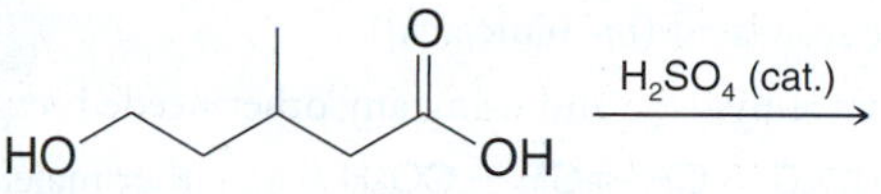

(e)

, AlCl₃

(d)

H₂SO₄ (cat.)

(b)

NH₂ (excess)

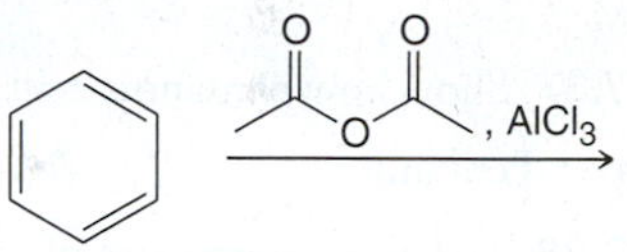

17.31 Write structural formulas for the major organic products from each of the following reactions.

(a)

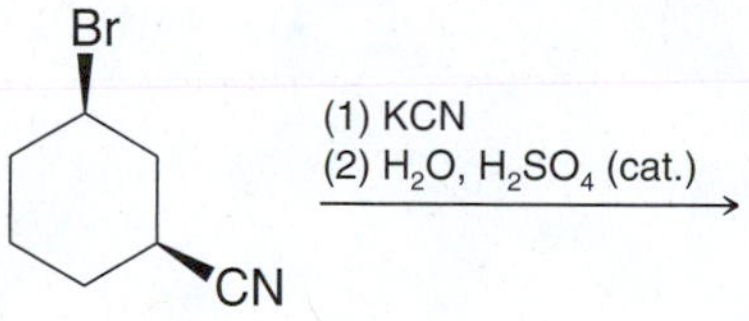

(1) KCN
(2) H₂O, H₂SO₄ (cat.)

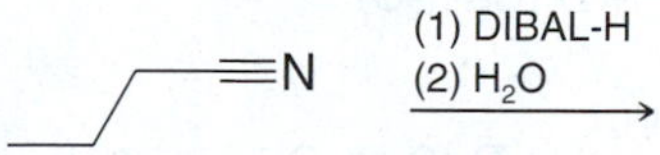

(c)

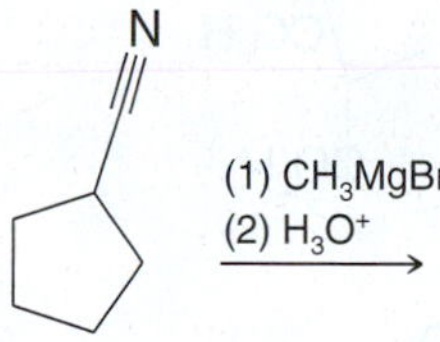

(1) CH₃MgBr
(2) H₃O⁺

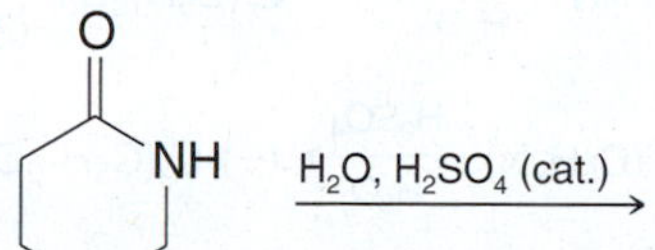

(e)

(1) CH₃CH₂MgBr (excess)
(2) H₃O⁺

(b)

(1) DIBAL-H
(2) H₂O

(d)

H₂O, H₂SO₄ (cat.)

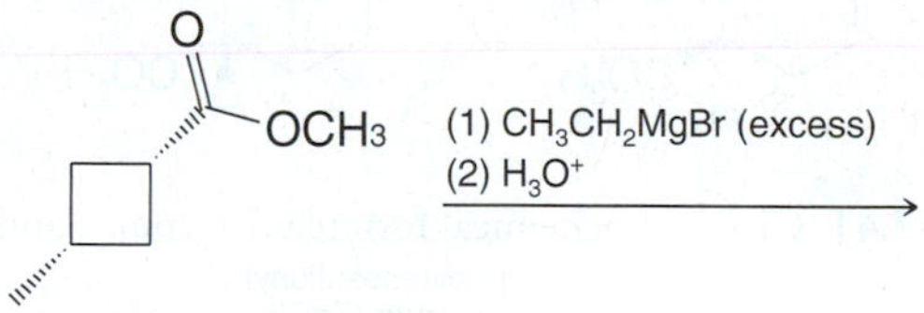

MECHANISMS

17.32 Write detailed mechanisms for the acidic and basic hydrolysis of propanamide.

17.33 Provide a detailed mechanism for each of the following reactions.

(a)

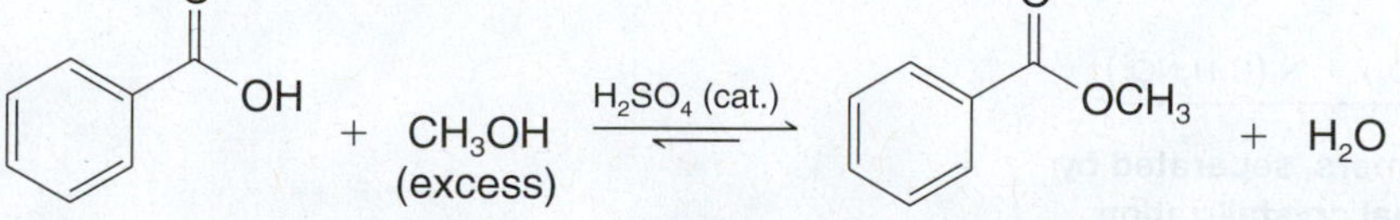

+ CH₃OH
(excess)

H₂SO₄ (cat.)

+ H₂O

(b)

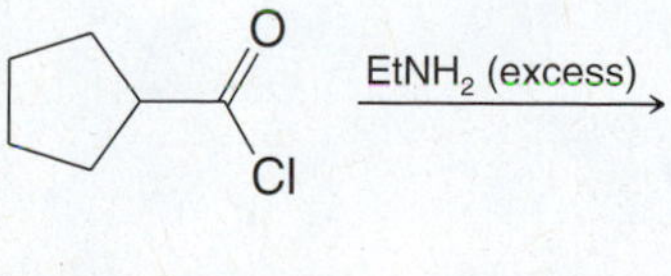

EtNH₂ (excess)

(c)

H₂O, H₂SO₄ (cat.)

17.34 On heating, *cis*-4-hydroxycyclohexanecarboxylic acid forms a lactone but *trans*-4-hydroxycyclohexanecarboxylic acid does not. Explain.

SYNTHESIS

17.35 Show how *p*-chlorotoluene could be converted to each of the following:

(a) *p*-Chlorobenzoic acid

(b)

(c) *p*-Chlorophenylacetic acid

(d)

17.36 Indicate the reagents needed for each of the following syntheses. More than one step may be needed.

(a)

(b)

17.37 Show how pentanoic acid can be prepared from each of the following:

(a) 1-Pentanol   (b) 1-Bromobutane (two ways)   (c) 5-Decene   (d) Pentanal

17.38 The active ingredient of the insect repellent Off is *N,N*-diethyl-*m*-toluamide, *m*-CH₃C₆H₄CON(CH₂CH₃)₂. Outline a synthesis of this compound starting with 3-methylbenzoic acid (*m*-toluic acid).

17.39 Starting with benzene and succinic anhydride and using any other needed reagents, outline a synthesis of 1-phenylnaphthalene.

17.40 Starting with either *cis*- or *trans*-HO₂C—CH=CH—CO₂H (i.e., either maleic or fumaric acid) and using any other needed compounds, outline syntheses of each of the following:

(a)   (b)   (c)

17.41 Give stereochemical formulas for compounds **A–Q**:

(a) (*R*)-(−)-2-Butanol $\xrightarrow[\text{pyridine}]{\text{p-toluenesulfonyl chloride (TsCl)}}$ **A** $\xrightarrow{\text{⁻C≡N}}$ **B** (C₅H₉N) $\xrightarrow[\text{H₂O}]{\text{H₂SO₄}}$ (+)-**C** (C₅H₁₀O₂) $\xrightarrow[\text{(2) H₂O}]{\text{(1) LiAlH₄}}$ (−)-**D** (C₅H₁₂O)

(b) (*R*)-(−)-2-Butanol $\xrightarrow[\text{pyridine}]{\text{PBr₃}}$ **E** (C₄H₉Br) $\xrightarrow{\text{⁻C≡N}}$ **F** (C₅H₉N) $\xrightarrow[\text{H₂O}]{\text{H₂SO₄}}$ (−)-**C** (C₅H₁₀O₂) $\xrightarrow[\text{(2) H₂O}]{\text{(1) LiAlH₄}}$ (+)-**D** (C₅H₁₂O)

(c) **A** $\xrightarrow{\text{CH₃CO₂⁻}}$ **G** (C₆H₁₂O₂) $\xrightarrow{\text{HO⁻}}$ (+)-**H** (C₄H₁₀O) + CH₃CO₂⁻

(d) (−)-**D** $\xrightarrow{\text{PBr₃}}$ **J** (C₅H₁₁Br) $\xrightarrow[\text{Et₂O}]{\text{Mg}}$ **K** (C₅H₁₁MgBr) $\xrightarrow[\text{(2) H₃O⁺}]{\text{(1) CO₂}}$ **L** (C₆H₁₂O₂)

(e) 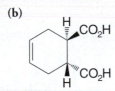 (*R*)-(+)-Glyceraldehyde $\xrightarrow{\text{HCN}}$ 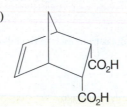 **M** (C₄H₇NO₃) + **N** (C₄H₇NO₃)

Diastereomers, separated by fractional crystallization

(f) **M** $\xrightarrow[\text{H₂O}]{\text{H₂SO₄}}$ **P** (C₄H₈O₅) $\xrightarrow[\text{HNO₃}]{\text{[O]}}$ *meso*-tartaric acid

(g) **N** $\xrightarrow[\text{H₂O}]{\text{H₂SO₄}}$ **Q** (C₄H₈O₅) $\xrightarrow[\text{HNO₃}]{\text{[O]}}$ (−)-tartaric acid

17.42 (*R*)-(+)-Glyceraldehyde can be transformed into (+)-malic acid by the following synthetic route. Give stereochemical structures for the products of each step.

$$(R)\text{-}(+)\text{-Glyceraldehyde} \xrightarrow[\text{oxidation}]{\text{Br}_2, \text{H}_2\text{O}} (-)\text{-glyceric acid} \xrightarrow{\text{PBr}_3}$$

$$(-)\text{-3-bromo-2-hydroxypropanoic acid} \xrightarrow{\text{NaCN}} \text{C}_4\text{H}_5\text{NO}_3 \xrightarrow[\text{heat}]{\text{H}_3\text{O}^+} (+)\text{-malic acid}$$

17.43 (*R*)-(+)-Glyceraldehyde can also be transformed into (−)-malic acid. This synthesis begins with the conversion of (*R*)-(+)-glyceraldehyde into (−)-tartaric acid, as shown in Problem 17.41, parts (e) and (g). Then (−)-tartaric acid is allowed to react with phosphorus tribromide in order to replace one alcoholic —OH group with —Br. This step takes place with inversion of configuration at the carbon that undergoes attack. Treating the product of this reaction with dimethyl sulfide produces (−)-malic acid. **(a)** Outline all steps in this synthesis by writing stereochemical structures for each intermediate. **(b)** The step in which (−)-tartaric acid is treated with phosphorus tribromide produces only one stereoisomer, even though there are two replaceable —OH groups. How is this possible? **(c)** Suppose that the step in which (−)-tartaric acid is treated with phosphorus tribromide had taken place with "mixed" stereochemistry, that is, with both inversion and retention at the carbon under attack. How many stereoisomers would have been produced? **(d)** What difference would this have made to the overall outcome of the synthesis?

17.44 Cantharidin is a powerful vesicant that can be isolated from dried beetles (*Cantharis vesicatoria,* or "Spanish fly"). Outlined here is the stereospecific synthesis of cantharidin reported by Gilbert Stork (Columbia University). Supply the missing reagents (a)–(n).

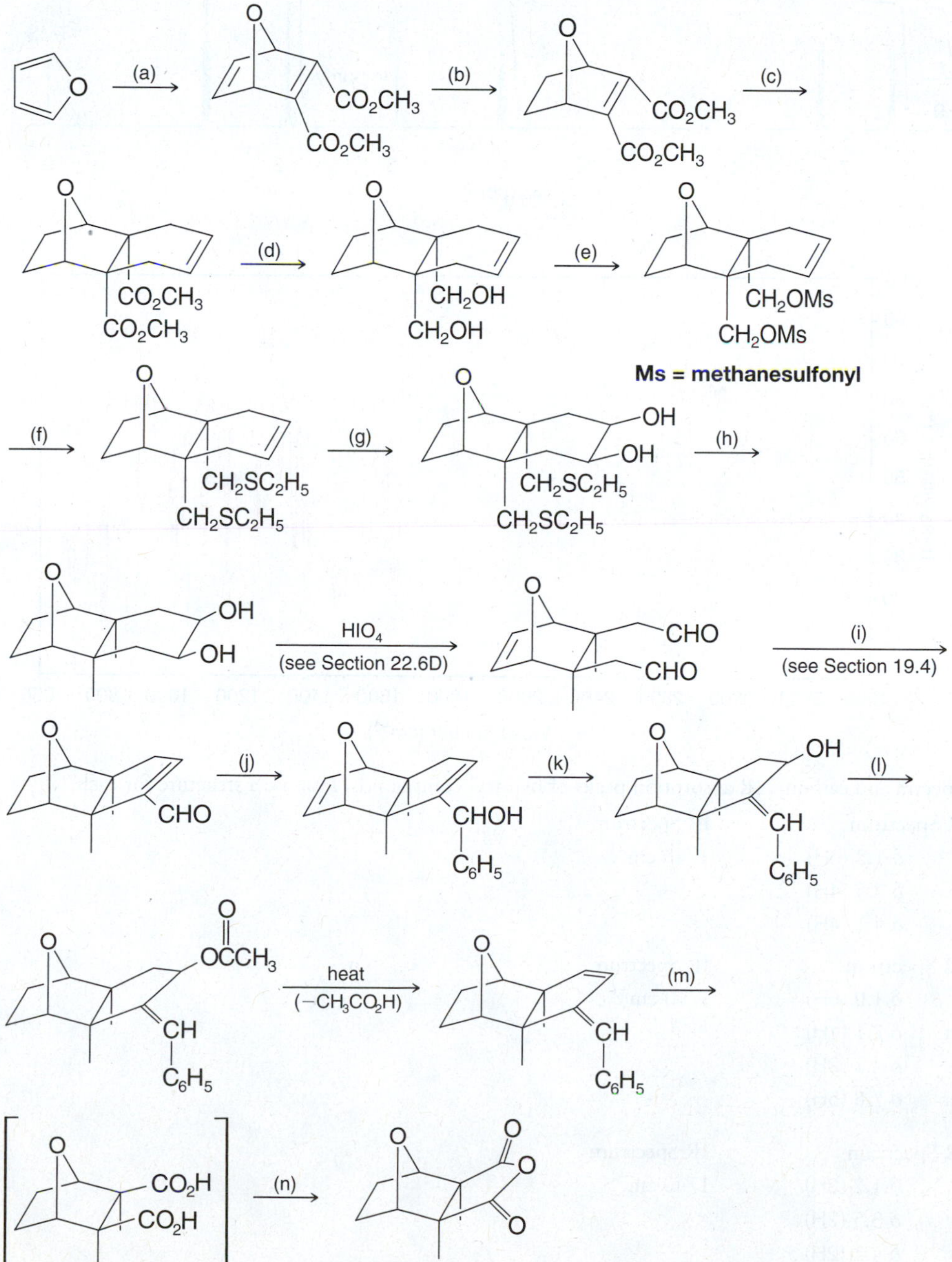

Cantharidin

SPECTROSCOPY

17.45 The IR and ^{1}H NMR spectra of phenacetin ($C_{10}H_{13}NO_2$) are given in Fig. 17.7. Phenacetin is an analgesic and antipyretic compound and was the P of A–P–C tablets (aspirin–phenacetin–caffeine). (Because of its toxicity, phenacetin is no longer used medically.) When phenacetin is heated with aqueous sodium hydroxide, it yields phenetidine ($C_8H_{11}NO$) and sodium acetate. Propose structures for phenacetin and phenetidine.

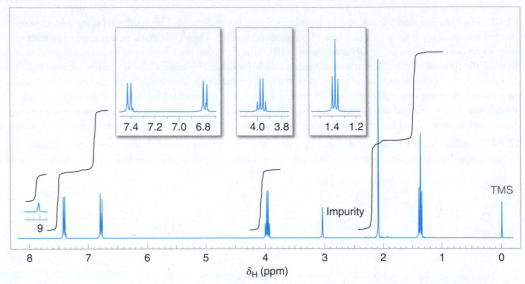

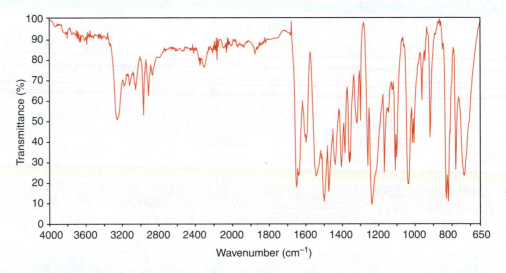

FIGURE 17.7 The 300-MHz ^{1}H NMR and IR spectra of phenacetin. Expansions of the ^{1}H NMR signals are shown in the offset plots.

17.46 Given here are the ^{1}H NMR spectra and carbonyl IR absorption peaks of five acyl compounds. Propose a structure for each.

(a) $C_8H_{14}O_4$

^{1}H NMR Spectrum		IR Spectrum
Triplet	δ 1.2 (6H)	1740 cm^{-1}
Singlet	δ 2.5 (4H)	
Quartet	δ 4.1 (4H)	

(b) $C_{11}H_{14}O_2$

^{1}H NMR Spectrum		IR Spectrum
Doublet	δ 1.0 (6H)	1720 cm^{-1}
Multiplet	δ 2.1 (1H)	
Doublet	δ 4.1 (2H)	
Multiplet	δ 7.8 (5H)	

(c) $C_{10}H_{12}O_2$

^{1}H NMR Spectrum		IR Spectrum
Triplet	δ 1.2 (3H)	1740 cm^{-1}
Singlet	δ 3.5 (2H)	
Quartet	δ 4.1 (2H)	
Multiplet	δ 7.3 (5H)	

(d) $C_2H_2Cl_2O_2$

¹H NMR Spectrum   IR Spectrum

Singlet   δ 6.0   Broad peak 2500–2700 cm⁻¹

Singlet   δ 11.70   1705 cm⁻¹

(e) $C_4H_7ClO_2$

¹H NMR Spectrum   IR Spectrum

Triplet   δ 1.3   1745 cm⁻¹

Singlet   δ 4.0

Quartet   δ 4.2

17.47 Compound **X** ($C_7H_{12}O_4$) is insoluble in aqueous sodium bicarbonate. The IR spectrum of **X** has a strong absorption peak near 1740 cm⁻¹, and its broadband proton-decoupled ¹³C spectrum is given in Fig. 17.8. Propose a structure for **X**.

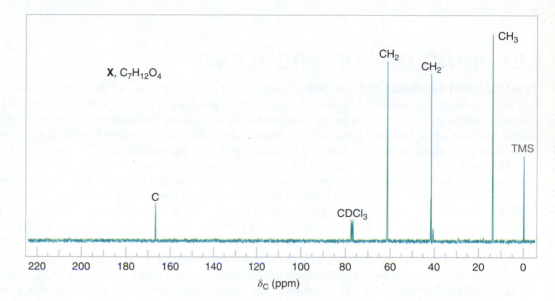

FIGURE 17.8 Broadband proton-decoupled ¹³C NMR spectrum of compound X, Problem 17.47. Information from the DEPT ¹³C NMR spectra is given above each peak.

17.48 Compound **Y** ($C_8H_4O_3$) dissolves slowly when warmed with aqueous sodium bicarbonate. The IR spectrum of **Y** has strong peaks at 1779 and at 1854 cm⁻¹. The broadband proton-decoupled ¹³C spectrum of **Y** exhibits signals at δ 125 (CH), 130 (C), 136 (CH), and 162 (C). Acidification of the bicarbonate solution of **Y** gave compound **Z**. The proton-decoupled ¹³C NMR spectrum of **Z** showed four signals. When **Y** was warmed in ethanol, compound **AA** was produced. The ¹³C NMR spectrum of **AA** displayed 10 signals. Propose structures for **Y**, **Z**, and **AA**.

CHALLENGE PROBLEMS

17.49 Ketene, $H_2C=C=O$, is an important industrial chemical. Predict the products that would be formed when ketene reacts with **(a)** ethanol, **(b)** acetic acid, and **(c)** ethylamine. (*Hint:* Markovnikov addition occurs.)

17.50 Two unsymmetrical anhydrides react with ethylamine as follows:

Explain the factors that might account for the formation of the products in each reaction.

17.51 Starting with 1-naphthol, suggest an alternative synthesis of the insecticide Sevin to the one given in Section 17.9A.

17.52 Suggest a synthesis of ibuprofen (Section 5.11) from benzene, employing **chloromethylation** as one step. Chloromethylation is a special case of the Friedel–Crafts reaction in which a mixture of HCHO and HCl, in the presence of $ZnCl_2$, introduces a —CH_2Cl group into an aromatic ring.

17.53 An alternative synthesis of ibuprofen is given below. Supply the structural formulas for compounds **A–D**:

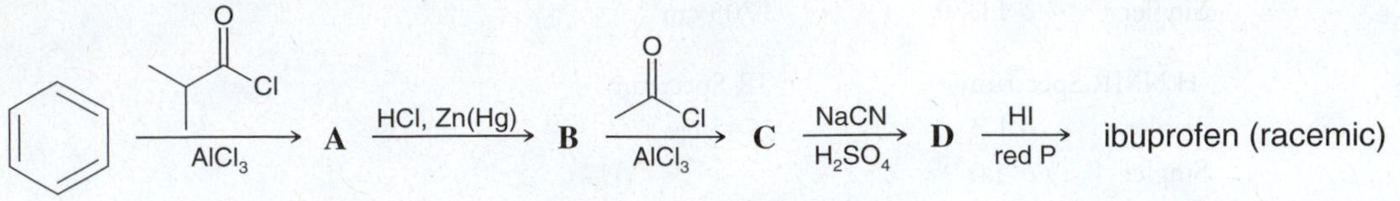

17.54 As a method for the synthesis of cinnamaldehyde (3-phenyl-2-propenal), a chemist treated 3-phenyl-2-propen-1-ol with $K_2Cr_2O_7$ in sulfuric acid. The product obtained from the reaction gave a signal at δ 164.5 in its ^{13}C NMR spectrum. Alternatively, when the chemist treated 3-phenyl-2-propen-1-ol with PCC in CH_2Cl_2, the ^{13}C NMR spectrum of the product displayed a signal at δ 193.8. (All other signals in the spectra of both compounds appeared at similar chemical shifts.) **(a)** Which reaction produced cinnamaldehyde? **(b)** What was the other product?

LEARNING GROUP PROBLEMS

The Chemical Synthesis of Peptides Carboxylic acids and acyl derivatives of the carboxyl functional group are very important in biochemistry. For example, the carboxylic acid functional group is present in the family of lipids called fatty acids. Lipids called glycerides contain the ester functional group, a derivative of carboxylic acids. Furthermore, the entire class of biopolymers called proteins contain repeating amide functional group linkages. Amides are also derivatives of carboxylic acids. Both laboratory and biochemical syntheses of proteins require reactions that involve substitution at activated acyl carbons.

This Learning Group Problem focuses on the chemical synthesis of small proteins, called peptides. The essence of peptide or protein synthesis is formation of the amide functional group by reaction of an activated carboxylic acid derivative with an amine.

First we shall consider reactions for traditional chemical synthesis of peptides and then we look at reactions used in automated solid-phase peptide synthesis. The method for solid-phase peptide synthesis was invented by R. B. Merrifield (Rockefeller University), for which he earned the 1984 Nobel Prize in Chemistry. Solid-phase peptide synthesis reactions are so reliable that they have been incorporated into machines called peptide synthesizers (Section 24.7D).

1. The first step in peptide synthesis is blocking (protection) of the amine functional group of an amino acid (a compound that contains both amine and carboxylic acid functional groups). Such a reaction is shown in Section 24.7C in the reaction between Ala (alanine) and benzyl chloroformate. The functional group formed in the structure labeled Z-Ala is called a carbamate (or urethane).

(Z is a benzyloxycarbonyl group, $C_6H_5CH_2O\overset{\overset{\displaystyle O}{\|}}{C}$ —).

(a) Write a detailed mechanism for formation of Z-Ala from Ala and benzyl chloroformate in the presence of hydroxide.

(b) In the reaction of part (a), why does the amino group act as the nucleophile preferentially over the carboxylate anion?

(c) Another widely used amino protecting group is the 9-fluorenylmethoxycarbonyl (Fmoc) group. Fmoc is the protecting group most often used in automated solid-phase peptide synthesis (see part 4 below). Write a detailed mechanism for formation of an Fmoc-protected amino acid under the conditions given in Section 24.7A.

2. The second step in the reactions of Section 24.7C is the formation of a mixed anhydride. Write a detailed mechanism for the reaction between Z-Ala and ethyl chloroformate ($ClCO_2C_2H_5$) in the presence of triethylamine to form the mixed anhydride. What is the purpose of this step?

3. The third step in the sequence of reactions in Section 24.7C is the one that actually joins the new amino acid (in this case leucine, abbreviated Leu) by another amide functional group. Write a detailed mechanism for this step (from the mixed anhydride of Z-Ala to Z-Ala-Leu). Show how CO_2 and ethanol are formed in the course of this mechanism.

4. A sequence of reactions commonly used for solid-phase peptide synthesis is shown in Section 24.7D.

(a) Write a detailed mechanism for step 1, in which diisopropylcarbodiimide is used to join the carboxyl group of the first amino acid (in Fmoc-protected form) to a hydroxyl group on the polymer solid support.

(b) Step 3 of the automated synthesis involves removal of the Fmoc group by reaction with piperidine (a reaction also shown in Section 24.7A). Write a detailed mechanism for this step.

CHAPTER

18

Reactions at the α Carbon of Carbonyl Compounds

ENOLS AND ENOLATES

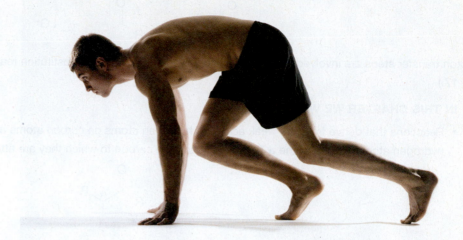

When we exercise vigorously, our bodies rely heavily on the metabolic process of glycolysis to derive energy from glucose. Glycolysis splits glucose into two three-carbon molecules. Only one of these three-carbon molecules (glyceraldehyde-3-phosphate, GAP) is directly capable of going further in the glycolytic pathway. The other three-carbon molecule (dihydroxyacetone-3-phosphate, DHAP) is not wasted, however. It is converted to a second molecule of GAP, via a type of intermediate that is key to our studies in this chapter—an enol (so named because the intermediate is an alkene alcohol). We shall learn about enols and enolates, their conjugate bases, in this chapter.

In Chapter 16, we saw how aldehydes and ketones can undergo nucleophilic addition at their carbonyl groups. For example:

Nucleophilic addition

PHOTO CREDIT: © Adam Burn/fStop/Age Fotostock America, Inc.

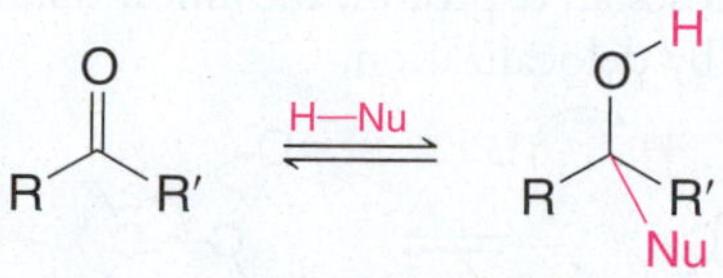

In Chapter 17 we saw how substitution could occur at a carbonyl group if a suitable leaving group is present. This type of reaction is called acyl substitution. For example:

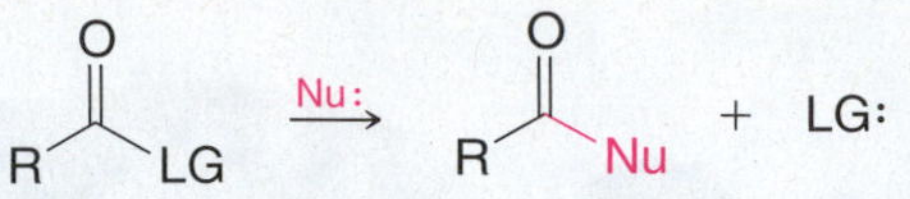

(Proton transfer steps are involved in some nucleophilic addition and acyl substitution reactions, as detailed in Chapters 16 and 17.)

IN THIS CHAPTER WE WILL CONSIDER:

- Reactions that derive from the weak acidity of hydrogen atoms on carbon atoms adjacent to α carbonyl group. These hydrogen atoms are called the α **hydrogens**, and the carbon to which they are attached is called the α **carbon**.

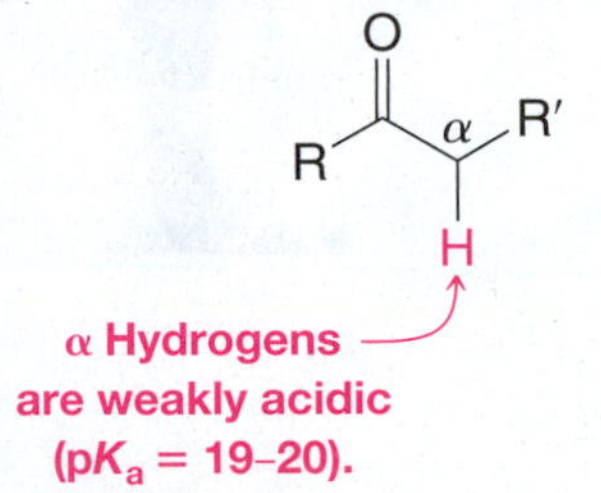

α **Hydrogens**
are weakly acidic
(pK_a = 19–20).

- The processes by which enols and enolates can be formed
- The concept of kinetic and thermodynamic deprotonations to generate different enolates from the same starting material
- Alkylations, acylations, and other electrophile additions to enols and enolates
- A special version of the same chemistry using the nitrogen analog of an enol—that is, an enamine

[**WHY** DO THESE TOPICS MATTER?] At the end of this chapter, we will show how the chemistry of enamines affords the ability to execute highly complex bond formations pertinent to the synthesis of complex, bioactive molecules, and how this chemistry has even been used to produce several tons of a highly valuable medicine.

18.1 THE ACIDITY OF THE α HYDROGENS OF CARBONYL COMPOUNDS: ENOLATE ANIONS

When we say that the α hydrogens of carbonyl compounds are acidic, *we mean that they are unusually acidic for hydrogen atoms attached to carbon.*

- The pK_a values for the α hydrogens of most simple aldehydes or ketones are of the order of 19–20.

This means that they are more acidic than hydrogen atoms of ethyne, pK_a = 25, and are far more acidic than the hydrogens of ethene (pK_a = 44) or of ethane (pK_a = 50).

The reasons for the unusual acidity of the α hydrogens of carbonyl compounds are straightforward.

- The carbonyl group is strongly electron withdrawing, and when a carbonyl compound loses an α proton, the anion that is produced, called an **enolate**, is stabilized by delocalization.

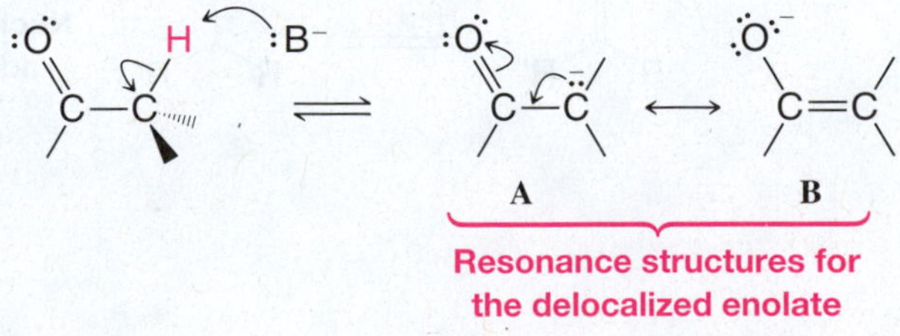

Resonance structures for
the delocalized enolate

Two resonance structures, **A** and **B**, can be written for the enolate. In structure **A** the negative charge is on carbon, and in structure **B** the negative charge is on oxygen. Both structures contribute to the hybrid. Although structure **A** is favored by the strength of its carbon–oxygen π bond relative to the weaker carbon–carbon π bond of **B**, structure **B** makes a greater contribution to the hybrid because oxygen, being highly electronegative, is better able to accommodate the negative charge. We can depict the enolate hybrid in the following way:

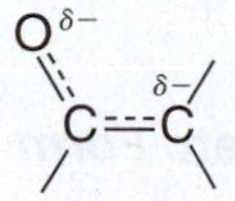

Hybrid resonance structure for an enolate

When this resonance-stabilized enolate accepts a proton, it can do so in either of two ways: It can accept the proton at carbon to form the original carbonyl compound in what is called the **keto form** or it may accept the proton at oxygen to form an **enol** (alk**en**e alco**hol**).

- The enolate is the conjugate base of both the enol and keto forms.

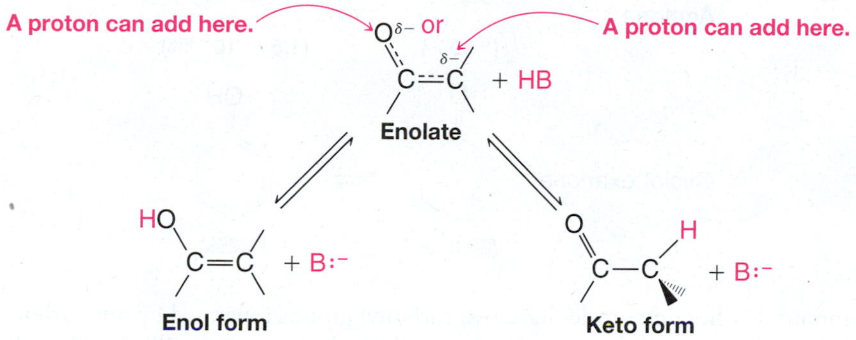

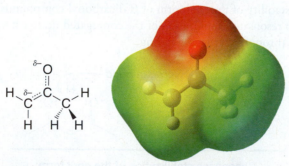

A proton can add here. ← O$^{\delta-}$ or → **A proton can add here.**

C==C + HB

Enolate

Enol form + B:$^-$ **Keto form** + B:$^-$

- Both of these reactions are reversible.

A calculated electrostatic potential map for the enolate of acetone is shown below. The map indicates approximately the outermost extent of electron density (the van der Waals surface) of the acetone enolate. Red color near the oxygen is consistent with oxygen being better able to stabilize the excess negative charge of the anion. Yellow at the carbon where the α hydrogen was removed indicates that some of the excess negative charge is localized there as well. These implications are parallel with the conclusions above about charge distribution in the hybrid based on delocalization and electronegativity effects.

Acetone enolate

18.2 KETO AND ENOL TAUTOMERS

The keto and enol forms of carbonyl compounds are constitutional isomers, but of a special type. Because they are easily interconverted by proton transfers in the presence of an acid or base, chemists use a special term to describe this type of constitutional isomerism.

- Interconvertible **keto and enol forms** are called **tautomers**, and their interconversion is called **tautomerization**.

Under most circumstances, we encounter keto–enol tautomers in a state of equilibrium. (The surfaces of ordinary laboratory glassware are able to catalyze the interconversion and establish the equilibrium.) For simple monocarbonyl compounds such as acetone and acetaldehyde, the amount of the enol form present at equilibrium is *very small*. In acetone it is much less than 1%; in acetaldehyde the enol concentration is too small to be detected. The greater stability of the following keto forms of monocarbonyl compounds can be related to the greater strength of the carbon–oxygen π bond compared to the carbon–carbon π bond (~364 versus ~ 250 kJ mol^{-1}):

Keto Form **Enol Form**

Acetaldehyde
(~100%) (extremely small)

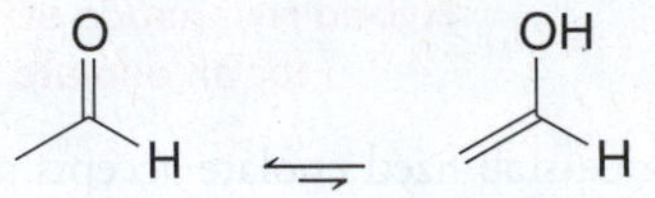

Acetone
(>99%) (1.5 × 10^{-4}%)

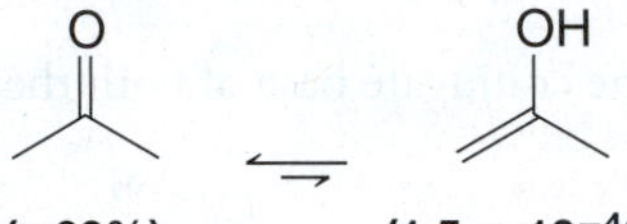

Cyclohexanone
(98.8%) (1.2%)

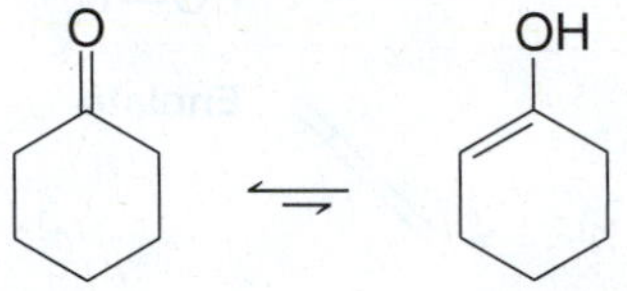

Helpful Hint

Keto–enol tautomers are not resonance structures. They are constitutional isomers in equilibrium (generally favoring the keto form).

In compounds whose molecules have two carbonyl groups separated by one carbon atom (called β-dicarbonyl compounds), the amount of enol present at equilibrium is far higher. For example, pentane-2,4-dione exists in the enol form to an extent of 76%:

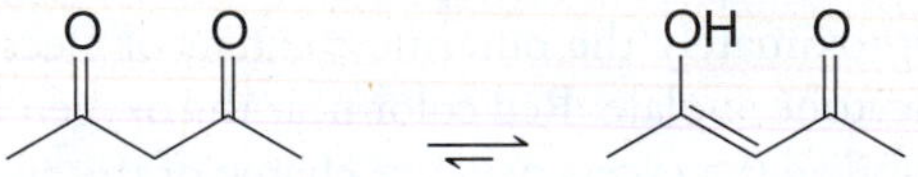

Pentane-2,4-dione (24%) Enol form (76%)

Helpful Hint

See "The Chemistry of... TIM (Triose Phosphate Isomerase) Recycles Carbon via an Enol" in *WileyPLUS* for more information relating to this chapter's opener about an important energy-yielding biochemical process.

- The greater stability of the enol form of β-dicarbonyl compounds can be attributed to resonance stabilization of the conjugated double bonds and (in a cyclic form) through hydrogen bonding.

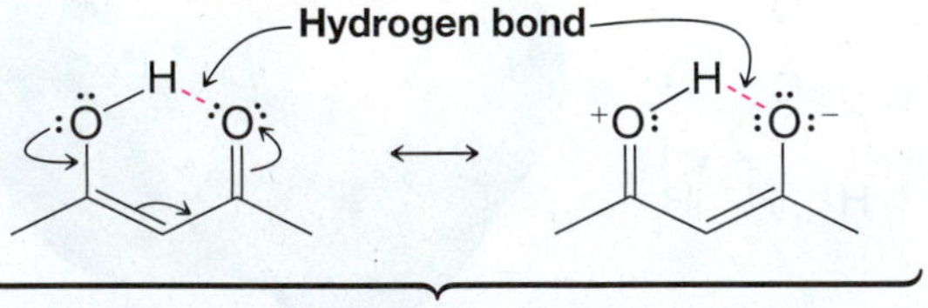

Hydrogen bond

Resonance stabilization of the enol form

••• SOLVED PROBLEM 18.1

Write bond-line structures for the keto and enol forms of 3-pentanone.

ANSWER:

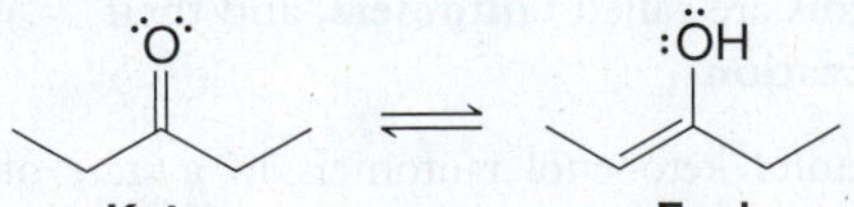

Keto Enol

For all practical purposes, the compound cyclohexa-2,4-dien-1-one exists totally in its enol form. Write the structure of cyclohexa-2,4-dien-1-one and of its enol form. What special factor accounts for the stability of the enol form?

18.3 REACTIONS VIA ENOLS AND ENOLATES

18.3A Racemization

When a solution of (R)-(+)-2-methyl-1-phenylbutan-1-one (see the following reaction) in aqueous ethanol is treated with either acids or bases, the solution gradually loses its optical activity. After a time, isolation of the ketone shows that it has been completely racemized. The (+) form of the ketone has been converted to an equimolar mixture of its enantiomers through its enol form.

Racemization via an enol

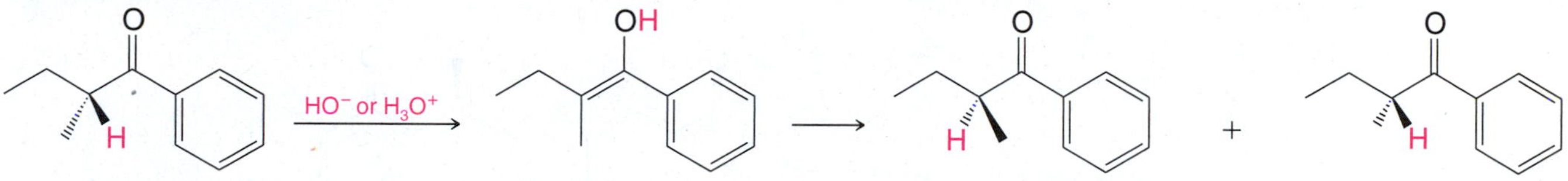

(R)-(+)-2-Methyl-1-
phenylbutan-1-one

Enol
(achiral)

(±)-2-Methyl-1-
phenylbutan-1-one
(racemic form)

- Racemization at an α carbon takes place in the presence of acids or bases because the keto form slowly but reversibly changes to its enol *and the enol is achiral.* When the **enol** reverts to the **keto form**, it can produce equal amounts of the two enantiomers.

A base catalyzes the formation of an enol through the intermediate formation of an **enolate** anion.

A MECHANISM FOR THE REACTION — Base-Catalyzed Enolization

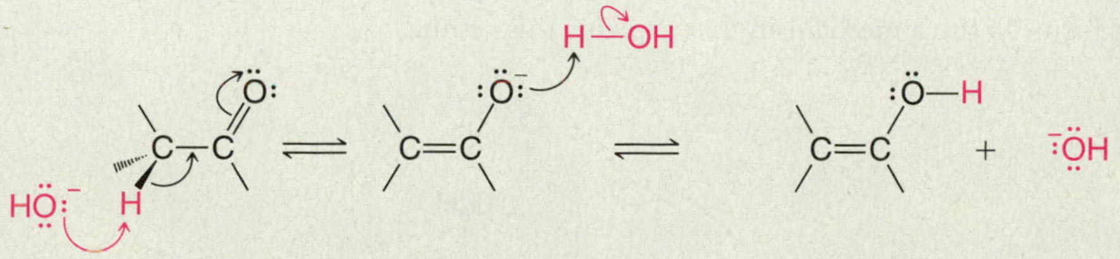

Enolate
(achiral)

Enol
(achiral)

An acid can catalyze enolization in the following way.

A MECHANISM FOR THE REACTION — Acid-Catalyzed Enolization

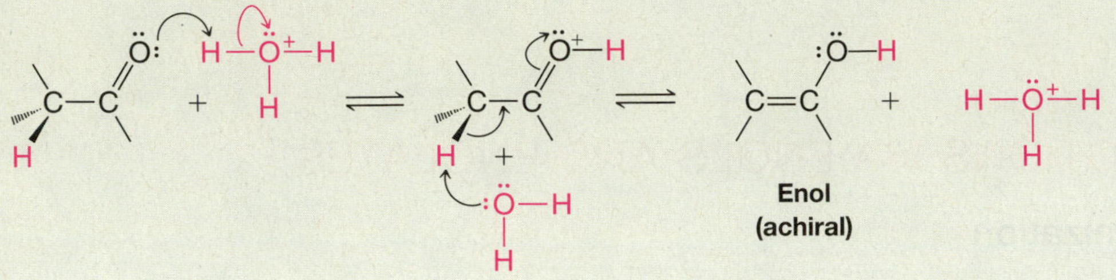

In acyclic ketones, the enol or enolate formed can be (E) or (Z). Protonation on one face of the (E) isomer and protonation on the same face of the (Z) isomer produces enantiomers.

● ● ●

PRACTICE PROBLEM 18.2 Would optically active ketones such as the following undergo acid- or base-catalyzed racemization? Explain your answer.

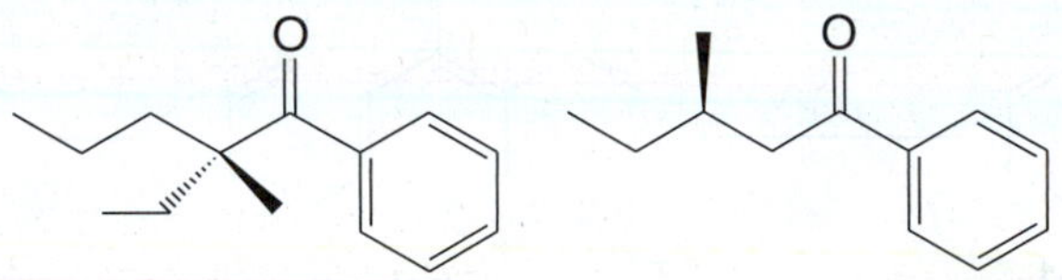

- Diastereomers that differ in configuration at only one of several chirality centers are sometimes called **epimers**.

Keto–enol tautomerization can sometimes be used to convert a less stable epimer to a more stable one. This equilibration process is an example of **epimerization**. An example is the epimerization of *cis*-decalone to *trans*-decalone:

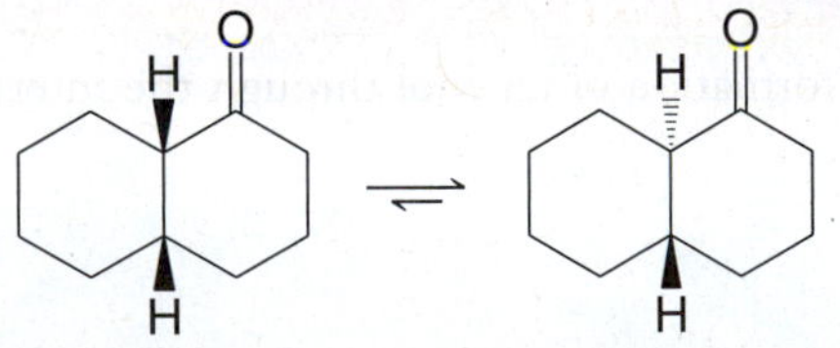

cis-Decalone   *trans*-Decalone

● ● ● **SOLVED PROBLEM 18.2**

Treating racemic 2-methyl-1-phenylbutan-1-one with **NaOD** in the presence of **D₂O** produces a deuterium-labeled compound as a racemic form. Write a mechanism that explains this result.

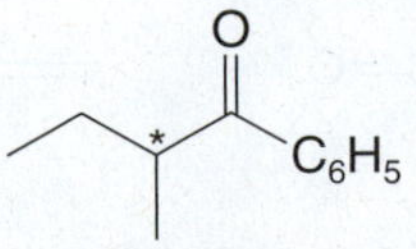

Racemic

STRATEGY AND ANSWER: Either enantiomer of the ketone can transfer an α proton to the ⁻OD ion to form an achiral enolate that can accept a deuteron to form a racemic mixture of the deuterium-labeled product.

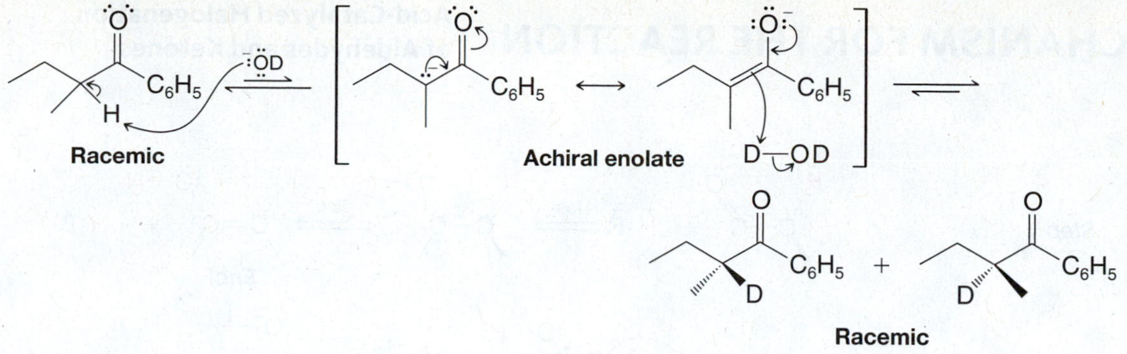
Racemic
Achiral enolate
Racemic

PRACTICE PROBLEM 18.3

Write a mechanism using sodium ethoxide in ethanol for the epimerization of *cis*-decalone to *trans*-decalone. Draw chair conformational structures that show why *trans*-decalone is more stable than *cis*-decalone. You may find it helpful to also examine handheld molecular models of *cis*- and *trans*-decalone.

18.3B Halogenation at the α Carbon

- Carbonyl compounds bearing an α hydrogen can undergo halogen substitution at the α carbon in the presence of acid or base.

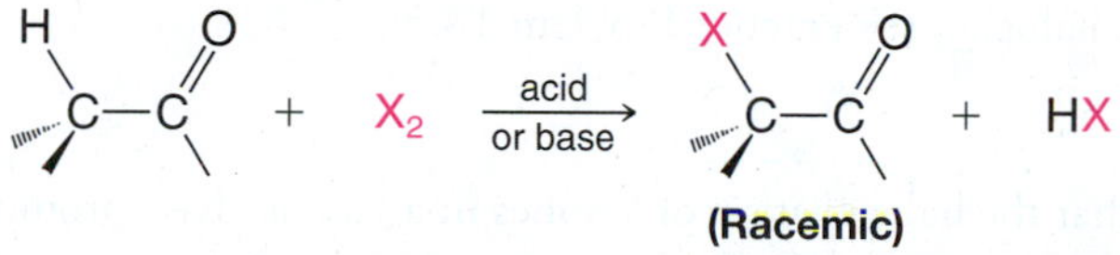
acid
or base
(Racemic)

Base-Promoted Halogenation In the presence of bases, halogenation takes place through the slow formation of an enolate anion or an enol followed by a rapid reaction of the enolate or enol with halogen.

A MECHANISM FOR THE REACTION — Base-Promoted Halogenation of Aldehydes and Ketones

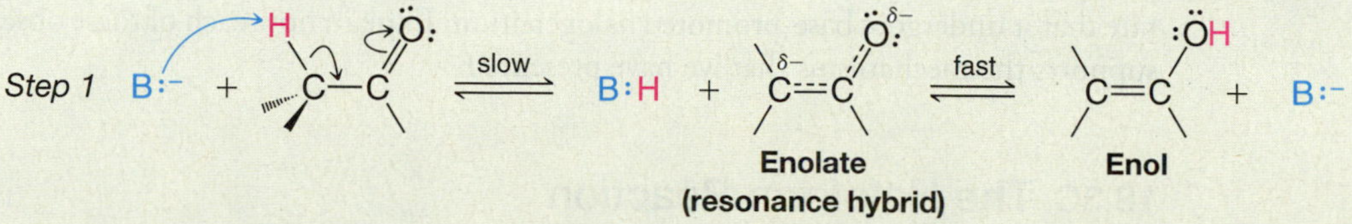
Step 1
slow
fast
Enolate
(resonance hybrid)
Enol

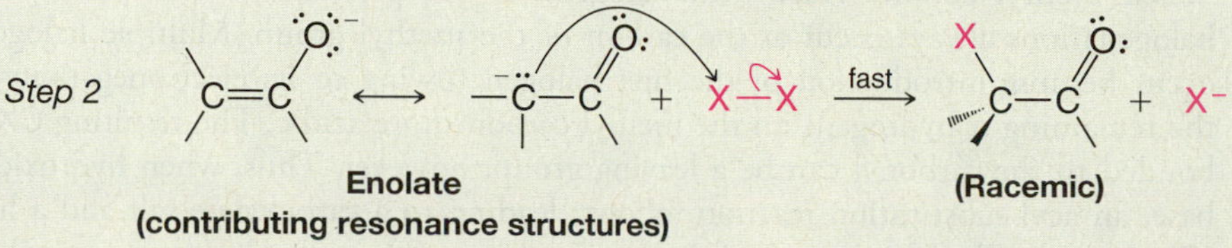
Step 2
fast
Enolate
(contributing resonance structures)
(Racemic)

As we shall see in Section 18.3C, multiple halogenations can occur.

Acid-Catalyzed Halogenation In the presence of acids, halogenation takes place through the slow formation of an enol followed by rapid reaction of the enol with the halogen.

[A MECHANISM FOR THE REACTION

Acid-Catalyzed Halogenation of Aldehydes and Ketones]

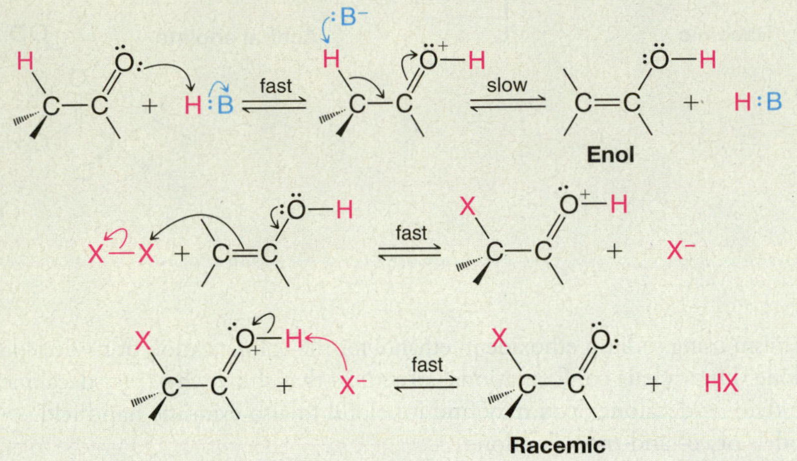

Step 1

Enol

Step 2

Step 3

Racemic

Part of the evidence that supports these mechanisms comes from studies of reaction kinetics. Both base-promoted and acid-catalyzed halogenations of ketones *show initial rates that are independent of the halogen concentration.* The mechanisms that we have written are in accord with this observation: in both instances the slow step of the mechanism occurs before the intervention of the halogen. (The initial rates are also independent of the nature of the halogen; see Practice Problem 18.5.)

PRACTICE PROBLEM 18.4 Why do we say that the halogenation of ketones in a base is "base promoted" rather than "base catalyzed"?

PRACTICE PROBLEM 18.5 Additional evidence for the halogenation mechanisms that we just presented comes from the following facts: **(a)** Optically active 2-methyl-1-phenylbutan-1-one undergoes acid-catalyzed racemization at a rate exactly equivalent to the rate at which it undergoes acid-catalyzed halogenation. **(b)** 2-Methyl-1-phenylbutan-1-one undergoes acid-catalyzed iodination at the same rate that it undergoes acid-catalyzed bromination. **(c)** 2-Methyl-1-phenylbutan-1-one undergoes base-catalyzed hydrogen–deuterium exchange at the same rate that it undergoes base-promoted halogenation. Explain how each of these observations supports the mechanisms that we have presented.

18.3C The Haloform Reaction

When methyl ketones react with halogens in the presence of excess base, multiple halogenations always occur at the carbon of the methyl group. Multiple halogenations occur because introduction of the first halogen (owing to its electronegativity) makes the remaining α hydrogens on the methyl carbon more acidic. The resulting CX_3 group bonded to the carbonyl can be a leaving group, however. Thus, when hydroxide is the base, an acyl substitution reaction follows, leading to a carboxylate salt and a haloform (CHX_3, e.g., chloroform, bromoform, or iodoform). The following is an example.

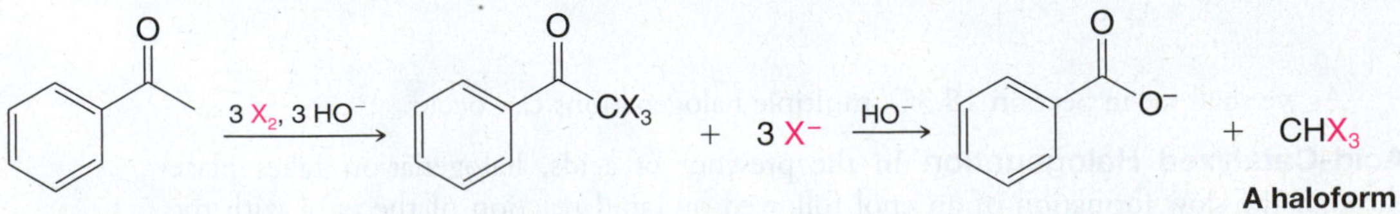

A haloform (X = Cl, Br, I)

The haloform reaction is one of the rare instances in which a carbanion acts as a leaving group. This occurs because the trihalomethyl anion is unusually stable; its negative charge is dispersed by the three electronegative halogen atoms (when X = Cl, the conjugate acid, $CHCl_3$, has $pK_a = 13.6$). In the last step, a proton transfer takes place between the carboxylic acid and the trihalomethyl anion.

The **haloform reaction** is synthetically useful as a means of converting methyl ketones to carboxylic acids. When the haloform reaction is used in synthesis, chlorine and bromine are most commonly used as the halogen component. Chloroform ($CHCl_3$) and bromoform ($CHBr_3$) are both liquids that are immiscible with water and are easily separated from the aqueous solution containing the carboxylate anion. When iodine is the halogen component, the bright yellow solid iodoform (CHI_3) results. This version is the basis of the iodoform classification test for methyl ketones and methyl secondary alcohols (which are oxidized to methyl ketones first under the reaction conditions):

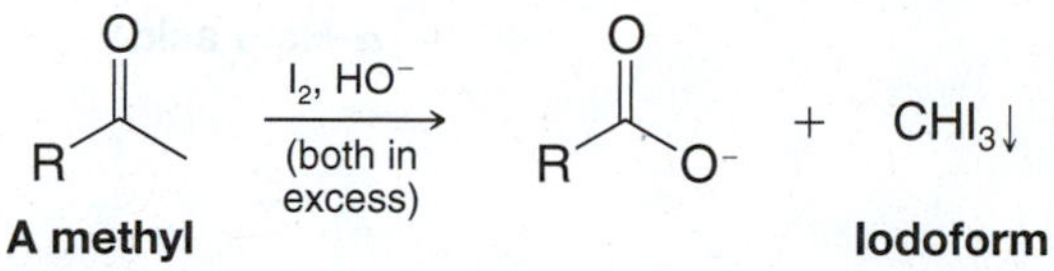

A methyl ketone $\xrightarrow[\substack{(both\ in \\ excess)}]{I_2,\ HO^-}$ + **CHI$_3$↓**

Iodoform (a yellow precipitate)

A MECHANISM FOR THE REACTION — The Haloform Reaction

Halogenation Step

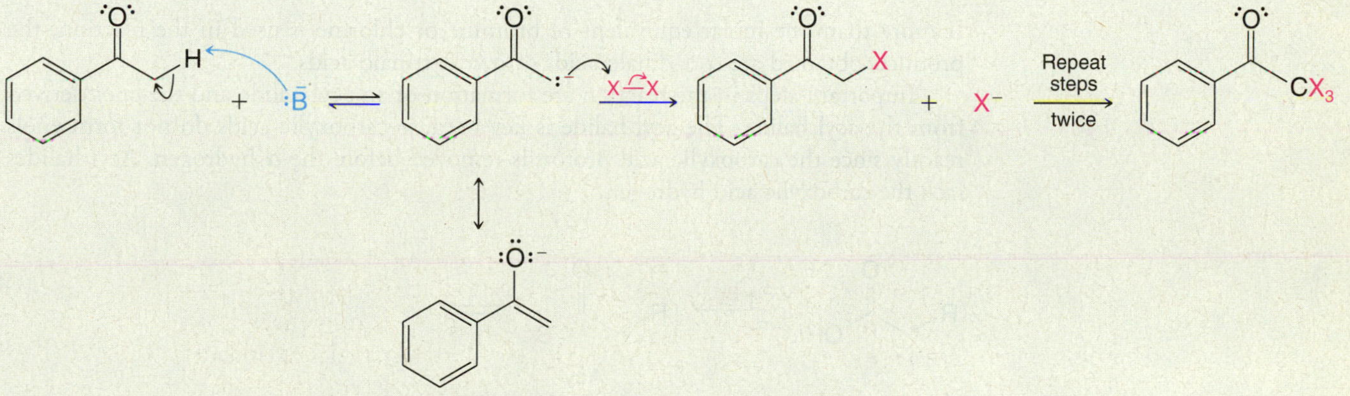

Enolate

Acyl Substitution Step

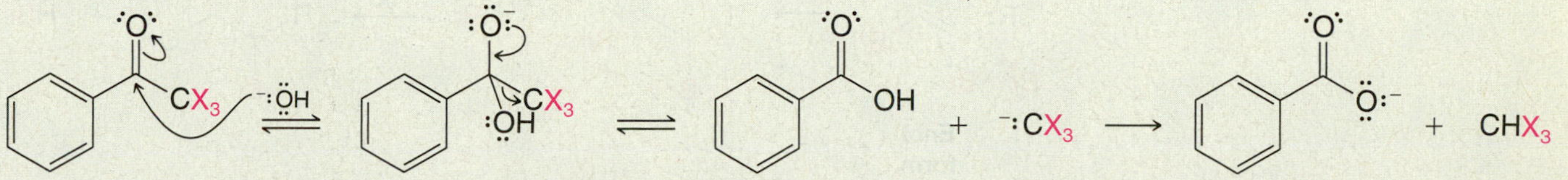

Carboxylate anion   **A haloform**

THE CHEMISTRY OF... Chloroform in Drinking Water

When water is chlorinated to purify it for public consumption, chloroform is produced from organic impurities in the water via the haloform reaction. (Many of these organic impurities are naturally occurring, such as humic substances.) The presence of chloroform in public water is of concern for water treatment plants and environmental officers, because chloroform is carcinogenic. Thus, the technology that solves one problem creates another. It is worth recalling, however, that before chlorination of water was introduced, thousands of people died in epidemics of diseases such as cholera and dysentery.

18.3D α-Halo Carboxylic Acids: The Hell–Volhard–Zelinski Reaction

Carboxylic acids bearing α **hydrogen** atoms react with bromine or chlorine in the presence of phosphorus (or a phosphorus halide) to give α-halo carboxylic acids through a reaction known as the Hell–Volhard–Zelinski (or HVZ) reaction.

General Reaction

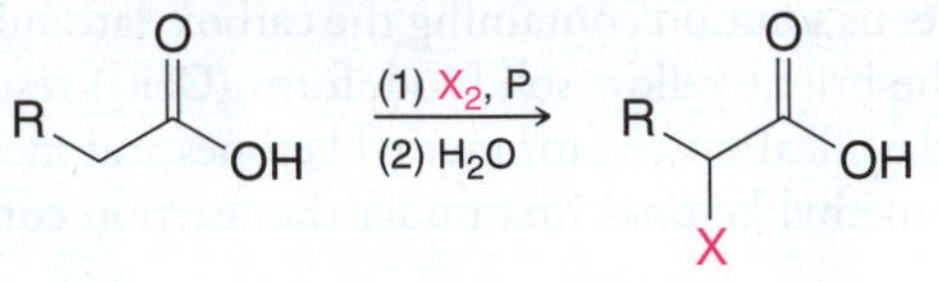

α-Halo acid

Specific Example

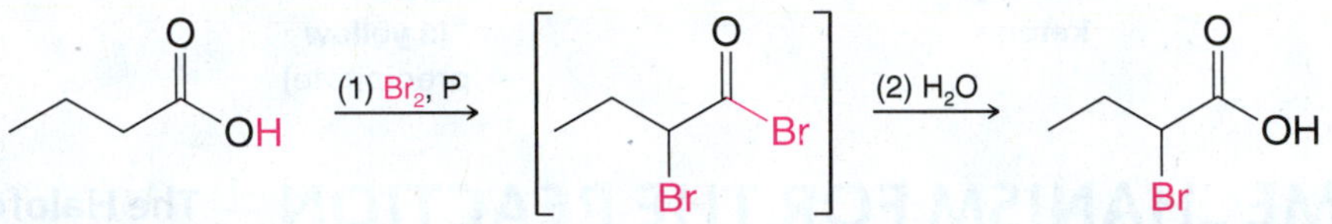

Butanoic acid

2-Bromobutanoic acid
(77%)

If more than one molar equivalent of bromine or chlorine is used in the reaction, the products obtained are α,α-dihalo acids or α,α,α-trihalo acids.

Important steps in the reaction are formation of an acyl halide and the enol derived from the acyl halide. The acyl halide is key because carboxylic acids do not form enols readily since the carboxylic acid proton is removed before the α hydrogen. Acyl halides lack the carboxylic acid hydrogen.

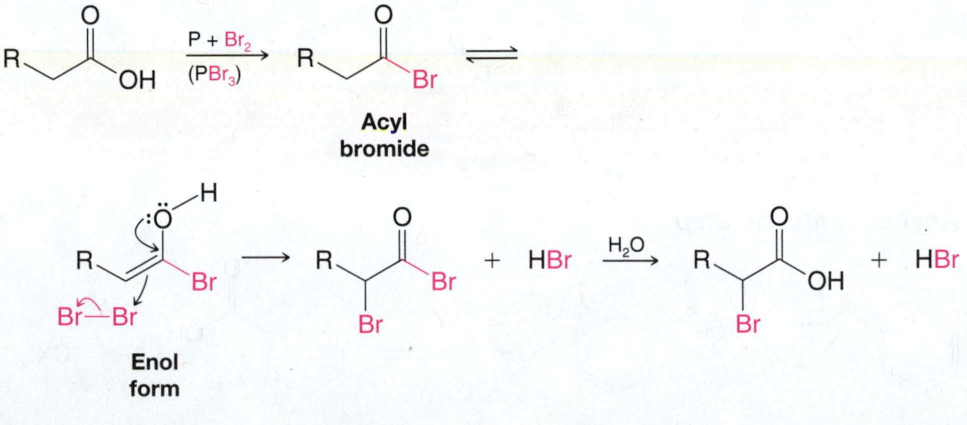

Acyl bromide

Enol form

An alternative method for α-halogenation has been developed by D. N. Harpp (McGill University). Acyl chlorides, formed *in situ* by the reaction of the carboxylic acid with $SOCl_2$, are treated with the appropriate *N*-halosuccinimide and a trace of HX to produce α-chloro and α-bromo acyl chlorides.

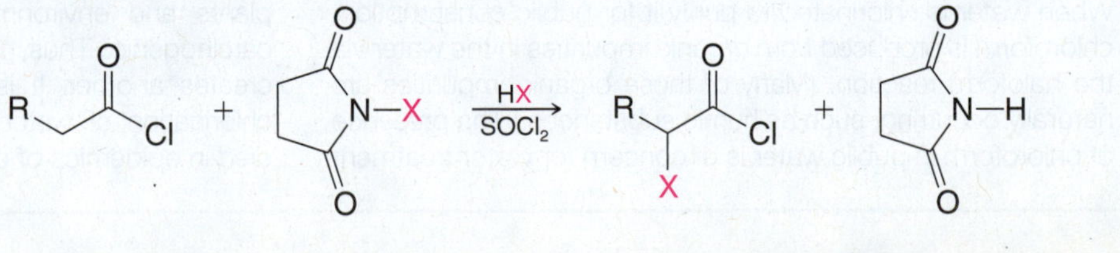

α-Iodo acyl chlorides can be obtained by using molecular iodine in a similar reaction.

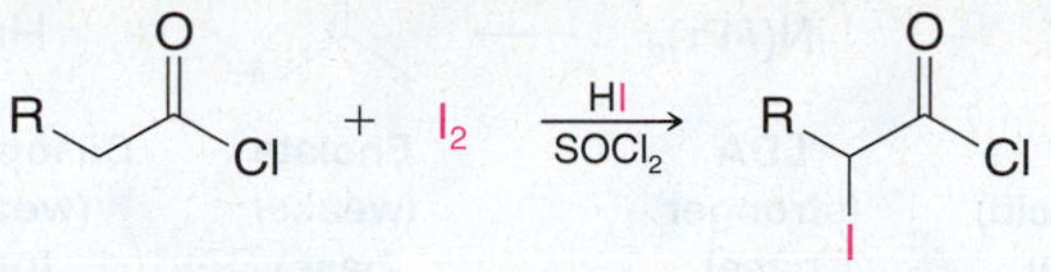

α-Halo acids are important synthetic intermediates because they are capable of reacting with a variety of nucleophiles:

Conversion to α-Hydroxy Acids

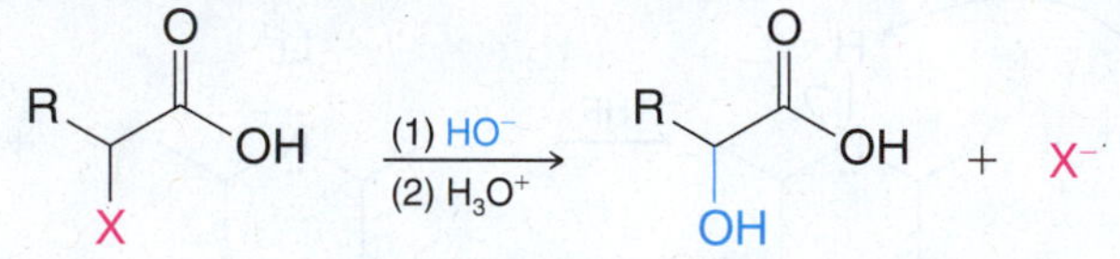

α–Halo acid α–Hydroxy acid

Specific Example

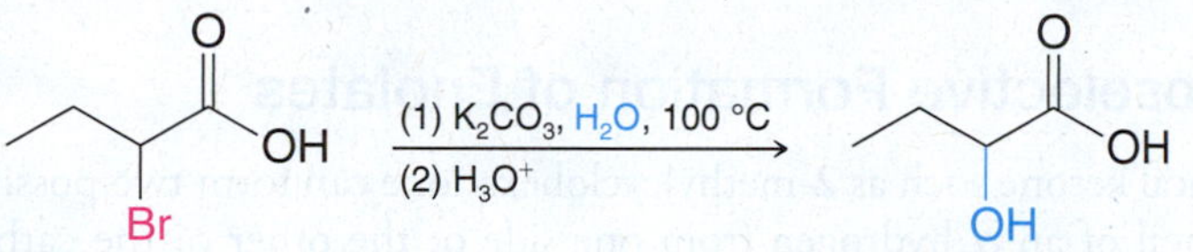

2-Hydroxybutanoic acid
(69%)

Conversion to α-Amino Acids

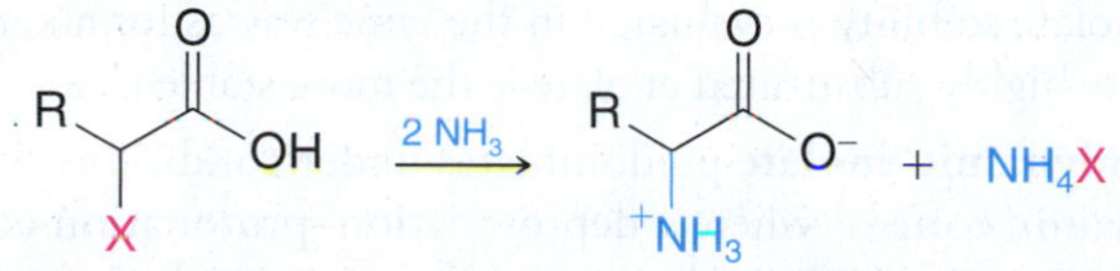

α–Halo acid α–Amino acid

Specific Example

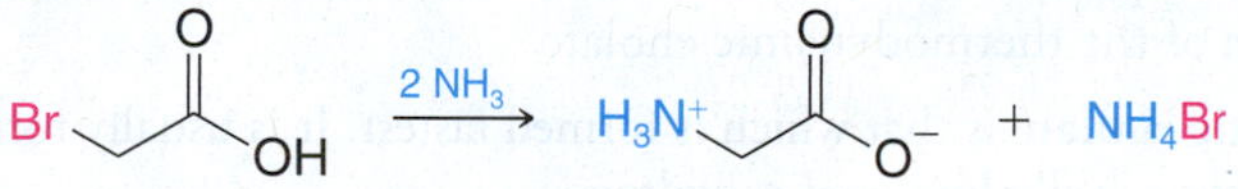

Aminoacetic acid
(glycine)
(60–64%)

18.4 LITHIUM ENOLATES

The position of the equilibrium by which an **enolate** forms depends on the strength of the base used. If the base employed is a weaker base than the enolate, then the equilibrium lies to the left. This is the case, for example, when a ketone is treated with sodium ethoxide in ethanol.

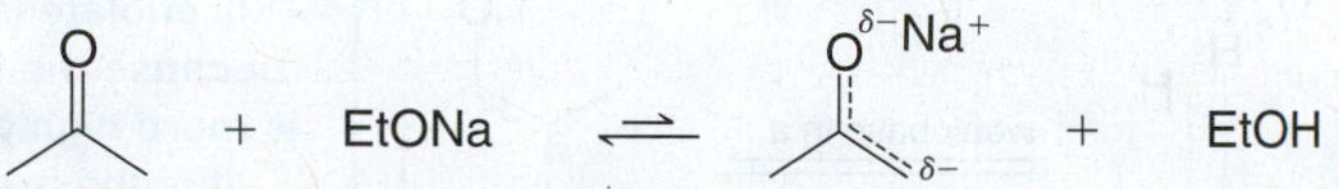

Weaker acid   Weaker base   Stronger base   Stronger acid
($pK_a = 19$)   ($pK_a = 16$)

On the other hand, if a very strong base is employed, the equilibrium lies far to the right. One very useful strong base for converting carbonyl compounds to enolates is **lithium diisopropylamide (LDA)** or $LiN(i\text{-}Pr)_2$:

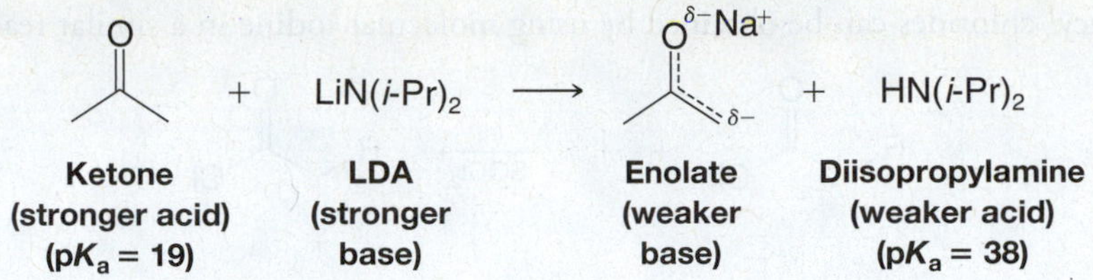

| Ketone (stronger acid) ($pK_a = 19$) | + | LiN(i-Pr)$_2$ LDA (stronger base) | $\longrightarrow$ | Enolate (weaker base) | + | HN(i-Pr)$_2$ Diisopropylamine (weaker acid) ($pK_a = 38$) |

- Lithium diisopropylamide (LDA) can be prepared by dissolving diisopropylamine in a solvent such as diethyl ether or THF and treating it with an alkyllithium:

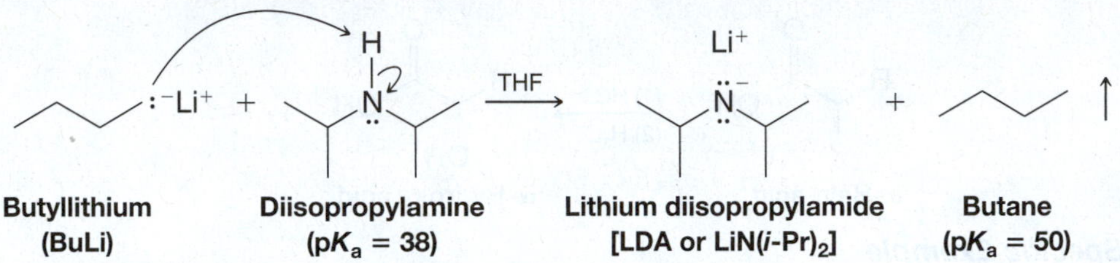

| Butyllithium (BuLi) | Diisopropylamine ($pK_a = 38$) | Lithium diisopropylamide [LDA or LiN(i-Pr)$_2$] | Butane ($pK_a = 50$) |

18.4A Regioselective Formation of Enolates

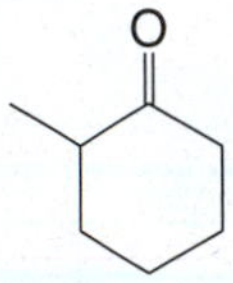

An unsymmetrical ketone such as 2-methylcyclohexanone can form two possible enolates, arising by removal of an α hydrogen from one side or the other of the carbonyl group. Which enolate predominates in the reaction depends on whether the enolate is formed under conditions that favor an acid–base equilibrium.

2-Methylcyclohexanone

- The **thermodynamic enolate** is that which is most stable among the possible enolates. Enolate stability is evaluated in the same way as for alkenes, meaning that the more highly substituted enolate is the more stable one.

- The **thermodynamic enolate** predominates under conditions of **thermodynamic control** where a deprotonation–protonation equilibrium allows interconversion among the possible enolates, such that eventually the more stable enolate exists in higher concentration. This is the case when the pK_a of the conjugate acid of the base is similar to the pK_a of the α hydrogen of the carbonyl compound. Use of hydroxide or an alkoxide in a protic solvent favors formation of the thermodynamic enolate.

- The **kinetic enolate** is that which is formed fastest. It is usually formed by removal of the least sterically hindered α hydrogen.

- The **kinetic enolate** predominates under conditions of **kinetic control** that do not favor equilibrium among the possible enolates. Use of a very strong and sterically hindered base in an aprotic solvent, such as LDA in tetrahydrofuran (THF) or dimethoxyethane (DME) favors formation of the kinetic enolate.

Conditions favoring formation of the thermodynamic and kinetic enolates from 2-methylcyclohexanone are illustrated below.

Formation of a Thermodynamic Enolate

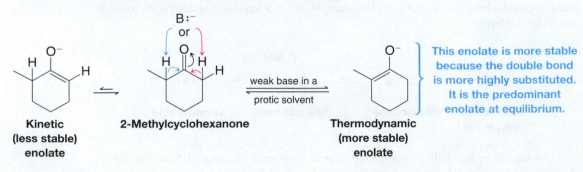

| Kinetic (less stable) enolate | 2-Methylcyclohexanone | Thermodynamic (more stable) enolate |

This enolate is more stable because the double bond is more highly substituted. It is the predominant enolate at equilibrium.

Formation of a Kinetic Enolate

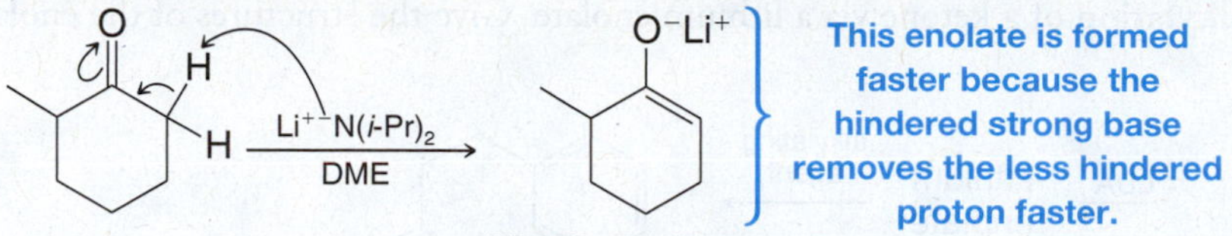

Kinetic enolate

This enolate is formed faster because the hindered strong base removes the less hindered proton faster.

18.4B Direct Alkylation of Ketones via Lithium Enolates

The formation of lithium enolates using lithium diisopropylamide furnishes a useful way of alkylating ketones in a regioselective way. For example, the lithium enolate formed from 2-methylcyclohexanone can be methylated or benzylated at the less hindered α carbon by allowing it to react with LDA followed by methyl iodide or benzyl bromide, respectively:

Helpful Hint

Alkylation of lithium enolates is a useful method for synthesis.

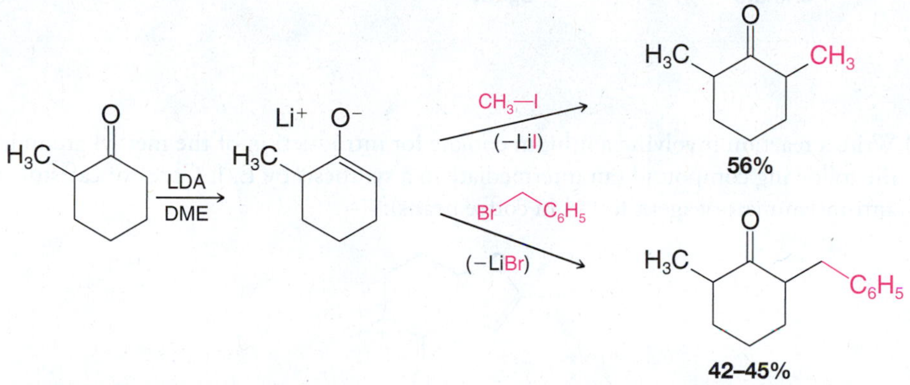

Alkylation reactions like these have an important limitation, however, because the reactions are S_N2 reactions, and also because enolates are strong bases.

- Successful alkylations occur only when primary alkyl, primary benzylic, and primary allylic halides are used. With secondary and tertiary halides, elimination becomes the main course of the reaction.

Helpful Hint

Proper choice of the alkylating agent is key to successful lithium enolate alkylation.

18.4C Direct Alkylation of Esters

Examples of the **direct alkylation** of esters are shown below. In the second example the ester is a lactone (Section 17.7C):

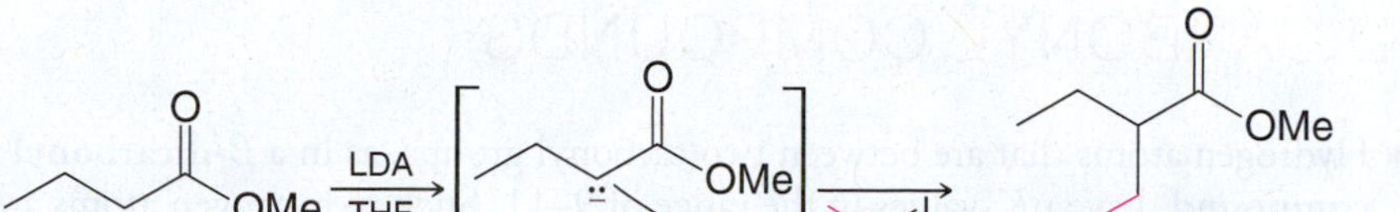

Methyl butanoate

Methyl 2-ethylbutanoate (96%)

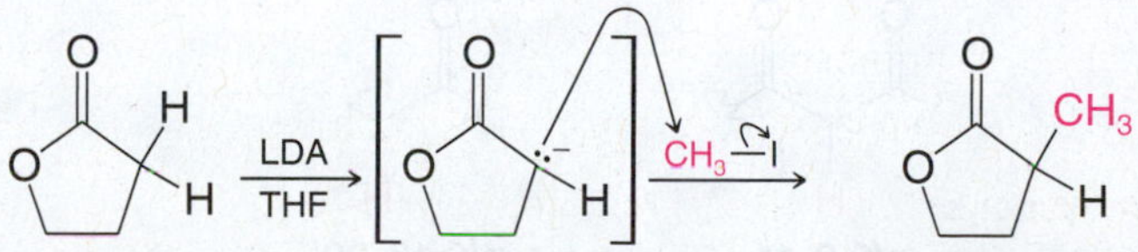

Butyrolactone

2-Methylbutyrolactone (88%)

••• SOLVED PROBLEM 18.3

The following synthesis illustrates the **alkylation** of a ketone via a lithium enolate. Give the structures of the enolate and the alkylating agent.

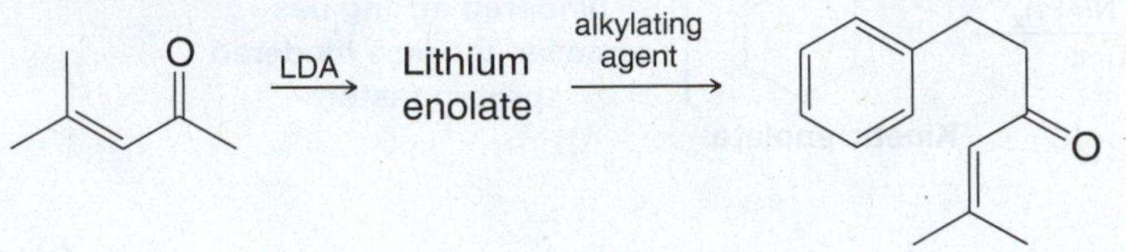

ANSWER:

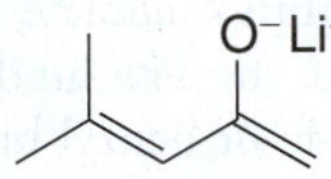

| **Lithium enolate** | **Alkylating agent** |

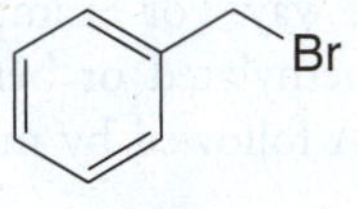

••• PRACTICE PROBLEM 18.6   **(a)** Write a reaction involving a lithium enolate for introduction of the methyl group in the following compound (an intermediate in a synthesis by E. J. Corey of cafestol, an anti-inflammatory agent found in coffee beans):

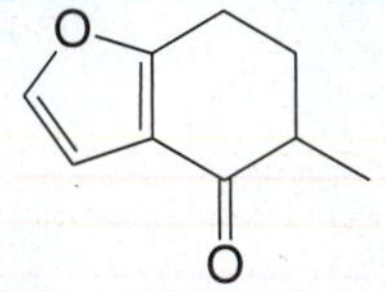

(b) Dienolates can be formed from β-keto esters using two equivalents of LDA. The di-enolate can then be alkylated selectively at the more basic of the two enolate carbons. Write a reaction for synthesis of the following compound using a dienolate and the appropriate alkyl halide:

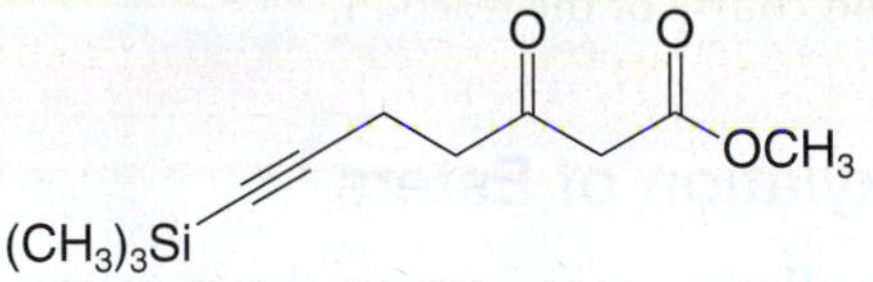

18.5 ENOLATES OF β-DICARBONYL COMPOUNDS

- Hydrogen atoms that are between two carbonyl groups, as in a β-**dicarbonyl compound**, have pK_a values in the range of 9–11. Such α-hydrogen atoms are much more acidic than α hydrogens adjacent to only one carbonyl group, which have pK_a values of 18–20.

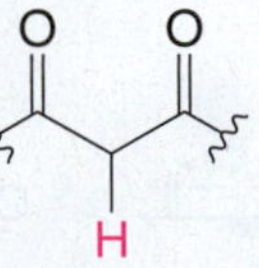

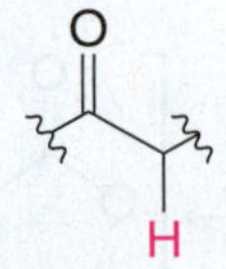

- A much weaker base than LDA, such as an alkoxide, can be used to form an enolate from a β-dicarbonyl compound.

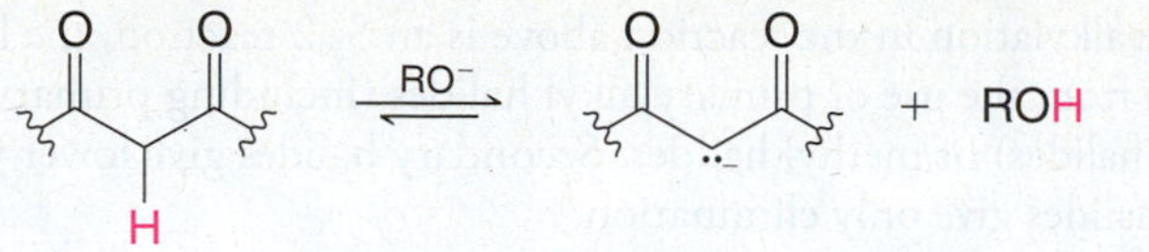

We can account for the greater acidity of β-dicarbonyl systems, as compared to single carbonyl systems, by delocalization of the negative charge to two oxygen atoms instead of one. We can represent this delocalization by drawing contributing resonance structures for a β-dicarbonyl enolate and its resonance hybrid:

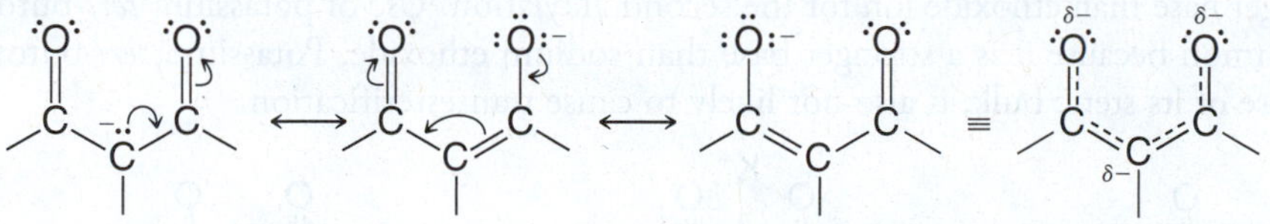

Contributing resonance structures   **Resonance hybrid**

We can visualize the enhanced charge delocalization of a β-dicarbonyl enolate by examining maps of electrostatic potential for enolates derived from pentane-2,4-dione and acetone. Here we see that the negative charge of the enolate from pentane-2,4-dione is associated substantially with the two oxygen atoms, as compared with the enolate from acetone, where significant negative charge in the enolate remains at the α-carbon atom:

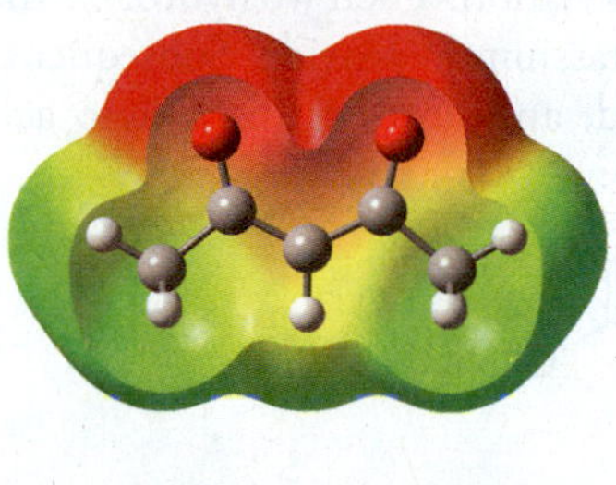

Pentane-2,4-dione enolate 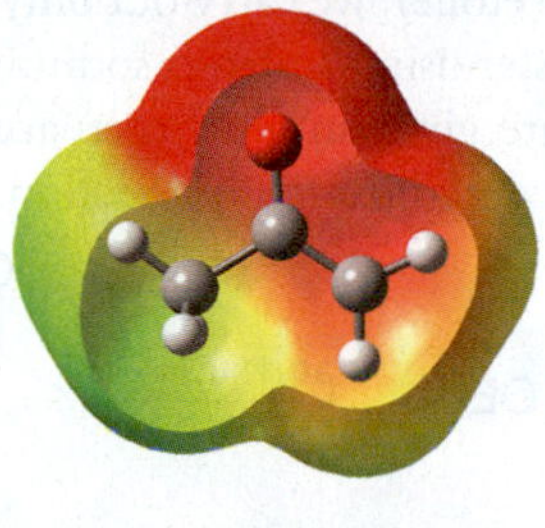   **Acetone enolate**

Two specific β-dicarbonyl compounds have had broad use in organic synthesis. These are acetoacetic ester (ethyl acetoacetate, ethyl 3-oxobutanoate), which can be used to make substituted acetone derivatives, and diethyl malonate (diethyl 1,3-propanedicarboxylic acid), which can be used to make substituted acetic acid derivatives. We shall consider syntheses involving ethyl acetoacetate and diethyl malonate in the upcoming sections of this chapter.

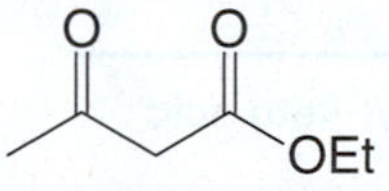

Acetoacetic ester
(ethyl acetoacetate;
ethyl 3-oxobutanoate)

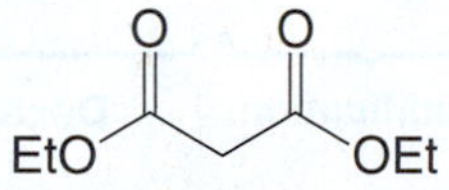

Diethyl malonate
(diethyl 1,3-propanedicarboxylic acid)

18.6 SYNTHESIS OF METHYL KETONES: THE ACETOACETIC ESTER SYNTHESIS

Acetoacetic ester, because it is a β-dicarbonyl compound, can easily be converted to an enolate using sodium ethoxide. We can then alkylate the resulting enolate (called sodio-acetoacetic ester) with an alkyl halide. This process is called an **acetoacetic ester synthesis**.

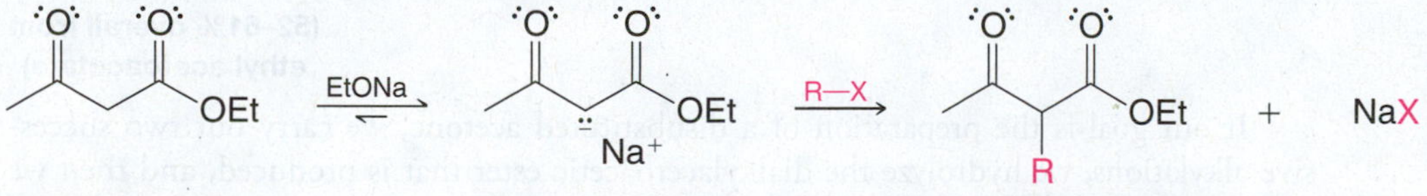

Acetoacetic
ester

Sodioacetoacetic
ester

Monoalkylacetoacetic
ester

- Since the alkylation in the reaction above is an S_N2 reaction, the best yields are obtained from the use of primary alkyl halides (including primary allylic and benzylic halides) or methyl halides. Secondary halides give lower yields, and tertiary halides give only elimination.

Dialkylation The monoalkylacetoacetic ester shown above still has one appreciably acidic hydrogen, and, if we desire, we can carry out a second alkylation. Because a monoalkylacetoacetic ester is somewhat less acidic than acetoacetic ester itself due to the electron-donating effect of the added alkyl group, it is usually helpful to use a stronger base than ethoxide ion for the second alkylation. Use of potassium *tert*-butoxide is common because it is a stronger base than sodium ethoxide. Potassium *tert*-butoxide, because of its steric bulk, is also not likely to cause transesterification.

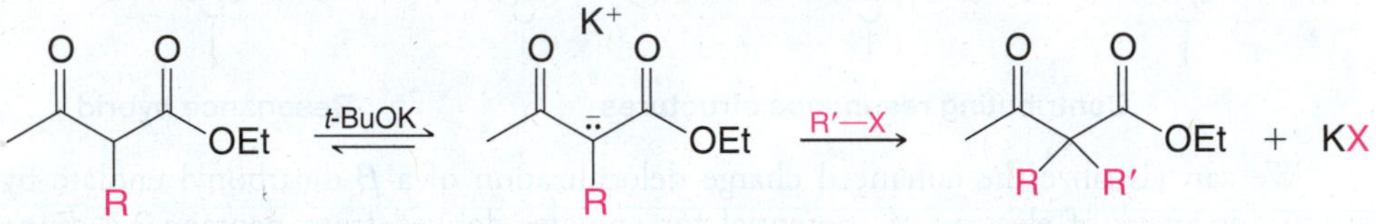

Monoalkylacetoacetic ester

Dialkylacetoacetic ester

Substituted Methyl Ketones To synthesize a monosubstituted methyl ketone (monosubstituted acetone), we carry out only one alkylation. Then we hydrolyze the monoalkylacetoacetic ester using aqueous sodium or potassium hydroxide. Subsequent acidification of the mixture gives an alkyl-acetoacetic acid, and heating this β-keto acid to 100 °C brings about decarboxylation (Section 17.10):

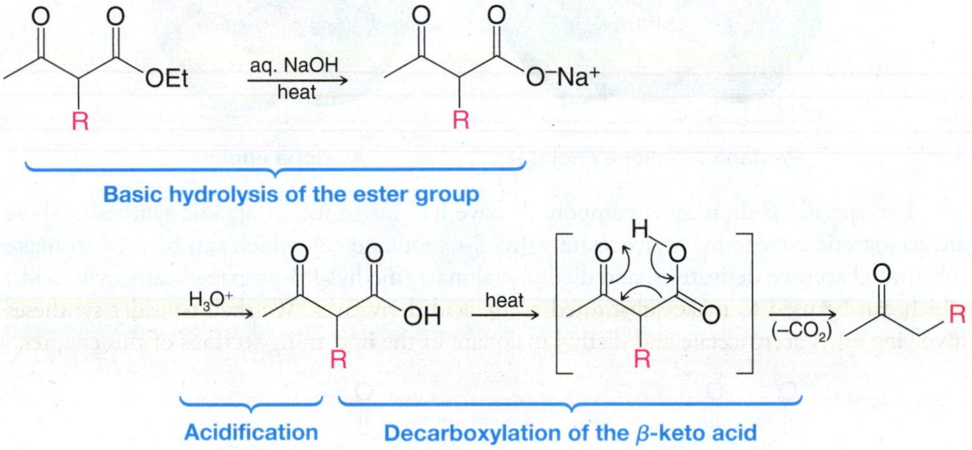

Basic hydrolysis of the ester group

Acidification   **Decarboxylation of the β-keto acid**

A specific example is the following synthesis of 2-heptanone:

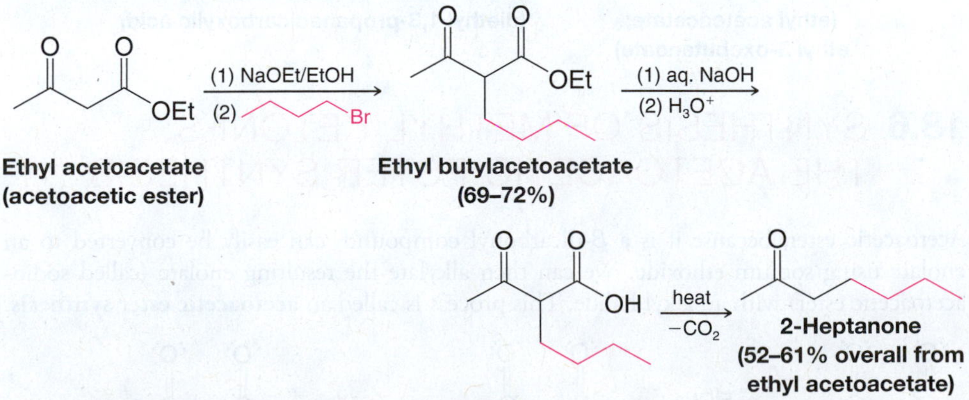

Ethyl acetoacetate (acetoacetic ester)

Ethyl butylacetoacetate (69–72%)

2-Heptanone (52–61% overall from ethyl acetoacetate)

If our goal is the preparation of a disubstituted acetone, we carry out two successive alkylations, we hydrolyze the dialkylacetoacetic ester that is produced, and then we decarboxylate the dialkylacetoacetic acid. An example of this procedure is the synthesis of 3-butyl-2-heptanone.

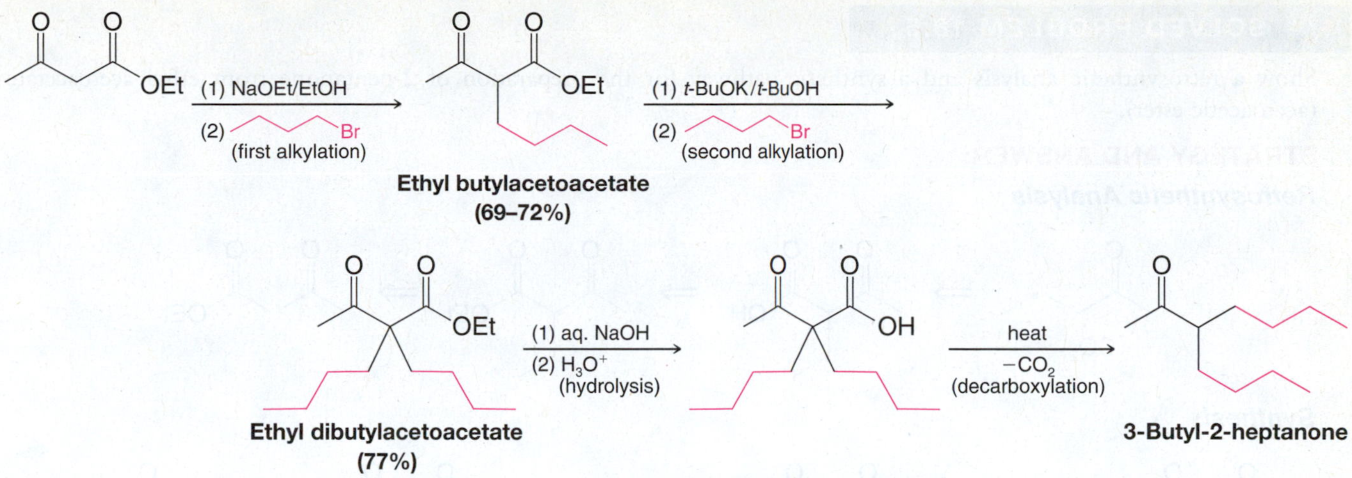

**Ethyl butylacetoacetate
(69–72%)**

**Ethyl dibutylacetoacetate
(77%)**

3-Butyl-2-heptanone

Although both alkylations in the example just given were carried out with the same alkyl halide, we could have used different alkyl halides if our synthesis had required it.

- As we have seen, ethyl acetoacetate is a useful reagent for the preparation of substituted acetones (methyl ketones) of the types shown below.

A monosubstituted acetone   **A disubstituted acetone**

- Ethyl acetoacetate therefore serves as the synthetic equivalent of the enolate from acetone shown below.

A **synthetic equivalent** is a reagent whose structure, when incorporated into a product, gives the appearance of having come from one type of precursor when as a reactant it actually had a different structural origin. Although it is possible to form the enolate of acetone, use of ethyl acetoacetate as a synthetic equivalent is often more convenient because its α hydrogens are so much more acidic ($pK_a = 9$–11) than those of acetone itself ($pK_a = $ 19–20). If we had wanted to use the acetone enolate directly, we would have had to use a much stronger base and other special conditions (e.g., a lithium enolate, Section 18.4).

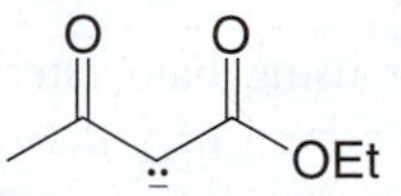

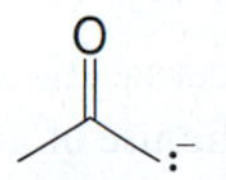

Ethyl acetoacetate ion   is the synthetic equivalent of   **Acetone enolate**

●●● SOLVED PROBLEM 18.4

Explain how compounds with the following general structure are formed as occasional side products of sodioacetoacetic ester alkylations.

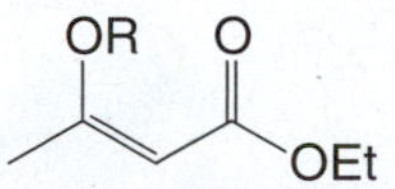

STRATEGY AND ANSWER: The partially negative oxygen atom of the sodioacetoacetic ester enolate acts as a nucleophile.

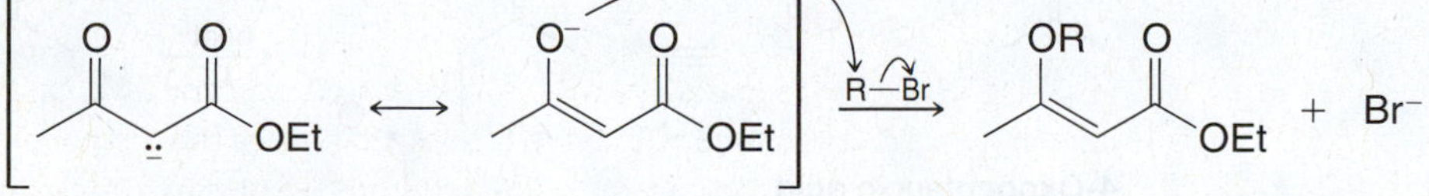

Show a retrosynthetic analysis and a synthetic pathway for the preparation of 2-pentanone from ethyl acetoacetate (acetoacetic ester).

STRATEGY AND ANSWER:

Retrosynthetic Analysis

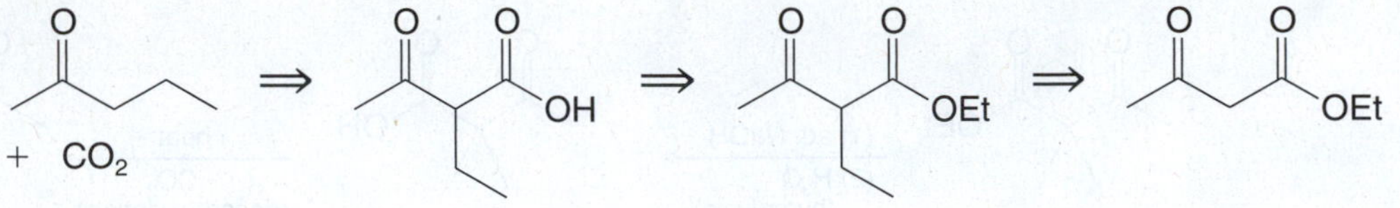

Synthesis

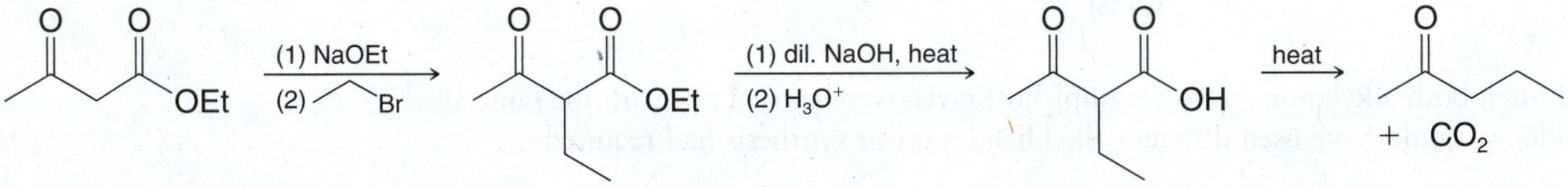

●●●

PRACTICE PROBLEM 18.7 Show how you would use the acetoacetic ester synthesis to prepare **(a)** 3-propyl-2-hexanone and **(b)** 4-phenyl-2-butanone.

●●●

PRACTICE PROBLEM 18.8 The acetoacetic ester synthesis generally gives best yields when primary halides are used in the alkylation step. Secondary halides give low yields, and tertiary halides give practically no alkylation product at all. **(a)** Explain. **(b)** What products would you expect from the reaction of sodioacetoacetic ester and *tert*-butyl bromide? **(c)** Bromobenzene cannot be used as an arylating agent in an acetoacetic ester synthesis in the manner we have just described. Why not?

The acetoacetic ester synthesis can also be carried out using halo esters and halo ketones. The use of an α-halo ester provides a convenient synthesis of γ-keto acids:

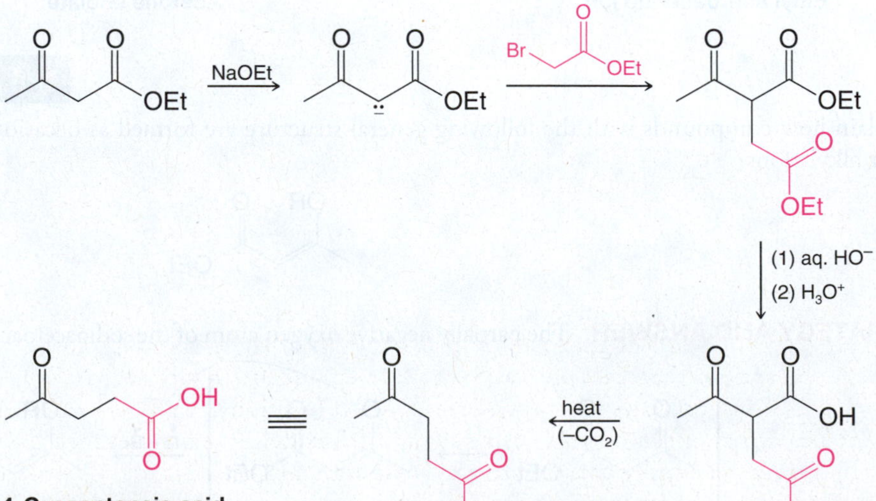

4-Oxopentanoic acid
(a γ-keto acid)

PRACTICE PROBLEM 18.9

In the synthesis of the keto acid just given, the dicarboxylic acid decarboxylates in a specific way; it gives

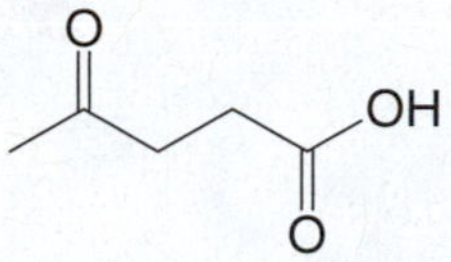

rather than

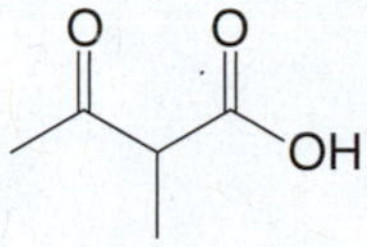

Explain.

The use of an α-halo ketone in an acetoacetic ester synthesis provides a general method for preparing γ-diketones:

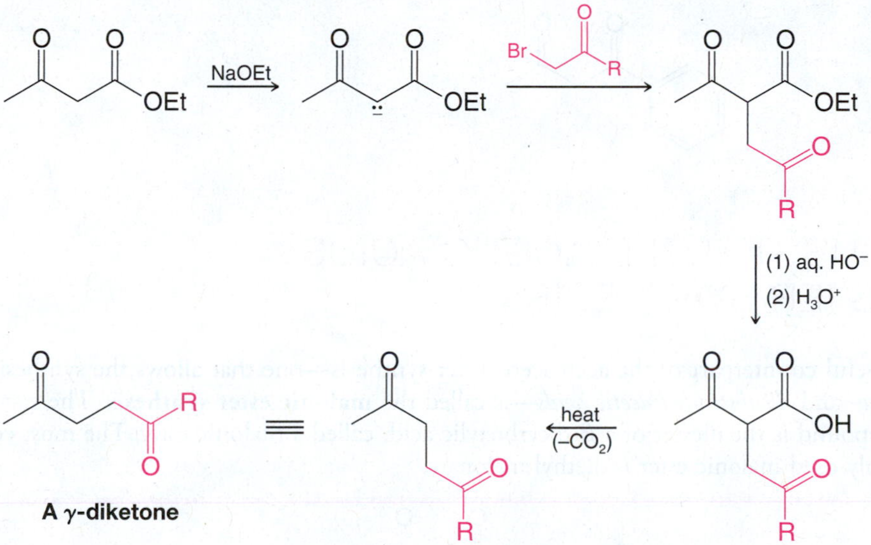

A γ-diketone

PRACTICE PROBLEM 18.10

How would you use the acetoacetic ester synthesis to prepare the following?

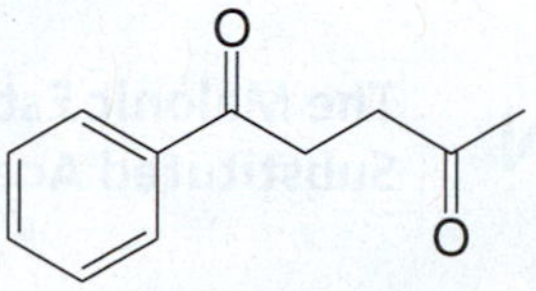

18.6A Acylation

Anions obtained from acetoacetic esters undergo acylation when they are treated with acyl chlorides or acid anhydrides. Because both of these acylating agents react with alcohols, acylation reactions cannot be carried out in ethanol and must be carried out in aprotic solvents such as DMF or DMSO (Section 6.13C). (If the reaction were to be carried out in ethanol, using sodium ethoxide, for example, then the acyl chloride would be rapidly converted to an ethyl ester and the ethoxide ion would be neutralized.) Sodium hydride can be used to generate the enolate ion in an aprotic solvent:

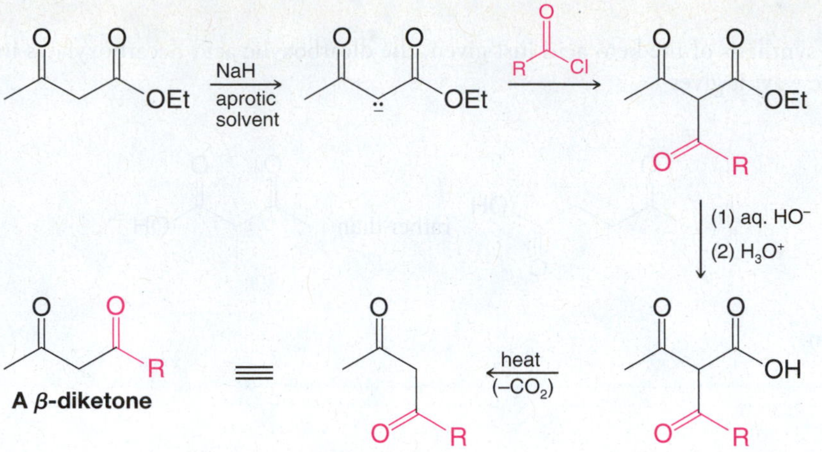

PRACTICE PROBLEM 18.11 How would you use the acetoacetic ester synthesis to prepare the following?

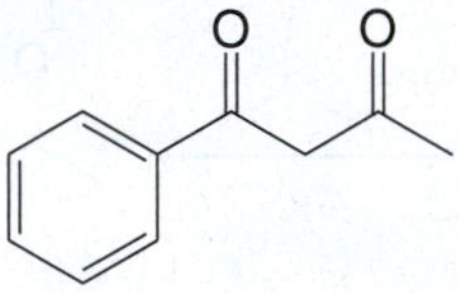

18.7 SYNTHESIS OF SUBSTITUTED ACETIC ACIDS: THE MALONIC ESTER SYNTHESIS

A useful counterpart of the acetoacetic ester synthesis—one that allows the synthesis of *mono-* and *disubstituted acetic acids*—is called the **malonic ester synthesis**. The starting compound is the diester of a β-dicarboxylic acid, called a malonic ester. The most commonly used malonic ester is diethyl malonate.

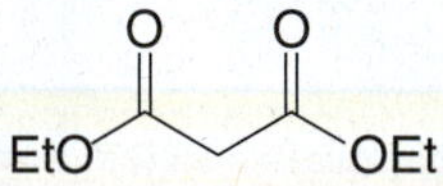

Diethyl malonate (a β-dicarboxylic acid ester)

We shall see by examining the following mechanism that the malonic ester synthesis resembles the acetoacetic ester synthesis in several respects.

[A MECHANISM FOR THE REACTION — The Malonic Ester Synthesis of Substituted Acetic Acids]

Step 1 *Diethyl malonate, the starting compound, forms a relatively stable enolate:*

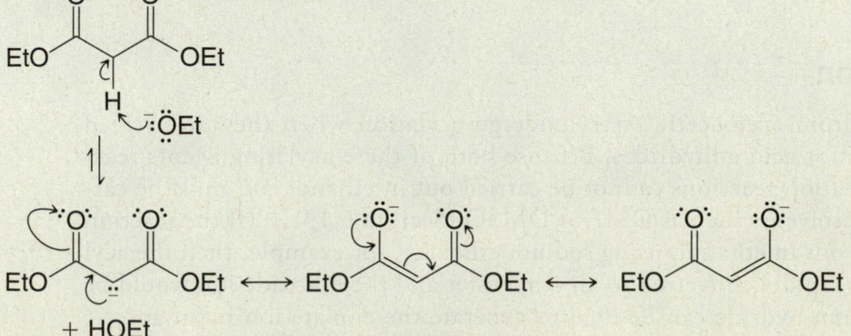

+ HOEt

Resonance-stabilized anion

Step 2   *This enolate can be alkylated in an S_N2 reaction,*

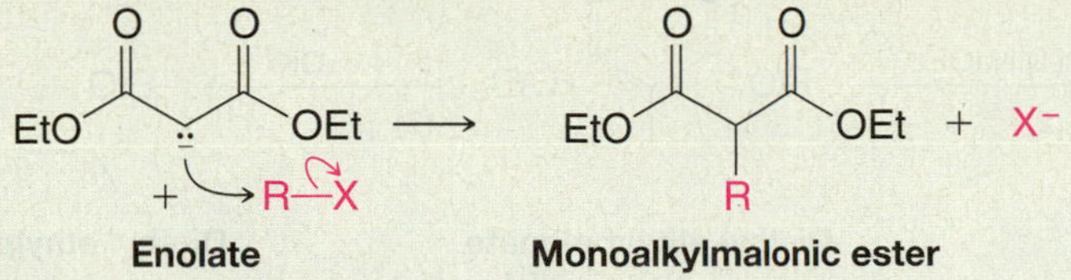

Enolate   **Monoalkylmalonic ester**

and the product can be alkylated again if our synthesis requires it:

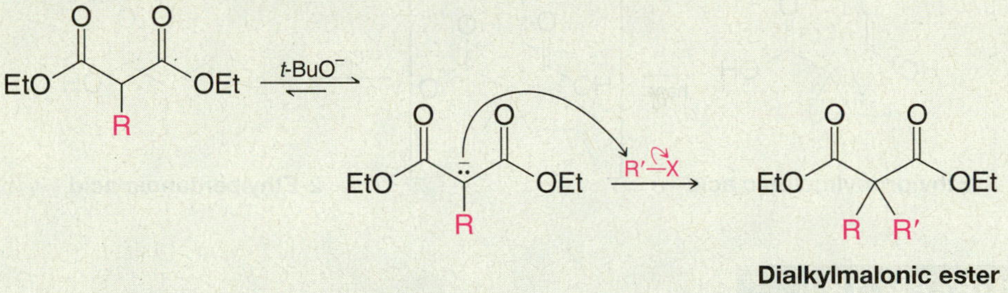

Dialkylmalonic ester

Step 3   *The mono- or dialkylmalonic ester can then be hydrolyzed to a mono- or dialkylmalonic acids decarboxylate readily. Decarboxylation gives a mono- or disubstituted acetic acid:*

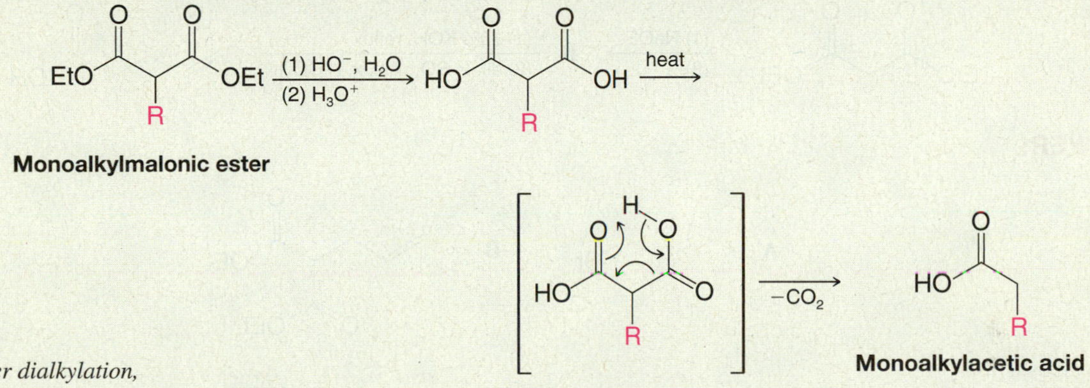

Monoalkylmalonic ester

Monoalkylacetic acid

or after dialkylation,

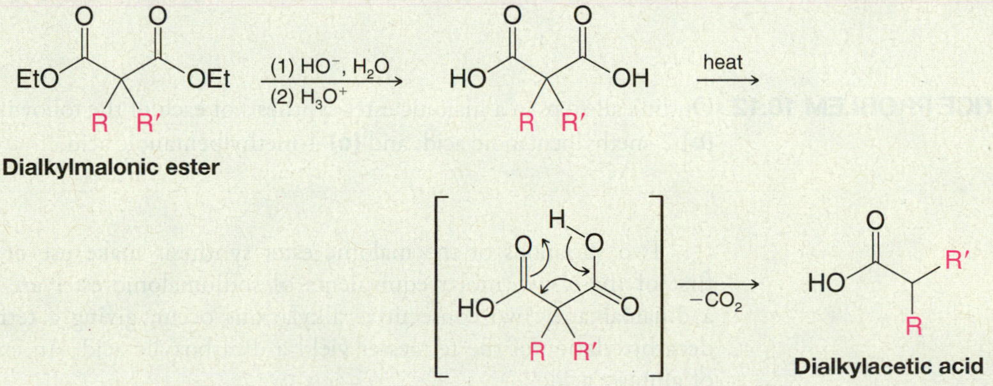

Dialkylmalonic ester

Dialkylacetic acid

     Two specific examples of the malonic ester synthesis are the syntheses of hexanoic acid and 2-ethylpentanoic acid that follow.

A Malonic Ester Synthesis of Hexanoic Acid

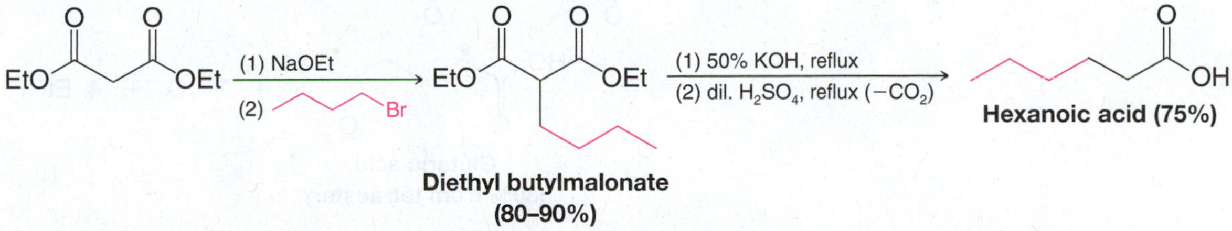

Diethyl butylmalonate
(80–90%)

Hexanoic acid (75%)

A Malonic Ester Synthesis of 2-Ethylpentanoic Acid

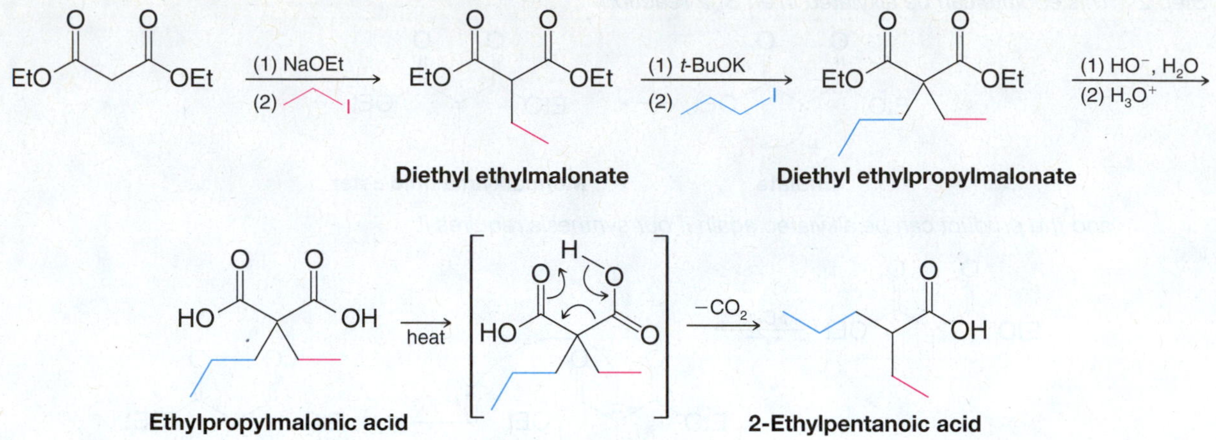

Diethyl ethylmalonate　　　　**Diethyl ethylpropylmalonate**

Ethylpropylmalonic acid　　　　　　　　　　**2-Ethylpentanoic acid**

●●● **SOLVED PROBLEM 18.6**

Provide structures for compounds **A** and **B** in the following synthesis.

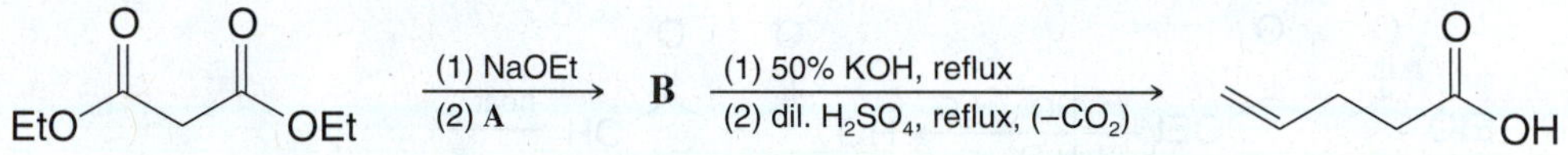

ANSWER:

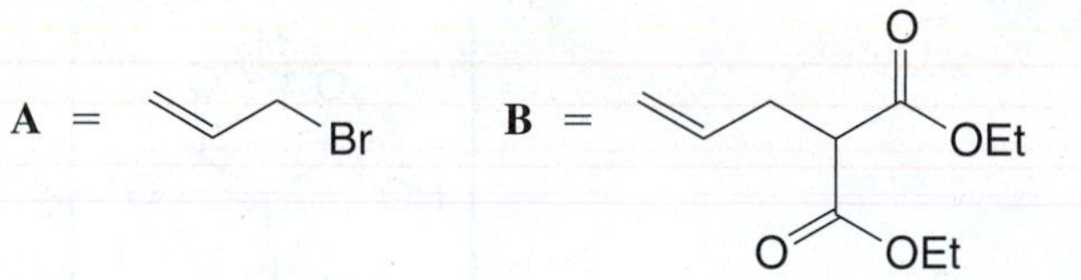

●●●

PRACTICE PROBLEM 18.12　Outline all steps in a malonic ester synthesis of each of the following: **(a)** pentanoic acid, **(b)** 2-methylpentanoic acid, and **(c)** 4-methylpentanoic acid.

Two variations of the malonic ester synthesis make use of dihaloalkanes. In the first of these, two molar equivalents of sodiomalonic ester are allowed to react with a dihaloalkane. Two consecutive alkylations occur, giving a tetraester; hydrolysis and decarboxylation of the tetraester yield a dicarboxylic acid. An example is the synthesis of glutaric acid:

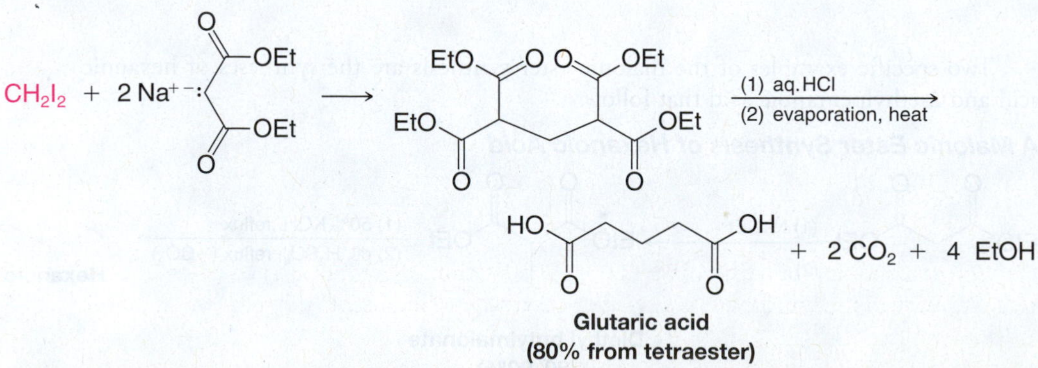

CH_2I_2 + 2 Na⁺⁻:

(1) aq. HCl
(2) evaporation, heat

+ 2 CO_2 + 4 EtOH

Glutaric acid
(80% from tetraester)

In a second variation, one molar equivalent of sodiomalonic ester is allowed to react with one molar equivalent of a dihaloalkane. This reaction gives a haloalkylmalonic ester, which, when treated with sodium ethoxide, undergoes an internal alkylation reaction. This method has been used to prepare three-, four-, five-, and six-membered rings. An example is the synthesis of cyclobutanecarboxylic acid:

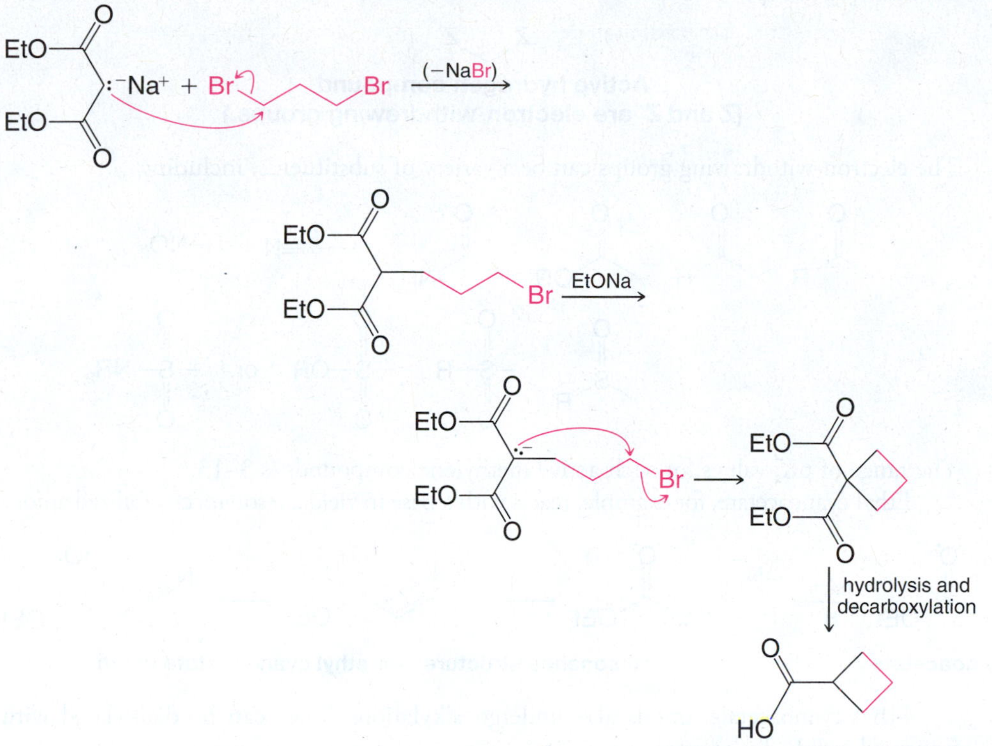

Cyclobutanecarboxylic acid

- As we have seen, the malonic ester synthesis is a useful method for preparing mono- and dialkylacetic acids:

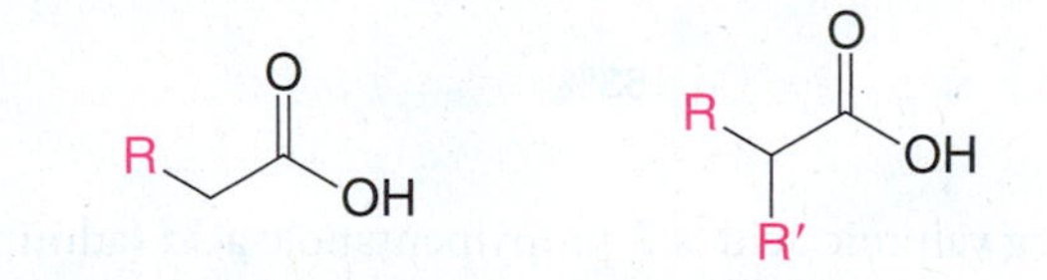

A monoalkylacetic acid   **A dialkylacetic acid**

Helpful Hint

The malonic ester synthesis is a tool for synthesizing substituted acetic acids.

- Thus, the malonic ester synthesis provides us with a synthetic equivalent of an ester enolate of acetic acid or acetic acid dianion.

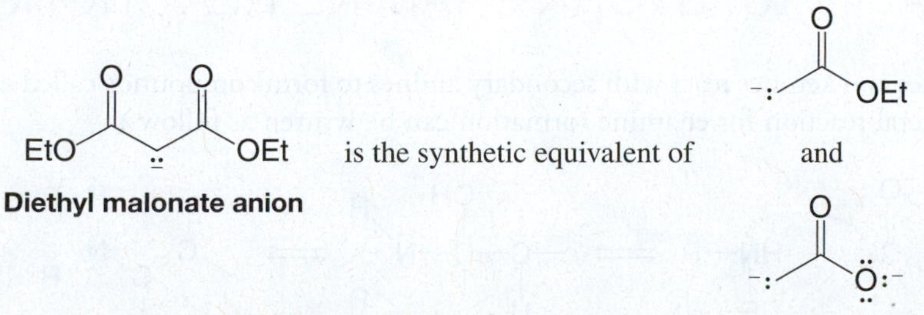

Diethyl malonate anion   is the synthetic equivalent of   and

Direct formation of such anions is possible (Section 18.4), but it is often more convenient to use diethyl malonate as a synthetic equivalent because its α hydrogens are more easily removed.

In Special Topic E (in *WileyPLUS*) we shall see biosynthetic equivalents of these anions.

18.8 FURTHER REACTIONS OF ACTIVE HYDROGEN COMPOUNDS

Because of the acidity of their methylene hydrogens malonic esters, acetoacetic esters, and similar compounds are often called **active hydrogen compounds** or active methylene compounds. Generally speaking, active hydrogen compounds have two electron-withdrawing groups attached to the same carbon atom:

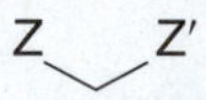

**Active hydrogen compound
(Z and Z′ are electron-withdrawing groups.)**

The electron-withdrawing groups can be a variety of substituents, including

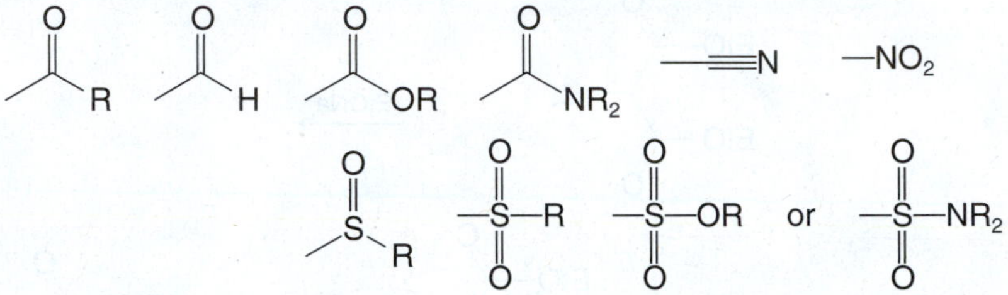

The range of pK_a values for such active methylene compounds is 3–13.
Ethyl cyanoacetate, for example, reacts with a base to yield a resonance-stabilized anion:

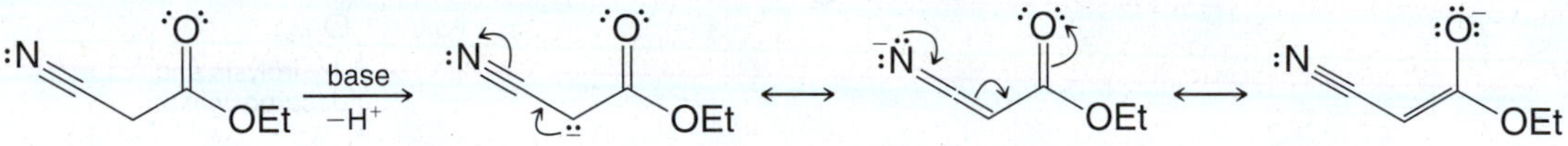

Ethyl cyanoacetate   **Resonance structures for ethyl cyanoacetate anion**

Ethyl cyanoacetate anions also undergo alkylations. They can be dialkylated with isopropyl iodide, for example:

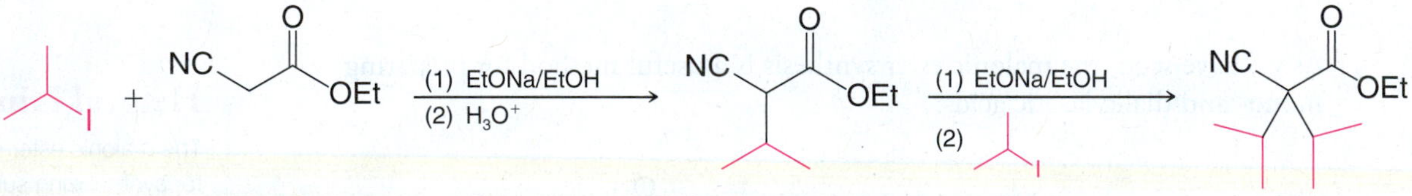

63%   **95%**

PRACTICE PROBLEM 18.13 The antiepileptic drug valproic acid is 2-propylpentanoic acid (administered as the sodium salt). One commercial synthesis of valproic acid begins with ethyl cyanoacetate. The penultimate step of this synthesis involves a decarboxylation, and the last step involves hydrolysis of a nitrile. Outline this synthesis.

18.9 SYNTHESIS OF ENAMINES: STORK ENAMINE REACTIONS

Aldehydes and ketones react with secondary amines to form compounds called **enamines**. The general reaction for enamine formation can be written as follows:

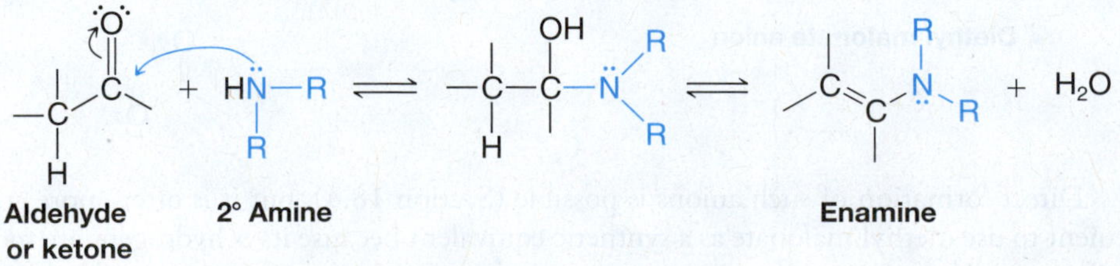

**Aldehyde
or ketone**   **2° Amine**   **Enamine**

See Section 16.8C for the mechanism of enamine formation.

Since enamine formation requires the loss of a molecule of water, enamine preparations are usually carried out in a way that allows water to be removed as an azeotrope or by a drying agent. This removal of water drives the reversible reaction to completion. Enamine formation is also catalyzed by the presence of a trace of an acid. The secondary amines most commonly used to prepare enamines are cyclic amines such as pyrrolidine, piperidine, and morpholine:

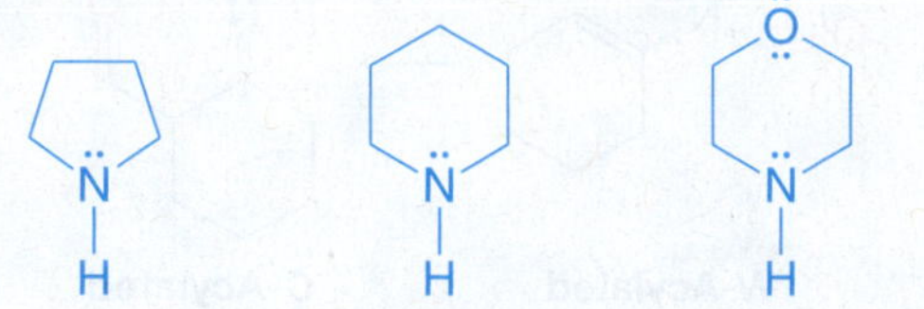

Pyrrolidine   **Piperidine**   **Morpholine**

Cyclohexanone, for example, reacts with pyrrolidine in the following way:

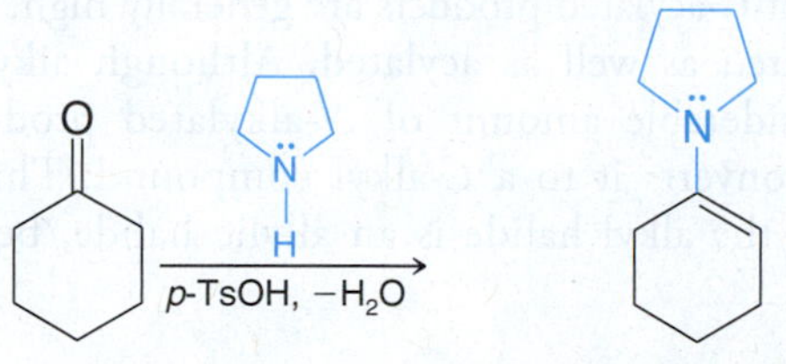

p-TsOH, $-H_2O$

N-(1-Cyclohexenyl)pyrrolidine
(an enamine)

Enamines are good nucleophiles. Examination of the resonance structures that follow show that we should expect enamines to have both a nucleophilic nitrogen and a *nucleophilic carbon*. A map of electrostatic potential highlights the nucleophilic region of an enamine.

Contribution to the hybrid made by this structure confers nucleophilicity on nitrogen.

Contribution to the hybrid made by this structure confers nucleophilicity on carbon and decreases the nucleophilicity of nitrogen.

A map of electrostatic potential for N-(1-cyclohexenyl)pyrrolidine shows the distribution of negative charge and the nucleophilic region of an enamine.

The nucleophilicity of the carbon of enamines makes them particularly useful reagents in organic synthesis because they can be **acylated, alkylated,** and used in **Michael additions** (see Section 19.7A). Enamines can be used as synthetic equivalents of aldehyde or ketone enolates because the alkene carbon of an enamine reacts the same way as does the α carbon of an aldehyde or ketone enolate and, after hydrolysis, the products are the same. Development of these techniques originated with the work of Gilbert Stork of Columbia University, and in his honor they have come to be known as **Stork enamine reactions**.

When an enamine reacts with an acyl halide or an acid anhydride, the product is the *C*-acylated compound. The iminium ion that forms hydrolyzes when water is added, and the overall reaction provides a synthesis of β-diketones:

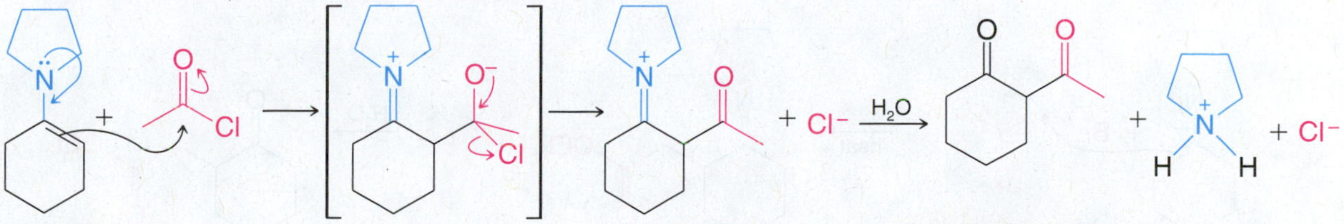

Helpful Hint

Enamines are the synthetic equivalents of aldehyde and ketone enolates.

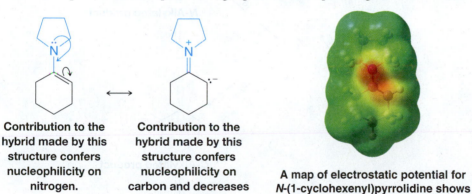

Iminium salt   **2-Acetylcyclohexanone**
(a β-diketone)

Although *N*-acylation may occur in this synthesis, the *N*-acyl product is unstable and can act as an acylating agent itself:

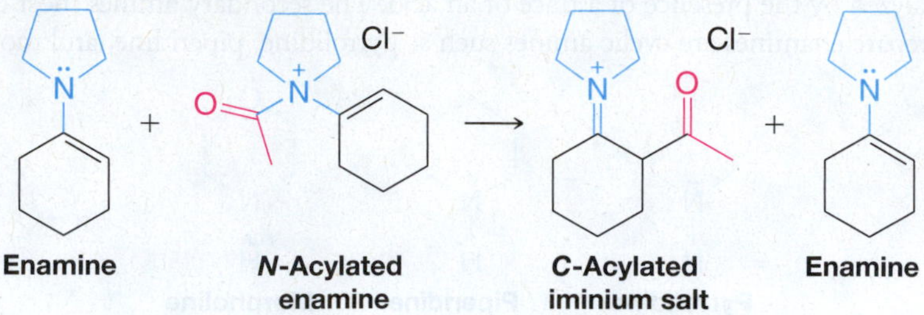

| **Enamine** | **N-Acylated enamine** | **C-Acylated iminium salt** | **Enamine** |

As a consequence, the yields of *C*-acylated products are generally high.

Enamines can be alkylated as well as acylated. Although alkylation may lead to the formation of a considerable amount of *N*-alkylated product, heating the *N*-alkylated product often converts it to a *C*-alkyl compound. This rearrangement is particularly favored when the alkyl halide is an allylic halide, benzylic halide, or α-haloacetic ester:

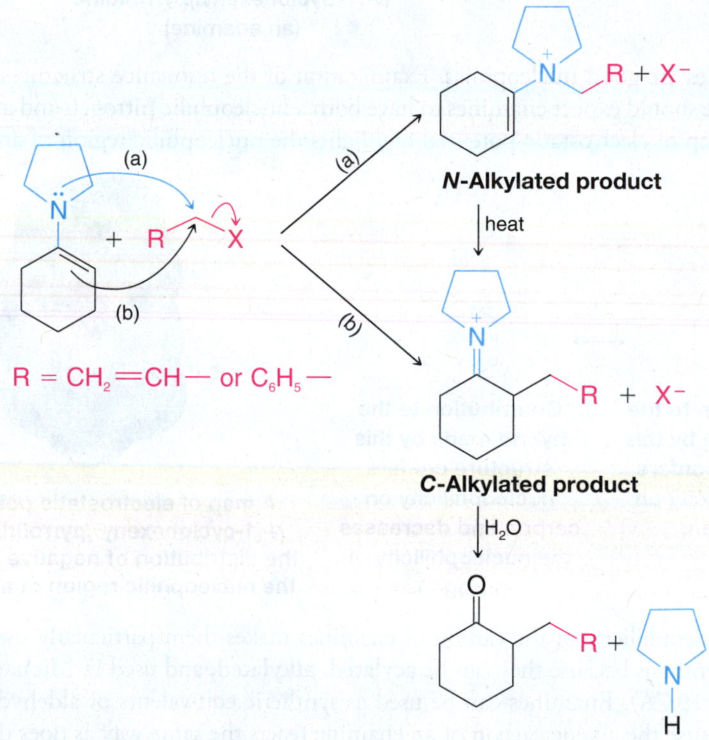

N-Alkylated product

heat

C-Alkylated product

H₂O

R = CH₂=CH— or C₆H₅—

Enamine alkylations are S_N2 reactions; therefore, when we choose our alkylating agents, we are usually restricted to the use of methyl, primary, allylic, and benzylic halides. α-Halo esters can also be used as the alkylating agents, and this reaction provides a convenient synthesis of β-keto esters:

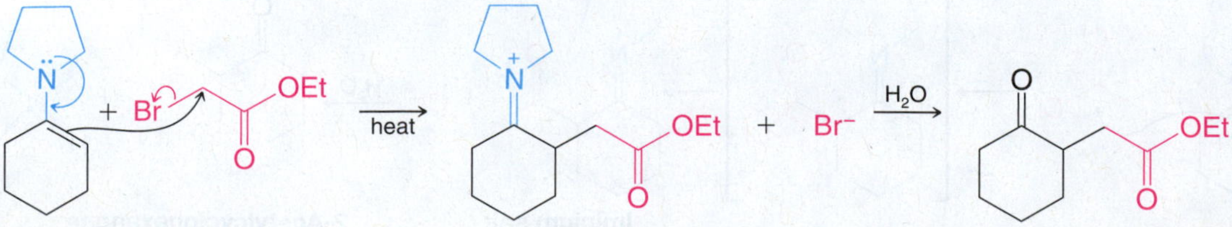

A γ-keto ester (75%)

Show how you could employ enamines in syntheses of the following compounds:

(a) 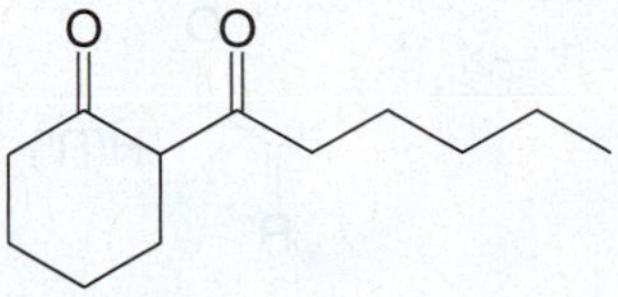　(b) 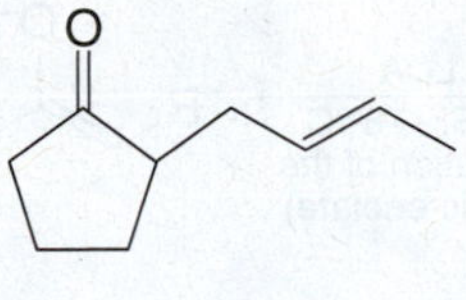　(c) 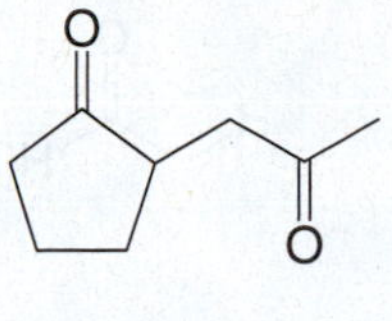　(d)

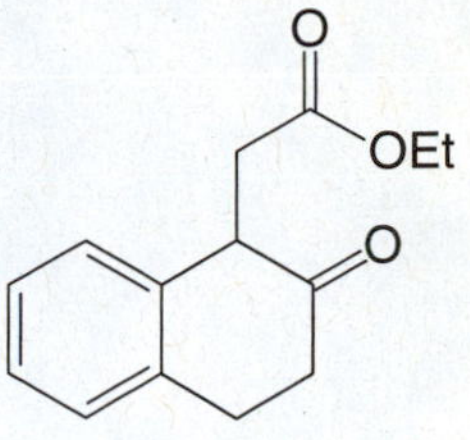

18.10 SUMMARY OF ENOLATE CHEMISTRY

1. *Formation of an Enolate* (Section 18.1)

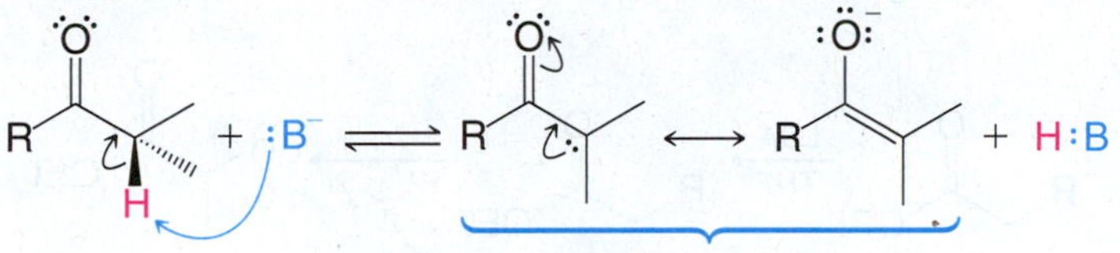

Resonance-stabilized enolate

$$:B = \overset{..}{\text{O}}\text{H}, \ \overset{..}{\text{O}}\text{R}, \text{ or } \overset{..}{\text{N}}(i\text{-Pr})_2 \quad \text{(Section 18.4)}$$

2. *Racemization* (Section 18.3A)

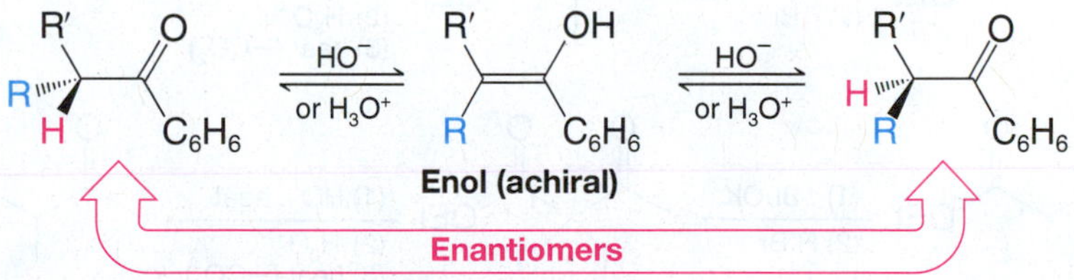

Enol (achiral)

Enantiomers

3. *Halogenation of Aldehydes and Ketones* (Sections 18.3B and 18.3C)

General Reaction

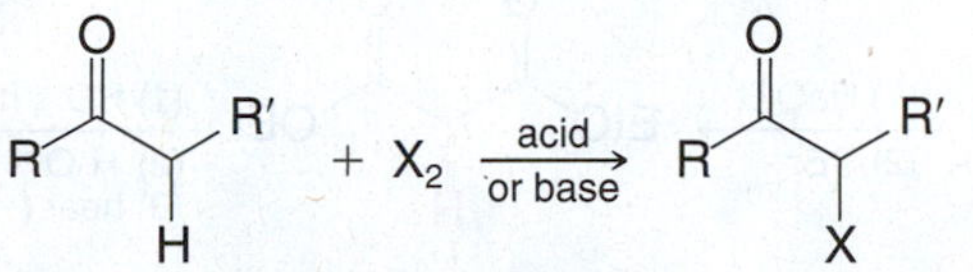

$$R \overset{O}{\underset{H}{\overset{|}{\text{—}}}} R' + X_2 \xrightarrow[\text{or base}]{\text{acid}} R \overset{O}{\underset{X}{\overset{|}{\text{—}}}} R'$$

Specific Example—Haloform Reaction

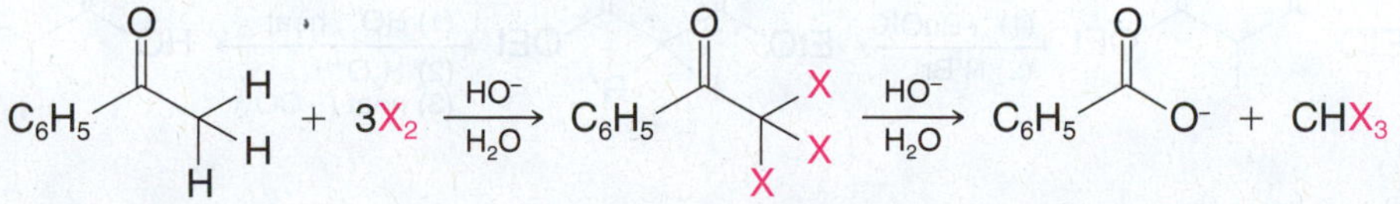

$$C_6H_5\text{—CO—CH}_3 + 3X_2 \xrightarrow[H_2O]{HO^-} C_6H_5\text{—CO—CX}_3 \xrightarrow[H_2O]{HO^-} C_6H_5\text{—CO—O}^- + CHX_3$$

4. *Halogenation of Carboxylic Acids: The HVZ Reaction* (Section 18.3D)

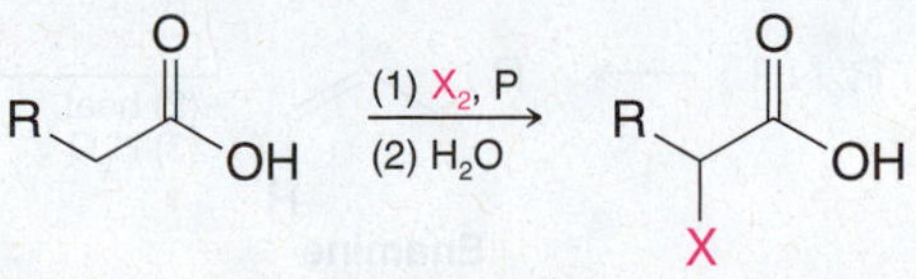

$$R\text{—CH}_2\text{—COOH} \xrightarrow[\text{(2) } H_2O]{\text{(1) } X_2, \text{ P}} R\text{—CHX—COOH}$$

5. *Direct Alkylation via Lithium Enolates* **(Section 18.4)**

General Reaction

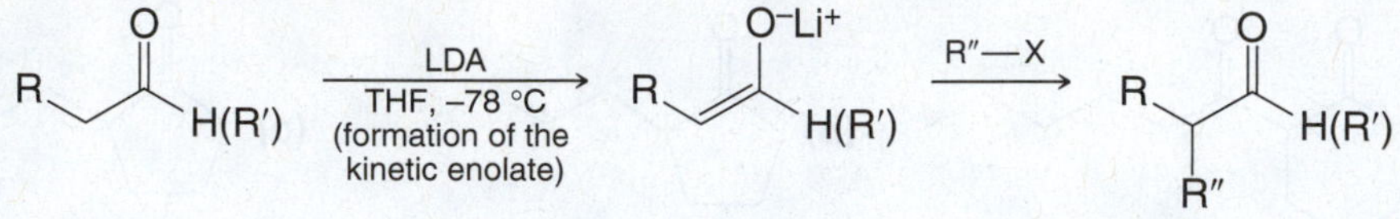

Specific Example

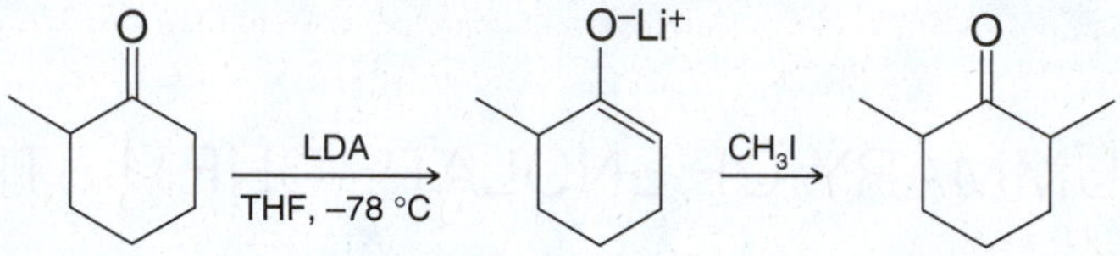

6. *Direct Alkylation of Esters* **(Section 18.4C)**

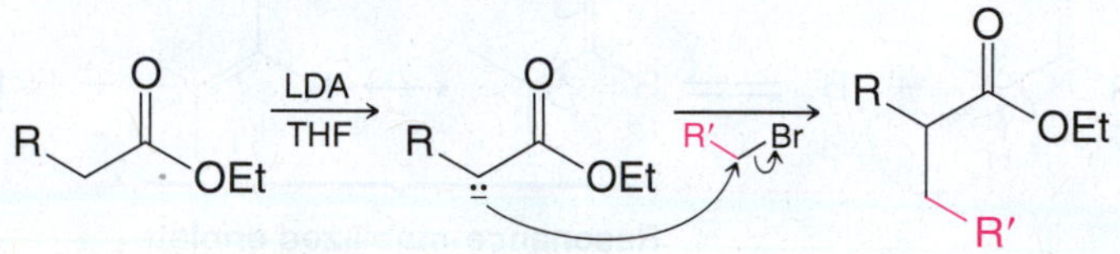

7. *Acetoacetic Ester Synthesis* **(Section 18.6)**

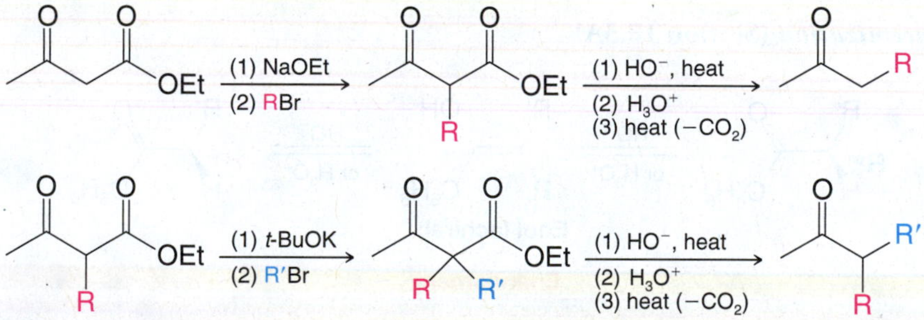

8. *Malonic Ester Synthesis* **(Section 18.7)**

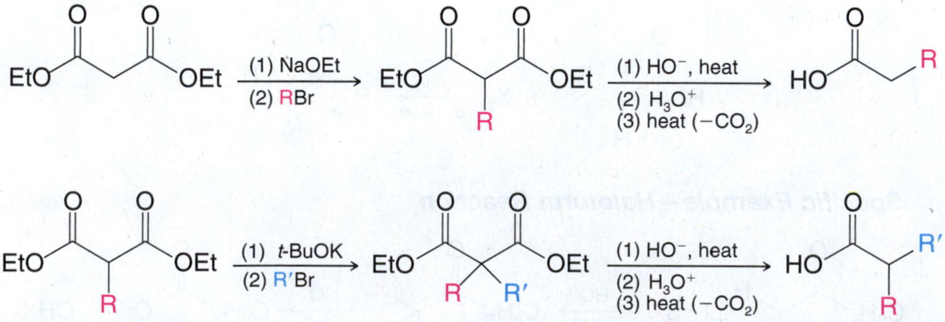

9. *Stork Enamine Reaction* **(Section 18.9)**

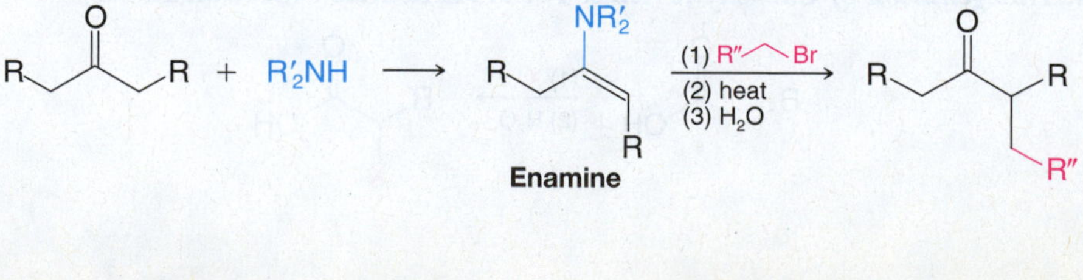

Enamine

[WHY Do These Topics Matter?

USING ENAMINE CHEMISTRY TO MAKE COMPLEXITY

The reactions that you have learned in this chapter are not just of academic interest; they are critical tools that make possible the syntheses of powerful pharmaceuticals and bioactive molecules, some even on ton scale! These reactions are significant because they constitute highly powerful methods for forming C—C bonds. Of the reactions you have seen thus far, though, perhaps the most versatile is the Stork enamine reaction. This general transformation was inspired by trying to copy mechanisms that nature uses for forming such C—C bonds. Since its initial discovery over half a century ago, the Stork enamine reaction has found countless applications. Here, we will mention four.

The first two reactions shown below highlight the merger of acid chlorides with enamines to make new C—C bonds (shown in red) along the lines presented in Section 18.9. One reason this transformation is of such importance is that so many functional groups can be contained within the reactants. As a result, the products possess most of the handles needed to form the final targets. Shown here are syntheses of jatrapholone A, which has antitumor properties, and epiibogamine, an alkaloid known to have value in fighting chemical addictions and cancer.

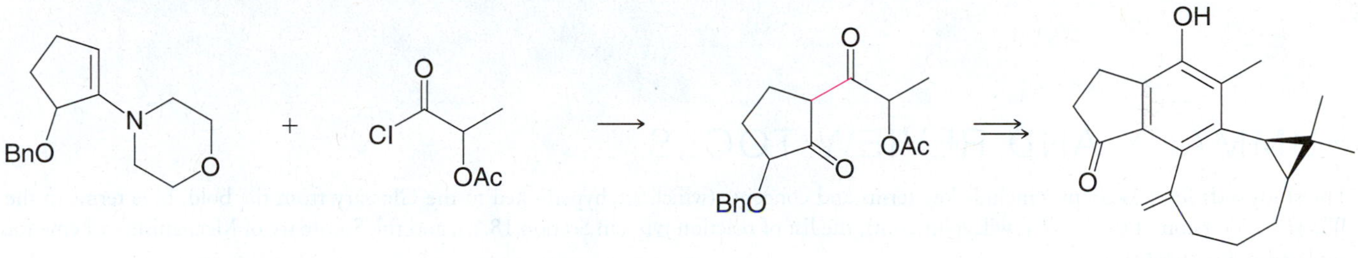

Jatrapholone A

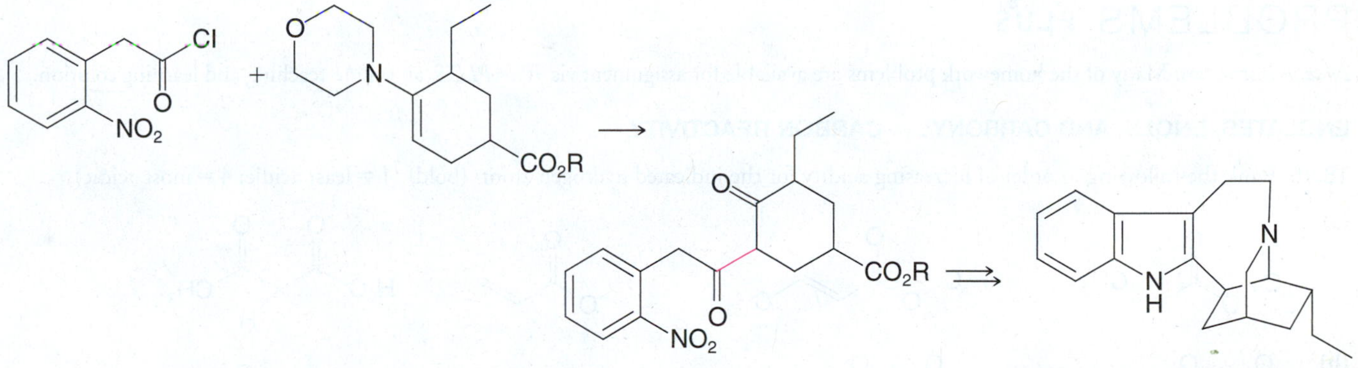

Epiibogamine

Enamine chemistry can also leave the nitrogen atom of the original enamine in the final product. Although understanding the specific examples shown below requires knowledge of some of the reactions found in the next chapter, we provide them now in the hope that they will build further appreciation for the power of the enamine functional group and its bond-forming coupling reactions. In the first case, it afforded a rapid synthesis of aspidospermine, a molecule with diuretic and respiratory stimulant activity; in the second, it provided a ton-scale synthesis of a novel drug candidate from AstraZeneca that has been evaluated in clinical trials to treat urinary incontinence.

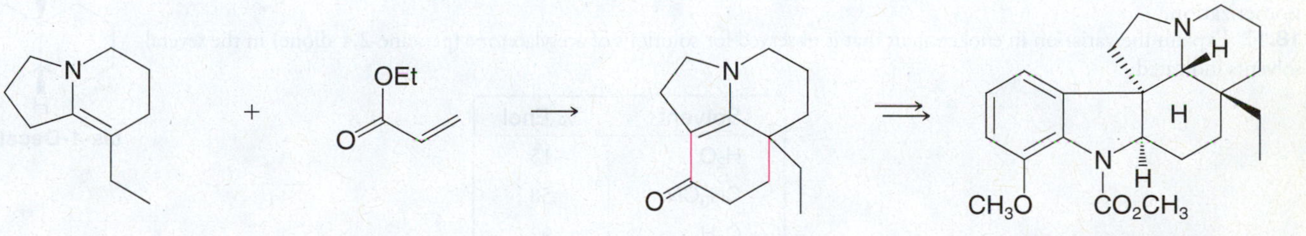

Aspidospermine

(continues on next page)

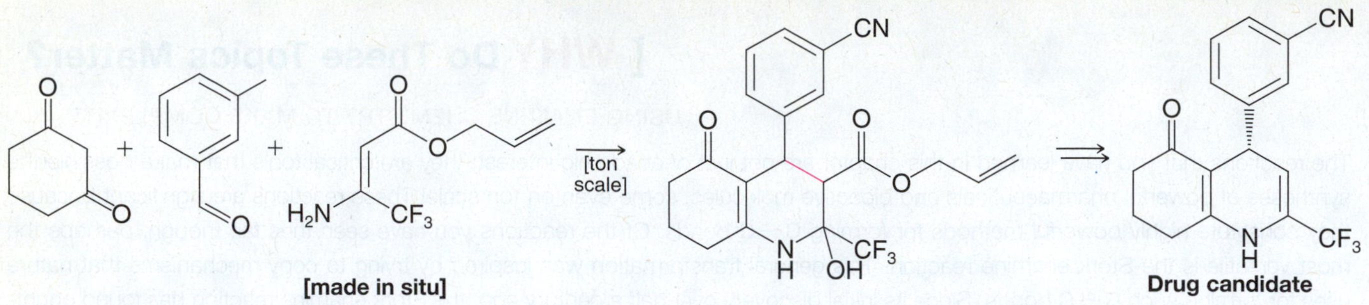

[made in situ]

[ton scale]

Drug candidate

To learn more about these topics, see:

1. Kürti, L.: Czakó, B. *Strategic Applications of Named Reactions in Organic Synthesis*. Elsevier: London **2005**, pp. 444–445.
2. Kuehne, M. E. "Application of enamines to syntheses of natural products and related compounds" in *Synthesis* **1970**, 510–537.
3. Hopes, P. A.; Parker, A. J.; Patel, I. "Development and optimization of an unsymmetrical Hantzsch reaction for plant-scale manufacture" in *Org. Proc. Res. Dev.* **2006**, *10*, 808–813.
4. Smith, III, A. B.; Liverton, N. J.; Hrib, N. J.; Sivaramakrishnan, H.; Winzenberg, K. "Total Synthesis of (+) Jatropholones A and B: Exploitation of the High-pressure Technique." *J. Am. Chem. Soc.* **1986**, *108*, 3040-3048.

SUMMARY AND REVIEW TOOLS

The study aids for this chapter include key terms and concepts (which are hyperlinked to the Glossary from the bold, blue terms in the *WileyPLUS* version of the book at wileyplus.com), the list of reaction types in Section 18.10, and the Summary of Mechanisms scheme for enolates and α-substitution.

PROBLEMS

Note to Instructors: Many of the homework problems are available for assignment via *WileyPLUS*, an online teaching and learning solution.

ENOLATES, ENOLS, AND CARBONYL α-CARBON REACTIVITY

18.15 Rank the following in order of increasing acidity for the indicated hydrogen atoms (bold) (1 = least acidic; 4 = most acidic).

(a)

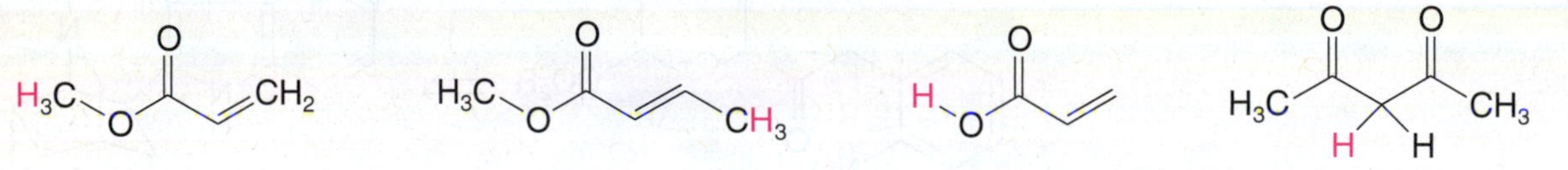

(b)

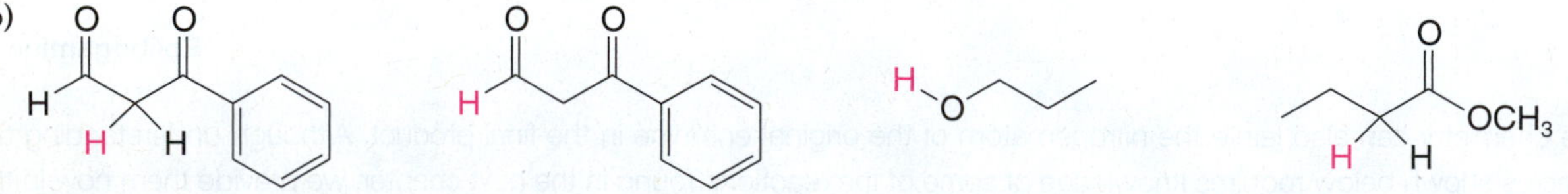

(c)

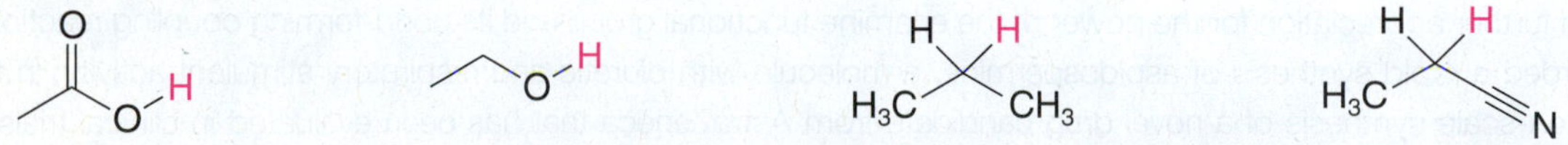

18.16 Treating a solution of *cis*-1-decalone with base causes an isomerization to take place. When the system reaches equilibrium, the solution is found to contain about 95% *trans*-1-decalone and about 5% *cis*-1-decalone. Explain this isomerization.

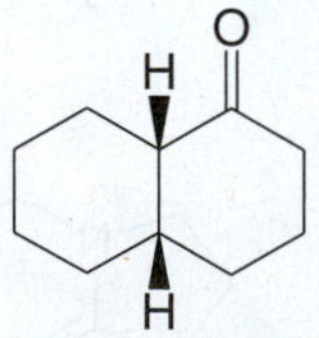

18.17 Explain the variation in enol content that is observed for solutions of acetylacetone (pentane-2,4-dione) in the several solvents indicated:

***cis*-1-Decalone**

Solvent	% Enol
H_2O	15
CH_3CN	58
C_6H_{14}	92
Gas phase	92

18.18 Provide a structural formula for the product from each of the following reactions.

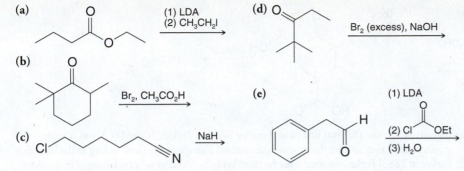

(a) (ethyl butanoate) → (1) LDA (2) CH₃CH₂I

(d) → Br₂ (excess), NaOH

(f) (acetophenone) → (1) I₂, NaOH (2) H₃O⁺

(b) (2,6,6-trimethylcyclohexanone) → Br₂, CH₃CO₂H

(e) (phenylacetaldehyde) → (1) LDA (2) Cl—C(=O)—OEt (3) H₂O

(c) Cl—(CH₂)₄—C≡N → NaH

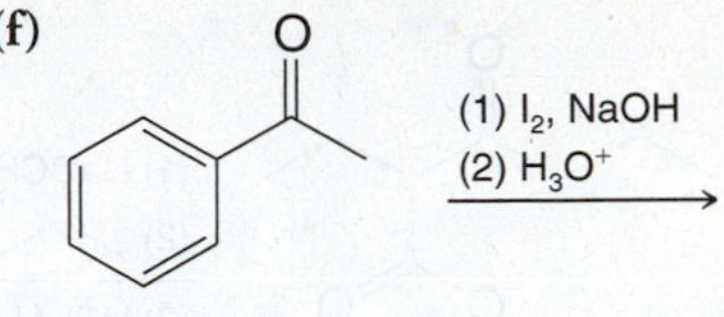

18.19 Write a stepwise mechanism for each of the following reactions.

(a) (cyclohexenone) → Br₂, CH₃CO₂H → (bromo cyclohexenone) + HBr

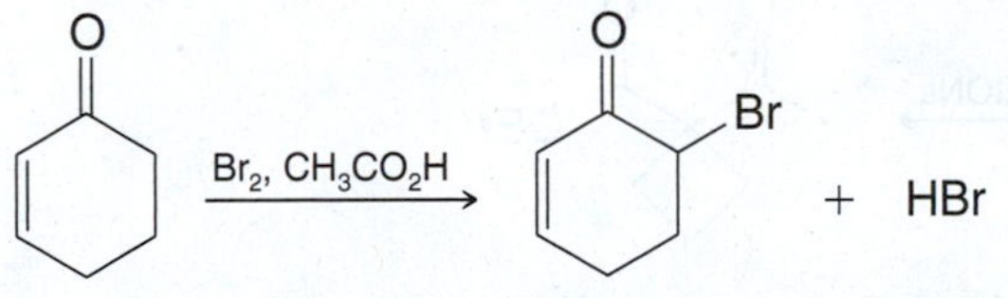

(b) (cyclohexyl methyl ketone) → excess I₂, NaOH → (cyclohexyl carboxylate sodium salt) O⁻Na⁺ + CHI₃

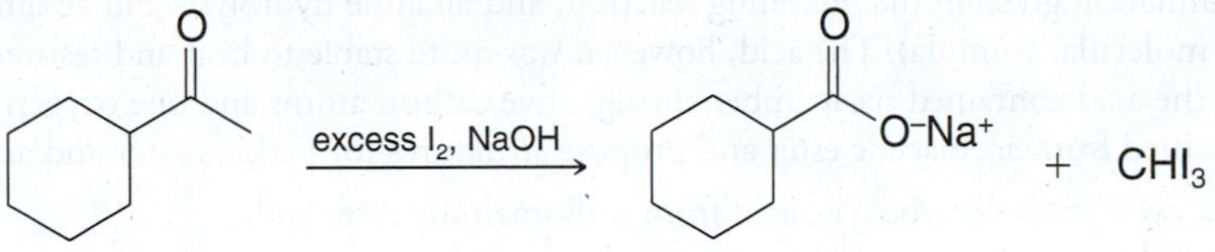

(c) (MeO vinyl ketone) → H₃O⁺ → (keto aldehyde) + MeOH

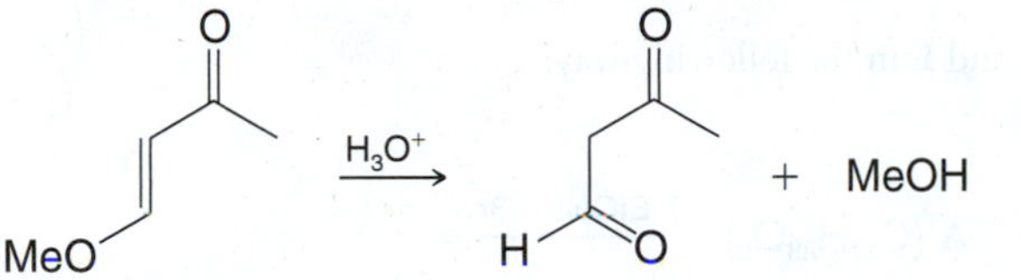

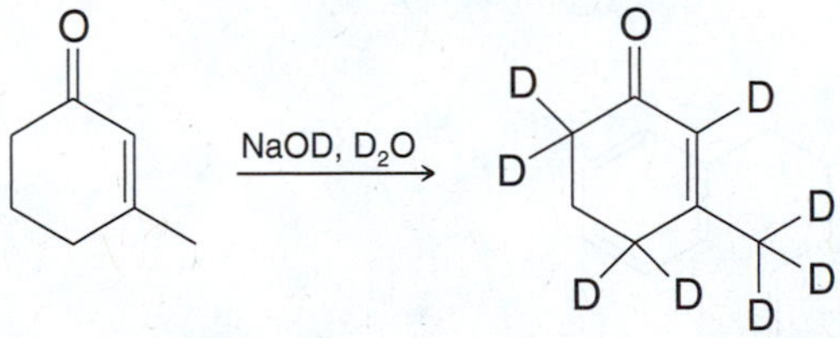

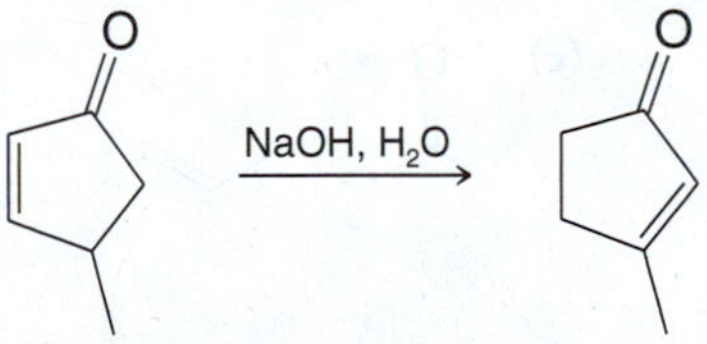

(d) (bicyclic diketone) → CH₃OH, H₂SO₄ ⇌ (cyclohexanone with CH₂CO₂CH₃) OCH₃

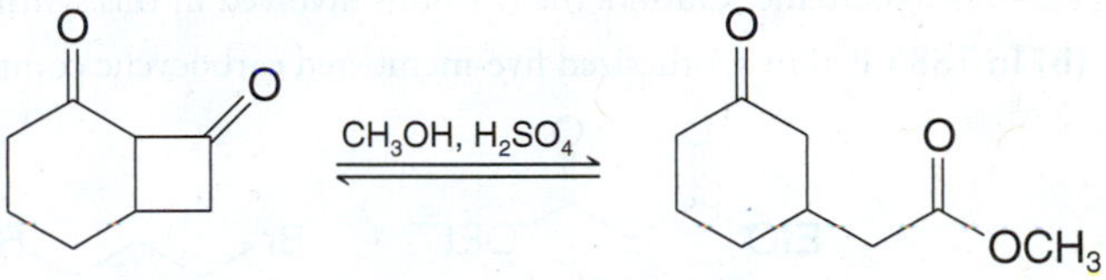

18.20 Write a stepwise mechanism for each of the following reactions.

(a) (4-methylenecyclohexanone) → NaOH, H₂O → (3-methylcyclohexenone)

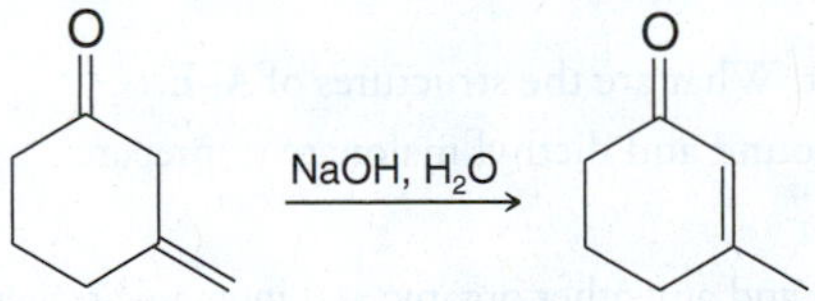

(b) (3-methylcyclohexenone) → NaOD, D₂O → (deuterated product with D labels)

(c) (4-methylcyclopentanone) → NaOH, H₂O → (3-methylcyclopentenone)

(d) (phenyl propenyl ketone) → (1) LDA (2) CH₃I → (phenyl pentenone) + (phenyl methyl butenone)

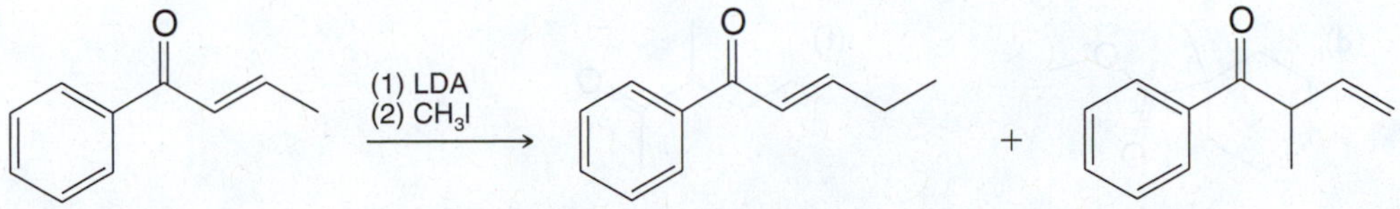

ACETOACETIC ESTER AND MALONIC ESTER SYNTHESES

18.21 Outline syntheses of each of the following from acetoacetic ester and any other required reagents:

(a) *tert*-Butyl methyl ketone   (c) 2,5-Hexanedione   (e) 2-Ethyl-1,3-butanediol

(b) 2-Hexanone   (d) 4-Hydroxypentanoic acid   (f) 1-Phenyl-1,3-butanediol

18.22 Outline syntheses of each of the following from diethyl malonate and any other required reagents:

(a) 2-Methylbutanoic acid

(b) 4-Methyl-1-pentanol

(c) 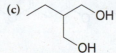 (branched diol with OH groups)

(d) HO—(CH₂)₄—OH

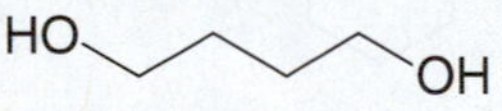

18.23 Provide a structural formula for the product from each of the following reactions.

(a)

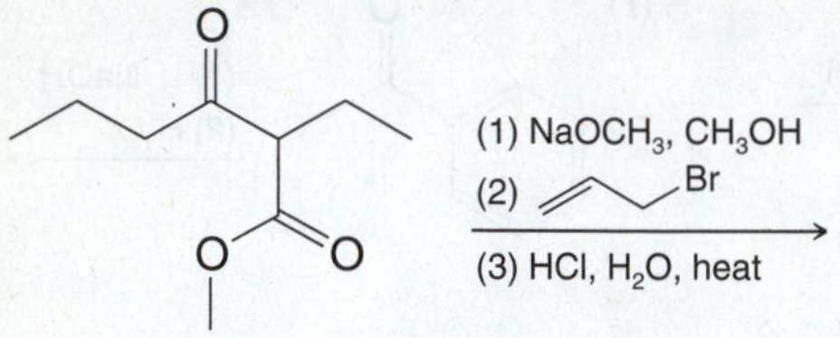

(1) NaOCH₃, CH₃OH
(2) [allyl]Br
(3) HCl, H₂O, heat

(b)

heat

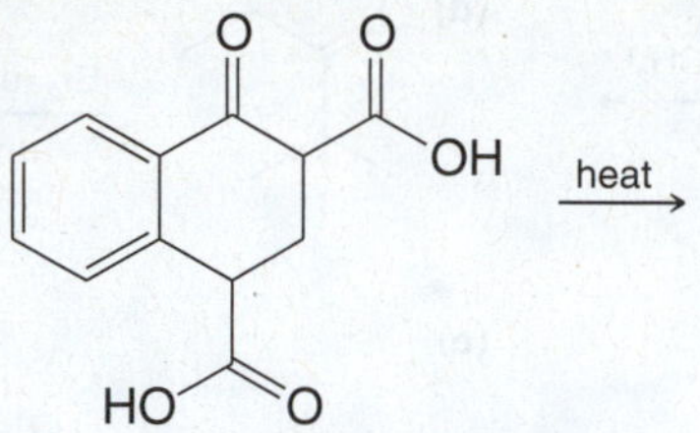

18.24 The synthesis of cyclobutanecarboxylic acid given in Section 18.7 was first carried out by William Perkin, Jr., in 1883, and it represented one of the first syntheses of an organic compound with a ring smaller than six carbon atoms. (There was a general feeling at the time that such compounds would be too unstable to exist.) Earlier in 1883, Perkin reported what he mistakenly believed to be a cyclobutane derivative obtained from the reaction of acetoacetic ester and 1,3-dibromopropane. The reaction that Perkin had expected to take place was the following:

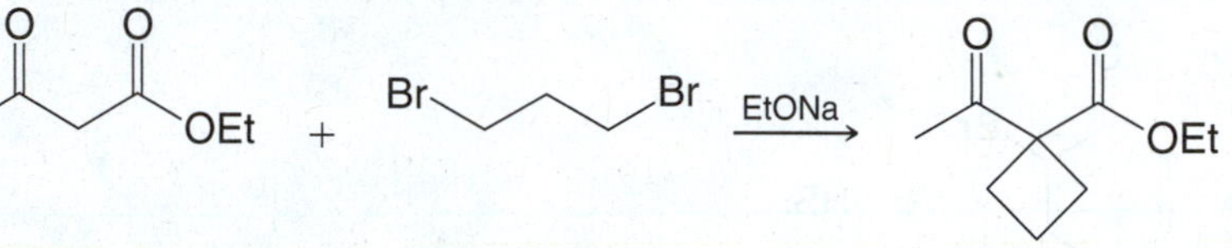

The molecular formula for his product agreed with the formulation given in the preceding reaction, and alkaline hydrolysis and acidification gave a nicely crystalline acid (also having the expected molecular formula). The acid, however, was quite stable to heat and resisted decarboxylation. Perkin later found that both the ester and the acid contained six-membered rings (five carbon atoms and one oxygen atom). Recall the charge distribution in the enolate ion obtained from acetoacetic ester and propose structures for Perkin's ester and acid.

18.25 (a) In 1884 Perkin achieved a successful synthesis of cyclopropanecarboxylic acid from sodiomalonic ester and 1,2-dibromoethane. Outline the reactions involved in this synthesis.

(b) In 1885 Perkin synthesized five-membered carbocyclic compounds **D** and **E** in the following way:

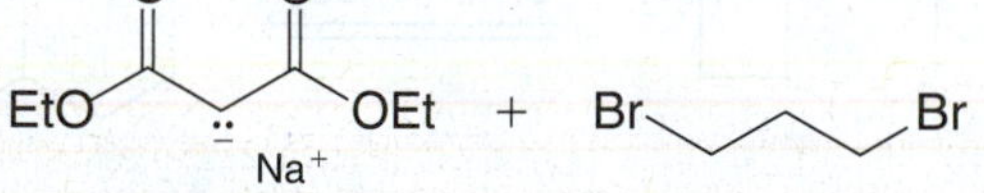

$$\text{B (C}_{17}\text{H}_{28}\text{O}_8) \xrightarrow[\text{(2) H}_3\text{O}^+]{\text{(1) HO}^-/\text{H}_2\text{O}} \text{C (C}_9\text{H}_{10}\text{O}_8) \xrightarrow{\text{heat}} \text{D (C}_7\text{H}_{10}\text{O}_4) + \text{E (C}_7\text{H}_{10}\text{O}_4)$$

where **D** and **E** are diastereomers; **D** can be resolved into enantiomeric forms while **E** cannot. What are the structures of **A–E**?

(c) Ten years later Perkin was able to synthesize 1,4-dibromobutane; he later used this compound and diethyl malonate to prepare cyclopentanecarboxylic acid. Show the reactions involved.

18.26 Synthesize each of the following compounds from diethyl malonate or ethyl acetoacetate and any other organic and inorganic reagents.

(a)

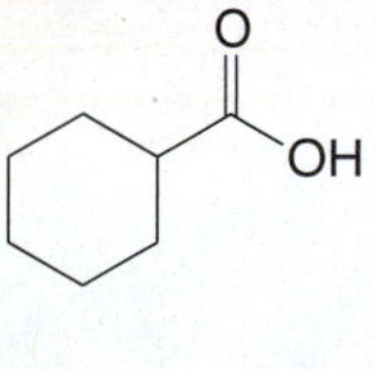

(c)

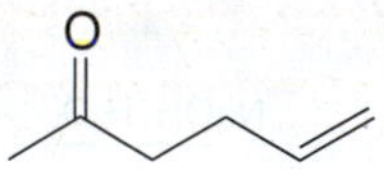

(e)

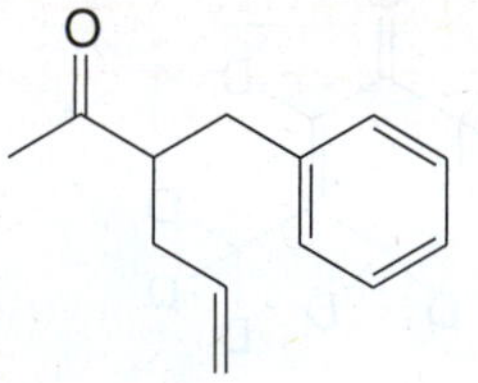

(b)

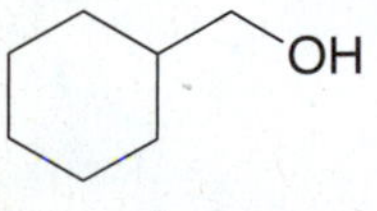

(d)

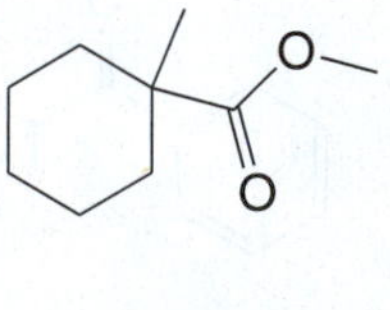

(f)

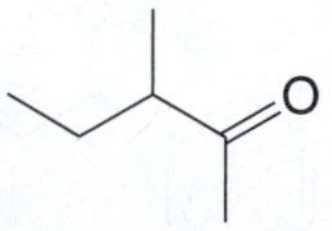

GENERAL PROBLEMS

18.27 Outline a reaction sequence for synthesis of each of the following compounds from the indicated starting material and any other organic or inorganic reagents needed.

(a)

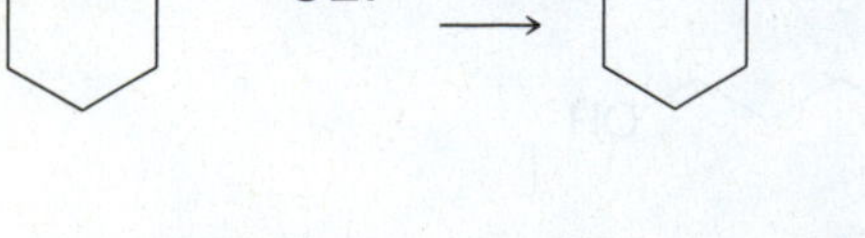

(b)

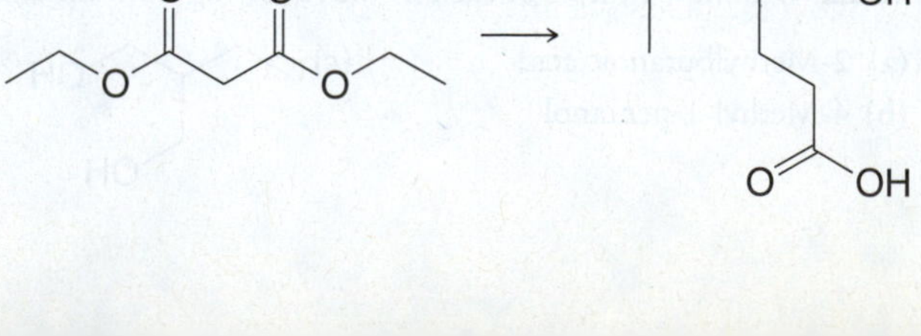

(c)

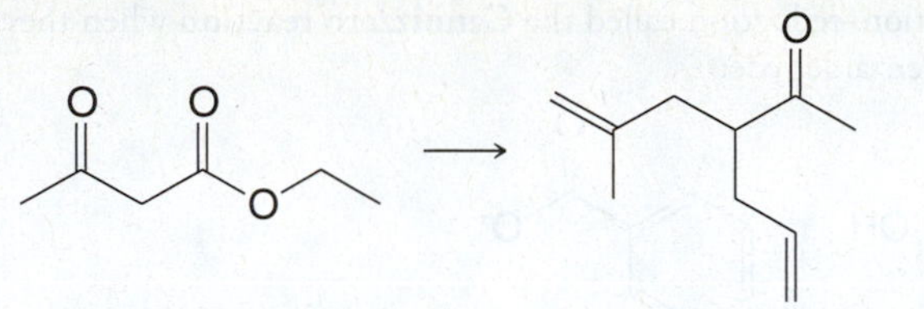

(d)

(e)

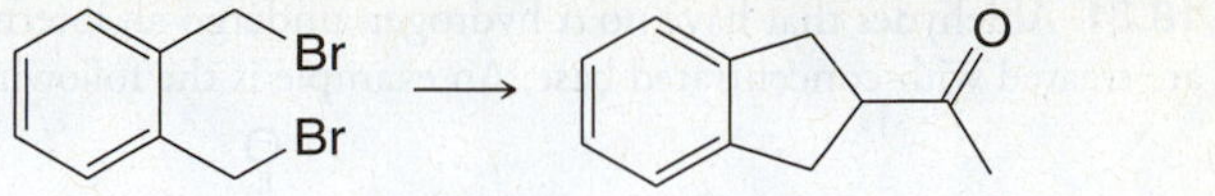

(f)

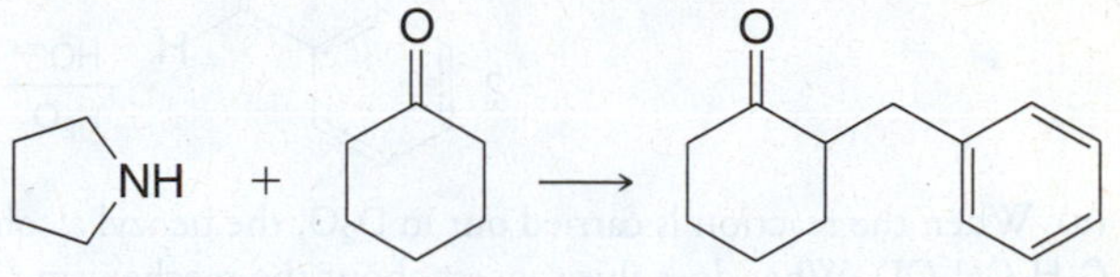

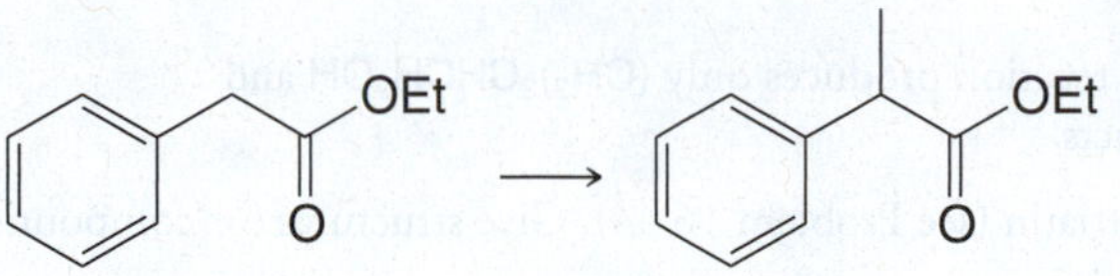

(g)

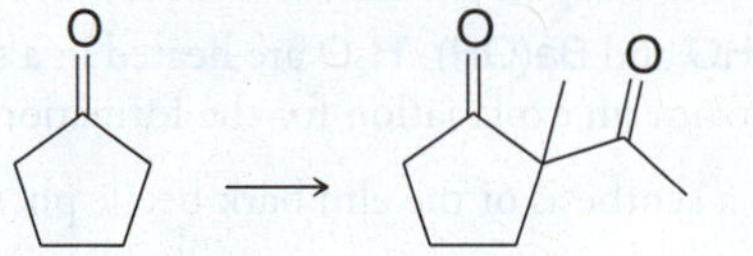

18.28 Linalool, a fragrant compound that can be isolated from a variety of plants, is 3,7-dimethyl-1,6-octadien-3-ol. Linalool is used in making perfumes, and it can be synthesized in the following way:

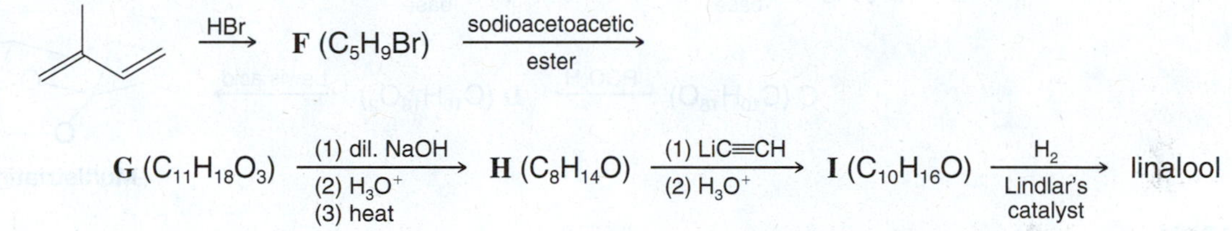

$$\text{(isoprene-like diene)} \xrightarrow{\text{HBr}} \mathbf{F}\ (C_5H_9Br) \xrightarrow{\substack{\text{sodioacetoacetic} \\ \text{ester}}}$$

$$\mathbf{G}\ (C_{11}H_{18}O_3) \xrightarrow[\substack{(2)\ H_3O^+ \\ (3)\ heat}]{(1)\ dil.\ NaOH} \mathbf{H}\ (C_8H_{14}O) \xrightarrow[(2)\ H_3O^+]{(1)\ LiC\equiv CH} \mathbf{I}\ (C_{10}H_{16}O) \xrightarrow[\substack{\text{Lindlar's} \\ \text{catalyst}}]{H_2} \text{linalool}$$

Outline the reactions involved. (*Hint*: Compound **F** is the more stable isomer capable of being produced in the first step.)

18.29 Compound **J**, a compound with two four-membered rings, has been synthesized by the following route. Outline the steps that are involved.

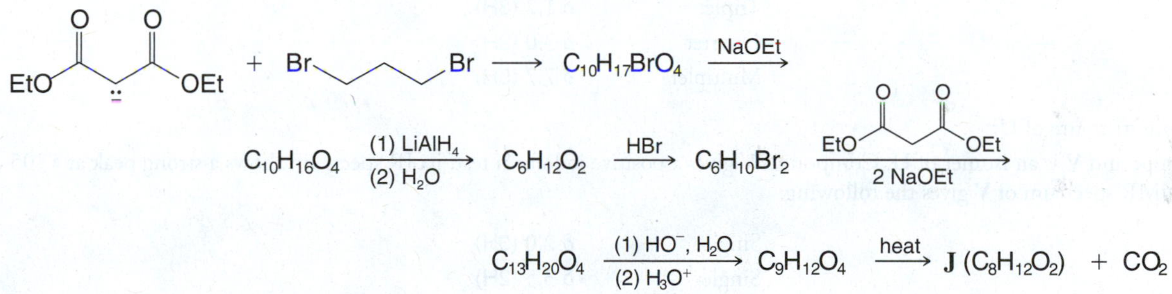

$$\text{(diethyl malonate)} + \text{Br}\!\!-\!\!\text{(chain)}\!\!-\!\!\text{Br} \longrightarrow C_{10}H_{17}BrO_4 \xrightarrow{\text{NaOEt}}$$

$$C_{10}H_{16}O_4 \xrightarrow[(2)\ H_2O]{(1)\ LiAlH_4} C_6H_{12}O_2 \xrightarrow{HBr} C_6H_{10}Br_2 \xrightarrow[2\ NaOEt]{\text{(diethyl malonate)}}$$

$$C_{13}H_{20}O_4 \xrightarrow[(2)\ H_3O^+]{(1)\ HO^-,\ H_2O} C_9H_{12}O_4 \xrightarrow{\text{heat}} \mathbf{J}\ (C_8H_{12}O_2)\ +\ CO_2$$

18.30 The Wittig reaction (Section 16.10) can be used in the synthesis of aldehydes, for example,

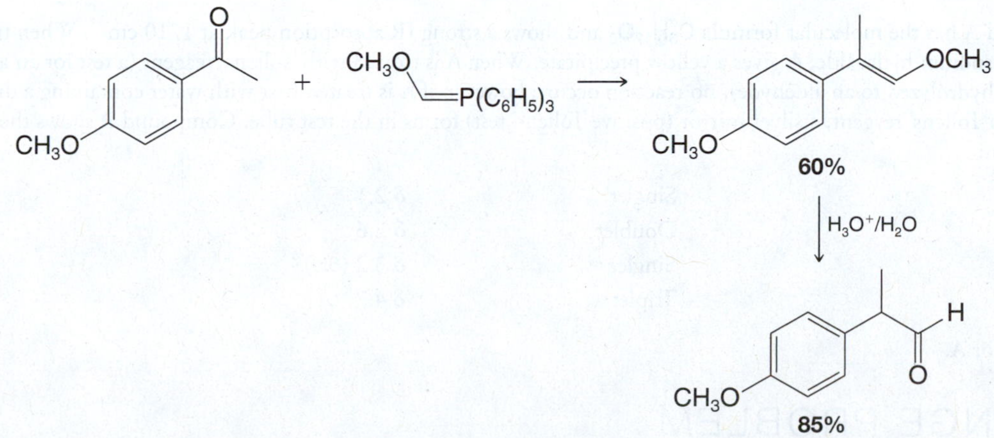

$$+\ CH_3O\text{—}CH\!=\!\!P(C_6H_5)_3 \longrightarrow \quad 60\%$$

$$\xrightarrow{H_3O^+/H_2O}$$

85%

(a) How would you prepare $CH_3OCH\!=\!\!P(C_6H_5)_3$?

(b) Show with a mechanism how the second reaction produces an aldehyde.

(c) How would you use this method to prepare 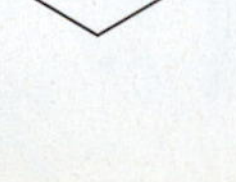 CHO from cyclohexanone?

18.31 Aldehydes that have no α hydrogen undergo an intermolecular oxidation–reduction called the **Cannizzaro reaction** when they are treated with concentrated base. An example is the following reaction of benzaldehyde:

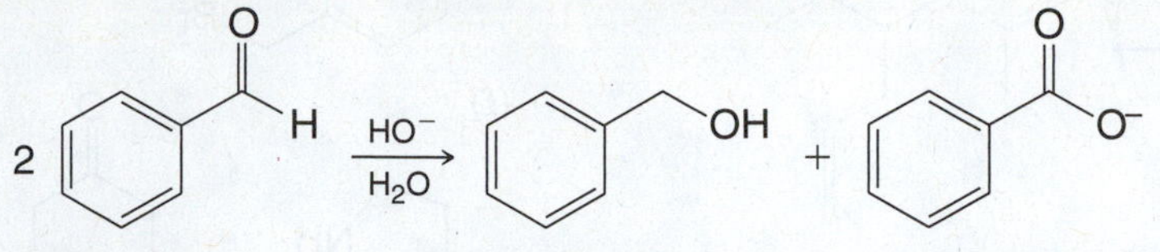

(a) When the reaction is carried out in D_2O, the benzyl alcohol that is isolated contains no deuterium bound to carbon. It is $C_6H_5CH_2OD$. What does this suggest about the mechanism for the reaction?

(b) When $(CH_3)_2CHCHO$ and $Ba(OH)_2/H_2O$ are heated in a sealed tube, the reaction produces only $(CH_3)_2CHCH_2OH$ and $[(CH_3)_2CHCO_2]_2Ba$. Provide an explanation for the formation of these products.

18.32 Shown below is a synthesis of the elm bark beetle pheromone, multistriatin (see Problem 16.44). Give structures for compounds **A**, **B**, **C**, and **D**.

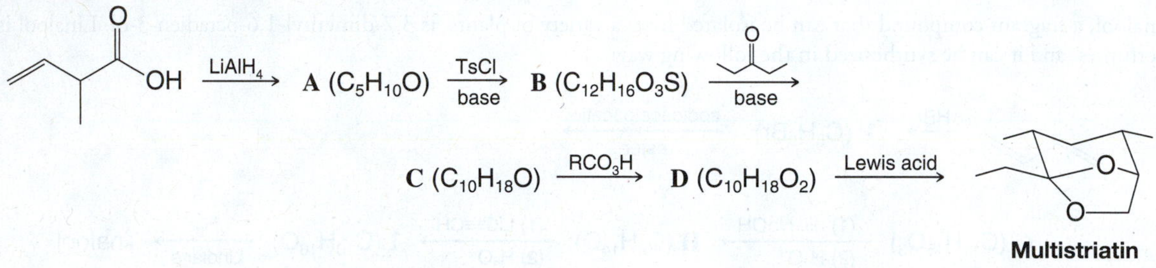

Multistriatin

SPECTROSCOPY

18.33 **(a)** A compound **U** ($C_9H_{10}O$) gives a negative iodoform test. The IR spectrum of **U** shows a strong absorption peak at 1690 cm^{-1}. The ^1H NMR spectrum of **U** gives the following:

Triplet	δ 1.2 (3H)	
Quartet	δ 3.0 (2H)	
Multiplet	δ 7.7 (5H)	

What is the structure of **U**?

(b) A compound **V** is an isomer of **U**. Compound **V** gives a positive iodoform test; its IR spectrum shows a strong peak at 1705 cm^{-1}. The ^1H NMR spectrum of **V** gives the following:

Singlet	δ 2.0 (3H)	
Singlet	δ 3.5 (2H)	
Multiplet	δ 7.1 (5H)	

What is the structure of **V**?

18.34 Compound **A** has the molecular formula $C_6H_{12}O_3$ and shows a strong IR absorption peak at 1710 cm^{-1}. When treated with iodine in aqueous sodium hydroxide, **A** gives a yellow precipitate. When **A** is treated with Tollens' reagent (a test for an aldehyde or a group that can be hydrolyzed to an aldehyde), no reaction occurs; however, if **A** is treated first with water containing a drop of sulfuric acid and then with Tollens' reagent, a silver mirror (positive Tollens' test) forms in the test tube. Compound **A** shows the following ^1H NMR spectrum:

Singlet	δ 2.1	
Doublet	δ 2.6	
Singlet	δ 3.2 (6H)	
Triplet	δ 4.7	

Write a structure for **A**.

CHALLENGE PROBLEM

18.35 The following is an example of a reaction sequence developed by Derin C. D'Amico and Michael E. Jung (UCLA) that results in enantiospecific formation of two new chirality centers and a carbon–carbon bond. The sequence includes a Horner–Wadsworth–Emmons reaction (Section 16.10B), a Sharpless asymmetric epoxidation (Section 11.13), and a novel rearrangement that ultimately leads to the product. Propose a mechanism for rearrangement of the epoxy alcohol under the conditions shown to form the aldol product. [*Hint*: The rearrangement can also be accomplished by preparing a trialkylsilyl ether from the epoxy alcohol in a separate reaction first and then treating the resulting silyl ether with a Lewis acid catalyst (e.g., BF_3).]

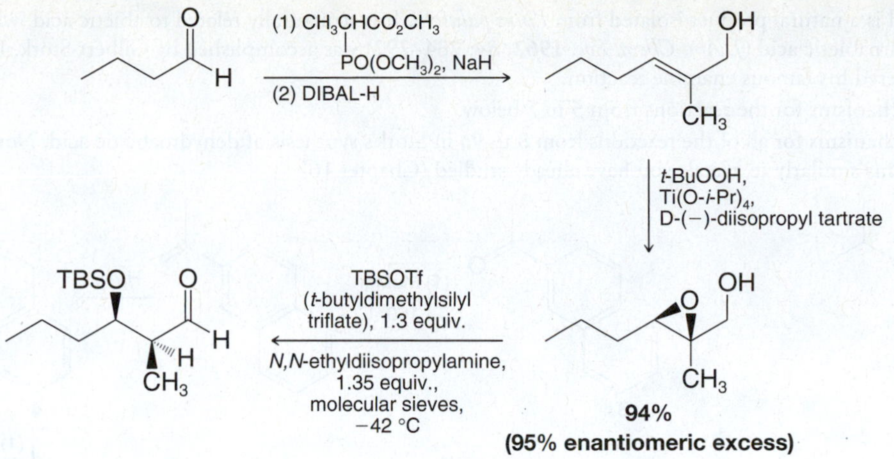

LEARNING GROUP PROBLEMS

β-CAROTENE, DEHYDROABEITIC ACID

1. β-Carotene is a highly conjugated hydrocarbon with an orange-red color. Its biosynthesis occurs via the isoprene pathway (Special Topic E in *WileyPLUS*), and it is found in, among other sources, pumpkins. One of the chemical syntheses of β-carotene was accomplished near the turn of the twentieth century by W. Ipatiew (*Ber.* **1901**, *34*, 594–596). The first few steps of this synthesis involve chemistry that should be familiar to you. Write mechanisms for all of the reactions from compounds **2** to **5**, and from **8** to **9** and **10**.

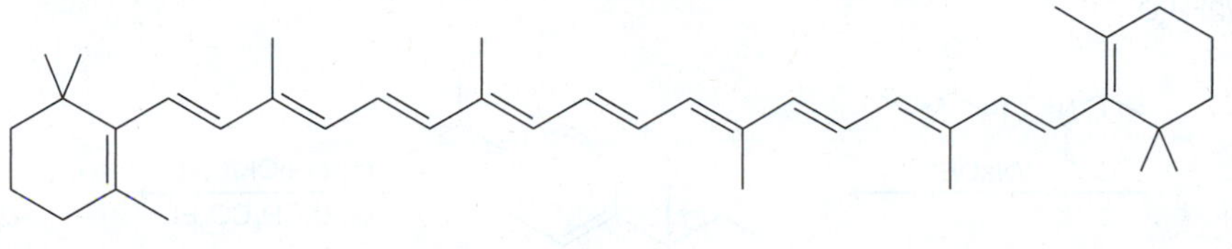

β-Carotene

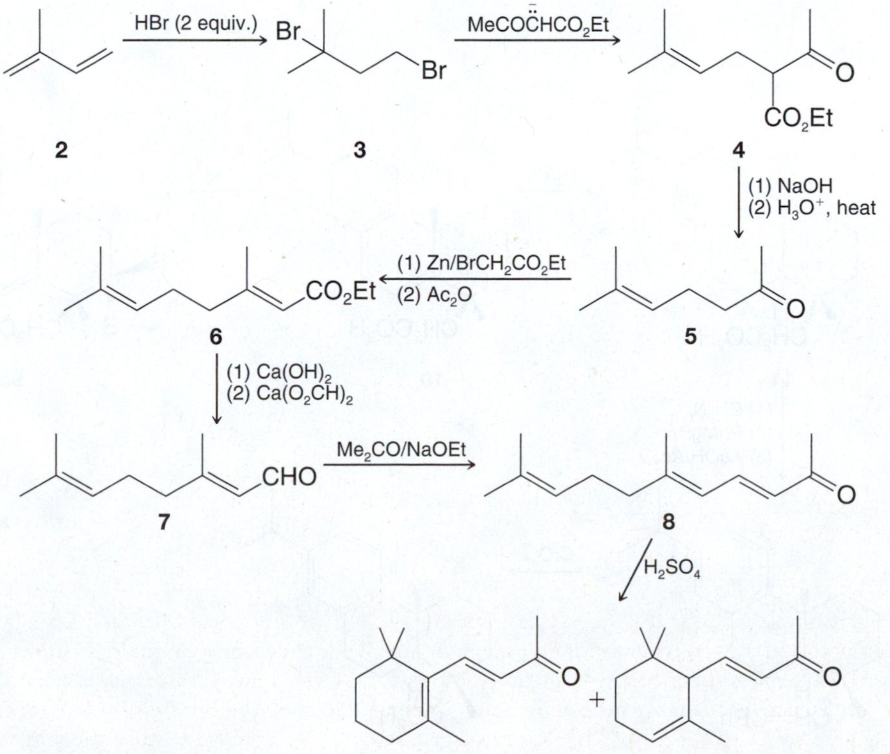

2. Dehydroabietic acid is a natural product isolated from *Pinus palustris*. It is structurally related to abietic acid, which comes from rosin. The synthesis of dehydroabietic acid (*J. Am. Chem. Soc.* **1962**, *84*, 284–292) was accomplished by Gilbert Stork. In the course of this synthesis, Stork discovered his famous enamine reaction.

(a) Write detailed mechanisms for the reactions from **5** to **7** below.

(b) Write detailed mechanisms for all of the reactions from **8** to **9a** in Stork's synthesis of dehydroabietic acid. Note that **9a** contains a dithioacetal, which forms similarly to acetals you have already studied (Chapter 16).

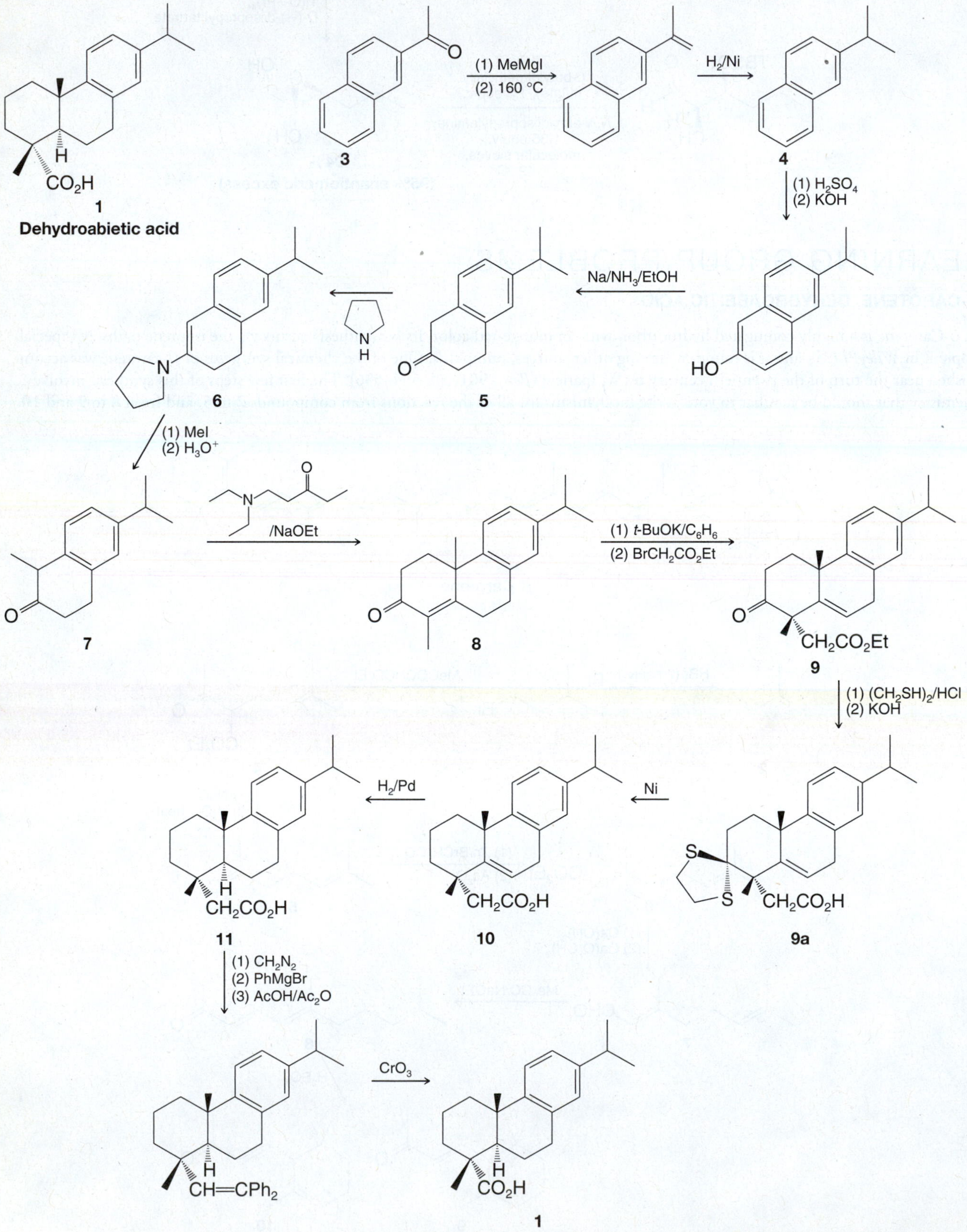

(Structures from Fleming, I., *Selected Organic Synthesis*, p. 76. Copyright John Wiley & Sons, Limited. Reproduced with permission.)

[S U M M A R Y   O F   M E C H A N I S M S]

Enolates: α-Substitution

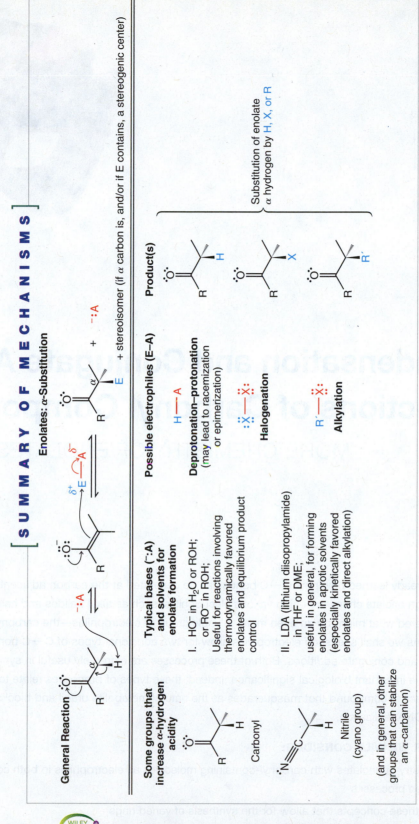

General Reaction

Some groups that increase α-hydrogen acidity

Carbonyl

Nitrile (cyano group)

(and in general, other groups that can stabilize an α-carbanion)

Typical bases (⁻:A) and solvents for enolate formation

I. HO⁻ in H_2O or ROH; or RO⁻ in ROH; Useful for reactions involving thermodynamically favored enolates and equilibrium product control

II. LDA (lithium diisopropylamide) in THF or DME; useful, in general, for forming enolates in aprotic solvents (especially kinetically favored enolates and direct alkylation)

Possible electrophiles (E–A)

H—A
Deprotonation–protonation (may lead to racemization or epimerization)

Halogenation

Alkylation

Product(s)

+ stereoisomer (if α carbon is, and/or if E contains, a stereogenic center)

Substitution of enolate α hydrogen by H, X, or R

WILEY PLUS See Special Topic C in *WileyPLUS*

CHAPTER
19

Condensation and Conjugate Addition Reactions of Carbonyl Compounds
MORE CHEMISTRY OF ENOLATES

We have already learned how new C—C bonds can be generated at the carbon adjacent to certain carbonyl functional groups through enolate chemistry using various electrophiles, such as alkyl halides and halogens. However, we have not yet considered what might be an even more valuable group of electrophiles—the carbonyl-containing molecules themselves. As we shall see, such electrophiles allow for two additional types of C—C bond-forming reactions: condensation reactions and conjugate additions. Both of these processes are extremely useful in synthesizing complex molecules, and they have important biological significance. Indeed, these types of processes relate to the cancer-fighting properties of 5-fluorouracil, a compound that masquerades as the natural metabolite uracil and blocks the biosynthesis of a compound needed for DNA replication.

IN THIS CHAPTER WE WILL CONSIDER:

- Additional chemistry of enolates with carbonyl-containing molecules as electrophiles in both condensation and conjugate addition processes
- Reactions using these concepts that allow for the synthesis of varied rings
- A special version of such reactions involving nitrogen that creates some unique carbonyl-containing amines

[WHY DO THESE TOPICS MATTER?] In "The Chemistry of... A Suicide Enzyme Substrate," we shall see how 5-fluorouracil works. Then, at the end of this chapter, we will show how the combination of several of these reactions in series, each setting up the next step like dominos falling in a row, can enable the one-pot preparation of a highly important alkaloid known as tropinone. Tropinone contains the core of several useful pharmaceuticals.

19.1 INTRODUCTION

In carbonyl **condensation reactions** the enolate or enol of one carbonyl compound reacts with the carbonyl group of another to join the two reactants. As part of the process, a new molecule that is derived from them "condenses" (forms). Often this molecule is that of an alcohol or water. The main types of condensation reactions we shall study are the **Claisen condensation** and the **aldol condensation**. Aldol condensations are preceded mechanistically by aldol additions, which we shall also study. The name **aldol** derives from the fact that **ald**ehyde and alco**hol** functional groups are present in the products of many aldol reactions.

An Example of a Claisen Condensation

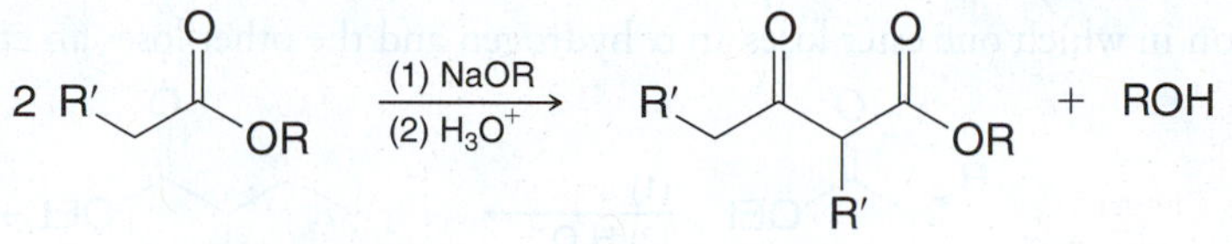

An Example of an Aldol Addition and Condensation

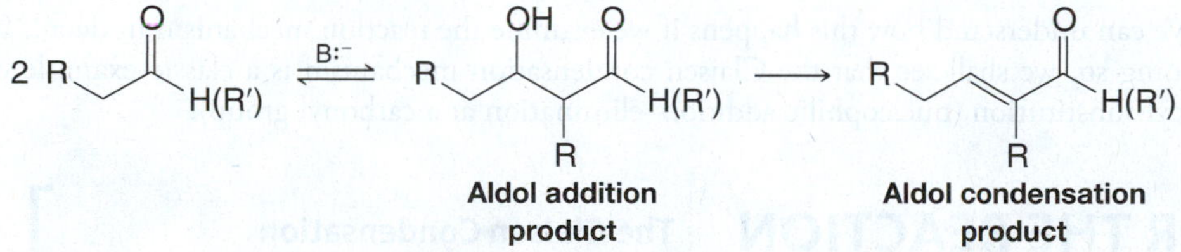

Conjugate addition reactions involve a nucleophile, which is often an enolate, adding to the β position of an α,β-unsaturated carbonyl compound. One of the most common conjugate addition reactions is the Michael addition. As we shall see, the aldol condensation provides a way to synthesize α,β-unsaturated carbonyl compounds that we can then use for subsequent conjugate addition reactions.

An Example of Conjugate Addition

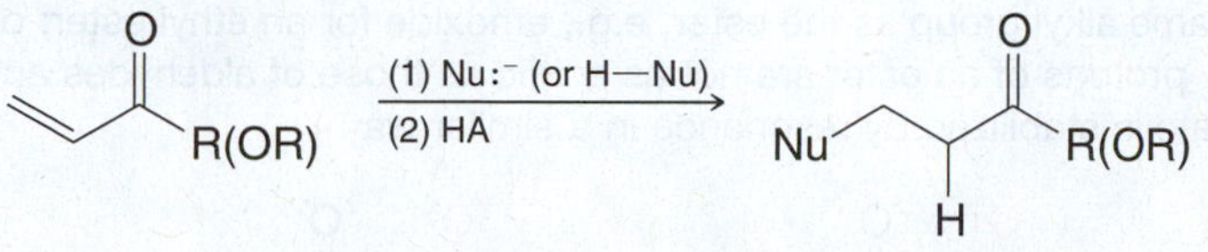

19.2 THE CLAISEN CONDENSATION: A SYNTHESIS OF β-KETO ESTERS

The Claisen condensation is a carbon–carbon bond-forming reaction that is useful for synthesizing β-keto esters. In Chapter 18 we saw how β-keto esters are useful in synthesis. In a Claisen condensation, the enolate of one ester molecule adds to the carbonyl group of another, resulting in an acyl substitution reaction that forms a β-keto ester and an alcohol molecule. The alcohol molecule that is formed derives from the alkoxyl group of the ester.

A classic example is the Claisen condensation by which ethyl acetoacetate (acetoacetic ester) can be synthesized.

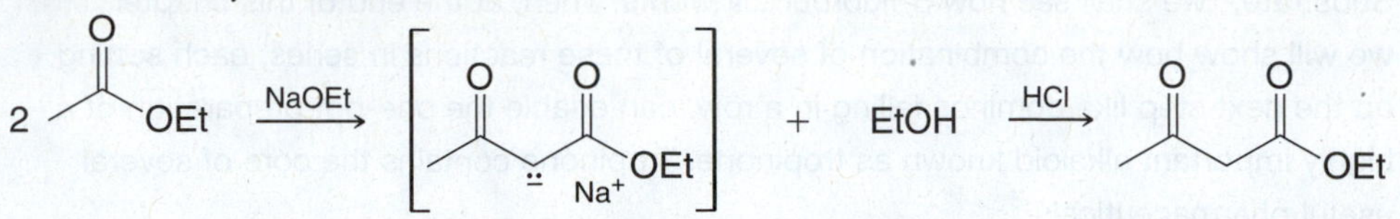

Sodioacetoacetic ester   (removed by distillation)   **Ethyl acetoacetate (acetoacetic ester) (76%)**

Another example is the Claisen condensation of two molecules of ethyl pentanoate, leading to ethyl 3-oxo-2-propylheptanoate.

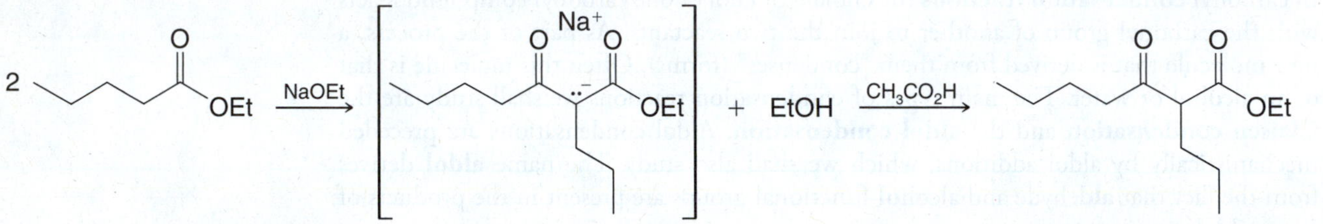

Ethyl pentanoate   **Ethyl 3-oxo-2-propylheptanoate (77%)**

If we look closely at these examples, we can see that, overall, both reactions involve a condensation in which one ester loses an α hydrogen and the other loses an ethoxide ion:

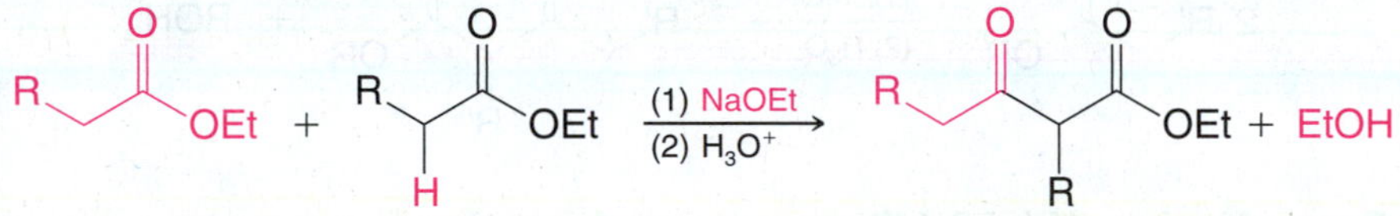

(R may also be H)   **A β-keto ester**

We can understand how this happens if we examine the reaction mechanism in detail. In doing so, we shall see that the Claisen condensation mechanism is a classic example of acyl substitution (nucleophilic addition–elimination at a carbonyl group).

[A MECHANISM FOR THE REACTION ── The Claisen Condensation]

Step 1

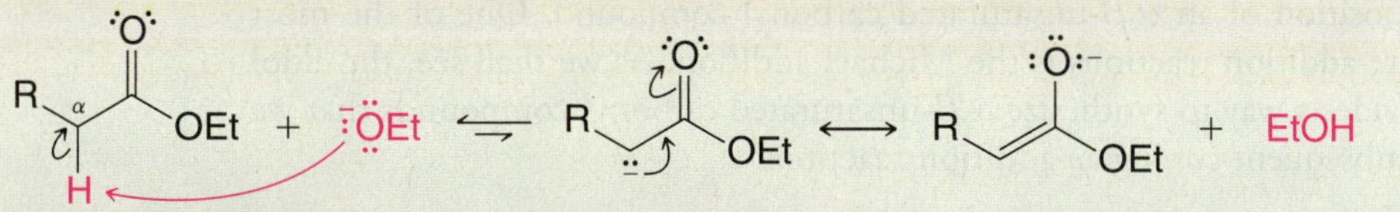

An alkoxide base removes an α proton from the ester, generating a nucleophilic enolate ion. (The alkoxide base used to form the enolate should have the same alkyl group as the ester, e.g., ethoxide for an ethyl ester; otherwise transesterification may occur.) Although the α protons of an ester are not as acidic as those of aldehydes and ketones, the resulting enolate is stabilized by resonance in a similar way.

Step 2

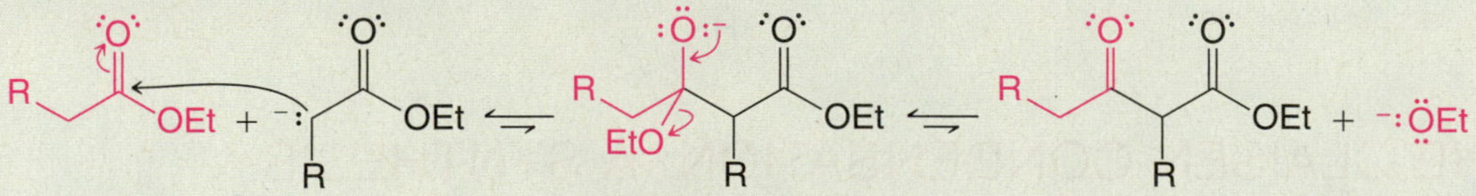

Nucleophilic addition   **Tetrahedral intermediate and elimination**

The enolate attacks the carbonyl carbon of another ester molecule, forming a tetrahedral intermediate. The tetrahedral intermediate expels an alkoxide ion, resulting in substitution of the alkoxide by the group derived from the enolate. The net result is nucleophilic addition–elimination at the ester carbonyl group. *The overall equilibrium for the process is unfavorable thus far, however,* but it is drawn toward the final product by removal of the acidic α hydrogen from the new β-dicarbonyl system.

Step 3

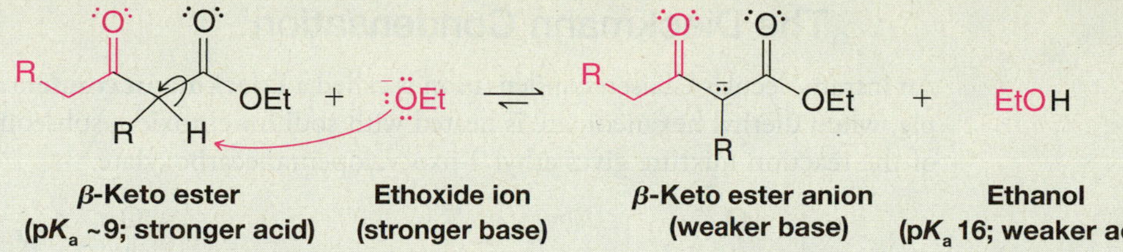

β-Keto ester   Ethoxide ion   β-Keto ester anion   Ethanol
(pK_a ~9; stronger acid)   (stronger base)   (weaker base)   (pK_a 16; weaker acid)

An alkoxide ion removes an α proton from the newly formed condensation product, resulting in a resonance stabilized β-keto ester ion. This step is highly favorable and draws the overall equilibrium toward product formation. The alcohol by-product (ethanol in this case) can be distilled from the reaction mixture as it forms, thereby further drawing the equilibrium toward the desired product.

Step 4

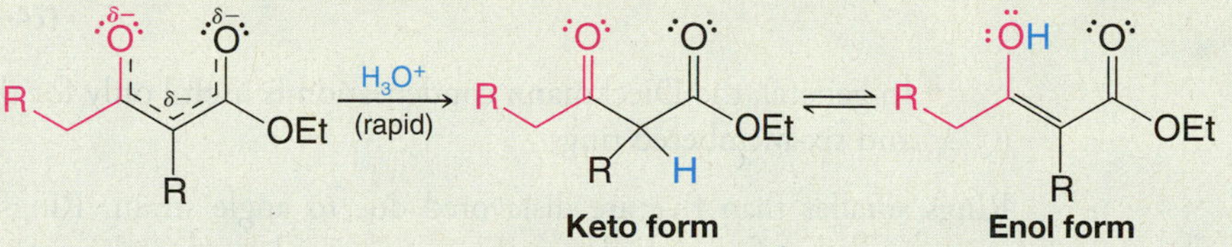

Keto form   Enol form

Addition of acid quenches the reaction by neutralizing the base and protonating the Claisen condensation product. The β-keto ester product exists as an equilibrium mixture of its keto and enol tautomers.

- When planning a reaction with an ester and an alkoxide ion it is important to use an alkoxide that has the same alkyl group as the alkoxyl group of the ester.

The alkoxyl group of the ester and the alkoxide must be the same so as to avoid trans-esterification (which occurs with alkoxides by the same mechanism as base-promoted ester hydrolysis; Section 17.7B). Ethyl esters and methyl esters, as it turns out, are the most common ester reactants in these types of syntheses. Therefore, we use sodium ethoxide when ethyl esters are involved and sodium methoxide when methyl esters are involved. There are some occasions when we shall choose to use other bases, but we shall discuss these later.

- Esters that have only one α hydrogen do not undergo the usual Claisen condensation.

An example of an ester that does not react in a normal Claisen condensation, because it has only one α hydrogen, is ethyl 2-methylpropanoate:

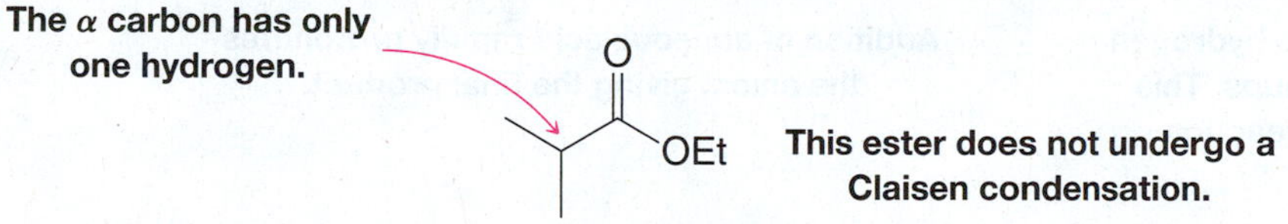

The α carbon has only one hydrogen.

This ester does not undergo a Claisen condensation.

Ethyl 2-methylpropanoate

- Inspection of the mechanism just given will make clear why this is so: an ester with only one α hydrogen will not have an acidic hydrogen when step 3 is reached, and step 3 provides the favorable equilibrium that ensures the success of the reaction.

In Section 19.2B we shall see how esters with only one α hydrogen can be converted to a β-keto ester by a method that uses a strong base.

PRACTICE PROBLEM 19.1

(a) Write a mechanism for all steps of the Claisen condensation that take place when ethyl propanoate reacts with ethoxide ion. **(b)** What products form when the reaction mixture is acidified?

PRACTICE PROBLEM 19.2

Since the products obtained from Claisen condensations are β-keto esters, subsequent hydrolysis and decarboxylation of these products give a general method for the synthesis of ketones. Show how you would employ this technique in a synthesis of 4-heptanone.

19.2A Intramolecular Claisen Condensations: The Dieckmann Condensation

An intramolecular Claisen condensation is called a **Dieckmann condensation**. For example, when diethyl hexanedioate is heated with sodium ethoxide, subsequent acidification of the reaction mixture gives ethyl 2-oxocyclopentanecarboxylate:

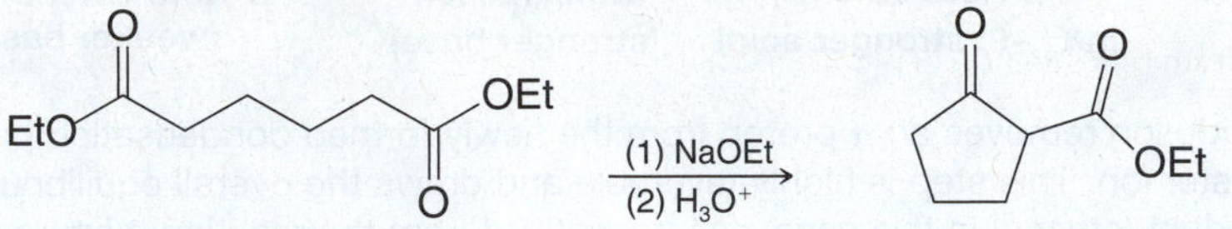

Diethyl hexanedioate (diethyl adipate)

$\xrightarrow{\text{(1) NaOEt} \atop \text{(2) } H_3O^+}$

Ethyl 2-oxocyclopentane-carboxylate (74–81%)

- In general, the Dieckmann condensation is useful only for the preparation of five- and six-membered rings.

Rings smaller than five are disfavored due to angle strain. Rings larger than seven are entropically less favorable due to the greater number of conformations available to a longer chain precursor, in which case intermolecular condensation begins to compete strongly.

[A MECHANISM FOR THE REACTION — The Dieckmann Condensation]

Ethoxide ion removes an α hydrogen.

The enolate ion attacks the carbonyl group at the other end of the chain.

An ethoxide ion is expelled.

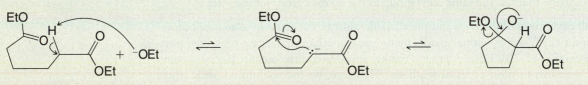

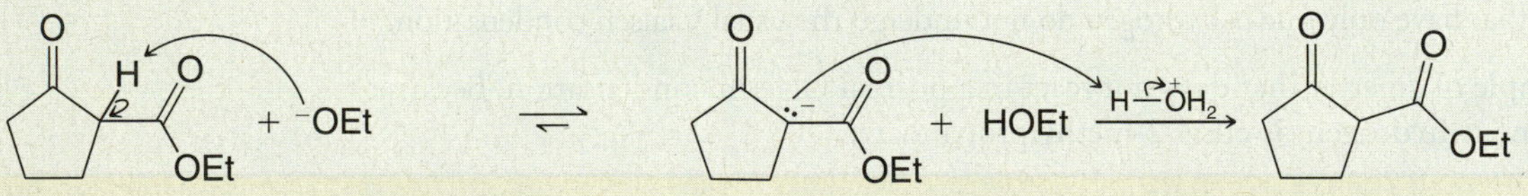

The ethoxide ion removes the acidic hydrogen located between two carbonyl groups. This favorable equilibrium drives the reaction.

Addition of aqueous acid rapidly protonates the anion, giving the final product.

PRACTICE PROBLEM 19.3 **(a)** What product would you expect from a Dieckmann condensation of diethyl heptanedioate? **(b)** Can you account for the fact that diethyl pentanedioate (diethyl glutarate) does not undergo a Dieckmann condensation?

19.2B Crossed Claisen Condensations

- Crossed Claisen condensations are possible **when one ester component has no α hydrogens** and, therefore, is unable to form an enolate ion and undergo self-condensation.

Ethyl benzoate, for example, condenses with ethyl acetate to give ethyl benzoylacetate:

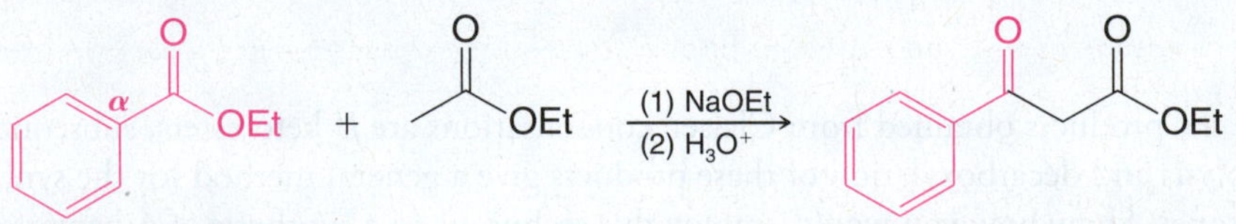

Ethyl benzoate (no α hydrogen) + $\xrightarrow{\text{(1) NaOEt} \atop \text{(2) } H_3O^+}$ **Ethyl benzoylacetate (60%)**

Ethyl phenylacetate condenses with diethyl carbonate to give diethyl phenylmalonate:

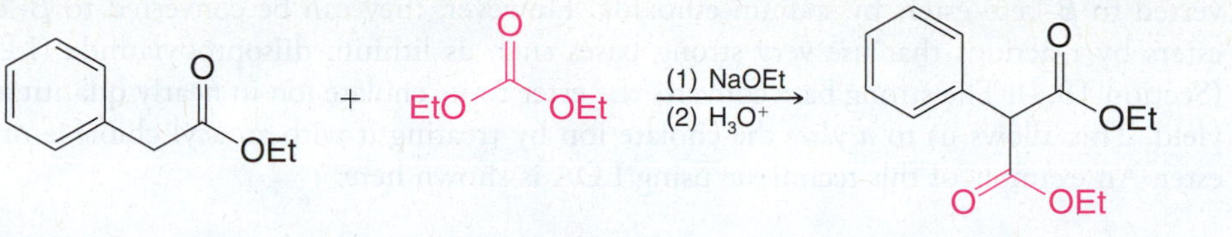

Ethyl phenylacetate   **Diethyl carbonate**   **Diethyl phenylmalonate**
                                  (no α carbon)   **(65%)**

●●● **SOLVED PROBLEM 19.1**

Write a mechanism for all of the steps in the Claisen condensation above between ethyl benzoate and ethyl acetate.

ANSWER: Ethyl benzoate contains no α hydrogens, so we begin by removing an α hydrogen from ethyl acetate to form an enolate.

Step 1

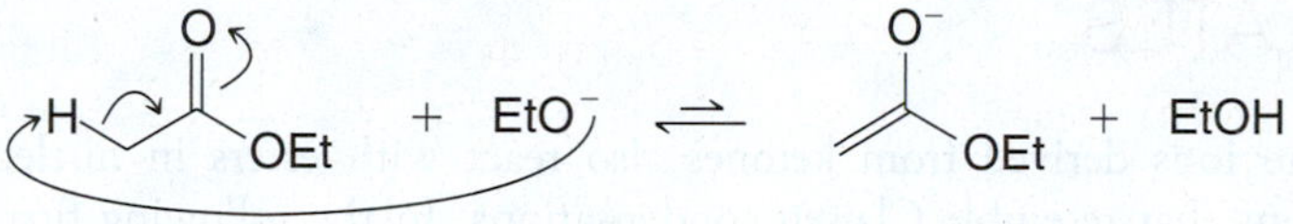

Step 2

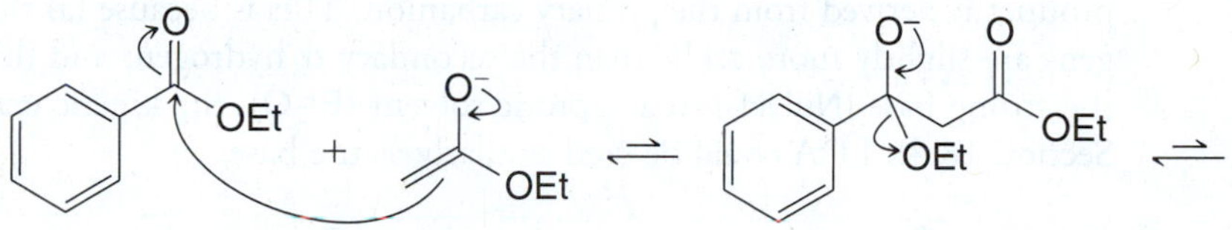

Step 3

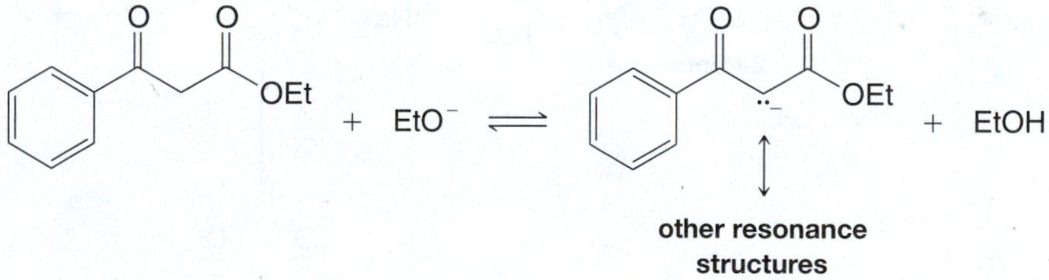

other resonance structures

Step 4

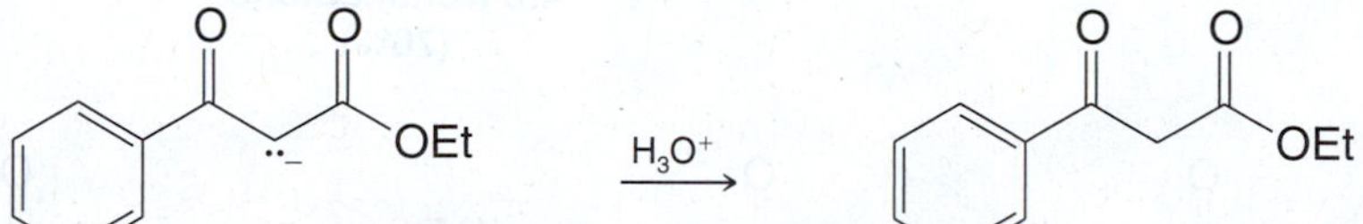

●●●

Write mechanisms that account for the products that are formed in the crossed Claisen condensation illustrated earlier between ethyl phenylacetate and diethyl carbonate.

PRACTICE PROBLEM 19.4

●●●

What products would you expect to obtain from each of the following crossed Claisen condensations?

PRACTICE PROBLEM 19.5

(a) Ethyl propanoate + diethyl oxalate $\xrightarrow[\text{(2) H}_3\text{O}^+]{\text{(1) NaOEt}}$

(b) Ethyl acetate + ethyl formate $\xrightarrow[\text{(2) H}_3\text{O}^+]{\text{(1) NaOEt}}$

As we learned earlier in this section, esters that have only one α hydrogen cannot be converted to β-keto esters by sodium ethoxide. However, they can be converted to β-keto esters by reactions that use very strong bases such as lithium diisopropylamide (LDA) (Section 18.4). The strong base converts the ester to its enolate ion in nearly quantitative yield. This allows us to *acylate* the enolate ion by treating it with an acyl chloride or an ester. An example of this technique using LDA is shown here:

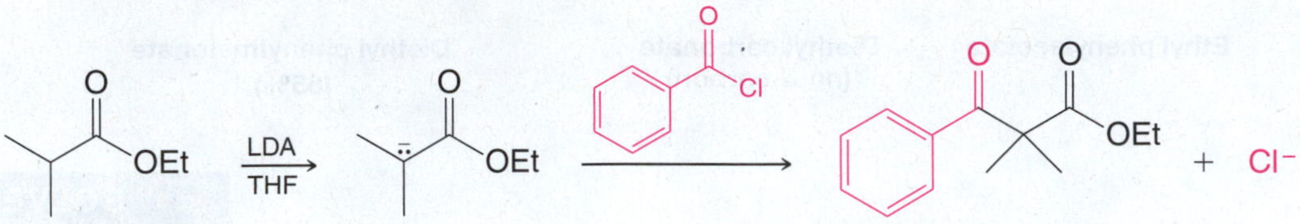

Ethyl 2,2-dimethyl-3-oxo-
3-phenylpropanoate

19.3 β-DICARBONYL COMPOUNDS BY ACYLATION OF KETONE ENOLATES

Enolate ions derived from ketones also react with esters in nucleophilic substitution reactions that resemble Claisen condensations. In the following first example, although two anions are possible from the reaction of the ketone with sodium amide, the major product is derived from the primary carbanion. This is because (a) the primary α hydrogens are slightly more acidic than the secondary α hydrogens and (b) in the presence of the strong base ($NaNH_2$) in an aprotic solvent (Et_2O), the kinetic enolate is formed (see Section 18.4). LDA could be used similarly as the base.

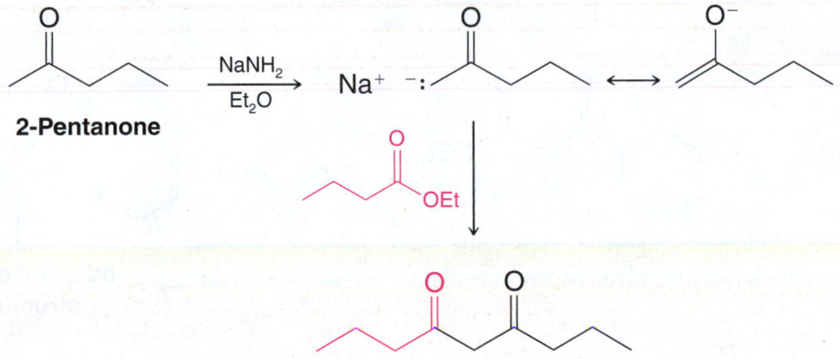

2-Pentanone

**4,6-Nonanedione
(76%)**

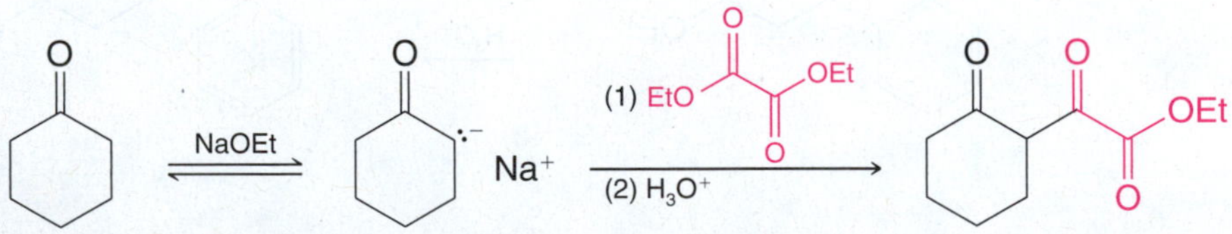

67%

● ● ● **SOLVED PROBLEM 19.2**

Keto esters are capable of undergoing cyclization reactions similar to the Dieckmann condensation. Write a mechanism for the following reaction.

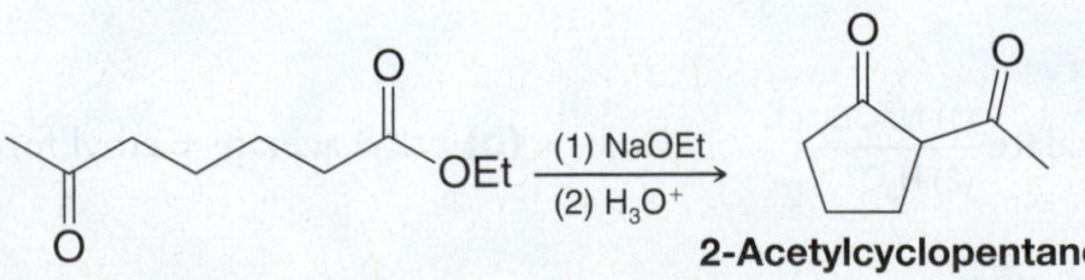

2-Acetylcyclopentanone

ANSWER:

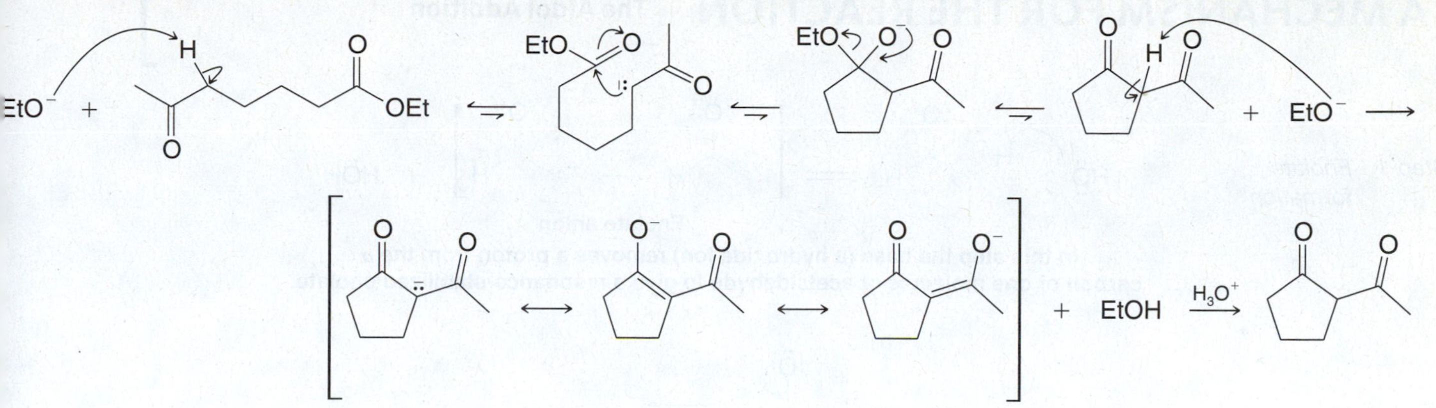

PRACTICE PROBLEM 19.6

Show how you might synthesize each of the following compounds using, as your starting materials, esters, ketones, acyl halides, and so on:

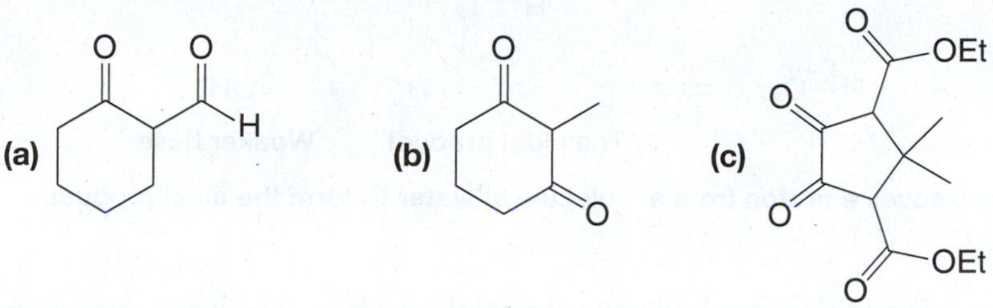

(a) (b) (c)

19.4 ALDOL REACTIONS: ADDITION OF ENOLATES AND ENOLS TO ALDEHYDES AND KETONES

- Aldol additions and aldol condensations together represent an important class of carbon–carbon bond-forming reaction.

An **aldol reaction** begins with addition of an enolate or enol to the carbonyl group of an aldehyde or ketone, leading to a β-hydroxy aldehyde or ketone as the initial product. A simple example is shown below, whereby two molecules of acetaldehyde (ethanal) react to form 3-hydroxybutanal. 3-Hydroxybutanal is an "aldol" because it contains both an **ald**ehyde and an alcoh**ol** functional group. Reactions of this general type are known as **aldol additions**.

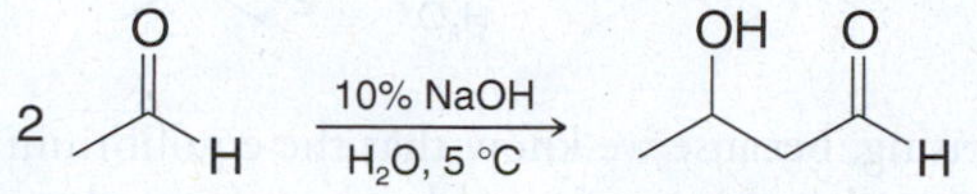

3-Hydroxybutanal
(50%)

As we shall see, the initial aldol addition product often dehydrates to form an α,β-unsaturated aldehyde or ketone. When this is the result, the overall reaction is an **aldol condensation**. First let us consider the mechanism of an aldol addition.

19.4A Aldol Addition Reactions

An aldol addition is an equilibrium reaction when it is conducted in a protic solvent with a base such as hydroxide or an alkoxide. The mechanism for an aldol addition involving an aldehyde is shown on the next page.

[A MECHANISM FOR THE REACTION — The Aldol Addition]

Step 1   Enolate formation

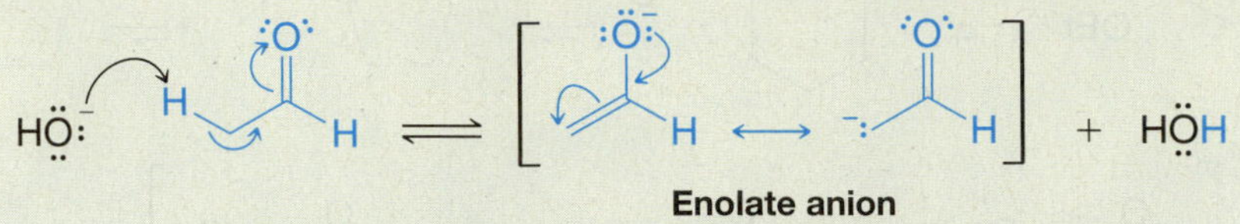

Enolate anion

In this step the base (a hydroxide ion) removes a proton from the α carbon of one molecule of acetaldehyde to give a resonance-stabilized enolate.

Step 2   Addition of the enolate

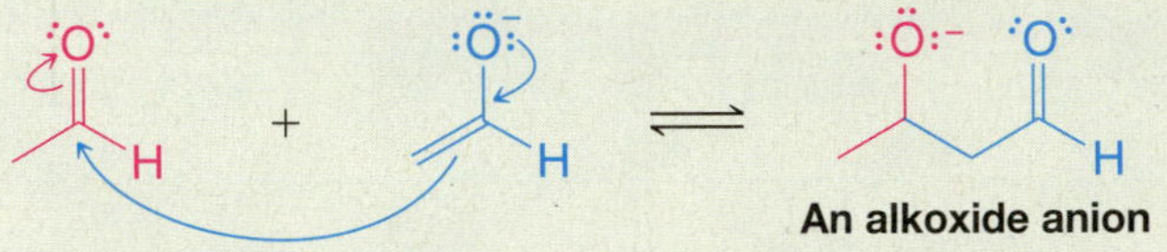

An alkoxide anion

The enolate then acts as a nucleophile and attacks the carbonyl carbon of a second molecule of acetaldehyde, producing an alkoxide anion.

Step 3   Protonation of the alkoxide

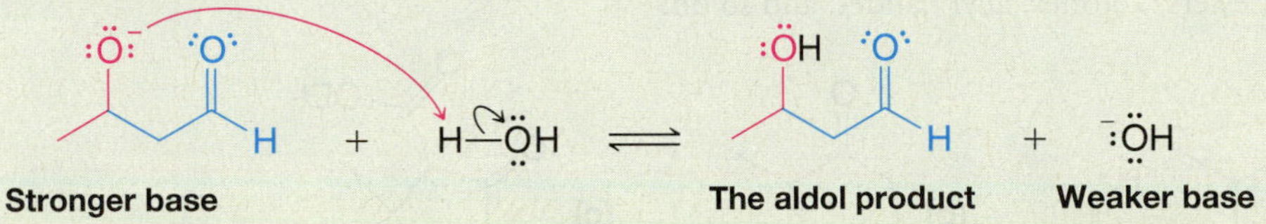

Stronger base   The aldol product   Weaker base

The alkoxide anion now removes a proton from a molecule of water to form the aldol product.

With ketones, the addition step leading to the aldol is unfavorable due to steric hindrance, and the equilibrium favors the aldol precursors rather than the addition product (Section 19.4B). However, as we shall see in Section 19.4C, dehydration of the aldol addition product can draw the equilibrium toward completion, whether the reactant is an aldehyde or a ketone. Enolate additions to both aldehydes and ketones are also feasible when a stronger base (such as LDA) is used in an aprotic solvent (Section 19.5B).

19.4B The Retro-Aldol Reaction

Because the steps in an aldol addition mechanism are readily reversible, a **retro-aldol reaction** can occur that converts a β-hydroxy aldehyde or ketone back to the precursors of an aldol addition. For example, when 4-hydroxy-4-methyl-2-pentanone is heated with hydroxide in water, the final equilibrium mixture consists primarily of acetone, the retro-aldol product.

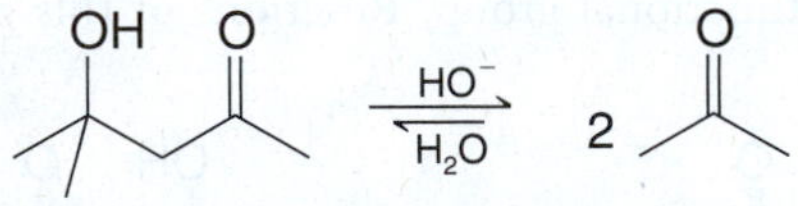

Helpful Hint

See "The Chemistry of... A Retro-Aldol Reaction in Glycolysis: Dividing Assets to Double the ATP Yield", page 870, for an important biochemical application that increases the energy yield from glucose.

This result is not surprising, because we know that the equilibrium for an aldol addition (the reverse of the reaction above) is not favorable when the enolate adds to a ketone. But, as mentioned earlier, dehydration of an aldol addition product can draw the equilibrium forward. We shall discuss the dehydration of aldols next (Section 19.4C).

●●● SOLVED PROBLEM 19.3

The carbon–carbon bond cleavage step in a retro-aldol reaction involves, under basic conditions, a leaving group that is an enolate, or under acidic conditions, an enol. Write a mechanism for the retro-aldol reaction of 4-hydroxy-4-methyl-2-pentanone under basic conditions (shown above).

STRATEGY AND ANSWER: Base removes the proton from the β-hydroxyl group, setting the stage for reversal of the aldol addition. As the alkoxide reverts to the carbonyl group, a carbon–carbon bond breaks with expulsion of the enolate as a leaving group. This liberates one of the original carbonyl molecules. Protonation of the enolate forms the other.

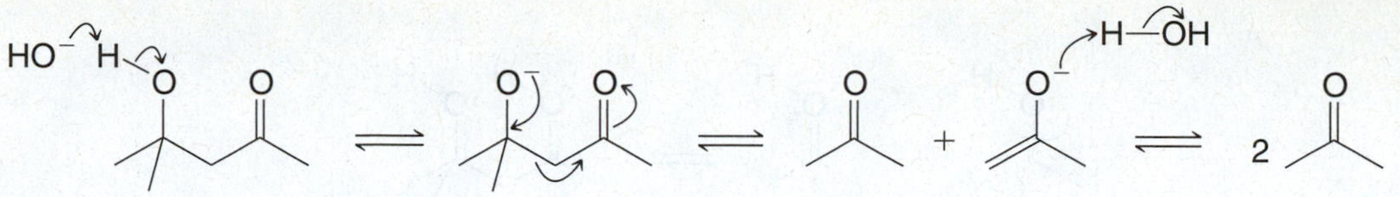

19.4C Aldol Condensation Reactions: Dehydration of the Aldol Addition Product

Dehydration of an aldol addition product leads to a conjugated α,β-unsaturated carbonyl system. The overall process is called an **aldol condensation**, and the product can be called an enal (alk*ene al*dehyde) or enone (alk*ene* ket*one*), depending on the carbonyl group in the product. The stability of the conjugated enal or enone system means that the dehydration equilibrium is essentially irreversible. For example, the aldol addition reaction that leads to 3-hydroxybutanal, shown in Section 19.4, dehydrates on heating to form 2-butenal. A mechanism for the dehydration is shown here.

[A MECHANISM FOR THE REACTION — Dehydration of the Aldol Addition Product]

The α hydrogens are acidic.

The double bonds of the alkene and carbonyl groups are conjugated, stabilizing the product.

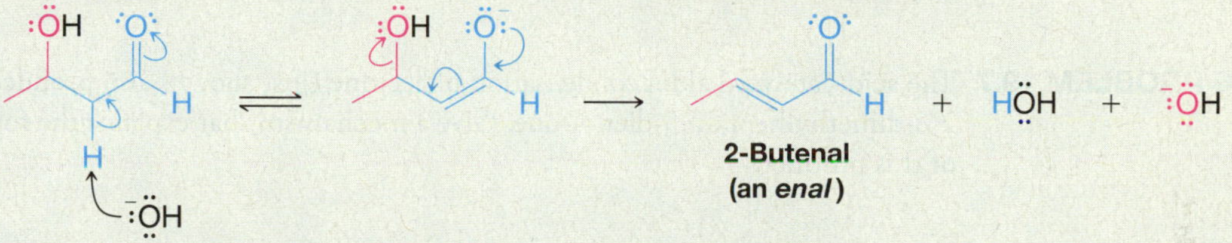

2-Butenal
(an *enal*)

Even though hydroxide is a leaving group in this reaction, the fact that each dehydrated molecule forms irreversibly, due to the stability from conjugation, draws the reaction forward.

19.4D Acid-Catalyzed Aldol Condensations

Aldol reactions can occur under acid catalysis, in which case the reaction generally leads to the α,β-unsaturated product by direct dehydration of the β-hydroxy aldol intermediate. This is one way by which ketones can successfully be utilized in an aldol reaction. The following is an example, in which acetone forms its aldol condensation product, 4-methylpent-3-en-2-one, on treatment with hydrogen chloride.

[A MECHANISM FOR THE REACTION — The Acid-Catalyzed Aldol Reaction]

Reaction

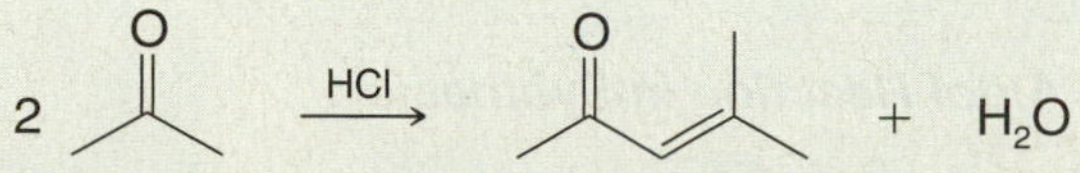

4-Methylpent-3-en-2-one

Mechanism

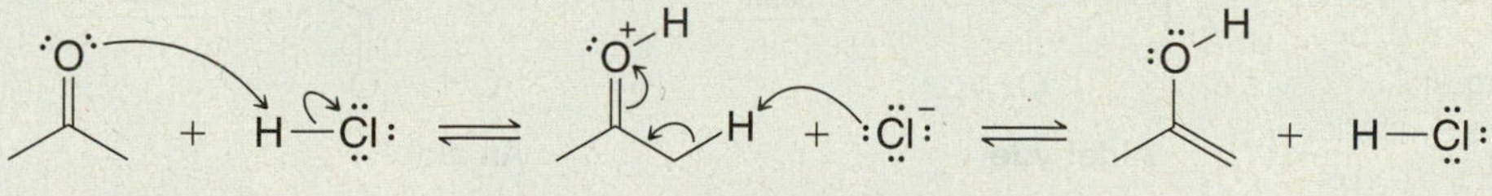

The mechanism begins with the acid-catalyzed formation of the enol.

(mechanism continues on the next page)

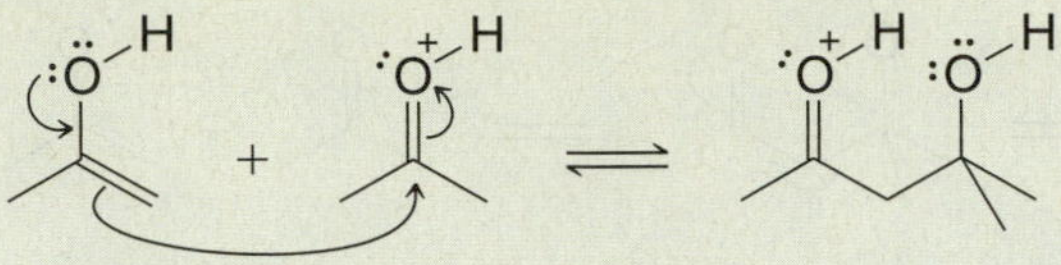

**Then the enol adds to the protonated carbonyl group
of another molecule of acetone.**

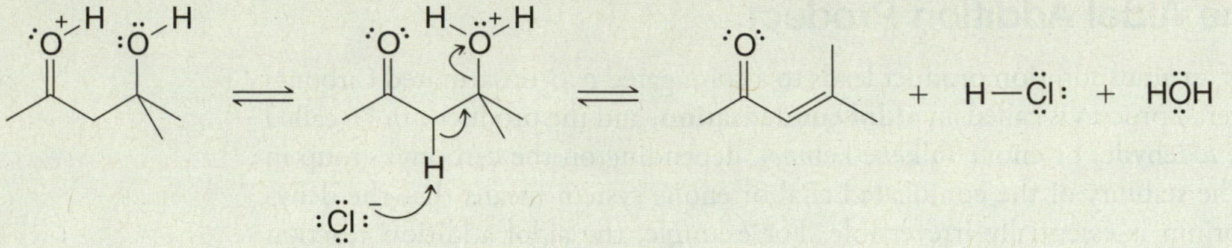

**Finally, proton transfers and dehydration lead
to the product.**

Acid catalysis can promote further reactions after the aldol condensation. An example is given in Practice Problem 19.8. Generally, it is more common in synthesis for an aldol reaction to be conducted under basic rather than acidic conditions.

PRACTICE PROBLEM 19.7 The acid-catalyzed aldol condensation of acetone (just shown) also produces some 2,6-dimethylhepta-2,5-dien-4-one. Give a mechanism that explains the formation of this product.

PRACTICE PROBLEM 19.8 Heating acetone with sulfuric acid leads to the formation of mesitylene (1,3,5-trimethylbenzene). Propose a mechanism for this reaction.

19.4E Synthetic Applications of Aldol Reactions

As we are beginning to see, aldol additions and aldol condensations are important methods for carbon–carbon bond formation. They also result in β-hydroxy and α,β-unsaturated carbonyl compounds that are themselves useful for further synthetic transformations. Some representative reactions are shown below.

The Aldol Reaction in Synthesis

Helpful Hint

The aldol reaction: a tool for synthesis. See also the Synthetic Connections review at the end of the chapter.

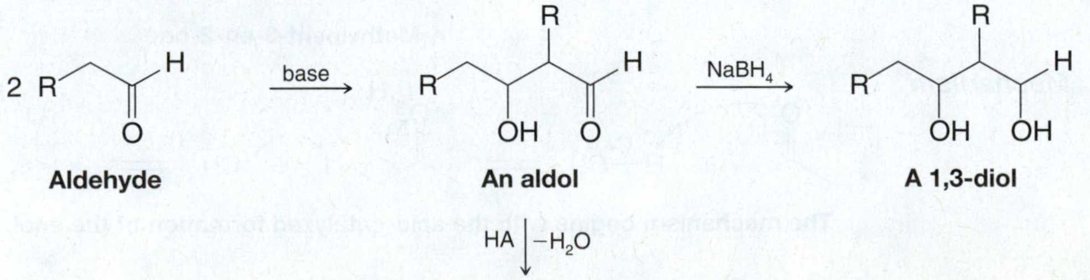

Aldehyde   An aldol   A 1,3-diol

$HA \downarrow -H_2O$

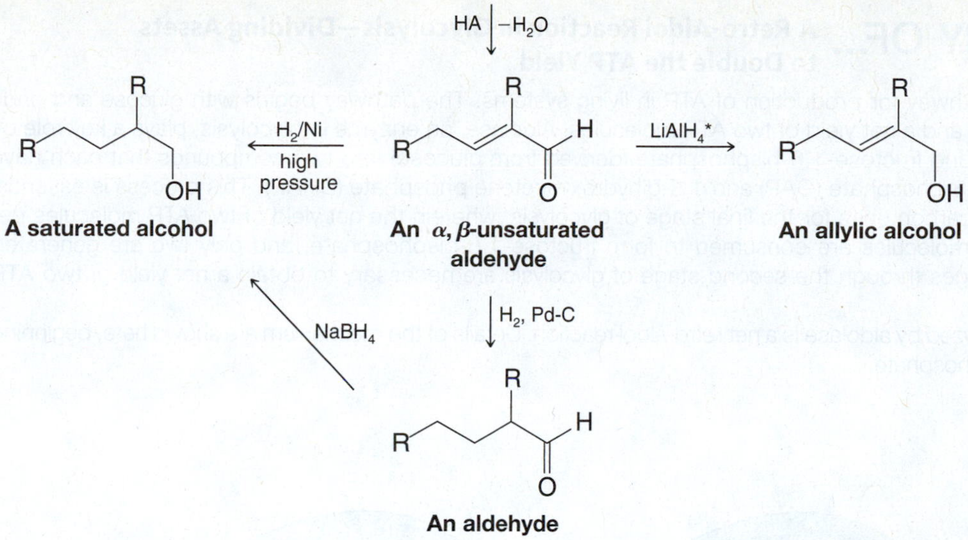

A saturated alcohol   **An α, β-unsaturated aldehyde**   **An allylic alcohol**

An aldehyde

●●● **SOLVED PROBLEM 19.4**

One industrial process for the synthesis of 1-butanol begins with ethanal. Show how this synthesis might be carried out.

STRATEGY AND ANSWER: Ethanal can be converted to an aldol via an aldol addition. Then, dehydration would produce 2-butenal, which can be hydrogenated to furnish 1-butanol.

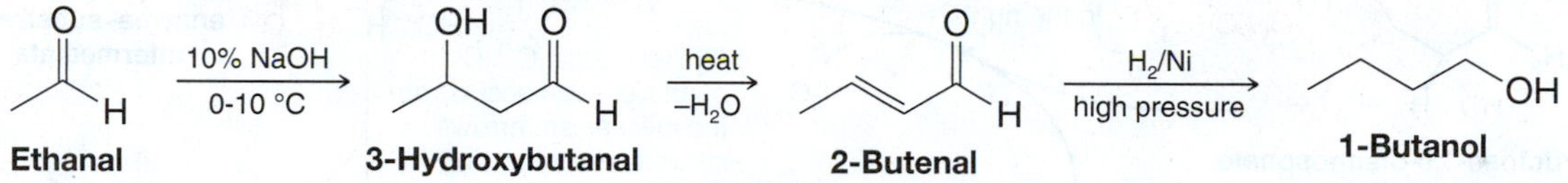

Ethanal   **3-Hydroxybutanal**   **2-Butenal**   **1-Butanol**

●●● **PRACTICE PROBLEM 19.9**

(a) Provide a mechanism for the aldol addition of propanal shown here.

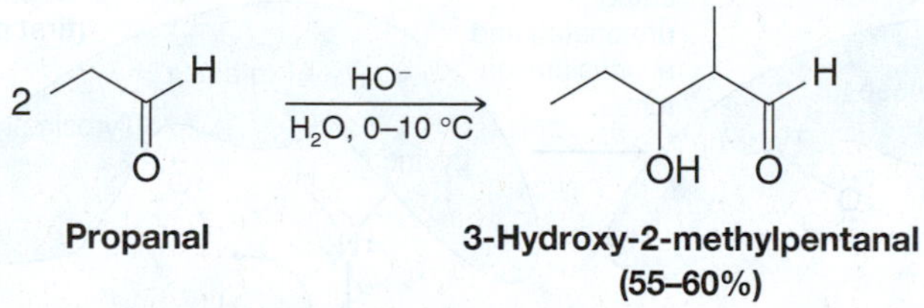

Propanal   **3-Hydroxy-2-methylpentanal**
(55–60%)

(b) How can you account for the fact that the product of the aldol addition is 3-hydroxy-2-methylpentanal and not 4-hydroxyhexanal?

(c) What products would be formed if the reaction mixture were heated?

●●● **PRACTICE PROBLEM 19.10**

Show how each of the following products could be synthesized from butanal:

(a) 2-Ethyl-3-hydroxyhexanal   **(c)** 2-Ethylhexan-1-ol
(b) 2-Ethylhex-2-en-1-ol   **(d)** 2-Ethylhexane-1,3-diol (the insect repellent "6–12")

Thus far we have only considered examples of aldol reactions where the reactant forms a product by dimerization. In the coming sections we shall discuss the use of aldol reactions to more generally prepare β-hydroxy and α,β-unsaturated carbonyl compounds. We shall then study reactions called conjugate addition reactions (Section 19.7), by which we can further build on the α,β-unsaturated carbonyl systems that result from aldol condensations.

*LiAlH₄ reduces the carbonyl group of α,β-unsaturated aldehydes and ketones cleanly. NaBH₄ often reduces the carbon–carbon double bond as well.

THE CHEMISTRY OF... A Retro-Aldol Reaction in Glycolysis—Dividing Assets to Double the ATP Yield

Glycolysis is a fundamental pathway for production of ATP in living systems. The pathway begins with glucose and ends with two molecules of pyruvate and a net yield of two ATP molecules. Aldolase, an enzyme in glycolysis, plays a key role by dividing the six-carbon compound fructose-1,6-bisphosphate (derived from glucose) into two compounds that each have three carbons, glyceraldehyde-3-phosphate (GAP) and 1,3-dihydroxyacetone phosphate (DHAP). This process is essential because it provides two three-carbon units for the final stage of glycolysis, wherein the net yield of two ATP molecules per glucose is realized. (Two ATP molecules are consumed to form fructose-1,6-bisphosphate, and only two are generated per pyruvate. Thus, two passages through the second stage of glycolysis are necessary to obtain a net yield of two ATP molecules per glucose.)

The cleavage reaction catalyzed by aldolase is a net retro-aldol reaction. Details of the mechanism are shown here, beginning at the left with fructose-1,6-bisphosphate.

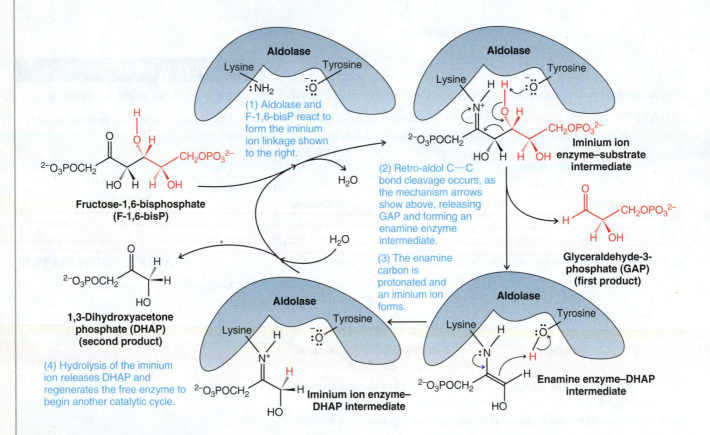

Two key intermediates in the aldolase mechanism involve functional groups that we have studied (Chapter 16)—an imine (protonated in the form of an iminium cation) and an enamine. In the mechanism of aldolase, an iminium cation acts as a sink for electron density during C—C bond cleavage (step 2), much like a carbonyl group does in a typical retro-aldol reaction. In this step the iminium cation is converted to an enamine, corresponding to the enolate or enol that is formed when a carbonyl group accepts electron density during C—C bond cleavage in an ordinary retro-aldol reaction. The enamine intermediate is then a source of an electron pair used to bond with a proton taken from the tyrosine hydroxyl at the aldolase active site (step 3). Finally, the resulting iminium group undergoes hydrolysis (step 4), freeing aldolase for another catalytic cycle and releasing DHAP, the second product of the retro-aldol reaction. Then, by a process catalyzed by the enzyme TIM (triose phosphate isomerase), DHAP undergoes isomerization to GAP for processing to pyruvate and synthesis of two more ATP molecules.

As we have seen with aldolase, imine and enamine functional groups have widespread roles in biological chemistry. Yet the functions of imines and enamines in biology are just as we would predict based on their native chemical reactivity.

19.5 CROSSED ALDOL CONDENSATIONS

An **aldol reaction** that starts with two different carbonyl compounds is called a **crossed aldol reaction**. Unless specific conditions are involved, a crossed aldol reaction can lead to a mixture of products from various pairings of the carbonyl reactants, as the following example illustrates with ethanal and propanal:

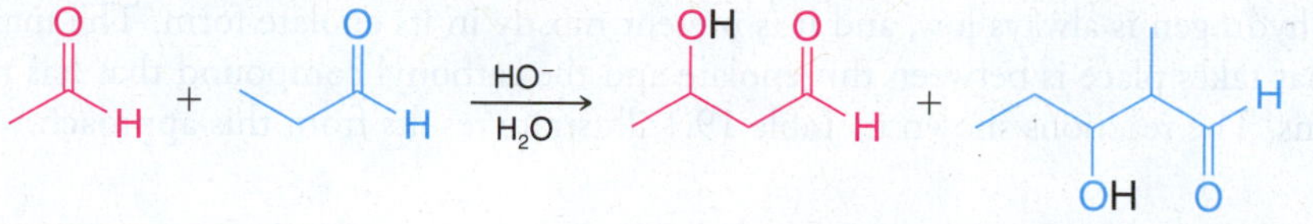

Ethanal   **Propanal**   **3-Hydroxybutanal** (from two molecules of ethanal)   **3-Hydroxy-2-methylpentanal** (from two molecules of propanal)

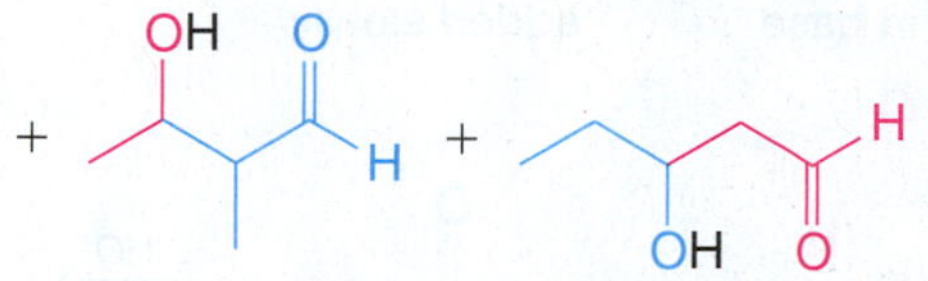

3-Hydroxy-2-methylbutanal   **3-Hydroxypentanal**
(from one molecule of ethanal and one molecule of propanal)

We shall therefore consider crossed aldol condensations by two general approaches that allow control over the distribution of products. The first approach hinges on structural factors of the carbonyl reactants and the role that favorable or unfavorable aldol addition equilibria play in determining the product distribution. In this approach relatively weak bases such as hydroxide or an alkoxide are used in a protic solvent such as water or an alcohol. The second approach, called a directed aldol reaction, involves use of a strong base such as LDA in an aprotic solvent. With a strong base, one reactant can be converted essentially completely to its enolate, which can then be allowed to react with the other carbonyl reactant.

●●● **SOLVED PROBLEM 19.5**

Show how each of the four products shown at the beginning of this section is formed in the crossed aldol addition between ethanal and propanal.

ANSWER: In the basic aqueous solution, four organic entities will initially be present: molecules of ethanal, molecules of propanal, enolate anions derived from ethanal, and enolate anions derived from propanal.

We have already seen (Section 19.4) how a molecule of ethanal can react with its enolate to form 3-hydroxybutanal (aldol). We have also seen (Practice Problem 19.9) how propanal can react with its enolate anion to form 3-hydroxy-2-methylpentanal. The other two products are formed as follows.

3-Hydroxy-2-methylbutanal results when the enolate of propanal reacts with ethanal.

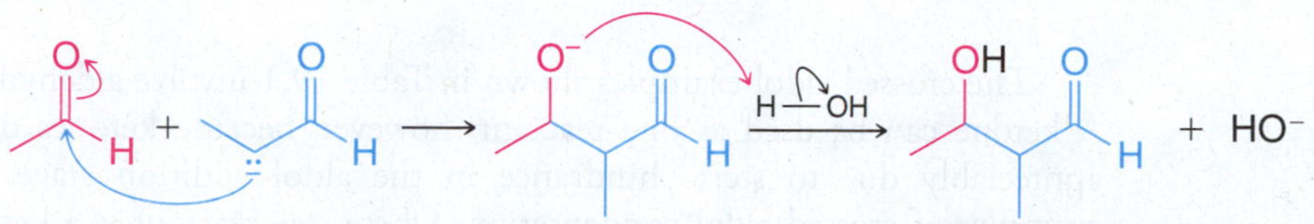

Ethanal   **Enolate of propanal**   **3-Hydroxy-2-methylbutanal**

And finally, 3-hydroxypentanal results when the enolate of ethanal reacts with propanal.

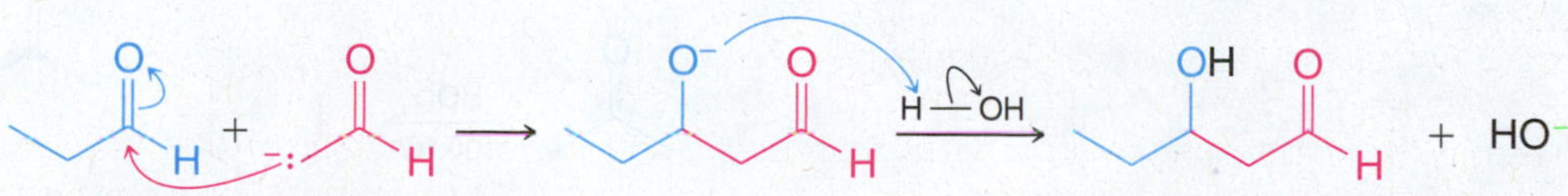

Propanal   **Enolate of ethanal**   **3-Hydroxypentanal**

19.5A Crossed Aldol Condensations Using Weak Bases

Crossed aldol reactions are possible with weak bases such as hydroxide or an alkoxide when one carbonyl reactant does not have an α hydrogen. A reactant without α hydrogens cannot self-condense because it cannot form an enolate. We avoid self-condensation of the other reactant, that which has an α hydrogen, by adding it slowly to a solution of the first reactant and the base. Under these conditions the concentration of the reactant with an α hydrogen is always low, and it is present mostly in its enolate form. The main reaction that takes place is between this enolate and the carbonyl compound that has no α hydrogens. The reactions shown in Table 19.1 illustrate results from this approach.

TABLE 19.1 CROSSED ALDOL REACTIONS

The reactant with no α hydrogen is placed in base	The reactant with an α hydrogen Is added slowly	Product

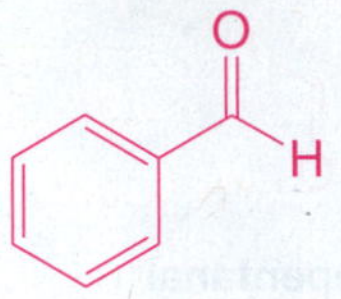

Benzaldehyde + Propanal $\xrightarrow[\text{10 °C}]{\text{HO}^-}$ 2-Methyl-3-phenyl-2-propenal (α-methylcinnamaldehyde) (68%)

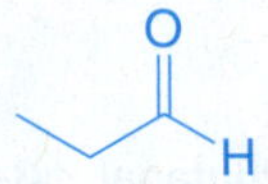

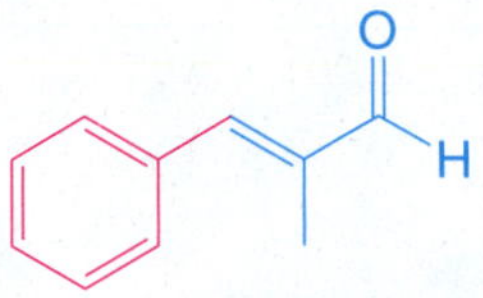

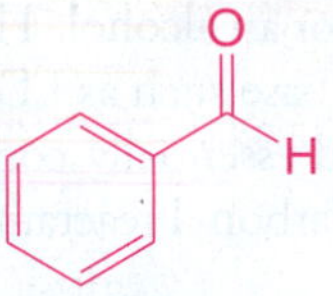

Benzaldehyde + Phenylacetaldehyde $\xrightarrow[\text{20 °C}]{\text{HO}^-}$ 2,3-Diphenyl-2-propenal

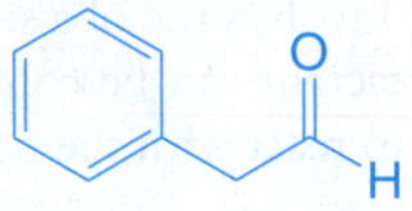

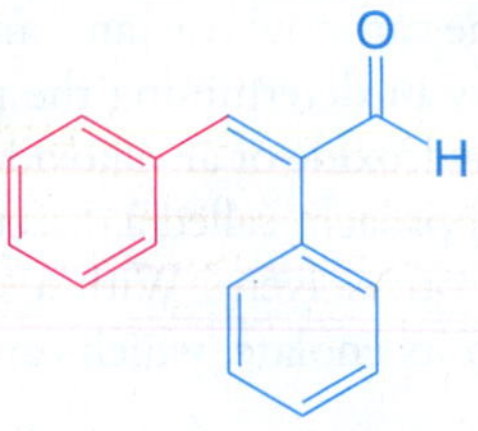

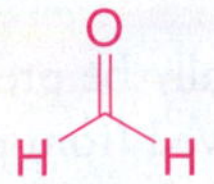

Formaldehyde + 2-Methylpropanal $\xrightarrow[\text{40 °C}]{\text{dilute Na}_2\text{CO}_3}$ 3-Hydroxy-2,2-dimethylpropanal (>64%)

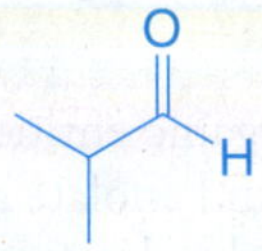

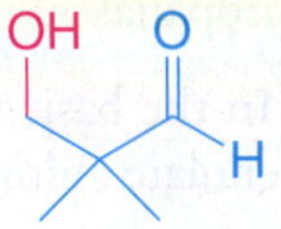

The crossed aldol examples shown in Table 19.1 involve aldehydes as both reactants. A ketone can be used as one reactant, however, because ketones do not self-condense appreciably due to steric hindrance in the aldol addition stage. The following are examples of crossed aldol condensations where one reactant is a ketone. Reactions such as these are sometimes called Claisen–Schmidt condensations. Schmidt discovered and Claisen developed this type of aldol reaction in the late 1800s.

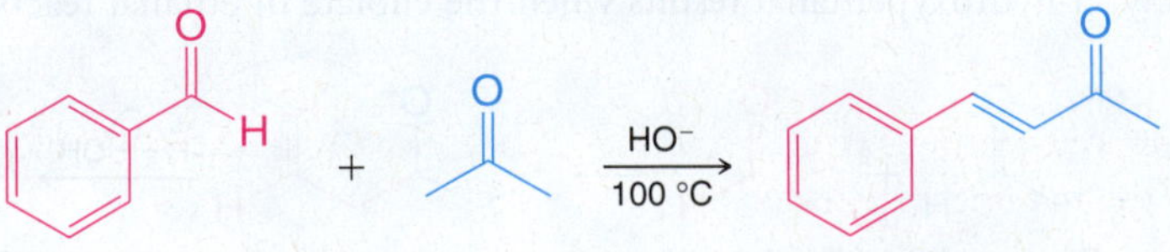

$\xrightarrow[\text{100 °C}]{\text{HO}^-}$ 4-Phenylbut-3-en-2-one (benzalacetone) (70%)

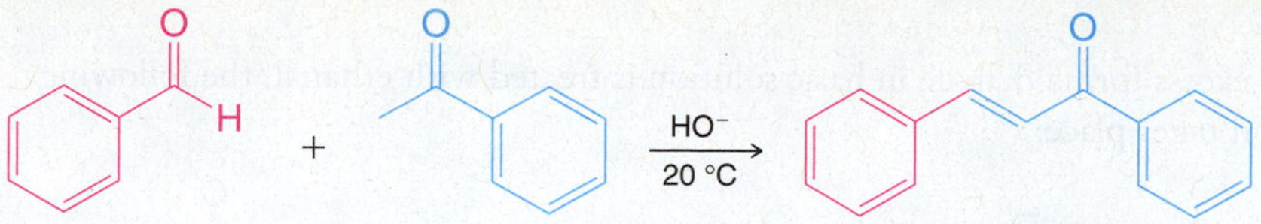

**1,3-Diphenylprop-2-en-1-one
(benzalacetophenone)
(85%)**

In these reactions, dehydration occurs readily because the double bond that forms is conjugated both with the carbonyl group and with the benzene ring. In general, dehydration of the aldol is especially favorable when it leads to extended conjugation.

As a further example, an important step in a commercial synthesis of vitamin A makes use of a crossed aldol condensation between geranial and acetone:

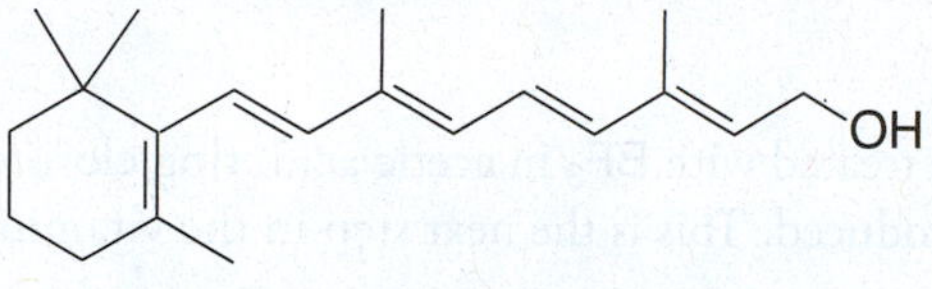

Vitamin A

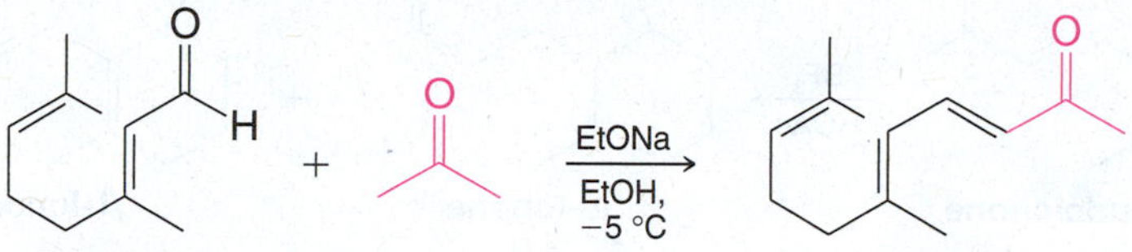

Helpful Hint

See "The Chemistry of... Antibody-catalyzed Aldol Condensations" in *WileyPLUS* for a method that uses the selectivity of antibodies to catalyze aldol reactions.

Geranial   **Pseudoionone
(49%)**

Geranial is a naturally occurring aldehyde that can be obtained from lemongrass oil. Its α hydrogen is *vinylic* and, therefore, not appreciably acidic. Notice, in this reaction, too, dehydration occurs readily because dehydration extends the conjugated system.

●●● **SOLVED PROBLEM 19.6**

Outlined below is a practical crossed aldol reaction that can be used for the synthesis of cinnamaldehyde (the essence of cinnamon, used in cooking). Provide the missing ingredients for this recipe.

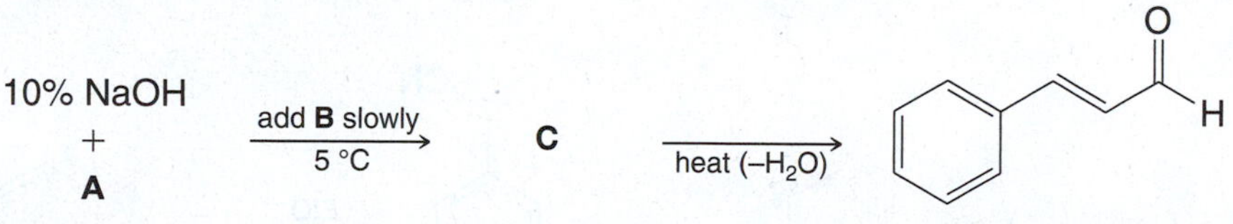

Cinnamaldehyde

STRATEGY AND ANSWER: Compound **A** is benzaldehyde, **B** is ethanal (acetaldehyde), and the intermediate **C** is shown below.

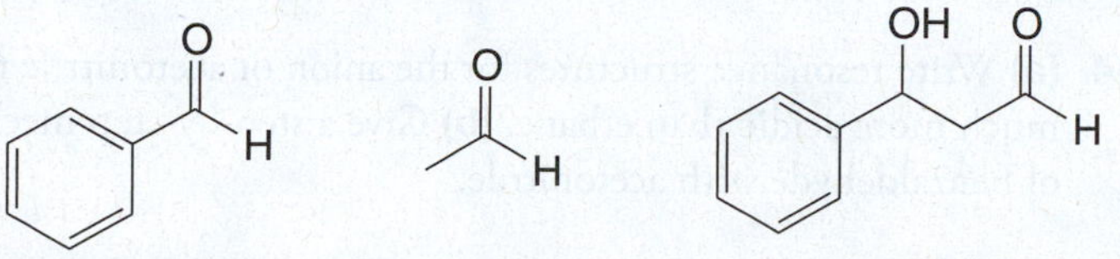

A   **B**   **C**

●●● **PRACTICE PROBLEM 19.11**

Outlined below is a synthesis of a compound used in perfumes, called lily aldehyde. Provide all of the missing structures.

$$p\text{-tert-Butylbenzyl alcohol} \xrightarrow[CH_2Cl_2]{PCC} C_{11}H_{14}O \xrightarrow[HO^-]{propanal} C_{14}H_{18}O \xrightarrow[CH_2Cl_2]{H_2, \text{ Pd-C}} \text{lily aldehyde } (C_{14}H_{20}O)$$

● ● ●

PRACTICE PROBLEM 19.12 When excess formaldehyde in basic solution is treated with ethanal, the following reaction takes place:

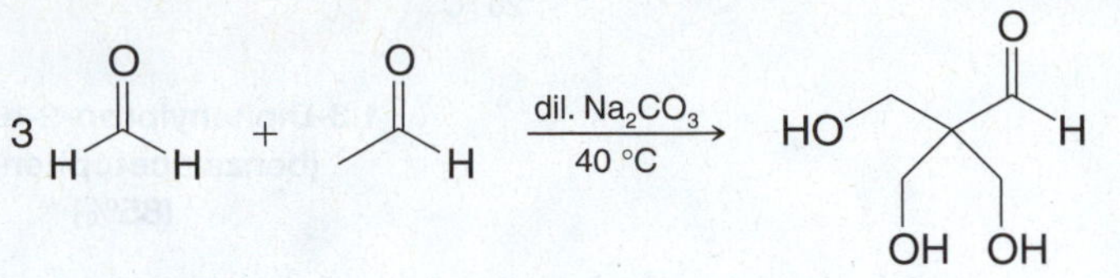

Write a mechanism that accounts for the formation of the product.

● ● ●

PRACTICE PROBLEM 19.13 When pseudoionone is treated with BF_3 in acetic acid, ring closure takes place and α- and β-ionone are produced. This is the next step in the vitamin A synthesis.

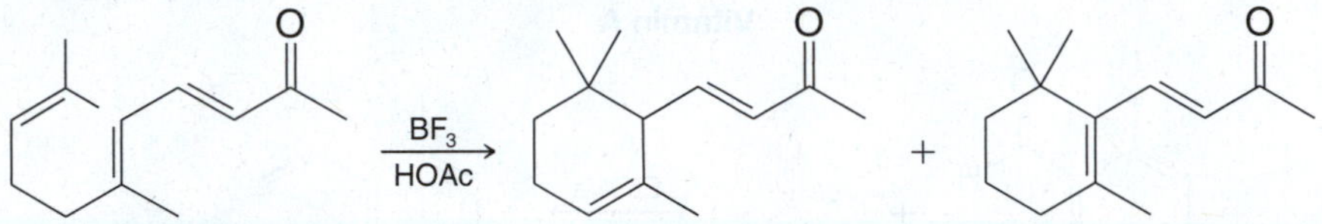

Pseudoionone α-**Ionone** β-**Ionone**

(a) Write mechanisms that explain the formation of α- and β-ionone.

(b) β-Ionone is the major product. How can you explain this?

(c) Which ionone would you expect to absorb at longer wavelengths in the UV–visible region? Why?

Nitriles with α hydrogens are also weakly acidic ($pK_a \cong 25$) and consequently these nitriles undergo condensations of the aldol type. An example is the condensation of benzaldehyde with phenylacetonitrile:

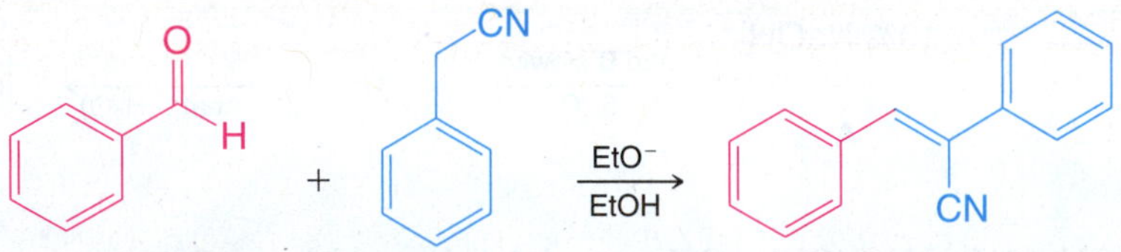

● ● ●

PRACTICE PROBLEM 19.14 **(a)** Write resonance structures for the anion of acetonitrile that account for its being much more acidic than ethane. **(b)** Give a step-by-step mechanism for the condensation of benzaldehyde with acetonitrile.

19.5B Crossed Aldol Condensations Using Strong Bases: Lithium Enolates and Directed Aldol Reactions

Helpful Hint

Lithium enolates are useful for crossed aldol syntheses.

One of the most effective and versatile ways to bring about a crossed aldol reaction is to use a lithium enolate obtained from a ketone as one component and an aldehyde or ketone as the other. An example of this approach, called a **directed aldol reaction**, is shown by the following mechanism.

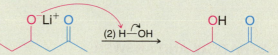

A MECHANISM FOR THE REACTION — A Directed Aldol Synthesis Using a Lithium Enolate

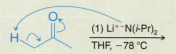

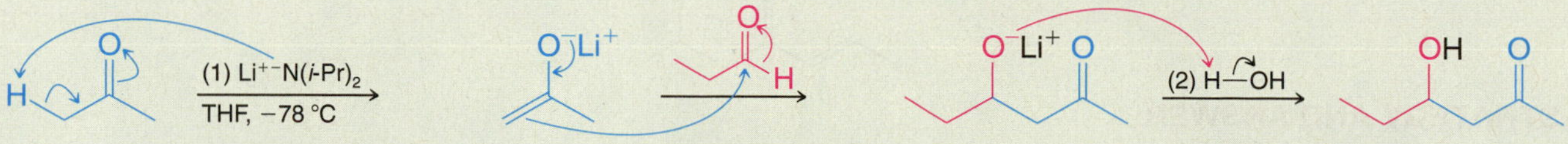

| The ketone is added to LiN(*i*-Pr)₂ (LDA), a strong base, which removes an α hydrogen from the ketone to produce an enolate. | The aldehyde is added and the enolate reacts with the aldehyde at its carbonyl carbon. | An acid–base reaction occurs when water is added at the end, protonating the lithium alkoxide. |

Regioselectivity can be achieved when unsymmetrical ketones are used in directed aldol reactions by generating the kinetic enolate using lithium diisopropylamide (LDA). This ensures production of the enolate in which the proton has been removed from the less substituted α carbon. The following is an example:

An Aldol Reaction via the Kinetic Enolate (Using LDA)

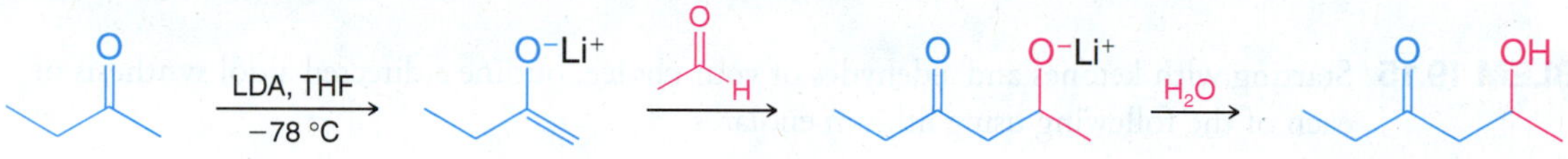

75%
A single crossed aldol product results.

If this aldol reaction had been carried out in the classic way (Section 19.5A) using hydroxide ion as the base, then at least two products would have been formed in significant amounts. Both the kinetic and thermodynamic enolates would have been formed from the ketone, and each of these would have added to the carbonyl carbon of the aldehyde:

An Aldol Reaction That Produces a Mixture via Both Kinetic and Thermodynamic Enolates (Using a Weaker Base under Protic Conditions)

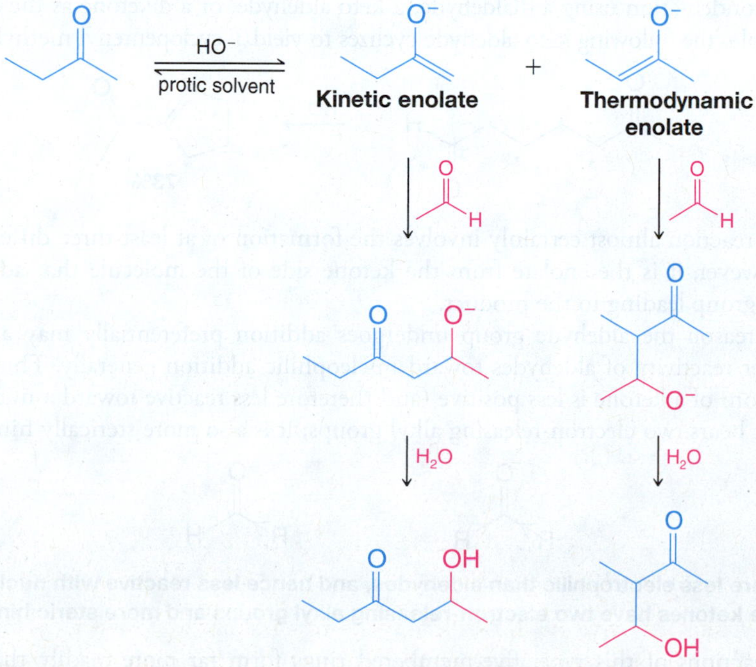

Kinetic enolate   +   Thermodynamic enolate

A mixture of crossed aldol products results.

●●● **SOLVED PROBLEM 19.7**

Outline a directed aldol synthesis of the following compound.

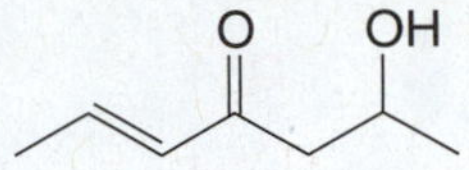

STRATEGY AND ANSWER:

Retrosynthetic Analysis

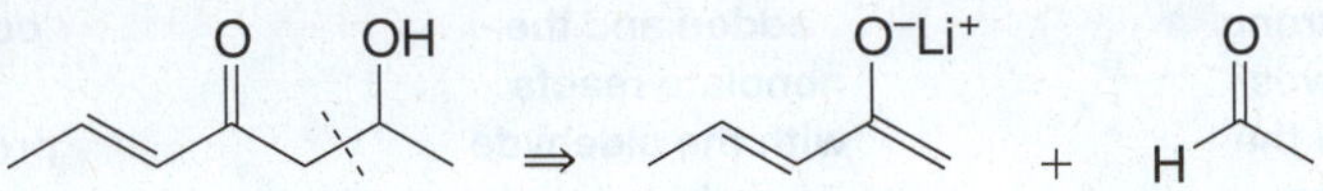

Synthesis

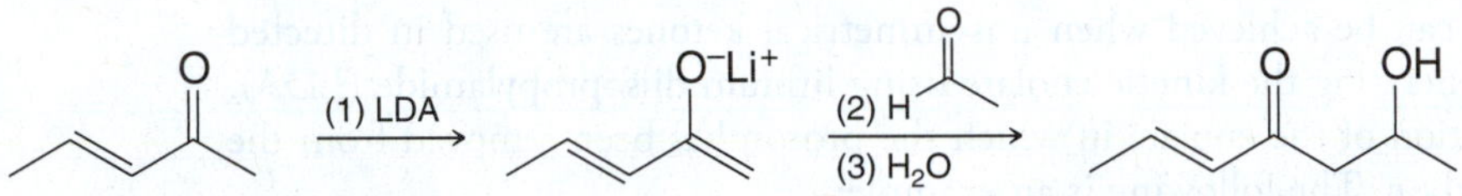

●●●

PRACTICE PROBLEM 19.15 Starting with ketones and aldehydes of your choice, outline a directed aldol synthesis of each of the following using lithium enolates:

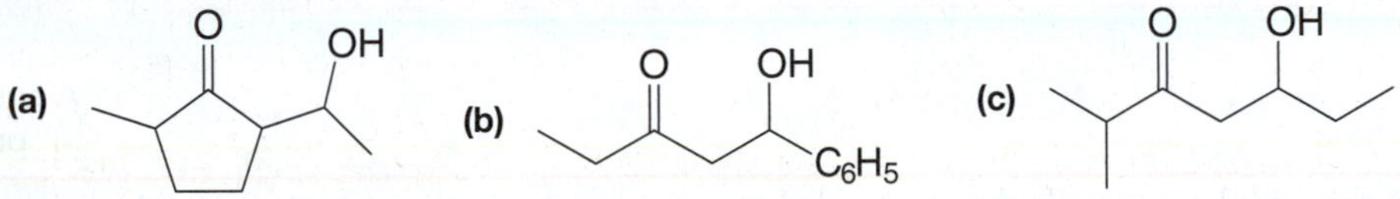

19.6 CYCLIZATIONS VIA ALDOL CONDENSATIONS

The aldol condensation also offers a convenient way to synthesize molecules with five- and six-membered rings (and sometimes even larger rings). This can be done by an intramolecular aldol condensation using a dialdehyde, a keto aldehyde, or a diketone as the substrate. For example, the following keto aldehyde cyclizes to yield 1-cyclopentenyl methyl ketone:

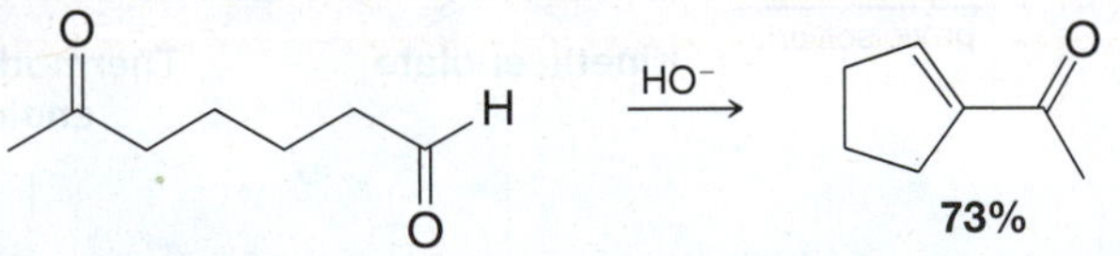

This reaction almost certainly involves the formation of at least three different enolates. However, it is the enolate from the ketone side of the molecule that adds to the aldehyde group leading to the product.

The reason the aldehyde group undergoes addition preferentially may arise from the greater reactivity of aldehydes toward nucleophilic addition generally. The carbonyl carbon atom of a ketone is less positive (and therefore less reactive toward a nucleophile) because it bears two electron-releasing alkyl groups; it is also more sterically hindered.

Ketones are less electrophilic than aldehydes, and hence less reactive with nucleophiles, because ketones have two electron-releasing alkyl groups and more steric hindrance.

In reactions of this type, five-membered rings form far more readily than seven-membered rings, and six-membered rings are more favorable than four- or eight-membered rings, when possible.

Helpful Hint

Selectivity in aldol cyclizations is influenced by carbonyl type and ring size.

A MECHANISM FOR THE REACTION — The Aldol Cyclization

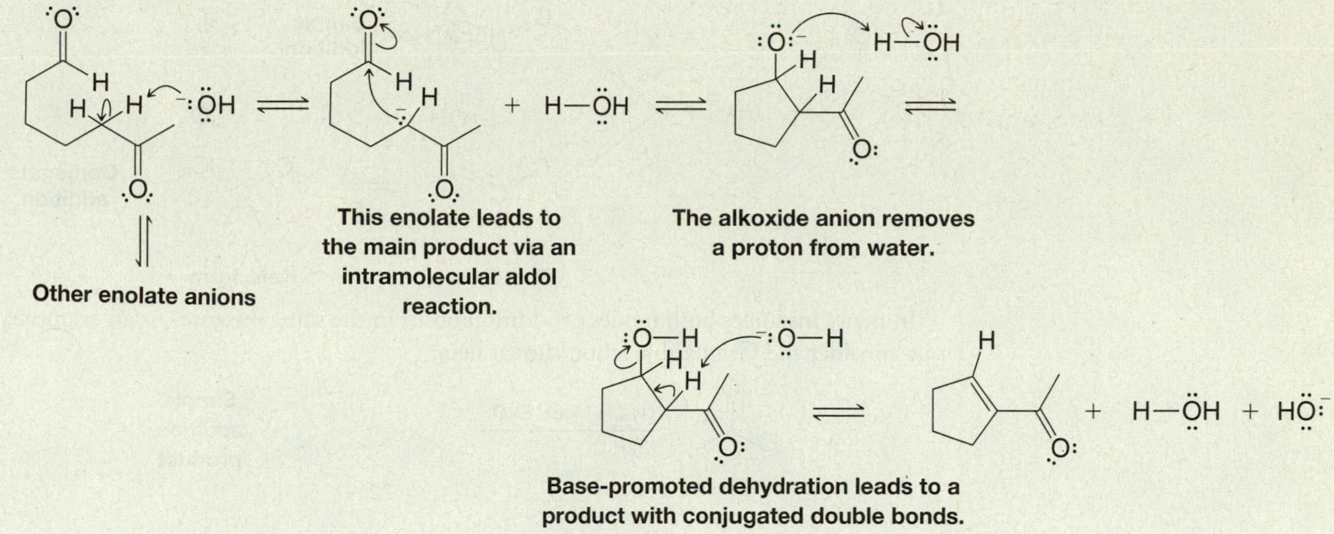

Other enolate anions

This enolate leads to the main product via an intramolecular aldol reaction.

The alkoxide anion removes a proton from water.

Base-promoted dehydration leads to a product with conjugated double bonds.

PRACTICE PROBLEM 19.16

Assuming that dehydration occurs, write the structures of the two other products that might have resulted from the aldol cyclization just given. (One of these products will have a five-membered ring and the other will have a seven-membered ring.)

PRACTICE PROBLEM 19.17

What starting compound would you use in an aldol cyclization to prepare each of the following?

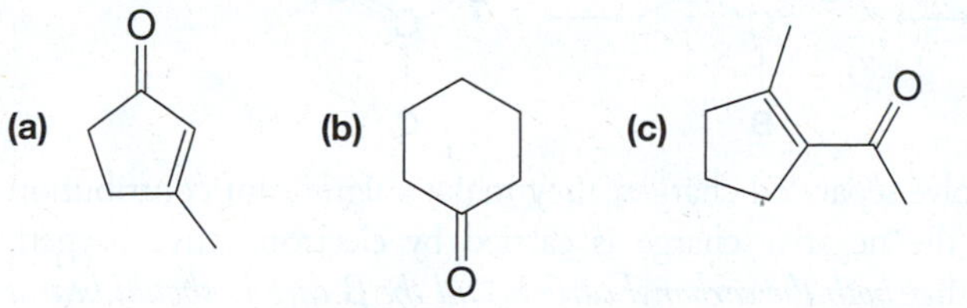

(a) (b) (c)

PRACTICE PROBLEM 19.18

What experimental conditions would favor the cyclization process in an intramolecular aldol reaction over intermolecular condensation?

19.7 ADDITIONS TO α,β-UNSATURATED ALDEHYDES AND KETONES

When α,β-unsaturated aldehydes and ketones react with nucleophilic reagents, they may do so in two ways. They may react by a **simple addition,** that is, one in which the nucleophile adds across the double bond of the carbonyl group, or they may react by a

conjugate addition. These two processes resemble the 1,2- and the 1,4-addition reactions of conjugated dienes (Section 13.10):

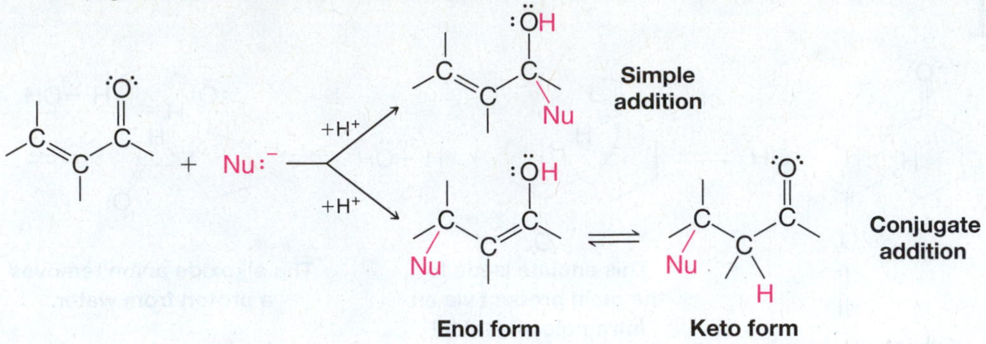

In many instances both modes of addition occur in the same mixture. As an example, let us consider the Grignard reaction shown here:

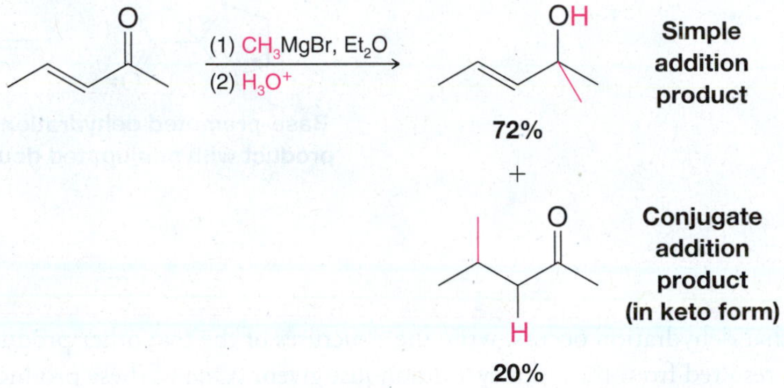

In this example we see that simple addition is favored, and this is generally the case with strong nucleophiles. Conjugate addition is favored when weaker nucleophiles are employed.

If we examine the resonance structures that contribute to the overall hybrid for an α,β-unsaturated aldehyde or ketone (see structures **A–C**), we shall be in a better position to understand these reactions:

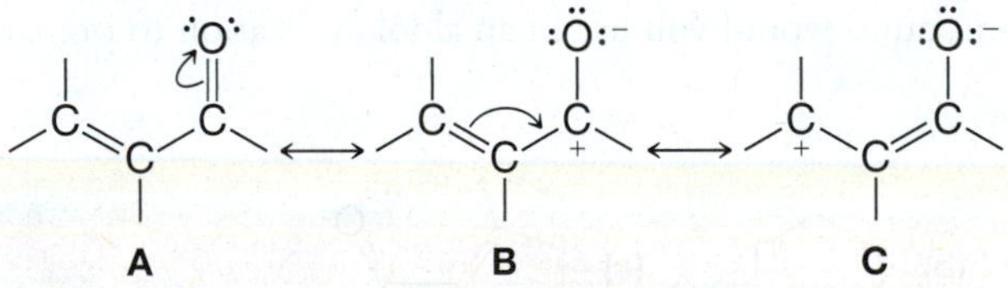

Although structures **B** and **C** involve separated charges, they make a significant contribution to the hybrid because, in each, the negative charge is carried by electronegative oxygen. Structures **B** and **C** also indicate that *both the carbonyl carbon and the β carbon should bear a partial positive charge*. They indicate that we should represent the hybrid in the following way:

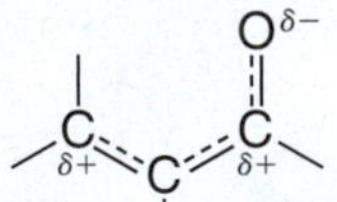

This structure tells us that we should expect a nucleophilic reagent to attack either the carbonyl carbon or the β carbon.

Almost every nucleophilic reagent that adds at the carbonyl carbon of a simple aldehyde or ketone is capable of adding at the β carbon of an α,β-unsaturated carbonyl compound. In many instances when weaker nucleophiles are used, conjugate addition is the major reaction path. Consider the following addition of hydrogen cyanide:

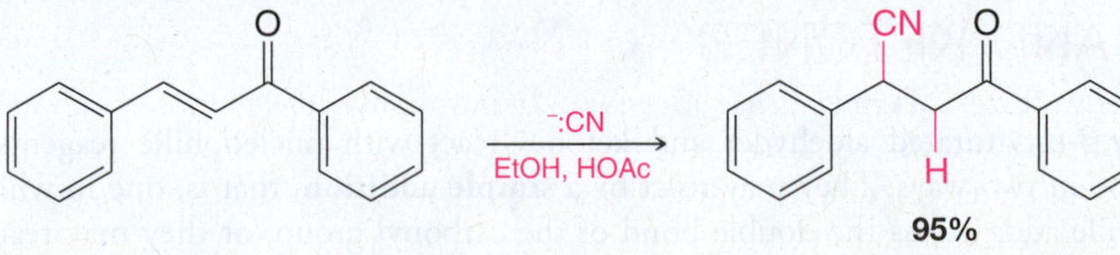

Helpful Hint

Note the influence of nucleophile strength on conjugate versus simple addition.

[A MECHANISM FOR THE REACTION ─ The Conjugate Addition of HCN]

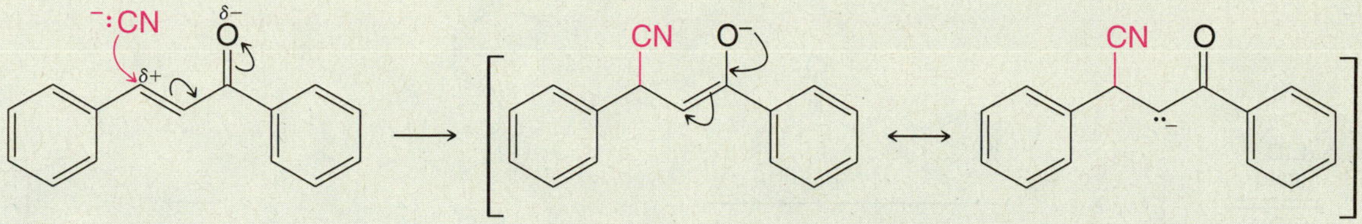

Enolate intermediate

Cyanide attacks at the partially positive β carbon. Then, the enolate intermediate accepts a proton in either of two ways:

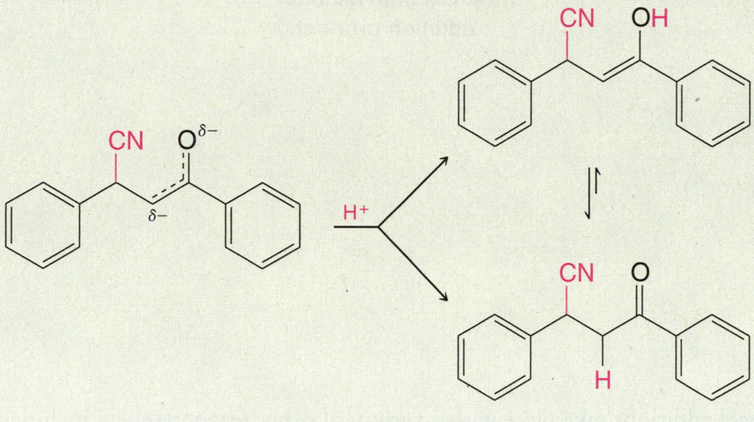

Enol form

Keto form

Another example of this type of addition is the following:

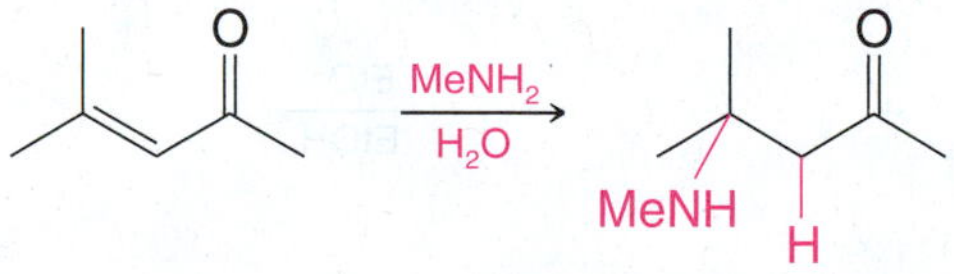

75%

[A MECHANISM FOR THE REACTION ─ The Conjugate Addition of an Amine]

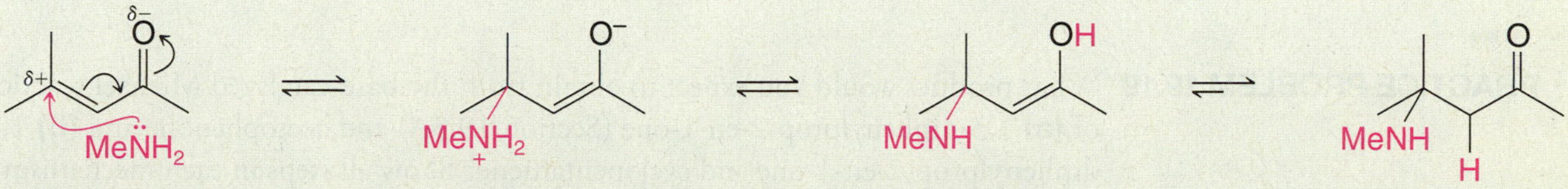

| **The nucleophile attacks the partially positive β carbon.** | **In two separate steps, a proton is lost from the nitrogen atom and a proton is gained at the oxygen.** | **Enol form** | **Keto form** |

We shall see examples of biochemically relevant conjugate additions in "The Chemistry of...Conjugate Additions to Activate Drugs" (see Section 19.7B) and in "The Chemistry of...A Suicide Enzyme Substrate" (Section 19.8).

19.7A Conjugate Additions of Enolates: Michael Additions

Conjugate additions of enolates to α,β-unsaturated carbonyl compounds are known generally as **Michael additions** (after their discovery, in 1887, by Arthur Michael, of Tufts University and later of Harvard University). The following mechanism box provides an example of a Michael addition.

[A MECHANISM FOR THE REACTION — The Michael Addition]

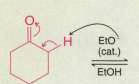

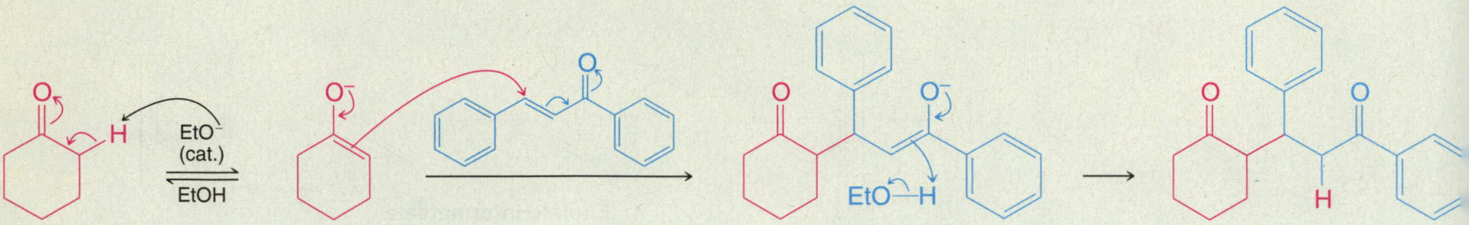

| **A base removes an α proton to form an enolate from one carbonyl reactant.** | **This enolate adds to the β carbon of the α,β-unsaturated carbonyl compound, forming a new carbon–carbon bond between them. As this bond is formed, electron density in the α,β-unsaturated compound shifts to its carbonyl oxygen, leading to a new enolate.** | **Protonation of the resulting enolate leads to the final Michael addition product.** |

Michael additions take place with a variety of other reagents; these include acetylenic esters and α,β-unsaturated nitriles:

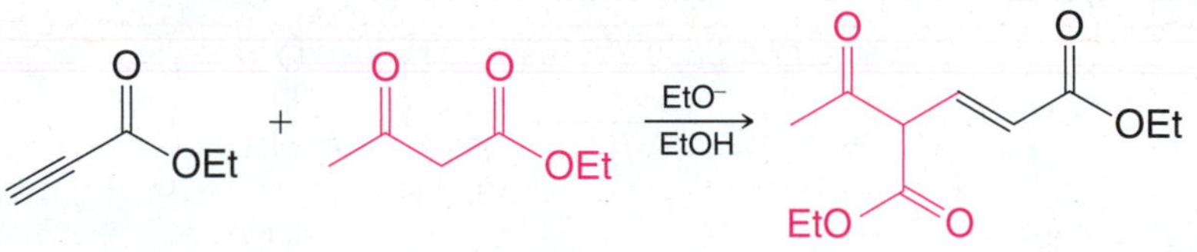

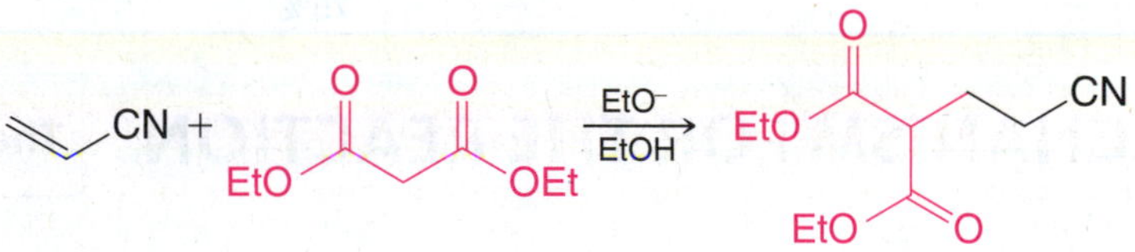

●●●

PRACTICE PROBLEM 19.19 What product would you expect to obtain from the base-catalyzed Michael reaction of **(a)** 1,3-diphenylprop-2-en-1-one (Section 19.5A) and acetophenone and **(b)** 1,3-diphenylprop-2-en-1-one and cyclopentadiene? Show all steps in each mechanism.

●●●

PRACTICE PROBLEM 19.20 When acrolein (propenal) reacts with hydrazine, the product is a dihydropyrazole:

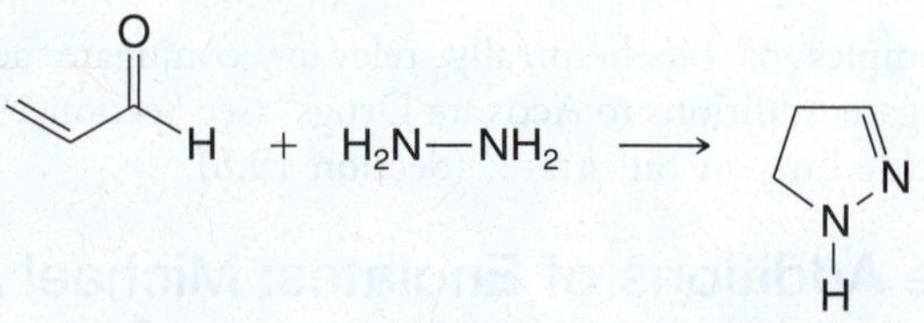

Acrolein   Hydrazine   A dihydropyrazole

Suggest a mechanism that explains this reaction.

Enamines can also be used in Michael additions. An example is the following:

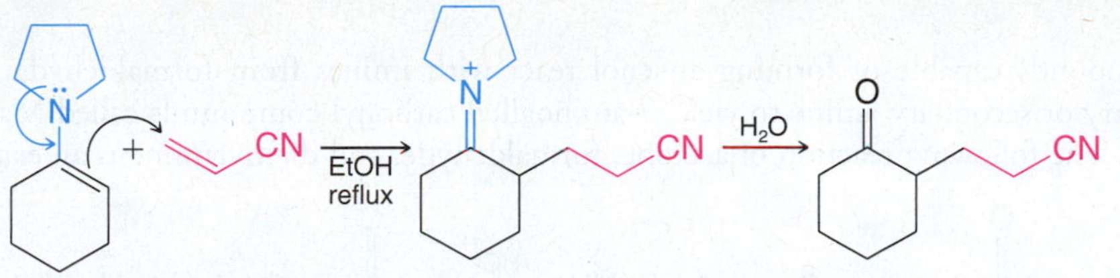

19.7B The Robinson Annulation

A Michael addition followed by a simple aldol condensation may be used to build one ring onto another. This procedure is known as the *Robinson annulation* (ring-forming) reaction, after the English chemist, Sir Robert Robinson, who won the Nobel Prize in Chemistry in 1947 for his research on naturally occurring compounds:

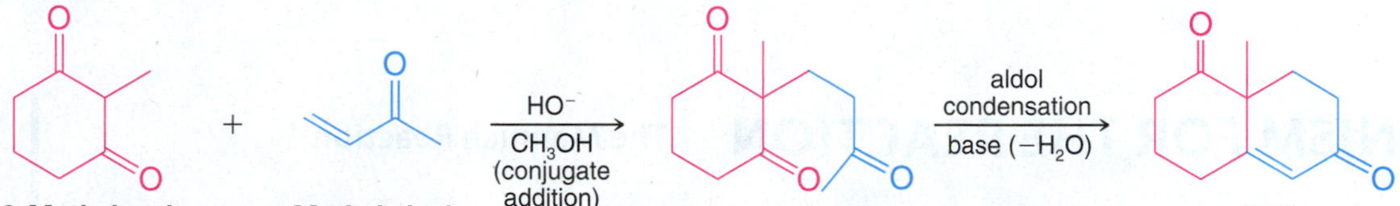

2-Methylcyclo-hexane-1,3-dione   **Methyl vinyl ketone**   **65%**

(a) Propose step-by-step mechanisms for both transformations of the Robinson annulation sequence just shown. **(b)** Would you expect 2-methylcyclohexane-1,3-dione to be more or less acidic than cyclohexanone? Explain your answer.

PRACTICE PROBLEM 19.21

THE CHEMISTRY OF... Conjugate Additions to Activate Drugs

At the end of Chapter 10, we considered the special reactivity of an antitumor antibiotic known as calicheamicin γ_1^I. There, we focused on how a chemical reaction transformed a stable enediyne into one capable of undergoing a Bergman cycloaromatization. Now that we have covered conjugate additions in Section 19.7, you can understand the reaction that started the process. It turns out that there are many situations where a conjugate, or Michael, addition can set a critical process in motion. Here we briefly present the story of the mitomycins, molecules from nature known to possess antitumor properties.

If any one of the three natural products denoted below (isomitomycin A, albomitomycin A, or mitomycin A) is simply dissolved in an alcohol solvent like methanol, it will rearrange into an equilibrium mixture that contains the other two materials; the favored compound is mitomycin A. The process for that equilibration is a series of Michael reactions and retro-Michael reactions as shown. All are potent compounds, but it is their ability to rearrange through such chemistry that is equally remarkable!

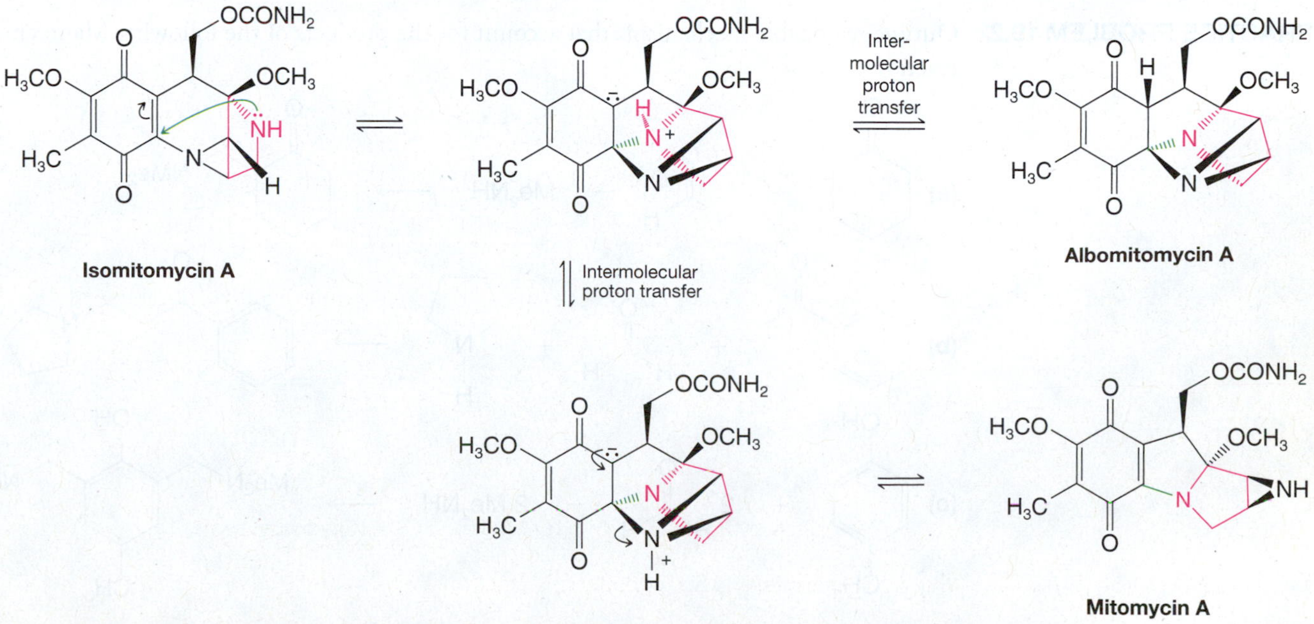

Isomitomycin A

Inter-molecular proton transfer

Albomitomycin A

Intermolecular proton transfer

Mitomycin A

19.8 THE MANNICH REACTION

Compounds capable of forming an enol react with imines from formaldehyde and a primary or secondary amine to yield β-aminoalkyl carbonyl compounds called Mannich bases. The following reaction of acetone, formaldehyde, and diethylamine is an example:

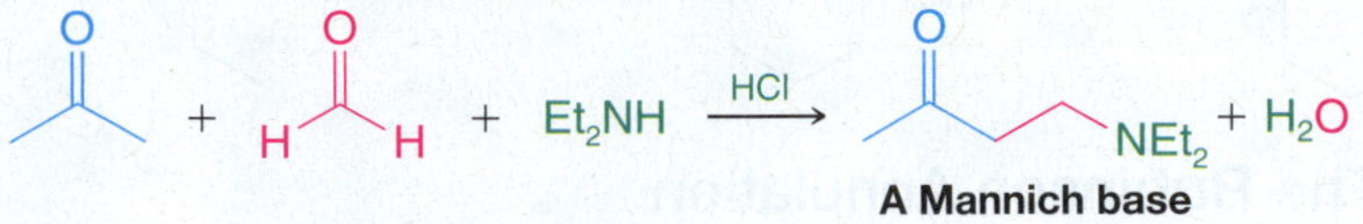

A Mannich base

The **Mannich reaction** apparently proceeds through a variety of mechanisms depending on the reactants and the conditions that are employed. The mechanism below appears to operate in neutral or acidic media. Note the aspects in common with imine formation and with reactions of enols and carbonyl groups.

[A MECHANISM FOR THE REACTION — The Mannich Reaction]

Step 1

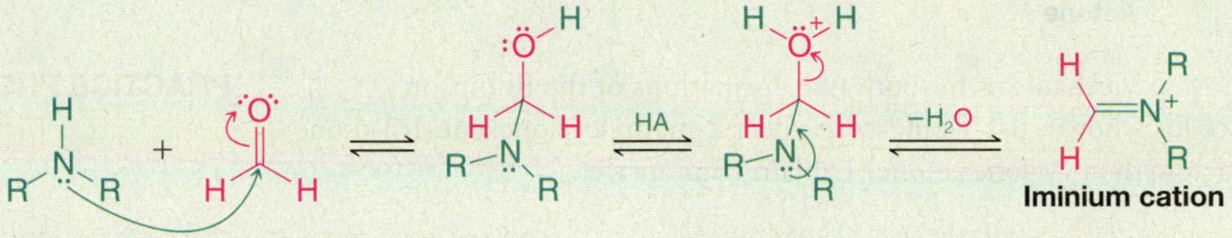

Reaction of the secondary amine with the aldehyde forms a hemiaminal.

The hemiaminal loses a molecule of water to form an iminium cation.

Step 2

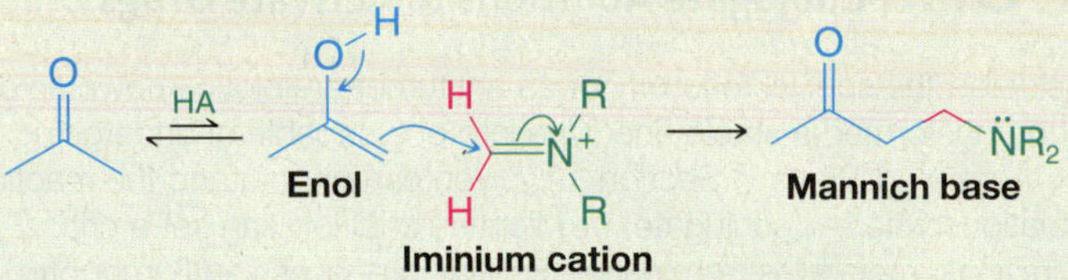

The enol form of the active hydrogen compound reacts with the iminium cation to form a β-aminocarbonyl compound (a Mannich base).

PRACTICE PROBLEM 19.22 Outline reasonable mechanisms that account for the products of the following Mannich reactions:

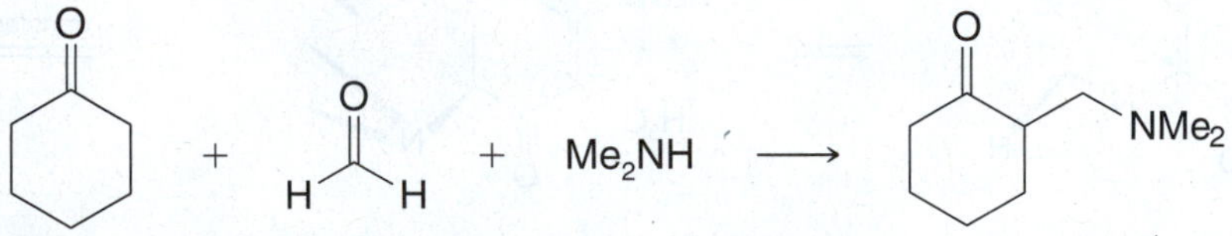

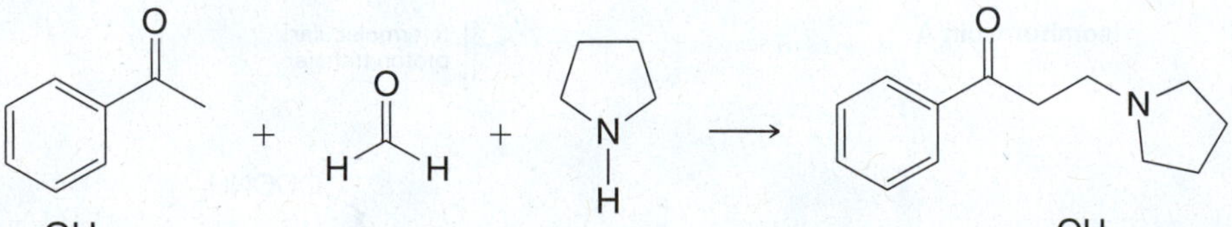

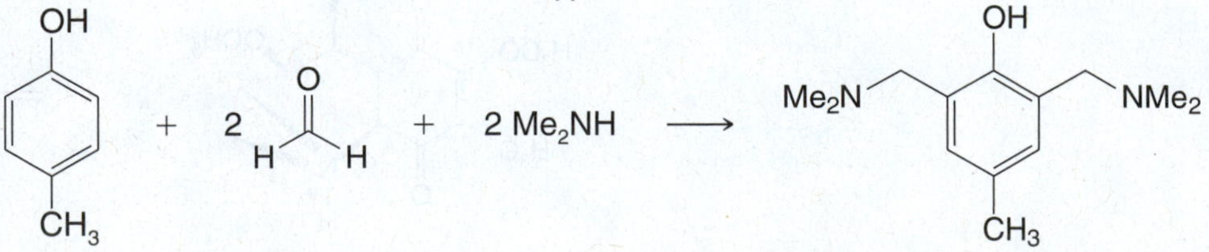

THE CHEMISTRY OF... A Suicide Enzyme Substrate

5-Fluorouracil is a chemical imposter for uracil and a potent clinical anticancer drug. This effect arises because 5-fluorouracil irreversibly destroys the ability of thymidylate synthase (an enzyme) to catalyze a key transformation needed for DNA synthesis. 5-Fluorouracil acts as a mechanism-based inhibitor (or suicide substrate) because it engages thymidylate synthase as though it were the normal substrate but then leads to self-destruction of the enzyme's activity by its own mechanistic pathway. The initial deception is possible because the fluorine atom in the inhibitor occupies roughly the same amount of space as the hydrogen atom does in the natural substrate. Disruption of the enzyme's mechanism occurs because a fluorine atom cannot be removed by a base in the way that is possible for a hydrogen atom to be removed.

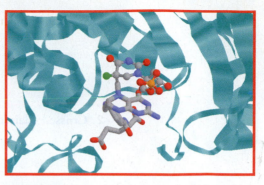

5-Fluorodeoxyuracil monophosphate covalently bound to tetrahydrofolate in thymidylate synthase, blocking the enzyme's catalytic activity.

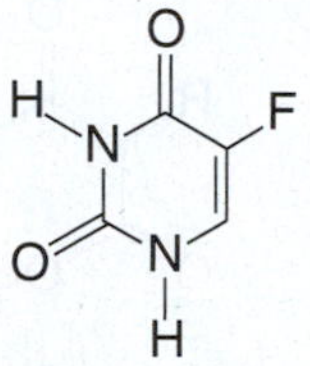

5-Fluorouracil

The mechanism of thymidylate synthase in both its normal mode and when it is about to be blocked by the inhibitor involves attack of an enolate ion on an iminium cation. This process is closely analogous to the Mannich reaction discussed in Section 19.8. The enolate ion in this attack arises by conjugate addition of a thiol group from thymidylate synthase to the α,β-unsaturated carbonyl group of the substrate. This process is analogous to the way an enolate intermediate occurs in a Michael addition. The iminium ion that is attacked in this process derives from the coenzyme N^5,N^{10}-methylenetetrahydrofolate (N^5,N^{10}-methylene-THF). Attack by the enolate in this step forms the bond that covalently links the substrate to the enzyme. It is this bond that cannot be broken when the fluorinated inhibitor is used. The mechanism of inhibition is shown at right.

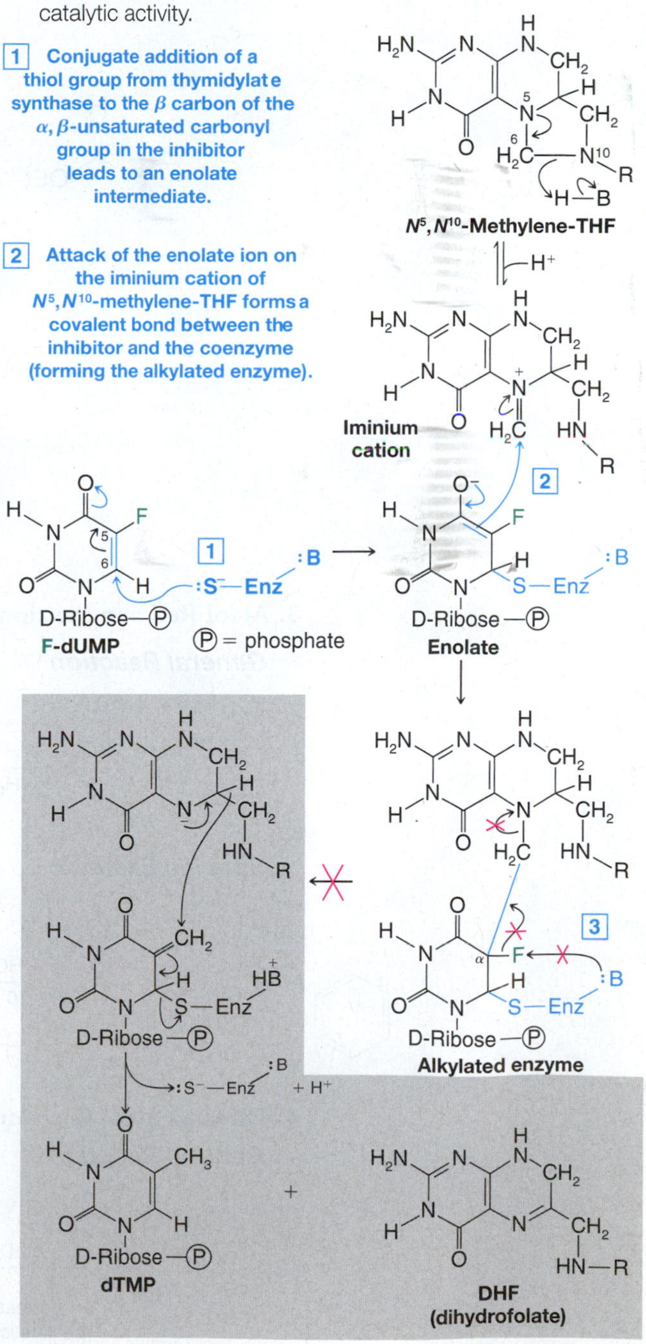

1 Conjugate addition of a thiol group from thymidylate synthase to the β carbon of the α,β-unsaturated carbonyl group in the inhibitor leads to an enolate intermediate.

2 Attack of the enolate ion on the iminium cation of N^5,N^{10}-methylene-THF forms a covalent bond between the inhibitor and the coenzyme (forming the alkylated enzyme).

N^5,N^{10}-**Methylene-THF**

Iminium cation

F-dUMP $\textcircled{P}$ = phosphate

Enolate

3 The next step in the normal mechanism would be an elimination reaction involving loss of a proton at the carbon α to the substrate's carbonyl group, releasing the tetrahydrofolate coenzyme as a leaving group. In the case of the fluorinated inhibitor, this step is not possible because a fluorine atom takes the place of the hydrogen atom needed for removal in the elimination. The enzyme cannot undergo the elimination reaction necessary to free it from the tetrahydrofolate coenzyme. These blocked steps are marked by cross-outs. Neither can the subsequent hydride transfer occur from the coenzyme to the substrate, which would complete formation of the methyl group and allow release of the product from the enzyme thiol group. These blocked steps are shown in the shaded area. The enzyme's activity is destroyed because it is irreversibly bonded to the inhibitor.

Alkylated enzyme

dTMP

DHF (dihydrofolate)

19.9 SUMMARY OF IMPORTANT REACTIONS

1. Claisen Condensation (Section 19.2):

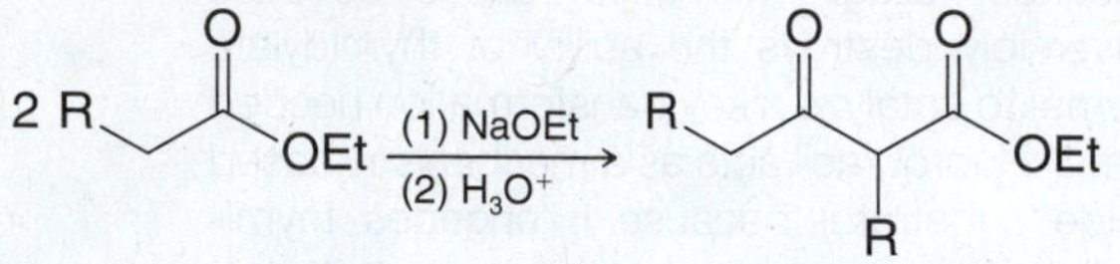

2. Crossed Claisen Condensation (Section 19.2B):

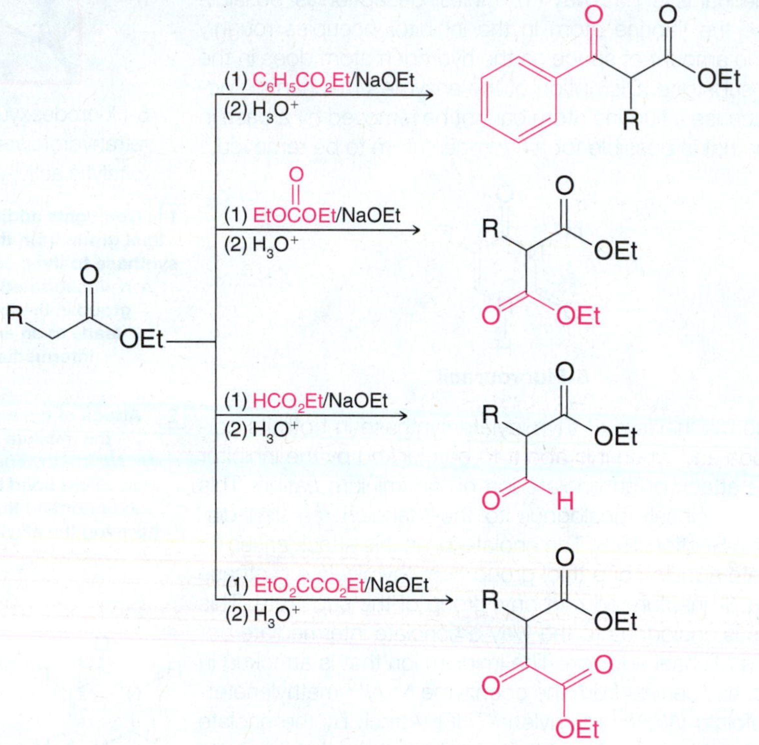

3. Aldol Reaction (Section 19.4)

General Reaction

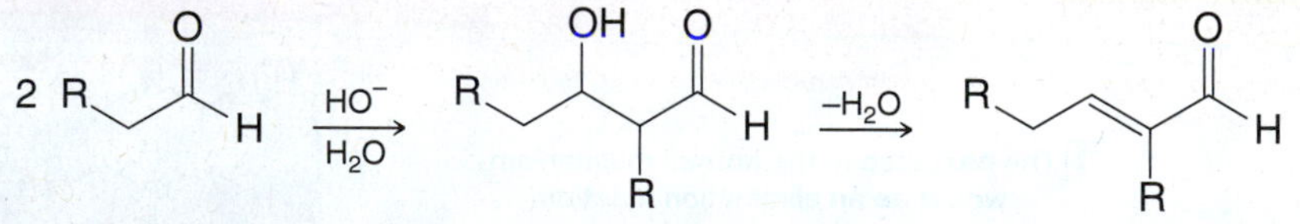

Specific Example

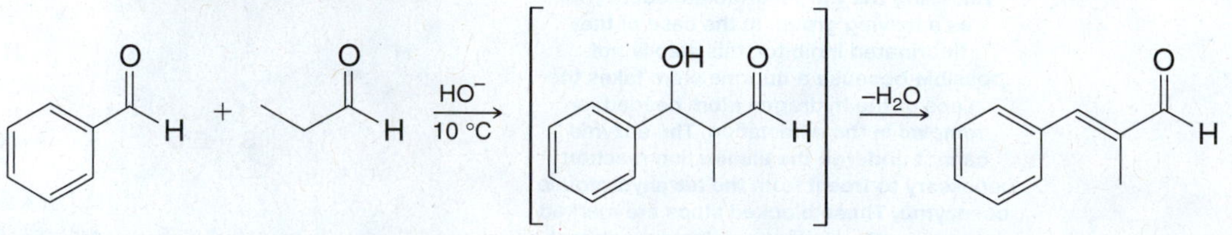

4. Directed Aldol Reactions via Lithium Enolates (Section 19.5B)

General Reaction

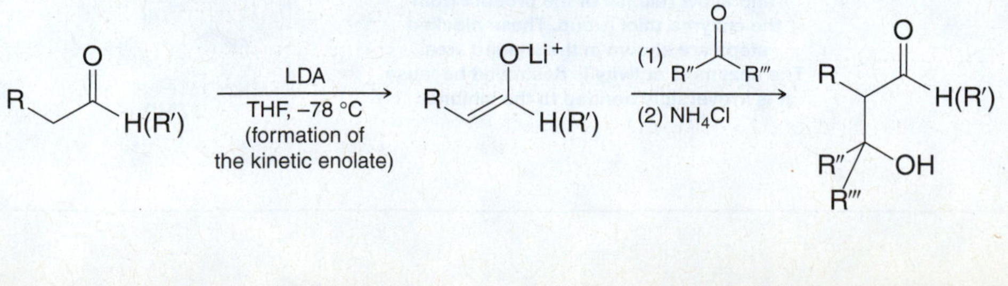

Specific Example

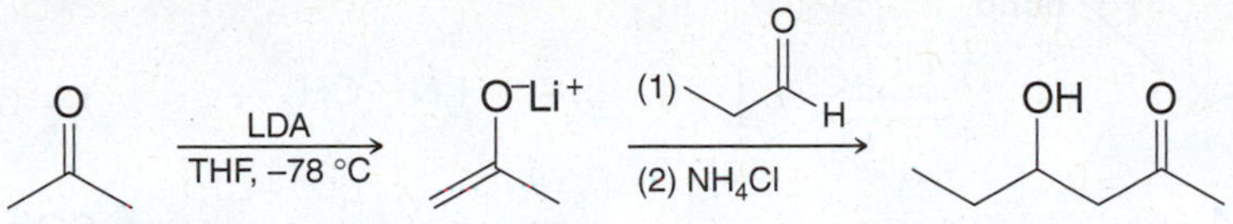

5. Conjugate Addition (Section 19.7)

General Example

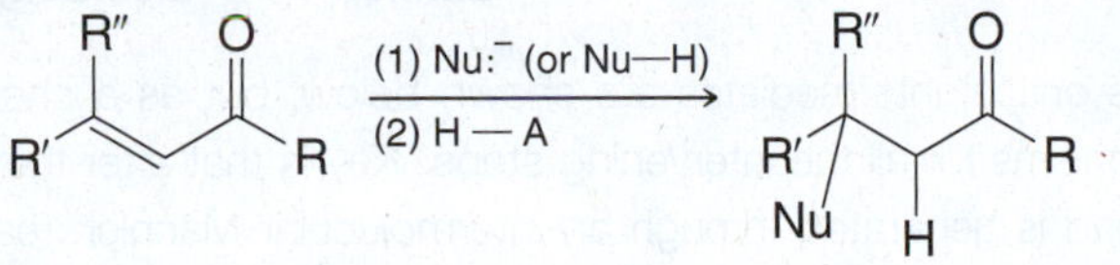

Nu:⁻ = ⁻CN ; an enolate (Michael addition); R‴MgBr
Nu—H = 1° or 2° amines; an enamine

Specific Example

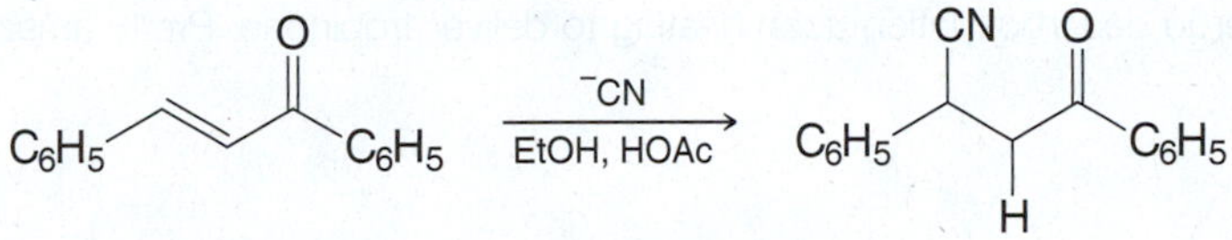

Specific Example (Michael Addition)

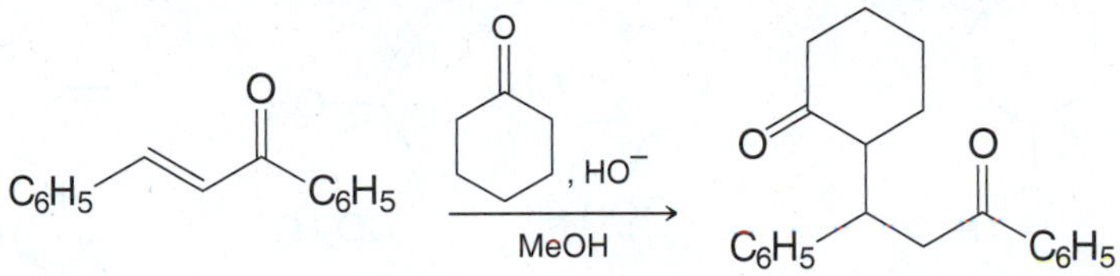

6. Mannich Reaction (Section 19.8):

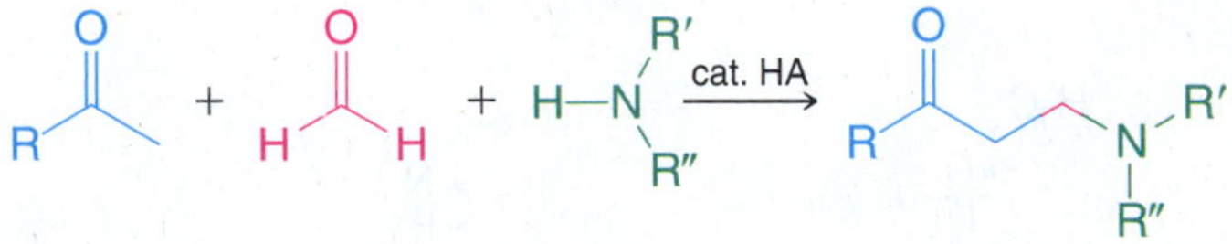

[WHY Do These Topics Matter?

PUTTING MULTIPLE REACTIONS TOGETHER IN ONE POT

Over the course of the past several chapters, you have had the chance to learn about several powerful tools in C—C bond construction using carbonyls and their derivatives as both nucleophiles and electrophiles. While these reactions are clearly powerful in their own right, when they are combined in series, they can deliver incredibly complex molecules all at once. Such processes are known as cascade, or domino, sequences in that each step sets the stage for the next event, all in the same reaction flask. Here we illustrate what is perhaps the earliest example of this concept as accomplished by Sir Robert Robinson (a future chemistry Nobel Laureate) during the middle of World War I (1917). His target was a natural product known as tropinone. This compound constitutes the core of a number of other bioactive substances, including cocaine and atropine. At that point in the war, atropine was desperately needed by soldiers at the front to combat poisoning from organophosphate nerve agents.

How could this complex bicyclic compound be synthesized efficiently? Using the positioning of the nitrogen atom relative to the ketone, Robinson believed that the entire molecule could potentially arise from a dialdehyde, methylamine, and acetone dicarboxylic acid in a single, one-pot transformation, as color-coded below. The key reactions in the actual union would be a series of carefully orchestrated iminium ion formations and Mannich reactions to make the new C—C bonds (colored in green), followed by carboxylic acid decarboxylations to complete the target.

(continues on next page)

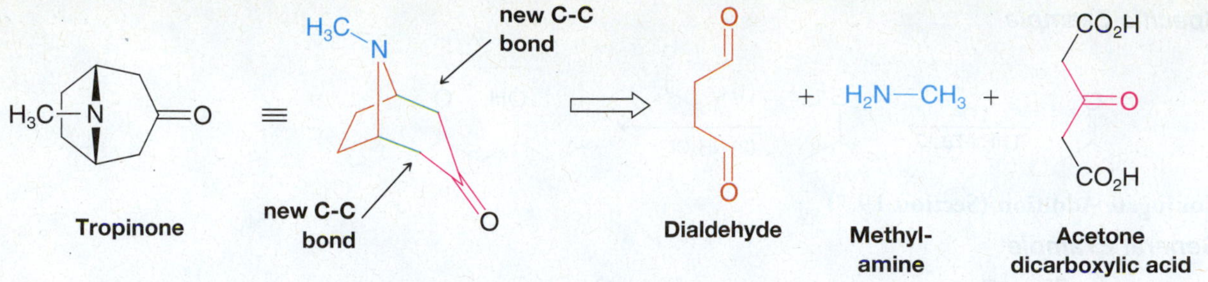

As shown below, that idea actually worked! Only the critical intermediates are shown below, but as a check of what you have learned so far you should be able to write the mechanisms for all the intervening steps. Key is that after the five-membered nitrogen-containing ring is formed, the first new C—C bond is generated through an intermolecular Mannich reaction. Because the reaction conditions are acidic, it is an enol tautomer that serves as the key nucleophile in this event; the two carboxylic acids attached to the acetone core of this piece aid in the ease of that tautomerization. Following acid-induced expulsion of the alcohol within the resultant aminal, a new iminium ion is generated. Once formed, an intramolecular Mannich reaction can then form the second C—C bond needed to complete the entire core of the target. Finally, the two carboxylic acids positioned strategically in a 1,3-fashion relative to the central ketone undergo decarboxylation upon heating to deliver tropinone. Pretty amazing what these reactions in series can accomplish!

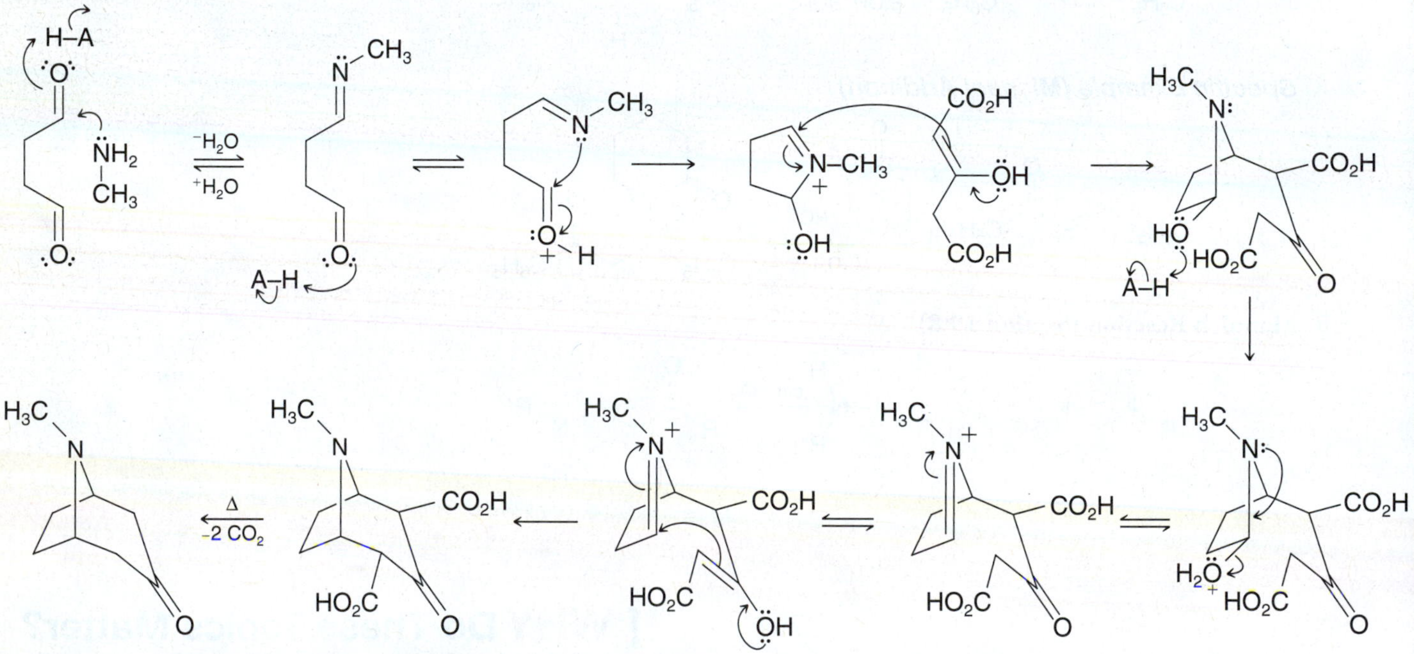

To learn more about these topics, see:
1. Nicolaou, K. C.; Montagnon, T. *Molecules that Changed the World*. Wiley-VCH: Weinheim, **2008**, p. 366.
2. Nicolaou, K. C.; Vourloumis, D.; Winssinger, N.; Baran, P.S. "The Art and Science of Total Synthesis at the Dawn of the Twenty-First Century" in *Angew. Chem. Int. Ed.* **2000**, *39*, 44–122.

SUMMARY AND REVIEW TOOLS

The study aids for this chapter include key terms and concepts (which are highlighted in bold, blue text within the chapter and defined in the Glossary (at the back of the book) and have hyperlinked definitions in the accompanying *WileyPLUS* course (www.wileyplus.com), the list of reaction types in Section 19.9, and the Summary of Mechanisms scheme for Enolate Reactions with Carbonyl Electrophiles and Synthetic Connections Involving Enolates.

PROBLEMS

Note to Instructors: Many of the homework problems are available for assignment via *WileyPLUS*, an online teaching and learning solution.

CLAISEN CONDENSATION REACTIONS

19.23 Write a structural formula for the product from each of the following reactions.

(a)

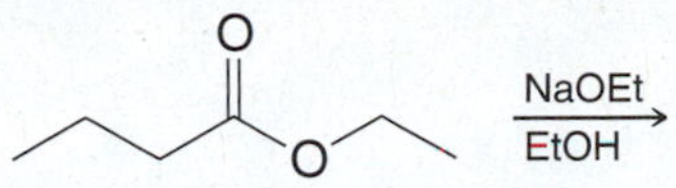
O
O
NaOEt
EtOH

(b)

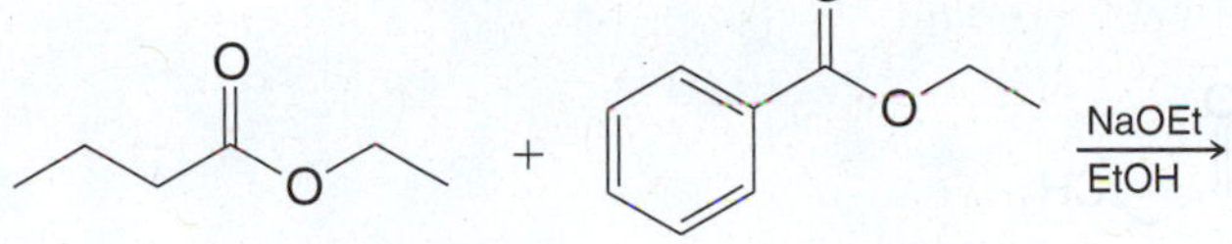
O
O
+
O
O
NaOEt
EtOH

(c)

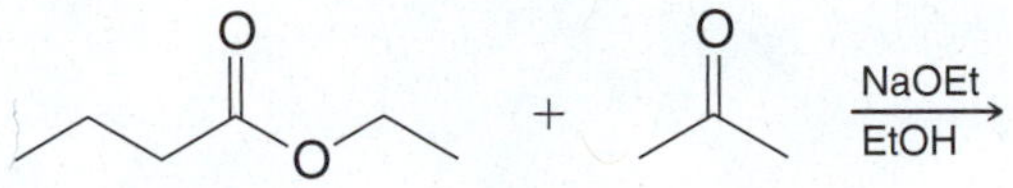
O
O
+
O
NaOEt
EtOH

(d)

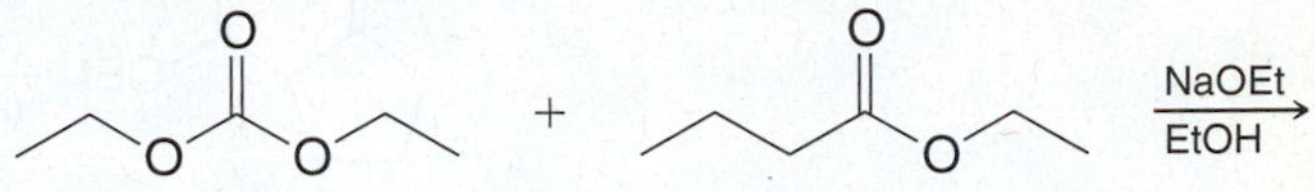
O
O
O
+
O
O
NaOEt
EtOH

(e)

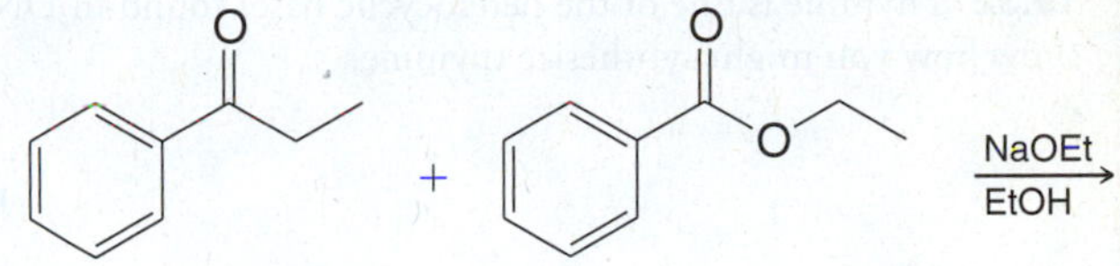
O
+
O
O
NaOEt
EtOH

(f)

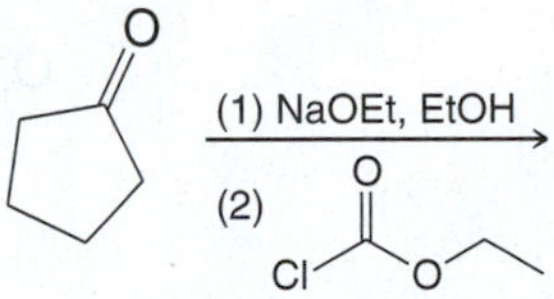
O
(1) NaOEt, EtOH
(2)
O
Cl
O

19.24 Show all steps in the following syntheses. You may use any other needed reagents but you should begin with the compound given.

(a)

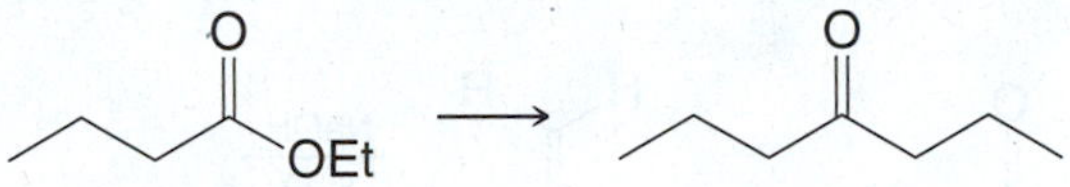
O
OEt
O

(b)

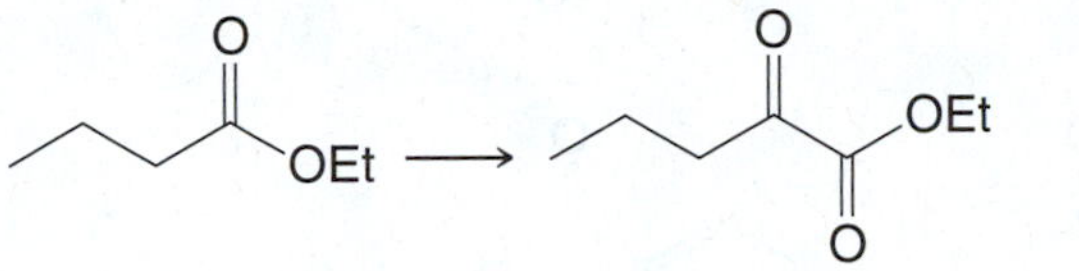
O
OEt
O
OEt
O

(c)

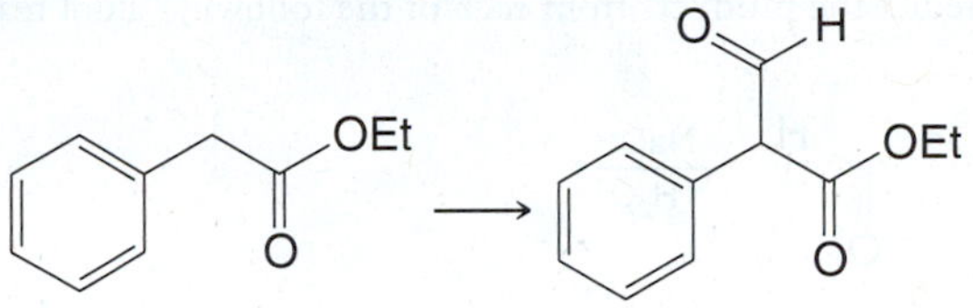
OEt
O
O
H
OEt
O

19.25 Provide the starting materials needed to synthesize each compound by acylation of an enolate.

(a)

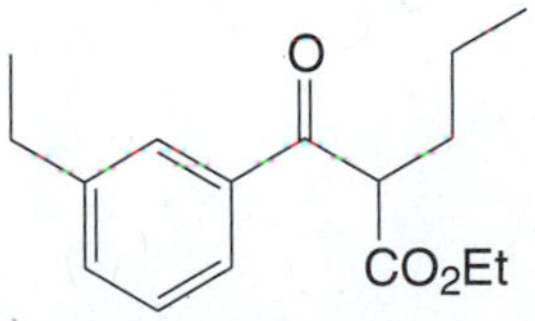
O
CO2Et

(b)

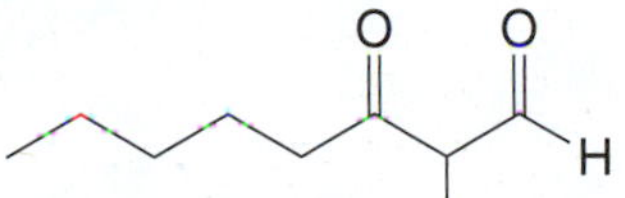
O
O
H

(c)

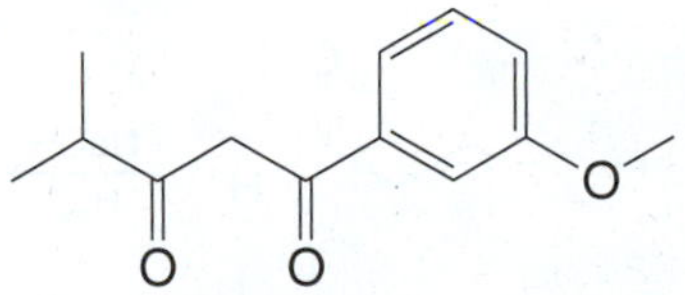
O
O
O

19.26 Write structural formulas for both of the possible products from the following Dieckmann condensation, and predict which one would likely predominate.

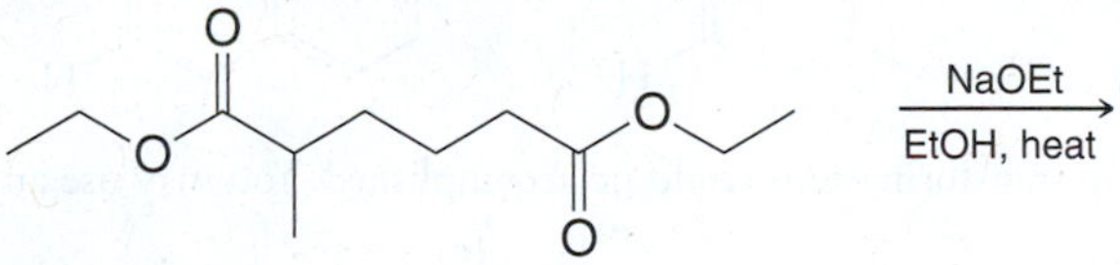
O
O
O
O
NaOEt
EtOH, heat

19.27 When a Dieckmann condensation is attempted with diethyl succinate, the product obtained has the molecular formula $C_{12}H_{16}O_6$. What is the structure of this compound?

19.28 Show how this diketone could be prepared by a condensation reaction:

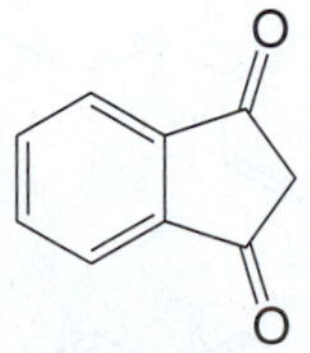
O
O

19.29 In contrast to the reaction with dilute alkali (Section 18.6), when concentrated solutions of NaOH are used, acetoacetic esters undergo cleavage as shown below.

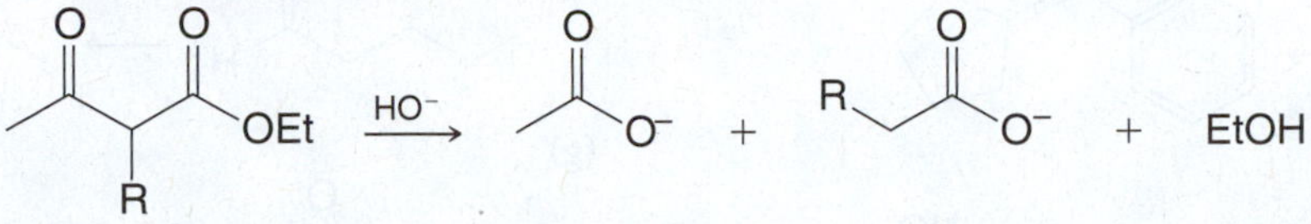
O
O
OEt
R
HO−
O
O−
+
R
O
O−
+
EtOH

Provide a mechanistic explanation for this outcome.

19.30 Write a detailed mechanism for the following reaction.

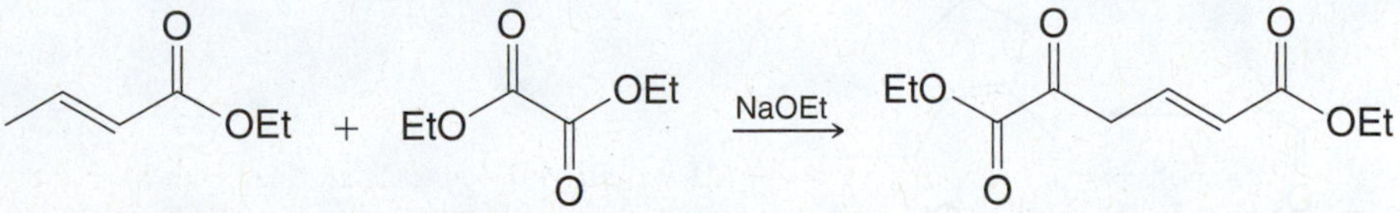
O
OEt
+
EtO
O
OEt
O
NaOEt
EtO
O
O
O
OEt

Ethyl crotonate **Diethyl oxalate**

19.31 In the presence of sodium ethoxide the following transformation occurs. Explain.

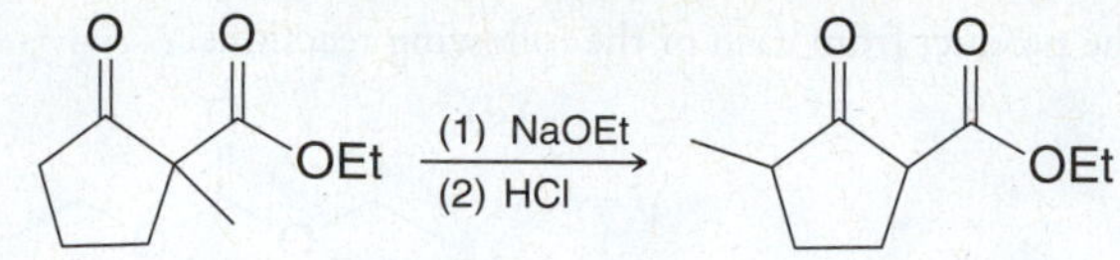

19.32 Thymine is one of the heterocyclic bases found in DNA. Starting with ethyl propanoate and using any other needed reagents, show how you might synthesize thymine.

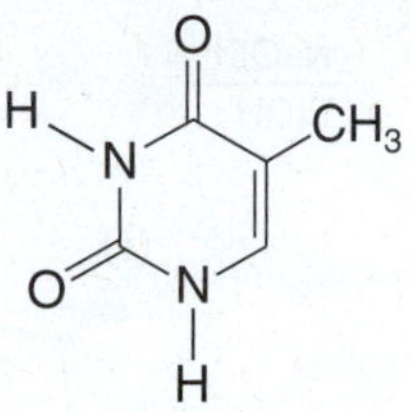

Thymine

ALDOL REACTIONS

19.33 Predict the products from each of the following aldol reactions.

(a)

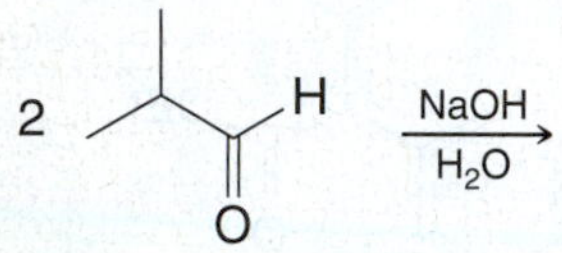

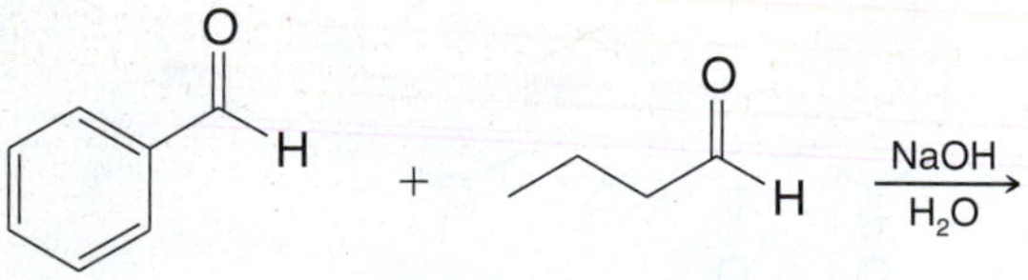

(d)

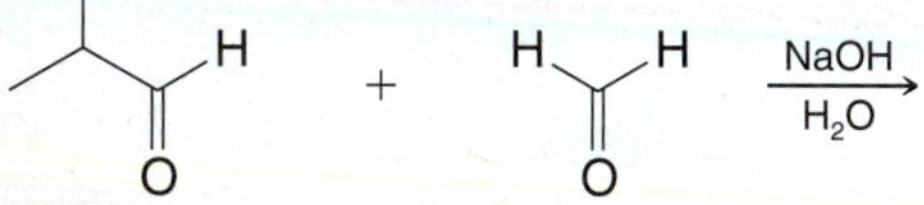

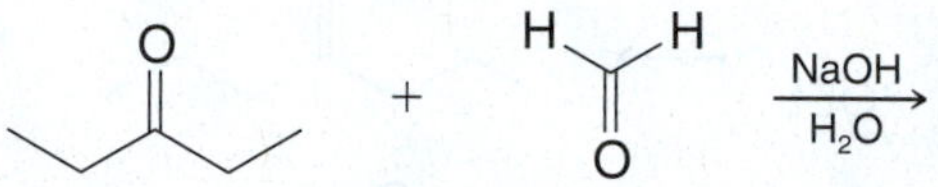

(b)

(e)

(c)

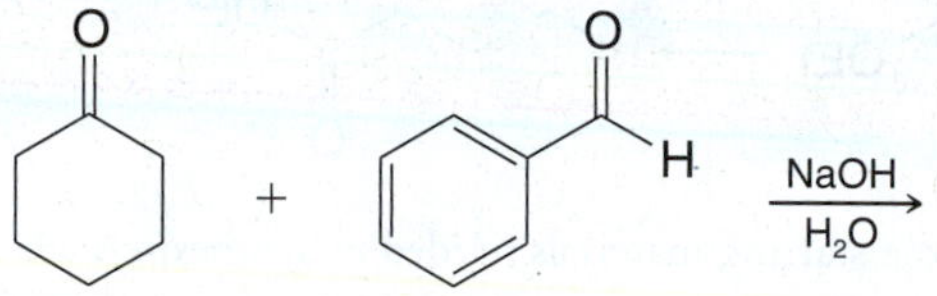

19.34 What four β-hydroxy aldehydes would be formed by a crossed aldol reaction between the following compounds?

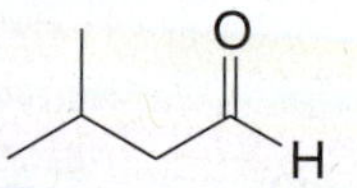

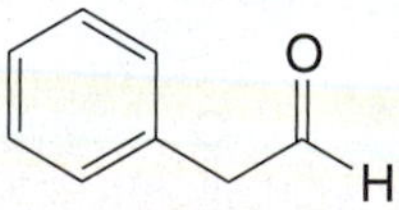

19.35 Show how each of the following transformations could be accomplished. You may use any other required reagents.

(a)

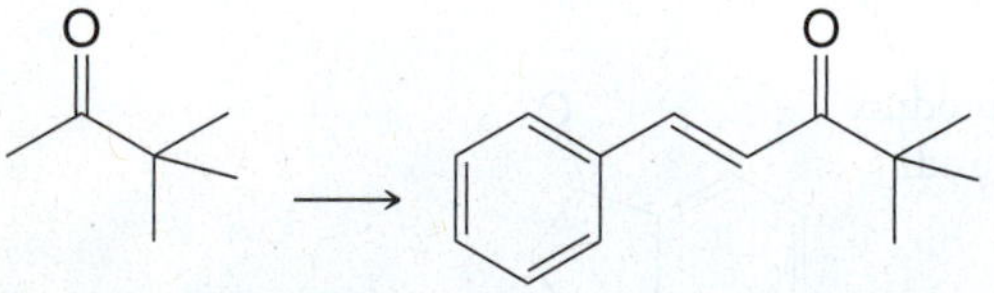

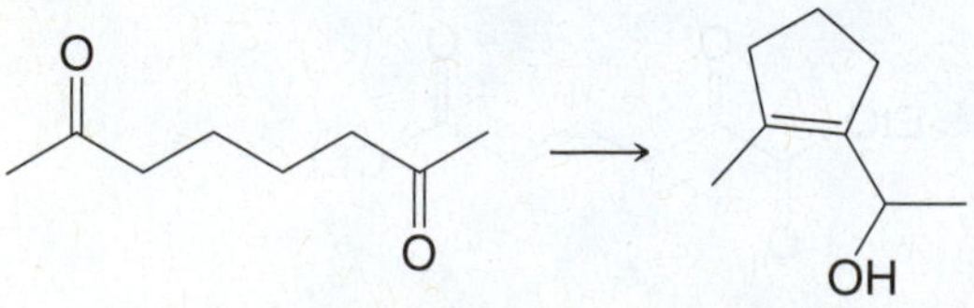

(e)

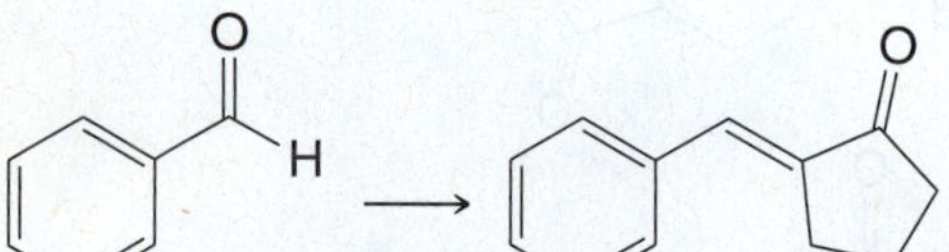

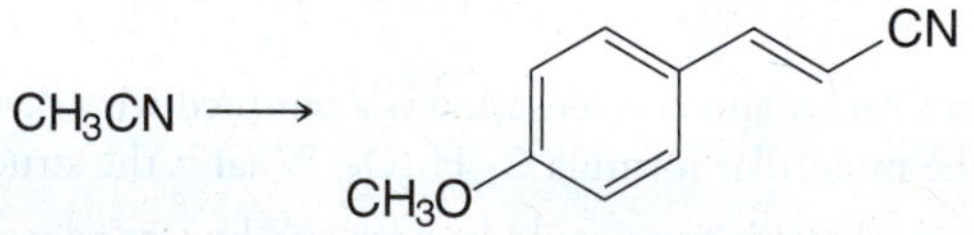

(b)

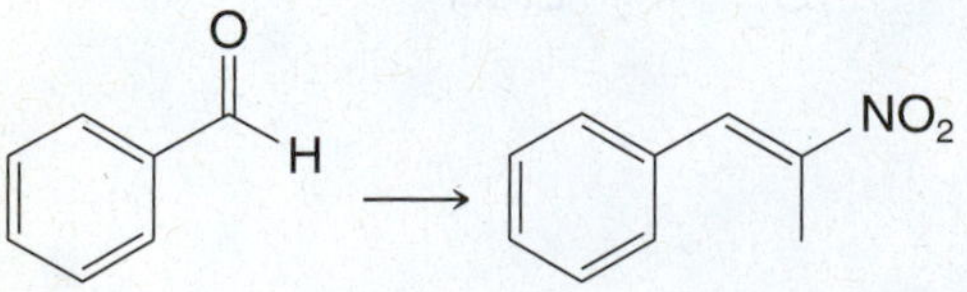

(f)

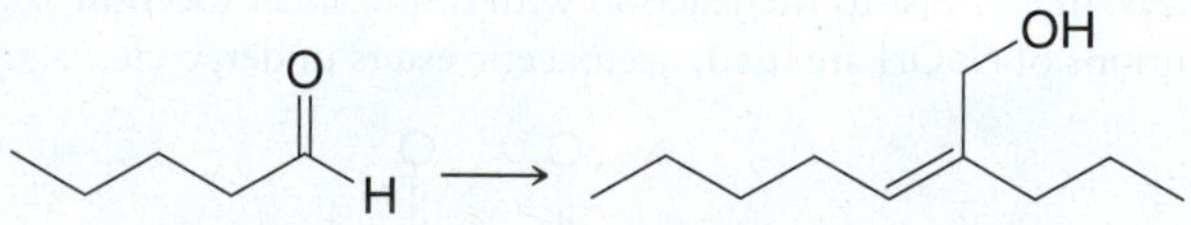

(c)

(g)

(d)

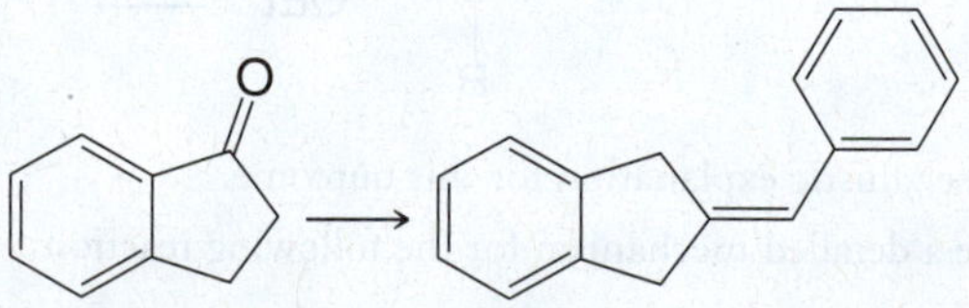

889

19.36 What starting materials are needed to synthesize each of the following compounds using an aldol reaction?

(a)

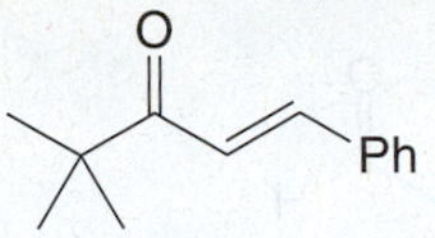

(c)

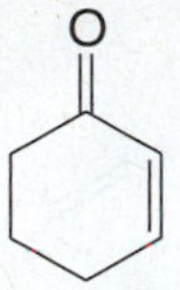

(e)

(g)

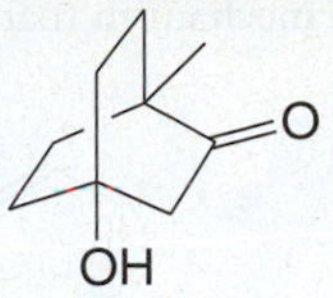

(b)

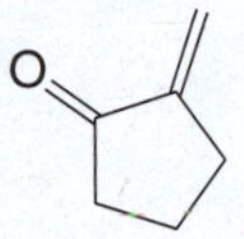

(d)

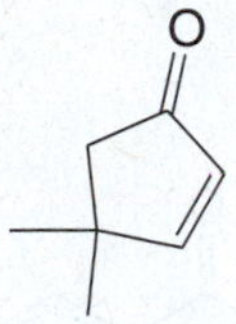

(f)

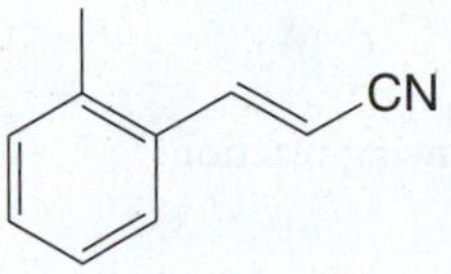

(h)

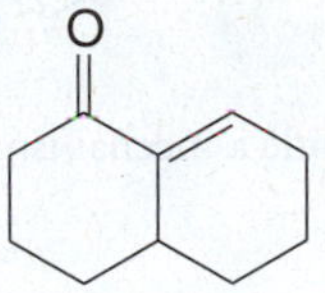

19.37 What reagents would you use to bring about each step of the following syntheses?

(a)

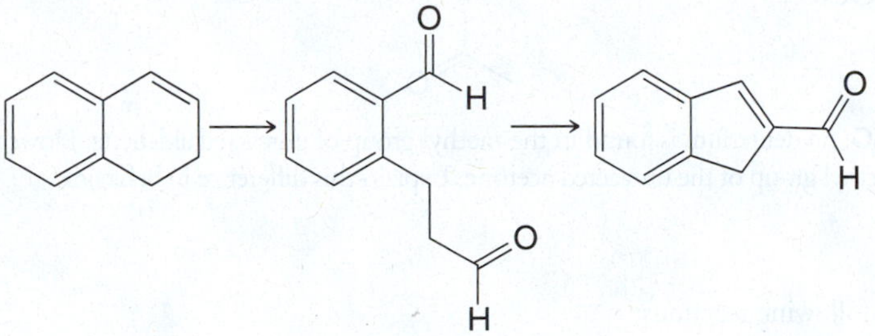

(b)

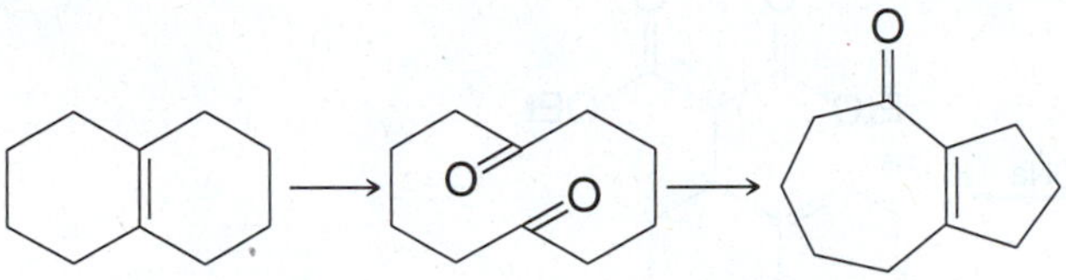

(c)

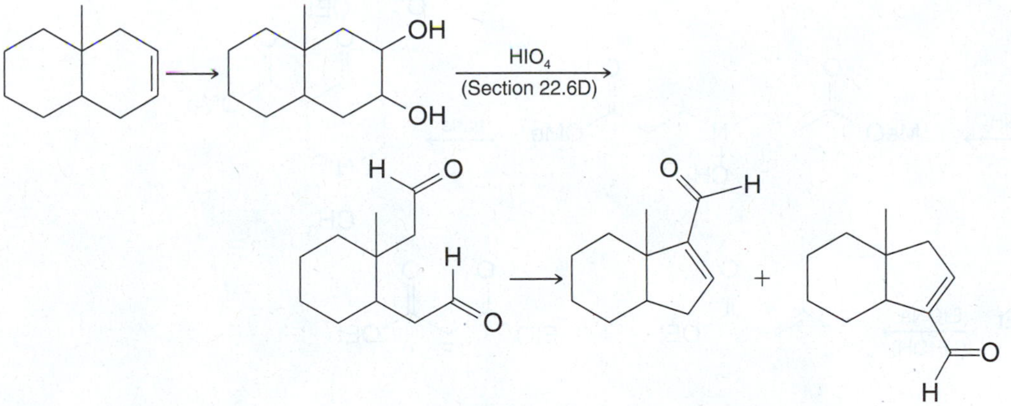

$$\xrightarrow{\text{HIO}_4}$$
(Section 22.6D)

(d)

19.38 The hydrogen atoms of the γ carbon of crotonaldehyde are appreciably acidic (p$K_a \cong 20$).

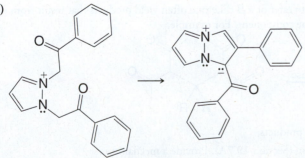

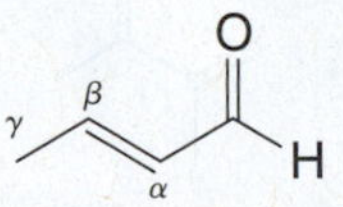

Crotonaldehyde

(a) Write resonance structures that will explain this fact.

(b) Write a mechanism that accounts for the following reaction:

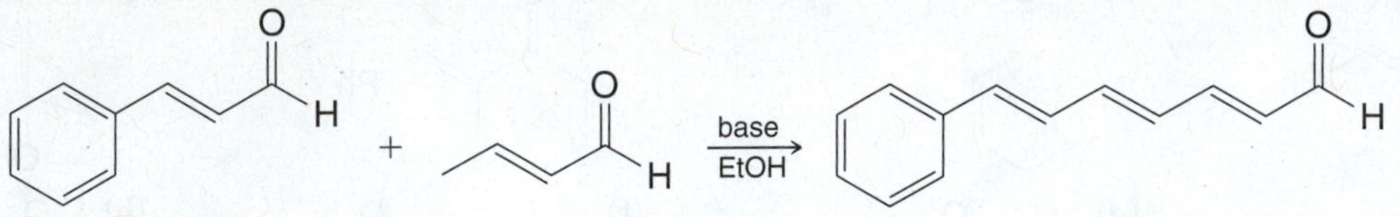

87%

19.39 Provide a mechanism for the following reaction.

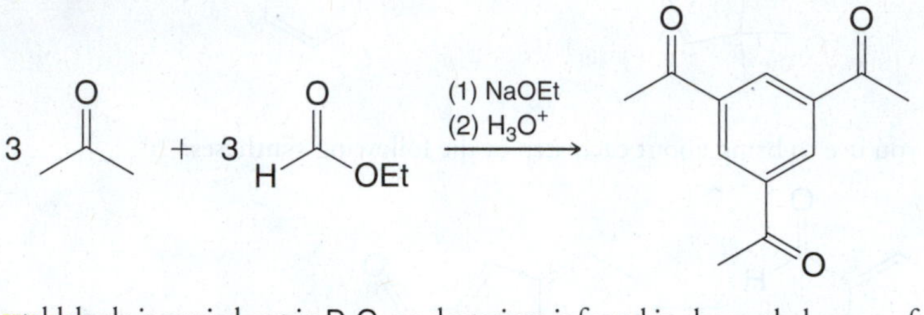

19.40 When the aldol reaction of acetaldehyde is carried out in D_2O, no deuterium is found in the methyl group of unreacted aldehyde. However, in the aldol reaction of acetone, deuterium is incorporated in the methyl group of the unreacted acetone. Explain this difference in behavior.

CONJUGATE ADDITION REACTIONS

19.41 Write mechanisms that account for the products of the following reactions:

(a)

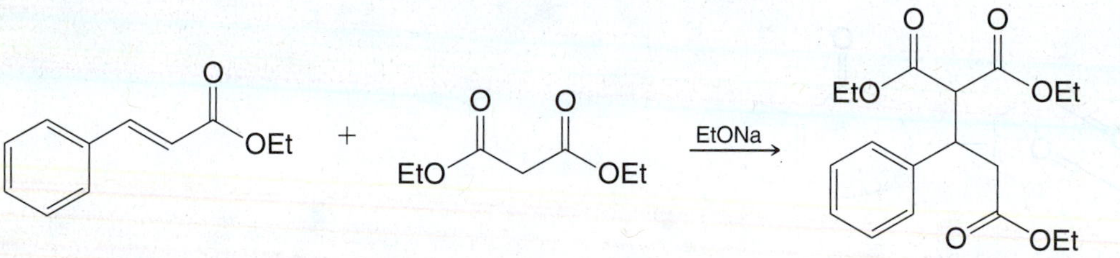

(b)

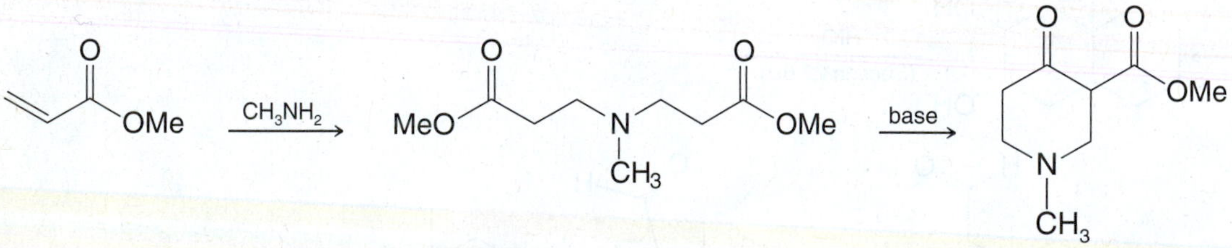

(c)

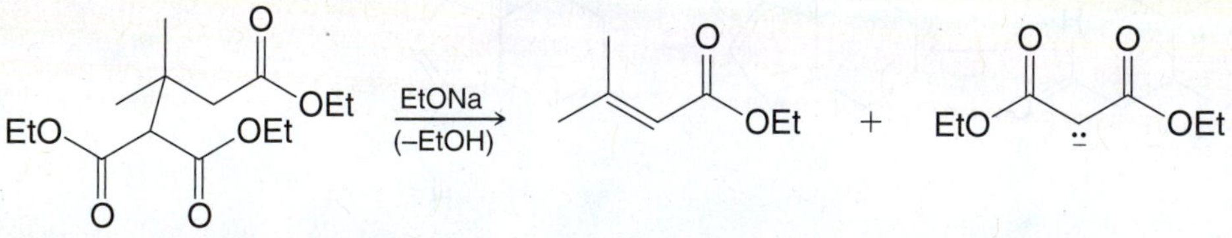

19.42 Condensations in which the active hydrogen compound is a β-keto ester or a β-diketone often yield products that result from one molecule of aldehyde or ketone and two molecules of the active methylene component. For example,

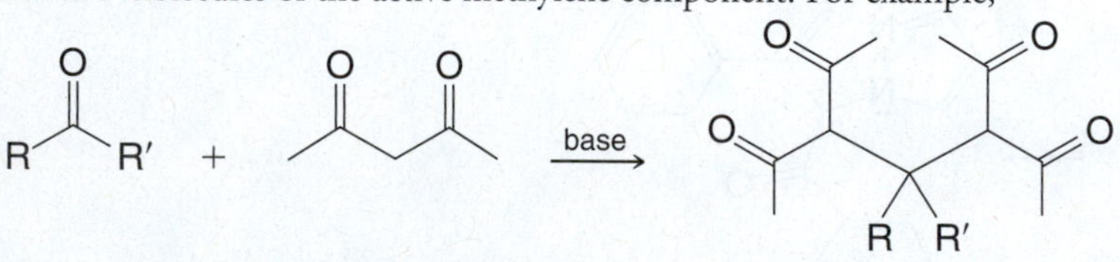

Suggest a reasonable mechanism that accounts for the formation of these products.

19.43 The following reaction illustrates the Robinson annulation reaction (Section 19.7A). Provide a mechanism.

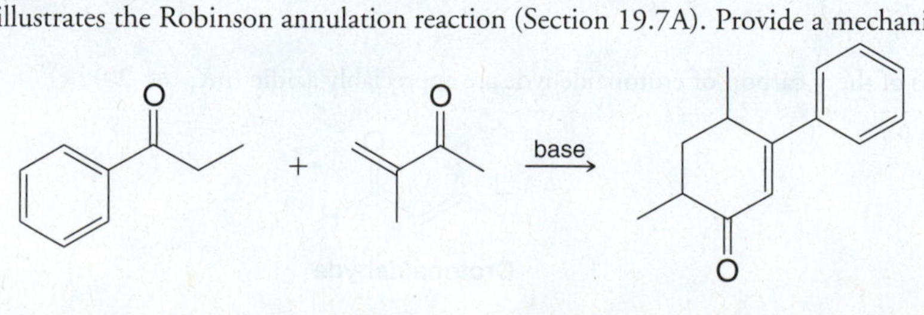

19.44 What is the structure of the *cyclic* compound that forms after the Michael addition of **1** to **2** in the presence of sodium ethoxide?

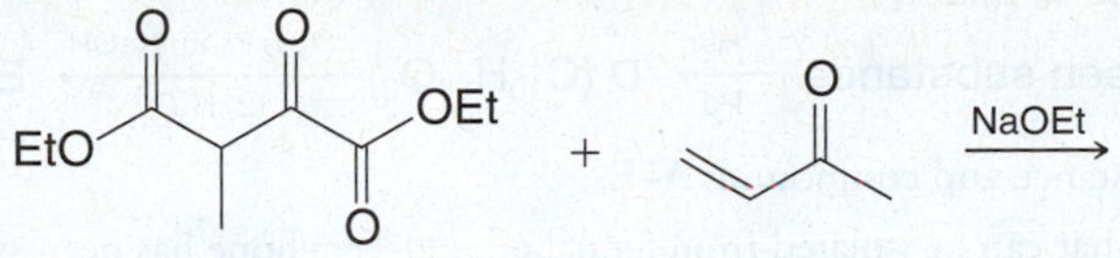

1   **2**

GENERAL PROBLEMS

19.45 Synthesize each compound starting from cyclopentanone.

(a)

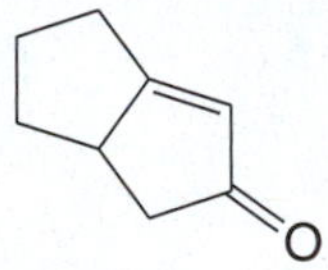

(b)

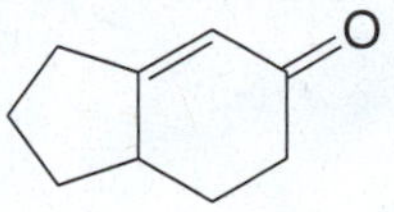

19.46 Provide a mechanism for the following reaction.

3 $\xrightarrow[\text{H}_2\text{O}]{\text{NaOH}}$

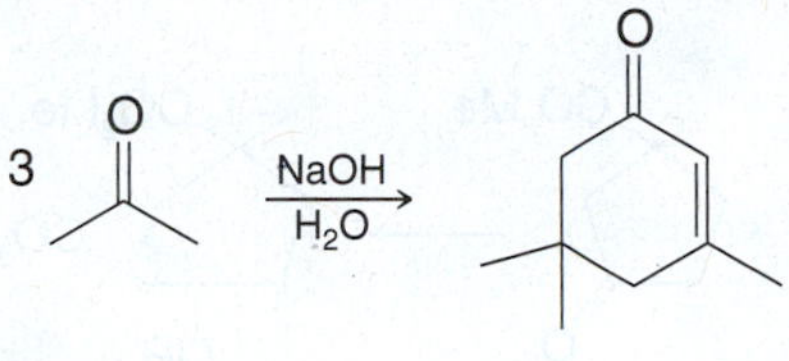

19.47 Predict the products of the following reactions.

(a)

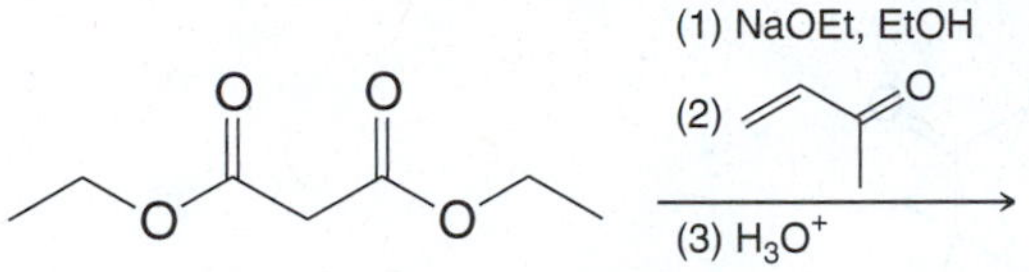

(1) NaOEt, EtOH
(2)
(3) H$_3$O$^+$

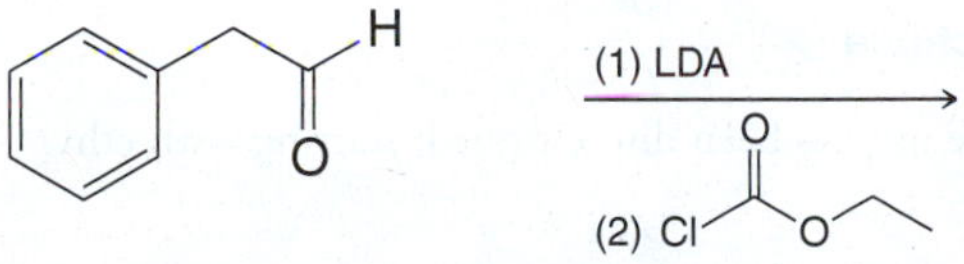

(c)

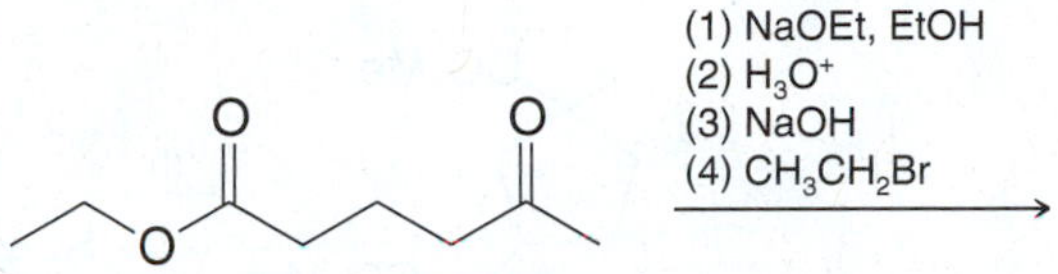

(1) NaOEt, EtOH
(2) H$_3$O$^+$
(3) NaOH
(4) CH$_3$CH$_2$Br

(b)

(1) LDA
(2)

(d)

(1) NaOEt, EtOH
(2)
(3) H$_3$O$^+$

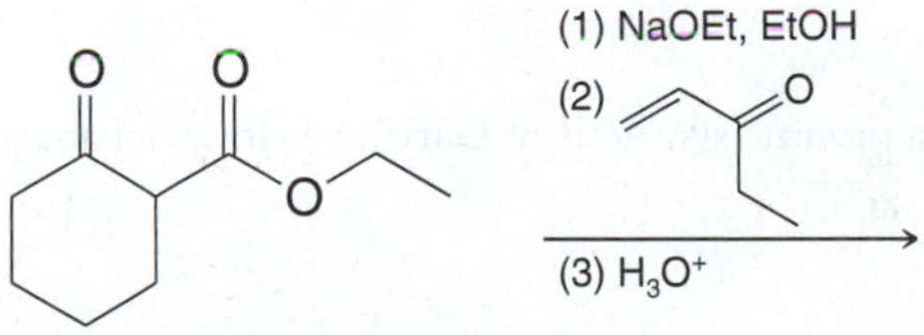

19.48 Predict the products from the following reactions.

(a)

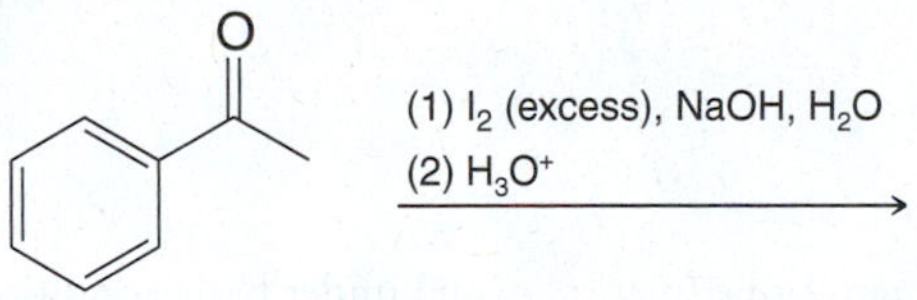

(1) I$_2$ (excess), NaOH, H$_2$O
(2) H$_3$O$^+$

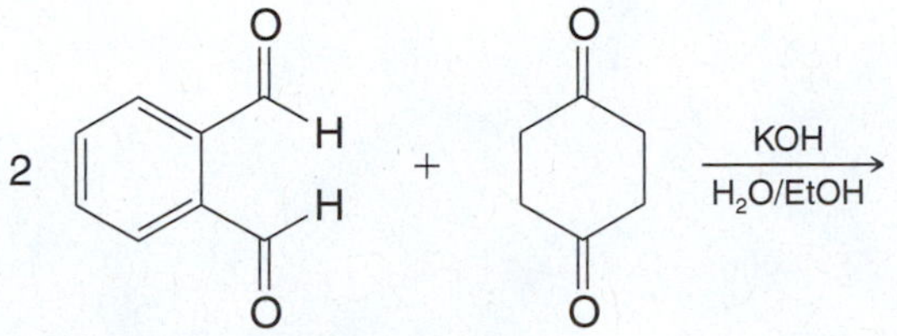

(c)

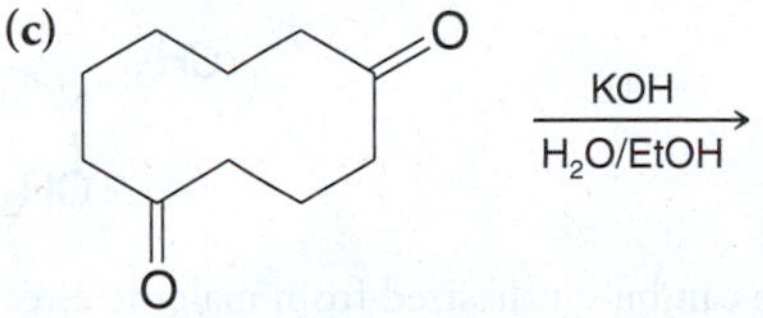

$\xrightarrow[\text{H}_2\text{O/EtOH}]{\text{KOH}}$

(b)

2 + $\xrightarrow[\text{H}_2\text{O/EtOH}]{\text{KOH}}$

(d)

(1) LDA (1.1 equiv)
(2)
(3) H$_3$O$^+$

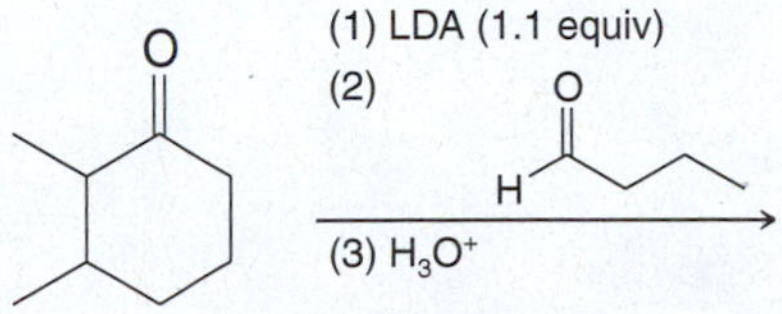

19.49 The mandibular glands of queen bees secrete a fluid that contains a remarkable compound known as "queen substance." When even an exceedingly small amount of the queen substance is transferred to worker bees, it inhibits the development of their ovaries and prevents the workers from bearing new queens. Queen substance, a monocarboxylic acid with the molecular formula $C_{10}H_{16}O_3$, has been synthesized by the following route:

Cycloheptanone $\xrightarrow[\text{(2) H}_3\text{O}^+]{\text{(1) CH}_3\text{MgI}}$ **A** $(C_8H_{16}O)$ $\xrightarrow{\text{HA, heat}}$ **B** (C_8H_{14}) $\xrightarrow[\text{(2) Me}_2\text{S}]{\text{(1) O}_3}$

C $(C_8H_{14}O_2)$ $\xrightarrow[\text{pyridine}]{}$ queen substance $(C_{10}H_{16}O_3)$

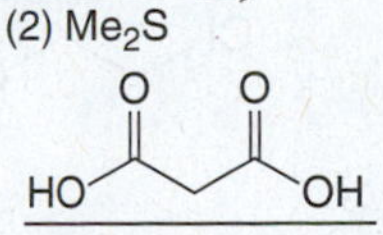

On catalytic hydrogenation, queen substance yields compound **D**, which, on treatment with iodine in sodium hydroxide and subsequent acidification, yields a dicarboxylic acid **E**; that is,

$$\text{Queen substance} \xrightarrow[\text{Pd}]{\text{H}_2} \textbf{D} \text{ (C}_{10}\text{H}_{18}\text{O}_3\text{)} \xrightarrow[\text{(2) H}_3\text{O}^+]{\text{(1) I}_2 \text{ in aq. NaOH}} \textbf{E} \text{ (C}_9\text{H}_{16}\text{O}_4\text{)}$$

Provide structures for the queen substance and compounds **A–E**.

19.50 (+)-Fenchone is a terpenoid that can be isolated from fennel oil. (±)-Fenchone has been synthesized through the following route. Supply the missing intermediates and reagents.

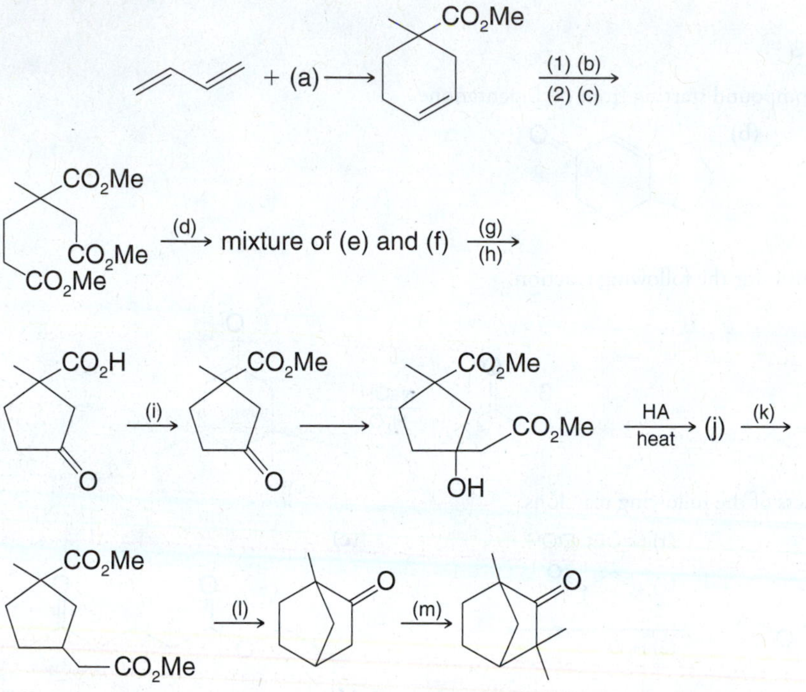

(±)-Fenchone

19.51 Outline a racemic synthesis of Darvon (below), an analgesic compound whose use has been discontinued, starting with ethyl phenyl ketone.

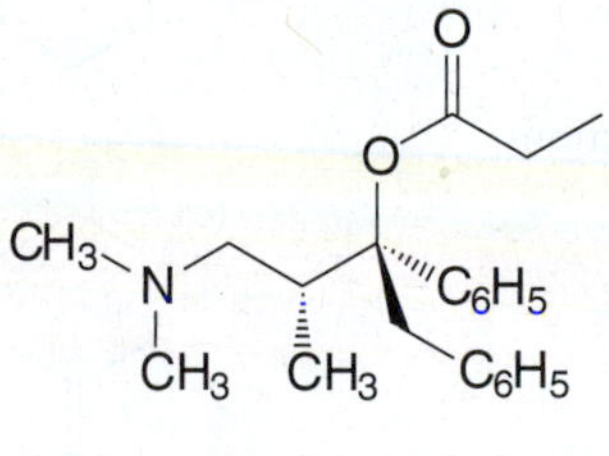

19.52 Show how dimedone can be synthesized from malonic ester and 4-methyl-3-penten-2-one (mesityl oxide) under basic conditions.

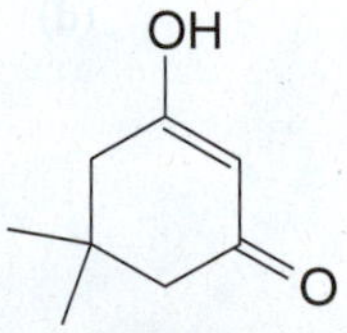

Dimedone

19.53 Write the mechanistic steps in the cyclization of ethyl phenylacetoacetate (ethyl 3-oxo-4-phenylbutanoate) in concentrated sulfuric acid to form naphthoresorcinol (1,3-naphthalenediol).

19.54 When an aldehyde or a ketone is condensed with ethyl α-chloroacetate in the presence of sodium ethoxide, the product is an α,β-epoxy ester called a *glycidic ester*. The synthesis is called the Darzens condensation.

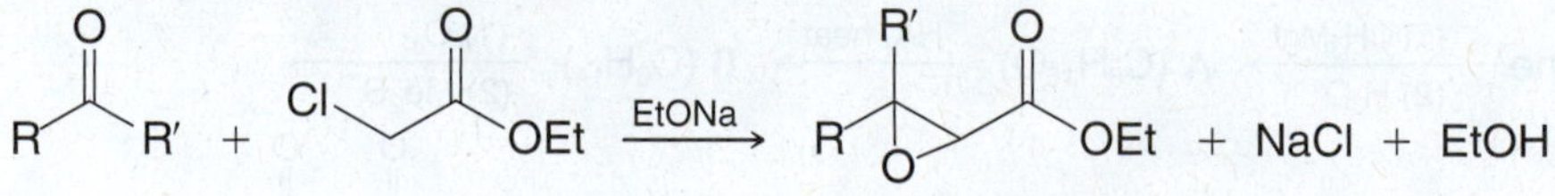

A glycidic ester

(a) Outline a reasonable mechanism for the Darzens condensation. (b) Hydrolysis of the epoxy ester leads to an epoxy acid that, on heating with pyridine, furnishes an aldehyde. What is happening here?

(c) Starting with β-ionone (Practice Problem 19.13), show how you might synthesize the following aldehyde. (This aldehyde is an intermediate in an industrial synthesis of vitamin A.)

19.55 The *Perkin condensation* is an aldol-type condensation in which an aromatic aldehyde (ArCHO) reacts with a carboxylic acid anhydride, $(RCH_2CO)_2O$, to give an α,β-unsaturated acid ($ArCH=CRCO_2H$). The catalyst that is usually employed is the potassium salt of the carboxylic acid (RCH_2CO_2K). (a) Outline the Perkin condensation that takes place when benzaldehyde reacts with propanoic anhydride in the presence of potassium propanoate. (b) How would you use a Perkin condensation to prepare *p*-chlorocinnamic acid, $p\text{-}ClC_6H_4CH=CHCO_2H$?

SPECTROSCOPY

19.56

(a) Infrared spectroscopy provides an easy method for deciding whether the product obtained from the addition of a Grignard reagent to an α,β-unsaturated ketone is the simple addition product or the conjugate addition product. Explain. (What peak or peaks would you look for?)

(b) How might you follow the rate of the following reaction using UV spectroscopy?

19.57 Allowing acetone to react with 2 molar equivalents of benzaldehyde in the presence of KOH in ethanol leads to the formation of compound **X**. The ^{13}C NMR spectrum of **X** is given in Fig. 19.1. Propose a structure for compound **X**.

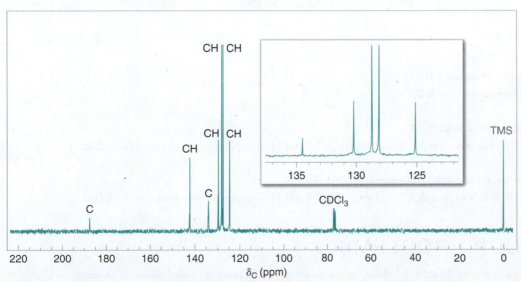

FIGURE 19.1 The broadband proton-decoupled ^{13}C NMR spectrum of compound X, Problem 19.57. Information from the DEPT ^{13}C NMR spectra is given above the peaks.

CHALLENGE PROBLEMS

19.58 Provide a mechanism for each of the following reactions.

(a)

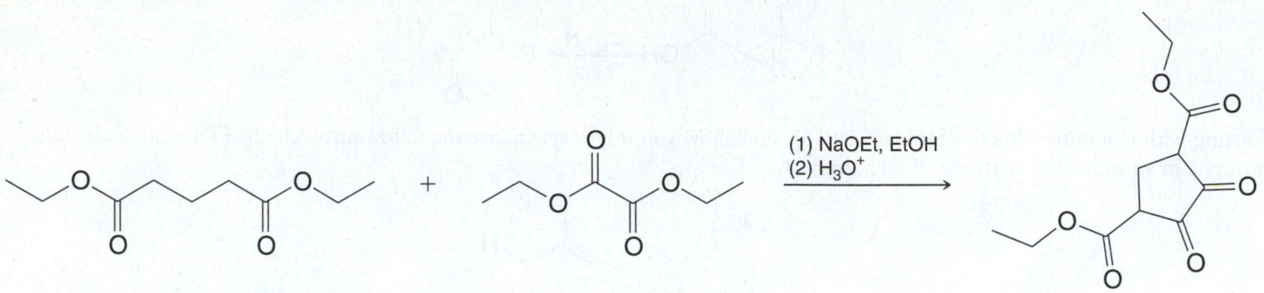

(b)

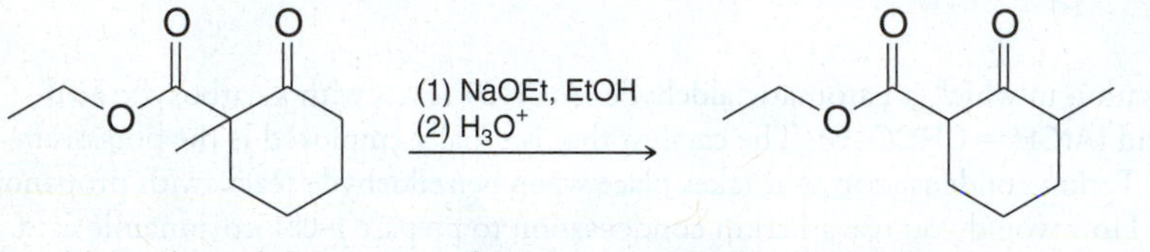

19.59 **(a)** Deduce the structure of product **A**, which is highly symmetrical:

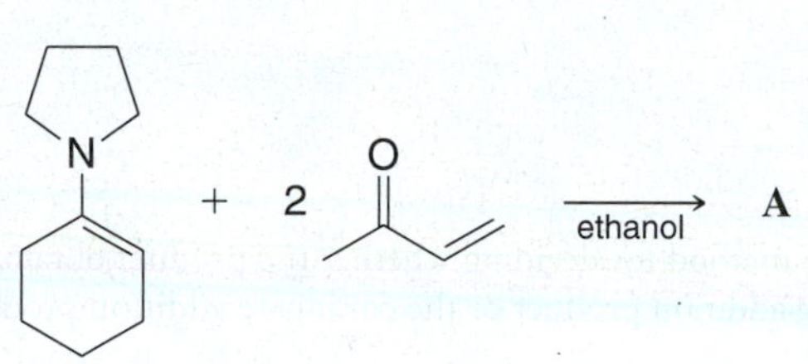

The following are selected spectral data for **A**:

MS (*m/z*): 220 (M$^{+\cdot}$)

IR (cm^{-1}): 2930, 2860, 1715

^{1}**H NMR** (δ): 1.25 (m), 1.29 (m), 1.76 (m), 1.77 (m), 2.14 (s), and 2.22 (t); (area ratios 2:1:2:1:2:2, respectively)

^{13}**C NMR** (δ): 23 (CH$_2$), 26 (CH$_2$), 27 (CH$_2$), 29 (C), 39 (CH), 41 (CH$_2$), 46 (CH$_2$), 208 (C)

(b) Write a mechanism that explains the formation of **A**.

19.60 Write the structures of the three products involved in this reaction sequence:

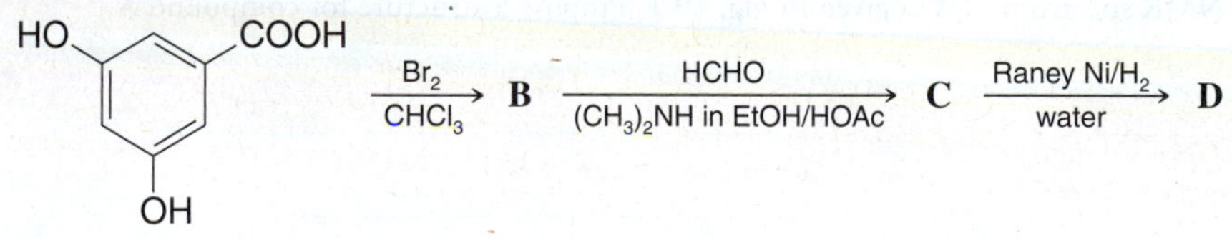

Spectral data for **B**:

MS (*m/z*): 314, 312, 310 (relative abundance 1:2:1)

^{1}**H NMR** (δ): only 6.80 (s) after treatment with D$_2$O

Data for **C**:

MS (*m/z*): 371, 369, 367 (relative abundance 1:2:1)

^{1}**H NMR** (δ): 2.48 (s) and 4.99 (s) in area ratio 3:1; broad singlets at 5.5 and 11 disappeared after treatment with D$_2$O.

Data for **D**:

MS (*m/z*): 369 (M$^{+\cdot}$−CH$_3$) [when studied as its tris(trimethylsilyl) derivative]

^{1}**H NMR** (δ): 2.16 (s) and 7.18 (s) in area ratio 3:2; broad singlets at 5.4 and 11 disappeared after treatment with D$_2$O.

LEARNING GROUP PROBLEMS

1. Lycopodine is a naturally occurring amine. As such, it belongs to the family of natural products called alkaloids. Its synthesis (*J. Am. Chem. Soc.* **1968**, *90*, 1647–1648) was accomplished by one of the great synthetic organic chemists of our time, Gilbert Stork (Columbia University). Write a detailed mechanism for all the steps that occur when **2** reacts with ethyl acetoacetate in the presence of ethoxide ion. Note that a necessary part of the mechanism will be a base-catalyzed isomerization (via a conjugated enolate) of the alkene in **2** to form the corresponding α,β-unsaturated ester.

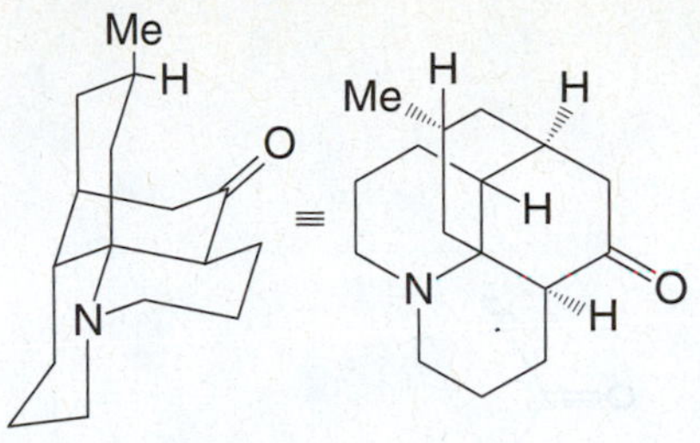

1
Lycopodine

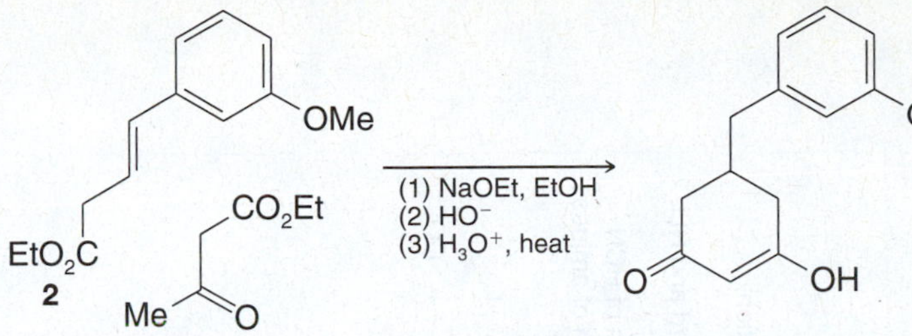

2

Reagents: (1) NaOEt, EtOH (2) HO⁻ (3) H₃O⁺, heat

3

2. Steroids are an extremely important class of natural and pharmaceutical compounds. Synthetic efforts directed toward steroids have been underway for many years and continue to be an area of important research. The synthesis of cholesterol by R. B. Woodward (Harvard University, recipient of the Nobel Prize in Chemistry for 1965) and co-workers represents a paramount accomplishment in steroid synthesis, and it is rich with examples of carbonyl chemistry and other reactions we have studied. Selected reactions from Woodward's cholesterol synthesis and the questions for this Learning Group Problem are shown in the *WileyPLUS* materials for this chapter. Access those materials online to complete this problem.

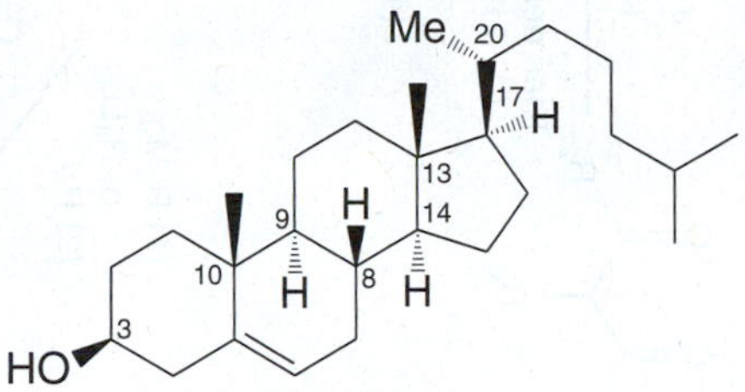

Cholesterol

[SUMMARY OF MECHANISMS]

Enolate Reactions with Carbonyl Electrophiles

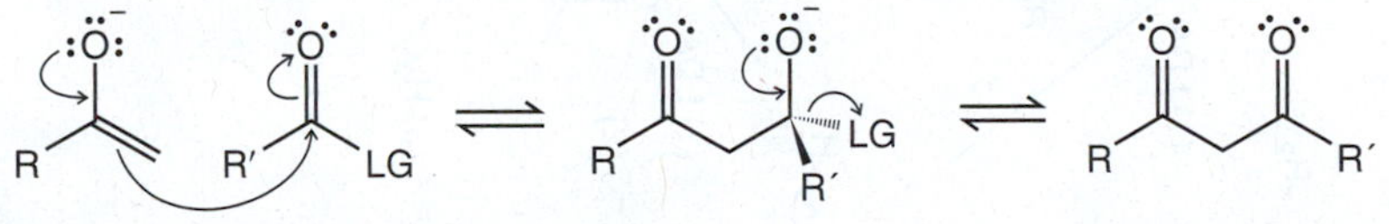

Acyl substitution (addition–elimination), e.g., Claisen condensation when LG = OR

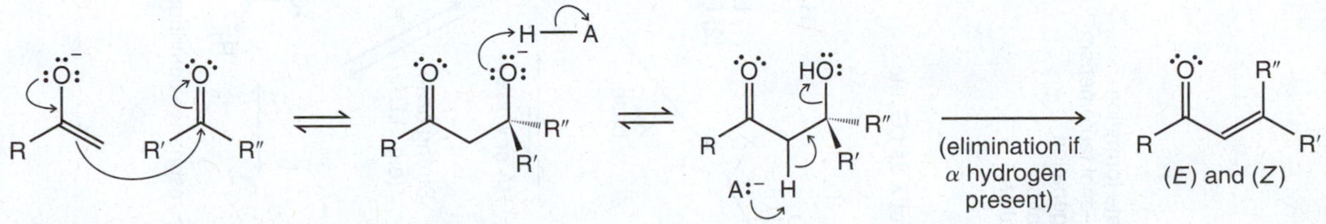

Aldol reactions (addition and condensation)

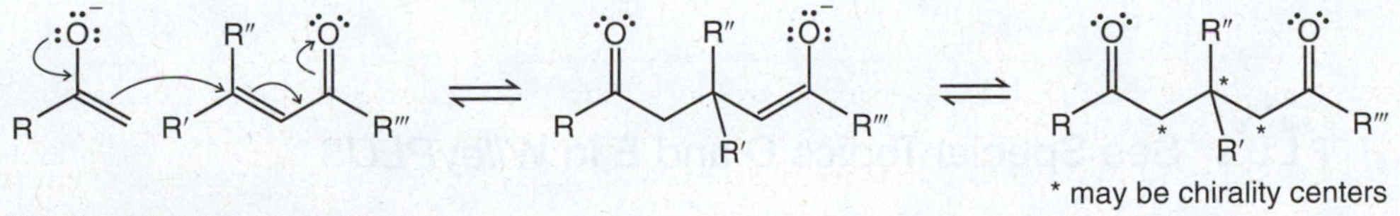

Michael (conjugate) addition

* may be chirality centers

[SYNTHETIC CONNECTIONS]

Some Synthetic Connections Involving Enolates

Enolates provide many ways to functionalize the α-carbon of a carbonyl compound. Most importantly, enolates provide ways to form new carbon–carbon bonds. Some of these synthetic connections are shown here. Previously studied reactions of carbonyl, alcohol, and alkene functional groups (e.g., reduction, oxidation, addition, substitution) lead to or from some of these pathways.

- Enolate formation
- Keto–enol tautomerism
- Halogenation
- Alkylation
- Acylation

- Claisen condensation
- Aldol reactions
- Addition of Grignard and RLi
- Michael addition
- Conjugate addition of HCN
- Conjugate addition of amines

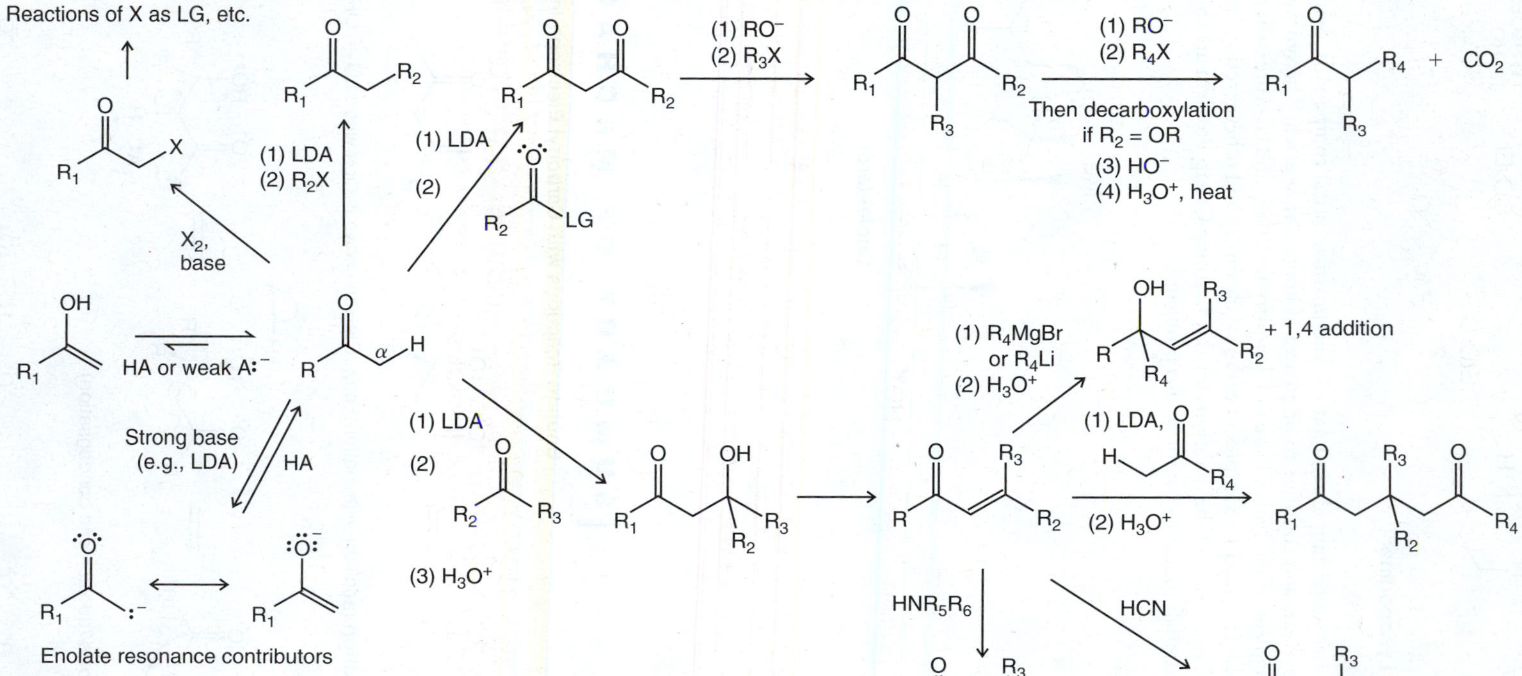

Reactions of X as LG, etc.

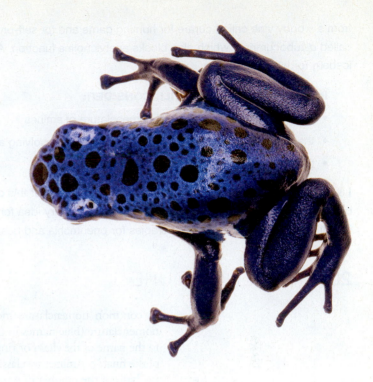

CHAPTER

20

Amines

Amine-containing compounds have an incredible range of biochemical properties. Some, like acetylcholine, act as neurotransmitters, control muscle function, enhance sensory perceptions, and sustain attention span. Others, however, can play far more sinister roles. Colombian poison dart frogs, for example, are tiny and beautiful, but they are also deadly. They produce a compound known as histrionicotoxin, an amine that causes paralysis and eventually death through suffocation. The respiratory muscles cease to function because acetylcholine cannot act, preventing it from initiating the electrical signaling that makes the muscles of our lungs function. Similarly, Amazon tribes have long used a mixture of compounds

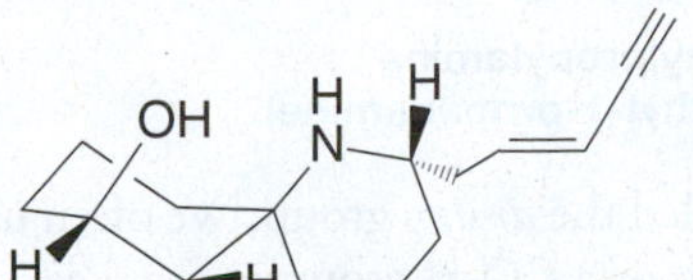

Histrionicotoxin

d-Tubocurarine chloride

Acetylcholine

PHOTO CREDIT: © Eric Isselée/iStockphoto

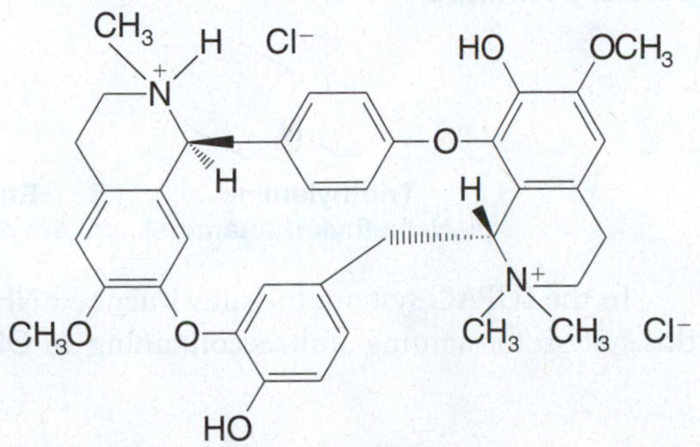

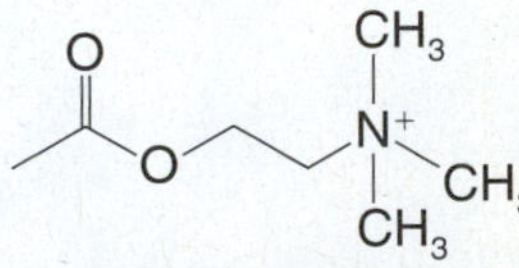

from a woody vine called curare for hunting game and for self-protection; this material includes another paralytic neurotoxin called *d*-tubocurarine, which also blocks acetylcholine function. As we shall see, these examples represent just the tip of the iceberg for what amines do.

IN THIS CHAPTER, WE WILL CONSIDER:

- the properties, structure, and nomenclature of amines
- the ability of amines to act as bases, salts, and resolving agents
- the synthesis and reactivity of amines

[WHY DO THESE TOPICS MATTER?]   At the end of this chapter we will show you how amine-containing compounds led not only to the genesis of a revolutionary idea for how small molecules can treat disease, but also to the identification of the world's first therapies for pneumonia and gastrointestinal infections.

20.1 NOMENCLATURE

In common nomenclature most primary amines are named as *alkylamines*. In systematic nomenclature (blue names in parentheses below) they are named by adding the suffix *-amine* to the name of the chain or ring system to which the NH_2 group is attached with replacement of the final *-e*. Amines are classified as being **primary** (**1°**), **secondary** (**2°**), or **tertiary** (**3°**) on the basis of the number of organic groups attached to the nitrogen (Section 2.8).

Primary Amines

$$CH_3\ddot{N}H_2$$

Methylamine
(methanamine)

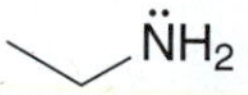

Ethylamine
(ethanamine)

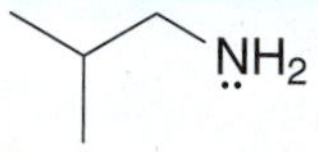

Isobutylamine
(2-methyl-1-propanamine)

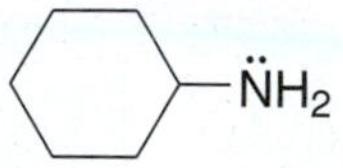

Cyclohexylamine
(cyclohexanamine)

Most **secondary** and **tertiary amines** are named in the same general way. In common nomenclature we either designate the organic groups individually if they are different or use the prefixes di- or tri- if they are the same. In systematic nomenclature we use the locant *N* to designate substituents attached to a nitrogen atom.

Secondary Amines

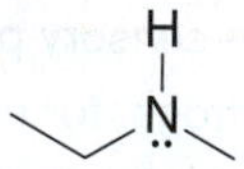

Ethylmethylamine
(*N*-methylethanamine)

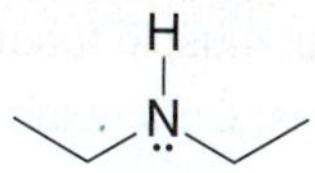

Diethylamine
(*N*-ethylethanamine)

Tertiary Amines

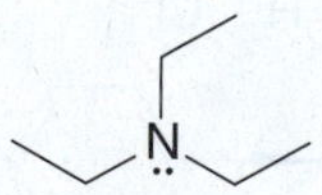

Triethylamine
(*N,N*-diethylethanamine)

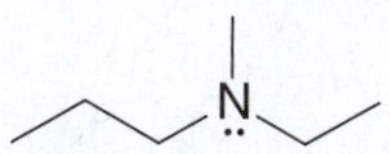

Ethylmethylpropylamine
(*N*-ethyl-*N*-methyl-1-propanamine)

In the IUPAC system, the substituent $-NH_2$ is called the *amino* group. We often use this system for naming amines containing an OH group or a CO_2H group:

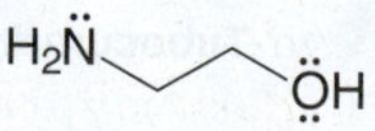

2-Aminoethanol

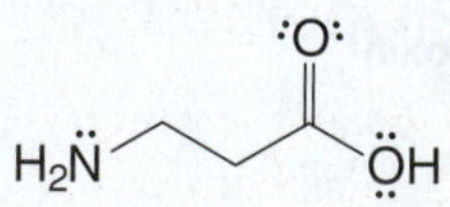

3-Aminopropanoic acid

20.1A Arylamines

Some common **arylamines** have the following names:

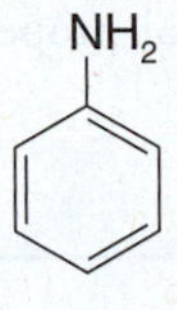

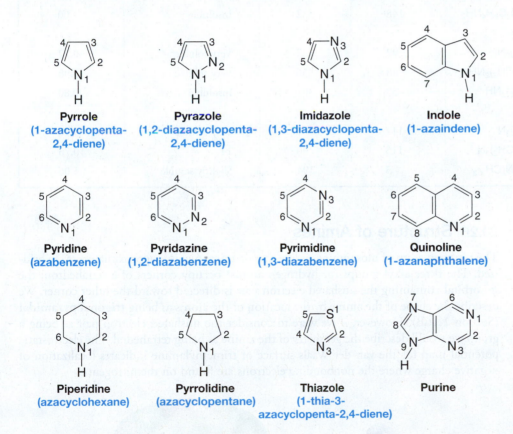

Aniline	**N-Methylaniline**	**p-Toluidine**	**p-Anisidine**
(benzenamine)	(N-methyl-benzenamine)	(4-methyl-benzenamine)	(4-methoxy-benzenamine)

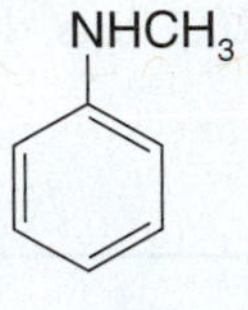

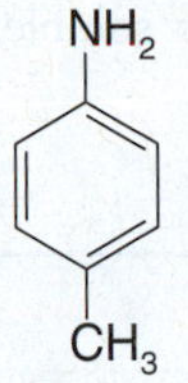

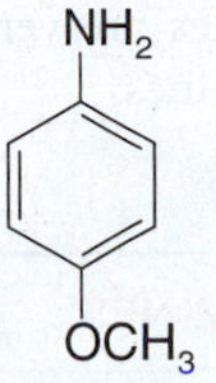

20.1B Heterocyclic Amines

The important **heterocyclic amines** all have common names. In systematic replacement nomenclature the prefixes *aza-, diaza-,* and *triaza-* are used to indicate that nitrogen atoms have replaced carbon atoms in the corresponding hydrocarbon. A nitrogen atom in the ring (or the highest atomic weight heteroatom, as in the case of thiazole) is designated position 1 and numbering proceeds to give the lowest overall set of locants to the heteroatoms:

Pyrrole (1-azacyclopenta-2,4-diene)   **Pyrazole** (1,2-diazacyclopenta-2,4-diene)   **Imidazole** (1,3-diazacyclopenta-2,4-diene)   **Indole** (1-azaindene)

Pyridine (azabenzene)   **Pyridazine** (1,2-diazabenzene)   **Pyrimidine** (1,3-diazabenzene)   **Quinoline** (1-azanaphthalene)

Piperidine (azacyclohexane)   **Pyrrolidine** (azacyclopentane)   **Thiazole** (1-thia-3-azacyclopenta-2,4-diene)   **Purine**

20.2 PHYSICAL PROPERTIES AND STRUCTURE OF AMINES

20.2A Physical Properties

Amines are moderately polar substances; they have boiling points that are higher than those of alkanes but generally lower than those of alcohols of comparable molecular weight. Molecules of primary and secondary amines can form strong hydrogen bonds to each other and to water. Molecules of tertiary amines cannot form hydrogen bonds to each other, but they can form hydrogen bonds to molecules of water or other hydroxylic solvents. As a result, tertiary amines generally boil at lower temperatures than primary and secondary amines of comparable molecular weight, but all low-molecular-weight amines are very water soluble. Table 20.1 lists the physical properties of some common amines.

TABLE 20.1 PHYSICAL PROPERTIES OF AMINES

Name	Structure	mp (°C)	bp (°C)	Water Solubility (25 °C) (g 100 mL^{-1})	pK$_a$ (aminium ion)
Primary Amines					
Methylamine	CH_3NH_2	−94	−6	Very soluble	10.64
Ethylamine	$CH_3CH_2NH_2$	−81	17	Very soluble	10.75
Isopropylamine	$(CH_3)_2CHNH_2$	−101	33	Very soluble	10.73
Cyclohexylamine	Cyclo-$C_6H_{11}NH_2$	−18	134	Slightly soluble	10.64
Benzylamine	$C_6H_5CH_2NH_2$	10	185	Slightly soluble	9.30
Aniline	$C_6H_5NH_2$	−6	184	3.7	4.58
4-Methylaniline	$4\text{-}CH_3C_6H_4NH_2$	44	200	Slightly soluble	5.08
4-Nitroaniline	$4\text{-}NO_2C_6H_4NH_2$	148	332	Insoluble	1.00
Secondary Amines					
Dimethylamine	$(CH_3)_2NH$	−92	7	Very soluble	10.72
Diethylamine	$(CH_3CH_2)_2NH$	−48	56	Very soluble	10.98
Diphenylamine	$(C_6H_5)_2NH$	53	302	Insoluble	0.80
Tertiary Amines					
Trimethylamine	$(CH_3)_3N$	−117	3	Very soluble	9.70
Triethylamine	$(CH_3CH_2)_3N$	−115	90	14	10.76
N,N-Dimethylaniline	$C_6H_5N(CH_3)_2$	3	194	Slightly soluble	5.06

20.2B Structure of Amines

The nitrogen atom of most amines is like that of ammonia; it is approximately sp^3 hybridized. The three alkyl groups (or hydrogen atoms) occupy corners of a tetrahedron; the sp^3 orbital containing the unshared electron pair is directed toward the other corner. We describe the shape of the amine by the location of the atoms as being **trigonal pyramidal** (Section 1.16B). However, if we were to consider the unshared electron pair as being a group we would describe the geometry of the amine as being tetrahedral. The electrostatic potential map for the van der Waals surface of trimethylamine indicates localization of negative charge where the nonbonding electrons are found on the nitrogen:

Structure of an amine.   A calculated structure for trimethylamine. The electrostatic potential map shows charge associated with the nitrogen unshared electron pair.

The bond angles are what one would expect of a tetrahedral structure; they are very close to 109.5°. The bond angles for trimethylamine, for example, are 108°.

If the alkyl groups of a tertiary amine are all different, the amine will be chiral. There will be two enantiomeric forms of the tertiary amine, and, theoretically, we ought to be able to resolve (separate) these enantiomers. In practice, however, resolution is usually impossible because the enantiomers interconvert rapidly:

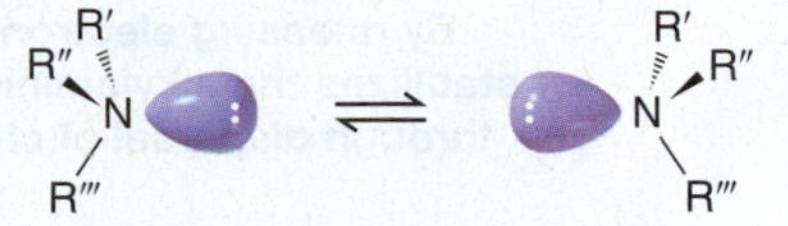

**Interconversion of amine
enantiomers**

This interconversion occurs through what is called a **pyramidal** or **nitrogen inversion**. The barrier to the interconversion is about 25 kJ mol^{-1} for most simple amines, low enough to occur readily at room temperature. In the transition state for the inversion, the nitrogen atom becomes sp^2 hybridized with the unshared electron pair occupying a p orbital.

Ammonium salts cannot undergo nitrogen inversion because they do not have an unshared pair. Therefore, those **quaternary ammonium salts** with four different groups are chiral and can be resolved into separate (relatively stable) enantiomers:

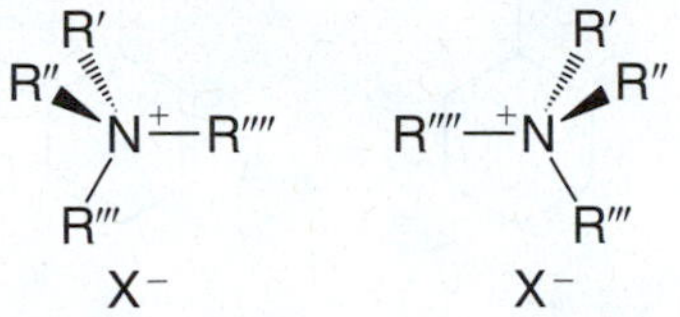

**Quaternary ammonium salts such as
these can be resolved.**

20.3 BASICITY OF AMINES: AMINE SALTS

- Amines are relatively weak bases. Most are stronger bases than water but are far weaker bases than hydroxide ions, alkoxide ions, and alkanide anions.

A convenient way to compare the **base strengths** of amines is to compare the pK_a values of their conjugate acids, the corresponding alkylaminium ions (Sections 3.6C and 20.3D).

$$\overset{+}{R}NH_3 + H_2O \rightleftharpoons RNH_2 + H_3O^+$$

$$K_a = \frac{[RNH_2][H_3O^+]}{[RNH_3^+]}$$

$$pK_a = -\log K_a$$

The equilibrium for an amine that is relatively more basic will lie more toward the left in the above chemical equation than for an amine that is less basic.

- The aminium ion of a more basic amine will have a larger pK_a than the aminium ion of a less basic amine.

When we compare aminium ion acidities in terms of this equilibrium, we see that most primary alkylaminium ions (RNH_3^+) are less acidic than ammonium ion (NH_4^+). In other words, primary alkylamines (RNH_2) are more basic than ammonia (NH_3):

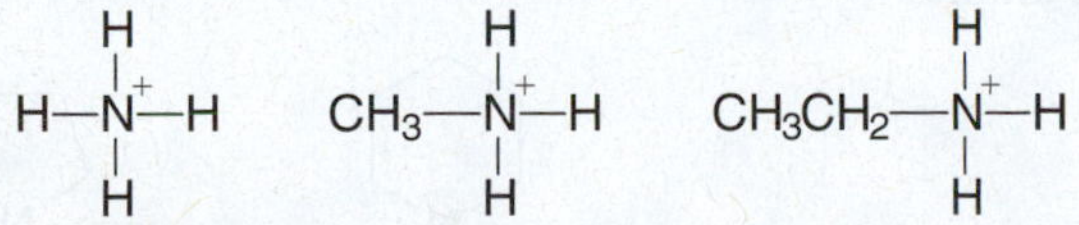

Aminium ion pK_a 9.26	10.64	10.75

We can account for this on the basis of the electron-releasing ability of an alkyl group. An alkyl group releases electrons, and it *stabilizes* the alkylaminium ion that results from the acid–base reaction *by dispersing its positive charge*. It stabilizes the alkylaminium ion to a greater extent than it stabilizes the amine:

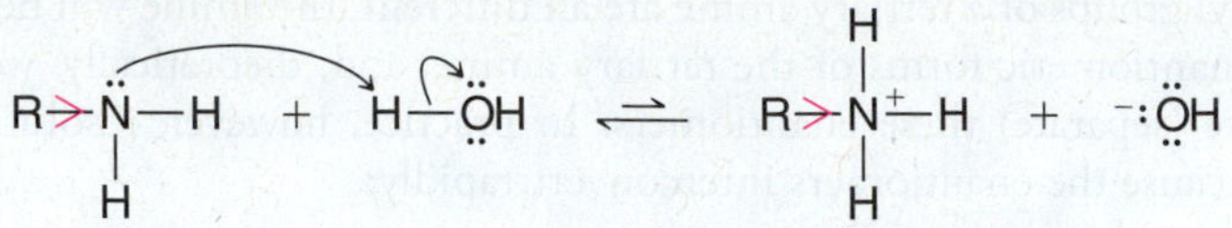

$$R \rightarrow \overset{H}{\underset{H}{N}} - H \ + \ H - \overset{..}{\underset{}{O}}H \ \rightleftharpoons \ R \rightarrow \overset{H}{\underset{H}{N}} - H \ + \ \ ^- \!:\! \overset{..}{\underset{}{O}}H$$

By releasing electrons, R $\rightarrow$ stabilizes the alkylaminium ion through dispersal of charge.

20.3A Basicity of Arylamines

• Aromatic amines are much weaker bases than alkylamines.

Considering amine basicity from the perspective of aminium ion acidity, when we examine the pK_a values of the conjugate acids of aromatic amines (e.g., aniline and 4-methylaniline) in Table 20.1, we see that they are much weaker bases than the nonaromatic amine, cyclohexylamine:

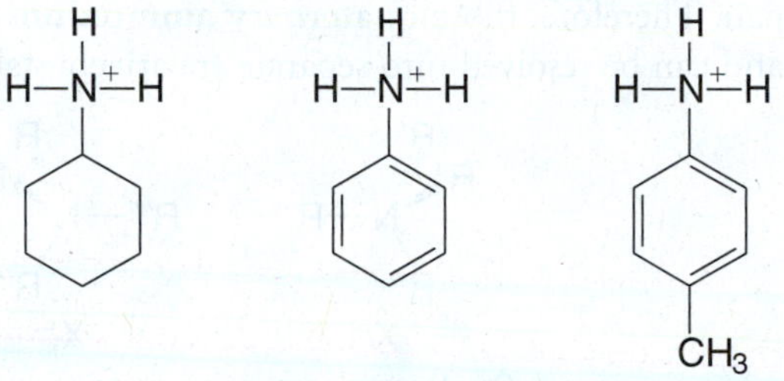

Aminium ion pK_a	10.64	4.58	5.08

We can account for this effect, in part, on the basis of resonance contributions to the overall hybrid of an arylamine. For aniline, the following contributors are important:

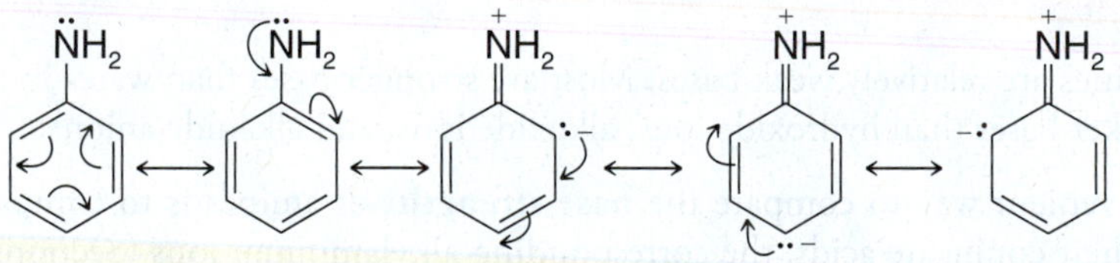

Structures **1** and **2** are the Kekulé structures that contribute to any benzene derivative. Structures **3–5**, however, *delocalize* the unshared electron pair of the nitrogen over the ortho and para positions of the ring. This delocalization of the electron pair makes it less available to a proton, and *delocalization of the electron pair stabilizes aniline*.

Another important effect in explaining the lower basicity of aromatic amines is the **electron-withdrawing effect of a phenyl group**. Because the carbon atoms of a phenyl group are sp^2 hybridized, they are more electronegative (and therefore more electron withdrawing) than the sp^3-hybridized carbon atoms of alkyl groups. We shall discuss this effect further in Section 21.5A.

20.3B Basicity of Heterocyclic Amines

Nonaromatic heterocyclic amines have basicities that are approximately the same as those of acyclic amines:

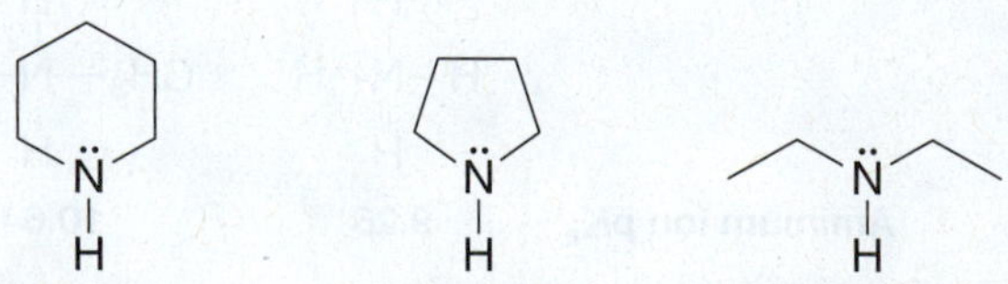

pK_a of corresponding aminium ion	Piperidine 11.20	Pyrrolidine 11.11	Diethylamine 10.98

In aqueous solution, aromatic heterocyclic amines such as pyridine, pyrimidine, and pyrrole are much weaker bases than nonaromatic amines or ammonia.

	Pyridine	Pyrimidine	Pyrrole	Quinoline
pK_a of corresponding aminium ion	5.23	2.70	0.40	4.5

20.3C Amines versus Amides

- Amides are far less basic than amines (even less basic than arylamines). The pK_a of the conjugate acid of a typical amide is about zero.

The lower basicity of amides when compared to amines can be understood in terms of resonance and inductive effects. An amide is stabilized by resonance involving the non-bonding pair of electrons on the nitrogen atom. However, an amide protonated on its nitrogen atom lacks this type of resonance stabilization. This is shown in the following resonance structures:

Amide

Larger resonance stabilization

N-Protonated Amide

Smaller resonance stabilization

However, a more important factor accounting for amides being weaker bases than amines is the powerful electron-withdrawing effect of the carbonyl group of the amide. This effect is illustrated by the electrostatic potential maps for ethylamine and acetamide shown in Fig. 20.1. Significant negative charge is localized at the position of the nonbonding electron pair in ethylamine (as indicated by the red color). In acetamide, however, less negative charge resides near the nitrogen than in ethylamine.

Comparing the following equilibria, the reaction with the amide lies more to the left than the corresponding reaction with an amine. This is consistent with the amine being a stronger base than an amide.

The nitrogen atoms of amides are so weakly basic that when an amide accepts a proton, it does so on its oxygen atom instead (see the mechanism for hydrolysis of an amide, Section 17.8F). Protonation on the oxygen atom occurs even though oxygen atoms (because of their greater electronegativity) are typically less basic than nitrogen atoms. Notice, however, that if an amide accepts a proton on its oxygen atom, resonance stabilization involving the nonbonding electron pair of the nitrogen atom is possible:

Ethylamine

Acetamide

FIGURE 20.1 Calculated electrostatic potential maps (calibrated to the same charge scale) for ethylamine and acetamide. The map for ethylamine shows localization of negative charge at the unshared electron pair of nitrogen. The map for acetamide shows most of the negative charge at its oxygen atom instead of at nitrogen, due to the electron-withdrawing effect of the carbonyl group.

20.3D Aminium Salts and Quaternary Ammonium Salts

When primary, secondary, and tertiary amines act as bases and react with acids, they form compounds called **aminium salts**. In an aminium salt the positively charged nitrogen atom is attached to at least one hydrogen atom:

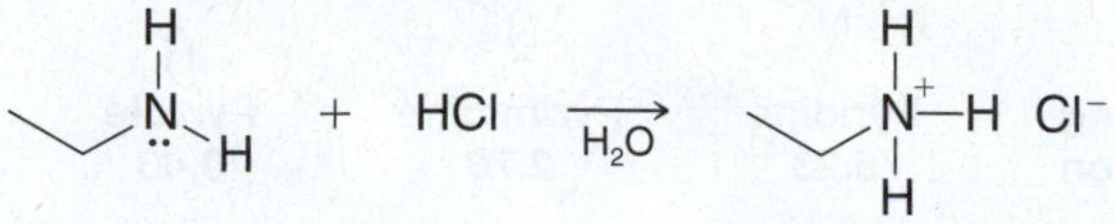

Ethylaminium chloride (an aminium salt)

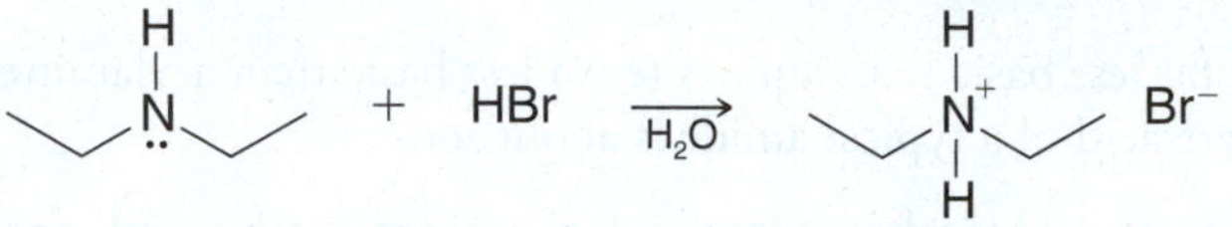

Diethylaminium bromide

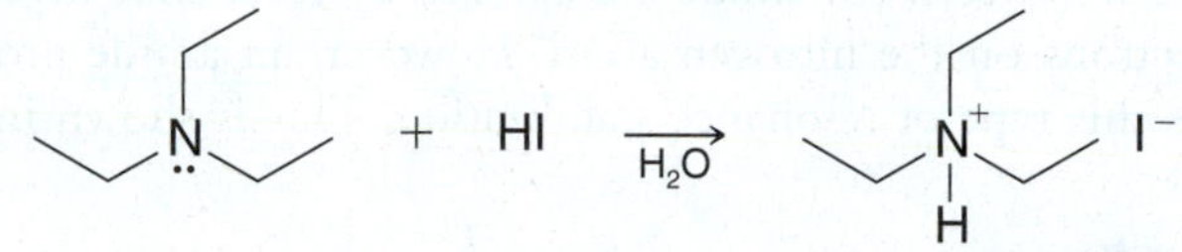

Triethylaminium iodide

When the central nitrogen atom of a compound is positively charged *but is not attached to a hydrogen atom*, the compound is called a **quaternary ammonium salt**. For example,

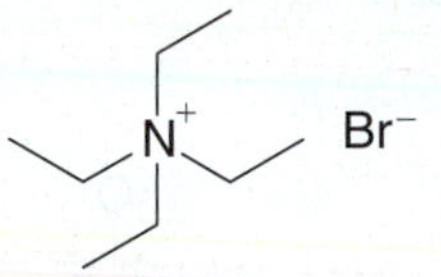

Tetraethylammonium bromide (a quaternary ammonium salt)

Quaternary ammonium halides—because they do not have an unshared electron pair on the nitrogen atom—cannot act as bases. Quaternary ammonium *hydroxides*, however, are strong bases. As solids, or in solution, they consist *entirely* of quaternary ammonium cations (R_4N^+) and hydroxide ions (HO^-); they are, therefore, strong bases—as strong as sodium or potassium hydroxide. Quaternary ammonium hydroxides react with acids to form quaternary ammonium salts:

$$(CH_3)_4 \overset{+}{N}OH^- + HCl \longrightarrow (CH_3)_4 \overset{+}{N}Cl^- + H_2O$$

In Section 20.12A we shall see how quaternary ammonium salts can be used to form alkenes by a reaction called the *Hofmann elimination*.

20.3E Solubility of Amines in Aqueous Acids

- Almost all alkylaminium chloride, bromide, iodide, and sulfate salts are soluble in water. Thus, primary, secondary, or tertiary amines that are not soluble in water will dissolve in dilute aqueous HCl, HBr, HI, and H_2SO_4.

Solubility in dilute acid provides a convenient chemical method for distinguishing amines from nonbasic compounds that are insoluble in water. Solubility in dilute acid also gives us a useful method for separating amines from nonbasic compounds that are insoluble in water. The amine can be extracted into aqueous acid (dilute HCl) and then recovered by making the aqueous solution basic and extracting the amine into ether or CH_2Cl_2.

Helpful Hint

You may make use of the basicity of amines in your organic chemistry laboratory work for the separation of compounds or for the characterization of unknowns.

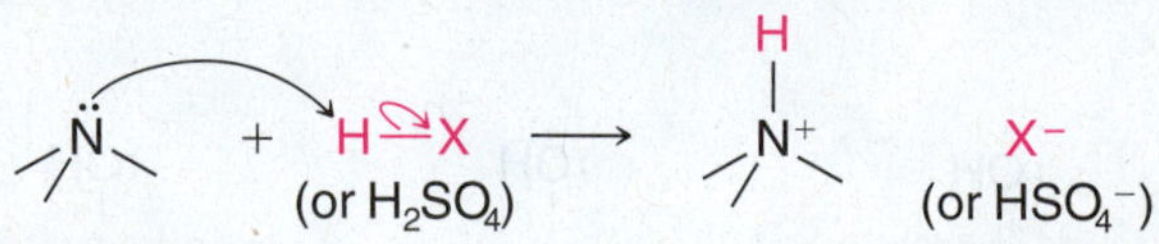

Water-insoluble amine + H—X (or H_2SO_4) ⟶ **Water-soluble aminium salt** X⁻ (or HSO_4^-)

Because amides are far less basic than amines, water-insoluble amides *do not dissolve* in dilute aqueous HCl, HBr, HI, or H_2SO_4:

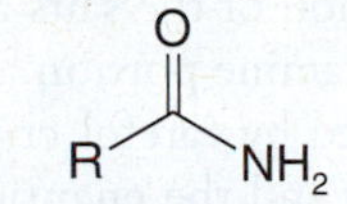

O
‖
R   NH₂

Water-insoluble amide
(not soluble in aqueous acids)

Outline a procedure for separating hexylamine from cyclohexane using dilute HCl, aqueous NaOH, and diethyl ether.

PRACTICE PROBLEM 20.1

Outline a procedure for separating a mixture of benzoic acid, 4-methylphenol, aniline, and benzene using acids, bases, and organic solvents.

PRACTICE PROBLEM 20.2

20.3F Amines as Resolving Agents

- Enantiomerically pure amines are often used to resolve racemic forms of acidic compounds by the formation of diastereomeric salts.

We can illustrate the principles involved in **resolution** by showing how a racemic form of an organic acid might be resolved (separated) into its enantiomers with the single enantiomer of an **amine as a resolving agent** (Fig. 20.2).

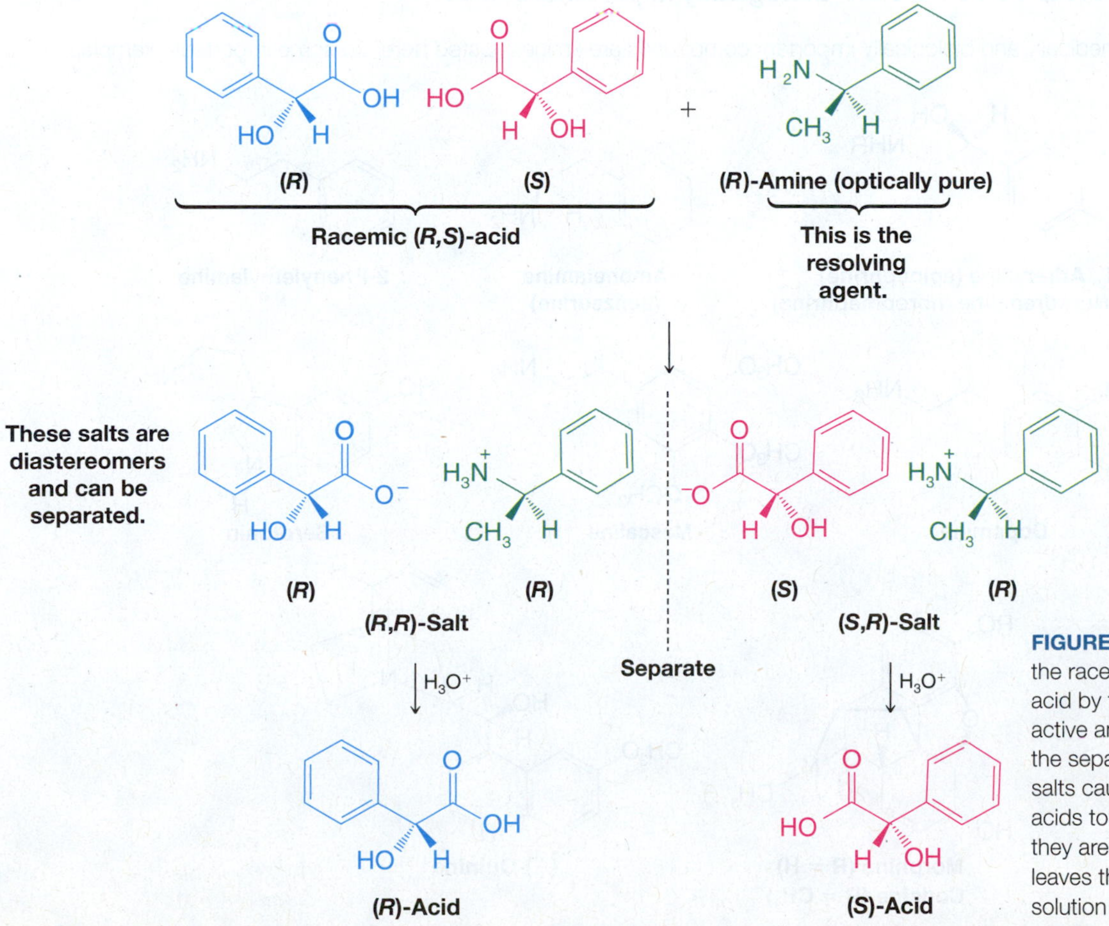

FIGURE 20.2 Resolution of the racemic form of an organic acid by the use of an optically active amine. Acidification of the separated diastereomeric salts causes the enantiomeric acids to precipitate (assuming they are insoluble in water) and leaves the resolving agent in solution as its conjugate acid.

Helpful Hint

See "The Chemistry of... HPLC Resolution of Enantiomers" in *WileyPLUS* for information about another technique for resolving enantiomers.

In this procedure the single enantiomer of an amine, (*R*)-1-phenylethylamine, is added to a solution of the racemic form of an acid. The salts that form are *diastereomers*. The chirality centers of the acid portion of the salts are enantiomerically related to each other, but the chirality centers of the amine portion are not. The diastereomers have different solubilities and can be separated by careful crystallization. The separated salts are then acidified with hydrochloric acid and the enantiomeric acids are obtained from the separate solutions. The amine remains in solution as its hydrochloride salt.

Single enantiomers that are employed as resolving agents are often readily available from natural sources. Because most of the chiral organic molecules that occur in living organisms are synthesized by enzymatically catalyzed reactions, most of them occur as single enantiomers. Naturally occurring optically active amines such as (−)-quinine (see "The Chemistry of... Biologically Important Amines" in this section), (−)-strychnine, and (−)-brucine are often employed as resolving agents for racemic acids. Acids such as (+)- or (−)-tartaric acid (Section 5.15A) are often used for resolving racemic bases.

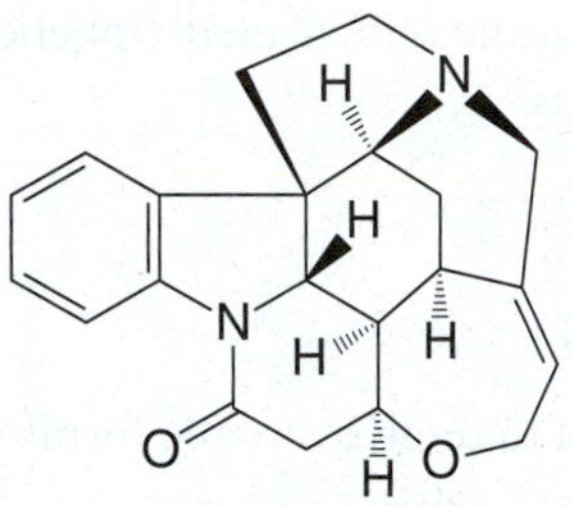

(−)-Strychnine

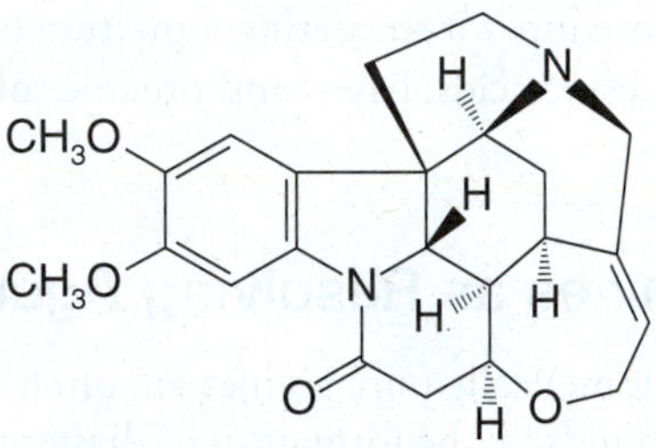

(−)-Brucine

THE CHEMISTRY OF... Biologically Important Amines

A large number of medically and biologically important compounds are amines. Listed here are some important examples:

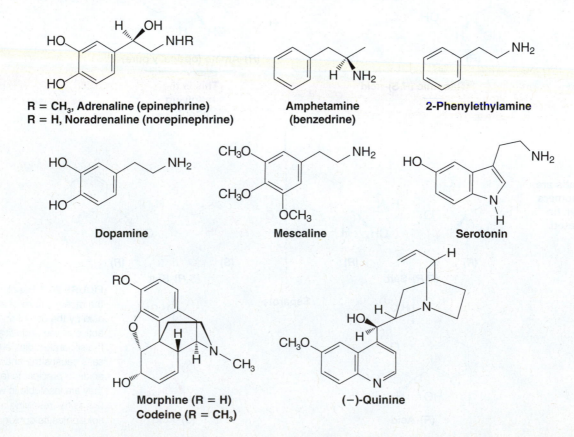

R = CH_3, Adrenaline (epinephrine)
R = H, Noradrenaline (norepinephrine)

Amphetamine (benzedrine)

2-Phenylethylamine

Dopamine

Mescaline

Serotonin

Morphine (R = H)
Codeine (R = CH_3)

(−)-Quinine

2-Phenylethylamines

Many phenylethylamine compounds have powerful physiological and psychological effects. Adrenaline and noradrenaline are two hormones secreted in the medulla of the adrenal gland. Released into the bloodstream when an animal senses danger, adrenaline causes an increase in blood pressure, a strengthening of the heart rate, and a widening of the passages of the lungs. All of these effects prepare the animal to fight or to flee. Noradrenaline also causes an increase in blood pressure, and it is involved in the transmission of impulses from the end of one nerve fiber to the next. Dopamine and serotonin are important neurotransmitters in the brain. Abnormalities in the level of dopamine in the brain are associated with many psychiatric disorders, including Parkinson's disease. Dopamine plays a pivotal role in the regulation and control of movement, motivation, and cognition. Serotonin is a compound of particular interest because it appears to be important in maintaining stable mental processes. It has been suggested that the mental disorder schizophrenia may be connected with abnormalities in the metabolism of serotonin.

Amphetamine (a powerful stimulant) and mescaline (a hallucinogen) have structures similar to those of serotonin, adrenaline, and noradrenaline. They are all derivatives of 2-phenylethylamine. (In serotonin the nitrogen is connected to the benzene ring to create a five-membered ring.) The structural similarities of these compounds must be related to their physiological and psychological effects because many other compounds with similar properties are also derivatives of 2-phenylethylamine. Examples (not shown) are *N*-methylamphetamine and LSD (lysergic acid diethylamide). Even morphine and codeine, two powerful analgesics, have a 2-phenylethylamine system as a part of their structures. [Morphine and codeine are examples of compounds called alkaloids (Special Topic F in *WileyPLUS*). Try to locate the 2-phenylethylamine system in their structures.]

Vitamins and Antihistamines

A number of amines are vitamins. These include nicotinic acid and nicotinamide, pyridoxine (vitamin B_6, see "The Chemistry of . . . Pyridoxal Phosphate" in *WileyPLUS* for Chapter 16), and thiamine chloride (vitamin B_1, see "The Chemistry of . . . Thiamine," in *WileyPLUS* for Chapter 17). Nicotine is a toxic alkaloid found in tobacco that makes smoking habit forming. Histamine, another toxic amine, is found bound to proteins in nearly all tissues of the body. Release of free histamine causes the symptoms associated with allergic reactions and the common cold. Chlorpheniramine, an "antihistamine," is an ingredient of many over-the-counter cold remedies.

Pyridoxine (vitamin B₆) **Thiamine chloride (vitamin B₁)** **Nicotine** **Nicotinic acid (niacin)**

Histamine **Chlorpheniramine** **Valium (diazepam)**

Tranquilizers

Valium (diazepam) is a widely prescribed tranquilizer. Chlordiazepoxide is a closely related compound. Phenobarbital (also see the model) is used to control epileptic seizures and as a sedative for insomnia and relief of anxiety.

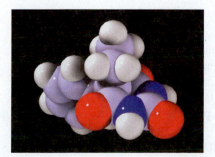

Phenobarbital.

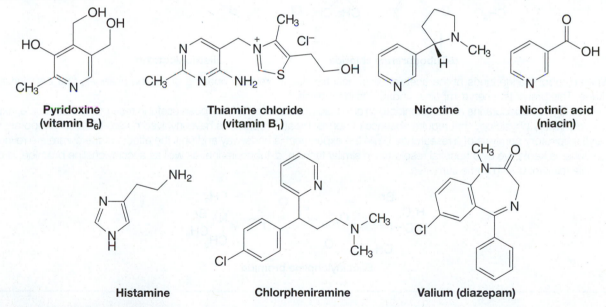

Chlordiazepoxide **Phenobarbital**

(continued on next page)

Neurotransmitters

Nerve cells interact with other nerve cells or with muscles at junctions, or gaps, called synapses. Nerve impulses are carried across the synaptic gap by chemical compounds called *neurotransmitters*. Acetylcholine (see the following reaction) is an important neurotransmitter at neuromuscular synapses called *cholinergic synapses*. Acetylcholine contains a quaternary ammonium group. Being small and ionic, acetylcholine is highly soluble in water and highly diffusible, qualities that suit its role as a neurotransmitter. Acetylcholine molecules are released by the presynaptic membrane in the neuron in packets of about 10^4 molecules. The packet of molecules then diffuses across the synaptic gap.

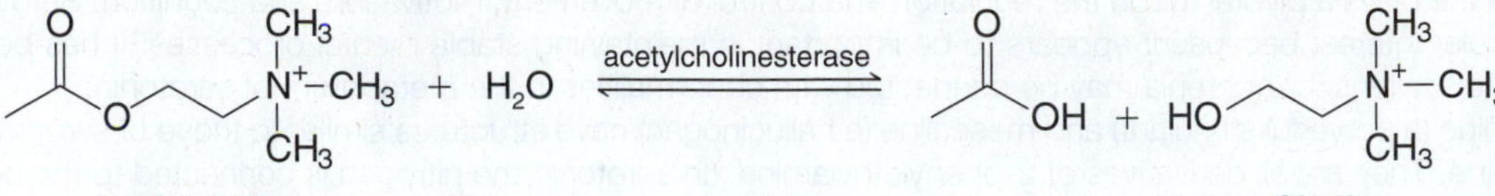

Acetylcholine   **Choline**

Having carried a nerve impulse across the synapse to the muscle where it triggers an electrical response, the acetylcholine molecules must be hydrolyzed (to choline) within a few milliseconds to allow the arrival of the next impulse. This hydrolysis is catalyzed by an enzyme of almost perfect efficiency called *acetylcholinesterase*.

The acetylcholine receptor on the postsynaptic membrane of muscle is the target for some of the most deadly neurotoxins, including *d*-tubocurarine and histrionicotoxin, shown here.

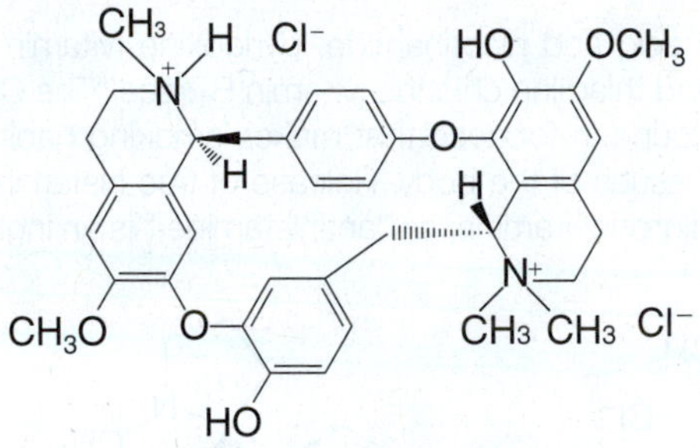

d-Tubocurarine chloride 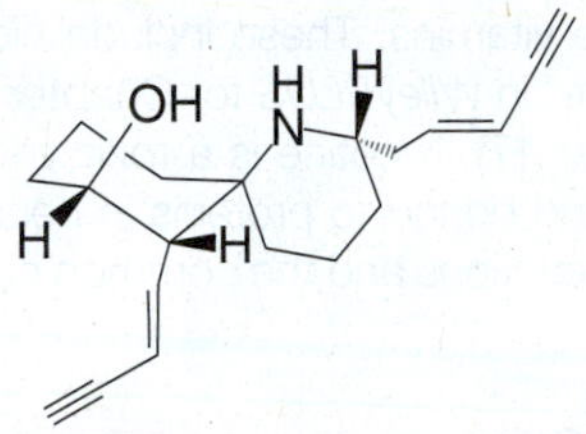   **Histrionicotoxin**

When *d*-tubocurarine binds at the acetylcholine receptor site, it prevents opening of the ion channels that depolarize the membrane. This prevents a nerve impulse, and results in paralysis.

Even though *d*-tubocurarine and histrionicotoxin are deadly poisons, both have been useful in research. For example, experiments in respiratory physiology that require absence of normal breathing patterns have involved curare-induced temporary (and voluntary!) respiratory paralysis of a researcher. While the experiment is underway and until the effects of the curare are reversed, the researcher is kept alive by a hospital respirator. In similar fashion, *d*-tubocurarine, as well as succinylcholine bromide, is used as a muscle relaxant during some surgeries.

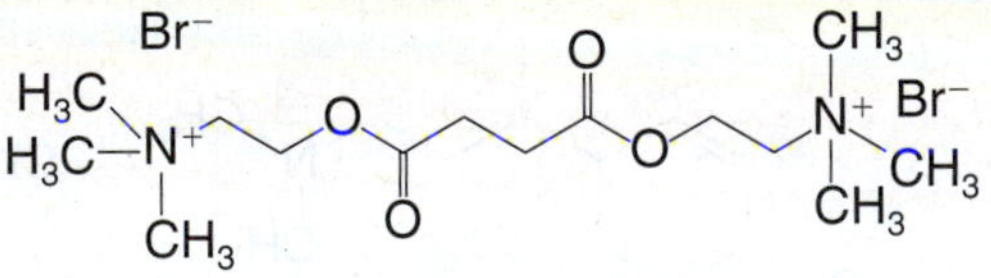

Succinylcholine bromide

20.4 PREPARATION OF AMINES

In this section we discuss a variety of ways to synthesize amines. Some of these methods will be new to you, while others are methods you have studied earlier in the context of related functional groups and reactions. Later, in Chapter 24, you will see how some of the methods presented here, as well as some others for asymmetric synthesis, can be used to synthesize α-amino acids, the building blocks of peptides and proteins.

20.4A Through Nucleophilic Substitution Reactions

Alkylation of Ammonia Salts of primary amines can be prepared from ammonia and alkyl halides by nucleophilic substitution reactions. Subsequent treatment of the resulting aminium salts with a base gives primary amines:

$$\overset{\cdot\cdot}{N}H_3 \;+\; R\overset{\frown}{-}X \;\longrightarrow\; R\overset{+}{-}NH_3\; X^- \;\xrightarrow{HO^-}\; RNH_2$$

- This method is of very limited synthetic application because multiple alkylations occur.

When ethyl bromide reacts with ammonia, for example, the ethylaminium bromide that is produced initially can react with ammonia to liberate ethylamine. Ethylamine can then compete with ammonia and react with ethyl bromide to give diethylaminium bromide. Repetitions of alkylation and proton transfer reactions ultimately produce some tertiary amines and even some quaternary ammonium salts if the alkyl halide is present in excess.

A MECHANISM FOR THE REACTION — Alkylation of NH_3

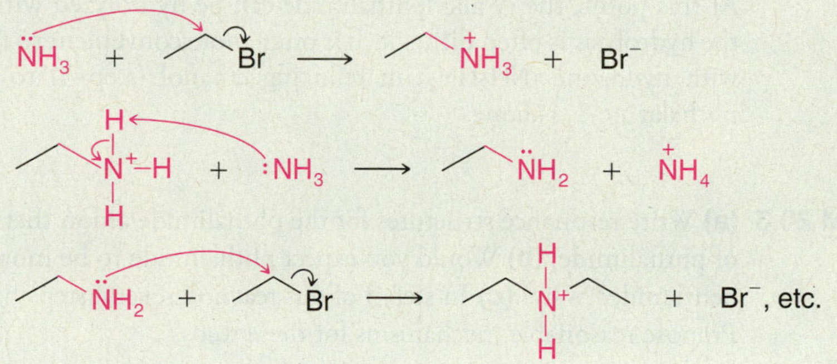

Multiple alkylations can be minimized by using a large excess of ammonia. (Why?) An example of this technique can be seen in the synthesis of alanine from 2-bromopropanoic acid:

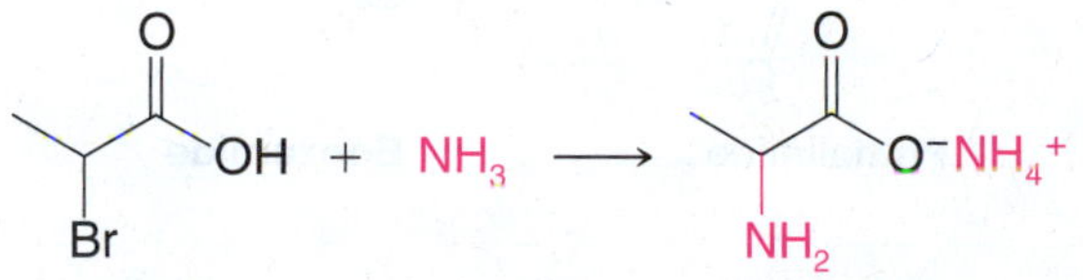

(1 mol) **(70 mol)** **Alanine (65–70%)**

Alkylation of Azide Ion and Reduction A much better method for preparing a primary amine from an alkyl halide is first to convert the alkyl halide to an alkyl azide ($R{-}N_3$) by a nucleophilic substitution reaction, then reduce the azide to a primary amine with lithium aluminum hydride.

$$R{-}X + :\overset{-}{\ddot{N}}{=}\overset{+}{N}{=}\overset{-}{\ddot{N}}: \xrightarrow[(-X^-)]{S_N2} R{-}\overset{-}{\ddot{N}}{=}\overset{+}{N}{=}\overset{-}{\ddot{N}}: \xrightarrow{LiAlH_4} RNH_2$$

Azide ion (a good nucleophile) **Alkyl azide**

A word of caution: Alkyl azides are explosive, and low-molecular-weight alkyl azides should not be isolated but should be kept in solution. Sodium azide is used in automotive airbags.

The Gabriel Synthesis Potassium phthalimide (see the following reaction) can also be used to prepare primary amines by a method known as the *Gabriel synthesis*. This synthesis also avoids the complications of multiple alkylations that occur when alkyl halides are treated with ammonia:

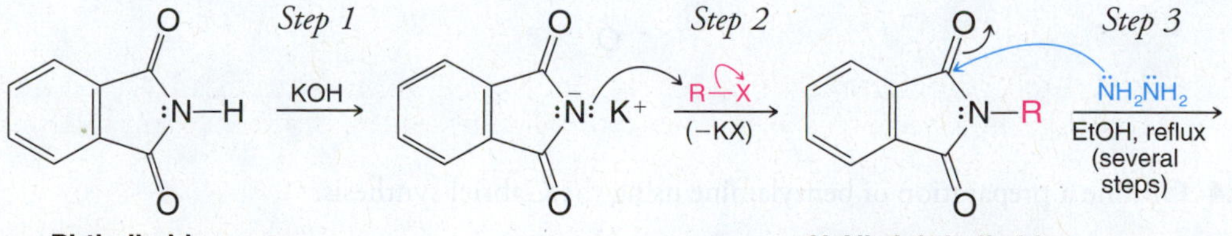

Phthalimide ***N*-Alkylphthalimide**

(continued from previous page)

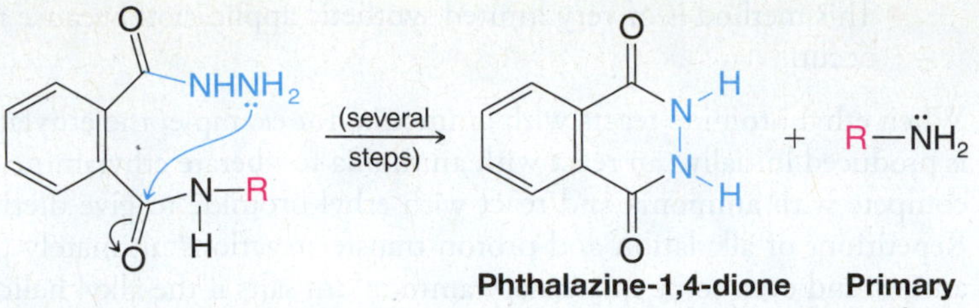

Phthalazine-1,4-dione   **Primary amine**

Phthalimide is quite acidic ($pK_a = 9$); it can be converted to potassium phthalimide by potassium hydroxide (step 1). The phthalimide anion is a strong nucleophile and (in step 2) it reacts with an alkyl halide by an S_N2 mechanism to give an *N*-alkylphthalimide. At this point, the *N*-alkylphthalimide can be hydrolyzed with aqueous acid or base, but the hydrolysis is often difficult. It is often more convenient to treat the *N*-alkylphthalimide with hydrazine (NH_2NH_2) in refluxing ethanol (step 3) to give a primary amine and phthalazine-1,4-dione.

PRACTICE PROBLEM 20.3 **(a)** Write resonance structures for the phthalimide anion that account for the acidity of phthalimide. **(b)** Would you expect phthalimide to be more or less acidic than benzamide? Why? **(c)** In step 3 of our reaction several steps have been omitted. Propose reasonable mechanisms for these steps.

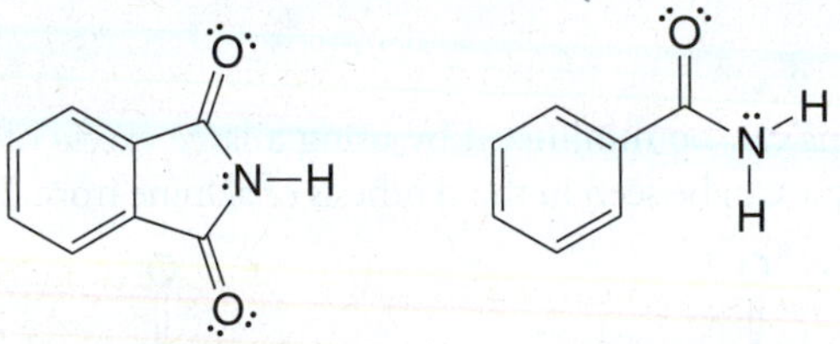

Phthalimide   **Benzamide**

Syntheses of amines using the Gabriel synthesis are, as we might expect, restricted to the use of methyl, primary, and secondary alkyl halides. The use of tertiary halides leads almost exclusively to eliminations.

••• SOLVED PROBLEM 20.1

Outline a synthesis of 4-methylpentanamine using the Gabriel synthesis.

ANSWER:

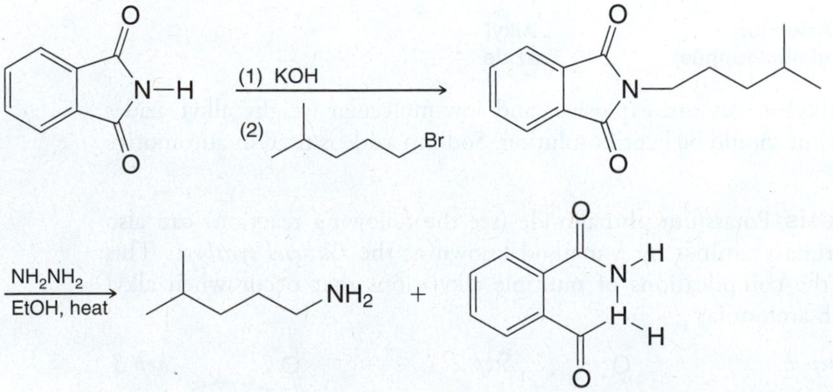

PRACTICE PROBLEM 20.4 Outline a preparation of benzylamine using the Gabriel synthesis.

Alkylation of Tertiary Amines Multiple alkylations are not a problem when tertiary amines are alkylated with methyl or primary halides. Reactions such as the following take place in good yield:

$$R_3N: \ + \ RCH_2-Br \xrightarrow{S_N2} R_3\overset{+}{N}-CH_2R \ + \ Br^-$$

20.4B Preparation of Aromatic Amines through Reduction of Nitro Compounds

The most widely used method for preparing aromatic amines involves nitration of the ring and subsequent reduction of the nitro group to an amino group:

$$Ar-H \xrightarrow[H_2SO_4]{HNO_3} Ar-NO_2 \xrightarrow{[H]} Ar-NH_2$$

We studied ring nitration in Chapter 15 and saw there that it is applicable to a wide variety of aromatic compounds. Reduction of the nitro group can also be carried out in a number of ways. The most frequently used methods employ catalytic hydrogenation, or treatment of the nitro compound with acid and iron. Zinc, tin, or a metal salt such as $SnCl_2$ can also be used. Overall, this is a $6e^-$ reduction.

General Reaction

$$Ar-NO_2 \xrightarrow[\text{or (1) Fe, HCl \ (2) HO}^-]{H_2, \text{ catalyst}} Ar-NH_2$$

Specific Example

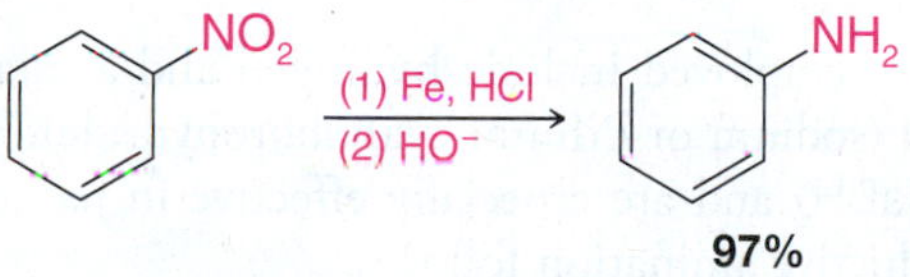

97%

20.4C Preparation of Primary, Secondary, and Tertiary Amines through Reductive Amination

Aldehydes and ketones can be converted to amines through catalytic or chemical reduction in the presence of ammonia or an amine. Primary, secondary, and tertiary amines can be prepared this way:

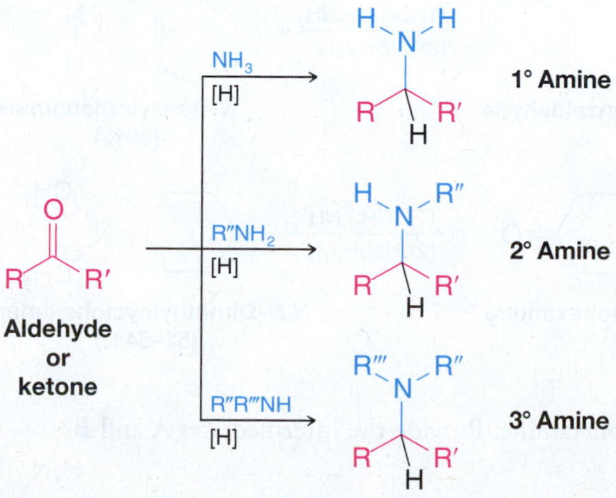

This process, called **reductive amination** of the aldehyde or ketone (or *reductive alkylation* of the amine), appears to proceed through the following general mechanism (illustrated with a 1° amine).

[A MECHANISM FOR THE REACTION] — Reductive Amination

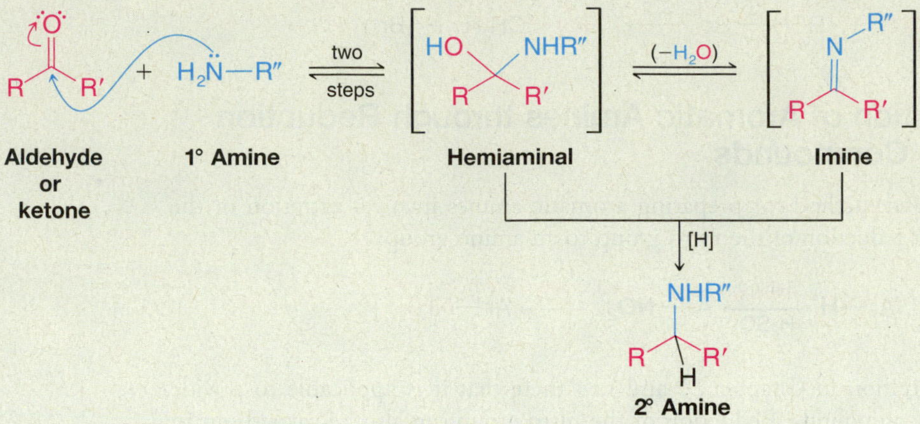

Aldehyde or ketone + 1° Amine ⇌ (two steps) [Hemiaminal] ⇌ (−H₂O) [Imine]

[H]

2° Amine

Helpful Hint

We saw the importance of imines in "The Chemistry of . . . Pyridoxal Phosphate" (vitamin B₆) in *WileyPLUS* for Chapter 16 (Section 16.8).

When ammonia or a primary amine is used, there are two possible pathways to the product: via an amino alcohol that is similar to a hemiacetal and is called a *hemiaminal* or via an imine (Section 16.8A). When secondary amines are used, an imine cannot form, and, therefore, the pathway is through the hemiaminal or through an iminium ion:

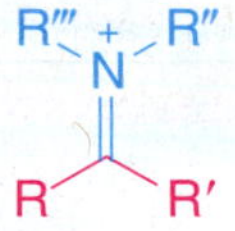

Iminium ion

The reducing agents employed include hydrogen and a catalyst (such as nickel) or NaBH₃CN or LiBH₃CN (sodium or lithium cyanoborohydride). The latter two reducing agents are similar to NaBH₄ and are especially effective in reductive aminations. Three specific examples of reductive amination follow:

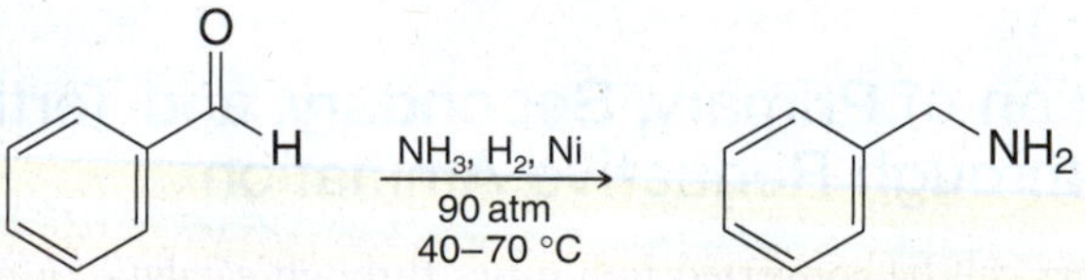

Benzaldehyde →(NH₃, H₂, Ni / 90 atm / 40–70 °C)→ **Benzylamine (89%)**

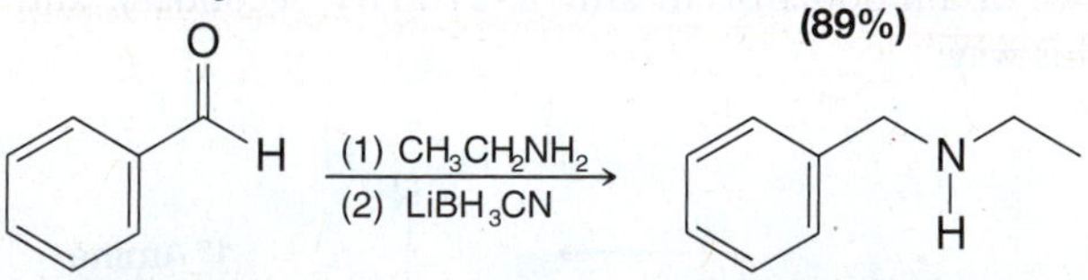

Benzaldehyde →((1) CH₃CH₂NH₂ / (2) LiBH₃CN)→ **N-Benzylethanamine (89%)**

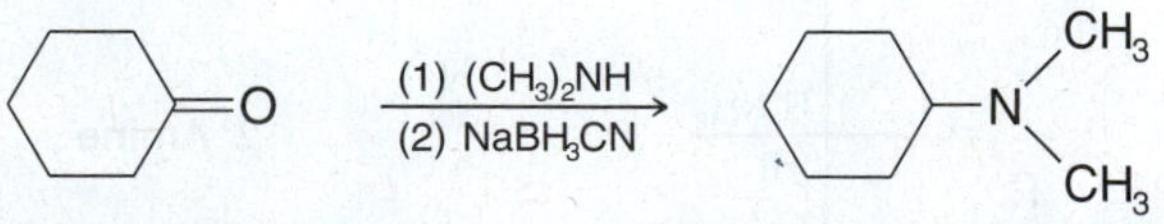

Cyclohexanone →((1) (CH₃)₂NH / (2) NaBH₃CN)→ **N,N-Dimethylcyclohexanamine (52–54%)**

●●● SOLVED PROBLEM 20.2

Outlined below is a synthesis of the stimulant amphetamine. Provide the intermediates **A** and **B**.

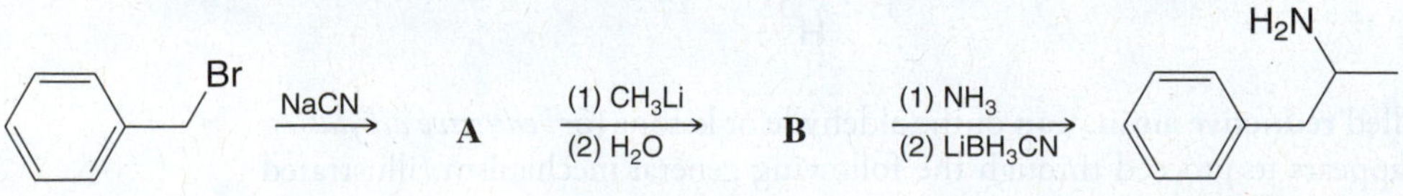

→(NaCN)→ **A** →((1) CH₃Li / (2) H₂O)→ **B** →((1) NH₃ / (2) LiBH₃CN)→ **Amphetamine**

ANSWER:

A = [structure: benzyl-CH₂-CN]   B = [structure: benzyl-CH₂-C(=O)-CH₃]

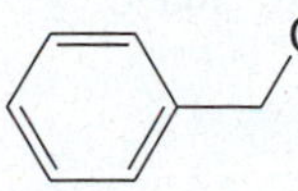

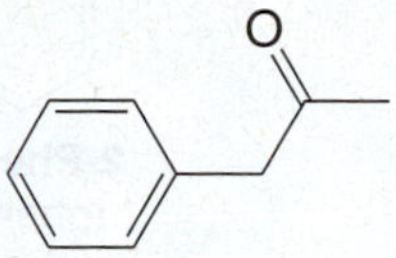

Show how you might prepare each of the following amines through reductive amination:

(a) 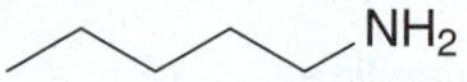 [structure with NH₂]   **(b)** 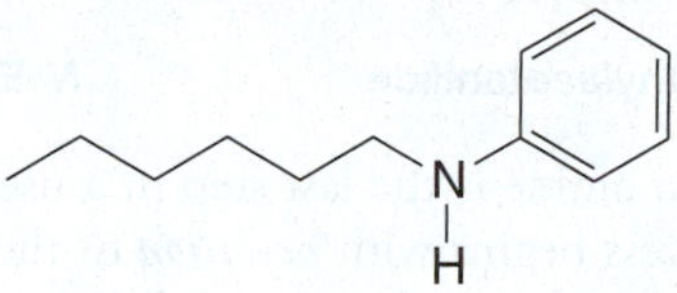 [structure with N–H attached to phenyl]   **(c)** 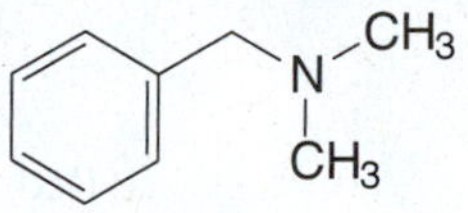 [structure: benzyl-N(CH₃)₂]

●●● SOLVED PROBLEM 20.3

Reductive amination of a ketone is almost always a better method for the synthesis of an amine of the type 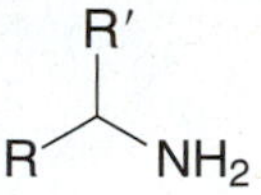 R–CH(R')–NH₂ than treatment of an alkyl halide with ammonia. Explain why this is true.

STRATEGY AND ANSWER: Consider the structure of the required alkyl halide. Reaction of a secondary halide with ammonia would inevitably be accompanied by considerable elimination, thereby decreasing the yield of the secondary amine. Multiple *N*-alkylations may also occur.

20.4D Preparation of Primary, Secondary, or Tertiary Amines through Reduction of Nitriles, Oximes, and Amides

Nitriles, oximes, and amides can be reduced to amines. Reduction of a nitrile or an oxime yields a primary amine; reduction of an amide can yield a primary, secondary, or tertiary amine:

$$R-C\equiv N \xrightarrow{[H]} RCH_2NH_2$$

Nitrile → **1° Amine**

> Nitriles can be prepared from alkyl halides and ⁻CN (Section 17.3) or from aldehydes and ketones as cyanohydrins (Section 16.9).

$$RCH=NOH \xrightarrow{[H]} RCH_2NH_2$$

Oxime → **1° Amine**

> Oximes can be prepared from aldehydes and ketones (Section 16.8B).

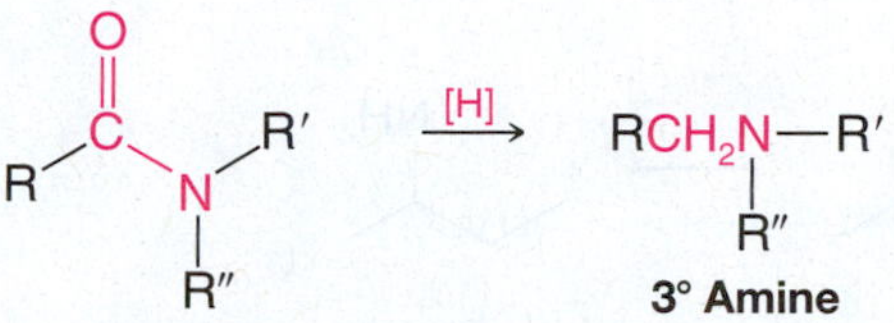

$$\underset{\text{Amide}}{R-\overset{O}{\overset{\|}{C}}-\underset{R''}{\overset{|}{N}}-R'} \xrightarrow{[H]} \underset{\text{3° Amine}}{RCH_2-\underset{R''}{\overset{|}{N}}-R'}$$

> Amides can be prepared from acid chlorides, acid anhydrides, and esters (Section 17.8).

(In the last example, if R′ = H and R″ = H, the product is a 1° amine; if only R′ = H, the product is a 2° amine.)

All of these reductions can be carried out with hydrogen and a catalyst or with LiAlH₄. Oximes are also conveniently reduced with sodium in ethanol.

Specific examples follow:

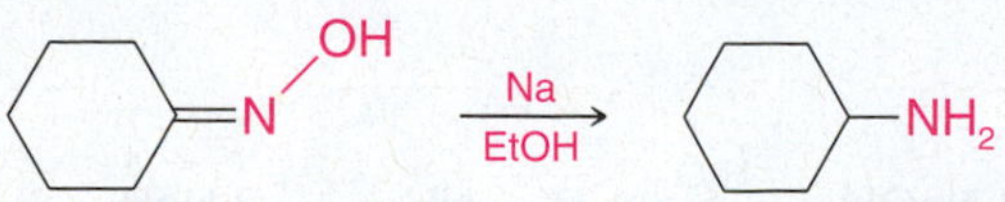

[structure: cyclohexanone oxime =N–OH] $\xrightarrow[\text{EtOH}]{\text{Na}}$ [structure: cyclohexyl–NH₂]

50–60%

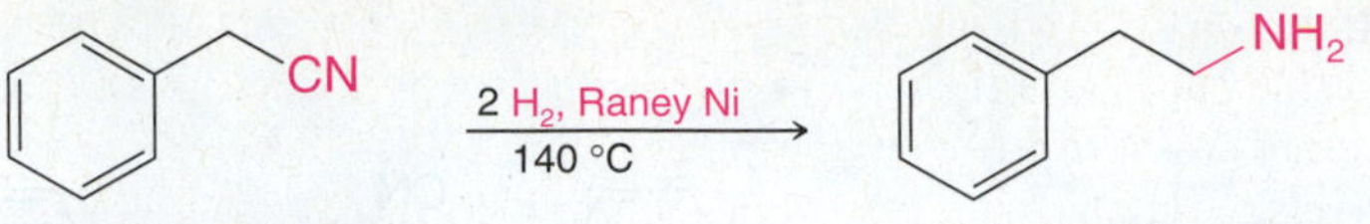

2-Phenylethanenitrile
(phenylacetonitrile)

$\xrightarrow[\text{140 °C}]{\text{2 H}_2\text{, Raney Ni}}$

2-Phenylethanamine
(71%)

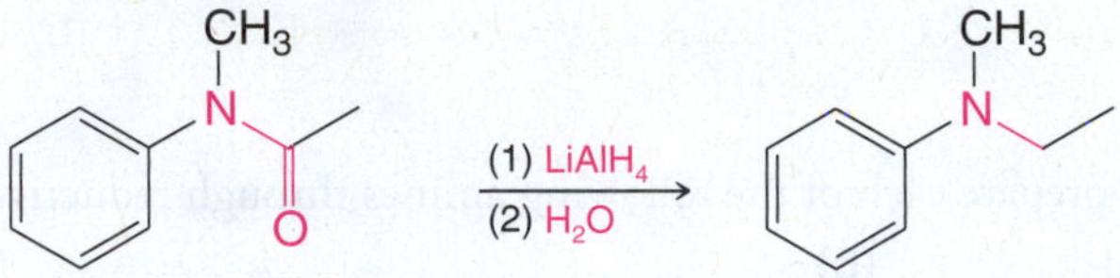

N-Methylacetanilide

$\xrightarrow[\text{(2) H}_2\text{O}]{\text{(1) LiAlH}_4}$

N-Ethyl-N-methylaniline

Reduction of an amide is the last step in a useful procedure for **monoalkylation of an amine**. The process begins with *acylation* of the amine using an acyl chloride or acid anhydride; then the amide is reduced with lithium aluminum hydride. For example,

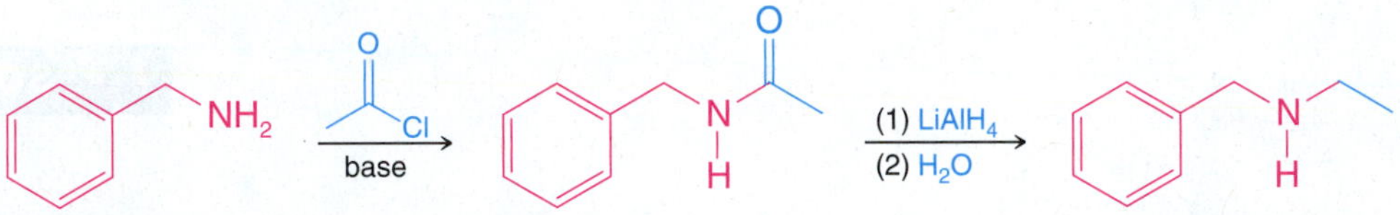

Benzylamine

$\xrightarrow{\text{base}}$

$\xrightarrow[\text{(2) H}_2\text{O}]{\text{(1) LiAlH}_4}$

Benzylethylamine

●●● SOLVED PROBLEM 20.4

Show how you might synthesize 2-propanamine from a three-carbon starting material that is a ketone, aldehyde, nitrile, or amide.

STRATEGY AND ANSWER: We begin by recognizing that 2-propanamine has a primary amine group bonded to a secondary carbon. Neither a three-carbon nitrile nor a three-carbon amide can lead to this structural unit from a C_3 starting material. An oxime can lead to the proper structure, but we must start with a three-carbon ketone rather than an aldehyde. Therefore, we choose propanone as our starting material, convert it to an oxime, and then reduce the oxime to an amine.

●●●

PRACTICE PROBLEM 20.6 Show how you might utilize the reduction of an amide, oxime, or nitrile to carry out each of the following transformations:

(a)

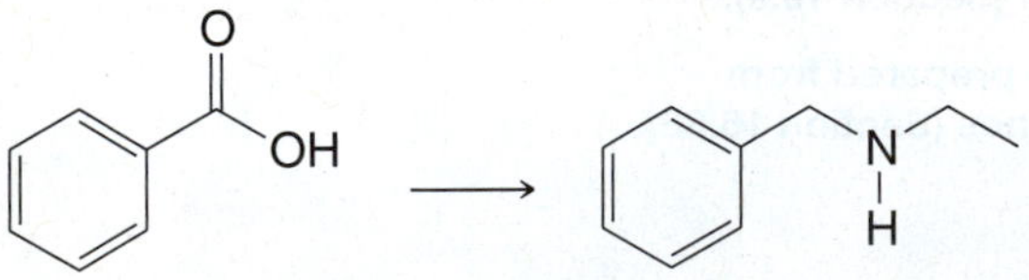

(b)

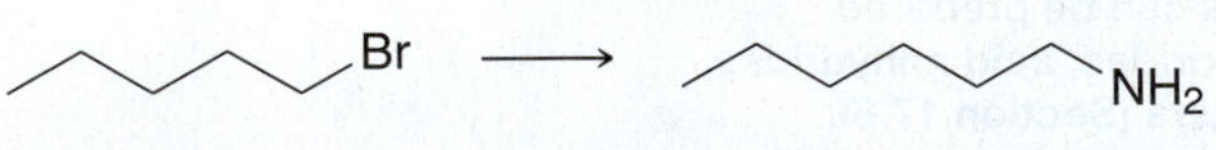

(c)

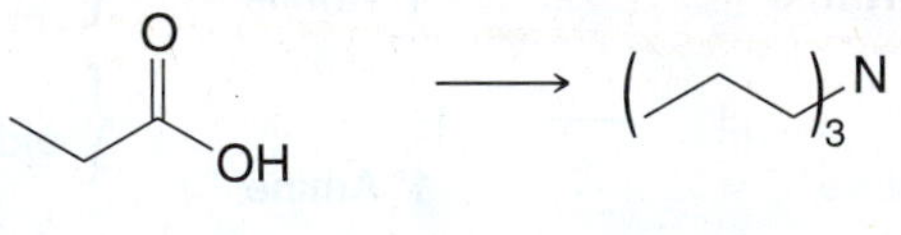

(d)

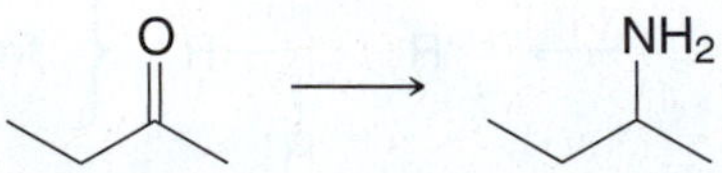

20.4E Preparation of Primary Amines through the Hofmann and Curtius Rearrangements

Hofmann Rearrangement Amides with no substituent on the nitrogen react with solutions of bromine or chlorine in sodium hydroxide to yield amines through loss of their carbonyl carbon by a reaction known as the *Hofmann rearrangement* or *Hofmann degradation*:

$+ \text{Br}_2 + \text{4 NaOH} \xrightarrow{\text{H}_2\text{O}} \text{R} - \text{NH}_2 + \text{2 NaBr} + \text{Na}_2\text{CO}_3 + \text{2 H}_2\text{O}$

From this equation we can see that the carbonyl carbon atom of the amide is lost (as CO_3^{2-}) and that the R group of the amide becomes attached to the nitrogen of the amine. Primary amines made this way are not contaminated by 2° or 3° amines.

The mechanism for this interesting reaction is shown in the following scheme. In the first two steps the amide undergoes a base-promoted bromination, in a manner analogous to the base-promoted halogenation of a ketone that we studied in Section 18.3B. (The electron-withdrawing acyl group of the amide makes the amido hydrogens much more acidic than those of an amine.) The *N*-bromo amide then reacts with hydroxide ion to produce an anion, which spontaneously rearranges with the loss of a bromide ion to produce an isocyanate (Section 17.9A). In the rearrangement the R— group migrates with its electrons from the acyl carbon to the nitrogen atom at the same time the bromide ion departs. The isocyanate that forms in the mixture is quickly hydrolyzed by the aqueous base to a carbamate ion, which undergoes spontaneous decarboxylation resulting in the formation of the amine.

[A MECHANISM FOR THE REACTION — The Hofmann Rearrangement]

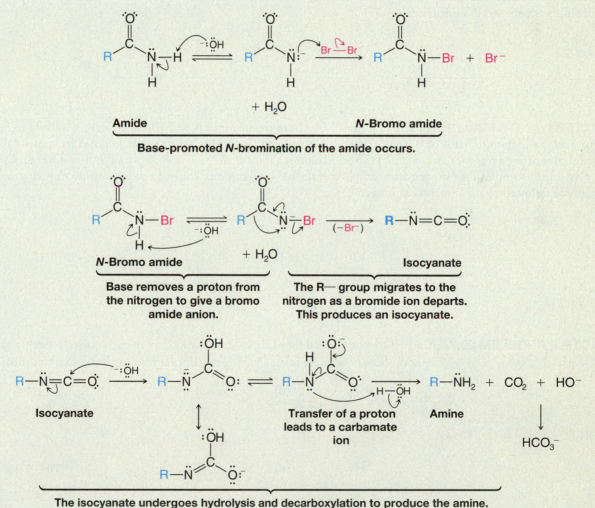

An examination of the first two steps of this mechanism shows that, initially, two hydrogen atoms must be present on the nitrogen of the amide for the reaction to occur. Consequently, the Hofmann rearrangement is limited to amides of the type $RCONH_2$.

Studies of the Hofmann rearrangement of optically active amides in which the chirality center is directly attached to the carbonyl group have shown that these reactions occur with *retention of configuration*. Thus, the R group migrates to nitrogen with its electrons, *but without inversion*.

Curtius Rearrangement The *Curtius rearrangement* is a rearrangement that occurs with acyl azides and yields a primary amine with loss of the acyl carbon. It resembles the Hofmann rearrangement in that an R— group migrates from the acyl carbon to the nitrogen atom as the leaving group departs. In this instance the leaving group is N_2 (the best of all possible leaving groups since it is highly stable, is virtually nonbasic, and being a gas, removes itself from the medium). Acyl azides are easily prepared by allowing acyl chlorides to react with sodium azide. Heating the acyl azide brings about the rearrangement; afterward, adding water causes hydrolysis and decarboxylation of the isocyanate:

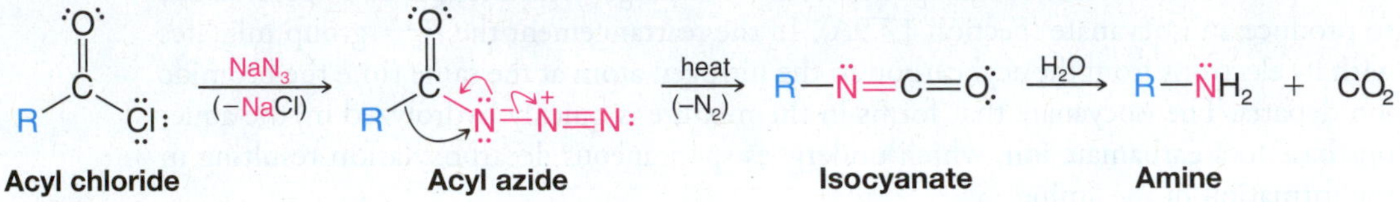

Acyl chloride   Acyl azide   Isocyanate   Amine

••• SOLVED PROBLEM 20.5

The reaction sequence below shows how a methyl group on a benzene ring can be replaced by an amino group. Supply the missing reagents and intermediates.

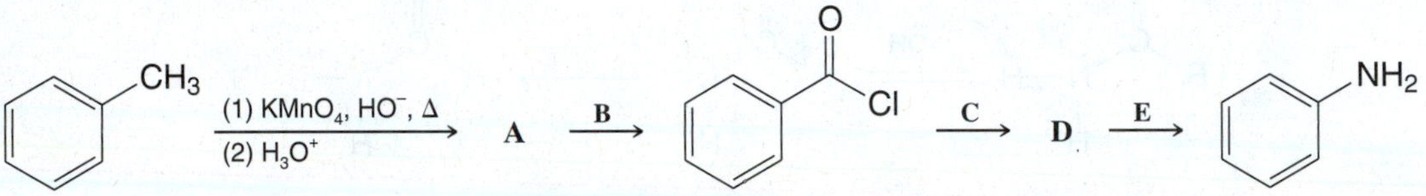

STRATEGY AND ANSWER: An acid chloride results from treatment of **A** with **B**. Therefore, **A** is likely to be a carboxylic acid, a conclusion that is consistent with the oxidizing conditions that led to formation of **A** from methylbenzene (toluene). **B** must be a reagent that can lead to an acid chloride. Thionyl chloride or PCl_5 would suffice. Overall, **C**, **D**, and **E** involve introduction of the nitrogen atom and loss of the carbonyl carbon. This sequence is consistent with preparation of an amide followed by a Hofmann rearrangement.

$A =$ (benzene ring with CO_2H) $B = SO_2Cl$ $C = NH_3$ $D =$ (benzene ring with $C(=O)NH_2$) $E = Br_2/HO^-$

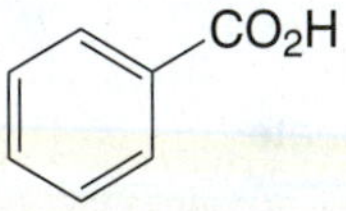

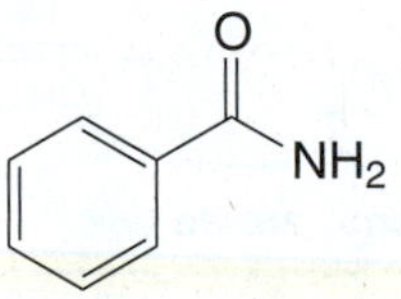

•••

PRACTICE PROBLEM 20.7 Using a different method for each part, but taking care in each case to select a *good* method, show how each of the following transformations might be accomplished:

(a)

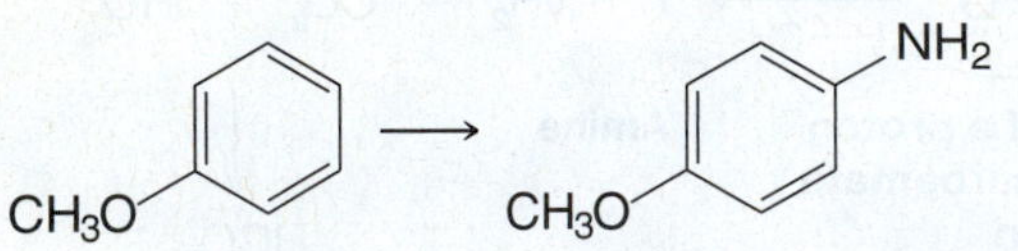

(b)

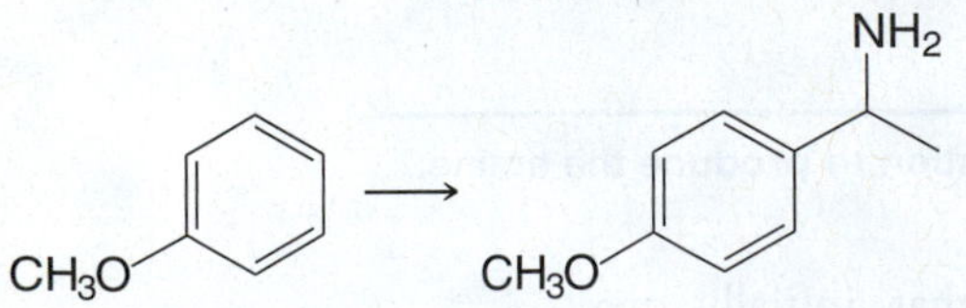

(c)

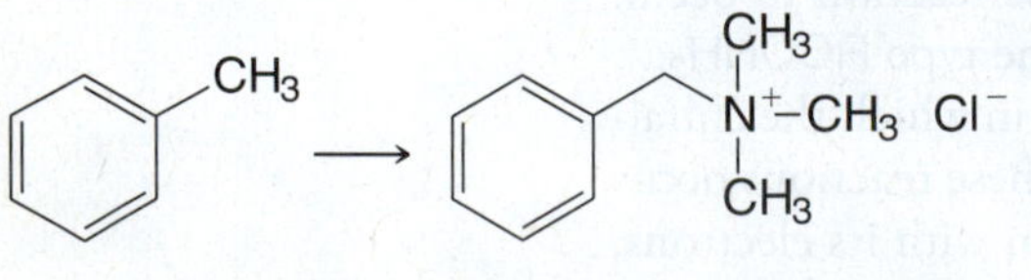

(d)

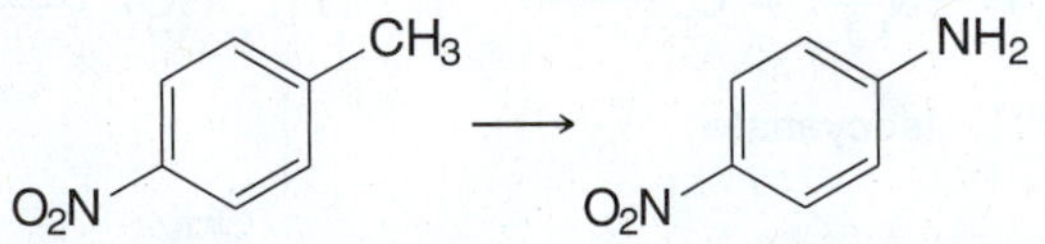

(e)

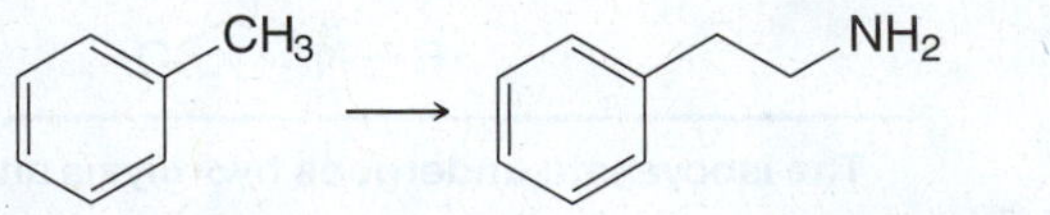

20.5 REACTIONS OF AMINES

We have encountered a number of important reactions of amines in earlier sections. In Section 20.3 we saw reactions in which primary, secondary, and tertiary amines act *as bases*. In Section 20.4 we saw their reactions as *nucleophiles* in *alkylation reactions*, and in Chapter 17 as *nucleophiles* in *acylation reactions*. In Chapter 15 we saw that an amino group on an aromatic ring acts as a powerful *activating group* and as an *ortho–para director*.

The feature of amines that underlies all of these reactions and that forms a basis for our understanding of most of the chemistry of amines is the ability of nitrogen to share an electron pair:

Acid–Base Reactions

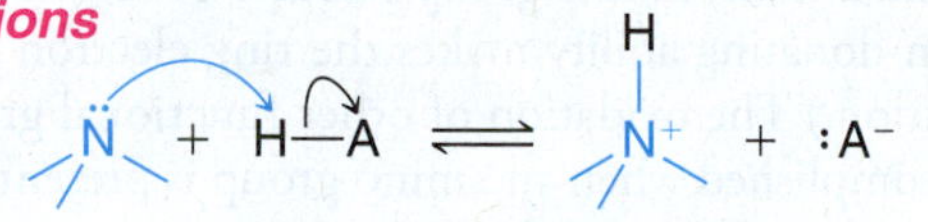

An amine acting as a base

Alkylation

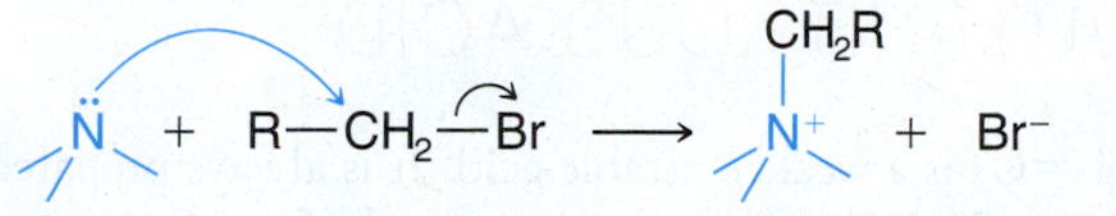

An amine acting as a nucleophile in an alkylation reaction

Acylation

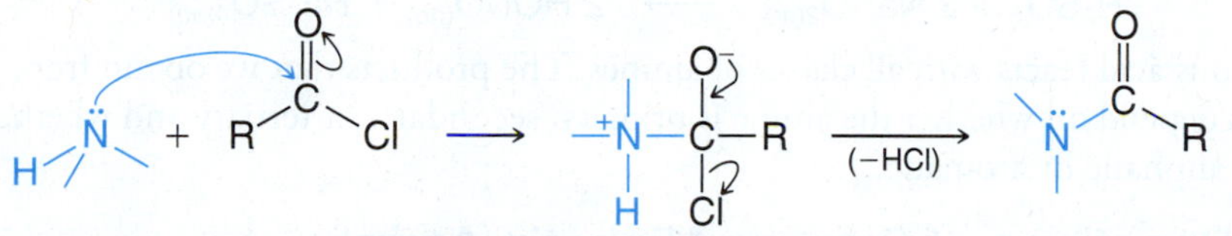

A primary or secondary amine acting as a nucleophile in an acylation reaction

In the preceding examples the amine acts as a nucleophile by donating its electron pair to an electrophilic reagent. In the following example, resonance contributions involving the nitrogen electron pair make *carbon* atoms nucleophilic:

Electrophilic Aromatic Substitution

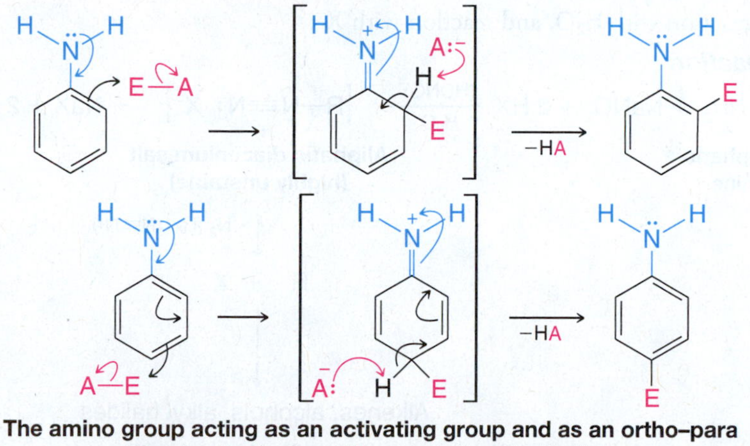

The amino group acting as an activating group and as an ortho–para director in electrophilic aromatic substitution

- - - ....

Review the chemistry of amines given in earlier sections and provide a specific example for each of the previously illustrated reactions.

PRACTICE PROBLEM 20.8

20.5A Oxidation of Amines

Primary and secondary aliphatic amines are subject to oxidation, although in most instances useful products are not obtained. Complicated side reactions often occur, causing the formation of complex mixtures.

Tertiary amines can be oxidized cleanly to tertiary amine oxides. This transformation can be brought about by using hydrogen peroxide or a peroxy acid:

$$R_3N: \xrightarrow{H_2O_2 \text{ or } RCOOH} R_3\overset{+}{N}-\overset{..}{\overset{..}{O}}:^-$$

A tertiary amine oxide

Tertiary amine oxides undergo a useful elimination reaction to be discussed in Section 20.12B.

Arylamines are very easily oxidized by a variety of reagents, including the oxygen in air. Oxidation is not confined to the amino group but also occurs in the ring. (The amino group through its electron-donating ability makes the ring electron rich and hence especially susceptible to oxidation.) The oxidation of other functional groups on an aromatic ring cannot usually be accomplished when an amino group is present on the ring, because oxidation of the ring takes place first.

20.6 REACTIONS OF AMINES WITH NITROUS ACID

Nitrous acid ($HO-N=O$) is a weak, unstable acid. It is always prepared *in situ*, usually by treating sodium nitrite ($NaNO_2$) with an aqueous solution of a strong acid:

$$HCl_{(aq)} + NaNO_{2(aq)} \longrightarrow HONO_{(aq)} + NaCl_{(aq)}$$
$$H_2SO_4 + 2\,NaNO_{2(aq)} \longrightarrow 2\,HONO_{(aq)} + Na_2SO_{4(aq)}$$

Nitrous acid reacts with all classes of amines. The products that we obtain from these reactions depend on whether the amine is primary, secondary, or tertiary and whether the amine is aliphatic or aromatic.

20.6A Reactions of Primary Aliphatic Amines with Nitrous Acid

Primary aliphatic amines react with nitrous acid through a reaction called *diazotization* to yield highly unstable aliphatic **diazonium salts**. Even at low temperatures, *aliphatic diazonium salts decompose spontaneously by losing nitrogen to form carbocations*. The carbocations go on to produce mixtures of alkenes, alcohols, and alkyl halides by removal of a proton, reaction with H_2O, and reaction with X^-:

General Reaction

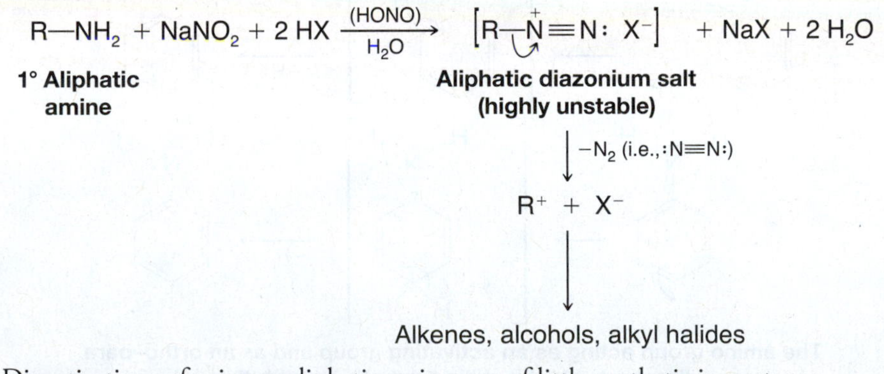

$$R-NH_2 + NaNO_2 + 2\,HX \xrightarrow[H_2O]{(HONO)} [R-\overset{+}{N}\equiv N: \; X^-] + NaX + 2\,H_2O$$

1° Aliphatic amine   **Aliphatic diazonium salt (highly unstable)**

$$\downarrow -N_2 \;(i.e., :N\equiv N:)$$

$$R^+ + X^-$$

$$\downarrow$$

Alkenes, alcohols, alkyl halides

- Diazotizations of primary aliphatic amines are of little synthetic importance because they yield such a complex mixture of products.

Diazotizations of primary aliphatic amines are used in some analytical procedures, however, because the evolution of nitrogen is quantitative. They can also be used to generate and thus study the behavior of carbocations in water, acetic acid, and other solvents.

20.6B Reactions of Primary Arylamines with Nitrous Acid

The most important reaction of amines with nitrous acid, by far, is the reaction of primary arylamines. We shall see why in Section 20.7.

- Primary arylamines react with nitrous acid to give arenediazonium salts.

Even though arenediazonium salts are unstable, they are still far more stable than aliphatic **diazonium salts**; they do not decompose at an appreciable rate in solution when the temperature of the reaction mixture is kept below 5 °C:

$$Ar—NH_2 \ + \ NaNO_2 \ + \ 2 HX \longrightarrow Ar—\overset{+}{N}\equiv N: \ X^- \ + \ NaX \ + \ 2 H_2O$$

Primary arylamine   **Arenediazonium salt**
                                         (stable if kept
                                         below 5 °C)

Diazotization of a primary amine takes place through a series of steps. In the presence of strong acid, nitrous acid dissociates to produce $^+$NO ions. These ions then react with the nitrogen of the amine to form an unstable *N*-nitrosoaminium ion as an intermediate. This intermediate then loses a proton to form an *N*-nitrosoamine, which, in turn, tautomerizes to a diazohydroxide in a reaction that is similar to keto–enol tautomerization. Then, in the presence of acid, the diazohydroxide loses water to form the diazonium ion.

Helpful Hint

Primary arylamines can be converted to aryl halides, nitriles, and phenols via aryl diazonium ions (Section 20.7).

[A MECHANISM FOR THE REACTION — Diazotization]

$$HO—\ddot{N}=O \ + \ H_3O^+ \ + \ A:^- \ \rightleftharpoons \ H_2\overset{+}{O}{-}\ddot{N}=O \ + \ H_2O \ \rightleftharpoons \ 2 H_2O \ + \ ^+\ddot{N}=O$$

1° Arylamine (or alkylamine) + $^+\ddot{N}=O$ $\longrightarrow$ *N*-Nitroso-aminium ion $\xrightarrow{-H_3O^+}$ *N*-Nitrosoamine

$$Ar—\ddot{N}—\ddot{N}=\overset{+}{O} \ \underset{+HA}{\overset{-HA}{\rightleftharpoons}} \ Ar—\ddot{N}=\ddot{N}—\ddot{O}H \ \underset{-HA}{\overset{+HA}{\rightleftharpoons}} \ Ar—\ddot{N}=\ddot{N}—\overset{+}{O}H_2 \ \rightleftharpoons$$

Diazohydroxide

$$Ar—\overset{+}{N}\equiv N: \longleftrightarrow Ar—\ddot{N}=\overset{+}{\overset{+}{N}} \ + \ H_2O$$

Diazonium ion

- Diazotization reactions of primary arylamines are of considerable synthetic importance because the diazonium group, $—\overset{+}{N}\equiv N:$ can be replaced by a variety of other functional groups.

We shall examine these reactions in Section 20.7.

THE CHEMISTRY OF... *N*-Nitrosoamines

N-Nitrosoamines are very powerful carcinogens which scientists fear may be present in many foods, especially in cooked meats that have been cured with sodium nitrite.

Sodium nitrite is added to many meats (e.g., bacon, ham, frankfurters, sausages, and corned beef) to inhibit the growth of *Clostridium botulinum* (the bacterium that produces botulinus toxin) and to keep red meats from turning brown. (Food poisoning by botulinus toxin is often fatal.) In the presence of acid or under the influence of heat, sodium nitrite reacts with amines always present in the meat to produce *N*-nitrosoamines. Cooked bacon, for example, has been shown to contain *N*-nitrosodimethylamine and *N*-nitrosopyrrolidine.

There is also concern that nitrites from food may produce nitrosoamines when they react with amines in the presence of the acid found in the stomach. In 1976, the FDA reduced the permissible amount of nitrite allowed in cured meats from 200 parts per million (ppm) to 50–125 ppm. Nitrites (and nitrates that can be converted to nitrites by bacteria) also occur naturally in many foods.

Cigarette smoke is known to contain *N*-nitrosodimethylamine. Someone smoking a pack of cigarettes a day inhales about 0.8 μg of *N*-nitrosodimethylamine, and even more has been shown to be present in the sidestream smoke.

20.6C Reactions of Secondary Amines with Nitrous Acid

Secondary amines—both aryl and alkyl—react with nitrous acid to yield *N*-nitrosoamines. *N*-Nitrosoamines usually separate from the reaction mixture as oily yellow liquids:

Specific Examples

$$(CH_3)_2\ddot{N}H \ + \ HCl \ + \ NaNO_2 \ \xrightarrow[H_2O]{(HONO)} \ (CH_3)_2\ddot{N}-\ddot{N}=O$$

Dimethylamine   **N-Nitrosodimethylamine**
                                                         (a yellow oil)

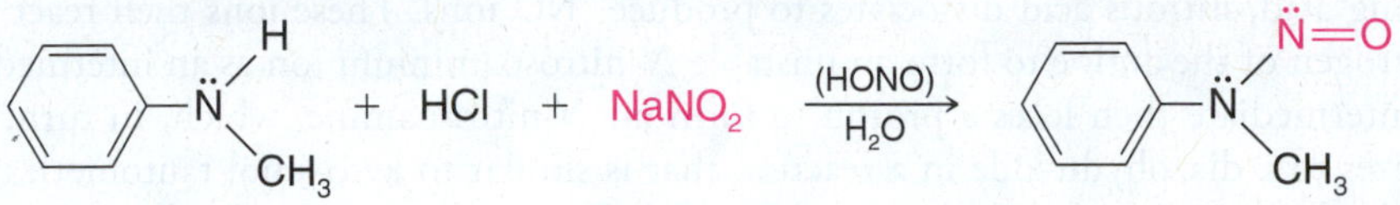

N-Methylaniline + HCl + NaNO$_2$ $\xrightarrow[H_2O]{(HONO)}$ N-Nitroso-*N*-methylaniline

N-Methylaniline   **N-Nitroso-*N*-methylaniline**
                                                         (87–93%)
                                                       (a yellow oil)

20.6D Reactions of Tertiary Amines with Nitrous Acid

When a tertiary aliphatic amine is mixed with nitrous acid, an equilibrium is established among the tertiary amine, its salt, and an *N*-nitrosoammonium compound:

$$2\ R_3N\text{:} \ + \ HX \ + \ NaNO_2 \ \rightleftharpoons \ R_3\overset{+}{N}H \ X^- \ + \ R_3\overset{+}{N}-\ddot{N}=O \ X^-$$

Tertiary aliphatic   **Amine salt**   **N-Nitrosoammonium**
amine   **compound**

Although *N*-nitrosoammonium compounds are stable at low temperatures, at higher temperatures and in aqueous acid they decompose to produce aldehydes or ketones. These reactions are of little synthetic importance, however.

Tertiary arylamines react with nitrous acid to form *C*-nitroso aromatic compounds. Nitrosation takes place almost exclusively at the para position if it is open and, if not, at the ortho position. The reaction (see Practice Problem 20.9) is another example of electrophilic aromatic substitution.

Specific Example

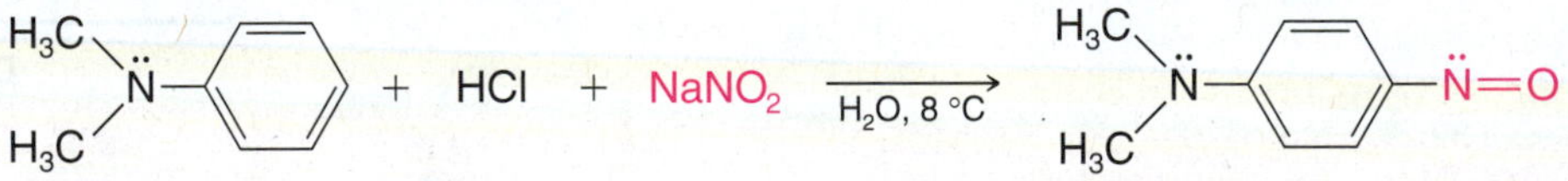

N,N-dimethylaniline + HCl + NaNO$_2$ $\xrightarrow{H_2O,\ 8\ °C}$ *p*-Nitroso-*N,N*-dimethylaniline

p-Nitroso-N,N-dimethylaniline
(80–90%)

●●●
PRACTICE PROBLEM 20.9 Para-nitrosation of *N,N*-dimethylaniline (*C*-nitrosation) is believed to take place through an electrophilic attack by $\overset{+}{N}O$ ions. **(a)** Show how $\overset{+}{N}O$ ions might be formed in an aqueous solution of NaNO$_2$ and HCl. **(b)** Write a mechanism for *p*-nitrosation of *N,N*-dimethylaniline. **(c)** Tertiary aromatic amines and phenols undergo *C*-nitrosation reactions, whereas most other benzene derivatives do not. How can you account for this difference?

20.7 REPLACEMENT REACTIONS OF ARENEDIAZONIUM SALTS

- Arenediazonium salts are highly useful intermediates in the synthesis of aromatic compounds, because the diazonium group can be replaced by any one of a number of other atoms or groups, including —F, —Cl, —Br, —I, —CN, —OH, and —H.

Diazonium salts are almost always prepared by diazotizing primary aromatic amines. Primary arylamines can be synthesized through reduction of nitro compounds that are readily available through direct nitration reactions.

20.7A Syntheses Using Diazonium Salts

Most arenediazonium salts are unstable at temperatures above 5–10 °C, and many explode when dry. Fortunately, however, most of the replacement reactions of diazonium salts do not require their isolation. We simply add another reagent (CuCl, CuBr, KI, etc.) to the mixture, gently warm the solution, and the replacement (accompanied by the evolution of nitrogen) takes place:

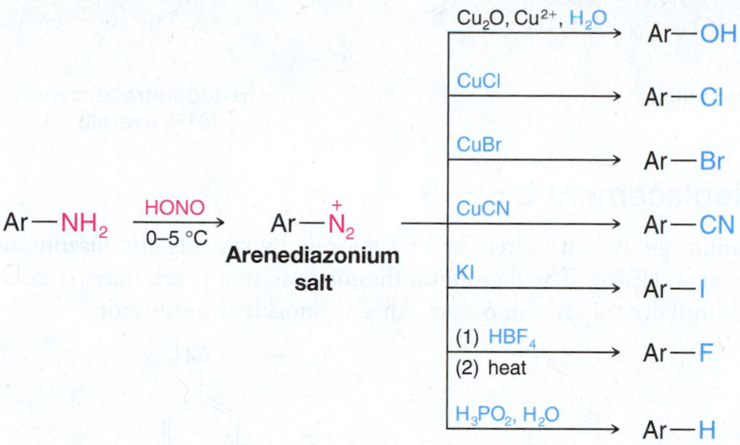

Only in the replacement of the diazonium group by —F need we isolate a diazonium salt. We do this by adding HBF_4 to the mixture, causing the sparingly soluble and reasonably stable arenediazonium fluoroborate, $ArN_2^+\ BF_4^-$, to precipitate.

20.7B The Sandmeyer Reaction: Replacement of the Diazonium Group by —Cl, —Br, or —CN

Arenediazonium salts react with cuprous chloride, cuprous bromide, and cuprous cyanide to give products in which the diazonium group has been replaced by —Cl, —Br, and —CN, respectively. These reactions are known generally as *Sandmeyer reactions*. Several specific examples follow. The mechanisms of these replacement reactions are not fully understood; the reactions appear to be radical in nature, not ionic.

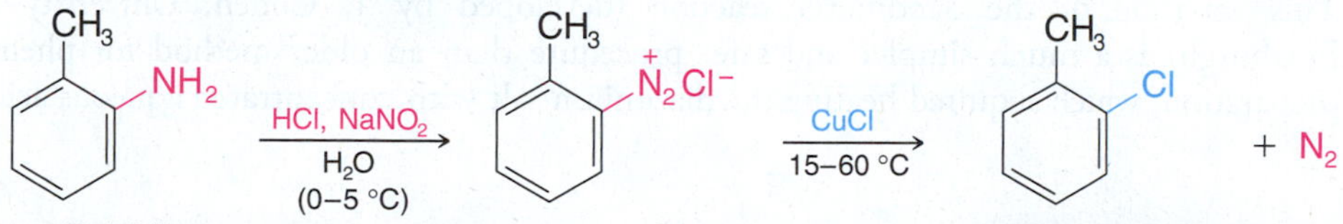

o-Toluidine **o-Chlorotoluene (74–79% overall)**

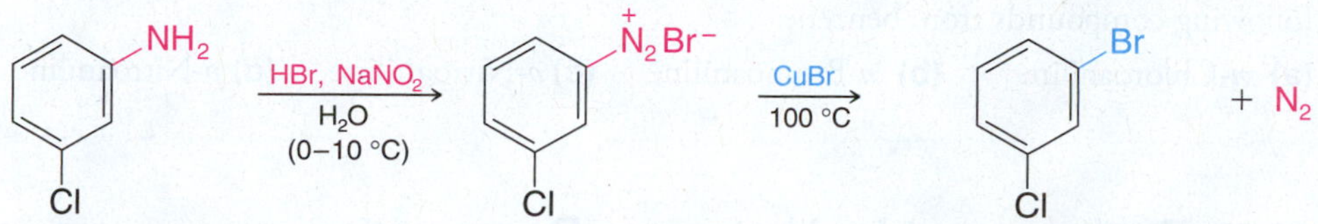

***m*-Chloroaniline** ***m*-Bromochlorobenzene (70% overall)**

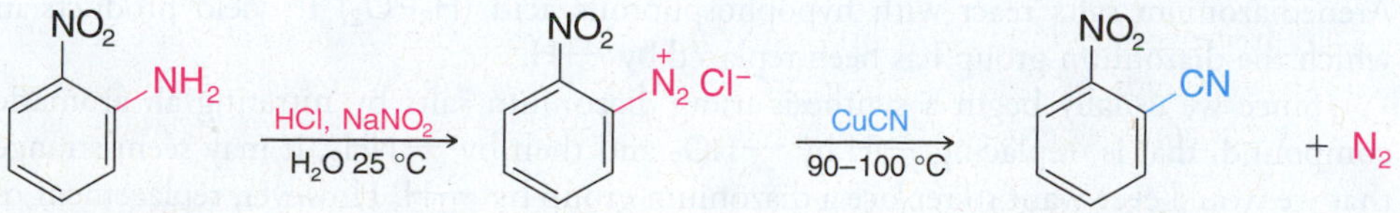

***o*-Nitroaniline** ***o*-Nitrobenzonitrile (65% overall)**

20.7C Replacement by —I

Arenediazonium salts react with potassium iodide to give products in which the diazonium group has been replaced by —I. An example is the synthesis of *p*-iodonitrobenzene:

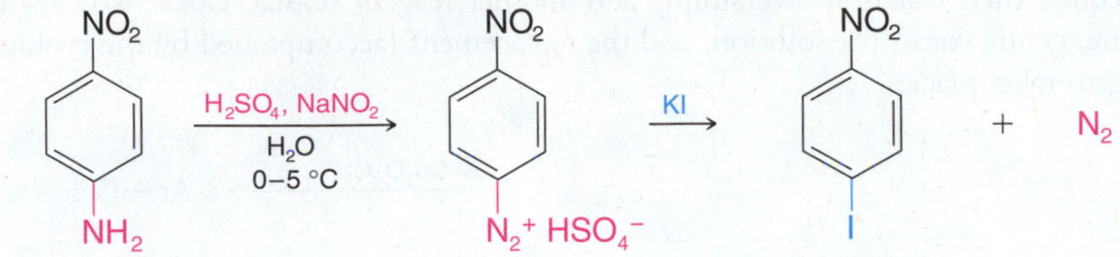

p-Nitroaniline

p-Iodonitrobenzene
(81% overall)

20.7D Replacement by —F

The diazonium group can be replaced by fluorine by treating the diazonium salt with fluoroboric acid (HBF$_4$). The diazonium fluoroborate that precipitates is isolated, dried, and heated until decomposition occurs. An aryl fluoride is produced:

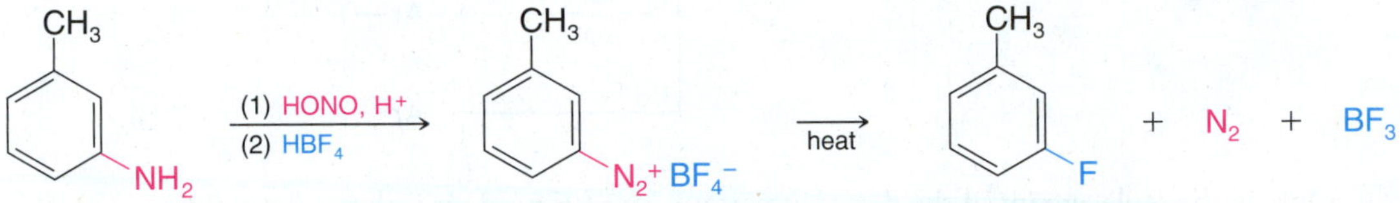

m-Toluidine

m-Toluenediazonium fluoroborate
(79%)

m-Fluorotoluene
(69%)

20.7E Replacement by —OH

The diazonium group can be replaced by a hydroxyl group by adding cuprous oxide to a dilute solution of the diazonium salt containing a large excess of cupric nitrate:

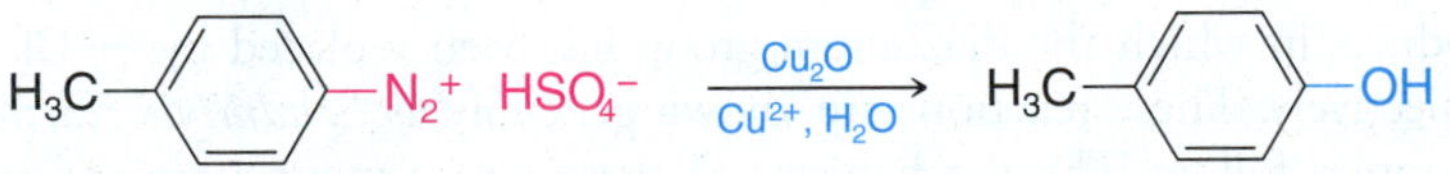

p-Toluenediazonium hydrogen sulfate

p-Cresol
(93%)

This variation of the Sandmeyer reaction (developed by T. Cohen, University of Pittsburgh) is a much simpler and safer procedure than an older method for phenol preparation, which required heating the diazonium salt with concentrated aqueous acid.

PRACTICE PROBLEM 20.10 In the preceding examples of diazonium reactions, we have illustrated syntheses beginning with the compounds (a)–(d) here. Show how you might prepare each of the following compounds from benzene:

(a) *m*-Chloroaniline   **(b)** *m*-Bromoaniline   **(c)** *o*-Nitroaniline   **(d)** *p*-Nitroaniline

20.7F Replacement by Hydrogen: Deamination by Diazotization

Arenediazonium salts react with hypophosphorous acid (H$_3$PO$_2$) to yield products in which the diazonium group has been replaced by —H.

Since we usually begin a synthesis using diazonium salts by nitrating an aromatic compound, that is, replacing —H by —NO$_2$ and then by —NH$_2$, it may seem strange that we would ever want to replace a diazonium group by —H. However, replacement of the diazonium group by —H can be a useful reaction. We can introduce an amino group into an aromatic ring to influence the orientation of a subsequent reaction. Later we can remove the amino group (i.e., carry out a *deamination*) by diazotizing it and treating the diazonium salt with H_3PO_2.

We can see an example of the usefulness of a deamination reaction in the following synthesis of *m*-bromotoluene.

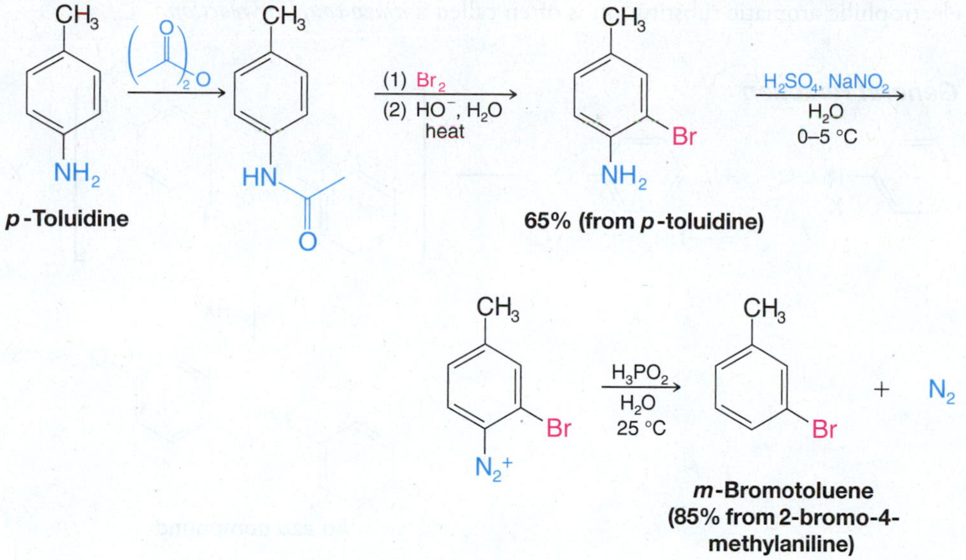

p-Toluidine   65% (from *p*-toluidine)

m-Bromotoluene
(85% from 2-bromo-4-
methylaniline)

We cannot prepare *m*-bromotoluene by direct bromination of toluene or by a Friedel–Crafts alkylation of bromobenzene because both reactions give *o*- and *p*-bromotoluene. (Both CH_3— and Br— are ortho–para directors.) However, if we begin with *p*-toluidine (prepared by nitrating toluene, separating the para isomer, and reducing the nitro group), we can carry out the sequence of reactions shown and obtain *m*-bromotoluene in good yield. The first step, synthesis of the *N*-acetyl derivative of *p*-toluidine, is done to reduce the activating effect of the amino group. (Otherwise both ortho positions would be brominated.) Later, the acetyl group is removed by hydrolysis.

●●● SOLVED PROBLEM 20.6

Suggest how you might modify the preceding synthesis in order to prepare 3,5-dibromotoluene.

STRATEGY AND ANSWER: An amino group is a stronger activating group than an amido group. If we brominate directly with the amino group present, rather than after converting the amine to an amide, we can brominate both ortho positions. We must also be sure to provide sufficient bromine.

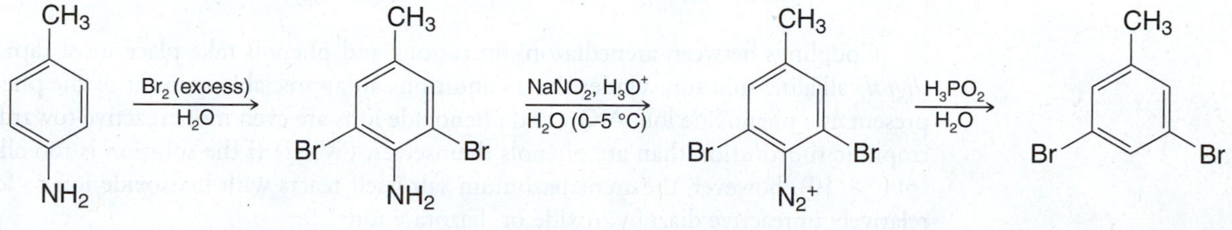

●●●

(a) In Section 20.7D we showed a synthesis of *m*-fluorotoluene starting with *m*-toluidine. How would you prepare *m*-toluidine from toluene? **(b)** How would you prepare *m*-chlorotoluene? **(c)** *m*-Bromotoluene? **(d)** *m*-Iodotoluene? **(e)** *m*-Tolunitrile (m-$CH_3C_6H_4CN$)? **(f)** *m*-Toluic acid?

PRACTICE PROBLEM 20.11

●●●

Starting with *p*-nitroaniline [Practice Problem 20.10(d)], show how you might synthesize 1,2,3-tribromobenzene.

PRACTICE PROBLEM 20.12

20.8 COUPLING REACTIONS OF ARENEDIAZONIUM SALTS

Arenediazonium ions are weak electrophiles; they react with highly reactive aromatic compounds—with phenols and tertiary arylamines—to yield *azo* compounds. This electrophilic aromatic substitution is often called a *diazo coupling reaction*.

General Reaction

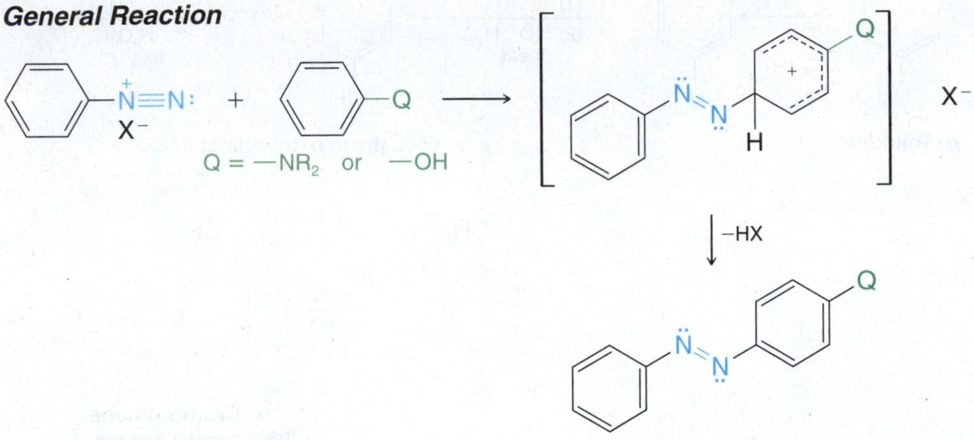

$Q = -NR_2$ or $-OH$

An azo compound

Specific Examples

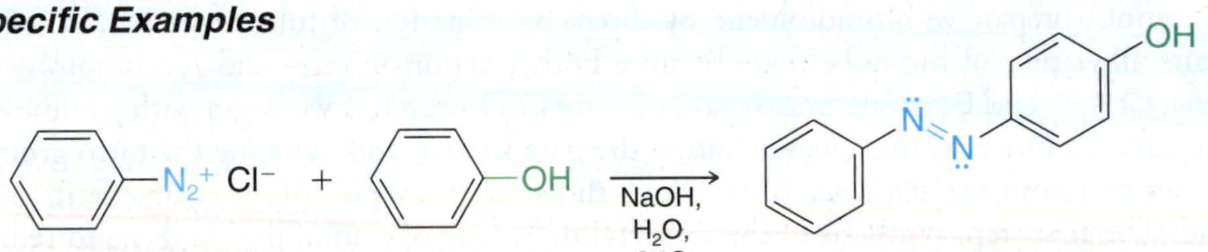

Benzenediazonium chloride + **Phenol** → $\xrightarrow[\substack{H_2O, \\ 0\,^\circ C}]{NaOH,}$ ***p*-(Phenylazo)phenol (orange solid)**

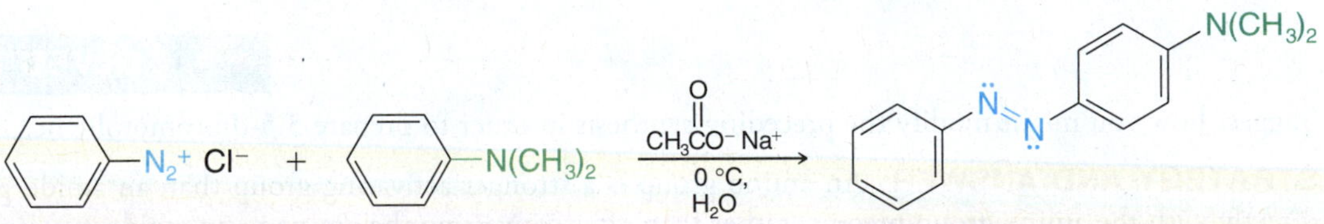

Benzenediazonium chloride + ***N,N*-Dimethylaniline** → $\xrightarrow[\substack{0\,^\circ C, \\ H_2O}]{CH_3CO^- Na^+}$ ***N,N*-Dimethyl-*p*-(phenylazo)aniline (yellow solid)**

Couplings between arenediazonium cations and phenols take place most rapidly in *slightly* alkaline solution. Under these conditions an appreciable amount of the phenol is present as a phenoxide ion, ArO$^-$, and phenoxide ions are even more reactive toward electrophilic substitution than are phenols themselves. (Why?) If the solution is too alkaline (pH > 10), however, the arenediazonium salt itself reacts with hydroxide ion to form a relatively unreactive diazohydroxide or diazotate ion:

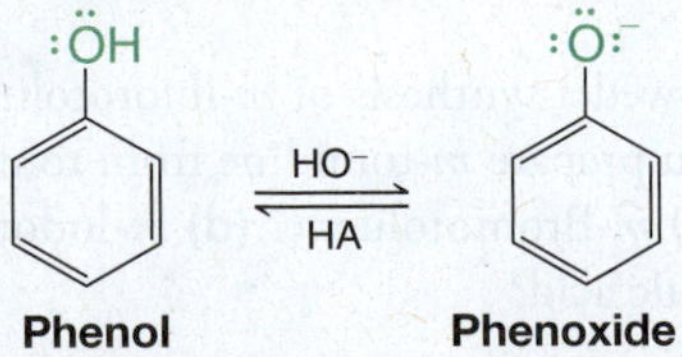

Phenol (couples slowly) $\underset{HA}{\overset{HO^-}{\rightleftharpoons}}$ **Phenoxide ion (couples rapidly)**

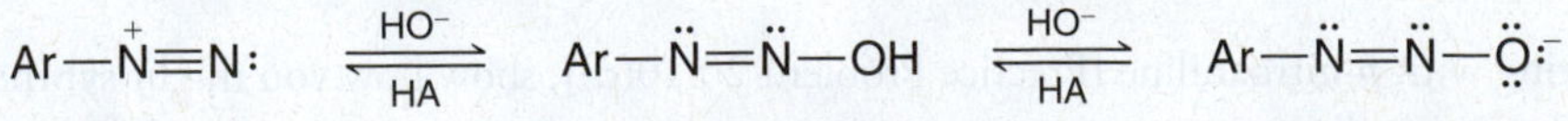

$Ar-\overset{+}{N}\equiv N:$ $\underset{HA}{\overset{HO^-}{\rightleftharpoons}}$ $Ar-\ddot{N}=\ddot{N}-OH$ $\underset{HA}{\overset{HO^-}{\rightleftharpoons}}$ $Ar-\ddot{N}=\ddot{N}-\ddot{\underset{..}{O}}:^-$

Arenediazonium ion (couples)   **Diazohydroxide (does not couple)**   **Diazotate ion (does not couple)**

Couplings between arenediazonium cations and amines take place most rapidly in slightly acidic solutions (pH 5–7). Under these conditions the concentration of the arenediazonium cation is at a maximum; at the same time an excessive amount of the amine has not been converted to an unreactive aminium salt:

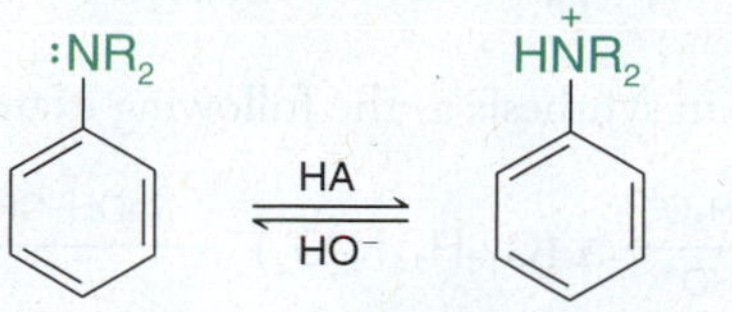

Amine (couples)   **Aminium salt** (does not couple)

If the pH of the solution is lower than 5, the rate of amine coupling is low.

With phenols and aniline derivatives, coupling takes place almost exclusively at the para position if it is open. If it is not, coupling takes place at the ortho position.

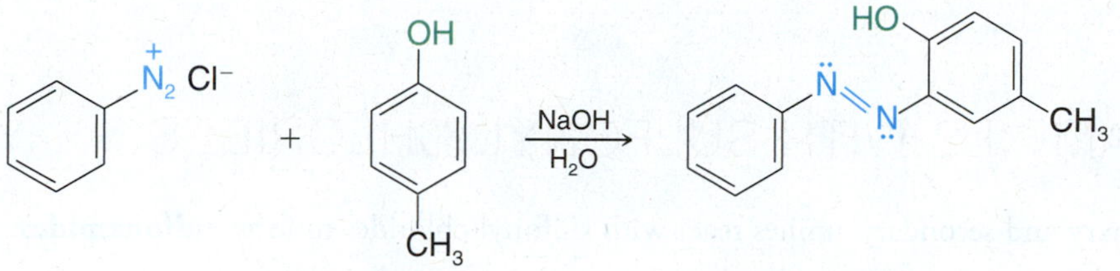

4-Methylphenol (p-cresol)   **4-Methyl-2-(phenylazo)phenol**

Azo compounds are usually intensely colored because the azo (diazenediyl) linkage, —N=N—, brings the two aromatic rings into conjugation. This gives an extended system of delocalized π electrons and allows absorption of light in the visible region. Azo compounds, because of their intense colors and because they can be synthesized from relatively inexpensive compounds, are used extensively as *dyes*.

Azo dyes almost always contain one or more —$SO_3^-Na^+$ groups to confer water solubility on the dye and assist in binding the dye to the surfaces of polar fibers (wool, cotton, or nylon). Many dyes are made by coupling reactions of naphthylamines and naphthols.

Orange II, a dye introduced in 1876, is made from 2-naphthol:

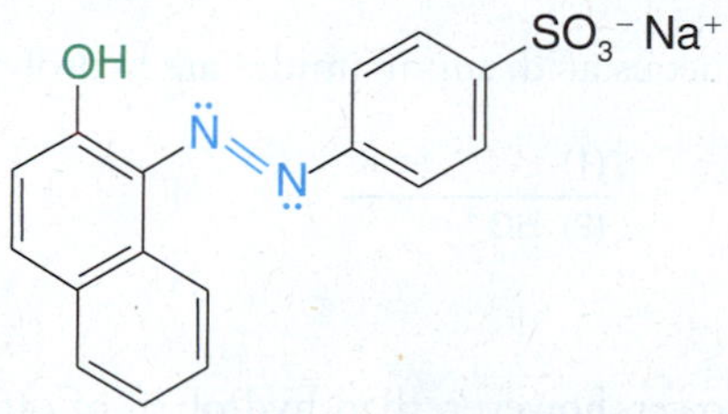

Orange II

Outline a synthesis of orange II from 2-naphthol and *p*-aminobenzenesulfonic acid.

PRACTICE PROBLEM 20.13

Butter yellow is a dye once used to color margarine. It has since been shown to be carcinogenic, and its use in food is no longer permitted. Outline a synthesis of butter yellow from benzene and *N,N*-dimethylaniline.

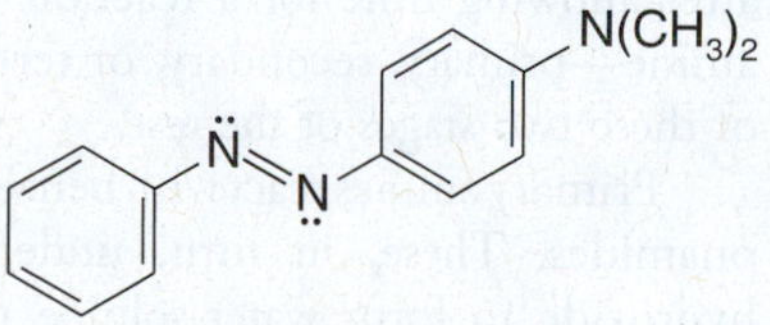

Butter yellow

PRACTICE PROBLEM 20.14

PRACTICE PROBLEM 20.15 Azo compounds can be reduced to amines by a variety of reagents including stannous chloride ($SnCl_2$):

$$Ar-N=N-Ar' \xrightarrow{SnCl_2} ArNH_2 + Ar'NH_2$$

This reduction can be useful in synthesis as the following example shows:

4-Ethoxyaniline $\xrightarrow[\text{(2) phenol, HO}^-]{\text{(1) HONO, H}_3\text{O}^+}$ **A** ($C_{14}H_{14}N_2O_2$) $\xrightarrow{\text{NaOH, CH}_3\text{CH}_2\text{Br}}$

B ($C_{16}H_{18}N_2O_2$) $\xrightarrow{SnCl_2}$

two molar equivalents of **C** ($C_8H_{11}NO$) $\xrightarrow{\text{acetic anhydride}}$ phenacetin ($C_{10}H_{13}NO_2$)

Give a structure for phenacetin and for the intermediates **A**, **B**, and **C**. (Phenacetin, formerly used as an analgesic, is also the subject of Problem 17.45.)

20.9 REACTIONS OF AMINES WITH SULFONYL CHLORIDES

Primary and secondary amines react with sulfonyl chlorides to form **sulfonamides**:

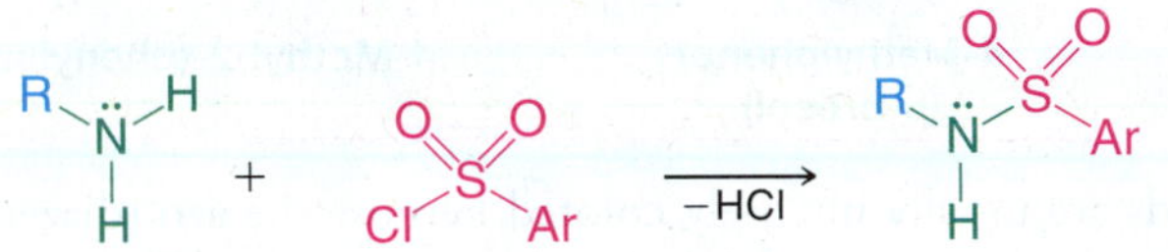

| 1° Amine | Sulfonyl chloride | | *N*-Substituted sulfonamide |

| 2° Amine | | | *N,N*-Disubstituted sulfonamide |

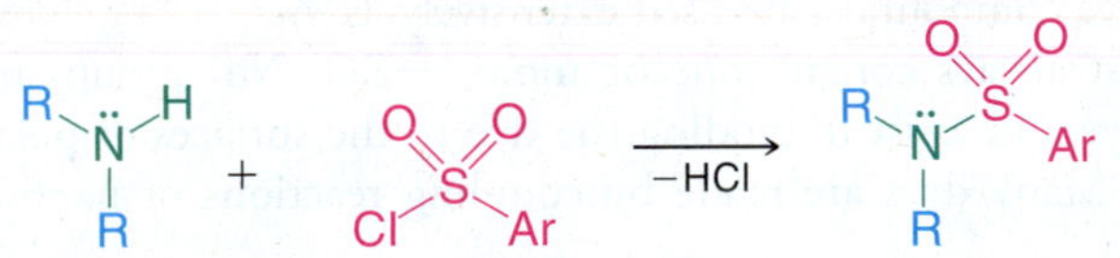

When heated with aqueous acid, sulfonamides are hydrolyzed to amines:

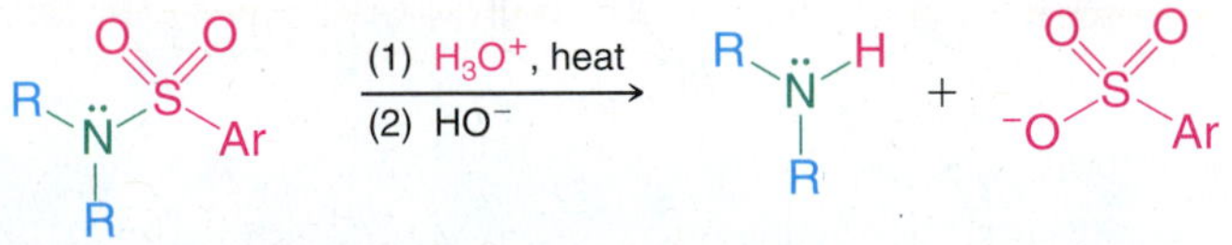

This hydrolysis is much slower, however, than hydrolysis of carboxamides.

20.9A The Hinsberg Test

- Sulfonamide formation is the basis for a chemical test, called the Hinsberg test, that can be used to demonstrate whether an amine is primary, secondary, or tertiary.

A Hinsberg test involves two steps. First, a mixture containing a small amount of the amine and benzenesulfonyl chloride is shaken with *excess* potassium hydroxide. Next, after allowing time for a reaction to take place, the mixture is acidified. Each type of amine—primary, secondary, or tertiary—gives a different set of *visible* results after each of these two stages of the test.

Primary amines react with benzenesulfonyl chloride to form *N*-substituted benzenesulfonamides. These, in turn, undergo acid–base reactions with the excess potassium hydroxide to form water-soluble potassium salts. (These reactions take place because the hydrogen attached to nitrogen is made acidic by the strongly electron-withdrawing —SO_2— group.) At this stage our test tube contains a clear solution. Acidification of this solution will, in the next stage, cause the water-insoluble *N*-substituted sulfonamide to precipitate:

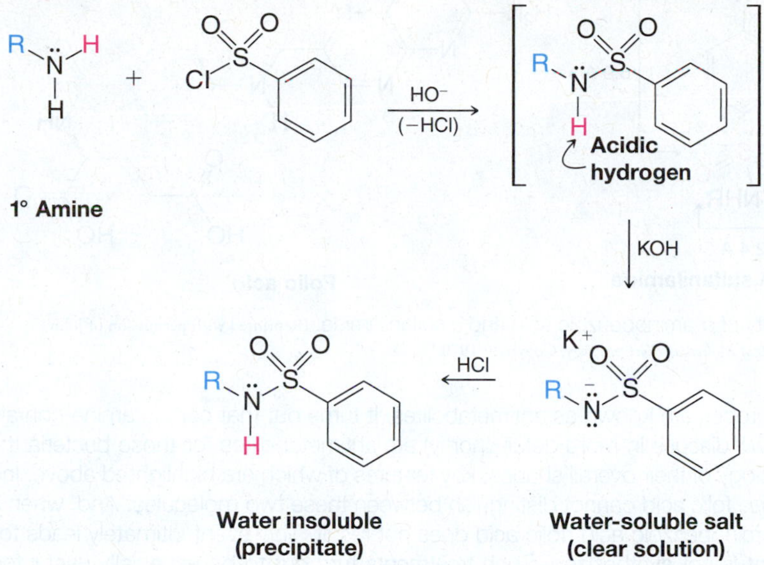

1° Amine

Acidic hydrogen

Water insoluble (precipitate)

Water-soluble salt (clear solution)

Secondary amines react with benzenesulfonyl chloride in aqueous potassium hydroxide to form insoluble *N,N*-disubstituted sulfonamides that precipitate after the first stage. *N,N*-Disubstituted sulfonamides do not dissolve in aqueous potassium hydroxide because they do not have an acidic hydrogen. Acidification of the mixture obtained from a secondary amine produces no visible result; the nonbasic *N,N*-disubstituted sulfonamide remains as a precipitate and no new precipitate forms:

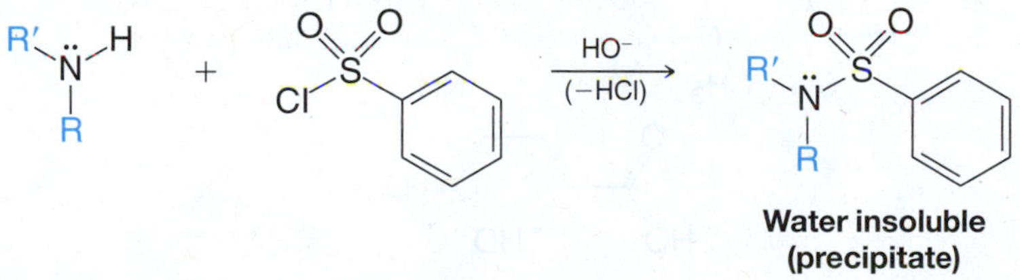

Water insoluble (precipitate)

If the amine is a tertiary amine and if it is water insoluble, no apparent change will take place in the mixture as we shake it with benzenesulfonyl chloride and aqueous KOH. When we acidify the mixture, the tertiary amine dissolves because it forms a water-soluble salt.

PRACTICE PROBLEM 20.16

An amine **A** has the molecular formula C_7H_9N. Compound **A** reacts with benzenesulfonyl chloride in aqueous potassium hydroxide to give a clear solution; acidification of the solution gives a precipitate. When **A** is treated with $NaNO_2$ and HCl at 0–5 °C, and then with 2-naphthol, an intensely colored compound is formed. Compound **A** gives a single strong IR absorption peak at 815 cm^{-1}. What is the structure of **A**?

PRACTICE PROBLEM 20.17

Sulfonamides of primary amines are often used to synthesize *pure* secondary amines. Suggest how this synthesis is carried out.

THE CHEMISTRY OF... Essential Nutrients and Antimetabolites

All higher animals and many microorganisms lack the biochemical ability to synthesize certain essential organic compounds. These essential nutrients include many amine-containing compounds, such as vitamins, certain amino acids, unsaturated carboxylic acids, components of DNA bases such as purines and pyrimidines. The aromatic amine *p*-aminobenzoic acid, for example, is an essential nutrient for many bacteria (see Figure 20.3). These microorganisms rely on *p*-aminobenzoic acid as a key starting material, along with several other compounds, to synthesize folic acid in enzymatically-controlled processes.

(continues on next page)

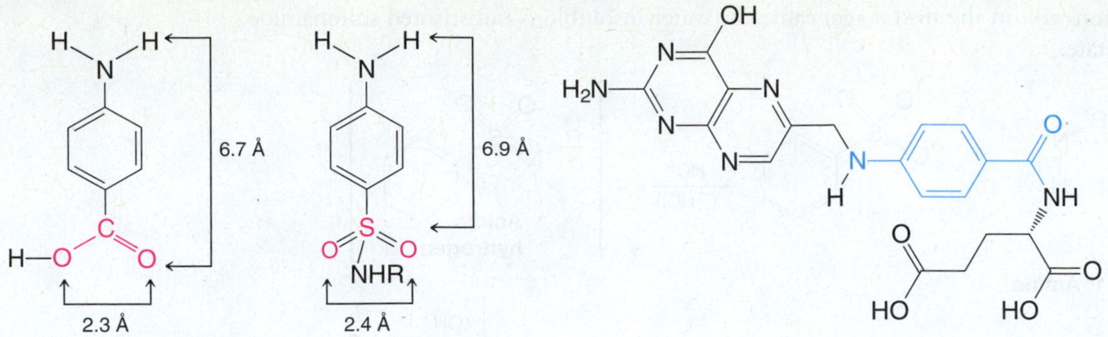

p-Aminobenzoic acid **A sulfanilamide** **Folic acid**

FIGURE 20.3 The structural similarity of *p*-aminobenzoic acid and a sulfanilamide. (Reprinted with permission of John Wiley and Sons, Inc. from Korolkovas, *Essentials of Molecular Pharmacology*, Copyright 1970.)

Chemicals that inhibit the growth of microbes are known as antimetabolites. It turns out that certain amine-containing molecules known as sulfanilamides (which we will discuss in more detail shortly) are antimetabolites for those bacteria that rely on *p*-aminobenzoic acid. The reason: the homology of their overall shapes, key features of which are highlighted above. Indeed, the enzymes that these bacteria use to synthesize folic acid cannot distinguish between these two molecules. And, when a sulfanilamide is used as a substrate instead of *p*-aminobenzoic acid, folic acid does not result. This event ultimately leads to bacterial death since enough of that essential nutrient is not synthesized. Such treatments turn out to be especially useful for humans because we derive our folic acid from dietary sources (folic acid is a vitamin). Thus, we do not have any enzymes that synthesize it from *p*-aminobenzoic acid and are, as a result, unaffected in any negative ways by a sulfanilamide therapy.

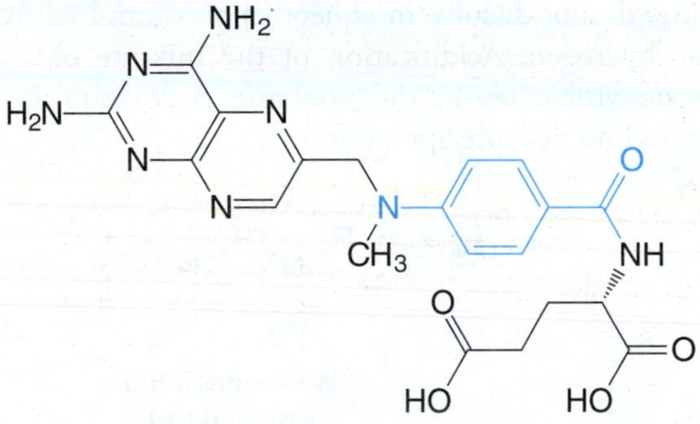

Methotrexate

Many other examples of this concept exist. A recent example is methotrexate, a derivative of folic acid that has been used successfully in treating certain carcinomas as well as rheumatoid arthritis. Just as in the case above, methotrexate, because of its resemblance to folic acid, can enter into some of the same chemical reactions as folic acid, but it cannot ultimately serve the same inherent biological function. Here that role is involvement in reactions critical to cellular division. Although methotrexate is toxic to all dividing cells, those cells that divide most rapidly—cancer cells—are most vulnerable to its effect.

20.10 SYNTHESIS OF SULFA DRUGS

Sulfanilamides (**sulfa drugs**) can be synthesized from aniline through the following sequence of reactions:

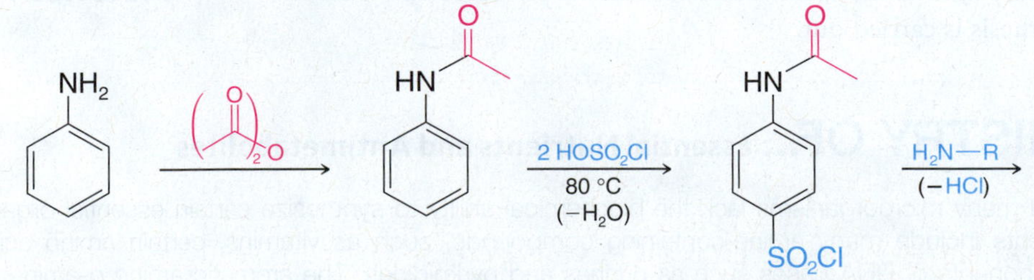

Aniline (1) **Acetanilide (2)** ***p*-Acetamidobenzenesulfonyl chloride (3)**

(continued on the next page)

(continued from previous page)

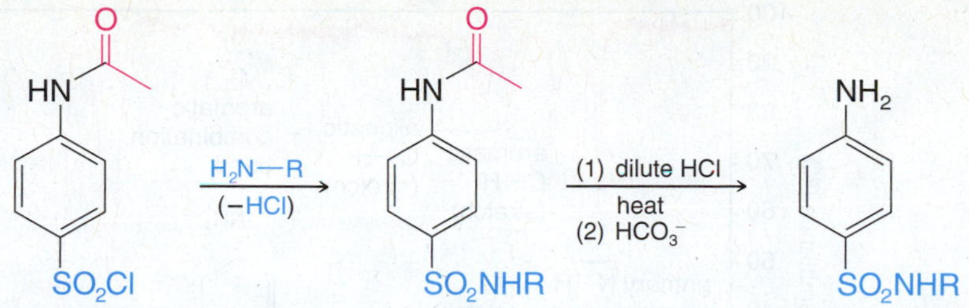

The scheme shows three structures. Structure **(3)** with an acetamido group (HN–C(=O)–CH₃) at top and SO₂Cl at bottom, labeled **p-Acetamidobenzene-sulfonyl chloride (3)**. An arrow with H_2N-R above and $(-HCl)$ below leads to structure **(4)** with HN–C(=O)–CH₃ at top and SO₂NHR at bottom. An arrow with (1) dilute HCl, heat, (2) HCO_3^- leads to structure **(5)** with NH_2 at top and SO₂NHR at bottom, labeled **A sulfanilamide (5)**.

Acetylation of aniline produces acetanilide (**2**) and protects the amino group from the reagent to be used next. Treatment of **2** with chlorosulfonic acid brings about an electrophilic aromatic substitution reaction and yields *p*-acetamidobenzene-sulfonyl chloride (**3**). Addition of ammonia or a primary amine gives the diamide, **4** (an amide of both a carboxylic acid and a sulfonic acid). Finally, refluxing **4** with dilute hydrochloric acid selectively hydrolyzes the carboxamide linkage and produces a sulfanilamide. (Hydrolysis of carboxamides is much more rapid than that of sulfonamides.)

(a) Starting with aniline and assuming that you have 2-aminothiazole available, show how you would synthesize sulfathiazole. **(b)** How would you convert sulfathiazole to succinylsulfathiazole?

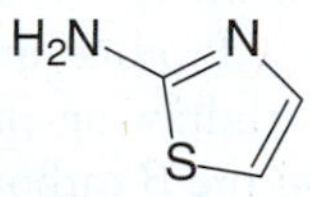

2-Aminothiazole

PRACTICE PROBLEM 20.18

20.11 ANALYSIS OF AMINES

20.11A Chemical Analysis

Amines are characterized by their basicity and, thus, by their ability to dissolve in dilute aqueous acid (Sections 20.3A, 20.3E). Moist pH paper can be used to test for the presence of an amine functional group in an unknown compound. If the compound is an amine, the pH paper shows the presence of a base. The unknown amine can then readily be classified as 1°, 2°, or 3° by IR spectroscopy (see below). Primary, secondary, and tertiary amines can also be distinguished from each other on the basis of the Hinsberg test (Section 20.9A). Primary aromatic amines are often detected through diazonium salt formation and subsequent coupling with 2-naphthol to form a brightly colored azo dye (Section 20.8).

20.11B Spectroscopic Analysis

Infrared Spectra Primary and secondary amines are characterized by IR absorption bands in the 3300–3555-cm^{-1} region that arise from N—H stretching vibrations. Primary amines give two bands in this region (see Fig. 20.4); secondary amines generally give only one. Tertiary amines, because they have no N—H group, do not absorb in this region. Absorption bands arising from C—N stretching vibrations of aliphatic amines occur in the 1020–1220-cm^{-1} region but are usually weak and difficult to identify. Aromatic amines generally give a strong C—N stretching band in the 1250–1360-cm^{-1} region. Figure 20.4 shows an annotated IR spectrum of 4-methylaniline.

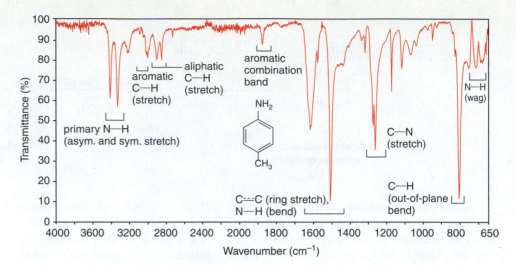

FIGURE 20.4 Annotated IR spectrum of 4-methylaniline.

¹H NMR Spectra Primary and secondary amines show N—H proton signals in the region δ 0.5–5. These signals are usually broad, and their exact position depends on the nature of the solvent, the purity of the sample, the concentration, and the temperature. Because of proton exchange, N—H protons are not usually coupled to protons on adjacent carbons. As such, they are difficult to identify and are best detected by proton counting or by adding a small amount of D_2O to the sample. Exchange of N—D deuterons for the N—H protons takes place, and the N—H signal disappears from the spectrum.

Protons on the α carbon of an aliphatic amine are deshielded by the electron-withdrawing effect of the nitrogen and absorb typically in the δ 2.2–2.9 region; protons on the β carbon are not deshielded as much and absorb in the range δ 1.0–1.7.

Figure 20.5 shows an annotated ¹H NMR spectrum of diisopropylamine.

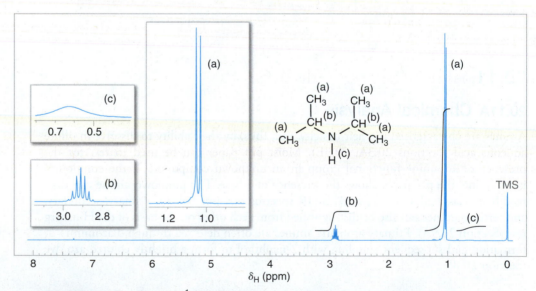

FIGURE 20.5 The 300-MHz ¹H NMR spectrum of diisopropylamine. Note the integral for the broad NH peak at approximately δ 0.7. Vertical expansions are not to scale.

¹³C NMR Spectra The α carbon of an aliphatic amine experiences deshielding by the electronegative nitrogen, and its absorption is shifted downfield, typically appearing at δ 30–60. The shift is not as great as for the α carbon of an alcohol (typically δ 50–75), however, because nitrogen is less electronegative than oxygen. The downfield shift is even less for the β carbon, and so on down the chain, as the chemical shifts of the carbons of pentyl amine show:

$$\underset{14.3 \quad 29.7 \quad 42.5}{23.0 \quad 34.0} NH_2$$

¹³C NMR chemical shifts (δ)

Mass Spectra of Amines   The molecular ion in the mass spectrum of an amine has an odd number mass (unless there is an even number of nitrogen atoms in the molecule). The peak for the molecular ion is usually strong for aromatic and cyclic aliphatic amines but weak for acyclic aliphatic amines. Cleavage between the α and β carbons of aliphatic amines is a common mode of fragmentation.

20.12 ELIMINATIONS INVOLVING AMMONIUM COMPOUNDS

20.12A The Hofmann Elimination

All of the eliminations that we have described so far have involved electrically neutral substrates. However, eliminations are known in which the substrate bears a positive charge. One of the most important of these is the E2-type elimination that takes place when a quaternary ammonium hydroxide is heated. The products are an alkene, water, and a tertiary amine:

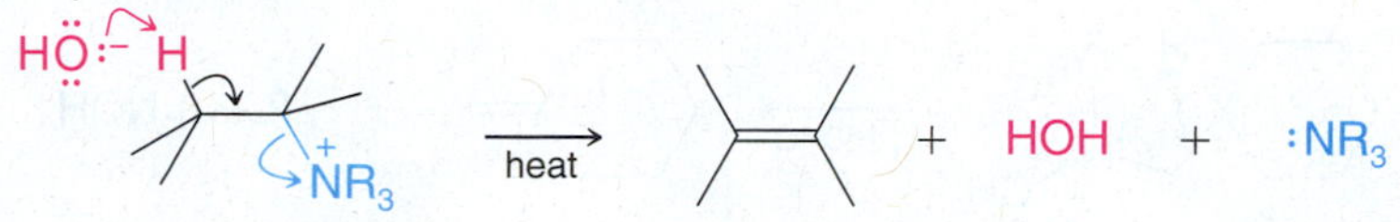

| A quaternary ammonium hydroxide | An alkene | A tertiary amine |

This reaction was discovered in 1851 by August W. von Hofmann and has since come to bear his name.

Quaternary ammonium hydroxides can be prepared from quaternary ammonium halides in aqueous solution through the use of silver oxide or an ion exchange resin:

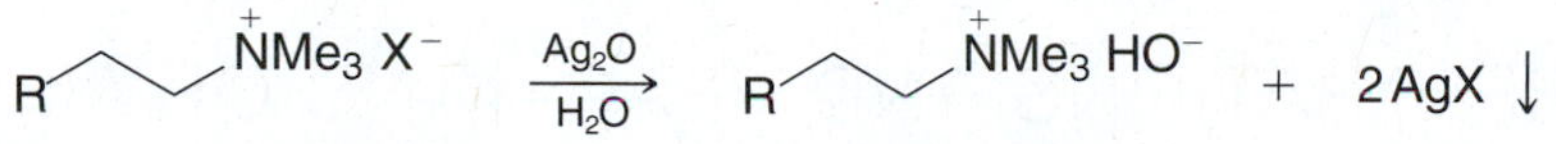

| A quaternary ammonium halide | A quaternary ammonium hydroxide |

Silver halide precipitates from the solution and can be removed by filtration. The quaternary ammonium hydroxide can then be obtained by evaporation of the water.

Although most eliminations involving neutral substrates tend to follow the *Zaitsev rule* (Section 7.6B), eliminations with charged substrates tend to follow what is called the **Hofmann rule** and *yield mainly the least substituted alkene*. We can see an example of this behavior if we compare the following reactions:

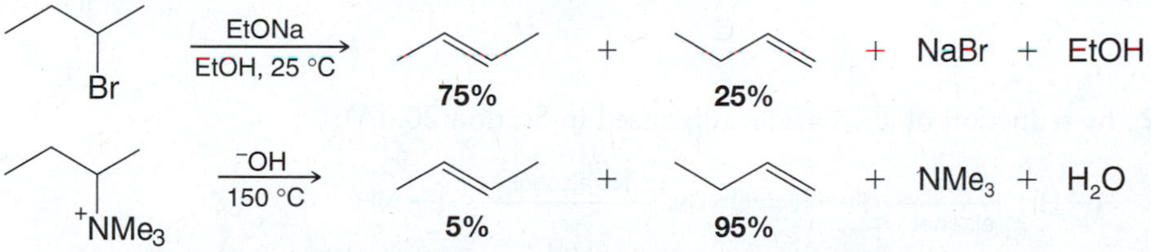

The precise mechanistic reasons for these differences are complex and are not yet fully understood. One possible explanation is that the transition states of elimination reactions with charged substrates have considerable carbanionic character. Therefore, these transition states show little resemblance to the final alkene product and are not stabilized appreciably by a developing double bond:

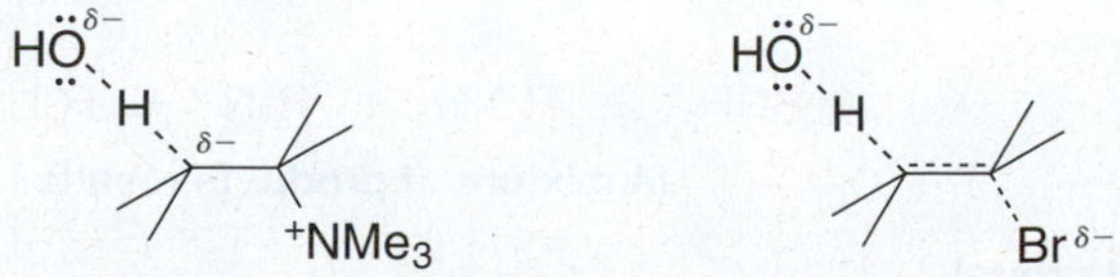

Carbanion-like transition state
(gives Hofmann orientation)   **Alkene-like transition state**
(gives Zaitsev orientation)

With a charged substrate, the base attacks the most acidic hydrogen instead. A primary hydrogen atom is more acidic because its carbon atom bears only one electron-releasing group.

20.12B The Cope Elimination

Tertiary amine oxides undergo the elimination of a dialkylhydroxylamine when they are heated. The reaction is called the Cope elimination, it is a syn elimination and proceeds through a cyclic transition state.

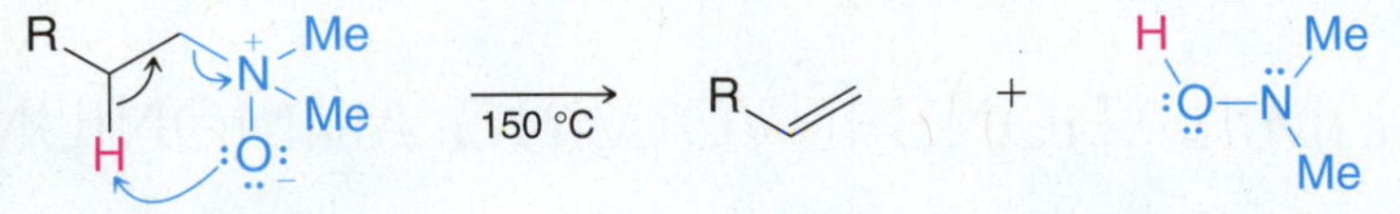

A tertiary amine oxide   **An alkene**   **N,N-Dimethylhydroxylamine**

Tertiary amine oxides are easily prepared by treating tertiary amines with hydrogen peroxide (Section 20.5A).

The Cope elimination is useful synthetically. Consider the following synthesis of methylenecyclohexane:

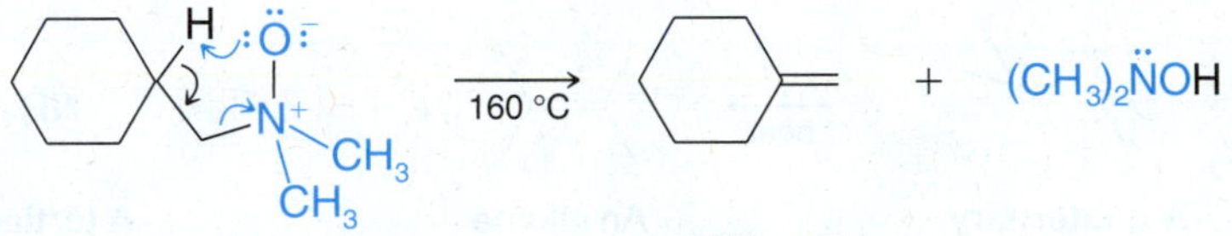

98%

20.13 SUMMARY OF PREPARATIONS AND REACTIONS OF AMINES

PREPARATION OF AMINES

1. Gabriel synthesis (discussed in Section 20.4A):

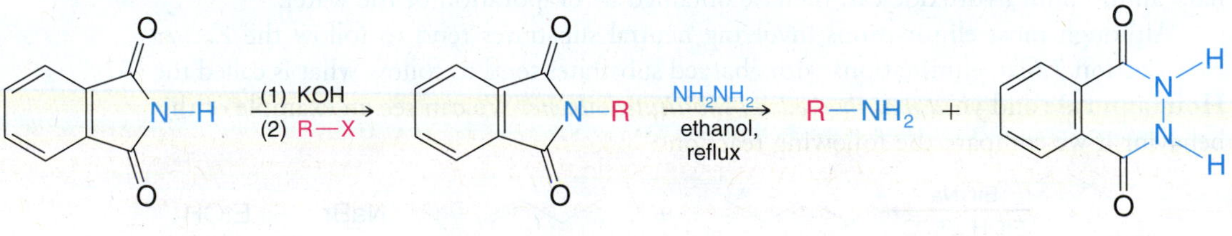

2. By reduction of alkyl azides (discussed in Section 20.4A):

$$R—Br \xrightarrow[\text{ethanol}]{NaN_3} R—N=\overset{+}{N}=\overset{-}{N} \xrightarrow[\substack{\text{or} \\ LiAlH_4}]{\text{Na/alcohol}} R—NH_2$$

3. By amination of alkyl halides (discussed in Section 20.4A):

$$R—Br + NH_3 \longrightarrow RNH_3^+Br^- + R_2NH_2^+Br^- + R_3NH^+Br^- + R_4N^+Br^-$$

$\downarrow HO^-$

$$RNH_2 + R_2NH + R_3N + R_4N^+OH^-$$

(A mixture of products results.)

(R = a 1° alkyl group)

4. By reduction of nitroarenes (discussed in Section 20.4B):

$$Ar—NO_2 \xrightarrow[\text{(1) Fe/HCl (2) NaOH}]{H_2, \text{ catalyst}} Ar—NH_2$$

5. By reductive amination (discussed in Section 20.4C):

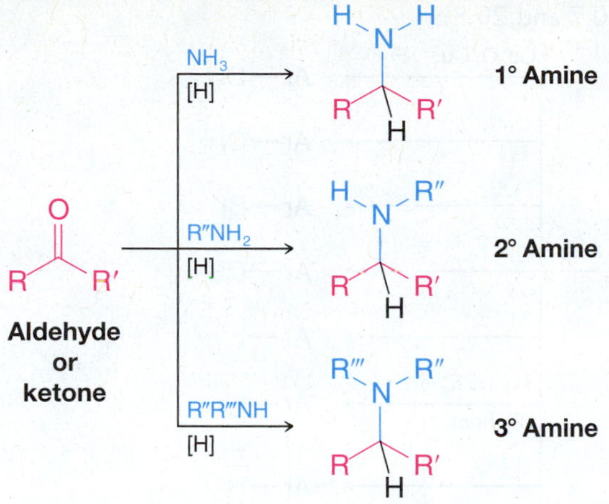

6. By reduction of nitriles, oximes, and amides (discussed in Section 20.4D):

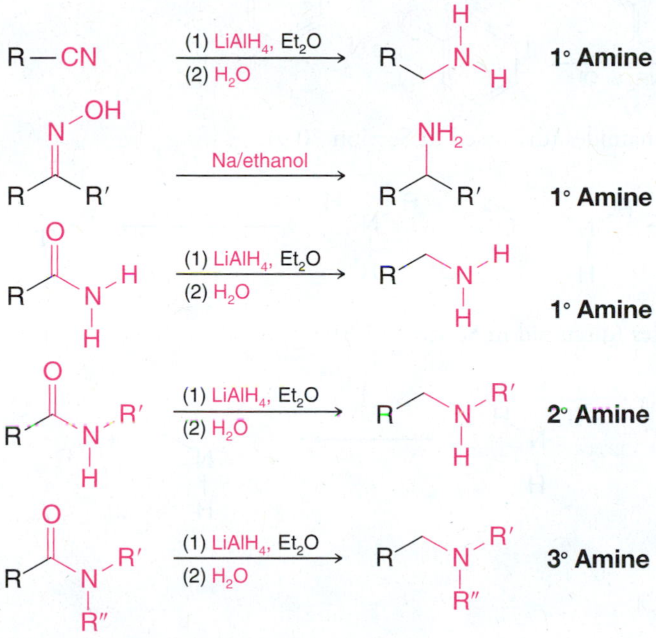

$$R—CN \quad \xrightarrow[\text{(2) } H_2O]{\text{(1) LiAlH}_4,\ Et_2O} \quad R \overset{}{\frown} \underset{H}{\overset{H}{N}} \qquad \textbf{1° Amine}$$

$$\text{oxime} \quad \xrightarrow{\text{Na/ethanol}} \quad R \overset{NH_2}{\underset{}{\diagup}} R' \qquad \textbf{1° Amine}$$

$$\text{amide} \quad \xrightarrow[\text{(2) } H_2O]{\text{(1) LiAlH}_4,\ Et_2O} \quad R \overset{}{\frown} \underset{H}{\overset{H}{N}} \qquad \textbf{1° Amine}$$

$$\text{amide} \quad \xrightarrow[\text{(2) } H_2O]{\text{(1) LiAlH}_4,\ Et_2O} \quad R \overset{}{\frown} \underset{H}{\overset{R'}{N}} \qquad \textbf{2° Amine}$$

$$\text{amide} \quad \xrightarrow[\text{(2) } H_2O]{\text{(1) LiAlH}_4,\ Et_2O} \quad R \overset{}{\frown} \underset{R''}{\overset{R'}{N}} \qquad \textbf{3° Amine}$$

7. Through the Hofmann and Curtius rearrangements (discussed in Section 20.4E):

Hofmann Rearrangement

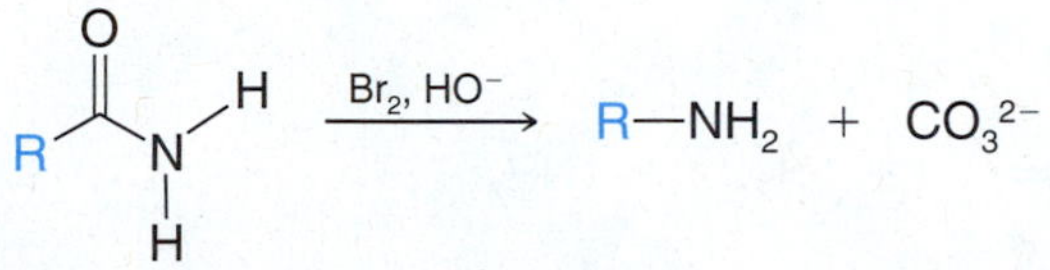

$$\text{amide} \quad \xrightarrow{Br_2,\ HO^-} \quad R—NH_2 \ + \ CO_3{}^{2-}$$

Curtius Rearrangement

$$\text{acyl chloride} \quad \xrightarrow[(-NaCl)]{NaN_3} \quad R \overset{O}{\underset{}{\diagdown}} N_3 \quad \xrightarrow[\text{heat}]{(-N_2)} \quad R—N{=}C{=}O \quad \xrightarrow{H_2O} \quad R—NH_2 \ + \ CO_2$$

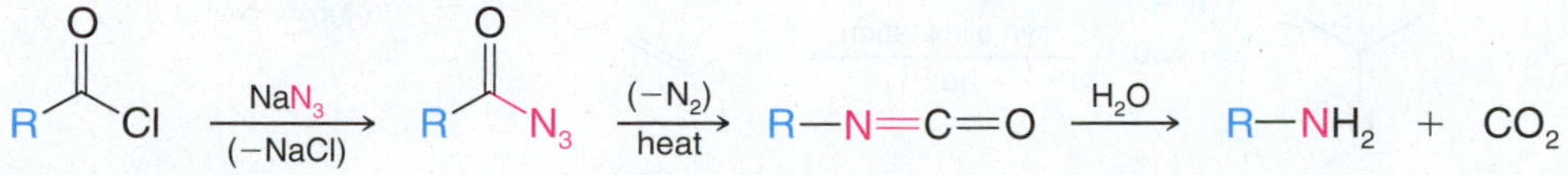

REACTIONS OF AMINES

1. As bases (discussed in Section 20.3):

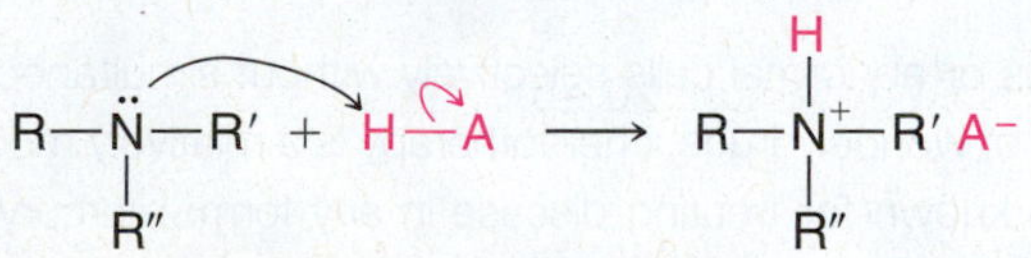

$$R—\ddot{N}—R' \ + \ H—A \ \longrightarrow \ R—\overset{+}{N}—R' \ A^-$$

(R, R′, and/or R″ may be alkyl, H, or Ar)

2. Diazotization of 1° arylamines and replacement of, or coupling with, the diazonium group (discussed in Sections 20.7 and 20.8):

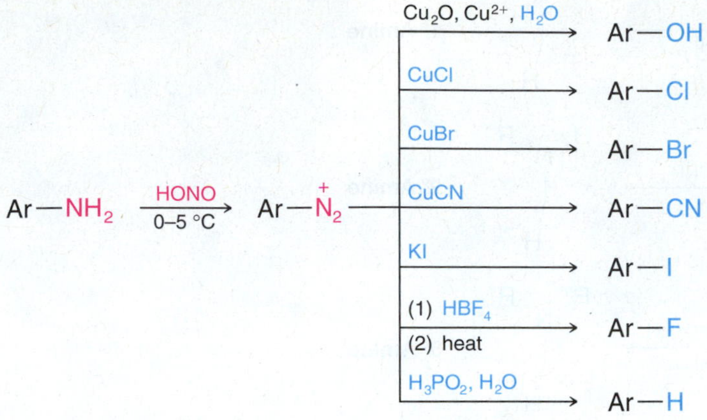
Ar—NH2
HONO
0–5 °C
Ar—N2+
Cu2O, Cu2+, H2O
Ar—OH
CuCl
Ar—Cl
CuBr
Ar—Br
CuCN
Ar—CN
KI
Ar—I
(1) HBF4
(2) heat
Ar—F
H3PO2, H2O
Ar—H

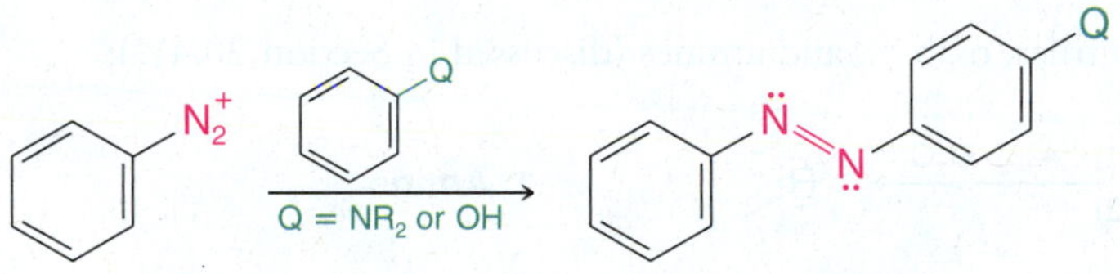
N2+
Q
Q = NR2 or OH
N
N
Q

3. Conversion to sulfonamides (discussed in Section 20.9):

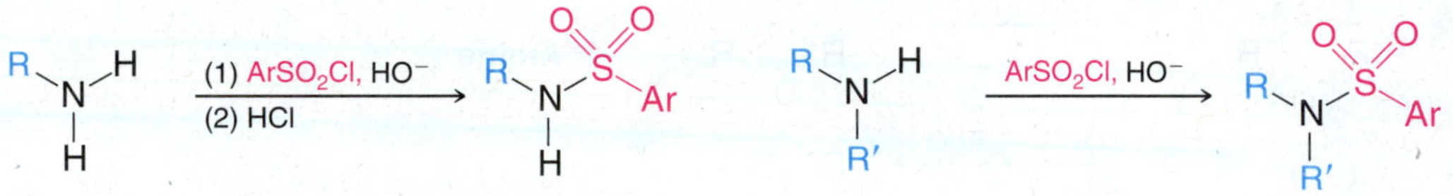
(1) ArSO2Cl, HO−
(2) HCl
ArSO2Cl, HO−
Ar

4. Conversion to amides (discussed in Section 17.8):

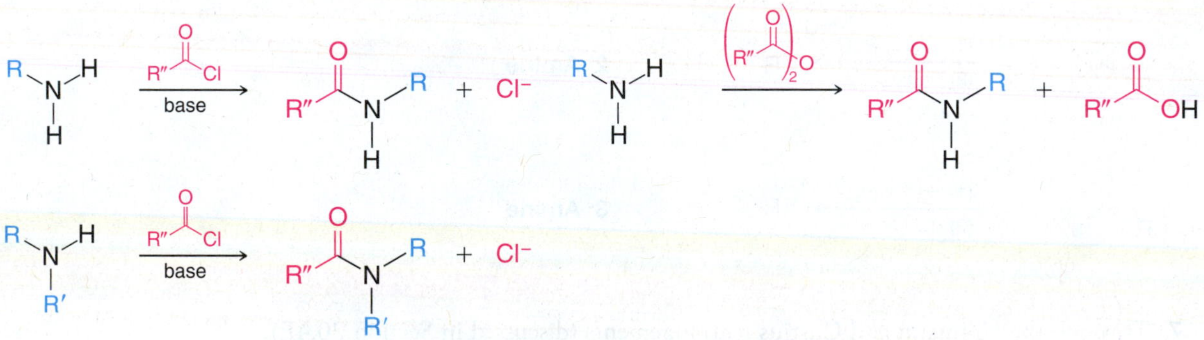
base
Cl−
OH

5. Hofmann and Cope eliminations (discussed in Section 20.12):

Hofmann Elimination

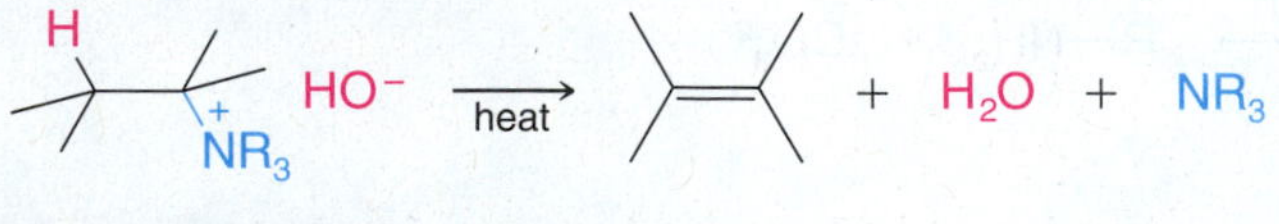
HO−
heat
+ H2O + NR3

Cope Elimination

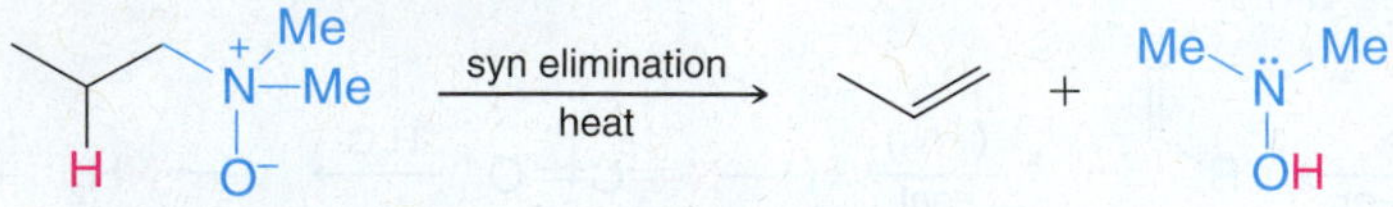
Me
N—Me
O−
syn elimination
heat
Me
Me
N
OH

WHY Do These Topics Matter?]

THE ORIGIN OF CHEMOTHERAPY AND SULFA DRUGS

Chemotherapy is defined as the use of chemical agents to destroy infectious or abnormal cells selectively without simultaneously destroying normal host cells. Although it may be difficult to believe in this age of wonder drugs, chemotherapy is a relatively modern phenomenon. Indeed, before 1900 there were only three specific remedies known for treating disease in any form: mercury (for syphilis, but with often disastrous results!), cinchona bark (i.e., quinine, for malaria), and ipecacuanha (for dysentery).

The term chemotherapy itself can be traced to a doctor named Paul Ehrlich. As a medical school student, Ehrlich had become impressed with the ability of certain dyes to stain tissues selectively. Believing that such staining was the result of a chemical reaction between the tissue and the dye, Ehrlich wondered whether it would be possible to identify dyes with selective affinities for microorganisms. He then hoped that he might be able to modify such dyes so that they could be specifically lethal to microorganisms but harmless to humans. He called such substances "magic bullets." In 1907, he discovered just such a substance in the form of a dye known as trypan red 1. This dye, which combated trypanosomiasis, led to his receipt of the 1908 Nobel Prize in Physiology or Medicine. In 1909, he followed up his initial discovery with a second magic bullet known as salvarsan, a remedy for syphilis, which contains aromatic amines in combination with arsenic atoms.

PAUL EHRLICH's work in chemotherapy led to his sharing one-half of the 1908 Nobel Prize in Physiology or Medicine with Ilya Mechnikov.

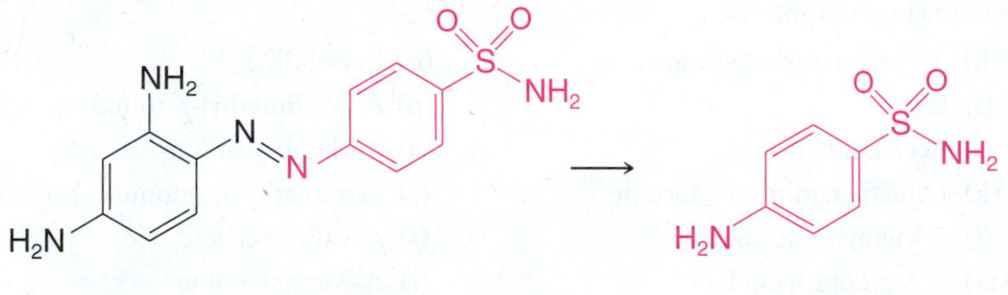

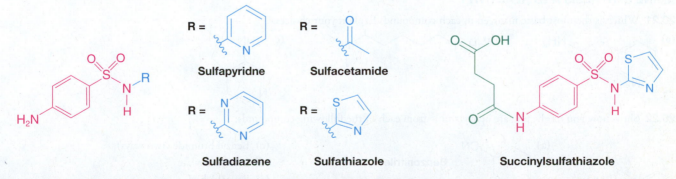

Protosil → **Sulfanilamide**

With these powerful proofs of principle in hand, for the next several decades Ehrlich and many other scientists tested tens of thousands of chemicals (not just dyes) looking for additional magic bullets. Unfortunately, very few were found to have any promising effects. Then, in 1935, the daughter of Gerhard Domagk, a doctor employed by a German dye manufacturer (I. G. Farbenindustrie), contracted a streptococcal infection from a pinprick. As she neared death, Domagk decided to give her an oral dose of a dye called prontosil, a substance his firm had developed. In tests with mice prontosil had inhibited the growth of streptococci. Within a short time the little girl recovered. Domagk's gamble not only saved his daughter's life, but it also initiated a new and spectacularly modern phase in modern chemotherapy and ultimately led to his receipt of the Nobel Prize in Physiology or Medicine (1939).

GERHARD DOMAGK won the 1939 Nobel Prize in Physiology or Medicine for discovering the antibacterial effects of prontosil.

In 1936, Ernest Fourneau of the Pasteur Institute in Paris demonstrated that (1) prontosil breaks down in the human body to produce sulfanilamide, and (2) sulfanilamide is the actual active agent against streptococci. Prontosil, therefore, is a prodrug because it is converted into the active compound in vivo. Fourneau's announcement of these results set in motion a search for other chemicals related to sulfanilamide that might have even better chemotherapeutic effects. Literally thousands of chemical variations were played on the sulfanilamide theme; its structure was varied in almost every imaginable way. At the end, the best therapeutic agents were obtained from compounds in which one hydrogen of the $-SO_2NH_2$ group was replaced by some other group, usually a heterocyclic ring (shown in blue in the following structures). Among the most successful variations were the compounds shown below; sulfanilamide itself, ultimately, proved too toxic for general use.

Sulfapyridne   **Sulfacetamide**

Sulfadiazene   **Sulfathiazole**   **Succinylsulfathiazole**

Sulfapyradine was shown to be effective against pneumonia in 1938; before that time, pneumonia epidemics had brought death to tens of thousands. Sulfacetamide was first used successfully in treating urinary tract infections in 1941. Succinylsulfathiazole and the related compound phthalylsulfathiazole were used as chemotherapeutics against infections of the gastrointestinal tract. Both compounds are hydrolyzed to sulfathiazole, a molecule that on its own saved the lives of countless soldiers in World War II.

SUMMARY AND REVIEW TOOLS

The study aids for this chapter include key terms and concepts (which are highlighted in bold, blue text within the chapter and defined in the glossary (at the back of the book) and have hyperlinked definitions in the accompanying *WileyPLUS* course (www.wileyplus.com), and the list of reaction types in Section 20.13.

PROBLEMS

Note to Instructors: Many of the homework problems are available for assignment via *WileyPLUS*, an online teaching and learning solution.

NOMENCLATURE

20.19 Write structural formulas for each of the following compounds:

(a) Benzylmethylamine
(b) Triisopropylamine
(c) *N*-Ethyl-*N*-methylaniline
(d) *m*-Toluidine
(e) 2-Methylpyrrole
(f) *N*-Ethylpiperidine
(g) *N*-Ethylpyridinium bromide
(h) 3-Pyridinecarboxylic acid
(i) Indole
(j) Acetanilide
(k) Dimethylaminium chloride
(l) 2-Methylimidazole
(m) 3-Aminopropan-1-ol
(n) Tetrapropylammonium chloride
(o) Pyrrolidine
(p) *N,N*-Dimethyl-*p*-toluidine
(q) 4-Methoxyaniline
(r) Tetramethylammonium hydroxide
(s) *p*-Aminobenzoic acid
(t) *N*-Methylaniline

20.20 Give common or systematic names for each of the following compounds:

(a)
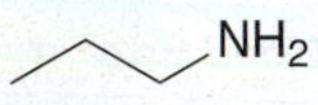

(b)
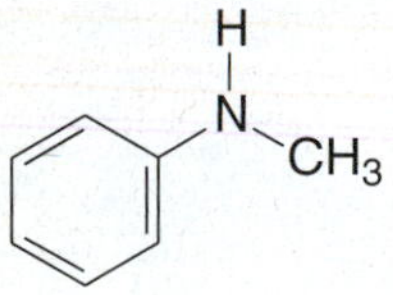

(c)
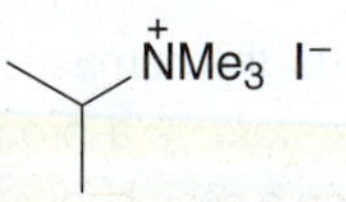

(d)
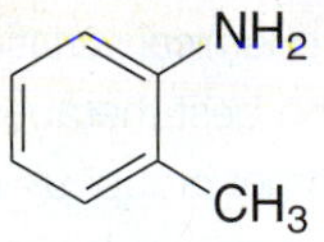

(e)
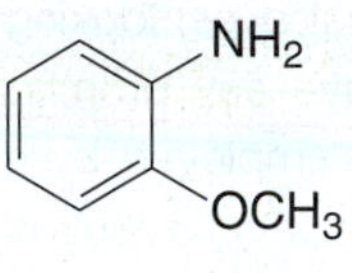

(f)
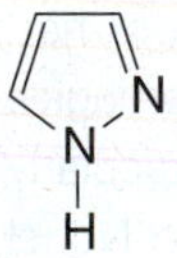

(g)
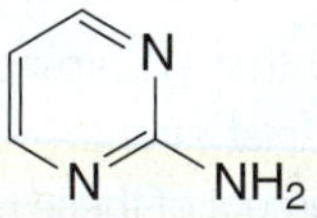

(h)
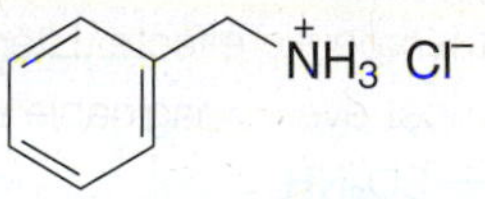

(i)
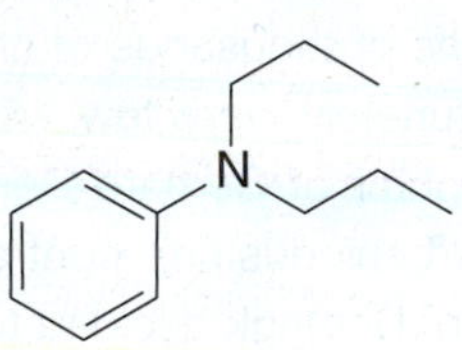

(j)
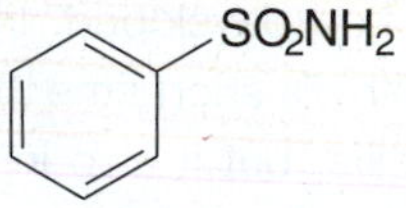

(k) $CH_3\overset{+}{N}H_3\ CH_3CO_2^-$

(l)
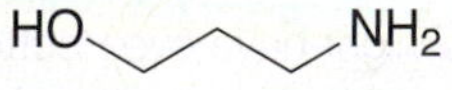

(m)
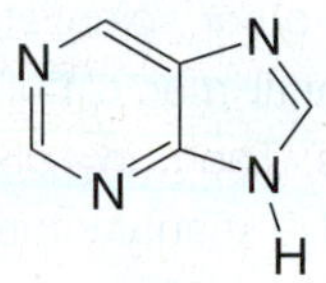

(n)
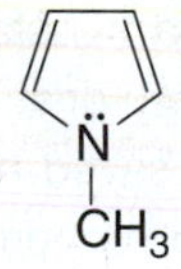

AMINE SYNTHESIS AND REACTIVITY

20.21 Which is the most basic nitrogen in each compound. Explain your choices.

(a)
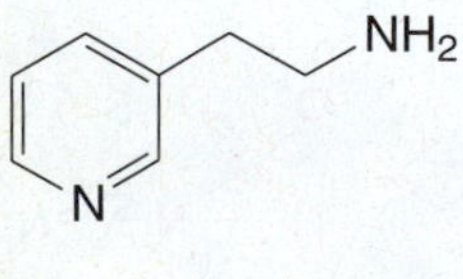

(b)
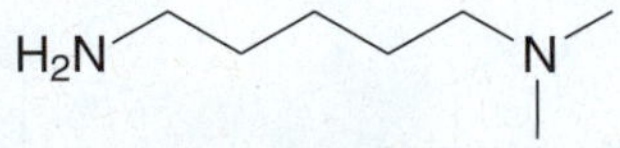

(c)
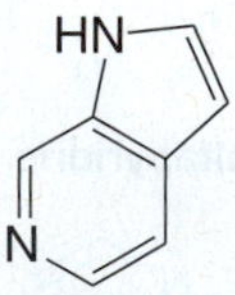

20.22 Show how you might prepare benzylamine from each of the following compounds:

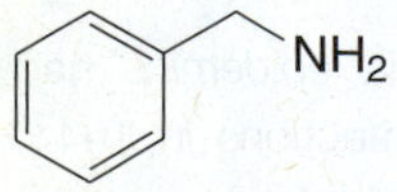

Benzylamine

(a)
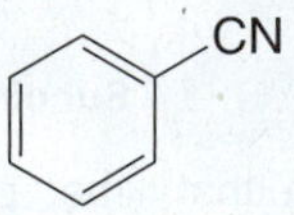

Benzonitrile

(b)
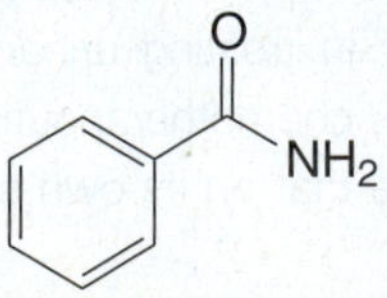

Benzamide

(c) Benzyl bromide (two ways)
(d) Benzyl tosylate
(e) Benzaldehyde
(f) Phenylnitromethane
(g)
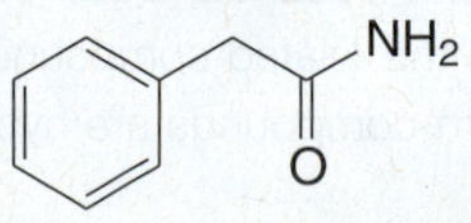

Phenylacetamide

20.23 Show how you might prepare aniline from each of the following compounds:

(a) Benzene **(b)** Bromobenzene **(c)** Benzamide

20.24 Show how you might synthesize each of the following compounds from 1-butanol:

(a) Butylamine (free of 2° and 3° amines) **(b)** Pentylamine **(c)** Propylamine **(d)** Butylmethylamine

20.25 Show how you might convert aniline into each of the following compounds. (You need not repeat steps carried out in earlier parts of this problem.)

(a) Acetanilide
(b) N-Phenylphthalimide
(c) p-Nitroaniline
(d) Sulfanilamide
(e) N,N-Dimethylaniline
(f) Fluorobenzene
(g) Chlorobenzene
(h) Bromobenzene
(i) Iodobenzene

(j) Benzonitrile
(k) Benzoic acid
(l) Phenol
(m) Benzene
(n)

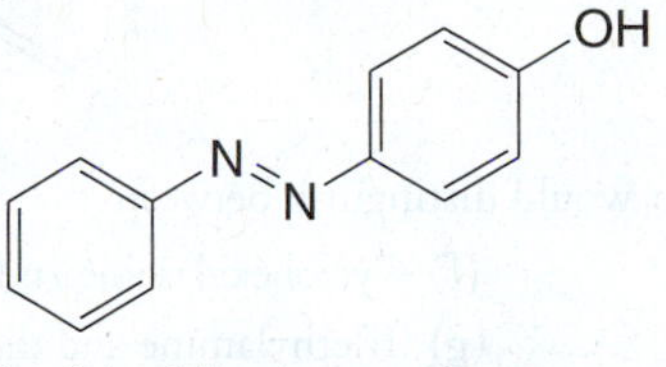

(o)

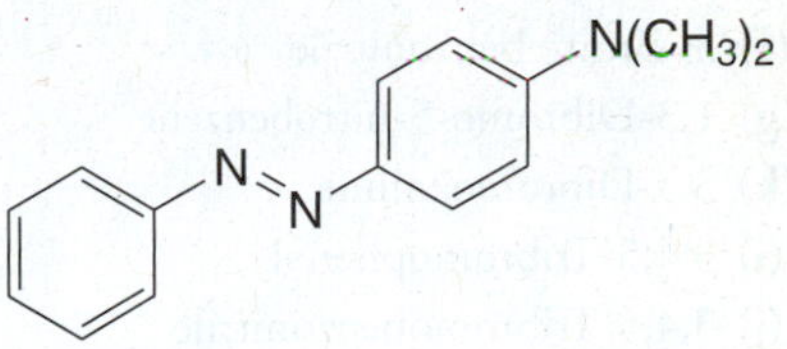

20.26 Provide the major organic product from each of the following reactions.

(a)

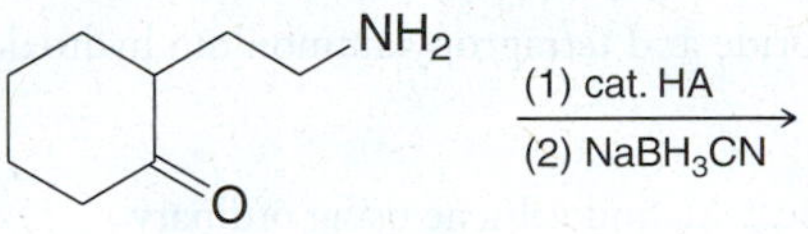

(1) cat. HA
(2) NaBH₃CN

(b)

(1) NaCN
(2) LiAlH₄
(3) H₃O⁺

(c)

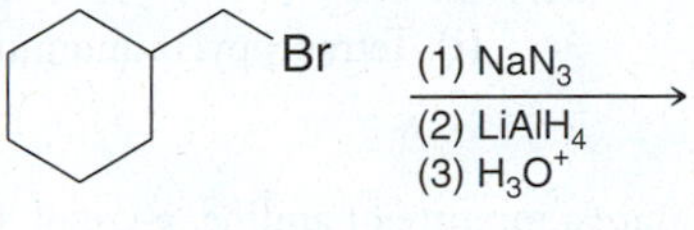

(1) NaN₃
(2) LiAlH₄
(3) H₃O⁺

(d)

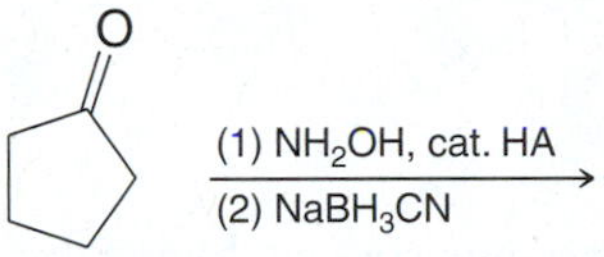

(1) NH₂OH, cat. HA
(2) NaBH₃CN

(e)

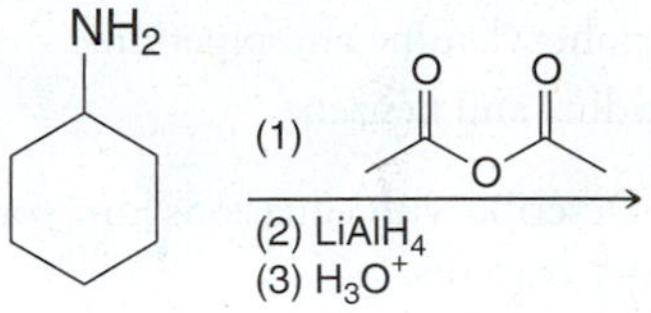

(1)
(2) LiAlH₄
(3) H₃O⁺

20.27 What products would you expect to be formed when each of the following amines reacts with aqueous sodium nitrite and hydrochloric acid?

(a) Propylamine **(b)** Dipropylamine **(c)** N-Propylaniline **(d)** N,N-Dipropylaniline **(e)** p-Propylaniline

20.28 **(a)** What products would you expect to be formed when each of the amines in the preceding problem reacts with benzenesulfonyl chloride and excess aqueous potassium hydroxide? **(b)** What would you observe in each reaction? **(c)** What would you observe when the resulting solution or mixture is acidified?

20.29 What product would you expect to obtain from each of the following reactions?

(a)

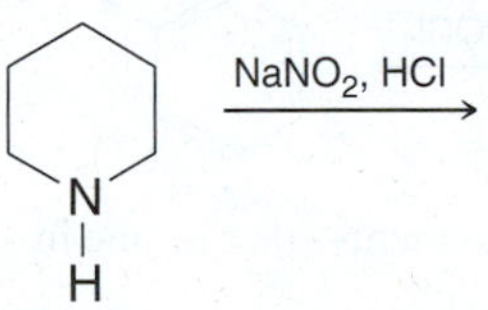

NaNO₂, HCl

(b)

aq. KOH

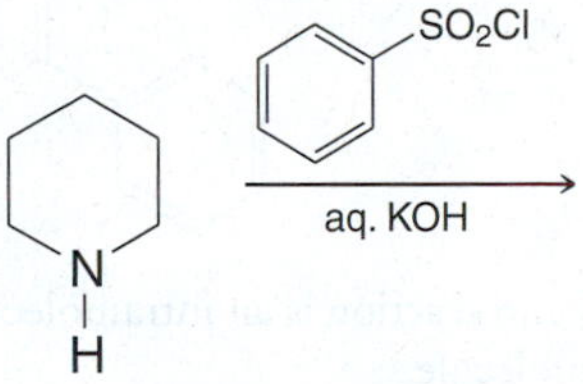

20.30 Give structures for the products of each of the following reactions:

(a)

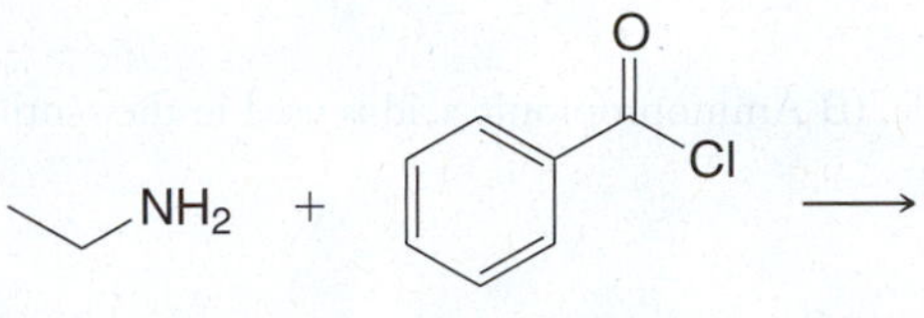

(b)

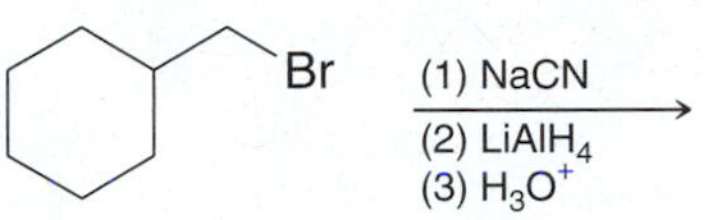

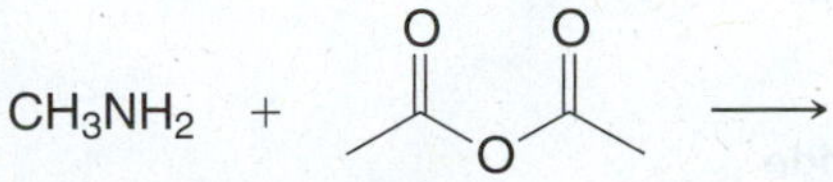

CH₃NH₂ +

(c)

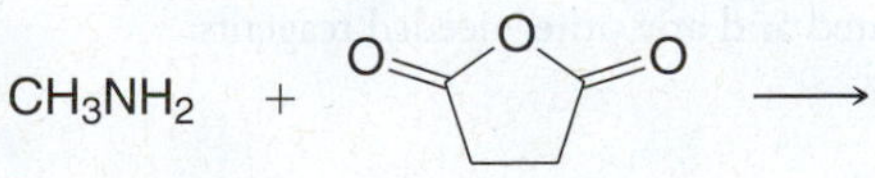

CH₃NH₂ +

(d) Product of (c) $\xrightarrow{\text{heat}}$

(e)

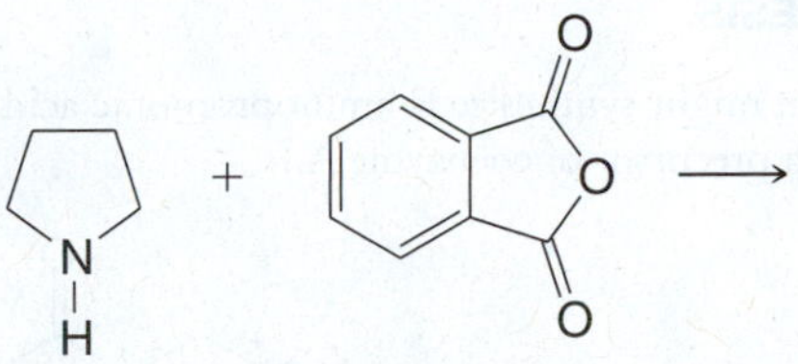

(f)

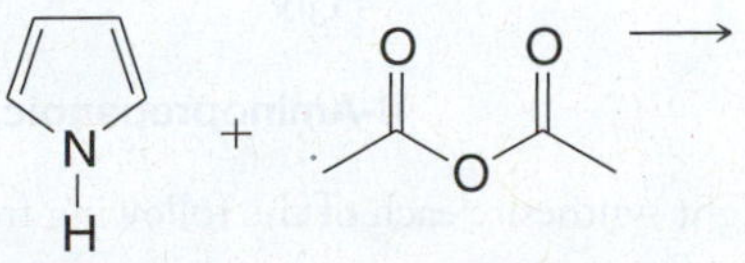

(g) Aniline + propanoyl chloride $\longrightarrow$

(h) Tetraethylammonium hydroxide $\xrightarrow{\text{heat}}$

(i) p-Toluidine + Br₂ (excess) $\xrightarrow{\text{H}_2\text{O}}$

20.31 Starting with benzene or toluene, outline a synthesis of each of the following compounds using diazonium salts as intermediates. (You need not repeat syntheses carried out in earlier parts of this problem.)

(a) *p*-Fluorotoluene
(b) *o*-Iodotoluene
(c) *p*-Cresol
(d) *m*-Dichlorobenzene
(e) *m*-C$_6$H$_4$(CN)$_2$
(f) *m*-Bromobenzonitrile
(g) 1,3-Dibromo-5-nitrobenzene
(h) 3,5-Dibromoaniline
(i) 3,4,5-Tribromophenol
(j) 3,4,5-Tribromobenzonitrile
(k) 2,6-Dibromobenzoic acid

(l) 1,3-Dibromo-2-iodobenzene
(m)

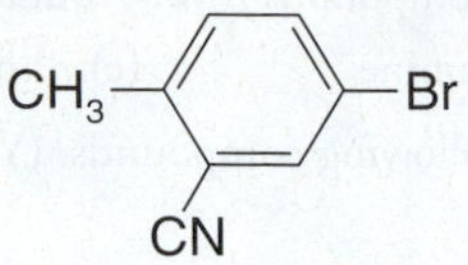

(n) CH$_3$

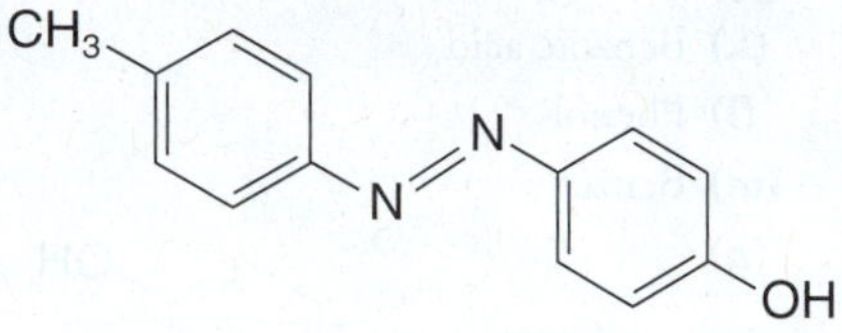

(o) CH$_3$

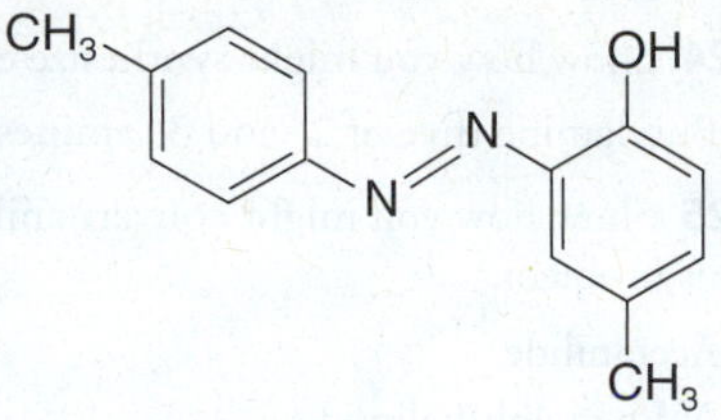

20.32 Write equations for simple chemical tests that would distinguish between

(a) Benzylamine and benzamide
(b) Allylamine and propylamine
(c) *p*-Toluidine and *N*-methylaniline
(d) Cyclohexylamine and piperidine
(e) Pyridine and benzene

(f) Cyclohexylamine and aniline
(g) Triethylamine and diethylamine
(h) Tripropylaminium chloride and tetrapropylammonium chloride
(i) Tetrapropylammonium chloride and tetrapropylammonium hydroxide

20.33 Describe with equations how you might separate a mixture of aniline, *p*-cresol, benzoic acid, and toluene using ordinary laboratory reagents.

MECHANISMS

20.34 Using reactions that we have studied in this chapter, propose a mechanism that accounts for the following reaction:

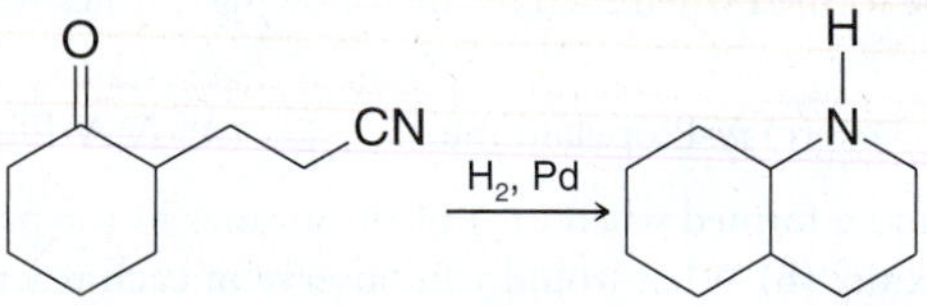

20.35 Provide a detailed mechanism for each of the following reactions.

(a)

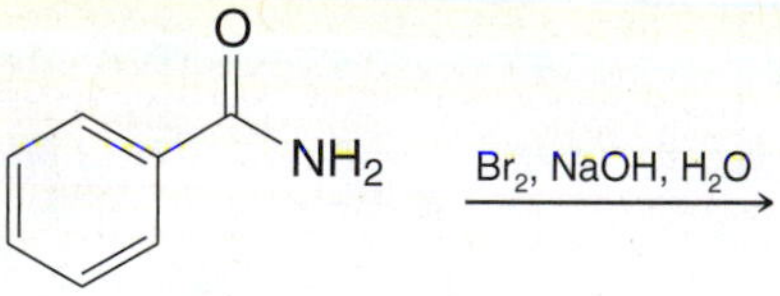

(b)

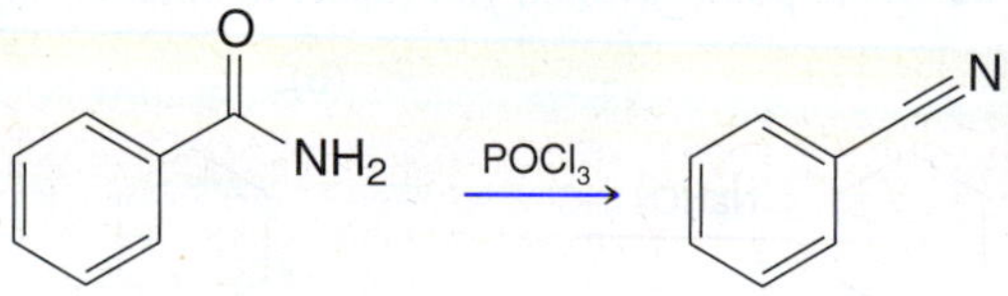

20.36 Suggest an experiment to test the proposition that the Hofmann reaction is an intramolecular rearrangement—that is, one in which the migrating R group never fully separates from the amide molecule.

GENERAL SYNTHESIS

20.37 Show how you might synthesize β-aminopropionic acid from succinic anhydride. (β-Aminopropionic acid is used in the synthesis of pantothenic acid, a precursor of coenzyme A.)

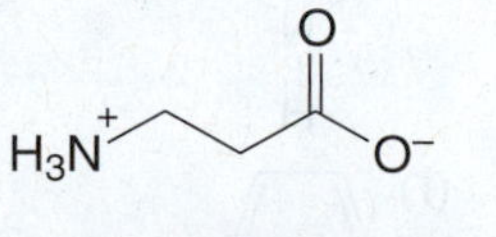

β-Aminopropanoic acid 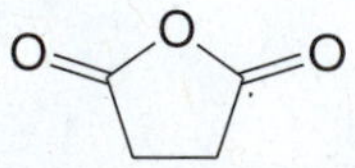   **Succinic anhydride**

20.38 Show how you might synthesize each of the following from the compounds indicated and any other needed reagents:

(a) Me$_3$N$^+$ ⌒⌒⌒ N$^+$Me$_3$ 2Br$^-$ from 1,10-decanediol

(b) Succinylcholine bromide (see "The Chemistry of... Biologically Important Amines" in Section 20.3) from succinic acid, 2-bromoethanol, and trimethylamine

20.39 A commercial synthesis of folic acid consists of heating the following three compounds with aqueous sodium bicarbonate. Propose reasonable mechanisms for the reactions that lead to folic acid. *Hint:* The first step involves formation of an imine between the lower right NH$_2$ group of the heterocyclic amine and the ketone.

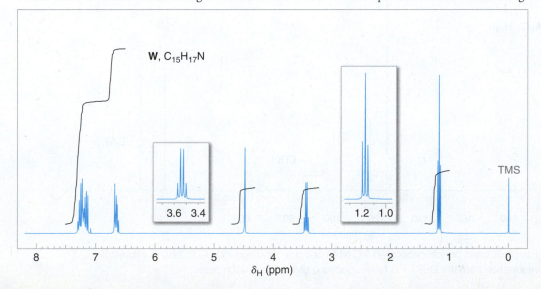

Folic acid
(~10%)

20.40 Give structures for compounds **R–W**:

N-Methylpiperidine $\xrightarrow{CH_3I}$ **R** ($C_7H_{16}NI$) $\xrightarrow[H_2O]{Ag_2O}$ **S** ($C_7H_{17}NO$) $\xrightarrow[heat]{(-H_2O)}$

T ($C_7H_{15}N$) $\xrightarrow{CH_3I}$ **U** ($C_5H_{18}NI$) $\xrightarrow[H_2O]{Ag_2O}$ **V** ($C_8H_{19}NO$) $\xrightarrow{heat}$ **W** (C_5H_8) + H_2O + (CH_3)$_3$N

20.41 Outline a synthesis of acetylcholine iodide using dimethylamine, oxirane, iodomethane, and acetyl chloride as starting materials.

Acetylcholine iodide

20.42 Ethanolamine, HOCH$_2$CH$_2$NH$_2$, and diethanolamine, (HOCH$_2$CH$_2$)$_2$NH, are used commercially to form emulsifying agents and to absorb acidic gases. Propose syntheses of these two compounds.

20.43 Diethylpropion (shown here) is a compound used in the treatment of anorexia. Propose a synthesis of diethylpropion starting with benzene and using any other needed reagents.

Diethylpropion

20.44 Using as starting materials 2-chloropropanoic acid, aniline, and 2-naphthol, propose a synthesis of naproanilide, a herbicide used in rice paddies in Asia:

Naproanilide

SPECTROSCOPY

20.45 When compound **W** (C$_{15}$H$_{17}$N) is treated with benzenesulfonyl chloride and aqueous potassium hydroxide, no apparent change occurs. Acidification of this mixture gives a clear solution. The ^{1}H NMR spectrum of **W** is shown in Fig. 20.6. Propose a structure for **W**.

FIGURE 20.6 The 300-MHz ^{1}H NMR spectrum of compound W, Problem 20.45. Expansions of the signals are shown in the offset plots.

20.46   Propose structures for compounds **X**, **Y**, and **Z**:

$$X \ (C_7H_7Br) \xrightarrow{\text{NaCN}} Y \ (C_8H_7N) \xrightarrow{\text{LiAlH}_4} Z \ (C_8H_{11}N)$$

The ^{1}H NMR spectrum of **X** gives two signals, a multiplet at δ 7.3 (5H) and a singlet at δ 4.25 (2H); the 680–840-cm^{-1} region of the IR spectrum of **X** shows peaks at 690 and 770 cm^{-1}. The ^{1}H NMR spectrum of **Y** is similar to that of **X**: multiplet at δ 7.3 (5H), singlet at δ 3.7 (2H). The ^{1}H NMR spectrum of **Z** is shown in Fig. 20.7.

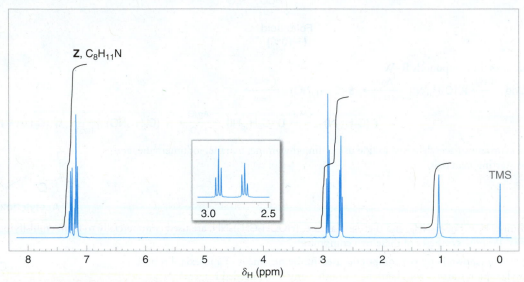

FIGURE 20.7 The 300-MHz ^{1}H NMR spectrum of compound Z, Problem 20.46. Expansion of the signals is shown in the offset plot.

20.47   Compound **A** ($C_{10}H_{15}N$) is soluble in dilute HCl. The IR absorption spectrum shows two bands in the 3300–3500-cm^{-1} region. The broadband proton-decoupled ^{13}C spectrum of **A** is given in Fig. 20.8. Propose a structure for **A**.

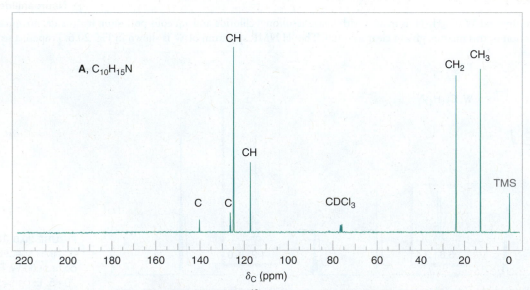

FIGURE 20.8 The broadband proton-decoupled ^{13}C NMR spectra of compounds A, B, and C, Problems 20.47–20.49. Information from the DEPT ^{13}C NMR spectra is given above each peak.

FIGURE 20.8 *(continued)*

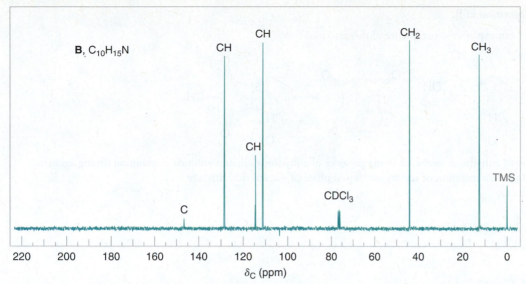

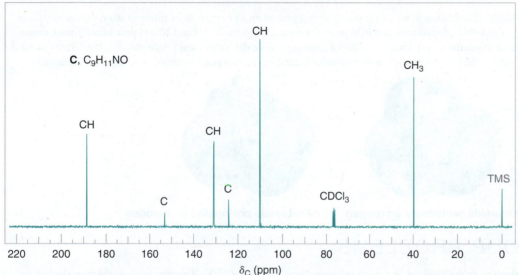

20.48 Compound **B**, an isomer of **A** (Problem 20.47), is also soluble in dilute HCl. The IR spectrum of **B** shows no bands in the 3300–3500-cm⁻¹ region. The broadband proton-decoupled ¹³C spectrum of **B** is given in Fig. 20.8. Propose a structure for **B**.

20.49 Compound **C** (C₉H₁₁NO) gives a positive Tollens' test (can be oxidized to a carboxylic acid) and is soluble in dilute HCl. The IR spectrum of **C** shows a strong band near 1695 cm⁻¹ but shows no bands in the 3300–3500-cm⁻¹ region. The broadband proton-decoupled ¹³C NMR spectrum of **C** is shown in Fig. 20.8. Propose a structure for **C**.

CHALLENGE PROBLEMS

20.50 When phenyl isothiocyanate, $C_6H_5N\!=\!C\!=\!S$, is reduced with lithium aluminum hydride, the product formed has these spectral data:

  MS (*m/z*): 107, 106

  IR (cm⁻¹): 3330 (sharp), 3050, 2815, 760, 700

  ¹H NMR (δ): 2.7 (s, 3H), 3.5 (broad, 1H), 6.6 (δ, 2H), 6.7 (t, 1H) 7.2 (t, 2H)

  ¹³C NMR (δ): 30 (CH₃), 112 (CH), 117 (CH), 129 (CH), 150 (C)

(a) What is the structure of the product?

(b) What is the structure that accounts for the 106 *m/z* peak and how is it formed? (It is an iminium ion.)

20.51 When *N,N′*-diphenylurea (**A**) is reacted with tosyl chloride in pyridine, it yields product **B**.

  The spectral data for **B** include:

  MS (*m/z*): 194 (M⁺)

  IR (cm⁻¹): 3060, 2130, 1590, 1490, 760, 700

  ¹H NMR (δ): only 6.9–7.4 (m)

  ¹³C NMR (δ): 122 (CH), 127 (CH), 130 (CH), 149 (C), and 163 (C)

(a) What is the structure of **B**?

(b) Write a mechanism for the formation of **B**.

20.52 Propose a mechanism that can explain the occurrence of this reaction:

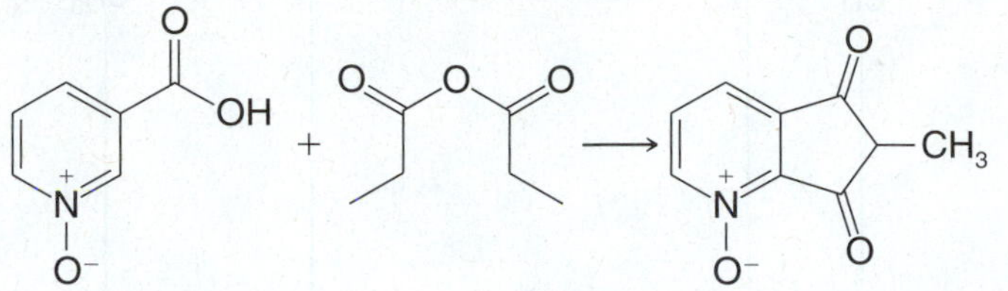

20.53 When acetone is treated with anhydrous ammonia in the presence of anhydrous calcium chloride (a common drying agent), crystalline product **C** is obtained on concentration of the organic liquid phase of the reaction mixture. These are spectral data for product **C**:

> **MS** (*m/z*): 155 (M$^{+\cdot}$), 140
> **IR** (cm^{-1}): 3350 (sharp), 2850–2960, 1705
> ^{1}**H NMR** (δ): 2.3 (s, 4H), 1.7 (1H; disappears in D$_2$O), and 1.2 (s, 12H)

(a) What is the structure of **C**?

(b) Propose a mechanism for the formation of **C**.

20.54 The difference in positive-charge distribution in an amide that accepts a proton on its oxygen or its nitrogen atom can be visualized with electrostatic potential maps. Consider the electrostatic potential maps for acetamide in its O—H and N—H protonated forms shown below. On the basis of the electrostatic potential maps, which protonated form appears to delocalize, and hence stabilize, the formal positive charge more effectively? Discuss your conclusion in terms of resonance contributors for the two possible protonated forms of acetamide.

Acetamide protonated on oxygen   **Acetamide protonated on nitrogen**

LEARNING GROUP PROBLEMS

1. Reserpine is a natural product belonging to the family of alkaloids (see Special Topic F in *WileyPLUS*). Reserpine was isolated from the Indian snakeroot *Rauwolfia serpentina*. Clinical applications of reserpine include treatment of hypertension and nervous and mental disorders. The synthesis of reserpine, which contains six chirality centers, was a landmark accomplishment reported by R. B. Woodward in 1955. Incorporated in the synthesis are several reactions involving amines and related nitrogen-containing functional groups, as we shall see on the following page.

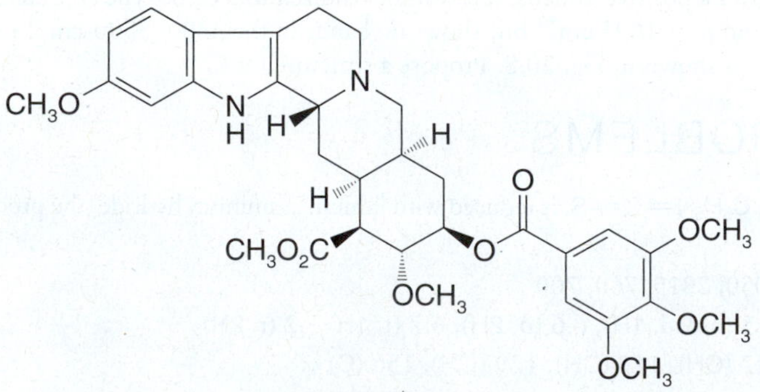

Reserpine

(a) The goal of the first two steps shown in the scheme on the following page, prior to formation of the amide, is preparation of a secondary amine. Draw the structure of the products labeled **A** and **B** from the first and second reactions, respectively. Write a mechanism for formation of **A**.

(b) The next sequence of reactions involves formation of a tertiary amine together with closure of a new ring. Write curved arrows to show how the amide functional group reacts with phosphorus oxychloride (POCl$_3$) to place the leaving group on the bracketed intermediate.

(c) The ring closure from the bracketed intermediate involves a type of electrophilic aromatic substitution reaction characteristic of indole rings. Identify the part of the structure that contains the indole ring. Write mechanism arrows to show how the nitrogen in the indole ring, via conjugation, can cause electrons from the adjacent carbon to attack an electrophile. In this case, the attack by the indole ring in the bracketed intermediate is an addition–elimination reaction, somewhat like reactions that occur at carbonyls bearing leaving groups.

Reserpine

2. (a) A student was given a mixture of two unknown compounds and asked to separate and identify them. One of the compounds was an amine and the other was a neutral compound (neither appreciably acidic nor basic). Describe how you would go about separating the unknown amine from the neutral compound using extraction techniques involving diethyl ether and solutions of aqueous 5% HCl and 5% NaHCO₃. The mixture as a whole was soluble in diethyl ether, but neither component was soluble in water at pH 7. Using R groups on a generic amine, write out the reactions for any acid–base steps you propose and explain why the compound of interest will be in the ether layer or aqueous layer at any given time during the process.

(b) Once the amine was successfully isolated and purified, it was reacted with benzenesulfonyl chloride in the presence of aqueous potassium hydroxide. The reaction led to a solution that on acidification produced a precipitate. The results just described constitute a test (Hinsberg's) for the class of an amine. What class of amine was the unknown compound: primary, secondary, or tertiary? Write the reactions involved for a generic amine of the class you believe this one to be.

(c) The unknown amine was then analyzed by IR, NMR, and MS. The following data were obtained. On the basis of this information, deduce the structure of the unknown amine. Assign the spectral data to specific aspects of the structure you propose for the amine.

  IR (cm^{-1}): 3360, 3280, 3020, 2962, 1604, 1450, 1368, 1021, 855, 763, 700, 538
  ¹H NMR (δ): 1.35 (d, 3H), 1.8 (bs, 2H), 4.1 (q, 1H), 7.3 (m, 5H)
  MS (*m/z*): 121, 120, 118, 106 (base peak), 79, 77, 51, 44, 42, 28, 18, 15

See Special Topic F in *WileyPLUS*

CHAPTER
21

Phenols and Aryl Halides

NUCLEOPHILIC AROMATIC SUBSTITUTION

Although we have already studied the chemistry of both alcohols and halides, when these functional groups are attached to benzene rings to make phenols and aryl halides, some unique chemical reactivity results. For instance, phenols take part in a powerful bond-forming process known as the Claisen rearrangement, while certain aryl halides can participate in *nucleophilic* aromatic substitution reactions. Phenols and aryl halides also have unique physical and biochemical properties. Some phenols act as hormones and play a role in chemical signaling, others are antioxidants and antibiotics, and still others are the blistering agents found in poison ivy (shown below). Aryl halides are also quite useful, although several, such as polychlorinated and polybrominated biphenyls, have been shown to be harmful to the environment.

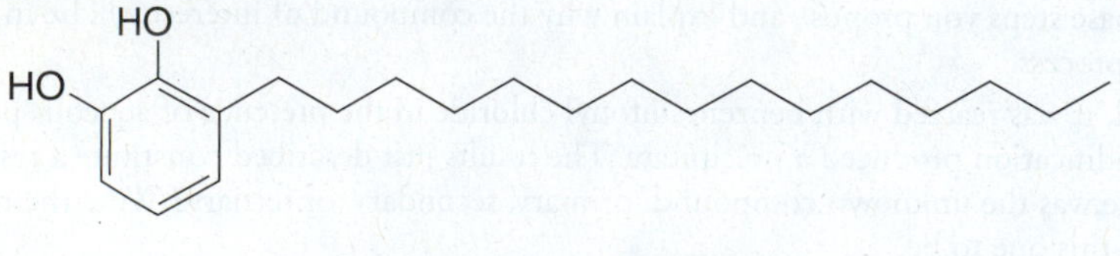

A poison ivy blistering agent

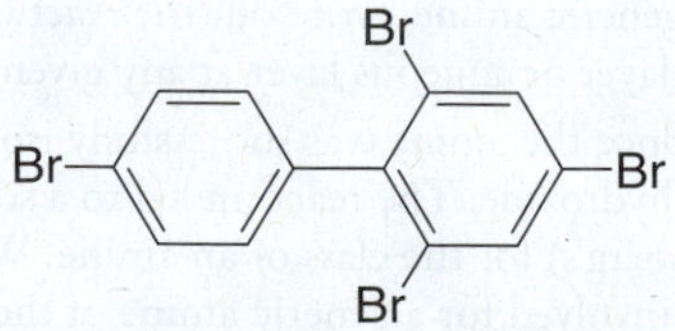

A polybrominated biphenyl

IN THIS CHAPTER WE WILL CONSIDER:

- the synthesis and properties of phenols
- the synthesis and properties of aryl halides
- powerful transformations using the functionality of phenols and aryl halides, such as the Claisen rearrangement and nucleophilic aromatic substitution

[WHY DO THESE TOPICS MATTER?] At the end of this chapter, we will show how subtle alterations in the structures of one specific group of phenols can turn compounds with a pleasant odor and taste profile into molecules that can both lead to the perception of pain as well as serve as analgesics.

21.1 STRUCTURE AND NOMENCLATURE OF PHENOLS

Compounds that have a hydroxyl group directly attached to a benzene ring are called **phenols**. Thus, **phenol** is the specific name for hydroxybenzene, and it is the general name for the family of compounds derived from hydroxybenzene:

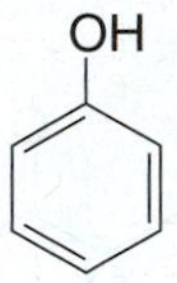

Phenol 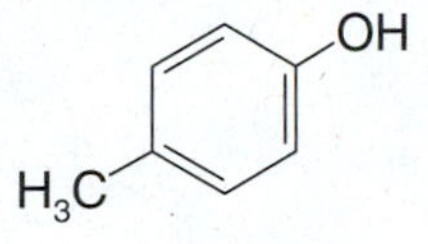   **4-Methylphenol**
                          (a phenol)

Compounds that have a hydroxyl group attached to a polycyclic benzenoid ring are chemically similar to phenols, but they are called **naphthols** and **phenanthrols**. For example:

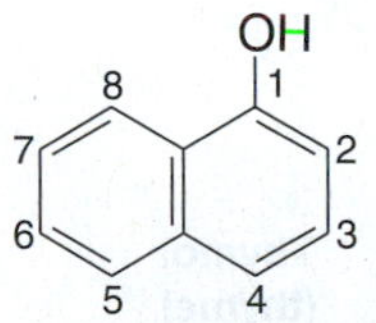

1-Naphthol   **2-Naphthol** 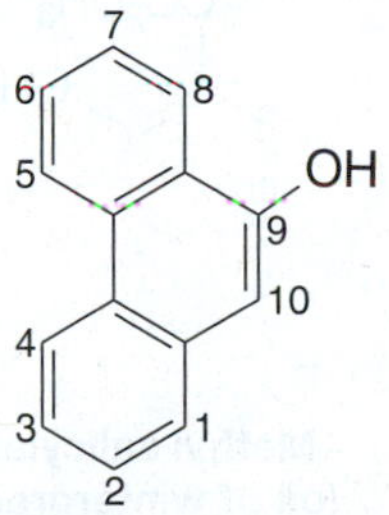   **9-Phenanthrol**
(α-naphthol)   **(β-naphthol)**

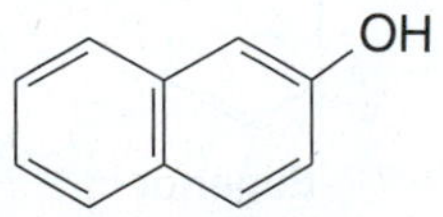

21.1A Nomenclature of Phenols

We studied the nomenclature of some phenols in Chapter 14. In many compounds *phenol* is the base name:

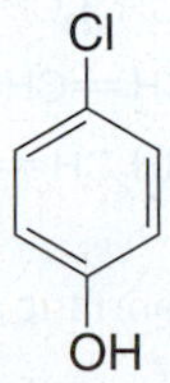

4-Chlorophenol   **2-Nitrophenol** 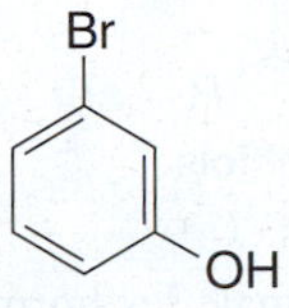   **3-Bromophenol**
(p-chlorophenol)   **(o-nitrophenol)**   **(m-bromophenol)**

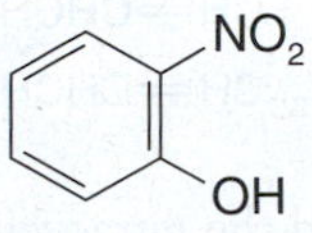

The methylphenols are commonly called *cresols*:

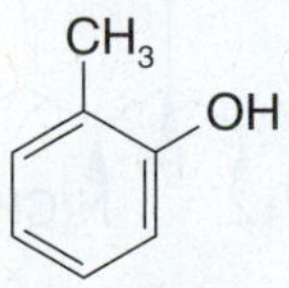

2-Methylphenol   **3-Methylphenol**   **4-Methylphenol**
(o-cresol)   **(m-cresol)**   **(p-cresol)**

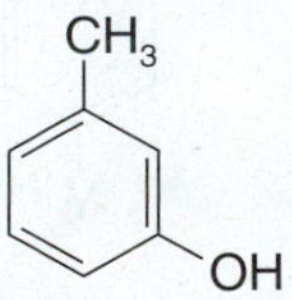

The benzenediols also have common names:

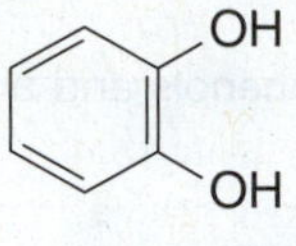

1,2-Benzenediol
(catechol)

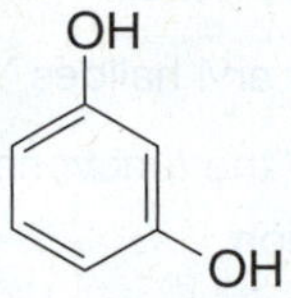

1,3-Benzenediol
(resorcinol)

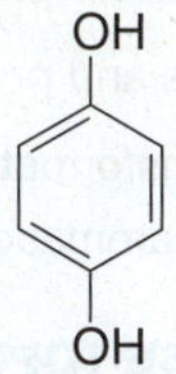

1,4-Benzenediol
(hydroquinone)

21.2 NATURALLY OCCURRING PHENOLS

Phenols and related compounds occur widely in nature. Tyrosine is an amino acid that occurs in proteins. Methyl salicylate is found in oil of wintergreen, eugenol is found in oil of cloves, and thymol is found in thyme.

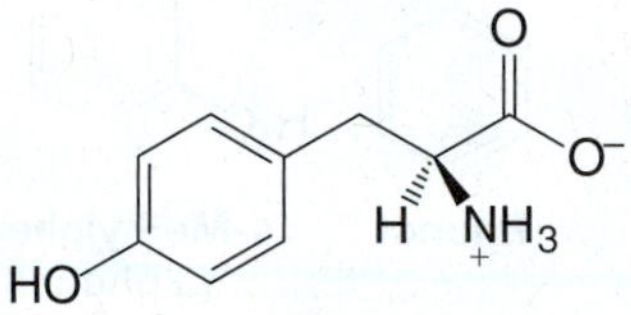

L-Tyrosine

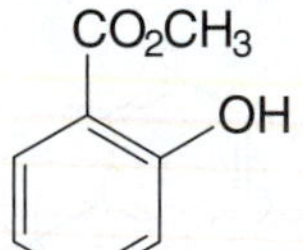

Methyl salicylate
(oil of wintergreen)

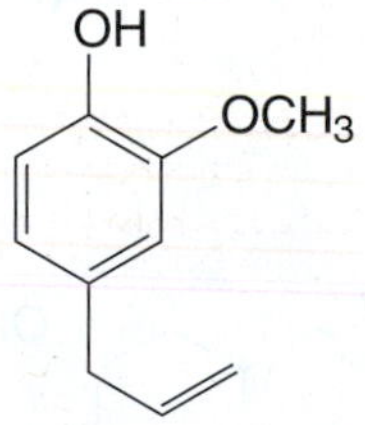

Eugenol
(oil of cloves)

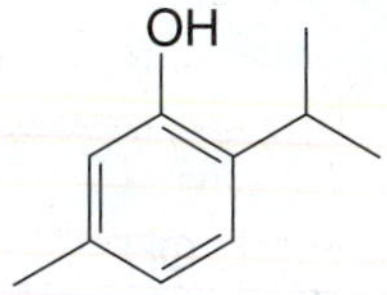

Thymol
(thyme)

The urushiols are blistering agents (vesicants) found in poison ivy.

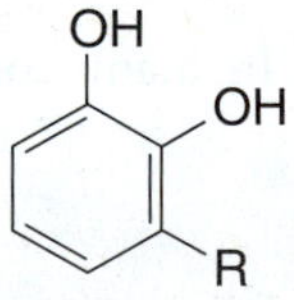

Urushiols

$R = -(CH_2)_{14}CH_3,$

$-(CH_2)_7CH=CH(CH_2)_5CH_3,$ or

$-(CH_2)_7CH=CHCH_2CH=CH(CH_2)_2CH_3,$ or

$-(CH_2)_7CH=CHCH_2CH=CHCH=CHCH_3$ or

$-(CH_2)_7CH=CHCH_2CH=CHCH_2CH=CH_2$

Estradiol is a female sex hormone, and the tetracyclines are important antibiotics.

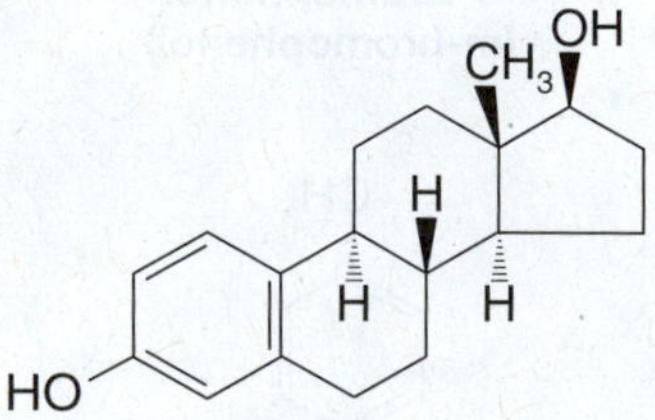

Estradiol

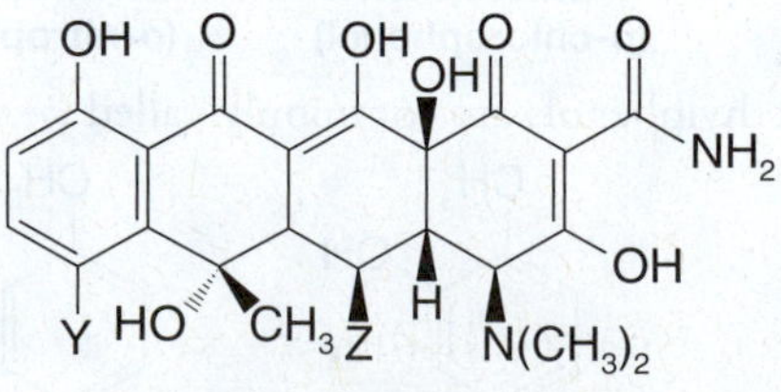

Tetracyclines
(Y = Cl, Z = H; Aureomycin)
(Y = H, Z = OH; Terramycin)

21.3 PHYSICAL PROPERTIES OF PHENOLS

The presence of hydroxyl groups in phenols means that phenols are like alcohols (Section 11.2) in some respects. For example, they are able to form strong intermolecular hydrogen bonds, and therefore have higher boiling points than hydrocarbons of the same molecular weight. Phenol (bp 182 °C) has a boiling point more than 70 °C higher than toluene (bp 110.6 °C), even though the two compounds have almost the same molecular weight. Phenols are also modestly soluble in water because of their ability to form strong hydrogen bonds with water molecules.

21.4 SYNTHESIS OF PHENOLS

21.4A Laboratory Synthesis

The most important laboratory synthesis of phenols is by hydrolysis of arenediazonium salts (Section 20.7E). This method is highly versatile, and the conditions required for the diazotization step and the hydrolysis step are mild. This means that other groups present on the ring are unlikely to be affected.

General Reaction

$$\text{Ar-NH}_2 \xrightarrow{\text{HONO}} \text{Ar-}\overset{+}{\text{N}}_2 \xrightarrow[\text{H}_2\text{O}]{\text{Cu}_2\text{O, Cu}^{2+}} \text{Ar-OH}$$

Specific Example

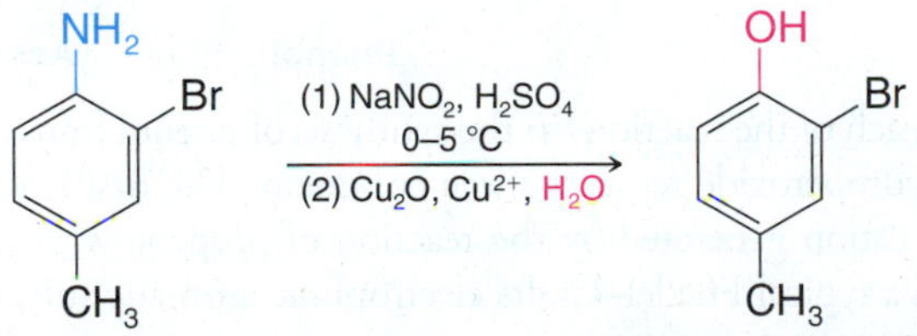

2-Bromo-4-methylphenol
(80–92%)

Reagents: (1) NaNO$_2$, H$_2$SO$_4$ 0–5 °C; (2) Cu$_2$O, Cu^{2+}, H$_2$O

21.4B Industrial Syntheses

Phenol is a highly important industrial chemical; it serves as the raw material for a large number of commercial products ranging from aspirin to a variety of plastics. Worldwide production of phenol (which in industry is sometimes called carbolic acid) is more than 3 million tons per year. Several methods have been used to synthesize phenol commercially.

1. **Hydrolysis of Chlorobenzene (Dow Process).** In this process chlorobenzene is heated at 350 °C (under high pressure) with aqueous sodium hydroxide. The reaction produces sodium phenoxide, which, on acidification, yields phenol. The mechanism for the reaction probably involves the formation of benzyne (Section 21.11B).

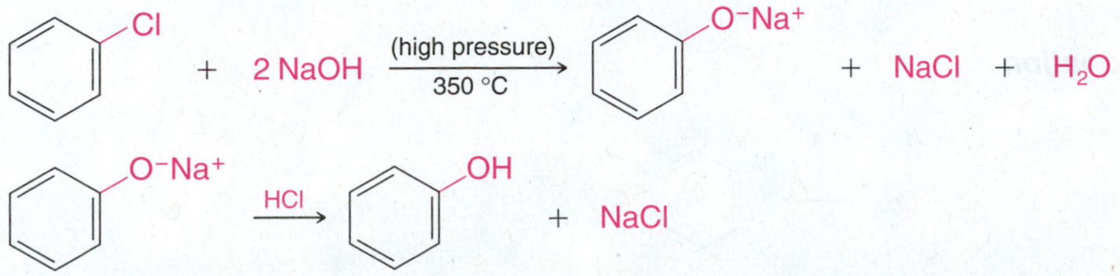

2. **From Cumene Hydroperoxide.** This process illustrates industrial chemistry at its best. Overall, it is a method for converting two relatively inexpensive organic compounds—benzene and propene—into two more valuable ones—phenol and acetone. The only other substance consumed in the process is oxygen from air. Most of the worldwide production of phenol is now based on this method. The synthesis

begins with the Friedel–Crafts alkylation of benzene with propene to produce cumene (isopropylbenzene):

Reaction 1

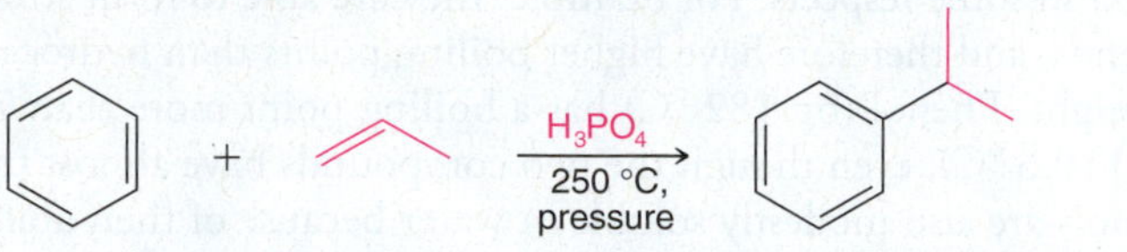

Then cumene is oxidized to cumene hydroperoxide:

Reaction 2

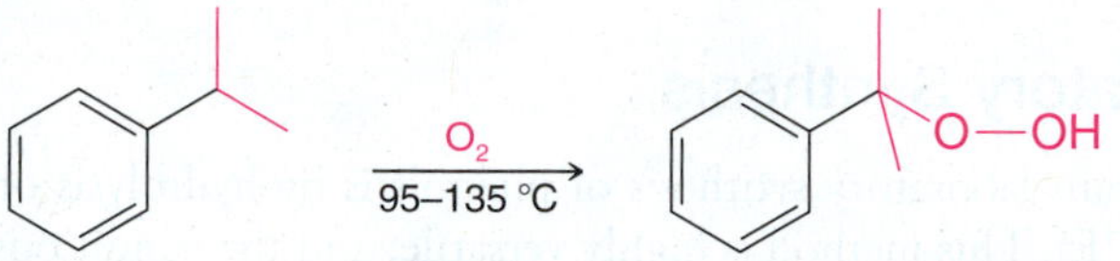

Cumene hydroperoxide

Finally, when treated with 10% sulfuric acid, cumene hydroperoxide undergoes a hydrolytic rearrangement that yields phenol and acetone:

Reaction 3

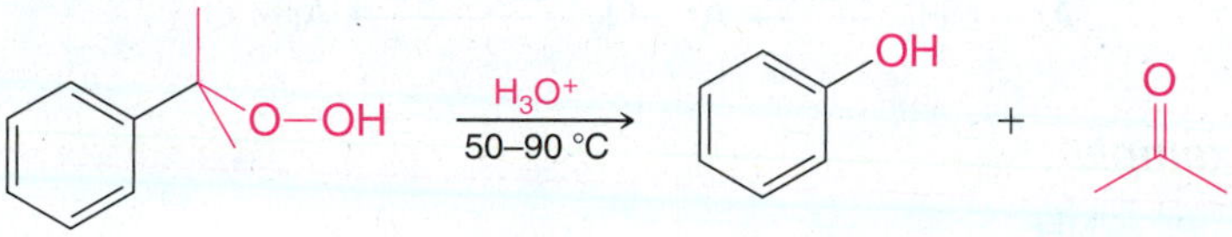

Phenol   **Acetone**

The mechanism of each of the reactions in the synthesis of phenol from benzene and propene via cumene hydroperoxide requires some comment. The first reaction is a familiar one. The isopropyl cation generated by the reaction of propene with the acid (H_3PO_4) alkylates benzene in a typical Friedel–Crafts electrophilic aromatic substitution:

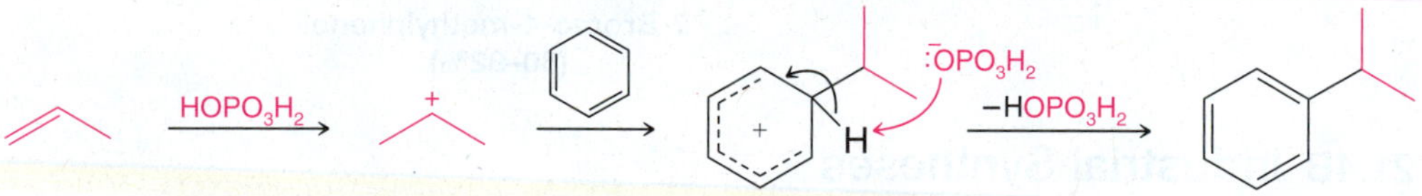

The second reaction is a radical chain reaction. A radical initiator abstracts the benzylic hydrogen atom of cumene, producing a 3° benzylic radical. Then a chain reaction with oxygen, which exists as a paramagnetic diradical in the ground state, produces cumene hydroperoxide:

Chain Initiation

Step 1

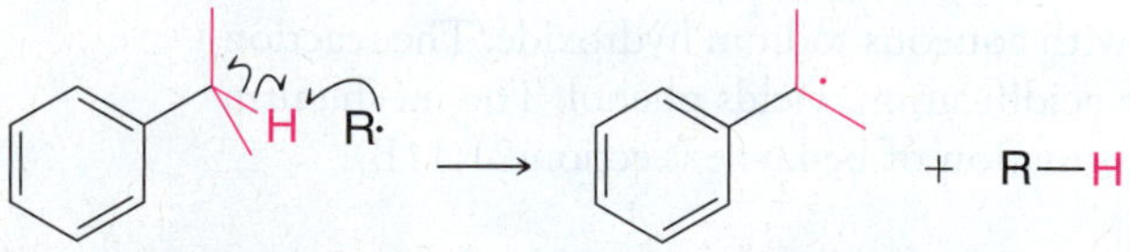

Chain Propagation

Step 2

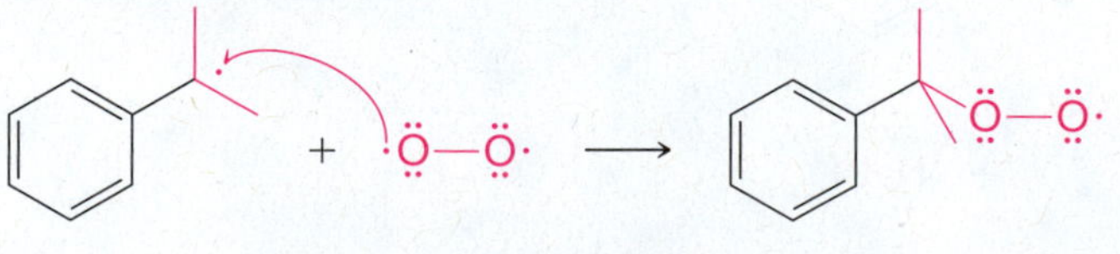

Step 3

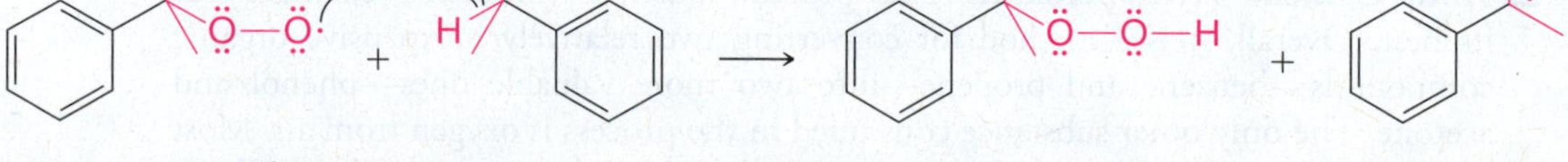

The reaction continues with steps 2, 3, 2, 3, and so on.

The third reaction—the hydrolytic rearrangement—resembles the carbocation rearrangements that we have studied before. In this instance, however, the rearrangement involves the migration of a phenyl group to *a cationic oxygen atom*. Phenyl groups have a much greater tendency to migrate to a cationic center than do methyl groups. The following equations show all the steps of the mechanism.

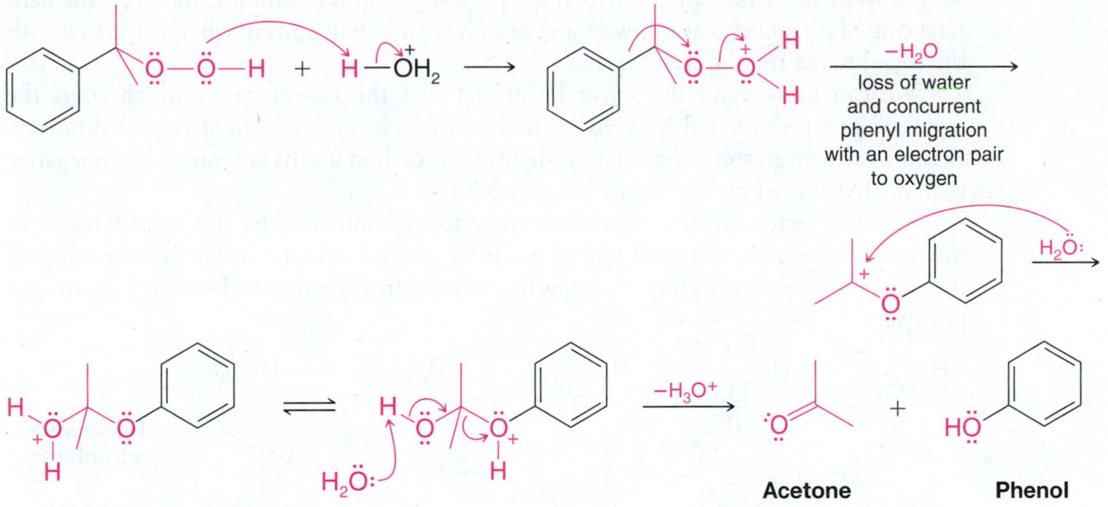

loss of water and concurrent phenyl migration with an electron pair to oxygen

Acetone **Phenol**

The second and third steps of the mechanism may actually take place at the same time; that is, the loss of H_2O and the migration of C_6H_5- may be concerted.

21.5 REACTIONS OF PHENOLS AS ACIDS

21.5A Strength of Phenols as Acids

Although phenols are structurally similar to alcohols, they are much stronger acids. The pK_a values of most alcohols are of the order of 18. However, as we see in Table 21.1, the pK_a values of phenols are smaller than 11.

TABLE 21.1 ACIDITY CONSTANTS OF PHENOLS

Name	pK_a (in H_2O at 25 °C)	Name	pK_a (in H_2O at 25 °C)
Phenol	9.89	3-Nitrophenol	8.28
2-Methylphenol	10.20	4-Nitrophenol	7.15
3-Methylphenol	10.01	2,4-Dinitrophenol	3.96
4-Methylphenol	10.17	2,4,6-Trinitrophenol (picric acid)	0.38
2-Chlorophenol	8.11		
3-Chlorophenol	8.80	1-Naphthol	9.31
4-Chlorophenol	9.20	2-Naphthol	9.55
2-Nitrophenol	7.17		

Let us compare two *superficially* similar compounds, cyclohexanol and phenol:

Cyclohexanol $pK_a = 18$ **Phenol** $pK_a = 9.89$

Although phenol is a weak acid when compared with a carboxylic acid such as acetic acid ($pK_a = 4.76$), phenol is a much stronger acid than cyclohexanol (by a factor of eight pK_a units).

Experimental and theoretical results have shown that the greater acidity of phenol owes itself primarily to an electrical charge distribution in phenol that causes the —OH oxygen to be more positive; therefore, the proton is held less strongly. In effect, the benzene ring of phenol acts as if it were an electron-withdrawing group when compared with the cyclohexane ring of cyclohexanol.

We can understand this effect by noting that the carbon atom which bears the hydroxyl group in phenol is sp^2 hybridized, whereas in cyclohexane it is sp^3 hybridized. Because of their greater s character, sp^2-hybridized carbon atoms are more electronegative than sp^3-hybridized carbon atoms (Section 3.8A).

Another factor influencing the electron distribution may be the contributions to the overall resonance hybrid of phenol made by structures **2–4**. Notice that the effect of these structures is to withdraw electrons from the hydroxyl group and to make the oxygen positive:

Resonance structures for phenol

| 1a | 1b | 2 | 3 | 4 |

An alternative explanation for the greater acidity of phenol relative to cyclohexanol can be based on similar resonance structures for the phenoxide ion. Unlike the structures for phenol, **2–4**, resonance structures for the phenoxide ion do not involve charge separation. According to resonance theory, such structures should stabilize the phenoxide ion more than structures **2–4** stabilize phenol. (No resonance structures can be written for cyclohexanol or its anion, of course.) Greater stabilization of the phenoxide ion (the conjugate base) than of phenol (the acid) has an acid-strengthening effect.

●●● SOLVED PROBLEM 21.1

Rank the following compounds in order of increasing acidity.

STRATEGY AND ANSWER: Alcohols are less acidic than phenols, and phenols are less acidic than carboxylic acids. An electron-withdrawing group increases the acidity of a phenol relative to phenol itself. Thus the order of increasing acidity among these examples is cyclohexanol < phenol < 4-nitrophenol < benzoic acid.

●●●

PRACTICE PROBLEM 21.1 If we examine Table 21.1, we find that the methylphenols (cresols) are less acidic than phenol itself. For example,

| Phenol | 4-Methylphenol |
| $pK_a = 9.89$ | $pK_a = 10.17$ |

This behavior is characteristic of phenols bearing electron-releasing groups. Provide an explanation.

If we examine Table 21.1, we see that phenols having electron-withdrawing groups (Cl— or O_2N—) attached to the benzene ring are more acidic than phenol itself. Account for this trend on the basis of resonance and inductive effects. Your answer should also explain the large acid-strengthening effect of nitro groups, an effect that makes 2,4,6-trinitrophenol (also called *picric acid*) so exceptionally acidic ($pK_a = 0.38$) that it is more acidic than acetic acid ($pK_a = 4.76$).

21.5B Distinguishing and Separating Phenols from Alcohols and Carboxylic Acids

Because phenols are more acidic than water, the following reaction goes essentially to completion and produces water-soluble sodium phenoxide:

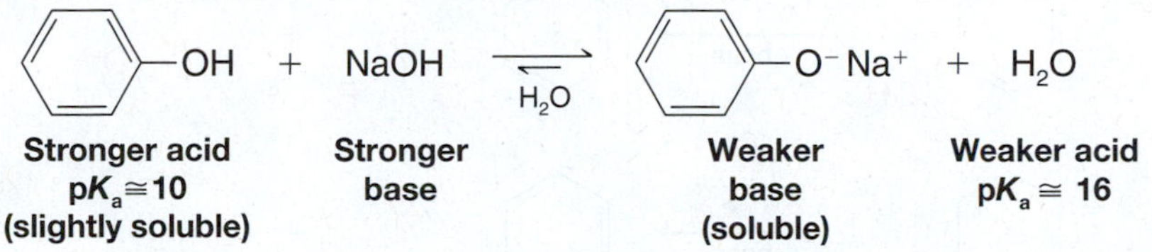

Stronger acid	Stronger	Weaker	Weaker acid
$pK_a \cong 10$	base	base	$pK_a \cong 16$
(slightly soluble)		(soluble)	

The corresponding reaction of 1-hexanol with aqueous sodium hydroxide does not occur to a significant extent because 1-hexanol is a weaker acid than water:

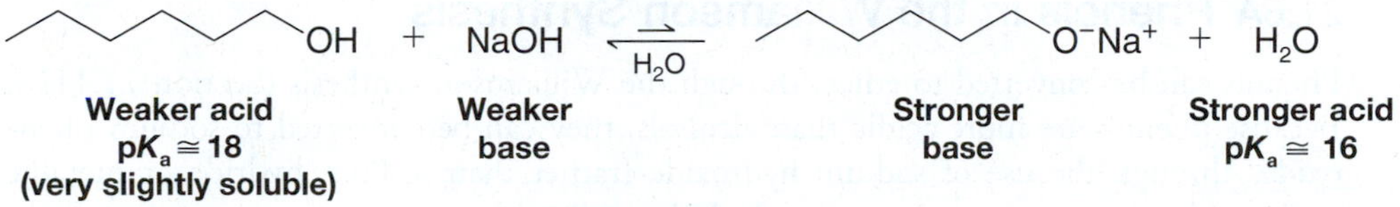

Weaker acid	Weaker	Stronger	Stronger acid
$pK_a \cong 18$	base	base	$pK_a \cong 16$
(very slightly soluble)			

The fact that phenols dissolve in aqueous sodium hydroxide, whereas most alcohols with six carbon atoms or more do not, gives us a convenient means for distinguishing and separating phenols from most alcohols. (Alcohols with five carbon atoms or fewer are quite soluble in water—some are infinitely so—and so they dissolve in aqueous sodium hydroxide even though they are not converted to sodium alkoxides in appreciable amounts.)

Most phenols, however, are not soluble in aqueous sodium bicarbonate ($NaHCO_3$), but carboxylic acids are soluble. Thus, aqueous $NaHCO_3$ provides a method for distinguishing and separating most phenols from carboxylic acids.

●●● SOLVED PROBLEM 21.2

Assume that each of the following mixtures was added to a flask or a separatory funnel that contained diethyl ether (as an organic solvent) and mixed well. In which layer (diethyl ether or water) would the organic compound predominate in each case, and in what form would it exist (in its neutral form or as its conjugate base)?

(a)

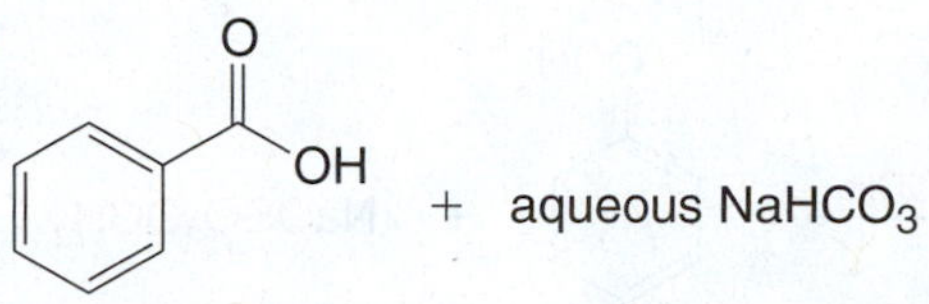

+ aqueous $NaHCO_3$

(c)

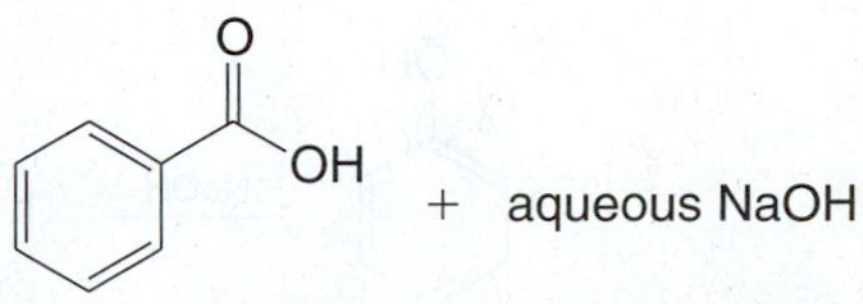

+ aqueous NaOH

(b)

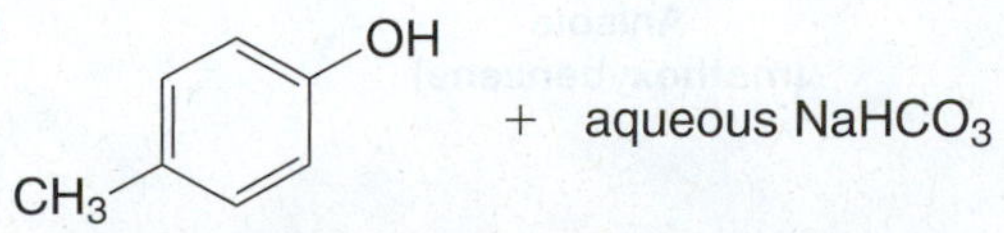

+ aqueous $NaHCO_3$

(d)

+ aqueous NaOH

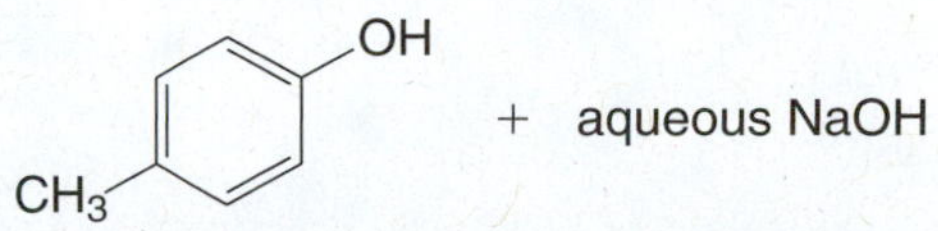

STRATEGY AND ANSWER: Sodium bicarbonate will remove a proton from a carboxylic acid to form a water-soluble carboxylate salt, but sodium bicarbonate will not remove a proton from a typical phenol. Sodium hydroxide will remove a proton from both a carboxylic acid and a phenol to form salts in each case. Thus, in **(a)** benzoic acid will be found in the water layer as its sodium salt, whereas in **(b)** 4-methylphenol will remain in its neutral form and be found predominantly in the ether layer. In **(c)** and **(d)** both benzoic acid and 4-methylphenol will be found in the aqueous layer as their corresponding salts.

PRACTICE PROBLEM 21.3 Your laboratory instructor gives you a mixture of 4-methylphenol, benzoic acid, and toluene. Assume that you have available common laboratory acids, bases, and solvents and explain how you would proceed to separate this mixture by making use of the solubility differences of its components.

21.6 OTHER REACTIONS OF THE O—H GROUP OF PHENOLS

Phenols react with carboxylic acid anhydrides and acid chlorides to form esters. These reactions are quite similar to those of alcohols (Section 17.7).

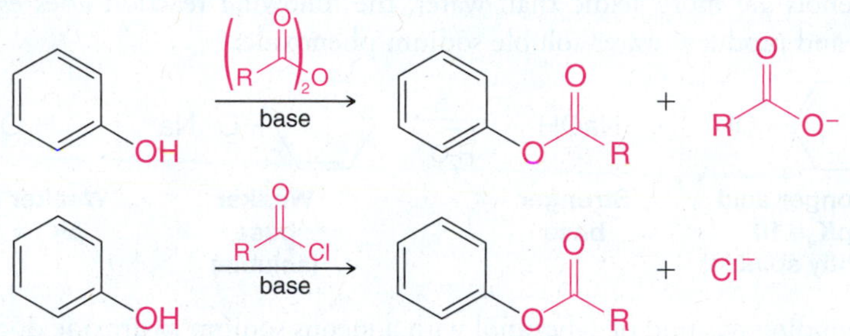

21.6A Phenols in the Williamson Synthesis

Phenols can be converted to ethers through the Williamson synthesis (Section 11.11B). Because phenols are more acidic than alcohols, they can be converted to sodium phenoxides through the use of sodium hydroxide (rather than sodium hydride or metallic sodium, the reagents used to convert alcohols to alkoxide ions).

General Reaction

$$\text{ArOH} \xrightarrow{\text{NaOH}} \text{ArO}^-\,\text{Na}^+ \xrightarrow[\substack{(\text{X = Cl, Br, I,}\\ \text{OSO}_2\text{OR}'\text{ or}\\ \text{OSO}_2\text{R}')}]{\text{R—X}} \text{ArOR} + \text{NaX}$$

Specific Examples

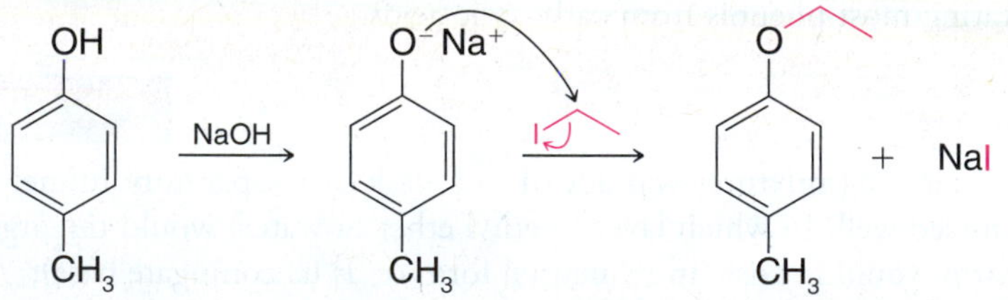

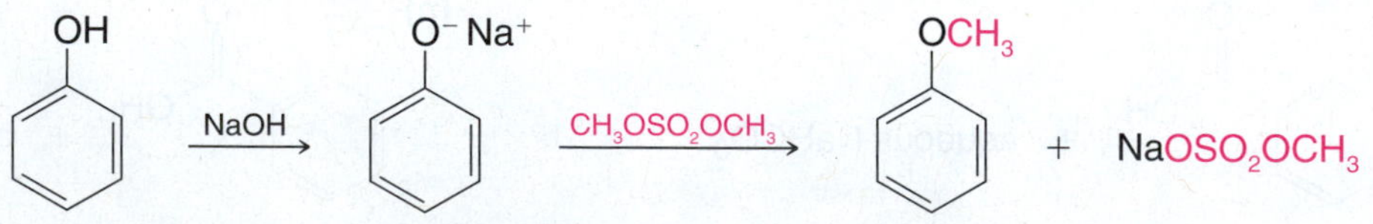

Anisole (methoxybenzene)

21.7 CLEAVAGE OF ALKYL ARYL ETHERS

We learned in Section 11.12A that when dialkyl ethers are heated with excess concentrated HBr or HI, the ethers are cleaved and alkyl halides are produced from both alkyl groups:

$$\text{R—O—R}' \xrightarrow[\text{heat}]{\text{concd HX}} \text{R—X} + \text{R}'\text{—X} + H_2O$$

When alkyl aryl ethers react with strong acids such as HI and HBr, the reaction produces an alkyl halide and a phenol. The phenol does not react further to produce an aryl halide because the phenol carbon–oxygen bond is very strong and because phenyl cations do not form readily.

General Reaction

$$Ar—O—R \xrightarrow[\text{heat}]{\text{concd HX}} Ar—OH \ + \ R—X$$

Specific Example

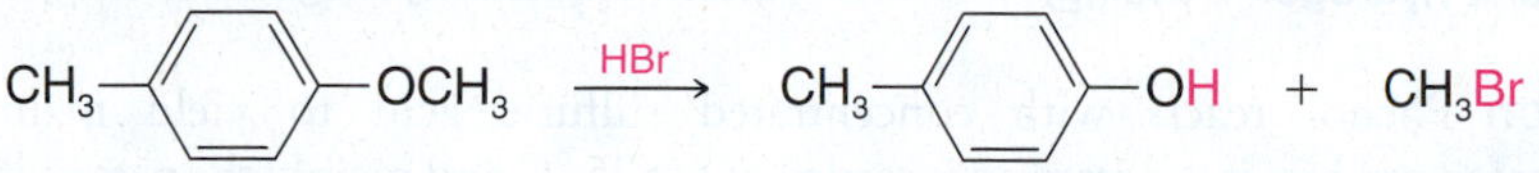

$$CH_3 \overbrace{\hspace{2cm}} —OCH_3 \xrightarrow{\text{HBr}} CH_3 \overbrace{\hspace{2cm}} —OH \ + \ CH_3Br$$

p-Methylanisole   4-Methylphenol   Methyl bromide

21.8 REACTIONS OF THE BENZENE RING OF PHENOLS

Bromination The hydroxyl group is a powerful activating group—and an ortho–para director—in **electrophilic aromatic substitutions**. Phenol itself reacts with bromine in aqueous solution to yield 2,4,6-tribromophenol in nearly quantitative yield. Note that a Lewis acid is not required for the bromination of this highly activated ring:

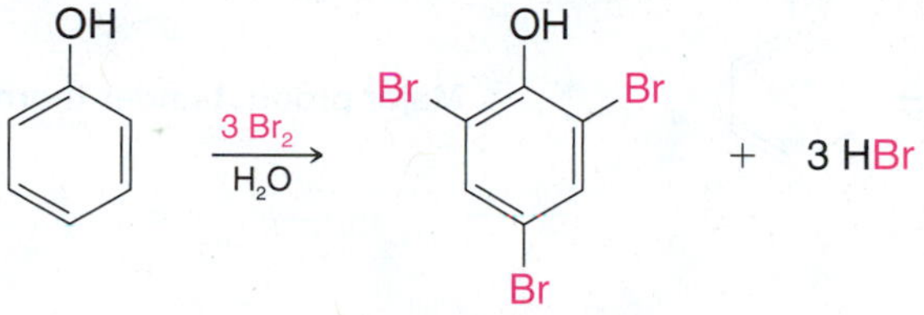

2,4,6-Tribromophenol
(~100%)

Monobromination of phenol can be achieved by carrying out the reaction in carbon disulfide at a low temperature, conditions that reduce the electrophilic reactivity of bromine. The major product is the para isomer:

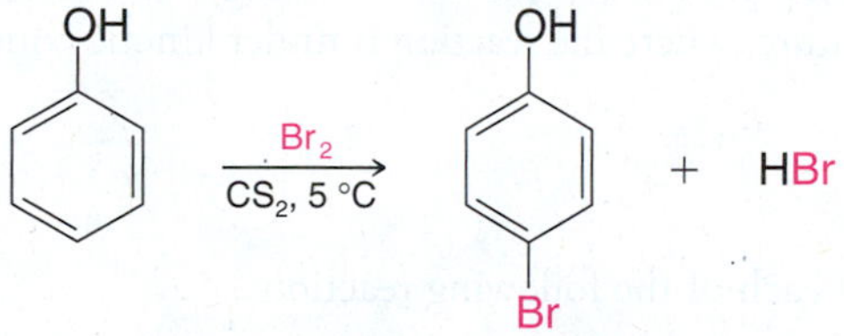

p-Bromophenol
(80–84%)

Nitration Phenol reacts with dilute nitric acid to yield a mixture of *o*- and *p*-nitrophenol:

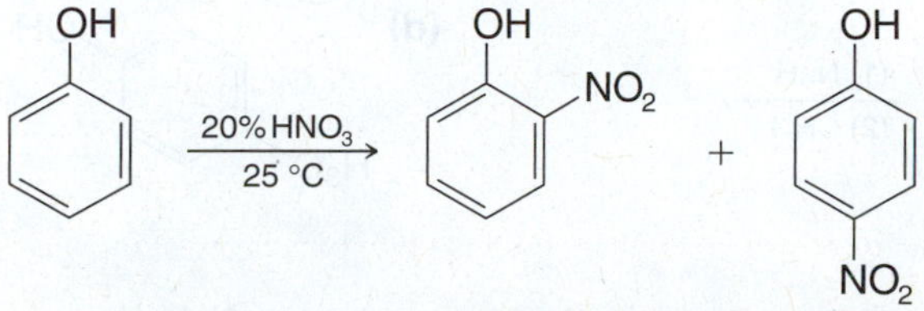

30–40%   15%

Although the yield is relatively low (because of oxidation of the ring), the ortho and para isomers can be separated by steam distillation. *o*-Nitrophenol is the more volatile isomer because its hydrogen bonding (see the following structures) is *intramolecular*. *p*-Nitrophenol is less volatile because *intermolecular* hydrogen bonding causes association among its molecules. Thus, *o*-nitrophenol passes over with the steam, and *p*-nitrophenol remains in the distillation flask.

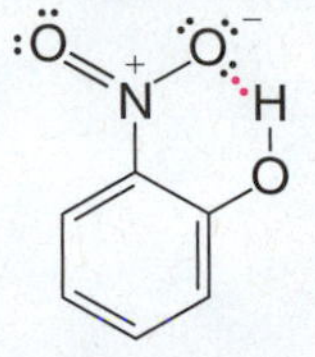

o-Nitrophenol
(more volatile because of intramolecular hydrogen bonding)

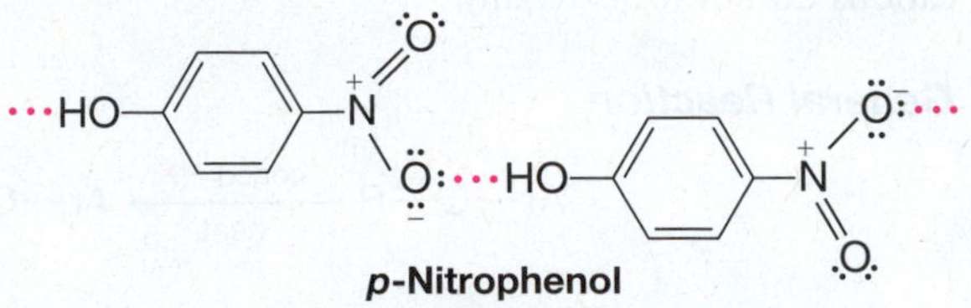

p-Nitrophenol
(less volatile because of intermolecular hydrogen bonding)

Sulfonation Phenol reacts with concentrated sulfuric acid to yield mainly the orthosulfonated product if the reaction is carried out at 25 °C and mainly the para-sulfonated product at 100 °C. This is another example of thermodynamic versus kinetic control of a reaction (Section 13.10A):

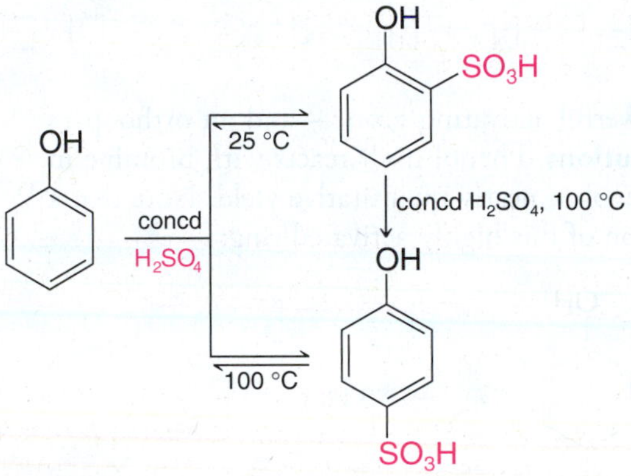

OH → (25 °C)

OH with SO$_3$H
Major product under kinetic control

concd H$_2$SO$_4$, 100 °C

OH → (100 °C)

OH with SO$_3$H (para)
Major product under thermodynamic control

●●● SOLVED PROBLEM 21.3

Consider the sulfonation reactions of phenol shown above. **(a)** Which sulfonation product is more stable, ortho or para? **(b)** For which sulfonation product is the free energy of activation lower?

ANSWER: (a) The para-sulfonated phenol is more stable. We know this because at the higher temperature, where the reaction is under equilibrium control, it is the major product. **(b)** The free energy of activation is lower for ortho substitution. We know this because at the lower temperature, where the reaction is under kinetic control, it is formed faster.

●●●

PRACTICE PROBLEM 21.4 Predict the products of each of the following reactions.

(a)

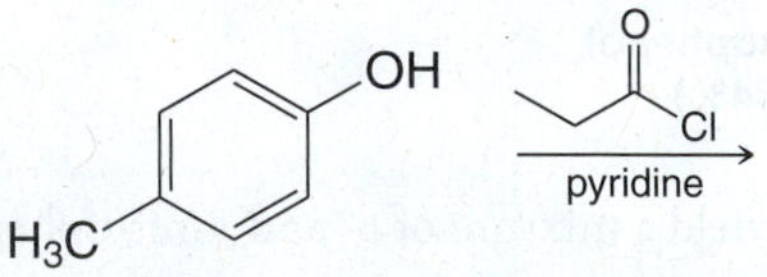

H$_3$C — (ring) — OH + (acyl chloride with Cl) → (pyridine)

(c)

(ring) — O — ethyl → (HBr)

(b)

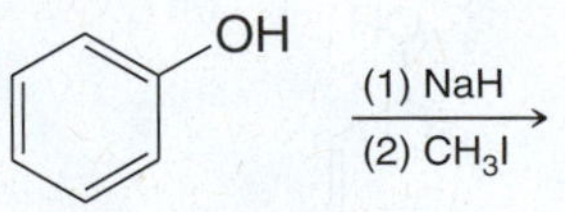

(ring) — OH → (1) NaH (2) CH$_3$I

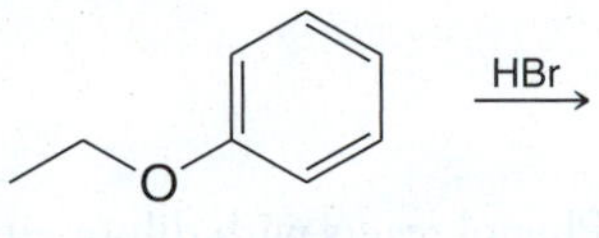

(d)

H$_3$C — (ring) — OH → (HNO$_3$, H$_2$SO$_4$)

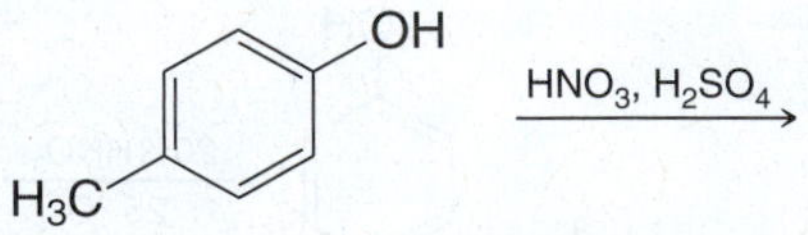

◆ THE CHEMISTRY OF... Polyketide Anticancer Antibiotic Biosynthesis

Doxorubicin (also known as adriamycin) is a highly potent anticancer drug that contains phenol functional groups. It is effective against many forms of cancer, including tumors of the ovaries, breast, bladder, and lung, as well as against Hodgkin's disease and other acute leukemias. Doxorubicin is a member of the anthracycline family of antibiotics.

Another member of the family is daunomycin. Both of these antibiotics are produced in strains of *Streptomyces* bacteria by a pathway called polyketide biosynthesis.

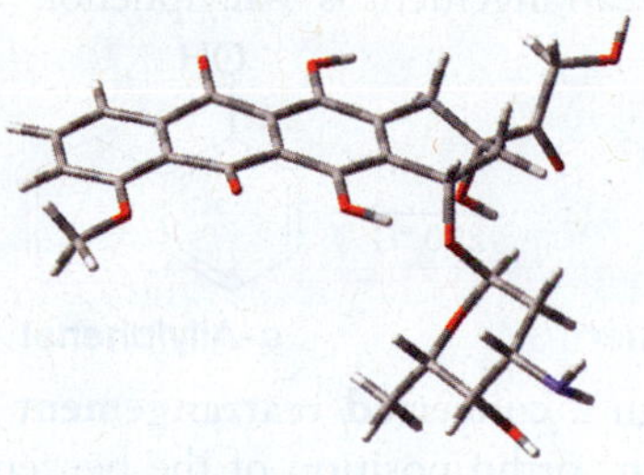

A molecular model of doxorubicin.

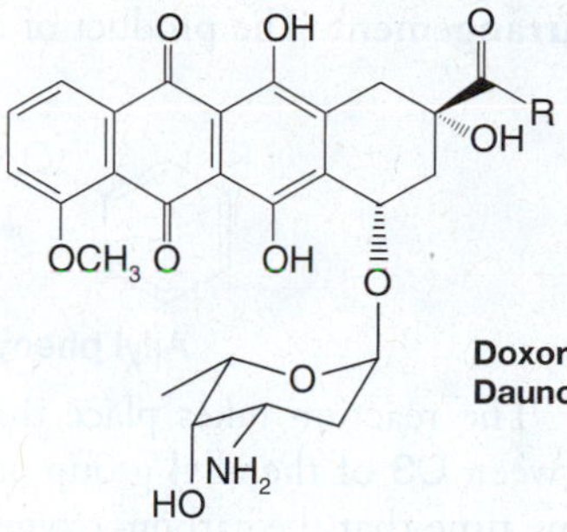

Doxorubicin (R = CH$_2$OH)
Daunomycin (R = CH$_3$)

Isotopic labeling experiments have shown that daunomycin is synthesized in *Streptomyces galilaeus* from a tetracyclic precursor called aklavinone. Aklavinone, in turn, is synthesized from acetate. When *S. galilaeus* is grown in a medium containing acetate labeled with carbon-13 and oxygen-18, the aklavinone produced has isotopic labels in the positions indicated below. Notice that oxygen atoms occur at alternate carbons in several places around the structure, consistent with the linking of acetate units in head-to-tail fashion. This is typical of aromatic polyketide biosynthesis.

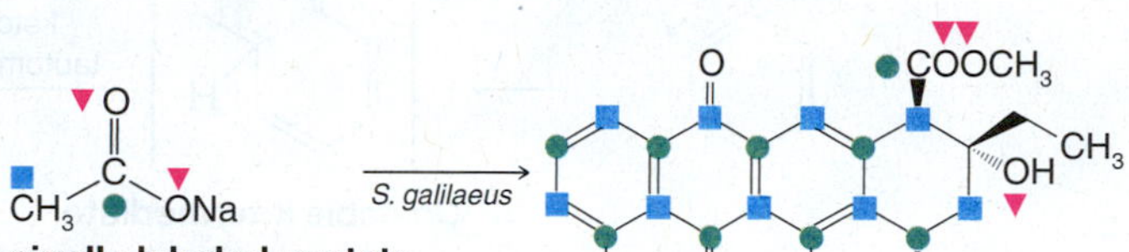

Isotopically labeled acetate
■, ● = ^{13}C labels
▼ = ^{18}O labels

S. galilaeus

Aklavinone

This and other information show that nine C$_2$ units from malonyl-coenzyme A and one C$_3$ unit from propionyl-coenzyme A condense to form the linear polyketide intermediate shown below. These units are joined by acylation reactions that are the biosynthetic equivalent of the *malonic ester synthesis* we studied in Section 18.7. These reactions are also similar to the acylation steps we saw in fatty acid biosynthesis (Special Topic E in *WileyPLUS*). Once formed, the linear polyketide cyclizes by enzymatic reactions akin to intramolecular *aldol additions and dehydrations* (Section 19.6). These steps form the tetracyclic core of aklavinone. Phenolic hydroxyl groups in aklavinone arise by enolization of ketone carbonyl groups present after the aldol condensation steps. Several other transformations ultimately lead to daunomycin:

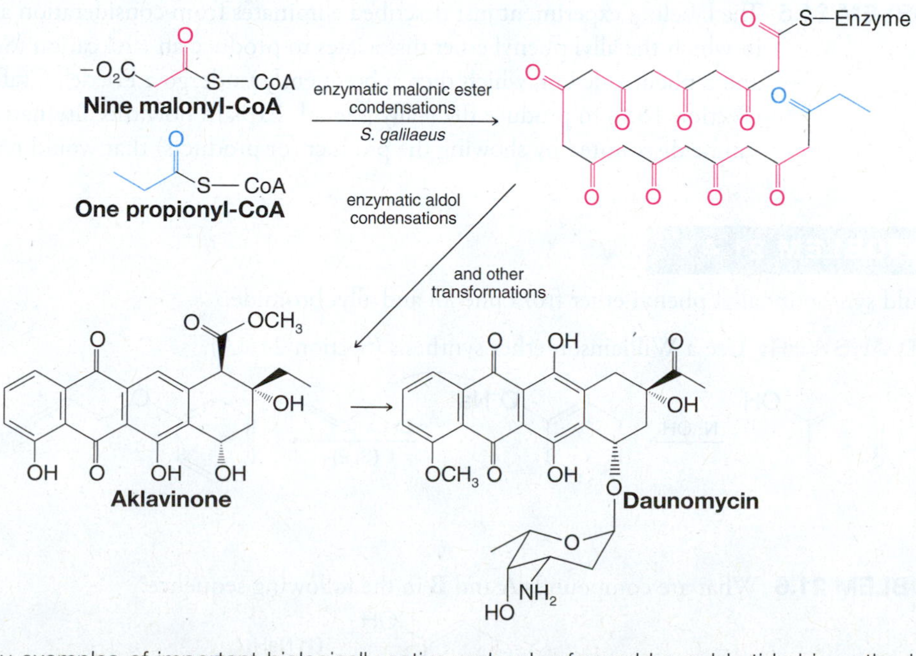

Nine malonyl-CoA

One propionyl-CoA

enzymatic malonic ester condensations
S. galilaeus

enzymatic aldol condensations

and other transformations

Aklavinone

Daunomycin

There are many examples of important biologically active molecules formed by polyketide biosynthesis. Aureomycin and terramycin (Section 21.2) are examples of other aromatic polyketide antibiotics. Erythromycin (Section 17.7C) and aflatoxin, a carcinogen (see "Why do these topics matter?" in Chapter 14), are polyketides from other pathways.

21.9 THE CLAISEN REARRANGEMENT

Heating allyl phenyl ether to 200 °C effects an intramolecular reaction called a **Claisen rearrangement**. The product of the rearrangement is *o*-allylphenol:

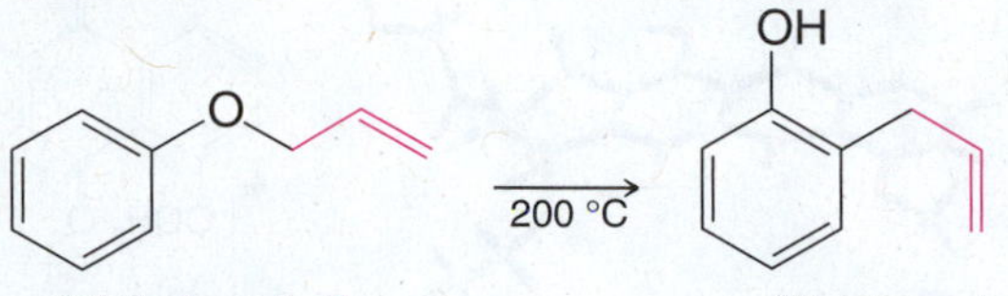

Allyl phenyl ether   **o-Allylphenol**

The reaction takes place through a **concerted rearrangement** in which the bond between C3 of the allyl group and the ortho position of the benzene ring forms at the same time that the carbon–oxygen bond of the allyl phenyl ether breaks. The product of this rearrangement is an unstable intermediate that, like the unstable intermediate in the Kolbe reaction (Section 21.8), undergoes a proton shift (a keto–enol tautomerization, see Section 18.2) that leads to the *o*-allylphenol:

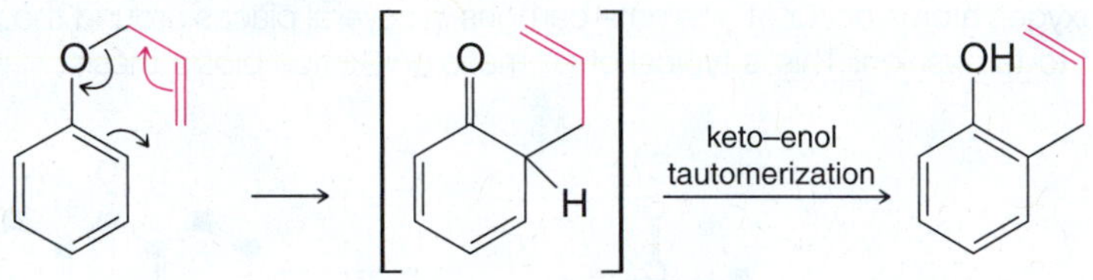

Unstable intermediate

That only C3 of the allyl group becomes bonded to the benzene ring was demonstrated by carrying out the rearrangement with allyl phenyl ether containing ^{14}C at C3. All of the product of this reaction had the labeled carbon atom bonded to the ring:

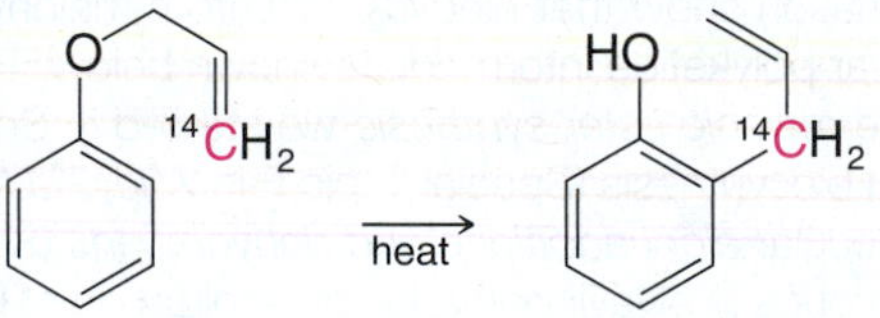

Only product

● ● ●

PRACTICE PROBLEM 21.5 The labeling experiment just described eliminates from consideration a mechanism in which the allyl phenyl ether dissociates to produce an allyl cation (Section 13.4) and a phenoxide ion, which then subsequently undergo a Friedel–Crafts alkylation (Section 15.6) to produce the *o*-allylphenol. Explain how this alternative mechanism can be discounted by showing the product (or products) that would result from it.

● ● ●   **SOLVED PROBLEM 21.4**

Show how you could synthesize allyl phenyl ether from phenol and allyl bromide.

STRATEGY AND ANSWER: Use a Williamson ether synthesis (Section 21.6A).

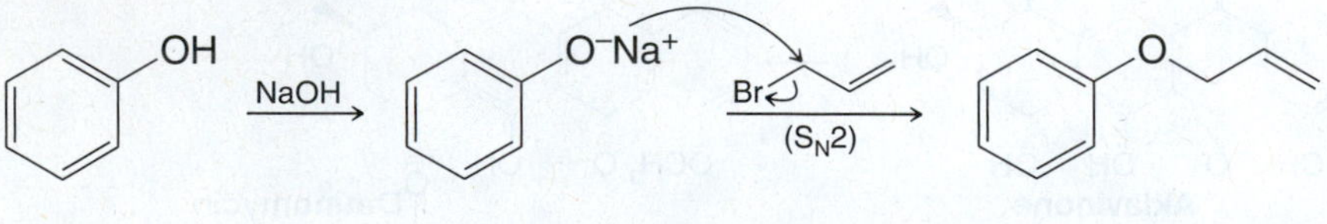

● ● ●

PRACTICE PROBLEM 21.6 What are compounds **A** and **B** in the following sequence?

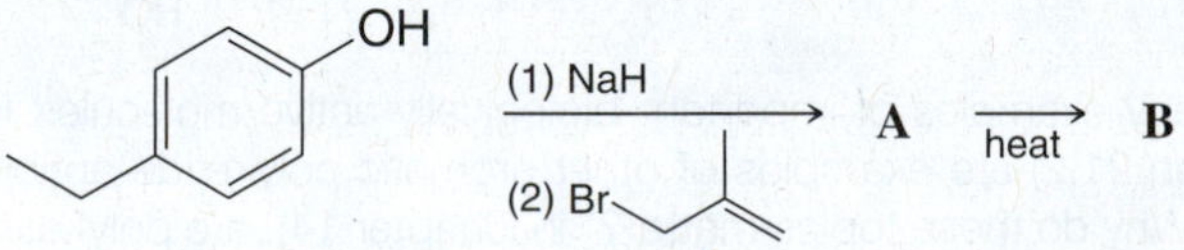

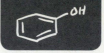

A Claisen rearrangement also takes place when allyl vinyl ethers are heated. For example,

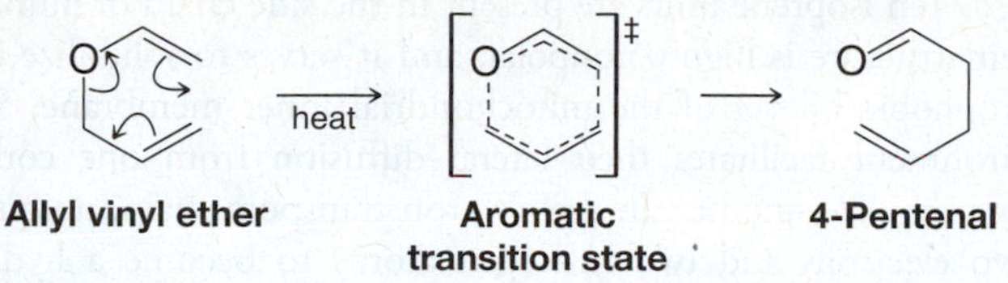

Allyl vinyl ether → (heat) → **Aromatic transition state** → **4-Pentenal**

The transition state for the Claisen rearrangement involves a cycle of six electrons. Having six electrons suggests that the transition state has aromatic character (Section 14.7). Other reactions of this general type are known, and they are called **pericyclic reactions**. Another similar pericyclic reaction is the **Cope rearrangement** shown here:

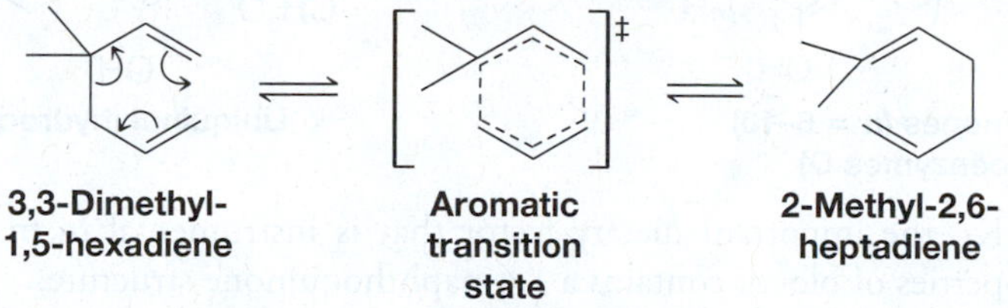

3,3-Dimethyl-1,5-hexadiene ⇌ **Aromatic transition state** ⇌ **2-Methyl-2,6-heptadiene**

The Diels–Alder reaction (Section 13.11) is also a pericyclic reaction. The transition state for the Diels–Alder reaction also involves six electrons:

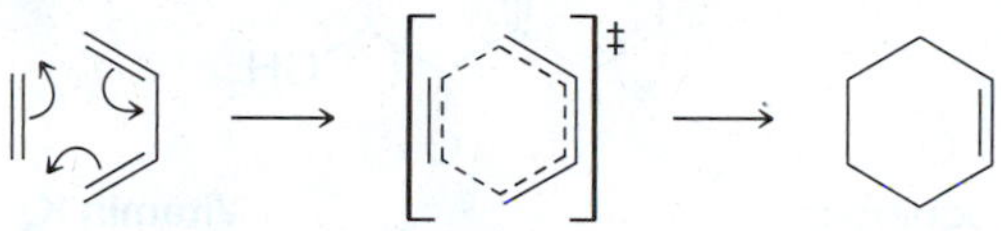

Aromatic transition state

The mechanism of the Diels–Alder reaction is discussed further in Special Topic H in *WileyPLUS*.

21.10 QUINONES

Oxidation of hydroquinone (1,4-benzenediol) produces a compound known as *p*-benzoquinone. The oxidation can be brought about by mild oxidizing agents, and, overall, the oxidation amounts to the removal of a pair of electrons ($2\ e^-$) and two protons from hydroquinone. (Another way of visualizing the oxidation is as the loss of a hydrogen molecule, $H\!:\!H$, making it a dehydrogenation.)

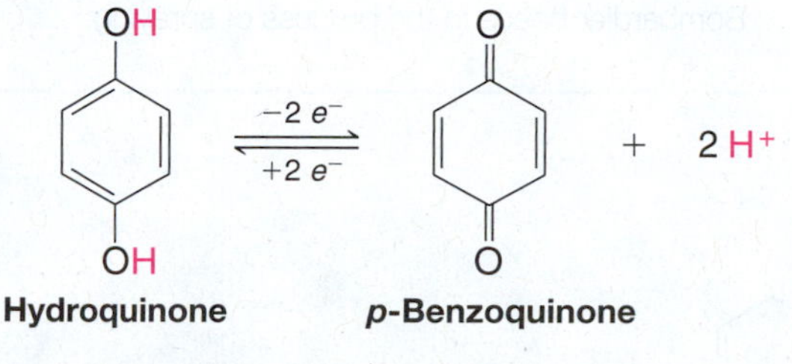

Hydroquinone $\underset{+2\ e^-}{\overset{-2\ e^-}{\rightleftharpoons}}$ **p-Benzoquinone** $+\ 2\ H^+$

This reaction is reversible; *p*-benzoquinone is easily reduced by mild reducing agents to hydroquinone.

Nature makes much use of this type of reversible oxidation–reduction to transport a pair of electrons from one substance to another in enzyme-catalyzed reactions. Important compounds in this respect are the compounds called **ubiquinones** (from *ubiquitous* + quinone—these quinones are found within the inner mitochondrial membrane of every living cell). Ubiquinones are also called coenzymes Q (CoQ).

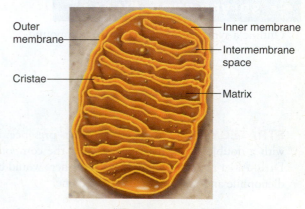

Cross section of a mitochondrion.

Outer membrane — Inner membrane — Intermembrane space — Cristae — Matrix

Ubiquinones have a long, isoprene-derived side chain (see Special Topic E in *WileyPLUS* and Section 23.3). Ten isoprene units are present in the side chain of human ubiquinones. This part of their structure is highly nonpolar, and it serves to solubilize the ubiquinones within the hydrophobic bilayer of the mitochondrial inner membrane. Solubility in the membrane environment facilitates their lateral diffusion from one component of the electron transport chain to another. In the electron transport chain, ubiquinones function by accepting two electrons and two hydrogen atoms to become a hydroquinone. The hydroquinone form carries the two electrons to the next acceptor in the chain:

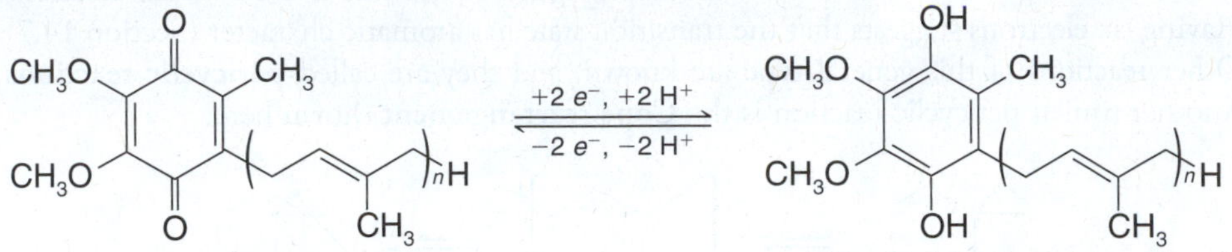

$$+2\ e^-,\ +2\ H^+ \qquad -2\ e^-,\ -2\ H^+$$

Ubiquinones (*n* = 6–10)
(coenzymes Q)

Ubiquinol (hydroquinone form)

Vitamin K_1, the important dietary factor that is instrumental in maintaining the coagulant properties of blood, contains a 1,4-naphthoquinone structure:

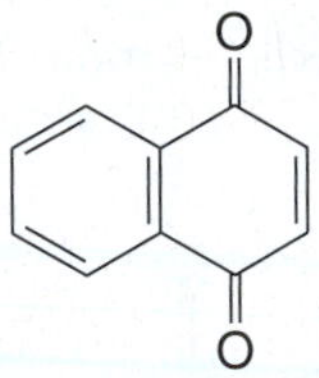

1,4-Naphthoquinone

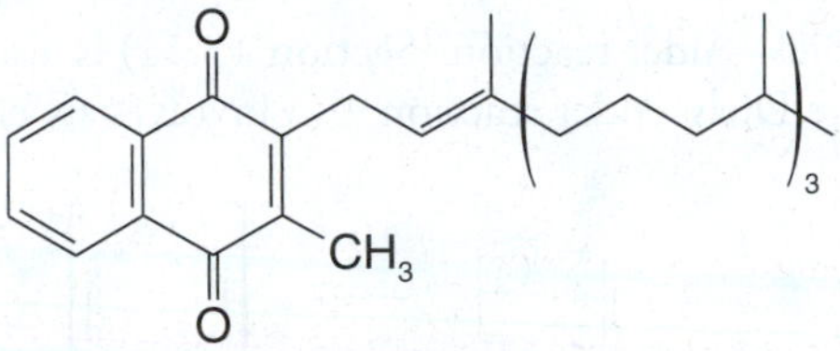

Vitamin K₁

THE CHEMISTRY OF... The Bombardier Beetle's Noxious Spray

The bombardier beetle defends itself by spraying a jet stream of hot (100 °C), noxious *p*-benzoquinones at an attacker. The beetle mixes *p*-hydroquinones and hydrogen peroxide from one abdominal reservoir with enzymes from another reservoir. The enzymes convert hydrogen peroxide to oxygen, which in turn oxidizes the *p*-hydroquinones to *p*-benzoquinones and explosively propels the irritating spray at the attacker. Photos by T. Eisner and D. Aneshansley (Cornell University) have shown that the amazing bombardier beetle can direct its spray in virtually any direction, even parallel over its back, to ward off a predator.

Thomas Eisner and Daniel Aneshansley, Cornell University

Bombardier beetle in the process of spraying.

●●● SOLVED PROBLEM 21.5

Outline a synthesis of the following compound.

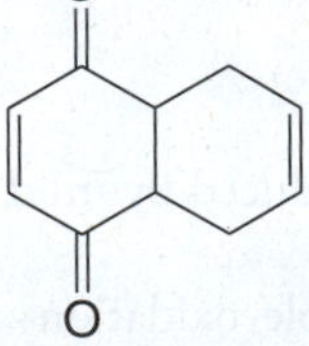

STRATEGY AND ANSWER: The presence of a cyclohexane ring with a double bond in it suggests that the compound could be made by a Diels–Alder reaction. Suitable reactants here would be *p*-benzoquinone as the dienophile and 1,3-butadiene as the diene.

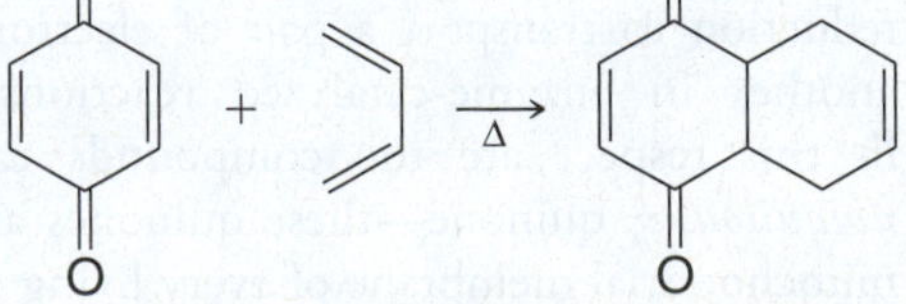

p-Benzoquinone and 1,4-naphthoquinone act as dienophiles in Diels–Alder reactions. Give the structures of the products of the following reactions:

PRACTICE PROBLEM 21.7

(a) 1,4-Naphthoquinone + butadiene   **(b)** *p*-Benzoquinone + cyclopentadiene

Outline a possible synthesis of the following compound.

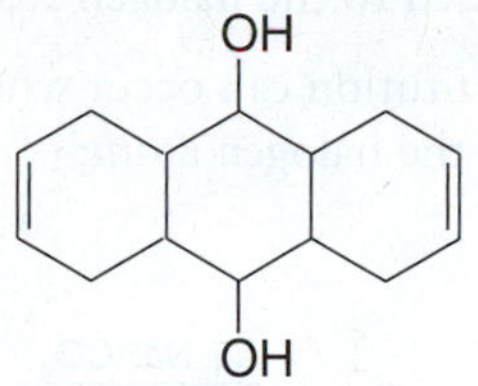

PRACTICE PROBLEM 21.8

21.11 ARYL HALIDES AND NUCLEOPHILIC AROMATIC SUBSTITUTION

- Simple aryl halides, like vinylic halides (Section 6.14A), are relatively unreactive toward nucleophilic substitution under conditions that would give rapid nucleophilic substitution with alkyl halides.

Chlorobenzene, for example, can be boiled with sodium hydroxide for days without producing a detectable amount of phenol (or sodium phenoxide).* Similarly, when vinyl chloride is heated with sodium hydroxide, no substitution occurs:

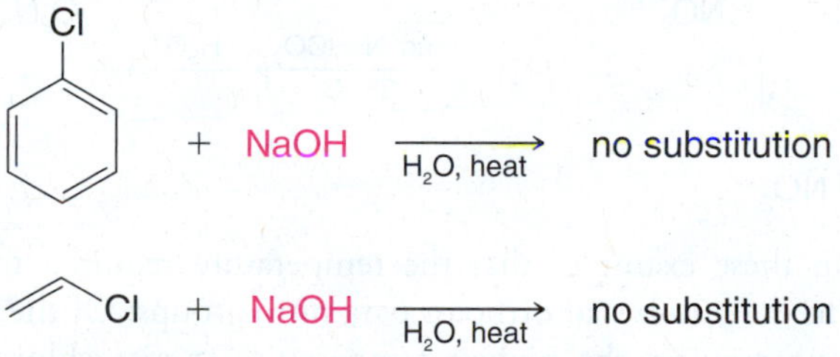

We can understand this lack of reactivity on the basis of several factors. The benzene ring of an aryl halide prevents back-side attack in an S_N2 reaction:

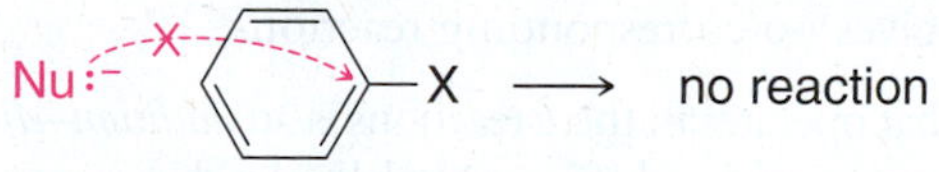

Phenyl cations are very unstable; thus S_N1 reactions do not occur. The carbon–halogen bonds of aryl (and vinylic) halides are shorter and stronger than those of alkyl, allylic, and benzylic halides. Stronger carbon–halogen bonds mean that bond breaking by either an S_N1 or S_N2 mechanism will require more energy.

Two effects make the carbon–halogen bonds of aryl and vinylic halides shorter and stronger: (1) The carbon of either type of halide is sp^2 hybridized, and therefore the electrons of the carbon orbital are closer to the nucleus than those of an sp^3-hybridized carbon. (2) Resonance of the type shown here strengthens the carbon–halogen bond by giving it *double-bond character*:

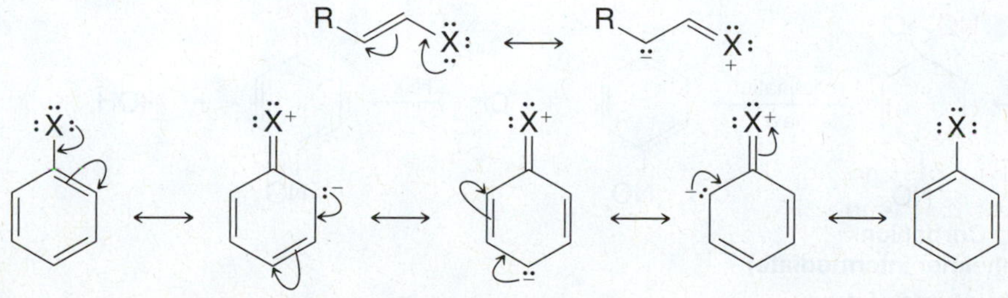

*The Dow process for making phenol by substitution (Section 21.4B) requires extremely high temperature and pressure to effect the reaction. These conditions are not practical in the laboratory.

Having said all this, we shall find in the next two subsections that *aryl halides can be remarkably reactive toward nucleophiles* if they bear certain substituents or when we allow them to react under the proper conditions.

21.11A Nucleophilic Aromatic Substitution by Addition–Elimination: The S$_N$Ar Mechanism

Nucleophilic substitution reactions of aryl halides *do* occur readily when an electronic factor makes the aryl carbon bonded to the halogen susceptible to nucleophilic attack.

● Nucleophilic aromatic substitution can occur when strong electron-withdrawing groups are ortho or para to the halogen atom:

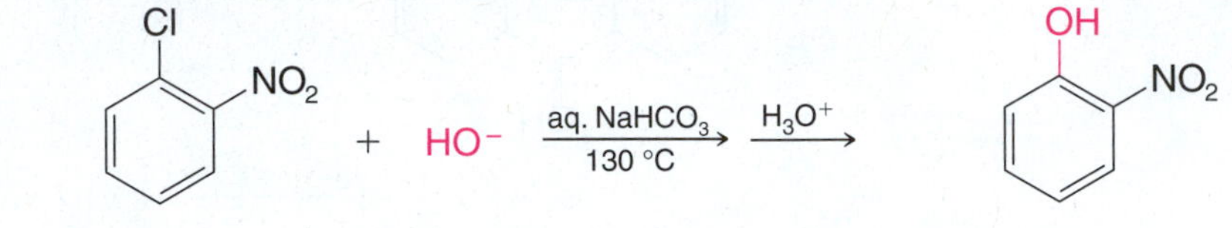

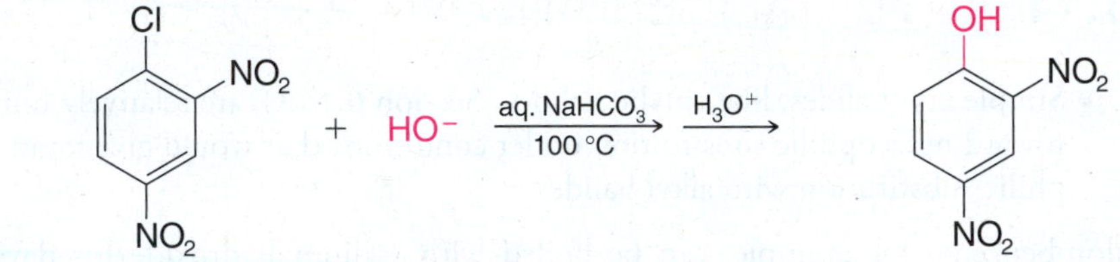

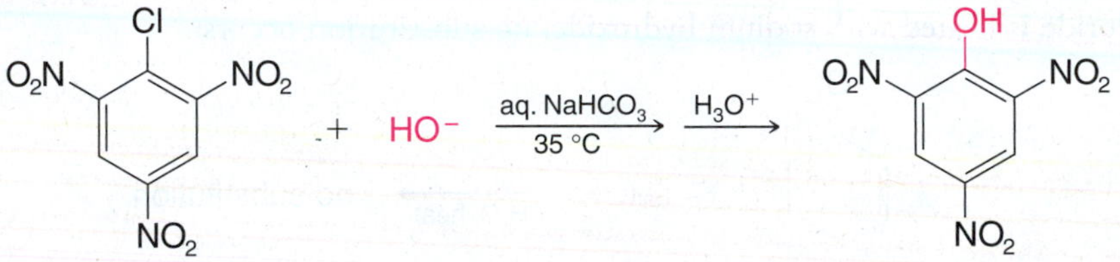

We also see in these examples that the temperature required to bring about the reaction is related to the number of ortho or para nitro groups. Of the three compounds, *o*-nitrochlorobenzene requires the highest temperature (*p*-nitrochlorobenzene reacts at 130 °C as well) and 2,4,6-trinitrochlorobenzene requires the lowest temperature.

A meta-nitro group does not produce a similar activating effect. For example, *m*-nitrochlorobenzene gives no corresponding reaction.

● The mechanism that operates in these reactions is an *addition–elimination* mechanism involving the formation of a *carbanion* with delocalized electrons, called a **Meisenheimer intermediate**. The process is called **nucleophilic aromatic substitution (S$_N$Ar)**.

In the first step of the following example, addition of a hydroxide ion to *p*-nitrochlorobenzene produces the carbanion; then elimination of a chloride ion yields the substitution product as the aromaticity of the ring is recovered.

A MECHANISM FOR THE REACTION — The S$_N$Ar Mechanism

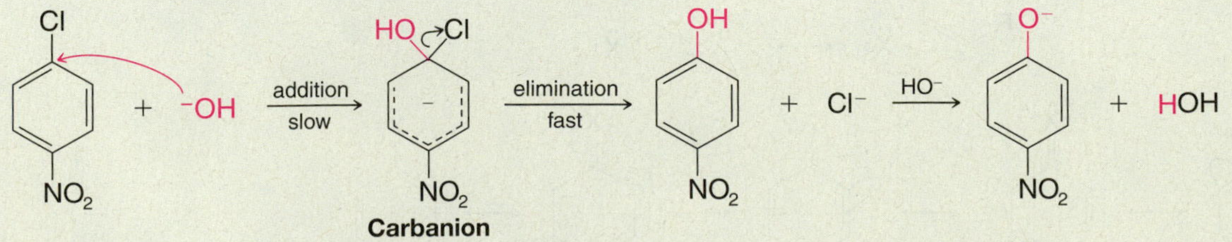

**Carbanion
(Meisenheimer intermediate)
Structures of the
contributing resonance
forms are shown further below.**

The carbanion is stabilized by *electron-withdrawing groups* in the positions ortho and para to the halogen atom. If we examine the following resonance structures for a Meisenheimer intermediate, we can see how:

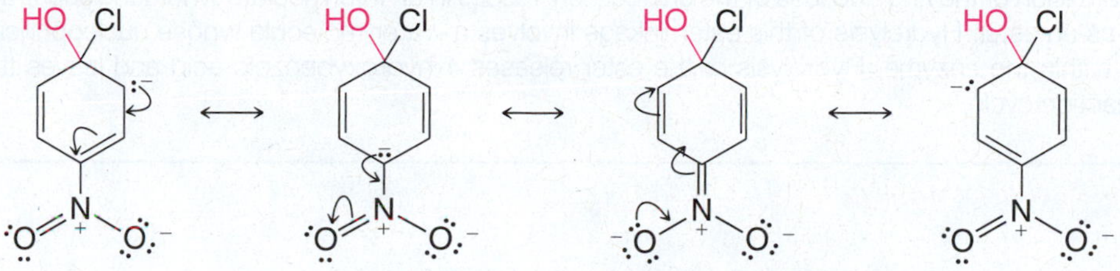

**Especially stable
(Negative charges
are both on oxygen atoms.)**

What is the product of the following reaction?

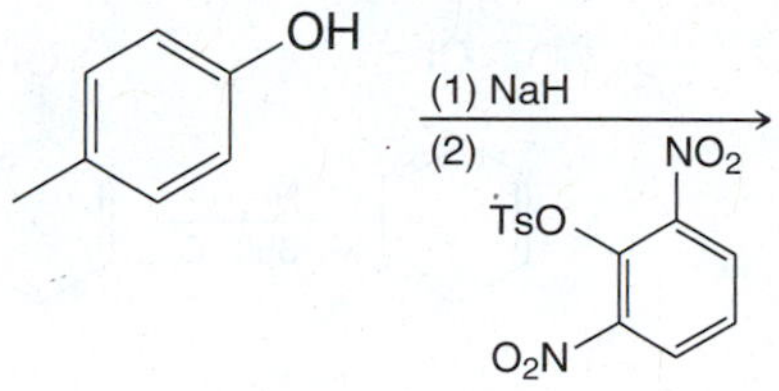

(1) NaH
(2)
TsO   NO$_2$
O$_2$N

STRATEGY AND ANSWER: NaH is a strong base that will convert 4-methylphenol to its phenoxide salt. 1-(*p*-Toluenesulfonyl)-2,6-dinitrobenzene contains both a good leaving group and two strong electron-withdrawing groups. Thus the likely reaction is a nucleophilic aromatic substitution (S$_N$Ar), leading to the following diaryl ether.

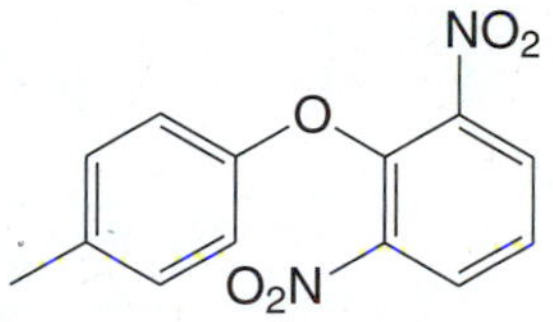

NO$_2$

O$_2$N

●●●

1-Fluoro-2,4-dinitrobenzene is highly reactive toward nucleophilic substitution through an S$_N$Ar mechanism. (In Section 24.5B we shall see how this reagent is used in the Sanger method for determining the structures of proteins.) What product would be formed when 1-fluoro-2,4-dinitrobenzene reacts with each of the following reagents?

(a) EtONa   **(b)** NH$_3$ **(c)** C$_6$H$_5$NH$_2$ **(d)** EtSNa

THE CHEMISTRY OF... Bacterial Dehalogenation of a PCB Derivative

Polychlorinated biphenyls (PCBs) are compounds that were once used in a variety of electrical devices, industrial applications, and polymers. Their use and production were banned in 1979, however, owing to the toxicity of PCBs and their tendency to accumulate in the food chain.

4-Chlorobenzoic acid is a degradation product of some PCBs. It is now known that certain bacteria are able to dehalogenate 4-chlorobenzoic acid by an enzymatic nucleophilic aromatic substitution reaction. The product is 4-hydroxybenzoic acid, and a mechanism for this enzyme-catalyzed process is shown here. The sequence begins with the thioester of 4-chlorobenzoic acid derived from coenzyme A (CoA):

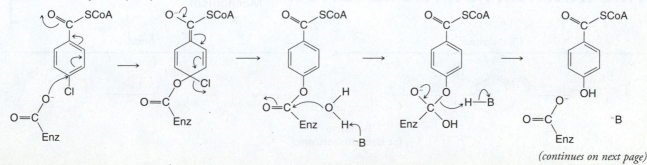

(continues on next page)

Some key features of this enzymatic S_NAr mechanism are the following. The nucleophile that attacks the chlorinated benzene ring is a carboxylate anion of the enzyme. When the carboxylate attacks, positively charged groups within the enzyme stabilize the additional electron density that develops in the thioester carbonyl group of the Meisenheimer intermediate. Collapse of the Meisenheimer intermediate, with rearomatization of the ring and loss of the chloride ion, results in an intermediate where the substrate is covalently bonded to the enzyme as an ester. Hydrolysis of this ester linkage involves a water molecule whose nucleophilicity has been enhanced by a basic site within the enzyme. Hydrolysis of the ester releases 4-hydroxybenzoic acid and leaves the enzyme ready to catalyze another reaction cycle.

21.11B Nucleophilic Aromatic Substitution through an Elimination–Addition Mechanism: Benzyne

Although aryl halides such as chlorobenzene and bromobenzene do not react with most nucleophiles under ordinary circumstances, they do react under highly forcing conditions. Chlorobenzene can be converted to phenol by heating it with aqueous sodium hydroxide in a pressurized reactor at 350 °C (Section 21.4B):

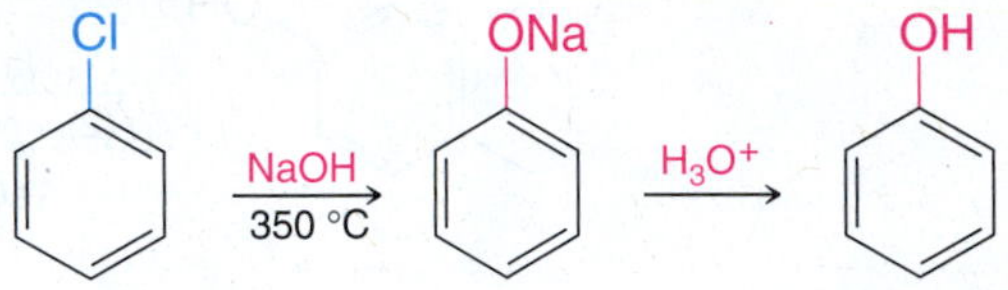

Bromobenzene reacts with the very powerful base, $^-NH_2$, in liquid ammonia:

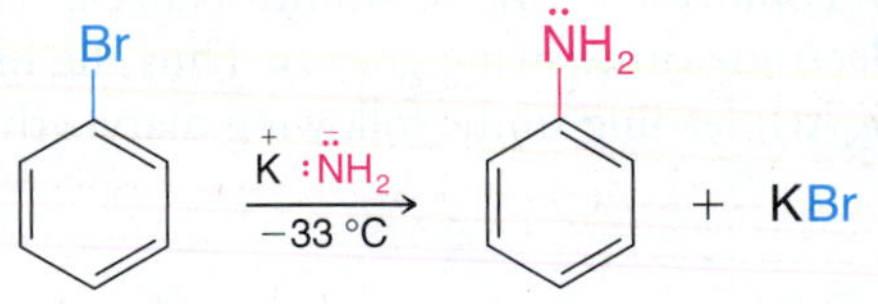

Aniline

- These reactions take place through an **elimination–addition mechanism** that involves the formation of a highly unstable intermediate called *benzyne* (or *dehydrobenzene*).

We can illustrate this mechanism with the reaction of bromobenzene and amide ion. In the first step (see the following mechanism), the amide ion initiates an elimination by abstracting one of the ortho protons because they are the most acidic. The negative charge that develops on the ortho carbon is stabilized by the inductive effect of the bromine. The anion then loses a bromide ion. This elimination produces the highly unstable, and thus highly reactive, **benzyne**. Benzyne then reacts with any available nucleophile (in this case, an amide ion) by a two-step addition reaction to produce aniline.

A MECHANISM FOR THE REACTION — The Benzyne Elimination–Addition Mechanism

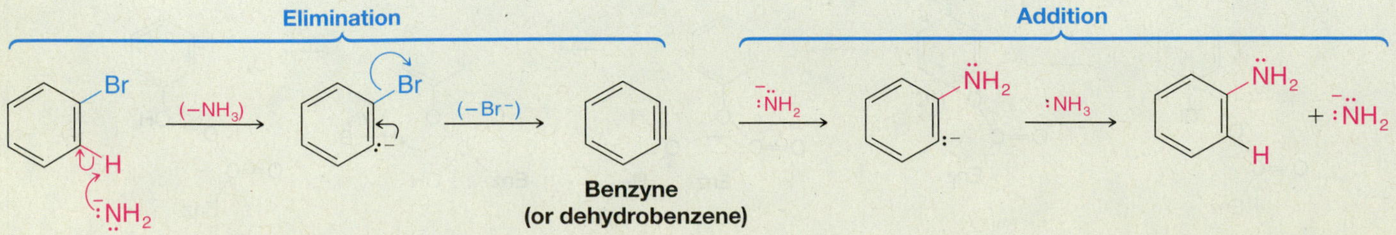

We can better understand the reactive and unstable nature of benzyne if we consider aspects of its electronic structure.

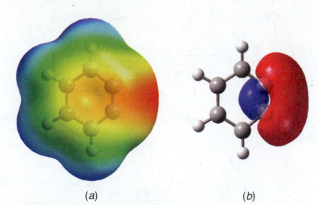

(a)   (b)

FIGURE 21.1 (a) A calculated electrostatic potential map for benzyne shows the relatively greater negative charge (in red) at the edge of the ring, corresponding to electron density from the additional π bond in benzyne. (b) A schematic representation of the molecular orbital associated with the additional π bond in benzyne. (Red and blue indicate orbital phase, not charge distribution.) Note that the orientation of this orbital is in the same plane as the ring and perpendicular to the axis of the aromatic π system.

The calculated electrostatic potential map for benzyne, shown in Fig. 21.1a, shows the relatively greater negative charge at the edge of the ring, corresponding to the electron density from the additional π bond in benzyne. Figure 21.1b shows a schematic representation of the orbital associated with the additional π bond. We can see from these models that the orbitals of the additional π bond in benzyne lie in the same plane as the ring, perpendicular to the axis of the aromatic π system. We can also see in Fig. 21.1 that, because the carbon ring is not a perfect hexagon as in benzene, there is angle strain in the structure of benzyne. The distance between the carbons of the additional π bond in benzyne is shorter than between the other carbons, and the bond angles of the ring are therefore distorted from their ideal values. The result is that benzyne is highly unstable and highly reactive. Consequently, benzyne has never been isolated as a pure substance, but it has been detected and trapped in various ways.

What, then, is some of the evidence for an elimination–addition mechanism involving benzyne in some nucleophilic aromatic substitutions?

The first piece of clear-cut evidence was an experiment done by J. D. Roberts (Section 9.10) in 1953—one that marked the beginning of benzyne chemistry. Roberts showed that when ^{14}C-labeled (C*) chlorobenzene is treated with amide ion in liquid ammonia, the aniline that is produced has the label equally divided between the 1 and 2 positions. This result is consistent with the following elimination–addition mechanism but is, of course, not at all consistent with a direct displacement or with an addition–elimination mechanism. (Why?)

Elimination   **Addition**   50%

An even more striking illustration can be seen in the following reaction. When the ortho derivative **1** is treated with sodium amide, the only organic product obtained is *m*-(trifluoromethyl)aniline:

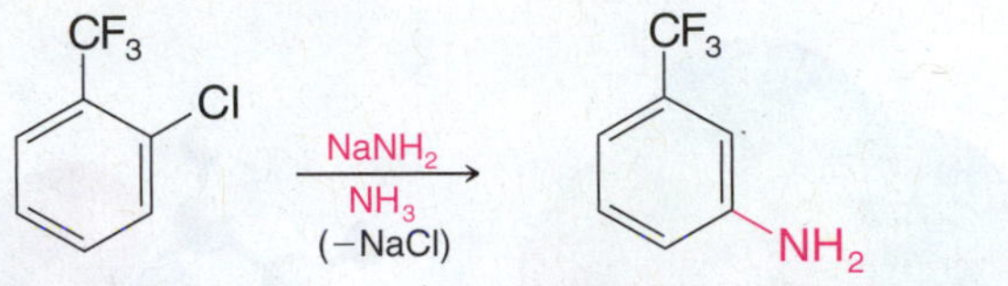

1 **m-(Trifluoromethyl)aniline**

This result can also be explained by an elimination–addition mechanism. The first step produces the benzyne **2**:

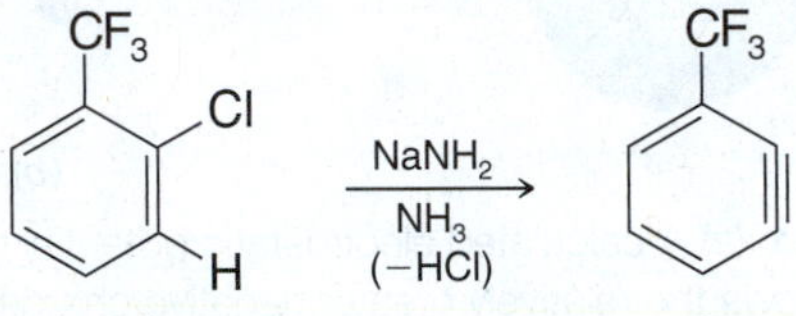

1 **2**

This benzyne then adds an amide ion in the way that produces the more stable carbanion **3** rather than the less stable carbanion **4**:

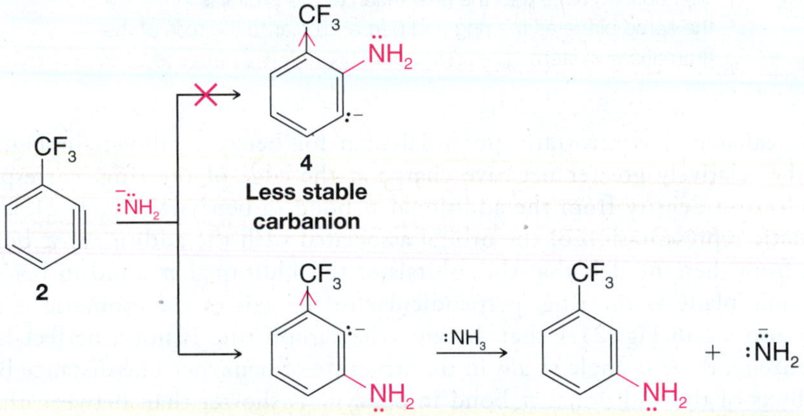

4
Less stable carbanion

2

3
More stable carbanion (The negative charge is closer to the electronegative trifluoromethyl group.)

Carbanion **3** then accepts a proton from ammonia to form *m*-(trifluoromethyl)aniline.

Carbanion **3** is more stable than **4** because the carbon atom bearing the negative charge is closer to the highly electronegative trifluoromethyl group. The trifluoromethyl group stabilizes the negative charge through its inductive effect. (Resonance effects are not important here because the sp^2 orbital that contains the electron pair does not overlap with the π orbitals of the aromatic system.)

Benzyne intermediates have been "trapped" through the use of Diels–Alder reactions. One convenient method for generating benzyne is the diazotization of anthranilic acid (2-aminobenzoic acid) followed by elimination of CO_2 and N_2:

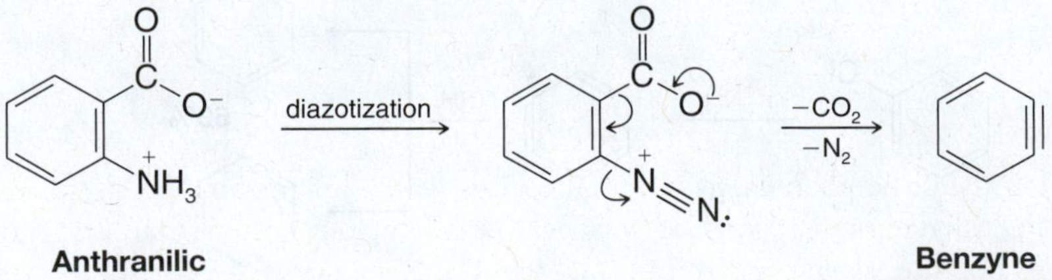

Anthranilic acid **Benzyne (trapped in situ)**

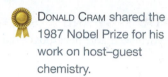

When benzyne is generated in the presence of the diene *furan*, the product is a Diels–Alder adduct:

Benzyne (generated by an elimination reaction) + **Furan** → **Diels–Alder adduct**

In a fascinating application of host–guest chemistry (an area founded by the late D. Cram, and for which he shared the Nobel Prize in Chemistry in 1987), benzyne itself has been trapped at very low temperature inside a molecular container called a hemicarcerand. Under these conditions, R. Warmuth and D. Cram found that the incarcerated benzyne was sufficiently stabilized for its ^{1}H and ^{13}C NMR spectra to be recorded (see Fig. 21.2), before it ultimately underwent a Diels–Alder reaction with the container molecule.

DONALD CRAM shared the 1987 Nobel Prize for his work on host–guest chemistry.

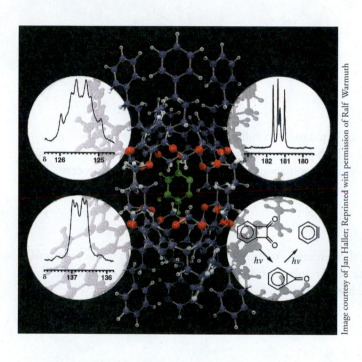

Image courtesy of Jan Haller; Reprinted with permission of Ralf Warmuth

FIGURE 21.2 A molecular graphic of benzyne (green) trapped in a hemicarcerand. Images of ^{13}C NMR data from benzyne and a reaction used to synthesize it are shown in the white circles.

21.11C Phenylation

Reactions involving benzyne can be useful for formation of a carbon–carbon bond to a phenyl group (a process called phenylation). For example, if acetoacetic ester is treated with bromobenzene and two molar equivalents of sodium amide, phenylation of ethyl acetoacetate occurs. The overall reaction is as follows:

$$\text{(acetoacetic ester)} + \text{(bromobenzene)} \xrightarrow[\text{liq. NH}_3]{2 \text{ NaNH}_2} \text{(phenylated product)}$$

Malonic esters can be phenylated in an analogous way. This process is a useful complement to the alkylation reactions of acetoacetic and malonic esters that we studied in Chapter 18 because, as you may recall, substrates like bromobenzene are not susceptible to S_N2 reactions [see Section 6.14A and Practice Problem 18.8(c)].

••• SOLVED PROBLEM 21.7

Outline a synthesis of phenylacetic acid from diethyl malonate.

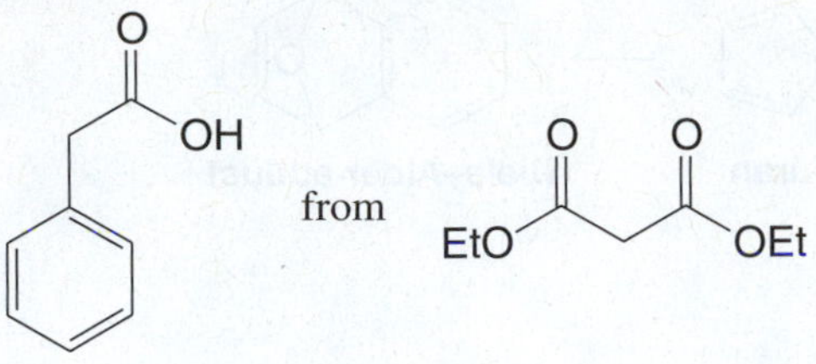

STRATEGY AND ANSWER: Diethyl malonate must first be substituted at the α carbon by a phenyl group, and then hydrolyzed and decarboxylated. Introduction of the phenyl group requires involvement of a benzyne intermediate.

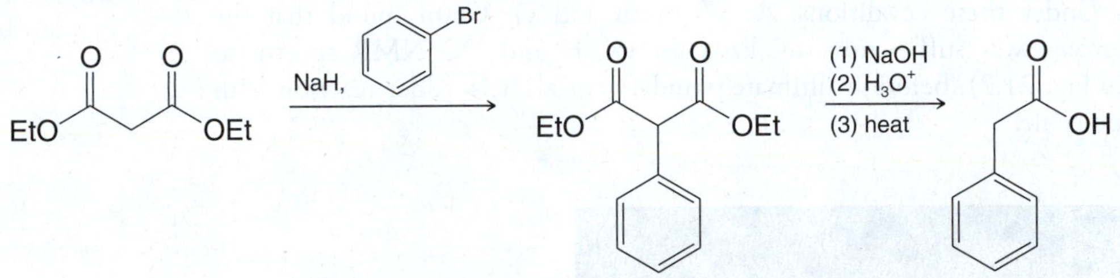

•••

PRACTICE PROBLEM 21.10 When *o*-chlorotoluene is subjected to the conditions used in the Dow process (i.e., aqueous NaOH at 350 °C at high pressure), the products of the reaction are *o*-cresol and *m*-cresol. What does this result suggest about the mechanism of the Dow process?

•••

PRACTICE PROBLEM 21.11 When 2-bromo-1,3-dimethylbenzene is treated with sodium amide in liquid ammonia, no substitution takes place. This result can be interpreted as providing evidence for the elimination–addition mechanism. Explain how this interpretation can be given.

•••

PRACTICE PROBLEM 21.12 **(a)** Outline a step-by-step mechanism for the phenylation of acetoacetic ester by bromobenzene and two molar equivalents of sodium amide. (Why are two molar equivalents of NaNH₂ necessary?) **(b)** What product would be obtained by hydrolysis and decarboxylation of the phenylated acetoacetic ester? **(c)** How would you prepare 2-phenylpropanoic acid from malonic ester?

21.12 SPECTROSCOPIC ANALYSIS OF PHENOLS AND ARYL HALIDES

Infrared Spectra Phenols show a characteristic absorption band (usually broad) arising from O—H stretching in the 3400–3600-cm^{-1} region. Phenols and aryl halides also show the characteristic absorptions that arise from their benzene rings (see Section 14.11C).

¹H NMR Spectra The hydroxylic proton of a phenol is more deshielded than that of an alcohol due to proximity of the benzene π electron ring current. The exact position of the O—H signal depends on the extent of hydrogen bonding and on whether the hydrogen bonding is *intermolecular* or *intramolecular*. The extent of intermolecular hydrogen bonding depends on the concentration of the phenol, and this strongly affects the position of the O—H signal. In phenol, itself, for example, the position of the O—H signal varies from δ 2.55 for pure phenol to δ 5.63 at 1% concentration in CCl_4. Phenols with strong intramolecular hydrogen bonding, such as salicylaldehyde, show O—H signals between δ 0.5 and δ 1.0, and the position of the signal varies only slightly with concentration. As with other protons that undergo exchange (Section 9.10), the identity of the O—H proton of a phenol can be determined by adding D_2O to the sample. The O—H proton undergoes rapid exchange with deuterium and the proton signal disappears. The aromatic protons of phenols and aryl halides give signals in the δ 7–9 region.

^{13}C NMR Spectra The carbon atoms of the aromatic ring of phenols and aryl halides appear in the region δ 135–170.

Mass Spectra Mass spectra of phenols often display a prominent molecular ion peak, $M^{\ddagger}$. Phenols that have a benzylic hydrogen produce an $M^{\ddagger} - 1$ peak that can be larger than the $M^{\ddagger}$ peak.

THE CHEMISTRY OF... Aryl Halides: Their Uses and Environmental Concerns

Aryl Halides as Insecticides

Insects, especially mosquitoes, fleas, and lice, have been responsible for innumerable human deaths throughout history. The bubonic plague or "black death" of medieval times that killed nearly one-third of Europe's population was borne by fleas. Malaria and yellow fever, diseases that were responsible for the loss of millions of lives in the twentieth century alone, are mosquito-borne diseases.

One compound widely known for its insecticidal properties and environmental effects is DDT [1,1,1-trichloro-2,2-bis(4-chlorophenyl)ethane].

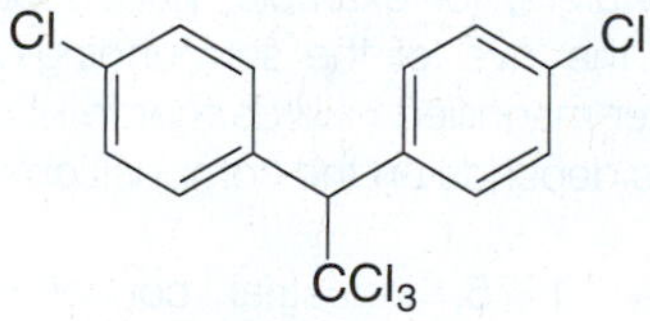

DDT
[1,1,1-trichloro-2,2-bis(4-chlorophenyl)ethane]

From the early 1940s through the early 1970s, when its use was banned in the United States, vast quantities of DDT were sprayed over many parts of the world in an effort to destroy insects. These efforts rid large areas of the world of disease-carrying insects, especially those responsible for malaria, yellow fever, sleeping sickness (caused by tsetse flies), and typhus. Though it has since resurged, by 1970, malaria had been largely eliminated from the developed world. According to estimates by the National Academy of Sciences, the use of DDT during that time had prevented more that 500 million deaths from malaria alone.

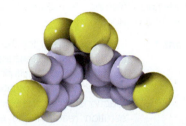

DDT

Eventually it began to become clear that the prodigious use of DDT had harmful side effects. Aryl halides are usually highly stable compounds that are only slowly destroyed by natural processes. As a result they remain in the environment for years; they are what we now call "persistent insecticides" or "hard insecticides." The U.S. Environmental Protection Agency banned the use of DDT beginning in 1973.

Aryl halides are also fat soluble and tend to accumulate in the fatty tissues of most animals. The food chain that runs from plankton to small fish to birds and to larger animals, including humans, tends to magnify the concentrations of aryl halides at each step.

The chlorohydrocarbon DDT is prepared from inexpensive starting materials, chlorobenzene and trichloroacetaldehyde. The reaction, shown here, is catalyzed by acid.

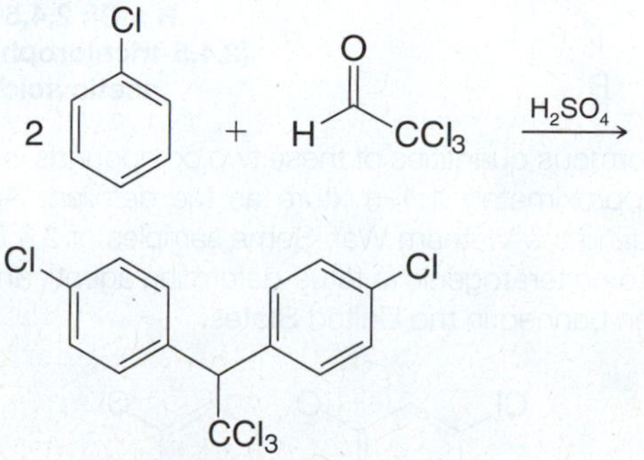

DDT
[1,1,1-trichloro-2,2-bis(4-chlorophenyl)ethane]

(continues on next page)

In nature the principal decomposition product of DDT is DDE.

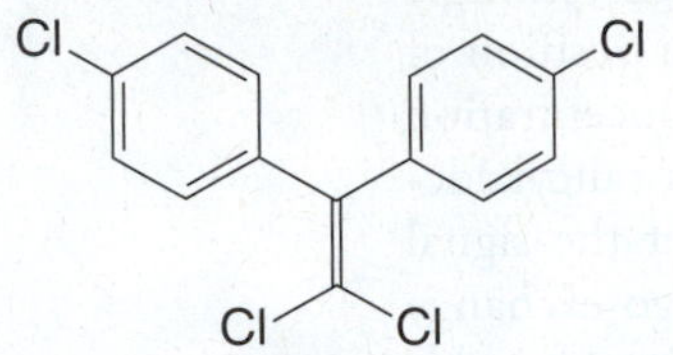

DDE
[1,1-dichloro-2,2-
bis(4-chlorophenyl)ethene]

Estimates indicate that nearly 1 billion pounds of DDT were spread throughout the world ecosystem. One pronounced environmental effect of DDE, after conversion from DDT, has been in its action on eggshell formation in many birds. DDE inhibits the enzyme *carbonic anhydrase* that controls the calcium supply for shell formation. As a consequence, the shells are often very fragile and do not survive to the time of hatching. During the late 1940s the populations of eagles, falcons, and hawks dropped dramatically. There can be little doubt that DDT was primarily responsible. DDE also accumulates in the fatty tissues of humans. Although humans appear to have a short-range tolerance to moderate DDE levels, the long-range effects are uncertain.

Study Problem 1 The mechanism for the formation of DDT from chlorobenzene and trichloroacetaldehyde in sulfuric acid involves two electrophilic aromatic substitution reactions. In the first electrophilic substitution reaction, the electrophile is protonated trichloroacetaldehyde. In the second, the electrophile is a carbocation. Propose a mechanism for the formation of DDT.

Study Problem 2 What kind of reaction is involved in the conversion of DDT to DDE?

Organic Halides as Herbicides

Other chlorinated organic compounds have been used extensively as herbicides. The following two examples are 2,4-D and 2,4,5-T.

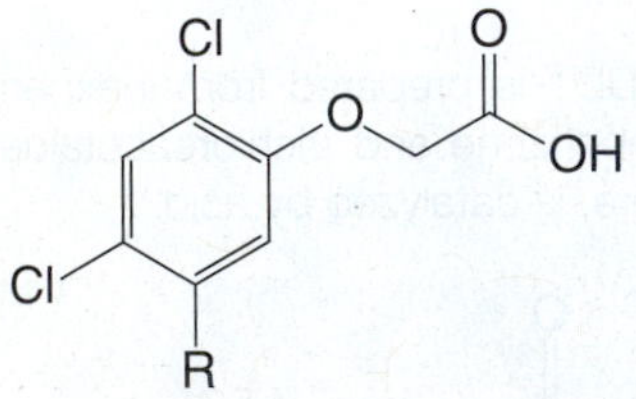

R = H; 2,4-D
(2,4-dichlorophenoxy-
acetic acid)

R = Cl; 2,4,5-T
(2,4,5-trichlorophenoxy-
acetic acid)

Enormous quantities of these two compounds were used in an approximately 1:1 mixture as the defoliant Agent Orange during the Vietnam War. Some samples of 2,4,5-T were shown to be teratogenic (a fetus-deforming agent), and its use has been banned in the United States.

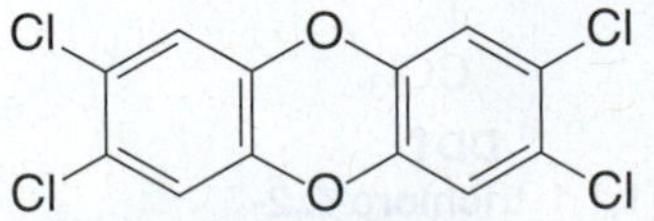

2,3,7,8-Tetrachlorodibenzodioxin
(also called TCDD)

This dioxin is also highly stable; it persists in the environment and because of its fat solubility can be passed up the food chain. In sublethal amounts it can cause a disfiguring skin disease called chloracne.

Polychlorinated Biphenyls (PCBs)

Mixtures of polychlorinated biphenyls have been produced and used commercially since 1929. In these mixtures, biphenyls with chlorine atoms at any of the numbered positions (see the following structure) may be present. In all, there are 210 possible compounds. A typical commercial mixture may contain as many as 50 different PCBs. Mixtures are usually classified on the basis of their chlorine content, and most industrial mixtures contain from 40 to 60% chlorine.

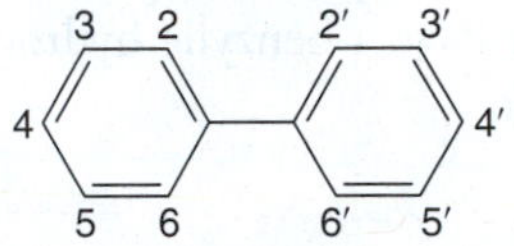

Biphenyl

Polychlorinated biphenyls have had a multitude of uses: as heat-exchange agents in transformers; in capacitors, thermostats, and hydraulic systems; as plasticizers in polystyrene coffee cups, frozen food bags, bread wrappers, and plastic liners for baby bottles. They have been used in printing inks, in carbonless carbon paper, and as waxes for making molds for metal castings. Between 1929 and 1972, about 500,000 metric tons of PCBs were manufactured.

Polychlorinated biphenyls are highly persistent in the environment, and, being fat soluble, tend to accumulate in the food chain. PCBs have been found in rainwater, in many species of fish, birds, and other animals (including polar bears) all over the globe, and in human tissue. Fish that feed in PCB-contaminated waters, for example, have PCB levels 1000–100,000 times the level of the surrounding water, and this amount is further magnified in birds that feed on the fish. The toxicity of PCBs depends on the composition of the individual mixture.

As late as 1975, industrial concerns were legally discharging PCBs into the Hudson River. In 1977, the EPA banned the direct discharge into waterways, and since 1979 their manufacture, processing, and distribution have been prohibited. In 2000 the EPA specified certain sections of the Hudson River for cleanup of PCBs. In 2009, a plan to decontaminate parts of the Hudson River by dredging was finally implemented. See "The Chemistry of...Bacterial Dehalogenation of a PCB Derivative" (Section 21.11A) for a potential method of PCB remediation.

Polybrominated Biphenyls and Biphenyl Ethers (PBBs and PBDEs)

As with polychlorinated biphenyls (PCBs), polybrominated aromatic compounds have been used in industry since the early twentieth century. The fire retardant properties of polybrominated and polychlorinated biphenyls and biphenyl ethers, for example, led to their use in building materials, furniture, clothing, and other consumer items. However, the 1970s discovery in Michigan of polybrominated biphenyls (PBBs) in feed for livestock, and subsequently in meat and dairy products, led to suspension of the use of PBBs in 1979.

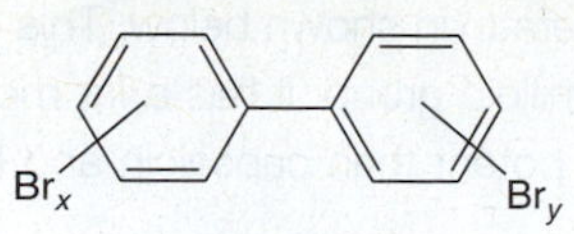

x + *y* = 1-10
Polybrominated biphenyls
(PBBs)

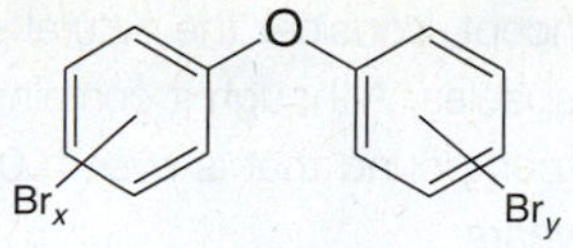

x + *y* = 1-10
Polybrominated diphenyl ethers
(PBDEs)

(*x* and *y* indicate the possibility of multiple bromine
substitution sites on each ring.)

Meanwhile, there is mounting concern about polybromo-diphenyl ethers (PBDEs). Although use of PBDEs could potentially save lives and property in their roles as flame retardants, these compounds are now widespread in the environment, and studies have led to significant concern about their toxicity to humans and other animals. As with PCBs, polybrominated biphenyls and polybrominated diphenyl ethers persist in the environment and accumulate in fatty biological tissues. PBDEs have been found in birds, fish, and breast milk. They are now banned in a number of areas.

[WHY Do These Topics Matter?

RELATIONSHIPS BETWEEN CHEMICAL STRUCTURE AND ACTIVITY

Knowing what we do about functional groups, it is not surprising that changing the structure of a molecule can lead to changes in activity. However, it is sometimes surprising how very small structural changes can lead to extreme alterations in activity, like the proverbial Dr. Jekyll turning into Mr. Hyde. There is at least one group of phenols found in nature that beautifully illustrates this idea. Drawn below is the molecule vanillin, the compound responsible for the wonderful smell and taste of vanilla. This compound is used in large quantities in the food and fragrance industries.

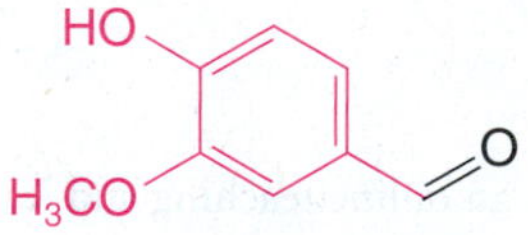

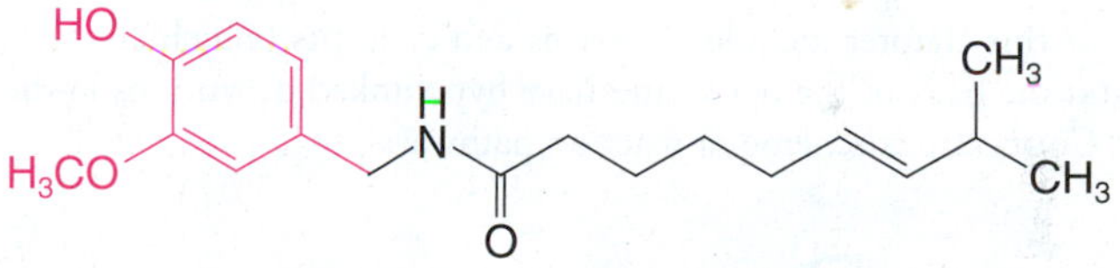

Capsaicin

Vanillin

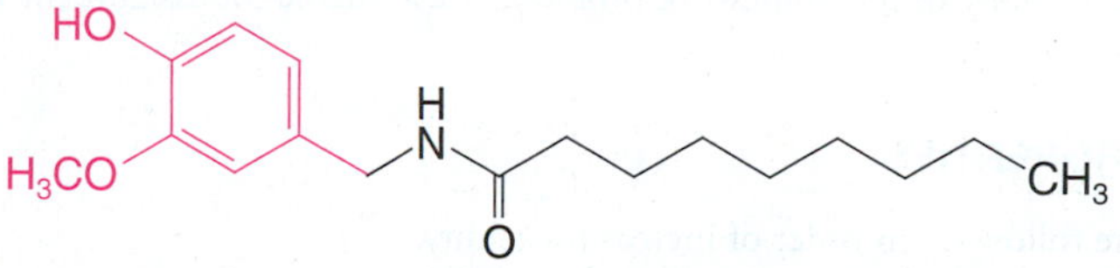

Nonivamide

If the aldehyde group of vanillin is changed to the amide-linked alkyl fragment of capsaicin, we then obtain a natural product found in many peppers. Instead of having a pleasant taste, this compound activates pain receptors in our mouths, sending signals to the brain that register as a sensation of heat. You have likely had this experience if you have ever eaten a jalapeño pepper. If you activate these pain receptors enough times, you can eventually destroy their efficacy, training your mouth to be able to tolerate more and more "heat." Capsaicin, though, is not all bad. In fact, applied to your skin (as the active ingredient in the medicine Capzacin), it can help to modulate pain by activating the pain receptors and preventing them from firing further, thus serving as an analgesic. Interestingly, subtle changes to the structure of capsaicin can diminish its impact. For instance, the natural product nonivamide, which is also found in peppers, is missing one of the terminal methyl groups and the double bond of capsaicin. These changes are small, but they are sufficient to decrease the "hotness" of the compound by nearly half. Nonivamide is still hot enough, though, that it has been used commercially as the active ingredient in some pepper sprays.

The key structural component that is consistent between all these molecules is the aryl ring, the phenol, and the methyl ether, which collectively is termed a vanilloid group (highlighted in magenta in each) and is recognized by a number of critical receptors throughout our bodies. Key to our perception of the resultant activity, be it a pleasant smell or pain, are the remaining atoms attached to the other side of the benzene ring.

(continues on next page)

As a final example of this concept, consider the natural product resiniferatoxin shown below. This compound comes from the latex of several flowering cactus species. Although it contains the same vanilloid group, it has a far more complex right-hand half. These structural changes yield a compound that is over 1000 times more potent than capsaicin and has been used as a natural analgesic for over two thousand years.

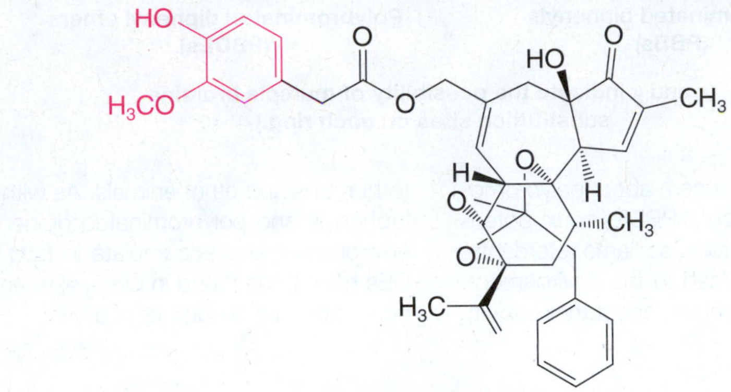

Resiniferatoxin

To learn more about these topics, see:

1. Walpole, C. S. J. et al. *Similarities and Differences in the Structure-Activity Relationships of Capsaicin and Resiniferatoxin Analogues.* J. Med. Chem. **1996**, 39, 2939-2952.
2. Nicolaou, K. C.; Montagnon, T. *Molecules that Changed the World.* Wiley-VCH: Weinheim, **2008**, p. 262.

SUMMARY AND REVIEW TOOLS

The study aids for this chapter include key terms and concepts (which are highlighted in bold, blue text within the chapter and defined in the Glossary (at the back of the book) and have hyperlinked definitions in the accompanying *WileyPLUS* course (www.wileyplus.com), and a Synthetic Connections scheme of reaction pathways.

PROBLEMS

Note to Instructors: Many of the homework problems are available for assignment via *WileyPLUS*, an online teaching and learning solution.

PHYSICAL PROPERTIES

21.13 Rank the following in order of increasing acidity.

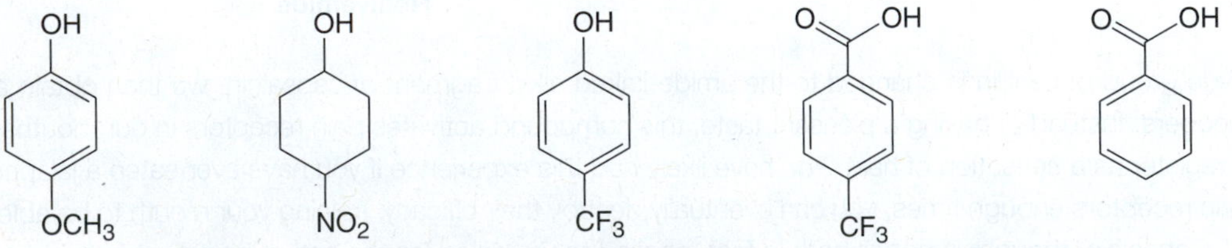

21.14 Without consulting tables, select the stronger acid from each of the following pairs:

(a) 4-Methylphenol and 4-fluorophenol
(b) 4-Methylphenol and 4-nitrophenol
(c) 4-Nitrophenol and 3-nitrophenol
(d) 4-Methylphenol and benzyl alcohol
(e) 4-Fluorophenol and 4-bromophenol

21.15 What products would be obtained from each of the following acid–base reactions?

(a) Sodium ethoxide in ethanol + phenol →
(b) Phenol + aqueous sodium hydroxide →
(c) Sodium phenoxide + aqueous hydrochloric acid →
(d) Sodium phenoxide + H_2O + CO_2 →

21.16 Describe a simple chemical test that could be used to distinguish between members of each of the following pairs of compounds:

(a) 4-Chlorophenol and 4-chloro-1-methylbenzene
(b) 4-Methylphenol and 4-methylbenzoic acid
(c) 4-Methylphenol and 2,4,6-trinitrophenol
(d) Ethyl phenyl ether and 4-ethylphenol

GENERAL REACTIONS

21.17 Complete the following equations:

(a) Phenol + Br_2 $\xrightarrow{5\,°C,\ CS_2}$

(b) Phenol + concd H_2SO_4 $\xrightarrow{25\,°C}$

(c) Phenol + concd H_2SO_4 $\xrightarrow{100\,°C}$

(d)  + *p*-toluenesulfonyl chloride $\xrightarrow{HO^-}$

(e) Phenol + Br_2 $\xrightarrow{H_2O}$

(f) Phenol + $\longrightarrow$

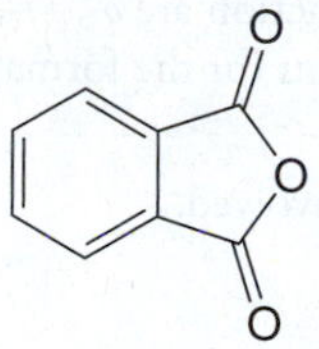

(g) *p*-Cresol + Br_2 $\xrightarrow{H_2O}$

(h) Phenol + 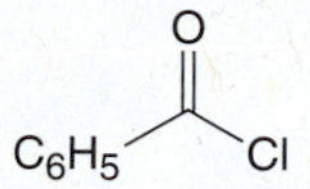 $\xrightarrow{base}$

(i) Phenol + 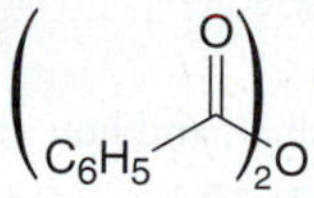 $\xrightarrow{base}$

(j) Phenol + NaH $\longrightarrow$

(k) Product of (j) + $CH_3OSO_2OCH_3$ $\longrightarrow$

(l) Product of (j) + CH_3I $\longrightarrow$

(m) Product of (j) + $C_6H_5CH_2Cl$ $\longrightarrow$

21.18 Predict the product of the following reactions.

(a) 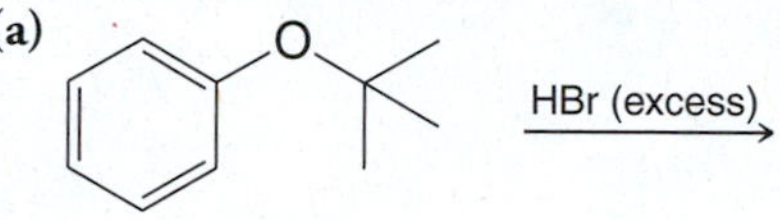 $\xrightarrow{HBr\ (excess)}$

(b) 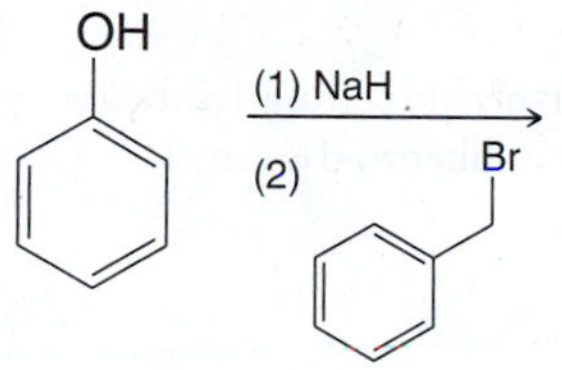 $\xrightarrow[(2)]{(1)\ NaH}$

(c) 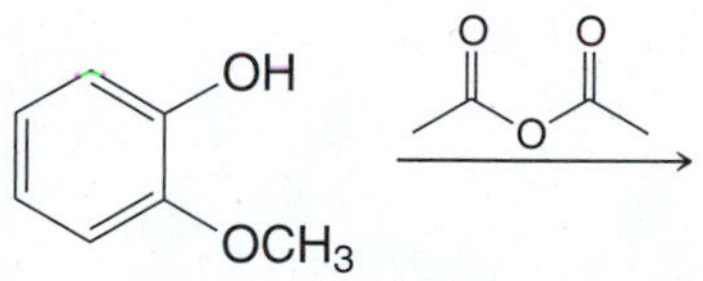 $\longrightarrow$

(d) 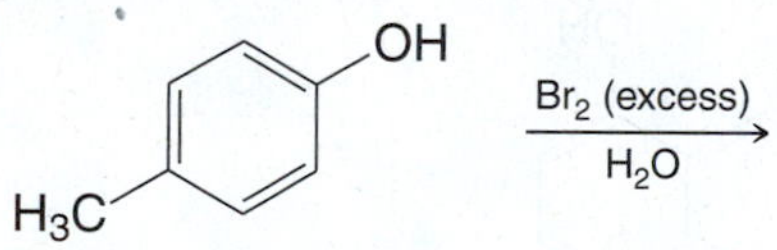 $\xrightarrow[H_2O]{Br_2\ (excess)}$

(e) $\xrightarrow{excess}$

(f) $\xrightarrow[(2)\ heat]{(1)\ NaH,\ \text{allyl bromide}}$

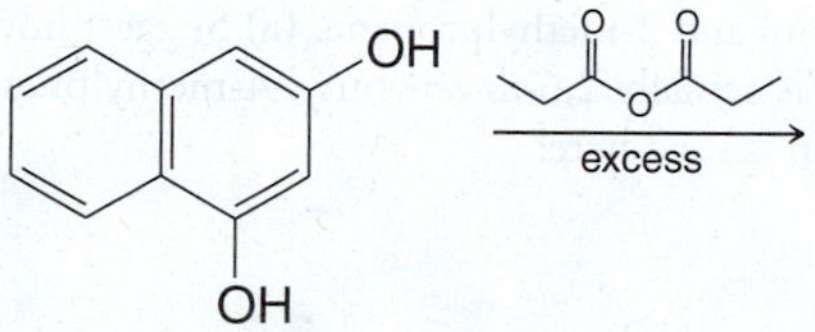

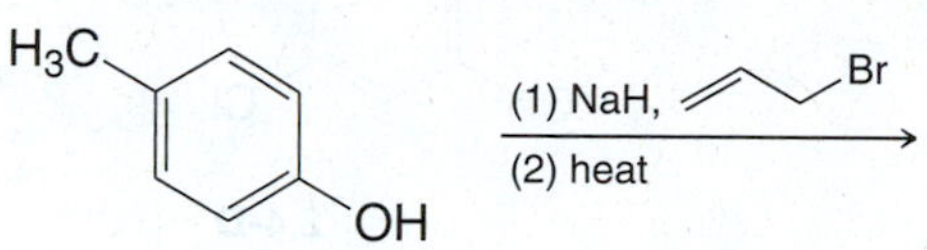

(g) 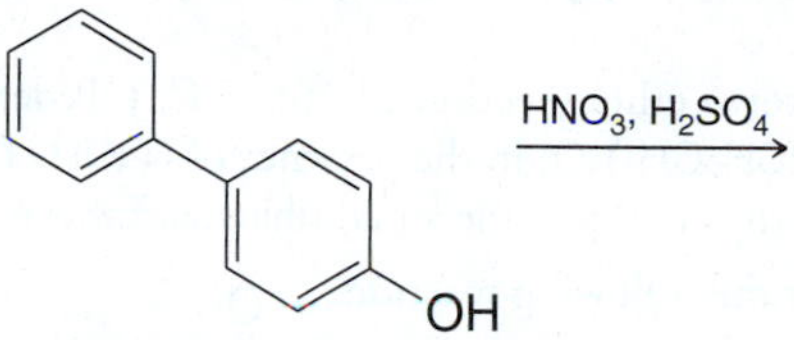 $\xrightarrow{HNO_3,\ H_2SO_4}$

(h) $\xrightarrow{NaNH_2,\ NH_3}$

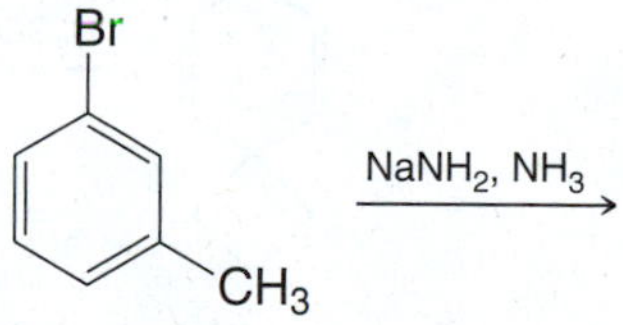

(i) 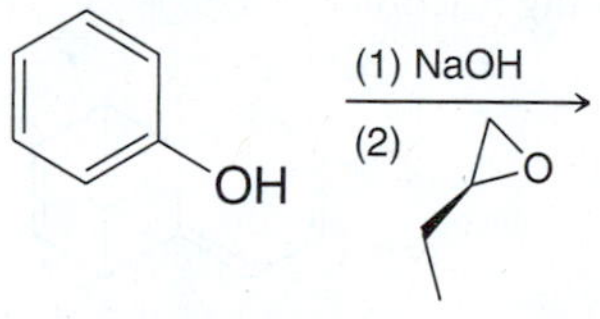 $\xrightarrow[(2)]{(1)\ NaOH}$

MECHANISMS AND SYNTHESIS

21.19 A synthesis of the β-receptor blocker called toliprolol begins with a reaction between 3-methylphenol and epichlorohydrin. The synthesis is outlined below. Give the structures of the intermediates and of toliprolol.

3-Methylphenol + 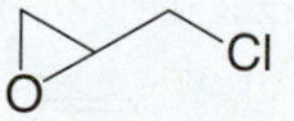 $\longrightarrow$ $C_{10}H_{13}O_2Cl$ $\xrightarrow{HO^-}$

Epichlorohydrin

$C_{10}H_{12}O_2$ $\xrightarrow{(CH_3)_2CHNH_2}$ toliprolol, $C_{13}H_{21}NO_2$

21.20 *p*-Chloronitrobenzene was allowed to react with sodium 2,6-di-*tert*-butylphenoxide with the intention of preparing the diphenyl ether **1**. The product was not **1**, but rather was an isomer of **1** that still possessed a phenolic hydroxyl group.

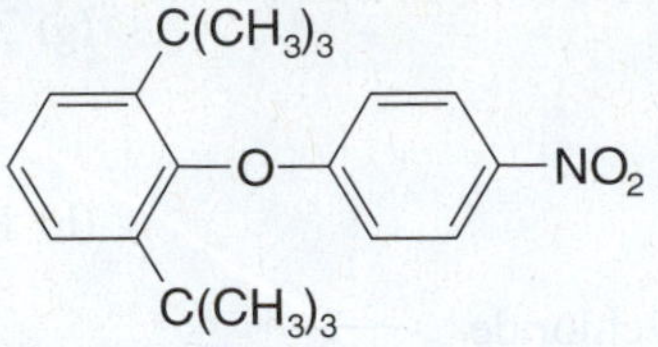

1

What was this product, and how can one account for its formation?

21.21 When *m*-chlorotoluene is treated with sodium amide in liquid ammonia, the products of the reaction are *o*-, *m*-, and *p*-toluidine (i.e., *o*-CH₃C₆H₄NH₂, *m*-CH₃C₆H₄NH₂, and *p*-CH₃C₆H₄NH₂). Propose plausible mechanisms that account for the formation of each product.

21.22 The herbicide **2,4-D** can be synthesized from phenol and chloroacetic acid. Outline the steps involved.

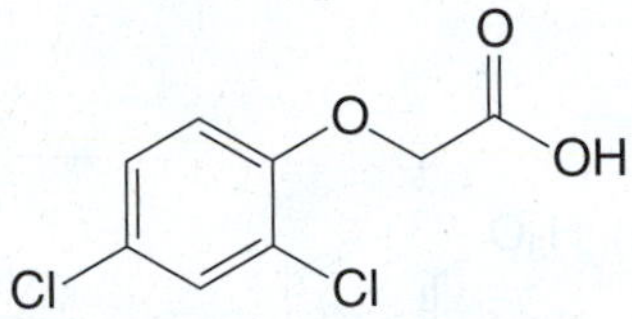

2,4-D
(2,4-dichlorophenoxyacetic acid)

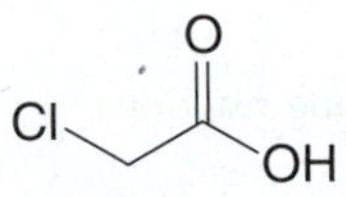

Chloroacetic acid

21.23 The first synthesis of a crown ether (Section 11.16) by C. J. Pedersen (of the DuPont Company) involved treating 1,2-benzenediol with di(2-chloroethyl) ether, (ClCH₂CH₂)₂O, in the presence of NaOH. The product was a compound called dibenzo-18-crown-6. Give the structure of dibenzo-18-crown-6 and provide a plausible mechanism for its formation.

21.24 Provide a mechanism for the following reaction.

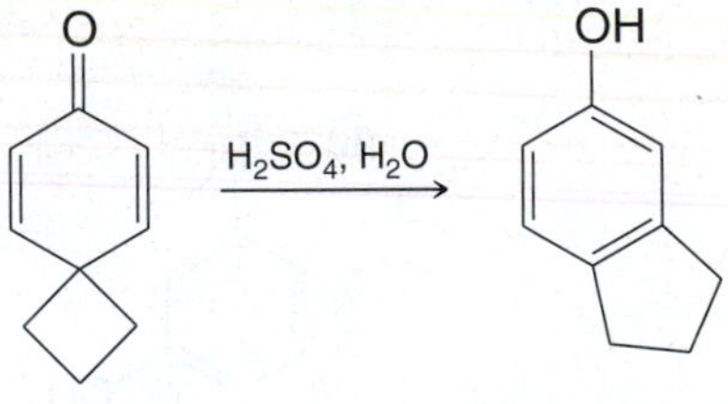

21.25 Provide a mechanism for the following reaction.

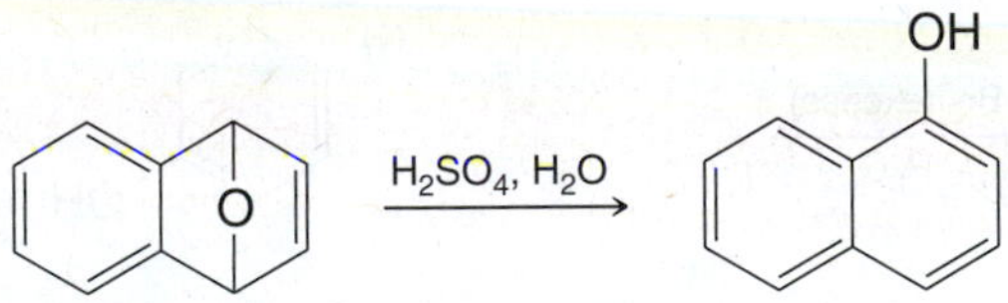

21.26 The widely used antioxidant and food preservative called **BHA** (**b**utylated **h**ydroxy**a**nisole) is actually a mixture of 2-*tert*-butyl-4-methoxyphenol and 3-*tert*-butyl-4-methoxyphenol. BHA is synthesized from *p*-methoxyphenol and 2-methylpropene. **(a)** Suggest how this is done. **(b)** Another widely used antioxidant is BHT (**b**utylated **h**ydroxy**t**oluene). BHT is actually 2,6-di-*tert*-butyl-4-methylphenol, and the raw materials used in its production are *p*-cresol and 2-methylpropene. What reaction is used here?

21.27 Provide a mechanism for the following reaction.

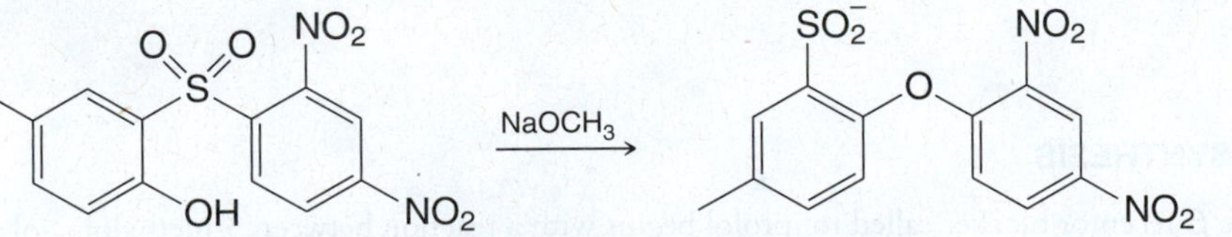

21.28 Account for the fact that the Dow process for the production of phenol produces both diphenyl ether (**1**) and 4-hydroxybiphenyl (**2**) as by-products:

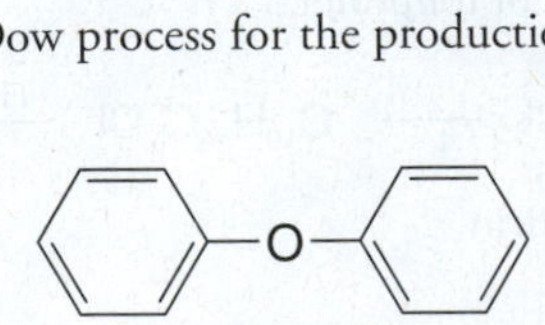

1 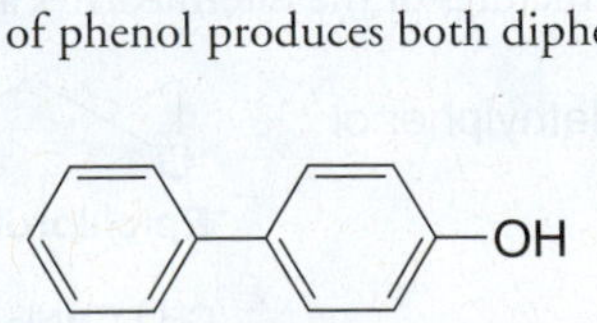   **2**

21.29 Predict the outcome of the following reactions:

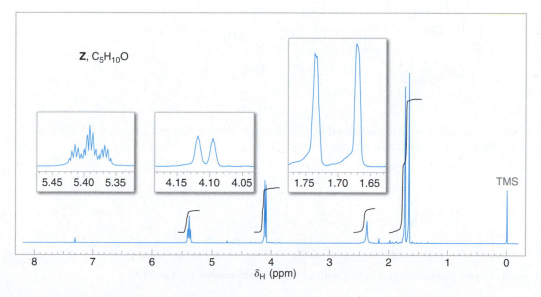

(a) (structure with CN, Cl) 2 equiv. KNH₂, liq. NH₃, −33 °C

(b) (structure) 2 equiv. NaNH₂, liq. NH₃, −33 °C

21.30 Explain how it is possible for 2,2′-dihydroxy-1,1′-binaphthyl (shown at right) to exist in enantiomeric forms.

21.31 Phenols are often effective antioxidants (see Problem 21.26 and "The Chemistry of…Antioxidants" in Section 10.11) because they are said to "trap" radicals. The trapping occurs when phenols react with highly reactive radicals to produce less reactive (more stable) phenolic radicals. **(a)** Show how phenol itself might react with an alkoxyl radical (RO·) in a hydrogen abstraction reaction involving the phenolic —OH. **(b)** Write resonance structures for the resulting radical that account for its being relatively unreactive.

SPECTROSCOPY

21.32 A compound **X** ($C_{10}H_{14}O$) dissolves in aqueous sodium hydroxide but is insoluble in aqueous sodium bicarbonate. Compound **X** reacts with bromine in water to yield a dibromo derivative, $C_{10}H_{12}Br_2O$. The 3000–4000-cm⁻¹ region of the IR spectrum of **X** shows a broad peak centered at 3250 cm⁻¹; the 680–840-cm⁻¹ region shows a strong peak at 830-cm⁻¹. The ¹H NMR spectrum of **X** gives the following: singlet at δ 1.3 (9H), singlet at δ 4.9 (1H), and multiplet at δ 7.0 (4H). What is the structure of **X**?

21.33 Compound **Z** ($C_5H_{10}O$) decolorizes bromine in carbon tetrachloride. The IR spectrum of **Z** shows a broad peak in the 3200–3600-cm⁻¹ region. The 300-MHz ¹H NMR spectrum of **Z** is given in Fig. 21.3. Propose a structure for **Z**.

Z, $C_5H_{10}O$

FIGURE 21.3 The 300-MHz ¹H NMR spectrum of compound Z (Problem 21.33). Expansions of the signals are shown in the offset plots.

CHALLENGE PROBLEMS

21.34 Explain why, in the case shown, the allyl group has migrated with no change having occurred in the position of the labeled carbon atom within the allyl group:

21.35 In protic solvents the naphthoxide ion (**I**) is alkylated primarily at position 1 (*C*-alkylation) whereas in polar aprotic solvents, such as DMF, the product is almost exclusively the result of a conventional Williamson ether synthesis (*O*-alkylation):

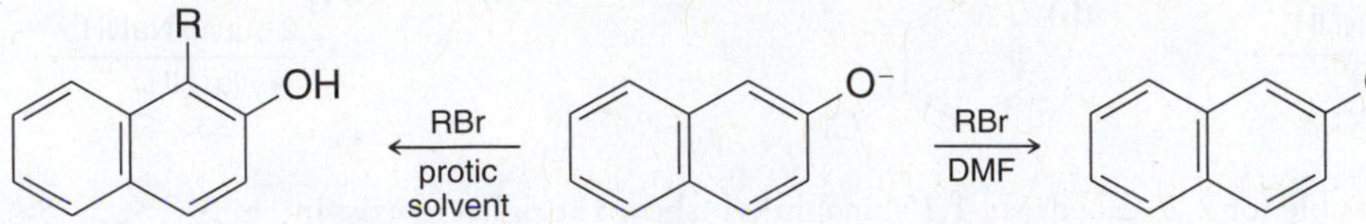

Why does the change in solvent make a difference?

21.36 In comparing nucleophilic aromatic substitution reactions that differ only in the identity of the halogen that is the leaving group in the substrate, it is found that the fluorinated substrate reacts faster than either of the cases where bromine or chlorine is the leaving group. Explain this behavior, which is contrary to the trend among the halogens as leaving groups in S_N1 and S_N2 reactions (in protic solvents).

21.37 In the case of halogen-substituted azulenes, a halogen atom on C6 can be displaced by nucleophiles while one on C1 is unreactive toward nucleophiles. Rationalize this difference in behavior.

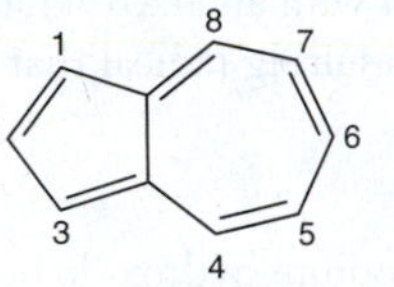

21.38 In the Sommelet–Hauser rearrangement, a benzyl quaternary ammonium salt reacts with a strong base to give a benzyl tertiary amine, as exemplified below:

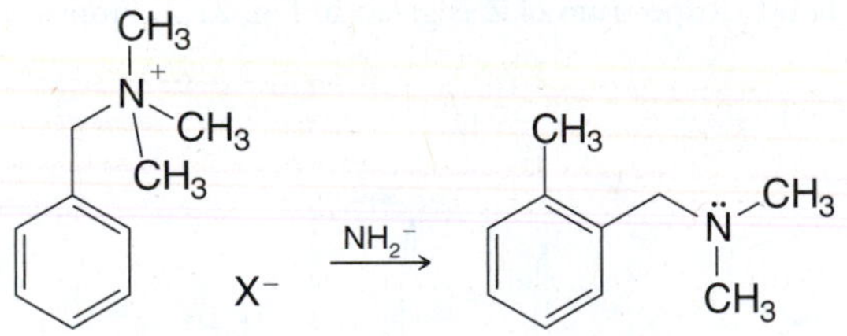

Suggest a mechanism for this rearrangement.

21.39 Hexachlorophene was a widely used germicide until it was banned in 1972 after tests showed that it caused brain damage in test animals. Suggest how this compound might be synthesized, starting with benzene.

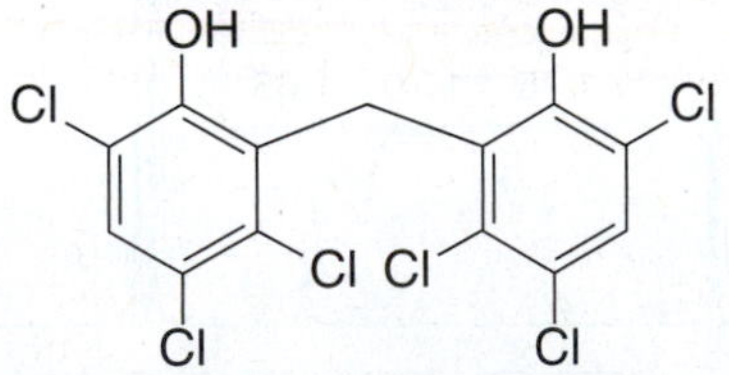

21.40 The Fries rearrangement occurs when a phenolic ester is heated with a Friedel–Crafts catalyst such as $AlCl_3$:

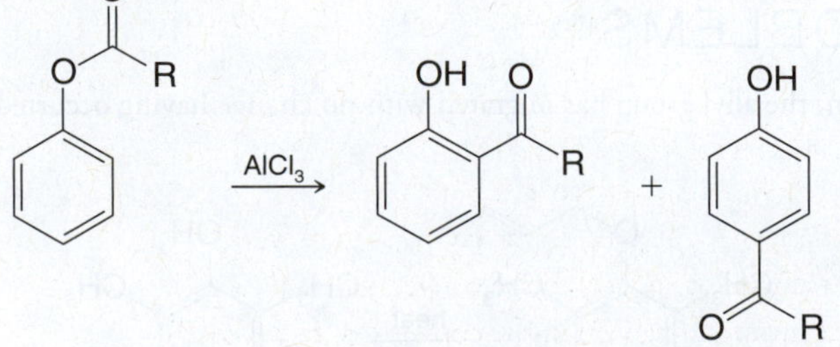

The reaction may produce both ortho and para acylated phenols, the former generally favored by high temperatures and the latter by low temperatures. **(a)** Suggest an experiment that might indicate whether the reaction is inter- or intramolecular. **(b)** Explain the temperature effect on product formation.

21.41 Compound **W** was isolated from a marine annelid commonly used in Japan as a fish bait, and it was shown to be the substance that gives this organism its observed toxicity to some insects that contact it.

> **MS** (*m/z*): 151 (relative abundance 0.09), 149 (M^{+}, rel. abund. 1.00), 148
>
> **IR** (cm^{-1}): 2960, 2850, 2775
>
> **^{1}H NMR** (δ): 2.3 (s, 6H), 2.6 (d, 4H), and 3.2 (pentet, 1H)
>
> **^{13}C NMR** (δ): 38 (CH$_3$), 43 (CH$_2$), and 75 (CH)

These reactions were used to obtain further information about the structure of **W**:

$$W \xrightarrow{\ NaBH_4\ } X \xrightarrow{\ C_6H_5COCl\ } Y \xrightarrow{\ Raney\ Ni\ } Z$$

Compound **X** had a new infrared band at 2570 cm^{-1} and these NMR data:

> **^{1}H NMR** (δ): 1.6 (t, 2H), 2.3 (s, 6H), 2.6 (m, 4H), and 3.2 (pentet, 1H)
>
> **^{13}C NMR** (δ): 28 (CH$_2$), 38 (CH$_3$), and 70 (CH)

Compound **Y** had these data:

> **IR** (cm^{-1}): 3050, 2960, 2850, 1700, 1610, 1500, 760, 690
>
> **^{1}H NMR** (δ): 2.3 (s, 6H), 2.9 (d, 4H), 3.0 (pentet, 1H), 7.4 (m, 4H), 7.6 (m, 2H), and 8.0 (m, 4H)
>
> **^{13}C NMR** (δ): 34 (CH$_2$), 39 (CH$_3$), 61 (CH), 128 (CH), 129 (CH), 134 (CH), 135 (C), and 187 (C)

Compound **Z** had

> **MS** (*m/z*): 87 (M^{+}), 86, 72
>
> **IR** (cm^{-1}): 2960, 2850, 1385, 1370, 1170
>
> **^{1}H NMR** (δ): 1.0 (d, 6H), 2.3 (s, 6H), and 3.0 (heptet, 1H)
>
> **^{13}C NMR** (δ): 21 (CH$_3$), 39 (CH$_3$), and 55 (CH)

What are the structures of **W** and of each of its reaction products **X**, **Y**, and **Z**?

21.42 Phenols generally are not changed on treatment with sodium borohydride followed by acidification to destroy the excess, unreacted hydride. For example, the 1,2-, 1,3-, and 1,4-benzenediols and 1,2,3-benzenetriol are unchanged under these conditions. However, 1,3,5-benzenetriol (phloroglucinol) gives a high yield of a product **A** that has these properties:

> **MS** (*m/z*): 110
>
> **IR** (cm^{-1}): 3250 (broad), 1613, 1485
>
> **^{1}H NMR** (δ in DMSO): 6.15 (m, 3H), 6.89 (t, 1H), and 9.12 (s, 2H)

(a) What is the structure of **A**?

(b) Suggest a mechanism by which the above reaction occurred. (1,3,5-Benzenetriol is known to have more tendency to exist in a keto tautomeric form than do simpler phenols.)

21.43 Open the molecular model file for benzyne and examine the following molecular orbitals: the LUMO (lowest unoccupied molecular orbital), the HOMO (highest occupied molecular orbital), the HOMO-1 (next lower energy orbital), the HOMO-2 (next lower in energy), and the HOMO-3 (next lower in energy). **(a)** Which orbital best represents the region where electrons of the additional π bond in benzyne would be found? **(b)** Which orbital would accept electrons from a Lewis base on nucleophilic addition to benzyne? **(c)** Which orbitals are associated with the six π electrons of the aromatic system? Recall that each molecular orbital can hold a maximum of two electrons.

LEARNING GROUP PROBLEMS

1. Thyroxine is a hormone produced by the thyroid gland that is involved in regulating metabolic activity. In a previous Learning Group Problem (Chapter 15) we considered reactions involved in a chemical synthesis of thyroxine. The following is a synthesis of optically pure thyroxine from the amino acid tyrosine (also see Problem 2, below). This synthesis proved to be useful on an industrial scale. (Scheme adapted from Fleming, I., *Selected Organic Syntheses*, pp. 31–33. © 1973 John Wiley & Sons, Limited. Reproduced with permission.)

(a) 1 to 2   What type of reaction is involved in the conversion of **1** to **2**? Write a detailed mechanism for this transformation. Explain why the nitro groups appear where they do in **2**.

(b) 2 to 3   (i) Write a detailed mechanism for step (1) in the conversion of **2** to **3**.

   (ii) Write a detailed mechanism for step (2) in the conversion of **2** to **3**.

   (iii) Write a detailed mechanism for step (3) in the conversion of **2** to **3**.

(c) 3 to 4   (i) What type of reaction mechanism is involved in the conversion of **3** to **4**?

   (ii) Write a detailed mechanism for the reaction from **3** to **4**. What key intermediate is involved?

(d) 5 to 6   Write a detailed mechanism for conversion of the methoxyl group of **5** to the phenolic hydroxyl of **6**.

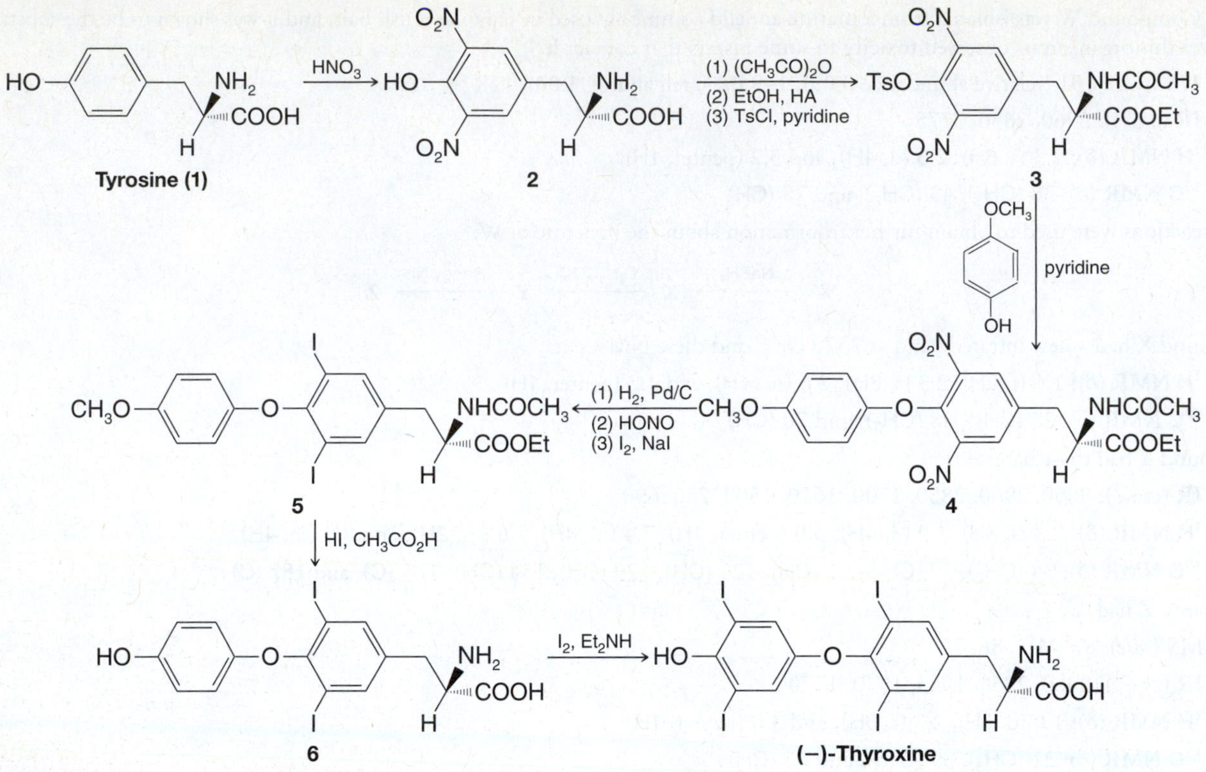

2. Tyrosine is an amino acid with a phenolic side chain. Biosynthesis in plants and microbes of tyrosine involves enzymatic conversion of chorismate to prephenate, below. Prephenate is then processed further to form tyrosine. These steps are shown here:

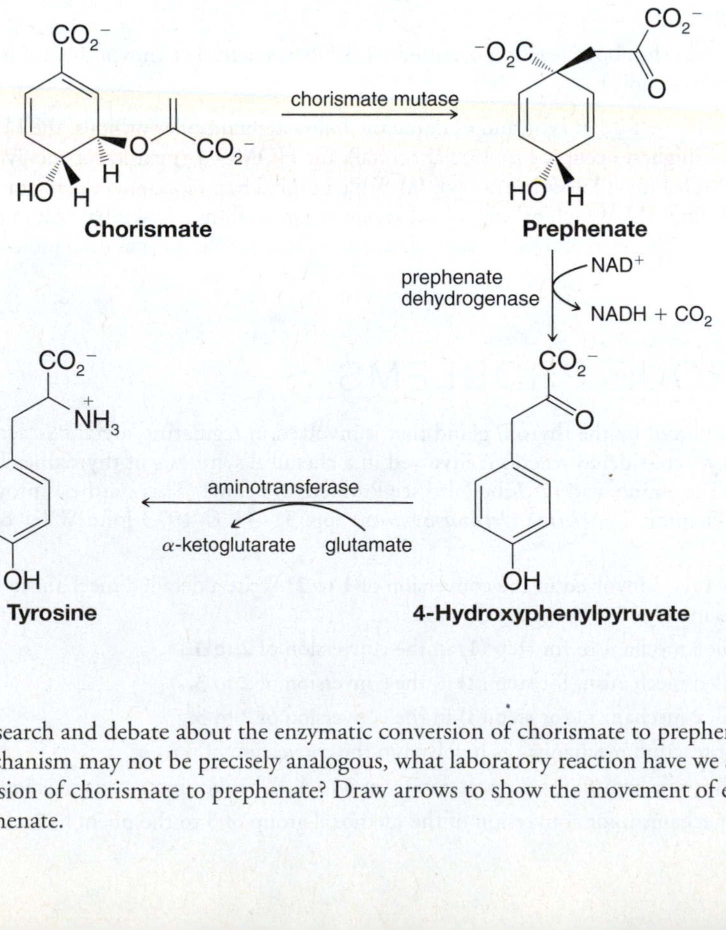

(a) There has been substantial research and debate about the enzymatic conversion of chorismate to prephenate by chorismate mutase. Although the enzymatic mechanism may not be precisely analogous, what laboratory reaction have we studied in this chapter that resembles the biochemical conversion of chorismate to prephenate? Draw arrows to show the movement of electrons involved in such a reaction from chorismate to prephenate.

(b) When the type of reaction you proposed above is applied in laboratory syntheses, it is generally the case that the reaction proceeds by a concerted chair conformation transition state. Five of the atoms of the chair are carbon and one is oxygen. In both the reactant and product, the chair has one bond missing, but at the point of the bond reorganization there is roughly concerted flow of electron density throughout the atoms involved in the chair. For the reactant shown below, draw the structure of the product and the associated chair conformation transition state for this type of reaction:

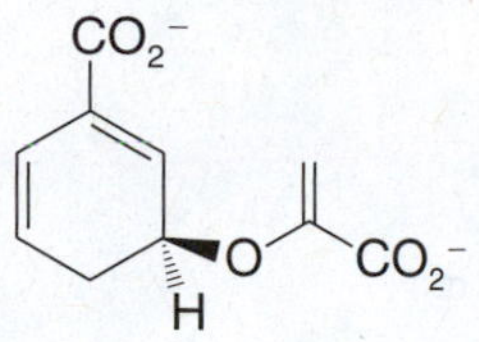

(c) Draw the structure of the nicotinamide ring of NAD^+ and draw mechanism arrows to show the decarboxylation of prephenate to 4-hydroxyphenylpyruvate with transfer of the hydride to NAD^+ (this is the type of process involved in the mechanism of prephenate dehydrogenase).

(d) Look up the structures of glutamate (glutamic acid) and α-ketoglutarate and consider the process of transamination involved in conversion of 4-hydroxyphenylpyruvate to tyrosine. Identify the source of the amino group in this transamination (i.e., what is the amino group "donor"?). What functional group is left after the amino group has been transferred from its donor? Propose a mechanism for this transamination. Note that the mechanism you propose will likely involve formation and hydrolysis of several imine intermediates— reactions similar to others we studied in Section 16.8.

CONCEPT MAP

Some Synthetic Connections of Phenols and Related Aromatic Compounds

Reactions of benzene related to phenol synthesis:
- Nitration and nitro-group reduction
- Friedel–Crafts alkylation
- Sulfonation
- Chlorination (X = Cl)

Synthesis of phenols:
- Laboratory: via diazonium salts
- Industrial synthesis of phenols
  (1) Via cumene hydroperoxide
  (2) Dow process (via benzyne)

Reactions of phenols:
- Acylation
- Alkylation (Williamson ether synthesis)
- Electrophilic aromatic substitution

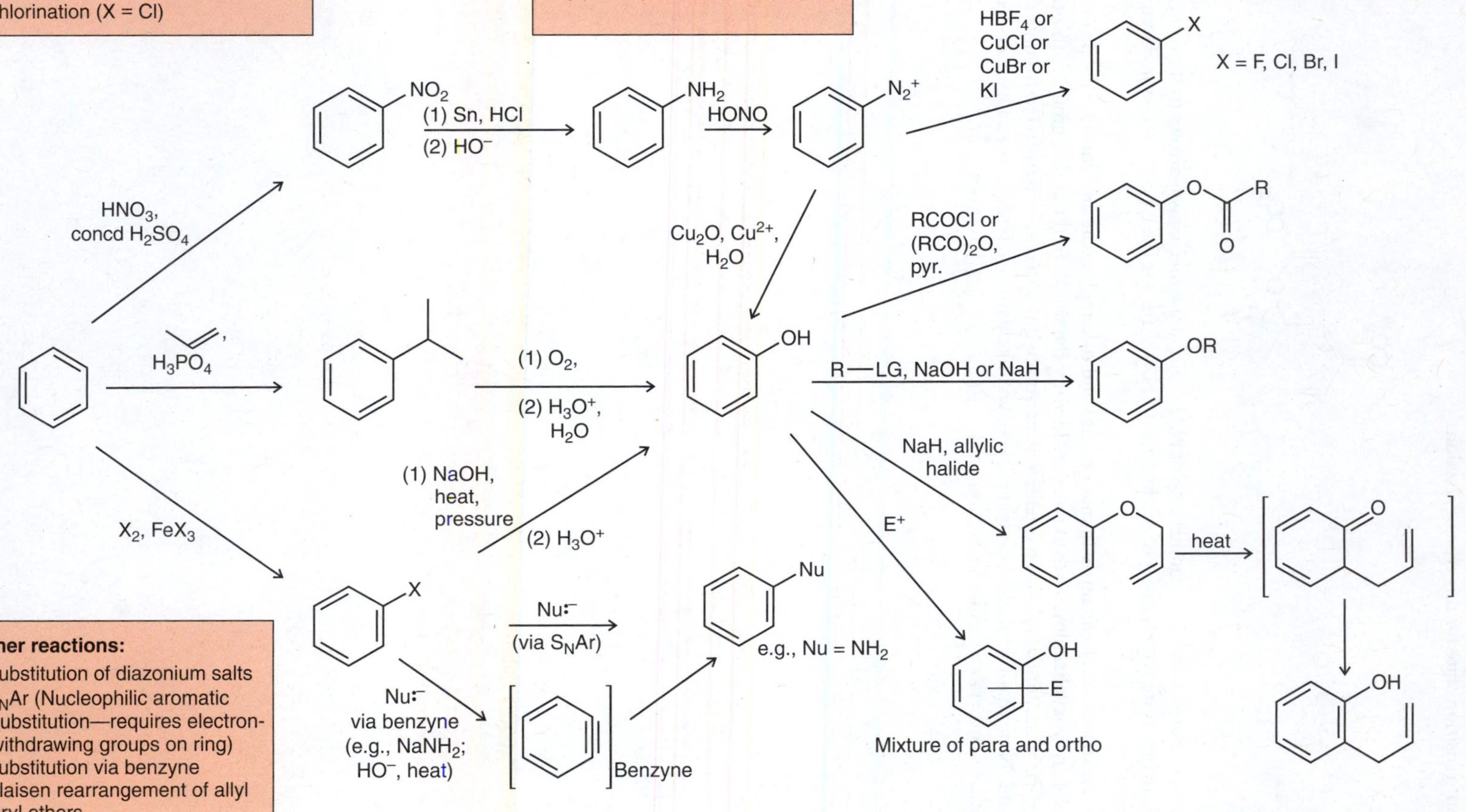
NO_2
(1) Sn, HCl
(2) HO^-
NH_2
HONO
N_2^+
HBF_4 or CuCl or CuBr or KI
X
X = F, Cl, Br, I
HNO_3, concd H_2SO_4
H_3PO_4
Cu_2O, Cu^{2+}, H_2O
RCOCl or $(RCO)_2O$, pyr.
O
R
(1) O_2,
(2) H_3O^+, H_2O
OH
R—LG, NaOH or NaH
OR
(1) NaOH, heat, pressure
(2) H_3O^+
NaH, allylic halide
X_2, FeX_3
E+
heat
Nu:−
(via S_NAr)
Nu
e.g., Nu = NH_2
E
Mixture of para and ortho
Nu:− via benzyne (e.g., $NaNH_2$; HO^-, heat)
Benzyne

Other reactions:
- Substitution of diazonium salts
- S_NAr (Nucleophilic aromatic substitution—requires electron-withdrawing groups on ring)
- Substitution via benzyne
- Claisen rearrangement of allyl aryl ethers

See Second Review Problem Set in *WileyPLUS*

Carbon–Carbon Bond–Forming and Other Reactions of Transition Metal Organometallic Compounds

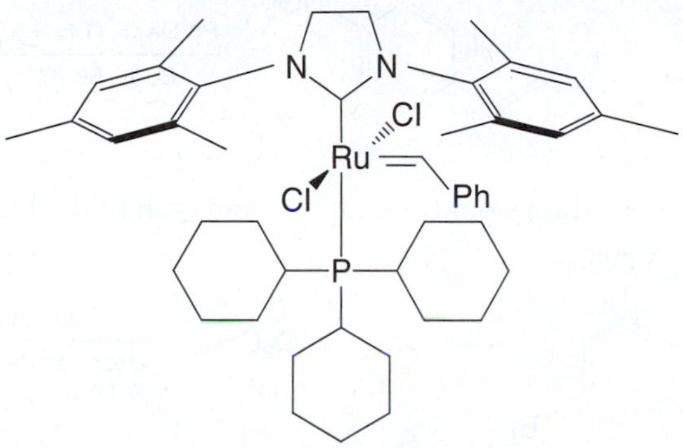

Second generation Grubbs olefin metathesis catalyst

A number of transition metal–catalyzed carbon–carbon bond-forming reactions have been developed into highly useful tools for organic synthesis. The great power of many transition metal–catalyzed reactions is that they provide ways to form bonds between groups for which there are very limited or perhaps no other carbon–carbon bond-forming reactions available. For example, using certain **transition metal catalysts** we can form bonds between alkenyl (vinyl) or aryl substrates and sp^2- or sp-hybridized carbons of other reactants. We shall provide examples of a few of these methods here, including the **Heck–Mizoroki reaction**, the **Suzuki–Miyaura coupling**, the **Stille coupling**, and the **Sonogashira coupling**. These reactions are types of **cross-coupling reactions**, whereby two reactants of appropriate structure are coupled by a new carbon–carbon σ bond.

Olefin metathesis is another reaction type, whereby the groups of two alkene reactants exchange position with each other. We shall discuss olefin metathesis reactions that are promoted by **Grubbs' catalyst**.

Another transition metal–catalyzed carbon–carbon bond-forming reaction we shall discuss is the **Corey–Posner, Whitesides–House reaction**. Using this reaction an alkyl halide can be coupled with the alkyl group from a **lithium dialkyl cuprate** reagent (often called a **Gilman reagent**). This reaction does not have a catalytic mechanism.

All of these reactions involve transition metals such as palladium, copper, and ruthenium, usually in complex with certain types of ligands. After we see the practical applications of these reactions for carbon–carbon bond formation, we shall consider some general aspects of transition metal complex structure and representative steps in the mechanisms of transition metal–catalyzed reactions. We shall consider as specific examples the mechanism for a transition metal–catalyzed hydrogenation using a rhodium complex called Wilkinson's catalyst, and the mechanism for the Heck–Mizoroki reaction.

G.1 CROSS-COUPLING REACTIONS CATALYZED BY TRANSITION METALS

G.1A The Heck–Mizoroki Reaction

ROBERT F. HECK shared the 2010 Nobel Prize in Chemistry with Ei-ichi Negishi and Akira Suzuki for development of palladium-catalyzed cross coupling reactions.

The Heck–Mizoroki reaction involves palladium-catalyzed coupling of an alkene with an alkenyl or aryl halide, leading to a substituted alkene. The alkene product is generally trans due to a 1,2-elimination step in the mechanism.

General Reaction

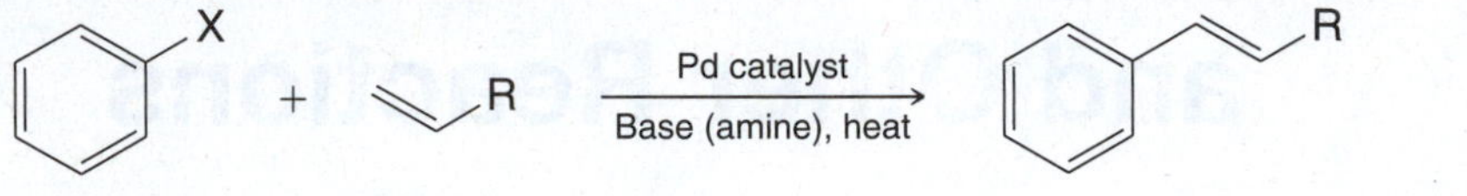

X = I, Br, Cl
(in order of relative reactivity)

Specific Example

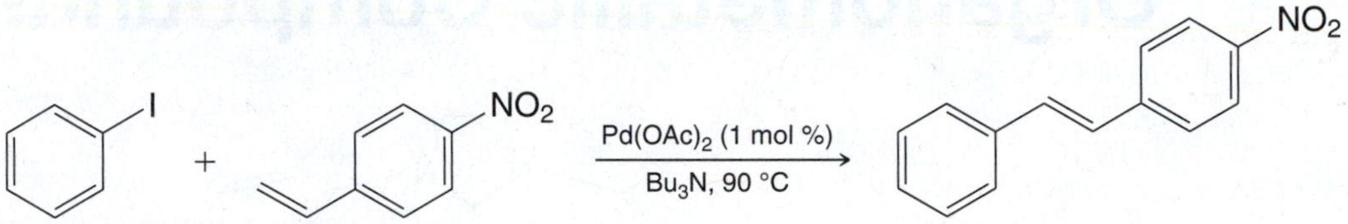

PRACTICE PROBLEM G.1 What product would you expect from each of the following reactions?

(a)

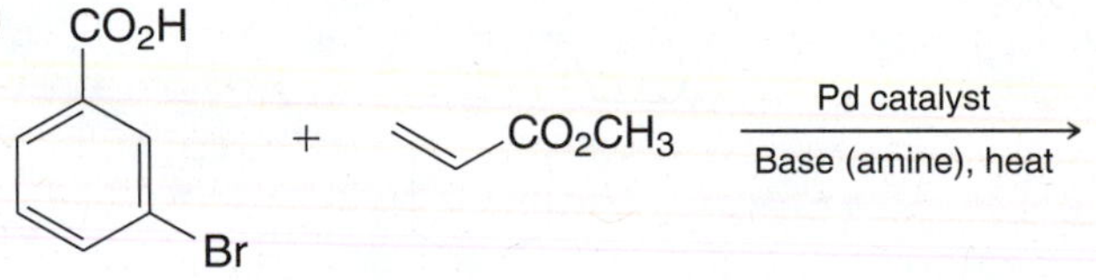

(b)

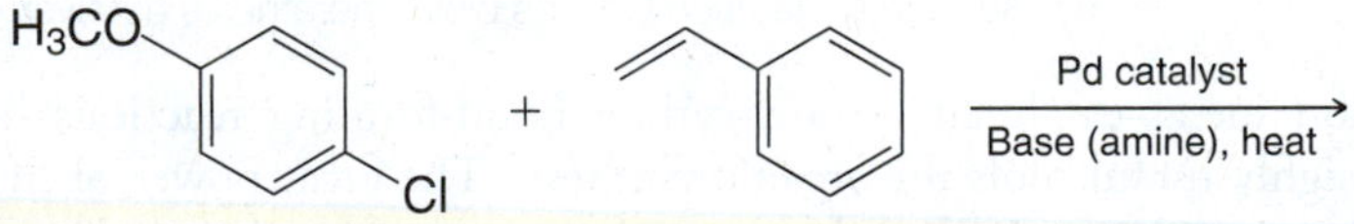

PRACTICE PROBLEM G.2 What starting materials could be used to synthesize each of the following compounds by a Heck–Mizoroki reaction?

(a)

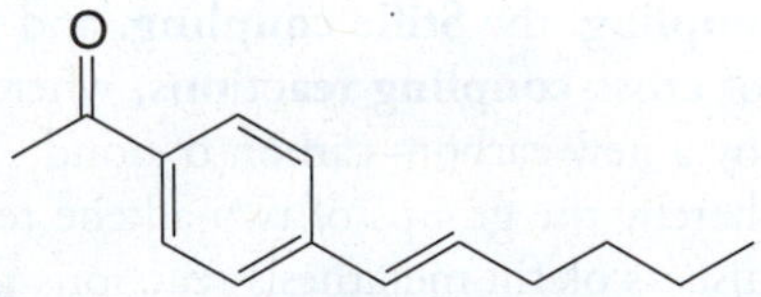

(b)

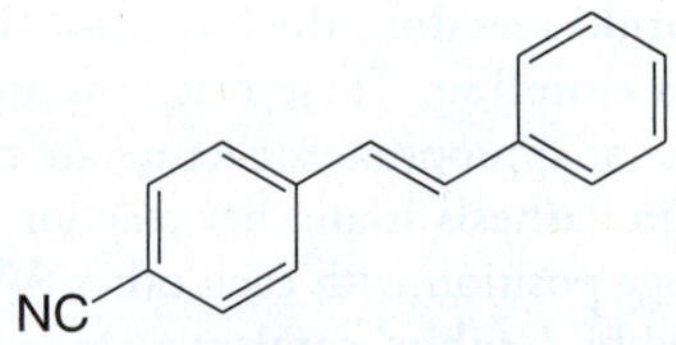

G.1B The Suzuki–Miyaura Coupling

The Suzuki–Miyaura coupling joins an alkenyl or aryl borate with an alkenyl or aryl halide in the presence of a palladium catalyst. The stereochemistry of alkenyl reactants is preserved in the coupling.

General Reactions

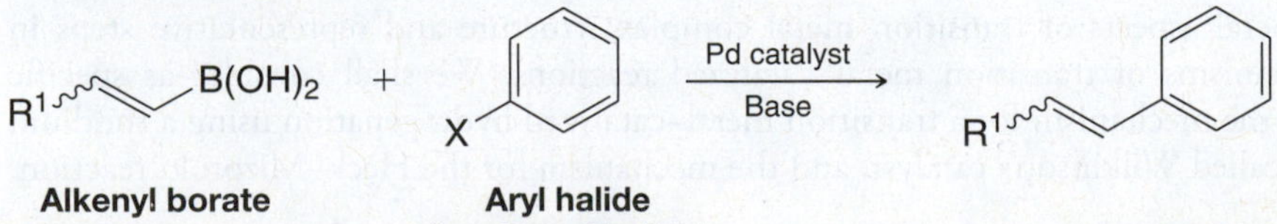

Alkenyl borate   **Aryl halide**
                        (or alkenyl halide)

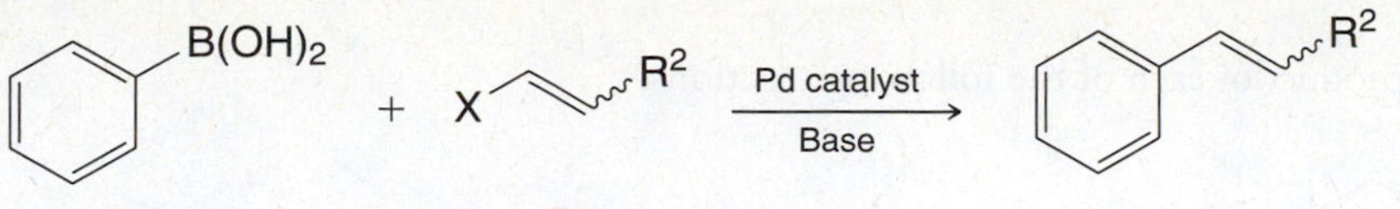

Aryl borate **Alkenyl halide (or aryl halide)**

Specific Example

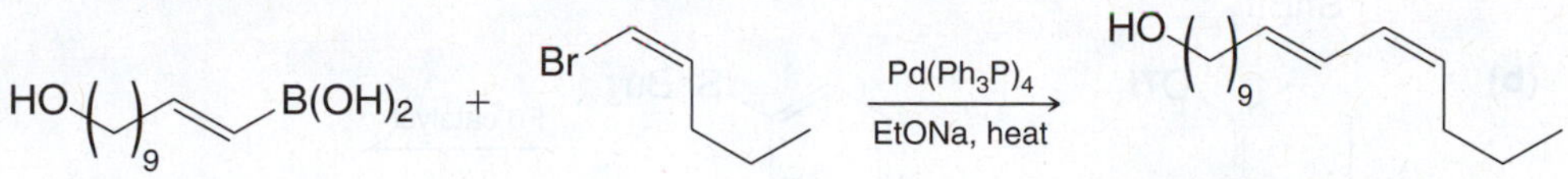

Bombykol (a sex hormone released by the female silkworm)

What is the product of the following Suzuki–Miyaura coupling?

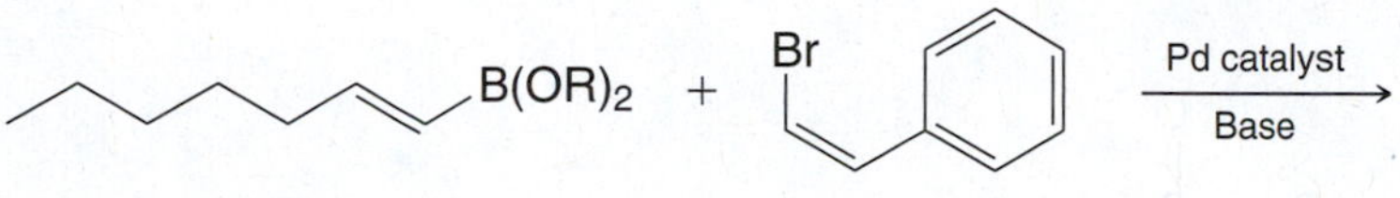

PRACTICE PROBLEM G.3

What starting materials could be used to synthesize the following compound by a Suzuki–Miyaura coupling?

PRACTICE PROBLEM G.4

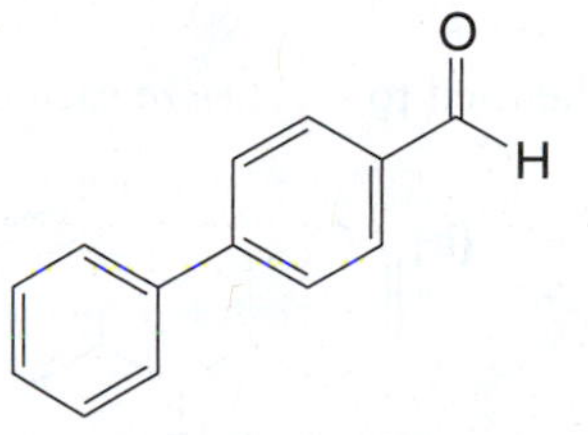

G.1C The Stille Coupling and Carbonylation

The Stille coupling is a cross-coupling reaction that involves an organotin reagent as one reactant. In the presence of appropriate palladium catalysts, alkenyl and aryl tin reactants can be coupled with alkenyl triflates, iodides, and bromides, as well as allylic chlorides and acid chlorides.

General Reaction

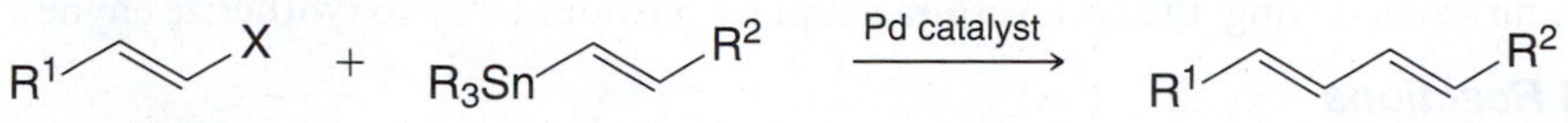

X = triflate, I, or Br

Specific Example

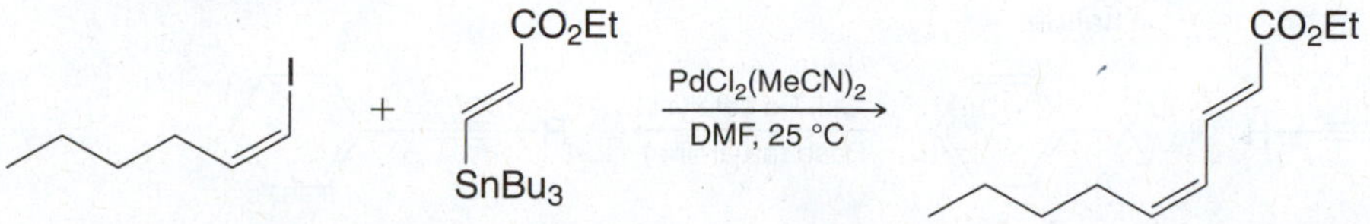

Ketones can be synthesized by a variation of the Stille coupling that involves coupling in the presence of carbon monoxide. The following reaction is an example.

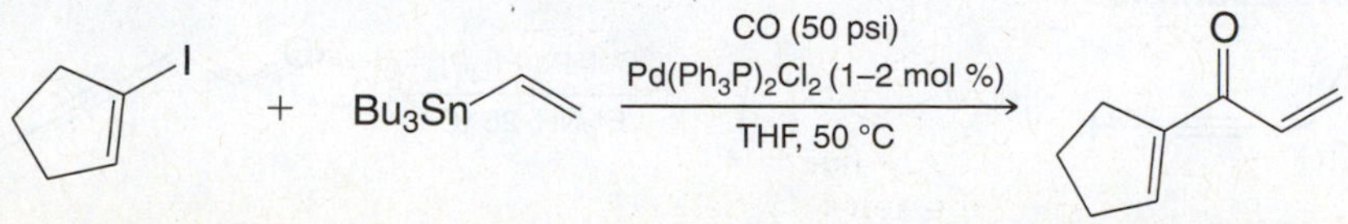

PRACTICE PROBLEM G.5 What is the product of each of the following reactions?

(a)

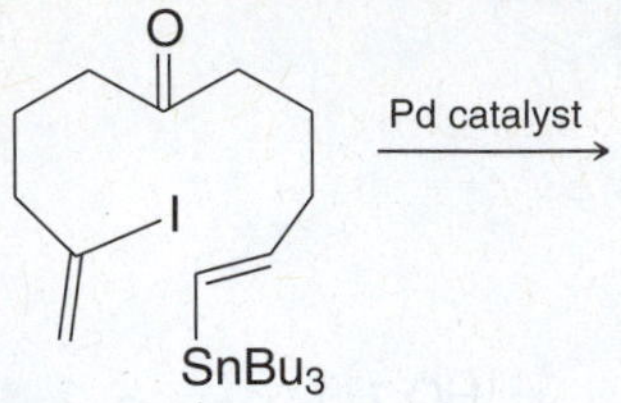

(b)

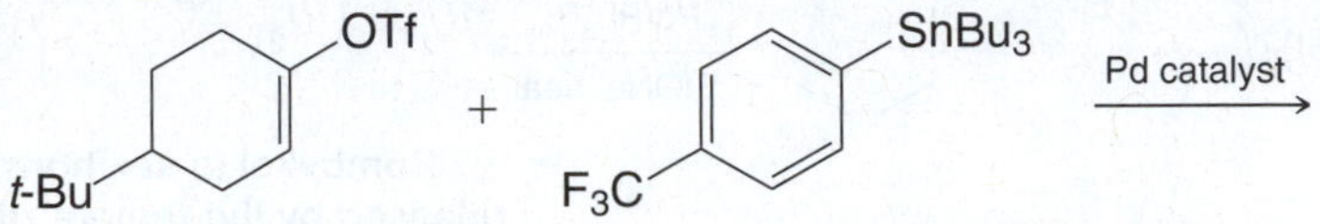

(c)

(d)

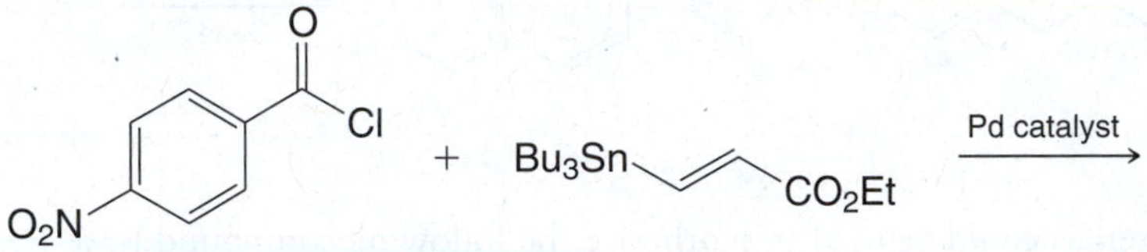

PRACTICE PROBLEM G.6 What starting materials could be used to synthesize each of the following compounds by a Stille coupling reaction?

(a)

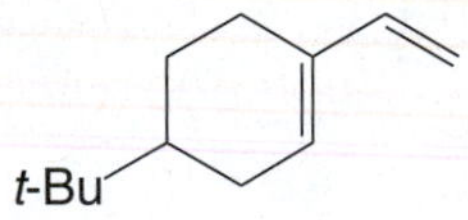

(b)

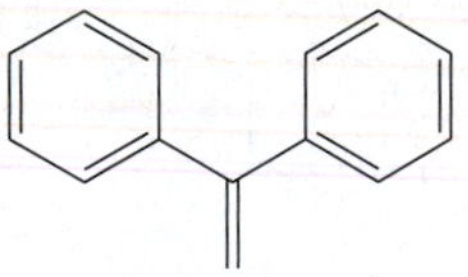

G.1D The Sonogashira Coupling

The Sonogashira coupling joins an alkyne with an alkenyl or aryl halide in the presence of catalytic palladium and copper. A copper alkynide is formed as an intermediate in the reaction. (When palladium is not used, the reaction is called the Stephens–Castro coupling, and it is not catalytic.) In addition to providing a method for joining an alkyne directly to an aromatic ring, the Sonogashira coupling provides a way to synthesize enynes.

General Reactions

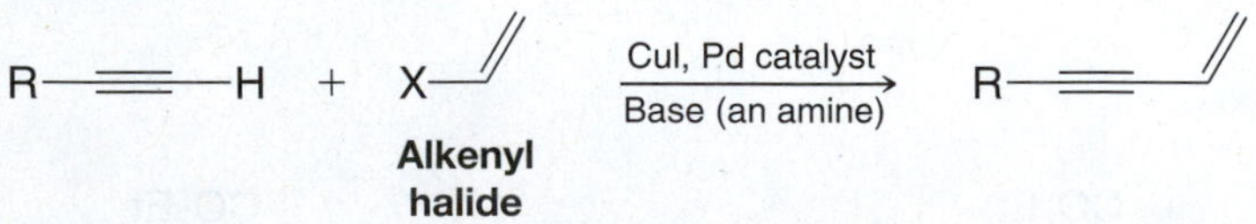

Alkenyl halide

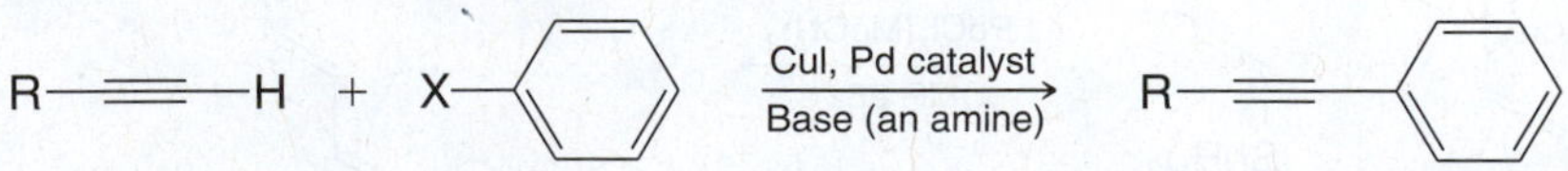

Aryl halide

Specific Example

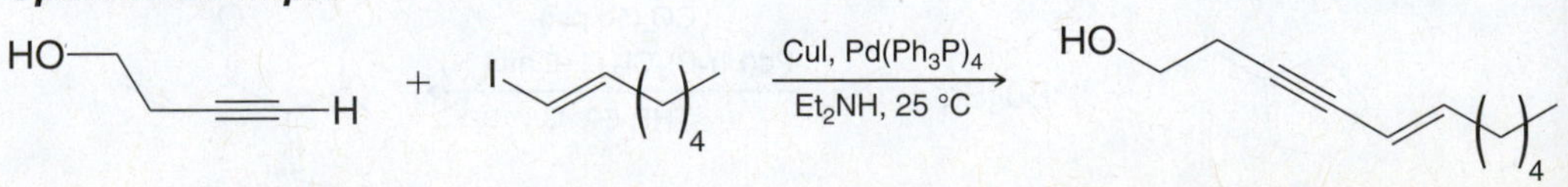

Provide the products of each of the following reactions.

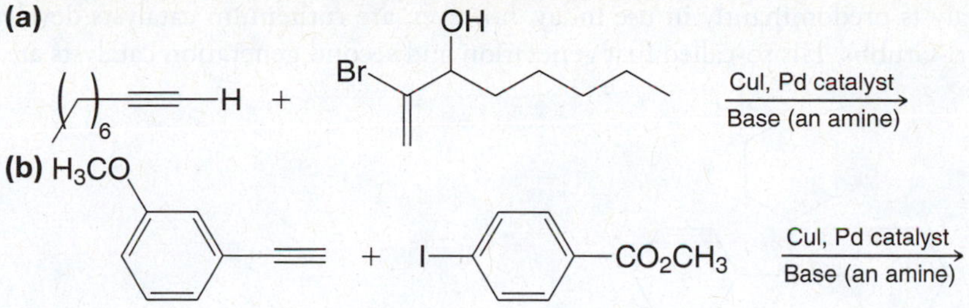

(a)

$$\frac{\text{CuI, Pd catalyst}}{\text{Base (an amine)}}$$

(b) H$_3$CO

$$\frac{\text{CuI, Pd catalyst}}{\text{Base (an amine)}}$$

PRACTICE PROBLEM G.7

What starting materials could be used to synthesize each of the following compounds by a Sonogashira coupling reaction?

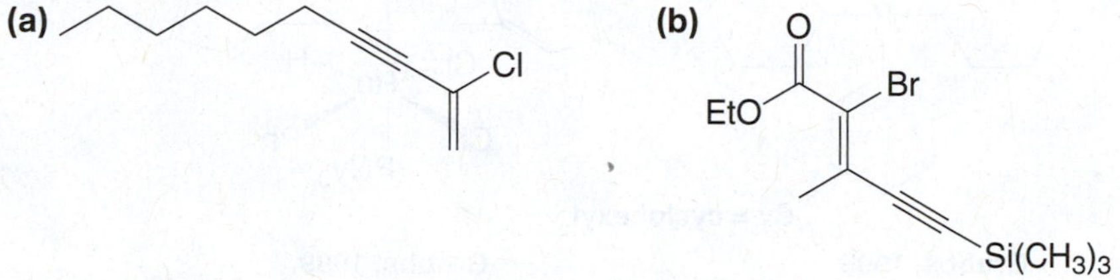

(a)

(b)

PRACTICE PROBLEM G.8

G.2 OLEFIN METATHESIS: RUTHENIUM CARBENE COMPLEXES AND GRUBBS' CATALYSTS

Pairs of alkene double bonds can trade ends with each other in a remarkable molecular "dance" called olefin metathesis (*meta*, Greek: to change; *thesis*, Greek: position).

The overall reaction is the following.

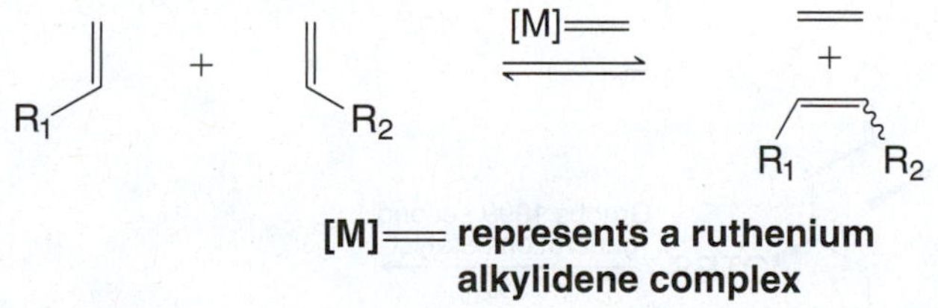

[M]═ **represents a ruthenium alkylidene complex**

The generally accepted catalytic cycle for this "change partners" dance was proposed by Yves Chauvin and is believed to involve metallocyclobutane intermediates that result from reaction of metal alkylidenes (also called metal carbenes) with alkenes. The catalysts themselves are metal alkylidenes, in fact. Chauvin's catalytic cycle for olefin metathesis is summarized here.

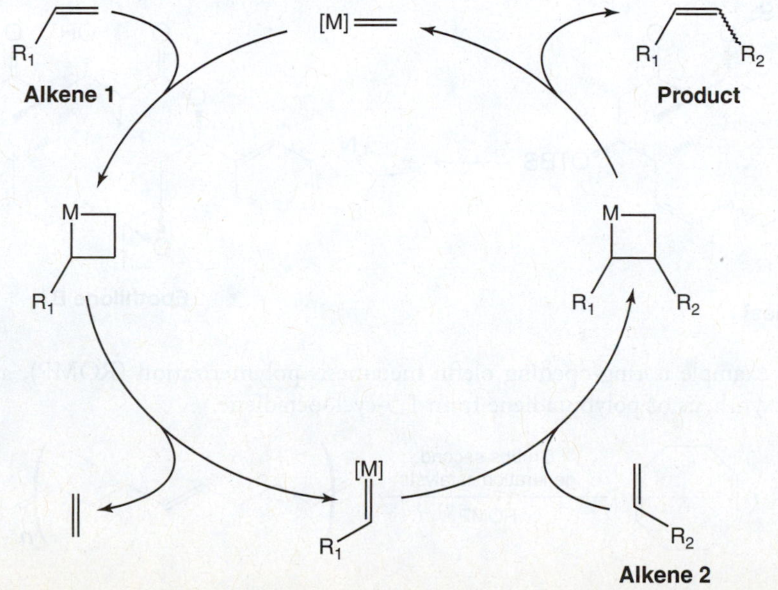

Richard Schrock investigated the properties of some of the first catalysts for olefin metathesis. His work included catalysts prepared from tantalum, titanium, and molybdenum. The catalysts predominantly in use today, however, are ruthenium catalysts developed by Robert Grubbs. His so-called first generation and second generation catalysts are shown here.

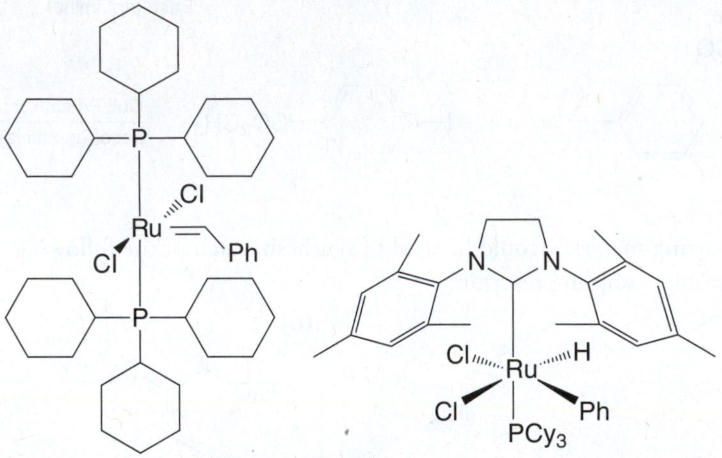

Cy = cyclohexyl

Grubbs, 1995
First generation Grubbs catalyst
(commercially available)

Grubbs, 1999
Second generation Grubbs catalyst
(commercially available)

 2005 Nobel Prize: Chauvin, Grubbs, and Schrock

Olefin metathesis has proved to be such a powerful tool for synthesis that the 2005 Nobel Prize in Chemistry was awarded to Chauvin, Grubbs, and Shrock for their work in this area. Uses include its application to several syntheses of anticancer agents in the epothilone family by Danishefsky, Nicolaou, Shinzer, Sinha, and others, as in the example shown here.

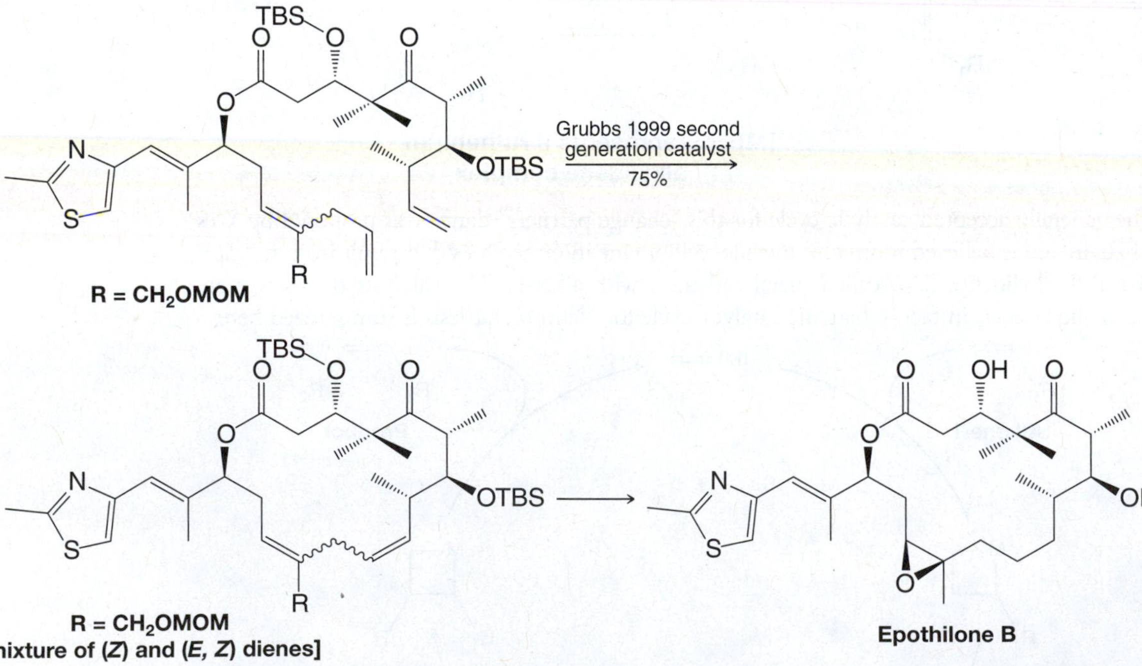

R = CH₂OMOM

Grubbs 1999 second generation catalyst
75%

R = CH₂OMOM
[mixture of (Z) and (E, Z) dienes]

Epothilone B

Another example is ring-opening olefin metathesis polymerization (ROMP), as can be used for synthesis of polybutadiene from 1,5-cyclooctadiene.

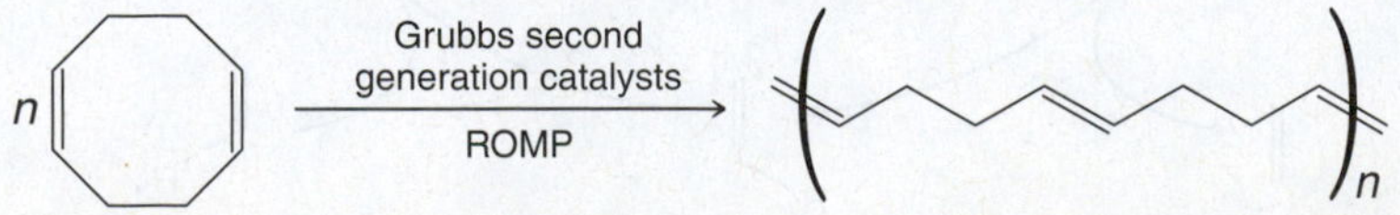

Grubbs second generation catalysts
ROMP

What products would form when each of the following compounds is treated with (PCy$_3$)$_2$Cl$_2$Ru=CHPh, one of Grubbs' catalysts?

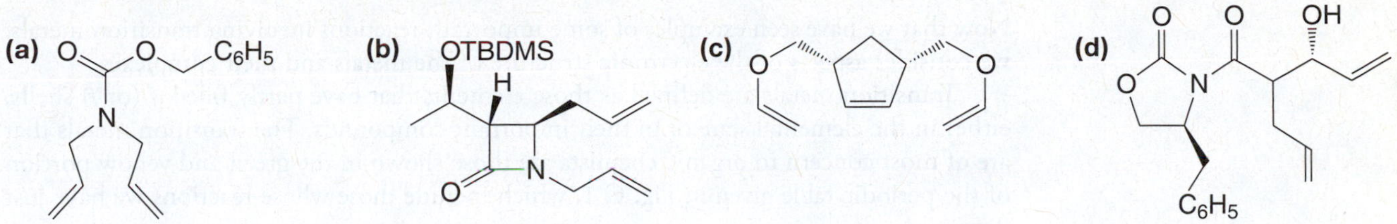

(a) O=C, O, C$_6$H$_5$, N

(b) OTBDMS, H, N, O

(c) O, O

(d) O, O, OH, N, O, C$_6$H$_5$

G.3 THE COREY–POSNER, WHITESIDES–HOUSE REACTION: USE OF LITHIUM DIALKYL CUPRATES (GILMAN REAGENTS) IN COUPLING REACTIONS

The Corey–Posner, Whitesides–House reaction involves the coupling of a lithium dialkyl-cuprate (called a Gilman reagent) with an alkyl, alkenyl, or aryl halide. The alkyl group of the lithium dialkylcuprate reagent may be primary, secondary, or tertiary. However, the halide with which the Gilman reagent couples must be a primary or cyclic secondary alkyl halide if it is not alkenyl or aryl.

General Reaction

$$R_2CuLi \quad + \quad R'-X \longrightarrow R-R' + RCu + LiX$$

A lithium dialkyl cuprate (a Gilman reagent) **Alkenyl, aryl, or 1° or cyclic 2° alkyl halide**

Specific Example

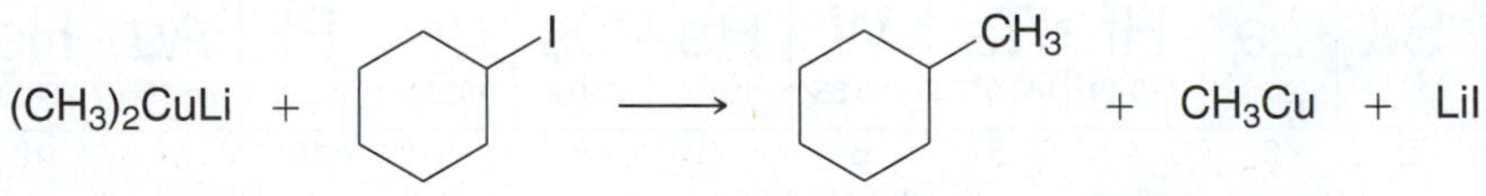

(CH$_3$)$_2$CuLi + [cyclohexyl iodide] ⟶ [methylcyclohexane] + CH$_3$Cu + LiI

Lithium dimethylcuprate **75%**

The required lithium dimethylcuprate (Gilman) reagent must be synthesized by a two-step process from the corresponding alkyl halide, as follows.

Synthesis of an Organolithium Compound
$$R-X \xrightarrow{2Li} R-Li + LiX$$

Synthesis of the Lithium Dialkylcuprate (Gilman) Reagent
$$2R-Li \xrightarrow{CuI} R_2CuLi + LiI$$

All of the reagents in a Corey–Posner, Whitesides–House reaction are consumed stoichiometrically. The mechanism does not involve a catalyst, as in the other reactions of transition metals that we have studied.

Show how 1-bromobutane could be converted to the Gilman reagent lithium dibutylcu-prate, and how you could use it to synthesize each of the following compounds.

(a)

(b)

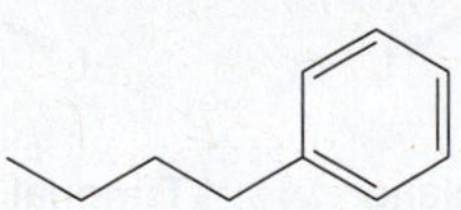

G.4 SOME BACKGROUND ON TRANSITION METAL ELEMENTS AND COMPLEXES

Now that we have seen examples of some important reactions involving transition metals, we consider aspects of the electronic structure of the metals and their complexes.

Transition metals are defined as those elements that have partly filled *d* (or *f*) shells, either in the elemental state or in their important compounds. The transition metals that are of most concern to organic chemists are those shown in the green and yellow portion of the periodic table given in Fig. G.1, which include those whose reactions we have just discussed.

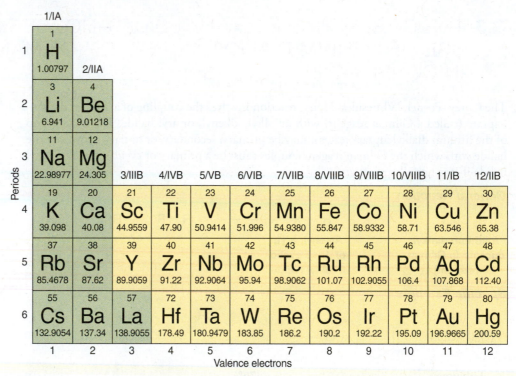

FIGURE G.1 Important transition elements are shown in the green and yellow portion of the periodic table. Given across the bottom is the total number of valence electrons (*s* and *d*) of each element.

Transition metals react with a variety of molecules or groups, called *ligands*, to form *transition metal complexes*. In forming a complex, the ligands donate electrons to vacant orbitals of the metal. The bonds between the ligand and the metal range from very weak to very strong. The bonds are covalent but often have considerable polar character.

Transition metal complexes can assume a variety of geometries depending on the metal and on the number of ligands around it. Rhodium, for example, can form complexes with four ligands in a configuration called *square planar*. On the other hand, rhodium can also form complexes with five or six ligands that are trigonal bipyramidal or octahedral. These typical shapes are shown below, with the letter L used to indicate a ligand.

Square planar rhodium complex   **Trigonal bipyramidal rhodium complex**   **Octahedral rhodium complex**

G.5 ELECTRON COUNTING IN METAL COMPLEXES

Transition metals are like the elements that we have studied earlier in that they are most stable when they have the electronic configuration of a noble gas. In addition to s and p orbitals, transition metals have five d orbitals (which can hold a total of 10 electrons). Therefore, the noble gas configuration for a transition metal is *18 electrons*, not 8 as with carbon, nitrogen, oxygen, and so on.

- When the metal of a transition metal complex has 18 valence electrons, it is said to be **coordinatively saturated**.*

To determine the valence electron count of a transition metal in a complex, we take the total number of valence electrons of the metal in the elemental state (see Fig. G.1) and subtract from this number the oxidation state of the metal in the complex. This gives us what is called the d electron count, d^n. The oxidation state of the metal is the charge that would be left on the metal if all the ligands (Table G.1) were removed.

$$d^n = \text{total number of valence electrons of the elemental metal} - \text{oxidation state of the metal in the complex}$$

Then to get the total valence electron count of the metal *in the complex*, we add to d^n the number of electrons donated by all of the ligands. Table G.1 gives the number of electrons donated by several of the most common ligands.

$$\text{total number of valence electrons of the metal in the complex} = d^n + \text{electrons donated by ligands}$$

Let us now work out the valence electron count of two examples.

TABLE G.1 COMMON LIGANDS IN TRANSITION METAL COMPLEXES[a]

Ligand	Count as	Number of Electrons Donated
Negatively charged ligands		
Hydride, H	$H{:}^-$	2
Alkanide, R	$R{:}^-$	2
Halide, X	$X{:}^-$	2
Allyl anion	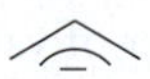	4
Cyclopentadienyl anion, Cp	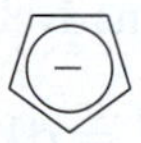	6
Electrically neutral ligands		
Carbonyl (carbon monoxide)	$:C{\equiv}O:$	2
Phosphine	$R_3P:$ or $Ph_3P:$	2
Alkene	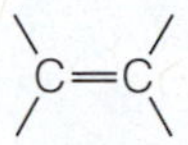	2
Diene	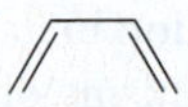	4
Benzene	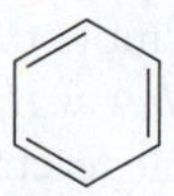	6

[a]Based on data obtained from the *Journal of Chemical Education*, Vol. 57, No. 1, 1980, pp. 170–175, copyright ©1980, Division of Chemical Education.

*We do not usually show the unshared electron pairs of a metal complex in our structures, because to do so would make the structure unnecessarily complicated.

Example A Consider iron pentacarbonyl, $Fe(CO)_5$, a toxic liquid that forms when finely divided iron reacts with carbon monoxide.

$$Fe \; + \; 5\,CO \longrightarrow Fe(CO)_5 \quad \text{or}$$

Iron pentacarbonyl

From Fig. G.1 we find that an iron atom in the elemental state has 8 valence electrons. We arrive at the oxidation state of iron in iron pentacarbonyl by noting that the charge on the complex as a whole is zero (it is not an ion), and that the charge on each CO ligand is also zero. Therefore, the iron is in the zero oxidation state.

Using these numbers, we can now calculate d^n and, from it, the total number of valence electrons of the iron in the complex.

$$d^n = 8 - 0 = 8$$

$$\begin{array}{c} \text{total number of} \\ \text{valence electrons} \end{array} = d^n + 5(CO) = 8 + 5(2) = 18$$

We find that the iron of $Fe(CO)_5$ has 18 valence electrons and is, therefore, coordinatively saturated.

Example B Consider the rhodium complex $Rh[(C_6H_5)_3P]_3H_2Cl$, a complex that, as we shall see later, is an intermediate in certain alkene hydrogenations.

$$L = Ph_3P \;[\text{i.e., } (C_6H_5)_3P]$$

The oxidation state of rhodium in the complex is +3. [The two hydrogen atoms and the chlorine are each counted as −1 (hydride and chloride, respectively), and the charge on each of the triphenylphosphine ligands is zero. Removing all the ligands would leave a Rh^{3+} ion.] From Fig. G.1 we find that, in the elemental state, rhodium has 9 valence electrons. We can now calculate d^n for the rhodium of the complex.

$$d^n = 9 - 3 = 6$$

Each of the six ligands of the complex donates two electrons to the rhodium in the complex, and, therefore, the total number of valence electrons of the rhodium is 18. The rhodium of $Rh[(C_6H_5)_3P]_3H_2Cl$ is coordinatively saturated.

$$\begin{array}{c} \text{total number of valence} \\ \text{electrons rhodium} \end{array} = d^n + 6(2) = 6 + 12 = 18$$

G.6 MECHANISTIC STEPS IN THE REACTIONS OF SOME TRANSITION METAL COMPLEXES

Much of the chemistry of organic transition metal compounds becomes more understandable if we are able to follow the mechanisms of the reactions that occur. These mechanisms, in most cases, amount to nothing more than a sequence of reactions, each of which represents *a fundamental reaction type that is characteristic of a transition metal complex.* Let us examine three of the fundamental reaction types now. In each instance we shall use steps that occur when an alkene is hydrogenated using a catalyst called Wilkinson's catalyst. In Section G.7 we shall examine the entire hydrogenation mechanism. In Section G.8 we shall see how similar types of steps are involved in the Heck–Mizoroki reaction.

1. **Ligand Dissociation–Association (Ligand Exchange).** A transition metal complex can lose a ligand (by dissociation) and combine with another ligand (by association). In the process it undergoes *ligand exchange.* For example, the rhodium complex that

we encountered in Example B above can react with an alkene (in this example, with ethene) as follows:

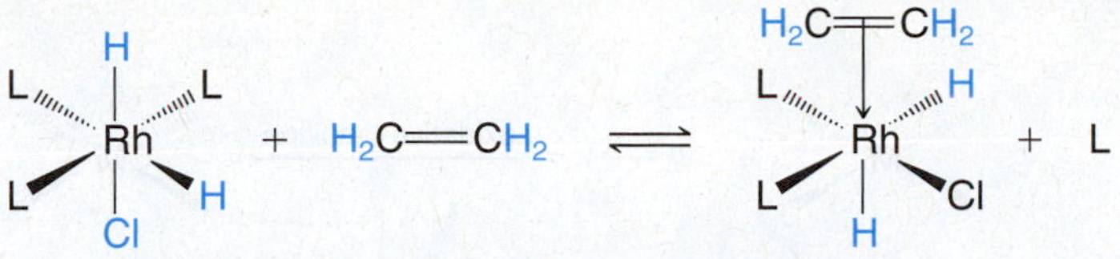

$$L = Ph_3P \; [\text{i.e.,} \; (C_6H_5)_3P]$$

Two steps are actually involved. In the first step, one of the triphenylphosphine ligands dissociates. This leads to a complex in which the rhodium has only 16 electrons and is, therefore, coordinatively *unsaturated*.

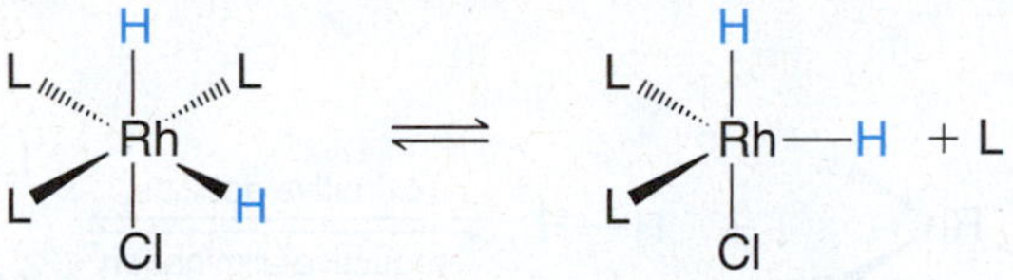

(18 electrons)   **(16 electrons)**

$$L = Ph_3P$$

In the second step, the rhodium associates with the alkene to become coordinatively saturated again.

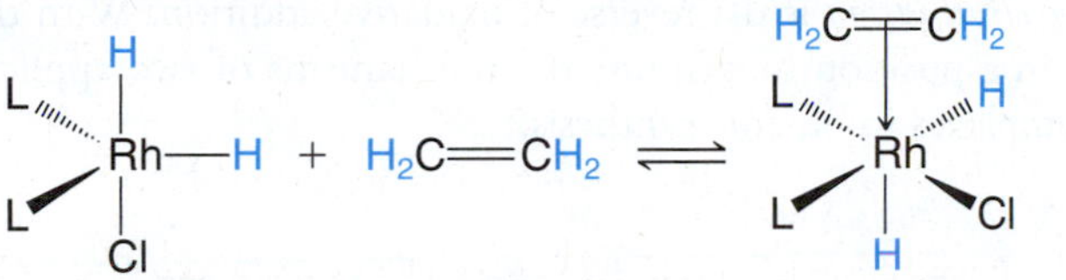

(16 electrons)   **(18 electrons)**

The complex between the rhodium and the alkene is called a π *complex*. In it, two electrons are donated by the alkene to the rhodium. Alkenes are often called π donors to distinguish them from σ donors such as $Ph_3P:$, Cl^-, and so on.

In a π complex such as the one just given, there is also a donation of electrons from a populated d orbital of the metal back to the vacant π^* orbital of the alkene. This kind of donation is called "back-bonding."

2. **Insertion–Deinsertion.** An unsaturated ligand such as an alkene can undergo *insertion* into a bond between the metal of a complex and a hydrogen or a carbon. These reactions are reversible, and the reverse reaction is called *deinsertion*.

The following is an example of insertion–deinsertion.

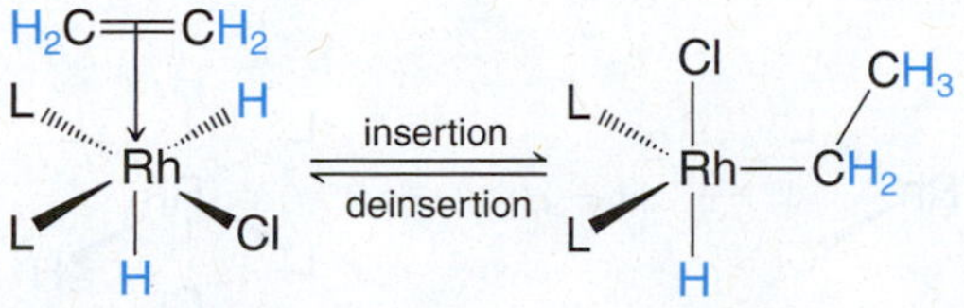

(18 electrons)   **(16 electrons)**

In this process, a π bond (between the rhodium and the alkene) and a σ bond (between the rhodium and the hydrogen) are exchanged for two new σ bonds (between rhodium and carbon, and between carbon and hydrogen). The valence electron count of the rhodium decreases from 18 to 16.

This insertion–deinsertion occurs in a stereospecific way, as a *syn addition* of the M—H unit to the alkene.

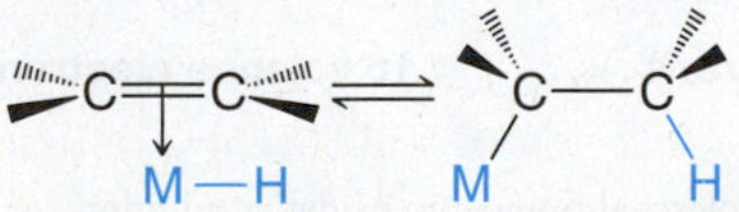

3. Oxidative Addition–Reductive Elimination. Coordinatively unsaturated metal complexes can undergo oxidative addition of a variety of substrates in the following way.*

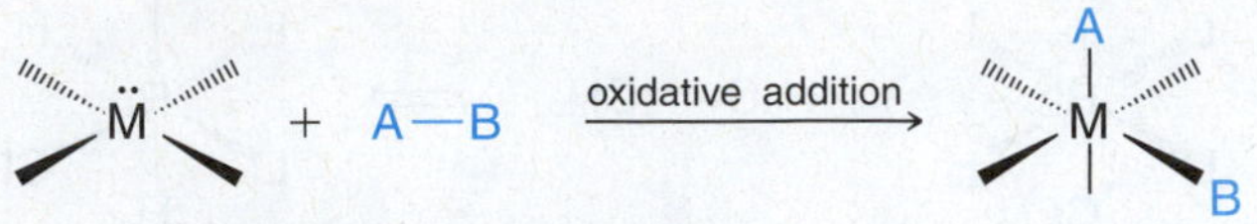

The substrate, A—B, can be H—H, H—X, R—X, RCO—H, RCO—X, and a number of other compounds.

In this type of oxidative addition, the metal of the complex undergoes an increase in the number of its valence electrons *and in its oxidation state*. Consider, as an example, the oxidative addition of hydrogen to the rhodium complex that follows ($L = Ph_3P$).

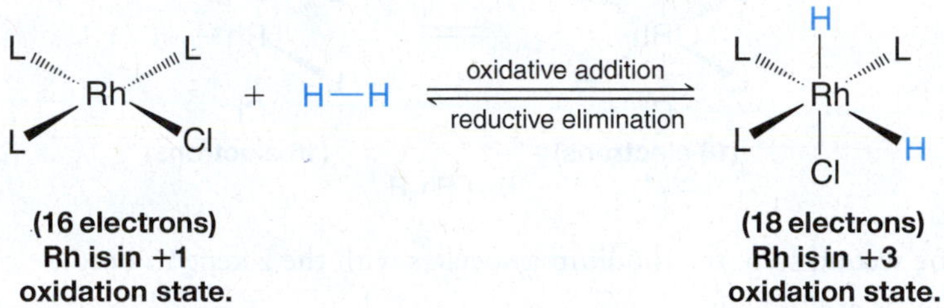

(16 electrons)
Rh is in +1
oxidation state.

(18 electrons)
Rh is in +3
oxidation state.

Reductive elimination is the reverse of oxidative addition. With this background, we are now in a position to examine the mechanisms of two applications of transition metal complexes in organic synthesis.

G.7 THE MECHANISM FOR A HOMOGENEOUS HYDROGENATION: WILKINSON'S CATALYST

The catalytic hydrogenations that we have examined in prior chapters have been heterogeneous processes. Two phases were involved: the solid phase of the catalyst (Pt, Pd, Ni, etc.), containing the adsorbed hydrogen, and the liquid phase of the solution, containing the unsaturated compound. In homogeneous hydrogenation using a transition metal complex such as $Rh[(C_6H_5)_3P]_3Cl$ (Wilkinson's catalyst), hydrogenation takes place *in a single phase*, i.e., in solution.

When Wilkinson's catalyst is used to carry out the hydrogenation of an alkene, the following steps take place ($L = Ph_3P$).

Step 1

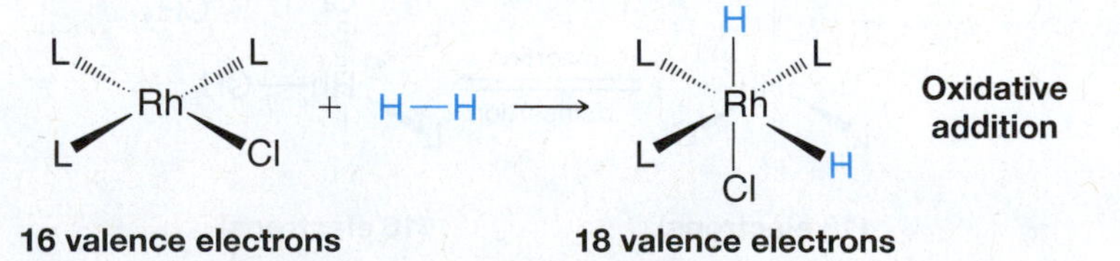

16 valence electrons

18 valence electrons

Oxidative addition

Step 2

18 valence electrons

16 valence electrons

Ligand dissociation

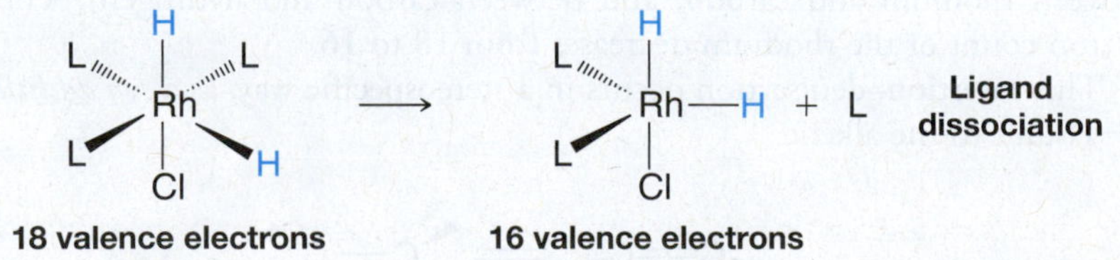

*Coordinatively saturated complexes also undergo oxidative addition.

Step 3

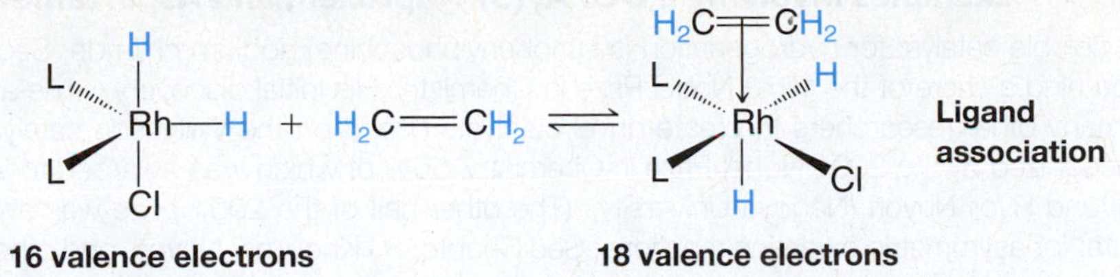

16 valence electrons   18 valence electrons

Ligand association

Step 4

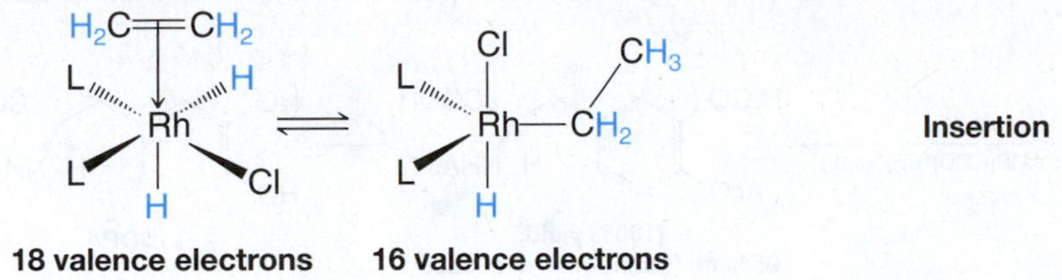

18 valence electrons   16 valence electrons

Insertion

Step 5

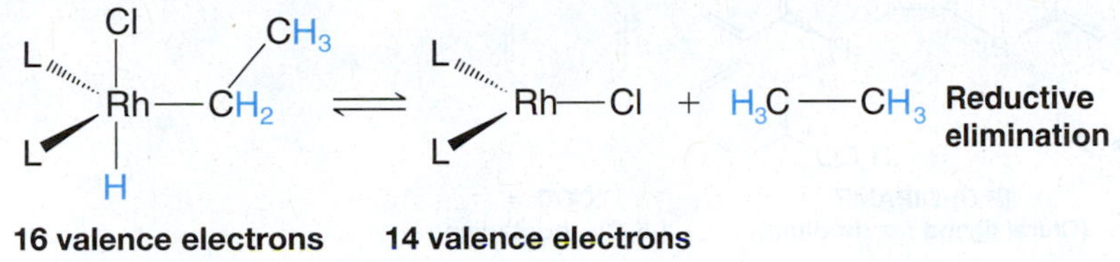

16 valence electrons   14 valence electrons

Reductive elimination

Step 6

14 valence electrons   16 valence electrons
(Cycle repeats from step 3.)

Oxidative addition

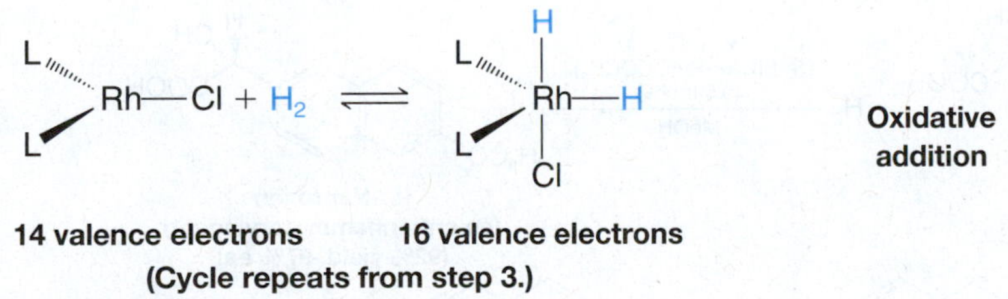

Step 6 regenerates the hydrogen-bearing rhodium complex and reaction with another molecule of the alkene begins at step 3.

Because the insertion step 4 and the reductive elimination step 5 are stereospecific, the net result of the hydrogenation using Wilkinson's catalyst is a *syn addition* of hydrogen to the alkene. The following example (with D_2 in place of H_2) illustrates this aspect.

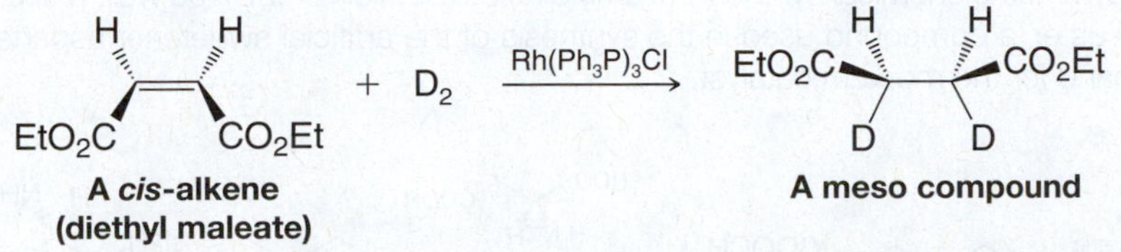

A *cis*-alkene
(diethyl maleate)

A meso compound

What product (or products) would be formed if the *trans*-alkene corresponding to the *cis*-alkene (see the previous reaction) had been hydrogenated with D_2 and Wilkinson's catalyst?

PRACTICE PROBLEM G.11

THE CHEMISTRY OF... Homogeneous Asymmetric Catalytic Hydrogenation: Examples Involving L-DOPA, (S)-Naproxen, and Aspartame

Development by Geoffrey Wilkinson of a soluble catalyst for hydrogenation [tris(triphenylphosphine)rhodium chloride, Section 7.13 and Special Topic G] led to Wilkinson's earning a share of the 1973 Nobel Prize in Chemistry. His initial discovery, while at Imperial College, University of London, inspired many other researchers to create novel catalysts based on the Wilkinson catalyst. Some of these researchers were themselves recognized by the 2001 Nobel Prize in Chemistry, 50% of which was awarded to William S. Knowles (Monsanto Corporation, retired) and Ryoji Noyori (Nagoya University). (The other half of the 2001 prize was awarded to K. B. Sharpless, Scripps Research Institute, for asymmetric oxidation reactions. See Chapter 8.) Knowles, Noyori, and others developed chiral catalysts for homogeneous hydrogenation that have proved extraordinarily useful for enantioselective syntheses ranging from small laboratory-scale reactions to industrial- (ton-) scale reactions. An important example is the method developed by Knowles and co-workers at Monsanto Corporation for synthesis of L-DOPA, a compound used in the treatment of Parkinson's disease:

Asymmetric Synthesis of L-DOPA

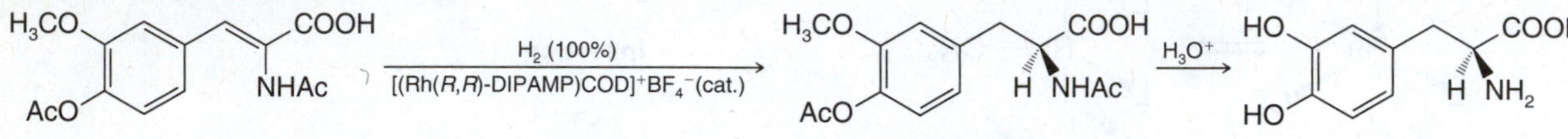

(100% yield, 95% ee [enantiomeric excess])

L-DOPA

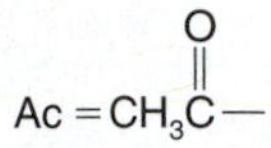

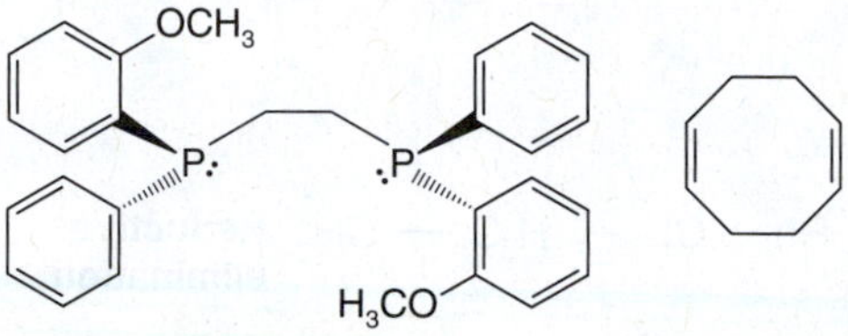

(*R*,*R*)-DIPAMP (Chiral ligand for rhodium)

COD = 1,5-Cyclooctadiene

Another example is synthesis of the over-the-counter analgesic (*S*)-naproxen using a BINAP rhodium catalyst developed by Noyori (Sections 5.11 and 5.18).

Asymmetric Synthesis of (S)-Naproxen

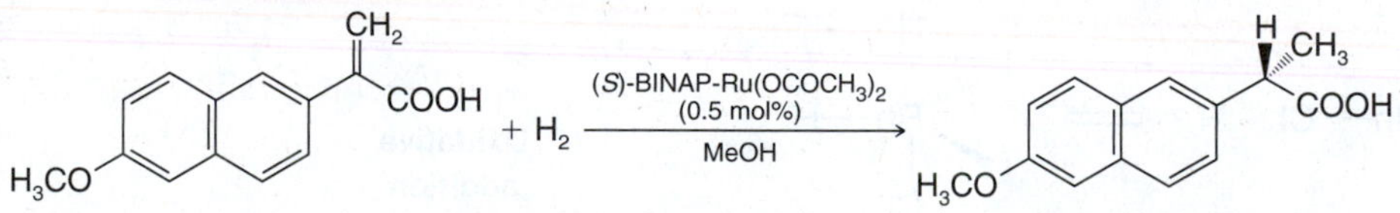

(S)-Naproxen (an anti-inflammatory agent) (92% yield, 97% ee)

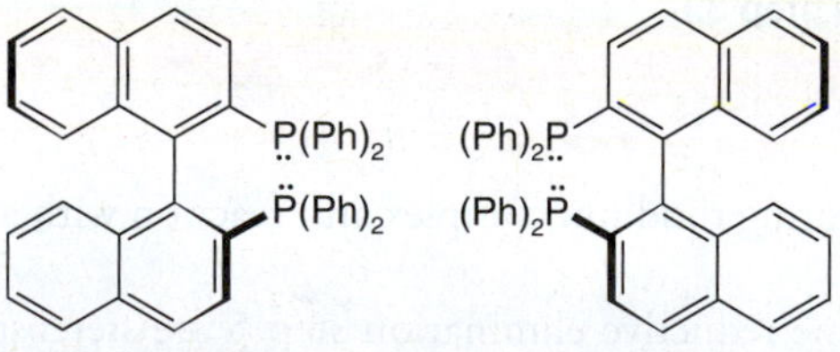

(S)-BINAP **(R)-BINAP**

(S)-BINAP and (R)-BINAP are chiral atropisomers (see Section 5.18).

Catalysts like these are important for asymmetric chemical synthesis of amino acids (Section 24.3D), as well. A final example is the synthesis of (*S*)-phenylalanine methyl ester, a compound used in the synthesis of the artificial sweetener aspartame. This preparation employs yet a different chiral ligand for the rhodium catalyst.

Asymmetric Synthesis of Aspartame

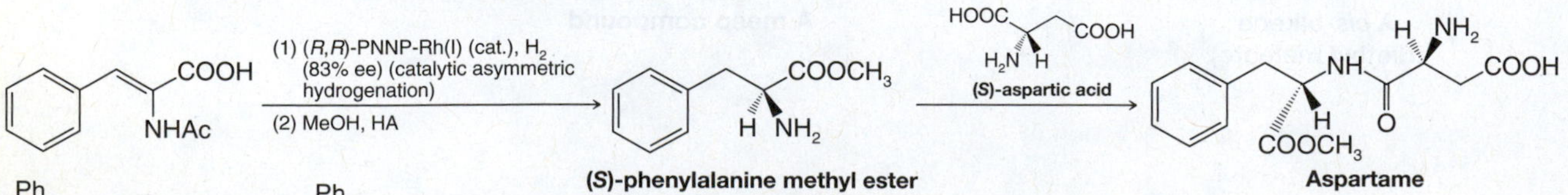

(S)-phenylalanine methyl ester (97% ee after recrystallization)

Aspartame

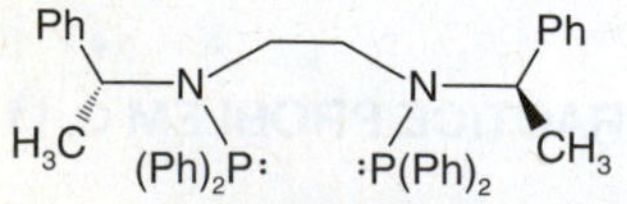

(*R*,*R*)-PNNP (Chiral ligand for rhodium)

The mechanism of homogeneous catalytic hydrogenation involves reactions characteristic of transition metal organometallic compounds. A general scheme for hydrogenation using Wilkinson's catalyst is shown here. We have seen structural details of the mechanism in Section G.7.

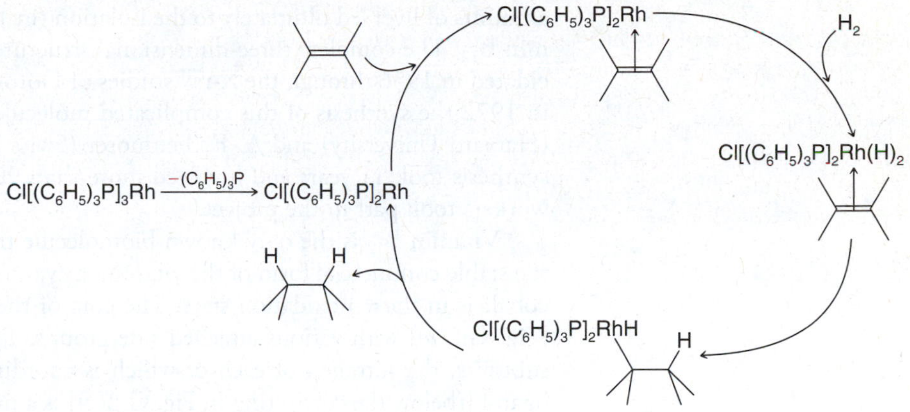

A general mechanism for the Wilkinson catalytic hydrogenation method, adapted with permission of John Wiley & Sons, Inc. from Noyori, *Asymmetric Catalysis in Organic Synthesis*, p. 17. Copyright 1994.

G.8 THE MECHANISM FOR AN EXAMPLE OF CROSS-COUPLING: THE HECK–MIZOROKI REACTION

Having seen steps such as oxidative addition, insertion, and reductive elimination in the context of transition metal–catalyzed hydrogenation using Wilkinson's catalyst, we can now see how these same types of mechanistic steps are involved in a mechanism proposed for the Heck–Mizoroki reaction. Aspects of the Heck–Mizoroki mechanism are similar to steps proposed for other cross-coupling reactions as well, although there are variations and certain steps that are specific to each, and not all of the steps below are involved or serve the same purpose in other cross-coupling reactions.

[**A MECHANISM FOR THE REACTION** — **The Heck–Mizoroki Reaction Using an Aryl Halide Substrate**]

General Reaction

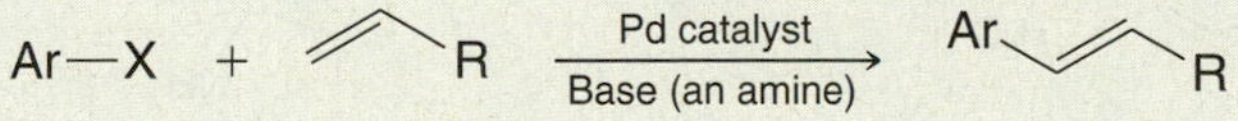

Mechanism

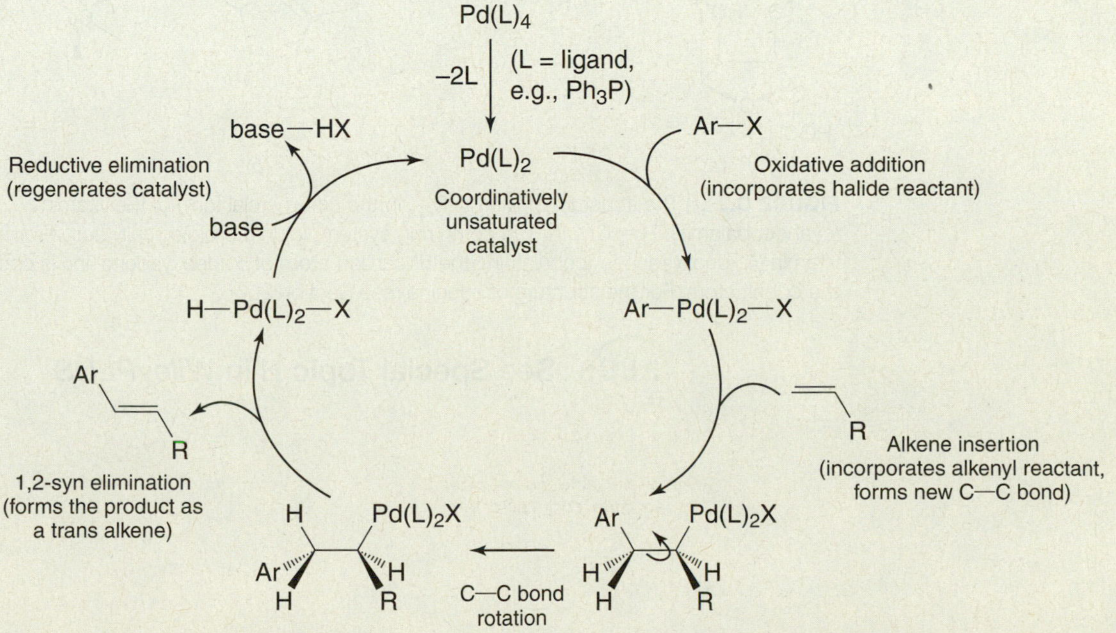

G.9 VITAMIN B$_{12}$: A TRANSITION METAL BIOMOLECULE

The discovery (in 1926) that pernicious anemia can be overcome by the ingestion of large amounts of liver led ultimately to the isolation (in 1948) of the curative factor, called vitamin B$_{12}$. The complete three-dimensional structure of vitamin B$_{12}$ [Fig. G.2(*a*)] was elucidated in 1956 through the X-ray studies of Dorothy Hodgkin (Nobel Prize, 1964), and in 1972 the synthesis of this complicated molecule was announced by R. B. Woodward (Harvard University) and A. Eschenmoser (Swiss Federal Institute of Technology). The synthesis took 11 years and involved more than 90 separate reactions. One hundred coworkers took part in the project.

Vitamin B$_{12}$ is the only known biomolecule that possesses a carbon–metal bond. In the stable commercial form of the vitamin, a cyano group is bonded to the cobalt, and the cobalt is in the +3 oxidation state. The core of the vitamin B$_{12}$ molecule is a *corrin ring* [Fig. G.2(*b*)] with various attached side groups. The corrin ring consists of four pyrrole subunits, the nitrogen of each of which is coordinated to the central cobalt. The sixth ligand [(below the corrin ring in Fig. G.2(*a*)] is a nitrogen of a heterocyclic group derived from 5,6-dimethylbenzimidazole.

The cobalt of vitamin B$_{12}$ can be reduced to a +2 or a +1 oxidation state. When the cobalt is in the +1 oxidation state, vitamin B$_{12}$ (called B$_{12s}$) becomes one of the most powerful nucleophiles known, being more nucleophilic than methanol by a factor of 10^{14}.

Acting as a nucleophile, vitamin B$_{12s}$ reacts with adenosine triphosphate (Fig. 22.2) to yield the biologically active form of the vitamin [Fig. G.2(*c*)].

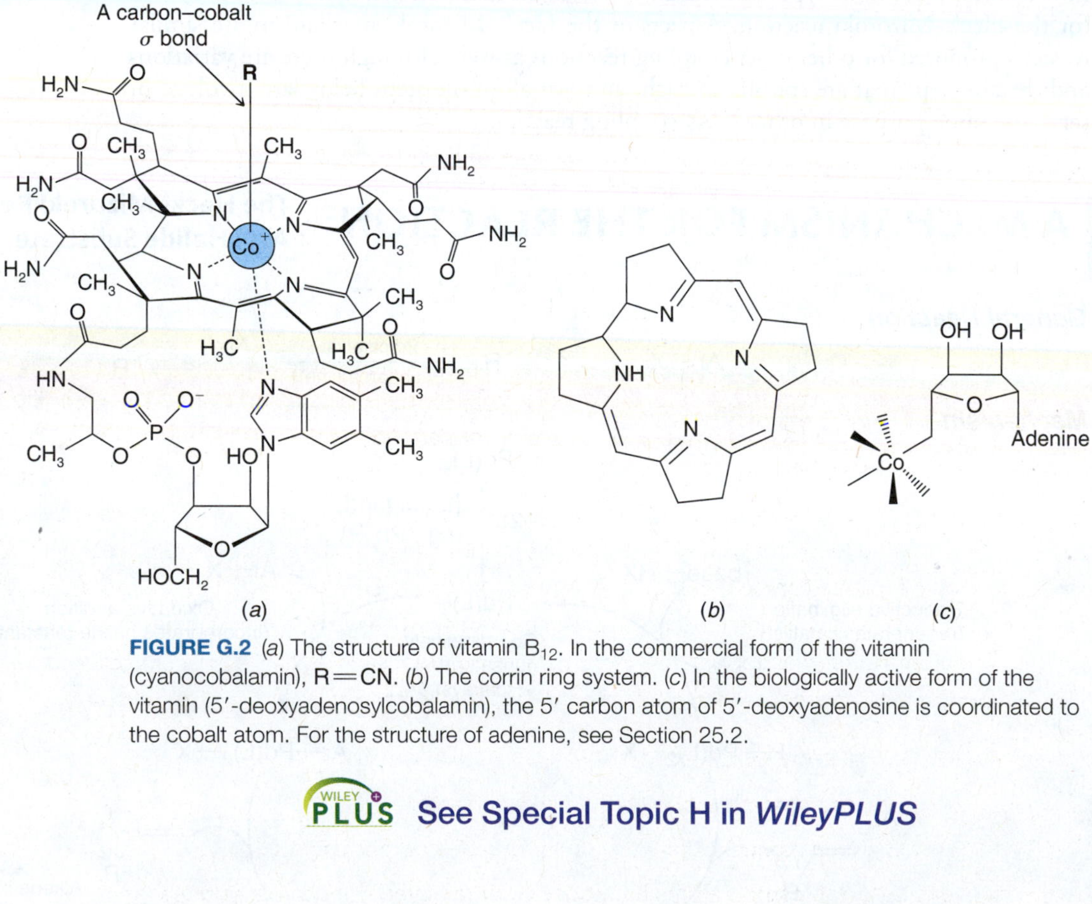

FIGURE G.2 (*a*) The structure of vitamin B$_{12}$. In the commercial form of the vitamin (cyanocobalamin), **R═CN**. (*b*) The corrin ring system. (*c*) In the biologically active form of the vitamin (5′-deoxyadenosylcobalamin), the 5′ carbon atom of 5′-deoxyadenosine is coordinated to the cobalt atom. For the structure of adenine, see Section 25.2.

PLUS   See Special Topic H in *WileyPLUS*

CHAPTER

22

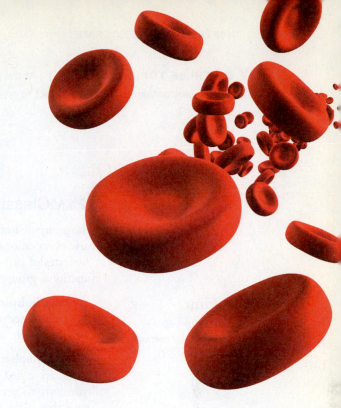

Carbohydrates

Molecules in which most carbon atoms formally have a molecule of water attached in the form of an H and an OH are known as carbohydrates, for hydrated carbon. They are also sometimes called saccharides. But what is most important about this significant group of organic compounds is that they come in many different forms and have an incredible range of properties. Nearly all carbohydrates, such as sucrose (normal table sugar), taste sweet, and are critical to our perception and enjoyment of the foods that we eat. Carbohydrates also serve as stores of chemical energy in our bodies, determine our blood type, and in plants can be united to make important fibers like cellulose and amylose. As we will see later in the chapter, they also can serve as critical molecules in the form of sialyl Lewis^x for the recognition and healing of traumatized tissue. Sometimes, atoms other than oxygen are part of carbohydrates, such as the nitrogen of amines; some of these materials, such as glucosamine, are believed to have the ability to modulate joint pain.

IN THIS CHAPTER WE WILL CONSIDER:

- the structures and properties of different carbohydrates
- reactions by which monosaccharides join to form di- and polysaccharides
- reactions by which carbon atoms are added to or removed from carbohydrates
- the functions of selected carbohydrates

[WHY DO THESE TOPICS MATTER?] At the end of this chapter we will show how chemists have used the structure of a unique glucose-containing natural product to treat diabetes, a disease characterized by having too much glucose in the bloodstream.

22.1 INTRODUCTION

22.1A Classification of Carbohydrates

The group of compounds known as carbohydrates received their general name because of early observations that they often have the formula $C_x(H_2O)_y$—that is, they appear to be "hydrates of carbon" as noted in the chapter opener. They are also characterized by the functional groups that they contain.

> **Helpful Hint**
>
> You may find it helpful now to review the chemistry of hemiacetals and acetals (Section 16.7).

- **Carbohydrates** are usually defined as polyhydroxy aldehydes and ketones or substances that hydrolyze to yield polyhydroxy aldehydes and ketones. They exist primarily in their hemiacetal or acetal forms (Section 16.7).

The simplest carbohydrates, those that cannot be hydrolyzed into simpler carbohydrates, are called **monosaccharides**. On a molecular basis, carbohydrates that undergo hydrolysis to produce only 2 molecules of monosaccharide are called **disaccharides**; those that yield 3 molecules of monosaccharide are called **trisaccharides**; and so on. (Carbohydrates that hydrolyze to yield 2–10 molecules of monosaccharide are sometimes called **oligosaccharides**.) Carbohydrates that yield a large number of molecules of monosaccharides (>10) are known as **polysaccharides**.

Maltose and sucrose are examples of disaccharides. On hydrolysis, 1 mol of maltose yields 2 mol of the monosaccharide glucose; sucrose undergoes hydrolysis to yield 1 mol of glucose and 1 mol of the monosaccharide fructose. Starch and cellulose are examples of polysaccharides; both are glucose polymers. Hydrolysis of either yields a large number of glucose units. The following shows these hydrolyses in a schematic way:

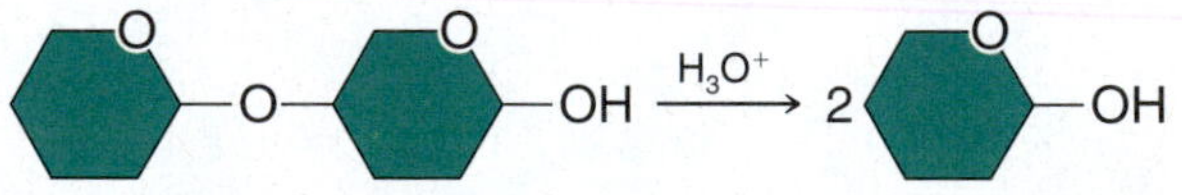

1 mol of maltose
A disaccharide

2 mol of glucose
A monosaccharide

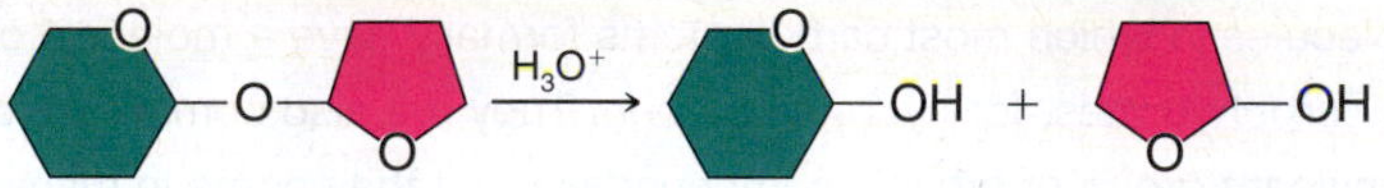

1 mol of sucrose
A disaccharide

1 mol of glucose + 1 mol of fructose
Monosaccharides

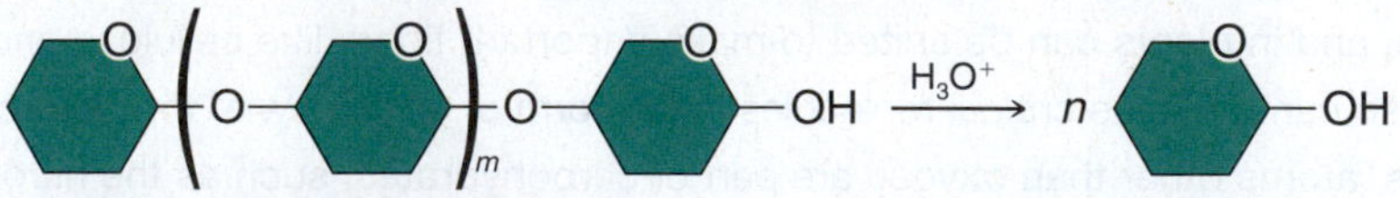

1 mol of starch or
1 mol of cellulose
Polysaccharides

many moles of glucose
Monosaccharides

Carbohydrates are the most abundant organic constituents of plants. They not only serve as an important source of chemical energy for living organisms (sugars and starches are important in this respect), but also in plants and in some animals they serve as important constituents of supporting tissues (this is the primary function of the cellulose found in wood, cotton, and flax, for example).

We encounter carbohydrates at almost every turn of our daily lives. The paper on which this book is printed is largely cellulose; so, too, is the cotton of our clothes and the wood of our houses. The flour from which we make bread is mainly starch, and starch is also a major constituent of many other foodstuffs, such as potatoes, rice, beans, corn, and

peas. Carbohydrates are central to metabolism, and they are important for cell recognition (see the chapter opening vignette and Section 22.16).

22.1B Photosynthesis and Carbohydrate Metabolism

Carbohydrates are synthesized in green plants by *photosynthesis*—a process that uses solar energy to reduce, or "fix," carbon dioxide. Photosynthesis in algae and higher plants occurs in cell organelles called chloroplasts. The overall equation for photosynthesis can be written as follows:

$$x\,CO_2 + y\,H_2O + \text{solar energy} \rightarrow C_x(H_2O)_y + x\,O_2$$
Carbohydrate

Schematic diagram of a chloroplast from corn. (Reprinted with permission of John Wiley & Sons, Inc., from Voet, D. and Voet, J. G., *Biochemistry*, Second Edition. © 1995 Voet, D. and Voet, J. G.)

Many individual enzyme-catalyzed reactions take place in the general photosynthetic process and not all are fully understood. We know, however, that photosynthesis begins with the absorption of light by the important green pigment of plants, chlorophyll (Fig. 22.1). The green color of chlorophyll and, therefore, its ability to absorb sunlight in the visible region are due primarily to its extended conjugated system. As photons of sunlight are trapped by chlorophyll, energy becomes available to the plant in a chemical form that can be used to carry out the reactions that reduce carbon dioxide to carbohydrates and oxidize water to oxygen.

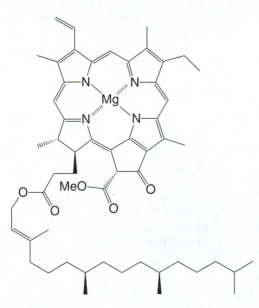

FIGURE 22.1 Chlorophyll a. [The structure of chlorophyll a was established largely through the work of H. Fischer (Munich), R. Willstätter (Munich), and J. B. Conant (Harvard). A synthesis of chlorophyll a from simple organic compounds was achieved by R. B. Woodward (Harvard) in 1960, who won the Nobel prize in 1965 for his outstanding contributions to synthetic organic chemistry.]

Carbohydrates act as a major chemical repository for solar energy. Their energy is released when animals or plants metabolize carbohydrates to carbon dioxide and water:

$$C_x(H_2O)_y + x\,O_2 \rightarrow x\,CO_2 + y\,H_2O + \text{energy}$$

The metabolism of carbohydrates also takes place through a series of enzyme-catalyzed reactions in which each energy-yielding step is an oxidation (or the consequence of an oxidation).

Although some of the energy released in the oxidation of carbohydrates is inevitably converted to heat, much of it is conserved in a new chemical form through reactions that are coupled to the synthesis of adenosine triphosphate (ATP) from adenosine diphosphate (ADP) and inorganic phosphate (P_i) (Fig. 22.2). The phosphoric anhydride bond that forms between the terminal phosphate group of ADP and the phosphate ion becomes another repository of chemical energy. Plants and animals can use the conserved energy of ATP (or very similar substances) to carry out all of their energy-requiring processes: the contraction of a muscle, the synthesis of a macromolecule, and so on. When the energy in ATP is used, a coupled reaction takes place in which ATP is hydrolyzed,

$$ATP + H_2O \rightarrow ADP + P_i + \text{energy}$$

or a new anhydride linkage is created,

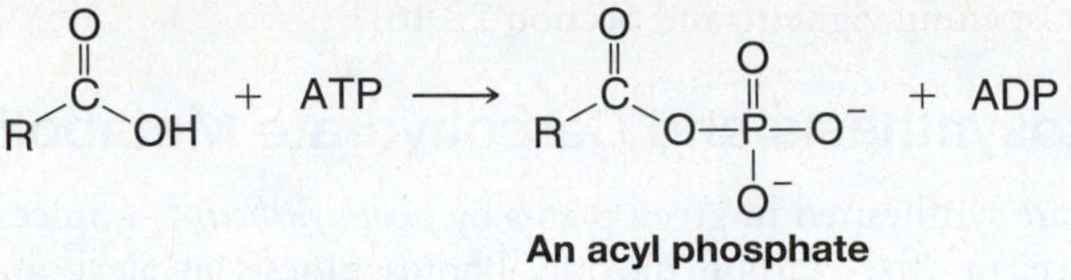

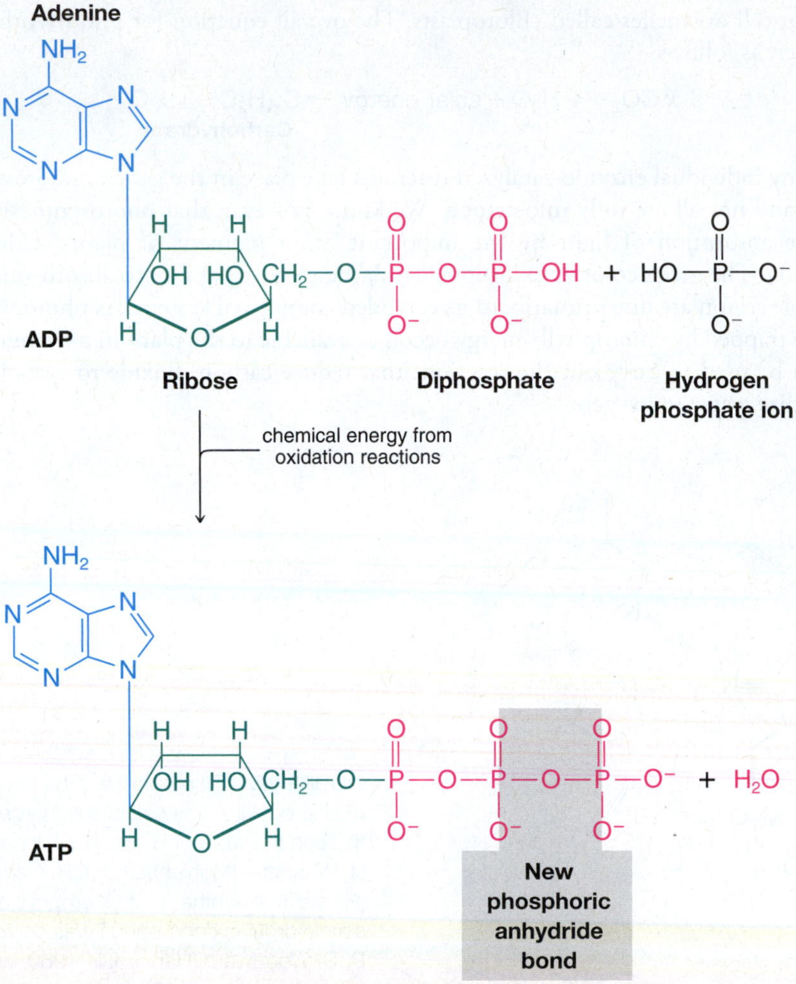

FIGURE 22.2 The synthesis of adenosine triphosphate (ATP) from adenosine diphosphate (ADP) and hydrogen phosphate ion. This reaction takes place in all living organisms, and adenosine triphosphate is the major compound into which the chemical energy released by biological oxidations is transformed.

22.2 MONOSACCHARIDES

22.2A Classification of Monosaccharides

Monosaccharides are classified according to (1) the number of carbon atoms present in the molecule and (2) whether they contain an aldehyde or keto group. Thus, a monosaccharide containing three carbon atoms is called a *triose*; one containing four carbon atoms is called a *tetrose*; one containing five carbon atoms is a *pentose*; and one containing six carbon atoms is a *hexose*. A monosaccharide containing an aldehyde group is called an **aldose**; one containing a keto group is called a **ketose**. These two classifications are frequently combined. A C_4 aldose, for example, is called an *aldotetrose*; a C_5 ketose is called a *ketopentose*.

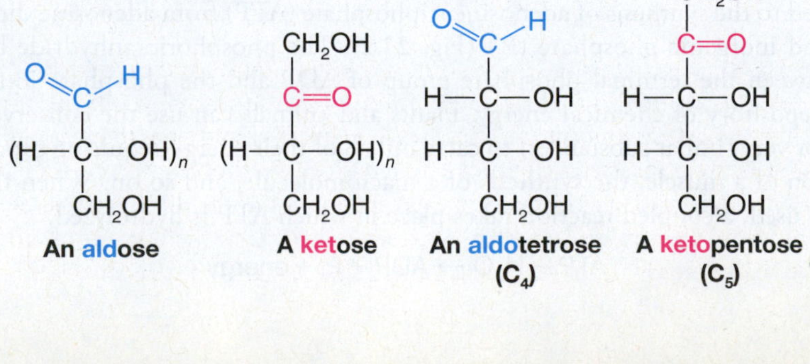

PRACTICE PROBLEM 22.1

How many chirality centers are contained in **(a)** the aldotetrose and **(b)** the ketopentose just given? **(c)** How many stereoisomers would you expect from each general structure?

22.2B D and L Designations of Monosaccharides

The simplest monosaccharides are the compounds glyceraldehyde and dihydroxyacetone (see the following structures). Of these two compounds, only glyceraldehyde contains a chirality center.

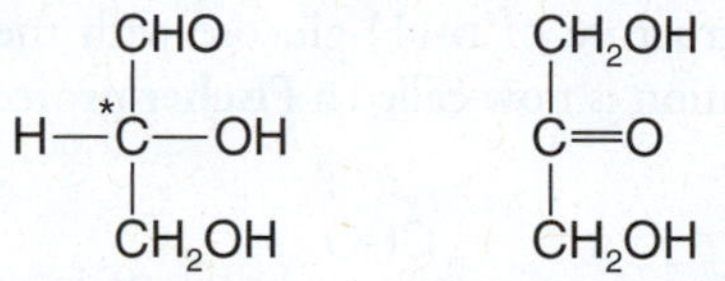

$$
\begin{array}{cc}
\text{CHO} & \text{CH}_2\text{OH} \\
| & | \\
\text{H—}\overset{*}{\text{C}}\text{—OH} & \text{C=O} \\
| & | \\
\text{CH}_2\text{OH} & \text{CH}_2\text{OH} \\
\textbf{Glyceraldehyde} & \textbf{Dihydroxyacetone} \\
\textbf{(an aldotriose)} & \textbf{(a ketotriose)}
\end{array}
$$

Glyceraldehyde exists, therefore, in two enantiomeric forms that are known to have the absolute configurations shown here:

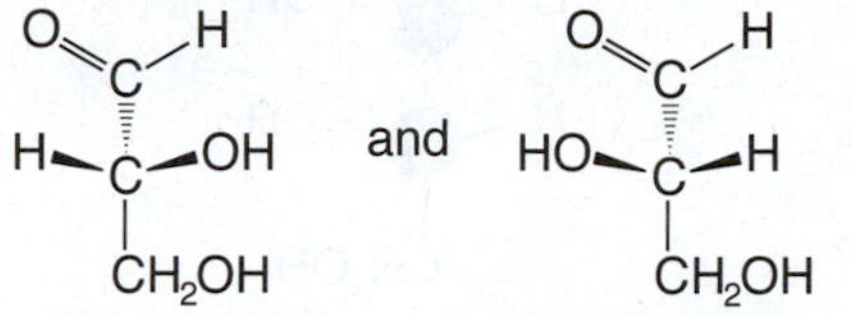

(+)-Glyceraldehyde   and   **(−)-Glyceraldehyde**

We saw in Section 5.7 that, according to the Cahn–Ingold–Prelog convention, (+)-glyceraldehyde should be designated (*R*)-(+)-glyceraldehyde and (−)-glyceraldehyde should be designated (*S*)-(−)-glyceraldehyde.

Early in the twentieth century, before the absolute configurations of any organic compounds were known, another system of stereochemical designations was introduced. According to this system (first suggested by M. A. Rosanoff of New York University in 1906), (+)-glyceraldehyde is designated D-(+)-glyceraldehyde and (−)-glyceraldehyde is designated L-(−)-glyceraldehyde. These two compounds, moreover, serve as configurational standards for all monosaccharides. A monosaccharide *whose highest numbered chirality center* (the penultimate carbon) has the same configuration as D-(+)-glyceraldehyde is designated as a D sugar; one whose highest numbered chirality center has the same configuration as L-glyceraldehyde is designated as an L sugar. By convention, acyclic forms of monosaccharides are drawn vertically with the aldehyde or keto group at or nearest the top. When drawn in this way, D sugars have the —OH on their penultimate carbon on the right:

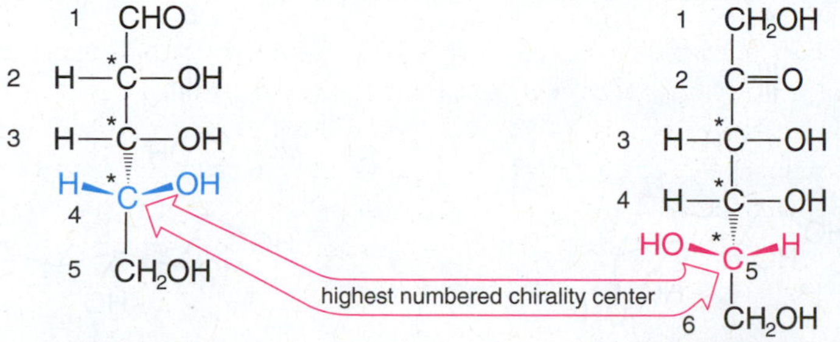

A D-aldopentose   highest numbered chirality center   **An L-ketohexose**

The **D and L nomenclature** designations are like (*R*) and (*S*) designations in that they are not necessarily related to the optical rotations of the sugars to which they are applied. Thus, one may encounter other sugars that are D-(+) or D-(−) and ones that are L-(+) or L-(−).

The D–L system of stereochemical designations is thoroughly entrenched in the literature of carbohydrate chemistry, and even though it has the disadvantage of specifying the configuration of only one chirality center—that of the highest numbered chirality center—we shall employ the D–L system in our designations of carbohydrates.

PRACTICE PROBLEM 22.2 Write three-dimensional formulas for each aldotetrose and ketopentose isomer in Practice Problem 22.1 and designate each as a D or L sugar.

22.2C Structural Formulas for Monosaccharides

Later in this chapter we shall see how the great carbohydrate chemist Emil Fischer* was able to establish the stereochemical configuration of the aldohexose D-(+)-glucose, the most abundant monosaccharide. In the meantime, however, we can use D-(+)-glucose as an example illustrating the various ways of representing the structures of monosaccharides.

Fischer represented the structure of D-(+)-glucose with the cross formulation **(1)** in Fig. 22.3. This type of formulation is now called a **Fischer projection** (Section 5.13) and is

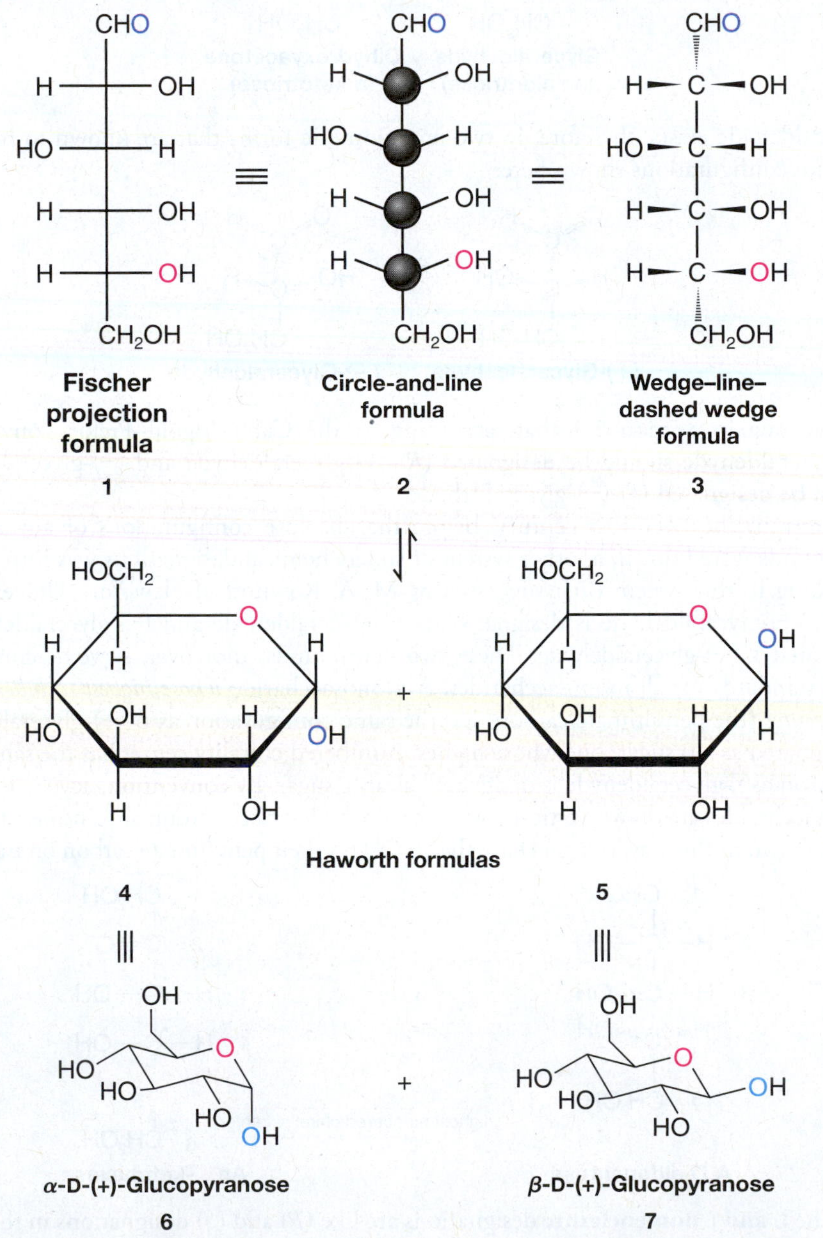

Fischer projection formula

1

Circle-and-line formula

2

Wedge–line–dashed wedge formula

3

Haworth formulas

4

+

5

α-D-(+)-Glucopyranose

6

+

β-D-(+)-Glucopyranose

7

FIGURE 22.3 Formulas 1–3 are used for the open-chain structure of D-(+)-glucose. Formulas 4–7 are used for the two cyclic hemiacetal forms of D-(+)-glucose.

*Emil Fischer (1852–1919) was professor of organic chemistry at the University of Berlin. In addition to monumental work in the field of carbohydrate chemistry, where Fischer and co-workers established the configuration of most of the monosaccharides, Fischer also made important contributions to studies of amino acids, proteins, purines, indoles, and stereochemistry generally. As a graduate student, Fischer discovered phenylhydrazine, a reagent that was highly important in his later work with carbohydrates. Fischer was the second recipient (in 1902) of the Nobel Prize in Chemistry.

still useful for carbohydrates. In Fischer projections, by convention, *horizontal lines project out toward the reader and vertical lines project behind the plane of the page. When we use Fischer projections, however, we must not* (in our mind's eye) *remove them from the plane of the page in order to test their superposability and we must not rotate them by 90°.* In terms of more familiar formulations, the Fischer projection translates into formulas **6** and **7**. In IUPAC nomenclature and with the Cahn–Ingold–Prelog system of stereochemical designations, the open-chain form of D-(+)-glucose is (2*R*,3*S*,4*R*,5*R*)-2,3,4,5,6-pentahydroxyhexanal.

The meaning of formulas **1**, **2**, and **3** can be seen best through the use of molecular models: we first construct a chain of six carbon atoms with the —CHO group at the top and a —CH$_2$OH group at the bottom. We then bring the CH$_2$OH group up behind the chain until it almost touches the —CHO group. Holding this model so that the —CHO and —CH$_2$OH groups are directed generally away from us, we then begin placing —H and —OH groups on each of the four remaining carbon atoms. The —OH group of C2 is placed on the right; that of C3 on the left; and those of C4 and C5 on the right.

Although many of the properties of D-(+)-glucose can be explained in terms of an open-chain structure (**1**, **2**, or **3**), a considerable body of evidence indicates that the open-chain structure exists, primarily, in equilibrium with two cyclic forms. These can be represented by structures **4** and **5** or **6** and **7**. The cyclic forms of D-(+)-glucose are **hemiacetals** formed by an intramolecular reaction of the —OH group at C5 with the aldehyde group (Fig. 22.4). Cyclization creates a new chirality center at C1, and this chirality center explains how two cyclic forms are possible. These two cyclic forms are *diastereomers* that differ only in the configuration of C1.

- In carbohydrate chemistry diastereomers differing only at the hemiacetal or acetal carbon are called **anomers**, and the hemiacetal or acetal carbon atom is called the **anomeric carbon** atom.

Structures **4** and **5** for the glucose anomers are called **Haworth formulas*** and, although they do not give an accurate picture of the shape of the six-membered ring, they have many practical uses. Figure 22.4 demonstrates how the representation of each chirality center of the open-chain form can be correlated with its representation in the Haworth formula.

Each glucose anomer is designated as an **α anomer** or a **β anomer** depending on the location of the —OH group of C1. When we draw the cyclic forms of a D sugar in the orientation shown in Figs. 22.3 or 22.4, the α anomer has the —OH trans to the —CH$_2$OH group and the β anomer has the —OH cis to the —CH$_2$OH group.

Studies of the structures of the cyclic hemiacetal forms of D-(+)-glucose using X-ray analysis have demonstrated that the actual conformations of the rings are the chair forms represented by conformational formulas **6** and **7** in Fig. 22.3. This shape is exactly what we would expect from our studies of the conformations of cyclohexane (Chapter 4), and it is especially interesting to notice that in the β anomer of D-glucose all of the large substituents, —OH and —CH$_2$OH, are equatorial. In the α anomer, the only bulky axial substituent is the —OH at C1.

It is convenient at times to represent the cyclic structures of a monosaccharide without specifying whether the configuration of the anomeric carbon atom is α or β. When we do this, we shall use formulas such as the following:

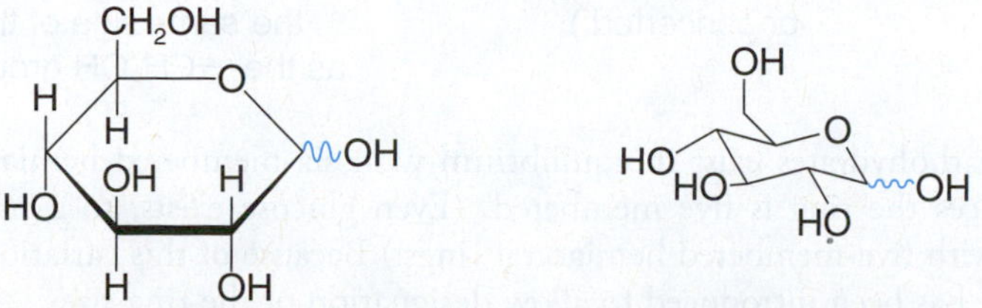

The symbol ∿ indicates α or β (three-dimensional view not specified).

*Haworth formulas are named after the English chemist W. N. Haworth (University of Birmingham), who, in 1926, along with E. L. Hirst, demonstrated that the cyclic form of glucose acetals consists of a six-membered ring. Haworth received the Nobel Prize for his work in carbohydrate chemistry in 1937. For an excellent discussion of Haworth formulas and their relation to open-chain forms, see "The Conversion of Open Chain Structures of Monosaccharides into the Corresponding Haworth Formulas," Wheeler, D. M. S., Wheeler, M. M., and Wheeler, T. S., *J. Chem. Educ.* **1982**, *59*, 969–970.

Helpful Hint

Use molecular models to help you learn to interpret Fischer projection formulas.

Helpful Hint

α and β also find common use in steroid nomenclature (Section 23.4A).

FIGURE 22.4 Haworth formulas for the cyclic hemiacetal forms of D-(+)-glucose and their relation to the open-chain polyhydroxy aldehyde structure. (Reprinted with permission of John Wiley & Sons, Inc., from Holum, J. R., *Organic Chemistry: A Brief Course*, p. 316. Copyright 1975.)

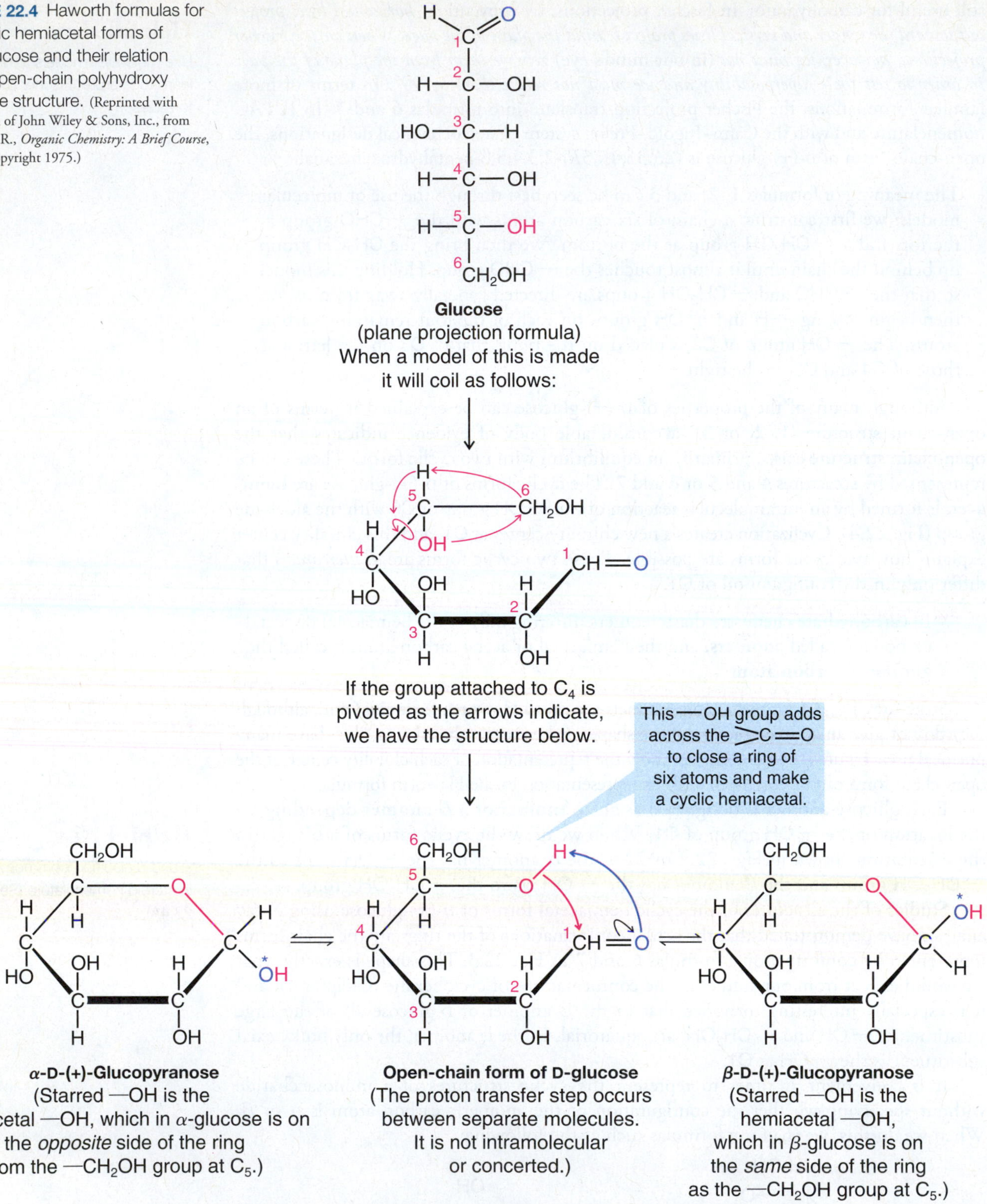

Not all carbohydrates exist in equilibrium with six-membered hemiacetal rings; in several instances the ring is five membered. (Even glucose exists, to a small extent, in equilibrium with five-membered hemiacetal rings.) Because of this variation, a system of nomenclature has been introduced to allow designation of the ring size.

● If the monosaccharide ring is six membered, the compound is called a **pyranose**; if the ring is five membered, the compound is designated as a **furanose**.

These names come from the names of the oxygen heterocycles *pyran* and *furan* + *ose*:

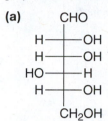

A pyran   **Furan**

Thus, the full name of compound **4** (or **6**) is α-D-(+)-glucopyranose, while that of **5** (or **7**) is β-D-(+)-glucopyranose.

PRACTICE PROBLEM 22.3

Draw the β-pyranose form of **(a)** in its lowest energy chair conformation, and a Fischer projection for **(b)**.

(a)

```
        CHO
   H ───── OH
   H ───── OH
  HO ───── H
   H ───── OH
       CH2OH
```

(b)

```
   OH  OH
          O
            OH
      HO
   HO      OH
```

22.3 MUTAROTATION

Part of the evidence for the cyclic hemiacetal structure for D-(+)-glucose comes from experiments in which both α and β forms have been isolated. Ordinary D-(+)-glucose has a melting point of 146 °C. However, when D-(+)-glucose is crystallized by evaporating an aqueous solution kept above 98 °C, a second form of D-(+)-glucose with a melting point of 150 °C can be obtained. When the optical rotations of these two forms are measured, they are found to be significantly different, but when an aqueous solution of either form is allowed to stand, its rotation changes. The specific rotation of one form decreases and the rotation of the other increases, *until both solutions show the same value*. A solution of ordinary D-(+)-glucose (mp 146 °C) has an initial specific rotation of +112, but, ultimately, the specific rotation of this solution falls to +52.7. A solution of the second form of D-(+)-glucose (mp 150 °C) has an initial specific rotation of +18.7, but, slowly, the specific rotation of this solution rises to +52.7.

- This change in specific rotation toward an equilibrium value is called **mutarotation**.

The explanation for this mutarotation lies in the existence of an equilibrium between the open-chain form of D-(+)-glucose and the α and β forms of the cyclic hemiacetals:

```
   OH                    O   H                    OH
          O          H ───── OH                          O
  HO                 HO ───── H             HO                  OH
   HO                 H ───── OH             HO
      HO              H ───── OH                HO
         OH                 ─── OH
```

α-D-(+)-Glucopyranose   Open-chain   β-D-(+)-Glucopyranose
(mp 146 °C; $[\alpha]_D^{25}$ = +112)   form of   (mp 150 °C; $[\alpha]_D^{25}$ = +18.7)
                                   D-(+)-glucose

X-ray analysis has confirmed that ordinary D-(+)-glucose has the α configuration at the anomeric carbon atom and that the higher melting form has the β configuration.

The concentration of open-chain D-(+)-glucose in solution at equilibrium is very small. Solutions of D-(+)-glucose give no observable UV or IR absorption band for a carbonyl group, and solutions of D-(+)-glucose give a negative test with Schiff's reagent— a special reagent that requires a relatively high concentration of a free aldehyde group (rather than a hemiacetal) in order to give a positive test.

Assuming that the concentration of the open-chain form is negligible, one can, by use of the specific rotations in the preceding figures, calculate the percentages of the α and β anomers present at equilibrium. These percentages, 36% α anomer and 64% β anomer, are in accord with a greater stability for β-D-(+)-glucopyranose. This preference is what we might expect on the basis of its having only equatorial groups:

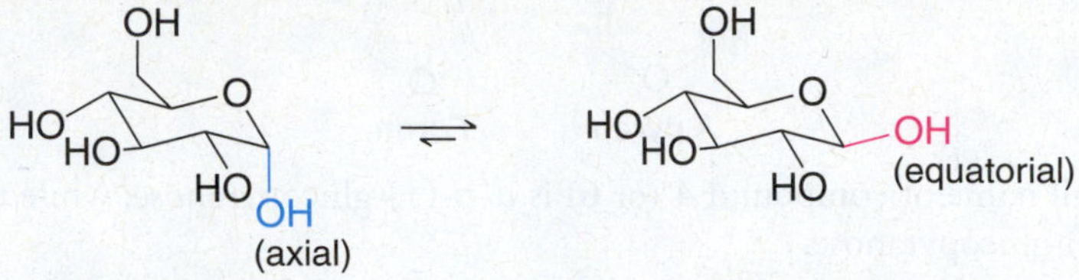

α-D-(+)-Glucopyranose
(36% at equilibrium)

β-D-(+)-Glucopyranose
(64% at equilibrium)

The β anomer of a pyranose is not always the more stable, however. With D-mannose, the equilibrium favors the α anomer, and this result is called an *anomeric effect*:

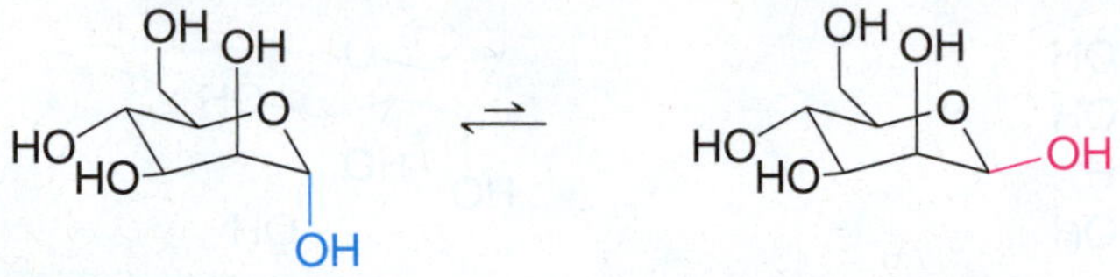

α-D-Mannopyranose
(69% at equilibrium)

β-D-Mannopyranose
(31% at equilibrium)

The anomeric effect is widely believed to be caused by hyperconjugation. An axially oriented orbital associated with nonbonding electrons of the ring oxygen can overlap with a σ^* orbital of the axial exocyclic C—O hemiacetal bond. This effect is similar to that which helps cause the lowest energy conformation of ethane to be the anti conformation (Section 4.8). An anomeric effect will frequently cause an electronegative substituent, such as a hydroxyl or alkoxyl group, to prefer the axial orientation.

22.4 GLYCOSIDE FORMATION

When a small amount of gaseous hydrogen chloride is passed into a solution of D-(+)-glucose in methanol, a reaction takes place that results in the formation of anomeric methyl *acetals*:

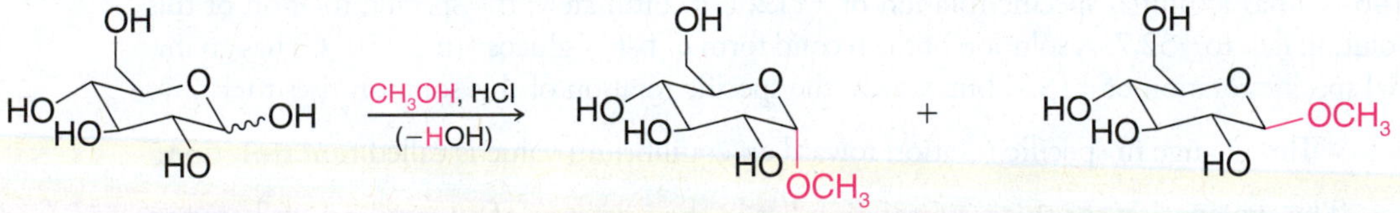

D-(+)-Glucose

Methyl α-D-glucopyranoside
(mp 165 °C; $[\alpha]_D^{25} = +158$)

Methyl β-D-glucopyranoside
(mp 107 °C; $[\alpha]_D^{25} = -33$)

- Carbohydrate acetals are generally called **glycosides** (see the following mechanism), and an acetal of glucose is called a *glucoside*. (Acetals of mannose are *mannosides*, acetals of fructose are *fructosides*, and so on.)

The methyl D-glucosides have been shown to have six-membered rings (Section 22.2C) so they are properly named methyl α-D-glucopyranoside and methyl β-D-glucopyranoside.
The mechanism for the formation of the methyl glucosides (starting arbitrarily with β-D-glucopyranose) is as follows:

[A MECHANISM FOR THE REACTION ┤ Formation of a Glycoside]

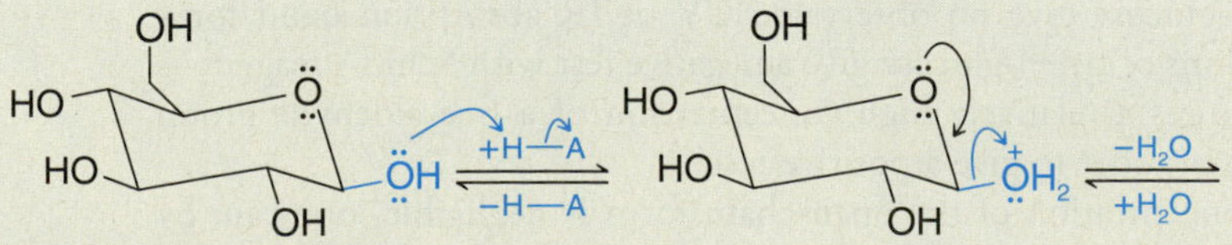

β-D-Glucopyranose

(continued on next page)

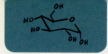

(continued from the previous page)

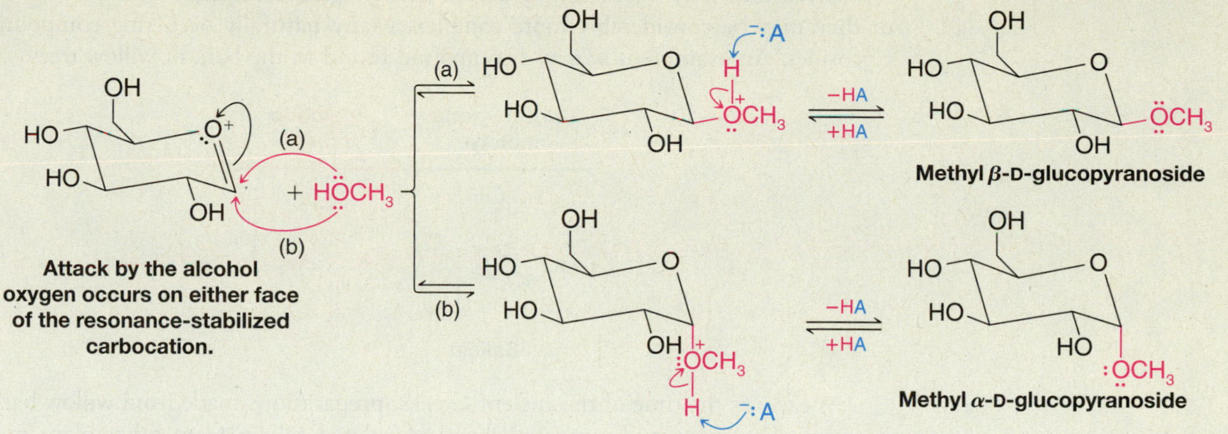

(a)

Methyl β-D-glucopyranoside

(b)

Methyl α-D-glucopyranoside

Attack by the alcohol
oxygen occurs on either face
of the resonance-stabilized
carbocation.

You should review the mechanism for acetal formation given in Section 16.7B and compare it with the steps given here. Notice, again, the important role played by the electron pair of the adjacent oxygen atom in stabilizing the carbocation that forms in the second step.

Glycosides are stable in basic solutions because they are acetals. In acidic solutions, however, glycosides undergo hydrolysis to produce a sugar and an alcohol (again, because they are acetals, Section 16.7B). The alcohol obtained by hydrolysis of a glycoside is known as an **aglycone**:

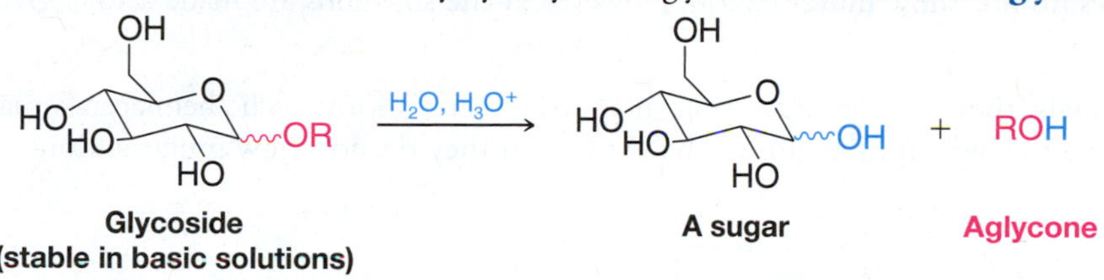

$$ \text{Glycoside} \xrightarrow{\text{H}_2\text{O, H}_3\text{O}^+} \text{A sugar} + \text{ROH} $$

Glycoside
(stable in basic solutions)   **A sugar**   **Aglycone**

For example, when an aqueous solution of methyl β-D-glucopyranoside is made acidic, the glycoside undergoes hydrolysis to produce D-glucose as a mixture of the two pyranose forms (in equilibrium with a small amount of the open-chain form).

A MECHANISM FOR THE REACTION — Hydrolysis of a Glycoside

Methyl β-D-glucopyranoside

(a)

βD-Glucopyranose

(b)

αD-Glucopyranose

(a)

(b)

Attack by water
occurs on either face of
the resonance-stabilized
carbocation.

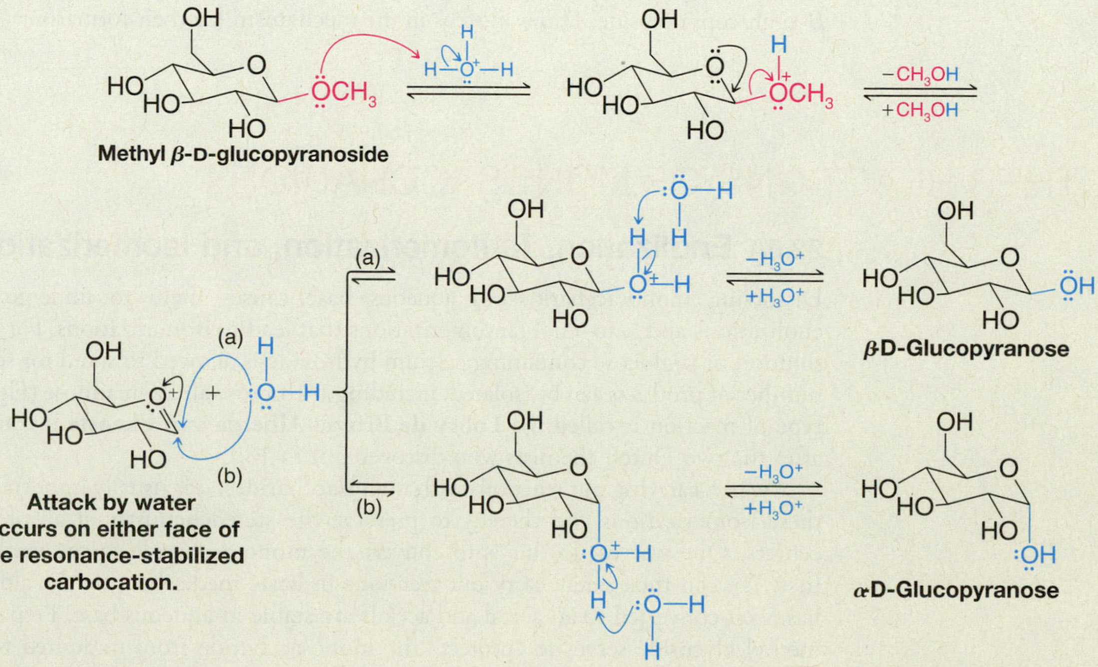

Glycosides may be as simple as the methyl glucosides that we have just studied or they may be considerably more complex. Many naturally occurring compounds are glycosides. An example is *salicin*, a compound found in the bark of willow trees:

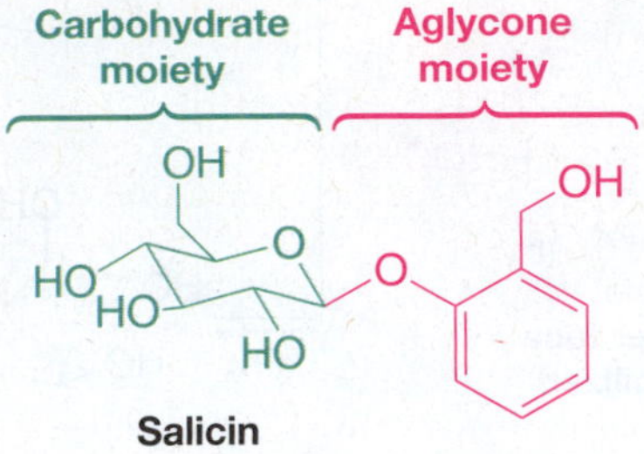

Salicin

As early as the time of the ancient Greeks, preparations made from willow bark were used in relieving pain. Eventually, chemists isolated salicin from other plant materials and were able to show that it was responsible for the analgesic effect of the willow bark preparations. Salicin can be converted to salicylic acid, which in turn can be converted into the most widely used modern analgesic, *aspirin* (Section 21.8).

●●● SOLVED PROBLEM 22.1

In neutral or basic solutions, glycosides do not show mutarotation. However, if the solutions are made acidic, glycosides show mutarotation. Explain.

ANSWER: Because glycosides are acetals, they undergo hydrolysis in aqueous acid to form cyclic hemiacetals that then undergo mutarotation. Acetals are stable to base, and therefore in basic solution they do not show mutarotation.

●●●

PRACTICE PROBLEM 22.4 **(a)** What products would be formed if salicin were treated with dilute aqueous HCl?
(b) Outline a mechanism for the reactions involved in their formation.

●●●

PRACTICE PROBLEM 22.5 How would you convert D-glucose to a mixture of ethyl α-D-glucopyranoside and ethyl β-D-glucopyranoside? Show all steps in the mechanism for their formation.

22.5 OTHER REACTIONS OF MONOSACCHARIDES

22.5A Enolization, Tautomerization, and Isomerization

Dissolving monosaccharides in aqueous base causes them to undergo a series of enolizations and keto–enol tautomerizations that lead to isomerizations. For example, if a solution of D-glucose containing calcium hydroxide is allowed to stand for several days, a number of products can be isolated, including D-fructose and D-mannose (Fig. 22.5). This type of reaction is called the **Lobry de Bruyn–Alberda van Ekenstein transformation** after the two Dutch chemists who discovered it in 1895.

When carrying out reactions with monosaccharides, it is usually important to prevent these isomerizations and thereby to preserve the stereochemistry at all of the chirality centers. One way to do this is to convert the monosaccharide to the methyl glycoside first. We can then safely carry out reactions in basic media because the aldehyde group has been converted to an acetal and acetals are stable in aqueous base. Preparation of the methyl glycoside serves to "protect" the monosaccharide from undesired reactions that could occur with the anomeric carbon in its hemiacetal form.

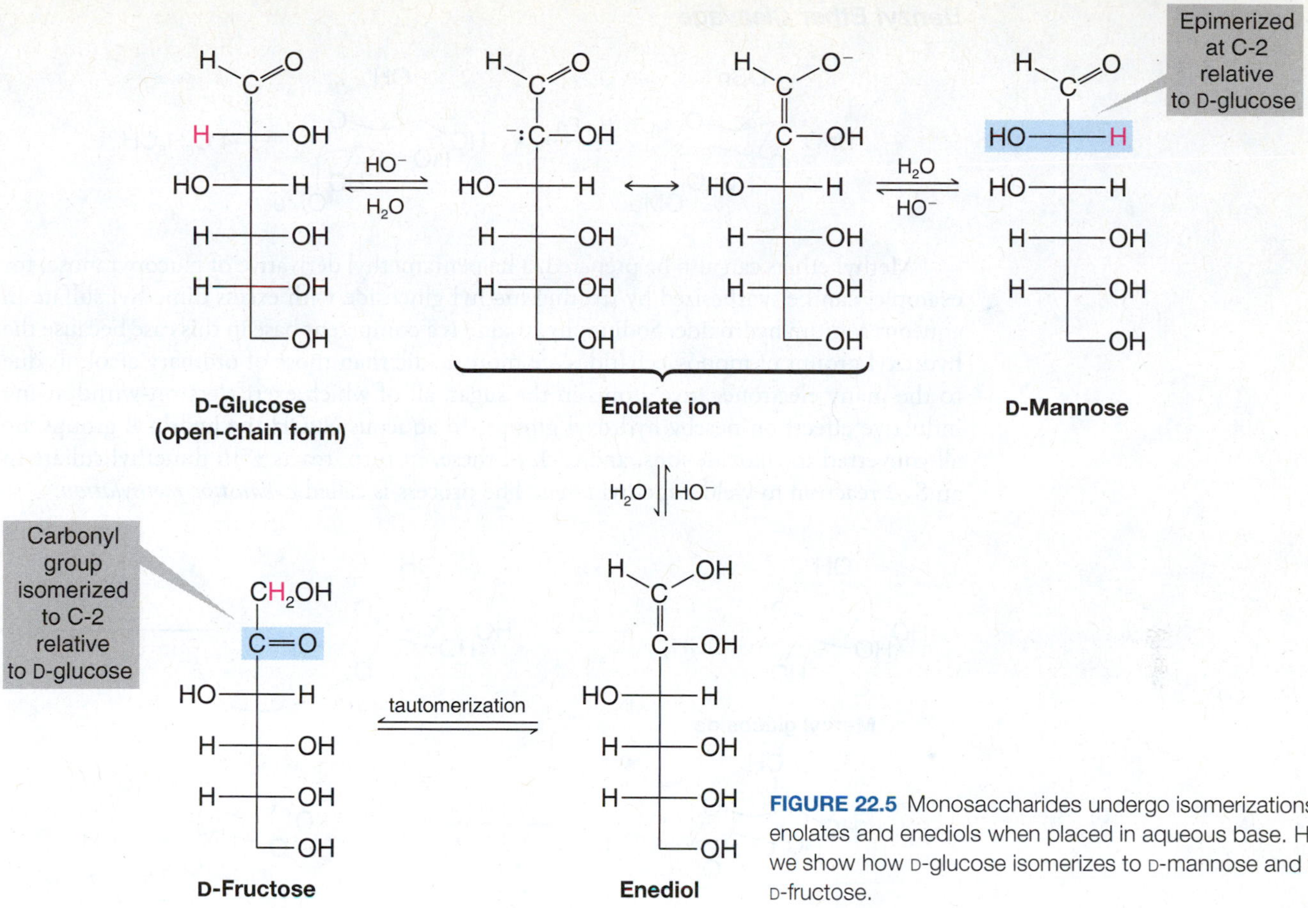

Epimerized at C-2 relative to D-glucose

D-Glucose (open-chain form)   **Enolate ion**   **D-Mannose**

Carbonyl group isomerized to C-2 relative to D-glucose

D-Fructose   tautomerization   **Enediol**

FIGURE 22.5 Monosaccharides undergo isomerizations via enolates and enediols when placed in aqueous base. Here we show how D-glucose isomerizes to D-mannose and to D-fructose.

22.5B Use of Protecting Groups in Carbohydrate Synthesis

Protecting groups are functional groups introduced selectively to block the reactivity of certain sites in a molecule while desired transformations are carried on elsewhere. After the desired transformations are accomplished, the protecting groups are removed. Laboratory reactions involving carbohydrates often require the use of protecting groups due to the multiple sites of reactivity present in carbohydrates. As we have just seen, formation of a glycoside (an acetal) can be used to prevent undesired reactions that would involve the anomeric carbon in its hemiacetal form. Common protecting groups for the alcohol functional groups in carbohydrates include ethers, esters, and acetals.

22.5C Formation of Ethers

- Hydroxyl groups of sugars can be converted to ethers using a base and an alkyl halide by a version of the Williamson ether synthesis (Section 11.11B).

Benzyl ethers are commonly used to protect hydroxyl groups in sugars. Benzyl halides are easily introduced because they are highly reactive in S_N2 reactions. Sodium or potassium hydride is typically used as the base in an aprotic solvent such as DMF or DMSO. The benzyl groups can later be easily removed by hydrogenolysis using a palladium catalyst.

Benzyl Ether Formation

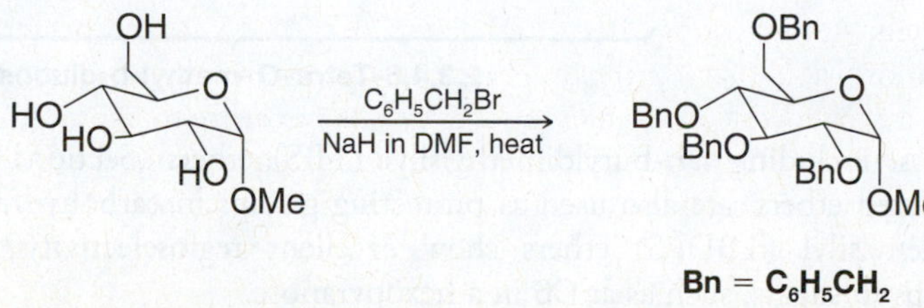

$$\xrightarrow[\text{NaH in DMF, heat}]{C_6H_5CH_2Br}$$

Bn = $C_6H_5CH_2$

Benzyl Ether Cleavage

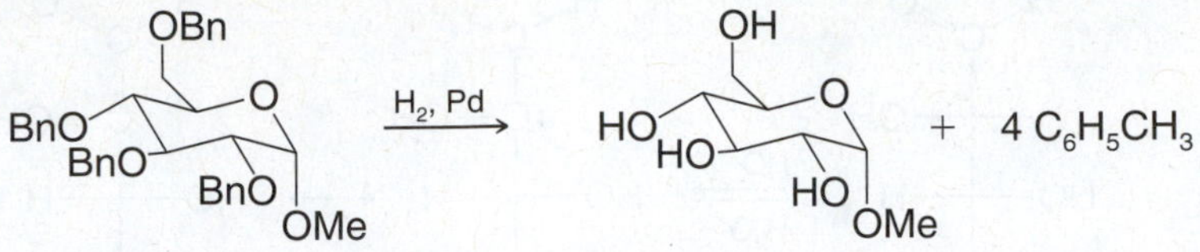

Methyl ethers can also be prepared. The pentamethyl derivative of glucopyranose, for example, can be synthesized by treating methyl glucoside with excess dimethyl sulfate in aqueous sodium hydroxide. Sodium hydroxide is a competent base in this case because the hydroxyl groups of monosaccharides are more acidic than those of ordinary alcohols due to the many electronegative atoms in the sugar, all of which exert electron-withdrawing inductive effects on nearby hydroxyl groups. In aqueous NaOH the hydroxyl groups are all converted to alkoxide ions, and each of these, in turn, reacts with dimethyl sulfate in an S_N2 reaction to yield a methyl ether. The process is called *exhaustive methylation*:

Methyl glucoside

Pentamethyl derivative

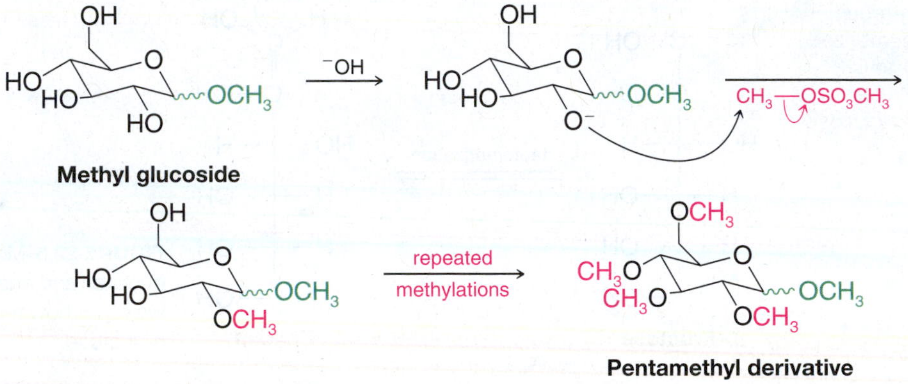

Although not often used as protecting groups for alcohols in carbohydrates, methyl ethers have been useful in the structure elucidation of sugars. For example, evidence for the pyranose form of glucose can be obtained by exhaustive methylation followed by aqueous hydrolysis of the acetal linkage. Because the C2, C3, C4, and C6 methoxy groups of the pentamethyl derivative are ethers, they are not affected by aqueous hydrolysis. (To cleave them requires heating with concentrated HBr or HI, Section 11.12.) The methoxyl group at C1, however, is part of an acetal linkage, and so it is labile under the conditions of aqueous hydrolysis. Hydrolysis of the pentamethyl derivative of glucose gives evidence that the C5 oxygen was the one involved in the cyclic hemiacetal form because in the open-chain form of the product (which is in equilibrium with the cyclic hemiacetal) it is the C5 oxygen that is not methylated:

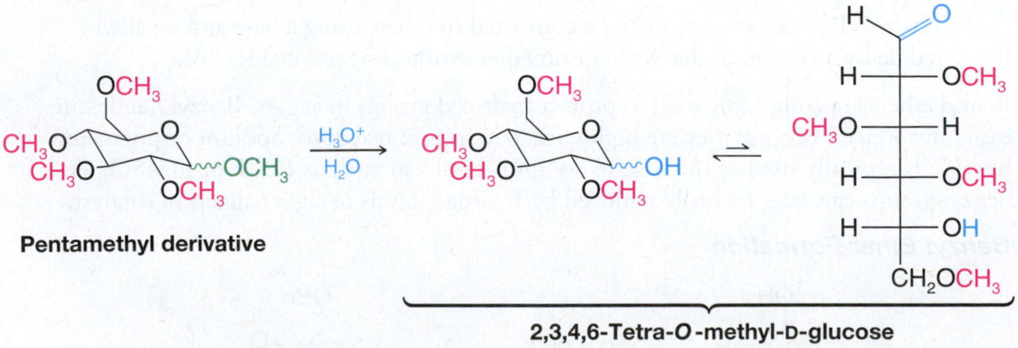

Pentamethyl derivative

2,3,4,6-Tetra-*O*-methyl-D-glucose

Silyl ethers, including *tert*-butyldimethylsilyl (TBS) ethers (Section 11.11E) and phenyl-substituted ethers, are also used as protecting groups in carbohydrate synthesis. *tert*-Butyldiphenylsilyl (TBDPS) ethers show excellent regioselectivity for primary hydroxyl groups in sugars, such as at C6 in a hexopyranose.

Regioselective TBDPS Ether Formation

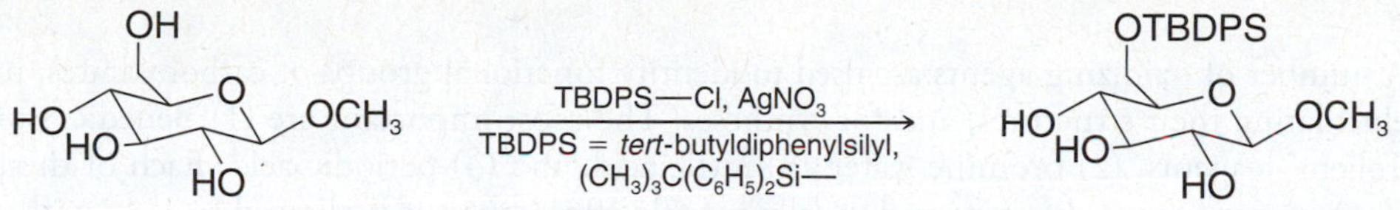

TBDPS—Cl, AgNO$_3$
TBDPS = *tert*-butyldiphenylsilyl,
(CH$_3$)$_3$C(C$_6$H$_5$)$_2$Si—

TBDPS Ether Cleavage

Bu$_4$N$^+$F$^-$
THF

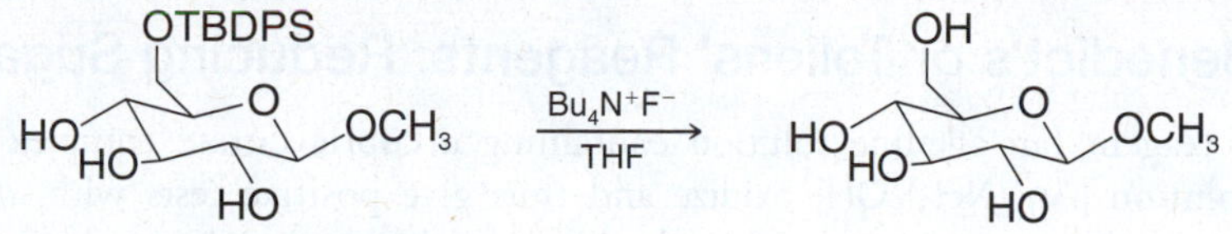

22.5D Conversion to Esters

Treating a monosaccharide with excess acetic anhydride and a weak base (such as pyridine or sodium acetate) converts all of the hydroxyl groups, including the anomeric hydroxyl, to ester groups. If the reaction is carried out at a low temperature (e.g., 0 °C), the reaction occurs stereospecifically; the α anomer gives the α-acetate and the β anomer gives the β-acetate. Acetate esters are common protecting groups for carbohydrate hydroxyls.

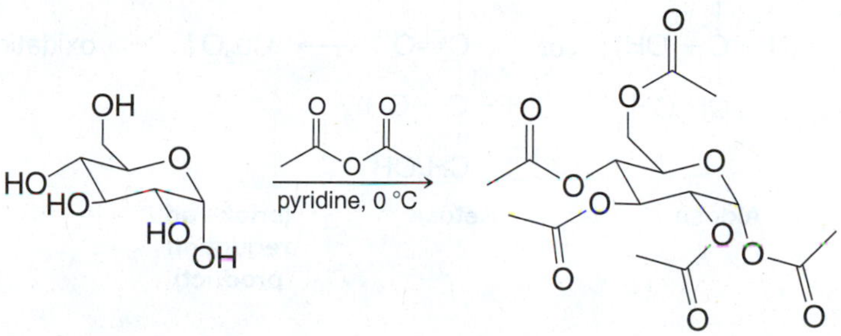

pyridine, 0 °C

22.5E Conversion to Cyclic Acetals

In Section 16.7B we learned that aldehydes and ketones react with open-chain 1,2-diols to produce **cyclic acetals**:

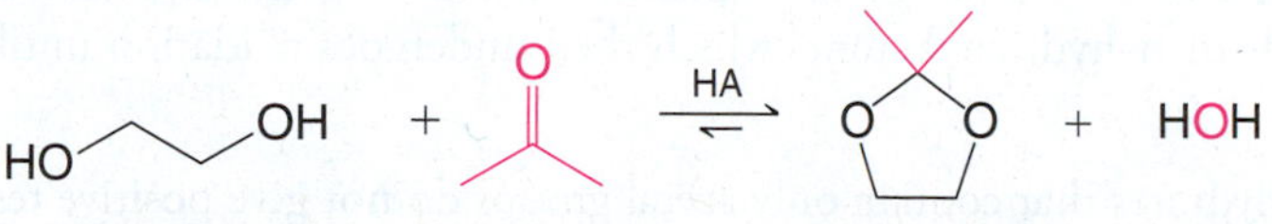

1,2-Diol   HA   **Cyclic acetal**   +   HOH

If the 1,2-diol is attached to a ring, as in a monosaccharide, **formation of the cyclic acetals occurs only when the vicinal hydroxyl groups are cis to each other.** For example, α-D-galactopyranose reacts with acetone in the following way:

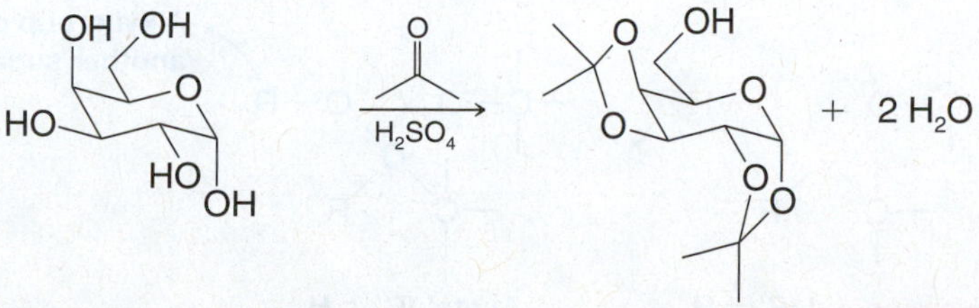

H$_2$SO$_4$ + 2 H$_2$O

Cyclic acetals are commonly used to protect vicinal cis hydroxyl groups of a sugar while reactions are carried out on other parts of the molecule. When acetals such as these are formed from acetone, they are called **acetonides**.

22.6 OXIDATION REACTIONS OF MONOSACCHARIDES

A number of oxidizing agents are used to identify functional groups of carbohydrates, in elucidating their structures, and for syntheses. The most important are (1) Benedict's or Tollens' reagents, (2) bromine water, (3) nitric acid, and (4) periodic acid. Each of these reagents produces a different and usually specific effect when it is allowed to react with a monosaccharide. We shall now examine what these effects are.

22.6A Benedict's or Tollens' Reagents: Reducing Sugars

Benedict's reagent (an alkaline solution containing a cupric citrate complex ion) and Tollens' solution $[Ag^+(NH_3)_2\bar{O}H]$ oxidize and thus give positive tests with *aldoses and ketoses*. The tests are positive even though aldoses and ketoses exist primarily as cyclic hemiacetals.

We studied the use of Tollens' silver mirror test in Section 16.13B. Benedict's solution and the related Fehling's solution (which contains a cupric tartrate complex ion) give brick-red precipitates of Cu_2O when they oxidize an aldose. [In alkaline solution ketoses are converted to aldoses (Section 22.5A), which are then oxidized by the cupric complexes.] Since the solutions of cupric tartrates and citrates are blue, the appearance of a brick-red precipitate is a vivid and unmistakable indication of a positive test.

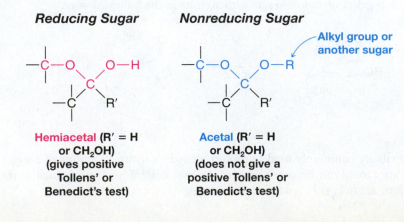

	O=C–H	CH₂OH		
Cu²⁺ (complex) +	(H–C–OH)ₙ or	C=O ⟶	Cu₂O↓ +	oxidation products
	CH₂OH	(H–C–OH)ₙ		
		CH₂OH		
Benedict's solution (blue)	**Aldose**	**Ketose**	**(brick-red reduction product)**	

- Sugars that give positive tests with Tollens' or Benedict's solutions are known as **reducing sugars**, and all carbohydrates that contain a *hemiacetal group* give positive tests.

In aqueous solution the hemiacetal form of sugars exists in equilibrium with relatively small, but not insignificant, concentrations of noncyclic aldehydes or α-hydroxy ketones. It is the latter two that undergo the oxidation, perturbing the equilibrium to produce more aldehyde or α-hydroxy ketone, which then undergoes oxidation until one reactant is exhausted.

- Carbohydrates that contain only acetal groups do not give positive tests with Benedict's or Tollens' solutions, and they are called *nonreducing sugars*.

Acetals do not exist in equilibrium with aldehydes or α-hydroxy ketones in the basic aqueous media of the test reagents.

Reducing Sugar   **Nonreducing Sugar**

Alkyl group or another sugar

Hemiacetal (R′ = H or CH₂OH) (gives positive Tollens' or Benedict's test)   **Acetal (R′ = H or CH₂OH)** (does not give a positive Tollens' or Benedict's test)

PRACTICE PROBLEM 22.6

How might you distinguish between α-D-glucopyranose (i.e., D-glucose) and methyl α-D-glucopyranoside?

Although Benedict's and Tollens' reagents have some use as diagnostic tools [Benedict's solution can be used in quantitative determinations of reducing sugars (reported as glucose) in blood or urine], neither of these reagents is useful as a preparative reagent in carbohydrate oxidations. Oxidations with both reagents take place in alkaline solution, *and in alkaline solutions sugars undergo a complex series of reactions that lead to isomerizations* (Section 22.5A).

22.6B Bromine Water: The Synthesis of Aldonic Acids

Monosaccharides do not undergo isomerization and fragmentation reactions in mildly acidic solution. Thus, a useful oxidizing reagent for preparative purposes is bromine in water (pH 6.0).

- Bromine water is a general reagent that selectively oxidizes the —CHO group to a —CO_2H group, thus converting an aldose to an **aldonic acid**:

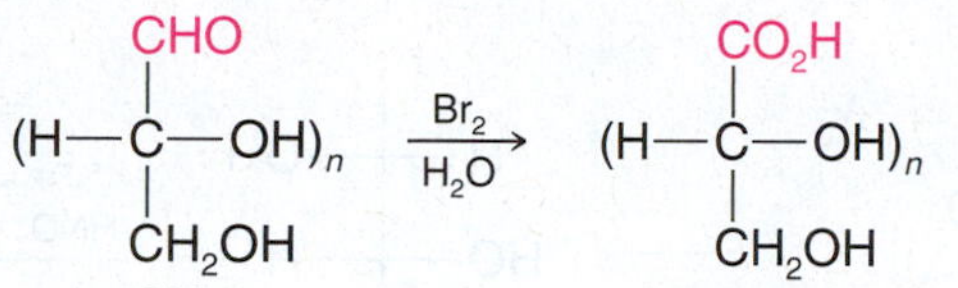

Aldose **Aldonic acid**

Experiments with aldopyranoses have shown that the actual course of the reaction is somewhat more complex than we have indicated. Bromine water specifically oxidizes the β anomer, and the initial product that forms is a *δ-aldonolactone*. This compound may then hydrolyze to an aldonic acid, and the aldonic acid may undergo a subsequent ring closure to form a *γ-aldonolactone:*

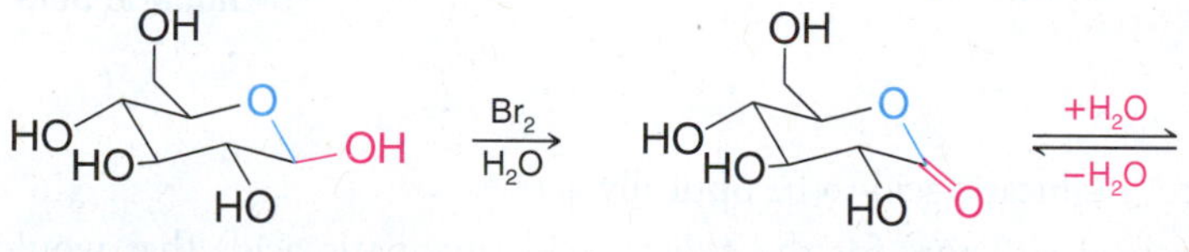

β-D-Glucopyranose **D-Glucono-δ-lactone**

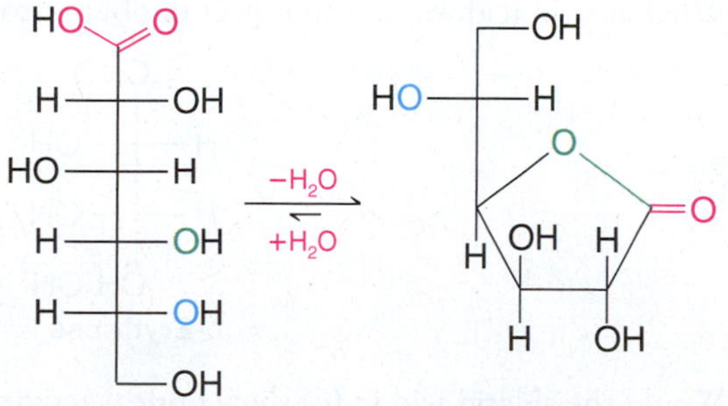

D-Gluconic acid **D-Gluconic-γ-lactone**

22.6C Nitric Acid Oxidation: Aldaric Acids

- Dilute nitric acid—a stronger oxidizing agent than bromine water—oxidizes both the —CHO group and the terminal —CH_2OH group of an aldose to —CO_2H groups, forming dicarboxylic acids are known as **aldaric acids**:

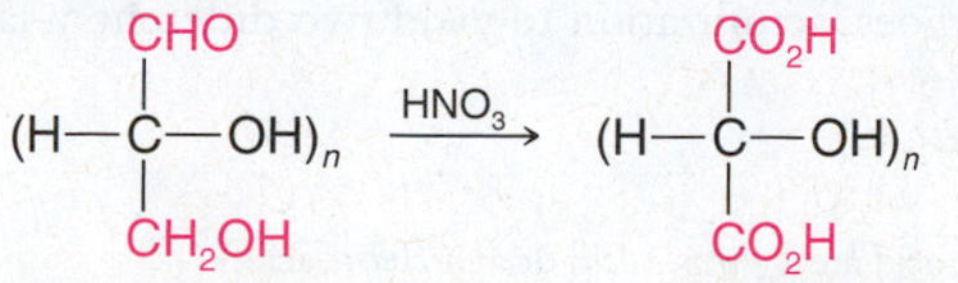

Aldose **Aldaric acid**

It is not known whether a lactone is an intermediate in the oxidation of an aldose to an aldaric acid; however, aldaric acids form γ- and δ-lactones readily:

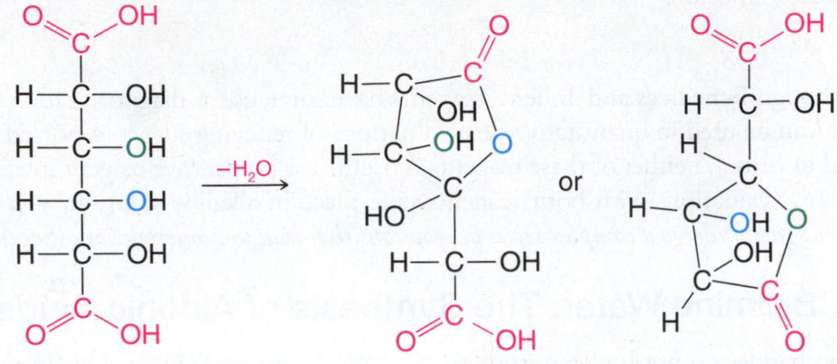

Aldaric acid (from an aldohexose) → $-H_2O$ → **γ-Lactones of an aldaric acid** or

The aldaric acid obtained from D-glucose is called D-glucaric acid*:

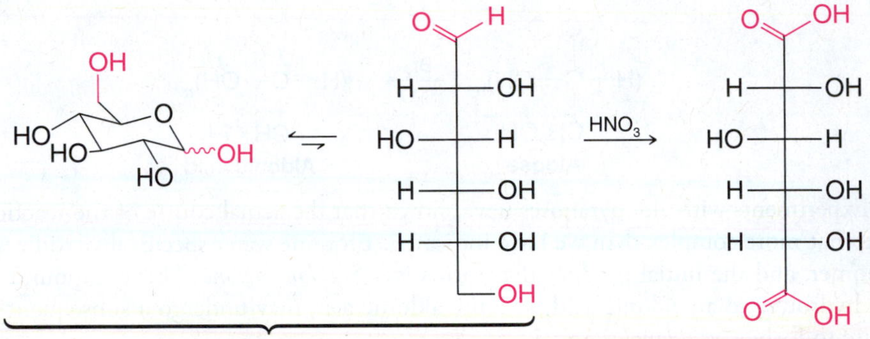

D-Glucose ⇌ → HNO_3 → **D-Glucaric acid**

• • •

PRACTICE PROBLEM 22.7

(a) Would you expect D-glucaric acid to be optically active?

(b) Write the open-chain structure for the aldaric acid (mannaric acid) that would be obtained by nitric acid oxidation of D-mannose.

(c) Would you expect mannaric acid to be optically active?

(d) What aldaric acid would you expect to obtain from D-erythrose?

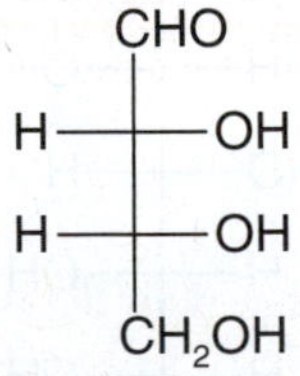

D-Erythrose

(e) Would the aldaric acid in (d) show optical activity?

(f) D-Threose, a diastereomer of D-erythrose, yields an optically active aldaric acid when it is subjected to nitric acid oxidation. Write Fischer projection formulas for D-threose and its nitric acid oxidation product.

(g) What are the names of the aldaric acids obtained from D-erythrose and D-threose?

• • •

PRACTICE PROBLEM 22.8

D-Glucaric acid undergoes lactonization to yield two different γ-lactones. What are their structures?

---

*Older terms for an aldaric acid are a *glycaric* acid or a *saccharic* acid.

22.6D Periodic Oxidations: Oxidative Cleavage of Polyhydroxy Compounds

- Compounds that have hydroxyl groups on adjacent atoms undergo oxidative cleavage when they are treated with aqueous periodic acid (HIO_4). The reaction breaks carbon–carbon bonds and produces carbonyl compounds (aldehydes, ketones, or acids).

The stoichiometry of oxidative cleavage by periodic acid is

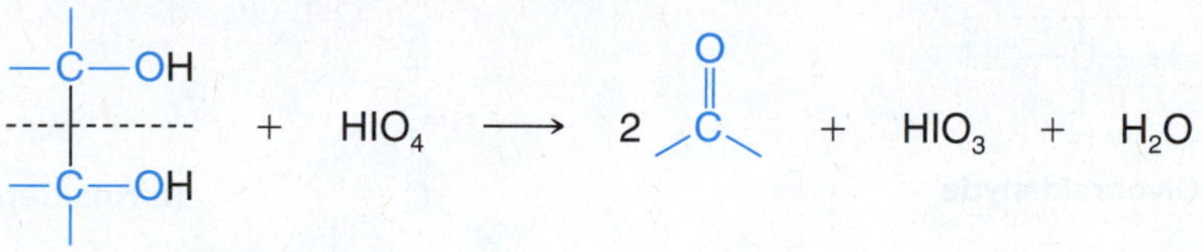

$$\begin{array}{c} -\!\!\!\overset{|}{\underset{|}{C}}\!\!-\!OH \\ ------- \\ -\!\!\!\overset{|}{\underset{|}{C}}\!\!-\!OH \end{array} + HIO_4 \longrightarrow 2 \begin{array}{c} O \\ \| \\ C \end{array} + HIO_3 + H_2O$$

Since the reaction usually takes place in quantitative yield, valuable information can often be gained by measuring the number of molar equivalents of periodic acid consumed in the reaction as well as by identifying the carbonyl products.

Periodate oxidations are thought to take place through a cyclic intermediate:

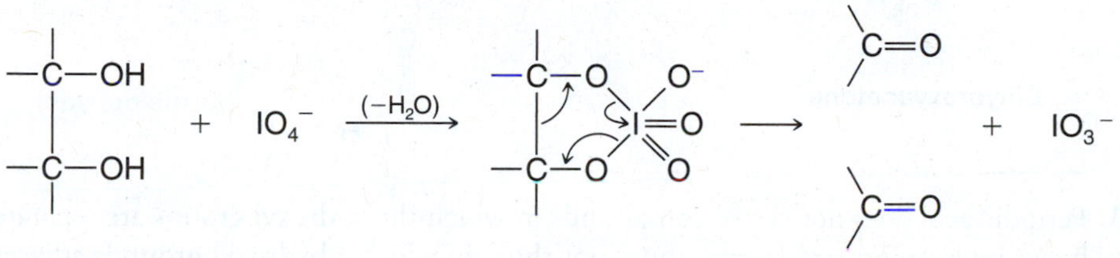

$$\begin{array}{c} -\!\!\overset{|}{C}\!\!-\!OH \\ \\ -\!\!\overset{|}{C}\!\!-\!OH \end{array} + IO_4^- \xrightarrow{(-H_2O)} \begin{array}{c} -\!\!\overset{|}{C}\!\!-\!O \quad O^- \\ \quad\quad\diagdown I \!=\! O \\ -\!\!\overset{|}{C}\!\!-\!O \quad O \end{array} \longrightarrow \begin{array}{c} \diagdown C\!=\!O \\ \diagup \\ \\ \diagdown C\!=\!O \\ \diagup \end{array} + IO_3^-$$

Before we discuss the use of periodic acid in carbohydrate chemistry, we should illustrate the course of the reaction with several simple examples. Notice in these periodate oxidations that *for every C—C bond broken, a C—O bond is formed at each carbon.*

1. When three or more —CHOH groups are contiguous, the internal ones are obtained as *formic acid*. Periodate oxidation of glycerol, for example, gives two molar equivalents of formaldehyde and one molar equivalent of formic acid:

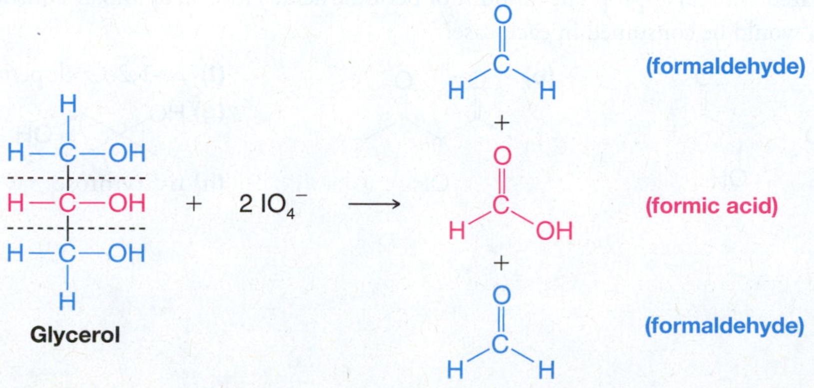

$$\begin{array}{c} H \\ | \\ H\!-\!C\!-\!OH \\ ------ \\ H\!-\!C\!-\!OH \\ ------ \\ H\!-\!C\!-\!OH \\ | \\ H \end{array} + 2\,IO_4^- \longrightarrow \begin{array}{c} O \\ \| \\ H\diagup C \diagdown H \quad \text{(formaldehyde)} \\ + \\ O \\ \| \\ H\diagup C \diagdown OH \quad \text{(formic acid)} \\ + \\ O \\ \| \\ H\diagup C \diagdown H \quad \text{(formaldehyde)} \end{array}$$

Glycerol

2. Oxidative cleavage also takes place when an —OH group is adjacent to the carbonyl group of an aldehyde or ketone (but not that of an acid or an ester). Glyceraldehyde yields two molar equivalents of formic acid and one molar equivalent of formaldehyde,

while dihydroxyacetone gives two molar equivalents of formaldehyde and one molar equivalent of carbon dioxide:

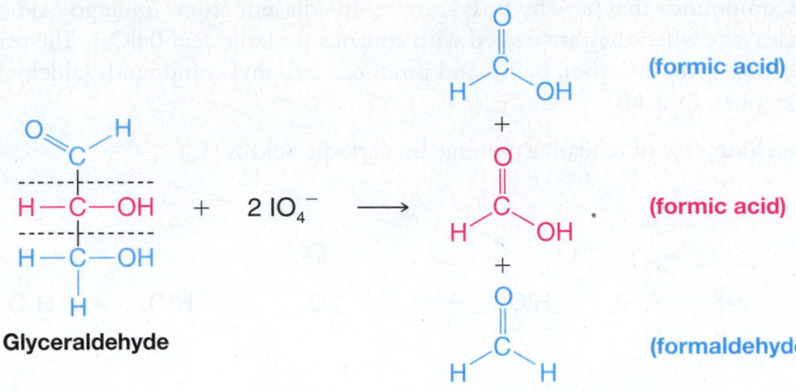

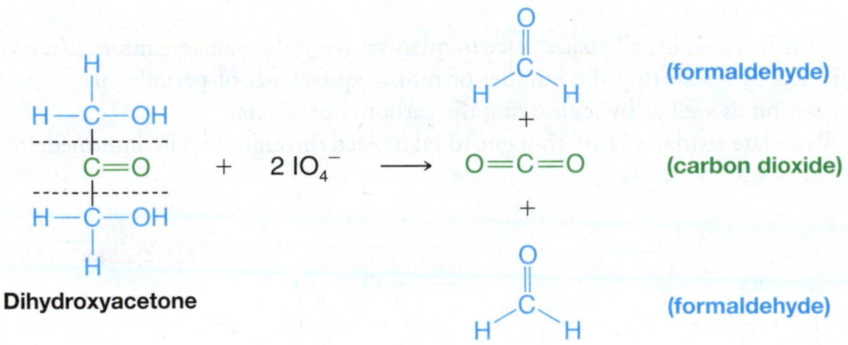

3. Periodic acid does not cleave compounds in which the hydroxyl groups are separated by an intervening $—CH_2—$ group, nor those in which a hydroxyl group is adjacent to an ether or acetal function:

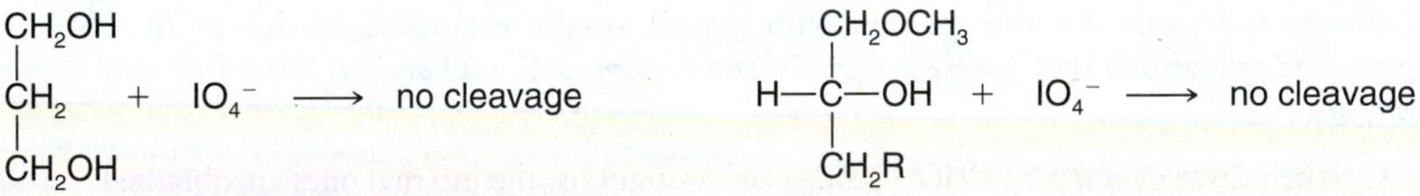

• • •

PRACTICE PROBLEM 22.9 What products would you expect to be formed when each of the following compounds is treated with an appropriate amount of periodic acid? How many molar equivalents of HIO_4 would be consumed in each case?

(a) 2,3-Butanediol

(b) 1,2,3-Butanetriol

(c)

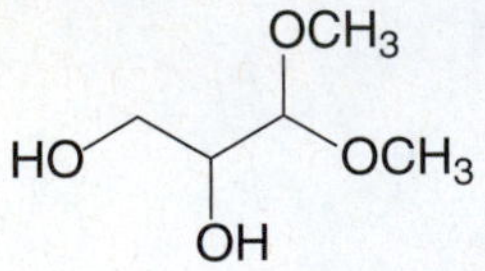

(d)

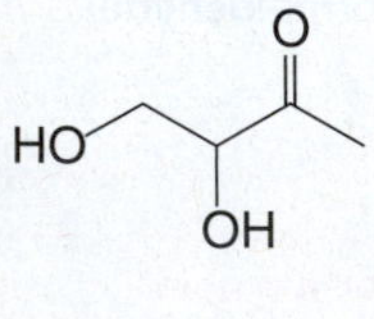

(e)

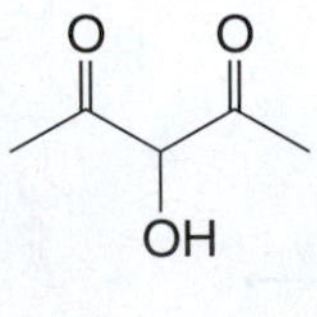

(f) *cis*-1,2-Cyclopentanediol

(g)

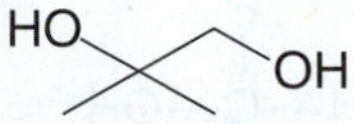

(h) D-Erythrose

• • •

PRACTICE PROBLEM 22.10 Show how periodic acid could be used to distinguish between an aldohexose and a ketohexose. What products would you obtain from each, and how many molar equivalents of HIO_4 would be consumed?

22.7 REDUCTION OF MONOSACCHARIDES: ALDITOLS

- Aldoses (and ketoses) can be reduced with sodium borohydride to give compounds called **alditols**:

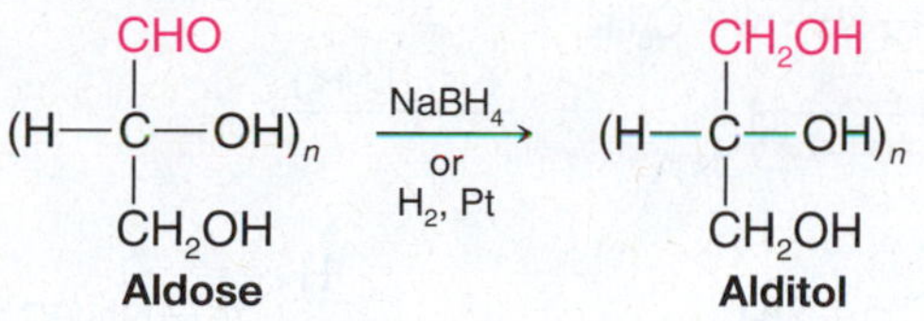
CHO
(H—C—OH)$_n$
CH$_2$OH
NaBH$_4$
or
H$_2$, Pt
CH$_2$OH
(H—C—OH)$_n$
CH$_2$OH
Aldose
Alditol

Reduction of D-glucose, for example, yields D-glucitol:

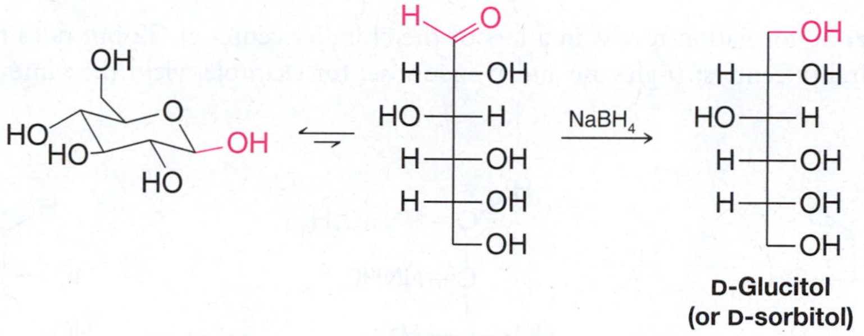
NaBH$_4$
D-Glucitol
(or D-sorbitol)

PRACTICE PROBLEM 22.11

(a) Would you expect D-glucitol to be optically active? **(b)** Write Fischer projection formulas for all of the D-aldohexoses that would yield *optically inactive alditols.*

22.8 REACTIONS OF MONOSACCHARIDES WITH PHENYLHYDRAZINE: OSAZONES

The aldehyde group of an aldose reacts with such carbonyl reagents as hydroxylamine and phenylhydrazine (Section 16.8B). With hydroxylamine, the product is the expected oxime. With enough phenylhydrazine, however, three molar equivalents of phenylhydrazine are consumed and a second phenylhydrazone group is introduced at C2. The product is called a *phenylosazone.* Phenylosazones crystallize readily (unlike sugars) and are useful derivatives for identifying sugars.

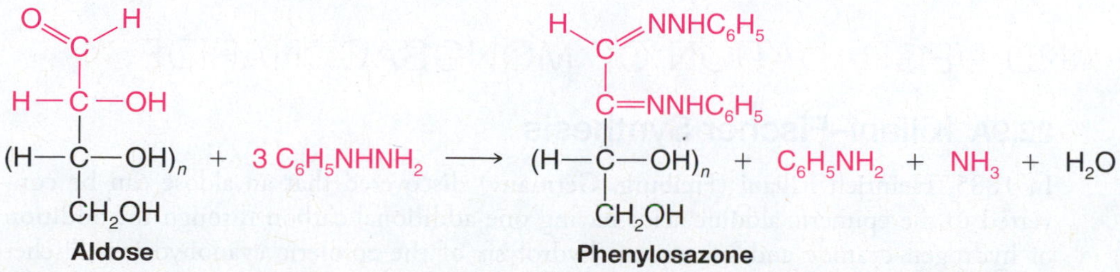
3 C$_6$H$_5$NHNH$_2$
NNHC$_6$H$_5$
C=NNHC$_6$H$_5$
(H—C—OH)$_n$
CH$_2$OH
C$_6$H$_5$NH$_2$ + NH$_3$ + H$_2$O
Aldose
Phenylosazone

The mechanism for osazone formation probably depends on a series of reactions in which >C=N— behaves very much like >C=O in giving a nitrogen version of an enol.

[A MECHANISM FOR THE REACTION — Phenylosazone Formation]

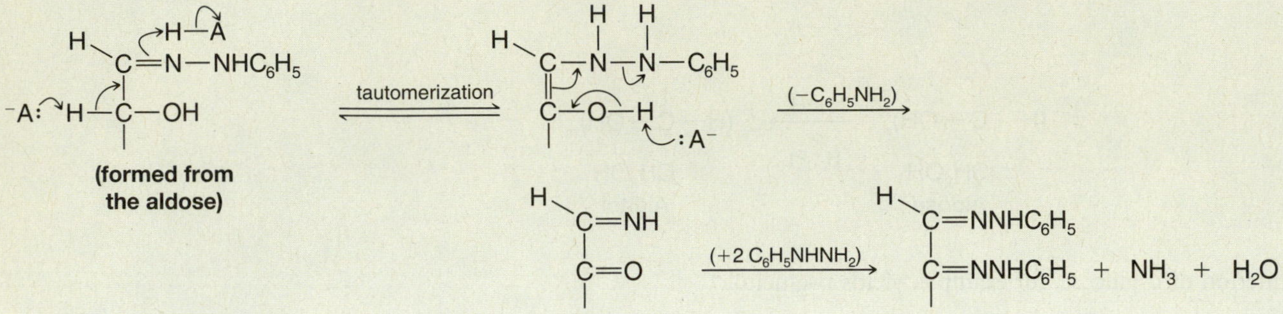

Osazone formation results in a loss of the chirality center at **C2** but does not affect other chirality centers; D-glucose and D-mannose, for example, yield the same phenylosazone:

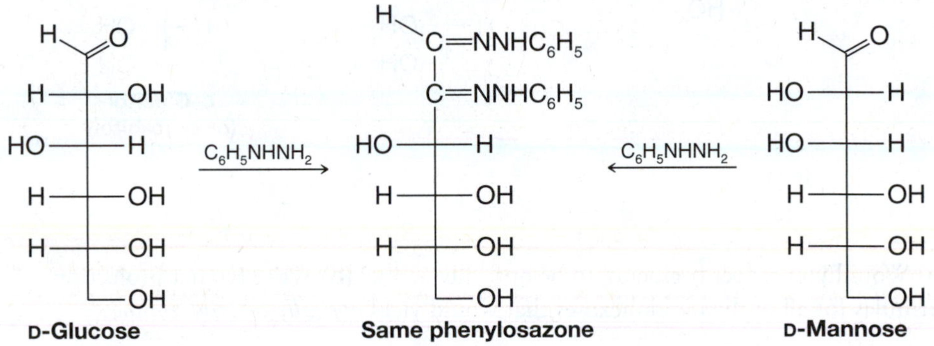

This experiment, first done by Emil Fischer, established that D-glucose and D-mannose have the same configurations about **C3, C4,** and **C5.** Diastereomeric aldoses that differ in configuration at only one carbon (such as D-glucose and D-mannose) are called epimers. In general, any pair of diastereomers that differ in configuration at only a single tetrahedral chirality center can be called **epimers**.

PRACTICE PROBLEM 22.12 Although D-fructose is not an epimer of D-glucose or D-mannose (D-fructose is a keto-hexose), all three yield the same phenylosazone. **(a)** Using Fischer projection formulas, write an equation for the reaction of fructose with phenylhydrazine. **(b)** What information about the stereochemistry of D-fructose does this experiment yield?

22.9 SYNTHESIS AND DEGRADATION OF MONOSACCHARIDES

22.9A Kiliani–Fischer Synthesis

In 1885, Heinrich Kiliani (Freiburg, Germany) discovered that an aldose can be converted to the epimeric aldonic acids having one additional carbon through the addition of hydrogen cyanide and subsequent hydrolysis of the epimeric cyanohydrins. Fischer later extended this method by showing that aldonolactones obtained from the aldonic acids can be reduced to aldoses. Today, this method for lengthening the carbon chain of an aldose is called the Kiliani–Fischer synthesis.

We can illustrate the Kiliani–Fischer synthesis with the synthesis of D-threose and D-erythrose (aldotetroses) from D-glyceraldehyde (an aldotriose) in Fig. 22.6.

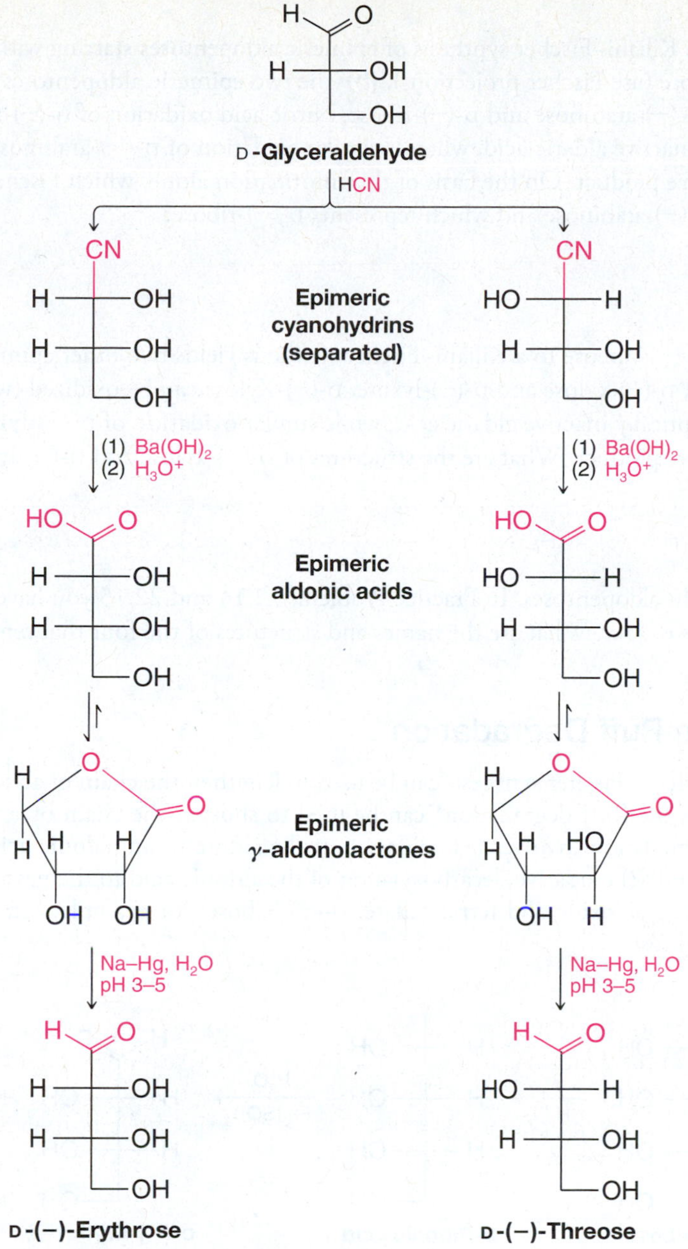

FIGURE 22.6 A Kiliani–Fischer synthesis of D-(−)-erythrose and D-(−)-threose from D-glyceraldehyde.

Addition of hydrogen cyanide to glyceraldehyde produces two epimeric cyanohydrins because the reaction creates a new chirality center. The cyanohydrins can be separated easily (since they are diastereomers), and each can be converted to an aldose through hydrolysis, acidification, lactonization, and reduction with Na–Hg at pH 3–5. One cyanohydrin ultimately yields D-(−)-erythrose and the other yields D-(−)-threose.

We can be sure that the aldotetroses that we obtain from this Kiliani–Fischer synthesis are both D sugars because the starting compound is D-glyceraldehyde and its chirality center is unaffected by the synthesis. On the basis of the Kiliani–Fischer synthesis, we cannot know just which aldotetrose has both —OH groups on the right and which has the top —OH on the left in the Fischer projection. However, if we oxidize both aldotetroses to aldaric acids, one [D-(−)-erythrose] will yield an *optically inactive* (meso) product while the other [D-(−)-threose] will yield a product that is *optically active* (see Practice Problem 22.7).

PRACTICE PROBLEM 22.13

(a) What are the structures of L-(+)-threose and L-(+)-erythrose? **(b)** What aldotriose would you use to prepare them in a Kiliani–Fischer synthesis?

PRACTICE PROBLEM 22.14 (a) Outline a Kiliani–Fischer synthesis of epimeric aldopentoses starting with D-(−)-erythrose (use Fischer projections). (b) The two epimeric aldopentoses that one obtains are D-(−)-arabinose and D-(−)-ribose. Nitric acid oxidation of D-(−)-ribose yields an optically inactive aldaric acid, whereas similar oxidation of D-(−)-arabinose yields an optically active product. On the basis of this information alone, which Fischer projection represents D-(−)-arabinose and which represents D-(−)-ribose?

PRACTICE PROBLEM 22.15 Subjecting D-(−)-threose to a Kiliani–Fischer synthesis yields two other epimeric aldopentoses, D-(+)-xylose and D-(−)-lyxose. D-(+)-Xylose can be oxidized (with nitric acid) to an optically inactive aldaric acid, while similar oxidation of D-(−)-lyxose gives an optically active product. What are the structures of D-(+)-xylose and D-(−)-lyxose?

PRACTICE PROBLEM 22.16 There are eight aldopentoses. In Practice Problems 22.14 and 22.15 you have arrived at the structures of four. What are the names and structures of the four that remain?

22.9B The Ruff Degradation

Just as the Kiliani–Fischer synthesis can be used to lengthen the chain of an aldose by one carbon atom, the Ruff degradation* can be used to shorten the chain by a similar unit. The Ruff degradation involves (1) oxidation of the aldose to an aldonic acid using bromine water and (2) oxidative decarboxylation of the aldonic acid to the next lower aldose using hydrogen peroxide and ferric sulfate. D-(−)-Ribose, for example, can be degraded to D-(−)-erythrose:

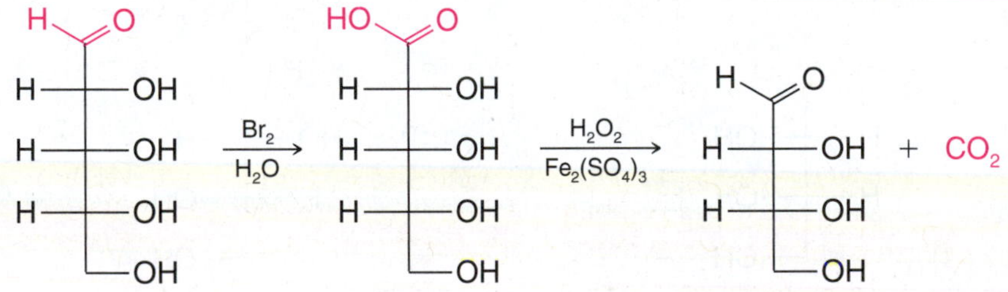

PRACTICE PROBLEM 22.17 The aldohexose D-(+)-galactose can be obtained by hydrolysis of *lactose*, a disaccharide found in milk. When D-(+)-galactose is treated with nitric acid, it yields an optically inactive aldaric acid. When D-(+)-galactose is subjected to Ruff degradation, it yields D-(−)-lyxose (see Practice Problem 22.15). Using only these data, write the Fischer projection formula for D-(+)-galactose.

22.10 THE D FAMILY OF ALDOSES

The Ruff degradation and the Kiliani–Fischer synthesis allow us to place all of the aldoses into families or "family trees" based on their relation to D- or L-glyceraldehyde. Such a tree is constructed in Fig. 22.7 and includes the structures of the D-aldohexoses, **1–8**.

- Most, but not all, of the naturally occurring aldoses belong to the D family, with D-(+)-glucose being by far the most common.

*Developed by Otto Ruff, 1871–1939, a German chemist.

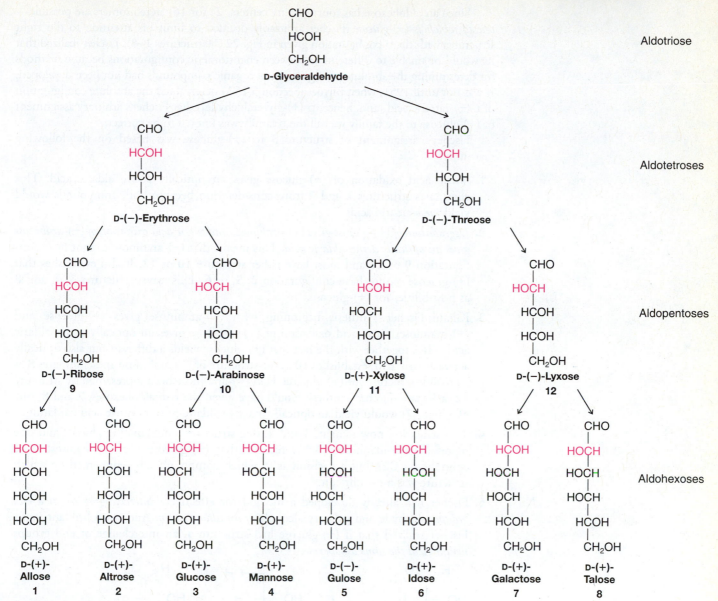

FIGURE 22.7 The D family of aldohexoses.

D-(+)-Galactose can be obtained from milk sugar (lactose), but L-(−)-galactose occurs in a polysaccharide obtained from the vineyard snail, *Helix pomatia*. L-(+)-Arabinose is found widely, but D-(−)-arabinose is scarce, being found only in certain bacteria and sponges. Threose, lyxose, gulose, and allose do not occur naturally, but one or both forms (D or L) of each have been synthesized.

22.11 FISCHER'S PROOF OF THE CONFIGURATION OF D-(+)-GLUCOSE

Emil Fischer began his work on the stereochemistry of (+)-glucose in 1888, only 12 years after van't Hoff and Le Bel had made their proposal concerning the tetrahedral structure of carbon. Only a small body of data was available to Fischer at the beginning. Only a few monosaccharides were known, including (+)-glucose, (+)-arabinose, and (+)-mannose. [(+)-Mannose had just been synthesized by Fischer.] The sugars (+)-glucose and (+)-mannose were known to be aldohexoses; (+)-arabinose was known to be an aldopentose.

Since an aldohexose has four chirality centers, 2^4 (or 16) stereoisomers are possible—*one of which is (+)-glucose.* Fischer arbitrarily decided to limit his attention to the eight structures with the D configuration given in Fig. 22.7 (structures **1–8**). Fischer realized that he would be unable to differentiate between enantiomeric configurations because methods for determining the absolute configuration of organic compounds had not been developed. It was not until 1951, when Bijvoet (Section 5.15A) determined the absolute configuration of L-(+)-tartaric acid [and, hence, D-(+)-glyceraldehyde], that Fischer's arbitrary assignment of (+)-glucose to the family we call the D family was known to be correct.

Fischer's assignment of structure **3** to (+)-glucose was based on the following reasoning:

1. Nitric acid oxidation of (+)-glucose gives an optically active aldaric acid. This eliminates structures **1** and **7** from consideration because both compounds would yield *meso*-aldaric acids.

2. *Degradation of (+)-glucose gives (−)-arabinose, and nitric acid oxidation of (−)-arabinose gives an optically active aldaric acid.* This means that (−)-arabinose cannot have configuration **9** or **11** and must have either structure **10** or **12**. It also establishes that (+)-glucose cannot have configuration **2**, **5**, or **6**. This leaves structures **3**, **4**, and **8** as possibilities for (+)-glucose.

3. Kiliani–Fischer synthesis beginning with (−)-arabinose gives (+)-glucose and (+)-mannose; nitric acid oxidation of (+)-mannose gives an optically active aldaric acid. This, together with the fact that (+)-glucose yields a different but also optically active aldaric acid, establishes **10** as the structure of (−)-arabinose and eliminates **8** as a possible structure for (+)-glucose. Had (−)-arabinose been represented by structure **12**, a Kiliani–Fischer synthesis would have given the two aldohexoses, 7 and 8, one of which (7) would yield an optically inactive aldaric acid on nitric acid oxidation.

4. Two structures now remain, **3** and **4**; one structure represents (+)-glucose and one represents (+)-mannose. Fischer realized that (+)-glucose and (+)-mannose were epimeric (at **C2**), but a decision as to which compound was represented by which structure was most difficult.

5. Fischer had already developed a method for effectively *interchanging the two end groups* (aldehyde and primary alcohol) *of an aldose chain.* And, with brilliant logic, Fischer realized that if (+)-glucose had structure **4**, an interchange of end groups *would yield the same aldohexose:*

(Recall that it is permissible to turn a Fischer projection 180° in the plane of the page.)

On the other hand, if (+)-glucose has structure **3**, *an end-group interchange will yield a different aldohexose,* **13**:

This new aldohexose, if it were formed, would be an L sugar and it would be the mirror reflection of D-gulose. Thus its name would be L-gulose.

Fischer carried out the end-group interchange starting with (+)-glucose and *the product was the new aldohexose* 13. This outcome proved that (+)-glucose has structure 3. It also established 4 as the structure for (+)-mannose, and it proved the structure of L-(+)-gulose as 13.

The procedure Fischer used for interchanging the ends of the (+)-glucose chain began with one of the γ-lactones of D-glucaric acid (see Practice Problem 22.8) and was carried out as follows:

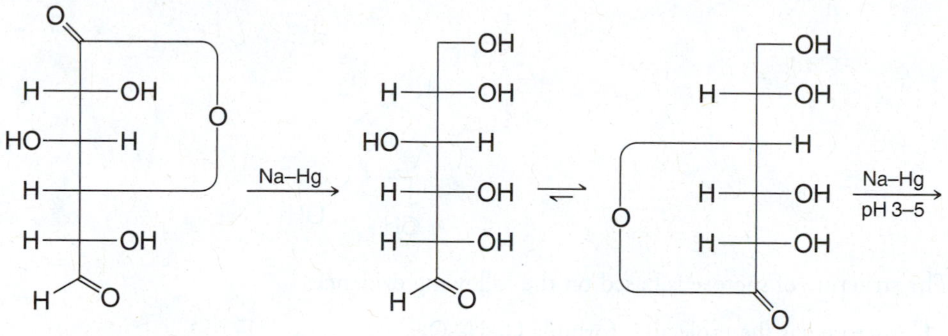

A γ-lactone of D-glucaric acid

L-Gulonic acid

A γ-aldonolactone

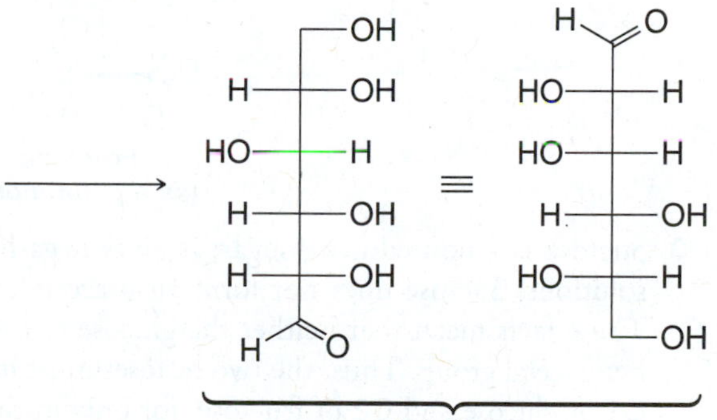

L-(+)-Gulose
13

Helpful Hint

See *WileyPLUS* for "The Chemistry of...Stereoselective Synthesis of all the L-Aldohexoses."

Notice in this synthesis that the second reduction with Na–Hg is carried out at pH 3–5. Under these conditions, reduction of the lactone yields an aldehyde and not a primary alcohol.

Fischer actually had to subject both γ-lactones of D-glucaric acid (Practice Problem 22.8) to the procedure just outlined. What product does the other γ-lactone yield?

PRACTICE PROBLEM 22.18

22.12 DISACCHARIDES

22.12A Sucrose

Ordinary table **sugar** is a **disaccharide** called *sucrose*. Sucrose, the most widely occurring disaccharide, is found in all photosynthetic plants and is obtained commercially from sugarcane or sugar beets. Sucrose has the structure shown in Fig. 22.8.

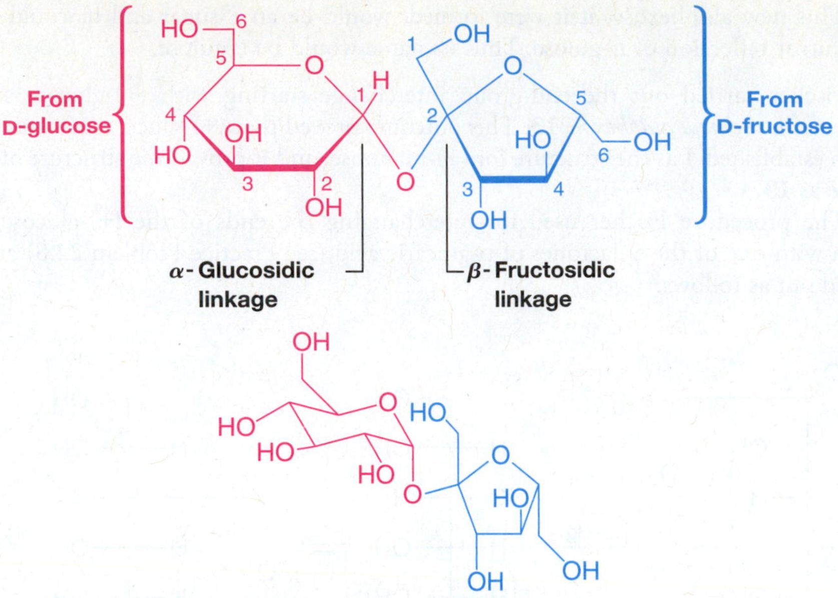

FIGURE 22.8 Two representations of the formula for (+)-sucrose (α-D-glucopyranosyl β-D-fructofuranoside).

The structure of sucrose is based on the following evidence:

1. Sucrose has the molecular formula $C_{12}H_{22}O_{11}$.

2. Acid-catalyzed hydrolysis of 1 mol of sucrose yields 1 mol of D-glucose and 1 mol of D-fructose.

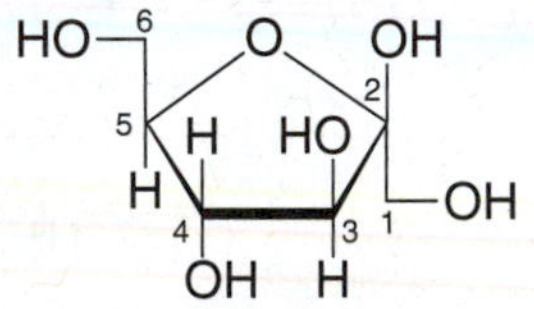

Fructose
(as a β-furanose)

3. Sucrose is a nonreducing sugar; it gives negative tests with Benedict's and Tollens' solutions. Sucrose does not form an osazone and does not undergo mutarotation. These facts mean that neither the glucose nor the fructose portion of sucrose has a hemiacetal group. Thus, the two hexoses must have a glycosidic linkage that involves C1 of glucose and C2 of fructose, for only in this way will both carbonyl groups be present as full acetals (i.e., as glycosides).

4. The stereochemistry of the glycosidic linkages can be inferred from experiments done with enzymes. Sucrose is hydrolyzed by an α-*glucosidase* obtained from yeast but not by β-glucosidase enzymes. This hydrolysis indicates *an α configuration at the glucoside portion.* Sucrose is also hydrolyzed by *sucrase,* an enzyme known to hydrolyze β-fructofuranosides but not α-fructofuranosides. This hydrolysis indicates a β *configuration at the fructoside portion.*

5. Methylation of sucrose gives an octamethyl derivative that, on hydrolysis, gives 2,3,4,6-tetra-*O*-methyl-D-glucose and 1,3,4,6-tetra-*O*-methyl-D-fructose. The identities of these two products demonstrate that the glucose portion is a *pyranoside* and that the fructose portion is a *furanoside.*

The structure of sucrose has been confirmed by X-ray analysis and by an unambiguous synthesis.

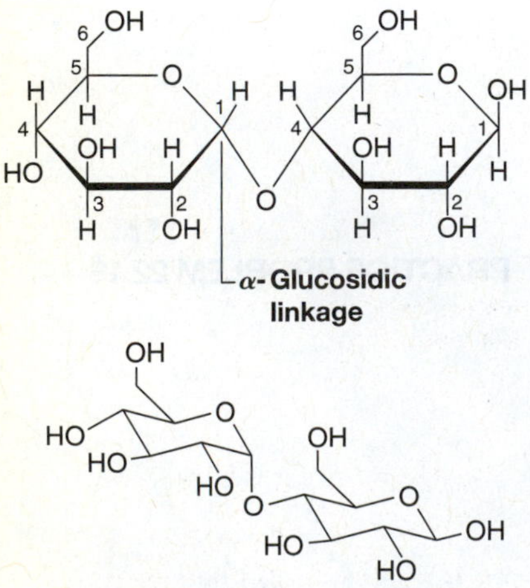

22.12B Maltose

When starch (Section 22.13A) is hydrolyzed by the enzyme *diastase,* one product is a disaccharide known as *maltose* (Fig. 22.9). The structure of maltose was deduced based on the following evidence:

1. When 1 mol of maltose is subjected to acid-catalyzed hydrolysis, it yields 2 mol of D-(+)-glucose.

α-Glucosidic linkage

FIGURE 22.9 Two representations of the structure of the β anomer of (+)-maltose, 4-*O*-(α-D-glucopyranosyl)-β-D-glucopyranose.

2. Unlike sucrose, *maltose is a reducing sugar*; it gives positive tests with Fehling's, Benedict's, and Tollens' solutions. Maltose also reacts with phenylhydrazine to form a monophenylosazone (i.e., it incorporates two molecules of phenylhydrazine).

3. Maltose exists in two anomeric forms: α-(+)-maltose, $[\alpha]_D^{25} = +168$, and β-(+)-maltose, $[\alpha]_D^{25} = +112$. The maltose anomers undergo mutarotation to yield an equilibrium mixture, $[\alpha]_D^{25} = +136$.

Facts 2 and 3 demonstrate that one of the glucose residues of maltose is present in a hemiacetal form; the other, therefore, must be present as a glucoside. The configuration of this glucosidic linkage can be inferred as α, because maltose is hydrolyzed by α-glucosidase enzymes and not by β-glucosidase enzymes.

4. Maltose reacts with bromine water to form a monocarboxylic acid, maltonic acid (Fig. 22.10*a*). This fact, too, is consistent with the presence of only one hemiacetal group.

5. Methylation of maltonic acid followed by hydrolysis gives 2,3,4,6-tetra-*O*-methyl-D-glucose and 2,3,5,6-tetra-*O*-methyl-D-gluconic acid. That the first product has a free

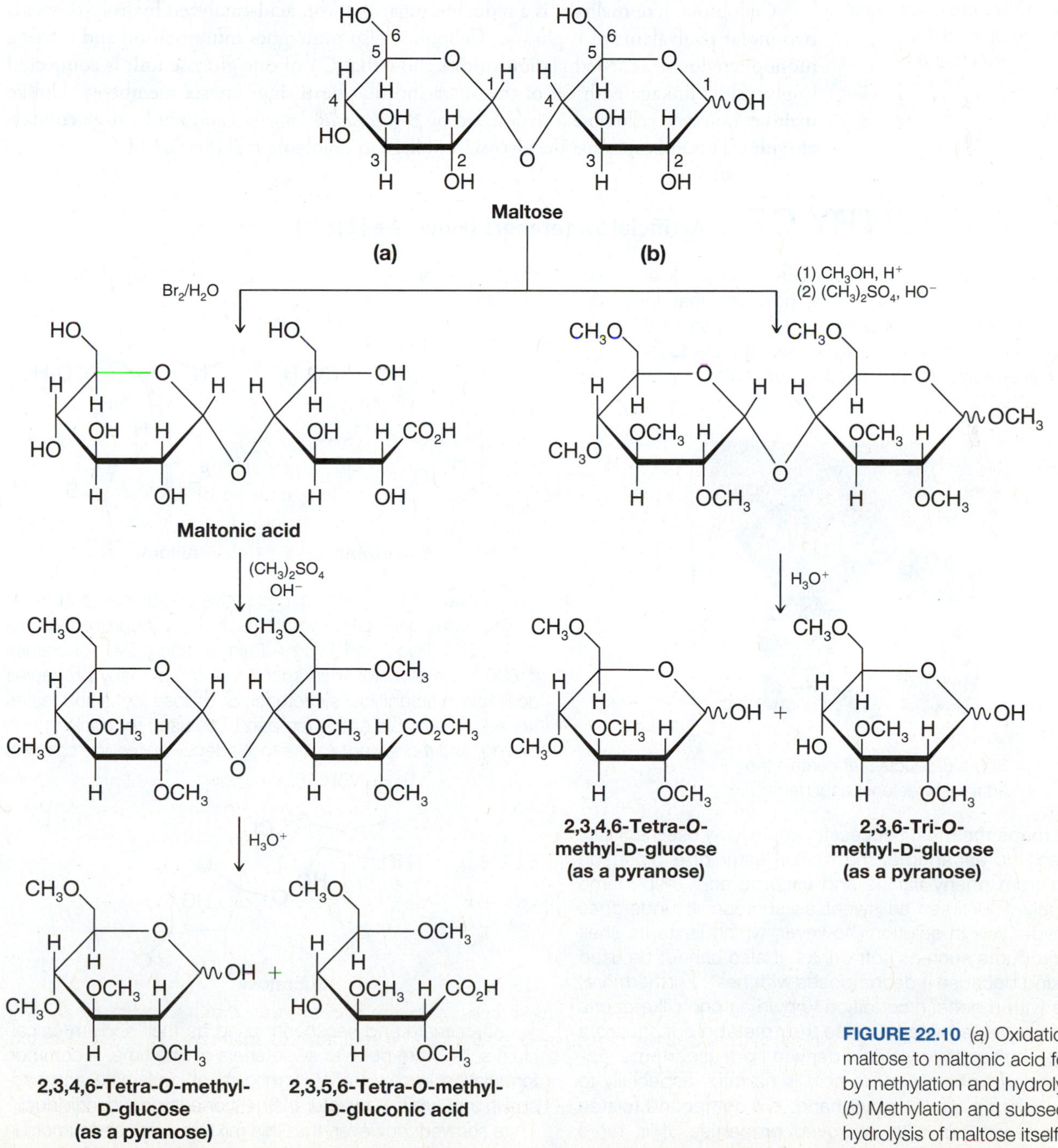

Maltose

(a)   **(b)**

Maltonic acid

2,3,4,6-Tetra-*O*-methyl-D-glucose (as a pyranose)

2,3,5,6-Tetra-*O*-methyl-D-gluconic acid

2,3,4,6-Tetra-*O*-methyl-D-glucose (as a pyranose)

2,3,6-Tri-*O*-methyl-D-glucose (as a pyranose)

FIGURE 22.10 (*a*) Oxidation of maltose to maltonic acid followed by methylation and hydrolysis. (*b*) Methylation and subsequent hydrolysis of maltose itself.

β-Glycosidic linkage

FIGURE 22.11 Two representations of the β anomer of cellobiose, 4-O-(β-D-glucopyranosyl)-β-D-glucopyranose.

—OH at C5 indicates that the nonreducing glucose portion is present as a pyranoside; that the second product, 2,3,5,6-tetra-O-methyl-D-gluconic acid, has a free —OH at C4 indicates that this position was involved in a glycosidic linkage with the nonreducing glucose. Only the size of the reducing glucose ring needs to be determined.

6. Methylation of maltose itself, followed by hydrolysis (Fig. 22.10*b*), gives 2,3,4,6-tetra-O-methyl-D-glucose and 2,3,6-tri-O-methyl-D-glucose. The free —OH at C5 in the latter product indicates that it must have been involved in the oxide ring and that the reducing glucose is present as a *pyranose*.

22.12C Cellobiose

Partial hydrolysis of cellulose (Section 22.13C) gives the disaccharide cellobiose ($C_{12}H_{22}O_{11}$) (Fig. 22.11). Cellobiose resembles maltose in every respect except one: the configuration of its glycosidic linkage.

Cellobiose, like maltose, is a reducing sugar that, on acid-catalyzed hydrolysis, yields two molar equivalents of D-glucose. Cellobiose also undergoes mutarotation and forms a monophenylosazone. Methylation studies show that C1 of one glucose unit is connected in glycosidic linkage with C4 of the other and that both rings are six membered. Unlike maltose, however, cellobiose is hydrolyzed by β-*glucosidase* enzymes and not by α-glucosidase enzymes: This indicates that the glycosidic linkage in cellobiose is β (Fig. 22.11).

⬡ THE CHEMISTRY OF... Artificial Sweeteners (How Sweet It Is)

Sucrose (table sugar) and fructose are the most common natural sweeteners. We all know, however, that they add to our calorie intake and promote tooth decay. For these reasons, many people find artificial sweeteners to be an attractive alternative to the natural and calorie-contributing counterparts.

Some products that contain the artificial sweetener aspartame.

Perhaps the most successful and widely used artificial sweetener is aspartame, the methyl ester of a dipeptide formed from phenylalanine and aspartic acid. Aspartame is roughly 100 times as sweet as sucrose. It undergoes slow hydrolysis in solution, however, which limits its shelf life in products such as soft drinks. It also cannot be used for baking because it decomposes with heat. Furthermore, people with a genetic condition known as phenylketonuria cannot use aspartame because their metabolism causes a buildup of phenylpyruvic acid derived from aspartame. Accumulation of phenylpyruvic acid is harmful, especially to infants. Alitame, on the other hand, is a compound related to aspartame, but with improved properties. It is more

stable than aspartame and roughly 2000 times as sweet as sucrose.

Aspartame

Alitame

Sucralose is a trichloro derivative of sucrose that is an artificial sweetener. Like aspartame, it is also approved for use by the U.S. Food and Drug Administration (FDA). Sucralose is 600 times sweeter than sucrose and has many properties desirable in an artificial sweetener. Sucralose looks and tastes like sugar, is stable at the temperatures used for cooking and baking, and it does not cause tooth decay or provide calories.

Sucralose

Cyclamate and saccharin, used as their sodium or calcium salts, were popular sweeteners at one time. A common formulation involved a 10:1 mixture of cyclamate and saccharin that proved sweeter than either compound individually. Tests showed, however, that this mixture produced tumors in animals, and the FDA subsequently banned it. Certain exclusions to the regulations nevertheless allow continued use of saccharin in some products.

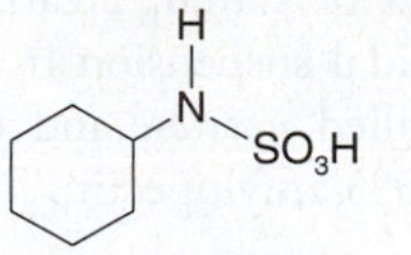

Cyclamate 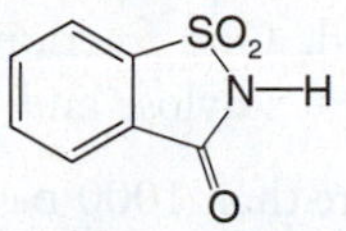   **Saccharin**

Many other compounds have potential as artificial sweeteners. For example, L sugars are also sweet, and they presumably would provide either zero or very few calories because our enzymes have evolved to selectively metabolize their enantiomers instead, the D sugars. Although sources of L sugars are rare in nature, all eight L-hexoses have been synthesized by S. Masamune and K. B. Sharpless using the Sharpless asymmetric epoxidation (Sections 11.13 and 22.11) and other enantioselective synthetic methods.

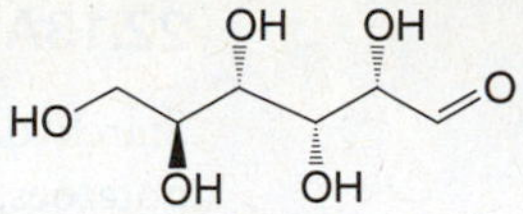

L-Glucose

Much of the research on sweeteners involves probing the structure of sweetness receptor sites. One model proposed for a sweetness receptor incorporates eight binding interactions that involve hydrogen bonding as well as van der Waals forces. Sucronic acid is a synthetic compound designed on the basis of this model. Sucronic acid is reported to be 200,000 times as sweet as sucrose.

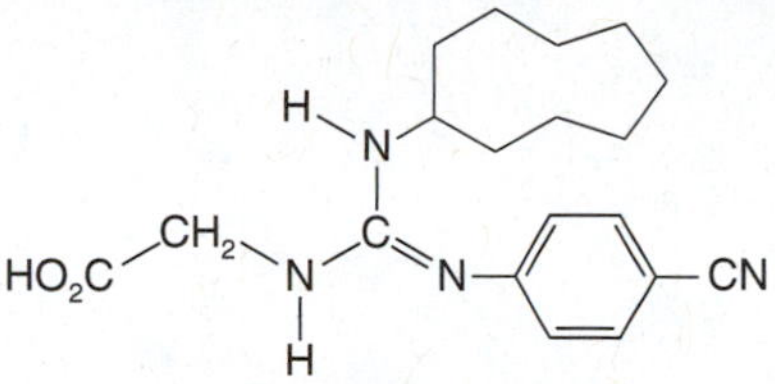

Sucronic acid

22.12D Lactose

Lactose (Fig. 22.12) is a disaccharide present in the milk of humans, cows, and almost all other mammals. Lactose is a reducing sugar that hydrolyzes to yield D-glucose and D-galactose; the glycosidic linkage is β.

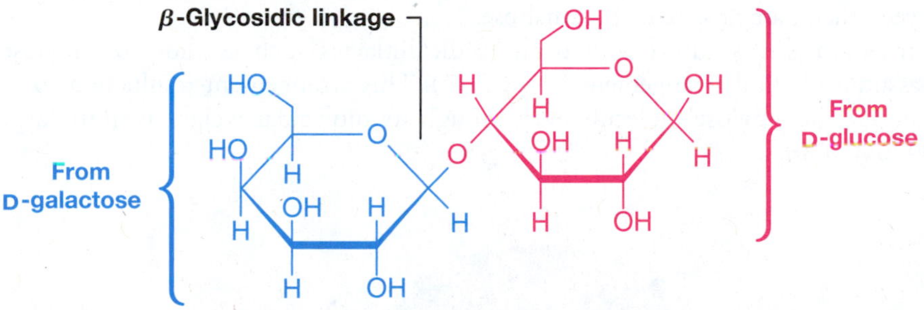

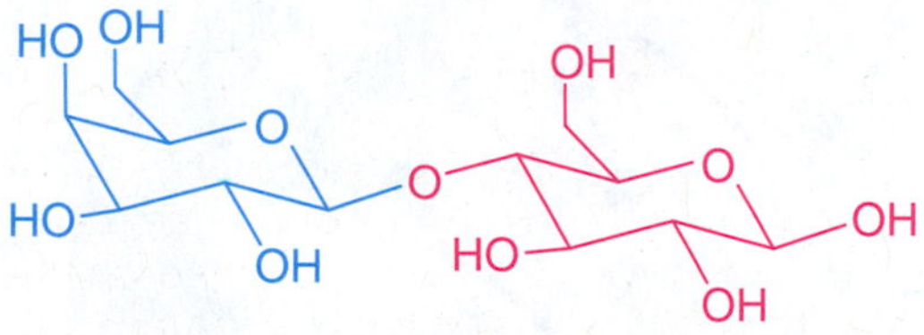

FIGURE 22.12 Two representations of the β anomer of lactose, 4-O-(β-D-galactopyranosyl)-β-D-glucopyranose.

22.13 POLYSACCHARIDES

- **Polysaccharides**, also known as **glycans**, consist of monosaccharides joined together by glycosidic linkages.

Polysaccharides that are polymers of a single monosaccharide are called **homopolysaccharides**; those made up of more than one type of monosaccharide are called **heteropolysaccharides**. Homopolysaccharides are also classified on the basis of their monosaccharide units. A homopolysaccharide consisting of glucose monomeric units is called a **glucan**; one consisting of galactose units is a **galactan**, and so on.

Three important polysaccharides, all of which are glucans, are starch, glycogen, and cellulose.

- Starch is the principal food reserve of plants, glycogen functions as a carbohydrate reserve for animals, and cellulose serves as structural material in plants.

As we examine the structures of these three polysaccharides, we shall be able to see how each is especially suited for its function.

22.13A Starch

Starch occurs as microscopic granules in the roots, tubers, and seeds of plants. Corn, potatoes, wheat, and rice are important commercial sources of starch. Heating starch with water causes the granules to swell and produce a colloidal suspension from which two major components can be isolated. One fraction is called *amylose* and the other *amylopectin*. Most starches yield 10–20% amylose and 80–90% amylopectin.

- Amylose typically consists of more than 1000 D-glucopyranoside units *connected in α linkages* between C1 of one unit and C4 of the next (Fig. 22.13).

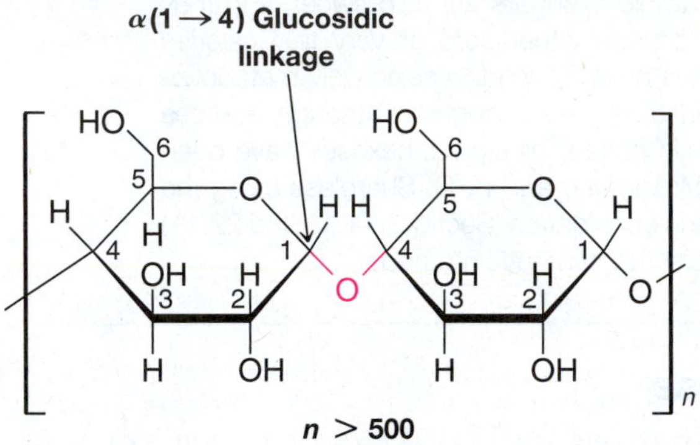

FIGURE 22.13 Partial structure of amylose, an unbranched polymer of D-glucose connected in $\alpha(1 \rightarrow 4)$ glycosidic linkages.

$\alpha(1 \rightarrow 4)$ Glucosidic linkage

$n > 500$

Thus, in the ring size of its glucose units and in the configuration of the glycosidic linkages between them, amylose resembles maltose.

Chains of D-glucose units with α-glycosidic linkages such as those of amylose tend to assume a helical arrangement (Fig. 22.14). This arrangement results in a compact shape for the amylose molecule even though its molecular weight is quite large (150,000–600,000).

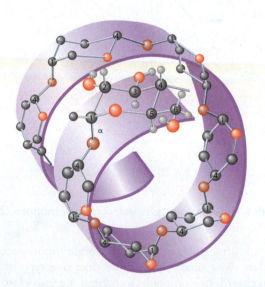

FIGURE 22.14 Amylose. The $\alpha(1 \rightarrow 4)$ linkages cause it to assume the shape of a left-handed helix.
(Illustration, Irving Geis. Image from the Irving Geis Collection, HHMI. Rights owned by Howard Hughes Medical Institute. Not to be reproduced without permission.)

- Amylopectin has a structure similar to that of amylose [i.e., $\alpha(1 \rightarrow 4)$ links], except that in amylopectin the chains are branched. Branching takes place between C6 of one glucose unit and C1 of another and occurs at intervals of 20–25 glucose units (Fig. 22.15).

Physical measurements indicate that amylopectin has a molecular weight of 1–6 million; thus amylopectin consists of hundreds of interconnecting chains of 20–25 glucose units each.

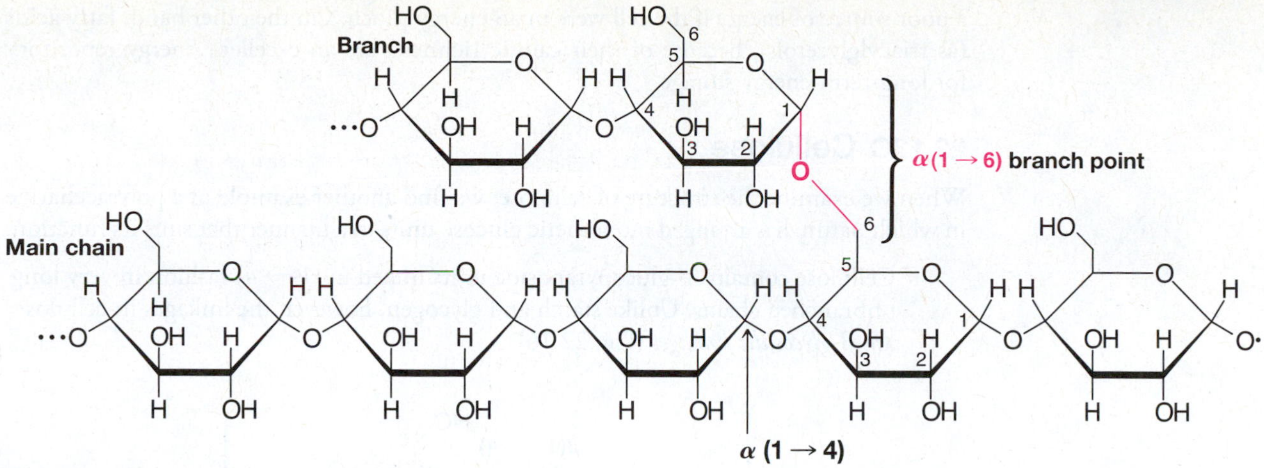

FIGURE 22.15 Partial structure of amylopectin.

22.13B Glycogen

- Glycogen has a structure very much like that of amylopectin; however, in glycogen the chains are much more highly branched.

Methylation and hydrolysis of glycogen indicate that there is one end group for every 10–12 glucose units; branches may occur as often as every 6 units. Glycogen has a very high molecular weight. Studies of glycogens isolated under conditions that minimize the likelihood of hydrolysis indicate molecular weights as high as 100 million.

The size and structure of glycogen beautifully suit its function as a reserve carbohydrate for animals. First, its size makes it too large to diffuse across cell membranes; thus, glycogen remains inside the cell, where it is needed as an energy source. Second, because glycogen incorporates tens of thousands of glucose units in a single molecule, it solves an important osmotic problem for the cell. Were so many glucose units present in the cell as individual molecules, the osmotic pressure within the cell would be enormous—so large that the cell membrane would almost certainly break.* Finally, the localization of glucose units within a large, highly branched structure simplifies one of the cell's logistical problems: that of having a ready source of glucose when cellular glucose concentrations are low and of being able to store glucose rapidly when cellular glucose concentrations are high. There are enzymes within the cell that catalyze the reactions by which glucose units are detached from (or attached to) glycogen. These enzymes operate at end groups by hydrolyzing (or forming) $\alpha(1 \rightarrow 4)$ glycosidic linkages. Because glycogen is so highly branched, a very large number of end groups is available at which these enzymes can operate. At the same time the overall concentration of glycogen (in moles per liter) is quite low because of its enormous molecular weight.

Amylopectin presumably serves a similar function in plants. The fact that amylopectin is less highly branched than glycogen is, however, not a serious disadvantage. Plants have a much lower metabolic rate than animals—and plants, of course, do not require sudden bursts of energy.

Animals store energy as fats (triacylglycerols) as well as glycogen. Fats, because they are more highly reduced, are capable of furnishing much more energy. The metabolism of a typical fatty acid, for example, liberates more than twice as much energy per carbon as glucose or glycogen. Why, then, we might ask, have two different energy repositories evolved? Glucose (from glycogen) is readily available and is highly water soluble.** Glucose, as a result, diffuses rapidly through the aqueous medium of the cell and serves as an ideal source of "ready energy." Long-chain fatty acids, by contrast, are almost insoluble in water, and their concentration inside the cell could never be very high. They would be

---

*The phenomenon of osmotic pressure occurs whenever two solutions of different concentrations are separated by a membrane that allows penetration (by osmosis) of the solvent but not of the solute. The osmotic pressure (π) on one side of the membrane is related to the number of moles of solute particles (n), the volume of the solution (V), and the gas constant times the absolute temperature (RT): $\pi V = nRT$.

**Glucose is actually liberated as glucose-6-phosphate (G6P), which is also water soluble.

a poor source of energy if the cell were in an energy pinch. On the other hand, fatty acids (as triacylglycerols), because of their caloric richness, are an excellent energy repository for long-term energy storage.

22.13C Cellulose

When we examine the structure of cellulose, we find another example of a polysaccharide in which nature has arranged monomeric glucose units in a manner that suits its function.

- Cellulose contains D-glucopyranoside units linked in (1 → 4) fashion in very long unbranched chains. Unlike starch and glycogen, however, the linkages in cellulose are *β-glycosidic linkages* (Fig. 22.16).

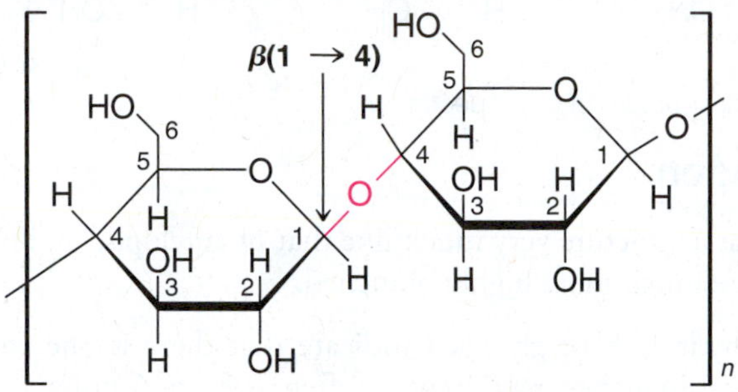

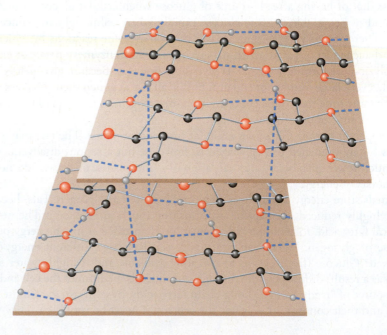

FIGURE 22.16 A portion of a cellulose chain. The glycosidic linkages are β(1 → 4).

The β-glycosidic linkages of cellulose make cellulose chains essentially linear; they do not tend to coil into helical structures as do glucose polymers when linked in an α(1 → 4) manner.

The linear arrangement of β-linked glucose units in cellulose presents a uniform distribution of —OH groups on the outside of each chain. When two or more cellulose chains make contact, the hydroxyl groups are ideally situated to "zip" the chains together by forming hydrogen bonds (Fig. 22.17). Zipping many cellulose chains together in this way gives a highly insoluble, rigid, and fibrous polymer that is ideal as cell-wall material for plants.

FIGURE 22.17 A proposed structure for cellulose. A fiber of cellulose may consist of about 40 parallel strands of glucose molecules linked in a β(1 → 4) fashion. Each glucose unit in a chain is turned over with respect to the preceding glucose unit and is held in this position by hydrogen bonds (dashed lines) between the chains. The glucan chains line up laterally to form sheets, and these sheets stack vertically so that they are staggered by one-half of a glucose unit. (Hydrogen atoms that do not participate in hydrogen bonding have been omitted for clarity.)
(Illustration, Irving Geis. Image from the Irving Geis Collection, HHMI. Rights owned by Howard Hughes Medical Institute. Not to be reproduced without permission.)

This special property of cellulose chains, we should emphasize, is not just a result of β(1 → 4) glycosidic linkages; it is also a consequence of the precise stereochemistry of D-glucose at each chirality center. Were D-galactose or D-allose units linked in a similar fashion, they almost certainly would not give rise to a polymer with properties like

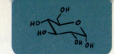

cellulose. Thus, we get another glimpse of why D-glucose occupies such a special position in the chemistry of plants and animals. Not only is it the most stable aldohexose (because it can exist in a chair conformation that allows all of its bulky groups to occupy equatorial positions), but its special stereochemistry also allows it to form helical structures when α linked as in starches, and rigid linear structures when β linked as in cellulose.

There is another interesting and important fact about cellulose: the digestive enzymes of humans cannot attack its $\beta(1 \rightarrow 4)$ linkages. Hence, cellulose cannot serve as a food source for humans, as can starch. Cows and termites, however, can use cellulose (of grass and wood) as a food source because symbiotic bacteria in their digestive systems furnish β-glucosidase enzymes.

Perhaps we should ask ourselves one other question: Why has D-(+)-glucose been selected for its special role rather than L-(−)-glucose, its mirror image? Here an answer cannot be given with any certainty. The selection of D-(+)-glucose may simply have been a random event early in the course of the evolution of enzyme catalysts. Once this selection was made, however, the chirality of the active sites of the enzymes involved would retain a bias toward D-(+)-glucose and against L-(−)-glucose (because of the improper fit of the latter). Once introduced, this bias would be perpetuated and extended to other catalysts.

Finally, when we speak about evolutionary selection of a particular molecule for a given function, we do not mean to imply that evolution operates on a molecular level. Evolution, of course, takes place at the level of organism populations, and molecules are selected only in the sense that their use gives the organism an increased likelihood of surviving and procreating.

22.13D Cellulose Derivatives

A number of derivatives of cellulose are used commercially. Most of these are compounds in which two or three of the free hydroxyl groups of each glucose unit have been converted to an ester or an ether. This conversion substantially alters the physical properties of the material, making it more soluble in organic solvents and allowing it to be made into fibers and films. Treating cellulose with acetic anhydride produces the triacetate known as "Arnel" or "acetate," used widely in the textile industry. Cellulose trinitrate, also called "gun cotton" or nitrocellulose, is used in explosives.

Rayon is made by treating cellulose (from cotton or wood pulp) with carbon disulfide in a basic solution. This reaction converts cellulose to a soluble xanthate:

$$\text{Cellulose—OH} \quad + \quad CS_2 \quad \xrightarrow{\text{NaOH}} \quad \text{cellulose—O—}\overset{\overset{\displaystyle S}{\|}}{C}\text{—S}^-\,Na^+$$

Cellulose xanthate

The solution of cellulose xanthate is then passed through a small orifice or slit into an acidic solution. This operation regenerates the —OH groups of cellulose, causing it to precipitate as a fiber or a sheet:

$$\text{Cellulose—O—}\overset{\overset{\displaystyle S}{\|}}{C}\text{—S}^-\,Na^+ \quad \xrightarrow{H_3O^+} \quad \text{cellulose—OH}$$

Rayon or cellophane

The fibers are *rayon*; the sheets, after softening with glycerol, are *cellophane*.

Cellophane on rollers at a manufacturing plant.

22.14 OTHER BIOLOGICALLY IMPORTANT SUGARS

Monosaccharide derivatives in which the —CH_2OH group at C6 has been specifically oxidized to a carboxyl group are called **uronic acids**. Their names are based on the monosaccharide from which they are derived. For example, specific oxidation of C6 of glucose to a carboxyl group converts *glucose* to **glucuronic acid**. In the same way, specific oxidation of C6 of *galactose* would yield **galacturonic acid**:

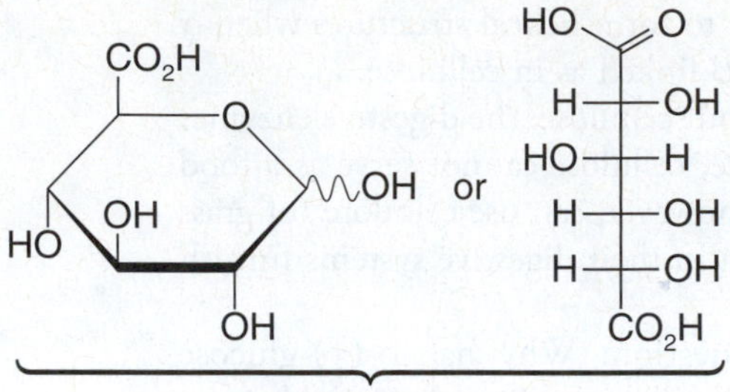

D-**Glucuronic acid** 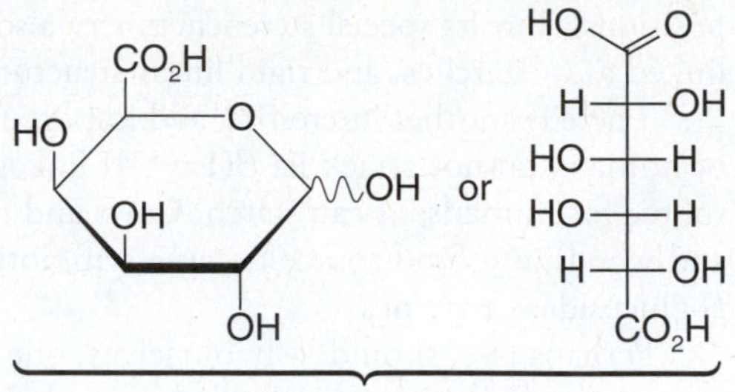   D-**Galacturonic acid**

PRACTICE PROBLEM 22.19 Direct oxidation of an aldose affects the aldehyde group first, converting it to a carboxylic acid (Section 22.6B), and most oxidizing agents that will attack 1° alcohol groups will also attack 2° alcohol groups. Clearly, then, a laboratory synthesis of a uronic acid from an aldose requires protecting these groups from oxidation. Keeping this in mind, suggest a method for carrying out a specific oxidation that would convert D-galactose to D-galacturonic acid. (*Hint*: See Section 22.5E.)

- Monosaccharides in which an —OH group has been replaced by —H are known as **deoxy sugars**.

The most important deoxy sugar, because it occurs in DNA, is **deoxyribose**. Other deoxy sugars that occur widely in polysaccharides are L-rhamnose and L-fucose:

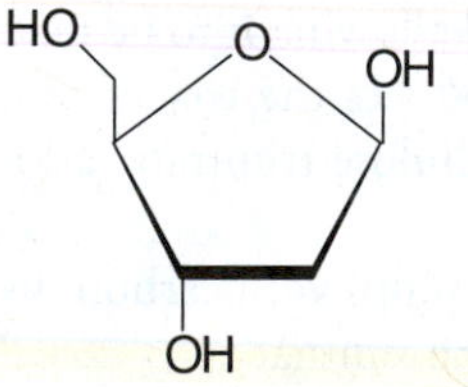

β-2-Deoxy-D-ribose 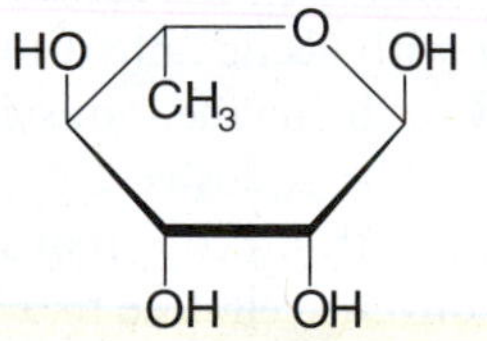   **α-L-Rhamnose**   **α-L-Fucose**
                          (6-deoxy-L-mannose)   (6-deoxy-L-galactose)

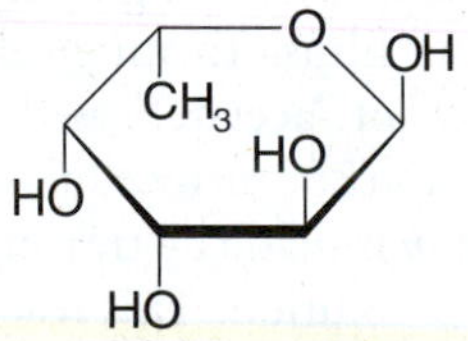

22.15 SUGARS THAT CONTAIN NITROGEN

22.15A Glycosylamines

A sugar in which an amino group replaces the anomeric —OH is called a glycosylamine. Examples are β-D-glucopyranosylamine and adenosine:

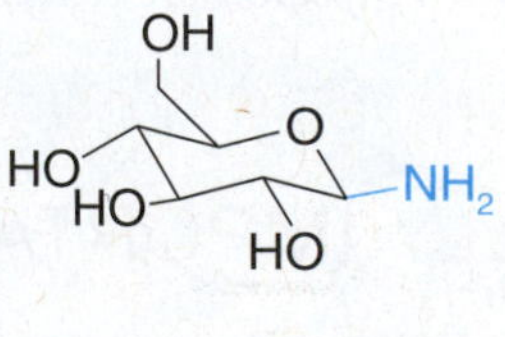

β-D-Glucopyranosylamine 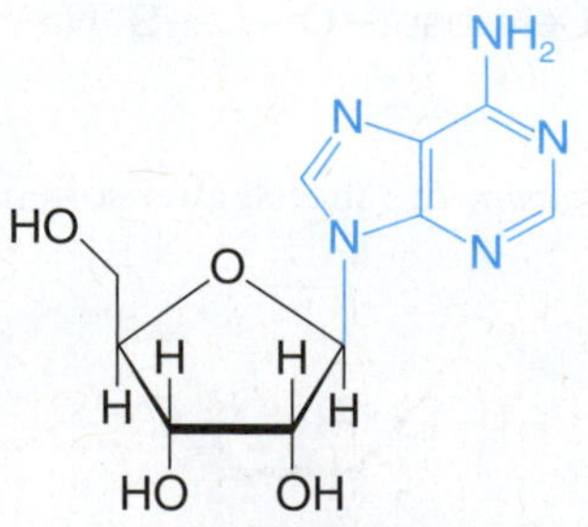   **Adenosine**

Adenosine is an example of a glycosylamine that is also called a nucleoside.

- **Nucleosides** are glycosylamines in which the amino component is a pyrimidine or a purine (Section 20.1B) and in which the sugar component is either D-ribose or 2-deoxy-D-ribose (i.e., D-ribose minus the oxygen at the 2 position).

Nucleosides are the important components of RNA (ribonucleic acid) and DNA (deoxyribonucleic acid). We shall describe their properties in detail in Section 25.2.

22.15B Amino Sugars

- A sugar in which an amino group replaces a nonanomeric —OH group is called an **amino sugar**.

D-**Glucosamine** is an example of an amino sugar. In many instances the amino group is acetylated as in *N*-**acetyl-D-glucosamine**. *N*-**Acetylmuramic acid** is an important component of bacterial cell walls (Section 24.10).

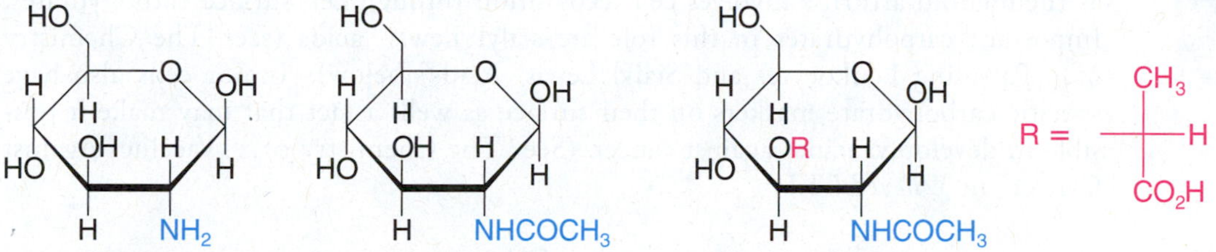

β-D-Glucosamine   β-*N*-Acetyl-D-glucosamine   β-*N*-Acetylmuramic acid
                    (NAG)   (NAM)

D-Glucosamine can be obtained by hydrolysis of **chitin**, a polysaccharide found in the shells of lobsters and crabs and in the external skeletons of insects and spiders. The amino group of D-glucosamine as it occurs in chitin, however, is acetylated; thus, the repeating unit is actually *N*-acetylglucosamine (Fig. 22.18). The glycosidic linkages in chitin are $\beta(1 \rightarrow 4)$. X-Ray analysis indicates that the structure of chitin is similar to that of cellulose.

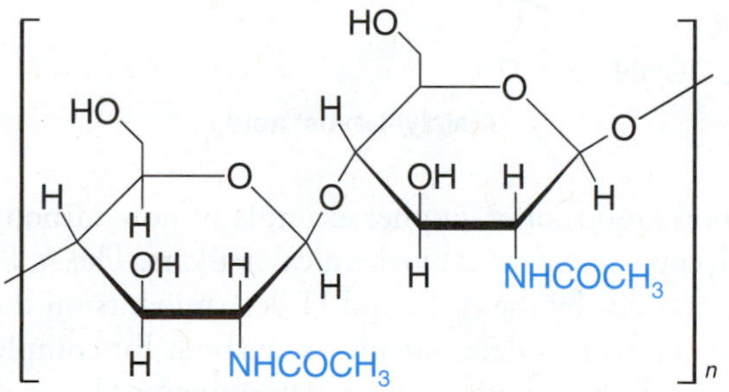

FIGURE 22.18 A partial structure of chitin. The repeating units are *N*-acetylglucosamines linked $\beta(1 \rightarrow 4)$.

D-Glucosamine can also be isolated from **heparin**, a sulfated polysaccharide that consists predominately of alternating units of D-glucuronate-2-sulfate and *N*-sulfo-D-glucosamine-6-sulfate (Fig. 22.19). Heparin occurs in intracellular granules of mast cells that line arterial walls, where, when released through injury, it inhibits the clotting of blood. Its purpose seems to be to prevent runaway clot formation. Heparin is widely used in medicine to prevent blood clotting in postsurgical patients.

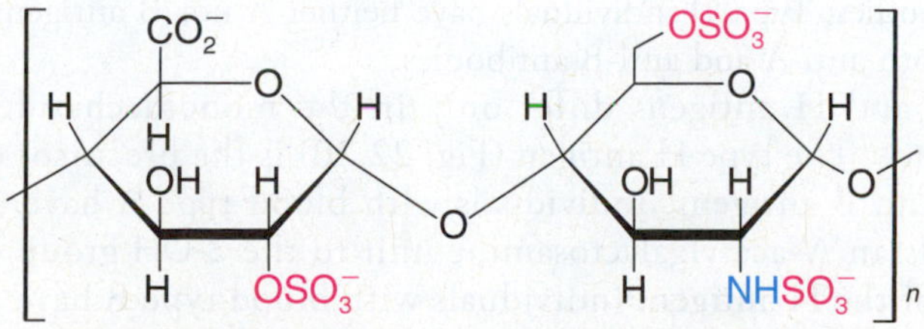

D-Glucuronate-2-sulfate   *N*-Sulfo-D-glucosamine-6-sulfate

FIGURE 22.19 A partial structure of heparin, a polysaccharide that prevents blood clotting.

22.16 GLYCOLIPIDS AND GLYCOPROTEINS OF THE CELL SURFACE: CELL RECOGNITION AND THE IMMUNE SYSTEM

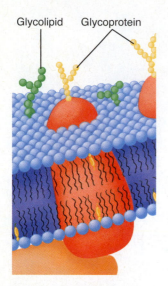

Glycolipid   Glycoprotein

(Reprinted with permission of John Wiley & Sons, Inc., from Voet, D., and Voet, J.G. *Biochemistry*, Second Edition © 1995 Voet, D., and Voet, J.G.)

Before 1960, it was thought that the biology of carbohydrates was rather uninteresting—that, in addition to being a kind of inert filler in cells, carbohydrates served only as an energy source and, in plants, as structural materials. Research has shown, however, that carbohydrates joined through glycosidic linkages to lipids (Chapter 23) and to proteins (Chapter 24), called **glycolipids** and **glycoproteins**, respectively, have functions that span the entire spectrum of activities in the cell. Indeed, most proteins are glycoproteins, of which the carbohydrate content can vary from less than 1% to greater than 90%.

Glycolipids and glycoproteins on the cell surface (Section 23.6A) are now known to be the agents by which cells interact with other cells and with invading bacteria and viruses. The immune system's role in healing and in autoimmune diseases such as rheumatoid arthritis involves cell recognition through cell surface carbohydrates. Important carbohydrates in this role are sialyl Lewis^x acids (see "The Chemistry of...Patroling Leukocytes and Sialyl Lewis^x Acids" below). Tumor cells also have specific carbohydrate markers on their surface as well, a fact that may make it possible to develop vaccines against cancer. (See "The Chemistry of...Vaccines Against Cancer" in *WileyPLUS*.)

A sialyl Lewis^x acid

The human blood groups offer another example of how carbohydrates, in the form of glycolipids and glycoproteins, act as biochemical markers. The A, B, and O blood types are determined, respectively, by the A, B, and H determinants on the blood cell surface. (The odd naming of the type O determinant came about for complicated historical reasons.) Type AB blood cells have both A and B determinants. These determinants are the carbohydrate portions of the A, B, and H **antigens**.

Antigens are characteristic chemical substances that cause the production of **antibodies** when injected into an animal. Each antibody can bind at least two of its corresponding antigen molecules, causing them to become linked. Linking of red blood cells causes them to agglutinate (clump together). In a transfusion this agglutination can lead to a fatal blockage of the blood vessels.

Individuals with type A antigens on their blood cells carry anti-B antibodies in their serum; those with type B antigens on their blood cells carry anti-A antibodies in their serum. Individuals with type AB cells have both A and B antigens but have neither anti-A nor anti-B antibodies. Type O individuals have neither A nor B antigens on their blood cells but have both anti-A and anti-B antibodies.

The A, B, and H antigens differ only in the monosaccharide units at their nonreducing ends. The type H antigen (Fig. 22.20) is the precursor oligosaccharide of the type A and B antigens. Individuals with blood type A have an enzyme that specifically adds an *N*-acetylgalactosamine unit to the **3-OH** group of the terminal galactose unit of the H antigen. Individuals with blood type B have an enzyme that specifically adds galactose instead. In individuals with type O blood, the enzyme is inactive.

Helpful Hint

See "The Chemistry of...Oligosaccharide Synthesis on a Solid Support–the Glycal Assembly Approach" in *WileyPLUS* regarding the synthesis of promising carbohydrate anticancer vaccines.

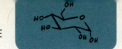

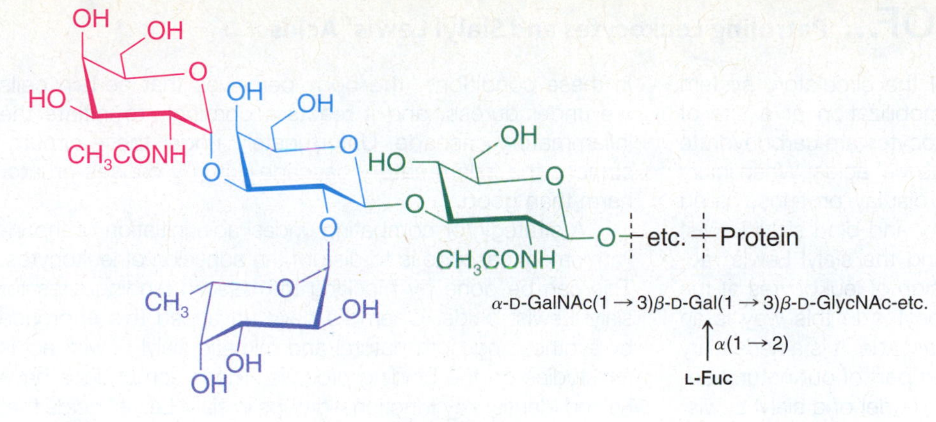

α-D-GalNAc(1 → 3)β-D-Gal(1 → 3)β-D-GlycNAc-etc.

↑α(1 → 2)

L-Fuc

Type A determinant

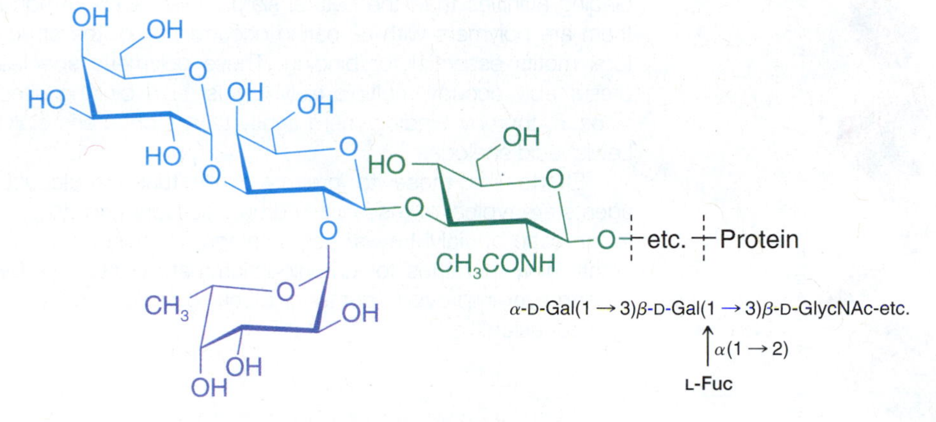

α-D-Gal(1 → 3)β-D-Gal(1 → 3)β-D-GlycNAc-etc.

↑α(1 → 2)

L-Fuc

Type B determinant

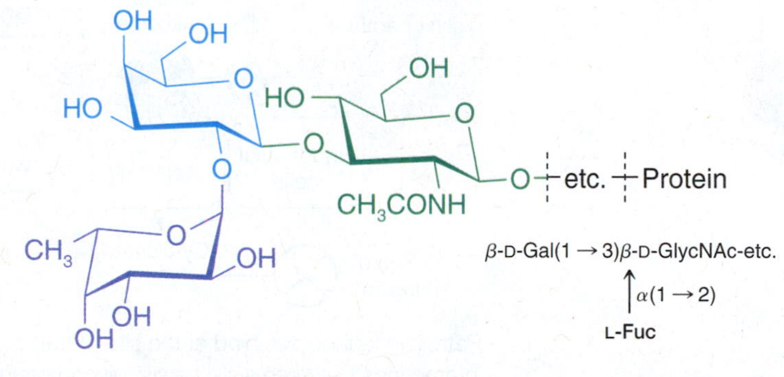

β-D-Gal(1 → 3)β-D-GlycNAc-etc.

↑α(1 → 2)

L-Fuc

Type H determinant

FIGURE 22.20 The terminal monosaccharides of the antigenic determinants for types A, B, and O blood. The type H determinant is present in individuals with blood type O and is the precursor of the type A and B determinants. These oligosaccharide antigens are attached to carrier lipid or protein molecules that are anchored in the red blood cell membrane (see Fig. 23.9 for a depiction of a cell membrane). Ac = acetyl, Gal = D-galactose, GalNAc = N-acetylgalactosamine, GlycNAc = N-acetylglucosamine, Fuc = fucose.

Antigen–antibody interactions like those that determine blood types are the basis of the immune system. These interactions often involve the chemical recognition of a glycolipid or glycoprotein in the antigen by a glycolipid or glycoprotein of the antibody. In "The Chemistry of... Antibody-Catalyzed Aldol Condensations" (in *WileyPLUS*, Chapter 19), however, we saw a different and emerging dimension of chemistry involving antibodies. We shall explore this topic further in the Chapter 24 opening vignette on designer catalysts and in "The Chemistry of... Some Catalytic Antibodies" (Section 24.12).

THE CHEMISTRY OF...   Patroling Leukocytes and Sialyl Lewisˣ Acids

White blood cells continually patrol the circulatory system and interstitial spaces, ready for mobilization at a site of trauma. The frontline scouts for leukocytes are carbohydrate groups on their surface called sialyl Lewisˣ acids. When injury occurs, cells at the site of trauma display proteins, called selectins, that signal the site of injury and bind sialyl Lewisˣ acids. Binding between selectins and the sialyl Lewisˣ acids on the leukocytes causes adhesion of leukocytes at the affected area. Recruitment of leukocytes in this way is an important step in the inflammatory cascade. It is a necessary part of the healing process as well as part of our natural defense against infection. A molecular model of a sialyl Lewisˣ acid is shown below, and its structural formula is given in Section 22.16.

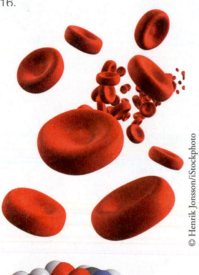

© Henrik Jonsson/iStockphoto

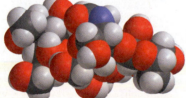

Sialyl Lewisˣ, a carbohydrate that is important in the recognition and healing of traumatized tissue.

There are some maladies, however, that result from the over-enthusiastic recruitment of leukocytes. Rheumatoid arthritis, strokes, and injuries related to perfusion during surgery and organ transplantation are a few examples.

In these conditions, the body perceives that certain cells are under duress, and it reacts accordingly to initiate the inflammatory cascade. Unfortunately, under these circumstances the inflammatory cascade actually causes greater harm than good.

A strategy for combating undesirable initiation of the inflammatory cascade is to disrupt the adhesion of leukocytes. This can be done by blocking the selectin binding sites for sialyl Lewisˣ acids. Chemists have advanced this approach by synthesizing both natural and mimetic sialyl Lewisˣ acids for studies on the binding process. These compounds have helped identify key functional groups in sialyl Lewisˣ acids that are required for recognition and binding. Chemists have even designed and synthesized novel compounds that have tighter binding affinities than the natural sialyl Lewisˣ acids. Among them are polymers with repeating occurrences of the structural motifs essential for binding. These polymeric species presumably occupy multiple sialyl Lewisˣ acid binding sites at once, thereby binding more tightly than monomeric sialyl Lewisˣ acid analogs.

Efforts like these to prepare finely tuned molecular agents are typical of research in drug discovery and design. In the case of sialyl Lewisˣ acid analogs, chemists hope to create new therapies for chronic inflammatory diseases by making ever-improved agents for blocking undesired leukocyte adhesion.

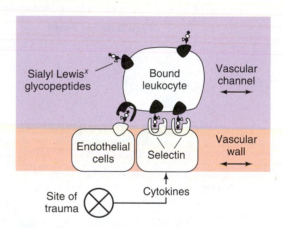

Patrolling leukocytes bind at the site of trauma by interactions between sialyl Lewisˣ glycoproteins on their surface and selectin proteins on the injured cell. (Reprinted with permission from Simanek, E. E.; McGarvey, G. J.; Jablonowski, J. A.; Wong, C. A., *Chemical Reviews*, *98*, p. 835, Figure 1, 1998. Copyright 1998 American Chemical Society.)

22.17 CARBOHYDRATE ANTIBIOTICS

One of the important discoveries in carbohydrate chemistry was the isolation (in 1944) of the carbohydrate antibiotic called *streptomycin*. Streptomycin disrupts bacterial protein synthesis. Its structure is made up of the following three subunits:

All three components are unusual: the amino sugar is based on L-glucose; streptose is a branched-chain monosaccharide; and streptidine is not a sugar at all, but a cyclohexane derivative called an amino cyclitol.

Other members of this family are antibiotics called kanamycins, neomycins, and gentamicins (not shown). All are based on an amino cyclitol linked to one or more amino sugars. The glycosidic linkage is nearly always α. These antibiotics are especially useful against bacteria that are resistant to penicillins.

22.18 SUMMARY OF REACTIONS OF CARBOHYDRATES

The reactions of carbohydrates, with few exceptions, are the reactions of functional groups that we have studied in earlier chapters, especially those of aldehydes, ketones, and alcohols. The most central reactions of carbohydrates are those of hemiacetal and acetal formation and hydrolysis. Hemiacetal groups form the pyranose and furanose rings in carbohydrates, and acetal groups form glycoside derivatives and join monosaccharides together to form di-, tri-, oligo-, and polysaccharides.

Other reactions of carbohydrates include those of alcohols, carboxylic acids, and their derivatives. Alkylation of carbohydrate hydroxyl groups leads to ethers. Acylation of their hydroxyl groups produces esters. Alkylation and acylation reactions are sometimes used to protect carbohydrate hydroxyl groups from reaction while a transformation occurs elsewhere. Hydrolysis reactions are involved in converting ester and lactone derivatives of carbohydrates back to their polyhydroxy form. Enolization of aldehydes and ketones leads to epimerization and interconversion of aldoses and ketoses. Addition reactions of aldehydes and ketones are useful, too, such as the addition of ammonia derivatives in osazone formation, and of cyanide in the Kiliani–Fischer synthesis. Hydrolysis of nitriles from the Kiliani–Fischer synthesis leads to carboxylic acids.

Oxidation and reduction reactions have their place in carbohydrate chemistry as well. Reduction reactions of aldehydes and ketones, such as borohydride reduction and catalytic hydrogenation, are used to convert aldoses and ketoses to alditols. Oxidation by Tollens' and Benedict's reagents is a test for the hemiacetal linkage in a sugar. Bromine water oxidizes the aldehyde group of an aldose to an aldonic acid. Nitric acid oxidizes both the aldehyde group and terminal hydroxymethyl group of an aldose to an aldaric acid (a dicarboxylic acid). And last, periodate cleavage of carbohydrates yields oxidized fragments that can be useful for structure elucidation.

WHY Do These Topics Matter?]

Chances are fairly high that you know someone with diabetes, given that it is estimated that at least 26 million people in the United States currently suffer from the condition, while over 50 million more are close to developing the disease. Diabetes is a metabolic disorder that is characterized by an individual having far too much of the carbohydrate glucose in his or her bloodstream—a problem that, if left untreated, can lead to a number of chronic problems such as kidney failure and cardiovascular disease.

Diabetes results because a critical protein known as insulin (see Section 24.6B), whose role is to regulate the overall amount of glucose in our systems by signaling cells to remove it from the bloodstream and store it as glycogen, either is no longer produced in sufficient amounts (affording what is known as Type 1, or juvenile, diabetes) or is no longer used effectively by cells to control glucose levels (affording what is known as Type 2, or adult-onset, diabetes). Either form results in a chronic need for treatment to control blood sugar at as normal a level as possible. For Type 1 patients, that goal often can be achieved simply with insulin treatments. However, for Type 2 patients, alternatives are often needed. Fortunately, there are several treatments available for these individuals, but most of these come with some undesired side effects, including too much glucose removal (leading to hypoglycemia) and/or unwanted weight gain.

Pharmaceutical companies throughout the world are currently working on therapies to counterbalance these side effects with Type 2 patients, and in several recent efforts, it has been a D-glucose-containing natural product that has been critical. That compound is phlorizin. This natural product is an inhibitor of several different types of sodium-dependent glucose transport systems (SDGT) found in cells. Some of these transporters, known as SDGT-1, are found throughout the body and play a role in controlling glucose uptake from our diets. If they are inhibited, then glucose from food will not enter the bloodstream. A second group, known as SDGT-2, is responsible for the reuptake of glucose filtered by our kidneys into our bloodstreams. If this group of transporters is inhibited, then that filtered glucose will be excreted instead in urine. Of the two, it is the second that many scientists believe would have a stronger impact on the disease if inhibited, with the hope that these compounds would not cause unwanted weight gain or hypoglycemia since they act by a different mechanism than other available therapies.

Phlorizin   Sergliflozin   Canagliflozin   Pfizer compound

Pleasingly, altering the structure of this natural product has led to new molecules such as sergliflozin that can selectively inhibit SDGT-2 in cellular assays. When dosed in humans, however, this and related molecules had to be abandoned in clinical trials because they were too easily degraded by glycosidases, enzymes that can cleave the glycosidic bond (Section 22.4) between the sugar portion and the aromatic domain of these pharmaceuticals into inactive molecules. However, if the carbohydrate backbone is changed to a glycosidic linkage based on carbon, not oxygen, then glycosidases cannot cleave the bond at that same position. As a result, new and longer-lived compounds have resulted such as canagliflozin and the unnamed molecule from Pfizer, both of which have been explored in advanced clinical trials and may provide new and highly needed therapies to treat the disease. If so, then it would be a carbohydrate-containing molecule that would be involved in controlling the levels of another key carbohydrate in our bodies.

To learn more about these topics, see:

1. V. Mascitti et al. "Discovery of a Clinical Candidate from the Structurally Unique Dioxa-bicyclo[3.2.1]octane Class of Sodium-Dependent Glucose Cotransporter 2 Inhibitors." *J. Med. Chem.* **2011**, *54*, 2952–2960.
2. E. C. Chao. "Canigliflozin." *Drugs of the Future* **2011**, *36*, 351–357.

SUMMARY AND REVIEW TOOLS

The study aids for this chapter include key terms and concepts (which are highlighted in bold, blue text within the chapter and defined in the Glossary (at the back of the book) and have hyperlinked definitions in the accompanying *WileyPLUS* course (www.wileyplus.com), and a summary of reactions involving monosaccharides.

PROBLEMS PLUS

Note to Instructors: Many of the homework problems are available for assignment via *WileyPLUS*, an online teaching and learning solution.

CARBOHYDRATE STRUCTURE AND REACTIONS

22.20 Give appropriate structural formulas to illustrate each of the following:

(a) An aldopentose   **(e)** An aldonic acid   **(i)** A furanose   **(m)** Epimers   **(q)** A polysaccharide

(b) A ketohexose   **(f)** An aldaric acid   **(j)** A reducing sugar   **(n)** Anomers   **(r)** A nonreducing sugar

(c) An L-monosaccharide   **(g)** An aldonolactone   **(k)** A pyranoside   **(o)** A phenylosazone

(d) A glycoside   **(h)** A pyranose   **(l)** A furanoside   **(p)** A disaccharide

22.21 Draw conformational formulas for each of the following: **(a)** α-D-allopyranose, **(b)** methyl β-D-allopyranoside, and **(c)** methyl 2,3,4,6-tetra-*O*-methyl-β-D-allopyranoside.

22.22 Draw structures for furanose and pyranose forms of D-ribose. Show how you could use periodate oxidation to distinguish between a methyl ribofuranoside and a methyl ribopyranoside.

22.23 One reference book lists D-mannose as being dextrorotatory; another lists it as being levorotatory. Both references are correct. Explain.

22.24 The starting material for a commercial synthesis of vitamin C is L-sorbose (see the following reaction); it can be synthesized from D-glucose through the following reaction sequence:

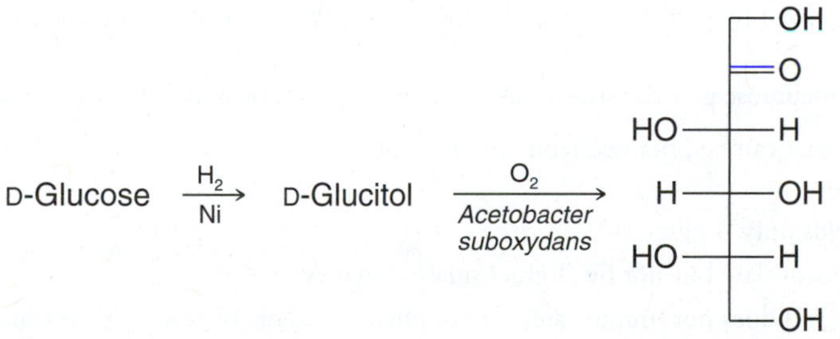

$$\text{D-Glucose} \xrightarrow[\text{Ni}]{\text{H}_2} \text{D-Glucitol} \xrightarrow[\substack{Acetobacter \\ suboxydans}]{\text{O}_2}$$

```
        ─── OH
        ═══ O
HO ───── H
H  ───── OH
HO ───── H
        ─── OH
      L-Sorbose
```

The second step of this sequence illustrates the use of a bacterial oxidation; the microorganism *A. suboxydans* accomplishes this step in 90% yield. The overall result of the synthesis is the transformation of a D-aldohexose (D-glucose) into an L-ketohexose (L-sorbose). What does this mean about the specificity of the bacterial oxidation?

22.25 What two aldoses would yield the same phenylosazone as L-sorbose (Problem 22.24)?

22.26 In addition to fructose (Practice Problem 22.12) and sorbose (Problem 22.24), there are two other 2-ketohexoses, *psicose* and *tagatose*. D-Psicose yields the same phenylosazone as D-allose (or D-altrose); D-tagatose yields the same osazone as D-galactose (or D-talose). What are the structures of D-psicose and D-tagatose?

22.27 **A**, **B**, and **C** are three aldohexoses. Compounds **A** and **B** yield the same optically active alditol when they are reduced with hydrogen and a catalyst; **A** and **B** yield different phenylosazones when treated with phenylhydrazine; **B** and **C** give the same phenylosazone but different alditols. Assuming that all are D sugars, give names and structures for **A**, **B**, and **C**.

22.28 Xylitol is a sweetener that is used in sugarless chewing gum. Starting with an appropriate monosaccharide, outline a possible synthesis of xylitol.

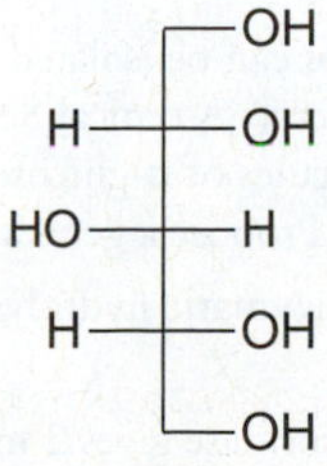

```
        ─── OH
H  ───── OH
HO ───── H
H  ───── OH
        ─── OH
       Xylitol
```

22.29 Although monosaccharides undergo complex isomerizations in base (see Section 22.5A), aldonic acids are epimerized specifically at C2 when they are heated with pyridine. Show how you could make use of this reaction in a synthesis of D-mannose from D-glucose.

22.30 The most stable conformation of most aldopyranoses is one in which the largest group, the —CH_2OH group, is equatorial. However, D-idopyranose exists primarily in a conformation with an axial —CH_2OH group. Write formulas for the two chair conformations of α-D-idopyranose (one with the —CH_2OH group axial and one with the —CH_2OH group equatorial) and provide an explanation.

STRUCTURE ELUCIDATION

22.31 **(a)** Heating D-altrose with dilute acid produces a nonreducing *anhydro sugar* ($C_6H_{10}O_5$). Methylation of the anhydro sugar followed by acid hydrolysis yields 2,3,4-tri-*O*-methyl-D-altrose. The formation of the anhydro sugar takes place through a chair conformation of β-D-altropyranose in which the —CH_2OH group is axial. What is the structure of the anhydro sugar, and how is it formed? **(b)** D-Glucose also forms an anhydro sugar but the conditions required are much more drastic than for the corresponding reaction of D-altrose. Explain.

22.32 Show how the following experimental evidence can be used to deduce the structure of lactose (Section 22.12D):

1. Acid hydrolysis of lactose ($C_{12}H_{22}O_{11}$) gives equimolar quantities of D-glucose and D-galactose. Lactose undergoes a similar hydrolysis in the presence of a β-*galactosidase*.

2. Lactose is a reducing sugar and forms a phenylosazone; it also undergoes mutarotation.

3. Oxidation of lactose with bromine water followed by hydrolysis with dilute acid gives D-galactose and D-gluconic acid.

4. Bromine water oxidation of lactose followed by methylation and hydrolysis gives 2,3,6-tri-*O*-methylgluconolactone and 2,3,4,6-tetra-*O*-methyl-D-galactose.

5. Methylation and hydrolysis of lactose give 2,3,6-tri-*O*-methyl-D-glucose and 2,3,4,6-tetra-*O*-methyl-D-galactose.

22.33 Deduce the structure of the disaccharide *melibiose* from the following data:

1. Melibiose is a reducing sugar that undergoes mutarotation and forms a phenylosazone.

2. Hydrolysis of melibiose with acid or with an α-*galactosidase* gives D-galactose and D-glucose.

3. Bromine water oxidation of melibiose gives *melibionic acid*. Hydrolysis of melibionic acid gives D-galactose and D-gluconic acid. Methylation of melibionic acid followed by hydrolysis gives 2,3,4,6-tetra-*O*-methyl-D-galactose and 2,3,4,5-tetra-*O*-methyl-D-gluconic acid.

4. Methylation and hydrolysis of melibiose give 2,3,4,6-tetra-*O*-methyl-D-galactose and 2,3,4-tri-*O*-methyl-D-glucose.

22.34 Trehalose is a disaccharide that can be obtained from yeasts, fungi, sea urchins, algae, and insects. Deduce the structure of trehalose from the following information:

1. Acid hydrolysis of trehalose yields only D-glucose.

2. Trehalose is hydrolyzed by α-glucosidase but not by β-glucosidase enzymes.

3. Trehalose is a nonreducing sugar; it does not mutarotate, form a phenylosazone, or react with bromine water.

4. Methylation of trehalose followed by hydrolysis yields two molar equivalents of 2,3,4,6-tetra-*O*-methyl-D-glucose.

22.35 Outline chemical tests that will distinguish between members of each of the following pairs:

(a) D-Glucose and D-glucitol   **(c)** D-Glucose and D-fructose   **(e)** Sucrose and maltose

(b) D-Glucitol and D-glucaric acid   **(d)** D-Glucose and D-galactose   **(f)** Maltose and maltonic acid

(g) Methyl β-D-glucopyranoside and 2,3,4,6-tetra-*O*-methyl-β-D-glucopyranose

(h) Methyl α-D-ribofuranoside (**I**) and methyl 2-deoxy-α-D-ribofuranoside (**II**):

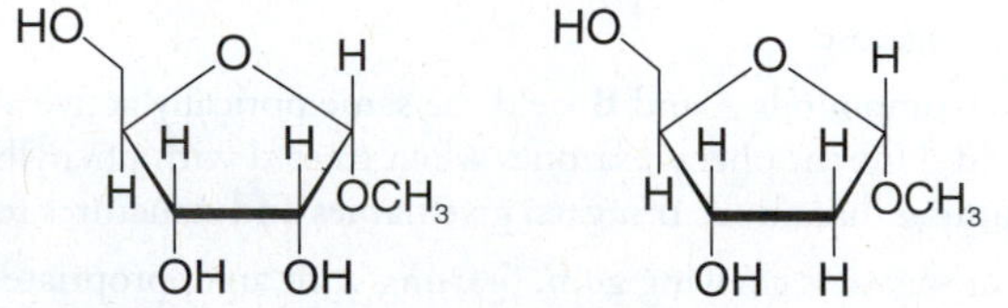

22.36 A group of oligosaccharides called *Schardinger dextrins* can be isolated from *Bacillus macerans* when the bacillus is grown on a medium rich in amylose. These oligosaccharides are all *nonreducing*. A typical Schardinger dextrin undergoes hydrolysis when treated with an acid or an α-glucosidase to yield six, seven, or eight molecules of D-glucose. Complete methylation of a Schardinger dextrin followed by acid hydrolysis yields only 2,3,6-tri-*O*-methyl-D-glucose. Propose a general structure for a Schardinger dextrin.

22.37 *Isomaltose* is a disaccharide that can be obtained by enzymatic hydrolysis of amylopectin. Deduce the structure of isomaltose from the following data:

1. Hydrolysis of 1 mol of isomaltose by acid or by an α-glucosidase gives 2 mol of D-glucose.

2. Isomaltose is a reducing sugar.

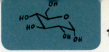

3. Isomaltose is oxidized by bromine water to isomaltonic acid. Methylation of isomaltonic acid and subsequent hydrolysis yields 2,3,4,6-tetra-*O*-methyl-D-glucose and 2,3,4,5-tetra-*O*-methyl-D-gluconic acid.

4. Methylation of isomaltose itself followed by hydrolysis gives 2,3,4,6-tetra-*O*-methyl-D-glucose and 2,3,4-tri-*O*-methyl-D-glucose.

22.38 *Stachyose* occurs in the roots of several species of plants. Deduce the structure of stachyose from the following data:

1. Acidic hydrolysis of 1 mol of stachyose yields 2 mol of D-galactose, 1 mol of D-glucose, and 1 mol of D-fructose.

2. Stachyose is a nonreducing sugar.

3. Treating stachyose with an α-galactosidase produces a mixture containing D-galactose, sucrose, and a nonreducing trisaccharide called *raffinose*.

4. Acidic hydrolysis of raffinose gives D-glucose, D-fructose, and D-galactose. Treating raffinose with an α-galactosidase yields D-galactose and sucrose. Treating raffinose with invertase (an enzyme that hydrolyzes sucrose) yields fructose and *melibiose* (see Problem 22.33).

5. Methylation of stachyose followed by hydrolysis yields 2,3,4,6-tetra-*O*-methyl-D-galactose, 2,3,4-tri-*O*-methyl-D-galactose, 2,3,4-tri-*O*-methyl-D-glucose, and 1,3,4,6-tetra-*O*-methyl-D-fructose.

SPECTROSCOPY

22.39 *Arbutin*, a compound that can be isolated from the leaves of barberry, cranberry, and pear trees, has the molecular formula $C_{12}H_{16}O_7$. When arbutin is treated with aqueous acid or with a β-glucosidase, the reaction produces D-glucose and a compound **X** with the molecular formula $C_6H_6O_2$. The ^1H NMR spectrum of compound **X** consists of two singlets, one at δ 6.8 (4H) and one at δ 7.9 (2H). Methylation of arbutin followed by acidic hydrolysis yields 2,3,4,6-tetra-*O*-methyl-D-glucose and a compound **Y** ($C_7H_8O_2$). Compound **Y** is soluble in dilute aqueous NaOH but is insoluble in aqueous $NaHCO_3$. The ^1H NMR spectrum of **Y** shows a singlet at δ 3.9 (3H), a singlet at δ 4.8 (1H), and a multiplet (that resembles a singlet) at δ 6.8 (4H). Treating compound **Y** with aqueous NaOH and $(CH_3)_2SO_4$ produces compound **Z** ($C_8H_{10}O_2$). The ^1H NMR spectrum of **Z** consists of two singlets, one at δ 3.75 (6H) and one at δ 6.8 (4H). Propose structures for arbutin and for compounds **X**, **Y**, and **Z**.

22.40 When subjected to a Ruff degradation, a D-aldopentose, **A**, is converted to an aldotetrose, **B**. When reduced with sodium borohydride, the aldotetrose **B** forms an optically active alditol. The ^{13}C NMR spectrum of this alditol displays only two signals. The alditol obtained by direct reduction of **A** with sodium borohydride is not optically active. When **A** is used as the starting material for a Kiliani–Fischer synthesis, two diastereomeric aldohexoses, **C** and **D**, are produced. On treatment with sodium borohydride, **C** leads to an alditol **E**, and **D** leads to **F**. The ^{13}C NMR spectrum of **E** consists of three signals; that of **F** consists of six. Propose structures for **A**–**F**.

22.41 Figure 22.21 shows the ^{13}C NMR spectrum for the product of the reaction of D-(+)-mannose with acetone containing a trace of acid. This compound is a mannofuranose with some hydroxyl groups protected as acetone acetals (as acetonides). Use the ^{13}C NMR spectrum to determine how many acetonide groups are present in the compound.

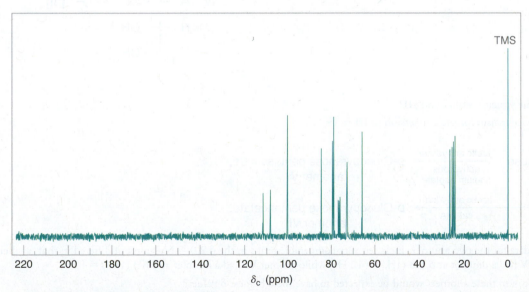

FIGURE 22.21 The broadband proton-decoupled ^{13}C NMR spectrum for the reaction product in Problem 22.41.

22.42 D-(+)-Mannose can be reduced with sodium borohydride to form D-mannitol. When D-mannitol is dissolved in acetone containing a trace amount of acid and the product of this reaction subsequently oxidized with $NaIO_4$, a compound whose ^{13}C NMR spectrum consists of six signals is produced. One of these signals is near δ 200. What is the structure of this compound?

CHALLENGE PROBLEMS

22.43 Of the two anomers of methyl 2,3-anhydro-D-ribofuranoside, **I**, the β form has a strikingly lower boiling point. Suggest an explanation using their structural formulas.

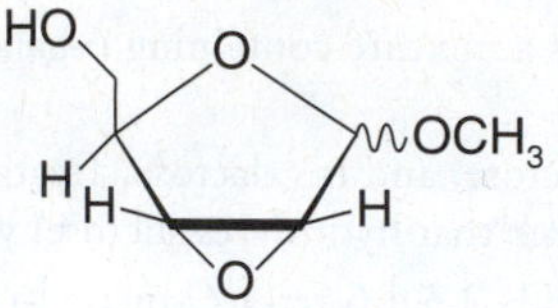

I

22.44 The following reaction sequence represents an elegant method of synthesis of 2-deoxy-D-ribose, **IV**, published by D. C. C. Smith in 1955:

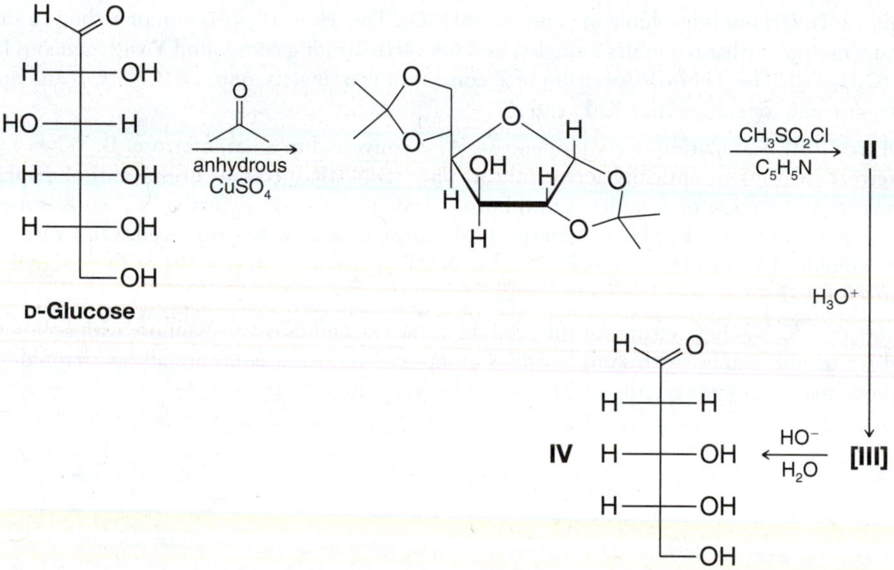

(a) What are the structures of **II** and **III**?

(b) Propose a mechanism for the conversion of **III** to **IV**.

22.45

D-Glucose —(acetic anhydride, anhydrous sodium acetate)→ D-Glucopyranose pentaacetate, anomer **V**

D-Glucose —(acetic anhydride, cat. HA)→ D-Glucopyranose pentaacetate, anomer **VI**

The ^{1}H NMR data for the two anomers included very comparable peaks in the δ 2.0–5.6 region but differed in that, as their highest δ peaks, anomer **V** had a doublet at δ 5.8 (1H, J = 12 Hz) while anomer **VI** had a doublet at δ 6.3 (1H, J = 4 Hz).

(a) Which proton in these anomers would be expected to have these highest δ values?

(b) Why do the signals for these protons appear as doublets?

(c) The relationship between the magnitude of the observed coupling constant and the dihedral angle (when measured using a Newman projection) between C—H bonds on the adjacent carbons of a C—C bond is given by the Karplus equation. It indicates that an axial–axial relationship results in a coupling constant of about 9 Hz (observed range is 8–14 Hz) and an equatorial–axial relationship results in a coupling constant of about 2 Hz (observed range is 1–7 Hz). Which of **V** and **VI** is the α anomer and which is the β anomer?

(d) Draw the most stable conformer for each of **V** and **VI**.

LEARNING GROUP PROBLEMS

1. (a) The members of one class of low-calorie sweeteners are called polyols. The chemical synthesis of one such polyol sweetener involves reduction of a certain disaccharide to a mixture of diastereomeric glycosides. The alcohol (actually polyol) portion of the diastereomeric glycosides derives from one of the sugar moieties in the original disaccharide. Exhaustive methylation of the sweetener (e.g., with dimethyl sulfate in the presence of hydroxide) followed by hydrolysis would be expected to produce 2,3,4,6-tetra-*O*-methyl-*α*-D-glucopyranose, 1,2,3,4,5-penta-*O*-methyl-D-sorbitol, and 1,2,3,4,5-penta-*O*-methyl-D-mannitol, in the ratio of 2:1:1. On the basis of this information, deduce the structure of the two disaccharide glycosides that make up the diastereomeric mixture in this polyol sweetener.

(b) Knowing that the mixture of two disaccharide glycosides in this sweetener results from reduction of a single disaccharide starting material (e.g., reduction by sodium borohydride), what would be the structure of the disaccharide *reactant* for the reduction step? Explain how reduction of this compound would produce the two glycosides.

(c) Write the lowest energy chair conformational structure for 2,3,4,6-tetra-*O*-methyl-*α*-D-glucopyranose.

2. Shikimic acid is a key biosynthetic intermediate in plants and microorganisms. In nature, shikimic acid is converted to chorismate, which is then converted to prephenate, ultimately leading to aromatic amino acids and other essential plant and microbial metabolites (see the Chapter 21 Learning Group problem). In the course of research on biosynthetic pathways involving shikimic acid, H. Floss (University of Washington) required shikimic acid labeled with ^{13}C to trace the destiny of the labeled carbon atoms in later biochemical transformations. To synthesize the labeled shikimic acid, Floss adapted a synthesis of optically active shikimic acid from D-mannose reported earlier by G. W. J. Fleet (Oxford University). This synthesis is a prime example of how natural sugars can be excellent chiral starting materials for the chemical synthesis of optically active target molecules. It is also an excellent example of classic reactions in carbohydrate chemistry. The Fleet–Floss synthesis of D-(−)-[1,7-^{13}C]-shikimic acid (**1**) from D-mannose is shown in Scheme 1.

(a) Comment on the several transformations that occur between D-mannose and **2**. What new functional groups are formed?

(b) What is accomplished in the steps from **2** to **3**, **3** to **4**, and **4** to **5**?

(c) Deduce the structure of compound **9** (a reagent used to convert **5** to **6**), knowing that it was a carbanion that displaced the trifluoromethanesulfonate (triflate) group of **5**. Note that it was compound **9** that brought with it the required ^{13}C atoms for the final product.

(d) Explain the transformation from **7** to **8**. Write out the structure of the compound in equilibrium with **7** that would be required for the process from **7** to **8** to occur. What is the name given to the reaction from this intermediate to **8**?

(e) Label the carbon atoms of D-mannose and **1** by number or letter so as to show which atoms in **1** came from which atoms of D-mannose.

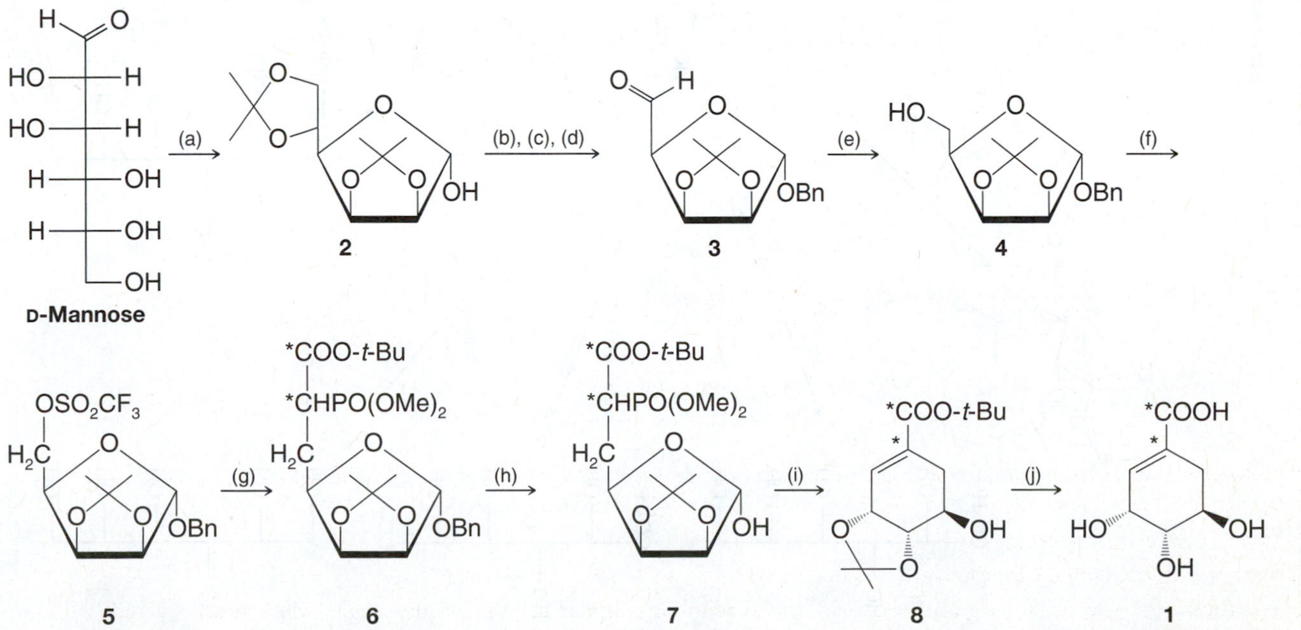

SCHEME 1 The synthesis of D-(−)-[1,7-^{13}C]-shikimic acid (1) by H. G. Floss, based on the route of Fleet et al. Conditions: (a) acetone, HA; (b) BnCl, NaH; (c) HCl, aq. MeOH; (d) NaIO$_4$; (e) NaBH$_4$; (f) (CF$_3$SO$_2$)$_2$O, pyridine; (g) **9**, NaH; (h) HCOO$^−$NH$_4^+$, Pd/C; (i) NaH; (j) 60% aq. CF$_3$COOH.

[S U M M A R Y A N D R E V I E W T O O L S]

A Summary of Reactions Involving Monosaccharides

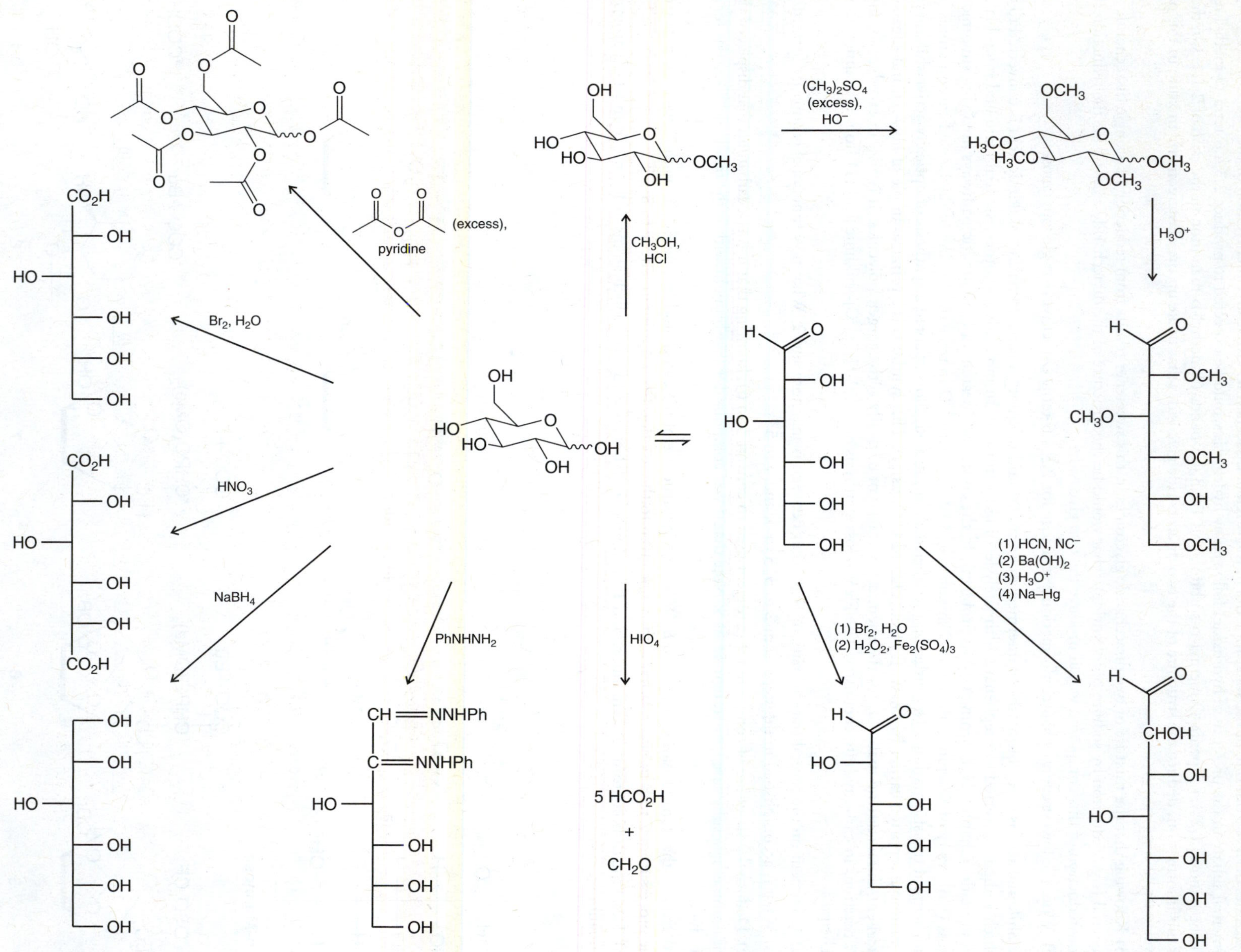

CHAPTER

23

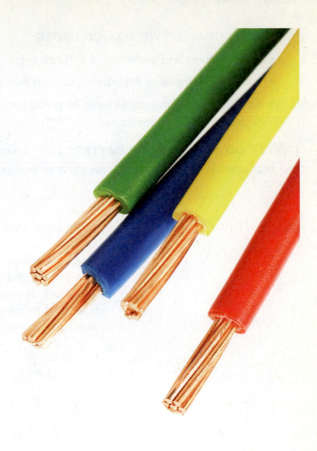

Lipids

I f you have ever worked with electrical wires, you know that a live bare wire will short circuit if it touches another conductor. To make sure that such a circuit does not do this, potentially causing a fire or an injury to an electrician, electrical wires are always insulated with a nonconducting material like plastic. Electrical signaling in our bodies occurs in much the same way through the connections between nerve cells. There, the insulation is provided by lipid-rich cells comprising the myelin sheath that wraps in layers encircling the long, thin nerve fibers called axons. In cross section, cells of the myelin sheath look much like the rings of a cut tree trunk, and by insulating the nerve cell axons the myelin sheath increases the overall speed of the electrical signals, or impulses, in the nervous system (where rates can reach as high as 100 meters per second). The myelin sheath is also critical for proper neurological functioning. For instance, too little sheathing of the nerves, a condition known as demyelination, can result from the autoimmune disorder multiple sclerosis; this condition usually leads to serious challenges in muscle movement, among other disorders. Too much of these lipids can cause problems as well, something encountered in Tay Sach's disease, an ailment fatal to children under the age of 3. As we shall see in this chapter, lipids play a number of varied biochemical roles, and they are often obtained from natural sources along with another special class of molecules called steroids that regulate a number of critical functions.

IN THIS CHAPTER WE WILL CONSIDER:

- the structures and properties of different lipids

- selected examples of important lipids and their functions

- how lipid-based molecules serve as precursors to a number of unique carbon frameworks, including steroids, waxes, and other signaling molecules

[**WHY** DO THESE TOPICS MATTER?]   At the end of this chapter we will show how one particularly unique steroid can both account for a classical Greek myth as well as hold promise as a potential new cancer therapy.

23.1 INTRODUCTION

Lipids are compounds of biological origin that dissolve in nonpolar solvents, such as chloroform and diethyl ether. The name lipid comes from the Greek word *lipos*, for fat. Unlike carbohydrates and proteins, which are defined in terms of their structures, lipids are defined by the physical operation that we use to isolate them. Not surprisingly, then, lipids include a variety of structural types. Examples are the following:

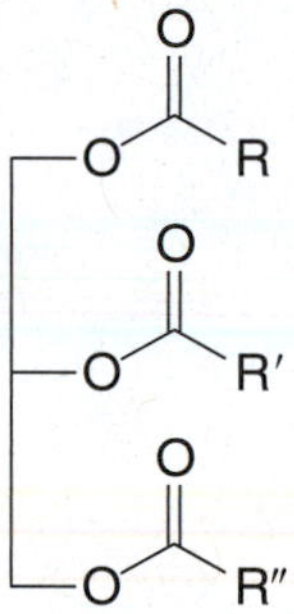

A fat or oil
(a triacylglycerol)

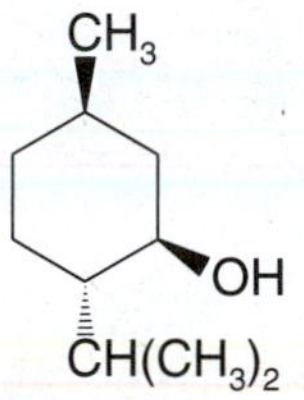

Menthol
(a terpenoid)

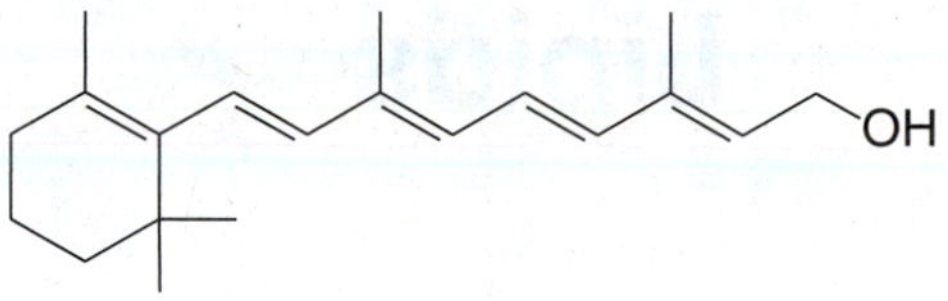

Vitamin A
(a terpenoid)

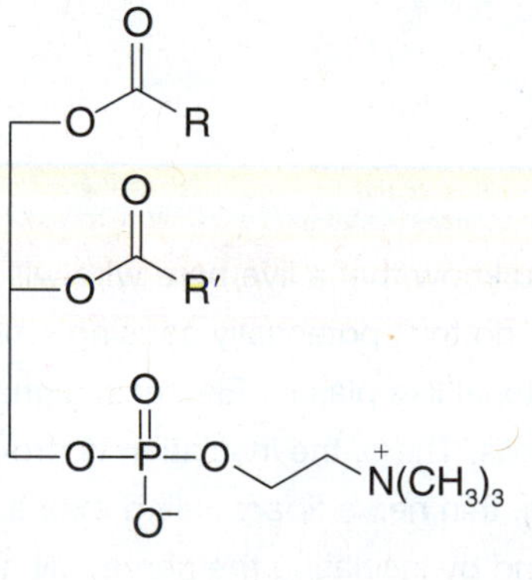

A lecithin
(a phosphatide)

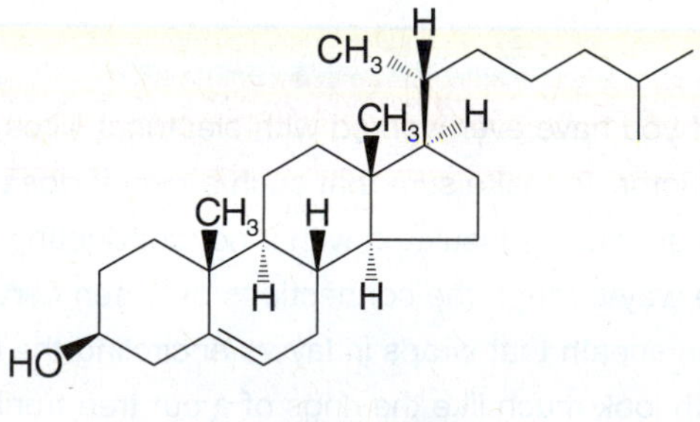

Cholesterol
(a steroid)

23.2 FATTY ACIDS AND TRIACYLGLYCEROLS

Only a small portion of the total lipid fraction obtained by extraction with a nonpolar solvent consists of long-chain carboxylic acids. Most of the carboxylic acids of biological origin are found as *esters of glycerol*, that is, as **triacylglycerols** (Fig. 23.1).*

Triacylglycerols are the oils of plants and the fats of animal origin. They include such common substances as peanut oil, soybean oil, corn oil, sunflower oil, butter, lard, and tallow.

---

*In the older literature triacylglycerols were referred to as triglycerides, or simply as glycerides. In IUPAC nomenclature, because they are esters of glycerol, they should be named as glyceryl trialkanoates, glyceryl trialkenoates, and so on.

- Triacylglycerols that are liquids at room temperature are generally called **oils**; those that are solids are called **fats**.

Triacylglycerols can be **simple triacylglycerols** in which all three acyl groups are the same. More commonly, however, the triacylglycerol is a **mixed triacylglycerol** in which the acyl groups are different.

- Hydrolysis of a fat or oil produces a mixture of **fatty acids**:

$$\text{A fat or oil} \xrightarrow[\text{(2) } H_3O^+]{\text{(1) } HO^- \text{ in } H_2O, \text{ heat}} \text{Glycerol} + \text{Fatty acids}$$

A fat or oil   **Glycerol**   **Fatty acids**

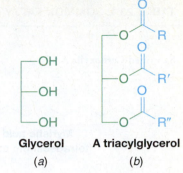

Glycerol   **A triacylglycerol**
    (*a*)   (*b*)

FIGURE 23.1 (*a*) Glycerol. (*b*) A triacylglycerol. The groups R, R′, and R″ are usually long-chain alkyl groups. R, R′, and R″ may also contain one or more carbon–carbon double bonds. In a triacylglycerol R, R′, and R″ may all be different.

- Most natural fatty acids have **unbranched chains** and, because they are synthesized from two-carbon units, **they have an even number of carbon atoms**.

Table 23.1 lists some of the most common fatty acids, and Table 23.2 gives the fatty acid composition of a number of common fats and oils. Notice that in the **unsaturated fatty acids** in Table 23.1 **the double bonds are all cis**. Many naturally occurring fatty acids contain two or three double bonds. The fats or oils that these come from are called **polyunsaturated fats** or oils. The first double bond of an unsaturated fatty acid commonly occurs between C9 and C10; the remaining double bonds tend to begin with C12 and C15 (as in linoleic acid and linolenic acid). The double bonds, therefore, *are not conjugated*. Triple bonds rarely occur in fatty acids.

The carbon chains of **saturated fatty acids** can adopt many conformations but tend to be fully extended because this minimizes steric repulsions between neighboring methylene groups.

- Saturated fatty acids pack efficiently into crystals, and because dispersion force attractions are large, they have relatively high melting points. The melting points increase with increasing molecular weight.
- The cis configuration of the double bond of an unsaturated fatty acid puts a rigid bend in the carbon chain that interferes with crystal packing, causing reduced dispersion force attractions between molecules. Unsaturated fatty acids, consequently, have lower melting points.

Fatty acids known as omega-3 fatty acids are those where the third to last carbon in the chain is part of a carbon–carbon double bond. Long-chain omega-3 fatty acids incorporated in the diet are believed to have beneficial effects in terms of reducing the risk of fatal heart attack and easing certain autoimmune diseases, including rheumatoid arthritis and psoriasis. Oil from fish such as tuna and salmon is a good source of omega-3 fatty acids, including the C_{22} omega-3 fatty acid docosahexaenoic acid [DHA, whose full IUPAC name is (4*Z*, 7*Z*,10*Z*,13*Z*,16*Z*,19*Z*)-4,7,10,13,16,19-docosahexaenoic acid]. DHA is also found in breast milk, gray matter of the brain, and retinal tissue.

**(4Z,7Z,10Z,13Z,16Z,19Z)-4,7,10,13,16,19-Docosahexaenoic acid
(DHA, an omega-3 fatty acid)**

What we have just said about the fatty acids applies to the triacylglycerols as well. Triacylglycerols made up of largely saturated fatty acids have high melting points and are solids at room temperature. They are what we call *fats*. Triacylglycerols with a high proportion of unsaturated and polyunsaturated fatty acids have lower melting points and most are *oils*.

Helpful Hint

We saw how fatty acids are biosynthesized in two-carbon units in Special Topic E (*WileyPLUS*).

A saturated triacylglycerol

TABLE 23.1 COMMON FATTY ACIDS

	mp (°C)
Saturated Carboxylic Acids	

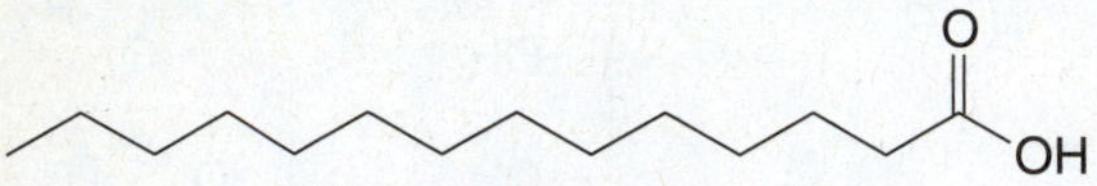

Myristic acid
(tetradecanoic acid) — 54

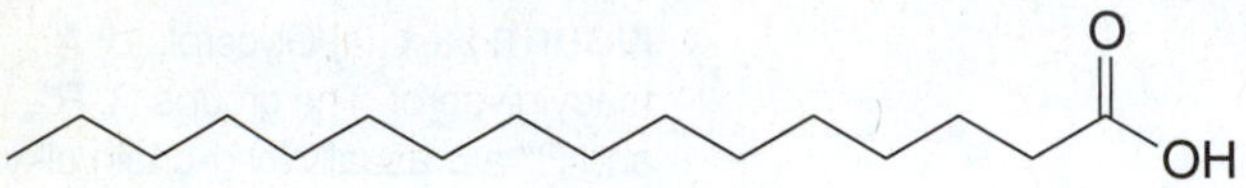

Palmitic acid
(hexadecanoic acid) — 63

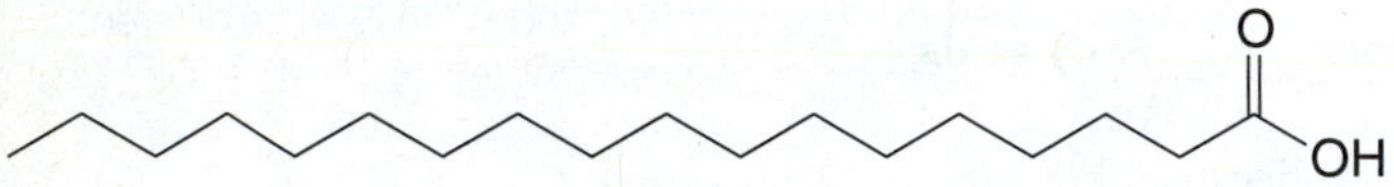

Stearic acid
(octadecanoic acid) — 70

Unsaturated Carboxylic Acids

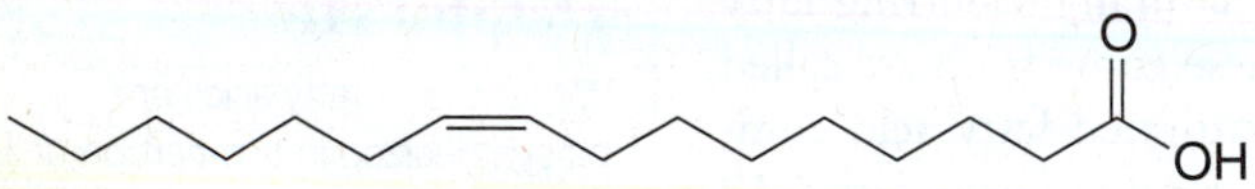

Palmitoleic acid
(*cis*-9-hexadecenoic acid) — 32

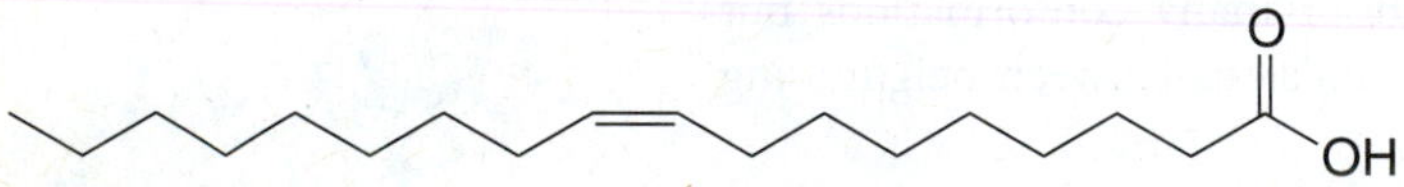

Oleic acid
(*cis*-9-octadecenoic acid) — 4

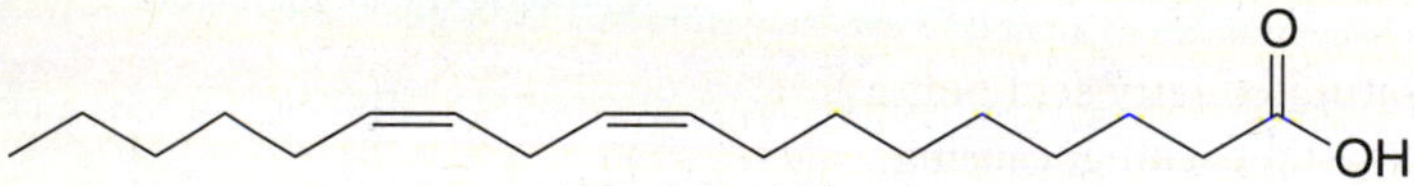

Linoleic acid
(*cis*,*cis*-9,12-octadecadienoic acid) — −5

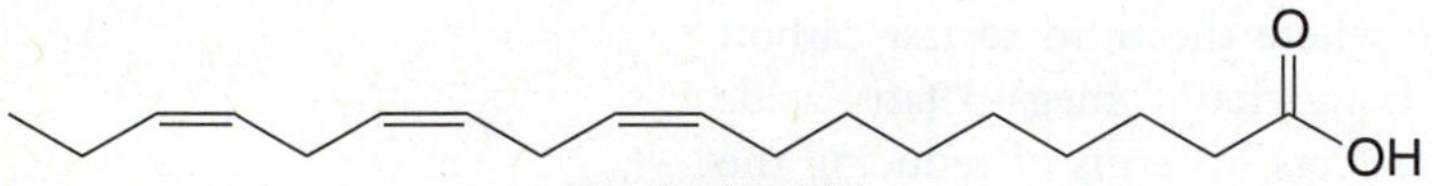

Linolenic acid
(*cis*,*cis*,*cis*-9,12,15-octadecatrienoic acid) — −11

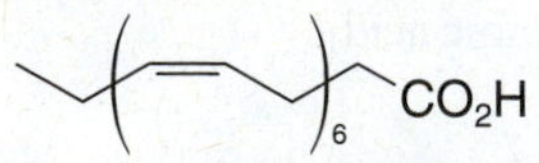

DHA, an omega-3 fatty acid
[(4Z,7Z,10Z,13Z,16Z,19Z)-4,7,10,13,16,19-docosahexaenoic acid] — −44

Arachidonic acid, an omega-6 fatty acid
[(5Z,8Z,11Z,14Z)-5,8,11,14-eicosatetraenoic acid] — −49

TABLE 23.2 FATTY ACID COMPOSITION OBTAINED BY HYDROLYSIS OF COMMON FATS AND OILS

	Average Composition of Fatty Acids (mol %)											
	Saturated								Unsaturated			
Fat or Oil	C_4 Butyric Acid	C_6 Caproic Acid	C_8 Caprylic Acid	C_{10} Capric Acid	C_{12} Lauric Acid	C_{14} Myristic Acid	C_{16} Palmitic Acid	C_{18} Stearic Acid	C_{16} Palmitoleic Acid	C_{18} Oleic Acid	C_{18} Linoleic Acid	C_{18} Linolenic Acid
Animal Fats												
Butter	3–4	1–2	0–1	2–3	2–5	8–15	25–29	9–12	4–6	18–33	2–4	
Lard						1–2	25–30	12–18	4–6	48–60	6–12	0–1
Beef tallow						2–5	24–34	15–30		35–45	1–3	0–1
Vegetable Oils												
Olive						0–1	5–15	1–4		67–84	8–12	
Peanut							7–12	2–6		30–60	20–38	
Corn						1–2	7–11	3–4	1–2	25–35	50–60	
Cottonseed						1–2	18–25	1–2	1–3	17–38	45–55	
Soybean						1–2	6–10	2–4		20–30	50–58	5–10
Linseed							4–7	2–4		14–30	14–25	45–60
Coconut		0–1	5–7	7–9	40–50	15–20	9–12	2–4	0–1	6–9	0–1	
Marine Oils												
Cod liver						5–7	8–10	0–1	18–22	27–33	27–32	

Reprinted with permission of John Wiley & Sons, Inc., from Holum, J. R., *Organic and Biological Chemistry*, p. 220. Copyright 1978.

Figure 23.2 shows how the introduction of a single cis double bond affects the shape of a triacylglycerol and how catalytic hydrogenation can be used to convert an unsaturated triacylglycerol into a saturated one.

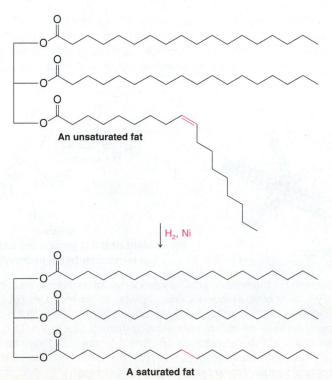

FIGURE 23.2 Two typical triacylglycerols, one unsaturated and one saturated. The cis double bond of the unsaturated triacylglycerol interferes with efficient crystal packing and causes an unsaturated fat to have a lower melting point. Hydrogenation of the double bond causes an unsaturated triacylglycerol to become saturated.

23.2A Hydrogenation of Triacylglycerols

Solid commercial cooking fats are manufactured by partial hydrogenation of vegetable oils. The result is the familiar "partially hydrogenated fat" present in so many prepared foods. Complete hydrogenation of the oil is avoided because a completely saturated triacylglycerol is very hard and brittle. Typically, the vegetable oil is hydrogenated until a semisolid of appealing consistency is obtained. One commercial advantage of partial hydrogenation is to give the fat a longer shelf life. Polyunsaturated oils tend to react by autoxidation (Section 10.12D), causing them to become rancid. One problem with partial hydrogenation, however, is that the catalyst isomerizes some of the unreacted double bonds from the natural cis arrangement to the unnatural trans arrangement, and there is accumulating evidence that trans fats are associated with an increased risk of cardiovascular disease.

23.2B Biological Functions of Triacylglycerols

The primary function of triacylglycerols in animals is as an energy reserve. When triacylglycerols are converted to carbon dioxide and water by biochemical reactions (i.e., when triacylglycerols are *metabolized*), they yield more than twice as many kilocalories per gram as do carbohydrates or proteins. This is largely because of the high proportion of carbon–hydrogen bonds per molecule.

In animals, specialized cells called **adipocytes** (fat cells) synthesize and store triacylglycerols. The tissue containing these cells, adipose tissue, is most abundant in the abdominal cavity and in the subcutaneous layer. Men have a fat content of about 21%, women about 26%. This fat content is sufficient to enable us to survive starvation for 2–3 months. By contrast, glycogen, our carbohydrate reserve, can provide only one day's energy need.

All of the saturated triacylglycerols of the body, and some of the unsaturated ones, can be synthesized from carbohydrates and proteins. Certain polyunsaturated fatty acids, however, are essential in the diets of higher animals.

The amount of fat in the diet, especially the proportion of saturated fat, has been a health concern for many years. There is compelling evidence that too much saturated fat in the diet is a factor in the development of heart disease and cancer.

THE CHEMISTRY OF... Olestra and Other Fat Substitutes

Olestra is a zero-calorie commercial fat substitute with the look and feel of natural fats. It is a synthetic compound whose structure involves a novel combination of natural components. The core of olestra is derived from sucrose, ordinary table sugar. Six to eight of the hydroxyl groups on the sucrose framework have long-chain carboxylic acids (fatty acids) appended to them by ester linkages. These fatty acids are from C_8 to C_{22} in length. In the industrial synthesis of olestra, these fatty acids derive from cottonseed or soybean oil.

(Illustration in center reprinted with permission from Doyle, E. Olestra? The Jury's Still Out. *Journal of Chemical Education*, Vol. 74, No. 4, 1997, pp. 370–372; © 1997, Division of Chemical Education, Inc. Copyright 1997 American Chemical Society.)

A food product made with olestra.

Olestra.

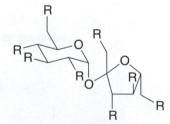

Olestra
Six to eight of the R groups are fatty acid esters, the remainder being hydroxyl groups.

The presence of fatty acid esters in olestra bestows on it the taste and culinary properties of an ordinary fat. Yet, olestra is not digestible like a typical fat. This is because the steric bulk of olestra renders it unacceptable to the enzymes that catalyze hydrolysis of ordinary fats. Olestra passes through the digestive tract unchanged and thereby adds no calories to the diet. As it does so, however, olestra associates with and carries away some of the lipid-soluble vitamins, namely, vitamins A, D, E, and K. Foods prepared with olestra are supplemented with these vitamins to compensate for any loss that may result from their extraction by olestra. Studies conducted since olestra's approval have demonstrated that people report no more bothersome digestive effects when eating Olean (the trademark name for olestra) snacks than they do when eating full-fat chips.

Many other fat substitutes have received consideration. Among these are polyglycerol esters, which presumably by their steric bulk would also be undigestible, like the polyester olestra. Another approach to low-calorie fats, already in commercial use, involves replacement of some long-chain carboxylic acids on the glycerol backbone with medium- or short-chain carboxylic acids (C_2 to C_4). These compounds provide fewer calories because each CH_2 group that is absent from the glycerol ester (as compared to long-chain fatty acids) reduces the amount of energy (calories) liberated when that compound is metabolized. The calorie content of a given glycerol ester can essentially be tailored to provide a desired calorie output, simply by adjusting the ratio of long-chain to medium- and short-chain carboxylic acids. Still other low-calorie fat substitutes are carbohydrate- and protein-based compounds. These materials act by generating a similar gustatory response to that of fat, but for various reasons produce fewer calories.

23.2C Saponification of Triacylglycerols

- **Saponification** is the alkaline hydrolysis of triacylglycerols, leading to glycerol and a mixture of salts of long-chain carboxylic acids:

$$\xrightarrow[\text{H}_2\text{O}]{3\ \text{NaOH}}$$

Glycerol   **Sodium carboxylates "soap"**

These salts of long-chain carboxylic acids are **soaps**, and this saponification reaction is the way most soaps are manufactured. Fats and oils are boiled in aqueous sodium hydroxide until hydrolysis is complete. Adding sodium chloride to the mixture then causes the soap to precipitate. (After the soap has been separated, glycerol can be isolated from the aqueous phase by distillation.) Crude soaps are usually purified by several reprecipitations. Perfumes can be added if a toilet soap is the desired product. Sand, sodium carbonate, and other fillers can be added to make a scouring soap, and air can be blown into the molten soap if the manufacturer wants to market a soap that floats.

The sodium salts of long-chain carboxylic acids (soaps) are almost completely miscible with water. However, they do not dissolve as we might expect, that is, as individual ions. Except in very dilute solutions, soaps exists as **micelles** (Fig. 23.3). Soap micelles

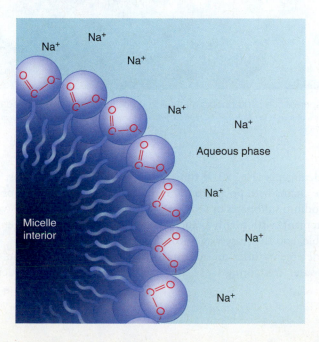

FIGURE 23.3 A portion of a soap micelle showing its interface with the polar dispersing medium.
(Reprinted with permission of John Wiley & Sons, Inc., from Karp, G., *Cell and Molecular Biology: Concepts and Experiments*, Fourth Edition, Copyright 1999.)

are usually spherical clusters of carboxylate anions that are dispersed throughout the aqueous phase. The carboxylate anions are packed together with their negatively charged (and thus, *polar*) carboxylate groups at the surface and with their nonpolar hydrocarbon chains on the interior. The sodium ions are scattered throughout the aqueous phase as individual solvated ions.

Micelle formation accounts for the fact that soaps dissolve in water. The nonpolar (and thus **hydrophobic**) alkyl chains of the soap remain in a nonpolar environment—in the interior of the micelle. The polar (and therefore **hydrophilic**) carboxylate groups are exposed to a polar environment—that of the aqueous phase. Because the surfaces of the micelles are negatively charged, individual micelles repel each other and remain dispersed throughout the aqueous phase.

Soaps serve their function as "dirt removers" in a similar way. Most dirt particles (e.g., on the skin) become surrounded by a layer of an oil or fat. Water molecules alone are unable to disperse these greasy globules because they are unable to penetrate the oily layer and separate the individual particles from each other or from the surface to which they are stuck. Soap solutions, however, *are* able to separate the individual particles because their hydrocarbon chains can "dissolve" in the oily layer (Fig. 23.4). As this happens, each individual particle develops an outer layer of carboxylate anions and presents the aqueous phase with a much more compatible exterior—a polar surface. The individual globules now repel each other and thus become dispersed throughout the aqueous phase. Shortly thereafter, they make their way down the drain.

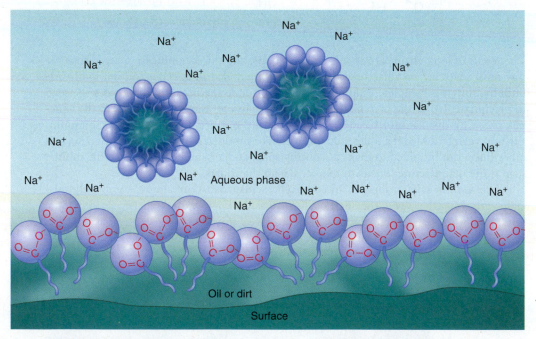

FIGURE 23.4 Dispersal of a hydrophobic material (e.g., oil, grease, or fat) by a soap.
(Adapted with permission of John Wiley & Sons, Inc., from Karp, G., *Cell and Molecular Biology: Concepts and Experiments*, Fourth Edition, Copyright 1999.)

Synthetic detergents (Fig. 23.5) function in the same way as soaps; they have long nonpolar alkane chains with polar groups at the end. The polar groups of most synthetic detergents are sodium sulfonates or sodium sulfates. (At one time, extensive use was made of synthetic detergents with highly branched alkyl groups. These detergents proved to be nonbiodegradable, and their use was discontinued.)

Synthetic detergents offer an advantage over soaps; they function well in "hard" water, that is, water containing Ca^{2+}, Fe^{2+}, Fe^{3+}, and Mg^{2+} ions. Calcium, iron, and magnesium salts of alkanesulfonates and alkyl hydrogen sulfates are largely water soluble, and thus synthetic detergents remain in solution. Soaps, by contrast, form precipitates—the ring around the bathtub—when they are used in hard water.

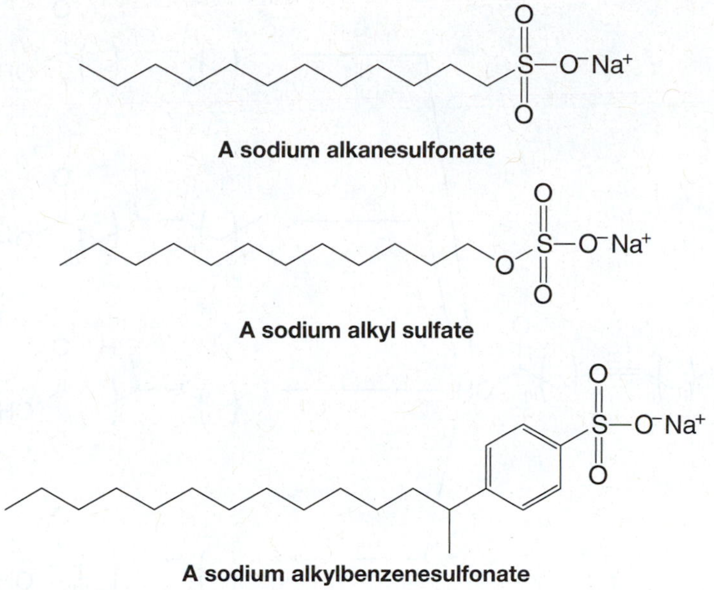

A sodium alkanesulfonate

A sodium alkyl sulfate

A sodium alkylbenzenesulfonate

FIGURE 23.5 Typical synthetic detergents.

23.2D Reactions of the Carboxyl Group of Fatty Acids

Fatty acids, as we might expect, undergo reactions typical of carboxylic acids (see Chapter 17). They react with $LiAlH_4$ to form alcohols, with alcohols and mineral acid to form esters, with bromine and phosphorus to form α-halo acids, and with thionyl chloride to form acyl chlorides:

> **Helpful Hint**
>
> The reactions presented in Sections 23.2D and 23.2E in the context of fatty acids are the same as those we studied in earlier chapters regarding carboxylic acids and alkenes.

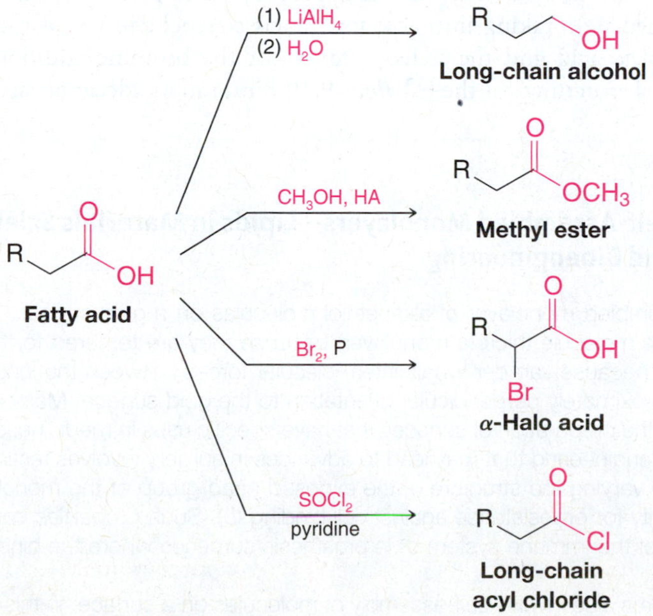

Fatty acid

(1) $LiAlH_4$
(2) H_2O → **Long-chain alcohol**

CH_3OH, HA → **Methyl ester**

Br_2, P → **α-Halo acid**

$SOCl_2$ / pyridine → **Long-chain acyl chloride**

23.2E Reactions of the Alkenyl Chain of Unsaturated Fatty Acids

The double bonds of the carbon chains of fatty acids undergo characteristic alkene addition reactions (see Chapters 7 and 8):

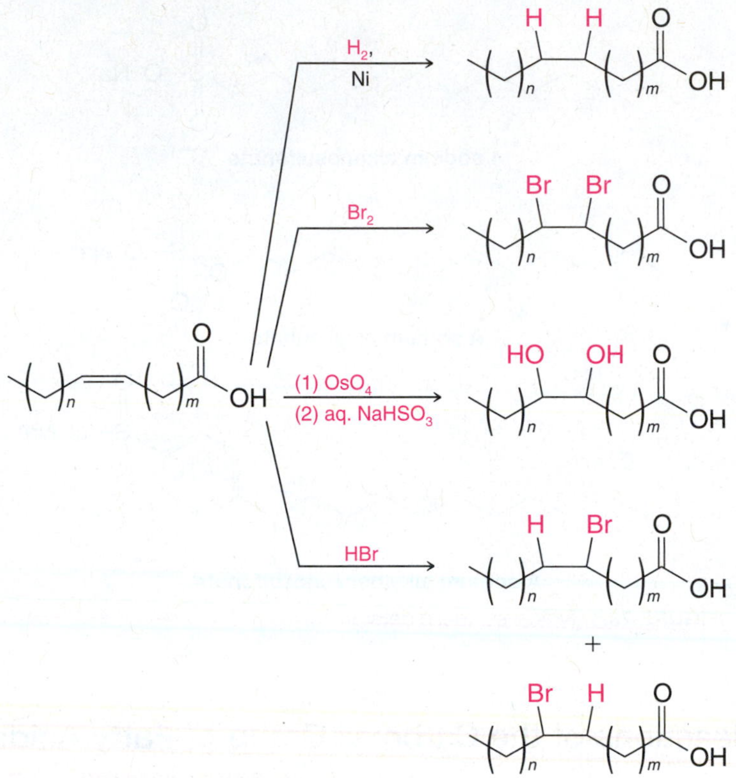

● ● ●

PRACTICE PROBLEM 23.1 **(a)** How many stereoisomers are possible for 9,10-dibromohexadecanoic acid?

(b) The addition of bromine to palmitoleic acid yields primarily one set of enantiomers, (±)-*threo*-9,10-dibromohexadecanoic acid. The addition of bromine is an anti addition to the double bond (i.e., it apparently takes place through a bromonium ion intermediate). Taking into account the cis stereochemistry of the double bond of palmitoleic acid and the stereochemistry of the bromine addition, write three-dimensional structures for the (±)-*threo*-9,10-dibromohexadecanoic acids.

THE CHEMISTRY OF... Self-Assembled Monolayers—Lipids in Materials Science and Bioengineering

The graphic shown below (*a*) depicts a self-assembled monolayer of alkanethiol molecules on a gold surface. The alkanethiol molecules spontaneously form a layer that is one molecule thick (a monolayer) because they are tethered to the gold surface at one end by a covalent bond to the metal and because van der Waals intermolecular forces between the long alkane chains cause them to align next to each other in an approximately perpendicular orientation to the gold surface. Many researchers are exploiting self-assembled monolayers (SAMs) for the preparation of surfaces that have specific uses in medicine, computing, and telecommunications. One example in biomedical engineering that may lead to advances in surgery involves testing cells for their response to SAMs with varying head groups. By varying the structure of the exposed head group of the monolayer, it may be possible to create materials that have either affinity for or resistance against cell binding (*b*). Such properties could be useful in organ transplants for inhibiting rejection by cells of the immune system or in prosthesis surgeries where the binding of tissue to the artificial device is desired.

Monolayers called Langmuir–Blodgett (LB) films also involve self-assembly of molecules on a surface. In this case, however, the molecules do not become covalently attached to the surface. These LB films are inherently less stable than covalently bonded monolayers, but they have characteristics that are useful for certain applications in nanotechnology. For example, an LB film made from phospholipid (Section 23.6) and catenane molecules was used in making the array of molecular switches we discussed in

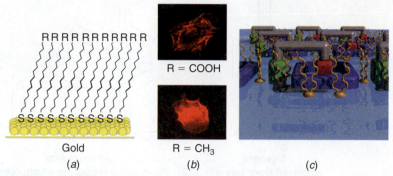

"The Chemistry of ... Nanoscale Motors and Molecular Switches" (Chapter 4). This LB monolayer (c) was formed at a water–air interface where the polar phosphate head groups of the phospholipid buried themselves in water and the hydrophobic carbon tails projected out into the air. Interspersed among them were the catenane molecules. In later steps, this monolayer was lifted from the water–air surface and transferred onto a solid gold surface.

R R R R R R R R R R R

S S S S S S S S S S S

Gold

(a)

R = COOH

R = CH₃

(b)

(c)

(a) A self-assembled monolayer of alkanethiol molecules on a gold surface (R = CH_3 or COOH). (b) Spreading of a Swiss 3T3 fibroblast cell plated on a COOH-terminated self-assembled monolayer (top) indicates effective signaling on the surface. The fibroblast cell on a CH_3-terminated monolayer (bottom) curls away from surface. The cells were stained with a rhodamine-tagged toxin that binds to filamentous actin and then were imaged under fluorescent light. (c) A Langmuir–Blodgett (LB) film formed from phospholipid molecules (golden color) and catenane molecules (purple and gray with green and red groups) at an air–water interface.

(Part (c) is reprint with permission from Pease, A. R.; Jeppensen, J. O.; Stoddart, J. F.; Luo, Y.; Colier, C. P.; Heath, J. R., *Accounts of Chemical Research*, **2001**, *34*, 433–444.)

23.3 TERPENES AND TERPENOIDS

People have isolated organic compounds from plants since antiquity. By gently heating or by steam distilling certain plant materials, one can obtain mixtures of odoriferous compounds known as **essential oils**. These compounds have had a variety of uses, particularly in early medicine and in the making of perfumes.

As the science of organic chemistry developed, chemists separated the various components of these mixtures and determined their molecular formulas and, later, their structural formulas. Even today these natural products offer challenging problems for chemists interested in structure determination and synthesis. Research in this area has also given us important information about the ways the plants themselves synthesize these compounds.

- Hydrocarbons known generally as **terpenes** and oxygen-containing compounds called **terpenoids** are the most important constituents of essential oils.

- Most terpenes have skeletons of 10, 15, 20, or 30 carbon atoms and are classified in the following way:

Number of Carbon Atoms	Class
10	Monoterpenes
15	Sesquiterpenes
20	Diterpenes
30	Triterpenes

- One can view terpenes as being built up from two or more C_5 units known as **isoprene units**. Isoprene is 2-methyl-1,3-butadiene.

Isoprene and the isoprene unit can be represented in various ways:

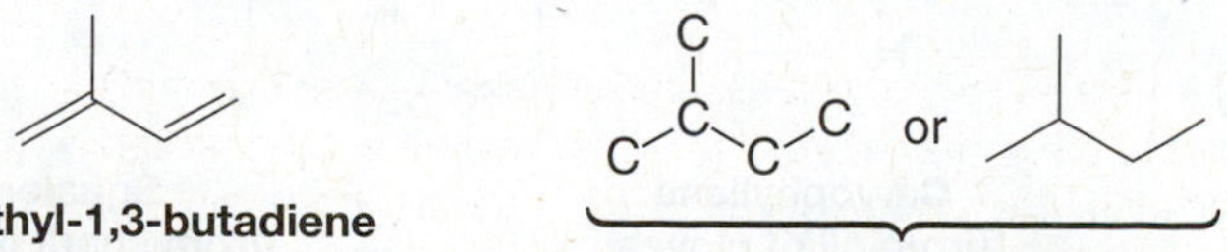

2-Methyl-1,3-butadiene (isoprene)

C—C—C—C   or
    |
    C

An isoprene unit

Helpful Hint

Terpene biosynthesis is described in Special Topic E (*WileyPLUS*).

We now know that plants do not synthesize terpenes from isoprene (see Special Topic E, *WileyPLUS*). However, recognition of the isoprene unit as a component of the structure of terpenes has been a great aid in elucidating their structures. We can see how if we examine the following structures:

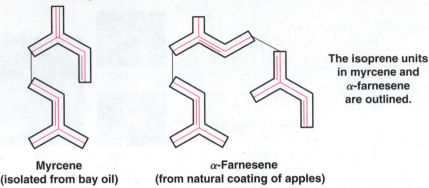

Myrcene
(isolated from bay oil)

α-Farnesene
(from natural coating of apples)

The isoprene units
in myrcene and
α-farnesene
are outlined.

By the outlines in the formulas above, we can see that the monoterpene (myrcene) has two isoprene units; the sesquiterpene (α-farnesene) has three. In both compounds the isoprene units are linked head to tail:

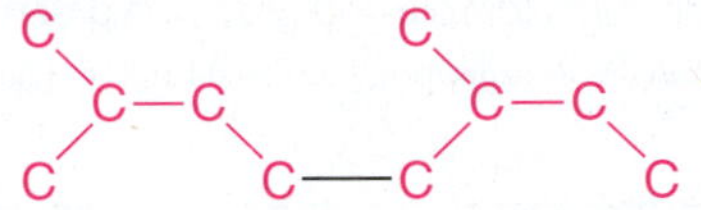

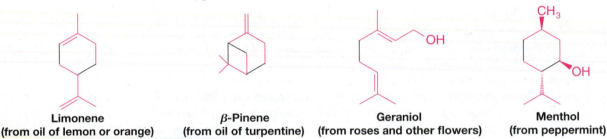

(head)   (tail) (head)   (tail)

Many terpenes also have isoprene units linked in rings, and others (terpenoids) contain oxygen:

Limonene
(from oil of lemon or orange)

β-Pinene
(from oil of turpentine)

Geraniol
(from roses and other flowers)

Menthol
(from peppermint)

●●● SOLVED PROBLEM 23.1

Hydrogenation of the sesquiterpene caryophyllene ($C_{15}H_{24}$) produces a compound with the molecular formula $C_{15}H_{28}$. What information does this provide about the structure of caryophyllene?

STRATEGY AND ANSWER: The molecular formula $C_{15}H_{24}$ gives an index of hydrogen deficiency (IHD) of 4 for caryophyllene. Its reaction with two molar equivalents of hydrogen suggests that caryophyllene has two double bonds or one triple bond, accounting for two of the four units of hydrogen deficiency. The remaining two units of hydrogen deficiency are due to rings. (The structure of caryophyllene is given in Practice Problem 23.2.)

●●●

PRACTICE PROBLEM 23.2 **(a)** Show the isoprene units in each of the following terpenes. **(b)** Classify each as a monoterpene, sesquiterpene, diterpene, and so on.

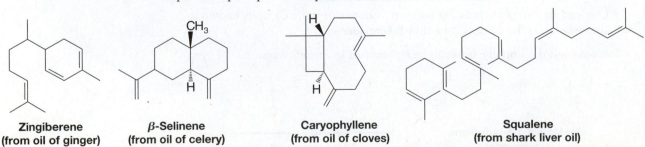

Zingiberene
(from oil of ginger)

β-Selinene
(from oil of celery)

Caryophyllene
(from oil of cloves)

Squalene
(from shark liver oil)

SOLVED PROBLEM 23.2

What products would you expect to obtain if caryophyllene were subjected to ozonolysis followed by workup with dimethyl sulfide?

ANSWER:

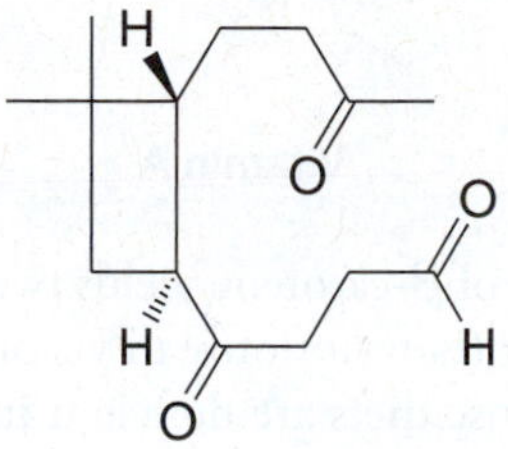

and HCHO (formaldehyde)

PRACTICE PROBLEM 23.3

What products would you expect to obtain if each of the following terpenes were subjected to ozonolysis and subsequent treatment with dimethyl sulfide?

(a) Myrcene **(b)** Limonene **(c)** α-Farnesene **(d)** Geraniol **(e)** Squalene

PRACTICE PROBLEM 23.4

Give structural formulas for the products that you would expect from the following reactions:

(a) β-Pinene $\xrightarrow{\text{KMnO}_4\text{, heat}}$

(b) Zingiberene $\xrightarrow{\text{H}_2\text{, Pt}}$

(c) Caryophyllene $\xrightarrow{\text{HCl}}$

(d) β-Selinene $\xrightarrow[\text{(2) H}_2\text{O}_2\text{, HO}^-]{\text{(1) BH}_3\text{:THF (2 equiv.)}}$

PRACTICE PROBLEM 23.5

What simple chemical test could you use to distinguish between geraniol and menthol?

The carotenes are tetraterpenes. They can be thought of as two diterpenes linked in tail-to-tail fashion:

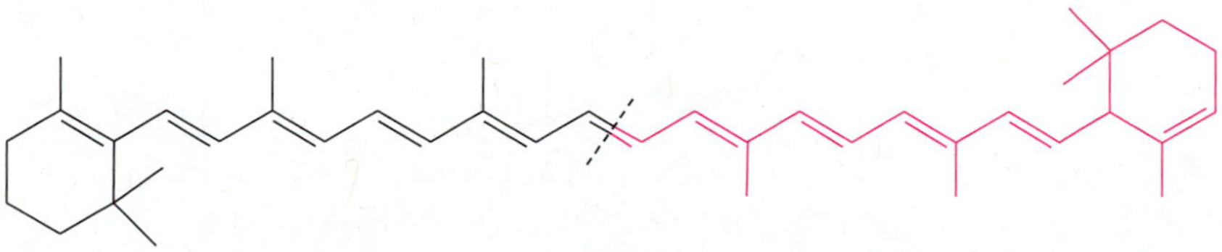

α-Carotene

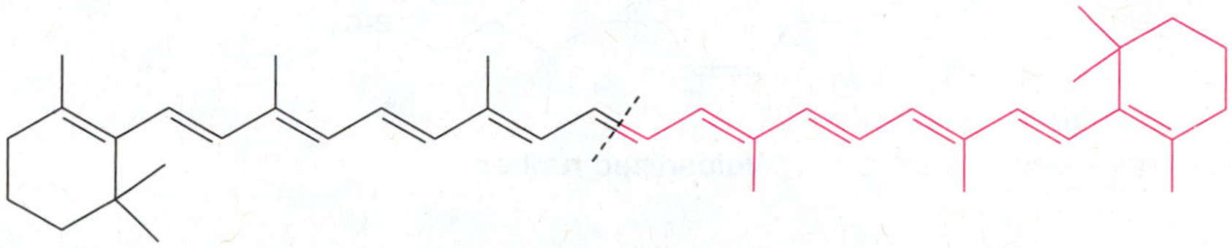

β-Carotene

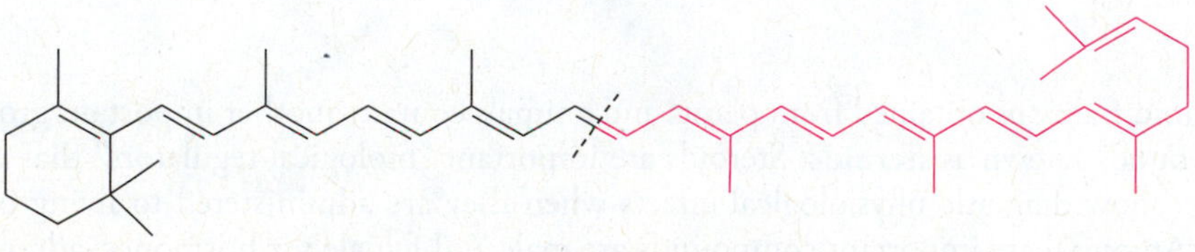

γ-Carotene

The carotenes are present in almost all green plants. In animals, all three carotenes serve as precursors for vitamin A, for they all can be converted to vitamin A by enzymes in the liver.

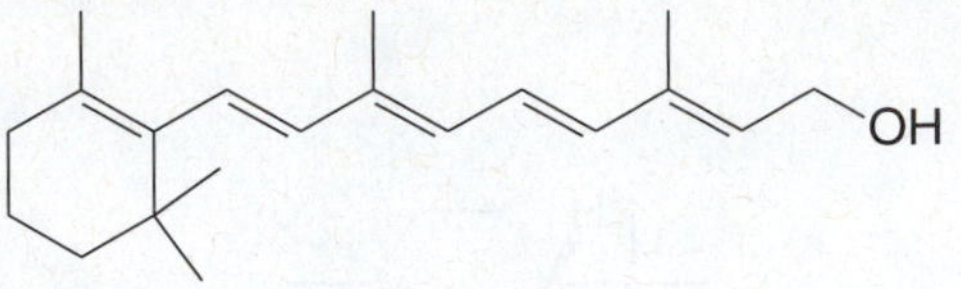

Vitamin A

In this conversion, one molecule of β-carotene yields two of vitamin A; α- and γ-carotene give only one. Vitamin A is important not only in vision but in many other ways as well. For example, young animals whose diets are deficient in vitamin A fail to grow. Vitamin A, β-carotene, and vitamin E ("The Chemistry of ... Antioxidants," Section 10.12) are also important lipid-soluble antioxidants.

23.3A Natural Rubber

Natural rubber can be viewed as a 1,4-addition polymer of isoprene. In fact, pyrolysis degrades natural rubber to isoprene. Pyrolysis (Greek: *pyros*, a fire, + *lysis*) is the heating of a substance in the absence of air until it decomposes. The isoprene units of natural rubber are all linked in a head-to-tail fashion, and all of the double bonds are cis:

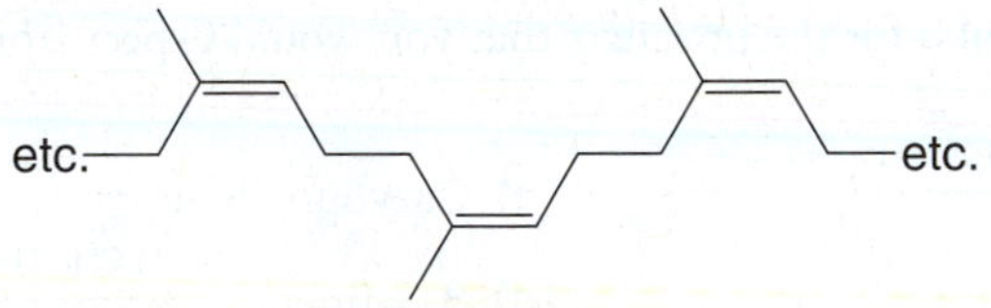

Natural rubber
(*cis*-1,4-polyisoprene)

Ziegler–Natta catalysts (see Special Topic B in *WileyPLUS*) make it possible to polymerize isoprene and obtain a synthetic product that is identical with the rubber obtained from natural sources.

Pure natural rubber is soft and tacky. To be useful, natural rubber has to be *vulcanized*. In vulcanization, natural rubber is heated with sulfur. A reaction takes place that produces cross-links between the *cis*-polyisoprene chains and makes the rubber much harder. Sulfur reacts both at the double bonds and at allylic hydrogen atoms:

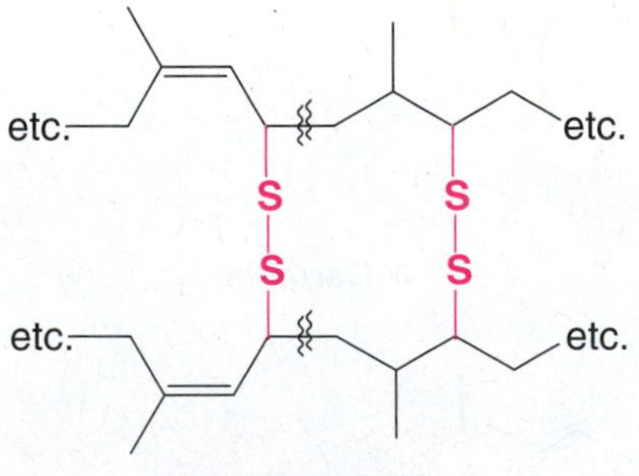

Vulcanized rubber

23.4 STEROIDS

The lipid fractions obtained from plants and animals contain another important group of compounds known as **steroids**. Steroids are important "biological regulators" that nearly always show dramatic physiological effects when they are administered to living organisms. Among these important compounds are male and female sex hormones, adrenocortical hormones, D vitamins, the bile acids, and certain cardiac poisons.

23.4A Structure and Systematic Nomenclature of Steroids

Steroids are derivatives of the following perhydrocyclopentanophenanthrene ring system:

The carbon atoms of this ring system are numbered as shown. The four rings are designated with letters. In most steroids the B,C and C,D ring junctions are trans. The A,B ring junction, however, may be either cis or trans, and this possibility gives rise to two general groups of steroids having the three-dimensional structures shown in Fig. 23.6.

The methyl groups that are attached at points of ring junction (i.e., those numbered 18 and 19) are called **angular methyl groups**, and they serve as important reference points for stereochemical designations. The angular methyl groups protrude above the general plane of the ring system when it is written in the manner shown in Fig. 23.6. By convention, other groups that lie on the same general side of the molecule as the angular methyl groups (i.e., on the top side) are designated β **substituents** (these are written with a solid wedge). Groups that lie generally on the bottom (i.e., are trans to the angular methyl groups) are designated α **substituents** (these are written with a dashed wedge). When α and β designations are applied to the hydrogen atom at position 5, the ring system in which the A,B ring junction is trans becomes the 5α series; the ring system in which the A,B ring junction is cis becomes the 5β series.

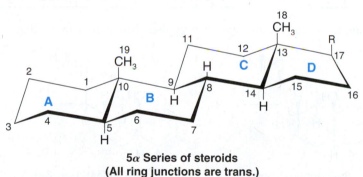

**5α Series of steroids
(All ring junctions are trans.)**

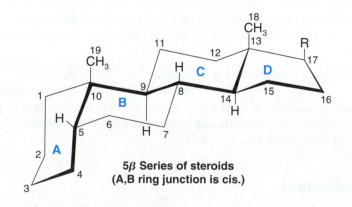

**5β Series of steroids
(A,B ring junction is cis.)**

FIGURE 23.6 The basic ring systems of the 5α and 5β series of steroids.

> ### Helpful Hint
> Build handheld molecular models of the 5α and 5β series of steroids and use them to explore the structures of steroids discussed in this chapter.

Draw the two basic ring systems given in Fig. 23.6 for the 5α and 5β series showing all hydrogen atoms of the cyclohexane rings. Label each hydrogen atom as to whether it is axial or equatorial.

PRACTICE PROBLEM 23.6

In systematic nomenclature the nature of the R group at position 17 determines (primarily) the base name of an individual steroid. These names are derived from the steroid hydrocarbon names given in Table 23.3.

TABLE 23.3 NAMES OF STEROID HYDROCARBONS

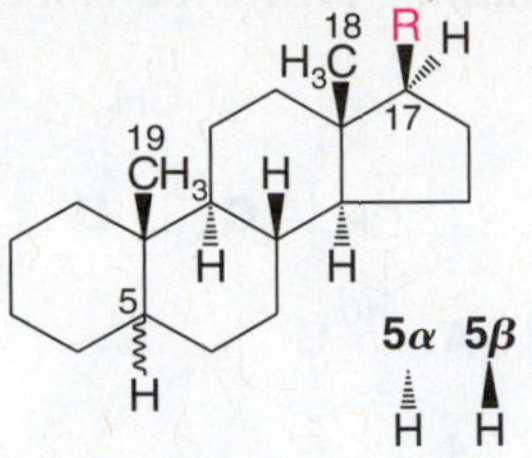

R	Name
H	Androstane
H (with —H also replacing —CH$_3$)	Estrane
 20 21	Pregnane
 21 20 22 23 24	Cholane
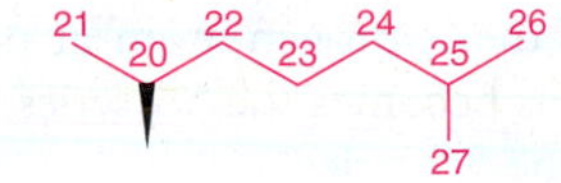 21 20 22 23 24 25 26 27	Cholestane

The following two examples illustrate the way these base names are used:

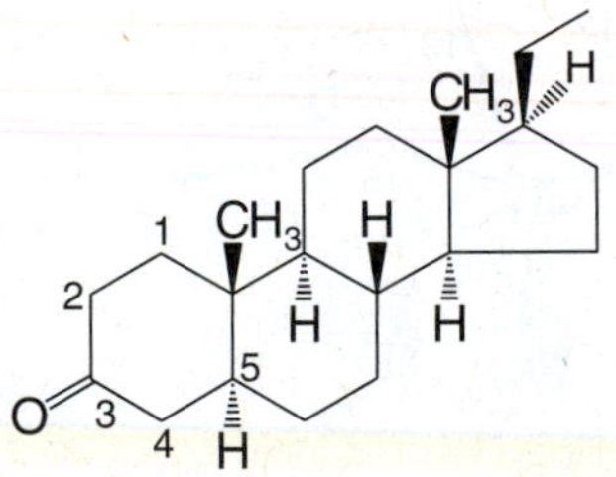

5α-Pregnan-3-one 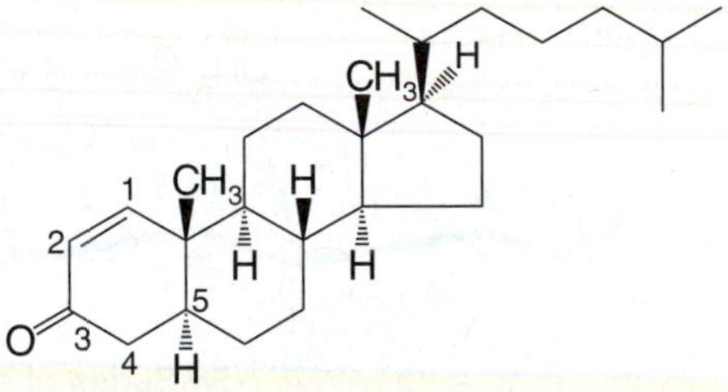   **5α-Cholest-1-en-3-one**

We shall see that many steroids also have common names and that the names of the steroid hydrocarbons given in Table 23.3 are derived from these common names.

• • •

PRACTICE PROBLEM 23.7 **(a)** Androsterone, a secondary male sex hormone, has the systematic name 3α-hydroxy-5α-androstan-17-one. Give a three-dimensional formula for androsterone.

(b) Norethynodrel, a synthetic steroid that has been widely used in oral contraceptives, has the systematic name 17α-ethynyl-17β-hydroxy-5(10)-estren-3-one. Give a three-dimensional formula for norethynodrel.

23.4B Cholesterol

Cholesterol, one of the most widely occurring steroids, can be isolated by extraction of nearly all animal tissues. Human gallstones are a particularly rich source.

Cholesterol was first isolated in 1770. In the 1920s, two German chemists, Adolf Windaus (University of Göttingen) and Heinrich Wieland (University of Munich), were responsible for outlining a structure for cholesterol; they received Nobel prizes for their work in 1927 and 1928.*

*The original cholesterol structure proposed by Windaus and Wieland was incorrect. This became evident in 1932 as a result of X-ray diffraction studies done by the British physicist J. D. Bernal. By the end of 1932, however, English scientists, and Wieland himself, using Bernal's results, were able to outline the correct structure of cholesterol.

Part of the difficulty in assigning an absolute structure to cholesterol is that cholesterol contains *eight* tetrahedral chirality centers. This feature means that 2^8, or 256, possible stereoisomeric forms of the basic structure are possible, *only one of which is cholesterol*:

5-Cholesten-3β-ol
(absolute configuration of cholesterol)

Helpful Hint

We saw how cholesterol is biosynthesized in "The Chemistry of...Cholesterol Biosynthesis" in *WileyPLUS* materials for Chapter 8.

Designate with asterisks the eight chirality centers of cholesterol.

PRACTICE PROBLEM 23.8

Cholesterol occurs widely in the human body, but not all of the biological functions of cholesterol are yet known. Cholesterol is known to serve as an intermediate in the biosynthesis of all of the steroids of the body. Cholesterol, therefore, is essential to life. We do not need to have cholesterol in our diet, however, because our body can synthesize all we need. When we ingest cholesterol, our body synthesizes less than if we ate none at all, but the total cholesterol is more than if we ate none at all. Far more cholesterol is present in the body than is necessary for steroid biosynthesis. High levels of blood cholesterol have been implicated in the development of arteriosclerosis (hardening of the arteries) and in heart attacks that occur when cholesterol-containing plaques block arteries of the heart. Considerable research is being carried out in the area of cholesterol metabolism with the hope of finding ways of minimizing cholesterol levels through the use of dietary adjustments or drugs.

It is important to note that, in common language, "cholesterol" does not necessarily refer only to the pure compound that chemists call cholesterol, but often refers instead to mixtures that contain cholesterol, other lipids, and proteins. These aggregates are called chylomicrons, high-density lipoproteins (HDLs), and low-density lipoproteins (LDLs). They have structures generally resembling globular micelles, and they are the vehicles by which cholesterol is transported through the aqueous environment of the body. Hydrophilic groups of their constituent proteins and phospholipids, and cholesterol hydroxyl substituents are oriented outward toward the water medium so as to facilitate transport of the lipids through the circulatory system. Chylomicrons transport dietary lipids from the intestines to the tissues. HDLs (the "good cholesterol") carry lipids from the tissues to the liver for degradation and excretion. LDL ("bad cholesterol") carries biosynthesized lipids from the liver to the tissues (see Fig. 23.7).

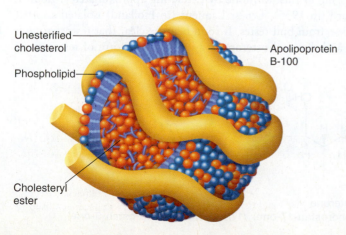

Unesterified cholesterol
Phospholipid
Apolipoprotein B-100
Cholesteryl ester

FIGURE 23.7 An LDL showing a core of cholesterol esters and a shell of phospholipids and unesterified cholesterol (hydroxyl groups exposed), wrapped in an apolipoprotein. The phospholipid head groups and hydrophilic residues of the protein support the water compatibility of the LDL particle.

(Reprinted with permission of John Wiley & Sons, Inc., from Voet, D. and Voet, J. G., *Biochemistry*, Second Edition. © 1995 Voet, D. and Voet, J. G.)

Certain compounds related to steroids and derived from plants are now known to lower total blood cholesterol when used in dietary forms approved by the FDA. Called phytostanols and phytosterols, these patented compounds act by inhibiting intestinal absorption of dietary cholesterol. They are marketed as food in the form of edible spreads. An example of a phytostanol is shown here.

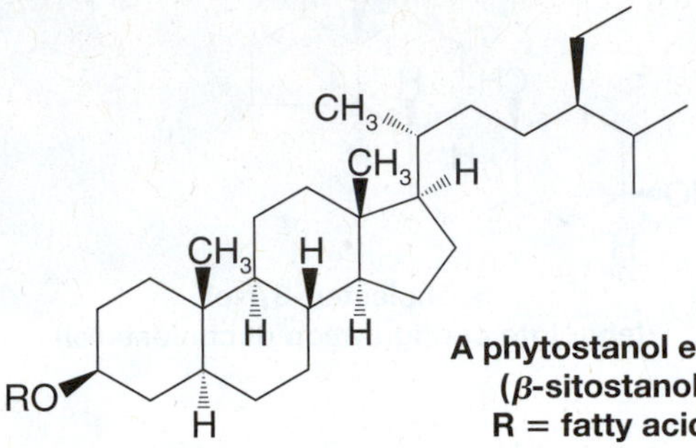

A phytostanol ester
(β-sitostanol,
R = fatty acid)

23.4C Sex Hormones

The sex hormones can be classified into three major groups: (1) the female sex hormones, or **estrogens**; (2) the male sex hormones, or **androgens**; and (3) the pregnancy hormones, or **progestins**.

The first sex hormone to be isolated was an estrogen, *estrone*. Working independently, Adolf Butenandt (in Germany at the University of Göttingen) and Edward Doisy (in the United States at St. Louis University) isolated estrone from the urine of pregnant women. They published their discoveries in 1929. Later, Doisy was able to isolate the much more potent estrogen, *estradiol*. In this research Doisy had to extract *4 tons* of sow ovaries in order to obtain just 12 mg of estradiol. Estradiol, it turns out, is the true female sex hormone, and estrone is a metabolized form of estradiol that is excreted.

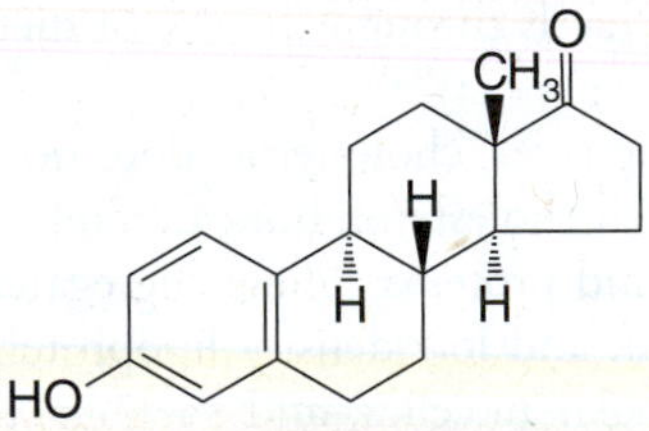

Estrone
[3-hydroxy-1,3,5(10)-estratrien-17-one]

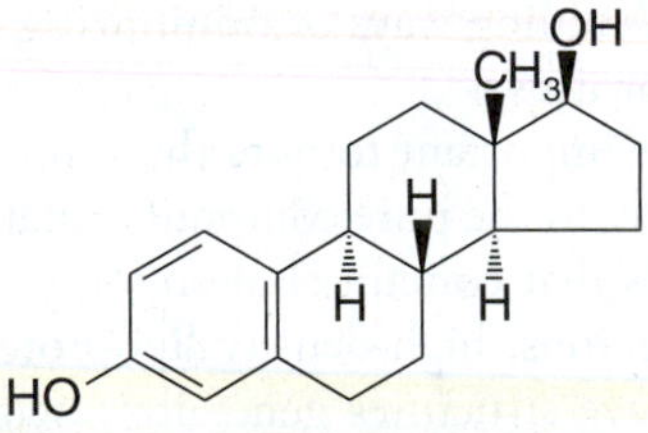

Estradiol
[1,3,5(10)-estratriene-3,17β-diol]

Estradiol is secreted by the ovaries and promotes the development of the secondary female characteristics that appear at the onset of puberty. Estrogens also stimulate the development of the mammary glands during pregnancy and induce estrus (heat) in animals.

In 1931, Butenandt and Kurt Tscherning isolated the first androgen, *androsterone*. They were able to obtain 15 mg of this hormone by extracting approximately 15,000 L of male urine. Soon afterward (in 1935), Ernest Laqueur (in Holland) isolated another male sex hormone, *testosterone*, from bull testes. It soon became clear that testosterone is the true male sex hormone and that androsterone is a metabolized form of testosterone that is excreted in the urine.

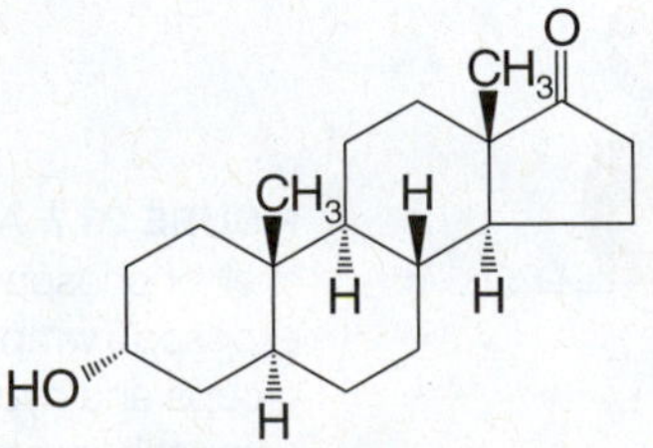

Androsterone
(3α-hydroxy-5α-androstan-17-one)

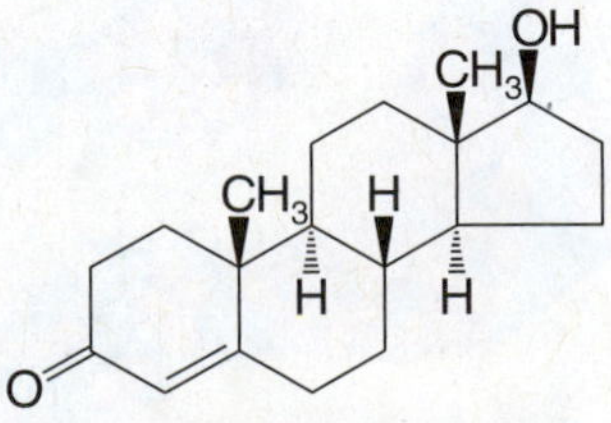

Testosterone
(17β-hydroxy-4-androsten-3-one)

Testosterone, secreted by the testes, is the hormone that promotes the development of secondary male characteristics: the growth of facial and body hair, the deepening of the voice, muscular development, and the maturation of the male sex organs.

Testosterone and estradiol, then, are the chemical compounds from which "maleness" and "femaleness" are derived. It is especially interesting to examine their structural formulas and see how very slightly these two compounds differ. Testosterone has an angular methyl group at the A,B ring junction that is missing in estradiol. Ring A of estradiol is a benzene ring and, as a result, estradiol is a phenol. Ring A of testosterone contains an α,β-unsaturated keto group.

● ● ●

PRACTICE PROBLEM 23.9

The estrogens (estrone and estradiol) are easily separated from the androgens (androsterone and testosterone) on the basis of one of their chemical properties. What is that property, and how could such a separation be accomplished?

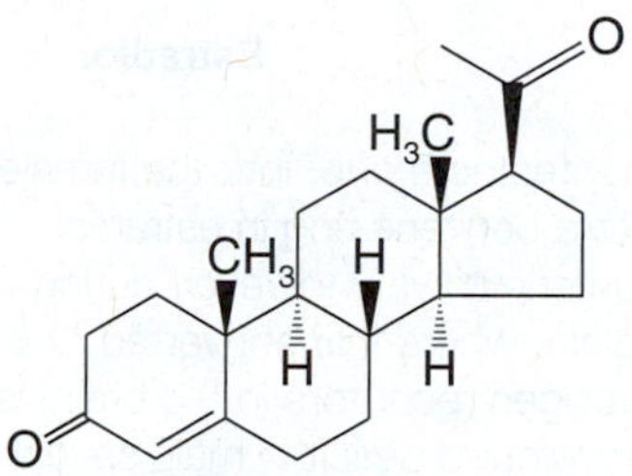

Progesterone
(4-pregnene-3,20-dione)

Progesterone is the most important *progestin* (pregnancy hormone). After ovulation occurs, the remnant of the ruptured ovarian follicle (called the *corpus luteum*) begins to secrete progesterone. This hormone prepares the lining of the uterus for implantation of the fertilized ovum, and continued progesterone secretion is necessary for the completion of pregnancy. (Progesterone is secreted by the placenta after secretion by the corpus luteum declines.)

Progesterone *also suppresses ovulation*, and it is the chemical agent that apparently accounts for the fact that pregnant women do not conceive again while pregnant. It was this observation that led to the search for synthetic progestins that could be used as oral contraceptives. (Progesterone itself requires very large doses to be effective in suppressing ovulation when taken orally because it is degraded in the intestinal tract.) A number of such compounds have been developed and are now widely used. In addition to norethynodrel (see Practice Problem 23.7), another widely used synthetic progestin is its double-bond isomer, *norethindrone*:

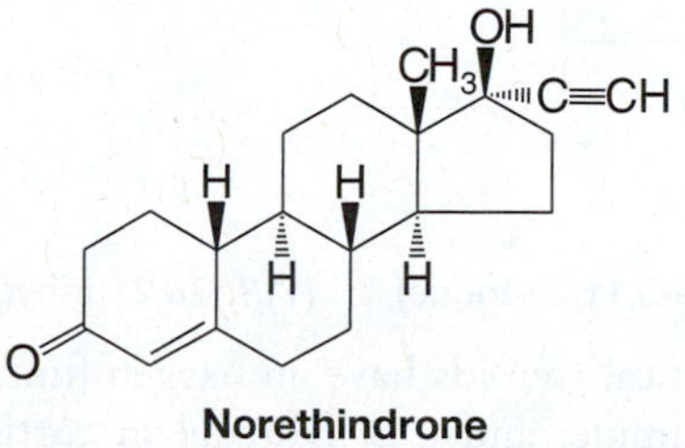

Norethindrone
(17α-ethynyl-17 β-hydroxy-4-estren-3-one)

Synthetic estrogens have also been developed, and these are often used in oral contraceptives in combination with synthetic progestins. A very potent synthetic estrogen is the compound called *ethynylestradiol* or *novestrol*:

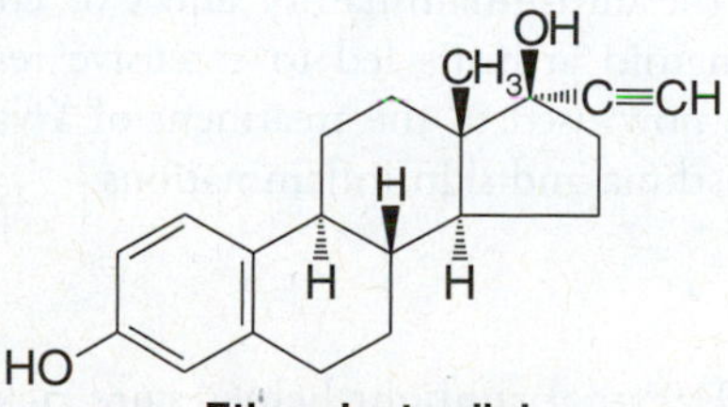

Ethynylestradiol
[17α-ethynyl-1,3,5(10)-estratriene-3,17β-diol]

THE CHEMISTRY OF... The Enzyme Aromatase

Look at the structures for testosterone and estradiol below. Testosterone is the primary male sex hormone, or **androgen**. It is the hormone that promotes the development of secondary male characteristics at puberty, such as muscular development and the maturation of the male sex organs. Estradiol is the primary **estrogen**. Estrogens promote the development of secondary female characteristics that occur at the onset of puberty and regulate the reproductive cycle. A significant molecular difference between the two hormones is the presence of a benzene ring in the female sex hormone.

Testosterone   Aromatase →   **Estradiol**

Aromatase is an enzyme that converts the male sex hormone, testosterone, into the female sex hormone, estradiol. In the course of this transformation, ring A of testosterone is converted to a benzene ring in estradiol.

Estrogen is essential for male behaviors. This may seem counterintuitive. However, during fetal development, testosterone produced in the male fetus interacts with aromatase in the fetal brain, where it is converted to estrogen. There is mounting evidence that this locally produced estrogen (which interacts with estrogen receptors in the brain) is responsible for male behavior. In fact, mutant male mice deficient in aromatase activity display a profound deficit in male sexual behavior.

In women who have not reached menopause the main source of estradiol is the ovaries. After menopause, aromatase turns testosterone, produced by the adrenal glands, into estradiol.

Certain breast cancers require estrogen to grow. Aromatase inhibitors, because they block the synthesis of estrogen, are a new class of drugs used in the treatment of breast cancer in postmenopausal women.

23.4D Adrenocortical Hormones

At least 28 different hormones have been isolated from the adrenal cortex, part of the adrenal glands that sit on top of the kidneys. Included in this group are the following two steroids:

Cortisone
(17α,21-dihydroxy-4-pregnene-3,11,20-trione)

Cortisol
(11β,17α,21-trihydroxy-4-pregnene-3,20-dione)

Most of the adrenocortical steroids have an oxygen function at position 11 (a keto group in cortisone, for example, and a β-hydroxyl in cortisol). Cortisol is the major hormone synthesized by the human adrenal cortex.

The adrenocortical steroids are apparently involved in the regulation of a large number of biological activities, including carbohydrate, protein, and lipid metabolism; water and electrolyte balance; and reactions to allergic and inflammatory phenomena. Recognition, in 1949, of the anti-inflammatory effect of cortisone and its usefulness in the treatment of rheumatoid arthritis led to extensive research in this area. Many 11-oxygenated steroids are now used in the treatment of a variety of disorders ranging from Addison's disease to asthma and skin inflammations.

23.4E D Vitamins

The demonstration, in 1919, that sunlight helped cure rickets—a childhood disease characterized by poor bone growth—began a search for a chemical explanation.

Subsequent investigations showed that D vitamins were involved, and eventually it became known that one of several D vitamins, called vitamin D_3, is the curative factor. Vitamin D_3 is formed in the skin from 7-dehydrocholesterol by two reactions. In the first reaction (below), ultraviolet light in the UV-B range (280–320 nm, which can penetrate the epidermal layer) brings about a 6-electron conrotatory electrocyclic reaction (see Special Topic H, *WileyPLUS*) to produce pre-vitamin D_3. Following this event, a spontaneous isomerization (by way of a [1,7] sigmatropic hydride shift) produces vitamin D_3 itself.

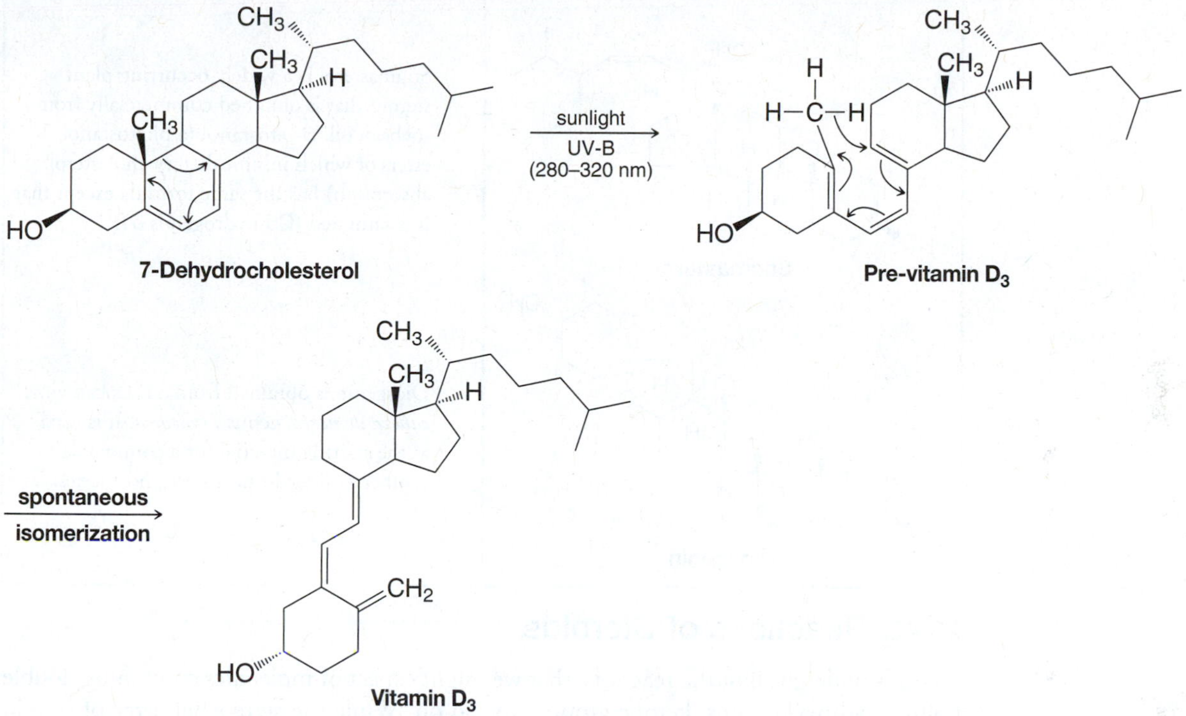

7-Dehydrocholesterol

Pre-vitamin D₃

Vitamin D₃

Vitamin D_3 is required for good health because it is essential in the process by which calcium (as Ca^{2+}) is absorbed from the intestine so as to allow for proper bone growth.

Various factors can cause a deficiency of sunlight and therefore of vitamin D_3, including one's geographic latitude and the season of the year. Sunlight levels are lower in extreme northern and southern latitudes, and are much lower in winter, so much so that for these conditions dietary guidelines in many countries call for supplemental D_3 for children and older persons. Other factors that can affect vitamin D_3 production in the skin are skin coloration, cloud cover, and the use of sunscreens.

23.4F Other Steroids

The structures, sources, and physiological properties of a number of other important steroids are given in Table 23.4.

TABLE 23.4 OTHER IMPORTANT STEROIDS

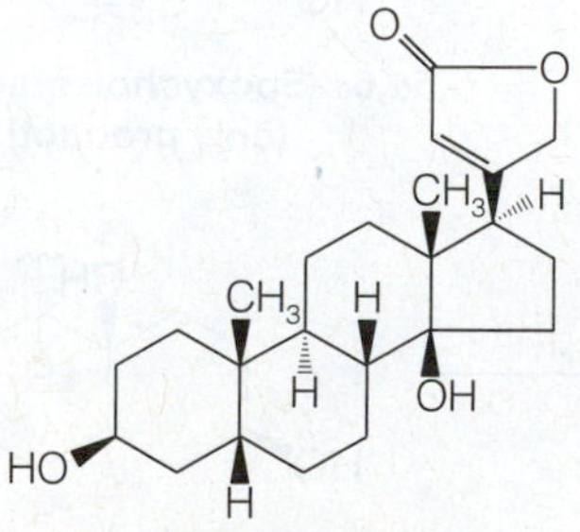

Digitoxigenin

Digitoxigenin is a cardiac aglycone that can be isolated by hydrolysis of digitalis, a pharmaceutical that has been used in treating heart disease since 1785. In digitalis, sugar molecules are joined in acetal linkages to the **3-OH** group of the steroid. In small doses digitalis strengthens the heart muscle; in larger doses it is a powerful heart poison. The aglycone has only about one-fortieth the activity of digitalis.

(continues on next page)

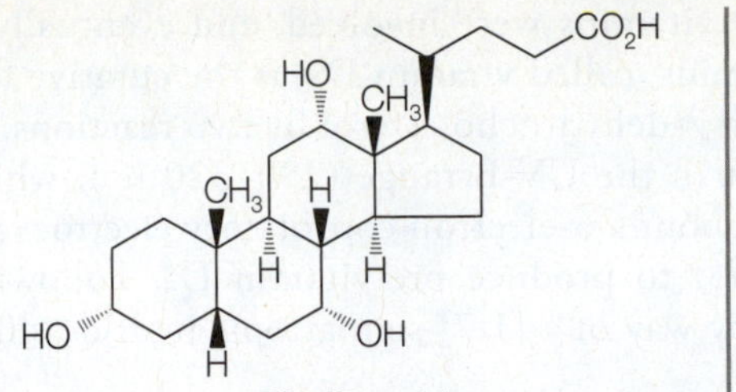

Cholic acid

Cholic acid is the most abundant acid obtained from the hydrolysis of human or ox bile. Bile is produced by the liver and stored in the gallbladder. When secreted into the small intestine, bile emulsifies lipids by acting as a soap. This action aids in the digestive process.

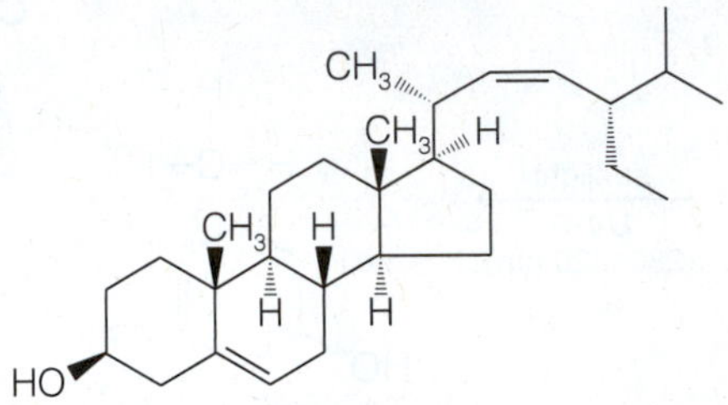

Stigmasterol

Stigmasterol is a widely occurring plant steroid that is obtained commercially from soybean oil. β-Sitostanol (a phytostanol, esters of which inhibit dietary cholesterol absorption) has the same formula except that it is saturated (C5 hydrogen is α).

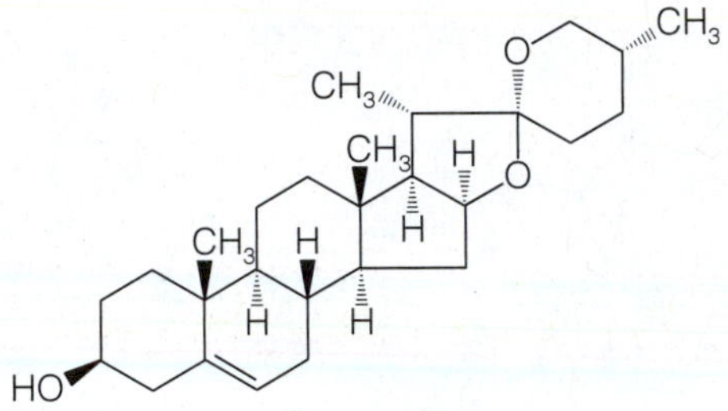

Diosgenin

Diosgenin is obtained from a Mexican vine, *cabeza de negro*, genus *Dioscorea*. It is used as the starting material for a commercial synthesis of cortisone and sex hormones.

23.4G Reactions of Steroids

Steroids undergo all of the reactions that we might expect of molecules containing double bonds, hydroxyl groups, ketone groups, and so on. While the stereochemistry of steroid reactions can be quite complex, it is often strongly influenced by the steric hindrance presented at the β face of the molecule by the angular methyl groups. Many reagents react preferentially at the relatively unhindered α face, especially when the reaction takes place at a functional group very near an angular methyl group and when the attacking reagent is bulky. Examples that illustrate this tendency are shown in the reactions below:

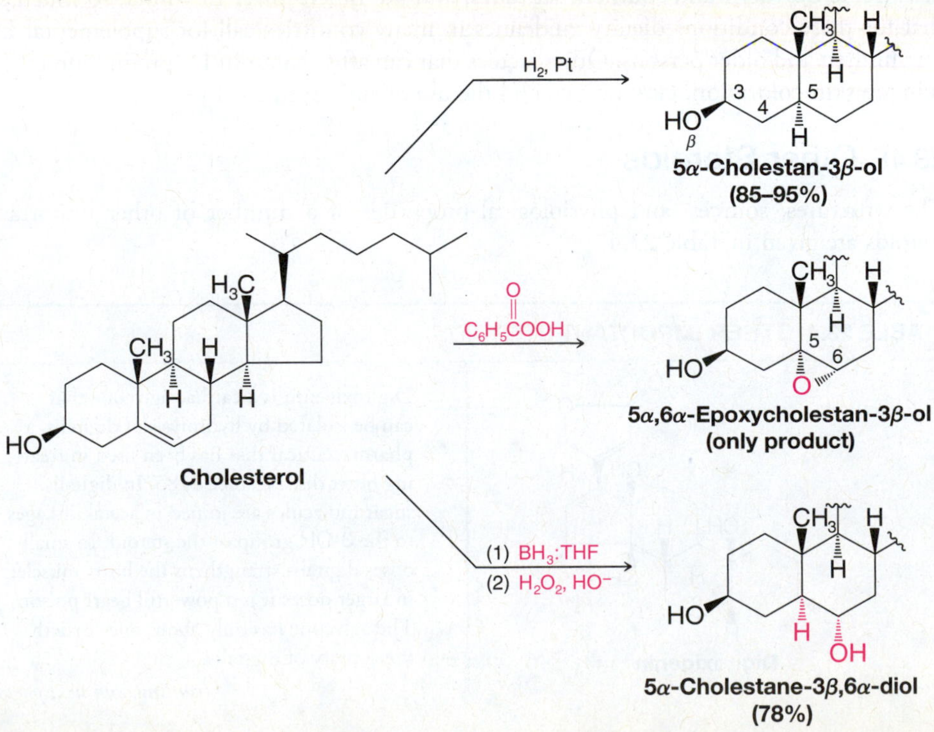

5α-Cholestan-3β-ol
(85–95%)

5α,6α-Epoxycholestan-3β-ol
(only product)

5α-Cholestane-3β,6α-diol
(78%)

When the epoxide ring of 5α,6α-epoxycholestan-3β-ol (see the following reaction) is opened, attack by chloride ion must occur from the β face, but it takes place at the more open 6 position. Notice that the 5 and 6 substituents in the product are *diaxial* (Section 8.13):

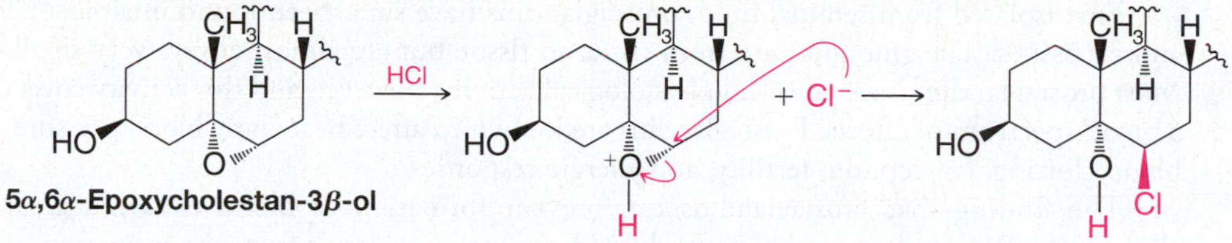

5α,6α-Epoxycholestan-3β-ol

Show how you might convert cholesterol into each of the following compounds:

PRACTICE PROBLEM 23.10

(a) 5α,6β-Dibromocholestan-3β-ol   **(c)** 5α-Cholestan-3-one   **(e)** 6β-Bromocholestane-3β,5α-diol

(b) Cholestane-3β,5α,6β-triol   **(d)** 6α-Deuterio-5α-cholestan-3β-ol

The relative openness of equatorial groups (when compared to axial groups) also influences the stereochemical course of steroid reactions. When 5α-cholestane-3β,7α-diol (see the following reaction) is treated with excess ethyl chloroformate (EtOCOCl), only the equatorial 3β-hydroxyl becomes esterified. The axial 7α-hydroxyl is unaffected by the reaction:

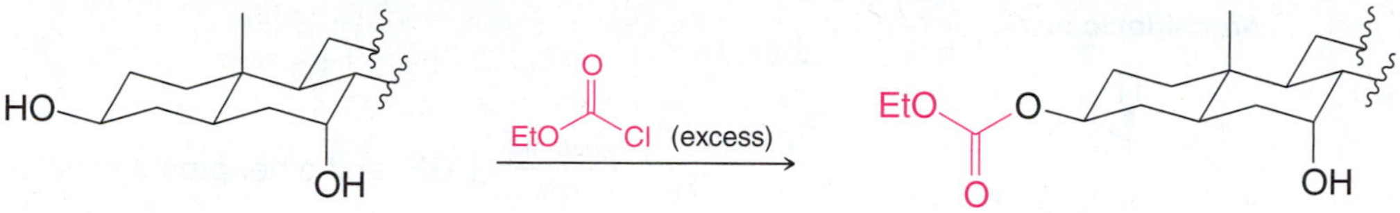

5α-Cholestane-3β,7α-diol   **(only product)**

By contrast, treating 5α-cholestane-3β,7β-diol with excess ethyl chloroformate esterifies both hydroxyl groups. In this instance, both groups are equatorial:

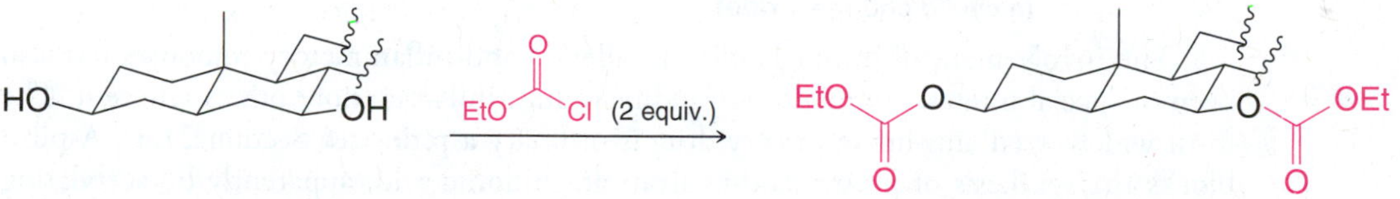

5α-Cholestane-3β,7β-diol   **(only product)**

23.5 PROSTAGLANDINS

One very active area of research has concerned a group of lipids called **prostaglandins**. Prostaglandins are C_{20} carboxylic acids that contain a five-membered ring, at least one double bond, and several oxygen-containing functional groups. Two of the most biologically active prostaglandins are prostaglandin E_2 and prostaglandin $F_{1\alpha}$:

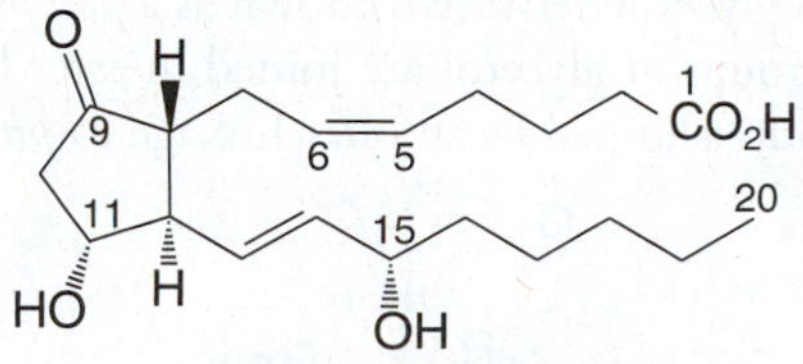

Prostaglandin E₂
(PGE₂)

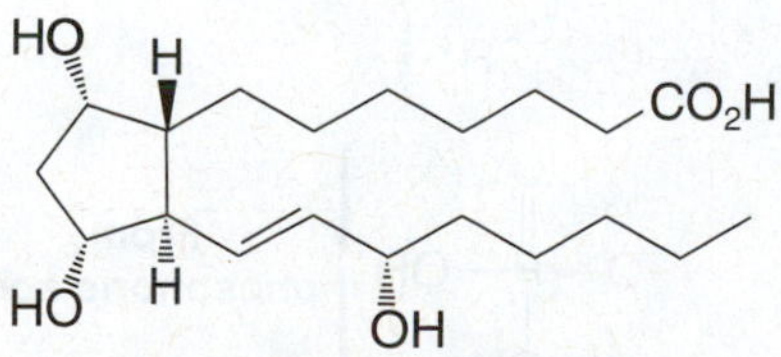

> ### Helpful Hint
>
> These names for the prostaglandins are abbreviated designations used by workers in the field; systematic names are seldom used for prostaglandins.

Prostaglandin F₁ₐ
(PGF₁ₐ)

The 1982 Nobel Prize in Physiology or Medicine was awarded to S. K. Bergström and B. I. Samuelsson (Karolinska Institute, Stockholm, Sweden) and to J. R. Vane (Wellcome Foundation, Beckenham, England) for their work on prostaglandins.

Prostaglandins of the E type have a carbonyl group at C9 and a hydroxyl group at C11; those of the F type have hydroxyl groups at both positions. Prostaglandins of the 2 series have a double bond between C5 and C6; in the 1 series this bond is a single bond.

First isolated from seminal fluid, prostaglandins have since been found in almost all animal tissues. The amounts vary from tissue to tissue but are almost always very small. Most prostaglandins have powerful physiological activity, however, and this activity covers a broad spectrum of effects. Prostaglandins are known to affect heart rate, blood pressure, blood clotting, conception, fertility, and allergic responses.

The finding that prostaglandins can prevent formation of blood clots has great clinical significance, because heart attacks and strokes often result from the formation of abnormal clots in blood vessels. An understanding of how prostaglandins affect the formation of clots may lead to the development of drugs to prevent heart attacks and strokes.

The biosynthesis of prostaglandins of the 2 series begins with a C_{20} polyenoic acid, arachidonic acid, an omega-6 fatty acid. (Synthesis of prostaglandins of the 1 series begins with a fatty acid with one fewer double bond.) The first step requires two molecules of oxygen and is catalyzed by an enzyme called *cyclooxygenase*:

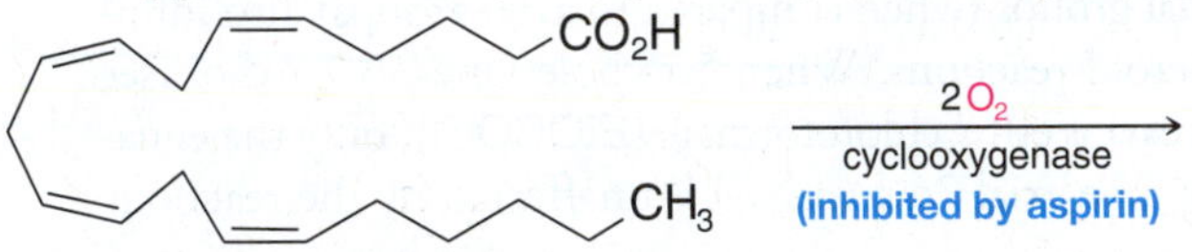

Arachidonic acid

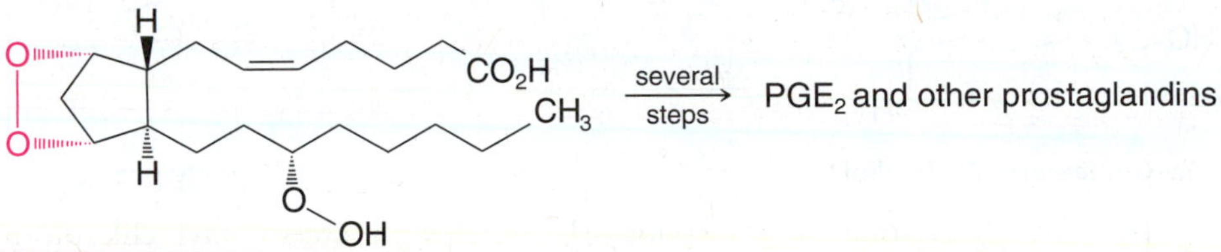

$\xrightarrow[\text{(inhibited by aspirin)}]{\substack{2 O_2 \\ \text{cyclooxygenase}}}$

PGG$_2$
(a cyclic endoperoxide)

$\xrightarrow[\text{steps}]{\text{several}}$ PGE$_2$ and other prostaglandins

The involvement of prostaglandins in allergic and inflammatory responses has also been of special interest. Some prostaglandins induce inflammation; others relieve it. The most widely used anti-inflammatory drug is ordinary aspirin (see Section 21.8). Aspirin blocks the synthesis of prostaglandins from arachidonic acid, apparently by acetylating the enzyme cyclooxygenase, thus rendering it inactive (see the previous reaction). This reaction may represent the origin of aspirin's anti-inflammatory properties. Another prostaglandin (PGE$_1$) is a potent fever-inducing agent (pyrogen), and aspirin's ability to reduce fever may also arise from its inhibition of prostaglandin synthesis.

23.6 PHOSPHOLIPIDS AND CELL MEMBRANES

Another large class of lipids are those called **phospholipids**. Most phospholipids are structurally derived from a glycerol derivative known as a *phosphatidic acid*. In a phosphatidic acid, two hydroxyl groups of glycerol are joined in ester linkages to fatty acids and one terminal hydroxyl group is joined in an ester linkage to *phosphoric acid*:

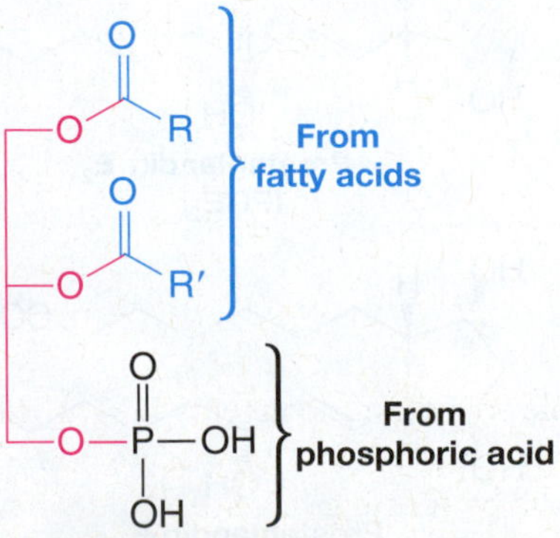

A phosphatidic acid
(a diacylglyceryl phosphate)

23.6A Phosphatides

In *phosphatides*, the phosphate group of a phosphatidic acid is bound through another phosphate ester linkage to one of the following nitrogen-containing compounds:

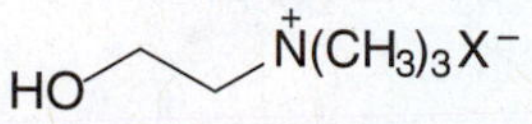

Choline 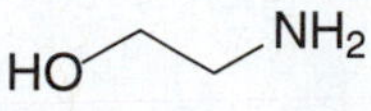   2-Aminoethanol (ethanolamine) 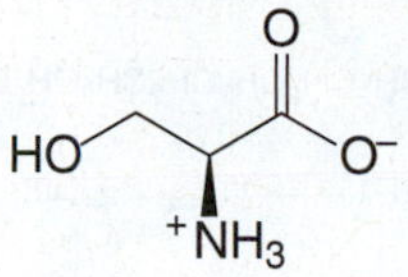   L-Serine

The most important phosphatides are the **lecithins, cephalins, phosphatidylserines**, and **plasmalogens** (a phosphatidyl derivative). Their general structures are shown in Table 23.5.

TABLE 23.5 PHOSPHATIDES

Lecithins

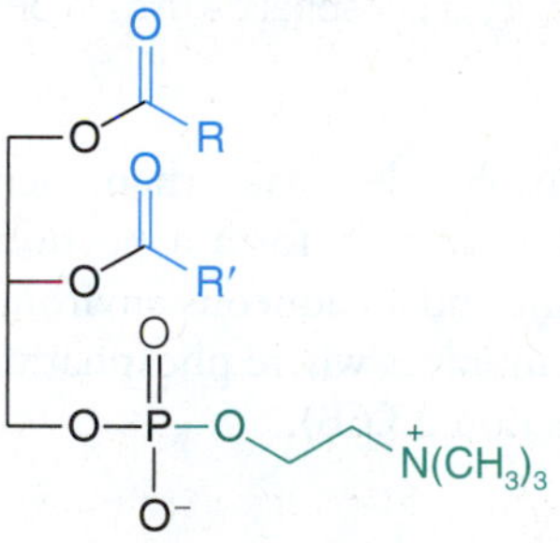

(from choline)
R is saturated and R′ is unsaturated.

Cephalins

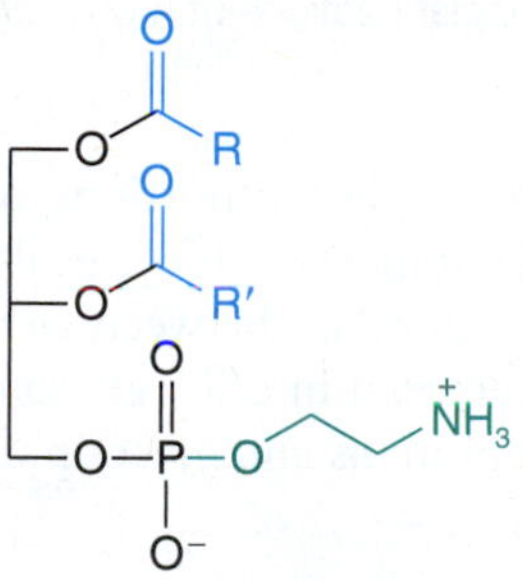

(from 2-aminoethanol)

Phosphatidylserines

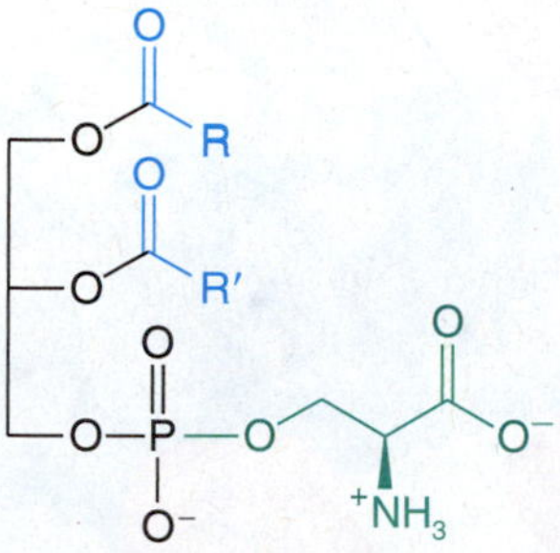

(from L-serine)
R is saturated and R′ is unsaturated.

Plasmalogens

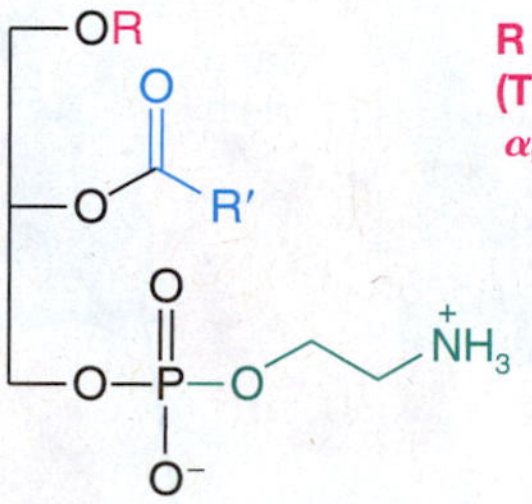

R is —CH=CH(CH$_2$)$_n$CH$_3$
(This linkage is that of an α,β-unsaturated ether.)

(from 2-aminoethanol) or
—OCH$_2$CH$_2$N$^+$(CH$_3$)$_3$ (from choline)
R′ is an unsaturated fatty acid.

Phosphatides resemble soaps and detergents in that they are molecules having both polar and nonpolar groups (Fig. 23.8a). Like soaps and detergents, too, phosphatides "dissolve" in aqueous media by forming micelles. There is evidence that in biological systems the preferred micelles consist of three-dimensional arrays of "stacked" bimolecular micelles (Fig. 23.8b) that are better described as **lipid bilayers**.

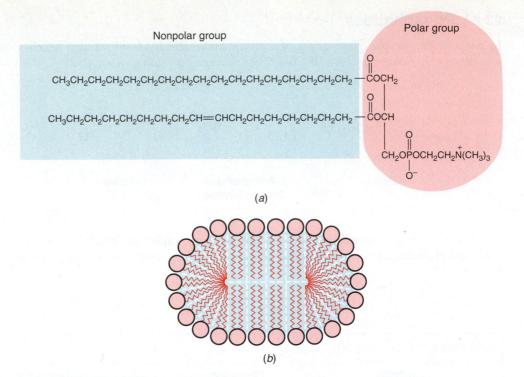

Nonpolar group

Polar group

$CH_3CH_2CH_2CH_2CH_2CH_2CH_2CH_2CH_2CH_2CH_2CH_2CH_2CH_2CH_2CH_2CH_2$ — COCH₂

$CH_3CH_2CH_2CH_2CH_2CH_2CH_2CH_2CH=CHCH_2CH_2CH_2CH_2CH_2CH_2CH_2$ — COCH

CH₂OPOCH₂CH₂N(CH₃)₃

(a)

(b)

FIGURE 23.8 (a) Polar and nonpolar sections of a phosphatide. (b) A phosphatide micelle or lipid bilayer.

The hydrophilic and hydrophobic portions of phosphatides make them perfectly suited for one of their most important biological functions: they form a portion of a structural unit that creates an interface between an organic and an aqueous environment. This structure (Fig. 23.9) is located in cell walls and membranes where phosphatides are often found associated with proteins and glycolipids (Section 23.6B).

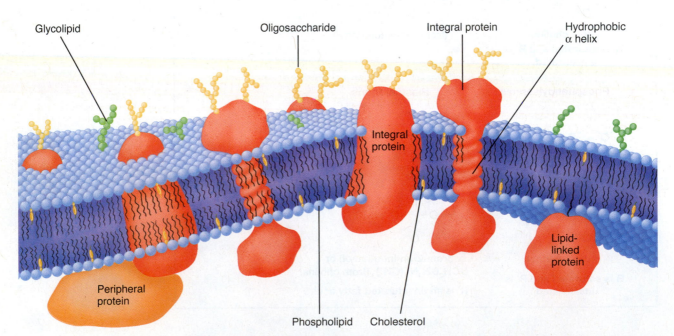

Glycolipid

Oligosaccharide

Integral protein

Hydrophobic α helix

Integral protein

Lipid-linked protein

Peripheral protein

Phospholipid   Cholesterol

FIGURE 23.9 A schematic diagram of a plasma membrane. Integral proteins (*red-orange*), shown for clarity in much greater proportion than they are found in actual biological membranes, and cholesterol (*yellow*) are embedded in a bilayer composed of phospholipids (*blue spheres with two wiggly tails*). The carbohydrate components of glycoproteins (*yellow beaded chains*) and glycolipids (*green beaded chains*) occur only on the external face of the membrane.
(Reprinted with permission of John Wiley & Sons, Inc., from Voet, D.; Voet, J. G.; Pratt, C., *Fundamentals of Biochemistry, Life at the Molecular Level*; © 1999 Voet, D. and Voet, J. G.)

PRACTICE PROBLEM 23.11

Under suitable conditions all of the ester (and ether) linkages of a phosphatide can be hydrolyzed. What organic compounds would you expect to obtain from the complete hydrolysis of (see Table 23.5) **(a)** a lecithin, **(b)** a cephalin, and **(c)** a choline-based plasmalogen? [*Note:* Pay particular attention to the fate of the α,β-unsaturated ether in part (c).]

THE CHEMISTRY OF... STEALTH® Liposomes for Drug Delivery

The anticancer drug Doxil (doxorubicin) has been packaged in STEALTH® liposomes that give each dose of the drug extended action in the body. During manufacture of the drug it is ensconced in microscopic bubbles (vesicles) formed by a phospholipid bilayer and then given a special coating that masks it from the immune system. Ordinarily, a foreign particle such as this would be attacked by cells of the immune system and degraded, but a veil of polyethylene glycol oligomers on the liposome surface masks it from detection. Because of this coating, the STEALTH® liposome circulates through the body and releases its therapeutic contents over a period of time significantly greater than the lifetime for circulation of the undisguised drug. Coatings like those used for STEALTH® liposomes may also be able to reduce the toxic side effects of some drugs. Furthermore, by attaching specific cell recognition "marker" molecules to the polymer, it may be possible to focus binding of the liposomes specifically to cells of a targeted tissue. One might be tempted to call a targeted liposome a "smart stealth liposome."

23.6B Derivatives of Sphingosine

Another important group of lipids is derived from **sphingosine**; the derivatives are called **sphingolipids**. Two sphingolipids, a typical *sphingomyelin* and a typical *cerebroside*, are shown in Fig. 23.10.

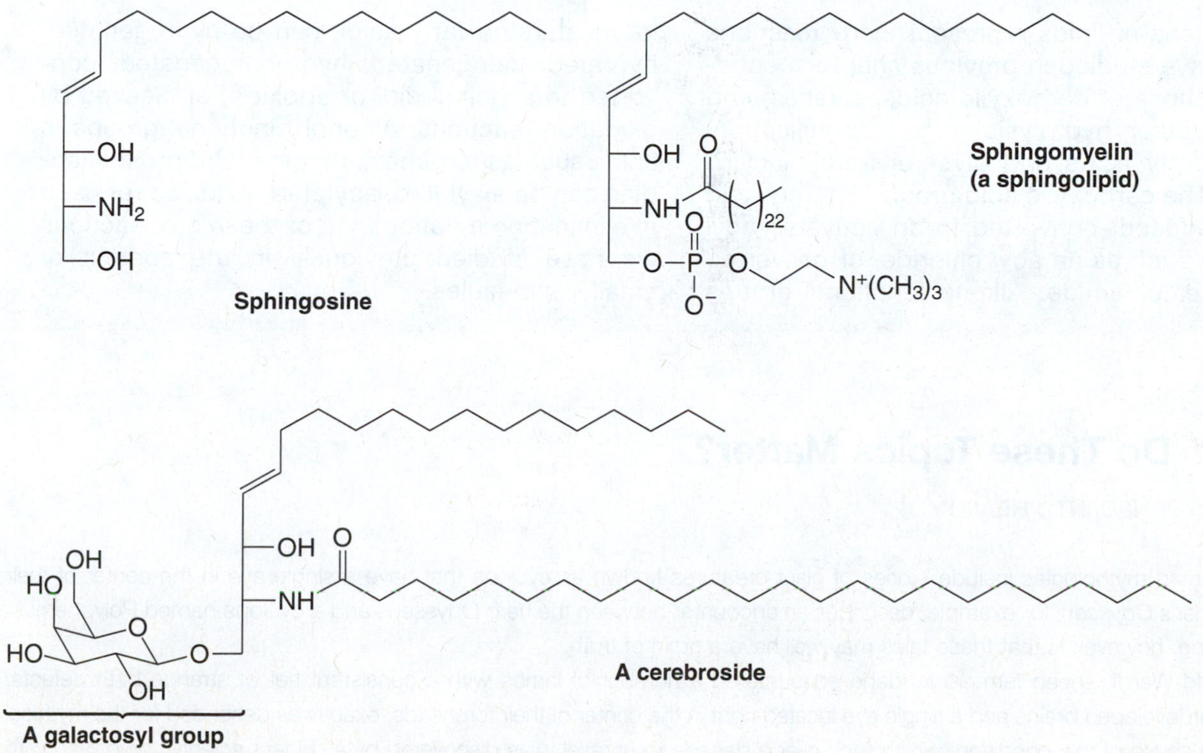

FIGURE 23.10 A sphingosine and two sphingolipids.

On hydrolysis, sphingomyelins yield sphingosine, choline, phosphoric acid, and a C_{24} fatty acid called lignoceric acid. In a sphingomyelin this last component is bound to the $-NH_2$ group of sphingosine. The sphingolipids do not yield glycerol when they are hydrolyzed.

The cerebroside shown in Fig. 23.10 is an example of a **glycolipid**. Glycolipids have a polar group that is contributed by a *carbohydrate*. They do not yield phosphoric acid or choline when they are hydrolyzed.

The sphingolipids, together with proteins and polysaccharides, make up **myelin**, the protective coating that encloses nerve fibers or **axons**. The axons of nerve cells carry electrical nerve impulses. Myelin has a function relative to the axon similar to that of the insulation on an ordinary electric wire (see the chapter opening vignette).

23.7 WAXES

Most **waxes** are esters of long-chain fatty acids and long-chain alcohols. Waxes are found as protective coatings on the skin, fur, and feathers of animals and on the leaves and fruits of plants. Several esters isolated from waxes are the following:

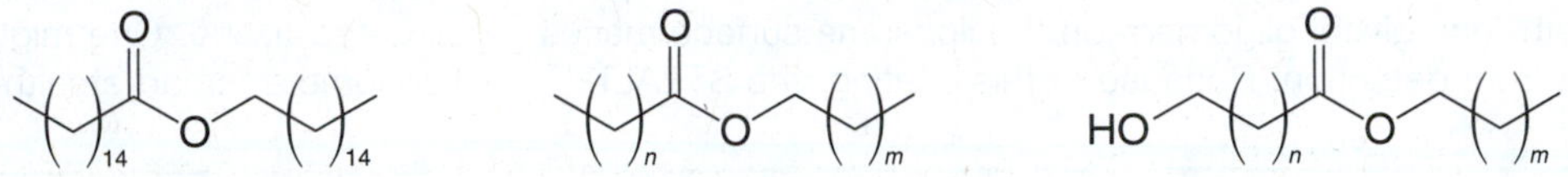

Cetyl palmitate
(from spermaceti)

$n = 24$ or 26; $m = 28$ or 30
(from beeswax)

$n = 16–28$; $m = 30$ or 32
(from carnauba wax)

[**SUMMARY OF REACTIONS OF LIPIDS**]

The reactions of lipids represent many reactions that we have studied in previous chapters, especially reactions of carboxylic acids, alkenes, and alcohols. Ester hydrolysis (e.g., saponification) liberates fatty acids and glycerol from triacylglycerols. The carboxylic acid group of a fatty acid can be reduced, converted to an activated acyl derivative such as an acyl chloride, or converted to an ester or amide. Alkene functional groups in unsaturated fatty acids can be hydrogenated, hydrated, halogenated, hydrohalogenated, converted to a vicinal diol or epoxide, or cleaved by oxidation reactions. Alcohol functional groups in lipids such as terpenes, steroids, and prostaglandins can be alkylated, acylated, oxidized, or used in elimination reactions. All of these are reactions we have studied previously in the context of smaller molecules.

[WHY Do These Topics Matter?

MYTHS TURNED INTO REALITY

Greek and Roman mythologies include stories of giant creatures known as cyclops that have a single eye in the center of their forehead. Homer's *Odyssey*, for example, describes an encounter between the hero Odysseus and a cyclops named Polyphemus. What is amazing, however, is that these tales may well have a grain of truth.

After World War II, sheep farmers in Idaho encountered a number of lambs with a consistent set of strange birth defects, including underdeveloped brains and a single eye located right in the center of their foreheads, exactly as described for the mythical creatures. The cause of this condition, which took over a decade to unravel, was discovered by a diligent scientist who lived with the sheep for a number of summers and cataloged their behaviors, particularly their diets. What proved key was the observation

that during periods of drought the grazing sheep moved higher into the hills and ate corn lilies instead of grass. These flowers, it turns out, produce the nitrogen-containing steroid shown below that is now named cyclopamine for its effects. Though seemingly harmless to adult sheep, it stunts the development of embryonic lambs and produces cyclops-like abnormalities. The effect is the same in other organisms as well. What is perhaps more amazing, however, is that not all the effects of this molecule are harmful. In fact, it may well be a future cancer therapy.

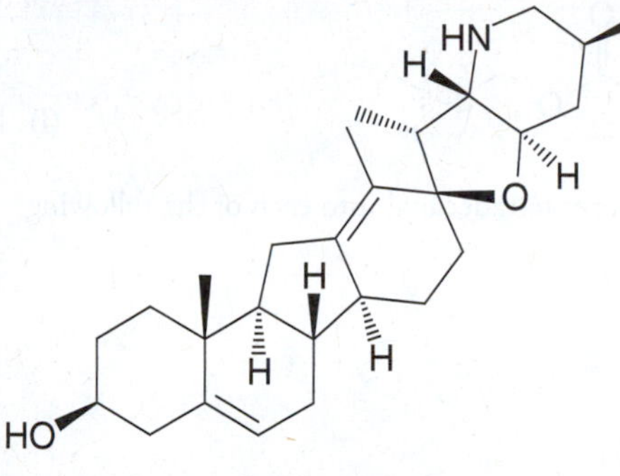

Cyclopamine

Starting in the late 1990s, scientists at a number of pharmaceutical, biotechnology, and academic laboratories determined just how cyclopamine impacts developing embryos. The compound acts on a critical signaling pathway called hedgehog, blocking its function and leading to abnormal development of the brain and other organs in the fetus. In adults, the hedgehog signaling pathway continues to play an important role, largely in controlling the division of adult stem cells for proper maintenance and regeneration of organ tissues. If the genes in the pathway become abnormal, many deadly cancers can result due to uncontrolled cellular division. It is this knowledge that led to the idea that cyclopamine could be a cancer therapy. Since it can block hedgehog functioning, it could potentially prevent cell division when the pathway is not operating normally. This theory is currently showing promise, with both cyclopamine and related analogs being able to combat pancreatic cancer and basal cell carcinoma in a number of human clinical trials. Thus, out of myth has come not only reality, but potentially an even more important discovery pertinent to treating a major human disease.

To learn more about these topics, see:

1. Heretsch, P.; Tzagkaroulaki, L.; Giannis, A. "Cyclopamine and Hedgehog Signaling: Chemistry, Biology, Medical Perspectives" in *Angew. Chem Int. Ed.* **2010**, *49*, 3418–3427.

SUMMARY AND REVIEW TOOLS

The study aids for this chapter include a narrative summary of reactions of lipids (after Section 23.7), and key terms and concepts, which are highlighted in bold, blue text within the chapter, defined in the Glossary at the back of the book, and which have hyperlinked definitions in the accompanying *WileyPLUS* course (www.wileyplus.com).

PROBLEMS

Note to Instructors: Many of the homework problems are available for assignment via *WileyPLUS*, an online teaching and learning solution.

GENERAL REACTIONS

23.12 How would you convert stearic acid, $CH_3(CH_2)_{16}CO_2H$, into each of the following?

(a) Ethyl stearate, 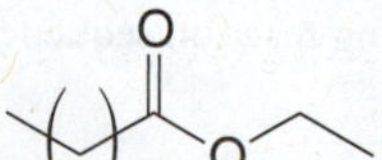 (two ways)

(b) *tert*-Butyl stearate,

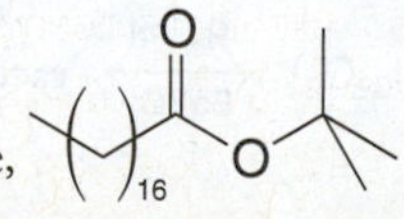

(c) Stearamide,

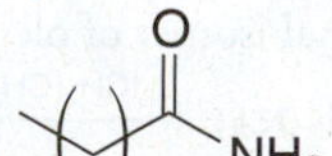

(d) *N,N*-Dimethylstearamide,

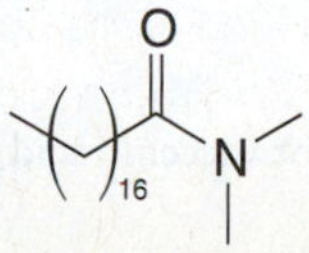

(e) Octadecylamine,

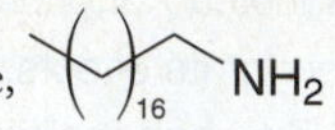

(f) Heptadecylamine,

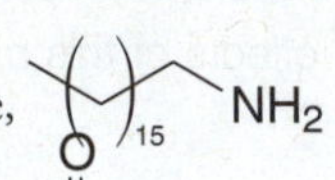

(g) Octadecanal,

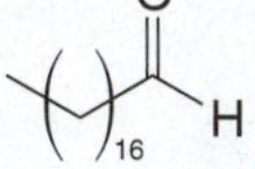

(h) Octadecyl stearate,

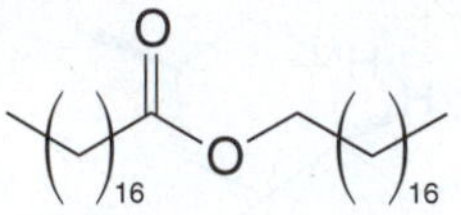

(i) 1-Octadecanol, 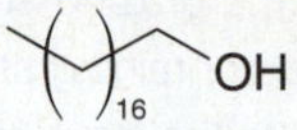 (two ways)

(j) 2-Nonadecanone,

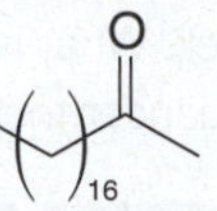

(k) 1-Bromooctadecane,

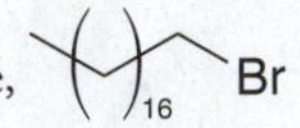

(l) Nonadecanoic acid,

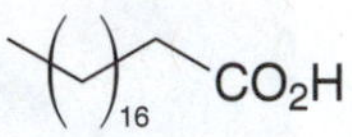

23.13 How would you transform tetradecanal into each of the following?

(a)

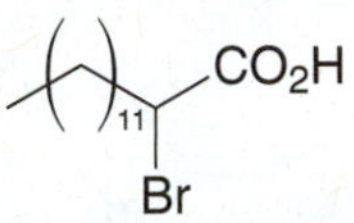

(b)

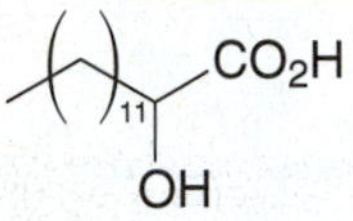

(c)

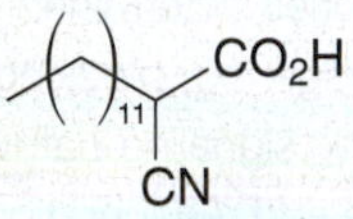

(d)

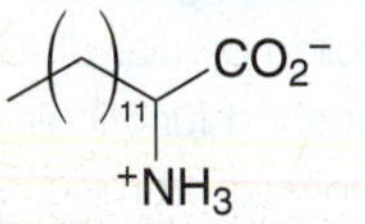

23.14 Using palmitoleic acid as an example and neglecting stereochemistry, illustrate each of the following reactions of the double bond:

(a) Addition of bromine   **(b)** Addition of hydrogen   **(c)** Hydroxylation   **(d)** Addition of HCl

23.15 When oleic acid is heated to 180–200 °C (in the presence of a small amount of selenium), an equilibrium is established between oleic acid (33%) and an isomeric compound called elaidic acid (67%). Suggest a possible structure for elaidic acid.

23.16 When limonene (Section 23.3) is heated strongly, it yields 2 mol of isoprene. What kind of reaction is involved here?

23.17 Gadoleic acid ($C_{20}H_{38}O_2$), a fatty acid that can be isolated from cod-liver oil, can be cleaved by hydroxylation and subsequent treatment with periodic acid to $CH_3(CH_2)_9CHO$ and $OHC(CH_2)_7CO_2H$. **(a)** What two stereoisomeric structures are possible for gadoleic acid? **(b)** What spectroscopic technique would make possible a decision as to the actual structure of gadoleic acid? **(c)** What peaks would you look for?

23.18 α-Phellandrene and β-phellandrene are isomeric compounds that are minor constituents of spearmint oil; they have the molecular formula $C_{10}H_{16}$. Each compound has a UV absorption maximum in the 230–270 nm range. On catalytic hydrogenation, each compound yields 1-isopropyl-4-methylcyclohexane. On vigorous oxidation with potassium permanganate, α-phellandrene yields

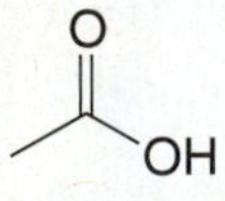

and

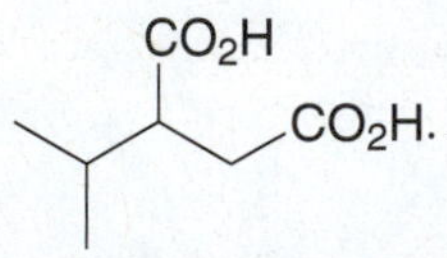

CO_2H. A similar oxidation of β-phellandrene yields

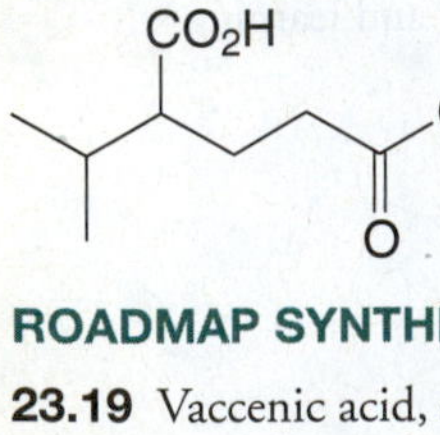

CO_2H as the only isolable product. Propose structures for α- and β-phellandrene.

ROADMAP SYNTHESES

23.19 Vaccenic acid, a constitutional isomer of oleic acid, has been synthesized through the following reaction sequence:

Propose a structure for vaccenic acid and for the intermediates **A–E**.

23.20 ω-Fluorooleic acid can be isolated from a shrub, *Dechapetalum toxicarium*, that grows in Africa. The compound is highly toxic to warm-blooded animals; it has found use as an arrow poison in tribal warfare, in poisoning enemy water supplies, and by witch doctors "for terrorizing the native population." Powdered fruit of the plant has been used as a rat poison; hence ω-fluorooleic acid has the common name "ratsbane." A synthesis of ω-fluorooleic acid is outlined here. Give structures for compounds **F–I**:

1-Bromo-8-fluorooctane + sodium acetylide $\longrightarrow$ **F** ($C_{10}H_{17}F$) $\xrightarrow[\text{(2) I(CH}_2)_7\text{Cl}]{\text{(1) NaNH}_2}$

G ($C_{17}H_{30}FCl$) $\xrightarrow{\text{NaCN}}$ **H** ($C_{18}H_{30}NF$) $\xrightarrow[\text{(2) H}_3\text{O}^+]{\text{(1) KOH}}$ **I** ($C_{18}H_{31}O_2F$) $\xrightarrow[\text{Ni}_2\text{B (P-2)}]{\text{H}_2}$

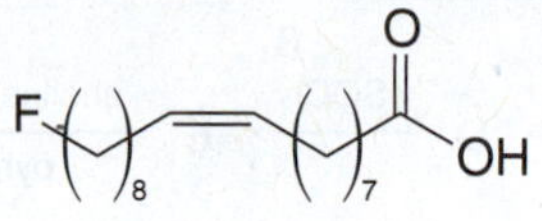

ω-Fluorooleic acid
(46% yield, overall)

23.21 Give formulas and names for compounds **A** and **B**:

5α-Cholest-2-ene $\xrightarrow{\text{C}_6\text{H}_5\text{COOH}}$ **A** (an epoxide) $\xrightarrow{\text{HBr}}$ **B**

(*Hint*: **B** is not the most stable stereoisomer.)

23.22 The initial steps of a laboratory synthesis of several prostaglandins reported by E. J. Corey (Section 7.15B) and co-workers in 1968 are outlined here. Supply each of the missing reagents:

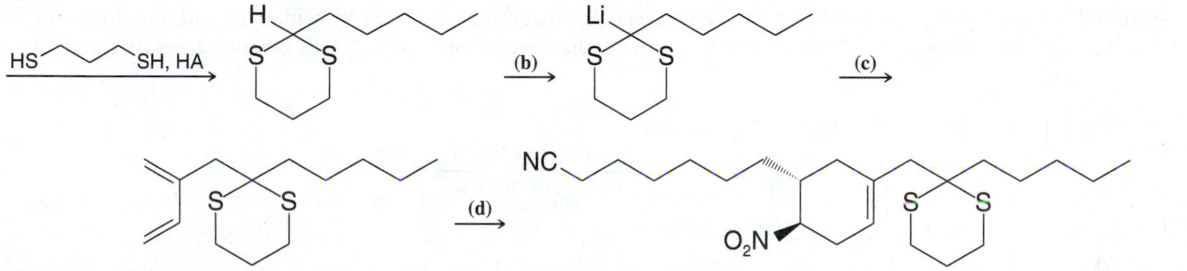

(e) The initial step in another prostaglandin synthesis is shown in the following reaction. What kind of reaction—and catalyst—is needed here?

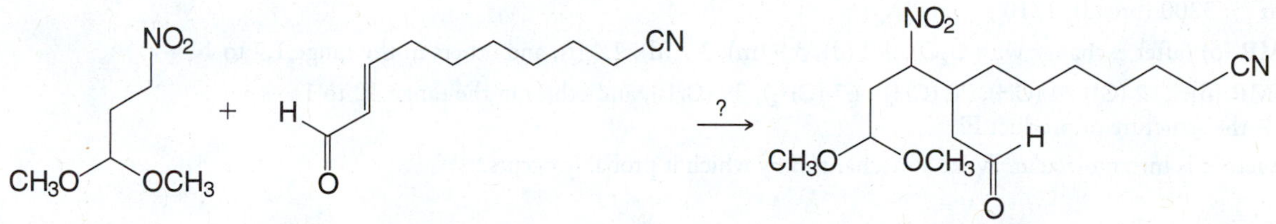

23.23 A useful synthesis of sesquiterpene ketones, called *cyperones*, was accomplished through a modification of the following Robinson annulation procedure (Section 19.7B). Write a mechanism that accounts for each step of this synthesis.

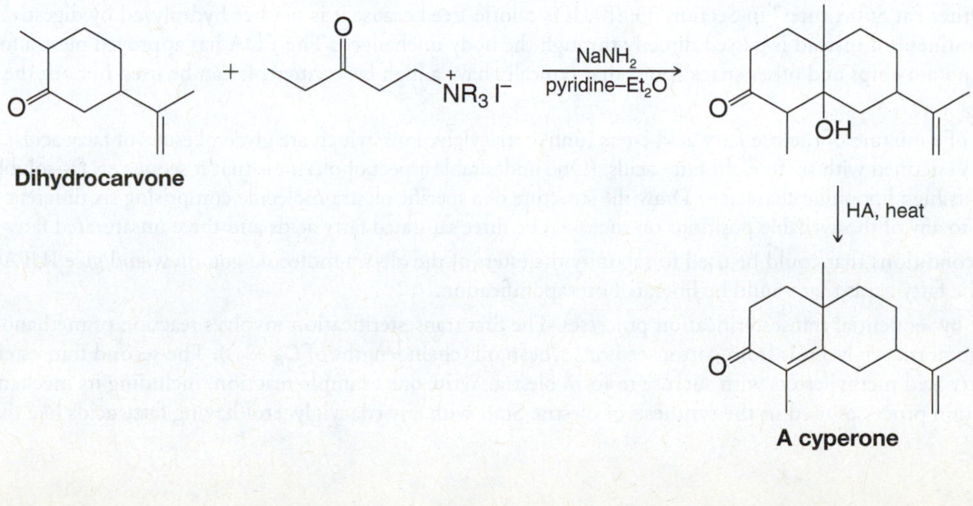

Dihydrocarvone

A cyperone

CHALLENGE PROBLEMS

23.24 A Hawaiian fish called the pahu or boxfish (*Ostracian lentiginosus*) secretes a toxin that kills other fish in its vicinity. The active agent in the secretion was named pahutoxin by P. J. Scheuer, and it was found by D. B. Boylan and Scheuer to contain an unusual combination of lipid moieties. To prove its structure, they synthesized it by this route:

Compound	Selected Infrared Absorption Bands (cm^{-1})
A	1725
B	3300 (broad), 1735
C	3300–2500 (broad), 1710
D	3000–2500 (broad), 1735, 1710
E	1800, 1735
Pahutoxin	1735

What are the structures of **A**, **C**, **D**, and **E** and of pahutoxin?

23.25 The reaction illustrated by the equation below is a very general one that can be catalyzed by acid, base, and some enzymes. It therefore needs to be taken into consideration when planning syntheses that involve esters of polyhydroxy substances like glycerol and sugars:

Spectral data for **F**:

MS (*m/z*): (after trimethylsilylation): 546, 531

IR (cm^{-1}): 3200 (broad), 1710

^{1}H NMR (δ) (after exchange with D_2O): 4.2 (d), 3.9 (m), 3.7 (d), 2.2 (t), and others in the range 1.7 to 1

^{13}C NMR (δ): 172 (C), 74 (CH), 70 (CH$_2$), 67 (CH$_2$), 39 (CH$_2$), and others in the range 32 to 14

(a) What is the structure of product **F**?

(b) The reaction is intramolecular. Write a mechanism by which it probably occurs.

LEARNING GROUP PROBLEMS

1. Olestra is a fat substitute patented by Procter and Gamble that mimics the taste and texture of triacylglycerols (see "The Chemistry of…Olestra and Other Fat Substitutes" in Section 23.2B). It is calorie-free because it is neither hydrolyzed by digestive enzymes nor absorbed by the intestines but instead is passed directly through the body unchanged. The FDA has approved olestra for use in a variety of foods, including potato chips and other snack foods that typically have a high fat content. It can be used in both the dough and the frying process.

(a) Olestra consists of a mixture of sucrose fatty acid esters (unlike triacylglycerols, which are glycerol esters of fatty acids). Each sucrose molecule in olestra is esterified with six to eight fatty acids. (One undesirable aspect of olestra is that it sequesters fat-soluble vitamins needed by the body, due to its high lipophilic character.) Draw the structure of a specific olestra molecule comprising six different naturally occurring fatty acids esterified to any of the available positions on sucrose. Use three saturated fatty acids and three unsaturated fatty acids.

(b) Write reaction conditions that could be used to saponify the esters of the olestra molecule you drew and give IUPAC and common names for each of the fatty acids that would be liberated on saponification.

(c) Olestra is made by sequential transesterification processes. The first transesterification involves reaction of methanol under basic conditions with natural triacylglycerols from cottonseed or soybean oil (chain lengths of C$_8$–C$_{22}$). The second transesterification involves reaction of these fatty acid methyl esters with sucrose to form olestra. Write one example reaction, including its mechanism, for each of these transesterification processes used in the synthesis of olestra. Start with any triacylglycerol having fatty acids like those incorporated into olestra.

2. The biosynthesis of fatty acids is accomplished two carbons at a time by an enzyme complex called fatty acid synthetase. The biochemical reactions involved in fatty acid synthesis are described in Special Topic E (*WileyPLUS*). Each of these biochemical reactions has a counterpart in synthetic reactions you have studied. Consider the biochemical reactions involved in adding each —CH_2CH_2— segment during fatty acid biosynthesis (those in Special Topic E that begin with acetyl-S-ACP and malonyl-S-ACP, and end with butyryl-S-ACP). Write laboratory synthetic reactions using reagents and conditions you have studied (not biosynthetic reactions) that would accomplish the same sequence of transformations (i.e., the condensation–decarboxylation, ketone reduction, dehydration, and alkene reduction steps).

3. A certain natural terpene produced peaks in its mass spectrum at *m/z* 204, 111, and 93 (among others). On the basis of this and the following information, elucidate the structure of this terpene. Justify each of your conclusions.

(a) Reaction of the unknown terpene with hydrogen in the presence of platinum under pressure results in a compound with molecular formula $C_{15}H_{30}$.

(b) Reaction of the terpene with ozone followed by dimethyl sulfide produces the following mixture of compounds (1 mol of each for each mole of the unknown terpene):

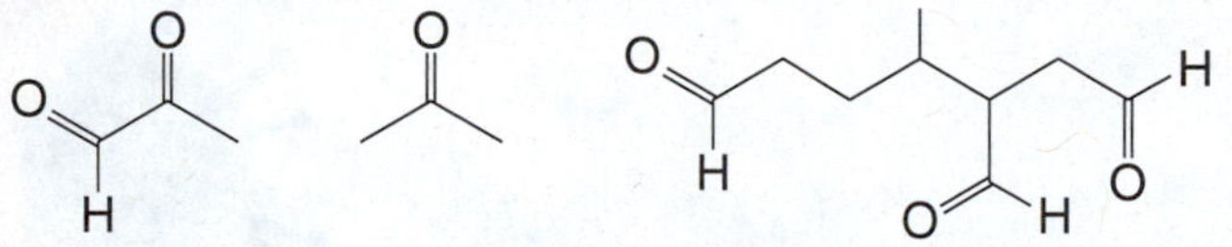

(c) After writing the structure of the unknown terpene, circle each of the isoprene units in this compound. To what class of terpenes does this compound belong (based on the number of carbons it contains)?

4. Draw the structure of a phospholipid (from any of the subclasses of phospholipids) that contains one saturated and one unsaturated fatty acid.

(a) Draw the structure of all of the products that would be formed from your phospholipid if it were subjected to complete hydrolysis (choose either acidic or basic conditions).

(b) Draw the structure of the product(s) that would be formed from reaction of the unsaturated fatty acid moiety of your phospholipid (assuming it had been released by hydrolysis from the phospholipid first) under each of the following conditions:

    (i) Br_2

    (ii) OsO_4, followed by $NaHSO_3$

    (iii) HBr

    (iv) Hot alkaline $KMnO_4$, followed by H_3O^+

    (v) $SOCl_2$, followed by excess CH_3NH_2

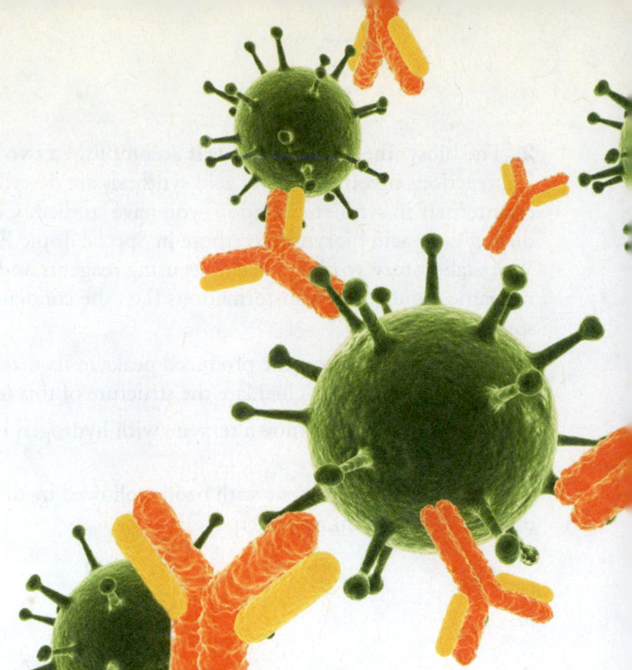

CHAPTER
24

Amino Acids and Proteins

mong the major classes of biomolecules, proteins arguably have the most diverse array of functions. As enzymes they serve as catalysts to affect chemical reactions; as antibodies they protect from disease; as molecules they form critical structures, including skin, hair, and nails; and as hormones they control many body functions, including metabolism, growth, and reproduction. As we shall see, not only do proteins come in all shapes and sizes, but each individual protein is also the product of an evolutionary process that has led to its specific properties and functions. By learning about the overall structure and function of proteins, chemists are now able to apply that knowledge to the development of some highly valuable, fully synthetic proteins of their own design. For instance, as we will see later in this chapter, lessons learned from the natural adaptability of antibodies generated by the immune system provided insights for the development of unnatural, synthetic analogs that can catalyze chemical reactions such as the Claisen rearrangement, aldol reactions, and the Diels–Alder reaction, as in the graphic shown to the right.

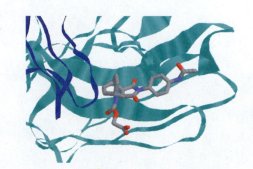

PHOTO CREDIT: Ingram Publishing/Getty Images, Inc. ILLUSTRATION CREDIT: PDB ID: 1A4K, http://www.pdb.org. Romesberg, F. E., Spiller, B., Schultz, P. G., Stevens, R. C. Immunological origins of binding and catalysis in a Diels–Alderase antibody. Science 279, pp. 1929–1933, 1998.

N THIS CHAPTER WE WILL CONSIDER:

- the structures and properties of amino acids that constitute proteins
- methods to determine the amino acid sequence of a given protein, as well as synthesize it
- the primary, secondary, tertiary, and quaternary structures of proteins
- selected examples of enzymes and their function

[**WHY** DO THESE TOPICS MATTER?]   Not only will we show you how some novel catalytic antibodies work, but also at the end of this chapter we will show how chemists have gone beyond the standard amino acids found in nature to create proteins in diverse cells that include many new and fully synthetic amino acids.

24.1 INTRODUCTION

The three groups of biological polymers are polysaccharides, proteins, and nucleic acids. We studied polysaccharides in Chapter 22 and saw that they function primarily as energy reserves, as biochemical labels on cell surfaces, and, in plants, as structural materials. When we study nucleic acids in Chapter 25, we shall find that they serve two major purposes: storage and transmission of information. Of the three groups of biopolymers, proteins have the most diverse functions. As enzymes and hormones, proteins catalyze and regulate the reactions that occur in the body; as muscles and tendons they provide the body with the means for movement; as skin and hair they give it an outer covering; as hemoglobin molecules they transfer all-important oxygen to its most remote corners; as antibodies they provide it with a means of protection against disease; and in combination with other substances in bone they provide it with structural support.

Given such diversity of functions, we should not be surprised to find that proteins come in all sizes and shapes. By the standard of most of the molecules we have studied, even small proteins have very high molecular weights. Lysozyme, an enzyme, is a relatively small protein and yet its molecular weight is 14,600. The molecular weights of most proteins are much larger. Their shapes cover a range from the globular proteins such as lysozyme and hemoglobin to the helical coils of α-keratin (hair, nails, and wool) and the pleated sheets of silk fibroin.

And yet, in spite of such diversity of size, shape, and function, all proteins have common features that allow us to deduce their structures and understand their properties. Later in this chapter we shall see how this is done.

- Proteins are **polyamides**, and their monomeric units are composed of about 20 different α-amino acids:

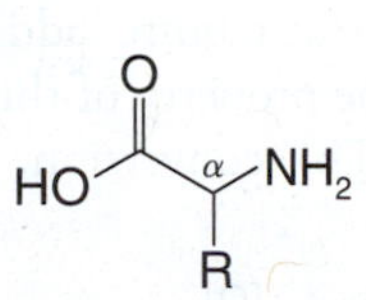

An α-amino acid

R is a side chain at the α carbon that determines the identity of the amino acid (Table 24.1).

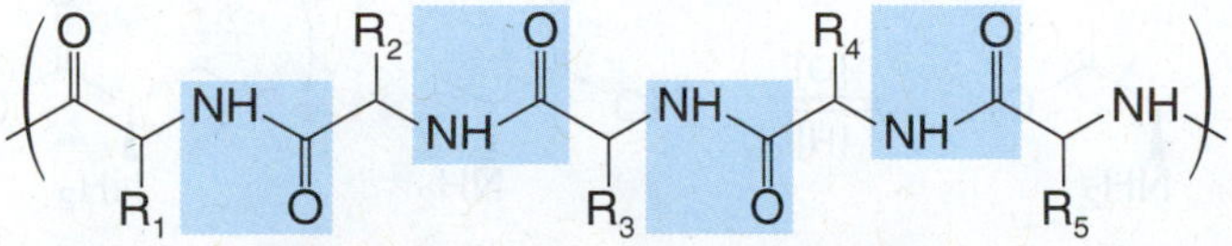

A portion of a protein molecule

Amide (peptide) linkages are shaded.
R₁ to R₅ may be any of the possible side chains.

- The exact sequence of the different α-amino acids along the protein chain is called the **primary structure** of the protein.

A protein's primary structure, as its name suggests, is of fundamental importance. For the protein to carry out its particular function, the primary structure must be correct. We shall see later that when the primary structure is correct, the protein's polyamide chain folds in particular ways to give it the shape it needs for its particular task.

- Folding of the polyamide chain gives rise to higher levels of complexity called the **secondary** and **tertiary structures** of the protein.
- **Quaternary structure** results when a protein contains an aggregate of more than one polyamide chain.
- Hydrolysis of proteins with acid or base yields a mixture of amino acids.

Although hydrolysis of naturally occurring proteins may yield as many as 22 different amino acids, the amino acids have an important structural feature in common: with the exception of glycine (whose molecules are achiral), almost all naturally occurring amino acids have the L configuration at the α carbon.* That is, they have the same relative configuration as L-glyceraldehyde:

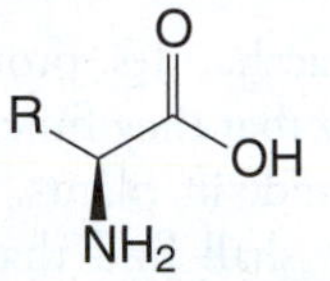

An L-α-amino acid
[usually an (S)-α-amino acid]

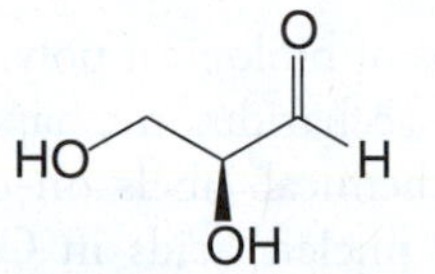

L-Glyceraldehyde
[(S)-glyceraldehyde]

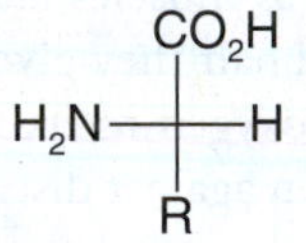

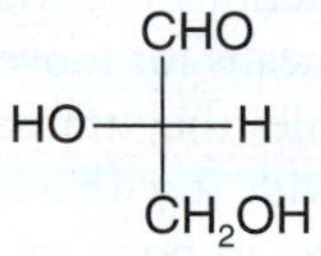

Fischer projections for an L-α-amino acid and L-glyceraldehyde

24.2 AMINO ACIDS

24.2A Structures and Names

- The 22 α-amino acids that can be obtained from proteins can be subdivided into three different groups on the basis of the structures of their side chains, R. These are given in Table 24.1.

Only 20 of the 22 α-amino acids in Table 24.1 are actually used by cells when they synthesize proteins. Two amino acids are synthesized after the polyamide chain is intact. Hydroxyproline (present mainly in collagen) is synthesized by oxidation of proline, and cystine (present in most proteins) is synthesized from cysteine.

The conversion of cysteine to cystine requires additional comment. The —SH group of cysteine makes cysteine a *thiol.* One property of thiols is that they can be converted to disulfides by mild oxidizing agents. This conversion, moreover, can be reversed by mild reducing agents:

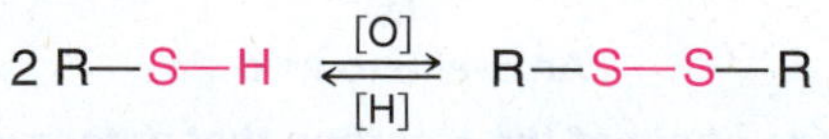

$$2\ R\!-\!S\!-\!H \underset{[H]}{\overset{[O]}{\rightleftharpoons}} R\!-\!S\!-\!S\!-\!R$$

Thiol   **Disulfide**

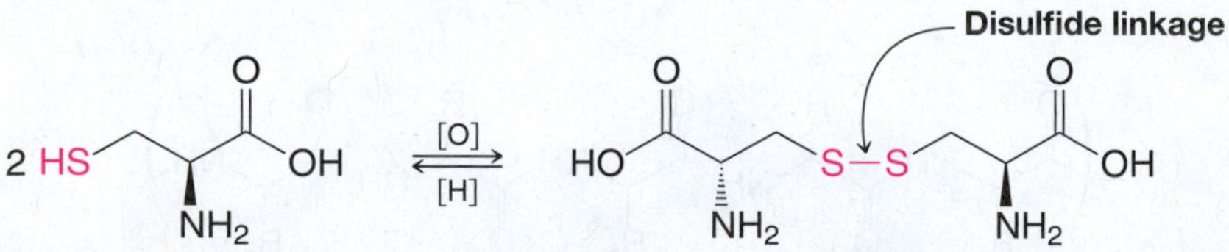

Cysteine   **Cystine**

We shall see later how the **disulfide linkage** between cysteine units in a protein chain contributes to the overall structure and shape of the protein.

*Some D-amino acids have been obtained from the material comprising the cell walls of bacteria and by hydrolysis of certain antibiotics.

TABLE 24.1 L-AMINO ACIDS FOUND IN PROTEINS

Structure	Name	Abbreviations[a]	pK_{a_1} α-CO$_2$H	pK_{a_2} α-NH$_3^+$	pK_{a_3} R group	pI
Neutral Amino Acids						
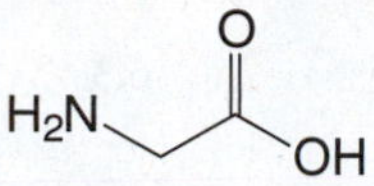	Glycine	G or Gly	2.3	9.6		6.0
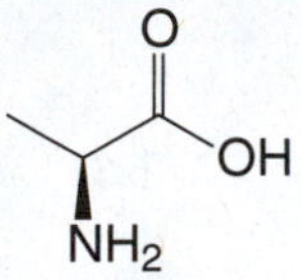	Alanine	A or Ala	2.3	9.7		6.0
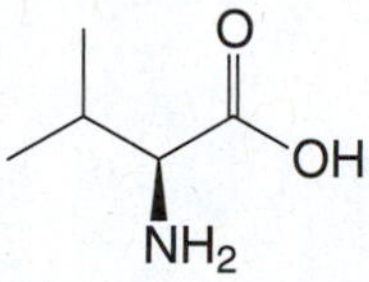	Valine[b]	V or Val	2.3	9.6		6.0
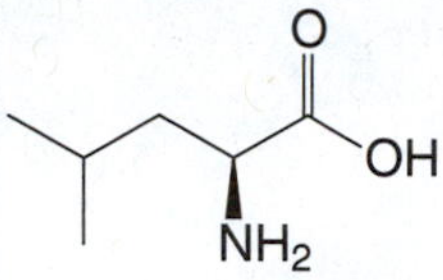	Leucine[b]	L or Leu	2.4	9.6		6.0
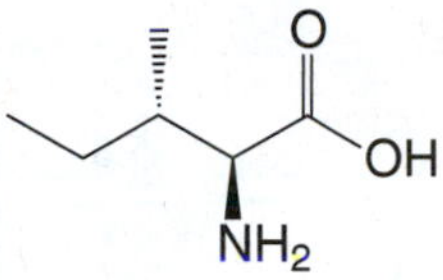	Isoleucine[b]	I or Ile	2.4	9.7		6.1
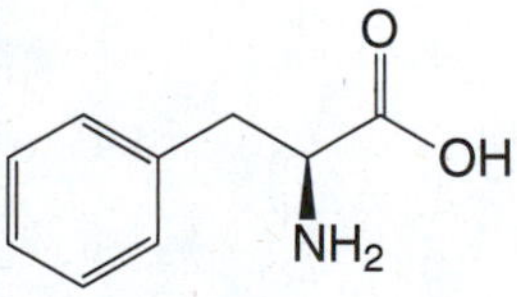	Phenylalanine[b]	F or Phe	1.8	9.1		5.5
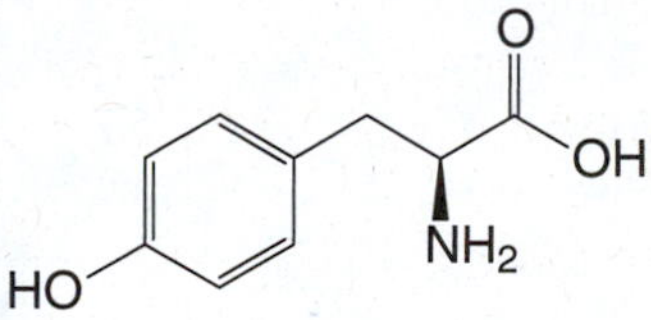	Tyrosine	Y or Tyr	2.2	9.1	10.1	5.7
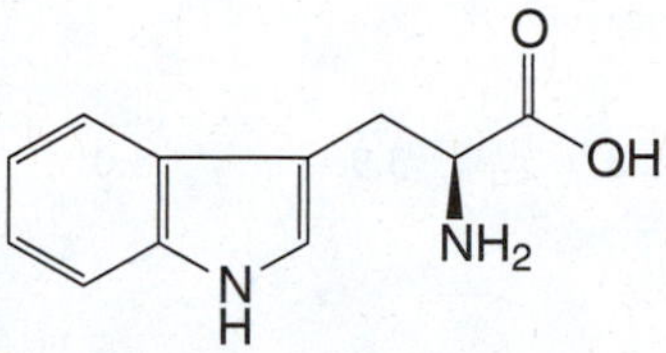	Tryptophan[b]	W or Trp	2.4	9.4		5.9
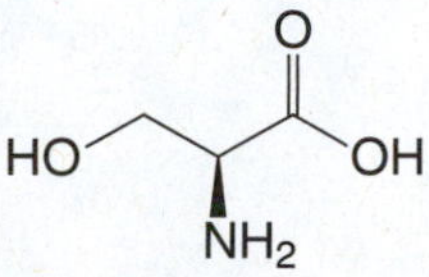	Serine	S or Ser	2.2	9.2		5.7
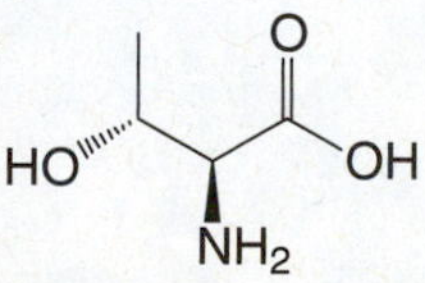	Threonine[b]	T or Thr	2.6	10.4		6.5

(continues on next page)

TABLE 24.1 CONTINUED

Structure	Name	Abbreviations[a]	pK_{a_1} α-CO$_2$H	pK_{a_2} α-NH$_3^+$	pK_{a_3} R group	pI
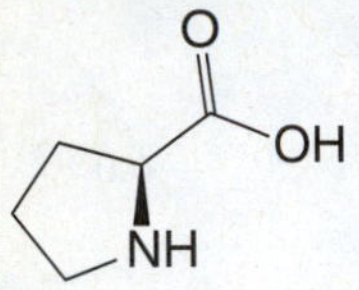	Proline	P or Pro	2.0	10.6		6.3
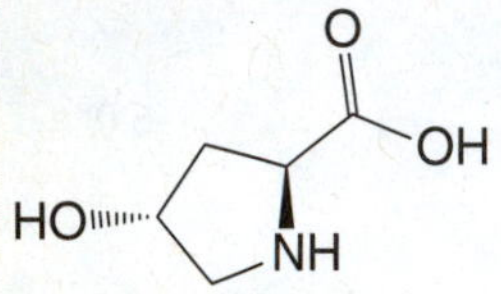	4-Hydroxyproline (cis and trans)	O or Hyp	1.9	9.7		6.3
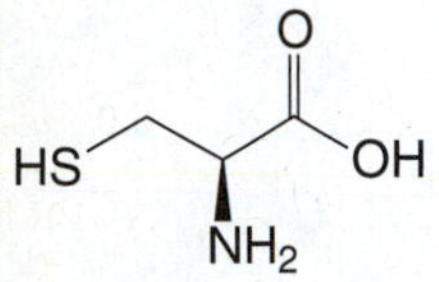	Cysteine	C or Cys	1.7	10.8	8.3	5.0
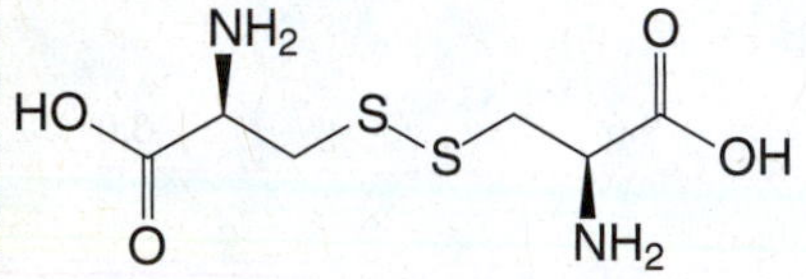	Cystine	Cys-Cys	1.6 2.3	7.9 9.9		5.1
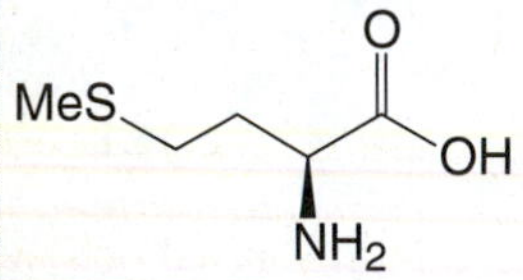	Methionine[b]	M or Met	2.3	9.2		5.8
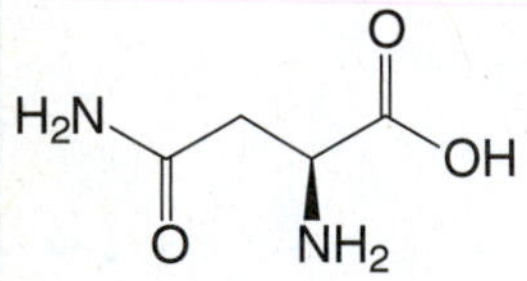	Asparagine	N or Asn	2.0	8.8		5.4
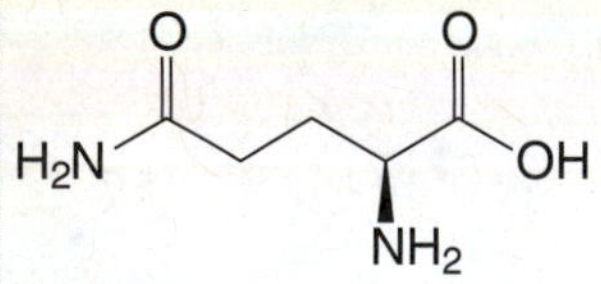	Glutamine	Q or Gln	2.2	9.1		5.7

Side Chains Containing an Acidic (Carboxyl) Group

Structure	Name	Abbreviations[a]	pK_{a_1} α-CO$_2$H	pK_{a_2} α-NH$_3^+$	pK_{a_3} R group	pI
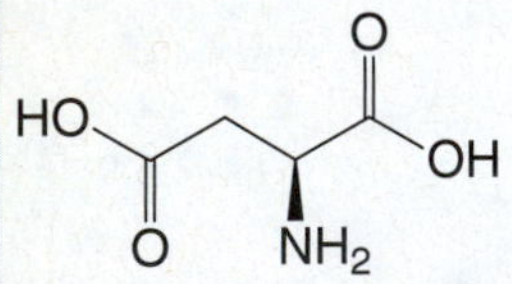	Aspartic acid	D or Asp	2.1	9.8	3.9	3.0
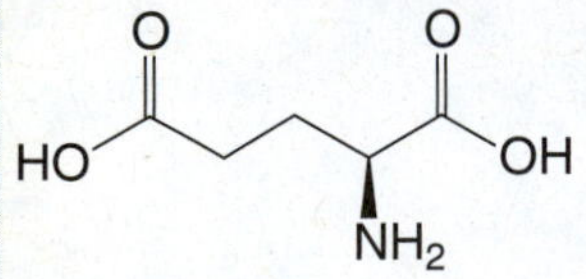	Glutamic acid	E or Glu	2.2	9.7	4.3	3.2

Side Chains Containing a Basic Group

Structure	Name	Abbreviations[a]	pK_{a_1} α-CO$_2$H	pK_{a_2} α-NH$_3^+$	pK_{a_3} R group	pI
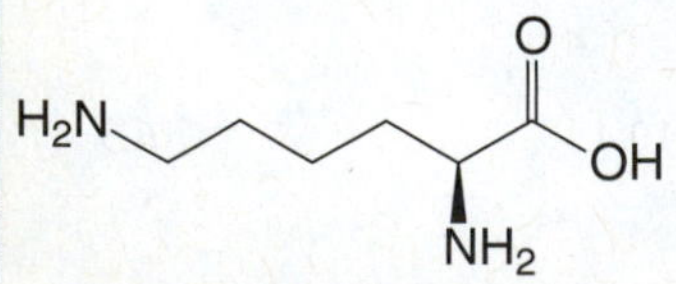	Lysine[b]	K or Lys	2.2	9.0	10.5[c]	9.8

TABLE 24.1 CONTINUED

Structure	Name	Abbreviations[a]	pK$_{a_1}$ α-CO$_2$H	pK$_{a_2}$ α-NH$_3^+$	pK$_{a_3}$ R group	pI
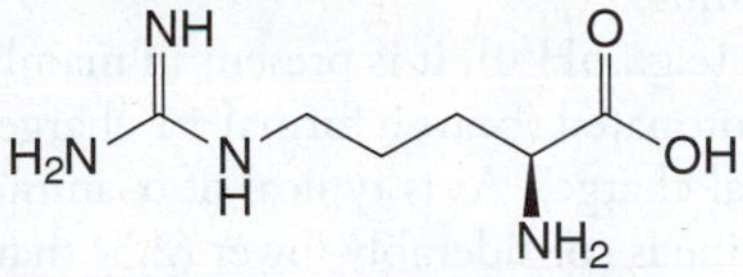	Arginine	R or Arg	2.2	9.0	12.5[c]	10.8
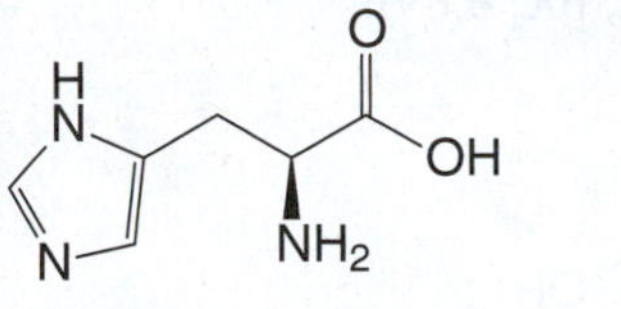	Histidine	H or His	1.8	9.2	6.0[c]	7.6

[a]Single-letter abbreviations are now the most commonly used form in current biochemical literature.

[b]An essential amino acid.

[c]pK$_a$ is of protonated amine of R group.

24.2B Essential Amino Acids

Amino acids can be synthesized by all living organisms, plants and animals. Many higher animals, however, are deficient in their ability to synthesize all of the amino acids they need for their proteins. Thus, these higher animals require certain amino acids as a part of their diet. For adult humans there are eight **essential amino acids**; these are identified in Table 24.1 by a footnote.

24.2C Amino Acids as Dipolar Ions

- Amino acids contain both a basic group (—NH$_2$) and an acidic group (—CO$_2$H).
- In the dry solid state, amino acids exist as **dipolar ions**, a form in which the carboxyl group is present as a carboxylate ion, —CO$_2^-$, and the amino group is present as an aminium ion, —NH$_3^+$. (Dipolar ions are also called **zwitterions**.)
- In aqueous solution, an equilibrium exists between the dipolar ion and the anionic and cationic forms of an amino acid.

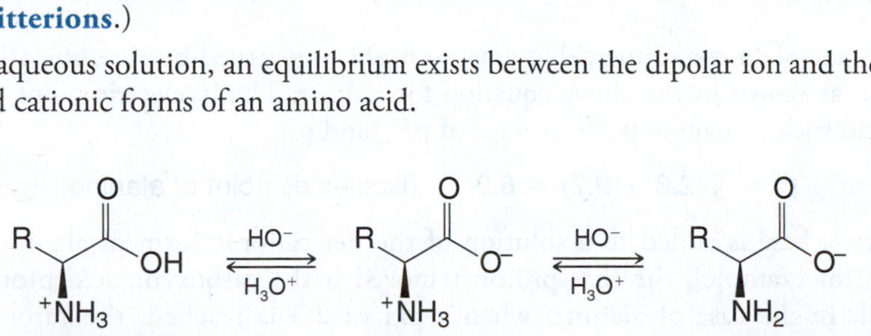

Cationic form (predominant in strongly acidic solutions, e.g., at pH 0) ⇌ HO$^-$ / H$_3$O$^+$ ⇌ **Dipolar ion** ⇌ HO$^-$ / H$_3$O$^+$ ⇌ **Anionic form** (predominant in strongly basic solutions, e.g., at pH 14)

The predominant form of the amino acid present in a solution depends on the pH of the solution and on the nature of the amino acid. In strongly acidic solutions all amino acids are present primarily as cations; in strongly basic solutions they are present as anions.

- The **isoelectric point (pI)** is the pH at which the concentration of the dipolar ion is at its maximum and the concentrations of the anions and cations are equal.

Each amino acid has a particular isoelectric point. These are given in Table 24.1. Proteins have isoelectric points as well. As we shall see later (Sections 24.13 and 24.14), this property of proteins is important for their separation and identification.

Let us consider first an amino acid with a side chain that contains neither acidic nor basic groups—an amino acid, for example, such as alanine.

If alanine is dissolved in a strongly acidic solution (e.g., pH 0), it is present in mainly a net cationic form. In this state the amino group is protonated (bears a formal +1 charge) and the carboxylic acid group is neutral (has no formal charge). As is typical of α-amino acids, the pK_a for the carboxylic acid hydrogen of alanine is considerably lower (2.3) than the pK_a of an ordinary carboxylic acid (e.g., propanoic acid, pK_a 4.89):

Cationic form of alanine   **Propanoic acid**
$pK_{a_1} = 2.3$ $pK_a = 4.89$

The reason for this enhanced acidity of the carboxyl group in an α-amino acid is the inductive effect of the neighboring aminium cation, which helps to stabilize the carboxylate anion formed when it loses a proton. Loss of a proton from the carboxyl group in a cationic α-amino acid leaves the molecule electrically neutral (in the form of a dipolar ion). This equilibrium is shown in the red-shaded portion of the equation below.

The protonated amino group of an α-amino acid is also acidic, but less so than the carboxylic acid group. The pK_a of the aminium group in alanine is 9.7. The equilibrium for loss of an aminium proton is shown in the blue-shaded portion of the equation below. The carboxylic acid proton is always lost before a proton from the aminium group in an α-amino acid.

Cationic form   **Dipolar ion**   **Anionic form**
$(pK_{a_1} = 2.3)$ $(pK_{a_2} = 9.7)$

The state of an α-amino acid at any given pH is governed by a combination of two equilibria, as shown in the above equation for alanine. The isoelectric point (pI) of an amino acid such as alanine is the average of pK_{a_1} and pK_{a_2}:

$$pI = \tfrac{1}{2}(2.3 + 9.7) = 6.0 \qquad \text{(isoelectric point of alanine)}$$

When a base is added to a solution of the net cationic form of alanine (initially at pH 0, for example), the first proton removed is the carboxylic acid proton, as we have said. In the case of alanine, when a pH of 2.3 is reached, the carboxylic acid proton will have been removed from half of the molecules. This pH represents the pK_a of the alanine carboxylic acid proton, as can be demonstrated using the **Henderson–Hasselbalch equation**.

- The **Henderson–Hasselbalch equation** shows that for an acid (HA) and its conjugate base (A$^-$) when [HA] = [A$^-$], then pH = pK_a.

$$pK_a = pH + \log \frac{[HA]}{[A^-]} \qquad \boxed{\textbf{Henderson–Hasselbalch equation}}$$

Therefore, when the acid is half neutralized,

$$[HA] = [A^-], \ \log \frac{[HA]}{[A^-]} = 0, \text{ and thus } pH = pK_a$$

As more base is added to this solution, alanine reaches its isoelectric point (pI), the pH at which all of alanine's carboxylic acid protons have been removed but not its aminium protons. The molecules are therefore electrically neutral (in their dipolar ion or zwitter ionic form) because the carboxylate group carries a −1 charge and the aminium group a +1 charge. The pI for alanine is 6.0.

Now, as we continue to add the base, protons from the aminium ions will begin to be removed, until at pH 9.7 half of the aminium groups will have lost a proton. This pH represents the pK_a of the aminium group. Finally, as more base is added, the remaining aminium protons will be lost until all of the alanine molecules have lost their aminium protons. At this point (e.g., pH 14) the molecules carry a net anionic charge from their carboxylate group. The amino groups are now electrically neutral.

Figure 24.1 shows a titration curve for these equilibria. The graph represents the change in pH as a function of the number of molar equivalents of base. Because alanine has two protons to lose in its net cationic form, when one molar equivalent of base has been added, the molecules will have each lost one proton and they will be electrically neutral (the dipolar ion or zwitterionic form).

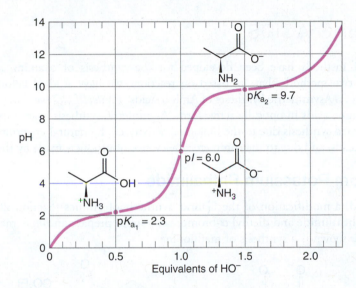

FIGURE 24.1 A titration curve for alanine.

If an amino acid contains a side chain that has an acidic or basic group, the equilibria become more complex. Consider lysine, for example, an amino acid that has an additional —NH_2 group on its ε carbon. In strongly acidic solution, lysine is present as a dication because both amino groups are protonated. The first proton to be lost as the pH is raised is a proton of the carboxyl group (pK_{a_1} = 2.2), the next is from the α-aminium group (pK_{a_2} = 9.0), and the last is from the ε-aminium group (pK_{a_3} = 10.5):

Dicationic form of lysine
(pK_{a_1} = 2.2)

Monocationic form
(pK_{a_2} = 9.0)

Dipolar ion
(pK_{a_3} = 10.5)

Anionic form

The isoelectric point of lysine is the average of pK_{a_2} (the monocation) and pK_{a_3} (the dipolar ion).

$$pI = \tfrac{1}{2}(9.01 + 0.5) = 9.8 \quad \text{(isoelectric point of lysine)}$$

PRACTICE PROBLEM 24.1 What form of glutamic acid would you expect to predominate in **(a)** strongly acidic solution, **(b)** strongly basic solution, and **(c)** at its isoelectric point ($pI = 3.2$)? **(d)** The isoelectric point of glutamine ($pI = 5.7$) is considerably higher than that of glutamic acid. Explain.

PRACTICE PROBLEM 24.2 The guanidino group $-NH-\overset{\overset{\displaystyle NH}{\|}}{C}-NH_2$ of arginine is one of the most strongly basic of all organic groups. Explain.

24.3 SYNTHESIS OF α-AMINO ACIDS

A variety of methods have been developed for the synthesis of α-amino acids. Here we describe two methods that are based on reactions we have studied before. In "The Chemistry of . . . Asymmetric Syntheses of Amino Acids" (*WileyPLUS*) we show methods to prepare α-amino acids in optically active form. Asymmetric synthesis is an important goal in α-amino acid synthesis due to the biological activity of the natural enantiomeric forms of α-amino acids, and due to the commercial relevance of products made by these routes.

24.3A From Potassium Phthalimide

This method, a modification of the Gabriel synthesis of amines (Section 20.4A), uses potassium phthalimide and diethyl α-bromomalonate to prepare an *imido* malonic ester. The example shown is a synthesis of methionine:

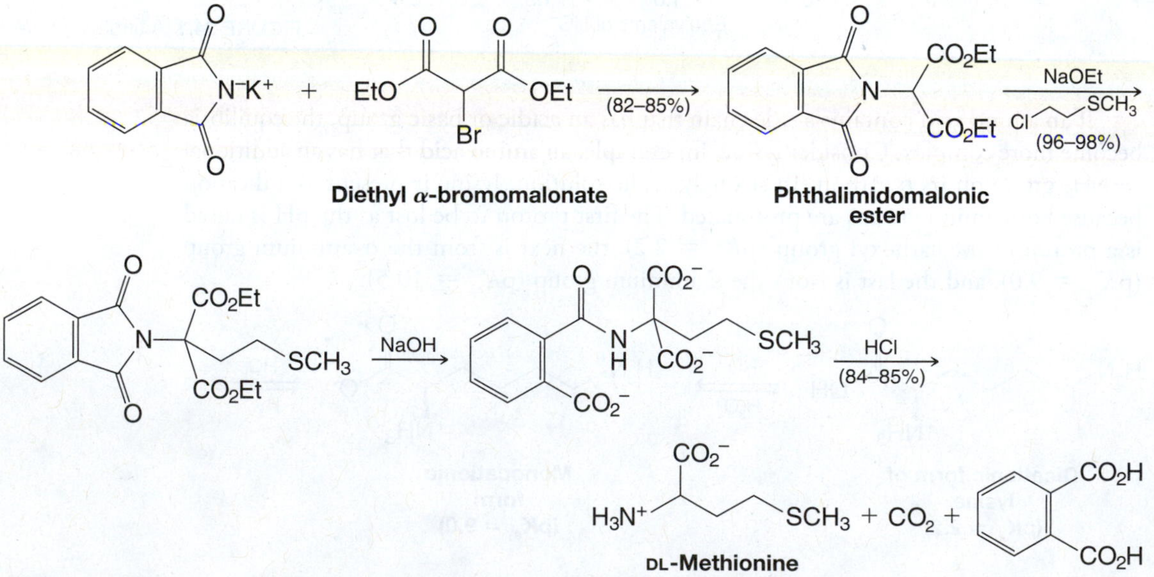

PRACTICE PROBLEM 24.3 Starting with diethyl α-bromomalonate and potassium phthalimide and using any other necessary reagents, show how you might synthesize **(a)** DL-leucine, **(b)** DL-alanine, and **(c)** DL-phenylalanine.

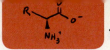

24.3B The Strecker Synthesis

Treating an aldehyde with ammonia and hydrogen cyanide produces an α-aminonitrile. Hydrolysis of the nitrile group (Section 17.3) of the α-aminonitrile converts the latter to an α-amino acid. This synthesis is called the Strecker synthesis:

α-Amino-nitrile

α-Amino acid

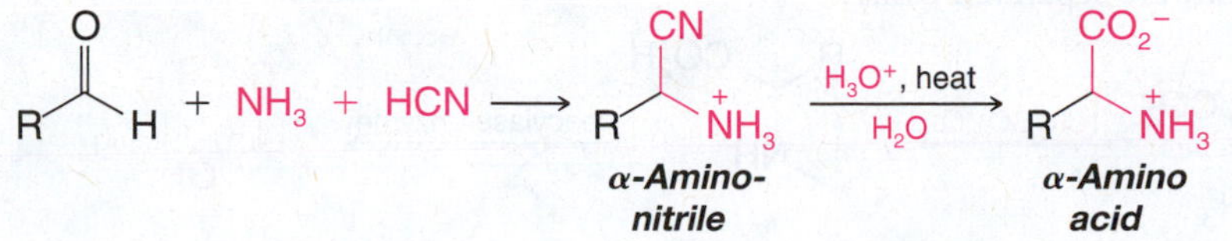

The first step of this synthesis probably involves the initial formation of an imine from the aldehyde and ammonia followed by the addition of hydrogen cyanide.

A MECHANISM FOR THE REACTION — Formation of an α-Aminonitrile during the Strecker Synthesis

Intermolecular proton transfer

Imine

α-Aminonitrile

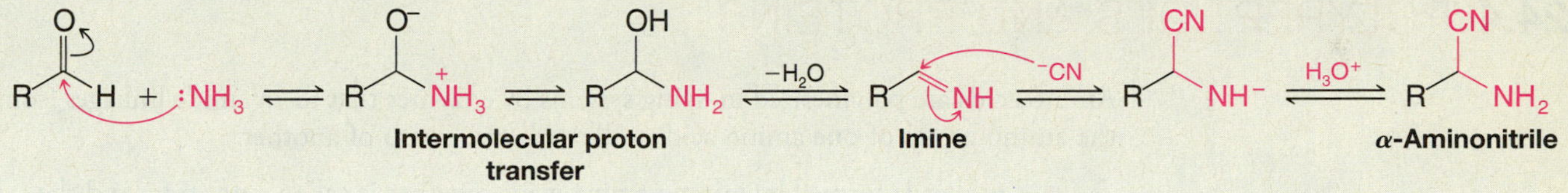

●●● SOLVED PROBLEM 24.1

Outline a Strecker synthesis of DL-tyrosine.

ANSWER:

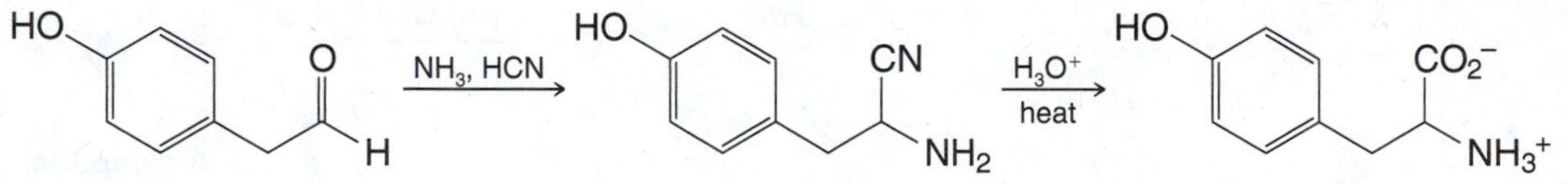

DL-Tyrosine

(a) Outline a Strecker synthesis of DL-phenylalanine. **(b)** DL-Methionine can also be synthesized by a Strecker synthesis. The required starting aldehyde can be prepared from acrolein (CH_2=CHCHO) and methanethiol (CH_3SH). Outline all steps in this synthesis of DL-methionine.

PRACTICE PROBLEM 24.4

24.3C Resolution of DL-Amino Acids

With the exception of glycine, which has no chirality center, the amino acids that are produced by the methods we have outlined are all produced as racemic forms. To obtain the naturally occurring L-amino acid, we must, of course, resolve the racemic form. This can be done in a variety of ways, including the methods outlined in Section 20.3F.

One especially interesting method for resolving amino acids is based on the use of enzymes called *deacylases*. These enzymes catalyze the hydrolysis of *N-acylamino acids* in living organisms. Since the active site of the enzyme is chiral, it hydrolyzes only *N*-acylamino acids of the L configuration. When it is exposed to a racemic mixture of *N*-acylamino acids, only the derivative of the L-amino acid is affected and the products, as a result, are separated easily:

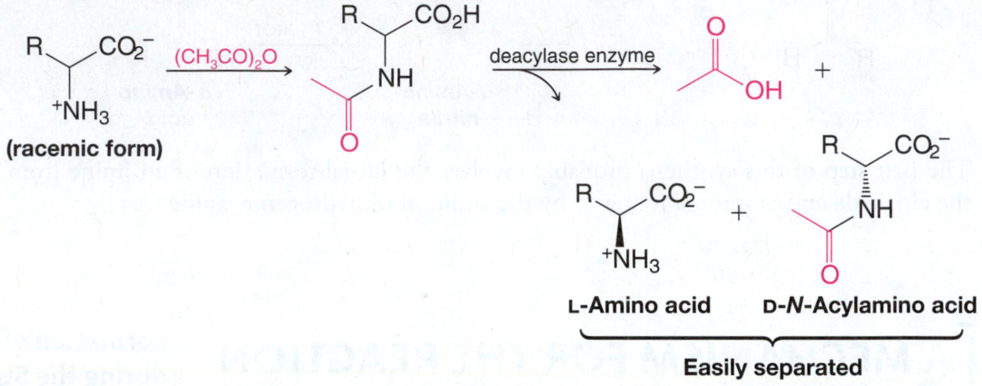

L-Amino acid   **D-*N*-Acylamino acid**

Easily separated

24.4 POLYPEPTIDES AND PROTEINS

Amino acids are polymerized in living systems by enzymes that form amide linkages from the amino group of one amino acid to the carboxyl group of another.

- A molecule formed by joining amino acids together is called a **peptide**, and the amide linkages in them are called **peptide bonds** or **peptide linkages**. Each amino acid in the peptide is called an **amino acid residue**.

Peptides that contain 2, 3, a few (3–10), or many amino acids are called **dipeptides**, **tripeptides**, **oligopeptides**, and **polypeptides**, respectively. **Proteins** are polypeptides consisting of one or more polypeptide chains.

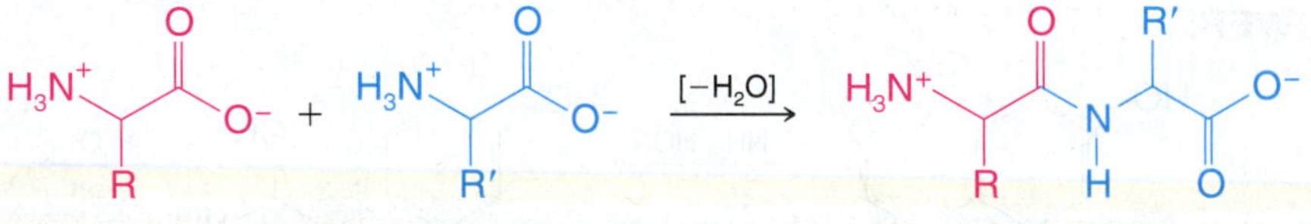

A dipeptide

Polypeptides are **linear polymers**. One end of a polypeptide chain terminates in an amino acid residue that has a free $—NH_3^+$ group; the other terminates in an amino acid residue with a free $—CO_2^-$ group. These two groups are called the **N-terminal** and the **C-terminal residues**, respectively:

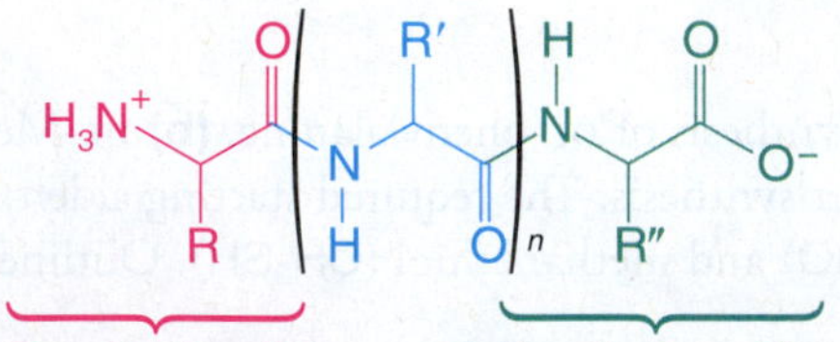

N-Terminal residue   **C-Terminal residue**

- By convention, we write peptide and protein structures with the N-terminal amino acid residue on the left and the C-terminal residue on the right:

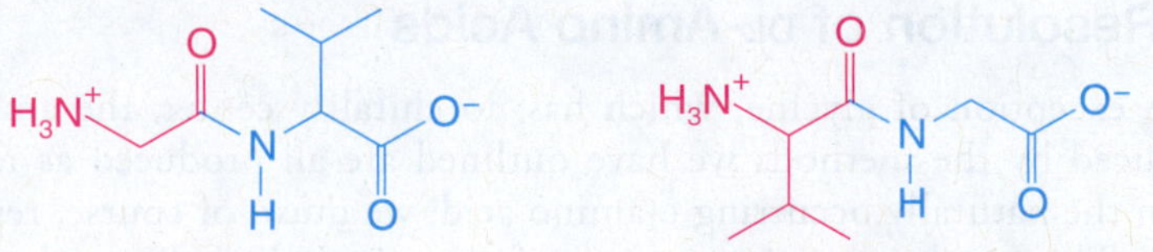

Glycyl valine   **Valyl glycine**
(GV)   **(VG)**

The tripeptide glycylvalylphenylalanine has the following structural formula:

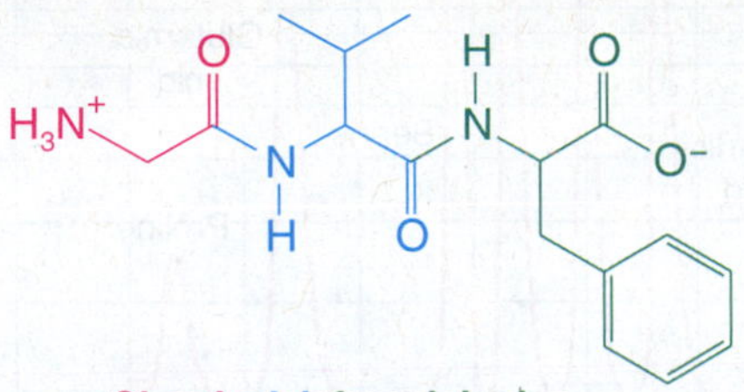

Glycylvalyl**phenylalanine**
(GVF)

It becomes a significant task to write a full structural formula for a polypeptide chain that contains any more than a few amino acid residues. In this situation, use of the one-letter abbreviations (Table 24.1) is the norm for showing the sequence of amino acids. Very short peptide sequences are sometimes still represented with the three-letter abbreviations (Table 24.1).

24.4A Hydrolysis

When a protein or polypeptide is refluxed with 6 M hydrochloric acid for 24 h, hydrolysis of all the amide linkages usually takes place, liberating its constituent amino acids as a mixture. Chromatographic separation and quantitative analysis of the resulting mixture can then be used to determine which amino acids composed the intact polypeptide and their relative amounts.

One chromatographic method for separation of a mixture of amino acids is based on the use of *cation-exchange resins* (Fig. 24.2), which are insoluble polymers containing sulfonate groups. If an acidic solution containing a mixture of amino acids is passed through a column packed with a cation-exchange resin, the amino acids will be adsorbed by the resin because of attractive forces between the negatively charged sulfonate groups and the positively charged amino acids. The strength of the adsorption varies with the basicity of the individual amino acids; those that are most basic are held most strongly. If the column is then washed with a buffered solution at a given pH, the individual amino acids move down the column at different rates and ultimately become separated. In an automated version of this analysis developed at Rockefeller University in 1950, the eluate is allowed to mix with **ninhydrin**, a reagent that reacts with most amino acids to give a derivative with an intense purple color (λ_{max} 570 nm). The amino acid analyzer is designed so that it can measure the absorbance of the eluate (at 570 nm) continuously and record this absorbance as a function of the volume of the effluent.

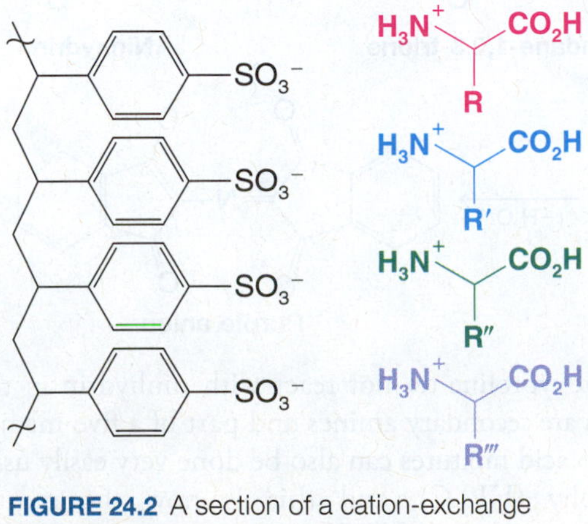

FIGURE 24.2 A section of a cation-exchange resin with adsorbed amino acids.

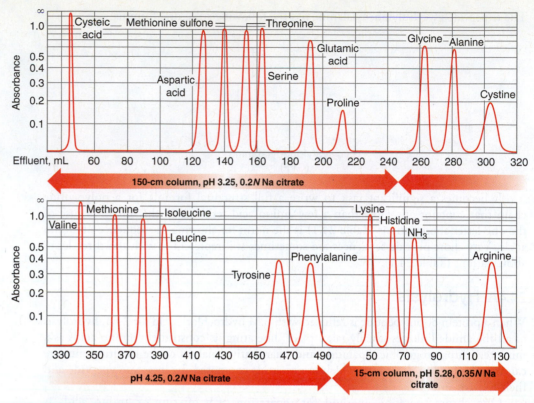

FIGURE 24.3 Typical result given by an automatic amino acid analyzer. (Adapted with permission from Spackman, D. H., Stein, W. H., and Moore, S., *Analytical Chemistry, 30(7)*, pp. 1190–1206, Figure 2, 1958. Copyright 1958 American Chemical Society.)

A typical graph obtained from an automatic amino acid analyzer is shown in Fig. 24.3. When the procedure is standardized, the positions of the peaks are characteristic of the individual amino acids, and the areas under the peaks correspond to their relative amounts.

Ninhydrin is the hydrate of indane-1,2,3-trione. With the exception of proline and hydroxyproline, all of the α-amino acids found in proteins react with ninhydrin to give the same intensely colored purple anion (λ_{max} 570 nm). We shall not go into the mechanism here, but notice that the only portion of the anion that is derived from the α-amino acid is the nitrogen:

Indane-1,2,3-trione   **Ninhydrin**

Purple anion

Proline and hydroxyproline do not react with ninhydrin in the same way because their α-amino groups are secondary amines and part of a five-membered ring.

Analysis of amino acid mixtures can also be done very easily using high-performance liquid chromatography (HPLC), and this is now the most common method. A cation-exchange resin is used for the column packing in some HPLC analyses

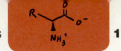

(see Section 24.14), while other analyses require hydrophobic (reversed-phase) column materials. Identification of amino acids separated by HPLC can be done by comparison with retention times of standard samples. Instruments that combine HPLC with mass spectrometry make direct identification possible (see Section 24.5E).

24.5 PRIMARY STRUCTURE OF POLYPEPTIDES AND PROTEINS

The sequence of amino acid residues in a polypeptide or protein is called its **primary structure**. A simple peptide composed of three amino acids (a tripeptide) can have 6 different amino acid sequences; a tetrapeptide can have as many as 24 different sequences. For a protein composed of 20 different amino acids in a single chain of 100 residues, there are $2^{100} = 1.27 \times 10^{130}$ possible peptide sequences, a number much greater than the number of atoms estimated to be in the universe (9×10^{78})! Clearly, one of the most important things to determine about a protein is the sequence of its amino acids. Fortunately, there are a variety of methods available to determine the sequence of amino acids in a polypeptide. We shall begin with **terminal residue analysis** techniques used to identify the N- and C-terminal amino acids.

24.5A Edman Degradation

The most widely used procedure for identifying the N-terminal amino acid in a peptide is the **Edman degradation** method (developed by Pehr Edman of the University of Lund, Sweden). Used repetitively, the Edman degradation method can be used to sequence peptides up to about 60 residues in length. The process works so well that machines called amino acid sequencers have been developed to carry out the Edman degradation process in automated cycles.

The chemistry of the Edman degradation is based on a labeling reaction between the N-terminal amino group and phenyl isothiocyanate, C_6H_5—N=C=S. Phenyl isothiocyanate reacts with the N-terminal amino group to form a phenylthiocarbamyl derivative, which is then cleaved from the peptide chain by acid. The result is an unstable anilinothioazolinone (ATZ), which rearranges to a stable phenylthiohydantoin (PTH) derivative of the amino acid. In the automated process, the PTH derivative is introduced directly to a high-performance liquid chromatograph and identified by comparison of its retention time with known amino acid PTH derivatives (Fig. 24.4). The cycle is then repeated for the next N-terminal amino acid. Automated peptide sequence analyzers can

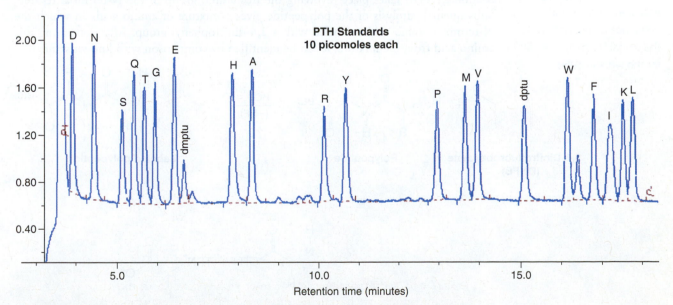

FIGURE 24.4 PTH amino acid standards run on a Procise instrument; see Table 24.1 for amino acid abbreviations. Peaks marked dmptu (dimethylphenylthiourea) and dptu (diphenylthiourea) represent side-reaction products of the Edman degradation. (*Copyright © 2012 Life Technologies Corporation. Used under permission.*)

perform a single iteration of the Edman degradation in approximately 30 min using only picomole amounts of the polypeptide sample.

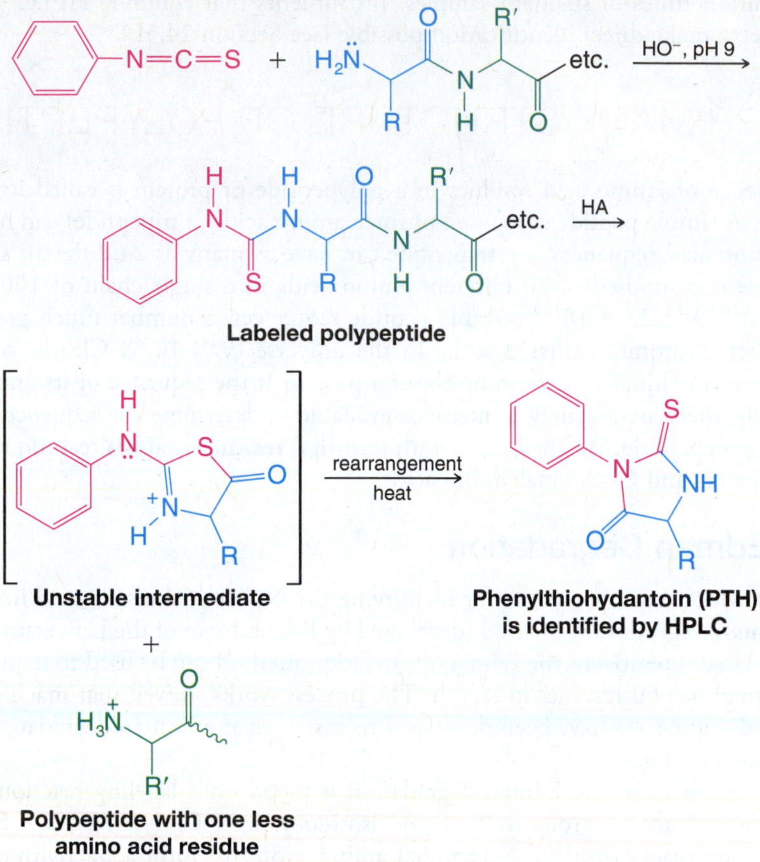

Labeled polypeptide

Unstable intermediate

Phenylthiohydantoin (PTH)
is identified by HPLC

+

Polypeptide with one less
amino acid residue

 This method was introduced by Frederick Sanger of Cambridge University in 1945. Sanger made extensive use of this procedure in his determination of the amino acid sequence of insulin and won the Nobel Prize in Chemistry for the work in 1958.

24.5B Sanger N-Terminal Analysis

Another method for sequence analysis is the **Sanger N-terminal analysis**, based on the use of 2,4-dinitrofluorobenzene (DNFB). When a polypeptide is treated with DNFB in mildly basic solution, a nucleophilic aromatic substitution reaction (S_NAr, Section 21.11A) takes place involving the free amino group of the N-terminal residue. Subsequent hydrolysis of the polypeptide gives a mixture of amino acids in which the N-terminal amino acid is labeled with a 2,4-dinitrophenyl group. After separating this amino acid from the mixture, it can be identified by comparison with known standards.

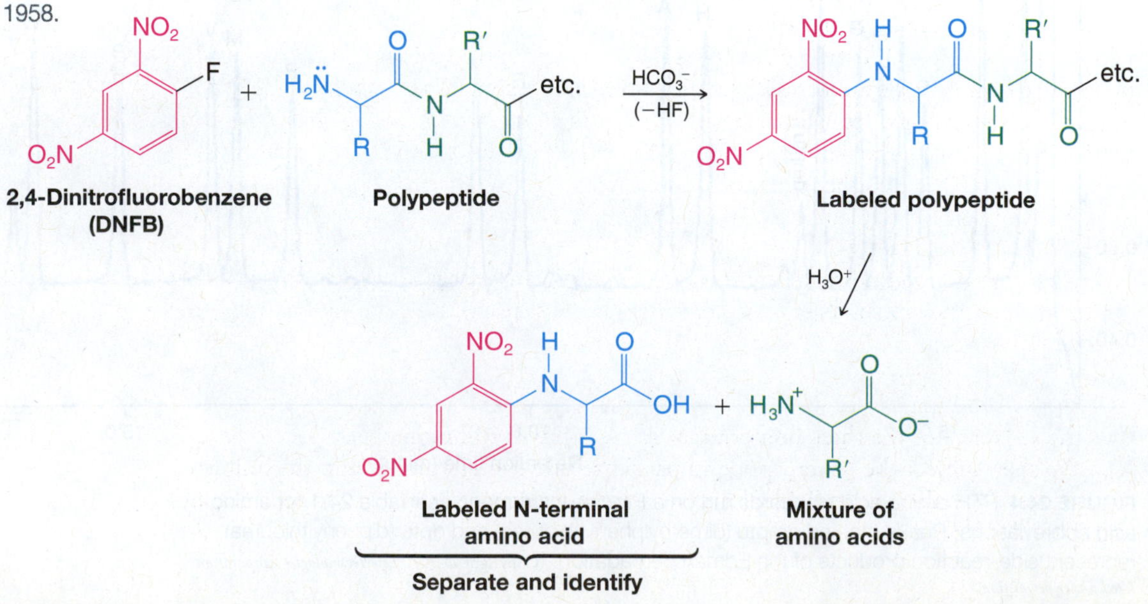

2,4-Dinitrofluorobenzene
(DNFB)

Polypeptide

Labeled polypeptide

Labeled N-terminal
amino acid

Mixture of
amino acids

Separate and identify

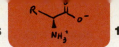

2,4-Dinitrofluorobenzene will react with any free amino group in a polypeptide, including the ε-amino group of lysine, and this fact complicates Sanger analyses. Only the N-terminal amino acid residue of a peptide will bear the 2,4-dinitrophenyl group at its α-amino group, however. Nevertheless, the Edman method of N-terminal analysis is much more widely used.

The electron-withdrawing property of the 2,4-dinitrophenyl group makes separation of the labeled amino acid very easy. Suggest how this is done.

PRACTICE PROBLEM 24.5

24.5C C-Terminal Analysis

C-Terminal residues can be identified through the use of digestive enzymes called *carboxypeptidases*. These enzymes specifically catalyze the hydrolysis of the amide bond of the amino acid residue containing a free —CO_2H group, liberating it as a free amino acid. A carboxypeptidase, however, will continue to attack the polypeptide chain that remains, successively lopping off C-terminal residues. As a consequence, it is necessary to follow the amino acids released as a function of time. The procedure can be applied to only a limited amino acid sequence for, at best, after a time the situation becomes too confused to sort out.

(a) Write a reaction showing how 2,4-dinitrofluorobenzene could be used to identify the N-terminal amino acid of VAG. **(b)** What products would you expect (after hydrolysis) when VKG is treated with 2,4-dinitrofluorobenzene?

PRACTICE PROBLEM 24.6

Write the reactions involved in a sequential Edman degradation of MIR.

PRACTICE PROBLEM 24.7

24.5D Complete Sequence Analysis

Sequential analysis using the Edman degradation or other methods becomes impractical with large proteins and polypeptides. Fortunately, there are techniques to cleave peptides into fragments that are of manageable size. **Partial hydrolysis** with dilute acid, for example, generates a family of peptides cleaved in random locations and with varying lengths. Sequencing these cleavage peptides and looking for points of overlap allows the sequence of the entire peptide to be pieced together.

Consider a simple example: we are given a pentapeptide known to contain valine (two residues), leucine (one residue), histidine (one residue), and phenylalanine (one residue), as determined by hydrolysis and automatic amino acid analysis. With this information we can write the "molecular formula" of the protein in the following way, using commas to indicate that the sequence is unknown:

2V, L, H, F

Then, let us assume that by using DNFB and carboxypeptidase we discover that valine and leucine are the N- and C-terminal residues, respectively. So far we know the following:

V (V, H, F) L

But the sequence of the three nonterminal amino acids is still unknown.

We then subject the pentapeptide to partial acid hydrolysis and obtain the following dipeptides. (We also get individual amino acids and larger pieces, i.e., tripeptides and tetrapeptides.)

VH + HV + VF + FL

The points of overlap of the dipeptides (i.e., H, V, and F) tell us that the original pentapeptide must have been the following:

VHVFL

Site-specific cleavage of peptide bonds is possible with enzymes and specialized reagents as well, and these methods are now more widely used than partial hydrolysis. For example, the enzyme trypsin preferentially catalyzes hydrolysis of peptide bonds on the C-terminal side of arginine and lysine. Chemical cleavage at specific sites can be done with cyanogen bromide (CNBr), which cleaves peptide bonds on the C-terminal side of methionine residues. Using these site-selective cleavage methods on separate samples of a given polypeptide results in fragments that have overlapping sequences. After sequencing the individual fragments, aligning them with each other on the basis of their overlapping sections results in a sequence for the intact protein.

24.5E Peptide Sequencing Using Mass Spectrometry and Sequence Databases

Other methods for determining the sequence of a polypeptide include mass spectrometry and comparison of partial peptide sequences with databases of known complete sequences.

Ladder Sequencing Mass spectrometry is especially powerful because sophisticated techniques allow mass analysis of proteins with very high precision. In one mass spectrometric method, called "ladder sequencing," an enzymatic digest is prepared that yields a mixture of peptide fragments that each differ in length by one amino acid residue (e.g., by use of carboxypeptidase). The digest is a family of peptides where each one is the result of cleavage of one successive residue from the chain. Mass spectrometric analysis of this mixture yields a family of peaks corresponding to the molecular weight of each peptide. Each peak in the spectrum differs from the next by the molecular weight of the amino acid that is the difference in their structures. With these data, one can ascend the ladder of peaks from the lowest weight fragment to the highest (or vice versa), "reading" the sequence of the peptide from the difference in mass between each peak. The difference in mass between each peptide fragment and the next represents the amino acid in that spot along the sequence, and hence an entire sequence can be read from the ladder of fragment masses. This technique has also been applied to the sequencing of oligonucleotides.

Tandem Mass Spectrometry (MS/MS) Random cleavage of a peptide, similar to that from partial hydrolysis with acid, can also be accomplished with mass spectrometry. An intact protein introduced into a mass spectrometer can be cleaved into smaller fragments by collision with gas molecules deliberately leaked into the mass spectrometer vacuum chamber (a technique called collision-induced dissociation, CID). These peptide fragments can be individually selected for analysis using a technique called tandem mass spectrometry (MS/MS). The mass spectra of these random fragments can be compared with mass spectra databases to determine the protein sequence.

Partial Hydrolysis and Sequence Comparison In some cases it is also possible to determine the sequence of an unknown polypeptide by sequencing just a few of its amino acids and comparing this partial sequence with the database of known sequences for complete polypeptides or proteins. This procedure works if the unknown peptide turns out to be one that has been studied previously. (Studies of the expression of known proteins is one dimension of the field of proteomics, Section 24.14.) Due to the many sequence permutations that are theoretically possible and the uniqueness of a given protein's structure, a sequence of just 10–25 peptide residues is usually sufficient to generate data that match only one or a small number of known polypeptides. The partial sequence can be determined by the Edman method or by mass spectrometry. For example, the enzyme lysozyme with 129 amino acid residues (see Section 24.10) can be identified based on the sequence of just its first 15 amino acid residues. Structure determination based on comparison of sequences with computerized databases is part of the burgeoning field of bioinformatics.

An analogous approach using databases is to infer the *DNA sequence* that codes for a partial peptide sequence and compare this DNA sequence with the database of known

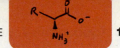

DNA sequences. If a satisfactory match is found, the remaining sequence of the polypeptide can be read from the DNA sequence using the genetic code (see Section 25.5). In addition, the inferred oligonucleotide sequence for the partial peptide can be synthesized chemically (see Section 25.7) and used as a probe to find the gene that codes for the protein. This technique is part of molecular biological methods used to clone and express large quantities of a protein of interest.

Glutathione is a tripeptide found in most living cells. Partial acid-catalyzed hydrolysis of glutathione yields two dipeptides, CG and one composed of E and C. When this second dipeptide was treated with DNFB, acid hydrolysis gave *N*-labeled glutamic acid. **(a)** On the basis of this information alone, what structures are possible for glutathione? **(b)** Synthetic experiments have shown that the second dipeptide has the following structure:

PRACTICE PROBLEM 24.8

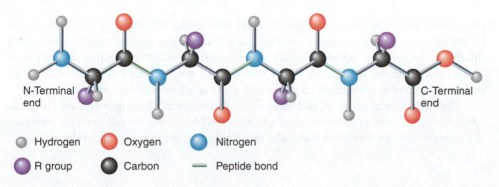

What is the structure of glutathione?

Give the amino acid sequence of the following polypeptides using only the data given by partial acidic hydrolysis:

PRACTICE PROBLEM 24.9

(a) S, O, P, T $\xrightarrow[\text{H}_2\text{O}]{\text{H}_3\text{O}^+}$ ST + TO + PS   **(b)** A, R, C, V, L $\xrightarrow[\text{H}_2\text{O}]{\text{H}_3\text{O}^+}$ AC + CR + RV + LA

24.6 EXAMPLES OF POLYPEPTIDE AND PROTEIN PRIMARY STRUCTURE

- The covalent structure of a protein or polypeptide is called its **primary structure** (Fig. 24.5).

N-Terminal end   C-Terminal end

- Hydrogen   - Oxygen   - Nitrogen
- R group   - Carbon   — Peptide bond

FIGURE 24.5 A representation of the primary structure of a tetrapeptide.

Using the techniques we described, chemists have had remarkable success in determining the primary structures of polypeptides and proteins. The compounds described in the following pages are important examples.

24.6A Oxytocin and Vasopressin

Oxytocin and vasopressin (Fig. 24.6) are two rather small polypeptides with strikingly similar structures (where oxytocin has leucine, vasopressin has arginine, and where oxytocin has isoleucine, vasopressin has phenylalanine). In spite of the similarity of their

Vincent du Vigneaud of Cornell Medical College synthesized oxytocin and vasopressin in 1953; he received the Nobel Prize in Chemistry in 1955.

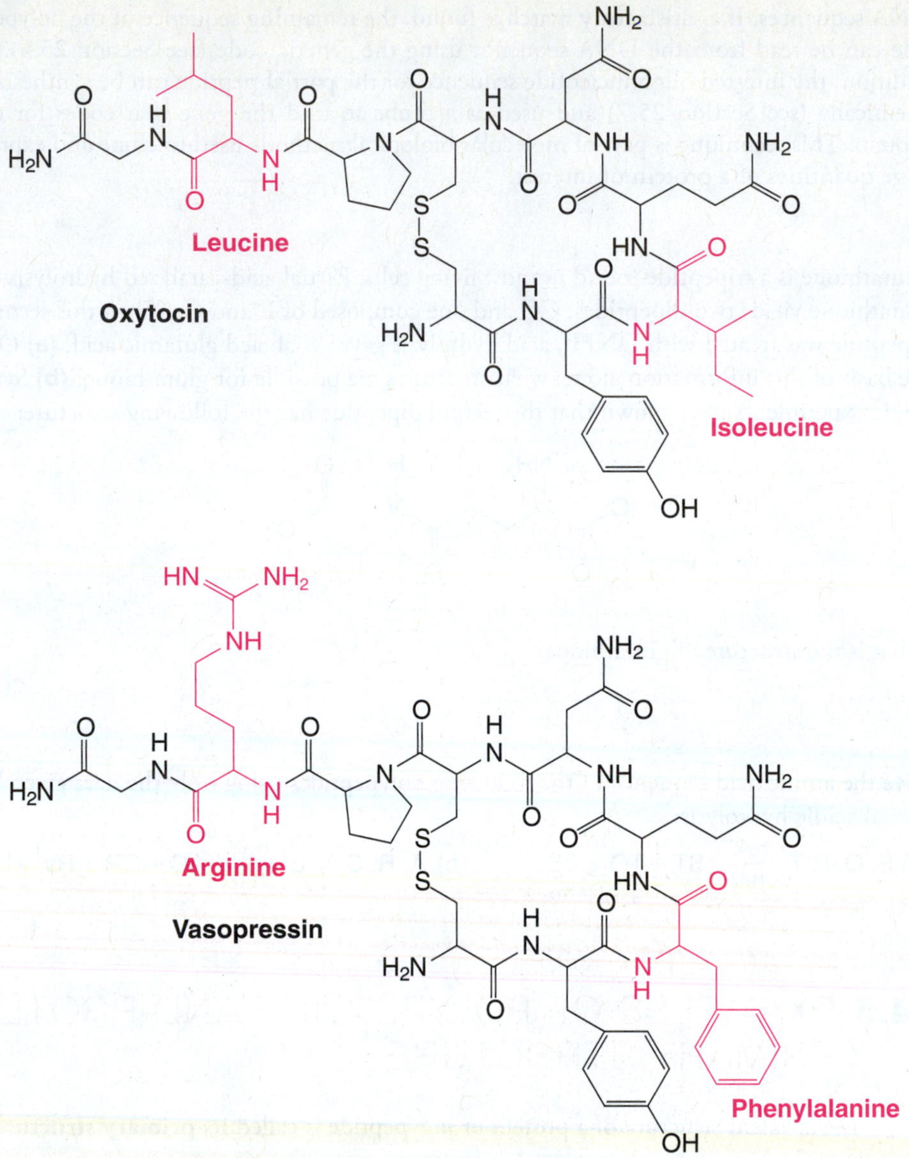

Leucine

Oxytocin

Isoleucine

Arginine

Vasopressin

Phenylalanine

FIGURE 24.6 The structures of oxytocin and vasopressin. Amino acid residues that differ between them are shown in red.

amino acid sequences, these two polypeptides have quite different physiological effects. Oxytocin occurs only in the female of a species and stimulates uterine contractions during childbirth. Vasopressin occurs in males and females; it causes contraction of peripheral blood vessels and an increase in blood pressure. Its major function, however, is as an *antidiuretic;* physiologists often refer to vasopressin as an *antidiuretic hormone.*

The structures of oxytocin and vasopressin also illustrate the importance of the disulfide linkage between cysteine residues (Section 24.2A) in the overall primary structure of a polypeptide. In these two molecules this disulfide linkage leads to a cyclic structure.

● ● ●

PRACTICE PROBLEM 24.10 Treating oxytocin with certain reducing agents (e.g., sodium in liquid ammonia) brings about a single chemical change that can be reversed by air oxidation. What chemical changes are involved?

24.6B Insulin

Insulin, a hormone secreted by the pancreas, regulates glucose metabolism. Insulin deficiency in humans is the major problem in diabetes mellitus.

A Chain

GIVEQCCASVCSLYQLENYCN

B Chain

FVNQHLCGSHLVEALYLVCGERGFFYTPKA

FIGURE 24.7 The amino acid sequence of bovine insulin. Lines between chains indicate disulfide linkages.

The amino acid sequence of bovine insulin (Fig. 24.7) was determined by Sanger in 1953 after 10 years of work. Bovine insulin has a total of 51 amino acid residues in two polypeptide chains, called the A and B chains. These chains are joined by two disulfide linkages. The A chain contains an additional disulfide linkage between cysteine residues at positions 6 and 11.

Human insulin differs from bovine insulin at only three amino acid residues: Threonine replaces alanine once in the A chain (residue 8) and once in the B chain (residue 30), and isoleucine replaces valine once in the A chain (residue 10). Insulins from most mammals have similar structures.

THE CHEMISTRY OF... Sickle-Cell Anemia

The genetically based disease sickle-cell anemia results from a single amino acid error in the β chain of hemoglobin. In normal hemoglobin, position 6 has a glutamic acid residue, whereas in sickle-cell hemoglobin position 6 is occupied by valine.

Red blood cells (erythrocytes) containing hemoglobin with this amino acid residue error tend to become crescent shaped ("sickle") when the partial pressure of oxygen is low, as it is in venous blood. These distorted cells are more difficult for the heart to pump through small capillaries. They may even block capillaries by clumping together; at other times the red cells may even split open. Children who inherit this genetic trait from both parents suffer from a severe form of the disease and usually do not live past the age of two. Children who inherit the disease from only one parent generally have a much milder form. Sickle-cell anemia arose among the populations of central and western Africa where, ironically, it may have had a beneficial effect. People with a mild form of the disease are far less susceptible to malaria than those with normal hemoglobin. Malaria, a disease caused by an infectious microorganism, is especially prevalent in central and western Africa. Mutational changes such as those that give rise to sickle-cell anemia are very common. Approximately 150 different types of mutant hemoglobin have been detected in humans; fortunately, most are harmless.

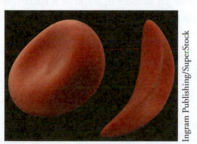

Ingram Publishing/SuperStock

Normal (left) and sickled (right) red blood cells viewed with a scanning electron microscope at 18,000× magnification.

24.6C Other Polypeptides and Proteins

Successful sequential analyses have now been achieved with thousands of other polypeptides and proteins, including the following:

1. **Bovine ribonuclease.** This enzyme, which catalyzes the hydrolysis of ribonucleic acid (Chapter 25), has a single chain of 124 amino acid residues and four intrachain disulfide linkages.

2. **Human hemoglobin.** There are four peptide chains in this important oxygen-carrying protein. Two identical α chains have 141 residues each, and two identical β chains have 146 residues each.

3. **Bovine trypsinogen and chymotrypsinogen.** These two digestive enzyme precursors have single chains of 229 and 245 residues, respectively.

4. **Gamma globulin.** This immunoprotein has a total of 1320 amino acid residues in four chains. Two chains have 214 residues each; the other two have 446 each.

5. **p53, an anticancer protein.** The protein called p53 (the p stands for protein), consisting of 393 amino acid residues, has a variety of cellular functions, but the most

important ones involve controlling the steps that lead to cell growth. It acts as a **tumor suppressor** by halting abnormal growth in normal cells, and by doing so it prevents cancer. Discovered in 1979, p53 was originally thought to be a protein synthesized by an oncogene (a gene that causes cancer). Research has shown, however, that the form of p53 originally thought to have this cancer-causing property was a mutant form of the normal protein. The unmutated (or *wild type*) p53 apparently coordinates a complex set of responses to changes in DNA that could otherwise lead to cancer. When p53 becomes mutated, it no longer provides the cell with its cancer-preventing role; it apparently does the opposite, by acting to increase abnormal growth.

More than half of the people diagnosed with cancer each year have a mutant form of p53 in their cancers. Different forms of cancer have been shown to result from different mutations in the protein, and the list of cancer types associated with mutant p53 includes cancers of most of the body parts: brain, breast, bladder, cervix, colon, liver, lung, ovary, pancreas, prostate, skin, stomach, and so on.

6. *Ras* **proteins.** *Ras* proteins are modified proteins associated with cell growth and the cell's response to insulin. They belong to a class of proteins called prenylated proteins, in which lipid groups derived from isoprenoid biosynthesis (Special Topic E, *WileyPLUS*) are appended as thioethers to C-terminal cysteine residues. Certain mutated forms of *ras* proteins cause oncogenic changes in various eukaryotic cell types. One effect of prenylation and other lipid modifications of proteins is to anchor these proteins to cellular membranes. Prenylation may also assist with molecular recognition of prenylated proteins by other proteins.*

24.7 POLYPEPTIDE AND PROTEIN SYNTHESIS

We saw in Chapter 17 that the synthesis of an amide linkage is a relatively simple one. We must first "activate" the carboxyl group of an acid by converting it to an anhydride or acid chloride and then allow it to react with an amine:

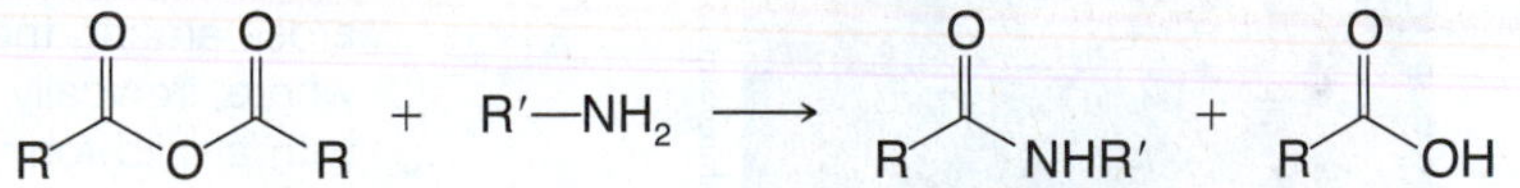

The problem becomes somewhat more complicated, however, when both the acid group and the amino group are present in the same molecule, as they are in an amino acid, and especially when our goal is the synthesis of a naturally occurring polyamide where the sequence of different amino acids is all important. Let us consider, as an example, the synthesis of the simple dipeptide alanylglycine, AG. We might first activate the carboxyl group of alanine by converting it to an acid chloride, and then we might allow it to react with glycine. Unfortunately, however, we cannot prevent alanyl chloride from reacting with itself. So our reaction would yield not only AG but also AA. It could also lead to AAA and AAG, and so on. The yield of our desired product would be low, and we would also have a difficult problem separating the dipeptides, tripeptides, and higher peptides.

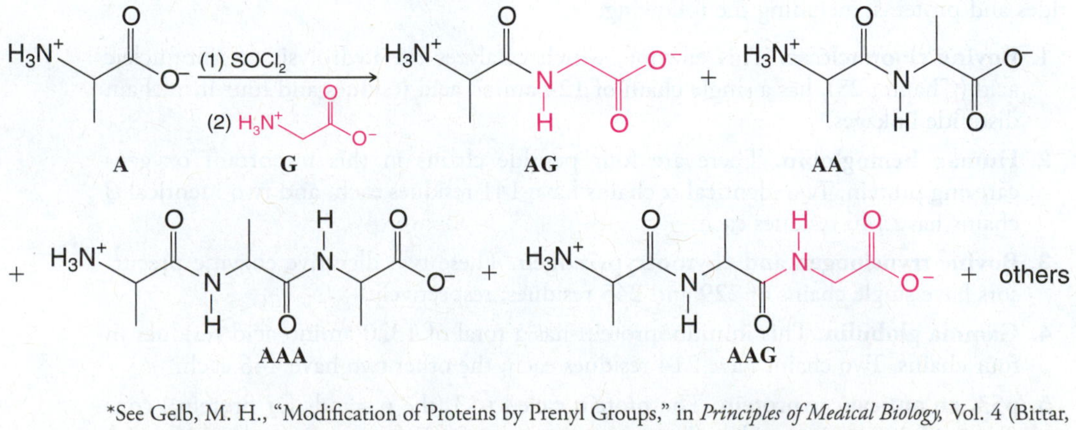

*See Gelb, M. H., "Modification of Proteins by Prenyl Groups," in *Principles of Medical Biology*, Vol. 4 (Bittar, E. E., and Bittar, N., eds.), JAI Press: Greenwich, CT, 1995; Chapter 14, pp. 323–333.

24.7A Protecting Groups

The solution to this problem is to "protect" the amino group of the first amino acid before we activate it and allow it to react with the second. By protecting the amino group, we mean that we must convert it to some other group of low nucleophilicity—*one that will not react with a reactive acyl derivative.* The **protecting group** must be carefully chosen because after we have synthesized the amide linkage between the first amino acid and the second, we will want to be able to remove the protecting group without disturbing the new amide bond.

A number of reagents have been developed to meet these requirements. Three that are often used are benzyl chloroformate, di-*tert*-butyl dicarbonate (sometimes abbreviated Boc$_2$O, where Boc stands for *tert*-butyloxycarbonyl), and 9-fluorenylmethyl chloroformate:

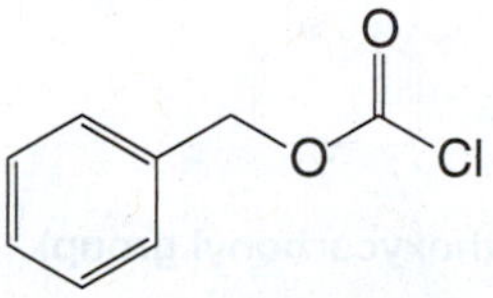

Benzyl chloroformate 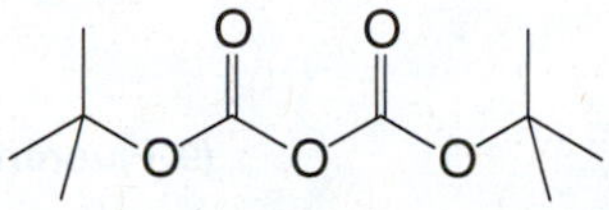   **Di-*tert*-butyl dicarbonate (Boc$_2$O)**   **9-Fluorenylmethyl chloroformate**

All three reagents react with the amine to block it from further acylation. These derivations, however, are types that allow removal of the protecting group under conditions that do not affect peptide bonds. The benzyloxycarbonyl group (abbreviated Z) can be removed with catalytic hydrogenation or cold HBr in acetic acid. The *tert*-butyloxycarbonyl group can be removed with trifluoroacetic acid (CF$_3$CO$_2$H) in acetic acid. The 9-fluorenylmethoxycarbonyl (Fmoc) group is stable under acid conditions but can be removed under mild basic conditions using piperidine (a secondary amine).

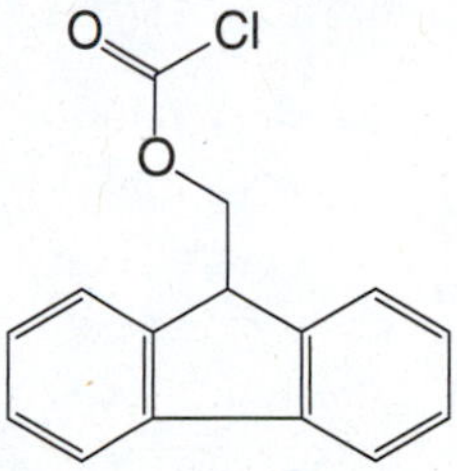

Benzyloxycarbonyl Group

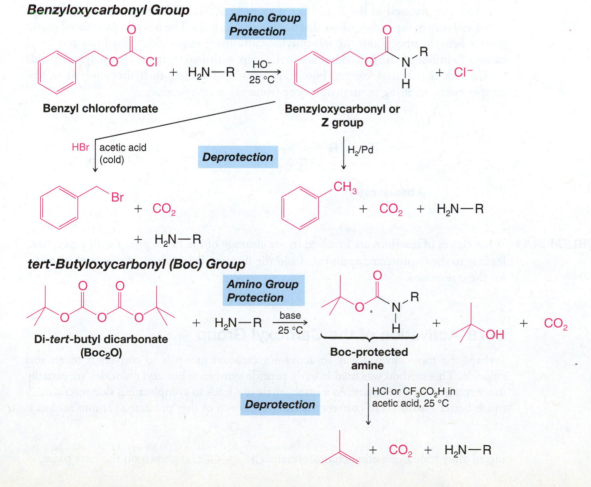

Benzyl chloroformate   **Benzyloxycarbonyl or Z group**

tert-Butyloxycarbonyl (Boc) Group

Di-*tert*-butyl dicarbonate (Boc$_2$O)   **Boc-protected amine**

9-Fluorenylmethoxycarbonyl Group

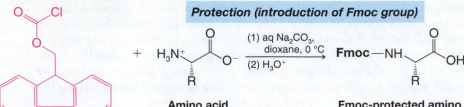

Protection (introduction of Fmoc group)

9-Fluorenylmethyl chloroformate + Amino acid (side chain protected in advance if necessary) $\xrightarrow[\text{(2) H}_3\text{O}^+]{\substack{\text{(1) aq Na}_2\text{CO}_3,\\ \text{dioxane, 0 °C}}}$ Fmoc-protected amino acid (stable in acid)

Fmoc = (9-Fluorenylmethoxycarbonyl group)

Deprotection (removal of Fmoc group)

Fmoc—NH ... OH + Piperidine (C$_5$H$_{11}$N) $\xrightarrow{\text{DMF}}$ Unprotected amino acid + CO$_2$ + (By-product)

The easy removal of the Z and Boc groups in acidic media results from the exceptional stability of the carbocations that are formed initially. The benzyloxycarbonyl group gives a benzyl carbocation; the *tert*-butyloxycarbonyl group yields, initially, a *tert*-butyl cation. Removal of the benzyloxycarbonyl group with hydrogen and a catalyst depends on the fact that benzyl–oxygen bonds are weak and subject to hydrogenolysis at low temperatures, resulting in methylbenzene (toluene) as one product:

A benzyl ester $\xrightarrow[\text{25 °C}]{\text{H}_2\text{, Pd}}$ CH$_3$ + HO ... R

PRACTICE PROBLEM 24.11   What classes of reactions are involved in the cleavage of the Fmoc group with piperidine, leading to the unprotected amino acid and the fluorene by-product? Write mechanisms for these reactions.

24.7B Activation of the Carboxyl Group

Perhaps the most obvious way to activate a carboxyl group is to convert it to an acyl chloride. This method was used in early peptide syntheses, but acyl chlorides are actually more reactive than necessary. As a result, their use leads to complicating side reactions. A much better method is to convert the carboxyl group of the "protected" amino acid to a

mixed anhydride using ethyl chloroformate, Cl—C(=O)—OEt, as shown on the next page:

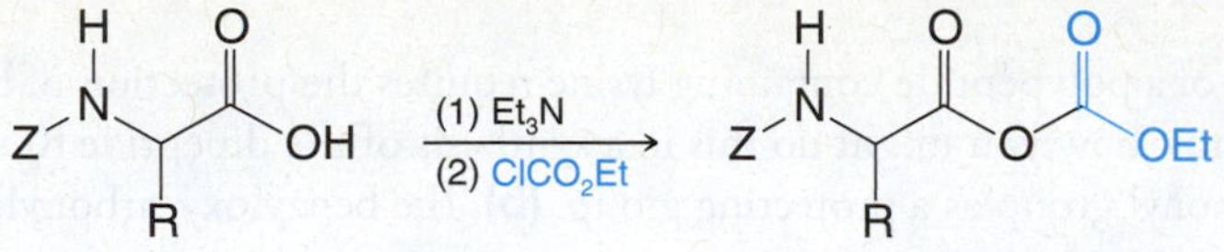

"Mixed anhydride"

The mixed anhydride can then be used to acylate another amino acid and form a peptide linkage:

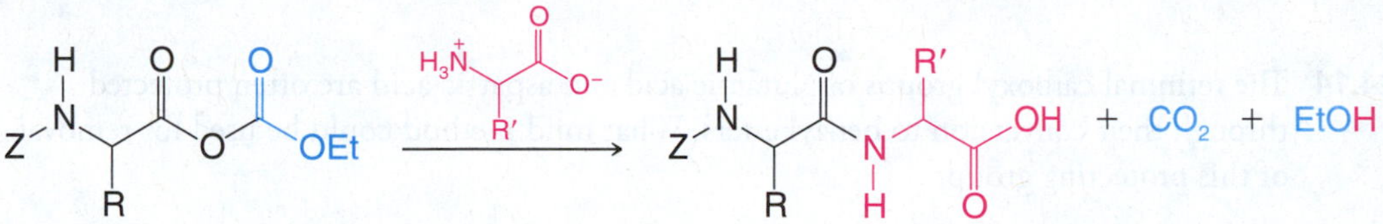

Diisopropylcarbodiimide and dicyclohexylcarbodiimide (Section 17.8E) can also be used to activate the carboxyl group of an amino acid. In Section 24.7D we shall see how diisopropylcarbodiimide is used in an automated peptide synthesis.

24.7C Peptide Synthesis

Let us examine now how we might use these reagents in the preparation of the simple dipeptide AL. The principles involved here can, of course, be extended to the synthesis of much longer polypeptide chains.

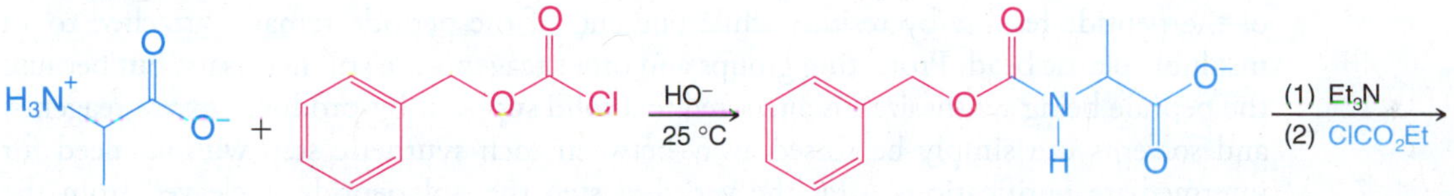

A   Benzyl chloroformate   Z-A

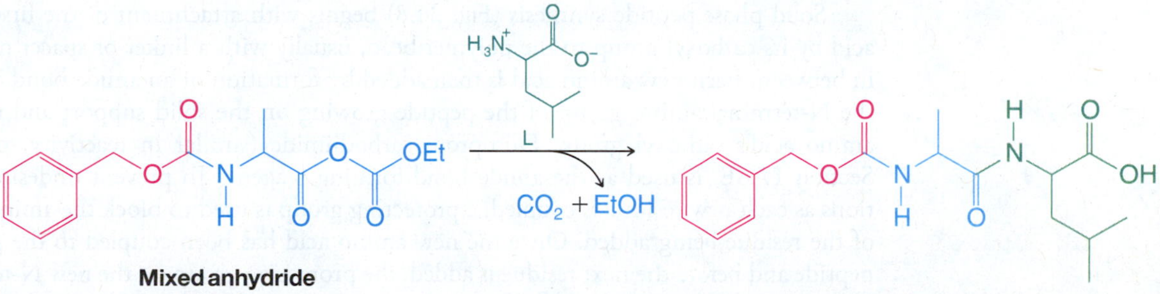

Mixed anhydride
of Z-A   Z-AL

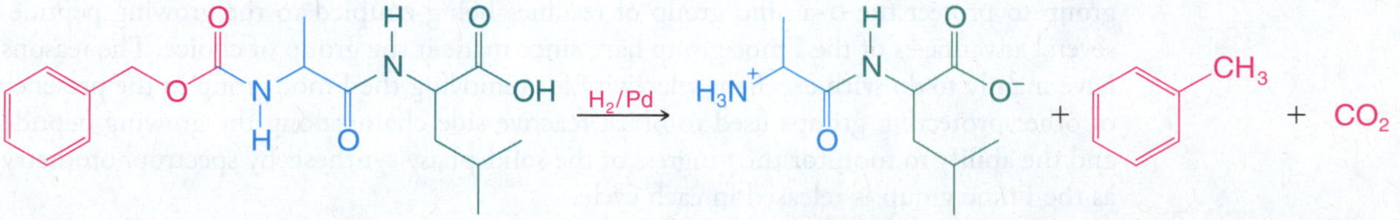

AL

Show all steps in the synthesis of GVA using the *tert*-butyloxycarbonyl (Boc) group as a protecting group.

PRACTICE PROBLEM 24.12

PRACTICE PROBLEM 24.13 The synthesis of a polypeptide containing lysine requires the protection of both amino groups. **(a)** Show how you might do this in a synthesis of the dipeptide **KI** using the benzyloxycarbonyl group as a protecting group. **(b)** The benzyloxycarbonyl group can

$$\begin{array}{c} NH \\ \parallel \end{array}$$

also be used to protect the guanidino group, —NHC—NH$_2$, of arginine. Show a synthesis of the dipeptide **RA**.

PRACTICE PROBLEM 24.14 The terminal carboxyl groups of glutamic acid and aspartic acid are often protected through their conversion to benzyl esters. What mild method could be used for removal of this protecting group?

24.7D Automated Peptide Synthesis

R. B. MERRIFIELD, received the 1984 Nobel Prize in Chemistry for development of a method for solid-phase peptide synthesis.

The methods that we have described thus far have been used to synthesize a number of polypeptides, including ones as large as insulin. They are extremely time-consuming and tedious, however. One must isolate the peptide and purify it by lengthy means at almost every stage. Furthermore, significant loss of the peptide can occur with each isolation and purification stage. The development of a procedure by R. B. Merrifield (Rockefeller University, dec. 2005) for automating this process was therefore a breakthrough in peptide synthesis. Merrifield's method, for which he received the 1984 Nobel Prize in Chemistry, is called **solid-phase peptide synthesis (SPPS),** and it hinges on synthesis of the peptide residue by residue while one end of the peptide remains attached to an insoluble plastic bead. Protecting groups and other reagents are still necessary, but because the peptide being synthesized is anchored to a solid support, by-products, excess reagents, and solvents can simply be rinsed away between each synthetic step without need for intermediate purification. After the very last step the polypeptide is cleaved from the polymer support and subjected to a final purification by HPLC. The method works so well that it was developed into an automated process.

Solid-phase peptide synthesis (Fig. 24.8) begins with attachment of the first amino acid by its carboxyl group to the polymer bead, usually with a linker or spacer molecule in between. Each new amino acid is then added by formation of an amide bond between the N-terminal amino group of the peptide growing on the solid support and the new amino acid's carboxyl group. Diisopropylcarbodiimide (similar in reactivity to DCC, Section 17.8E) is used as the amide bond-forming reagent. To prevent undesired reactions as each new residue is coupled, a protecting group is used to block the amino group of the residue being added. Once the new amino acid has been coupled to the growing peptide and before the next residue is added, the protecting group on the new N-terminus is removed, making the peptide ready to begin the next cycle of amide bond formation.

Although Merrifield's initial method for solid-phase peptide synthesis used the Boc group to protect the α-amino group of residues being coupled to the growing peptide, several advantages of the Fmoc group have since made it the group of choice. The reasons have mainly to do with excellent selectivity for removing the Fmoc group in the presence of other protecting groups used to block reactive side chains along the growing peptide and the ability to monitor the progress of the solid-phase synthesis by spectrophotometry as the Fmoc group is released in each cycle.

Let us discuss the choice of protecting groups further. As noted (Section 24.7A), *basic conditions* (piperidine in DMF) are used to remove the Fmoc group. On the other hand, protecting groups for the side chains of the peptide residues are generally blocked with *acid-labile* moieties. The base-labile Fmoc groups and acid-labile side-chain protecting groups are said to be **orthogonal protecting groups** because one set of protecting groups is stable under conditions for removal of the other, and vice versa. Another advantage of Fmoc as compared to Boc groups for protecting the α-amino group of each new residue

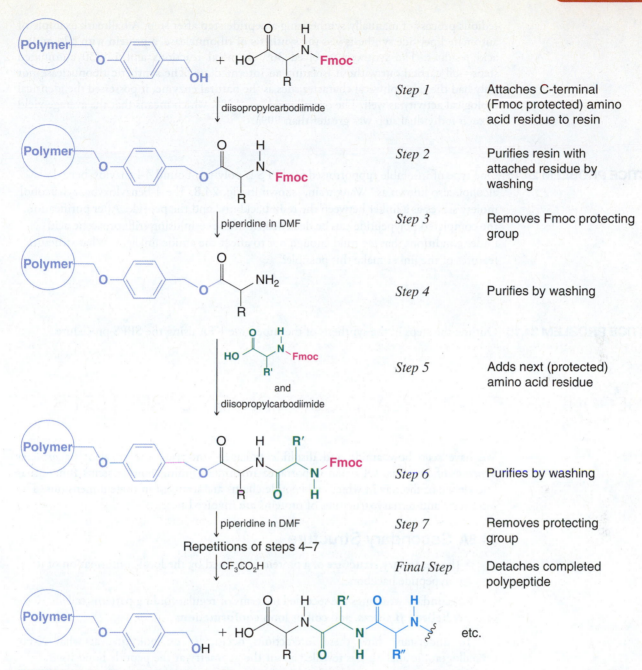

FIGURE 24.8 A method for automated solid-phase peptide synthesis.

Step 1	Attaches C-terminal (Fmoc protected) amino acid residue to resin
Step 2	Purifies resin with attached residue by washing
Step 3	Removes Fmoc protecting group
Step 4	Purifies by washing
Step 5	Adds next (protected) amino acid residue
Step 6	Purifies by washing
Step 7	Removes protecting group
Final Step	Detaches completed polypeptide

is that repetitive application of the acidic conditions to remove Boc groups from each new residue slowly sabotages the synthesis by prematurely cleaving some peptide molecules from the solid support and deprotecting some of the side chains. The basic conditions for Fmoc removal avoid these problematic side reactions.

- The great advantage of solid-phase peptide synthesis is that purification of the peptide at each stage involves simply rinsing the beads of the solid support to wash away excess reagent, by-products, and solvents.

- Furthermore, having the peptide attached to a tangible solid during the synthesis allows all of the steps in the synthesis to be carried out by a machine in repeated cycles.

Automated peptide synthesizers are available that can complete one cycle in 40 min and carry out 45 cycles of unattended operation. Though not as efficient as protein synthesis in the body, where enzymes directed by DNA can catalyze assembly of a protein with 150 amino acids in about 1 min, automated peptide synthesis is a far cry from the tedious process of manually synthesizing a peptide step after step. A hallmark example of automated peptide synthesis was the synthesis of ribonuclease, a protein with 124 amino acid residues. The synthesis involved 369 chemical reactions and 11,930 automated steps—all carried out without isolating an intermediate. The synthetic ribonuclease not only had the same physical characteristics as the natural enzyme, it possessed the identical biological activity as well. The overall yield was 17%, which means that the average yield of each individual step was greater than 99%.

PRACTICE PROBLEM 24.15 One type of insoluble support used for SPPS is polymer-bound 4-benzyloxybenzyl alcohol, also known as "Wang resin," shown in Fig. 24.8. The 4-benzyloxybenzyl alcohol moiety serves as a linker between the resin backbone and the peptide. After purification, the completed polypeptide can be detached from the resin using trifluoroacetic acid under conditions that are mild enough not to affect the amide linkages. What structural features of the linker make this possible?

PRACTICE PROBLEM 24.16 Outline the steps in the synthesis of the tripeptide KFA using the SPPS procedure.

24.8 SECONDARY, TERTIARY, AND QUATERNARY STRUCTURES OF PROTEINS

We have seen how amide and disulfide linkages constitute the covalent or *primary structure* of proteins. Of equal importance in understanding how proteins function is knowledge of the way in which the peptide chains are arranged in three dimensions. The secondary and tertiary structures of proteins are involved here.

24.8A Secondary Structure

- The **secondary structure** of a protein is defined by the local conformation of its polypeptide backbone.
- Secondary structures are specified in terms of regular folding patterns called **α helices, β sheets**, and **coil** or **loop conformations**.

To understand how these interactions occur, let us look first at what X-ray crystallographic analysis has revealed about the geometry at the peptide bond itself.

- Peptide bonds tend to assume a geometry such that six atoms of the amide linkage are coplanar (Fig. 24.9).

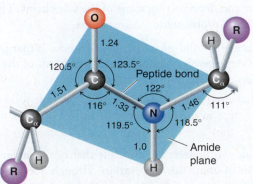

FIGURE 24.9 The geometry and bond lengths (in angstroms, Å) of the peptide linkage. The six enclosed atoms tend to be coplanar and assume a "transoid" arrangement.
(Reprinted with permission of John Wiley & Sons, Inc., from Voet, D. and Voet, J. G., *Biochemistry*, Second Edition. © 1995 Voet, D. and Voet, J. G.)

trans-Peptide group

The carbon–nitrogen bond of the amide linkage is unusually short, indicating that resonance contributions of the type shown here are important:

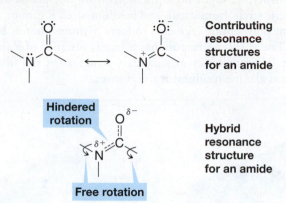

Contributing resonance structures for an amide

Hindered rotation

Hybrid resonance structure for an amide

Free rotation

- The amide carbon–nitrogen bond, consequently, has considerable double-bond character (~ 40%), and rotations of groups about this bond are severely hindered.
- Rotations of groups attached to the amide nitrogen and the carbonyl carbon are relatively free, however, and these rotations allow peptide chains to form different conformations.

A transoid arrangement of groups around the relatively rigid amide bond would cause the side-chain R groups to alternate from side to side of a single fully extended peptide chain:

Two American scientists, Linus Pauling and Robert B. Corey, were pioneers in the X-ray analysis of proteins. Beginning in 1939, Pauling and Corey initiated a long series of studies of the conformations of peptide chains. At first, they used crystals of single amino acids, then dipeptides and tripeptides, and so on. Moving on to larger and larger molecules and using the precisely constructed molecular models, they were able to understand the secondary structures of proteins for the first time. Pauling won the 1954 Nobel Prize in Chemistry and the 1962 Nobel Peace Prize.

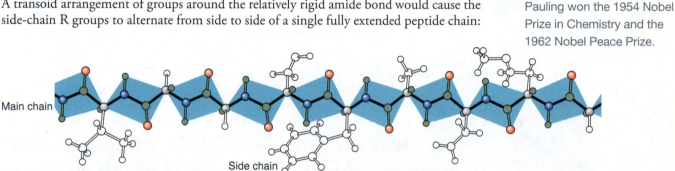

Main chain

Side chain

Calculations show that such a polypeptide chain would have a repeat distance (i.e., distance between alternating units) of 7.2 Å.

Fully extended polypeptide chains could hypothetically form a flat-sheet structure, with each alternating amino acid in each chain forming two hydrogen bonds with an amino acid in the adjacent chain:

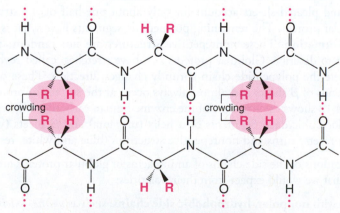

Hypothetical flat-sheet structure
(not formed because of steric hindrance)

However, this structure does not exist in naturally occurring proteins because of the crowding that would exist between R groups. If such a structure did exist, it would have the same repeat distance as the fully extended peptide chain, that is, 7.2 Å.

● Many proteins incorporate a **β sheet** or **β configuration** (Fig. 24.10).

In a β sheet structure, slight bond rotations from one planar amide group to the next relieve the steric strain from small- and medium-sized R groups. This allows amide groups on adjacent polypeptide segments to form hydrogen bonds between the chains (see Fig. 24.10). The β sheet structure has a repeat distance of 7.0 Å between amide groups in a chain. The predominant secondary structure in silk fibroin (48% glycine and 38% serine and alanine residues) is the β sheet.

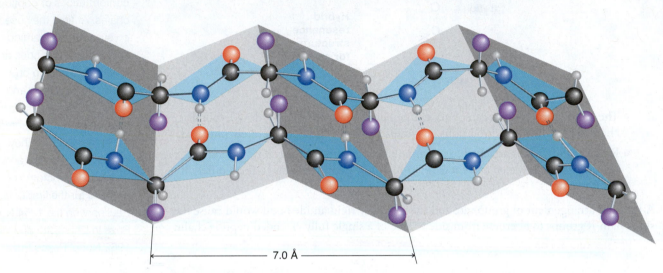

|← 7.0 Å →|

FIGURE 24.10 The β sheet or β configuration of a protein. (Illustration, Irving Geis. Image from the Irving Geis Collection, HHMI. Rights owned by Howard Hughes Medical Institute. Not to be reproduced without permission.)

FIGURE 24.11 A representation of the α-helical structure of a polypeptide. Hydrogen bonds are denoted by dashed lines. (Illustration, Irving Geis. Image from the Irving Geis Collection, HHMI. Rights owned by Howard Hughes Medical Institute. Not to be reproduced without permission.)

● The **α helix** is also a very important secondary structure in proteins (Fig. 24.11).

The α helix of a polypeptide is right-handed with 3.6 amino acid residues per turn. Each amide group in the chain has a hydrogen bond to an amide group at a distance of three amino acid residues in either direction, and the R groups all extend away from the axis of the helix. The repeat distance of the α helix is 5.4 Å.

The α-helical structure is found in many proteins; it is the predominant structure of the polypeptide chains of fibrous proteins such as *myosin*, the protein of muscle, and of *α-keratin*, the protein of hair, unstretched wool, and nails.

Helices and pleated sheets account for only about one-half of the structure of the average globular protein. The remaining polypeptide segments have what is called a **coil** or **loop conformation**. These nonrepetitive structures are not random; they are just more difficult to describe. Globular proteins also have stretches, called **reverse turns** or **β bends**, where the polypeptide chain abruptly changes direction. These often connect successive strands of β sheets and almost always occur at the surface of proteins.

Figure 24.12 shows the structure of the enzyme human carbonic anhydrase, based on X-ray crystallographic data. Segments of α helix (magenta) and β sheets (yellow) intervene between reverse turns and nonrepetitive structures (blue and white, respectively).

● The locations of the side chains of amino acids of globular proteins are usually those that we would expect from their polarities:

1. Residues with **nonpolar, hydrophobic side chains**, such as *valine, leucine, isoleucine, methionine, and phenylalanine,* are almost always found in the interior of the protein, out of contact with the aqueous solvent. (These hydrophobic interactions are largely responsible for the tertiary structure of proteins that we discuss in Section 24.8B.)

2. Side chains of **polar residues with positive or negative charges**, such as *arginine, lysine, aspartic acid, and glutamic acid,* are usually on the surface of the protein in contact with the aqueous solvent.

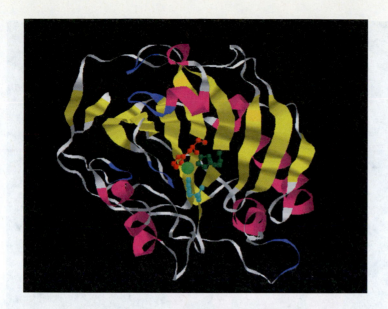

FIGURE 24.12 The structure of the enzyme human carbonic anhydrase, based on X-ray crystallographic data. Alpha helices are shown in magenta and strands of β sheets are yellow. Turns are shown in blue and random coils are white. The side chains of three histidine residues (shown in red, green, and cyan) coordinate with a zinc atom (light green). Not obvious from this image is the interesting fact that the C-terminus is tucked through a loop of the polypeptide chain, making carbonic anhydrase a rare example of a native protein in which the polypeptide chain forms a knot. (PDB ID CA2, *http://www.pdb.org*. Eriksson, Jones, Liljas, *Proteins: Structure, Function and Genetics*, Volume 4, Issue 4, 1988, pp. 274–282.)

3. **Uncharged polar side chains**, such as those of *serine, threonine, asparagine, glutamine, tyrosine, and tryptophan,* are most often found on the surface, but some of these are found in the interior as well. When they are found in the interior, they are virtually all hydrogen bonded to other similar residues. Hydrogen bonding apparently helps neutralize the polarity of these groups.

Certain peptide chains assume what is called a **random coil arrangement**, a structure that is flexible, changing, and statistically random. Synthetic polylysine, for example, exists as a random coil and does not normally form an α helix. At pH 7, the ε-amino groups of the lysine residues are positively charged, and, as a result, repulsive forces between them are so large that they overcome any stabilization that would be gained through hydrogen bond formation of an α helix. At pH 12, however, the ε-amino groups are uncharged and polylysine spontaneously forms an α helix.

The presence of proline or hydroxyproline residues in polypeptide chains produces another striking effect: because the nitrogen atoms of these amino acids are part of five-membered rings, the groups attached by the nitrogen–α carbon bond cannot rotate enough to allow an α-helical structure. Wherever proline or hydroxyproline occur in a peptide chain, their presence causes a kink or bend and interrupts the α helix.

24.8B Tertiary Structure

- The **tertiary structure** of a protein is the overall three-dimensional shape that arises from all of the secondary structures of its polypeptide chain.

Proteins typically have either **globular** or **fibrous** tertiary structures. These tertiary structures do not occur randomly. Under the proper environmental conditions the tertiary structure of a protein occurs in one particular way—a way that is characteristic of that particular protein and one that is often highly important to its function.

Various forces are involved in stabilizing tertiary structures, including the disulfide bonds of the primary structure.

- One characteristic of most proteins is that the folding takes place in such a way as to expose the maximum number of polar (hydrophilic) groups to the aqueous environment and enclose a maximum number of nonpolar (hydrophobic) groups within its interior.

The soluble globular proteins tend to be much more highly folded than fibrous proteins. Myoglobin (Fig. 24.13) is an example of a globular protein. However, fibrous proteins also have a tertiary structure; the α-helical strands of α-keratin, for example, are wound together into a "superhelix." The superhelix makes one complete turn for each 35 turns of the α helix. The tertiary structure does not end here, however. Even the superhelices can be wound together to give a ropelike structure of seven strands.

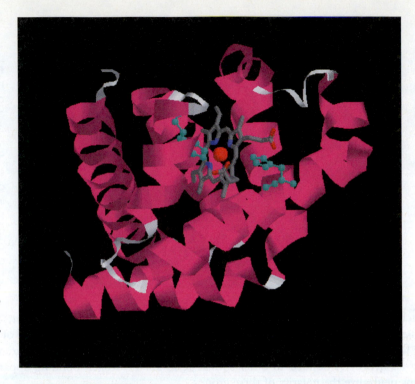

FIGURE 24.13 The three-dimensional structure of myoglobin. The heme ring is shown in gray. The iron atom is shown as a red sphere, and the histidine side chains that coordinate with the iron are shown in cyan. (PDB ID 1MBD, *http://www.pdb.org*. Phillips, S.E., Schoenberg, B.P. Neutron diffraction reveals oxygen-histidine hydrogen bond in oxymyoglobin. *Nature* 292, pp. 81–82, 1981.)

24.8C Quaternary Structure

Many proteins exist as stable and ordered noncovalent aggregates of more than one polypeptide chain. The overall structure of a protein having multiple subunits is called its **quaternary structure**. The quaternary structure of hemoglobin, for example, involves four subunits (see Section 24.12).

24.9 INTRODUCTION TO ENZYMES

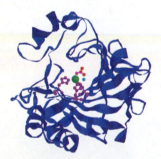

Carbonic anhydrase

Carbonic anhydrase is an enzyme that catalyzes the following reaction: $H_2O + CO_2 \rightleftharpoons H_2CO_3$. (PDB ID CA2, *http://www.pdb.org*. Eriksson, Jones, Liljas, *Proteins: Structure, Function and Genetics*, Volume 4, Issue 4, 1988, pp. 274–282.)

- The reactions of cellular metabolism are mediated by remarkable biological catalysts called **enzymes**.

Enzymes have the ability to bring about vast increases in the rates of reactions; in most instances, the rates of enzyme-catalyzed reactions are faster than those of uncatalyzed reactions by factors of 10^6–10^{12}. For living organisms, rate enhancements of this magnitude are important because they permit reactions to take place at reasonable rates, even under the mild conditions that exist in living cells (i.e., approximately neutral pH and a temperature of about 35 °C).

- Enzymes show remarkable **specificity** for their substrates and for formation of specific products.

The specificity of enzymes is far greater than that shown by most chemical catalysts. In the enzymatic synthesis of proteins, for example (through reactions that take place on ribosomes, Section 25.5E), polypeptides consisting of well over 1000 amino acid residues are synthesized virtually without error. It was Emil Fischer's discovery, in 1894, of the ability of enzymes to distinguish between α- and β-glycosidic linkages (Section 22.12) that led him to formulate his **lock-and-key hypothesis** for enzyme specificity.

- According to the **lock-and-key hypothesis**, the specificity of an enzyme (the lock) and its substrate (the key) comes from their geometrically complementary shapes.
- In an enzyme-catalyzed reaction, the enzyme and the substrate combine to form an **enzyme–substrate complex**.
- Formation of the enzyme–substrate complex often induces a conformational change in the enzyme called an **induced fit** that allows it to bind the substrate more effectively.

Binding of the substrate can cause certain of its bonds to become strained, and therefore more easily broken. The product of the reaction usually has a different shape from the substrate, and this altered shape, or in some instances the intervention of another molecule, causes the complex to dissociate. The enzyme can then accept another molecule of the substrate, and the whole process is repeated:

Enzyme + substrate $\rightleftharpoons$ enzyme–substrate complex $\rightleftharpoons$ enzyme + product

- The place where a substrate binds to an enzyme and where the reaction takes place is called the **active site**.

The noncovalent forces that bind the substrate to the active site are the same forces that account for the conformations of proteins: dispersion forces, electrostatic forces, hydrogen bonding, and hydrophobic interactions. The amino acids located in the active site are arranged so that they can interact specifically with the substrate.

- Reactions catalyzed by enzymes are **stereospecific** because enzymes are chiral.

The specificity of enzymes arises in the way enzymes bind their substrates. An α-glycosidase will only bind the α stereoisomeric form of a glycoside, not the β form. Enzymes that metabolize sugars bind only D sugars; enzymes that synthesize most proteins bind only L amino acids; and so on.

Although enzymes catalyze reactions stereospecifically, they often vary considerably in what is called their **geometric specificity**. By geometric specificity, we mean a specificity that is related to the identities of the chemical groups of the substrates. Some enzymes will accept only one compound as their substrate. Others, however, will accept a range of compounds with similar groups. Carboxypeptidase A, for example, will hydrolyze the C-terminal peptide from all polypeptides as long as the penultimate residue is not arginine, lysine, or proline and as long as the next preceding residue is not proline. Chymotrypsin, a digestive enzyme that catalyzes the hydrolysis of peptide bonds, will also catalyze the hydrolysis of esters. We shall consider its mechanism of hydrolysis in Section 24.11.

Peptide

Ester

- A compound that can negatively alter the activity of an enzyme is called an **inhibitor**.

A **competitive inhibitor** is a compound that competes directly with the substrate for the active site. We learned in Section 20.9, for example, that sulfanilamide is a competitive inhibitor for a bacterial enzyme that incorporates p-aminobenzoic acid into folic acid.

Some enzymes require the presence of a **cofactor**. The cofactor may be a metal ion as, for example, the zinc atom of human carbonic anhydrase (see the Chemistry of…box, Section 24.10 and Fig. 24.12). Others may require the presence of an organic molecule, such as NAD^+ (Section 14.10), called a **coenzyme**. Coenzymes become chemically changed in the course of the enzymatic reaction. NAD^+ becomes converted to **NADH**. In some enzymes the cofactor is permanently bound to the enzyme, in which case it is called a **prosthetic group**.

Many of the water-soluble vitamins are the precursors of coenzymes. Niacin (nicotinic acid) is a precursor of NAD^+, for example. Pantothenic acid is a precursor of coenzyme A.

Niacin   **Pantothenic acid**

Certain RNA molecules, called ribozymes, can also act as enzymes. The 1989 Nobel Prize in Chemistry went to Sidney Altman (Yale University) and to Thomas R. Cech (University of Colorado, Boulder) for the discovery of ribozymes.

Helpful Hint

In *WileyPLUS* we have highlighted several coenzymes because they are the "organic chemistry machinery" of some enzymes. For example, see "The Chemistry of ... Pyridoxal Phosphate" and "The Chemistry of ... Thiamine."

24.10 LYSOZYME: MODE OF ACTION OF AN ENZYME

Lysozyme is an enzyme that breaches the cell wall of gram-positive bacteria by hydrolyzing specific acetal linkages in the cell's peptidoglycan polymer, causing lysis and cell death. We shall discuss the mechanism of this reaction below, but first let us consider the structure of lysozyme. The primary structure of lysozyme is shown in Figure 24.14.

Lysozyme's secondary structure includes α-helices at residues 5–15, 24–34, and 88–96; β-sheet involving residues 41–45 and 50–54; and a hairpin turn at residues 46–49. The remaining polypeptide segments of lysozyme have coil or loop formations. Glu-35 and Asp-52 are the amino acid residues directly involved in the hydrolysis reaction catalyzed by lysozyme. A three-dimensional structure of lysozyme is shown in Fig. 24.15. The amino acid residues responsible for its catalytic activity are highlighted in ball-and-stick format (Glu-35 and Asp-52 to the left).

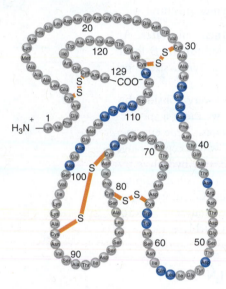

FIGURE 24.14 The primary structure of hen egg white lysozyme. The amino acids that line the substrate-binding pocket are shown in blue. (Reprinted with permission of John Wiley & Sons, Inc. from Voet, D. and Voet, J. G. *Biochemistry*, Second Edition. © 1995 Voet, D. and Voet, J. G.)

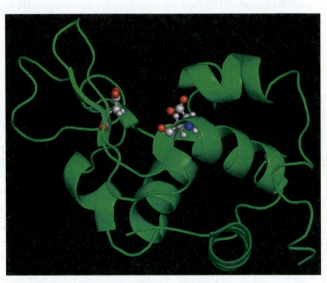

FIGURE 24.15 A ribbon diagram of lysozyme highlighting aspartic acid 52 (left) and glutamic acid 35 (right) in ball-and-stick format. (PDB ID: 1AZF, *http://www.pdb.org*. Lim, K., Nadarajah A., Forsythe, E. L., Pusey, M. L. Locations of bromide ions in tetragonal lysozyme crystals. *Acta Crystallogr.*, Sect. D, 54, pp. 899–904, 1998.)

As mentioned, lysozyme hydrolyzes glycosidic linkages in the peptidoglycan polymer of gram-positive bacterial cell walls. The structure of an oligosaccharide similar to the polysaccharide found in bacterial cell walls is shown in Fig. 24.16. *N*-Acetylglucosamine (NAG) and *N*-acetylmuramic acid (NAM) form alternating repeat units in this polysaccharide.

Lysozyme selectively binds a six-unit segment of the peptidoglycan polymer and hydrolyzes specifically the acetal linkage between rings D and E shown in Fig. 24.16 (NAM and NAG units, respectively).

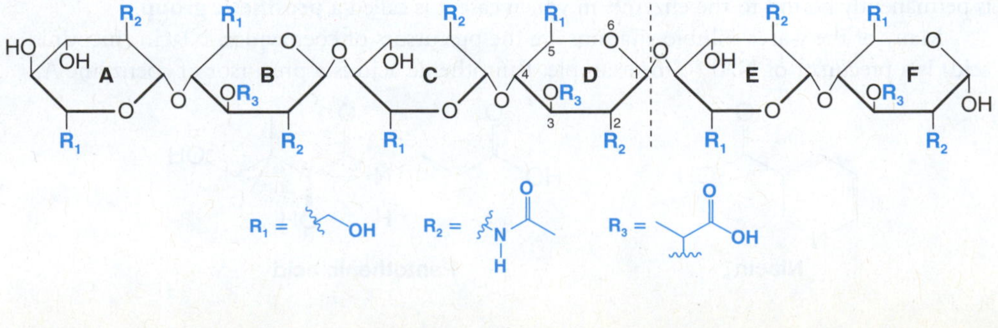

FIGURE 24.16 A hexasaccharide that has the same general structure as the cell wall polysaccharide on which lysozyme acts. Two different amino sugars are present: rings A, C, and E are derived from a monosaccharide called *N*-acetylglucosamine; rings B, D, and F are derived from a monosaccharide called *N*-acetylmuramic acid. When lysozyme acts on this oligosaccharide, hydrolysis takes place and results in cleavage at the glycosidic linkage between rings D and E.

The overall reaction catalyzed by lysozyme is as follows:

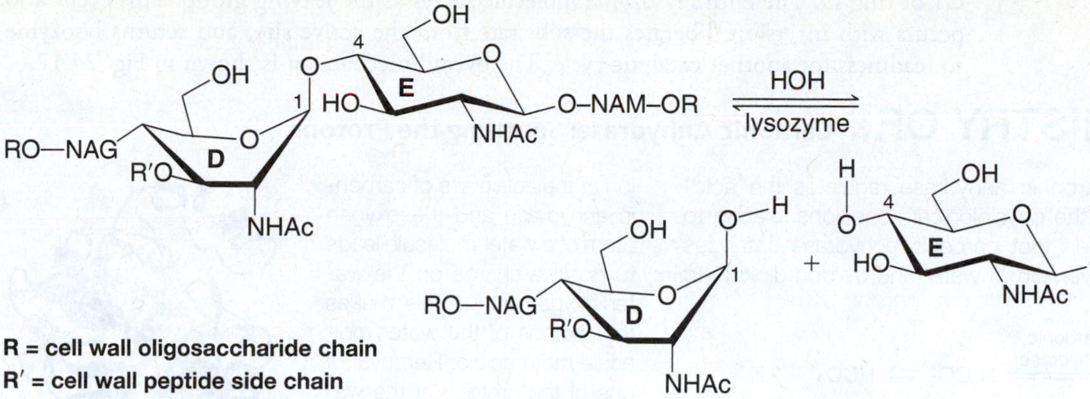

R = cell wall oligosaccharide chain
R′ = cell wall peptide side chain

Lysozyme binds the cell wall substrate in a cleft within its tertiary structure, such that the Glu-35 residue is close to the substrate on one side and Asp-52 is close on the other. Both amino acid residues are positioned in a way that facilitates reaction with the D–E glycosidic linkage of the polysaccharide.

Strong evidence from mass spectrometry suggests that the mechanism of lysozyme involves sequential S_N2 reactions and a covalent enzyme–substrate intermediate (based on work by Stephen Withers and colleagues at the University of British Columbia and elsewhere). Asp-52 acts as the nucleophile in the first step, covalently bonding the substrate to the enzyme. A water molecule acts as a nucleophile in the second step to complete the formation of product and free the substrate from the active site. In both steps, Glu-35 serves as a general acid–base catalyst. The details are as follows.

As lysozyme binds the substrate, the active site cleft closes slightly and C1 of ring D in the oligosaccharide substrate moves downward. The carboxylate group of Asp-52 attacks C1 of ring D from below (Figure 24.17), displacing the ring E C4 oxygen as a leaving group. The ring E C4 oxygen departs as a neutral species because it is protonated concurrently by the carboxylic acid group of Glu-35. The transition state for this S_N2 reaction is presumed to be the point at which ring D is nearly flat during the boat to chair conformational change. This step occurs with inversion, as expected for an S_N2 reaction, and leaves one part of the substrate covalently bound to the enzyme.

In the second step, a water molecule, now in the site formerly occupied by ring E, attacks C1 and displaces the carboxylate group of Asp-52 as a leaving group. The Glu-35

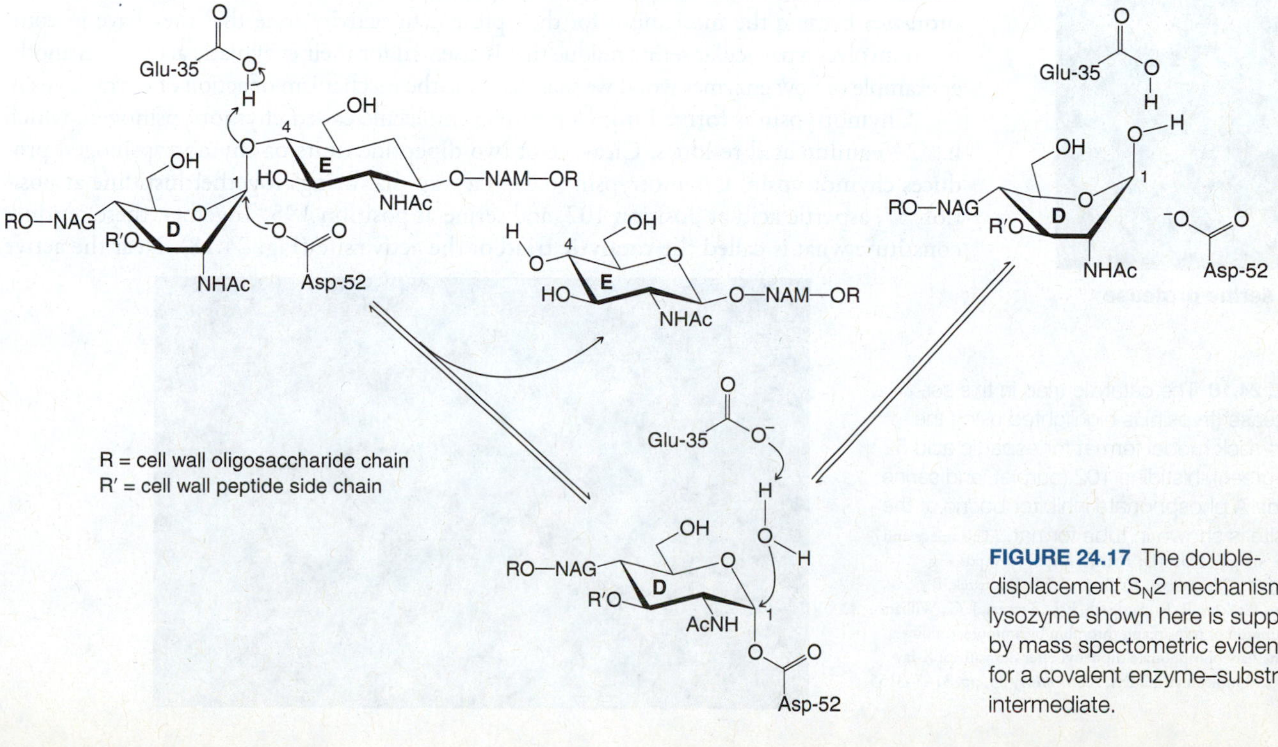

R = cell wall oligosaccharide chain
R′ = cell wall peptide side chain

FIGURE 24.17 The double-displacement S_N2 mechanism for lysozyme shown here is supported by mass spectometric evidence for a covalent enzyme–substrate intermediate.

anion assists as a base by removing a proton from the water molecule as it bonds with C1 of ring D. The entire lysozyme molecule serves as the leaving group. This event also occurs with inversion, liberates the substrate from the active site, and returns lysozyme to readiness for another catalytic cycle. The overall mechanism is shown in Fig. 24.17.

◆ THE CHEMISTRY OF... Carbonic Anhydrase: Shuttling the Protons

An enzyme called carbonic anhydrase regulates the acidity (pH) of blood and the physiological conditions relating to blood pH. The reaction that carbonic anhydrase catalyzes is the equilibrium conversion of water and carbon dioxide to carbonic acid (H_2CO_3).

$$H_2O + CO_2 \underset{}{\overset{\text{carbonic}}{\underset{\text{anhydrase}}{\rightleftharpoons}}} H_2CO_3 \rightleftharpoons HCO_3^- + H^+$$

The rate at which one breathes, for example, is influenced by one's relative blood acidity. Mountain climbers going to high elevations sometimes take a drug called Diamox (acetazolamide) to prevent altitude sickness. Diamox inhibits carbonic anhydrase, and this, in turn, increases blood acidity. This increased blood acidity stimulates breathing and thereby decreases the likelihood of altitude sickness.

Carbonic anhydrase consists of a chain of 260 amino acids that naturally folds into a specific globular shape. Included in its structure is a cleft or pocket, the active site, where the reactants are converted to products. The protein chain of carbonic anhydrase is shown here as a blue ribbon.

At the active site of carbonic anhydrase a water molecule loses a proton to form a hydroxide (HO$^-$) ion. This proton is removed by a part of carbonic anhydrase that acts as a base. Ordinarily the proton of a water molecule is not very acidic. However, the Lewis acid–base interaction between a zinc cat-

ion at the active site of carbonic anhydrase and the oxygen atom of a water molecule leads to positive charge on the water oxygen atom. This makes the protons of the water molecule more acidic. Removal of one of the protons of the water molecule forms hydroxide, which reacts with a carbon dioxide molecule at the active site to form HCO_3^- (hydrogen carbonate, or bicarbonate). In the structure of carbonic anhydrase shown here (based on X-ray crystallographic data), a bicarbonate ion at the active site is shown in red, the zinc cation at the active site is green, a water molecule is shown in blue, and the basic sites that coordinate with the zinc cation (as Lewis bases) or remove the proton from water to form hydroxide (as Brønsted–Lowry bases) are magenta (these bases are nitrogen atoms from histidine imidazole rings). No hydrogen atoms are shown in any of these species. As you can see, a remarkable orchestration of Lewis and Brønsted–Lowry acid–base reactions is involved in catalysis by carbonic anhydrase.

Carbonic anhydrase

(PDB ID CA2, *http://www.pdb.org.* Eriksson, Jones, Liljas, *Proteins: Structure, Function and Genetics*, Volume 4, Issue 4, pp. 274–282, 1988.)

24.11 SERINE PROTEASES

Chymotrypsin, trypsin, and elastin are digestive enzymes secreted by the pancreas into the small intestine to catalyze the hydrolysis of peptide bonds. These enzymes are all called **serine proteases** because the mechanism for their proteolytic activity (one that they have in common) involves a particular serine residue that is essential for their enzymatic activity. As another example of how enzymes work, we shall examine the mechanism of action of chymotrypsin.

Chymotrypsin is formed from a precursor molecule called chymotrypsinogen, which has 245 amino acid residues. Cleavage of two dipeptide units of chymotrypsinogen produces chymotrypsin. Chymotrypsin folds in a way that brings together histidine at position 57, aspartic acid at position 102, and serine at position 195. Together, these residues constitute what is called the **catalytic triad** of the active site (Fig. 24.18). Near the active

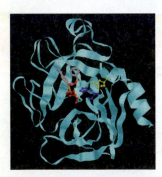

A serine protease

FIGURE 24.18 The catalytic triad in this serine protease (trypsin) is highlighted using the ball-and-stick model format for aspartic acid 52 (yellow-green), histidine 102 (purple), and serine 195 (red). A phosphonate inhibitor bound at the active site is shown in tube format. (This image and that in the margin, PDB ID: 1MAX, http://www.pdb.org. Bertrand, J. A., Oleksyszyn, J., Kam, C. M., Boduszek, B., Presnell, S., Plaskon, R. R., Suddath, F. L., Powers, J. C., Williams, L. D., Inhibition of trypsin and thrombin by amino(4-amidinophenyl)methanephosphonate diphenyl ester derivatives: X-ray structures and molecular models. *Biochemistry* 35, pp. 3147–3155, 1996.)

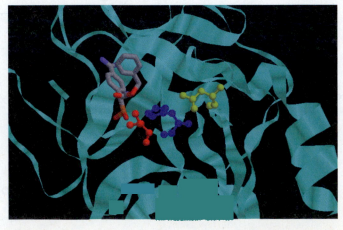

site is a hydrophobic binding site, a slotlike pocket that preferentially accommodates the nonpolar side chains of Phe, Tyr, and Trp.

After chymotrypsin has bound its protein substrate, the serine residue at position 195 is ideally situated to attack the acyl carbon of the peptide bond (Fig. 24.19). This serine residue is made more nucleophilic by transferring its proton to the imidazole nitrogen of the histidine residue at position 57. The imidazolium ion that is formed is stabilized by the polarizing effect of the carboxylate ion of the aspartic acid residue at position 102. (Neutron diffraction studies, which show the positions of hydrogen atoms, confirm that the carboxylate ion remains as a carboxylate ion throughout and does not actually accept a proton from the imidazole.) Nucleophilic attack by the serine leads to an acylated serine through a tetrahedral intermediate. The new N-terminal end of the cleaved polypeptide chain diffuses away and is replaced by a water molecule.

Regeneration of the active site of chymotrypsin is shown in Fig. 24.20. In this process water acts as the nucleophile and, in a series of steps analogous to those in

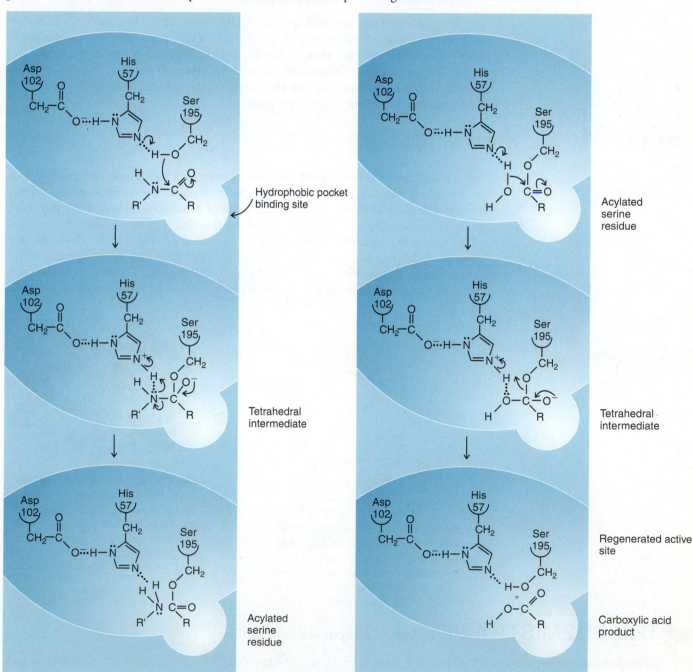

FIGURE 24.19 The catalytic triad of chymotrypsin causes cleavage of a peptide bond by acylation of serine residue 195 of chymotrypsin. Near the active site is a hydrophobic binding site that accommodates nonpolar side chains of the protein.

FIGURE 24.20 Regeneration of the active site of chymotrypsin. Water causes hydrolysis of the acyl–serine bond.

Fig. 24.19, hydrolyzes the acyl–serine bond. The enzyme is now ready to repeat the whole process.

There is much evidence for this mechanism that, for reasons of space, we shall have to ignore. One bit of evidence deserves mention, however. There are compounds such as **diisopropylphosphofluoridate (DIPF)** that irreversibly inhibit serine proteases. It has been shown that they do this by reacting only with Ser 195:

Ser195 — OH + *i*-Pr—O—P(=O)(F)(O—*i*-Pr) ⟶ Ser195 — O—P(=O)(O—*i*-Pr)(O—*i*-Pr)

Diisopropylphospho-fluoridate (DIPF)   **DIP-enzyme**

Recognition of the inactivating effect of DIPF came about as a result of the discovery that DIPF and related compounds are powerful **nerve poisons**. (They are the "nerve gases" of military use, even though they are liquids dispersed as fine droplets, and not gases.) Diisopropylphosphofluoridate inactivates **acetylcholinesterase** (Section 20.3) by reacting with it in the same way that it does with chymotrypsin. Acetylcholinesterase is a **serine esterase** rather than a serine protease.

24.12 HEMOGLOBIN: A CONJUGATED PROTEIN

Some proteins, called **conjugated proteins**, contain as a part of their structure a non-protein group called a **prosthetic group.** An example is the oxygen-carrying protein hemoglobin. Each of the four polypeptide chains of hemoglobin is bound to a prosthetic group called *heme* (Fig. 24.21). The four polypeptide chains of hemoglobin are wound in such a way as to give hemoglobin a roughly spherical shape (Fig. 24.22). Moreover, each heme group lies in a crevice with the hydrophobic vinyl groups of its porphyrin structure surrounded by hydrophobic side chains of amino acid residues. The two propanoate side chains of heme lie near positively charged amino groups of lysine and arginine residues.

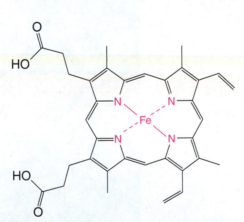

FIGURE 24.21 The structure of heme, the prosthetic group of hemoglobin. Heme has a structure similar to that of chlorophyll (Fig. 22.1) in that each is derived from the heterocyclic ring, porphyrin. The iron of heme is in the ferrous (2+) oxidation state.

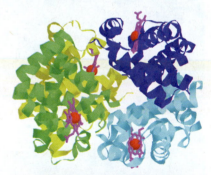

FIGURE 24.22 Hemoglobin. The two α subunits of hemoglobin are shown in blue and green. The two β subunits are shown in yellow and cyan. The four heme groups are shown in purple, and their iron atoms are in red. (PDB ID: IOUU, *http://www.pdb.org*. Tame, J. R., Wilson, J. C., Weber, R. E. The crystal structures of trout Hb I in the deoxy and carbonmonoxy forms. *J. Mol. Biol.* Volume 259, Issue 4, pp. 749–760, 1996.)

THE CHEMISTRY OF... Some Catalytic Antibodies

Antibodies are chemical warriors of the immune system. Each antibody is a protein produced specifically in response to an invading chemical species (e.g., molecules on the surface of a virus or pollen grain). The purpose of antibodies is to bind with these foreign agents and cause their removal from the organism. The binding of each antibody with its target (the antigen) is usually highly specific.

One way that *catalytic* antibodies have been produced is by prompting an immune response to a chemical species resembling the transition state for a reaction. According to this idea, if an antibody is created that preferentially binds with a stable molecule that has a transition state-like structure, other molecules that are capable of reaction *through* this transition state should, in principle, react faster as a result of binding with the antibody. (By facilitating association of the reactants and favoring formation of the transition state structure, the antibody acts in a way similar to an enzyme.) In stunning fashion, precisely this strategy has worked to generate catalytic antibodies for certain Diels–Alder reactions, Claisen rearrangements, and ester hydrolyses. Chemists have synthesized stable molecules that resemble transition states for these reactions, allowed antibodies to be generated against these molecules (called haptens), and then isolated the resulting antibodies. The antibodies thus produced are catalysts when actual substrate molecules are provided.

The following are examples of haptens used as transition state analogs to elicit catalytic antibodies for a Claisen rearrangement, hydrolysis of a carbonate, and a Diels–Alder

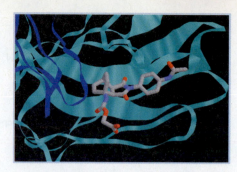

A hapten related to the Diels–Alder adduct from cyclohexadiene and maleimide, bound within a Diels–Alderase catalytic antibody. (PDB ID: 1A4K, http://www.pdb.org. Romesberg, F. E., Spiller, B., Schultz, P. G., Stevens, R. C. Immunological origins of binding and catalysis in a Diels–Alderase antibody. *Science* **279**, pp. 1929–1933, 1998.)

reaction. The reaction catalyzed by the antibody generated from each hapten is shown as well.

Claisen Rearrangement

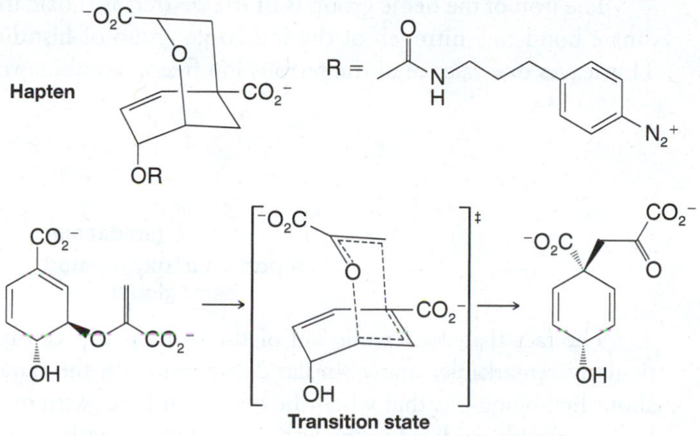

Hapten

Transition state

Carbonate Hydrolysis

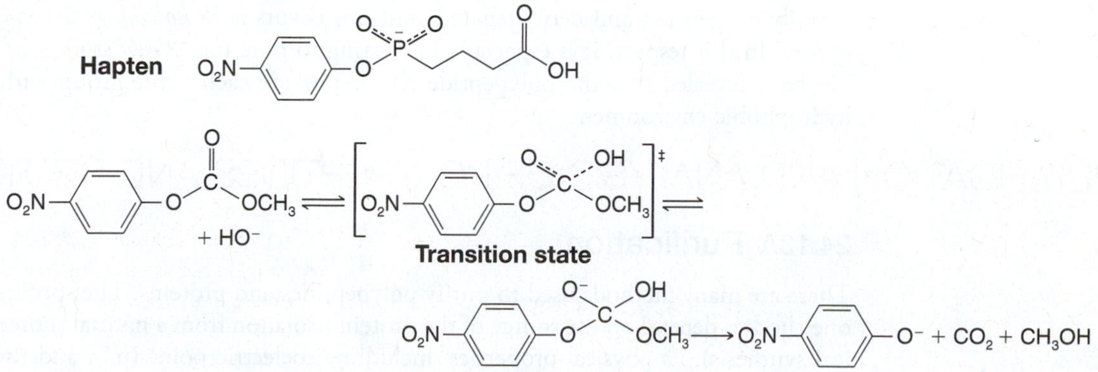

Hapten

Transition state

Diels–Alder Reaction

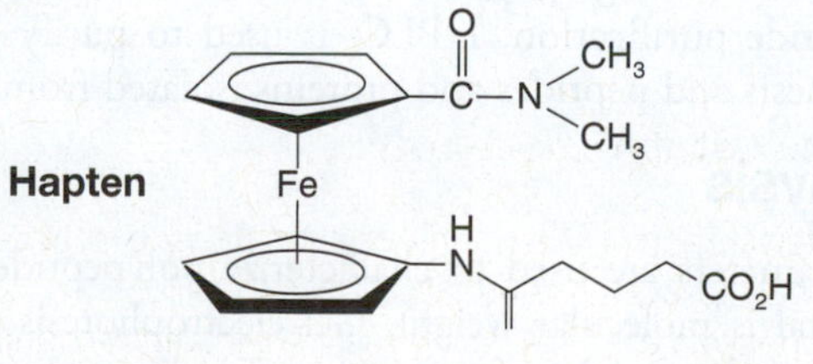

Hapten

(continues on next page)

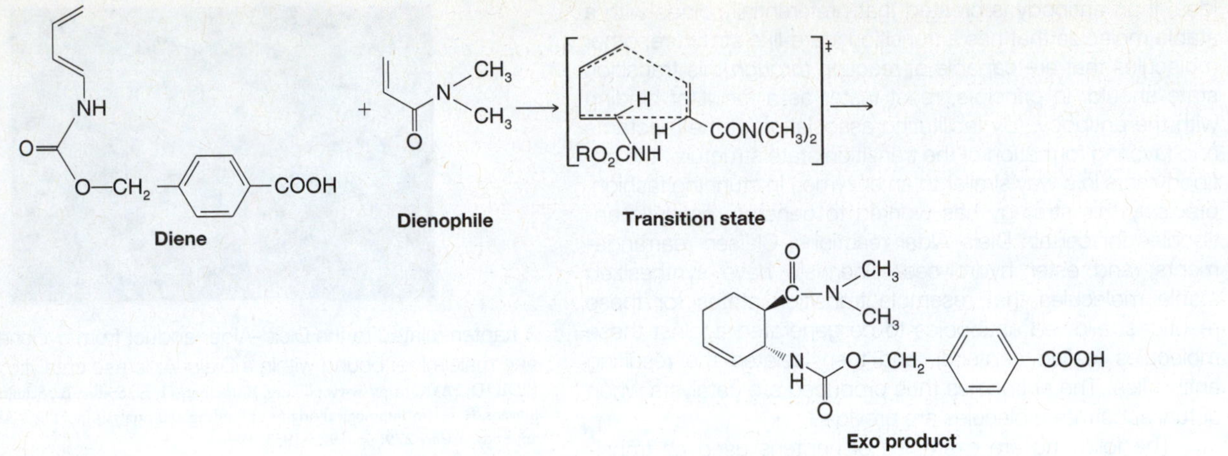

Diene   **Dienophile**   **Transition state**

Exo product

This marriage of enzymology and immunology, resulting in chemical offspring, is just one area of exciting research at the interface of chemistry and biology.

The iron of the heme group is in the 2+ (ferrous) oxidation state and it forms a coordinate bond to a nitrogen of the imidazole group of histidine of the polypeptide chain. This leaves one valence of the ferrous ion free to combine with oxygen as follows:

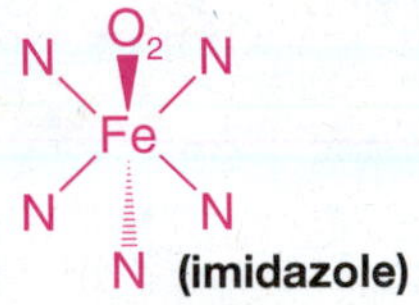

(imidazole)
A portion of oxygenated
hemoglobin

The fact that the ferrous ion of the heme group combines with oxygen is not particularly remarkable; many similar compounds do the same thing. What is remarkable about hemoglobin is that when the heme combines with oxygen the ferrous ion does not become readily oxidized to the ferric state. Studies with model heme compounds in water, for example, show that they undergo a rapid combination with oxygen but they also undergo a rapid oxidation of the iron from Fe^{2+} to Fe^{3+}. When these same compounds are embedded in the hydrophobic environment of a polystyrene resin, however, the iron is easily oxygenated and deoxygenated, and this occurs *with no change in oxidation state of iron*. In this respect, it is especially interesting to note that X-ray studies of hemoglobin have revealed that the polypeptide chains provide each heme group with a similar hydrophobic environment.

24.13 PURIFICATION AND ANALYSIS OF POLYPEPTIDES AND PROTEINS

24.13A Purification

There are many methods used to purify polypeptides and proteins. The specific methods one chooses depend on the source of the protein (isolation from a natural source or chemical synthesis), its physical properties, including isoelectric point (pI), and the quantity of the protein on hand. Initial purification methods may involve precipitation, various forms of column chromatography, and electrophoresis. Perhaps the most important final method for peptide purification, HPLC, is used to purify both peptides generated by automated synthesis and peptides and proteins isolated from nature.

24.13B Analysis

A variety of parameters are used to characterize polypeptides and proteins. One of the most fundamental is molecular weight. Gel electrophoresis can be used to measure the approximate molecular weight of a protein. Gel electrophoresis involves migration of a peptide or protein dissolved in a buffer through a porous polymer gel under the influence of a high-voltage electric field. The buffer used (typically about pH 9) imparts an overall negative charge to the protein such that the protein migrates toward the positively charged terminal. Migration rate depends on the overall charge and size of the protein as well as the average pore size of the gel. The molecular weight of the protein is inferred by comparing the distance traveled through the gel by the protein of interest with the migration distance of proteins with known molecular weights used as internal standards. The version of this technique called SDS–PAGE (sodium dodecyl sulfate–polyacrylamide gel electrophoresis) allows protein molecular weight determinations with an accuracy of about 5–10%.

Mass spectrometry can be used to determine a peptide's molecular weight with very high accuracy and precision. Earlier we discussed mass spectrometry in the context of protein sequencing. Now we shall consider the practical aspects of how molecules with very high molecular weight, such as proteins, can be transferred to the gas phase for mass spectrometric analysis. This is necessary, of course, whether the analysis regards peptide sequencing or full molecular analysis. Small organic molecules, as we discussed in Chapter 9, can be vaporized simply with high vacuum and heat. High-molecular-weight species cannot be transferred to the gas phase solely with heat and vacuum. Fortunately, very effective techniques have been developed for generating gas-phase ions of large molecules without destruction of the sample.

One ionization method is electrospray ionization (ESI, Fig. 24.23), whereby a solution of a peptide (or other analyte) in a volatile solvent containing a trace of acid is sprayed through a high-voltage nozzle into the vacuum chamber of a mass spectrometer. The acid in the solvent generates ions by protonating Lewis basic sites within the analyte. Peptides are typically protonated multiple times. Once injected through the high-voltage nozzle into the vacuum chamber, solvent molecules evaporate from the analyte ions (Fig. 24.23a), and the ions are drawn into the mass analyzer (Fig. 24.23b). The mass analyzer detects the analyte ions according to their time of flight, and registers their mass-to-charge ratio (m/z) (Fig. 24.23c). Each peak displayed in the mass spectrum represents the molecular weight of an ion divided by the number of positive charges it carries. From

One-quarter of the 2002 Nobel Prize in Chemistry was awarded to John B. Fenn for his development of ESI mass spectrometry. Another quarter of the prize was awarded to Koichi Tanaku for discoveries that led to matrix-assisted laser desorption ionization (MALDI, see below).

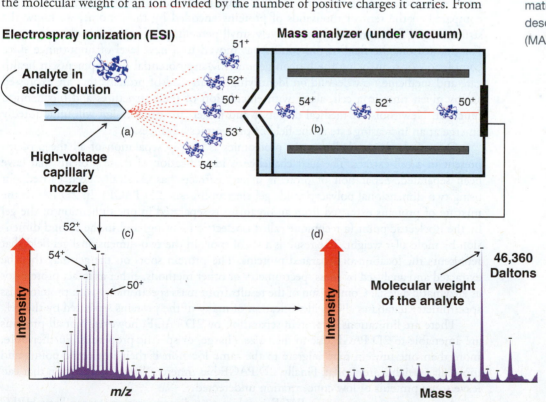

FIGURE 24.23 Electrospray ionization (ESI) mass spectrometry. (a) Analyte ions, protonated multiple times by an acidic solvent system, are sprayed through a high-voltage nozzle into a vacuum chamber (diagram is not to scale). Molecules of the solvent evaporate. The mutiply charged analyte ions are drawn into the mass analyzer. (b) The analyte ions are separated and detected in the mass analyzer. (c) The family of detected ions is displayed in a spectrum according to m/z ratio. (d) Computerized deconvolution of the m/z peak series leads to the molecular weight of the analyte.

this series of *m/z* peaks, the molecular weight of the analyte is calculated by a computerized process called deconvolution. An example of a deconvoluted spectrum, indicating a molecular weight of 46,360 atomic mass units (daltons), is shown in Fig. 24.23*d*.

If fragmentation of the analyte molecules is desired, it can be caused by collision-induced dissociation (CID, Section 24.5E). In this case, tandem mass spectrometry is necessary because the first mass analyzer in the system is used to select fragments of the peptide from CID based on their overall mass, while the second mass analyzer in the system records the spectrum of the selected peptide fragment. Multiple fragments from the CID procedure can be analyzed this way. The final spectrum for each peptide fragment selected has the typical appearance of a family of ions, as shown below.

Mass spectrometry with electrospray ionization (ESI-MS) is especially powerful when combined with HPLC because the two techniques can be used in tandem. With such an instrument the effluent from the HPLC is introduced directly into an ESI mass spectrometer. Thus, chromatographic separation of peptides in a mixture and direct structural information about each of them are possible using this technique.

Another method for ionization of nonvolatile molecules is MALDI (matrix-assisted laser desorption ionization, Section 9.18A). Energy from laser bombardment of a sample adsorbed in a solid chemical matrix leads to generation of gas-phase ions that are detected by the mass spectrometer. Both MALDI and ESI are common ionization techniques for the analysis of biopolymers.

24.14 PROTEOMICS

Proteomics and genomics are two fields that have blossomed in recent years. **Proteomics** has to do with the study of all proteins that are expressed in a cell at a given time. **Genomics** (Sections 25.1 and 25.9) focuses on the study of the complete set of genetic instructions in an organism. While the genome holds the instructions for making proteins, it is proteins that carry out the vast majority of functions in living systems. Yet, compared to the tens of thousands of proteins encoded by the genome, we know the structure and function of only a relatively small percentage of proteins in the proteome. For this reason, the field of proteomics has moved to a new level of importance since completion of sequencing the human genome. Many potential developments in health care and medicine now depend on identifying the myriad of proteins that are expressed at any given time in a cell, along with elucidation of their structures and biochemical function. New tools for medical diagnosis and targets for drug design will undoubtedly emerge at an increasing rate as the field of proteomics advances.

One of the basic challenges in proteomics is simply separation of all the proteins present in a cell extract. The next challenge is identification of those proteins that have been separated. Separation of proteins in cell extracts has classically been carried out using two-dimensional polyacrylamide gel electrophoresis (2D PAGE). In 2D PAGE the mixture of proteins extracted from an organism is separated in one dimension of the gel by the isoelectric point (a technique called isoelectric focusing) and in the second dimension by molecular weight. The result is a set of spots in the two-dimensional gel field that represents the location of separated proteins. The protein spots on the gel may then be extracted and analyzed by mass spectrometry or other methods, either as intact proteins or as enzymatic digests. Comparison of the results from mass spectrometry with protein mass spectrometry databases allows identification of many of the proteins separated by the gel.

There are limitations to protein separation by 2D PAGE, however. Not all proteins are amenable to 2D PAGE due to their size, charge, or specific properties. Furthermore, more than one protein may migrate to the same location if their isoelectric points and molecular weights are similar. Finally, 2D PAGE has inherent limits of detection that can leave some proteins of low concentration undetected.

An improvement over 2D PAGE involves two-dimensional microcapillary HPLC coupled with mass spectrometry (see Fig. 24.24). In this technique, called MudPIT (multidimensional protein identification technology, developed by John Yates and co-workers at The Scripps Research Institute), a microcapillary HPLC column is used that has been packed first with a strong cation-exchange resin and then a reversed-phase (hydrophobic) material. The two packing materials used in sequence and with different resolving

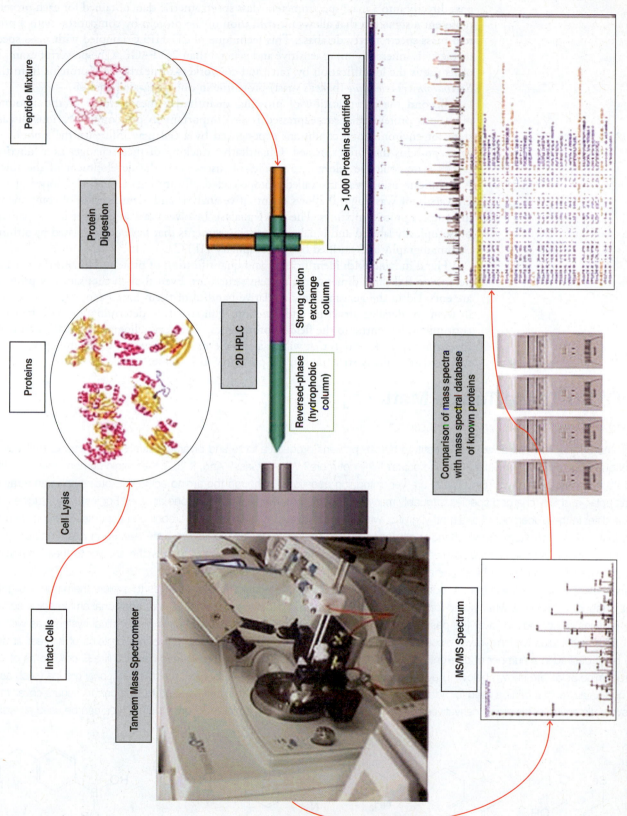

FIGURE 24.24 The high-throughput multidimensional protein identification technology (MudPIT) process. (Reprinted from *International Journal for Parasitology*, December 4, 32(13), Carucci, D.J.; Yates, J. R. 3rd; Florens, L.; Exploring the Proteome of Plasmodium, pp. 1539–1542, Copyright 2002, with permission from Elsevier.)

properties represent the two-dimensional aspect of this technique. A peptide mixture is introduced to the microcapillary column and eluted with pH and solvent gradients over a sequence of automated steps. As the separated peptides are eluted from the column they pass directly into a mass spectrometer. Mass spectrometric data obtained for each protein represent a signature that allows identification of the protein by comparison with a protein mass spectrometry database. This technique of 2D HPLC coupled with mass spectrometry is inherently more sensitive and general than 2D PAGE. One powerful example of its use is the identification by Yates and co-workers of nearly 1500 proteins from the *Saccharomyces cerevisiae* (baker's yeast) proteome in one integrated analysis.

Beyond the identification of proteins, quantitative measurement of the amounts of various proteins that are expressed is also important in proteomics. Various disease states or environmental conditions experienced by a cell may influence the amount of some proteins that are expressed. Quantitative tracking of these changes as a function of cell state could be relevant to studies of disease and the development of therapies. A technique using reagents called isotope-coded affinity tags (ICAT, developed at the University of Washington) allows quantitative analysis and identification of components in complex protein mixtures. The ICAT analysis involves mass spectrometric comparison of isotopically labeled and unlabeled protein segments that have been isolated by affinity chromatography and purified by microcapillary HPLC.

Hand in hand with identification and quantification of proteins remains the need to determine full three-dimensional protein structures. Even though thousands of proteins are encoded in the genome, only a relative handful of them have been studied in depth in terms of detailed structure and function. Full structure determination will therefore continue to be central to the field of proteomics. X-ray crystallography, NMR, and mass spectrometry are key tools that will be applied ever more fervently as the quest intensifies to elucidate as many structures in the proteome as possible.

WHY Do These Topics Matter?]

LIFE WITH MORE THAN 20 AMINO ACIDS!?

Although the 20 major amino acids that constitute human proteins produce a large and diverse selection of functional molecules, a natural question might be why just 20 amino acids? Why not more? Why not less? And, if 20 is the magic number, then why the specific 20 amino acids that we use? These are very intriguing questions. Although the amino acids in proteins span an array of nonpolar, polar, and fully charged species, arguably many of them are not very different from one another. For instance, leucine and isoleucine differ in the placement of one methyl group, while valine and leucine or serine and threonine or asparagine and glutamine differ by just one carbon atom. Small changes can often lead to profound differences, as we have learned in past chapters, but these acids do not quite seem to be as diverse as they could be. Ultimately, we may never know the answer to these intriguing questions, since they are largely philosophical.

Nevertheless, current knowledge of biochemical systems and synthetic techniques may allow us to explore them from a slightly different perspective. Namely, what would life look like with an expanded genetic code—that is, with additional amino acids added into the proteins of life. Indeed, over the past few years chemists have been able to utilize native biochemical systems as well as evolved tRNA molecules (which we will discuss in Chapter 25) to load many unique amino acids into proteins of interest at any specific point desired in a number of different cells, including those of yeast, some mammals, and bacteria like *E. coli*. Some of the unnatural amino acids are shown below. They include ones with unique metals (like selenium), reactive functional groups (such as a ketone and an azide) that can be used for additional chemistry, and a boronic acid that can be used to bind certain sugars covalently. These synthetic amino acids are all derivatives of phenylalanine, but many other amino acid parent structures can be used as well.

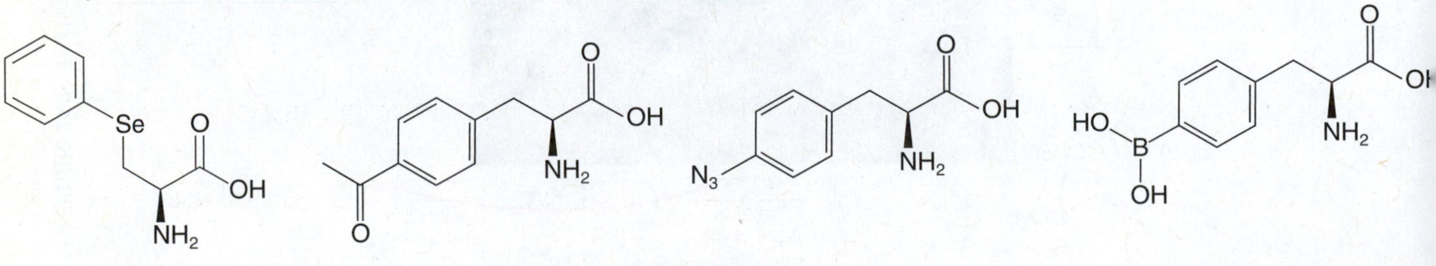

Unique metal; thiophilic | **Contains reactive groups to label proteins** | **Can bind certain carbohydrates**

These modifications allow the alteration of an amino acid at the active site of an enzyme, thereby stopping or changing its function. This event makes it possible to determine the potential mechanism of action of the enzyme and/or to observe downstream effects as a result of the alteration. In other cases, a reactive group has allowed chemists to perform labeling experiments and watch what a protein might accomplish in response to an external stimulus. Whether life forms can be generated that have the full machinery to incorporate such additional amino acids on a prolonged basis remains to be seen, but for sure, the ability to load such unnatural amino acids into proteins has led to a wealth of new and valuable knowledge. Indeed, without this capability, in some cases there would be no other way to examine the function of certain proteins.

To learn more about these topics, see:

1. Xie, J.; Schultz, P.G. "Adding amino acids to the genetic repertoire" in *Current Opinion in Chemical Biology* **2005**, *9*, 548–554.
2. Wang, Q; Parrish, A.R.; Wang, L. "Expanding the genetic code for biological studies." *Chem. Biol.* **2009**, *16*, 323–336.
3. Wang, L.; Brock, A.; Herberich, B.; Schultz, P. G. "Expanding the Genetic Code of Escherichia coli". *Science* **2001**, *292*, 498–500.
4. Brustad, E.; Bushey, M. L.; Lee, J. W.; Groff, D.; Liu, W.; Schultz, P. G. "A Genetically Encoded Boronate Amino Acid." *Angew. Chem. Int. Ed.* **2008**, *47*, 8220–8223.

SUMMARY AND REVIEW TOOLS

The study aids for this chapter include key terms and concepts, which are highlighted in bold, blue text within the chapter, defined in the Glossary at the back of the book, and which have hyperlinked definitions in the accompanying *WileyPLUS* course (www.wileyplus.com).

PROBLEMS PLUS

Note to Instructors: Many of the homework problems are available for assignment via *WileyPLUS*, an online teaching and learning solution.

STRUCTURE AND REACTIVITY

24.17 **(a)** Which amino acids in Table 24.1 have more than one chirality center?
(b) Write Fischer projections for the isomers of each of these amino acids that would have the L configuration at the α carbon.
(c) What kind of isomers have you drawn in each case?

24.18 **(a)** What product would you expect to obtain from treating tyrosine with excess bromine water?
(b) What product would you expect to be formed in the reaction of phenylalanine with ethanol in the presence of hydrogen chloride?
(c) What product would you expect from the reaction of alanine and benzoyl chloride in aqueous base?

24.19 **(a)** On the basis of the following sequence of reactions, Emil Fischer was able to show that L-(−)-serine and L-(+)-alanine have the same configuration. Write Fischer projections for the intermediates **A–C**:

$$(-)\text{-Serine} \xrightarrow[\text{CH}_3\text{OH}]{\text{HCl}} \mathbf{A}\ (\text{C}_4\text{H}_{10}\text{ClNO}_3) \xrightarrow{\text{PCl}_5} \mathbf{B}\ (\text{C}_4\text{H}_9\text{Cl}_2\text{NO}_2) \xrightarrow[\text{(2) HO}^-]{\text{(1) H}_3\text{O}^+,\ \text{H}_2\text{O, heat}} \mathbf{C}\ (\text{C}_3\text{H}_6\text{ClNO}_2) \xrightarrow[\text{dilute H}_3\text{O}^+]{\text{Na–Hg}} \text{L-(+)-alanine}$$

(b) The configuration of L-(+)-cysteine can be related to that of L-(−)-serine through the following reactions. Write Fischer projections for **D** and **E**:

$$\mathbf{B}\ (\text{from part a}) \xrightarrow{\text{HO}^-} \mathbf{D}\ (\text{C}_4\text{H}_8\text{ClNO}_2) \xrightarrow{\text{NaSH}} \mathbf{E}\ (\text{C}_4\text{H}_9\text{NO}_2\text{S}) \xrightarrow[\text{(2) HO}^-]{\text{(1) H}_3\text{O}^+,\ \text{H}_2\text{O, heat}} \text{L-(+)-cysteine}$$

(c) The configuration of L-(−)-asparagine can be related to that of L-(−)-serine in the following way. What is the structure of **F**?

$$\text{L-(−)-Asparagine} \xrightarrow[\substack{\text{Hofmann} \\ \text{rearrangement}}]{\text{NaOBr/HO}^-} \mathbf{F}\ (\text{C}_3\text{H}_7\text{N}_2\text{O}_2)$$

$$\mathbf{C}\ (\text{from part a}) \xrightarrow{\text{NH}_3} \quad \uparrow$$

24.20 **(a)** DL-Glutamic acid has been synthesized from diethyl acetamidomalonate in the following way. Outline the reactions involved.

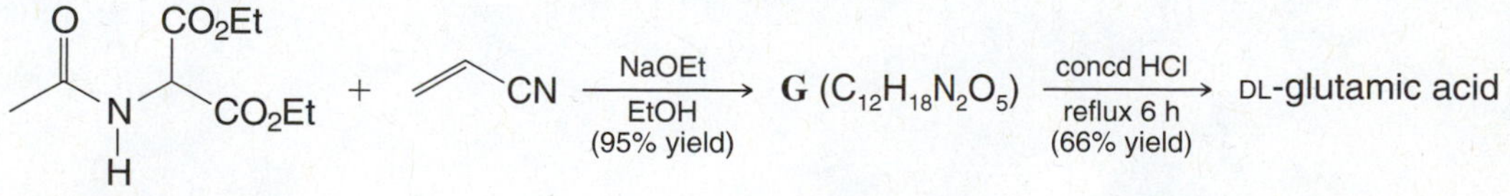

Diethyl acetamido-malonate $+$ (acrylonitrile) $\xrightarrow[\substack{\text{EtOH} \\ \text{(95% yield)}}]{\text{NaOEt}} \mathbf{G}\ (\text{C}_{12}\text{H}_{18}\text{N}_2\text{O}_5) \xrightarrow[\substack{\text{reflux 6 h} \\ \text{(66% yield)}}]{\text{concd HCl}} \text{DL-glutamic acid}$

**Diethyl acetamido-
malonate**

(b) Compound **G** has also been used to prepare the amino acid DL-ornithine through the following route. Outline the reactions involved here.

$$\text{G } (C_{12}H_{18}N_2O_5) \xrightarrow[\substack{68\ ^\circ C,\ 1000\ psi \\ (90\%\ yield)}]{H_2,\ Ni} \text{H } (C_{10}H_{16}N_2O_4,\ a\ \delta\text{-lactam}) \xrightarrow[\substack{reflux\ 4\ h \\ (97\%\ yield)}]{concd\ HCl} \begin{array}{c} \text{DL-ornithine hydrochloride} \\ (C_5H_{13}ClN_2O_2) \end{array}$$

(L-Ornithine is a naturally occurring amino acid but does not occur in proteins. In one metabolic pathway L-ornithine serves as a precursor for L-arginine.)

24.21 Synthetic polyglutamic acid exists as an α helix in solution at pH 2–3. When the pH of such a solution is gradually raised through the addition of a base, a dramatic change in optical rotation takes place at pH 5. This change has been associated with the unfolding of the α helix and the formation of a random coil. What structural feature of polyglutamic acid and what chemical change can you suggest as an explanation of this transformation?

PEPTIDE SEQUENCING

24.22 Bradykinin is a nonapeptide released by blood plasma globulins in response to a wasp sting. It is a very potent pain-causing agent. Its constituent amino acids are 2R, G, 2F, 3P, S. The use of 2,4-dinitrofluorobenzene and carboxypeptidase shows that both terminal residues are arginine. Partial acid hydrolysis of bradykinin gives the following di- and tripeptides:

FS + PGF + PP + SPF + FR + RP

What is the amino acid sequence of bradykinin?

24.23 Complete hydrolysis of a heptapeptide showed that it has the following constituent amino acids:

2A, E, L, K, F, V

Deduce the amino acid sequence of this heptapeptide from the following data.

1. Treatment of the heptapeptide with 2,4-dinitrofluorobenzene followed by incomplete hydrolysis gave, among other products: valine labeled at the α-amino group, lysine labeled at the ε-amino group, and a dipeptide, DNP—VL (DNP = 2,4-dinitrophenyl-).

2. Hydrolysis of the heptapeptide with carboxypeptidase gave an initial high concentration of alanine, followed by a rising concentration of glutamic acid.

3. Partial enzymatic hydrolysis of the heptapeptide gave a dipeptide **(A)** and a tripeptide **(B)**.

(a) Treatment of **A** with 2,4-dinitrofluorobenzene followed by hydrolysis gave DNP-labeled leucine and lysine labeled only at the ε-amino group.

(b) Complete hydrolysis of **B** gave phenylalanine, glutamic acid, and alanine. When **B** was allowed to react with carboxypeptidase, the solution showed an initial high concentration of glutamic acid. Treatment of **B** with 2,4-dinitrofluorobenzene followed by hydrolysis gave labeled phenylalanine.

CHALLENGE PROBLEM

24.24 Part of the evidence for restricted rotation about the carbon–nitrogen bond in a peptide linkage (see Section 24.8A) comes from ^1H NMR studies done with simple amides. For example, at room temperature the ^1H NMR spectrum of N,N-dimethylformamide, $(CH_3)_2NCHO$, shows a doublet at δ 2.80 (3H), a doublet at δ 2.95 (3H), and a multiplet at δ 8.05 (1H). When the spectrum is determined at lower magnetic field strength the doublets are found to have shifted so that the distance (in hertz) that separates one doublet from the other is smaller. When the temperature at which the spectrum is determined is raised, the doublets persist until a temperature of 111 °C is reached; then the doublets coalesce to become a single signal. Explain in detail how these observations are consistent with the existence of a relatively large barrier to rotation about the carbon–nitrogen bond of DMF.

LEARNING GROUP PROBLEMS

1. The enzyme lysozyme and its mechanism are described in Section 24.10. Using the information presented there (and perhaps with additional information from a biochemistry textbook), prepare notes for a class presentation on the mechanism of lysozyme.

2. Chymotrypsin is a member of the serine protease class of enzymes. Its mechanism of action is described in Section 24.11. Using the information presented there (and perhaps supplemented by information from a biochemistry textbook), prepare notes for a class presentation on the mechanism of chymotrypsin. Consider especially the role of the "catalytic triad" with regard to acid–base catalysis and the relative propensity of various groups to act as nucleophiles or leaving groups.

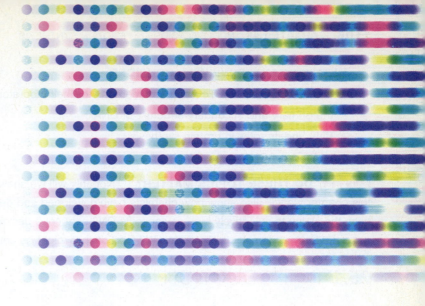

Nucleic Acids and Protein Synthesis

Chemistry is the central science because it is involved in every aspect of life. Much of what we have learned about organic chemistry thus far is related to how things work chemically, how diseases can be treated at the molecular level with small molecules, and how we can create new compounds and materials that improve our daily lives. One of the most interesting of the many applications of organic chemistry is its ability to solve critical challenges in identification through DNA matching. Studying the structure of genes and DNA, scientists can determine genetic relationships between different species (and hence the course of evolution) or between people. They can also identify the remains of individuals through DNA matching, a valuable tool if there are no other physical means to make such an identification. In fact, DNA, the genetic material, is the key to all this work. DNA is the chemical fingerprint in every tissue of every individual. With the use of chemistry involving fluorescent dyes, radioactive isotopes, enzymes, gel electrophoresis, and a process called the polymerase chain reaction (PCR) that earned its inventor the 1993 Nobel Prize in Chemistry, it is now easy to synthesize millions of copies of DNA from a single molecule of DNA, as well as to sequence it rapidly and conveniently. To understand just how this amazing process works, we need to understand this final class of biomolecules in much more detail.

A guanine-cytosine base pair

PHOTO CREDIT: PASIEKA/Science Photo Library/Getty Images, Inc.

IN THIS CHAPTER WE WILL CONSIDER:

- the structures of nucleic acids and methods for their laboratory synthesis
- the primary and secondary structure of DNA
- RNA and its roles in protein synthesis
- methods of DNA sequencing

[WHY DO THESE TOPICS MATTER?] Not only will we show you how the PCR reaction works, but at the end of this chapter we will also show how chemists have developed and designed small molecules that, with hydrogen bonding, have the ability to selectively bind any specific DNA sequence desired. Through this technique, chemists can potentially target a drug molecule selectively to any DNA portion that might be critical to treating disease.

25.1 INTRODUCTION

Deoxyribonucleic acid (DNA) and **ribonucleic acid (RNA)** are molecules that carry genetic information in cells. DNA is the molecular archive of instructions for protein synthesis. RNA molecules transcribe and translate the information from DNA for the mechanics of protein synthesis. The storage of genetic information, its passage from generation to generation, and the use of genetic information to create the working parts of the cell all depend on the molecular structures of DNA and RNA. For these reasons, we shall focus our attention on the structures and properties of these **nucleic acids** and of their components, nucleotides and nucleosides.

DNA is a biological polymer composed of two molecular strands held together by hydrogen bonds. Its overall structure is that of a twisted ladder with a backbone of alternating sugar and phosphate units and rungs made of hydrogen-bonded pairs of heterocyclic amine bases (Fig. 25.1). DNA molecules are very long polymers. If the DNA from a

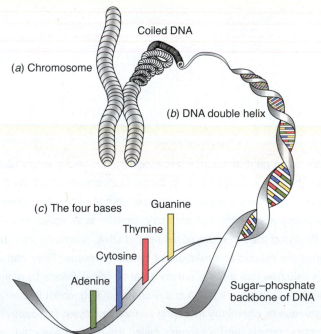

FIGURE 25.1 The basics of genetics. Each cell in the human body (except red blood cells) contains 23 pairs of chromosomes. Chromosomes are inherited: each parent contributes one chromosome per pair to their children. (a) Each chromosome is made up of a tightly coiled strand of DNA. The structure of DNA in its uncoiled state reveals (b) the familiar double-helix shape. If we picture DNA as a twisted ladder, the sides, made of sugar and phosphate molecules, are connected by (c) rungs made of heterocyclic amine bases. DNA has four, and only four, bases—adenine (A), thymine (T), guanine (G), and cytosine (C)—that form interlocking pairs. The order of the bases along the length of the ladder is called the DNA sequence. Within the overall sequence are genes, which encode the structure of proteins. (*Science and Technology Review*, November 1996, "The Human Genome Project," https://www.llnl.gov/str/Ashworth.html. Credit must be given to Linda Ashworth, the University of California, Lawrence Livermore National Laboratory, and the Department of Energy under whose auspices the work was performed, when this information or a reproduction of it is used.)

single human cell were extracted and laid straight end-to-end, it would be roughly a meter long. To package DNA into the microscopic container of a cell's nucleus, however, it is supercoiled and bundled into the 23 pairs of chromosomes with which we are familiar from electron micrographs.

Four types of heterocyclic bases are involved in the rungs of the DNA ladder, and it is the sequence of these bases that carries the information for protein synthesis. Human DNA consists of approximately 3 billion base pairs. In an effort that marks a milestone in the history of science, a working draft of the sequence of the 3 billion base pairs in the human genome was announced in 2000. A final version was announced in 2003, the 50th anniversary of the structure determination of DNA by Watson and Crick.

- Each section of DNA that codes for a given protein is called a **gene**.
- The set of all genetic information coded by DNA in an organism is its **genome**.

There are approximately 30,000–35,000 genes in the human genome. The set of all proteins encoded within the genome of an organism and expressed at any given time is called its **proteome** (Section 24.14). Some scientists estimate there could be up to one million different proteins in the cells of our various tissues—a number much greater than the number of genes in the genome due to gene splicing during protein expression and post-translational protein modification.

Hopes are very high that, having sequenced the human genome, knowledge of it will bring increased identification of genes related to disease states (Fig. 25.2) and study of these genes and the proteins encoded by them will yield a myriad of benefits for human health and longevity. Determining the structure of all of the proteins encoded in the genome, learning their functions, and creating molecular therapeutics based on this rapidly expanding store of knowledge are some of the key research challenges that lie ahead.

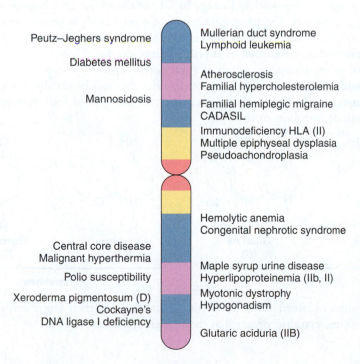

FIGURE 25.2 Schematic map of the location of genes for diseases on chromosome 19. *(From Dept. of Energy Joint Genome Institute Website. (http://www.jgi.doe.gov/whoweare/). Credit to the University of California, Lawrence Livermore National Laboratory, and the Department of Energy under whose auspices the work was performed.)*

Let us begin with a study of the structures of nucleic acids. Each of their monomer units contains a cyclic amine base, a carbohydrate group, and a phosphate ester.

25.2 NUCLEOTIDES AND NUCLEOSIDES

Mild degradations of nucleic acids yield monomeric units called **nucleotides**. A general formula for a nucleotide and the specific structure of one called adenylic acid are shown in Fig. 25.3.

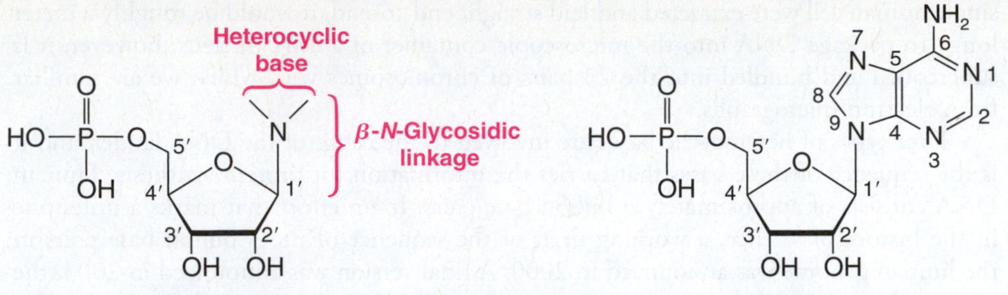

A nucleotide

(a)

Adenylic acid

(b)

FIGURE 25.3 (a) General structure of a nucleotide obtained from RNA. The heterocyclic base is a purine or pyrimidine. *In nucleotides obtained from DNA, the sugar component is 2′-deoxy-D-ribose; that is, the* — *OH at position 2′ is replaced by* — *H.* The phosphate group of the nucleotide is shown attached at **C5′**; it may instead be attached at **C3′**. In DNA and RNA a phosphodiester linkage joins **C5′** of one nucleotide to **C3′** of another. The heterocyclic base is always attached through a β-*N*-glycosidic linkage at **C1′**. (b) Adenylic acid, a typical nucleotide.

Complete hydrolysis of a nucleotide furnishes:

1. A heterocyclic base from either the purine or pyrimidine family.

2. A five-carbon monosaccharide that is either D-ribose or 2-deoxy-D-ribose.

3. A phosphate ion.

The central portion of the nucleotide is the monosaccharide, and it is always present as a five-membered ring, that is, as a furanoside. The heterocyclic base of a nucleotide is attached through an *N*-glycosidic linkage to C1′ of the ribose or deoxyribose unit, and this linkage is always β. The phosphate group of a nucleotide is present as a phosphate ester and may be attached at C5′ or C3′. (In nucleotides, the carbon atoms of the monosaccharide portion are designated with primed numbers, i.e., 1′, 2′, 3′, etc.)

Removal of the phosphate group of a nucleotide converts it to a compound known as a **nucleoside** (Section 22.15A). The nucleosides that can be obtained from DNA all contain 2-deoxy-D-ribose as their sugar component and one of four heterocyclic bases: adenine, guanine, cytosine, or thymine:

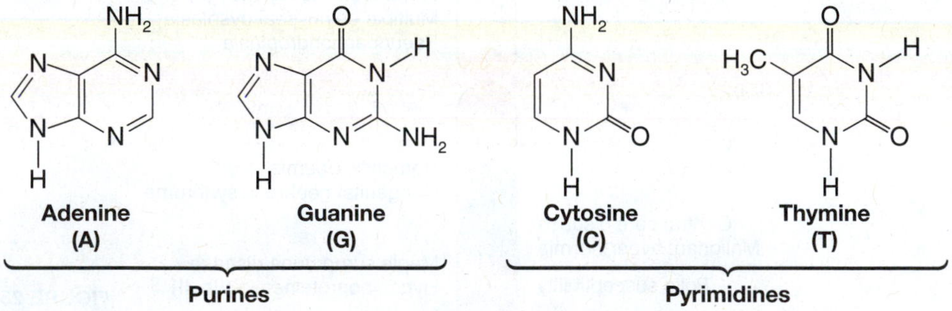

| **Adenine** (A) | **Guanine** (G) | **Cytosine** (C) | **Thymine** (T) |

Purines   **Pyrimidines**

The nucleosides obtained from RNA contain D-ribose as their sugar component and adenine, guanine, cytosine, or uracil as their heterocyclic base.

Uracil replaces thymine in an RNA nucleoside (or nucleotide). Some nucleosides obtained from specialized forms of RNA may also contain other, but similar, purines and pyrimidines.

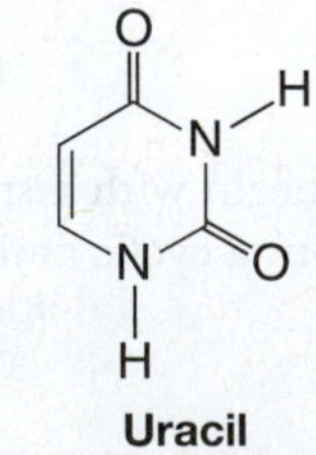

Uracil (a pyrimidine)

The heterocyclic bases obtained from nucleosides are capable of existing in more than one tautomeric form. The forms that we have shown are the predominant forms that the bases assume when they are present in nucleic acids.

Adenine

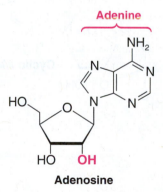

2'-Deoxyadenosine

Guanine

2'-Deoxyguanosine

Cytosine

2'-Deoxycytidine

Thymine

2'-Deoxythymidine

FIGURE 25.4 Nucleosides that can be obtained from DNA. DNA is 2'-deoxy at the position where the blue shaded box is shown. RNA (see Fig 25.5) has hydroxyl groups at that location. RNA has a hydrogen where there is a methyl group in thymine, which in RNA makes the base uracil (and the nucleoside uridine).

The names and structures of the nucleosides found in DNA are shown in Fig. 25.4; those found in RNA are given in Fig. 25.5.

Adenine

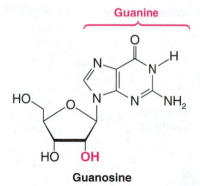

Adenosine

Guanine

Guanosine

Cytosine

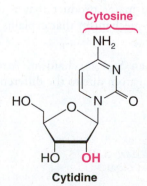

Cytidine

Uracil

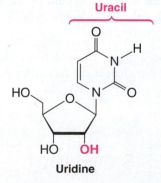

Uridine

FIGURE 25.5 Nucleosides that can be obtained from RNA. DNA (see Fig 25.4) has hydrogen atoms where the red hydroxyl groups of ribose are shown (DNA is 2'-deoxy with respect to its ribose moiety).

Write the structures of other tautomeric forms of adenine, guanine, cytosine, thymine, and uracil.

PRACTICE PROBLEM 25.1

● ● ●

PRACTICE PROBLEM 25.2 The nucleosides shown in Figs. 25.4 and 25.5 are stable in dilute base. In dilute acid, however, they undergo rapid hydrolysis yielding a sugar (deoxyribose or ribose) and a heterocyclic base.

(a) What structural feature of the nucleoside accounts for this behavior?

(b) Propose a reasonable mechanism for the hydrolysis.

Nucleotides are named in several ways. Adenylic acid (Fig. 25.3), for example, is usually called AMP, for adenosine monophosphate. The position of the phosphate group is sometimes explicitly noted by use of the names adenosine 5′-monophosphate or 5′-adenylic acid. Uridylic acid is usually called UMP, for uridine monophosphate, although it can also be called uridine 5′-monophosphate or 5′-uridylic acid. If a nucleotide is present as a diphosphate or triphosphate, the names are adjusted accordingly, such as ADP for adenosine diphosphate or GTP for guanosine triphosphate.

Nucleosides and nucleotides are found in places other than as part of the structure of DNA and RNA. We have seen, for example, that adenosine units are part of the structures of two important coenzymes, NADH and coenzyme A. The 5′-triphosphate of adenosine is, of course, the important energy source, ATP (Section 22.1B). The compound called 3′,5′-cyclic adenylic acid (or cyclic AMP) (Fig. 25.6) is an important regulator of hormone activity. Cells synthesize this compound from ATP through the action of an enzyme, *adenylate cyclase*. In the laboratory, 3′,5′-cyclic adenylic acid can be prepared through dehydration of 5′-adenylic acid with dicyclohexylcarbodiimide.

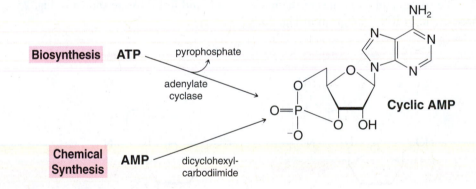

FIGURE 25.6 3′,5′-Cyclic adenylic acid (cyclic AMP) and its biosynthesis and laboratory synthesis.

● ● ● **SOLVED PROBLEM 25.1**

When 3′,5′-cyclic adenylic acid is treated with aqueous sodium hydroxide, the major product that is obtained is 3′-adenylic acid (adenosine 3′-phosphate) rather than 5′-adenylic acid. Suggest a mechanism that explains the course of this reaction.

STRATEGY AND ANSWER: The reaction appears to take place through an S_N2 mechanism. Attack occurs preferentially at the primary 5′-carbon atom rather than at the secondary 3′-carbon atom due to the difference in steric hindrance.

25.3 LABORATORY SYNTHESIS OF NUCLEOSIDES AND NUCLEOTIDES

A variety of methods have been developed for the chemical synthesis of nucleosides from the constituent sugars and bases or their precursors. The following is an example of a *silyl–Hilbert–Johnson nucleosidation*, where a benzoyl protected sugar (D-ribose) reacts in the presence of tin(IV) chloride with an *N*-benzoyl protected base (cytidine) that is protected further by *in situ* silylation.* The trimethylsilyl protecting groups for the base are introduced using *N,O*-bis(trimethylsilyl)acetamide (BSA) and they are removed with aqueous acid in the second step. The result is a protected form of the nucleoside cytosine, from which the benzoyl groups can be removed with ease using a base:

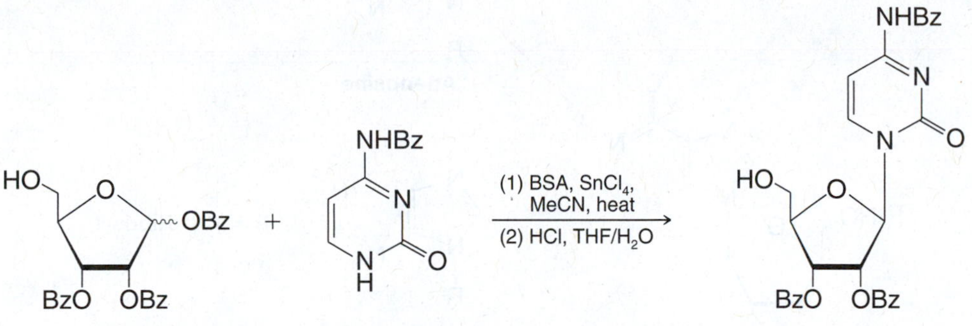

D-Ribose (Bz protected) + ***N*⁴-Benzoylcytosine** → **Cytidine** (Bz protected)

(1) BSA, SnCl₄, MeCN, heat
(2) HCl, THF/H₂O

Bz = C₆H₅CO (benzoyl)

BSA = CH₃C[=NSi(CH₃)₃]OSi(CH₃)₃,
[*N,O*-bis(trimethylsilyl)acetamide, a reagent for
trimethylsilylation of nucleophilic oxygen and nitrogen atoms]

Another technique involves formation of the heterocyclic base on a protected ribosylamine derivative:

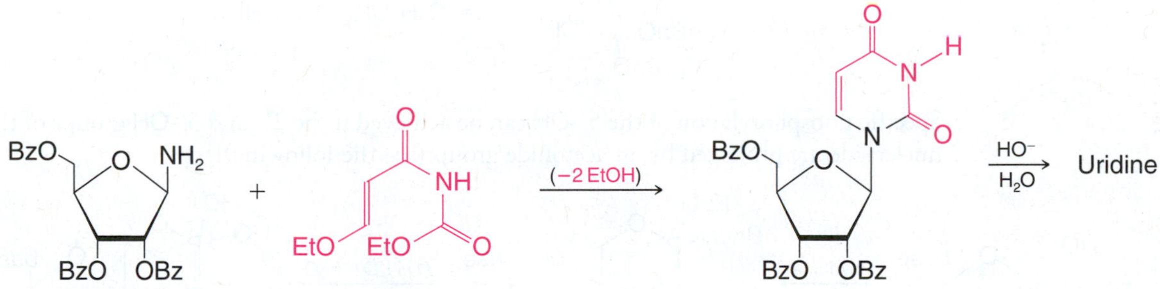

2,3,5-Tri-*O*-benzoyl-β-D-ribofuranosylamine + **β-Ethoxy-*N*-ethoxycarbonylacrylamide** →(−2 EtOH)→ →(HO⁻/H₂O)→ Uridine

- - - ......

PRACTICE PROBLEM 25.3

Basing your answer on reactions that you have seen before, propose a likely mechanism for the condensation reaction in the first step of the preceding uridine synthesis.

A third technique involves the synthesis of a nucleoside with a substituent in the heterocyclic ring that can be replaced with other groups. This method has been used extensively to synthesize unusual nucleosides that do not necessarily occur naturally. The

---

*These conditions were applied using L-ribose in a synthesis of the unnatural enantiomer of RNA (Pitsch, S. An efficient synthesis of enantiomeric ribonucleic acids from D-glucose. *Helv. Chim. Acta* **1997**, *80*, 2286–2314). The protected enantiomeric cytidine was produced in 94% yield by the above reaction. After adjusting protecting groups, solid-phase oligonucleotide synthesis methods (Section 25.7) were used with this compound and the other three nucleotide monomers (also derived from L-ribose) for preparation of the unnatural RNA enantiomer. See also Vorbrüggen, H.; Ruh-Pohlenz, C., *Handbook of Nucleoside Synthesis*; Wiley: Hoboken, NJ, 2001.

following example makes use of a 6-chloropurine derivative obtained from the appropriate ribofuranosyl chloride and chloromercuripurine:

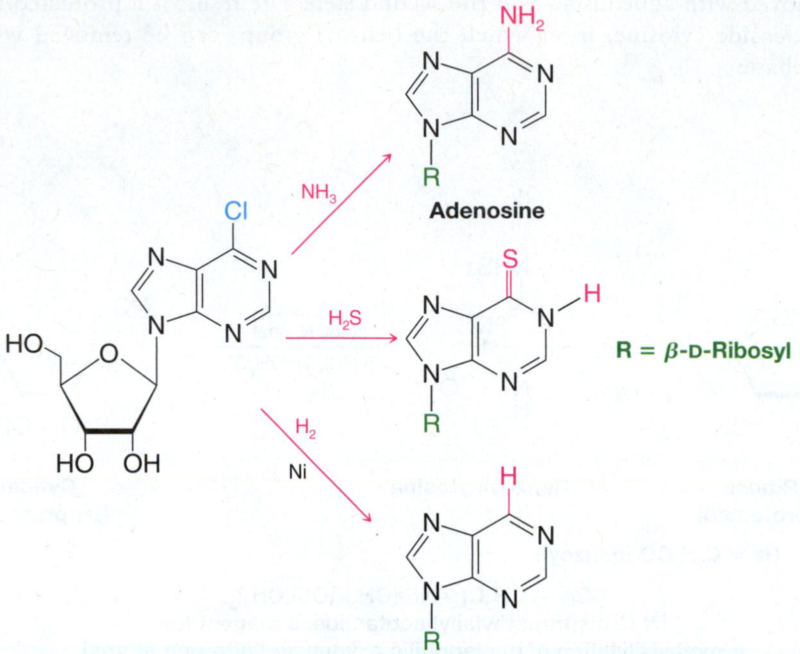

Numerous phosphorylating agents have been used to convert nucleosides to nucleotides. One of the most useful is dibenzyl phosphochloridate:

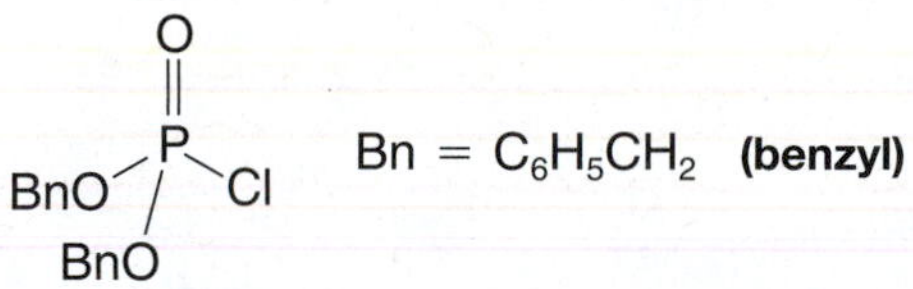

Bn = C₆H₅CH₂ **(benzyl)**

Specific phosphorylation of the 5'-OH can be achieved if the 2'- and 3'-OH groups of the nucleoside are protected by an acetonide group (see the following):

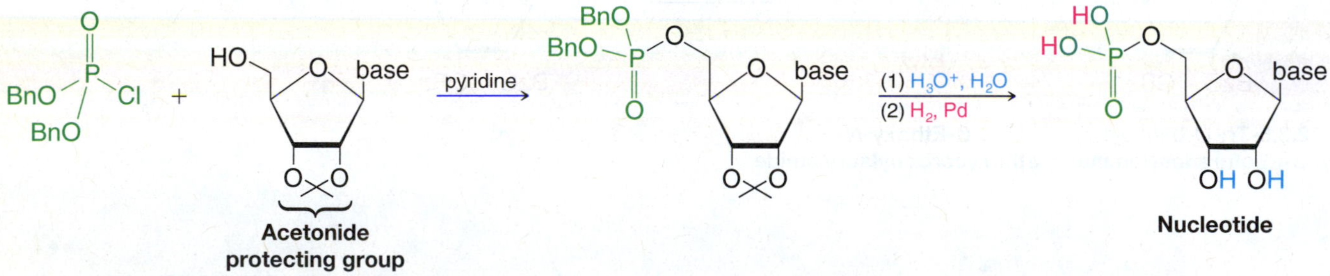

Mild acid-catalyzed hydrolysis removes the acetonide group, and hydrogenolysis cleaves the benzyl phosphate bonds.

●●●

PRACTICE PROBLEM 25.4 **(a)** What kind of linkage is involved in the acetonide group of the protected nucleoside, and why is it susceptible to mild acid-catalyzed hydrolysis? **(b)** How might such a protecting group be installed?

●●●

PRACTICE PROBLEM 25.5 The following reaction scheme is from a synthesis of cordycepin (a nucleoside antibiotic) and the first synthesis of 2'-deoxyadenosine (reported in 1958 by C. D. Anderson, L. Goodman, and B. R. Baker, Stanford Research Institute):

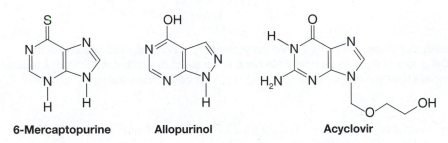

(a) What is the structure of cordycepin? (**I** and **II** are isomers.)

(b) Propose a mechanism that explains the formation of **II**.

25.3A Medical Applications

In the early 1950s, Gertrude Elion and George Hitchings (of the Wellcome Research Laboratories) discovered that 6-mercaptopurine had antitumor and antileukemic properties. This discovery led to the development of other purine derivatives and related compounds, including nucleosides, of considerable medical importance. Three examples are the following:

6-Mercaptopurine   **Allopurinol**   **Acyclovir**

ELION AND HITCHINGS shared the 1988 Nobel Prize in Physiology or Medicine for their work in the development of chemotherapeutic agents derived from purines.

6-Mercaptopurine is used in combination with other chemotherapeutic agents to treat acute leukemia in children, and almost 80% of the children treated are now cured. Allopurinol, another purine derivative, is a standard therapy for the treatment of gout. Acyclovir, a nucleoside that lacks two carbon atoms of its ribose ring, is highly effective in treating diseases caused by certain herpes viruses, including *herpes simplex* type 1 (fever blisters), type 2 (genital herpes), and varicella-zoster (shingles).

25.4 DEOXYRIBONUCLEIC ACID: DNA

25.4A Primary Structure

Nucleotides bear the same relation to a nucleic acid that amino acids do to a protein: they are its monomeric units. The connecting links in proteins are amide groups; in nucleic acids they are phosphate ester linkages. Phosphate esters link the 3'-OH of one ribose (or deoxyribose) with the 5'-OH of another. This makes the nucleic acid a long unbranched chain with a "backbone" of sugar and phosphate units with heterocyclic bases protruding

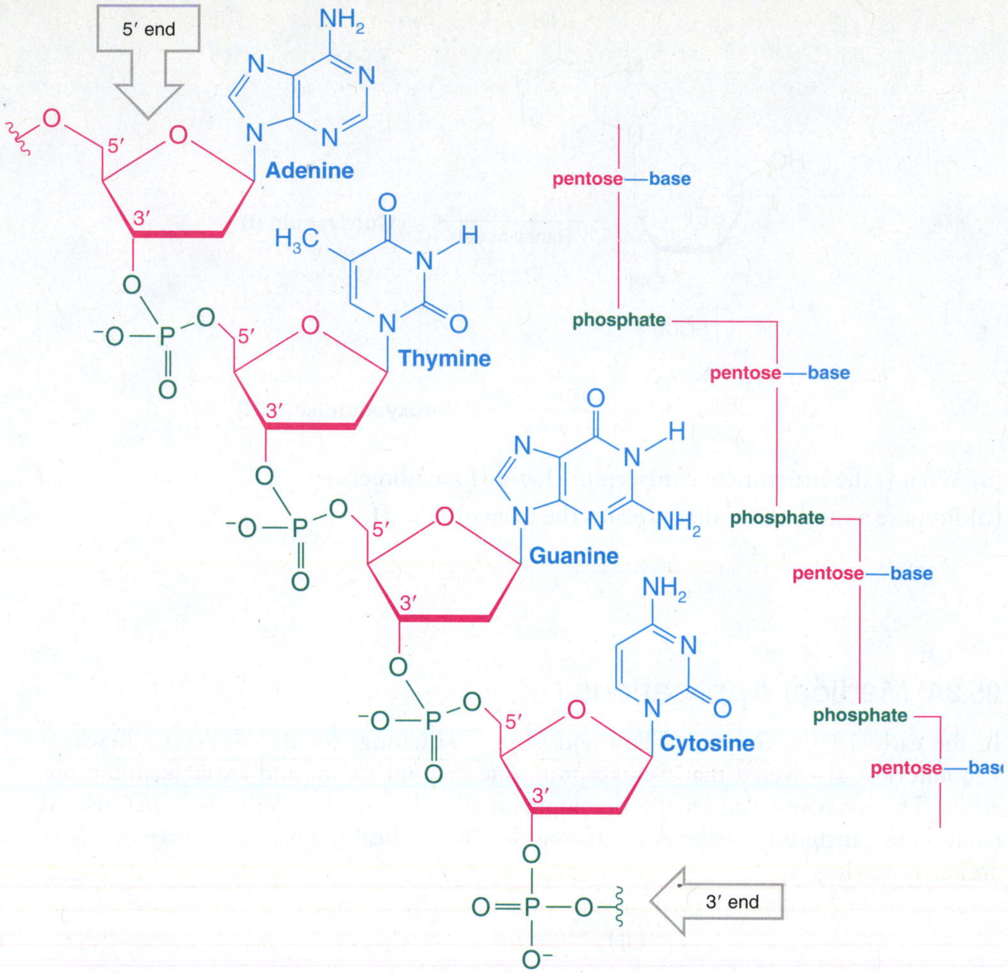

FIGURE 25.7 A segment of one DNA chain showing how phosphate ester groups link the 3'- and 5'-OH groups of deoxyribose units. RNA has a similar structure with two exceptions: a hydroxyl replaces a hydrogen atom at the 2' position of each ribose unit and uracil replaces thymine.

from the chain at regular intervals (Fig. 25.7). We would indicate the direction of the bases in Fig. 25.7 in the following way:

$$5' \leftarrow A—T—G—C \rightarrow 3'$$

It is, as we shall see, the **base sequence** along the chain of DNA that contains the encoded genetic information. The sequence of bases can be determined using enzymatic methods and chromatography (Section 25.6).

25.4B Secondary Structure

"I cannot help wondering whether some day an enthusiastic scientist will christen his newborn twins Adenine and Thymine."
F. H. C. Crick*

It was the now-classic proposal of James Watson and Francis Crick (made in 1953 and verified shortly thereafter through the X-ray analysis by Maurice Wilkins) that gave a model for the secondary structure of DNA. This work earned Crick, Watson, and Wilkins the 1962 Nobel Prize in Physiology or Medicine. Many believe that Rosalind Franklin, whose X-ray data was also key to solving the structure of DNA, should have shared the Nobel Prize, but her death from cancer in 1958 precluded it. The secondary structure of DNA is especially important because it enables us to understand how genetic information is preserved, how it can be passed on during the process of cell division, and how it can be transcribed to provide a template for protein synthesis.

*Taken from Crick, F. H. C., The structure of the hereditary material. *Sci. Am.* **1954**, *191*(10), 20, 54–61.

Of prime importance to Watson and Crick's proposal was an earlier observation (made in the late 1940s) by Erwin Chargaff that certain regularities can be seen in the percentages of heterocyclic bases obtained from the DNA of a variety of species. Table 25.1 gives results that are typical of those that can be obtained.

TABLE 25.1 DNA COMPOSITION OF VARIOUS SPECIES

Species	\multicolumn{9}{c}{Base Proportions (mol %)}
	G	A	C	T	$\dfrac{G + A}{C + T}$	$\dfrac{A + T}{G + C}$	$\dfrac{A}{T}$	$\dfrac{G}{C}$
Sarcina lutea	37.1	13.4	37.1	12.4	1.02	0.35	1.08	1.00
Escherichia coli K12	24.9	26.0	25.2	23.9	1.08	1.00	1.09	0.99
Wheat germ	22.7	27.3	22.8[a]	27.1	1.00	1.19	1.01	1.00
Bovine thymus	21.5	28.2	22.5[a]	27.8	0.96	1.27	1.01	0.96
Staphylococcus aureus	21.0	30.8	19.0	29.2	1.11	1.50	1.05	1.11
Human thymus	19.9	30.9	19.8	29.4	1.01	1.52	1.05	1.01
Human liver	19.5	30.3	19.9	30.3	0.98	1.54	1.00	0.98

[a]Cytosine + methylcytosine.

Source: Reproduced with permission of The McGraw-Hill Companies, Inc. from Smith, E.L.; Hill, R.L.; Jehman, I.R.; Lefkowitz, R.J.; Handler, P.; and White, A., *Principles of Biochemistry: General Aspects*, 7th edition, ©1982.

Chargaff pointed out that for all species examined:

1. The total mole percentage of purines is approximately equal to that of the pyrimidines, that is, (%G + %A)/(%C + %T) $\cong$ 1.

2. The mole percentage of adenine is nearly equal to that of thymine (i.e., %A/%T $\cong$ 1), and the mole percentage of guanine is nearly equal to that of cytosine (i.e., %G/%C $\cong$ 1).

Chargaff also noted that the ratio which varies from species to species is the ratio (%A + %T)/(%G + %C). He noted, moreover, that whereas this ratio is characteristic of the DNA of a given species, it is the same for DNA obtained from different tissues of the same animal and does not vary appreciably with the age or conditions of growth of individual organisms within the same species.

Watson and Crick also had X-ray data that gave them the bond lengths and angles of the purine and pyrimidine rings of model compounds. In addition, they had data from Franklin and Wilkins that indicated a repeat distance of 34 Å in DNA.

Reasoning from these data, Watson and Crick proposed a double helix as a model for the secondary structure of DNA. According to this model, two nucleic acid chains are held together by hydrogen bonds between base pairs on opposite strands. This double chain is wound into a helix with both chains sharing the same axis. The base pairs are on the inside of the helix, and the sugar–phosphate backbone is on the outside (Fig. 25.8). The pitch of the helix is such that 10 successive nucleotide pairs give rise to one complete turn in 34 Å (the repeat distance). The exterior width of the spiral is about 20 Å, and the internal distance between 1′ positions of ribose units on opposite chains is about 11 Å.

Using molecular-scale models, Watson and Crick observed that the internal distance of the double helix is such that it allows only a purine–pyrimidine type of hydrogen bonding between base pairs. Purine–purine base pairs do not occur because they would be too large to fit, and pyrimidine–pyrimidine base pairs do not occur because they would be too far apart to form effective hydrogen bonds.

Helpful Hint

The use of models was critical to Watson and Crick in their Nobel prize-winning work on the three-dimensional structure of DNA.

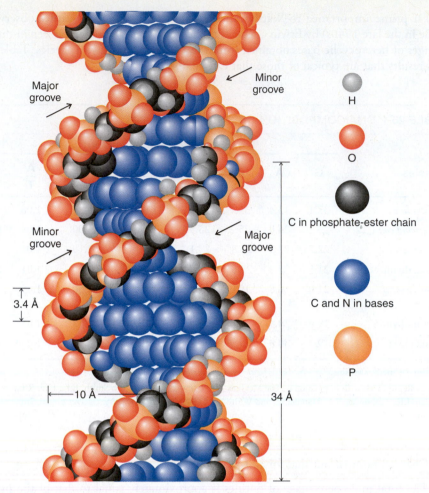

FIGURE 25.8 A molecular model of a portion of the DNA double helix. (Reprinted with permission of The McGraw-Hill Companies from Neal, L., *Chemistry and Biochemistry: A Comprehensive Introduction*, © 1971.)

Watson and Crick went one crucial step further in their proposal. Assuming that the oxygen-containing heterocyclic bases existed in keto forms, they argued that base pairing through hydrogen bonds can occur in only a specific way: adenine (A) with thymine (T) and cytosine (C) with guanine (G). Dimensions of the pairs and electrostatic potential maps for the individual bases are shown in Fig. 25.9.

Adenine Pairs with Thymine   **Guanine Pairs with Cytosine**

Thymine   Adenine   **Cytosine   Guanine**

Specific base pairing of this kind is consistent with Chargaff's finding that %A/%T ≅ 1 and %G/%C ≅ 1.

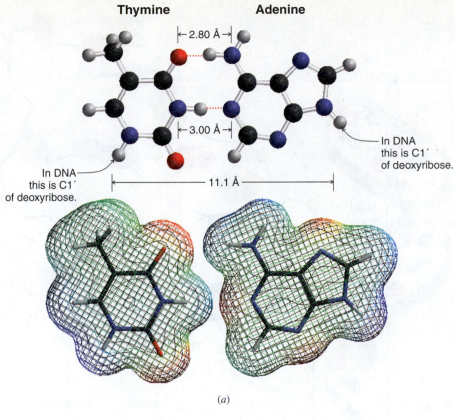

(a)

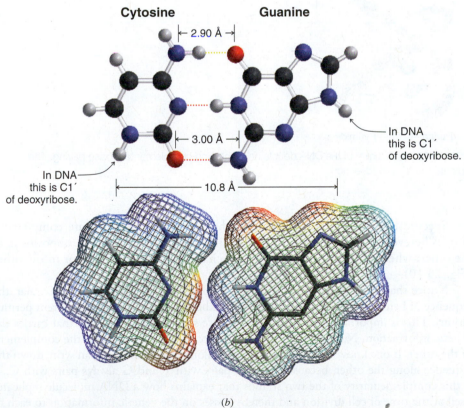

(b)

FIGURE 25.9 Base pairing of adenine with thymine (a) and cytosine with guanine (b). The dimensions of the thymine–adenine and cytosine–guanine hydrogen-bonded pairs are such that they allow the formation of strong hydrogen bonds and also allow the base pairs to fit inside the two phosphate–ribose chains of the double helix. Electrostatic potential maps calculated for the individual bases show the complementary distribution of charges that leads to hydrogen bonding. (Ball and stick models reprinted from Archives of *Biochemistry and Biophysics*, **65**, Pauling, I., Corey, R., p. 164–181, 1956. Copyright 1956, with permission from Elsevier.)

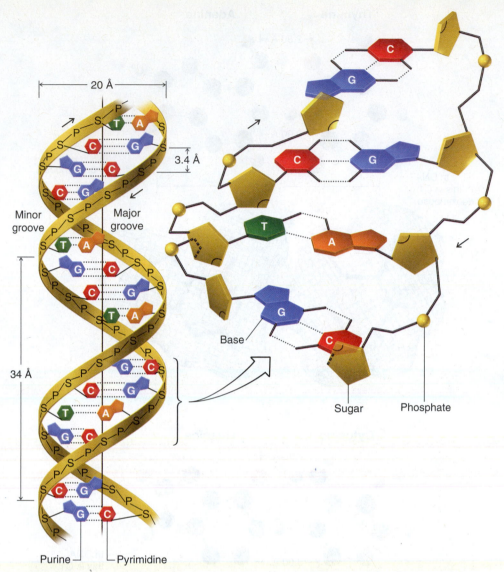

FIGURE 25.10 Diagram of the DNA double helix showing complementary base pairing. The arrows indicate the 3′ → 5′ direction.

Specific base pairing also means that the two chains of DNA are complementary. Wherever adenine appears in one chain, thymine must appear opposite it in the other; wherever cytosine appears in one chain, guanine must appear in the other (Fig. 25.10).

Notice that while the sugar–phosphate backbone of DNA is completely regular, the sequence of heterocyclic base pairs along the backbone can assume many different permutations. This is important because it is the precise sequence of base pairs that carries the genetic information. Notice, too, that one chain of the double strand is the complement of the other. If one knows the sequence of bases along one chain, one can write down the sequence along the other, because A always pairs with T and G always pairs with C. It is this complementarity of the two strands that explains how a DNA molecule replicates itself at the time of cell division and thereby passes on the genetic information to each of the two daughter cells.

25.4C Replication of DNA

Just prior to cell division the double strand of DNA begins to unwind. Complementary strands are formed along each chain (Fig. 25.11). Each chain acts, in effect, as a template

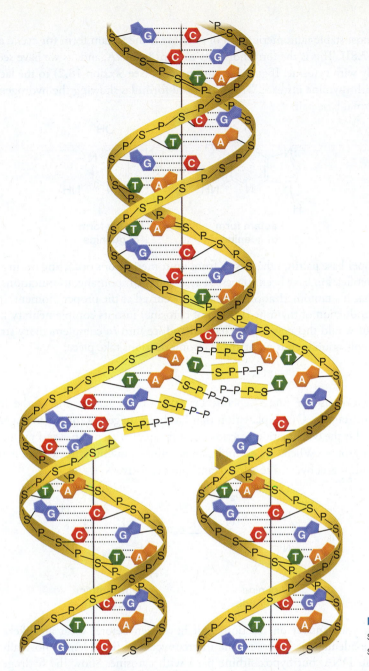

FIGURE 25.11 Replication of DNA. The double strand unwinds from one end and complementary strands are formed along each chain.

for the formation of its complement. When unwinding and **replication** are complete, there are two identical DNA molecules where only one had existed before. These two molecules can then be passed on, one to each daughter cell.

PRACTICE PROBLEM 25.6

(a) There are approximately 3 billion base pairs in the DNA of a single human cell. Assuming that this DNA exists as a double helix, calculate the length of all the DNA contained in a human cell. **(b)** The weight of DNA in a single human cell is 6×10^{-12} g. Assuming that Earth's population is about 6.5 billion, we can conclude that all of the genetic information that gave rise to all human beings now alive was once contained in the DNA of a corresponding number of fertilized ova. What is the total weight of DNA in this many ova? (The volume that this DNA would occupy is approximately that of a raindrop, yet if the individual molecules were laid end-to-end, they would stretch to the moon and back almost eight times.)

PRACTICE PROBLEM 25.7 **(a)** The most stable tautomeric form of guanine is the lactam form (or cyclic amide, see Section 17.8I). This is the form normally present in DNA, and, as we have seen, it pairs specifically with cytosine. If guanine tautomerizes (see Section 18.2) to the lactim form, it pairs with thymine instead. Write structural formulas showing the hydrogen bonds in this abnormal base pair.

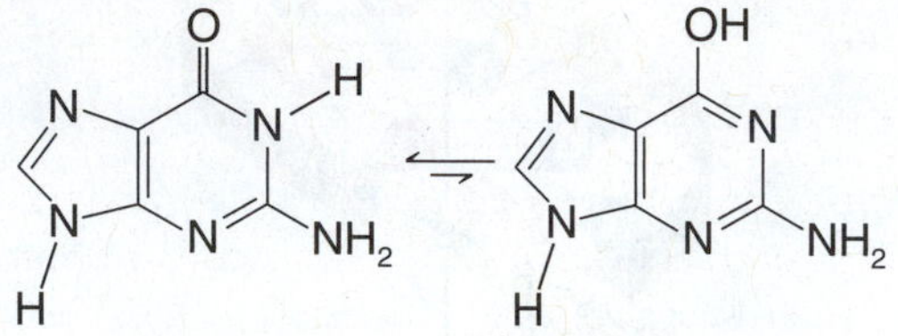

Lactam form of guanine

Lactim form of guanine

(b) Improper base pairings that result from tautomerizations occurring during the process of DNA replication have been suggested as a source of spontaneous mutations. We saw in part (a) that if a tautomerization of guanine occurred at the proper moment, it could lead to the introduction of thymine (instead of cytosine) into its complementary DNA chain. What error would this new DNA chain introduce into *its* complementary strand during the next replication even if no further tautomerizations take place?

PRACTICE PROBLEM 25.8 Mutations can also be caused chemically, and nitrous acid is one of the most potent chemical **mutagens**. One explanation that has been suggested for the mutagenic effect of nitrous acid is the deamination reactions that it causes with purines and pyrimidines bearing amino groups. When, for example, an adenine-containing nucleotide is treated with nitrous acid, it is converted to a hypoxanthine derivative:

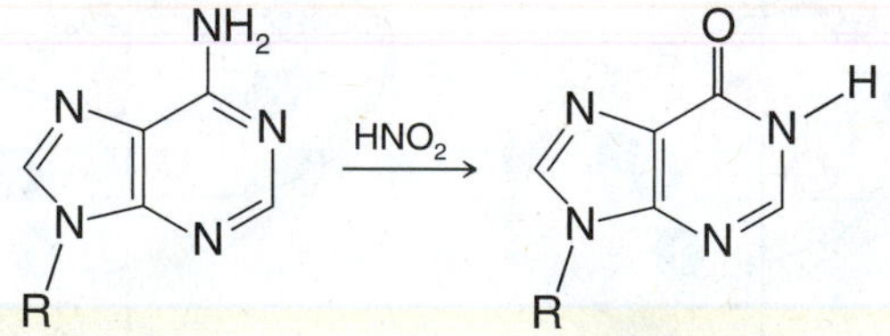

Adenine nucleotide

HNO_2

Hypoxanthine nucleotide

(a) Basing your answer on reactions you have seen before, what are likely intermediates in the adenine $\rightarrow$ hypoxanthine interconversion? **(b)** Adenine normally pairs with thymine in DNA, but hypoxanthine pairs with cytosine. Show the hydrogen bonds of a hypoxanthine–cytosine base pair. **(c)** Show what errors an adenine $\rightarrow$ hypoxanthine interconversion would generate in DNA through two replications.

25.5 RNA AND PROTEIN SYNTHESIS

Soon after the Watson–Crick hypothesis was published, scientists began to extend it to yield what Crick called "the central dogma of molecular genetics." This dogma stated that genetic information flows as follows:

$$DNA \rightarrow RNA \rightarrow protein$$

The synthesis of protein is, of course, all important to a cell's function because proteins (as enzymes) catalyze its reactions. Even the very primitive cells of bacteria require as many as 3000 different enzymes. This means that the DNA molecules of these cells must contain a corresponding number of genes to direct the synthesis of these proteins. A **gene** is that segment of the DNA molecule that contains the information necessary to direct the synthesis of one protein (or one polypeptide).

DNA is found primarily in the nucleus of eukaryotic cells. Protein synthesis takes place primarily in that part of the cell called the *cytoplasm*. Protein synthesis requires that two major processes take place; the first occurs in the cell nucleus, the second in the cytoplasm. The first is **transcription**, a process in which the genetic message is transcribed onto a form of RNA called messenger RNA (mRNA). The second process involves two other forms of RNA, called ribosomal RNA (rRNA) and transfer RNA (tRNA).

There are viruses, called retroviruses, in which information flows from RNA to DNA. The virus that causes AIDS is a retrovirus.

25.5A Messenger RNA Synthesis—Transcription

The events leading to protein synthesis begin in the cell nucleus with the synthesis of mRNA. Part of the DNA double helix unwinds sufficiently to expose on a single chain a portion corresponding to at least one gene. Ribonucleotides, present in the cell nucleus, assemble along the exposed DNA chain by pairing with the bases of DNA. The pairing patterns are the same as those in DNA with the exception that in RNA uracil replaces thymine. The ribonucleotide units of mRNA are joined into a chain by an enzyme called *RNA polymerase*. This process is illustrated in Fig. 25.12.

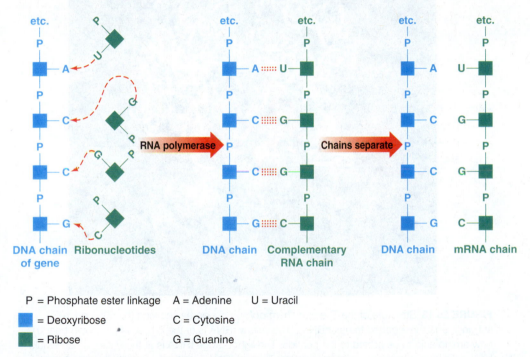

P = Phosphate ester linkage   A = Adenine   U = Uracil

■ = Deoxyribose   C = Cytosine

■ = Ribose   G = Guanine

FIGURE 25.12 Transcription of the genetic code from DNA to mRNA.

PRACTICE PROBLEM 25.9

Write structural formulas showing how the keto form of uracil (Section 25.2) in mRNA can pair with adenine in DNA through hydrogen bond formation.

Most eukaryotic genes contain segments of DNA that are not actually used when a protein is expressed, even though they are transcribed into the initial mRNA. These segments are called **introns**, or intervening sequences. The segments of DNA within a gene that are expressed are called **exons**, or expressed sequences. Each gene usually contains a number of introns and exons. After the mRNA is transcribed from DNA, the introns in the mRNA are removed and the exons are spliced together.

After mRNA has been synthesized and processed in the cell nucleus to remove the introns, it migrates into the cytoplasm where, as we shall see, it acts as a template for protein synthesis.

25.5B Ribosomes—rRNA

Protein synthesis is catalyzed by ribosomes in the cytoplasm. Ribosomes (Fig. 25.13) are ribonucleoproteins, comprised of approximately two-thirds RNA and one-third protein. They have a very high molecular weight (about 2.6×10^6). The RNA component is present in two subunits, called the 50S and 30S subunits (classified according to their sedimentation behavior during ultracentrifugation*). The 50S subunit is roughly twice the molecular weight of the 30S subunit. Binding of RNA with mRNA is mediated by the 30S subunit. The 50S subunit carries the catalytic activity for translation that joins one amino acid by an amide bond to the next. In addition to the rRNA subunits there are approximately 30–35 proteins tightly bound to the ribosome, the entire structure resembling an exquisite three-dimensional jigsaw puzzle of RNA and protein. The mechanism for ribosome-catalyzed amide bond formation is discussed below.

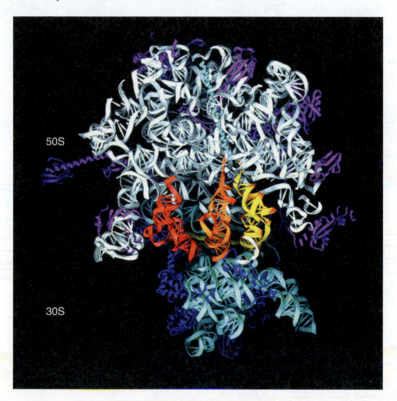

FIGURE 25.13 Structure of the *Thermus thermophilus* ribosome showing the 50S and 30S subunits and three bound transfer RNAs. The yellow tRNA is at the A site, which would bear the new amino acid to be added to the peptide. The light orange tRNA is at the P site, which would be the tRNA that bears the growing peptide. The red tRNA is at the E site, which is the "empty" tRNA after it has transferred the peptide chain to the new amino acid. *(Courtesy of Harry Noller, University of California, Santa Cruz.)*

Ribosomes, as reaction catalysts, are most appropriately classified as **ribozymes** rather than enzymes, because it is RNA that catalyzes the peptide bond formation during protein synthesis and not the protein subunits of the ribosome. The mechanism for peptide bond formation catalyzed by the 50S ribosome subunit (Fig. 25.14), proposed by Moore and co-workers based on X-ray crystal structures, suggests that attack by the α-amino group is facilitated by acid–base catalysis involving nucleotide residues along the 50S ribosome subunit chain, specifically a nearby adenine group. Full or partial removal of a proton from the α-amino group of the amino acid by N3 of the adenine group imparts greater nucleophilicity to the amino nitrogen, facilitating its attack on the acyl carbon of the adjacent peptide–tRNA moiety. A tetrahedral intermediate is formed, which collapses to form the new amide bond with release of the tRNA that had been joined to the peptide. Other moieties in the 50S ribosome

*S stands for svedberg unit; it is used in describing the behavior of proteins in an ultracentrifuge.

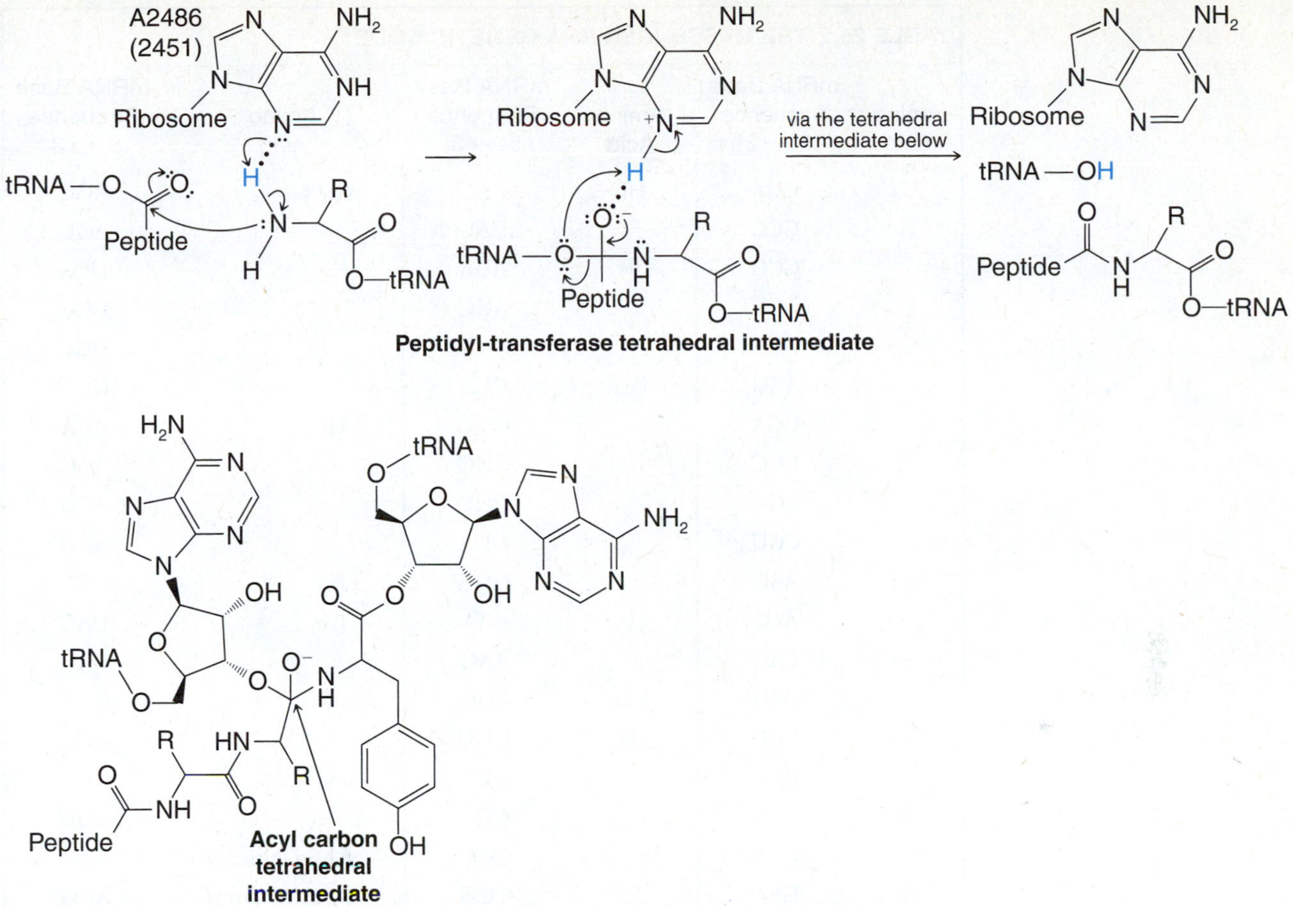
A2486
(2451)
Ribosome
tRNA—O
Peptide
R
O—tRNA
via the tetrahedral intermediate below
tRNA—OH
Peptidyl-transferase tetrahedral intermediate
tRNA
NH2
OH
Acyl carbon tetrahedral intermediate

FIGURE 25.14 A mechanism for peptide bond formation catalyzed by the 50S subunit of the ribosome (as proposed by Moore and co-workers). The new amide bond in the growing peptide chain is formed by attack of the α-amino group in the new amino acid, brought to the A site of the ribosome by its tRNA, on the acyl carbon linkage of the peptide held at the P site by its tRNA. Acid–base catalysis by groups in the ribosome facilitate the reaction. (Reprinted with permission from Nissen et al., The structural basis of ribosome activity in peptide bond synthesis. SCIENCE 289:920–930 (2000). Reprinted with permission from AAAS. Also reprinted from Monro, R. E., and Marker, K. A., Ribosome-catalysed reaction of puromycin with a formylmethionine containing oligonucleotide, *J. Mol. Biol.* 25 pp. 347–350. Copyright 1967, with permission of Elsevier.)

subunit are believed to help stabilize the transfer of charge that occurs as N3 of the adenyl group accepts the proton from the α-amino group of the new amino acid (see Problem 25.16).

25.5C Transfer RNA

Transfer RNA has a very low molecular weight when compared to those of mRNA and rRNA. Transfer RNA, consequently, is much more soluble than mRNA or rRNA and is sometimes referred to as soluble RNA. The function of tRNA is to transport amino acids to specific areas on the mRNA bound to the ribosome. There are, therefore, many forms of tRNA, more than one for each of the 20 amino acids that is incorporated into proteins, including the redundancies in the **genetic code** (see Table 25.2).*

The structures of most tRNAs have been determined. They are composed of a relatively small number of nucleotide units (70–90 units) folded into several loops or

*Although proteins are composed of 22 different amino acids, protein synthesis requires only 20. Proline is converted to hydroxyproline and cysteine is converted to cystine after synthesis of the polypeptide chain has taken place.

TABLE 25.2 THE MESSENGER RNA GENETIC CODE

Amino Acid	mRNA Base Sequence 5′ → 3′	Amino Acid	mRNA Base Sequence 5′ → 3′	Amino Acid	mRNA Base Sequence 5′ → 3′
Ala	GCA	His	CAC	Ser	AGC
	GCC		CAU		AGU
	GCG	Ile	AUA		UCA
	GCU		AUC		UCG
Arg	AGA		AUU		UCC
	AGG	Leu	CUA		UCU
	CGA		CUC	Thr	ACA
	CGC		CUG		ACC
	CGG		CUU		ACG
	CGU		UUA		ACU
Asn	AAC		UUG	Trp	UGG
	AAU	Lys	AAA	Tyr	UAC
Asp	GAC		AAG		UAU
	GAU	Met	AUG	Val	GUA
Cys	UGC	Phe	UUU		GUG
	UGU		UUC		GUC
Gln	CAA	Pro	CCA		GUU
	CAG		CCC	Chain initiation	
Glu	GAA		CCG	fMet (*N*-formyl-methionine)	AUG
	GAG		CCU	Chain termination	UAA
Gly	GGA				UAG
	GGC				UGA
	GGG				
	GGU				

arms through base pairing along the chain (Fig. 25.15). One arm always terminates in the sequence cytosine–cytosine–adenine (CCA). It is to this arm that a specific amino acid becomes attached *through an ester* linkage to the 3′-OH of the terminal adenosine. This attachment reaction is catalyzed by an enzyme that is specific for the tRNA and for the amino acid. The specificity may grow out of the enzyme's ability to recognize base sequences along other arms of the tRNA.

At the loop of still another arm is a specific sequence of bases, called the **anticodon**. The anticodon is highly important because it allows the tRNA to bind with a specific site—called the **codon**—of mRNA. The order in which amino acids are brought by their tRNA units to the mRNA strand is determined by the sequence of codons. This sequence, therefore, constitutes a genetic message. Individual units of that message (the individual words, each corresponding to an amino acid) are triplets of nucleotides.

25.5D The Genetic Code

The triplets of nucleotides (the codons) on mRNA are the genetic code (see Table 25.2). The code must be in the form of three bases, not one or two, because there are 20 different amino acids used in protein synthesis but there are only four different bases in mRNA. If only two bases were used, there would be only 4^2, or 16, possible combinations, a number too small to accommodate all of the possible amino acids. However, with a three-base code, 4^3, or 64, different sequences are possible. This is far more than

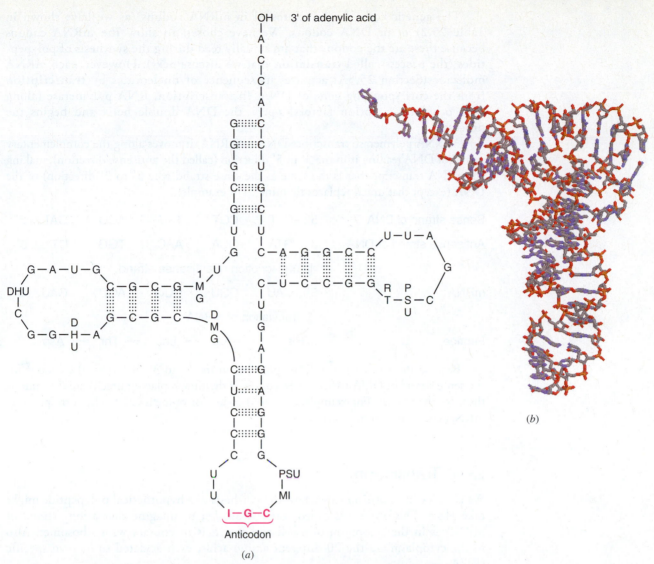

FIGURE 25.15 (a) Structure of a tRNA isolated from yeast that has the specific function of transfer-ring alanine residues. Transfer RNAs often contain unusual nucleosides. PSU = pseudouridine, RT = ribothymidine, MI = 1-methylinosine, I = inosine, DMG = N^2-methylguanosine, DHU = 4,5-dihydrouridine, 1MG = 1-methylguanosine. (b) The X-ray crystal structure of a phenylalanine–tRNA from yeast. (For part b, Protein Data Bank PDB ID: 4TNA. *http://www.pdb.or* Reprinted from Hingerty, E., Brown, R.S., Jack, A., Further refinement of the structure of yeast tRNA$_{Phe}$, *J. Mol. Biol.* **124**, p. 523. Copyright 1978, with permission of Elsevier.)

are needed, and it allows for multiple ways of specifying an amino acid. It also allows for sequences that punctuate protein synthesis, sequences that say, in effect, "start here" and "end here."

Both methionine (Met) and *N*-formylmethionine (fMet) have the same mRNA code (AUG); however, *N*-formylmethionine is carried by a different tRNA from that which carries methionine. *N*-Formylmethionine appears to be the first amino acid incorporated into the chain of proteins in bacteria, and the tRNA that carries fMet appears to be the punctuation mark that says "start here." Before the polypeptide synthesis is complete, *N*-formylmethionine is removed from the protein chain by an enzymatic hydrolysis.

N-Formylmethionine (fMet)

The genetic code can be expressed in mRNA codons (as we have shown in Table 25.2) or in DNA codons. We have chosen to show the mRNA codons because these are the codons that are actually read during the synthesis of polypeptides (the process called **translation** that we discuss next). However, each mRNA molecule (Section 25.5A) acquires its sequence of nucleotides by **transcription** from the corresponding gene of DNA. In transcription, RNA polymerase (along with other transcription factors) opens the DNA double helix and begins the process.

As RNA polymerase transcribes DNA to mRNA, it moves along the complementary strand of DNA reading it in the 3′ to 5′ direction (called the antisense direction), making an mRNA transcript that is the same as the sense strand (the 5′ to 3′ direction) of the DNA (except that uracil replaces thymine). For example:

Sense strand of DNA	5′... CAT	CGT	TTG	ACC	GAT ... 3′
Antisense strand of DNA	3′... GTA	GCA	AAC	TGG	CTA ... 5′

⇓ Transcription of antisense strand

mRNA	5′... CAU	CGU	UUG	ACC	GAU ... 3′

⇓ Translation of mRNA

Peptide	... His —	Arg —	Leu —	Thr —	Asp ...

Because the synthesis of mRNA proceeds in the 5′ to 3′ direction, the codons for the sense strand of DNA (with the exception of thymine replacing uracil) are the same as those for the mRNA. For example, one DNA codon for valine is GTA. The corresponding mRNA codon for valine is GUA.

25.5E Translation

We are now in a position to see how the synthesis of a hypothetical polypeptide might take place. This process is called **translation**. Let us imagine that a long strand of mRNA is in the cytoplasm of a cell and that it is in contact with ribosomes. Also in the cytoplasm are the 20 different amino acids, each acylated to its own specific tRNA.

As shown in Fig. 25.16, a tRNA bearing fMet uses its anticodon to associate with the proper codon (AUG) on that portion of mRNA that is in contact with a ribosome. The next triplet of bases on the mRNA chain in this figure is AAA; this is the codon that specifies lysine. A lysyl-tRNA with the matching anticodon UUU attaches itself to this site. The two amino acids, fMet and Lys, are now in the proper position for the 50S ribosome subunit to catalyze the formation of an amide bond between them, as shown in Fig. 25.16 (by the mechanism in Fig. 25.14). After this happens, the ribosome moves down the chain so that it is in contact with the next codon. This one, GUA, specifies valine. A tRNA bearing valine (and with the proper anticodon) binds itself to this site. Another peptide bond–forming reaction takes place attaching valine to the polypeptide chain. Then the whole process repeats itself again and again. The ribosome moves along the mRNA chain, other tRNAs move up with their amino acids, new peptide bonds are formed, and the polypeptide chain grows. At some point an enzymatic reaction removes fMet from the beginning of the chain. Finally, when the chain is the proper length, the ribosome reaches a punctuation mark, UAA, saying "stop here." The ribosome separates from the mRNA chain and so, too, does the protein.

Even before the polypeptide chain is fully grown, it begins to form its own specific secondary and tertiary structure. This happens because its primary structure is correct—its amino acids are ordered in just the right way. Hydrogen bonds form, giving rise to specific segments of α helix, pleated sheet, and coil or loop. Then the whole chain folds and bends; enzymes install disulfide linkages, so that when the chain is fully grown, the whole protein has just the shape it needs to do its job. (Predicting 2° and 3° protein

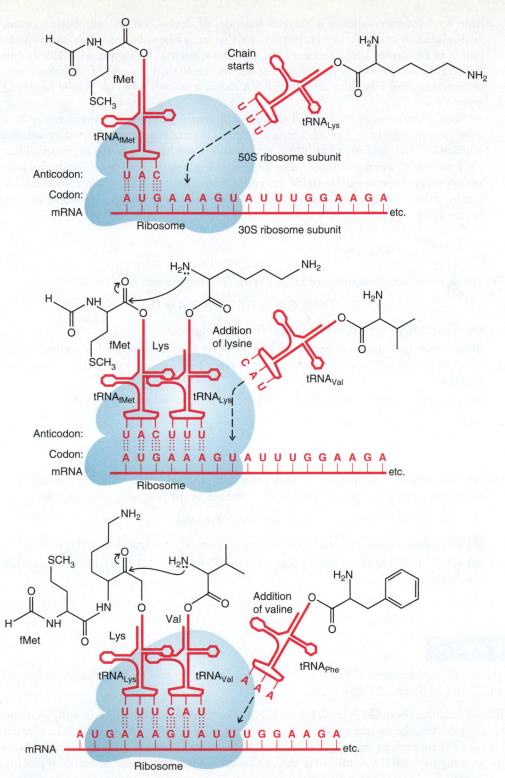

FIGURE 25.16 Step-by-step growth of a polypeptide chain with mRNA acting as a template. Transfer RNAs carry amino acid residues to the site of mRNA that is in contact with a ribosome. Codon–anticodon pairing occurs between mRNA and RNA at the ribosomal surface. An enzymatic reaction joins the amino acid residues through an amide linkage. After the first amide bond is formed, the ribosome moves to the next codon on mRNA. A new tRNA arrives, pairs, and transfers its amino acid residue to the growing peptide chain, and so on.

structure from amino acid sequence, however, remains a critical problem in structural biochemistry.)

In the meantime, other ribosomes nearer the beginning of the mRNA chain are already moving along, each one synthesizing another molecule of the polypeptide. The time required to synthesize a protein depends, of course, on the number of amino acid residues it contains, but indications are that each ribosome can cause 150 peptide bonds to be formed each minute. Thus, a protein, such as lysozyme, with 129 amino acid residues requires less than a minute for its synthesis. However, if four ribosomes are working their way along a single mRNA chain, a protein molecule can be produced every 13 s.

But why, we might ask, is all this protein synthesis necessary—particularly in a fully grown organism? The answer is that proteins are not permanent; they are not synthesized once and then left intact in the cell for the lifetime of the organism. They are synthesized when and where they are needed. Then they are taken apart, back to amino acids; enzymes disassemble enzymes. Some amino acids are metabolized for energy; others—new ones—come in from the food that is eaten, and the whole process begins again.

●●●

PRACTICE PROBLEM 25.10 The sense strand of a segment of DNA has the following sequence of bases:

5′...T G G G G G T T T T A C A G C...3′

(a) What mRNA sequence would result from this segment?

(b) Assume that the first base in this mRNA is the beginning of a codon. What order of amino acids would be translated into a polypeptide synthesized along this segment?

(c) Give anticodons for each tRNA associated with the translation in part (b).

●●●

PRACTICE PROBLEM 25.11 **(a)** Using the first codon given for each amino acid in Table 25.2, write the base sequence of mRNA that would translate the synthesis of the following pentapeptide:

Arg · Ile · Cys · Tyr · Val

(b) What base sequence in the DNA sense strand would correspond with this mRNA?

(c) What anticodons would appear in the tRNAs involved in the pentapeptide synthesis?

●●● **SOLVED PROBLEM 25.2**

Explain how an error of a single base in each strand of DNA could bring about the amino acid error that causes sickle-cell anemia (see "The Chemistry of..." box in Section 24.6B).

STRATEGY AND ANSWER: A change from GAA to GTA in DNA would lead to a change in mRNA from GAA to GUA (see Table 25.2). This change would result in the glutamic acid residue at position 6 in normal hemoglobin becoming valine (as it is in persons with sickle-cell anemia). Alternatively, a change from GAG to GTG in DNA would lead to a change in mRNA from GAG to GUG that would also result in valine replacing glutamic acid.

25.6 DETERMINING THE BASE SEQUENCE OF DNA: THE CHAIN-TERMINATING (DIDEOXYNUCLEOTIDE) METHOD

Certain aspects of the strategy used to sequence DNA resemble the methods used to sequence proteins. Both types of molecules require methods amenable to lengthy polymers, but in the case of DNA, a single DNA molecule is so long that it is absolutely necessary to cleave it into smaller, manageable fragments. Another similarity between DNA and proteins is that small sets of molecular building blocks comprise the structures of each, but in the case of DNA, only four nucleotide monomer units are involved instead of the 20 amino acid building blocks used to synthesize proteins. Finally, both proteins and nucleic acids are charged molecules that can be separated on the basis of size and charge using chromatography.

The first part of the process is accomplished by using enzymes called **restriction endonucleases**. These enzymes cleave double-stranded DNA at specific base sequences. Several hundred restriction endonucleases are now known. One, for example, called *Alu*I, cleaves the sequence AGCT between G and C. Another, called *Eco*R1, cleaves GAATTC between G and A. Most of the sites recognized by restriction enzymes have sequences of base pairs with the same order in both strands when read from the 5′ direction to the 3′ direction. For example,

$$5' \leftarrow G-A-A-T-T-C \rightarrow 3'$$
$$3' \leftarrow C-T-T-A-A-G \rightarrow 5'$$

Such sequences are known as **palindromes**. (Palindromes are words or sentences that read the same forward or backward. Examples are "radar" and "Madam, I'm Adam.")

Sequencing of the fragments (often called restriction fragments) can be done chemically or with the aid of enzymes. The first chemical method was introduced by A. Maxam and W. Gilbert (both of Harvard University); the **chain-terminating (dideoxynucleotide) method** was introduced in the same year by F. Sanger (Cambridge University). Essentially all DNA sequencing is currently done using an automated version of the chain-terminating method, which involves enzymatic reactions and 2′,3′-dideoxynucleotides.

25.6A DNA Sequencing by the Chain-Terminating (Dideoxynucleotide) Method

The chain-terminating method for sequencing DNA involves replicating DNA in a way that generates a family of partial copies that differ in length by one base pair. These partial copies of the parent DNA are separated according to length, and the terminal base on each strand is detected by a covalently attached fluorescent marker.

The mixture of partial copies of the target DNA is made by "poisoning" a replication reaction with a low concentration of unnatural nucleotides. The unnatural terminating nucleotides are 2′,3′-dideoxy analogues of the four natural nucleotides. Lacking the 3′-hydroxyl, each 2′,3′-dideoxynucleotide incorporated is incapable of forming a phosphodiester bond between its 3′ carbon and the next nucleotide that would be needed to continue the polymerization, and hence the chain terminates. Because a low concentration of the dideoxynucleotides is used, only occasionally is a dideoxynucleotide incorporated at random into the growing chains, and thus DNA molecules of essentially all different lengths are synthesized from the parent DNA.

Each terminating dideoxynucleotide is labeled with a fluorescent dye that gives a specific color depending on the base carried by that terminating nucleotide. (An alternate method is to label the *primer*, a short oligonucleotide sequence used to initiate replication of the specific DNA, with specific fluorescent dyes, instead of the dideoxynucleotide terminators, but the general method is the same.) One of the dye systems in use (patented by ABI) consists of a donor chromophore that is initially excited by the laser and which then transfers its energy to an acceptor moiety which produces the observed fluorescence. The donor is tethered to the dideoxynucleotide by a short linker.

GILBERT AND SANGER shared the Nobel Prize in Chemistry in 1980 with Paul Berg for their work on nucleic acids. Sanger (Section 24.5B), who pioneered the sequencing of proteins, had won an earlier Nobel Prize in 1958 for the determination of the structure of insulin.

A 2'3'-Dideoxynucleotide, Linker, and Fluorescent Dye Moiety Like Those Used in Fluorescence-Tagged Dideoxynucleotide DNA Sequencing Reactions

2',3'-Dideoxycytosine triphosphate

Lack of 3'-hydroxyl causes chain termination.

Linker between nucleotide and dye

Fluorescence donor moiety

Fluorescence acceptor moiety

Donor–acceptor fluorescent dye

The replication reaction used to generate the partial DNA copies is similar but not identical to the polymerase chain reaction (PCR) method (Section 25.8). In the dideoxy sequencing method only one primer sequence of DNA is used, and hence only one strand of the DNA is copied, whereas in the PCR, two primers are used and both strands are copied simultaneously. Furthermore, in sequencing reactions the chains are deliberately terminated by addition of the dideoxy nucleotides.

Capillary electrophoresis is the method most commonly used to separate the mixture of partial DNAs that results from a sequencing reaction. Capillary electrophoresis separates the DNAs on the basis of size and charge, allowing nucleotides that differ by only one base length to be resolved. Computerized acquisition of fluorescence data as the differently terminated DNAs pass the detector generates a four-color chromatogram, wherein each consecutive peak represents a DNA molecule one nucleotide longer than the previous one. The color of each peak represents the terminating nucleotide in that molecule. Since each of the four types of dideoxy terminating bases fluoresces a different color, the sequence of nucleotides in the DNA can be read directly. An example of sequence data from this kind of system is shown in Fig. 25.17.

Use of automated methods for DNA sequencing represents an exponential increase in speed over manual methods employing vertical slab polyacrylamide gel electrophoresis (Fig. 25.18). Only a few thousand bases per day (at most) could be sequenced by a person using the manual method. Now it is possible for a single machine running parallel and continuous analyses to sequence almost 3 million bases per day using automated capillary electrophoresis and laser fluorescence detection. As an added benefit, the ease of DNA sequencing often makes it easier to determine the sequence of a protein by the sequence of all or part of its corresponding gene, rather than by sequencing the protein itself (see Section 24.5).

The development of high-throughput methods for sequencing DNA is largely responsible for the remarkable success achieved in the Human Genome Project. Sequencing the 3 billion base pairs in the human genome could never have been completed before 2003 and the 50th anniversary of Watson and Crick's elucidation of the structure of DNA had high-throughput sequencing methods not come into existence.

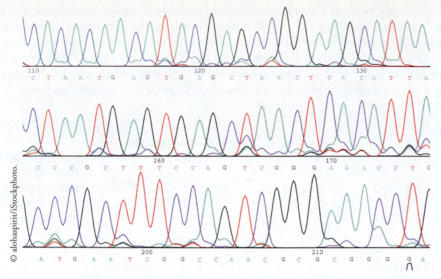

FIGURE 25.17 Example of data from an automated DNA sequencer.

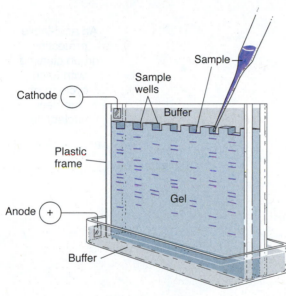

FIGURE 25.18 An apparatus for gel electrophoresis. Samples are applied in the slots at the top of the gel. Application of a voltage difference causes the samples to move. The samples move in parallel lanes. (Reprinted with permission of John Wiley & Sons, Inc., from Voet, D. and Voet, J. G., *Biochemistry*, Second Edition. © 1995 Voet, D. and Voet, J. G.)

25.7 LABORATORY SYNTHESIS OF OLIGONUCLEOTIDES

Synthetic oligonucleotides are needed for a variety of purposes. One of the most important and common uses of synthetic oligonucleotides is as primers for nucleic acid sequencing and for PCR (Section 25.8). Another important application is in the research and development of **antisense oligonucleotides**, which hold potential as therapies for a variety of diseases. An antisense oligonucleotide is one that has a sequence complementary to the coding sequence in a DNA or RNA molecule. Synthetic oligonucleotides that bind tightly to DNA or mRNA sequences from a virus, bacterium, or other disease condition may be able to shut down expression of the target protein associated with those conditions. For example, if the sense portion of DNA in a gene reads

A—G—A—C—C—G—T—G—G

the antisense oligonucleotide would read

T—C—T—G—G—C—A—C—C

The ability to deactivate specific genes in this way holds great medical promise. Many viruses and bacteria, during their life cycles, use a method like this to regulate some of their own genes. The hope, therefore, is to synthesize antisense oligonucleotides that will seek out and destroy viruses in a person's cells by binding with crucial sequences of the viral DNA or RNA. Synthesis of such oligonucleotides is an active area of research today and is directed at many viral diseases, including AIDS, as well as lung and other forms of cancer.

Current methods for **oligonucleotide synthesis** are similar to those used to synthesize proteins, including the use of automated solid-phase techniques (Section 24.7D). A suitably protected nucleotide is attached to a solid phase called a "controlled pore glass," or CPG (Fig. 25.19), through a linkage that can ultimately be cleaved. The next protected nucleotide in the form of a **phosphoramidite** is added, and coupling is brought about by a coupling agent, usually 1,2,3,4-tetrazole. The phosphite triester that results from the

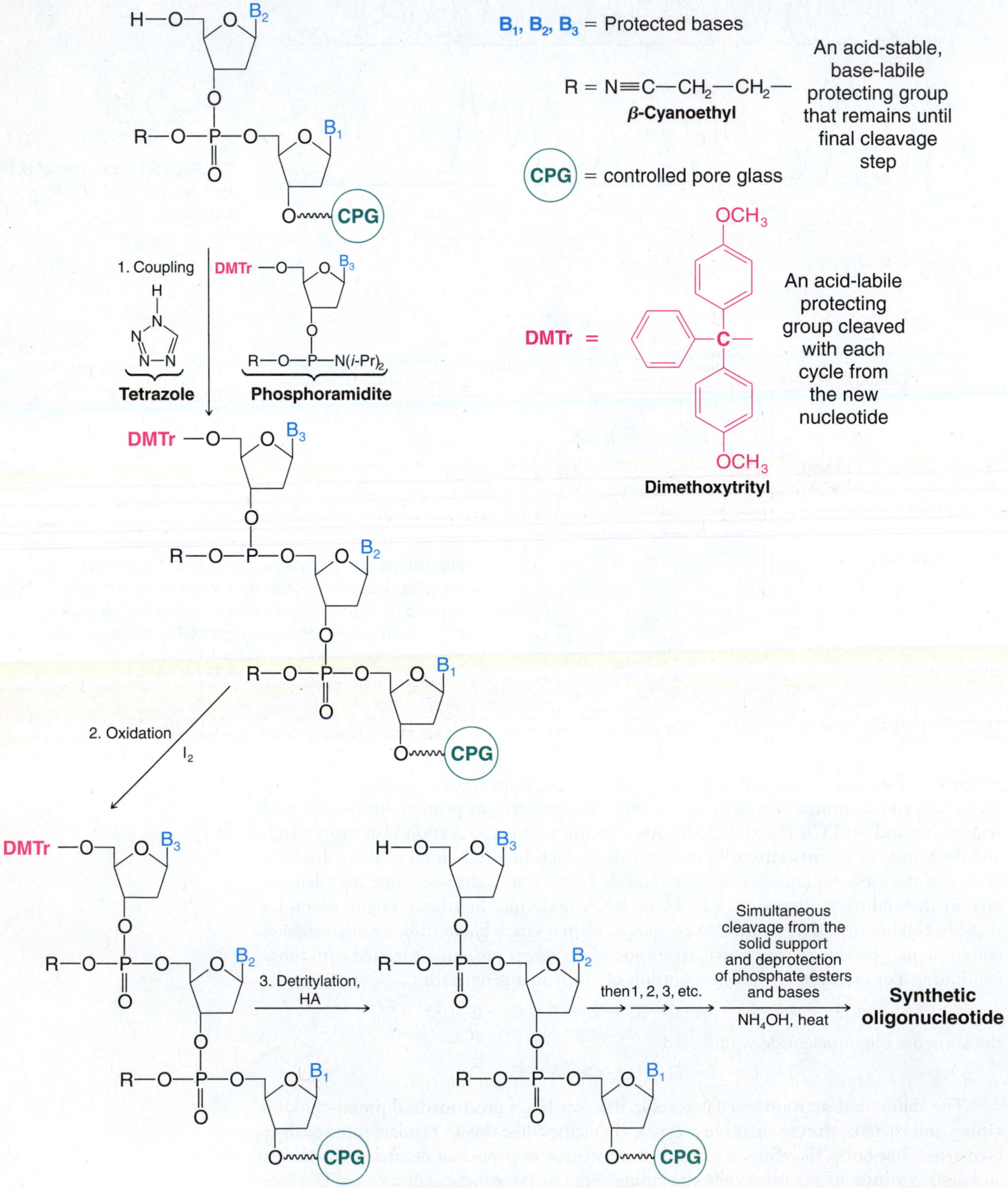

B_1, B_2, B_3 = Protected bases

$R = N{\equiv}C{-}CH_2{-}CH_2{-}$
β-Cyanoethyl

An acid-stable, base-labile protecting group that remains until final cleavage step

CPG = controlled pore glass

DMTr =

An acid-labile protecting group cleaved with each cycle from the new nucleotide

Dimethoxytrityl

FIGURE 25.19 The steps involved in automated synthesis of oligonucleotides using the phosphoramidite coupling method.

coupling is oxidized to phosphate triester with iodine, producing a chain that has been lengthened by one nucleotide. The **dimethoxytrityl (DMTr)** group used to protect the 5′ end of the added nucleotide is removed by treatment with acid, and the steps **coupling, oxidation, detritylation**, as shown in Figure 25.19, are repeated. (All the steps are carried out in nonaqueous solvents.) With automatic synthesizers the process can be repeated at least 50 times and the time for a complete cycle is 40 min or less. The synthesis is monitored by spectrophotometric detection of the dimethoxytrityl cation as it is released in each cycle (much like the monitoring of Fmoc release in solid-phase peptide synthesis). After the desired oligonucleotide has been synthesized, it is released from the solid support and the various protecting groups, including those on the bases, are removed.

25.8 THE POLYMERASE CHAIN REACTION

The **polymerase chain reaction (PCR)** is an extraordinarily simple and effective method for exponentially multiplying (amplifying) the number of copies of a DNA molecule. Beginning with even just a single molecule of DNA, the PCR can generate 100 billion copies in a single afternoon. The reaction is easy to carry out: It requires only a miniscule sample of the target DNA (picogram quantities are sufficient), a supply of nucleotide triphosphate reagents and primers to build the new DNA, DNA polymerase to catalyze the reaction, and a device called a thermal cycler to control the reaction temperature and automatically repeat the reaction. The PCR has had a major effect on molecular biology. Perhaps its most important role has been in the sequencing of the human genome (Sections 25.6 and 25.9), but now virtually every aspect of research involving DNA involves the PCR at some point.

One of the original aims in developing the PCR was to use it in increasing the speed and effectiveness of prenatal diagnosis of sickle-cell anemia (Section 24.6B). It is now being applied to the prenatal diagnosis of a number of other genetic diseases, including muscular dystrophy and cystic fibrosis. Among infectious diseases, the PCR has been used to detect cytomegalovirus and the viruses that cause AIDS, certain cervical carcinomas, hepatitis, measles, and Epstein–Barr disease.

The PCR is a mainstay in forensic sciences as well, where it may be used to copy DNA from a trace sample of blood or semen or a hair left at the scene of a crime. It is also used in evolutionary biology and anthropology, where the DNA of interest may come from a 40,000-year-old woolly mammoth or the tissue of a mummy. It is also used to match families with lost relatives (see the chapter opening vignette). There is almost no area with biological significance that does not in some way have application for use of the PCR reaction.

The PCR was invented by Kary B. Mullis and developed by him and his co-workers at Cetus Corporation. It makes use of the enzyme DNA polymerase, discovered in 1955 by Arthur Kornberg and associates at Stanford University. In living cells, DNA polymerases help repair and replicate DNA. The PCR makes use of a particular property of DNA polymerases: their ability to attach additional nucleotides to a short oligonucleotide "primer" when the primer is bound to a complementary strand of DNA called a template. The nucleotides are attached at the 3′ end of the primer, and the nucleotide that the polymerase attaches will be the one that is complementary to the base in the adjacent position on the template strand. If the adjacent template nucleotide is G, the polymerase adds C to the primer; if the adjacent template nucleotide is A, then the polymerase adds T, and so on. Polymerase repeats this process again and again as long as the requisite nucleotides (as triphosphates) are present in the solution, until it reaches the 5′ end of the template.

Figure 25.20 shows one PCR cycle. The target DNA, a supply of nucleotide triphosphate monomers, DNA polymerase, and the appropriate oligonucleotide primers (one primer sequence for each 5′ to 3′ direction of the target double-stranded DNA) are added to a small reaction vessel. The mixture is briefly heated to approximately 90 °C to separate the DNA strands (denaturation); it is cooled to 50–60 °C to allow the primer sequences and DNA polymerase to bind to each of the separated strands (annealing); and it is warmed to about 70 °C to extend each strand by polymerase-catalyzed condensation of nucleotide triphosphate monomers complementary to the parent DNA strand. Another cycle of the PCR begins by heating to separate the new collection of DNA molecules into single strands, cooling for the annealing step, and so on.

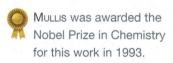

MULLIS was awarded the Nobel Prize in Chemistry for this work in 1993.

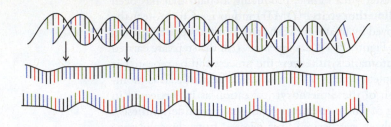

30–40 cycles of 3 steps:

Step 1: Denaturation of double-stranded DNA to single strands.
1 minute at approximately 90 °C

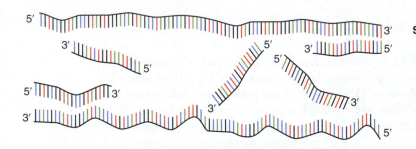

Step 2: Annealing of primers to each single-stranded DNA. Primers are needed with sequences complementary to both single strands.
45 seconds at 50–60 °C

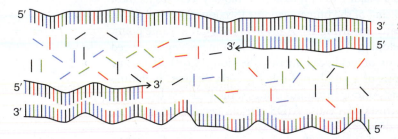

Step 3: Extension of the parent DNA strands with nucleotide triphosphate monomers from the reaction mixture.
2 minutes at approximately 70 °C

FIGURE 25.20 One cycle of the PCR. Heating separates the strands of DNA of the target to give two single-stranded templates. Primers, designated to complement the nucleotide sequences flanking the targets, anneal to each strand. DNA polymerase, in the presence of nucleotide triphosphates, catalyzes the synthesis of two pieces of DNA, each identical to the original target DNA. *(Used with permisson from Andy Vierstraete, University of Ghent.)*

Each cycle, taking only a few minutes, doubles the amount of target DNA that existed prior to that step (Fig. 25.21). The result is an exponential increase in the amount of DNA over time. After n cycles, the DNA will have been replicated 2^n times—after 10 cycles there is roughly 1000 times as much DNA; after 20 cycles roughly 1 million times as much; and so on. Thermal cycling machines can carry out approximately 20 PCR cycles per hour, resulting in billions of DNA copies over a single afternoon.

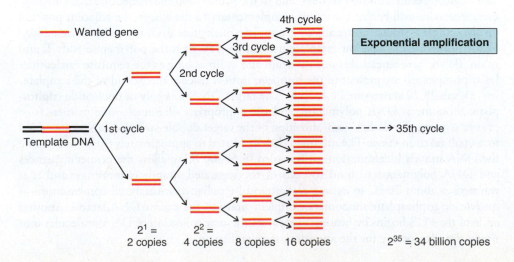

FIGURE 25.21 Each cycle of the PCR doubles the number of copies of the target area. *(Used with permisson from Andy Vierstraete, University of Ghent.)*

Each application of PCR requires primers that are 10–20 nucleotides in length and whose sequences are complementary to short, conveniently located sequences flanking the target DNA sequence. The primer sequence is also chosen so that it is near sites that are cleavable with restriction enzymes. Once a researcher determines what primer sequence is needed, the primers are usually purchased from commercial suppliers who synthesize them on request using solid-phase oligonucleotide synthesis methods like that described in Section 25.7.

As an intriguing adjunct to the PCR story, it turns out that cross-fertilization between disparate research fields greatly assisted development of current PCR methods. In particular, the discovery of extremozymes, which are enzymes from organisms that live in high-temperature environments, has been of great use. DNA polymerases now typically used in PCR are heat-stable forms derived from thermophilic bacteria. Polymerases such as Taq polymerase, from the bacterium *Thermus aquaticus*, found in places such as geyser hot springs, and Vent$_R$^{TM}, from bacteria living near deep-sea thermal vents, are used. Use of extremozyme polymerases facilitates PCR by allowing elevated temperatures to be used for the DNA melting step without having to worry about denaturing the polymerase enzyme at the same time. All materials can therefore be present in the reaction mixture throughout the entire process. Furthermore, use of a higher temperature during the chain extension also leads to faster reaction rates. (See "The Chemistry of... Stereoselective Reductions of Carbonyl Groups," Section 12.3C, for another example of the use of high-temperature enzymes.)

Simon Terry/Photo Researchers, Inc.

Thermophilic bacteria, growing in hot springs like these at Yellowstone National Park, produce heat-stable enzymes called extremozymes that have proved useful for a variety of chemical processes.

25.9 SEQUENCING OF THE HUMAN GENOME: AN INSTRUCTION BOOK FOR THE MOLECULES OF LIFE

The announcement by scientists from the public Human Genome Project and Celera Genomics Company in June 2000 that sequencing of the approximately 3 billion base pairs in the human genome was complete marked the achievement of one of the most important and ambitious scientific endeavors ever undertaken. To accomplish this feat, data were pooled from thousands of scientists working around the world using tools including PCR (Section 25.8), dideoxynucleotide sequencing reactions (Section 25.6), capillary electrophoresis, laser-induced fluorescence, and supercomputers. What was ultimately produced is a transcript of our chromosomes that could be called an instruction book for the molecules of life.

But what do the instructions in the genome say? How can we best make use of the molecular instructions for life? Of the roughly 35,000 genes in our DNA, the function of only a small percentage of genes is understood. Discovering genes that can be used to benefit our human condition and the chemical means to turn them on or off presents some of the greatest opportunities and challenges for scientists of today and the future. Sequencing the genome was only the beginning of the story.

As the story unfolds, chemists will continue to add to the molecular archive of compounds used to probe our DNA. DNA microchips, with 10,000 or more short diagnostic sequences of DNA chemically bonded to their surface in predefined arrays, will be used to test DNA samples for thousands of possible genetic conditions in a single assay. With the map of our genome in hand, great libraries of potential drugs will be tested against genetic targets to discover more molecules that either promote or inhibit expression of key gene products. Sequencing of the genome will also accelerate development of molecules that interact with proteins, the products of gene expression. Knowledge of

"This structure has novel features, which are of considerable biological importance." James Watson, one of the scientists who determined the structure of DNA.

the genome sequence will expedite identification of the genes coding for interesting proteins, thus allowing these proteins to be expressed in virtually limitless quantities. With an ample supply of target proteins available, the challenges of solving three-dimensional protein structures and understanding their functions will also be overcome more easily. Optimization of the structures of small organic molecules that interact with proteins will also occur more rapidly because the protein targets for these molecules will be available faster and in greater quantity. There is no doubt that the pace of research to develop new and useful organic molecules for interaction with gene and protein targets will increase dramatically now that the genome has been sequenced. The potential to use our chemical creativity in the fields of genomics and proteomics is immense.

WHY Do These Topics Matter?]

SELECTIVELY TARGETING A DNA SEQUENCE

Just as specific hydrogen bonds are the basis for the base-pairing of A with T and C with G in a DNA molecule, hydrogen bonds can also be used to bind small molecules to specific DNA sequences. The result is often a significant biochemical outcome. For instance, the unique cyclopropane-containing natural product (+)-CC-1065 possesses helicity that allows it to line up with the edge of the DNA minor groove. Once it encounters a domain rich in AT sequences, it can serve as an electrophile in an S_N2 reaction that leads to alkylation of an adenine residue as shown below. As a result, the cell is eventually destroyed. Thus, (+)-CC-1065 can serve as an antitumor agent if the target is a cancerous cell.

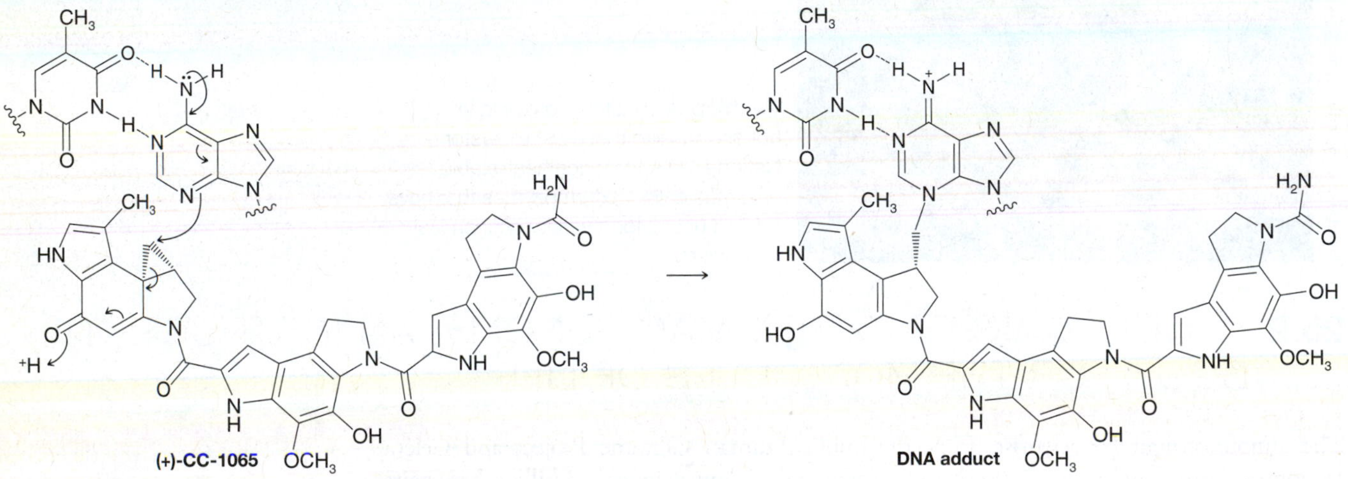

(+)-CC-1065   DNA adduct

This mechanism of action is not unique. In fact, there are several natural products that can similarly "read" the edge of the minor groove of DNA through hydrogen bond interactions. One such compound is distamycin A, which, as shown below, contains three peptide bond–linked pyrroles that can also target AT-rich regions of DNA.

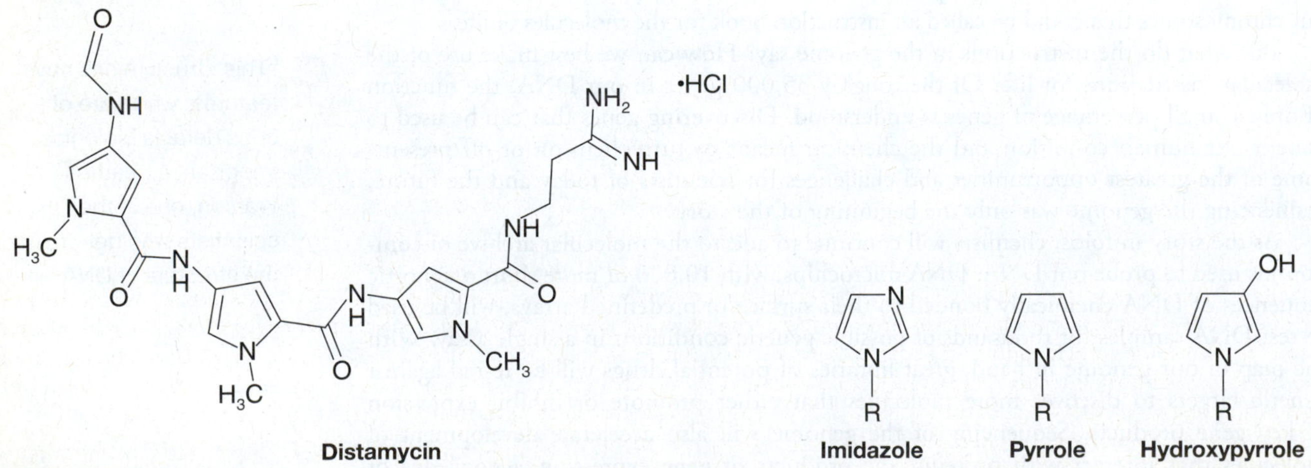

Distamycin   **Imidazole**   **Pyrrole**   **Hydroxypyrrole**

This molecule, however, has proved highly significant in that it served as the main inspiration for chemists to go beyond natural products and develop a set of molecules with the power to literally read, or differentiate, not only AT-rich sequences, but any specific DNA target sequence desired. The leader of these efforts has been Peter Dervan of the California Institute of Technology. Over a period spanning two decades, his team developed a group of molecules reminiscent of distamycin that contain pyrrole, imidazole, and hydroxypyrrole rings in two separate domains linked by a flexible tether. These compounds can read any of the main DNA sequences as shown on the previous page based on the use of A, C, G, and T as the purines and pyrimidines comprising the backbone structure, and you can learn more about their specific structures in reference 2 cited below.

To give a sense of the significance of this finding, a DNA segment containing 8 base pairs has 32,896 different possible sequences. Rather than having to identify thousands of different and distinct solutions for selectively targeting such an array sequences, this solution provides a common, predictable system that can be easily tailored to target any of these possible sequences at will simply by changing the positioning of these three heterocyclic systems within the two arms of the molecules. Current work is directed toward determining whether drugs can be combined with such sequence-specific molecules to provide novel treatments. While time will tell if new medicines will result, for now it is satisfying to see how natural products, first chemical principles like hydrogen bonding, and thoughtful molecular design can be combined to do something that even nature does not appear to be able to accomplish!

To learn more about these topics, see:

1. Boger, D. L.; Johnson, D. S. "CC-1065 and the duocarmycins: Unraveling the keys to a new class of naturally derived DNA alkylating agents." *Proc. Natl. Acad. Sci. USA* **1995**, *92*, 3642–3649.
2. Dervan, P. B. "Molecular recognition of DNA by small molecules." *Bioorg. Med. Chem.* **2001**, *9*, 2215–2135.

SUMMARY AND REVIEW TOOLS

The study aids for this chapter include key terms and concepts, which are highlighted in bold, blue text within the chapter, defined in the Glossary at the back of the book, and which have hyperlinked definitions in the accompanying *WileyPLUS* course (www.wileyplus.com).

PROBLEMS

Note to Instructors: Many of the homework problems are available for assignment via *WileyPLUS*, an online teaching and learning solution.

NUCLEIC ACID STRUCTURE

25.12 Write the structure of the RNA dinucleotide G–C in which G has a free 5′-hydroxyl group and C has a free 3′-hydroxyl group.

25.13 Write the structure of the DNA dinucleotide T–A in which T has a free 5′-hydroxyl group and A has a free 3′-hydroxyl group.

MECHANISMS

25.14 The example of a silyl–Hilbert–Johnson nucleosidation reaction in Section 25.3 is presumed to involve an intermediate ribosyl cation that is stabilized by intramolecular interactions involving the **C2** benzoyl group. This intermediate blocks attack by the heterocyclic base from the α face of the ribose ring but allows attack on the β face, as required for formation of the desired product. Propose a structure for the ribosyl cation intermediate that explains the stereoselective bonding of the base.

25.15 (a) Mitomycin is a clinically used antitumor antibiotic that acts by disrupting DNA synthesis through covalent bond-forming reactions with deoxyguanosine in DNA. Maria Tomasz (Hunter College) and others have shown that alkylation of DNA by mitomycin occurs by a complex series of mechanistic steps. The process begins with reduction of the quinone ring in mitomycin to its hydroquinone form, followed by elimination of methanol from the adjacent ring to form an intermediate called leuco-aziridinomitosene. One of the paths by which leuco-aziridinomitosene alkylates DNA involves protonation and opening of the three-membered aziridine ring, resulting in an intermediate cation that is resonance stabilized by the hydroquinone group. Attack on the cation by N2 of a deoxyguanosine residue leads to a monoalkylated DNA product, as shown in the scheme. Write a detailed mechanism to show how the ring opening might occur, including resonance forms for the cation intermediate, followed by nucleophilic attack by DNA. (Intra- or interstrand cross-linking of DNA can further occur by reaction of another deoxyguanosine residue to displace the carbamoyl group of the initial mitosene–DNA monoadduct. A cross-linked adduct is also shown.) **(b)** 1-Dihydromitosene A is sometimes formed from the cation intermediate in part (a) by loss of a proton and tautomerization. Propose a detailed mechanism for the formation of 1-dihydromitosene A from the resonance-stabilized cation of part (a).

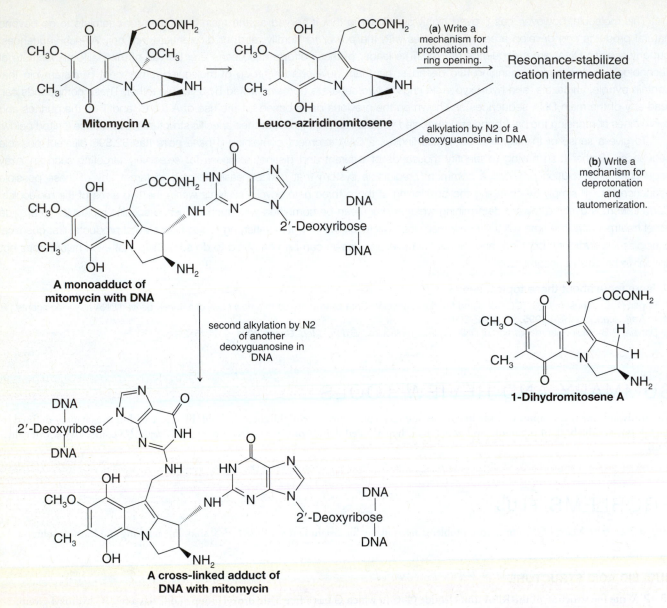

25.16 As described in Section 25.5B, acid–base catalysis is believed to be the mechanism by which ribosomes catalyze the formation of peptide bonds in the process of protein translation. Key to this proposal is assistance by the N3 nitrogen (highlighted in the scheme on the next page) of a nearby adenine in the ribosome for the removal of a proton from the α-amino group of the amino acid adding to the growing peptide chain (Fig. 25.14). The ability of this adenine group to remove the proton is, in turn, apparently facilitated by relay of charge made possible by other nearby groups in the ribosome. The constellation of these groups is shown in the scheme. Draw mechanism arrows to show formation of a resonance contributor wherein the adenine group could carry a formal negative charge, thereby facilitating its removal of the α-amino proton of the amino acid. (The true electronic structure of these groups is not accurately represented by any single resonance contributor, of course. A hybrid of the contributing resonance structures weighted according to stability would best reflect the true structure.)

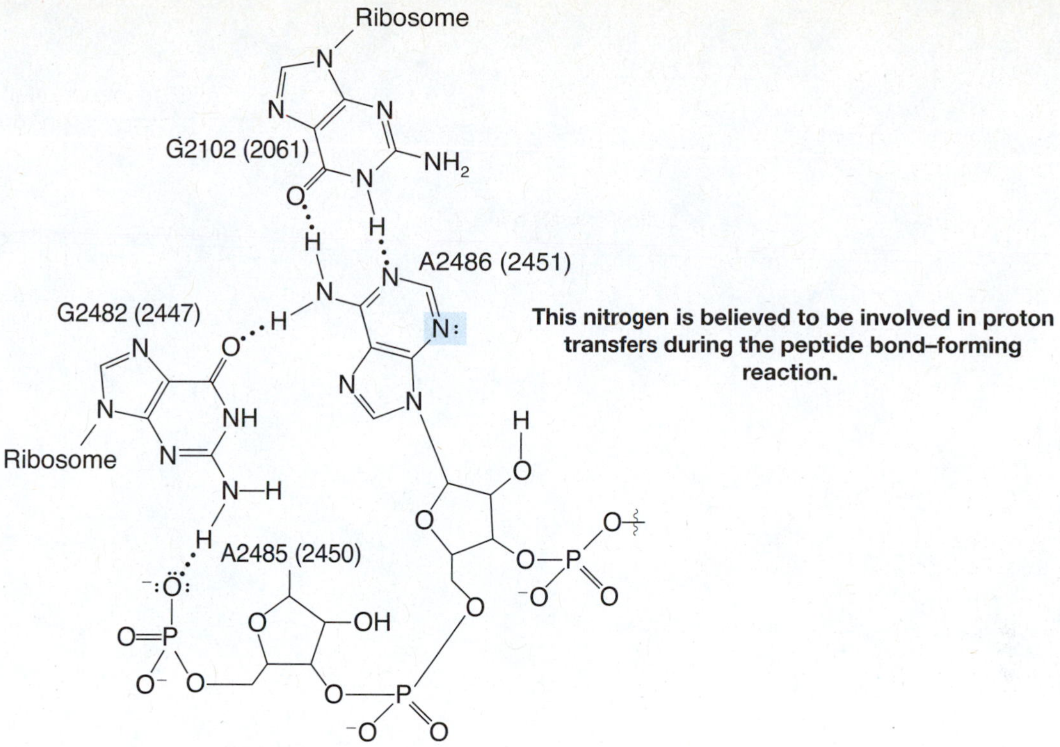

This nitrogen is believed to be involved in proton transfers during the peptide bond–forming reaction.

G2102 (2061)

G2482 (2447)

A2486 (2451)

Ribosome

A2485 (2450)

LEARNING GROUP PROBLEM

Research suggests that expression of certain genes is controlled by conversion of some cytosine bases in the genome to 5-methylcytosine by an enzyme called DNA methyltransferase. Cytosine methylation may be a means by which some genes are turned off as cells differentiate during growth and development. It may also play a role in some cancer processes and in defending the genome from foreign DNA such as viral genes. Measuring the level of methylation in DNA is an important analytical process. One method for measuring cytosine methylation is known as methylation-specific PCR. This technique requires that all unmethylated cytosines in a sample of DNA be converted to uracil by deamination of the C4 amino group in the unmethylated cytosines. This is accomplished by treating the DNA with sodium bisulfite ($NaHSO_3$) to form a bisulfite addition product with its unmethylated cytosine residues. The cytosine sulfonates that result are then subjected to hydrolysis conditions that convert the C4 amino group to a carbonyl group, resulting in uracil sulfonate. Finally, treatment with base causes elimination of the sulfonate group to produce uracil. The modified DNA is then amplified by PCR using primers designed to distinguish DNA with methylated cytosine from cytosine-to-uracil converted bases.

Write detailed mechanisms for the reactions used to convert cytosine to uracil by the above sequence of steps.

[ANSWERS TO SELECTED PROBLEMS]

CHAPTER 1

1.15 (a) and (d); (b) and (e); and (c) and (f).

1.27 (a), (c), (d), (f), (g), and (h) have tetrahedral geometry; (b) is linear; (e) is trigonal planar.

1.35 (a), (g), (i), (l), represent different compounds that are not isomeric; (b–e), (h), (j), (m), (n), (o) represent the same compound; (f), (k), (p) represent constitutional isomers.

1.42 (a) The structures differ in the positions of the nuclei. (b) The anions are resonance structures.

1.44 (a) A negative charge; (b) a negative charge; (c) trigonal pyramidal.

CHAPTER 2

2.11 (c) Propyl bromide; (d) isopropyl fluoride; (e) phenyl iodide.

2.14 (a) (b)
(e) diisopropyl ether.

2.25 (a) (b)
(c)

2.29 (a) ketone; (c) 2° alcohol; (e) 2° alcohol.

2.30 (a) 3 alkene groups, and a 2° alcohol; (c) phenyl and 1° amine; (e) phenyl, ester and 3° amine; (g) alkene and 2 ester groups.

2.35 (f)

2.53 Ester

CHAPTER 3

3.3 (a), (c), (d), and (f) are Lewis bases; (b) and (e) are Lewis acids.

3.5 (a) $[H_3O^+] = [HCO_2^-] = .0042\ M$; (b) Ionization $= 4.2\%$.

3.6 (a) $pK_a = 7$; (b) $pK_a = -0.7$; (c) Because the acid with a $pK_a = 5$ has a larger K_a, it is the stronger acid.

3.8 The pK_a of the methylaminium ion is equal to 10.6 (Section 3.6B). Because the pK_a of the anilinium ion is equal to 4.6, the anilinium ion is a stronger acid than the methylaminium ion, and aniline ($C_6H_5NH_2$) is a weaker base than methylamine (CH_3NH_2).

3.14 (a) $CHCl_2CO_2H$ would be the stronger acid because the electron-withdrawing inductive effect of two chlorine atoms would make its hydroxyl proton more positive. (c) CH_2FCO_2H would be the stronger acid because a fluorine atom is more electronegative than a bromine atom and would be more electron withdrawing.

3.28 (a) $pK_a = 3.752$; (b) $K_a = 10^{-13}$.

CHAPTER 4

4.8 (a) (1,1-dimethylethyl)cyclopentane or *tert*-butyl-cyclopentane; (c) butylcyclohexane; (e) 2-chlorocyclopentanol.

4.9 (a) 2-Chlorobicyclo[1.1.0]butane; (c) bicyclo[2.1.1]hexane; (e) 2-methylbicyclo[2.2.2]octane.

4.10 (a) *trans*-3-Heptene; (c) 4-ethyl-2-methyl-1-hexene

4.11

(a)

(c)

(e)

(g)   (i)

4.12 1-Hexyne, 2-Hexyne, 3-Hexyne, 4-Methyl-1-pentyne, 4-Methyl-2-pentyne, 3,3-Dimethyl-1-butyne

(R)-3-Methyl-1-pentyne   **(S)-3-Methyl-1-pentyne**

4.24 (a) 5-ethyl-7-isopropyl-2,3-dimethyldecane; (c) 4-bromo-6-chloro-3-methyloctane; (e) 2-Bromobicyclo[3.3.1]nonane; (g) 5,6-dimethyl-2-heptene

4.39 (a) Pentane would boil higher because its chain is unbranched. (c) 2-Chloropropane because it is more polar and has a higher molecular weight. (e) CH_3COCH_3 because it is more polar.

4.43

(a)

More stable conformation because both alkyl groups are equatorial

(b)

More stable because larger group is equatorial

(c)

More stable conformation because both alkyl groups are equatorial

(d)

More stable because larger group is equatorial

CHAPTER 5

5.1 (a) achiral; (c) chiral; (e) chiral.

5.2 (a) Yes; (c) no.

5.3 (a) They are the same. (b) They are enantiomers.

5.7 The following possess a plane of symmetry and are, therefore, achiral: screwdriver, baseball bat, hammer.

5.11

(a) $-Cl > -SH > -OH > -H$

(c) $-OH > -CHO > -CH_3 > -H$

(e) $-OCH_3 > -N(CH_3)_2 > -CH_3 > -H$

5.13 (a) enantiomers; (c) enantiomers.

5.19 (a) diastereomers; (c) no; (e) no.

5.21 (a) represents **A**; (b) represents **C**; (c) represents **B**.

5.23 **B** (2S,3S)-2,3-Dibromobutane; **C** (2R,3S)-2,3-Dibromobutane.

5.40 (a) same; (c) diastereomers; (e) same; (g) diastereomers; (i) same; (k) diastereomers; (m) diastereomers; (o) diastereomers; (q) same.

CHAPTER 6

6.6 (a) The reaction is S_N2 and, therefore, occurs with inversion of configuration. Consequently, the configuration of (+)-2-chlorobutane is opposite [i.e., (S)] to that of (−)-2-butanol [i.e., (R)]. (b) The configuration of (−)-2-iodobutane is (R).

6.14 Protic solvents are formic acid, formamide, ammonia, and ethylene glycol. The others are aprotic.

6.16 (a) CH_3O^-; (c) $(CH_3)_3P$.

6.20 (a) 1-Bromopropane would react more rapidly, because, being a primary halide, it is less hindered. (c) 1-Chlorobutane, because the carbon bearing the leaving group is less hindered than in 1-chloro-2-methylpropane. (e) 1-Chlorohexane because it is a primary halide. Phenyl halides are unreactive in S_N2 reactions.

6.21 (a) Reaction (1) because ethoxide ion is a stronger nucleophile than ethanol; (c) reaction (2) because triphenylphosphine, $(C_6H_5)_3P$, is a stronger nucleophile than triphenylamine. (Phosphorus atoms are larger than nitrogen atoms.)

6.22 (a) Reaction (2) because bromide ion is a better leaving group than chloride ion; (c) reaction (2) because the concentration of the substrate is twice that of reaction (1).

6.26 The better yield is obtained by using the secondary halide, 1-bromo-1-phenylethane, because the desired reaction is E2. Using the primary halide will result in substantial S_N2 reaction as well, producing the alcohol instead of the desired alkene.

6.38 (a) You should use a strong base, such as RO^-, at a higher temperature to bring about an E2 reaction. (b) Here we want an S_N1 reaction. We use ethanol as the solvent *and as the nucleophile*, and we carry out the reaction at a low temperature so that elimination is minimized.

CHAPTER 7

7.4 (a) 2,3-Dimethyl-2-butene would be the more stable because the double bond is tetrasubstituted. (c) *cis*-3-Hexene would be more stable because its double bond is disubstituted.

7.7 (a)

(trisubstituted, more stable) + **(monosubstituted, less stable)**

Major product   **Minor product**

7.25 (a) We designate the position of the double bond by using the *lower* of the two numbers of the doubly bonded carbon atoms, and the chain is numbered from the end nearer the double bond. The correct name is *trans*-2-pentene. (c) We use the lower number of the two doubly bonded carbon atoms to designate the position of the double bond. The correct name is 1-methylcyclohexene.

7.26

(a)   (c)

(e)   (g)

7.28 (a) (E)-3,5-Dimethyl-2-hexene; (c) 6-methyl-3-heptyne; (e) (3Z,5R)-5-chloro-3-hepten-6-yne.

7.43 Only the deuterium atom can assume the anti coplanar orientation necessary for an E2 reaction to occur.

CHAPTER 8

8.1 2-Bromo-1-iodopropane.

8.8 The order reflects the relative ease with which these alkenes accept a proton and form a carbocation. 2-Methylpropene reacts fastest because it leads to a 3° cation; ethene reacts slowest because it leads to a 1° cation.

8.25 By converting the 3-hexyne to *cis*-3-hexene using H_2/Ni_2B (P-2).

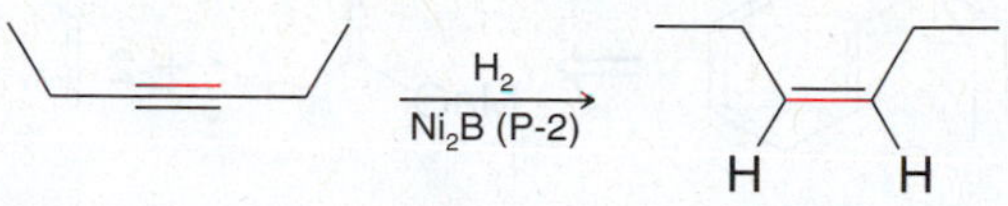

Then, addition of bromine to *cis*-3-hexene will yield (3*R*,4*R*), and (3*S*,4*S*)-3,4-dibromohexane as a racemic form.

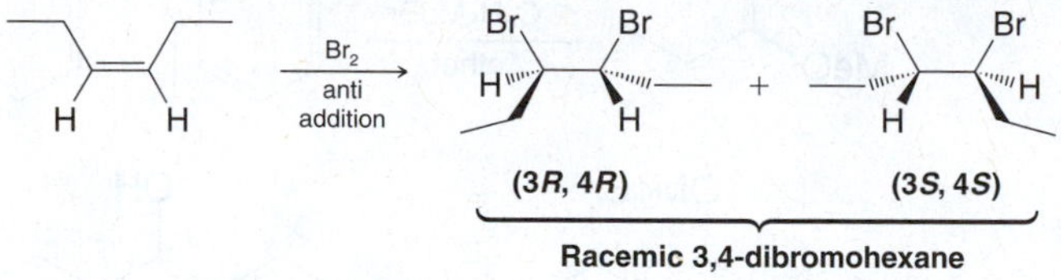

(3*R*, 4*R*) + (3*S*, 4*S*)

Racemic 3,4-dibromohexane

8.26 (a) 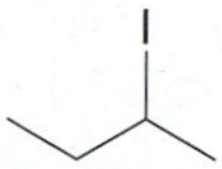   (b)    (e) 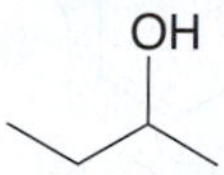 OH

(i) 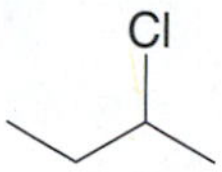 Cl   (j) 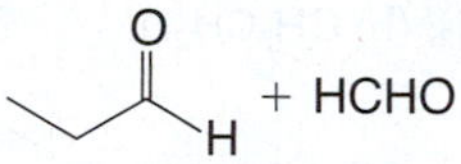 O + HCHO

8.29 (a) 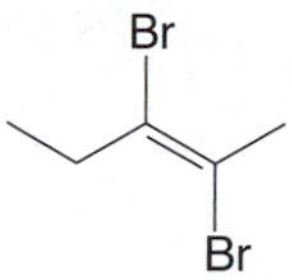 Br   (c) 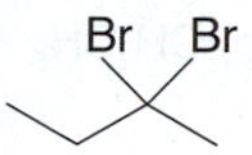 Br Br   (e)

Br

8.33 (a) 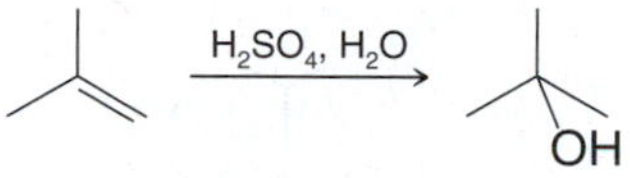 $\xrightarrow{H_2SO_4, H_2O}$ OH

(c) 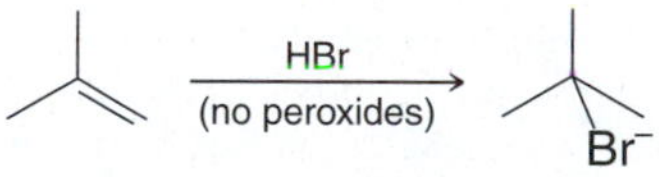 $\xrightarrow[\text{(no peroxides)}]{HBr}$ Br^-

(d) 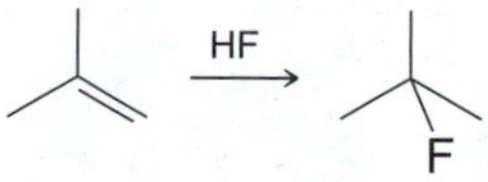 $\xrightarrow{HF}$ F

8.34 (a) 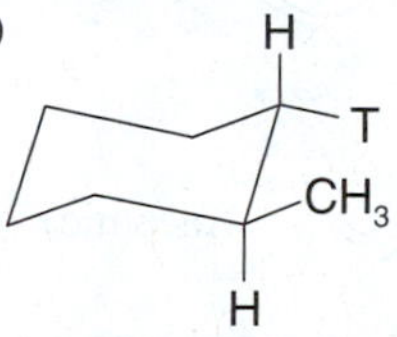 H T CH3 H

(c) 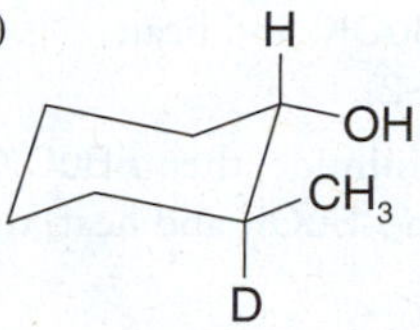 H OH CH3 D

8.64 H CH3 $\xrightarrow{H_2, Pt}$

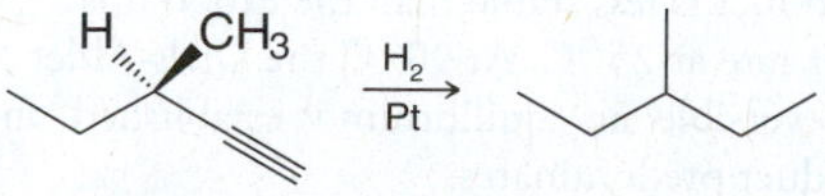

D   **E**

CHAPTER 9

9.4 (a) One; (b) two; (c) two; (d) one; (e) two; (f) two.

9.9 A doublet (3H) at relatively higher frequency; a quartet (1H) at relatively lower frequency.

9.10 A, CH_3CHICH_3; B, CH_3CHCl_2; C, $CH_2ClCH_2CH_2Cl$

9.25 G, Br

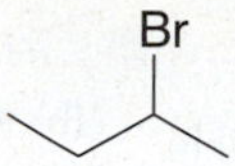

H, Br Br

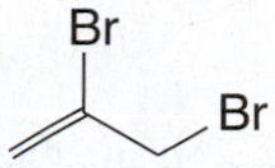

9.28 Q is bicyclo[2.2.1]hepta-2,5-diene.
R is bicyclo[2.2.1]heptane.

9.39 E is phenylacetylene.

CHAPTER 10

10.1

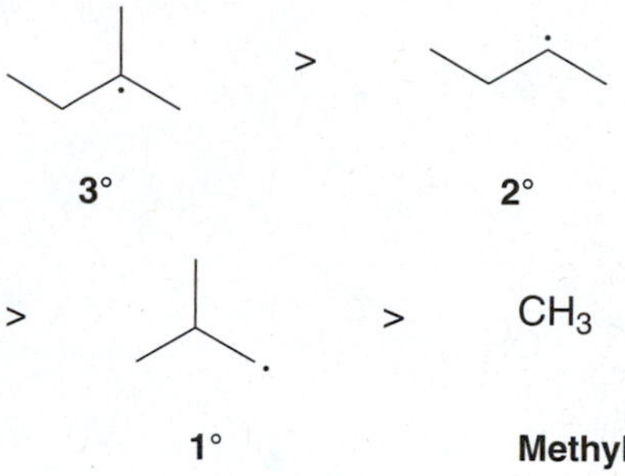

3°   >   **2°**

>   **1°**   > CH_3 **Methyl**

10.8 (a) Cyclopentane; (b) 2,2,3,3-tetramethylbutane

10.9 (a)

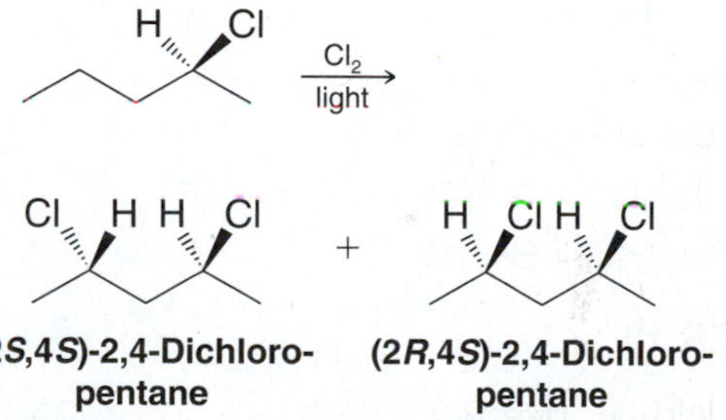

H Cl $\xrightarrow[\text{light}]{Cl_2}$

Cl H H Cl + H Cl H Cl

(2S,4S)-2,4-Dichloro-pentane   **(2R,4S)-2,4-Dichloro-pentane**

(c) No, (2*R*,4*S*)-2,4-dichloropentane is achiral because it is a meso compound. (It has a plane of symmetry passing through C3.)

(e) Yes, by fractional distillation or by gas–liquid chromatography. (Diastereomers have different physical properties. Therefore, the two isomers would have different vapor pressures.)

10.10 (a) The only fractions that would contain chiral molecules (as enantiomers) would be those containing 1-chloro-2-methylbutane and the two diastereomers of 2-chloro-3-methylbutane. These fractions would not show optical activity, however, because they would contain racemic forms of the enantiomers.

(b) Yes, the fractions containing 1-chloro-2-methylbutane and the two containing the 2-chloro-3-methylbutane diastereomers.

10.25

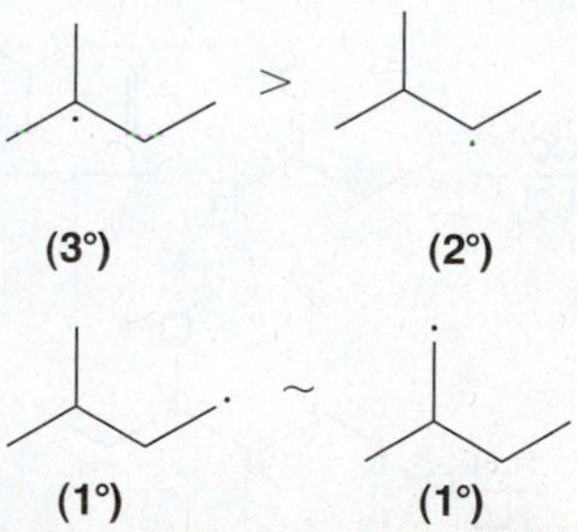

>

(3°)   **(2°)**

>   ~

(1°)   **(1°)**

CHAPTER 11

11.3 (a) 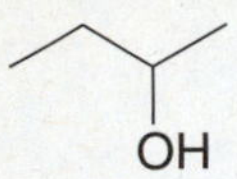 [structure] (b) 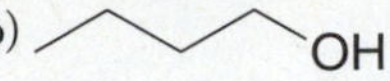 [structure]

(b) 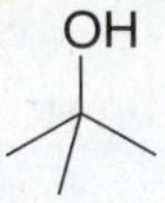 [structure]

(c) [structure]

$$\Rightarrow \text{MeO}\text{(ester)} + 2\ \text{(phenyl)MgBr}$$

11.10 Use an alcohol containing labeled oxygen. If all of the labeled oxygen appears in the sulfonate ester, then it can be concluded that the alcohol C—O bond does not break during the reaction.

11.25 (a) 3,3-Dimethyl-1-butanol; (c) 2-methyl-1,4-butanediol; (e) 1-methyl-2-cyclopenten-1-ol.

11.26 (a) 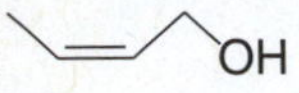 [structure]

(c) [structure]

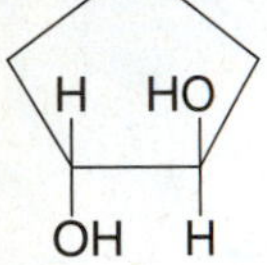

(e) 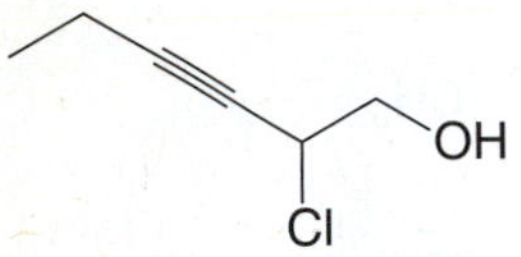 [structure]

(g) 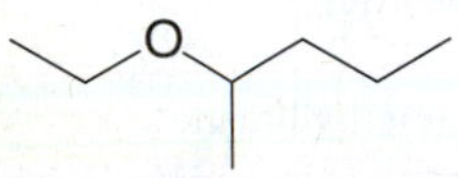 [structure]

(i) 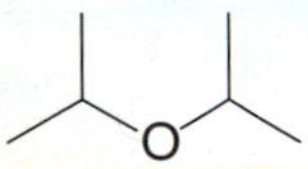 [structure]

11.33 (a) CH_3Br + [structure]

(c)  [structure]

$$\text{MeO}\text{(ester)} \xrightarrow[\text{ether}]{2\ C_6H_5MgBr} \text{(OMgBr structure)}$$

$$\text{(OMgBr structure)} \xrightarrow[\text{H}_2\text{O}]{\text{NH}_4^+} \text{(OH structure)}$$

CHAPTER 12

12.3 (a) $LiAlH_4$; (c) $NaBH_4$

12.4 (a) 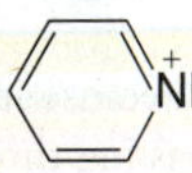 [structure] $\overset{+}{N}HCrO_3Cl^-$ (PCC)/CH_2Cl_2

(c) H_2CrO_4/acetone

12.10
(a) [structure] $\Rightarrow$ [structure] +

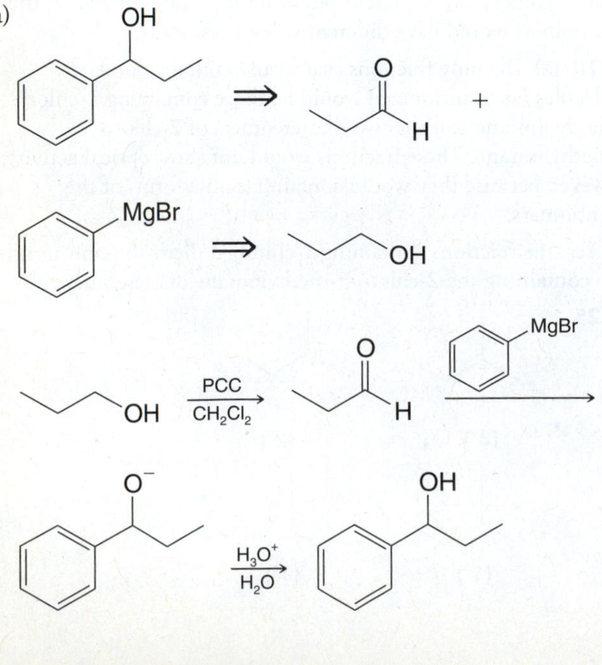

[structure] $\Rightarrow$ [structure]

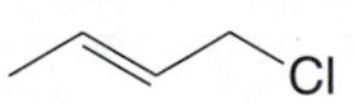

[structure] $\xrightarrow[CH_2Cl_2]{PCC}$ [structure] + [structure (MgBr)] $\longrightarrow$

[structure] $\xrightarrow[H_2O]{H_3O^+}$ [structure]

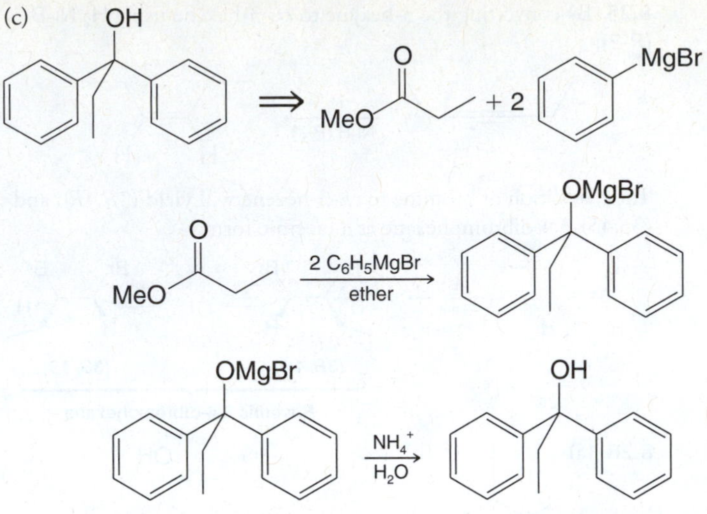

12.11 (a) CH_3CH_3; (b) CH_3CH_2D;

(c) 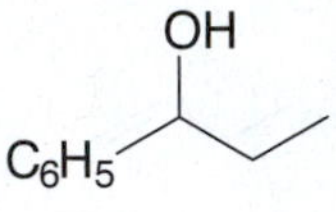 C_6H_5 [structure]   **(g)** CH_3CH_3 + [structure]

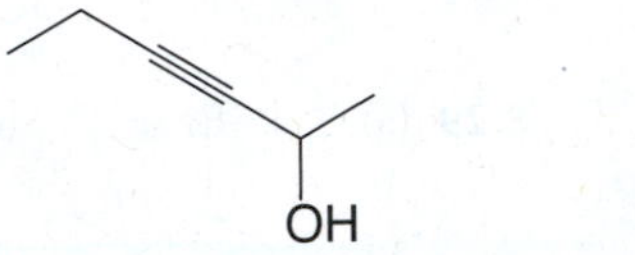

12.12 (a) 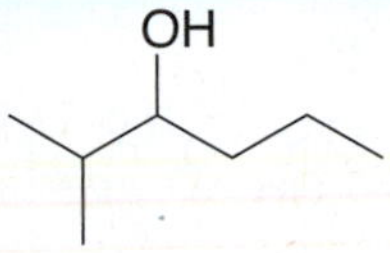 [structure]   (b) 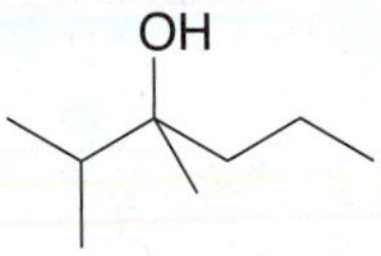 [structure]

(e)  [structure]

CHAPTER 13

13.2 (a) 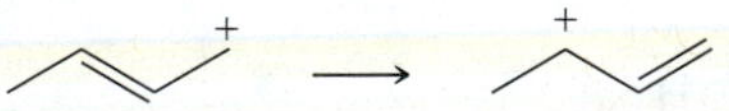 [structure] + [structure] $\longrightarrow$ [structure] +

(c) [structure] and [structure] (racemic)

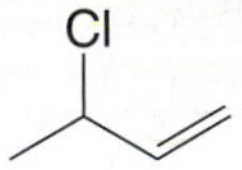

13.6 (b) 1,4-Cyclohexadiene and 1,4-pentadiene are isolated dienes.

13.18 (a) 1,4-Dibromobutane + t-BuOK, and heat;
(g) $HC{\equiv}CCH{=}CH_2$ + H_2, Ni_2B (P-2).

13.22 (a) 1-Butene + N-bromosuccinimide, then t-BuOK and heat; (e) cyclopentane + Br_2, hv, then t-BuOK and heat, then N-bromosuccinimide.

13.45 The endo adduct is less stable than the exo, but is produced at a faster rate at 25 °C. At 90 °C the Diels-Alder reaction becomes reversible; an equilibrium is established, and the more stable exo adduct predominates.

CHAPTER 14

14.1 (a) 4-Bromobenzoic acid (or p-bromobenzoic acid)
(b) 2-Benzyl-1.3-cyclohexadiene
(c) (2-chloro-2-pentyl) benzene
(d) Phenyl propyl ether

14.7 (a)

(b)

Tropylium bromide

These results suggest that the bonding in tropylium bromide is ionic; that is, it consists of a positive tropylium ion and a negative bromide ion.

14.9 The cyclopropenyl cation.

14.15 A, *o*-bromotoluene; **B**, *p*-bromotoluene; **C**, *m*-bromotoluene; **D**, benzyl bromide.

14.23 Hückel's rule should apply to both pentalene and heptalene. Pentalene's antiaromaticity can be attributed to its having 8 π electrons. Heptalene's lack of aromaticity can be attributed to its having 12 π electrons. Neither 8 nor 12 is a Hückel number.

14.25 The bridging —CH$_2$— group causes the 10 π electron ring system to become planar. This allows the ring to become aromatic.

14.28 (a) The cyclononatetraenyl anion, with 10 π electrons, obeys Hückel's rule.

14.31 A,   **B**,   **C**,

14.33 F,

CHAPTER 15

15.6 If the methyl group had no directive effect on the incoming electrophile, we would expect to obtain the products in purely statistical amounts. Since there are two ortho hydrogen atoms, two meta hydrogen atoms, and one para hydrogen, we would expect to get 40% ortho (2/5), 40% meta (2/5), and 20% para (1/5). Thus, we would expect that only 60% of the mixture of mononitrotoluenes would have the nitro group in the ortho or para position. And, we would expect to obtain 40% of *m*-nitrotoluene. In actuality, we get 96% of combined *o*- and *p*-nitrotoluene and only 4% *m*-nitrotoluene. This result shows the ortho–para directive effect of the methyl group.

15.9 (b) Structures such as the following compete with the benzene ring for the oxygen electrons, making them less available to the benzene ring.

(d) Structures such as the following compete with the benzene ring for the nitrogen electrons, making them less available to the benzene ring.

15.32 (a)

(c)

15.33 (a)

(c)

(e)

(g)

CHAPTER 16

16.2 (a) 1-Pentanol; (c) pentanal; (e) benzyl alcohol.

16.6 A hydride ion.

16.17 (b) CH$_3$CH$_2$Br + (C$_6$H$_5$)$_3$P, then strong base, then C$_6$H$_5$COCH$_3$; (d) CH$_3$I + (C$_6$H$_5$)$_3$P, then strong base, then cyclopentanone; (f) CH$_2$=CHCH$_2$Br + (C$_6$H$_5$)$_3$P, then strong base, then C$_6$H$_5$CHO.

16.23 (a) CH$_3$CH$_2$CH$_2$OH; (c) CH$_3$CH$_2$CH$_2$OH (h) CH$_3$CH$_2$CH=CHCH$_3$; (j) CH$_3$CH$_2$CO$_2^-$NH$_4^+$ + Ag↓ (l) CH$_3$CH$_2$CH=NNHCONH$_2$; (n) CH$_3$CH$_2$CO$_2^-$

16.49 X is

16.50 Y is 1-phenyl-2-butanone; **Z** is 4-phenyl-2-butanone.

CHAPTER 17

17.3 (a) CH_2FCO_2H; (c) $CH_3CH_2CHFCO_2H$;

(e)

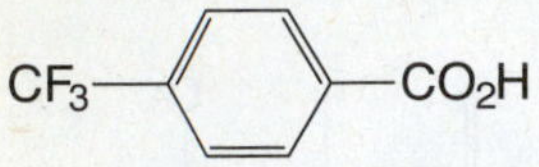

CF_3—⬡—CO_2H

17.6 (a) $C_6H_5CH_2Br$ + Mg in diethyl ether, then CO_2, then H_3O^+; (c) $CH_2\!=\!CHCH_2Br$ + Mg in diethyl ether, then CO_2, then H_3O^+.

17.7 (a), (c), and (e).

17.9 In the carboxyl group of benzoic acid.

17.14 (a) $(CH_3)_3CCO_2H$ + $SOCl_2$, then NH_3, then P_4O_{10}, heat;

(b)

17.22 (a)
    + HCl

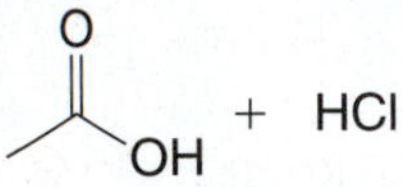

(b)

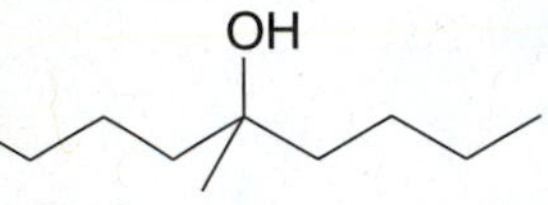

(c) 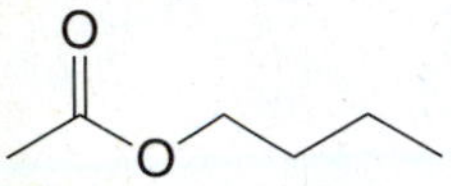   **(d)**

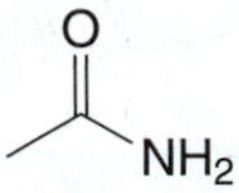

(e)
    +

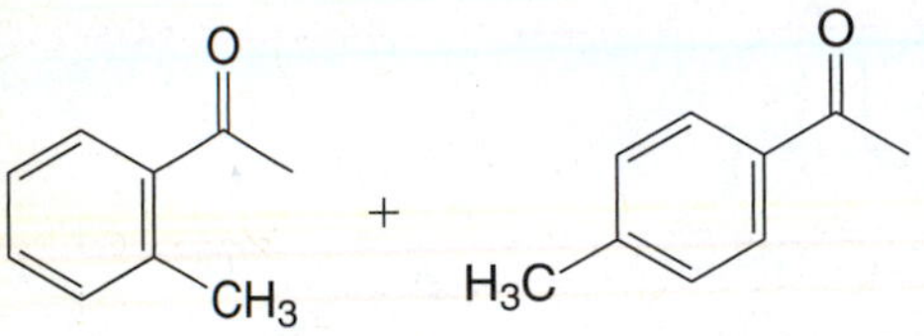

(f) 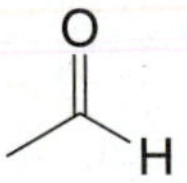   **(g)**

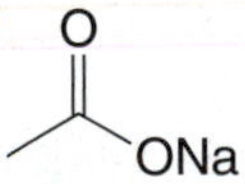

(h)
    +

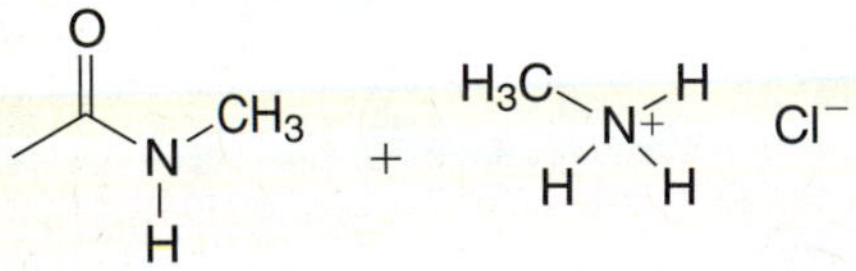

(i)
    +

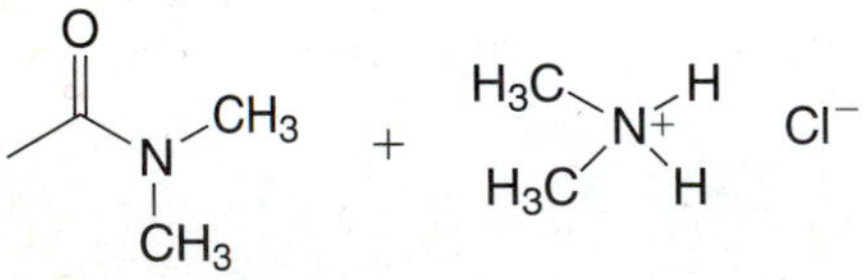

(j) 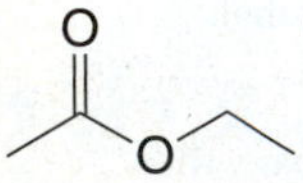   **(k)**

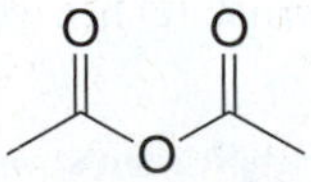

(l)

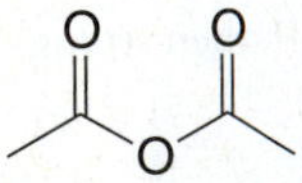

17.46 (a) Diethyl succinate; (c) ethyl phenylacetate; (e) ethyl chloroacetate.

17.47 **X** is diethyl malonate.

CHAPTER 18

18.1 The enol form is phenol. It is especially stable because it is aromatic.

18.2 No. 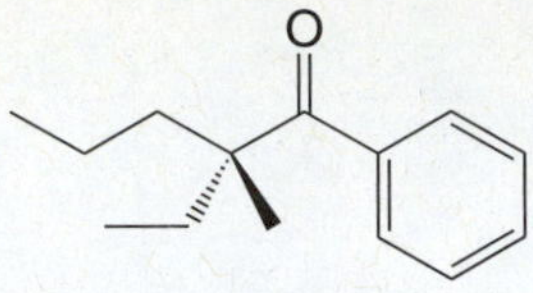
    does not have a hydrogen

attached to its α-carbon atom (which is a chirality center) and thus enol formation involving the chirality center is not possible.

With   the chirality center is a β carbon

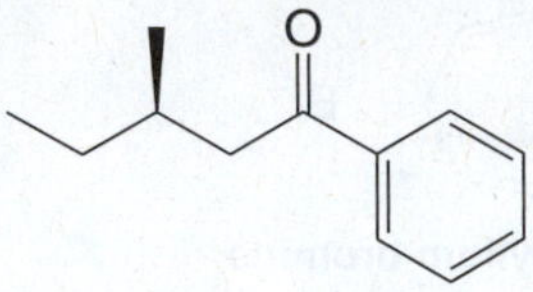

and thus enol formation does not affect it.

18.5 Base is consumed as the reaction takes place. A catalyst, by definition, is not consumed.

18.8 (a) Reactivity is the same as with any S_N2 reaction. With primary halides substitution is highly favored, with secondary halides elimination competes with substitution, and with tertiary halides elimination is the exclusive course of the reaction. (b) Acetoacetic ester and 2-methylpropene. (c) Bromobenzene is unreactive toward nucleophilic substitution.

18.10 Working backward

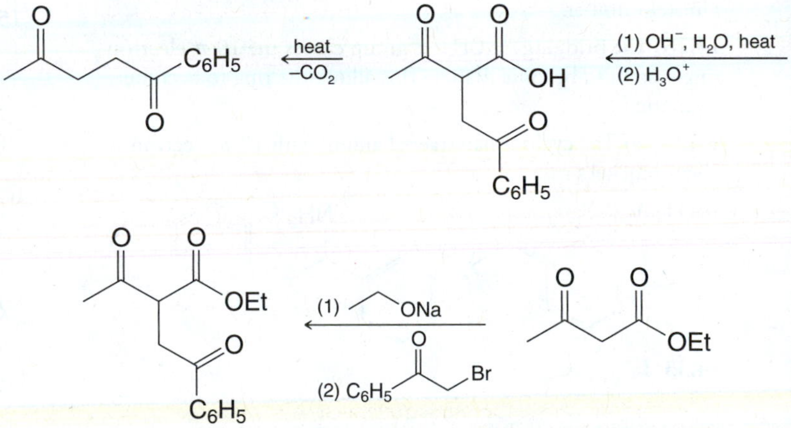

18.17 In a polar solvent, such as water, the keto form is stabilized by solvation. When the interaction with the solvent becomes minimal, the enol form achieves stability by internal hydrogen bonding.

18.25 (b) **D** is racemic *trans*-1,2-cyclopentanedicarboxylic acid, **E** is *cis*-1,2-cyclopentanedicarboxylic acid, a meso compound.

CHAPTER 19

19.3 (a)
    OEt

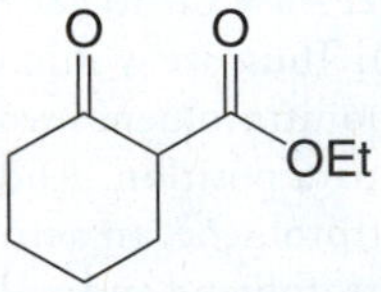

(b) To undergo a Dieckmann condensation, diethyl 1,5-pentanedioate would have to form a highly strained four-membered ring.

19.5 (a)
    EtO   OEt

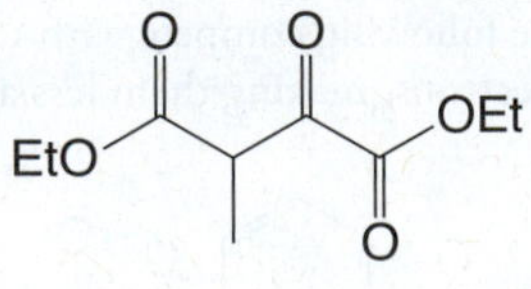

(b)
    H   OEt

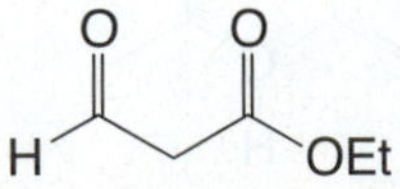

19.11

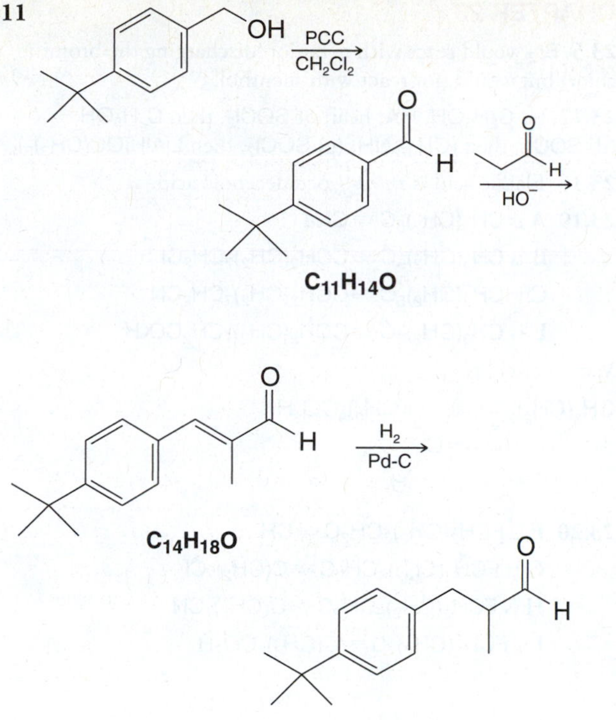

[structure] + OH $\xrightarrow[\text{CH}_2\text{Cl}_2]{\text{PCC}}$ [aldehyde structures] $\xrightarrow{\text{HO}^-}$

$C_{11}H_{14}O$

[structure] $\xrightarrow[\text{Pd-C}]{\text{H}_2}$

$C_{14}H_{18}O$

Lily aldehyde ($C_{14}H_{20}O$)

19.17

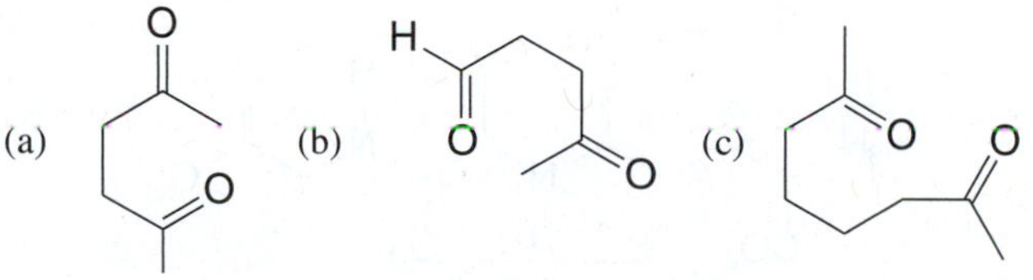

(a) [structure] (b) [structure] (c) [structure]

Notice that starting compounds are drawn so as to indicate which atoms are involved in the cyclization reaction.

19.19

(a)

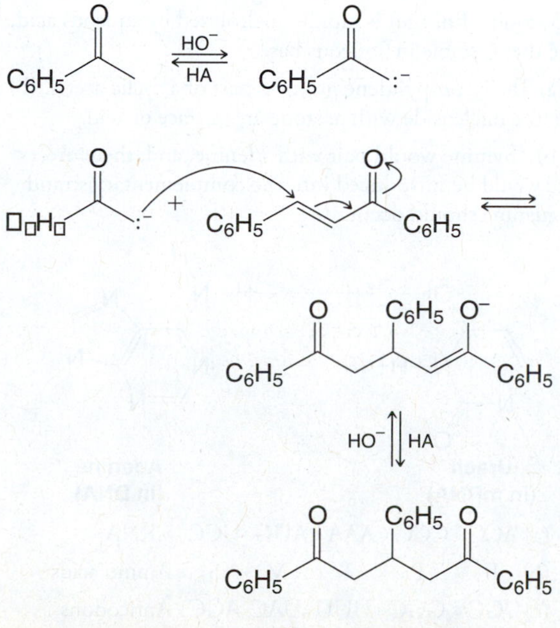

C_6H_5—[structure] $\xrightleftharpoons[\text{HA}]{\text{HO}^-}$ C_6H_5—[structure] :$^-$

[structure] + C_6H_5—[structure]—C_6H_5 $\rightleftharpoons$

C_6H_5—[structure]—C_6H_5

$\xrightleftharpoons[\text{HA}]{\text{HO}^-}$

C_6H_5—[structure]—C_6H_5

(b) H H [structure] $\xrightleftharpoons[\text{HA}]{:\text{A}^-}$ [structure]

[structure] + C_6H_5—[structure]—C_6H_5 $\rightleftharpoons$

C_6H_5—[structure]—C_6H_5 $\xrightleftharpoons[:\text{A}^-]{\text{HA}}$ C_6H_5—[structure]—C_6H_5

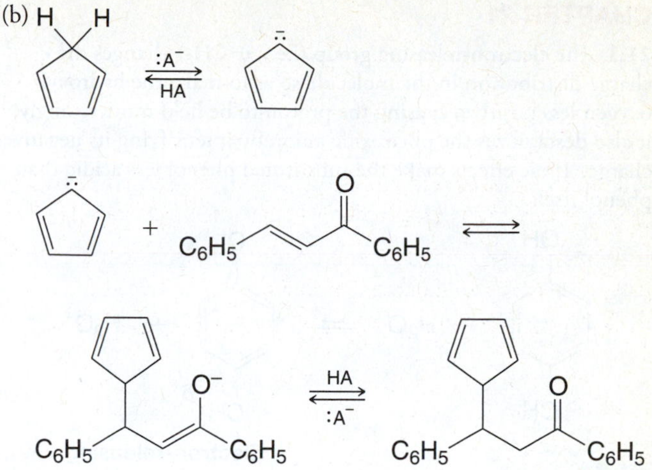

19.50 (a) $CH_2{=}C(CH_3)CO_2CH_3$; (b) $KMnO_4$, HO^-; H_3O^+;
(c) CH_3OH, HA; (d) CH_3ONa, then H_3O^+

(e) and (f) [structure with CO_2CH_3, CO_2CH_3] and

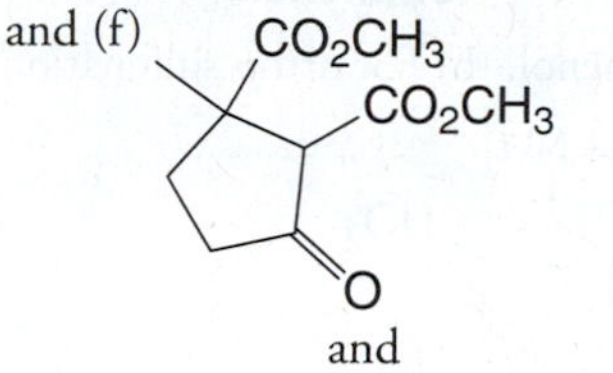

[structure with CO_2CH_3, CH_3O_2C]

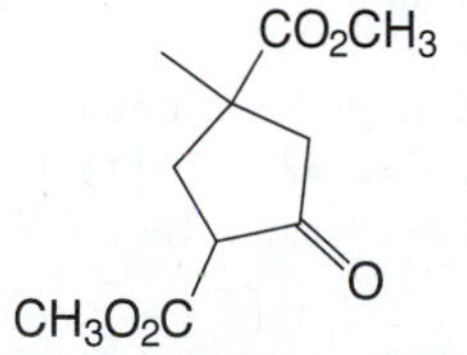

(g) HO^-, H_2O, then H_3O^+; (h) heat ($-CO_2$); (i) CH_3OH, HA;

(j) [structure with CO_2CH_3, $CHCO_2CH_3$]

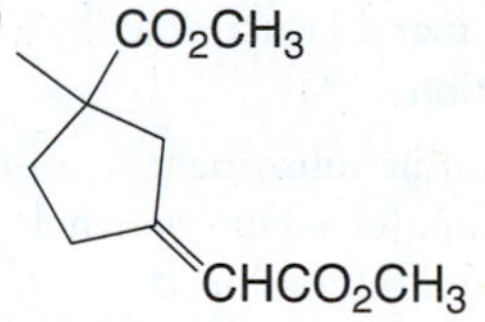

(k) H_2, Pt; (m) CH_3ONa, then H_3O^+; (n) 2 $NaNH_2$ + 2 CH_3I

CHAPTER 20

20.5 (a) $CH_3(CH_2)_3CHO + NH_3 \xrightarrow[\text{LiBH}_3\text{CN}]{\text{H}_2,\ \text{Ni}} CH_3(CH_2)_3CH_2NH_2$
(c) $CH_3(CH_2)_4CHO + C_6H_5NH_2 \longrightarrow$

$CH_3(CH_2)_4CH_2NHC_6H_5$

20.6 The reaction of a secondary halide with ammonia is almost always accompanied by some elimination.

20.7 (a) Methoxybenzene + HNO_3 + H_2SO_4, then Fe + HCl;
(b) Methoxybenzene + CH_3COCl + $AlCl_3$, then NH_3 + H_2 + Ni;
(c) toluene + Cl_2 and light, then $(CH_3)_3N$; (d) p-nitrotoluene + $KMnO_4$ + HO^-, then H_3O^+, then $SOCl_2$ followed by NH_3, then NaOBr (Br_2 in NaOH); (e) toluene + N-bromosuccinimide then KCN, then $LiAlH_4$.

20.12 p-Nitroaniline + Br_2 + Fe, followed by $H_2SO_4/NaNO_2$ followed by CuBr, then H2/Pt, then $H_2SO_4/NaNO_2$ followed by H_3PO_2.

20.45 **W** is N-benzyl-N-ethylaniline.

CHAPTER 21

21.1 The electron-releasing group (i.e., —CH$_3$) changes the charge distribution in the molecule so as to make the hydroxyl oxygen less positive, causing the proton to be held more strongly; it also destabilizes the phenoxide anion by intensifying its negative charge. These effects make the substituted phenol less acidic than phenol itself.

Electron-releasing — CH$_3$ destabilizes the anion more than the acid. pKa is larger than for phenol.

21.4 (a) The para-sulfonated phenol. (b) For ortho sulfonation.

21.9 (a)

(b)

21.10 That *o*-chlorotoluene leads to the formation of two products (*o*-cresol and *m*-cresol), when submitted to the conditions used in the Dow process, suggests that an elimination-addition mechanism takes place.

21.11 2-Bromo-1,3-dimethylbenzene, because it has no *o*-hydrogen atom, cannot undergo an elimination. Its lack of reactivity toward sodium amide in liquid ammonia suggests that those compounds (e.g., bromobenzene) that do react, react by a mechanism that begins with an elimination.

21.14 (a) 4-Fluorophenol because a fluorine substituent is more electron withdrawing than a methyl group. (e) 4-Fluorophenol because fluorine is more electronegative than bromine.

21.16 (a) 4-Chlorophenol will dissolve in aqueous NaOH; 4-chloro-1-methylbenzene will not. (c) Phenyl vinyl ether will react with bromine by addition (thus decolorizing the solution); ethyl phenyl ether will not. (e) 4-Ethylphenol will dissolve in aqueous NaOH; ethyl phenyl ether will not.

CHAPTER 22

22.1 (a) Two; (b) two; (c) four.

22.5 Acid catalyzes hydrolysis of the glycosidic (acetal) group.

22.9 (a) 2 CH$_3$CHO, one molar equivalent HIO$_4$; (b) HCHO + HCO$_2$H + CH$_3$CHO, two molar equivalents HIO$_4$; (c) HCHO + OHCCH(OCH$_3$)$_2$, one molar equivalent HIO$_4$; (d) HCHO + HCO$_2$H + CH$_3$CO$_2$H, two molar equivalents HIO$_4$; (e) 2 CH$_3$CO$_2$H + HCO$_2$H, two molar equivalents HIO$_4$

22.18 D-(+)-Glucose.

22.23 One anomeric form of D-mannose is dextrorotatory ($[\alpha]_D = +29.3$), the other is levorotatory ($[\alpha]_D = -17.0$).

22.24 The microorganism selectively oxidizes the —CHOH group of D-glucitol that corresponds to C5 of D-glucose.

22.27 **A** is D-altrose; **B** is D-talose, **C** is D-galactose

CHAPTER 23

23.5 Br$_2$ would react with geraniol (discharging the bromine color) but would not react with menthol.

23.12 (a) C$_2$H$_5$OH, HA, heat; or SOCl$_2$, then C$_2$H$_5$OH; (d) SOCl$_2$, then (CH$_3$)$_2$NH; (g) SOCl$_2$, then LiAlH[OC(CH$_3$)$_3$]$_3$

23.15 Elaidic acid is *trans*-9-octadecenoic acid.

23.19 **A** is CH$_3$(CH$_2$)$_5$C≡CNa

   B is CH$_3$(CH$_2$)$_5$C≡CCH$_2$(CH$_2$)$_7$CH$_2$Cl

   C is CH$_3$(CH$_2$)$_5$C≡CCH$_2$(CH$_2$)$_7$CH$_2$CN

   E is CH$_3$(CH$_2$)$_5$C≡CCH$_2$(CH$_2$)$_7$CH$_2$CO$_2$H

Vaccenic acid is

23.20 **F** is FCH$_2$(CH$_2$)$_6$CH$_2$C≡CH

   G is FCH$_2$(CH$_2$)$_6$CH$_2$C≡C(CH$_2$)$_7$Cl

   H is FCH$_2$(CH$_2$)$_6$CH$_2$C≡C(CH$_2$)$_7$CN

   I is FCH$_2$(CH$_2$)$_7$C≡C(CH$_2$)$_7$CO$_2$H

CHAPTER 24

24.5 The labeled amino acid no longer has a basic —NH$_2$ group; it is, therefore, insoluble in aqueous acid.

24.8 Glutathione is

24.22 Arg·Pro·Pro·Gly·Phe·Ser·Pro·Phe·Arg

24.23 Val·Leu·Lys·Phe·Ala·Glu·Ala

CHAPTER 25

25.2 (a) The nucleosides have an *N*-glycosidic linkage that (like an *O*-glycosidic linkage) is rapidly hydrolyzed by aqueous acid, but one that is stable in aqueous base.

25.4 (a) The isopropylidene group is part of a cyclic acetal. (b) By treating the nucleoside with acetone and a trace of acid.

25.7 (b) Thymine would pair with adenine, and, therefore, adenine would be introduced into the complementary strand where guanine should occur.

25.9

Uracil (in mRNA)   **Adenine (in DNA)**

25.10

(a)	ACC	CCC	AAA	AUG	UCC	mRNA
(b)	T	P	K	M	S	Amino acids
(c)	UGC	GGC	UUU	UAC	AGC	Anticodons

A

Absolute configuration (Section 5.15A): The actual arrangement of groups in a molecule. The absolute configuration of a molecule can be determined by X-ray analysis or by relating the configuration of a molecule, using reactions of known stereochemistry, to another molecule whose absolute configuration is known.

Absorption spectrum (Section 13.8B): A plot of the wavelength (λ) of a region of the spectrum versus the absorbance (A) at each wavelength. The absorbance at a particular wavelength (A_λ) is defined by the equation $A_\lambda = \log(I_R/I_S)$, where I_R is the intensity of the reference beam and I_S is the intensity of the sample beam.

Acetal (Section 16.7B): A functional group, consisting of a carbon bonded to alkoxy groups [i.e., $RCH(OR')_2$ or $R_2C(OR')_2$], derived by adding 2 molar equivalents of an alcohol to an aldehyde or ketone. An acetal synthesized from a ketone is sometimes called a ketal.

Acetoacetic ester synthesis (Section 18.6): A sequence of reactions involving removal of the α-hydrogen of ethyl 3-oxobutanoate (ethyl acetoacetate, also called "acetoacetic ester"), creating a resonance-stabilized anion which then can serve as a nucleophile in an S_N2 reaction. The α-carbon can be substituted twice; the ester functionality can be converted into α carboxylic acid which, after decarboxylation, yields a substituted ketone.

Acetonide (Section 22.5E): A cyclic acetal formed from acetone.

Acetylene (Sections 1.14, 7.1, and 7.11): A common name for ethyne.

Acetylenic hydrogen atom (Sections 4.6, and 7.9): A hydrogen atom attached to a carbon atom that is bonded to another carbon atom by a triple bond.

Achiral molecule (Sections 5.3 and 5.4): A molecule that is superposable on its mirror image. Achiral molecules lack handedness and are incapable of existing as a pair of enantiomers.

Acid strength (Section 3.5): The strength of an acid is related to its acidity constant, K_a or to its pK_a. The larger the value of its K_a or the smaller the value of its pK_a, the stronger is the acid.

Acidity constant, K_a (Section 3.5A): An equilibrium constant related to the strength of an acid. For the reaction,

$$HA + H_2O \rightleftharpoons H_3O^+ + A^-$$

$$K_a = \frac{[H_3O^+][A^-]}{[HA]}$$

Activating group (Sections 15.10, 15.10D): A group that when present on a benzene ring causes the ring to be more reactive in electrophilic substitution than benzene itself.

Activation energy, E_{act} (See **Energy of activation** and Section 10.5A)

Active hydrogen compounds or *active methylene compounds* (Section 18.8): Compounds in which two electron-withdrawing groups are attached to the same carbon atom (a methylene or methane carbon). The electron-withdrawing groups enhance the acidity of the hydrogens on carbon; these hydrogens are easily removed, creating a resonance-stabilized nucleophilic anion.

Active site (Section 24.9): The location in an enzyme where a substrate binds.

Acylation (Section 15.7): The introduction of an acyl group into a molecule.

Acyl compounds (Section 17.1): A compound containing the group $(R-C=O)-$, usually derived from a carboxylic acid, such as an ester, acid halide (acyl halide), amide, or carboxylic acid anhydride.

Acyl group (Section 15.7): The general name for groups with the structure $RCO-$ or $ArCO-$.

Acyl halide (Section 15.7): Also called an *acid halide*. A general name for compounds with the structure RCOX or ArCOX.

Acylium ion (Sections 9.16C and 15.7): The resonance-stabilized cation:

$$R-\overset{+}{C}=\overset{..}{\overset{..}{O}} \longleftrightarrow R-C\equiv\overset{+}{\overset{..}{O}}$$

Acyl transfer reactions (Section 17.4): A reaction in which a new acyl compound is formed by a nucleophilic addition-elimination reaction at a carbonyl carbon bearing a leaving group.

Addition polymer (Section 10.11 and Special Topic B in *WileyPLUS*): A polymer that results from a stepwise addition of monomers to a chain (usually through a chain reaction) with no loss of other atoms or molecules in the process. Also called a chain-growth polymer.

Addition reaction (Sections Chapter 8 intro, 8.1–8.9, 8.11, 8.12, 12.1A, 16.6B, and 17.4): A reaction that results in an increase in the number of groups attached to a pair of atoms joined by a double or triple bond. An addition reaction is the opposite of an elimination reaction.

Adduct (Section 13.10): The product formed by a Diels-Alder [4 + 2] cycloaddition reaction, so called because two compounds (a *diene* and a *dienophile*) are added together to form the product.

Aglycone (Section 22.4): The alcohol obtained by hydrolysis of a glycoside.

Aldaric acid (Section 22.6C): An α,ω-dicarboxylic acid that results from oxidation of the aldehyde group and the terminal 1° alcohol group of an aldose.

Alditol (Section 22.7): The alcohol that results from the reduction of the aldehyde or keto group of an aldose or ketose.

Aldol (Section 19.4): A common name for 3-hydoxybutanal, which contains both *ald*ehyde and an alco*hol* functional groups. Aldol is formed from the *aldol reaction* (see below) of ethanal (acetaldehyde) with itself.

Aldol additions (Section 19.4): See **Aldol reaction** and **Aldol condensation**.

Aldol condensation (Sections 19.1 and 19.4): An aldol reaction that forms an α,β-unsaturated product by dehydration of the β-hydroxy aldehyde or ketone aldol product.

Aldol reactions (Sections 19.4–19.6): Reactions in which the enol or enolate ion of an aldehyde or ketone reacts with the carbonyl group of the same or a different aldehyde or ketone, creating a β-hydroxy aldehyde or ketone and a new carbon-carbon σ-bond.

Aldonic acid (Section 22.6C): A monocarboxylic acid that results from oxidation of the aldehyde group of an aldose.

Aliphatic compound (Section 14.1): A nonaromatic compound such as an alkane, cycloalkane, alkene, or alkyne.

Alkaloid (Special Topic F in *WileyPLUS*): A naturally occurring basic compound that contains an amino group. Most alkaloids have profound physiological effects.

Alkanes (Sections 2.1, 2.1A, 4.1–4.3, 4.7, and 4.16): Hydrocarbons having only single (σ) bonds between carbon atoms. Acyclic alkanes have the general formula C_nH_{2n+2}. Monocyclic alkanes have the general formula of C_nH_{2n}. Alkanes are said to be "saturated" because C—C single bonds cannot react to add hydrogen to the molecule.

Alkanide (Section 7.8A): An alkyl anion, R:⁻, or alkyl species that reacts as though it were an alkyl anion.

Alkenes (Sections 2.1, 2.1B, 4.1, and 4.5): Hydrocarbons having at least one double bond between carbon atoms. Acyclic alkenes have the general formula C_nH_{2n}. Monocyclic alkenes have the general formula of C_nH_{2n-2}. Alkenes are said to be "unsaturated" because their C=C double bonds can react to add hydrogen to the molecule, yielding an alkane.

Alkenyl halides (Section 6.1): An organic halide in which the halogen atom is bonded to an alkene carbon.

Alkylation (Sections 7.11A, 15.6, and 18.4C): The introduction of an alkyl group into a molecule.

Alkyl group (See **R**) (Sections 2.4A and 4.3A): The designation given to a fragment of a molecule hypothetically derived from an alkane by removing a hydrogen atom. Alkyl group names end in "yl." Example: the methyl group, CH_3—, is derived from methane, CH_4.

Alkyl halide (Section 6.1): An organic halide in which the halogen atom is bonded to an alkyl carbon.

Alkynes (Sections 2.1, 2.1C, 4.1, and 4.6): Hydrocarbons having at least one triple bond between carbon atoms. Acyclic alkynes have the general formula C_nH_{2n-2}. Monocyclic alkynes have the general formula of C_nH_{2n+4}. Alkynes are said to be "unsaturated" because C≡C triple bonds can react to add two molecules of hydrogen to the molecule, yielding an alkane.

Allyl group (Section 4.5): The CH_2—$CHCH_2$— group.

Allylic carbocation (Sections 13.1, 13.9, and 15.15): A substructure involving a three-carbon delocalized carbocation in which the positively charged carbon is adjacent to a carbon-carbon double bond in each of two contributing resonance structures.

Allylic group (Section 10.8): An atom or group that is bonded to an sp^3-hybridized carbon adjacent to an alkene double bond.

Allylic position (Section 10.8): The location of a group that is bonded to an sp^3-hybridized carbon adjacent to an alkene double bond.

Allylic substitution (Section 10.8): The replacement of a group at an allylic position.

Allyl (propenyl cation) (Section 13.3): The carbocation formally related to propene by removal of a proton from its methyl group. The two contributing resonance structures of the delocalized carbocation each include a positive charge on a carbon adjacent to the double bond, such that a *p* orbital on each of the three carbons overlaps to delocalize positive charge to each end of the allyl system.

Allyl radical (Sections 10.8A and 13.3): The radical formally related to propene by removal of a hydrogen atom from its methyl group. The two contributing resonance structures of the delocalized radical each include an unpaired electron on a carbon adjacent to the double bond, such that a *p* orbital on each of the three carbons overlaps to delocalize the radical to each end of the allyl system, in which the radical carbon is adjacent to a carbon-carbon double bond.

Alpha (α) anomer (Section 22.2C): In the standard Haworth formula representation for a D-hexopyranose, the α anomer has the hemiacetal hydroxyl or acetal alkoxyl group trans to C6. Similar usage applies to other carbohydrate forms regarding the stereochemical relationship of the anomeric hydroxyl or alkoxyl group and the configuration at the carbon bearing the ring oxygen that forms the hemiacetal or acetal.

Alpha (α) carbon (Section 18.1): A carbon adjacent to a carbonyl (C=O) group.

Alpha (α) helix (Section 24.8A): A secondary structure in proteins where the polypeptide chain is coiled in a right-handed helix.

Alpha (α) hydrogens (Sections 18.1 and 18.5D): A hydrogen atom bonded to an α carbon. These hydrogens are significantly more acidic than the typical alkane hydrogen.

Aminium salt (Section 20.3D): The product of the reaction of an amine, acting as a Bronsted-Lowry base, with an acid. The amine can be primary, secondary, or tertiary. The positively charged nitrogen in an aminium salt is attached to at least one hydrogen atom. (An ammonium salt has no hydrogen atoms bonded directly to the nitrogen.)

Amino acid residue (Section 24.4): An amino acid that is part of a peptide.

Angle strain (Section 4.10): The increased potential energy of a molecule (usually a cyclic one) caused by deformation of a bond angle away from its lowest energy value.

Annulene (Section 14.7B): Monocyclic hydrocarbon that can be represented by a structure having alternating single and double bonds. The ring size of an annulene is represented by a number in brackets, e.g., benzene is [6]annulene and cyclooctatetraene is [8]annulene.

Anomeric carbon (Section 22.2C): The hemiacetal or acetal carbon in the cyclic form of a carbohydrate. The anomeric carbon can have either the α or β stereochemical configuration (using carbohydrate nomenclature), resulting in diastereomeric forms of the carbohydrate called anomers (α-anomers and β-anomers). Anomers differ *only* in the stereochemistry at the anomeric carbon.

Anomers (Section 22.2C): A term used in carbohydrate chemistry. Anomers are diastereomers that differ only in configuration at the acetal or hemiacetal carbon of a sugar in its cyclic form.

Anti 1,2-dihydroxylation (Section 11.15): The installation of hydroxyl groups at adjacent carbons and on opposite faces of an alkene, often accomplished by ring-opening of an epoxide.

Anti addition (Sections 7.13A, 7.14B, and 8.11A): An addition that places the parts of the adding reagent on opposite faces of the reactant.

Antiaromatic compound (Section 14.7E): A cyclic conjugated system whose π electron energy is greater than that of the corresponding open-chain compound.

Antibonding molecular orbital (antibonding MO) (Sections 1.11, 1.13, and 1.15): A molecular orbital whose energy is higher than that of the isolated atomic orbitals from which it is constructed. Electrons in an antibonding molecular orbital destabilize the bond between the atoms that the orbital encompasses.

Anticodon (Section 25.5C): A sequence of three bases on transfer RNA (tRNA) that associates with a codon of messenger RNA (mRNA).

Anti conformation (Section 4.9): An anti conformation of butane, for example, has the methyl groups at an angle of 180° to each other:

Anti
conformation
of butane

Anti coplanar (Section 7.6D): The relative position of two groups that have a 180° dihedral angle between them.

anti-Markovnikov addition (Sections 8.2D, 8.6–8.9, 8.18, and 10.10): An addition reaction where the hydrogen atom of a reagent becomes bonded to an alkene or alkyne at the carbon having the fewer hydrogen atoms initially. This orientation is the opposite of that predicted by Markovnikov's rule.

Arenium ion (Section 15.2): A general name for the cyclohexadienyl carbocations that form as intermediates in electrophilic aromatic substitution reactions.

Aromatic compound (Sections 2.1, 2.1D, 14.1–14.8, and 14.11): A cyclic conjugated unsaturated molecule or ion that is stabilized by π electron delocalization. Aromatic compounds are characterized by having large resonance energies, by reacting by substitution rather than addition, and by deshielding of protons exterior to the ring in their ^{1}H NMR spectra caused by the presence of an induced ring current.

Aromatic ions (Section 14.7D): Cations and anions that fulfill the criteria for aromaticity (planarity, electron delocalization, and a Hückel number of π-electrons) and thus have additional (aromatic) stability.

Arylamines (Section 20.1A): A compound in which the carbon of an aromatic ring bears the amine nitrogen atom. Aryl amines can be primary, secondary, or tertiary.

Aryl halide (Sections 2.5 and 6.1): An organic halide in which the halogen atom is attached to an aromatic ring, such as a benzene ring.

Atactic polymer (Special Topic B.1 in *WileyPLUS*): A polymer in which the configuration at the stereogenic centers along the chain is random.

Atomic orbital (AO) (Sections 1.10, 1.11, and 1.15): A volume of space about the nucleus of an atom where there is a high probability of finding an electron. An atomic orbital can be described mathematically by its wave function. Atomic orbitals have characteristic quantum numbers; the *principal quantum number, n,* is related to the energy of the electron in an atomic orbital and can have the values 1, 2, 3,.... The *azimuthal quantum number, l,* determines the angular momentum of the electron that results from its motion around the nucleus, and can have the values 0, 1, 2,..., $(n-1)$. The *magnetic quantum number, m,* determines the orientation in space of the angular momentum and can have values from $+l$ to $-l$. The *spin quantum number, s,* specifies the intrinsic angular momentum of an electron and can have the values of $+\frac{1}{2}$ and $-\frac{1}{2}$ only.

Atropisomers (Section 5.18): Conformational isomers that are stable, isolable compounds.

Aufbau principle (Section 1.10A): A principle that guides us in assigning electrons to orbitals of an atom or molecule in its lowest energy state or ground state. The aufbau principle states that electrons are added so that orbitals of lowest energy are filled first.

Autoxidation (Section 10.12C): The reaction of an organic compound with oxygen to form a hydroperoxide.

Axial bond (Section 4.12): The six bonds of a cyclohexane ring (below) that are perpendicular to the general plane of the ring, and that alternate up and down around the ring.

B

Base peak (Section 9.13): The most intense peak in a mass spectrum.

Base strength (Sections 3.5C and 20.3): The strength of a base is inversely related to the strength of its conjugate acid; the weaker the conjugate acid, the stronger is the base. In other words, if the conjugate acid has a large pK_a, the base will be strong.

Benzene (Section 2.1D): The prototypical aromatic compound having the formula C_6H_6. Aromatic compounds are planar, cyclic, and contain $4n + 2$ π electrons *delocalized* in contiguous fashion about a ring of electron density in the molecule. Electron delocalization gives aromatic compounds a high degree of stability.

Benzenoid aromatic compound (Section 14.8A): An aromatic compound whose molecules have one or more benzene rings.

Benzyl group (Sections 2.4B and 10.9): The $C_6H_5CH_2$— group.

Benzylic carbocation (Section 15.15): A carbocation located adjacent to a benzene ring.

Benzylic cation (Section 15.12A): A carbocation where the positive charge is on a carbon bonded to a benzene ring. The positive charge is delocalized into the benzene ring through conjugation, resulting in a relatively stable carbocation.

Benzylic position (Section 10.9): The location of a group that is bonded to an sp^3-hybridized carbon adjacent to a benzene ring.

Benzylic radical (Section 15.12A): The radical comprised of a methylene (CH_2) group bonded to a benzene ring, wherein the unpaired electron is *delocalized* over the methylene group and the ring. As a highly *conjugated* system, the benzylic radical has greatly enhanced stability.

Benzylic substituent (Sections 15.12A): Refers to a substituent on a carbon atom adjacent to a benzene ring.

Benzyne (Section 21.11B): An unstable, highly reactive intermediate consisting of a benzene ring with an additional bond resulting from sideways overlap of sp^2 orbitals on adjacent atoms of the ring.

Beta (β) anomer (Section 22.2C): In the standard Haworth formula representation for a D-hexopyranose, the β anomer has the hemiacetal hydroxyl or acetal alkoxy group cis to C6. Similar usage applies to other carbohydrate forms regarding the stereochemical relationship of the anomeric hydroxyl or alkoxyl group and the configuration at the carbon bearing the ring oxygen that forms the hemiacetal or acetal.

Beta (β)-carbonyl compound (Section 18.5C): A compound having two carbonyl groups separated by an intervening carbon atom.

Beta (β)-pleated sheet (Section 24.8A): A type of protein secondary structure involving alignment of two polypeptide regions alongside each other through hydrogen bonding of their amide groups.

Bicyclic compounds (Section 4.4B): Compounds with two fused or bridged rings.

Bimolecular reaction (Section 6.5B): A reaction whose rate-determining step involves two initially separate species.

Boat conformation (Section 4.11): A conformation of cyclohexane that resembles a boat and that has eclipsed bonds along its two sides:

It is of higher energy than the chair conformation.

Boiling point (Sections 2.13A and 2.13C): The temperature at which the vapor pressure of a liquid is equal to the pressure above the surface of the liquid.

Bond angle (Section 1.7A): The angle between two bonds originating at the same atom.

Bond dissociation energy (See **Homolytic bond dissociation energy** and Section 10.2)

Bonding molecular orbital (bonding MO) (Sections 1.11, 1 .12, and 1.15): The energy of a bonding molecular orbital is lower than the energy of the isolated atomic orbitals from which it arises. When electrons occupy a bonding molecular orbital they help hold together the atoms that the molecular orbital encompasses.

Bond length (Sections 1.11 and 1.14A): The equilibrium distance between two bonded atoms or groups.

Bond-line formula (Section 1.7C): A formula that shows the carbon skeleton of a molecule with lines. The number of hydrogen atoms necessary to fulfill each carbon's valence is assumed to be present but not written in. Other atoms (e.g., O, Cl, N) are written in.

Broadband (BB) proton decoupling (see **Proton decoupling**) (Section 9.11B): A method of eliminating carbon-proton coupling by irradiating the sample with a wide-frequency ("broadband") energy input in the frequencies in which protons absorb energy. This energy input causes the protons to remain in the high energy state, eliminating coupling with carbon nuclei.

Bromohydrin (Section 8.13): A compound bearing a bromine atom and a hydroxyl group on adjacent (vicinal) carbons.

Bromonium ion (Section 8.11A): An ion containing a positive bromine atom bonded to two carbon atoms.

Brønsted–Lowry theory of acid–base (Section 3.1A): An acid is a substance that can donate (or lose) a proton; a base is a substance that can accept (or remove) a proton. The *conjugate acid* of a base is the molecule or ion that forms when a base accepts a proton. The *conjugate base* of an acid is the molecule or ion that forms when an acid loses its proton.

C

Carbanion (Sections 3.4 and 12.1A): A chemical species in which a carbon atom bears a formal negative charge.

Carbene (Section 8.14): An uncharged species in which a carbon atom is divalent. The species :CH₂, called methylene, is a carbene.

Carbenoid (Section 8.14C): A carbene-like species. A species such as the reagent formed when diiodomethane reacts with a zinc–copper couple. This reagent, called the Simmons–Smith reagent, reacts with alkenes to add methylene to the double bond in a stereospecific way.

Carbocation (Sections 3.4, 6.11, and 6.12): A chemical species in which a trivalent carbon atom bears a formal positive charge.

Carbohydrate (Section 22.1A): A group of naturally occurring compounds that are usually defined as polyhydroxyaldehydes or polyhydroxyketones, or as substances that undergo hydrolysis to yield such compounds. In actuality, the aldehyde and ketone groups of carbohydrates are often present as hemiacetals and acetals. The name comes from the fact that many carbohydrates possess the empirical formula $C_x(H_2O)_y$.

Carbon-carbon double bond (Section 1.3B): A bond between two carbon atoms comprised of four electrons; two of the electrons are in a sigma bond and two of the electrons are in a pi bond.

Carbon-carbon single bond (Section 1.3B): A bond between two carbon atoms comprised of two electrons shared in a sigma bond.

Carbon-carbon triple bond (Section 1.3B): A bond between two carbon atoms comprised of six electrons; two of the electrons are in a sigma bond and four of the electrons are as pairs in each of two pi bonds.

Carbon-13 NMR spectroscopy (Section 9.11): NMR spectroscopy applied to carbon. Carbon-13 is NMR active, whereas carbon-12 is not and therefore cannot be studied by NMR. Only 1.1% of all naturally occurring carbon is carbon-13.

Carbonyl group (Section 16.1): A functional group consisting of a carbon atom doubly bonded to an oxygen atom. The carbonyl group is found in aldehydes, ketones, esters, anhydrides, amides, acyl halides, and so on. Collectively these compounds are referred to as carbonyl compounds.

Carboxylic acid derivatives (Section 17.1): Acyl compounds that can be synthesized from a carboxylic acid or another carboxylic acid derivative. Examples include esters, amides, acid halides, anhydrides, etc.

CFC (see **Freon**): A chlorofluorocarbon.

Chain-growth polymer (See **Addition polymer** and Special Topic B in *WileyPLUS*): Polymers (macromolecules with repeating units) formed by adding subunits (called *monomers*) repeatedly to form a chain.

Chain reaction (Sections 10.4 and 10.10): A reaction that proceeds by a sequential, stepwise mechanism, in which each step generates the reactive intermediate that causes the next step to occur. Chain reactions have *chain-initiating steps, chain-propagating steps*, and *chain-terminating steps*.

Chain-terminating (dideoxynucleotide) method (Section 25.6): A method of sequencing DNA that involves replicating DNA in a way that generates a family of partial copies, each differing in length by one base pair and containing a nucleotide-specific fluorescent tag on the terminal base. The partial copies of the parent DNA are separated by length, usually using capillary electrophoresis, and the terminal base on each strand is identified by the covalently attached fluorescent marker.

Chair conformation (Section 4.11): The all-staggered conformation of cyclohexane that has no angle strain or torsional strain and is, therefore, the lowest energy conformation:

Chemical exchange (Section 9.10): In the context of NMR, transfer of protons bonded to heteroatoms from one molecule to another, broadening their signal and eliminating spin-spin coupling.

Chemical shift, δ (Sections 9.2A, 9.7, and 9.11C): The position in an NMR spectrum, relative to a reference compound, at which a nucleus absorbs. The reference compound most often used is tetramethylsilane (TMS), and its absorption point is arbitrarily designated zero. The chemical shift of a given nucleus is proportional to the strength of the magnetic field of the spectrometer. The chemical shift in delta units, δ, is determined by dividing the observed shift from TMS in hertz multiplied by 10^6 by the operating frequency of the spectrometer in hertz.

Chirality (Sections 5.1, 5.4, and 5.6): The property of having handedness.

Chirality center (Sections 5.4 and 5.17): An atom bearing groups of such nature that an interchange of any two groups will produce a stereoisomer.

Chiral molecule (Sections 5.3 and 5.12): A molecule that is not superposable on its mirror image. Chiral molecules have handedness and are capable of existing as a pair of enantiomers.

Chlorination (Sections 8.12, 10.3B, 10.4, and 10.5): A reaction in which one or more chlorine atoms are introduced into a molecule.

Chlorohydrin (Section 8.13): A compound bearing a chlorine atom and a hydroxyl group on adjacent (vicinal) carbons.

Cis–trans isomers (Sections 1.13B, 4.13, and 7.2): Diastereomers that differ in their stereochemistry at adjacent atoms of a double bond or on different atoms of a ring. Cis groups are on the same side of a double bond or ring. Trans groups are on opposite sides of a double bond or ring.

Claisen condensation (Section 19.1): A reaction in which an enolate anion from one ester attacks the carbonyl function of another ester, forming a new carbon-carbon σ-bond. A tetrahedral intermediate is involved that, with expulsion of an alkoxyl group, collapses to a β-ketoester. The two esters are said to *"condense"* into a larger product with loss of an alcohol molecule.

Claisen rearrangement (Section 21.9): A [3,3] sigmatropic rearrangement reaction involving an allyl vinyl ether, in which the allyl group of migrates to the other end of the vinyl system, with bond reorganization leading to a γ,δ-unsaturated carbonyl compound.

Codon (Section 25.5C): A sequence of three bases on messenger RNA (mRNA) that contains the genetic information for one amino acid. The codon associates, by hydrogen bonding, with an anticodon of a transfer RNA (tRNA) that carries the particular amino acid for protein synthesis on the ribosome.

Coenzyme (Section 24.9): A small organic molecule that participates in the mechanism of an enzyme and which is bound at the active site of the enzyme.

Cofactor (Section 24.9): A metal ion or organic molecule whose presence is required in order for an enzyme to function.

Concerted reaction (Section 6.6): A reaction where bond forming and bond breaking occur simultaneously (in concert) through a single transition state.

Condensation polymer (See **Step-growth polymer**, Section 17.12, and Special Topic C in *WileyPLUS*): A polymer produced when bifunctional monomers (or potentially bifunctional monomers) react with each other through the intermolecular elimination of water or an alcohol. Polyesters, polyamides, and polyurethanes are all condensation polymers.

Condensation reaction (Section 19.1): A reaction in which molecules become joined through the intermolecular elimination of water or an alcohol.

Condensed structural formula (Section 1.7B): A chemical formula written using letters of the elemental symbols for the atoms involved, listed in sequence for the connections of the central chain of atoms and without showing the bonds between them. In organic compounds, all of the substituent atoms that are bonded to a given carbon atom are written immediately after the symbol for that carbon atom, then the next carbon atom in the chain is written, and so on.

Configuration (Sections 5.7, 5.15, and 6.8): The particular arrangement of atoms (or groups) in space that is characteristic of a given stereoisomer.

Conformation (Section 4.8): A particular temporary orientation of a molecule that results from rotations about its single bonds.

Conformational analysis (Sections 4.8, 4.9, 4.11, and 4.12): An analysis of the energy changes that a molecule undergoes as its groups undergo rotation (sometimes only partial) about the single bonds that join them.

Conformational stereoisomers (Section 4.9A): Stereoisomers differing in space only due to rotations about single (σ) bonds.

Conformations of cyclohexane (Sections 4.11 and 4.13): Rotations about the carbon-carbon single bonds of cyclohexane can produce different conformations which are interconvertible. The most important are the chair conformation, the boat conformation, and the twist conformation.

Conformer (Section 4.8): A particular staggered conformation of a molecule.

Conjugate acid (Section 3.1A): The molecule or ion that forms when a base accepts a proton.

Conjugate addition (Sections 19.1 and 19.7): A form of nucleophilic addition to an α,β-unsaturated carbonyl compound in which the nucleophile adds to the β carbon. Also called Michael addition.

Conjugate base (Sections 3.1A and 3.5C): The molecule or ion that forms when an acid loses its proton.

Conjugated protein (Section 24.12): A protein that contains a nonprotein group (called a prosthetic group) as part of its structure.

Connectivity (Sections 1.6 and 1.7A): The sequence, or order, in which the atoms of a molecule are attached to each other.

Constitutional isomers (Sections 1.6, 4.2, and 5.2A): Compounds that have the same molecular formula but that differ in their connectivity (i.e., molecules that have the same molecular formula but have their atoms connected in different ways).

Coplanar (Section 7.6D): A conformation in which vicinal groups lie in the same plane.

Copolymer (Special Topic B in *WileyPLUS*): A polymer synthesized by polymerizing two monomers.

COSY (Correlation Spectroscopy) (Section 9.12): A two-dimensional NMR method that displays coupling relationships between protons in a molecule.

Coupling (Section 9.2C): In NMR, the splitting of the energy levels of a nucleus under observation by the energy levels of nearby NMR-active nuclei, causing characteristic splitting patterns for the signal of the nucleus being observed. The signal from an NMR-active nucleus will be split into $(2nI + 1)$ peaks, where n = the number of equivalent neighboring magnetic nuclei and I = the spin quantum number. For hydrogen ($I = 1/2$) this rule devolves to $(n + 1)$, where n = the number of equivalent neighboring hydrogen nuclei.

Coupling constant, J_{ab} (Section 9.9C): The separation in frequency units (hertz) of the peaks of a multiplet caused by spin–spin coupling between atoms a and b.

Covalent bond (Section 1.3B): The type of bond that results when atoms share electrons.

Cracking (Section 4.1A): A process used in the petroleum industry for breaking down the molecules of larger alkanes into smaller ones. Cracking may be accomplished with heat (thermal cracking), or with a catalyst (catalytic cracking).

Crossed-aldol reaction (Section 19.5): An aldol reaction involving two different aldehyde or ketone reactants. If both aldol reactants have α hydrogens, four products can result. Crossed aldol reactions are synthetically useful when one reactant has no α hydrogens, such that it can serve only as an electrophile that is subject to attack by the enolate from the other reactant.

Crown ether (Section 11.16): Cyclic polyethers that have the ability to form complexes with metal ions. Crown ethers are named as *x*-crown-*y* where *x* is the total number of atoms in the ring and *y* is the number of oxygen atoms in the ring.

Curved arrows (Sections 1.8, 3.2, and 10.1): Curved arrows show the direction of electron flow in a reaction mechanism. They point from the source of an electron pair to the atom receiving the pair. Double-barbed curved arrows are used to indicate the movement of a pair of electrons; single-barbed curved arrows are used to indicate the movement of a single electron. Curved arrows are never used to show the movement of atoms.

Cyanohydrin (Sections 16.9 and 17.3): A functional group consisting of a carbon atom bonded to a cyano group and to a hydroxyl group, i.e., RHC(OH)(CN) or R₂C(OH)(CN), derived by adding HCN to an aldehyde or ketone.

1,4-Cycloaddition (Section 13.10): A ring-forming reaction where new bonds are formed to the first and fourth atoms of a molecular moiety, as at the ends of a 1,3-diene in a Diels-Alder reaction.

Cycloaddition (Section 13.10): A reaction, like the Diels–Alder reaction, in which two connected groups add to the end of a π system to generate a new ring. Also called 1,4-cycloaddition.

Cycloalkanes (Sections 4.1, 4.4, 4.7, 4.10, and 4.11): Alkanes in which some or all of the carbon atoms are arranged in a ring. Saturated cycloalkanes have the general formula C_nH_{2n}.

D

1,3-Diaxial interaction (Section 4.12): The interaction between two axial groups that are on adjacent carbon atoms.

1,2-Dihydroxylation (Section 8.15): The installation of hydroxyl groups on adjacent carbons, such as by the reaction of OsO_4 or $KMnO_4$ with an alkene.

D and L nomenclature (Section 22.2B): A method for designating the configuration of monosaccharides and other compounds in which the reference compound is (+)- or (−)-glyceraldehyde. According to this system, (+)-glyceraldehyde is designated D-(+)-glyceraldehyde and (−)-glyceraldehyde is designated L-(−)-glyceraldehyde. Therefore, a monosaccharide whose highest numbered stereogenic center has the same general configuration as D-(+)-glyceraldehyde is designated a D-sugar; one whose highest numbered stereogenic center has the same general configuration as L-(+)-glyceraldehyde is designated an L-sugar.

Dash structural formulas (Sections 1.3B and 1.7A): Structural formulas in which atom symbols are drawn and a line or "dash" represents each pair of electrons (a covalent bond). These formulas show connectivities between atoms but do not represent the true geometries of the species.

Deactivating group (Sections 15.10, 15.10E, 15.10F, and 15.11A): A group that when present on a benzene ring causes the ring to be less reactive in electrophilic substitution than benzene itself.

Debye (Section 2.2): The unit in which dipole moments are stated. One debye, D, equals 1×10^{-18} esu cm.

Decarboxylation (Section 17.10): A reaction whereby a carboxylic acid loses CO_2.

Degenerate orbitals (Section 1.10A): Orbitals of equal energy. For example, the three $2p$ orbitals are degenerate.

Dehydration (Sections 7.7 and 7.8): An elimination that involves the loss of a molecule of water from the substrate.

Dehydrohalogenation (Sections 6.15A and 7.6): An elimination reaction that results in the loss of HX from adjacent carbons of the substrate and the formation of a π bond.

Delocalization effect (Sections 3.10A and 6.11B): The dispersal of electrons (or of electrical charge). Delocalization of charge always stabilizes a system.

Deoxyribonucleic acid (DNA) (Sections 25.1 and 25.4A): One of the two molecules (the other is RNA) that carry genetic information in cells. Two molecular strands held together by hydrogen bonds give DNA a "twisted ladder"-like structure, with four types of heterocyclic bases (adenine, cytosine, thymine, and guanine) making up the "rungs" of the ladder.

DEPT ¹³C NMR spectra (Section 9.11D): Distortionless enhanced polarization transfer (DEPT) ¹³C NMR spectra indicate how many hydrogen atoms are bonded to a given carbon atom.

Deshielded (Section 9.7): See Shielding.

Dextrorotatory (Section 5.8B): A compound that rotates plane-polarized light clockwise.

Diastereomers (Section 5.2C): Stereoisomers that are not mirror images of each other.

Diastereoselective reaction (see **Stereoselective reaction** and Sections 5.10B and 12.3D)

Diastereotopic hydrogens (or *ligands*) (Section 9.8B): If replacement of each of two hydrogens (or ligands) by the same groups yields compounds that are diastereomers, the two hydrogen atoms (or ligands) are said to be diastereotopic.

Diazonium salts (Sections 20.6A, 20.6B, and 20.7): Salts synthesized from the reaction of primary amines with nitrous acid. Diazonium salts have the structure $[R-N\equiv N]^+ X^-$. Diazonium salts of primary aliphatic amines are unstable and decompose rapidly; those from primary aromatic amines decompose slowly when cold, and are useful in the synthesis of substituted aromatics and *azo* compounds.

Dieckmann condensation (Section 19.2A): An intramolecular Claisen condensation of a diester; the enolate from one ester group attacks the carbonyl of another ester function in the same molecule, leading to a cyclic product.

Dielectric constant (Section 6.13C): A measure of a solvent's ability to insulate opposite charges from each other. The dielectric constant of a solvent roughly measures its polarity. Solvents with high dielectric constants are better solvents for ions than are solvents with low dielectric constants.

Diels-Alder reaction (Section 13.10): In general terms, a reaction between a conjugated diene (a 4-π-electron system) and a compound containing a double bond (a 2-π-electron system), called a dienophile, to form a cyclohexene ring.

Diene (Section 13.10): A molecule containing two double bonds (*di* = two, *ene* = alk*ene* or double bonds). In a Diels-Alder reaction, a *conjugated* diene in the *s-cis* conformation reacts with a dienophile.

Dienophile (Section 13.10): The diene-seeking component of a Diels–Alder reaction.

Dihedral angle (Sections 4.8A and 9.9D): See Fig. 4.4. The angle between two atoms (or groups) bonded to adjacent atoms, when viewed as a projection down the bond between the adjacent atoms.

Dihydroxylation (Section 8.15): A process by which a starting material is converted into a product containing adjacent alcohol functionalities (called a "1,2-diol" or "glycol").

Dipeptide (Section 24.4): A peptide comprised of two amino acids.

Dipolar ion (Section 24.2C): The charge-separated form of an amino acid that results from the transfer of a proton from a carboxyl group to a basic group.

Dipole–dipole force (Section 2.13B): An interaction between molecules having permanent dipole moments.

Dipole moment, μ (Section 2.2): A physical property associated with a polar molecule that can be measured experimentally. It is defined as the product of the charge in electrostatic units (esu) and the distance that separates them in centimeters: $\mu = e \times d$.

Direct alkylation (Section 18.4C): A synthetic process in which the α-hydrogen of an ester is removed by a strong, bulky base such as LDA, creating a resonance-stabilized anion which will act as a nucleophile in an S_N2 reaction.

Directed aldol reaction (Section 19.5B): A crossed aldol reaction in which the desired enolate anion is generated first and rapidly using a strong base (e.g., LDA) after which the carbonyl reactant to be attacked by the enolate is added. If both a *kinetic enolate anion* and a *thermodynamic enolate anion* are possible, this process favors generation of the kinetic enolate anion.

Disaccharide (Sections 22.1A and 22.12): A carbohydrate that, on a molecular basis, undergoes hydrolytic cleavage to yield two molecules of a monosaccharide.

Dispersion force (or *London force*) (Sections 2.13B and 4.12B): Weak forces that act between nonpolar molecules or between parts of the same molecule. Bringing two groups (or molecules) together first results in an attractive force between them because a temporary unsymmetrical distribution of electrons in one group induces an opposite polarity in the other. When groups are brought closer than

their *van der Waals radii*, the force between them becomes repulsive because their electron clouds begin to interpenetrate each other.

Distortionless enhanced polarization transfer (DEPT) spectra (Section 9.11D): A technique in ^{13}C NMR spectroscopy by which the number of hydrogens at each carbon, e.g., C, CH, CH_2, and CH_3 can be determined.

Disulfide linkage (Section 24.2A): A sulfur-sulfur single bond in a peptide or protein formed by an oxidative reaction between the thiol groups of two cysteine amino acid residues.

Double bonds (Sections 1.4A and 1.13A): Bonds composed of four electrons: two electrons in a sigma (σ) bond and two electrons in a pi (π) bond.

Doublet (Section 9.2C): An NMR signal comprised of two peaks with equal intensity, caused by signal splitting from one neighboring NMR-active nucleus.

Downfield (Section 9.3): Any area or signal in an NMR spectrum that is to the left relative to another. (See **Upfield** for comparison.) A signal that is downfield of another occurs at higher frequency (and higher δ and ppm values) than the other signal.

E

E1 reaction (Sections 6.15C, 6.17, and 6.18B): A unimolecular elimination in which, in a slow, rate-determining step, a leaving group departs from the substrate to form a carbocation. The carbocation then in a fast step loses a proton with the resulting formation of a π bond.

E2 reaction (Sections 6.15C, 6.16, and 6.18B): A bimolecular 1,2 elimination in which, in a single step, a base removes a proton and a leaving group departs from the substrate, resulting in the formation of a π bond.

Eclipsed conformation (Section 4.8A): A temporary orientation of groups around two atoms joined by a single bond such that the groups directly oppose each other.

An eclipsed conformation

Edman degradation (Section 24.5A): A method for determining the *N*-terminal amino acid in a peptide. The peptide is treated with phenylisothiocyanate ($C_6H_5-N=C=S$), which reacts with the *N*-terminal residue to form a derivative that is then cleaved from the peptide with acid and identified. Automated sequencers use the Edman degradation method.

Electromagnetic spectrum (Section 13.8A): The full range of energies propagated by wave fluctuations in an electromagnetic field.

Electron density surface (Section 1.12B): An electron density surface shows points in space that happen to have the same electron density. An electron density surface can be calculated for any chosen value of electron density. A "high" electron density surface (also called a "bond" electron density surface) shows the *core* of electron density around each atomic nucleus and regions where neighboring atoms share electrons (bonding regions). A "low" electron density surface roughly shows the *outline* of a molecule's electron cloud. This surface gives information about molecular shape and volume, and usually looks the same as a van der Waals or space-filling model of the molecule. (Contributed by Alan Shusterman, Reed College, and Warren Hehre, Wavefunction, Inc.)

Electronegativity (Sections 1.3A and 2.2): A measure of the ability of an atom to attract electrons it is sharing with another and thereby polarize the bond.

Electron impact (EI) (Sections 9.14 and 9.16A): A method of ion formation in mass spectrometry whereby the sample to be analyzed (analyte) is placed in a high vacuum and, when in the gas phase, bombarded with a beam of high-energy electrons. A valence electron is displaced by the impact of the electron beam, yielding a species called the *molecular ion* (if there has been no fragmentation), with a +1 charge and an unshared electron (a radical cation).

Electron probability density (Section 1.10): The likelihood of finding an electron in a given volume of space. If the electron probability density is large, then the probability of finding an electron in a given volume of space is high, and the corresponding volume of space defines an orbital.

Electrophile (Sections 3.4A and 8.1A): A Lewis acid, an electron-pair acceptor, an electron-seeking reagent.

Electrophilic aromatic substitutions (Sections 15.1, 15.2, and 21.8): A reaction of aromatic compounds in which an *electrophile* ("electron-seeker" – a positive ion or other electron-deficient species with a full or large partial positive charge) replaces a hydrogen bonded to the carbon of an aromatic ring.

Electrophoresis (Section 25.6A): A technique for separating charged molecules based on their different mobilities in an electric field.

Electrospray ionization (ESI) (Section 9.19): A method of ion formation in mass spectrometry whereby a solution of the sample to be analyzed (analyte) is sprayed into the vacuum chamber of the mass spectrometer from the tip of a high-voltage needle, imparting charge to the mixture. Evaporation of the solvent in the vacuum chamber yields charged species of the analyte; some of which may have charges greater than +1. A family of *m/z* peaks unique to the formula weight of the analyte results, from which the formula weight itself can be calculated by computer.

Elimination-addition (via benzyne) (Section 21.11B): A substitution reaction in which a base, under highly forcing conditions, deprotonates an aromatic carbon that is adjacent to a carbon bearing a leaving group. Loss of the leaving group and overlap of the adjacent *p* orbitals creates a species, called *benzyne*, with a π-bond in the plane of the ring (separate from the aromatic π-system). Attack by a nucleophile on this π-bond followed by protonation yields a substituted aromatic compound.

Elimination reaction (Sections 3.1, 6.15–6.17, 7.5, 7.7): A reaction that results in the loss of two groups from the substrate and the formation of a π bond. The most common elimination is a 1,2 elimination or β elimination, in which the two groups are lost from adjacent atoms.

Enamines (Sections 16.8 and 18.9): An *enamine* group consists of an amine function bonded to the sp^2 carbon of an alkene.

Enantiomeric excess (or *enantiomeric purity*) (Section 5.9B): A percentage calculated for a mixture of enantiomers by dividing the moles of one enantiomer minus the moles of the other enantiomer by the moles of both enantiomers and multiplying by 100. The enantiomeric excess equals the percentage optical purity.

Enantiomers (Sections 5.2C, 5.3, 5.7, 5.8, and 5.16): Stereoisomers that are mirror images of each other.

Enantioselective reaction (see **Stereoselective reaction** and Sections 5.10B and 12.3D)

Enantiotopic hydrogens (or *ligands*) (Section 9.8B): If replacement of each of two hydrogens (or ligands) by the same group yields compounds that are enantiomers, the two hydrogen atoms (or ligands) are said to be enantiotopic.

Endergonic reaction (Section 6.7): A reaction that proceeds with a positive free-energy change.

Endo group (Section 13.10B): A group on a bicyclic compound that is on the same side (syn) as the longest bridge in the compound.

Endothermic reaction (Section 3.8A): A reaction that absorbs heat. For an endothermic reaction D*H*° is positive.

Energy (Section 3.8): Energy is the capacity to do work.

Energy of activation, E_{act} (Section 10.5A): A measure of the difference in potential energy between the reactants and the transition state of a reaction. It is related to, but not the same as, the free energy of activation, $\Delta G^{\ddagger}$.

Enol (Section 18.1): An alkene alcohol, where the hydroxyl group is bonded to an alkene carbon. A generally minor tautomeric equilibrium contributor to the keto form of a carbonyl group that has at least one alpha hydrogen.

Enolate (Sections 18.1, 18.3, and 18.4): The delocalized anion formed when an enol loses its hydroxylic proton or when the carbonyl tautomer that is in equilibrium with the enol loses an α proton.

Enthalpy change (Sections 3.8A, 3.9, and 3.16): Also called the heat of reaction. The *standard enthalpy change*, $\Delta H°$, is the change in enthalpy after a system in its standard state has undergone a transformation to another system, also in its standard state. For a reaction, $\Delta H°$ is a measure of the difference in the total bond energy of the reactants and products. It is one way of expressing the change in potential energy of molecules as they undergo reaction. The enthalpy change is related to the free-energy change, $\Delta G°$, and to the entropy change, $\Delta S°$, through the expression:

$$\Delta H° = \Delta G° + T\Delta S°$$

Entropy change (Section 3.9): The standard entropy change, $\Delta S°$, is the change in entropy between two systems in their standard states. Entropy changes have to do with changes in the relative order of a system. The more random a system is, the greater is its entropy. When a system becomes more disorderly its entropy change is positive.

Enzyme (Section 24.9): A protein or polypeptide that is a catalyst for biochemical reactions.

Enzyme-substrate complex (Section 24.9): The species formed when a substrate (reactant) binds at the active site of an enzyme.

Epimers, epimerization (Sections 18.3A and 22.8): Diastereomers that differ in configuration at only a single tetrahedral chirality center. Epimerization is the interconversion of epimers.

Epoxidation (Section 11.13A): The process of synthesizing an epoxide. Peroxycarboxylic acids (RCO_3H) are reagents commonly used for epoxidation.

Epoxide (Sections 11.13 and 11.14): An oxirane. A three-membered ring containing one oxygen and two carbon atoms.

Equatorial bond (Section 4.12): The six bonds of a cyclohexane ring that lie generally around the "equator" of the molecule:

Equilibrium constant, K_{eq} (Section 3.5A): A constant that expresses the position of an equilibrium. The equilibrium constant is calculated by multiplying the molar concentrations of the products together and then dividing this number by the number obtained by multiplying together the molar concentrations of the reactants.

Equilibrium control (see **Thermodynamic control**)

Essential amino acid (Section 24.2B) An amino acid that cannot be synthesized by the body and must be ingested as part of the diet. For adult humans there are eight essential amino acids ($RCH(NH_2)$ CO_2H): valine (R = isopropyl), Leucine (R = isobutyl), isoleucine (R = *sec*-butyl), phenylalanine (R = benzyl), threonine (R = 1-hydroxyethyl), methionine (R = 2-(methylthio)ethyl), lysine (R = 4-aminobutyl), and tryptophen (R = 3-methyleneindole).

Essential oil (Section 23.3): A volatile odoriferous compound obtained by steam distillation of plant material.

Esterification (Section 17.7A): The synthesis of an ester, usually involving reactions of carboxylic acids, acid chlorides or acid anhydrides with alcohols.

Exchangeable protons (Section 9.10): Protons that can be transferred rapidly from one molecule to another. These protons are often attached to electronegative elements such as oxygen or nitrogen.

Exergonic reaction (Section 6.7): A reaction that proceeds with a negative free-energy change.

Exo group (Section 13.10B): A group on a bicyclic compound that is on the opposite side (anti) to the longest bridge in the compound.

Exon (Section 25.5A): Short for "expressed sequence," an exon is a segment of DNA that is used when a protein is expressed. (See **Intron**).

Exothermic reaction (Section 3.8A): A reaction that evolves heat. For an exothermic reaction, $\Delta H°$ is negative.

(E)–(Z) system (Section 7.2): A system for designating the stereochemistry of alkene diastereomers based on the priorities of groups in the Cahn–Ingold–Prelog convention. An E isomer has the highest priority groups on opposites sides of the double bond, a Z isomer has the highest priority groups on the same side of the double bond.

F

Fat (Section 23.2): A triacylglycerol. The triester of glycerol with carboxylic acids.

Fatty acid (Section 23.2): A long-chained carboxylic acid (usually with an even number of carbon atoms) that is isolated by the hydrolysis of a fat.

Fischer projection (Sections 5.13 and 22.2C): A two-dimensional formula for representing the configuration of a chiral molecule. By convention, Fischer projection formulas are written with the main carbon chain extending from top to bottom with all groups eclipsed. Vertical lines represent bonds that project behind the plane of the page (or that lie in it). Horizontal lines represent bonds that project out of the plane of the page.

Fischer   Wedge-dashed
projection   wedge formula

Formal charge (Section 1.5): The difference between the number of electrons assigned to an atom in a molecule and the number of electrons it has in its outer shell in its elemental state. Formal charge can be calculated using the formula: $F = Z − S/2 − U$, where F is the formal charge, Z is the group number of the atom (i.e., the number of electrons the atom has in its outer shell in its elemental state), S is the number of electrons the atom is sharing with other atoms, and U is the number of unshared electrons the atom possesses.

Fourier transform NMR (Section 9.5): An NMR method in which a pulse of energy in the radiofrequency region of the electromagnetic spectrum is applied to nuclei whose nuclear magnetic moment is precessing about the axis of a magnetic field. This pulse of energy causes the nuclear magnetic moment to "tip" toward the xy plane. The component of the nuclear magnetic moment in the x–y plane generates ("induces") a radiofrequency signal, which is detected by the instrument. As nuclei relax to their ground states this signal decays over time; this time-dependent signal is called a "Free Induction Decay" (FID) curve. A mathematical operation (a Fourier transform) converts time-dependent data into frequency-dependent data–the NMR signal.

Fragmentation (Section 9.16): Cleavage of a chemical species by the breaking of covalent bonds, as in the formation of fragments during mass spectrometric analysis.

Free energy of activation, $\Delta G^{\ddagger}$ (Section 6.7): The difference in free energy between the transition state and the reactants.

Free-energy change (Section 3.9): The *standard free-energy change*, $\Delta G°$, is the change in free energy between two systems in their standard states. At constant temperature, $\Delta G° = \Delta H° − T\Delta S° = −RT \ln K_{eq}$, where $\Delta H°$ is the standard enthalpy change, $\delta S°$ is the standard entropy change, and K_{eq} is the equilibrium constant. A negative value of $\Delta G°$ for a reaction means that the formation of products is favored when the reaction reaches equilibrium.

Free-energy diagram (Section 6.7): A plot of free-energy changes that take place during a reaction versus the reaction coordinate. It displays free-energy changes as a function of changes in bond orders and distances as reactants proceed through the transition state to become products.

Freon (Section 10.12D): A chlorofluorocarbon or CFC.

Frequency, ν (Sections 2.15 and 13.8A): The number of full cycles of a wave that pass a given point in each second.

Friedel-Crafts acylation (Sections 15.6 and 15.7): Installation of an acyl group on a benzene ring by electrophilic aromatic substitution using an acylium ion as the electrophile (generated in situ using a Lewis acid).

Friedel-Crafts alkylation (Section 15.6): Installation of an alkyl group on a benzene ring by electrophilic aromatic substitution using an alkyl carbocation as the electrophile (generated in situ using a Lewis acid).

Fullerenes (Section 14.8C): Cagelike aromatic molecules with the geometry of a truncated icosahedron (or geodesic dome). The structures are composed of a network of pentagons and hexagons. Each carbon is sp^2 hybridized; the remaining electron at each carbon is delocalized into a system of molecular orbitals that gives the *whole molecule* aromatic character.

Functional class nomenclature (Section 4.3E): A system for naming compounds that uses two or more words to describe the compound. The final word corresponds to the functional group present; the preceding words, usually listed in alphabetical order, describe the remainder of the molecule. Examples are methyl alcohol, ethyl methyl ether, and ethyl bromide.

Functional group (Sections 2.2 and 2.4): The particular group of atoms in a molecule that primarily determines how the molecule reacts.

Functional group interconversion (Section 6.14): A process that converts one functional group into another.

Furanose (Section 22.2C): A sugar in which the cyclic acetal or hemiacetal ring is five membered.

G

Gauche conformation (Section 4.9): A gauche conformation of butane, for example, has the methyl groups at an angle of 60° to each other:

CH$_3$

CH$_3$

**A gauche
conformation
of butane**

GC/MS analysis (Section 9.18): An analytical method that couples a gas chromatograph (GC) with a mass spectrometer (MS). The GC separates the components of a mixture to be analyzed by sweeping the compounds, in the gas phase, through a column containing an adsorbant called a *stationary phase*. The gaseous molecules will cling to the surface of the stationary phase (be *adsorbed*) with different strengths. Those molecules that cling (adsorb) weakly will pass through the column quickly; those that *adsorb* more strongly will pass through the column more slowly. The separated components of the mixture are then introduced into the mass spectrometer, where they are analyzed.

***gem*-Dihalide** (Section 7.10A): A general term for a molecule or group containing two halogen atoms bonded to the same carbon.

Geminal (*gem*-) substituents (Section 7.10A): Substituents that are on the same atom.

Gene (Section 25.1): A section of DNA that codes for a given protein.

Genetic code (Sections 25.5C and 25.5D): The correspondence of specific three-base sequences in mRNA (codons) that each code for a specific amino acid. Each codon pairs with the anticodon of a specific tRNA, which in turn carries the corresponding amino acid.

Genome (Sections 25.1 and 25.9): The set of all genetic information coded by DNA in an organism.

Genomics (Section 24.14): The study of the complete set of genetic instructions in an organism.

Glycan (See **Polysaccharide** and Section 22.13): An alternate term for a polysaccharide; monosaccharies joined together by glycosidic linkages.

Glycol (Sections 4.3F and 8.15): A diol.

Glycolipids (Section 22.16): Carbohydrates joined through glycosidic linkages to lipids.

Glycoproteins (Section 22.16): Carbohydrates joined through glycosidic linkages to proteins.

Glycoside (Section 22.4): A cyclic mixed acetal of a sugar with an alcohol.

Grignard reagent (Section 12.6B): An organomagnesium halide, usually written RMgX.

Ground state (Section 1.12): The lowest electronic energy state of an atom or molecule.

H

^{1}H — ^{1}H correlation spectroscopy (COSY) (Section 9.12): A two-dimensional NMR method used to display the coupling between hydrogen atoms.

Haloform reaction (Section 18.3C): A reaction specific to methyl ketones. In the presence of base multiple halogenations occur at the carbon of the methyl group; excess base leads to acyl substitution of the trihalomethyl group, resulting in a carboxylate anion and a *haloform* (CHX$_3$).

Halogenation (Sections 10.3–10.5 and 10.8A): A reaction in which one or more halogen atoms are introduced into a molecule.

Halohydrin (Section 8.13): A compound bearing a halogen atom and a hydroxyl group on adjacent (vicinal) carbons.

Halonium ion (Section 8.11A): An ion containing a positive halogen atom bonded to two carbon atoms.

Hammond–Leffler postulate (Section 6.13A): A postulate stating that the structure and geometry of the transition state of a given step will show a greater resemblance to the reactants or products of that step depending on which is closer to the transition state in energy. This means that the transition state of an endothermic step will resemble the products of that step more than the reactants, whereas the transition state of an exothermic step will resemble the reactants of that step more than the products.

Heat of hydrogenation (Section 7.3A): The standard enthalpy change that accompanies the hydrogenation of 1 mol of a compound to form a particular product.

Heisenberg uncertainty principle (Section 1.11): A fundamental principle that states that both the position and momentum of an electron (or of any object) cannot be exactly measured simultaneously.

Hemiacetal (Sections 16.7A and 22.2C): A functional group, consisting of an sp^3 carbon atom bearing both an alkoxyl group and a hydroxyl group [i.e., RCH(OH)(OR′) or R$_2$C(OH)(OR′)].

Hemiketal (See **Hemiacetal** and Section 16.7A)

Henderson-Hasselbalch equation (Sect. 24.2C): The Henderson-Hasselbalch equation ($pK_a = pH + log[HA]/[A-]$) shows that when the concentration of an acid and its conjugate base are equal, the pH of the solution equals the pK_a of the acid.

Hertz (Hz) (Sections 9.6A, 9.9C, and 13.8A): The frequency of a wave. Now used instead of the equivalent cycles per second (cps).

Heteroatom (Section 2.1): Atoms such as oxygen, nitrogen, sulfur and the halogens that form bonds to carbon and have unshared pairs of electrons.

Heterocyclic amines (Section 20.1B): A secondary or tertiary amine in which the nitrogen group is part of a carbon-based ring.

Heterocyclic compound (Sections 14.9): A compound whose molecules have a ring containing an element other than carbon.

Heterogeneous catalysis (Sections 7.12 and 7.14A): Catalytic reactions in which the catalyst is insoluble in the reaction mixture.

Heterolysis (Section 3.4): The cleavage of a covalent bond so that one fragment departs with both of the electrons of the covalent bond that joined them. Heterolysis of a bond normally produces positive and negative ions.

Heteronuclear correlation spectroscopy (HETCOR or C-H HETCOR) (Section 9.12): A two-dimensional NMR method used to display the coupling between hydrogens and the carbons to which they are attached.

Heterotopic (chemically nonequivalent atoms) (Section 9.8A): Atoms in a molecule where replacement of one or the other leads to a new compound. Heterotopic atoms are not chemical shift equivalent in NMR spectroscopy.

Hofmann rule (Sections 7.6C and 20.12A): When an elimination yields the alkene with the less substituted double bond, it is said to follow the Hofmann rule.

HOMO (Sections 3.3A, 6.6, and 13.8C): The highest occupied molecular orbital.

Homogeneous catalysis (Section 7.12): Catalytic reactions in which the catalyst is soluble in the reaction mixture.

Homologous series (Section 4.7): A series of compounds in which each member differs from the next member by a constant unit.

Homolysis (Section 10.1): The cleavage of a covalent bond so that each fragment departs with one of the electrons of the covalent bond that joined them.

Homolytic bond dissociation energy, $DH°$ (Section 10.2): The enthalpy change that accompanies the homolytic cleavage of a covalent bond.

Homotopic (chemically equivalent) atoms (Section 9.8A): Atoms in a molecule where replacement of one or another results in the same compound. Homotopic atoms are chemical shift equivalent in NMR spectroscopy.

Hückel's rule (Section 14.7): A rule stating that planar monocyclic rings with $(4n + 2)$ delocalized π electrons (i.e., with 2, 6, 10, 14,..., delocalized π electrons) will be aromatic.

Hund's rule (Section 1.10A): A rule used in applying the aufbau principle. When orbitals are of equal energy (i.e., when they are degenerate), electrons are added to each orbital with their spins unpaired, until each degenerate orbital contains one electron. Then electrons are added to the orbitals so that the spins are paired.

Hybrid atomic orbitals (Sections 1.12 and 1.15): An orbital that results from the mathematical combination of pure atomic orbitals, such as the combination of pure s and p orbitals in varying proportions to form hybrids such as sp^3, sp^2, and sp orbitals.

Hydration (Sections 8.4–8.9 and 11.4): The addition of water to a molecule, such as the addition of water to an alkene to form an alcohol.

Hydrazone (Section 16.8B): An imine in which an amino group ($-NH_2$, $-NHR$, $-NR_2$) is bonded to the nitrogen atom.

Hydride (Section 7.8A): A hydrogen anion, $H:^-$ Hydrogen with a filled 1s shell (containing two electrons) and negative charge.

Hydride ion (Section 12.1A): The anionic form of hydrogen; a proton with two electrons.

Hydroboration (Sections 8.6, 8.7, and 11.4): The addition of a boron hydride (either BH_3 or an alkylborane) to a multiple bond.

Hydrocarbon (Section 2.2): A molecular containing only carbon and hydrogen atoms.

Hydrogen abstraction (Section 10.1B): The process by which a species with an unshared electron (a radical) removes a hydrogen atom from another species, breaking the bond to the hydrogen homolytically.

Hydrogenation (Sections 4.16, 7.3A, and 7.13–7.15): A reaction in which hydrogen adds to a double or triple bond. Hydrogenation is often accomplished through the use of a metal catalyst such as platinum, palladium, rhodium, or ruthenium.

Hydrogen bond (Sections 2.13B, 2.13E, 2.13F, and 2.14): A strong dipole–dipole interaction (4–38 kJ mol^{-1}) that occurs between hydrogen atoms bonded to small strongly electronegative atoms (O, N, or F) and the nonbonding electron pairs on other such electronegative atoms.

Hydrophilic group (Sections 2.13D and 23.2C): A polar group that seeks an aqueous environment.

Hydrophobic group (See also **Lipophilic group**) (Sections 2.13D and 23.2C): A nonpolar group that avoids an aqueous surrounding and seeks a nonpolar environment.

Hyperconjugation (Sections 4.8B and 6.11B): Electron delocalization (via orbital overlap) from a filled bonding orbital to an adjacent unfilled orbital. Hyperconjugation generally has a stabilizing effect.

I

Imines (Section 16.8): A structure with a carbon-nitrogen double bond. If the groups bonded to carbon are not the same, (E) and (Z) isomers are possible.

Index of hydrogen deficiency (Section 4.17): The index of hydrogen deficiency (or IHD) equals the number of pairs of hydrogen atoms that must be subtracted from the molecular formula of the corresponding alkane to give the molecular formula of the compound under consideration.

Induced fit hypothesis (Section 24.9): An hypothesis regarding enzyme reactivity whereby formation of the enzyme-substrate complex causes conformational changes in the enzyme that facilitate conversion of the substrate to product.

Inductive effect (Sections 3.7B, 3.10B, and 15.11B): An intrinsic electron-attracting or -releasing effect that results from a nearby dipole in the molecule and that is transmitted through space and through the bonds of a molecule.

Infrared (IR) spectroscopy (Section 2.15): A type of optical spectroscopy that measures the absorption of infrared radiation. Infrared spectroscopy provides structural information about functional groups present in the compound being analyzed.

Inhibitor (Section 24.9): A compound that can negatively alter the activity of an enzyme.

Integration (Section 9.2B): A numerical value representing the relative area under a signal in an NMR spectrum. In ^{1}H NMR, the integration value is proportional to the number of hydrogens producing a given signal.

Intermediate (Sections 3 intro, 6.10, and 6.11): A transient species that exists between reactants and products in a state corresponding to a local energy minimum on a potential energy diagram.

Intermolecular forces (Sections 2.13B and 2.13F): Also known as van der Waals forces. Forces that act between molecules because of permanent (or temporary) electron distributions. Intermolecular forces can be attractive or repulsive. Dipole-dipole forces (including hydrogen bonds) and dispersion forces (also called London forces), are intermolecular forces of the van der Waal type.

Intron (Section 25.5A): Short for "intervening sequence," an intron is a segments of DNA that is not actually used when a protein is expressed, even though it is transcribed into the initial mRNA.

Inversion of configuration (Sections 6.6 and 6.14): At a tetrahedral atom, the process whereby one group is replaced by another bonded 180° opposite to the original group. The other groups at the tetrahedral atom "turn inside out" (shift) in the same way that an umbrella "turns inside out." When a chirality center undergoes configuration inversion, its (R,S) designation may switch, depending on the relative Cahn-Ingold-Prelog priorities of the groups before and after the reaction.

Ion (Sections 1.3A and 3.1A): A chemical species that bears an electrical charge.

Ion–dipole force (Section 2.13D): The interaction of an ion with a permanent dipole. Such interactions (resulting in solvation) occur between ions and the molecules of polar solvents.

Ionic bond (Section 1.3A): A bond formed by the transfer of electrons from one atom to another resulting in the creation of oppositely charged ions.

Ionic reaction (Sections 3.1B and 10.1): A reaction involving ions as reactants, intermediates, or products. Ionic reactions occur through the heterolysis of covalent bonds.

Ion-ion forces (Section 2.13A): Strong electrostatic forces of attraction between ions of opposite charges. These forces hold ions together in a crystal lattice.

Ionization (Section 9.14): Conversion of neutral molecules to ions (charged species).

Isoelectric point (pI) (Section 24.2C): The pH at which the number of positive and negative charges on an amino acid or protein are equal.

Isomers (Sections 1.6 and 5.2A): Different molecules that have the same molecular formula.

Isoprene unit (Section 23.3): A name for the structural unit found in all terpenes:

Isotactic polymer (Special Topic B.1 in *WileyPLUS*): A polymer in which the configuration at each stereogenic center along the chain is the same.

Isotopes (Section 1.2A): Atoms that have the same number of protons in their nuclei but have differing atomic masses because their nuclei have different numbers of neutrons.

IUPAC system (Section 4.3): (also called the "systematic nomenclature") A set of nomenclature rules overseen by the International Union of Pure and Applied Chemistry (IUPAC) that allows every compound to be assigned an unambiguous name.

K

Karplus correlation (Section 9.9D): An empirical correlation between the magnitude of an NMR coupling constant and the dihedral angle between two coupled protons. The dihedral angles derived in this manner can provide information about molecular geometries.

Kekulé structure (Sections 2.1D and 14.4): A structure in which lines are used to represent bonds. The Kekulé structure for benzene is a hexagon of carbon atoms with alternating single and double bonds around the ring, and with one hydrogen atom attached to each carbon.

Ketal (See **Acetal** and Section 16.7B)

Keto and enol forms (Sections 18.1–18.3): Tautomeric forms of a compound related by a common resonance-stabilized intermediate. An *enol* structure consists of an alcohol functionality bonded to the sp^2 carbon of an alkene. Shifting the hydroxyl proton to the alkene and creation of a carbon-oxygen π-bond results in the *keto* form of the species.

Ketose (Section 22.2A): A monosaccharide containing a ketone group or a hemiacetal or acetal derived from it.

Kinetic control (Sections 7.6B, 13.9A, and 18.4A): A principle stating that when the ratio of products of a reaction is determined by relative rates of reaction, the most abundant product will be the one that is formed fastest. Also called rate control.

Kinetic energy (Section 3.8): Energy that results from the motion of an object. Kinetic energy $(KE) = \frac{1}{2}mv^2$, where m is the mass of the object and v is its velocity.

Kinetic enolate (Section 18.4A): In a situation in which more than one enolate anion can be formed, the *kinetic enolate anion* is that which is formed most rapidly. This is usually the enolate anion with the less substituted double bond; the decrease in steric hindrance permits more rapid deprotonation by the base. A kinetic enolate anion is formed predominantly under conditions that do not permit the establishment of an equilibrium.

Kinetic product (Section 13.9): The product formed fastest when multiple products are possible; the product formed via the lowest energy of activation pathway.

Kinetic resolution (Section 5.10B): A process in which the rate of a reaction with one enantiomer is different than with the other, leading to a preponderance of one product stereoisomer. This process is said to be "stereoselective" in that it leads to the preferential formation of one stereoisomer over other stereoisomers that could possibly be formed.

Kinetics (Section 6.5): A term that refers to rates of reactions.

L

Lactam (Section 17.8I): A cyclic amide.

Lactone (Section 17.7C): A cyclic ester.

LCAO (linear combination of atomic orbitals, Section 1.11): A mathematical method for arriving at wave functions for molecular obitals that involves adding or subtracting wave functions for atomic orbitals.

Leaving group (Sections 6.2, 6.4, and 6.13E): The substituent that departs from the substrate in a nucleophilic substitution reaction.

Leveling effect of a solvent (Section 3.14): An effect that restricts the use of certain solvents with strong acids and bases. In principle, no acid stronger than the conjugate acid of a particular solvent can exist to an appreciable extent in that solvent, and no base stronger than the conjugate base of the solvent can exist to an appreciable extent in that solvent.

Levorotatory (Section 5.8B): A compound that rotates plane-polarized light in a counterclockwise direction.

Lewis acid–base theory (Section 3.3): An acid is an electron pair acceptor, and a base is an electron pair donor.

Lewis structure (or *electron-dot structure*) (Sections 1.3B, 1.4, and 1.5): A representation of a molecule showing electron pairs as a pair of dots or as a dash.

Lipid (Section 23.1): A substance of biological origin that is soluble in nonpolar solvents. Lipids include fatty acids, triacylglycerols (fats and oils), steroids, prostaglandins, terpenes and terpenoids, and waxes.

Lipid bilayers (Section 23.6A): A two-layer noncovalent molecular assembly comprised primarily of phospholipids. The hydrophobic phospholipid "tail" groups of each layer orient toward each other in the center of the two-layered structure due to attractive dispersion forces. The hydrophilic "head" groups of the lipids orient toward the aqueous exterior of the bilayer. Lipid bilayers are important in biological systems such as cell membranes.

Lipophilic group (See also **Hydrophobic group**) (Sections 2.13D and 23.2C): A nonpolar group that avoids an aqueous surrounding and seeks a nonpolar environment.

Lithium diisopropylamide (LDA) (Section 18.4): $(i\text{-}C_3H_7)_2N^-Li^+$ The lithium salt of diisopropylamine. A strong base used to form *lithium enolates* from carbonyl compounds.

Lock-and-key hypothesis (Section 24.9): An hypothesis that explains enzyme specificity on the basis of complementary geometry between the enzyme (the "lock") and the substrate (the "key"), such that their shapes "fit together" correctly for a reaction to occur.

LUMO (Sections 3.3A and 13.8C): The lowest unoccupied molecular orbital.

M

Macromolecule (Section 10.11): A very large molecule.

Magnetic resonance imaging (MRI) (Section 9.12B): A technique based on NMR spectroscopy that is used in medicine.

Malonic ester synthesis (Section 18.7): A reaction in which the α-hydrogen of diethyl propanedioate (diethyl malonate, also called "malonic ester") is removed, creating a resonance-stabilized anion which can serve as a nucleophile in an S_N2 reaction. The α-carbon can be substituted twice; the ester functionalities can be converted into a carboxylic acid which, after decarboxylation, will yield a substituted ketone.

Mannich reaction (Section 19.8): The reaction of an enol with an iminium cation (formed from the reaction of a primary or secondary amine with formaldehyde) to yield a β-aminoalkyl carbonyl compound.

Markovnikov's rule (Sections 8.2B and 8.18): A rule for predicting the regiochemistry of electrophilic additions to alkenes and alkynes that can be stated in various ways. As originally stated (in 1870) by Vladimir Markovnikov, the rule provides that "if an unsymmetrical alkene combines with a hydrogen halide, the halide ion adds to the carbon with the fewer hydrogen atoms." More commonly the rule has been stated in reverse: that in the addition of HX to an alkene or alkyne the hydrogen atom adds to the carbon atom that already has the greater number of hydrogen atoms. A modern expression of Markovnikov's rule is: *In the ionic addition of an unsymmetrical reagent to a multiple bond, the positive portion of the reagent (the electrophile) attaches itself to a carbon atom of the reagent in the way that leads to the formation of the more stable intermediate carbocation.*

Mass spectrometry (MS) (Section 9.13): A technique, useful in structure elucidation, that involves the generation of ions from a molecule, the sorting and detecting of the ions, and the display of the result in terms of the mass/charge ratio and relative amount of each ion.

Matrix-assisted laser desorption-ionization (MALDI) (Section 9.19): A method in mass spectrometry for ionizing analytes that do not ionize well by electrospray ionization. The analyte is mixed with low molecular weight organic molecules that can absorb energy from a laser and then transfer this energy to the analyte, producing +1 ions which are then analyzed by the mass spectrometer.

Mechanism (See **Reaction mechanism**)

Melting Point (Section 2.13A): The temperature at which an equilibrium exists between a well-ordered crystalline substance and the more random liquid state. It reflects the energy needed to overcome the attractive forces between the units (ions, molecules) that comprise the crystal lattice.

Meso compound (Section 5.12B): An optically inactive compound whose molecules are achiral even though they contain tetrahedral atoms with four different attached groups.

Mesylate (Section 11.10): A methanesulfonate ester. Methanesulfonate esters are compounds that contain the CH_3SO_3— group, i.e., CH_3SO_3R.

Meta directors (Section 15.10B): An electron-withdrawing group on an aromatic ring. The major product of electrophilic aromatic substitution on a ring bearing a meta-directing group will have the newly substituted electrophile located meta to the substituent.

Methanide (Section 7.8A): A methyl anion, $-\!:\!CH3$, or methyl species that reacts as though it were a methyl anion.

Methylene (Section 8.14A): The carbene with the formula $:CH_2$.

Methylene group (Section 2.4B): The —CH_2— group.

Micelle (Section 23.2C): A spherical cluster of ions in aqueous solution (such as those from a soap) in which the nonpolar groups are in the interior and the ionic (or polar) groups are at the surface.

Michael addition (See **Conjugate addition** and Sections 18.9 and 19.7): A reaction between an active hydrogen compound and an α,β-unsaturated carbonyl compound. The attack by the anion of the active hydrogen compound takes place at the β-carbon of the α,β-unsaturated carbonyl compound. A Michael addition is a type of conjugate addition.

Molar absorptivity, ε (Section 13.8B): A proportionality constant that relates the observed absorbance (A) at a particular wavelength (λ) to the molar concentration of the sample (C) and the length (l) (in centimeters) of the path of the light beam through the sample cell:

$$\varepsilon = A/C \times l$$

Molecular formula (Section 1.6): A formula that gives the total number of each kind of atom in a molecule. The molecular formula is a whole number multiple of the empirical formula. For example the molecular formula for benzene is C_6H_6; the empirical formula is CH.

Molecular ion (Sections 9.14, 9.15, and 9.17): The cation produced in a mass spectrometer when one electron is dislodged from the parent molecule, symbolized $M^{\ddot{+}}$.

Molecularity (Section 6.5B): The number of species involved in a single step of a reaction (usually the rate-determining step).

Molecular orbital (MO) (Sections 1.11 and 1.15): Orbitals that encompass more than one atom of a molecule. When atomic orbitals combine to form molecular orbitals, the number of molecular orbitals that results always equals the number of atomic orbitals that combine.

Molecule (Section 1.3B): An electrically neutral chemical entity that consists of two or more bonded atoms.

Monomer (Section 10.11): The simple starting compound from which a polymer is made. For example, the polymer polyethylene is made from the monomer ethylene.

Monosaccharide (Sections 22.1A and 22.2): The simplest type of carbohydrate, one that does not undergo hydrolytic cleavage to a simpler carbohydrate.

Mutarotation (Section 22.3): The spontaneous change that takes place in the optical rotation of α and β anomers of a sugar when they are dissolved in water. The optical rotations of the sugars change until they reach the same value.

N

Nanotube (Section 14.8C): A tubular structure with walls resembling fused benzene rings, capped by half of a "buckyball" (buckminsterfullerene) at each end. The entire structure exhibits aromatic character.

Newman projection formula (Section 4.8A): A means of representing the spatial relationships of groups attached to two atoms of a molecule. In writing a Newman projection formula we imagine ourselves viewing the molecule from one end directly along the bond axis joining the two atoms. Bonds that are attached to the front atom are shown as radiating from the center of a circle; those attached to the rear atom are shown as radiating from the edge of the circle:

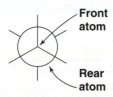

Front atom

Rear atom

N-nitrosoamines (Section 20.6C): Amines bearing an $N{=}O$ on the nitrogen, such as $R{-}NH{-}N{=}O$ or $Ar{-}NH{-}N{=}O$. Often referred to as "nitrosamines" in the popular press. N-nitrosoamines are very powerful carcinogens.

Node (Section 1.15): A place where a wave function (ψ) is equal to zero. The greater the number of nodes in an orbital, the greater is the energy of the orbital.

Nonbenzenoid aromatic compound (Section 14.8B): An aromatic compound, such as azulene, that does not contain benzene rings.

Nuclear magnetic resonance (NMR) spectroscopy (Sections 9.2 and 9.11A): A spectroscopic method for measuring the absorption of radio frequency radiation by certain nuclei when the nuclei are in a strong magnetic field. The most important NMR spectra for organic chemists are ^{1}H NMR spectra and ^{13}C NMR spectra. These two types of spectra provide structural information about the carbon framework of the molecule, and about the number and environment of hydrogen atoms attached to each carbon atom.

Nucleic acids (Sections 25.1 and 25.4A): Biological polymers of nucleotides. DNA and RNA are, respectively, nucleic acids that preserve and transcribe hereditary information within cells.

Nucleophile (Sections 3.4A, 6.2, 6.3, and 6.13B): A Lewis base, an electron pair donor that seeks a positive center in a molecule.

Nucleophilic addition-elimination (Section 17.4): Addition of a nucleophile to a carbonyl (or other trigonal) carbon, yielding a tetrahedral intermediate, followed by elimination of a leaving group to yield a trigonal planar product.

Nucleophilic addition to the carbonyl carbon (Sections 12.1A and 16.6): A reaction in which a *nucleophile* (an electron-pair

donor) forms a bond to the carbon of a *carbonyl* ($C{=}O$) group. To avoid violating the octet rule, the electrons of the carbon-oxygen π-bond shift to the oxygen, resulting in a four-coordinate (tetrahedral) carbon.

Nucleophilic aromatic substitution (Section 21.11A): A substitution reaction in which a nucleophile attacks an aromatic ring bearing strongly electron-withdrawing groups in ortho and/or para positions relative to the site of attack and the leaving group. This step is an addition reaction that yields and aryl carbanion (called a Meisenheimer Complex) which is stabilized by the electron-withdrawing groups on the ring. Loss of the leaving group in an elimination step regenerates the aromatic system, yielding a substituted aromatic compound by what was, overall, an addition-elimination process.

Nucleophilicity (Section 6.13B): The relative reactivity of a nucleophile in an S_N2 reaction as measured by relative rates of reaction.

Nucleophilic substitution reaction (Section 6.2): A reaction initiated by a nucleophile (a neutral or negative species with an unshared electron pair) in which the nucleophile reacts with a substrate to replace a substituent (called the leaving group) that departs with an unshared electron pair.

Nucleoside (Sections 22.15A, 25.2, and 25.3): A five-carbon monosaccharide bonded at the 1′ position to a purine or pyrimidine.

Nucleotide (Sections 25.2 and 25.3): A five-carbon monosaccharide bonded at the 1′ position to a purine or pyrimidine and at the 3′ or 5′ position to a phosphate group.

O

Octet rule (Sections 1.3 and 1.4A): An empirical rule stating that atoms not having the electronic configuration of a noble gas tend to react by either transferring electrons or sharing electrons so as to achieve the valence electron configuration (i. e., eight electrons) of a noble gas.

Oil (Section 23.2): A triacylglycerol (see below) that is liquid at room temperature.

Olefin (Section 7.1): An old name for an alkene.

Oligonucleotide synthesis (Section 25.7): Synthesis of specific sequence of nucleotides, often by automated solid-phase techniques, in which the nucleotide chain is built up by adding a protected nucleotide in the form of a phosphoramidite to a protected nucleotide linked to a solid phase, (usually a "controlled pore glass") in the presence of a coupling agent. The phosphite triester product is oxidized to a phosphate triester with iodine, producing a chain that has been lengthened by one nucleotide. The protecting group is then removed, and the steps (coupling, oxidation, deprotection) are repeated. After the desired oligonucleotide has been synthesized it is cleaved from the solid support and the remaining protecting groups removed.

Oligopeptide (Section 24.4): A peptide comprised of 3–10 amino acids.

Oligosaccharides (Section 22.1A): A carbohydrate that hydrolyzes to yield 2–10 monosaccharide molecules.

Optically active compound (Sections 5.8 and 5.9): A compound that rotates the plane of polarization of plane-polarized light.

Optical purity (Section 5.9B): A percentage calculated for a mixture of enantiomers by dividing the observed specific rotation for the mixture by the specific rotation of the pure enantiomer and multiplying by 100. The optical purity equals the enantiomeric purity or enantiomeric excess.

Orbital (Section 1.10): A volume of space in which there is a high probability of finding an electron. Orbitals are described mathematically by the squaring of wave functions, and each orbital has a characteristic energy. An orbital can hold two electrons when their spins are paired.

Orbital hybridization (Section 1.12): A mathematical (and theoretical) mixing of two or more atomic orbitals to give the same number of new orbitals, called *hybrid orbitals*, each of which has some of the character of the original atomic orbitals.

Organometallic compound (Section 12.5): A compound that contains a carbon–metal bond.

Orthogonal protecting groups (Section 24.7D): Protecting groups in which one set of protecting groups is stable under conditions for removal of the other, and vice versa.

Ortho-para directors (Section 15.10B): An electron-donating group on an aromatic ring. The major product of electrophilic aromatic substitution on a ring bearing such a group will have the newly substituted electrophile located ortho and/or para to the ortho-para-directing group.

Osazone (Section 22.8): A 1,2-bisarylhydrazone formed by reaction of an aldose or ketose with three molar equivalents of an arylhydrazone. Most common are phenylosazones, formed by reaction with phenylhydrazine, and 2,4-dinitrophenylhydrazones.

Oxidation (Sections 12.2 and 12.4): A reaction that increases the oxidation state of atoms in a molecule or ion. For an organic substrate, oxidation usually involves increasing its oxygen content or decreasing its hydrogen content. Oxidation also accompanies any reaction in which a less electronegative substituent is replaced by a more electronegative one.

Oxidative cleavage (Sections 8.16 and 8.19): A reaction in which the carbon-carbon double bond of an alkene or alkyne is both cleaved and oxidized, yielding compounds with carbon-oxygen double bonds.

Oxidizing agent (Section 12.2): A chemical species that causes another chemical species to become oxidized (lose electrons, or gain bonds to more electronegative elements, often losing bonds to hydrogen in the process). The oxidizing agent is reduced in this process.

Oxime (Section 16.8B): An imine in which a hydroxyl group is bonded to the nitrogen atom.

Oxirane (See **Epoxide** and Section 11.13)

Oxonium ion (Sections 3.12 and 11.12): A chemical species with an oxygen atom that bears a formal positive charge.

Oxonium salt (Section 11.12): A salt in which the cation is a species containing a positively charged oxygen.

Oxymercuration (Sections 8.5 and 11.4): The addition of —OH and —HgO$_2$CR to a multiple bond.

Oxymercuration-demercuration (Sections 8.5 and 11.4): A two-step process for adding the elements of water (H and OH) to a double bond in a Markovnikov orientation without rearrangements. An alkene reacts with mercuric acetate (or trifluoroacetate), forming a bridged mercurinium ion. Water preferentially attacks the more substituted side of the bridged ion, breaking the bridge and resulting, after loss of a proton, in an alcohol. Reduction with NaBH$_4$ replaces the mercury group with a hydrogen atom, yielding the final product.

Ozonolysis (Sections 8.16B and 8.19): The oxidative cleavage of a multiple bond using O$_3$ (ozone). The reaction leads to the formation of a cyclic compound called an *ozonide*, which is then reduced to carbonyl compounds by treatment with dimethyl sulfide (Me$_2$S) or zinc and acetic acid.

P

p orbitals (Section 1.10): A set of three degenerate (equal energy) atomic orbitals shaped like two tangent spheres with a nodal plane at the nucleus. For *p* orbitals of second row elements, the principal quantum number, *n* (see **Atomic orbital**), is 2; the azimuthal quantum number, *l*, is 1; and the magnetic quantum numbers, *m*, are +1, 0, or −1.

Paraffin (Section 4.15): An old name for an alkane.

Partial hydrolysis (Section 24.5D): Random cleavage of a polypeptide with dilute acid, resulting in a family of peptides of varying lengths that can be more easily sequenced than the parent polypeptide. Once each fragment peptide is sequenced, the areas of overlap indicate the sequence of the initial peptide.

Pauli exclusion principle (Section 1.10A): A principle that states that no two electrons of an atom or molecule may have the same set of four quantum numbers. It means that only two electrons can occupy the same orbital, and then only when their spin quantum numbers are opposite. When this is true, we say that the spins of the electrons are paired.

Peptide (Section 24.4): A molecule comprised of amino acids bonded via amide linkages.

Peptide bond, peptide linkage (Section 24.4): The amide linkage between amino acids in a peptide.

Peracid (See **Peroxy acid,** Section 11.13A)

Periplanar (See **Coplanar,** Section 7.6D)

Peroxide (Section 10.1A): A compound with an oxygen–oxygen single bond.

Peroxy acid (Section 11.13A): An acid with the general formula RCO$_3$H, containing an oxygen–oxygen single bond.

Phase sign (Section 1.9): Signs, either + or −, that are characteristic of all equations that describe the amplitudes of waves.

Phase transfer catalysis (Section 11.16): A reaction using a reagent that transports an ion from an aqueous phase into a nonpolar phase where reaction takes place more rapidly. Tetraalkylammonium ions and crown ethers are phase-transfer catalysts.

Phenyl halide (Section 6.1): An organic halide in which the halogen atom is bonded to a benzene ring. A phenyl halide is a specific type of aryl halide (Section 6.1).

Phospholipid (Section 23.6): Compound that is structurally derived from *phosphatidic acid*. Phosphatidic acids are derivatives of glycerol in which two hydroxyl groups are joined to fatty acids, and one terminal hydroxyl group is joined in an ester linkage to phosphoric acid. In a phospholipid the phosphate group of the phosphatidic acid is joined in ester linkage to a nitrogen-containing compound such as choline, 2-aminoethanol, or L-serine.

Physical property (Section 2.13): Properties of a substance, such as melting point and boiling point, that relate to physical (as opposed to chemical) changes in the substance.

Pi (π) bond (Section 1.13): A bond formed when electrons occupy a bonding π molecular orbital (i.e., the lower energy molecular orbital that results from overlap of parallel *p* orbitals on adjacent atoms).

Pi (π) molecular orbital (Section 1.13): A molecular orbital formed when parallel *p* orbitals on adjacent atoms overlap. Pi molecular orbitals may be *bonding* (*p* lobes of the same phase sign overlap) or *antibonding* (*p* orbitals of opposite phase sign overlap).

p*K*a (Section 3.5B): The pK_a is the negative logarithm of the acidity constant, K_a. p$K_a = -\log K_a$.

Plane-polarized light (Section 5.8A): Light in which the oscillations of the electrical field occur only in one plane.

Plane of symmetry (Sections 5.6 and 5.12A): An imaginary plane that bisects a molecule in a way such that the two halves of the molecule are mirror images of each other. Any molecule with a plane of symmetry will be achiral.

Polar aprotic solvent (Section 6.13C): A polar solvent that does not have a hydrogen atom attached to an electronegative element. Polar aprotic solvents do *not* hydrogen bond with a Lewis base (e.g., a nucleophile).

Polar covalent bond (Section 2.2): A covalent bond in which the electrons are not equally shared because of differing electronegativities of the bonded atoms.

Polarimeter (Section 5.8B): A device used for measuring optical activity.

Polarizability (Section 6.13C): The susceptibility of the electron cloud of an uncharged molecule to distortion by the influence of an electric charge.

Polar molecule (Section 2.3): A molecule with a dipole moment.

Polar protic solvent (Section 6.13D): A polar solvent that has at least one hydrogen atom bonded to an electronegative element. These hydrogen atoms of the solvent can form hydrogen bonds with a Lewis base (e.g., a nucleophile).

Polymer (Section 10.11): A large molecule made up of many repeating subunits. For example, the polymer polyethylene is made up of the repeating subunit $-(CH_2CH_2)_n-$.

Polymerase chain reaction (PCR) (Section 25.8): A method for multiplying (amplifying) the number of copies of a DNA molecule. The reaction uses DNA polymerase enzymes to attach additional nucleotides to a short oligonucleotide "primer" that is bound to a complementary strand of DNA called a "template." The nucleotide that the polymerases attach are those that are complementary to the base in the adjacent position on the template strand. Each cycle doubles the amount of target DNA that existed prior to the reaction step, yielding an exponential increase in the amount of DNA over time.

Polymerizations (Section 10.11): Reactions in which individual subunits (called *monomers*) are joined together to form long-chain macromolecules.

Polypeptide (Section 24.4): A peptide comprised of many (>10) amino acids.

Polysaccharide (Sections 22.1A and 22.13): A carbohydrate that, on a molecular basis, undergoes hydrolytic cleavage to yield many molecules of a monosaccharide. Also called a glycan.

Polyunsaturated fatty acid/ester (Section 23.2): A fatty acid or ester of a fatty acid whose carbon chain contain two or more double bonds.

Potential energy (Section 3.8): Potential energy is stored energy; it exists when attractive or repulsive forces exist between objects.

Potential energy diagram (Section 4.8B); A graphical plot of the potential energy changes that occurs as molecules (or atoms) react (or interact). Potential energy is plotted on the vertical axis, and the progress of the reaction on the horizontal axis

Primary carbon atom (Section 2.5): A carbon atom that has only one other carbon atom attached to it.

Primary structure (Sections 24.1, 24.5, and 24.6): The covalent structure of a polypeptide or protein. This structure is determined, in large part, by determining the sequence of amino acids in the protein.

Prochiral center (Section 12.3D): A group is prochiral if replacement of one of two identical groups at a tetrahedral atom, or if addition of a group to a trigonal planar atom, leads to a new chirality center. At a tetrahedral atom where there are two identical groups, the identical groups can be designated pro-*R* and pro-*S* depending on what configuration would result when it is imagined that each is replaced by a group of next higher priority (but not higher than another existing group).

Prostaglandins (Section 23.5): Natural C_{20} carboxylic acids that contain a five-membered ring, at least one double bond, and several oxygen-containing functional groups. Prostaglandins mediate a variety of physiological processes.

Prosthetic group (Sections 24.9 and 24.12): An enzyme cofactor that is permanently bound to the enzyme.

Protecting group (Sections 11.11D, 11.11E, 12.9, 15.5, 16.7C, and 24.7A): A group that is introduced into a molecule to protect a sensitive group from reaction while a reaction is carried out at some other location in the molecule. Later, the protecting group is removed. Also called blocking group. (See also **Orthogonal protecting group**.)

Protein (Section 24.4): A large biological polymer of α-amino acids joined by amide linkages.

Proteome Proteome (Sections 25.1 and 25.9): The set of all proteins encoded within the genome of an organism and expressed at any given time.

Proteomics (Section 24.14): The study of all proteins that are expressed in a cell at a given time.

Protic solvent (Sections 3.11, 6.13C, and 6.13D): A solvent whose molecules have a hydrogen atom attached to a strongly electronegative element such as oxygen or nitrogen. Molecules of a protic solvent can therefore form hydrogen bonds to unshared electron pairs of oxygen or nitrogen atoms of solute molecules or ions, thereby stabilizing them. Water, methanol, ethanol, formic acid, and acetic acid are typical protic solvents.

Proton decoupling (Section 9.11B): An electronic technique used in ^{13}C NMR spectroscopy that allows decoupling of spin–spin interactions between ^{13}C nuclei and ^1H nuclei. In spectra obtained in this mode of operation all carbon resonances appear as singlets.

Psi (ψ) function (See **Wave function** and Section 1.9)

Pyranose (Section 22.2C): A sugar in which the cyclic acetal or hemiacetal ring is six membered.

Q

Quartet (Section 9.2): An NMR signal comprised of four peaks in a 1:3:3:1 area ratio, caused by signal splitting from three neighboring NMR-active spin 1/2 nuclei.

Quaternary ammonium salt (Sections 20.2B and 20.3D): Ionic compounds in which a nitrogen bears four organic groups and a positive charge, paired with a counterion.

Quaternary structure (Sections 24.1 and 24.8C): The overall structure of a protein having multiple subunits (non-covalent aggregates of more than one polypeptide chain). Each subunit has a primary, secondary, and tertiary structure of its own.

R

R (Section 2.4A): A symbol used to designate an alkyl group. Oftentimes it is taken to symbolize any organic group.

R,S-System (Section 5.7): A method for designating the configuration of tetrahedral chirality centers.

Racemic form (*racemate* or *racemic mixture*) (Sections 5.9A, 5.9B, and 5.10A): An equimolar mixture of enantiomers. A racemic form is optically inactive.

Racemization (Section 6.12A): A reaction that transforms an optically active compound into a racemic form is said to proceed with racemization. Racemization takes place whenever a reaction causes chiral molecules to be converted to an achiral intermediate.

Radical addition to alkenes (Section 10.10): A process by which an atom with an unshared electron, such as a bromine atom, adds to an alkene with homolytic cleavage of the π-bond and formation of a σ-bond from the radical to the carbon; the resulting carbon radical then continues the chain reaction to product the final product plus another species with an unshared electron.

Radical cation (Section 9.14): A chemical species containing an unshared electron and a positive charge.

Radical (or *free radical*) (Sections 10.1, 10.6, and 10.7): An uncharged chemical species that contains an unpaired electron.

Radical halogenation (Section 10.3): Substitution of a hydrogen by a halogen through a radical reaction mechanism.

Radical reaction (Section 10.1B): A reaction involving radicals. Homolysis of covalent bonds occurs in radical reactions.

Random coil arrangement (Section 24.8): A type of protein secondary structure that is flexible, changing, and statistically random in its conformations.

Rate control (See **Kinetic control**)

Rate-determining step (Section 6.9A): If a reaction takes place in a series of steps, and if the first step is intrinsically slower than all of the others, then the rate of the overall reaction will be the same as (will be determined by) the rate of this slow step.

Reaction coordinate (Section 6.7): The abscissa in a potential energy diagram that represents the progress of the reaction. It represents the changes in bond orders and bond distances that must take place as reactants are converted to products.

Reaction mechanism (Sections 3 intro and 3.13): A step-by-step description of the events that are postulated to take place at the molecular level as reactants are converted to products. A mechanism will include a description of all intermediates and transition states. Any mechanism proposed for a reaction must be consistent with all experimental data obtained for the reaction.

Rearrangement (Sections 3.1, 7.8A, and 7.8B): A reaction that results in a product with the same atoms present but a different carbon skeleton from the reactant. The type of rearrangement called a 1,2 shift involves the migration of an organic group (with its electrons) from one atom to the atom next to it.

Reducing agent (Sections 12.2 and 12.3A): A chemical species that causes another chemical species to become reduced (to gain electrons, or to lose bonds to electronegative elements, often gaining bonds to hydrogen in the process). The reducing agent is oxidized in this process.

Reducing sugar (Section 22.6A): Sugars that reduce Tollens' or Benedict's reagents. All sugars that contain hemiacetal or hemiketal groups (and therefore are in equilibrium with aldehydes or α-hydroxyketones) are reducing sugars. Sugars in which only acetal or ketal groups are present are nonreducing sugars.

Reduction (Sections 12.2 and 12.3): A reaction that lowers the oxidation state of atoms in a molecule or ion. Reduction of an organic compound usually involves increasing its hydrogen content or decreasing its oxygen content. Reduction also accompanies any reaction that results in replacement of a more electronegative substituent by a less electronegative one.

Reductive amination (Section 20.4C): A method for synthesizing primary, secondary, or tertiary amines in which an aldehyde or ketone is treated with a primary or secondary amine to produce an imine (when primary amines are used) or an iminium ion (when secondary amines are used), followed by reduction to produce an amine product.

Regioselective reaction (Sections 8.2C and 8.18): A reaction that yields only one (or a predominance of one) constitutional isomer as the product when two or more constitutional isomers are possible products.

Relative configuration (Section 5.15A): The relationship between the configurations of two chiral molecules. Molecules are said to have the same relative configuration when similar or identical groups in each occupy the same position in space. The configurations of molecules can be related to each other through reactions of known stereochemistry, for example, through reactions that cause no bonds to a stereogenic center to be broken.

Replication (Section 25.4C): A process in which DNA unwinds, allowing each chain to act as a template for the formation of its complement, producing two identical DNA molecules from one original molecule.

Resolution (Sections 5.16B and 20.3F): The process by which the enantiomers of a racemic form are separated.

Resonance (Sections 3.10A, 13.4, and 15.11B): An effect by which a substituent exerts either an electron-releasing or electron-withdrawing effect through the π system of the molecule.

Resonance energy (Section 14.5): An energy of stabilization that represents the difference in energy between the actual compound and that calculated for a single resonance structure. The resonance energy arises from delocalization of electrons in a conjugated system.

Resonance structures (or *resonance contributors*) (Sections 1.8, 1.8A, 13.2B, and 13.4A): Lewis structures that differ from one another only in the position of their electrons. A single resonance structure will not adequately represent a molecule. The molecule is better represented as a *hybrid* of all of the resonance structures.

Restriction endonucleases (Section 25.6): Enzymes that cleave double-stranded DNA at specific base sequences.

Retro-aldol reaction (Section 19.4B): Aldol reactions are reversible; under certain conditions an aldol product will revert to its aldol reaction precursors. This process is called a *retro-aldol reaction*.

Retrosynthetic analysis (Section 7.15B): A method for planning syntheses that involves reasoning backward from the target molecule through various levels of precursors and thus finally to the starting materials.

Ribonucleic acid (RNA) (Sections 25.1 and 25.5): One of the two classes of molecules (the other is DNA) that carry genetic information in cells. RNA molecules transcribe and translate the information from DNA for the mechanics of protein synthesis.

Ribozyme (Section 25.5B): A ribonucleic acid that acts as a reaction catalyst.

Ring flip (Section 4.12): The change in a cyclohexane ring (resulting from partial bond rotations) that converts one ring conformation to another. A chair–chair ring flip converts any equatorial substitutent to an axial substituent and vice versa.

Ring strain (Section 4.10): The increased potential energy of the cyclic form of a molecule (usually measured by heats of combustion) when compared to its acyclic form.

S

1,2 Shift (Section 7.8A): The migration of a chemical bond with its attached group from one atom to an adjacent atom.

S_N1 reaction (Sections 6.9, 6.10, 6.12, 6.13, and 6.18B): Literally, substitution nucleophilic unimolecular. A multistep nucleophilic substitution in which the leaving group departs in a unimolecular step before the attack of the nucleophile. The rate equation is first order in substrate but zero order in the attacking nucleophile.

S_N2 reaction (Sections 6.5B, 6.6–6.8, 6.13, and 6.18A): Literally, substitution nucleophilic bimolecular. A bimolecular nucleophilic substitution reaction that takes place in a single step. A nucleophile attacks a carbon bearing a leaving group from the back side, causing an inversion of configuration at this carbon and displacement of the leaving group.

Salt (Section 1.3A): The product of a reaction between an acid and a base. Salts are ionic compounds composed of oppositely charged ions.

Sanger *N*-terminal analysis (Section 24.5B): A method for determining the *N*-terminal amino acid residue of a peptide by its S_NAr (nucleophilic aromatic substitution) reaction with dinitrofluorobenzene, followed by peptide hydrolysis and comparison of the product with known standards.

Saponification (Sections 17.7B and 23.2C): Base-promoted hydrolysis of an ester.

Saturated compound (Sections 2.1, 7.12, and 23.2): A compound that does not contain any multiple bonds.

Saturated fatty acids (Section 23.2): Fatty acids that contain no carbon-carbon double bonds.

Sawhorse formula (Section 4.8A): A chemical formula that depicts the spatial relationships of groups in a molecule in a way similar to dash-wedge formulas.

Secondary amine (Section 20.1): A derivative of ammonia in which there are two carbons bonded to a nitrogen atom. Secondary amines have a formula R_2NH, where the R groups can be the same or different.

Secondary carbon (Section 2.5): A carbon atom that has two other carbon atoms attached to it.

Secondary structure (Sections 24.1 and 24.8A): The local conformation of a polypeptide backbone. These local conformations are specified in terms of regular folding patterns such as pleated sheets, α helixes, and turns.

Shielding and deshielding (Section 9.7): Effects observed in NMR spectra caused by the circulation of sigma and pi electrons within the molecule. Shielding causes signals to appear at lower frequencies (upfield), deshielding causes signals to appear at higher frequencies (downfield).

Sigma (σ) bond (Section 1.12A): A single bond. A bond formed when electrons occupy the bonding σ orbital formed by the end-on overlap of atomic orbitals (or hybrid orbitals) on adjacent atoms. In a sigma bond the electron density has circular symmetry when viewed along the bond axis.

Sigma (σ) orbital (Section 1.13): A molecular orbital formed by end-on overlap of orbitals (or lobes of orbitals) on adjacent atoms. Sigma orbitals may be *bonding* (orbitals or lobes of the same phase sign overlap) or *antibonding* (orbitals or lobes of opposite phase sign overlap).

Signal splitting (Sections 9.2C and 9.9): Splitting of an NMR signal into multiple peaks, in patterns such as doublets, triplets, quartets, etc., caused by interactions of the energy levels of the magnetic nucleus under observation with the energy levels of nearby magnetic nuclei.

Silyl ether (silylation) (Section 11.11E): Conversion of an alcohol, R—OH, to a silyl ether (usually of the form R—O—SiR'_3, where the groups on silicon may be the same or different). Silyl ethers are used as protecting groups for the alcohol functionality.

Single bond (Section 1.12A): A bond between two atoms comprised of two electrons shared in a sigma bond.

Singlet (Section 9.2C): An NMR signal with only a single, unsplit peak.

Site-specific cleavage (Section 24.5D): A method of cleaving peptides at specific, known sites using enzymes and specialized reagents. For example, the enzyme trypsin preferentially catalyzes hydrolysis of peptide bonds on the C-terminal side of arginine and lysine. Other bonds in the peptide are not cleaved by this reagent.

Solid-phase peptide synthesis (SPPS) (Section 24.7D): A method of peptide synthesis in which the peptide is synthesized on a solid support, one amino acid residue at a time. The first amino acid of the peptide is bonded as an ester between its carboxylic acid group and a hydroxyl of the solid support (a polymer bead). This is then treated with a solution of the second amino acid and appropriate coupling reagents, creating a dipeptide. Excess reagents, byproducts, etc. are washed away. Further linkages are synthesized in the same manner. The last step of the synthesis is cleavage of the peptide from the solid support and purification.

Solubility (Section 2.13D): The extent to which a given solute dissolves in a given solvent, usually expressed as a weight per unit volume (e.g., grams per 100 mL).

Solvent effect (Sections 6.13C and 6.13D): An effect on relative rates of reaction caused by the solvent. For example, the use of a polar solvent will increase the rate of reaction of an alkyl halide in an S_N1 reaction.

Solvolysis (Section 6.12B): Literally, cleavage by the solvent. A nucleophilic substitution reaction in which the nucleophile is a molecule of the solvent.

***s* orbital** (Section 1.10): A spherical atomic orbital. For *s* orbitals the azimuthal quantum number $l = 0$ (See **Atomic orbital**).

Specific rotation (Section 5.8C): A physical constant calculated from the observed rotation of a compound using the following equation:

$$[\alpha]_D = \frac{\alpha}{c \times l}$$

where α is the observed rotation using the D line of a sodium lamp, c is the concentration of the solution or the density of a neat liquid in grams per milliliter, and l is the length of the tube in decimeters.

Spectroscopy (Section 9.1): The study of the interaction of energy with matter. Energy can be absorbed, transmitted, emitted or cause a chemical change (break bonds) when applied to matter. Among other uses, spectroscopy can be used to probe molecular structure.

Spin decoupling (Section 9.10): An effect that causes spin–spin splitting not to be observed in NMR spectra.

Spin–spin splitting (Section 9.9): An effect observed in NMR spectra. Spin–spin splittings result in a signal appearing as a multiplet (i.e., doublet, triplet, quartet, etc.) and are caused by magnetic couplings of the nucleus being observed with nuclei of nearby atoms.

Splitting tree diagrams (Section 9.9B): A method of illustrating the NMR signal splittings in a molecule by drawing "branches" from the original signal. The distance between the branches is proportional to the magnitude of the coupling constant. This type of analysis is especially useful when multiple splittings (splitting of already split signals) occur due to coupling with non-equivalent protons.

***sp* orbital** (Section 1.14): A hybrid orbital that is derived by mathematically combining one *s* atomic orbital and one *p* atomic orbital. Two *sp* hybrid orbitals are obtained by this process, and they are oriented in opposite directions with an angle of 180° between them.

***sp*² orbital** (Section 1.13): A hybrid orbital that is derived by mathematically combining one *s* atomic orbital and two *p* atomic orbitals. Three *sp*² hybrid orbitals are obtained by this process, and they are directed toward the corners of an equilateral triangle with angles of 120° between them.

***sp*³ orbital** (Section 1.12A): A hybrid orbital that is derived by mathematically combining one *s* atomic orbital and three *p* atomic orbitals. Four *sp*³ hybrid orbitals are obtained by this process, and they are directed toward the corners of a regular tetrahedron with angles of 109.5° between them.

Staggered conformation (Section 4.8A): A temporary orientation of groups around two atoms joined by a single bond such that the bonds of the back atom exactly bisect the angles formed by the bonds of the front atom when shown in a Newman projection formula:

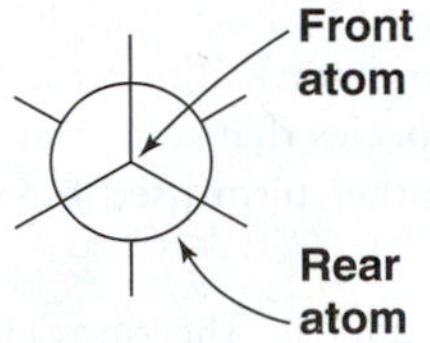

Front atom

Rear atom

A staggered conformation

Step-growth polymer (See also **Condendsation polymer**, Section 17.12 and Special Topic C in *WileyPLUS*): A polymer produced when bifunctional monomers (or potentially bifunctional monomers) react with each other through the intermolecular elimination of water or an alcohol. Polyesters, polyamides, and polyurethanes are all step-growth (condensation) polymers

Stereochemistry (Sections 5.2B, 6.8, and 6.14): Chemical studies that take into account the spatial aspects of molecules.

Stereogenic carbon (Section 5.4): A single tetrahedral carbon with four different groups attached to it. Also called an *asymmetric carbon, a stereocenter, or a chirality center*. The last usage is preferred.

Stereogenic center (Section 5.4): When the exchange of two groups bonded to the same atom produces stereoisomers, the atom is said to be a stereogenic atom, or stereogenic center.

Stereoisomers (Sections 1.13B, 4.9A, 4.13, 5.2B, and 5.14): Compounds with the same molecular formula that differ only in the arrangement of their atoms in space. Stereoisomers have the same connectivity and, therefore, are not constitutional isomers. Stereoisomers are classified further as being either enantiomers or diastereomers.

Stereoselective reaction (Sections 5.10B, 8.21C, and 12.3D): In reactions where chirality centers are altered or created, a stereoselective reaction produces a preponderance of one stereoisomer. Furthermore, a stereoselective reaction can be either enantioselective, in which case the reaction produces a preponderance of one enantiomer, or diastereoselective, in which case the reaction produces a preponderance of one diastereomer.

Stereospecific reaction (Sections 8.12 and 8.20C): A reaction in which a particular stereoisomeric form of the reactant reacts in such a way that it leads to a specific stereoisomeric form of the product.

Steric effect (Section 6.13A): An effect on relative reaction rates caused by the space-filling properties of those parts of a molecule attached at or near the reacting site.

Steric hindrance (Sections 4.8B and 6.13A): An effect on relative reaction rates caused when the spatial arrangement of atoms or groups at or near the reacting site hinders or retards a reaction.

Steroid (Section 23.4): Steroids are lipids that are derived from the following perhydrocyclopentanophenanthrene ring system:

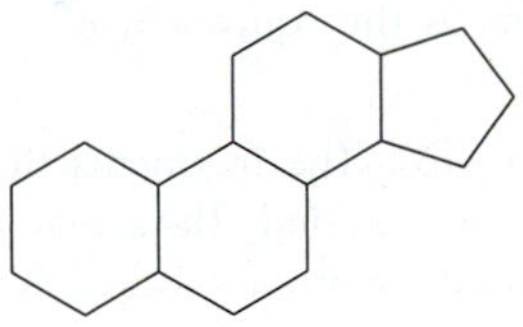

Structural formula (Section 1.7): A formula that shows how the atoms of a molecule are attached to each other.

Substituent effect (Sections 3.10D and 15.11F): An effect on the rate of reaction (or on the equilibrium constant) caused by the replacement of a hydrogen atom by another atom or group. Substituent effects include those effects caused by the size of the atom or group, called steric effects, and those effects caused by the ability of the group to release or withdraw electrons, called electronic effects. Electronic effects are further classified as being inductive effects or resonance effects.

Substitution reaction (Sections 3.13, 6.2, 10.3, 15.1, and 17.4): A reaction in which one group replaces another in a molecule.

Substitutive nomenclature (Section 4.3F): A system for naming compounds in which each atom or group, called a substituent, is cited as a prefix or suffix to a parent compound. In the IUPAC system only one group may be cited as a suffix. Locants (usually numbers) are used to tell where the group occurs.

Substrate (Sections 6.2 and 24.9): The molecule or ion that undergoes reaction.

Sugar (Section 22.12A): A carbohydrate.

Sulfa drugs (Section 20.10): Sulfonamide antibacterial agents, most of which have the general structure p-$H_2NC_6H_4SO_2NHR$. Sulfa drugs act as *antimetabolites* (they inhibit the growth of microbes) by inhibiting the enzymatic steps that are involved in the synthesis of folic acid; when deprived of folic acid, the microorganism dies.

Sulfonamides (Section 20.9): An amide derivative of a sulfonic acid, usually made by the reaction of ammonia, or a primary or secondary amine, with a sulfonyl chloride, resulting in compounds having the general formulas $R'SO_2NH_2$, $R'SO_2NHR$, or $R'SO_2NR_2$, respectively.

Sulfonate ester (Section 11.10): A compound with the formula $ROSO_2R'$ and considered to be derivatives of sulfonic acids, $HOSO_2R'$. Sulfonate esters are used in organic synthesis because of the excellent leaving group ability of the fragment $^-OSO_2R'$.

Superposable (Sections 1.13B and 5.1): Two objects are superposable if, when one object is placed on top of the other, all parts of each coincide. To be superposable is different than to be superimposable. Any two objects can be superimposed simply by putting one object on top of the other, whether or not all parts coincide. The condition of superposability must be met for two things to be identical.

Syn addition (Sections 7.13A and 8.15A): An addition that places both parts of the adding reagent on the same face of the reactant.

Syn coplanar (Section 7.6D): The relative position of two groups that have a 0° degree dihedral angle between them.

Syn dihydroxylation (Section 8.16A): An oxidation reaction in which an alkene reacts to become a 1,2-diol (also called a *glycol*) with the newly bonded hydroxyl groups added to the same face of the alkene.

Syndiotactic polymer (Special Topic B.1 in *WileyPLUS*): A polymer in which the configuration at the stereogenic centers along the chain alternate regularly: (*R*), (*S*), (*R*), (*S*), etc.

Synthetic equivalent (Sections 8.20B, 18.6, and 18.7): A compound that functions as the equivalent of a molecular fragment needed in a synthesis.

Synthon (Section 8.20B): The fragments that result (on paper) from the disconnection of a bond. The actual reagent that will, in a synthetic step, provide the synthon is called the *synthetic equivalent*.

T

Tautomerization (Section 18.2): An isomerization by which tautomers are rapidly interconverted, as in keto-enol tautomerization.

Tautomers (Section 18.2): Constitutional isomers that are easily interconverted. Keto and enol tautomers, for example, are rapidly interconverted in the presence of acids and bases.

Terminal residue analysis (Section 24.5): Methods used to determine the sequence of amino acids in a peptide by reactions involving the *N*- and *C*-terminal residues.

Terpene (Section 23.3): Terpenes are lipids that have a structure that can be derived on paper by linking isoprene units.

Terpenoids (Section 23.3): Oxygen-containing derivatives of terpenes.

Tertiary amine (Section 20.1): A derivative of ammonia in which there are three carbons bonded to a nitrogen atom. Tertiary amines have a formula R_3N where the R groups can be the same or different.

Tertiary carbon (Section 2.5): A carbon atom that has three other carbon atoms attached to it.

Tertiary structure (Sections 24.1 and 24.8B): The three dimensional shape of a protein that arises from folding of its polypeptide chains superimposed on its α helices and pleated sheets.

Tetrahedral intermediate (Section 17.4): A species created by the attack of a nucleophile on a trigonal carbon atom. In the case of a carbonyl group, as the electrons of the nucleophile form a bond to the carbonyl carbon the electrons of the carbon-oxygen π-bond shift to the oxygen. The carbon of the carbonyl group becomes four-coordinate (tetrahedral), while the oxygen gains an electron-pair and becomes negatively charged.

Thermodynamic control (Section 18.4A): A principle stating that the ratio of products of a reaction that reaches equilibrium is determined by the relative stabilities of the products (as measured by their standard free energies, $\Delta G°$). The most abundant product will be the one that is the most stable. Also called equilibrium control.

Thermodynamic enolate (Section 18.4A): In a situation in which more than one enolate anion can be formed, the *thermodynamic enolate* is the more stable of the possible enolate anions—usually the enolate with the more substituted double bond. A thermodynamic enolate is formed predominantly under conditions that permit the establishment of an equilibrium.

Thermodynamic or equilibrium product (Section 13.9A): When multiple products are possible, the product formed that is most stable; sometimes formed via a reversible, equilibrium process.

Torsional barrier (Section 4.8B): The barrier to rotation of groups joined by a single bond caused by repulsions between the aligned electron pairs in the eclipsed form.

Torsional strain (Sections 4.9 and 4.10): The strain associated with an eclipsed conformation of a molecule; it is caused by repulsions between the aligned electron pairs of the eclipsed bonds.

Tosylate (Section 11.10): A *p*-toluenesulfonate ester, which is a compound that contains the p-$CH_3C_6H_4SO_3$— group, i.e., p-$CH_3C_6H_4SO_3R$

Transcription (Section 25.5): Synthesis of a messenger RNA (mRNA) molecule that is complimentary to a section of DNA that carries genetic information.

Transesterification (Section 17.7A): A reaction involving the exchange of the alkoxyl portion of an ester for a different alkoxyl group, resulting in a new ester.

Transition state (Sections 6.6, 6.7, and 6.10): A state on a potential energy diagram corresponding to an energy maximum (i.e., characterized by having higher potential energy than immediately adjacent states). The term transition state is also used to refer to the species that occurs at this state of maximum potential energy; another term used for this species is *the activated complex*.

Translation (Section 25.5E): The ribosomal synthesis of a polypeptide using an mRNA template.

Triacylglycerols (Section 23.2): An ester of glycerol (glycerin) in which all three of the hydroxyl groups are esterified.

Triflate (Section 11.10): A methanesulfonate ester, which is a compound that contains the CH_3SO_3— group, i.e., p-CH_3SO_3R

Tripeptide (Section 24.4): A peptide comprised of three amino acids.

Triple bonds (Section 1.3B): Bonds comprised of one sigma (σ) bond and two pi (π) bonds.

Triplet (Section 9.2C): An NMR signal comprised of three peaks in a 1:2:1 area ratio, caused by signal splitting from two neighboring NMR-active spin 1/2 nuclei.

Trisaccharides (Section 22.1A): A carbohydrate that, when hydrolyzed, yields three monosaccharide molecules.

Two-dimensional (2D) NMR (Section 9.12): NMR techniques such as COSY and HETCOR that correlate one property (e.g., coupling), or type of nucleus, with another. (See **COSY** and **HETCOR**.)

U

Ultraviolet–visible (UV–Vis) spectroscopy (Section 13.8): A type of optical spectroscopy that measures the absorption of light in the visible and ultraviolet regions of the spectrum. Visible–UV spectra primarily provide structural information about the kind and extent of conjugation of multiple bonds in the compound being analyzed.

Unimolecular reaction (Section 6.9): A reaction whose rate-determining step involves only one species.

Unsaturated compound (Sections 2.1, 7.13, and 23.2): A compound that contains multiple bonds.

Unsaturated fatty acids (Section 23.2): Fatty acids that contain at least one carbon-carbon double bond.

Upfield (Section 9.3): Any area or signal in an NMR spectrum that is to the right relative to another. (See **Downfield** for comparison.) A signal that is upfield of another occurs at lower frequency (and lower δ and ppm values) than the other signal.

V

Valence shell (Section 1.3): The outermost shell of electrons in an atom.

***vic*-Dihalide** (Section 7.10): A general term for a molecule having halogen atoms bonded to each of two adjacent carbons.

Vicinal coupling (Sections 9.9 and 9.12A): The splitting of an NMR signal caused by hydrogen atoms on adjacent carbons. (See also **Coupling** and **Signal Splitting**.)

Vicinal (*vic*-) substituents (Section 7.10): Substituents that are on adjacent atoms.

Vinyl group (Sections 4.5 and 6.1): The CH_2—CH— group.

VSEPR model (valence shell electron pair repulsion) (Section 1.16): A method of predicting the geometry at a covalently bonded atom by considering the optimum geometric separation between groups of bonding and non-bonding electrons around the atom

W

Wave function (or ψ **function**) (Section 1.9): A mathematical expression derived from *quantum mechanics* corresponding to an energy state for an electron, i.e., for an orbital. The square of the ψ function, ψ^2, gives the probability of finding the electron in a particular place in space.

Wavelength, λ (Sections 2.15 and 13.8A): The distance between consecutive crests (or troughs) of a wave.

Wavenumber, $\bar{v}$ (Section 2.15): A way to express the frequency of a wave. The wavenumber is the number of waves per centimeter, expressed as cm^{-1}.

Waxes (Section 23.7): Esters of long-chain fatty acids and long-chain alcohols.

Williamson ether synthesis (Section 11.11B): The synthesis of an ether by the S_N2 reaction of an alkoxide ion with a substrate bearing a suitable leaving group (often a halide, sulfonate, or sulfate).

Y

Ylide (Section 16.10): An electrically neutral molecule that has a negative carbon with an unshared electron pair adjacent to a positive heteroatom.

Z

Zaitsev's rule (Sections 7.6B and 7.8A): A rule stating that an elimination will give as the major product the most stable alkene (i.e., the alkene with the most highly substituted double bond).

Zwitterion (See **Dipolar ion** and Section 24.2C): Another name for a dipolar ion.

[INDEX]

A

Absolute configuration, 228–230

Absorption spectrum, 599

Acetaldehyde, 10, 72, 76, 78, 372, 723

Acetaldehyde enolate, 53
  physical properties, 10

Acetals, 738–740

Acetic acid, 73–74, 76, 78, 113, 118–119, 126, 128, 145, 772
  physical properties, 78
  and pK_a, 114

Acetoacetic ester synthesis, 835–840
  acylation, 839–840
  dialkylation, 836
  substituted methyl ketones, 836–837

Acetone, 14, 57, 72, 79, 720

Acetonides, 993, 1023

Acetonitrile, 75

Acetyl-coenzyme A, 784

Acetyl group, 722

Acetylcholine, 897–898, 908

Acetylcholinesterase, 908, 1096

Acetylenes, 7, 57, 159, 159fn, 292, 326
  structure of, 40–42

Acetylenic hydrogen atom, 159, 313
  of terminal alkynes, substitution of, 313–314

Achiral molecules, 192, 195, 198, 212, 471

Acid anhydrides, reactions of, 810–811

Acid-base reactions, 105–109
  of amines, 917
  curved arrows, 107
  mechanism for, 107–108

Acid-catalyzed aldol condensations, 867–868

Acid-catalyzed aldol enolization, 826

Acid-catalyzed dehydration, of alcohols, 303–309

Acid-catalyzed esterification, 790–792

Acid-catalyzed halogenation, of aldehydes and ketones, 828

Acid-catalyzed hemiacetal formation, 736

Acid-catalyzed hydration, of alkenes, 354, 505–506

Acid chlorides, See Acyl chlorides

Acid derivatives, synthesis of, 786

Acid strength, 113

Acid–base reactions, 120–123
  acids and bases in water, 106–107

Brønsted–Lowry acids and bases, 105–106
  opposite charges attract, 135
  predicting the outcome of, 118–120
  and the synthesis of deuterium and tritium-labeled compounds, 134–135
  water solubility as the result of salt formation, 119–120

Acidic hydrolysis of a nitrile, 801

Acidity:
  effect of solvent on, 130
  effect of the solvent on, 130
  hybridization, 122–123
  inductive effects, 123
  relationships between structure and, 120–123

Acidity constant (K_a), 113–114

Acids:
  alcohols as, 509
  Brønsted–Lowry, 105–106, 137
  Lewis, 102–104
  in nonaqueous solutions, 133–134
  relative strength of, 115
  in water, 106

Acne medications, chemistry of, 459

Acrylonitrile, anionic polymerization of, 486

Actin, 166

Activating groups, 685–686
  meta directors, 685–686
  ortho–para directors, 685

Activation energies, 471

Active hydrogen compounds, 844

Active methylene compounds, 844

Active site, 1091
  of enzyme, 1070

Acyclovir, 1113

Acyl chlorides (acid chlorides), 679, 725–726, 777, 785–786, 786–788
  aldehydes by reduction of, 725–727
  esters from, 791
  reactions of, 787–788, 810
  synthesis of, 786–787
    using thionyl chloride, 787

Acyl compounds:
  relative reactivity of, 785–786
  spectroscopic properties of, 779–781

Acyl groups, 678

Acyl halide, 678

Acyl substitution, 771, 784–786, 822
  by nucleophilic addition–elimination, 784–786

Acyl transfer reactions, 784

Acylation, 839–840

Acylation reaction, 678

Acylium ions, 437

Adamantane, 180

Addition polymers, 483, 808

Addition reaction, 337–390
  of alkenes, 338–340

Adduct, 608

Adenosine diphosphate (ADP), 431

Adenosine triphosphate (ATP), 273, 431, 981–982

Adenylate cyclase, 1110

Adipic acid, 776

Adipocytes, 1032

Adrenaline, 273, 906

Adrenocortical hormones, 1046

Adriamycin, See Doxorubicin

Aggregation compounds, 161

Aglycone, 989–990

Aklavinone, 955

Alanine, 909, 1063, 1066
  isolectric point of, 1066
  titration curve for, 1067

Albrecht, Walther, 618

Albuterol, 501

Alcohol dehydrogenase, 548

Alcohols, 55, 67–68, 130, See also Primary alcohols; Secondary alcohols; Tertiary alcohols
  acid-catalyzed dehydration of, 303–309
  as acids, 509
  addition of:
    acetals, 738–740
    hemiacetals, 735–736
    thioacetals, 741
  alcohol carbon atom, 499
  from alkenes through hydroboration–oxidation, 349–352
  from alkenes through oxymercuration–demercuration, 349–350
  boiling points, 501
  conversion of, into a mesylate, 516
  conversion of, into alkyl halides, 510
  dehydration of, 296–297
    acid-catalyzed, 303–308

SEE TABLE 2.7 FOR A TABLE OF IR FREQUENCIES

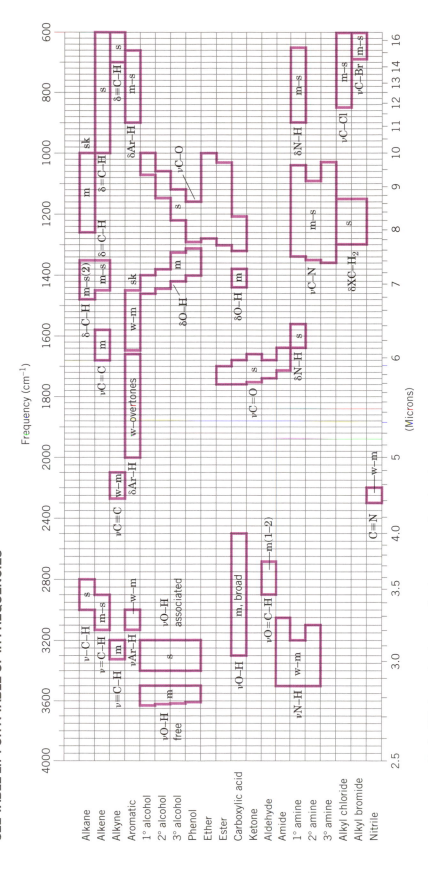

Typical IR absorption frequencies for common functional groups.
Absorptions are as follows: ν = stretching; δ = bending; w = weak; m = medium; s = strong; sk = skeletal
From *Multiscale Organic Chemistry: A Problem-Solving Approach* by John W. Lehman © 2002.
Reprinted by permission of Pearson Education, Inc., Upper Saddle River, NJ.

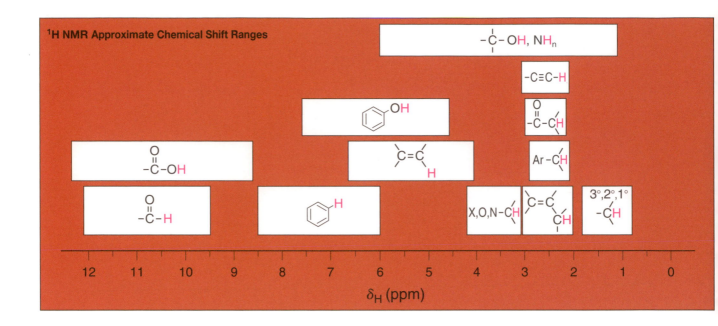

¹H NMR Approximate Chemical Shift Ranges

δ_H (ppm)

TABLE 9.1 APPROXIMATE PROTON CHEMICAL SHIFTS

Type of Proton	Chemical Shift (δ, ppm)	Type of Proton	Chemical Shift (δ, ppm)
1° Alkyl, RCH_3	0.8–1.2	Alkyl bromide, RCH_2Br	3.4–3.6
2° Alkyl, RCH_2R	1.2–1.5	Alkyl chloride, RCH_2Cl	3.6–3.8
3° Alkyl, R_3CH	1.4–1.8	Vinylic, $R_2C{=}CH_2$	4.6–5.0
Allylic, $R_2C{=}C{-}CH_3$ $\;\;\;\;\;\;\;\;\;\;\;R$	1.6–1.9	Vinylic, $R_2C{=}CH$ $\;\;\;\;\;\;\;\;\;\;\;R$	5.2–5.7
Ketone, $RCCH_3$ $\;\;\;\;\;\;\;\;\;\;\parallel$ $\;\;\;\;\;\;\;\;\;\;O$	2.1–2.6	Aromatic, ArH	6.0–8.5
Benzylic, $ArCH_3$	2.2–2.5	Aldehyde, RCH $\;\;\;\;\;\;\;\;\;\parallel$ $\;\;\;\;\;\;\;\;\;O$	9.5–10.5
Acetylenic, $RC{\equiv}CH$	2.5–3.1	Alcohol hydroxyl, ROH	0.5–6.0[a]
Alkyl iodide, RCH_2I	3.1–3.3	Amino, $R{-}NH_2$	1.0–5.0[a]
Ether, $ROCH_2R$	3.3–3.9	Phenolic, ArOH	4.5–7.7[a]
Alcohol, $HOCH_2R$	3.3–4.0	Carboxylic, $RCOH$ $\;\;\;\;\;\;\;\;\;\;\;\parallel$ $\;\;\;\;\;\;\;\;\;\;\;O$	10–13[a]

[a]The chemical shifts of these protons vary in different solvents and with temperature and concentration.